Signal Processing and Machine Learning Theory

AF764267

Signal Processing and Machine Learning Theory

Edited by

Paulo S.R. Diniz

ELSEVIER

ACADEMIC PRESS
An imprint of Elsevier

Academic Press is an imprint of Elsevier
125 London Wall, London EC2Y 5AS, United Kingdom
525 B Street, Suite 1650, San Diego, CA 92101, United States
50 Hampshire Street, 5th Floor, Cambridge, MA 02139, United States
The Boulevard, Langford Lane, Kidlington, Oxford OX5 1GB, United Kingdom

Copyright © 2024 Elsevier Inc. All rights reserved.

MATLAB® is a trademark of The MathWorks, Inc. and is used with permission.
The MathWorks does not warrant the accuracy of the text or exercises in this book.
This book's use or discussion of MATLAB® software or related products does not constitute endorsement or sponsorship by The MathWorks of a particular pedagogical approach or particular use of the MATLAB® software.

No part of this publication may be reproduced or transmitted in any form or by any means, electronic or mechanical, including photocopying, recording, or any information storage and retrieval system, without permission in writing from the publisher. Details on how to seek permission, further information about the Publisher's permissions policies and our arrangements with organizations such as the Copyright Clearance Center and the Copyright Licensing Agency, can be found at our website: www.elsevier.com/permissions.

This book and the individual contributions contained in it are protected under copyright by the Publisher (other than as may be noted herein).

Notices

Knowledge and best practice in this field are constantly changing. As new research and experience broaden our understanding, changes in research methods, professional practices, or medical treatment may become necessary.

Practitioners and researchers must always rely on their own experience and knowledge in evaluating and using any information, methods, compounds, or experiments described herein. In using such information or methods they should be mindful of their own safety and the safety of others, including parties for whom they have a professional responsibility.

To the fullest extent of the law, neither the Publisher nor the authors, contributors, or editors, assume any liability for any injury and/or damage to persons or property as a matter of products liability, negligence or otherwise, or from any use or operation of any methods, products, instructions, or ideas contained in the material herein.

ISBN: 978-0-323-91772-8

For information on all Academic Press publications
visit our website at https://www.elsevier.com/books-and-journals

Publisher: Mara E. Conner
Acquisitions Editor: Tim Pitts
Editorial Project Manager: Zsereena Rose Mampusti
Production Project Manager: Kamesh R
Cover Designer: Vicky Pearson Esser

Typeset by VTeX

Printed in India

Last digit is the print number: 9 8 7 6 5 4 3 2 1

ELSEVIER Book Aid International
Working together
to grow libraries in
developing countries
www.elsevier.com • www.bookaid.org

Love, Solidarity, and Dedication are the legacies of the swift comet. You arrived, enchanted everyone, and left before anyone could reach your spiritual goodness. The divinely bright light waned, transforming the days into dark and lonely nights, but her legacy lives on in those who met someone as beautiful as her.

Contents

List of contributors xxiii
Contributors xxvii
Signal processing and machine learning theory xlvii

CHAPTER 1 Introduction to signal processing and machine learning theory **1**
Marcele O.K. Mendonça, Isabela F. Apolinário, and Paulo S.R. Diniz

1.1 Introduction 1
1.2 Continuous-time signals and systems 2
1.3 Discrete-time signals and systems 3
1.4 Random signals and stochastic processes 5
1.5 Sampling and quantization 6
1.6 FIR and IIR filter design 8
1.7 Digital filter structures and implementations 10
1.8 Multirate signal processing 12
1.9 Filter banks and transform design 14
1.10 Discrete multiscale and transforms 18
1.11 Frames 19
1.12 Parameter estimation 20
1.13 Adaptive filtering 21
1.14 Machine learning: review and trends 22
1.15 Signal processing over graphs 24
1.16 Tensor methods in deep learning 25
1.17 Nonconvex graph learning: sparsity, heavy tails, and clustering 27
1.18 Dictionaries in machine learning 30
1.19 Closing comments 31
References 33

CHAPTER 2 Continuous-time signals and systems **35**
José A. Apolinário Jr. and Carla L. Pagliari

Glossary 35
Nomenclature 35
2.1 Introduction 36
2.2 Continuous-time systems 38
2.3 Differential equations 44
2.4 Laplace transform: definition and properties 48
2.5 Transfer function and stability 62
2.6 Frequency response 66
2.7 The Fourier series and the Fourier transform 72

2.8 Conclusion and future trends 82
Relevant websites 83
References 84

CHAPTER 3 Discrete-time signals and systems **85**
Leonardo G. Baltar and Josef A. Nossek

3.1 Introduction 85
3.2 Discrete-time signals: sequences 85
3.3 Discrete-time systems 87
3.3.1 Classification 88
3.4 Linear time-invariant systems 90
3.4.1 State-space description 90
3.4.2 Transfer function of a discrete-time LTI system 94
3.4.3 Finite duration impulse response systems 96
3.4.4 Infinite duration impulse response systems 98
3.4.5 Observability and controllability 101
3.4.6 Stability 103
3.5 Discrete-time signals and systems with MATLAB® 109
3.5.1 Discrete-time signals 109
3.5.2 Discrete-time systems representation and implementation 113
3.6 Conclusions 118
Notation 119
References 121

CHAPTER 4 Random signals and stochastic processes **123**
Luiz Wagner Pereira Biscainho and Eduardo Alves da Silva

Foreword 123
Acknowledgment 123
4.1 Introduction 123
4.2 Probability 124
4.2.1 Joint, conditional, and total probability – Bayes' rule 126
4.2.2 Probabilistic independence 128
4.2.3 Combined experiments – Bernoulli trials 129
4.3 Random variable 129
4.3.1 Probability distributions 131
4.3.2 Usual distributions 134
4.3.3 Conditional distribution 137
4.3.4 Statistical moments 138
4.3.5 Transformation of random variables 141
4.3.6 Multiple random variable distributions 142
4.3.7 Statistically independent random variables 146
4.3.8 Joint statistical moments 146

4.3.9 Central limit theorem 148
4.3.10 Multivariate Gaussian distribution 149
4.3.11 Transformation of vector random variables 149
4.3.12 Complex random variables 150
4.3.13 Application: estimators 151
4.3.14 Application: feature selection 153
4.4 Random process 154
4.4.1 Distributions of random processes and sequences 154
4.4.2 Statistical independence 155
4.4.3 First- and second-order moments for random processes and sequences 155
4.4.4 Stationarity 157
4.4.5 Properties of correlation functions for WSS processes and sequences 158
4.4.6 Time averages of random processes and sequences 159
4.4.7 Ergodicity 161
4.4.8 An encompassing example 162
4.4.9 Gaussian processes and sequences 165
4.4.10 Poisson random process 166
4.4.11 Complex random processes and sequences 166
4.4.12 Markov chains 167
4.4.13 Spectral description of random processes and sequences 169
4.4.14 White and colored noise 172
4.4.15 Applications: modulation, "bandpass," and band-limited processes, and sampling 174
4.4.16 Processing of random processes and sequences 175
4.4.17 Application: characterization of LTI systems 179
4.4.18 Modeling of bandpass WSS random processes 180
4.4.19 Statistical modeling of signals: random sequence as the output of an LTI system 181
4.5 Afterword 183
References 183
CHAPTER 5 Sampling and quantization **185**
Håkan Johansson
5.1 Introduction 185
5.1.1 Scope and prerequisites 185
5.1.2 Chapter outline 186
5.1.3 Recent and current trends 186
5.2 Preliminaries 187
5.2.1 Classification of signals 187
5.2.2 Discrete-time signals – sequences 188

5.2.3 Sampling of continuous-time signals 189
5.2.4 Classification of systems 190
5.2.5 Digital signal processing of analog signals 190
5.2.6 Digital communication – analog signal processing of digital signals 194
5.3 Sampling of deterministic signals 194
5.3.1 Uniform sampling 194
5.3.2 Poisson's summation formula 195
5.3.3 The sampling theorem 196
5.3.4 Antialiasing filter 199
5.3.5 Reconstruction 200
5.3.6 Distortion caused by undersampling 208
5.3.7 Distortion measure for energy signals 209
5.3.8 Bandpass sampling 211
5.4 Sampling of stochastic processes 212
5.4.1 Uniform sampling 212
5.4.2 Reconstruction of stochastic processes 214
5.5 Nonuniform sampling and generalizations 220
5.5.1 Time-interleaved ADCs 221
5.5.2 Problem formulation 221
5.5.3 Reconstruction 223
5.5.4 Reconstruction using a time-varying FIR system 223
5.5.5 Error metrics 225
5.5.6 Reconstruction based on least-squares design 227
5.5.7 Reconstruction using a polynomial-based approach 229
5.5.8 Performance discussion 230
5.6 Quantization 234
5.6.1 Quantization errors in A/D conversion 236
5.6.2 Round-off errors 239
5.7 Oversampling techniques and MIMO systems 244
5.7.1 Oversampled A/D converters 244
5.7.2 Oversampled D/A converters 248
5.7.3 Multiple-input multiple-output systems 250
5.8 Discrete-time modeling of mixed-signal systems 252
5.8.1 Input/output relations 255
5.8.2 Determining $G(z)$ 256
5.8.3 Stability 257
5.8.4 Frequency responses $H(j\Omega)$ and $G(e^{j\Omega T})$ 259
5.8.5 Example 12: continuous-time-input $\Sigma\Delta$-modulator 259
5.8.6 Example 13: single feedback loop system 260
References 263

CHAPTER 6 Digital filter structures and their implementation 267
Lars Wanhammar and Yajun Yu
6.1 Introduction 267
6.1.1 Nonrecursive FIR filters 267
6.1.2 Analog filters 268
6.1.3 Wave digital filters 268
6.1.4 Implementation of digital filters 269
6.1.5 MATLAB® toolbox 269
6.2 Properties of digital filters 270
6.2.1 Design target 271
6.3 Synthesis of digital FIR filters 272
6.3.1 REMEZ_FIR 276
6.3.2 Notch FIR filters 277
6.3.3 Half-band FIR filters 278
6.3.4 Complementary FIR filters 278
6.3.5 Minimum-phase FIR filters 279
6.3.6 Miscellaneous FIR filters 279
6.4 FIR structures 280
6.4.1 Direct-form FIR structures 280
6.4.2 Transposed direct form 281
6.4.3 Linear-phase FIR structures 281
6.4.4 Half-band FIR filters 282
6.4.5 Delay-complementary FIR structure 283
6.4.6 Cascade-form FIR structures 284
6.4.7 Lattice FIR structures 285
6.4.8 Recursive frequency-sampling structures 285
6.4.9 Multirate FIR filters 285
6.5 Frequency response masking filters 285
6.5.1 Basic FRM structure 286
6.5.2 Multistage FRM structure 293
6.5.3 Half-band filter 294
6.5.4 Hilbert transformer 301
6.5.5 Decimator and interpolator 301
6.5.6 Filter banks 308
6.5.7 Cosine-modulated transmultiplexer 309
6.5.8 FRM structure for recursive filters 310
6.5.9 2D FRM structure 313
6.5.10 Summary 315
6.6 The analog approximation problem 316
6.6.1 Typical requirements 317
6.6.2 Standard low-pass approximations 317

6.6.3 Comparison of the standard approximations ... 318
6.6.4 Filters with constant pole radius ... 319
6.6.5 Frequency transformations ... 321
6.7 Doubly resistively terminated lossless networks ... 321
6.7.1 Maximal power transfer ... 321
6.7.2 Reflection function ... 322
6.7.3 Element sensitivity ... 322
6.7.4 Errors in the elements in doubly terminated filters ... 324
6.7.5 Filters with diminishing ripple ... 326
6.8 Design of doubly resistively terminated analog filters ... 327
6.8.1 Ladder structures ... 327
6.8.2 Analog lattice structures ... 330
6.8.3 Realization of lossless one-ports ... 332
6.8.4 Commensurate length transmission line networks ... 336
6.8.5 Richards' variable ... 337
6.8.6 Unit elements ... 337
6.9 Design and realization of IIR filters ... 339
6.10 Wave digital filters ... 339
6.10.1 Wave descriptions ... 341
6.10.2 Wave-flow building blocks ... 342
6.10.3 Interconnection networks ... 346
6.10.4 Adaptor transformations ... 354
6.10.5 Resonance circuits ... 355
6.10.6 Parasitic oscillations ... 358
6.11 Ladder wave digital filters ... 359
6.12 Design of lattice wave digital filters ... 361
6.12.1 Bandpass lattice filters ... 363
6.12.2 Lattice wave digital filter with a constant pole radius ... 366
6.12.3 Lattice filters with a pure delay branch ... 371
6.13 Circulator-tree wave digital filters ... 374
6.14 Numerically equivalent state-space realization of wave digital filters ... 377
6.15 Computational properties of filter algorithms ... 378
6.15.1 Latency and throughput ... 378
6.15.2 Maximal sample rate ... 378
6.15.3 Cyclic scheduling ... 379
6.16 Architecture ... 380
6.17 Arithmetic operations ... 381
6.17.1 Addition and subtraction ... 381
6.17.2 Multiplication ... 384
6.18 Composite arithmetic operations ... 389
6.18.1 Multiple-constant multiplication ... 389

6.18.2 Distributed arithmetic 390
6.19 Power reduction techniques 393
6.19.1 Power supply voltage scaling 396
6.19.2 Overall strategy 396
References 397

CHAPTER 7 Multirate signal processing for software radio architectures 403
Fred Harris, Elettra Venosa, Xiaofei Chen, Richard Bell, Srivatsan Rajagopal, Radhika Mathuria, and Dinesh Bharadia

7.1 Introduction 404
7.2 The sampling process and the "resampling" process 407
7.3 Digital filters 413
7.4 Windowing 423
7.5 Basics on multirate filters 427
7.6 From single-channel down converter to standard down converter channelizer 428
7.6.1 From single channel up converter to standard up converter channelizer 438
7.7 Modifications of the standard down converter channelizer – M:2 down converter channelizer 445
7.7.1 Modifications of the standard up converter channelizer – 2:M up converter channelizer 454
7.8 Preliminaries on software-defined radios 460
7.9 Proposed architectures for software radios 462
7.9.1 Proposed digital down converter architecture 464
7.9.2 Digital down converter simulation results 470
7.9.3 Proposed up converter architecture 474
7.9.4 Digital up converter simulation results 477
7.10 Case study: enabling automated signal detection, segregation, and classification 479
7.10.1 Signal detection and segregation using the channelizer 481
7.10.2 Signal classification 485
7.11 Closing comments 492
Glossary 493
References 493

CHAPTER 8 Modern transform design for practical audio/image/video coding applications 495
Trac D. Tran

8.1 Introduction 495
8.2 Background and fundamentals 497
8.2.1 Notation 497
8.2.2 Transform fundamentals 498
8.2.3 Optimal orthogonal transform 499
8.2.4 Popular transforms in signal processing: DFT, WHT, DCT 499

8.2.5 The filter bank connection 500
8.2.6 The lapped transform connection 502
8.3 Design strategy 503
8.3.1 Desirable transform properties 504
8.4 Approximation approach via direct scaling 505
8.4.1 H.264 4 × 4 transform design 506
8.4.2 Integer DCT design via the principle of dyadic symmetry 507
8.4.3 Direct scaling of a rotation angle 507
8.5 Approximation approach via structural design 508
8.5.1 Lifting step 509
8.5.2 Lifting-based approximation 511
8.5.3 Lossless color transform design 512
8.5.4 Integer DCT design 512
8.5.5 Lifting for complex-coefficient transforms 515
8.6 Wavelet filter design via spectral factorization 516
8.6.1 Wavelets and filter banks 516
8.6.2 Spectral factorization 517
8.6.3 Lifting and the wavelet transform 521
8.7 Higher-order design approach via optimization 522
8.7.1 General modular construction 523
8.7.2 Higher-order design with pre-/postprocessing operators 524
8.7.3 Adaptive decomposition design 529
8.7.4 Modulated design 531
8.8 Conclusion 534
References 534

CHAPTER 9 Data representation: from multiscale transforms to neural networks 537
Yufang Bao and Hamid Krim

9.1 Introduction 537
9.1.1 An overview of multiscale transforms 537
9.1.2 Signal representation in functional bases 540
9.1.3 The continuous Fourier transform 541
9.1.4 Fourier series 542
9.1.5 The discrete Fourier transform 542
9.1.6 The windowed Fourier transform 543
9.2 Wavelets: a multiscale analysis tool 546
9.2.1 The wavelet transform 546
9.2.2 Inverting the wavelet transform 547
9.2.3 Wavelets and multiresolution analysis 549
9.2.4 Multiresolution applications in estimation problems 559

9.3 Curvelets and their applications ... 567
9.3.1 The continuous curvelet transform ... 569
9.3.2 The discrete curvelet transform ... 576
9.3.3 Applications ... 585
9.4 Contourlets and their applications ... 586
9.4.1 The contourlet transform ... 586
9.4.2 Laplacian pyramid frames ... 590
9.4.3 Directional filter banks ... 594
9.4.4 Contourlet filter bank ... 601
9.5 Shearlets and their applications ... 605
9.5.1 Composite wavelets ... 605
9.5.2 Shearlet atoms in the spatial plane ... 606
9.5.3 Shearlet atoms in the frequency domain ... 607
9.5.4 The continuous shearlet transform ... 608
9.5.5 Discrete shearlet atoms ... 613
9.5.6 Numerical implementation of the shearlet transform ... 615
9.5.7 Applications ... 618
9.6 Incorporating wavelets into neural networks ... 625
9.6.1 Using wavelets in neural networks for signal processing ... 626
9.6.2 Using wavelets in neural networks for image processing ... 638
Appendix 9.A ... 644
9.A.1 The z-transform ... 644
References ... 644

CHAPTER 10 Frames in signal processing ... 651
Lisandro Lovisolo, Eduardo A.B. da Silva, and Allan Freitas da Silva

10.1 Introduction ... 651
10.1.1 Notation ... 652
10.2 Basic concepts ... 652
10.2.1 The dual frame ... 654
10.2.2 Signal analysis and synthesis using frames ... 655
10.3 Relevant definitions ... 656
10.3.1 The frame operator ... 657
10.3.2 The inverse frame ... 659
10.3.3 Characterization of frames: basic property ... 661
10.4 Some computational remarks ... 664
10.4.1 Frames in discrete spaces ... 664
10.4.2 Finite vector spaces ... 665
10.5 Construction of frames from a prototype signal ... 667
10.5.1 Translation, modulation, and dilation operators ... 668
10.5.2 Common frame constructions ... 668

10.5.3 Frames of translates 669
10.5.4 Gabor frames 670
10.5.5 Wavelet frames 672
10.5.6 Finite vector spaces 673
10.6 Some remarks and highlights on applications 673
10.6.1 Signal analysis 673
10.6.2 Robust data transmission 674
10.6.3 Gaborgram 676
10.6.4 Inverse Gabor frame 677
10.6.5 Gabor frames in discrete spaces 679
10.6.6 Fast analysis and synthesis operators for Gabor frames 680
10.6.7 Time-frequency content analysis using frame expansions 681
10.7 Conclusion 684
References 684

CHAPTER 11 Parametric estimation 689
Ryan M. Corey, Suleyman Serdar Kozat, and Andrew C. Singer
11.1 Introduction 689
11.1.1 What is parametric estimation? 689
11.1.2 Parametric models for signals and systems 690
11.2 Preliminaries 692
11.2.1 Signals 692
11.2.2 Data-driven framework 693
11.2.3 Stochastic model and criteria 695
11.3 Parametric models for linear time-invariant systems 696
11.3.1 Autoregressive models 697
11.3.2 Moving average models 699
11.3.3 Autoregressive moving average models 701
11.3.4 Parametric modeling for system function approximation 702
11.4 Joint process estimation and sequential modeling 707
11.4.1 Batch joint process estimation 708
11.4.2 Segmented least-squares 710
11.4.3 Recursive estimation with RLS and LMS 711
11.5 Model order estimation 714
References 716

CHAPTER 12 Adaptive filters 717
Vítor H. Nascimento and Magno T.M. Silva
12.1 Introduction 717
12.1.1 Motivation – acoustic echo cancelation 718
12.1.2 A quick tour of adaptive filtering 722
12.1.3 Applications 740

12.2 Optimum filtering ... 745
12.2.1 Linear least-mean squares estimation ... 746
12.2.2 Complex variables and multichannel filtering ... 760
12.3 Stochastic algorithms ... 773
12.3.1 LMS algorithm ... 775
12.3.2 Normalized LMS algorithm ... 778
12.3.3 RLS algorithm ... 781
12.3.4 Comparing convergence rates ... 787
12.4 Statistical analysis ... 790
12.4.1 Data model ... 792
12.4.2 Relating the autocorrelation matrix of the weight error vector to the EMSE and the MSD ... 794
12.4.3 Statistical analysis of the LMS algorithm ... 796
12.4.4 A unified statistical analysis ... 801
12.5 Extensions and current research ... 818
12.5.1 Finite precision arithmetic ... 819
12.5.2 Regularization ... 830
12.5.3 Variable step size ... 832
12.5.4 Non-Gaussian noise and robust filters ... 833
12.5.5 Blind equalization ... 835
12.5.6 Subband and transform-domain adaptive filters ... 837
12.5.7 Affine projections algorithm ... 837
12.5.8 Cooperative estimation ... 839
12.5.9 Adaptive IIR filters ... 846
12.5.10 Set membership and projections onto convex sets ... 846
12.5.11 Adaptive filters with constraints ... 847
12.5.12 Reduced-rank adaptive filters ... 848
12.5.13 Kernel adaptive filtering ... 848
References ... 851

CHAPTER 13 Machine learning ... 869
Marcele O.K. Mendonça, Sergio L. Netto, Paulo S.R. Diniz, and Sergios Theodoridis

13.1 Introduction ... 869
13.2 Learning concepts ... 870
13.2.1 Data concepts ... 871
13.2.2 Kinds of learning ... 873
13.2.3 Underfitting and overfitting ... 874
13.3 Unsupervised learning ... 876
13.3.1 Introduction ... 876
13.3.2 K-Means algorithm ... 877

13.3.3 Self-organizing map ... 877
13.3.4 Mean-shift clustering ... 878
13.3.5 MNIST example ... 878
13.4 Supervised learning ... 879
13.4.1 Perceptron ... 879
13.4.2 Fully connected neural networks ... 883
13.4.3 Regression ... 887
13.4.4 A regression example ... 889
13.4.5 Classification ... 892
13.4.6 A classification example ... 892
13.4.7 EMNIST letter dataset ... 894
13.4.8 Dropout ... 896
13.4.9 Number of flops ... 897
13.4.10 Convolutional neural networks ... 898
13.4.11 Recurrent neural networks ... 899
13.4.12 Support vector machines ... 908
13.5 Ensemble learning ... 921
13.5.1 Random forest ... 921
13.5.2 Boosting ... 923
13.6 Deep learning ... 924
13.6.1 Large dataset corpora ... 924
13.6.2 Nonlinear activation function ... 926
13.6.3 Network weight initialization ... 926
13.6.4 Loss function ... 927
13.6.5 Learning algorithm ... 928
13.6.6 Network regularization ... 930
13.6.7 Deep learning frameworks ... 931
13.6.8 Autoencoders ... 931
13.6.9 Adversarial training ... 932
13.7 CNN visualization ... 937
13.7.1 Saliency map (vanilla and guided backpropagation) ... 937
13.8 Deep reinforcement learning ... 938
13.9 Current trends ... 941
13.9.1 Transfer learning ... 942
13.9.2 Learning on graphs ... 942
13.9.3 Interpretable artificial intelligence ... 942
13.9.4 Federated learning ... 943
13.9.5 Causal machine learning ... 946
13.10 Concluding remarks ... 949
13.11 Appendix: RNN's gradient derivations ... 949
13.11.1 Gradient derivations of the standard RNN ... 949

13.11.2 Gradient derivations for the LSTM scheme 952
References 956

CHAPTER 14 A primer on graph signal processing 961
Wallace Alves Martins, Juliano Bandeira Lima, Cédric Richard, and Symeon Chatzinotas

14.1 The case for GSP 961
14.2 Fundamentals of graph theory 963
14.3 Graph signals and systems 965
14.3.1 Graph signals 965
14.3.2 Inferring graphs 967
14.3.3 Graph shift operators 968
14.4 Graph Fourier transform 969
14.4.1 Adjacency-based graph Fourier transform 971
14.4.2 Laplacian-based graph Fourier transform 975
14.5 Graph filtering 976
14.5.1 Convolution and filtering 978
14.5.2 Graph filter design 982
14.6 Down- and upsampling graph signals 985
14.6.1 Band-limited graph signals 986
14.6.2 Approximately band-limited GSs 990
14.6.3 Optimal sampling strategies 990
14.6.4 Interpolation of band-limited GSs 992
14.7 Examples and applications 994
14.7.1 On-line filtering of network signals 994
14.7.2 Image processing 996
14.7.3 3D point cloud processing 998
14.7.4 Spatiotemporal analysis and visualization 999
14.8 A short history of GSP 1001
References 1003

CHAPTER 15 Tensor methods in deep learning 1009
Yannis Panagakis, Jean Kossaifi, Grigorios G. Chrysos, James Oldfield, Taylor Patti, Mihalis A. Nicolaou, Anima Anandkumar, and Stefanos Zafeiriou

15.1 Introduction 1009
15.2 Preliminaries on matrices and tensors 1011
15.2.1 Notation 1011
15.2.2 Transforming tensors into matrices and vectors 1012
15.2.3 Matrix and tensor products 1013
15.2.4 Tensor diagrams 1014
15.3 Tensor decomposition 1015

15.3.1 Canonical-polyadic decomposition 1016
15.3.2 Tucker decomposition 1017
15.3.3 Tensor-train 1018
15.3.4 Tensor regression 1019
15.4 Tensor methods in deep learning architectures 1020
15.4.1 Parameterizing fully connected layers 1020
15.4.2 Tensor contraction layers 1020
15.4.3 Tensor regression layers 1022
15.4.4 Parameterizing convolutional layers 1023
15.4.5 Tensor structures in polynomial networks and attention mechanisms ... 1027
15.4.6 Tensor structures in generative adversarial networks 1029
15.4.7 Robustness of deep networks 1032
15.4.8 Applications 1033
15.5 Tensor methods in quantum machine learning 1034
15.5.1 Quantum mechanics 1034
15.5.2 Matrix product states for quantum simulation 1035
15.5.3 Factorized quantum circuit simulation with TensorLy-Quantum 1036
15.6 Software 1037
15.7 Limitations and guidelines 1038
15.7.1 Choosing the appropriate decomposition 1039
15.7.2 Rank selection 1039
15.7.3 Training pitfalls and numerical issues 1040
References 1040

CHAPTER 16 Nonconvex graph learning: sparsity, heavy tails, and clustering. 1049
José Vinícius de M. Cardoso, Jiaxi Ying, and Daniel P. Palomar
16.1 Introduction 1049
16.1.1 Learning undirected graphs 1050
16.1.2 Majorization-minimization: a brief visit 1052
16.1.3 Alternating direction method of multipliers: a brief visit 1053
16.2 Sparse graphs 1053
16.3 Heavy-tail graphs 1057
16.4 Clustering 1061
16.4.1 Soft-clustering via bipartite graphs 1066
16.5 Conclusion 1068
Acknowledgments 1069
References 1069

CHAPTER 17 Dictionaries in machine learning 1073
Kenneth Kreutz-Delgado, Bhaskar Rao, Igor Fedorov, and Srinjoy Das
17.1 Data-driven AI via dictionary learning for sparse signal processing 1073
17.2 Sparsity and sparse representations 1075

17.2.1 *"Something rather than anything"* – sparse structure in the natural world 1075
17.2.2 Sparsity: "*meaning*" versus "*information as choice of possibilities*" 1076
17.2.3 The linear sparse generative model 1077
17.2.4 Sufficient conditions for a unique sparse solution 1079
17.2.5 Sparsity classes 1080
17.2.6 Bases, frames, dictionaries, and sparsity classes 1081
17.3 Sparse signal processing 1082
17.3.1 Measures of sparsity and diversity 1082
17.3.2 Sparsity-inducing optimizations of diversity measures 1083
17.3.3 The stochastic model and sparse Bayesian learning 1084
17.4 Dictionary learning I – basic models and algorithms 1086
17.4.1 Dictionary learning from observational data – the setup 1087
17.4.2 Dictionary learning as a generalization of K-means clustering and vector quantization 1090
17.4.3 Dictionary learning as a generalization of PCA and FA 1096
17.4.4 Dictionary learning using the Method of Directions 1097
17.4.5 Dictionary learning using K-SVD 1099
17.5 Dictionary learning II – the hierarchical/empirical Bayes approach 1101
17.5.1 Hierarchical model for sparse Bayesian learning – Type I versus Type II estimation 1101
17.5.2 Dictionary learning in the SBL framework 1107
17.5.3 Information theory, generative models, and "Sparseland" 1110
17.5.4 Dictionary learning for conceptual understanding of the world 1114
17.6 Nonnegative matrix factorization in dictionary learning 1116
17.7 Dictionaries, data manifold learning, and geometric multiresolution analysis ... 1119
17.8 Hard clustering and classification in dictionary learning 1122
17.8.1 Clustering 1122
17.8.2 Classification using residual errors 1125
17.9 Multilayer dictionary learning and classification 1127
17.10 Kernel dictionary learning 1131
17.11 Conclusion 1136
Appendix 17.A Derivation and properties of the K-SVD algorithm 1137
Appendix 17.B Derivation of the SBL EM update equation 1142
Appendix 17.C SBL dictionary learning algorithm 1142
Appendix 17.D Mathematical background for kernelizing dictionary learning 1146
References 1151

Index 1161

List of contributors

Wallace Alves Martins
Interdisciplinary Centre for Security, Reliability and Trust (SnT), University of Luxembourg, Luxembourg, Luxembourg

Anima Anandkumar
NVIDIA, Santa Clara, CA, United States
Caltech, Pasadena, CA, United States

Isabela F. Apolinário
Universidade Federal do Rio de Janeiro, Rio de Janeiro, Brazil

José A. Apolinário Jr.
Military Institute of Engineering (IME), Department of Electrical Engineering (SE/3), Rio de Janeiro, RJ, Brazil

Leonardo G. Baltar
Institute for Circuit Theory and Signal Processing, Technische Universität München, München, Germany

Juliano Bandeira Lima
Department of Electronics and Systems, Federal University of Pernambuco (UFPE), Recife, Brazil

Yufang Bao
Department of Mathematics and Computer Science, UNC Fayetteville State University, Fayetteville, NC, United States

Richard Bell
University of California, San Diego, Department of Electrical and Computer Engineering, San Diego, CA, United States

Dinesh Bharadia
University of California, San Diego, Department of Electrical and Computer Engineering, San Diego, CA, United States

Luiz Wagner Pereira Biscainho
Federal University of Rio de Janeiro (UFRJ), Rio de Janeiro, Brazil

Symeon Chatzinotas
Interdisciplinary Centre for Security, Reliability and Trust (SnT), University of Luxembourg, Luxembourg, Luxembourg

Xiaofei Chen
Meta Platforms, Inc., Menlo Park, CA, United States

Grigorios G. Chrysos
Department of Electrical Engineering, Ecole Polytechnique Federale de Lausanne (EPFL), Lausanne, Switzerland

Ryan M. Corey
University of Illinois, Urbana, IL, United States

Srinjoy Das
West Virginia University, Morgantown, WV, United States

Allan Freitas da Silva
SMT – DEL/PEE – COPPE/UFRJ, Rio de Janeiro, Brazil

Eduardo A.B. da Silva
SMT – DEL/PEE – COPPE/UFRJ, Rio de Janeiro, Brazil

Eduardo Alves da Silva
Federal University of Rio de Janeiro (UFRJ), Rio de Janeiro, Brazil

José Vinícius de M. Cardoso
The Hong Kong University of Science and Technology, Hong Kong SAR, China

Paulo S.R. Diniz
Universidade Federal do Rio de Janeiro, Rio de Janeiro, Brazil
Program of Electrical Engineering and Department of Electronics & Computer Engineering, COPPE/Poli/Universidade Federal do Rio de Janeiro, Rio de Janeiro, Brazil

Igor Fedorov
Meta, Menlo Park, CA, United States

Fred Harris
University of California, San Diego, Department of Electrical and Computer Engineering, San Diego, CA, United States

Håkan Johansson
Division of Communication Systems, Department of Electrical Engineering, Linköping University, Linköping, Sweden

Jean Kossaifi
NVIDIA, Santa Clara, CA, United States

Suleyman Serdar Kozat
Bilkent University, Ankara, Turkey

Kenneth Kreutz-Delgado
University of California San Diego, San Diego, CA, United States
Pattern Computer Inc., Friday Harbor, WA, United States

Hamid Krim

Electrical and Computer Engineering Department, North Carolina State University, Raleigh, NC, United States

Lisandro Lovisolo

PROSAICO – DETEL/PEL – UERJ, Rio de Janeiro, Brazil

Radhika Mathuria

University of California, San Diego, Department of Electrical and Computer Engineering, San Diego, CA, United States

Marcele O.K. Mendonça

Universidade Federal do Rio de Janeiro, Rio de Janeiro, Brazil

Program of Electrical Engineering and Department of Electronics & Computer Engineering, COPPE/Poli/Universidade Federal do Rio de Janeiro, Rio de Janeiro, Brazil

Vítor H. Nascimento

Dept. of Electronic Systems Engineering, University of São Paulo, São Paulo, SP, Brazil

Sergio L. Netto

Program of Electrical Engineering and Department of Electronics & Computer Engineering, COPPE/Poli/Universidade Federal do Rio de Janeiro, Rio de Janeiro, Brazil

Mihalis A. Nicolaou

Computation-based Science and Technology Research Center at The Cyprus Institute, Nicosia, Cyprus

Josef A. Nossek

Institute for Circuit Theory and Signal Processing, Technische Universität München, München, Germany

James Oldfield

Queen Mary University of London, London, United Kingdom

Carla L. Pagliari

Military Institute of Engineering (IME), Department of Electrical Engineering (SE/3), Rio de Janeiro, RJ, Brazil

Daniel P. Palomar

The Hong Kong University of Science and Technology, Hong Kong SAR, China

Yannis Panagakis

Department of Informatics and Telecommunications, National and Kapodistrian University of Athens, Athens, Greece

Taylor Patti

NVIDIA, Santa Clara, CA, United States

Srivatsan Rajagopal
University of California, San Diego, Department of Electrical and Computer Engineering, San Diego, CA, United States

Bhaskar Rao
University of California San Diego, San Diego, CA, United States

Cédric Richard
Department of Electrical Engineering, University Côte d'Azur, Nice, France

Magno T.M. Silva
Dept. of Electronic Systems Engineering, University of São Paulo, São Paulo, SP, Brazil

Andrew C. Singer
University of Illinois, Urbana, IL, United States

Sergios Theodoridis
Department of Informatics and Telecommunications, National and Kapodistrian University of Athens, Athens, Greece
Electronic Systems Department, Aalborg University, Aalborg, Denmark

Trac D. Tran
Department of Electrical and Computer Engineering, The Johns Hopkins University, Baltimore, MD, United States

Elettra Venosa
Space Micro, Inc., San Diego, CA, United States

Lars Wanhammar
Department of Electrical Engineering, Linköping University, Linköping, Sweden

Jiaxi Ying
The Hong Kong University of Science and Technology, Hong Kong SAR, China

Yajun Yu
Department of Electronic and Electrical Engineering, Southern University of Science and Technology, Shenzhen, Guangdong Province, PR China

Stefanos Zafeiriou
Department of Computing, Imperial College London, London, United Kingdom

Contributors

Chapter 1

Marcele O. K. de Mendonça was born in Mesquita, Brazil. She received her degree in Telecommunications Engineering from the Fluminense Federal University (UFF) in 2016. In addition, she received her MSc and PhD degrees from the Signal Multimedia and Telecommunications Laboratory (SMT) at the Federal University of Rio de Janeiro (UFRJ) in 2018 and 2022, respectively. She won 2nd place in the UFF Vasconcellos Torres Award of Scientific Initiation with the project OFDM Systems in Software Gnuradio using USRP in 2014. Moreover, she received the Best Demo Award at the Brazilian Telecom Symposium with the project FM transmission and reception using software-defined radios in 2019. She was a PhD visiting scholar in the Swiss Government Excellence Scholarships program with the LTS4 Research Group in Ecole Polytechnique Fédérale de Lausanne (EPFL), Switzerland, in 2021–2022, supervised by Prof. Pascal Frossard. She joined the Interdisciplinary Centre for Security, Reliability, and Trust (SnT) at the University of Luxembourg as a Research Associate in 2022. Her research interests mainly include signal processing, OFDM systems with reduced redundancy, massive MIMO systems with low-complexity antenna selection to reduce transmission power, and machine learning for wireless communications, including robust systems based on adversarial learning.

Isabela Ferrão Apolinário was born in Brasília, Brazil. She studied Electronics Engineering at UFRJ and obtained her MSc degree in Electrical Engineering from COPPE/UFRJ. She was a member of the Audio Processing Group (GPA) from SMT for eight years. Her main interest is in digital audio signal processing, more specifically in time-frequency representation of audio and music signals.

Paulo S. R. Diniz was born in Niterói, Brazil. He received his degree in Electronics Engineering (Cum Laude) from UFRJ, Rio de Janeiro, Brazil, in 1978, his MSc degree in electrical engineering from COPPE/UFRJ, Rio de Janeiro, Brazil, in 1981, and his PhD degree in electrical engineering from Concordia University, Montreal, QC, Canada, in 1984. He is with both the Program of Electrical Engineering of COPPE, and the Department of Electronics and Computer Engineering of the Polytechnic School, Universidade Federal do Rio de Janeiro (UFRJ). He has authored or coauthored refereed papers in journals and conference papers in some of these areas and authored the following textbooks: *Online Learning and Adaptive Filters* (Cambridge, U.K.: Cambridge University Press, 2022, with M. L. Campos, W. A. Martins, M. V. S. Lima, and J. A. Apolinário Jr.), *Online Component Analysis, Architectures and Applications* (Boston-Delft, NOW Publishers, 2022, with J. B. O. Souza Filho, L.-D. Van, and T.-P. Jung), *Adaptive Filtering: Algorithms and Practical Implementation* (NY: Springer, Fifth Edition, 2020), and *Digital Signal Processing: System Analysis and Design* (Cambridge, U.K.: Cambridge University Press, Second Edition, 2010, with E. A. B. da Silva and S. L. Netto). He has coauthored the monograph *Block Transceivers: OFDM and Beyond* (New York, NY, USA: Morgan and Claypool, 2012, with W. A. Martins and M. V. S. Lima). His teaching and research interests include analog and digital signal processing, adaptive signal processing, digital communications, wireless communications, multirate systems, stochastic processes, and electronic circuits. Dr. Diniz is a Life Fellow of the IEEE and Fellow of the European Association for Signal Processing (EURASIP). He is a Member of the National Academy of Engineering and the Brazilian Academy of Science. He was the recipient of the 2014 Charles Desoer Technical Achievement Award of the IEEE Circuits and Systems Society. He was also the recipient of some Best Paper awards from several conferences and an IEEE journal. He was an Associate Editor for *IEEE Transactions on Circuits and Systems II: Analog and Digital Signal Processing* from 1996 to 1999, *Circuits, Systems, and Signal Processing* from 1998 to 2002, and *IEEE Transactions on Signal Processing* from 1999 to 2002. From 2000 to 2001, he was a Distinguished Lecturer of the IEEE Circuits and Systems Society. In 2004, he was a Distinguished Lecturer of the IEEE Signal Processing Society.

Chapter 2

José Antonio Apolinário Junior graduated from the Military Academy of Agulhas Negras (AMAN), Brazil, in 1981. He received his BSc degree from the Military Institute of Engineering (IME), Brazil, in 1988, his MSc degree from the University of Brasília (UnB), Brazil, in 1993, and his DSc degree from COPPE/UFRJ, Rio de Janeiro, Brazil, in 1998, all in electrical engineering. He is currently an Associate Professor at the Department of Electrical Engineering, IME, where he has served as Head of the Department and Vice-Rector of Study and Research. He was a Visiting Professor at the *Escuela Politécnica del Ejército* (ESPE), Quito, Ecuador, from 1999 to 2000 and a Visiting Researcher and twice a Visiting Professor at the Helsinki University of Technology (HUT, today Aalto University), Finland, in 1997, 2004, and 2006, respectively. His research interests comprise many aspects of linear and nonlinear digital signal processing, including adaptive filtering, acoustics, and array processing. He has organized and has been the first Chair of the Rio de Janeiro Chapter of the IEEE Communications Society. He has edited the book "*QRDRLS Adaptive Filtering*" (Springer, 2009) and recently coauthored the book "*Online Learning and Adaptive Filters*" (Cambridge University Press, 2022). He served as the Finance Chair of IEEE ISCAS 2011 (Rio de Janeiro, Brazil, May 2011) and is a senior member of the IEEE and the Brazilian Telecommunications Society (SBrT).

Carla Liberal Pagliari received her PhD degree in Electronic Systems Engineering from the University of Essex, UK, in 2000. In 1983 and 1985 she worked for TV Globo, Rio de Janeiro, Brazil. From 1986 to 1992 she was a Researcher at the Instituto de Pesquisa e Desenvolvimento, Rio de Janeiro, Brazil. Since 1993 she has been with the Department of Electrical Engineering at the Military Institute of Engineering (IME), Rio de Janeiro, Brazil. Her main research activities and interests are focused on light field signal processing and coding, image and video processing, and computer vision. She is a member of the ISO/IEC JTC1/SC29/WG1 (JPEG) standardization committee and the Chair of the Ad-hoc Group on JPEG Pleno Light Field.

Chapter 3

Leonardo Gomes Baltar received his B.Sc. and M.Sc. degrees in Electrical Engineering from the Federal University of Rio de Janeiro (UFRJ), Brazil, in 2004 and 2006, respectively. He received the Dr.-Ing. degree from the Technical University of Munich (TUM), Germany, in 2017. From 2006 to 2015, he was the Chair for Circuit Theory and Signal Processing at TUM as a research and teaching associate. His research was in the area of digital signal processing techniques applied to wired and wireless communications systems including multi-rate systems, filter banks, block transforms, digital filter design, and efficient processing structures. He was involved in the European FP7 projects PHYDYAS and EMPHATIC and in projects with industry partners. In 2015 he joined Intel Germany to get involved in the EC 5G-PPP H2020 projects Flex5Gware and FANTASTIC5G, where he mainly contributed to the topic of 5G waveforms and their computational complexity. Since 2018, Leonardo has been actively contributing to the 5G Automotive Association (5GAA) and ETSI Intelligence Transportation Systems (ITS) on diverse topics related to connected vehicles and smart road infrastructure. In 5GAA, he currently holds the position of Working Group Chair. He has contributed to 3GPP RAN1 in the past and, more recently to the EC Horizon Project on 6G Hexa-X on topics related to the physical layer and AI/ML for wireless networks. Leonardo is a Senior Member of the IEEE and a member of the German Association of electrical engineers (VDE).

Josef A. Nossek (Life Fellow, IEEE) received the Dipl.-Ing. and Dr.Techn. degrees in electrical engineering from the Vienna University of Technology, Vienna, Austria, in 1974 and 1980, respectively. In 1974, he joined Siemens AG, Munich, Germany as a member of the technical staff. In 1978, he became a Supervisor and from 1980 to 1987, he was the Head of the Department. In 1987, he was promoted to the Head of all radio systems design. Since 1989, he has been a Full Professor in circuit theory and signal processing with the Munich University of Technology, where he teaches undergraduate and graduate courses on circuit and systems theory and signal processing and leads research on signal processing algorithms for communications, especially multiantenna systems. He was the President-Elect, the President, and the Past President of the IEEE Circuits and Systems Society in 2001, 2002, and 2003, respectively. He was the Vice President of Verband der Elektrotechnik (VDE), Elektronik und Informationstechnik e.V., in 2005 and 2006, the President of VDE in 2007 and 2008, and the Vice President in 2009 and 2010. His awards include the ITG Best Paper Award in 1988, the Mannesmann Mobilfunk (now Vodafone) Innovations Award in 1998, and the Award for Excellence in Teaching from the Bavarian Ministry for Science, Research and Art in 1998. From the IEEE Circuits and Systems Society, he received the Golden Jubilee Medal for Outstanding Contributions to the Society in 1999 and the Education Award in 2008. In 2008, he also received the Order of Merit of the Federal Republic of Germany and in 2009, he was an Elected Member of the German National Academy of Engineering Sciences (acatech). In 2013, he received an Honorary Doctorate and the Ring of Honor from VDE in 2014. From 2016 to 2019, he was a Full Professor at the Federal University of Ceará, Brazil.

Chapter 4

Luiz W. P. Biscainho was born in Rio de Janeiro, Brazil, in 1962. He received his degree in Electronic Engineering (magna cum laude) from the EE (now Poli) at UFRJ, Brazil, in 1985, and his MSc and DSc degrees in Electrical Engineering from COPPE at UFRJ in 1990 and 2000, respectively. Having worked in the telecommunications industry between 1985 and 1993, Dr. Biscainho is now Associate Professor at the Department of Electronic and Computer Engineering (DEL) of Poli and COPPE's Electrical Engineering Program (PEE) (of which he was Academic Coordinator in 2010/2011 and 2017/2018), at UFRJ. His research area is digital audio processing. He is currently a member of the IEEE, the Audio Engineering Society (AES), SBrT, and the Brazilian Computer Society (SBC).

Eduardo Alves da Silva was born in Rio de Janeiro, Brazil, in 1996. He received his degree in Electronic Engineering (Cum Laude) from the Escola Politécnica at UFRJ, Brazil, in 2020, and his MSc degree in Electrical Engineering from COPPE at UFRJ in 2022. He currently works at Halliburton (Brazil) and collaborates with his former research group at UFRJ. His research area is array signal processing and acoustics. He is currently a member of the IEEE, AES, and SBrT.

Chapter 5

Håkan Johansson received his MSc degree in computer science and his Licentiate, Doctoral, and Docent degrees in Electronics Systems from Linkoping University, Sweden, in 1995, 1997, 1998, and 2001, respectively. During 1998 and 1999 he held a postdoctoral position at Signal Processing Laboratory, Tampere University of Technology, Finland. He is currently Professor in Electronics Systems at the Department of Electrical Engineering of Linkoping University. His research encompasses design and implementation of efficient and flexible signal processing systems, mainly for communication applications. During the past decade, he has developed many different signal processing algorithms for various purposes, including filtering, sampling rate conversion, signal reconstruction, and parameter estimation. He has developed new estimation and compensation algorithms for errors in analog circuits such as compensation of mismatch errors in time-interleaved analog-to-digital converters and mixers. He is one of the founders of the company Signal Processing Devices Sweden AB, which sells this type of advanced signal processing apparatuses. He is the author or coauthor of four books and some 160 international journal and conference papers. He is the coauthor of three papers that have received best paper awards and he has authored one invited paper in *IEEE Transactions* and four invited chapters. He has served as Associate Editor for *IEEE Trans. on Circuits and Systems I* and *II*, *IEEE Trans. Signal Processing*, and *IEEE Signal Processing Letters*, and he is currently an Area Editor of the Elsevier *Digital Signal Processing* journal and a member of the IEEE Int. Symp. Circuits. Syst. DSP track committee.

Chapter 6

Lars Wanhammar was born in Vansbro, Sweden, on August 19, 1944. He has received the following degrees: Teknisk magister (teknisk fysik) in 1970, civilingenjör in 1980, teknisk doktor in 1981, and docent in 1986, in electrical engineering from Linköping University, Sweden. During 1964–1970 he worked at Televerket (Royal Swedish Telephone Board), Division of Communication, and during 1970–1971 as a Lecturer at the technical college in Norrköping. Since 1971 he has worked at Linksöping University, Department of Electrical Engineering, Division of Applied Electronics, as Assistant, then Research Assistant, from 1982 as Associate Professor (universitetslektor), and from 1997 as full Professor and Head of the Division of Electronics Systems, at the Department of Electrical Engineering, Linköping University, Sweden. Since 2011 he is Professor Emeritus. In addition, from 1995 to 2004 he worked as Adjunct Professor at the Norwegian Institute of Technology (NTNU) at the departments of Physical Electronics and Telecommunications. His research interests are primary theory and design of digital signal processing and telecommunication systems, particularly analog and digital filters, and discrete transforms as well as computational properties of digital signal processing algorithms, bit-serial, digit-serial, and distributed arithmetic, CAD tools, and globally asynchronous locally synchronous techniques for ULSI. He is the author or coauthor of five books on analog and digital filters, one on parallel processing in industrial real-time applications, and one on integrated digital signal processing circuits.

Ya Jun Yu received her BSc and MEng degrees in biomedical engineering from Zhejiang University, Hangzhou, China, in 1994 and 1997, respectively, and her PhD degree in electrical and computer engineering from the National University of Singapore, Singapore, in 2004. From 1997 to 1998, she was a Teaching Assistant at Zhejiang University. She joined the Department of Electrical and Computer Engineering, National University of Singapore as a Post Master Fellow in 1998 and remained in the same department as a Research Engineer until 2004. She joined the Temasek Laboratories at Nanyang Technological University as a Research Fellow in 2004. Since 2005, she has been with the School of Electrical and Electronic Engineering, Nanyang Technological University, Singapore, where she is currently an Assistant Professor. Her research interests include digital signal processing and VLSI circuits and systems design. She has served as an associate editor for *Circuits Systems and Signal Processing* and *IEEE Transactions on Circuits and Systems II* since 2009 and 2010, respectively.

Chapter 7

Fred Harris. Professor harris is at the University of California San Diego where he teaches and conducts research on Digital Signal Processing and Communication Systems. He holds 40 patents on digital receiver and DSP technology and lectures throughout the world on DSP applications. He consults for organizations requiring high performance, cost effective DSP solutions.

He has written some 260 journal and conference papers, the most well-known being his (9700 citations) 1978 paper "On the use of Windows for Harmonic Analysis with the Discrete Fourier Transform". He is the author of the book **Multirate Signal Processing for Communication Systems** and has contributed to several other DSP books. His special areas include Polyphase Filter Banks, Physical Layer Modem design, Synchronizing Digital Modems and Spectral Estimation.

He was the Technical and General Chair respectively of the 1990 and 1991 Asilomar Conference on Signals, Systems, and Computers, was Technical Chair of the 2003 Software Defined Radio Conference, of the 2006 Wireless Personal Multimedia Conference, of the DSP-2009, DSP-2013 Conferences and of the SDR-WinnComm 2015 Conference. He became a Fellow of the IEEE in 2003, cited for contributions of DSP to communications systems. In 2006 he received the Software Defined Radio Forum's "Industry Achievement Award". He received the DSP-2018 conference's commemorative plaque with the citation: *We wish to recognize and pay tribute to fred harris for his pioneering contributions to digital signal processing algorithmic design and implementation, and his visionary and distinguished service to the Signal Processing Community.*

The spelling of his name with all lower case letters is a source of distress for typists and spell checkers. A child at heart, he collects toy trains, old slide-rules, and vintage gyroscopes.

Elettra Venosa received her "Laurea" (BS/MS) degree (summa cum laude) in Electrical Engineering in January 2007 from Seconda Università degli Studi di Napoli, Italy. From January 2007 to November 2007 she was a researcher at the Italian National Inter-University Consortium for Telecommunications. In January 2011, she received her PhD in Telecommunication/DSP from Seconda Università degli Studi di Napoli. From June 2008 to September 2008 she worked as a project manager for Kiranet – ICT Research Center – to develop an advanced radio identification system for avionics, in collaboration with the Italian Center for Aerospace Research (CIRA). From April 2009 to September 2009 she worked as associate researcher at the Communications and Signal Processing Laboratory (CSPL) in the Department of Electrical and Computer Engineering at Drexel University, Philadelphia, where she worked on sparse sampling techniques for software defined radio receivers. From January 2011 to July 2012 she worked as principal system engineer at IQ-Analog, CA, developing algorithms for digital correction in TI-ADCs. From August 2012 to December 2013 she was associate researcher at Qualcomm, CA. Her aim was to improve the current commercial modem architectures. Currently, she is working, as a postdoctoral researcher, on multirate signal processing for software defined radios in the Department of Electrical Engineering at San Diego State University, San Diego, CA, USA, where she also teaches graduate and undergraduate courses. She is also working as a software defined radio engineer at Space Micro, CA. She is the author of more than 40 scientific publications on software defined radios and of the book "*Software Radio: Sampling Rate Selection Design and Synchronization*."

Xiaofei Chen received his Bachelor's degree in Electrical Engineering in June 2006 from Xi'an University of Posts & Telecommunications, China. In December 2008, he received his Master's degree at the Electrical & Computer Engineering department, San Diego State University, USA. In March 2009, he joined the Joint Doctoral Program between San Diego State University and the University of California, San Diego. His current research interests are in the area of multirate signal processing and software defined radio.

Richard Bell received the B.S. in Physics and Mathematics from SUNY Binghamton in 2005 and M.S. in Electrical Engineering from SUNY Buffalo in 2009. Richard is currently an Electrical and Computer Engineering Ph.D. candidate at UC San Diego with a research focus on detecting and localizing unknown RF emitters. Richard enjoys seeing ideas progress from infancy to instantiation.

From 2009 to 2021 he was with the Naval Information Warfare Center (NIWC) Pacific in San Diego, CA where he contributed to the modernization of the Talon IV explosive ordinance removal robot, the modernization of the Link 16 tactical communications waveform, as well as leading work on the passive localization of RF emitters. In 2022 he joined JASR Systems in San Diego where he focuses on the development of a state-of-the-art wideband spectrum sensing platform.

Srivatsan Rajagopal completed the PhD in theoretical physics from MIT in 2019. Following that, he has had post-doctoral appointments in Physics at UIUC, and in Electrical Engineering at UCSD, where he is situated currently. He is broadly interested in applying machine learning and statistical signal processing methods to problems in wireless communication. Currently he is working on detecting and characterizing RF anomalies.

Radhika Mathuria received her Bachelors in Technology in Electronics and Communication Engineering from the National Institute of Technology, Surat, India in 2021. She is currently a Masters student at UC San Diego majoring in Signal and Image Processing.

Dinesh Bharadia is associate professor in ECE deparment at the University of California San Diego since July 2022. He received early promotion to tenured professorship, held Assistant Professorship for brief four years from 2018 – 2022. He received his Ph.D. from Stanford University in 2016 and was a Postdoctoral Associate at MIT. His research interests include advancing the theory and design of modern wireless communication, wireless sensing, and sensor design with applications to privacy, security, robotics, health, and everyday life. Much of his research has inspired new research areas for border communities: communication theory, circuits, RFIC, and robotics. Much of his research work has been translated into the startup and commercial products (Haila, Kumu Networks, Totemic Labs).

Specifically, he built a prototype of a radio that invalidated a long-held assumption in wireless that radios cannot transmit and receive simultaneously on the same frequency, which inspired research on this topic from different communities (communication theory to RFIC). From 2013 to 2015, he worked to commercialize his research on full-duplex radios, building a product that underwent successful field trials at Tier 1 network providers worldwide like Deutsche Telekom and SK Telecom. This product is currently under deployment. He serves as a technical advisor for multiple startups.

In recognition of his work, Dinesh was named to Forbes 30 under 30 for the science category worldwide list. Dinesh was named a Marconi Young Scholar for outstanding wireless research and was awarded the Michael Dukakis Leadership award. He was also named one of the top 35 Innovators under 35 worldwide by MIT Technology Review in 2016.

Chapter 8

Trac D. Tran received his BS and MS degrees from the Massachusetts Institute of Technology, Cambridge, in 1993 and 1994, respectively, and his PhD degree from the University of Wisconsin, Madison, in 1998, all in electrical engineering. In July 1998, he joined the Department of Electrical and Computer Engineering, Johns Hopkins University, Baltimore, MD, where he currently holds the rank of Professor. His research interests are in the field of signal processing, particularly in sparse representation, sparse recovery, sampling, multirate systems, filter banks, transforms, wavelets, and their applications in signal analysis, compression, processing, and communications. His pioneering research on integer-coefficient transforms and pre-/postfiltering operators has been adopted as critical components of Microsoft Windows Media Video 9 and JPEG XR—the latest international still-image compression standard ISO/IEC 29199–2. He is currently a regular consultant for the US Army Research Laboratory, Adelphi, MD. He was the codirector (with Prof. J. L. Prince) of the 33rd Annual Conference on Information Sciences and Systems (CISS'99), Baltimore, in March 1999. In the summer of 2002, he was an ASEE/ONR Summer Faculty Research Fellow at the Naval Air Warfare Center Weapons Division (NAWCWD), China Lake, CA. He has served as Associate Editor of *IEEE Transactions on Signal Processing* as well as *IEEE Transactions on Image Processing*. He was a former member of the IEEE Technical Committee on Signal Processing Theory and Methods

(SPTM TC) and is a current member of the IEEE Image Video and Multidimensional Signal Processing (IVMSP) Technical Committee. He received the NSF CAREER award in 2001, the William H. Huggins Excellence in Teaching Award from Johns Hopkins University in 2007, and the Capers and Marion McDonald Award for Excellence in Mentoring and Advising in 2009.

Chapter 9

Yufang Bao has been an Assistant Professor in the Department of Mathematics and Computer Science at UNC Fayetteville State University (UNCFSU) since 2007. She is also a scholar of the Center of Defense and Homeland Security (CDHS) at UNCFSU. She received her first PhD degree in probability/statistics from Beijing Normal University (BNU), Beijing, China, and her second PhD degree in Electrical Engineering from North Carolina State University (NCSU), Raleigh, NC. Her research interest was in probability/statistics. She subsequently directed her research into the area of applying mathematics in signal/image processing and analysis. Her contributions in signal/image processing included algorithm development that bridged stochastic diffusion and multiscale wavelet theory with scale space analysis methods for image denoising and segmentation. Between 2002 and 2007, she worked at the VA Center, UCSF, CA and then at the University of Miami, School of Medicine, FL, both as a research scientist focusing on statistical image reconstruction in frequency domain with MR spectroscopy imaging and parallel MR image reconstruction using mathematical modeling. Currently, her research interests are in applying mathematics to statistical digital signal/image processing and analysis, mathematical modeling, and their applications.

Hamid Krim received his degrees in Electrical and Computer Engineering from the University of Washington and Northeastern University. He was a Member of Technical Staff at AT&T Bell Labs, where he has conducted research and development in the areas of telephony and digital communication systems/subsystems. Following an NSF postdoctoral fellowship at Foreign Centers of Excellence, LSS/University of Orsay, Paris, France, he joined the Laboratory for Information and Decision Systems, Massachusetts Institute of Technology, Cambridge, MA as a Research Scientist, where he was performing and supervising research. He is presently Professor of Electrical Engineering in the Electrical and Computer Engineering Department, North Carolina State University, Raleigh, leading the Vision, Information and Statistical Signal Theories and Applications group. His research interests are in statistical signal and image analysis and mathematical modeling with a keen emphasis on applied problems in classification and recognition using geometric and topological tools.

Chapter 10

Lisandro Lovisolo was born in Neuquen, Argentina, but considers himself Brazilian. He received his Electronics Engineering degree from the Universidade Federal do Rio de Janeiro (UFRJ), his MSc degree (in 2001), and his DSc degree, all in Electrical Engineering, from COPPE/UFRJ. Since 2003 he has been with the Department of Electronics and Communications Engineering (the undergraduate department), UERJ. He has also been with the Postgraduate in Electronics Program, since 2008. His research interests lie in the fields of digital signal and image processing and communications.

Eduardo A. B. da Silva was born in Rio de Janeiro, Brazil. He received his degree in Electronics Engineering from the Instituto Militar de Engenharia (IME), Brazil, in 1984, his MSc degree in Electrical Engineering from COPPE/UFRJ in 1990, and his PhD degree in Electronics from the University of Essex, England, in 1995. In 1987 and 1988 he worked at the Department of Electrical Engineering at Instituto Militar de Engenharia, Rio de Janeiro, Brazil. Since 1989 he has worked at the Department of Electronics Engineering (the undergraduate department), UFRJ. He has also been with the Department of Electrical Engineering (the graduate studies department), COPPE/UFRJ, since 1996. His research interests lie in the fields of digital signal and image processing, especially signal compression, digital television, wavelet transforms, mathematical morphology, and applications to telecommunications.

Allan Freitas da Silva was born in Rio de Janeiro, Brazil. He received his Engineering degree in Electronic and Computer Engineering from the UFRJ in 2013, his MSc degree in Electrical Engineering from COPPE/UFRJ in 2015, and his DSc at the same institution in 2019. Between November 2017 and October 2018, he received a Doctoral Exchange Program grant (SWE) from CNPq, where he integrated the Integration: from Material to Systems (IMS) laboratory at the University of Bordeaux. Since 2022, he is a postdoctoral researcher at Institute of Systems and Robotics in the University of Coimbra (ISR-UC). His research interests include the areas of computer vision and video and image processing.

Chapter 11

Ryan M. Corey received the B.S.E. degree in Electrical Engineering summa cum laude from Princeton University and the M.S. and Ph.D. degrees in Electrical and Computer Engineering from the University of Illinois Urbana-Champaign, where he also completed a postdoctoral fellowship. While in Urbana, he helped to found and lead the Illinois Augmented Listening Laboratory, an interdisciplinary research group that has trained dozens of students in audio-related technologies. He currently holds a dual appointment as Assistant Professor in Electrical and Computer Engineering at the University of Illinois Chicago and Research Scientist at the Discovery Partners

Institute, a Chicago-based innovation hub operated by the University of Illinois System and dedicated to equitable economic development. A hearing aid user for most of his life, Professor Corey's research focuses on audio and acoustic signal processing for assistive and augmentative hearing applications. His research emphasizes spatial signal processing, including microphone arrays, distributed sensor networks, and wireless audio systems. He is the recipient of the NSF Graduate Research Fellowship, the Intelligence Community Postdoctoral Research Fellowship, the Microsoft Research Dissertation Grant, the Microsoft AI for Accessibility Grant, and the inaugural ORISE Future of Science postdoctoral award. He is the author of a high-impact study on the acoustic effects of face masks during the COVID-19 pandemic and has also received best-paper awards for his work on deformable microphone arrays and dynamic range compression algorithms.

Suleyman Serdar Kozat received his BS degree with full scholarship and high honors from Bilkent University, Turkey. He received his MS and PhD degrees in Electrical and Computer Engineering from the University of Illinois at Urbana Champaign, Urbana, IL, in 2001 and 2004, respectively. After graduation, he joined IBM Research, T.J. Watson Research Center, Yorktown, NY as a Research Staff Member in the Pervasive Speech Technologies Group, where he focused on problems related to statistical signal processing and machine learning. While doing his PhD, he was also working as a Research Associate at Microsoft Research, Redmond, WA, in the Cryptography and Anti-Piracy Group. He holds several patent applications for his works performed in IBM Research and Microsoft Research. Currently, he is an Assistant Professor at the Electrical and Electronics Engineering department, Koc University, Turkey. He coauthored more than 50 papers in refereed high-impact journals and conference proceedings and has several patent applications. Overall, his research interests include intelligent systems, adaptive filtering for smart data analytics, online learning, and machine learning algorithms for signal processing. He has been serving as an Associate Editor for *IEEE Transactions on Signal Processing* and he is a Senior Member of the IEEE. He has been awarded the IBM Faculty Award by IBM Research in 2011, the Outstanding Faculty Award by Koc University in 2011, the Outstanding Young Researcher Award by the Turkish National Academy of Sciences in 2010, and the ODTU Prof. Dr. Mustafa N. Parlar Research Encouragement Award in 2011, and he holds the Career Award from the Scientific Research Council of Turkey, 2009. He has won several scholarships and medals in international and national science and math competitions.

Andrew C. Singer received his SB, SM, and PhD degrees, all in Electrical Engineering and Computer Science, from the Massachusetts Institute of Technology. Since 1998, he has worked for the Department of Electrical and Computer Engineering at the University of Illinois at Urbana-Champaign, where he is currently a Professor in the Electrical and Computer Engineering department and the Coordinated Science Laboratory. During the academic year 1996, he was a Postdoctoral Research Affiliate in the Research Laboratory of Electronics at MIT. From 1996 to 1998, he was a Research Scientist at Sanders, a Lockheed Martin Company in Manchester, New Hampshire, where he designed algorithms, architectures, and systems for a variety of DOD applications. His research interests include signal processing and communication systems. He was a Hughes Aircraft Masters Fellow and was the recipient of the Harold L. Hazen Memorial Award for excellence in teaching in 1991. In 2000, he received the National Science Foundation CAREER Award, in 2001 he received the Xerox Faculty Research Award, and in 2002 he was named a Willett Faculty Scholar. He has served as an Associate Editor for *IEEE Transactions on Signal Processing* and is a member of the MIT Educational Council and of Eta Kappa Nu and Tau Beta Pi. He is a Fellow of the IEEE. In 2005, he was appointed as the Director of the Technology Entrepreneur Center (TEC) in the College of Engineering. He also cofounded Intersymbol Communications, Inc., a venture-funded fabless semiconductor IC company, based in Champaign Illinois. A developer of signal processing enhanced chips for ultra-high speed optical communications systems, Intersymbol was acquired by Finisar Corporation (NASD:FNSR) in 2007. He serves on the board of directors of a number of technology companies and as an expert witness for electronics, communications, and circuit-related technologies.

Chapter 12

Vítor H. Nascimento was born in São Paulo, Brazil. He obtained his BS and MS degrees in Electrical Engineering from the University of São Paulo in 1989 and 1992, respectively, and his PhD degree from the University of California, Los Angeles, in 1999. From 1990 to 1994 he was a Lecturer at the University of São Paulo, and in 1999 he joined the faculty at the same school, where he is now an Associate Professor. One of his papers received the 2002 IEEE SPS Best Paper Award. He served as an Associate Editor for *IEEE Signal Processing Letters* from 2003 to 2005, for *IEEE Transactions on Signal Processing* from 2005 to 2008, and for the *EURASIP Journal on Advances in Signal Processing* from 2006 to 2009, and he was a member of the IEEE-SPS Signal Processing Theory and Methods Technical Committee from 2007 to 2012. Since 2010 he is the chair of the São Paulo SPS Chapter. His research interests include signal processing theory and applications, robust and nonlinear estimation, and applied linear algebra.

Magno T. M. Silva was born in São Sebastião do Paraíso, Brazil. He received his BS, MS, and PhD degrees, all in electrical engineering, from Escola Politécnica, University of São Paulo, São Paulo, Brazil, in 1999, 2001, and 2005, respectively. From February 2005 to July 2006, he was an Assistant Professor at Mackenzie Presbyterian University, São Paulo, Brazil. Since August 2006, he has been with the Department of Electronic Systems Engineering at Escola Politécnica, University of São Paulo, where he is currently an Assistant Professor. From January to July 2012, he worked as a Postdoctoral Researcher at Universidad Carlos III de Madrid, Leganés, Spain. His research interests include linear and nonlinear adaptive filtering.

Chapter 13

Marcele O. K. Mendonça (See Chapter 1)

Sergio L. Netto was born in Rio de Janeiro, Brazil. He received his BSc degree (cum laude) from UFRJ, Brazil, in 1991, his MSc degree from COPPE/UFRJ in 1992, and his PhD degree from the

University of Victoria, BC, Canada, in 1996, all in electrical engineering. Since 1997, he has been with the Department of Electronics and Computer Engineering, Poli/UFRJ. He is the coauthor (with P. S. R. Diniz and E. A. B. da Silva) of *Digital Signal Processing: System Analysis and Design* (Cambridge University Press, 2nd ed., 2010) and (with L. P. Cinelli, M. A. Marins, and E. A. B. da Silva) of *Variational Methods for Machine Learning with Applications to Deep Networks* (Springer, 2021). His research and teaching interests lie in the areas of digital signal processing, adaptive filtering, speech processing, information theory, computer vision, and machine learning.

Paulo S. R. Diniz (See Chapter 1)

Sergios Theodoridis currently serves as a Distinguished Professor at Aalborg University, Denmark. He is also Professor Emeritus of Signal Processing and Machine Learning with the Department of Informatics and Telecommunications of the National and Kapodistrian University of Athens, Greece. His research interests lie in the areas of Machine Learning and Signal Processing.

He is the author of the book "*Machine Learning: A Bayesian and Optimization Perspective*," Academic Press, 2nd Ed., 2020, coauthor of the best-selling book "*Pattern Recognition*," Academic Press, 4th ed. 2009, coauthor of the book "*Introduction to Pattern Recognition: A MATLAB Approach*," Academic Press, 2010, and coeditor of the book "*Efficient Algorithms for Signal Processing and System Identification*," Prentice Hall 1993.

He is coauthor of seven papers that have received Best Paper Awards, including the 2014 IEEE Signal Processing Magazine Best Paper Award and the 2009 IEEE Computational Intelligence Society Transactions on Neural Networks Outstanding Paper Award.

He is the recipient of the 2021 IEEE Signal Processing Society (SPS) Norbert Wiener Award, which is the IEEE SP Society's highest honor, the 2017 EURASIP Athanasios Papoulis Award, the 2014 IEEE SPS Carl Friedrich Gauss Education Award, and the 2014 EURASIP Meritorious Service Award.

He currently serves as Chairman of the IEEE SP Society Awards Committee. He has served as Vice President for the IEEE Signal Processing Society and as President of EURASIP. He has served as Editor-in-Chief for *IEEE Transactions on Signal Processing*.

He is Fellow of IET, a Corresponding Fellow of the Royal Society of Edinburgh (RSE), a Fellow of EURASIP, and a Life Fellow of IEEE.

Chapter 14

Wallace A. Martins received his PhD degree in electrical engineering from UFRJ, Rio de Janeiro, Brazil, in 2011. He was a Research Visitor at the University of Notre Dame, USA, in 2008, Université de Lille 1, France, in 2016, and Universidad de Alcalá, Spain, in 2018. He was an Associate Professor at the Department of Electronics and Computer Engineering (DEL/Poli) and the Electrical Engineering Program (PEE/COPPE), UFRJ, from 2013 to 2022. He was Academic Coordinator and Deputy Department Chairman (DEL/Poli), UFRJ, from 2016 to 2017. He is currently a Research Scientist working with SnT, University of Luxembourg. He is a member (Associate Editor) of the Editorial Boards of *IEEE Signal Processing Letters* and the *EURASIP Journal on Advances in Signal Processing*. He was the recipient of the Best Student Paper Award from EURASIP at EUSIPCO-2009, Glasgow, Scotland, the 2011 Best Brazilian D.Sc. Dissertation Award from Capes, and the Best Paper Award at SBrT-2020, Florianópolis, Brazil. His research interests include digital signal processing and telecommunications, with focus on equalization and beamforming/precoding for terrestrial and nonterrestrial (satellite) wireless communications.

Juliano B. Lima was born in Brazil. He received his MSc and PhD degrees in electrical engineering from the Federal University of Pernambuco (UFPE), Brazil, in 2004 and 2008, respectively. He is an IEEE Senior Member, and since 2015 he has been a Research Productivity Fellow awarded by the Conselho Nacional de Desenvolvimento Científico e Tecnológico. He is currently an Associate Professor at the Department of Electronics and Systems, UFPE. His main research interests include discrete and number-theoretic transforms and their applications in digital signal processing, communications, and cryptography.

Cédric Richard is a Full Professor at Université Côte d'Azur, France, and holder of the Senior Chair "Smart cities and Secure territories" at the Interdisciplinary Institute for Artificial Intelligence (3IA). His research interests include statistical signal processing and machine learning. He is the author of over 350 publications. He has been the Director-at-Large of the Region Europe, Middle East, Africa of the IEEE Signal Processing Society (IEEE-SPS), and a Member of the Board of Governors of the IEEE-SPS. He is currently the Chair of SPTM TC of the IEEE-SPS.

Symeon Chatzinotas has been a Visiting Professor at the University of Parma, Italy, and was involved in numerous research and development projects for NCSR Demokritos, CERTH Hellas and CCSR, and the University of Surrey. He is currently a Full Professor and the Head of the SIGCOM Research Group, SnT, University of Luxembourg. He has coauthored more than 700 technical papers in refereed international journals, conferences, and scientific books. He is coordinating the research activities on communications and networking across a group of 80 researchers, acting as a PI for more than 40 projects, and a main representative for 3GPP, ETSI, and DVB. He was a corecipient of the 2014 IEEE Distinguished Contributions to Satellite Communications Award and the Best Paper Awards at WCNC, 5GWF, EURASIP JWCN, CROWNCOM, and ICSSC. He is also serving in the editorial board for *IEEE Transactions on Communications*, *IEEE Open Journal of Vehicular Technology*, and the *International Journal of Satellite Communications and Networking*.

Chapter 15

Yannis Panagakis is an Associate Professor of machine learning and the Director of the MSc programme in Language Technology in the Department of Informatics and Telecommunications at the National and Kapodistrian University of Athens, Greece. Previously, he held positions at the Samsung AI Research Centre in Cambridge and Imperial College London, UK, where he is now an Honorary Research Fellow. He obtained his PhD and MSc degrees from Aristotle University of Thessaloniki and his BSc degree from the University of Athens. His research focuses on artificial intelligence (*AI*) algorithms for information processing tasks arising in a broad spectrum of application domains. He is particularly interested in mathematical aspects of deep learning related to generalization, compression, robustness, and interpretation and their interface with tensor methods. Yannis has published over 90 articles on the above topics in leading journals (e.g., *IEEE TPAMI*, *IJCV*, *JMLR*) and top-tier conferences (e.g., CVPR, NeurIPS, ICML, ICCV), and he has received several research grants and awards.

Jean Kossaifi is a Senior Research Scientist at NVIDIA. Prior to this, he was a Research Scientist at the Samsung AI Center in Cambridge. He is a leading expert on tensor methods for machine learning and deep learning and a major contributor to facial affect estimation in naturalistic conditions, a field which bridges the gap between computer vision and machine learning. He created TensorLy to make tensor methods accessible for all. Jean received his PhD and MSc degrees from Imperial College London, where he worked with Prof. Maja Pantic. He also holds a French Engineering Diploma/MSc in Applied Mathematics, Computing and Finance and obtained a BSc degree in advanced mathematics in parallel.

Grigorios G. Chrysos is a postdoctoral researcher at EPFL following the completion of his PhD at Imperial College London. His research interests focus on reliable machine learning, and in particular on comprehending the inductive bias and out-of-distribution performance of deep networks. His recent work has been published in top-tier conferences (NeurIPS, ICLR, CVPR, ICML) and prestigious journals (*T-PAMI*, *IJCV*, *T-IP*). Grigorios has co-organized several workshops and tutorials (CVPR, ICCV, AAAI), while he has been recognized as an outstanding reviewer in journals and conferences (NeurIPS'22, ICML'21/22, ICLR'22).

James Oldfield is currently a PhD student at Queen Mary University of London. Previously, he was a research intern at The Cyprus Institute. He has published at some of the most authoritative venues and journals in the fields of machine learning and computer vision, including *ICLR*, *CVPR*, *IJCV*, and the *Proceedings of the IEEE*. His recent research focuses on controllable and interpretable generative models of visual data.

Taylor Lee Patti is a Research Scientist at NVIDIA working on Quantum Algorithms and Simulation. She holds received her PhD and Masters degrees in Theoretical Physics from Harvard University, where she was both an NSF-GRFP and a Goldhaber recipient. Taylor graduated from Chapman University with degrees in Physics and Computational Science, Mathematics, and Spanish.

Mihalis A. Nicolaou is currently an Assistant Professor at the Computation-based Science and Technology Center at the Cyprus Institute. He received his BSc degree from the University of Athens, Greece, and his MSc and PhD degrees from the Department of Computing, Imperial College London, UK, where he was also postdoctoral researcher. His research interests lie in machine learning, computer vision, and signal processing, focusing on the analysis and interpretation of multimodal, high-dimensional data, in a wide range of interdisciplinary applications. He has published over 70 research articles and has received several best paper awards.

Anima Anandkumar is a Bren Professor at Caltech and Director of ML Research at NVIDIA. She was previously a Principal Scientist at Amazon Web Services. She has received several honors such as an Alfred. P. Sloan Fellowship, an NSF Career Award, Young investigator awards from DoD, and Faculty Fellowships from Microsoft, Google, Facebook, and Adobe. She is part of the World Economic Forum's Expert Network. She is passionate about designing principled AI algorithms and applying them in interdisciplinary applications. Her research focus is on unsupervised AI, optimization, and tensor methods.

Stefanos Zafeiriou is currently a Professor of machine learning and computer vision at the Department of Computing, Imperial College London, London, UK. From 2016 to 2020, he was a Distinguishing Research Fellow at the University of Oulu, Oulu, Finland, under the Finnish Distinguishing Professor Program. He cofounded and exited two startups, including Facesoft and Ariel Artificial Intelligence. He has coauthored more than 80 journal articles, mainly on novel machine learning methodologies applied to various domains published in the most prestigious journals in his field of research, including *IEEE Transactions on Pattern Analysis and Machine Intelligence* and the *International Journal of Computer Vision*, and many articles in top conferences, including Computer Vision and Pattern Recognition (CVPR), the International Conference on Computer Vision (ICCV), the European Conference on Computer Vision (ECCV), and the International Conference on Machine Learning (ICML). He has more than 17K+ citations to his work and an h-index of 60. Prof. Zafeiriou is an Engineering and Physical Sciences Research Council (EPSRC) Early Career Research Fellow. His students are frequent recipients of very prestigious and highly competitive fellowships, such as the Google Fellowship, the Intel Fellowship, and the Qualcomm Fellowship. He was a recipient of the Prestigious Junior Research Fellowships from Imperial College London in 2011, the President's Medal for Excellence in Research Supervision for 2016, the Google Faculty Research Awards, and the

Amazon Web Services (AWS) Machine Learning Research Award. He was the General Chair of the British Machine Vision Conference (BMVC) in 2017. He is currently the Area Chair in the top venues of his field. He was an Associate Editor and the Guest Editor in more than eight journals. He was the Guest Editor of more than 10 journal special issues and co-organized more than 20 workshops or special sessions on specialized computer vision and machine learning topics in top venues, including CVPR, the International Conference on Automatic Face and Gesture Recognition (FG), ICCV, ECCV, and Neural Information Processing Systems (NeurIPS).

Chapter 16

José Vinícius de Miranda Cardoso was born in Campina Grande, Brazil. He received his Bachelor degree in Electrical Engineering from the Universidade Federal de Campina Grande in 2019. Since then, he has been pursuing his PhD degree in Electronic and Computer Engineering at the Hong Kong University of Science and Technology (HKUST) in Clear Water Bay, Hong Kong SAR China. Previously, he was a scientific software engineering intern for one year at the National Aeronautics and Space Administration (NASA) Ames Research Center, in Mountain View, California, where he worked for the K2 mission as part of the Guest Observer Office. Prior to that, he was a software developer for AstroPy as part of the Google Summer of Code program. He spent a summer as a guest researcher in the Center for Nanoscale Science and Technology (CNST) at the National Institute of Standards and Technology (NIST) in Gaithersburg, Maryland. He was a recipient of the NeurIPS 2022 Scholars Award. His research interests include statistical graph learning and optimization algorithms.

Jiaxi Ying received his PhD degree from the Department of Electronic and Computer Engineering, the HKUST, Hong Kong, in 2022. He is currently a postdoctoral fellow at the Department of Mathematics, the HKUST. He was the recipient of the HKUST RedBird PhD Scholarship Program. His research interests are mainly in the intersection of optimization, machine learning, signal processing, and statistics.

Daniel P. Palomar (S'99-M'03-SM'08-F'12) received his Electrical Engineering and PhD degrees from the Technical University of Catalonia (UPC), Barcelona, Spain, in 1998 and 2003, respectively, and was a Fulbright Scholar at Princeton University during 2004–2006. He is a Professor in the Department of Electronic & Computer Engineering and in the Department of Industrial Engineering & Decision Analytics at HKUST, Hong Kong, which he joined in 2006. He had previously held several research appointments, namely, at King's College London (KCL), London, UK; Stanford University, Stanford, CA; the Telecommunications Technological Center of Catalonia (CTTC), Barcelona, Spain; the Royal Institute of Technology (KTH), Stockholm, Sweden; University of Rome "La Sapienza," Rome, Italy; and Princeton University, Princeton, NJ. His current research interests include applications of optimization theory, graph methods, and signal processing in financial systems and big data analytics.

Dr. Palomar is an IEEE Fellow, a recipient of a 2004/06 Fulbright Research Fellowship, the 2004, 2015, and 2020 (coauthor) Young Author Best Paper Awards by the IEEE Signal Processing Society, the 2015-16 HKUST Excellence Research Award, the 2002/03 Best Ph.D. Prize in Information Technologies and Communications by the Technical University of Catalonia (UPC), the 2002/03 Rosina

Ribalta first prize for the Best Doctoral Thesis in Information Technologies and Communications by the Epson Foundation, and the 2004 prize for the best Doctoral Thesis in Advanced Mobile Communications by Vodafone Foundation and COIT. He has been a Guest Editor of the *IEEE Journal of Selected Topics in Signal Processing* 2016 Special Issue on "Financial Signal Processing and Machine Learning for Electronic Trading," an Associate Editor of IEEE Transactions on Information Theory and of IEEE Transactions on Signal Processing, a Guest Editor of the IEEE Signal Processing Magazine 2010 Special Issue on "Convex Optimization for Signal Processing," the *IEEE Journal on Selected Areas in Communications* 2008 Special Issue on "Game Theory in Communication Systems," and the *IEEE Journal on Selected Areas in Communications* 2007 Special Issue on "Optimization of MIMO Transceivers for Realistic Communication Networks."

Chapter 17

Srinjoy Das received his BE degree (Hons.) in electrical and electronics engineering, his MSc degree (Hons.) in physics from the Birla Institute of Technology and Science, Pilani, an MS degree in electrical engineering from the University of California at Irvine, and an MS degree in statistics and his PhD degree in electrical engineering from the University of California at San Diego. He has previously worked for Qualcomm Inc. and Atheros Communications, where he led the development and design of embedded low-power devices for several real-time applications, including 3G and 4G cellular communications and Global Positioning System technologies. He is currently an Assistant Professor of Data Science at the School of Mathematical and Data Sciences, West Virginia University. His research interests and publications are focused on algorithms for predictive inference on time series and random fields, novel distance measures for accurate inference on time series, and generative deep learning algorithms for efficient implementation on resource-constrained edge computing platforms.

Kenneth Kreutz-Delgado is an Emeritus Professor at the University of California San Diego (UCSD) and Senior Technical Fellow at Pattern Computer Inc. After receiving an MS in Physics and a PhD in Engineering Systems Science from UCSD, he worked for five years as a researcher in intelligent robotics at the NASA/Caltech Jet Propulsion Laboratory. Thereafter he joined the faculty at UCSD engaging in research in a variety of areas including nonlinear robot kinematics and control; sparse signal processing, compressive sensing and dictionary learning; communication systems; EEG and fMRI brain imaging; computational aspects of stochastic spiking neural networks; and, most recently, implementation of machine learning algorithms for efficient IoT and edge-of-the-cloud applications. Currently he is engaged with the development of explainable machine learning algorithms ("XAI"), an interdisciplinary endeavor that lies at the interface of mathematical statistics; differential and Riemannian geometry; algebraic and differential topology; algebraic and spectral graph theory; machine learning; and statistical physics. Prof. Kreutz-Delgado is a Life Fellow of the IEEE.

Igor Federov is an AI Research Scientist at Meta. From 2018 to 2022, he was a member of the ARM Machine Learning Research Laboratory. He received his PhD in 2018 in the Electrical and Computer Engineering department at the UCSD, focusing on Bayesian methods for dictionary learning and sparse signal recovery. Igor received his MS and BS degrees in Electrical and Computer Engineering at the University of Illinois Urbana-Champaign in 2014 and 2012, respectively.

Bhaskar D. Rao received his BTech degree in electronics and electrical communication engineering from the Indian Institute of Technology, Kharagpur, India, in 1979, and his MS and PhD degrees from the University of Southern California, Los Angeles, in 1981 and 1983, respectively. Since 1983, he has been with the University of California at San Diego, La Jolla, where he is currently a Distinguished Professor in the Electrical and Computer Engineering department. He is the holder of the Ericsson endowed chair in Wireless Access Networks and was the Director of the Center for Wireless Communications (2008–2011). Prof. Rao's interests are in the areas of digital signal processing, estimation theory, and optimization theory, with applications to digital communications, speech signal processing, and biomedical signal processing. For his contributions to the statistical analysis of subspace algorithms for harmonic retrieval, Prof. Rao was elected fellow of the IEEE in 2000. His work has received several paper awards: the 2013 best paper award in fall 2013, the IEEE Vehicular Technology Conference for the paper "Multicell Random Beamforming with CDF-based Scheduling: Exact Rate and Scaling Laws," by Yichao Huang and Bhaskar D Rao; the 2012 Signal Processing Society (SPS) best paper award for the paper "An Empirical Bayesian Strategy for Solving the Simultaneous Sparse Approximation Problem," by David P. Wipf and Bhaskar D. Rao published in *IEEE Transaction on Signal Processing*, Volume: 55, No. 7, July 2007; and the 2008 Stephen O. Rice Prize paper award in the field of communication systems for the paper "Network Duality for Multiuser MIMO Beamforming Networks and Applications," by B. Song, R. L. Cruz, and B. D. Rao, that appeared in *IEEE Transactions on Communications*, Vol. 55, No. 3, March 2007, pp. 618–630 (http://www.comsoc.org/awards/rice.html); among others. Prof. Rao is also the recipient of the 2016 IEEE Signal Processing Society Technical Achievement Award.

Signal processing and machine learning theory

Signal processing and machine learning are vital areas of knowledge that find applications in virtually all aspects of modern life. Indeed, human beings have been employing signal processing tools for centuries without being aware of them. In recent years the affordable availability of computational and storage resources, in addition to theoretical advances, allowed a booming activity in both related areas. In present days the younger generation might not be able to understand how one can live without carrying a mobile phone, travel long distances without an almost self-piloted airplane, explore other planets without human presence, and utilize a medical facility without a wide range of diagnostic and intervention equipment.

Signal processing consists of mapping or transforming information-bearing signals into another form of signals at the output, aiming at some application benefits. This mapping defines a continuous or analog system if it involves functions representing the input and output signals. On the other hand, the system is discrete or digital if sequences of numbers represent its input and output signals.

Signal processing theory includes a wide range of tools which have found applications in uncountable areas, such as bioengineering, communications, control, surveillance, environment monitoring, oceanography, and astronomy, to mention a few. Many of these applications benefited from machine learning theory by solving some unprecedented problems and enhancing existing solutions. Machine learning basically consists of fitting models to the provided data. There are many algorithms to perform the desired fitting, but it is worth mentioning that data quality is essential for a successful outcome. In addition, the book includes many topics on machine learning, highlighting some connections with many classical methods related to learning from data. Indeed, machine learning is a framework for many disciplines dealing with tools to learn from data and perform modeling tasks such as prediction and classification.

This book includes an overview of classical signal processing and machine learning tools, whose basic concepts are encountered in numerous textbooks available in the literature, as well as topics usually not covered in the books available in the market. As a result, we believe that students and professionals can benefit from reading many parts of this book, if not all, in order to deepen and widen their current knowledge as well as to start exploiting new ideas. Another objective of this book is to offer a wide range of self-contained topics in signal processing and machine learning which can be used as complementary to existing standard courses and, above all, serve as the basis for advanced courses. The breadth of subjects discussed provides ample coverage, so that many configurations are possible for an entire course on special topics in signal processing and machine learning according to the selection of chapters.

The broad scope of topics is inspired by the fact that several solutions for practical problems start by acquiring some continuous-time signals, followed by digitizing and storing them to perform further processing. The set of tasks may entail signal filtering, modeling, prediction, and classification, among others.

The theories of signal processing and machine learning have been developing for over five decades at a fast pace. The following chapters will cover many of the classical tools widely employed in applications that we perceive in our daily life through the use of mobile phones, media players, medical equipment, transportation, and so on. These chapters will also cover recent advances and describe many new tools that can potentially be utilized in new applications and further investigated. The authors of the chapters are frequent and vital contributors to the fields of signal processing and machine learning theory. The goal of each chapter is to provide an overview and a brief tutorial of important topics pertaining to the signal processing and machine learning theory, including key references for further studies. The topics included in this book are the following:

- Chapter 1 provides a potpourri of the matters covered in the book.
- Chapter 2 covers the basic concepts of continuous-time signals and systems, highlighting the main tools that can be employed to analyze and design such systems. Several examples are included to illustrate the use of the tools in a whole continuous-time environment. The content of this chapter is also essential to connect the continuous-time and the discrete-time systems.
- Chapter 3 addresses discrete-time signals and systems, emphasizing the essential analysis tools that will be crucial to understand some more advanced techniques presented in the forthcoming chapters. This chapter shows how the powerful state-space representation of discrete-time systems can be employed as analysis and design tool.
- Chapter 4 on random signals and stochastic processes comprehensively describes the basic concepts required to deal with random signals. It starts with the concept of probability, useful to model chance experiments, whose outcomes give rise to the random variable. Next, the time-domain signals representing nonstatic outcomes are known as stochastic processes, where their full definition is provided. Finally, the chapter describes the tools to model the interactions between random signals and linear time-invariant systems.
- Chapter 5 covers sampling and quantization, where the basic concepts of properly sampling a continuous-time signal and its representation as a sequence of discrete-valued samples are addressed in detail. The chapter discusses a wide range of topics rarely found in a single textbook, such as uniform sampling and reconstruction of deterministic signals, the extension to sampling and reconstruction of stochastic processes, and time-interleaved analog-to-digital (A/D) converters (ADCs) for high-speed A/D conversion. In addition, topics related to the correction of analog channel mismatches, principles of quantization, and oversampled ADCs and digital-to-analog converters (DACs) are briefly discussed. The chapter introduces the more advanced method for discrete-time modeling of mixed-signal systems employed in $\Sigma\Delta$-modulator-based ADCs. This chapter also includes the basic concepts of compressive sampling, featuring many practical issues related to this important field.
- Chapter 6 takes us on a tour of the design of finite length impulse response (FIR) and infinite length impulse response (IIR) transfer functions of fixed filters satisfying prescribed specifications, explaining their fundamental realizations, and exploring more sophisticated realizations. For IIR filters, the chapter introduces the concept of wave digital filters concisely and clearly, motivating its conception from the analog filter structures. The resulting IIR filter realizations keep the low-sensitivity properties of their analog originators. For FIR filter structures, the chapter covers the frequency masking approach that exploits redundancy in the impulse response of typical FIR filters with high selectivity, aiming to reduce the computational complexity. The chapter also explains in detail how to implement digital filters efficiently in specific hardware by adequately scheduling the arithmetic operations.

- Chapter 7 on multirate signal processing for software radio architecture applies several concepts presented in the previous chapters as tools to develop and conceive the implementation of radio defined by software, a subject of great interest in modern communications systems. Software-defined radio entails implementing radio functionality into software, resulting in flexible radio systems whose main objective is to provide multiservice, multistandard, multiband features, all reconfigurable by software. From the signal processing perspective, several basic concepts of multirate systems such as interpolation, decimation, polyphase decomposition, and transmultiplexing play a central role. The chapter reviews the required concepts of signal processing theory and proposes a software-based radio architecture. Chapter 7 uses multirate signal processing and transmultiplexers to discuss a design procedure to be applied in communications, including the application of machine learning to signal detection, segregation, and classification.
- Chapter 8 on modern transform design for practical audio/image/video coding applications addresses the design of several application-oriented transform methods. Most signal processing textbooks present the classical transforms requiring a large number of multiplications; however, the demand for low-power multimedia platforms requires the development of computationally efficient transforms for coding applications. This chapter has the unique feature of presenting systematic procedures to design these transforms while illustrating their practical use in standard codecs.
- Chapter 9 entitled "Discrete multiscale transforms in signal processing" presents a deeper high-level exposition of the theoretical frameworks of the classical wavelets and the anisotropic wavelets. The aim is to enable the reader with enough knowledge of the wavelet-based tools in order to exploit their potential for applications in information sciences even further. Indeed, the material covered in this chapter is unique in the sense that every discrete multiscale transform is linked to a potential application. This chapter includes the current and important issue of incorporating wavelet to neural networks providing some potential application in machine learning.
- Chapter 10 on frames discusses the use of overcomplete signal representations employing frames and their duals, frame operators, inverse frames, and frame bounds. The signal analysis and synthesis using frame representation are also addressed in detail. In particular, the chapter's emphasis is on frames generated from a fixed prototype signal by using translations, modulations, and dilations and analyzing frames of translates, Gabor frames, and wavelet frames. The chapter also presents signal analysis techniques based on the Gaborgram and time-frequency analysis using frames and the matching pursuit algorithm.
- Chapter 11 on parametric estimation exploits the key engineering concept related to modeling signals and systems so that the model provides an understanding of the underlying phenomena and possibly their control. This chapter emphasizes the parametric models leading to simple estimation of the parameters while exposing the main characteristics of the signals or systems under study. By employing both statistical and deterministic formulations, the chapter explains how to generate autoregressive and moving average models, which are then applied to solve problems related to spectrum estimation, prediction, and filtering.
- Chapter 12 on adaptive filters presents a comprehensive description of the current trends as well as some open problems in this area. The chapter discusses the basic building blocks utilized in adaptive filtering setups as well as its typical applications. Then, the chapter discusses the optimal solutions followed by the derivation of the main algorithms. The authors confront for the first time two standard approaches to access the performance of the adaptive filtering algorithms. The chapter closes by discussing some current research topics and open problems for further investigation.

- Chapter 13 on machine learning describes a wide range of tools currently utilized to learn from data. In particular, the chapter overviews the established learning concepts and describes classical tools to perform unsupervised and supervised learning. In addition, the idea of deep learning is highlighted along with some of its structural variations. Finally, the chapter also addresses more recent techniques such as adversarial training and federated learning and includes many illustrative examples. The chapter aims to set the stage for the reader to acquire a broad view of the current machine learning techniques and delve into the details addressed in the remaining chapters of the book.
- Chapter 14 on signal processing over graphs covers an emerging field where data are collected in the vertices of a graph, forming a finite set of samples, with each vertex providing a sample at a given time. The generic data representation originating from graphs departs from the usual regular structures inherent to standard signal processing, such as filtering, translation, and modulation. In this field it is challenging to generalize many of these tools to allow their application to signals from graphs. Many existing problems are topology-dependent, where the graph representation is a natural solution. Among these problems, we can list transportation networks, infectious disease spreading, security systems, human migration, and many others. The graph represents the topology of the considered underlying space from where high-dimensional data are linked to the graph's nodes and edges. The chapter includes the basic concepts of signal processing on graphs and discusses some open problems in this new area.
- Chapter 15 on tensor methods in deep learning addresses the representation of multidimensional arrays using tensors, widely employed in deep neural network algorithms. The tensors allow for a sparse representation of arrays with multiple dimensions, leading to efficient network implementation. The chapter consists of an overview of tensors and tensor methods in the context of deep learning.
- Chapter 16 on nonconvex graph learning addresses an overview of recent advancements in the area of learning on undirected and weighted graphs from data. The chapter describes the practical requirements needed from such models with often appearances in machine learning problems. The requirements include imposing sparsity and handling data with outliers or heavy tails, which are part of tasks such as clustering and graphical-variational autoencoding, etc. The algorithms presented in the chapter are readily available for applications.
- Chapter 17 on dictionaries in machine learning deals with tools to improve the ability to learn and extract meaningful sparse and compact latent signals and feature representations from large datasets originating from natural signals. The chapter provides an overview of both deterministic and statistical dictionary learning algorithms. The chapter includes a discussion about the relationship of learned dictionary-based signal processing and detection with contemporary machine learning techniques.

In summary, this book shows that numerous signal processing and machine learning theory tools, which are apparently unrelated, inspire new ideas for further advancements. Indeed, this cross-fertilization is a powerful enabler to allow intuitive interpretation of many successful machine learning solutions and induce new endeavors. A current challenge in learning from data is using some signal processing tools to close the gap between machine learning and natural intelligence.

CHAPTER

Introduction to signal processing and machine learning theory

1

Marcele O.K. Mendonça, Isabela F. Apolinário, and Paulo S.R. Diniz
Universidade Federal do Rio de Janeiro, Rio de Janeiro, Brazil

1.1 Introduction

Signal processing is a crucial area of knowledge that finds applications in virtually all aspects of modern life. Indeed, human beings are employing signal processing tools for centuries without realizing it [3]. In present days the younger generation might not be able to understand how one can live without carrying a mobile phone, travel long distances without an almost self-piloted airplane, explore other planets without human presence, and utilize a medical facility without a wide range of diagnostic and intervention equipment.

Signal processing consists of mapping or transforming information-bearing signals into another form of signals at the output, aiming at some application benefits. This mapping defines a continuous or analog system that involves functions representing the input and output signals. On the other hand, the system is discrete or digital if its input and output signals are represented by sequences of numbers.

Machine learning is an expression that is used in many disciplines dealing with tools to learn from data and perform modeling tasks such as prediction and classification. Machine learning theory enables the solution of unprecedented problems and also enhances existing solutions. Machine learning basically consists of fitting models to the provided data. There are many machine learning algorithms available. However, the quality of the training data plays a crucial role in the algorithm's success.

Signal processing and machine learning theories are very rich and have been finding applications in uncountable areas, among which we can mention bioengineering, communications, control, surveillance, environment monitoring, oceanography, astronomy, and image classification, to mention a few. In particular, the enormous advance in digital integrated circuit technology, responsible for the computing and information revolutions that we are so used to today, enabled the widespread use of digital signal processing (DSP) and machine learning-based systems. These systems allowed advances in fields such as speech, audio, image, video, automatic translation, self-driving cars, recommendation systems, and spam filters, as well as several applications of wireless communications and digital control.

Many practical problems start by acquiring continuous-time signals, followed by digitizing and storing them to perform further processing, eventually including data mining. This chapter is an overview of the classical signal processing and machine learning theories whose details are encountered in numerous textbooks and monographs, such as [2,4–10,12,15–17,19,21–25,29–31,33,35,36]. The topics briefly treated in this chapter and detailed in the following chapters entail the description of many signal processing and machine learning tools and how they intermingle. As a result, the interested reader can benefit from reading many of these chapters, if not all, in order to deepen and widen their current knowledge as well as to start exploiting new ideas.

Signal Processing and Machine Learning Theory. https://doi.org/10.1016/B978-0-32-391772-8.00007-7
Copyright © 2024 Elsevier Inc. All rights reserved.

1.2 Continuous-time signals and systems

The processing of signals starts by accessing the type of information we want to deal with. Many signals originating from nature are continuous in time, and as such, the processing involved must include analog signal acquisition systems followed by the implementation of a continuous-time system if the processing remains analog all the way. Alternatively, assuming the original continuous-time signal contains limited spectral information, we can sample it in order to generate a sequence representing the continuous-time signal unambiguously.[1] In this latter form one can benefit from the advanced digital integrated circuit technology to process the signal. However, many real-life applications still include continuous-time signal processing, at least at the acquisition and actuator phases of the processing.

A continuous-time system maps an analog input signal represented by $x(t)$ into an output signal represented by $y(t)$, as depicted in Fig. 1.1. A general representation of this mapping is

$$y(t) = \mathcal{H}\{x(t)\},$$

where $\mathcal{H}\{\cdot\}$ denotes the operation performed by the continuous-time system. If the system is linear and time-invariant (LTI), as will be described in Chapter 2, there are several tools to describe the mapping features of the system. Typically, a general description of the systems consists of a differential equation which in turn can be solved and analyzed by employing the Laplace transform. A key feature of the LTI systems is their full representation through their impulse responses. The behavior of a nonlinear system is more complicated to analyze since its impulse response cannot fully characterize it.

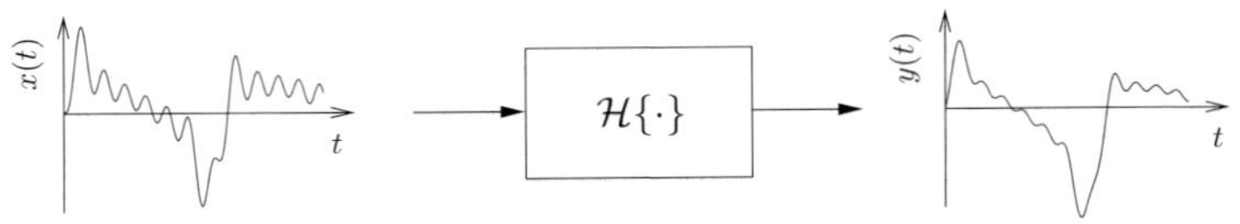

FIGURE 1.1

Continuous-time system.

Other important tools to deal with periodic and nonperiodic continuous-time signals are the Fourier series and the Fourier transform, respectively. As will be discussed in Chapter 2, the Fourier and Laplace transforms are closely related, both being essential tools to describe the behavior of continuous-time signals when applied to LTI systems.

The Laplace transform is suitable to represent a time-domain nonperiodic function of a continuous and real (time) variable, resulting in a frequency-domain nonperiodic function of a continuous and complex frequency variable. That is,

$$X(s) = \int_{-\infty}^{\infty} x(t)e^{-st}\,dt \quad \Longleftrightarrow \quad x(t) = \frac{1}{2\pi}e^{\sigma t}\int_{-\infty}^{\infty} X(\sigma + j\omega)e^{j\omega t}\,d\omega.$$

[1] As will be seen in Chapter 5, we can completely reconstruct a continuous-time band-limited signal $x(t)$ from its sampled version $x(n)$ if the sampling frequency is chosen correctly.

The Fourier transform is employed to represent a time-domain nonperiodic function of a continuous and real (time) variable with a frequency-domain nonperiodic function of a continuous and imaginary frequency variable. The Fourier transform is described by

$$X(\Omega) = \int_{-\infty}^{\infty} x(t)\mathrm{e}^{-\mathrm{j}\Omega t}\,\mathrm{d}t \iff x(t) = \frac{1}{2\pi}\int_{-\infty}^{\infty} X(\Omega)\mathrm{e}^{\mathrm{j}\Omega t}\,\mathrm{d}\Omega.$$

The representation of a time-domain periodic function of a continuous and real (time) variable is performed by the Fourier series. In the frequency domain, the representation consists of a nonperiodic function of a discrete and integer frequency variable, described as follows:

$$X(k) = \frac{1}{T}\int_{0}^{T} x(t)\mathrm{e}^{-\mathrm{j}(2\pi/T)kt}\,\mathrm{d}t \iff x(t) = \sum_{k=-\infty}^{\infty} X(k)\mathrm{e}^{\mathrm{j}(2\pi/T)kt}.$$

1.3 Discrete-time signals and systems

A discrete-time signal is represented by a sequence of numbers and can be denoted as $x(n)$ with $n \in \mathbb{Z}$, where $\mathbb{Z}$ is the set of integer numbers. The samples $x(n)$ might represent numerically the amplitude of a continuous-time signal sample at every T seconds or can originate from discrete information. In many applications, the sampling period T represents the time interval between two acquired samples of the continuous-time signal. However, in other situations, it might represent the distance between two sensors of two pixels of an image, or the separation between two antennas, to mention a few examples.

Discrete-time systems map input sequences into output sequences. A general representation of this mapping is

$$y(n) = \mathcal{H}\{x(n)\},$$

where $\mathcal{H}\{\cdot\}$ denotes the operation performed by the discrete-time system. Fig. 1.2 depicts the input-to-output mapping of sequences. According to the properties of $\mathcal{H}\{\cdot\}$, the discrete-time system might be LTI. This way, it benefits from a number of analysis and design tools, such as frequency-domain representations. However, there are many applications employing nonlinear, time-varying, and even noncausal systems [4,10,20,22–25].

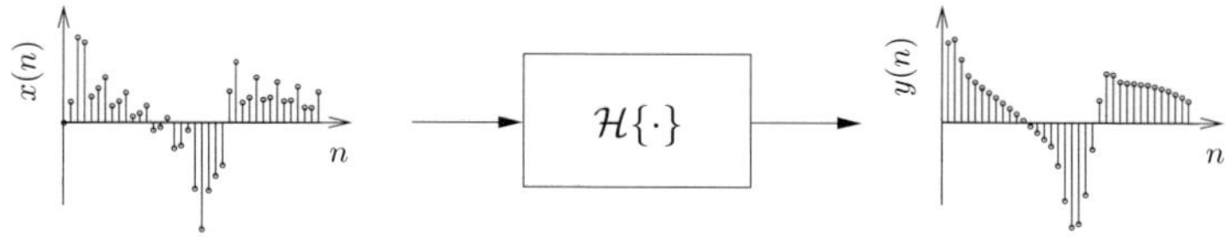

FIGURE 1.2

Discrete-time signal representation.

If the system is LTI, as will be described in Chapter 2, there are several tools to describe the mapping features of the system. Typically, a general description of discrete-time systems through a difference equation can be solved and analyzed by employing the z-transform. Also, a discrete-time LTI system is fully described by its impulse response, whereas a nonlinear system cannot be fully characterized by its impulse response.

The z-transform is the key tool to represent a time-domain nonperiodic function of a discrete and integer variable through a frequency-domain nonperiodic function of a continuous and complex frequency variable. That is,

$$X(z) = \sum_{n=-\infty}^{\infty} x(n)z^{-n} \quad \Longleftrightarrow \quad x(n) = \frac{1}{2\pi \mathrm{j}} \oint_C X(z)z^{n-1}\,\mathrm{d}z.$$

The discrete-time Fourier transform (DTFT) represents a time-domain nonperiodic function of a discrete and integer variable by a periodic function of a continuous frequency variable in the frequency domain as follows:

$$X(\mathrm{e}^{\mathrm{j}\omega}) = \sum_{n=-\infty}^{\infty} x(n)\mathrm{e}^{-\mathrm{j}\omega n} \quad \Longleftrightarrow \quad x(n) = \frac{1}{2\pi} \int_{-\infty}^{\infty} X(\mathrm{e}^{\mathrm{j}\omega})\mathrm{e}^{\mathrm{j}\omega n}\,\mathrm{d}\omega.$$

Finally, the discrete Fourier transform (DFT) is the right tool to represent a time-domain periodic function of a discrete and integer variable through a periodic function of a discrete and integer frequency variable in the frequency domain. The DFT is defined as

$$X(k) = \sum_{n=0}^{N-1} x(n)\mathrm{e}^{-\mathrm{j}(2\pi/N)kn} \quad \Longleftrightarrow \quad x(n) = \frac{1}{N} \sum_{k=0}^{N-1} X(k)\mathrm{e}^{\mathrm{j}(2\pi/N)kn}.$$

The DFT plays a key role in DSP since it represents a finite length sequence in the time domain through a finite length sequence in the frequency domain. Since both domains utilize sequences, this feature makes the DFT a natural choice for the time-frequency representation of information in a digital computer. A little mental exercise allows us to infer that the DFT is the perfect representation in the frequency domain of a finite length sequence in the time domain if we interpret the latter as a period of a periodic infinite length sequence. In addition, by employing appropriate zero-padding in two finite length sequences, the product of their DFTs represents the DFT of their linear convolution (see Chapter 3), turning the DFT into a valuable tool for LTI system implementation. There is a plethora of applications for the DFT in signal processing and communications; some of them can be accessed in the references [10,19].

Often, whenever a DSP system is LTI, it can be described through a difference equation as follows:

$$\sum_{i=0}^{N} a_i y(n-i) + \sum_{l=0}^{M} b_l x(n-l) = 0.$$

The above description is suitable for implementation in digital computers. The difference equation represents a causal LTI system if its auxiliary conditions correspond to its initial conditions, and those are zeros [10]. Assuming $a_0 = 1$, the above equation can be rewritten as

$$y(n) = -\sum_{i=1}^{N} a_i y(n-i) + \sum_{l=0}^{M} b_l x(n-l).$$

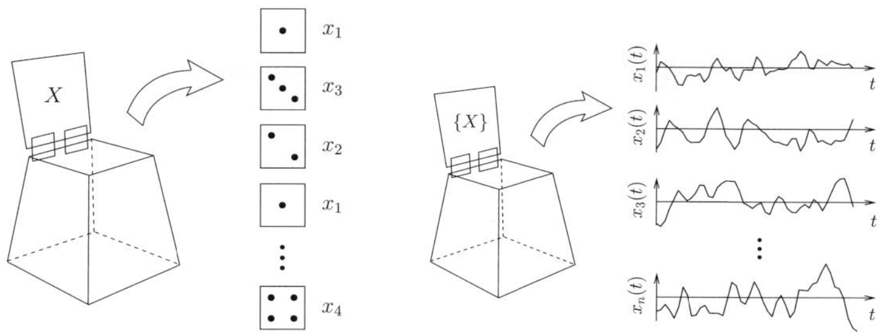

(a) Random variable. (b) Stochastic process.

FIGURE 1.3

Examples of a random variable and a stochastic process.

This class of systems has infinite impulse response (IIR) (i.e., $y(n) \neq 0$ when $n \to \infty$), and as such, it is called an IIR system or filter.

The nonrecursive system generates the output signal from past input samples, that is,

$$y(n) = \sum_{l=0}^{M} b_l x(n-l).$$

In this case, the resulting system has a finite impulse response (FIR) and is known as FIR system or filter.

1.4 Random signals and stochastic processes

In most practical applications, we have to deal with signals that cannot be described by a function of time since their waveforms follow a random pattern [26,28]. Take, for example, the thermal noise inherent to any material. Despite the lack of exact knowledge of the signal values, it is possible to analyze and extract information the signals contain by employing the mathematical tools available to deal with random signals. Chapter 4 will present a compact and yet clarifying review of random signals and stochastic processes, which are crucial to understanding several concepts that will be discussed in the more advanced chapters.

The theory starts by defining a random variable as a mapping of the result of a random process experiment onto a set of numbers. A random variable X is a function that assigns a number x to every outcome of a random process experiment. An example is given in Fig. 1.3(a), in which each outcome of a throw of a dice, numbers 1–6, corresponds to a determined value, x_1–x_6, respectively.

The stochastic process is a rule to describe the time evolution of the random variable depending on the random process experiment, whereas the set of all experimental outcomes is its domain known as the ensemble. An example is given in Fig. 1.3(b), in which at any time instant t_0, the set formed by the output samples of the stochastic process $\{X\}$, $\{x_1(t_0), x_2(t_0), x_3(t_0), \ldots, x_n(t_0)\}$, represents a random variable. A single outcome $x_i(t)$ of $\{X\}$ is a random signal. Since random signals do not have a precise description of their waveforms, we have to characterize them either via measured statistics or through a probabilistic model. Generally, the first- and second-order statistics (mean and variance, respectively) are sufficient for characterizing the stochastic process, mainly because these statistics are suitable for measurements.

As an illustration, Fig. 1.4 shows a random signal as an input of a high-pass filter. We can observe that the output signal is still random but clearly shows a faster changing behavior, given that the high-pass filter attenuated the low-frequency contents of the input signal.

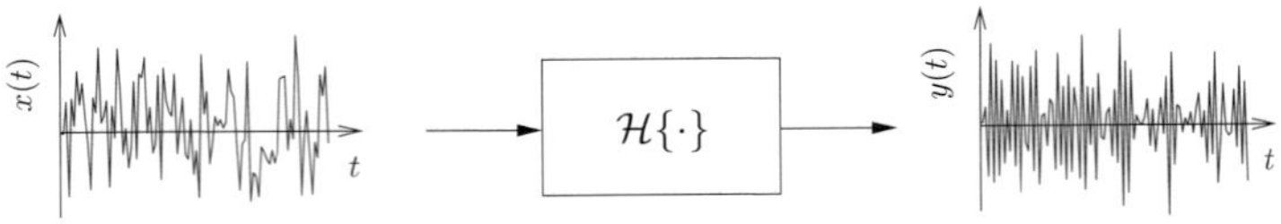

FIGURE 1.4

Filtering of a random signal.

1.5 Sampling and quantization

A DSP system whose signals originated from continuous-time sources includes several building blocks, namely, an analog-to-digital (A/D) converter, a DSP system, a digital-to-analog (D/A) converter, and a low-pass filter. Fig. 1.5 illustrates a typical DSP setup, where:

- The A/D converter produces a set of samples in equally spaced time intervals, which may retain the information of the continuous-time signal in the case the latter is band-limited. These samples are converted into a numeric representation in order to be applied to the DSP system.
- The DSP performs the desired mapping between the input and output sequences.
- The D/A converter produces a set of equally spaced-in-time analog samples representing the DSP system output.
- The low-pass filter interpolates the analog samples to produce a smooth continuous-time signal.

Chapter 5 will discuss in detail the conditions which a continuous-time signal must satisfy so that its sampled version retains the information of the original signal, dictated by the sampling theorem. This theorem determines that a band-limited continuous-time signal can be theoretically recovered from its sampled version by filtering the sequence with an analog filter with a prescribed frequency response.

It is also important to mention that while processing the signals in the digital domain, these are subject to quantization errors, such as round-off errors originating from internal arithmetic operations performed in the signals, deviations in the filter response due to the finite word length representation of the multiplier coefficients inherent to the signal processing operation, and errors due to the representation of the acquired continuous-time signals with a set of discrete levels.

The actual implementation of a DSP system might rely on general-purpose digital machines, where the user writes a computer program to implement the DSP tasks. This strategy allows fast prototyping and testing. Another way is to use special-purpose commercially available CPUs, known as digital sig-

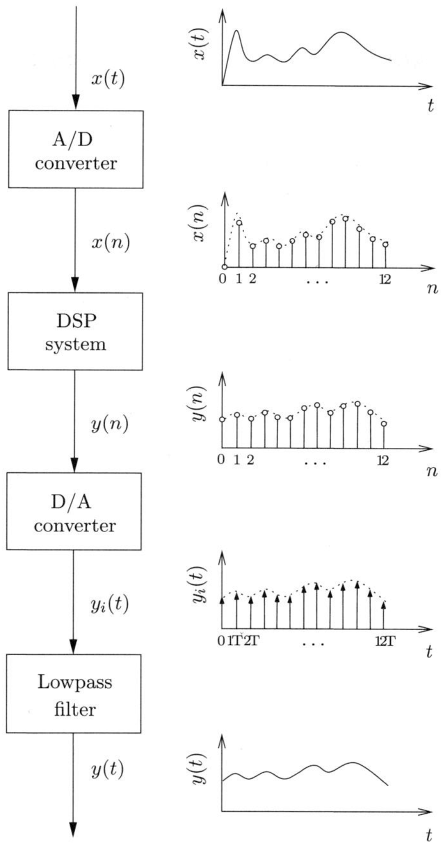

FIGURE 1.5

A DSP system.

nal processors, that are capable of implementing sum of product operations in a very efficient manner. Yet another approach is to employ special-purpose hardware tailored for the given application.

Fig. 1.6 depicts a continuous-time signal, its digitized version, and its recovered continuous-time representation. We can note from Figs. 1.6(a) to 1.6(f) the effects of a low sampling frequency and a small number of quantization steps on the A/D and D/A processes.

Fig. 1.7 shows a detailed example of the steps entailing an A/D and a D/A conversion, where in Fig. 1.7(a) a continuous-time signal is depicted along with its equally spaced samples. These samples are then quantized, as illustrated in Fig. 1.7(b). Assuming the quantized samples are converted into an analog signal through a zero-order hold, the output of the D/A converter becomes as illustrated in Fig. 1.7(c). The sampled-and-held continuous-time signal is low-pass filtered in order to recover the original continuous-time signal. As can be observed in Fig. 1.7(d), the recovered signal resembles the

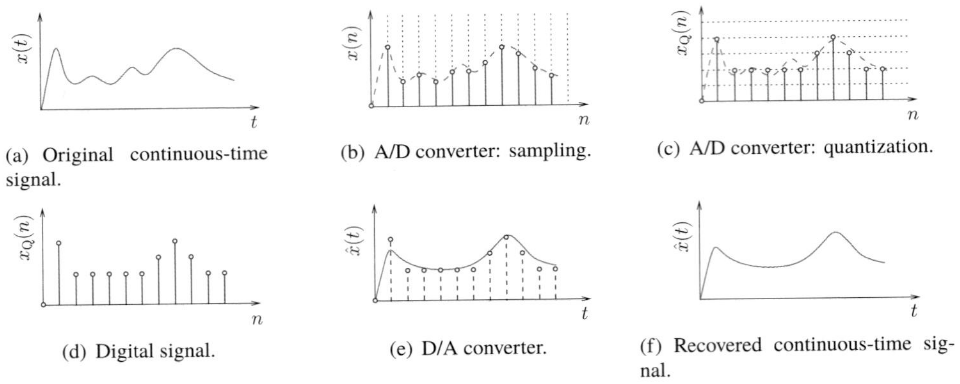

FIGURE 1.6

A digital signal generation.

original where the difference originates from the quantization effects and the nonideal low-pass filter at the D/A converter output.

1.6 FIR and IIR filter design

The general form of an FIR transfer function is given by

$$H(z) = \sum_{l=0}^{M} b_l z^{-l} = H_0 z^{-M} \prod_{l=0}^{M} (z - z_l).$$

Since all the poles of the FIR transfer function are placed at $z = 0$, it is always stable. As a result, the transfer function above will always represent a stable filter. The main feature of the FIR filters is the possibility of designing structurally induced linear-phase filters. On the other hand, an FIR filter requires a higher number of multiplications, additions, and storage elements when compared to their IIR counterparts in order to satisfy a prescribed specification for the magnitude response.

There are many methods to design an FIR filter, ranging from the simple ones, such as the window and frequency sampling methods, to more efficient and sophisticated ones that rely on some kind of optimization method [1,11,18]. The simple methods lead to solutions that are far from optimum in most practical applications, in the sense that they either require higher order or cannot satisfy prescribed specifications. A popular optimization solution is the minimax method based on the Remez exchange algorithm, which leads to transfer functions with a minimum order to meet prescribed specifications.

With the growing computational power, it is possible to design flexible FIR filters by utilizing weighted least-squares (WLS) solutions. These filters are particularly suitable for multirate systems since the resulting transfer functions can have decreasing energy in their stopband [10]. The WLS

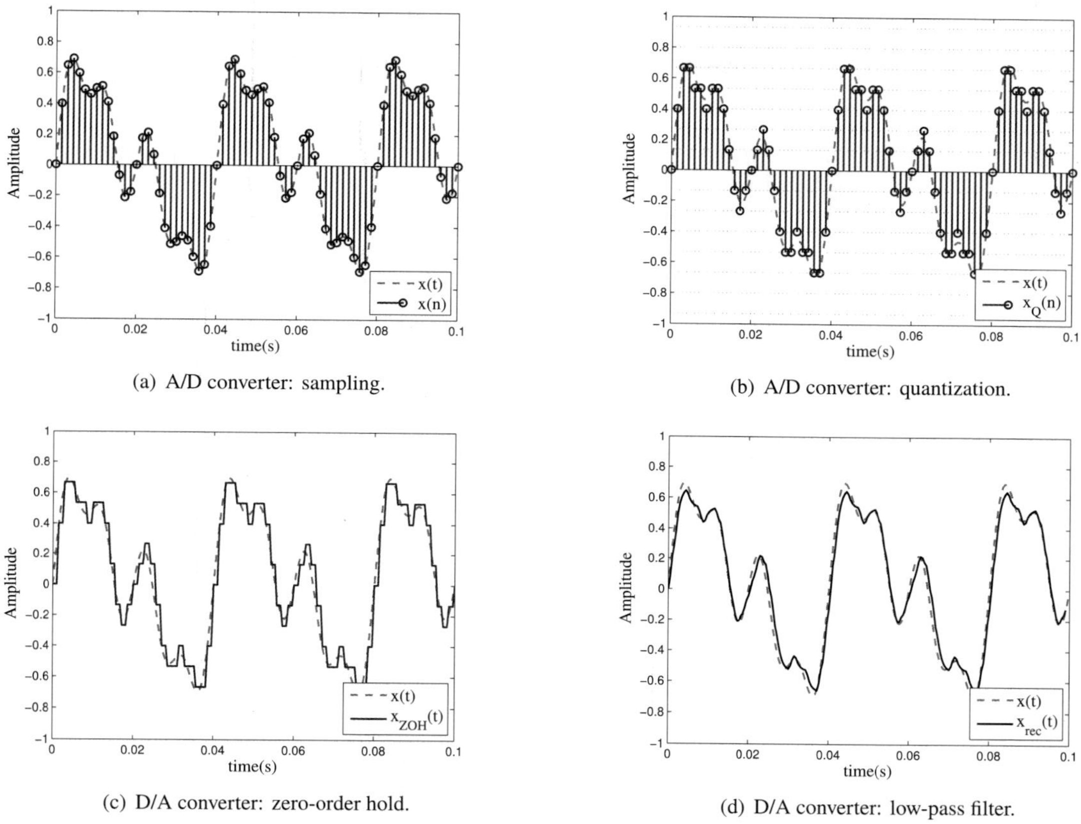

(a) A/D converter: sampling.

(b) A/D converter: quantization.

(c) D/A converter: zero-order hold.

(d) D/A converter: low-pass filter.

FIGURE 1.7

A real digital signal generation.

method enables the design of filters satisfying prescribed specifications, such as the maximum deviation in the passband and part of the stopband, while minimizing the energy in a range of frequencies in the stopband. As can be observed in Fig. 1.8, for a given filter order, the minimax design leads to lower maximum stopband attenuation, whereas the WLS solution leads to lower stopband energy. Note that there are solutions in between where some maximum values of stopband ripples can be minimized while the energy of the remaining ones is also minimized. Therefore, it is possible to observe that the WLS solution does not satisfy the prescribed specifications, given that it has the same filter order as the minimax solution.

The typical transfer function of an IIR filter is given by

$$H(z) = \frac{Y(z)}{X(z)} = \frac{\sum_{l=0}^{M} b_l z^{-l}}{1 + \sum_{i=1}^{N} a_i z^{-i}},$$

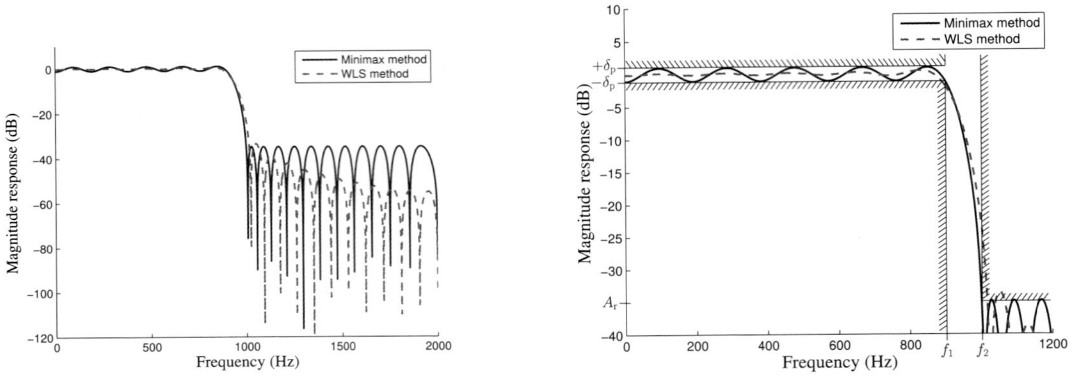

(a) Magnitude response.

(b) Specifications for a low-pass filter using minimax design.

FIGURE 1.8

Typical magnitude responses of FIR filters: minimax and WLS.

which can be rewritten in a form explicitly showing the poles' positions as

$$H(z) = H_0 \frac{\prod_{l=0}^{M}(1 - z^{-1} z_l)}{\prod_{i=0}^{N}(1 - z^{-1} p_i)} = H_0 z^{N-M} \frac{\prod_{l=0}^{M}(z - z_l)}{\prod_{i=0}^{N}(z - p_i)}.$$

IIR filters are usually designed by using well-established analog filter approximations, such as the Butterworth, Chebyshev, and elliptic methods (Fig. 1.9). The prototype analog transfer function is then transformed into a digital transfer function by using an adequate transformation method. The most widely used transformation methods are bilinear and the impulse invariance methods [10].

1.7 Digital filter structures and implementations

From a closer examination of the FIR and IIR transfer functions, one can infer that they can be realized with three basic operators: adder, multiplier, and delay (represented by z^{-1}).

For example, Fig. 1.10 shows a linear-phase FIR filter realization of odd order. Linear-phase filters have symmetric or antisymmetric impulse response and the symmetry should be exploited in order to minimize the number of multipliers in the filter implementation, as illustrated in Fig. 1.10.

Fig. 1.11 depicts a possible direct-form realization of an IIR transfer function. This realization requires the minimum number of multiplications, adders, and delays to implement the desired IIR transfer function.

There are many alternative structures to implement the FIR and IIR filters. For example, IIR transfer functions can be implemented as a product or as a summation of lower-order IIR structures by starting

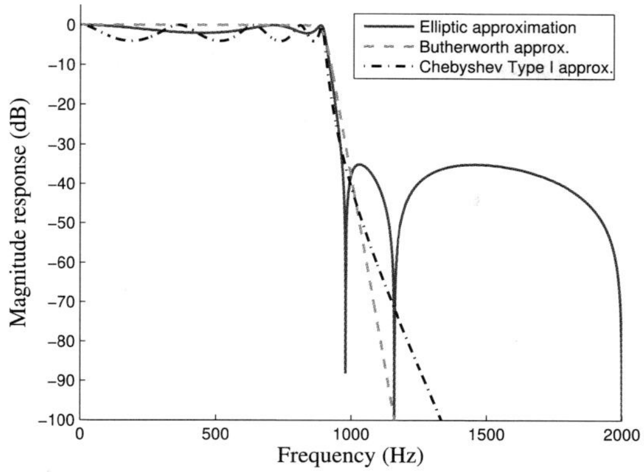

FIGURE 1.9

Typical magnitude response of an IIR filter.

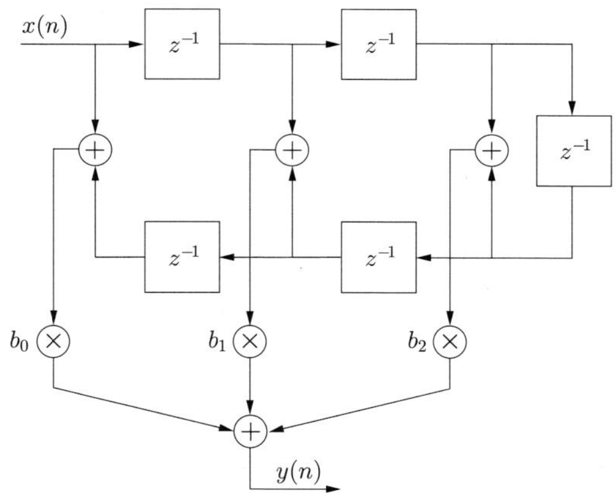

FIGURE 1.10

Odd-order linear-phase FIR filter structure with $M = 5$.

with their description as

$$\begin{aligned} H(z) &= \prod_{k=1}^{m} \frac{\gamma_{0k} + \gamma_{1k} z^{-1} + \gamma_{2k} z^{-2}}{1 + m_{1k} z^{-1} + m_{2k} z^{-2}} \\ &= h_0^p + \sum_{k=1}^{m} \frac{\gamma_{1k}^p z + \gamma_{2k}^p}{z^2 + m_{1k} z + m_{2k}}, \end{aligned}$$

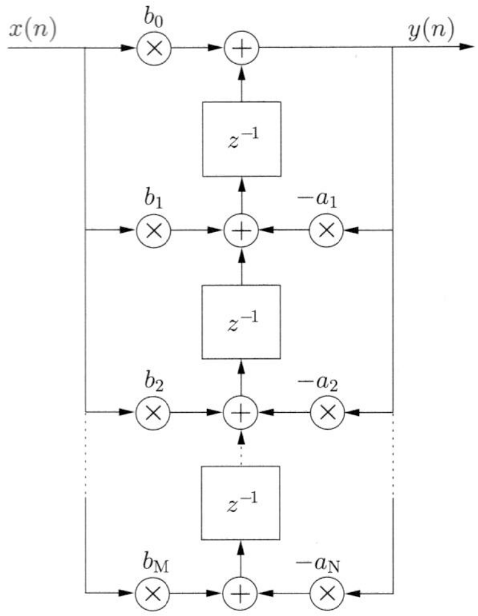

FIGURE 1.11

Direct-form IIR filter structure.

respectively. The right choice for a filter realization must take into consideration some issues, such as modularity, quantization effects, design simplicity, and power consumption [10]. Chapter 6 will address advanced methods for FIR and IIR filter realization and implementation, where elegant and creative solutions are introduced in a clear manner. That chapter will also discuss strategies to implement digital filters efficiently.

1.8 Multirate signal processing

In DSP, it is easy to change the sampling rate of the underlying sequences by a ratio of integer values. This feature is highly desirable whenever we want to merge information of systems employing distinct sampling rates. In these cases, those rates should be made compatible [6,33,36].

Systems that internally employ multiple sampling rates are collectively referred to as multirate systems. In most cases, these systems are time-varying or, at most, periodically time-invariant.

The basic operations of the multirate systems are decimation and interpolation. The proper composition of these operations allows arbitrary rational sampling rate changes. If this sampling rate change is arbitrary and nonrational, the only solution is to recover the band-limited continuous-time signal $x(t)$ from its samples $x(m)$ and then resample it with a different sampling rate, thus generating a distinct discrete-time signal $\overline{x}(n)$.

If we assume that a discrete-time signal $x(m)$ was generated from an continuous-time signal $y(t)$ with sampling period T_1, so that $x(m) = y(mT_1)$, with $m = \ldots, 0, 1, 2 \ldots$, the sampling theorem requires that $y(t)$ should be band-limited to the range $[-\frac{\pi}{T_1}, \frac{\pi}{T_1}]$. The sampled continuous-time signal is given by

$$y'(t) = \sum_{m=-\infty}^{\infty} x(m)\delta(t - mT_1).$$

The spectrum of the sampled signal is periodic with period $\frac{2\pi}{T_1}$. The original continuous-time signal $y(t)$ can be recovered from $y'(t)$ through an ideal low-pass filter. This interpolation filter, whose impulse response is denoted as $h(t)$, has an ideal frequency response $H(e^{j\omega})$ as follows:

$$H(e^{j\omega}) = \begin{cases} 1, & \omega \in [-\frac{\pi}{T_1}, \frac{\pi}{T_1}], \\ 0, & \text{otherwise}, \end{cases}$$

so that

$$y(t) = y'(t) * h(t) = \frac{1}{T_1} \sum_{m=-\infty}^{\infty} x(m)\text{sinc}\frac{\pi}{T_1}(t - mT_1).$$

By resampling $y(t)$ with period T_2, we obtain the discrete-time signal in the new desired sampling rate as follows:

$$\overline{x}(n) = \frac{1}{T_1} \sum_{m=-\infty}^{\infty} x(m)\text{sinc}\frac{\pi}{T_1}(nT_2 - mT_1),$$

where $\overline{x}(n) = y(nT_2)$, with $n = \ldots, 0, 1, 2 \ldots$.

In this general equation governing sampling rate changes, there is no restriction on the values of T_1 and T_2. However, if $T_2 > T_1$ and aliasing is to be avoided, the interpolation filter must have a zero gain for $\omega \notin [-\frac{\pi}{T_2}, \frac{\pi}{T_2}]$.

In the case the sampling rate change corresponds to a ratio of integer numbers, all we need is simple decimation and interpolation operations. The decimation operation, or downsampling, consists of retaining every Mth sample of a given sequence $x(m)$. Since we are disposing of samples from the original sequence, either the signal is sufficiently limited in band or the signal has to be filtered by a low-pass filter so that the decimated signal retains the useful information of the original signal. Decimation is a time-varying operation, since, if $x(m)$ is shifted by m_0, the output signal will, in general, differ from the unshifted output shifted by m_0. Indeed, decimation is a periodically time-invariant operation, which consists of an extra degree of freedom with respect to the traditional time-invariant systems.

The interpolation, or upsampling, of a discrete-time signal $x(m)$ by a factor of L consists of inserting $L - 1$ zeros between each of its samples. In the frequency domain, the spectrum of the upsampled signal is periodically repeated. Given that the spectrum of the original discrete-time signal is periodic with period 2π, the interpolated signal will have period $\frac{2\pi}{L}$. Therefore, in order to obtain a smooth interpolated version of $x(m)$, the spectrum of the interpolated signal must have the same shape as the spectrum of $x(m)$. This can be obtained by filtering out the repetitions of the spectra beyond $[-\frac{\pi}{L}, \frac{\pi}{L}]$. Thus, the upsampling operation is generally followed by a low-pass filter. The interpolation is only periodically time-invariant and does not entail any loss of information.

Fig. 1.12 illustrates discrete-time signal decimation and interpolation operations. As can be observed in Fig. 1.12(a), the decimation factor of two widens the spectrum of the sampled signal in comparison to the new sampling rate. Fig. 1.12(b) depicts the effect of increasing the sampling rate of a given sequence by two, where it can be seen that the frequency contents of the original signal repeat twice as often and appear narrower in the new sampling rate.

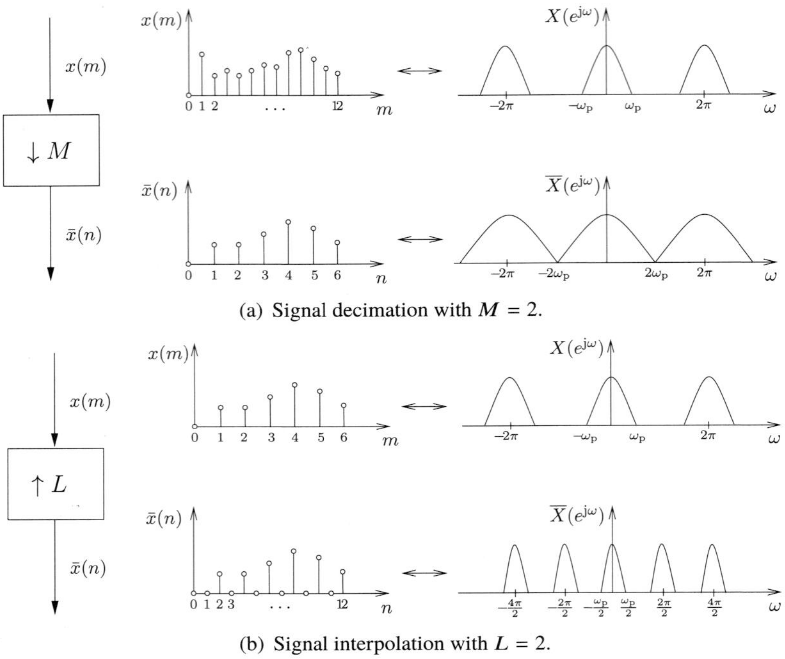

FIGURE 1.12

Signal decimation and interpolation.

Chapter 8 will further discuss the theory and practice of multirate signal processing. This chapter will show how sophisticated real-life signal processing problems can be addressed, starting from the basic theory. In particular, the authors address the design of flexible communications systems incorporating several advanced signal processing tools [9,32,34,37].

1.9 Filter banks and transform design

In many applications, it is desirable to decompose a wide-band discrete-time signal into several nonoverlapping narrow-band subsignals in order to be transmitted, quantized, or stored. In this case, each narrow-band channel can reduce its sampling rate, since the subband signals have lower bandwidth than their originator signal. The signal processing tool that performs these tasks is the analysis filter bank. The analysis filters, represented by the transfer functions $H_i(z)$, for $i = 0, 1, \ldots, M - 1$, consist of a low-pass filter $H_0(z)$, bandpass filters $H_i(z)$, for $i = 1, 2, \ldots, M - 2$, and a high-pass filter

$H_{M-1}(z)$. Ideally, these filters have nonoverlapping passbands, whereas their spectrum combination should cover the overall spectrum of the input signal. The outputs of the analysis filters, denoted as $x_i(n)$, for $i = 0, 1, \ldots, M-1$, have together M times the number of samples of the original signal $x(n)$. However, since each subband signal occupies a spectrum band M times narrower than the input signal, they can be decimated so that the number of samples in the subbands does not increase. Indeed, if the input signal is uniformly split into subbands, we can decimate each $x_i(n)$ by a factor of L smaller than or equal to M without generating unrecoverable aliasing effects. Fig. 1.13 shows the general configuration of a maximally (or critically) decimated analysis filter, where the decimation factor is equal to the number of subbands, i.e., $L = M$.

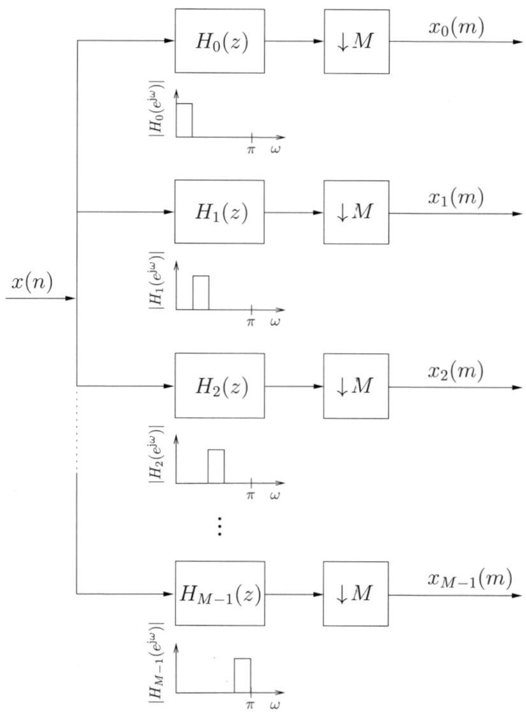

FIGURE 1.13

Analysis filter bank.

Whenever the input signal is split into subbands and decimated, if $L \leq M$, it is always possible to recover the input signal by properly designing the analysis filters in conjunction with the synthesis filters $G_i(z)$, for $i = 0, 1, \ldots, M-1$. The synthesis filters are placed after interpolators, and they have the task of smoothing the upsampled signals. Fig. 1.14 illustrates the general configuration of the synthesis filter bank. A key feature of the synthesis filter bank is to cancel out or reduce the aliasing effects.

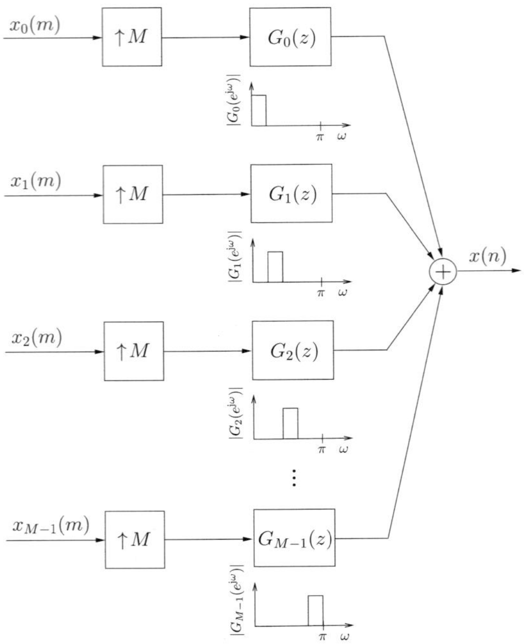

FIGURE 1.14

Synthesis filter bank.

If the subband signals are not modified, the filter bank output $y(n)$ can be a delayed copy signal of $x(n)$. The cascade connection of the analysis and synthesis filter banks satisfying this condition is called a perfect reconstruction filter bank.

The cascade of an M-channel synthesis filter bank with an M-channel analysis filter bank gives rise to a transmultiplex system. The transmultiplex model is widely used in communications to model multiuser, multicarrier, and multiaccess systems [9], [37], [32]. Chapter 7 will utilize multirate signal processing and transmultiplexers to discuss a design procedure to be applied in communications, also providing an illustrative example related to the application of machine learning to signal detection, segregation, and classification.

Chapter 8 will present several methods for designing the analysis filters $H_i(z)$ and the synthesis filters $G_i(z)$ such that perfect reconstruction is achieved and other features are also met. This chapter will focus on the connections between modern transforms and filter bank design.

Fig. 1.15 shows a four-band filter bank design and its effect on a composed signal. Fig. 1.16 illustrates how a four-band synthesis filter bank performs the recomposition of the input signal starting from the decimated subband signals. It is worth mentioning that the filter banks, when the subbands are divided into octave bands, give rise to the discrete-time wavelet series, which is part of the discussion in the following section.

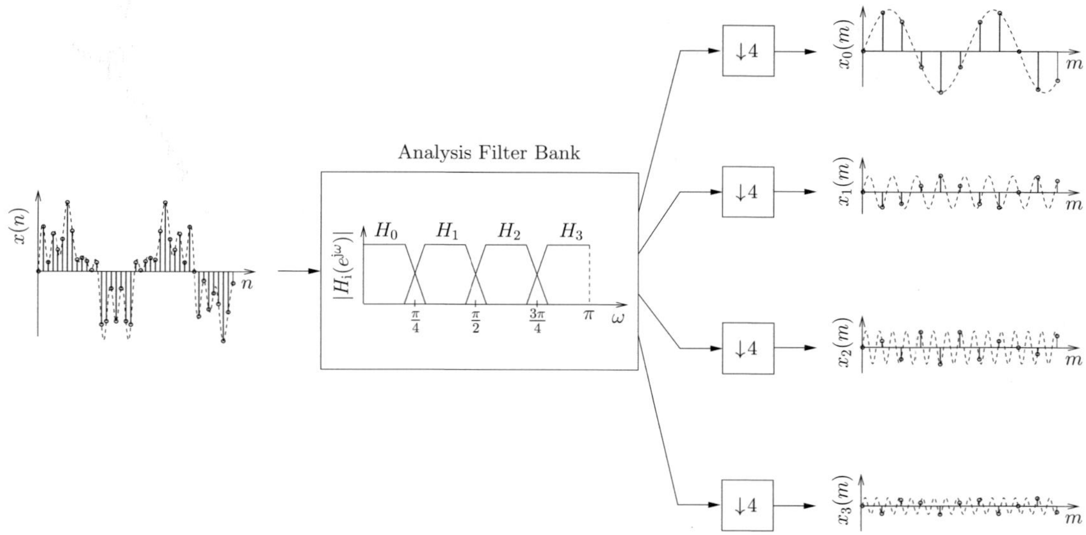

FIGURE 1.15

Analysis filter bank design.

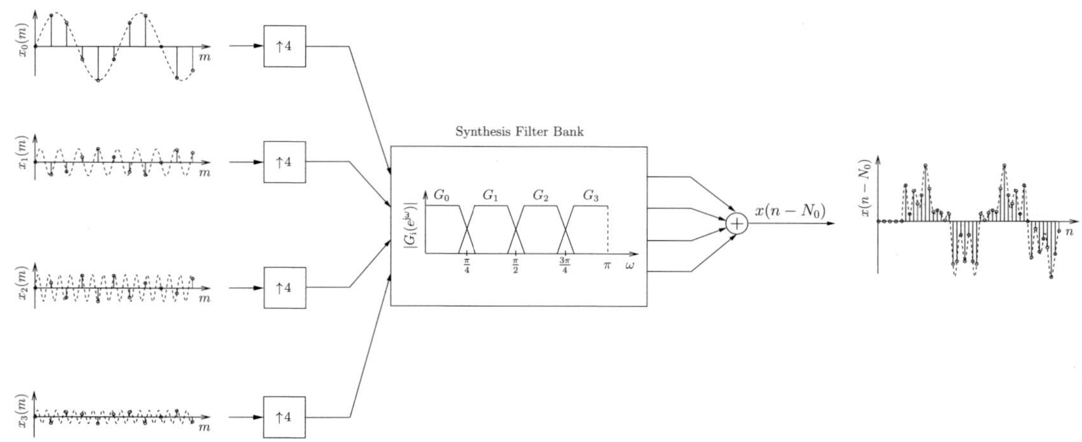

FIGURE 1.16

Synthesis filter bank design.

1.10 Discrete multiscale and transforms

Discrete-time signals and systems can be characterized in the frequency domain by their Fourier transforms. One of the main advantages of discrete-time signals is that they can be processed and represented by digital computers. However, when we examine the definition of the DTFT,

$$X(\mathrm{e}^{\mathrm{j}\omega}) = \sum_{n=-\infty}^{\infty} x(n)\mathrm{e}^{-\mathrm{j}\omega n},$$

we notice that such a characterization in the frequency domain depends on the continuous variable ω. This implies that the Fourier transform, as it is, is not suitable for the processing of discrete-time signals in digital computers. We need a transform depending on a discrete-frequency variable that, if possible, preserves the handy interpretations of the Fourier-based tools, which retain important information. This can be obtained from the Fourier transform itself in a very simple way, by sampling uniformly the continuous-frequency variable ω as long as the input sequence has a finite length. Otherwise, the time-domain signal will be affected by aliasing. Using this strategy, it is possible to obtain a mapping of a signal depending on a discrete-time variable n to a transform depending on a discrete-frequency variable k, leading to another interpretation for the DFT. Unfortunately, the DFT has its limitations in representing a more general class of signals, as will be discussed next.

The wavelet transform of a function belonging to $\mathcal{L}^2\{\mathbb{R}\}$, the space of the square-integrable functions, is the function decomposition in a basis composed of expansions, compressions, and translations of a single mother function $\psi(t)$, called wavelet.

The wavelet transform can be defined as

$$x(t) = \sum_{m=-\infty}^{\infty} \sum_{n=-\infty}^{\infty} c_{m,n}\overline{\psi}_{m,n}(t),$$

$$c_{m,n} = \int_{-\infty}^{\infty} x(t)\psi^*_{m,n}(t)\,\mathrm{d}t,$$

where

$$\psi_{m,n}(t) = 2^{-m/2}\psi(2^{-m}t - n),$$
$$\overline{\psi}_{m,n}(t) = 2^{-m/2}\overline{\psi}(2^{-m}t - n).$$

This pair of equations defines a biorthogonal wavelet transform which is characterized by two wavelets: the analysis wavelet, $\psi(t)$, and the synthesis wavelet, $\overline{\psi}(t)$. Any function $x(t) \in \mathcal{L}^2\{\mathbb{R}\}$ can be decomposed as a linear combination of contractions, expansions, and translations of the synthesis wavelet $\overline{\psi}(t)$. The expansion weights can be computed via the inner product of $x(t)$ with expansions, contractions, and translations of the analysis wavelet $\psi(t)$.

Functions $\psi_{m,n}(t)$ do not comprise an orthogonal set, so neither do the functions $\overline{\psi}_{m,n}(t)$. However, functions $\psi_{m,n}(t)$ are orthogonal to $\overline{\psi}_{m,n}(t)$ so that

$$\langle\psi_{m,n}(t), \overline{\psi}_{k,l}(t)\rangle = \delta(m-k)\delta(n-l).$$

A byproduct of the development of the wavelet transforms is the so-called wavelet series, allowing the representation of signals in the time and frequency domains through sequences. The cascade of two-band filter banks can produce many alternative maximally decimated decompositions. The hierarchical decomposition achieving an octave-band decomposition is of particular interest, in which only the low-pass band is further decomposed. This configuration gives rise to the widely used wavelets series.

As discussed in Chapter 10, many transform-based tools in signal processing theory are available to meet the requirements of different applications. The classical Fourier-based transforms have been used for centuries, but their efficiency in dealing with nonstationary signals is limited. With wavelet series, for example, it is possible to obtain good resolution in time and frequency domains, unlike the Fourier-based analysis. Therefore, Chapter 10 will include a myriad of transform solutions tailored for distinct practical situations. The chapter concludes with a detailed discussion about the incorporation of wavelets to neural networks, illustrating how much a robust data analysis tool enhances the modeling capability of neural networks providing to machine learning applications.

Fig. 1.17 illustrates a wavelet series representation of a composed signal, where the input signal is decomposed by an octave-band filter bank representing a wavelet series.

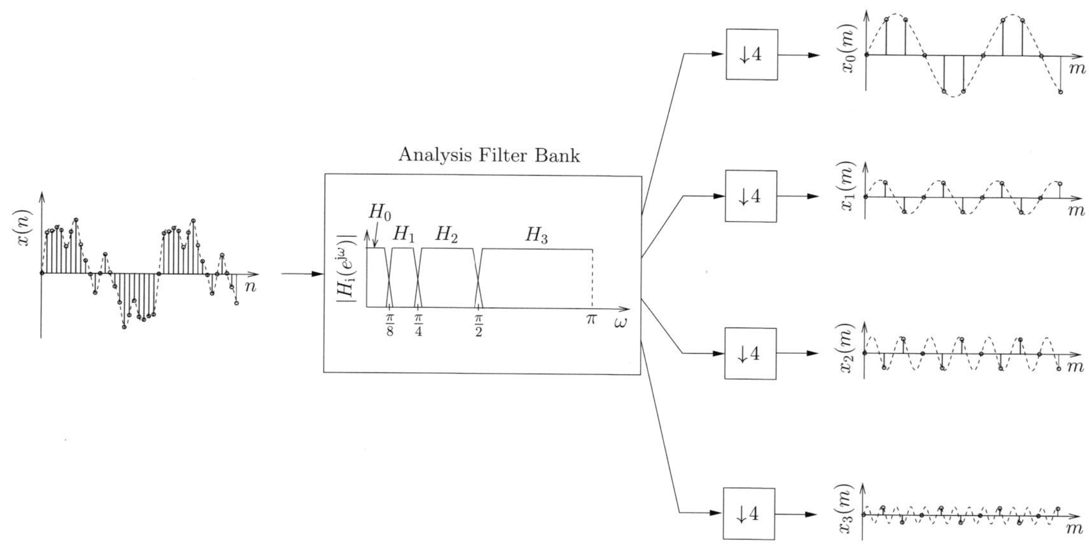

FIGURE 1.17

Wavelet series of a composed sinusoidal signal.

1.11 Frames

Frames consist of a set of vectors that enables a stable representation of a signal starting from an analysis operation whereby a set of coefficients is generated from the signal projection with the frame vectors and a synthesis operation where the synthesis vectors are linearly combined to approximate the original signal. Chapter 12 will introduce the idea of frames by discussing the concepts of overcomplete representations, frames and their duals, frame operators, inverse frames, and frame bounds. In particular, it will show how to implement signal analysis and synthesis employing frames generated from a fixed prototype signal by using translations, modulations, and dilations and analyzes frames of translates, Gabor frames, and wavelet frames. The frames representation is usually redundant by requiring more coefficients than the minimum, where redundancy is measured by frame bounds [19]. The end result of employing the framework of frames is a set of signal processing tools such as the Gaborgram, time-frequency analysis, and the matching pursuit algorithm.

Fig. 1.18 shows the representation of a composed signal into frames. As can be observed, the representation of $x(n)$ by $x_1(n)$ and $x_0(n)$ is redundant, since $x_1(n)$ has the same rate as the input signal $x(n)$, whereas $x_0(n)$ has half the rate.

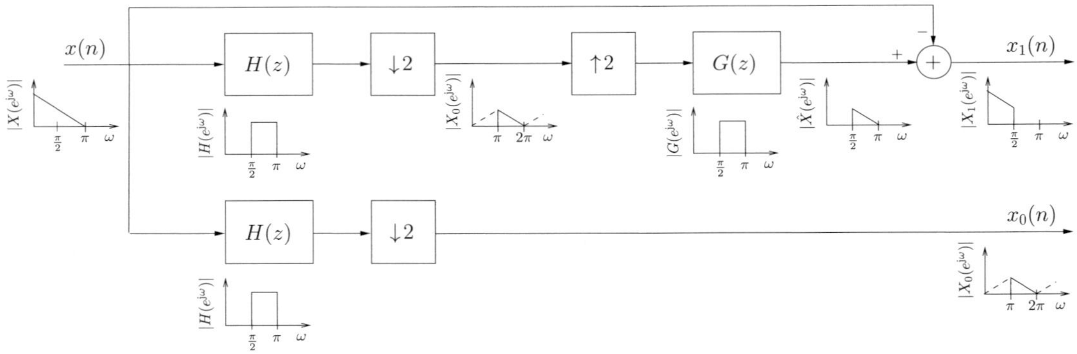

FIGURE 1.18

Frame representation.

1.12 Parameter estimation

In many signal processing applications, it is crucial to estimate the power spectral density (PSD) of a given discrete-time signal. Examples of the use of PSD are vocal-track modeling, radar systems, antenna arrays, sonar systems, and synthesis of speech and music, just to name a few. Usually, the first step in PSD estimation consists of estimating the autocorrelation function associated with the given data, followed by a Fourier transform to obtain the desired spectral description of the process.

In the literature, we can find a large number of sources dealing with algorithms for performing spectral estimation. However, the alternative solutions differ in their computational complexity, precision, frequency resolution, or other statistical aspects. In general, spectral estimation methods can be classified as nonparametric or parametric methods. The nonparametric methods do not prescribe any particular structure for the given data. In contrast, parametric schemes assume that the provided process follows the pattern of a prescribed model characterized by a specific set of parameters [17].

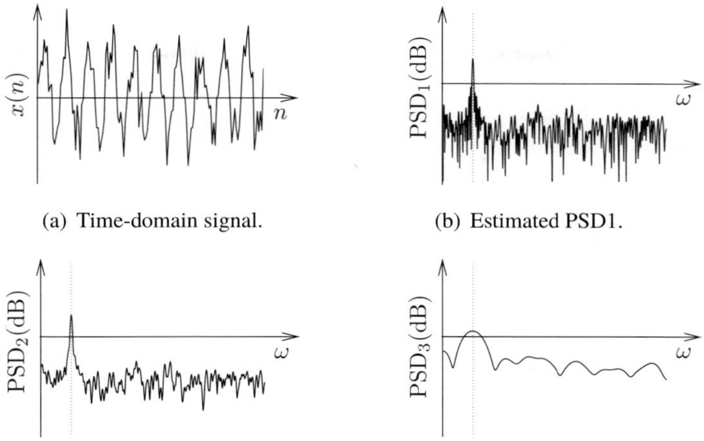

(a) Time-domain signal. (b) Estimated PSD1. (c) Estimated PSD2. (d) Estimated PSD3.

FIGURE 1.19

Spectral estimation.

Since the parametric approaches are simpler and more accurate, they will be emphasized in Chapter 12. However, parametric methods rely on some (a priori) information regarding the problem at hand.

Fig. 1.19 shows a noisy signal record and the estimated PSDs utilizing distinct methods. As can be observed, some of the resulting PSDs expose the presence of a sinusoid, not clearly observed in the time-domain signal.

1.13 Adaptive filtering

In many practical situations, we can extract information from a signal by comparing it to another signal or a reference. An adaptive filter is a natural solution for cases where the specifications are neither known nor time-invariant. The coefficients of these filters are data-driven, and the resulting processing task does not meet the homogeneity and additive conditions of linear filtering. An adaptive filter is also time-varying since its parameters are continually changing in order to minimize a cost function.

Adaptive filters are self-designing and perform the approximation step on-line. Starting from an initial condition, the adaptive filter searches for an optimal solution in an iterative format. Since adaptive filters are nonlinear filters, the analysis of their performance behavior requires the solution of time-domain equations, usually involving the second-order statistics of their internal and external quantities. Although analyzing an adaptive filter is more complicated than analyzing fixed filters, it is a simple solution to deal with unknown stationary and nonstationary environments. On the other hand, since the adaptive filters are self-designing filters, their design can be considered less involved than those of standard fixed digital filters [8]. Important topics related to adaptive filtering are neural networks and machine learning [13], [31].

Fig. 1.20 depicts the general setup of an adaptive-filtering environment, where, for a given iteration n, an input signal $x(n)$ excites the adaptive filter which will produce an output signal $y(n)$, which is then compared with a reference signal $d(n)$. From this comparison, an error signal $e(n)$ is formed according to $d(n) - y(n)$. The error signal is the argument of a cost (or objective) function whose minimization is achieved by properly adapting the adaptive filter coefficients. Fig. 1.21 shows a signal enhancement where it is recognized in the error signal that the input signal contained a sinusoidal component. In Fig. 1.21(a) it is difficult to observe the presence of a sinusoidal signal, whereas the error signal clearly shows a nearly sinusoidal behavior.

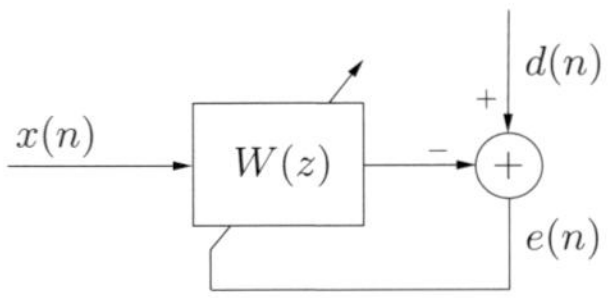

FIGURE 1.20

General adaptive filtering configuration.

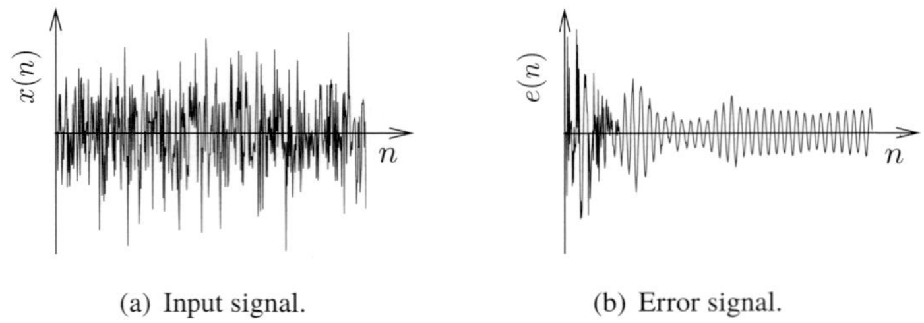

FIGURE 1.21

An adaptive filtering application.

Chapter 12 will present a comprehensive overview related to adaptive signal processing that unifies several concepts that go beyond the concepts covered in textbooks, such as [7,8,14,27]. The adaptive filters belong to the class of data-learning algorithms.

1.14 Machine learning: review and trends

Machine learning has increasingly been finding applications in many areas pertaining to the realm of artificial intelligence. However, there is room for further development, since in many situations, natural intelligence excels machine learning solutions relying on data. Despite its limitations, machine learning has achieved sounding advances in critical areas, highlighting solutions for prediction and classification in a wide variety of fields.

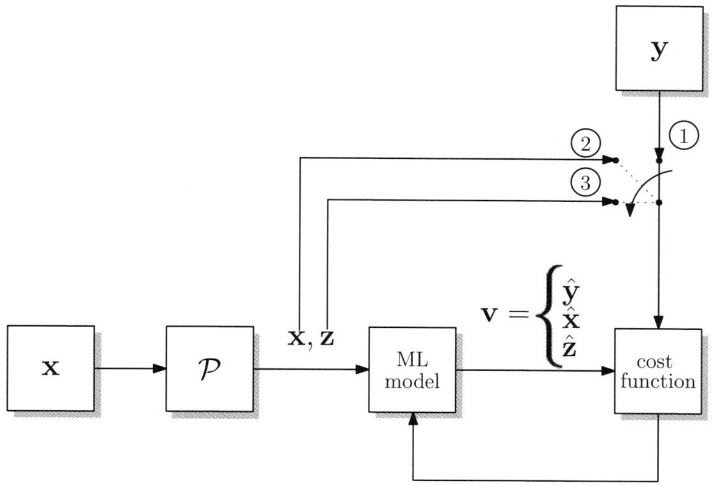

FIGURE 1.22

Using raw data $\mathbf{x}$ to train a machine learning model under different learning techniques.

Machine learning algorithms are usually divided into four categories: supervised, unsupervised, self-supervised, and semisupervised. A dataset having observations or inputs and matching labels or outputs is necessary for supervised machine learning models. By learning from this labeled dataset, they can forecast the result from unknown input data. Fig. 1.22 illustrates supervised learning when the switch is in position 1, and the process P rule is to multiply the input by the identity matrix. In this case, the machine learning model learns the relationship between the input data $\mathbf{x}$ and its label $\mathbf{y}$, and produces output $\mathbf{v} = \hat{\mathbf{y}}$. Unsupervised machine learning models, on the other hand, require observations. Finding latent patterns in unlabeled input data that could be utilized to categorize previously unknown input data is the goal of unsupervised learning. The switch in Fig. 1.22 is in position 2 in unsupervised mode, and the process $\mathcal{P}$ also multiplies the input by the identity matrix. The model is then trained by minimizing the cost function that considers the output $\mathbf{v} = \hat{\mathbf{x}}$ and the input data $\mathbf{x}$. Since neither technique uses annotations to learn, self-supervised learning can be viewed as a specific case of unsupervised learning. Self-supervised learning produces pseudolabels $\mathbf{z}$ by transforming the input data with $\mathcal{P}$, as shown in Fig. 1.22. With the switch in position 3, the model learns by minimizing the cost function between $\mathbf{v} = \hat{\mathbf{z}}$ and the pseudolabels $\mathbf{z}$. By using both labeled and unlabeled training data, semisupervised learning falls between supervised and unsupervised learning. Unlabeled data are used in semisupervised models to reorder the priority of hypotheses derived from labeled data. In this case, the switch in Fig. 1.22 might operate in position 1 or 3. For example, in reinforcement learning with ϵ-greed policy, the switch is in position 3, and P obtains a random label when ϵ is high. A memory replay stores the generated labels. Therefore, when ϵ is low, the switch moves to position 1, and the labels in memory replay are used to train the machine learning model.

This section provides a couple of examples illustrating the power of machine learning solutions to stimulate the readers to widen their knowledge in the chapters ahead.

For instance, given a training dataset with labels, artificial neural networks, or simply neural networks, are effective at learning complicated tasks in supervised learning. Neural networks are interconnected nodes organized in layers that mimic the human brain. The connections between the nodes serve as the neural network's weights and the nodes themselves can represent the input, a bias, an activation function, or the output. The error backpropagation algorithm is used by the neural network to compare input data with its label and update its weights as input data are fed into it. After presenting, for example, some images of different animals with correspondent labels during training, the model is able to classify an unseen panda image with 84.283555% of accuracy, as indicated by the left-hand entry of Fig. 1.23.

Although neural networks excel at tackling difficult problems in a variety of domains, they are incredibly weak against adversarial examples. Adversarial examples are perturbed versions of the original samples that are invisible to the human eye, but are powerful enough to cause the model to confidently misclassify the sample. For instance, by adding a small perturbation η, the model might misclassify a panda as an indri with high confidence, as illustrated in Fig. 1.23. Approaches like adversarial training can be employed to mitigate the effects of these malicious attacks, as will be further discussed in Chapter 13.

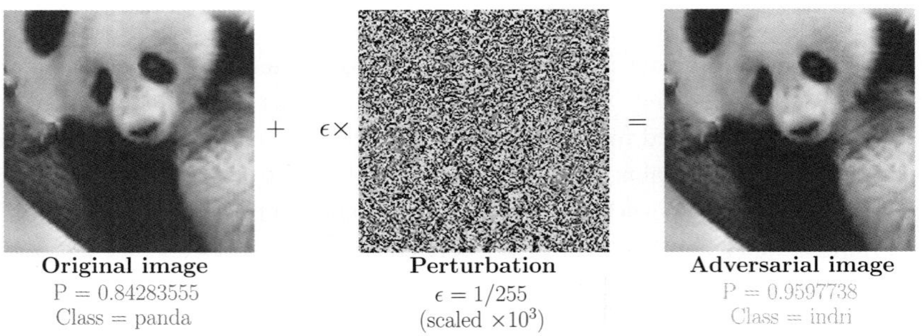

FIGURE 1.23

Example of adversarial image generation.

1.15 Signal processing over graphs

Large-scale interacting systems, including climate, social, and financial networks, have made a vast amount of data available. In those applications, the data or signals can be represented by structures known as graphs, which are composed of vertices and edges. For example, the graph's vertices could represent certain sensors positioned at a specific geographic location to monitor a climate variable, while the edges could reflect the distance among them. In this sense, graph signal processing (GSP) is a promising study area to deal with these signals that lie over irregular domains since many classical DSP methods are inappropriate for analyzing them. For instance, graph spectral analysis is regarded as the analog of Fourier analysis in DSP. Hence, well-known operations like translation, convolution, modulation, and filtering can be comparably described on the graph. Moreover, the eigenvalues and eigenvectors of particular graph-related matrices, such as the adjacency matrix and the Laplacian matrix, are connected to a spectrum in spectral graph theory. For example, the eigenvalues of the Laplacian matrix, λ_l, correspond to the frequencies, and the eigenvectors, $\mathbf{v}_l$, correspond to the complex exponentials in traditional DSP. In this way, the quicker the eigenvector $\mathbf{v}_l$ changes values, the higher the corresponding frequency λ_l.

Analyzing how frequently the signal changes polarity is one method of determining the signal frequency. This can be accomplished by counting the number of edges that connect samples of different polarities. To illustrate that, we highlight in red (mid gray in print version) the edges of the graph in Fig. 1.24(a) that connect a positive sample to a negative one for some eigenvectors of the Laplacian operator in Fig. 1.24(b–f). Observe that the number of red edges (mid gray in print version) increases for higher eigenvalues, indicating that the corresponding eigenvectors change values more often, as expected. This confirms that the eigenvalues play the role of frequencies in graph signals.

Chapter 14 introduces the fundamental concepts of GSP, starting with the GSP building blocks, which are surrogates of the classical signal processing network structures, along with some practical applications.

1.16 Tensor methods in deep learning

Deep neural networks include a large number of parameters requiring a considerable amount of storage elements as well as high computational complexity. As a result, it is crucial that such complex networks benefit from sparse representation employing tensors enabling multidimensional data and mappings representations. Furthermore, many sophisticated learning architectures entail multidimensional blocks of parameters, which higher-order tensors can efficiently represent.

This section highlights the tensors as a multidimensional generalization of matrices and exemplifies how tensors can be used to represent and analyze information hidden in multidimensional data. Tensor decompositions are critical to the reduction in the number of unknown parameters of deep models and further mitigate the curse of dimensionality. Chapter 15 addresses the fundamentals of tensors and their decompositions and discusses the application of tensor methods in emerging fields.

Fig. 1.25 shows an example of a tensor applied to the MNIST dataset. The resulting tensor can be represented by $\mathcal{X} \in \mathbb{R}^{I_1 \times I_2 \times I_3}$, in which I_1 is the height, I_2 is the width, and I_3 is the number of images.

It is possible to reduce the tensor's high dimensionality, which produces compact representations and lowers computational complexity. To do so, we need a model $\mathcal{M}$ that approximates $\mathcal{X} \in \mathbb{R}^{I_1 \times I_2 \times I_3}$ such that

$$\mathcal{X} \approx \mathcal{M} = \sum_{r=1}^{R} \mathbf{u}_r^{(1)} \circ \mathbf{u}_r^{(2)} \circ \mathbf{u}_r^{(3)}, \tag{1.1}$$

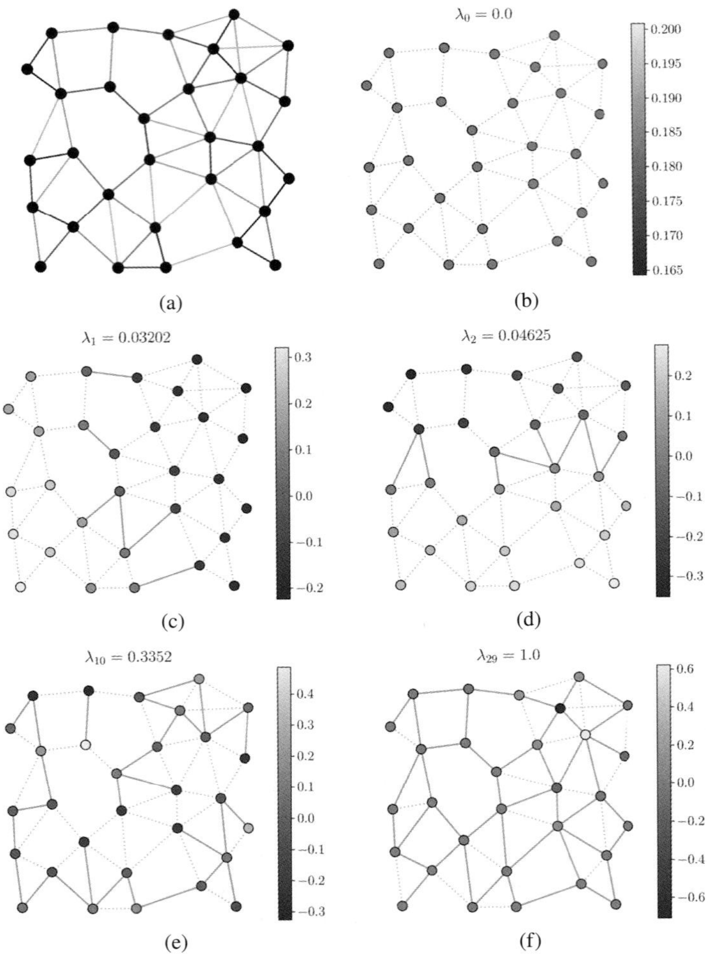

FIGURE 1.24

Edges connecting samples with different polarities for Laplacian eigenvectors on a randomly generated graph.

where the third-order tensor $\mathcal{M}$ is a sum of rank-1 tensors $u_r^{(n)}, r = 1, \cdots R$. These vectors can be combined into three matrices, each of which has the format

$$\mathbf{U}^{(n)} = \left[\mathbf{u}_1^{(n)} \quad \mathbf{u}_2^{(n)} \quad \mathbf{u}_R^{(n)}\right]. \tag{1.2}$$

Often referred to as rank (not to be confused with matrix rank), R is the data's reduced dimension. By using, for instance, $R = 20$, we obtain the sparse matrices $\mathbf{U}^{(1)}$, $\mathbf{U}^{(2)}$, and $\mathbf{U}^{(3)}$ illustrated in Fig. 1.26. In this way, the samples in the second row of Fig. 1.27 are the results of reconstruction using equa-

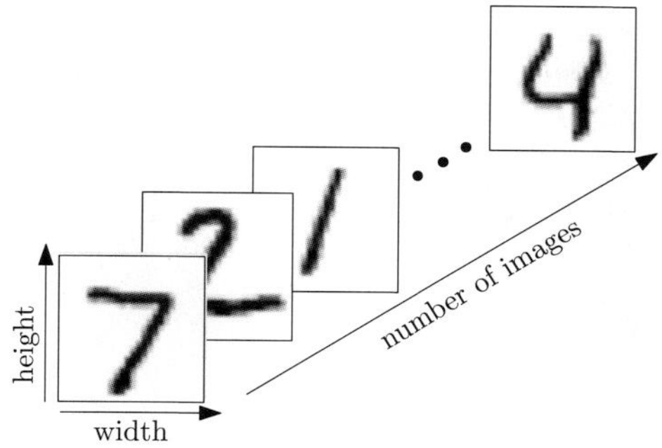

FIGURE 1.25

MNIST dataset as a tensor.

tion (1.1). We can see that the reconstruction was successful by comparing the first and second rows of samples in Fig. 1.27.

1.17 Nonconvex graph learning: sparsity, heavy tails, and clustering

This section illustrates some tools to promote sparsity and to deal with data outliers of heavy-tailed distributions, relying on the graph formulation. This graph learning framework naturally represents the connectivity experienced by devices that are the sources of big datasets. For instance, typical graph learning methods might rely on cost functions represented by convex or nonconvex formulations.

Graph learning algorithms aim to learn a Laplacian matrix, or equivalently an adjacency matrix relying on data. The problem is solved using an optimization formulation involving the penalized maximum likelihood estimator of the Laplacian-constrained precision matrix, utilizing the observed data.

This book introduces many optimization formulations for connected sparse graphs, heavy-tailed distributions, and k-component graphs widely employed in machine learning clustering. Several optimization methods are described to solve nonconvex problems found in many machine learning applications.

For instance, consider a sensor array in a forest collecting temperature measurements as illustrated in Fig. 1.28.

The sensors are jointly utilized to estimate a variable of interest $\mathbf{x} \in \mathbb{R}^{N\times 1}$. From their M linear measurements,

$$y_m = \mathbf{a}_m^{\mathrm{T}}\mathbf{x} + n_m \text{ for } m = 1, \cdots M, \tag{1.3}$$

where $\mathbf{a} \in \mathbb{R}^{N\times 1}$ characterizes the measurement and n_m represents noise, we can obtain the maximum likelihood estimate of $\mathbf{x}$. Since some of the sensors are closer to the fire, their contribution to the final estimation is more important. Our goal here is to obtain this subset of sensors that is more informative to estimate $\mathbf{x}$. By following one of the criteria to select the antennas based on minimizing a function of

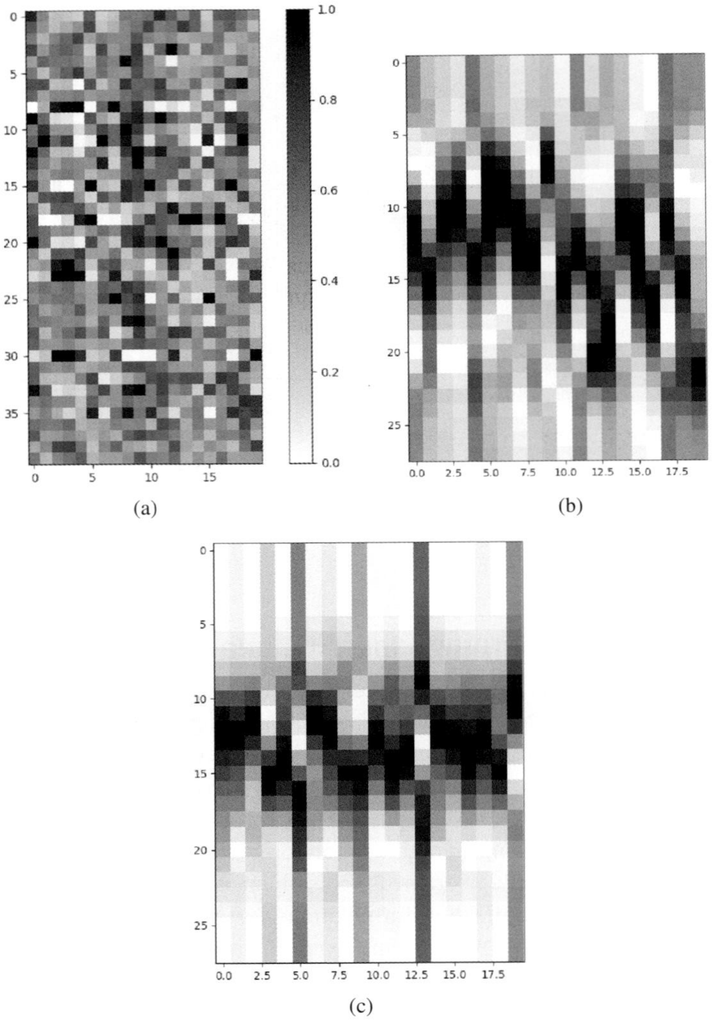

FIGURE 1.26

The decomposition gives us the matrices (a) $\mathbf{U}^{(1)}$, (b) $\mathbf{U}^{(2)}$, and (c) $\mathbf{U}^{(3)}$.

the estimation error, we end up with the following optimization problem:

$$\begin{aligned} &\max_{\mathbf{z}\in\mathbb{R}^{M\times 1}} \quad \operatorname{logdet}\left(\sum_{m=1}^{M} z_m \mathbf{a}_m \mathbf{a}_m^{\mathrm{T}}\right), \\ &\text{s.t.} \quad \|\mathbf{z}\|_0 = S, \end{aligned} \tag{1.4}$$

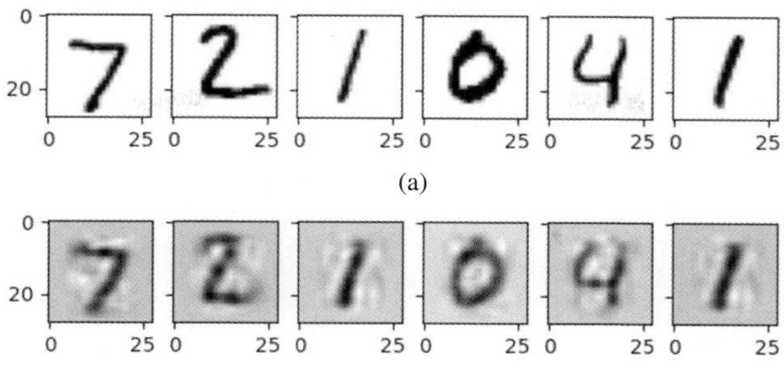

FIGURE 1.27

(a) Samples from original tensor $\mathcal{X}$ and (b) reconstructed samples from approximated tensor $\mathcal{M}$.

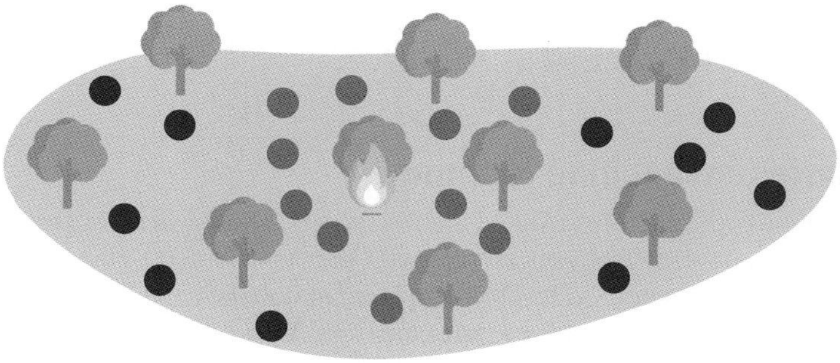

FIGURE 1.28

Sensors in a forest to estimate a variable of interest $\mathbf{x}$ from their temperature measurements y_m, $m = 1 \cdots M$.

where S is the number of selected sensors. However, the optimization problem in equation (1.4) is not convex. A possible way to solve the problem in equation (1.4) is to replace the l_0-norm constraint by convex constraints as in

$$\begin{aligned} \max_{\mathbf{z} \in \mathbb{R}^{M \times 1}} \quad & \text{logdet}\left(\sum_{m=1}^{M} z_m \mathbf{a}_m \mathbf{a}_m^{\mathrm{T}} \right), \\ \text{s.t.} \quad & \mathbf{1}^{\mathrm{T}} \mathbf{z} = S, \; 0 \leq z_m \leq 1, \; m = 1, \cdots M. \end{aligned} \tag{1.5}$$

For instance, we can solve the problem in equation (1.5) for $S = 10$ selected sensors and $M = 20$ measurements and obtain the selection vector $\mathbf{z}$ illustrated in Fig. 1.29. Observe that the S greatest z-values correspond to the most significant columns in matrix $\mathbf{A} = [\mathbf{a}_1, \cdots \mathbf{a}_M]$, also shown in Fig. 1.29.

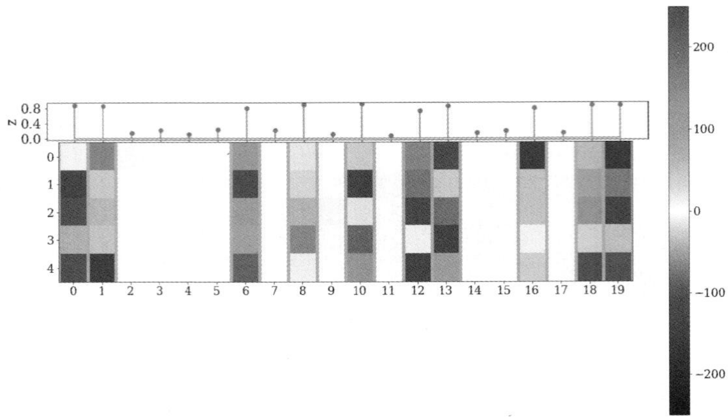

FIGURE 1.29

Matrix of measurements **A** with columns representing sensors. The selected sensors are highlighted in green (mid gray in print version), and they have the greatest z-values.

1.18 Dictionaries in machine learning

Machine learning, and more generally artificial intelligence, is a field where one tries, in some sense, to mimic how living beings develop from interacting with the environment. Therefore, the success of an machine learning algorithm depends on its ability to learn and extract sparse and compact latent signal and feature representations from measured natural datasets. This section briefly discusses the benefits of dictionary learning to aid the solution to machine learning problems.

Chapter 17 suggests easy ways to combine dictionary learning with, for instance, a neural network classifier. The idea is to perform a preprocessing step aiming at generating a dictionary-based sparse feature extraction. Then, with the sparse codes and dictionary elements, the network ahead can adapt its weights, capitalizing on the sparse codes provided by the dictionary. The connections between sparse coding and some types of neural networks also improve the interpretation of how the networks learn from data and the resulting representation.

A dictionary-based sparse feature extraction preprocessing stage can be added to the neural network as a technique to combine dictionary learning with a classifier. First, the dictionary **D** and the sparse codes $\mathbf{x}_l$ are learned aiming to minimize the loss function

$$\mathcal{L} = \frac{1}{l}\sum_{l=1}^{L}\frac{1}{2}\|\mathbf{y}_l - \mathbf{D}\mathbf{x}_l\|_2^2 + \lambda\|\mathbf{x}_l\|_1, \tag{1.6}$$

where $l = 1 \cdots L$, L is the number of data samples, and $\mathbf{y}_l$ is the lth labeled data sample.

Take a labeled dataset of faces $\mathbf{y}_l$ and their corresponding one-hot-encoded IDs $\mathbf{z}_l$ as an illustration. Some samples of this dataset are shown in Fig. 1.30(a). We then solve equation (1.6) and compute the dictionary **D** with 100 atoms and the sparse codes $\mathbf{x}_l$. Fig. 1.30(b) shows some of the dictionary components that can be reorganized as images. Observe how some features, like glasses, are highlighted by the atoms. Fig. 1.31 depicts the matrix with the columns representing the sparse codes. As expected, the majority of the sparse codes' elements are zero. Each labeled data $\mathbf{y}_l \in \mathbb{R}^{4096\times 1}$ is then replaced by its 100-sparse representation vector $\mathbf{x}_l \in \mathbb{R}^{100\times 1}$. The obtained sparse dataset is then used to train a simple neural network with only one hidden layer by forcing the output of the softmax function to match to the one-hot class encoding vectors $\mathbf{z}_l$. Fig. 1.32 shows that under this basic setup, validation accuracy can be maintained to around 80% even when using a condensed version of the input images.

(a)

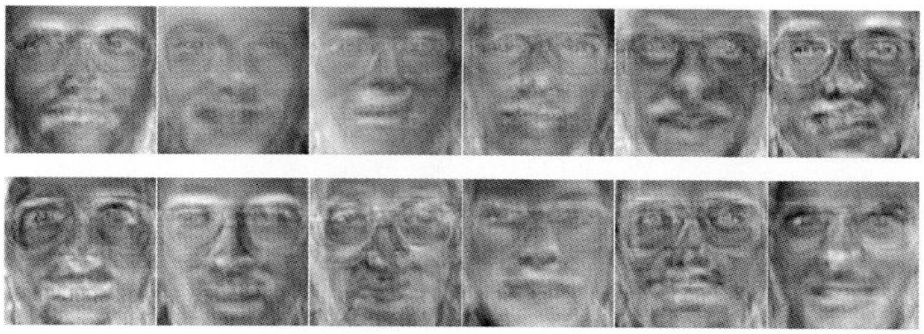

(b)

FIGURE 1.30

(a) Samples from the original dataset and (b) samples from the obtained dictionary with 100 atoms.

1.19 Closing comments

The theory of signal processing and machine learning has been developing for over five decades at a fast pace. The following chapters will cover many of the classical tools widely employed in applications that we perceive in our daily life through the use of mobile phones, media players, medical equipment, transportation, drones, robots, and so on. These chapters will also cover recent advances and describe many new tools that can potentially be utilized in new applications and further investigated. The authors of the chapters are frequent and vital contributors to the field of signal processing and machine learning theory. The goal of each chapter is to provide an overview and a brief tutorial on essential topics pertaining to signal processing and machine learning theory, including critical references for further studies. In addition, in the present chapter we attempted to describe the context in which each family of tools is employed.

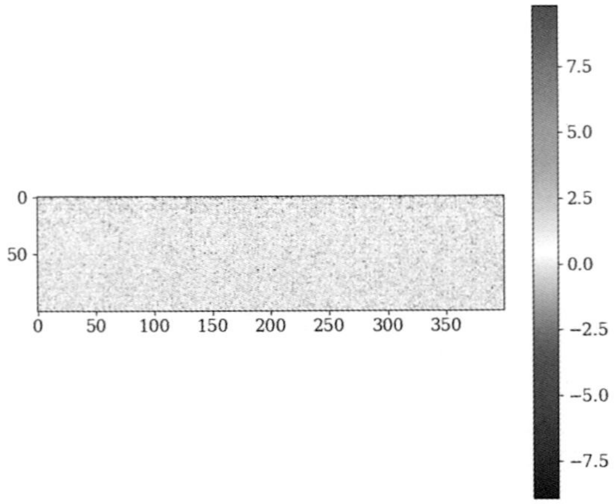

FIGURE 1.31

k-Sparse vectors obtained by solving equation (1.6) with $k = 100$ atoms and $L = 400$ images.

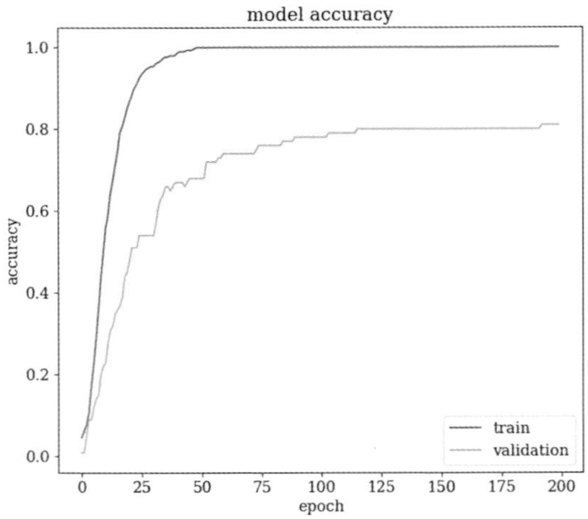

FIGURE 1.32

Accuracy obtained during training and validation.

The rest of this book is organized into 16 more chapters covering in more depth the subjects pertaining to signal processing and machine learning theory outlined in the present chapter. Using a bird's-eye view, we can infer that the contents ahead represent the many possible stages of acquiring data for several purposes. Chapters 2, 3, and 4 deal with descriptions of continuous and discrete signals and their basic manipulation that could be the dataset's source for further processing. Chapter 5 discusses issues related to sampling and quantization, which is part of preparing the data to be adequately represented for filtering, as discussed in Chapter 6. Filters are part of the mapping required to expose data or signal features, including sampling rate changes, widening the set of methods for analysis and systems design, as seen in Chapter 7. Then, with the previous tools, Chapters 8, 9, and 10 discuss many data-mapping methods that are instrumental in analyzing the data as the final target or performing a preprocessing step for further learning. In sequence, the subjects of Chapters 11 and 12 belong to the class of algorithms that allows for learning model parameters from data. Chapter 13 presents several alternatives for machine learning, broadening the power to learn from large datasets. Furthermore, Chapters 14 and 16 discuss methods to learn from data related to the nodes of a graph, where analysis, design, and learning tools are applied to data defined on graph structures. Finally, Chapters 15 and 17 deal with the techniques required to learn features from large datasets with sparse representations and utilize adequate data preprocessing. The cross-fertilization between these areas widens our ability to interpret the mechanisms behind the data-learning properties of the algorithms, network structures, and applications. Pick your choices and enjoy the ride!

References

[1] A. Antoniou, W.-S. Lu, Practical Optimization: Algorithms and Engineering Applications, 2nd edition, Springer, New York, NY, 2021.

[2] Y.S. Abu-Mostafa, M. Magdon-Ismail, H.-T. Lin, Learning from Data, AMLbook.com, Pasadena, CA, 2012.

[3] A. Antoniou, On the roots of digital signal processing: Part I and II, IEEE Circuits and Systems Magazine 7 (1) (2007) 8–18 and 7 (4) (2007) 8–19.

[4] A. Antoniou, Digital Filters: Analysis, Design, and Signal Processing Applications, McGraw-Hill, New York, NY, 2018.

[5] C.M. Bishop, Pattern Recognition and Machine Learning, Springer, New York, NY, 2006.

[6] R.E. Crochiere, L.R. Rabiner, Multirate Digital Signal Processing, Prentice-Hall, Englewood Cliffs, NJ, 1983.

[7] P.S.R. Diniz, M.L.R. de Campos, W.A. Martins, M.V.S. Lima, J.A. Apolinário Jr., Online Learning and Adaptive Filters, Cambridge University Press, Cambridge, UK, 2022.

[8] P.S.R. Diniz, Adaptive Filtering: Algorithms and Practical Implementation, 5th edition, Springer, New York, NY, 2020.

[9] P.S.R. Diniz, W.A. Martins, M.V.S. Lima, Block Transceivers: OFDM and Beyond, Springer, New York, NY, 2012.

[10] P.S.R. Diniz, E.A. Silva, S.L. Netto, Digital Signal Processing: System Analysis and Design, 2nd edition, Cambridge University Press, Cambridge, UK, 2010.

[11] R. Fletcher, Practical Methods of Optimization, John Wiley & Sons, Chichester, UK, 1980.

[12] I. Goodfellow, Y. Bengio, A. Courville, Deep Learning, MIT Press, Cambridge, MA, 2016.

[13] S. Haykin, Neural Networks and Learning Machines, 3rd edition, Prentice-Hall, Englewood Cliffs, NJ, 2008.

[14] S. Haykin, Adaptive Filter Theory, 5th edition, Prentice-Hall, Englewood Cliffs, NJ, 2014.

[15] A. Hyvärinen, J. Karhunen, E. Oja, Independent Component Analysis, Wiley Interscience, New York, NY, 2001.

[16] T. Hastie, R. Tibshirani, J. Friedman, Elements of Statistical Learning: Data Mining, Inference, and Prediction, Springer, New York, NY, 2016.

[17] S.M. Kay, Modern Spectral Estimation: Theory and Application, Prentice-Hall, Englewood Cliffs, NJ, 1988.

[18] D.G. Luenberger, Introduction to Linear and Nonlinear Programming, 2nd edition, Addison-Wesley, Boston, MA, 1984.

[19] S. Mallat, A Wavelet Tour of Signal Processing: The Sparse Way, 3rd edition, Academic Press, Burlington, MA, 2009.

[20] S.K. Mitra, Digital Signal Processing: A Computer-Based Approach, 4th edition, McGraw-Hill, New York, NY, 2010.
[21] K.P. Murphy, Machine Learning: A Probabilistic Perspective, MIT Press, Cambridge, MA, 2012.
[22] A.V. Oppenheim, R.W. Schaffer, Discrete-Time Signal Processing, 3rd edition, Prentice-Hall, Englewood Cliffs, NJ, 2010.
[23] A.V. Oppenheim, A.S. Willsky, A.H. Nawab, Signals and Systems, 2nd edition, Prentice-Hall, Englewood Cliffs, NJ, 1997.
[24] J.G. Proakis, D.G. Manolakis, Digital Signal Processing: Principles, Algorithms, and Applications, 4th edition, Prentice-Hall, Englewood Cliffs, NJ, 2006.
[25] J.G. Proakis, D.G. Manolakis, Applied Digital Signal Processing, Cambridge University Press, Cambridge, UK, 2011.
[26] A. Papoulis, S. Unnikrishna Pillai, Probability, Random Variables, and Stochastic Processes, 4th edition, McGraw-Hill, New York, NY, 2002.
[27] A.H. Sayed, Adaptive Filters, John Wiley & Sons, Hoboken, NJ, 2008.
[28] J.J. Shynk, Probability, Random Variables, and Random Processes: Theory and Signal Processing Applications, John Wiley & Sons, Hoboken, NJ, 2013.
[29] J.B.O. Souza Filho, L.-D. Van, T.-P. Jung, P.S.R. Diniz, Online component analysis, architectures and applications, Foundations and Trends in Signal Processing 16 (3–4) (2022) 224–429, https://doi.org/10.1561/2000000112, NOW Publishers, Boston, MA.
[30] B. Schölkopf, A.J. Smola, Learning with Kernels: State Vector Machines, Regularization, Optimization, and Beyond, MIT Press, Cambridge, MA, 2002.
[31] S. Theodoridis, Machine Learning: A Bayesian and Optimization Perspective, 2nd edition, Academic Press, London, UK, 2020.
[32] D. Tse, P. Viswanath, Fundamentals of Wireless Communication, Cambridge University Press, Cambridge, UK, 2005.
[33] P.P. Vaidyanathan, Multirate Systems and Filter Banks, Prentice-Hall, Englewood Cliffs, NJ, 1993.
[34] H.L. Van Trees, Optimum Array Processing, John Wiley & Sons, New York, NY, 2002.
[35] V.N. Vapnik, Statistical Learning Theory, Wiley Interscience, New Delhi, India, 1998.
[36] M. Vetterli, J. Kovačević, Wavelets and Subband Coding, Prentice-Hall, Englewood Cliffs, NJ, 1995.
[37] P.P. Vaidyanathan, S.-M. Phoong, Y.-P. Lin, Signal Processing and Optimization for Transceiver Systems, Cambridge University Press, Cambridge, UK, 2010.

CHAPTER

Continuous-time signals and systems

2

José A. Apolinário Jr.[a] **and Carla L. Pagliari**[a]

Military Institute of Engineering (IME), Department of Electrical Engineering (SE/3), Rio de Janeiro, RJ, Brazil

2.1 Introduction

Most signals present in our daily lives are continuous in time such as music and speech. A signal is a function of an independent variable, usually an observation measured from the real world such as position, depth, temperature, pressure, or time. Continuous-time signals have a continuous independent variable. The velocity of a car could be considered a continuous-time signal if, at any time t, the velocity $v(t)$ could be defined. A continuous-time signal with a continuous amplitude is usually called an analog signal, speech signals being a typical example.

Signals convey information and are generated by electronic or natural means such as when someone talks or plays a musical instrument. The goal of signal processing is to extract the useful information embedded in a signal.

Electronics for audio and last-generation mobile phones rely on universal concepts associated with the flow of information to make these devices fully functional. Therefore, a system designer, in order to have the necessary understanding of advanced topics, needs to master basic signals and systems theory. Systems theory could be defined as the study of the relation between input and output signals.

As characterizing the complete input/output properties of a system through measurement is, in general, impossible, the idea is to infer the system response to nonmeasured inputs based on a limited number of measurements. System design is a challenging task. However, several systems can be accurately modeled as linear systems. Hence, the designer has to create an expected, or predictable, relationship between the input and the output that is always the same (time-invariant). In addition, for a range of possible input types, the system should generate a predictable output in accordance with the input/output relationship.

A continuous-time signal is defined on the continuum of time values as the one depicted in Fig. 2.1, $x(t)$ for $t \in \mathrm{R}$. This signal is actually periodic, the period being equal to one, and the figure shows the interval of two periods.

Although signals are real mathematical functions, some transformations can produce complex signals that have both real and imaginary parts. Therefore, throughout this book, complex-domain equivalents of real signals will certainly appear.

[a] Authors thank Prof. Ney Bruno (*in memoriam*) for his kind and competent review of this chapter.

Signal Processing and Machine Learning Theory. https://doi.org/10.1016/B978-0-32-391772-8.00008-9
Copyright © 2024 Elsevier Inc. All rights reserved.

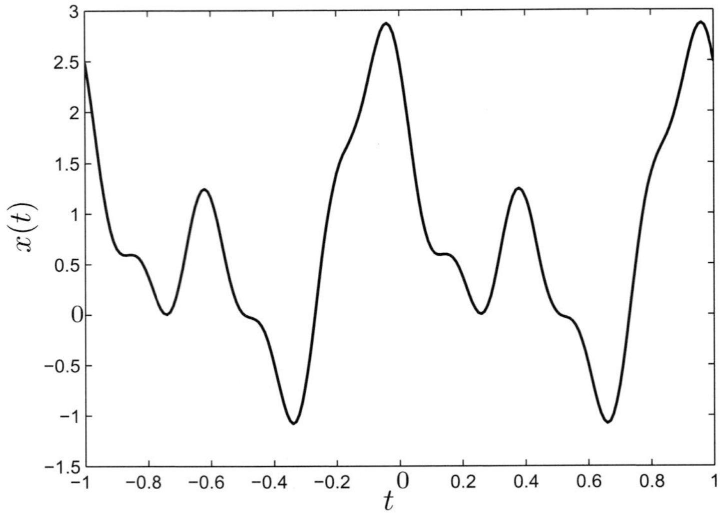

FIGURE 2.1

Example of a continuous-time signal.

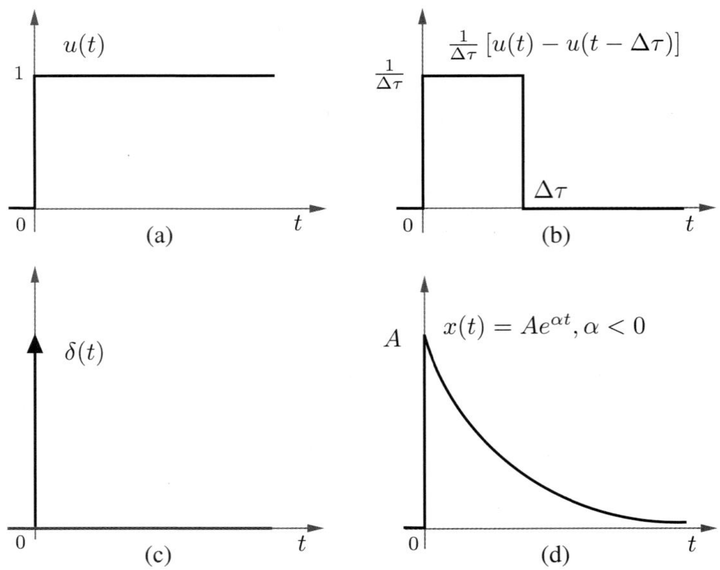

FIGURE 2.2

Basic continuous-time signals: (a) unit-step signal $u(t)$, (b) unit-area pulse, (c) impulse signal (unit-area pulse when $\Delta\tau \to 0$), and (d) exponential signal.

Elementary continuous-time signals are the basis to build more intricate signals. The unit-step signal, $u(t)$, displayed in Fig. 2.2(a) is equal to 1 for all time greater than or equal to zero and is equal to zero for all time less than zero. A step of height K can be made with $\mathrm{K}u(t)$.

A unit-area pulse is pictured in Fig. 2.2(b); as $\Delta\tau$ gets smaller, the pulse gets higher and narrower with a constant area of one (unit area). We can see the impulse signal in Fig. 2.2(c) as a limiting case of the pulse signal when the pulse is infinitely narrow. The impulse response will help to estimate how the system will respond to other possible *stimuli*, and we will further explore this topic in the next section.

Exponential signals are important for signals and systems as eigensignals of linear time-invariant (LTI) systems. In Fig. 2.2(d), for A and α being real numbers, α negative, the signal is a decaying exponential. When α is a complex number, the signal is called a complex exponential signal. The periodic signals sine and cosine are used for modeling the interaction of signals and systems, and the usage of complex exponentials to manipulate sinusoidal functions turns trigonometry into elementary arithmetic and algebra.

2.2 Continuous-time systems

We start this section by providing a simplified classification of continuous-time systems in order to focus on our main interest: an LTI system.

Why are linearity and time-invariance important? In general, real-world systems can be successfully modeled by theoretical LTI systems; in an electrical circuit, for instance, a system can be designed using a well-developed linear theory (valid for LTI systems) and be implemented using real components such as resistors, inductors, and capacitors. Even though the physical components, strictly speaking, do not comply with linearity and time-invariance for any input signals, the circuit will work quite well under reasonable conditions (input voltages constrained to the linear working ranges of the components).

A system can be modeled as a function that transforms, or maps, the input signal into a new output signal. Let a continuous-time system be represented as in Fig. 2.3, where $x(t)$ is the input signal and $y(t)$, its output signal, corresponds to a transformation applied to the input signal: $y(t) = \mathcal{T}\{x(t)\}$.

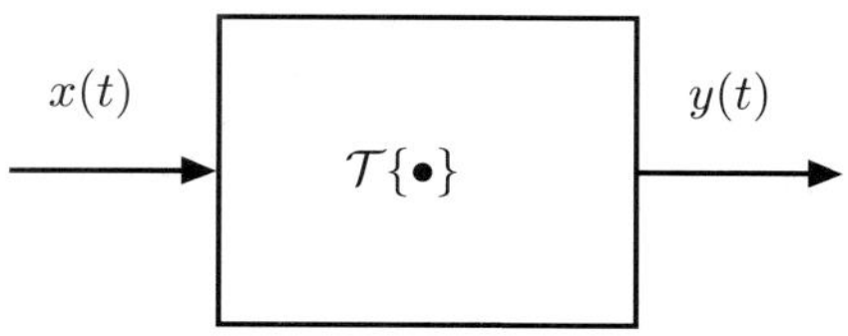

FIGURE 2.3

A generic continuous-time system representing the output signal as a transformation applied to the input signal: $y(t) = \mathcal{T}\{x(t)\}$.

The two following properties define an LTI system.

Linearity: A continuous-time system is said to be linear if a linear combination of input signals, when applied to this system, produces an output that corresponds to a linear combination of individual outputs, i.e.,

$$\begin{aligned} y(t) &= \mathcal{T}\{a_1x_1(t) + a_2x_2(t) + \cdots\} \\ &= a_1\mathcal{T}\{x_1(t)\} + a_2\mathcal{T}\{x_2(t)\} + \cdots \\ &= a_1y_1(t) + a_2y_2(t) + \cdots, \end{aligned} \tag{2.1}$$

where $y_i(t) = \mathcal{T}\{x_i(t)\}$.

Time-invariance: A continuous-time system is said to be time-invariant when the output signal, $y(t)$, corresponding to an input signal $x(t)$, will be a delayed version of the original output, $y(t - t_0)$, whenever the input is delayed accordingly, i.e.,

$$\begin{aligned} &\text{if} && y(t) = \mathcal{T}\{x(t)\}, \\ &\text{then} && y(t - t_0) = \mathcal{T}\{x(t - t_0)\}. \end{aligned}$$

The transformation caused in the input signal by an LTI system may be represented in a number of ways: with a set of differential equations, with the help of a set of state variables, with the aid of the concept of the impulse response, or in a transformed domain.

Let the circuit in Fig. 2.4 be an example of a continuous-time system where the input $x(t)$ corresponds to the voltage between terminals $\mathcal{A}$ and $\mathcal{B}$ while its output $y(t)$ is described by the voltage between terminals $\mathcal{C}$ and $\mathcal{D}$.

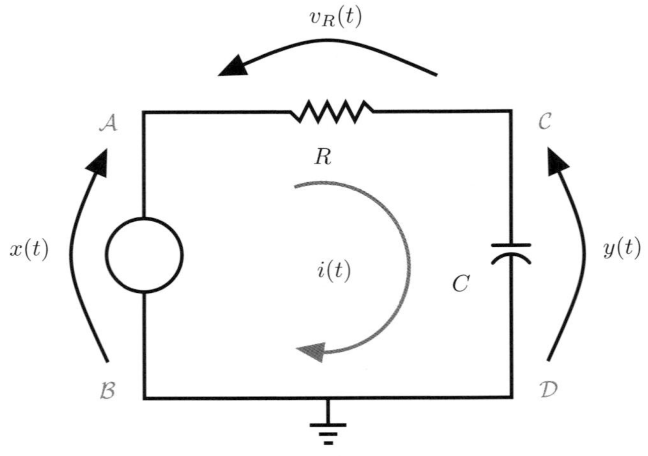

FIGURE 2.4

An example of an electric circuit representing a continuous-time system with input $x(t)$ and output $y(t)$.

We know for this circuit, according to Kirchhoff's voltage law (KVL), that $x(t) = y(t) + v_R(t)$ and that the current in the capacitor, $i(t)$, corresponds to $C dy(t)/dt$. Therefore, using Ohm's law, we obtain $v_R(t) = Ri(t) = RC dy(t)/dt$, and we can write

$$RC\frac{dy(t)}{dt} + y(t) = x(t), \tag{2.2}$$

which is an input–output (external) representation of a continuous-time system given as a differential equation.

Another representation of a linear system, widely used in control theory, is the state-space representation. This representation will be presented in the next section since it is better explored for systems having a higher-order differential equation.

In order to find an explicit expression for $y(t)$ as a function of $x(t)$, we need to solve the differential equation. The complete solution of this system, as shall be seen in the following section, is given by the sum of an homogeneous solution (zero-input solution or natural response) and a particular solution (zero-state solution or forced solution):

$$y(t) = \underbrace{y_h(t)}_{\text{homogeneous solution}} + \underbrace{y_p(t)}_{\text{particular solution}}. \tag{2.3}$$

The homogeneous solution, in our example, is obtained from (2.2) by setting $RDdy_h(t)/dt + y_h(t) = 0$. Since this solution and its derivatives must comply with the homogeneous differential equation, its usual choice is an exponential function such as Ke^{st}, s being, in general, a complex number. Replacing this general expression in the homogeneous differential equation, we obtain $RCs + 1 = 0$ (characteristic equation) and, therefore, $s = -1/RC$, such that the homogeneous solution becomes

$$y_h(t) = Ke^{-\frac{t}{RC}}, \tag{2.4}$$

where K is a constant.

The particular solution is usually of the same form as the forcing function (input voltage $x(t)$ in our example) and it must comply with the differential equation without an arbitrary constant. In our example, if $x(t) = Me^{-mt}$, $y_p(t)$ would be $\frac{M}{1-mRC}e^{-mt}$.

If we set $x(t) = u(t)$, the step function which can be seen as a DC voltage of 1 V switched at time instant $t = 0$, the forced solution is equal to one when $t \geq 0$. For this particular input $x(t) = u(t)$, the output signal is given by $y(t) = (Ke^{-\frac{t}{RC}} + 1)u(t)$. Constant K is obtained with the knowledge of the initial charge of the capacitor (*initial conditions*); assuming it is not charged ($v_c(t) = 0,\ t \leq 0$), we have

$$y(t) = (1 - e^{-\frac{t}{RC}})u(t), \tag{2.5}$$

which can be observed in Fig. 2.5.

Note that, for this case where $x(t) = u(t)$, the output $y(t)$ is known as the *step response*, i.e., $r(t) = y(t)\,|_{x(t)=u(t)}$. Moreover, since the impulse $\delta(t)$ corresponds to the derivative of $u(t)$, $\delta(t) = du(t)/dt$, and the system is LTI, we may write $dr(t)/dt = d\mathcal{T}\{u(t)\}/dt$, which corresponds to $\mathcal{T}\{du(t)/dt\}$, the transformation applied to the impulse signal leading to

$$h(t) = \frac{dr(t)}{dt}, \tag{2.6}$$

where $h(t)$ is known as the *impulse response* or $h(t) = y(t)\,|_{x(t)=\delta(t)}$.

The impulse response is, particularly, an important characterization of a linear system. It will help to estimate how the system will respond to other *stimuli*. In order to show the relevance of this, let us

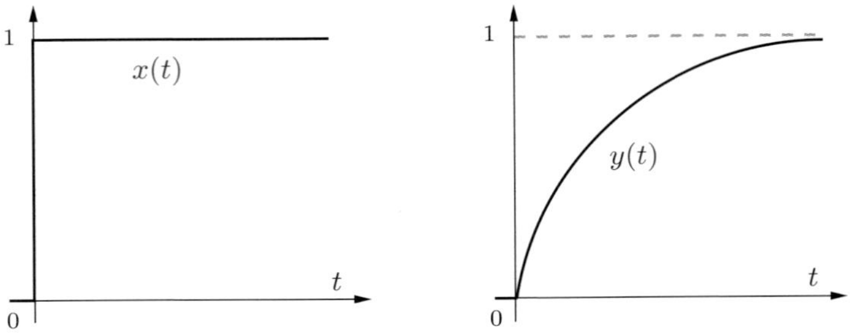

FIGURE 2.5

Input and output signals of the system depicted in Fig. 2.4. Click here to watch the animation.

define the unit impulse as

$$\delta(t) = \lim_{\Delta\tau \to 0} \frac{1}{\Delta\tau} [u(t) - u(t - \Delta\tau)], \tag{2.7}$$

such that we may see an input signal $x(t)$ as in Fig. 2.6:

$$x(t) \approx \sum_{k=-\infty}^{\infty} x(k\Delta\tau) \underbrace{\left[\frac{u(t - k\Delta\tau) - u(t - k\Delta\tau - \Delta\tau)}{\Delta\tau} \right]}_{\delta(t-k\Delta\tau) \text{ when } \Delta\tau \to 0} \Delta\tau. \tag{2.8}$$

The idea is to show that any signal, e.g., $x(t)$, can be expressed as a sum of scaled and shifted impulse functions.

In (2.8), making $\Delta\tau \to 0$ and representing the resulting continuous time by τ instead of $k\Delta\tau$, we can use the integral instead of the summation and find

$$x(t) = \int_{-\infty}^{\infty} x(\tau)\delta(t - \tau)d\tau. \tag{2.9}$$

In an LTI system, using the previous expression as an input signal to compute the output $y(t) = \mathcal{T}\{x(t)\}$, we obtain $\int_{-\infty}^{\infty} x(\tau)\mathcal{T}\{\delta(t-\tau)\}d\tau$ such that the output signal can be written as

$$\boxed{y(t) = \int_{-\infty}^{\infty} x(\tau)h(t - \tau)d\tau.} \tag{2.10}$$

This expression is known as the *convolution* integral between the input signal and the system impulse response, and is represented as

$$y(t) = x(t) * h(t). \tag{2.11}$$

Please note that the output signal provided by the convolution integral corresponds to the zero-state solution.

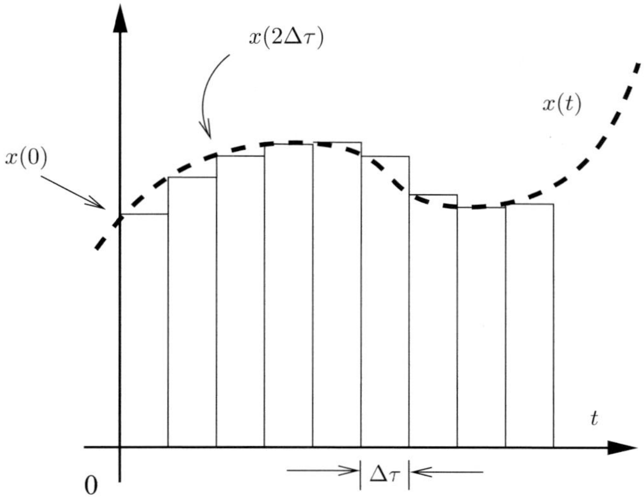

FIGURE 2.6

A signal $x(t)$ can be obtained from the pulses if we make $\Delta\tau \to 0$.

From Fig. 2.6, another approximation for $x(t)$ can be devised, leading to another expression relating input and output signals. At a certain instant $t = \tau$, let the angle of a tangent line to the curve of $x(t)$ be θ such that $\tan\theta = \frac{dx(\tau)}{d\tau}$. Assuming a very small $\Delta\tau$, this tangent can be approximated by $\frac{\Delta x(\tau)}{\Delta\tau}$ such that

$$\Delta x(\tau) = \frac{dx(\tau)}{d\tau}\Delta\tau. \tag{2.12}$$

Knowing each increment at every instant $t = k\Delta\tau$, we can visualize an approximation for $x(t)$ as a sum of shifted and weighted step functions $u(t - k\Delta\tau)$, i.e.,

$$x(t) \approx \sum_{k=-\infty}^{\infty} \Delta x(k\Delta\tau)u(t - k\Delta\tau) = \sum_{k=-\infty}^{\infty} \frac{dx(k\Delta\tau)}{d\tau}u(t - k\Delta\tau)\Delta\tau. \tag{2.13}$$

From the previous expression, if we once more, as in (2.9), make $\Delta\tau \to 0$ and represent the resulting continuous time by τ instead of $k\Delta\tau$, we can drop the summation and use the integral to obtain

$$x(t) = \int_{-\infty}^{\infty} \frac{dx(\tau)}{d\tau}u(t - \tau)d\tau. \tag{2.14}$$

Assuming, again, that the system is LTI, we use the last expression as an input signal to compute the output $y(t) = \mathcal{T}\{x(t)\}$, obtaining $\int_{-\infty}^{\infty} \frac{dx(\tau)}{d\tau}\mathcal{T}\{u(t - \tau)\}d\tau$, which can be written as

$$y(t) = \int_{-\infty}^{\infty} \frac{dx(\tau)}{d\tau}r(t - \tau)d\tau \tag{2.15}$$

or, $r(t)$ being the step response, as

$$y(t) = \frac{dx(t)}{dt} * r(t). \tag{2.16}$$

In our example from Fig. 2.4, taking the derivative of $r(t) = (1 - e^{-t/RC})u(t)$, we obtain the impulse response (the computation of this derivative is somehow tricky for we must bear in mind that the voltage in the terminals of a capacitor cannot present discontinuities, just as in the case of the current through an inductor) as

$$h(t) = \frac{1}{RC} e^{-\frac{t}{RC}} u(t). \tag{2.17}$$

In order to have a graphical interpretation of the convolution, we make $x(t) = u(t)$ in (2.11) and use

$$r(t) = h(t) * u(t) = \int_{-\infty}^{\infty} h(\tau) u(t - \tau) d\tau; \tag{2.18}$$

we then compute this integral, as indicated in Fig. 2.7, in two parts:

$$\begin{aligned} \text{for} \quad t < 0 : r(t) &= 0, \\ \text{for} \quad t > 0 : r(t) &= \int_0^t h(\tau) d\tau = \frac{1}{RC} \int_0^t e^{-\frac{\tau}{RC}} d\tau. \end{aligned} \tag{2.19}$$

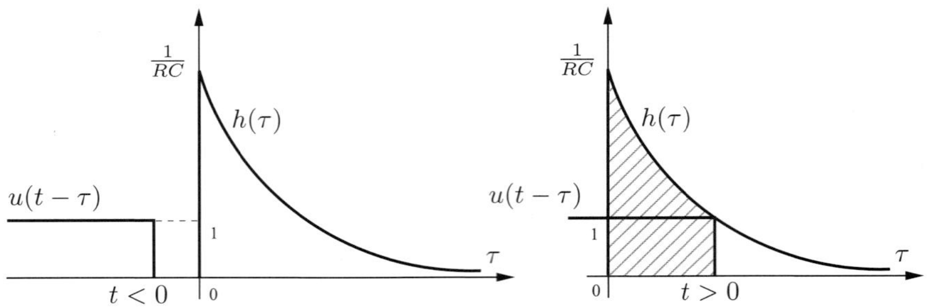

FIGURE 2.7

Graphical interpretation of the convolution integral. Click here to watch an animation.

Finally, from (2.19), we obtain $r(t) = (1 - e^{-t/RC})u(t)$, as previously known.

A few mathematical properties of the convolution are listed in the following:

- commutative: $x(t) * h(t) = h(t) * x(t)$,
- associative: $[x(t) * h_1(t)] * h_2(t) = x(t) * [h_1(t) * h_2(t)]$,
- distributive: $x(t) * [h_1(t) + h_2(t)] = x(t) * h_1(t) + x(t) * h_2(t)$,
- identity element of convolution: $x(t) * \delta(t) = x(t)$.

The properties of convolution can be used to analyze different system combinations. For example, if two systems with impulse responses $h_1(t)$ and $h_2(t)$ are cascaded, the whole cascade system will present the impulse response $h_1(t) * h_2(t)$. The commutative property allows the order of the systems of the cascade combination to be changed without affecting the whole system's response.

Two other important system properties are causability and stability.

Causability: A causal system, also known as nonanticipative system, is a system whose output $y(t)$ depends only on the current and past (not future) information about $x(t)$. A noncausal, or anticipative, system is actually not feasible to be implemented in real life. For example, $y(t) = x(t+1)$ is noncausal, whereas $y(t) = x(t-1)$ is causal.

With respect to the impulse response, it is also worth mentioning that, for a causal system, $h(t) = 0$ for $t < 0$.

Stability: A system is stable if and only if every bounded input produces a bounded output, i.e., if $|x(t)| < B_x$, then $|y(t)| < B_y$ for all values of t.

More about stability will be addressed in a forthcoming section, after the concept of poles and zeros is introduced.

The classic texts for the subjects discussed in this section include [1], [2], [3], [4], and [5], while a more recent discussion, with examples, on the Dirac delta function is found in [6].

2.3 Differential equations

In many engineering applications, the behavior of a system is described by a differential equation. A differential equation is merely an equation with a derivative of at least one of its variables. Two simple examples follow:

$$a\frac{d^2x}{dt^2} + b\frac{dx}{dt} + c = \sin(t) \text{ and} \tag{2.20}$$

$$\frac{\partial^2 z}{\partial x^2} + \frac{\partial^2 z}{\partial y^2} = x^2 + y^2, \tag{2.21}$$

where a, b, and c are constants.

In (2.20), we observe an *ordinary differential equation*, i.e., there is only one independent variable and $x = f(t)$, t being the independent variable. On the other hand, in (2.21), we have a *partial differential equation*, where $z = f(x, y)$, x and y being two independent variables. Both examples have *order 2* (highest derivative appearing in the equation) and *degree 1* (power of the highest derivative term).

This section deals with the solution of differential equations usually employed to represent the mathematical relationship between input and output signals of a linear system. We are therefore most interested in linear differential equations with constant coefficients having the following general form:

$$a_o y + a_1\frac{dy}{dt} \cdots + a_N\frac{d^N y}{dt^N} = b_0 x + b_1\frac{dx}{dt} + \cdots + b_M\frac{d^M x}{dt^M}. \tag{2.22}$$

The expression on the right side of (2.22) is known as a *forcing function*; we could name it $f(t)$. When $f(t) = 0$, the differential equation is known as *homogeneous*, while a nonzero forcing function corresponds to a nonhomogeneous differential equation such as in

$$a_o y + a_1 \frac{dy}{dt} \cdots + a_N \frac{d^N y}{dt^N} = f(t). \tag{2.23}$$

As mentioned in the previous section, the general solution of a differential equation as the one in (2.22) is given by the sum of two expressions: $y_H(t)$, the solution of the associated homogeneous differential equation

$$a_o y + a_1 \frac{dy}{dt} \cdots + a_N \frac{d^N y}{dt^N} = 0, \tag{2.24}$$

and a particular solution of the nonhomogeneous equation, $y_P(t)$, such that

$$y(t) = y_H(t) + y_P(t). \tag{2.25}$$

Solution of the homogeneous equation: The natural response of a linear system is given by $y_H(t)$, the solution of (2.24). Due to the fact that a linear combination of $y_H(t)$ and its derivatives must be equal to zero in order to comply with (2.24), it may be postulated that the nontrivial solution has the form

$$y_H(t) = e^{st}, \tag{2.26}$$

where s is a constant to be determined.

Replacing the assumed solution in (2.24), we obtain, after simplification, the characteristic (or auxiliary) equation

$$a_0 + a_1 s + \cdots + a_N s^N = 0. \tag{2.27}$$

Assuming N distinct roots of the polynomial in (2.27), s_1 to s_N, the homogeneous solution is obtained as a linear combination of N exponentials as in

$$y_H(t) = k_1 e^{s_1 t} + k_2 e^{s_2 t} + \cdots + k_N e^{s_N t}. \tag{2.28}$$

Two special cases follow.

(1) Nondistinct roots: If, for instance, $s_1 = s_2$, it is possible to show that $e^{s_1 t}$ and $te^{s_2 t}$ are independent solutions leading to

$$y_H(t) = k_1 e^{s_1 t} + k_2 t e^{s_2 t} + \cdots.$$

(2) Characteristic equation with complex roots: It is known that for real coefficients (a_0 to a_N), all complex roots will occur in complex conjugate pairs such as $s_1 = \alpha + j\beta$ and $s_2 = \alpha - j\beta$, α and β being real numbers. Hence, the solution would be

$$\begin{aligned} y_H(t) &= k_1 e^{(\alpha + j\beta)t} + k_2 e^{(\alpha - j\beta)t} + \cdots \\ &= [(k_1 + k_2)\cos\beta t + j(k_1 - k_2)\sin\beta t]e^{\alpha t} + \cdots \\ &= (k_3 \cos\beta t + k_4 \sin\beta t)e^{\alpha t} + \cdots, \end{aligned}$$

k_3 and k_4 being real numbers if we make $k_1 = k_2^*$. In that case, we will have

$$y_H(t) = k_5 \cos(\beta t - \theta)e^{\alpha t} + \cdots,$$

with $k_5 = \sqrt{k_3^2 + k_4^2}$ and $\theta = \arctan \frac{k_4}{k_3}$ (check the signs of k_3 and k_4 to determine the correct quadrant of θ). Also note from the last expression that in order to have a stable system (bounded output), α should be negative (causing a decaying exponential).

Solution of the nonhomogeneous equation: A (any) forced solution of the nonhomogeneous equation must satisfy (2.23) containing no arbitrary constant. A usual way to find the forced solution is employing the so-called method of undetermined coefficients: it consists in estimating a general form for $y_P(t)$ from $f(t)$, the forcing function. The coefficients $(A_0, A_1, \cdots)$, as seen in Table 2.1, which shows the most common assumed solutions for each forced function, are to be determined in order to comply with the nonhomogeneous equation. A special case is treated slightly differently: when a term of $f(t)$ corresponds to a term of $y_H(t)$, the corresponding term in $y_P(t)$ must be multiplied by t. Also, when we find nondistinct roots of the characteristic equation (assume multiplicity m as an example), the corresponding term in $y_P(t)$ shall be multiplied by t^m.

Table 2.1 Assumed particular solutions to common forcing functions.

Forcing function $f(t)$	Assumed $y_P(t)$
K	A_0
Kt	$A_0 t + A_1$
Kt^n	$A_0 t^n + A_1 t^{n-1} + \cdots + A_{n-1}t + A_n$
$Ke^{-\alpha t}$	$A_0 e^{-\alpha t}$
$\sin \Omega t$ or $\cos \Omega t$	$A_0 \sin \Omega t + A_1 \cos \Omega t$
$e^{-\alpha t} \sin \Omega t$ or $e^{-\alpha t} \cos \Omega t$	$e^{-\alpha t}(A_0 \sin \Omega t + A_1 \cos \Omega t)$

An *example* of a second-order ordinary differential equation (ODE) with constant coefficients is considered as follows:

$$LC\frac{d^2 v_o(t)}{dt^2} + RC\frac{dv_o(t)}{dt} + v_o(t) = v_i(t), \tag{2.29}$$

where $R = 2.0$ (in $\mathbf{\Omega}$), $L = 2.0$ (in **H**), $C = 1/2$ (in **F**), and $v_i(t) = e^{-t}$ (in **V**).

This equation describes the input × output relationship of the electrical circuit shown in Fig. 2.8 for $t > 0$.

Replacing the values of the components and the input voltage, the ODE becomes

$$\frac{d^2 v_o(t)}{dt^2} + \frac{dv_o(t)}{dt} + v_o(t) = \underbrace{e^{-t}}_{f(t)}. \tag{2.30}$$

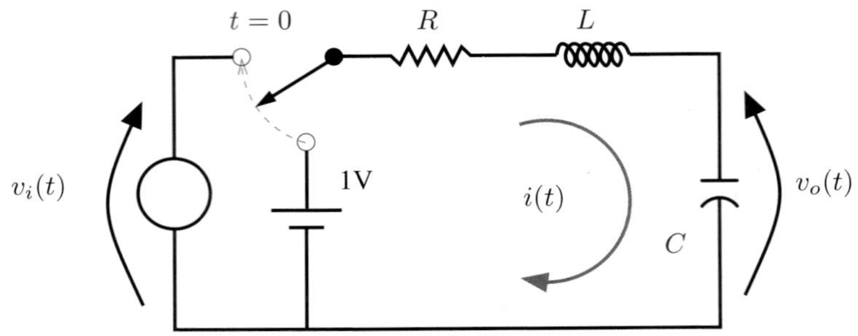

FIGURE 2.8

RLC series circuit with a voltage source as input and the voltage in the capacitor as output.

The associated characteristic equation $s^2 + s + 1 = 0$ has roots $s = \frac{-1 \pm j\sqrt{3}}{2}$ leading to an homogeneous solution given by

$$\begin{aligned} v_{oH}(t) &= k_1 \cos\left(\frac{\sqrt{3}}{2}t\right) e^{-t/2} + k_2 \sin\left(\frac{\sqrt{3}}{2}t\right) e^{-t/2} \\ &= k_3 \cos\left(\frac{\sqrt{3}}{2}t + k_4\right) e^{-t/2}. \end{aligned} \tag{2.31}$$

Next, from the forcing function $f(t) = e^{-t}$, we assume $v_{oP}(t) = k_5 e^{-t}$ and replace it in (2.30), resulting in $k_5 = 1$ such that

$$\begin{aligned} v_o(t) &= k_1 \cos\left(\frac{\sqrt{3}}{2}t\right) e^{-t/2} + k_2 \sin\left(\frac{\sqrt{3}}{2}t\right) e^{-t/2} + e^{-t} \\ &= k_3 \cos\left(\frac{\sqrt{3}t}{2} + k_4\right) e^{-t/2} + e^{-t}, \end{aligned} \tag{2.32}$$

where k_1 and k_2 (or k_3 and k_4) are constants to be obtained from previous knowledge of the physical system, the RLC circuit in this example. This knowledge comes as the initial conditions: an N-order differential equation having N constants and requiring N initial conditions.

Initial conditions: Usually, the differential equation describing the behavior of an electrical circuit is valid for any time $t > 0$ (instant 0 assumed the initial reference in time); the initial conditions correspond to the solution (and its $N - 1$ derivatives) at $t = 0^+$. *In the absence of pulses, the voltage at the terminals of a capacitance and the current through an inductance cannot vary instantaneously* and must be the same value at $t = 0^-$ and $t = 0^+$.

In the case of our example, based on the fact that there is a key switching $v_i(t)$ to the RLC series at $t = 0$, we can say that the voltage across the inductor is $v_o(0^-) = 1$ (the capacitor assumed charged

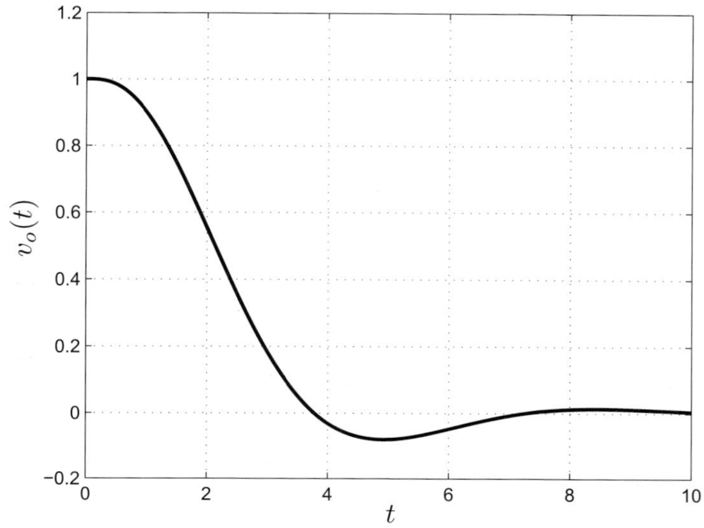

FIGURE 2.9

Output voltage as a function of time, $v_o(t)$, for the circuit in Fig. 2.8. Click here to watch an animation.

with the DC voltage). Therefore, since the voltage across C and the current through L do not alter instantaneously, we know that $v_o(0^+) = 1$ and $i(0^+) = 0$. Since we know that $i(t) = C\frac{dv_o(t)}{dt}$, we have $\left.\frac{dv_0(t)}{dt}\right|_{t=0^+} = 0$. With these initial conditions and from (2.32), we find $k_1 = 0$ and $k_2 = \frac{2\sqrt{3}}{3}$.

Finally, the general solution is given as

$$v_o(t) = 1.1547\sin(0.866t)e^{-t/2} + e^{-t}. \tag{2.33}$$

Fig. 2.9 shows $v_o(t)$ for $0 \leq t \leq 10$. An easy way to obtain this result with MATLAB® is

```
> y=dsolve('D2y+Dy+y=exp(-t)','y(0)=1','Dy(0)=0');
> ezplot(y,[0 10])
```

To end this section, we represent this example using the state-space approach. We first rewrite (2.29) using the first and second derivatives of $v_o(t)$ as $\dot{v}$ and $\ddot{v}$, respectively, and also $u = v_i(t)$ as in

$$LC\ddot{v} + RC\dot{v} + v = u. \tag{2.34}$$

In this representation, we define a state vector $\mathbf{x} = [v \;\; \dot{v}]^T$ and its derivative $\dot{\mathbf{x}} = [\dot{v} \;\; \ddot{v}]^T$, and from these definitions we write the *state equation* and the *output equation*:

$$\begin{cases} \dot{\mathbf{x}} &= \mathbf{A}\mathbf{x} + \mathbf{B}u = \begin{bmatrix} 0 & 1 \\ \frac{-1}{LC} & \frac{-R}{L} \end{bmatrix} \begin{bmatrix} v \\ \dot{v} \end{bmatrix} + \begin{bmatrix} 0 \\ \frac{1}{LC} \end{bmatrix} v_i(t), \\ y &= \mathbf{C}\mathbf{x} + \mathbf{D}u = [1 \;\; 0] \begin{bmatrix} v \\ \dot{v} \end{bmatrix} + 0 v_i(t), \end{cases} \tag{2.35}$$

where **A** is known as the *state matrix*, **B** as the *input matrix*, and **C** as the *output matrix*, and **D** (equal to zero in this example) would be the *feedthrough* (or *feedforward*) matrix.

For further reading, we suggest [7], [8], [9], and [10].

2.4 Laplace transform: definition and properties

The Laplace transform [11] is named after the French mathematician Pierre-Simon Laplace. It is an important tool for solving differential equations, and it is very useful for designing and analyzing linear systems [12].

The Laplace transform is a mathematical operation that maps, or transforms, a variable (or function) from its original domain into the Laplace domain, or s domain. This transform produces a time-domain response to transitioning inputs, whenever time-domain behavior is more interesting than frequency-domain behavior. When solving engineering problems one has to model a physical phenomenon that is dependent on the rates of change of a function (e.g., the velocity of a car as mentioned in Section 2.1). Hence, calculus and differential equations (Section 2.3) that model the phenomenon are the natural candidates to be the mathematical tools. However, calculus solves (ordinary) differential equations provided the functions are continuous and with continuous derivatives. In addition, engineering problems have often to deal with impulsive, nonperiodic, or piecewise-defined input signals.

The Fourier transform, to be addressed in Section 2.7, is also an important tool for signal analysis, as well as for linear filter design. However, while the unit-step function (Fig. 2.2(a)), discontinuous at time $t = 0$, has a Laplace transform, its forward Fourier integral does not converge. The Laplace transform is particularly useful for input terms that are impulsive, nonperiodic, or piecewise-defined [4].

The Laplace transform maps the time domain into the s-domain, with $s = \sigma + j\Omega$, converting integral and differential equations into algebraic equations. The function is mapped (transformed) to the s-domain, eliminating all the derivatives. Hence, solving the equation becomes simple algebra in the s-domain and the result is transformed back to the time domain. The Laplace transform converts a time-domain 1D signal, $x(t)$, into a complex representation, $X(s)$, defined over a complex plane (s-plane). The complex plane is spanned by the variables σ (real axis) and Ω (imaginary axis) [5].

The *two-sided* (or bilateral) *Laplace transform* of a signal $x(t)$ is the function $\mathcal{L}\{x(t)\}$ defined by

$$\boxed{X(s) = \mathcal{L}\{x(t)\} = \int_{-\infty}^{\infty} x(t)e^{-st}\,dt.} \tag{2.36}$$

The notation $X(s) = \mathcal{L}\{x(t)\}$ indicates that $X(s)$ is the Laplace transform of $x(t)$. Conversely, the notation $x(t) = \mathcal{L}^{-1}\{X(s)\}$ denotes that $x(t)$ is the inverse Laplace transform of $X(s)$. This relationship is expressed with the notation $x(t) \overset{\mathcal{L}}{\longleftrightarrow} X(s)$.

As $e^{-st} = e^{-(\sigma+j\Omega)t} = e^{-\sigma t}(\cos \Omega t - j \sin \Omega t)$, Eq. (2.36) can be rewritten as

$$\mathcal{L}\{x(t)\} = X(s)|_{s=\sigma+j\Omega} = \int_{-\infty}^{\infty} x(t)e^{-\sigma t} \cos \Omega t \, dt - j \int_{-\infty}^{\infty} x(t)e^{-\sigma t} \sin \Omega t \, dt. \tag{2.37}$$

This way, one can identify that the Laplace transform real part ($\text{Re}(X(s))$ represents the contribution of the (combined) exponential and cosine terms to $x(t)$, while its imaginary part ($\text{Im}(X(s))$ represents the contribution of the (combined) exponential and sine terms to $x(t)$. As the term e^{-st} is an eigenfunction, we can state that the Laplace transform represents time-domain signals in the s-domain as weighted combinations of eigensignals.

For those signals equal to zero for $t < 0$, the limits on the integral are changed for the *one-sided* (or unilateral) *Laplace transform*:

$$\boxed{X(s) = \mathcal{L}\{x(t)\} = \int_{0^-}^{\infty} x(t)e^{-st} dt,} \tag{2.38}$$

with $x(t) = 0$ for $t < 0^-$, in order to deal with signals that present singularities at the origin, i.e., at $t = 0$.

As the interest will be in signals defined for $t \geq 0$, let us find the one-sided Laplace transform of $x(t) = e^t u(t)$:

$$\mathcal{L}\{x(t)\} = X(s) = \int_{0^-}^{\infty} e^t e^{-st} dt = \int_{0^-}^{\infty} e^{(1-s)t} dt = \frac{1}{1-s} e^{(1-s)t} \Big|_{0^-}^{\infty} = \frac{1}{s-1}. \tag{2.39}$$

The integral, in (2.39), defining $\mathcal{L}\{x(t)\}$ is true if $e^{(1-s)t} \to 0$ as $t \to \infty$, $\forall\, s \in \mathcal{C}$ with $\text{Re}(s) > 1$, which is the region of convergence (ROC) of $X(s)$.

As the analysis of convergence and the conditions that guarantee the existence of the Laplace integral are beyond the scope of this chapter, please refer to [5] and [4].

For a given $x(t)$, the integral may converge for some values of σ, $\text{Re}(s = \sigma + j\Omega)$, but not for others. So, we have to guarantee the existence of the integral, i.e., $x(t)e^{-\sigma t}$ has to be absolutely integrable. The ROC of the integral in the complex s-plane should be specified for each transform $\mathcal{L}\{x(t)\}$, which exists if and only if the argument s is inside the ROC. The transformed signal, $X(s)$, will be well defined for a range of values in the s-domain that is the ROC, which is always given in association with the transform itself.

When $\sigma = 0$, so that $s = j\Omega$, the Laplace transform reverts to the Fourier transform, i.e., $x(t)$ has a Fourier transform if the ROC of the Laplace transform in the s-plane includes the imaginary axis.

For finite duration signals that are absolutely integrable, the ROC contains the entire s-plane. As $\mathcal{L}\{x(t)\} = X(s)$ cannot uniquely define $x(t)$, $X(s)$ and the ROC are required. If we wish to find the Laplace transform of a one-sided real exponential function $x(t) = e^{at}u(t)$, $a \in \mathcal{R}$, given by

$$x(t) = \begin{cases} 0, & t \leq 0, \\ e^{at}, & t > 0, \end{cases} \tag{2.40}$$

we have

$$X(s) = \int_{0^-}^{\infty} e^{-(s-a)t} dt = -\frac{1}{s-a} e^{-(s-a)t} \Big|_{0^-}^{\infty} = \frac{1}{s-a}. \tag{2.41}$$

The integral of (2.41) converges if $\sigma > a$, and the ROC is the region of the s-plane to the right of $\sigma = a$, as pictured in Fig. 2.10. If $\sigma < a$ the integral does not converge as $X(s) \to \infty$, and if $\sigma = a$ we cannot determine $X(s)$.

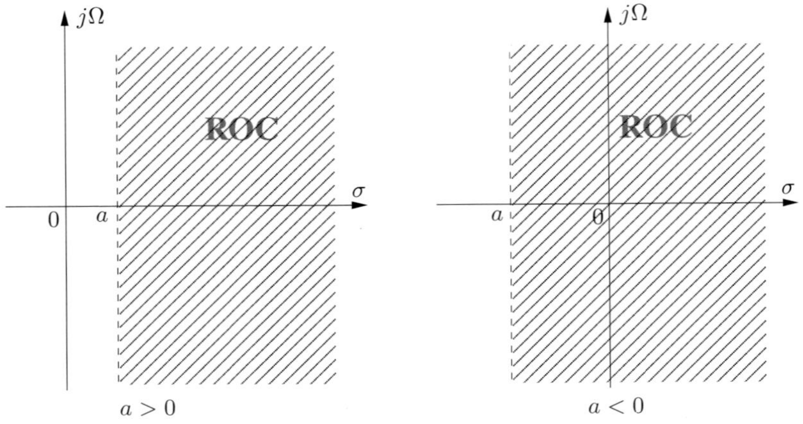

FIGURE 2.10

Region of convergence (ROC) on the s-plane for the signal defined by Eq. (2.40). Note that for a stable signal ($a < 0$), the ROC contains the vertical axis $s = j\Omega$.

In other words, if a signal $x(t)$ is nonzero only for $t \geq 0$, the ROC of its Laplace transform lies to the right-hand side of its poles (please refer to Section 2.5). Additionally, the ROC does not contain any pole. Poles are points where the Laplace transform reaches an infinite value in the s-plane (e.g., $s = a$ in Fig. 2.10).

An example of the two-sided Laplace transform of a right-sided exponential function $x(t) = e^{-at}u(t),\ a \in \mathcal{R}$, is given by

$$\begin{aligned} X(s) &= \int_{-\infty}^{\infty} e^{-at} u(t) e^{-st}\, dt \\ &= \int_{0^-}^{\infty} e^{-(s+a)t}\, dt \\ &= \int_{0^-}^{\infty} e^{-(\sigma+a)t} e^{-j\Omega t}\, dt \\ &= -\frac{1}{s+a} e^{-(s+a)t} \Big|_{0^-}^{\infty} = \frac{1}{s+a}. \end{aligned} \tag{2.42}$$

As the term $e^{-j\Omega t}$ is sinusoidal, only $\sigma = \text{Re}(s)$ is important, i.e., the integral given by (2.42), when t tends to infinity, converges if $e^{-(\sigma+a)t}$ is finite ($\sigma > -a$) in order for the exponential function to decay. The integral of Eq. (2.42) converges if $\text{Re}(s+1) > 0$, i.e. if $\text{Re}(s) > -a$, and the ROC is the region of the s-plane to the right of $\sigma = -a$, as pictured in the left side of Fig. 2.11. The ROC of the Laplace transform of a right-sided signal is to the right of its rightmost pole.

If we wish to find the two-sided Laplace transform of a left-sided signal, $x(t) = -e^{-at}u(-t),\ a \in \mathcal{R}$, we have

$$\begin{aligned} X(s) &= -\int_{-\infty}^{\infty} e^{-at}u(-t)e^{-st}dt \\ &= -\int_{-\infty}^{0^-} e^{-(s+a)t}dt \\ &= \frac{1}{s+a}e^{-(s+a)t}\bigg|_{\infty}^{0^-} = \frac{1}{s+a}. \end{aligned} \tag{2.43}$$

The integral of Eq. (2.43) converges if $\text{Re}(s) < -a$, and the ROC is the region of the s-plane to the left of $\sigma = -a$, as pictured in the right side of Fig. 2.11. The ROC of the Laplace transform of a left-sided signal is to the left of its leftmost pole.

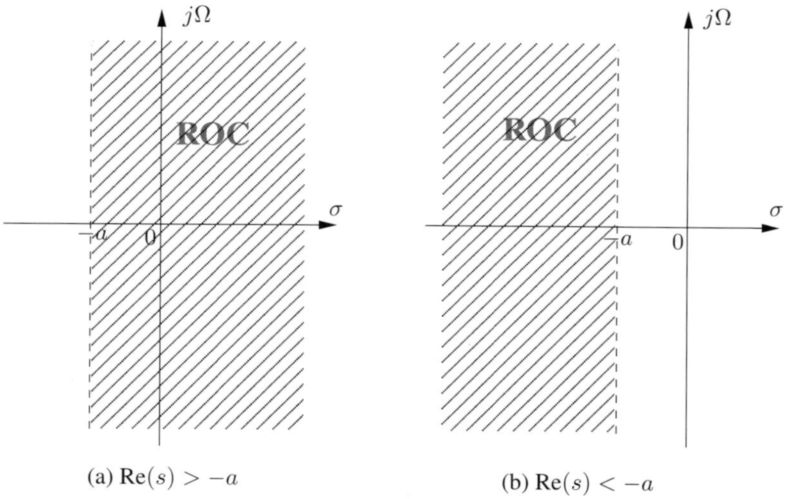

FIGURE 2.11

Regions of convergence (ROCs) on the s-plane: (a) for right-sided signals, $\text{Re}(s) > -a$, and (b) for left-sided signals, $\text{Re}(s) < -a$.

For causal systems, where $h(t) = 0,\ t < 0$, and right-sided signals, where $x(t) = 0,\ t < 0$, the unilateral (one-sided) and bilateral (two-sided) transforms are equal. However, it is important to stress that some properties change, such as the differentiation property reduced to $sX(s)$ for the bilateral case. Conversely, other properties, such as convolution, hold as is, provided the system is causal and the input starts at $t = 0^-$.

Common Laplace transform pairs [13] are summarized in Table 2.2.

Table 2.2 Laplace transform pairs.

$x(t) = \mathcal{L}^{-1}\{X(s)\}$	$X(s) = \mathcal{L}\{x(t)\}$	ROC
$\delta(t)$	1	for all s
$\delta(t-\tau),\ \tau > 0$	$e^{-s\tau}$	for all s
$u(t)$	$\frac{1}{s}$	$\mathrm{Re}(s) > 0$
$-u(-t)$	$\frac{1}{s}$	$\mathrm{Re}(s) < 0$
$t\,u(t)$	$\frac{1}{s^2}$	$\mathrm{Re}(s) > 0$
$t^n\,u(t)$	$\frac{n!}{s^{n+1}},\ n = 1, 2, \ldots$	$\mathrm{Re}(s) > 0$
$e^{-at}\,u(t)$	$\frac{1}{s+a}$	$\mathrm{Re}(s) > -\mathrm{Re}(a)$
$-e^{-at}\,u(-t)$	$\frac{1}{s+a}$	$\mathrm{Re}(s) < -\mathrm{Re}(a)$
$t^n e^{-at}\,u(t)$	$\frac{n!}{(s+a)^{n+1}},\ n = 1, 2, \ldots$	$\mathrm{Re}(s) > -\mathrm{Re}(a)$
$\sin \Omega t\,u(t)$	$\frac{\Omega}{s^2+\Omega^2}$	$\mathrm{Re}(s) > 0$
$\cos \Omega t\,u(t)$	$\frac{s}{s^2+\Omega^2}$	$\mathrm{Re}(s) > 0$
$e^{-at}\sin \Omega t\,u(t)$	$\frac{\Omega}{(s+a)^2+\Omega^2}$	$\mathrm{Re}(s) > -\mathrm{Re}(a)$
$e^{-at}\cos \Omega t\,u(t)$	$\frac{s+a}{(s+a)^2+\Omega^2}$	$\mathrm{Re}(s) > -\mathrm{Re}(a)$
$\sin(\Omega t + \theta)\,u(t)$	$\frac{s\sin\theta + \Omega\cos\theta}{s^2+\Omega^2}$	$\mathrm{Re}(s) > 0$
$\cos(\Omega t + \theta)\,u(t)$	$\frac{s\cos\theta - \Omega\sin\theta}{s^2+\Omega^2}$	$\mathrm{Re}(s) > 0$
$e^{-at}(\Omega\cos\Omega t - a\sin\Omega t)\,u(t)$	$\frac{\Omega s}{(s+a)^2+\Omega^2}$	$\mathrm{Re}(s) > -\mathrm{Re}(a)$

Recalling Fig. 2.3, where $x(t)$ is the input to a linear system with impulse response $h(t)$, to obtain the output $y(t)$ one could convolve $x(t)$ with $h(t)$ (see (2.11)). However, if $X(s)$ and $Y(s)$ are the associated Laplace transforms, the (computationally demanding) operation of convolution in the time domain is mapped into a (simple) operation of multiplication in the s-domain:

$$y(t) = h(t) * x(t) \stackrel{\mathcal{L}}{\longleftrightarrow} Y(s) = H(s)X(s). \tag{2.44}$$

Eq. (2.44) shows that convolution in the time domain is equivalent to multiplication in the Laplace domain. In the following, some properties of the Laplace transform disclose the symmetry between operations in the time and s-domains. Note that due to the practical interest on causal signals and systems, we focus on the properties of the unilateral (also known as ordinary or one-sided) Laplace transform.

Linearity: If $x_1(t) \stackrel{\mathcal{L}}{\longleftrightarrow} X_1(s)$, ROC $= R_1$, and $x_2(t) \stackrel{\mathcal{L}}{\longleftrightarrow} X_2(s)$, ROC $= R_2$, then

$$\mathcal{L}\{a_1x_1(t)+a_2x_2(t)\} = a_1\mathcal{L}\{x_1(t)\}+a_2\mathcal{L}\{x_2(t)\} = a_1X_1(s)+a_2X_2(s), \tag{2.45}$$

with a_1 and a_2 being constants and ROC $\supseteq R_1 \cap R_2$. The Laplace transform is a linear operation.

Time shifting: If $x(t) \stackrel{\mathcal{L}}{\longleftrightarrow} X(s)$, ROC $= R$, then $x(t-t_0) \stackrel{\mathcal{L}}{\longleftrightarrow} e^{-st_0}X(s)$, $\forall t_0 \in \mathcal{R}$, ROC $= R$. We have

$$\begin{aligned}\int_{0^-}^{\infty} x(t-t_0)e^{-st}dt &= \int_{0^-}^{\infty} x(\tau)e^{-s(\tau+t_0)}d\tau \\ &= e^{-st_0}\int_{0^-}^{\infty} x(\tau)e^{-s(\tau)}d\tau \\ &= e^{-st_0}X(s),\end{aligned} \tag{2.46}$$

where $t-t_0=\tau$. Time shifting (or time delay) in the time domain is equivalent to modulation (alteration of the magnitude and phase) in the s-domain.

Exponential scaling (frequency shifting): If $x(t) \stackrel{\mathcal{L}}{\longleftrightarrow} X(s)$, ROC $= R$, then $e^{at}x(t) \stackrel{\mathcal{L}}{\longleftrightarrow} X(s-a)$, ROC $= R + \mathrm{Re}(a)$, $\forall a \in \mathcal{R}$ or $\mathcal{C}$. We have

$$\mathcal{L}\{e^{at}x(t)\} = \int_{0^-}^{\infty} e^{at}x(t)e^{-st}dt = \int_{0^-}^{\infty} x(t)e^{-(s-a)t}dt = X(s-a). \tag{2.47}$$

Modulation in the time domain is equivalent to shifting in the s-domain.

Convolution: If $x_1(t) \stackrel{\mathcal{L}}{\longleftrightarrow} X_1(s)$, ROC $= R_1$, and $x_2(t) \stackrel{\mathcal{L}}{\longleftrightarrow} X_2(s)$, ROC $= R_2$, then $x_1(t) * x_2(t) \stackrel{\mathcal{L}}{\longleftrightarrow} X_1(s)X_2(s)$, ROC $\supseteq R_1 \cap R_2$:

$$\begin{aligned}\int_{0^-}^{\infty} [x_1(t)*x_2(t)]e^{-st}dt &= \int_{0^-}^{\infty}\left[\int_{0^-}^{\infty} x_1(\tau)x_2(t-\tau)d\tau\right]e^{-st}dt \\ &= \int_{0^-}^{\infty}\int_{0^-}^{\infty} x_1(\tau)x_2(t-\tau)e^{-s(t-\tau+\tau)}d\tau dt \\ &= \int_{0^-}^{\infty} x_1(\tau)e^{-s\tau}d\tau \int_{0^-}^{\infty} x_2(\upsilon)e^{-s\upsilon}d\upsilon \\ &= X_1(s)X_2(s),\end{aligned} \tag{2.48}$$

where in the inner integral $\upsilon = 1-\tau$. Convolution in the time domain is equivalent to multiplication in the s-domain.

Differentiation in time: If $x(t) \stackrel{\mathcal{L}}{\longleftrightarrow} X(s)$, ROC $= R$, then $\dot{x}(t) \stackrel{\mathcal{L}}{\longleftrightarrow} sX(s) - x(0^-)$, ROC $\supseteq R$. Integrating by parts, we have

$$\int_{0^-}^{\infty} \frac{dx(t)}{dt}e^{-st}dt = x(t)e^{-st}\Big|_{0^-}^{\infty} - \int_{0^-}^{\infty} x(t)(-s)e^{-st}dt$$

$$
\begin{aligned}
&= 0 - x(0^-) + s\int_{0^-}^{\infty} x(t)e^{-st}\,dt \\
&= sX(s) - x(0^-), \qquad (2.49)
\end{aligned}
$$

where any jump from $t = 0^-$ to $t = 0^+$ is considered. In other words, the Laplace transform of a derivative of a function is a combination of the transform of the function, multiplied by s, and its initial value. This property is quite useful as a differential equation in time can be turned into an algebraic equation in the Laplace domain, where it can be solved and mapped back into the time domain (inverse Laplace transform).

The initial- and final-value theorems show that the initial and final values of a signal in the time domain can be obtained from its Laplace transform without knowing its expression in the time domain.

Initial-value theorem: Assuming a strictly proper (having the degree of the numerator less than the degree of the denominator) LTI system that generated the signal $x(t)$, we have

$$\lim_{s\to\infty} sX(s) = \lim_{t\to 0^+} x(t). \qquad (2.50)$$

From the differentiation in time property, we have $\dot{x}(t) \leftrightarrow sX(s) - x(0^-)$. By taking the limit when $s \to \infty$ of the first expression of (2.49), we find

$$
\begin{aligned}
\lim_{s\to\infty}\int_{0^-}^{\infty}\frac{dx(t)}{dt}e^{-st}\,dt &= \lim_{s\to\infty}\left(\int_{0^-}^{0^+}\frac{dx(t)}{dt}e^{0}\,dt + \int_{0^+}^{\infty}\frac{dx(t)}{dt}e^{-st}\,dt\right) \\
&= \lim_{t\to 0^+} x(t) - \lim_{t\to 0^-} x(t) + 0 \\
&= \lim_{t\to 0^+} x(t) - \lim_{t\to 0^-} x(t). \qquad (2.51)
\end{aligned}
$$

If we take the limit of s to ∞ in the result of (2.49), we have

$$\lim_{s\to\infty}(sX(s) - x(0^-)) = \lim_{s\to\infty} sX(s) - \lim_{t\to 0^-} x(t). \qquad (2.52)$$

Then we can equal the results of (2.52) and (2.51) and obtain

$$\lim_{s\to\infty} sX(s) - \lim_{t\to 0^-} x(t) = \lim_{t\to 0^+} x(t) - \lim_{t\to 0^-} x(t). \qquad (2.53)$$

As the right-hand side of (2.53) is obtained taking the limit when $s \to \infty$ of the result of (2.49), we can see from (2.51) and (2.53) that

$$\lim_{s\to\infty} sX(s) = \lim_{t\to 0^+} x(t). \qquad (2.54)$$

The initial-value theorem provides the behavior of a signal in the time domain for small time intervals, i.e., for $t \to 0^+$. In other words, it determines the initial values of a function in time from its expression in the s-domain, which is particularly useful in circuits and systems.

Final-value theorem: Considering that the LTI system that generated the signal $x(t)$ is stable, we have

$$\lim_{s\to 0} sX(s) = \lim_{t\to\infty} x(t). \qquad (2.55)$$

From the differentiation in time property, we have $\dot{x}(t) \overset{\mathcal{L}}{\longleftrightarrow} sX(s) - x(0^-)$. By taking the limit when $s \to 0$ of the first expression of (2.49), we have

$$\begin{aligned} \lim_{s\to 0} \int_{0^-}^{\infty} \frac{dx(t)}{dt} e^{-st} dt &= \int_{0^-}^{\infty} \frac{dx(t)}{dt} dt \\ &= \lim_{t\to\infty} x(t) - \lim_{t\to 0^-} x(t). \end{aligned} \tag{2.56}$$

If we take the limit when $s \to 0$ of the last expression of (2.49), we get

$$\lim_{s\to 0}(sX(s) - x(0^-)) = \lim_{s\to 0} sX(s) - \lim_{t\to 0^-} x(t); \tag{2.57}$$

then, we can write

$$\lim_{s\to 0} sX(s) - \lim_{t\to 0^-} x(t) = \lim_{t\to\infty} x(t) - \lim_{t\to 0^-} x(t). \tag{2.58}$$

As the right-hand side of (2.58) is obtained taking the limit when $s \to 0$ of the result of (2.49), we can see from (2.57) and (2.58) that

$$\lim_{s\to 0} sX(s) = \lim_{t\to\infty} x(t). \tag{2.59}$$

The final-value theorem provides the behavior of a signal in the time domain for large time intervals, i.e., for $t \to \infty$. In other words, it obtains the final value of a function in time, assuming it is stable and well defined when $t \to \infty$, from its expression in the s-domain. An LTI system, as will be seen in the next section, is considered stable if all of its poles lie within the left side of the s-plane. As $t \to \infty$, $x(t)$ must reach a steady value, thus it is not possible to apply the final-value theorem to signals such as sine, cosine, or ramp.

A list of one-sided Laplace transform properties is summarized in Table 2.3.

The inverse Laplace transform is given by

$$\mathcal{L}^{-1}\{X(s)\} = x(t) = \int_{\sigma - j\infty}^{\sigma + j\infty} X(s) e^{st} ds. \tag{2.60}$$

The integration is performed along a line, parallel to the imaginary axis $(\sigma - j\infty, \sigma + j\infty)$, that lies in the ROC. However, the inverse transform can be calculated using partial fractions expansion with the method of residues. In this method, the s-domain signal $X(s)$ is decomposed into partial fractions, thus expressing $X(s)$ as a sum of simpler rational functions. Hence, as each term in the partial fraction is expected to have a known inverse transform, each transform may be obtained from a table like Table 2.2. Please recall that the ROC of a signal that is nonzero for $t \geq 0$ is located to the right-hand side of its poles. Conversely, for a signal that is nonzero for $t \leq 0$, the ROC of its Laplace transform lies to the left-hand side of its poles. In other words, if $X(s) = 1/(s + a)$, $a \in \mathcal{R}$, its inverse Laplace transform is given as follows:

$$\mathcal{L}^{-1}\left\{\frac{1}{s+a}\right\} = \begin{cases} -e^{-at}, & \mathrm{Re}(s) < -a, \\ e^{-at}, & \mathrm{Re}(s) > -a. \end{cases} \tag{2.61}$$

Table 2.3 Properties of (one-sided) Laplace transform.

Property	Time domain	s-domain	ROC
	$x(t)$	$X(s)$	R
	$x_1(t)$	$X_1(s)$	R_1
	$x_2(t)$	$X_2(s)$	R_2
Linearity	$a_1x_1(t)+a_2x_2(t)$	$a_1X_1(s)+a_2X_2(s)$	$\supseteq R_1\cap R_2$
Scalar multiplication	$ax(t)$	$aX(s)$	R
Scaling	$x(at)$	$\frac{1}{a}X\left(\frac{s}{a}\right)$	$\frac{R}{\lvert a\rvert}$
Time shifting	$x(t-t_0)$	$e^{-st_0}X(s)$	R
Exponential scaling	$e^{at}x(t)$	$X(s-a)$	$R+\mathrm{Re}(a)$
Differentiation	$\frac{dx(t)}{dt}$	$sX(s)-x(0^-)$	$\supseteq R$
	$\frac{d^2x(t)}{dt^2}$	$s^2X(s)-sx(0^-)-\dot{x}(0^-)$	
	$\frac{d^nx(t)}{dt^n}$	$s^nX(s)-\sum_{k=1}^{n}s^{n-k}x^{(k-1)}(0^-)$	
Differentiation (in the s-domain)	$-tx(t)$	$\frac{dX(s)}{ds}$	R
Convolution	$x_1(t)*x_2(t)$	$X_1(s)X_2(s)$	$\supseteq R_1\cap R_2$
Convolution (in the s-domain)	$x_1(t)x_2(t)$	$\frac{1}{2\pi j}X_1(s)*X_2(s)$	$\supseteq R_1\cap R_2$

In practice, the inverse Laplace transform is found recursing to tables of transform pairs (e.g., Table 2.2). Given a function $X(s)$ in the s-domain and an ROC, its inverse Laplace transform is given by $\mathcal{L}^{-1}\{X(s)\}=x(t)$ such that $X(s)=\mathcal{L}\{x(t)\},\ s\in\mathrm{ROC}$.

An immediate application of the Laplace transform is in circuit analysis. Assuming that all initial conditions are equal to zero, i.e., there is no initial charge on the capacitor, the response $y(t)$ of the circuit with input given by $x(t)=u(t)$ displayed by Fig. 2.4 could be obtained in the Laplace domain to be further transformed into the time domain.

First, the circuit elements are transformed from the time domain into the s-domain, thus creating an s-domain equivalent circuit. Hence, the RLC elements in the time domain and s-domain are:

Resistor (voltage-current relationship):

$$v(t)=Ri(t)\xleftrightarrow{\mathcal{L}}V(s)=RI(s). \tag{2.62}$$

The equivalent circuit is depicted in Fig. 2.12.

Inductor (initial current $i(0^-)$):

$$v(t)=L\frac{di(t)}{dt}\xleftrightarrow{\mathcal{L}}V(s)=sLI(s)-Li(0^-); \tag{2.63}$$

applying the differentiation property (Table 2.3) leads to

$$I(s) = \frac{1}{sL}V(s) + \frac{i(0^-)}{s}. \tag{2.64}$$

The equivalent circuit is depicted in Fig. 2.13. The inductor, L, is an impedance sL in the s-domain in series with a voltage source, $Li(0^-)$, or in parallel with a current source, $\frac{i(0^-)}{s}$.

Capacitor (initial voltage $v(0^-)$):

$$i(t) = C\frac{dv(t)}{dt} \xleftrightarrow{\mathcal{L}} I(s) = sCV(s) - v(0^-); \tag{2.65}$$

applying the differentiation property (Table 2.3) leads to

$$V(s) = \frac{I(s)}{sC} + \frac{v(0^-)}{s}. \tag{2.66}$$

The capacitor, C, is an impedance $\frac{1}{sC}$ in the s-domain in series with a voltage source, $\frac{v(0^-)}{s}$, or in parallel with a current source, $Cv(0^-)$. The voltage across the capacitor in the time domain corresponds to the voltage across both the capacitor and the voltage source in the frequency domain. The equivalent circuit is depicted in Fig. 2.14.

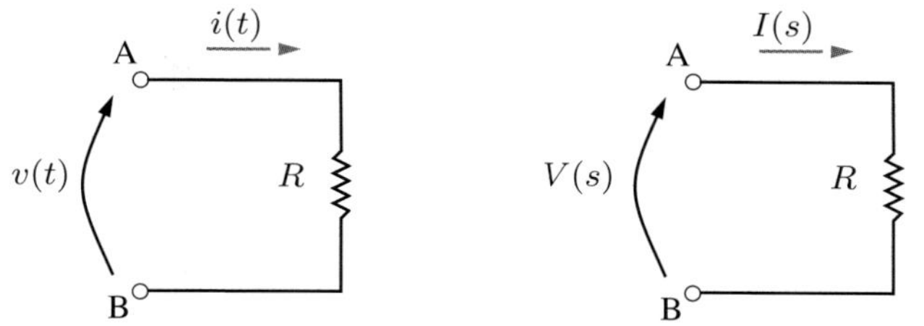

FIGURE 2.12

Equivalent circuit of a resistor in the Laplace domain.

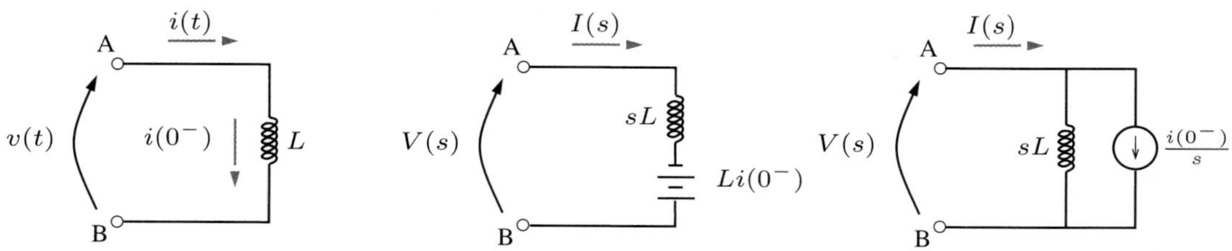

FIGURE 2.13

Equivalent circuit of an inductor in the Laplace domain.

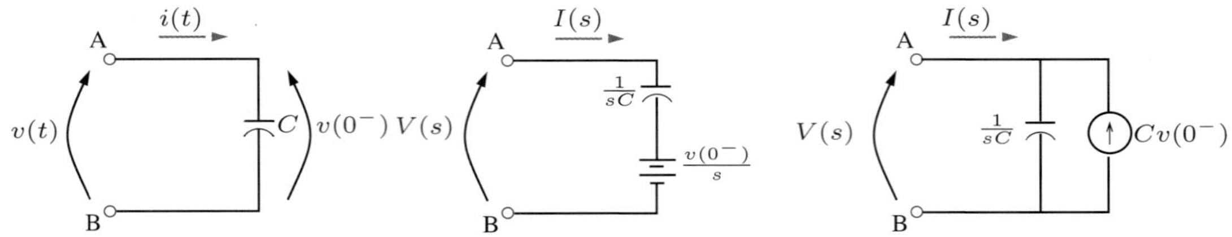

FIGURE 2.14

Equivalent circuit of a capacitor in the Laplace domain.

As an example, the system's response given by the voltage across the capacitor, $y(t)$, depicted in Fig. 2.4, is obtained in the Laplace domain as follows:

$$X(s) = \frac{1}{s} = I(s)\left(R + \frac{1}{sC}\right). \tag{2.67}$$

From (2.65), we have

$$Y(s) = \frac{I(s)}{sC} \tag{2.68}$$

with

$$I(s) = \frac{X(s)}{R + \frac{1}{sC}}. \tag{2.69}$$

Substituting (2.69) in (2.68) and applying the inverse Laplace transform (Table 2.2) and the linearity property (Table 2.3), we get

$$\begin{aligned} \mathcal{L}^{-1}\{Y(s)\} &= \mathcal{L}^{-1}\left\{\frac{1}{s} - \frac{1}{s + \frac{1}{RC}}\right\}, \\ y(t) &= u(t) - e^{-t\frac{1}{RC}}u(t), \end{aligned} \tag{2.70}$$

which is the same result presented in (2.5).

Another example where the Laplace transform is useful is the RLC circuit displayed by Fig. 2.8, where the desired response is the voltage across the capacitor C, $v_o(t) = v_C(t)$. Note that the initial conditions are part of the transform, as well as the transient and steady-state responses. Given the input signal $v_i(t) = e^{-t}u(t)$ with initial conditions $v_o(0^-) = 1$, $\dot{v_o}(0^-) = 0$, and $i_o(0^-) = 0$, applying KVL to the circuit, we have

$$v_i(t) = v_R(t) + v_L(t) + v_C(t). \tag{2.71}$$

Directly substituting (2.62), (2.63), and (2.65) in (2.71) and applying the Laplace transform to the input signal, $v_i(t) = e^{-t}u(t)$, we get

$$\begin{aligned} \frac{1}{s+1} &= \left(RI(s) + sLI(s) - Li_o(0^-) + \frac{1}{sC}I(s) + \frac{v_o(0^-)}{s}\right) \\ &= I(s)\left(R + sL + \frac{1}{sC}\right) - Li_o(0^-) + \frac{v_o(0^-)}{s} \\ &= \frac{I(s)}{sC}\left(s^2LC + sRC + 1\right) - Li_o(0^-) + \frac{v_o(0^-)}{s}. \end{aligned} \tag{2.72}$$

From (2.65), $\frac{I(s)}{sC} = V_o(s) - \frac{v_o(0^-)}{s}$, we have

$$\begin{aligned} \frac{1}{s+1} &= (V_o(s) - \frac{v_o(0^-)}{s})\left(s^2LC + sRC + 1\right) - Li_o(0^-)\frac{v_o(0^-)}{s} \\ &= V_o(s)\left(s^2LC + sRC + 1\right) - \frac{v_o(0^-)}{s}\left(s^2LC + sRC + 1\right) \\ &\quad - Li_o(0^-) + \frac{v_o(0^-)}{s}. \end{aligned} \tag{2.73}$$

Substituting the values of $R = 2.0$ (in **Ω**), $L = 2.0$ (in **H**), and $C = 1/2$ (in **F**), the equation becomes

$$\begin{aligned} \frac{1}{s+1} &= V_o(s)\left(s^2 + s + 1\right) - \frac{v_o(0^-)}{s}\left(s^2 + s + 1\right) \\ &\quad - i_o(0^-) + \frac{v_o(0^-)}{s}; \end{aligned} \tag{2.74}$$

considering that the initial conditions are $v_0(0^-) = 1$, $\dot{v_o}(0^-) = 0$, and $i_o(0^-) = 0$, we get

$$\begin{aligned} \frac{1}{s+1} &= V_o(s)\left(s^2 + s + 1\right) - \frac{1}{s}\left(s^2 + s + 1\right) - 0 + \frac{1}{s} \\ &= V_o(s)\left(s^2 + s + 1\right) - s - 1 - \frac{1}{s} - 0 + \frac{1}{s} \\ &= V_o(s)\left(s^2 + s + 1\right) - s - 1, \\ V_o(s) &= \left(\frac{1}{s+1} + s + 1\right)\left(\frac{1}{s^2 + s + 1}\right) \\ &= \left(\frac{s^2 + 2s + 2}{s+1}\right)\left(\frac{1}{s^2 + s + 1}\right). \end{aligned} \tag{2.75}$$

After decomposing (2.75) into partial fractions and finding the poles and residues, the inverse Laplace transform is applied in order to find the expression of $v_o(t)$:

$$V_o(s) = \frac{1}{(s+1)} - j\frac{\frac{\sqrt{3}}{3}}{(s + \frac{1}{2} - j\frac{\sqrt{3}}{2})} + j\frac{\frac{\sqrt{3}}{3}}{(s + \frac{1}{2} + j\frac{\sqrt{3}}{2})}, \tag{2.76}$$

$$\mathcal{L}^{-1}\{V_o(s)\} = \mathcal{L}^{-1}\left\{\frac{1}{(s+1)} - j\frac{\frac{\sqrt{3}}{3}}{(s+\frac{1}{2}-j\frac{\sqrt{3}}{2})} + j\frac{\frac{\sqrt{3}}{3}}{(s+\frac{1}{2}+j\frac{\sqrt{3}}{2})}\right\}. \tag{2.77}$$

Applying the linearity property (Table 2.3) and recursing to the Laplace transform pair table (Table 2.2), we get

$$v_o(t) = e^{-t} - j\frac{\sqrt{3}}{3}e^{-\frac{t}{2}+jt\frac{\sqrt{3}}{2}} + j\frac{\sqrt{3}}{3}e^{-\frac{t}{2}-jt\frac{\sqrt{3}}{2}}, \tag{2.78}$$

considering that the inverse Laplace transforms of $\frac{1}{s+a}$, for $\text{Re}(s) > -\text{Re}(a)$, given by Table 2.2, with $a = \frac{1}{2} - j\frac{\sqrt{3}}{2}$ and $a = \frac{1}{2} + j\frac{\sqrt{3}}{2}$, are

$$\begin{aligned}
\mathcal{L}^{-1}\left\{\frac{1}{(s+\frac{1}{2}-j\frac{\sqrt{3}}{2})}\right\} &= e^{-(\frac{1}{2}-j\frac{\sqrt{3}}{2})t} \text{ and} \\
\mathcal{L}^{-1}\left\{\frac{1}{(s+\frac{1}{2}+j\frac{\sqrt{3}}{2})}\right\} &= e^{-(\frac{1}{2}+j\frac{\sqrt{3}}{2})t},
\end{aligned}$$

respectively.

Algebraically manipulating (2.78) and substituting the terms of complex exponentials by trigonometric identities, we have (for $t \geq 0$)

$$\begin{aligned}
v_o(t) &= 2\frac{\sqrt{3}}{2}e^{-\frac{t}{2}}\sin(\frac{\sqrt{3}}{2}) + e^{-t} \\
&= 1.1547\sin(0.866t)e^{-\frac{t}{2}} + e^{-t}.
\end{aligned} \tag{2.79}$$

The solution given by (2.79) is the same as that given by (2.33). However, in the solution obtained using the Laplace transform, the initial conditions were part of the transform. We could also have applied the Laplace transform directly to the ODE (2.30) assisted by Tables 2.3 and 2.2, as follows:

$$\mathcal{L}\{e^{-t}\} = \mathcal{L}\left\{\frac{d^2v_o(t)}{dt^2} + \frac{dv_o(t)}{dt} + v_o(t)\right\}, \tag{2.80}$$

where

$$\begin{aligned}
\mathcal{L}\left\{\frac{d^2v_o(t)}{dt^2}\right\} &= s^2V_o(s) - sv_o(0^-) - \dot{v}_o(0^-), \\
\mathcal{L}\left\{\frac{dv_o(t)}{dt}\right\} &= sV_o(s) - v_o(0^-), \\
\mathcal{L}\{v_o(t)\} &= V_o(s), \text{ and} \\
\mathcal{L}\{e^{-t}\} &= \frac{1}{s+1}.
\end{aligned} \tag{2.81}$$

Hence, substituting the expressions from (2.81) into (2.80) and considering the initial conditions, we obtain the same final expression presented in (2.75):

$$\begin{aligned}\frac{1}{s+1} &= V_o(s)\left(s^2+s+1\right) - s v_o(0^-) - \dot{v}_o(0^-) - v_o(0^-)\\ &= V_o(s)\left(s^2+s+1\right) - s - 0 - 1\\ &= V_o(s)\left(s^2+s+1\right) - s - 1,\\ V_o(s) &= \left(\frac{1}{s+1}+s+1\right)\left(\frac{1}{s^2+s+1}\right)\\ &= \left(\frac{s^2+2s+2}{s+1}\right)\left(\frac{1}{s^2+s+1}\right). \end{aligned} \quad (2.82)$$

In Section 2.2 the importance of LTI systems was introduced. The convolution integral in (2.9), which expresses the input signal, $x(t)$, as a sum of scaled and shifted impulse functions, uses delayed unit impulses as its basic signals. Therefore, an LTI system response is the same linear combination of the responses to the basic inputs.

Considering that complex exponentials, such as e^{-st}, are eigensignals of LTI systems, Eq. (2.36) is defined if one uses e^{-st} as a basis for the set of all input functions in a linear combination made of infinite terms (i.e., an integral). As the signal is being decomposed in basic inputs (complex exponentials), the LTI system's response could be characterized by weighting factors applied to each component in that representation.

In Section 2.7, the Fourier transform is introduced and it is shown that the Fourier integral does not converge for a large class of signals. The Fourier transform may be considered a subset of the Laplace transform, or the Laplace transform could be considered a generalization (or expansion) of the Fourier transform. The Laplace integral term e^{-st}, from (2.36), forces the product $x(t)e^{-st}$ to zero as time t increases. Therefore, it could be regarded as the exponential weighting term that provides convergence to functions for which the Fourier integral does not converge.

Following the concept of representing signals as a linear combination of eigensignals, instead of choosing $e^{-(\sigma+j\Omega)t}$ as the eigensignals, one could choose $e^{-j\Omega t}$ as the eigensignals. The latter representation leads to the Fourier transform equation, and any LTI system's response could be characterized by the amplitude scaling applied to each of the basic inputs $e^{-j\Omega t}$.

The Laplace transform and its applications are discussed in greater detail in [5], [11], [12], [4], [2], and [9].

2.5 Transfer function and stability

The Laplace transform, seen in the previous section, is an important tool to solve differential equations and therefore to obtain, once given the input, the output of an LTI system. For an LTI system, the Laplace transform of its mathematical representation, given as a differential equation – with constant coefficients and null initial conditions – as in (2.22) or, equivalently, given by a convolution integral as in (2.10), leads to the concept of *transfer function*, the ratio

$$\boxed{H(s) = \frac{Y(s)}{X(s)}} \tag{2.83}$$

between the Laplace transform of the output signal and the Laplace transform of the input signal. This representation of an LTI system also corresponds to the Laplace transform of its impulse response, i.e., $H(s) = \mathcal{L}\{h(t)\}$.

Assuming that all initial conditions are equal to zero, solving a system using the concept of transfer function is usually easier for the convolution integral is replaced by a multiplication in the transformed domain:

$$Y(s) = \mathcal{L}\{h(t) * x(t)\} = H(s)X(s) \tag{2.84}$$

such that

$$y(t) = \mathcal{L}^{-1}\{H(s)X(s)\}, \tag{2.85}$$

the zero-state response of this system.

A simple example is given from the circuit in Fig. 2.4, where

$$H(s) = \frac{1/RC}{s + 1/RC}.$$

If $x(t) = u(t)$, then $X(s) = 1/s$ and

$$\begin{aligned} Y(s) &= H(s)X(s) = \frac{1/RC}{(s + 1/RC)s} \\ &= \frac{1}{s} - \frac{1}{s + 1/RC}. \end{aligned}$$

In this case, $y(t) = \mathcal{L}^{-1}\{Y(s)\}$ corresponds to

$$y(t) = u(t) - e^{-\frac{1}{RC}t}u(t),$$

which is the same solution given in (2.5).

Given (2.10) and provided that all initial conditions are null, the transfer function $H(s)$ corresponds to a ratio of two polynomials in s:

$$H(s) = \frac{P(s)}{Q(s)}. \tag{2.86}$$

The roots of the numerator $P(s)$ are named *zeros*, while the roots of the denominator $Q(s)$ are known as *poles*. Both are usually represented in the complex plane $s = \sigma + j\Omega$ and this plot is referred to as *pole-zero plot* or *pole-zero diagram*.

We use the ODE (2.29) to provide an example:

$$H(s) = \frac{V_o(s)}{V_i(s)} = \frac{1}{LCs^2 + RCs + 1}. \tag{2.87}$$

Using the same values, $R = 2.0$ (in $\mathbf{\Omega}$), $L = 2.0$ (in **H**), and $C = 1/2$ (in **F**), we obtain the poles (roots of the characteristic equation) $s = \frac{-1 \pm j\sqrt{3}}{2}$ as seen in Fig. 2.15.

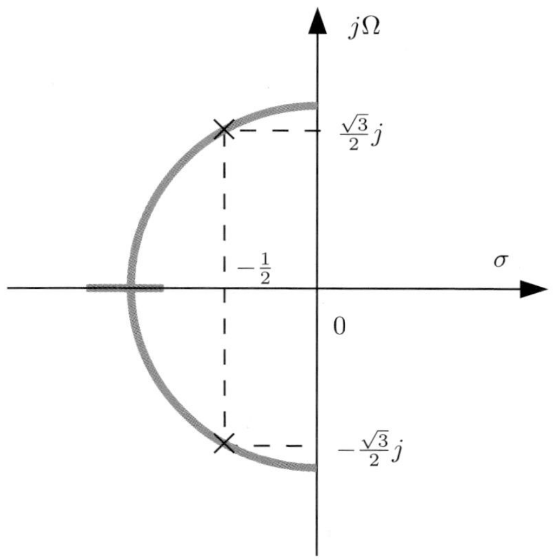

FIGURE 2.15

An example of a pole-zero diagram of the transfer function in (2.87) for $R = 2.0$ (in $\mathbf{\Omega}$), $L = 2.0$ (in **H**), and $C = 1/2$ (in **F**). The gray curve corresponds to the root locus of the poles when R varies from $R = 0$ (poles at $\pm j$) to $R = 4.1$ (poles at -1.25 and -0.8). Click here to watch an animation.

For an LTI system, a complex exponential e^{st} can be considered an eigensignal:

$$\begin{aligned} x(t) = e^{st} \longrightarrow \quad y(t) &= e^{st} * h(t) = \int_{-\infty}^{\infty} h(\tau) e^{s(t-\tau)} d\tau \\ &= e^{st} \underbrace{\int_{-\infty}^{\infty} h(\tau) e^{-s\tau} d\tau}_{H(s)} \\ &= H(s) e^{st} = \frac{P(s)}{Q(s)} e^{st}, \end{aligned} \tag{2.88}$$

which is valid only for those values of s where $H(s)$ exists.

If there is no common root between $P(s)$ and $Q(s)$, the denominator of $H(s)$ corresponds to the characteristic polynomial and, therefore, poles p_i of an LTI system correspond to their *natural frequencies*.

As mentioned in Section 2.2, the *stability* of a system, in a *bounded-input bounded-output* (BIBO) sense, implies that if $|x(t)| < B_x$, then $|y(t)| < B_y$, B_x and $B_y < \infty$, for all values of t. This corresponds to its input response being absolutely integrable, i.e.,

$$\int_{-\infty}^{\infty} |h(t)|dt = \|h(t)\|_1 < \infty. \tag{2.89}$$

As seen previously, the natural response of a linear system corresponds to a linear combination of complex exponentials $e^{p_i t}$. Let us assume, for convenience, that $H(s)$ has N distinct poles and that the order of $P(s)$ is lower than the order of $Q(s)$; in that case, we can write

$$h(t) = \mathcal{L}^{-1}\left\{\frac{P(s)}{\prod_{i=1}^{N}(s-p_i)}\right\} = \sum_{i=1}^{N} A_i e^{p_i t} u(t), \tag{2.90}$$

where constants A_i are obtained from simple partial fraction expansion.

In order for the system to be stable, each exponential should tend to zero as t tends to infinity. Considering a complex pole $p_i = \sigma_i + j\Omega_i$, $\lim_{t\to\infty} |h(t)| \to 0$ should imply that $\lim_{t\to\infty} |e^{(\sigma_i + j\Omega_i)t}| = \lim_{t\to\infty} |e^{\sigma_i t}| = 0$ or $\sigma_i < 0$; that is, the real part of p_i should be negative.

While the location of the *zeros* of $H(s)$ is irrelevant for the stability of an LTI system, their *poles* are of paramount importance. We summarize this relationship in the following.

1. A causal LTI system is BIBO stable if and only if all poles of its transfer function have negative real part (belong to the left half of the s-plane).
2. A causal LTI system is unstable if and only if at least one of the following conditions occur: one pole of $H(s)$ has a positive real part and repeated poles of $H(s)$ have real part equal to zero (belong to the imaginary axis of the s-plane).
3. A causal LTI system is marginally stable if and only if there are no poles on the right half of the s-plane and nonrepeated poles occur on its imaginary axis.

We can also state system stability from the ROC of $H(s)$: an LTI system is stable if and only if the ROC of $H(s)$ includes the imaginary axis ($s = j\Omega$). Fig. 2.16 depicts examples of the impulse response for BIBO stable, marginally stable, and unstable systems.

In this section, we have basically addressed BIBO stability, which, although not sufficient for asymptotic stability (a system can be BIBO stable without being stable when initial conditions are not null), is usually employed for linear systems. A more thorough stability analysis could be carried out by using Lyapunov criteria [14].

The Laplace transform applied to the input response of the differential equation provides the transfer function whose poles (roots of its denominator) determine the system stability: a causal LTI system is said to be stable when all poles lie in the left half of the complex s-plane. In this case, the system is also known as asymptotically stable, for its output always tends to decrease, not presenting permanent oscillation.

Whenever distinct poles have their real part equal to zero (poles on the imaginary axis), permanent oscillation will occur and the system is marginally stable; the output signal does not decay nor grows indefinitely. Fig. 2.16 shows the impulse response growing over time when two repeated poles are located on the imaginary axis. Although not shown in Fig. 2.16, decaying exponentials of oscillation will appear when poles are real and negative, while growing exponentials will appear in the case of real positive poles.

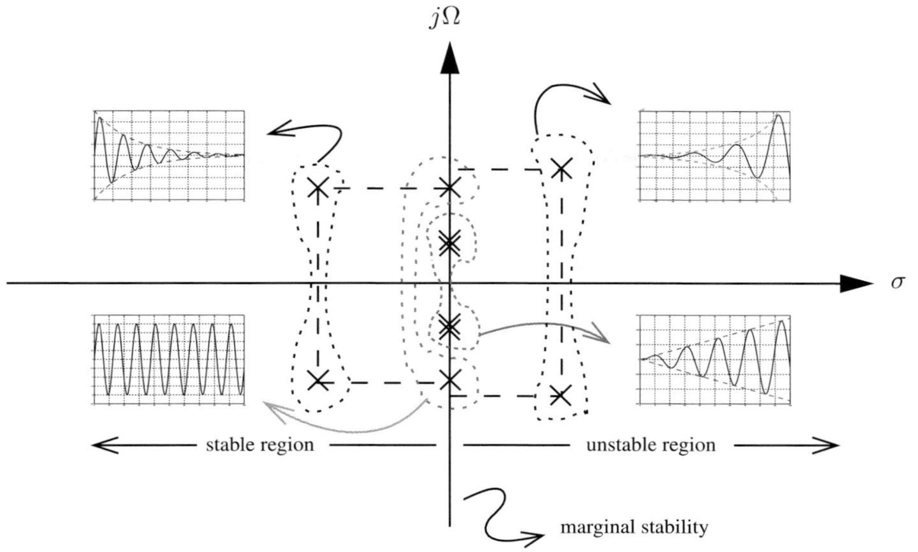

FIGURE 2.16

Pole location on the s-plane and system stability: examples of impulse responses of causal LTI systems. Visit the "exploring the s-plane" website.

Although we have presented the key concepts of stability related to a linear system, much more could be said about stability theory. Hence, further reading is encouraged: [2] and [14].

2.6 Frequency response

We have mentioned in the previous section that the complex exponential e^{st} is an eigensignal of a continuous-time linear system. A particular choice of s is $j\Omega_0 = j2\pi f_0$, i.e., the input signal

$$x(t) = e^{j\Omega_0 t} = \cos(\Omega_0 t) + j\sin(\Omega_0 t) \tag{2.91}$$

has a single frequency and the output, from (2.88), is

$$y(t) = e^{j\Omega_0 t} \underbrace{\int_{-\infty}^{\infty} h(\tau)e^{-j\Omega_0 \tau} d\tau}_{H(j\Omega_0)} . \tag{2.92}$$

With $e^{j\Omega_0 t}$ being an eigensignal, $H(j\Omega_0)$ in (2.92) corresponds to its eigenvalue. Allowing a variable frequency $\Omega = 2\pi f$ instead of a particular value $\Omega_o = 2\pi f_0$, we define the *frequency response* of a linear system having impulse response $h(t)$ as

$$\boxed{H(j\Omega) = \int_{-\infty}^{\infty} h(\tau)e^{-j\Omega\tau}d\tau.} \tag{2.93}$$

We note that the frequency response corresponds to the Laplace transform of $h(t)$ on a specific region of the s-plane, the vertical (or imaginary) axis $s = j\Omega$:

$$H(j\Omega) = H(s)|_{s=j\Omega}. \tag{2.94}$$

It is clear from (2.93) that the frequency response $H(j\Omega)$ of a linear system is a complex function of Ω. Therefore, it can be represented in its polar form as

$$H(j\Omega) = |H(j\Omega)|\, e^{\angle H(j\Omega)}. \tag{2.95}$$

As an example, we use the RLC series circuit given in Fig. 2.17 with $L = 1.0$ (in **H**), $C = 1.0$ (in **F**), and R varying from $0.2\boldsymbol{\Omega}$ to $1.2\boldsymbol{\Omega}$ (in $\boldsymbol{\Omega}$).

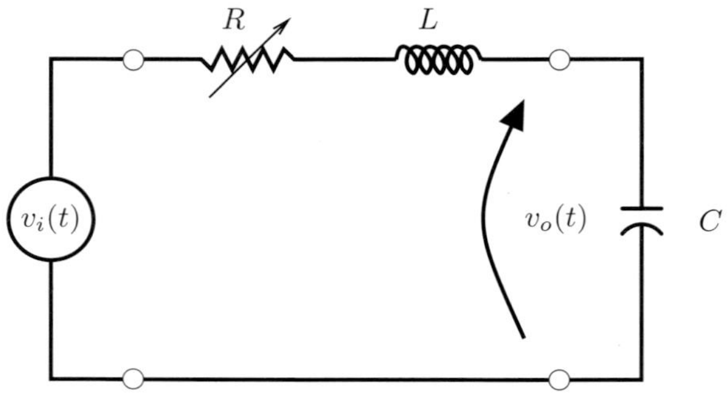

FIGURE 2.17

RLC series circuit: the transfer function is $H(s) = \frac{V_o(s)}{V_i(s)}$, $V_o(s) = \mathcal{L}\{v_0(t)\}$, $V_i(s) = \mathcal{L}\{v_i(t)\}$, and the frequency response is given by $H(j\Omega) = H(s)|_{s=j\Omega}$.

The frequency response, magnitude or absolute value (in dB), and argument or phase (in radians) are depicted in Fig. 2.18. For this example, we consider the voltage applied to the circuit as input and the voltage measured at the capacitor as output. This is worth mentioning since the output could be, for instance, the current in the circuit or the voltage across the inductor.

In low frequencies, the capacitor tends to become an open circuit such that $v_o(t) = v_i(t)$ and the gain tends to 1 or 0 dB. On the other hand, as the frequency Ω goes towards infinity, the capacitor tends to become a short circuit such that $v_o(t)$ goes to zero, gain in dB tending to minus infinity. The circuit behaves like a low-pass filter allowing low frequencies to the output while blocking high frequencies. Specially, when R has low values (when tending to zero), we observe a peak in $|H(j\Omega)|$

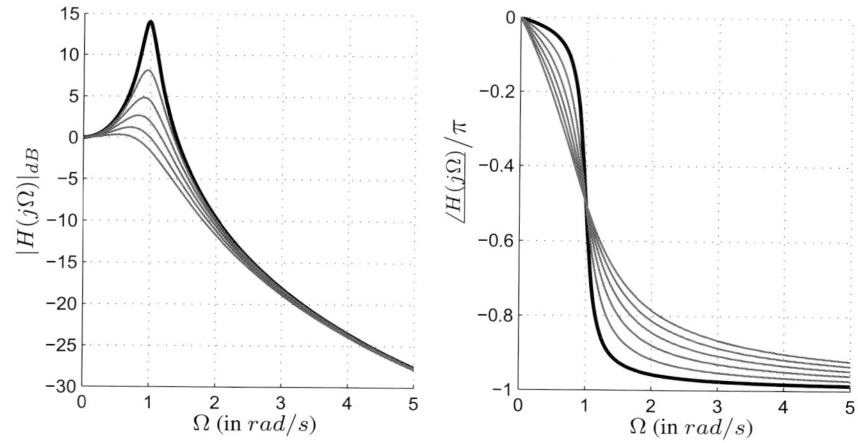

FIGURE 2.18

Frequency response in magnitude ($|H(j\Omega)|$ in dB) and normalized phase (argument of $H(j\Omega)$ divided by π, i.e., -1 corresponds to $-\pi$) for the circuit in Fig. 2.17 with $L = 1$ (in **H**), $C = 1$ (in **F**), and R varying from 0.2 (in **Ω**) (the highest peak in magnitude is highlighted) to 1.2 (in **Ω**). Download the MATLAB code.

at $\Omega \approx \Omega_0 = 1$ rad/s. This *frequency*, $\Omega_0 = 2\pi f_0$, is termed the resonance frequency and corresponds to the absolute value of the pole responsible for this *oscillation*.

Let us write the frequency response from $H(s)$, as given in (2.87):

$$H(j\Omega) = H(s)|_{s=j\Omega} = \frac{1}{(1-\Omega^2 LC) + j\Omega RC}. \tag{2.96}$$

We note that when $R \to 0$, the peak amplitude of $|H(j\Omega)|$ occurs for $1 - \Omega^2 LC \to 0$ (it tends to infinity as R goes to zero), i.e., the resonance frequency corresponds to $\Omega_0 = \frac{1}{\sqrt{LC}}$.

The impulse response and the pole diagram are shown in Fig. 2.19; from this figure, we observe the oscillatory behavior of the circuit as R tends to zero in both the time domain and the s-domain (poles approaching the vertical axis). The impulse response and the pole locations for the minimum value of the resistance, $R = 0.2$ (in **Ω**), are highlighted in this figure.

Before providing more details about the magnitude plot of the frequency response as a function of Ω, let us show the (zero-state) response of a linear system to a real single-frequency excitation $x(t) = A\cos(\Omega t + \theta)$. We know that the complex exponential is an eigensignal to a linear system such that, making $s = j\Omega$,

$$y(t)|_{x(t)=e^{j\Omega t}} = H(j\Omega)e^{j\Omega t}. \tag{2.97}$$

Since we can write $A\cos(\Omega t + \theta)$ as $A(e^{j(\Omega t+\theta)} + e^{-j(\Omega t+\theta)})/2$, the output is given as

$$y(t) = \frac{A}{2}e^{j\theta}e^{j\Omega t}H(j\Omega) + \frac{A}{2}e^{-j\theta}e^{-j\Omega t}H(-j\Omega). \tag{2.98}$$

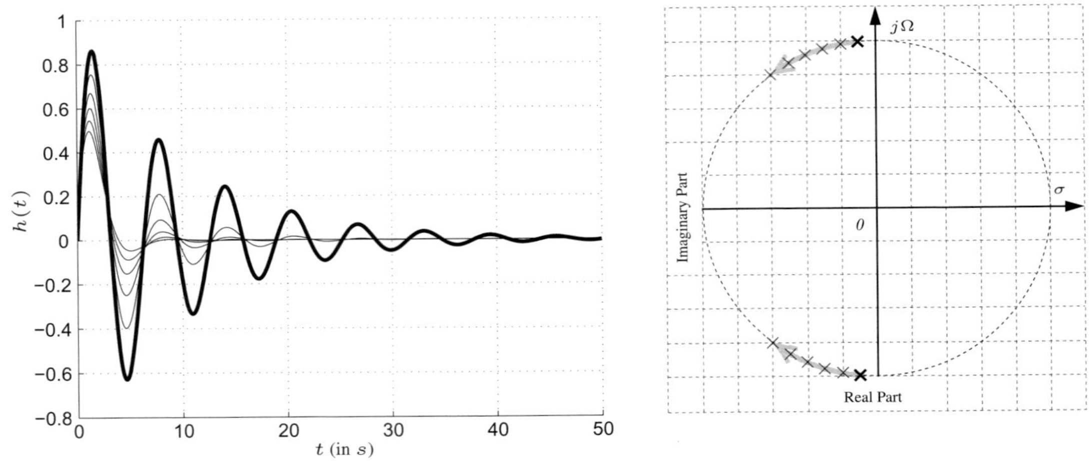

FIGURE 2.19

Impulse responses and pole locations in the s-plane (six different values of R) for the circuit in Fig. 2.17. Note that for lower values of R, $h(t)$ tends to oscillate and the poles are closer to the vertical axis (marginal stability region). Download the MATLAB code.

Assuming that $h(t)$ is real, it is possible to ensure that all poles of $H(s)$ will occur in complex conjugate pairs, leading (as shall be addressed in the next section) to $H^*(j\Omega) = H(-j\Omega)$. This conjugate symmetry property of the frequency response, i.e., $|H(j\Omega)| = |H(-j\Omega)|$ and $\angle H(j\Omega) = -\angle H(-j\Omega)$, when applied in (2.98), results in

$$\boxed{y(t)|_{x(t)=A\cos(\Omega t+\theta)} = |H(j\Omega)| A\cos(\Omega t + \theta + \angle H(j\Omega)).} \tag{2.99}$$

The previous expression is also valid as regime solution when the input is a sinusoid that is nonzero for $t \geq 0$, such as $x(t) = A\cos(\Omega t + \theta)u(t)$.

Now, back to our example of a frequency response given in (2.96), let us express its magnitude squared:

$$|H(j\Omega)|^2 = \frac{1}{1 + \Omega^2(R^2C^2 - 2LC) + \Omega^4 L^2 C^2}. \tag{2.100}$$

It is easy to see that when Ω tends to zero, the magnitude squared tends to one, while when Ω tends to infinity, the dominant term becomes $\Omega^4 L^2 C^2$:

(a) $\Omega \to 0 \Rightarrow |H(j\Omega)| \to 1$ or 0_{dB},
(b) $\Omega \to \infty \Rightarrow |H(j\Omega)| \to \frac{1}{\Omega^2 LC}$ or, in dB, $-40\log(\Omega/\Omega_0)$, $\Omega_0 = \frac{1}{\sqrt{LC}}$.

These two regions of $|H(j\Omega)|$, close to zero and tending to infinity, could be visualized by lines in a semilog plot; the frequency axis, due to its large range of values, is plotted using a logarithmic scale.

Particularly, when $\Omega \to \infty$, the approximation in (b) tells us that in an interval from Ω_{ref} to $10\Omega_{ref}$, $\Omega_{ref} \gg \Omega_0$ (the resonance frequency), we have an attenuation of 40 dB, as shown in Fig. 2.20.

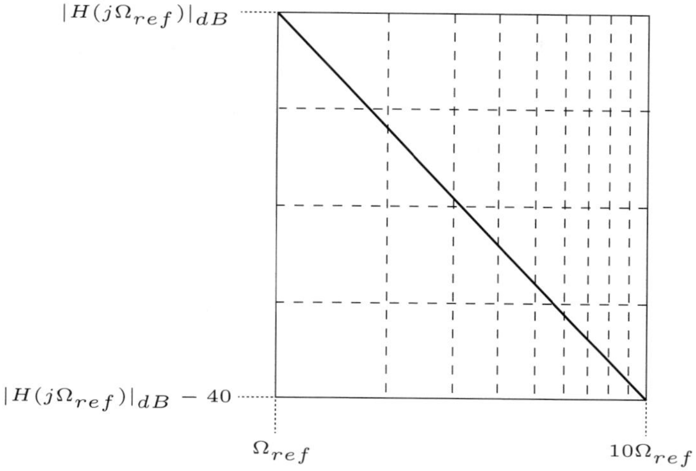

FIGURE 2.20

Asymptotical decay of 40 dB per decade.

A plot showing the magnitude in dB and the phase of $H(j\Omega)$ as a function of $\Omega = 2\pi f$, plotted in a logarithmic frequency scale, is known as *Bode plot* or *Bode diagram*. A Bode diagram may be sketched from lines (asymptotes) which are drawn from the structure of $H(s)$, its poles and zeros. For the case of a transfer function with two conjugate complex roots, we have

$$H(s) = \frac{\Omega_0^2}{s^2 + 2\zeta\Omega_0 s + \Omega_0^2}, \tag{2.101}$$

where, in our example in (2.87), $\Omega_0 = \frac{1}{\sqrt{LC}}$ and $\zeta = \frac{R}{2}\sqrt{\frac{C}{L}}$. Also note that (in order to have two conjugate complex roots) $0 < \zeta < 1$.

The magnitude of the frequency response in dB is given as

$$|H(j\Omega)|_{dB} = -10\log\left(\left(1 - \left(\frac{\Omega}{\Omega_0}\right)^2\right)^2 + \left(2\zeta\frac{\Omega}{\Omega_0}\right)^2\right). \tag{2.102}$$

When $\Omega = \Omega_0$, $|H(j\Omega)| = -20\log(2\zeta)$; this value is a good approximation for the peak magnitude (see Fig. 2.21) when ζ tends to zero (we have an error lower than 0.5% when $\zeta < 0.1$). The peak actually occurs in $\Omega = \Omega_0\sqrt{1 - 2\zeta^2}$, having a peak height equal to $\frac{1}{2\zeta\sqrt{1-\zeta^2}}$; we could also add that the peak is basically observable only when $0 < \zeta < 0.5$. The Bode plot of (2.87) with $R = 0.1$ (in $\mathbf{\Omega}$), $L = 1.0$ (in **H**), and $C = 1.0$ (in **F**) showing the (low-frequency and high-frequency) asymptotes is depicted in Fig. 2.21.

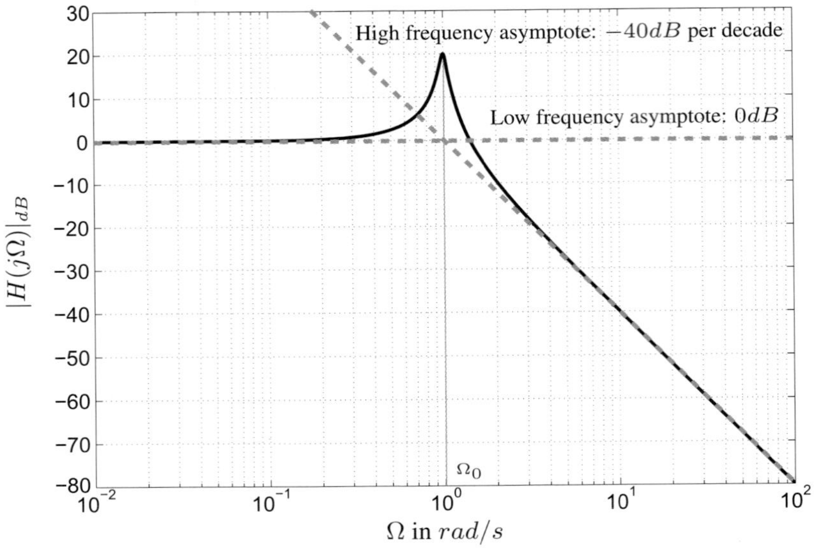

FIGURE 2.21

Example of Bode plot for $H(s) = \frac{\Omega_o^2}{s^2 + 2\zeta\Omega_0 s + \Omega_o^2}$ with $\Omega_0 = 1$ and $\zeta = 0.05$. Note that for this case, the approximated height of the peak $-20\log(2\zeta) = 20$ dB works well.

Another example with first-order poles and zeros will end our discussion on Bode plots. Let the transfer function $H(s)$ with poles at -1 and $-10{,}000$ and zeros at -10 and -1000 be represented as follows:

$$\begin{aligned} H(s) &= \frac{s^2 + 1.01s + 10{,}000}{s^2 + 10{,}001s + 10{,}000} \\ &= \frac{(s+1000)(s+10)}{(s+10{,}000)(s+1)} \\ &= \frac{\left(1+\frac{s}{1000}\right)\left(1+\frac{s}{10}\right)}{\left(1+\frac{s}{10{,}000}\right)(1+s)}. \end{aligned} \tag{2.103}$$

The magnitude in dB of its frequency response could then be written as

$$\begin{aligned} |H(j\Omega)|_{dB} &= 20\log\left|1+j\frac{\Omega}{1000}\right| + 20\log\left|1+j\frac{\Omega}{10}\right| \\ &\quad -20\log\left|1+j\frac{\Omega}{10{,}000}\right| - 20\log|1+j\Omega|. \end{aligned} \tag{2.104}$$

The first term, as in the previous example, can be checked for low and high frequencies:

(a) $\Omega \ll 1000 \Rightarrow 0_{dB}$,
(b) $\Omega \gg 1000 \Rightarrow$ increases 20 dB per *decade* (from Ω_{ref} to $10\Omega_{ref}$) or, equivalently, 6 dB per *octave* (from Ω_{ref} to $2\Omega_{ref}$).

When $\Omega = 1000$, the value of this term is $10\log(2) = 3.0103$ dB, which is the usual error at the these frequencies (*cutoff frequencies*).

Table 2.4 Asymptotes for the Bode plot of Fig. 2.22.

Ω	Event	asymptotic $\|H(j\Omega)\|_{dB}$
1 rad/s	New asymptote starts (−20 dB per decade)	0 dB
10 rad/s	New asymptote starts (+20 dB per decade)	−20 dB
1000 rad/s	New asymptote starts (+20 dB per decade)	−20 dB
10,000 rad/s	New asymptote starts (−20 dB per decade)	0 dB

We could, from the analysis of the first term, extend the results to all cutoff frequencies as shown in Table 2.4. The resulting asymptotic sketch and real curves are shown in Fig. 2.22, where we observe the 3-dB errors at the cutoff frequencies.

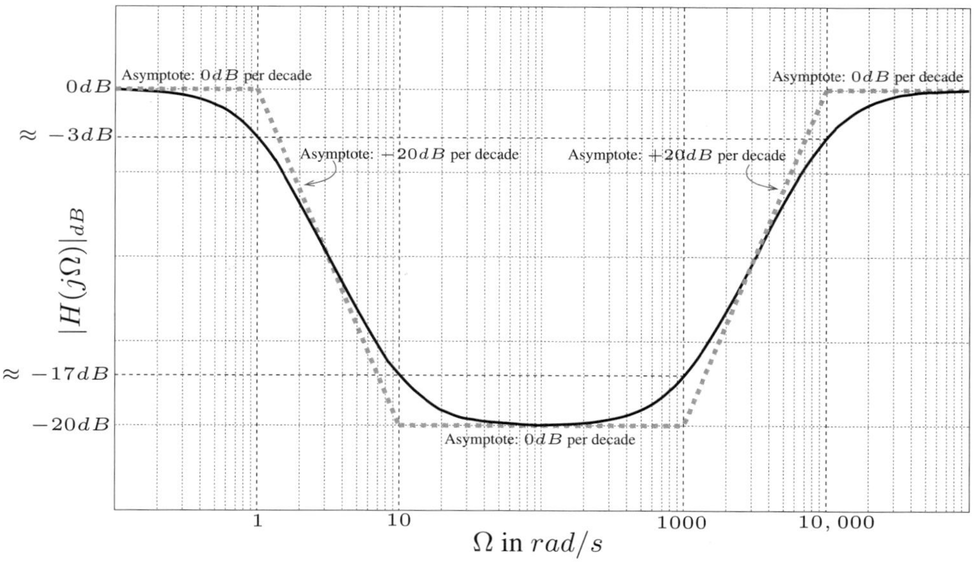

FIGURE 2.22

Bode magnitude (dB) plot for $H(s) = \frac{(1+\frac{s}{1000})(1+\frac{s}{10})}{\left(1+\frac{s}{10,000}\right)(1+\frac{s}{1})}$. Download the MATLAB code.

A final observation regards the phase of the frequency response: similarly to the magnitude, a sketch of the phase can also be obtained from the poles and zeros of $H(s)$:

$$\angle H(j\Omega) \;=\; \angle 1+j\Omega/1000 + \angle 1+j\Omega/10 - \angle 1+j\Omega/10{,}000 - \angle 1+j\Omega. \tag{2.105}$$

The first term on the right-hand side of (2.105), $\angle 1+j\Omega/1000 = \arctan(\Omega/1000)$, can be checked for low and high frequencies:

(a) $\Omega \ll 1000 \Rightarrow \arctan(\Omega/1000) \approx 0$,
(b) $\Omega \gg 1000 \Rightarrow \arctan(\Omega/1000) \approx 90$ degrees.

Considering only this zero (corresponding cutoff frequency $\Omega_0 = 1000$ rad/s), one may assume $0.1\Omega_0$ (100 rad/s) and $10\Omega_0$ (10,000 rad/s) sufficiently small and sufficiently large such that an asymptote could be drawn between points $(10^2, 0$ degrees) and $(10^4, 90$ degrees) as in Fig. 2.23. For a pole, the asymptote is mirrored with respect to 0 degrees.

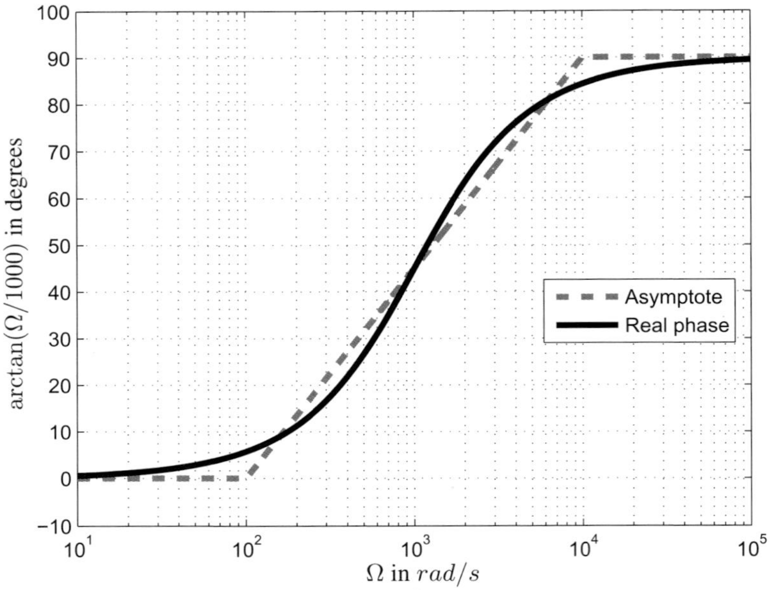

FIGURE 2.23

Asymptotical behavior of the phase for a single *zero.*

The asymptotic sketch and real phase curves for (2.105) are shown in Fig. 2.24; note that we have two negative slope asymptotes, starting in $\Omega = 10^{-1}$ rad/s and $\Omega = 10^3$ rad/s, and two positive slope asymptotes, starting at 10^0 rad/s and 10^2 rad/s.

The topics discussed in this section are found in the following books: [15], [16], [9], [3], and [1].

2.7 The Fourier series and the Fourier transform

In electronic circuits we often deal with periodic and nonperiodic signals. In addition, circuits could be driven by nonsinusoidal functions. Hence, methods that decompose continuous-time periodic (nonsi-

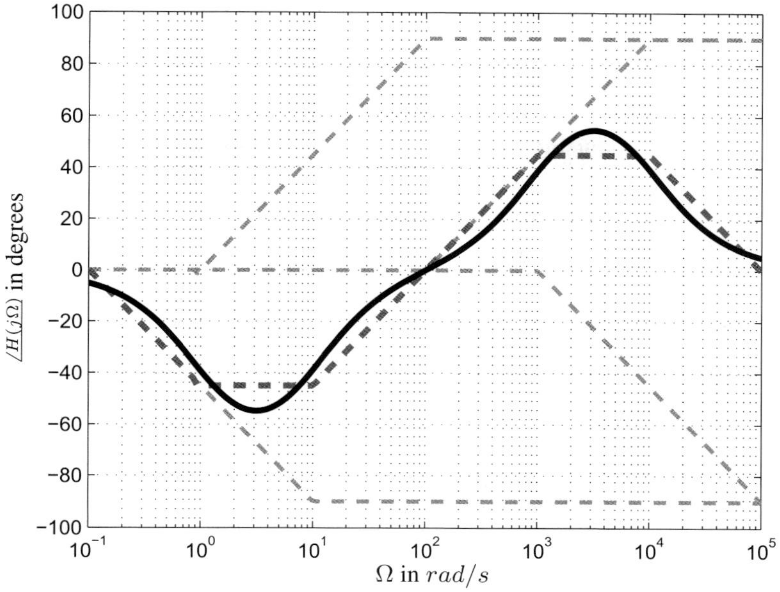

FIGURE 2.24

Bode phase plot for $H(s) = \frac{(1+\frac{s}{1000})(1+\frac{s}{10})}{\left(1+\frac{s}{10{,}000}\right)(1+\frac{s}{1})}$. Lighter gray dashed lines represent the asymptotes and the darker dashed line corresponds to the asymptotic sketch (sum of the asymptotes); the real phase is the continuous curve in black. Download the MATLAB code.

nusoidal) and nonperiodic signals into contributions from sinusoidal ($\sin(\Omega t)$ and $\cos(\Omega t)$) signals are extremely valuable for electronic system analysis; this is due to the fact that an LTI system only changes the amplitude and phase of a real sinusoid. Several properties make sinusoids an ideal choice to be the elementary basis functions for signal analysis and synthesis. For example, Eq. (2.99) shows that the LTI system changed the amplitude to $|H(j\Omega)|A$ and the phase to $\theta + \angle H(j\Omega)$ of the given sinusoidal input, $x(t) = A\cos(\Omega t + \theta)$. In addition, as sine and cosine are mutually orthogonal, sinusoidal basis functions are independent and can be processed independently. Usually, any sinusoidal function, such as the one mentioned above, can be expressed as a linear combination of sine and cosine functions: $x(t) = C\cos(\Omega t) - D\sin(\Omega t)$. If a signal corresponding to a sum of sinusoids,

$$x(t) = A_1\cos(\Omega_1 t + \theta_1) + A_2\cos(\Omega_2 t + \theta_2), \tag{2.106}$$

is fed to an LTI system, its output would be given as

$$y(t) = A_1|H(j\Omega_1)|\cos\left(\Omega_1 t + \theta_1 + \angle H(j\Omega_1)\right) + A_2|H(j\Omega_2)|\cos\left(\Omega_2 t + \theta_2 + \angle H(j\Omega_2)\right). \tag{2.107}$$

A periodic signal, $x(t) = x(t + nT)$, $\forall n = \pm 1, \pm 2, \pm 3, \ldots$, with period T, may contain components at frequencies $\Omega = 2\pi n/T$, where the fundamental frequency is given by $\Omega_0 = 2\pi f_0$, $f_0 = 1/T$. The frequencies $n\Omega_0$ are the harmonics, and any pair of signals are harmonically related if their periods are related by an integer ratio. In general, the resulting signal from a sum of harmonically related waveforms is also periodic, and its repetition period is equal to the fundamental period. The harmonic components exist only at discrete frequencies.

Consider that we add the harmonically related signals, $x_n(t)$, with $\Omega_0 = 2\pi/T$, over $T = 1$ such that $x(t) = \frac{1}{2} + \frac{2}{\pi}\sin(2\pi t) + \frac{2}{3\pi}\sin(6\pi t) + \frac{2}{5\pi}\sin(10\pi t) + \frac{2}{7\pi}\sin(14\pi t) + \cdots$, and get $x_N(t)$ as the result when $1 \le N \le 5$ and when $N \to \infty$, meaning that infinite terms are added. We have

$$x_N(t) = \frac{1}{2} + \sum_{n=1}^{N} x_n(t) = \frac{1}{2} + \frac{2}{\pi}\sum_{n=1}^{N} \frac{1}{2n-1}\sin\left(2\pi(2n-1)t\right). \tag{2.108}$$

The resulting waveforms, $x_N(t)$ for $1 \le N \le 5$ and $x_N(t)$ when N is large, i.e., $N \to \infty$, are displayed in Fig. 2.25. It is clear that as the number of terms becomes infinite in (2.108) the result converges at every value of the period T (to the square wave) except at the discontinuities. Actually, the ringing effect at the discontinuities (not shown in Fig. 2.25 for the curve with the large value of N) never dies out as N becomes larger, yet reaching a finite limit (Gibbs phenomenon).

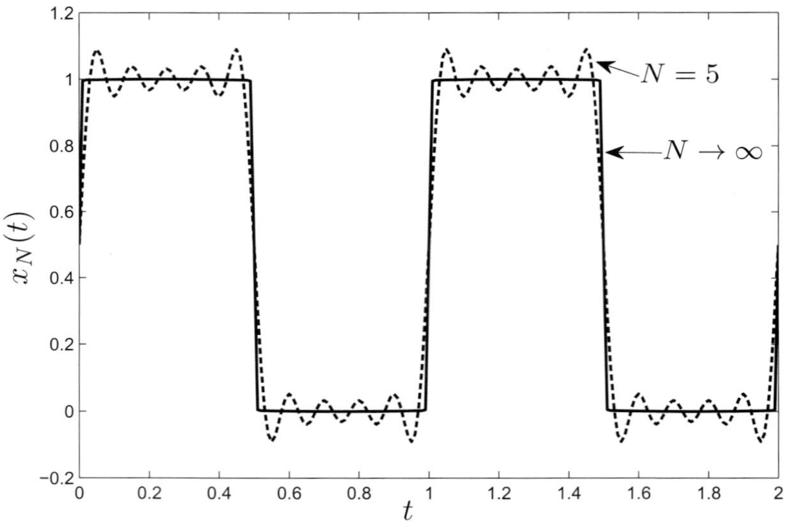

FIGURE 2.25

Synthesized square wave from Eq. (2.108).

Hence, for a period T, for a fundamental frequency Ω_0 with frequencies that are integer multiples of the fundamental frequency such that $n\Omega_0$, $\forall n \in \mathcal{Z}$ (the set of all integers), the representation of the harmonic components, $x_n(t)$, of a signal $x(t)$ can be written as

$$\begin{aligned} x_n(t) &= A_n \sin(n\Omega_0 t + \theta_n) \\ &= a_n \cos(n\Omega_0 t) + b_n \sin(n\Omega_0 t) \\ &= \frac{1}{2}(a_n - jb_n)e^{jn\Omega_0 t} + \frac{1}{2}(a_n + jb_n)e^{-jn\Omega_0 t}. \end{aligned} \tag{2.109}$$

Eq. (2.109) presents three equivalent forms to express the harmonic components $x_n(t)$: the first one is written as a sine function with amplitude A_n and phase θ_n; the second one is written as a sum of sine and cosine functions with real coefficients a_n and b_n, respectively; and the last form writes $x_n(t)$ in terms of complex exponentials.

The derivation, providing proofs, of the expressions for computing the coefficients in a Fourier series is beyond the scope of this chapter. Hence, we simply state, without proof, that if $x(t)$ is periodic with period T and fundamental frequency Ω_0, any arbitrary real periodic signal $x(t)$ could be represented by the infinite summation of harmonically related sinusoidal components, known as the Fourier series:

$$\begin{aligned} x(t) &= \frac{1}{2}a_0 + \sum_{n=1}^{\infty} A_n \sin(n\Omega_0 t + \theta_n) \\ &= \frac{1}{2}a_0 + \sum_{n=1}^{\infty} (a_n \cos(n\Omega_0 t) + b_n \sin(n\Omega_0 t)), \end{aligned} \tag{2.110}$$

where for $n = 0$, we have the term $\frac{a_0}{2}$ corresponding to the mean value of the signal, with a frequency of 0 Hz, known as the DC value (or level), and the other terms are as follows:

$$\begin{aligned} a_n &= A_n \sin(\theta_n), \\ b_n &= A_n \cos(\theta_n), \\ A_n &= \sqrt{a_n^2 + b_n^2}, \\ \theta_n &= \arctan\left(\frac{a_n}{b_n}\right). \end{aligned} \tag{2.111}$$

In the last form presented in (2.109), we substitute

$$\begin{aligned} c_n &= \frac{1}{2}(a_n - jb_n), \\ c_{-n} &= \frac{1}{2}(a_n + jb_n) \end{aligned} \tag{2.112}$$

to have a third representation of the Fourier series, given by

$$\begin{aligned} x(t) &= c_0 + \sum_{n=1}^{\infty} c_n e^{jn\Omega_0 t} + \sum_{n=-1}^{-\infty} c_n e^{jn\Omega_0 t} \\ &= \sum_{n=-\infty}^{\infty} c_n e^{jn\Omega_0 t}, \end{aligned} \tag{2.113}$$

where c_0 is real and c_n, $n \neq 0$, are complex coefficients. The coefficients of terms for positive (c_n) and negative (c_{-n}) values of n are complex conjugates ($c_{-n} = c_n^*$). This Fourier series form demands the summation to be performed over all $n \in \mathcal{Z}$, as the continuous-time complex sinusoids, $e^{jn\Omega_0 t}$, present an infinite number of terms (specific frequencies $n\Omega_0$). Moreover, for every integer value of n, $e^{jn\Omega_0 t} = \cos(n\Omega_0 t) + j\sin(n\Omega_0 t)$ is periodic with $T = 2\pi$. Hence, the periodic signal $x(t)$ must have $T = 2\pi$ in order to compute the Fourier series expansion. However, the analysis of the Fourier cosine series of a function x in the interval $[0, \pi]$ is equivalent to the analysis of its extension first to $[-\pi, \pi]$ as an even function and then to all components with period $T = 2\pi$.

The complex exponential representation of the Fourier series given by (2.113) generates a *two-sided spectrum*, as the summation requires all $n \in \mathcal{Z}$. The first and second representations of the Fourier series presented in (2.110) produce a *one-sided spectrum*, as the sum is computed only for values of $n > 0$.

Eqs. (2.110) and (2.113) are called Fourier synthesis equations as it is possible to synthesize any periodic (time) signal, $x(t)$, from an infinite set of harmonically related signals.

If $x(t)$ is periodic with fundamental frequency Ω_0 and period T, the coefficients of the three Fourier series equivalent representations are given by

$$c_n = \frac{1}{T}\int_{t_1}^{t_1+T} x(t)e^{-jn\Omega_0 t}dt, \tag{2.114}$$

and by manipulating (2.112) we obtain

$$\begin{aligned} a_n &= \frac{2}{T}\int_{t_1}^{t_1+T} x(t)\cos(n\Omega_0 t)dt, \\ b_n &= \frac{2}{T}\int_{t_1}^{t_1+T} x(t)\sin(n\Omega_0 t)dt, \end{aligned} \tag{2.115}$$

where the integrals in (2.114) and (2.115) are obtained over any interval that corresponds to the period T, with initial time t_1 for the integration arbitrarily defined.

Eqs. (2.114) and (2.115) are known as Fourier analysis or Fourier decomposition as the signal is being decomposed into its spectral components (basic inputs).

The continuous-time Fourier series expands any continuous-time periodic signal into a series of sine waves. The Fourier analysis methods are named after the French mathematician Jean-Baptiste Joseph Fourier. In general, almost any periodic signal can be written as an infinite sum of complex exponential functions (or real sinusoidal functions) and thus be represented as a Fourier series. Therefore, finding the response to a Fourier series means finding the response to any periodic function and its effect on its spectrum. When computing trigonometric coefficients, one could multiply a random signal by sinusoids of different frequencies to disclose all the (hidden) frequency components of the signal. In this way, it becomes easier to isolate the signal of the television station we want to watch from the others that are being simultaneously received by our television set.

In addition, the multiplication of a signal by $e^{j\Omega_0 t}$ means that the signal is being frequency shifted to Ω_0, which is equivalent to the Frequency shift Laplace property (Table 2.3).

Consider $x(t) = 3/4 + \cos(\Omega_0 t) + \cos(2\Omega_0 t + \pi/3) + 1/3\cos(5\Omega_0 t + \pi/4)$, where $\Omega_0 = 2\pi$, $3/4$ is the DC level, and we have three nonnull harmonics ($n = 1, 2,$ and 5) with the time-domain representation displayed in Fig. 2.1. The signal has several (hidden) frequencies, and no clue is provided when visually inspecting Fig. 2.1. The Fourier series representations are very helpful in this case, showing the amplitudes and phases of each harmonic in the frequency domain.

The exponential form, recursing to trigonometric identities, is given by

$$x(t) = \underbrace{\frac{3}{4}}_{c_0} + \left(\underbrace{\frac{1}{2}}_{c_1} e^{j\Omega_0 t} + \underbrace{\frac{1}{2}}_{c_{-1}} e^{-j\Omega_0 t} \right) + \left(\underbrace{\frac{1}{2} e^{j\frac{\pi}{3}}}_{c_2} e^{j2\Omega_0 t} + \underbrace{\frac{1}{2} e^{-j\frac{\pi}{3}}}_{c_{-2}} e^{-j2\Omega_0 t} \right) + \left(\underbrace{\frac{1}{6} e^{j\frac{\pi}{4}}}_{c_5} e^{j5\Omega_0 t} + \underbrace{\frac{1}{6} e^{-j\frac{\pi}{4}}}_{c_{-5}} e^{-j5\Omega_0 t} \right). \tag{2.116}$$

The complex-valued coefficient c_n conveys both amplitude ($|c_n|$) and phase ($\angle c_n$) of the frequency content of the signal $x(t)$ at each value $n\Omega_0$ rad/s as pictured in Fig. 2.26 (all other terms are null).

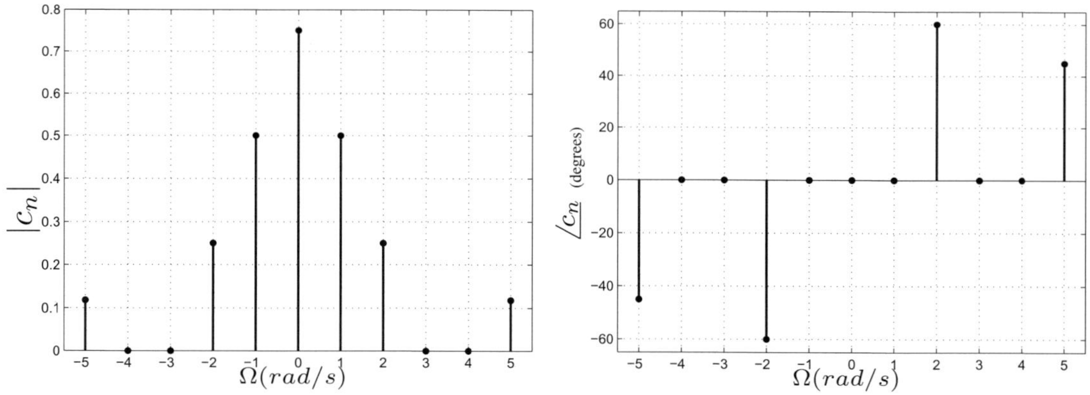

FIGURE 2.26

Two-sided spectrum of periodic signal $x(t)$ depicted in Fig. 2.1.

The sinusoidal basis functions of the Fourier series are smooth and infinitely differentiable but they only represent periodic signals. The Fourier representation for a nonperiodic (time-domain) signal has to treat the frequency spectrum of nonperiodic signals as a continuous function of frequency. This representation could be obtained by considering a nonperiodic signal as a special case of a periodic signal with an infinite period. The idea is that if the period of a signal is infinite ($T \to \infty$), then it is never repeated. Hence, it is nonperiodic. As $T \to \infty$, $\Omega_0 \to 0$, thus decreasing the spacing between adjacent line spectra (representing each harmonic contribution in the spectrum) towards a continuous representation of frequency.

The generalization to nonperiodic signals is known as the *Fourier transform* and is given by (2.117). It is analogous to the Fourier series analysis equation (2.114). It expresses the time-domain function

$x(t)$ as a continuous function of frequency $X(j\Omega)$ defined as follows:

$$\boxed{X(j\Omega) = \mathcal{F}\{x(t)\} = \int_{-\infty}^{\infty} x(t)e^{-j\Omega t}dt.} \tag{2.117}$$

The inverse Fourier transform is analogous to the Fourier series synthesis equation (2.113). It obtains the time-domain signal, $x(t)$, from its frequency-domain representation $X(j\Omega)$:

$$x(t) = \mathcal{F}^{-1}\{X(j\Omega)\} = \frac{1}{2\pi}\int_{-\infty}^{\infty} X(j\Omega)e^{j\Omega t}d\Omega. \tag{2.118}$$

The notation $X(j\Omega) = \mathcal{F}\{x(t)\}$ denotes that $X(j\Omega)$ is the Fourier transform of $x(t)$. Conversely, $x(t) = \mathcal{F}^{-1}\{X(j\Omega)\}$ denotes that $x(t)$ is the inverse Fourier transform of $X(j\Omega)$. This relationship is expressed with $x(t) \xleftrightarrow{\mathcal{F}} X(j\Omega)$.

The Fourier series coefficients have distinct units of amplitude, whereas $X(j\Omega)$ has units of amplitude density. The magnitude of $X(j\Omega)$ is termed as the *magnitude spectrum* ($|X(j\Omega)|$) and its phase as the *phase spectrum* ($\angle X(j\Omega)$).

Following the concept of representing signals as a linear combination of eigensignals, instead of choosing $e^{-(\sigma+j\Omega)t}$ as the eigensignals, one could choose $e^{-j\Omega t}$ as the eigensignals. The latter representation leads to the Fourier transform equation, and any LTI system's response could be characterized by the amplitude scaling applied to each of the basic inputs $e^{-j\Omega t}$.

This transform is very useful as it produces the frequency complex content of a signal, $X(j\Omega)$, from its temporal representation, $x(t)$. Moreover, the mapping provides a representation of the signal as a combination of complex sinusoids (or exponentials) for frequency-domain behavior analysis, where the (computationally consuming) computation of the convolution in the time domain is replaced by the (simple) operation of multiplication in the frequency domain.

Nevertheless, functions such as the unit step (Fig. 2.2(a)), discontinuous at time $t = 0$, do not have forward Fourier transforms as their integrals do not converge.

An analysis of convergence is beyond the scope of this chapter; hence, we assume that the existence of the Fourier transform is ensured by three (modified) Dirichlet conditions (for nonperiodic signals), requiring the temporal signal, $x(t)$, to have a finite number of discontinuities, with the size of each discontinuity being finite; to have a finite number of local maxima and minima; and to be absolutely integrable, i.e.,

$$\int_{-\infty}^{\infty} |x(t)|dt < \infty. \tag{2.119}$$

These are sufficient conditions, although not strictly necessary as several common signals, such as the unit step, are not absolutely integrable. Nevertheless, we can define a transform pair that satisfies the Fourier transform properties by the usage of impulses when dealing with this class of signals.

The Fourier transform converts a 1D time-domain signal, $x(t)$, into its 1D complex spectrum $X(j\Omega)$ representation in the frequency domain. In Section 2.4, the Laplace transform maps a time-domain 1D signal, $x(t)$, to a complex representation, $X(s)$, defined over a complex plane (s-plane), spanned by its two variables σ and Ω, as $s = \sigma + j\Omega$. The Laplace and Fourier transforms are closely related. When $s = j\Omega$, i.e., $\sigma = 0$, the Laplace integral becomes the Fourier transform. Referring to Fig. 2.10

(Section 2.4), as the imaginary axis ($j\Omega$) lies within the ROC, the Fourier transform exists only if $a < 0$. For a certain class of signals, there is a simple relationship between the Fourier and Laplace transforms given by

$$\mathcal{F}\{x(t)\} = \mathcal{L}\{x(t)\}|_{s=j\Omega}. \tag{2.120}$$

The Laplace transform for $x(t) = e^{-at}u(t)$, $a > 0$, is $X(s) = 1/(s+a)$ for $\sigma > -a$ (Table 2.2); $X(s)$ has a pole at $s = -a$, which lies at the left half s-plane. The ROC is located at the right side of $-a$ (right side of Fig. 2.11), thus containing the imaginary axis ($j\Omega$) of the s-plane. Hence, the signal $x(t)$ has a Fourier transform given by $X(j\Omega) = 1/(j\Omega + a)$. However, the Fourier transform does not converge for $\mathrm{Re}(s) < -a$ because $x(t)$ is not absolutely integrable, whereas $x(t)$ has a Laplace transform with the ROC pictured in the left side of Fig. 2.11.

Common Fourier transform pairs are summarized in Table 2.5.

Table 2.5 Fourier transform pairs.

$x(t) = \mathcal{F}^{-1}\{X(j\Omega)\}$	$X(j\Omega) = \mathcal{F}\{x(t)\}$
$\delta(t)$	1
1	$2\pi\delta(\Omega)$
$u(t)$	$\pi\delta(\Omega) + \frac{1}{j\Omega}$
$\lvert t\rvert$	$-\frac{2}{\Omega^2}$
e^{-at}	$\frac{1}{a + j\Omega}$
$e^{-a\lvert t\rvert}$, $a > 0$	$\frac{2a}{a^2 + \Omega^2}$
$e^{j\Omega_0 t}$	$2\pi\delta(\Omega - \Omega_0)$
$sgn(t)$	$\frac{2}{j\Omega}$
$\sin\Omega_0 t$	$-j\pi[\delta(\Omega - \Omega_0) + \delta(\Omega + \Omega_0)]$
$\cos\Omega_0 t$	$\pi[\delta(\Omega - \Omega_0) + \delta(\Omega + \Omega_0)]$

As the Laplace transform, the Fourier transform presents a set of properties. A comprehensive description of the properties of the Fourier transform is beyond the intent of this chapter, and the properties we examine are due to their importance in the study of linear system analysis, as is the case of linear electronic circuits. Much of the usefulness of the Fourier transform arises from its properties, which have important interpretations in the context of continuous-time signal processing. Understanding the relationship between time and frequency domains (symmetry between operations) is a key point in the study (and usage) of the Fourier transform.

Time and frequency scaling: If $x(t) \overset{\mathcal{F}}{\longleftrightarrow} X(j\Omega)$, then $x(at) \overset{\mathcal{F}}{\longleftrightarrow} \frac{1}{|a|}X\left(j\frac{\Omega}{a}\right)$. With $a \neq 0 \in \mathcal{R}$,

$$\mathcal{F}\{x(at)\} = \int_{-\infty}^{\infty} x(at)e^{-j\Omega t}\,dt. \tag{2.121}$$

Substituting $u = at$ we get

$$\mathcal{F}\{x(at)\} = \begin{cases} \int_{-\infty}^{\infty} x(u)e^{-j\frac{u}{a}\Omega}\frac{du}{a}, & a > 0, \\ \int_{\infty}^{-\infty} x(u)e^{-j\frac{u}{a}\Omega}\frac{du}{a}, & a < 0. \end{cases} \tag{2.122}$$

The two previous expressions can be combined in one single expression given by

$$\begin{aligned} \mathcal{F}\{x(at)\} &= \frac{1}{|a|}\int_{-\infty}^{\infty} x(u)e^{-j\frac{\Omega}{a}u}\,du \\ &= \frac{1}{|a|}X\left(j\frac{\Omega}{a}\right), \end{aligned} \tag{2.123}$$

implying that a linear expansion of the time axis in the time domain leads to linear compression of the frequency axis in the frequency domain, and vice versa. This property indicates an important trade-off between the two domains, as narrow time-domain signals have wide Fourier representations (narrow pulses have wide bandwidth).

Symmetry: The symmetry properties are quite useful as for a real-valued, a real-valued and even, a real-valued and odd, and an imaginary temporal signal, $x(t)$, we have the following relations:

$$x(t) \text{ real} \quad \stackrel{\mathcal{F}}{\longleftrightarrow} \quad X^*(j\Omega) = X(-j\Omega),$$

$$x(t) \text{ real and even} \quad \stackrel{\mathcal{F}}{\longleftrightarrow} \quad \text{Im}(X(j\Omega)) = 0,$$

$$x(t) \text{ real and odd} \quad \stackrel{\mathcal{F}}{\longleftrightarrow} \quad \text{Re}(X(j\Omega)) = 0,$$

$$x(t) \text{ imaginary} \quad \stackrel{\mathcal{F}}{\longleftrightarrow} \quad X^*(j\Omega) = -X(-j\Omega).$$

These properties are important as for a real-valued function in time its Fourier transform is conjugate symmetric, $X^*(j\Omega) = X(-j\Omega)$. This means that its real part (the magnitude) is an even function of frequency and that its imaginary part (phase) is an odd function of frequency. Therefore, when we obtain the Fourier transform of a real-valued time function, it is only necessary to display the transform for positive values of Ω. If the signal is even, i.e., $x(t) = x(-t)$, it has a Fourier transform that is a purely real function, $X(j\Omega) = |X(j\Omega)|$. Conversely, the transform of an odd function is an purely imaginary function.

Furthermore, other properties deserve to be highlighted. We start with the frequency shifting property presented in Table 2.6. As its representation in the frequency domain indicates, $X(j(\Omega - \Omega_0))$, a multiplication by $e^{j\Omega_0 t}$ in the time domain modulates the signal $x(t)$ onto a different frequency. As most modulation and demodulation techniques involve multiplication, it is important to stress the importance of the convolution property (Table 2.6). A simple operation of multiplication in time becomes a computationally intensive operation of convolution in frequency. Conversely, a convolution operation in the time domain is much easier to analyze in the frequency domain (multiplication in the frequency domain). Hence, as the convolution in the time domain defines the output of a time-invariant filter to a given input, the analysis and design of filters are typically performed in the frequency domain.

Another important relation is Parseval's relationship. It indicates that the information (energy) contained in the time domain is preserved when representing the signal in the frequency domain. If the power associated with a (complex) signal $x(t)$ is $|x(t)|^2$, Parseval's relationship states that the signal is represented equivalently in either the time or frequency domain without loss or gain of energy:

$$\int_{-\infty}^{\infty} |x(t)|^2 dt = \frac{1}{2\pi}\int_{-\infty}^{\infty} |X(j\Omega)|^2 d\Omega. \tag{2.124}$$

Hence, we can compute average power in either the time or frequency domain. The term $|X(j\Omega)|^2$ is known as the *energy spectrum* or *energy density spectrum* of the signal and shows how the energy of $x(t)$ is distributed across the spectrum.

A list of Fourier transform properties is presented in Table 2.6.

Table 2.6 Properties of the Fourier transform.

Property	time domain	frequency domain
Linearity	$a_1x_1(t) + a_2x_2(t)$	$a_1X_1(j\Omega) + a_2X_2(j\Omega)$
Scaling	$x(at)$	$\frac{1}{\lvert a\rvert}X(j\frac{\Omega}{a})$
Time shifting	$x(t - t_0)$	$e^{-j\Omega t_0}X(j\Omega)$
Frequency shifting	$e^{j\Omega_0 t}x(t)$	$X(j(\Omega - \Omega_0))$
Differentiation in time	$\frac{dx(t)}{dt}$	$j\Omega X(j\Omega)$
	$\frac{d^nx(t)}{dt}$	$(j\Omega)^nX(j\Omega)$
Integration in time	$\int_{-\infty}^{t} x(\tau)\,d\tau$	$\frac{1}{j\Omega}X(j\Omega) + \pi X(0)\delta(\Omega)$
Differentiation in frequency	$-jtx(t)$	$\frac{dX(j\Omega)}{d\Omega}$
Convolution in time	$x_1(t) * x_2(t)$	$X_1(j\Omega)X_2(j\Omega)$
Convolution in frequency	$x_1(t)x_2(t)$	$\frac{1}{2\pi}X_1(j\Omega) * X_2(j\Omega)$

One more time, we take, as an example, the RLC circuit displayed in Fig. 2.8, where the desired response is the voltage, $v_o(t)$, across the capacitor C, $v_C(t)$. If we use the Fourier transform, assisted by its differentiation property (Table 2.6), to solve the ODE in (2.30) we have

$$\mathcal{F}\{e^{-t}\} = V_i(j\Omega) = \mathcal{F}\left\{LC\frac{d^2v_o(t)}{dt^2} + RC\frac{dv_o(t)}{dt} + v_o(t)\right\} \tag{2.125}$$

and we get the following result:

$$V_i(j\Omega) = -LC\Omega^2V_o(j\Omega) + jRC\Omega V_o(j\Omega) + V_o(j\Omega), \tag{2.126}$$

where the output in the frequency domain, $V_o(j\Omega)$, is given by

$$V_o(j\Omega) = \frac{V_i(j\Omega)}{(1 - LC\Omega^2) + jRC\Omega}, \tag{2.127}$$

which exhibits a resonant behavior exactly as in (2.96). Also note that the frequency response, defined in (2.93), corresponds to the Fourier transform of the impulse response.

Fourier methods and applications are presented in [1], [4], [17], [18], and [5].

2.8 Conclusion and future trends

This chapter provided an introductory overview on the general concepts of signals, systems, and analysis tools used in the continuous-time domain.

We believe that the knowledge of continuous-time signals and systems, even for a book directed towards methods used in the discrete-time domain, is extremely important. While exploring their differences, the reader can enhance the understanding of natural phenomena.

A few papers discuss the relevance of teaching (analog) circuits and systems, as well their importance in engineering [19].

Analog circuit engineers have to deal with the entire circuit, and many parameters must be considered in their design. In order to simplify circuit models, designers have to make approximations about the analog components, which requires expertise and years of education.

This chapter dealt with traditional electrical components and their fundamental variables: resistors having a direct relationship between voltage and current ($v = Ri$, R being the resistance), inductors relating voltage and current with its flux ($d\phi = vdt$ and $d\phi = Ldi$, L being the inductance), and capacitors relating voltage and current with charge ($dq = Cdv$ and $dq = idt$, C being the capacitance). After first theorized in the early 1970s, the *memristor* regained attention in the media in 2008 when a first real implementation was announced: the fourth fundamental circuit element relating charge and flux as in its axiomatic definition, $d\phi = Mdq$ (M the *memristance*) [20]. The use of this component in usual electronic gadgets as well as teaching its theory even in undergraduate courses seems to be an upcoming technology trend.

In addition, most electronic devices have to interface with the external, analog, world, where data conversion is needed at the input and output sides. Analog technologies are used along with digital technologies, spanning from human interface components to analog circuits in wireless components. Furthermore, analog design is also present in the mixed-signal integrated circuits, and it is becoming increasingly critical due to high-speed circuitry [21]. Analog circuits are also used to process signals in very high frequency ranges, such as microwave and radiofrequency ranges. In addition, electronic digital circuits are supplied with power, and the trend is to reduce power consumption of digital circuitry. Hence, knowledge of analog circuits is required if an engineer is designing a high-performance digital circuit.

We have tried to condense in few pages all pertinent information regarding analog signals and systems in order to facilitate complete understanding of the forthcoming chapters. Nevertheless, we have not exhausted the description of all details. For interested readers, we outline in the following suggested references for further reading. We hope you enjoy the rest of the book.

Relevant websites

http://www.britannica.com/EBchecked/topic/330320/Pierre-Simon-marquis-de-Laplace/
(article from the Encyclopædia Britannica about Laplace)

http://www.genealogy.ams.org/id.php?id=108295
(American Mathematical Society Genealogy Project, Pierre-Simon Laplace)

https://www.academie-francaise.fr/les-immortels/pierre-simon-de-laplace?fauteuil=8&election=11-04-1816
(mention to Pierre-Simon de Laplace, Grand officier de la Légion d'honneur – Académie Française)

http://www.britannica.com/EBchecked/topic/215097/Joseph-Baron-Fourier/
(article from the Encyclopædia Britannica about Fourier)

http://www.genealogy.ams.org/id.php?id=17981
(American Mathematical Society Genealogy Project, Jean-Baptiste Joseph Fourier)

http://www.ieeeghn.org/
(IEEE Global History Network)

Glossary

BIBO	BIBO refers to a certain class of stability known as *bounded-input bounded-output*
Convolution	Convolution is an operation between two functions: the output (zero-state solution) of a linear and time-invariant system corresponds to the convolution integral between the input signal and the impulse response, i.e., $y(t) = \int_{-\infty}^{\infty} x(\tau)h(t-\tau)d\tau$
Frequency response	The frequency response of a linear system corresponds to the Fourier transform of its impulse response: $H(j\Omega) = \int_{-\infty}^{\infty} h(\tau)e^{-j\Omega\tau}d\tau$
Impulse signal	$\delta(t)$ is a pulse of infinite amplitude and infinitesimal time characterized by being equal to zero when $t \neq 0$ and $\int_{-\infty}^{\infty} \delta(t)dt = 1$
Impulse response	$h(t)$ is the output of a linear and time-invariant system when the input corresponds to the impulse $\delta(t)$
LTI	Linear and time-invariant system: when the output of a continuous-time system corresponds to a linear operation on the input signal while not depending on a particular instant but remaining the same as time goes by
ODE	Ordinary differential equation: a differential equation having only one independent variable
RLC	RLC is usually related to a circuit with resistor (R), inductor (L), and capacitor (C)
ROC	The region of convergence of the Laplace transform of a given signal $x(t)$: in the Laplace transform domain, it corresponds to the set of values of the complex variable s that ensures the existence of $\mathcal{L}\{x(t)\}$
Transfer function	A transfer function is the representation in the s-domain of the input–output relation for a linear system; it corresponds to the Laplace transform of the impulse response, i.e., $H(s) = \frac{Y(s)}{X(s)} = \mathcal{L}\{h(t)\}$
Unit-step signal	$u(t)$ is equal to 0 when $t < 0$ and equal to 1 when $t > 0$, and it relates to the impulse signal as follows: $u(t) = \int_{-\infty}^{t} \delta(t)dt$

Nomenclature

C	Capacitance, in Faraday (F)
f	Frequency, in cycles per second or Hertz (Hz)
L	Inductance, in Henry (H)
R	Resistance, in Ohms (**Ω**)
t	Time, in seconds (s)

$x(t)$	Input signal of a given continuous-time system, expressed in Volt (V) when it corresponds to an input voltage; $y(t)$ is usually used as the output signal
Ω	Angular frequency, in radians per second (rad/s) (it is sometimes referred to as *frequency* although corresponding to $2\pi f$)

References

[1] A. Oppenheim, A. Willsky, S. Nawab, Signals and Systems, 2nd edition, Prentice-Hall Signal Processing Series, Prentice Hall, 1996.
[2] B.P. Lathi, Linear Systems and Signals, 2nd edition, Oxford University Press, 2004.
[3] B.P. Lathi, Signal Processing and Linear Systems, 2nd edition, Oxford University Press, Incorporated, 2009.
[4] S. Haykin, B. Veen, Signals and Systems, 2nd edition, John Wiley & Sons, 2003.
[5] B. Girod, R. Rabenstein, A. Stenger, Signals and Systems, 1st edition, John Wiley & Sons, 2001.
[6] C. Candan, Proper definition and handling of Dirac delta functions [lecture notes], IEEE Signal Processing Magazine 38 (3) (2021) 186–203, https://doi.org/10.1109/MSP.2021.3055025.
[7] W. Boyce, R. DiPrima, Elementary Differential Equations, 9th edition, John Wiley & Sons, USA, 2008.
[8] A.C. Fischer-Cripps, The Mathematics Companion: Mathematical Methods for Physicists and Engineers, Taylor & Francis, New York, USA, 2005.
[9] R. Dorf, J. Svoboda, Introduction to Electric Circuits, 8th edition, John Wiley & Sons, USA, 2010.
[10] G. Strang, Computational Science and Engineering, Wellesley-Cambridge Press, Massachusetts, USA, 2007.
[11] W. Thomson, Laplace Transformation, 2nd edition, Prentice-Hall Electrical Engineering Series, Prentice-Hall Inc., 1960.
[12] J. Holbrook, Laplace Transforms for Electronic Engineers, 2nd edition, International Series of Monographs on Electronics and Instrumentation, Pergamon Press, 1966.
[13] F. Oberhettinger, L. Badii, Tables of Laplace Transforms, Grundlehren der mathematischen wissenschaften, Springer-Verlag, 1973.
[14] L. Padulo, M.A. Arbib, System Theory: a Unified State-Space Approach to Continuous and Discrete Systems, W. B. Saunders Company, Philadelphia, USA, 1974.
[15] C. Close, Analysis of Linear Circuits, international edition, Harcourt Publishers Ltd., California, USA, 1974.
[16] C. Desoer, E. Kuh, Basic Circuit Theory, Electrical and Electronic Engineering Series, McGraw Hill International Editions, New York, USA, 1969.
[17] R. Bracewell, The Fourier Transform and Its Applications, 3rd edition, McGraw-Hill Series in Electrical and Computer Engineering, McGraw Hill, 2000.
[18] A. Papoulis, The Fourier Integral and Its Applications, McGraw-Hill Electronic Sciences Series, McGraw-Hill, 1962.
[19] P. Diniz, Teaching circuits, systems, and signal processing, in: Circuits and Systems, 1998. ISCAS'98. Proceedings of the 1998 IEEE International Symposium on, vol. 1, 1998, pp. 428–431.
[20] G.E. Pazienza, J. Albo-Canals, Teaching memristors to EE undergraduate students [class notes], IEEE Circuits and Systems Magazine 11 (4) (2011) 36–44.
[21] R. Gray, History and trends in analog circuit challenges at the system level, in: Electronic System Design (ISED), 2010 International Symposium on, 2010, p. 24.

CHAPTER

Discrete-time signals and systems

3

Leonardo G. Baltar and Josef A. Nossek

Institute for Circuit Theory and Signal Processing, Technische Universität München, München, Germany

3.1 Introduction

Discrete-time systems are signal processing entities that process discrete-time signals, i.e., sequences of signal values that are generally obtained as equidistant samples of continuous-time waveforms along the time axis. Usually a clock signal will determine the period T (sampling interval) in which the input signal values enter the system and the output samples leave the system. The interval T also determines the cycle in which the internal signal values within the system are processed. Typical representatives of this class of systems are the analog discrete-time signal processing systems (switched capacitor circuits, charge-coupled devices [CCDs], bucket-brigade devices [BBDs]) as well as in omnipresent digital systems. In the case of the latter, the sequence value for each sampling interval will also be discretized, i.e., it will be represented with a finite set of amplitudes, usually in a binary finite word length representation. Chapter 5 covers the topic of sampling and quantization of continuous-time signals.

It is worth mentioning that in the case of multirate discrete-time systems, the time interval in which the internal and possibly the output signals are processed will usually be different from the input period T. The topic of multirate discrete-time systems is covered in Chapter 9.

It is important to emphasize that the discussion in this chapter considers the special case of discrete-time signals with unquantized values. This means that the amplitude of all the signals and the value of the parameters (coefficients) that describe the discrete-time systems are represented with infinite word length. This is an abstraction from real-world digital signals and systems, where the finite word length representations of both signals and parameters have to be taken into account to fully cover all important phenomena, such as limit cycles. Design of digital filters and their implementation will be discussed in Chapter 7.

3.2 Discrete-time signals: sequences

Discrete-time signals can arise naturally in situations where the data to be processed are inherently discrete in time, like in financial or social sciences, economy, etc. But more generally, the physical quantities that a discrete-time signal represents are obtained via sensors and evolve continuously over time, for example, voice, temperature, voltages, currents, or scattering variables. Nevertheless, inside a sampling interval of T seconds, the continuous-time signal can be completely characterized by means

Signal Processing and Machine Learning Theory. https://doi.org/10.1016/B978-0-32-391772-8.00009-0
Copyright © 2024 Elsevier Inc. All rights reserved.

of a single value or sample, as depicted in Fig. 3.1. In this case, the continuous-time signals are first sampled or discretized over the time axis and possibly also discretized or quantized in the amplitude domain.

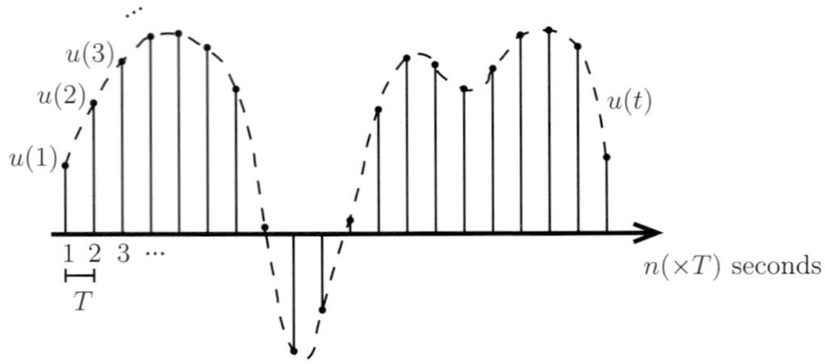

FIGURE 3.1

Discrete-time signal as a sampled version of a continuous-time waveform.

Still images are usually described as 2D discrete signals, where the two dimensions are spatial coordinates and pixels are finite, discrete quantities representing intensities either of a gray level or multiple color levels. For videos, a third dimension is added, for example, to represent the time axis typically in the form of a frame index or time stamp. Beyond images and videos, there are also discrete-time multidimensional signals which are obtained by simultaneous sampling of the output of multiple sensors such as antenna or microphone arrays, for example. Some examples of applications which involve multidimensional discrete-time signals are multiantenna wireless communications, e.g., local WLAN/WiFi, and cellular or multisensor radars, e.g., automotive radars. The representation and analysis of multidimensional signals is beyond the scope of this chapter.

Important and useful examples of discrete-time signals that are fundamental in digital signal processing and broadly used in practice will be addressed in the following paragraphs.

Analogously to the continuous-time unit impulse, known as the Dirac delta function, we can also define a discrete-time unit impulse, known as the Kronecker delta, as

$$\delta(n) = \begin{cases} 1, & n = 0, \\ 0, & \text{else}, \end{cases} \tag{3.1}$$

where n is an integer number. The discrete-time unit impulse is graphically shown in Fig. 3.2. Any arbitrary discrete-time sequence can be represented as a sum of weighted and delayed unit impulses. Also analogous to the continuous-time systems, the input–output behavior of a discrete-time system can also be described by the impulse response, as we will see in detail in Section 3.4.3.

Another commonly used test signal is a sequence where the samples are equidistantly taken from a sinusoid,

$$u(n) = U \sin(nT\omega + \varphi), \tag{3.2}$$

where U and φ are real-valued constants and ω is the frequency of the sinusoid in rad/s. In some cases, it is important to analyze a system for a certain frequency range (frequency-domain analysis)

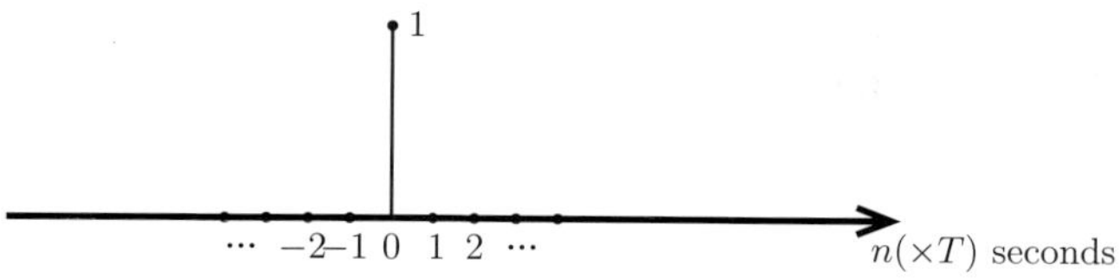

FIGURE 3.2

Discrete-time unit impulse.

and in this case a sinusoidal signal with varying frequency can be applied at its input. An example of a discrete-time sinusoid and the corresponding continuous-time waveform are depicted in Fig. 3.3.

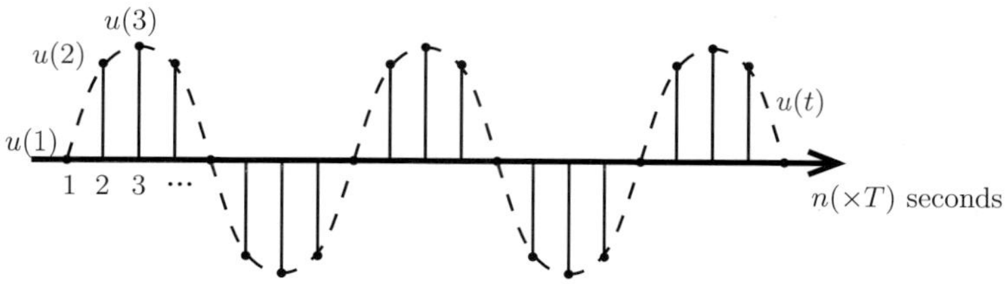

FIGURE 3.3

Discrete-time sinusoid and corresponding continuous-time sinusoid waveform.

We can also define a random sequence, an example being the case where the samples are distributed according to a normal distribution with zero mean and given variance σ^2 and the individual samples are statistically independent, i.e.,

$$\mathrm{E}\left[u(n)u(n-k)\right] = \begin{cases} \sigma^2, & k = 0, \\ 0, & \text{else.} \end{cases} \tag{3.3}$$

This is the discrete equivalent of the white Gaussian noise (WGN) and it is useful if the behavior of the discrete-time system is to be analyzed for all frequencies at once. Also in the case where some noise sources are to be included during the analysis or modeling of a discrete-time system implementation, this random sequence is the most used one.

3.3 Discrete-time systems

Mathematically, a discrete-time system with discrete-time input sequence $u(n)$ and output sequence $y(n)$ is an operator $\mathcal{T}$ that performs the mapping

$$y(n) = \mathcal{T}\{u(n)\}. \tag{3.4}$$

It is important to note that this mapping is applied to the entire input sequence $u(n)$, beginning with some starting instant in the far distant past, including the present instant n, and up to a far future instant, to obtain the present output $y(n)$.

3.3.1 Classification

There are many different classes of discrete-time systems. They can be categorized according to various important properties and criteria that will allow us to analyze and design systems using specific mathematical tools. All classes and properties explained here find an equivalent for continuous-time systems.

Memoryless systems

One sample of the output sequence $y(n)$ of a memoryless system for a specific index n_0 is a function exclusively of the input value $u(n_0)$. The output depends neither on any past or future value of the input ..., $u(n_0+1), u(n_0-1), u(n_0-2)$, ... nor on any past or future value of the output ..., $y(n_0+1), y(n_0-1), y(n_0-2)$, ..., i.e.,

$$y(n_0) = \mathcal{T}\{u(n_0)\}. \tag{3.5}$$

Dynamic systems

In contrast, a memory or dynamic system has its output as a function of at least one past value of the output or of the input,

$$y(n) = \mathcal{T}\{u(n), u(n-1), ...u(n-N), y(n-1), ...y(n-N)\}. \tag{3.6}$$

The positive integer constant N is usually called the degree or order of the system. We have assumed here, without loss of generality, that the same number of past inputs and outputs influence a certain output sample.

Linear systems

If the system is a linear system, the superposition principle should hold. Let us consider an input u_1 for the system. Then the output is

$$y_1(n) = \mathcal{T}\{u_1(n)\}. \tag{3.7}$$

Similarly, with another input u_2, the output is

$$y_2(n) = \mathcal{T}\{u_2(n)\}. \tag{3.8}$$

The superposition principle tells us that if each input is scaled by any real-valued constant α_i, and moreover, the sum of the inputs is employed as a new input, then for the new output, we have

$$\begin{aligned}\mathcal{T}\{\alpha_1 u_1(n) + \alpha_2 u_2(n)\} &= \mathcal{T}\{\alpha_1 u_1(n)\} + \mathcal{T}\{\alpha_2 u_2(n)\} \\ &= \alpha_1 \mathcal{T}\{u_1(n)\} + \alpha_2 \mathcal{T}\{u_2(n)\} = \alpha_1 y_1(n) + \alpha_2 y_2(n).\end{aligned} \tag{3.9}$$

This can be extended for any number of different inputs and any value of the scaling factors. One of the simplest and probably most popular linear dynamic systems is the recurrence used to generate

Fibonacci sequences [1]:

$$y(n) = u(n) + y(n-1) + y(n-2), \tag{3.10}$$

where $u(n) = \delta(n)$.

Nonlinear systems

In the case of nonlinear systems, the superposition principle does not apply. A good example is the famous logistic map [2] or discrete logistic equation, which arise in many contexts in the biological, economical, and social sciences:

$$y(n) = u(n) + \alpha y(n-1)(1 - y(n-1)), \tag{3.11}$$

where α is a positive real-valued number, $u(n) = \gamma\delta(n)$, and γ represents an initial ratio of population to a maximum population. The logistic map is a discrete-time demographic model analogous to the continuous-time logistic equation [3] and is a simple example of how chaotic behavior can arise. In this chapter, we are mainly interested in linear systems, since there exist many well-established analysis tools and they already cover a broad range of practical and useful systems.

Time-invariant systems

If the samples of the input sequence are delayed by a constant n_1 and the resulting output samples observed are delayed by the same constant, a discrete-time system is called time-invariant. Mathematically, this means

$$y(n - n_1) = \mathcal{T}\{u(n - n_1)\}. \tag{3.12}$$

Both the Fibonacci recurrence and the logistic map are good examples of time-invariant systems.

Time-variant systems

In the case of time-variant systems, internal parameters of the system, like multiplier coefficients, can also vary with time. This is the case for the so-called adaptive systems or adaptive filters (see Chapter 11 for more details). A simple example of a time-variant system is the amplitude modulation

$$y(n) = \sin(nT\omega + \varphi)u(n), \tag{3.13}$$

which is very useful for a broad range of wireless and wired communications systems. There exists a special case of time-variant systems where the internal multiplier coefficients does not necessarily vary with time, but the sampling rate is changed within the system. Those systems are called multirate systems and they frequently present a cyclical or periodically time-varying output. This means that for a certain multiple delay of the input signal, the output sequence will have the same samples as for the nondelayed input.

Causal systems

In a causal system, the output samples only depend on the actual or past samples of the input sequence and on past samples of the output itself. If the system is noncausal, this means that the output sample depends on an input value that has not yet entered the system. The definition of dynamic systems in (3.6) already considered a causal case. Noncausal systems are not relevant for almost all real-world practical applications.

3.4 Linear time-invariant systems

This is probably the most technically important class of discrete-time systems, for which a rich mathematical framework exists.

Linear time-invariant (LTI) continuous-time systems in the electrical/electronic domain can be built with memoryless (resistive) elements such as resistors, independent and controlled sources, and memory-possessing (reactive) elements, for example, capacitors or inductors. In LTI discrete-time systems, we also have memoryless and memory-possessing elements, such as:

- Memoryless: adders and multipliers, signal sources and sinks
- Memory-possessing: delay elements.

Symbolic representations of these elements are depicted in Fig. 3.4. It is important to note that in this context, nothing is said about how these elements are realized – neither with which electronic components nor with which technology.

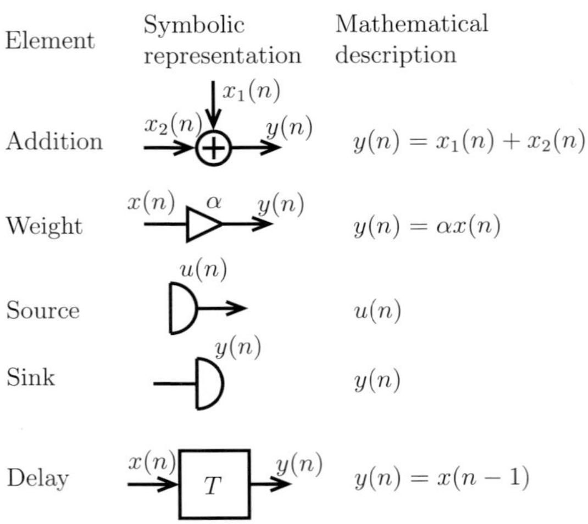

FIGURE 3.4

Basic building blocks of a discrete-time system.

3.4.1 State-space description

Any LTI system of order N can be structured into a memoryless linear $(2N+2)$ terminal network with N delay elements, a single input, and a single output. Although the system can easily extend to multiple inputs and/or multiple outputs, we will consider only single-input single-output (SISO) systems, with which we can study all important phenomena.

Because everything within the LTI memoryless box in Fig. 3.5 can only perform weighted sums, the $(N+1)$-dimensional input vector (state vector $\boldsymbol{x}(n)$ stacked with scalar input $u(n)$) is mapped onto the $(N+1)$-dimensional output vector (state vector $\boldsymbol{x}(n+1)$ stacked with scalar output $y(n)$) through

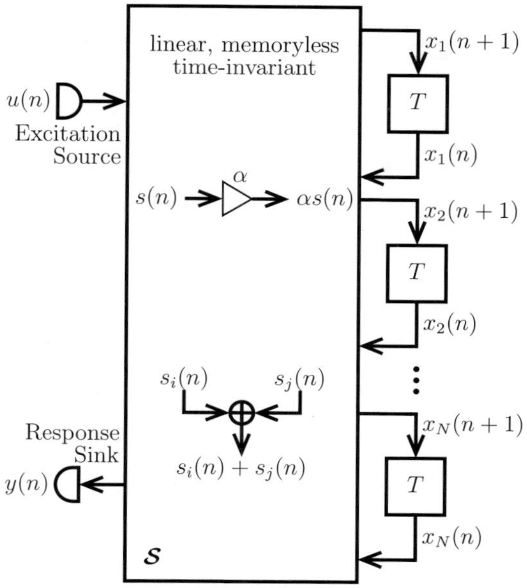

FIGURE 3.5

Discrete-time system.

multiplication with the $(N+1)\times(N+1)$ matrix $\boldsymbol{\mathcal{S}}$:

$$\begin{bmatrix} x_1(n+1) \\ x_2(n+1) \\ \vdots \\ x_N(n+1) \\ y(n) \end{bmatrix} = \left[\begin{array}{c|c} \boldsymbol{A} & \boldsymbol{b} \\ \hline \boldsymbol{c}^{\mathrm{T}} & d \end{array}\right] \begin{bmatrix} x_1(n) \\ x_2(n) \\ \vdots \\ x_N(n) \\ u(n) \end{bmatrix} = \boldsymbol{\mathcal{S}} \cdot \begin{bmatrix} \boldsymbol{x}(n) \\ u(n) \end{bmatrix}. \tag{3.14}$$

The matrix $\boldsymbol{\mathcal{S}}$ can obviously be partitioned into the well-known state-space description

$$\begin{aligned} \boldsymbol{x}(n+1) &= \boldsymbol{A}\boldsymbol{x}(n) + \boldsymbol{b}u(n), \\ y(n) &= \boldsymbol{c}^{\mathrm{T}}\boldsymbol{x}(n) + du(n), \end{aligned} \tag{3.15}$$

where $\boldsymbol{A} \in \mathcal{R}^{N\times N}$, $\boldsymbol{b} \in \mathcal{R}^N$, $\boldsymbol{c}^{\mathrm{T}} \in \mathcal{R}^{1\times N}$, $d \in \mathcal{R}$, $\boldsymbol{x}(n)$, $\boldsymbol{x}(n+1) \in \mathcal{R}^N$, and $u(n)$, $y(n) \in \mathcal{R}$.

Now we can take (3.15) and refine the block diagram depicted in Fig. 3.5 to show Fig. 3.6, which is the general state-space realization. The structure in Fig. 3.6 is still rather general, because for each element of $\boldsymbol{A}$, $\boldsymbol{b}$, $\boldsymbol{c}$, and d, a multiplication has to be performed, especially if we assume all of their elements with nontrivial values. The total number of multiplications is in this case $N(N+2)+1$. But, as we will see later, $\boldsymbol{A}$, $\boldsymbol{b}$, and $\boldsymbol{c}$ are not uniquely determined by a given input–output mapping.

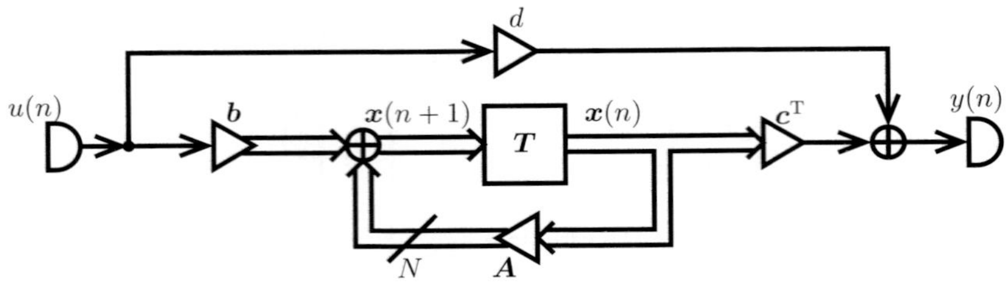

FIGURE 3.6

General state-space realization of a discrete system.

Consequently there is room for optimization, e.g., to reduce the number of actual multiplications to be performed per output sample.

Let us compute the output signal $y(n)$ using an initial state vector $\boldsymbol{x}(0)$ and the input signal sequence from the time instant $n = 0$ onwards

$$\begin{aligned} \boldsymbol{x}(n) &= \boldsymbol{A}\boldsymbol{x}(n-1) + \boldsymbol{b}u(n-1) \\ &= \boldsymbol{A}(\boldsymbol{A}\boldsymbol{x}(n-2) + \boldsymbol{b}u(n-2)) + \boldsymbol{b}u(n-1) \\ &= \dots \\ &= \boldsymbol{A}^n \boldsymbol{x}(0) + \sum_{k=1}^{n} \boldsymbol{A}^{k-1} \boldsymbol{b}u(n-k), \end{aligned} \tag{3.16}$$

$$y(n) = \underbrace{\boldsymbol{c}^{\mathrm{T}} \boldsymbol{A}^n \boldsymbol{x}(0)}_{\text{zero-input response}} + \underbrace{\sum_{k=1}^{n} (\boldsymbol{c}^{\mathrm{T}} \boldsymbol{A}^{k-1} \boldsymbol{b}u(n-k)) + du(n)}_{\text{zero-state response}}. \tag{3.17}$$

From this general response we can also derive the so-called impulse response, for which we set $\boldsymbol{x}(0) = \boldsymbol{0}$ and $u(n) = \delta(n)$, the unit impulse which was defined in (3.1). This leads to

$$y(n) = h(n) = \begin{cases} d & n = 0, \\ \boldsymbol{c}^{\mathrm{T}} \boldsymbol{A}^{n-1} \boldsymbol{b} & n \geq 1. \end{cases} \tag{3.18}$$

By making use of the impulse response $h(n)$, we can reformulate the zero-state response with a general input sequence $u(n)$:

$$\begin{aligned} y(n) &= \sum_{k=1}^{n} (\boldsymbol{c}^{\mathrm{T}} \boldsymbol{A}^{k-1} \boldsymbol{b}u(n-k)) + du(n) \\ &= \sum_{k=0}^{n} h(n)u(n-k) = h(n) * u(n), \end{aligned} \tag{3.19}$$

which is the so-called convolution sum. This summation formulation corresponds to the so-called convolution integral for continuous-time signals and systems.

Now it is time to show that we can generate a whole class of equivalent systems (equivalent in the sense that they have the same zero-state response and the same input–output mapping, respectively) with the aid of a so-called similarity transform.

With a nonsingular $N \times N$ matrix $\boldsymbol{T}$ we can define a new state vector as the linear transformation of the original one as $\boldsymbol{x}'(n) = \boldsymbol{T}^{-1}\boldsymbol{x}(n)$. Then we can rewrite the state-space equations with the new state vector $\boldsymbol{x}'(n)$ as

$$\begin{aligned} \boldsymbol{T}\boldsymbol{x}'(n+1) &= \boldsymbol{A}\boldsymbol{T}\boldsymbol{x}'(n) + \boldsymbol{b}u(n), \\ y(n) &= \boldsymbol{c}^{\mathrm{T}}\boldsymbol{T}\boldsymbol{x}'(n) + du(n), \end{aligned} \tag{3.20}$$

and by multiplying the first equation with $\boldsymbol{T}^{-1}$ from the left side we get

$$\begin{aligned} \boldsymbol{x}'(n+1) &= \boldsymbol{T}^{-1}\boldsymbol{A}\boldsymbol{T}\boldsymbol{x}'(n) + \boldsymbol{T}^{-1}\boldsymbol{b}u(n), \\ y(n) &= \boldsymbol{c}^{\mathrm{T}}\boldsymbol{T}\boldsymbol{x}'(n) + du(n). \end{aligned} \tag{3.21}$$

The new state-space representation can now be formulated as

$$\begin{aligned} \boldsymbol{x}'(n+1) &= \boldsymbol{A}'\boldsymbol{x}'(n) + \boldsymbol{b}'u(n), \\ y(n) &= \boldsymbol{c}'^{\mathrm{T}}\boldsymbol{x}(n) + d'u(n), \end{aligned} \tag{3.22}$$

where the equations

$$\boldsymbol{A}' = \boldsymbol{T}^{-1}\boldsymbol{A}\boldsymbol{T}, \quad \boldsymbol{b}' = \boldsymbol{T}^{-1}\boldsymbol{b}, \quad \boldsymbol{c}'^{\mathrm{T}} = \boldsymbol{c}^{\mathrm{T}}\boldsymbol{T}, \text{ and } d' = d \tag{3.23}$$

hold.

By inserting $\boldsymbol{A}'$, $\boldsymbol{b}'$, $\boldsymbol{c}'$, and d' into the zero-state response, we see that this response is invariant to the above transformation.

Depending on the choice of the transformation matrix, the matrix $\boldsymbol{A}'$ can have different structures or forms. A particularly interesting form is the so-called normal form, where the state matrix is diagonalized, allowing a much lower complexity than a dense matrix. First, we apply an eigenvalue decomposition (EVD) [4] to $\boldsymbol{A}$:

$$\begin{aligned} &\boldsymbol{A} = \boldsymbol{Q}\boldsymbol{\Lambda}\boldsymbol{Q}^{-1} && \text{EVD of } \boldsymbol{A}, \\ &\boldsymbol{Q} = [\boldsymbol{q}_1, \boldsymbol{q}_2, \ldots, \boldsymbol{q}_n] && \text{modal matrix with the eigenvectors } \boldsymbol{q}_k, \\ &\boldsymbol{\Lambda} = \operatorname{diag}(\lambda_k)_{k=1}^{n} && \text{eigenvalue } \lambda_k \text{ with } \boldsymbol{A}\boldsymbol{q}_k = \lambda_k \boldsymbol{q}_k, \end{aligned}$$

where we have assumed that $\boldsymbol{A}$ is diagonalizable and we do not need to resort to the Jordan form. The transformation matrix and the new state vector are $\boldsymbol{T} = \boldsymbol{Q}$ and $\boldsymbol{\xi}(n) = \boldsymbol{Q}^{-1}\boldsymbol{x}(n)$. The system can then be described by

$$\begin{aligned} \boldsymbol{\xi}(n+1) &= \boldsymbol{\Lambda}\boldsymbol{\xi}(n) + \boldsymbol{Q}^{-1}\boldsymbol{b}u(n), \\ y(n) &= \boldsymbol{c}^{\mathrm{T}}\boldsymbol{Q}\boldsymbol{\xi}(n) + du(n), \end{aligned} \tag{3.24}$$

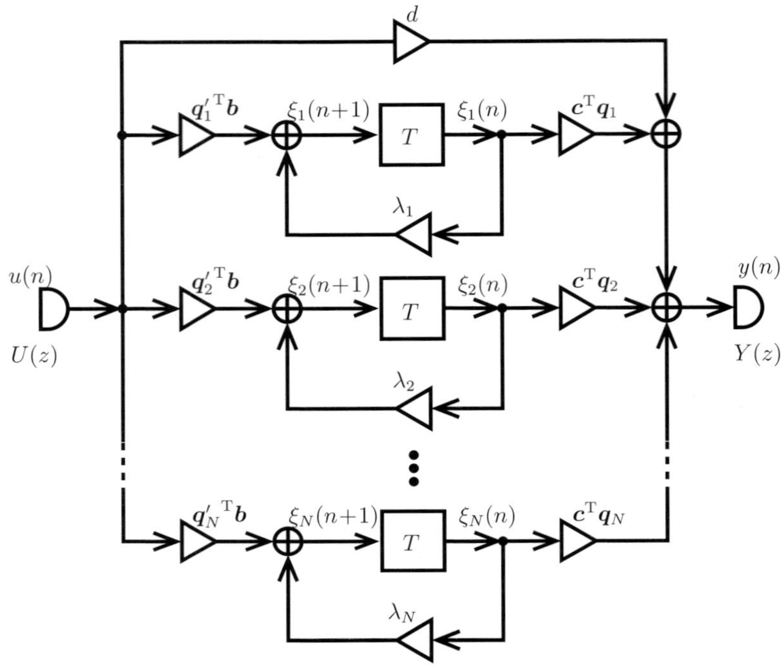

FIGURE 3.7

Normal form for real-valued eigenvalues.

which leads to the new specific implementation of Fig. 3.7. With this new implementation, the number of coefficients, and consequently the number of multiplications, is reduced to $3N + 1$.

If some of the eigenvalues are complex-valued (if this is the case they will always come in complex conjugate pairs provided $\boldsymbol{A}$ is real-valued), we always will merge the two corresponding first-order systems into one second-order system with real-valued multiplier coefficients only. This is equivalent to having a block-diagonal state matrix (with 2-by-2 blocks) instead of a diagonal one.

Now, we can compute the transfer function by transforming (3.15) to the frequency domain with the $\mathcal{Z}$-transform.

3.4.2 Transfer function of a discrete-time LTI system

Similar to continuous-time systems we can also analyze a discrete-time system with the help of a functional transform to make a transition to the frequency domain. However, we have only sequences of signal values in the time domain and not continuous waveforms. Therefore, we have to apply the $\mathcal{Z}$-transform [5]:

$$\mathcal{Z}\{u(n)\} = \sum_{n=-\infty}^{\infty} u(n)z^{-n} = U(z),$$

$$\mathcal{Z}\{y(n)\} = \sum_{n=-\infty}^{\infty} y(n)z^{-n} = Y(z),$$
$$\mathcal{Z}\{\boldsymbol{x}(n)\} = \sum_{n=-\infty}^{\infty} \boldsymbol{x}(n)z^{-n} = \boldsymbol{X}(z). \tag{3.25}$$

The delay in the time domain is represented in the frequency domain by a multiplication by z^{-1}:

$$\mathcal{Z}\{\boldsymbol{x}(n+1)\} = \sum_{n=-\infty}^{\infty} \boldsymbol{x}(n+1)z^{-n} = z \sum_{n=-\infty}^{\infty} \boldsymbol{x}(n+1)z^{-(n+1)}$$
$$\overset{n'=n+1}{=} z \sum_{n'=-\infty}^{\infty} \boldsymbol{x}[n']z^{-n'} = z\mathcal{Z}\{\boldsymbol{x}(n)\} = z\boldsymbol{X}(z). \tag{3.26}$$

Now we can transform the state-space equations to the frequency domain:

$$z\boldsymbol{X}(z) = \boldsymbol{A}\boldsymbol{X}(z) + \boldsymbol{b}U(z),$$
$$Y(z) = \boldsymbol{c}^{\mathrm{T}}\boldsymbol{X}(z) + dU(z), \tag{3.27}$$

and after we eliminate the state vector $\boldsymbol{X}(z)$, we obtain the transfer function

$$H(z) = \frac{Y(z)}{U(z)} = \boldsymbol{c}^{\mathrm{T}}\left(z\mathbf{1} - \boldsymbol{A}\right)^{-1}\boldsymbol{b} + d. \tag{3.28}$$

Since

$$z = \mathrm{e}^{pT} \quad \text{and hence for technical frequencies} \quad z|_{p=\mathrm{j}\omega} = \mathrm{e}^{\mathrm{j}\omega T} \tag{3.29}$$

holds, $H(\mathrm{e}^{\mathrm{j}\omega T})$ is obviously a periodical function of ω resp. f with frequency period $2\pi/T$ resp. $1/T$.

The question that arises here is: What type of function is (3.28)? To answer that we have to consider the expression $(z\mathbf{1} - \boldsymbol{A})^{-1}$ more closely. We assume of course that $(z\mathbf{1} - \boldsymbol{A})^{-1}$ is nonsingular and therefore the inverse exists (this means that the inversion will only be performed for the values of z at which it is possible). The inverse reads

$$(z\mathbf{1} - \boldsymbol{A})^{-1} = \frac{\operatorname{adj}(z\mathbf{1} - \boldsymbol{A})}{\det(z\mathbf{1} - \boldsymbol{A})}, \tag{3.30}$$

where the determinant is a polynomial in z of degree N, also called the characteristic polynomial

$$\det(z\mathbf{1} - \boldsymbol{A}) = z^N + \alpha_1 z^{N-1} + \cdots + \alpha_{N-1}z + \alpha_N, \quad \alpha_k \in \mathcal{R}, \tag{3.31}$$

and the elements of the adjugate matrix are the transposed cofactors of the matrix. The cofactors on the other hand are subdeterminants of $(z\mathbf{1} - \boldsymbol{A})$ and therefore polynomials of at most degree $(N-1)$.

The numerator of $\boldsymbol{c}^{\mathrm{T}}(z\mathbf{1} - \boldsymbol{A})^{-1}\boldsymbol{b}$ is hence a linear combination of cofactors, i.e., a linear combination of polynomials of degree $(N-1)$ and therefore a polynomial of at most degree $(N-1)$. If we apply this result in Eq. (3.28), we will obtain a rational fractional function of z,

$$H(z) = \frac{\beta_0 z^N + \beta_1 z^{N-1} + \cdots + \beta_{N-1} z + \beta_N}{z^N + \alpha_1 z^{N-1} + \cdots + \alpha_{N-1} z + \alpha_N}, \tag{3.32}$$

where $\beta_0 \neq 0$ only for $d \neq 0$, since $\beta_0 = d$.

We usually divide the nominator and the denominator by z^n and obtain in the numerator and in the denominator polynomials in z^{-1}:

$$H(z) = \frac{\beta_N z^{-N} + \beta_{N-1} z^{-(N-1)} + \cdots + \beta_1 z^{-1} + \beta_0}{\alpha_N z^{-N} + \alpha_{N-1} z^{-(N-1)} + \cdots + \alpha_1 z^{-1} + 1} = \frac{\sum_{k=0}^{N} \beta_k z^{-k}}{1 + \sum_{k=1}^{N} \alpha_k z^{-k}} = \frac{Y(z)}{U(z)}. \tag{3.33}$$

By multiplying both sides of (3.33) with the denominator we get

$$Y(z)\left(1 + \sum_{k=1}^{N} \alpha_k z^{-k}\right) = U(z) \sum_{k=0}^{N} \beta_k z^{-k}. \tag{3.34}$$

Now we go back to the time domain with help of the inverse $\mathcal{Z}$-transform and obtain a global difference equation:

$$y(n) + \sum_{k=1}^{N} \alpha_k y(n-k) = \sum_{k=0}^{N} \beta_k u(n-k),$$

$$y(n) = \sum_{k=0}^{N} \beta_k u(n-k) - \sum_{k=1}^{N} \alpha_k y(n-k). \tag{3.35}$$

The difference equation is equivalent to the description of a discrete-time system like a differential equation is for the description of a continuous-time system.

The representation of Eq. (3.35) leads us to an important special case, for which we have no equivalent in continuous-time systems built with lumped elements: the so-called finite impulse response (FIR) systems.

3.4.3 Finite duration impulse response systems

Those are systems with, as the name says, an FIR, i.e., the α_i's in (3.35) are equal to zero such that

$$y(n) = \sum_{k=0}^{N} \beta_k u(n-k) \tag{3.36}$$

and with a unit impulse as excitation $u(n) = \delta(n)$ we get

$$y(n) = h(k) = \sum_{k=0}^{N} \beta_k \delta(n-k) = \begin{cases} 0, & n < 0, \\ \beta_n, & 0 \leq n \leq N, \\ 0, & n > N. \end{cases} \tag{3.37}$$

We can see that the coefficients of the polynomial transfer function then directly give values of the impulse response so that it takes us directly to an implementation (Fig. 3.8). Note that the two structures

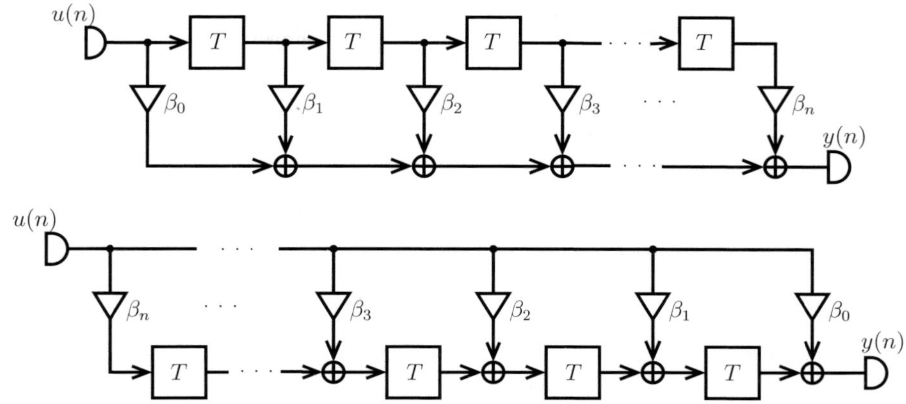

FIGURE 3.8

Two FIR realizations interconvertible through the principle of flow reversal.

are interconvertible if the principle of flow reversal is applied to the block diagram.

Let us now write down the state-space description for the uppermost structure in Fig. 3.8. By calling the input of the delay elements $x_k(n+1)$ and their outputs $x_k(n)$ indexing them from the leftmost to the rightmost we can see that

$$\begin{aligned} x_1(k+1) &= u(n), \\ x_2(k+1) &= x_1(n), \\ \vdots &= \vdots \\ x_N(k+1) &= x_{N-1}(n), \end{aligned} \tag{3.38}$$

and for the output

$$y(n) = \beta_1 x_1(n) + \beta_2 x_2(n) + \cdots + \beta_N x_N(n) + \beta_0 u(n) \tag{3.39}$$

holds. In matrix vector notation we get

$$\begin{bmatrix} x_1(n+1) \\ x_2(n+1) \\ \vdots \\ x_N(n+1) \end{bmatrix} = \begin{bmatrix} 0 & 0 & \cdots & 0 \\ 1 & 0 & \cdots & 0 \\ \vdots & \vdots & \ddots & \vdots \\ 0 & \cdots & 1 & 0 \end{bmatrix} \begin{bmatrix} x_1(n) \\ x_2(n) \\ \vdots \\ x_N(n) \end{bmatrix} + \begin{bmatrix} 1 \\ 0 \\ \vdots \\ 0 \end{bmatrix} u(n), \tag{3.40}$$

$$y(n) = \begin{bmatrix} \beta_1 & \beta_2 & \cdots & \beta_N \end{bmatrix} \boldsymbol{x}(n) + \beta_0 u(n). \tag{3.41}$$

It can be easily demonstrated that the second structure in Fig. 3.8 is obtained by a transposition of the state-space description of the first structure; the matrix and vectors for it are then given by

$$A' = A^{\mathrm{T}}, \quad b' = c, \quad c'^{\mathrm{T}} = b^{\mathrm{T}}, \quad d' = d. \tag{3.42}$$

Other usual names for FIR systems are:

- nonrecursive systems,
- moving average (MA),
- transversal filter.

3.4.4 Infinite duration impulse response systems

On the other hand, if the α_i's are different from zero, then the impulse response will have infinite duration, as the name says. We can again clearly derive a realization structure from the corresponding time-domain description, in this case from Eq. (3.35) (see Fig. 3.9). This per se inefficient realization

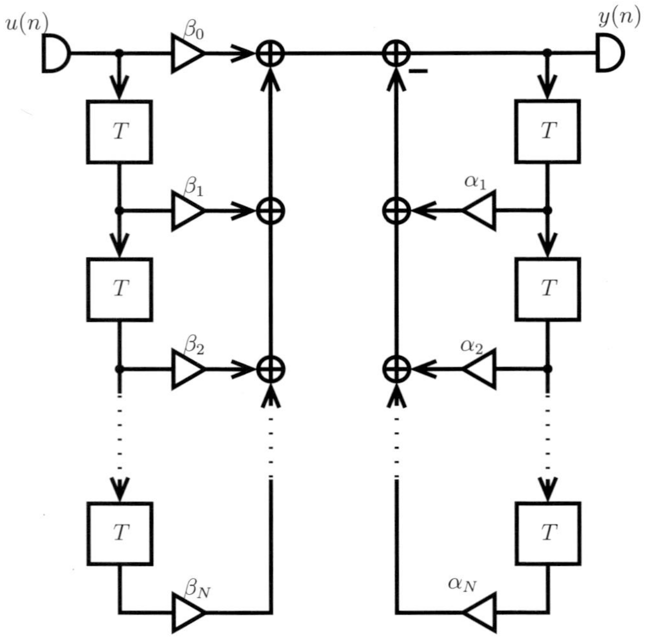

FIGURE 3.9

IIR realization according to Eq. (3.35).

(two times the number of necessary delay elements) can be transformed into the well-known direct form through simple rearranging of the two blocks like in Fig. 3.10. This is called a canonical realization because it possesses a minimum number of delay elements.

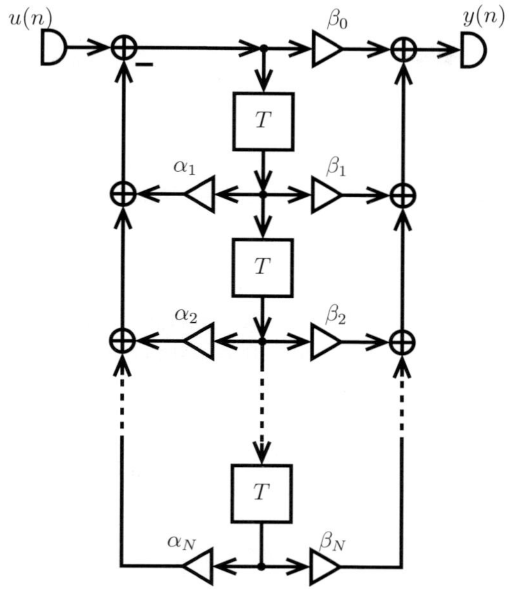

FIGURE 3.10

Canonical direct realization of an IIR system.

Now we can set up the state-space equations for the infinite impulse response (IIR) system structure of Fig. 3.10. As already mentioned we label the output of the delays the states $x_k(n)$ and their inputs $x_k(n+1)$. By numbering from the topmost to the lowest delay we can see that

$$x_1(n+1) = -\alpha_1 x_1(n) - \alpha_2 x_2(n) - \dots - \alpha_N x_N(n) + u(n) \tag{3.43}$$

holds and from the second until the Nth state we get

$$\begin{aligned} x_2(n+1) &= x_1(n), \\ x_3(n+1) &= x_2(n), \\ \vdots &= \vdots \\ x_N(n+1) &= x_{N-1}(n). \end{aligned} \tag{3.44}$$

For the output we can see that

$$\begin{aligned} y(n) &= \beta_0 x_1(n+1) + \beta_1 x_1(n) + \beta_2 x_2(n) + \dots + \beta_N x_N(n) \\ &= (\beta_1 - \beta_0\alpha_1)x_1(n) + (\beta_2 - \beta_0\alpha_2)x_2(n) + \dots + (\beta_N - \beta_0\alpha_N)x_N(n) + \beta_0 u(n), \end{aligned} \tag{3.45}$$

where we have used the definition of $x_1(n+1)$ of (3.43). In matrix vector notation we get

$$\begin{bmatrix} x_1(n+1) \\ x_2(n+1) \\ \vdots \\ x_N(n+1) \end{bmatrix} = \begin{bmatrix} -\alpha_1 & -\alpha_2 & \cdots & -\alpha_N \\ 1 & 0 & \cdots & 0 \\ \vdots & \vdots & \ddots & \vdots \\ 0 & \cdots & 1 & 0 \end{bmatrix} \begin{bmatrix} x_1(n) \\ x_2(n) \\ \vdots \\ x_N(n) \end{bmatrix} + \begin{bmatrix} 1 \\ 0 \\ \vdots \\ 0 \end{bmatrix} u(n), \tag{3.46}$$

$$y(n) = \begin{bmatrix} \beta_1 - \beta_0\alpha_1, & \beta_2 - \beta_0\alpha_2, & \cdots & \beta_N - \beta_0\alpha_N \end{bmatrix} \boldsymbol{x}(n) + \beta_0 u(n), \tag{3.47}$$

where because of its particular structure, $\boldsymbol{A}$ is called a companion matrix.

If we apply the principle of flow reversal to the diagram of Fig. 3.10 we will end up with the realization shown in Fig. 3.11. This is again equivalent to transformation of the state-space description

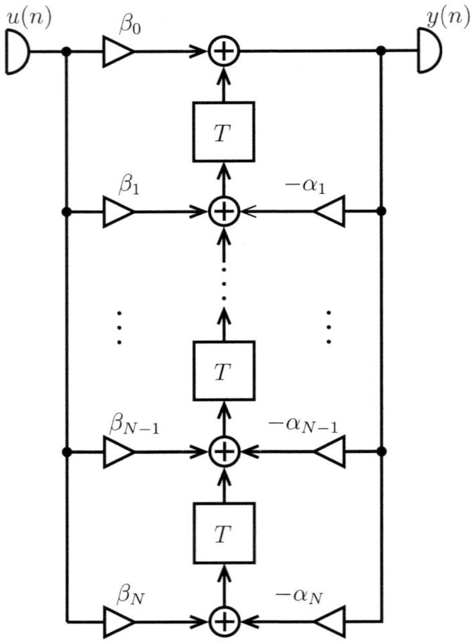

FIGURE 3.11

Transposed realization of the system from Fig. 3.10.

into the transposed system, i.e., the transformed system is represented by

$$\boldsymbol{A}' = \boldsymbol{A}^{\mathrm{T}}, \quad \boldsymbol{b}' = \boldsymbol{c}, \quad \boldsymbol{c}'^{\mathrm{T}} = \boldsymbol{b}^{\mathrm{T}}, \quad d' = d. \tag{3.48}$$

If we apply these definitions to (3.28) we can see that

$$H'(z) = \boldsymbol{c}'^{\mathrm{T}} \left(z\mathbf{1} - \boldsymbol{A}'\right)^{-1} \boldsymbol{b}' + d' = \boldsymbol{b}^{\mathrm{T}} \left(z\mathbf{1} - \boldsymbol{A}^{T}\right)^{-1} \boldsymbol{c} + d = (H'(z))^{\mathrm{T}}$$

$$= \boldsymbol{c}^{\mathrm{T}} \left(\left(z\mathbf{1} - \boldsymbol{A}^{T} \right)^{-1} \right)^{\mathrm{T}} (\boldsymbol{b}^{\mathrm{T}})^{\mathrm{T}} + d = \boldsymbol{c}^{\mathrm{T}} (z\mathbf{1} - \mathbf{A})^{-1} \boldsymbol{b} + d = H(z). \tag{3.49}$$

This means that the transposed realization has the same transfer function as the original system.

Other usual names for IIR systems are:

- recursive systems,
- autoregressive moving average (ARMA),
- systems with feedback.

3.4.5 Observability and controllability

As we have seen in the previous section, every LTI system can be completely characterized by a state-space description

$$\boldsymbol{x}(n+1) = \boldsymbol{A}\boldsymbol{x}(n) + \boldsymbol{b}u(n), \tag{3.50}$$

$$y(n) = \boldsymbol{c}^{\mathrm{T}}\boldsymbol{x}(n) + du(n), \tag{3.51}$$

from which we can uniquely derive the corresponding transfer function

$$H(z) = \boldsymbol{c}^{\mathrm{T}} (z\mathbf{1} - \mathbf{A})^{-1} \boldsymbol{b} + d. \tag{3.52}$$

It is important to note that although the way from a given state-space description to the input–output transfer function is unique, the same is not true for the other direction. For one given transfer function there are infinitely many different state-space descriptions and realizations, each having different specific properties, but all with the very same input–output relation! One of those properties is the stability of the so-called LTI system.

An often applied stability criterion is to check whether the zeros of the denominator polynomial of the transfer function $H(z)$ lie within the unit circle of the z-plane or, equivalently, the impulse response

$$h(n) = \mathcal{Z}^{-1}\{H(z)\} \tag{3.53}$$

decays over time. In the following, we will see that this criterion does not provide us with complete information about the stability of an LTI system, because stability depends on the actual implementation, which can be better reflected in detail via a state-space description and not via the transfer function or impulse response descriptions. To get a clearer view, we have to introduce the concepts of observability and controllability, and based on these, we can define different types of stability criteria.

Let us begin with the definition of observability. Starting with the state-space description (3.50), we assume that the excitation $u(n) = 0$ for $n \geq 0$ and an initial state $\boldsymbol{x}(0) \neq \mathbf{0}$. We then compute the system output in the time domain:

$$\begin{aligned} y(0) &= \boldsymbol{c}^{\mathrm{T}}\boldsymbol{x}(0), \\ y(1) &= \boldsymbol{c}^{\mathrm{T}}\boldsymbol{x}(1) = \boldsymbol{c}^{\mathrm{T}}(\boldsymbol{A}\boldsymbol{x}(0)), \\ &\vdots \end{aligned}$$

$$y(N-1) = \boldsymbol{c}^{\mathrm{T}} \boldsymbol{A}^{N-1} \boldsymbol{x}(0). \tag{3.54}$$

Next, we stack these subsequent N output values in one vector:

$$\boldsymbol{y}(0) = \begin{bmatrix} y(0) \\ y(1) \\ y(2) \\ \vdots \\ y(N-1) \end{bmatrix} = \begin{bmatrix} \boldsymbol{c}^{\mathrm{T}} \\ \boldsymbol{c}^{\mathrm{T}}\boldsymbol{A} \\ \boldsymbol{c}^{\mathrm{T}}\boldsymbol{A}^2 \\ \vdots \\ \boldsymbol{c}^{\mathrm{T}}\boldsymbol{A}^{N-1} \end{bmatrix} \boldsymbol{x}(0) = \mathcal{O}\left(\boldsymbol{c}^{\mathrm{T}}, \boldsymbol{A}\right)\boldsymbol{x}(0). \tag{3.55}$$

From (3.55) we see that this output vector $\boldsymbol{y}(0)$ is given by the product of the so-called observability matrix $\mathcal{O}\left(\boldsymbol{c}^{\mathrm{T}}, \boldsymbol{A}\right)$, which is completely determined by $\boldsymbol{A}$ and $\boldsymbol{c}^{\mathrm{T}}$ and the system state at time instant $n = 0$: $\boldsymbol{x}(0)$. Now, if the observability matrix is invertible, we can compute the state vector

$$\boldsymbol{x}(0) = \mathcal{O}\left(\boldsymbol{c}^{\mathrm{T}}, \boldsymbol{A}\right)^{-1} \boldsymbol{y}(0). \tag{3.56}$$

If this is possible, the system (3.50) is completely observable.

Next we define controllability in a corresponding way by assuming an initial zero state $\boldsymbol{x}(0) = \boldsymbol{0}$ and let us look for a sequence $u(n)$, $n \geq 0$, to drive the system into some desired state $\boldsymbol{x}(k)$. We look at the evolution of the state as controlled by the excitation:

$$\begin{aligned}
\boldsymbol{x}(1) &= \boldsymbol{b}u(0), \\
\boldsymbol{x}(2) &= \boldsymbol{A}\boldsymbol{x}(1) + \boldsymbol{b}u(1) = \boldsymbol{A}\boldsymbol{b}u(0) + \boldsymbol{b}u(1), \\
\boldsymbol{x}(3) &= \boldsymbol{A}\boldsymbol{x}(2) + \boldsymbol{b}u(2) = \boldsymbol{A}^2\boldsymbol{b}u(0) + \boldsymbol{A}\boldsymbol{b}u(1) + \boldsymbol{b}u(2), \\
&\vdots \\
\boldsymbol{x}(N) &= \boldsymbol{A}^{N-1}\boldsymbol{b}u(0) + \boldsymbol{A}^{N-2}\boldsymbol{b}u(1) + \cdots + \boldsymbol{A}^2\boldsymbol{b}u(N-3) + \boldsymbol{A}\boldsymbol{b}u(N-2) + \boldsymbol{b}u(N-1).
\end{aligned} \tag{3.57}$$

This can be put together as

$$\boldsymbol{x}(N) = \left[\boldsymbol{b}, \boldsymbol{A}\boldsymbol{b}, \boldsymbol{A}^2\boldsymbol{b}, \ldots, \boldsymbol{A}^{N-1}\boldsymbol{b}\right] \begin{bmatrix} u(N-1) \\ u(N-2) \\ u(N-3) \\ \vdots \\ u(0) \end{bmatrix} = \mathcal{C}(\boldsymbol{b}, \boldsymbol{A})\boldsymbol{u}(N), \tag{3.58}$$

where $\mathcal{C}(\boldsymbol{b}, \boldsymbol{A})$ is the so-called controllability matrix and $\boldsymbol{u}(N)$ is the vector, in which the subsequent excitation samples have been stacked. If the controllability matrix is invertible, we can solve for the excitation

$$\boldsymbol{u}(N) = \mathcal{C}(\boldsymbol{b}, \boldsymbol{A})^{-1} \boldsymbol{x}(N), \tag{3.59}$$

which will steer the system from the zero initial state into a specified desired state $\boldsymbol{x}(N)$. If this is possible, the system (3.50) is completely controllable.

3.4.6 Stability

Based on the concepts of observability and controllability, we can now define the following concepts of stability:

- internal stability,
- output stability,
- input stability,
- input–output stability.

We will start with *internal stability* by requiring that the Euclidean norm of the state vector has to decay over time from any initial value with zero excitation:

$$\forall \boldsymbol{x}(0) \in \mathcal{R}^N \quad \text{and} \quad u(n) = 0, \quad \text{for} \quad n \geq 0: \quad \lim_{n\to\infty} \|\boldsymbol{x}(n)\|_2 = 0. \tag{3.60}$$

This is equivalent to requiring the state vector to converge to the zero vector.

By looking at the evolution of the state vector over time,

$$\begin{aligned} \boldsymbol{x}(1) &= \boldsymbol{A}\boldsymbol{x}(0), \\ \boldsymbol{x}(2) &= \boldsymbol{A}\boldsymbol{x}(1) = \boldsymbol{A}^2\boldsymbol{x}(0), \\ &\vdots \\ \boldsymbol{x}(n) &= \boldsymbol{A}^n\boldsymbol{x}(0), \end{aligned} \tag{3.61}$$

and making use of the EVD of $\boldsymbol{A} = \boldsymbol{Q}\boldsymbol{\Lambda}\boldsymbol{Q}^{-1}$, $\boldsymbol{\Lambda} = \text{diag}\,\{\lambda_i\}_{i=1}^N$, we get

$$\boldsymbol{Q}^{-1}\boldsymbol{x}(n) = \boldsymbol{\Lambda}^n \boldsymbol{Q}^{-1}\boldsymbol{x}(0). \tag{3.62}$$

From (3.62) we can conclude that the vector $\boldsymbol{Q}^{-1}\boldsymbol{x}(n)$ will converge to the zero vector, i.e., $\lim_{n\to\infty} \|\boldsymbol{Q}^{-1}\boldsymbol{x}(n)\|_2 = 0$ if and only if all eigenvalues λ_i of $\boldsymbol{A}$ are smaller than one in magnitude:

$$|\lambda_i| < 1 \Longleftrightarrow \lim_{n\to\infty} \|\boldsymbol{Q}^{-1}\boldsymbol{x}(n)\|_2 = 0 \Longleftrightarrow \lim_{n\to\infty} \|\boldsymbol{x}(n)\|_2 = 0. \tag{3.63}$$

This is equivalent to saying that the poles of the transfer function are localized inside the unity circle, since the eigenvalues of $\boldsymbol{A}$ equal the poles of the transfer function. In the above derivation, we have assumed that $\boldsymbol{A}$ is a diagonalizable matrix. If this is not the case because of multiple eigenvalues, we have to turn to the so-called Jordan form. Although this is slightly more complicated, it leads to the same result.

A somewhat weaker stability criterion is to check only if the system output decays to zero from any initial state and zero excitation:

$$\forall \boldsymbol{x}(0) \in \mathcal{R}^N \quad \text{and} \quad u(n) = 0 \quad \text{for} \quad n \geq 0: \quad \lim_{n\to\infty} |y(n)|^2 = 0. \tag{3.64}$$

Computing the output we get

$$y(n) = \boldsymbol{c}^{\mathrm{T}}\boldsymbol{x}(n) = \boldsymbol{c}^{\mathrm{T}}\boldsymbol{A}^n\boldsymbol{x}(0) = \boldsymbol{c}^{\mathrm{T}}\boldsymbol{Q}\boldsymbol{\Lambda}^n\boldsymbol{Q}^{-1}\boldsymbol{x}(0) = \boldsymbol{\eta}^{\mathrm{T}}\boldsymbol{\Lambda}^n\left(\boldsymbol{Q}^{-1}\boldsymbol{x}(0)\right)$$

$$= \begin{bmatrix}\eta_1 & \eta_2 & \cdots & \eta_N\end{bmatrix}\begin{bmatrix}\lambda_1^n & 0 & \cdots & 0 \\ 0 & \lambda_2^n & \cdots & 0 \\ \vdots & \vdots & \ddots & \vdots \\ 0 & 0 & \cdots & \lambda_N^n\end{bmatrix}\left(\boldsymbol{Q}^{-1}\boldsymbol{x}(0)\right) = \sum_{i=1}^{N}\eta_i\,\lambda_i^n\left(\boldsymbol{Q}^{-1}\boldsymbol{x}(0)\right)_i, \tag{3.65}$$

where $\boldsymbol{\eta}^{\mathrm{T}} = \boldsymbol{c}^{\mathrm{T}}\boldsymbol{Q}$ is the transformed output vector. If an entry η_i is equal to zero, then the eigenmode (normal mode) λ_i will not contribute to the system output and, therefore, a nondecaying eigenmode $|\lambda_i| \geq 1$ will not violate the *output stability* criterion. This can only happen if the observability matrix is not full rank. From (3.55) we have

$$\boldsymbol{y}(0) = \mathcal{O}\left(\boldsymbol{c}^{\mathrm{T}}, \boldsymbol{A}\right)\boldsymbol{x}(0) = \mathcal{O}\left(\boldsymbol{c}^{\mathrm{T}}, \boldsymbol{A}\right)\boldsymbol{Q}\left(\boldsymbol{Q}^{-1}\boldsymbol{x}(0)\right) = \mathcal{O}\left(\boldsymbol{\eta}^{\mathrm{T}}, \boldsymbol{\Lambda}\right)\left(\boldsymbol{Q}^{-1}\boldsymbol{x}(0)\right)$$

$$= \begin{bmatrix}\eta_1 & \eta_2 & \cdots & \eta_N \\ \eta_1\lambda_1 & \eta_2\lambda_2 & \cdots & \eta_N\lambda_N \\ \eta_1\lambda_1^2 & \eta_2\lambda_2^2 & \cdots & \eta_N\lambda_N^2 \\ \vdots & \vdots & \ddots & \vdots \\ \eta_1\lambda_1^{N-1} & \eta_2\lambda_2^{N-1} & \cdots & \eta_N\lambda_N^{N-1}\end{bmatrix}\left(\boldsymbol{Q}^{-1}\boldsymbol{x}(0)\right) \tag{3.66}$$

and that $\mathcal{O}\left(\boldsymbol{\eta}^{\mathrm{T}}, \boldsymbol{\Lambda}\right)$ is rank-deficient if at least one η_i is zero, and since $\boldsymbol{Q}$ is full rank, $\mathcal{O}\left(\boldsymbol{c}^{\mathrm{T}}, \boldsymbol{A}\right)$ must also be rank-deficient. Therefore, a system can be output stable without being internally stable, if it is not completely observable.

A complementary criterion is that of *input stability*, requiring the state vector to converge to the zero vector, but not from any arbitrary initial state, but only from states which can be controlled from the system input. Now we start with an initial state $\boldsymbol{x}(N)$, which can be generated from an input sequence (see (3.58))

$$\boldsymbol{x}(N) = \mathcal{C}(\boldsymbol{b}, \boldsymbol{A})\boldsymbol{u}(N). \tag{3.67}$$

The transformed state $\boldsymbol{Q}^{-1}\boldsymbol{x}(N)$ then reads

$$\boldsymbol{Q}^{-1}\boldsymbol{x}(N) = \boldsymbol{Q}^{-1}\mathcal{C}(\boldsymbol{b}, \boldsymbol{A})\boldsymbol{u}(N) = \mathcal{C}(\boldsymbol{\mu}, \boldsymbol{\Lambda})\boldsymbol{u}(N) \tag{3.68}$$

with the transformed input vector $\boldsymbol{\mu} = \boldsymbol{Q}^{-1}\boldsymbol{b}$. After the input $\boldsymbol{u}(N)$ has produced the state $\boldsymbol{x}(N)$ it is switched off, i.e., $u(n) = 0$ for $n \geq N$.

We ask whether the state vector can converge to the zero vector although the system may not be internally stable. The answer is given by the structure of

$$\mathcal{C}(\boldsymbol{\mu}, \boldsymbol{\Lambda}) = \begin{bmatrix}\mu_1 & \mu_2 & \cdots & \mu_N \\ \mu_1\lambda_1 & \mu_2\lambda_2 & \cdots & \mu_N\lambda_N \\ \vdots & \vdots & \ddots & \vdots \\ \mu_1\lambda_1^{N-1} & \mu_2\lambda_2^{N-1} & \cdots & \mu_N\lambda_N^{N-1}\end{bmatrix} = \boldsymbol{Q}^{-1}\mathcal{C}(\boldsymbol{b}, \boldsymbol{A}). \tag{3.69}$$

If some $\mu_i = 0$, then $\boldsymbol{Q}^{-1}\boldsymbol{x}(0)$ will converge to the zero vector, even if the corresponding $|\lambda_i| \geq 1$, because this nondecaying eigenmode cannot be excited from the regular input. But this is only possible if $\boldsymbol{C}(\boldsymbol{b}, \boldsymbol{A})$ is not full rank and the system is, therefore, not completely controllable.

Finally, we look at the concept of the commonly used *input–output stability*, i.e., we excite the system with zero initial state with a unit pulse and require the output to converge to zero over time. In this setting the state and the output evolve as

$$\boldsymbol{x}(n) = \boldsymbol{A}^n \boldsymbol{b}, \tag{3.70}$$

$$y(n) = \boldsymbol{c}^{\mathrm{T}} \boldsymbol{A}^n \boldsymbol{b} = \boldsymbol{c}^{\mathrm{T}} \boldsymbol{Q} \boldsymbol{\Lambda}^n \boldsymbol{Q}^{-1} \boldsymbol{b} = \boldsymbol{\eta}^{\mathrm{T}} \boldsymbol{\Lambda}^n \boldsymbol{\mu} = \sum_{i=1}^{N} \eta_i \mu_i \lambda_i^n. \tag{3.71}$$

The output will converge to zero even if there are nondecaying eigenmodes λ_i, as long as for every such λ_i there is an $\eta_i = 0$ or a $\mu_i = 0$, or both are zero. We see that a system could be input–output stable, although it is neither internally stable nor input or output stable.

Only for systems which are completely observable and controllable, the four different stability criteria coincide.

Example: The four stability concepts will now be illustrated with an example: the Cascade Integrator Comb (CIC) filter [6]. These filters are attractive in situations where either a narrow-band signal should be extracted from a wide-band source or a wide-band signal should be constructed from a narrow-band source. The first situation is associated with the concept of sampling rate decrease or decimation and the second one leads to the concept of sampling rate increase or interpolation. CIC filters are very efficient from a hardware point of view, because they need no multipliers. In the first case, i.e., decimation, the integrators come first, and then the sampling rate is decreased by a factor of R, followed by a subsequent cascade of comb filters (Fig. 3.12(a)), while in the second case, i.e., interpolation, the comb filters come first, and then the sampling rate increase is followed by a subsequent cascade of integrators (Fig. 3.12(b)).

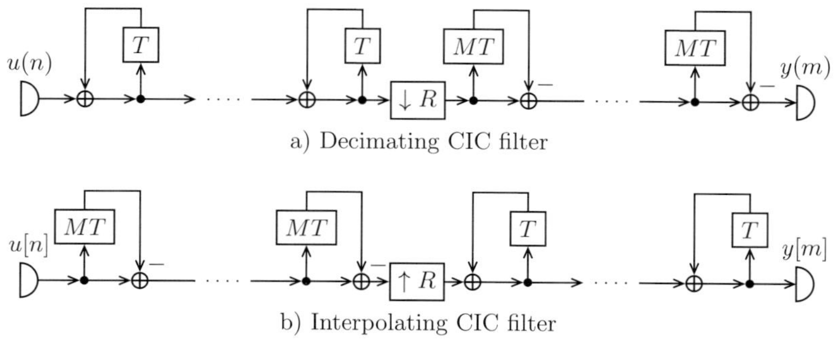

FIGURE 3.12

Multiplierless decimator and integrator.

The stability problem in these CIC filters stems from the integrators, which obviously exhibit a nondecaying eigenmode (pole with unit magnitude). To analyze the stability of such CIC filters we set the sampling rate decrease/increase factor $R = 1$ and the number of integrator and comb filter stages

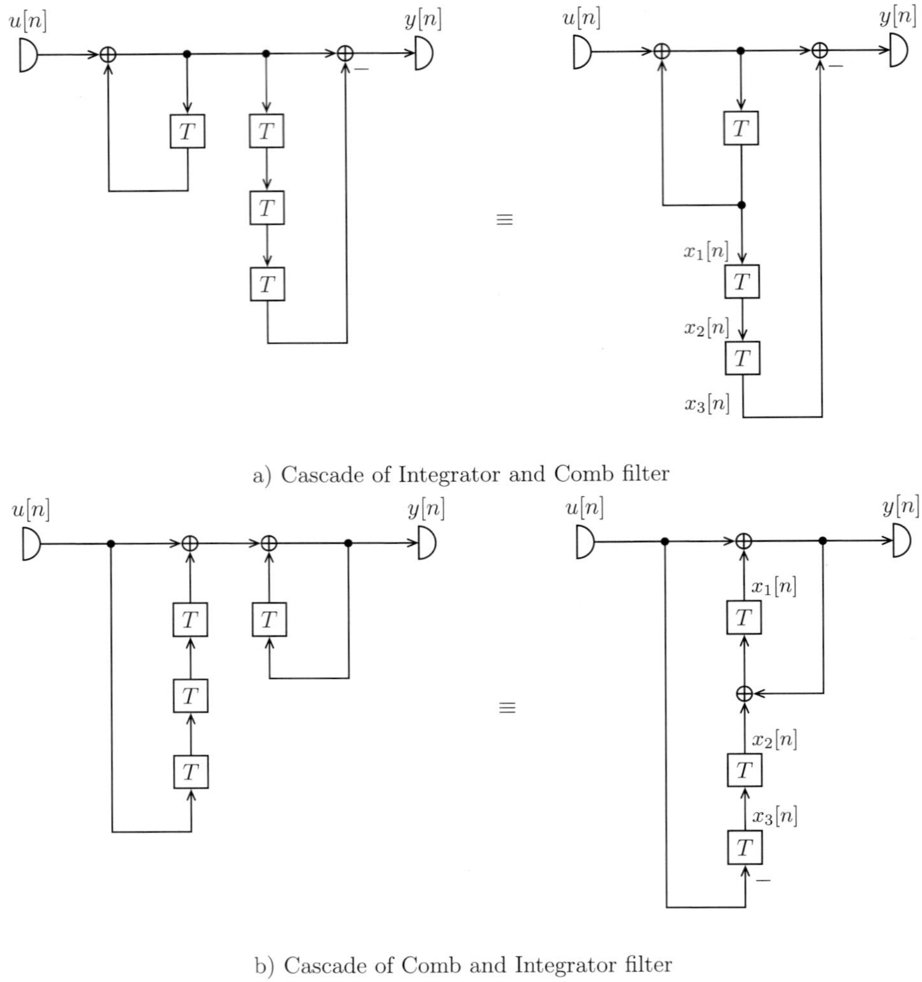

FIGURE 3.13

CIC filters without sampling rate alteration with $R = 1$, $L = 1$, and $M = 3$.

to $L = 1$ without loss of generality. In addition, we assume a differential delay of $M = 3$ samples per stage. This leads us to the following two implementations to be analyzed, which are shown in Fig. 3.13(a and b).

Let us start with an analysis of the filter in Fig. 3.13(a), the state-space description of which reads

$$\boldsymbol{A}_1 = \begin{bmatrix} 1 & 0 & 0 \\ 1 & 0 & 0 \\ 0 & 1 & 0 \end{bmatrix}, \ \boldsymbol{b}_1 = \begin{bmatrix} 1 \\ 0 \\ 0 \end{bmatrix}, \ \boldsymbol{c}_1 = \begin{bmatrix} 1 \\ 0 \\ -1 \end{bmatrix}, \ d_1 = 1 \tag{3.72}$$

and leads to the transfer function

$$H_1(z^{-1}) = \frac{1 - z^{-3}}{1 - z^{-1}} = 1 + z^{-1} + z^{-2}. \tag{3.73}$$

The observability matrix

$$\mathcal{O}\left(\boldsymbol{c}_1^{\mathrm{T}}, \boldsymbol{A}_1\right) = \begin{bmatrix} 1 & 0 & -1 \\ 1 & -1 & 0 \\ 0 & 0 & 0 \end{bmatrix}, \quad \det\left(\mathcal{O}\left(\boldsymbol{c}_1^{\mathrm{T}}, \boldsymbol{A}_1\right)\right) = 0 \tag{3.74}$$

is obviously rank-deficient and the system is therefore not completely observable.

Since the state matrix $\boldsymbol{A}$ is not diagonalizable, we have to turn to a Jordan form

$$\boldsymbol{A}_1 = \boldsymbol{Q}_1 \boldsymbol{J} \boldsymbol{Q}_1^{-1} = \begin{bmatrix} 0 & 0 & \frac{1}{\sqrt{3}} \\ 0 & 1 & \frac{1}{\sqrt{3}} \\ 1 & 0 & \frac{1}{\sqrt{3}} \end{bmatrix} \begin{bmatrix} 0 & 1 & 0 \\ 0 & 0 & 1 \\ 0 & 0 & 1 \end{bmatrix} \begin{bmatrix} -1 & 0 & 1 \\ -1 & 1 & 0 \\ \sqrt{3} & 0 & 0 \end{bmatrix}, \tag{3.75}$$

which shows the three eigenvalues are $\lambda_1 = 0$, $\lambda_2 = 0$, and $\lambda_3 = 1$. Not all eigenvalues are less than one in magnitude. Therefore, the system of Fig. 3.13(a) is not internally stable. But transforming the observability matrix according to (3.66), we get

$$\mathcal{O}\left(\boldsymbol{\eta}^{\mathrm{T}}, \boldsymbol{J}\right) = \mathcal{O}\left(\boldsymbol{c}_1^{\mathrm{T}}, \boldsymbol{A}_1\right) \boldsymbol{Q}_1 = \begin{bmatrix} 1 & 0 & 0 \\ 1 & 0 & 0 \\ 0 & 1 & 0 \end{bmatrix} \begin{bmatrix} 0 & 0 & \frac{1}{\sqrt{3}} \\ 0 & 1 & \frac{1}{\sqrt{3}} \\ 1 & 0 & \frac{1}{\sqrt{3}} \end{bmatrix} = \begin{bmatrix} -1 & 0 & 0 \\ 0 & -1 & 0 \\ 0 & 0 & 0 \end{bmatrix} \tag{3.76}$$

with the third component η_3 of the transformed output vector $\boldsymbol{\eta}^{\mathrm{T}} = \boldsymbol{c}_1^{\mathrm{T}} \boldsymbol{Q}_1$ equal to zero. This clearly means that the nonvanishing third eigenmode is not observable at the system output. Therefore, the system is output stable, i.e., the output converges in the limit $\lim_{n \to \infty} y(n) = 0$, although the state vector does not converge to the zero vector. But the controllability matrix

$$\mathcal{C}\left(\boldsymbol{b}_1, \boldsymbol{A}_1\right) = \begin{bmatrix} 1 & 1 & 1 \\ 1 & 1 & 1 \\ 0 & 0 & 1 \end{bmatrix}, \quad \det\left(\mathcal{C}\left(\boldsymbol{b}_1, \boldsymbol{A}_1\right)\right) = 1 \tag{3.77}$$

is full rank, i.e., the system is completely controllable, and an arbitrary input sequence may excite the nondecaying eigenmode. Therefore, the system is not input stable.

Now let us analyze the filter in Fig. 3.13(b), which has the following state-space description:

$$\boldsymbol{A}_2 = \begin{bmatrix} 1 & 1 & 0 \\ 0 & 0 & 1 \\ 0 & 0 & 0 \end{bmatrix}, \quad \boldsymbol{b}_2 = \begin{bmatrix} 1 \\ 0 \\ -1 \end{bmatrix}, \quad \boldsymbol{c}_2 = \begin{bmatrix} 1 \\ 0 \\ 0 \end{bmatrix}, \quad d_2 = 1 \tag{3.78}$$

and the same transfer function as the previous filter from Fig. 3.13(a). It is interesting to note that the second filter can be obtained from the first one by applying transposition:

$$A_2 = A_1^{\mathrm{T}}, \quad b_2 = c_1, \quad c_2 = b_1, \quad d_2 = d_1. \tag{3.79}$$

The controllability matrix of the second system

$$\mathcal{C}(b_2, A_2) = \begin{bmatrix} 1 & 1 & 0 \\ 0 & -1 & 0 \\ -1 & 0 & 0 \end{bmatrix}, \quad \det(\mathcal{C}(b_2, A_2)) = 0 \tag{3.80}$$

is obviously rank-deficient and the system is therefore not completely controllable.

Since the state matrix $A_2 = A_1^{\mathrm{T}}$ is not diagonalizable we again have to turn to a Jordan form:

$$A_2 = Q_2 J Q_2^{-1} = \begin{bmatrix} \frac{1}{\sqrt{2}} & \frac{1}{\sqrt{2}} & 1 \\ -\frac{1}{\sqrt{2}} & 0 & 0 \\ 0 & -\frac{1}{\sqrt{2}} & 0 \end{bmatrix} \begin{bmatrix} 0 & 1 & 0 \\ 0 & 0 & 1 \\ 0 & 0 & 1 \end{bmatrix} \begin{bmatrix} 0 & -\sqrt{2} & 0 \\ 0 & 0 & -\sqrt{2} \\ 1 & 1 & 1 \end{bmatrix}, \tag{3.81}$$

which again shows obviously the same eigenvalues $\lambda_1 = 0$, $\lambda_2 = 0$, and $\lambda_3 = 1$. The second system is again not internally stable. Transforming the above controllability matrix according to (3.68) we get

$$\mathcal{C}(\mu, J) = Q_2^{-1}\mathcal{C}(b_2, A_2) = \begin{bmatrix} 0 & -\sqrt{2} & 0 \\ 0 & 0 & -\sqrt{2} \\ 1 & 1 & 1 \end{bmatrix} \begin{bmatrix} 1 & 1 & 0 \\ 0 & -1 & 0 \\ -1 & 0 & 0 \end{bmatrix} = \begin{bmatrix} 0 & \sqrt{2} & 0 \\ \sqrt{2} & 0 & 0 \\ 0 & 0 & 0 \end{bmatrix} \tag{3.82}$$

with the third component μ_3 of the transformed input vector $\mu = Q_2^{-1} b_2$ being equal to zero. This shows that the nondecaying third eigenmode will not be excited by any possible input signal. Therefore, the system is input stable, although it is not internally stable.

The observability matrix of the second system

$$\mathcal{O}\left(c_2^{\mathrm{T}}, A_2\right) = \begin{bmatrix} 1 & 1 & 0 \\ 1 & 1 & 0 \\ 1 & 1 & 1 \end{bmatrix}, \quad \det(\mathcal{O}(b_2, A_2)) = 1 \tag{3.83}$$

is full rank and the second system is completely observable. Therefore, the nondecaying eigenmode, although not excited by any input sequence, is observable at the output. This can always happen because of an unfavorable initial state accidentally occurring in the switch-on transient of the power supply or because of some disturbance entering the system not through the regular input path. Therefore, the system is not output stable.

But both systems are input–output stable, because both have the same impulse response with finite duration.

This example shows that the standard input–output stability criterion, i.e., requiring a decaying impulse response, does not always tell the whole story. Judging whether a system is stable or not requires detailed knowledge about the system structure, or in other words, the realization, and not only about input–output mapping.

3.5 Discrete-time signals and systems with MATLAB®

In this section we will provide some examples how to generate discrete-time signals in MATLAB and how to represent and implement basic discrete-time systems. We assume here a basic MATLAB installation. The commands and programs shown here are not unique in the way they perform the analysis and implement discrete-time signals and systems. Our objective here is to give an introduction with very simple commands and functions. With this introduction the reader will get more familiar with this powerful tool that is widely employed to analyze, simulate, and implement digital signal processing systems. If one has access to the full palette of MATLAB toolboxes, like e.g., the Signal Processing Toolbox, Control System Toolbox, Communications System Toolbox, and DSP System Toolbox, many of the scripts included here can be substituted by functions encountered in those libraries.

3.5.1 Discrete-time signals

Discrete-time signals are defined in MATLAB as 1D arrays, i.e., as vectors. They can be row or column vectors and their entries can be either real- or complex-valued. The signals we will generate and analyze in this section are always row vectors and their entries are always real-valued. The length of the vectors will be defined according to the time span in which the system is supposed to be analyzed.

Unit impulse

The unit impulse $\delta(n)$ can be generated with the command

```
L = 100;
delta = [1 zeros(1,L-1)];
```

where `L` is the length of the desired input and the function `zeros(1,L-1)` generates a row vector with `L-1` zeros.

The unit impulse can be further weighted and delayed. For example, $2 * \delta(n-5)$ can be generated by

```
L = 100;
N = 5;
delta = [zeros(1,N) 2 zeros(1,L-1-N)];
```

where we have assumed a delay of five samples or sampling periods.

The length of the vector containing only a unit impulse depends on the objective of the analysis. For example, if the impulse response of a discrete-time system is to be studied, there should be enough elements in the input array so that a significant part of the impulse response is contained in the output vector.

Sinusoid

A sinusoid can be generated with the help of the function `sin`. Below is an example of a sinusoid where all its parameters are first defined.

```
fs = 1000;                  % Sampling frequency in Hz
T = 1/fs;                   % Sampling period
```

```
n = 1:100;                  % Time indices
U = 0.5;                    % Amplitude of the sinusoid
fc = 20;                    % Sinusoid frequency in Hz
omega = 2*pi*fc;            % Sinusoid angular frequency in rad/s
phi = 0;                    % phase in radians
x = U*sin(omega*n*T+phi);
```

We can see that the comments after each command explain what the parameter means.

If we would like to graphically represent the sinusoid, we could use

```
stem(n*T,x);
xlabel('Time in seconds');
ylabel('Amplitude');
```

and we get the plot in Fig. 3.14, where we can see two periods of the sinusoid. An alternative to `stem` is

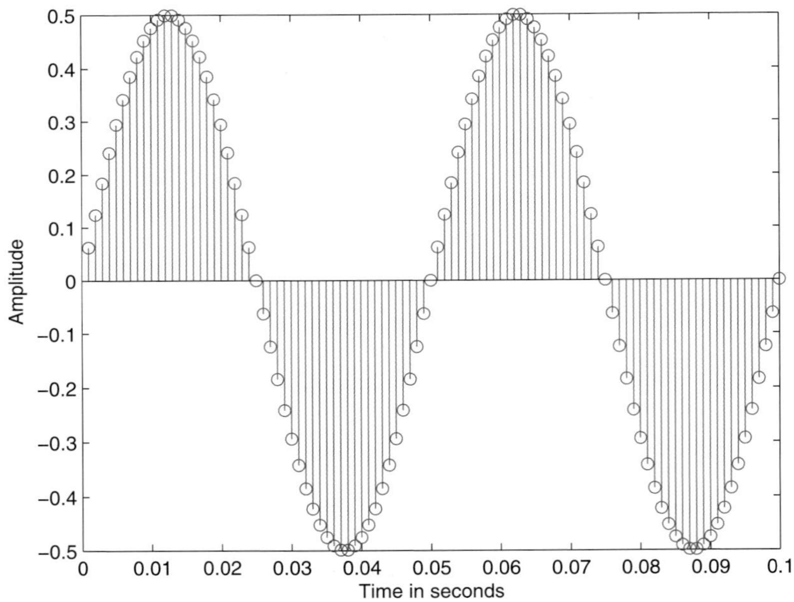

FIGURE 3.14

Discrete sinusoid with $f_c = 20$ Hz and $f_s = 1$ kHz plotted in MATLAB.

the command `plot`. If we employ it using the same parameters and the same syntax we obtain Fig. 3.15, where it should be noted that MATLAB graphically connects the amplitude samples with a straight line and as a consequence the discrete-time sinusoid looks like a continuous-time one. It is important to keep in mind that the true discrete-time sinusoid is only defined for certain time instants and the amplitude between two true samples in Fig. 3.15 is only an approximation of the amplitude of the equivalent continuous-time sinusoid.

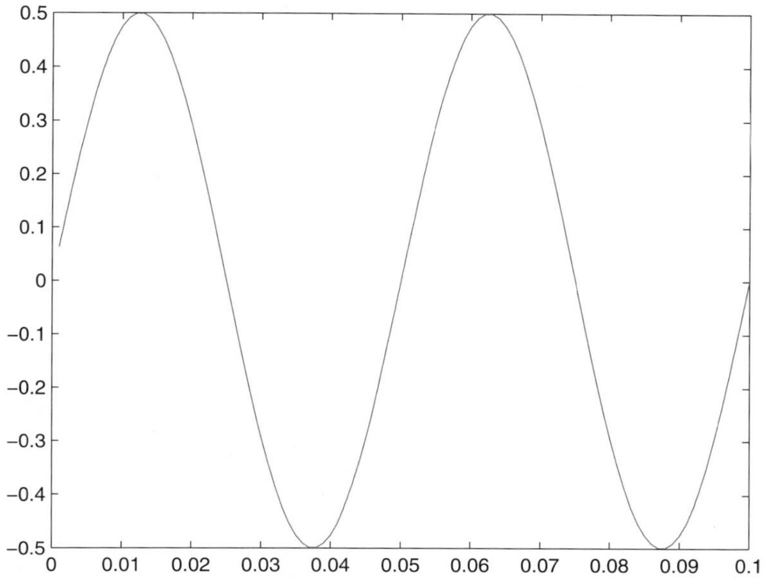

FIGURE 3.15

Discrete-time sinusoid represented in MATLAB with the command `plot`.

As an exercise the reader could play around with the parameters of the sinusoid. For example, he or she could change the phase ϕ to $\pi/2$ and plot the sinusoid again, try to use the function `cos` instead of `sin`, set ϕ to 0 again, and compare with the previous plots. Change the sinusoid frequency until you reach the Nyquist frequency, which is 500 Hz for the example above. Compare the use of the command `plot` for different sampling rates and different sinusoid frequencies to see how the approximation of the continuous-time sinusoid gets worse.

White Gaussian noise

To generate a WGN signal we have to employ the function `randn`, which generates pseudorandom numbers following the normal or Gaussian distribution with zero mean and unit variance. By using

```
L = 100;
sigma2 = 0.5;                % Variance of the WGN
u = sqrt(sigma2)*randn(1,L);
```

we generate a WGN vector with zero mean and variance $\sigma^2 = 0.5$.

Elaborated signal model

Many operations can be further performed with the generated signals. They can be, for example, added to each other or multiplied by a constant or by another signal. As a last example let us generate a typical signal model used in communications systems. Let us consider an input signal composed by a sum of weighted and delayed unit impulses. This could be a data-carrying signal generated at the transmitter.

Then we multiply it by a sinusoid, which could represent a modulation to a higher frequency, for example the radio frequency (RF), to allow the propagation of the signal as an electromagnetic wave in a physical medium between transmitter and receiver. Finally, we add WGN that represents the thermal noise generated by the analog components used in the transmitter and receiver.

The program to generate this signal is the following:

```
L = 100;
u_in = [0.5*ones(1,L/4) -0.5*ones(1,L/4) 0.5*ones(1,L/4) -0.5*ones(1,L/4)];
                                  % Sum of weighted and delayed unit impulses
fs = 1000;                        % Sampling frequency in Hz
T = 1/fs;                         % Sampling period
n = 1:L;                          % Time indices
U = 1;                            % Amplitude of the sinusoid
fc = 50;                          % Sinusoid frequency in Hz
omega = 2*pi*fc;                  % Sinusoid frequency in rad/s
phi = 0;                          % Phase in radians
u_sin = U*sin(omega*n*T+phi);     % Sinusoid
sigma2 = 0.02;                    % Variance of the white Gaussian noise
u_awgn = sqrt(sigma2)*randn(1,L); % White Gaussian noise vector
u_mod = u_in.*u_sin + u_awgn;     % Modulation and noise addition
```

In Fig. 3.16 we can see the signal that is composed of a sum of weighted and delayed unit impulses.

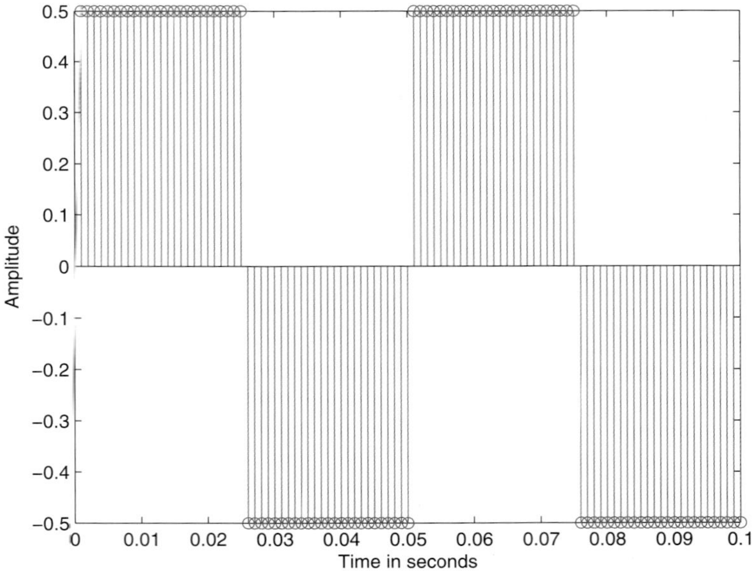

FIGURE 3.16

Weighted and delayed sum of unit impulses `u_in`.

In Fig. 3.17 the modulated sinusoid is depicted.

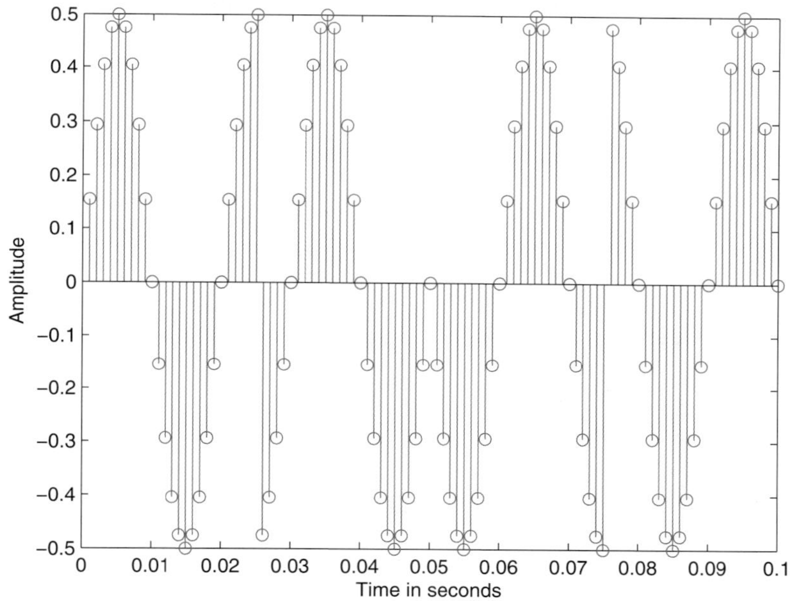

FIGURE 3.17

Modulated sinusoid `u_in*u_sin`.

We can see in Fig. 3.18 the modulated sinusoid with the additive WGN.

3.5.2 Discrete-time systems representation and implementation

There are different ways to represent a discrete-time system in MATLAB, like, for example, the space-state description or the transfer function. To start we revisit the elaborated signal model introduced in the last section. If we would like to, at least approximately, recover the input signal represented by the sum of delayed and weighted unit impulses, we should first demodulate it by multiplying it by a sinusoid with the same frequency as the one used to modulate

```
u_demod=u_mod.*u_sin;
```

We will then obtain the sequence shown in Fig. 3.19.

The only problem with the new sequence is that not only an approximation of the desired signal is obtained, but also another modulated version of it, but this time with a sinusoid with double the original modulation frequency, and we still have the additive noise. If the noise variance is low enough we are still able to approximately recover the original signal without any further processing. But first we have to eliminate the modulated signal component, and for this we will apply a discrete-time filter specifically designed for this task. We assume here that the filter is given and we only care about how to

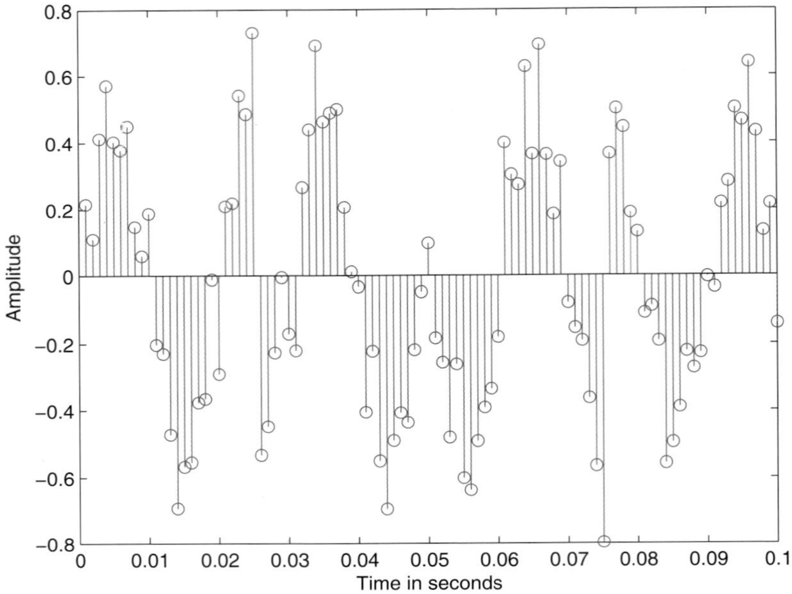

FIGURE 3.18

Modulated sinusoid with additive white Gaussian noise `u_mod`.

represent and implement it. In [7] and the references therein many methods for the design of discrete-time filters can be encountered.

Let us say that a state-space description of the filter is already known and is defined as

```
A = [0.6969    0.8263   -0.6089    0.0111;
          1         0         0        0;
          0         1         0        0;
          0         0         1        0];
b = [1;
     0;
     0;
     0];
c = [-0.4272    1.0189   -0.1298    0.0160];
d = [0.9738];
```

where we can identify the companion matrix structure of **A**.

To compute the output of the filter given an initial state and the input signal, the state equations can be directly employed and iteratively calculated as

```
L=100;
u= [1 zeros(1,L-1)]; % Input signal, here a unit impulse
```

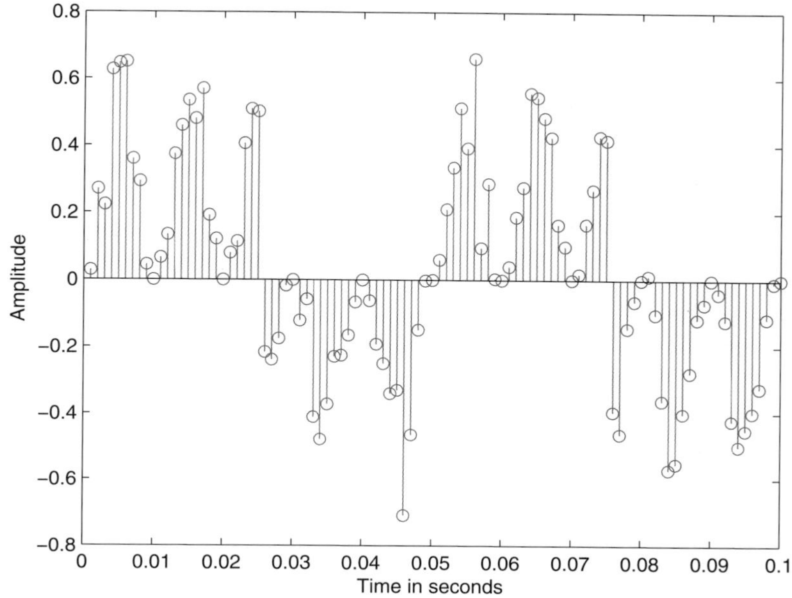

FIGURE 3.19

Demodulated signal `u_demod`.

```
x= [0; 0; 0; 0];        % Initial state vector
y=zeros(1,L);           % Initialization of the output vector
for n=1:L
  y(n)=c*x+d*u(n);
  x = A*x+b*u(n);
end
```

where we can see that the values of the state vector are not stored for all time instants. If one is interested in looking at the evolution of internal states, a matrix could be defined with so many rows as the number of states and so many columns as the length of the input/output signal, and the state vector for each time instant can be saved in its columns.

If we substitute the unit impulse by the demodulated signal as the input `u = u_demod`, we obtain the signal depicted in Fig. 3.20. One can see that we do not obtain exactly the input signal `u_in` but an approximation. Particularly in communications systems this is usually the case, where it is important to recover the information contained in `u_in`, even if the waveform has been distorted, and not the waveform itself.

But if one would like to implement Eq. (3.17) the program becomes more complicated, as shown below:

```
L=100;
u= [1 zeros(1,L-1)];             % Input signal, here a unit impulse
```

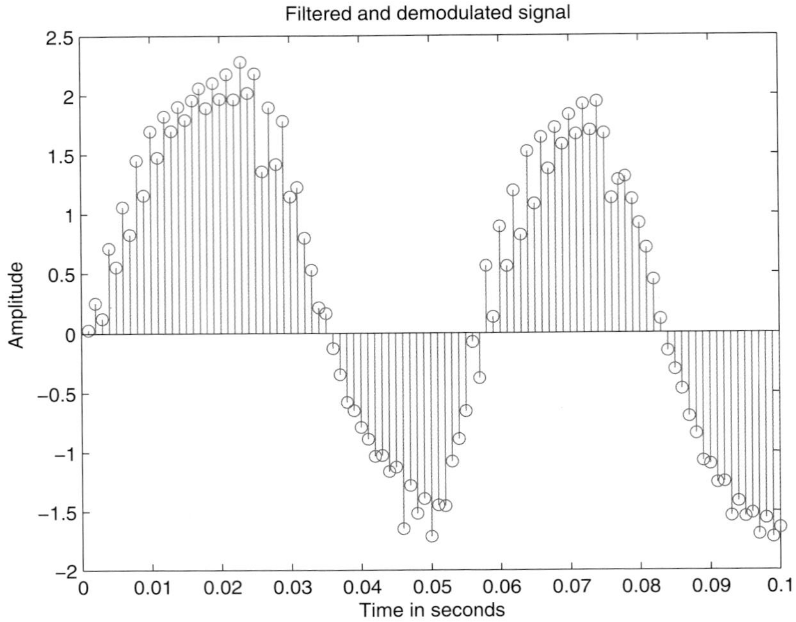

FIGURE 3.20

Demodulated and filtered signal u_demod.

```
x_0= [0; 0; 0; 0];              % Initial state vector
y=zeros(1,L);                   % Allocation of the output vector
cAb_array=zeros(1,L);           % Allocation of the array c*A^(k-1)*b
A_power=eye(4);                 % Allocation of the powers of A matrix
cAb_array(1)=c*b;               % First element of the array cAb_array
y(1)=d*u(1);                    % Output at sample n=0
for n=2:L
  cAb_array(n)=c*A_power*b;     % Inclusion of a new element in cAb_array
  y(n)=c*(A*A_power)*x_0+sum(cAb_array(1:n).*u(n:-1:1))+d*u(n);
                                % Output
  A_power=A*A_power;            % New power of A
end
```

In both programs presented above the impulse response can be obtained. Its first 100 samples are plotted in Fig. 3.21.

The similarity transformation can also be easily implemented. Let us consider the transformation into the normal form. The code to perform it is

```
[Q,Lambda] = eig(A); % Eigenvalue decomposition:
                     %  Lambda -> Eigenvalues in main diagonal
```

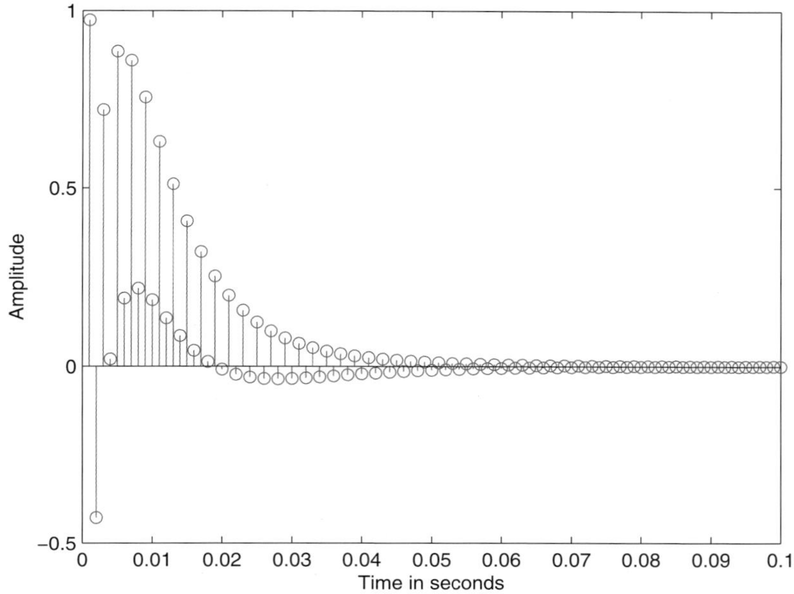

FIGURE 3.21

First 100 samples of the impulse response.

```
                    %  Q -> Eigenvectors in the columns
T = Q;
A_norm = inv(T)*A*T;
b_norm = inv(T)*b;
c_norm = c*T;
d_norm = d;
```

The resulting state-space description for our example is

```
A_norm = [-0.9242        0         0         0
               0    0.8057         0         0
               0         0    0.7968         0
               0         0         0    0.0186];
b_norm =  [0.6380;
          126.6454;
          127.3681;
            1.7322];
c_norm = [ -0.7502    0.2283   -0.2268    0.0140];
d_norm = 0.9738;
```

From the state-space representation it is possible to calculate the coefficients of the numerator and of the denominator polynomials of the transfer function with help of the Faddeev–LeVerrier algorithm [8]:

```
beta = [d zeros(1,3)];
alpha = [1 zeros(1,3)];
F = eye(4);
for i=1:4
  alpha(i+1) = -(1/i)*trace(A*F);
  beta(i+1) = c*F*b+d*alpha(i+1);
  F = A*F+alpha(i+1)*eye(4);
end
```

where the arrays `beta` and `alpha` contain the coefficients of the nominator and denominator polynomials.

With the coefficients of the transfer function it is also possible to calculate the output given the input of the system and some past values of the output by using the difference equation that can be implemented as

```
L=100;
u=[1 zeros(1,L-1)];
y=[zeros(1,L)];          % We assume a relaxed initial condition
y(1) = beta(1).*u(1);    % Transition phase or first initial output
y(2) = sum(beta(1:2).*u(2:-1:1))-alpha(2).*y(1);
y(3) = sum(beta(1:3).*u(3:-1:1))-sum(alpha(2:3).*y(2:-1:1));
y(4) = sum(beta(1:4).*u(4:-1:1))-sum(alpha(2:4).*y(3:-1:1));
for n=5:L
 y(n)=sum(beta.*u(n:-1:n-4))-sum(alpha(2:end).*y(n-1:-1:n-4));
end
```

As we saw before this is equivalent to a state-space realization in direct form.

3.6 Conclusions

In this chapter we have introduced an important and fundamental topic in electrical engineering that provides the basics for digital signal processing. We started with the presentation of some frequently used discrete-time signals. Then we showed how to classify discrete-time systems and started the study of the widely employed class of LTI systems. We showed the most basic way of representing LTI systems such that not only the input–output behavior is described, but also important internal signals, the so-called state-space representation. We have seen that the state-space representation is not unique and, consequently, there is room for optimization in the way discrete-time systems are realized. After that, the transfer function was derived from the state-space description, and the two main classes of discrete-time systems, namely FIR and IIR, were introduced. We formulated the concepts of controllability and observability and showed how they can be used to evaluate the stability of the system by also considering its internal signals. We finally gave some examples how to implement and analyze simple discrete-time signals and systems with MATLAB.

Notation

Throughout this chapter, the following notations was used:

$\mathcal{R}$ Set of real numbers
$\mathcal{C}$ Set of complex numbers
$\mathcal{R}^m$ Real Euclidean space of dimension m
$\mathcal{H}$ Denotes a Hilbert space
$.^T$ Matrix transpose
$.^H$ Hermitian operation for a matrix
$.^*$ Complex conjugation
$\boldsymbol{x}$ Bold letter denotes a vector
x Letter in normal font denotes a scalar
$\boldsymbol{X}$ Capital bold letter denotes a matrix
x_i or $x(i)$ The ith component of vector $\boldsymbol{x}$
$\boldsymbol{X}(n)$ or $\boldsymbol{x}(n)$ or $x_i(n)$ or $x(n)$ Time dependence on the time index n, depending on if it is a matrix, vector, vector component, or scalar
$\nabla f(\boldsymbol{x})$ or $\nabla_{\boldsymbol{x}} f(\boldsymbol{x})$ or $\frac{\partial f(\boldsymbol{x})}{\partial \boldsymbol{x}}$ The gradient of a function
$X(i, j)$ or X_{ij} The (i, j) element of matrix $\boldsymbol{X}$
$X_{ij}(n)$ Time dependence on the time index n of the (i, j) element of a matrix $\boldsymbol{X}$
$\mathrm{E}[\cdot]$ Denotes expectation
$\hat{x}$ Denotes the estimate of a variable, x

In case the authors wish to distinguish the random variable from its corresponding values that occur from the experiments, the following notation is proposed. Very often, authors use bold font to denote the random variable and nonbold for the values that result from the experiment. However, we have adopted the use of bold fonts for the vectors.

$\underline{\boldsymbol{X}}$ Denotes a random matrix
$\underline{\boldsymbol{x}}$ Denotes a random vector
$\underline{x}$ Denotes a random scalar

$\boldsymbol{x} \perp \boldsymbol{y}$ Denotes orthogonal vectors
$|x|$ and $|z|$ Absolute values for a real number x and a complex number z
$||\boldsymbol{x}||$ or $||\boldsymbol{x}||_2$ Denotes Euclidean norm
$||\boldsymbol{A}||_F$ Denotes Frobenius norm
$||\boldsymbol{x}||_p$ Denotes l_p-norm
$\mathrm{Re}(x)$ Denotes the real part of a complex number
$\mathrm{Im}(x)$ Denotes the imaginary part of a complex number
I_m or I Denotes the identity matrix of dimension $n \times n$
$\mathrm{vec}(\boldsymbol{A})$ Column vector formed by stacking the columns of $\boldsymbol{A}$
$\mathrm{diag}(\boldsymbol{A})$ Column vector with the diagonal entries of $\boldsymbol{A}$
$\mathrm{rank}(\boldsymbol{A})$ Rank of matrix $\boldsymbol{A}$
$\boldsymbol{A} \otimes \boldsymbol{B}$ The Kronecker product of two matrices
$\det(\cdot)$ The determinant of a matrix

$\mathrm{Tr}(\cdot)$ The trace of a matrix
$\lambda_i,\ i = 1, 2, \ldots, m$ The eigenvalues of an $m \times m$ matrix
$R(\boldsymbol{A})$ Range space of $\boldsymbol{A}$
$N(\boldsymbol{A})$ Null space of $\boldsymbol{A}$
$P(\cdot)$ Probability of a discrete event
$p(\cdot)$ Probability density function of a random variable
$\boldsymbol{R}$ or $\boldsymbol{R}_x$ The autocorrelation matrix of a random vector $\underline{\boldsymbol{x}}$
$\boldsymbol{\Sigma}$ or $\boldsymbol{\Sigma}_x$ The covariance matrix of a random vector $\underline{\boldsymbol{x}}$
σ^2 or σ_x^2 The variance of the random variable $\underline{x}$
$\log a$ The logarithm of a relative to base 10
$\ln a$ The natural logarithm of a
$\exp(\cdot)$ The exponential function
$\boldsymbol{A} > \boldsymbol{B}$ Means that $\boldsymbol{A} - \boldsymbol{B}$ is positive definite
$\boldsymbol{A} \geq \boldsymbol{B}$ Means that $\boldsymbol{A} - \boldsymbol{B}$ is positive semidefinite
$X(z)$ The z-transform of sequence $x(n)$
$X(e^{j\omega})$ The Fourier transform of a sequence $x(n)$
$X(e^{j\Omega T})$ The Fourier transform of a sampled sequence $x(nT)$, with sampling period T
$X(j\Omega)$ The Fourier transform of a function $x(t)$
$X(m)$ or X_m The discrete Fourier transform of $x(n)$
$F(\Omega_x, \Omega_y)$ The Fourier transform of $f(x, y)$
$F(\omega_x, \omega_y)$ or $F(\omega_1, \omega_2)$, etc. The Fourier transform of $f(n, m)$
$F(k, l)$ The discrete Fourier transform of the sequence $f(n, m)$
$X(s)$ The Laplace transform of a function $x(t)$
Notation of a matrix A

$$\boldsymbol{A} = \begin{bmatrix} A(1,1) & A(1,2) & A(1,3) \\ A(2,1) & A(2,2) & A(2,3) \\ A(3,1) & A(3,2) & A(3,3) \end{bmatrix}$$

Notation of a vector $\boldsymbol{x}$

$$\boldsymbol{x} = [x_1, x_2, \ldots, x_m]^T = \begin{bmatrix} x_1 \\ x_2 \\ \vdots \\ x_m \end{bmatrix}$$

It will also be nice to make an effort to use the following symbols:

- For input signals: $u(n)$ or $x(n)$.
- For the output signal: $y(n)$ and $d(n)$ for the desired response.
- For the impulse response of a system: $h(n)$ or $w(n)$ or $g(n)$ and similarly for the point spread function, $h(n, m)$ or $w(n, m)$ or $g(n, m)$. Similar notation for analog systems, i.e., $h(t)$ or $h(x, y)$.
- $e(n)$ for an error estimation sequence, $e_a(n)$ for an a posteriori error sequence, and $e_p(n)$ for an a priori error sequence.

References

[1] Wikipedia, Fibonacci number — Wikipedia, the free encyclopedia, 2012 [online; accessed 16 March 2012].

[2] Robert M. May, Simple mathematical models with very complicated dynamics, Nature 261 (5560) (June 1976) 459–467.

[3] Pierre-François Verhulst, Notice sur la loi que la population poursuit dans son accroissement, Correspondance mathématique et physique 10 (1838) 113–121.

[4] Gilbert Strang, Linear Algebra and Its Applications, 3rd ed., Brooks Cole, February 1988.

[5] E.I. Jury, Theory and Application of the z-Transform Method, John Wiley, New York, USA, 1964.

[6] E.B. Hogenauer, An economical class of digital filters for decimation and interpolation, IEEE Transactions on Acoustics, Speech and Signal Processing 29 (2) (April 1981) 155–162.

[7] P.S.R. Diniz, E.A.B. da Silva, S. Lima Netto, Digital Signal Processing: System Analysis and Design, second edition, Cambridge University Press, Cambridge, UK, 2010.

[8] R.A. Roberts, C.T. Mullis, Digital Signal Processing, Addison-Wesley Series in Electrical Engineering, vol. 1, Addison-Wesley, 1987.

CHAPTER

Random signals and stochastic processes

4

Luiz Wagner Pereira Biscainho and Eduardo Alves da Silva

Federal University of Rio de Janeiro (UFRJ), Rio de Janeiro, Brazil

Foreword

This chapter is a new edition of *Random signals and stochastic processes*, which first appeared in [1]. Compared to the previous manuscript, in the following content the reader will find better-structured examples, a few corrected mistakes, and important changes in convention and notation aimed at greater cohesion and clarity, which led to a complete revision of the formulas of the final two-thirds of the chapter. Additionally, supplemental material, such as course slides and exercises, are provided by the authors on a companion web page.

Acknowledgment

The authors would like to thank Dr. Paulo A. A. Esquef for the careful review and Leonardo de O. Nunes for kindly preparing the illustrations of the first version of this text.

4.1 Introduction

Probability is an abstract concept useful to model chance experiments. The definition of a numerical representation for the outcomes of such experiments (the random variable) is essential to build a complete and general framework for probabilistic models. These models can be extended to nonstatic outcomes in the form of time signals,[1] leading to the so-called stochastic process, which can evolve along continuous or discrete time. Its complete description is usually too complicated to be applied to practical situations. Fortunately, a well-accepted set of simplifying properties like stationarity and ergodicity allows the modeling of many problems in areas as diverse as biology, economics, and communications.

[1] Of course, other independent variables (e.g., space) can substitute for time in this context.

Signal Processing and Machine Learning Theory. https://doi.org/10.1016/B978-0-32-391772-8.00010-7
Copyright © 2024 Elsevier Inc. All rights reserved.

There are plenty of books on random processes,[2] each one following some preferred order and notation, but covering essentially the same topics. This chapter is just one more attempt to present the subject in a compact manner. It is structured in the usual way: probability is first introduced and then described through random variables; within this framework, stochastic processes are presented and then associated with processing systems. Due to space limitations, proofs are avoided and examples are kept at a minimum. No attempt was made to cover many families of probability distributions, for example. The preferred path was to clearly and unambiguously define the concepts and entities associated with the subject and whenever possible give them simple and intuitive interpretations. Even risking to seem redundant, the authors decided to explicitly duplicate the formulations related to random processes and sequences (i.e., continuous-time and discrete-time random processes, respectively); the idea was to provide always a direct response to a consulting reader, instead of suggesting modifications in the given expressions.

Writing technical material always poses a difficult problem: which level of detail and depth will make the text useful? Our goal was to make the text approachable for undergraduate students and provide a consistent reference for more advanced students or even researchers (re)visiting random processes. The authors tried to be especially careful with the notation consistence throughout the chapter in order to avoid confusion and ambiguity (which may easily occur in advanced texts). The choices of covered topics and order of presentation reflect several years of teaching the subject, and obviously match those of some preferred books. A selected list of didactic references on statistics [4,3], random variables and stochastic processes [11,12,10,15], and applications [14,9] is given at the end of the chapter. We hope you enjoy the reading.

4.2 Probability

Probabilistic models are useful to describe and study phenomena that cannot be precisely predicted. To establish a precise framework for the concept of probability, one should define a chance *experiment*, of which each *trial* yields an *outcome* s. The set of all possible outcomes is the *sample space* S. In this context, one speaks of the *probability* that one trial of the experiment yields an outcome s that belongs to a desired set $A \subset S$ (when one says the *event* A has occurred), as illustrated in Fig. 4.1.

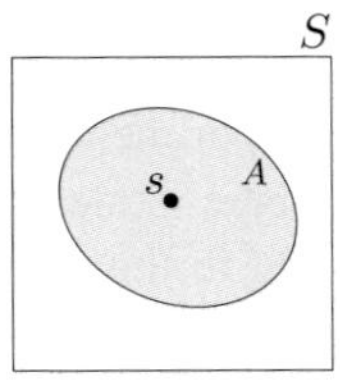

FIGURE 4.1

Sample space S, event A, and outcome s.

[2] Even if the literature sometimes prefers to employ "stochastic" for processes and "random" for signals, the expression "random processes" has become common and will be used throughout the chapter.

There are two different views of probability: the subjectivist and the objectivist. For subjectivists, the probability measures someone's degree of belief in the occurrence of a given event, while for objectivists it results from concrete reality. Polemics aside, objectivism can be more rewarding didactically.

From the objectivist (or frequentist) viewpoint, one of the possible ways to define probability is the so-called *classical* or *a priori* approach: given an experiment whose possible outcomes s are equally likely, the probability of an event A is defined as the ratio between the number $N(A)$ of acceptable outcomes (elements of A) and the number $N(S)$ of possible outcomes (elements of S):

$$P(A) = \frac{N(A)}{N(S)}. \tag{4.1}$$

The experiment of flipping a fair coin, with $S = \{\text{head, tail}\}$, is an example in which the probabilities of both individual outcomes can be theoretically set: $P(\{\text{head}\}) = P(\{\text{tail}\}) = 0.5$.

Another way to define probability is the *relative frequency* or *a posteriori* approach: the probability that the event A occurs after one trial of a given experiment can be obtained by taking the limit of the ratio between the number n_A of successes (i.e., occurrences of A) and the number n of experiment trials, when the repeats go to infinity:

$$P(A) = \lim_{n \to \infty} \frac{n_A}{n}. \tag{4.2}$$

In the ideal coin flip experiment, one is expected to find equal probabilities for head and tail. On the other hand, this pragmatic approach allows modeling the nonideal case, provided the experiment can be repeated.

A pure *mathematical* definition can provide a sufficiently general framework to encompass every conceptual choice: the *axiomatic* approach develops a complete probability theory from three *axioms*:

1. $P(A) \geq 0$;
2. $P(S) = 1$;
3. given A_m, $1 \leq m \leq M$, such that $A_m \cap A_{m'} = \emptyset \ \forall \ m' \neq m$, $1 \leq m' \leq M$, we have $P\left(\bigcup_{m=1}^{M} A_m\right) = \sum_{m=1}^{M} P(A_m)$.

Referring to an experiment with sample space S:

- The events A_m, $1 \leq m \leq M$, are said to be *mutually exclusive* when the occurrence of one prevents the occurrence of the others. They are the subject of the third axiom.
- The *complement* $\overline{A}$ of a given event A, illustrated in Fig. 4.2, is determined by the nonoccurrence of A, i.e., $\overline{A} = S - A$. From this definition, $P(\overline{A}) = 1 - P(A)$. Complementary events are also mutually exclusive.
- An event $A = \emptyset$ is called *impossible*. From this definition, $P(A) = 0$.
- An event $A = S$ is called *certain*. From this definition, $P(A) = 1$.

It should be emphasized that all events A related to a given experiment are completely determined by the sample space S, since by definition $A \subset S$. Therefore, a set B of outcomes not in S is mapped to

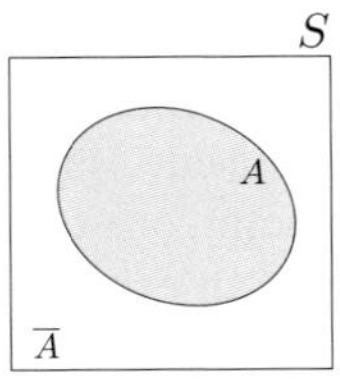

FIGURE 4.2

Event A and its complement $\overline{A}$.

an event $B \cap S = \emptyset$. For instance, for the experiment of rolling a fair die, $S = \{1, 2, 3, 4, 5, 6\}$; the event corresponding to $B = \{7\}$ (showing a 7) is $B \cap S = \emptyset$.

According to the experiment, sample spaces may be *countable* or *uncountable*.[3] For example, the sample space of the coin flip experiment is countable. On the other hand, the sample space of the experiment that consists in sampling with no preference any real number from the interval $S = (0, 100]$ is uncountable. Given an individual outcome $s \in S$, defining $A = \{s\}$,

$$P(A) \begin{cases} \neq 0, & \text{if S is countable,} \\ = 0, & \text{if S is uncountable.} \end{cases} \tag{4.3}$$

As a consequence, one should not be surprised to find an event $A \neq \emptyset$ with $P(A) = 0$ or an event $A \neq S$ with $P(A) = 1$. In the $(0, 100]$ interval sampling experiment, $A = \{50\} \neq \emptyset$ has $P(A) = 0$ and $\overline{A} \neq S$ has $P(\overline{A}) = 1$.

4.2.1 Joint, conditional, and total probability – Bayes' rule

The *joint probability* of a set of events A_m, $1 \leq m \leq M$, is the probability of their simultaneous occurrence $\bigcap_{m=1}^{M} A_m$. Referring to Fig. 4.3, given two events A and B, their joint probability can be found as

$$P(A \cap B) = P(A) + P(B) - P(A \cup B). \tag{4.4}$$

By rewriting this equation as

$$P(A \cup B) = P(A) + P(B) - P(A \cap B), \tag{4.5}$$

one finds an intuitive result: the term $P(A \cap B)$ is included twice in $P(A) + P(B)$, and should thus be discounted. Moreover, when A and B are mutually exclusive, we arrive at the third axiom. In the die experiment, defining $A = \{s \in S | s \text{ is even}\} = \{2, 4, 6\}$ and $B = \{s \in S | s > 3\} = \{4, 5, 6\}$, we have $P(A) = 1/2$, $P(B) = 1/2$, $P(A \cap B) = P(\{4, 6\}) = 1/3$, and $P(A \cup B) = P(\{2, 4, 5, 6\}) = 2/3$.

[3] One could think of mixed cases, but this discussion is better conveyed in the random variable framework, which comprises every possible mapping from the original experiment to a subset of the real numbers.

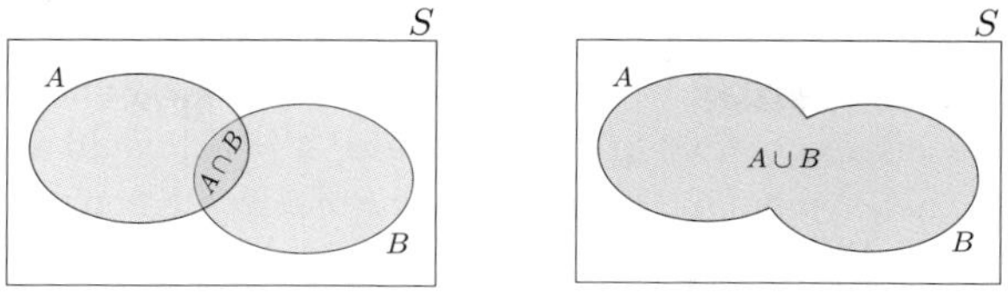

FIGURE 4.3

$A \cap B$ and $A \cup B$.

The *conditional probability* $P(A|B)$ is the probability of event A conditioned to the occurrence of event B, and can be computed as

$$P(A|B) = \frac{P(A \cap B)}{P(B)}, \quad P(B) > 0. \tag{4.6}$$

The value of $P(A \cap B)$ accounts for the uncertainty of both A and B; the term $P(B)$ discounts the uncertainty of B, since it is certain in this context. In fact, the conditioning event B is the new (reduced) sample space for the experiment. By rewriting this equation as

$$P(A \cap B) = P(A|B)P(B), \tag{4.7}$$

one gets another interpretation: the joint probability of A and B combines the uncertainty of B with the uncertainty of A when B is known to occur. Using the example in the last paragraph, $P(A|B) = [P(\{4, 6\})$ in sample space $B] = 2/3$.

Since $P(A \cap B) = P(B \cap A) = P(B|A)P(A)$, the *Bayes Rule* follows straightforwardly:

$$P(A|B) = \frac{P(B|A)P(A)}{P(B)}. \tag{4.8}$$

This formula allows computing one conditional probability $P(A|B)$ from its reverse $P(B|A)$. Using again the example in the last paragraph, one would arrive at the same result for $P(A|B)$ using $P(B|A) = [P(\{4, 6\})$ in sample space $A] = 2/3$ in the last equation.

If the sample space S is partitioned into M disjoint sets A_m, $1 \le m \le M$, such that $A_m \cap A_{m'} = \emptyset \ \forall \ m' \neq m, 1 \le m' \le M$, then any event $B \subset S$ can be written as $B = \bigcup_{m=1}^{M} (B \cap A_m)$. Since $B \cap A_m$, $1 \le m \le M$, are disjoint sets,

$$P(B) = \sum_{m=1}^{M} P(B|A_m)P(A_m), \tag{4.9}$$

which is called the *total probability* of B.

For a single event of interest A, for example, the application of Eq. (4.9) to Eq. (4.8) yields

$$P(A|B) = \frac{P(B|A)P(A)}{P(B|A)P(A) + P(B|\overline{A})P(\overline{A})}. \tag{4.10}$$

Within the Bayes context, $P(A)$ and $P(A|B)$ are usually known as the *a priori* and *a posteriori* probabilities of A, respectively; $P(B|A)$ is called *likelihood* in the estimation context and *transition probability* in the communications context. In the latter case, $A = \{a\}$ could refer to a symbol a sent by the transmitter and $B = \{b\}$ to a symbol b recognized by the receiver. Knowing $P(B|A)$, the probability of recognizing b when a is sent (which models the communication channel), and $P(A)$, the *a priori* probability of the transmitter to send a, allows to compute $P(A|B)$, the probability of a having been sent given that b has been recognized. In the case of binary communication, we could partition the sample space either in the events TX0 (0 transmitted) and TX1 (1 transmitted) or in the events RX0 (0 recognized) and RX1 (1 recognized), as illustrated in Fig. 4.4; the event "error" would be $E = (\text{TX0} \cap \text{RX1}) \cup (\text{TX1} \cap \text{RX0})$.

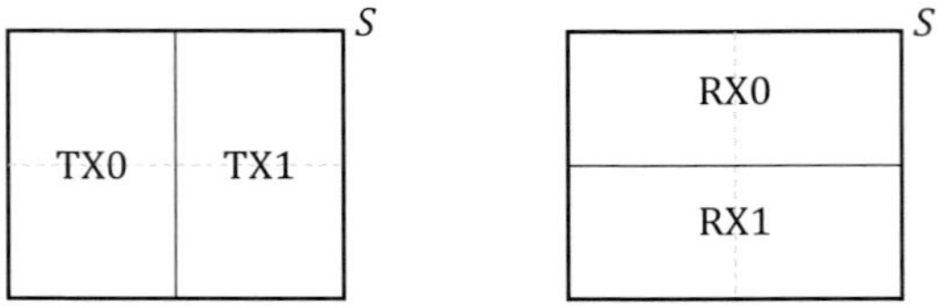

FIGURE 4.4

Binary communication example: partitions of S regarding transmission and reception.

4.2.2 Probabilistic independence

The events A_m, $1 \leq m \leq M$, are said to be *mutually independent* when the occurrence of one does not affect the occurrence of any combination of the others.

For two events A and B, three equivalent tests can be employed: they are independent if and only if:

1. $P(A|B) = P(A)$,
2. $P(B|A) = P(B)$, or
3. $P(A \cap B) = P(A)P(B)$.

The first two conditions follow directly from the definition of independence. Using any of them in Eq. (4.6) one arrives at the third one. The reader is invited to return to Eq. (4.7) and conclude that B and $A|B$ are independent events. Consider the experiment of rolling a special die with the numbers 1, 2, and 3 stamped in black on three faces and stamped in red on the other three; the events $A = \{1\}$ and $B = \{s \in S | s \text{ is a red number}\}$ are mutually independent.

Algorithmically, the best choice for testing the mutual independence of more than two events is using the third condition for every combination of $2, 3, \ldots, M$ events among $A_1, A_2, \ldots, A_M$.

As a final observation, mutually exclusive events are not mutually independent; on the contrary, one could say they are maximally dependent, since the occurrence of one precludes the occurrence of the other.

4.2.3 Combined experiments – Bernoulli trials

The theory we have discussed so far can be applied when more than one experiment is performed at a time. The generalization to the *multiple experiment* case can be easily done by using Cartesian products.

Consider the experiments E_m with their respective sample spaces S_m, $1 \le m \le M$. Define the combined experiment E such that each of its trials is composed of one trial of each E_m. An outcome of E is the M-tuple $(s_1, s_2, \ldots, s_M)$, where $s_m \in S_m$, $1 \le m \le M$, and the sample space of E can be written as $S = S_1 \times S_2 \times \cdots \times S_M$. Correspondingly, any event of E can be expressed as unions and intersections of sets in the form $A = A_1 \times A_2 \times \cdots \times A_M$, where A_m is a properly chosen event of E_m, $1 \le m \le M$.

In the special case when the subexperiments are mutually independent, i.e., the outcomes of one do not affect the outcomes of the others, we have $P(A) = \prod_{m=1}^{M} A_m$. However, this is not the general case. Consider the experiment of randomly selecting a card from a 52-card deck, repeated twice: if the first card drawn is replaced, the subexperiments are independent; if not, the first outcome affects the second experiment. For example, for $A = \{(\text{ace of spades, ace of spades})\}$, $P(A) = (1/52)^2$ if the first card is replaced, and $P(A) = 0$ if not.

At this point, an interesting counting experiment (called *Bernoulli trials*) can be defined. Take a random experiment E with sample space S and a desired event A (success) with $P(A) = p$, which also defines $\overline{A}$ (failure) with $P(\overline{A}) = 1 - p$. What is the probability of getting exactly k successes in N independent repeats of E? The solution can be easily found by noticing that the desired result is composed of k successes and $N - k$ failures, which may occur in $\binom{N}{k}$ different orders. Then,

$$P(k \text{ successes after } N \text{ trials}) = \binom{N}{k} p^k (1-p)^{N-k}. \tag{4.11}$$

When $N \to \infty$, $p \to 0$, and $Np \to$ a constant,

$$P(k \text{ successes after } N \text{ trials}) \to \frac{(Np)^k \mathrm{e}^{-Np}}{k!}. \tag{4.12}$$

Returning to the card deck experiment (with replacement), the probability of selecting exactly two aces of spades after 300 repeats is approximately 5.09%, from Eq. (4.11); Eq. (4.12) provides the approximate value 5.20%. In a binary communication system where 0 and 1 are randomly transmitted with equal probabilities, the probability that exactly 2 bits out of 3 transmitted bits are equal to 1 is 0.25%; this result can be easily checked by inspection of Fig. 4.5.

4.3 Random variable

Mapping each outcome of a random experiment to a real number provides a different framework for the study of probabilistic models, amenable to simple interpretation and easy mathematical manipulation. This mapping is performed by the so-called random variable.

Given a random experiment with sample space S, a *random variable* is any function $X(s)$ that maps each $s \in S$ into some $x \in \mathbb{R}$ (see Fig. 4.6). The image of this transformation with domain S and

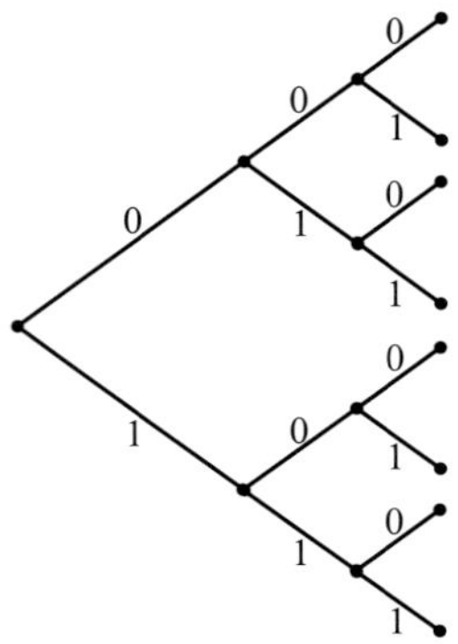

FIGURE 4.5

Three consecutive bits sent through a binary communication system.

codomain $\mathbb{R}$, which results from the convenient choice of the mapping function, can be seen as the sample space of X; any event A of X can be described as a subset of $\mathbb{R}$; and each mapped outcome x is called a sample of X. The following conditions should be satisfied by a random variable X:

1. $\{X \leq x\}$ is an event $\forall x \in \mathbb{R}$;
2. $P(\{X = -\infty\}) = P(\{X = \infty\}) = 0$.

We will see later that these conditions allow the proper definition of the cumulative probability distribution function (CDF) of X.

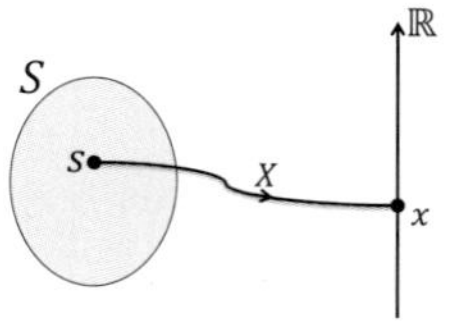

FIGURE 4.6

Sample space S mapped into $\mathbb{R}$ via random variable $X(s)$.

As seen in Section 4.2, sample spaces (and events) can be countable or uncountable, according to the nature of the random experiment's individual outcomes. After mapping into subsets of $\mathbb{R}$, countable events remain countable – e.g., one could associate with the coin flip experiment a random variable V such that $V(\text{head}) = 0$ and $V(\text{tail}) = 1$. On the other hand, uncountable events may or may not remain uncountable. Consider the following four distinct definitions of a random variable I associated with the $(0, 100]$ interval experiment defined immediately before Eq. (4.3):

1. $I_1(s) = s$;
2. $I_2(s) = \begin{cases} -1, & s \leq 50, \\ 1, & s > 50; \end{cases}$

3. $I_3(s) = \begin{cases} s, & s \leq 50, \\ 100, & s > 50; \end{cases}$
4. $I_4(s) = \min[s, 50]$.

The sample space of:

1. I_1, $(0, 100]$, is uncountable;
2. I_2, $\{-1, 1\}$, is countable;
3. I_3, $(0, 50] \cup \{100\}$, is part uncountable, part countable;
4. I_4, $(0, 50]$, even though it is an interval, can be considered as part uncountable, part countable.

The classification of sample spaces as countable or uncountable leads directly to the classification of random variables as discrete, continuous, or mixed. A *discrete* random variable D has a countable sample space S_D and has $P(\{D = d\}) > 0,\ \forall\ d \in S_D$ – this is the case of V and I_2 defined above. A *continuous* random variable C has an uncountable sample space S_D and (to avoid ambiguity) has $P(\{C = c\}) = 0,\ \forall\ c \in S_C$ – this is the case of I_1 defined above. A *mixed* variable M has a sample space S_D composed by the union of real intervals with continuously distributed probabilities, within which $P(\{M = m\}) = 0$, and discrete real values with finite probabilities $P(\{M = m\}) > 0$ – this is the case of I_3 and I_4 defined above. Specifically, since $P(0 < I_4 < 50) = P(I_4 = 50) = 50\%$, S_{I_4} should rather be treated as part uncountable, part countable than as uncountable.

4.3.1 Probability distributions

From the conditions that must be satisfied by any random variable X, an overall description of its probability distribution can be provided by the so-called *cumulative probability distribution function* (shortened to *CDF*),

$$F_X(x) = P(\{X \leq x\}). \tag{4.13}$$

Since $P(X = -\infty) = 0$, $F_X(-\infty) = 0$. It is obvious that $F_X(\infty) = 1$; but since $P(\{X = \infty\}) = 0$, $P(\{X < \infty\}) = 1$. Moreover, $0 \leq F_X(x) \leq 1$ and $F_X(x)$ is a nondecreasing function of x, i.e., $F_X(x_1) \leq F_X(x_2)$ if $x_1 < x_2$. Also, $F_X(x^+) = F_X(x)$, i.e., $F_X(x)$ is continuous from the right; this will be important in the treatment of discrete random variables. A typical CDF is depicted in Fig. 4.7. One can use the CDF to calculate probabilities by noticing that

$$P(\{x_1 < X \leq x_2\}) = F_X(x_2) - F_X(x_1). \tag{4.14}$$

For the random variable X whose CDF is described in Fig. 4.7, one can easily check by inspection that $P\left(X \leq \frac{1}{2}\right) = 50\%$.

The CDF of the random variable I_1 associated above with the $(0, 100]$ interval sampling experiment is

$$F_{I_1}(i_1) = \begin{cases} 0, & i_1 \leq 0, \\ \frac{i_1}{100}, & 0 < i_1 \leq 100, \\ 1, & i_1 > 100. \end{cases} \tag{4.15}$$

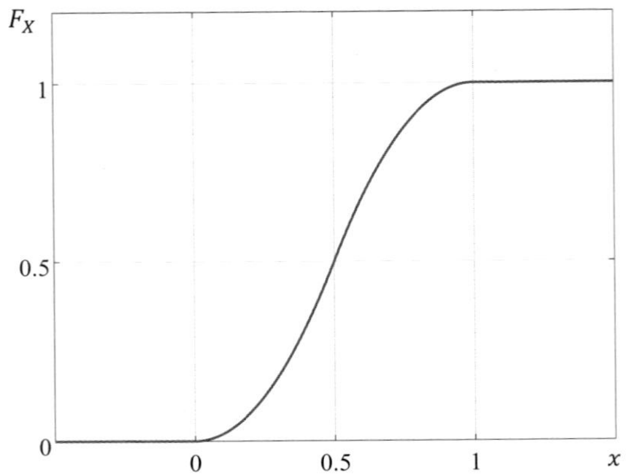

FIGURE 4.7

Example of a cumulative probability distribution function.

The CDF of the random variable V associated above with the coin flip experiment is

$$F_V(v) = \begin{cases} 0, & v < 0, \\ 0.5, & 0 \le v < 1, \\ 1, & v \ge 1. \end{cases} \tag{4.16}$$

Given a random variable $\underline{x}$, any single value x_0 such that $P(\{X = x_0\}) = p > 0$ contributes a step[4] of amplitude p to $F_X(x)$. Then, it is easy to conclude that for a discrete random variable D with $S_D = \{\dots, d_{i-1}, d_i, d_{i+1}, \dots\}$,

$$F_D(d) = \sum_i P(\{D = d_i\})u(d - d_i). \tag{4.17}$$

An even more informative function that can be derived from the CDF to describe the probability distribution of a random variable X is the so-called *probability density function* (shortened to *PDF*),

$$f_X(x) = \frac{\mathrm{d}F_X(x)}{\mathrm{d}x}. \tag{4.18}$$

Since $F_X(x)$ is a nondecreasing function of x, it follows that $f_X(x) \ge 0$. From the definition,

$$F_X(x) = \int_{-\infty}^{x^+} f_X(\tilde{x})\mathrm{d}\tilde{x}. \tag{4.19}$$

[4] Unit step function: $u(x) = \begin{cases} 0, & x < 0, \\ 1, & x \ge 0. \end{cases}$

Then,

$$\int_{-\infty}^{\infty} f_X(x)\mathrm{d}x = 1. \tag{4.20}$$

The PDF corresponding to the CDF shown in Fig. 4.7 is depicted in Fig. 4.8.

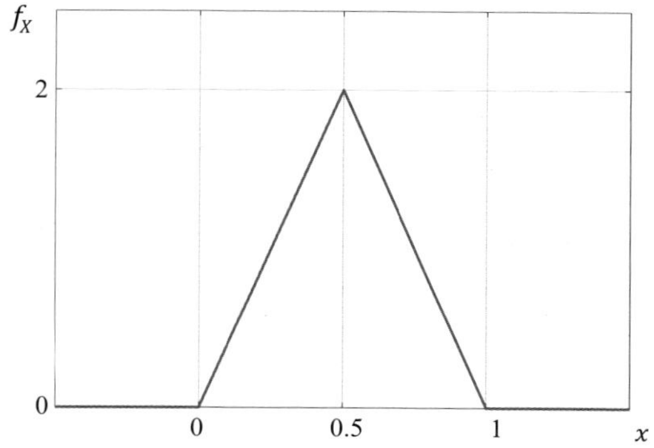

FIGURE 4.8

Example of a probability density function.

One can use the PDF to calculate probabilities by noticing that

$$P(\{x_1 < X \leq x_2\}) = \int_{x_1^+}^{x_2^+} f_X(x)\mathrm{d}x. \tag{4.21}$$

Again, for the random variable X whose distribution is described in Fig. 4.8, one can easily check by inspection that $\left(X \leq \frac{1}{2}\right) = 50\%$.

The PDF of the random variable I_1 associated above with the (0, 100] interval sampling experiment is

$$f_{I_1}(i_1) = \begin{cases} \frac{1}{100}, & 0 < i_1 \leq 100, \\ 0, & \text{otherwise}. \end{cases} \tag{4.22}$$

The PDF of the random variable V associated above with the coin flip experiment is

$$f_V(v) = \begin{cases} \infty, & v = 0 \text{ or } v = 1, \\ 0, & \text{otherwise}, \end{cases} \tag{4.23}$$

which is not well defined. But coherently with what has been seen for the CDF, given a random variable X, any single value x_0 such that $P(\{X = x_0\}) = p > 0$ contributes an impulse[5] of area p to $f_X(x)$. Then,

[5] Unit impulse distribution, or Dirac delta: For $x \in \mathbb{R}$, $\delta(x) = 0\ \forall x \neq 0$ and $\int_{-\infty}^{\infty} \delta(x)\mathrm{d}x = 1$.

it is easy to conclude that for a discrete random variable D with $S_D = \{\ldots, d_{i-1}, d_i, d_{i+1}, \ldots\}$,

$$f_D(d) = \sum_i P(\{D = d_i\})\delta(d - d_i). \tag{4.24}$$

In particular, for the coin flip experiment the PDF is $f_V(v) = 0.5\delta(v) + 0.5\delta(v-1)$.

In the case of discrete random variables, in order to avoid the impulses in the PDF, one can operate directly on the so-called *mass probability function*[6]:

$$P_D(d) = P(\{D = d\}). \tag{4.25}$$

In this case,

$$F_D(d) = \sum_{i \mid d_i \leq d} P_D(d_i). \tag{4.26}$$

This chapter favors an integrate framework for continuous and discrete variables, based on CDFs and PDFs.

It is usual in the literature referring (for short) to the CDF as "the distribution" and to the PDF as "the density" of the random variable. This text avoids this loose terminology, since the word "distribution" better applies to the overall probabilistic behavior of the random variable, both in the form of a CDF or a PDF.

4.3.2 Usual distributions

The simplest continuous distribution is the so-called *uniform* distribution. A random variable X is said to be uniformly distributed between a and $b > a$ if its PDF is

$$f_X(x) = \begin{cases} \dfrac{1}{b-a}, & a \leq x \leq b, \\ 0, & \text{otherwise}, \end{cases} \tag{4.27}$$

also depicted in Fig. 4.9. Notice that the inclusion or not of the interval bounds is unimportant here, since the variable is continuous. The error produced by uniform quantization of real numbers is an example of a uniform random variable.

Perhaps the most recurrent continuous distribution is the so-called *Gaussian* (or *normal*) distribution. A Gaussian random variable X is described by the PDF

$$f_X(x) = \frac{1}{\sqrt{2\pi\sigma_X^2}} \mathrm{e}^{-\dfrac{(x-\overline{X})^2}{2\sigma_X^2}}. \tag{4.28}$$

As seen in Fig. 4.10, this function is symmetrical around $\overline{X}$, with spread controlled by σ_X. These parameters, respectively called statistical mean and standard deviation of X, will be precisely defined

6 This unusual notation is employed here for the sake of compatibility with the other definitions.

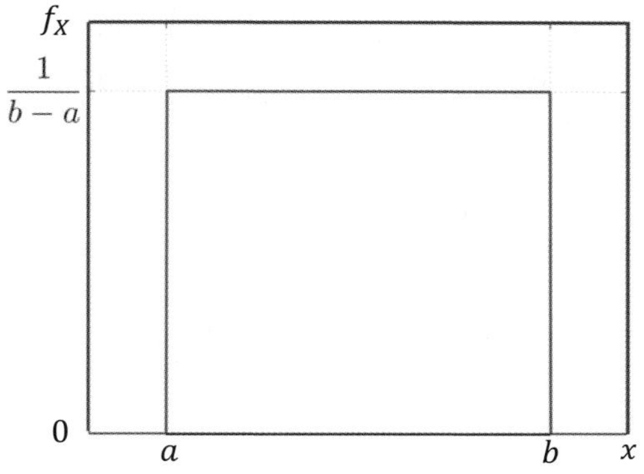

FIGURE 4.9

Uniform probability distribution.

in Section 4.3.4. The Gaussian distribution arises from the combination of several independent random phenomena, and is often associated with noise models.

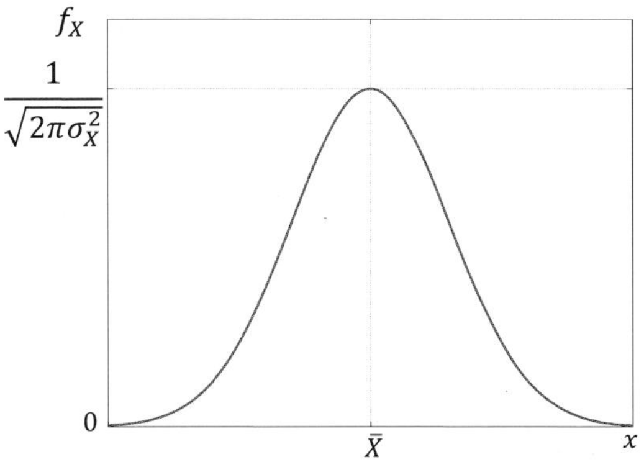

FIGURE 4.10

Gaussian probability distribution.

There is no closed expression for the Gaussian CDF, which is usually tabulated for a normalized Gaussian random variable $\tilde{X}$ with $\overline{\tilde{X}} = 0$ and $\sigma_{\tilde{X}}^2 = 1$ such that

$$f_{\tilde{X}}(\tilde{x}) = \frac{1}{\sqrt{2\pi}} e^{-\frac{\tilde{x}^2}{2}}. \tag{4.29}$$

In order to compute $p = P(\{x_1 < X \leq x_2\})$ for a Gaussian random variable X, one can build an auxiliary variable $\tilde{X} = \dfrac{X - \overline{X}}{\sigma_X^2}$ such that $p = P(\{\tilde{x}_1 < \tilde{X} \leq \tilde{x}_2\}) = F_{\tilde{X}}(\tilde{x}_2) - F_{\tilde{X}}(\tilde{x}_1)$, which can then be approximated by tabulated values of $F_{\tilde{X}_1}(\tilde{x})$.

Example: The values r of a 10% resistor series produced by a component industry can be modeled by a Gaussian random variable R with $\sigma_R = \dfrac{1}{30}\overline{R}$. What is the probability of producing a resistor within $\pm 1\%$ of $\overline{R}$?

Solution: The normalized counterpart of R is $\tilde{R} = \dfrac{R - \overline{R}}{\frac{1}{30}\overline{R}}$. Then,

$$\begin{aligned} P(\{0.99\overline{R} \leq R \leq 1.01\overline{R}\}) &= P(\{-0.3 \leq \tilde{R} \leq 0.3\}) \\ &= F_{\tilde{R}}(0.3) - F_{\tilde{R}}(-0.3) \approx 0.2358. \end{aligned} \tag{4.30}$$

Notice once more that since the variable is continuous, the inclusion or not of the interval bounds has no influence on the result.

In Section 4.2.3, an important discrete random variable was implicitly defined. A random variable X that follows the so-called *binomial* distribution is described by

$$f_X(x) = \sum_{k=0}^{N} \binom{N}{k} p^k (1-p)^{N-k} \delta(x-k), \tag{4.31}$$

and counts the number of occurrences of an event A which has probability p, after N independent runs of a random experiment. Games of chance are related to binomial random variables.

The so-called *Poisson* distribution, described by

$$f_X(x) = \mathrm{e}^{-\lambda T} \sum_{k=0}^{\infty} \frac{(\lambda T)^k}{k!} \delta(x-k), \tag{4.32}$$

counts the number of occurrences of an event A that follows a mean rate of λ occurrences per unit time, during the time interval T. Traffic studies are related to Poisson random variables.

It is simple to derive the Poisson distribution from the binomial distribution. If an event A may occur with no preference anytime during a time interval $(t_0, t_0 + T_0)$, the probability that A occurs within the time interval $[t, t+T] \subset (t_0, t_0 + T_0)$ is $\dfrac{T}{T_0}$. If A occurs N times in $(t_0, t_0 + T_0)$, the probability that it falls exactly k times in $[t, t+T] \subset (t_0, t_0 + T_0)$ is $\binom{N}{k} p^k (1-p)^{N-k}$. If $T_0 \to \infty$, $p \to 0$; if A follows a mean rate of $\lambda = \dfrac{N}{T_0}$ occurrences per unit time, then $N \to \infty$ and the probability of exact k occurrences in a time interval of duration T becomes $\mathrm{e}^{-\lambda T} \sum_{k=0}^{\infty} \dfrac{(\lambda T)^k}{k!}$, where λT substituted for Np. If a central office receives 100 telephone calls per minute in the mean, the probability that more than one call arrives during 1 second is $1 - \mathrm{e}^{-\frac{100}{60}1}\left(1 + \frac{100}{60}1\right) \approx 49.6\%$.

Due to space restrictions, this chapter does not detail other random distributions, which can be easily found in the literature.

4.3.3 Conditional distribution

In many instances, one is interested in studying the behavior of a given random variable X under some constraints. A *conditional distribution* can be built to this effect, if the event $B \subset \mathbb{R}$ summarizes those constraints. The corresponding CDF of X conditioned to B can be computed as

$$F_X(x|B) = P(\{X \leq x|B\}) = \frac{P(\{X \leq x\} \cap B)}{P(B)}. \tag{4.33}$$

The related PDF is simply given by

$$f_X(x|B) = \frac{\mathrm{d}F_X(x|B)}{\mathrm{d}x}. \tag{4.34}$$

Conditional probabilities can be straightforwardly computed from conditional CDFs or PDFs.

When the conditioning event is an interval $B = (a, b]$, it can be easily deduced that

$$F_X(x|B) = \begin{cases} 0, & x \leq a, \\ \dfrac{F_X(x) - F_X(a)}{F_X(b) - F_X(a)}, & a < x \leq b, \\ 1, & x > b \end{cases} \tag{4.35}$$

and

$$f_X(x|B) = \begin{cases} \dfrac{f_X(x)}{\displaystyle\int_a^b f_X(x)\mathrm{d}x}, & a < x \leq b, \\ 0, & \text{otherwise}. \end{cases} \tag{4.36}$$

Both sentences in Eq. (4.36) can be easily interpreted:

- Within $(a, b]$, the conditional PDF has the same shape as the original one; the normalization factor $\int_a^b f_X(x)\mathrm{d}x$ ensures $P((a, b]) = 100\%$ in the restricted sample space $(a, b]$.
- By definition, there is null probability of getting x outside $(a, b]$.

As an example, the random variable H defined by the nonnegative outcomes of a normalized Gaussian variable follows the PDF

$$f_H(h) = \sqrt{\frac{2}{\pi}}\mathrm{e}^{-\frac{h^2}{2}}u[h]. \tag{4.37}$$

The results shown in this section can be promptly generalized to any conditioning event.

4.3.4 Statistical moments

The concept of mean does not need any special introduction: the single value that substituted for each member in a set of numbers produces the same total. In the context of probability, following a frequentist path, one could define the arithmetic mean of infinite samples of a random variable X as its *statistical mean* $\overline{X}$ or its *expected value* $\mathrm{E}[X]$.

Recall the random variable V associated with the fair coin flip experiment, with $P_V(0) = P_V(1) = 0.5$. After infinite repeats of the experiment, one gets 50% of heads ($V = 0$) and 50% of tails ($V = 1$); then, the mean outcome will be[7] $E[V] = 0.5$. If another variable V' is associated with an unfair coin with probabilities $P_{V'}(0) = 0.4$ and $P_{V'}(1) = 0.6$, the same reasoning leads to $\mathrm{E}[V'] = 0.6$. Instead of averaging infinite outcomes, just summing the possible values of the random variable weighted by their respective probabilities also yields its statistical mean. Thus we can state that for any discrete random variable D with sample space $S = \{\ldots, d_{i-1}, d_i, d_{i+1}, \ldots\}$,

$$\mathrm{E}[D] = \sum_i d_i P_D(d_i). \tag{4.38}$$

This result can be generalized. Given a continuous random variable X, the probability of drawing a value in the interval $\mathrm{d}x$ around x_0 is given by $f_X(x_0)\mathrm{d}x$. The weighted sum of every $x \in \mathbb{R}$ is simply

$$\mathrm{E}[X] = \int_{-\infty}^{\infty} x f_X(x)\mathrm{d}x. \tag{4.39}$$

By substituting the PDF of a discrete variable D (see Eq. (4.24)) into this expression, one arrives at Eq. (4.38). Then, Eq. (4.39) is the analytic expression for the expected value of any random variable X.

Suppose another random variable $g(X)$ is built as a function of X. Since the probability of getting the value x_0 is the same as getting the respective $g(x_0)$, we can deduce that

$$\mathrm{E}[g(X)] = \int_{-\infty}^{\infty} g(x) f_X(x)\mathrm{d}x. \tag{4.40}$$

A complete family of measures based on expected values can be associated with a random variable. The so-called nth-order *moment* of X (about the origin) is defined as

$$m_n = \mathrm{E}[X^n],\ n \in \mathbb{Z}. \tag{4.41}$$

The first two moments of X are:

- $m_1 = \overline{X}$, i.e., the *mean* of X, given by the centroid of $f_X(x)$;
- $m_2 = \overline{X^2}$, i.e., the *mean square value* of X.

A modified family of parameters can be formed by computing the moments about the mean. The so-called nth-order *central moment* of X is defined as

$$\mu_n = \mathrm{E}[(X - \overline{X})^n],\ n \in \mathbb{Z}. \tag{4.42}$$

[7] Note that the expected value of a random variable is not necessarily meaningful for the associated experiment.

Subtracting the statistical mean from a random variable can be interpreted as disregarding its "deterministic" part (represented by its statistical mean). We describe three special cases:

- $\mu_2 = \sigma_X^2$, known as the *variance* of X, measures the spread of $f_X(x)$ around $\overline{X}$. The so-called *standard deviation* $\sigma_X = \sqrt{\mu_2}$ is a convenient measure with the same dimension of X.
- μ_3, whose standardized version $\frac{\mu_3}{\sigma_X^3}$ is the so-called *skewness* of X, measures the asymmetry of $f_X(x)$.
- μ_4, whose standardized version[8] minus three $(\frac{\mu_4}{\sigma_X^4} - 3)$ is the so-called *kurtosis* of X, measures the peakedness of $f_X(x)$. One can say the distributions are measured against the Gaussian, which has a null kurtosis.

Of course, analogous definitions apply to the moments computed over conditional distributions.

A useful expression, which will be recalled in the context of random processes, relates three of the measures defined above:

$$\sigma_X^2 = \overline{X^2} - \overline{X}^2. \tag{4.43}$$

Rewritten as

$$\overline{X^2} = \overline{X}^2 + \sigma_X^2, \tag{4.44}$$

it allows to split the overall "intensity" of X, measured by its mean square value, into a deterministic part, represented by $\overline{X}^2$, and a random part, represented by σ_X^2.

As an example, consider a discrete random variable D distributed as shown in Table 4.1. Their respective parameters are:

Table 4.1 Statistical distribution of *D*.

d	p_D
−1	0.2
0	0.5
1	0.3

- mean $\overline{D} = 0.1$, indicating the PDF of D is shifted to the right of the origin;
- mean square value $\overline{D^2} = 1$, which measures the intensity of D;
- variance $\sigma_D^2 = 0.49$, which measures the random part of the intensity of D;
- standard deviation $\sigma_D = 0.7$, which measures the spread of D around its mean;
- skewness $\frac{\mu_3}{\sigma_D^3} \approx -0.14$, indicating the PDF of D is left-tailed;
- kurtosis $\frac{\mu_4}{\sigma_D^4} - 3 \approx -0.96$, indicating the PDF of D is less peaky than the PDF of a Gaussian variable.

[8] The classical definition does not subtract three.

At this point, we can discuss two simple transformations of a random variable X whose effects can be summarized by low-order moments:

- A random variable $Y = X + x_0$ can be formed by adding a fixed offset x_0 to each sample of X. As a consequence of this operation, the new PDF is a shifted version of the original one: $f_Y(y) = f_X(y - x_0)$, thus adding x_0 to the mean of X: $\overline{Y} = \overline{X} + x_0$.
- A random variable $Y = \alpha X$ can be formed by scaling by α each sample of X. As a consequence of this operation, the new PDF is a scaled version of the original one: $f_Y(y) = \frac{1}{|\alpha|} f_X\left(\frac{y}{\alpha}\right)$, thus scaling by α the standard deviation of X: $\sigma_Y = \alpha\sigma_X$.

Such transformations do not change the shape (and therefore the type) of the original distribution. In particular, one can generate:

- a zero-mean version of X by making $Y = X - \overline{X}$, which disregards the deterministic part of X;
- a unit-standard deviation version of X by making $Y = \frac{X}{\sigma_X}$, which enforces a standard statistical variability to X;
- a normalized version of X by making $Y = \frac{X - \overline{X}}{\sigma_X}$, which combines both effects.

We already defined a normalized Gaussian distribution in Section 4.3.2. The normalized version $\tilde{D}$ of the random variable D of the last example would be distributed as shown in Table 4.2.

Table 4.2 Statistical distribution of $\tilde{D}$.

$\tilde{d}$	$P_{\tilde{D}}$
−11/7	0.2
−1/7	0.5
9/7	0.3

These expectation-based parameters are mainly important because they provide a partial description of an underlying distribution without the need to resource to the PDF. In a practical situation in which only a few samples of a random variable are available, as opposed to its PDF and related moments, it is easier to get more reliable estimates for the latter (especially low-order ones) than for the PDF itself. The same rationale applies to the use of certain auxiliary inequalities that avoid the direct computation of probabilities on a random variable (which otherwise would require the knowledge or estimation of its PDF) by providing upper bounds for them based on low-order moments. Two such inequalities are:

- Markov's inequality for a nonnegative random variable X:
  $P(\{X \geq a\}) \leq \frac{\overline{X}}{a}, \quad \forall a > 0;$
- Chebyshev's inequality for a random variable X:
  $P(\{|X - \overline{X}| \geq a\}) \leq \frac{\sigma_X^2}{a^2}, \quad \forall a > 0.$

The derivation of Markov's inequality is quite simple:

$$\begin{aligned} P\{X \geq a\} &= \int_a^{\infty} f_X(x)\mathrm{d}x = \int_{-\infty}^{\infty} f_X(x)u(x-a)\mathrm{d}x \\ &\leq \int_{-\infty}^{\infty} \frac{x}{a} f_X(x)u(x-a)\mathrm{d}x \leq \int_{-\infty}^{\infty} \frac{x}{a} f_X(x)\mathrm{d}x = \frac{\mathrm{E}[X]}{a}. \end{aligned} \tag{4.45}$$

Chebyshev's inequality follows directly by substituting $(X-\overline{X})^2 \geq a^2$ for $X \geq a$ in Markov's inequality. As an example, the probability of getting a sample $g > 3$ from the normalized Gaussian random variable G is $P(\{G > 3\}) \approx 0.0013$; Chebyshev's inequality predicts $P(\{G > 3\}) \leq \frac{1}{9}$, an upper bound almost 100 times greater than the actual probability.

Two representations of the distribution of a random variable in alternative domains provide interesting links with its statistical moments. The so-called *characteristic function* of a random variable X is defined as

$$\Phi_X(\omega) = \mathrm{E}[\mathrm{e}^{\mathrm{j}\omega X}], \ \omega \in \mathbb{R}, \tag{4.46}$$

and interrelates the moments m_n by successive differentiations:

$$m_n = (-\mathrm{j})^n \left. \frac{\mathrm{d}^n \Phi_X(\omega)}{\mathrm{d}\omega^n} \right|_{\omega=0}. \tag{4.47}$$

The so-called *moment-generating function* of a random variable X is defined as

$$M_X(\nu) = \mathrm{E}[\mathrm{e}^{\nu X}], \ \nu \in \mathbb{R}, \tag{4.48}$$

and provides an even more direct way to compute m_n:

$$m_n = \left. \frac{\mathrm{d}^n M_X(\nu)}{\mathrm{d}\nu^n} \right|_{\nu=0}. \tag{4.49}$$

4.3.5 Transformation of random variables

Consider the problem of describing a random variable $Y = g(X)$ obtained by transformation of another random variable X, as illustrated in Fig. 4.11. By definition,

$$F_Y(y) = P(\{Y \leq y\}) = P(\{g(X) \leq y\}). \tag{4.50}$$

If X is a continuous random variable and $g(\cdot)$ is a differentiable function, the sentence $g(X) \leq y$ is equivalent to a set of sentences in the form $h_1(y) < x \leq h_2(y)$. Since $P(\{h_1(y) < x \leq h_2(y)\}) = F_X(h_2(y)) - F_X(h_1(y))$, $F_Y(y)$ can be expressed as a function of $F_X(x)$.

Fortunately, an intuitive formula expresses $f_Y(y)$ as a function of $f_X(x)$:

$$f_Y(y) = \sum_n \frac{f_X(x_n)}{\left| \frac{\mathrm{d}g(x)}{\mathrm{d}x} \right| \Big|_{x=x_n}}, \ \text{with } y = g(x_n). \tag{4.51}$$

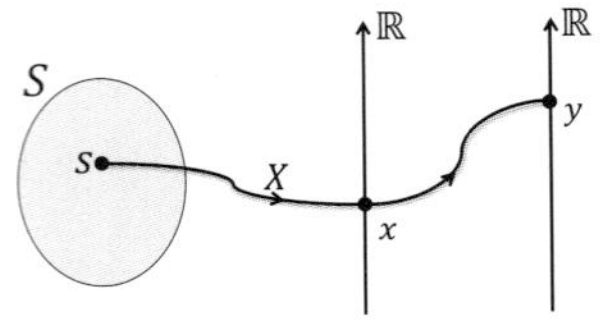

FIGURE 4.11

Random variable X mapped into random variable Y.

Given that the transformation $g(\cdot)$ is not necessarily monotonic, x_n are all possible values mapped to y; therefore, their contributions must be summed up. It is reasonable that $f_Y(y)$ must be directly proportional to $f_X(x)$: the more frequent a given x_0, the more frequent its respective $y_0 = g(x_0)$. In turn, the term $\left|\frac{\mathrm{d}g(x)}{\mathrm{d}x}\right|$ accounts for the distortion imposed to $f_X(x)$ by $g(x)$. For example, if this transformation is almost constant in a given region, both if increasing or decreasing, then $\left|\frac{\mathrm{d}g(x)}{\mathrm{d}x}\right|$ in the denominator will be close to zero; this just reflects the fact that a wide range of X values will be mapped into a narrow range of values of Y, which will then become denser than X in that region.

As an example, consider that a continuous random variable X is transformed into a new variable $Y = X^2$. For the CDF of Y,

$$Y \le y \Rightarrow X^2 \le y \Rightarrow -\sqrt{y} \le X \le \sqrt{y} \Rightarrow F_Y(y) = F_X(\sqrt{y}) - F_X(-\sqrt{y}). \tag{4.52}$$

By Eq. (4.51), or by differentiation of Eq. (4.52), one arrives at the PDF of Y:

$$f_Y(y) = \frac{f_X(\sqrt{y}) + f_X(-\sqrt{y})}{\sqrt{y}}. \tag{4.53}$$

The case when real intervals of X are mapped into single values of Y requires an additional care to handle the nonzero individual probabilities of the resulting values of the new variable. However, the case of a discrete random variable X with $S = \{\ldots, x_{i-1}, x_i, x_{i+1}, \ldots\}$ being transformed is trivial:

$$f_Y(y) = \sum_i P_Y(y_i)\delta(y - y_i), \text{ with } y_i = g(x_{i_n}) \text{ and } P_Y(y_i) = \sum_n P_X(x_{i_n}). \tag{4.54}$$

Again, x_{i_n} are all possible values mapped to y_i.

An interesting application of transformations of random variables is to obtain samples y of a given random variable Y from samples x of another random variable X, both with known distributions. Assume that $Y = g(X)$ exists such that $F_Y(y = g(x)) = F_X(x)$. Then, $y = F_Y^{-1} F_X(x)$, which requires only the invertibility of $F_Y(y)$.

4.3.6 Multiple random variable distributions

A single random experiment E (composed or not) may give rise to a set of N random variables if each individual outcome $s \in S$ is mapped into several real values $x_1, x_2, \ldots, x_N$. Consider, for example, the random experiment of sampling the climate conditions at every point on the globe; the daily mean temperature T and relative air humidity H can be considered as two random variables that serve as numerical summarizing measures. One must generalize the probability distribution descriptions to cope with this multiple random variable situation: it should be clear that the set of N individual probability distribution descriptions, one for each distinct random variable, provides less information than their joint probability distribution description, since the former cannot convey information about their mutual influences.

We start by defining a *multiple* or *vector random variable* as the function $\mathbf{X}$ that maps $s \in S$ into $\mathbf{x} \in \mathbb{R}^N$, such that

$$\mathbf{X} = \begin{bmatrix} X_1 \\ X_2 \\ \vdots \\ X_N \end{bmatrix} \tag{4.55}$$

and

$$\mathbf{x} = \begin{bmatrix} x_1 \\ x_2 \\ \vdots \\ x_N \end{bmatrix}. \tag{4.56}$$

Note that $x_1, x_2, \ldots, x_N$ are jointly sampled from $X_1, X_2, \ldots, X_N$.

The *joint cumulative probability distribution function* (*joint CDF*) of $X_1, X_2, \ldots, X_N$, or simply the CDF of $\boldsymbol{X}$, can be defined as

$$\begin{aligned} F_{\mathbf{X}}(\mathbf{x}) &= F_{X_1, X_2, \ldots, X_N}(x_1, x_2, \ldots, x_N) \\ &= P(\{X_1 \le x_1\} \cap \{X_2 \le x_2\} \cap \cdots \cap \{X_N \le x_N\}). \end{aligned} \tag{4.57}$$

The following relevant properties of the joint CDF can be easily deduced:

- $F_{\mathbf{X}}(\ldots, x_{n-1}, -\infty, x_{n+1}, \ldots) = 0$, since $P(\{X_n \le -\infty\}) = 0$ for any $1 \le n \le N$;
- $F_{\mathbf{X}}(\infty, \infty, \ldots, \infty) = 1$, since this condition encompasses the complete S;
- $0 \le F_{\mathbf{X}}(\mathbf{x}) \le 1$ and $F_{\mathbf{X}}(\mathbf{x})$ is a nondecreasing function of x_n, $1 \le n \le N$, by construction.

Define $\mathbf{X}_{\mathrm{i}}$ containing N_{i} variables of interest among the random variables in $\mathbf{X}$ and leave the remaining $N - N_{\mathrm{i}}$ *nuisance* variables in $\mathbf{X}_{\mathrm{n}}$. The so-called *marginal CDF* of $X_{\mathrm{i}_1}, X_{\mathrm{i}_2}, \ldots, X_{\mathrm{i}_{N_\mathrm{i}}}$ separately describes these variables' distribution from the knowledge of the CDF of $\mathbf{X}$:

$$F_{\mathbf{X}_\mathrm{i}}(\mathbf{x}_\mathrm{i}) = F_{\mathbf{X}}(\mathbf{x})|_{x_{\mathrm{n}_1} = x_{\mathrm{n}_2} = \cdots = x_{\mathrm{n}_{N-N_\mathrm{i}}} = \infty}. \tag{4.58}$$

One says the variables $X_{\mathrm{n}_1}, X_{\mathrm{n}_2}, \ldots, X_{\mathrm{n}_{N-N_\mathrm{i}}}$ have been *marginalized* out: the condition $X_{\mathrm{n}_m} \le \infty$ simply means they must be in the sample space, thus one does not need to care about them anymore.

The CDF of a discrete random variable $\mathbf{D}$ with $S_\mathbf{D} = \{\ldots, \mathbf{d}_{i-1}, \mathbf{d}_i, \mathbf{d}_{i+1}, \ldots\}$, analogously to the single-variable case, is composed of steps at the admissible N-tuples:

$$F_\mathbf{D}(\mathbf{d}) = \sum_i P_\mathbf{D}(\mathbf{d}_i)u(d_1 - d_{1i})u(d_2 - d_{2i})\ldots u(d_N - d_{Ni}). \tag{4.59}$$

The *joint probability density function* of $X_1, X_2, \ldots, X_N$, or simply the PDF of $\mathbf{X}$, can be defined as

$$f_\mathbf{X}(\mathbf{x}) = f_{X_1,X_2,\ldots,X_N}(x_1, x_2, \ldots, x_N) = \frac{\partial^N F_\mathbf{X}(\mathbf{x})}{\partial x_1 \partial x_2 \ldots \partial x_N}. \tag{4.60}$$

Since $F_\mathbf{X}(\mathbf{x})$ is a nondecreasing function of x_n, $1 \le n \le N$, it follows that $f_\mathbf{X}(\mathbf{x}) \ge 0$. From the definition,

$$F_\mathbf{X}(\mathbf{x}) = \int_{-\infty}^{x_N} \cdots \int_{-\infty}^{x_2} \int_{-\infty}^{x_1} f_\mathbf{X}(\tilde{\mathbf{x}}) \mathrm{d}\tilde{x}_1 \mathrm{d}\tilde{x}_2 \cdots \mathrm{d}\tilde{x}_N. \tag{4.61}$$

Then,

$$\int_{-\infty}^{\infty} \cdots \int_{-\infty}^{\infty} \int_{-\infty}^{\infty} f_\mathbf{X}(\mathbf{x}) \mathrm{d}x_1 \mathrm{d}x_2 \cdots \mathrm{d}x_N = 1. \tag{4.62}$$

The probability of any event $B \in \mathbb{R}^N$ can be computed as

$$P(B) = \int_{\mathbf{x} \in B} f_\mathbf{X}(\mathbf{x}) \mathrm{d}x_1 \mathrm{d}x_2 \cdots \mathrm{d}x_N. \tag{4.63}$$

Once more we can marginalize the nuisance variables $X_{\mathrm{n}_1}, X_{\mathrm{n}_2}, \ldots, X_{\mathrm{n}_{N-N_\mathrm{i}}}$ to obtain the *marginal PDF* of $\mathbf{X}_\mathrm{i}$ from the PDF of $\mathbf{X}$:

$$f_{\mathbf{X}_\mathrm{i}}(\mathbf{x}_\mathrm{i}) = \int_{-\infty}^{\infty} \cdots \int_{-\infty}^{\infty} \int_{-\infty}^{\infty} f_\mathbf{X}(\mathbf{x}) \mathrm{d}x_{\mathrm{n}_1} \mathrm{d}x_{\mathrm{n}_2} \cdots \mathrm{d}x_{\mathrm{n}_{N-N_\mathrm{i}}}. \tag{4.64}$$

Note that if $\mathbf{X}_\mathrm{i}$ and $\mathbf{X}_\mathrm{n}$ are statistically dependent, the marginalization of $\mathbf{X}_\mathrm{n}$ does not "eliminate" (only "hides") the effect of $\mathbf{X}_\mathrm{n}$ on $\mathbf{X}_\mathrm{i}$: the integration is performed over $\mathbf{X}$, which describes their mutual influences.

For a discrete random variable $\mathbf{D}$ with $S_\mathbf{D} = \{\ldots, \mathbf{d}_{i-1}, \mathbf{d}_i, \mathbf{d}_{i+1}, \ldots\}$, the PDF consists of impulses at the admissible N-tuples:

$$f_\mathbf{D}(\mathbf{d}) = \sum_i P_\mathbf{D}(\mathbf{d}_i)\delta(d_1 - d_{1i})\delta(d_2 - d_{2i})\ldots\delta(d_N - d_{Ni}). \tag{4.65}$$

Suppose, for example, that we want to find the joint and marginal CDFs and PDFs of two random variables X and Y jointly uniformly distributed in the region $0 \le y \le x \le 1$, such that

$$f_{X,Y}(x, y) = \begin{cases} 2, & 0 \le y \le x \le 1, \\ 0, & \text{otherwise}. \end{cases} \tag{4.66}$$

The marginalization of Y yields

$$f_X(x) = \int_{-\infty}^{\infty} f_{X,Y}(x,y)\mathrm{d}y = \int_0^x 2\mathrm{d}y = \begin{cases} 2x, & 0 \le x \le 1, \\ 0, & \text{otherwise.} \end{cases} \tag{4.67}$$

The marginalization of X yields

$$f_Y(y) = \int_{-\infty}^{\infty} f_{X,Y}(x,y)\mathrm{d}x = \int_y^1 2\mathrm{d}x = \begin{cases} 2(1-y), & 0 \le y \le 1, \\ 0, & \text{otherwise.} \end{cases} \tag{4.68}$$

By definition,

$$F_{X,Y}(x,y) = \int_{-\infty}^{y}\int_{-\infty}^{x} f_{X,Y}(\tilde{x},\tilde{y})\mathrm{d}\tilde{x}\mathrm{d}\tilde{y} = \begin{cases} 0, & x < 0 \text{ or } y < 0, \\ y(2x-y), & 0 \le y \le x \le 1, \\ x^2, & 0 \le x \le 1 \text{ and } y > x, \\ y(2-y), & 0 \le y \le 1 \text{ and } x > 1, \\ 1, & x > 1 \text{ and } y > 1. \end{cases} \tag{4.69}$$

The reader is invited to sketch the admissible region $0 \le y \le x \le 1$ on the (x, y)-plane and try to solve Eq. (4.69) by visual inspection. The marginal CDF of X is given by

$$F_X(x) = F_{X,Y}(x,\infty) = \begin{cases} 0, & x < 0, \\ x^2, & 0 \le x \le 1, \\ 1, & x > 1, \end{cases} \tag{4.70}$$

which could also have been obtained by direct integration of $f_X(x)$. The marginal CDF of Y is given by

$$F_Y(y) = F_{X,Y}(\infty,y) = \begin{cases} 0, & y < 0, \\ y(2-y), & 0 \le y \le 1, \\ 1, & y > 1, \end{cases} \tag{4.71}$$

which could also have been obtained by direct integration of $f_Y(y)$.

Similarly to the univariate case discussed in Section 4.3.3, the *conditional distribution* of a vector random variable $\mathbf{X}$ restricted by a conditioning event $B \in \mathbb{R}^N$ can be described by its respective CDF

$$F_{\mathbf{X}}(\mathbf{x}|B) = \frac{P(\{X_1 \le x_1\} \cap \{X_2 \le x_2\} \cap \cdots \cap \{X_N \le x_N\} \cap B)}{P(B)} \tag{4.72}$$

and PDF

$$f_{\mathbf{X}}(\mathbf{x}|B) = \frac{\partial^N F_{\mathbf{X}}(\mathbf{x}|B)}{\partial x_1 \partial x_2 \cdots \partial x_N}. \tag{4.73}$$

A special case arises when B imposes a point conditioning: Define $\mathbf{X}_B$ containing N_{B} variables among those in $\mathbf{X}$, such that $B = \{\mathbf{X}_B = \mathbf{x}_B\}$. It can be shown that

$$f_{\mathbf{X}}(\mathbf{x}|B) = \frac{f_{\mathbf{X}}(\mathbf{x})}{f_{\mathbf{X}_B}(\mathbf{x}_B)}, \tag{4.74}$$

an intuitive result in light of Eq. (4.6), to which one may directly refer in the case of discrete random variables. Returning to the last example, the point-conditioned PDFs can be shown to be

$$f_X(x|Y=y) = \begin{cases} \dfrac{1}{1-y}, & 0 \leq y \leq x \leq 1, \\ 0, & \text{otherwise} \end{cases} \tag{4.75}$$

and

$$f_Y(y|X=x) = \begin{cases} \dfrac{1}{x}, & 0 \leq y \leq x \leq 1, \\ 0, & \text{otherwise.} \end{cases} \tag{4.76}$$

4.3.7 Statistically independent random variables

Based on the concept of statistical independence, discussed in Section 4.2.2, the mutual *statistical independence* of a set of random variables means that the probabilistic behavior of each one is not affected by the probabilistic behavior of the others. Formally, the random variables $X_1, X_2, \ldots, X_N$ are independent when any of the following conditions is fulfilled:

- $F_{\mathbf{X}}(\mathbf{x}) = \prod_{n=1}^{N} F_{X_n}(x_n)$,
- $f_{\mathbf{X}}(\mathbf{x}) = \prod_{n=1}^{N} f_{X_n}(x_n)$,
- $F_{X_n}(x_n|\text{any condition on the remaining variables}) = F_{X_n}(x_n), \forall n = 1, \ldots, N$, or
- $f_{X_n}(x_n|\text{any condition on the remaining variables}) = f_{X_n}(x_n), \forall n = 1, \ldots, N$.

Returning to the example developed in Section 4.3.6, x and y are clearly statistically dependent. However, if they were jointly uniformly distributed in the region defined by $0 \leq x \leq 1,\ 0 \leq y \leq 1$, they would be statistically independent – this verification is left to the reader.

The PDF of a random variable $Y = \sum_{n=1}^{N} X_n$ composed of the sum of N independent variables X_n can be computed as[9]

$$f_Y(y) = (f_{X_1} * f_{X_2} * \cdots * f_{X_N})(y). \tag{4.77}$$

4.3.8 Joint statistical moments

The definition of *statistical mean* can be directly extended to a vector random variable $\mathbf{X}$ with known PDF. The *expected value* of a real scalar function $g(\mathbf{X})$ is given by

---

[9] Convolution integral: $(a * b)(y) = \int_{-\infty}^{\infty} a(\tilde{y})b(y - \tilde{y})\mathrm{d}\tilde{y}$.

$$\mathrm{E}[g(\mathbf{X})] = \int_{-\infty}^{\infty} \cdots \int_{-\infty}^{\infty} \int_{-\infty}^{\infty} g(\mathbf{x}) f_{\mathbf{X}}(\mathbf{x}) \mathrm{d}x_1 \mathrm{d}x_2 \cdots \mathrm{d}x_N. \tag{4.78}$$

We can analogously proceed to the definition of N-variable joint statistical moments. However, due to the more specific usefulness of the $N > 2$ cases, only 2-variable moments will be presented.

An $(n+k)$th-order *joint moment* of X and Y (about the origin) has the general form

$$m_{nk} = \mathrm{E}[X^n Y^k], \ n, k \in \mathbb{Z}. \tag{4.79}$$

The most important case corresponds to $m = n = 1$: $m_{11} = R_{XY}$ is called the *statistical correlation* between X and Y. From a frequentist perspective, R_{XY} is the average of infinite products of samples x and y jointly drawn from $f_{X,Y}(x, y)$, which links the correlation to an inner product between the random variables. Indeed, when $R_{XY} = 0$, X and Y are called *orthogonal*. One can say the correlation quantifies the overall linear relationship between the two variables. Moreover, when $R_{XY} = \overline{X}\,\overline{Y}$, X and Y are called mutually *uncorrelated*; this "separability" in the mean is weaker than the distribution separability implied by the independence. In fact, if X and Y are independent,

$$\begin{aligned} R_{XY} &= \int_{-\infty}^{\infty} \int_{-\infty}^{\infty} xy f_{X,Y}(x, y) \mathrm{d}x \mathrm{d}y = \int_{-\infty}^{\infty} \int_{-\infty}^{\infty} xy f_X(x) f_Y(y) \mathrm{d}x \mathrm{d}y \\ &= \int_{-\infty}^{\infty} x f_X(x) \mathrm{d}x \int_{-\infty}^{\infty} y f_Y(y) \mathrm{d}y = \overline{X}\,\overline{Y}, \end{aligned} \tag{4.80}$$

i.e., they are also uncorrelated. The converse is not necessarily true.

An $(n+k)$th-order *joint central moment* of X and Y, computed about the mean, has the general form

$$\mu_{nk} = \mathrm{E}[(X - \overline{X})^n (Y - \overline{Y})^k)], \ n, k \in \mathbb{Z}. \tag{4.81}$$

Once more, the case $m = n = 1$ is specially important: $\mu_{11} = C_{XY}$ is called the *statistical covariance* between X and Y, which quantifies the linear relationship between their random parts. Since[10] $C_{XY} = R_{XY} - \overline{X}\,\overline{Y}$, X and Y are uncorrelated when $C_{XY} = 0$. If $C_{XY} > 0$, one says X and Y are positively correlated, i.e., the variations of their statistical samples tend to occur in the same direction; if $C_{XY} < 0$, one says X and Y are negatively correlated, i.e., the variations of their statistical samples tend to occur in opposite directions. For example, the age and the annual medical expenses of an individual are expected to be positively correlated random variables. A normalized covariance can be computed as the correlation between the normalized versions of X and Y:

$$\rho_{XY} = \frac{C_{XY}}{\sigma_X \sigma_Y} = \mathrm{E}\left[\frac{X - \overline{X}}{\sigma_X} \cdot \frac{Y - \overline{Y}}{\sigma_Y}\right], \tag{4.82}$$

known as the *correlation coefficient* between X and Y, where $-1 \leq \rho_{XY} \leq 1$. It can be interpreted as the percentage of correlation between x and y. Recalling the inner product interpretation, the correlation coefficient can be seen as the cosine of the angle between the statistical variations of the two random variables.

[10] Eq. (4.43) can be seen as the next expression computed for $X = Y$, since $R_{XX} = \overline{X^2}$ and $C_{XX} = \sigma_X^2$.

Consider, for example, the discrete variables D_1 and D_2 jointly distributed as shown in Table 4.3. Their joint second-order parameters are:

Table 4.3 Statistical distribution of *D*.

d_1	d_2	P_{D_1,D_2}
−1	1	0.3
1	−1	0.6
1	1	0.1

- correlation $R_{XY} = -0.80$, indicating D_1 and D_2 are not orthogonal;
- covariance $C_{XY} = -0.72$, indicating D_1 and D_2 tend to evolve in opposite directions;
- correlation coefficient $\rho_{XY} \approx -0.80$, indicating this negative correlation is relatively strong.

Returning to N-dimensional random variables, the *characteristic function* of a vector random variable $\mathbf{X}$ is defined as

$$\Phi_{\mathbf{X}}(\omega_1, \omega_2, \ldots, \omega_N) = \mathrm{E}[\mathrm{e}^{\mathrm{j}(\omega_1 X_1 + \omega_2 X_2 + \cdots \omega_N X_N)}], \ \omega_1, \omega_2, \ldots, \omega_N \in \mathbb{R}, \tag{4.83}$$

and interrelates the moments of order $m_{n_1,n_2,\ldots,n_N}$ by

$$m_{n_1,n_2,\ldots,n_N} = (-\mathrm{j})^{n_1+n_2+\cdots+n_N} \times \left. \frac{\partial^{n_1+n_2+\cdots+n_N} \Phi_{\mathbf{X}}(\omega_1, \omega_2, \ldots, \omega_N)}{\partial \omega_1^{n_1} \omega_2^{n_2} \cdots \partial \omega_N^{n_N}} \right|_{\omega_1=\omega_2=\cdots=\omega_N=0}. \tag{4.84}$$

The *moment-generating function* of a vector random variable $\mathbf{X}$ is defined as

$$M_{\mathbf{X}}(\nu_1, \nu_2, \ldots, \nu_N) = \mathrm{E}[\mathrm{e}^{\nu_1 X_1 + \nu_2 X_2 + \cdots + \nu_N X_N}], \ \nu_1, \nu_2, \ldots, \nu_N \in \mathbb{R}, \tag{4.85}$$

and allows the computation of $m_{n_1,n_2,\ldots,n_N}$ as

$$m_{n_1,n_2,\ldots,n_N} = \left. \frac{\partial^{n_1+n_2+\cdots+n_N} M_{\mathbf{X}}(\nu_1, \nu_2, \ldots, \nu_N)}{\partial \nu_1^{n_1} \nu_2^{n_2} \cdots \partial \nu_N^{n_N}} \right|_{\nu_1=\nu_2=\cdots=\nu_N=0}. \tag{4.86}$$

4.3.9 Central limit theorem

A result that partially justifies the ubiquitous use of Gaussian models, the *central limit theorem* (CLT), states that (under mild conditions in practice) the distribution of a sum $Y = \sum_{n=1}^{N} X_n$ of N independent variables X_n approaches a Gaussian distribution as $N \to \infty$. Having completely omitted the proof of the CLT, we can at least recall that in this case, $f_Y(y) = \lim_{N\to\infty} f_Y(y) = (f_{X_1} * f_{X_2} * \cdots * f_{X_N})(y)$ (see Eq. (4.77)). Of course, $\overline{Y} = \sum_{n=1}^{N} \overline{X_n}$, and by the independence property, $\sigma_Y^2 = \sum_{n=1}^{N} \sigma_{X_n}^2$.

Gaussian approximations for finite N models are useful for sufficiently high N, but due to the shape of the Gaussian distribution, the approximation error grows as y moves away from $\overline{Y}$.

The reader is invited to verify the validity of the approximation provided by the CLT for successive sums of independent X_n uniformly distributed in [0, 1].

Interestingly, the CLT also applies to the sum of discrete distributions: even if $f_Y(y)$ remains impulsive, the shape of $F_Y(y)$ approaches the shape of the CDF of a Gaussian random variable as N grows.

4.3.10 Multivariate Gaussian distribution

The PDF of an N-dimensional Gaussian variable is defined as

$$f_{\mathbf{X}}(\mathbf{x}) = \frac{1}{\sqrt{(2\pi)^N \det(\boldsymbol{C}_{\mathbf{X}})}} \mathrm{e}^{-\frac{(\mathbf{x}-\overline{\mathbf{X}})^{\mathrm{T}} \boldsymbol{C}_{\mathbf{X}}^{-1} (\mathbf{x}-\overline{\mathbf{X}})}{2}}, \quad (4.87)$$

where $\overline{\mathbf{X}}$ is an $N \times 1$ vector with elements $\{\overline{\mathbf{X}}\}_n = \overline{X_n},\ 1 \le n \le N$, and $\boldsymbol{C}_{\mathbf{X}}$ is an $N \times N$ matrix with elements $\{\boldsymbol{C}_{\mathbf{X}}\}_{mn} = C_{X_m X_n},\ 1 \le m,\ n \le N$, such that any related marginal distribution remains Gaussian. As a consequence, by Eq. (4.74), any conditional distribution $f_{\mathbf{X}}(\mathbf{x}|B)$ with point conditioning B is also Gaussian.

By this definition, we can see that the Gaussian distribution is completely defined by its first- and second-order moments, which means that if N jointly distributed random variables are known to be Gaussian, estimating their *mean vector* $\overline{\mathbf{X}}$ and their *covariance matrix* $\boldsymbol{C}_{\mathbf{X}}$ is equivalent to estimating their overall PDF. Moreover, if X_m and X_n are mutually uncorrelated, $\forall m \neq n$, then $\boldsymbol{C}_{\mathbf{X}}$ is diagonal, and the joint PDF becomes $f_{\mathbf{X}}(\mathbf{x}) = \prod_{n=1}^{N} f_{X_n}(x_n)$, i.e., X_n will be mutually independent. This is a strong result inherent to Gaussian distributions.

4.3.11 Transformation of vector random variables

The treatment of a general multiple random variable $\mathbf{Y} = \mathbf{T}(\mathbf{X})$ resulting from the application of the transformation $\mathbf{T}(\cdot)$ to the multiple variable $\mathbf{X}$ always starts by enforcing $F_{\mathbf{X}}(\mathbf{x}) = F_{\mathbf{Y}}(\mathbf{y})$, with $\mathbf{y} = \mathbf{T}(\mathbf{x})$.

The special case when $\mathbf{T}$ is invertible, i.e., $\mathbf{X} = \mathbf{T}^{-1}(\mathbf{Y})$ (or $X_n = {T_n}^{-1}(\mathbf{Y}),\ 1 \le n \le N$), follows a closed expression:

$$f_{\mathbf{Y}}(\mathbf{y}) = f_{\mathbf{X}}(\mathbf{x} = \mathbf{T}^{-1}(\mathbf{y}))|J|, \quad (4.88)$$

where

$$J = \det\left(\begin{bmatrix} \frac{\partial T_1^{-1}(\mathbf{y})}{\partial y_1} & \frac{\partial T_1^{-1}(\mathbf{y})}{\partial y_2} & \cdots & \frac{\partial T_1^{-1}(\mathbf{y})}{\partial y_N} \\ \frac{\partial T_2^{-1}(\mathbf{y})}{\partial y_1} & \frac{\partial T_2^{-1}(\mathbf{y})}{\partial y_2} & \cdots & \frac{\partial T_2^{-1}(\mathbf{y})}{\partial y_N} \\ \vdots & \vdots & \ddots & \vdots \\ \frac{\partial T_N^{-1}(\mathbf{y})}{\partial y_1} & \frac{\partial T_N^{-1}(\mathbf{y})}{\partial y_2} & \cdots & \frac{\partial T_N^{-1}(\mathbf{y})}{\partial y_N} \end{bmatrix}\right). \quad (4.89)$$

The reader should notice that this expression with $N = 1$ reduces to Eq. (4.51) particularized to the invertible case.

Given the multiple random variable $\mathbf{X}$ with known mean vector $\overline{\mathbf{X}}$ and covariance matrix $\boldsymbol{C}_{\mathbf{X}}$, its linear transformation $\mathbf{Y} = \boldsymbol{T}\mathbf{X}$ will have:

- a mean vector $\overline{\mathbf{Y}} = \boldsymbol{T}\overline{\mathbf{X}}$,
- a covariance matrix $\boldsymbol{C}_{\mathbf{Y}} = \boldsymbol{T}\boldsymbol{C}_{\mathbf{X}}\boldsymbol{T}^{\mathrm{T}}$.

It is possible to show that the linear transformation $\mathbf{Y}$ of an N-dimensional Gaussian random variable $\mathbf{X}$ is also Gaussian. Therefore, in this case the two expressions above completely determine the PDF of the transformed variable.

The generation of samples that follow a desired multivariate probabilistic distribution from samples that follow another known multivariate probabilistic distribution applies the same reasoning followed in the univariate case to the multivariate framework. The reader is invited to show that given the pair x_1, x_2 of samples from the random variables X_1, X_2, jointly uniform in the region $0 \leq x \leq 1,\ 0 \leq y \leq 1$, it is possible to generate the pair y_1, y_2 of samples from the random variables Y_1, Y_2, jointly Gaussian with the desired parameters $\overline{Y_1},\ \sigma^2_{Y_1},\ \overline{Y_2},\ \sigma^2_{Y_2}$, and $\rho_{Y_1Y_2}$, by making

$$\begin{cases} y_1 = \overline{Y_1} + \sigma_{Y_1}\sqrt{-2\ln x_1}\cos(2\pi x_2), \\ y_2 = \overline{Y_2} + \rho_{Y_1Y_2}\sigma_{Y_2}\sqrt{-2\ln x_1}\cos(2\pi x_2) + \sigma_{Y_2}\sqrt{1-\rho^2_{Y_1Y_2}}\sqrt{-2\ln x_1}\sin(2\pi x_2). \end{cases} \tag{4.90}$$

4.3.12 Complex random variables

At first sight, defining a complex random variable Z may seem impossible, since the seminal event $\{Z \leq z\}$ makes no sense. However, in the vector random variable framework, this issue can be circumvented if one jointly describes the real and imaginary parts (or magnitude and phase) of Z.

The single complex random variable $Z = X + \mathrm{j}Y,\ \ x, y \in \mathbb{R}$, is completely represented by $f_{X,Y}(x, y)$, which allows one to compute the expected value of a scalar function $g(Z)$ as $\mathrm{E}[g(Z)] = \int_{-\infty}^{\infty}\int_{-\infty}^{\infty} g(z) f_{X,Y}(x, y)\mathrm{d}x\mathrm{d}y$. Moreover, we can devise general definitions for the mean and variance of Z as, respectively:

- $\overline{Z} = \overline{X} + \mathrm{j}\overline{Y}$,
- $\sigma^2_Z = \mathrm{E}[|Z - \overline{Z}|^2] = \sigma^2_X + \sigma^2_Y$, which measures the spread of Z about its mean in the complex plan.

An N-dimensional complex random variable $\mathbf{Z} = \mathbf{X} + \mathrm{j}\mathbf{Y}$ can be easily handled through a properly defined joint distribution, e.g., $f_{\mathbf{X},\mathbf{Y}}(\mathbf{x}, \mathbf{y})$. The individual variables $Z_n,\ 1 \leq n \leq N$, will be independent when

$$f_{\mathbf{X},\mathbf{Y}}(\mathbf{x}, \mathbf{y}) = \prod_{n=1}^{N} f_{X_n,Y_n}(x_n, y_n). \tag{4.91}$$

The following definitions must be generalized to cope with complex random variables:

- correlation: $R_{Z_1Z_2} = \mathrm{E}[Z_1 Z_2^*]$;
- covariance: $C_{Z_1Z_2} = \mathrm{E}[(Z_1 - \overline{Z_1})(Z_2 - \overline{Z_2})^*]$.

Now, $C_{Z_1Z_2} = R_{Z_1Z_2} - \overline{Z_1}\,\overline{Z_2}^*$. As before, Z_1 and Z_2 will be:

- orthogonal when $R_{Z_1Z_2} = 0$;
- uncorrelated when $C_{Z_1Z_2} = 0$.

4.3.13 Application: estimators

One of the most encompassing applications of vector random variables is the so-called *parameter estimation*, per se an area supported by its own theory. The typical framework comprises using a finite set of measured data from a given phenomenon to estimate the parameters[11] of its underlying model.

A set $X_1, X_2, \ldots, X_N$ of N independent identically distributed (i.i.d.) random variables such that

$$f_{\mathbf{X}}(\mathbf{x}) = \prod_{n=1}^{N} f_X(x_n) \tag{4.92}$$

can describe an N-size *random sample* of a population modeled by $f_X(x)$. Any function $g(\mathbf{X})$ (called a *statistic* of $\mathbf{X}$) can constitute a *point estimator* $\hat{\Theta}$ of some parameter θ such that given an N-size sample $x_1, x_2, \ldots, x_N$ of $\mathbf{X}$, one can compute an estimate $\hat{\theta} = g(\mathbf{x})$ of θ.

A classical example of point estimator for a fixed parameter is the so-called *sample mean*, which estimates $\theta = \overline{X}$ by the arithmetic mean of the available samples x_n, $n = 1, 2, \ldots, N$:

$$\hat{\overline{x}}_N = \frac{1}{N}\sum_{n=1}^{N} x_n. \tag{4.93}$$

Defining a desired estimator is not always a trivial task as in the previous example. In general, resorting to a proper analytical criterion is necessary. Suppose we are given the so-called likelihood function of θ, $L(\theta) = f_{\mathbf{X}}(\mathbf{x}|\theta)$, which provides a probabilistic model for $\mathbf{X}$ as an explicit function of θ. Given the sample $\mathbf{x}$, the so-called *maximum likelihood* (ML) estimator $\hat{\Theta}_{\text{ML}}$ computes an estimate of θ by direct maximization of $L(\theta)$. This operation is meant to find the value of θ that would make the available sample $\mathbf{x}$ the most probable.

Sometimes the parameter θ itself can be a sample of a random variable Θ described by an a priori PDF $f_\Theta(\theta)$. For example, suppose one wants to estimate the mean value of a shipment of components received from a given factory; since the components may come from several production units, we can think of a statistical model for their nominal values. In such situations, any reliable information on the parameter distribution can be taken into account to better tune the estimator formulation; these so-called *Bayesian estimators* rely on the a posteriori PDF of θ:

$$f_\Theta(\theta|\mathbf{x}) = \frac{f_{\mathbf{X},\Theta}(\mathbf{x},\theta)}{f_{\mathbf{X}}(\mathbf{x})} \propto L(\theta) f_\Theta(\theta). \tag{4.94}$$

Given a sample $\mathbf{x}$, the so-called *maximum a posteriori* (MAP) estimator $\hat{\Theta}_{\text{MAP}}$ computes an estimate of θ as the mode of $f_\Theta(\theta|\mathbf{x})$. This choice is meant to find the most probable θ that would have produced

[11] We will consider only scalar parameters, without loss of generality.

the available data **x**. Several applications favor the *posterior mean* estimator, which for a given sample **x** computes $\hat{\theta}_{\text{PM}} = \text{E}[\Theta|\mathbf{x}]$, over the MAP estimator.

The quality of an estimator $\hat{\Theta}$ can be assessed through some of its statistical properties.[12] An overall measure of the estimator performance is its *mean square error* $\text{MSE}(\hat{\Theta}) = \text{E}[(\theta - \hat{\Theta})^2]$, which can be decomposed in two parts:

$$\text{MSE}(\hat{\Theta}) = b^2(\hat{\Theta}) + \sigma^2_{\hat{\Theta}}, \tag{4.95}$$

where $b(\hat{\Theta}) = \text{E}[\theta - \hat{\Theta}]$ is the estimator *bias* and $\sigma^2_{\hat{\Theta}}$ is its *variance*. The bias measures the deterministic part of the error, i.e., how much the estimates deviate from the target in the mean, thus providing an *accuracy* measure: the smaller its bias, the more accurate the estimator. The variance, in turn, measures the random part of the error, i.e., how much the estimates spread about their mean, thus providing a *precision* measure: the smaller its variance, the more precise the estimator. Another property attributable to an estimator is *consistence*[13]: $\hat{\Theta}$ is said to be consistent when $\lim_{N\to\infty} P(|\hat{\Theta} - \theta| \geq \epsilon) = 0,\ \forall \epsilon > 0$. Using the Chebyshev inequality (see Section 4.3.4), one finds that an unbiased estimator with $\lim_{N\to\infty} \sigma^2_{\hat{\Theta}} = 0$ is consistent. It can be shown that the sample mean defined in Eq. (4.93) has $b(\hat{\overline{X}}_N) = 0$ (thus is unbiased) and $\sigma^2_{\hat{\overline{X}}_N} = \frac{\sigma^2_X}{N}$ (thus is consistent). The similarly built estimator for σ^2_X, the *unbiased sample variance*, computes

$$\widehat{\sigma^2_X} = \frac{1}{N-1}\sum_{n=1}^{N}\left(x_n - \hat{\overline{X}}_N\right)^2. \tag{4.96}$$

The reader is invited to prove that this estimator is unbiased, while its more intuitive form with denominator N instead of $N-1$ is not.

One could argue how reliable a point estimation is. In fact, if the range of θ is continuous, one can deduce that $P(\{\hat{\Theta} = \theta\}) = 0$. However, perhaps we would feel more confident if the output of an estimator were something like "$\theta_1 \leq \theta \leq \theta_2$ with probability p." The confidence limits of this *interval estimate* are θ_1 and θ_2 and p is the confidence.

Example: A resistance meter uses a batch of 10 measurements and outputs the sample mean $\hat{\overline{R}}_{10}$. Assuming that all measurements are i.i.d. and that they can be described by a unit-variance Gaussian random variable centered at the nominal resistance value $\overline{R}$, determine the 95% confidence interval for the estimates.

Solution: Evidently, $\hat{\overline{R}}_{10}$ is a Gaussian random variable with $\text{E}[\hat{\overline{R}}_{10}] = \overline{R}$ and $\sigma^2_{\hat{\overline{R}}_{10}} = 0.1$. We should find a such that $P(\{\hat{\overline{R}}_{10} - a \leq \overline{R} \leq \hat{\overline{R}}_{10} + a\}) = 0.95$. For the associated normalized Gaussian random variable, this is equivalent to $P\left(\left\{-1.96 \leq \frac{\overline{R} - \hat{\overline{R}}_{10}}{1/\sqrt{10}} \leq 1.96\right\}\right) \approx 0.95 \Rightarrow a \approx 0.62$.

[12] For the sake of simplicity, we refer only to the estimation of a fixed parameter θ.

[13] There are other definitions of consistence.

4.3.14 Application: feature selection

An important use of estimators is in the process of feature selection, which is widely used in machine learning systems. Consider, for example, classification machines, whose purpose is to use a set of characteristics of an entity (e.g., height, weight, and glucose level of a human, morphological characteristics of a plant, etc.) to classify it. These models are trained in order to take an input vector of features $\mathbf{x}$ and identify the "class" y of that input.[14]

When modeling the input as a sample of a vector random variable $\mathbf{X}$, it is easy to see that high correlation between its entries is undesirable. If two different components X_1 and X_2 of the random vector are such that $|\rho_{X_1 X_2}| \approx 1$, this means that knowledge of one of the variables is virtually sufficient to determine the other. This, combined with the fact that dealing with a high number of dimensions is detrimental to machine learning systems, makes a tool to identify correlations between features necessary.

The covariance matrix $\boldsymbol{C}_{\mathbf{XX}}$ presented above is a strong candidate for this task, as its off-diagonal terms represent cross-covariances between different features. Since the input distribution $f_{\mathbf{X}}(\mathbf{x})$ is often unknown, an estimate $\widehat{\boldsymbol{C}}_{\mathbf{XX}}$ is used instead. Suppose we have N samples of the input vector $\mathbf{X}$, given by $\mathbf{x_n} \in \mathbb{R}^d$, where d is the number of features. Assume, without loss of generality, that all components of the input vector are zero-mean, unit-variance random variables,[15] so that the off-diagonal terms of $\boldsymbol{C}_{\mathbf{XX}}$ become correlation coefficients. An estimation of the covariance matrix is given by

$$\widehat{\boldsymbol{C}}_{\mathbf{XX}} = \frac{1}{N}\sum_{n=1}^{N} \mathbf{x}_n \mathbf{x}_n^{\mathrm{T}} = \frac{1}{N}\boldsymbol{D}_{\mathbf{X}}^{\mathrm{T}} \boldsymbol{D}_{\mathbf{X}}, \tag{4.97}$$

where $\boldsymbol{D}_{\mathbf{X}}$ is the *data matrix*, given by

$$\boldsymbol{D}_{\mathbf{X}} = \begin{bmatrix} \mathbf{x}_1^{\mathrm{T}} \\ \mathbf{x}_2^{\mathrm{T}} \\ \vdots \\ \mathbf{x}_N^{\mathrm{T}} \end{bmatrix}. \tag{4.98}$$

An inspection of $\widehat{\boldsymbol{C}}_{\mathbf{XX}}$ makes it possible to identify highly correlated features (i.e., above a given threshold) and remove redundant ones, producing a reduced input vector, better suited to the problem. As only the input vectors themselves are used to make the selection, this type of procedure is called an unsupervised feature selection. Alternatively, one can take into account the correlation between each input characteristic and the desired result, which is called supervised feature selection.

[14] Although we are only considering numerical features, one could talk about categorical features as well.

[15] A simple normalization of the original samples based on estimations of their means and variances can approximate this condition.

4.4 Random process

If the outcomes s of a given random experiment vary with time t, each one can be mapped by some $X(s(t))$ into a real function $x(t)$. This is a direct generalization of the random variable. As a result, we get an infinite *ensemble* of time-varying *sample functions* or *realizations* $x(t)$ which form the so-called *stochastic* or *random process* $X(t)$. For a fixed t_0, $X(t_0)$ reduces to a single random variable, thus a random process can be seen as a time-ordered multiple random variable. An example of a random process is the simultaneous observation of air humidity at every point of the globe, $H(t)$: each realization $h(t)$ describes the humidity variation in time at a randomly chosen place on earth, whereas the random variable $H(t_0)$ describes the distribution of air humidity over the earth at instant t_0. A similar construction in the discrete-time n domain would produce $X[n]$ composed of realizations $x[n]$. If the hourly measures of air humidity substitute for its continuous measure in the former example, we get a new random process $H[n]$[16]

As seen, there are *continuous-time* and *discrete-time* random processes. Since random variables have already been classified as continuous or discrete, one analogously refers to *continuous-valued* and *discrete-valued* random processes. The combination of the two classifications can become ambiguous or awkwardly lengthy; whenever this complete categorization is necessary, one favors the nomenclature *random process* for the continuous-time case and *random sequence* for the discrete-time case, using the words "continuous" and "discrete" only with reference to amplitude values. In the former examples, the continuous random process $H(t)$ and random sequence $H[n]$ would model the idealized actual air humidity as a physical variable, whereas the practical digitized measures of air humidity would be modeled by their discrete counterparts.

4.4.1 Distributions of random processes and sequences

One could think of the complete description of a random process or sequence as a joint CDF (or PDF) encompassing every possible instant of time t or n, respectively, which is obviously impractical. However, their partial representation by Lth-order distributions (employing a slightly modified notation to include the reference to time) can be useful. The CDF of a random process $X(t)$ can be defined as

$$\begin{aligned} F_X(x_1, x_2, \ldots, x_L; t_1, t_2, \ldots, t_L) \\ = P(\{X(t_1) \leq x_1, X(t_2) \leq x_2 \ldots, X(t_L) \leq x_L\}), \end{aligned} \tag{4.99}$$

with an associated PDF

$$f_X(x_1, x_2, \ldots, x_L; t_1, t_2, \ldots, t_L) = \frac{\partial^L F_X(x_1, x_2, \ldots, x_L; t_1, \ldots, t_L)}{\partial x_1 \partial x_2 \cdots \partial x_L}. \tag{4.100}$$

The CDF of a random sequence $X[n]$ can be defined as

$$\begin{aligned} F_X(x_1, x_2, \ldots, x_L; n_1, n_2, \ldots, n_L) \\ = P(\{X[n_1] \leq x_1, X[n_2] \leq x_2 \ldots, X[n_L] \leq x_L\}), \end{aligned} \tag{4.101}$$

16 In this text, for completeness we opted for always explicitly writing both formulations, for processes and sequences.

with an associated PDF

$$f_X(x_1, x_2, \ldots, x_L; n_1, n_2, \ldots, n_L) = \frac{\partial^L F_X(x_1, x_2, \ldots, x_L; n_1, n_2, \ldots, n_L)}{\partial x_1 \partial x_2 \cdots \partial x_L}. \tag{4.102}$$

This convention can be easily extended to the joint distributions of M random processes $X_m(t),\ m = 1, 2, \ldots, M$ (sequences $X_m[n],\ m = 1, 2, \ldots, M$), as will be seen in the next subsection. The notation for point-conditioned distributions can also be promptly inferred (see Section 4.4.12).

4.4.2 Statistical independence

As with random variables, the mutual independence of random processes or sequences is tested by the complete separability of their respective distributions. The random processes $X_m(t),\ m = 1, 2, \ldots, M$, are independent when

$$\begin{aligned} &f_{X_1,X_2,\ldots,X_M}(x_{11}, x_{12}, \ldots, x_{1L_1}, x_{21}, x_{22}, \ldots, x_{2L_2}, \ldots, x_{M1}, x_{M2}, \ldots, x_{ML_M};\\ &\qquad t_{11}, t_{12}, \ldots, t_{1L_1}, t_{21}, t_{22}, \ldots, t_{2L_2}, \ldots, t_{M1}, t_{M2}, \ldots, t_{ML_M})\\ &\qquad = \prod_{m=1}^{M} f_{X_m}(x_{m1}, x_{m2}, \ldots, x_{mL_m}; t_{m1}, t_{m2}, \ldots, t_{mL_m}) \end{aligned} \tag{4.103}$$

for any choice of time instants. Similarly, the random sequences $X_m[n],\ m = 1, 2, \ldots, M$, are independent when

$$\begin{aligned} &f_{X_1,X_2,\ldots,X_M}(x_{11}, x_{12}, \ldots, x_{1L_1}, x_{21}, x_{22}, \ldots, x_{2L_2}, \ldots, x_{M1}, x_{M2}, \ldots, x_{ML_M};\\ &\qquad n_{11}, n_{12}, \ldots, n_{1L_1}, n_{21}, n_{22}, \ldots, n_{2L_2}, \ldots, n_{M1}, n_{M2}, \ldots, n_{ML_M})\\ &\qquad = \prod_{m=1}^{M} f_{X_m}(x_{m1}, x_{m2}, \ldots, x_{mL_m}; n_{m1}, n_{m2}, \ldots, n_{mL_m}) \end{aligned} \tag{4.104}$$

for any choice of time instants.

4.4.3 First- and second-order moments for random processes and sequences

It is not necessary to redefine moments in the context of random processes and sequences, since we will always be dealing with a set of random variables as in Section 4.3.8. But it is useful to revisit the first- and second-order cases with a slightly modified notation to include time information.

For a random process $X(t)$, the following definitions apply:

- the *mean* $\overline{X}(t) = \mathrm{E}[X(t)]$;
- the *mean square value* or *mean instantaneous power* $\overline{X^2}(t) = \mathrm{E}[X^2(t)]$;
- the *variance* $\sigma_X^2(t) = \mathrm{E}[(X(t) - \overline{X}(t))^2]$;
- the *autocorrelation* $R_{XX}(t_1, t_2) = \mathrm{E}[X(t_1)X(t_2)]$;
- the *autocovariance* $C_{XX}(t_1, t_2) = \mathrm{E}[(X(t_1) - \overline{X}(t_1))(X(t_2) - \overline{X}(t_2))]$.

As seen for random variables,

$$\sigma_X^2(t) = \overline{X^2}(t) - (\overline{X}(t))^2 \Rightarrow \overline{X^2}(t) = (\overline{X}(t))^2 + \sigma_X^2(t), \tag{4.105}$$

which means that the mean instantaneous power of the process is the sum of a deterministic parcel with a random parcel.

Analogously, for a random sequence $X[n]$, the following definitions apply:

- the *mean* $\overline{X}[n] = \mathrm{E}[X[n]]$;
- the *mean square value* or *mean instantaneous power* $\overline{X^2}[n] = \mathrm{E}[X^2[n]]$;
- the *variance* $\sigma_X^2[n] = \mathrm{E}[(X[n] - \overline{X}[n])^2]$;
- the *autocorrelation* $R_{XX}[n_1, n_2] = \mathrm{E}[X[n_1]X[n_2]]$;
- the *autocovariance* $C_{XX}[n_1, n_2] = \mathrm{E}[(X[n_1] - \overline{X}[n_1])(X[n_2] - \overline{X}[n_2])]$.

Again,

$$\sigma_X^2[n] = \overline{X^2}[n] - (\overline{X}[n])^2 \Rightarrow \overline{X^2}[n] = (\overline{X}[n])^2 + \sigma_X^2[n], \tag{4.106}$$

i.e., the mean instantaneous power of the sequence is the sum of a deterministic parcel with a random parcel.

Given two random processes $X(t)$ and $Y(t)$, one defines:

- the *cross-correlation* $R_{XY}(t_1, t_2) = \mathrm{E}[X(t_1)Y(t_2)]$;
- the *cross-covariance* $C_{XY}(t_1, t_2) = \mathrm{E}[(X(t_1) - \overline{X}(t_1))(Y(t_2) - \overline{Y}(t_2))]$.

The processes $X(t)$ and $Y(t)$ are said to be mutually:

- *orthogonal* when $R_{XY}(t_1, t_2) = 0, \ \forall t_1, t_2$;
- *uncorrelated* when $C_{XY}(t_1, t_2) = 0$, which is the same as $R_{XY}(t_1, t_2) = \overline{X}(t_1)\overline{Y}(t_2) \ \forall t_1, t_2$.

The mean instantaneous power of $(X(t) + Y(t))$ is given by

$$\mathrm{E}[(X(t) + Y(t))^2] = \overline{X^2}(t) + 2\mathrm{E}[X(t)Y(t)] + \overline{Y^2}(t), \tag{4.107}$$

which suggests the definition of $\mathrm{E}[X(t)Y(t)]$ as the *mean instantaneous cross-power* between $X(t)$ and $Y(t)$.

Analogously, given two random sequences $X[n]$ and $Y[n]$, one defines:

- the *cross-correlation* $R_{XY}[n_1, n_2] = \mathrm{E}[X[n_1]Y[n_2]]$;
- the *cross-covariance* $C_{XY}[n_1, n_2] = \mathrm{E}[(X[n_1] - \overline{X}[n_1])(Y[n_2] - \overline{Y}[n_2])]$.

The sequences $X[n]$ and $Y[n]$ are said to be mutually:

- *orthogonal* when $R_{XY}[n_1, n_2] = 0, \ \forall n_1, n_2$;
- *uncorrelated* when $C_{XY}[n_1, n_2] = 0$, which is the same as $R_{XY}[n_1, n_2] = \overline{X}[n_1]\overline{Y}[n_2] \ \forall n_1, n_2$.

The mean instantaneous power of $(X[n] + Y[n])$ is given by

$$\mathrm{E}[(X[n] + Y[n])^2] = \overline{X^2}[n] + 2\mathrm{E}[X[n]Y[n]] + \overline{Y^2}[n], \tag{4.108}$$

which suggests the definition of E$[X[n]Y[n]]$ as the *mean instantaneous cross-power* between $X[n]$ and $Y[n]$.

The interpretation of this "cross-power" is not difficult: it accounts for the constructive or destructive interaction of the two involved processes (sequences) as dictated by their mutual correlation.

We will dedicate some space later to the properties of second-order moments.

4.4.4 Stationarity

The stationarity of random processes and sequences refers to the time-invariance of their statistical properties, which makes their treatment considerably easier.

The *strict-sense* is the strongest class of stationarity. A strict-sense stationary (SSS) random process $X(t)$ (sequence $X[n]$) bears exactly the same statistical properties as $X(t+\Delta t)$, $\forall \Delta t \in \mathbb{R}$ ($X[n+\Delta n]$, $\forall \Delta n \in \mathbb{Z}$), which means that its associated joint distribution for any set of time instants does not change when they are all shifted by the same time lag. A random process (sequence) in which each realization is a constant sampled from some statistical distribution in time is SSS.

Under this perspective, we can define that a random process $X(t)$ is *Lth-order* stationary when

$$\begin{aligned} &F_X(x_1, x_2, \ldots, x_L; t_1, t_2, \ldots, t_L) \\ &\qquad = F_X(x_1, x_2, \ldots, x_L; t_1+\Delta t, t_2+\Delta t, \ldots, t_L+\Delta t) \end{aligned} \tag{4.109}$$

or, alternatively,

$$\begin{aligned} &f_X(x_1, x_2, \ldots, x_L; t_1, t_2, \ldots, t_L) \\ &\qquad = f_X(x_1, x_2, \ldots, x_L; t_1+\Delta t, t_2+\Delta t, \ldots, t_L+\Delta t). \end{aligned} \tag{4.110}$$

Analogously, a random sequence $X[n]$ is *Lth-order* stationary when

$$\begin{aligned} &F_X(x_1, x_2, \ldots, x_L; n_1, n_2, \ldots, n_L) \\ &\qquad = F_X(x_1, x_2, \ldots, x_L; n_1+\Delta n, n_2+\Delta n, \ldots, n_L+\Delta n) \end{aligned} \tag{4.111}$$

or, alternatively,

$$\begin{aligned} &f_X(x_1, x_2, \ldots, x_L; n_1, n_2, \ldots, n_L) \\ &\qquad = f_X(x_1, x_2, \ldots, x_L; n_1+\Delta n, n_2+\Delta n, \ldots, n_L+\Delta n). \end{aligned} \tag{4.112}$$

We can say that an SSS random process or sequence is stationary for any order L. Moreover, Lth-order stationarity implies $(L-1)$th-order stationarity, by marginalization of both sides of any of the four equations above.

First-order stationarity says the statistical distribution at any individual instant of time t (n) is the same, i.e., $f_X(x;t)$ does not depend on t ($f_X(x;n)$ does not depend on n). A natural consequence is the time-invariance of its statistical mean: $\overline{X}(t) = \overline{X}$ ($\overline{X}[n] = \overline{X}$). Second-order stationarity says the joint statistical distribution at any two time instants t and $t-\tau$ for a fixed τ (n and $n-k$ for a fixed k) is the same. A natural consequence is the time-invariance of its autocorrelation, which can depend only on the time lag between the two time instants: $R_{XX}(t, t-\tau) = R_{XX}(\tau)$ ($R_{XX}[n, n-k] = R_{XX}[k]$).

A weaker class of stationarity called *wide-sense* is much used in practice for its simple testability. A random process $X(t)$ (sequence $X[n]$) is said to be wide-sense stationary (WSS) when its mean and autocorrelation are time-invariant. Such conditions do not imply stationarity of any order, although second-order stationarity implies wide-sense stationarity. As an extension of such definition, one says that two processes $X(t)$ and $Y(t)$ (sequences $X[n]$ and $Y[n]$) are *jointly WSS* when they are individually WSS and their cross-correlation is time-invariant, i.e., can depend only on the time lag between the two time instants: $R_{XY}(t, t-\tau) = R_{XY}(\tau)$ ($R_{XY}[n, n-k] = R_{XY}[k]$).

4.4.5 Properties of correlation functions for WSS processes and sequences

Wide-sense stationarity in random processes and sequences induces several interesting properties, some of which are listed below. For processes,

- $|R_{XX}(\tau)| \leq R_{XX}(0)$;
- $R_{XX}(-\tau) = R_{XX}(\tau)$;
- if $\lim\limits_{|\tau|\to\infty} R_{XX}(\tau) = B$, then $B = \overline{X}^2$;
- if $X(t)$ has a periodic component, then $R_{XX}(\tau)$ has the same periodic component;
- $R_{XY}(-\tau) = R_{YX}(\tau)$;
- $R_{XY}(\tau) \leq \sqrt{R_{XX}(0)R_{YY}(0)}$;
- $R_{XY}(\tau) \leq \dfrac{R_{XX}(0) + R_{YY}(0)}{2}$.

For sequences,

- $|R_{XX}[k]| \leq R_{XX}[0]$;
- $R_{XX}[-k] = R_{XX}[k]$;
- if $\lim\limits_{|k|\to\infty} R_{XX}[k] = B$, then $B = \overline{X}^2$;
- if $X[n]$ has a periodic component, then $R_{XX}[k]$ has the same periodic component;
- $R_{XY}[-k] = R_{YX}[k]$;
- $R_{XY}[k] \leq \sqrt{R_{XX}[0]R_{YY}[0]}$;
- $R_{XY}[k] \leq \dfrac{R_{XX}[0] + R_{YY}[0]}{2}$.

Even if the properties are not proven here, some interesting observations can be traced about them:

- The statistical behavior of the process (sequence) at a given instant of time is maximally correlated with itself, an intuitive result.
- The autocorrelation commutes.
- The statistical behaviors of a WSS process (sequence) with no periodic components at two time instants separated by $\tau \to \infty$ ($k \to \infty$) are uncorrelated. Then, the autocorrelation tends to the product of two (identical) statistical means.
- Periodic components in the process (sequence) cause the correlation maximum (and surrounding behavior) to repeat at each fundamental period.
- The arithmetic and geometric mean properties are in a certain sense linked to the first property. The second property is more stringent than the first.

4.4.6 Time averages of random processes and sequences

The proper use of random processes and sequences to model practical problems depends on the careful understanding of both their statistical and time variations. The main difference between these two contexts, which must be kept in mind, is the fact that time is inexorably ordered, thus allowing the inclusion in the model of some deterministic (predictable) time structure. When we examine the statistical autocorrelation between two time instants of a random process or sequence, we learn how the statistical samples taken at those instants follow each other. In a WSS process or sequence, this measure for different lags can convey an idea of statistical periodicity.[17] But what if one is concerned with the time structure and characteristics of each (or some) individual realization(s) of a given process or sequence? The use of time averages can provide this complementary description.

The time average of a given continuous-time function $f(t)$ can be defined as

$$\mathrm{A}[f(t)] = \lim_{T\to\infty} \frac{1}{2T}\int_{-T}^{T} f(t)\mathrm{d}t. \tag{4.113}$$

Given the random process $X(t)$, the *time average* of any of its realizations $x(t)$ is

$$\overline{\chi} = \mathrm{A}[x(t)]. \tag{4.114}$$

The *average power* of $x(t)$ is

$$\mathcal{p}_{xx} = \mathrm{A}[x^2(t)]. \tag{4.115}$$

The *time autocorrelation* of $x(t)$ for a lag τ is defined as

$$\mathcal{r}_{xx}(\tau) = \mathrm{A}[x(t)x(t-\tau)]. \tag{4.116}$$

The *time cross-correlation* between the realizations $x(t)$ of process $X(t)$ and $y(t)$ of process $Y(t)$ for a lag τ is defined as

$$\mathcal{r}_{xy}(\tau) = \mathrm{A}[x(t)y(t-\tau)]. \tag{4.117}$$

As in the statistical case, an *average cross-power* between $x(t)$ and $y(t)$, which accounts for their mutual constructive or destructive interferences due to their time structure, can be defined as

$$\mathcal{p}_{xy} = \mathrm{A}[x(t)y(t)]. \tag{4.118}$$

Computed for the complete ensemble, such measures produce the random variables $\overline{\mathcal{X}}$, $\mathcal{P}_{xx}$, $\mathcal{R}_{xx}(\tau)$, $\mathcal{R}_{xy}(\tau)$, and $\mathcal{P}_{xy}$, respectively. Under mild convergence conditions,

$$\mathrm{E}[\overline{\mathcal{X}}] = \mathrm{E}[\mathrm{A}[X(t)]] = \mathrm{A}[\mathrm{E}[X(t)]] = \mathrm{A}[\overline{X}(t)], \tag{4.119}$$

$$\mathrm{E}[\mathcal{P}_{xx}] = \mathrm{E}[\mathrm{A}[X^2(t)]] = \mathrm{A}[\mathrm{E}[X^2(t)]] = \mathrm{A}[\overline{X^2}(t)], \tag{4.120}$$

$$\mathrm{E}[\mathcal{R}_{xx}(\tau)] = \mathrm{E}[\mathrm{A}[X(t)X(t-\tau)]] = \mathrm{A}[\mathrm{E}[X(t)X(t-\tau)]] = \mathrm{A}[R_{XX}(t,t-\tau)], \tag{4.121}$$

[17] We have already addressed this fact when discussing the properties of second-order moments in Section 4.4.5.

$$\mathrm{E}[\mathcal{R}_{xy}(\tau)] = \mathrm{E}[\mathrm{A}[X(t)Y(t-\tau)]] = \mathrm{A}[\mathrm{E}[X(t)Y(t-\tau)]] = \mathrm{A}[R_{XY}(t, t-\tau)], \tag{4.122}$$

and

$$\mathrm{E}[\mathcal{P}_{xy}] = \mathrm{E}[\mathrm{A}[X(t)Y(t)]] = \mathrm{A}[\mathrm{E}[X(t)Y(t)]], \tag{4.123}$$

which are respectively the overall mean value, mean power, and autocorrelation (for lag τ) of process $X(t)$ and the cross-correlation (for lag τ) and mean cross-power between processes $X(t)$ and $Y(t)$.

The time average of a given discrete-time function $f[n]$ can be defined as

$$\mathrm{A}[f[n]] = \lim_{N\to\infty} \frac{1}{2N+1} \sum_{n=-N}^{N} f[n]. \tag{4.124}$$

Given the random sequence $X[n]$, the *time average* of any of its realizations $x[n]$ is

$$\overline{x} = \mathrm{A}[x[n]]. \tag{4.125}$$

The *average power* of $x[n]$ is

$$p_{xx} = \mathrm{A}[x^2[n]]. \tag{4.126}$$

The *time autocorrelation* of $x[n]$ for lag k is defined as

$$r_{xx}[k] = \mathrm{A}[x[n]x[n-k]]. \tag{4.127}$$

The *time cross-correlation* between the realizations $x[n]$ of sequence $X[n]$ and $y[n]$ of sequence $Y[n]$ for lag k is defined as

$$r_{xy}[k] = \mathrm{A}[x[n]y[n-k]]. \tag{4.128}$$

As in the statistical case, an *average cross-power* between $x[n]$ and $y[n]$, which accounts for their mutual constructive or destructive interferences due to their time structure, can be defined as

$$p_{xy} = \mathrm{A}[x[n]y[n]]. \tag{4.129}$$

Computed for the complete ensemble, such measures produce the random variables $\overline{X}$, $\mathcal{P}_{xx}$, $\mathcal{R}_{xx}[k]$, $\mathcal{R}_{xy}[k]$, and $\mathcal{P}_{xy}$, respectively. Under mild convergence conditions,

$$\mathrm{E}[\overline{X}] = \mathrm{E}[\mathrm{A}[X[n]]] = \mathrm{A}[\mathrm{E}[X[n]]] = \mathrm{A}[\overline{X}[n]], \tag{4.130}$$

$$\mathrm{E}[\mathcal{P}_{xx}] = \mathrm{E}[\mathrm{A}[X^2[n]]] = \mathrm{A}[\mathrm{E}[X^2[n]]] = \mathrm{A}[\overline{X^2}[n]], \tag{4.131}$$

$$\mathrm{E}[\mathcal{R}_{xx}[k]] = \mathrm{E}[\mathrm{A}[X[n]X[n-k]]] = \mathrm{A}[\mathrm{E}[X[n]X[n-k]]] = \mathrm{A}[R_{XX}[n, n-k]], \tag{4.132}$$

$$\mathrm{E}[\mathcal{R}_{xy}[k]] = \mathrm{E}[\mathrm{A}[X[n]Y[n-k]]] = \mathrm{A}[\mathrm{E}[X[n]Y[n-k]]] = \mathrm{A}[R_{XY}[n, n-k]], \tag{4.133}$$

and

$$\mathrm{E}[\mathcal{P}_{xy}] = \mathrm{E}[\mathrm{A}[X[n]Y[n]]] = \mathrm{A}[\mathrm{E}[X[n]Y[n]]], \tag{4.134}$$

which are respectively the overall mean value, mean power, and autocorrelation (for lag k) of sequence $X[n]$ and the cross-correlation (for lag k) and mean cross-power between sequences $X[n]$ and $Y[n]$.

As an example, consider that a deterministic signal $s(t) \neq 0$ (with $\mathrm{A}[s(t)] = 0$) is additively contaminated by random noise $n(t)$ that can be modeled as a realization of the WSS random process $N(t)$ (with $\overline{N}(t) = 0$ and $\overline{N^2}(t) = \sigma_N^2$), thus yielding the random signal $x(t) = s(t) + n(t)$. The corresponding process $X(t) = s(t) + N(t)$ is obviously not WSS, since $\overline{X}(t) = s(t)$. Furthermore, (following the nomenclature we have adopted) we can compute its:

- mean instantaneous power $\overline{X^2}(t) = s^2(t) + \sigma_N^2$;
- mean power $\mathcal{P}_{xx} = \mathrm{A}[s^2(t)] + \mathrm{A}[n^2(t)]$;
- overall mean power $E[\mathcal{P}_{xx}] = \mathrm{A}[s^2(t)] + \sigma_N^2$.

Defining the signal-to-noise ratio (SNR) as the ratio between signal and noise powers, we can conclude that $\mathrm{SNR}_X = \dfrac{\mathrm{A}[s^2(t)]}{\sigma_N^2}$.

Now, suppose that two differently contaminated versions of $s(t)$ are available, $x_1(t) = s(t) + n_1(t)$ and $x_2(t) = s(t) + n_2(t)$, and that the underlying noise processes $N_1(t)$ and $N_2(t)$ are jointly WSS and uncorrelated, with $\overline{N_1}(t) = \overline{N_2}(t) = 0$, $\overline{N_1^2}(t) = \sigma_{N_1}^2$, and $\overline{N_2^2}(t) = \sigma_{N_2}^2$. If an average signal $x_{\mathrm{m}}(t) = \dfrac{x_1(t) + x_2(t)}{2}$ is computed, we can guarantee $\mathrm{SNR}_{X_{\mathrm{m}}} > \mathrm{SNR}_{X_1}$ and $\mathrm{SNR}_{X_{\mathrm{m}}} > \mathrm{SNR}_{X_2}$ (i.e., noise is reduced) as long as $\dfrac{1}{3} < \dfrac{\sigma_{N_1}^2}{\sigma_{N_2}^2} < 3$.

4.4.7 Ergodicity

Ergodicity is a property that allows interchanging statistic and temporal characteristics of some random processes (sequences) – which are then called ergodic. There are several levels of ergodicity, some of which are discussed below.

A random process $X(t)$ with constant statistical mean is said to be mean-ergodic when any time average $\overline{\mathcal{X}}$ is equal to $\overline{X}$ with probability 1, which requires $\sigma_{\overline{\mathcal{X}}}^2 = 0$. If $X(t)$ is WSS, a necessary and sufficient condition for mean-ergodicity is

$$\lim_{T\to\infty} \frac{1}{2T} \int_{-2T}^{2T} C_{XX}(\tau)\left(1 - \frac{|\tau|}{2T}\right) \mathrm{d}\tau = 0. \tag{4.135}$$

A random process $X(t)$ with time-invariant autocorrelation is said to be autocorrelation-ergodic when any time autocorrelation $r_{xx}(\tau)$ is equal to $R_{XX}(\tau)$ with probability 1, which requires $\sigma_{\mathcal{R}_{xx}(\tau)}^2 = 0$. Two processes $X(t)$ and $Y(t)$ with time-invariant cross-correlation are cross-correlation-ergodic when any time cross-correlation $r_{xy}(\tau)$ is equal to $R_{XY}(\tau)$ with probability 1, which requires $\sigma_{\mathcal{R}_{xy}(\tau)}^2 = 0$. Conditions for correlation-ergodicity of random processes involve fourth-order moments.

A random sequence $X[n]$ with constant statistical mean is said to be mean-ergodic when any time average $\overline{\mathcal{X}}$ is equal to $\overline{X}$ with probability 1, which requires $\sigma_{\overline{\mathcal{X}}}^2 = 0$. If $X[n]$ is WSS, a necessary and sufficient condition for mean-ergodicity is

$$\lim_{M\to\infty} \frac{1}{2M+1} \sum_{k=-2M}^{2M} C_{XX}[k]\left(1 - \frac{|k|}{2M+1}\right) = 0. \tag{4.136}$$

A random sequence $X[n]$ with time-invariant autocorrelation is said to be autocorrelation-ergodic when any time autocorrelation $\mathcal{r}_{xx}[k]$ is equal to $R_{XX}[k]$ with probability 1, which requires $\sigma^2_{\mathcal{R}_{xx}[k]} = 0$. Two sequences $X[n]$ and $Y[n]$ with time-invariant cross-correlation are cross-correlation-ergodic when any time cross-correlation $\mathcal{r}_{xy}[k]$ is equal to $R_{XY}[k]$ with probability 1, which requires $\sigma^2_{\mathcal{R}_{xy}[k]} = 0$. Conditions for correlation-ergodicity of random sequences involve fourth-order moments.

A process (sequence) that is mean- and autocorrelation-ergodic is called *wide-sense ergodic*. Two wide-sense ergodic processes (sequences) that are cross-correlation-ergodic are called *jointly wide-sense ergodic*.

A process (sequence) is *distribution-ergodic* when it is ergodic for every moment.

The SSS random process (sequence) formed by random constant realizations is not ergodic in any sense. From Eqs. (4.135) and (4.136), a WSS process $X(t)$ (sequence $X[n]$), statistically uncorrelated at any different time instants t_1 and t_2 (n_1 and n_2), i.e., with $C_{XX}(t_1 - t_2) = \delta(t_1 - t_2)$ ($C_{XX}[n_1 - n_2] = \delta[n_1 - n_2]$), is mean-ergodic.

In practical real-life situations, quite often just a single realization of an underlying random process (sequence) is available. In such cases, if the latter is known to be ergodic, we can make use of the complete powerful statistical modeling framework described in this chapter. But how can one guarantee the property is satisfied by a process (sequence) of which a single sample function is known? This is not so stringent a requirement: in fact, one needs just to be sure that there can be an ergodic process (sequence) of which that time function may be a realization. Strictly speaking, a given music recording $s(t)$ additively contaminated by background noise $d(t)$ cannot be modeled as a realization of a mean-ergodic process,[18] since each member of the ensemble would combine the same $s(t)$ with a different $d(t)$. The random process which describes the noisy signal can be written as $X(t) = s(t) + D(t)$; thus, $\overline{X(t)} = s(t)$, while in practice we know that $\overline{\mathcal{x}} = 0$.

4.4.8 An encompassing example

Define the random sequences

$$X_1[n] = A_1 \cos(\Omega_0 n + \Phi_1) \tag{4.137}$$

and

$$X_2[n] = A_2 \sin(\Omega_0 n + \Phi_2) \tag{4.138}$$

such that:

- A_1 is a random variable with mean $\overline{A_1}$ and variance $\sigma^2_{A_1}$;
- A_2 is a random variable with mean $\overline{A_2}$ and variance $\sigma^2_{A_2}$;
- $(-\pi \leq \Omega_0 < \pi)$ rad/sample is a real constant;
- Φ_1 is a random variable uniformly distributed between $-\pi$ and π rad;
- Φ_2 is a random variable uniformly distributed between $-\pi$ and π rad;
- A_1, A_2, Φ_1, and Φ_2 are mutually independent.

[18] Even though experts do it without reservation...

Compute their time and statistical means, mean powers, autocorrelations, cross-correlations, and mean cross-powers and discuss their correlation, orthogonality, stationarity, and ergodicity.

Solution For the time averages,

$$\overline{x_1} = \mathrm{A}[x_1[n]] = a_1 \mathrm{A}[\cos(\Omega_0 n + \phi_1)] = 0, \tag{4.139}$$

$$\overline{x_2} = \mathrm{A}[x_2[n]] = a_2 \mathrm{A}[\sin(\Omega_0 n + \phi_2)] = 0. \tag{4.140}$$

We have

$$\begin{aligned}\mathcal{R}_{x_1x_1}[k] &= \mathrm{A}[x_1[n]x_1[n-k]] \\ &= a_1^2 \mathrm{A}[\cos(\Omega_0 n + \phi_1)\cos(\Omega_0 n - \Omega_0 k + \phi_1)] \\ &= a_1^2 \mathrm{A}\left[\frac{\cos(2\Omega_0 n - \Omega_0 k + 2\phi_1) + \cos(\Omega_0 k)}{2}\right] \\ &= \frac{a_1^2}{2}\cos(\Omega_0 k).\end{aligned}$$

Then,

$$\mathcal{P}_{x_1x_1} = \mathcal{R}_{x_1x_1}[0] = \frac{a_1^2}{2}. \tag{4.141}$$

We have

$$\begin{aligned}\mathcal{R}_{x_2x_2}[k] &= \mathrm{A}[x_2[n]x_2[n-k]] \\ &= a_2^2 \mathrm{A}[\sin(\Omega_0 n + \phi_2)\sin(\Omega_0 n - \Omega_0 k + \phi_2)] \\ &= a_2^2 \mathrm{A}\left[\frac{\cos(\Omega_0 k) - \cos(2\Omega_0 n - \Omega_0 k + 2\phi_2)}{2}\right] \\ &= \frac{a_2^2}{2}\cos(\Omega_0 k).\end{aligned} \tag{4.142}$$

Then,

$$\mathcal{P}_{x_2x_2} = \mathcal{R}_{x_2x_2}[0] = \frac{a_2^2}{2}. \tag{4.143}$$

We have

$$\begin{aligned}\mathcal{R}_{x_1x_2}[k] &= \mathrm{A}[x_1[n]x_2[n-k]] \\ &= a_1a_2 \mathrm{A}[\cos(\Omega_0 n + \phi_1)\sin(\Omega_0 n - \Omega_0 k + \phi_2)] \\ &= a_1a_2 \mathrm{A}\left[\frac{\sin(\phi_2 - \phi_1 - \Omega_0 k) + \sin(2\Omega_0 n - \Omega_0 k + \phi_1 + \phi_2)}{2}\right] \\ &= -\frac{a_1a_2}{2}\sin(\Omega_0 k + \phi_1 - \phi_2).\end{aligned} \tag{4.144}$$

Then,

$$\mathcal{P}_{x_1x_2} = \mathcal{R}_{x_1x_2}[0] = \frac{a_1a_2}{2}\sin(\phi_2 - \phi_1). \tag{4.145}$$

For the expected values,

$$\begin{aligned}\overline{X_1} &= \mathrm{E}\left[X_1[n]\right] = \mathrm{E}\left[A_1\right]\mathrm{E}\left[\cos\left(\Omega_0 n + \Phi_1\right)\right] \\ &= \overline{A_1}\int_{-\pi}^{\pi}\frac{1}{2\pi}\cos(\Omega_0 n + \phi_1)\mathrm{d}\phi_1 = \overline{A_1}.0 = 0, \\ \overline{X_2} &= \mathrm{E}\left[X_2[n]\right] = \mathrm{E}\left[A_2\right]\mathrm{E}\left[\sin\left(\Omega_0 n + \Phi_2\right)\right] \\ &= \overline{A_2}\int_{-\pi}^{\pi}\frac{1}{2\pi}\sin(\Omega_0 n + \phi_2)\mathrm{d}\phi_2 = \overline{A_2}.0 = 0.\end{aligned}$$

We have

$$\begin{aligned}R_{X_1X_1}[n, n-k] &= \mathrm{E}\left[X_1[n]X_1[n-k]\right] \\ &= \mathrm{E}\left[A_1^2\right]\mathrm{E}\left[\cos\left(\Omega_0 n + \Phi_1\right)\cos\left(\Omega_0 n - \Omega_0 k + \Phi_1\right)\right] \\ &= \left(\overline{A_1}^2 + \sigma_{A_1}^2\right)\mathrm{E}\left[\frac{\cos\left(2\Omega_0 n - \Omega_0 k + 2\Phi_1\right) + \cos(\Omega_0 k)}{2}\right] \\ &= \frac{\overline{A_1}^2 + \sigma_{A_1}^2}{2}\cos(\Omega_0 k).\end{aligned}$$

Then,

$$\overline{X_1^2} = R_{X_1X_1}[n, n] = \frac{\overline{A_1}^2 + \sigma_{A_1}^2}{2}. \tag{4.146}$$

We have

$$\begin{aligned}R_{X_2X_2}[n, n-k] &= \mathrm{E}\left[X_2[n]X_2[n-k]\right] \\ &= \mathrm{E}\left[A_2^2\right]\mathrm{E}\left[\sin\left(\Omega_0 n + \Phi_2\right)\sin\left(\Omega_0 n - \Omega_0 k + \Phi_2\right)\right] \\ &= \left(\overline{A_2}^2 + \sigma_{A_2}^2\right)\mathrm{E}\left[\frac{\cos(\Omega_0 k) - \cos\left(2\Omega_0 n - \Omega_0 k + 2\Phi_2\right)}{2}\right] \\ &= \frac{\overline{A_2}^2 + \sigma_{A_2}^2}{2}\cos(\Omega_0 k).\end{aligned}$$

Then,

$$\overline{X_2^2} = R_{X_2X_2}[n, n] = \frac{\overline{A_2}^2 + \sigma_{A_2}^2}{2}. \tag{4.147}$$

We have

$$R_{X_1X_2}[n, n-k] = \mathrm{E}\left[X_1[n]X_2[n-k]\right] \tag{4.148}$$

$$= \mathrm{E}\left[A_1\right]\mathrm{E}\left[A_2\right]\mathrm{E}\left[\cos\left(\Omega_0 n + \Phi_1\right)\right]\mathrm{E}\left[\sin\left(\Omega_0 n - \Omega_0 k + \Phi_2\right)\right]$$

$$= \overline{A_1}\,\overline{A_2}.0.0 = 0. \tag{4.149}$$

We conclude the following:

- Since $\overline{X_1}$ and $R_{X_1X_1}[n, n-k]$ do not depend on n, $X_1[n]$ is WSS.
- Since $\overline{X_2}$ and $R_{X_2X_2}[n, n-k]$ do not depend on n, $X_2[n]$ is WSS.
- Since $X_1[n]$ is WSS, $X_2[n]$ is WSS, and since $R_{X_1X_2}[n, n-k]$ does not depend on n, $X_1[n]$ and $X_2[n]$ are jointly WSS.
- Since $R_{X_1X_2}[n, n-k] = \overline{X_1}\,\overline{X_2}$, $X_1[n]$ and $X_2[n]$ are uncorrelated.
- Since $R_{X_1X_2}[n, n-k] = 0$, $X_1[n]$ and $X_2[n]$ are orthogonal.
- Since $\overline{X_1} = \overline{\chi_1}$, $X_1[n]$ is mean-ergodic.
- Since $\overline{X_2} = \overline{\chi_2}$, $X_2[n]$ is mean-ergodic.

4.4.9 Gaussian processes and sequences

A special case of random processes (sequences) are the Gaussian-distributed ones. The corresponding Lth-order PDFs can be written (see Section 4.3.10) as follows:

- For processes they can be written as

$$f_X(x_1, x_2, \ldots, x_L; t_1, t_2, \ldots, t_L) = \frac{1}{\sqrt{(2\pi)^L \det(\boldsymbol{C}_{\mathbf{X}})}} e^{-\frac{(\mathbf{x}-\overline{\mathbf{X}})^{\mathrm{T}} \boldsymbol{C}_{\mathbf{X}}^{-1} (\mathbf{x}-\overline{\mathbf{X}})}{2}}, \tag{4.150}$$

  where $\overline{\mathbf{X}}$ is the $L \times 1$ mean vector with elements $\{\overline{\mathbf{X}}\}_l = \overline{X}(t_l)$, for $1 \le l \le L$, and $\boldsymbol{C}_{\mathbf{X}}$ is the $L \times L$ covariance matrix with elements $\{\boldsymbol{C}_{\mathbf{X}}\}_{l_1 l_2} = C_{XX}(t_{l_1}, t_{l_2})$, for $1 \le l_1, l_2 \le L$.
- For sequences they can be written as

$$f_X(x_1, x_2, \ldots, x_L; n_1, n_2, \ldots, n_L) = \frac{1}{\sqrt{(2\pi)^L \det(\boldsymbol{C}_{\mathbf{X}})}} e^{-\frac{(\mathbf{x}-\overline{\mathbf{X}})^{\mathrm{T}} \boldsymbol{C}_{\mathbf{X}}^{-1} (\mathbf{x}-\overline{\mathbf{X}})}{2}}, \tag{4.151}$$

  where $\overline{\mathbf{X}}$ is the $L \times 1$ mean vector with elements $\{\overline{\mathbf{X}}\}_l = \overline{X}[n_l]$, for $1 \le l \le L$, and $\boldsymbol{C}_{\mathbf{X}}$ is the $L \times L$ covariance matrix with elements $\{\boldsymbol{C}_{\mathbf{X}}\}_{l_1 l_2} = C_{XX}[n_{l_1}, n_{l_2}]$, for $1 \le l_1, l_2 \le L$.

From the definition above we have the following:

- A WSS Gaussian random process (sequence) is SSS: Since the PDF of a Gaussian process (sequence) is entirely described by first- and second-order moments, if these moments do not depend on t (n), the PDF itself does not depend on t (n).
- Uncorrelated Gaussian processes (sequences) are independent: A joint PDF of two uncorrelated Gaussian processes (sequences) can be easily factorized as the product of their individual PDFs, since the covariance matrix becomes block-diagonal.

These two strong properties make Gaussian models mathematically simpler to handle, and they make the strict-sense stationarity and independence conditions easier to meet in practice.

The reader is invited to show that a WSS Gaussian random process (sequence) whose square autocovariance is absolutely integrable in τ (summable in k) is ergodic.

4.4.10 Poisson random process

Poisson is an example of a discrete random process that we have already defined in Section 4.3.2. Each realization $x(t)$ of $X(t)$ counts the occurrences along time of a certain event whose mean occurrence rate is λ per time unit, provided that:

- there are no simultaneous occurrences;
- occurrence times are independent.

It can model the entry of clients in a store, for example. By convention:

- $X(0)=0$, i.e., the count starts at $t=0$;
- $X(t_0>0)$ provides the count between $t=0$ and $t=t_0$;
- $X(t_0<0)=-X(t_0>0)$.

Time origin can be shifted if necessary.

From Eq. (4.32), the first-order PDF of a Poisson process is given by

$$f_X(x;t)=\sum_{k=0}^{\infty}\frac{(\lambda t)^k \mathrm{e}^{-\lambda t}}{k!}\delta(x-k). \tag{4.152}$$

Applying the definition iteratively, one can show that for $t_1\le t_2\le\cdots\le t_L$, the corresponding Lth-order PDF is

$$f_X(x_1,x_2,\ldots,x_L;t_1,t_2,\ldots,t_L)$$
$$=\begin{cases}\displaystyle\sum_{k_1=0}^{\infty}\sum_{k_2=0}^{\infty}\cdots\sum_{k_L=0}^{\infty}\frac{(\lambda t_1)^{k_1}}{k_1!}\prod_{l=2}^{L}\frac{[\lambda(t_l-t_{l-1})]^{k_l-k_{l-1}}}{(k_l-k_{l-1})!} & \\ \quad\times\,\mathrm{e}^{-\lambda t_L}\delta(x_1-k_1)\delta(x_2-k_2)\cdots\delta(x_L-k_L), & k_1\le k_2\le\cdots\le k_L,\\ 0, & \text{otherwise.}\end{cases} \tag{4.153}$$

4.4.11 Complex random processes and sequences

Analogously to what has been done for random variables, real random processes and sequences can be generalized to handle the complex case.

A complex random process can be described as $Z(t)=X(t)+\mathrm{j}Y(t)$, where $X(t)$ and $Y(t)$ are real random processes. It is stationary in some sense as long as $X(t)$ and $Y(t)$ are jointly stationary in that sense. The remaining definitions related to process stationarity are kept the same.

For a complex random process $Z(t)$, the following definitions apply:

- the *mean* $\overline{Z}(t)=\overline{X}(t)+\mathrm{j}\overline{Y}(t)$;
- the *mean instantaneous power* $\overline{|Z|^2}(t)=\overline{|X|^2}(t)+\overline{|Y|^2}(t)$;
- the *variance* $\sigma_Z^2(t)=\mathrm{E}[(|Z(t)-\overline{Z}(t)|^2]=\sigma_X^2(t)+\sigma_Y^2(t)$;
- the *autocorrelation* $R_{ZZ}(t_1,t_2)=\mathrm{E}[Z(t_1)Z^*(t_2)]$;
- the *autocovariance* $C_{ZZ}(t_1,t_2)=\mathrm{E}[(Z(t_1)-\overline{Z}(t_1))(Z(t_2)-\overline{Z}(t_2))^*]$.

Given two random processes $Z_1(t)$ and $Z_2(t)$, one defines:

- the *mean instantaneous cross-power* $\mathrm{E}[Z_1(t)Z_2^*(t)]$;
- the *cross-correlation* $R_{Z_1Z_2}(t_1,t_2)=\mathrm{E}[Z_1(t_1)Z_2^*(t_2)]$;
- the *cross-covariance* $C_{Z_1Z_2}(t_1,t_2)=\mathrm{E}[(Z_1(t_1)-\overline{Z_1}(t_1))(Z_2(t_2)-\overline{Z_2}(t_2))^*]$.

The processes $Z_1(t)$ and $Z_2(t)$ are said to be mutually:

- *orthogonal* when $R_{Z_1Z_2}(t_1,t_2)=0,\ \forall t_1,t_2$;
- *uncorrelated* when $C_{Z_1Z_2}(t_1,t_2)=0$, which is the same as $R_{Z_1Z_2}(t_1,t_2)=\overline{Z_1}(t_1)\,\overline{Z_2}^*(t_2)\ \forall t_1,t_2$.

For a random sequence $Z[n]$, the following definitions apply:

- the *mean* $\overline{Z}[n]=\overline{X}[n]+\mathrm{j}\overline{Y}[n]$;
- the *mean instantaneous power* $\overline{|Z|^2[n]}=\overline{|X|^2[n]}+\overline{|Y|^2[n]}$;
- the *variance* $\sigma_Z^2[n]=\mathrm{E}[(|Z[n]-\overline{Z}[n]|^2]=\sigma_X^2[n]+\sigma_Y^2[n]$;
- the *autocorrelation* $R_{ZZ}[n_1,n_2]=\mathrm{E}[Z[n_1]Z^*[n_2]]$;
- the *autocovariance* $C_{ZZ}[n_1,n_2]=\mathrm{E}[(Z[n_1]-\overline{Z}[n_1])(Z[n_2]-\overline{Z}[n_2])^*]$.

Given two random processes $Z_1[n]$ and $Z_2[n]$, one defines:

- the *mean instantaneous cross-power* $\mathrm{E}[Z_1[n]Z_2^*[n]]$;
- the *cross-correlation* $R_{Z_1Z_2}[n_1,n_2]=\mathrm{E}[Z_1[n_1]Z_2^*[n_2]]$;
- the *cross-covariance* $C_{Z_1Z_2}[n_1,n_2]=\mathrm{E}[(Z_1[n_1]-\overline{Z_1}[n_1])(Z_2[n_2]-\overline{Z_2}[n_2])^*]$.

The processes $Z_1[n]$ and $Z_2[n]$ are said to be mutually:

- *orthogonal* when $R_{Z_1Z_2}[n_1,n_2]=0,\ \forall n_1,n_2$;
- *uncorrelated* when $C_{Z_1Z_2}[n_1,n_2]=0$, which is the same as $R_{Z_1Z_2}[n_1,n_2]=\overline{Z_1}[n_1]\,\overline{Z_2}^*[n_2]\ \forall n_1,n_2$.

4.4.12 Markov chains

The simplest random processes (sequences) are those whose statistics at a given instant are independent from every other instant: a first-order distribution suffices to describe them. A less trivial model is found when the statistical time interdependency assumes a special recursive behavior such that the knowledge of a past conditioning state summarizes all previous history of the process (sequence). For the so-called *Markov* random process (sequence), the knowledge of the past does not affect the expectation of the future when the present is known.

Mathematically, a random process $X(t)$ is said to be a Markov process when for $t_{n-1}<t_n$,

$$f_X(x_{t_n};t_n|x_t;\text{all } t\le t_{n-1})=f_X(x_{t_n};t_n|x_{t_{n-1}};t_{n-1}), \tag{4.154}$$

or for $t_1<t_2<\cdots<t_{n-1}<t_n$,

$$f_X(x_{t_n};t_n|x_{t_{n-1}},\ldots,x_{t_2},x_{t_1};t_{n-1},\ldots,t_2,t_1)=f_X(x_{t_n};t_n|x_{t_{n-1}};t_{n-1}). \tag{4.155}$$

We will restrict ourselves to the discrete-time case. A random sequence $X[n]$ is said to be a Markov sequence when

$$f_X(x_n;n|x_{n-1},\ldots,x_1,x_0;n-1,\ldots,1,0)=f_X(x_n;n|x_{n-1};n-1), \tag{4.156}$$

which can be seen as a *transition PDF* (also known as *evolution model*). From the definition, one arrives at the *chain rule*

$$f_X(x_0, x_1, \ldots, x_n; 0, 1, \ldots, n) \\ = f_X(x_n; n|x_{n-1}; n-1) \cdots f_X(x_1; 1|x_0; 0) f_X(x_0; 0), \tag{4.157}$$

which means that the overall Markov sequence can be statistically described for $n \geq 0$ from the knowledge of its distribution at $n = 0$ and its subsequent transition distributions. Some interesting properties can be deduced from the expressions above:

- Since

$$f_X(x_n; n|x_{n+1}, x_{n+2} \ldots, x_{n-k}; n+1, n+2, \ldots, n-k) = f_X(x_n; n|x_{n+1}; n+1),$$

  a time-reversed Markov sequence is also a Markov sequence.
- For $n_1 < n_2 < n_3$,

$$f_X(x_{n_1}, x_{n_3}; n_1, n_3|x_{n_2}; n_2) = f_X(x_{n_3}; n_3|x_{n_2}; n_2)\, f_X(x_{n_1}; n_1|x_{n_2}; n_2).$$

A very important property of Markov sequences is the so-called *Chapman–Kolmogorov equation*: for $n_1 < n_2 < n_3$,

$$f_X(x_{n_3}; n_3|x_{n_1}; n_1) = \int_{-\infty}^{\infty} f_X(x_{n_3}; n_3|x_{n_2}; n_2)\, f_X(x_{n_2}; n_2|x_{n_1}; n_1) \mathrm{d}x_{n_2}, \tag{4.158}$$

which provides a recursive way to compute arbitrary transition PDFs.

We can say a Markov sequence $X[n]$ is *stationary* when $f_X(x_n; n)$ and $f_X(x_n; n|x_{n-1}; n-1)$ are shift-invariant. In this case, the overall sequence can be obtained from $f_X(x_1, x_2; 1, 2)$. A less trivial (and more useful) model is provided by *homogeneous* Markov sequences, which are characterized only by a shift-invariant transition distribution; they are not stationary in general, but can be asymptotically stationary (i.e., for $n \to \infty$) under certain conditions.

Discrete Markov processes and sequences are called *Markov chains*, which can assume a countable number of random states a_i described by their *state probabilities* ($P(X[n] = a_i) = p_i[n]$ for sequences) and *transition probabilities* ($P(X[n_2] = a_j|X[n_1] = a_i) = \Pi_{ij}[n_1, n_2]$, $n_1 < n_2$ for sequences). Discrete-time Markov chains satisfy the following properties, for $n_1 < n_2 < n_3$:

- $\sum_j \Pi_{ij}[n_1, n_2] = 1$, which totals all possible ways to leave state a_i in n_1;
- $\sum_i p_i[n_1]\,\Pi_{ij}[n_1, n_2] = p_j[n_2]$, which totals all possible ways to arrive at state a_j in n_2;
- $\Pi_{il}[n_1, n_3] = \sum_j \Pi_{ij}[n_1, n_2]\,\Pi_{jl}[n_2, n_3]$ (*Chapman–Kolmogorov equation*).

If the chain has a finite number of states, a matrix notation can be employed. The state probability vector $\mathbf{p}[n]$, with elements $\{\mathbf{p}_i[n]\} = p_i[n]$, and the transition probability matrix $\mathbf{\Pi}[n1, n2]$, with elements $\{\mathbf{\Pi}_{ij}[n1, n2]\} = \Pi_{ij}[n1, n2]$, are related by $\mathbf{p}[n_2] = \mathbf{\Pi}^{\mathrm{T}}[n_1, n_2]\mathbf{p}[n_1]$.

The special class of homogeneous Markov chains satisfies the following additional properties:

- $\Pi_{ij}[n_1, n_2] = \Pi_{ij}[k],\ k = n_2 - n_1$;
- $\Pi_{ij}[k_2 + k_1] = \sum_l \Pi_{il}[k_1]\Pi_{lj}[k_2]$.

Accordingly, the transition probability matrix becomes $\mathbf{\Pi}[k]$. Defining $\mathbf{\Pi}[1] = \mathbf{\Pi}$,

$$\mathbf{p}[n] = \mathbf{\Pi}^{\mathrm{T}}\mathbf{p}[n-1] = (\mathbf{\Pi}^{\mathrm{T}})^n\mathbf{p}[0], \tag{4.159}$$

i.e., $\mathbf{\Pi}[n] = \mathbf{\Pi}^n$. When asymptotic stationarity is reachable, one can find the *steady-state probability vector* $\mathbf{p} = \mathbf{p}[\infty]$ such that $\mathbf{p} = \mathbf{\Pi}^{\mathrm{T}}\mathbf{p}$.

Consider, for example, the homogeneous Markov chain $X[n]$ depicted in Fig. 4.12, with states 1 and 2, state probabilities $P(X[n] = 1) = p_1[n]$ and $P(X[n] = 2) = p_2[n]$, and transition probabilities $P(X[n] = 2|X[n-1] = 1) = \Pi_{12}[1] = a$ and $P(X[n] = 1|X[n-1] = 2) = \Pi_{21}[1] = b$. Its corresponding one-sample transition matrix is

$$\mathbf{\Pi}[1] = \mathbf{\Pi} = \begin{bmatrix} 1-a & a \\ b & 1-b \end{bmatrix}, \tag{4.160}$$

such that

$$\mathbf{\Pi}[k] = \mathbf{\Pi}^k = \begin{bmatrix} \frac{b+a(1-a-b)^k}{a+b} & \frac{a-a(1-a-b)^k}{a+b} \\ \frac{b-b(1-a-b)^k}{a+b} & \frac{a+b(1-a-b)^k}{a+b} \end{bmatrix}. \tag{4.161}$$

It can also be shown that the chain reaches the steady-state probability vector

$$\lim_{n\to\infty} \mathbf{p}[n] = \lim_{n\to\infty} \begin{bmatrix} p_1[n] \\ p_2[n] \end{bmatrix} = \begin{bmatrix} \frac{b}{a+b} \\ \frac{a}{a+b} \end{bmatrix}. \tag{4.162}$$

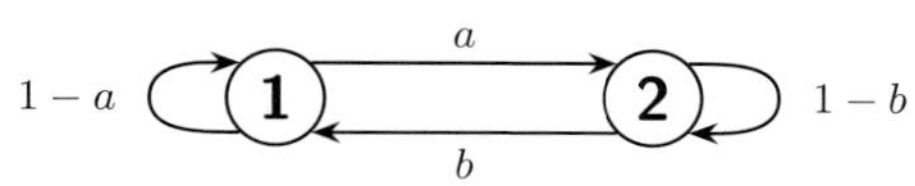

FIGURE 4.12

Homogeneous Markov chain with two states.

4.4.13 Spectral description of random processes and sequences

At first sight, the direct conversion of each realization $x(t)$ ($x[n]$) to the frequency domain seems the easiest way to spectrally characterize a random process $X(t)$ (sequence $X[n]$). However, it is difficult to guarantee the existence of the Fourier transform

$$X(\mathrm{j}\omega) = \mathcal{F}[x(t)] \triangleq \int_{-\infty}^{\infty} x(t)\mathrm{e}^{-\mathrm{j}\omega t}\mathrm{d}t \tag{4.163}$$

(in the case of random processes) or

$$X(e^{j\Omega}) = \mathcal{F}[x[n]] \triangleq \sum_{n=-\infty}^{\infty} x[n]e^{-j\Omega n} \tag{4.164}$$

(in the case of random sequences) for every realization. And even if possible, we would get a random spectrum of limited applicability.

In Section 4.4.5, we found that the autocorrelation conveys information about every sinusoidal component found in the process (sequence). A correct interpretation of the Fourier transform[19] suggests that the autocorrelation in fact conveys information about the overall spectrum of the process (sequence). Furthermore, it is a better-behaved function than the individual realizations, and thus amenable to be Fourier-transformed.

In Section 4.4.6, Eq. (4.120), we defined the overall mean power of the random process $X(t)$, which for the general complex case becomes

$$\overline{\mathcal{P}_{xx}} = \mathrm{A}[\mathrm{E}[|X^2(t)|]]. \tag{4.165}$$

It can be shown that

$$\overline{\mathcal{P}_{xx}} = \frac{1}{2\pi}\int_{-\infty}^{\infty} \lim_{T\to\infty} \frac{\mathrm{E}[|X_T(\mathrm{j}\omega)|^2]}{2T} \mathrm{d}\omega, \tag{4.166}$$

where

$$X_T(\mathrm{j}\omega) = \mathcal{F}[X_T(t)] \tag{4.167}$$

and

$$x_T(t) = \begin{cases} x(t), & -T < t < T, \\ 0, & \text{otherwise.} \end{cases} \tag{4.168}$$

We can then define a *power spectral density*

$$S_{XX}(\mathrm{j}\omega) = \lim_{T\to\infty} \frac{\mathrm{E}[|X_T(\mathrm{j}\omega)|^2]}{2T} \tag{4.169}$$

such that $\overline{\mathcal{P}_{xx}} = \frac{1}{2\pi}\int_{-\infty}^{\infty} S_{XX}(\mathrm{j}\omega)\mathrm{d}\omega$. Some additional algebraic manipulation yields

$$S_{XX}(\mathrm{j}\omega) = \mathcal{F}[\mathrm{A}[R_{XX}(t, t-\tau)]], \tag{4.170}$$

which directly relates the autocorrelation to the power spectral density of the process, as predicted. From the expressions above, $S_{XX}(\mathrm{j}\omega)$ is a nonnegative real function of ω. Furthermore, if $X(t)$ is real, then $S_{XX}(\mathrm{j}\omega)$ is an even function of ω.

In Section 4.4.6, Eq. (4.123), we also defined the overall mean cross-power between the processes $X(t)$ and $Y(t)$, which for the general complex case becomes

$$\overline{\mathcal{P}_{xy}} = \mathrm{A}[\mathrm{E}[X(t)Y^*(t)]]. \tag{4.171}$$

19 The Fourier transform represents a time signal as a linear combination of continuously distributed sinusoids.

Following the same steps as above, we arrive at

$$\overline{\mathcal{P}_{xy}} = \frac{1}{2\pi}\int_{-\infty}^{\infty} \lim_{T\to\infty} \frac{\mathrm{E}[X_T(\mathrm{j}\omega)Y_T^*(\mathrm{j}\omega)]}{2T}\mathrm{d}\omega, \tag{4.172}$$

where $X_T(\mathrm{j}\omega)$ is defined as before,

$$Y_T(\mathrm{j}\omega) = \mathcal{F}[Y_T(t)], \tag{4.173}$$

and

$$y_T(t) = \begin{cases} y(t), & -T < t < T, \\ 0, & \text{otherwise.} \end{cases} \tag{4.174}$$

We can then define the corresponding *cross-power spectral density*

$$S_{XY}(\mathrm{j}\omega) = \lim_{T\to\infty} \frac{\mathrm{E}[X_T(\mathrm{j}\omega)Y_T^*(\mathrm{j}\omega)]}{2T} \tag{4.175}$$

such that $\overline{\mathcal{P}_{xy}} = \frac{1}{2\pi}\int_{-\infty}^{\infty} S_{XY}(\mathrm{j}\omega)\mathrm{d}\omega$. In terms of the cross-correlation, the cross-power spectral density can be written as

$$S_{XY}(\mathrm{j}\omega) = \mathcal{F}[\mathrm{A}[R_{XY}(t, t-\tau)]]. \tag{4.176}$$

As expected, the cross-power density of orthogonal processes is zero. From the expressions above, $S_{XY}(\mathrm{j}\omega) = S_{YX}^*(\mathrm{j}\omega)$. Furthermore, if $X(t)$ and $Y(t)$ are real, then the real part of $S_{XY}(\mathrm{j}\omega)$ is even and the imaginary part of $S_{XY}(\mathrm{j}\omega)$ is odd.

It should be noted that in the WSS case, $\overline{\mathcal{P}_{xx}} = \mathrm{E}[|X^2(t)|]$, $\overline{\mathcal{P}_{xy}} = \mathrm{E}[X(t)Y^*(t)]$, $S_{XX}(\mathrm{j}\omega) = \mathcal{F}[R_{XX}(\tau)]$, and $S_{XY}(\mathrm{j}\omega) = \mathcal{F}[R_{XY}(\tau)]$.

A similar development can be done for random sequences. In Section 4.4.6, Eq. (4.131), we defined the overall mean power of the random sequence $X[n]$, which for the general complex case becomes

$$\overline{\mathcal{P}_{xx}} = \mathrm{A}[\mathrm{E}[|X^2[n]|]]. \tag{4.177}$$

It can be shown that

$$\overline{\mathcal{P}_{xx}} = \frac{1}{2\pi}\int_{-\pi}^{\pi} \lim_{N\to\infty} \frac{\mathrm{E}[|X_N(\mathrm{e}^{\mathrm{j}\Omega})|^2]}{2N+1}\mathrm{d}\Omega, \tag{4.178}$$

where

$$X_N(\mathrm{e}^{\mathrm{j}\Omega}) = \mathcal{F}[X_N[n]] \tag{4.179}$$

and

$$x_N[n] = \begin{cases} x[n], & -N \le n \le N, \\ 0, & \text{otherwise.} \end{cases} \tag{4.180}$$

We can then define a *power spectral density*

$$S_{XX}(\mathrm{e}^{\mathrm{j}\Omega}) = \lim_{N\to\infty} \frac{\mathrm{E}[|X_N(\mathrm{e}^{\mathrm{j}\Omega})|^2]}{2N+1} \tag{4.181}$$

such that $\overline{\mathcal{P}_{xx}} = \frac{1}{2\pi}\int_{-\pi}^{\pi} S_{XX}(e^{j\Omega})d\Omega$. Some additional algebraic manipulation yields

$$S_{XX}(e^{j\Omega}) = \mathcal{F}[A[R_{XX}[n, n-k]]], \tag{4.182}$$

which directly relates the autocorrelation to the power spectral density of the random sequence, as expected. From the expressions above, $S_{XX}(e^{j\Omega})$ is a nonnegative real function of Ω. Furthermore, if $X[n]$ is real, then $S_{XX}(e^{j\Omega})$ is an even function of Ω.

In Section 4.4.6, Eq. (4.134), we also defined the overall mean cross-power between the random sequences $X[n]$ and $Y[n]$, which for the general complex case becomes

$$\overline{\mathcal{P}_{xy}} = A[E[X[n]Y^*[n]]]. \tag{4.183}$$

Following the same steps as above, we arrive at

$$\overline{\mathcal{P}_{xy}} = \frac{1}{2\pi}\int_{-\pi}^{\pi} \lim_{N\to\infty} \frac{E[X_N(e^{j\Omega})Y_N^*(e^{j\Omega})]}{2N+1} d\Omega, \tag{4.184}$$

where $X_N(e^{j\Omega})$ is defined as before,

$$Y_N(e^{j\Omega}) = \mathcal{F}[Y_N[n]], \tag{4.185}$$

and

$$y_N[n] = \begin{cases} y[n], & -N \le n \le N, \\ 0, & \text{otherwise}. \end{cases} \tag{4.186}$$

We can then define the corresponding *cross-power spectral density*

$$S_{XY}(e^{j\Omega}) = \lim_{N\to\infty} \frac{E[X_N(e^{j\Omega})Y_N^*(e^{j\Omega})]}{2N+1} \tag{4.187}$$

such that $\overline{\mathcal{P}_{xy}} = \frac{1}{2\pi}\int_{-\pi}^{\pi} S_{XY}(e^{j\Omega})d\Omega$. In terms of the cross-correlation, the cross-power spectral density can be written as

$$S_{XY}(e^{j\Omega}) = \mathcal{F}[A[R_{XY}[n, n-k]]]. \tag{4.188}$$

As expected, the cross-power density of orthogonal processes is zero. From the expressions above, $S_{XY}(e^{j\Omega}) = S_{YX}^*(e^{j\Omega})$. Furthermore, if $X[n]$ and $Y[n]$ are real, then the real part of $S_{XY}(e^{j\Omega})$ is even and the imaginary part of $S_{XY}(e^{j\Omega})$ is odd.

It should be noted that in the WSS case, $\overline{\mathcal{P}_{xx}} = E[|X^2[n]|]$, $\overline{\mathcal{P}_{xy}} = E[X[n]Y^*[n]]$, $S_{XX}(e^{j\Omega}) = \mathcal{F}[R_{XX}[k]]$, and $S_{XY}(e^{j\Omega}) = \mathcal{F}[R_{XY}[k]]$.

4.4.14 White and colored noise

One usually refers to a disturbing signal $d(t)$ ($d[n]$) as noise. Since noise signals are typically unpredictable, they are preferably modeled as realizations of some noise random process $D(t)$ (sequence $D[n]$).

A very special case is the so-called *white noise* (by analogy with white light), which is a uniform combination of all frequencies.

Continuous-time white noise $w(t)$ is sampled from a random process $W(t)$ characterized by the following properties:

- zero mean: $\overline{W(t)} = 0$;
- noncorrelation between distinct time instants: $R_{WW}(t, t-\tau) = 0, \quad \tau \neq 0$;
- constant power spectral density: $S_{WW}(\mathrm{j}\omega) = S_{WW}, \ \forall \omega$;
- infinite overall mean power: $\overline{\mathcal{P}_{xx}} = \infty$.

From the last property, continuous-time white noise is not physically realizable. Furthermore, WSS continuous-time white noise has $R_{WW}(\tau) = S_{WW}\delta(\tau)$.

Discrete-time white noise $w[n]$ is sampled from a random sequence $W[n]$ characterized by the following properties:

- zero mean: $\overline{W[n]} = 0$;
- noncorrelation between distinct time instants: $R_{WW}[n, n-k] = 0, \ k \neq 0$;
- constant power spectral density: $S_{WW}(\mathrm{e}^{\mathrm{j}\Omega}) = \mathrm{A}[\sigma_W^2[n]], \ \forall \Omega$;
- overall mean power $\overline{\mathcal{P}_{xx}} = \mathrm{A}[\sigma_W^2[n]]$.

For WSS discrete-time white noise, $S_{WW}(\mathrm{e}^{\mathrm{j}\Omega}) = \sigma_W^2$, $R_{WW}[k] = \sigma_W^2 \delta[k]$, and $\overline{\mathcal{P}_{xx}} = \sigma_W^2$.

Unless stated otherwise, white noise is implicitly assumed to be generated by a WSS random process (sequence). Note, however, that the sequence $d[n] = w(n)\cos(\Omega_0 n)$, where $0 < \Omega_0 < \pi$ rad and $W(n)$ is WSS white noise with unit variance, for example, satisfies all conditions to be called white noise, even if $R_{DD}[n, n-k] = \delta[k]\cos(\Omega_0 n)$.

A common (more stringent) model of white noise imposes that values at different time instants of the underlying process (sequence) be statistically i.i.d.

Any random noise whose associated power spectral density is not constant is said to be *colored noise*. One can find in the literature several identified colored noises (pink, gray, etc.), each one with a prespecified spectral behavior. It should also be noted that any band-limited approximation of white noise loses its noncorrelation property. Consider $d(t)$ generated by a WSS random process $D(t)$ such that

$$S_{DD}(\mathrm{j}\omega) = \begin{cases} \dfrac{P\pi}{W}, & -W < \omega < W, \\ 0, & \text{otherwise}. \end{cases} \tag{4.189}$$

It is common practice to call $d(t)$ "white noise" of bandwidth W and overall mean power P. Since its autocorrelation is

$$R_{DD}(\tau) = \frac{P \sin(W\tau)}{W\tau}, \tag{4.190}$$

strictly speaking one cannot say $d(t)$ is white noise.

4.4.15 Applications: modulation, "bandpass," and band-limited processes, and sampling

An interesting application of the spectral description of random processes is modeling of amplitude modulation (AM). Assuming a carrier signal $c(t) = A_0 \cos(\omega_0 t)$ modulated by a real random signal $m(t)$ sampled from a process $M(t)$, the resulting modulated random process $X(t) = M(t)A_0 \cos(\omega_0 t)$ has autocorrelation

$$R_{XX}(t, t-\tau) = \frac{A_0^2}{2} R_{MM}(t, t-\tau)[\cos(\omega_0\tau) + \cos(2\omega_0 t - \omega_0\tau)]. \tag{4.191}$$

If $M(t)$ is WSS, then

$$A[R_{XX}(t, t-\tau)] = \frac{A_0^2}{2} R_{MM}(\tau)\cos(\omega_0\tau), \tag{4.192}$$

and the corresponding power spectral density is

$$S_{XX}(\mathrm{j}\omega) = \frac{A_0^2}{4}[S_{MM}(\mathrm{j}(\omega - \omega_0)) + S_{MM}(\mathrm{j}(\omega + \omega_0))]. \tag{4.193}$$

We conclude that the AM constitution of $X(t)$ carries through its autocorrelation, which provides a simple statistical model for AM.

Quite often we find ourselves dealing with *band-limited* random signals. In the AM discussion above, typically $M(t)$ has bandwidth $W \ll \omega_0$. For a random process $X(t)$ whose spectrum is at least concentrated around $\omega = 0$ (*baseband* process), we can attribute it a *root mean squared* (RMS) *bandwidth* W_{RMS} such that

$$W_{\mathrm{RMS}}^2 = \frac{\displaystyle\int_{-\infty}^{\infty} \omega^2 S_{XX}(\mathrm{j}\omega)\mathrm{d}\omega}{\displaystyle\int_{-\infty}^{\infty} S_{XX}(\mathrm{j}\omega)\mathrm{d}\omega}. \tag{4.194}$$

If the spectrum is concentrated around a centroid

$$\overline{\omega}_0 = \frac{\displaystyle\int_{0}^{\infty} \omega S_{XX}(\mathrm{j}\omega)\mathrm{d}\omega}{\displaystyle\int_{0}^{\infty} S_{XX}(\mathrm{j}\omega)\mathrm{d}\omega}, \tag{4.195}$$

W_{RMS} is given by

$$\left(\frac{W_{\mathrm{RMS}}}{2}\right)^2 = \frac{\displaystyle\int_{0}^{\infty} (\omega - \overline{\omega}_0)^2 S_{XX}(\mathrm{j}\omega)\mathrm{d}\omega}{\displaystyle\int_{0}^{\infty} S_{XX}(\mathrm{j}\omega)\mathrm{d}\omega}. \tag{4.196}$$

If the process bandwidth excludes $\omega = 0$, it is called[20] a "*bandpass*" process.

It can be shown that if a band-limited baseband WSS random process $X(t)$ with bandwidth W (i.e., such that $S_{XX}(j\omega) = 0$ for $|\omega| > W$) is sampled at rate $\omega_s > 2W$ to generate the random sequence

$$X[n] = X\left(n\frac{2\pi}{\omega_s}\right), \tag{4.197}$$

then $X(t)$ can be recovered from $X[n]$ with zero mean squared error. This is the stochastic version of the *Nyquist criterion* for lossless sampling.

4.4.16 Processing of random processes and sequences

Statistical signal modeling is often employed in the context of signal processing, and thus the interaction between random signals and processing systems calls for a systematic approach. In this section, we will restrict ourselves to *linear time-invariant* (LTI) systems.

A system defined by the operation $y = L[x]$ between input x and output y is called *linear* if any linear combination of m inputs produces as output the same linear combination of their m corresponding outputs:

- For a continuous-time system, $L\left[\sum_{m=1}^{M} \alpha_m x_m(t)\right] = \sum_{m=1}^{M} \alpha_m L[x_m(t)], \quad \alpha_m \in \mathbb{C}$.
- For a discrete-time system, $L\left[\sum_{m=1}^{M} \alpha_m x_m[n]\right] = \sum_{m=1}^{M} \alpha_m L[x_m[n]], \quad \alpha_m \in \mathbb{C}$.

For a linear system, one can define an *impulse response* h as the system output to a unit impulse δ applied to the system input at a given time instant:

- For a continuous-time system, a unit impulse $\delta(\cdot)$ applied at instant τ produces an impulse response $h_\tau(t) = L[\delta(t-\tau)]$.
- For a discrete-time system, a unit impulse[21] $\delta[\cdot]$ applied at instant k produces an impulse response $h_k[n] = L[\delta[n-k]]$.

A system is called *time-invariant* if a given input x applied at different time instants produces the same output y:

- For a continuous-time system, if $L[x(t)] = y(t)$, then $L[x(t-\tau)] = y(t-\tau)]\ \forall\tau$; thus, $h_\tau(t) = h_0(t-\tau) \triangleq h(t-\tau)$.
- For a discrete-time system, if $L[x[n]] = y[n]$, then $L[x[n-k]] = y[n-k]]\ \forall k$; thus, $h_k[n] = h_0[n-k] \triangleq h[n-k]$.

The output of an LTI system to any input x is the *convolution* $x * h$ between the input and the impulse response h of the system:

---

[20] With considerable looseness of nomenclature.

[21] Unit impulse function, or Kronecker delta: For $x \in \mathbb{Z}$, $\delta[x] = 0\ \forall x \neq 0$ and $\delta[0] = 1$.

- For a continuous-time system, $L[x(t)] = \int_{-\infty}^{\infty} x(\tau)h(t-\tau)\mathrm{d}\tau \triangleq (x * h)(t)$.
- For a discrete-time system, $L[x[n]] = \sum_{k=-\infty}^{\infty} x[k]h[n-k] \triangleq (x * h)[n]$.

In the frequency domain, the output Y is related to the input X by the system *frequency response* H:

- For a continuous-time system, $Y(\mathrm{j}\omega) = X(\mathrm{j}\omega)H(\mathrm{j}\omega)$, $H(\mathrm{j}\omega) = \mathcal{F}[h(t)]$.
- For a discrete-time system, $Y(\mathrm{e}^{\mathrm{j}\Omega}) = X(\mathrm{e}^{\mathrm{j}\Omega})H(\mathrm{e}^{\mathrm{j}\Omega})$, $H(\mathrm{e}^{\mathrm{j}\Omega}) = \mathcal{F}[h[n]]$.

The product XH brings the concept of *filtering*: H defines how much of each frequency component of input X passes to the output Y. From now on, when referring to the processing of a random process (sequence) by a system we imply that each process realization is filtered by that system.

The filtering of a complex random process $X(t)$ by an LTI system with a complex impulse response $h(t)$ results in another random process $Y(t)$ such that $Y(t) = (X * h)(t)$. Assuming $X(t)$ is WSS, then $Y(t)$ will be also WSS and[22]:

- $\overline{Y} = \overline{X}H(\mathrm{j}0)$;
- $R_{XY}(\tau) = R^*_{YX}(-\tau) = (R_{XX} * h^*_-)(\tau)$;
- $R_{YY}(\tau) = (R_{XY} * h)(\tau) = (R_{XX} * h * h^*_-)(\tau)$;
- $S_{XY}(\mathrm{j}\omega) = S^*_{YX}(\mathrm{j}\omega) = S_{XX}(\mathrm{j}\omega)H^*(\mathrm{j}\omega)$;
- $S_{YY}(\mathrm{j}\omega) = S_{XY}(\mathrm{j}\omega)H(\mathrm{j}\omega) = S_{XX}(\mathrm{j}\omega)|H(\mathrm{j}\omega)|^2$.

The filtering of a complex random sequence $X[n]$ by an LTI system with a complex impulse response $h[n]$ results in another random sequence $Y[n]$ such that $Y[n] = (X * h)[n]$. Assuming $X[n]$ is WSS, then $Y[n]$ will be also WSS and[23]:

- $\overline{Y} = \overline{X}H(\mathrm{e}^{\mathrm{j}0})$;
- $R_{XY}[k] = R^*_{YX}[-k] = (R_{XX} * h^*_-)[k]$;
- $R_{YY}[k] = (R_{XY} * h)[k] = (R_{XX} * h * h^*_-)[k]$;
- $S_{XY}(\mathrm{e}^{\mathrm{j}\Omega}) = S^*_{YX}(\mathrm{e}^{\mathrm{j}\Omega}) = S_{XX}(\mathrm{e}^{\mathrm{j}\Omega})H^*(\mathrm{e}^{\mathrm{j}\Omega})$;
- $S_{YY}(\mathrm{e}^{\mathrm{j}\Omega}) = S_{XY}(\mathrm{e}^{\mathrm{j}\Omega})H(\mathrm{e}^{\mathrm{j}\Omega}) = S_{XX}(\mathrm{e}^{\mathrm{j}\Omega})|H(\mathrm{e}^{\mathrm{j}\Omega})|^2$.

Among the above results, one of the most important is the effect of filtering on the power spectral density S_{XX} of the input WSS random process (sequence): it is multiplied by the squared magnitude $|H|^2$ of the frequency response of the LTI system to produce the power spectral density S_{YY} of the output WSS random process (sequence). As a direct consequence, any WSS random process (sequence) with known power spectral density S can be modeled by the output of an LTI system with squared magnitude response $|H|^2 \propto S$ whose input is white noise.

In order to qualitatively illustrate this filtering effect:

- Fig. 4.13 shows 200 samples of white noise $w[n]$;

[22] Note that we used for notation $f_-(t) \triangleq f(-t)$.

[23] Note that we used for notation $f_-[n] \triangleq f[-n]$.

- Fig. 4.14 shows 200 samples of $l[n]$, a slowly evolving signal obtained by low-pass filtering of $w[n]$ to 20% of its original bandwidth;
- Fig. 4.15 shows 200 samples of $b[n]$, an almost sinusoidal signal obtained by bandpass filtering of $w[n]$ to the central 4% of its original bandwidth.

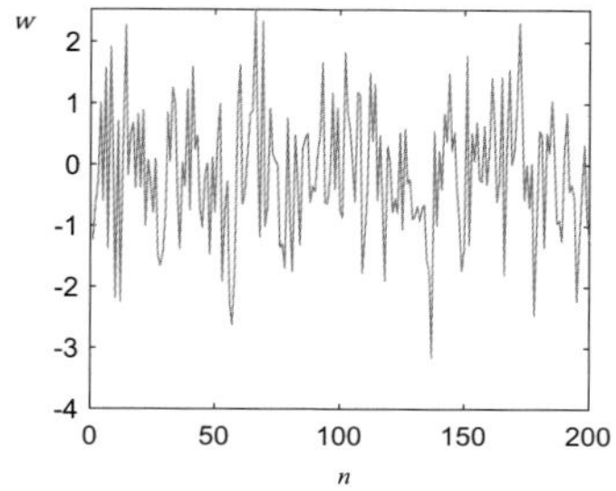

FIGURE 4.13

White noise.

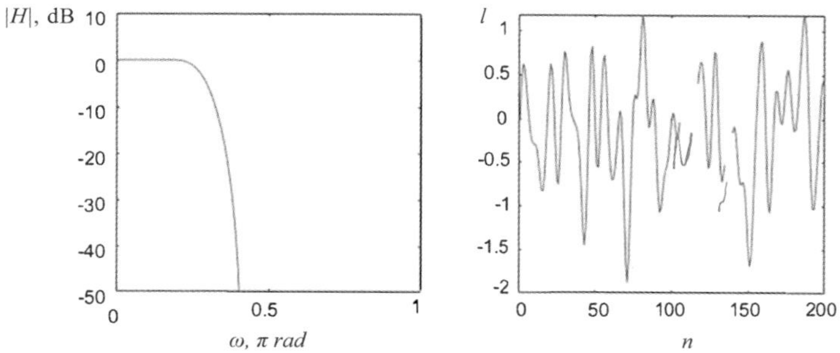

FIGURE 4.14

Realization of a "low-pass" sequence.

From the area of *optimum filtering*, whose overall goal is finding the best filter to perform a defined task, we can bring an important application of the concepts presented in this section. The general discrete *Wiener filter* H, illustrated in Fig. 4.16, is the LTI filter which minimizes the mean quadratic value $\mathrm{E}[|e^2|[n]]$ of the error $e[n] = d[n] - y[n]$ between the desired signal $d[n]$ and the filtered version $y[n]$ of the input signal $x[n]$. Assuming that $x[n]$ and $d[n]$ are realizations of two jointly WSS random sequences $X[n]$ and $D[n]$, if the optimum filter is not constrained to be causal, one finds that it must satisfy the equation

$$(h * R_{XX})[k] = R_{DX}[k], \tag{4.198}$$

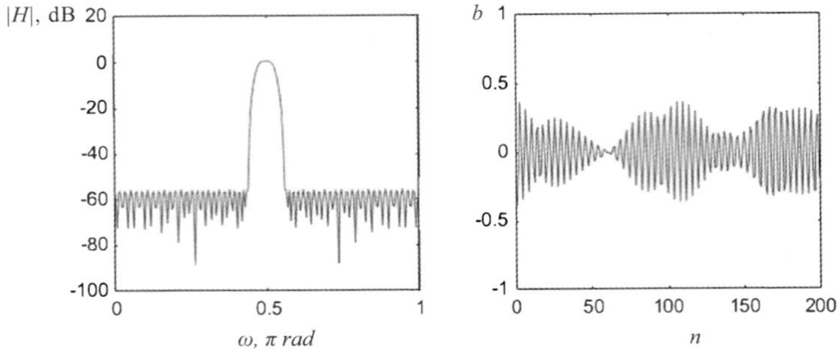

FIGURE 4.15

Realization of a "bandpass" sequence.

i.e., its frequency response is

$$H(e^{j\Omega}) = \frac{S_{DX}(e^{j\Omega})}{S_{XX}(e^{j\Omega})}. \tag{4.199}$$

If one needs a causal finite impulse response (FIR) solution, a different but equally simple solution can be found for $h[n]$.

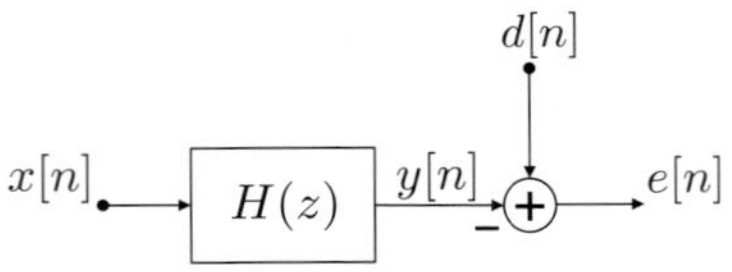

FIGURE 4.16

General discrete Wiener filter.

We can solve a very simple example where we compute an optimum zero-order predictor for a given signal of which a noisy version is available. In this case, $H(e^{j\Omega}) = c$ (a simple gain to be determined), the input $x[n] = s[n] + v[n]$ (signal $s[n]$ additively corrupted by uncorrelated noise $v[n]$), and $d[n] = s[n+1]$. The solution is

$$c = \frac{R_{SS}[1]}{R_{SS}[0] + R_{VV}[0]}, \tag{4.200}$$

which yields the minimum mean quadratic error

$$\mathrm{E}[e^2[n]] = \frac{R_{SS}^2[0] + R_{SS}[0]R_{VV}[0] - R_{SS}^2[1]}{R_{SS}[0] + R_{VV}[0]}. \tag{4.201}$$

Note that if no noise is present,

$$c = \frac{R_{SS}[1]}{R_{SS}[0]} \tag{4.202}$$

and

$$\mathrm{E}[e^2[n]] = \frac{R_{SS}^2[0] - R_{SS}^2[1]}{R_{SS}[0]}. \tag{4.203}$$

The reader should be aware that this example studies a theoretical probabilistic model, which may therefore depend on several parameters which probably would not be available in practice and should rather be estimated. In fact, it points out the solution a practical system should pursue.

The Wiener filter is a kind of benchmark for adaptive filters [5], which optimize themselves recursively.

4.4.17 Application: characterization of LTI systems

A direct way to find the impulse response $h(t)$ of an LTI system is to apply white noise $W(t)$ with power spectral density $S_{WW}(\omega) = S_{WW}\ \forall\omega$ to the system input, which yields $R_{YX}(\tau) = S_{WW}h(\tau)$. From an estimate $\widehat{R_{YX}}(\tau)$ of the cross-correlation, one can obtain an estimate of the desired impulse response: $\hat{h}(t) \approx \widehat{R_{YX}}(t)$. In practice, assuming ergodicity, the following approximations are employed:

- $R_{XX}(\tau) \approx \delta(\tau)$;
- $\widehat{R_{YX}}(\tau) = \widehat{\mathcal{R}_{yx}}(\tau) = \frac{1}{2T}\int_{-T}^{T} y(t)x(t-\tau)\mathrm{d}t$ for sufficiently long T.

Furthermore, the operations are often performed in the discrete-time domain. In order to exemplify this procedure, a simulation was carried. With the goal of identifying the impulse response $h[n]$ of the system described by the difference equation $y[n] = -0.155y[n-1] + 0.36y[n-2] - 0.06469y[n-3] + x[n] + 0.5x[n-1]$, we used 100,000 samples of white Gaussian noise.[24] Each output allows us to calculate $\hat{\mathcal{R}}_{XY}$. Fig. 4.17 shows a comparison of the actual impulse response and the estimated value by the aforementioned method.

The effective bandwidth of an LTI system with frequency response $H(\mathrm{j}\omega)$ can be estimated by its *noise bandwidth* W_{w}. For a low-pass system, the idea is to identify the frequency W_{w} such that both $H(\mathrm{j}\omega)$ and an ideal low-pass filter with cutoff frequency W_{w} and passband gain $|H(0)|$ would deliver equal output powers when subjected to the same input white noise process. A straightforward algebraic development yields

$$W_{\mathrm{w}} = \frac{\int_0^\infty |H(\mathrm{j}\omega)|^2 \mathrm{d}\omega}{|H(0)|^2}. \tag{4.204}$$

For a bandpass system with frequency response centered at ω_0, the idea is to identify the frequency bandwidth Ww such that both $H(\mathrm{j}\omega)$ and an ideal bandpass filter with passband of width $[\omega_0 - \frac{W_{\mathrm{w}}}{2}, \omega_0 + \frac{W_{\mathrm{w}}}{2}]$ and gain $|H(\mathrm{j}\omega_0)|$ would deliver the same output powers when subjected to the same input white noise process. In this case,

$$W_{\mathrm{w}} = \frac{\int_0^\infty |H(\mathrm{j}\omega)|^2 \mathrm{d}\omega}{|H(\omega_0)|^2}. \tag{4.205}$$

[24] Defining the number of samples needed to characterize an unknown LTI system is a challenge in itself.

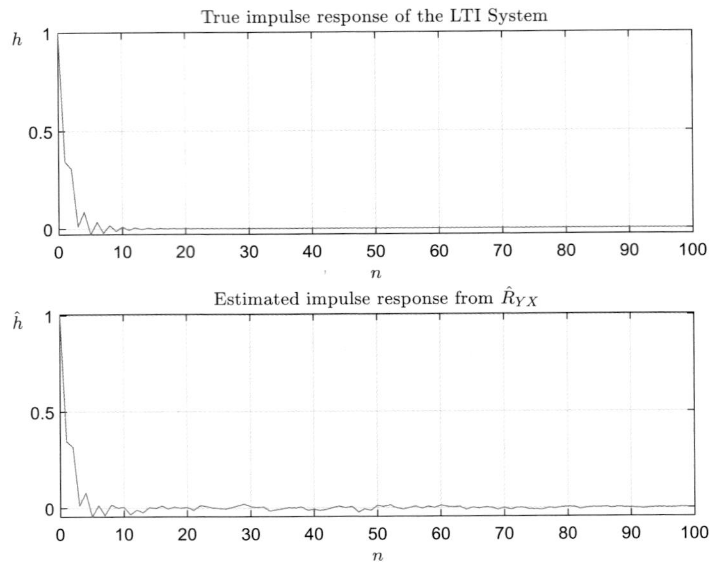

FIGURE 4.17

True system impulse response and estimation via cross-correlation.

Illustrations of this concept are given in Fig. 4.18 for a sixth-order low-pass filter and a 12th-order bandpass filter.

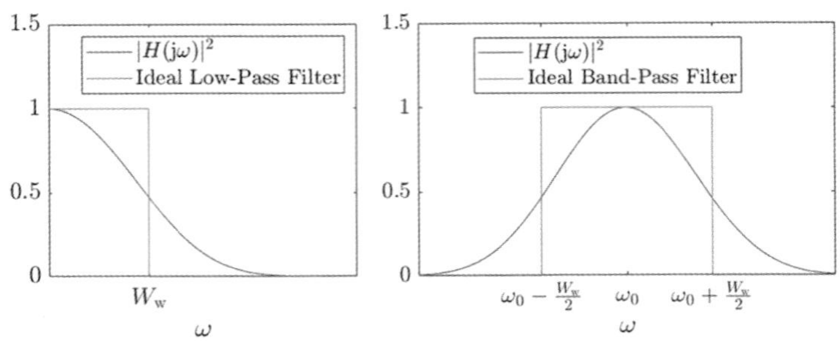

FIGURE 4.18

Equivalent noise bandwidth: ideal filter modeling of a low-pass system (on the left) and of a bandpass system (on the right).

4.4.18 Modeling of bandpass WSS random processes

Any real "bandpass" random process (see Section 4.4.15) at center frequency ω_0 can be expressed as $N(t) = M(t)\cos[\omega_0 t + \Theta(t)]$, where the envelope $M(t)$ and the phase $\Theta(t)$ are both baseband random processes. We can write

$$N(t) = N_x(t)\cos(\omega_0 t) - N_y(t)\sin(\omega_0 t), \tag{4.206}$$

where $N_x(t) = M(t)\cos(\Theta(t))$ and $N_y(t) = M(t)\sin(\Theta(t))$ are the quadrature components of $N(t)$. In the typical situation where $N(t)$ is a zero-mean WSS random process with autocorrelation $R_{NN}(\tau)$, we can explicitly find the first- and second-order moments of the quadrature components that validate the model. First define the *Hilbert transform* of a given signal $n(t)$ as $\hat{n}(t) = (n * h)(t)$, where

$$H(\mathrm{j}\omega) = \mathcal{F}[h(t)] = \begin{cases} -\mathrm{j}, & \omega > 0, \\ \mathrm{j}, & \omega < 0. \end{cases} \tag{4.207}$$

From $R_{NN}(\tau)$, we can find $R_{\hat{N}N}(\tau)$ and proceed to write:

- $\overline{N_x} = \overline{N_y} = 0$;
- $R_{N_xN_x}(\tau) = R_{N_yN_y}(\tau) = R_{NN}(\tau)\cos(\omega_0\tau) + R_{\hat{N}N}(\tau)\sin(\omega_0\tau)$;
- $R_{N_xN_y}(\tau) = -R_{N_yN_x}(\tau) = R_{NN}(\tau)\sin(\omega_0\tau) - R_{\hat{N}N}(\tau)\cos(\omega_0\tau)$.

In the frequency domain, if we define

$$S^+_{NN}(\mathrm{j}\omega) \triangleq \begin{cases} S_{NN}(\mathrm{j}\omega), & \omega > 0, \\ 0, & \omega \le 0, \end{cases} \tag{4.208}$$

then:

- $S_{N_xN_x}(\mathrm{j}\omega) = S_{N_yN_y}(\mathrm{j}\omega) = S^+_{NN}(\mathrm{j}(\omega + \omega_0)) + S^+_{NN}(\mathrm{j}(-\omega + \omega_0))$;
- $S_{N_xN_y}(\mathrm{j}\omega) = -S_{N_yN_x}(\mathrm{j}\omega) = \mathrm{j}[S^+_{NN}(\mathrm{j}(-\omega + \omega_0)) - S^+_{NN}(\mathrm{j}(\omega + \omega_0))]$.

In the particular case where $N(t)$ is Gaussian, it can be shown that $N_x(t)$ and $N_y(t)$ are Gaussian, and thus completely defined by their first- and second-order moments above.

4.4.19 Statistical modeling of signals: random sequence as the output of an LTI system

A discrete-time signal $x[n]$ with spectrum $X(\mathrm{e}^{\mathrm{j}\Omega})$ can be modeled as the output of an LTI system whose frequency response is equal to $X(\mathrm{e}^{\mathrm{j}\Omega})$ when excited by a unit impulse $\delta[n]$. In this case, the impulse response of the system is expected to be $h[n] = x[n]$.

In the context of statistical models, an analogous model can be built: now, we look for an LTI system which produces as its output the WSS random sequence $X[n]$ with power spectral density $S_{XX}(\mathrm{e}^{\mathrm{j}\Omega})$, when excited by a unit-variance random sequence $W[n]$ of white noise. As before, the frequency response of the system alone is expected to shape the output spectrum, yet this time in the mean power sense, i.e., the system must be such that $|H(\mathrm{e}^{\mathrm{j}\Omega})|^2 = S_{XX}(\mathrm{e}^{\mathrm{j}\Omega})$.

Assume the overall modeling system is described by the difference equation

$$x[n] + \sum_{k=1}^{p} a_p[k]x[n-k] = \sum_{k=0}^{q} b_q[k]w[n-k], \tag{4.209}$$

where $w[n]$ is white noise with $S_{WW}(e^{j\Omega}) = 1$. The corresponding transfer function is

$$H(z) = \frac{B_q(z)}{A_p(z)} = \frac{\sum_{k=0}^{q} b_q[k]z^{-k}}{1 + \sum_{k=1}^{p} a_p[k]z^{-k}}. \tag{4.210}$$

The output $X[n]$ of this general system model with $W[n]$ at its input is called *autoregressive moving average*[25] (ARMA) process of order (p, q). It can be described by the following difference equation (known as the *Yule–Walker equation*) in terms of its autocorrelation:

$$R_{XX}[k] + \sum_{l=1}^{p} a_p[l] R_{XX}[k-l] = \begin{cases} \sum_{l=0}^{q-k} b_q[l+k]h^*[l], & 0 \le k \le q, \\ 0, & k > q. \end{cases} \tag{4.211}$$

The values of $R_{XX}[k]$ for $k < 0$ can be easily found by symmetry.

If one restricts the system to be FIR (i.e., to have a finite duration impulse response), then $p = 0$, i.e., $a_p[k] = 0$ for $1 \le k \le p$, and

$$H(z) = \sum_{k=0}^{q} b_q[k]z^{-k}. \tag{4.212}$$

The output of this "all-zero"[26] system model with $W[n]$ at its input is called a *moving average* (MA) process of order q. Its autocorrelation can be found to be

$$R_{XX}[k] = \sum_{l=0}^{q-|k|} b_q[l+|k|]b_q^*[l]. \tag{4.213}$$

In this case, $R_{XX}[k] = 0$ for $k < -q$ or $k > q$. This model is more suited for modeling notches in the random sequence power spectrum.

If one restricts the system to be "all-pole",[27] then $q = 0$, i.e., $b_q[k] = 0$ for $1 \le k \le p$, and

$$H(z) = \frac{b[0]}{1 + \sum_{k=1}^{p} a_p[k]z^{-k}}. \tag{4.214}$$

---

25 Autoregressive for its output feedback, moving average for its weighted sum of past input samples.

26 This is misnamed, since the poles of the system are at the origin. To be precise, the zeros are the only responsible for shaping the magnitude frequency response of the system.

27 This is also misnamed, since the zeros of the system are at the origin. To be precise, the poles are the only responsible for shaping the magnitude frequency response of the system.

The output of this system model with $W[n]$ at its input is called an *autoregressive* (AR) process of order p. Its autocorrelation follows the following difference equation:

$$R_{XX}[k] + \sum_{l=1}^{p} a_p[l] R_{XX}[k-l] = |b[0]|^2 \delta[k]. \qquad (4.215)$$

Again, the values of $R_{XX}[k]$ for $k < 0$ can be easily found by symmetry. This model is more suited for modeling peaks in the random sequence power spectrum, which is typical of quasiperiodic signals (as audio signals, for example). It should be noted that, differently from the ARMA and MA cases, the equation for the AR process autocorrelation is linear in the system coefficients, which makes their estimation easier. If $R_{XX}[k]$ is known for $0 \leq k \leq p$, recalling that $R_{XX}[-k] = R^*_{XX}[k]$, one can:

- solve the pth-order linear equation system obtained by substituting $k = 1, 2, \ldots, p$ in Eq. (4.215) to find $a_p[k]$, $1 \leq k \leq p$;
- compute $|b[0]|$ from Eq. (4.215) with $k = 0$.

All-pole modeling of audio signals is extensively used in applications like speech coding [13] and audio restoration [7], for example.

4.5 Afterword

Spanning fields as diverse as biology, economics, and telecommunications, it is impossible to list all areas that make extensive use of probabilistic models.

In particular, digital signal processing [6] is going through a real revolution thanks to the growing use of advanced machine learning [2] methods, previously impractical due to their high data, storage, and computational requirements. Benefiting from deep learning [8], black box type models are being massively adopted in various industries.

The most virtuous objective of our times is to combine the highest quality of life with the most efficient use of natural resources. In this context, exercising maximum control over processing means and tools becomes essential. Since the theoretical basis behind computational intelligence is inherently probabilistic, a solid background in the subject of this chapter provides the necessary tools to that end.

Thanks for getting this far.

References

[1] L.W.P. Biscainho, Random signals and stochastic processes, in: P.S. Diniz, J.A. Suykens, R. Chellappa, S. Theodoridis (Eds.), Academic Press Library in Signal Processing: Volume 1, in: Academic Press Library in Signal Processing, vol. 1, Academic Press, Waltham, 2014, pp. 113–168.

[2] C.M. Bishop, Pattern Recognition and Machine Learning, Springer, New York, 2006.

[3] G. Casella, R.L. Berger, Statistical Inference, 2nd ed., Duxbury, Pacific Grove, 2001.

[4] R. Deep, Probability and Statistics, Academic Press, Burlington, 2006.

[5] P.S.R. Diniz, Adaptive Filtering: Algorithms and Practical Implementation, 5th ed., Springer, Boston, 2020.

[6] P.S.R. Diniz, E.A.B. da Silva, S.L. Netto, Digital Signal Processing: Systems Analysis and Design, 2nd ed., Cambridge University Press, New York, 2010.
[7] S.J. Godsill, J.W. Rayner, Digital Audio Restoration: A Statistical Model Based Approach, Springer, London, 1998.
[8] I. Goodfellow, Y. Bengio, A. Courville, Deep Learning, MIT, Cambridge, 2016.
[9] M.H. Hayes, Statistical Digital Signal Processing and Modeling, Wiley, Hoboken, 1996.
[10] S. Miller, D. Childers, Probability and Random Processes, 2nd ed., Academic Press, Waltham, 2012.
[11] A. Papoulis, P. Unnikrishna, Probability, Random Variables and Stochastic Processes, 4th ed., McGraw-Hill, New York, 2002.
[12] P.Z. Peebles Jr., Probability, Random Variables and Random Signal Principles, 4th ed., McGraw-Hill, New York, 2001.
[13] L.R. Rabiner, R.W. Schafer, Theory and Applications of Digital Speech Processing, Pearson, Upper Saddle River, 2010.
[14] K.S. Shanmugan, A.M. Breipohl, Random Signals: Detection, Estimation and Data Analysis, Wiley, Hoboken, 1988.
[15] C.W. Therrien, M. Tummala, Probability and Random Processes for Electrical and Computer Engineers, 2nd ed., CRC Press, Boca Raton, 2012.

CHAPTER

5 Sampling and quantization

Håkan Johansson
Division of Communication Systems, Department of Electrical Engineering, Linköping University, Linköping, Sweden

5.1 Introduction

The shift from analog signal processing (ASP) to digital signal processing (DSP) lies behind the tremendous development of information processing systems that are used today in practically all areas one can think of. For example, DSP has enabled the fast growth of mobile communication systems and has paved the way for the development of new sophisticated medical aids, just to mention a couple of important applications. The foremost advantages of DSP over ASP are its robustness and that it can perform functions with arbitrary precision. This is because arithmetic operations in the digital domain can be done with as small numerical errors as desired by simply increasing the resolution, i.e., by allocating more bits to the operands. This can always be done as the resolution in DSP systems is independent of physical variations, like temperature changes, processing inaccuracies, etc., in contrast to ASP systems, which are highly dependent on such variations. The trend during the past decades has therefore been to use as little ASP as possible and to replace analog functions with the corresponding digital functions. This trend has been sped up by the fast development of integrated circuit techniques and progress in the implementation of digital systems. Specifically, the introduction of cost-efficient signal processors and so-called application-specific integrated circuits has enabled low-cost implementation and manufacturing of digital systems. The real world is however analog by nature, which means that the need for ASP cannot be eliminated completely. In particular, it will always be necessary to use analog-to-digital converters (abbreviated A/D converters or ADCs) and digital-to-analog converters (abbreviated D/A converters or DACs) for interfacing the two realms.

Evidently, ADCs and DACs are crucial components in virtually all practical applications today. The fundamental signal processing operations in these components are sampling and quantization (in ADCs) and signal reconstruction (in DACs). This chapter considers the basic principles of these operations as detailed below.

5.1.1 Scope and prerequisites

The theory of sampling has a long history [23,38,47,62] and a large literature body; see for example [17,51,53] and references therein. In this chapter, we are only able to cover some parts of this field. The emphasis is on basic principles and analysis methods for uniform sampling and quantization, which are the most common sampling and quantization schemes used in conventional ADCs and DACs. Furthermore, we will only consider linear models of the sampling and quantization processes. Practical ADCs and DACs exhibit (small) nonlinearities that also need be analyzed, but this topic is beyond the scope of

Signal Processing and Machine Learning Theory. https://doi.org/10.1016/B978-0-32-391772-8.00011-9
Copyright © 2024 Elsevier Inc. All rights reserved.

this chapter. In other words, this chapter concentrates on the underlying principles of uniform sampling and quantization rather than fine implementation details. For such details, we refer to books on ADCs and DACs [54]. Furthermore, we take an "engineering approach" in that we concentrate on explaining the concepts that are sufficient for understanding and analyzing practical signals and systems. Some of the fine mathematical details are therefore sometimes left out.

It is assumed that the reader has a basic knowledge of continuous-time and discrete-time signals and systems as well as the Fourier, Laplace, and z-transforms. Some of the basics regarding signals and systems are however briefly recapitulated in Section 5.2. Further, for a complete understanding of Section 5.4, the reader must be familiar with stochastic processes. Most of the material in the other sections can however be understood without reading Section 5.4.

5.1.2 Chapter outline

After this introductory section, some basic concepts are given in Section 5.2. Then, Section 5.3 discusses the basics of uniform sampling and reconstruction of deterministic signals. The content of this section is typically also found in introductory books on signal processing [34,39]. Section 5.4 considers the extension to sampling and reconstruction of stochastic processes (also called random processes). In introductory books on signal processing, this is typically treated very briefly or not at all.

Section 5.5 considers an application referred to as time-interleaved ADCs, which is a scheme that has been introduced for high-speed A/D conversion [3]. The section discusses the principles and signal reconstruction methods that are required in this scheme due to analog channel mismatches. Without correction, this scheme corresponds to nonuniform sampling (or generalizations), but the overall goal, with correction, is to achieve uniform sampling.

Section 5.6 covers the principles of quantization, both in ADCs and internally in digital systems, as they are closely related. In Section 5.7, oversampled ADCs and DACs are briefly discussed. Oversampling techniques are used in practice to, e.g., increase the performance and relax the requirements of the analog components. This chapter also includes a discussion on the use of few-bit ADCs in massive multiple-input multiple-output (MIMO) systems. Finally, Section 5.8 presents a method for discrete-time modeling of mixed-signal systems. This is useful in systems with feedback, like so-called $\Sigma\Delta$-modulator-based ADCs. Such converters can reach a very high resolution using oversampling together with noise shaping [37,43].

Finally, it is pointed out that, throughout the chapter, real-valued signals are primarily used in the examples. However, the basic principles to be presented are valid for complex signals as well.

5.1.3 Recent and current trends

The research and development of data converters have been quite intensive for several decades, both in academia and industry. This interest is foreseen to continue for many years to come as the requirements on signal processing systems and thus the converters continuously increase. At the same time, it is also desirable to decrease the size and power consumption of the converters. To develop energy-efficient high-performance devices, it is therefore vital to continue to conduct research on new principles of data conversion as well their practical implementations.

There are different paths one can take to increase the performance. One way is to make use of parallel converters like time-interleaved ADCs [3] (to be discussed in Section 5.5) and/or parallel $\Sigma\Delta$-ADCs [22]. Such converters have great potential but due to analog channel mismatches they also introduce errors that must be compensated for. A recent trend here has been "digitally assisted analog circuits," where digital circuitry is added to correct for analog errors [36]. With the increasing computational capacity of digital systems, one can expect that this trend will continue. The digital parts may however be computationally excessive, and it is therefore vital to devise new algorithms to reduce their complexity.

Another recent trend is that of sub-Nyquist sampling covering so-called sparse sampling and compressed sampling (also called compressed sensing) [4,33,48]. This is different from the traditional Nyquist-sampled systems in that additional knowledge of the input signal is taken into account. Specifically, if the signal is sparse in some sense, one can reduce the sampling rate and thereby relax the requirements on the converters. One scenario is for applications where the signals are locally narrow-band (in time) but globally wide-band. Conventionally, high-speed converters are then needed. Utilizing the sparse or compressed sampling paradigm, the sampling rate can be substantially reduced.

Finally, in communication systems, massive MIMO systems have emerged as key enablers for 5G systems and beyond [24,30,50]. To reduce the cost of such systems it is desired to make use of few-bit ADCs, and low-precision analog components in general.

5.2 Preliminaries

5.2.1 Classification of signals

We communicate by conveying information (messages), like speech and images. In some contexts, one wishes to transmit information from one place to another, like in telephone systems. In other cases, it may be desired to store the information for later use, e.g., music stored on CDs.

Information can be transmitted using information bearing signals. A signal is a quantity that varies with time or other independent variables. Usually, the information must be converted to a desired signal type. For example, when we speak in a microphone, the generated sound is converted to an electrical voltage that varies with time. The sound can then be recovered by connecting the voltage to a loudspeaker. (In this context, however, we usually do not consider the sound as being recovered since it, in principle, is reconstructed at the same time as it is generated.) The same information can be represented in many different ways. Different types of signals can in other words contain exactly the same information. For example, speech and music can be represented in analog form on LPs and in digital form on CDs. The corresponding signals are then analog and digital, respectively.

In mathematical form, a signal is represented as a function (or sequence) of one or several independent variables. The signal is a 1D signal if it is a function of one variable and a multidimensional signal if it is a function of several variables. This chapter considers 1D signals only. An example of such signals is a speech signal. It is customary to let the independent variable represent time, although it may represent whatever we wish, like temperature, pressure, etc. The time variable in the representation of a signal can be either continuous or discrete (quantized). The signal is a continuous-time signal if the variable is continuous and a discrete-time signal if the variable is discrete. A continuous-time signal is thus defined for all time instances (except possibly at discontinuities), whereas a discrete-time signal is defined only at discrete time instances. We also distinguish between signals that can take on continuous signal values and those that only can take on discrete signal values. Consequently, signals can be divided into four different categories according to Fig. 5.1.

In the literature, a continuous-time signal with continuous signal values is often referred to as an analog signal, whereas a discrete-time signal with discrete signal values is called a digital signal. How-

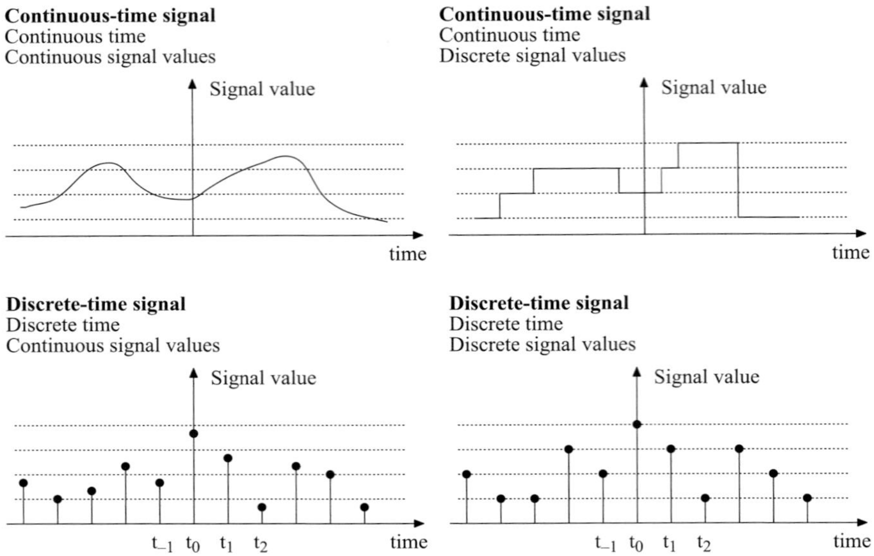

FIGURE 5.1

Different types of signals.

ever, in some (earlier) literature, "analog" and "digital" are only related to the signal values, i.e., the signals to the left in Fig. 5.1 are then analog, whereas the signals to the right in the same figure are digital. In this book, we assume hereafter that analog signals are continuous-time signals whereas digital signals are discrete-time signals.

5.2.2 Discrete-time signals – sequences

A continuous-time signal can mathematically be represented by a continuous function $x(t)$, where t is a continuous variable. A discrete-time signal cannot be represented in this way since it is only defined at discrete instances. To handle discrete-time signals mathematically, one therefore makes use of sequences of numbers.

As one can see from Fig. 5.1, the signal values of a discrete-time signal constitute a sequence of numbers. In mathematical form, a sequence of numbers can be written as

$$\{x(n)\}, \quad n = \ldots, -2, -1, 0, 1, 2, \ldots, \tag{5.1}$$

where $\{x(n)\}$ denotes the sequence and $x(n)$ denotes the value of the nth number in the sequence. The values $x(n)$ are usually called samples because they are often generated through sampling of continuous-time signals (see Section 5.2.3). The notation of the sequence is thus $\{x(n)\}$, but when no misunderstandings can occur we write "the sequence $x(n)$" for the sake of simplicity. A sequence can be represented graphically as illustrated in Fig. 5.2. The horizontal axis is drawn as a continuous line but it is important to remember that the sequence is only defined for discrete values.

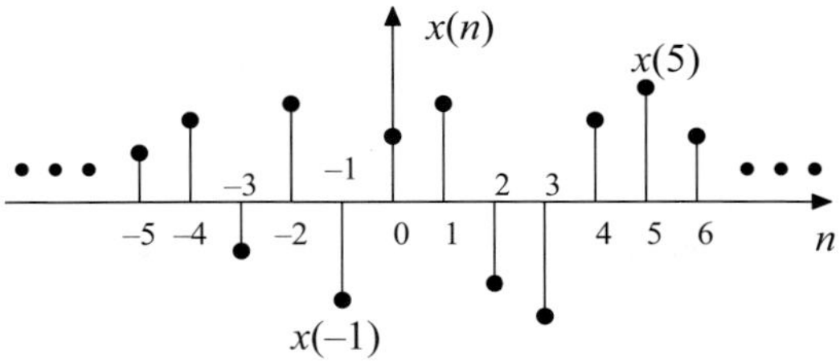

FIGURE 5.2

Graphical representation of a sequence $x(n)$.

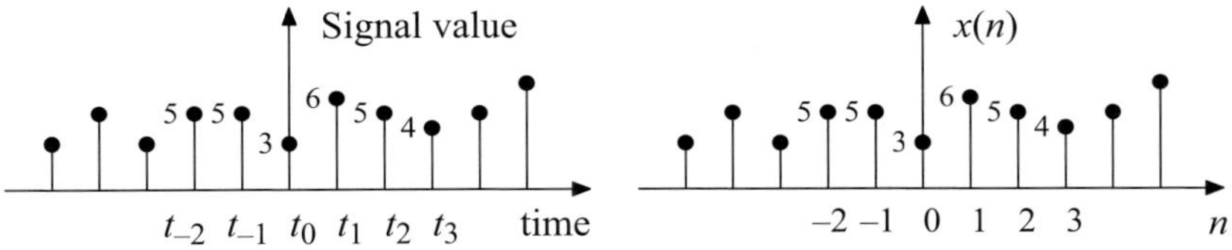

FIGURE 5.3

A discrete-time signal and its representation in the form of a sequence $x(n)$.

When a sequence represents a discrete-time signal, the samples $x(n)$ equal the signal values of the signal at the time instances $t = t_n$, as illustrated in Fig. 5.3. Since sequences often represent discrete-time signals, we frequently write "the discrete-time signal $x(n)$" instead of "the sequence $x(n)$." This is in accordance with the continuous-time case, where one often writes "the continuous-time signal $x(t)$" instead of "the function $x(t)$."

5.2.3 Sampling of continuous-time signals

Discrete-time signals are often generated through sampling (measurement) of continuous-time signals. It is common to use the same time interval T between the sampling instances. The sampling then takes place at the time instances $t = nT$, as illustrated in Fig. 5.4. The time interval is called the sampling period, whereas $f_s = 1/T$ is referred to as the sampling frequency, which thus denotes the number of sampling instances per second. This type of sampling is called uniform sampling. If $x_a(t)$ denotes the continuous-time signal,[1] the corresponding discrete-time signal $x(n)$ becomes $x(n) = x_a(nT)$. Instead of using the notation $x(n)$ one may thus use $x_a(nT)$ to stress that the signal values are obtained via uniform sampling of a continuous-time signal. Throughout this chapter, we mainly use $x(n)$ for the sake of simplicity.

[1] The subscript "a" in $x_a(t)$ stands for analog. We use the notation $x_a(t)$ instead of $x_c(t)$, which would be natural when we deal with continuous-time signals. This is to avoid confusion with frequency edges Ω_c, etc., where "c" stands for cutoff.

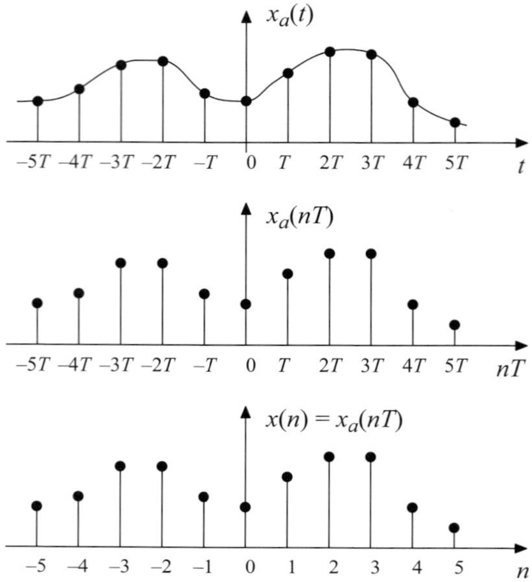

FIGURE 5.4

Sampling of a continuous-time signal.

5.2.4 Classification of systems

Systems can be classified in accordance with their input and output signal types. As we saw in the previous section, we can divide signals into four different categories. Hence, one may also divide systems into four different classes, as seen in Fig. 5.5. A continuous-time system is a system whose input and output signals are continuous-time signals, whereas a discrete-time system is a system whose input and output signals are discrete-time signals. Mathematically, the role of a discrete-time system is thus to generate an output sequence given an input sequence. Therefore, instead of the terms "input signal" and "output signal," we often use the terms "input sequence" and "output sequence." A digital system is a discrete-time system in which the samples can only take on discrete values. Today, signal processing systems usually contain both continuous-time and discrete-time parts.

5.2.5 Digital signal processing of analog signals

The principle of DSP of analog signals is illustrated in Figs. 5.6–5.8. The first step is to convert the signal from analog to digital form. This is carried out by an ADC which takes care of both sampling and quantization. The digital signal thus obtained can then be processed by a digital system which takes as input a sequence of samples, $x_Q(n)$, and generates an output sequence, $y_Q(n)$ (Q indicates quantized signal values). Finally, reconstruction takes place in order to go back to a continuous-time representation. This is performed by a DAC followed by an analog reconstruction filter.

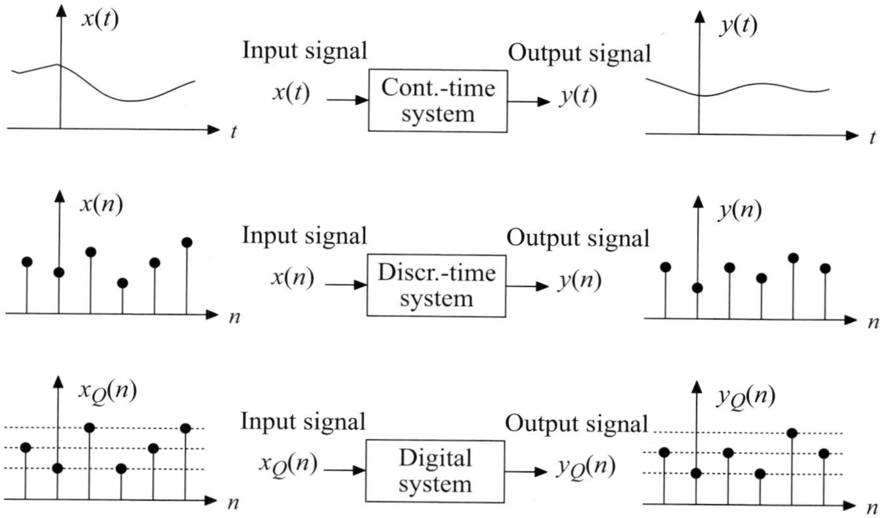

FIGURE 5.5

Three different types of systems.

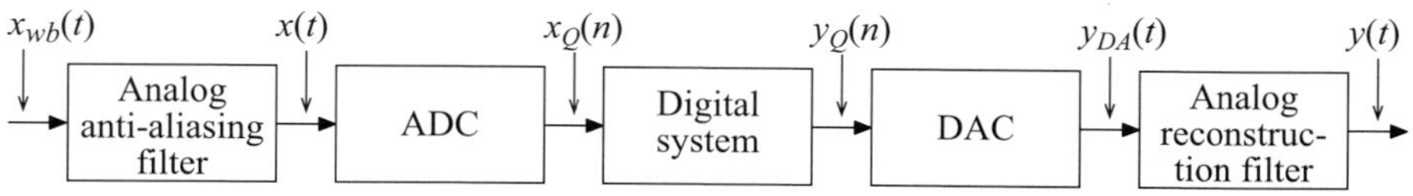

FIGURE 5.6

Digital signal processing of analog signals.

To be able to perform A/D and D/A conversion without errors, the original signal must be band-limited. Before the A/D conversion, the signal is therefore band-limited with the aid of an analog antialiasing filter (wb in $x_{wb}(t)$ in Fig. 5.6 stands for wide-band). A/D conversion is carried out by first sampling and then quantizing the signal (see Fig. 5.7(a)). In practice, the sampling is performed by a sample-and-hold (S/H) circuit. An S/H circuit samples an analog signal at the time instances $t = nT$ and holds this sample value until the next sampling instance. The output signal from an ideal S/H circuit is a new analog signal that ideally is piecewise constant. The function of the S/H circuit can be seen in Fig. 5.8(a and b). The next step in the A/D conversion is to quantize the sample values using a quantizer. The input signal of the quantizer is thus the output signal from the S/H circuit. The input to the quantizer needs to be kept constant long enough for the quantizer to produce a correct result, which is why an S/H circuit is needed.

Quantization amounts to representing each sample in a binary form with a finite word length. There exist many different types of binary representations. One commonly used binary form is the two's complement representation. A number x, which for the sake of simplicity is assumed here to lie between

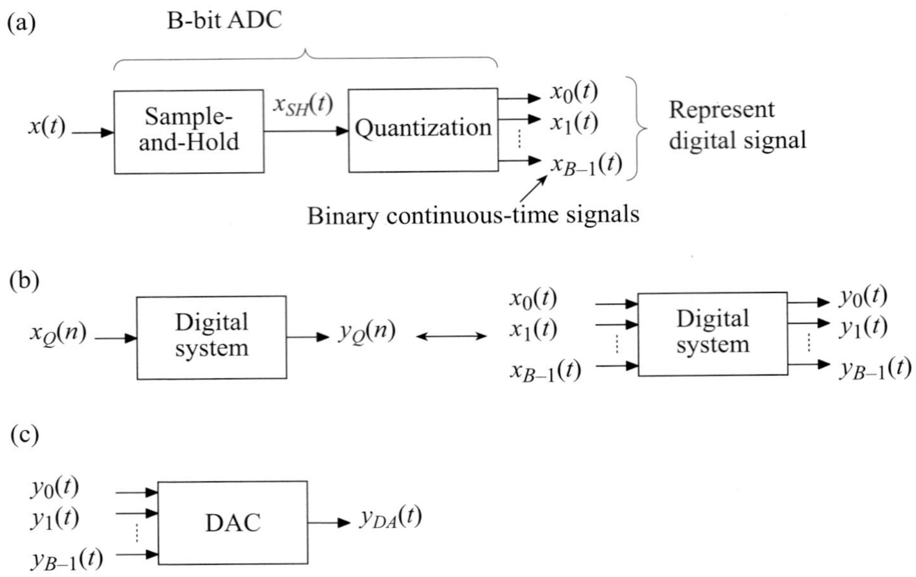

FIGURE 5.7

ADC, digital system, and DAC.

-1 and 1, is then represented as

$$x = -x_0 + \sum_{i=1}^{B-1} 2^{-i} x_i, \tag{5.2}$$

where x_i, $i = 0, 1, \ldots, B-1$, can take on the values zero and one. The number of bits is thus B. The bits from a quantizer can be delivered in series or in parallel. In the latter case, the output signal from a B-bit quantizer consists in practice of B continuous-time signals which together represent the different sample values (see Fig. 5.7(a)). In the ideal case, these continuous-time signals are binary, that is, they can only take on two different values, usually zero and one. The principle can be seen in Fig. 5.8(c and d). In practice, the signals must hold the values zero or one long enough so one can store them in registers, etc. Furthermore, it takes a certain time to perform the quantizations, which means that a certain amount of delay will be introduced. The output signal from the quantizer is the desired digital signal that can be processed by a digital system. In a practical implementation, the sample values are represented in the same form as the quantizer output. The output signal from a digital system is thus represented in the same form as its input signal (see Fig. 5.8(e and f)).

A digital signal can be converted to an analog signal using a DAC followed by an analog reconstruction filter. The input signal to the DAC is a digital signal represented in the same form as the output signal from the ADC, i.e., the output from the quantizer. The output signal from the DAC is a continuous-time signal that is piecewise constant with a limited number of levels (see Fig. 5.8(g)). In other words, it is of the same type as the output signal from the S/H circuit with the difference that

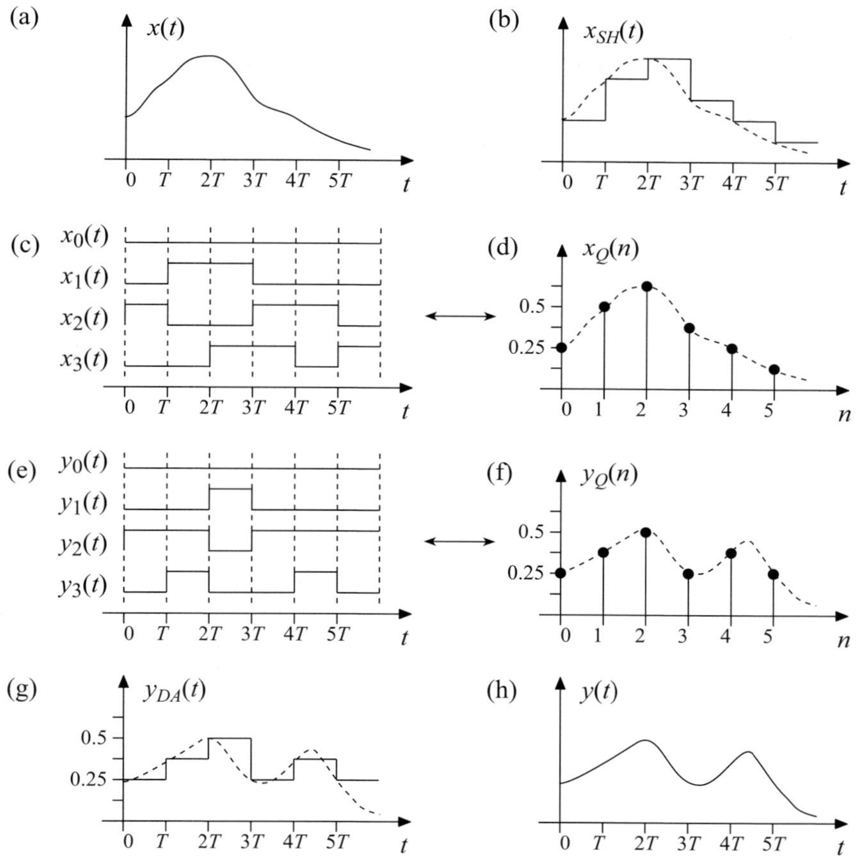

FIGURE 5.8

Typical signals in digital signal processing of analog signals.

the DAC output signal takes on discrete signal values, whereas the S/H circuit output signal takes on continuous signal values. Finally, to obtain the desired analog signal, the output signal from the DAC must be filtered. The role of this filter is to "smooth out" the piecewise constant signal from the DAC so as to obtain the desired signal.

As evident from above, sampling and reconstruction as well as quantization play fundamental roles in DSP. The following sections treat the basic principles of these operations. We will only deal with relations between continuous-time signals and their discrete-time counterparts. Mathematically, this amounts to studying relations between functions $x_a(t)$ and the corresponding sequences $x(n) = x_a(nT)$ as well as their transform–domain relations. We will not discuss the physical representation of the signals seen in Figs. 5.7 and 5.8.

FIGURE 5.9

Simple model of a digital communication system.

5.2.6 Digital communication – analog signal processing of digital signals

In digital communication systems, the situation is opposite to that in the previous section in that ASP is now used for processing digital signals. Fig. 5.9 shows a simple model of a digital communication system. The original digital information is contained in $x_Q(n)$. This information is D/A converted before being transmitted over an analog channel. Finally, on the receiver side, the signal is A/D converted, generating $y_Q(n)$. It is desired to have $y_Q(n) = x_Q(n)$, i.e., to receive the information sent. In practice, due to many different error sources in the communication channel, advanced DSP methods are required to transmit information successfully. Regarding the sampling and reconstruction, one can make use of the same analysis methods here as for the system depicted earlier in Fig. 5.7.

5.3 Sampling of deterministic signals

Discrete-time signals are often generated through sampling of continuous-time signals. It is therefore of interest to investigate the relations between the Fourier transform of a discrete-time signal and that of the underlying continuous-time signal. This relation is given by the so-called Poisson summation formula, which leads us to the sampling theorem. This theorem states that a continuous-time signal can be sampled and perfectly reconstructed provided that the signal is band-limited and the sampling frequency exceeds twice the bandwidth. If the theorem is fulfilled, the original signal can be retained via an ideal pulse amplitude modulator. This section treats basic concepts and theory on sampling and reconstruction. In practice, one can neither fulfill the sampling theorem nor perform the reconstruction perfectly because ideal pulse amplitude modulators (PAMs) cannot be implemented. Some of these practical aspects are included in the section. It is also noted that this section deals with deterministic signals. The extension to stochastic signals is treated in Section 5.4.

5.3.1 Uniform sampling

Discrete-time signals are typically generated through sampling (measurement) of continuous-time signals. Most signal processing applications are based on uniform sampling, which means that the time interval between two consecutive sampling instances is constant. The sampling then takes place at the time instances $t = nT$, $n = ..., -2, -1, 0, 1, 2, ...$, as illustrated in Fig. 5.10. The time interval T is called the sampling period, whereas $f_s = 1/T$ is referred to as the sampling frequency, which thus denotes the number of sampling instances per second. If $x_a(t)$ denotes the continuous-time signal, the corresponding discrete-time signal, represented by the sequence $x(n)$, becomes

$$x(n) = x_a(nT). \tag{5.3}$$

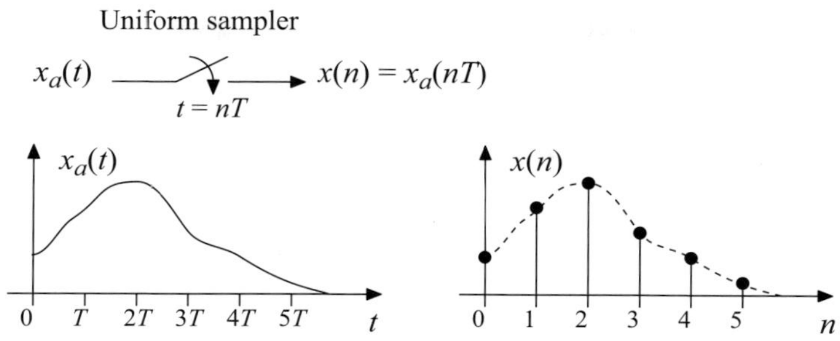

FIGURE 5.10

Uniform sampling.

5.3.2 Poisson's summation formula

This section derives a relation between the Fourier transforms of $x(n)$ and $x_a(t)$. Assume that the continuous-time signal $x_a(t)$ has the Fourier transform $X_a(j\Omega)$. We can then express $x_a(t)$ in terms of its inverse Fourier transform according to

$$x_a(t) = \frac{1}{2\pi}\int_{-\infty}^{\infty} X_a(j\Omega)e^{j\Omega t}d\Omega. \tag{5.4}$$

In particular, we have for $t = nT$

$$x_a(nT) = \frac{1}{2\pi}\int_{-\infty}^{\infty} X_a(j\Omega)e^{j\Omega nT}d\Omega, \tag{5.5}$$

which can be rewritten as

$$\begin{aligned} x_a(nT) &= \ldots + \frac{1}{2\pi}\int_{-3\pi/T}^{-\pi/T} X_a(j\Omega)e^{j\Omega nT}d\Omega + \frac{1}{2\pi}\int_{-\pi/T}^{\pi/T} X_a(j\Omega)e^{j\Omega nT}d\Omega \\ &\quad + \frac{1}{2\pi}\int_{\pi/T}^{3\pi/T} X_a(j\Omega)e^{j\Omega nT}d\Omega + \ldots. \end{aligned} \tag{5.6}$$

By making the variable substitutions $\Omega \to \Omega - 2\pi k/T$, k integer, utilizing $e^{-j2\pi kn} = 1$, and changing the order between summation and integration, we obtain

$$\begin{aligned} x_a(nT) &= \sum_{k=-\infty}^{\infty} \frac{1}{2\pi}\int_{-\pi/T}^{\pi/T} X_a\left(j\Omega - j\frac{2\pi k}{T}\right)e^{j(\Omega - \frac{2\pi k}{T})nT}d\Omega \\ &= \frac{1}{2\pi}\int_{-\pi/T}^{\pi/T} \sum_{k=-\infty}^{\infty} X_a\left(j\Omega - j\frac{2\pi k}{T}\right)e^{j\Omega Tn}d\Omega. \end{aligned} \tag{5.7}$$

With ΩT as integration variable we get

$$x_a(nT) = \frac{1}{2\pi}\int_{-\pi}^{\pi}\frac{1}{T}\sum_{k=-\infty}^{\infty} X_a\left(j\Omega - j\frac{2\pi k}{T}\right)e^{j\Omega T n}d(\Omega T). \tag{5.8}$$

Further, we know that a discrete-time signal $x(n)$ can be expressed in terms of its inverse Fourier transform according to

$$x(n) = \frac{1}{2\pi}\int_{-\pi}^{\pi} X(e^{j\Omega T})e^{j\Omega T n}d(\Omega T). \tag{5.9}$$

Thus, if we let $x(n) = x_a(nT)$ and utilize the uniqueness of the Fourier transform, we obtain the following relation from (5.8) and (5.9):

$$X(e^{j\Omega T}) = \frac{1}{T}\sum_{k=-\infty}^{\infty} X_a\left(j\Omega - j\frac{2\pi k}{T}\right). \tag{5.10}$$

Eq. (5.10) is referred to as Poisson's summation formula[2] and gives the relation between the Fourier transform of a continuous-time signal and the Fourier transform of the discrete-time signal that is generated by sampling the continuous-time signal uniformly with the sampling period T (sampling frequency $f_s = 1/T$) [40]. Poisson's summation formula states that the spectrum of the discrete-time signal is obtained by summing the spectrum of the continuous-time signal and its shifted versions, weighted by $1/T$. This is illustrated in Fig. 5.11 in the following example.

5.3.2.1 *Example 1: illustration of Poisson's summation formula*

Fig. 5.11(a) shows the spectrum of a band-limited continuous-time signal. Fig. 5.11(b–d) shows the spectra of three different discrete-time signals generated by sampling the continuous-time signal with sampling frequencies of 100 Hz, 50 Hz, and 40 Hz, i.e., $T = 1/100$ s, $1/50$ s, and $1/40$ s, respectively. Studying Fig. 5.11, the following observations can be made. In the first two cases, the partial spectra do not overlap with each other (Fig. 5.11(b and c)). This means that the shape of $X_a(j\Omega)$ is preserved and it is thereby possible in principle to reconstruct $x_a(t)$ from $x(n)$, as we shall see later. In the third case, the shape of $X_a(j\Omega)$ is deteriorated since the partial spectra to some extent overlap with each other (Fig. 5.11(d)). In this case, we can no longer perfectly reconstruct $x_a(t)$ from $x(n)$. What distinguishes the three sequences is that they have been obtained using three different sampling frequencies. If the continuous-time signal is band-limited, it can thus be sampled and reconstructed perfectly provided that the sampling frequency is sufficiently high. This is summarized below in the sampling theorem.

5.3.3 The sampling theorem

As illustrated above, a continuous-time signal can be sampled and reconstructed from the generated samples if the so-called sampling theorem is fulfilled.

[2] If the sampled signal is discontinuous at a certain time instant and the signal is sampled at this instant, then the sample value must be chosen as the average of the left and right limits in order to make (5.10) valid.

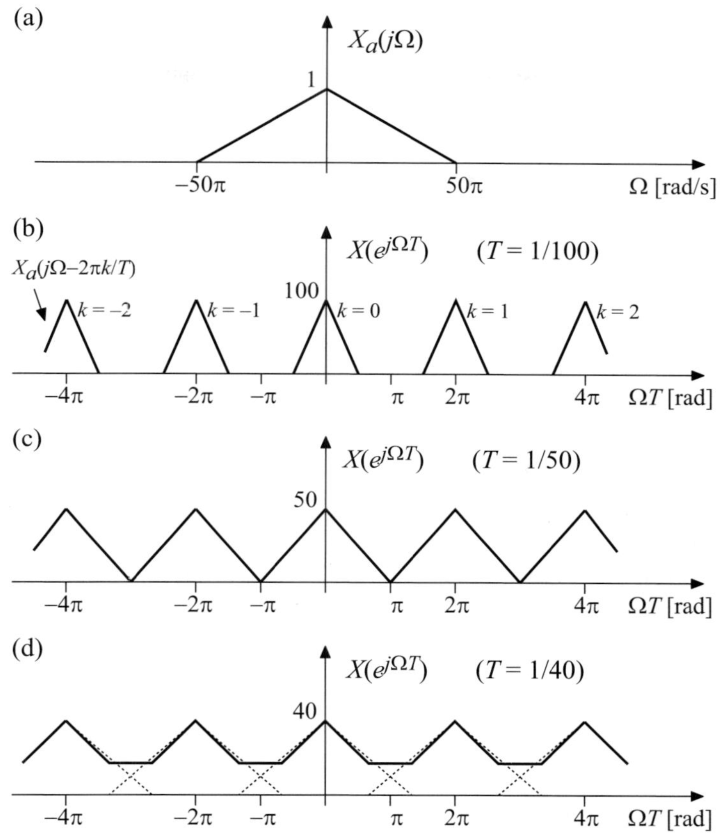

FIGURE 5.11

Spectra for a continuous-time signal and the corresponding sequences generated via sampling with three different sampling frequencies (Example 1).

Sampling theorem. *If a continuous-time signal $x_a(t)$ is band-limited to $\Omega = \Omega_0$ ($f = f_0$), i.e.,*

$$X_a(j\Omega) = 0, \quad |\Omega| > \Omega_0 = 2\pi f_0, \tag{5.11}$$

then $x_a(t)$ can be recovered from the samples $x(n) = x_a(nT)$ provided that

$$f_s = \frac{1}{T} > 2f_0. \tag{5.12}$$

Thus, if (5.11) and (5.12) are satisfied, the continuous-time signal $x_a(t)$ can be recovered from $x(n) = x_a(nT)$. One can understand this by studying Fig. 5.12 (and also Fig. 5.11). If the partial spectra in Poisson's summation formula do not overlap each other, the shape of $X_a(j\Omega)$ is preserved and it is

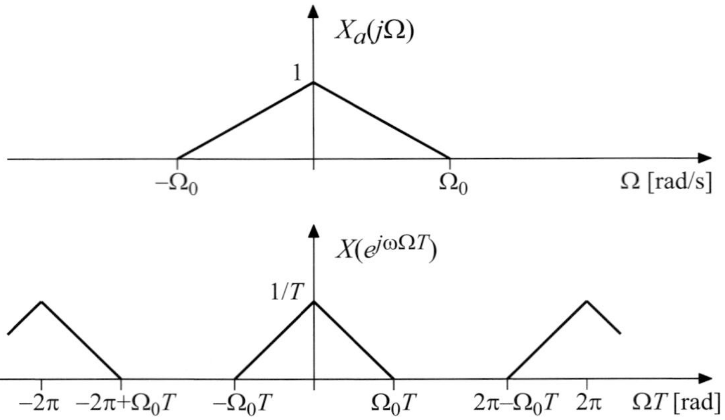

FIGURE 5.12

Illustration of the sampling theorem.

thereby possible, in principle, to reconstruct $x_a(t)$ from $x(n)$. From Fig. 5.12, we see that overlap of spectra is avoided if

$$2\pi - \Omega_0 T > \Omega_0 T \Leftrightarrow \Omega_0 T < \pi. \tag{5.13}$$

Since $\Omega = 2\pi f$ and $f_s = 1/T$, (5.13) is equivalent to

$$2\pi f_0 T < \pi \Leftrightarrow f_s = \frac{1}{T} > 2 f_0, \tag{5.14}$$

which is the same as (5.12). The frequency $f_s/2$ is called the Nyquist frequency, whereas f_s is referred to as the Nyquist sampling frequency. Further, the frequency band $\Omega \in [-\pi f_s, \pi f_s]$ (thus $f \in [-f_s/2, f_s/2]$) is referred to as the first Nyquist band.

The sampling theorem states that the sampling frequency f_s must be strictly greater than $2 f_0$ in order to be able to recover the signal. It should be mentioned however that for some classes of signals, like energy signals, one may select f_s equal to $2 f_0$ and still reconstruct the signals (in a certain sense). For other classes of signals, like sinusoidal signals, it is necessary to fulfill $f_s > 2 f_0$. This is because for a sinusoidal signal with frequency f_0, one may for example sample the zero-crossings. That is, if f_s equals $2 f_0$, one may obtain a sequence with zero-valued samples, which means that the continuous-time signal cannot be reconstructed.

If $f_s \le 2 f_0$, the sampling theorem is not fulfilled, in which case the signal is undersampled. This means that the partial spectra will overlap, as discussed earlier and illustrated in Fig. 5.11(d). This in turn implies that the shape of the continuous-time signal's spectrum is not preserved, which means that the signal cannot be reconstructed exactly. The error that is then introduced is called aliasing distortion. The reason for this name is that frequency components above the Nyquist frequency are aliased to the baseband $\Omega T \in [-\pi, \pi]$ (see Fig. 5.11(d)). In practice, one often chooses a higher sampling frequency than that required to fulfill the sampling theorem, the reason being that one can thereby reduce the requirements on the analog components that are used in the A/D and D/A conversions. In such a case, the signal is oversampled. If $f_s = M2f_0$, one has M times oversampling. Oversampled A/D and D/A converters are discussed in Section 5.7.

5.3.3.1 Example 2: illustration of aliasing

To further illustrate that the sampling theorem must be fulfilled in order to sample and reconstruct a signal, we consider stationary sinusoidal signals. Let $x_{a1}(t)$ and $x_{a2}(t)$ be two stationary sinusoidal signals according to

$$x_{a1}(t) = \sin(\Omega_0 t) = \sin(2\pi f_0 t) \tag{5.15}$$

and

$$x_{a2}(t) = -\sin(3\Omega_0 t) = -\sin(6\pi f_0 t). \tag{5.16}$$

We now form the discrete-time signals $x_1(n)$ and $x_2(n)$ by sampling $x_{a1}(t)$ and $x_{a2}(t)$, respectively, with a sampling frequency of $f_s = 4f_0$, i.e., a sampling period of $T = 1/(4f_0)$. This gives us

$$x_1(n) = x_{a1}(nT) = x_{a1}(0.25n/f_0) = \sin(0.5\pi n) \tag{5.17}$$

and

$$x_2(n) = x_{a2}(nT) = x_{a2}(0.25n/f_0) = -\sin(1.5\pi n). \tag{5.18}$$

Since $\sin(a) = \sin(a + 2\pi n)$ for all integers n and $\sin(a) = -\sin(-a)$, we can rewrite $x_2(n)$ as

$$\begin{aligned} x_2(n) &= -\sin(1.5\pi n) = -\sin(1.5\pi n - 2\pi n) \\ &= -\sin(-0.5\pi n) = \sin(0.5\pi n) = x_1(n). \end{aligned} \tag{5.19}$$

That is, $x_2(n)$ equals $x_1(n)$. The reason for this is that the sampling theorem is not fulfilled for $x_{a2}(t)$. This means that the spectrum of $x_{a2}(t)$ is aliased into the band $\Omega T \in [-\pi, \pi]$, as illustrated in Fig. 5.13. If the sampling theorem is not met, we can thus obtain the same sequence by sampling different signals.

5.3.4 Antialiasing filter

A condition for sampling and perfect reconstruction (PR) of a continuous-time signal is that it is strictly band-limited. In practice, continuous-time signals are not strictly band-limited, which means that we will always have aliasing distortion since the partial spectra then overlap each other. This in turn implies that we can never reconstruct the signal perfectly, but instead we obtain a distorted signal. To obtain an acceptable level of distortion, one usually has to filter the signal before the sampling takes place, as seen in Fig. 5.14(a). In this context, the filter $H(s)$ is called antialiasing filter because it is used to reduce the aliasing distortion.

Apparently, the antialiasing filter affects the signal that is to be sampled. However, in many cases, the information in the signal lies in the low-frequency region, whereas the high-frequency band only contains undesired frequency components like noise, etc. If the antialiasing filter is an ideal low-pass filter, the information is unaffected, whereas the undesired components are removed. This is illustrated in Fig. 5.14(b and c). In practice, one can not implement ideal filters, which means that one should always take into account a certain amount of distortion. More about distortion is given in Section 5.3.6.

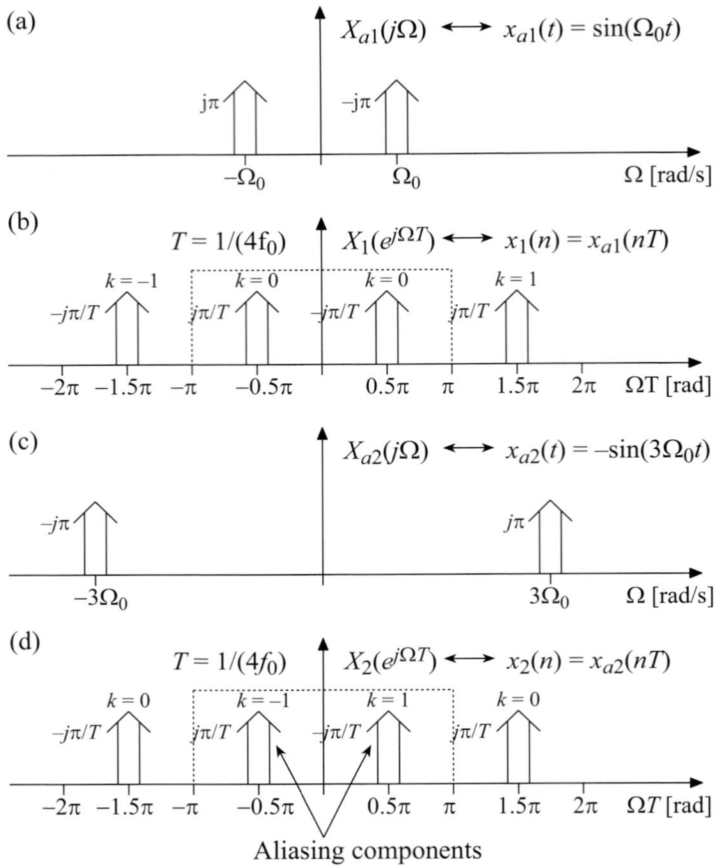

FIGURE 5.13

Illustration of aliasing (Example 2).

5.3.5 Reconstruction

Reconstruction refers to the process that forms a continuous-time signal from a discrete-time signal. The reconstruction can be done with the aid of pulse amplitude modulation. The system that performs the pulse amplitude modulation is called PAM. The sampling and reconstruction can be represented as in Fig. 5.15. We denote the reconstructed signal as $x_r(t)$ to distinguish it from the original signal $x_a(t)$.

The PAM forms the output signal $x_r(t)$ through multiplication of the samples $x(n)$ by the pulse (signal) $p(t)$ and its shifted versions $p(t–nT)$. The output signal is given by

$$x_r(t) = \sum_{n=-\infty}^{\infty} x(n)p(t-nT). \tag{5.20}$$

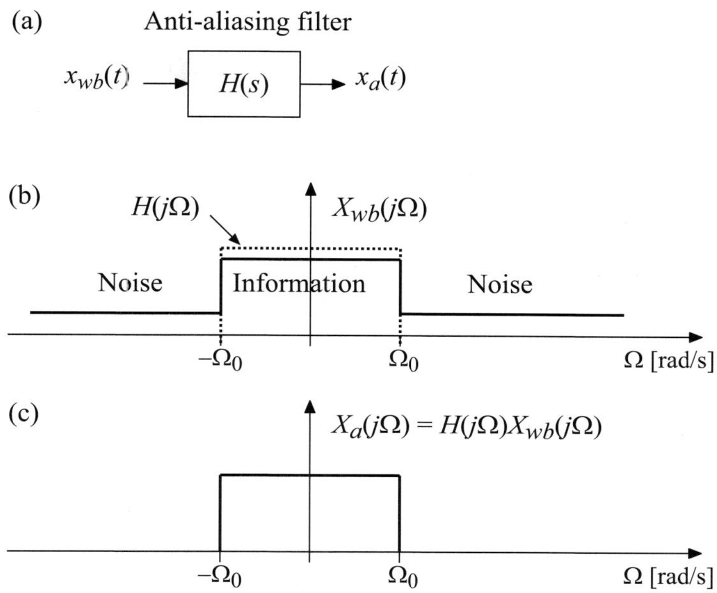

FIGURE 5.14

Band limitation using an antialiasing filter (wb stands for wide-band).

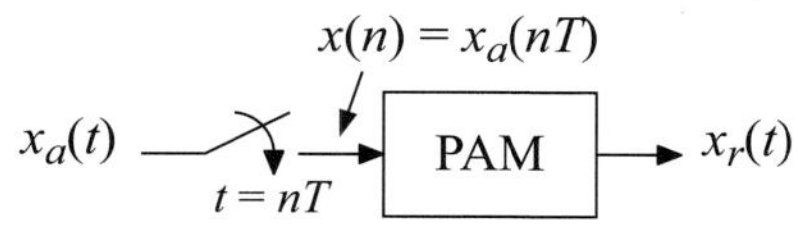

FIGURE 5.15

Sampling and reconstruction using a PAM.

Fig. 5.16 shows an example of reconstruction according to (5.20).

5.3.5.1 *Ideal reconstruction*

When the sampling theorem is fulfilled, it is in principle possible to obtain $x_r(t) = x_a(t)$ by properly selecting $p(t)$. If the original signal $x_a(t)$ is perfectly reconstructed from the samples $x(n) = x_a(nT)$, we have ideal reconstruction. As will be clear below, it is however not possible to perform ideal reconstruction in practice.

To see how to choose $p(t)$ in order to obtain $x_r(t) = x_a(t)$, we proceed as follows. If we utilize the expression for the reconstructed signal $x_r(t)$ in (5.20), its Fourier transform can be written as

$$X_r(j\Omega) = \int_{-\infty}^{\infty} x_r(t)e^{-j\Omega t}dt = \int_{-\infty}^{\infty} \sum_{n=-\infty}^{\infty} x(n)p(t-nT)e^{-j\Omega t}dt$$

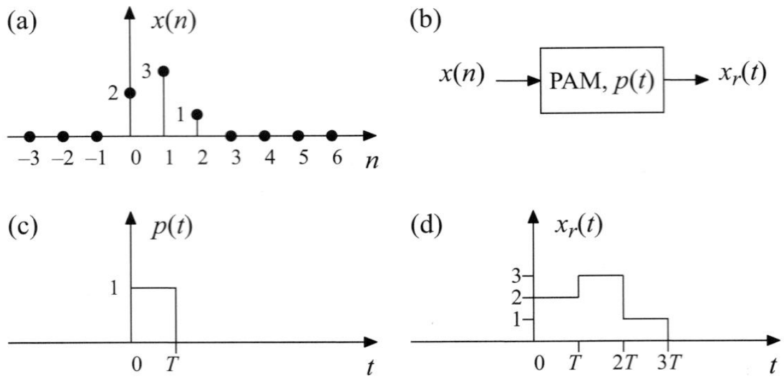

FIGURE 5.16

Example of reconstruction according to (5.20).

$$= \sum_{n=-\infty}^{\infty} x(n) \int_{-\infty}^{\infty} p(t-nT)e^{-j\Omega t}dt, \tag{5.21}$$

assuming that integration and summation can be interchanged. The variable substitution $v = t-nT$ in (5.21) yields

$$\begin{aligned} X_r(j\Omega) &= \sum_{n=-\infty}^{\infty} x(n) \int_{-\infty}^{\infty} p(v)e^{-j\Omega(v+nT)}dv \\ &= \left(\sum_{n=-\infty}^{\infty} x(n)e^{-j\Omega Tn}\right)\left(\int_{-\infty}^{\infty} p(v)e^{-j\Omega v}dv\right) \\ &= X(e^{j\Omega T})P(j\Omega). \end{aligned} \tag{5.22}$$

We also know that $X(e^{j\Omega T})$ is related to $X_a(j\Omega)$ via Poisson's summation formula. The Fourier transform of the reconstructed signal can therefore be related to the Fourier transform of the original signal according to

$$X_r(j\Omega) = P(j\Omega)X(e^{j\Omega T}) = P(j\Omega)\frac{1}{T}\sum_{k=-\infty}^{\infty} X_a\left(j\Omega - j\frac{2\pi k}{T}\right). \tag{5.23}$$

If the sampling theorem is fulfilled, the partial spectra $X_a\left(j\Omega - j2\pi k/T\right)$ do not overlap with each other, as illustrated in Fig. 5.17. We can therefore obtain $X_r(j\Omega) = X_a(j\Omega)$ by letting $P(j\Omega)$ be an ideal low-pass filter according to

$$P(j\Omega) = \begin{cases} T, & |\Omega| \le \Omega_c < \pi/T, \\ 0, & |\Omega| > \Omega_c, \end{cases} \tag{5.24}$$

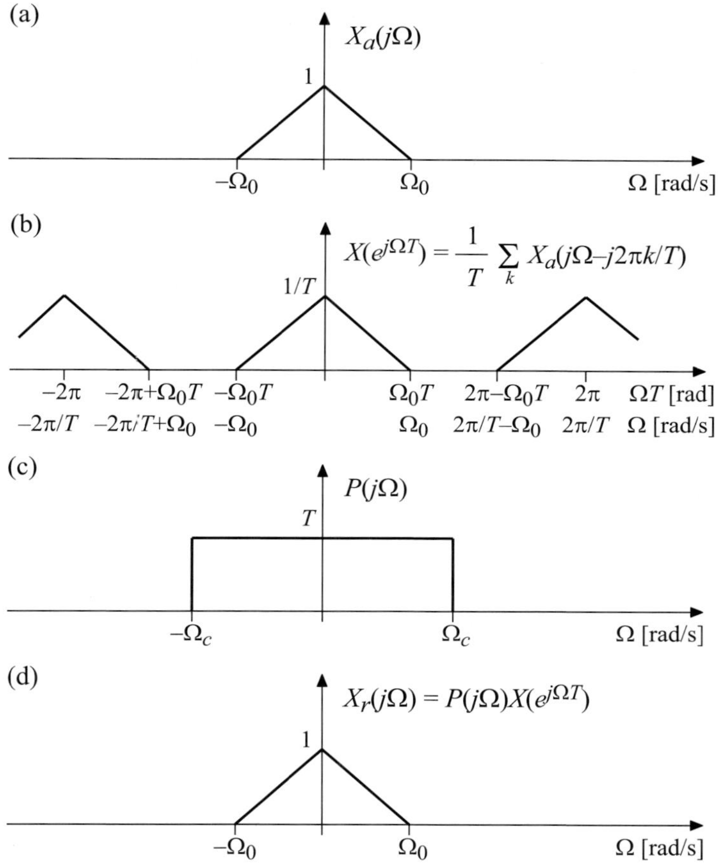

FIGURE 5.17

Illustration of ideal reconstruction.

with $\Omega_0 < \Omega_c \leq \pi/T$, since $P(j\Omega)$ then leaves $X_a(j\Omega)$ unaffected and removes the remaining partial spectra. This is also illustrated in Fig. 5.17. We then obtain (utilizing the uniqueness of the Fourier transform)

$$X_r(j\Omega) = X_a(j\Omega) \Leftrightarrow x_r(t) = x_a(t). \tag{5.25}$$

That is, the original signal $x_a(t)$ is reconstructed from the samples $x(n)$.

The pulse $p(t)$ is obtained by taking the inverse Fourier transform of $P(j\Omega)$ in (5.24). For example, with $\Omega_c = \pi/T$, we get

$$p(t) = \frac{\sin(\pi t/T)}{\pi t/T}, \tag{5.26}$$

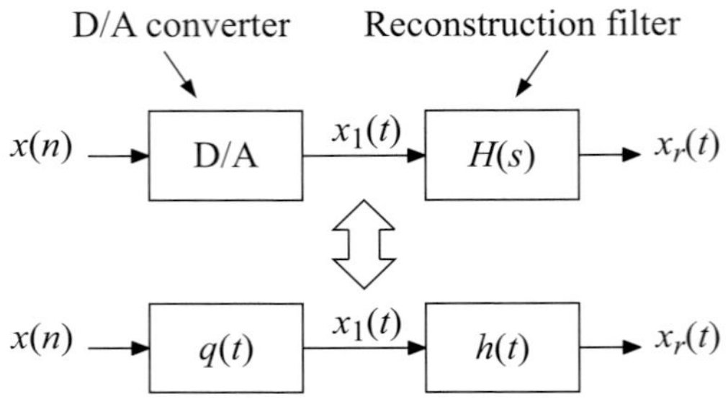

FIGURE 5.18

Illustration of ideal reconstruction.

which is a two-sided function and thus not realizable in practice by causal systems.

5.3.5.2 Reconstruction using a D/A converter and an analog reconstruction filter

In practice, the reconstruction is performed using a D/A converter followed by an analog reconstruction filter as depicted in Fig. 5.18. The D/A converter is in principle a PAM and can therefore be described by a pulse in accordance with the previous section. If we denote this pulse as $q(t)$ and let $x_1(t)$ denote the output from the D/A converter, we get

$$X_1(j\Omega) = Q(j\Omega)X(e^{j\Omega T}). \tag{5.27}$$

Furthermore, we have

$$X_r(j\Omega) = H(j\Omega)X_1(j\Omega). \tag{5.28}$$

Combining (5.27) and (5.28) gives us

$$X_r(j\Omega) = Q(j\Omega)H(j\Omega)X(e^{j\Omega T}). \tag{5.29}$$

We can now write $X_r(j\Omega)$ as

$$X_r(j\Omega) = P(j\Omega)X(e^{j\Omega T}), \tag{5.30}$$

where

$$P(j\Omega) = Q(j\Omega)H(j\Omega). \tag{5.31}$$

We know that if $P(j\Omega)$ is chosen as in (5.24), then $x_r(t) = x_a(t)$. Hence, also a scheme with a D/A converter followed by a reconstruction filter according to Fig. 5.18 achieves ideal reconstruction provided that $Q(j\Omega)H(j\Omega)$ satisfies

$$Q(j\Omega)H(j\Omega) = \begin{cases} T, & |\Omega| \leq \Omega_c < \pi/T, \\ 0, & |\Omega| > \Omega_c, \end{cases} \tag{5.32}$$

where $\Omega_0 < \Omega_c \leq \pi/T$. Dividing the reconstruction into two steps gives us more degrees of freedom. In particular, one can thereby let $q(t)$ be a simple pulse since the only requirement is that the product

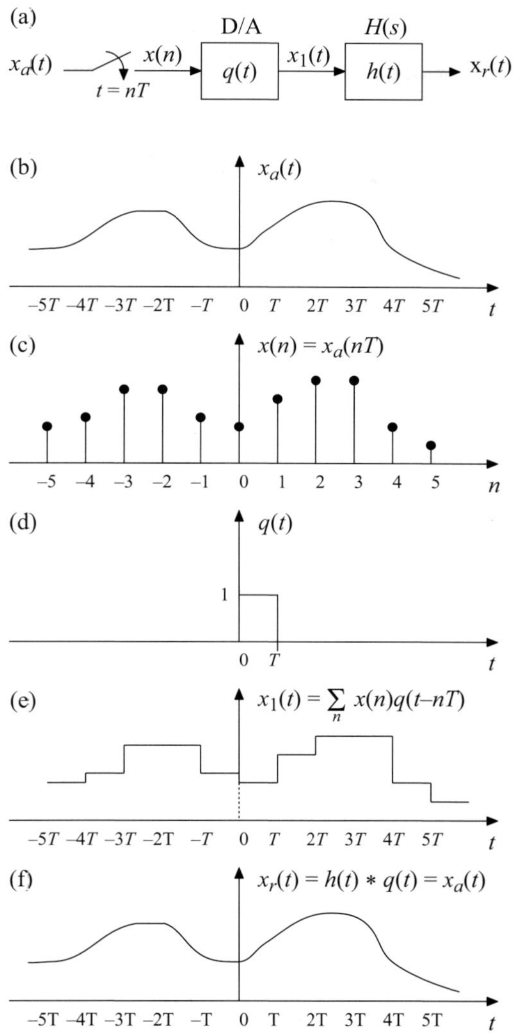

FIGURE 5.19

Illustration of reconstruction using a D/A converter and an analog filter.

$Q(j\Omega)H(j\Omega)$ must meet the right-hand side of (5.32). In principle, the actual shapes of the individual responses $Q(j\Omega)$ and $H(j\Omega)$ are therefore arbitrary. In practice, it is much easier to generate simple pulses that are subsequently filtered through a conventional analog filter, instead of generating more complicated pulses directly, like the sinc function in (5.26).

A common type of D/A converter is of the type zero-order hold. In this case, $q(t)$ is a square pulse according to Fig. 5.19(d). This pulse can mathematically be expressed as

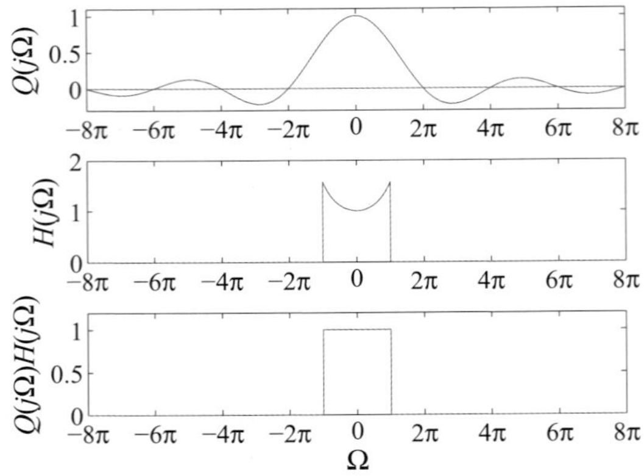

FIGURE 5.20

Frequency responses in ideal reconstruction using a D/A converter followed by an analog reconstruction filter ($T = 1$).

$$q(t) = \begin{cases} 1, & 0 < t < T, \\ 0, & t < 0,\ t > T. \end{cases} \tag{5.33}$$

Its Fourier transform is

$$Q(j\Omega) = e^{-j\Omega T/2}\frac{\sin(\Omega T/2)}{\Omega/2}. \tag{5.34}$$

Hence, the Fourier transform $Q(j\Omega)$ is a sinc function with zero-crossings at $\pm 2\pi k/T$, for all integers k. We note that $Q(j\Omega)H(j\Omega)$ satisfies (5.32) if the frequency response $H(j\Omega)$ is selected as

$$H(j\Omega) = \begin{cases} T/Q(j\Omega), & |\Omega| \le \Omega_c < \pi/T, \\ 0, & |\Omega| > \Omega_c. \end{cases} \tag{5.35}$$

For low frequencies, $H(j\Omega)$ is thus

$$H(j\Omega) = e^{j\Omega T/2}\frac{\Omega/2}{\sin(\Omega T/2)}, \quad |\Omega| \le \Omega_c < \pi/T. \tag{5.36}$$

When $\Omega_c = \pi/T$, the frequency response is as shown in Fig. 5.20. The response of this filter is discontinuous at $\Omega_c = \pi/T$ and it can therefore not be implemented in practice. Furthermore, it would be very costly to approximate this function, since a high filter order would be required in order to avoid a large distortion.

To get around these problems in practical implementations, one uses a certain amount of oversampling. This means that the sampling frequency is higher than the sampling theorem requires. One can thereby let the reconstruction filter have a transition band as shown in Fig. 5.21. The higher is the sampling frequency, the wider the transition band can be. The filter requirements are thus relaxed when the sampling frequency is increased. Another advantage of using oversampling is that the quantization

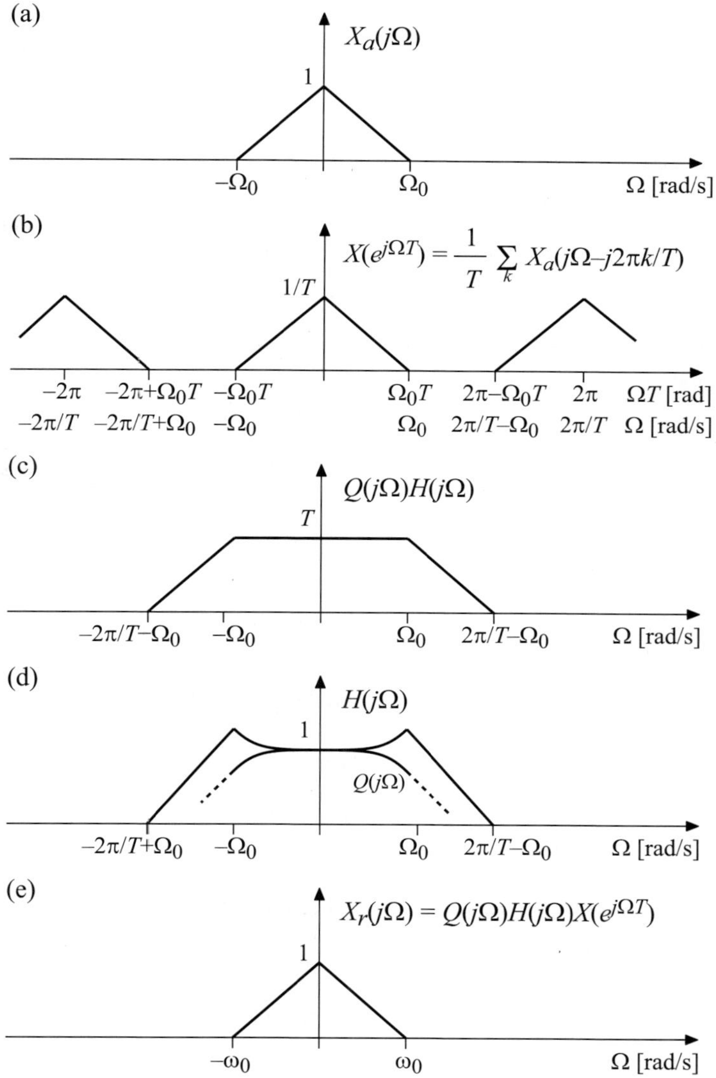

FIGURE 5.21

Typical spectra using oversampling and reconstruction via a D/A converter followed by an analog reconstruction filter.

noise that is introduced in the A/D conversion can be reduced via digital filtering. Oversampling will be discussed in Section 5.7. An oversampled A/D converter generates more samples than is actually required to reconstruct the signal. To make sure that the subsequent digital system does not have to work at higher data rates than necessary, one reduces the data rate by using decimation [52]. This can be done without losing information since the signal has been oversampled. In other words, there is a redundancy in the corresponding discrete-time signal.

5.3.6 Distortion caused by undersampling

In practice, a signal is never strictly band-limited, which means that one will always undersample the signal and thus cause a certain amount of aliasing distortion. The sampling frequency must be high enough to ensure that the reconstructed signal does not differ too much from the original one. The following example illustrates how the choice of sampling frequency affects the reconstructed signal.

5.3.6.1 *Example 3: illustration of distortion due to undersampling*

Let $x_a(t)$ be a continuous-time signal according to

$$x_a(t) = e^{-a|t|}, \quad a > 0. \tag{5.37}$$

This signal has a real Fourier transform given by

$$X_a(j\Omega) = \frac{2a}{\Omega^2 + a^2}. \tag{5.38}$$

Clearly, $X_a(j\Omega)$ is not band-limited. We now sample $x_a(t)$ with sampling frequency $f_s = 1/T$, which gives us the discrete-time signal $x(n)$ according to

$$x(n) = x_a(nT) = e^{-a|nT|} = \left(e^{-aT}\right)^{|n|}. \tag{5.39}$$

The Fourier transform of $x(n)$ is

$$X(e^{j\Omega T}) = \frac{1 - e^{-2aT}}{1 - 2e^{-aT}\cos(\Omega T) + e^{-2aT}}. \tag{5.40}$$

When $X_a(j\Omega)$ is real, so is $X(e^{j\Omega T})$, since the two transforms can be related to each other via Poisson's summation formula. Finally, we perform reconstruction using an ideal PAM, where $P(j\Omega)$ is given by

$$P(j\Omega) = \begin{cases} T, & |\Omega| \leq \pi/T, \\ 0, & |\Omega| > \pi/T. \end{cases} \tag{5.41}$$

This gives us the signal $x_r(t)$, whose Fourier transform is

$$X_r(j\Omega) = \begin{cases} TX(e^{j\Omega T}), & |\Omega| \leq \pi/T, \\ 0, & |\Omega| > \pi/T, \end{cases} \tag{5.42}$$

where $X(e^{j\Omega T})$ is given by (5.40). Figs. 5.22 and 5.23 show the spectra $X_a(j\Omega)$ and $X(e^{j\Omega T})$ as well as the signals $x_a(t)$ and $x_r(t)$. In Fig. 5.22, $T = 1/2$, whereas $T = 1/20$ in Fig. 5.23. For $T = 1/2$, the distortion is clearly visible, in both the time and frequency domains. If we instead choose $T = 1/20$, i.e., we increase the sampling frequency by a factor of 10, the distortion is no longer visually noticeable. An acceptable level of distortion can thus be obtained by selecting a high enough sampling frequency.

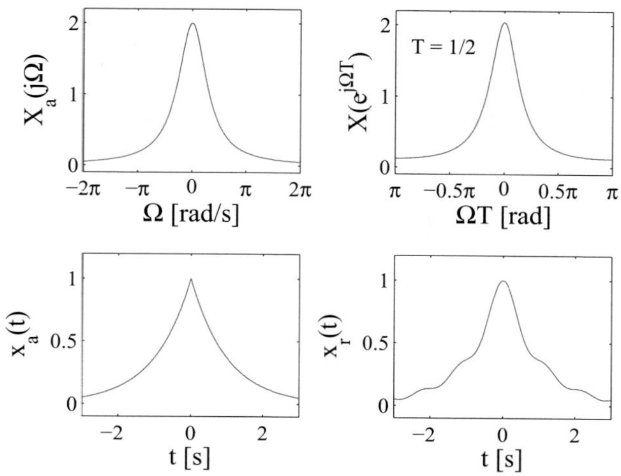

FIGURE 5.22

Spectra and signals in Example 3 for $T = 1/2$.

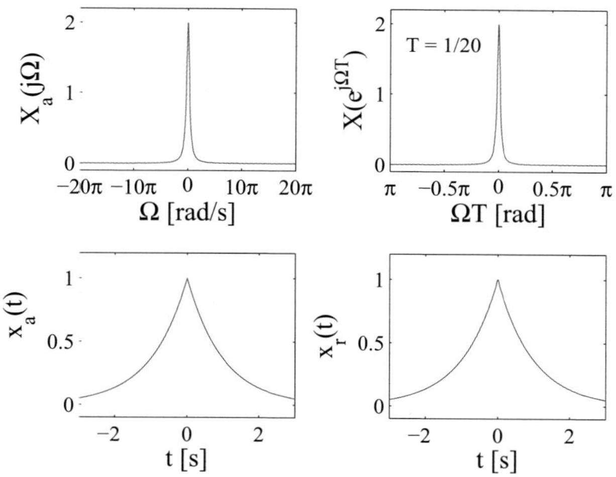

FIGURE 5.23

Spectra and signals in Example 3 for $T = 1/20$.

5.3.7 Distortion measure for energy signals

When the sampling theorem is not satisfied, the reconstructed signal $x_r(t)$ will differ from the original signal $x_a(t)$. We thereby get an error $e(t)$ according to

$$e(t) = x_r(t) - x_a(t). \tag{5.43}$$

For energy signals, a common measure of the "size" of the distortion is the ratio between the error energy and the signal energy, whereby one computes E_e/E_x, where E_x and E_e denote the energies of $x_a(t)$ and $e(t)$, respectively. These energies are given by

$$E_x = \int_{-\infty}^{\infty} x_a^2(t)dt = \frac{1}{2\pi}\int_{-\infty}^{\infty} |X_a(j\Omega)|^2 d\Omega \tag{5.44}$$

and

$$E_e = \int_{-\infty}^{\infty} e^2(t)dt = \frac{1}{2\pi}\int_{-\infty}^{\infty} |E(j\Omega)|^2 d\Omega, \tag{5.45}$$

where we have utilized Parseval's formula. If we now make use of (5.43) we can rewrite (5.45) as

$$E_e = \frac{1}{2\pi}\int_{-\infty}^{\infty} |X_r(j\Omega) - X_a(j\Omega)|^2 d\Omega. \tag{5.46}$$

When the reconstruction is done with an ideal PAM, $X_r(j\Omega)$ is given by

$$X_r(j\Omega) = \begin{cases} TX(e^{j\Omega T}), & |\Omega| \leq \pi/T, \\ 0, & |\Omega| > \pi/T, \end{cases} \tag{5.47}$$

where

$$TX(e^{j\Omega T}) = \sum_{k=-\infty}^{\infty} X_a\left(j\Omega - j\frac{2\pi k}{T}\right). \tag{5.48}$$

The energy E_e can in this case be expressed as

$$E_e = \frac{1}{2\pi}\int_{-\pi/T}^{\pi/T} |TX(e^{j\Omega T}) - X_a(j\Omega)|^2 d\Omega + \frac{1}{2\pi}\int_{|\Omega|\geq\pi/T} |X_a(j\Omega)|^2 d\Omega \tag{5.49}$$

or, alternatively,

$$E_e = \frac{1}{2\pi}\int_{-\pi/T}^{\pi/T} \left|\sum_{k\neq 0} X_a\left(j\Omega - j\frac{2\pi k}{T}\right)\right|^2 d\Omega + \frac{1}{2\pi}\int_{|\Omega|\geq\pi/T} |X_a(j\Omega)|^2 d\Omega. \tag{5.50}$$

The energies E_x and E_e are often difficult to compute analytically. In such cases, one has to use numerical computations.

5.3.7.1 Example 4: illustration of distortion measure

We consider the same signals as in Example 3 of Section 5.3.6.1 with $a = 1$. The energies E_x and E_e can in this case be computed as

$$E_x = \frac{1}{2\pi}\int_{-\infty}^{\infty} \left(\frac{2}{\Omega^2+1}\right)^2 d\Omega \tag{5.51}$$

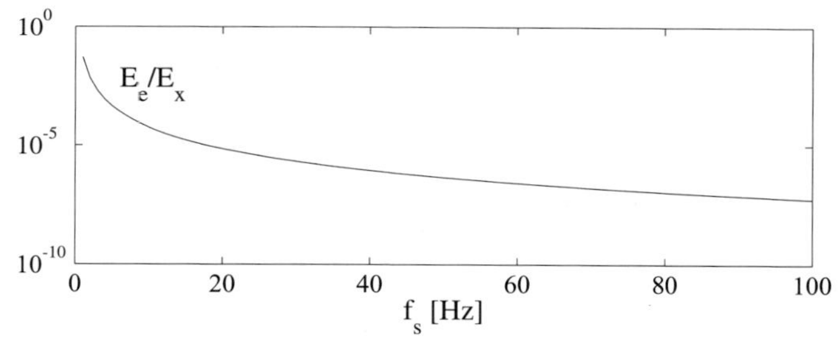

FIGURE 5.24

Distortion as a function of the sampling frequency in Example 4.

and

$$E_e = \frac{1}{2\pi}\int_{-\pi/T}^{\pi/T}\left(\frac{1-e^{-2aT}}{1-2e^{-aT}\cos(\Omega T)+e^{-2aT}}-\frac{2}{\Omega^2+1}\right)^2 d\Omega$$
$$+\frac{1}{2\pi}\int_{|\Omega|\geq\pi/T}\left(\frac{2}{\Omega^2+1}\right)^2 d\Omega, \quad (5.52)$$

respectively. Since the integrands are symmetric here, we can alternatively express E_x and E_e according to

$$E_x = \frac{1}{\pi}\int_0^{\infty}\left(\frac{2}{\Omega^2+1}\right)^2 d\Omega \quad (5.53)$$

and

$$E_e = \frac{1}{\pi}\int_0^{\pi/T}\left(\frac{1-e^{-2aT}}{1-2e^{-aT}\cos(\Omega T)+e^{-2aT}}-\frac{2}{\Omega^2+1}\right)^2 d\Omega$$
$$+\frac{1}{\pi}\int_{\pi/T}^{\infty}\left(\frac{2}{\Omega^2+1}\right)^2 d\Omega, \quad (5.54)$$

respectively. The integral in (5.53) can easily be computed analytically, whereby one obtains $E_x = 1$. The second integral in (5.54) can be similarly solved. For the first integral in (5.54), it is difficult to find a closed-form expression and we therefore compute it numerically. Fig. 5.24 plots the ratio E_e/E_x for different sampling frequencies $f_s = 1/T$. (Here, E_e/E_x equals E_e since $E_x = 1$.) As expected, we see that the distortion reduces when the sampling frequency is increased. The example in Section 5.3.6.1 considered the two cases $f_s = 2$ and $f_s = 20$. In these cases, the distortion figures are about 6.86×10^{-3} and 5.30×10^{-6}, respectively.

5.3.8 Bandpass sampling

We have so far considered low-pass signals, meaning that the signal spectrum $X_a(j\Omega)$ is contained in the low-frequency region $\Omega \in [-\Omega_0, \Omega_0]$. When the sampling frequency satisfies $f_s > 2f_0 = \Omega_0/\pi$,

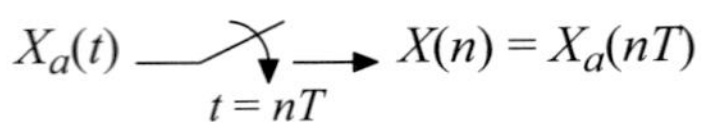

FIGURE 5.25

Uniform sampling of a stochastic process.

sampling and ideal reconstruction can theoretically be achieved. For a given sampling frequency, $f_s = 1/T$, the signal is in other words restricted to the frequency region $\Omega \in [-\pi/T, \pi/T]$. This frequency region is referred to as the first Nyquist band (or Nyquist zone). The theory can then be extended to bandpass signals for which the spectrum is contained in the bandpass frequency region $\Omega \in [-N\pi/T, -(N-1)\pi/T] \cup [(N-1)\pi/T, N\pi/T]$, which is called the Nth Nyquist band. Using again Poisson's summation formula (and following, e.g., the example in Section 5.3.2.1) it is then readily shown that the partial spectra do not overlap, which means that the signal is not aliased and thus theoretically can be reconstructed without errors. In the reconstruction, the pulse $p(t)$ must here correspond to a bandpass filter instead of a low-pass filter.

5.4 Sampling of stochastic processes

This section extends the sampling theory from deterministic signals to wide-sense stationary (WSS) stochastic processes.

5.4.1 Uniform sampling

Let $X_a(t)$ be a continuous-time WSS stochastic process with mean m_{X_a}, autocorrelation function $r_{X_aX_a}(\tau)$, and power spectrum $R_{X_aX_a}(j\Omega)$. We now form $X(n)$ by sampling $X_a(t)$ uniformly according to

$$X(n) = X_a(nT). \tag{5.55}$$

Graphically, we represent uniform sampling of a stochastic process as in Fig. 5.25.

As shown below, $X(n)$ is also a WSS process whose mean m_X, autocorrelation sequence $r_{XX}(k)$, and power spectrum $R_{XX}(e^{j\Omega T})$ are given by

$$m_X = m_{X_a}, \tag{5.56}$$

$$r_{XX}(k) = r_{X_aX_a}(kT), \tag{5.57}$$

and

$$R_{XX}(e^{j\Omega T}) = \frac{1}{T} \sum_{p=-\infty}^{\infty} R_{X_aX_a}\left(j\Omega - j\frac{2\pi p}{T}\right). \tag{5.58}$$

Eq. (5.58) is the same relation as in Poisson's summation formula for deterministic signals. One can therefore understand that the following sampling theorem can be formulated for WSS processes.

Sampling theorem for WSS processes. *If a continuous-time WSS process $X_a(t)$ is band-limited to $\Omega = \Omega_0$ ($f = f_0$), i.e., if*

$$R_{X_aX_a}(j\Omega) = 0, \quad |\Omega| > \Omega_0 = 2\pi f_0, \tag{5.59}$$

then $X_a(t)$ can be recovered from the samples $X(n) = X_a(nT)$ if

$$f_s = \frac{1}{T} > 2f_0. \tag{5.60}$$

Here, recovered means that we can form a new process $X_r(t)$ from $X(n)$, such that $X_r(t)$ equals $X_a(t)$ in the sense that $E\{[X_r(t)-X_a(t)]^2\} = 0$, where E denotes expectation.

Eqs. (5.56) and (5.57) follow immediately from their definitions according to

$$m_X = E\{X(n)\} = E\{X_a(nT)\} = m_{X_a} \tag{5.61}$$

and

$$\begin{aligned} r_{XX}(k) &= E\{X(n)X(n-k)\} \\ &= E\{X_a(nT)X_a(nT-kT)\} = r_{X_aX_a}(kT), \end{aligned} \tag{5.62}$$

respectively. This shows that $X(n)$ is a WSS process since both the mean and autocorrelation sequence are independent of n. As for (5.58), it is first noted that the autocorrelation function can be written in terms of its inverse Fourier transform according to

$$r_{X_aX_a}(\tau) = \frac{1}{2\pi}\int_{-\infty}^{\infty} R_{X_aX_a}(j\Omega)e^{j\Omega\tau}d\Omega. \tag{5.63}$$

In particular, one obtains for $\tau = kT$

$$r_{X_aX_a}(kT) = \frac{1}{2\pi}\int_{-\infty}^{\infty} R_{X_aX_a}(j\Omega)e^{j\Omega Tk}d\Omega. \tag{5.64}$$

By dividing the infinite interval integral into an infinite number of finite interval integrals, in the same way as we did in Section 5.3.2 when deriving Poisson's summation formula, we obtain

$$r_{X_aX_a}(kT) = \frac{1}{2\pi}\int_{-\pi}^{\pi} \frac{1}{T}\sum_{p=-\infty}^{\infty} R_{X_aX_a}\left(j\Omega - j\frac{2\pi p}{T}\right)e^{j\Omega Tk}d(\Omega T). \tag{5.65}$$

We also know that the autocorrelation sequence can be expressed in terms of its inverse Fourier transform according to

$$r_{XX}(k) = \frac{1}{2\pi}\int_{-\infty}^{\infty} R_{XX}(e^{j\Omega T})e^{j\Omega Tk}d(\Omega T). \tag{5.66}$$

Since $r_{XX}(k) = r_{X_aX_a}(kT)$ and the Fourier transform is unique, the right-hand sides in the two equations above must be equal, which finally gives us (5.58).

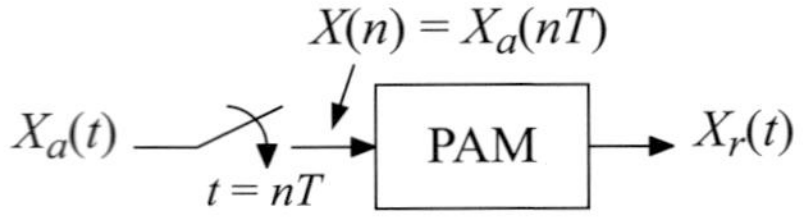

FIGURE 5.26

Sampling and reconstruction of a stochastic process using a PAM.

5.4.2 Reconstruction of stochastic processes

Reconstruction of stochastic processes can be performed with a PAM according to Fig. 5.26. The reconstructed process $X_r(t)$ is given by

$$X_r(t) = \sum_{n=-\infty}^{\infty} X(n)p(t - nT). \tag{5.67}$$

One can show that $X_r(t)$ is equal to $X_a(t)$ if the sampling theorem is fulfilled and $p(t)$ is chosen as

$$p(t) = \frac{\sin(\pi t/T)}{\pi t/T} = \text{sinc}(t/T). \tag{5.68}$$

This is in analogy with ideal reconstruction of deterministic signals. Note again that equality holds in the sense that $E[X_r(t)–X_a(t)]^2 = 0$. However, in practice, one cannot have strictly band-limited processes, which introduces aliasing distortion in the sampling process. Furthermore, errors are introduced in the reconstruction, since only approximate sinc functions can be generated. The reconstructed process will therefore differ from the original process.

An additional problem here is that $X_r(t)$ generally will not be a WSS process, which complicates the analysis of distortion, etc. To get around this dilemma, the sampling and reconstruction in (5.55) and (5.67) are replaced with

$$X(n) = X_a(nT + \Psi) \tag{5.69}$$

and

$$X_r(t) = \sum_{n=-\infty}^{\infty} X(n)p(t - nT - \Psi), \tag{5.70}$$

respectively, where Ψ is a stochastic variable that is uniformly distributed in the interval $(0, T)$. In practice, this means that each realization of the sampling and reconstruction process incorporates a stochastic delay Ψ. However, the sampling and reconstruction are still synchronized, since they incorporate the same delay Ψ. The purpose of introducing the delay Ψ is that the reconstructed process $X_r(t)$ then becomes WSS. The stochastic delay Ψ introduced in (5.69) does not have any implications as to stationarity and ensemble averages, that is, $X(n)$ is still WSS and (5.56)–(5.58) still hold. Furthermore, as shown below, the same stochastic delay Ψ introduced in (5.70) makes $X_r(t)$ a WSS process with the following mean m_{X_r}, autocorrelation function $r_{X_rX_r}(\tau)$, and power spectrum $R_{X_rX_r}(j\Omega)$:

$$m_{X_r} = \frac{1}{T} m_X P(0), \tag{5.71}$$

$$r_{X_rX_r}(\tau) = \frac{1}{T}\sum_{k=-\infty}^{\infty} r_{XX}(k)\int_{-\infty}^{\infty} p(u)p(u-\tau+kT)du, \tag{5.72}$$

and

$$R_{X_rX_r}(j\Omega) = \frac{1}{T}|P(j\Omega)|^2 R_{XX}(e^{j\Omega T}). \tag{5.73}$$

Moreover, as is also shown below, the reconstruction error $E\{[X_r(t)–X_a(t)]^2\}$ is given by

$$\begin{aligned} E\{[X_r(t)–X_a(t)]^2\} &= \frac{1}{2\pi T^2}\int_{-\infty}^{\infty}|P(j\Omega)-T|^2 R_{X_aX_a}(j\Omega)d\Omega \\ &+\frac{1}{2\pi T^2}\int_{-\infty}^{\infty}\sum_{p\neq 0}|P(j\Omega)|^2 R_{X_aX_a}\left(j\Omega - j\frac{2\pi p}{T}\right)d\Omega. \end{aligned} \tag{5.74}$$

From this equation, we see that if $p(t)$ is ideal according to (5.68), in which case $P(j\Omega)$ is an ideal low-pass filter with passband up to π/T and gain T, then the reconstruction error becomes

$$\begin{aligned} E\{[X_r(t)–X_a(t)]^2\} &= \frac{1}{2\pi}\int_{|\Omega|\geq\pi/T} R_{X_aX_a}(j\Omega)d\Omega \\ &+\frac{1}{2\pi}\int_{-\pi/T}^{\pi/T}\sum_{p\neq 0} R_{X_aX_a}\left(j\Omega - j\frac{2\pi p}{T}\right)d\Omega, \end{aligned} \tag{5.75}$$

which can be rewritten as

$$E\{[X_r(t)–X_a(t)]^2\} = \frac{2}{\pi}\int_{\pi/T}^{\infty} R_{X_aX_a}(j\Omega)d\Omega. \tag{5.76}$$

If, in addition, $X_a(t)$ is band-limited to π/T, i.e., $R_{X_aX_a}(j\Omega)=0$ for $|\Omega|\geq\pi/T$, it follows immediately from (5.76) that $E[X_r(t)–X_a(t)]^2=0$, in which case PR is achieved.

We now show (5.71)–(5.73). The mean of $X_r(t)$ becomes

$$\begin{aligned} m_{X_r} &= E\{X_r(t)\} = E\left\{\sum_{n=-\infty}^{\infty} X(n)p(t-nT-\Psi)\right\} \\ &= \sum_{n=-\infty}^{\infty} E\{X(n)p(t-nT-\Psi)\} \\ &= \sum_{n=-\infty}^{\infty} E\{X(n)\}E\{p(t-nT-\Psi)\}, \end{aligned} \tag{5.77}$$

where we have utilized that $X(n)$ and $p(t-nT-\Psi)$ are uncorrelated. Further, we have $E\{X(n)\}=m_X$ and

$$E\{p(t-nT-\Psi)\} = \frac{1}{T}\int_0^T p(t-nT-\Psi)d\Psi = \frac{1}{T}\int_{t-nT-T}^{t-nT} p(v)dv, \tag{5.78}$$

which gives us

$$m_{X_r} = \frac{1}{T} m_X \sum_{n=-\infty}^{\infty} \int_{t-nT-T}^{t-nT} p(v) dv = \frac{1}{T} m_X \int_{-\infty}^{\infty} p(v) dv = \frac{1}{T} m_X P(0). \tag{5.79}$$

We see that the mean is independent of t.

The autocorrelation function for $X_r(t)$ becomes

$$\begin{aligned}
r_{X_r X_r}(\tau) &= E\{X_r(t) X_r(t-\tau)\} \\
&= E\left\{ \sum_{n=-\infty}^{\infty} X(n) p(t-nT-\Psi) \sum_{m=-\infty}^{\infty} X(m) p(t-\tau-mT-\Psi) \right\} \\
&= E\left\{ \sum_{n=-\infty}^{\infty} \sum_{m=-\infty}^{\infty} X(n) X(m) p(t-nT-\Psi) p(t-\tau-mT-\Psi) \right\} \\
&= \sum_{n=-\infty}^{\infty} \sum_{m=-\infty}^{\infty} E\{X(n) X(m)\} \\
&\quad \times E\{p(t-nT-\Psi) p(t-\tau-mT-\Psi)\} \\
&= \sum_{n=-\infty}^{\infty} \sum_{m=-\infty}^{\infty} r_{XX}(n-m) \\
&\quad \times \frac{1}{T} \int_0^T p(t-nT-\Psi) p(t-\tau-mT-\Psi) d\Psi.
\end{aligned} \tag{5.80}$$

The variable substitutions $k = n–m$ and $u = t–nT–\Psi$ yield

$$\begin{aligned}
r_{X_r X_r}(\tau) &= \sum_{n=-\infty}^{\infty} \sum_{k=-\infty}^{\infty} r_{XX}(k) \frac{1}{T} \int_{t-nT-T}^{t-nT} p(u) p(u-\tau+kT) du \\
&= \frac{1}{T} \sum_{k=-\infty}^{\infty} r_{XX}(k) \int_{-\infty}^{\infty} p(u) p(u-\tau+kT) du.
\end{aligned} \tag{5.81}$$

We see that also the autocorrelation function is independent of t. Hence, $X_r(t)$ is a WSS process.

The power spectrum of $X_r(t)$ becomes

$$\begin{aligned}
R_{X_r X_r}(\tau) &= \int_{-\infty}^{\infty} r_{X_r X_r}(\tau) e^{-j\Omega\tau} d\tau \\
&= \int_{-\infty}^{\infty} \left(\frac{1}{T} \sum_{k=-\infty}^{\infty} r_{XX}(k) \int_{-\infty}^{\infty} p(u) p(u-\tau+kT) du \right) e^{-j\Omega\tau} d\tau \\
&= \frac{1}{T} \sum_{k=-\infty}^{\infty} r_{XX}(k) \int_{-\infty}^{\infty} p(u) \int_{-\infty}^{\infty} p(u-\tau+kT) e^{-j\Omega\tau} d\tau du.
\end{aligned} \tag{5.82}$$

The variable substitution $v = u - \tau + kT$ gives us

$$\begin{aligned}
R_{X_rX_r}(\tau) &= \frac{1}{T}\sum_{k=-\infty}^{\infty} r_{XX}(k)\int_{-\infty}^{\infty} p(u)\int_{-\infty}^{\infty} p(v)e^{-j\Omega(u-v+kT)}dvdu \\
&= \frac{1}{T}\sum_{k=-\infty}^{\infty} r_{XX}(k)e^{-j\Omega Tk}\int_{-\infty}^{\infty} p(u)e^{-j\Omega u}du\int_{-\infty}^{\infty} p(v)e^{j\Omega v}dvdu \\
&= \frac{1}{T}R_{XX}(e^{j\Omega T})P(j\Omega)P^{\star}(j\Omega) \\
&= \frac{1}{T}|P(j\Omega)|^2 R_{XX}(e^{j\Omega T}).
\end{aligned} \tag{5.83}$$

The last two equalities hold when $p(t)$ is a real function.

Next, (5.74) is shown. To this end, the reconstruction error is first expanded as

$$E\left\{[X_r(t) - X_a(t)]^2\right\} = E\left\{X_r^2(t)\right\} + E\left\{X_a^2(t)\right\} - 2E\left\{X_r(t)X_a(t)\right\}. \tag{5.84}$$

The first term in (5.84) can be computed as

$$\begin{aligned}
E\left\{X_r^2(t)\right\} &= \frac{1}{2\pi}\int_{-\infty}^{\infty} R_{X_rX_r}(j\Omega)d\Omega = \frac{1}{2\pi}\int_{-\infty}^{\infty}\frac{1}{T}|P(j\Omega)|^2 R_{XX}(e^{j\Omega T})d\Omega \\
&= \frac{1}{2\pi T^2}\int_{-\infty}^{\infty}|P(j\Omega)|^2\sum_{p=-\infty}^{\infty} R_{X_aX_a}\left(j\Omega - j\frac{2\pi p}{T}\right)d\Omega.
\end{aligned} \tag{5.85}$$

The second term in (5.84) is simply

$$E\left\{X_a^2(t)\right\} = \frac{1}{2\pi}\int_{-\infty}^{\infty} R_{X_aX_a}(j\Omega)d\Omega. \tag{5.86}$$

In the third term of (5.84) we have

$$\begin{aligned}
E\left\{X_r(t)X_a(t)\right\} &= E\left\{\sum_{n=-\infty}^{\infty} X_a(nT+\Psi)p(t-nT-\Psi)X_a(t)\right\} \\
&= E\left\{\sum_{n=-\infty}^{\infty} r_{X_aX_a}(t-nT-\Psi)p(t-nT-\Psi)\right\},
\end{aligned} \tag{5.87}$$

assuming that $X_a(t)$ and Ψ are uncorrelated. Since it is assumed that Ψ is uniformly distributed on $[0,\ T]$, it follows from (5.87) that

$$E\left\{X_r(t)X_a(t)\right\} = \sum_{n=-\infty}^{\infty}\frac{1}{T}\int_{0}^{T} r_{X_aX_a}(t-nT-\Psi)p(t-nT-\Psi)d\Psi. \tag{5.88}$$

The variable substitution $u=t–nT–\Psi$ now yields

$$
\begin{aligned}
E\{X_r(t)X_a(t)\} &= \frac{1}{T}\sum_{n=-\infty}^{\infty}\int_{t-nT-T}^{t-nT} r_{X_aX_a}(u)p(u)du \\
&= \frac{1}{T}\int_{-\infty}^{\infty} r_{X_aX_a}(u)p(u)du. \qquad (5.89)
\end{aligned}
$$

Using Parseval's relation

$$
\int_{-\infty}^{\infty} x(t)y^{\star}(t)dt = \frac{1}{2\pi}\int_{-\infty}^{\infty} X(j\Omega)Y^{\star}(j\Omega)d\Omega \qquad (5.90)
$$

together with the fact that $p(u)$ is real, i.e., $p(u)=p^{\star}(u)$, (5.89) can equivalently be written as

$$
E\{X_r(t)X_a(t)\} = \frac{1}{2\pi T}\int_{-\infty}^{\infty} R_{X_aX_a}(j\Omega)P^{\star}(j\Omega)d\Omega. \qquad (5.91)
$$

Making use of the facts that $R_{X_aX_a}(j\Omega)=R_{X_aX_a}(-j\Omega)$ and $P^{\star}(j\Omega)=P(-j\Omega)$, (5.91) can equivalently be written as

$$
E\{X_r(t)X_a(t)\} = \frac{1}{2\pi T}\int_{-\infty}^{\infty} R_{X_aX_a}(j\Omega)P(j\Omega)d\Omega. \qquad (5.92)
$$

Inserting (5.85), (5.86), (5.91), and (5.92) into (5.84), we finally obtain

$$
\begin{aligned}
E[X_r(t)–X_a(t)]^2 &= \frac{1}{2\pi T^2}\int_{-\infty}^{\infty} |P(j\Omega)|^2 \sum_{p=-\infty}^{\infty} R_{X_aX_a}\left(j\Omega - j\frac{2\pi p}{T}\right)d\Omega \\
&+ \frac{1}{2\pi}\int_{-\infty}^{\infty} R_{X_aX_a}(j\Omega)d\Omega \\
&- \frac{1}{2\pi T}\int_{-\infty}^{\infty} R_{X_aX_a}(j\Omega)P^{\star}(j\Omega)d\Omega \\
&- \frac{1}{2\pi T}\int_{-\infty}^{\infty} R_{X_aX_a}(j\Omega)P(j\Omega)d\Omega \\
&= \frac{1}{2\pi T^2}\int_{-\infty}^{\infty} |P(j\Omega)-T|^2 R_{X_aX_a}(j\Omega)d\Omega \\
&+ \frac{1}{2\pi T^2}\int_{-\infty}^{\infty} \sum_{p\neq 0} |P(j\Omega)|^2 R_{X_aX_a}\left(j\Omega - j\frac{2\pi p}{T}\right)d\Omega. \qquad (5.93)
\end{aligned}
$$

5.4.2.1 *Example 5: computation of reconstruction error power*

Let $X_a(t)$ be the output of an analog filter $H(s)$ for the input $X_{wb}(t)$. Assume that $X_{wb}(t)$ is white noise with zero mean and power spectrum $\sigma^2_{X_{wb}}$ and that $H(s)$ is a first-order low-pass filter according to

$$
H(s) = \frac{\Omega_0}{s+\Omega_0}. \qquad (5.94)
$$

We now form $X(n)$ and $X_r(t)$ through uniform sampling of $X_a(t)$ and reconstruction with a PAM according to (5.69) and (5.70), respectively. We assume that $P(j\Omega)$ is ideal and given by (5.68). We now wish to compute the average powers of $X_a(t)$, $X(n)$, and $X_r(t)$. These quantities are obtained by integrating the corresponding power spectra.

The power spectrum of $X_a(t)$ is

$$R_{X_aX_a}(j\Omega) = |H(j\Omega)|^2 R_{X_{wb}X_{wb}}(j\Omega), \tag{5.95}$$

where

$$R_{X_{wb}X_{wb}}(j\Omega) = \sigma^2_{X_{wb}} \tag{5.96}$$

and

$$|H(j\Omega)|^2 = \frac{\Omega_0^2}{\Omega^2 + \Omega_0^2}. \tag{5.97}$$

The power of $X_a(t)$ therefore becomes

$$\begin{aligned} E\left\{X_a^2(t)\right\} &= \frac{1}{2\pi}\int_{-\infty}^{\infty} R_{X_aX_a}(j\Omega)d\Omega = \frac{1}{2\pi}\int_{-\infty}^{\infty} |H(j\Omega)|^2 R_{X_{wb}X_{wb}}(j\Omega)d\Omega \\ &= \frac{1}{2\pi}\int_{-\infty}^{\infty} \frac{\Omega_0^2}{\Omega^2+\Omega_0^2}\sigma^2_{X_{wb}} d\Omega = \frac{\Omega_0\sigma^2_{X_{wb}}}{2\pi}[\arctan(\Omega/\Omega_0)]_{-\infty}^{\infty} \\ &= \frac{\Omega_0\sigma^2_{X_{wb}}}{2}. \end{aligned} \tag{5.98}$$

The power of $X(n)$ is the same as that of $X_a(t)$ because

$$E\left\{X^2(n)\right\} = r_{XX}(0) = r_{X_aX_a}(0) = E\left\{X_a^2(t)\right\}. \tag{5.99}$$

The power spectrum of $X_r(t)$ is

$$R_{X_rX_r}(j\Omega) = \frac{1}{T}|P(j\Omega)|^2 R_{XX}(e^{j\Omega T}), \tag{5.100}$$

where

$$R_{XX}(e^{j\Omega T}) = \frac{1}{T}\sum_{p=-\infty}^{\infty} R_{X_aX_a}\left(j\Omega - j\frac{2\pi p}{T}\right). \tag{5.101}$$

When $p(t)$ is given by (5.68), one obtains

$$P(j\Omega) = \begin{cases} T, & |\Omega| \le \pi/T, \\ 0, & |\Omega| > \pi/T. \end{cases} \tag{5.102}$$

The power of $X_r(t)$ thus becomes

$$E\left\{X_r^2(t)\right\} = \frac{1}{2\pi}\int_{-\infty}^{\infty} R_{X_rX_r}(j\Omega)d\Omega = \frac{1}{2\pi}\int_{-\pi/T}^{\pi/T} T R_{XX}(e^{j\Omega T})d\Omega$$

$$
\begin{aligned}
&= \frac{1}{2\pi}\int_{-\pi/T}^{\pi/T}\sum_{p=-\infty}^{\infty} R_{X_aX_a}\left(j\Omega - j\frac{2\pi p}{T}\right)d\Omega \\
&= \frac{1}{2\pi}\sum_{p=-\infty}^{\infty}\int_{-\pi/T}^{\pi/T} R_{X_aX_a}\left(j\Omega - j\frac{2\pi p}{T}\right)d\Omega. \qquad (5.103)
\end{aligned}
$$

The variable substitution $j\Omega - j2\pi p/T \to j\Omega$ finally gives us

$$
\begin{aligned}
E\left\{X_r^2(t)\right\} &= \frac{1}{2\pi}\sum_{p=-\infty}^{\infty}\int_{-\pi/T-j2\pi p/T}^{\pi/T-j2\pi p/T} R_{X_aX_a}(j\Omega)d\Omega \\
&= \frac{1}{2\pi}\int_{-\infty}^{\infty} R_{X_aX_a}(j\Omega)d\Omega = E\left\{X_a^2(t)\right\}. \qquad (5.104)
\end{aligned}
$$

That is, the power of $X_r(t)$ is the same as that of $X_a(t)$. We thus have

$$
E\left\{X_a^2(t)\right\} = E\left\{X^2(n)\right\} = E\left\{X_r^2(t)\right\}. \qquad (5.105)
$$

Finally, it is observed that the reconstruction error is

$$
\begin{aligned}
E\left\{[X_r(t) - X_a(t)]^2\right\} &= \frac{2}{\pi}\int_{\pi/T}^{\infty} R_{X_aX_a}(j\Omega)d\Omega \\
&= \frac{2}{\pi}\int_{\pi/T}^{\infty} |H(j\Omega)|^2 R_{X_{wb}X_{wb}}(j\Omega)d\Omega \\
&= \frac{2}{\pi}\int_{\pi/T}^{\infty} \frac{\Omega_0^2}{\Omega^2+\Omega_0^2}\sigma_{X_{wb}}^2 d\Omega \\
&= \frac{2\Omega_0\sigma_{X_{wb}}^2}{\pi}[\arctan(\Omega/\Omega_0)]_{\pi/T}^{\infty} \\
&= \frac{2\Omega_0\sigma_{X_{wb}}^2}{\pi}(\pi/2 - \arctan[\pi/(\Omega_0 T)]). \qquad (5.106)
\end{aligned}
$$

We see that with a fixed Ω_0, $E\left\{[X_r(t) - X_a(t)]^2\right\} \to 0$ when $T \to 0$, i.e., we can make the reconstruction error as small as desired by a proper selection of the sampling period T (sampling frequency $f_s = 1/T$).

5.5 Nonuniform sampling and generalizations

This section considers the extension to nonuniform sampling [29] and generalizations. This appears in many different forms and contexts and it is beyond the scope of this section to present a comprehensive survey. Instead, we will concentrate on one particular application, namely M-channel time-interleaved A/D converters, which in the simplest case corresponds to an M-periodic nonuniform sampling grid

[10,19,21]. In this application, the sampling grid is close to a uniform grid and, thus, the obtained nonuniform-sampling sequence is close to the desired uniform-sampling sequence. Nevertheless, reconstruction is required to recover the uniform-sampling sequence from the nonuniform-sampling sequence. The generalized case is an extension in the sense that the M channels experience different and general frequency responses.

5.5.1 Time-interleaved ADCs

Time interleaving of multiple parallel ADCs is a technique to increase the effective sampling rate of the overall converter. Using an M-channel time-interleaved ADC, the effective sampling rate is increased by a factor of M. Unfortunately, the effective resolution of the individual channel converters is not maintained in the overall converter because of channel mismatch errors. It is therefore necessary to compensate for these errors in order to restore the resolution [56,57]. The errors can broadly be divided into linear and nonlinear mismatch errors [57,58]. This section only deals with linear mismatch errors, in which case the channels are modeled as linear systems, thus having specific frequency responses. There are also static offset mismatch errors present but they are signal-independent and straightforward to compensate for, and therefore not explicitly included in the formulas to be presented in this section. However, as noted in Section 5.5.5.1, one can make use of the same frequency-domain expression as that used for relating the input and output signals.

Up to a certain resolution, one can assume that the channel frequency responses have frequency-independent magnitude and phase delay responses, which corresponds to static gain and linear-phase (time-skew) mismatch errors. Without gain errors, this case corresponds to nonuniform sampling and the problem is then to recover the uniform-sampling sequence from the nonuniform-sampling sequence. Numerous papers have addressed this problem over the last decades; see for example [7–9,13–15,56] and references therein. However, to reach a very high resolution for high-speed conversion, one needs to extend the channel model to general frequency responses, thus with frequency-dependent magnitude and phase delay responses [49,57]. In this case, one has to compensate for these frequency response mismatch errors, not only static gain and linear-phase parts which correspond to approximations of the true responses. Recent papers have considered the more general problem [18,20,26,31,32,42,44–46,49, 59] which corresponds to the "generalizations" mentioned above.

Before proceeding, it is also noted that calibration of time-interleaved ADCs requires estimation of and compensation for the channel frequency response mismatches. This section only discusses the compensation which requires that accurate models of the channel frequency responses are available. However, the estimation benefits from efficient compensation techniques, for example when the calibration is done through simultaneous estimation and compensation by minimizing an appropriate cost measure [46]. The compensation techniques to be discussed here can thus be used either after the channel frequency responses have been estimated or as a part of a calibration technique.

5.5.2 Problem formulation

The point of departure is that we have a continuous-time signal $x_a(t)$ band-limited to $\Omega_0 < \pi/T$, which means that the Nyquist criterion for uniform sampling with a sampling frequency of $1/T$ without aliasing is fulfilled, that is, sampling according to $x(n) = x_{\mathrm{a}}(nT)$ does not introduce aliasing and we have in the frequency domain

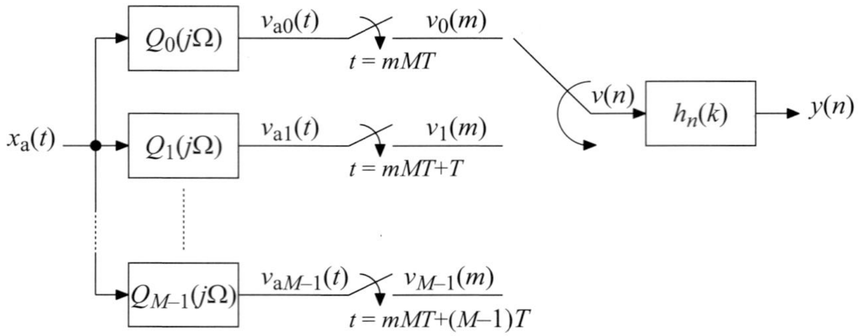

FIGURE 5.27

M-channel time-interleaved ADC with different channel frequency responses $Q_n(j\Omega)$, $n = 0, 1, \ldots, M-1$, and an M-periodic time-varying compensation system with impulse response $h_n(k) = h_{n+M}(k)$.

$$X(e^{j\Omega T}) = \frac{1}{T} X_a(j\Omega), \quad \Omega T \in [-\pi, \pi], \tag{5.107}$$

where $X(e^{j\Omega T})$ and $X_a(j\Omega)$ denote the Fourier transforms of the sequence $x(n)$ and the continuous-time signal $x_a(t)$, respectively. This means that $x_a(t)$ can be recovered from $x(n)$. It is generally advisable to make Ω_0 correspond to some 80%–90% of the overall Nyquist band as the computational complexity of the compensation system becomes prohibitive when it approaches 100%, as exemplified in Section 5.5.5.2. However, normally this does not impose additional band limitations as a certain amount of oversampling is required anyhow to get reasonable requirements on the analog antialiasing filter that precedes the ADC, in accordance with Fig. 5.39 in Section 5.7.1, which depicts a typical system for conversion of an analog signal into its digital representation.

In an M-channel time-interleaved ADC, without correction, we do not obtain the desired uniform sequence $x(n)$ but instead another sequence, say $v(n)$, through interleaving of M subsequences, as seen in Fig. 5.27. This can equivalently be viewed as if $v(n)$ is obtained by applying $x_a(t)$ to a time-varying continuous-time system and then sampling the output at $t = nT$. Utilizing (5.107), the sequence $v(n)$ can then be expressed via the inverse Fourier transform as

$$v(n) = \frac{1}{2\pi} \int_{-\Omega_0 T}^{\Omega_0 T} Q_n(j\Omega) X(e^{j\Omega T}) e^{j\Omega T n} d(\Omega T), \tag{5.108}$$

where $Q_n(j\Omega) = Q_{n+M}(j\Omega)$ is an M-periodic time-varying system frequency response and $Q_n(j\Omega)$, $n = 0, 1, \ldots, M-1$, constitute the M ADC channel frequency responses. For example, with static gain constants, g_n, and static time skews, d_n (given as a percentage of the sampling period $T = 1/f_s$), the channel frequency responses are modeled as $Q_n(j\Omega) = g_n e^{j\Omega T d_n}$. With $g_n = 1$, this corresponds to $v(n) = x_a(nT + d_n T)$ and thus a nonuniform sampling and reconstruction problem. It is also noted that the model above holds also in the more general case when $M \to \infty$, thus also for single-channel ADCs with a time-varying frequency response.

The problem to be solved can now be formulated as follows: Given the sequence $v(n)$ in (5.108), form a new sequence $y(n)$ that should approximate the sequence $x(n) = x_a(nT)$ as closely as desired. This problem is considered in Section 5.5.4.

5.5.2.1 Relaxed problem

The overall output $y(n)$ should ideally approximate the sequence $x(n) = x_a(nT)$. This requires that all channel frequency responses $Q_n(j\Omega)$ are known. In practice, it is customary to determine the ratios between the channel frequency responses and then match the channels. Typically, one chooses the first channel as a reference and then matches all other channels to this reference channel. In the reconstruction, this means that we set $Q_0(j\Omega) = 1$ and divide the remaining $Q_n(j\Omega)$ by the actual $Q_0(j\Omega)$. In the estimation, the frequency response ratios are obtained directly so we will never have available $Q_n(j\Omega)$ but only an estimation of $Q_n(j\Omega)/Q_0(j\Omega)$. After the compensation, the overall ADC will then experience a linear distortion, $Q_0(j\Omega)$, because we then have $Y(e^{j\Omega T}) = Q_0(j\Omega)X(e^{j\Omega T})$. In other words, the linear distortion after compensation is determined by the frequency response of the reference ADC. In the compensation, this also means that the samples from the reference channel are taken as they are and only the other channels' samples are corrected. Finally, it is noted that in the overall system where the A/D converter is to be used, the linear distortion needs to be equalized but this problem exists for all types of A/D converters. In a larger composite system, like a communication system, a single equalizer can be used to equalize all linear distortion at the same time, emanating from the filters, A/D converters, communication channel, etc.

5.5.3 Reconstruction

Regardless whether the sequence $x(n)$ has been obtained through uniform sampling of $x_a(t)$ or through nonuniform sampling and/or its generalization (like in time-interleaved ADCs considered here), it is often desired to reconstruct $x_a(t)$ from the generated sequence of samples $x(n)$. Thus, in the generalized case, it is desired to recover $x_a(t)$ from the sequence $v(n)$. This can, in principle, be done in two different ways. The first way is to reconstruct $x_a(t)$ directly from $v(n)$ through analog reconstruction functions. Although it is known how to do this in principle (see, e.g., [17,29,41,63]), problems arise when it comes to practical implementations. In particular, it is very difficult to practically implement analog functions with high precision. It is therefore desired to use the second way, which is to perform the reconstruction in the digital domain, i.e., to first recover $x(n) = x_a(nT)$. One then needs only one conventional DAC and analog filter to obtain $x_a(t)$. Such components are much easier to implement than components that are to approximate complicated analog functions. Recovery of $x(n)$ is also of interest even if $x_a(t)$ is not to be reconstructed. For example, in receivers in digital communication systems, $x(n)$ is the desired result. Sections 5.5.4–5.5.7 give more details regarding the reconstruction of $x(n)$ from $v(n)$ in time-interleaved ADCs.

5.5.4 Reconstruction using a time-varying FIR system

A time-interleaved ADC corresponds to a time-varying system, and the compensation thus corresponds to inverting such a system. For an M-channel system, this amounts to determining an M-periodic time-varying discrete-time system characterized by an M-periodic impulse response, say $h_n(k) = h_{n+M}(k)$, or, equivalently, M time-invariant impulse responses $h_n(k)$, $n = 0, 1, \ldots, M-1$.

Using an Nth-order M-periodic time-varying finite length impulse response (FIR) system, the reconstructed output sequence, $y(n)$, is given by the convolution sum as

$$y(n) = \sum_{k=-N/2}^{N/2} v(n-k)h_n(k), \tag{5.109}$$

where $h_n(k) = h_{n+M}(k)$ denotes the M-periodic impulse response. Thus, to correct each output sample, one needs to compute $N+1$ multiplications and N additions. It is convenient here to make use of a noncausal system, which means that the impulse response is assumed to be centered around $k=0$, which corresponds to zero delay. A causal system is then obtained by introducing a delay of $N/2$ samples into the system. It is noted that we have assumed here that the order N of the system is even, to keep the notation simple. Odd-order systems can be used as well after some minor appropriate modifications.

To determine the impulse response coefficients, it is convenient to express the output $y(n)$ in the time-frequency domain [19], instead of the time domain given in (5.109). This will be done here in terms of an M-periodic time-frequency function $A_n(j\Omega) = A_{n+M}(j\Omega)$, or, equivalently, M frequency functions $A_n(j\Omega), n = 0, 1, \ldots, M-1$. To this end, we insert (5.108) into (5.109) and interchange the summation and integration, and we obtain

$$y(n) = \frac{1}{2\pi} \int_{-\Omega_0 T}^{\Omega_0 T} A_n(j\Omega) X(e^{j\Omega T}) e^{j\Omega T n} d(\Omega T), \tag{5.110}$$

where

$$A_n(j\Omega) = \sum_{k=-N/2}^{N/2} h_n(k) Q_{n-k}(j\Omega) e^{-j\Omega T k}. \tag{5.111}$$

Further, we can write the desired sequence $x(n)$ in terms of the inverse Fourier transform according to

$$x(n) = \frac{1}{2\pi} \int_{-\Omega_0 T}^{\Omega_0 T} X(e^{j\Omega T}) e^{j\Omega T n} d(\Omega T). \tag{5.112}$$

Comparing (5.110) and (5.112), it is seen that PR is obtained if

$$A_n(j\Omega) = 1, \quad \Omega \in [-\Omega_0, \Omega_0], \tag{5.113}$$

for $n = 0, 1, \ldots, M-1$, because then $y(n) = x(n)$ for all n as $A_n(j\Omega)$ is M-periodic. Thus, if all $A_n(j\Omega) = 1$, we get the desired result $x(n) = x_a(nT)$. The problem is thus to determine the M impulse responses $h_n(k)$, $n = 0, 1, \ldots, M-1$, so that $A_n(j\Omega)$ approximate unity. If the channel frequency responses $Q_n(j\Omega)$ are known, this can be done optimally in, for example, least-squares or minimax senses. In practice, one does not explicitly estimate $Q_n(j\Omega)$ but instead $Q_n(j\Omega)/Q_0(j\Omega)$ if $Q_0(j\Omega)$ is the reference channel, in accordance with the discussion presented in Section 5.5.2.1. The ratios $Q_n(j\Omega)/Q_0(j\Omega)$ are then used in the design, which corresponds to selecting $h_0(k) = \delta(k)$ (unit impulse). Therefore, it is sufficient to determine the $M-1$ impulse responses $h_n(k)$ for $n = 1, 2, \ldots, M-1$, so that the corresponding $A_n(j\Omega)$ approximate unity. After compensation with these filters, the overall ADC will experience the actual linear distortion $Q_0(j\Omega)$.

5.5.5 Error metrics

A common metric of ADCs is the signal-to-noise ratio (SNR). In this case, the aim is to minimize $A_n(j\Omega) - 1$ in the least-squares sense, which means that the quantities P_n given by

$$P_n = \frac{1}{2\pi} \int_{-\Omega_0 T}^{\Omega_0 T} |A_n(j\Omega) - 1|^2 d(\Omega T) \tag{5.114}$$

are minimized, either separately or simultaneously (e.g., by minimizing the mean of P_n). The minimization of P_n corresponds to the minimization of the expected error $E\{[y(n) - x(n)]^2\}$ when the input signal is a stochastic process with a constant power spectrum in the region $[-\Omega_0, \Omega_0]$ [19].

Another common metric of ADCs is the spurious-free dynamic range. In this case, it is appropriate to write the input–output relation in the frequency domain in terms of a distortion function $V_0(e^{j\Omega T})$ and $M-1$ aliasing functions $V_m(e^{j\Omega T})$, $m = 1, 2, \ldots, M-1$. This is similar to frequency-domain input–output relations for multirate filter banks [52]. One can then write the output Fourier transform as

$$Y(e^{j\Omega T}) = V_0(e^{j\Omega T}) X(e^{j\Omega T}) + \sum_{m=1}^{M-1} V_m(e^{j\Omega T}) X(e^{j(\Omega T - 2\pi m/M)}), \tag{5.115}$$

where

$$V_0(e^{j\Omega T}) = \frac{1}{M} \sum_{n=0}^{M-1} B_n(e^{j\Omega T}) \tag{5.116}$$

and

$$V_m(e^{j\Omega T}) = \frac{1}{M} \sum_{n=0}^{M-1} e^{-j2\pi mn/M} B_n(e^{j(\Omega T - 2\pi m/M)}), \tag{5.117}$$

with $B_n(e^{j\Omega T})$ being the 2π-periodic extensions of $A_n(j\Omega)$ in (5.111). That is, $B_n(e^{j\Omega T}) = A_n(j\Omega)$ for $-\pi \leq \Omega T \leq \pi$. The distortion function corresponds to a regular frequency response seen from the input to the output, whereas the aliasing functions give the size of the aliased (frequency shifted) versions of the input. Specifically, a frequency component at $\Omega = \Omega_0$ is affected by the modulus and phase of the distortion frequency response $V_0(e^{j\Omega_0 T})$, and we also get aliased versions at the "digital frequencies" $\Omega_m T = \Omega_0 T + 2\pi m/M$, $m = 1, 2, \ldots, M-1$, with amplitudes determined by $V_m(e^{j(\Omega_0 T + 2\pi m/M)})$. Note also that for real signals, which always have frequency components at $\Omega_0 T$ and $-\Omega_0 T$ simultaneously, aliased versions also appear at $-\Omega_0 T + 2\pi m/M$, $m = 1, 2, \ldots, M-1$.

Ideally, the distortion function should equal unity, whereas the aliasing functions should be zero, in which case we have PR, because then we obtain $Y(e^{j\Omega T}) = X(e^{j\Omega T})$ and thus $y(n) = x(n)$. In practice, we can only approximate PR. Typically, one then aims at minimizing two tolerances, say δ_d and δ_a, between unity and zero. One can then solve the problem in several slightly different ways. One typical way is to minimize δ subject to the constraints

$$|V_0(e^{j\Omega T}) - 1| \leq \delta, \quad \Omega T \in [-\Omega_0 T, \Omega_0 T], \tag{5.118}$$

and

$$|V_m(e^{j\Omega T})| \leq \delta(\delta_a/\delta_d), \quad \Omega T \in \Omega_m T, \tag{5.119}$$

for $m = 1, 2, \ldots, M-1$, where

$$\Omega_m T = [-\Omega_0 T + \frac{2\pi m}{M}, \ \Omega_0 T + \frac{2\pi m}{M}]. \tag{5.120}$$

Clearly, $|V_0(e^{j\Omega T})| \leq \delta_d$ and $|V_m(e^{j\Omega T})| \leq \delta_a$ if $\delta \leq \delta_d$. It is noted that δ_d and δ_a are related through

$$|\delta_d| \leq (M-1)|\delta_a| + |A_n(j\Omega) - 1|_{\min}, \tag{5.121}$$

which means that it is usually sufficient to ensure that the aliasing is suppressed, as this implies a small distortion as well. Specifically, for the relaxed problem, (5.121) reduces to $|\delta_d| \leq (M-1)|\delta_a|$ provided $|V_0(e^{j\Omega T}) - 1|$ in (5.118) is replaced with $|V_0(e^{j\Omega T}) - B_q(e^{j\Omega T})|$, q denoting the reference channel. The relation above is a consequence of the following equation which can be derived from (5.117):

$$\sum_{m=0}^{M-1} e^{j2\pi mq/M} V_m(e^{j(\Omega T + 2\pi m/M)}) = B_q(e^{j\Omega T}). \tag{5.122}$$

However, in a practical system, we cannot minimize δ directly as we do not have access to the distortion and aliasing functions. Instead, δ is reduced indirectly to an acceptable level through some estimation and correction scheme.

5.5.5.1 DC offset

It is finally noted in this section that the representation in (5.115) can be used for DC offset mismatch as well, if we let X represent a DC signal, because it can be seen as a DC input that is time-interleaved and amplified differently in the different channels. In accordance with the discussion on aliasing above, it is seen that offset mismatches give rise to tones at $\Omega_m T = 2\pi m/M$, $m = 0, 1, \ldots, M-1$, with amplitudes determined by the frequency-independent V_m, as B_n in this case are frequency-independent constants equaling the channel offset values. Specifically, with all B_n being equal, say $B_n = c$, we obtain from (5.117) that $V_0 = c$ and $V_m = 0$ for $m = 1, 2, \ldots, M-1$, as we in this case have a regular DC offset.

5.5.5.2 Example 6: effect of oversampling

A good approximation can generally only be achieved if the input signal is band-limited in accordance with the Nyquist theorem. It is generally advisable to use at least 10% oversampling; otherwise the complexity of the reconstructor may become prohibitive. However, if low complexity is not crucial, the amount of oversampling can be reduced. The increase in complexity with decreasing oversampling is illustrated in Figs. 5.28 and 5.29, which plot the system (filter) order versus time skew in a two-channel time-interleaved ADC for a reconstruction error (P_1 in (5.114)) of -60, -80, and -100 dB. It is seen that the order is roughly doubled when the don't-care band between the upper band edge and the Nyquist frequency $f_s/2$ is halved (which occurs when we go from 80% to 90%, etc.). However, it is also seen that for small time skews and moderate reconstruction errors, one may increase the bandwidth, as the order in such cases is relatively low. Recall that for a reconstruction system of order

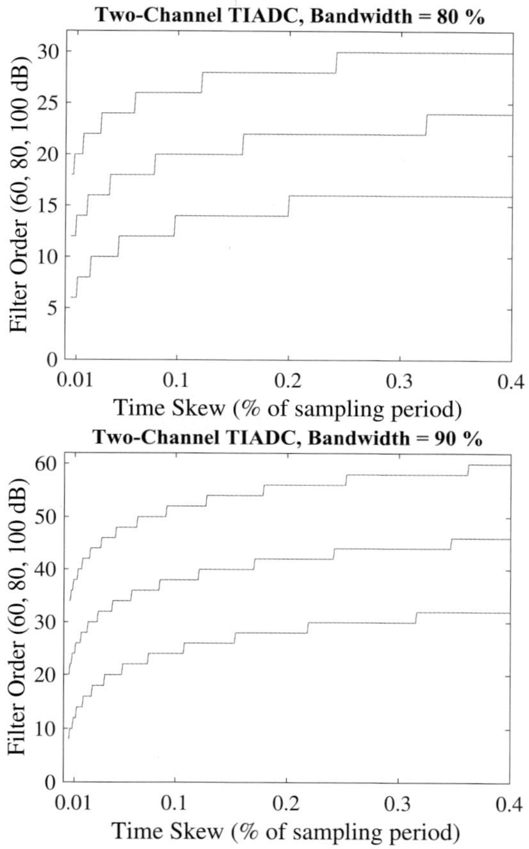

FIGURE 5.28

System (filter) order versus time skew in percent of the sampling period $T = 1/f_s$ in a two-channel time-interleaved ADC for 80% and 90% of the Nyquist band.

N, in accordance with the convolution sum in (5.109), one needs to compute $N + 1$ multiplications and N additions to correct each output sample.

5.5.6 Reconstruction based on least-squares design

From the previous section, we see that the overall compensation system can be designed by minimizing the quantities P_n, $n = 0, 1, \ldots, M - 1$, in (5.114). The quantities P_n can be minimized either separately or simultaneously. In the former case, the corresponding M impulse responses $h_n(k)$, $n = 0, 1, \ldots, M - 1$, can be computed separately as [20]

$$\mathbf{h}_n = -0.5\mathbf{S}_n^{-1}\mathbf{c}_n, \tag{5.123}$$

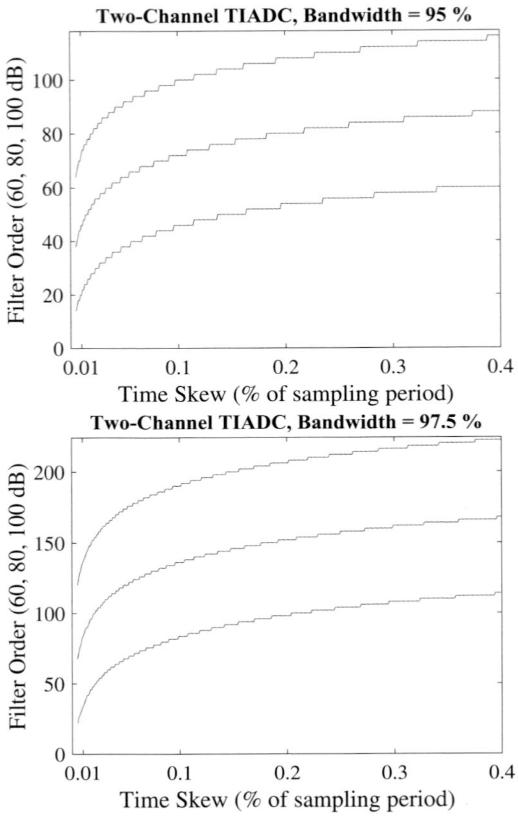

FIGURE 5.29

System (filter) order versus time skew in percent of the sampling period $T = 1/f_s$ in a two-channel time-interleaved ADC for 95% and 97.5% of the Nyquist band.

with $c_{n,k}$, $k = -N, -N+1, \ldots, N$, being

$$\begin{aligned} c_{n,k} &= -\frac{1}{\pi}\int_{-\Omega_0 T}^{\Omega_0 T} |Q_{n-k}(j\Omega)| \times \\ &\cos\big(\Omega T k - \arg\{Q_{n-k}(j\Omega)\}\big) d(\Omega T), \end{aligned} \tag{5.124}$$

$\mathbf{S}_n$ being $(N+1)\times(N+1)$ symmetric and positive definite matrices with entries $s_{n,kp}$, $k, p = -N/2, -N+1, \ldots, N/2$, given by

$$\begin{aligned} s_{n,kp} &= \frac{1}{2\pi}\int_{-\Omega_0 T}^{\Omega_0 T} |Q_{n-k}(j\Omega)||Q_{n-p}(j\Omega)| \times \\ &\cos\big(\Omega T(p-k) + \arg\{Q_{n-k}(j\Omega)\} - \arg\{Q_{n-p}(j\Omega)\}\big) d(\Omega T), \end{aligned} \tag{5.125}$$

and the constant C being $C = \Omega_0 T/\pi$. In order to compute $\mathbf{h}_n$ in (5.123), it is necessary to compute the integrals in (5.124) and (5.125), which generally has to be done numerically.

Above, the overall compensation system is designed by determining the M filter impulse responses $h_n(k)$ through M separate matrix inversions ($M-1$ in the practical relaxed case), where the size of the matrices is $(N+1)\times(N+1)$, where $N+1$ is the filter impulse response length [20,26]. In the alternative case, corresponding to simultaneous design, the impulse responses $h_n(k)$ are designed simultaneously. This was done in [44], which poses the design problem in terms of M synthesis filters that are designed simultaneously by inverting one matrix of size $M(N+1)\times M(N+1)$. Whereas the separate-design technique yields optimum filters with respect to the reconstruction error in the band of interest, the technique in [44] may give somewhat better results as to distortion and aliasing, because it includes those quantities in the overall error cost measure. The main advantage of the separate-design approach is that the design problem comprises M smaller subproblems instead of one big problem. It may therefore be more numerically robust and simpler to design and implement.

In addition to the abovementioned design techniques [20,26,44], there are a few other related techniques [26,31,32,49]. In [49], separately designed synthesis filters were obtained through windowing techniques, which, however, are known to result in suboptimum filters. A somewhat different compensation technique that also utilizes separately designed filters was proposed in [31], [32], but that technique needs additional cosine and sine modulators, which increases the implementation cost of the compensation system. Finally, it is noted that one may alternatively use numerical optimization in accordance with [19], but this is costly and therefore less suitable for on-line design applications.

For all methods described above, new values of the impulse responses $h_n(k)$ must be computed whenever the frequency responses $Q_n(j\Omega)$ are changed, which occur in a practical time-interleaved ADC due to temperature variations, etc. This requires recomputation of the filter coefficients using windowing techniques (including inverse Fourier transforms), matrix inversions, or numerical optimization. This may be acceptable for off-line design, which can be adopted, for example, in spectrum analysis applications. In real-time low-power applications, on the other hand, these on-line redesign procedures can become too costly and time consuming to implement. The next section discusses a compensation structure for which on-line design is not needed. This structure was introduced in [18]. A similar structure has been reported in [59].

5.5.7 Reconstruction using a polynomial-based approach

As discussed in the previous section, the reconstruction can in principle be done through the time-varying system $h_n(k)$. For an Nth-order reconstructor, each system $h_n(k)$ has then $N+1$ free parameters (multipliers) that must be estimated and implemented, and thus $(M-1)(N+1)$ in total for a practical M-channel system (in the relaxed case). This becomes difficult from the estimation point of view and expensive to implement except for small values of M.

Depending on the behavior of the channel responses $Q_n(j\Omega)$, the number of free parameters may be reduced by modeling them as Pth-order polynomials in $j\Omega$ according to

$$Q_n(j\Omega) = \varepsilon_n^{(0)}[1 + \sum_{p=1}^{P} \varepsilon_n^{(p)}(j\Omega T)^p], \tag{5.126}$$

where $\varepsilon_n^{(0)}$ is a gain constant (ideally equal to unity) and the remaining $\varepsilon_n^{(p)}$ are constants (ideally equal to zero) used for modeling the frequency dependency. This does not impose a fundamental restriction, as it is known that any frequency response can be approximated as closely as desired by a properly chosen polynomial and polynomial order. For example, when the system has frequency-independent time skews, $d_n T$, $Q_n(j\Omega)$ can be written as

$$Q_n(j\Omega) = \varepsilon_n^{(0)} e^{j\Omega T d_k} \approx \varepsilon_n^{(0)}[1 + \sum_{p=1}^{P} d_n^p (j\Omega T)^p / p!], \tag{5.127}$$

from which we obtain $\varepsilon_n^{(p)} = d_n^p / p!$. In this case it suffices to estimate two parameters per channel (three, including offset errors).

A general structure for the reconstruction based on the model in (5.126) is seen in Fig. 5.30. This structure makes use of K sets of fixed subfilters $G_p(z)$, $p = 1, 2, \ldots, P$, that approximate the pth-order differentiators in (5.126). Fig. 5.30 shows a noncausal filter structure for convenience. The corresponding causal structure is obtained by propagating D delay elements into each vertical branch and replacing $\varepsilon_n^{(p)}$ with $\varepsilon_{n-D}^{(p)}$, where D is the (possibly approximate) delay of the subfilters, which can be either linear-phase finite length impulse response (FIR) subfilters or approximately linear-phase FIR or infinite length impulse response (IIR) subfilters. Using Nth-order linear-phase FIR subfilters, $D = N/2$. The remaining errors in the output $y_K(n)$ of the K-stage structure in Fig. 5.30 are of orders $(\Omega_0 T \varepsilon_n^{(p)})^{K+1}$. This means that one can reach any desired error level by increasing the number of stages provided $|\Omega_0 T \varepsilon_n^{(p)}|$ is small, which is the case in practical time-interleaved ADCs. The error will in practice decrease with K to a level that is determined by the channel model accuracy, i.e., by the distances between $Q_n(j\Omega)$ and the actual frequency responses.

5.5.8 Performance discussion

The performance of the overall compensated ADC is naturally bounded by the performance of the individual converters. In this section, we only consider the compensation of the linear mismatch errors. Ultimately, the overall ADC's performance is therefore determined by the nonlinear distortion of the individual converters. In practice, the compensation structure presented above has virtually no effect on the nonlinear distortion. That is, the undesired frequency components caused by nonlinear distortion pass the compensation structure more or less unaffected. This is because the nonlinear distortion components, which are small themselves, pass a time-varying system that approximates a pure delay provided $\varepsilon_n^{(p)}$, $p = 1, 2, \ldots, P$, are small, which is the case in practice. This will be illustrated below in an example. It should be noted though that the time interleaving introduces nonlinear-distortion frequency components that are not present in the individual converters, the total nonlinear-distortion power is virtually the same [57,58].

5.5.8.1 Example 7: frequency response mismatch correction

To illustrate the theory above, an example is provided. It is assumed here that the number of channels is $M = 4$, the bandwidth is $\Omega_0 T = 0.9\pi$, and the four channel frequency responses $Q_n(j\Omega)$,

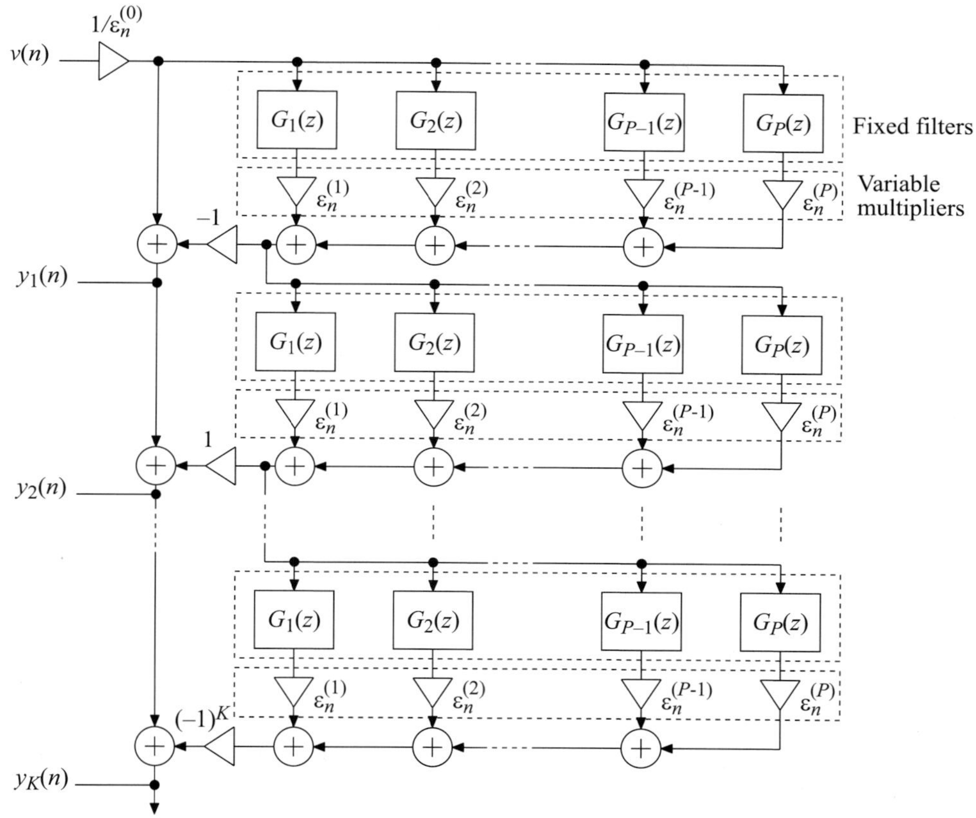

FIGURE 5.30

Reconstruction based on the model in (5.126).

$n = 0, 1, 2, 3$, are modeled as

$$Q_n(j\Omega) = \frac{e^{j\Omega T d_n}}{1 + j\frac{\Omega}{\Omega_c}(1 + \Delta_n)} = \frac{e^{j\Omega T d_n}}{1 + j\Omega T \frac{f_s}{2\pi f_c}(1 + \Delta_n)}, \tag{5.128}$$

where $f_s = 1/T$ denotes the sampling frequency, whereas f_c and Ω_c denote the 3-dB cutoff frequency and angular frequency, respectively, and d_n and Δ_n correspond to analog matching errors. Here, we have chosen $f_s/f_c = 1/2$, which corresponds to a bandwidth of about two times the overall Nyquist band or, equivalently, eight times the individual channel ADC Nyquist band. Further, we have set $d_0 = -0.02$, $d_1 = 0.02$, $d_2 = -0.01$, $d_3 = 0.01$, $\Delta_0 = -0.005$, $\Delta_1 = 0.005$, $\Delta_2 = -0.004$, and $\Delta_3 = 0.004$. The term in the numerator in (5.128) models static aperture delay mismatches, whereas the denominator models the S/H circuit [49,57,58]. The values for these parameters correspond to time and bandwidth mismatch spurs of some -30 and -60 dB, respectively, which is realistic at least for well-matched

ADCs operating in their lower Nyquist bands. Hence, for a 10-bit resolution, one may ignore the bandwidth mismatch.

We assume that the zeroth channel is the reference channel. Consequently, we need to find the polynomial models of the rational functions $Q_n(j\Omega)/Q_0(j\Omega)$, $n = 0, 1, 2, 3$. Including terms up to third order, we obtain from (5.128)

$$\frac{Q_n(j\Omega)}{Q_0(j\Omega)} = 1 + j\Omega T\varepsilon_n^{(1)} + (j\Omega T)^2\varepsilon_n^{(2)} + \ldots + (j\Omega T)^3\varepsilon_n^{(3)}, \tag{5.129}$$

where

$$\varepsilon_n^{(1)} = d_n - d_0 + \Delta_0 - \Delta_n, \tag{5.130}$$

$$\varepsilon_n^{(2)} = \frac{(d_n - d_0)^2}{2} - \Delta_n(\Delta_0 - \Delta_n) - (d_n - d_0)(\Delta_0 - \Delta_n), \tag{5.131}$$

and

$$\varepsilon_n^{(3)} = \frac{(d_n - d_0)^3}{6} + \Delta_n^2(\Delta_0 - \Delta_n) - (d_n - d_0)\Delta_n(\Delta_0 - \Delta_n) + \frac{(d_n - d_0)^2}{2}(\Delta_0 - \Delta_n). \tag{5.132}$$

Here, we can compute the different parameters $\varepsilon_n^{(p)}$ needed for the compensation, as we have assumed that we know the channel frequency responses. In practice, these parameters are obtained through some estimation technique for this purpose. It is noted that $\varepsilon_0^{(1)} = \varepsilon_0^{(2)} = \varepsilon_0^{(3)} = 0$, as we have assumed that the zeroth channel is the reference, i.e., the samples from this channel are taken directly without compensation.

We have applied a 16-bit multisine input to this four-channel system. Without compensation, the output spectrum becomes as seen in Fig. 5.31, which reveals large aliasing spurs. Fig. 5.31 also plots the output spectra after compensation using one, two, and three stages. For each plot, an 8192-point discrete Fourier transform (DFT) and a Blackman–Harris window have been used. The signal contains 45 unity-amplitude sinusoidal signals with frequencies at $\Omega T = m \times 0.02\pi - 0.00477\pi$, $m = 1, 2, \ldots, 45$. The frequencies have been chosen so that the aliasing spurs appear approximately in the middle of two adjacent fundamental tones, to clearly show the aliasing, and so that the sampling is noncoherent to avoid periodic quantization errors. Recall that coherent sampling corresponds to frequencies coinciding with the DFT frequencies $2\pi k/N$, $k = 0, 1, \ldots, N-1$, for an N-point DFT.

From Fig. 5.31, it is seen that after three stages the errors have been reduced to a level that corresponds to some 16 bits. This reveals that any desired error level can be reached in the frequency band $\Omega \in [-\Omega_0, \Omega_0]$ by increasing the number of stages, provided the model order is high enough. In this example, the third-order model in (5.129) is required to reach 16 bits. Using instead a second-order model, we end up with spurs of some -80 dB for $K \geq 2$, which corresponds to 13–14 bits. Hence, the approximation of Eq. (5.128) by a numerator polynomial of a certain order imposes a performance bound that cannot be surpassed by increasing K alone. In other words, to reach the desired result, one must select both the model order and the number of stages appropriately. It is also noted that one stage suffices to reach some -60 dB, thus about 10 bits.

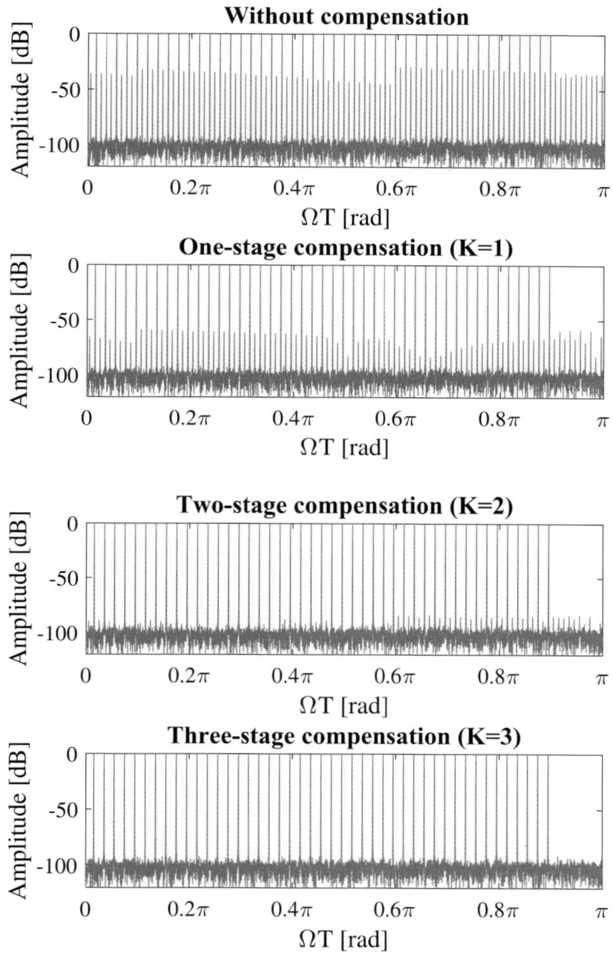

FIGURE 5.31

Spectra in Example 7.

5.5.8.2 Example 8: frequency response mismatch correction in the presence of nonlinearities

To illustrate that the compensation has virtually no effect on nonlinear distortion, as discussed previously, we have applied a 13-bit two-tone input including the distortion term $0.0001x^2(n) + 0.0002x^3(n)$ and plotted the output spectra in Fig. 5.32 for the uncompensated system and compensated system with $K = 3$. From the figure, it is seen that the two large unwanted spurs (of some -40 dB) caused by the interleaving have been suppressed, whereas the smaller nonlinear-distortion

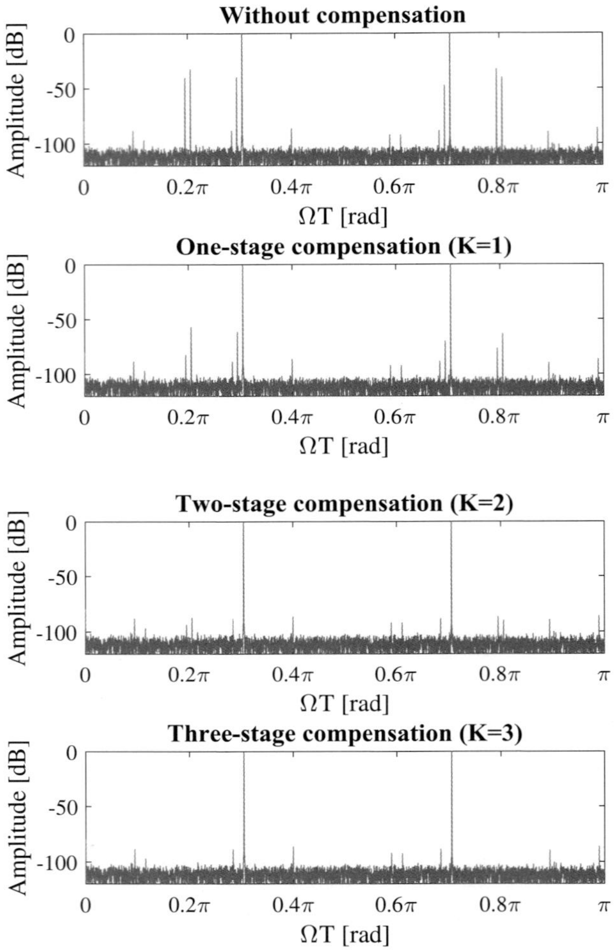

FIGURE 5.32

Spectra in Example 8.

spurs (of some −90 to −85 dB) are practically unaffected by the compensation. It is also noted that the nonlinearity terms can be suppressed by applying appropriate linearization techniques [5,27,60].

5.6 Quantization

In ideal linear systems, all signal values are represented with real numbers and all computations are done with infinite precision. In a practical implementation, one can however only use finite precision

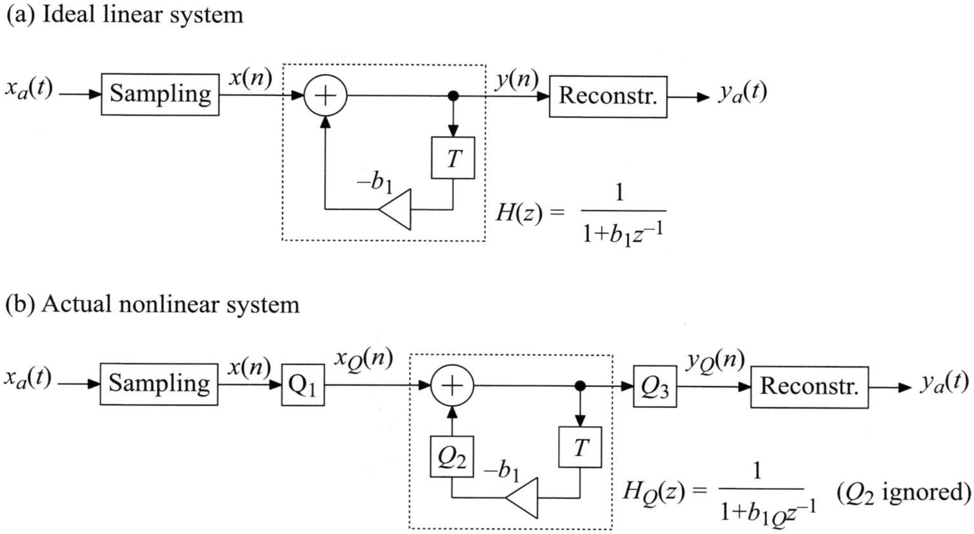

FIGURE 5.33

(a) Ideal linear system. (b) Actual nonlinear system in a practical implementation.

for both the input and output signal values and the internal signal values. Hence, instead of the ideal system in Fig. 5.33(a), we will in practice implement the one shown in Fig. 5.33(b). The boxes Q (with additional indices) represent quantizers. The function of a quantizer Q depends on the type of arithmetic that is to be used in the implementation. In general-purpose computers, one uses floating-point arithmetic. However, for many fixed functions (like filters), the large number range provided by floating-point arithmetic is not needed and thus becomes unnecessarily expensive to implement. In very large-scale integration (VLSI) implementations, one therefore usually employs fixed-point arithmetic in order to minimize the implementation cost. Henceforth, we assume that fixed-point arithmetic is used. One may further choose between several different formats like sign and magnitude, one's complement, two's complement, etc. Here, we consider the two's complement format, which is commonly used in VLSI implementations.

Using two's complement arithmetic, a quantized number x_Q within the number range $-X_m \leq x_Q \leq X_m(1 - Q)$ is represented by the B-bit fractional binary number x_B given by

$$x_B = -x_0 + \sum_{i=1}^{B-1} 2^{-i} x_i, \tag{5.133}$$

where x_i, $i = 0, 1, ..., B{-}1$, are either zero or one and x_B is related to x_Q according to

$$x_Q = X_m x_B. \tag{5.134}$$

The quantity x_0 is referred to as the sign bit, which is zero for nonnegative numbers and one for negative numbers. Furthermore, it is seen that $-1 \leq x_B \leq 1 - Q$. Hence, the quantity X_m is a scaling constant that is used for relating numbers within the range $-1 \leq x_B \leq 1 - Q$ to numbers within the range $-X_m \leq x_Q \leq X_m(1 - Q)$. For example, in the context of A/D conversion, we can view X_m as the full-scale amplitude of the A/D converter. Another role of X_m is to enable representation of numbers larger than one in magnitude. In this case, $X_m = 2^P$, for some integer P, which can be interpreted as if the binary point has been moved to the right. Often, this scaling constant is not explicitly implemented but instead implicit.

The quantizers that inevitably are present introduce a number of errors that need to be analyzed when designing and implementing digital systems. The errors can be divided into four broad categories as discussed below. It is beyond the scope of this section to discuss all of these issues in detail. The subsequent subsections concentrate on quantization errors in the A/D conversion process and the related round-off errors in digital systems. For a comprehensive treatment of quantization effects, we refer in particular to the digital filter literature [1,16,61].

- Quantization errors in the A/D conversion – A/D conversion consists of sampling followed by quantization of the sample values. Thereby, a quantization error is introduced in the sampled signal.
- Overflow errors – Overflow errors occur when the available number range is exceeded. This may give rise to large sustaining errors referred to as parasitic large-scale oscillations. Digital systems must be designed so that such errors are suppressed once the disturbance that caused them vanishes. This is done by scaling the signal levels appropriately, not only at inputs and outputs but also inside the systems.
- Round-off errors – In digital systems, the results of arithmetic operations usually need to be rounded (truncated). This gives rise to round-off errors at the output of the systems.
- Coefficient errors – The coefficients of a digital system must be represented with finite precision. This gives rise to a static error in the corresponding transfer function, frequency response, etc.

5.6.1 Quantization errors in A/D conversion

A/D conversion consists of sampling followed by quantization according to Fig. 5.34. In the sampling process, the samples $x(n)$ are obtained as $x(n) = x_a(nT)$. Quantization means that each sample is represented in binary form with finite word length. This gives rise to a quantization error, i.e., an error in the representation of $x_a(nT)$, according to

$$e(n) = x_Q(n) - x(n). \tag{5.135}$$

This is also illustrated in Fig. 5.34. The size of the quantization error depends on the types of binary representation and quantizer that are used. It is common to use fixed-point two's complement representation and uniform quantization, in which case $x(n)$ is rounded to the nearest number in the number representation.

If we assume that two's complement representation is used and that the number range for the quantizer is $-X_m \leq x_Q \leq X_m(1 - Q)$, where Q is the quantization step, $x_Q(n)$ can be written for a B-bit quantizer as

$$x_Q(n) = X_m x_B(n), \tag{5.136}$$

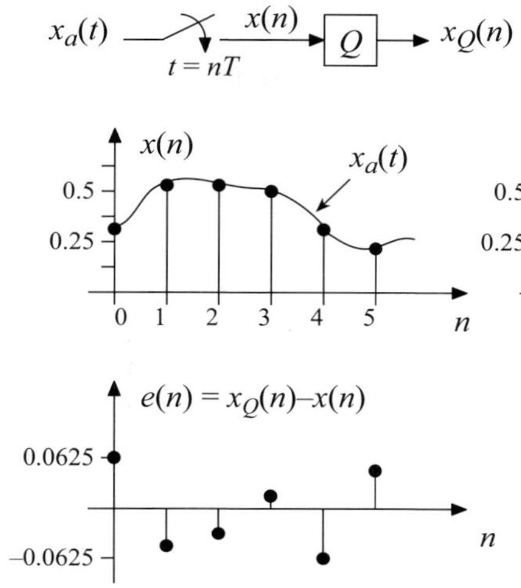

FIGURE 5.34

A/D conversion – sampling and quantization.

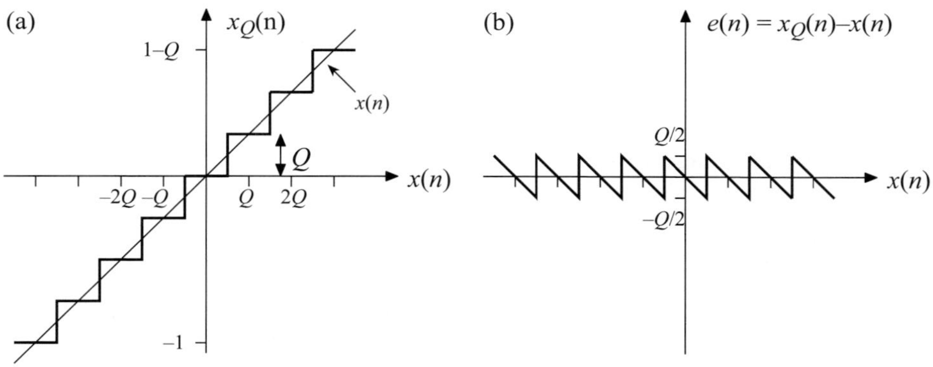

FIGURE 5.35

(a) Uniform quantization with $X_m = 1$ and $B = 3$ bits. (b) Quantization error.

with

$$x_B(n) = -x_0(n) + \sum_{i=1}^{B-1} 2^{-i} x_i(n), \tag{5.137}$$

where $x_i(n)$, $i = 0, 1, \ldots, B-1$, are either zero or one. When the samples $x_Q(n)$ are represented with B bits, they can take on 2^B different values, as illustrated in Fig. 5.35 for a uniform quantizer, assuming

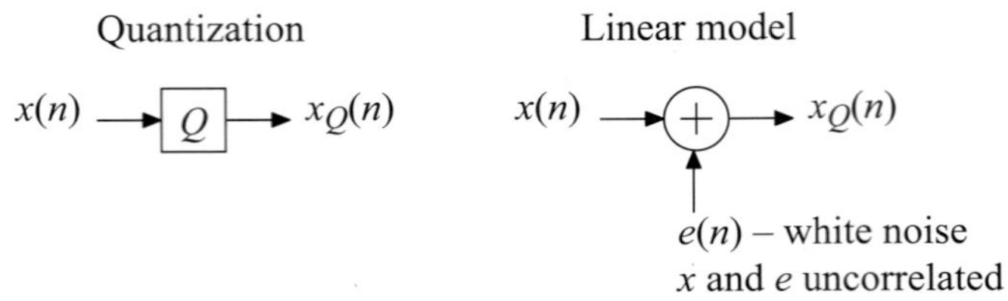

FIGURE 5.36

Linear model of quantization.

for simplicity that $X_m = 1$. The quantization step Q is given by

$$Q = X_m 2^{-(B-1)}. \tag{5.138}$$

Quantization is a nonlinear operation which in general is much more difficult to analyze than a linear operation. To analyze the effects of the quantization errors, one therefore makes use of a linear model of the quantization according to Fig. 5.36. In the linear model, $x_Q(n)$ is obtained by adding $x(n)$ and an error sequence $e(n)$,

$$x_Q(n) = x(n) + e(n), \tag{5.139}$$

where it is assumed that $x(n)$ and $e(n)$ are uncorrelated. Further, $e(n)$ is assumed to be uniformly distributed white noise. That is, one understands that the samples $e(n)$ are uncorrelated and the errors that $e(n)$ can take on occur with equal probability. This model is appropriate when $x_a(t)$ varies irregularly, since the quantization errors then vary more or less randomly and independently of previous errors.

Using uniform quantization, we have

$$|e(n)| \le Q/2, \tag{5.140}$$

where Q is the quantization step according to (5.138). Each sample $e(n)$ is therefore assumed to have the probability density function

$$f(e) = \begin{cases} 1/Q, & |e| \le Q/2, \\ 0, & |e| > Q/2. \end{cases} \tag{5.141}$$

The mean value and variance of $e(n)$ become

$$m_e = 0 \tag{5.142}$$

and

$$\sigma_e^2 = \frac{Q^2}{12} = X_m^2 \frac{2^{-2(B-1)}}{12}, \tag{5.143}$$

respectively.

A measure that is commonly used for comparing the (useful) signal power and noise power is the SNR, which is defined as

$$\mathrm{SNR} = 10 \times \log_{10}\left(\frac{\text{signal power}}{\text{noise power}}\right). \tag{5.144}$$

Since here the noise power is the variance of $e(n)$ (due to $m_e = 0$), i.e., σ_e^2 in (5.143), we can write the SNR as

$$\mathrm{SNR} = 10 \times \log_{10}(\text{signal power}) - 10 \times \log_{10}\left(X_m^2 \frac{2^{-2(B-1)}}{12}\right), \tag{5.145}$$

which alternatively can be written as

$$\mathrm{SNR} = 10 \times \log_{10}(\text{signal power}) + 4.77 + 6.02B - 20 \times \log_{10}(X_m). \tag{5.146}$$

We note that the SNR increases by 6 dB for each added bit. For a full-scale sinusoidal signal with amplitude X_m, the signal power is $X_m^2/2$, which corresponds to $20 \times \log_{10}(X_m) - 3.01$ dB. In this case, the SNR is

$$\mathrm{SNR} = 1.76 + 6.02B. \tag{5.147}$$

For example, in a CD player, $B = 16$ bits, in which case the SNR becomes 98.1 dB.

5.6.2 Round-off errors

This section considers round-off errors (or round-off noise) in digital systems. The round-off noise is computed under the assumption that the quantization errors can be modeled as white noise, which is appropriate in many practical situations. It is further assumed that the digital system operates under normal conditions, which means that overflows or other abnormal disturbances do not occur. In other words, we assume a linear model of the system incorporating one input signal and one or several noise sources. The problem is then to determine the round-off noise power and SNR at the output of the system. To this end, one makes use of the theory of linear systems excited with white stochastic processes. We will explain the principle by means of examples, where we assume $X_m = 1$ for simplicity, but without loss of generality.

5.6.2.1 Example 9: quantization and round-off noise in an IIR filter

Consider a system represented by the structure in Fig. 5.37(a). First, sampling and quantization take place, with the quantization step Q_1, i.e., A/D conversion with B_1 bits, where $Q_1 = 2^{-(B_1-1)}$. The output from the quantization is $x_Q(n)$, which is subsequently filtered. The filter is realized by a first-order direct-form recursive IIR filter structure. In an implementation of this filter, the result of the arithmetic operations must be rounded or truncated at some point(s) inside the filter. Otherwise, the word length would increase in each iteration of the filter algorithm, since it is recursive. Here, only

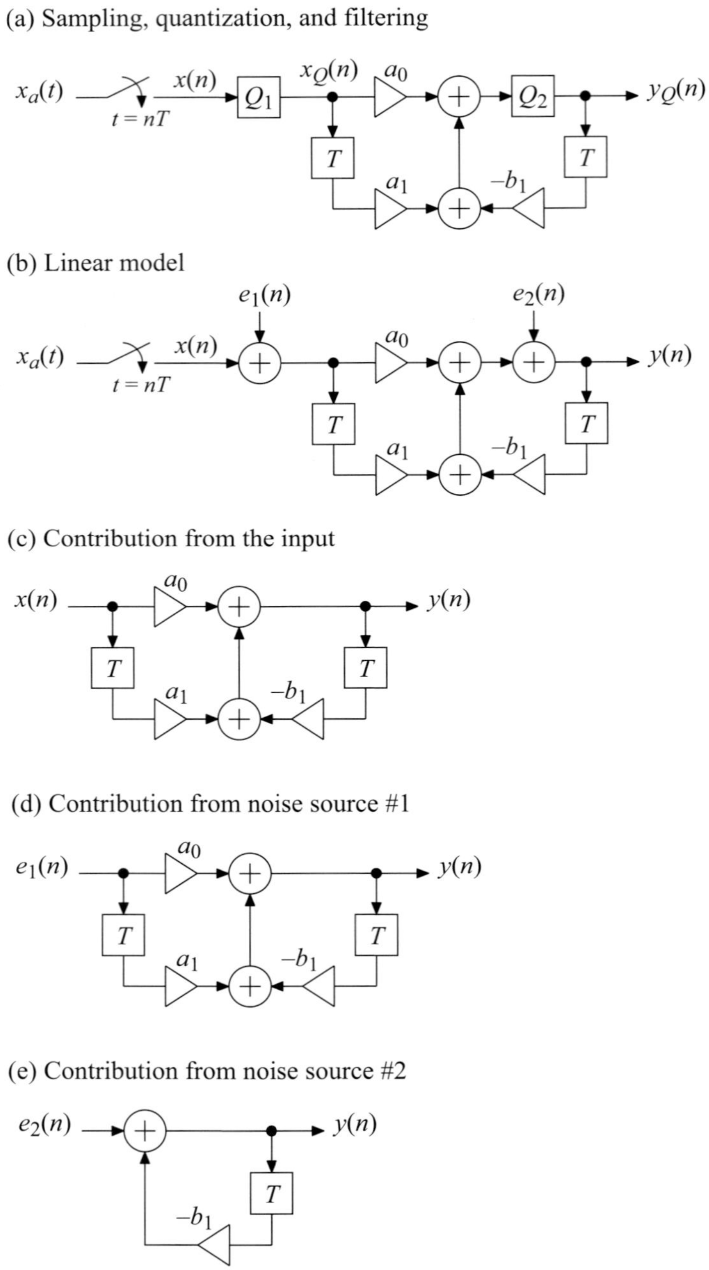

FIGURE 5.37

Signal-flow graphs used for computation of the round-off noise in Example 9.

one quantizer with the quantization step $Q_2 = 2^{-(B_2-1)}$ is used, and it is placed after the uppermost addition.[3]

We now wish to analyze the effects at the output of the filter caused by quantization errors. In order to separate the origins of the errors, we distinguish between the errors emanating from the A/D conversion, here referred to as quantization noise, and the errors coming from the quantizations of the arithmetic operations inside the filter, usually referred to as round-off noise. To be able to analyze the different errors, we make use of a linear model of the quantizations, resulting in the structure shown in Fig. 5.37(b). The error sequences $e_1(n)$ and $e_2(n)$ are assumed to be white noise with zero mean and variances σ_{e1}^2 and σ_{e2}^2, respectively.[4] These error sequences are often called noise sources.

Since we use linear models, we can consider the input and noise sources one at a time according to Fig. 5.37(c–e). The total output round-off noise variance is then computed as the sum

$$\sigma_y^2 = \sigma_{y1}^2 + \sigma_{y2}^2, \tag{5.148}$$

where

$$\sigma_{y1}^2 = \sigma_{e1}^2 \sum_{n=-\infty}^{\infty} h_1^2(n) \tag{5.149}$$

is the output quantization noise variance and

$$\sigma_{y2}^2 = \sigma_{e2}^2 \sum_{n=-\infty}^{\infty} h_2^2(n) \tag{5.150}$$

is the output round-off noise variance. Here, $h_1(n)$ and $h_2(n)$ are the impulse responses as seen from the respective noise source to the output. They can be obtained through inverse transformation of the corresponding transfer functions $H_1(z)$ and $H_2(z)$.

Here, the transfer function from $e_1(n)$ to $y(n)$ is the same as that from $x(n)$ to $y(n)$ and given by

$$H_1(z) = \frac{a_0 + a_1 z^{-1}}{1 + b_1 z^{-1}}, \tag{5.151}$$

whereas the transfer function from $e_2(n)$ to $y(n)$ is

$$H_2(z) = \frac{1}{1 + b_1 z^{-1}}. \tag{5.152}$$

In the causal-filter case, the region of convergence is $|z| > b_1$ for both of these transfer functions. We then get

$$h_1(n) = a_0(-b_1)^n u(n) + a_1(-b_1)^{n-1} u(n-1) \tag{5.153}$$

[3] Therefore, in this case the results of the remaining operations must be represented with more bits. But this is only one option: for example, one may instead introduce quantizers after each operation. This will increase the noise at the output of the filter, but, on the other hand, one may then use a shorter word length inside the filter. It is generally not obvious which alternative one should use in order to attain the cheapest implementation. It has to be investigated in each specific case.

[4] When quantizing previously quantized numbers, as is the case at the output of Q_2, one should actually use a discrete rectangular probability function. The difference is however negligible, except when only a few bits are discarded.

and

$$h_2(n) = (-b_1)^n u(n), \tag{5.154}$$

respectively, where $u(n)$ denotes the unit-step sequence. This gives us the corresponding noise gains

$$\sum_{n=-\infty}^{\infty} h_1^2(n) = a_0^2 + \left(a_0 - \frac{a_1}{b_1}\right)^2 \sum_{n=1}^{\infty} b_1^{2n} = \frac{a_0^2 + a_1^2 - 2a_0a_1b_1}{1-b_1^2} \tag{5.155}$$

and

$$\sum_{n=-\infty}^{\infty} h_2^2(n) = \sum_{n=0}^{\infty} b_1^{2n} = \frac{1}{1-b_1^2}, \tag{5.156}$$

respectively.

If the quantizations Q_1 and Q_2 correspond to B_1 and B_2 bits, respectively, the variances σ_{e1}^2 and σ_{e2}^2 become

$$\sigma_{e1}^2 = \frac{Q_1^2}{12}, \quad \sigma_{e2}^2 = \frac{Q_2^2}{12}. \tag{5.157}$$

This gives us the variances σ_{y1}^2 and σ_{y2}^2 according to

$$\sigma_{y1}^2 = \sigma_{e1}^2 \sum_{n=-\infty}^{\infty} h_1^2(n) = \frac{Q_1^2}{12} \frac{a_0^2 + a_1^2 - 2a_0a_1b_1}{1-b_1^2} \tag{5.158}$$

and

$$\sigma_{y2}^2 = \sigma_{e2}^2 \sum_{n=-\infty}^{\infty} h_2^2(n) = \frac{Q_2^2}{12} \frac{1}{1-b_1^2}, \tag{5.159}$$

respectively. Finally, we obtain

$$\sigma_y^2 = \sigma_{y1}^2 + \sigma_{y2}^2 = \frac{Q_1^2}{12} \frac{a_0^2 + a_1^2 - 2a_0a_1b_1}{1-b_1^2} + \frac{Q_2^2}{12} \frac{1}{1-b_1^2}, \tag{5.160}$$

which can be rewritten as

$$\sigma_y^2 = \frac{Q_1^2}{12} \frac{1}{1-b_1^2} \left(a_0^2 + a_1^2 - 2a_0a_1b_1 + \frac{Q_2^2}{Q_1^2}\right). \tag{5.161}$$

From the equation above, we see that the contribution from the quantizer Q_2 can be neglected if we choose Q_2 sufficiently small as compared to Q_1. That is, by increasing the number of bits within the filter, the round-off noise can be made negligible compared to the quantization noise. This shows that digital systems can be implemented with as high resolution as desired. However, an increased internal word length also results in a more expensive implementation. One should therefore not use longer word lengths than necessary to meet the specification at hand, typically given in terms of an SNR required.

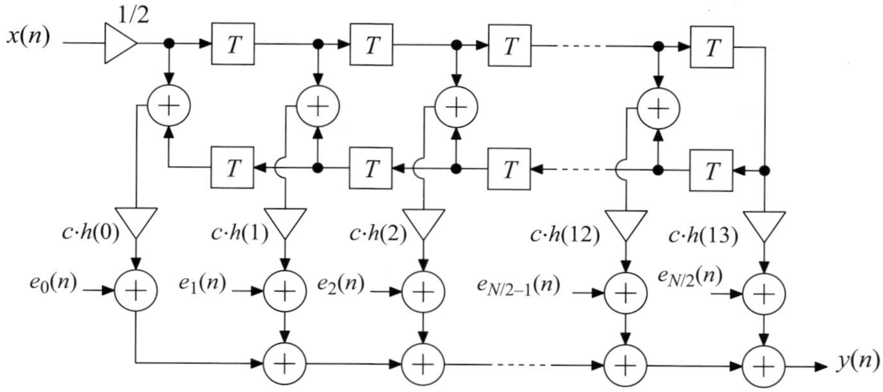

FIGURE 5.38

Linear model of a linear-phase FIR filter with quantizers (Example 10).

It should also be noted that the digital signal that at the end is to be D/A converted must be quantized to B_1 bits if we want to use the same number of bits as used in the A/D converter. This will give rise to additional round-off noise, equally large as in the A/D conversion, which in turn degrades the resolution by half a bit. One can alleviate this problem by using oversampling techniques.

5.6.2.2 Example 10: quantization and round-off noise in an FIR filter

In this example, we compute the round-off noise at the output of a linear-phase direct-form FIR filter structure and the increase in SNR that is achieved by using L_2-norm scaling instead of safe scaling.

Assuming that quantization takes place after each multiplication, the linear model of the whole filter is as shown in Fig. 5.38. Apparently, the errors pass directly to the output, i.e., the transfer function from each noise source to the filter's output is equal to unity. Assuming that the errors are uncorrelated white noise sources, the total output noise variance equals the sum of the individual variances. As a consequence, if all noise sources have the same variance, say σ_e^2, the total output variance for an Nth-order filter, for N even, becomes

$$\sigma_y^2 = (N/2 + 1)\sigma_e^2. \tag{5.162}$$

With $N = 26$, as in this example, we thus have

$$\sigma_y^2 = 14\sigma_e^2. \tag{5.163}$$

This corresponds to a degradation of almost 2 bits as compared to the case where one quantizer is placed at the output. That is, the outputs of quantizers placed inside the filter need to be 2 bits longer than the output of a single quantizer located at the output, in order to achieve (roughly) the same round-off noise. On the other hand, this means that we can use adders with shorter word lengths, which reduces the implementation cost. The alternative of using one quantizer at the output requires the use of full-length adders inside the filter, which increases the implementation cost.

Next, we consider the increase in SNR that is achieved by using L_2-norm scaling instead of safe scaling in the filter. Scaling of signal levels is used to avoid or reduce the probability of overflows, but it may also be used to increase the SNR. In the FIR filter in Fig. 5.38, the only node that needs to be scaled (assuming a scaled input) is the output, which is scaled by multiplying all impulse response values $h(n)$ by the scaling coefficient c. In this filter, the output SNR is affected by the scaling because the output round-off noise variance is independent of the multiplier coefficient values, whereas the output signal power is scaled with c. With safe scaling, c is chosen as

$$c = \frac{1}{\sum_{n=-\infty}^{\infty} |h(n)|}. \tag{5.164}$$

Using the convolution sum and the triangle inequality, one can show that c according to (5.164) guarantees no overflow. On the other hand, it also results in a lower SNR. The SNR can be increased by using a larger value of c, but at the cost of accepting the occurrence of overflow with a certain probability. With L_2-norm scaling, c is chosen as

$$c = \frac{1}{\sqrt{\sum_{n=-\infty}^{\infty} |h(n)|^2}}. \tag{5.165}$$

Using L_2-norm scaling for white noise Gaussian input signals, the probability of overflow at the output will be the same as the probability of overflow at the input.

Therefore, there is a trade-off between high SNR and probability of overflow. In this example, we obtain $c = 1.238477$ using safe scaling and $c = 3.330192$ using L_2-norm scaling. The increase in SNR that is achieved by using L_2-norm scaling instead of safe scaling is therefore

$$10 \times \log_{10}(3.330192^2) - 10 \times \log_{10}(1.238477) \approx 8.59 \text{ [dB]}, \tag{5.166}$$

which roughly corresponds to some 1.5 bits. That is, using L_2-norm scaling instead of safe scaling, we can implement the filter using 1.5 bits shorter data word length, which in practice means 1 bit or 2 bits shorter word length.

5.7 Oversampling techniques and MIMO systems

This section discusses oversampling techniques which are used in A/D and D/A conversions for relaxing the requirements on the analog filters and quantizers. It also discusses quantization in MIMO systems.

5.7.1 Oversampled A/D converters

The principle of an oversampled A/D converter is shown in Fig. 5.39. To elaborate on this principle, consider the typical spectra shown in Fig. 5.40. Assume that we have a signal $x_{wb}(t)$ containing information in the baseband up to π/T and undesired frequency components above π/T, like wide-band noise (Fig. 5.40(a)). The information can, in principle, be sampled and reconstructed without errors if

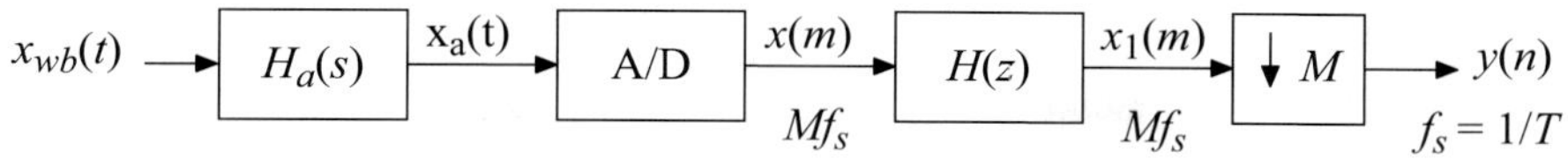

FIGURE 5.39

Oversampled A/D converter.

the sampling frequency is $f_s = 1/T$. However, due to the wide-band noise, the sampling will also introduce aliasing distortion. Theoretically, aliasing distortion can be avoided by band-limiting the input signal using an ideal analog low-pass filter with passband up to π/T, but such a filter cannot be realized in practice. Hence, we must let the filter have a transition band. The wider this band is, the simpler it becomes to implement the filter. On the other hand, we must then increase the sampling frequency so as to avoid aliasing distortion.

Assume that we use M times oversampling, i.e., that we choose a sampling frequency f_{s1} according to

$$f_{s1} = M f_s. \tag{5.167}$$

This corresponds to a sampling period T_1 according to

$$T_1 = T/M. \tag{5.168}$$

We can now avoid aliasing distortion by selecting the stopband edge of the antialiasing filter $H_a(s)$ appropriately. It suffices to ensure that aliasing into the baseband is avoided since the information is then unaffected (see Fig. 5.40(c)). Frequency components that are aliased into the remaining frequency region are allowed since they can be eliminated by the digital filter $H(z)$ that succeeds the A/D converter (see Fig. 5.40(d)). We see that the stopband edge of the analog filter must satisfy

$$2\pi - \Omega_{as} T_1 > \pi/M \Longleftrightarrow \Omega_{as} < 2\pi M f_s - \pi f_s \tag{5.169}$$

in order to avoid aliasing distortion. In practice, the distortion is suppressed to a level determined by the attenuation of the analog antialiasing filter. To make sure that one can reduce the distortion to any desired level, (5.169) must be satisfied, as the distortion is then determined by the filter's stopband attenuation.

In addition, one has to take into account that the digital filter has a transition band as well, which means that the overall design is somewhat more complicated than indicated above. Specifically, to avoid aliasing into the baseband, the information must therefore be limited to an angle frequency below π/T. If the information is limited to, say Ω_c, then the transition band of the digital filter ranges from $\Omega_c T_1$ to π/M.

We see that $x_1(m)$ corresponds to the information oversampled by a factor of M. The A/D converter thus generates M times more samples per time unit than actually needed for reconstruction of the information. To make sure that the succeeding digital system does not have to work at a higher data rate than necessary, one therefore reduces the sampling rate through decimation (downsampling) by a factor of M. Since we have M times oversampling after $H(z)$, we can do this without losing information. When M is an integer, the decimation is done by simply extracting only every Mth sample in the output

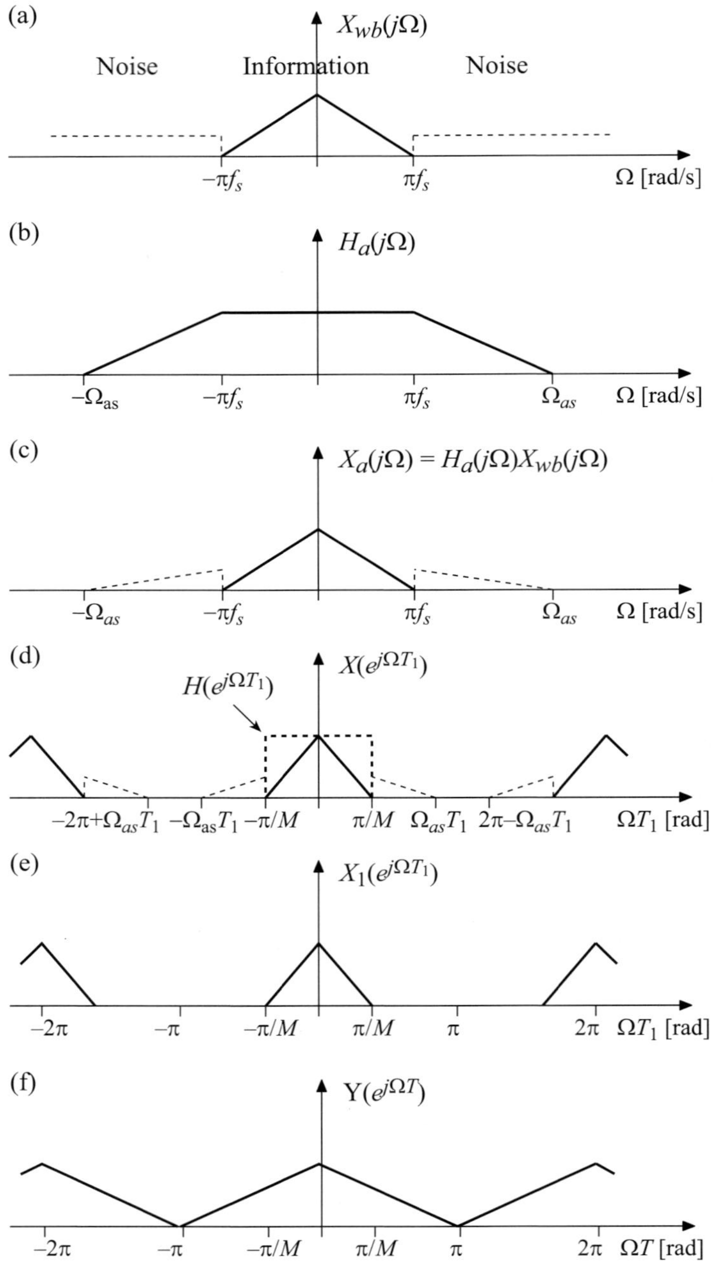

FIGURE 5.40

Conceptual spectra in an oversampled A/D converter.

sequence of $H(z)$ [6,11,52]. This is done by a downsampler, represented by $\downarrow M$. (For a noninteger M, more elaborate schemes must be employed.) The result is the sequence $y(n)$ corresponding to the information sampled with the frequency $f_s = 1/T$ (see Fig. 5.40(f)).

One motivation for using oversampling is, as mentioned earlier, that one in this way can relax the requirements on the analog antialiasing filter. Another reason is that oversampling enables the use of an A/D converter with lower resolution than the desired one, i.e., with fewer bits than the desired number of bits. The reason is that the digital filter removes a large part of the noise power while leaving the information virtually unaffected. This means that the SNR of $y(n)$ is higher than that of $x(m)$. To elaborate on this point, say that $x(m)$ is quantized to B_1 bits (and again assuming a maximum signal value of $X_m = 1$ for simplicity). We then know that the noise variance is

$$\sigma_e^2 = \frac{2^{-2(B_1-1)}}{12}. \tag{5.170}$$

Furthermore, we know that the noise variance at the output of the filter $H(z)$, say σ_h^2, is given by

$$\sigma_h^2 = \sigma_e^2 \frac{1}{2\pi} \int_{-\pi}^{\pi} |H(e^{j\Omega T})|^2 d(\Omega T). \tag{5.171}$$

If we now assume for the sake of simplicity that $H(z)$ is an ideal unity-gain low-pass filter with passband up to π/M, we get

$$\sigma_h^2 = \sigma_e^2 \frac{1}{2\pi} \int_{-\pi}^{\pi} |H(e^{j\Omega T})|^2 d(\Omega T) = \frac{1}{2\pi} \int_{-\pi/M}^{\pi/M} |H(e^{j\Omega T})|^2 d(\Omega T) = \frac{\sigma_e^2}{M}. \tag{5.172}$$

Since $H(z)$ leaves the information unaffected, the SNR is M times higher at the output of the filter than at the input of the filter, i.e., the output of the A/D converter. We can therefore obtain a desired resolution of B bits by using oversampling and an A/D converter with a lower resolution of B_1 bits. From the equations above, we obtain

$$\frac{2^{-2(B_1-1)}}{12M} = \frac{2^{-2(B-1)}}{12}, \tag{5.173}$$

which gives us

$$B_1 = B - 0.5\log_2(M). \tag{5.174}$$

We see that for each doubling of the sampling frequency, we gain half a bit in resolution. By using oversampling techniques, one can thus use A/D converters with lower resolution than the desired one. In the extreme case, the so-called 1-bit converters are employed. For such an A/D converter, one must however use a very high oversampling factor if a high resolution is desired. This causes problems in wide-band applications since the sampling frequency then becomes too high to handle. By utilizing the so-called $\Sigma\Delta$-modulators, the oversampling factor required to achieve a certain resolution can be reduced [37]. The basic principle is to shape the noise so that most of its energy is located in the high-frequency region (for a low-pass converter). In this way, one gains more than a factor of M in noise reduction when using a low-pass filter with a passband width of π/M. It is beyond the scope of this chapter to discuss $\Sigma\Delta$-converters in detail. However, Section 5.8 deals with modeling of mixed-signal

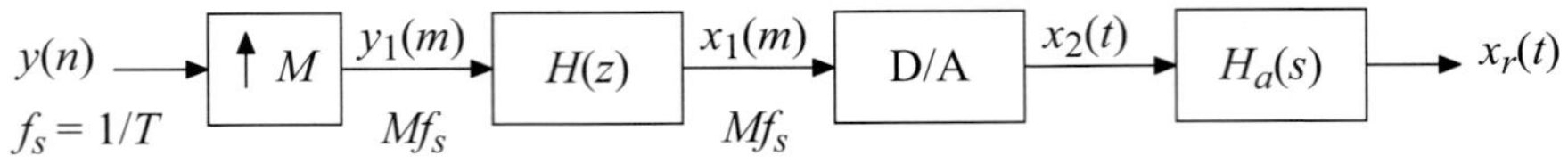

FIGURE 5.41

Oversampled D/A converter.

systems, which is useful in the analysis of such converters. That section will also provide an example of second-order $\Sigma\Delta$-modulators.

5.7.2 Oversampled D/A converters

As discussed in the previous section, the requirements on the analog antialiasing filters can be relaxed by using oversampled A/D converters. The requirements on the analog reconstruction filter can likewise be relaxed by using oversampled D/A converters according to Fig. 5.41.

We start with the sequence $y(n)$ generated by the oversampled A/D converter in Fig. 5.39. To illustrate the principle we use again the spectra in Fig. 5.40. The spectrum of $y(n)$ is then as shown in Fig. 5.42(a). The first step in an oversampled D/A converter is digital interpolation by a factor of M. This is done through an upsampler, represented by $\uparrow M$, followed by a filter $H(z)$ [6,11,52]. The result is the sequence $x_1(m)$, whose spectrum is shown in Fig. 5.42(c). We see that this sequence has the same spectrum as $x_1(m)$ in the oversampled A/D converter (see Figs. 5.39 and 5.40). In other words, it corresponds to the information oversampled by a factor of M. In the next step, reconstruction is performed through a D/A converter followed by a reconstruction filter. The D/A converter works at the higher sampling frequency Mf_s. As in Section 5.3.5.2, the D/A converter is here described by the pulse $q(t)$ according to Fig. 5.19(d). Ideally, $Q(j\Omega)H_a(j\Omega)$ and $H_a(j\Omega)$ are then as shown in Fig. 5.42(d and e), respectively. We see that the stopband edge of the analog reconstruction filter must satisfy

$$\Omega_{as} < 2\pi M f_s - \pi f_s, \tag{5.175}$$

which is the same requirement as that of the antialiasing filter in an oversampled A/D converter. By increasing M, we thus relax the filter requirements since a wider transition band is then allowed.

In practice, one can let the digital filter equalize the deviation of $Q(j\Omega)$, which is needed since $Q(j\Omega)$ is not constant in the baseband. That is, $H(z)$ is designed so that

$$|H(e^{j\Omega T_1})Q(j\Omega)| \approx \text{constant}, \quad \text{for } |\Omega| < \pi f_s. \tag{5.176}$$

As analog reconstruction filter, one can then use a conventional low-pass filter with (approximately) constant gain in both the passband and stopband. In that way, the requirements on this filter are further relaxed. Instead, the requirements on the digital filter have become more severe, but it is much easier to implement digital filters with stringent requirements than analog filters.

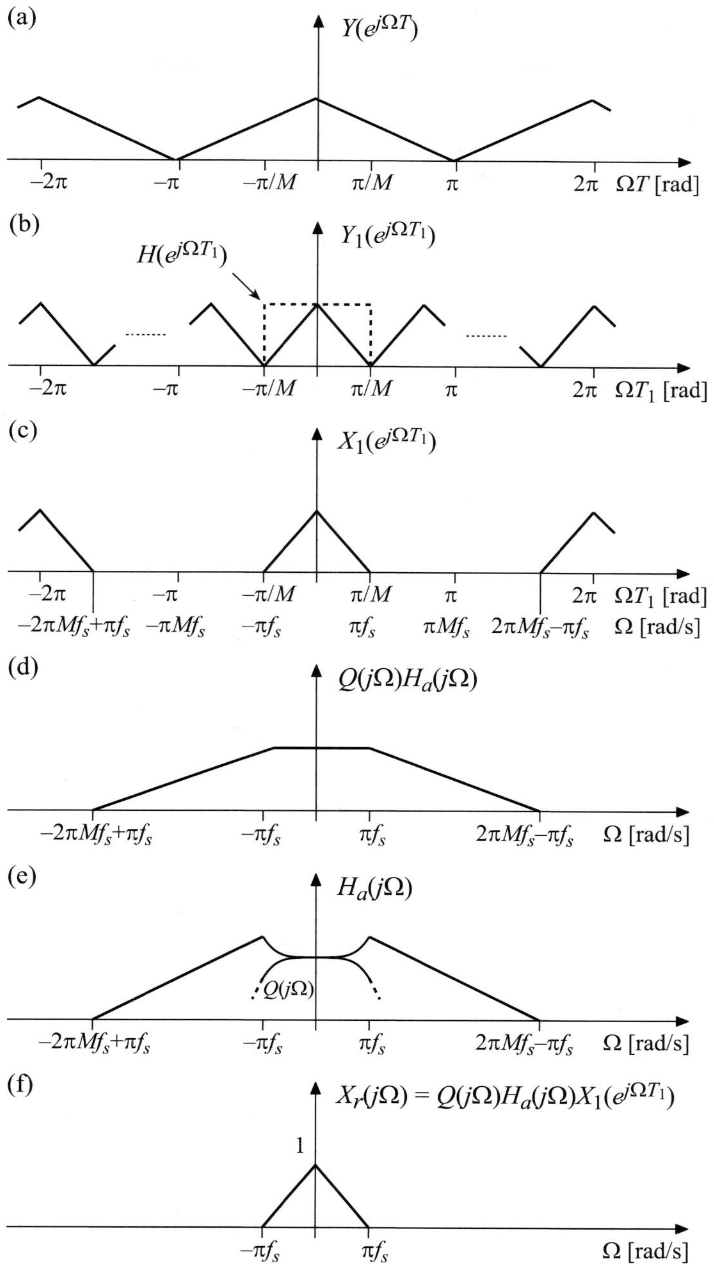

FIGURE 5.42

Conceptual spectra in an oversampled D/A converter.

5.7.3 Multiple-input multiple-output systems

In communication systems, multiple antennas can be used at the transmitters and/or receivers in order to increase the data transfer capacity. In the general case, a multiple-antenna system corresponds to a MIMO system. In particular, massive MIMO systems have been researched extensively during the past two decades and have emerged as key enablers for 5G systems and beyond [24,30,50]. Massive MIMO systems enable transmission of data to/from multiple users using the same time-frequency resources. This is achieved by designing the system so that signals add up constructively at an intended user whereas they add up destructively (cancel out) for all the other users. Similar techniques (constructive and destructive interference) are used in beamforming networks, which correspond to a multiple-input single-output (MISO) system at the receiver and a single-input multiple-output (SIMO) system at the transmitter [55].

Focusing now on receivers in MIMO and MISO systems, there is a need to convert the received signals to digital form before their further processing and final combination into the output signal(s). This can be carried out using an RF chain for each antenna. However, this solution may become very costly when the number of antennas is large. For such cases, it is therefore crucial to use as simple RF chains as possible. As ADCs constitute a large part of the total power consumption, the use of few-bit and 1-bit conversions has been studied in these contexts. As MIMO and MISO systems rely on constructive and destructive interference, there is also an averaging effect on the quantization noise, which may enable ADCs with few bits. It has been argued that the use of 1-bit ADCs is feasible, but under assumptions that may not hold in practice [25,35]. One issue is that when the number of antennas increases, the quantization errors may become correlated instead of uncorrelated, which eventually implies that there is no improved averaging effect. This needs to be taken into account to properly analyze the effect of using ADCs with few bits. In practice, there are also stronger signals (blockers) that interfere with weaker signals, which makes it harder to identify the weaker signal with few-bit ADCs [50].

5.7.3.1 Example 11: correlated channel quantization errors

To illustrate the effect of correlation, consider the case where a complex baseband signal $x(n)$ is transmitted through M channels and experience a random phase rotation in each channel (which approximately corresponds to random delays in each channel for narrow-band signals and line-of-sight transmission). Then, on the receiver side (after estimating the phase rotations), an estimation of $x(n)$, say $y(n)$, is formed as

$$\begin{aligned} y(n) &= \frac{1}{M}\sum_{m=0}^{M-1}\left(x(n)e^{j\alpha_m}+e_m(n)\right)e^{-j\alpha_m} \\ &= x(n)+\frac{1}{M}\sum_{m=0}^{M-1}e_m(n)e^{-j\alpha_m}, \end{aligned} \tag{5.177}$$

where α_m denote the phase rotations and $e_m(n)$ denote the quantization errors, given by the difference between the quantized value, $[x(n)e^{j\alpha_m}]_Q$, and the unquantized value, $x(n)e^{j\alpha_m}$, i.e., $e_m(n) = [x(n)e^{j\alpha_m}]_Q - x(n)e^{j\alpha_m}$. If $x(n)$ and $e_m(n)$ are uncorrelated, the expected average power of $y(n)$ becomes

$$P_y = P_x + P_e, \tag{5.178}$$

where

$$P_x = E\left\{|x(n)|^2\right\}. \tag{5.179}$$

Assuming that all $e_m(n)$ have the same quantization power, say P_N, and that $e_m(n)$ and $e_p(n)$ are uncorrelated for $m \neq p$, the expected average power, say $P_e(M)$, of the total quantization error, $e(n) = \frac{1}{M}\sum_{m=0}^{M-1} e_m(n)e^{-j\alpha_m}$, becomes

$$\begin{aligned} P_e(M) &= E\left\{|e(n)|^2\right\} \\ &= E\left\{e(n)e^*(n)\right\} \\ &= E\left\{\left(\frac{1}{M}\sum_{m=0}^{M-1} e_m(n)e^{-j\alpha_m}\right)\left(\frac{1}{M}\sum_{p=0}^{M-1} e_p^*(n)e^{j\alpha_p}\right)\right\} \\ &= \frac{1}{M^2}\sum_{m=0}^{M-1} E\left\{|e_m(n)|^2\right\} \\ &= \frac{P_N}{M}. \end{aligned} \tag{5.180}$$

As seen, with uncorrelated noise sources, the total noise power reduces with M in the same manner as when we use oversampling by M. However, this will only hold true as long as the channel quantization errors are uncorrelated. Especially with few bits in the ADC, the channel quantization errors tend to become correlated, even for relatively small values of M. This is because when M increases, more phase rotation values become closer to each other. Then, using few bits when α_m and α_p have approximately the same value implies that the quantization errors when quantizing $x(n)e^{j\alpha_m}$ and $x(n)e^{j\alpha_p}$ will have a large common (correlated) part and smaller uncorrelated parts. As a consequence, little or no further noise reduction will be achieved when M increases beyond a certain value.

As an illustration, for 1–10-bit quantizations, Fig. 5.43 plots the $\text{SNR} = 10\log_{10}(P_x/P_e)$ as a function of M, using a nonperiodic complex sinusoidal sequence $x(n) = \exp(j0.5\sqrt{3}\pi n)$ (thus with $P_x = 1$) and random phase rotation values, generated from a random variable uniformly distributed between 0 and 2π. As a reference, the figure also plots the SNR in the ideal case (dashed curves), where the channel quantization errors are uncorrelated, in which case the total error power is as given by Eq. (5.180). Also recall from Section 5.7.1 that doubling M results in a 3-dB SNR increase in the ideal case with uncorrelated errors.

As seen in Fig. 5.43, beyond a certain value of M, depending on the number of bits, practically no further SNR increase will be achieved when M increases. For the extreme case of 1 bit, the SNR saturates at some 18 dB and it begins to deviate from the ideal uncorrelated case already for $M = 2$. For 9 and 10 bits, the solid and corresponding dashed curves are practically the same, that is, there is no (or little) correlation in these cases for $M = 256 = 2^8$ and below. This illustrates that the errors are (approximately) uncorrelated as long as M is (approximately) smaller than the quantization step $2^{(B-1)}$, B denoting the number of bits.

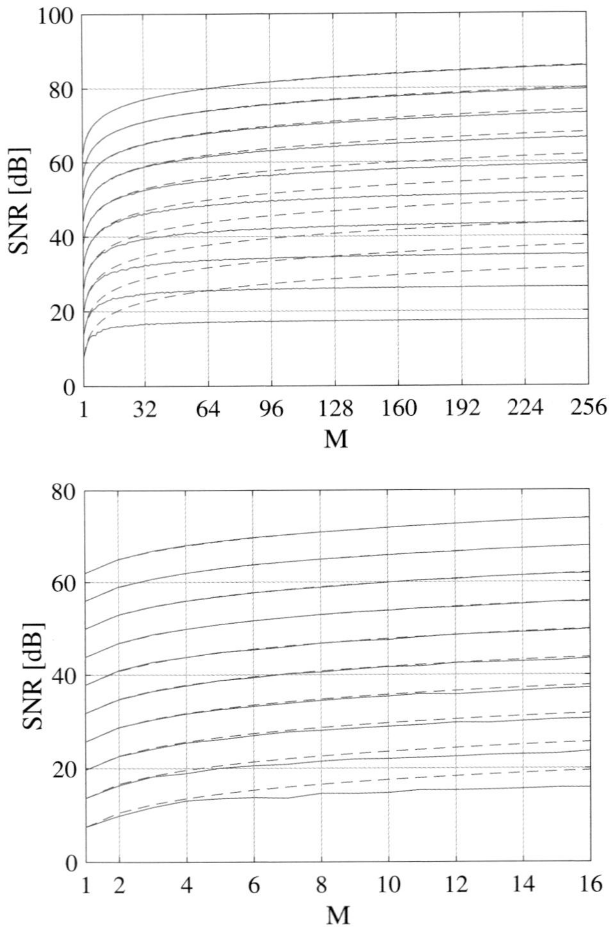

FIGURE 5.43

SNR as a function of the number of channels M using 1–10 bits (increasing SNR with the number of bits). The lower plot is a zoom-in of the upper plot. The dashed curves correspond to the ideal uncorrelated case. The difference between a solid curve and the corresponding dashed curve is thus the SNR loss due to correlation between the channel quantization errors.

5.8 Discrete-time modeling of mixed-signal systems

Many systems today are mixed-signal systems conducting both continuous-time and discrete-time signal processing. From the analysis point of view, it is often advantageous to use a discrete-time model of the mixed-signal system. This section presents an analysis method for linear time-invariant (LTI) systems containing sampling and reconstruction inside a feedback loop [28]. Using this method, the

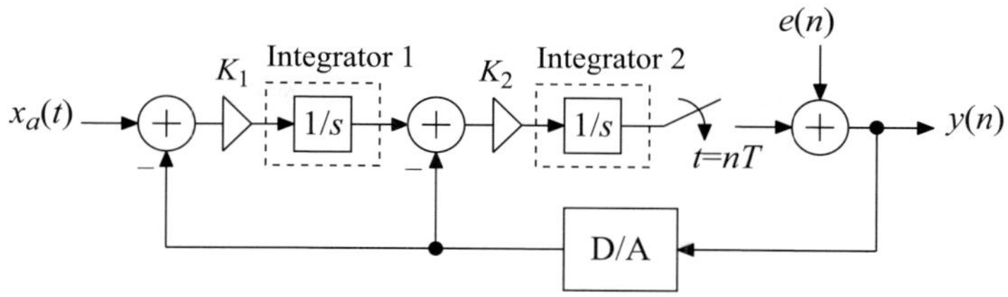

FIGURE 5.44

Linear model of a second-order double feedback loop continuous-time-input $\Sigma\Delta$-modulator.

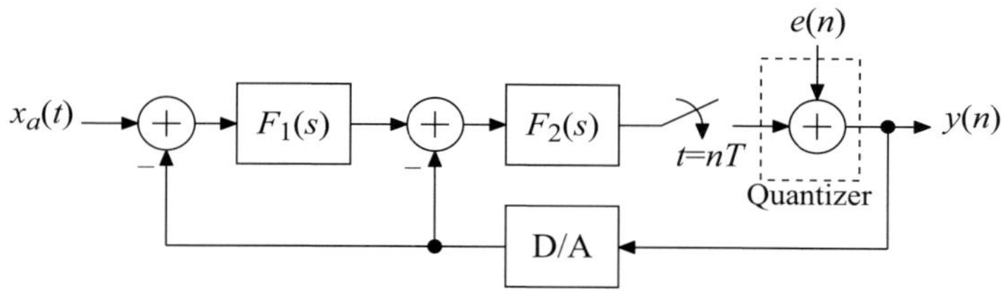

FIGURE 5.45

Generalization of the scheme in Fig. 5.44.

input–output stability as well as the frequency responses of such mixed systems can easily be analyzed. One application is ADCs realized with $\Sigma\Delta$-modulators with continuous-time input signals. An example is seen in Fig. 5.44, which is a linear model of a double feedback loop continuous-time-input $\Sigma\Delta$-modulator. Fig. 5.45 shows a generalization where $F_0(s)$ and $F_1(s)$ are transfer functions of general loop filters. However, the analysis method that will be presented here is applicable to any LTI system that can be represented by the general structure shown in Fig. 5.46, where N is a two-input/one-output system and where the single output is uniformly sampled and reconstructed through pulse amplitude modulation before being fed back again to N. Clearly, the structures of Figs. 5.44 and 5.45 belong to the class of structures that can be represented by the general structure of Fig. 5.46.

It is of practical interest to analyze systems such as that of Fig. 5.44 with respect to stability as well as frequency responses as seen from $x_a(t)$ and $e(n)$ to the output $y(n)$. Here, $e(n)$ models quantization errors that occur in the quantizer that follows the sampler. For continuous-time-input $\Sigma\Delta$-modulators, stability and frequency response analyses are often performed by using the so-called "sampled-data equivalent" discrete-time-input $\Sigma\Delta$-modulator seen in Fig. 5.47 [2,37,43], where integrators are replaced with accumulators. However, these "sampled-data equivalents" are only approximations of the actual behavior of the continuous-time-input $\Sigma\Delta$-modulator, as will be exemplified later in this section. That is, they do not correctly describe the properties of the continuous-time-input circuit.

To get a correct description, the system at hand can instead be represented by the system in Fig. 5.48, where the continuous-time input signal $x_a(t)$ is prefiltered through a continuous-time system G_p before being sampled uniformly. The resulting sampled signal and the system output signal are then used as input signals to a linear discrete-time system G_s, usually represented in state-space form [12,37]. It

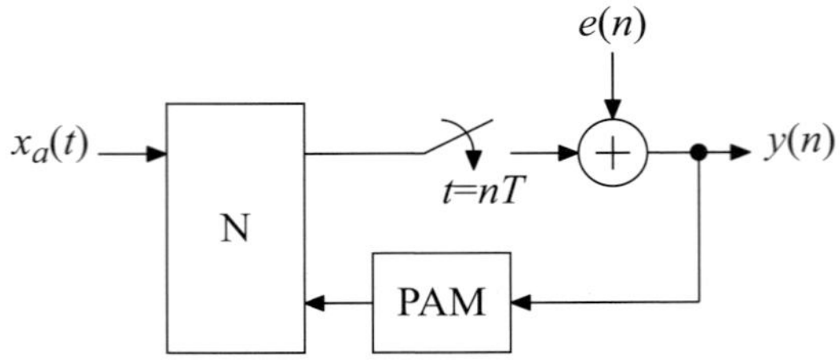

FIGURE 5.46

General hybrid continuous-time/discrete-time system setup with a feedback loop.

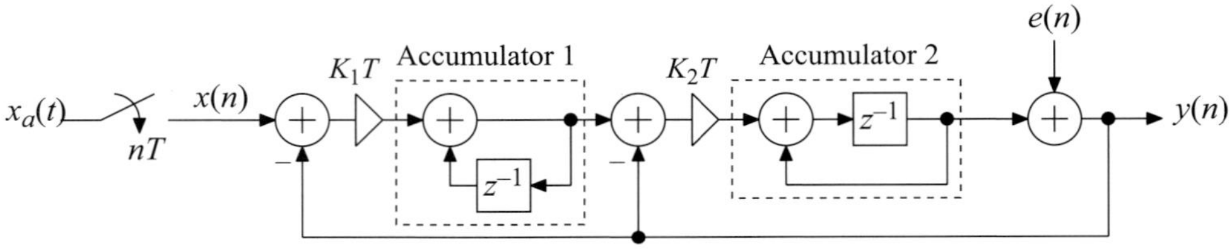

FIGURE 5.47

Linear model of a second-order discrete-time-input ΣΔ-modulator.

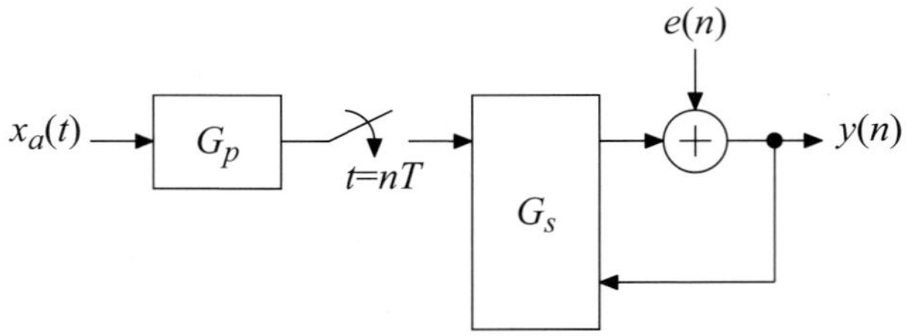

FIGURE 5.48

State-space representation of the system in Fig. 5.46.

is then required to determine a corresponding input–output difference equation using state variable representation. To that end, a state matrix, A, of G_s is needed, requiring the computation of the power series

$$A = e^{A_c} = \sum_{k=0}^{\infty} \frac{1}{k!} A_c^k, \tag{5.181}$$

where A_c is the state matrix of the system N in Fig. 5.46, as well as the inversion of two particular matrices. This method gives a correct description of the circuit at hand but requires the computation

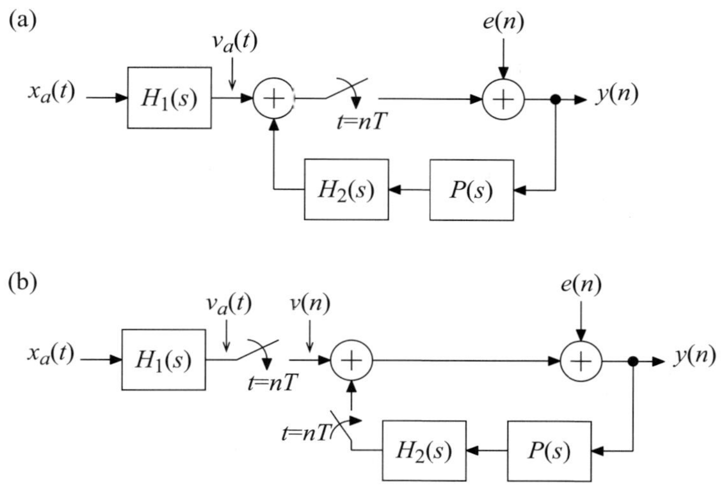

FIGURE 5.49

Two equivalent models for the system in Fig. 5.46.

of (5.181), which can be a nontrivial task, and the abovementioned matrices often are singular. In fact, this is always the case when $F_0(s)$ and $F_1(s)$ represent ideal integrators $1/s$.

In this section, we discuss an alternative method for the analysis of the external stability as well as the frequency responses from the inputs $x_a(t)$ and $e(n)$ to the output $y(n)$ of the general system in Fig. 5.48 [28]. This is done by using the models shown in Fig. 5.49. It is readily shown that all LTI systems belonging to the class of systems that can be represented by the system in Fig. 5.46 can be represented as in Fig. 5.49. It should however be noted that the systems in Fig. 5.49 are only used for input/output analysis purposes and do not capture the internal properties of the underlying system of interest. From the input/output stability and frequency response points of view, the systems in Fig. 5.49 are equivalent to their underlying system in Fig. 5.46.

5.8.1 Input/output relations

For mixed-signal systems, direct use of transfer functions is less appropriate as they depend on two variables (s and z). Therefore, the analysis becomes somewhat more involved than for purely discrete-time or continuous-time systems. The mixed-signal systems are analyzed here by using equivalent discrete-time systems that produce the same output samples as the actual mixed-signal system.

To be precise, stability is analyzed by first determining $G(z)$ in the z-transform relation between $y(n)$, $v(n)$, and $e(n)$ given as

$$Y(z) = G(z)V(z) + G(z)E(z). \qquad (5.182)$$

Stability is then investigated by considering the poles of $G(z)$ and the boundedness of the resulting output signal. As to the frequency responses, as seen from $x_a(t)$ and $e(n)$ to the output $y(n)$, it is appropriate to express the output Fourier transform as

$$Y(e^{j\Omega T}) = H(j\Omega)\frac{1}{T}X_a(j\Omega) + G(e^{j\Omega T})E(e^{j\Omega T}), \tag{5.183}$$

where $H(j\Omega)$ represents the signal frequency response, whereas $G(e^{j\Omega T})$ represents the noise frequency response. The latter is immediately obtained from the noise transfer function $G(z)$ evaluated for $z = e^{j\Omega T}$.

In order to arrive at (5.182) and (5.183), the models of Fig. 5.49 are used to represent the system in Fig. 5.46. In Fig. 5.49, $P(s)$ represents pulse amplitude modulation.

5.8.2 Determining $G(z)$

We will make use of the generalized Poisson summation formula that relates the Laplace transform of a signal $x_a(t)$ with the corresponding z-transform of its uniformly sampled version $x(n) = x_a(nT)$ according to

$$X(e^{sT}) = \frac{1}{T}\sum_{p=-\infty}^{\infty} X_a\left(s - j\frac{2\pi p}{T}\right) \tag{5.184}$$

with the z-transform evaluated for $z = e^{sT}$. Furthermore, we will make use of the following equation, which relates the output $x_r(t)$ of the PAM $p(t)$ with its input $x(n)$:

$$X_r(s) = P(s)X(e^{sT}). \tag{5.185}$$

Using the above relations, the z-transform of the output signal $y(n)$ in Fig. 5.49(a) is easily derived as

$$\begin{aligned} Y(e^{sT}) &= E(e^{sT}) \\ &+ \frac{1}{T}\sum_{p=-\infty}^{\infty} H_1\left(s - j\frac{2\pi p}{T}\right) X_a\left(s - j\frac{2\pi p}{T}\right) \\ &- Y(e^{sT})\frac{1}{T}\sum_{p=-\infty}^{\infty} H_2\left(s - j\frac{2\pi p}{T}\right) P\left(s - j\frac{2\pi p}{T}\right), \end{aligned} \tag{5.186}$$

where $E(z)$ is the z-transform of $e(n)$ and $P(s)$ is defined by the PAM used. Here, a rectangular pulse $p(t)$ is assumed according to

$$p(t) = \begin{cases} 1, & 0 \le t \le T, \\ 0, & \text{otherwise}, \end{cases} \tag{5.187}$$

which is commonly used in D/A conversion. This gives

$$P(s) = \frac{1 - e^{-sT}}{s}. \tag{5.188}$$

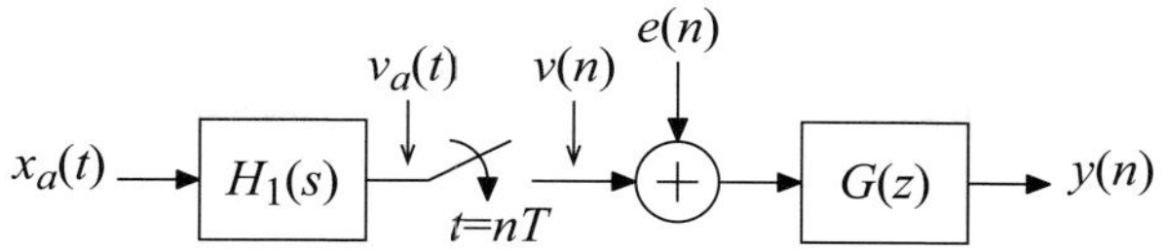

FIGURE 5.50

Representation of the system in Fig. 5.46 in view of (5.182).

Eq. (5.186) can be rearranged according to (5.182) with

$$V(e^{sT}) = \frac{1}{T}\sum_{p=-\infty}^{\infty} H_1\left(s - j\frac{2\pi p}{T}\right) X_a\left(s - j\frac{2\pi p}{T}\right), \tag{5.189}$$

which corresponds to the z-transform of $v(n)$ in Fig. 5.49(b), i.e., a uniformly sampled version of $v_a(t)$, and

$$G(z) = \frac{1}{1 + L(z)} = \frac{N(z)}{D(z)} \tag{5.190}$$

with

$$L(e^{sT}) = \frac{1}{T}\sum_{p=-\infty}^{\infty} H_2\left(s - j\frac{2\pi p}{T}\right) P\left(s - j\frac{2\pi p}{T}\right). \tag{5.191}$$

Eq. (5.182) can be further interpreted as shown in Fig. 5.50.

In Fig. 5.49, $H_1(s)$ models the output signal dependency of the input signal, whereas $H_2(s)$ describes the behavior of the feedback signal. Eqs. (5.182) and (5.189)–(5.191) do not give us suitable expressions for the stability analysis. In order to express $L(z)$ in a more suitable form for this purpose, we make use of the observation that $L(z)$ is the z-transform of the sequence $l(n) = l_a(nT)$, i.e., a uniformly sampled version of the signal $l_a(t)$. Furthermore, from (5.191), it is seen that $l_a(t)$ is the inverse Laplace transform of $H_2(s)P(s)$. Hence, a more convenient representation of $L(z)$ and thereby $G(z)$ can be found using the following procedure.

Starting with $H_2(s)P(s)$, its inverse Laplace transform gives $l_a(t)$, from which $l(n) = l_a(nT)$ follows directly. A closed-form expression for $L(z)$ is then obtained through the computation of the z-transform of $l(n)$. The approach is applicable for general $H_2(s)$ and $P(s)$, and is straightforward when $H_2(s)$ and $P(s)$ are rational functions in s.

5.8.3 Stability

In order to investigate the stability, it is convenient to first consider the system in Fig. 5.51(a), where the numerator and denominator polynomials of $G(z)$ in Fig. 5.50, i.e., $N(z)$ and $D(z)$, respectively, have been shown explicitly. The system of Fig. 5.46 will be bounded-input bounded-output (BIBO) stable if the following two conditions are satisfied:

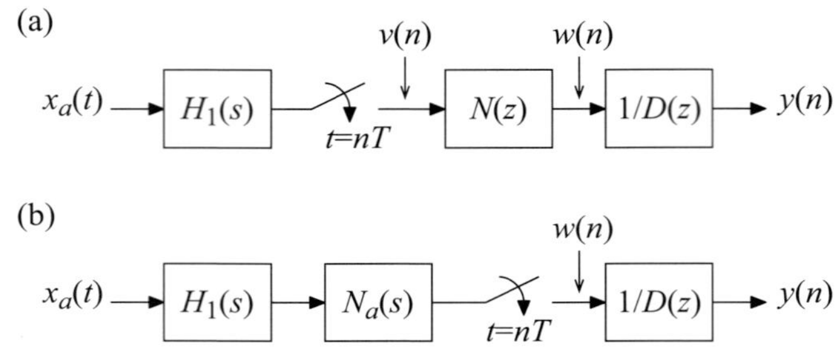

FIGURE 5.51

Representations of the system in Fig. 5.50. They are useful in the stability analysis.

1. The poles of $G(z)$, i.e., the roots of $D(z)$, lie inside the unit circle.
2. The signal $w(n)$ in Fig. 5.51 is bounded.

The first condition above is easily checked once $G(z)$ is available. The second condition requires an analysis of $H_1(s)$ and $N(z)$. We consider the following two different cases, which are of practical interest.

Case 1: $H_1(s)$ is stable. This gives directly that the second condition above is satisfied, and BIBO stability is then ensured by also fulfilling the first condition.

Case 2: $H_1(s)$ can be factored according to

$$H_1(s) = A(s)\frac{1}{s^{Q_1}}\prod_{q=1}^{Q_2}\frac{1}{s+p_q}, \tag{5.192}$$

where $A(s)$ is stable and p_q are imaginary roots. Then, $H_1(s)$ contains Q_1 poles at $s=0$ and Q_2 poles at $s=-p_q$. In order to investigate stability in this case, it is convenient to consider the system shown in Fig. 5.51(b), where

$$N_a(s) = N(e^{sT}). \tag{5.193}$$

Since $N(z)$ is of the form

$$N(z) = \sum_{k=0}^{K} a_k z^{-k}, \tag{5.194}$$

the signal $w(n)$ of Fig. 5.51(a) is identical to that in Fig. 5.51(b). This follows from Poisson's summation formula together with the fact that $N_a(s)$ above is periodic. It then remains to investigate if $H_1(s)N_a(s)$ is stable, i.e., if it exists for all $\Re\{s\} \geq 0$, assuming a causal system, and if $G(z)$ (or, equivalently, $1/D(z)$) is stable.

5.8.4 Frequency responses $H(j\Omega)$ and $G(e^{j\Omega T})$

The frequency response from $e(n)$ to the output $y(n)$, i.e., $G(e^{j\Omega T})$, is directly obtained from $G(z)$ in (5.182) if it exists for $z = e^{j\Omega T}$. It is then given by

$$G(e^{j\Omega T}) = \frac{N(e^{j\Omega T})}{D(e^{j\Omega T})} = \frac{1}{1 + \frac{1}{T}\sum_{p-\infty}^{\infty} H_1\left(j\Omega - j\frac{2\pi p}{T}\right) P\left(j\Omega - j\frac{2\pi p}{T}\right)}. \tag{5.195}$$

As to the signal frequency response, $H(j\Omega)$, it is first noted from (5.182) and (5.183) that $H(j\Omega)\frac{1}{T}X_a(j\Omega) = G(e^{j\Omega T})V(e^{j\Omega T})$, where

$$V(e^{j\Omega T}) = \frac{1}{T}\sum_{p-\infty}^{\infty} H_1\left(j\Omega - j\frac{2\pi p}{T}\right) X_a\left(j\Omega - j\frac{2\pi p}{T}\right). \tag{5.196}$$

Assume now that the input signal $x_a(t)$ is band-limited according to the Nyquist criterion for sampling with the sampling period T, i.e.,

$$X_a(j\Omega) = 0, \quad |\Omega| \geq \pi/T. \tag{5.197}$$

In this case only the term for $p = 0$ in (5.196) is nonzero in the interval $-\pi \leq \Omega T \leq \pi$. Hence, $Y(e^{j\Omega T})$ can be written as in (5.183) with

$$H(j\Omega) = H_1(j\Omega)G(e^{j\Omega T}). \tag{5.198}$$

5.8.5 Example 12: continuous-time-input $\Sigma\Delta$-modulator

Consider the second-order continuous-time-input $\Sigma\Delta$-modulator shown in Fig. 5.44. This is a special case of the system of Fig. 5.45 with $F_1(s)$ and $F_2(s)$ being ideal integrators, i.e.,

$$F_1(s) = \frac{K_1}{s}, \quad F_2(s) = \frac{K_2}{s}. \tag{5.199}$$

In this case, one obtains that $H_1(s)$ and $H_2(s)$ are given by

$$H_1(s) = F_1(s)F_2(s), \quad H_2(s) = [1 + F_1(s)]F_2(s). \tag{5.200}$$

From this equation, we obtain

$$L_a(s) = H_2(s)P(s) = \left(\frac{K_1}{s^2} + \frac{K_1K_2}{s^3}\right)(1 - e^{-sT}). \tag{5.201}$$

Taking the inverse Laplace transform, we get

$$l_a(t) = K_1[tu(t) - (t-T)u(t-T)] + \frac{K_1K_2}{2}[t^2u(t) - (t-T)^2u(t-T)]. \tag{5.202}$$

Sampling $l_a(t)$ at $t = nT$ then generates

$$l(n) = l_a(nT) = K_1T[n \times u(n) - (n-1) \times u(n-1)] + \frac{K_0K_1T^2}{2}[n^2 \times u(n) - (n-1)^2 \times u(n-1)], \tag{5.203}$$

where $u(n)$ is the unit-step sequence. Then, taking the z-transform of $l(n)$ gives

$$L(z) = \frac{K_1Tz^{-1}}{1-z^{-1}}\left[1 + \frac{K_0T}{2}\frac{1+z^{-1}}{1-z^{-1}}\right], \tag{5.204}$$

whereby

$$G(z) = \frac{N(z)}{D(z)} = \frac{1}{1+L(z)} = \frac{(1-z^{-1})^2}{1+az^{-1}+bz^{-2}}, \tag{5.205}$$

where

$$a = K_2T + 0.5K_1K_2T^2 - 2 \tag{5.206}$$

and

$$b = 1 - K_2T + 0.5K_1K_2T^2. \tag{5.207}$$

Furthermore, with $F_0(s)$ and $F_1(s)$ as in (5.199), we get

$$H_1(s)N_a(s) = \frac{K_0K_2}{s^2}(1-e^{-sT})^2, \tag{5.208}$$

which is a stable system because the corresponding impulse response is finite. Hence, stability is in this case solely determined by the stability of $G(z)$ given by (5.205). For example, with $K_0 = K_1 = K$, it is straightforward to show that stability is ensured by constraining K according to $0 < K < 2/T$. Using instead the often employed and so-called "sampled-data equivalents," where integrators $1/s$ are replaced by the accumulators $1/(1-z^{-1})$ and $z^{-1}/(1-z^{-1})$, one obtains $0 < K < 1.236/T$. Hence, the accumulator-based structures are not appropriate models.

Furthermore, from (5.198), (5.200), and (5.205), we obtain the frequency response

$$H(j\Omega) = \frac{K_0K_1}{(j\Omega)^2}\frac{(1-e^{-j\Omega T})^2}{1+ae^{-j\Omega T}+be^{-j2\Omega T}}. \tag{5.209}$$

This expression differs from that obtained via the accumulator-based structures. Also the noise transfer functions are different. Again, this means that such structures are not appropriate. Figs. 5.52 and 5.53 illustrate these differences by plotting the modulus of the signal and noise transfer functions in the two cases.

5.8.6 Example 13: single feedback loop system

Consider the scheme in Fig. 5.54 with $F(s)$ being a second-order transfer function according to

$$F(s) = \frac{K}{s(s+p)} \tag{5.210}$$

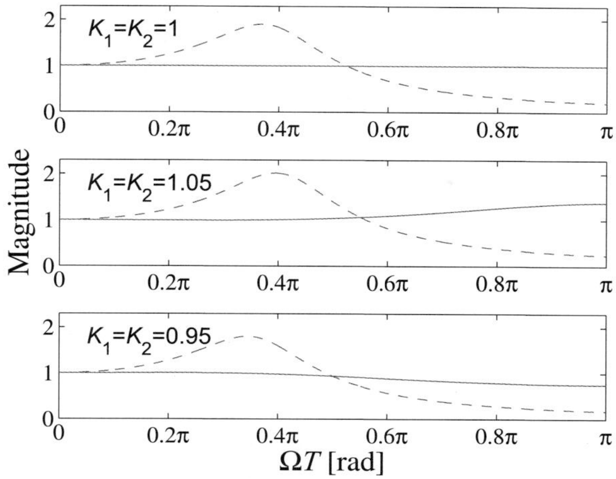

FIGURE 5.52

Modulus of the signal frequency responses for second-order continuous-time-input (dashed) and discrete-time-input ΣΔ-modulators (solid).

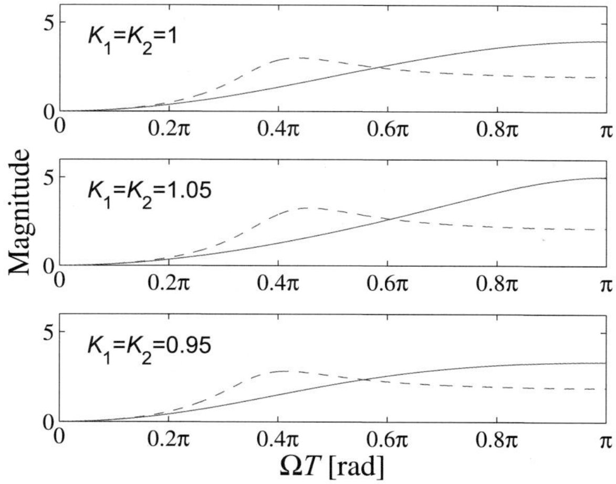

FIGURE 5.53

Modulus of the noise frequency responses for second-order continuous-time-input (dashed) and discrete-time-input ΣΔ-modulators (solid).

with a real-valued pole at $s = -p$, with $p > 0$ assumed. For the equivalent representation of Fig. 5.49, $H_1(s) = H_2(s) = F(s)$. Thus, $H_1(s)$ follows (5.192) as $H_1(s) = A(s)/s$ with $A(s) = K/(s + p)$.

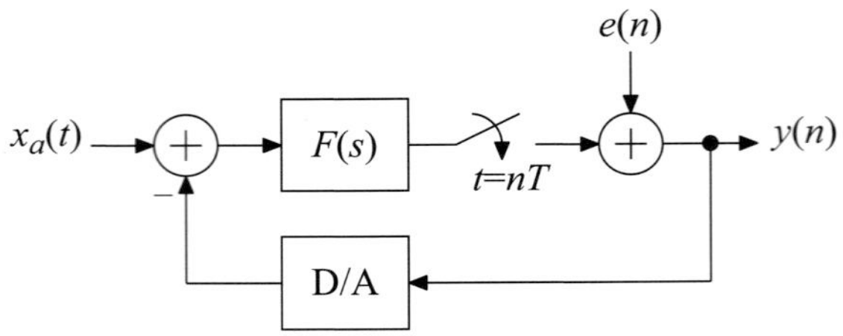

FIGURE 5.54

System with a single feedback loop.

We first investigate the stability. We note that Case 2 applies here since $H_1(s)$ has a pole at $s = 0$. To derive $G(z)$, we first note that $H_1(s)P(s)$, using partial fraction expansion, becomes

$$H_1(s)P(s) = \left(\frac{K/p}{s^2} - \frac{K/p^2}{s} + \frac{K/p^2}{s+p}\right)(1 - e^{-sT}). \tag{5.211}$$

Hence, $l(n) = l_a(nT)$ is given by

$$l(n) = l_{a0}(nT) - l_{a0}(nT - T) \tag{5.212}$$

with $l_{a0}(t)$ being

$$l_{a0}(t) = \left(\frac{K}{p}t - \frac{K}{p^2} + \frac{K}{p^2}e^{-pt}\right)u(t). \tag{5.213}$$

Hence, taking the z-transform of $l(n)$, we obtain

$$L(z) = \frac{KT}{p(z-1)} - \frac{K}{p^2} + \frac{K}{p^2}\frac{z-1}{z - e^{-pT}}. \tag{5.214}$$

Finally, this gives us

$$G(z) = \frac{1}{1 + L(z)} = \frac{(1 - z^{-1})(1 - z^{-1}e^{-pT})}{1 + az^{-1} + bz^{-2}}, \tag{5.215}$$

where

$$a = \frac{KT}{p} - (1 + e^{-pT}) - \frac{K}{p^2}(1 - e^{-pT}) \tag{5.216}$$

and

$$b = \frac{K}{p^2}(1 - e^{-pT}) + \left(1 - \frac{KT}{p}\right)e^{-pT}. \tag{5.217}$$

We note that

$$H_1(s)N_a(s) = \frac{A(s)(1 - e^{sT})(1 - e^{-(s+p)T})}{s}, \tag{5.218}$$

which corresponds to a stable system, because it consists of (readily shown) stable subsystems. Hence, the stability is given by the poles of $G(z)$, which are most easily found numerically.

Further, we get the frequency responses

$$G(e^{j\Omega T}) = \frac{(1 - e^{-j\Omega T})(1 - e^{-j\Omega T}e^{-pT})}{1 + ae^{-j\Omega T} + be^{-j2\Omega T}} \tag{5.219}$$

and

$$\begin{aligned} H(j\Omega) &= H_1(j\Omega)G(e^{j\Omega T}) \\ &= \frac{K}{j\Omega(j\Omega + p)} \frac{(1 - e^{-j\Omega T})(1 - e^{-j\Omega T}e^{-pT})}{1 + ae^{-j\Omega T} + be^{-j2\Omega T}}. \end{aligned} \tag{5.220}$$

References

[1] A. Antoniou, Digital Filters: Analysis, Design, and Applications, 2nd ed., McGraw-Hill, 1993.

[2] S.H. Ardalan, J.J. Paulos, An analysis of nonlinear behavior in delta-sigma modulators, IEEE Trans. Circuits Syst. CAS–34 (6) (June 1987) 593–603.

[3] W.C. Black, D.A. Hodges, Time interleaved converter arrays, IEEE J. Solid-State Circuits SC–15 (6) (Dec. 1980) 1022–1029.

[4] E.J. Candes, M.B. Wakin, An introduction to compressive sampling, IEEE Signal Process. Mag. 25 (2) (Mar. 2008) 21–30.

[5] H.-W. Chen, Modeling and identification of parallel nonlinear systems: structural classification and parameter estimation methods, IEEE Proc. 83 (1) (Jan. 1995) 39–66.

[6] R.E. Crochiere, L.R. Rabiner, Multirate Digital Signal Processing, Prentice Hall, NJ, 1983.

[7] V. Divi, G. Wornell, Scalable blind calibration of timing skew in high-resolution time-interleaved ADCs, in: Proc. IEEE Int. Symp. Circuits Syst., Kos, Greece, May 21–24, 2006.

[8] M. El-Chammas, B. Murmann, General analysis on the impact of phase-skew mismatch in time-interleaved ADCs, in: Proc. IEEE Int. Symp. Circuits Syst., Seattle, USA, May 18–21, 2008.

[9] J. Elbornsson, F. Gustafsson, J.E. Eklund, Blind equalization of time errors in a time-interleaved ADC system, IEEE Trans. Signal Process. 53 (4) (2005) 1413.

[10] Y.C. Eldar, A.V. Oppenheim, Filterbank reconstruction of bandlimited signals from nonuniform and generalized samples, IEEE Trans. Signal Process. 48 (10) (2000) 2864.

[11] N.J. Fliege, Multirate Digital Signal Processing, Wiley, New York, 1994.

[12] H. Freeman, Discrete-Time Systems, John Wiley, 1965.

[13] A. Haftbaradaran, K.W. Martin, A background sample-time error calibration technique using random data for wide-band high-resolution time-interleaved ADCs, IEEE Trans. Circuits Syst. II 55 (3) (Mar. 2008) 234–238.

[14] S. Huang, B.C. Levy, Adaptive blind calibration of timing offset and gain mismatch for two-channel time-interleaved ADCs, IEEE Trans. Circuits Syst. I 53 (6) (June 2006) 1278–1288.

[15] S. Huang, B.C. Levy, Blind calibration of timing offsets for four-channel time-interleaved ADCs, IEEE Trans. Circuits Syst. I 54 (4) (Apr. 2007) 863–876.

[16] L.B. Jackson, Digital Filters and Signal Processing, 3rd ed., Kluwer Academic Publishers, 1996.

[17] A.J. Jerri, The Shannon sampling theorem–its various extensions and applications: a tutorial review, IEEE Proc. 65 (11) (Nov. 1977) 1565–1596.

[18] H. Johansson, A polynomial-based time-varying filter structure for the compensation of frequency-response mismatch errors in time-interleaved ADCs: Special issue on DSP techniques for RF/analog circuit impairments, IEEE J. Sel. Top. Signal Process. 3 (3) (June 2009) 384–396.

[19] H. Johansson, P. Löwenborg, Reconstruction of nonuniformly sampled bandlimited signals by means of time-varying discrete-time FIR filters, J. Appl. Signal Process. 2006 (2006) 64185, Special Issue on Frames and Overcomplete Representations in Signal Processing, Communications, and Information Theory.

[20] H. Johansson, P. Löwenborg, A least-squares filter design technique for the compensation of frequency-response mismatch errors in time-interleaved A/D converters, IEEE Trans. Circuits Syst. II, Express Briefs 55 (11) (2008) 1154–1158.

[21] H. Johansson, P. Löwenborg, K. Vengattaramane, Reconstruction of M-periodic nonuniformly sampled signals using multivariate polynomial impulse response time-varying FIR filters, in: Proc. XII European Signal Processing Conf., Florence, Italy, Sept. 2006.

[22] R. Khoini-Poorfard, L.B. Lim, D.A. Johns, Time-interleaved oversampling A/D converters: theory and practice, IEEE Trans. Circuits Syst. II, Analog Digit. Signal Process. 44 (8) (1997) 634–645.

[23] V.A. Kotelnikov, On the transmission capacity of "ether" and wire in electrocommunications, Izd. Red. Upr. Svyazi RKKA (Moscow) (1933).

[24] E.G. Larsson, O. Edfors, F. Tufvesson, T.L. Marzetta, Massive MIMO for next generation wireless systems, IEEE Commun. Mag. 52 (2) (Feb. 2014) 186–195.

[25] Y. Li, C. Tao, G. Seco-Granados, A. Mezghani, A.L. Swindlehurst, Channel estimation and performance analysis of one-bit massive MIMO systems, IEEE Trans. Signal Process. 65 (15) (Aug. 1 2017) 4075–4089.

[26] Y.C. Lim, Y.X. Zou, J.W. Lee, S.C. Chan, Time-interleaved analog-to-digital converter compensation using multichannel filters, IEEE Trans. Circuits Syst. I, Regul. Pap. 56 (10) (Oct. 2009) 2234–2247.

[27] X. Liu, H. Xu, H. Johansson, Y. Wang, N. Li, Correlation-based calibration for nonlinearity mismatches in dual-channel TIADCs, IEEE Trans. Circuits Syst. II, Express Briefs 67 (3) (Mar. 2019) 585–589.

[28] P. Löwenborg, H. Johansson, Analysis of continuous-time-input $\Sigma\Delta$ A/D modulators and their generalizations, in: Proc. European Conf. Circuit Theory Design, Krakow, Poland, Sept. 1–4, 2003.

[29] F. Marvasti, Nonuniform Sampling: Theory and Practice, Kluwer, NY, 2001.

[30] T.L. Marzetta, E.G. Larsson, H. Yang, H.Q. Ngo, Fundamentals of Massive MIMO, Cambridge Univ. Press, Cambridge, U.K., 2016.

[31] S. Mendel, C. Vogel, A compensation method for magnitude response mismatches in two-channel time-interleaved analog-to-digital converters, in: Proc. IEEE Int. Conf. Electronics, Circuits, Syst., Nice, France, Dec. 2006.

[32] S. Mendel, C. Vogel, On the compensation of magnitude response mismatches in M-channel time-interleaved ADCs, in: Proc. IEEE Int. Symp. Circuits, Syst., New Orleans, USA, May 2007, pp. 3375–3378.

[33] M. Mishali, Y.C. Eldar, Sub-Nyqust sampling: bridging theory and practice, IEEE Signal Process. Mag. 28 (6) (2011) 98–124.

[34] S.K. Mitra, Digital Signal Processing: A Computer-Based Approach, McGraw-Hill, 2006.

[35] C. Mollen, J. Choi, E.G. Larsson, R.W. Heath, Uplink performance of wideband massive MIMO with one-bit ADCs, IEEE Trans. Wirel. Commun. 16 (1) (Jan. 2017) 87–100.

[36] B. Murmann, Trends in low-power, digitally assisted A/D conversion, IEICE Trans. Electron. E93–C (6) (June 2010) 718–729.

[37] S.R. Norsworthy, R. Schreier, G.C. Temes, Delta-Sigma Data Converters: Theory, Design, and Simulation, IEEE Press, 1997.

[38] H. Nyquist, Certain topics in telegraph transmission theory, Trans. Am. Inst. Electr. Eng. 47 (1928).

[39] A.V. Oppenheim, R.W. Schafer, Discrete-Time Signal Processing, Prentice Hall, 1989.

[40] A. Papoulis, The Fourier Integral and Its Applications, McGraw-Hill, 1962.

[41] A. Papoulis, Generalized sampling expansion, IEEE Trans. Circuits, Syst. CAS–24 (11) (Nov. 1977) 652–654.

[42] P. Satarzadeh, B.C. Levy, P.J. Hurst, Bandwidth mismatch correction for a two-channel time-interleaved A/D converter, in: Proc. IEEE Int. Symp. Circuits Syst., New Orleans, USA, May 2007.

[43] R. Schreier, G.C. Temes, Understanding Delta-Sigma Data Converters, John Wiley and Sons, NJ, 2005.

[44] M. Seo, M.J.W. Rodwell, U. Madhow, Comprehensive digital correction of mismatch errors for a 400-msamples/s 80-dB SFDR time-interleaved analog-to-digital converter, IEEE Trans. Microw. Theory Tech. 53 (3) (Mar. 2005) 1072–1082.
[45] M. Seo, M.J.W. Rodwell, U. Madhow, Generalized blind mismatch correction for a two-channel time-interleaved ADC: analytic approach, in: Proc. IEEE Int. Symp. Circuits Syst., New Orleans, USA, May 27–30, 2007.
[46] M. Seo, M.J.W. Rodwell, U. Madhow, Generalized blind mismatch correction for two-channel time-interleaved A-to-D converters, in: Proc. IEEE Int. Conf. Acoustics, Speech, Signal Processing, Hawaii, USA, Apr. 2007.
[47] C.E. Shannon, Communication in the presence of noise, Proc. IRE 37 (1949) 10–21.
[48] J.A. Tropp, J.N. Laska, M.F. Duarte, J.K. Romberg, R.G. Baraniuk, Beyond Nyquist: efficient sampling of sparse bandlimited signals, IEEE Trans. Inf. Theory 56 (1) (Jan. 2010) 520–544.
[49] T. Tsai, P.J. Hurst, S.H. Lewis, Bandwidth mismatch and its correction in time-interleaved analog-to-digital converters, IEEE Trans. Circuits Syst. II 53 (10) (Oct. 2006) 1133–1137.
[50] A.B. Ucuncu, E. Björnson, H. Johansson, A.Ö. Yilmaz, E.G. Larsson, Performance analysis of quantized uplink massive MIMO-OFDM with oversampling under adjacent channel interference, IEEE Trans. Commun. 68 (2) (Feb. 2020) 871–886.
[51] M. Unser, Sampling – 50 years after Shannon, IEEE Proc. 88 (4) (Apr. 2000) 569–587.
[52] P.P. Vaidyanathan, Multirate Systems and Filter Banks, Prentice Hall, 1993.
[53] P.P. Vaidyanathan, Generalizations of the sampling theorem: seven decades after Nyquist, IEEE Trans. Circuits, Syst. I 48 (9) (Sept. 2001) 1094–1109.
[54] R.J. van de Plassche, CMOS Integrated Analog-to-Digital and Digital-to-Analog Converters, Kluwer Academic Publ., 2003.
[55] B.D. Van Veen, K.M. Buckley, Beamforming: a versatile approach to spatial filtering, IEEE ASSP Mag. 5 (2) (1988) 4–24.
[56] C. Vogel, The impact of combined channel mismatch effects in time-interleaved ADCs, IEEE Trans. Instrum. Meas. 55 (1) (Feb. 2005) 415–427.
[57] C. Vogel, Modeling, Identification, and Compensation of Channel Mismatch Errors in Time-Interleaved Analog-to-Digital Converters, Diss., Graz Univ., 2005.
[58] C. Vogel, G. Kubin, Modeling of time-interleaved ADCs with nonlinear hybrid filter banks, AEÜ, Int. J. Electron. Commun. 59 (5) (July 2005) 288–296.
[59] C. Vogel, S. Mendel, A flexible and scalable structure to compensate frequency response mismatches in time-interleaved ADCs, IEEE Trans. Circuits Syst. I, Regul. Pap. 56 (11) (Nov. 2009) 2463–2475.
[60] Y. Wang, H. Johansson, H. Xu, Z. Sun, Joint blind calibration for mixed mismatches in two-channel time-interleaved ADCs, IEEE Trans. Circuits Syst. I, Regul. Pap. 62 (6) (June 2015) 1508–1517.
[61] L. Wanhammar, H. Johansson, Digital Filters Using Matlab, Linköping University, 2011.
[62] E.T. Whittaker, On the functions which are represented by the expansion of interpolating theory, Proc. R. Soc. Edinb. 35 (1915) 181–194.
[63] J.L. Yen, On nonuniform sampling of bandlimited signals, IRE Trans. Circuit Theory CT–3 (1956).

CHAPTER

Digital filter structures and their implementation

6

Lars Wanhammar[a] **and Yajun Yu**[b]

[a]*Department of Electrical Engineering, Linköping University, Linköping, Sweden*

[b]*Department of Electronic and Electrical Engineering, Southern University of Science and Technology, Shenzhen, Guangdong Province, PR China*

6.1 Introduction

Digital filtering is used in most communication applications because of the inherent flexibility and reliability of digital processing. Other important factors that support this trend are the rapid development in integrated circuit technology and the associated design tools that simplify and reduce the cost of implementing digital signal processing systems and digital filters in particular. Of interest here are frequency-selective digital filters, for example, low-pass and bandpass filters.

Typically we specify the filter requirements in terms of a behavioral description, i.e., output versus input signals, for example, transfer functions, frequency responses, impulse responses, or poles and zeros. Once these are determined, we need to determine a suitable structural description, i.e., an algorithm that realizes the proper behavior. Furthermore, this algorithm must be robust under finite length arithmetic operations. We will later discuss two main classes of such robust algorithms, i.e., nonrecursive finite length impulse response (FIR) filters and infinite length impulse response (IIR) filters, namely, wave digital filters.

6.1.1 Nonrecursive FIR filters

In Section 6.2, we will discuss the properties of digital filters and our design target. We are mainly interested in designing digital filters with low power consumption, which we assume are implemented using an field-programmable gate array (FPGA).

In Section 6.3, we will discuss computer-aided approaches to synthesize nonrecursive FIR filters. Typically, we will try to find a behavioral description in terms of frequency response, impulse response, or pole-zero positions. In Section 6.4, we discuss structures that can efficiently realize FIR filters. However, these structures are not well suited for filters with narrow transition bands. In Section 6.5, we discuss efficient structures suited for FIR and IIR filters with very narrow transition bands. We also extend the frequency response masking (FRM) technique to realizations of half-band filters, Hilbert transformers, multirate filters, filter banks, transmultiplexers (TMUXs), and 2D filters.

Signal Processing and Machine Learning Theory. https://doi.org/10.1016/B978-0-32-391772-8.00012-0
Copyright © 2024 Elsevier Inc. All rights reserved.

6.1.2 Analog filters

Analog filters are often used as a first step to design digital filters. One reason is that the poles and zeros can be mapped using the bilinear transformation to a corresponding digital filter. The bilinear transformation [1,2,3,4] between the s- and z-domains is

$$s = \frac{2}{T}\frac{z-1}{z+1}. \tag{6.1}$$

A more important fact is that we know how to design and realize an analog circuit with very low (optimal) passband sensitivity to errors in the circuit elements. A doubly resistively terminated lossless circuit, i.e., a voltage source in series with a source resistor and loaded with a load resistor that is designed for maximal power transfer between the source and load, will have low passband sensitivity. The concept of maximal power transfer is essential. The port voltage and port current characterize the power delivered into a port. It takes two variables to represent power. Hence, the corresponding digital filter will also require two variables. We use incident and reflected waves in wave digital filters, and we obtain a corresponding pseudopower concept.

In Section 6.6, we discuss the synthesis of analog filters using the toolbox [5]. Optimal high-pass filters can be designed by frequency transformation of an optimal low-pass filter. Near-optimal bandpass and bandstop filters can also be designed by frequency transformations. In Section 6.7, we discuss the properties and design of low-sensitivity analog structures that are realized using lumped elements (*LCs*) and lossless, commensurate length transmission line networks. We will show that these networks can be mapped to a similar network consisting of resistors, inductors, and capacitors. The latter network can be designed using classical design tools.

6.1.3 Wave digital filters

Wave digital filters are derived from an analog reference filter that consists of commensurate length transmission lines and resistors. A wave digital filter structure inherits its reference filter's low-sensitivity properties, being designed for maximum power transfer in the passband. Moreover, the concept of pseudopower in the digital domain allows us to suppress parasitic oscillations. Hence, a wave digital filter cannot support parasitic oscillations, which is very important under finite length arithmetic operations.

The overall design procedure is illustrated in Fig. 6.1.

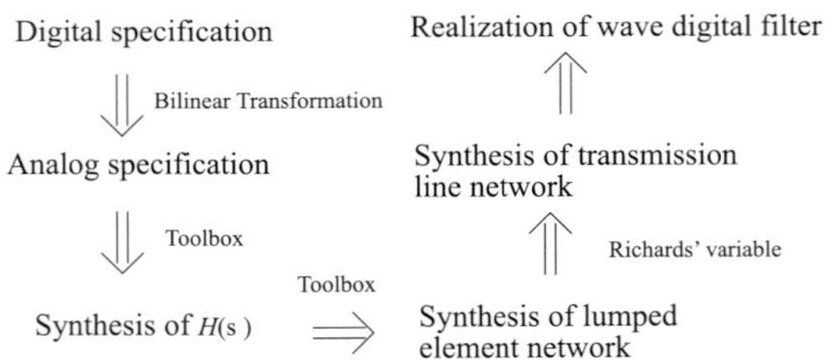

FIGURE 6.1

Design procedure for wave digital filters.

- The specification of the digital filters is mapped using the bilinear transformation and possible using a frequency transformation to a corresponding analog low-pass filter.
- The analog transfer function, $H(s)$ (frequency response), is determined using the toolbox.
- Next, the transfer function is realized using an *LC* filter structure, e.g., a ladder or lattice structure. We will later show that a properly designed *LC* filter will have optimal element sensitivity in the passband. Unfortunately, this structure is not possible to simulate in the z-domain because delay-free loops occur.
- We use an analogy (Richards' variable) between an *LC* network and a commensurate length transmission line network to circumvent this problem. The latter contain sources, resistors, and lossless transmission lines of equal length. The transmission line network has a periodic frequency response, and there is a one-to-one mapping between the continuous-time transmission line network and its digital counterpart. We will later discuss this in more detail.
- The digital algorithm is obtained using a wave description, i.e., wave digital filters [6,7,8], corresponding to a linear combination of voltages and current in the analog reference network. This allows us to simulate power in the analog filter, which we later will show is a necessity to obtain a low-sensitivity algorithm.

We will describe various steps in the design process, illustrated in Fig. 6.1, and the designer's options to minimize the implementation cost. First, we discuss in Section 6.6 the analog approximation problem and show that classical design recommendations are not always the best. In Section 6.7, we discuss the realization of the analog transfer function using a doubly resistively terminated network. We also discuss what the designer can do in order to minimize the element sensitivity. We also discuss the realization of commensurate length transmission line filters. In Sections 6.8.1 and 6.8.2, we discuss two of the most common analog filter networks and their properties.

In Section 6.10, we discuss the design of wave digital filters, their building blocks, and their properties. In Section 6.11, we discuss the design of ladder wave digital filters. In Section 6.12, we discuss the design of lattice wave digital filters. In addition, we compare two types of wave digital filter realizations using an example. In Section 6.13, we discuss the design of wave digital filter structures that are based on a factorization of the scattering matrix. These filters are similar to lattice structures. In the following section, we discus a numerically equivalent state-space realization of the previous structures.

6.1.4 Implementation of digital filters

In Section 6.15, we discuss some properties of digital filter algorithms from a computational point of view. In Section 6.16, we discuss computational architectures. In Section 6.17, we discuss arithmetic operations. In Section 6.18, we discuss the computation of sum-of-products (SOP). In Section 6.19, we discuss power reduction techniques [9].

6.1.5 MATLAB® toolbox

We will here use a toolbox for the design of both analog and digital filters [5]. This toolbox is also valid for use in the book *Analog Filters Using MATLAB* [10], except that _S or _Z has been added to the function names to differentiate between analog and digital functions, e.g., CA_POLES_S.m and CA_POLES_Z.m find the poles and zeros for an analog and digital Cauer filter, respectively. The toolbox includes a large number of MATLAB scripts. A key feature is that the MATLAB scripts only perform a single task. Hence, these scripts do not interact and can therefore be combined and modified by the user to solve many different design tasks.

6.2 Properties of digital filters

The following properties are characteristic of digital filters.

Flexibility

The frequency response can easily and quickly be changed. This means that a digital filter implementation can be time-shared between several input signals and act as several filters. Furthermore, a digital filter can be multiplexed to simultaneously act as several filters with one input signal and realize, for example, a filter bank. The high flexibility is explored in adaptive filters. This feature will not be explored in this chapter.

Special transfer functions

Discrete-time and digital filters can realize special transfer functions that are not realizable with continuous-time *LC* filters, for example, filters with exact linear phase response.

Coefficient sensitivity

Digital filters are not affected by temperature variations, variations in the power supply voltage, stray capacitances, etc., which are significant problems in analog filters. This is because the properties of a digital filter are determined by the numerical operations on numbers and are not dependent on any tolerances of electrical components. For the same reason, there is no aging or drift in digital filters. The independence of element sensitivity leads to high flexibility, miniaturization, and high signal quality.

Reproducibility

Exact reproducible filters can be manufactured. This is also a consequence of the fact that the filter properties are only dependent on the numerical operation and not on the tolerances of the electrical components. Thus, tolerance problems as such do not exist in digital filters. However, we will see that the sensitivity of a filter structure to errors in the coefficient directly affects the required signal word length and implementation cost. Furthermore, no trimming is necessary, which leads to low cost. Digital filters can be manufactured with, in principle, arbitrary precision, linearity, and dynamic range, however, at an increasing cost.

Sample frequency range

Digital filters can be made to operate over a wide range of frequencies, from zero up to several hundred MHz, depending on the filter complexity and the technology used for the implementation. However, the cost in terms of power consumption and the number of gates in the integrated circuits increases with the filter order and the sampling frequency. The maximal sample rate of a nonrecursive algorithm is, in principle, unlimited. However, the amount of circuitry and the inherent speed of the used CMOS technology becomes a limit. Recursive algorithms have an inherent limit on the throughput, which will be further discussed in Section 6.15.2

Power consumption

Digital filters generally have relatively high power consumption, but filters with low power consumption can be implemented for applications with low sampling frequencies. However, as the CMOS geometries are reduced, the power consumption is also reduced. Therefore, the power consumption compared with a corresponding analog filter counterpart is often lower for a digital filter.

Implementation techniques

Digital filters can be implemented using several techniques. At low sample rates and arithmetic workloads, typically below a few MHz, standard signal processors can be used while application-specific or algorithm-specific digital signal processors can be used up to a few tenths of MHz. For high sample rates, specialized implementation techniques are required [11,4].

6.2.1 Design target

The design should meet the requirements with minimal cost. The cost to be minimized is typically composed of several parameters, e.g., power consumption, chip area, and design effort. To complicate the issue even further, the designer has many alternative design options available. Hence, filter design is a challenging task, and it typically involves several conflicting optimization problems.

Power consumption

Until recently, throughput and chip area were the primary metrics for evaluating digital signal processing (DSP) algorithms and their implementations. However, with the advancement of CMOS technology, it is no longer a significant problem to achieve high enough throughput. As a result, the cost of the chip area can usually be neglected, except in very high-volume products. Today, the focus should be on design simplicity and low power consumption since many implementations are battery-powered. Too high power dissipation may require more expensive packaging and cooling. Further, there is a trend to use more sophisticated and complex DSP, increasing the power dissipation.

Chip area

The required chip area is essential since it directly affects the cost of manufacturing the filter. Moreover, the leakage current will be proportional to the chip area. A designer can minimize the required chip area using filter structures with few arithmetic operations per sample, simplified arithmetic operations, and small memory requirements. We are interested in structures with low element sensitivity, i.e., where a small error in the coefficients causes only a small error in the frequency response. Hence, it is possible to select simple coefficient values that deviate slightly from their ideal values and still meet the requirements on the frequency response. We will later discuss filter structures with very low coefficient sensitivities to reduce the required chip area and efficient schemes to realize filter coefficients.

The signal word length generally affects the execution time and the required chip area for the arithmetic operations. Jackson [12,4] has shown that the round-off noise is bounded from below by the coefficient sensitivities, i.e.,

$$\sigma^2 \geq \sum_i \sigma_{0i}^2 \left\| \frac{\partial \left| H(e^{j\omega T}) \right|}{\partial c_i} \right\|_p^2, \tag{6.2}$$

where $p = 1$ or $p = 2$, σ_{0i}^2 is the variance of the round-off noise sources, and the sum is over all noise sources due to rounding at the outputs of the coefficients c_i. Hence, the right-hand side of (6.2) is small for a low-sensitivity structure and results in low round-off noise at the output of the structure. The round-off noise will be low, even though there is no proof for it. A structure with high sensitivity will generate a large round-off noise, and it will require a large signal word length. To summarize, the element sensitivity affects both the required signal and coefficient word lengths and, hence, the overall cost.

Robustness

Recursive algorithms require that the signal word length is rounded or truncated at least once in every recursive loop that contains at least one multiplication. Otherwise, the signal word length would increase for each iteration. Rounding/truncation and overflow correction of the available signal range are nonlinear operations that will, in most cases, cause parasitic oscillations. As a result, the output of the filter will contain a periodic or even a chaotic signal component that is not due to the input signal [13,4]. This unwanted distortion can be substantial. Of course, this is highly undesirable and should be avoided. Fig. 6.2 shows an example of the input and output of a recursive loop when the input signal overflows the signal range just before it becomes zero. The output, which is multiplied with 512 to more clearly show its behavior, enters an unacceptable periodic oscillation.

Nonrecursive FIR filter structures cannot support parasitic oscillations, while most recursive IIR structures will, however, support parasitic oscillations. Therefore we will focus on nonrecursive FIR and a class of filter structures, i.e., wave digital filters, which will suppress such undesirable oscillations.

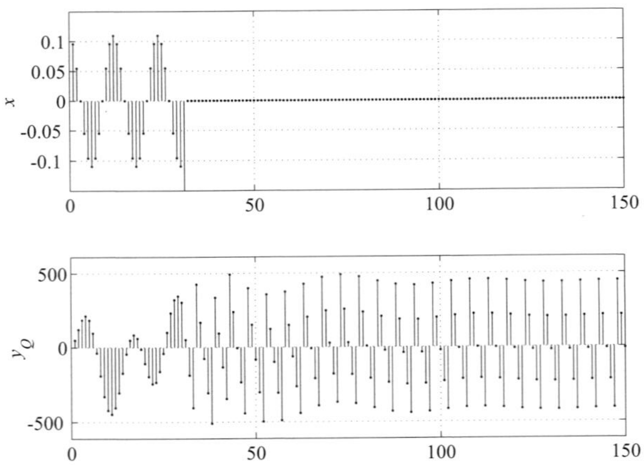

FIGURE 6.2

Persistent parasitic overflow oscillation.

6.3 Synthesis of digital FIR filters

Digital filters can be categorized into FIR and IIR filters. Advantages of FIR filters over IIR filters are that they can be guaranteed to be stable and may have a linear-phase response. Further, they require generally shorter signal word length than the corresponding IIR filters. However, FIR filters usually require much higher orders than IIR filters for the same magnitude specification. In addition, they often introduce a large group delay, which makes them unsuitable in many applications. In addition, FIR filters are well suited for many multirate filter realizations. In addition, an advantage is that FIR filters are always stable, except when used inside a recursive loop.

The synthesis of digital filters can be performed along several different routes. If the filter has an FIR, the frequency response is typically designed directly in the z-domain using computer-based optimization methods. Section 6.3 will discuss classical FIR filters and associated structures, suitable for cases when the transition band is not too small. Unfortunately, for requirements with a narrow transition band, the filter order and arithmetic workload become very large for FIR filters. In such cases, we instead recommend recursive structures (IIR), frequency response masking techniques, which are discussed in Section 6.5, or multirate techniques [14,15,4].

The most common method to design a linear-phase FIR filter that satisfies a specification of the magnitude response is to use a numeric optimization procedure to determine the impulse response $h(n)$ in (6.3). A widely used algorithm for designing linear-phase FIR filters with multiple passbands and stopbands is that of McClellan, Parks, and Rabiner [16,17,18,5]. We will use the function REMEZ_FIR to design FIR filters, an improved version based on the Remez algorithm.

The impulse response $h(n)$ of a causal FIR filter of order N (length $L = N+1$) is nonzero for $0 \leq n \leq L-1$ and zero for all other values of n, as illustrated in Fig. 6.3.

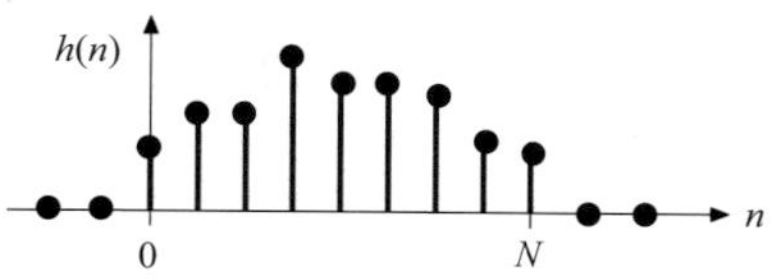

FIGURE 6.3

Impulse response of a causal FIR filter of order N (length $L = N+1$).

The transfer function and frequency response of an Nth-order FIR filter is

$$H(z) = \sum_{n=0}^{N} h(n) z^{-n}. \tag{6.3}$$

The transfer function poles can only be placed at the origin of the z-plane for nonrecursive FIR filters. The zeros can be placed anywhere in the z-plane, but they are usually located on the unit circle or complex conjugate pairs mirrored in the unit circle.

Of primary interest are FIR filters with linear phase for which the impulse response exhibits symmetry or antisymmetry. Linear-phase FIR filters are widely used in digital communication systems, speech and image processing systems, spectral analysis, and particularly applications where nonlinear phase distortion cannot be tolerated.

A linear-phase FIR filter is obtained by letting the impulse response exhibit symmetry around $n = N/2$ or antisymmetry around $n = N/2$, i.e.,

$$h(n) = \pm h(N - n) \tag{6.4}$$

for $n = 0,1,..., N$. Since the order N is either even or odd, there are four different types of FIR filters with linear phases. These are denoted as Type I, II, III, and IV linear-phase FIR filters.

$$\begin{array}{llll} \text{Type I:} & h(n) = h(N-n), & N & \text{even,} \\ \text{Type II:} & h(n) = h(N-n), & N & \text{odd,} \\ \text{Type III:} & h(n) = -h(N-n), & N & \text{even,} \\ \text{Type IV:} & h(n) = -h(N-n), & N & \text{odd.} \end{array}$$

Fig. 6.4 shows typical impulse responses for the different cases. Note that the center value $h(N/2)$ is always equal to zero for Type III filters. Also, when N is even, the point of symmetry is an integer corresponding to one of the sample values. When N is odd, the point of symmetry lies between two sample values.

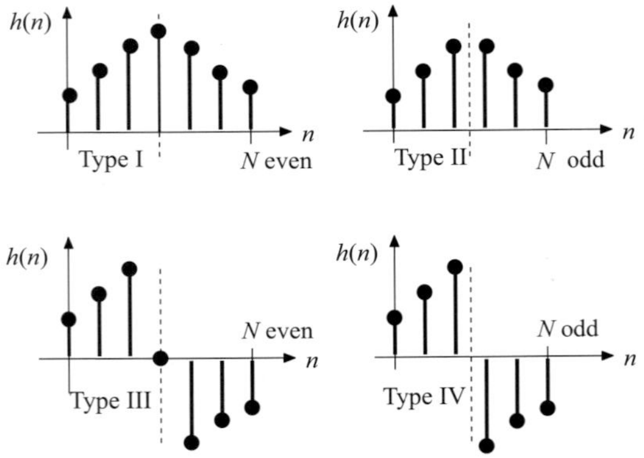

FIGURE 6.4

Impulse responses for the four types of linear-phase FIR filters.

Fig. 6.5 shows a typical specification of the zero-phase response for a low-pass linear-phase FIR filter. Several approximate expressions for the required filter order for equiripple, linear-phase FIR filters have been determined empirically.

The deviation in the attenuation in the passband is

$$A_{max} = 20\,log(\frac{1+\delta_c}{1-\delta_c}), \tag{6.5}$$

and the minimum stopband attenuation is

$$A_{min} = 20 log(\frac{1+\delta_c}{\delta_s}). \tag{6.6}$$

One simple estimate of the required filter order for a linear-phase low-pass (or high-pass) filter that meets the specification shown in Fig. 6.5 is

$$N \approx \frac{-4\pi}{3}\frac{log\,(13\delta_c\delta_s)}{\Delta\omega T}, \tag{6.7}$$

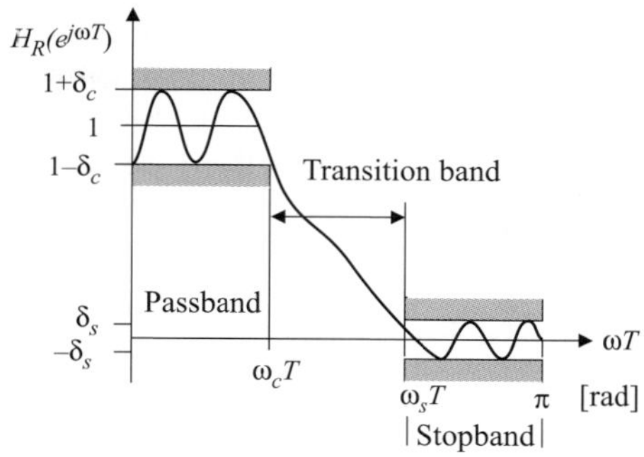

FIGURE 6.5

Specification of FIR filters.

where $\Delta\omega T = \omega_s T - \omega_c T$ is the transition band. Note that the required filter order will be high for filters with narrow transition bands.

Here we are concerned with linear-phase FIR filters that are frequency-selective, e.g., low-pass, high-pass, bandpass, and bandstop filters. When designing such filters, it is convenient to consider the zero-phase frequency response, $H_R\left(e^{j\omega T}\right)$, which is defined by

$$H\left(e^{j\omega T}\right) = H_R(e^{j\omega T})e^{j\phi(\omega T)}. \tag{6.8}$$

Note that $\left|H\left(e^{j\omega T}\right)\right| = \left|H_R(e^{j\omega T})\right|$.

The specification of a low-pass filter is commonly given as

$$\begin{aligned} 1-\delta_c \le \left|H(e^{j\omega T})\right| &\le 1+\delta_c, & \omega T \in [0,\ \omega_c T], \\ \left|H(e^{j\omega T})\right| &\le \delta_s, & \omega T \in [\omega_s T,\ \pi]. \end{aligned} \tag{6.9}$$

For linear-phase FIR filters, the specification of (6.9) can be stated with the aid of $H_R\left(e^{j\omega T}\right)$ according to

$$\begin{aligned} 1-\delta_c \le H_R\left(e^{j\omega T}\right) &\le 1+\delta_c, & \omega T \in [0,\ \omega_c T], \\ -\delta_s \le H_R\left(e^{j\omega T}\right) &\le \delta_s, & \omega T \in [\omega_s T,\ \pi]. \end{aligned} \tag{6.10}$$

The REMEZ_FIR program will find a unique impulse response that minimizes a weighted error function. The algorithm solves the approximation problem

$$minimize\,(E_\infty) = max\left|E\left(e^{j\omega T}\right)\right|,\ \omega T \in \Omega, \tag{6.11}$$

where $E\left(e^{j\omega T}\right)$ is the weighted (Chebyshev) error function given by

$$E\left(e^{j\omega T}\right) = W\left(e^{j\omega T}\right)\left[H_R\left(e^{j\omega T}\right) - D\left(e^{j\omega T}\right)\right],\ \omega T \in \Omega, \tag{6.12}$$

with Ω being the union of the passband and stopband regions. In practice, Ω is a dense set of frequency samples taken from the passbands and stopbands, including the band edges. It should be stressed that all functions involved here, $H_R\left(e^{j\omega T}\right)$, $E\left(e^{j\omega T}\right)$, $D\left(e^{j\omega T}\right)$, and $W\left(e^{j\omega T}\right)$ (nonnegative function), are real functions of ωT. The desired function $D\left(e^{j\omega T}\right)$ is the one to be approximated by $H_R\left(e^{j\omega T}\right)$. For example, $D\left(e^{j\omega T}\right)$ is for the standard low-pass filter

$$D\left(e^{j\omega T}\right) = \begin{cases} 1 & \omega T \in [0,\ \omega_c T], \\ 0 & \omega T \in [\omega_s T,\ \pi]. \end{cases} \tag{6.13}$$

That is, $D\left(e^{j\omega T}\right)$ is for conventional filters equal to one in the passbands and zero in the stopbands. The weighting function $W\left(e^{j\omega T}\right)$ specifies the cost of the deviation from the desired function. The use of the weighting function makes it possible to obtain different deviations in different frequency bands. The larger the weighting function is, the smaller is the ripple. For example, for a standard low-pass filter, $W\left(e^{j\omega T}\right)$ is usually given as

$$W\left(e^{j\omega T}\right) = \begin{cases} 1 & \omega T \in [0,\ \omega_c T], \\ \delta_c/\delta_s & \omega T \in [\omega_s T,\ \pi]. \end{cases} \tag{6.14}$$

The specification of (6.10) is then satisfied if $E_\infty \leq \delta_c$. The resulting filter is optimized in the Chebyshev (minimax) sense. MATLAB programs for designing FIR filters are found [5]. The design of special cases of linear-phase FIR filters and nonlinear phase FIR filters is discussed in detail [4]. The following MATLAB script can be used to compute the impulse response of a linear-phase FIR filter [5,4].

6.3.1 REMEZ_FIR

The function REMEZ_FIR [18,5], which is used to design linear-phase FIR filters, computes the unique impulse response of an Nth-order linear-phase, equiripple FIR filter. The function REMEZ_FIR

[h, Err] = REMEZ_FIR(N, Be, D, W, Ftype, LGRID)

is called using the following parameters: h is the desired impulse response and Err is the maximal deviation in the error function. N is the filter order (positive integer), e.g., N = 32. Be is the band edges, e.g., Be = [0 0.2 0.5 1]$\times\pi$ for a low-pass filter with $\omega_c T = 0.2\pi$ rad and $\omega_s T = 0.5\pi$ rad. Both the beginning and the end of the band must be included, i.e., 0 and π rad. Thus, the length of Be is even. D is the desired values at the band edges, e.g., D = [1 1 0 0] for a low-pass filter. W is the weighting of the bands, e.g., W = [δ_c δ_s] for low-pass filters. Ftype is m, d, and h for multibands, differentiators, and Hilbert transformers, respectively. LGRID is typically set to {128}.

6.3.1.1 Example

Synthesize a linear-phase equiripple FIR filter that satisfies the following low-pass specifications: $\omega_s T = 0.2$ rad, $\omega_s T = 0.5\pi$ rad, $\delta_c = 0.01$, and $\delta_s = 0.001$ (low-pass). The required filter order is estimated to $N = 19$ using the function LPHASE_LP_FIR_ORDER. We, however, select $N = 20$ to meet the requirements. From Eqs. (6.5) and (6.6) we get $A_{maxreq} = 0.087296$ dB and $A_{minreq} = 60.0864$ dB. The linear-phase equiripple FIR filter is synthesized using the program [5] shown in Listing 6.1.

Listing 6.1: Program design of linear-phase equiripple FIR filters.

```
dc = 0.01; ds = 0.001;
wTedges = [0.2 0.5]*pi; % Band edges
b = [1 0];          % Gains in the bands
d = [dc ds]; % Acceptable deviations
Amaxreq = 20*log10(1+dc/(1-dc))
Aminreq = 20*log10((1+dc)/ds)
% Estimate filter order
N, Be, D, W] = L_PHASE_LP_FIR_ORDER(wTedges, d);
W = [1 0.5*dc/ds] % Yields an acceptable solution
N     % Estimated order = 19
N = N+1 % Minimum order 20 to meet the criteria
Ftype = 'm';
[h, Err] = REMEZ_FIR(N, Be, D, W, Ftype, {128});
deltac = Err/W(1);   % Resulting passband and
deltas = Err/W(2);   % Stopband ripples
hn = h/(1+deltac);   % Make |H]max = 1
fprintf('Err= %2.8f \n', Err);
Amax = 20*log10((1+deltac)/(1-deltac))
Amin = 20*log10((1+deltac)/deltas)
fprintf('Amax = %2.6f dB \n', Amax)
fprintf('Amin = %2.4f dB \n', Amin);
wT = linspace(0, pi, 1001); % wT axis 0 to  rad
Mag = H_2_MAG(freqz(hn,1,wT)); % Magnitude in dB
opt = 'Mag';
 PLOT_MAG_PZ_h_FIR(wT,Mag,hn,Be,D,d,opt)
```

We get

Amax = 0.079552 dB,

Amin = 60.8030 dB.

The required filter order that is obtained from L_PHASE_LP_FIR_ORDER is only an estimation, and it is, in this case, too low. Therefore, it is necessary to try different filter orders to find an order that satisfies the specification. See [4] for a practical scheme. Fig. 6.6 shows the magnitude response and the specification, pole and zeros, and the impulse response. Note that there are 20 poles at $z = 0$. The filter has linear-phase response. Hence, the impulse response is symmetric for an even-order FIR filter.

6.3.2 Notch FIR filters

Narrow-stopband FIR filters are frequently used to remove a single-frequency component from the signal spectrum, often referred to as notch filters. The minimax-based approximation used in REMEZ_FIR has some inherent limitations, i.e., the solution may fail in the transition region and for notch filters in particular. Nevertheless, maximally flat notch FIR filters can be designed using the

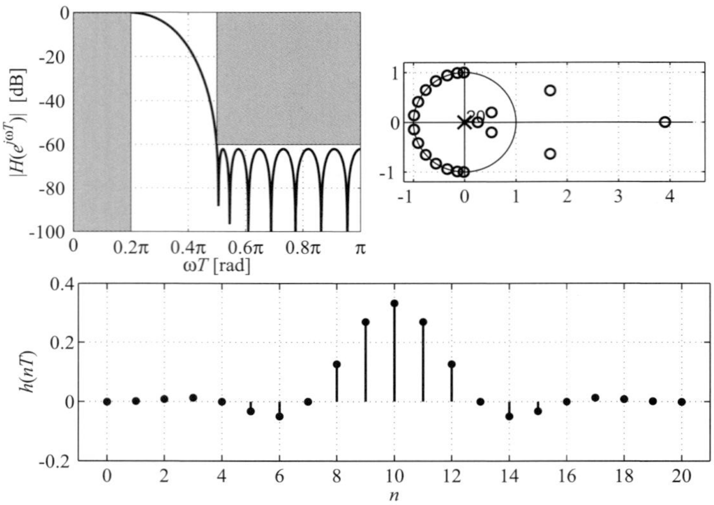

FIGURE 6.6

Magnitude, pole-zeros, and impulse response for a 20th-order linear-phase FIR filter.

function MF_NOTCH_FIR. DC notch filters are used to remove a DC component from the signal spectrum. An efficient algorithm for designing equiripple DC notch FIR filters of even order is the function DC_NOTCH_FIR.

6.3.3 Half-band FIR filters

Many digital filter structures can exploit the fact that a large number of coefficient values are zero. Thus, the required number of arithmetic operations can be reduced since it is unnecessary to perform multiplications by zero coefficients. One such case is a filter with a 3-dB cutoff angle of π/2 rad, referred to as half-band filters since the bandwidth is half the whole frequency band. We will discuss half-band FIR filters in Section 6.4.4.

6.3.4 Complementary FIR filters

A pair of complementary filters are used in many applications, for example, in low-sensitivity filter structures and filter banks. Several types of complementary filter pairs are discussed in [4]. In addition, in many applications, the need arises to split the input signal into two or more frequency bands.

For a delay-complementary filter pair, $H(z)$ and $H_c(z)$, we have

$$H(z) + H_c(z) = z^{-N/2}, \tag{6.15}$$

where N is an even integer.

Note that $H(e^{jwT}) = H_c(e^{jwT}) = 0.5$, i.e., the attenuation is 6.02 dB at the crossover angle ωT. The corresponding impulse responses are

$$h(n) + h_c(n) = \delta(n - N/2). \tag{6.16}$$

We will later show that these filters, in some cases, can be effectively realized by a slight modification of an even-order FIR filter of Type I.

For a power-complementary filter pair, we have

$$|H(e^{jwT})|^2 + |H_c(e^{jwT})|^2 = 1. \tag{6.17}$$

Complementary half-band filters are particularly useful since the saving in the computational workload is significant.

6.3.5 Minimum-phase FIR filters

One of the advantages of FIR filters compared to IIR filters is attaining an exact linear-phase response. However, the linear-phase characteristic implies symmetry or antisymmetry in the impulse response that constrains *N*/2 coefficients to be the same. Hence, only *N*/2+1 of the remaining coefficients are freely assignable (assuming *N* is even). In a nonlinear-phase filter, all *N* +1 coefficients are free parameters. The design of general nonlinear-phase FIR filters is discussed in [4].

A linear-phase response is often desirable because it introduces no phase distortion, but it has the disadvantage that the overall delay is usually considerable. A minimum-phase FIR filter might prove useful if a linear-phase response is not as critical as the overall delay. For each linear-phase filter, there exists a corresponding minimum-phase filter. The function MIN_PHASE_ LP_FIR_ORDER can be used to estimate the order of a minimum-phase FIR filter. We have

$$N_{MP} = N_{LP} + \frac{2\pi}{3}\frac{log(13\delta c)}{\Delta\omega T}. \tag{6.18}$$

The linear-phase filter may be up to twice as long as the corresponding minimum-phase filter for the same magnitude specification. A minimum-phase FIR filter has all its zeros inside or on the unit circle and can be designed using MIN_PHASE_ LP_FIR.

6.3.6 Miscellaneous FIR filters

Two commonly used FIR filters are differentiators and Hilbert transformers, which also can be designed using REMEZ_FIR. Linear-phase FIR filters can also be designed using linear programming (linprog.m) and quadratic programming techniques (quadprog.m). Several solved design cases for these filters are given in [4].

Both analog and digital filters can be transformed into a new filter with different frequency responses. For example, a low-pass filter, usually called the reference filter, can be transformed into high-pass, bandpass, and bandstop filters. Frequency transformation can be applied to both FIR and IIR filters. There are two main techniques for frequency transformations: direct frequency transformation and structural frequency transformation. The direct frequency transformation is carried out by mapping the poles and zeros of the reference filter to new positions to obtain the new filter.

A structural frequency transformation is similar, but instead of mapping the poles and zeros, the delay elements in the reference filter structure are replaced by a subnetwork corresponding to the mapping. However, there are severe restrictions on the subnetwork because the resulting filter structure may have delay-free loops. Structural transformations can be applied to both FIR and IIR reference filters [4].

6.4 FIR structures

An FIR filter can be realized using both recursive or nonrecursive algorithms. The former, however, suffer from several drawbacks and should rarely be used in practice. An exception is so-called running-sum filters, which often are used in decimation filters for $\sum\Delta$-converters.

Nonrecursive filters are always stable and cannot sustain parasitic oscillations, except when the filters are a part of a recursive loop. They generate little round-off noise and can therefore use shorter signal word lengths. However, they usually require a large number of arithmetic operations and a large amount of memory. Therefore great efforts have been directed towards developing methods to reduce the arithmetic workload.

This can be done by reducing the number of multiplications per sample or by simplifying the coefficient values so that the multiplications become simpler to implement. The latter approach results in what is referred to as multiplier-free structures since all multiplications have been replaced with addition/subtraction and shift operations. That is, the coefficients are selected among a restricted set of values, i.e., as a sum-of-powers-of-two (SPOT), $c = \pm 2^{\pm n_1} \pm 2^{\pm n_2} \pm \ldots \pm 2^{\pm n_m}$. Most FIR filters can be realized using no more than four powers of two. Today the arithmetic workload for a properly designed FIR structure is still relatively large, but chip area and power consumption are becoming more significant.

6.4.1 Direct-form FIR structures

There exist several structures that are of interest for the realization of FIR filters. One simple structure is the direct-form FIR, which has relatively low element sensitivity, i.e., the deviation in the frequency response due to a small change in any of the coefficients, is relatively small. This structure has many arithmetic operations and is therefore only suitable for relatively short FIR filters. The round-off noise variance is $(N+1)\,Q^2/12$ if all products are rounded. For filters with very narrow transition bands, we recommend frequency response masking structures, which are discussed later.

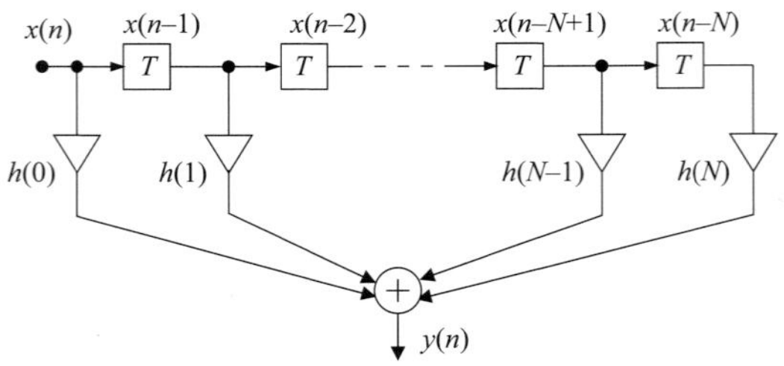

FIGURE 6.7

Direct-form FIR structure (transversal filter).

6.4.1.1 *Adder tree*

The magnitude and word length of the impulse response values of a low-pass filter is approximately sin(x)/x. Hence, the magnitudes of the tail values are relatively small, while the center value is the largest. It is, therefore, advantageous to pairwise add small values using adders with short word lengths.

The partial sums, which have increasing word lengths, can then be added so that successive values have about the same word length [4]. This scheme tends to reduce the required hardware requirement. An implementation of the adder tree in FPGAs or ASICs may result in different wire lengths and capacitive loads, and the adder tree may become irregular. This requires load balancing by sizing the gates to achieve high speed and avoid glitches that waste energy. We usually realize FIR structures using adder trees, but for reasons of simplicity, we will in figures use a simplified signal-flow representation compared to the one used in Fig. 6.7. A detailed discussion of the optimization of the adder tree may distract the reader from the issues at hand. However, an adder tree should, if appropriate, be used in the final realization.

6.4.2 Transposed direct form

The transposition theorem [19] can generate new structures that realize the same transfer function as the original filter structure. For example, the transposed direct-form FIR structure, shown in Fig. 6.8, is derived from the direct-form structure. The throughput is higher for the transposed form because the longest critical path is through only one multiplier and an adder, but the capacitive load at the input node is larger since all multipliers load this node. Therefore, the transposed structure is often preferred for an FPGA implementation.

The direct-form structures require, in general, $N+1$ multiplications, which may become a too large workload for high-order FIR filters. However, the required amount of hardware can be significantly reduced since many multiplications are performed with the same input value or the transposed case where many signals are multiplied and then added. The internal signal level increases from left to right as more and more internal signals are added. Also, in this structure, the round-off noise is $(N+1)\,Q^2/12$ if all products are rounded.

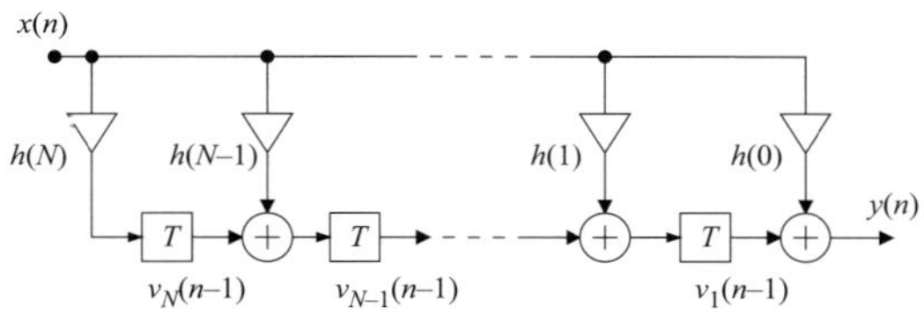

FIGURE 6.8

Transposed direct-form FIR structure.

6.4.3 Linear-phase FIR structures

A major reason why FIR filters are used in practice is that they can realize an exact linear-phase response. Linear-phase implies that the impulse response is either symmetric or antisymmetric. Thus, an advantage of linear-phase FIR filters is that the number of multiplications can be reduced by exploiting the symmetry (or antisymmetry) in the impulse response, as illustrated in Fig. 6.9. This structure is called a direct-form linear-phase FIR structure since the phase response is independent of the coefficient values.

The number of multiplications for the direct-form linear-phase structure is $(N+1)/2$ for N = odd and $N/2+1$ for N = even. The number of multiplications is significantly smaller than for the more

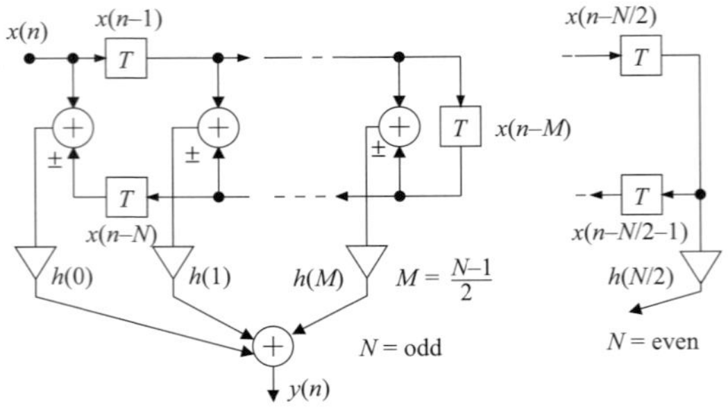

FIGURE 6.9

Direct-form linear-phase FIR structure.

general structures in Figs. 6.7 and 6.8, whereas the number of additions is the same. Subtractors are used instead of adders if the impulse response is antisymmetric. The signal levels at the inputs of the multipliers are twice as large as the input signal. Hence, the input signal should be divided by two, and the filter coefficients should be scaled so that a proper out signal level is obtained. The round-off noise is only $(N+1)\,Q^2/24$ and $(N+2)\,Q^2/24$ for $N=$ odd and even, respectively, independently of the filter coefficients.

A significant drawback is that the group delay for linear-phase FIR filters is often too large to be useful in many applications. Even-order FIR filters are often preferred since the group delay is an integer multiple of the sample period.

The number of arithmetic operations is further reduced in linear-phase, half-band filters. Each zero-valued coefficient makes one multiplication and one addition redundant. The number of actual multiplications in a linear-phase, half-band filter is only $(N+6)/4$. The order is always $N = 4m+2$ for some integer m. A half-band filter is a particular case of the Nyquist filters.

6.4.4 Half-band FIR filters

Many digital filter structures can exploit the fact that a large number of coefficient values are zero. Therefore the required number of arithmetic operations can be reduced since it is unnecessary to perform multiplications by zero coefficients. One such case is a filter, with a 3-dB cutoff angle of π/2 rad, referred to as half-band filters since the bandwidth is half the whole frequency band [4]. Every even sample in the impulse response corresponding to the zero-phase frequency response of a half-band filter is zero, except $h_R(0)$.

The zero-phase response of even-order half-band filters is antisymmetric around π/2 rad, i.e.,

$$H_R\left(e^{j\omega T}\right) = 1 - H_R\left(e^{j(\pi-\omega T)}\right), \tag{6.19}$$

and we have

$$\omega_c T + \omega_s T = \pi \text{ rad}, \tag{6.20}$$

and the passband and stopband deviations are $\delta_c = \delta_s$. Hence, if high stopband attenuation is required, then the passband has a small ripple, and vice versa. Hence, the smallest of the ripple factors determines the filter order.

Only even-order (N = even) half-band, low-pass FIR filters are of interest since the coefficients are nonzero in odd-order filters. Hence, only filter orders with $N = 4n+2$, where n is a positive integer, are useful since for an even-order filter, every other, except for the first and the last coefficients, is zero. Thus, the number of multiplications per sample in a linear-phase, half-band filter is only $(N+6)/4$. Even-order half-band low-pass filters are advantageous since they use half as many arithmetic operations as general FIR filters. Thus, they are used in filter banks, wavelets, and multirate filters, e.g., interpolating or decimating the sampling frequency by a factor of two.

6.4.5 Delay-complementary FIR structure

Even more dramatic reductions of the required arithmetic operations are possible when a pair of delay-complementary FIR filters is needed. We illustrate the basic ideas by example. The delay-complementary filter $H_c(z)$ to an even-order linear-phase filter of Type I or II can be obtained from the direct-form linear-phase structure by subtracting the ordinary filter output from the delayed input value, as shown in Fig. 6.10.

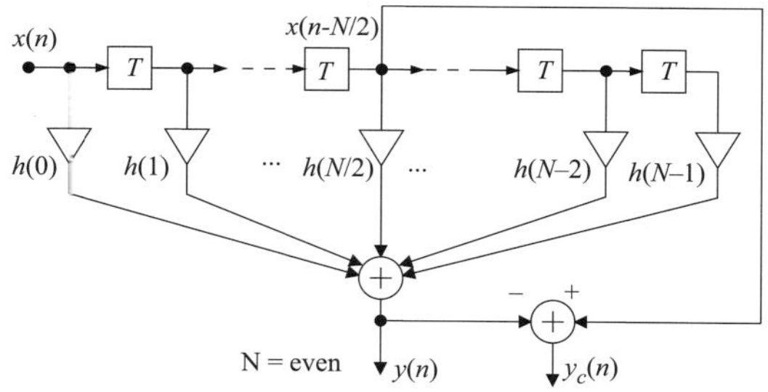

FIGURE 6.10

Delay-complementary FIR filter structure with N = even.

Odd-order delay-complementary FIR filters are not feasible since the symmetry axis, as shown in Fig. 6.4, does not coincide with the impulse response. The cost of realizing the delay-complementary filter seems to be only one subtraction. However, the passband requirement on the FIR filter realizing $H(z)$ must usually be increased significantly to meet the stopband requirements on the complementary filter, $H_c(z)$. Fortunately, a reduction of the passband ripple requires only a slight increase in the filter order (see (6.7)). Furthermore, only even-order filters are useful. Hence, the arithmetic workload is significantly smaller than that for two separate filters. This technique can, of course, also be applied to even-order linear-phase FIR filter structures, with either symmetric or antisymmetric impulse response and their transposes. Complementary half-band FIR filters are particularly useful since the saving in arithmetic workload is substantial.

In many applications, the need arises to split the input signal into two or more frequency bands. For example, in certain transmission systems for speech, the speech signal is partitioned into several frequency bands using a filter bank. A filter bank is a set of bandpass filters with staggered center frequencies to cover the whole frequency range. The first and last filters are low-pass and high-pass filters, respectively. The filtered signals are then processed individually to reduce the number of bits that have to be transmitted to the receiver, where the frequency components are added to an intelligible speech signal.

A particular case of band-splitting filters, a low-pass and high-pass filter pair, is obtained by imposing the following symmetry constraints. Let $H(z)$ be an even-order FIR filter (N even). The delay-complementary transfer function $H_c(z)$ is defined by

$$\left| H\left(e^{j\omega T}\right) + H_c\left(e^{j\omega T}\right)\right| = 1. \tag{6.21}$$

This requirement can be satisfied by two FIR filters that are related according to

$$H(z) + H_c(z) = z^{-N/2}. \tag{6.22}$$

Hence, the two transfer functions are complementary. A signal is split into two parts that, if added, combine into the original signal except for a delay corresponding to the group delay of the filters. Note that the attenuation is 6.02 dB at the crossover frequency. The output of the complementary filter $H_c(z)$ is obtained, as shown in Fig. 6.10, by subtracting the ordinary filter output from the central value $x(n–N/2)$, which significantly reduces the arithmetic workload. However, the complementary filter is not obtained for free because if the stopband attenuation is large, then the passband ripple of the ordinary filter must be very small. Hence, the filter order must be increased. The complementary FIR filters considered here cannot be of odd order since the center value of the impulse response, in that case, is not available.

6.4.6 Cascade-form FIR structures

High-order FIR and IIR filters may be realized in cascade form, i.e., as a cascade of several lower-order subfilters, as shown in Fig. 6.11 for an FIR filter. The overall transfer function is

$$H(z) = \prod_{k=1}^{M} H_k(z), \tag{6.23}$$

where the subfilters $H_k(z)$ usually have orders one and two. The cascading of two minimum-phase filters yields an overall minimum-phase filter. Moreover, cascading several linear-phase filters also results in an overall linear-phase filter.

The benefit of cascading several filters is that the stopband attenuations of the filters add. Hence, each subfilter needs only to contribute with a small fraction of the overall stopband attenuation, and they can be simplified. As a result, the coefficients in the subfilters need not be as accurate, i.e., the coefficient sensitivity in the stopband is small. However, the benefit obtained in the stopband sensitivity is offset by increased sensitivity in the passband since errors in the lower-order filters add. Furthermore, the cascade form suffers from a decrease in the dynamic signal range. Hence, the internal signal word length must be increased. The ordering of the sections is essential since it affects the dynamic signal range significantly. There are $M!$ possible ways to order the sections for an FIR filter realized with M sections.

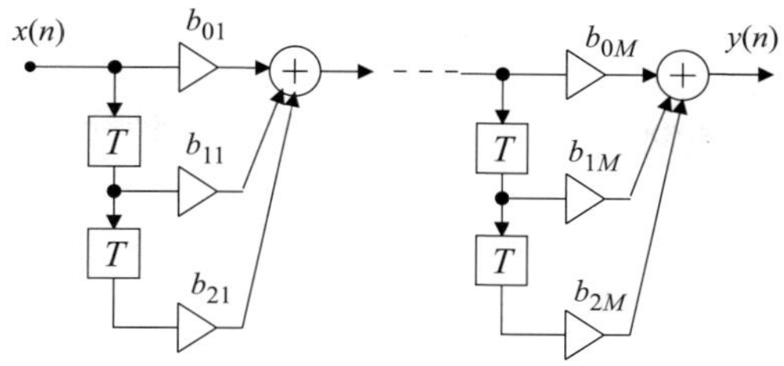

FIGURE 6.11

FIR filter in cascade form.

A minimum-phase filter can be partitioned into a linear-phase filter and a filter with a nonlinear phase in cascade form [4]. This allows the minimum-phase filter to be realized using the linear-phase FIR structure.

6.4.7 Lattice FIR structures

Lattice FIR structures are used in a wide range of applications, for example, speech and image processing, adaptive filters, and filter banks [20]. The lattice structure can be of FIR or IIR type. Note, however, that the lattice structures discussed in this section are a different type of structure compared to the lattice wave digital filter structures discussed in Section 6.12, although confusingly, they have the name lattice in common.

6.4.8 Recursive frequency-sampling structures

Traditionally, most textbooks [21,22,3] contain a discussion of so-called frequency-sampling FIR filter structures. These realizations are recursive algorithms that rely on pole-zero canceling techniques. Although they may seem attractive, they are not recommended due to their high coefficient sensitivity, low dynamic range, and severe stability problems [4].

6.4.9 Multirate FIR filters

Multirate filters are efficient in terms of arithmetic workload since a large part of the arithmetic operations can be performed at a slower rate [14,15,4]. Polyphase structures are efficient for decimation and interpolation since redundant arithmetic operations are avoided. FIR and IIR filters with narrow transition bands can be realized by methods based on decimation–filtering–interpolation schemes [4]. Common to these techniques is that the effective number of arithmetic operations per sample is reduced.

6.5 Frequency response masking filters

For given passband ripple and stopband attenuation, the filter order as well as the number of nontrivial coefficients of a direct-form FIR filter is according to (6.7) inversely proportional to the transition width. Therefore, sharp filters suffer from high computational complexity due to the high filter orders. Lim [23] proposed an FRM FIR filter structure in 1986 for the design of sharp FIR filters with low computational complexity. The structure was thereafter modified and improved by many researchers to achieve even further computation savings, and has been extended to the design of various types of filters, such as half-band filters, multirate filters, recursive filters, and 2D filters.

6.5.1 Basic FRM structure

Just as the name implies, FRM uses masking filters to obtain the desired frequency responses. The overall structure of an FRM is a subfilter network, consisting of three subfilters, as shown in Fig. 6.12. The three subfilters are a prototype band edge shaping filter (or simply called prototype filter) with z-transform transfer function $F(z)$ and two masking filters with z-transform transfer functions $G_1(z)$ and $G_2(z)$, respectively; their corresponding frequency responses are denoted as $F(e^{j\omega T})$, $G_1(e^{j\omega T})$, and $G_2(e^{j\omega T})$, respectively. A fourth filter function is realized by using a delay-complementary counterpart.

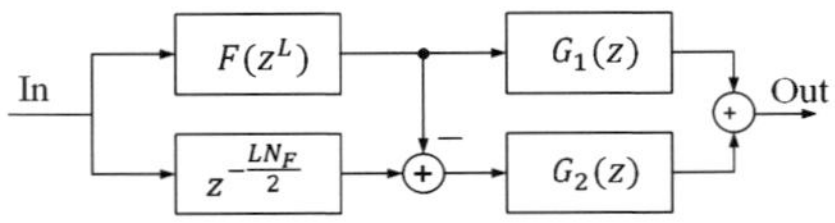

FIGURE 6.12

Structure of the frequency response masking filter.

In FRM structure, every delay of the prototype band edge shaping filter $F(z)$ is replaced by L delays. The frequency response of the resulting filter, $F(z^L)$, is a frequency-compressed (by a factor of L) and periodic version of $F(z)$, as shown in Fig. 6.13(a and b). For this reason, the number of delays replacing one delay in the impulse response of the band edge shaping prototype filter, L, is referred to as the compression factor of the FRM filter.

As a result of the delay element replacement, the transition band of $F(z)$ is mapped to L transition bands with transition width shrunk by a factor of L. The complementary filter of $F(z^L)$ is obtained by subtracting $F(z^L)$ from a pure delay term, $z^{-LN_F/2}$, where N_F is the order of $F(z)$. To avoid a half delay, N_F is chosen to be even. Thus, the entire frequency region from DC to Nyquist frequency is decomposed into L bands; $F(z^L)$ and its complement hold alternate bands respectively, as shown in Fig. 6.13(b).

Two masking filters $G_1(z)$ and $G_2(z)$ are used to filter (i.e., mask) the outputs of $F(z^L)$ and its complement. The desired frequency bands of $F(z^L)$ and its complement are kept and then combined to obtain the final frequency response.

The overall z-transform transfer function of the FRM structure, $H(z)$, is thus given by

$$H(z) = F(z^L)G_1(z) + \left[z^{-LN_F/2} - F(z^L)\right]G_2(z). \tag{6.24}$$

In a synthesis problem, the passband and stopband edges of the overall filter $H(z)$, denoted as $\omega_c T$ and $\omega_s T$, respectively, are given, whereas the passband and stopband edges of the prototype filter $F(z)$, denoted as $\omega_c^F T$ and $\omega_s^F T$, have to be determined.

From Fig. 6.13(b), it can be seen that the transition band of the overall filter $H(z)$ may be generated either from $F(z^L)$ or from its complement. We denote the two cases as Case A and Case B. The filter $F(z^L)$ (or its complement) is thus called periodic band edge shaping filter, or simply periodic filter or band edge shaping filter.

Throughout this section, we use low-pass filters as examples (except in sections for filter banks) to illustrate the FRM technique. The extension of the technique to other frequency-selective filters is straightforward.

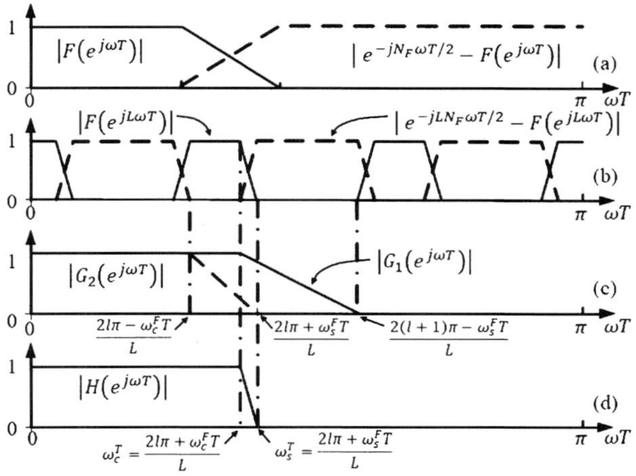

FIGURE 6.13

Frequency response of FRM subfilters of Case A.

6.5.1.1 Case A

In Case A, the transition band of $H(z)$ is formed by the lth transition band of $F(z^L)$, as shown in Fig. 6.13, i.e.,

$$\omega_c T = \frac{2l\pi + \omega_c^F T}{L} \quad \text{and} \quad \omega_s T = \frac{2l\pi + \omega_s^F T}{L}, \tag{6.25}$$

respectively. For a low-pass filter $H(z)$, the band edge of the prototype filter $F(z)$ must have $0 < \omega_c^F T < \omega_s^F T < \pi$, and meanwhile, we have $\omega_c T < \omega_s T$ and l must be an integer. According to (6.25), there is only one value of l that can meet these requirements if there is a solution, i.e., we must select $l = \lfloor L\omega_c T/2\pi \rfloor$, where $\lfloor x \rfloor$ is the largest integer less than or equal to x.

Thus, for Case A, we have

$$\omega_c^F T = L\omega_c T - 2l\pi, \quad \omega_s^F T = L\omega_s T - 2l\pi \quad \text{for} \quad l = \lfloor L\omega_c T/2\pi \rfloor. \tag{6.26}$$

6.5.1.2 Case B

In Case B, the transition band of $H(z)$ is formed by the $(l-1)$th transition band of $F(z^L)$'s complement, as shown in Fig. 6.14, i.e.,

$$\omega_c T = \frac{2l\pi - \omega_s^F T}{L} \quad \text{and} \quad \omega_s T = \frac{2l\pi - \omega_c^F T}{L}, \tag{6.27}$$

respectively. For the similar requirements as in Case A, the only value of l if there is a solution is $l = \lceil L\omega_s T/2\pi \rceil$, where $\lceil x \rceil$ is the smallest integer larger than or equal to x. Thus, for Case B, we have

$$\omega_c^F T = 2l\pi - L\omega_s T, \quad \omega_s^F T = 2l\pi - L\omega_c T \quad \text{for} \quad l = \lceil L\omega_s T/2\pi \rceil. \tag{6.28}$$

Once the band edges of the prototype filter $F(z)$ are determined, for a given L, the band edges of the masking filters $G_1(z)$ and $G_2(z)$ may be chosen to keep the desired bands to obtain the overall responses. The band edge values of the masking filters are shown in Figs. 6.13(c) and 6.14(c), respectively, for Case A and Case B.

Overall, Fig. 6.13 shows a Case A example, whereas Fig. 6.14 shows a Case B example.

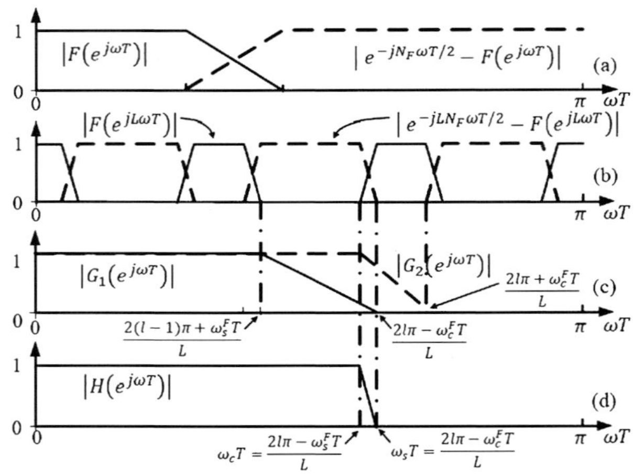

FIGURE 6.14

Frequency response of FRM subfilters of Case B.

6.5.1.3 Conditions for feasible FRM solutions

The conditions derived for Case A and Case B in Sections 6.5.1.1 and 6.5.1.2 assume that a solution is available for a given set of $\omega_c T$, $\omega_s T$, and L. It is possible that neither Case A nor Case B generates a feasible solution for a combination of $\omega_c T$, $\omega_s T$, and L.

First, obviously only when $L > 2\pi/(\omega_c T)$ and $L > 2\pi/(\pi - \omega_s T)$ may the FRM technique be considered; otherwise, the passband of the filter is too narrow or too wide to be obtained by combining bands from both the band edge shaping filter and its complement.

Besides the above condition, from (6.25), (6.27), the fact that $0 < \omega_c^F T < \omega_s^F T < \pi$, and $\omega_c T < \omega_s T$, for a low-pass design, we can further find that L must satisfy

$$\left\lceil \frac{L\omega_c T}{\pi} \right\rceil \neq \left\lfloor \frac{L\omega_s T}{\pi} \right\rfloor, \tag{6.29}$$

so that either Case A or Case B, but not both, leads to a feasible solution. When (6.29) is violated, no feasible solution exists. Such requirements were first presented in [24], where two separate conditions were derived. Eq. (6.29) is a compact expression of those two conditions.

6.5.1.4 Selection of L

Since the transition width of the prototype filter $F(z)$ is $L(\omega_s T - \omega_c T)$ for given $\omega_c T$ and $\omega_s T$ and the sum of the transition width of $G_1(z)$ and $G_2(z)$ is $2\pi/L$, the complexity of $F(z)$ decreases with increasing L, while the complexity of $G_1(z)$ and $G_2(z)$ increases with increasing L. It has been shown that the minimum arithmetic complexity in terms of the number of multipliers is achieved by selecting

$$L = \frac{1}{\sqrt{2\beta(\omega_s T - \omega_c T)/\pi}}, \tag{6.30}$$

where $\beta = 1$ if the subfilters are optimized separately [25] and $\beta = 0.6$ if the subfilters are optimized jointly [26].

6.5.1.5 Computational complexity

The basic FRM structure utilizing the complementary periodic filter pairs and masking filters synthesizes sharp filters with low arithmetic complexity. When

$$\omega_s T - \omega_c T < 0.126\pi \text{ rad},$$

the arithmetic complexity saving is achieved by using the FRM structure, rather than the direct-form FIR filters [27]. In a typical design with $\omega_s T - \omega_c T < 0.002\pi$ rad, the FRM structure reduces the number of multipliers by a factor of 10, and the price paid for the arithmetic saving is a less than 5% increase in the overall filter order [26].

6.5.1.6 Related structure

A special case of FRM filter, named interpolated FIR (IFIR) filter [28], was earlier developed by Neuvo et al. IFIR consists of the upper branch of the FRM structure and can therefore only be used to design narrow-band and wide-band sharp FIR filters.

Some modified FRM structures were proposed in [29,30,27] for further complexity reductions mainly of the masking filters.

6.5.1.7 Design procedure

For a given set of low-pass filter specifications, including passband edge, stopband edge, passband ripple, and stopband ripple, $\omega_c T$, $\omega_s T$, δ_c, and δ_s, respectively, a general design procedure, where all the subfilters are optimized separately, is given as follows:

Step 1. Determine the optimum L that minimizes the filter complexity according to (6.30). If the selected L violates (6.29), increase or decrease L by 1.

Step 2. Determine the passband and stopband edges of the prototype band edge shaping filter according to (6.26) or (6.28).

Step 3. Determine the passband and stopband edges of the masking filters according to the parameters given in Fig. 6.13(c) or Fig. 6.14(c).

Step 4. Design the prototype band edge shaping filter and masking filters with 50% ripple margins.

Step 5. Verify the frequency response of the resulting FRM filter. If the specification has been satisfied, then the design is successful; otherwise, increase the ripple margin and go to **Step 4**.

6.5.1.8 Design example

Use the FRM technique to design an FIR low-pass filter with passband edge, stopband edge, passband ripple, and stopband ripple of 0.398π rad, 0.4π rad, 0.01, and 0.001, respectively.

We get a design of Case A with a prototype filter with $L = 16$. The orders of the three subfilters are 186, 78, and 124, respectively. It is advantageous to make $G_1(z)$ and $G_2(z)$ either both even or both odd in order to use the multiple-constant implementation technique [4]. Here we increase the order of $G_2(z)$ to 124. Note that our selection of filter orders, ripples, and band edges is not necessarily the best. A standard realization of this filter specification using direct form requires a filter order of about 2542. Therefore, the number of nonzero coefficients in the FRM technique is about 17.1% of that required in the direct form. If joint optimization techniques such as that in [26] are used, the value of the optimum L may be selected differently, and the number of nonzero coefficients can be further reduced.

The MATLAB code shown in Listing 6.2 implements the above design procedure.

Listing 6.2: Program for designing arbitrary bandwidth FRM IIR/FIR filters.

```
wcT = 0.398*pi;   wsT = 0.4*pi;
dc = 0.01;   ds = 0.001; d = [dc, ds];
% Conventional direct form
[N,Be,D,W] = HERRMANN_LP_FIR_ORDER([wcT wsT],d); N
% Determine the optimum L to minimize the complexity
L = round(1/sqrt(2*(wsT-wcT)/pi)); % beta = 1
% Determine Case, band edges of the filters
[Case,wcFT,wsFT,G1Edge,G2Edge] = FRM_CASES(wcT,wsT,L); Case
while Case == 'C'
    L = L + 1;
    [Case,wcFT,wsFT,G1Edge,G2Edge] = FRM_CASES(wcT,wsT,L);
end; L
% Prototype filter
if strcmp(Case, 'B')
    d = [ds, dc]/2;  % allow a design margin
else
    d = [dc, ds]/2;
end;
[N_F,Be,D,W] = HERRMANN_LP_FIR_ORDER([wcFT wsFT],d);
N_F = N_F + mod(N_F,2) + 2  % make the order even;
                            % add 2 to meet spec
[f,Err] = REMEZ_FIR(N_F,Be,D,W,'m');
% Masking filters
[N_G1,Be,D,W] = HERRMANN_LP_FIR_ORDER([G1Edge(2) ...
    G1Edge(3)],d);
N_G1 = N_G1 + mod(N_G1,2)
```

```
[g1,ErrG1] = REMEZ_FIR(N_G1,Be,D,W,'m'); ErrG1
[N_G2,Be,D,W] = HERRMANN_LP_FIR_ORDER([G2Edge(2) ...
    G2Edge(3)],d);
N_G2 = N_G2 + mod(N_G2,2)
[g2,ErrG2] = REMEZ_FIR(N_G2,Be,D,W,'m'); ErrG2
% Equal group delays
g1 = [zeros(1,(N_G2-N_G1)/2) g1 zeros(1,(N_G2-N_G1)/2)];
% Verify the frequency response of the FRM fitler
TN = N_F*L*16;
LwT = [linspace(0,L*pi,TN)]; wT = [linspace(0,pi,TN)];
FL = freqz(f,1,LwT);
G1 = freqz(g1,1,wT);  G2 = freqz(g2,1,wT);
figure(1); subplot(2,1,1)
PLOT_MAG_Z_dB(wT,abs(FL),pi,90,' '), hold on
PLOT_MAG_Z_dB(wT,abs(G1),pi,90,'F^L, G_1'), hold off
subplot(2,1,2)
PLOT_MAG_Z_dB(wT,1-abs(FL),pi,90,' '), hold on
PLOT_MAG_Z_dB(wT,G2,pi,90,'F_c^L, G_2'), hold off
figure(2), subplot(2, 1,1)
PLOT_MAG_Z_dB(wT,abs(FL).*abs(G1),pi,90,'F^L G_1')
subplot(2,1,2)
PLOT_MAG_Z_dB(wT,(1-abs(FL)).*abs(G2),pi,90,'F_c^L G_2')
figure(3), subplot(2,1,1)
PLOT_MAG_Z_dB(wT,abs(FL).*abs(G1)+(1-abs(FL)) ...
    .*abs(G2),pi,90,'H')
```

In the above code, the function FRM_CASES is a rewritten form of the function FRM_CASE available in Toolbox [5], to make the parameters of the function consistent with the notations used in this section. The function FRM_CASES is shown in Listing 6.3.

Listing 6.3: Program for the function FRM_CASES.

```
function [Case, wcFT, wsFT, G1Edge, G2Edge] =...
        FRM_CASES(wcT,wsT, L)
% Determines the Case A or B for FRM filters.
% Determines the band edges for the prototype F(z).
% Determines the band edges for the masking filters.
% Determine the band edges of the prototype F(z)
l = floor(wcT*L/(2*pi));     l2pi = l*2*pi;
wcFT = L*wcT-l2pi;     wsFT = L*wsT-l2pi;
if (wcFT<wsFT) & (0<wcFT) & (wsFT<pi & l>0)
        Case = 'A';
else
        l = ceil(L*wsT/(2*pi)); l2pi = l*2*pi;
        wcFT = l2pi-L*wsT; wsFT = l2pi-L*wcT;
```

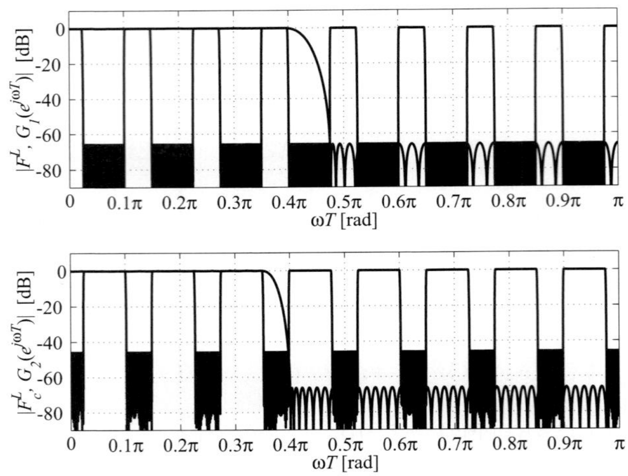

FIGURE 6.15

Magnitude response of FRM subfilters of Example 6.5.1.8.

```
        if (wcFT<wsFT) & (0<wcFT) & (wsFT<pi)
                Case = 'B';
        else
                Case = 'C';
        end;
end
% Determine the band edges of the masking filters
if (Case == 'A') % wcT_G1 = wcT and wsT_G2 = wsT
        wcT_G1=(12pi+wcFT)/L; wsT_G1=(12pi+2*pi-wsFT)/L;
        wcT_G2=(12pi-wcFT)/L; wsT_G2=(12pi+wsFT)/L;
else
        (Case == 'B') % wcT_G2 =  wcT and wsT_G1 = wsT
        wcT_G1=(12pi -2*pi+wsFT)/L; wsT_G1=(12pi-wcFT)/L;
        wcT_G2=(12pi-wsFT)/L;       wsT_G2=(12pi+wcFT)/L;
end;
if (Case ~= 'C')
        G1Edge = [0 wcT_G1 wsT_G1 pi];
        G2Edge = [0 wcT_G2 wsT_G2 pi];
else
        error('No feasible solution for the selected L');
end;
```

Fig. 6.15 shows the magnitude responses of the obtained band edge shaping filters and masking filters. The upper plot shows the magnitude responses of the filter $F(e^{jL\omega})$ and $G_1(e^{j\omega})$, while the lower plot shows that of $1 - F(e^{jL\omega})$ and $G_2(e^{j\omega})$. These two plots show that 4.5 and 5 passbands of

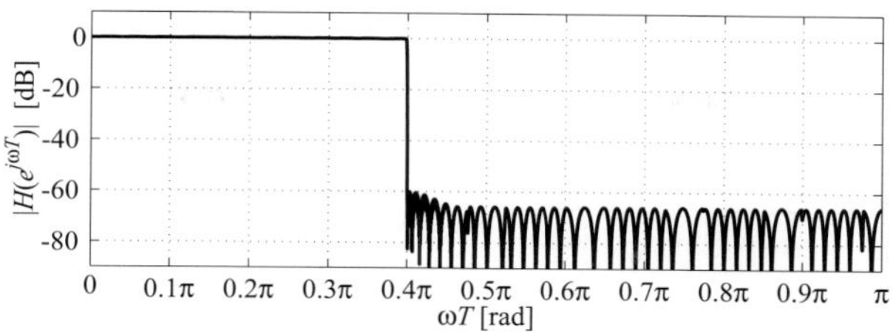

FIGURE 6.16

Magnitude response of the overall filter of Example 6.5.1.8.

the periodic filter pairs are attenuated, respectively. Finally, the remaining passbands are added to form the overall magnitude response of the FRM filter, which is shown in Fig. 6.16.

6.5.2 Multistage FRM structure

In the above basic FRM structure, if the order of $F(z)$ is too high due to a sharp transition width of $F(z)$, the FRM technique can be used recursively to form a multistage FRM structure to reduce the complexity.

In a K-stage FRM structure, the z-transform transfer function of the overall system $H(z)$ is recursively constructed as

$$H(z) = F^0(z),$$
$$F^{k-1}(z) = F^k(z^{L_k})G_1^k(z) + \left[z^{-\frac{L_k N_F^k}{2}} - F^k(z^{L_k})\right] G_2^k(z),$$

for $k = 1, 2, \ldots, K$, where N_F^k is the filter order of the kth stage prototype band edge shaping filter $F^k(z)$ and $G_1^k(z)$ and $G_2^k(z)$ are the kth stage masking filters, with orders of N_1^k and N_2^k, respectively. An example of a three-stage FRM structure is shown in Fig. 6.17.

It was discovered theoretically in [25] that the optimum number of stages to achieve the minimum arithmetic complexity is independent of the band ripple requirements, but solely determined by the transition width. The optimum value of the number of stage K satisfies the constraint

$$\beta_b(K) \le \omega_s T - \omega_c T \le \beta_b(K-1), \tag{6.31}$$

where $\beta_b(K) \cong \frac{1}{4}\left(\frac{K+1}{K+2}\right)^{(K+1)(K+2)}$. When a filter is constructed in FRM structure with the optimum number of stages K_{opt} satisfying the constraint in (6.31), the compression factors of L_k for $k = 1, 2, \ldots, K$ tend to be equal, and are given by

$$\left(\frac{K_{opt}+1}{K_{opt}}\right)^{K_{opt}} \le L_k \le \left(\frac{K_{opt}+2}{K_{opt}+1}\right)^{K_{opt}+2}. \tag{6.32}$$

It is interesting to note that both the lower bound and upper bound in (6.32) approach e ($\cong 2.7$, the base of the natural logarithm) when K is large.

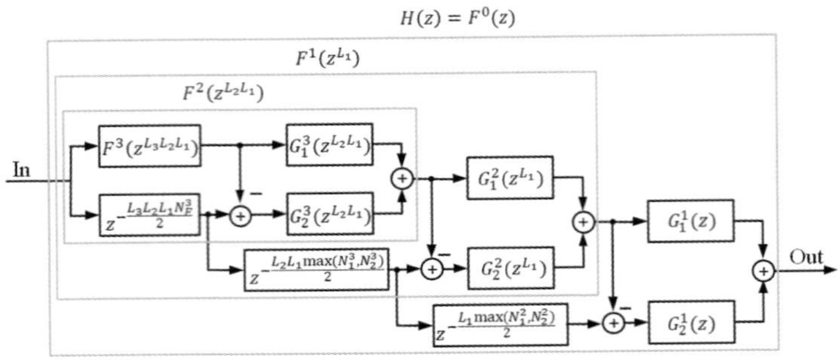

FIGURE 6.17

A three-stage FRM structure.

Practical design experience showed that for a given filter transition width and a given stage K, the minimum arithmetic complexity is achieved when [26]

$$L_k = \left\lceil \frac{1}{\sqrt[(K+1)]{2\beta(\omega_s T - \omega_c T)/\pi}} \right\rceil, \tag{6.33}$$

for $k = 1, 2, \ldots, K$, where $\beta = 1$ if the subfilters are optimized separately [25] and $\beta = 0.6$ if the subfilters are optimized jointly [26].

Considering both the theoretic analysis in [25] and the practical design experience in [26], the overall arithmetic complexity of a given filter specification is in proximity to the minimum if K in (6.33) is chosen such that L_k is 2 or 3.

6.5.2.1 Exercise

Use a two-stage FRM technique to design the FIR low-pass filter with specifications given in Example 6.5.1.8. Compare the numbers of nonzero coefficients required and the overall delays of the filters designed by the two techniques.

6.5.3 Half-band filter

When synthesizing half-band filters using FRM, one has to ensure that the alternate coefficient values are trivial except that the center coefficient has a value of 0.5 [31].

In order to simplify notations, in this section, as well as in Sections 6.5.4 and 6.5.5.2, zero-phase filters are used during derivations. The causality of the filter can be easily restored by delaying the impulse response of the zero-phase filters by appropriate numbers of delays.

6.5.3.1 Structure

Consider a zero-phase half-band prototype band edge shaping filter of order $N_F = 4N - 2$ with passband edge $\omega_c^F T$. Its transfer function $F(z)$ is given by

$$F(z) = \frac{1}{2} + F_o(z), \tag{6.34}$$

where $F_o(z)$ consists of only odd-power terms of z. A half-band FRM filter $H(z)$ can be constructed from this prototype filter by

$$H(z) = \left[\frac{1}{2} + F_o(z^L)\right] G_1(z) + \left[\frac{1}{2} - F_o(z^L)\right] G_2(z) \tag{6.35}$$

for odd L, where $G_1(z)$ and $G_2(z)$ are even-order low-pass masking filters. The parities of L and filter orders are chosen to ensure the resultant overall filter $H(z)$ is a half-band filter.

For FRM half-band filters, there are also two cases where L is an integer of $4i + 1$ or $4i - 1$ for positive i, denoted as Case A and Case B, respectively.

In Case A, a transition band of $\frac{1}{2} + F_o(z^L)$ is centered at $\frac{\pi}{2}$, and in Case B, that of $\frac{1}{2} - F_o(z^L)$ is centered at $\frac{\pi}{2}$ as desired. In either case, for given passband and stopband edges $\omega_c T$, $\omega_s T$, and given L, we have

$$\omega_c T = \frac{(L-1)\pi + 2\omega_c^F T}{2L} = \frac{(L+1)\pi - 2\omega_s^F T}{2L}. \tag{6.36}$$

Thus,

$$\omega_c^F T = \frac{2L\omega_c T - (L-1)\pi}{2} \quad \text{and} \quad \omega_s^F T = \frac{(L+1)\pi - 2L\omega_c T}{2}. \tag{6.37}$$

The derived $\omega_c^F T$ and $\omega_s^F T$ of the prototype half-band filter are consistent with that of the original FRM structure given in (6.26), where $l = \left\lfloor \frac{L\omega_c T}{2\pi} \right\rfloor = \frac{L-1}{4}$ for Case A, and that given in (6.28), where $l = \left\lceil \frac{L\omega_s T}{2\pi} \right\rceil = \frac{L+1}{4}$ for Case B.

Fig. 6.18 shows the frequency response of $F(z^L)$, $G_1(z)$, and $G_2(z)$ of Case A. The frequency response plot for Case B is similar to that of Fig. 6.18 except that the frequency responses of $F(z^L)$, $G_1(z)$, and $G_2(z)$ are replaced by those of $1 - F(z^L)$, $G_2(z)$, and $G_1(z)$, respectively.

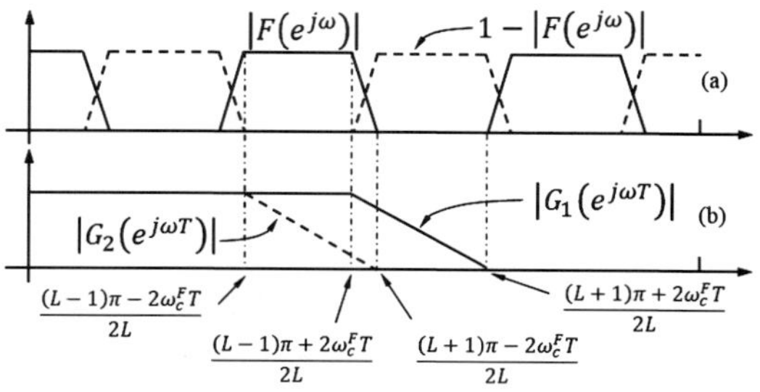

FIGURE 6.18

Frequency responses of $F(z^L)$, $1 - F(z^L)$, $G_1(z)$, and $G_2(z)$ for a Case A half-band filter.

From the band edges of masking filters indicated in Fig. 6.18, it seems that $G_1(z)$ may be related with $G_2(z)$ by $G_2(z) = 1 - G_1(-z)$ if the ripple requirements of the two filters are set the same. Thus,

(6.35) may be written as

$$H(z) = \left[\frac{1}{2} + F_o(z^L)\right] G_1(z) + \left[\frac{1}{2} - F_o(z^L)\right] [1 - G_1(-z)]. \tag{6.38}$$

Let $G_o(z)$ consist of the odd-power terms of z and let $G_e(z)$ consist of the even-power terms of z of the masking filter $G_1(z)$. Then $G_1(z) = G_o(z) + G_e(z)$, and we have

$$H(z) = \frac{1}{2} + G_o(z) + F_o(z^L)[2G_e(z) - 1]. \tag{6.39}$$

Thus, $H(z)$ consists of a constant term $\frac{1}{2}$ and terms with odd power of z, being a half-band filter. The filter structure is shown in Fig. 6.19.

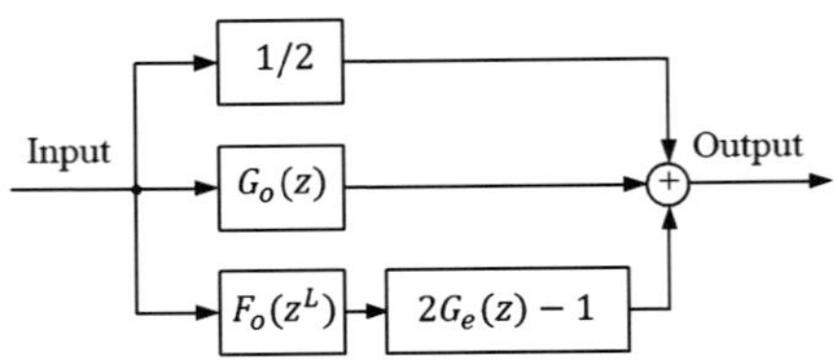

FIGURE 6.19

Structure for synthesizing a half-band filter.

6.5.3.2 Design procedure

In FRM half-band filter design, no closed-form formula to estimate the optimal compression factor L is available. A straightforward way to find an optimal L value is to traverse all odd integer L values within a certain range, estimate the filter length for each subfilter, and select the L value that results in the minimum number of multipliers required in the implementation. The general design procedure is given as follows:

Step 1. Traverse odd integer L values from 3 to a reasonably large number. For each L, repeat **Step 1.1** to **Step 1.3**.
- **Step 1.1** Determine the passband and stopband edges of the prototype band edge shaping filter according to (6.37).
- **Step 1.2** Determine the passband and stopband edges of the masking filters according to the parameters given in Fig. 6.18.
- **Step 1.3** Estimate the filter orders of $F(z)$ and $G_1(z)$.

Step 2. Find the value of L that achieves the minimum arithmetic complexity in **Step 1**.

Step 3. Design the prototype band edge shaping filter $F(z)$ and masking filters $G_1(z)$ with 50% ripple margins.

Step 4. Decompose $F(z)$ to $F_o(z)$ and $G_1(z)$ to $G_o(z)$ and $G_e(z)$.

Step 5. Verify the frequency response of the resulting FRM filter. If the specification has been satisfied, then the design is successful; otherwise, increase the ripple margin and go to **Step 3**.

6.5.3.3 Design example

Use the FRM technique to design a linear-phase half-band FIR filter with transition width 0.002π rad and peak ripple 0.001.

According to the design procedure, we get a design of $L = 15$ that minimizes the number of multipliers required. The filter orders of $F(z)$ and $G_1(z)$ are 258 and 110, respectively. A standard realization of this filter specification using direct form requires a filter order of about 3430. Therefore, the number of nonzero coefficients in the FRM technique is about 14.2% of that required in the direct form.

The MATLAB code shown in Listing 6.4 implements the above design procedure.

Listing 6.4: Program for the design of low-pass FRM half-band filters.

```
% use more stringent band edge to ensure
% the ripple at the edge
wcT = (0.5-0.0019/2)*pi;    wsT =(0.5+0.0019/2)*pi;
dc = 0.001;  ds = 0.001; d = [dc, ds]
% Conventional direct form
[N,Be,D,W] = HERRMANN_LP_FIR_ORDER([wcT wsT], d); N
% Determine the optimum L to minimize the complexity
d = d/2; % allow a design margin
BestTL = N;  BestL = 1;
for L = 3:2:39
    % Determine prototype filter and masking filter edges
    [Case,wcFT,wsFT,G1Edge,G2Edge] = FRM_CASES(wcT,wsT,L);
    % Estimate total number of multipliers
    [N_F,Be,D,W] = HERRMANN_LP_FIR_ORDER([wcFT wsFT],d);
    N_F = N_F-mod(N_F,4)+2; % make the order 4i-2
    [N_G1,Be,D,W] = HERRMANN_LP_FIR_ORDER([G1Edge(2) ...
        G1Edge(3)], d);
    if(ceil(N_F/4)+ceil(N_G1/2)<BestTL)
        BestTL = ceil(N_F/4)+ceil(N_G1/2)+2; BestL = L;
    end;
end;  BestL
[Case,wcFT,wsFT,G1Edge,G2Edge] = FRM_CASES(wcT,wsT,BestL);
% Prototype filter
[N_F,Be,D,W] = HERRMANN_LP_FIR_ORDER([wcFT wsFT],d);
N_F = N_F-mod(N_F,4)+2 % make the order 4i-2;
[f,Err] = REMEZ_FIR(N_F,Be,D,W,'m');
fo = [f(1:(end-1)/2) 0 f((end+3)/2:end)];
% Masking filter
[N_G1,Be,D,W] = HERRMANN_LP_FIR_ORDER([G1Edge(2) ...
    G1Edge(3)],d);
N_G1 = N_G1+mod(N_G1,2) % make the order even;
[g1,ErrG1] = REMEZ_FIR(N_G1,Be,D,W,'m'); ErrG1
% find Go(z), Ge(z) and G(-z)
```

```
go = g1; ge = g1; glm = g1;
go((end+1)/2:2:end) = 0; go((end+1)/2:-2:1) = 0;
ge((end+3)/2:2:end) = 0; ge((end+3)/2:-2:1) = 0;
glm((end+3)/2:2:end) = -glm((end+3)/2:2:end);
glm((end-1)/2:-2:1) = -glm((end-1)/2:-2:1);
% Verify the magnitude response
TN = N_F*BestL*16;
wT = [linspace(0, pi, TN)];
FoL = FREQZ_SYMMETRY(fo, BestL, wT);
FL = FREQZ_SYMMETRY(f, BestL, wT);
Go = FREQZ_SYMMETRY(go, 1, wT);
Ge = FREQZ_SYMMETRY(ge, 1, wT);
G1 = FREQZ_SYMMETRY(g1, 1, wT);
Glm = FREQZ_SYMMETRY(glm, 1, wT);
% Plot according to 1/2+Glo(z)+FoL(z^L).*(2*Gle(z)-1)
figure(2), subplot(2,1,1)
PLOT_MAG_Z_LS(wT,FoL,-1.1,0.6,' '); hold on;
PLOT_MAG_Z_LS(wT,2*Ge-1,-1.1,1.1,'FoL,2*Ge-1'); hold off;
subplot(2,1,2)
PLOT_MAG_Z_LS(wT,FoL.*(2*Ge-1),-.6,.6,'FoL.*(2*Gle-1)');
figure(3)
subplot(2,1,1)
PLOT_MAG_Z_LS(wT,Go,-0.6,0.6,'Go');
subplot(2,1,2)
PLOT_MAG_Z_LS(wT,1/2+Go+FoL.*(2*Ge-1),-0.1,1.1,'H');
```

In the above code, the function FREQZ_SYMMETRY, instead of function freqz as in Listing 6.2, is used to compute the zero-phase response of symmetric FIR filters. The program of the function is given in Listing 6.5.

Listing 6.5: Program for the function FREQZ_SYMMETRY.

```
function freq = FREQZ_SYMMETRY(h, L, w)
% h: impulse response of the symmetric FIR filter
% L: each delay is replaced by L delays
% w: frequency grid points
N = length(h);
tol = 10^(-5);
% Determine the symmetric type
if rem(N,2)==1
    if (abs(real(h(1:(N-1)/2))-real(h(N:-1:(N+3)/2)))<tol ...
            & abs(imag(h(1:(N-1)/2))-imag(h(N:-1:(N+3)/2)))<tol)
          type = 1;
    else if (abs(real(h(1:(N-1)/2))+real(h(N:-1:(N+3)/2)))<tol ...
                & abs(imag(h(1:(N-1)/2))+imag(h(N:-1:(N+3)/2)))<tol)
```

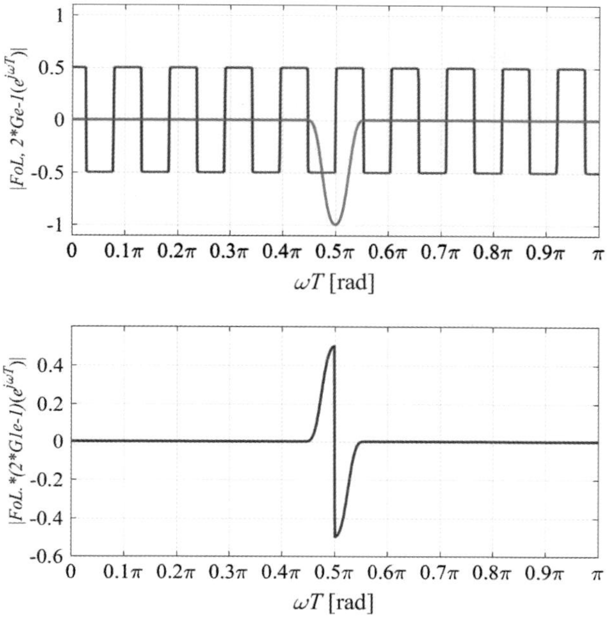

FIGURE 6.20

Zero-phase responses of $F_o(z^L)$ and $2G_e(z) - 1$ (top) and their product (bottom) for Example 6.5.3.3.

```
            type = 3;
        else error('Not symmetric filter');
        end;
    end;  hn = (N+1)/2;
else
    if (abs(real(h(1:N/2))-real(h(N:-1:N/2+1)))<tol ...
      & abs(imag(h(1:N/2))-imag(h(N:-1:N/2+1)))<tol)
        type = 2;
    else if (abs(real(h(1:N/2))+real(h(N:-1:N/2+1)))<tol ...
            & abs(imag(h(1:N/2))+imag(h(N:-1:N/2+1)))<tol)
             type = 4;
        else error('Not symmetric filter');
             end;
    end;  hn = N/2;
end;
i=hn:-1:1;
switch type
case 1, exp = cos(L*(hn-i')*w);
        a = [h(hn) 2.*h(hn-1:-1:1)];
```

```
case 2, exp = cos(L*(hn-i'+0.5)*w);
        a = [2.*h(hn:-1:1)];
case 3, exp = 1i*sin(L*(hn-i')*w);
        a = [h(hn) 2.*h(hn-1:-1:1)];
case 4, exp = 1i*sin(L*(hn-i'+0.5)*w);
        a = [2.*h(hn:-1:1)];
end;
freq = a*exp;
```

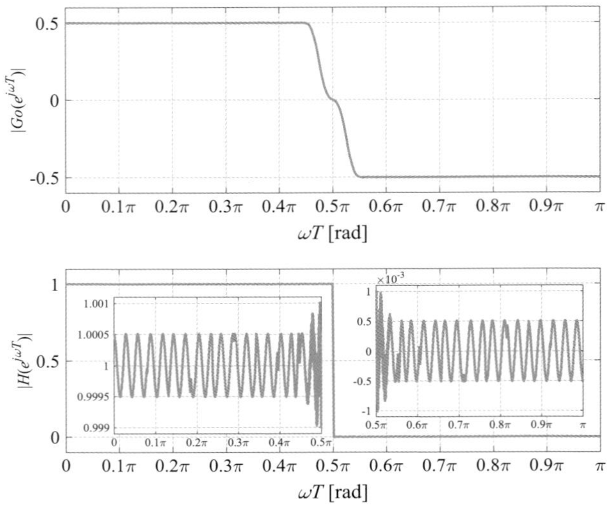

FIGURE 6.21

Zero-phase responses of $G_o(z)$ (top) and the overall filter $H(z)$ (bottom) for Example 6.5.3.3.

The zero-phase responses of the subfilters in the structure shown in Fig. 6.19 and that of the overall filter are shown in Figs. 6.20 and 6.21.

In Fig. 6.20, the top subplot shows the zero-phase responses of $F_o(z^L)$ and $2G_e(z) - 1$, and the bottom subplot shows that of their product. The top subplot of Fig. 6.21 is the zero-phase response of $G_e(z)$, and the bottom one is that of the overall filter.

It is interesting to note that the response of $F_o(z^L)[2G_e(z) - 1]$ forms a bulge and a plunge around the band edges of the overall filter. The bulge adds to the response of $G_o(z)$ in the passband, boosting it to unit, and the plunge cancels the response of $G_o(z)$ in the stopband, nullifying it to zero. In such an exquisite way, the final response is achieved. That is also the reason that zero-phase responses have to be computed in the program to realize the boosting and nullifying.

Zoom-in views of the passband ripple and stopband ripple of the overall filter are also shown in Fig. 6.21. Note that for the conciseness of the program in Listing 6.4, the code to show the zoom-in views is not listed.

The zoom-in views show that the designed filter meets the given specification. However, it is also noted that the ripples in the frequency range closer to the transition band are larger than the ripples in the other frequency ranges. This phenomenon is common for all FRM structures if subfilters are optimized separately without special processes. To achieve equiripple filters, which may reduce the overall arithmetic complexity, higher weights may be imposed to the frequency ranges closer to the transition band during the optimization of the subfilters.

6.5.4 Hilbert transformer

A Hilbert transformer may be derived from a unity half-band filter by subtracting the constant 1/2 from its transfer function, modulating the remaining coefficients by $e^{jn\pi/2}$, and then scale the response by 2. Thus, the transfer function of the Hilbert transformer $H_H(z)$ can be written from (6.39) to [32]

$$H_H(z) = 2G_o(jz) + 2F_o(j^L z^L)[2G_e(jz) - 1]. \tag{6.40}$$

The structure of the Hilbert transformer is shown in Fig. 6.22.

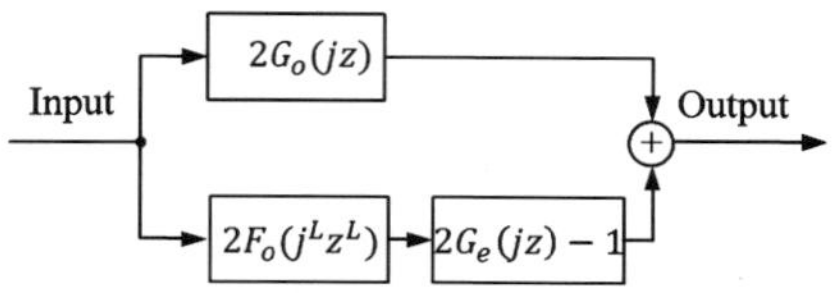

FIGURE 6.22

Structure for synthesizing a Hilbert transformer.

6.5.4.1 Exercise

Use the subfilters designed in Example 6.5.3.3 to construct a Hilbert transformer. Plot the zero-phase response of each subfilter in the structure shown in Fig. 6.22, as well as that of the overall filter. Without checking the plots, what are the expected transition bandwidth and passband ripple of the resultant Hilbert transformer? Verify your answer by the plots.

6.5.5 Decimator and interpolator

In classical DSP systems, all algorithms operate with the same sampling frequency. In multirate systems, the sampling rate is changed during different stages of the signal processing. These changes of the sampling frequency are done by decimators and interpolators, which often are realized using a polyphase structure. Multirate techniques are today used in many DSP applications, such as communications, image processing, digital audio, and multimedia. In many cases, multiple sampling frequencies are used to simplify and reduce the computational complexity and to achieve performances that are difficult by traditional approaches. We recommend Refs. [14,4] for further study of the design and implementation of decimators and interpolators.

6.5.5.1 Difficulties of FRM multirate filters

To synthesize a low-pass filter with a bandwidth of approximately π/M as required in the interpolator and decimator, there are two cases: (1) the transition band does not include π/M; and (2) the transition band includes π/M.

In a polyphase implementation of interpolator and decimator with a sampling rate change factor of M, a "factor of M" reduction in computational complexity may be achieved in general. However, if the filter is realized as a cascade of two subfilters (as in the upper branch or lower branch of the FRM structure), generally only one of the two filters may achieve the "factor of M" reduction in computational complexity.

A straightforward way to circumvent the above difficulty is to make the output of the first filter in the cascade nonzero only at intervals of M. This can be achieved by selecting L as an integer multiple of M in the original FRM structure shown in Fig. 6.12.

In such a strategy, the original FRM technique may be directly used for the first case, where the transition band does not include π/M to achieve the "factor of M" reduction in computational complexity.

However, in the second case, where the transition band includes π/M, since $\omega_c T < \pi/M < \omega_s T$, we have

$$\frac{L\omega_c T}{\pi} < \frac{L\pi/M}{\pi} < \frac{L\omega_s T}{\pi}, \tag{6.41}$$

where $\frac{L\pi/M}{\pi}$ is an integer for L being an integer multiple of M, leading to $\left\lceil \frac{L\omega_c T}{\pi} \right\rceil = \left\lfloor \frac{L\omega_s T}{\pi} \right\rfloor$. This violates the constraints on the selection of the compression factor L for the original FRM structure given in (6.29). Therefore, L cannot be selected as an integer multiple of M to make the output of the first filter nonzero only at intervals of M. Section 6.5.5.2 describes a variant of the FRM structure to design an interpolator or decimator with transition band including π/M [24].

Moreover, the filter designed in Section 6.5.5.2 is not an Mth band filter. An Mth band filter has impulse response values being zero for $n = rM$, $n \neq 0$ for all integers r. Two more variants of FRM structure are described in Section 6.5.5.5 to design an Mth band filter. In this technique, L does not have to be an integer multiple of M, and the transition band of the filter designed can either include or not include π/M. When the transition band includes π/M, the filter designed may be an Mth band filter if the prototype band edge shaping filter is an Mth band filter.

6.5.5.2 A variant of the FRM structure

A variant of the FRM structure proposed in [24] achieves the "factor of M reduction" in a multirate filter with sampling rate change factor M when the transition band is centered at π/M. The following is an illustration for an interpolator with an upsampling rate M. The same technique applies to decimators as well. When considering the zero-phase response, the variant of the FRM is constructed from

$$H(z) = \frac{1}{2}G_s(z) + F_o(z^L)G_d(z), \tag{6.42}$$

where

$$G_s(z) = G_1(z) + G_2(z), \tag{6.43}$$

$$G_d(z) = G_1(z) - G_2(z). \tag{6.44}$$

In (6.42)–(6.44), $F_o(z)$ is given in the condition of (6.34) for a half-band prototype band edge shaping filter $F(z)$; $G_1(z)$ and $G_2(z)$ are the masking filters as that in Fig. 6.12. The factor M interpolation

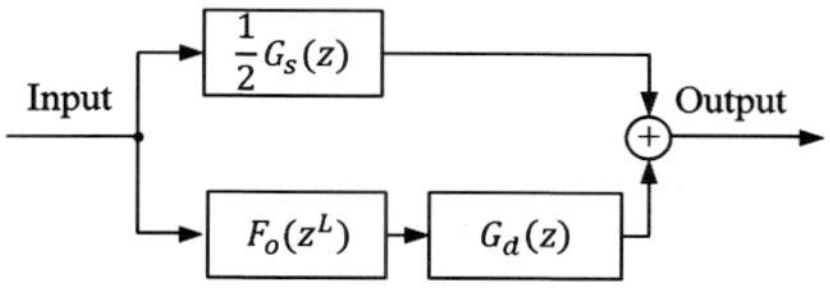

FIGURE 6.23

FRM structure for the synthesis of decimation and interpolation filters.

or decimation filter constructed in such a manner has a transition band centered at π/M. The filter structure is shown in Fig. 6.23.

Selection of L: To ensure that the output of $F_o(z^L)$ has nonzero values only at intervals of M, for M even, $2L$ has to be an odd multiple of M, while for M odd, $2L$ has to be an even multiple of M.

Even M case: For the case of M even, the band edge selection of the three subfilters $F(z)$, $G_1(z)$, and $G_2(z)$ is the same as that of the basic FRM structure described in Section 6.5.1.

Odd M case: However, with the selection of L for M odd, the condition $\lfloor L\omega_c T/\pi \rfloor \neq \lceil L\omega_s T/\pi \rceil$ is violated. This difficulty can be alleviated as follows: $F(z)$ is replaced by $F(ze^{-j\frac{\pi}{2L}})$, and $G_1(z)$ and $G_2(z)$ are frequency response shifted versions of a prototype masking filter $G(z)$. The frequency responses of $F(z^L)$, $F((ze^{-j\frac{\pi}{2L}})^L)$, $G(z)$, $G_1(z)$, and $G_2(z)$ are shown in Fig. 6.24.

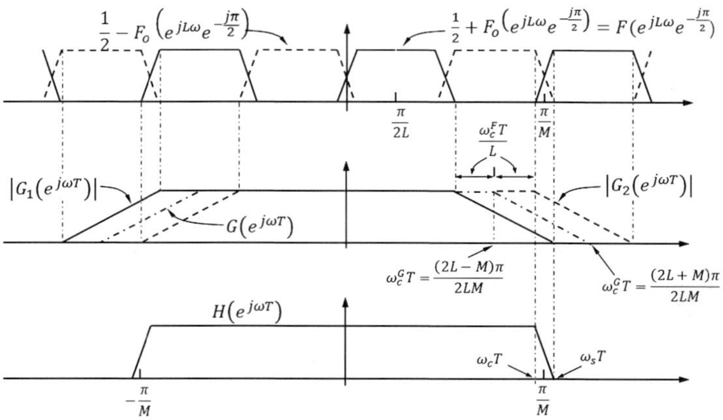

FIGURE 6.24

The magnitude responses of $F((ze^{-j\frac{\pi}{2L}})^L)$, $1 - F((ze^{-j\frac{\pi}{2L}})^L)$, $G(z)$, $G_1(z)$, $G_2(z)$, and the overall filter $H(z)$.

As $F(z)$ is a half-band filter, its band edges obviously are

$$\omega_c^F T = \frac{\pi}{2} - \frac{\omega_s T - \omega_c T}{2}L \quad \text{and} \quad \omega_s^F T = \frac{\pi}{2} + \frac{\omega_s T - \omega_c T}{2}L. \tag{6.45}$$

The band edges of the prototype masking filter $G(z)$ are

$$\omega_c^G T = \frac{(2L - M)\pi}{2LM} \quad \text{and} \quad \omega_s^G T = \frac{(2L + M)\pi}{2LM}. \tag{6.46}$$

Masking filters $G_1(z)$ and $G_2(z)$ are obtained from $G(z)$ by shifting its frequency response by $\frac{\omega_c^F T}{L}$ to the left and right, respectively, where $\omega_c^F T$ is the passband edge of $F(z)$. The frequency responses of $G_1(z)$ and $G_2(z)$ thus are complex conjugate symmetrical to each other.

Therefore, we have

$$G_s(z) = G_1(z) + G_2(z) = 2\mathscr{R}\{G(z)\} \tag{6.47}$$

and

$$G_d(z) = G_1(z) - G_2(z) = 2\mathscr{I}\{G(z)\}. \tag{6.48}$$

Polyphase implementation: In implementation, both $G_s(z)$ and $F_o(z^L)$ can be efficiently decomposed into M polyphase components using polyphase structures since the inputs are nonzero only at intervals of M. For $G_d(z)$, besides the fact that the output of $F_o(z^L)$ is nonzero at intervals of M samples, only the first polyphase component of $F_o(z^L)$ has nonzero coefficients, because it is a function of z^{-2L} and $2L$ is an integer multiple of M. Therefore, $G_d(z)$ may be efficiently decomposed into M polyphase components as well.

Optimum L: When the coefficient symmetry and zero values of half-band filters are considered, the optimum L may be selected as follows.

For M even, we have

$$L = [L_e + \text{mod}(L_e, 2) - 1]\frac{M}{2}, \tag{6.49}$$

where $L_e = \left\lfloor \frac{1}{M\sqrt{2(\omega_s T - \omega_c T)/\pi}} + 0.5 \right\rfloor$ and mod(x, n) stands for x modulo n. For M odd, we have

$$L = [L_o + \text{mod}(L_o, 2)]M, \tag{6.50}$$

where $L_o = \left\lfloor \frac{1}{2M\sqrt{2(\omega_s T - \omega_c T)/\pi}} + 0.5 \right\rfloor$. The modulus operation in (6.49) and (6.50) is to ensure $2L$ is an odd multiple and an even multiple of M, respectively.

6.5.5.3 Design example

Using the FRM technique, design a "factor of 7" interpolation filter with a very narrow transition width of 0.001π rad. The transition band is centered at $\frac{\pi}{7}$ rad; the passband and stopband peak ripple magnitude is 0.001. The estimated order of an optimum direct-form seventh band filter meeting this specification is 6512. Thus the computational rate is 931 ($= 6512/7$) multiplications per output sample when implemented using the conventional polyphase technique.

For the FRM technique presented above, L is chosen to be 14 according to (6.50). The orders of the prototype band edge shaping filter and the prototype masking filter are 522 and 104, respectively. The computational rate is reduced to 68 ($= 38 + 15 + 15$) multiplications per output sample when subfilters are implemented in the polyphase technique, i.e., a reduction factor of more than 13. Note that the

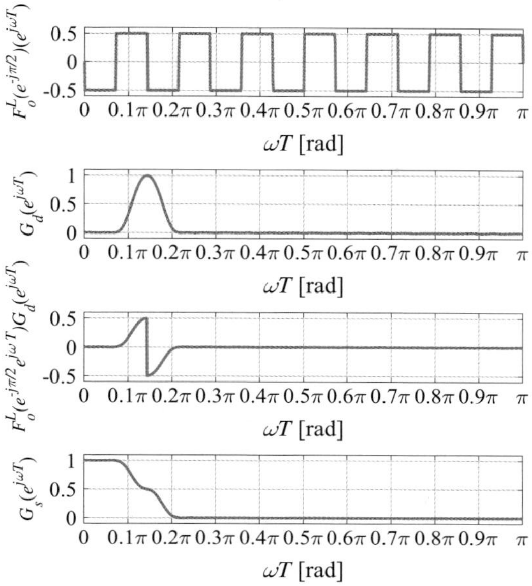

FIGURE 6.25

The zero-phase responses of subfilters of $F_o((ze^{-j\frac{\pi}{2L}})^L)$, $G_d(z)$, $F_o((ze^{-j\frac{\pi}{2L}})^L) \cdot G_d(z)$, and $G_s(z)$, from top to bottom.

coefficient symmetry is not considered when computing the arithmetic complexity for either the direct form or the FRM structure.

The program for the design of interpolation filters with an odd interpolation factor M is shown in Listing 6.6.

Listing 6.6: Program for the design of interpolation filters with an odd interpolation factor M.

```
M = 7; Wt = 0.001*pi; % Transition width
wcT = pi/M-Wt/2; wsT = pi/M+Wt/2;
dc = 0.004; ds = 0.004; d = [dc, ds];
% Conventional direct form
[N,Be,D,W] = HERRMANN_LP_FIR_ORDER([wcT wsT],d); N
% Determine L when coe. symmetry is considered
temp = max(floor(1/(2*M*sqrt(2*M*Wt/(2*pi)))+0.5),1);
L = (temp+mod(temp, 2))*M; L
% Determine the band edges of subfilters
[wcFT,wsFT,GEdge] = FRM_INTERP_ODDL_EDGES(wcT, ...
      wsT,L,M);
d = d/2;
[N_F,Be,D,W] = HERRMANN_LP_FIR_ORDER([wcFT wsFT],d);
```

```
N_F = N_F-mod(N_F,4)+2 % make the order 4i-2;
% Alternative way to design half-band filter
N_F = N_F/2;
[f,Err] = REMEZ_FIR(N_F,[Be(1);Be(2)*2;Be(4); ...
      Be(4)],D/2,W,'m');
fo = UP_SAMPLE(f,2);
foexp = [-(length(fo)-1)/2:(length(fo)-1)/2];
foj = fo.*(-1i).^foexp;
% Masking filter
[N_G,Be,D,W] = HERRMANN_LP_FIR_ORDER([GEdge(2) ...
      GEdge(3)],d);
N_G = N_G+mod(N_G,2) % make the order even;
[g,ErrG] = REMEZ_FIR(N_G,Be,D,W,'m'); ErrG
gexp = [-(length(g)-1)/2:(length(g)-1)/2];
gs = g.*cos(wcFT*gexp/L);
gd = 2*j*g.*sin(wcFT*gexp/L);
% Verify the frequency response
TN = N_F*L*16;
wT = [linspace(0,pi,TN)];
FoL = FREQZ_SYMMETRY(fo,L,wT);
FojL = FREQZ_SYMMETRY(foj,L,wT);
Gs = FREQZ_SYMMETRY(gs,1,wT);
Gd = FREQZ_SYMMETRY(gd,1,wT);
figure(1), subplot(4,1,1)
PLOT_MAG_Z_LS(wT,FojL,-0.6,0.6,'F_o^L(e^{-j\pi/2})');
subplot(4,1,2)
PLOT_MAG_Z_LS(wT, Gd, -0.1, 1.1, 'G_d');
subplot(4,1,3)
PLOT_MAG_Z_LS(wT,FojL.*Gd,-0.6,0.6, ...
      'F_o^L(e^{-j\pi/2}e^{j\omega T})G_d');
subplot(4,1,4)
PLOT_MAG_Z_LS(wT, Gs, -0.1, 1.1, 'G_s');
figure(2)
PLOT_MAG_Z_LS(wT, Gs+FojL.*Gd, -0.1, 1.1, 'H');
```

Note that in the above code, the half-band filter $F(z)$ is designed through a wide-band filter, whose coefficients are the alternate nonzero (and noncenter coefficient) values of the half-band, i.e., by inserting the center coefficient (which is 0.5) and alternate zeros to the wide-band low-pass filter, the required half-band filter is obtained. This is because for certain half-band filter specifications, the REMEZ_FIR function may not converge.

In addition, the function FRM_INTERP_ODDL_EDGES in the above code is to obtain the band edges of subfilters $F(z)$ and $G(z)$. This function can be easily developed according to (6.45) and (6.46), and is omitted here.

Fig. 6.25 shows the zero-phase responses of the subfilters in the FRM interpolation filter structure. From top to bottom, they are the responses of filters of $F_o((ze^{-j\frac{\pi}{2L}})^L)$, $G_d(z)$, $F_o((ze^{-j\frac{\pi}{2L}})^L) \cdot G_d(z)$,

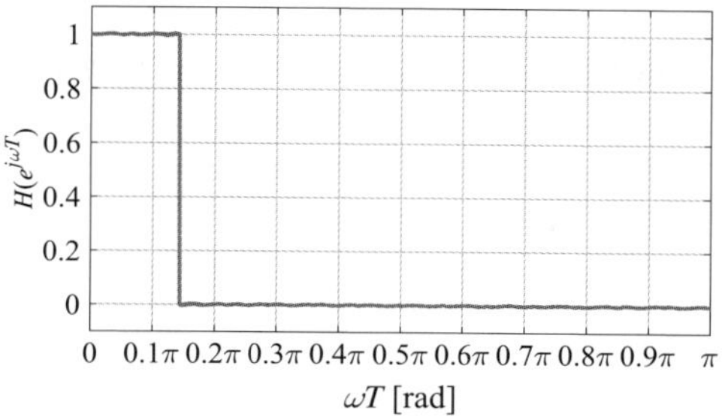

FIGURE 6.26

The zero-phase response of the overall interpolation filter $H(z)$.

and $G_s(z)$. It is noted that $F_o((ze^{-j\frac{\pi}{2L}})^L) \cdot G_d(z)$ has a similar response as that of $F_o(z^L)[2G_e(z) - 1]$ in the FRM half-band structure. Their roles are also similar. The responses of $F_o((ze^{-j\frac{\pi}{2L}})^L) \cdot G_d(z)$ and $G_s(z)$ are further added to achieve the overall response, shown in Fig. 6.26.

6.5.5.4 Exercise

Develop a program to design an interpolation filter with an even interpolation factor $M = 8$ using the FRM technique. The transition width and ripple requirements are the same as those in Example 6.5.5.3. Find the arithmetic saving of the design compared with the direct-form FIR filter.

6.5.5.5 Two more variants of FRM structures

In [33], two alternative FRM structures are proposed for the design of filters for interpolation and decimation by a factor of M. In this approach, additional constraints are imposed on the masking filters in the original FRM structure shown in Fig. 6.12, such that

$$G_2(z) = G_1(z) - \frac{M}{M-1} G_{10}(z^M) + \frac{1}{M-1} \tag{6.51}$$

or

$$G_1(z) = G_2(z) - \frac{M}{M-1} G_{20}(z^M) + \frac{1}{M-1}, \tag{6.52}$$

where $G_{10}(z)$ and $G_{20}(z)$ are the 0th polyphase components in the polyphase representations of $G_1(z)$ and $G_2(z)$, respectively. The relations in (6.51) and (6.52) specify the Class I and Class II FRM interpolator/decimator structures. Both classes of FRM structures can be used to design low-pass filters with a bandwidth of approximately π/M. The transition band may or may not include π/M, depending on the prototype band edge shaping filter. When the prototype band edge shaping filter $F(z)$ is constrained to an Mth band filter, the resultant overall filter $H(z)$ is an Mth band filter as well.

The relations given in (6.51) and (6.52) impose further constraints on the selection of l and L in (6.26) and (6.28). Nevertheless, the available choices are still much more than that available in the variant structure described in Section 6.5.5.2 (where $2L$ must be an even or odd integer multiple of M), and are sufficient to obtain arbitrary frequency selectivity in the filter design and meanwhile achieve complexity reduction.

When the masking filters are related to each other as those shown in (6.51) and (6.52), both the band edge shaping filter and the masking filters can be decomposed into a polyphase structure such that all filters are working at the lower sampling rate involved in the interpolation and decimation. Moreover, all the polyphase components of the masking filters share a common part. Utilizing this fact, more efficient interpolator and decimator structures may be developed.

It is interesting to note that both (6.51) and (6.52) are reduced to $G_2(z) = 1 - G_1(-z)$ when $M = 2$; this relation is the same as that in the half-band FRM structure given in (6.38). In fact, when the transition band is restricted to include π/M, the proposed approach is a generalization of that in Section 6.5.3 and an Mth band filter is obtained.

Note that although the transition band of the interpolation and decimation filters designed in Section 6.5.5.2 is centered at π/M, the resultant filter is not an Mth band filter.

The two variants of FRM structures presented in this section may achieve lower arithmetic complexity than that described in Section 6.5.5.2 when interpolation and decimation filters are designed, attributed to one or more of the following three merits: (1) the filter may be constrained to an Mth band filter; (2) more choices of L are available; and (3) all the polyphase components of the masking filters share a common part.

6.5.5.6 An alternative recursive structure

Besides the above three FRM interpolation/decimation filter structures, a factor of two interpolation and decimation filters are developed in [34], based on a recursive filter structure introduced in Section 6.5.8.

6.5.6 Filter banks

A two-channel maximally decimated filter bank utilizing the FRM structure is proposed in [35]. In the diagram of a two-channel filter bank shown in Fig. 6.27, each analysis and synthesis filter of the proposed technique is realized through a modified FRM structure with the same prototype band edge shaping filter $F(z)$. The z-transform transfer functions of the analysis and synthesis filters are given by

$$H_{ak}(z) = F(z^L)G_{k1}(z) + F_c(z^L)G_{k2}(z),$$
$$H_{sk}(z) = F(z^L)G_{k1}(z) - F_c(z^L)G_{k2}(z),$$

for $k = 0, 1$.

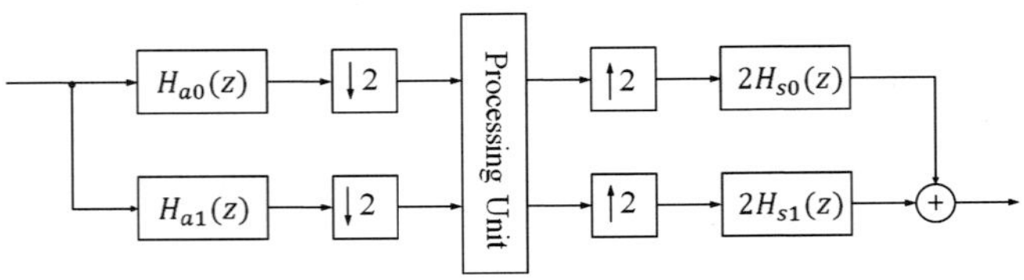

FIGURE 6.27

Two-channel maximally decimated filter bank.

In this FRM structure, the original magnitude complementary filter of the prototype band edge shaping filter, $1 - F(z)$, is modified to an approximately power complementary filter, i.e., $F_c(z)$ is related to $F(z)$ in a form of

$$F_c(z) = F(-z),$$

and the filter order is constrained to odd order. In addition, the compression factor L is selected to be odd and the masking filters are of even order.

While $G_{01}(z)$ and $G_{02}(z)$ are linear-phase low-pass filters, $G_{11}(z)$ and $G_{12}(z)$ are their frequency shifted versions, obtained by

$$\left|G_{11}\left(e^{j\omega T}\right)\right| = \left|G_{02}\left(e^{j(\omega T-\pi)}\right)\right| \quad \text{and} \quad \left|G_{12}\left(e^{j\omega T}\right)\right| = \left|G_{01}\left(e^{j(\omega T-\pi)}\right)\right|, \tag{6.53}$$

being linear-phase high-pass filters.

To satisfy the relations in (6.53), there are two ways to relate the masking filters. In the first method, we select

$$G_{11}(z) = -G_{02}(-z) \quad \text{and} \quad G_{12}(z) = G_{01}(-z),$$

and the filter bank thus constructed is aliasing-free. In the second method, we select

$$G_{11}(z) = G_{02}(-z) \quad \text{and} \quad G_{12}(z) = G_{01}(-z).$$

In such a way, the filter bank constructed suffers some aliasing errors, but the arithmetic complexity may be reduced to achieve the same distortion errors as with the aliasing-free structure.

This technique is further extended to modulated M-channel filter banks by using the analysis and synthesis filters in the upper branch of the two-channel filter bank as the prototype filters for modulation [36]. Together with a cosine modulation block and a sine modulation block, a near-perfect reconstructed M-channel maximally decimated filter bank is developed.

Other FRM structures for filter banks are mainly developed for two-channel filter banks, including nonuniform filter banks with rational sampling factor [37] and multiplet perfect reconstruction filter banks [38]. In addition, a nonmultirate fast filter bank is developed in [39] to realize the function of sliding fast Fourier transform filter bank with low computational complexity, but achieving high frequency selectivity and passband performance.

6.5.7 Cosine-modulated transmultiplexer

Due to the same constraints on the selection of the compression factor L as that for decimation and interpolation filters, the prototype band edge shaping filter for an M-band TMUX is not realizable if L is an integer multiple of M. Unlike the variant structure described in Section 6.5.5.2, an implementation of an M-band cosine-modulated TMUX (CMT) [40] for

$$L = 2K_aM + \frac{M}{K_b}, \tag{6.54}$$

where K_a is a nonnegative integer and K_b is a positive integer, is proposed.

If only the upper branch in the original FRM structure is used as the prototype filter for the CMT, the transfer function for the analysis filter becomes

$$H_m(z) = \sum_{n=0}^{N} c_{m,n} \left(f^L * g_1 \right)(n) z^{-n}, \tag{6.55}$$

where $c_{m,n} = 2\cos\left[\frac{(2m+1)(n-\frac{N}{2})}{2M} + (-1)^m \frac{\pi}{4}\right]$ for $m = 0, 1, \ldots, M-1$ and $n = 0, 1, \ldots, N-1$ and the term $(f^L * g_1)(n)$ denotes the convolution between the impulse responses of $F(z^L)$ and $G_1(z)$. Thus, $F(z^L)$ can be decomposed into $Q = 2K_b$ polyphase components assuming that $N_F = QK_c - 1$ with K_c being a positive integer, and $G_1(z)$ can be decomposed to $2M$ polyphase components assuming that $N_1 = 2K_d M - 1$ with K_d being a positive integer. Under the relations of L and M in (6.54) and $c_{m,(n+2kM)} = (-1)^k c_{m,n}$, the overall polyphase representation of $H_m(z)$ is given by

$$H_m(z) = \sum_{q=0}^{Q-1} \left[z^{-Lq} E_F^q(-z^{LQ}) \times \sum_{p=0}^{2M-1} c_{m,\left(n+\frac{M}{K_b}q\right)} z^{-p} E_G^p(-z^{2M}) \right],$$

where $E_F^q(z)$ is the qth polyphase component of the prototype band edge shaping filter $F(z)$, given by

$$E_F^q(z) = \sum_{k=0}^{K_c} (-1)^{K_a q} f(kQ+q) z^{-k},$$

and $E_G^p(z)$ is the pth polyphase component of the masking filter $G_1(z)$ given by

$$E_G^p(z) = \sum_{k=0}^{K_d-1} g_1(2kM+p) z^{-k}.$$

If both branches in the FRM design are used, this decomposition can be applied to the lower branch, and by enforcing both masking filters to have the same order $N_1 = N_2$, the responses of the two branches can be added together just before the modulation stage.

6.5.8 FRM structure for recursive filters

While FRM structures were originally and mainly developed for the design of sharp FIR filters to reduce computational complexity, a hybrid IIR and FIR filter structure was proposed for the design of high-speed recursive digital filters [41,4].

In the proposed structures, the prototype band edge shaping filters are power-complementary IIR filters, denoted as $F(z)$ and $F_c(z)$, realized as a parallel connection of two all-pass filters, and the masking filters are linear-phase FIR filters. The prototype band edge shaping filters are expressed as

$$F(z) = \frac{F_1(z) - F_2(z)}{2}$$

and

$$F_c(z) = \frac{F_1(z) + F_2(z)}{2},$$

where $F_1(z)$ and $F_2(z)$ are even- and odd-order stable all-pass filters. The overall filter order is always odd for low-pass filters. When the magnitude responses of the power-complementary IIR filter pair $F(z)$ and $F_c(z)$ are bounded by unity, the overall transfer function is expressed as

$$H(z) = F(z^L)G_2(z) + F_c(z^L)G_1(z),$$

where $G_1(z)$ and $G_2(z)$ are the linear-phase FIR masking filters. The structure of the overall filter is shown in Fig. 6.28.

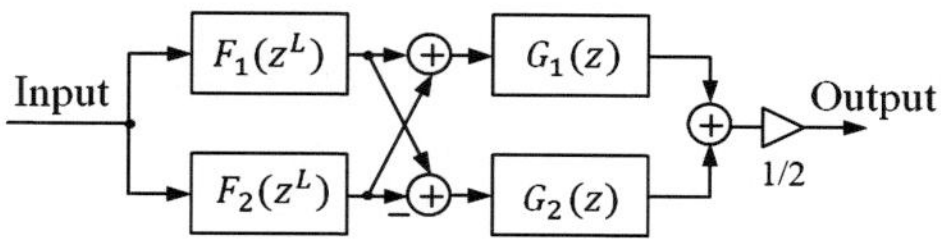

FIGURE 6.28

The overall structure of the recursive filter.

In this structure, the all-pass filters $F_1(z)$ and $F_2(z)$ can be realized in many different ways. The one used here is the wave digital filter, which makes it possible to maintain stability under finite arithmetic conditions. In this case, the IIR filter $F(z)$ corresponds to the well-known lattice wave digital filters. Power-complementary wave digital filters will be further discussed in Section 6.10.

During the optimization of the IIR prototype band edge shaping filter $F(z)$, a prototype Cauer filter is first designed, and the coefficients of the two all-pass filters $F_1(z)$ and $F_2(z)$ are computed from the poles of the Cauer filter.

Since the maximum sample frequency of recursive filters is determined by the ratio of the number of delay elements and the operational latency in the critical loop of the filter realization (see (6.130)), this structure introduced L delay elements in its critical loop, resulting in an L-fold increase of the maximal sampling frequency. The increased maximal sampling frequency can be used to reduce the power consumption using voltage scaling techniques [42].

Also for this structure we get two cases depending on L. For $L = 5, 9, 13, \ldots$, we get Case A and the resulting transition band is determined by $F(z^L)$. For $L = 3, 7, 11, \ldots$, we get Case B and the transition band is determined by $F_c(z^L)$. The complexity of $F(z)$ decreases with increasing L, while the complexity of $G_1(z)$ and $G_2(z)$ increases with increasing L.

6.5.8.1 Design example

Consider a low-pass filter meeting the following specifications: $\omega_c T = 0.494\pi$ rad, $\omega_s T = 0.506\pi$ rad, $A_{max} = 0.2$ dB, $A_{min} = 60$ dB. The transition band is very small.

The band edges are selected according to Section 6.5.1. We use the program shown in Listing 6.7 to design the filters. We select the filters $F_1(z)$ and $F_2(z)$ of orders 4 and 5 and the corresponding periodic band edge shaping lattice filters with compress factor $L = 5$. In this case, we get a Case A design. A corresponding program for Case B and FRM filters using only IIR structure is found in [4].

We select the passband ripple of the prototype filter to a small fraction of A_{max}. In this case, we select A_IIR = 0.01 dB. We get the following orders of the prototype Cauer filter and the masking filters: $N_F = 9$, $N_{G_1} = 25$, $N_{G_2} = 25$. We may use the function REMEZ_FIR to design the FIR filters $G_1(z)$ and $G_2(z)$ as two independent low-pass filters. Here, however, we use an optimization routine FRM_IIR_FIR that is based on a modified version of the Remez exchange algorithm to synthesize the FIR filters. This routine tries to utilize the different requirements in the periodic band edge shaping lattice filters. Note that the optimization may take a few minutes, and if a feasible solution is not found, then the program results in an error.

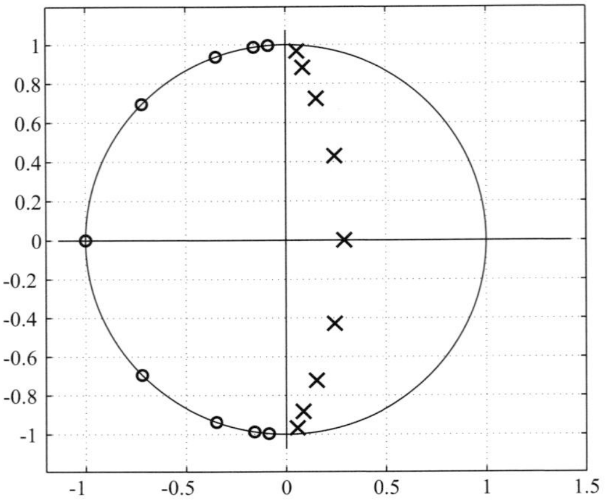

FIGURE 6.29

Poles and zeros for the ninth-order prototype Cauer filter.

Listing 6.7: Program for the design of arbitrary bandwidth FRM IIR/FIR filters.

```
wcT = 0.494*pi; wsT = 0.506*pi;
Amax = 0.2; Amin = 60; L = 5;
[Case,wcFT,wsFT,G2e,G1e] = FRM_CASES(wcT,wsT,L);
Case
% Select A_IIR of prototype Cauer filter << Amax
A_IIR = 0.01; % Typically select Amax/10 or less
N = CA_ORDER_Z(wcFT,wsFT,A_IIR,Amin), NF = ceil(N);
NF = NF+1-rem(NF,2) % Make the Cauer filter odd
% Find the maximal stopband att. of the Cauer filter
Amin_Max = CA_Amin_Z(NF,wcFT,wsFT,A_IIR)
Amin_comp = -10*log10(1-10^(-Amin_Max/10));
[F,Z,R_ZEROS,P,wsTnew] = CA_POLES_Z(wcFT,wsFT,...
    A_IIR,Amin_Max,NF);
Delta_p = 1-10^(-Amax/10); Delta_s = 1/10^(Amin/10);
figure(1), PLOT_PZ_Z(Z,P)
```

```
% Estimate the order of the two masking filters
AmaxFir = Amax-A_IIR; x = 10^(AmaxFir/20);
dc = (x-1)/(x+1); ds =(1+dc)/10^(Amin/20);
[NG1,Be,D,W] = HERRMANN_LP_FIR_ORDER([G1e(2), ...
    G1e(3)],[dc,ds]); NG1 = NG1+mod(NG1+1,2)
[NG2,Be,D,W] = HERRMANN_LP_FIR_ORDER([G2e(2), ...
   G2e(3)],[dc,ds]); NG2 = NG2+mod(NG2+1,2)
[g2,g1] = FRM_IIR_FIR(G2e,G1e,Delta_p,Delta_s, ...
    P,L,NG2,NG1);
LwT = linspace(0,L*pi,1000); wT = linspace(0,pi,1000);
% Determine the magnitude responses of the IIR filter
[F2num,F2den,F1num,F1den] = SORT_LATTICE_POLES_LP_Z(P);
[F2,F1] = LATTICE_2_H_Z(F2num,F2den,F1num,F1den,LwT);
 FM = (F1-F2)/2; FcM = (F1+F2)/2;
H_G2 = ZEROAMW(g2,wT); H_G1 = ZEROAMW(g1,wT);
figure(2), subplot(2,1,1)
PLOT_MAGNITUDE_Z_dB(wT,FM,pi,80,'-'); hold on
PLOT_MAGNITUDE_Z_dB(wT,H_G2,pi,80,'H_l_o_w_e_r_B')
 subplot(2,1,2)
PLOT_MAGNITUDE_Z_dB(wT,FcM,pi,80,'-'), hold on
PLOT_MAGNITUDE_Z_dB(wT,H_G1,pi,80,'H_u_p_p_e_r_B')
figure(3)
H = FM.*H_G2 + FcM.*H_G1; % The overall filter
PLOT_MAGNITUDE_Z_dB(wT,H,pi,80,'H'), hold on
axes('position',[0.240 .5 0.25 0.25]);
wT = linspace(0,wcT,1000);
[F2, F1] = LATTICE_2_H_Z(F2num,F2den,F1num,F1den,L*wT);
FM = (F1-F2)/2; FcM = (F1+F2)/2;
H_G2 = ZEROAMW(g2,wT); H_G1 = ZEROAMW(g1,wT);
H = FM.*H_G2+FcM.*H_G1;
PLOT_MAGNITUDE_Z_dB2(wT,H,wcT,0.2,'H')
```

The pole-zero configuration of the prototype Cauer filter is shown in Fig. 6.29, i.e., an odd-order lattice filter with a transition band equal to $L(\omega_s T - \omega_c T)$ rad. In this case, $L = 5$ and the order is L times smaller compared to that of a direct filter design. However, Cauer filters are not very sensitive to the width of the transition band. Thus, we do not get a significant reduction in the filter order.

The magnitude responses for the lower and upper branches are shown in Fig. 6.30. The overall magnitude and passband responses for the FRM filter are shown in Fig. 6.31.

6.5.9 2D FRM structure

The extension of the FRM structure to 2D sharp filters is not straightforward. The main difficulty lies in the generation of complementary band edge shaping filters, whose frequency responses of the compressed passbands should be separated by stopbands and be able to be masked by the subsequent masking filters.

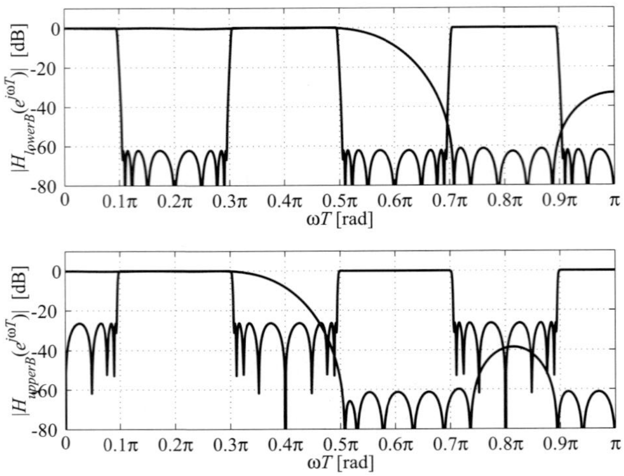

FIGURE 6.30

Magnitude responses for the periodic band edge shaping filters and the masking filters.

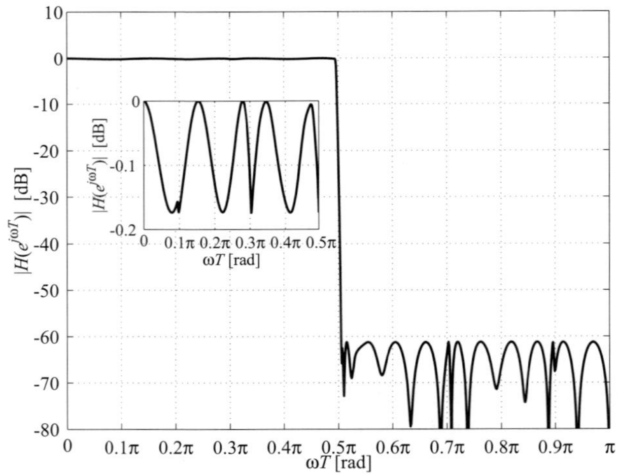

FIGURE 6.31

Overall magnitude response for FRM IIR/FIR filters.

A 2D diamond-shaped filter using FRM structure was proposed in [43]. The idea is to divide the frequency spectrum into four suitably chosen complementary components R_k for $k = 1, 2, 3$, and 4, as shown in Fig. 6.32, in the 2D region $\left[-\frac{\pi}{L}, \frac{\pi}{L}\right]^2$, for a compression factor of L. These regions are also defined for the periodic repetitions of the 2D region $\left[-\frac{\pi}{L}, \frac{\pi}{L}\right]^2$.

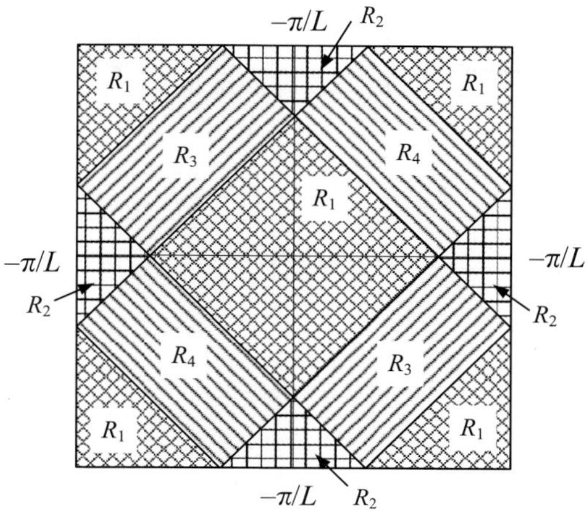

FIGURE 6.32

The four frequency regions for the complementary components of the 2D FRM filter.

These four complementary regions are obtained by four filters $\hat{F}_k(z_1, z_2)$ for $k = 1, 2, 3$, and 4. To make the four filters complementary, an error filter, denoted as $F_e(z_1, z_2)$, is introduced, such that

$$\sum_{k=1}^{4} \hat{F}_k(z_1, z_2) + F_e(z_1, z_2) = 1. \tag{6.56}$$

Thus, the four prototype band edge shaping filters, denoted as $F_k(z_1, z_2)$, are given by

$$F_k(z_1, z_2) = \hat{F}_k(z_1, z_2) + \frac{F_e(z_1, z_2)}{4}.$$

Each delay in each dimension of the four prototype band edge shaping filters is replaced by L delays, and four complementary band edge shaping filters $F_k(z_1^L, z_2^L)$ are obtained. The outputs of $F_k(z_1^L, z_2^L)$ are then filtered by respective masking filters $G_k(z_1, z_2)$, and the resulting outputs are summed to form the final output. The generation of the complementary band edge shaping filters and the overall 2D diamond-shaped FRM filter structure are shown in Fig. 6.33.

Thus, based on the same principle of 1D FRM filters, the specifications of the prototype band edge shaping filters and masking filters can be derived for the design procedure.

6.5.10 Summary

As one of the most successful low-computational complexity digital filter structures, the basic FRM filter consists of an elegant filter network with two branches comprising three subfilters and realizes frequency responses owning a sharp transition band. The price paid for the high computational saving is a small increase in the overall filter order. Further developments on the basic FRM structure have

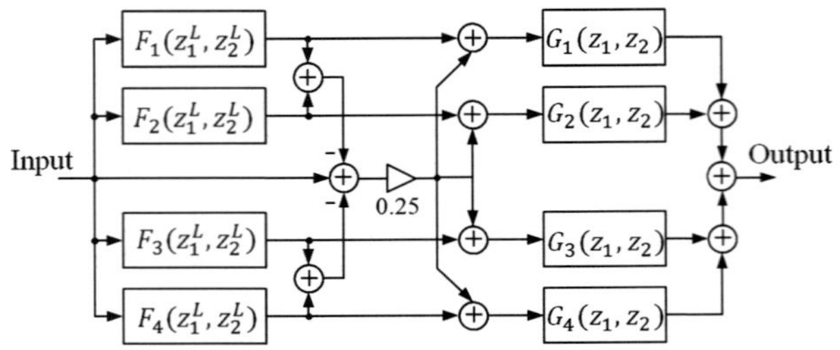

FIGURE 6.33

The overall filter structure of the 2D diamond-shaped FRM filter.

extended the technique to various areas of digital filters, including but not limited to recursive filters, multirate filters, filter banks, and 2D filters.

6.6 The analog approximation problem

Frequency-selective filters are specified in the frequency domain in terms of an acceptable deviation from the desired behavior of the magnitude or attenuation function. Fig. 6.34 shows a typical attenuation specification for an analog low-pass filter. The variation (ripple) in the attenuation (loss) function in the passband may not be larger than A_{max}, and the attenuation in the stopband may not be smaller than A_{min}. Usually, the acceptable tolerances are piecewise constant. During the early stages of the filter design process, it is convenient to use a normalized filter with unity gain, i.e., the minimum attenuation is normalized to 0 dB. The filter may be provided with the proper gain in the later stages of the design process. The passband is from 0 to ω_c rad/s, the transition band is between ω_c and ω_s, and the stopband is from ω_s to ∞. No requirement is placed on the attenuation in the transition band.

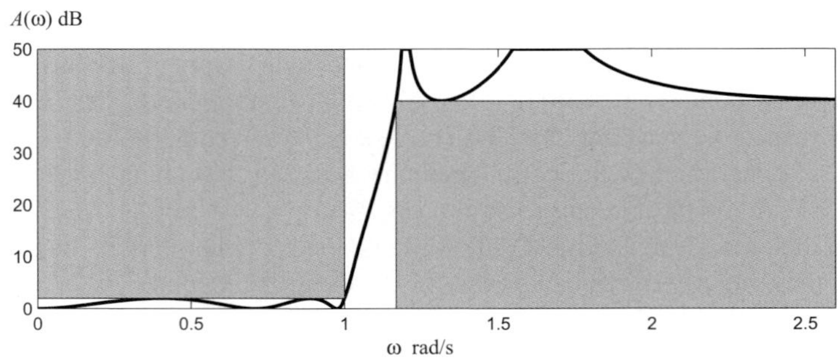

FIGURE 6.34

Attenuation specification for an analog low-pass filter, $A_{max} = 2$ dB, $A_{min} = 40$ dB, $\omega_c = 1$ rad/s, and $\omega_s = 1.3$ rad/s.

6.6.1 Typical requirements

Passband: The frequency band occupied by the wanted signal is referred to as the passband. In this band, the ideal requirement for the filter is to provide constant loss and constant group delay so that the wanted signal will be transmitted with no distortion. The passband frequency and the acceptable tolerance A_{max} for an analog low-pass filter span from 0 to ω_c rad/s.

Stopband: The frequency bands occupied by the unwanted signals are referred to as the stopbands. In these bands, the specification merely requires the attenuation relative to the lower limit set for the stopband to be equal to or greater than some minimum amount. The stopband begins at ω_s for an analog low-pass filter and extends towards infinity. For a digital low-pass filter, the stopband begins at $\omega_s T$ and extends to π rad. It is usually of no interest whether the actual filter loss in the stopband exceeds the specified amount by 1 dB or even 20 dB. Hence, the stopband requirement has only a lower limit, while the passband has both an upper and lower limit.

Transition band: The transition band is from ω_c to ω_s. There are no requirements for attenuation in the transition band. The loss function is a continuous function of frequency and cannot have any jump discontinuities. For this reason, there must always be some interval separating the edge of the passband from the edge of the stopband, in which the loss can start from the low value in the passband to that required in the stopband. The bandwidth allocated to the transition band is one of the main factors determining the order of the filter needed to meet the specification. Moreover, as the width of the transition band is decreased, not only do the filter order and complexity increase, but it also becomes more challenging to meet the specification for the passband. Narrower transition bands mean that the attenuation has to change more rapidly near the passband edges, and this causes the passband response to be more sensitive to component losses and component tolerances.

Phase response: A linear-phase or near-linear-phase response is desirable or even required in many filter applications. However, this is usually specified in terms of the group delay.

Group delay: For the same reason that one cannot get precisely constant loss over a finite band, it is also not possible to obtain a precisely constant group delay over a finite band with *LC* filters. However, it is possible to precisely get a linear phase with filters built with distributed elements and digital FIR filters. We will later show that the group delay is an important factor in determining the sensitivity in the passband for error in the circuit elements.

Impulse response: The impulse response is mainly used as a theoretical concept, and the impulse response is rarely specified. However, in some communication applications, it may be required that the impulse response is zero at specific time instances.

Step response: In most applications, the frequency domain characteristics such as attenuation and group delay requirements must be met, but in some cases, additional requirements in the time domain are used, for example, step response and intersymbol interference requirements.

6.6.2 Standard low-pass approximations

Many filter approximations have been developed to meet different requirements, particularly for analog filters. The main work has focused on approximations of low-pass filters since high-pass, bandpass, and bandstop filters can be obtained from low-pass filters through frequency transformations [44,45,10]. It is also possible to use these results to design digital filters. The classical low-pass filter approximations (standard approximations) are obtained by exploiting the available tolerances in the passband and stopband. We may require that the magnitude function is maximally flat at $\omega = 0$ and/or $\omega = \infty$. We may alternatively require that the tolerance band is fully exploited by requiring that the magnitude function has an equiripple. Hence, combining these requirements, we get four cases. These four cases and some related approximations can be designed by using the functions in [5].

Butterworth: The magnitude function is maximally flat at the origin, $\omega = 0$, and monotonically decreasing in both the passband and the stopband. The group delay and variation within the passband are large. This approximation requires a larger filter order to meet a given specification than the filter approximations discussed below. We use for the synthesis BW_ORDER_S.m and BW_POLES_S.m.

Chebyshev I: The magnitude function has an equal ripple in the passband and decreases monotonically in the stopband. The group delay and the group delay variation within the passband are somewhat less than for a Butterworth approximation that meets the same attenuation requirement. A lower filter order is required compared to the Butterworth approximation. We use for the synthesis CH_ORDER_S.m and CH_I_POLES_S.m.

Chebyshev II (inverse Chebyshev): The magnitude function is maximally flat at the origin, decreases monotonically in the passband, and has an equal ripple in the stopband. The group delay is the smallest of the four approximations, and it has the smallest variation within the passband. The same filter order is required for the Chebyshev I approximation. We use for the synthesis CH_ORDER_S.m and CH_II_POLES_S.m.

Cauer: The magnitude function has an equal ripple in both the passband and the stopband. The group delay and its variation are almost as low as for the Chebyshev II approximation. The Cauer filter, also called an elliptic filter, has the smallest order to meet a given magnitude specification. We use for the synthesis CA_ORDER_S.m and CA_POLES_S.m.

6.6.3 Comparison of the standard approximations

Comparing the standard approximations Butterworth, Chebyshev I, Chebyshev II, and Cauer filters, we find that the two latter have lower variations in the group delay. In the literature, it is often claimed that Cauer filters have a larger variation in the group delay than Butterworth filters and that this is a valid reason for using the Butterworth approximation. The mistake in this argument is that the two filter approximations are compared using the same filter order. This claim is obviously incorrect, which is evident in the following example, since Cauer filters can handle a considerably stricter requirement on the magnitude function. In addition, the step response for a Cauer filter is also better. The differences between Chebyshev II and Cauer filters are, however, relatively small; the former has a somewhat smaller group delay, but the order is larger.

6.6.3.1 Example

Compare the Butterworth, Chebyshev I, Chebyshev II, and Cauer approximations, which meet the same standard LP specification: $A_{max} = 0.043648$ dB, ($\rho = 10\%$), $A_{min} = 60$ dB, $\omega_c = 1$ rad/s, and $\omega_s = 2.5$ rad/s.

As will be further discussed in Section 6.7.4, A_{max} has been chosen very small in order to achieve low passband sensitivity, and we may use components with large tolerances ε at the expense of a slightly higher filter order.

We get the following filter orders with the four standard approximations:

Butterworth: $N_B = 10.046 ==> N_B = 11$,

Chebyshev I and Chebyshev II: $N_C = 6.3176 ==> N_C = 7$,

Cauer: $N_{Ca} = 4.687 ==> N_{Ca} = 5$.

Note the large difference between the required orders for the standard approximations that meet the same requirement. The difference tends to increase if the transition band is reduced. Fig. 6.35 shows the attenuation for the four approximations. The allowed passband ripple is very small, and we are only interested in that the requirement is met and not in detailed variation inside the passband.

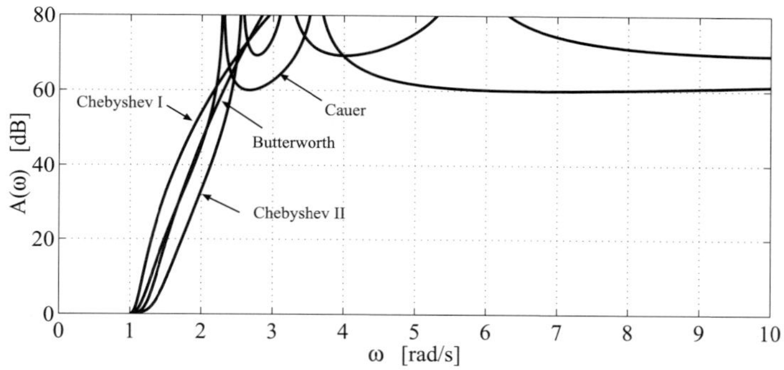

FIGURE 6.35

Attenuation for the four standard approximations.

The attenuation in the transition band and the stopband varies between the different filters. The Chebyshev II filter has a more gradual transition between the passband and stopband compared with the Chebyshev I filter, even though they have the same order. The attenuation approaches, in this case, i.e., odd-order filters, infinity for all of the filters.

Fig. 6.36 shows the corresponding group delays. The difference in the group delay is large between the different approximations, and the group delay maximum lies above the passband edge $\omega_c = 1$ rad/s. The Butterworth and Chebyshev I filters have a larger group delay in the passband, while the Chebyshev II and Cauer filters have considerably smaller group delay, and the difference between the latter two is relatively small. In the literature, it is commonly claimed that the Butterworth filter has the best group delay properties. This is obviously incorrect and is based on an unfair comparison between the standard approximations of the same order. According to Fig. 6.36, Chebyshev II and Cauer filters have considerably lower group delays.

The element sensitivity for an *LC* filter is proportional to the group delay. The group delays at the passband edge and the passband variations are shown in Table 6.1. Note that the Chebyshev II approximation has the smallest variation. In Section 6.7.4, we discuss the effect of the group delay on the sensitivity of a class of filter structures.

6.6.4 Filters with constant pole radius

Filters with a constant pole radius are particularly important since their digital counterpart can easily be designed and implemented. Note that in an *N*th-order filter, there are $2N+1$ degrees of freedom. Here we

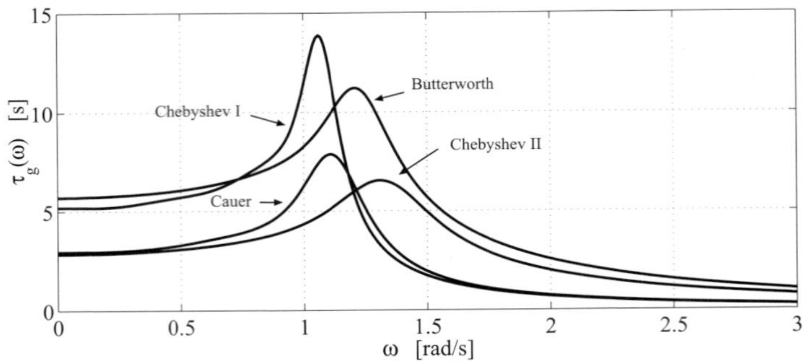

FIGURE 6.36

Group delays for the four approximations.

Table 6.1 Group delays.

	$\tau_g(\omega_c)$	$\tau_g(\omega_c) - \tau_g(0)$
Butterworth	8.27503	2.57287
Chebyshev I	11.4526	6.25390
Chebyshev II	4.36458	1.523593
Cauer	6.27324	3.31994

have restricted the poles to have the same radius. Hence, there are only N+1 degrees of freedom left once the pole radius has been set. That is the problem of finding simple digital filter coefficients and being significantly simplified. The characteristic functions for these filters obey the following bireciprocal property:

$$C(s) = \frac{1}{C(\frac{\omega_0}{s})}, \tag{6.57}$$

where

$$|H(j\omega)|^2 = \frac{1}{1+\varepsilon^2|C(j\omega)|^2}. \tag{6.58}$$

The magnitude function has odd symmetry around $\omega_0 = \sqrt{\omega_c \omega_s}$. The symmetry constraint in (6.114) forces the ripples in the passband and stopband of the filter to be the same. Hence, a filter with large stopband attenuation will have a very small passband attenuation. We have

$$A_{max} = 10log(\frac{10^{0.1A_{min}}}{10^{0.1A_{min}}-1}). \tag{6.59}$$

For example, with A_{min} = 60 dB we get A_{max} = 4.343 10^{-6} dB. Only Butterworth and Cauer filters can have a constant pole radius and obey the symmetry constraint in (6.57). We are mainly interested in the latter, which can be designed using the functions CA_CONST_R_ORDER_S.m and CA_CONST_R_POLES_S.m.

6.6.5 Frequency transformations

Optimal high-pass filters can be derived from an optimal low-pass filter using frequency transformations [44,45,10,4]. However, the frequency transformations yield suboptimal bandpass and bandstop filters. Optimal bandpass and stopband filters require more advanced methods [44,45,10,4,5].

6.7 Doubly resistively terminated lossless networks

Passive *LC* filters belong to the oldest implementation technologies, but they still play an essential role since they are used as prototypes to design advanced frequency-selective filters. However, a drawback with *LC* filters is that it is difficult to integrate resistors and coils with sufficiently high quality in integrated circuit technology. *LC* filters are, for this reason, not well suited for systems that are implemented in an integrated circuit.

A more important aspect is that *LC* filters are used as a basis for realizing high-performance frequency-selective filters. This is the case for the design of mechanical, active, discrete-time, and *SC* filters and digital filters. The main reason is that the magnitude function for a well-designed *LC* filter has low sensitivity in the passband for variations in the element values in the lossless network.

6.7.1 Maximal power transfer

To explain the excellent passband sensitivity properties of correctly designed *LC* filters, we first consider the power transferred from the source to the load in the circuit shown in Fig. 6.37. A lossless reciprocal network (two-port) can be realized using only lossless circuit elements, e.g., inductors, capacitors, transformers, gyrators, and lossless transmission lines. However, these filters are often referred to as *LC* filters.

The maximal power that can be transferred to the load, which occurs when the input impedance $Z(j\omega)$ to the lossless two-port equals R_s, is

$$P_{Lmax} = \frac{|V_{in}|^2}{4R_s}. \tag{6.60}$$

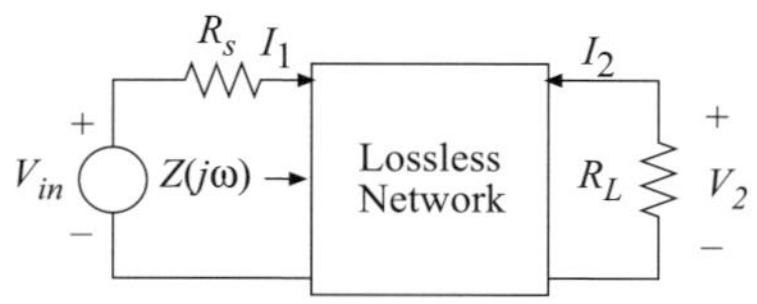

FIGURE 6.37

Doubly resistively terminated lossless network.

The ratio of the output power and the maximal output power is

$$\frac{P_{out}}{P_{outmax}} = \frac{4R_s}{R_L}\left|\frac{V_{out}(j\omega)}{V_{in}(j\omega)}\right|^2 \leq 1, \tag{6.61}$$

where $V_{in}(j\omega)$ is a sinusoidal input signal. An important observation is that the power that the signal source can deliver to the load is limited. The upper bound for the *maximal power transferred* explains why filter structures can have low element sensitivity. Note that a singly terminated network has no such limit, and the passband sensitivity is much higher.

We define the frequency response as the ratio between input and output voltages, i.e., the relation between signal quantities and corresponding physical signal carrier, according to

$$H(j\omega) = \sqrt{\frac{4R_s}{R_L}} \frac{V_{out}(j\omega)}{V_{in}(j\omega)}. \tag{6.62}$$

It is convenient to normalize the magnitude response to $|H(j\omega)| \leq 1$.

6.7.2 Reflection function

Note that the signal source shown in Fig. 6.37 does not deliver maximal power for all frequencies since the input impedance to the reactance network is in general not equal to R_s. This can be interpreted as a fraction of the maximally available power being reflected back to the source. The relationship between the power that is absorbed in R_L and the power that is reflected back to the source is illustrated in Fig. 6.38, i.e., we have

$$\frac{P_L}{P_{Lmax}} + \frac{P_r}{P_{Lmax}} = 1. \tag{6.63}$$

Feldtkeller's equation, which is based on the assumption that the source delivers constant power, relates the power delivered to the load and the power reflected back to the source using the reflection function, $\rho(j\omega)$. We have

$$\frac{4R_s}{R_L}|H(j\omega)|^2 + |\rho(j\omega)|^2 = 1. \tag{6.64}$$

The reflection function is defined as

$$\rho(j\omega) = \frac{Z(j\omega) - R_s}{Z(j\omega) + R_s}, \tag{6.65}$$

where $Z(j\omega)$ is the input impedance to the lossless network in Fig. 6.37. The magnitude of the reflection function will be small in the passband. We will later show that the reflection function is a key to designing low-sensitivity filter structures. Note that frequencies of maximum power transferred are the same at the zeros of the reflection function.

6.7.3 Element sensitivity

A measure of sensitivity to a parameter x is the relative sensitivity of the magnitude function

$$S_x^{|H(j\omega)|} = \frac{\dfrac{\partial|H(j\omega)|}{|H(j\omega)|}}{\dfrac{\partial x}{x}}. \tag{6.66}$$

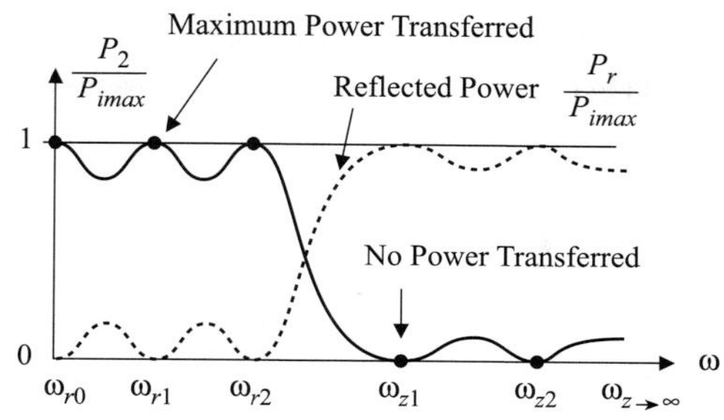

FIGURE 6.38

Transferred and reflected power.

It is difficult to find a simple and good measure of how the attenuation changes when several circuit elements vary at the same time. The reason for this is that the influence of errors in different element values interacts. In fact, for a doubly resistively terminated reactance network, we will demonstrate that they tend to cancel. We shall therefore use and interpret sensitivity measures according to (6.66) with care. Furthermore, it is not easy to compare different filter structures fairly.

The sensitivity of, for example, the magnitude function to a circuit element, x, is a function of the angular frequency. The sensitivity in the passband can be determined from the derivative of Feldtkeller's equation to an arbitrary circuit element, x. We get

$$S_x^{|H(j\omega)|} = -\frac{R_L}{4R_s}\left|\frac{\rho(j\omega)}{H(j\omega)}\right| S_x^{|\rho(j\omega)|}. \tag{6.67}$$

For a doubly resistively terminated LC filter, we have

$$|\rho(j\omega)| = \sqrt{10^{0.1A_{max}} - 1}, \tag{6.68}$$

where A_{max} is the acceptable ripple in the passband. Hence, the magnitude of the reflection function, as well as the sensitivity will be small in the passband if A_{max} is small.

Alfred Fettweis showed in 1960 that the sensitivity becomes minimal if the filter is designed for maximal power transfer at several angular frequencies in the passband. At these angular frequencies, the reflection function $\rho\,(j\omega)$ is zero, since input impedance to the network in Fig. 6.37 is $Z\,(j\omega) = R_s$. The sensitivity at these frequencies is, therefore, according to (6.67), zero. If A_{max} is small, both the reflection coefficient according to (6.65) and the magnitude of the reflection function $|\rho\,(j\omega)|$ according to (6.68) will be small throughout the passband. Hence, the sensitivity will be small. If the ripple is decreased in the passband, the sensitivity is also decreased.

A doubly resistively terminated LC filter with low element sensitivity can also be realized through the following reasoning. Irrespective of if the element value is increased or decreased from its nominal value, P_L will decrease since $P_L = P_{Lmax}$ for the nominal element value, since the derivative is zero where the magnitude function has a maximum, i.e., for $\omega = \omega_k$ with nominal element values. If there are many angular frequencies, ω_k, with maximal power transfer, the sensitivity will be low throughout the passband. This line of reasoning is referred to as Fettweis–Orchard's argument [46,10].

6.7.4 Errors in the elements in doubly terminated filters

Fig. 6.39 shows the typical deviation in the attenuation for doubly resistively terminated *LC* filters due to errors in the reactive elements. The sensitivities to R_s and R_L are proportional to the reflection function. Hence, the sensitivities are small in the passband since $|\rho(j\omega)| \ll 1$ and equal zero for the frequencies at maximal power transfer. In addition, the errors will essentially appear as a small change in the gain of the filter and not affect the frequency selectivity.

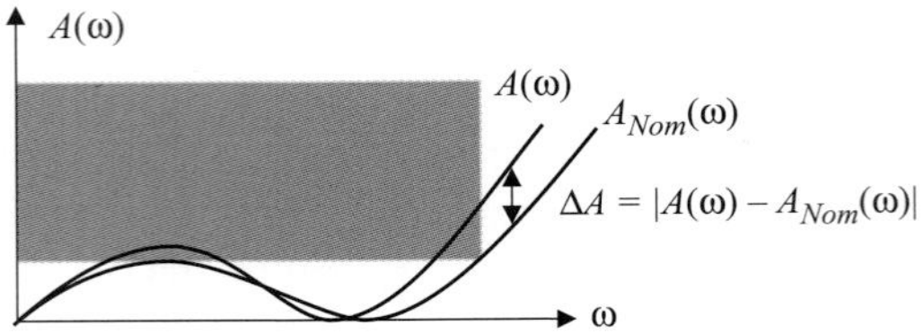

FIGURE 6.39

Deviation in the attenuation due to element errors.

It can be shown that the deviation in the passband attenuation, as shown in Fig. 6.39, for a doubly resistively terminated filter [47,4] is

$$\Delta A(\omega) \leq 4.343\varepsilon \frac{|\rho(j\omega)|}{|H(j\omega)|^2}\omega\tau_g(\omega), \tag{6.69}$$

where $\varepsilon = |\Delta L/L| = |\Delta C/C|$ represent the uniformly distributed errors in the inductors and the capacitors, i.e., $(1-\varepsilon)L \leq L \leq (1+\varepsilon)L$, etc. It can be shown that $\Delta A(\omega)$ is proportional to the electric and magnetic energy stored in the capacitors and inductors and that (6.69) holds for commensurate length transmission line filters [47]. The deviation will, according to (6.69), be largest for frequencies where $\omega\tau_g(\omega)$ is largest since the reflection function, $|\rho(j\omega)|$, is small and $|H(j\omega)| \lesssim 1$ in the passband. Hence, a doubly resistively terminated filter with 3-dB ripple in the passband is significantly more sensitive to element errors than a filter with a smaller passband ripple, e.g., 0.01 dB. Thus, it is often better to design a filter with a small ripple at the expense of a slightly higher filter order. Note that (6.69) is not valid for singly resistively terminated filters. The effect on the attenuation of lossy reactive elements is discussed in [45,10]. However, this is not an issue for the wave digital filters.

6.7.4.1 Example

Consider a fifth-order Chebyshev I filter with $A_{max} = 0.5$ dB, $A_{min} = 42$ dB, $\omega_c = 1$ rad/s, and $\omega_s = 2$ rad/s. We assume that the components' errors are uniformly distributed with $\varepsilon = \pm 1\%$.

Fig. 6.40 shows the deviations in the attenuation in the passband and stopband and the bound on the deviation according to (6.69).

As shown, a significant number of filters do not meet the specification since the design margin is very small. Therefore, in practice, a design margin should be allocated to the two bands and to the band edges [48,10,4]. The deviation increases towards the passband edge while it is insignificant at low frequencies. Moreover, the sensitivity at the band edge is large, and the cutoff frequency is sensitive to component errors.

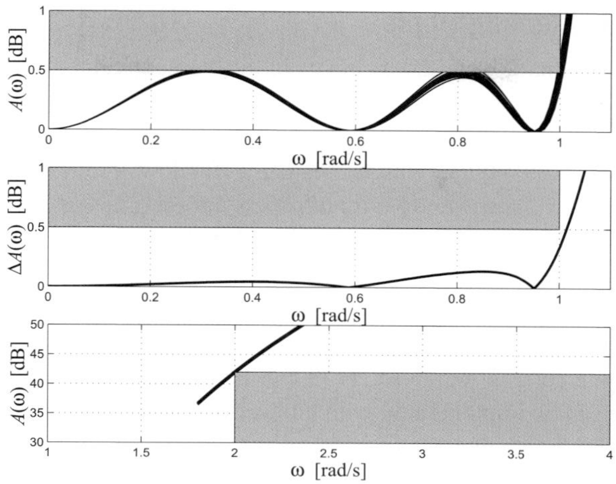

FIGURE 6.40

Deviation in the passband attenuation (top), bound on the attenuation (middle), and deviation in the stopband attenuation (bottom).

The maximum deviation due to component errors for the filters in Example 6.6.3.1 is shown in Table 6.2.

Table 6.2 Maximum passband deviation of the attenuation.

	$\Delta A\ (\omega)$ according to (6.69)
Butterworth	3.63010 ε dB
Chebyshev I	5.02407 ε dB
Chebyshev II	1.91466 ε dB
Cauer	2.75195 ε dB

Using a Chebyshev II or Cauer filter instead of a Butterworth filter reduces, in the case, the group delay with a factor of 1.89 and 1.31, respectively. Hence, the component tolerances can be increased with the same factor. An additional improvement of factor 2 to 3 can be obtained using a diminishing ripple approximation that allocates a larger design margin and reduces the sensitivity at the upper part of the passband. Analog components with large tolerances are considerably cheaper than those with small tolerances. In the corresponding wave digital filter, the multiplication with the filter coefficients becomes simpler and dissipates less power. In addition, the number of components is fewer, 9 and 7, compared to 13 for Butterworth. Therefore it is essential to use an approximation with small group delay. Cauer is often the preferred approximation since the required order is significantly lower than for Chebyshev II, and the group delay is almost as low.

The conclusion is that Cauer is the best approximation in most cases, i.e., when we have requirements on both the attenuation and group delay. In addition, the Cauer approximation yields an *LC* filter with fewer components and almost as low sensitivity to errors in the element values as Chebyshev II filters. Finally, we find that classical design recommendations are not always the best.

6.7.5 Filters with diminishing ripple

Due to deviations in the attenuation caused by errors in the element values, a part of the allowed ripple in the passband, A_{max}, must be reserved to allow for errors in the component values. Therefore, the filter must be synthesized with a design margin, i.e., with a ripple, which is less than required by the application. According to (6.69), the deviation is smaller for low frequencies and increases towards the passband edge. In practice, however, to simplify the synthesis, the design margin for the standard approximations is distributed evenly in the passband, even though the margin will not be exploited for lower frequencies. In order to exploit the allowed passband ripple better, we may let the reflection function $|\rho(j\omega)|$ of the synthesized filter decrease at the same rate as the other factors in $\Delta A(\omega)$ increase so that $A(\omega) + \Delta A(\omega) \leq A_{max}$, as shown in Fig. 6.40. The passband ripple will decay towards the passband edge, and the corresponding *LC* filter can be implemented with components with larger tolerances, i.e., the filter can be implemented with a lower overall cost. The group delay of a filter with diminishing ripples is slightly smaller than for the original filter. An additional advantage is that this will reduce the thermal noise as well [49]. The acceptable range for selecting the filter coefficients in the corresponding wave digital filter will be larger.

6.7.5.1 Example

Consider the seventh-order Cauer filter with $A_{max} = 0.177287$ dB, $\rho = 20\%$, $A_{min} = 60$ dB, $\omega_c = 1$ rad/s, and $\omega_s = 1.5$ rad/s shown in Fig. 6.41. The "minimum" filter order is $N = 5.979$. By increasing the order to $N = 7$, the error tolerance can be increased to $\epsilon = 0.043$. Hence, the room for the deviation in the circuit elements is significantly increased and this allows the multiplier coefficients in the corresponding digital filter to be more roughly quantized.

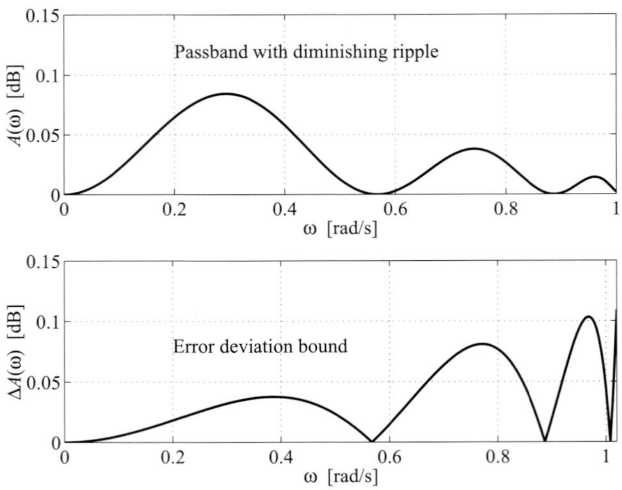

FIGURE 6.41

Passband for a seventh-order Cauer filter with diminishing ripple (top) and error bound (bottom).

Fig. 6.42 shows the overall passband deviation bound and the overall attenuation. We get $A_{max} = 0.11717$ dB and $A_{min} = 64.8222$ dB.

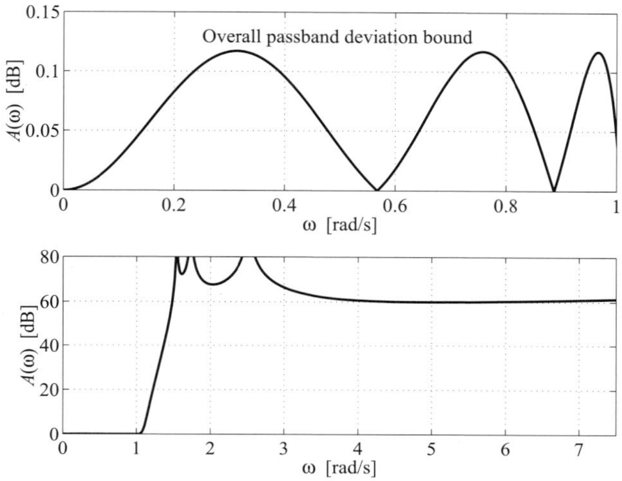

FIGURE 6.42

Overall passband deviation bound for a seventh-order Cauer filter with diminishing ripple, $A(\omega) + \Delta A(\omega) \leq A_{max}$ (top), and overall attenuation (bottom).

6.8 Design of doubly resistively terminated analog filters

Instead of using expensive components with low tolerances, we can compensate for an increase in the tolerance factor, ε, i.e., using components with larger tolerances, using some or all of the following possible trade-offs:

- Use a doubly resistively terminated reactance network that is designed for maximal power transfer, i.e., (6.69) is valid.
- Reduce $|\rho(j\omega)|$ by reducing the passband ripple, A_{max}, of the filter more than required by the application. However, this requires that the filter order is increased. We may use a few more but cheaper components to reduce the overall cost of the implementation.
- Use an approximation with low group delay, i.e., Chebyshev II and Cauer filters are preferred over Butterworth and Chebyshev I filters.
- Use an approximation with a diminishing ripple.

6.8.1 Ladder structures

LC ladder structures can be used to realize filters with transfer functions with zeros on the $j\omega$-axis. Ladder structures cannot realize zeros in the right-hand side of the s-plane, i.e., only minimum-phase transfer functions can be realized. Doubly resistively terminated ladder structures have low element sensitivity in the passband, as discussed above, and relatively small sensitivity in the stopband.

6.8.1.1 Structures for low-pass filters

Figs. 6.43 and 6.44 show ladder structures of T and π types that can realize low-pass filters without finite zeros, respectively. T and π type refer to the leftmost stage of the ladder. These ladder structures can realize Butterworth and Chebyshev I low-pass filters.

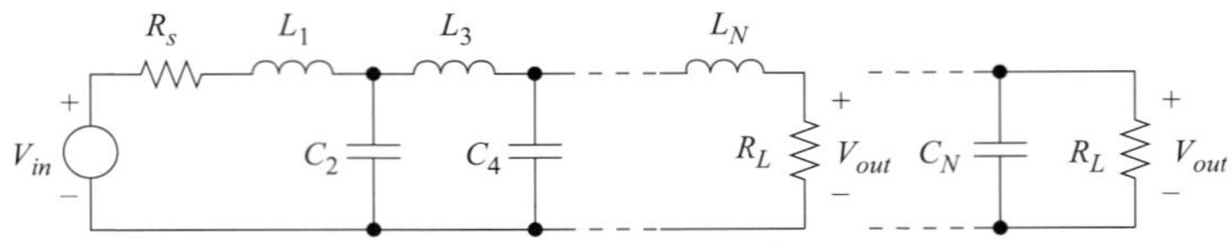

FIGURE 6.43

Nth-order T ladder structure for low-pass filters without finite zeros.

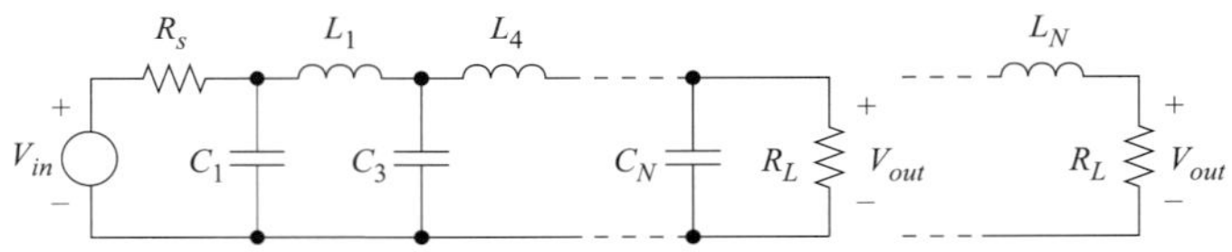

FIGURE 6.44

Nth-order π ladder structure for low-pass filters without finite zeros.

Figs. 6.45 and 6.46 show ladder structures that can realize transfer functions with finite zeros, e.g., Chebyshev II and Cauer filters.

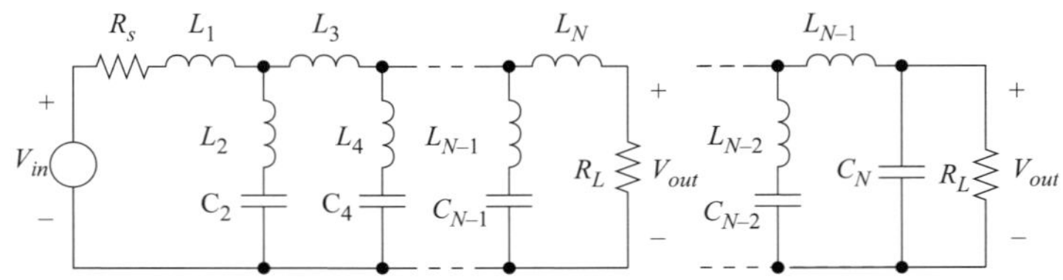

FIGURE 6.45

Nth-order T ladder structure with finite zeros.

There are two main methods for creating transmission zeros. The best method is to use series and parallel resonance circuits in the shunt and series arms in a ladder structure, respectively. The transmission zeros are created by reflecting the signal at the resonance circuits back to the source. The stopband sensitivity will be low if the transmission zeros are created by reflection. A zero pair on the $j\omega$-axis is determined by only two elements. The passband sensitivity and the ratio of the largest and smallest

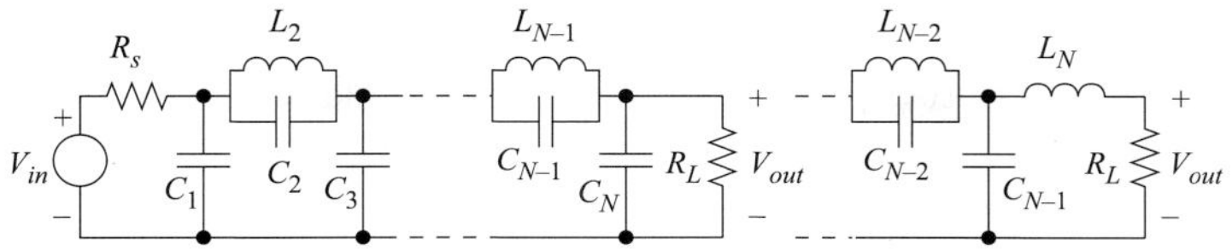

FIGURE 6.46

*N*th-order π ladder structure with finite zeros.

element values depend on the ordering of the transmission zeros. For low-pass filters, minimal sensitivity is obtained if the transmission zeros closest to the passband edge are positioned in the center of the ladder structure. The positioning of the zeros is very essential in narrow-band bandpass filters to optimize the dynamic signal range.

In lattice filters, which will be discussed in Section 6.8.2, the transmission zeros are created by a cancelation. In fact, lattice filters have low passband sensitivities but very poor stopband sensitivities. Therefore, the lattice structure is useful for piezoelectric and ceramic resonator filters: Lattice reference filters are suitable for wave digital filters since the element errors corresponding to binary filter coefficients are fixed.

We conclude that creating transmission zeros by using reflection at the shunt or series arms is a superior method compared to signal cancelation or dissipation.

A singly resistively terminated reactive network does not adhere to Feldtkeller's equation, and the power transferred is either zero or upwards unbounded. Furthermore, it is well known that singly terminated filters are much more sensitive than doubly terminated filters. Hence, singly terminated filters are not recommended as reference filters.

6.8.1.2 Design of ladder structures

Doubly terminated ladder structures are generally designed by using the insertion loss method. This method involves long, complicated, and numerically ill-conditioned calculations, which must be done by a computer. Two exceptions are the Butterworth and Chebyshev I low-pass filters for which analytical solutions have been found. See Refs. [10,4] for details on the synthesis of ladder structures as well as the design of high-pass, bandpass, and bandstop *LC* filers. Efficient MATLAB functions for the synthesis of ladder filters have been implemented in the toolbox [5]. For example, the MATLAB script in Listing 6.8 determines the component values in a T ladder.

Listing 6.8: Program for designing ladder structures of Cauer type.

```
Wc = 20000; Ws = 74000; r = 0.15; Amax = -10*log10(1-r^2);
Amin = 40; Rs = 50; RL = 50;
Ladder = 0; % 1 for T ladder and 0 for a pi ladder
N = CA_ORDER_S(Wc, Ws, Amax, Amin);
N = 5;          % Select an integer filter order
[G, Z, R_ZEROS, P, Wsnew] = CA_POLES_S(Wc,Ws,Amax,Amin);
[L, C, Rs, RL, W0, K] = CA_LADDER(G, Z, R_ZEROS,...
                P, Wc, Ws, RS, RL, Ladder);
```

6.8.1.3 Example

Determine the element values in a Cauer filter with $\omega_c = 20$ krad/s, $\omega_s = 74$ krad/s, and $\rho = 15\%$, which correspond to $A_{max} = -10\log(1-\rho^2)$, $A_{min} = 40$ dB. Use $R_s = R_L = 50\ \Omega$.

We use the program in Listing 6.8.

We get:

$C_1 = 1.0058733\ \mu\text{F}$,
$L_2 = 2.9753089$ mH, $C_2 = 0.1776245\ \mu\text{F}$, $\omega_{02} = 43.499325$ krad/s,
$C_3 = 1.58170022\ \mu\text{F}$,
$L_4 = 2.1813867$ mH, $C_4 = 0.5301041\ \mu\text{F}$, $\omega_{04} = 29.407160$ krad/s,
$C_5 = 0.7623550\ \mu\text{F}$.

6.8.2 Analog lattice structures

Analog lattice structures can be used to derive digital filter structures with low round-off noise, a high degree of parallelism, and modular circuitry that is suitable for digital implementation. In addition, the problems associated with unstable and inaccurate element values are not present in digital filters since the coefficients are fixed binary values.

Fig. 6.47 shows a symmetric analog lattice filter with the lossless reactances X_1 and X_2 and $R_s = R_L$. An analog lattice filter is in practice unusable, because of its high sensitivity in the stopband, except if the reactances are realized by highly stable and accurate components, for example, quartz and ceramic resonators. In fact, due to the high stopband sensitivity, the lattice structure is often used as a measuring device. However, lattice filters, which are designed for maximal power transfer, have low coefficient sensitivity in the passband.

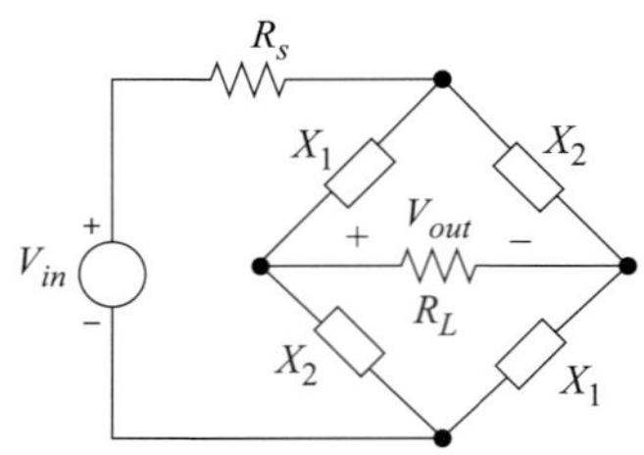

FIGURE 6.47

Analog lattice filter.

The transfer function for a lattice filter is

$$H(s) = \frac{(R/X_1+1)(R/X_2+1)}{R/X_1 - R/X_2}. \tag{6.70}$$

6.8.2.1 Wave description of two-ports

Instead of using voltages and currents to describe electrical networks, it is convenient to use a linear combination. One such linear combination is voltage waves. An impedance Z is usually described by the ratio $Z = V_1/I_1$, but it can also be uniquely described using voltage waves, i.e.,

$$\begin{cases} A_1 = V_1 + RI_1, \\ B_1 = V_1 - RI_1, \end{cases} \tag{6.71}$$

where R is a positive constant. Other possibilities are, for example, current waves and power waves [7,4]. The latter is often used to describe distributed networks.

Consider the lattice structure shown in Fig. 6.48, where we have added a second optional input signal source.

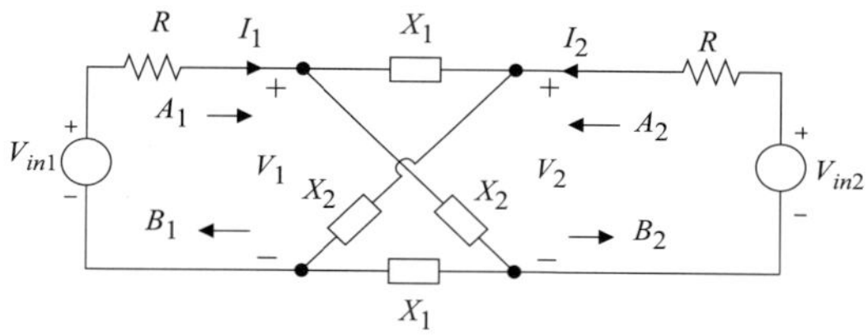

FIGURE 6.48

Symmetric analog lattice filter.

The lossless reciprocal two-port of the lattice structure can be described by the incident (A_1 and A_2) and reflected (B_1 and B_2) voltage waves

$$\begin{cases} A_1 = V_1 + RI_1 = V_{in1}, \\ B_1 = V_1 - RI_1 \end{cases} \tag{6.72}$$

and

$$\begin{cases} A_2 = V_2 + RI_2 = V_{in2}, \\ B_2 = V_2 - RI_2. \end{cases} \tag{6.73}$$

We define the (voltage wave) scattering matrix by

$$\begin{bmatrix} B_1 \\ B_2 \end{bmatrix} = \begin{bmatrix} s_{11} & s_{12} \\ s_{21} & s_{22} \end{bmatrix} \begin{bmatrix} A_1 \\ A_2 \end{bmatrix} = S \begin{bmatrix} A_1 \\ A_2 \end{bmatrix}. \tag{6.74}$$

The scattering matrix for a reciprocal two-port is symmetric if $s_{11} = s_{22}$ and $s_{12} = s_{21}$ and antimetric if $s_{11} = -s_{22}$ and $s_{12} = s_{21}$. Here we are interested in the symmetric case.

We get after elimination of voltages and currents in (6.73) through (6.74)

$$\begin{cases} 2B_1 = S_1 (A_1 - A_2) + S_2 (A_1 + A_2), \\ 2B_2 = S_1 (A_2 - A_1) + S_2 (A_1 + A_2), \end{cases} \tag{6.75}$$

where

$$S_1 = \frac{X_1 - R}{X_1 + R}, \tag{6.76}$$

$$S_2 = \frac{X_2 - R}{X_2 + R} \tag{6.77}$$

are the *reflectance functions* for X_1 and X_2, respectively. Note that S_1 and S_2 are all-pass functions if X_1 and X_2 are reactances. We can rewrite (6.75) as

$$\begin{cases} B_1 = \frac{S_2+S_1}{2} A_1 + \frac{S_2-S_1}{2} A_2, \\ B_2 = \frac{S_2-S_1}{2} A_1 + \frac{S_2+S_1}{2} A_2, \end{cases} \tag{6.78}$$

from which it follows that

$$s_{11} = \frac{S_2 + S_1}{2}, \tag{6.79}$$

$$s_{21} = \frac{S_2 - S_1}{2}. \tag{6.80}$$

The scattering parameter $s_{21}(s)$ corresponds to the transfer function $H(s)$ and $s_{11}(s)$ corresponds to the reflection function $\rho(s)$. Feldtkeller's equation can be expressed in terms of the scattering parameters s_{11} and s_{21} as

$$|s_{11}|^2 + |s_{21}|^2 = 1. \tag{6.81}$$

Hence, s_{11} and s_{21} are power-complementary transfer functions. Fig. 6.49 shows the wave-flow diagram for the lattice filter. Note that the filter consists of two all-pass filters in parallel and that the two outputs are obtained by adding and subtracting the outputs of the all-pass filters. The normal output is B_2. Only odd-order low-pass filters can be realized using a lattice structure, i.e., s_{21} is odd. Hence, S_1 and S_2 are odd (even) and even (odd), respectively. See [50,51,4,5] on how to select S_1 and S_2 for more general cases.

6.8.3 Realization of lossless one-ports

The reflectance functions S_1 and S_2 correspond according to (6.76) and (6.77) to the reactances X_1 and X_2. A reactance $X(s)$ can be realized in several canonic ways; see [10] for more design details. For example, Foster I and II realizations are shown in Fig. 6.50. Alternative realizations are Cauer I and II ladders, which are shown in Fig. 6.51.

6.8.3.1 *Richards' structures*

An arbitrary Nth-order reactance $X(s)$ can be realized by N lossless commensurate length transmission lines using a Richards' structure, which is shown in Fig. 6.52. The structure is terminated with either a short circuit or an open circuit.

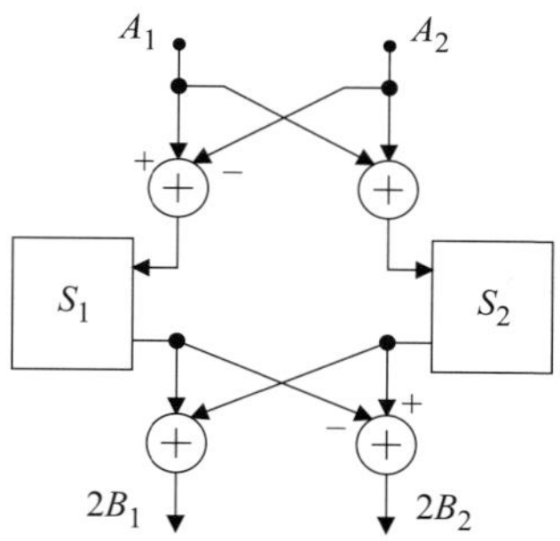

FIGURE 6.49

Wave-flow diagram for a lattice filter.

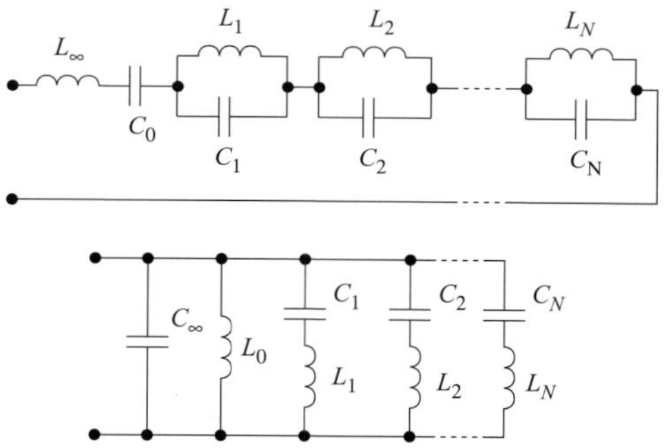

FIGURE 6.50

Foster I (upper) and II (lower) structures.

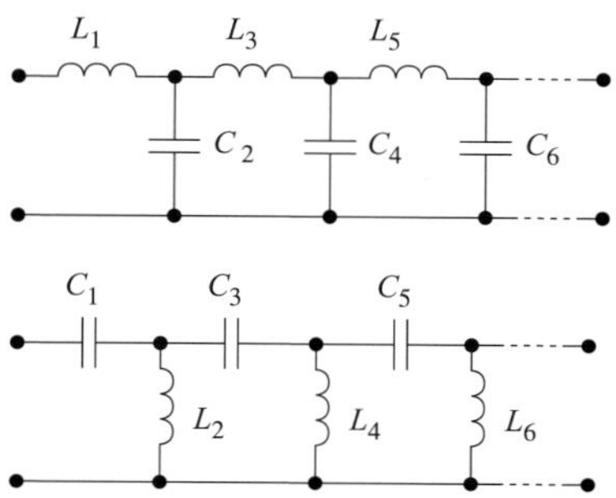

FIGURE 6.51

Cauer I (upper) and II (lower) structures.

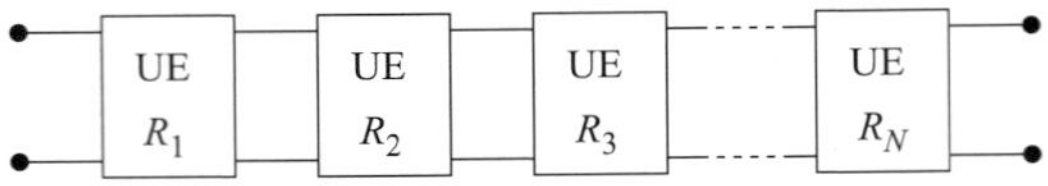

FIGURE 6.52

Realization of an *N*th-order reactance.

6.8.3.2 Example

Determine the element values in an analog lattice structure with the reactances realized by Cauer I ladder or Richards' structures when $A_{max} = 0.2$ dB, $A_{min} = 60$ dB, $\omega_c = 1$ rad/s, $\omega_s = 2$ rad/s, and $R = 1\ \Omega$ using the program shown in Listing 6.9.

We find that a fifth-order Cauer filter meets the requirements. The gain constant, zeros, and poles are

$G = 0.008905489,$
$Z = \pm\, 3.114263i, \pm\, 2.00868528i, \infty,$
$P = -0.367886 \pm 0.6978531i, -0.118784 \pm 1.044093i, -0.5071094.$

Listing 6.9: Program for the design of lattice structures.

```
Wc = 1; Ws = 2; Amax = 0.2;  Amin = 60;
N = CA_ORDER_S(Wc, Ws, Amax, Amin)
N = 5;
[G,Z,R_ZEROS,P,wsnew] = CA_POLES_S(Wc,Ws,Amax,Amin,N)
[S1num,S1den,S2num,S2den] = SORT_LATTICE_POLES_LP_S(P);
figure(1), subplot(1,2,1), xmax = 1; xmin = -2; ymax = 2;
PLOT_PZ_S(-S1den, S1den, Wc, Ws, xmin, xmax, ymax);
subplot(1,2,2)
PLOT_PZ_S(-S2den, S2den, Wc, Ws, xmin, xmax, ymax);
figure(2), subplot('position', [0.08 0.4 0.90 0.5]);
omega = linspace(0, 10, 1001);
Hodd = PZ_2_FREQ_S(0.5,S1num, S1den, omega);
Heven = PZ_2_FREQ_S(0.5, S2num, S2den, omega);
Att = MAG_2_ATT(Heven-Hodd);
PLOT_ATT_S(omega, Att),axis([0, 10, 0, 80]);
G = 1;
[Xnum, Xden] = S_2_Z(G, S1num, S1den);
[L, C, K] = X_CAUER_I(Xnum, Xden)
[Xnum, Xden] = S_2_Z(G, S2num, S2den);
[L, C, K] = X_CAUER_I(Xnum, Xden)
```

The scheme of splitting an odd-order low-pass transfer function into two all-pass sections is as follows:

- Assign the real pole to S_1 and remove it from the set of poles. Note that only odd-order low-pass and high-pass filters and even-order bandpass and stopband filters are feasible.

- Next, assign to S_2 the pole pair with the smallest angle between the negative real axis and the vector from the origin to the positive pole and remove the pole pair from the set of poles.
- Continue in the same way by alternating assigning the remaining pole pair with the smallest angle to S_1 and S_2 until all poles have been assigned.
- The two all-pass filters are obtained by adding the zeros according to $s_z \leftarrow -s_p$, that is, each pole has a corresponding zero but in the right-hand side half-plane. This is done in line 5 by the function SORT_LATTICE_POLES_LP_S.
- Determine the reflectance functions from the all-pass functions.
- Finally, select suitable realizations of the two reactances.

The corresponding pole-zero configurations are shown in Fig. 6.53.

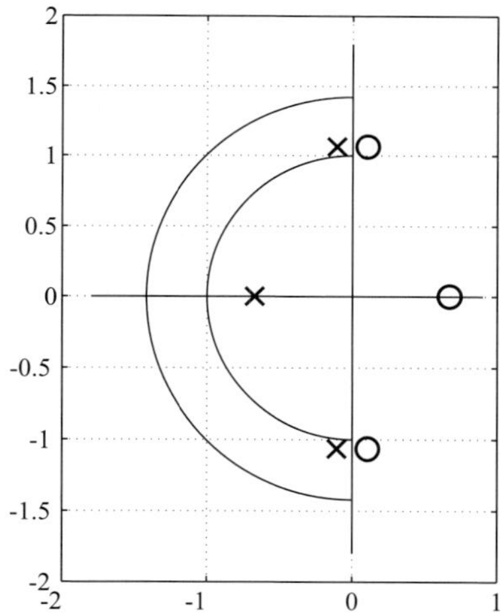

FIGURE 6.53

Poles and zeros for S_1 (odd) and S_2 (even).

In lines 18 and 20, the reactances X1 and X2 corresponding to the reflectances S_1 and S_2 are computed, and in lines 19 and 21 the element values are computed. Note that in lines 19 and 21, an infinite capacitor represents a short circuit, while K = 2 or 9 represents series inductors and shunt capacitors, respectively. See X_CAUER_I for details.

We get for X1 $L_1 = 1.34286$, $C_2 = 1.575207$, $L_3 = 0.8442389$.

We get for X2 $L_1 = 1.359117$, $C_2 = 1.182268$.

Fig. 6.54 shows the corresponding ladder structures.

The reactances in the analog lattice structure can also be realized by Richards' structures. We therefore replace lines 18 and 20 in Listing 6.9 with

```
18 [R, K, RL] = RICHARDS_REACTANCE_S(Xnum, Xden)
20 [R, K, RL] = RICHARDS_REACTANCE_S(Xnum, Xden)
```

which yields

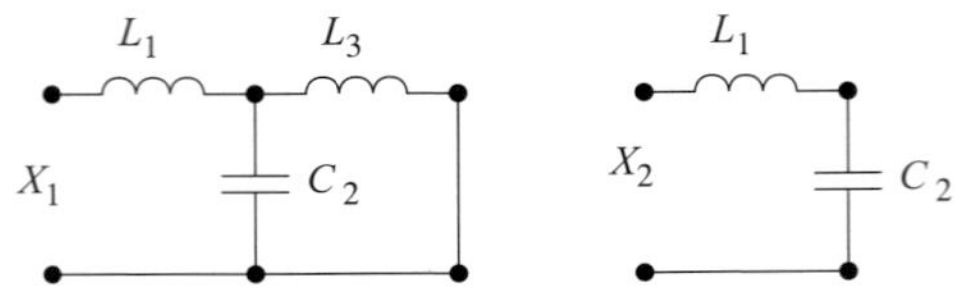

FIGURE 6.54

Cauer I ladders for S_1 (odd) and S_2 (even).

$$R_1 = 1.705221,\ R_2 = 0.2353787,\ R_3 = 0.246502, \text{ and } R_L = 0 \text{ (termination)}$$

and

$$R_1 = 2.2049488,\ R_2 = 1.372225, \text{ and } R_L = \text{Inf (termination).}$$

We may mix different realizations of the reactances and Foster, Cauer, and Richards' structures. Hence, there exist a larger number of realization alternatives.

6.8.4 Commensurate length transmission line networks

In a commensurate length transmission line network, all transmission lines have a common electrical propagation time. Commensurate length transmission line networks constitute a particular case of distributed element networks that can easily be designed by mapping to a lumped element structure, which can be designed using common analytical or computer-aided tools. This mapping is based on a one-to-one mapping between the LC network and the transmission line network expressed in terms of a complex frequency variable, Richards' variable, Ψ. Lossless commensurate length transmission lines can be used to build low-sensitivity filters.

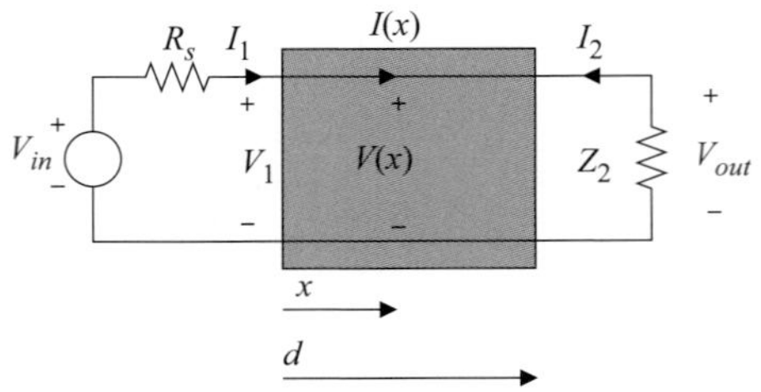

FIGURE 6.55

Transmission line.

A commensurate length transmission line with length d is illustrated in Fig. 6.55. The voltage and current at various positions along the line can be described by an incident and a reflected wave. Under stationary conditions, we have

$$\begin{cases} V(x) = Ae^{-\gamma x} + Be^{\gamma x}, \\ I(x) = \frac{1}{Z_0}\left(Ae^{-\gamma x} - Be^{\gamma x}\right), \end{cases} \tag{6.82}$$

where A and B are constants which can be determined by the boundary conditions. For example, $V_1 = V(0)$ and $I_1 = I(0)$. Furthermore, we have

$$\gamma^2 = (r + j\omega l)(g + j\omega c) \tag{6.83}$$

and

$$Z_0^2 = \frac{r + j\omega l}{g + j\omega c}, \tag{6.84}$$

where r, l, g, and c are the primary constants for the resistance, inductance, conductance per unit length, and capacitance per unit length, respectively, γ is the propagation constant, and Z_0 is the characteristic impedance, which is a real-valued positive constant, $Z_0 = R = \sqrt{\frac{l}{c}}$, for lossless transmission lines and is therefore sometimes called characteristic resistance.

6.8.5 Richards' variable

Commensurate length transmission line filters constitute a particular case of distributed element networks that can easily be designed by mapping them to an LC structure. This mapping involves Richards' variable, which is defined as

$$\Psi = \frac{s^{sT} - 1}{s^{sT} + 1} = \tanh\left(\frac{sT}{2}\right), \tag{6.85}$$

where $\Psi = \Sigma + j\Omega$ and T is the propagation time for a wave through the transmission line and back. Richards' variable is a dimensionless complex-valued variable. The real frequencies in the s- and Ψ-domains are related by

$$\Omega = \tan\left(\frac{\omega T}{2}\right). \tag{6.86}$$

Note the similarity between the bilinear transformation and Richards' variable. Commensurate length transmission line filters have the same periodic frequency responses as digital filters. Substituting Richards' variable into the chain matrix for a transmission line, we get [4]

$$K = \frac{1}{\sqrt{1 - \Psi^2}} \begin{bmatrix} 1 & Z_0\Psi \\ \frac{\Psi}{Z_0} & 1 \end{bmatrix}. \tag{6.87}$$

The matrix in (6.87) has element values that are rational functions in Richards' variable, except for the square-root factor. However, this factor can be handled separately during the synthesis. The synthesis procedures (programs) used for LC design can therefore be used with small modifications to synthesize commensurate length transmission line filters.

6.8.6 Unit elements

The transmission line filters of interest are, with a few exceptions, built using only one-ports. At this stage, it is, therefore, interesting to study the input impedance of the one-port shown in Fig. 6.56.

A lossless transmission line described by (6.87) is called a unit element. The input impedance to a unit element, which is terminated by the impedance Z_L, can be derived from (6.87).

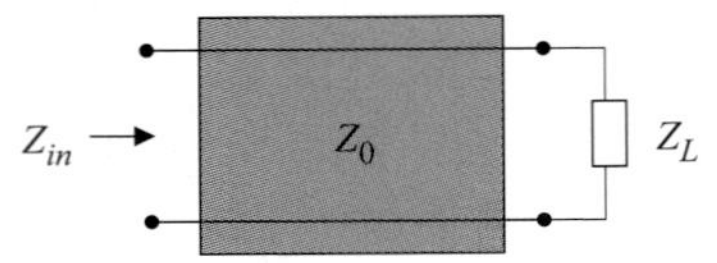

FIGURE 6.56

Terminated transmission line. UE, unit element.

We get the input impedance to a transmission line, with characteristic impedance Z_0 and loaded with an impedance Z_L, as

$$Z_{in}(\Psi) = \frac{Z_L + Z_0\Psi}{Z_0 + Z_L\Psi}. \tag{6.88}$$

We are interested in the input impedance of a lossless transmission line with characteristic impedance $Z_0 = R$ that is terminated by an impedance in the following three cases.

6.8.6.1 $Z_L = Z_0$ *(matched termination)*

For the case where $Z_L = Z_0$, we have according to (6.88)

$$Z_{in} = R. \tag{6.89}$$

Hence, if we match the unit element and the load, an incident wave to the load will not be reflected.

6.8.6.2 $Z_L = \infty$ *(open-ended)*

We get

$$Z_{in} = \frac{R}{\Psi}. \tag{6.90}$$

Hence, an open-ended unit element can be interpreted as a new kind of capacitor, i.e., a Ψ-plane capacitor with the value 1/R.

6.8.6.3 $Z_L = 0$ *(short-circuited)*

A short-circuited unit element has the input impedance

$$Z_{in} = R\Psi, \tag{6.91}$$

which can be interpreted as a Ψ-plane inductor with the value R.

6.9 Design and realization of IIR filters

In the case of IIR filters, there are many design options available. For complex and nonstandard requirements, the frequency response is often determined directly in the z-domain using computer-based optimization methods [4,5].

For frequency-selective filters with standard requirements, i.e., filters with a passband that approximate a constant gain in the passband, it is favorable to employ the vast available knowledge of the design of analog filters. There are several advantages of this approach, which are illustrated in Fig. 6.1.

- Accurate numerical algorithms have been developed for analog filters [44,52,10,5].
- Classical analog attenuation approximations can be converted using the bilinear transformation to their digital counterparts. The bilinear transformation [1,2,3,4,5] between the s- and z-domains is

$$s = \frac{2}{T}\frac{z-1}{z+1}. \tag{6.92}$$

Optimal high-pass filters can be designed by frequency transformation of an optimal low-pass filter. Near-optimal bandpass and bandstop filters can also be designed by frequency transformations. However, we will take an indirect route that will yield a low-sensitivity digital filter algorithm.

- Once the analog transfer function, $H(s)$, is determined, it is realized using an LC filter structure, e.g., a ladder structure. We will later show that a properly designed LC filter will have optimal element sensitivity in the passband. Unfortunately, this structure is not possible to simulate in the z-domain because delay-free loops occur.
- We use an analogy between an LC network and a commensurate length transmission line network to circumvent this problem. The latter contains sources, resistors, and lossless transmission lines of equal length. The transmission line network has a periodic frequency response, and there is a one-to-one mapping between the continuous-time transmission line network and its digital counterpart. We will later discuss this in more detail.
- The digital algorithm is obtained using a wave description, i.e., wave digital filters [53,8,7], consisting of a linear combination of voltages and current in the analog network. This allows us to simulate power in the analog filter, which we later will show is a necessity to obtain a low-sensitivity algorithm.
- Moreover, the concept of pseudopower in the digital domain allows us to suppress parasitic oscillations and obtain a robust realization.

Therefore, we will describe various steps in the design process, illustrated in Fig. 6.1, and options the designer has to minimize the implementation cost.

6.10 Wave digital filters

A wave digital filter is derived from an analog reference filter. The relations between the wave digital filter, the corresponding reference filter, i.e., the transmission line filter, and the LC filter are summarized in Fig. 6.57. The lumped element filter aims to simplify the design process using conventional LC filter theory and design tools. In principle, all commensurate length transmission line networks can be used as reference filters for wave digital filters. Furthermore, wave digital filters representing classical filter structures are also viable since all LC networks can be uniquely mapped onto commensurate length transmission line networks using Richards' variable.

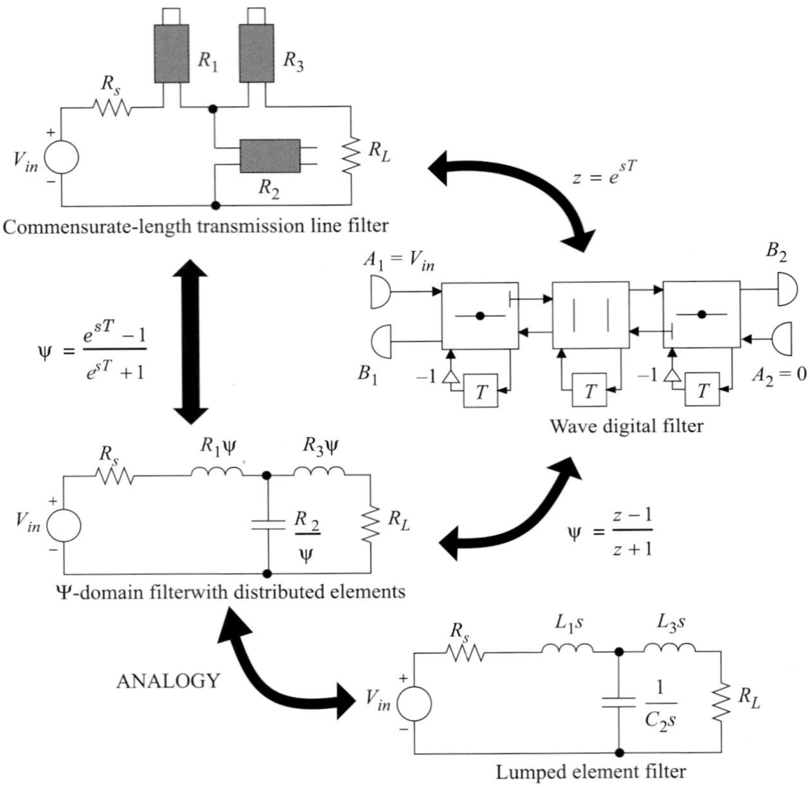

FIGURE 6.57

Summary of the design process for wave digital filters.

Wave digital filters constitute a wide class of digital IIR filters, and in some particular cases wave digital FIR filters. Wave digital filters have many advantageous properties, and they are therefore the recommended type of IIR filters. A wave digital filter is derived from an analog filter, referred to as its reference filter. Due to this relationship between the wave digital filter and the corresponding analog reference filter, wave digital filter structures inherit many fundamental properties from their analog counterparts. Of primary interest are favorable stability properties and low sensitivity to variations in element values. Furthermore, the approximation problem for the wave digital filter can be carried out in the analog domain using well-known design programs.

In order to retain the low element sensitivity and robustness of the analog reference filter, it is necessary to adhere to the maximal power transfer principle, which was discussed in Section 6.7.1.

Furthermore, it is necessary to use two variables to represent power or energy, e.g., voltage and current [54]. Thus the signal-flow graphs discussed in Section 6.4 cannot represent power.

Alfred Fettweis has developed a comprehensive theory, called wave digital filter theory, which solves these problems [6,7,8]. This theory is not only a digital filter design theory; it is a general theory describing the relations between the distributed element and discrete-time networks. Furthermore, it inherently contains an energy concept, which is essential for stability analysis of nonlinear digital networks and for retaining the low-sensitivity properties of the reference filter.

An essential property of wave digital filters is the guaranteed stability, which is inherited from the reference filter. In practice, the inductors in an *LC* ladder filter are nonlinear. Such nonlinearities may cause and sustain parasitic oscillations. However, in the passive *LC* filter, such oscillations are attenuated since the filter dissipates signal power. Hence, any oscillation will eventually vanish. This property is retained in the wave digital filter.

Wave digital filters, particularly wave digital lattice and circulator-tree filters, are suitable for high-speed applications. They are modular and possess a high degree of parallelism, making them easy to implement in hardware.

However, specific reference structures result in wave digital filter algorithms that are not sequentially computable because the wave-flow graph contains delay-free loops. Therefore, one of the main issues is to avoid these delay-free loops. There are three main approaches to avoid delay-free loops in wave digital filters [7,8,10]. Naturally, combinations of these methods can also be used. The main approaches to avoid delay-free loops are:

1. using cascaded transmission lines, so-called Richards' structures, and certain types of circulator filters,
2. introducing transmission lines between cascaded two-ports, and
3. using reflection-free ports.

6.10.1 Wave descriptions

It is necessary to use two variables to represent power [54,7]. We usually use voltages and currents in the analog domain, but we may also use any linear combination of voltage and current to describe a network [7,8]. A network can be described using incident and reflected waves instead of voltages and currents. Scattering parameter formalism has been used in microwave theory to describe networks with distributed circuit elements.

6.10.1.1 Voltage waves

The one-port network shown in Fig. 6.58 can be described by the incident and reflected waves instead of voltages and currents.

The voltage waves are defined as

$$\begin{cases} A = V + RI, \\ B = V - RI, \end{cases} \tag{6.93}$$

where A is the incident wave, B is the reflected wave, and R is a positive real constant, called port resistance. The port resistance corresponds to the characteristic impedance in a transmission line.

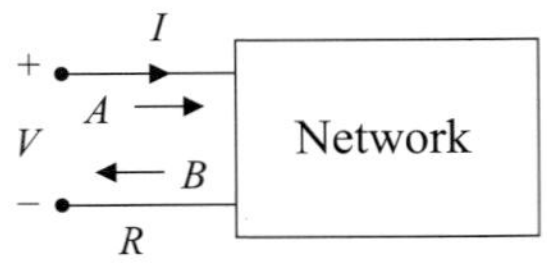

FIGURE 6.58

Incident and reflected waves into a port with a port resistance, R.

6.10.1.2 Current waves

Similarly, current waves can be defined, but they result in the same digital filter structure as voltage waves if we instead use the dual reference filter. Since the current wave description does not provide any additional structures, we will not discuss current waves further. There are also simple relations between wave digital filters derived using voltage, current, and power waves [8].

6.10.1.3 Power waves

In microwave theory, so-called power waves are used:

$$\begin{cases} A = \frac{V}{\sqrt{R}} + \sqrt{R}I, \\ B = \frac{V}{\sqrt{R}} - \sqrt{R}I. \end{cases} \tag{6.94}$$

The name "power waves" stems from the fact that their squared values have the dimension of power. However, we will mainly use voltage waves to provide simpler digital realizations than a power wave description.

6.10.1.4 Reflectance function

A one-port can be described by the reflectance function, which is defined by

$$S = \frac{B}{A}. \tag{6.95}$$

The reflectance function serves a similar purpose as impedances when voltages and currents to describe a network. For example, the reflectance for an impedance Z is

$$S = \frac{Z - R}{Z + R}. \tag{6.96}$$

The reflectance is an all-pass function for a reactance.

6.10.2 Wave-flow building blocks

The basic building blocks for the reference filter are unit elements that are either open- or short-circuited at the far end. The frequency response of such a unit element filter is periodic with a period of $2\pi/\tau$, i.e., the same as for a digital filter. The signals and components of the unit element filter can be mapped to a digital filter by sampling with the sample period, $T = \tau$, where τ is the round-trip for a wave through the unit element. We will derive the wave-flow equivalents to some common circuit elements and interconnection networks in the following sections.

6.10.2.1 Circuit elements

In this section, we derive the wave-flow representation for some basic circuit elements.

6.10.2.2 Open-ended unit element

The input impedance to an open-circuited lossless unit element with $Z_0 = R$ (a Ψ-plane capacitor) is given in (6.90). The reflectance is

$$S(\Psi) = \frac{Z_{in} - R}{Z_{in} + R} = \frac{1 - \Psi}{1 + \Psi} = e^{-sT} \tag{6.97}$$

and

$$S(z) = z^{-1}. \tag{6.98}$$

Fig. 6.59 shows the wave-flow of an open-ended, lossless transmission line and its wave-flow equivalent.

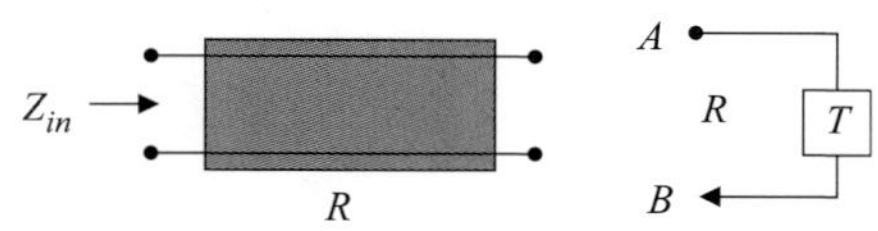

FIGURE 6.59

Open-ended lossless transmission line and its wave-flow equivalent.

6.10.2.3 Short-circuited unit element

The input impedance to a short-circuited, lossless unit element with $Z_0 = R$ (a Ψ-plane inductor) is given by (6.91). The reflectance is

$$S(\Psi) = \frac{Z_{in} - R}{Z_{in} + R} = \frac{\Psi - 1}{\Psi + 1} = -e^{-s\tau} \tag{6.99}$$

and

$$S(z) = -z^{-1}. \tag{6.100}$$

Fig. 6.60 shows the short-circuited lossless transmission line and its wave-flow equivalent.

For the sake of simplicity, we will not always distinguish between wave-flow graphs in the time and frequency domains. Hence, the wave variables in the figures are denoted A and B, but we will use the notation $a(n)$ and $b(n)$ when convenient.

An open-ended unit element corresponds to a pure delay, while a short-circuited unit element corresponds to a delay and a 180-degree phase shift.

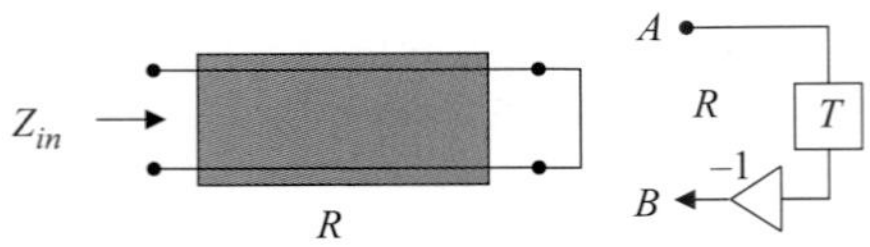

FIGURE 6.60

Short-circuited lossless transmission line and its wave-flow equivalent.

6.10.2.4 Matched termination

The reflectance for a lossless unit element terminated at the far end with a resistor with $Z_0 = R$ (matched termination) is

$$S(\Psi) = 0. \tag{6.101}$$

Hence, an input signal to the unit element is not reflected back to the input.

6.10.2.5 Resistive load

The reference filters of interest are, for sensitivity reasons, doubly resistively terminated. Thus, the load resistor and its corresponding wave-flow equivalent, a wave sink, are shown to the right in Fig. 6.61.

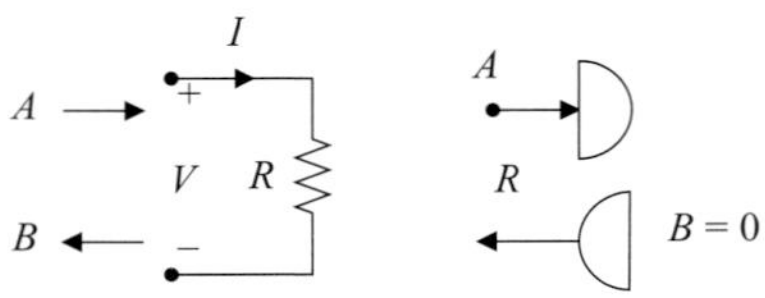

FIGURE 6.61

Wave-flow equivalent for a resistor, R.

6.10.2.6 Short circuit

For a short circuit, we have $V = 0$, which yields

$$S(\Psi) = -1 \tag{6.102}$$

and $B = -A$. Note that this holds independently of the port resistance. The corresponding wave-flow graph is shown to the right in Fig. 6.62.

6.10.2.7 Open circuit

For an open circuit, we have $I = 0$, which yields $B = A$. This result holds independently of the port resistance R. An open circuit has the reflectance

$$S(\Psi) = 1. \tag{6.103}$$

The wave-flow graph for an open circuit is shown to the right in Fig. 6.63.

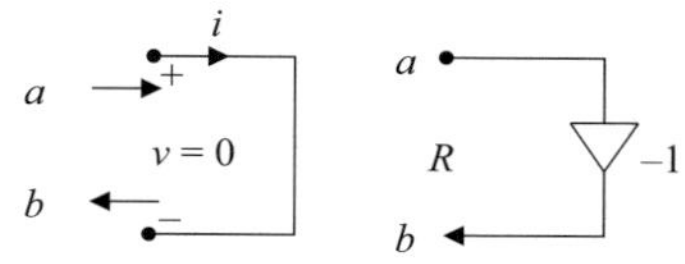

FIGURE 6.62

Wave-flow equivalent for a short-circuit.

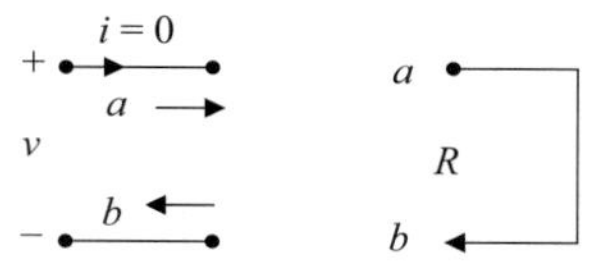

FIGURE 6.63

Wave-flow equivalent for an open-circuit.

6.10.2.8 *Voltage signal source*

For a signal source with a source resistance, we have $V_{in} = V - RI$, which yields $B = V_{in}$. The source does not reflect the incident wave since the source resistance equals the port resistance. The corresponding wave-flow graph is shown to the right in Fig. 6.64.

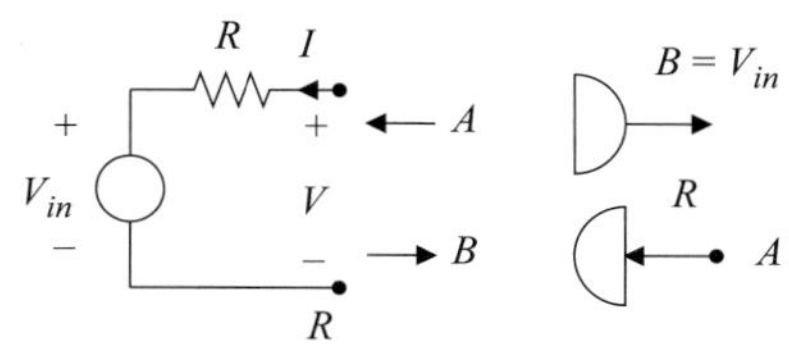

FIGURE 6.64

Wave-flow equivalent for a voltage source with source resistance.

6.10.2.9 *Circulator*

The symbol for a three-port called circulator and its corresponding wave-flow graph is shown in Fig. 6.65. The name comes from the fact that an incident wave is "circulated" to the next port, as shown in Fig. 6.65. The incident wave to port 1 is reflected to port 2, an incident wave to port 2 is reflected to port 3, and so on. Hence, the circulator "circulates" the incident waves to the next port in order. Circulators can be realized in the microwave range by using ferrites. Circulators are used, for example, to direct the signals from the transmitter to the antenna and from the antenna to the receiver in a radar. The circulator is a useful component that is inexpensive to implement in the wave digital domain.

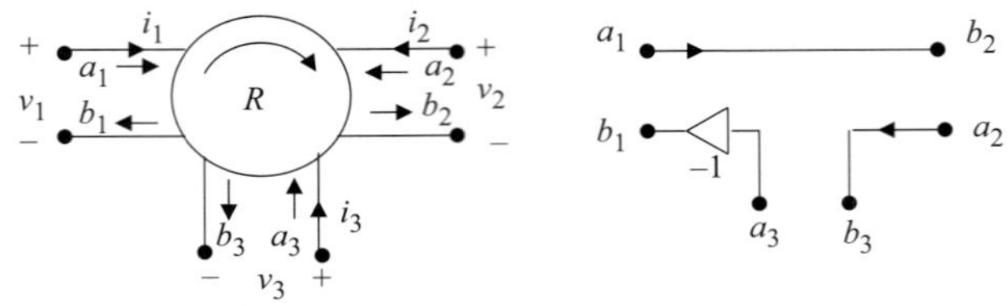

FIGURE 6.65

Three-port circulator and the corresponding wave-flow graph.

6.10.3 Interconnection networks

In order to interconnect different wave-flow graphs, it is necessary to obey Kirchhoff's laws at the interconnection. Generally, the incident waves are partially transmitted and reflected at the connection point, as illustrated in Fig. 6.66. A wave-flow graph describes the transmission and reflection at the connection port called an adaptor. There are several types of adaptors corresponding to different types of interconnection networks [55]. It can be shown that any interconnection network can be decomposed into a set of interconnected two-port and three-port adaptors. Since the interconnection network is lossless, the corresponding adaptor network will also be lossless.

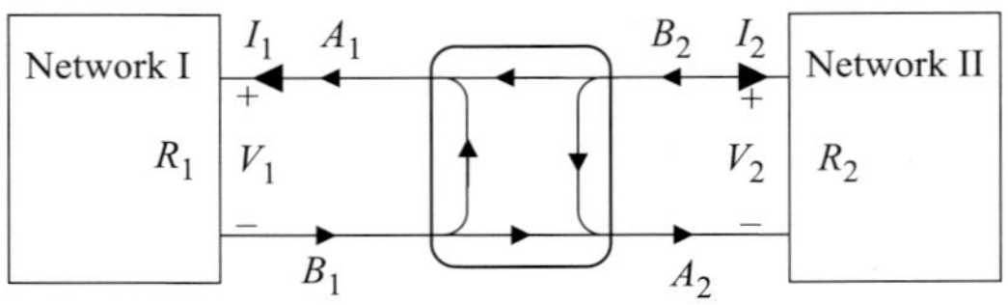

FIGURE 6.66

Connection of two ports.

6.10.3.1 Symmetric two-port adaptor

Fig. 6.67 shows the symbol for the symmetric two-port adaptor. The incident and reflected waves for the two-port are

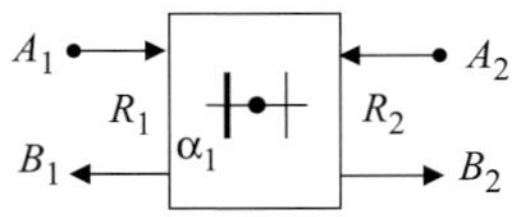

FIGURE 6.67

Symmetric two-port adaptor.

$$\begin{cases} A_1 = V_1 + RI_1, \\ B_1 = V_1 - RI_1 \end{cases} \tag{6.104}$$

and

$$\begin{cases} A_2 = V_2 + RI_2, \\ B_2 = V_2 - RI_2. \end{cases} \tag{6.105}$$

At the point of interconnection, we have according to Kirchhoff's current and voltage laws

$$\begin{cases} I_1 = -I_2, \\ V_1 = V_2. \end{cases} \tag{6.106}$$

By elimination of the voltages and currents, we get the following relations between the incident and reflected waves for the symmetric two-port adaptor:

$$\begin{cases} B_1 = A_2 + \alpha\,(A_2 - A_1), \\ B_2 = A_1 + \alpha\,(A_2 - A_1) \end{cases} \tag{6.107}$$

and

$$\alpha = \frac{R_1 - R_2}{R_1 + R_2}, \tag{6.108}$$

which are illustrated by the wave-flow graph in Fig. 6.68.

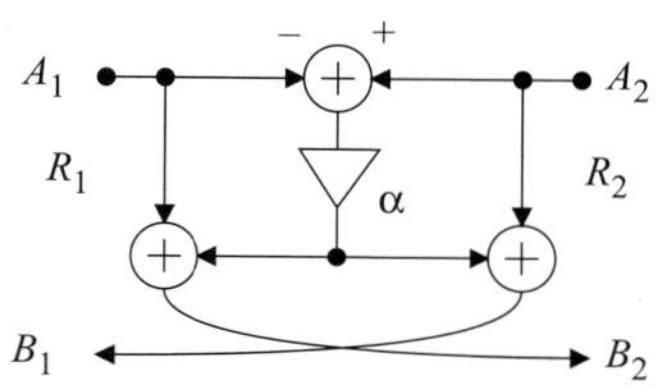

FIGURE 6.68

Symmetric two-port adaptor.

Note that $|\alpha| \leq 1$. The adaptor coefficient α is usually written on the side corresponding to port 1, or one of the parallel lines is thicker. Thus, as can be seen, the wave-flow graph is almost symmetric.

Note that $\alpha = 0$ for $R_1 = R_2$, the adaptor degenerates to a direct connection of the two ports, and the incident waves are not reflected at the point of interconnection. For $R_2 = 0$ we get $\alpha = 1$ and the incident wave at port 1 is reflected and multiplied by -1, while for $R_2 = \infty$ we get $\alpha = -1$ and the incident wave at port 1 is reflected without a change of sign.

Several different adaptor types exist that correspond to interconnections between circuit elements [55]. Symmetric adaptors are commonly used in the first step in the design of wave digital filters. In subsequent design steps, these adaptors can be transformed into other types to optimize the dynamic signal range.

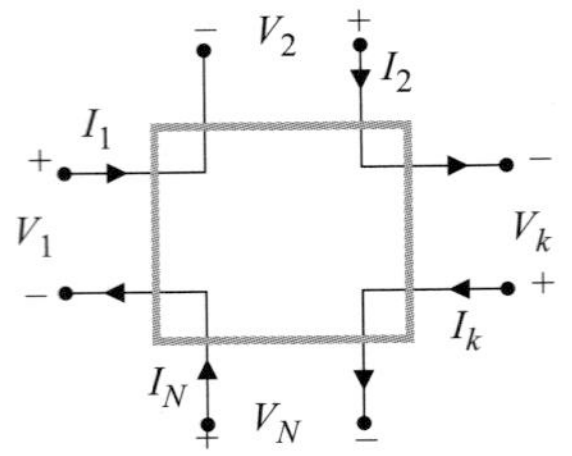

FIGURE 6.69

Series connected ports.

6.10.3.2 Series adaptors

Consider the network topology shown in Fig. 6.69 with N ports, where the currents that flow through the ports are the same. Hence, the ports are connected in series. We have

$$\begin{cases} I_1 = I_2 = \ldots = I_N, \\ V_1 + V_2 + \ldots + V_N = 0. \end{cases} \tag{6.109}$$

After elimination of voltages and currents, we get

$$\begin{cases} B_k = A_k - \alpha_k A_0, \\ A_0 = \sum_{k=1}^{N} A_k, \end{cases} \tag{6.110}$$

where

$$\alpha_k = \frac{2R_k}{\sum_{n=1}^{N} R_n}.$$

The symbol used for an N-port series adaptor is shown in Fig. 6.70. It is straightforward to show that the sum of the adaptor coefficients is

$$\sum_{k=1}^{N} \alpha_k = 2. \tag{6.111}$$

Hence, the adaptor coefficients are linearly dependent, and one of the coefficients can therefore be eliminated, i.e., expressed in terms of the others. A port for which the adaptor coefficient has been eliminated is called a dependent port. It is favorable to select the port with the largest adaptor coefficient as the dependent port. The number of different adaptor coefficients can be reduced further if some of the port resistances are the same.

6.10.3.3 Two-port series adaptor

Fig. 6.71 shows the symbol for the two-port series adaptor. In this case, we have only two adaptor coefficients, and one of them can be eliminated since they are linearly dependent. The resulting wave-flow graph, where α_2 has been eliminated, is shown in Fig. 6.72. The remaining adaptor coefficient

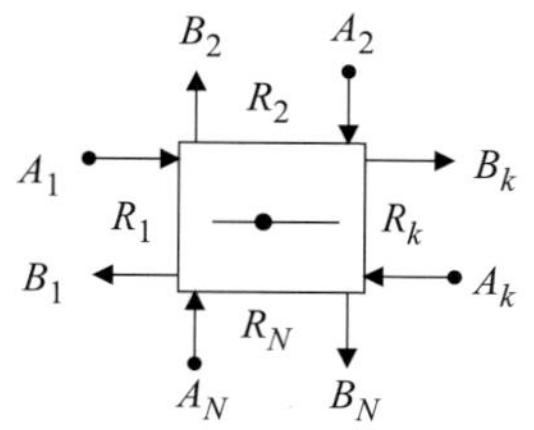

FIGURE 6.70

Series adaptor.

is

$$\alpha_1 = \frac{2R_1}{R_1 + R_2} \tag{6.112}$$

and $0 \leq \alpha_1 \leq 2$.

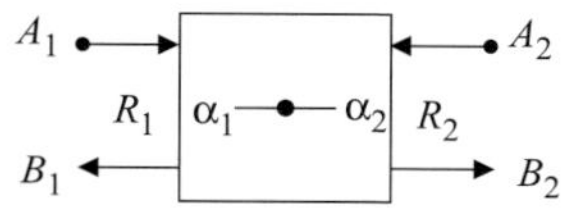

FIGURE 6.71

Two-port series adaptor.

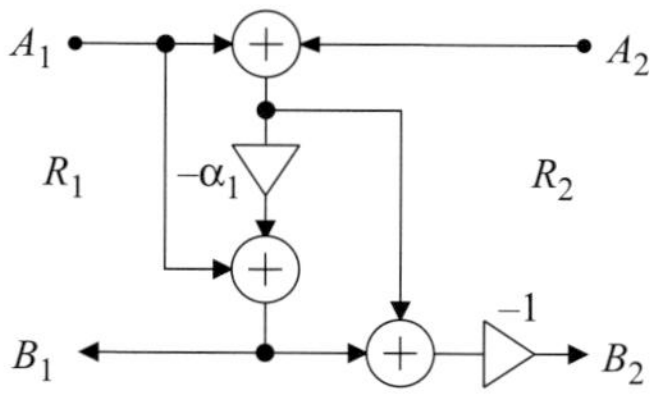

FIGURE 6.72

Wave-flow graphs for a two-port series adaptor.

6.10.3.4 Three-port series adaptor

A symbol for the three-port series adaptor is shown in Fig. 6.73, and Fig. 6.74 depicts the corresponding realization of the three-port series adaptor, where adaptor coefficient α_3 has been eliminated. Hence, port 3 depends on the adaptor coefficients for the two other ports.

This three-port series adaptor requires two multiplications and six additions. However, it is often favorable since it requires multiplications with shorter coefficient word length to eliminate the adaptor coefficient with the largest magnitude. Hence, the corresponding port is select as the dependent port.

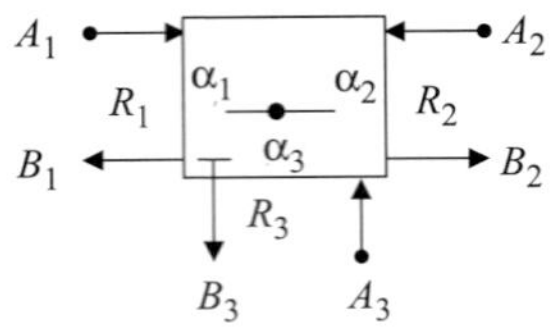

FIGURE 6.73

Three-port series adaptor.

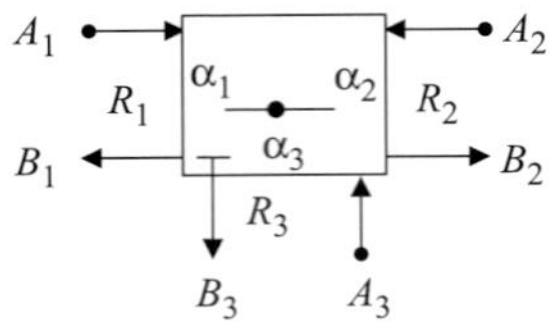

FIGURE 6.74

Three-port series adaptor with port 3 as the dependent port.

6.10.3.5 Parallel adaptors

If all ports have the same voltage, as illustrated in Fig. 6.75, then the ports are connected in parallel. By using the definition of parallel connection of the ports, we get

$$\begin{cases} V_1 = V_2 = \ldots = V_N, \\ I_1 + I_2 + \ldots + I_N = 0. \end{cases} \tag{6.113}$$

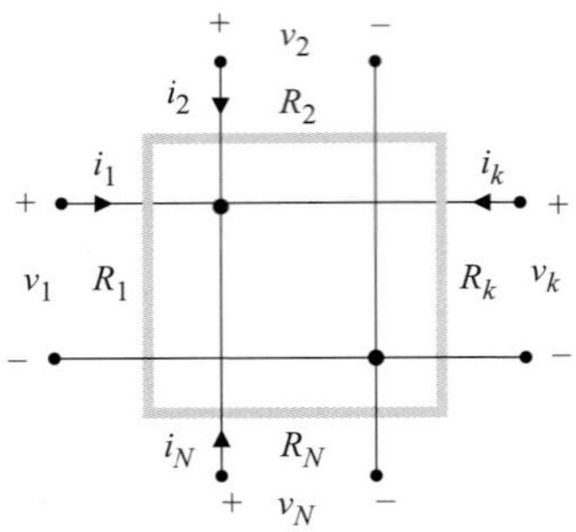

FIGURE 6.75

Parallel connection of ports.

By elimination of voltages and currents, we get

$$\begin{cases} B_k = A_0 - A_k, \\ A_0 = \sum_{k=1}^{N} \alpha_k A_k, \end{cases} \tag{6.114}$$

where

$$\alpha_k = \frac{2G_k}{\sum_{n=1}^{N} G_n}$$

and $G_k = 1/R_k$. The symbol for an N-port parallel adaptor is shown in Fig. 6.76. Also, for the parallel adaptor, we have

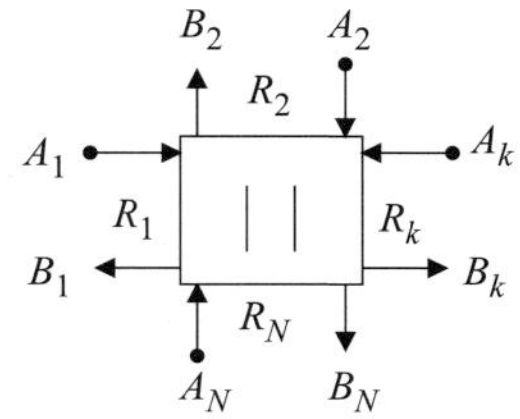

FIGURE 6.76

N-port parallel adaptor.

$$\sum_{k=1}^{N} \alpha_k = 2. \tag{6.115}$$

Hence, one of the adaptor coefficients can be expressed in terms of the others. The number of multiplications can be reduced further if some of the port conductances are equal.

6.10.3.6 Two-port parallel adaptor

Fig. 6.77 shows the symbol for a two-port parallel adaptor, and the corresponding wave-flow graph is shown in Fig. 6.78.

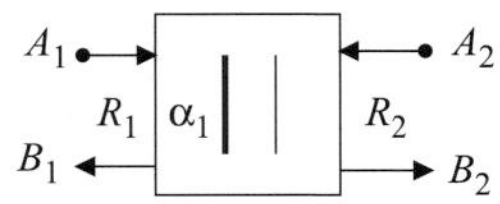

FIGURE 6.77

Two-port parallel adaptor.

The sum of the adaptor coefficients in an adaptor is always equal to 2, except for the symmetric two-port adaptor. Hence, one of the coefficients can be expressed in terms of the others and can therefore be eliminated.

The adaptor coefficient in Fig. 6.78 is

$$\alpha_1 = \frac{2G_1}{G_1 + G_2} \tag{6.116}$$

and $0 \leq \alpha_1 \leq 2$.

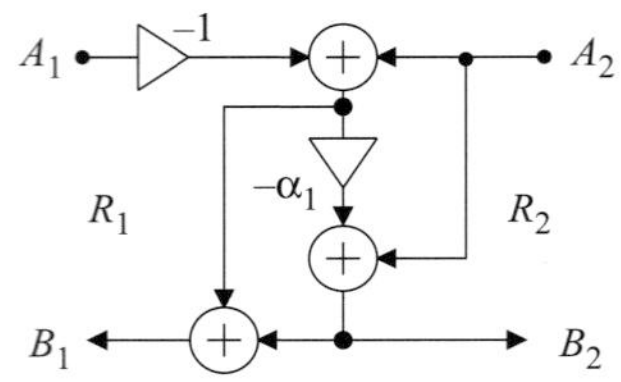

FIGURE 6.78

Wave-flow graph for two-port parallel adaptor.

6.10.3.7 Three-port parallel adaptor

A general three-port parallel adaptor requires two multiplications and six additions. The symbol for the three-port parallel adaptor is shown in Fig. 6.79.

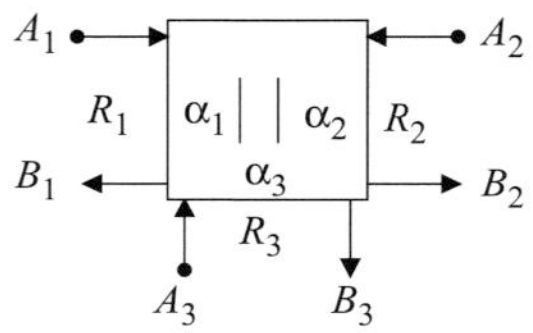

FIGURE 6.79

Three-port parallel adaptor.

Fig. 6.80 shows the wave-flow graph for the corresponding three-port parallel adaptor with port 3 as the dependent port.

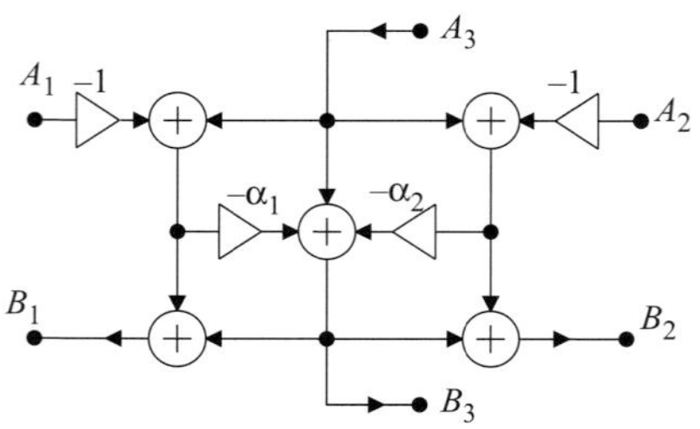

FIGURE 6.80

Three-port parallel adaptor with port 3 as dependent port.

6.10.3.8 Direct interconnection of adaptors

Arbitrary interconnection networks can be built using only two- and three-port series and parallel adaptors. However, in some cases also four-port adaptors are used. Unfortunately, delay-free loops will, in general, occur if two adaptors are connected, as illustrated in Fig. 6.81. The port resistances of the two connected ports must, of course, be equal to satisfy the Kirchhoff equations. However, this problem can be solved by a judicious choice of the port resistances in one of the connected adaptors.

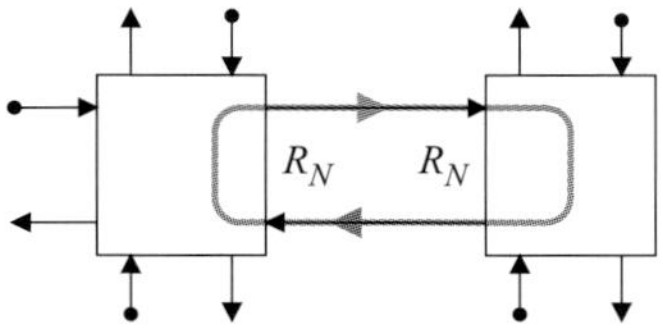

FIGURE 6.81

Potentially delay-free loop.

The port resistance must be selected so that one of the adaptor coefficients corresponding to the connected ports becomes equal to unity. We select, for example, $\alpha_N = 1$, which for the series adaptor is equivalent to

$$R_N = \sum_{k=1}^{N-1} R_k \tag{6.117}$$

and we get from (6.110)

$$B_N = -\sum_{k=1}^{N-1} A_k. \tag{6.118}$$

Hence, the reflected wave B_N is independent of the incident wave A_N and the delay-free loop is broken. For a parallel adaptor, we have

$$G_N = \sum_{k=1}^{N-1} G_k \tag{6.119}$$

and we get from (6.114)

$$B_N = \sum_{k=1}^{N-1} \alpha_k A_k. \tag{6.120}$$

In both cases, the expression for B_N is independent of A_N, i.e., a direct path from A_N to B_N in the wave-flow graph does not exist. Hence, the two ports can be connected without a delay-free loop occurring. There are no restrictions on the adjacent adaptor. The port with the coefficient equal to unity is called a reflection-free port. Since $\alpha_N = 1$, one of the remaining coefficients can be eliminated. Fig. 6.82 shows a three-port series adaptor with $\alpha_3 = 1$ and port 2 as the dependent port.

Note the nonexistent path between A_3 and B_3. A three-port parallel adaptor with $\alpha_2 = 1$ and port 3 as the dependent port is shown in Fig. 6.83. We denote a reflection-free port with the symbol shown in Fig. 6.84.

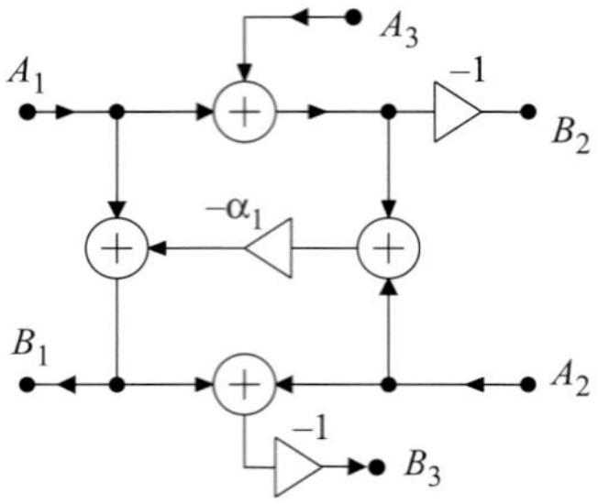

FIGURE 6.82

Series adaptor with port 3 and port 2 as dependent and reflection-free ports, respectively.

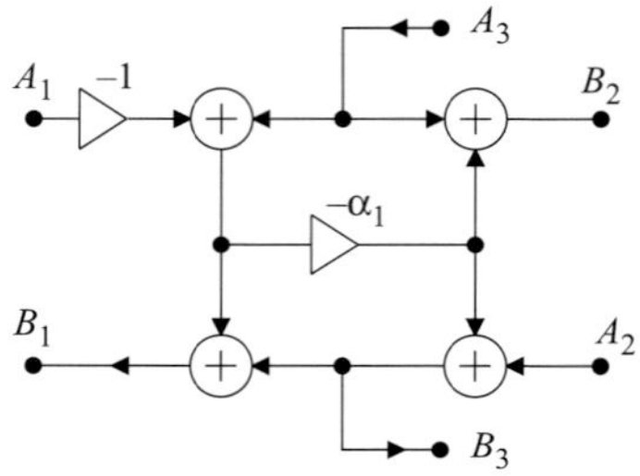

FIGURE 6.83

Parallel adaptor with port 3 as the dependent port and port 2 is reflection-free.

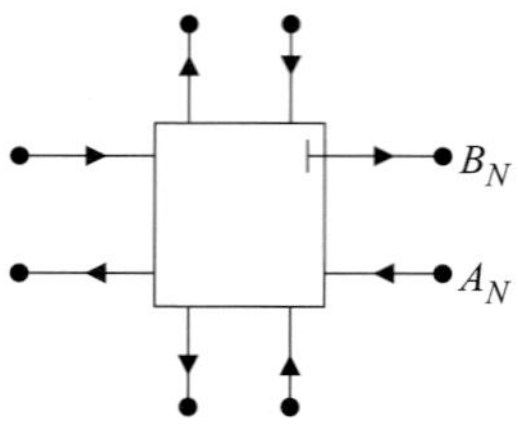

FIGURE 6.84

Adaptor with reflection-free port.

6.10.4 Adaptor transformations

Adaptor transformations, i.e., replacing series adaptors with parallel adaptors and vice versa, yet with the same adaptor coefficients, can improve the dynamic signal range, lower round-off noise, and improve the computational properties of the algorithm [55]. Replacing a series adaptor with a parallel adaptor in the reference filter corresponds to replacing an impedance in series with a shunt impedance embedded between two gyrators [10]. Thus, a series adaptor is replaced by a parallel adaptor with a pair of inverse multipliers at each port and an inverter at each output terminal, as shown in Fig. 6.85.

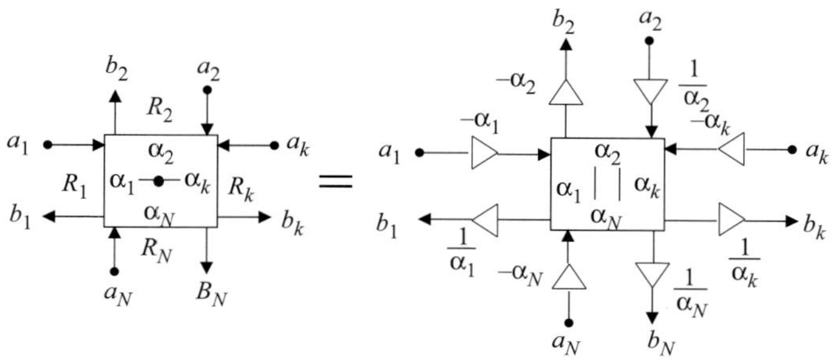

FIGURE 6.85

N-port series adaptor and its equivalent parallel adaptor.

Fig. 6.86 shows an N-port parallel adaptor and its equivalent series adaptor. A pair of inverse multipliers can be eliminated or replaced with any other pair of inverse multipliers, thereby changing the signal levels in the parallel adaptor network. Note that a pair of inverse multipliers correspond to a transformer in the reference filter. The equivalence transformations change neither the adaptor coefficients nor the transfer functions except for a possible change in the gain. Note, however, that inserting a pair of inverse multipliers requires that the coefficients are of the form $c = 2^{\pm n}$, where n is an integer, since the product $c(1/c)$, where both factors are in binary representation, must be equal to 1. Thus, a network with n adaptors has $2n$ equivalent realizations, which differ with respect to round-off noise, overflow probability, and dynamic signal range.

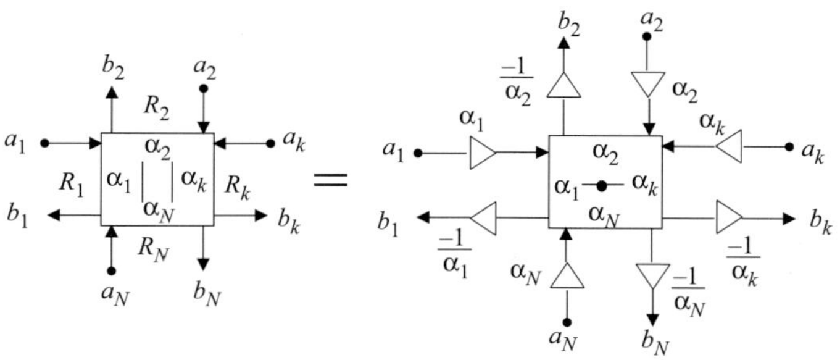

FIGURE 6.86

N-port parallel adaptor and its equivalent series adaptor.

6.10.5 Resonance circuits

First- and second-order resonance circuits play an important role as building blocks for frequency-selective filters. We are, in practice, only interested in low-order resonance circuits, i.e., first- and second-order circuits. In this section, we discuss the design and properties of some resonance circuits and their wave digital counterparts.

6.10.5.1 First-order circuits

A first-order wave digital circuit can be derived from a Ψ-plane capacitor or inductor using series, parallel, or symmetric adaptors. Note that a Ψ-plane capacitor represents an open-ended unit element while a Ψ-plane inductor represents a short-circuited unit element. Fig. 6.87 shows a first-order circuit corresponding to (a) an inductor and (b) a capacitor realized using the symmetric adaptor.

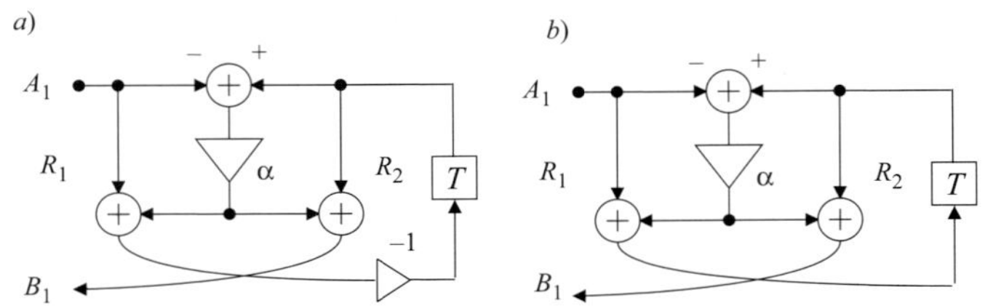

FIGURE 6.87

First-order section with (a) inductor and (b) capacitor.

The reflectance function for the inductor, $R_2\Psi$, is

$$S_L(z) = -\frac{\alpha z + 1}{z + \alpha}. \tag{6.121}$$

The reflectance function for the capacitor, R_2/Ψ, is

$$S_C(z) = \frac{1 - \alpha z}{z - \alpha}, \tag{6.122}$$

where α is given by (6.112). Alternatively, first-order circuits corresponding to an inductor or a capacitor can be derived using series or parallel adaptors [55,8,4].

6.10.5.2 Second-order series resonance circuit

Fig. 6.88 shows a second-order series resonance circuit and its wave digital realization using a three-port series adaptor. The reflectance function is

$$S = \frac{B_1}{A_1} = \frac{(1 - \alpha_1) z^2 - (2\alpha_2 + \alpha_1 - 2) z + 1}{z^2 - (2\alpha_2 + \alpha_1 - 2) z + 1 - \alpha_1}. \tag{6.123}$$

Note that the reflection function is an all-pass function.

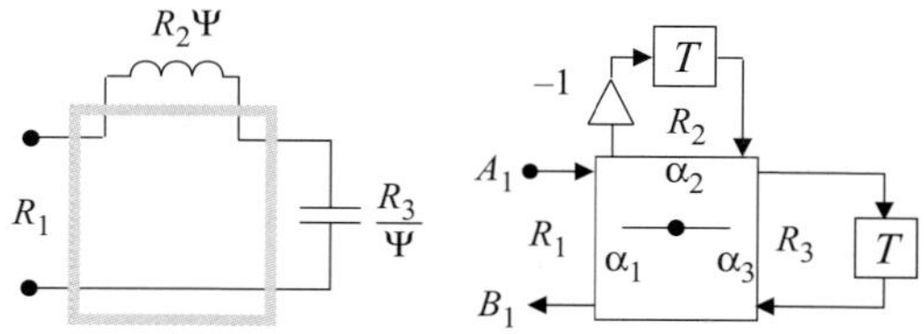

FIGURE 6.88

Realization of a series resonance circuit.

6.10.5.3 Second-order parallel resonance circuit

Fig. 6.89 shows a parallel resonance circuit and its wave digital realization using a three-port parallel adaptor. The reflectance function is

$$S = -\frac{(1-\alpha_1)\,z^2 + (2\alpha_3 + \alpha_1 - 2)\,z + 1}{z^2 + (2\alpha_3 + \alpha_1 - 2)\,z + 1 - \alpha_1}. \tag{6.124}$$

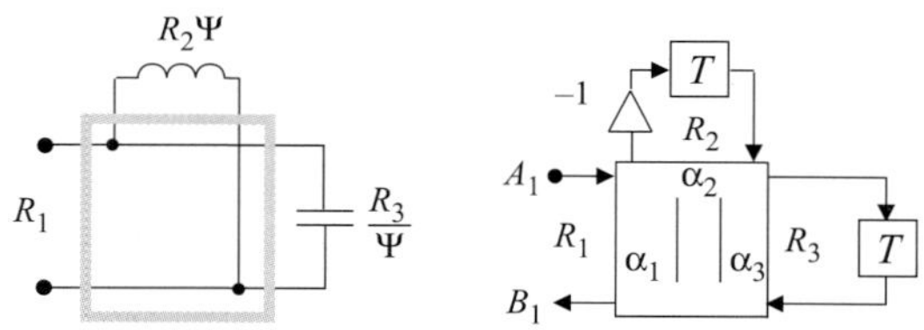

FIGURE 6.89

Parallel resonance circuit and its wave digital filter counterpart.

The pole density is the same as for the series resonance circuit [4].

6.10.5.4 Second-order Richards' structures

As discussed in Section 6.8.3, a Richards' structure that is either open-circuited or short-circuited at the far end can realize an arbitrary reactance. Fig. 6.90 shows a second-order Richards' structure and its wave digital counterpart. It can be shown by using Kuroda–Levy identities that this Richards' structure corresponds to a series resonance circuit in the Ψ-domain [8,4].

The reflectance function is

$$S(z) = \frac{-\alpha_1 z^2 + (\alpha_1 - 1)\,\alpha_2 z + 1}{z^2 + (\alpha_1 - 1)\,\alpha_2 z - \alpha_1}. \tag{6.125}$$

A Richard's parallel resonance circuit has the same reflectance function, except that α_1 has changed sign. Hence, the two structures have the same pole density.

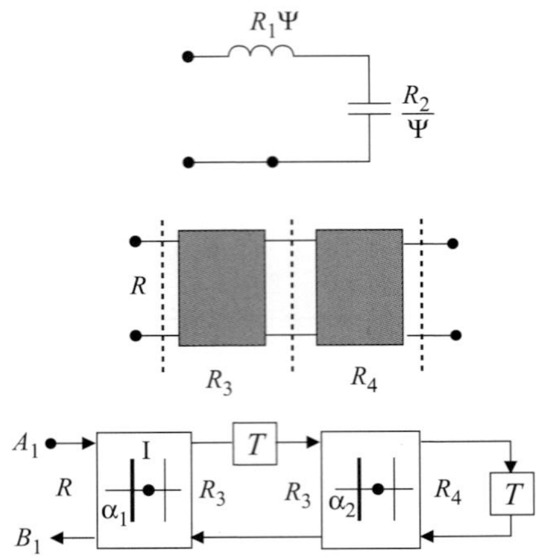

FIGURE 6.90

Richards' structure equivalent to second-order series resonance circuit with corresponding wave-flow graph.

6.10.6 Parasitic oscillations

Parasitic oscillations can occur only in recursive structures. Nonrecursive digital filters can have parasitic oscillations only if they are used inside a closed loop. Nonrecursive FIR filters are, therefore, robust filter structures that do not support parasitic oscillation. Among the IIR filter structures, wave digital filters and certain related state-space structures are of major interest as they can be designed to suppress parasitic oscillations. A. Fettweis has shown that overflow oscillations can be suppressed entirely in wave digital filters by placing appropriate restrictions on the overflow characteristic [56,57,58,8,59]. To show that a properly designed wave digital filter suppresses any parasitic oscillation, we use the concept of pseudopower, which corresponds to power in analog networks. Note that the pseudopower concept is defined only for wave digital filters and not for arbitrary filter structures.

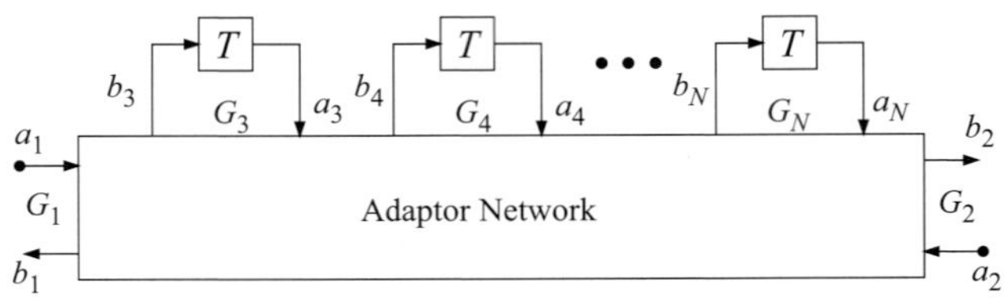

FIGURE 6.91

Wave digital filter.

The instantaneous pseudopower entering the adaptor network shown in Fig. 6.91 is defined as

$$p(n) = \sum_{k=1}^{N} G_k \left(a_k(n)^2 - b_k(n)^2 \right). \tag{6.126}$$

Ideally, $p(n) = 0$, since the adaptors are lossless and an arbitrary network of adaptors is also lossless.

Now, to suppress a disturbance (error signal component) caused by a nonlinearity, it is sufficient to introduce losses in each port of the network such that the nonlinearities correspond to energy losses. This can be accomplished by making each term $a_k(n)^2 - b_k(n)^2 > 0$, i.e., each recursive loop will be lossy. Note that all recursive loops must contain at least one delay element for a sequentially computable algorithm. Any parasitic oscillation will, therefore, by necessity appear at one or several of the ports with delay elements. Hence, a parasitic oscillation, which, of course, has finite pseudoenergy and is not supported from an external signal source, will decay to zero since it will dissipate energy in the lossy ports of the adaptor network. Lossy ports can be obtained by quantizing the reflected waves such that their magnitudes are consistently decreased. Hence, for each nonzero reflected wave, we require

$$\begin{cases} \left|b(n)_Q\right| < b(n)_{exact}, \\ b(n)_Q\, b(n)_{exact} \geq 0, \end{cases} \tag{6.127}$$

where the second constraint is to ensure that the signals have the same sign.

In order to suppress parasitic oscillations, all of the output signals from the adaptors should be sign-magnitude truncated. It can be shown that the sign-magnitude truncation can be substituted by truncation of the signals after the multiplications inside the adaptors and by adding a correcting term at the outputs of the adaptors [4]. This implies considerable hardware simplifications. In practice, it seems to be sufficient that the correcting terms are added only at those ports to which delays are connected.

6.11 Ladder wave digital filters

Doubly resistively terminated ladder structures, designed for maximal power transfer in the passband, are characterized by low element sensitivity. A corresponding wave digital filter inherits the sensitivity properties and the stability properties of the analog reference filter. The positions of the transmission zeros are also relatively insensitive since they are realized by reflections in the resonance circuits. This is valid for filters that approximate a constant passband gain. However, it is not true for, for example, a filter that approximates a differentiator.

There are several alternative ladder design techniques:

- Unit elements can be inserted between the branches in the ladder structure to avoid delay-free loops [6,60,8,7,4]. These filters are of less importance.
- Reference filters of Richards' type, i.e., cascaded transmission lines, may be used to realize low-pass and high-pass filters and a restricted set of bandpass and stopband filters [4]. Also these filters are of less importance.
- A commonly used wave digital filter class is derived from a singly resistively terminated Richards' structure [61,62,63,64,4]. The corresponding wave digital filter is used to realize the denominator. The numerator is realized by injecting the input signal into several nodes in the recursive wave digital filter structure, using a linear combination of the signal in several nodes, or a combination thereof. A singly resistively terminated filter has very high element sensitivity [10], and they are not

recommended for use as frequency-selective filters. Moreover, these structures suffer from significant variations in the internal signal levels, and they also generate considerable round-off noise. The design of the ladder wave digital filters is discussed in detail in Refs. [4,5].

- Here we recommend wave digital filters derived from a doubly resistively terminated ladder, and we use the technique with reflection-free ports to obtain a wave digital filter without a delay-free loop.
- Alternatively, we may select a wave digital filter derived from a doubly resistively terminated lattice filter and use reflection-free ports or a cascade of all-pass sections to obtain a wave digital filter without a delay-free loop.

6.11.0.1 Example

Consider a low-pass filter with the following specifications: $A_{max} = 0.1$ dB, $A_{min} = 65$ dB, $\omega_c T = 0.55\pi$ rad, and $\omega_s T = 0.6\pi$ rad. The minimum order is $N_{min} = 8.499$, which we increase to $N = 9$.

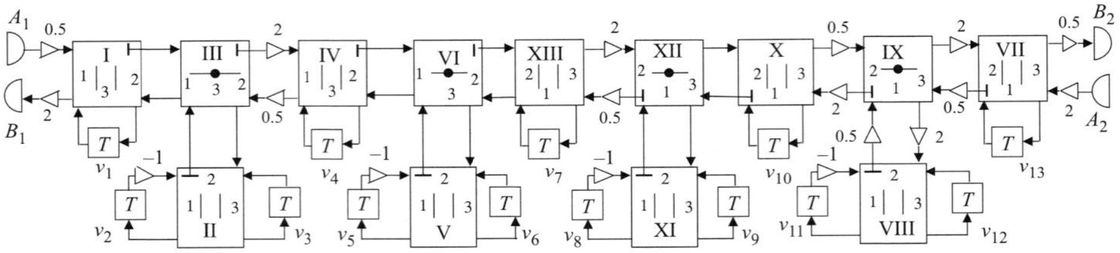

FIGURE 6.92

Ninth-order ladder wave digital filter.

The ladder structure can be scaled by inserting transformers between adjacent adaptors. Each transformer allows one node in an adaptor to be scaled. The adaptors with a reflection-free port have only one critical node, i.e., the input to the multiplier. Hence, all critical nodes can be scaled, except one node in adaptor XIII, which has two critical nodes. The L_2-norm scaled structure is shown in Fig. 6.92. Note that filters should be scaled to fairly compare the sensitivity properties, i.e., they must have about the same dynamic signal range.

In order to illustrate and make a crude comparison of the coefficient sensitivity of a π-ladder and lattice filter with cascaded all-pass sections, we generate 100 frequency responses where we randomly vary the coefficients in the range -2^{-8} to 2^{-8} around their nominal values. The overall deviation in the attenuation is shown in Fig. 6.93.

The deviation is most significant close to the stopband edge. We will later see that the deviation in the positions of the zeros is smaller for the ladder structure compared to a corresponding lattice structure. The deviation in the transmission zeros is relatively small because a transmission zero is determined by only one coefficient. For example, consider the structure shown in Fig. 6.92. Port 2 of adaptor II is reflection-free and hence $\alpha_2 = 1$, and therefore we have $\alpha_1 + \alpha_3 = 1$. Thus, the transmission zero, which is restricted to remain on the unit circle, is determined by only a single coefficient and the coefficient word length determines the accuracy. The deviation in the attenuation in the passband is shown in Fig. 6.94.

The variation increases towards the passband edge, which is in agreement with (6.69).

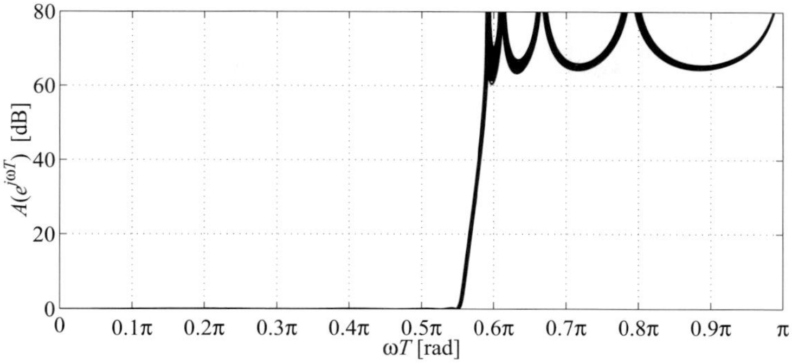

FIGURE 6.93

Variation in the attenuation for the ninth-order ladder filter.

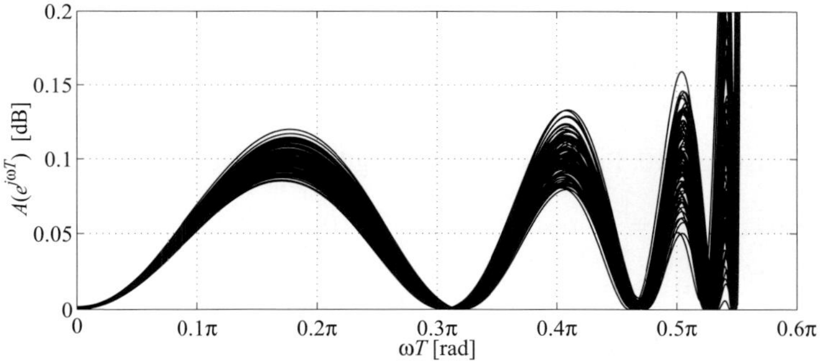

FIGURE 6.94

Passband variation in the attenuation for the ninth-order ladder filter.

A realization, possibly with a somewhat shorter coefficient word length, may have been found if the filter instead was synthesized with a diminishing passband ripple. Note that it is most likely that a set of coefficients can be found with a shorter word length than 9 bits. Typically, the coefficients can be represented using only two or three nonzero bits, i.e., requiring only one or two adders. Fourteen multipliers are required for this ladder structure. Wave digital ladder structures appear to yield complicated wave-flow graphs that may make them challenging to implement efficiently. However, a detailed analysis shows that this is not the case [4].

6.12 Design of lattice wave digital filters

Lattice wave digital filters are derived from an analog bridge structure and have very high stopband sensitivity. Different wave digital lattice structures are obtained depending on how the lattice branches are realized. For example, we may use Forster and Cauer networks or Richards' structures. A realization consisting of a set of interconnected three-port circulators loaded with first- and second-order reactances has become popular, mainly since they are easy to design and consist of simple all-pass sections [65, 66,67,4].

The transfer function for a low-pass lattice wave digital filter is

$$H(z) = \frac{1}{2}(S_2 - S_1),$$

where S_2 and S_1 are the reflectances of the even and odd branches, respectively. The power-complementary transfer function is

$$H(z) = \frac{1}{2}(S_2 + S_1).$$

6.12.0.1 Example

Fig. 6.95 shows a ninth-order lattice wave digital filter with the same specification as in Example 6.11.0.1, where the two all-pass branches have been realized by a circulator structure [10] loaded with first- and second-order sections of the types shown in Fig. 6.90. A program for computation of the adaptor coefficients is shown in Listing 6.10.

Listing 6.10: Program for designing lattice wave digital filters.

```
WcT = 0.55*pi; WsT = 0.6*pi; Amax = 0.1; Amin = 65;
N = CA_ORDER_Z(WcT, WsT, Amax, Amin)
N = 9;            % We MUST select an odd order
[G,Z,R_ZEROS,P,WsnewT] = CA_POLES_Z(WcT,WsT,Amax,Amin,N);
% Compute the adaptor coefficients
[Godd,Podd,Geven,Peven] = SORT_LATTICE_POLES_LP_Z(P);
[alfa_odd,alfa_even] = CASCADE_RICHARDS_SECTIONS(Podd,Peven)
alfa_even0 = alfa_even; alfa_odd0 = alfa_odd;
figure(1), subplot('position', [0.1 0.4 0.88 0.5]);
delta = 2^-14;  wT = linspace(0, pi, 1001);
for nn = 1:100
    alfa_odd = alfa_odd0 +delta*(rand(1,length(alfa_odd))-0.5);
    alfa_even = alfa_even0 + delta*(rand(1,length(alfa_even))-0.5);
    [S1, S2] = LATTICE_RICHARDS_H(alfa_odd, alfa_even, wT);
    Att = MAG_2_ATT(0.5*(S2-S1));
    PLOT_ATT_Z(wT, Att, pi, 80), hold on
end
figure(2),  subplot('position', [0.1 0.4 0.88 0.5]);
delta = 2^-7;  wT = linspace(0, 0.6*pi, 1001);
for nn = 1:100
    alfa_odd = alfa_odd0 +delta*(rand(1,length(alfa_odd))-0.5);
    alfa_even = alfa_even0 + delta*(rand(1,length(alfa_even))-0.5);
    [S1, S2] = LATTICE_RICHARDS_H(alfa_odd, alfa_even, wT);
```

```
    Att = MAG_2_ATT(0.5*(S2-S1));
    PLOT_ATT_Z(wT, Att, 0.6*pi,0.2), hold on
end
```

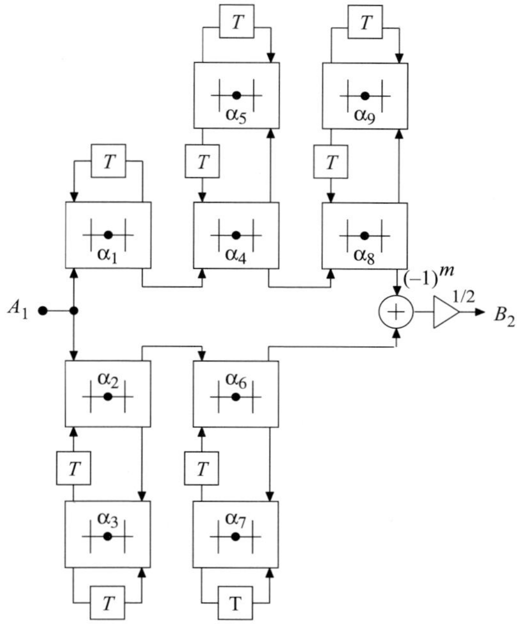

FIGURE 6.95

Ninth-order lattice wave digital filter.

If the order of the odd branch is larger than the order of the even branch, then $m =$ odd, otherwise $m =$ even [4].

To get a similar deviation in the passband and stopband attenuation as in Figs. 6.96 and 6.97 for the ladder structure, we restrict the error range to -2^{-15} to 2^{-15}, which is about $2^7 = 128$ times smaller. The resulting variation in the attenuation is shown in Fig. 6.96. It is evident that the stopband sensitivity is much higher than the sensitivity for the corresponding ladder structure and that the precision in the adaptor coefficients is not enough. The passband variation when the error range is decreased to -2^{-8} to 2^{-8} is shown in Fig. 6.97.

6.12.1 Bandpass lattice filters

Bandpass and bandstop filters can be designed by first designing a low-pass filter that is transformed into the desired frequency response. However, lattice low-pass filters must have odd order. The obtained bandpass and bandstop filters are restricted to $N = 2N_{LP}$, i.e., $N = 2, 6, 10, 14, \ldots$ Hence, the obtained filter may often be highly overdesigned. We have three bands for a bandpass filter, i.e., stopband, passband, stopband. The denominator is a Hurwitz polynomial of even order, N. The difference

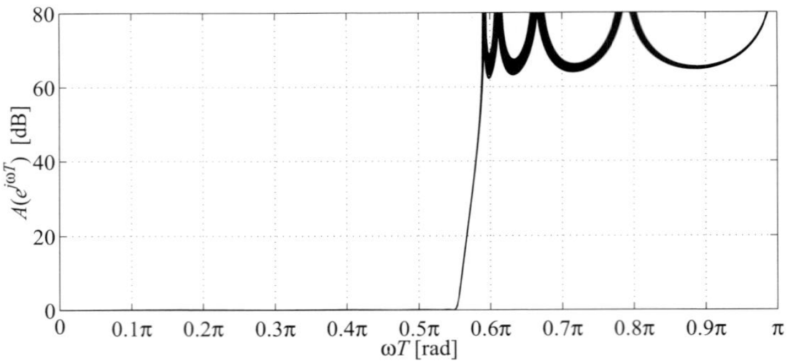

FIGURE 6.96

Variation in the attenuation for the ninth-order lattice filter.

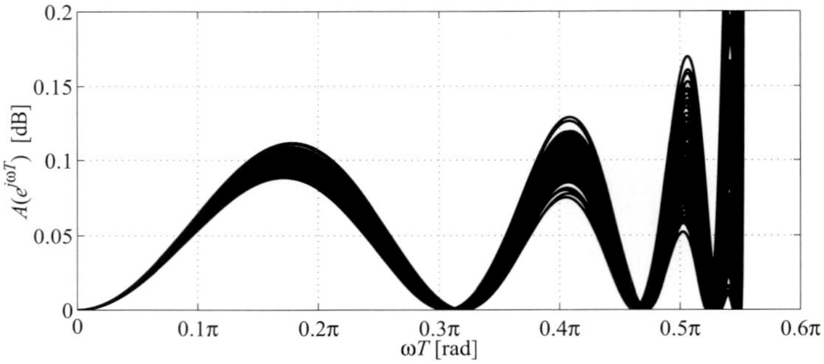

FIGURE 6.97

Passband variation in the attenuation for the ninth-order lattice filter.

in the order of S_1 and S_2 is an even number, and the two branches are added. The proper ordering of the poles and zeros is determined using the function SORT_LATTICE_POLES_BP_Z for both bandpass and bandstop filters.

6.12.1.1 Example

Consider the bandpass filter with $\omega_{c1}T = 0.4\pi$ rad, $\omega_{c2}T = 0.5\pi$ rad, $\omega_{s1}T = 0.2\pi$ rad, $\omega_{s2}T = 0.7\pi$ rad, $A_{max} = 0.1$ dB, and $A_{min} = 60$ dB. We use the program shown in Listing 6.11.

Listing 6.11: Program for the design of bandpass lattice wave digital filters with cascaded Richards' sections.

```
wc1T = 0.4*pi; wc2T = 0.5*pi; ws1T = 0.2*pi; ws2T = 0.7*pi;
Amax = 0.1; Amin = 60; T = 1;
% Requirements for the analog prototype low-pass filter
Wac1 = (2/T)*tan(wc1T/2);     Wac2 = (2/T)*tan(wc2T/2);
```

```
Was1 = (2/T)*tan(ws1T/2);        Was2 = (2/T)*tan(ws2T/2)
Was2 = Wac1*Wac2/Was1, % Band edges for the  analog filter
Omegac = Wac2 - Wac1; Omegas = Was2 - Was1;
% Design of the analog prototype low-pass filter (Cauer)
N = CA_ORDER_S(Omegac, Omegas, Amax, Amin)
N = 5;          % We must select an integer order
% Design of the analog prototype low-pass filter (Cauer)
[Ga,Za,R_ZEROS,Pa,Wsnew] = CA_POLES_S(Omegac,Omegas,Amax,Amin,N);
% Transformation the analog filter into a low-pass digital filter
[G, Z, P] = PZ_2_PZ_Z(Ga, Za, Pa, T);
% Transformation the low-pass filter into a digital bandpass
% filter
[Gbp, Zbp, Pbp] = PZ_2_BP_Z(G,Z,P,wc1T,wc2T,Omegac*T)
figure(1)
PLOT_PZ_Z(Zbp, Pbp)
[Zodd, Podd, Xeven, Peven] = SORT_LATTICE_POLES_BP_Z(Pbp);
figure(2),  subplot('position', [0.1 0.4 0.88 0.5]);
wT = linspace(0, pi, 1001);
[alfa_odd, alfa_even] = CASCADE_RICHARDS_SECTIONS(Podd, Peven)
[S1, S2] = LATTICE_RICHARDS_H(alfa_odd, alfa_even, wT);
Att = MAG_2_ATT(0.5*(S2-S1));
PLOT_ATT_Z(wT, Att, pi, 80)
```

We obtain the poles and zeros shown in Fig. 6.98. As shown below, the function SORT_LATTICE_POLES_BP_Z assigns every other pole pair to S_1 and S_2.

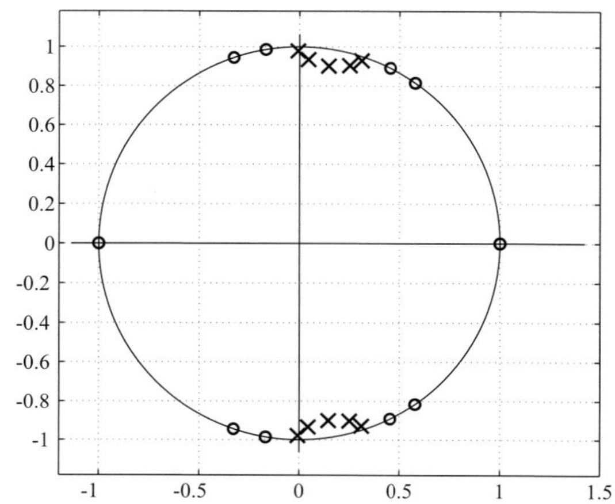

FIGURE 6.98

Poles and zeros for the bandpass lattice filter.

$$G_{bp} = 0.001540573$$

Zbp	Pbp	
±1.000000	0.1451952 ± 0.9013162i	S_2
−0.3274959 ± 0.9448525i	0.0443251 ± 0.9343518i	S_1
0.5780166 ± 0.81602500i	0.2495567 ± 0.9037655i	S_1
−0.1668641 ± 0.9859799i	−0.007322 ± 0.9785812i	S_2
0.4525461 ± 0.89174100i	0.3094194 ± 0.9295458i	S_2

The difference in order between the two branches is two. We may use second-order Richards' structure with symmetric two-port adaptors for the sections. Note that the attenuation, shown in Fig. 6.99 is symmetric around the center frequency. Hence, this filter can be realized using a symmetric ladder or a lattice structure. Multiband symmetrical lattice wave digital filters can be designed by using the function `SYM_WDF.m` [4].

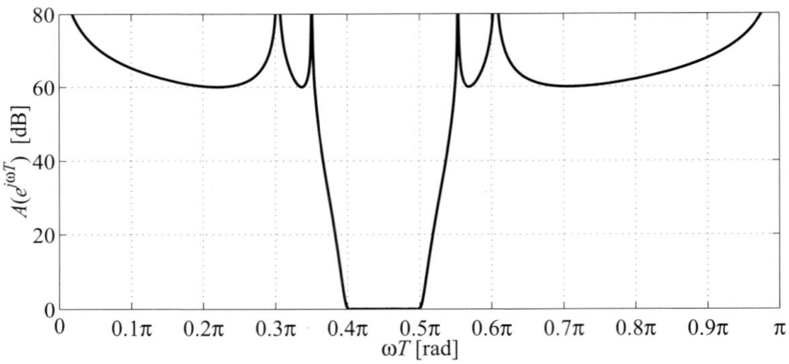

FIGURE 6.99

Attenuation for the bandpass lattice filter.

6.12.2 Lattice wave digital filter with a constant pole radius

Here we consider the design of odd-order low-pass lattice wave digital filters based on a reference filter with a constant pole radius. We assume that the transfer function is realized with a doubly resistive terminated lossless reciprocal network. Only Butterworth and Cauer filters with a constant pole radius are feasible. The magnitude response, $|s_{21}|$, of the corresponding wave digital filter is antisymmetric. Furthermore, if the passband edge is $\omega_c T$, then the stopband edge is $\omega_s T = \pi - \omega_c T$, and the relationship between the passband and stopband ripples is $\varepsilon_p = 1/\varepsilon_s$, where $\varepsilon_s = \sqrt{10^{0.1A_{min}} - 1}$. Hence, for reasonable stopband attenuations, the passband ripple will be very small, and the element sensitivity will be very small.

Note that $(N-1)/2$ adaptor coefficients in the second-order sections in an odd-order low-pass filter can be made equal. Hence, the dimensions of the parameter space are significantly constrained, and the complexity of finding quantized adaptor coefficients is thereby significantly reduced. Moreover, these adaptor coefficients can often be realized using trivial coefficients requiring only a few powers-of-two coefficients.

Note that the bilinear transformation is a conformal mapping between the s- and z-planes. Hence, circles and straight lines are mapped to circles or straight lines. The poles of the analog filter with a constant pole radius are on a circle with the pole radius. The corresponding poles in the wave digital filter are also on a circle with a center in the real axis in the z-domain [4]. The coefficient α_2 in the second-order all-pass section shown in Fig. 6.90 corresponds to $\alpha_2 = -(r_p T)^2$. Hence, α_2 will be a constant for all second-order sections. We may select the pole radius so that α_2 is simple to realize. The function `mincost`(n) determines the minimum number of add/subs required to implement a positive integer using less than five add/subs. This issue will be further discussed in Section 6.17.2.3.

Note that the coefficient α_2 is independent of the filter order and transition band. It only depends on the normalized bandwidth. Hence, it is possible to use the design margin to adjust the normalized bandwidth so that a_2 can be realized using only a few powers-of-two coefficients. Moreover, it is possible to realize a filter with variable bandwidth by merely changing α_2. We demonstrate the design process for digital Cauer filters with a constant pole radius by using an example.

6.12.2.1 Example

Consider the following specifications: $\omega_c T = 0.25\pi$ rad, $\omega_s T = 0.45\pi$ rad, and $A_{min} = 50$ dB. We use for the design the program shown in Listing 6.12. First, we synthesize an analog Cauer filter with a constant pole radius in lines 2 through 5 and map this filter to its digital counterpart in lines 6 through 8. We select, by experience, the coefficient word length of $W_c = 6$ bit, and we select the adaptor coefficient, which corresponds to the constant pole radius, to $a_2 = 34/64$, which can be realized using only one addition. In lines 18 through 43, we loop through a set of coefficients and evaluate the corresponding magnitude functions, while we use as cost function the sum of the number of additions required for their realization. There are only $(N+1)/2$ free coefficients when we have selected $a_3 = a_5 = a_7 = 34/64$. Thus, the search space is significantly reduced.

Listing 6.12: Program for the design of the lattice wave digital filter constant pole radius.

```
1 wcT = 0.25*pi;wsT = 0.45*pi; Amin = 50;T = 1;
2 wc = (2/T)*tan(wcT/2); ws = (2/T)*tan(wsT/2);
3 N = CA_CONST_R_ORDER_S(wc, ws, Amin)
4 N = 7
5 [Ga,Za,R_ZEROS,Pa,wsnew] = CA_CONST_R_POLES_S(wc,ws,Amin,N);
6 [G, Z, P] = PZ_2_PZ_Z(Ga, Za, Pa, 1);
7 [S1num, S1den, S2num, S2den] = SORT_LATTICE_POLES_LP_Z(P);
8 [a_odd,a_even] = CASCADE_RICHARDS_SECTIONS(S1den,S2den)
9 wTpassb = linspace(0,wcT,100); wTstopb = linspace(wsT,pi,100);
10 load mincost; % Number of add/sub for an integer
11 WX = 64;         % Coefficient word length: 2^6 = 64
12 TempCost = 100;
13 a_odd = round(a_odd*WX);  % a_odd(3) = mrpT2 = (-rpT)^2
14 a_even = round(a_even*WX); % a_even(2) = mrpT2;
15 N0 = 5; N1 = 5; Fail = 1; disp(['Searching, Please wait'])
16 mrpT2 = 34;  % Select the constant pole radius
```

```
C_odd(3) = mrpT2; C_even(2) = mrpT2; C_even(4) = mrpT2;
for k0 = -N0:N1
    C_odd(1) = a_odd(1)+k0;
for k1 = -N0:N1
     C_odd(2) = a_odd(2)+k1;
     Cost_odd = sum(mincost(abs(C_odd)));
for k2 = -N0:N1
    C_even(1) = a_even(1)+k2;
for k3 = -N0:N1
    C_even(3) = a_even(3)+k3;
    Cost = Cost_odd+sum(mincost(abs(C_even)));
if Cost < TempCost
    [S1, S2] = LATTICE_RICHARDS_H(C_odd/WX,...
            C_even/WX, wTstopb);
    Att = MAG_2_ATT(0.5*(S2-S1));
if min(Att) >= Amin
    [S1, S2] = LATTICE_RICHARDS_H(C_odd/WX,...
       C_even/WX, wTpassb);
    Att = MAG_2_ATT(0.5*(S2-S1));
    a_odd0 = C_odd; a_even0 = C_even;
    TempCost = Cost
    Fail = 0;
end
end
end
end
end
end
end
if Fail == 1, disp(['Failed']), break, end
figure(2); subplot('position', [0.1 0.4 0.88 0.5]);
wT = linspace(0, pi, 1001);
a_odd = a_odd0, a_even = a_even0, TempCost, WX
[S1,S2] = LATTICE_RICHARDS_H(a_odd/WX, a_even/WX,wT);
Att = MAG_2_ATT(0.5*(S2 - S1));
PLOT_ATT_Z(wT,Att,pi,70)
figure(3); PLOT_PZ_Z(Z, P)
```

We get $N = 6.309$ and select $N = 7$. The adaptor coefficients in the two branches are

a_odd = 0.2935188, -0.52011888. **0.54047406**
a_even = -0.2233453, **0.54047406**, -0.8342365, **0.54047406**

Searching yields TempCost = 7 (seven add/subs) and

a_odd = (16 –31 **34**)/64

a_even = (–12 **34** –51 **34**)/64

Note that there are three identical coefficients, 34/64, corresponding to the constant pole radius, i.e., $-(r_p T)^2$. There are often several coefficient sets with the same adder cost. Here the realization cost is only seven additions for the multiplications. There are seven adaptors with $7 \times 3 = 21$ additional structural adders. Hence, the realization cost of the multiplications is typically far less than the cost for the structural adders. In the literature, it is often assumed that multiplications dominate the implementation cost. This is, however, not the case when multiple-constant multiplication (MCM) techniques are employed.

If the order of the even branch is larger than the order of the odd branch, then $m =$ even, i.e., the outputs from the two branches are added.

6.12.2.2 Half-band lattice wave digital filter

The half-band IIR filters are a particular case of transfer functions based on reference filters with a constant pole radius. Half-band wave digital filters are often referred to as bireciprocal filters, although they are only a particular case [4]. The poles for a half-band filter lie on the imaginary axis. Thus, every odd adaptor coefficient in Fig. 6.95 becomes zero. The real pole lies at $z = 0$, and the odd-order section becomes a delay element. The number of adaptors and multipliers is only $(N-1)/2$, and the number of adders is $3(N-1)/2+1$, where N is odd for low-pass and high-pass filters.

The transfer function for a half-band lattice wave digital filter is

$$H(z) = \frac{1}{2}(S_2(z^2) - z^{-1}S(z^2)).$$

6.12.2.3 Example

Synthesize a half-band lattice wave digital filter with the following specifications: $\omega_s T = 0.54\pi$ rad and $A_{min} = 60$ dB. Note that the asymmetry requires that $\omega_c T + \omega_s T = \pi$ rad. An 11th-order filter meets the specification. In this case, we require 5 adaptors and 16 structural add/subs. The number of required add/subs can be determined by appropriately modifying the program in Listing 6.13.

The half-band lattice wave digital filter is realized using the structure shown in Fig. 6.100. If the order of the even branch is larger than the order of the odd branch, then $m =$ even, i.e., the outputs from the two branches are added.

Listing 6.13: Program for the design of half-band lattice wave digital filters.

```
wsT = 0.54*pi; Amin = 60;
N = CA_ORDER_HB_Z(wsT, Amin)
N = 11;
N = CA_ORDER_HB_Z(wsT, Amin)
% Compute the adaptor coefficients
[G,Z,P,wcT,wsT,Amax,Amin] = CA_POLES_HB_Z(wsT,Amin,N);
[S1num, S1den, S2num, S2den] = ...
      SORT_LATTICE_POLES_HB_Z(P);
[alfa_odd,alfa_even] = ...
      CASCADE_RICHARDS_SECTIONS(S1den,S2den);
```

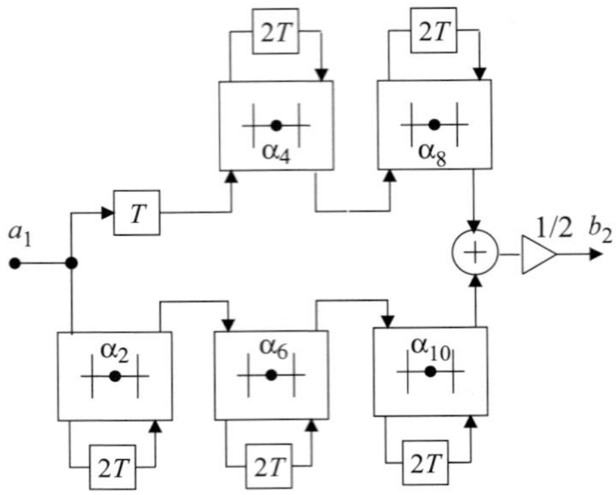

FIGURE 6.100

11th-order half-band lattice wave digital filter.

```
wT = linspace(0, pi, 1001);
[S1,S2] = LATTICE_RICHARDS_H(alfa_odd, alfa_even, wT);
Att = MAG_2_ATT(0.5*(S2 - S1));
figure(1);
subplot('position', [0.1 0.4 0.88 0.5]);
PLOT_ATT_Z(wT, Att, pi, 100), hold on
AttC = MAG_2_ATT(0.5*(S2 + S1));
plot(wT, AttC,'--','linewidth',2);
axis([0 pi 0 100]); grid on, hold off
figure(2);
subplot('position',[0.1 0.4 0.88 0.5]);
PLOT_ATT_Z(wT, Att, pi, 3*10^-6), hold on
plot(wT, AttC,'--', 'linewidth', 2)
```

The adaptor coefficients for the branches are:

odd branch:
$\alpha_1 = 0$,
$\alpha_4 = -0.315306$, $\alpha_5 = 0$,
$\alpha_8 = -0.759827$, $\alpha_9 = 0$,

even branch:
$\alpha_2 = -0.0936040$, $\alpha_3 = 0$,
$\alpha_6 = -0.5575275$, $\alpha_7 = 0$,
$\alpha_{10} = 0.9226919$, $\alpha_{11} = 0$.

All poles lie on the imaginary axis in the z-plane and alternate between the upper and lower branches of the lattice filter and $(N+1)/2$ adaptor coefficients are zero and do not require any multiplications. In addition, $3(N+1)/2$ structural adders are no longer required.

The attenuation of the low-pass filter and the complementary high-pass filter is shown in Fig. 6.101. Note that the passband ripple is extremely small. The symmetry between the two responses is perfect and remains symmetric independent of coefficient quantization. The crossover angle for a half-band lattice wave digital filter pair occurs at π/2 rad.

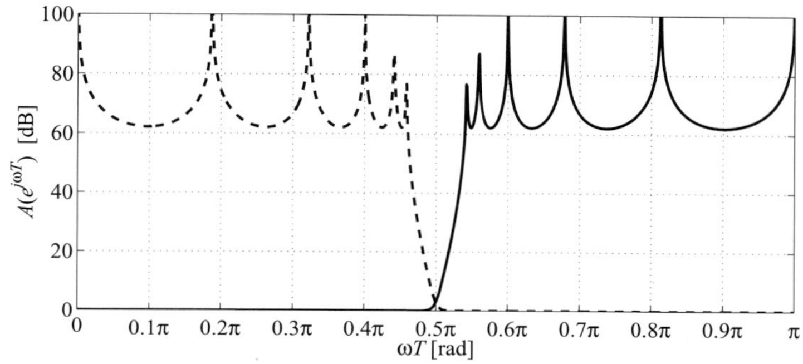

FIGURE 6.101

11th-order half-band lattice wave digital filter.

6.12.3 Lattice filters with a pure delay branch

One way to obtain a lattice wave digital filter with an approximately linear phase in the passband is to constrain one of the all-pass branches to a pure delay [68]. For example, the reflectance $S_2(z)$ corresponds to a cascade of M unit elements, all with port resistance equal to R. The transfer function is in this case

$$H(z) = \frac{1}{2}(z^{-M} - S_1(z)).$$

For low-pass and high-pass filters, the order of $S_1(z)$ must be $M+1$ or $M-1$. Hence, $N = 2M \pm 1$. It is beneficial to select the order of $S_1(z)$ as M+1, i.e., M = even [68]. These filters require much higher orders than their nonlinear-phase counterparts, especially when the transition band is narrow, but they are advantageous compared with FIR filters. There exist no closed-form solutions for these filters. Therefore, numerical optimization must be used, e.g., L_PHASE_LATTICE_WDF.m. This algorithm yields a filter with equiripple attenuation in the passband and stopband. Designed examples are found in [4].

6.12.3.1 Half-band lattice filter with a pure delay

Consider a half-band lattice wave digital filter with an approximately linear phase in the passband using a branch with a pure delay. The transfer function can, in this case, be written as

$$H(z) = \frac{1}{2}(S_2(z) - z^{-M})),$$

where M = odd. For a half-band lattice wave digital filter, all poles lie on the imaginary axis. For a half-band, linear-phase lattice wave digital filter, this is generally not the case. The poles can occur in pairs on the real or imaginary axis or as a quadruple of poles for these filters. Therefore, we use a cascade of sections of the type shown in Fig. 6.102.

For the synthesis of half-band, linear-phase lattice wave digital filters, there exist no closed-form solutions. Therefore, numerical optimization algorithms must be used. The approximately linear-phase half-band filters require much higher orders than their nonlinear-phase counterparts, especially when the transition band is narrow, but they are often advantageous compared with half-band FIR filters.

6.12.3.2 Example

Synthesize a 19th-order half-band linear-phase lattice wave digital filter having passband and stopband edges at $\omega_c T = 0.4\pi$ rad, $\omega_s T = 0.56\pi$ rad, and $A_{min} = 50$ dB. A_{max} will be very small, but here the function L_PHASE_IIR_HB unnecessarily requires it to be specified.

The filter is synthesized using the program shown in Listing 6.14.

Listing 6.14: Program for the design of half-band lattice wave digital filters with a pure delay branch.

```
wcT = 0.4*pi;wsT = 0.56*pi; Amin = 50; Amax = 0.1;
N = 19;
[G,Z,P,alfa] = L_PHASE_IIR_HB(wcT,wsT,Amax,Amin,N);
M = (N-1)/2;
figure(1), PLOT_PZ_Z(Z, P)
figure(2), subplot('position',[0.1 0.4 0.88 0.5]);
wT = linspace(0, pi,1001);
[S1, S2] = L_PHASE_LATTICE_RICHARDS_HB(alfa,M,wT);
Att = H_2_ATT(0.5*(S2 - S1));
subplot('position', [0.08 0.4 0.90 0.5]);
PLOT_ATT_Z(wT, Att, pi, 80), hold on
axes('position',[0.20 0.52 0.3 0.2]);
wTc = linspace(0, wcT,1000);
[S1, S2] = L_PHASE_LATTICE_RICHARDS_HB(alfa,M,wTc);
Att = H_2_ATT(0.5*(S2 - S1));
PLOT_ATT_Z_LS(wTc, Att, 3*10^-5)
figure(3), subplot(3,1,1);
[S1, S2] = L_PHASE_LATTICE_RICHARDS_HB(alfa,M,wTc);
H = 0.5*(S2 - S1);
Ph_error = unwrap(angle(H) + 9*wTc);
PLOT_DELTA_PHASE_Z(wTc, Ph_error,-3*10^-3,3*10^-3)
subplot(3,1,2)
Taug = PZ_2_TG_Z(G, Z, P, wTc);
PLOT_TG_Z(wTc, Taug, 14.4, 14.6)
```

The adaptor coefficients are

$$\alpha_1 = -0.7655339,\ a_2 = -0.09483945,\ a_3 = 0.4531888,\ a_4 = -0.1186705,\ a_5 = -0.1952483.$$

Note the different indexing of the adaptor coefficients. The realization of the filter is shown in Fig. 6.102.

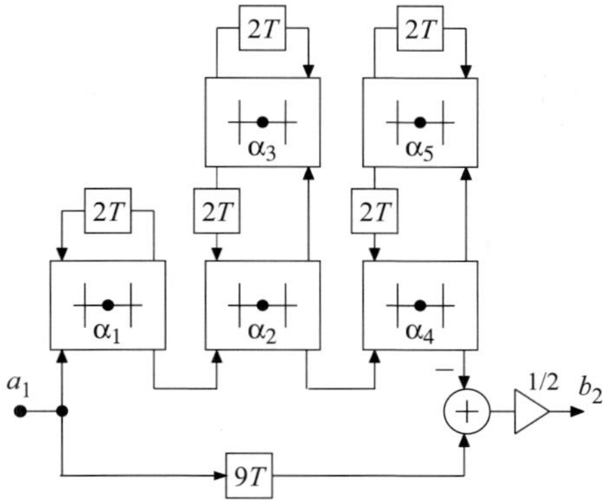

FIGURE 6.102

Half-band lattice filter with a pure delay.

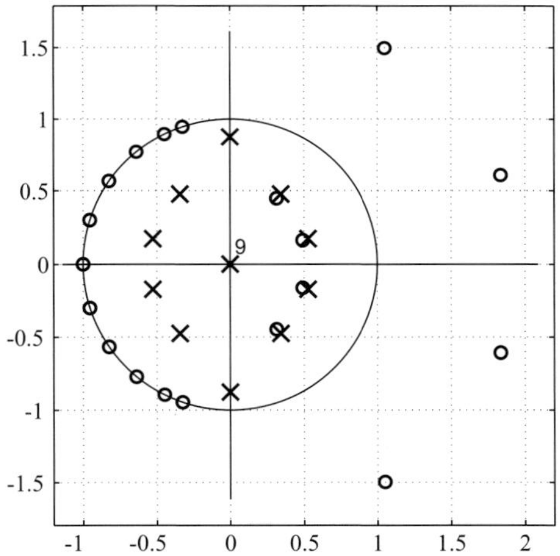

FIGURE 6.103

Poles and zeros for the half-band lattice filter with a pure delay.

If the order of the odd branch is larger than the order of the even branch, then m = odd, i.e., the output from the odd branch is subtracted.

The poles and zeros are shown in Fig. 6.103. The attenuation is shown in Fig. 6.104, whereas the phase error and group delay are shown in Fig. 6.105. As mentioned earlier, the passband ripple of a half-band lattice wave digital filter is very small if the stopband attenuation is relatively high. In the linear-phase case, this automatically leads to small phase error and group delay variations in the passband, as shown in Fig. 6.105.

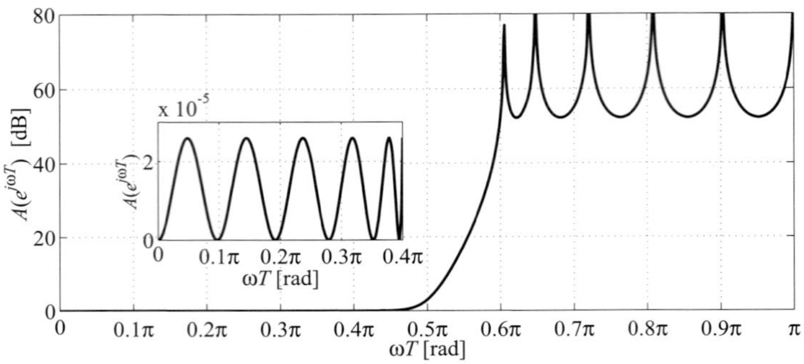

FIGURE 6.104

Attenuation for the half-band lattice filter with a pure delay.

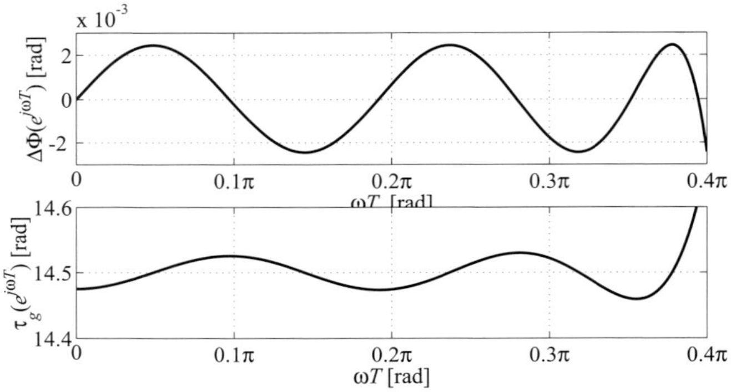

FIGURE 6.105

Phase error and group delay for the half-band lattice filter with a pure delay.

6.13 Circulator-tree wave digital filters

The reactance two-port that is shown in Fig. 6.37 can be described by the scattering matrix, S, given in (6.74). If the scattering matrix is symmetric, it can be factorized into a set of smaller elementary scattering matrices, S_i, corresponding to first- and second-order reactances. That is, for a tree with four two-ports,

$$S = S_1 S_2 S_3 S_4.$$

This corresponds to an analog structure involving circulator-trees, as shown in Fig. 6.106, where the lossless two-ports are elementary reactance two-ports.

Factorization of the scattering matrix is always possible for symmetric scattering matrices with $s_{11} = s_{22}$ and $s_{12} = s_{21}$. Hence, circulator-tree and lattice wave digital filters are directly related since they realize the same symmetric scattering matrices. The structure is a doubly resistively terminated

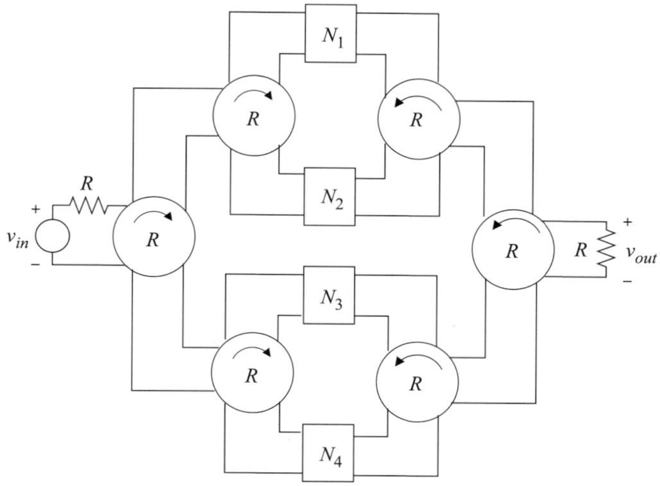

FIGURE 6.106

Circulator-tree structure.

reactance network that can be designed for maximum power transfer and, therefore, exhibits all the well-known favorable properties of such networks. Moreover, wave digital circulator-tree structures exhibit well-known properties with respect to finite word length effects, and they have a high degree of modularity.

The corresponding wave digital structure is shown in Fig. 6.107. Note that there are only feedforward waves from the inputs to the outputs. The only recursive parts are within the elementary scattering matrices. The two outputs are power-complementary, i.e., they obey Feldtkeller's equation. For an odd-order filter, one section is a first-order all-pass section, and it can be realized by using a three-port adaptor. The remaining all-pass sections require four-port adaptors, e.g., the four-port adaptor shown in Fig. 6.108. Note that if the sections are constrained to two-ports with a single transmission line, then the circulator-tree becomes a wave digital FIR filter.

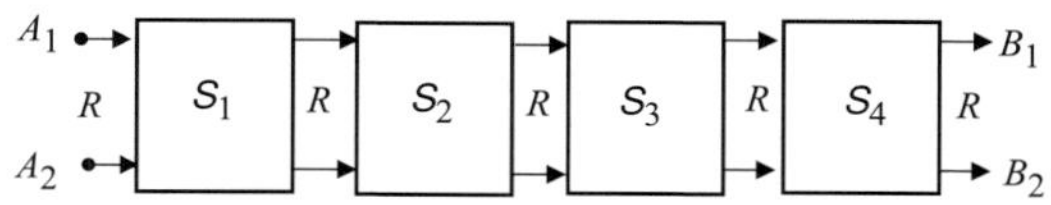

FIGURE 6.107

Wave digital circulator-tree structure.

Symmetric circulator-tree structures require the same number of coefficients as lattice wave digital filters, i.e., N coefficients for an Nth-order filter. Antisymmetric two-ports cannot be realized with a canonic number of multiplications. Hence, we do not discuss this case any further. Circulator-tree structures and lattice wave digital filters are restricted to odd-order low-pass and high-pass filters and even-order bandpass and bandstop filters.

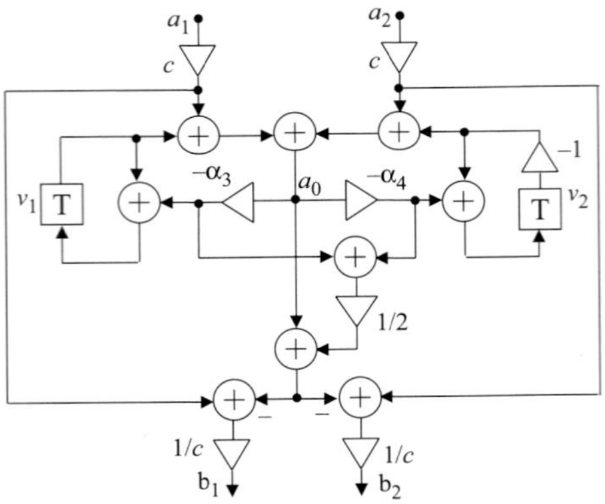

FIGURE 6.108

Four-port adaptor.

6.13.0.1 Example

Synthesize a circulator-tree wave digital filter with $\omega_c T = 0.3\pi$ rad, $\omega_s T = 0.4\pi$ rad, $A_{max} = 0.1$ dB, and $A_{min} = 60$ dB.

We compute the adaptor coefficients using the program shown in Listing 6.15.

Listing 6.15: Program for the design of circulator-tree wave digital filters.

```
wcT = 0.3*pi;wsT = 0.4*pi;Amax = 0.1;Amin = 60;
N = CA_ORDER_Z(wcT, wsT, Amax, Amin);
N = 7;          % We MUST select an odd order
[G,Z,R_ZEROS,P,wsnewT] =...
    CA_POLES_Z(wcT,wsT,Amax,Amin,N);
alfa = CT_TREE_WDF_LP(P)
wT = linspace(0, pi, 512);
[s11, s21] = CT_TREE_WDF_LP_H(alfa, wT);
Att = H_2_ATT(s21);
subplot('position', [0.1 0.4 0.88 0.5]);
PLOT_ATT_Z(wT, Att, pi, 80)
```

We get

S_i	Sign	$\alpha_1 = \alpha_2$	α_3	α_4
1	1	0.19636589	1.607268	-
2	1	0.249462	0.1647667	1.3363089
3	−1	0.1305664	0.3168790	1.421988
4	1	0.0383347	0.4151295	1.508200

A negative sign indicates that the corresponding section should have a sign inverter at a_2 and b_2 as shown below. The resulting realization of the circulator-tree structure is shown in Fig. 6.109.

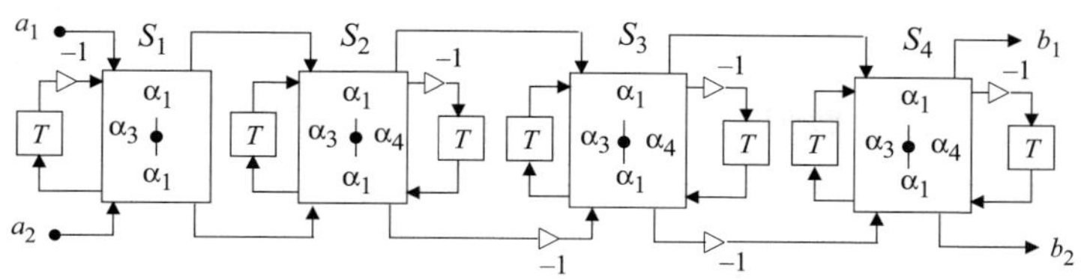

FIGURE 6.109

Circulator-tree structure.

Circulator-tree and lattice structures are closely related, and they have similar sensitivity properties. Design examples for circulator-tree filters with constant pole radius, half-band circulator-tree, band-pass, and stopband filters are found in [4].

6.14 Numerically equivalent state-space realization of wave digital filters

For a given linear signal-flow graph with input x, internal states v, and output y, we can write the state-space representation as

$$\begin{cases} V(n+1) = AV(n) + Bx(n), \\ y(n) = CV(n) + Dx(n), \end{cases} \tag{6.128}$$

where $V(n)$ is a vector containing the current states in the delay elements. We can rearrange these equations to obtain a more regular form suitable for realization using matrix-vector operations:

$$\begin{pmatrix} V(n+1) \\ y(n) \end{pmatrix} = \begin{bmatrix} B & A \\ D & C \end{bmatrix} \begin{pmatrix} x(n) \\ V(n) \end{pmatrix} = G \begin{pmatrix} x(n) \\ V(n) \end{pmatrix}. \tag{6.129}$$

Hence, we can realize the matrix-vector multiplication with the matrix G. We will later discuss the realization and implementation of such matrix operations.

In order to exactly represent a wave digital filter structure, we need to observe some constraints.

- First, the matrices A, B, C, and D are computed exactly from the wave-flow graph using quantized coefficients. Hence, the state-space representation is numerically equivalent to the original algorithm.
- Second, quantization of the internal signal word lengths is performed after the matrix multiplication, i.e., the signal b_k in Fig. 6.91 according to the constraint in (6.127). Hence, any parasitic oscillation will be suppressed.

The numerically equivalent state-space structure inherits the sensitivity properties and the robustness of the original wave digital filter [69,70,4]. Any wave digital filter structure can be realized using a numerically equivalent state-space realization. The state-space realization is obtained by computing the transmittances between the state variables (delay elements) and the input and output exactly. They have finite word length since the adaptor coefficients in the wave digital filter have finite word length. However, the word length becomes, in general, longer, but this drawback may be alleviated by using multiple-constant techniques or distributed arithmetic and a suitable wave digital structure [69,70,4]. Nevertheless, this approach has several advantages.

- It retains the robustness and sensitivity properties of the wave digital filter.
- The quantization nodes, i.e., in front of the delay elements, can be optimally scaled. Hence, shorter signal word lengths are sufficient.
- It can be implemented to achieve a maximally fast sample rate [4]. Any excess speed can be used to reduce power consumption by reducing the power supply voltage. This issue will be discussed later.
- It can be implemented using bit-parallel, digit-serial, and bit-serial arithmetic, based on multiple-constant techniques, as well as using distributed arithmetic, which is discussed later.

Consider a wave digital ladder filter. The elements in the matrix A will have a relatively long word lengths since the paths between the delay elements pass through several adaptors. Furthermore, few elements will be zero. Hence, ladder structures are not the best candidates for this realization technique.

Consider a lattice wave digital filter structure. It has two separate branches, with $(N-1)/2$ and $(N+1)/2$ delay elements that do not interact. Hence, the required coefficient word lengths will be shorter. Moreover, the A matrix has only $N^2/2$ nonzero elements. Lattice wave digital filters are therefore better candidates.

Circulator-tree structures yield a triangular A matrix and are therefore also a good candidate. In addition, it may be advantageous to introduce pipelining to reduce the lengths of the delay-to-delay paths.

We find that a numerically equivalent realization of a wave digital filter may in many cases be efficient if multiple-constant techniques or distributed arithmetic is employed instead of a direct realization of the wave-flow graph.

6.15 Computational properties of filter algorithms

In order to compare and evaluate different filter structures, we need to discuss a few basic properties of algorithms and their implementation.

6.15.1 Latency and throughput

We define *latency* as the time to generate an output value from the corresponding input value for an arithmetic operation, as illustrated in Fig. 6.110. The throughput is defined as the reciprocal of the time between successive outputs. In sequential algorithms, it is possible to increase the throughput of an algorithm, for example, by using pipelining [11].

6.15.2 Maximal sample rate

The maximal sample rate of a recursive algorithm is determined only by its recursive loops [71,11]. Nonrecursive parts of the signal-flow graph, e.g., input and output branches, generally do not limit the

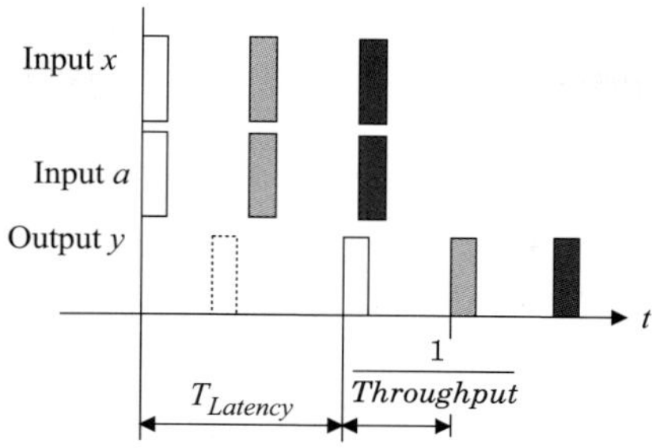

FIGURE 6.110

Latency and throughput.

sample rate, but additional delay elements may have to be introduced into the nonrecursive branches [11]. The maximal sample frequency for an iterative recursive algorithm, described by a fully specified signal-flow graph [11,4], is

$$f_{max} = \min_i \left\{ \frac{N_i}{T_{op}} \right\}, \tag{6.130}$$

where T_{op} is the total latency due to the arithmetic operations and N_i is the number of delay elements in the directed loop i. We will later define the latency for different arithmetic operations. The loop with the lowest f_{max} is called the critical loop. The minimum sample period is also referred to as the iteration period bound. It is, of course, not possible to improve the iteration period bound for a given algorithm. However, a new algorithm with a higher bound can sometimes be derived from the original algorithm using algorithm transformations [72,11,73]. We recognize from (6.130) that there are several possibilities to improve the bound:

- Reduce the operation latency, T_{op}, in the critical loop.
- Introduce additional delay elements into the loop without changing the behavior of the filter.
- Remove superfluous arithmetic or logic operations from the loop.

6.15.3 Cyclic scheduling

In order to achieve a maximally fast implementation that achieves the bound in (6.130), we must generally connect m computation graphs representing the operations within a single sample interval [11], as illustrated in Fig. 6.111.

This will result in a periodic scheduling formulation that allows a schedule period that is m times longer than the sample interval. This cyclic formulation will, under certain conditions, result in a maximally fast, periodic schedule. Unfortunately, it is not possible to determine the best choice of m in the general case. In order to attain the minimum sample period, it is necessary to perform cyclic scheduling of the operations belonging to several successive sample intervals if:

- the latency for a processing element (PE) is longer than T_{min} or
- the critical loop(s) contain(s) more than one delay element.

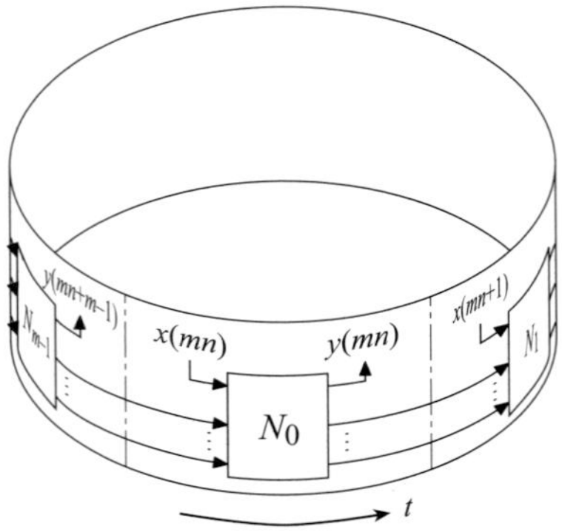

FIGURE 6.111

Cyclic scheduling.

Generally, the critical loop should be at least as long as the longest latency for any PEs in the loop. Bit-serial and digit-serial operations with latencies longer than the minimal sample period can be completed within the scheduling period. The minimum number of sample periods for the schedule is

$$m = \left\lceil \frac{max(T_{latency,i})}{T_{min}} \right\rceil, \tag{6.131}$$

where $T_{latency,i}$ is the total latency of operations in the loop i. The cyclic scheduling formulation makes it possible to find a resource-optimal schedule with a minimum number of concurrent operations [11,4].

6.16 Architecture

A significant problem when implementing an algorithm is to decide upon a suitable hardware architecture. For example, we may elect to use one or several PEs and one or several memories or memory ports and various communication networks between PEs and memories [11]. Typically the design target is to minimize the implementation cost, e.g., chip area and power consumption, while meeting the throughput requirement. Usually, there is no benefit in having a higher throughput in DSP applications than required.

We recommend that the arithmetic operations are scheduled cyclically to attain a maximally fast implementation, as illustrated in Fig. 6.111. A maximally fast schedule and the corresponding implementation are obtained using an isomorphic mapping between the arithmetic operations and the PEs [74,75,76]. This allows the hardware to operate with a low power supply voltage and low power consumption. Moreover, the communication network becomes simple and static. There are graphic-based methods to determine the number of independent memories or memory ports required [11].

It has been shown that a maximally fast implementation is obtainable using a numerically equivalent state-space representation [69,4] and distributed arithmetic [70,77]. These techniques can efficiently be implemented as a matrix-vector operation. Moreover, a numerically equivalent state-space representation of a wave digital filter can be scaled optimally, i.e., it uses the available signal range efficiently. Moreover, it will generate low round-off noise and will therefore require a shorter signal word length [69,70,11,4].

6.17 Arithmetic operations

Here we are mainly interested in the arithmetic operations: addition, subtraction, and multiplication, as well as SOP. There exist many alternative arithmetic algorithms to perform these operations. Consider, for example, addition. Addition of two numbers, independently of their sign, can be performed using a single circuit if two's-complement representation is used. Moreover, changing the circuit for addition into a subtractor requires only that the bits of one of the inputs are inverted. An advantage of the two's-complement number representation is that several numbers can be added, with possible intermediate overflows, if we can guarantee that the result is within the correct number range. This situation frequently occurs in digital filter algorithms. A drawback is, however, that the carry propagation limits the latency. Several techniques to speed up the computation of the carries and carry-free number representations have been proposed [78].

6.17.1 Addition and subtraction

Adding two or more numbers is in many ways the most fundamental arithmetic operation since most other operations are based on addition. The operands of concern here are either two's-complement or unsigned representation. Most DSP applications use fractional arithmetic instead of integer arithmetic [11,4]. The sum of two W-bit numbers is a $(W+1)$-bit number, while the product of two binary W-bit numbers is a $2W$-bit number. In many cases, and always in recursive algorithms, the resulting number needs to be quantized to a W-bit number. Hence, the question is which bits of the result are to be retained. In fractional arithmetic, two's-complement numbers are interpreted as being in the range $[-1, 1[$, i.e.,

$$x = -x_0 + \sum_{i=1}^{W-1} x_i 2^{-i}. \tag{6.132}$$

Therefore, the most significant part of a product is contained in the bits $(x_0, x_1, ...)$. We use the graphic representation shown in Fig. 6.112 to represent the operands and the sum bits with the most significant bit to the left.

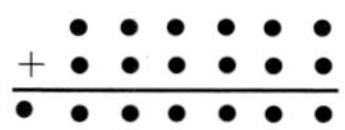

FIGURE 6.112

Illustration of addition of two binary numbers.

6.17.1.1 Ripple-carry addition

A straightforward way to add two numbers is to sequentially add the two bits of the same significance and the carry from the previous stage using a full adder (FA) and propagate the carry bit to the next stage. This is called ripple-carry addition and is illustrated in Fig. 6.113. Note that the operation propagates from right to left and that only one FA has correct inputs at a given moment. Hence, the circuit performs addition sequentially, one bit at a time, even though it is referred to as a parallel adder. This type of adder can add both unsigned and two's-complement numbers.

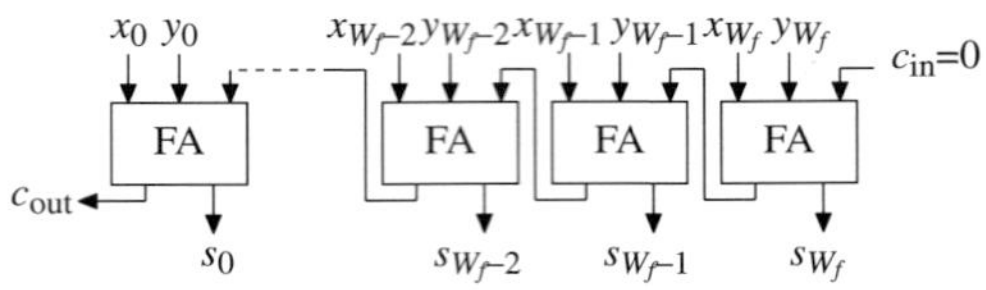

FIGURE 6.113

Ripple-carry adder.

The major drawback with the ripple-carry adder is that the latency is proportional to the word length. Also, typically the ripple-carry adder will produce many glitches since the FAs perform switching operations without having correct carries. This situation is improved if the delay for the carry bit is smaller than that for the sum bit [79].

However, due to the simple design, the energy per computation is still reasonable small [80]. Alternatively, all pairs of two bits of the same significance can be added simultaneously, and then the carries are added using some efficient scheme. Many adder schemes have been proposed. For more details, we refer to [80]. For carry-save addition we refer to [81,82]. It is also possible to perform addition in constant time using redundant number systems such as signed-digit or carry-save representations. An alternative is to use residue number systems, which break the carry chain into several shorter chains [83].

6.17.1.2 Bit-serial addition and subtraction

In bit-serial arithmetic, the inputs x and y are normally processed with the least significant bit first. Bit-serial numbers in two's-complement representation can be added or subtracted with the circuits shown in Fig. 6.114.

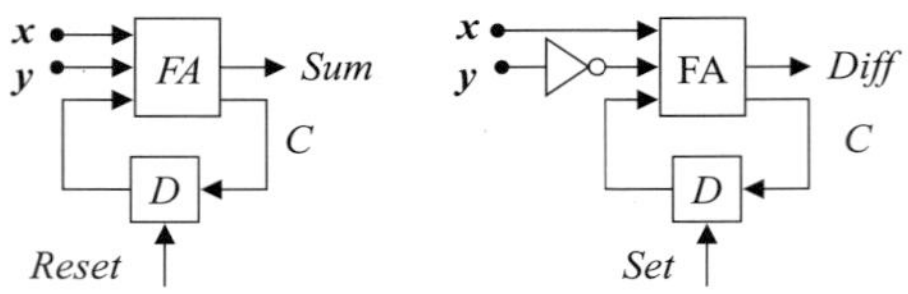

FIGURE 6.114

Bit-serial adder and subtracter.

Since the carries are saved from one bit position to the next, the circuits are called carry-save adder and carry-save subtractor, respectively. At the start of the addition, the D flip-flop is reset (set) for the adder (subtractor).

Fig. 6.115 shows two latency models for bit-serial adders. The leftmost is suitable for static CMOS logic and the rightmost is suitable for dynamic (clocked) logic styles. The time to add two W-bit numbers is W or W+1 clock cycles. Note that the first bit of the sum, delayed by the latency given in Fig. 6.117, can directly be fed to any subsequent bit-serial PE. Thus, the throughput of a complex computation may not necessarily be determined by the long addition time [11,4].

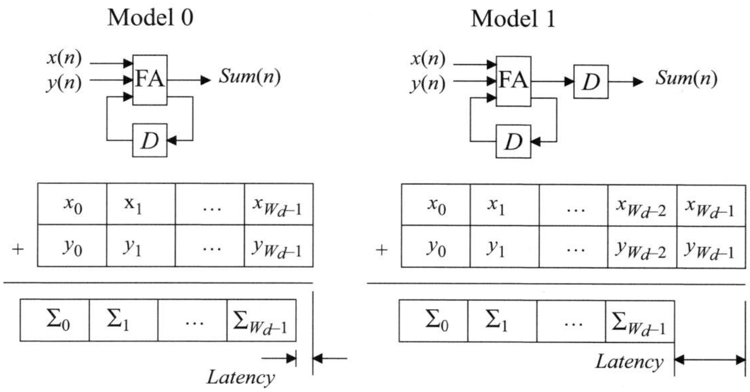

FIGURE 6.115

Latency models for a bit-serial adder.

Comparing the ripple-carry adder and the bit-serial adder, we note that both operate sequentially with one bit addition at a time. In the ripple-carry adder, each bit addition is mapped to its private FA, while in the bit-serial adder, all bit additions are mapped to a time-multiplexed FA. Hence, the difference in throughput is mainly due to the time lost in the D flip-flop. The chip area for bit-serial circuits is small, and consequently, the wiring is also small, but standard D elements have large power consumptions.

6.17.1.3 Digit-serial addition and subtraction

From the speed and power consumption points of view, it may sometimes be advantageous to process several bits simultaneously using so-called digit-serial processing [84,85,86,87,88,89,90]. The number of bits processed in a clock cycle is referred to as the digit size. Fig. 6.116 shows a d-bit digital-serial adder.

The flip-flop transmits the carry between successive digits. In the case of subtraction of a two's-complement number, the negative value is instead added by inverting the bits and setting the carry flip-flop. Most of the principles for bit-serial arithmetic can easily be extended to digit-serial arithmetic. An advantage of bit-serial and digit-serial arithmetic is that less chip area is required, and therefore the equivalent switched capacitance and leakage current is low [91]. Furthermore, the difference in throughput between a conventional word level and a digit-serial implementation is not great [92]. Therefore, both bit-serial and digit-serial arithmetic circuits are suitable for implementations in FPGA circuits.

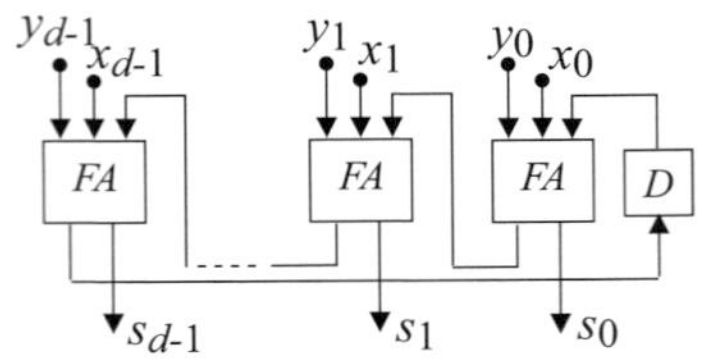

FIGURE 6.116

Digit-serial adder with digit size *d*.

6.17.2 Multiplication

Bit-parallel multiplication of two numbers can be divided into two main steps:

- partial product generation that determines the bits to be added and
- accumulation of the partial products.

Two main approaches to perform the accumulation have been proposed, namely array and tree type accumulation [78]. In addition, bit-serial and digit-serial multiplications are usually based on a shift-and-add approach [11,93]. However, we will focus on techniques that use addition and subtraction as generic arithmetic operations.

6.17.2.1 Serial/parallel multipliers

Two latency models for a serial/parallel multiplier are shown in Fig. 6.117. Denoting the number of fractional bits of the coefficient by W_{cf}, the latencies are W_{cf} for latency model 0 and $W_{cf} + 1$ for latency model 1. The corresponding latency models for digit-serial arithmetic are defined in the same manner.

In model 0, which corresponds to a static CMOS logic style without pipelining of the gates, the latency is equal to the gate delay of an FA. In model 1, which corresponds to a dynamic CMOS logic style or a static CMOS logic style with pipelining at the gate level, the FA is followed by a D flip-flop, making the latency equal to one clock cycle. Model 1 generally results in faster bit-serial implementation due to the shorter logic paths between the flip-flops in successive operations.

To identify the critical loop in the filter, the total latency of the operations inside the loops must be determined. The total latency of several sequential arithmetic operations depends on how they are realized at the logic level because the clock frequency is determined by the longest propagation path between any two registers. The propagation paths do not only depend on the internal circuitry in an arithmetic unit; they also depend on the paths between units. Hence, it may be efficient to use pipelining inside the critical loop to avoid long propagation paths between arithmetic units.

6.17.2.2 Example

Consider a lattice wave digital filter. The critical loop in a first-order all-pass section is indicated in Fig. 6.118 [75,76].

This loop, containing two additions, one multiplication with α_0, and one delay element, has the minimal sample period given by

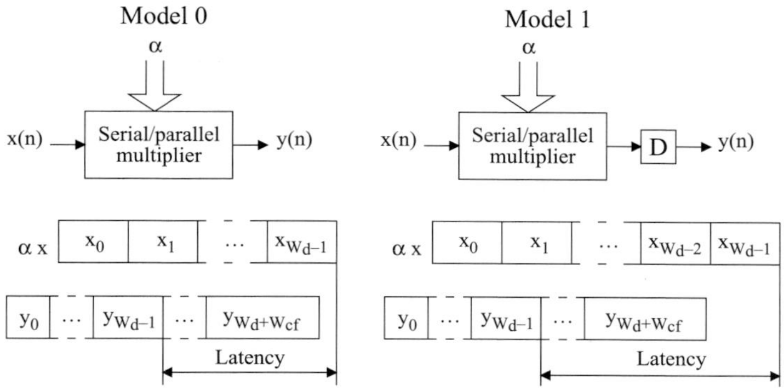

FIGURE 6.117

Latency models for a serial/parallel multiplier.

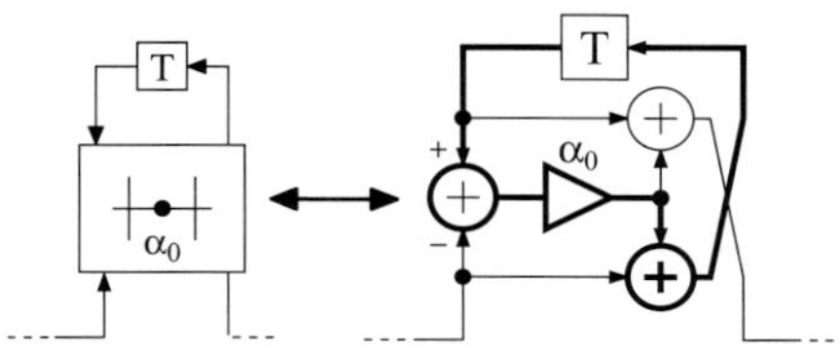

FIGURE 6.118

First-order all-pass section.

$$T_{min1} = \frac{T_{\alpha_0} + 2T_{add}}{1}. \tag{6.133}$$

In Fig. 6.119, the critical loop in a second-order all-pass section is indicated. For this loop, consisting of four additions, two multiplications, and one delay element, the corresponding minimal sample period becomes

$$T_{min2} = \frac{T_{\alpha i} + T_{\alpha i-1} + 4T_{add}}{1}. \tag{6.134}$$

The section with the lowest throughput then bounds the minimal sample period of the filter, i.e., the loop i that yields T_{min}, since each section has to operate with the same throughput.

Fig. 6.120 shows the cyclic scheduling of the operations belonging to m sample periods for a first-order all-pass section. The shaded areas indicate execution time for the operations, and darker shaded areas indicate their latency. Scheduling of second-order all-pass sections can be found in [11,4]. Since a bit-serial processing element only processes one bit in each clock cycle, a complete arithmetic operation requires at least W_d clock cycles. Therefore, for the minimal sample period of T_{min} clock cycles, where $T_{min} < W_d$, operations belonging to m sample periods need to be included in the schedule. It can be shown that m must be selected as

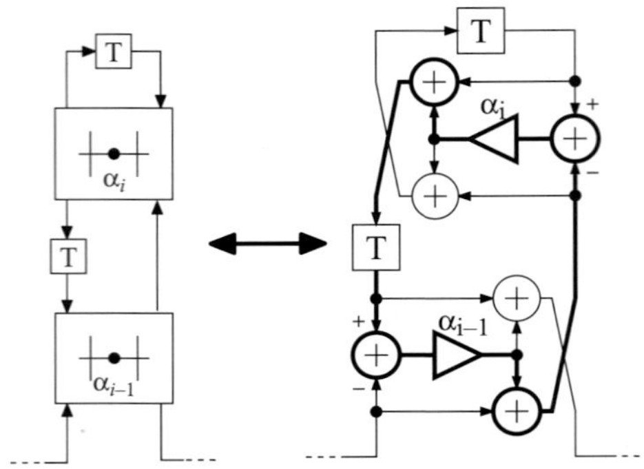

FIGURE 6.119

The critical loop in a second-order all-pass section.

$$m \geq \left\lceil \frac{W_{cf} + W_d + m_{n0}}{T_{min}} \right\rceil, \tag{6.135}$$

where m_{n0} is 0 for model 0 and 1 for model 1, to allow an arithmetic operation requiring $W_{cf} + W_d + m_{n0}$ clock cycles to be completed within the scheduling period. Thus, the scheduling period for a maximally fast schedule becomes mT_{min}. The implementation of the complete filter is illustrated in Fig. 6.121.

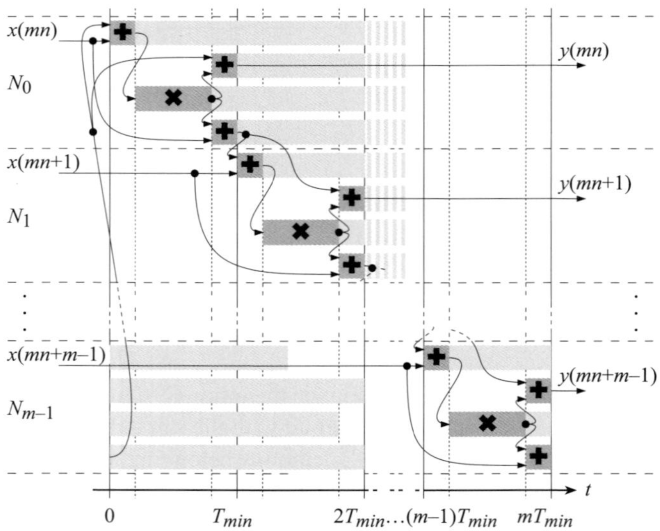

FIGURE 6.120

Maximally fast schedule for a first-order all-pass section.

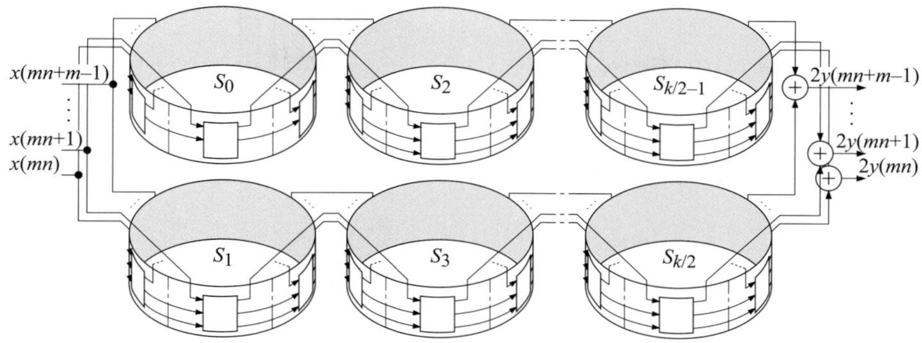

FIGURE 6.121

Complete lattice wave digital filter.

6.17.2.3 Constant multiplication

General multipliers that can execute arbitrary multiplications of two numbers with word lengths W_d and W_c are costly. Typically the coefficient word length is shorter than the signal word length, i.e., $W_c < W_d$. In many cases, we may assign a dedicated multiplier to each multiplication and simplify the implementation by specializing the multipliers for each particular coefficient value. Here we assume that the filter is implemented using FPGA or a standard cell approach. We have full freedom to design a specialized multiplier that is optimized for a specific coefficient. In this case, a significant reduction in the implementation cost, i.e., chip area or power consumption, can be achieved. In addition, in most cases, the throughput can also be increased.

Multiplication with a power-of-two can be realized by a simple shifting operation using two's-complement representation. For example, multiplication with a factor $2^n > 1$, $n > 0$, involves shifting the signal word to the left, e.g.,

$$2^2(x_0.x_1x_2x_3) = x_0x_1.x_2x_300,$$

where x_0 is the sign bit and (.) denotes the binary point. Zeros replace the bits to the left. Multiplication with a factor $2^n < 1$, $n < 0$, involves shifting the signal word to the right and copying the sign bit, e.g.,

$$2^{-2}(x_0.x_1x_2x_3) = x_0.x_0\mathbf{x_0}\mathbf{x_1}x_2x_3.$$

This property of two's-complement representation allows us to simplify the implementation of multipliers with constant coefficients. Any binary number can be split into a SPOT. Multiplication can then be implemented by a shift-and-add approach. It is advantageous to minimize the number of nonzero bits in the coefficients to significantly reduce the cost of the multiplier. The number of adders/subtractors equals the number of nonzero bits minus one [94,11,4]. Of particular interest are coefficients that can be implemented with a minimum number of shift-and-add operations.

Consider the multiplication with the integer 23. It is often simpler to view the multiplication in terms of integers rather than factional numbers. We can, at the end, change the position of the binary point.

We have

$$23_{10} \cdot x = (010111)_2 \cdot x = 2^4 x + 2^2 x + 2x + x = x{<<}4 + x{<<}2 + x{<<}1 + x,$$

where we have used the in the literature commonly used notation for left-shift operator. The number of adders equals the number of nonzero bits minus one; in this case, $4 - 1 = 3$.

We may also use a signed-digit representation. We have

$$23_{10} \bullet x = (01100\overline{1})_{SD} \bullet x = 2^4 x + 2^3 x - x,$$

which requires only two adders/subtractors; $\overline{1}$ denotes -1. Moreover, we may realize the multiplications by noting that $23_{10} \bullet x = 2^3(2x + x) - x$, which also requires two adders/subtractors. We conclude that there may exist several alternative realizations of a constant multiplication [78,4].

Extensive efforts have been made to find those multiplier algorithms that minimize the number of adders/subtractors required for performing the fixed coefficient multiplication. The assumption is that the number of adders represents the required chip area. However, it is often also important to consider throughput and power consumption. The longest propagation path limits the throughput, but this limit can be alleviated using pipelining. On the other hand, the dynamic power consumption, which depends on switching and glitches inside the adder network, tends to be high if the longest propagation path is long. Significant efforts have therefore been directed towards finding compromises between chip area, throughput, and power consumption.

Fig. 6.122 shows all possible graphical representations of a network consisting of adders/subtractors and shifts to realize a single multiplication with one, two, and three adders. The nodes with two incident branches represent a single addition/subtraction, while nodes with three incident branches represent two additions/subtractions. The branches have either transmittance equal to 1 or a power of two. For example, the first graph with only a single adder may realize a multiplication by 9 when one of the branches has transmittance 1 and the other 8, i.e., three left-shifts. There exist 11 and 34 graphs with four and five adders/subtractors, respectively. Only five adders/subtractors are required to realize all coefficients with word length less than 20 bits. These numbers are stored in the file `mincost.mat`. If the filter structure has low coefficient sensitivity, the coefficients can be quantized to have many zero-bits. Typically, only a few powers-of-two, i.e., less than five bits, are required. If the distance between the two furthest nonzero bits is large, then the coefficient sensitivity is also large.

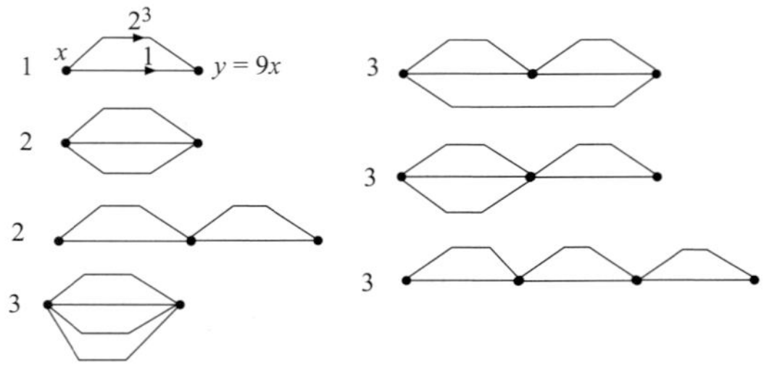

FIGURE 6.122

Graph representation of multiplication with one, two, or three adders/subtractors.

6.18 Composite arithmetic operations

This section discusses the two main techniques to simplify the hardware implementation of MCM and SOP. Fig. 6.123(a) shows a multiple-constant multiplication structure. Typically, the coefficients a_k are fixed and x_k are variable, but there are several common cases where both a_k and x_k are variable, for example, computation of correlation, which requires a general multiplier.

The arithmetic structure shown in Fig. 6.123(b) consists of a set of SOP, i.e.,

$$y = \sum_{k=1}^{N} a_k x_k. \tag{6.136}$$

The multiple-constant multiplication structure can be derived from a transposed version of the SOP by reversing the directions of the signal flow graphs. These two cases often appear as substructures in digital filter algorithms.

Distributed arithmetic is an efficient procedure for computing SOP between a fixed and a variable signal vector. The basic principle is owed to Croisier et al. [95]. Distributed arithmetic is suitable for the implementation of relatively short SOP, discrete cosine transforms, and complex multiplications [11,4].

6.18.1 Multiple-constant multiplication

It should be noted that by transposing an MCM structure, an SOP is obtained. If an optimal solution is obtained for one of these cases, the same solution, but transposed, is valid for the other case. There are two main routes to find efficient realizations of SOP using only shifts and adder/subtractor structures. The first approach is to find recurring bit patterns in the coefficient representation, which need to be computed only once and then reused. A problem with this approach is that it depends on the representation of the coefficients. This approach is referred to as subexpression sharing [96].

MCM techniques [78,11,4] are efficient methods to simplify multipliers with fixed coefficients and reduce power consumption. Single multipliers and substructures of the type shown in Fig. 6.123 can be simplified significantly by using MCM techniques [97,98,99,100,101,102]. Several effective algorithms have been proposed for the MCM case that avoids number representation dependency [103,104]. Lower bounds for MCM problems have been presented in [105]. Moreover, using linear programming, filter coefficient optimization and MCM implementation can be jointly addressed [106,107].

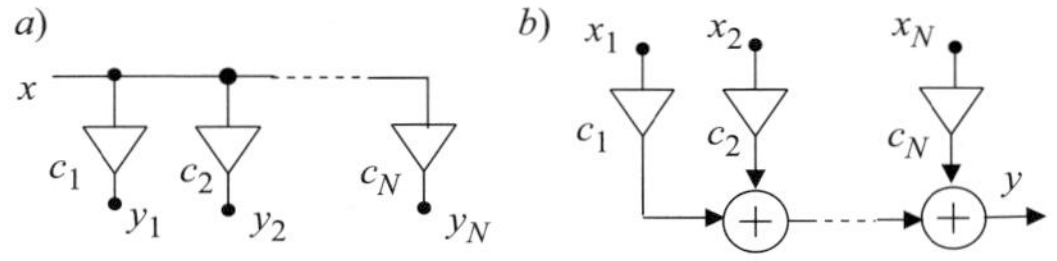

FIGURE 6.123

Multiple-constant multiplication and sum-of-products.

Fig. 6.124 shows a typical example of an MCM network. The basic idea is to reduce the number of basic operations (i.e., add/sub-and-shift operations) by factoring out common factors among the coefficients c_i by first generating a set of common subexpressions, d_1, d_2, d_3. In the next step, the simple coefficients are multiplied and added/subtracted, generating new subexpressions, and so on [82,108].

For MCM cases with many coefficients, for example, for long FIR filters, the average increase of the number of adders/subtractors approaches one, increasing the products' length by one. This is because most subexpressions have already been generated, and only a single addition/subtraction is needed to generate another coefficient. Thus, the problem of finding optimal solutions to the MCM problem is NP-complete. Therefore, several heuristic algorithms have been derived; see, e.g., [97,99,100,109,110, 111,108].

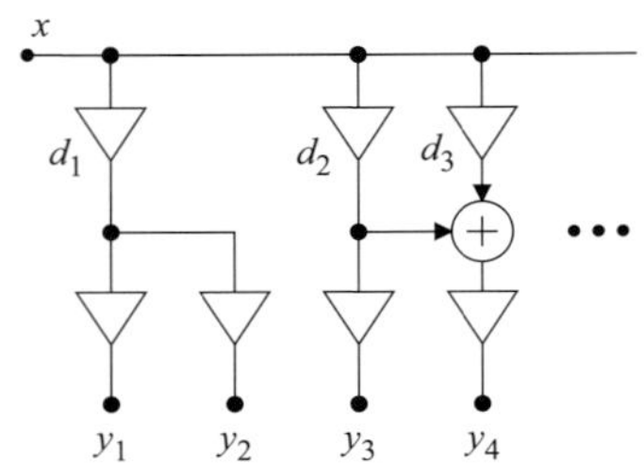

FIGURE 6.124

MCM network.

Fig. 6.125 shows a graph that simultaneously realizes the multiplication with 7, 11, and 106. Only three adders/subtractors and some shift operations are required. These techniques are efficient for implementing bit-parallel, digit-serial, and bit-serial processing elements [11,4].

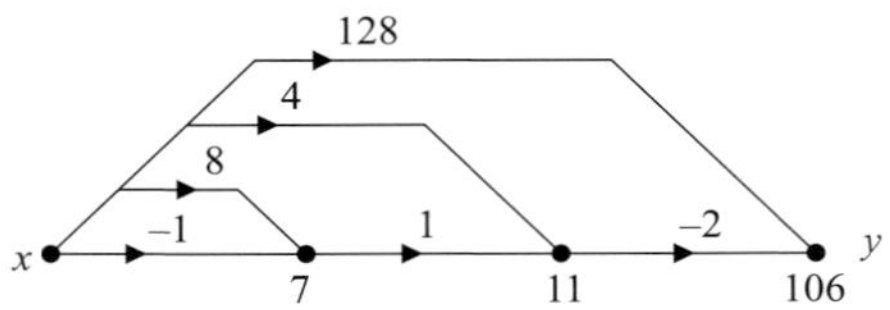

FIGURE 6.125

MCM example.

According to Fig. 6.111, a realization will consist of a set of MCM structures where the input to a network is taken from a register and the output is to be stored into other registers.

6.18.2 Distributed arithmetic

Consider the SOP in (6.136) and assume that the coefficients a_k, $k = 1, 2, ..., N$, are fixed. Two's-complement representation is used for both coefficients and signal. The inputs are scaled so that $|x_k| \leq 1$. The SOP can be rewritten as

$$y=\sum_{k=1}^{N} a_k\left(-x_{k0}+\sum_{i=1}^{W_d-1} x_{ki}2^{-i}\right), \tag{6.137}$$

where x_{ki} is the ith bit in x_k. By interchanging the order of the two sums, we get

$$y=-\sum_{k=1}^{N} a_k x_{k0}+\sum_{i=1}^{W_d-1}\sum_{k=1}^{N}(a_k x_{ki})\,2^{-i}, \tag{6.138}$$

which can be written as

$$y=-F_0\left(x_{10},x_{20},\ldots,x_{N0}\right)+\sum_{i=1}^{W_d-1} F_i\left(x_{1i},x_{2i},\ldots,x_{Ni}\right)2^{-i}, \tag{6.139}$$

where

$$F_i\left(x_{1i},x_{2i},\ldots,x_{Ni}\right)=\sum_{k=1}^{N} a_k x_{ki}. \tag{6.140}$$

F is a function of N binary variables, the kth variable being the ith bit in the signal, x_k. Since F_i can take on only a finite number of values, 2^N, it can be precomputed and stored in a read-only memory (ROM). Using Horner's method to evaluate a polynomial for $x = 0.5$, Eq. (6.139) can be rewritten as

$$y=\left(\left(\left(\left((0+F_{W_d-1})\,2^{-1}+\ldots+F_2\right)2^{-1}+F_1\right)2^{-1}-F_0\right)\right). \tag{6.141}$$

Fig. 6.126 shows a block diagram for computing an SOP according to (6.141).

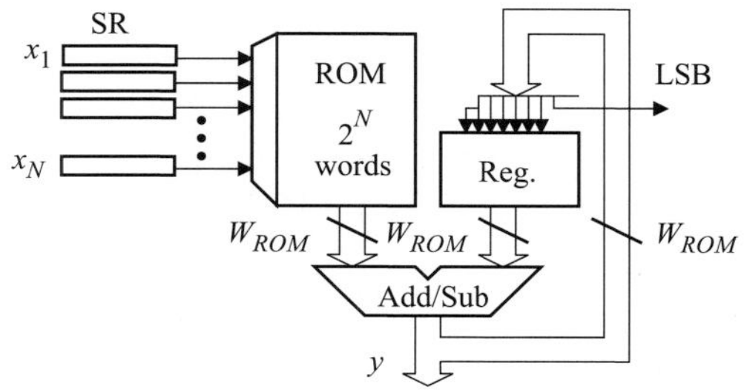

FIGURE 6.126

Block diagram for distributed arithmetic.

Inputs $x_1, x_2, \ldots, x_N$ are shifted out bit-serially from the shift registers with the least significant bit first. The bits x_{ki} are used as an address to the ROM storing the look-up table. Computation of the SOP starts by adding F_{W_d-1} to the initially cleared accumulator register, REG. In the next clock cycle, outputs from the shift registers address F_{W_d-2}, which is added to the value in the accumulator register. The result is divided by 2 and stored back into REG. After $W_d - 1$ clock cycles, F_0 is subtracted from the value in the accumulator register. The computation time is W_d clock cycles. The required word length in the ROM, W_{ROM}, depends on F_{imax} with the largest magnitude and the coefficient word length, W_c, and $W_{ROM} = \lceil W_c + \log_2(N) \rceil$.

The hardware cost is similar to that of a multiplier. In fact, most multiplier structures, e.g., serial/parallel, array, and tree-structure multipliers, can be used for distributed arithmetic. Distributed arithmetic can be used to implement first- and second-order direct-form I sections, FIR filters, and wave digital filters [112,95,113,114,4].

6.18.2.1 *Implementation of FIR filters using distributed arithmetic*

Fig. 6.127 shows an implementation of an 11th-order linear-phase FIR filter. $N/2$ bit-serial adders (subtractors) are used to sum the symmetrically placed values in the delay line. This reduces the number of terms in the SOP. Only 64 words are required, whereas $2^{12} = 4096$ words are required for the general case, e.g., a nonlinear-phase FIR filter. For higher-order FIR filters, the reduction in the number of terms by 50% is significant. Further, the logic circuitry has been pipelined by introducing D flip-flops between the adders (subtractors) and the ROM and between the ROM and the shift-and-add accumulator.

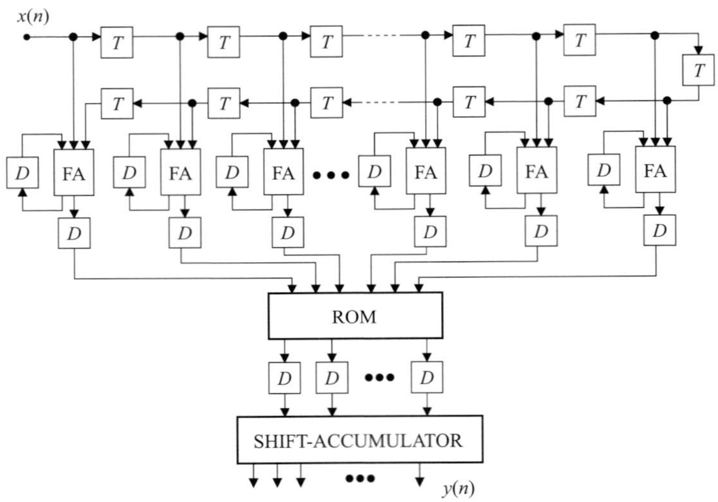

FIGURE 6.127

11th-order linear-phase FIR filter.

The number of words in the ROM is 2^N, where N is the number of terms in the SOP. The chip area for the ROM is small for SOP with up to 5 to 6 terms. The basic approach is useful for up to 10 to 11 terms. However, SOPs containing many terms can be partitioned into a number of smaller SOPs, which can be computed and summed by using either distributed arithmetic or an adder tree.

A complete SOP PE and several similar types of PEs, for example, adaptors and butterflies, can be based on distributed arithmetic. The main parts of an SOP PE are the shift-accumulator, the coefficient ROM with decoder, and the control circuits. Overflow and quantization circuitry are also necessary, but it is often advantageous to move parts of this circuitry to the serial/parallel converters used in the interconnection network [11].

6.18.2.2 Memory reduction techniques

The amount of memory required becomes very large for long SOP. There are mainly two ways to reduce the memory requirements, namely, partitioning and coding of input signals. The two methods can be applied at the same time to obtain a very small amount of memory.

One way to reduce the overall memory requirement is to partition the memory into smaller pieces that are added before the shift-accumulator, as shown in Fig. 6.128.

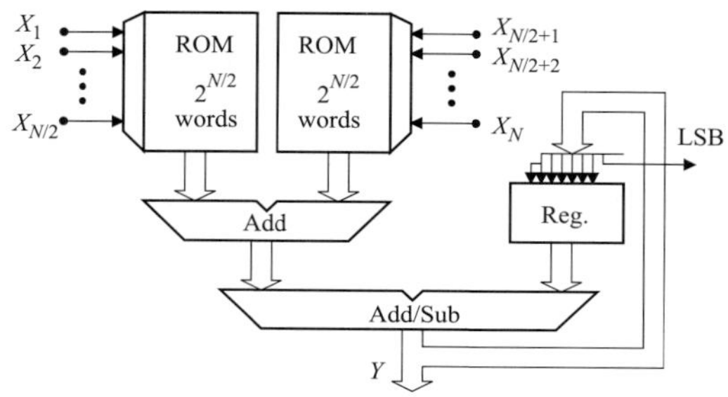

FIGURE 6.128

Reducing the memory by partitioning.

The amount of memory is reduced from 2^N words to $2 \cdot 2^{\frac{N}{2}}$ words if the original memory is partitioned into two parts. For example, for $N = 10$, we go from $2^{10} = 1024$ words to $2 \cdot 2^5 = 64$ words. Hence, this approach reduces the memory significantly at the cost of an additional adder. In addition, the partitioning may, in some cases, be done so that long and short values are stored in different partitions. Note that depending on the values in the ROM, it is often favorable from both speed and area points of view to implement a ROM by logic gates. Large ROMs tend to be slow.

The second approach is based on a special coding of the inputs, x_i, which results in a symmetric ROM content [95,11]. Memory size can be halved by using the ingenious scheme [95] based on the identity

$$x = \frac{1}{2}\left(x - (-x)\right). \tag{6.142}$$

This approach is also used for the realization of a complex multiplier using only two shift-accumulators and a ROM [11,4]

6.19 Power reduction techniques

Static CMOS technology is currently and will continue to be the dominating implementation technology. However, the feature size is rapidly shrinking towards 7 nm and below.[1] Note that the size of viruses is in the range of 10 to 400 nm. Circuits implemented using modern CMOS processes are no

[1] State-of-the-art 2021: 300-mm wafers with a transistor gate length of 2 nm.

longer characterized by the number of transistors on the chip. Instead, the chip area, overall throughput, and power consumption are more appropriate metrics. Here, we assume the digital filter is operated continuously, and therefore the leakage and short-circuit currents can be neglected [42]. The power dissipation associated with switching in CMOS circuits is approximately given by

$$P = \alpha f_{CL} C_L V_{DD} \Delta V_{DD}, \tag{6.143}$$

where α is the activity factor, i.e., the average number of transitions from 0 to 1 for the equivalent electrical node per clock cycle, f_{CL} is the clock frequency, C_L is the equivalent switched capacitance of the circuit, V_{DD} is the power supply voltage, and ΔV_{DD} is the voltage swing. In most cases, we use CMOS circuits with full swing, i.e., $\Delta V_{DD} = V_{DD}$. Reducing any of the factors in (6.143) can reduce power consumption. Below we briefly review various techniques that can be applied to reduce the power consumption in digital filters and are useful in many other DSP algorithms [9].

The *activity factor* α can be reduced in several ways. Some techniques exploit the known statistical properties of the signals. For example, operations to be executed on an adder or multiplier can be scheduled with respect to the correlation between successive operands so that the carry propagation and glitches are minimized. For example, only the first FA (LSB) in a bit-parallel ripple-carry adder has correct inputs and performs a valid addition in the first moment. The other FAs have carry-ins that are not yet determined; hence, they may perform invalid additions that consume energy. In the next moment, the second FA has correct inputs and performs a valid addition while the first FA is idle. In each instance, the remaining FAs may change their states and increase power dissipation. The bit-parallel ripple-carry adder performs a bit-by-bit addition sequentially, while in bit-serial arithmetic, a bit-by-bit addition is performed sequentially on a single FA. The execution time for a bit-parallel addition using a ripple-carry adder is almost the same as for a bit-serial addition if the safety margins needed for the flip-flops that separate two successive bit additions are neglected. Techniques to speed up the propagation of the carries favor, however, bit-parallel adders at the expense of increased power consumption.

Glitches: A redundant number representation [78] is another technique to avoid long carry propagation and glitches, which are costly in terms of power consumption. Glitches in logic circuits occur when the difference in arrival times of two logic signals to a gate is so large that false transitions occur and change the states of some electrical nodes. Slowing down the signal paths that are too fast so that the signals arrive simultaneously can minimize this effect and reduce power consumption by up to 40% in a bit-parallel multiplier.

In CMOS technologies with small feature sizes, the switched capacitance associated with the devices is relatively small compared to interwire capacitance, as illustrated in Fig. 6.129. The interwire capacitance can be about six times larger than the wire-substrate capacitance. For long buses, it is therefore important to order the transmission of successive signals on the bus so that the charging/discharging of the wire capacitors is minimized. A useful measure to minimize is the sum of the Hamming distances of the transmitted words. For example, in speech signals, it may be advantageous to use sign-magnitude representation on long buses since most samples have small magnitudes, which causes only the least significant bits to vary, while in two's-complement representation, almost all bits change when a sample value changes sign.

The *clock frequency*, f_{CL}, of the circuitry should of course not be reduced, since this would reduce the computational work performed by the circuitry, and the inherent processing capacity of the circuitry

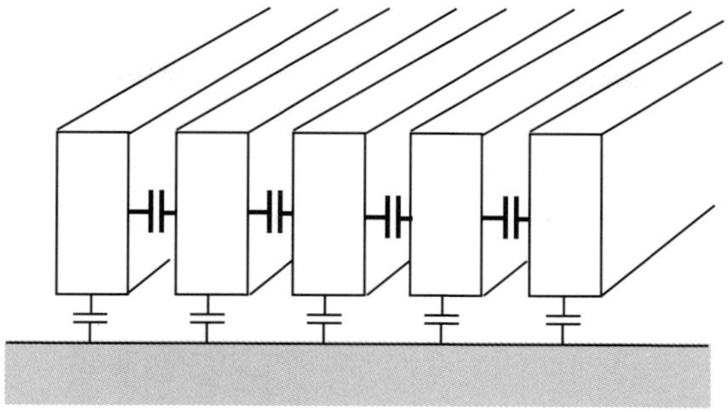

FIGURE 6.129

Interwire and wire-to-substrate capacitance model.

would not be fully utilized. Generally, it is efficient to use minimum-size transistors throughout the integrated circuit, except for a few critical paths where slightly larger and faster transistors are used. This strategy leads to a significant increase in the maximal clock frequency at the expense of a small increase in the switched capacitance.

If a long logic path requires gates with extra drive capacity in order to meet the throughput requirement, it may often be more efficient to break the long path into smaller paths to reduce the chip area and power consumption.

The required clock frequency for the algorithm may, however, be reduced at the expense of increased chip area by exploiting computational parallelism of the algorithm by mapping the arithmetic operations to several PEs or by using multirate techniques in which parts of the digital filter operate at lower sample rates. Parts of the algorithm that operate at a lower rate can often be multiplexed onto a single hardware structure in order to save chip area.

The equivalent *switched capacitance*, C_L, can be made small by using a combination of different techniques. On the algorithm level, most techniques focus on using filter structures with a small number of arithmetic operations per sample. Multirate filter structures are typical examples where a significant reduction in arithmetic workload can be obtained.

Low-sensitivity structures require shorter signal word lengths and shorter and more favorable coefficients that allow fixed multipliers to be simplified [11]. Canonic signed-digit code (CSDC) [78,11] and multiple-constant techniques are efficient techniques to simplify the realization of fixed multipliers [97,99,100,101,108,115,116,105]. The trade-off between the two extremes, bit-parallel and bit-serial arithmetic, i.e., digit-serial arithmetic, is a promising route to lower power consumption and reduce the chip area [11]. As mentioned above, the transmission of signals over long buses is costly in terms of time and power consumption. Architectures with local computations are therefore preferred. Digit-serial PEs yield smaller chip area and, hence, shorter communication distances and are a good compromise between a large number of glitches in bit-parallel arithmetic and many clock cycles per operation of the flip-flops in bit-serial arithmetic. At the logic level, the use of efficient, low-power CMOS circuits and clocking techniques is important [42].

6.19.1 Power supply voltage scaling

The power consumption can be significantly reduced by using a lower power supply voltage. Unfortunately, then the maximal clock frequency of the CMOS circuit is reduced [42] according to

$$f_{CL} = \frac{k\,(V_{DD} - V_T)^{\beta}}{V_{DD}}, \tag{6.144}$$

where k is a process-dependent constant, β is a constant slightly larger than one in modern CMOS processes, and V_T is the threshold voltage of a MOSFET. However, the reduction in throughput can often be compensated in parallel algorithms by exploiting more computational parallelism [11]. From (6.144), it is evident that a CMOS process with a low threshold voltage is preferred from the point of view of speed. In practice, many circuits are therefore manufactured with a process with two threshold voltages. Transistors with low threshold voltages are used for high-speed circuits, while transistors with high threshold voltages are used for low-speed circuits, since the latter have smaller leakage currents.

Many approaches to reduce power consumption are based on various techniques to decrease the arithmetic workload. This can be done at the algorithmic level by using efficient, low-sensitivity structures that allow few and simple coefficients and at the hardware level using simplified multiplier realizations. Usually, the power consumption, due to memory, communication, and control, is neglected in these techniques, although their contribution may be significant. Polyphase structures and frequency masking techniques are commonly used to reduce the arithmetic workload.

Another approach is based on power supply voltage scaling, where the implementation is operated with a low supply voltage. Pipelining and interleaving do not apply to recursive algorithms. However, pipelining inside recursive loops can be performed at the logic level. Therefore, it is important to exploit all parallelism of the recursive algorithm to achieve higher speed, which allows a reduction in the power supply voltage.

6.19.2 Overall strategy

We propose the following strategy to design recursive digital filters with low power consumption:

- Select an inherently fast and low-sensitivity filter algorithm, e.g., ladder or lattice wave digital filters, possibly combined with frequency masking techniques or multirate techniques.
- Use an approximation with a diminishing ripple, which reduces the passband sensitivity in the wave digital filters, and use discrete optimization techniques to reduce the operation latencies in the critical loop.
- Use pipelining to split the nonrecursive parts into smaller pieces so that T_{min} can be achieved.
- Find a maximally fast and resource-minimal schedule.
- Use an isomorphic mapping to an optimal hardware structure with bit- or digit-serial PEs.
- Use CMOS logic circuits that yield bit- or digit-serial PEs with low power consumption, especially for the D flip-flops in the bit-serial case.
- Convert any excess circuit speed to reduced power consumption by reducing the power supply voltage.

References

[1] A. Antoniou, Digital Filters, Analysis, Design and Applications, McGraw-Hill, 2000.

[2] M.G. Bellanger, G. Bonnerot, M. Coudreuse, Digital filtering by polyphase network: application to sample-rate alteration and filter banks, IEEE Trans. Acoust. Speech Signal Process. ASSP-24 (2) (Apr. 1976) 109–114.

[3] E.C. Ifeachor, B.W. Jervis, Digital Signal Processing: A Practical Approach, Addison-Wesley, 2002.

[4] L. Wanhammar, T. Saramäki, Digital Filters Using MATLAB, Springer, 2020.

[5] L. Wanhammar, Toolbox: Digital Filters Using MATLAB, Springer, 2020, https://www.springer.com/inbook/9783030240622.

[6] A. Fettweis, Some principles of designing digital filters imitating classical filter structures, IEEE Trans. Circuit Theory CT–18 (2) (March 1971) 314–316.

[7] A. Fettweis, Digital filter structures related to classical filter networks, Arch. Elektron. Übertrag.tech. 25 (2) (Feb. 1971) 79–89.

[8] A. Fettweis, Wave digital filters: theory and practice, Proc. IEEE 2 (2) (Feb. 1986) 270–327.

[9] O. Gustafsson, L. Wanhammar, Some issues in low power arithmetic for fixed-function dsp, in: Proc. Nat. Conf. Radio Science (RVK), Stockholm, Sweden, June 10–13, 2002, pp. 473–477.

[10] L. Wanhammar, Analog Filters Using MATLAB, Springer, 2009.

[11] L. Wanhammar, DSP Integrated Circuits, Academic Press, 1999.

[12] L.B. Jackson, Roundoff noise bounds derived from coefficient sensitivities for digital filters, IEEE Trans. Circuits Syst. CAS-23 (8) (Aug. 1976) 481–485.

[13] L.O. Chua, T. Lin, Chaos in digital filters, IEEE Trans. Circuits Syst. CAS-35 (June 1988) 648–658.

[14] E.G. Jovanovic-Dolecek, Multirate Systems: Design and Applications, Idea Group Publ., Hersey, USA, 2001.

[15] L. Milic, Multirate Filtering for Digital Signal Processing: MATLAB Applications, Information Science Reference, 2009.

[16] J.H. McClellan, T.W. Parks, L.R. Rabiner, A computer program for designing optimum FIR linear phase digital filters, IEEE Trans. Audio Electroacoust. AU-21 (Dec. 1973) 506–526.

[17] L.R. Rabiner, B. Gold, Theory and Application of Digital Signal Processing, Prentice Hall, Englewood Cliffs, N.J., 1975.

[18] M. Ahsan, T. Saramäki, A MATLAB based optimum multibands FIR filters design program following the original idea of the Remez multiple exchange algorithm, in: Proc. IEEE Intern. Symp. on Circuits and Systems, ISCAS-11, Rio de Janeiro, Brazil, May 15–18, 2011, pp. 137–140.

[19] R.E. Crochiere, L.R. Rabiner, Multirate Digital Signal Processing, Prentice Hall, NJ, 1983.

[20] N.J. Fliege, Multirate Digital Signal Processing, John Wiley and Sons, New York, 1994.

[21] A.V. Oppenheim, R.W. Schafer, Discrete-Time Signal Processing, Prentice Hall, 1989.

[22] J.G. Proakis, D.G. Manolakis, Digital Signal Processing, Principles, Algorithms, and Applications, third ed., Prentice Hall, 1996.

[23] Y.C. Lim, Frequency-response masking approach for the synthesis of sharp linear phase digital filters, IEEE Trans. Circuits Syst. CAS-33 (4) (Apr. 1986) 357–364.

[24] Y.C. Lim, R. Yang, On the synthesis of very sharp decimators and interpolators using the frequency-response masking technique, IEEE Trans. Signal Process. SP-53 (4) (April 2005) 1387–1397.

[25] Y.C. Lim, Y. Lian, The optimum design of one- and two-dimensional FIR filters using the frequency response masking technique, IEEE Trans. Circuits Syst. II CAS-II-40 (Feb. 1993) 88–95.

[26] J. Yli-Kaakinen, T. Saramäki, An efficient algorithm for the optimization of FIR filters synthesized using the multistage frequency-response, Circuits Syst. Signal Process. 30 (1) (2011) 157–183.

[27] Y. Lian, C.Z. Yang, Complexity reduction by decoupling the masking filters from the bandedge shaping filter in the FRM technique, Circuits Syst. Signal Process. 22 (2) (2003) 115–135.

[28] Y. Neuvo, D. Cheng-Yu, S.K. Mitra, Interpolated finite impulse response filters, IEEE Trans. Acoust. Speech Signal Process. ASSP-32 (3) (June 1984) 563–570.
[29] Y.C. Lim, Y. Lian, Frequency-response masking approach for digital filter design: complexity reduction via masking filter factorization, IEEE Trans. Circuits Syst. II CAS-II-41 (Aug. 1994) 518–525.
[30] O. Gustafsson, H. Johansson, L. Wanhammar, Single filter frequency-response masking FIR filters, J. Circuits Syst. Comput. 12 (5) (2003) 601–630.
[31] T. Saramäki, Y.C. Lim, R. Yang, The synthesis of half-band filter using frequency-response masking technique, IEEE Trans. Circuits Syst. II CAS-II-42 (1) (Jan. 1995) 58–60.
[32] Y.C. Lim, Y.J. Yu, Synthesis of very sharp Hilbert transformer using the frequency-response masking technique, IEEE Trans. Signal Process. SP-53 (7) (July 2005) 2595–2597.
[33] H. Johansson, Two classes of frequency-response masking linear-phase FIR filters for interpolation and decimation, Circuits Syst. Signal Process. 25 (2) (April 2006) 175–200, Special issue on Computationally Efficient Digital Filters: Design and Applications.
[34] H. Johansson, L. Wanhammar, Filter structures composed of allpass and FIR filters for interpolation and decimation by a factor of two, IEEE Trans. Circuits Syst. II CAS-II-46 (7) (July 1999) 896–905.
[35] H. Johansson, T. Saramäki, Two-channel FIR filter banks utilizing the frequency-response masking approach, Circuits Syst. Signal Process. 22 (2) (March 2003) 157–192.
[36] L. Rosenbaum, P. Löwenborg, H. Johansson, An approach for synthesis of modulated M-channel filter banks utilizing the frequency-response masking technique, EURASIP J. Adv. Signal Process. 2007 (1) (Jan. 2007), Special Issue on Multirate Systems and Applications.
[37] R. Bregovic, Y.C. Lim, T. Saramäki, Frequency-response masking-based design of nearly perfect-reconstruction two-channel FIR filterbanks with rational sampling factors, IEEE Trans. Circuits Syst. I CAS-I-55 (7) (Aug. 2008) 2002–2012.
[38] K.M. Tsui, S.C. Chan, Y.C. Lim, Design of multi-plet perfect reconstruction filter banks using frequency-response masking technique, IEEE Trans. Circuits Syst. I CAS-I-55 (9) (Sept. 2008) 2707–2715.
[39] Y.C. Lim, B. Farhang-Boroujeny, Fast filter bank (FFB), IEEE Trans. Circuits Syst. II CAS-II-39 (5) (May 1992) 316–318.
[40] P.S.R. Diniz, L.C.R. de Barcellos, S.L. Netto, Design of high-resolution cosine-modulated transmultiplexers with sharp transition band, IEEE Trans. Signal Process. SP-52 (5) (May 2004) 1278–1288.
[41] H. Johansson, L. Wanhammar, High-speed recursive digital filters based on the frequency-response masking approach, IEEE Trans. Circuits Syst. II CAS-II-47 (1) (Jan. 2000) 48–61.
[42] A.P. Chandrakasan, R.W. Brodersen, Low Power Digital CMOS Design, Kluwer, 1995.
[43] Y.C. Lim, S.H. Low, Frequency-response masking approach for the synthesis of sharp two-dimensional diamond-shaped filters, IEEE Trans. Circuits Syst. II CAS-II-45 (12) (Dec. 1998) 1573–1584.
[44] R.W. Daniels, Approximation Methods for Electronic Filter Design, McGraw-Hill, New York, 1974.
[45] A.S. Sedra, P.O. Brackett, Filter Theory and Design: Active and Passive, Pitman, London, 1978.
[46] H.J. Orchard, Inductorless filters, Electron. Lett. 2 (June 1966) 224–225.
[47] L. Wanhammar, A bound on the passband deviation for symmetric and antimetric commensurate transmission line filters, http://citeseerx.ist.psu.edu, 1991.
[48] M.D. Lutovac, D.V.Tosic, B.L. Evans, Advanced filter design, in: Proc. 34th Asilomar Conf. on Signals, Systems and Computers, Pacific Grove, CA, Nov. 2–5, 1997, pp. 710–715.
[49] G. Groenewold, Noise and group delay in active filters, IEEE Trans. Circuits Syst. I CAS-I-54 (7) (July 2007) 1471–1480.
[50] M. Yaseen, Robust simultaneous amplitude and phase approximations oriented to bandpass wave digital lattice filters, Int. J. Circuit Theory Appl. 26 (1998) 179–189.
[51] M. Yaseen, On the design of multiband transmission functions synthesized by one wave digital lattice structure, Int. J. Circuit Theory Appl. (Dec. 2011).
[52] M.D. Lutovac, D.V. Tosic, B.L. Evan, Filter Design for Signal Processing Using MATLAB and Mathematica, Prentice Hall, 2001.

[53] A. Fettweis, On the connection between multiplier word length limitation and roundoff noise in digital filters, IEEE Trans. Circuit Theory CT-19 (5) (Sept. 1972) 486–491.
[54] J.A. Nossek, M.T. Ivrla, On the relation of circuit theory and signals, systems and communications, in: Proc. IEEE Intern. Symp. on Circuits and Systems, ISCAS-11, Rio de Janeiro, Brazil, May 15–18, 2011, pp. 603–604.
[55] A. Fettweis, K. Meerkötter, On adaptors for wave digital filters, IEEE Trans. Acoust. Speech Signal Process. ASSP–23 (6) (Dec. 1975) 516–525.
[56] A. Fettweis, Pseudopassivity, sensitivity, and stability of wave digital filters, IEEE Trans. Circuit Theory CT-19 (6) (Nov. 1972) 668–673.
[57] A. Fettweis, K. Meerkötter, Suppression of parasitic oscillations in wave digital filters, in: Proc. IEEE Intern. Symp. on Circuits and Systems, ISCAS-74, San Francisco, April 1974, pp. 682–686.
[58] A. Fettweis, K. Meerkötter, Suppression of parasitic oscillations in wave digital filters, IEEE Trans. Circuits Syst. CAS–22 (3) (March 1975) 239–246.
[59] A. Fettweis, Passivity and losslessness in digital filtering, Arch. Elektron. Übertrag.tech. 42 (1) (Jan./Feb. 1988) 1–8.
[60] A. Fettweis, Scattering properties of wave digital filters, in: Proc. Florence Sem. on Digital Filtering, Florence, Italy, Sept. 1972, pp. 1–8.
[61] A.H. Gray, J.D. Markel, Digital lattice and ladder filter synthesis, IEEE Trans. Audio Electroacoust. AU-21 (6) (Dec. 1973) 491–500.
[62] A.H. Gray, J.D. Markel, A normalized digital filter structure, IEEE Trans. Acoust. Speech Signal Process. ASSP-23 (3) (June 1975) 268–277.
[63] A.H. Gray, J.D. Markel, Roundoff noise characteristics of a class of orthogonal polynomial structures, IEEE Trans. Acoust. Speech Signal Process. ASSP-23 (Oct. 1975) 473–486.
[64] A.H. Gray, J.D. Markel, Fixed-point implementation algorithms for a class of orthogonal polynomial filter structures, IEEE Trans. Acoust. Speech Signal Process. ASSP-23 (5) (Oct. 1975) 486–494.
[65] A.N. Willson, H.J. Orchard, Insights into digital filters made as a sum of two allpass functions, IEEE Trans. Circuits Syst. I CAS-I-42 (3) (March 1995) 129–137.
[66] V.I. Anzova, J. Yli-Kaakinen, T. Saramäki, An algorithm for the design of multiplierless IIR filters as a parallel connection of two all-pass filters, in: Proc. IEEE Asia Pacific Conf. on Circuits and Systems, APCCAS, Singapore, Dec. 2006, pp. 744–747.
[67] B. Jaworski, T. Saramäki, Linear phase IIR filters composed of two parallel allpass sections, in: IEEE Intern. Symp. on Circuits and Systems, ISCAS-94, vol. 2, London, May 1994, pp. 537–540.
[68] M. Renfors, T. Saramäki, A class of approximately linear phase digital filters composed of allpass subfilters, in: IEEE Intern Symp. on Circuits and Systems, ISCAS 86, San Jose, CA, May 1986, pp. 678–681.
[69] L. Wanhammar, Implementation of wave digital filters using vector-multipliers, in: Proc. First European Signal Processing Conf., EUSIPCO-80, Lausanne, Switzerland, Sept. 1980, pp. 21–26.
[70] L. Wanhammar, Implementation of wave digital filters using distributed arithmetic, Signal Process. 2 (3) (July 1980) 253–260.
[71] M. Renfors, Y. Neuvo, The maximal sampling rate of digital filters under hardware speed constraints, IEEE Trans. Circuits Syst. CAS-28 (3) (March 1981) 196–202.
[72] K. Palmkvist, M. Vesterbacka, L. Wanhammar, Arithmetic transformations for fast bit-serial VLSI implementations of recursive algorithms, in: Proc. Nordic Signal Processing Symp., NORSIG'96, Espoo, Finland, Sept. 24–27, 1996, pp. 391–394.
[73] H. Ohlsson, O. Gustafsson, L. Wanhammar, Arithmetic transformations for increased maximal sample rate of bit-parallel bireciprocal lattice wave digital filters, in: Proc. IEEE Int. Symp. Circuits Syst., Sydney, Australia, May 6–9, 2001.
[74] M. Vesterbacka, K. Palmkvist, P. Sandberg, L. Wanhammar, Implementation of fast bit-serial lattice wave digital filters, in: Proc. IEEE Intern. Symp. on Circuits and Systems, ISCAS'94, vol. 2, London, May 30–June 1, 1994, pp. 113–116.

[75] M. Vesterbacka, K. Palmkvist, L. Wanhammar, On implementation of fast, bit-serial loops, in: Proc. IEEE 1996 Midwest Symp. on Circuits and Systems, vol. 1, Ames, Iowa, Aug. 18–21, 1996, pp. 190–193.
[76] M. Vesterbacka, K. Palmkvist, L. Wanhammar, Maximally fast, bit-serial lattice wave digital filters, in: Proc. IEEE DSP Workshop 96, Loen, Norway, Sept. 2–4, 1996, pp. 207–210.
[77] O. Gustafsson, L. Wanhammar, Maximally fast scheduling of bit-serial lattice wave digital filters using three-port adaptor allpass sections, in: Proc. IEEE Nordic Signal Processing Symp., Kolmården, Sweden, June 13–15, 2000, pp. 441–444.
[78] I. Koren, Computer Arithmetic Algorithms, Prentice-Hall, 1993.
[79] K. Johansson, O. Gustafsson, L. Wanhammar, Power estimation for ripple-carry adders with correlated input data, in: Proc. IEEE Intern. Workshop on Power and Timing Modeling, Optimization and Simulation, Santorini, Greece, Sept. 15–17, 2004.
[80] R. Zimmermann, Binary adder architectures for cell-based VLSI and their synthesis, PhD Thesis, Swiss Federal Institute of Technology (ETH), Zurich, Hartung-Gorre Verlag, 1998.
[81] O. Gustafsson, L. Wanhammar, Low-complexity constant multiplication using carry-save arithmetic for high-speed digital filters, in: Int. Symp. Image, Signal Processing, Analysis, Istanbul, Turkey, Sept. 2007, pp. 27–29.
[82] O. Gustafsson, A.G. Dempster, L. Wanhammar, Multiplier blocks using carry-save adders, in: Proc. IEEE Intern. Symp. Circuits Systems, Vancouver, Canada, May 23–26, 2004.
[83] V.G. Oklobdzija, D. Villeger, S.S. Liu, A method for speed optimized partial product reduction and generation of fast parallel multipliers using an algorithmic approach, IEEE Trans. Comput. C-45 (3) (March 1996).
[84] M.J. Irwin, R.M. Owens, Design issues in digit serial signal processors, in: Proc. IEEE Intern. Symp. on Circuits and Systems, vol. 1, May 8–11, 1989, pp. 441–444.
[85] H. Lee, G.E. Sobelman, Digit-serial reconfigurable FPGA logic block architecture, in: Proc. IEEE Workshop on Signal Processing Systems, SIPS 98, Oct. 8–10, 1998, pp. 469–478.
[86] K.K. Parhi, A systematic approach for design of digit-serial signal processing architectures, IEEE Trans. Circuits Syst. CAS-38 (4) (April 1991) 358–375.
[87] H. Lee, G.E. Sobelman, Digit-serial DSP library for optimized FPGA configuration, in: Proc. IEEE Symp. on FPGAs for Custom Computing Machines, April 15–17, 1998, pp. 322–323.
[88] H. Suzuki, Y.-N. Chang, K.K. Parhi, Performance tradeoffs in digit-serial DSP systems, in: Proc. Thirty-Second Asilomar Conf., vol. 2, Pacific Grove, CA, Nov. 1–4, 1998, pp. 1225–1229.
[89] Y.-N. Chang, J.H. Satyanarayana, K.K. Parhi, Systematic design of high-speed and low-power digit-serial multipliers, IEEE Trans. Circuits Syst. II CAS-II-45 (12) (Dec. 1998) 1585–1596.
[90] Y.-N. Chang, J.H. Satyanarayana, K.K. Parhi, Low-power digit-serial multipliers, in: Proc. IEEE Intern. Symp. on Circuits and Systems, ISCAS'97, vol. 3, June 9–12, 1997, pp. 2164–2167.
[91] P. Nilsson, Arithmetic and architectural design to reduce leakage in nano-scale digital circuits, in: Proc. 18th European Conference on Circuit Design, ECCTD 07, Seville, Spain, 2007, pp. 373–375.
[92] A.N. Willson, H.J. Orchard, A design method for half-band FIR filters, IEEE Trans. Circuits Syst. I CAS-I-45 (1) (Jan. 1999) 95–101.
[93] K. Johansson, O. Gustafsson, L. Wanhammar, Low-complexity bit-serial constant-coefficient multipliers, in: Proc. IEEE Int. Symp. Circuits Syst., vol. 3, Vancouver, Canada, May 23–26, 2004, pp. 649–652.
[94] C.-H. Chang, O. Gustafsson, A.-P. Vinod, M. Faust, Shift-add circuits for multiplications, in: P.K. Meher, T. Soouraitis (Eds.), Arithmetic Circuits for DSP Applications, Wiley-IEEE Press, 2017.
[95] A. Croisier, D.J. Esteban, M.E. Levilion, V. Rizo, Digital filter for PCM encoded signals, U.S. Patent 3777130, Dec. 4, 1973.
[96] O. Gustafsson, L. Wanhammar, Basic arithmetic circuits, in: P.K. Meher, T. Soouraitis (Eds.), Arithmetic Circuits for DSP Applications, Wiley-IEEE Press, 2017.
[97] D.R. Bull, D.H. Horrocks, Primitive operator digital filters, IEE Proc. G 138 (3) (June 1991) 401–412.

[98] A.G. Dempster, M.D. Macleod, O. Gustafsson, Comparison of graphical and sub-expression elimination methods for design of efficient multipliers, in: Proc. Asilomar Conf. Signals Syst. Comp., Monterey, CA, Nov. 2004, pp. 7–10.

[99] A.G. Dempster, M.D. Macleod, Constant integer multiplication using minimum adders, IEE Proc., Circuits Devices Syst. 141 (5) (Oct. 1994) 407–413.

[100] A.G. Dempster, M.D. Macleod, Use of minimum-adder multiplier blocks in FIR digital filters, IEEE Trans. Circuits Syst. II CAS-II-42 (9) (Sept. 1995) 569–577.

[101] R.I. Hartley, Subexpression sharing in filters using canonic signed digit multipliers, IEEE Trans. Circuits Syst. II CAS-II-43 (10) (Oct. 1996) 677–688.

[102] M. Potkonjak, M.B. Srivastava, A.P. Chandrakasan, Multiple constant multiplications: efficient and versatile framework and algorithms for exploring common subexpression elimination, IEEE Trans. Comput.-Aided Des. CAD-15 (2) (Feb. 1996) 151–165.

[103] O. Gustafsson, A difference based adder graph heuristic for multiple constant multiplication problems, in: Proc. IEEE Int. Symp. Circuits Syst., New Orleans, LA, May 27–30, 2007, pp. 1097–1100.

[104] Y. Voronenko, M. Püschel, Multiplierless multiple constant multiplication, ACM Trans. Algorithms 3 (2) (May 2007).

[105] O. Gustafsson, Lower bounds for constant multiplication problems, IEEE Trans. Circuits Syst. II CAS-II-54 (11) (Nov. 2007) 974–978.

[106] Y.J. Yu, Y.C. Lim, Design of linear phase FIR filters in subexpression space using mixed integer linear programming, IEEE Trans. Circuits Syst. I 54 (10) (Oct. 2007) 2330–2338.

[107] D. Shi, Y.J. Yu, Design of linear phase FIR filters with high probability of achieving minimum number of adders, IEEE Trans. Circuits Syst. I 58 (1) (Jan. 2011) 126–136.

[108] M. Martinez-Peiro, E. Boemo, L. Wanhammar, Design of high speed multiplierless filters using a nonrecursive signed common subexpression algorithm, IEEE Trans. Circuits Syst. II CAS-II-49 (3) (March 2002) 196–203.

[109] K. Johansson, O. Gustafsson, L. Wanhammar, Multiple constant multiplication for digit-serial implementation of low power FIR filters, WSEAS Trans. Circuits Syst. 5 (7) (July 2006) 1001–1008.

[110] K. Johansson, O. Gustafsson, L.S. DeBrunner, L. Wanhammar, Minimum adder depth multiple constant multiplication algorithm for low power FIR filters, in: Proc. IEEE Int. Symp. Circuits Syst., Rio de Janeiro, Brazil, May 15–18, 2011.

[111] M. Martinez-Peiro, L. Wanhammar, High-speed low-complexity FIR filter using multiplier block reduction and polyphase decomposition, in: Proc. IEEE Int. Symp. Circuits Syst., vol. 3, Geneva, Switzerland, May 2000, pp. 367–370.

[112] M. Büttner, H.W. Schüssler, On structures for the implementation of the distributed arithmetic, Nachrichtentechn. Z. 29 (6) (1976) 472–477.

[113] A. Peled, B. Liu, Digital Signal Processing: Theory, Design and Implementation, John Wiley and Sons, 1976.

[114] S. White, On applications of distributed arithmetic to digital signal processing, a tutorial review, IEEE ASSP Mag. (July 1989) 4–19.

[115] O. Gustafsson, A.G. Dempster, On the use of multiple constant multiplication in polyphase FIR filters and filter banks, in: Proc. Nordic Signal Processing Symp., vol. 3, Espoo, Finland, June 9–11, 2004, pp. 53–56.

[116] O. Gustafsson, H. Johansson, L. Wanhammar, MILP design of frequency-response masking FIR filters with few SPT terms, in: Proc. Int. Symp. Control, Communications, Signal Processing, Hammamet, Tunisia, March 21–24, 2004.

CHAPTER

Multirate signal processing for software radio architectures

7

Fred Harris[a], Elettra Venosa[b], Xiaofei Chen[c], Richard Bell[a], Srivatsan Rajagopal[a], Radhika Mathuria[a], and Dinesh Bharadia[a]

[a] *University of California, San Diego, Department of Electrical and Computer Engineering, San Diego, CA, United States*

[b] *Space Micro, Inc., San Diego, CA, United States*

[c] *Meta Platforms, Inc., Menlo Park, CA, United States*

After the introductory sections on resampling theory and basics of digital filters, the architectures of up and down converter channelizers are presented. Those paragraphs are followed by the main core of this chapter in which the authors present novel digital up converter (DUC) and digital down converter (DDC) architectures for software-defined radios, which are based on variations of standard polyphase channelizers.

The proposed DUC is able to simultaneously up convert multiple signals, having arbitrary bandwidths, to arbitrarily located center frequencies. On the other side of the communication chain, the proposed DDC is able to simultaneously down convert multiple received signals with arbitrary bandwidth and arbitrarily located in the frequency domain. Both proposed structures avoid the digital data section replication, which is necessary in the current digital transmitters and receivers, when multiple signals have to be handled. Due to the inverse discrete Fourier transform (IDFT), performed using an inverse fast Fourier transform (IFFT) algorithm, embedded in the channelizers, the proposed architectures are very efficient in terms of total workload.

This chapter is structured into two main parts, which are composed, in total, of 10 sections. Its structure is described in Fig. 7.1. The first part provides preliminary concepts on the sampling process, the resampling process of a digital signal, digital filters, multirate structures, and standard M-path polyphase up and down converter channelizers, as well as their modified versions that represent the core of the structures we present in the second part of this chapter. These are well-known topics in the digital signal processing (DSP) area and they represent the necessary background for people who want to start learning about software radios. The second part of this chapter goes through new frontiers of the research presenting the novel transmitting and receiving designs for software radios and demonstrating, while explaining the way in which they work, the reasons that make them the perfect candidates for the upcoming cognitive radios.

Signal Processing and Machine Learning Theory. https://doi.org/10.1016/B978-0-32-391772-8.00013-2
Copyright © 2024 Elsevier Inc. All rights reserved.

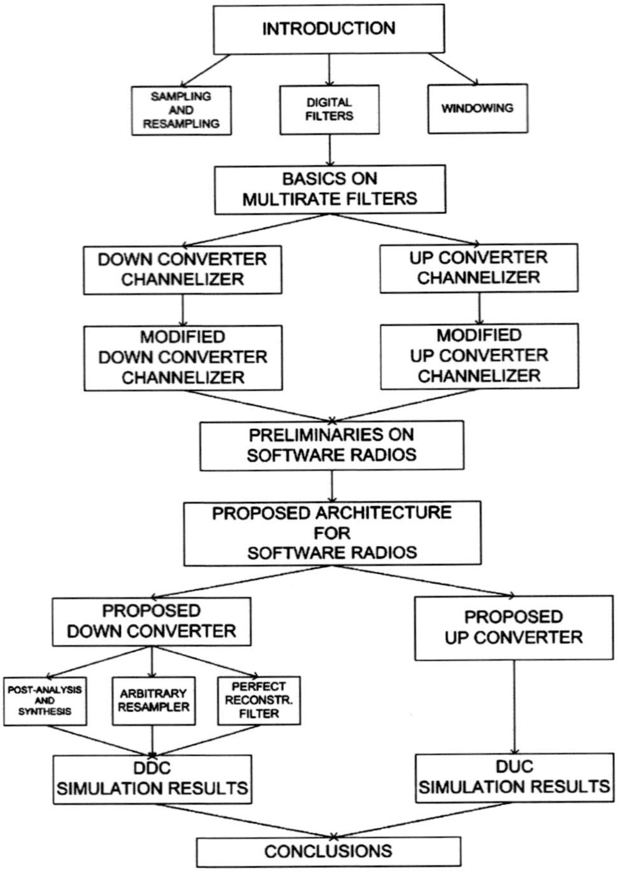

FIGURE 7.1

Structure of the chapter.

7.1 Introduction

Signal processing is the art of representing, manipulating, and transforming wave shapes and the information content that they carry by means of hardware and/or software devices. Until the 1960s, almost entirely continuous-time, analog technology was used for performing signal processing. The evolution of digital systems along with the development of important algorithms such as the well-known fast Fourier transform (FFT), by Cooley and Tukey in 1965, caused a major shift to digital technologies giving rise to new DSP architectures and techniques. The key difference between analog processing techniques and digital processing techniques is that while the first process analog signals, which are continuous functions of time, the second process sequences of values. The sequences of values, called digital signals, are series of quantized samples (discrete-time signal values) of analog signals.

The link between the analog signals and their sampled versions is provided by the sampling process. When sampling is properly applied it is possible to reconstruct the continuous-time signal from its samples while preserving its information content. Thus the selection of the sampling rate is crucial: an insufficient sampling frequency causes, surely, irrecoverable loss of information, while a more than necessary sampling rate causes useless overload on the digital systems.

"Multirate DSP" refers to the operation of changing the sample rate of digital signals, one or multiple times, while processing them. The sampling rate changes can occur at a single or multiple locations in the processing architecture. When the multirate processing applied to the digital signal is a filtering step, then the digital filter is named multirate filter; a multirate filter is a digital filter that operates with one or more sample rate changes embedded in the signal processing architecture. Selecting the most appropriate signal sampling rate at different stages of the processing architecture, rather than having a single (higher) sampling rate, enhances the performance of the system while reducing its implementation costs. However, the analysis and design of multirate systems could result complicated because of the fact that they are time-varying systems. It is interesting to know that the first multirate filters and systems were developed in the context of control systems. The pioneer papers on this topic were published in the second half of the 1950s [13–15]. Soon the idea spilled over to the areas of speech, audio, image processing [20,25], and communication systems [8]. Today, multirate structures seem to be the optimum candidates for cognitive and software-defined radio [4,6,7,12,24], which represents the frontier of innovation for the upcoming communication systems.

One of the well-known techniques that find its most efficient application when embedded in multirate structures is the polyphase decomposition of a prototype filter. The polyphase networks, of generic order M, originated in the late 1970s from the works by Bellanger et al. [1,2]. The term polyphase is the aggregation of two words: poly, which derives from the ancient Greek word *polys*, which means many, and phase. When applied to an M-path partitioned filter these two words underline the fact that on each path, each aliased filter spectral component experiences a unique phase rotation due to both their center frequencies and the time delays, which are different in each path because of the way in which the filter has been partitioned. When all the paths are summed together, the undesired spectral components, having phases with opposite polarity, cancel each other while only the desired components, which experience the same phase on all the arms of the partitions, constructively add up. By applying appropriate phase rotators to each path we can arbitrarily change the phases of the spectral components, selecting the spectral component that survives as a consequence of the summation. It is interesting to know that the first applications of the polyphase networks were in the areas of real-time implementation of decimation and interpolation filters, fractional sampling rate-changing devices, and uniform DFT filter banks, as well as perfect reconstruction analysis/synthesis systems. Today polyphase filter banks, embedded in multirate structures, are used in many modern DSP applications as well as in communication systems where they represent, according to the authors' belief, one of the most efficient options for designing software-based radio.

The processing architecture that represents the starting point for the derivation of the standard polyphase down converter channelizer is the single-channel DDC. In a digital radio receiver this engine performs the operations of filtering, frequency translation, and resampling on the intermediate frequency (IF) signals. The resampling is a downsampling operation to make the signal sampling rate commensurate to its new reduced bandwidth. When the output sampling rate is selected to be an integer multiple of the signal's center frequency, the signal spectrum is shifted, by aliasing, to baseband and the complex heterodyne defaults to unity value disappearing from the processing path. The two remaining operations of filtering and downsampling are usually performed in a single processing architecture (multirate filter). After exchanging the positions of the filter and resampler, by applying polyphase decomposition, we achieve the standard M-path polyphase down converter channelizer which shifts a single narrow-band channel to baseband while reducing its sampling rate. By following a similar procedure the standard up converter channelizer can also be obtained. It performs the operation of up converting while interpolating a single narrow-band channel.

When the polyphase channelizer is used in this fashion the complex phase rotators can be applied to each arm to select, by coherent summation, a desired channel arbitrarily located in the frequency domain. Moreover, the most interesting and efficient application of this engine is in the multichannel scenario. When the IDFT block is embedded, the polyphase channelizer acquires the capability to simultaneously up and down convert, by aliasing, multiple equally spaced narrow-band channels having equal bandwidths. Due to its computational efficiency, the polyphase channelizer, and its modified versions, represents the best candidate for building up new flexible multichannel digital radio architectures [18] like software-defined radio promises to be.

Software-defined radio represents one of the most important emerging technologies for the future of wireless communication services. By moving radio functionality into software, it promises to give flexible radio systems that are multiservice, multistandard, multiband, reconfigurable, and reprogrammable by software. The goal of software-defined radio is to solve the issue of optimizing the use of the radio spectrum that is becoming more and more pressing because of the growing deployment of new wireless devices and applications [17,19,21,26].

In the second part of this chapter we address the issue of designing software radio architectures for both the transmitter and the receiver. The novel structures are based on modified versions of the standard polyphase channelizer, which provides the system with the capability to optimally adapt its operating parameters according to the surrounding radio environment [3,12,16,23]. This implies, on the transmitter side, the capability of the radio to detect the available spectral holes [5,22] in the spanned frequency range and to dynamically use them for sending signals having different bandwidths at randomly located center frequencies. The signals could originate from different information sources or they could be spectral partitions of one signal, fragmented because of the unavailability of free space in the radio spectrum or for protecting it from unknown and undesired detections.

The core of the proposed transmitter is a synthesis channelizer [11]. It is a variant of the standard M-path polyphase up converter channelizer [7,8] that is able to perform 2-to-M upsampling while shifting, by aliasing, all the baseband channels to the desired center frequencies. Input signals with bandwidths wider than the synthesis channelizer bandwidth are preprocessed through small down converter channelizers that disassemble their bandwidths into reduced-bandwidth subchannels. The proposed transmitter avoids the need to replicate the sampled data section when multiple signals have to be simultaneously transmitted and it also allows partitioning of the signal spectra, when necessary, before transmitting them. Those spectral fragments will be reassembled in the receiver after being shifted to baseband without any loss of energy because perfect reconstruction filters are used as low-pass prototype in the channelizer polyphase decompositions.

On the other side of the communication chain, a cognitive receiver has to be able to simultaneously detect multiple signals, recognize, when necessary, all their spectral partitions, and filter, down convert, and recompose them without energy losses [12] independently of their bandwidths or center frequencies. An analysis channelizer is the key element of the proposed receiver [12]. This engine is able to perform M-to-2 downsampling while simultaneously demodulating, by aliasing, all the received signal spectra having arbitrary bandwidths residing at arbitrary center frequencies [6]. Postprocessing up converter channelizers are used for reassembling, from the analysis channelizer baseline channels, signal bandwidths wider than the analysis channelizer channel bandwidth.

In the transmitter, complex frequency rotators apply the appropriate frequency offsets for arbitrary center frequency positioning of the spectra. When the frequency rotators are placed in the proposed receiver, they are responsible for perfect DC alignment of the down converted signals.

This chapter is composed of 10 main sections. The first four sections are dedicated to the basics of DSP, multirate signal processing, and polyphase channelizers. These sections contain preliminary concepts necessary for completely understanding the last sections of this work in which actual research issues on software radio architecture design are discussed and innovative results are presented. In particular, Section 7.2 recalls basic concepts on the resampling process of a discrete-time signal as opposed to the sampling process of an analog continuous-time signal. Section 7.3 introduces the readers to digital filters providing preliminaries on their design techniques. Section 7.5 presents the standard single-path architectures for downsampling and upsampling digital signals. Section 7.6 introduces the standard version of the M-path polyphase down and up converter channelizers. Both of them are derived, step by step, by the single-path structures that represent the current state of the technology. In Section 7.7 we present the modified versions of these engines. These are the structures that give form to the proposed digital down and up converters for software-defined radios, which are presented in Section 7.9, which also provides the simulation results. Section 7.8 introduces the readers to the concept of software radio, while Section 7.11 gives concluding remarks along with suggestions for future research works in this area.

7.2 The sampling process and the "resampling" process

In the previous section we mentioned that the sampling process is the link between the continuous-time world and the discrete-time world. By sampling a continuous-time signal we achieve its discrete-time representation. When the sampling frequency is properly selected, the sampling process becomes invertible and it is possible to reshift from the discrete-time representation to the continuous-time representation of a signal while maintaining its information content during the process.

It is well known that when the signal is band-limited and has no frequency components above f_{MAX}, it can be uniquely described by its samples taken uniformly at frequency

$$f_s \geq 2 f_{\mathrm{MAX}}. \tag{7.1}$$

Eq. (7.1) is known as the Nyquist sampling criterion (or uniform sampling theorem), and the sampling frequency $f_s = 2 f_{\mathrm{MAX}}$ is called the Nyquist sampling rate. This theorem represents a theoretically sufficient condition for reconstructing the analog signal from its uniformly spaced samples.

Even though sampling is practically implemented in a different way, it is convenient, in order to facilitate the learning process, to represent it as a product of the analog waveform, $x(t)$, with a periodic train of unit impulse functions defined as

$$x_\delta(t) = \sum_{n=-\infty}^{\infty} \delta(t - nT_s),$$

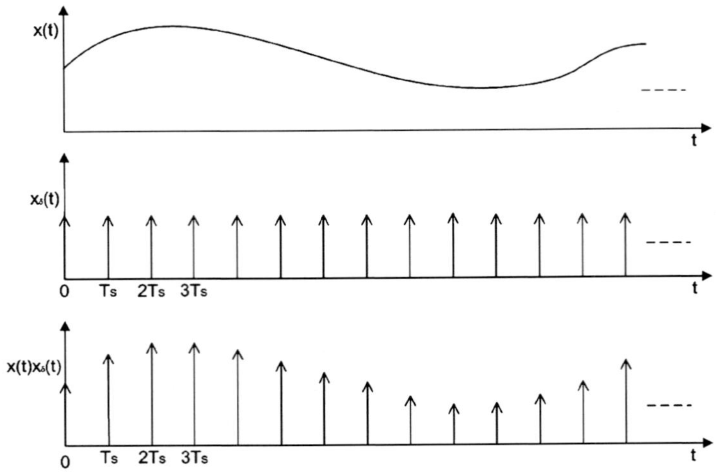

FIGURE 7.2

Sampling process as a product of the analog waveform, $x(t)$, with a periodic train of unit impulse functions $x_\delta(t)$.

where $T_s = 1/f_s$ is the sampling period. By using the shifting property of the impulse function we obtain

$$x_s(t) = x(t)x_\delta(t) = x(t)\sum_{n=-\infty}^{\infty}\delta(t-nT_s) = \sum_{n=-\infty}^{\infty}x(nT_s)\delta(t-nT_s). \tag{7.2}$$

A pictorial description of Eq. (7.2) is given in Fig. 7.2.

To facilitate the readers' understanding of the uniform sampling effects on the band-limited continuous-time signal, $x(t)$, we shift from the time domain to the frequency domain, where we are allowed to use the properties of the Fourier transform. The product of two functions in time domain becomes the convolution of their Fourier transforms in the frequency domain. Thus, if $X(f)$ is the Fourier transform of $x(t)$ and $X_\delta(f)$ is the Fourier transform of $x_\delta(t)$, then the Fourier transform of $x_s(t)$ is

$$X_s(f) = X(f) * X_\delta(f),$$

where $*$ indicates linear convolution and

$$X_\delta(f) = \frac{1}{T_s}\sum_{k=-\infty}^{\infty}\delta(f-kf_s).$$

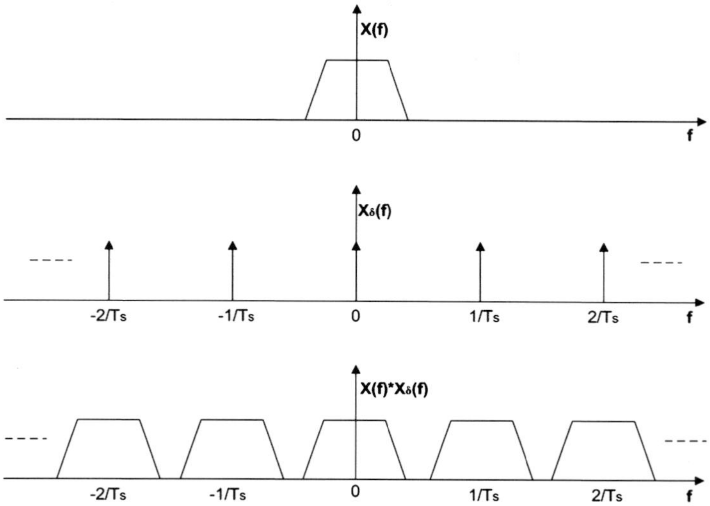

FIGURE 7.3

Sampling process in the frequency domain.

Note that the Fourier transform of an impulse train is another impulse train with the values of the periods reciprocally related to each other. Then in the frequency domain Eq. (7.2) becomes

$$X_s(f) = X(f) * X_\delta(f) = X(f) * \frac{1}{T_s} \sum_{k=-\infty}^{\infty} \delta(f - kf_s) = \frac{1}{T_s} \sum_{k=-\infty}^{\infty} X(f - kf_s), \tag{7.3}$$

whose pictorial view is shown in Fig. 7.3.

From Eq. (7.3) we conclude that the spectrum of the sampled signal, in the original signal bandwidth ($[-f_s/2, f_s/2]$), is the same as the continuous-time one (except for a scale factor, $\frac{1}{T_s}$); however, as a consequence of the sampling process, this spectrum periodically repeats itself with a period of $f_s = \frac{1}{T_s}$. We can easily recognize that it should be possible to recover the original spectrum (associated to the spectral replica which resides in $[-f_s/2, f_s/2]$) by filtering the periodically sampled signal spectrum with an appropriate low-pass filter.

Note from Eq. (7.3) that the spacing, f_s, between the signal replicas is the reciprocal of the sampling period T_s. A small sampling period corresponds to a large space between the spectral replicas.

A fundamental observation, regarding the selection of the uniform sampling frequency, needs to be made at this point: When the sampling rate is selected to be less than the maximum frequency component of the band-limited signal ($f_s < 2f_{\text{MAX}}$), the periodic spectral replicas overlap each other. The amount of overlap depends of the selected sampling frequency: the smaller the sampling frequency, the larger the amount of overlap experienced by the replicas. This phenomenon is well known as aliasing. When aliasing occurs it is impossible to recover the analog signal from its samples. When $f_s = 2f_{\text{MAX}}$, the spectral replicas touch each other without overlapping and it is theoretically (but not practically) possible to recover the analog signal from its samples; however, a filter with an infinite number of taps would be required. As a matter of practical consideration, we need to specify here that the signals (and filters) of interest are never perfectly band-limited and some amount of aliasing always occurs as an effect of sampling; however, some techniques can be used to limit this, making it less harmful.

After the sampling has been applied, the amplitude of each sample is one from an infinite set of possible values. This is the reason why the samples are not compatible with a digital system yet. A digital system can, in fact, only deal with a finite number of values. The digital samples need to be quantized before being sent to the digital data system. The quantization process limits the amplitude of the samples to a finite set of values. After the quantization process the signal can still be recovered; however, some additional imprecision is added to it. The amount of imprecision depends of the quantization levels used in the process, and it has the same effect on the signal as white noise. This is the reason why it is referred to as quantization noise. The sampling of a continuous-time signal and the quantization of its discrete-time samples are both performed with devices called analog-to-digital converters (ADCs).

The selection of the appropriate sampling rate is a fundamental issue faced when designing digital communication systems. In order to preserve the signal, the Nyquist criterion must always be satisfied. Also, it is not difficult to find tasks for which, at some points in the digital data section of the transmitter and the receiver, it is recommended to have more than two samples per signal bandwidth. On the other hand, large sample rates cause an increase in the total workload of the system. Thus an appropriate sampling rate must be selected according to both necessities: to have the required number of samples and to minimize the total workload. Most likely the optimum sampling frequency changes at different points in the digital architecture. It would be a huge advantage to have the option of changing the signal sampling rate at different parts in the system while processing, thus optimizing the number of samples as a function of the requirements. The process of changing the sampling rate of a digital signal

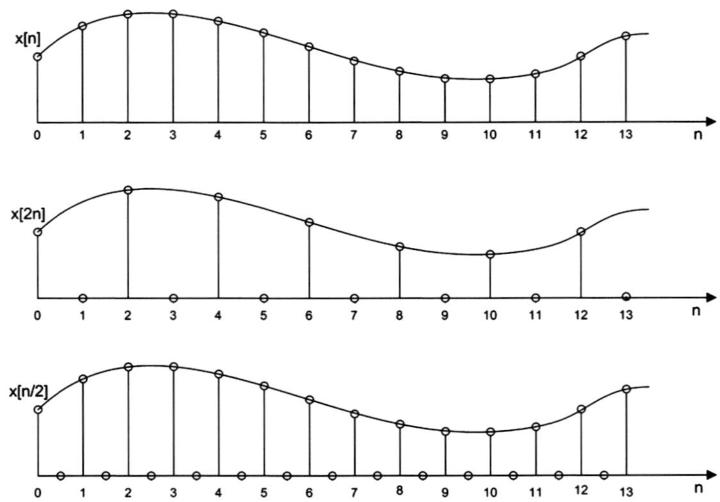

FIGURE 7.4

Resampling by zeroing sample values and by inserting zero-valued samples.

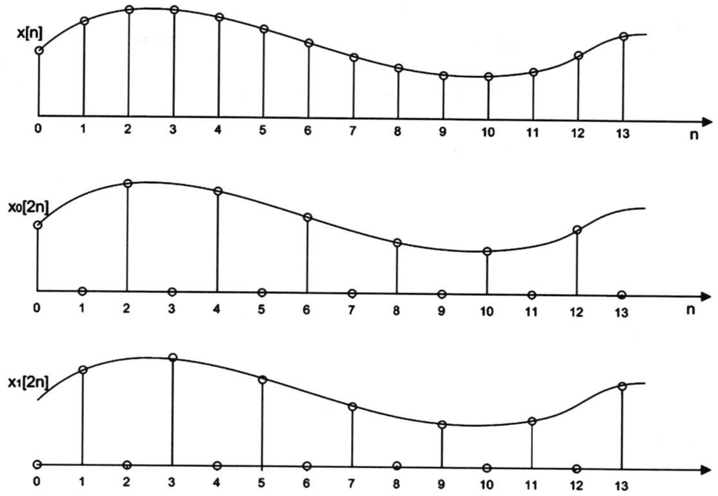

FIGURE 7.5

Two examples of 2-to-1 downsampling of a time series with different time offsets.

is referred to as resampling process. The resampling process is the key concept in multirate signal processing.

After this brief discussion about the sampling process of a continuous-time signal, we address now the process of resampling an already sampled signal. Note that when a continuous-time signal is sampled, there are no restrictions on the sample rate or the phase of the sample clock relative to the time base of the continuous-time signal. On the other hand, when we resample an already sampled signal, the output sample locations are intimately related to the input sample positions. A resampled time series contains samples of the original input time series separated by a set of zero-valued samples. The zero-valued time samples can be the result of setting a subset of input sample values to zero or the result of inserting zeros between existing input sample values. Both options are shown in Fig. 7.4. In the first example shown in this figure, the input sequence is resampled 2-to-l, keeping every second input sample starting at sample index 0 while zeroing the interim samples. In the second example, the input sequence is resampled 1-to-2, keeping every input sample but inserting a zero-valued sample between each pair of input samples. These two processes are called downsampling and upsampling, respectively.

The nonzero-valued samples of two sequences having the same sample rate can occur at different time indices (i.e., the same sequence can have different initial time offset), as shown in Fig. 7.5, in which the two 2-to-1 downsampled sequences, $s_2(n)$ and $s_2(n-1)$, have different starting time index, 0 and 1, which gives equal magnitude spectral components with different phase profiles. This is explicitly shown in Figs. 7.6 and 7.7, in which the time domain and frequency domain views of the resampled sequences $s_5(n)$ and $s_5(n-1)$ are depicted, respectively. Remember, in fact, the time shifting property of the discrete Fourier transform (DFT) for which a shift in time domain is equivalent to a linear phase shift in the frequency domain. The different phase profiles play a central role in multirate signal processing.

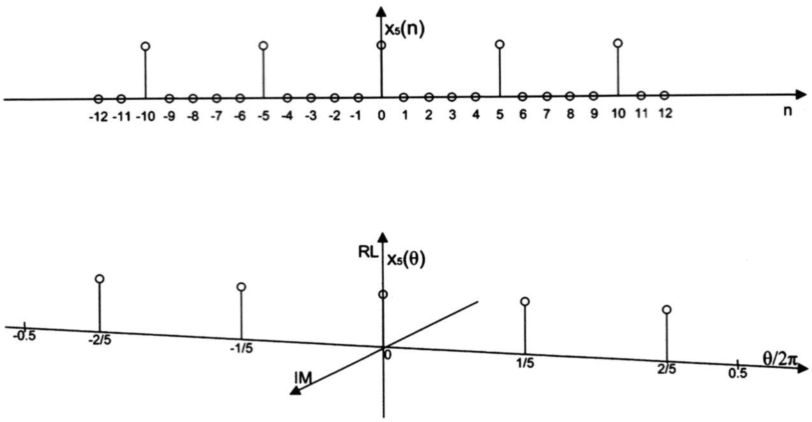

FIGURE 7.6

Sampling sequence $s_5(n)$ in the time and frequency domains.

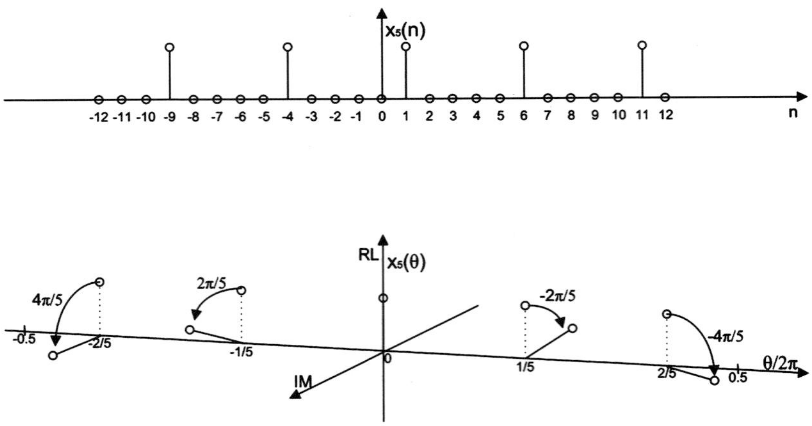

FIGURE 7.7

Sampling sequence $s_5(n-1)$ in the time and frequency domains.

The M zero-valued samples inserted between the signal samples create M periodic replicas of the original spectrum at the frequency locations k/M, with $k = 0, \ldots, M-1$ in the normalized domain. This observation suggests that resampling can be used to affect translation of spectral bands, up and down conversion, without the use of sample data heterodynes.

In the following paragraphs we clarify this concept showing how to embed the spectral translation of narrow-band signals in resampling filters and describe the process as aliasing.

Finally, we wish to mention that the resampling process can be applied to a time series or to the impulse response of a filter, which, of course, is simply another time series. When the resampling process is applied to a filter, the architecture of the filter changes considerably and the filter is called multirate filter.

7.3 Digital filters

Filtering is the practice of processing signals which results in some changes of their spectral contents. Usually the change implies a reduction in or filtering out of some undesired input spectral components. A filter allows certain frequencies to pass while attenuating others. While analog filters operate on continuous-time signals, digital filters operate on sequences of discrete sampled values (see Fig. 7.8). Although digital filters perform many of the same functions as analog filters, they are different!

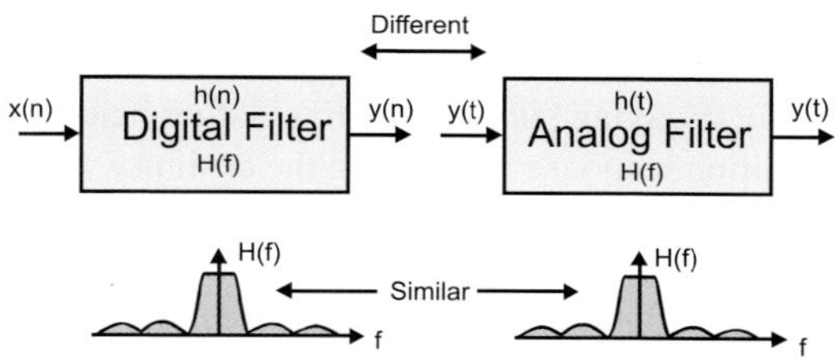

FIGURE 7.8

Digital and analog filters are different even though they perform similar tasks.

The two classes of filters share many of the same or analogous properties. In the standard order of our educational process we first learn about analog filters and later about digital filters. To ease the entry into the digital domain, we emphasize the similarities of the two classes of filters. This order is due to the fact that an analog filter is less abstract than a digital filter. We can touch the components of an analog filter: capacitors, inductors, resistors, operational amplifiers, and wires. On the other hand a digital filter does not have the same physical form because a digital filter is merely a set of instructions that perform arithmetic operations on an array of numbers. The operations can be weighted sums or inner products (see Fig. 7.9). It is convenient to visualize the array as a list of uniformly spaced sample values of an analog waveform. The instructions to perform these operations can reside as software in a computer or microprocessor or as firmware in a dedicated collection of interconnected hardware elements.

Let us examine the similarities that are emphasized when we learn about digital filters with some level of prior familiarity with analog filters and the tools to describe them. Many analog filters are formed by interconnections of lumped linear components modeled as ideal capacitors, inductors, and resistors while others are formed by interconnections of resistors, capacitors, and operational amplifiers. Sampled data filters are formed by interconnections of registers, adders, and multipliers. Most analog filters are designed to satisfy relationships defined by linear time-invariant differential equations. The differential equations are recursive, which means the initial condition time domain response, called the homogeneous or undriven response, is a weighted sum of exponentially decaying sinusoids. Most

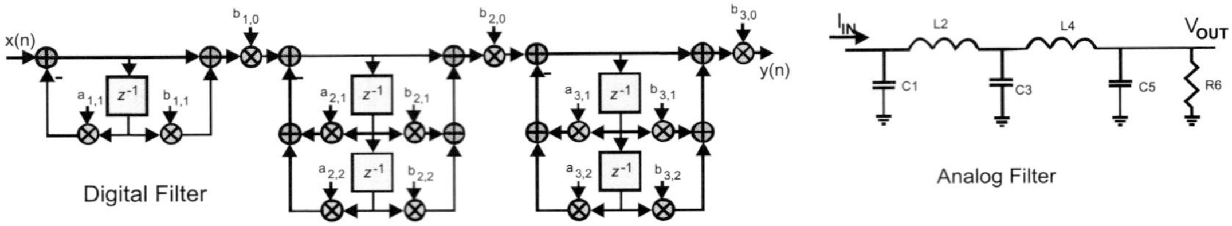

FIGURE 7.9

Digital filter: registers, adders, and multipliers. Analog filter: capacitor, inductors, and resistors.

sampled data filters are designed to satisfy relationships defined by linear time-invariant difference equations. Here we find the first major distinction between analog and digital filters: The difference equations for the sampled data filters can be recursive or nonrecursive. When the difference equations are selected to be recursive, the initial condition response is, as it was for the analog filter, samples of a weighted sum of exponentially decaying sinusoids. On the other hand, when the difference equation is nonrecursive, the initial condition response is anything the designer wants it to be and is limited only by her imagination but is usually designed to satisfy some frequency domain specifications.

When we compare the two filter classes, analog and digital, we call attention to the similarities of the tools we use to describe, analyze, and design them. They are described by differential and difference equations which are weighted sums of signal derivates or weighted sums of delayed sequences. They both have linear operators or transforms, the Laplace transform $L\{h(t)\} = H(s)$ and the z-transform $z\{h(n)\} = H(z)$, which offer us insight into the internal structure of the differential or difference equations (see Fig. 7.10). They both have operator descriptions that perform the equivalent functions. These are the integral operator, denoted by s^{-1}, and the delay operator, denoted by z^{-1}, in which the system memories or state of the analog and digital filters reside, respectively. Both systems have transfer functions $H(s)$ and $H(z)$, ratios of polynomials in the operator variables s and z. The roots of the denominator polynomial, the operator version of the characteristic equation, are the filter poles. These poles describe the filter modes, the exponentially damped sinusoids, of the recursive structure. The roots of the numerator polynomial are the filter zeros. The zeros describe how the filter internal mode responses are connected to the filter input and output ports. They tell us the amplitude of each response component (called residues to impress us) in the output signal's initial condition response. The transforms of the two filter classes have companion transforms, the Fourier transform $H(\omega)$ and the sampled data Fourier series $H(\theta)$, which describe, when they exist, the frequency domain behavior, the sinusoidal steady-state gain and phase, of the filters. It is remarkable how many similarities there are in the structure, the tools, and the time and frequency responses of the two types of filters. It is no wonder we emphasize their similarities.

We spend a great deal of effort examining recursive digital filters because of the close relationships with their analog counterparts. There are digital filters that are mappings of the traditional analog filters. These include maximally flat Butterworth filters, equal ripple passband Tchebychev-I filters, equal ripple stopband Tchebychev-II filters, and both equal ripple passband and equal ripple stopband elliptic filters. The spectra of these infinite impulse response (IIR) filters are shown in Fig. 7.11. All DSP filter design programs offer the DSP equivalents of their analog recursive filter counterparts.

$$H(z) = \sum_{n=0}^{\infty} h(n) z^{-n} \qquad H(s) = \int_0^{\infty} h(t)\, e^{-st} dt$$

$$H(z) = b_0 \frac{z^3 + b_1 z^2 + b_2 z^1 + b_3}{z^3 + a_1 z^2 + a_2 z^1 + a_3} \qquad H(s) = b_0 \frac{s^3 + b_1 s^2 + b_2 s^1 + b_3}{s^3 + a_1 s^2 + a_2 s^1 + a_3}$$

$$H(z) = b_0 \frac{(z - z_1)(z - z_2)(z - z_3)}{(z - p_1)(z - p_2)(z - p_3)} \qquad H(s) = b_0 \frac{(s - z_1)(s - z_2)(s - z_3)}{(s - p_1)(s - p_2)(s - p_3)}$$

$$H(z) = c_0 + c_1 \frac{z}{z - p_1} + c_2 \frac{z}{z - p_2} + c_2 \frac{z}{z - p_2} \qquad H(s) = c_0 + c_1 \frac{p_1}{s - p_1} + c_2 \frac{p_2}{s - p_2} + c_2 \frac{p_2}{s - p_2}$$

$$h(n) = c_0 + c_1 p_1^n + c_2 p_2^n + c_3 p_3^n \qquad h(t) = c_0 \delta(t) + c_1 e^{-p_1 t} + c_2 e^{-p_2 t} + c_3 e^{-p_3 t}$$

FIGURE 7.10

Similarity of system equations for digital and analog filters.

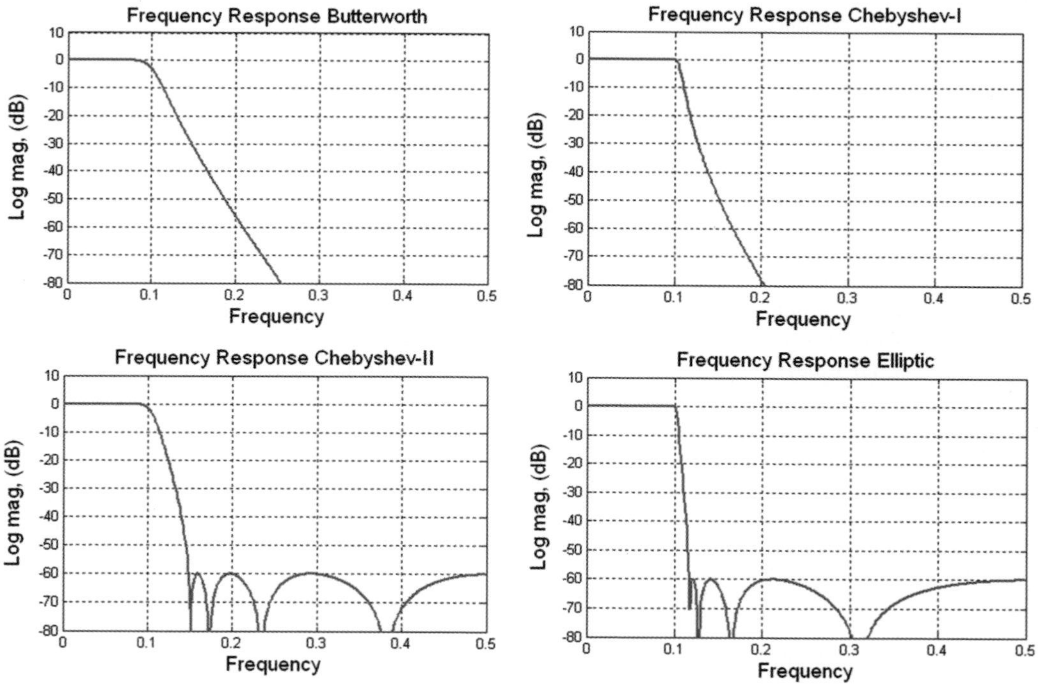

FIGURE 7.11

Frequency response of digital IIR filter realizations of standard analog filters.

In retrospect, emphasizing similarities between the analog filter and the equivalent digital was a poor decision. We make this claim for a number of reasons. The first is that there are limited counterparts in the analog domain to the nonrecursive filter of the digital domain. This is because the primary element of the nonrecursive filter is the pure delay, represented by the delay operator z^{-1}. We cannot form a pure delay response, represented by the analog delay operator e^{-sT}, in the analog domain as the solution of a linear differential equation. To obtain pure delay, we require a partial differential equation which offers the wave equation whose solution is a propagating wave which we use to convert distance to time delay. The propagation velocity of electromagnetic waves is too high to be of practical use in most analog filter designs but the velocity of sound waves in crystal structures enables an important class of analog filters known as acoustic surface wave (SAW) devices. Your cell phone contains one or more such filters. In general, most of us have limited familiarity with nonrecursive analog filters so our first introduction to the properties and design techniques for finite duration impulse response (FIR) filters is given in our DSP classes.

The second reason for not emphasizing the similarities between analog and digital filters is that we start believing the statement that "a digital filter is the same as an analog filter" and that all the properties of one are embedded in the other. That seemed like an okay perspective the first 10 or 20 times we heard that claim. The problem is, it is not true! The digital filter has a resource the analog filter does not have! The digital filter can perform tasks the analog filter cannot do! What is that resource? It is the sampling clock! We can do applied magic in the sampled data domain by manipulating the sample clock. There is no equivalent magic available in the analog filter domain! Analog designers shake their heads in disbelief when they see what a multirate filter can do. Actually, digital designers also shake their heads in disbelief when they learn that manipulating the clock can accomplish such amazing things.

We might ask, how can the sampling clock affect the performance, or more so, enhance the capabilities of a digital filter? Great question! The answer, my friend, is written in the way we define the frequency of a sampled data signal. Let us review how frequency as we understand it in the analog world is converted to frequency in the digital world. We now present a concise review of complex and real sinusoids with careful attention paid to their arguments.

The exponential function e^x has the interesting property that it satisfies the differential equation shown in Eq. (7.4). This means the exponential function replicates under the differential operator, an important property inherited by real and complex sinusoids. It is easy to verify that the Taylor series shown in Eq. (7.5) satisfies Eq. (7.4). When we replace the argument "x" of Eq. (7.5) with the argument "$j\theta$" we obtain the series shown in Eq. (7.6), and when we gather the terms corresponding to the even and odd powers respectively of the argument "$j\theta$" we obtain the partitioned series shown in Eq. (7.7). We recognize that the two Taylor series of the partitioned Eq. (7.7) are the Taylor series of the cosine and sine which we show explicitly in Eq. (7.8). We note that the arguments of both the real exponential and the complex exponential must be dimensionless; otherwise, we would not be able to sum the successive powers of the series. We also note that the units of the argument "θ" are radians, a dimensionless unit formed by taking the ratio of the arc length of a circle to the radius of the circle. When we replace the "θ" argument of Eq. (7.8) with a time-varying angle "$\theta(t)$" we obtain the time function shown in Eq. (7.9). The simplest time-varying angle is one that changes linearly with time, $\theta(t) = \omega_c t$, which when substituted in Eq. (7.9) gives Eq. (7.10). This discussion leads us to the next statement. Since the argument of a sinusoid has the dimensionless units radians and "t" has units of seconds, the unit of ω_c must be rad/s, a velocity. In other words, the frequency ω_c is the time derivative of the linearly increasing phase angle $\omega_c t$:

$$\frac{d}{dx} f(x) = f(x), \tag{7.4}$$

$$e^x = \sum_{k=0}^{\infty} \frac{x^k}{k!} = 1 + x + \frac{x^2}{2} + \frac{x^3}{6} + \frac{x^4}{24} + \frac{x^5}{120} + \cdots, \tag{7.5}$$

$$e^{j\theta} = \sum_{k=0}^{\infty} \frac{(j\theta)^k}{k!} = 1 + j\theta + \frac{(j\theta)^2}{2} + \frac{(j\theta)^3}{6} + \frac{(j\theta)^4}{24} + \cdots, \tag{7.6}$$

$$e^{j\theta} = \left[1 - \frac{\theta^2}{2} + \frac{\theta^4}{24} - \frac{\theta^6}{720} + \cdots\right] + j\left[\theta - \frac{\theta^3}{6} + \frac{\theta^5}{120} - \frac{\theta^7}{5040} + \cdots\right], \tag{7.7}$$

$$e^{j\theta} = \cos(\theta) + j\sin(\theta), \tag{7.8}$$

$$s(t) = e^{j\theta(t)} = \cos(\theta(t)) + j\sin(\theta(t)), \tag{7.9}$$

$$s(t) = e^{j\omega_c t} = \cos(\omega_c t) + j\sin(\omega_c t). \tag{7.10}$$

We now examine the phase argument of the sampled data sinusoid. The successive samples of a sampled complex sinusoid are formed as shown in Eq. (7.11) by replacing the continuous argument t with the sample time positions nT, as done in Eq. (7.12). We now note that while the argument of the sampled data sinusoid is dimensionless, the independent variable is no longer "t" with units of seconds but rather "n" with units of sample. This is consistent with the dimension of "T," which of course is s/sample. Consequently, the product $\omega_c T$ has units of (rad/s) · (s/smpl) or rad/smpl; thus, digital frequency is given in rad/smpl or emphatically the angle change per sample! A more useful version of Eq. (7.12) is obtained by replacing ω_c with $2\pi f_c$ as done in Eq. (7.13) and then replacing T with $1/f_s$ as done in Eq. (7.14). We finally replace $2\pi f_c/f_s$ with θ_c, as done in Eq. (7.15). Here we clearly see that the parameter θ_c has units of rad/smpl, which describes the fraction of the circle the argument traverses per successive sample. We have

$$s(n) = s(t)|_{t=nT} = \cos(\omega_c t)|_{t=nT} + j\sin(\omega_c t)|_{t=nT}, \tag{7.11}$$

$$s(n) = \cos(\omega_c Tn) + j\sin(\omega_c Tn), \tag{7.12}$$

$$s(n) = \cos(2\pi f_c Tn) + j\sin(2\pi f_c Tn), \tag{7.13}$$

$$s(n) = \cos\left(2\pi\frac{f_c}{f_s}n\right) + j\sin\left(2\pi\frac{f_c}{f_s}n\right), \tag{7.14}$$

$$s(n) = \cos(\theta_c n) + j\sin(\theta_c n). \tag{7.15}$$

Now suppose we have a sampled sinusoid with sample rate eight times the sinusoid's center frequency, so that there are eight samples per cycle. This is equivalent to the sampled sinusoid having a digital frequency of $2\pi/8$ rad/smpl. We can visualize a spinning phasor rotating 1/8 of the way around the unit circle per sample. In Fig. 7.12, subplots 1 and 2 show 21 samples of this sinusoid and its spectrum. Note the pair of spectral lines located at ± 0.125 on the normalized frequency axis. We can reduce the sample rate of this time series by taking every other sample. The spinning rate for this new sequence is $2 \cdot 2\pi/8$ or $2\pi/4$. The downsampled sequence has doubled its digital frequency. The newly sampled sequence and its spectrum are shown in Fig. 7.12, subplots 3 and 4. Note that the pair of spectral lines are now located at ± 0.25 on the normalized frequency axis. We can further reduce the sample rate by again taking every other sample (or every fourth sample of the original sequence). The spinning rate for this new sequence is $4 \cdot 2\pi/8$ or $2\pi/2$. The downsampled sequence has again doubled its digital frequency. The newly sampled sequence and its spectrum are shown in Fig. 7.12, subplots 5 and 6. Note

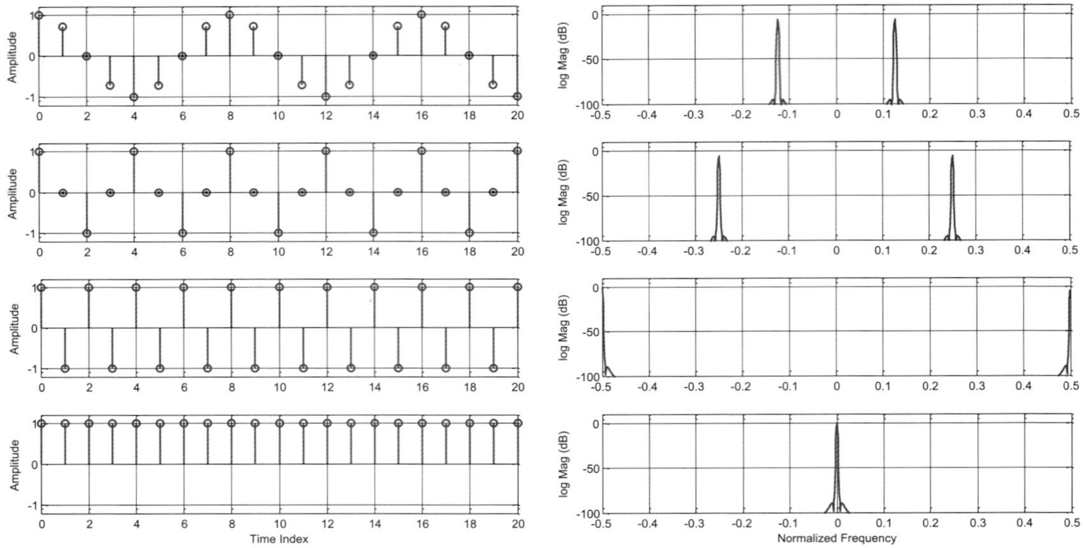

FIGURE 7.12

Time series and spectra of sinusoids aliased to new digital frequencies as a result of downsampling.

that the pair of spectral lines are now located at ± 0.5 on the normalized frequency axis. We once again reduce the sample rate by taking every other sample (or every eighth sample of the original sequence). The spinning rate for this new sequence is $8 \cdot 2\pi/8$ or 2π, but since a 2π rotation is equivalent to no rotation, the spinning phasor appears to be stationary or rotating at 0 rad/smpl. The newly sampled sequence and its spectrum are shown in Fig. 7.12, subplots 7 and 8. Note that the pair of spectral lines are now located at ± 1, which is equivalent to 0 on the normalized frequency axis. Frequency shifts due to sample rate changes are attributed to aliasing. Aliasing enables us to move the spectrum from one frequency location to another location without the need for the complex mixers we normally use in a DDC to move a spectral span to baseband.

Speaking of the standard DDC, Fig. 7.13 presents the primary signal flow blocks based on the DSP emulation of the conventional analog receiver. Fig. 7.14 shows the spectra that will be observed at successive output ports of the DDC. We can follow the signal transformations of the DDC by following the spectra in the following description. The top figure shows the spectrum at the output of the ADC. The second figure shows the spectrum at the output of the quadrature mixer. Here we see that the input spectrum has been shifted so that the selected channel band resides at zero frequency. The third figure shows the output of the low-pass filter pair. Here we see that a significant fraction of the input bandwidth has been rejected by the stopband of the low-pass filter. Finally, the last figure shows the spectrum after the M-to-1 downsampling that reduced the sample rate in proportion to the reduction in bandwidth performed by the filter. We will always reduce the sample rate if we reduce the bandwidth. We gain no performance advantage by oversatisfying the Nyquist criterion, but incur significant penalties by operating the DSP processors at higher than necessary sample rates.

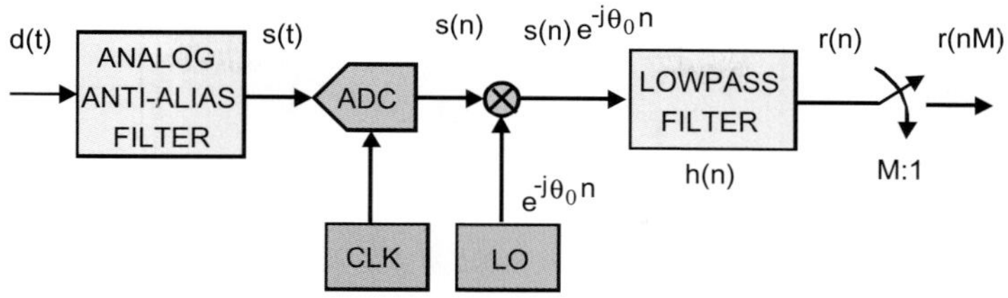

FIGURE 7.13

Building blocks of a conventional digital down converter.

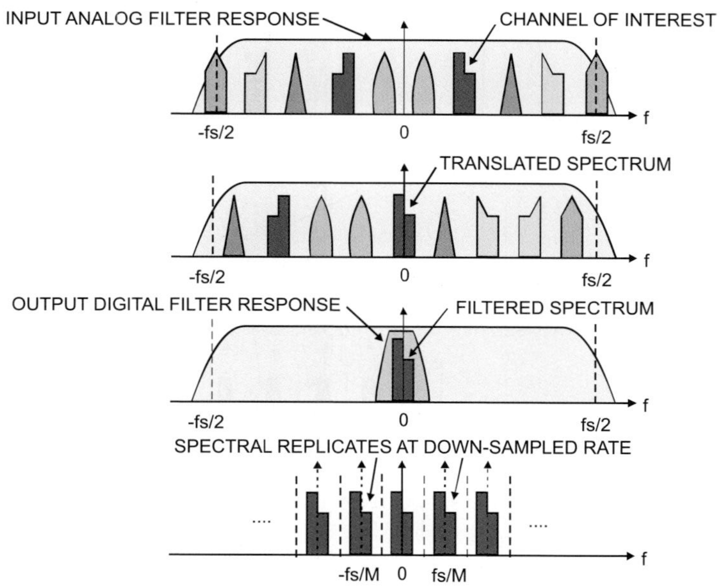

FIGURE 7.14

Spectra at consecutive points in a conventional digital down converter.

Fig. 7.15 presents the primary signal flow blocks based on aliasing the selected bandpass span to baseband by reducing the output sample rate of the bandpass filter. Fig. 7.16 shows the spectra that will be observed at successive output ports of the aliasing DDC. We can follow the signal transformations of the aliasing DDC by following the spectra in the following description. The top figure shows the spectrum at the output of the ADC. The second figure shows the spectrum of the complex bandpass filter and the output spectrum from this filter. The bandpass filter is formed by up converting, by a complex heterodyne, the coefficients of the prototype low-pass filter. We note that this filter does not have a mirror image. Analog designers sigh when they see this digital option. The third figure shows the output of the bandpass filter pair following the M-to-1 downsampling. If the center frequency of

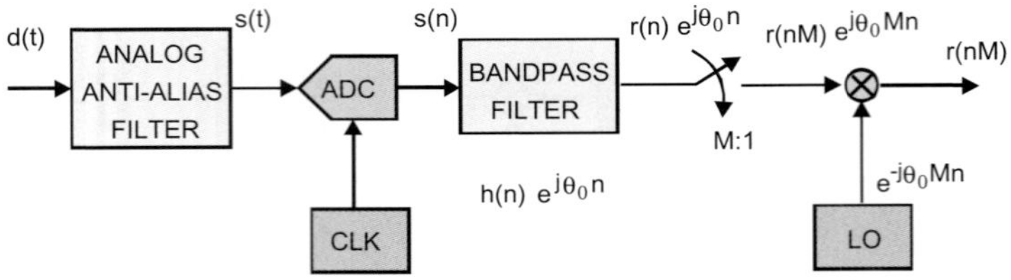

FIGURE 7.15

Building blocks of aliasing a digital down converter with M-to-1 downsampler.

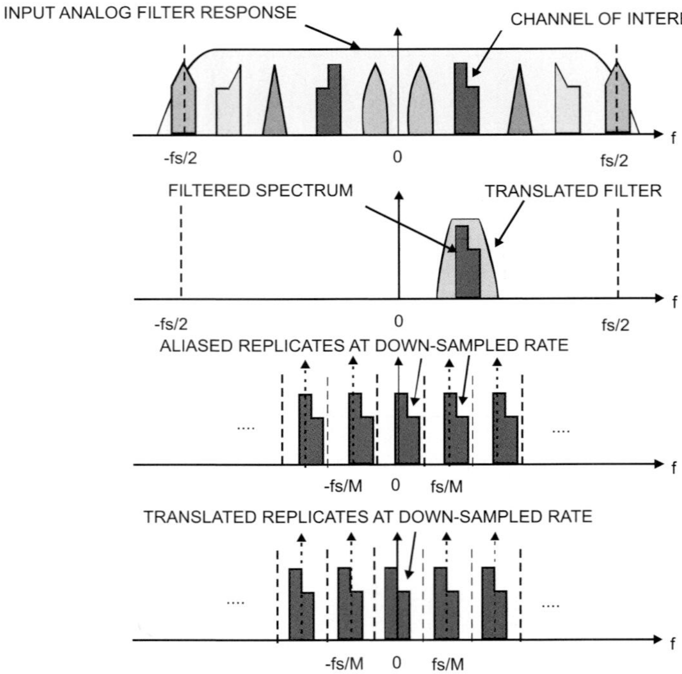

FIGURE 7.16

Spectra at consecutive points in aliasing digital down converter with downsampling.

the band is a multiple of the output sample rate, the aliased band resides at baseband. For the example shown here, the center frequency was Δf above $k\ f_s/M$, and since all multiples of f_s/M alias to baseband, our selected band has aliased to Δf above zero frequency. Finally, the last figure shows the spectrum after the output heterodyne from Δf to baseband. Note that the heterodyne is applied at the low output rate rather than at the high input rate as was done in Fig. 7.13.

The last part of this discussion deals with the equivalency of cascade operators. Here is the core idea. Suppose we have a cascade of two filters, say a low-pass filter followed by a derivative filter, as shown

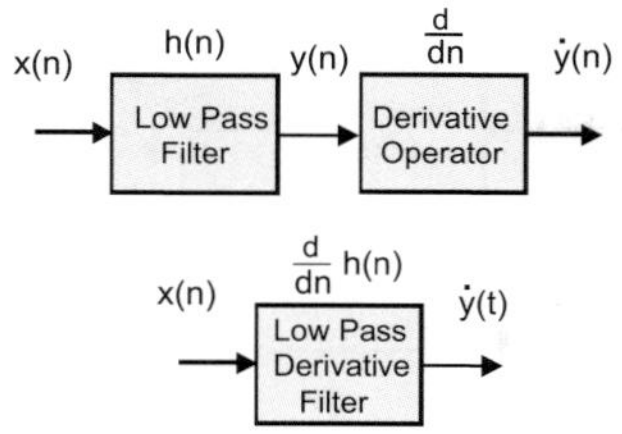

FIGURE 7.17

A cascade of two filters is the same as a single filter formed by a composite filter.

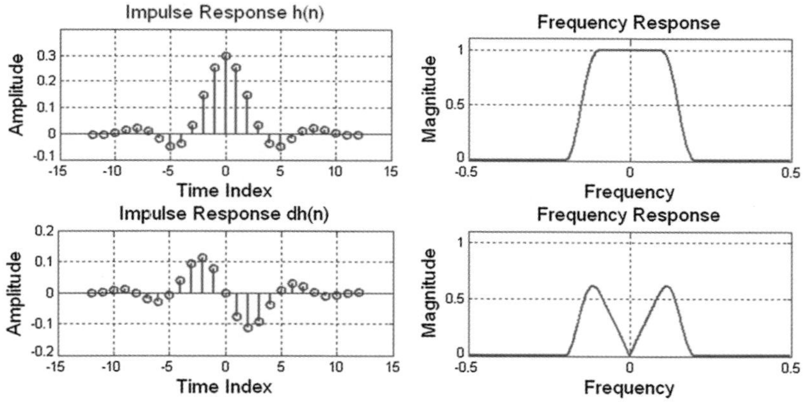

FIGURE 7.18

The spectrum of cascade filters is the same as the spectrum of a single filter containing combined responses.

in Fig. 7.17. The filters are linear operators that commute and are distributive. We could reorder the two filters and have the same output from their cascade or we could apply the derivative filter operator to the low-pass filter to form a composite filter and obtain the same output as from the cascade. This equivalency is shown in Fig. 7.18. The important idea here is that we can filter the input signal and then apply an operator to the filter output, or we can apply the operator to the filter and have the altered filter perform both operators to the input signal simultaneously.

Here comes the punchline! Let us examine the aliasing DDC of Fig. 7.15, which contains a bandpass filter followed by an M-to-1 resampler. This section is redrawn in the upper half of Fig. 7.19. If we think about it, this is almost silly: Here we compute one output sample for each input sample and then discard $M-1$ of these samples in the M-to-1 downsampler. Following the lead of the cascade linear operators, the filter and resampler, being equivalent to a composite filter containing the resampler applied to the filter, we replace the pair of operators with the composite operator, as shown in the lower half of Fig. 7.19.

Embedding the M-to-1 resampler in the bandpass filter is accomplished by the block diagram shown in Fig. 7.20. This filter accepts M inputs, one for each path, and computes one output. Thus we do not

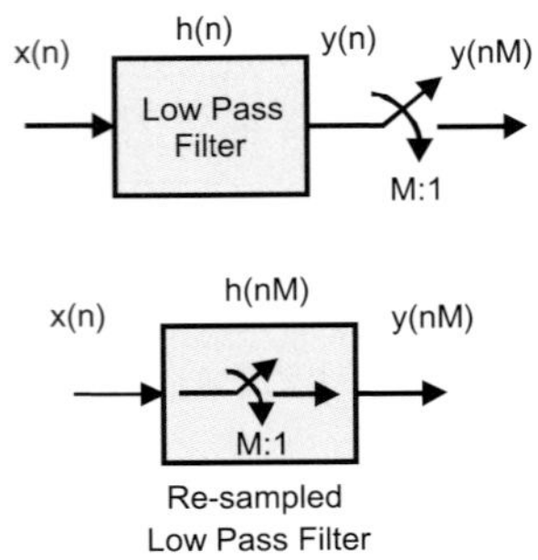

FIGURE 7.19

The cascade of a filter and M-to-1 resampler is equivalent to that of an M-to-1 resample filter.

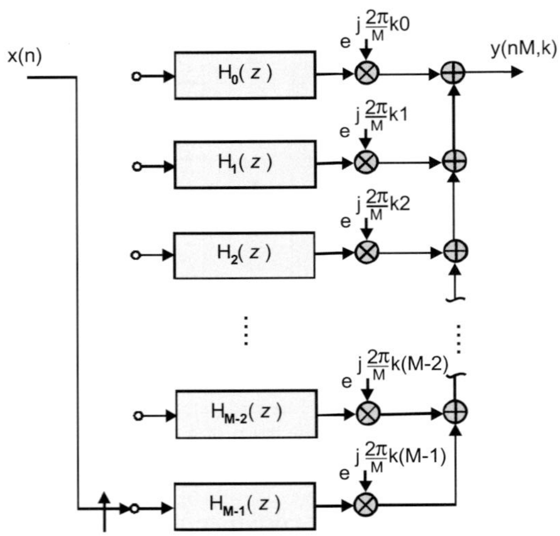

FIGURE 7.20

An M-path M-to-1 resample bandpass filter performs aliased digital down conversion.

waste operations computing output samples scheduled to be discarded. An interesting note here is that the complex rotators applied to the prototype low-pass filter in Fig. 7.15 have been factored out of each arm and are applied once at the output of each path. This means that if the input data are real, the samples are not made complex until they leave the filter. This represents a 2-to-1 savings over the architecture of Fig. 7.13, in which the samples are made complex on the way into the filter, which means there are two filters, one for each path of the complex heterodyne. The alias-based DDS with resampler embedded in the filter is not a bad architecture! It computes output samples at the output rate and it only uses one partitioned filter rather than two. "Not bad" is an understatement! The cherry on top of this dessert is the rotator vector applied to the output of the path filters. This rotator can extract any band from the aliased frequency spans that have been aliased to baseband by the resampling operator. If it can extract any band, then we have the option of extracting more than one band, and in fact we can extract every band from the output of the partitioned filter. That is amazing! This filter can service more than one output channel simultaneously; all it needs is additional rotator vectors. As we will see shortly, the single filter structure can service all the channels that alias to baseband and the most efficient bank of rotators is implemented by an M-point IFFT. There is more to come! Be sure to read on! We have only scratched the surface and the body of material related to multirate filters is filled with many wonderful properties and applications.

7.4 Windowing

As specified in the previous section, digital filters can be classified in many ways, starting with the most general frequency domain characteristics such as low-pass, high-pass, bandpass, and bandstop and finishing with secondary characteristics such as uniform and nonuniform group delay. We now also know that an important classification is based on the filter's architectural structure with a primary consideration being that of FIR and IIR filter. Further subclassifications, such as canonic forms, cascade forms, and lattice forms, are primarily driven by consideration of the sensitivity to finite arithmetic, memory requirements, the ability to perform pipeline arithmetic, and hardware constraints.

The choice to perform a given filtering task with a recursive or a nonrecursive filter is driven by a number of system considerations, including processing resources, clock speed, and various filter specifications. Performance specifications, which include the operating sample rate, passband and stopband edges, passband ripple, and out-of-band attenuation, all interact to determine the complexity required of the digital filter.

In filters with infinite impulse response, each output sample depends on previous input samples and on previous filter output samples. That is the reason why their practical implementation always requires a feedback loop. Thus, like all feedback-based architectures, IIR filters are sensitive to input perturbations that could cause instability and infinite oscillations at the output. However, IIR filters are usually very efficient and require far few multiplications than FIR filters per output sample. Note that an IIR filter could have an infinite sequence of output samples even if the input samples become all zeros. It is this characteristic which gives them their name.

FIR filters use only current and past input samples and none of the filter's previous output samples to obtain a current output sample value (remember that they are also sometimes referred to as nonrecursive filters). Given a finite duration of the input signal, the FIR filter will always have a finite duration of nonzero output samples, and this is the characteristic which gives them their name.

The procedure by which FIR filters calculate the output samples is the convolution of the input sequence with the filter impulse response:

$$y(n) = \sum_{k=0}^{M-1} h(k)x(n-k), \tag{7.16}$$

where $x(n)$ is the input sequence, $h(n)$ is the filter impulse response, and M is the number of filter taps. The convolution is nothing but a series of multiplications followed by the addition of the products, while the impulse response of a filter is nothing but what the name tells us: It is the time-domain view

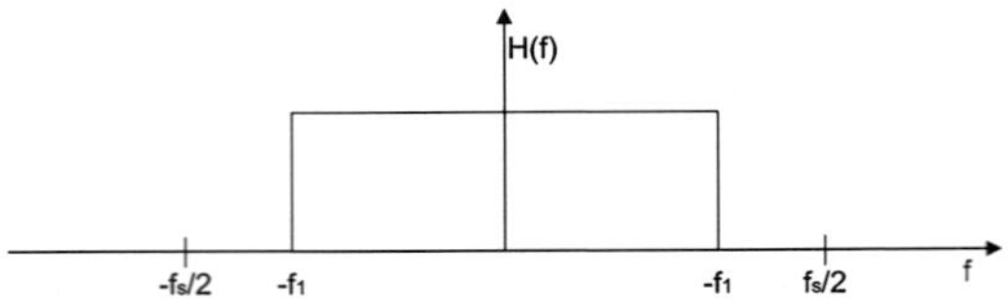

FIGURE 7.21

Frequency response of a prototype ideal low-pass filter.

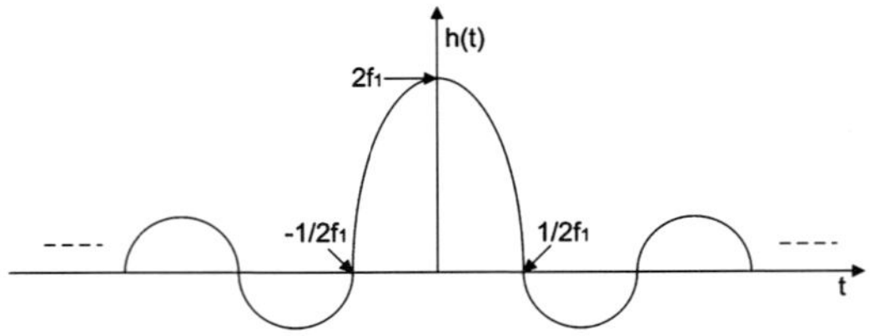

FIGURE 7.22

Impulse response of a prototype ideal low-pass filter.

of the filter's output values when the input is an impulse (a single unity-valued sample preceded and followed by zero-valued samples).

In the following paragraphs we introduce the basic window design method for FIR filters. Because low-pass filtering is the most common filtering task, we will introduce the window design method for low-pass FIR filters. However, the relationships between filter length and filter specifications and the interactions between filter parameters remain valid for other filter types. Many other techniques can also be used to design FIR filters. At the end of this section we will introduce the Remez algorithm, which is one of the most widely used filter design technique, as well as one of its modified versions, which allows us to achieve $1/f$ type decay for the out-of-band side lobes.

The frequency response of a prototype low-pass filter is shown in Fig. 7.21. The passband is seen to be characterized by an ideal rectangle with unity gain between the frequencies $\pm f_1$ Hz and zero gain elsewhere.

An attractive feature of the ideal low-pass filter $H(f)$ as a prototype is that from its closed-form inverse Fourier transform, we achieve the exact expression for its impulse response $h(t)$:

$$h(t) = 2f_1 \frac{\sin(2\pi f_1 t)}{2\pi f_1 t}. \tag{7.17}$$

The argument of the $\sin(x)/x$ function is always the product of 2π, half the spectral support $(2f_1)/2$, and the independent variable t. The numerator is periodic and becomes zero when the argument is a multiple of π. The impulse response of the prototype filter is shown in Fig. 7.22. The $\sin(x)/x$ filter shown in Fig. 7.22 is a continuous function which we have to sample to obtain the prototype sampled data impulse response. To preserve the filter gain during the sampling process we scale the sampled function by the sample rate $1/f_s$. The problem with the sample set of the prototype filter is that the number of samples is unbounded and the filter is noncausal. If we had a finite number of samples, we could delay the response to make it causal and solve the problem. Then our first task is to reduce the unbounded set of filter coefficients to a finite set. The process of pruning an infinite sequence to a finite sequence is called windowing. In this process, a new limited sequence is formed as the product of the finite sequence $w(n)$ and the infinite sequence, as shown in Eq. (7.18), where the $*$ operator is the standard MATLAB® point-by-point multiplication:

$$h_w(n) = w(n)*h(n). \tag{7.18}$$

Applying the properties of the Fourier transforms, we obtain the expression for the spectrum of the windowed impulse response, which is the circular convolution of the Fourier transform of $h(n)$ and $w(n)$.

The symmetric rectangle we selected as our window abruptly turns off the coefficient set at its boundaries. The sampled rectangle weighting function has a spectrum described by the Dirichlet kernel, which is the periodic extension of the transform of a continuous-time rectangle function. The convolution between the spectra of the prototype filter with the Dirichlet kernel forms the spectrum of the rectangle windowed filter coefficient set. The contribution to the corresponding output spectrum is seen to be the stopband ripple, the transition bandwidth, and the passband ripple. The passband and stopband ripples are due to the side lobes of the Dirichlet kernel moving through the passband of the prototype filter, while the transition bandwidth is due to the main lobe of the kernel moving from the stopband to the passband of the prototype. A property of the Fourier series is that their truncated version forms a new series exhibiting the minimum mean square (MMS) approximation to the original function. We thus note that the set of coefficients obtained by a rectangle window exhibits the minimum mean square approximation to the prototype frequency response. The problem with MMS approximations in numerical analysis is that there is no mechanism to control the location or value of the error maxima. The local maximum errors are attributed to the Gibbs phenomena, the failure of the series to converge in the neighborhood of a discontinuity. These errors can be objectionably large. A process must now be invoked to control the objectionably high side lobe levels. We have two ways to approach the problem. We can redefine the frequency response of the prototype filter so that the amplitude discontinuities are replaced with a specified tapering in the transition bandwidth. In this process, we trade off the transition bandwidth for side lobe control. Equivalently, knowing that the objectionable stopband side lobes are caused by the side lobes in the spectrum of the window, we can replace the rectangle window with other even symmetric functions with reduced amplitude side lobe levels. The two techniques, side lobe control and transition bandwidth control, are tightly coupled. The easiest way to visualize control of the side lobes is by destructive cancellation between the spectral side lobes of the Dirichlet kernel associated with the rectangle and the spectral side lobes of translated and scaled versions of the same kernel. The cost we incur to obtain reduced side lobe levels is an increase in main lobe bandwidth. Remembering that the window's two-sided main lobe width is an upper bound to the filter's transition bandwidth, we can estimate the transition bandwidth of a filter required to obtain a specified side lobe level. The primary reason we examined windows and their spectral description as weighted Dirichlet kernels was to develop a sense of how we trade the main lobe width of the window for its side lobe levels and in turn filter transition bandwidth and side lobe levels. Some windows perform this trade-off of bandwidth for side lobe level very efficiently while others do not.

The Kaiser–Bessel window is very effective, while the triangle (or Fejer) window is not. The Kaiser–Bessel window is in fact a family of windows parameterized over the time–bandwidth product, β, of the window. The main lobe width increases with β, while the peak side lobe level decreases with β. The Kaiser–Bessel window is a standard option in filter design packages such as MATLAB and QED-2000.

Other filter design techniques include the Remez algorithm [8], sometimes also referred to as the Parks–McClellan (P-M), the McClellan–Parks–Rabiner (MPR), the Equiripple, or the multiple exchange algorithm, which found large acceptance in practice. It is a very versatile algorithm capable of designing FIR filters with various frequency responses, including multiple passband and stopband frequency responses with independent control of ripple levels in the multiple bands. The desired passband cutoff frequencies, the frequencies where the attenuated bands begin, and the desired passband and stopband ripple are given as input parameters to the software which will generate N time domain filter coefficients. Here, N is the minimum number of taps required for the desired filter response and it is selected by using the Harris approximation or the Hermann approximation (in MATLAB). The problem with the Remez algorithm is that it shows equiripple side lobes, which is not a desirable characteristic. We would like to have a filter frequency response which has a $1/f$ out-of-band decay rate rather than exhibiting equiripple. The main reasons for desiring that characteristic are related to system performance. We often build systems comprised of a digital filter and a resampling switch. Here the digital filter reduces the bandwidth and is followed by a resampling switch that reduces the output sample commensurate with the reduced output bandwidth. When the filter output is resampled, the low-level energy residing in the out-of-band spectral region aliases back into the filter passband. When the reduction in sample rate is large, there are multiple spectral regions that alias or fold into the passband. For instance, in a 16-to-1 reduction in sample rate, there are 15 spectral regions that fold into the passband. The energy in these bands is additive and if the spectral density in each band is equal, as it is in an equiripple design, the folded energy level is increased by a factor of sqrt(15). The second reason for which we may prefer FIR filters with $1/f$ side lobe attenuation as opposed to uniform side lobes is finite arithmetic. A filter is defined by its coefficient set and an approximation to this filter is realized by a set of quantized coefficients. Given two filter sets $h(n)$ and $g(n)$, the first with equiripple side lobes and the second with $1/f$ side-lobes, we form two new sets, $h_Q(n)$ and $g_Q(n)$, by quantizing their coefficients. The quantization process is performed in two steps. First, we rescale the filters by dividing by the peak coefficient. Second, we represent the coefficients with a fixed number of bits to obtain the quantized approximations. The zeros of an FIR filter residing on the unit circle perform the task of holding down the frequency response in the stopband. The interval between the zeros contains the spectral side lobes. When the interval between adjacent zeros is reduced, the amplitude of the side lobe between them is reduced, and when the interval between adjacent zeros is increased, the amplitude of the side lobe between them is increased. The zeros of the filters are the roots of the polynomials $H(z)$ and $G(z)$. The roots of the polynomials formed by the quantized set of coefficients differ from the roots of the nonquantized polynomials. For small changes in coefficient size, the roots exhibit small displacements along the unit circle from their nominal positions. The amplitude of some of the side lobes must increase due to this root shift. In the equiripple design, the initial side lobes exactly meet the designed side lobe level with no margin for side lobe increase due to root shift caused by coefficient quantization. On the other hand, the filter with $1/f$ side lobe levels has plenty of margin for side lobe increase due to root shift caused by coefficient quantization.

More details on the modified Remez algorithm can be found in [8].

7.5 Basics on multirate filters

Multirate filters are digital filters that contain a mechanism to increase or decrease the sample rate while processing input sampled signals. The simplest multirate filter performs integer upsampling of 1-to-M or integer downsampling of Q-to-1. By extension, a multirate filter can employ both upsampling and downsampling in the same process to effect a rational ratio sample rate change of M-to-Q. More sophisticated techniques exist to perform arbitrary and perhaps slowly time-varying sample rate changes. The integers M and Q may be selected to be the same so that there is no sample rate change between input and output but rather an arbitrary time shift or phase offset between input and output sample positions of the complex envelope. The sample rate change can occur at a single location in the processing chain or can be distributed over several subsections.

Conceptually, the process of downsampling can be visualized as a two-step process, as indicated in Fig. 7.23. There are three distinct signals associated with this procedure. The process starts with an input series $x(n)$ that is processed by a filter $h(n)$ to obtain the output sequence $y(n)$ with reduced bandwidth. The sample rate of the output sequence is then reduced Q-to-1 to a rate commensurate with the reduced signal bandwidth. In reality the processes of bandwidth reduction and sample rate reduction are merged in a single process called multirate filtering. The bandwidth reduction performed by the digital filter can be a low-pass or a bandpass process.

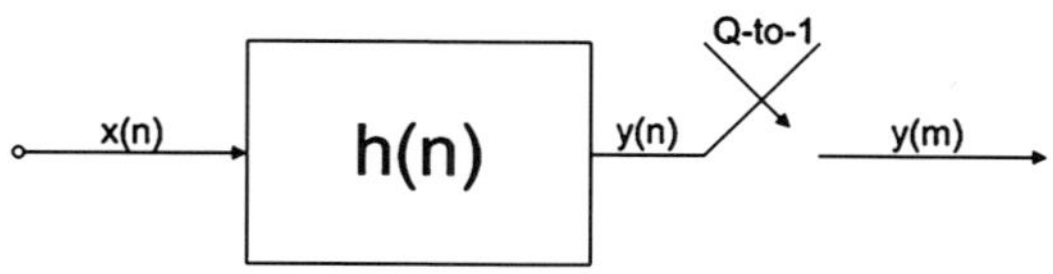

FIGURE 7.23

Downsampling process; filtering and sample rate reduction.

In a dual way, the process of upsampling can be visualized as a two-step process, as indicated in Fig. 7.24. Here too there are three distinct time series. The process starts by increasing the sample rate of an input series $x(n)$ by resampling it 1-to-M. The zero-packed time series with M-fold replication of the input spectrum is processed by a filter $h(n)$ to reject the spectral replicas and output the sequence $y(m)$ with the same spectrum as the input sequence but sampled at an M times higher sample rate. In reality the processes of sample rate increase and selected bandwidth rejection are also merged in a single process again called multirate filtering.

The presented structures are the most basic forms of multirate filters; in the following we present more complicated multirate architecture models. In particular, we focus on the polyphase decomposition of a prototype filter which is very efficient when embedded in multirate structures. We also propose the derivation, step by step, of both the polyphase down converter and up converter channelizers. These are the two standard engines that we will modify for designing a novel polyphase channelizer that better fits the needs of the future software-defined radio.

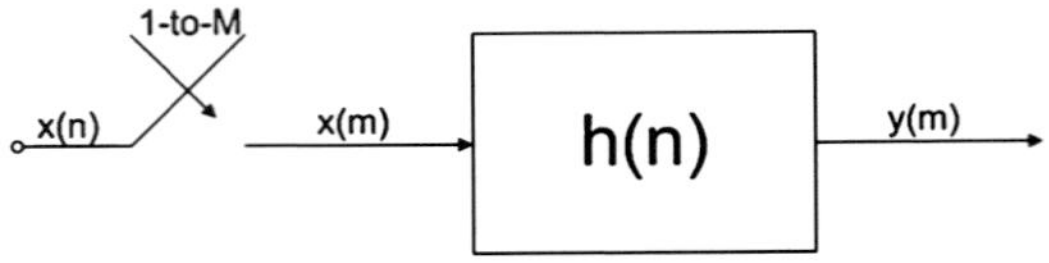

FIGURE 7.24

Upsampling process; sample rate increasing and filtering.

7.6 From single-channel down converter to standard down converter channelizer

The derivation of the standard polyphase channelizer begins with the issue of down converting a single frequency band, or channel, located in a multichannel frequency division multiplexed (FDM) input signal whose spectrum is composed of a set of M equally spaced, equal bandwidth channels, as shown in Fig. 7.25. Note that this signal has been band-limited by analog filters and has been sampled at a sufficiently high sample rate to satisfy the Nyquist criterion for the full FDM bandwidth.

We have many options available for down converting a single channel. The standard processing chain for accomplishing this task is shown in Fig. 7.26. This structure performs the standard operations of down converting a selected frequency band with a complex heterodyne, low-pass filtering to reduce the output signal bandwidth to the channel bandwidth, and downsampling to a reduced rate commensurate with the reduced bandwidth. The structure of this processor is seen to be a digital signal processor implementation of a prototype analog I-Q down converter. We mention that the downsampler is commonly referred to as a decimator, a term that means to destroy one tenth. Since nothing is destroyed and nothing happens in tenths, we prefer, and will continue to use, the more descriptive name, downsampler.

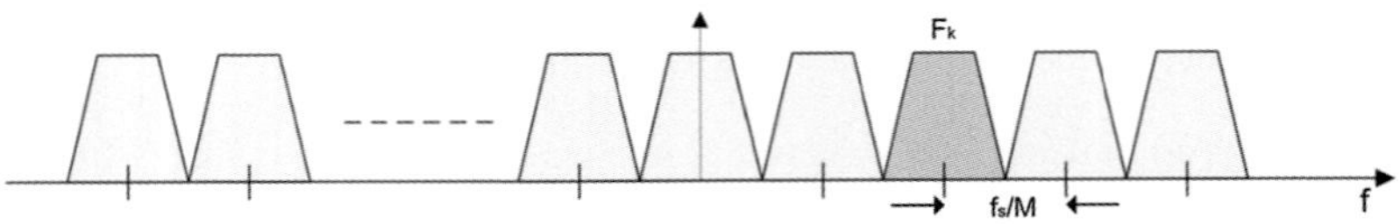

FIGURE 7.25

Spectrum of multichannel input signal, processing task: extract complex envelope of selected channel.

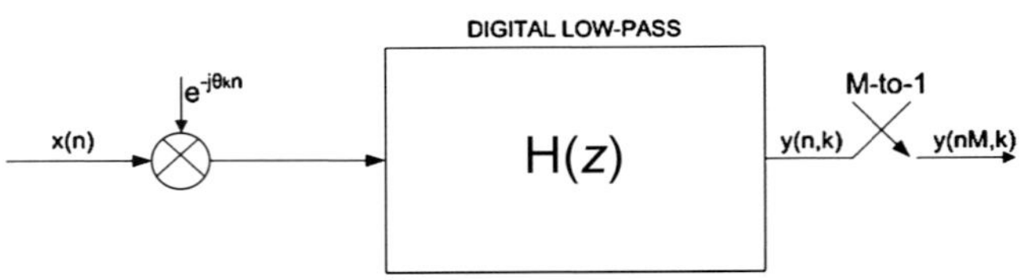

FIGURE 7.26

Standard single-channel down converter.

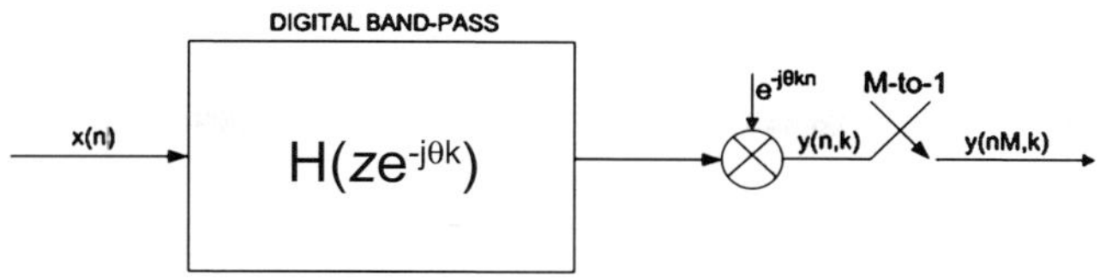

FIGURE 7.27

Bandpass filter version of single-channel down converter.

The output data from the complex mixer are complex; hence, the data are represented by two time series, $I(n)$ and $Q(n)$. The filter with real impulse response $h(n)$ is implemented as two identical filters, each processing one of the quadrature time series. The convolution process between a signal and a filter is often simply performed by multiplication and sum operations between signal data samples and filter coefficients extracted from two sets of addressed memory registers. In this form of the filter, one register set contains the data samples while the other contains the coefficients that define the filter impulse response. By using the equivalency theorem, which states that *the operations of down conversion followed by a low-pass filter are equivalent to the operations of bandpass filtering followed by a down conversion*, we can exchange the positions of the filter and of the complex heterodyne, achieving the block diagram shown in Fig. 7.27.

Note here that the up converted filter, $h(n)\ \exp(j\theta_k n)$, is complex and as such its spectrum resides only on the positive frequency axis without a negative frequency image. This is not a common structure for an analog prototype because of the difficulty of forming a pair of analog quadrature filters exhibiting a 90-degree phase difference across the filter bandwidth. The closest equivalent structure in the analog world is the filter pair used in image-reject mixers, and even there the phase relationship is maintained by a pair of complex heterodynes.

Applying the transformation suggested by the equivalency theorem to an analog prototype system does not make sense since it doubles the required hardware. We would have to replace a complex scalar heterodyne (two mixers) and a pair of low-pass filters with a pair of bandpass filters, containing twice the number of reactive components, and a full complex heterodyne (four mixers). If it makes no sense to use this relationship in the analog domain, why does it make sense in the digital world? The answer is found in the fact that we define a digital filter as a set of weights stored in the coefficient memory. Thus, in the digital world, we incur no cost in replacing the pair of low-pass filters $h(n)$ required in the first option with the pair of bandpass filters $h(n)\cos(n\theta_k)$ and $h(n)\sin(n\theta_k)$ required for the second one. We accomplish this task by a simple download to the coefficient memory.

An interesting historical perspective is worth noting here. In the early days of wireless communication, radio tuning was accomplished by sliding a narrow-band filter to the center frequency of the desired channel to be extracted from the FDM input signal. These radios were known as tuned radio frequency (TRF) receivers. The numerous knobs on the face of early radios adjusted reactive components of the amplifier chain to accomplish the tuning process. Besides the difficulty in aligning multiple tuned stages, shifting the center frequency of the amplifier chain often initiated undesired oscillation due to the parasitic coupling between the components in the radio. Edwin Howard Armstrong, an early radio pioneer, suggested moving the selected channel to the fixed frequency filter rather than moving the filter to the selected channel. This is known as the *superheterodyne principle*, a process invented by

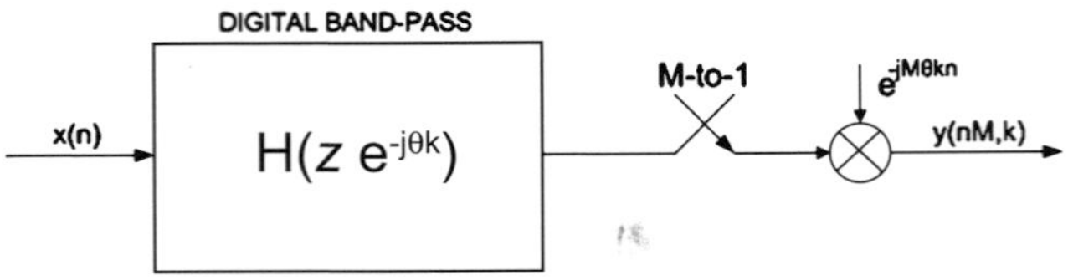

FIGURE 7.28

Downsampled bandpass down converter.

Armstrong in 1918 and quickly adopted by David Sarnoff of the Radio Corporation of America (RCA) in 1924. Acquiring exclusive rights to Armstrong's single-tuning dial radio invention ensured the commercial dominance of RCA in radio broadcasting as well the demise of hundreds of manufacturers of TRF radio receivers. It seems we have come full circle. We inherited from Armstrong a directive to move the desired spectrum to the filter and we have readily applied this legacy to DSP-based processing. We are now proposing that, under appropriate conditions, it makes more sense to move the filter to the selected spectral region. We still have to justify the full complex heterodyne required for the down conversion at the filter output rather than at the filter input, which is done in the subsequent paragraphs.

Examining Fig. 7.27, we note that following the output down conversion, we perform a sample rate reduction in which we retain one sample out of M samples. Because we find it useless to down convert the samples we discard in the next downsample operation, we move the downsampler before the down converter, producing the demodulator scheme shown in Fig. 7.28. We note in this figure that also the time series of the complex sinusoid has been downsampled. The rotation rates of the sampled complex sinusoid are θ_k and $M\theta_k$ rad per sample at the input and output of the M-to-1 resampler, respectively. This change in observed rotation rate is due to aliasing. When aliased, a sinusoid at one frequency or phase slope appears at another phase slope due to the resampling.

We now invoke a constraint on the sampled data center frequency of the down converted channel. We choose center frequencies θ_k, which will alias to DC as a result of downsampling to $M\theta_k$. This condition is ensured if $M\theta_k$ is congruent to 2π, which occurs when $M\theta_k = k2\pi$, or more specifically, when $\theta_k = k2\pi/M$. The modification to Fig. 7.28 to reflect this provision is seen in Fig. 7.29. The constraint that the center frequencies be limited to integer multiples of the output sample rate ensures aliasing to baseband by the sample rate change. When a channel aliases to baseband because of the resampling operation, the corresponding resampled heterodyne defaults to a unity-valued scalar, which consequently is removed from the signal processing path.

Note that if the center frequency of the aliased signal is offset by Δ_θ rad/smpl from a multiple of the output sample rate, the aliased signal will reside at an offset of Δ_θ rad/smpl from zero frequency at baseband and a complex heterodyne, or baseband converter, is needed to shift the signal by the residual Δ_θ offset. This baseband mixer operates at the output sample rate rather than at the input sample rate for a conventional down converter. We can consider this required final mixing operation as a postconversion task and allocate it to the next processing block.

The spectral effect of the signal processing performed by the structure in Fig. 7.29 is shown in Fig. 7.30. The savings realized by this form of down conversion is due to the fact that we no longer require a quadrature oscillator or the pair of input mixers to effect the required frequency translation.

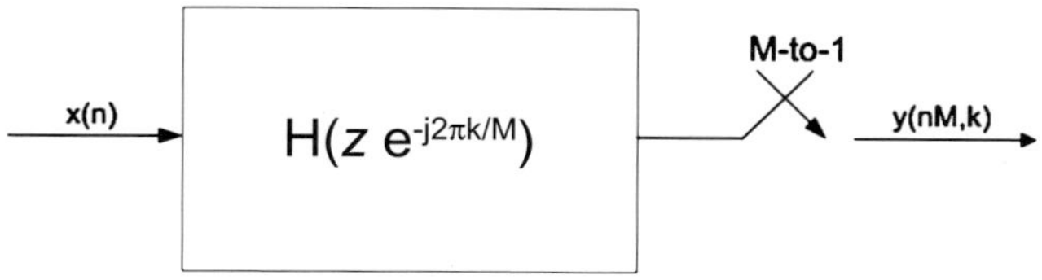

FIGURE 7.29

Bandpass down converter aliased to baseband by downsampler.

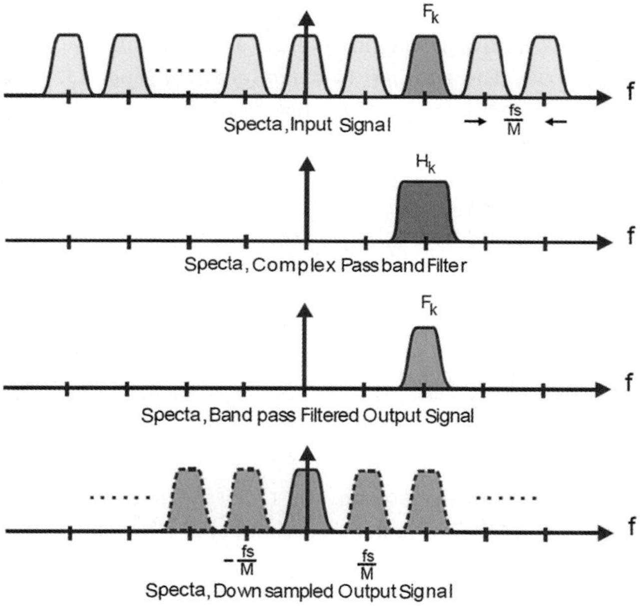

FIGURE 7.30

Spectrum, downsampled output signal.

Applying again the idea that it is useless to filter those samples that will be discarded by the downsampler, we apply the noble identity to exchange the operations of filtering and downsampling. The noble identity states that *a filter processing every Mth input sample followed by an output M-to-1 downsampler is the same as an input M-to-1 downsampler followed by a filter processing every input sample*. Its interpretation is that the M-to-1 downsampled time series from a filter processing every Mth input sample presents the same output by first downsampling the input by M-to-1 to discard the samples not used by the filter when computing the retained output samples and then operating the filter on only the retained input samples. The noble identity works because samples of delay at the input clock rate is the same interval as one-sample delay at the output clock rate.

At this point, with only a filter followed by a resampler, we can apply the polyphase decomposition to the filter and separate it into M parallel paths for achieving the standard M-path polyphase down converter channelizer. In order to facilitate understanding by the reader, we first perform the polyphase decomposition of a low-pass prototype filter and then we extend the results to the bandpass filter case. Eq. (7.19) represents the z-transform of a digital low-pass prototype filter $h(n)$:

$$H(z) = \sum_{n=0}^{N-1} h(n)z^{-n} = h(0) + h(1)z^{-1} + h(2)z^{-2} + \cdots + h(N-1)z^{-(N-1)}. \tag{7.19}$$

In order to achieve the polyphase partition of $H(z)$, we rewrite the sum in Eq. (7.19), as shown in Eq. (7.20), partitioning the 1D array of weights in a 2D array. In this mapping we load an array by columns but process it by rows. The partition forms columns of length M containing M successive terms in the original array, and continues to form adjacent M-length columns until we account for all the elements of the original 1D array:

$$H(z) = \begin{array}{c} h(0) + h(M+0)z^{-M} + h(2M+0)z^{-2M} + \cdots \\ h(1)z^{-1} + h(M+1)z^{-(M+1)} + h(2M+1)z^{-(2M+1)} + \cdots \\ h(2)z^{-2} + h(M+2)z^{-(M+2)} + h(2M+2)z^{-(2M+2)} + \cdots \\ \vdots \\ h(M-1)z^{-(M-1)} + h(2M-1)z^{-(2M-1)} + h(3M-1)z^{-(3M-1)} + \cdots \end{array} \tag{7.20}$$

We note that the first row of the 2D array is a polynomial in z^M, which we will denote $H_0(z^M)$, a notation to be interpreted as an addressing scheme to start at index 0 and increment in strides of length M. The second row of the same array, while not a polynomial in z^M, is made into one by factoring the common term z^{-1} and then identifying this row as $z^{-1}H_1(z^M)$. It is easy to see that each row of the 2D array described by Eq. (7.8) can be written as $z^{-r}H_r(z^M)$, achieving the more compact form as shown in Eq. (7.9), where r is the row index which is coincident with the path index

$$H(z) = \sum_{r=0}^{M-1} z^{-r} H_r\left(z^M\right) = \sum_{r=0}^{M-1} z^{-r} \sum_{n=0}^{\frac{N}{M}-1} h(r+nM)z^{-nM}. \tag{7.21}$$

The block diagram depicting the M-path polyphase decomposition of the resampled low-pass filter of Eq. (7.21) is depicted in Fig. 7.31.

At this point, in the structure depicted in Fig. 7.31, we can pull the resampler on the left side of the adder and downsample the separate filter outputs, performing the sum only for the retained filter output samples. With the resamplers at the output of each filter we can invoke again the noble identity and place them at the input of each filter.

The resamplers operate synchronously, all closing at the same clock cycle. When the switches are closed, the signal delivered to the filter on the top path is the current input sample. The signal delivered

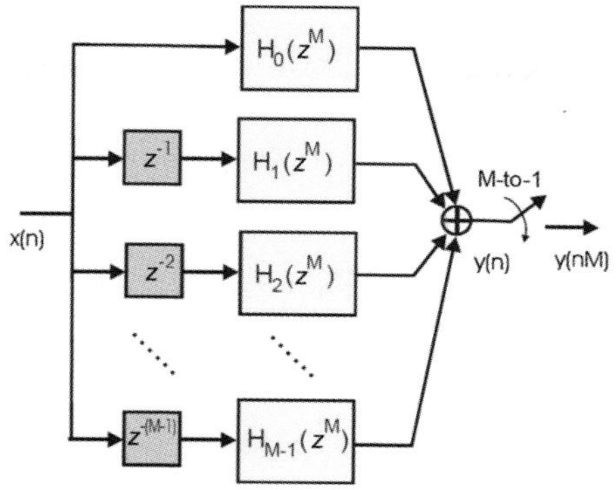

FIGURE 7.31

M-path partition of prototype low-pass filter with output resampler.

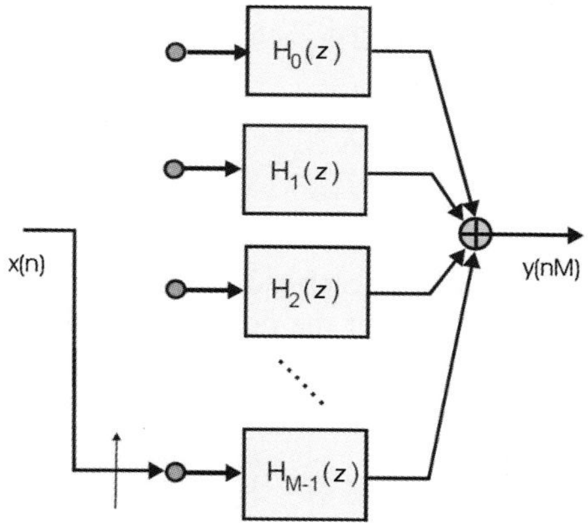

FIGURE 7.32

M-path partition of prototype low-pass filter with input delays and *M*-to-1 resamples replaced by input commutator.

to the filter one path down is the content of the one-stage delay line, which, of course, is the previous input sample. Similarly, as we traverse the successive paths of the M-path partition, we find upon switch closure that the kth path receives a data sample delivered k samples ago. We conclude that the interaction of the delay lines in each path with the set of synchronous switches can be likened to an input commutator that delivers successive samples to successive arms of the M-path filter. This interpretation is shown in Fig. 7.32, which depicts the final structure of the polyphase decomposition of a low-pass prototype filter.

At this point we can easily achieve the bandpass polyphase partition from the low-pass partition by applying the frequency translation property of the z-transform that states the following.

If

$$H(z) = h(0) + h(1)z^{-1} + h(2)z^{-2} + \cdots + h(N-1)z^{N-1} = \sum_{n=0}^{N-1} h(n)z^{-n}$$

and

$$\begin{aligned} G(z) &= h(0) + h(1)e^{j\theta}z^{-1} + h(2)e^{j2\theta}z^{-2} + \cdots + h(N-1)e^{j(N-1)\theta}z^{N-1} \\ &= h(0) + h(1)\left[e^{-j\theta}z\right]^{-1} + h(2)\left[e^{-j\theta}z\right]^{-2} + \cdots + h(N-1)\left[e^{-j(N-1)\theta}z\right]^{-(N-1)} \\ &= \sum_{n=0}^{N-1} h(n)\left[e^{-j\theta}z\right]^{-n}, \end{aligned}$$

then

$$G(z) = H(z)|_{z=e^{-j\theta}z} = H\left(e^{j\theta}z\right).$$

By using this property and replacing in Eq. (7.21) each z^{-r} with $z^{-r}e^{jr\theta}$, where $\theta = k\frac{2\pi}{M}$, we achieve

$$H(ze^{-j(2\pi k/M)}) = \sum_{r=0}^{M-1} z^{-r}e^{jr(2\pi k/M)}H_r\left(z^M\right). \tag{7.22}$$

The complex scalars $e^{jr\left(\frac{2\pi k}{M}\right)}$ attached to each path of the M-path filter can be placed anywhere along the path. We choose to place them after the downsampled path filter segments $H_r(z^M)$. This change is shown in Fig. 7.33, which depicts the standard polyphase filter bank used for down converting a single channel. The computation of the time series obtained from the output summation in

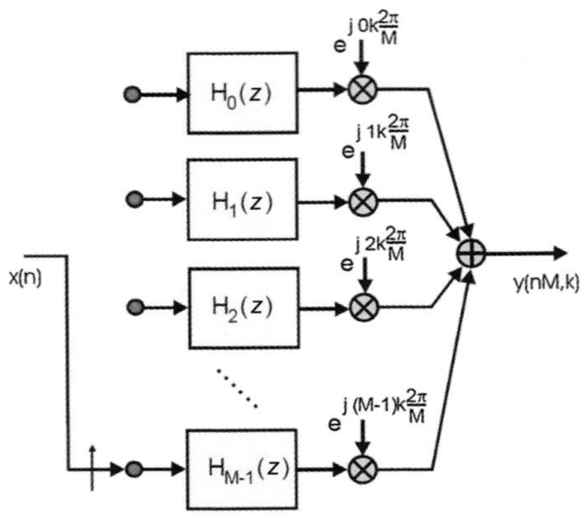

FIGURE 7.33

Resampling M-path down converter.

Fig. 7.33 is shown in Eq. (7.23). Here, the argument nM reflects the downsampling operation, which increments through the time index in strides of length M, delivering every Mth sample of the original output series. The variable $y_r(nM)$ is the nMth sample from the filter segment in the rth path, and $y(nM, k)$ is the nMth time sample of the time series from the rth center frequency. Remember that the down converted center frequencies located at integer multiples of the output sample frequency alias to zero frequency after the resampling operation. Note that the output $y(nM, k)$ is computed as a phase coherent summation of the output series $y_r(nM)$. This phase coherent sum is, in fact, an IDFT of the M-path outputs, which can be likened to beamforming the output of the path filters:

$$y(nM, k) = \sum_{r=1}^{M-1} y_r(nM)e^{j(2\pi/M)rk}. \tag{7.23}$$

The beamforming perspective offers interesting insight into the operation of the resampled down converter system we have just examined. The reasoning proceeds as follows. The commutator delivering consecutive samples to the M input ports of the M-path filter performs a downsampling operation. Each port of the M-path filter receives data at 1/Mth of the input rate. The downsampling causes the M-to-1 spectral folding, effectively translating the M multiples of the output sample rate to baseband. The alias terms in each path of the M-path filter exhibit unique phase profiles due to their distinct center frequencies and the time offsets of the different downsampled time series delivered to each port. These time offsets are, in fact, the input delays shown in Fig. 7.31 and Eq. (7.3). Each of the aliased center frequencies experiences a phase shift shown in Eq. (7.24) equal to the product of its center frequency and the path time delay:

$$\varphi(r, k) = w_k \Delta T_r = 2\pi \frac{f_s}{M} krT_s = 2\pi \frac{f_s}{M} kr \frac{1}{f_s} = \frac{2\pi}{M} kr. \tag{7.24}$$

The phase shifters of the IDFT perform phase coherent summations, very much like that performed in narrow-band beamforming, extracting from the myriad of aliased time series the alias with the particular matching phase profile. This phase-sensitive summation aligns contributions from the desired alias to realize the processing gain of the coherent sum while the remaining alias terms, which exhibit rotation rates corresponding to the M roots of unity, are destructively canceled in the summation. The inputs to the M-path filter are not narrow-band, and phase shift alone is insufficient to cause the destructive cancellation over the full bandwidth of the undesired spectral contributions. Continuing with our beamforming perspective, to successfully separate signals with unique phase profiles due to the input commutator delays, we must perform the equivalent of time-delay beamforming. The M-path filters obtained by M-to-1 downsampling of the prototype low-pass filter supply the required time delays. The M-path filters are approximations to all-pass filters, exhibiting, over the channel bandwidth, equal ripple approximation to unity gain and the set of linear phase shifts that provide the time delays required for the time-delay beamforming task. The filter achieves this property by virtue of the way it is partitioned. Each of the M-path filters, $h_r(n)$, for instance, with weights $h_r(r + nM)$, is formed by starting with an initial offset of r samples and then incrementing in strides of M samples. The initial offsets, unique to each path, are the source of the different linear phase-shift profiles. It is because of the different linear phase profiles that the filter partition is known as *polyphase* filter. A useful perspective is that the phase rotators following the filters perform phase alignment of the band center for each aliased spectral

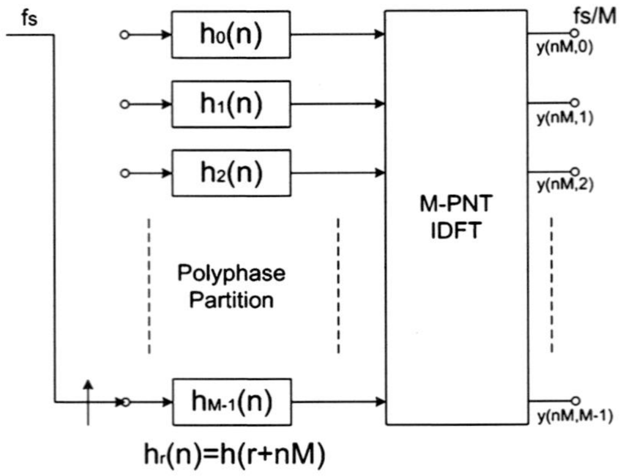

FIGURE 7.34

Standard polyphase down converter channelizer: commutator, *M*-path polyphase filter, and *M*-point IFFT.

band, while the polyphase filters perform the required differential phase shift across these same channel bandwidths.

When the polyphase filter is used to down convert and downsample a single channel, the phase rotators are implemented as external complex products following each path filter. When a small number of channels are being down converted and downsampled, appropriate sets of phase rotators can be applied to the filter stage outputs and summed to form each channel output. Therefore, the most advantageous applications of that structure are those in which the number of channels to be down converted becomes sufficiently large, which means on the order of $\log_2(N)$. Since the phase rotators following the polyphase filter stages are the same as the phase rotators of an IDFT, we can use the IDFT to simultaneously apply the phase shifters for all the channels we wish to extract from the aliased signal set. This is reminiscent of phased-array beamforming. For computational efficiency, the IFFT algorithm implements the IDFT.

The complete structure of a standard M-path polyphase down converter channelizer is shown in Fig. 7.34.

To summarize, in this structure the commutator performs an input sample rate reduction by commutating successive input samples to selected paths of the M-path filter. Sample rate reduction occurring prior to any signal processing causes spectral regions residing at multiples of the output sample rate to alias to baseband. The partitioned M-path filter performs the task of aligning the time origins of the offset sampled data sequences delivered by the input commutator to a single common output time origin. This is accomplished by the all-pass characteristics of the M-path filter sections that apply the required differential time delay to the individual input time series. The IDFT performs the equivalent of a beamforming operation, the coherent summation of the time-aligned signals at each output port with selected phase profiles. The phase coherent summation of the outputs of the M-path filters separates the various aliases residing in each path by constructively summing the selected aliased frequency components located in each path, while simultaneously destructively canceling the remaining aliased spectral components.

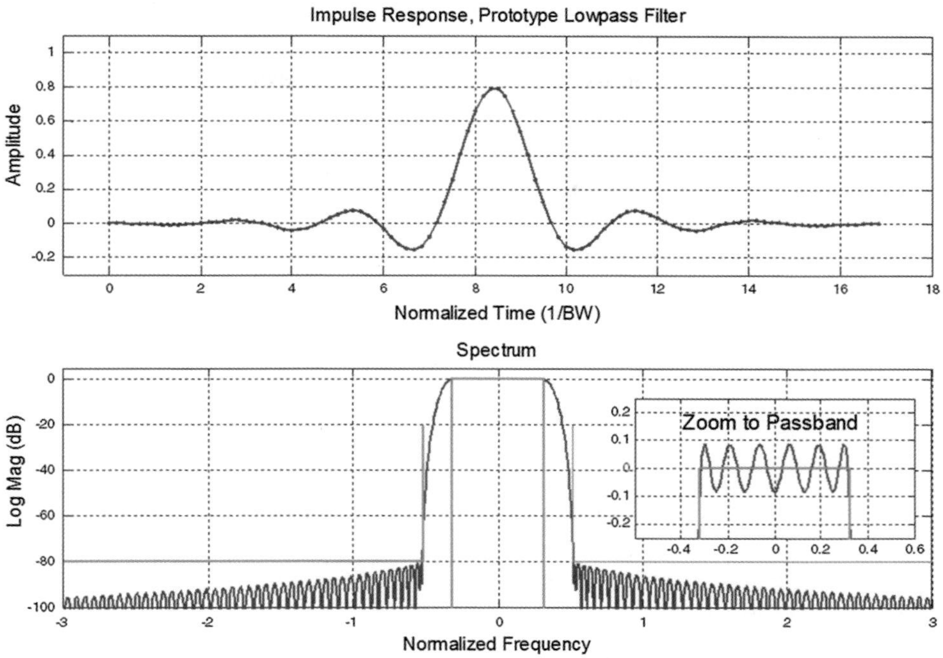

FIGURE 7.35

Impulse response and frequency response of prototype low-pass filter.

We conclude the reasoning on the standard M-path down converter channelizer by stating that it simultaneously performs the following three basic operations: sample rate reduction, due to the input commutator; bandwidth reduction, due to the M-path partitioned filter weights; and Nyquist zone selection, due to the IFFT block.

In the following, in order to facilitate understanding by the reader, we show some figures, which are results of MATLAB simulations, to support the theoretical reasoning of the previous paragraphs. The example concerns a six-path down converter channelizer that performs 6-to-1 downsampling. In particular, Fig. 7.35 shows the impulse response and the frequency response of the low-pass prototype filter used for the polyphase partition.

Fig. 7.36 shows the impulse responses of the zero-packed filter on each arm of the six-path polyphase partition, while Fig. 7.37 shows the spectra corresponding to the filters of Fig. 7.36. It is evident from this figure that the zero-packing process has the effect of producing spectral copies of the filter frequency response.

In Fig. 7.38 we show the phase of each spectral copy of the zero-packed prototype filter on each arm of the partition. In this figure the phases of the spectral copies of each arm are overlaid and each line of the plot corresponds to a specific arm in the partition.

In Fig. 7.39 the same phase profiles of Fig. 7.38 are detrended, zoomed, and made causal.

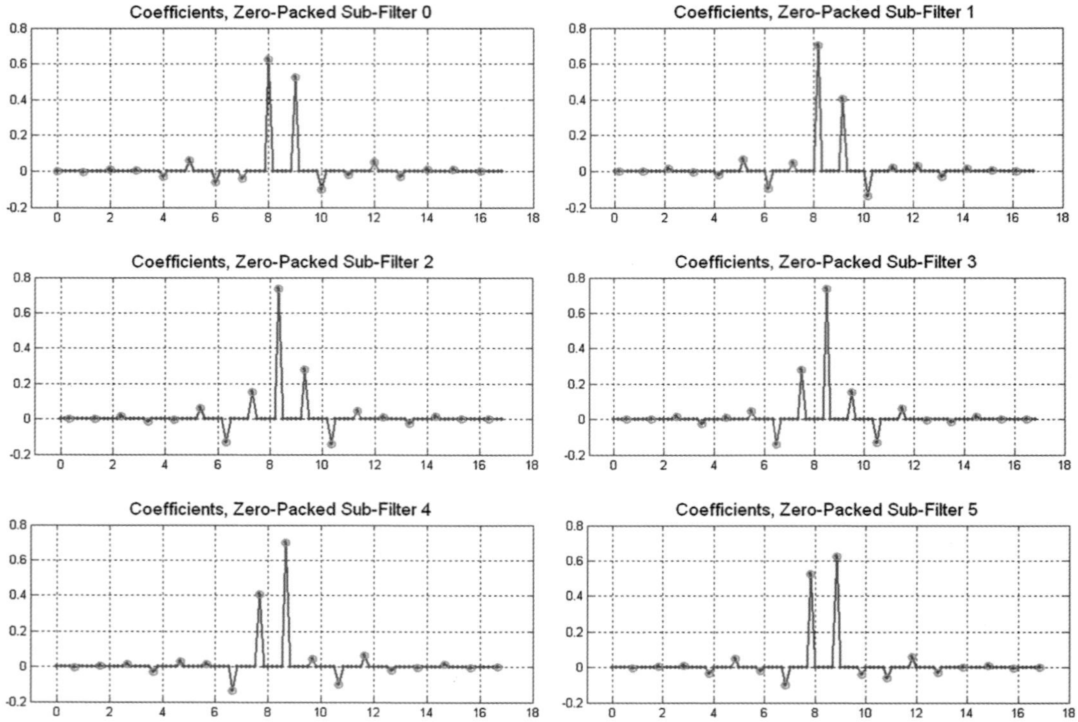

FIGURE 7.36

Impulse response of six-path polyphase partition prior to 6-to-1 resampling.

In Fig. 7.40 the 3D view of the filter spectral copies, with their phase profiles, is shown for each path, while in Fig. 7.41, the final result of the channelization process is shown, again in a 3D fashion.

7.6.1 From single channel up converter to standard up converter channelizer

The standard M-path up converter channelizer that performs 1-to-M upsampling while up converting the input time series is derived from the basic structure, shown in Fig. 7.11, of a single-channel interpolator consisting of an upsampler followed by an appropriate low-pass filter. In this configuration, the upsampler converts the Nyquist interval, which is the observable frequency span, from the input sample rate to a span M times wider, which is the output sample rate. The 1-to-M upsampler zero-packs the input time series, effectively decreasing the distance between the input samples without modifying the spectral content of the series. The wider Nyquist interval, spanning M input Nyquist intervals, presents M spectral copies of the input spectrum to the low-pass filter. The amplitude of each of the M copies is $1/M$th of the amplitude of the input signal spectrum. When a low-pass filter is scaled so that the peak coefficient is unity, the filter exhibits a processing gain inversely proportional to its fractional bandwidth. Thus, as the filter eliminates the $M-1$ spectral copies, reducing the bandwidth by a factor of

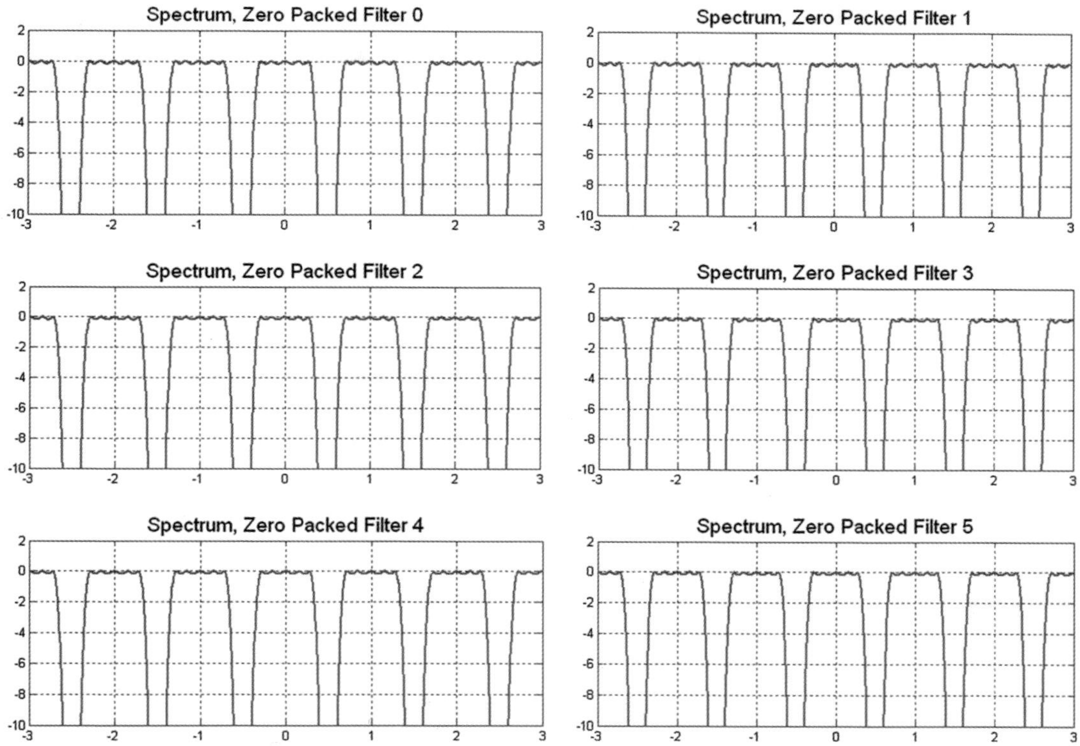

FIGURE 7.37

Frequency response of six-path polyphase partition prior to 6-to-1 resampling.

M, the filter gain precisely compensates for the attenuation of the input spectra due to the zero-packing of the input series.

We start the modifications of this basic architecture by observing that the zero-valued samples of the zero-packed input time series do not contribute to the weighted sums formed at the filter output. Since they do not contribute, there is no need to perform the product and sum from the input data registers containing the known zero-valued samples. Since only the nonzero-packed samples contribute to the filter output, we can track their location in the filter and perform the weighted sum only from the register locations containing these samples. These locations are separated by M sample values, and their position shifts through the filter as each new zero-valued input is presented to the input of the filter. Keeping track of the coefficient stride and the position of each coefficient set is automatically performed by the polyphase partition of the filter which is represented in Eq. (7.25). This equation describes the filter as a sum of successively delayed subfilters with coefficients separated by the stride of M samples. Eq. (7.26) is a compact representation of Eq. (7.25) where the rth stage $H_r(z^M)$ of the polyphase filter

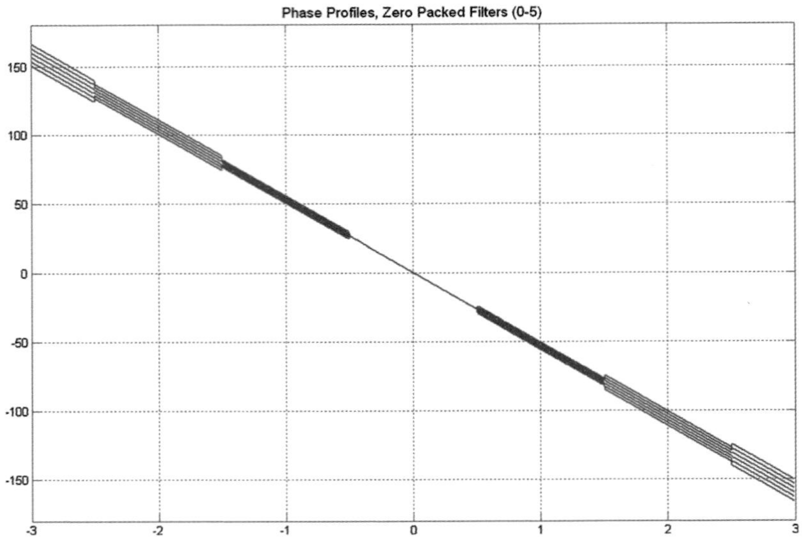

FIGURE 7.38

Overlaid phase response of six-path polyphase partition prior to 6-to-1 resampling.

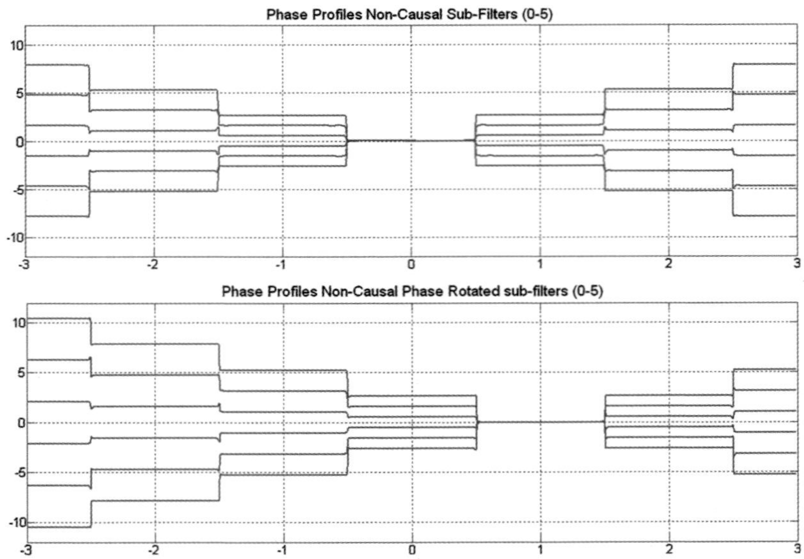

FIGURE 7.39

Detrended overlaid phase response; six-path partition prior to 6-to-1 resampling.

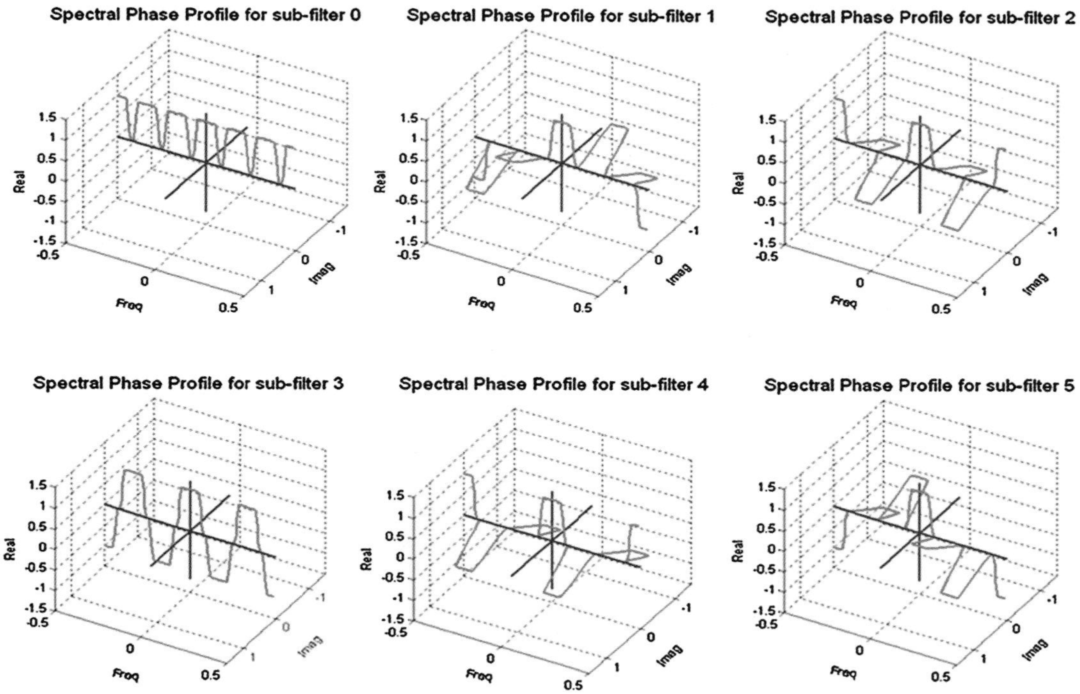

FIGURE 7.40

3D Paddle-wheel phase profiles; six-path partition prior to 6-to-1 resampling.

is formed by the coefficient set that starts at index r and increments in steps of length M:

$$H(z) = \sum_{r-0}^{M-1} z^{-r} \sum_{n=0}^{\left(\frac{N}{M}\right)-1} h(r+nM)z^{-nM}, \tag{7.25}$$

$$H(z) = \sum_{r-0}^{M-1} z^{-r} H_r\left(z^M\right). \tag{7.26}$$

The pictorial form of the filter in the M-path polyphase partition is shown in Fig. 7.42. This structure enables the application of the noble identity in which we slide the resampler through the filter and replace the M units of delay at the output clock rate with one unit of delay at the input clock rate. Note that the resampler cannot slide through the delays z^{-r} following each filter segment H_r.

The resamplers following the separate filter stages upsample each time series by a factor of M, and the delays in each arm shift each resulting time series at different time increments so that only one nonzero time sample is presented to the summing junction at each output time. Thus, rather than performing the sum with multiple zeros, we can simply point to the arm that sequentially supplies the

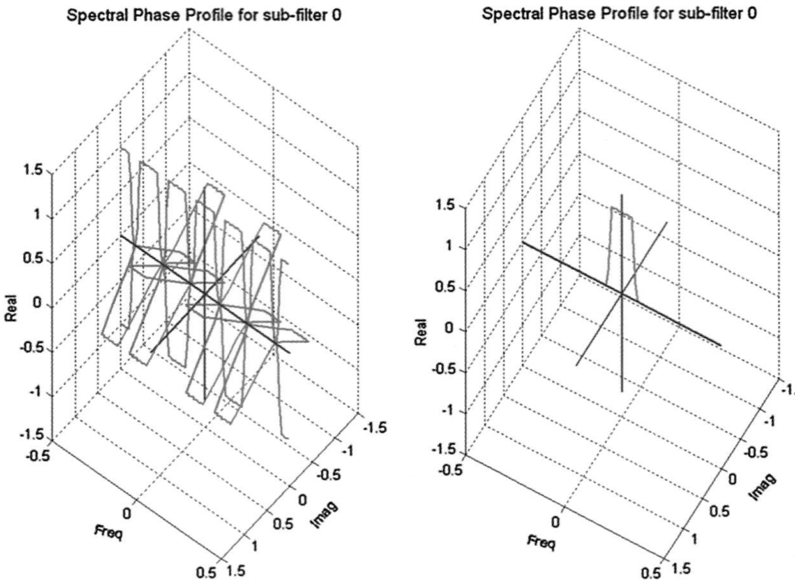

FIGURE 7.41

Overlaid 3D paddle-wheel phase profiles; six-path partition prior to 6-to-1 resampling.

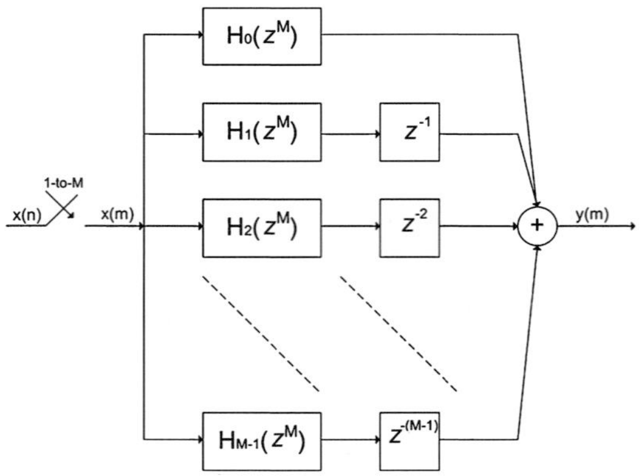

FIGURE 7.42

Initial structure of 1-to-M polyphase interpolator.

nonzero samples. The output commutator, shown in Fig. 7.43, performs this selective access. Fig. 7.42 represents the standard polyphase interpolator or upsampler. Due to the low-pass nature of the filter on which the polyphase decomposition has been applied, this structure does not perform the up conversion

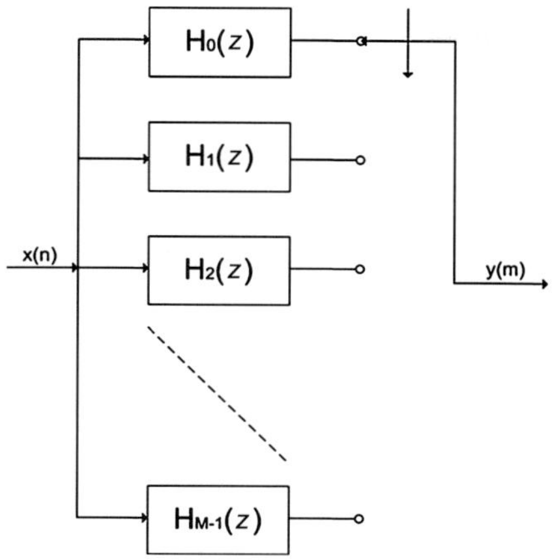

FIGURE 7.43

Standard structure of polyphase interpolator.

of the interpolated signal yet. Its output is the same low-pass input spectrum with a sample rate that is M times higher than the input sample rate.

We recall here that the purpose of the interpolation process is to increase the input sample rate while translating the input spectrum to a higher carrier frequency. If used in a communication system, such a structure directly translates the input signal to an IF and then outputs the digital IF signal via a single DAC. This option reduces the system costs by using a single DAC and a bandpass filter to replace the standard baseband process requiring matched DACs, matched low-pass filters, matched balanced mixers, and a quadrature oscillator to form the IF band.

In the structure of Fig. 7.24, the 1-to-M interpolation process zero-packed the input data giving us access to M spectral copies of the input time series. The spectral copies reside at multiples of the input sample rate. The low-pass filter rejects the spectral copies, retrieving only the baseband copy centered at DC that is then sent through a DUC for translation to the desired center frequency. It is possible to perform the spectral translation as a part of the interpolation process when the desired center frequency coincides with one of the multiples of the input sample rate. Rather than extracting the spectral copy at baseband from the replicated set of spectra and then translating it by means of a complex heterodyne, we can directly extract one of the spectral copies by using a bandpass filter as opposed to a low-pass filter. The bandpass filter is simply an up converted version of the low-pass filter $h(n)$ with weights shown in Eq. (7.27). Here the center frequency is interpreted as the kth multiple of $\frac{1}{M}$ of the output sample rate. We have

$$g(n,k) = h(n)\exp\left(j\frac{2\pi}{M}kn\right). \tag{7.27}$$

The z-transform of the bandpass filter $g(n,k)$ is

$$G_k(z) = \sum_{n=0}^{N-1} h(n) \exp\left(j\frac{2\pi}{M}kn\right) z^{-n}. \tag{7.28}$$

The z-transform of the polyphase decomposition of this filter is shown in Eq. (7.29). Here we see that the length M stride in coefficient index due to the 1-to-M resampling aliases the phase rotators in the polyphase filter stages to DC and hence has no effect on the polyphase weights. The phase rotator does, however, have a contribution related to the delay associated with each arm of the partition. The output commutator process absorbs the delays, while the phase rotators are applied to each arm to obtain the spectral translation as a part of the interpolation. We have

$$\begin{aligned} G_k(z) &= \sum_{r=0}^{M-1} z^{-r} \exp\left(j\frac{2\pi}{M}rk\right) \sum_{n=0}^{\frac{N}{M}-1} h(r+nM) \exp\left(j\frac{2\pi}{M}nMk\right) z^{-nM} \\ &= \sum_{r=0}^{M-1} z^{-r} \exp\left(j\frac{2\pi}{M}rk\right) \sum_{n=0}^{\frac{N}{M}-1} h(r+nM) z^{-nM}. \end{aligned} \tag{7.29}$$

The polyphase filter that accomplishes the task of simultaneous interpolation and up conversion to the kth Nyquist zone is shown in Fig. 7.44. This engine upsamples and up converts, by aliasing, a single narrow-band channel to a particular center frequency which is a multiple of the input sample rate. By noting, as we did in the previous section of this chapter for the dual down converter channelizer, that the phase rotators applied in each arm are the coefficients of an IDFT, it is easy to generalize this structure to the standard multichannel polyphase up converter channelizer by applying an IDFT block at the input of the filter bank.

Fig. 7.45 shows the complete structure of the standard M-path polyphase up converter channelizer. It is composed of an M-point IDFT block, an M-path partitioned filter, and an output commutator. In this structure, we enter the M-channel process at the IDFT block and leave the process by the output commutator. The M-point IDFT performs two simultaneous tasks: an initial upsampling of 1-to-M forming an M-length vector for each input sample $x(n,k)$ and further imparting a complex phase rotation of k cycles in M samples on the upsampled output vector. It generates a weighted sum of complex vectors containing an integer number of cycles per M-length vector. The polyphase filter forms a sequence of column coefficient-weighted MATLAB dot-multiply versions of these complex spinning vectors. The sum of these columns, formed by the set of inner products in the polyphase partitioned filter, is the shaped version of the up converted M-length vector output from the IFFT. On each output port of this engine we find the input baseband channel aliased to the specific Nyquist zone with a new sampling rate that is commensurate with its reduced bandwidth.

A closing comment on both the standard polyphase up converter and downconverter channelizer is that the operations of sampling rate change, spectral shaping, and Nyquist zone selection are completely independent of each other. The channel bandwidth, the channel spacing, and the output sampling rate do not necessarily have to be equal, but they can all be modified according to the applications.

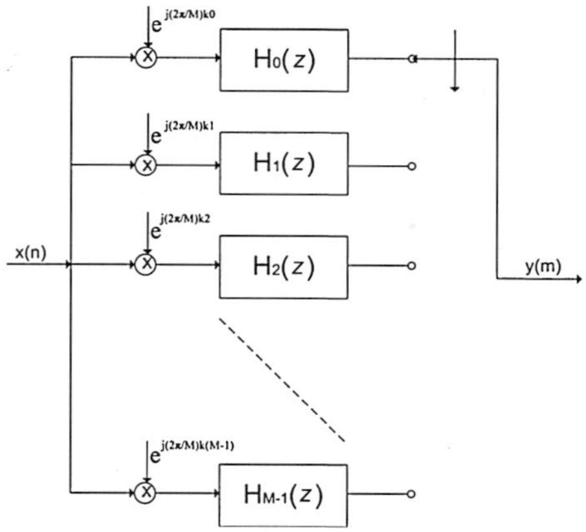

FIGURE 7.44

Structure of 1-to-*M* polyphase interpolator with phase rotators for selecting the spectrum centered in the *M*th Nyquist zone.

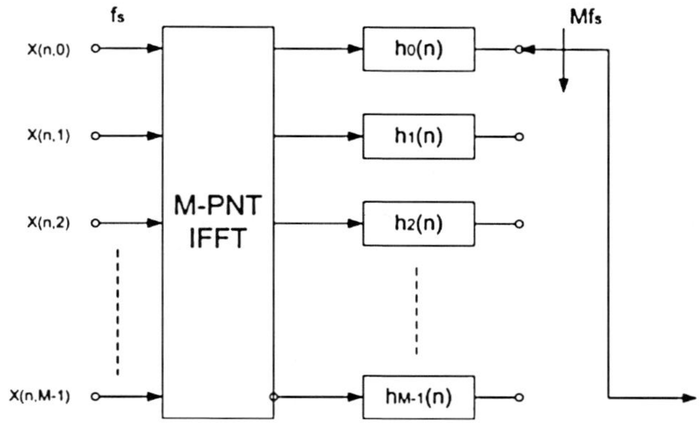

FIGURE 7.45

Polyphase up converter channelizer: *M*-point IFFT, *M*-path polyphase filter, and commutator.

7.7 Modifications of the standard down converter channelizer – *M*:2 down converter channelizer

In the standard polyphase down converter channelizer shown in Fig. 7.34, the channel bandwidth f_{BW}, the channel spacing f_{Δ}, and the sampling frequency f_s are fixed to be equal.

This configuration could represent a good choice when the polyphase channelizer is used for communication systems. In these kinds of applications, in fact, an output sample rate matching the channel spacing is sufficient to avoid adjacent channel crosstalk since the two-sided bandwidth of each channel is less than the channel spacing. An example of a signal that would require this mode of operation is the quadrature amplitude modulation (QAM) channels of a digital cable system. In North America, the channels are separated by 6 MHz centers and operate with square-root cosine tapered Nyquist-shaped spectra with 18% or 12% excess bandwidth, at symbol rates of approximately 5.0 MHz. The minimum sample rate required of a cable channelizer to satisfy the Nyquist criterion would be 6.0 MHz (the European cable plants have a channel spacing of 8.0 MHz and symbol rates of 7.0 MHz). The actual sample rate would likely be selected as a multiple of the symbol rate rather than as a multiple of the channel spacing. Systems that channelize and form samples of the Nyquist-shaped spectrum often present the sampled data to an interpolator to resample the time series collected at a bandwidth-related Nyquist rate to a rate offering two samples per symbol or twice the symbol rate. For the TV example just cited, the 6 Ms/s, 5 MHz symbol signal would have to be resampled by 5/3 to obtain the desired 10 Ms/s. This task is done quite regularly in the communication receivers and it may represent a significant computational burden. It would be appropriate if we could avoid the external interpolation process by modifying the design of the standard polyphase channelizer for directly providing us the appropriate output sampling frequency.

Many other applications desire channelizers in which the output sampling frequency is not equal to the channel spacing and the channel bandwidth. The design of a software-defined radio receiver and transmitter is only one of them. Another can be found in [7].

We concluded the previous section mentioning that the channel spacing, the channel bandwidth, and the output sampling frequency of the polyphase channelizer are completely independent of each other and that they can be arbitrarily selected based on the application. Fig. 7.46 shows some possible options in which the channel bandwidth, the output sampling rate, and the channel spacing of the channelizer are not equal. In this chapter, for the purpose of designing flexible radio architectures, the third option, when the selected low-pass prototype filter is a Nyquist filter (perfect reconstruction filter), results to be the most interesting one.

In this section of the chapter we derive the reconfigured version of the standard down converter channelizer that is able to perform the sample rate change from the input rate f_s to the output sampling rate $2f_s/M$ maintaining both the channel spacing and the channel bandwidth equal to f_s. Similar reasoning can be applied for implementing different selections.

Note that the standard down converter channelizer is critically sampled when the channel spacing and the channel bandwidth are both equal to f_s and when the output sample rate is also equal to f_s/M. This particular choice, in fact, causes the transition band edges of the channelizer filters to alias onto themselves, which would prevent us from further processing the signals when they are arbitrarily located in the frequency domain, which is the case for software radio. This problem can be visualized in the upper example depicted in Fig. 7.46.

For the record, recall that a polyphase filter bank can be operated with an output sample rate which can be any rational ratio of the input sample rate. With minor modifications the filter can be operated with totally arbitrary ratios between input and output sample rates. This is true for the sample rate reduction imbedded in a polyphase receiver as well as for the sample rate increase embedded in a polyphase modulator.

We have control over the output sampling rate of the down converter channelizer by means of the input commutator that delivers input data samples to the polyphase stages. We normally deliver M successive input samples to the M-path filter starting at port $M-1$ and progress up the stack to port 0, and by doing so we deliver M inputs per output, which means performing an M-to-1 downsampling operation. To obtain the desired $(M/2)$-to-1 downsampling, we have to modify the input commutator in

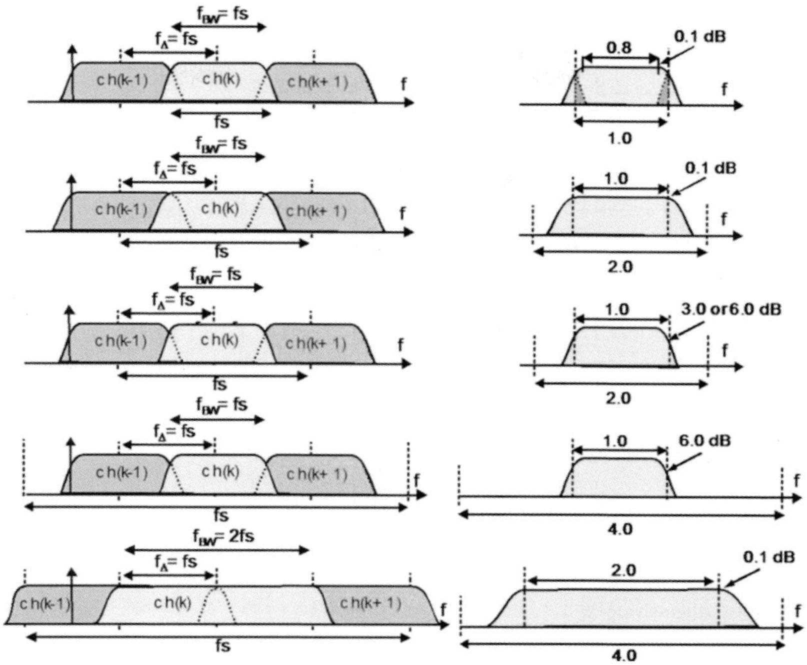

FIGURE 7.46

Some possible channel width, channel spacing, and output sampling frequency.

a way that it delivers $M/2$ successive input samples starting at port $(M/2) - 1$ and progressing up the stack to port 0. We develop and illustrate the modifications with the aid of Figs. 7.47, 7.48, 7.49, and 7.50.

Fig. 7.47 represents the polyphase M-path filter partition shown in Eq. (7.9) with an $M/2$-to-1 rather than the conventional M-to-1 downsample operation after the output summing junction. We need it in order to perform the desired sampling rate change.

In Fig. 7.48 we apply the noble identity to the polyphase paths by pulling the $M/2$-to-1 downsampler through the path filters which converts the polynomials in z^M operating at the high input rate to polynomials in z^2 operating at the lower output rate.

Fig. 7.49 shows the second application of the noble identity in which we again take the $M/2$-to-1 downsamplers through the $z^{-M/2}$ parts of the input path delays for the paths in the second or bottom half of the path set.

In Fig. 7.50 the $M/2$-to-1 downsamplers switch and their delays are replaced with a two-pronged commutator that delivers the same sample values to path inputs with the same path delay. Here we also merged the z^{-1} delays in the lower half of the filter bank with their path filters.

Fig. 7.51 shows and compares the block diagrams of the path filters in the upper and lower halves of this modified polyphase partition.

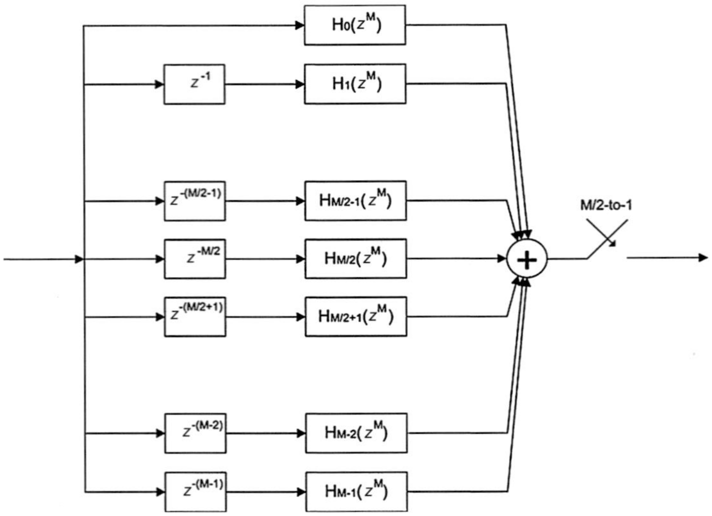

FIGURE 7.47

M-path filter and *M*/2:1 downsampler.

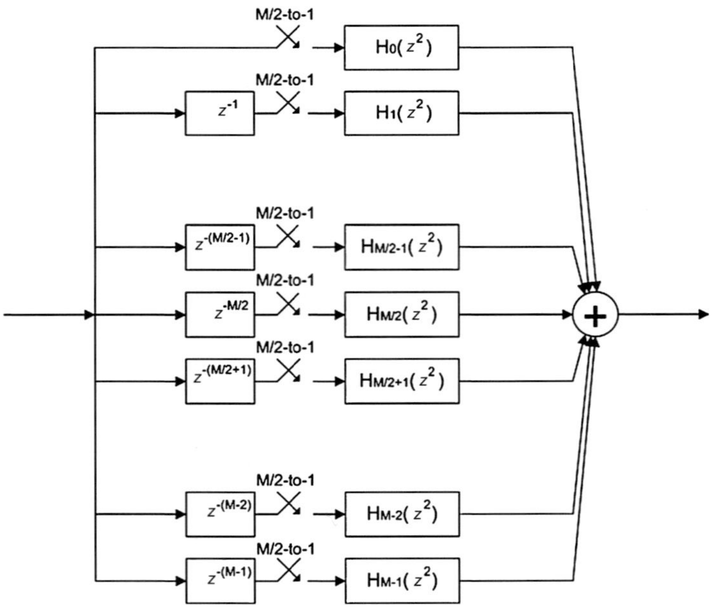

FIGURE 7.48

Noble identity applied to *M*-path filters.

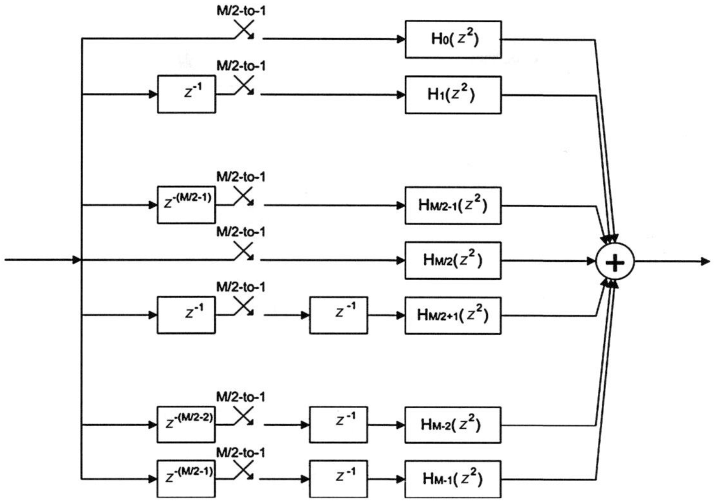

FIGURE 7.49

Noble identity applied to delays.

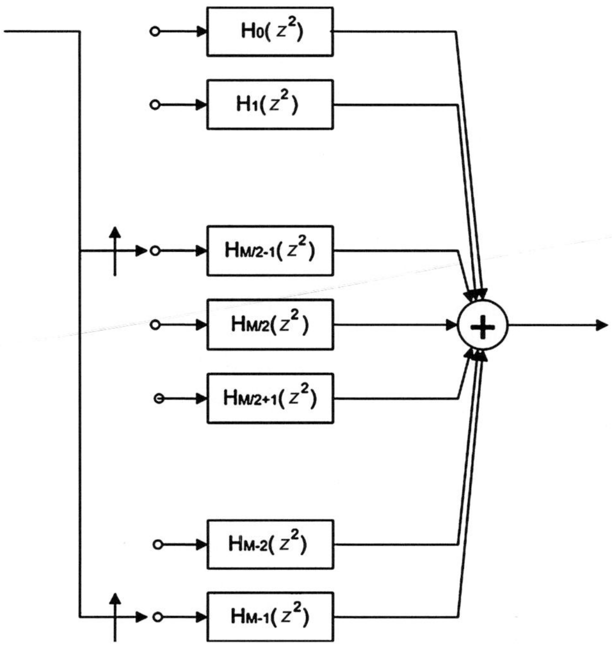

FIGURE 7.50

Path delays replaced by input commutator.

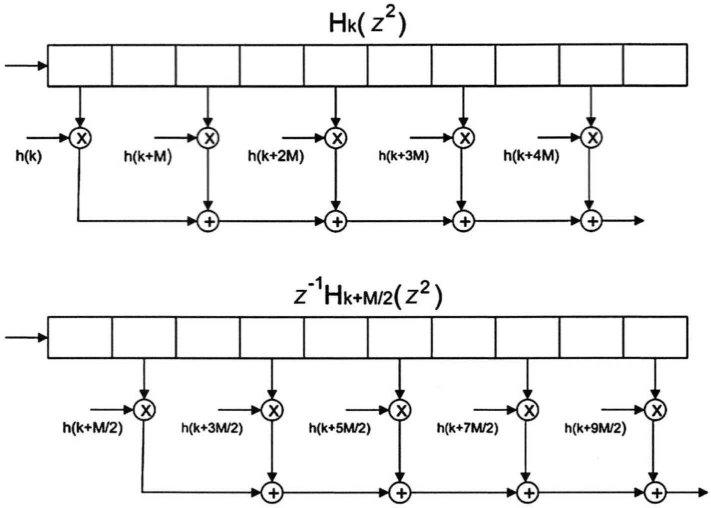

FIGURE 7.51

M-path filters with and without extra delays.

When the input commutator is so designed, the $M/2$ addresses to which the new $M/2$ input samples are delivered have to be first vacated by their former contents, the $M/2$ previous input samples. All the samples in the 2D filter undergo a serpentine shift of $M/2$ samples with the $M/2$ samples in the bottom half of the first column sliding into the $M/2$ top addresses of the second column while the $M/2$ samples in the top half of the second column slide into the $M/2$ addresses in the bottom half of the second column, and so on. This is equivalent to performing a linear shift through the prototype 1D filter prior to the polyphase partition. In reality, we do not perform the serpentine shift but rather perform an addressing manipulation that swaps two memory banks. This is shown in Fig. 7.52.

After each $M/2$-point data sequence is delivered to the partitioned M-stage polyphase filter, in the standard channelizer configuration, the outputs of the M stages are computed and conditioned for delivery to the M-point IFFT. What we need to do at this point is the time alignment of the shifting time origin of the input samples in the M-path filter with the stationary time origin of the phase rotator outputs of the IFFT. We can understand the problem by visualizing, as shown in Fig. 7.53, a single cycle of a sine wave extending over M samples being inserted in the input data register, the first column of the polyphase filter in segments of length $M/2$. We can assume that the data in the first $M/2$ addresses are phase-aligned with the first $M/2$ samples of a single cycle of the sine wave offered by the IFFT.

When the second $M/2$ input samples are delivered to the input data register, the first $M/2$ input samples shift to the second half of the M-length array. Its original starting point is now at address $M/2$ but the IFFT's origin still resides at address 0. The shift of the origin causes the input sine wave in the register to have the opposing phase of the sine wave formed by the IFFT; in fact the data shifting into the polyphase filter stages causes a frequency-dependent phase shift of the form shown in Eq. (7.30). The time delay due to shifting is nT, where n is the number of samples and T is the time interval between samples. The frequencies of interest are integer multiples k of $1/M$th of the sample rate $2\pi/T$.

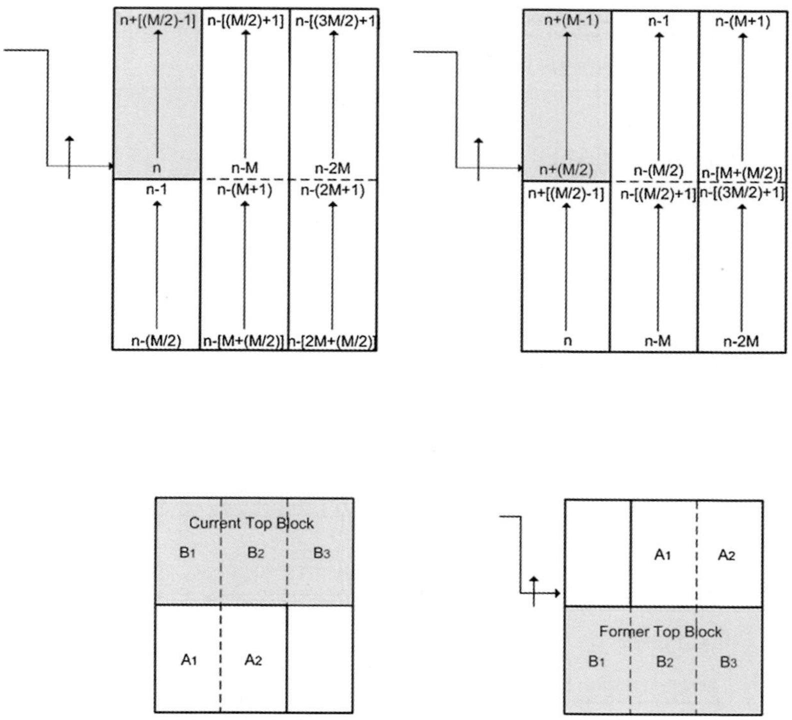

FIGURE 7.52

Data memory loading for successive $M/2$-point sequences in an M-stage polyphase channelizer.

Substituting these terms in Eq. (7.30) and canceling terms, we obtain the frequency-dependent phase shift shown in Eq. (7.31). From this relationship we see that for time shifts n equal to multiples of M, as demonstrated in Eq. (7.32), the phase shift is a multiple of 2π and contributes zero offset to the spectra observed at the output of the IFFT. The M-sample time shift is the time shift applied to the data in the normal use of the polyphase filter. Now suppose that the time shift is $M/2$ time samples. When substituted in Eq. (7.31), we find, as shown in Eq. (7.33), a frequency-dependent phase shift of $k\pi$, from which we conclude that odd-indexed frequency terms experience a phase shift of π rad for each successive $N/2$ shift of input data. We have

$$\theta(\omega) = \Delta t \cdot \omega, \tag{7.30}$$

$$\theta(\omega_k) = nT \cdot k\frac{1}{M}\frac{2\pi}{T} = \frac{nk}{M}2\pi, \tag{7.31}$$

$$\theta(\omega_k)|_{n=M} = \frac{nk}{M}2\pi\bigg|_{n=M} = 2\pi k, \tag{7.32}$$

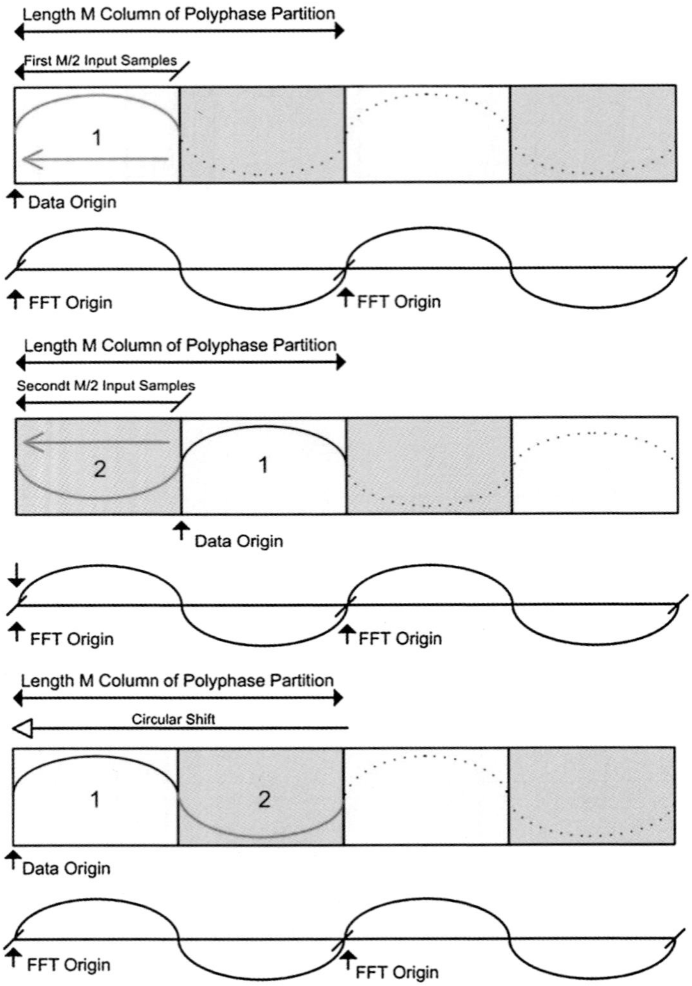

FIGURE 7.53

Illustration of phase reversal of an M-point sinusoid input to an $M/2$-path polyphase filter.

$$\theta(\omega_k)|_{n=M/2} = \left.\frac{nk}{M}2\pi\right|_{n=M/2} = \pi k. \tag{7.33}$$

This π-rad phase shift is due to the fact that the odd-indexed frequencies alias to the half sample rate when the input signal is downsampled by $M/2$. What we are observing is the sinusoids with an odd number of cycles in the length M array alias to the half sample rate when downsampled $M/2$-to-1. Note that when downsampled $M/2$-to-1, the sinusoids with an even number of cycles in the length M array

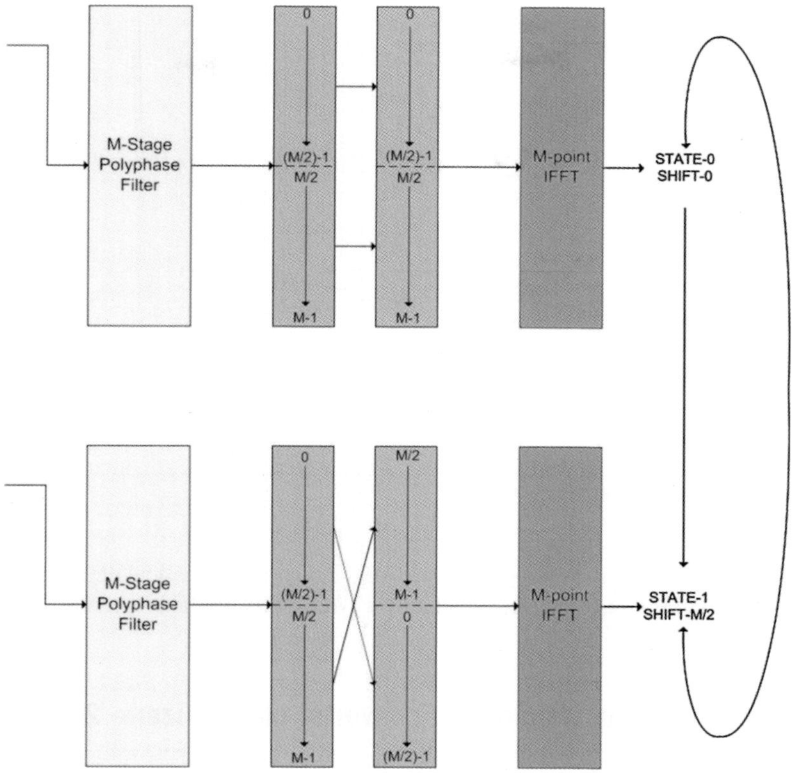

FIGURE 7.54

Cyclic shift of input data to IFFT to absorb phase shift due to linear time shift of data through polyphase filter.

alias to DC. We can compensate for the alternating signs in successive output samples by applying the appropriate phase correction to the spectral data as we extract successive time samples from the odd-indexed frequency bins of the IFFT. The phase correction here is trivial, but for other downsampling ratios, the residual phase correction would require a complex multiplication at each transform output port. Alternatively, since time delay imposes a frequency-dependent phase shift, we can use time shifts to cancel the frequency-dependent phase shifts. We accomplish this by applying a circular time shift of $N/2$ samples to the vector of samples prior to their presentation to the IFFT. As in the case of the serpentine shift of the input data, the circular shift of the polyphase filter output data is implemented as an address-manipulated data swap. This data swap occurs on alternate input cycles and a simple two-state machine determines for which input cycle the output data swap is applied. This is shown in Fig. 7.54.

The complete structure of the modified version of the M-to-2 down converter channelizer with the input data buffer and the circular data buffer is shown in Fig. 7.55.

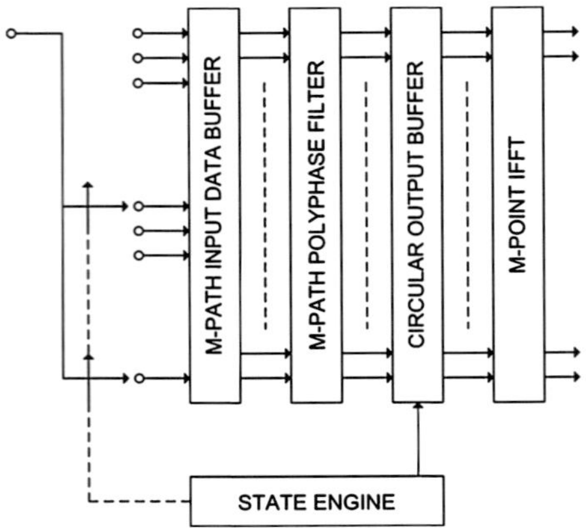

FIGURE 7.55

M-to-2 modified down converter channelizer.

For brevity of notation, we avoid reporting here all the dual mathematical derivations that led to the final structure of the modified up converter channelizer. We briefly explain its block diagram in the next subsection.

7.7.1 Modifications of the standard up converter channelizer – 2:*M* up converter channelizer

Dual reasoning drives us to the reconfigured version of the M-path polyphase up converter channelizer that is able to perform the sample rate change from $2f_s/M$ to f_s maintaining both the channel spacing and the channel bandwidth equal to f_s. The choice of using a 2-to-M upsampler avoids the difficulty of having the sample rate that precisely matches the two-sided bandwidth of the input signals as well as permitting a shorter length prototype channelizer filter due to an allowable wider transition bandwidth. In this chapter we briefly derive its block diagram (Fig. 7.45). More details on this structure can be found in [6–8]. We develop and illustrate the modifications with the aid of Figs. 7.56, 7.57, 7.58, 7.59, 7.60, and 7.61. In these figures we do not consider the IFFT block because for now it does not affect our reasoning. Moreover, we will introduce it later motivating the reason for which it is important to include a circular buffer in the design of the modified up converter channelizer.

Fig. 7.56 presents the structure of the M-path filter implementation of the polyphase partition shown in Eq. (7.7). Note the 1-to-$M/2$ upsample operation at the input port normally described as the zero-packing process. In Fig. 7.57 we apply the noble identity to the polyphase paths and pull the 1-to-$M/2$ upsampler through the path filters which convert the polynomials in z^M operating at the high output rate to polynomials in z^2 operating at the low input rate. Note that the paths are now polynomials in z^2 rather than z^M, as is the normal mode that we identify as the dual of the maximally decimated filter bank.

Fig. 7.58 shows the second application of the noble identity in which we again take the 1-to-$M/2$ upsampler through the $z^{-M/2}$ part of the output path delays for the paths in the second or bottom half

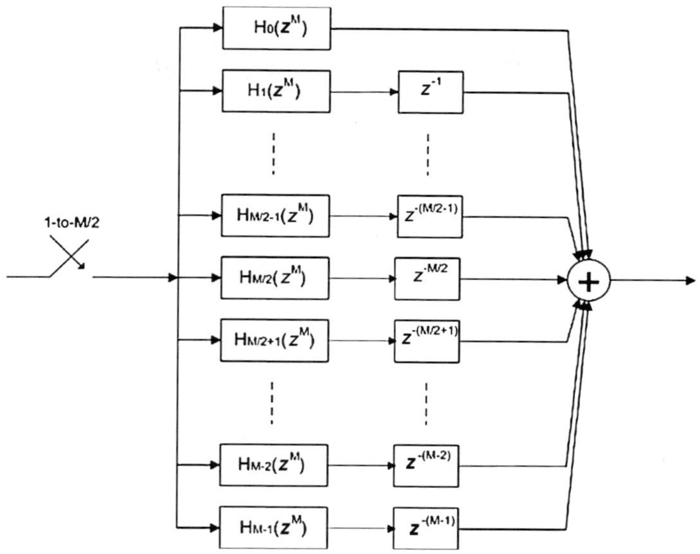

FIGURE 7.56

M-path polyphase filter.

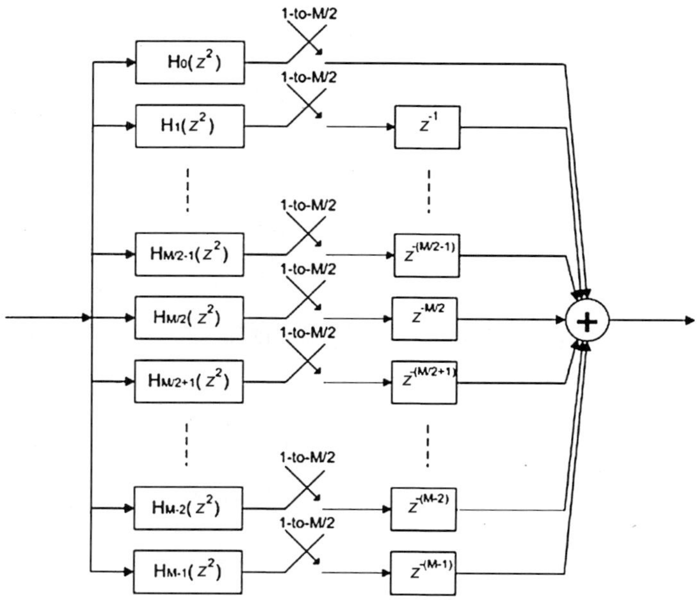

FIGURE 7.57

Noble identity applied to *M*-path filter.

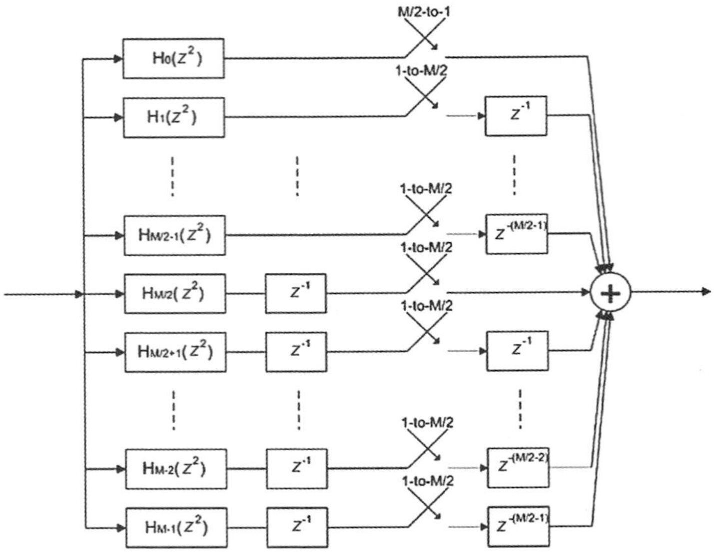

FIGURE 7.58

Noble identity applied to delays.

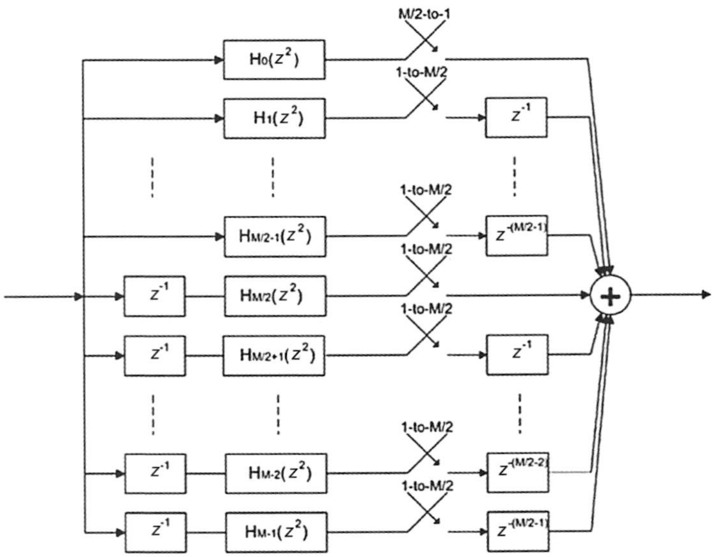

FIGURE 7.59

Interchange unit delay and path filters.

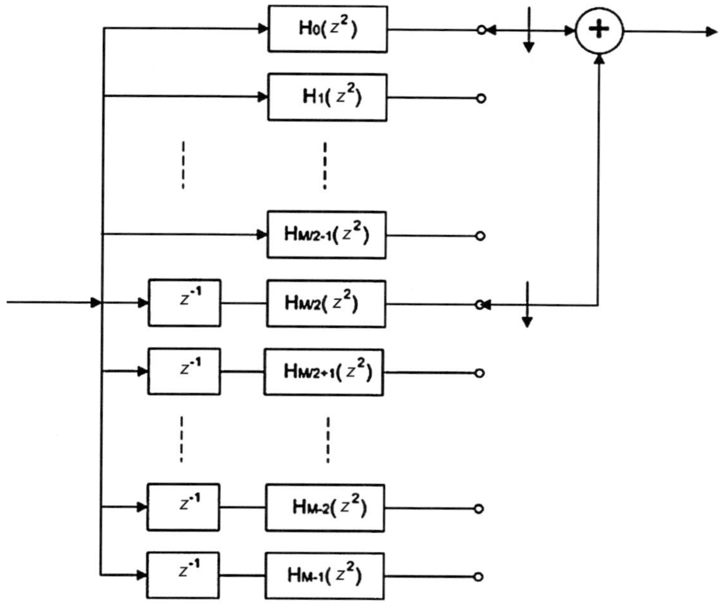

FIGURE 7.60

Insert commutator, add the same delay paths.

of the path set. The resultant delay, z^{-1}, now operating at the input clock rate, is then interchanged with its path filter, as shown in Fig. 7.59.

In Fig. 7.60 the 1-to-$M/2$ upsamplers switch and their delays are replaced with a pair of commutators that add the path outputs with the same path delay. Finally, in Fig. 7.61 we fold the single delay in front of the lower set of path filters into the path filter. After having changed the output sampling rate of the polyphase channelizer, the final modification we need to introduce is the time alignment of the phase rotators from the input IFFT and the shifted time origin in the M-path filter.

From Fig. 7.62, we note the locations of the nonzero coefficients in the polynomials in z^2 and conclude that the input sine wave only contributes to the output time samples in the path filters located in the upper half of the path set. When the next scaled sine wave output from the IFFT is inserted in the first column of the path filter memory, the previously scaled sine wave is shifted one column in the memory and it will now contribute to the output time samples from the filter paths in the lower half of the path filter set. The problem is that the sine wave samples in the lower half of the path filter set have opposite polarity of the sine wave samples in the upper half of the path filter set. The samples from the two half filter sets are added, and to do so they must have the same phase. We note that this alternating sign effect only occurs for the odd-indexed IFFT frequency bins, which have an odd number of cycles per interval.

An alternative description of the sign reversal is that in the $M/2$ resampling of the modified M-path filter the even-indexed frequencies alias to multiples of the input sample rate and the odd-indexed frequencies alias to odd multiples of the half sample rate.

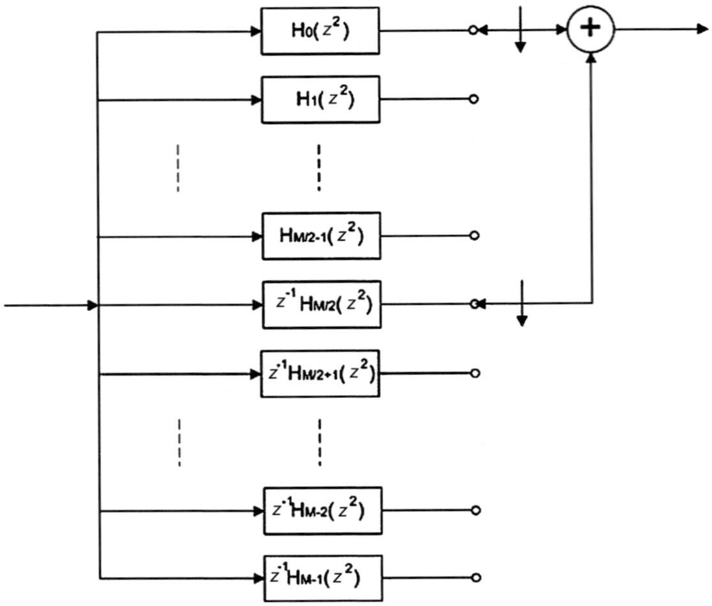

FIGURE 7.61

Fold unit delays into path filters.

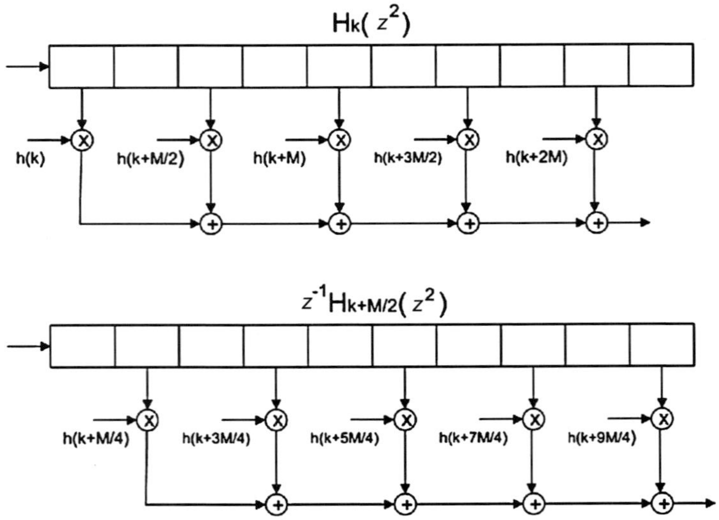

FIGURE 7.62

Comparison of path filter polynomials in z^2 with and without the additional input delay.

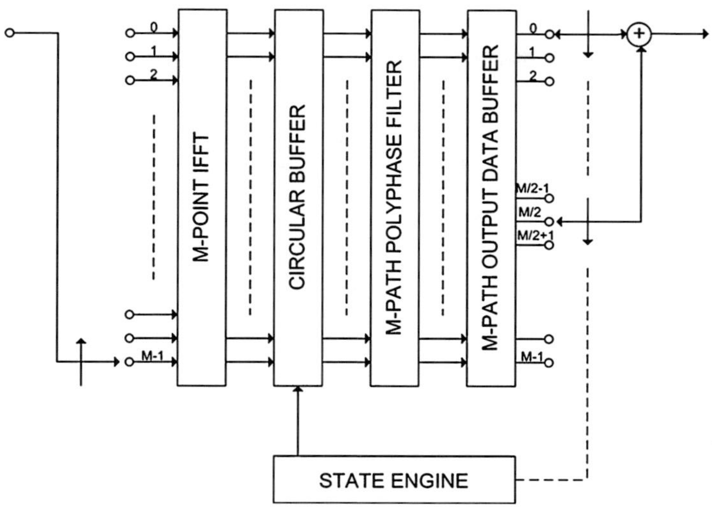

FIGURE 7.63

2-to-*M* modified up converter channelizer.

We recall here that there are two methods for performing the phase alignment of the successive output vectors from the IFFT. In the first method we simply invert the input phase of successive input samples to the odd-indexed IFFT bins. In the second method, recognizing that equivalency of phase shift and time delay for sinusoids on alternate outputs from the IFFT we apply an $M/2$ circular shift to its output buffer prior to delivering the phase-aligned vector to the path filter memory. This end around shift of the output buffer occurs during the data transfer in memory and requires no additional manipulation of the time samples.

Fig. 7.63 shows the complete block diagram of the modified M-path up converter channelizer. In this engine, the input commutator delivers the samples to the M-point IFFT. It applies the complex phase rotation to the separate baseband input signals and performs the initial 1-to-M upsampling of the input samples. The circular output buffer performs the correct data offset of the two $M/2$-point halves of the IFFT output vector to maintain phase alignment with the $M/2$ channelizer output vector. The complex sinusoid output by the IFFT always defines its time origin as the initial sample of its output vector. The output of the polyphase filter exhibits a time origin that shifts due to the $M/2$ time sample shift embedded in the output commutator. The $M/2$ time sample shift of the output time series causes sinusoids with an odd number of cycles in the length M array to alternate sign on successive shifts. The alternating sign is the reason that the odd-indexed frequency bins up convert to a frequency $k + N/2$ rather than frequency index k. Rather than reverse phase alternate input samples to the odd-indexed IFFT bins, we perform an $M/2$ point circular shift of alternate M-length vectors from the IFFT for applying the correct phase alignment to all frequencies simultaneously.

The polyphase down converter and up converter channelizers shown in Figs. 7.55 and 7.63, respectively, are the key elements of the proposed receiver and transmitter structures that we present in the following sections. Because of the signal processing tasks they handle, in the following, we refer to them as analysis and synthesis channelizers, respectively.

7.8 Preliminaries on software-defined radios

The 20th century saw the explosion of hardware-defined radio as a means of communicating all forms of data, audible and visual information, over vast distances. These radios have little or no software control. Their structures are fixed in accordance with the applications; the signal modulation formats and the carrier frequencies and bandwidths are only some of the factors that dictate the radio structures. The smallest change to one of these parameters could imply a replacement of the entire radio system. A consequence of this is, for example, the fact that a television receiver purchased in France does not work in England. The reason, of course, is that the different geographical regions employ different modulation standards for analog TV as well as for digital TV. Then, the citizens cannot use the same TV for receiving signals in both countries; they need to buy a new television for each country in which they decide to live. Sometimes, even if the communication devices are designed for the same application purposes and they work in the same geographical area, they are not able to communicate between each other. One of the most evident examples of this is that the city police car radio cannot communicate with the city fire truck radio or with the local hospital ambulance radio, even if they have the common purpose of helping and supporting the citizen. Also, the city fire truck radio cannot communicate with the county fire truck radio or with the radios of the fire truck operated by the adjacent city, or by the state park service, or the international airport. None of these services can communicate with the National Guard, or with the local Port Authority, or with the local Navy base, or the local Coast Guard base, or the US Border Patrol, or the US Customs Service. In a hardware-defined radio, if we decide to change one of the parameters of the transmitted signal, like bandwidth or carrier frequency (for example because the carrier frequency we want to use is the only one available at that particular moment), we need to change the transmitter. On the other side, every time we want to receive a signal having a different bandwidth or center frequency, we need to change the receiver. Hardware-defined transmitter and receiver devices are not flexible at all; we must modify their structure every time we change even one of the transmitting and receiving signal parameters.

In 1991, Joe Mitola coined the term software-defined radio. It referred to a class of reprogrammable (and reconfigurable) devices. At that time it was not clear at which level the digitization should occur to define a radio as software but the concept sounded pretty interesting, and the dream of building a completely reconfigurable radio device involved scientists from all over the world. Today the exact definition of software-defined radio is still controversial, and no consensus exists about the level of reconfigurability needed to qualify a radio as software.

Fig. 7.64 shows, in a simple block diagram, all the possible places in which the digitization can occur in a radio receiver. The exact dual block diagram can be portrayed for the radio transmitter. Current radios, often referred to as digital but sometimes referred to as software-defined radios (depending on their particular structure), after shifting the signals to IF, digitize them and assign all the remaining tasks to a digital signal processor. One of the main reasons for shifting the signals to IF, before digitizing them, is to reduce their maximum frequency so that a smaller number of samples can be taken for preserving the information content.

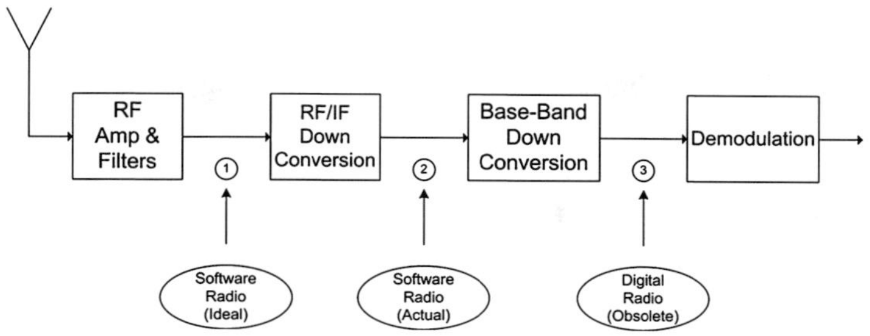

FIGURE 7.64

Simple block diagram indicating all the possible places in which the digitization can occur in a radio receiver.

Implementation of ideal software radios requires digitization at the antenna, allowing complete flexibility in the digital domain. Then it requires both the design of a flexible and efficient DSP-based structure and the design of a completely flexible radio frequency front-end for handling a wide range of carrier frequencies, bandwidths, and modulation formats. These issues have not been exploited yet in the commercial systems due to technology limitations and cost considerations.

As pointed out in the previous section, in a current software-defined radio receiver the signals are digitized in IF bands. The receiver employs a superheterodyne frequency down conversion, in which the radio frequency signals are picked up by the antenna along with other spurious, unwanted signals (noise and interferences), filtered, amplified with a low noise amplifier, and mixed with a local oscillator to shift it to IF. Depending on the application, the number of stages of this operation may vary. Digitizing the signal in the IF range eliminates the last analog stage in the conventional hardware-defined radios in which problems like carrier offset and imaging are encountered. When sampled, digital IF signals give spectral replicas that can be placed accurately near the baseband frequency, allowing frequency translation and digitization to be carried out simultaneously. Digital filtering and sample rate conversion are often needed to interface the output of the ADC to the processing hardware to implement the receiver. Likewise, on the transmitter side, digital filtering and sample rate conversion are often necessary to interface the digital hardware that creates the modulated waveforms to the DAC. DSP is usually performed in radio devices using field programmable gate arrays (FPGAs) or application-specific integrated circuits (ASICs).

Even if the dream of building a universal radio is still far away, current software-defined radio architectures are quite flexible, in the sense that they usually down convert to IF a collection of signals, and after sampling they are able to shift these signals to baseband via software. Changes in the signal bandwidths and center frequencies are made by changing some parameters of the digital data section. The flexibility of such a structure can be improved by moving the ADC closest to the receiver antenna. By digitizing the signals immediately after (and before in the transmitter) the antenna, which is the ideal software radio case, the down conversion (and up conversion) processes are performed completely in software and the radio acquires the capability of changing its personality, possibly in real time, guaranteeing a desired quality of service (QoS). The digitization after the receiver antenna, in fact, allows service providers to upgrade the infrastructure and new market services quickly. It promises multifunctionality, global mobility, ease of manufacture, compactness, and power efficiency. The flexibility in hardware architectures combined with flexibility in software architectures, through the implementation of techniques such as object-oriented programming, can provide software radio also with the ability to seamlessly integrate itself into multiple networks with widely different air and data interfaces.

7.9 Proposed architectures for software radios

A digital transmitter accepts binary input sequences and outputs radio frequency amplitude and phase-modulated wave shapes. The DSP part of this process starts by accepting b-bit words from a binary source at input symbol rate. These words address a look-up table that outputs gray-coded ordered pairs, i.e., I-Q constellation points, that control the amplitude and phase of the modulated carrier. The I-Q pair is input to DSP-based shaping filters that form 1-to-4 upsampled time series designed to control the wave shape and limit the baseband modulation bandwidth. The time series from the shaping filter are further upsampled by a pair of interpolating filters to obtain a wider spectral interval between spectral replicas of the sampled data. The interpolated data are then heterodyned by a DUC to a convenient digital IF and then moved from the sampled data domain to the continuous analog domain by a DAC and an analog IF filter. Further analog processing performs the spectral transformations required to couple the wave shape to the channel.

On the other side of the communication chain, a digital receiver has to perform the tasks of filtering, spectral translation, and analog-to-digital conversion to reverse the dual tasks performed at the transmitter. The receiver must also perform a number of other tasks absent in the transmitter for estimating the unknown parameters of the received signal such as amplitude, frequency, and timing alignment. It samples the output of the analog IF filter and down converts the IF centered signal to baseband with a DDC. The baseband signal is downsampled by a decimating filter and finally processed in the matched filter to maximize the signal-to-noise ratio (SNR) of the samples presented to the detector. The DSP portion of this receiver includes carrier alignment, timing recovery, channel equalization, automatic gain control, SNR estimation, signal detection, and interference suppression blocks. In order to suppress the undesired artifacts introduced by the analog components, the receiver also incorporates a number of DSP compensating blocks.

Fig. 7.65 shows the first-tier processing block diagrams of a typical digital transmitter–receiver chain. The depicted transmitter is designed for sending, one per time, signals having fixed bandwidths at precisely located center frequencies. On the other hand, the depicted receiver is also designed for down converting to baseband fixed-bandwidth signals having fixed center frequencies. Therefore, the goal of a cognitive radio is quite different: the cognitive transmitter and receiver should be completely flexible. The transmitter has, in fact, to be able to simultaneously up convert, at desired center frequencies, which are selected based on the temporary availability of the radio spectrum, a collection of signals having arbitrary bandwidths. It should also be able to decompose a signal in spectral fragments when a sufficiently wide portion of the radio spectrum is not available for transmission. Of course no energy loss has to occur in the partitioning process. A cognitive receiver has to be able to simultaneously down convert the transmitted signal spectra wherever they are positioned in the frequency domain and whatever bandwidths they have. Also, it should have the capability of recognizing spectral segments belonging to the same information signal and recompose them, after the baseband translation, without energy losses. It could be possible to reach this goal by using the current technology but the cost would be very high: For every signal we want to transmit we have to replicate the digital data section of both the transmitter and the receiver. The more signals we have to simultaneously up and down convert, the

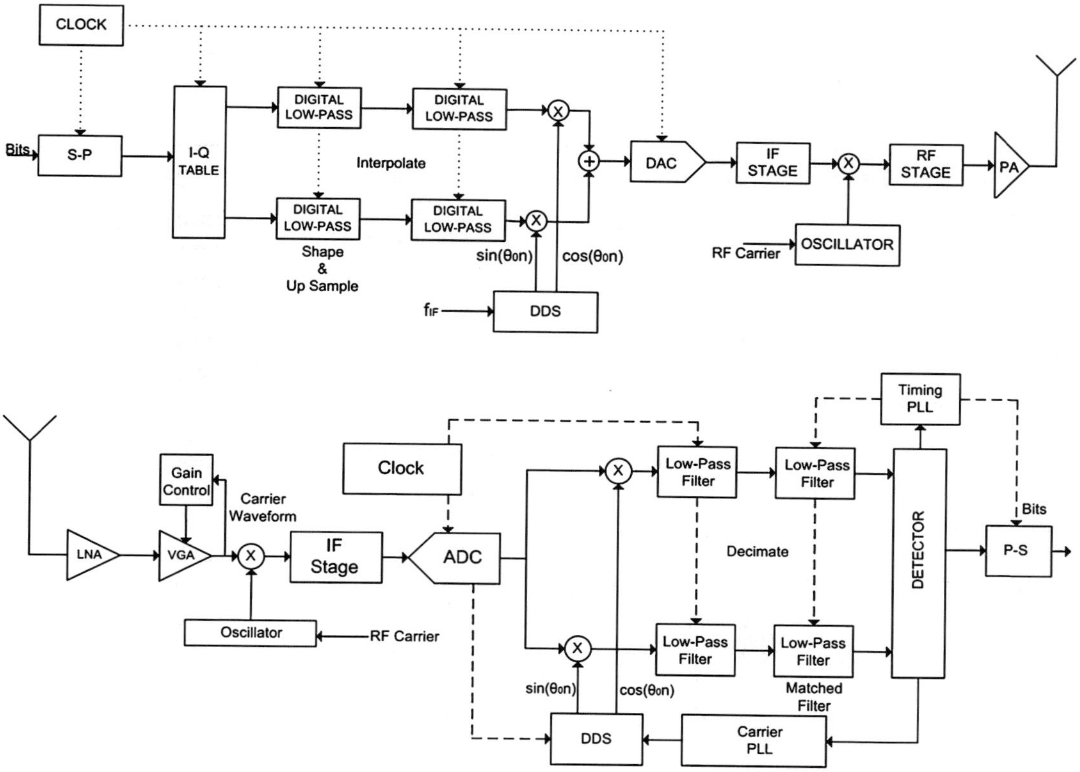

FIGURE 7.65

Block diagram of primary signal processing tasks in a typical transmitter and receiver.

more sampled data sections we need to implement! Also, it is not possible, using the current technology, to fragment a single signal, transmit its spectral partitions using the temporary available spectral holes, and perfectly recombine them at the receiver. Today the transmission happens under the constraint that the available space in the radio spectrum is large enough to accommodate the entire bandwidth of the signal to be transmitted.

In the next sections of this chapter, we present novel channelizers to simultaneously up and down convert arbitrary bandwidth signals randomly located in the frequency domain. Differently from the previous contents of this document, from this point onwards the presented material reproduces the results of the authors' research on software-defined radio design, thus it is new. The proposed architectures can be used for the transmitter and for the receiver devices of a cognitive radio. By using them we avoid the need to replicate the sampled data sections for simultaneously transmitting and receiving more signals with arbitrary bandwidths over multiple desired center frequencies. We also gain the capability to partition the signal spectra before transmitting them and to perfectly reassemble them in the receiver after the baseband shifting.

The core of the proposed transmitter structure is the variant of the standard M-path polyphase up converter channelizer that is able to perform 2-to-M upsampling while shifting, by aliasing, all the channels to desired center frequencies. This has been presented in Section 7.5 of this same chapter. In the following we use the term synthesis channelizer to refer to this structure. This name has been given for the tasks that this structure accomplishes when embedded in a software-defined radio transmitter. When the input signals have bandwidths wider than the synthesis channelizer bandwidth, they are preprocessed through small down converter channelizers that disassemble the signals' bandwidths into reduced bandwidth subchannels which are matched to the baseline synthesis channelizer bandwidths. Complex sampled frequency rotators are used to offset the signal spectra by arbitrary fractions of the sample frequency before processing them through the channelizers. This is necessary for the completely arbitrary center frequency positioning of the signals.

On the other side the variation of the standard M-path down converter channelizer, presented in Section 7.6, represents the key element of the proposed receiver. It is able to perform M-to-2 downsampling while simultaneously down converting, by aliasing, all the received signal spectra having arbitrary bandwidths. In the following, we refer to this structure as analysis channelizer. Postprocessing channelizers, which are a smaller version of the synthesis channelizer composing the transmitter, are used for reassembling, from the analysis channelizer baseline channels, signal bandwidths wider than the analysis channelizer channel bandwidth. Possible residual frequency offsets can be easily solved with the aid of a digital complex heterodyne while a postanalysis block performs further channelization, when it is required, for separating signal spectra falling in the same channelizer channel after the analysis processing.

7.9.1 Proposed digital down converter architecture

The standard M-path polyphase down converter channelizer is able to shift to baseband, by aliasing, the input signals that are exactly located on the channel's center frequencies, kf_c, where $k = 0, 1, \ldots, M-1$. In this engine the center frequencies of the channels are integer multiples of the channelizer output sampling frequency as well as the channelizer channel bandwidth.

We recall here that the standard M-path downsampling channelizer simultaneously down converts, by aliasing, from fixed center frequencies to baseband, M bands narrower than the channelizer channel bandwidth. If $kf_c \pm \Delta f_k$ are the center frequencies of the input spectra, where $\pm\Delta f_k$ are the arbitrary frequency offsets from the channel center frequencies, kf_c, after the down conversion process the output spectra will be shifted by the same offset from DC, i.e., the down converted signals will be centered at the frequency locations $\pm\Delta f_k$ because the polyphase down converter channelizer only compensates the kf_c frequency terms. It is unable to compensate the frequency offsets, $\pm\Delta f_k$, that are responsible for the arbitrary center frequency positioning of the input signals. The frequency offset compensation is one of the motivations for which we need to add further processing blocks in the digital down converter design when it is intended to be used in a software radio receiver. Other issues, like receiver resolution, baseband reassembly of wide bandwidths, and arbitrary interpolation of the baseband shifted signals, need to be considered too. All these issues are addressed and solved in the following paragraphs.

We also recall here that the standard M-to-1 down converter channelizer with the channel spacing and channel bandwidth equal to the output sampling frequency is critically sampled (see upper plot of Fig. 7.29). In order to use it in a cognitive radio receiver, we need to modify it, redesigning the channel spacing, channel bandwidth, and output sampling frequency. We presented one of these modifications

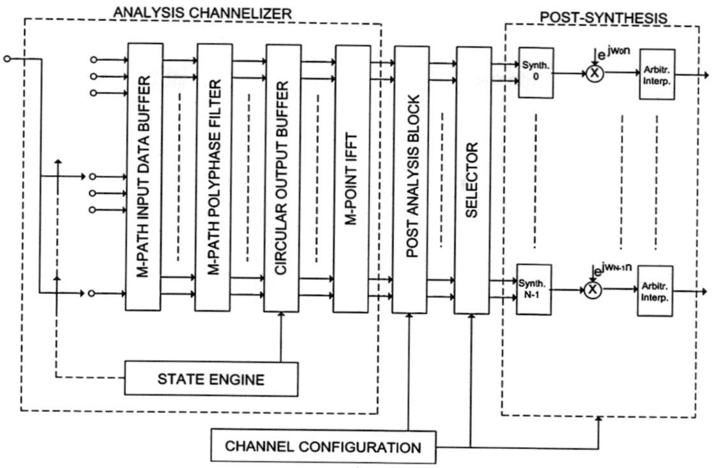

FIGURE 7.66

Block diagram of the proposed down converter; analysis down converter channelizer, postanalysis block, synthesis up converter channelizers, complex frequency rotators, and arbitrary interpolators.

in Section 7.5, where the M-to-2 down converter channelizer was derived. Also the selection of the low-pass prototype filter is an important issue that has to be addressed in order to avoid energy losses while the signal is processed.

Fig. 7.66 shows the complete block diagram of the proposed down converting chain. The input signal is processed, at first, by an analysis channelizer which is the modified version of the standard M-to-1 down converter channelizer presented in Section 7.5. By performing an $M/2$-to-1 downsampling of the input time series, this engine simultaneously shifts all the aliased channels to baseband presenting an output sampling rate that is twice the sample rate of the standard M-to-1 channelizer. Nyquist filters are used, as low-pass prototype, for avoiding energy losses while processing the signals. They are briefly described in the next subsections. The interested reader can find more details on Nyquist filter design in [8].

Note that if more signals or even spectral fragments belonging to different signals are channelized in the same channel, further processing is needed to separate them at the output of the analysis down converter. The extra filtering process is performed in the postanalysis block, which is connected to a channel configuration block that provides the receiver with the necessary information about the input signal bandwidths and center frequencies. Different options can be implemented to simplify the design of the postanalysis block. One of the possible options is to decrease the bandwidth of the perfect reconstruction prototype low-pass filter and to increase the number of points of the IFFT block in the analysis channelizer. By designing the channel spacing and bandwidth to accommodate the most likely expected signal width we minimize the possibility of having more than one signal in each channel. It is also possible to modify the analysis channelizer for having a more convenient frequency positioning of the aliased channels. The optimal choice, of course, depends on the receiver application and on the workload requirements.

At the output of the analysis channelizer, most of the signals with bandwidths narrower than the channel bandwidth are already down converted by means of the channelization process. However, the spectra wider than the channel bandwidth, or the spectra that, as a consequence of the arbitrary center frequency positioning, are processed by two or more adjacent channels, have been fragmented and their segments have all been aliased to the first Nyquist zone; these segments need to be recomposed before being translated to DC. The recomposition of spectral fragments is the task performed by the synthesis channelizers. They have been presented in Section 7.7.1, and their block diagram is depicted in Fig. 7.59. These are small 2-to-P_n polyphase up converters in which the IFFT size, P_n, is properly chosen in order to span the bandwidth of the nth received signal spectrum. We summarize the reason for including the synthesizers in our design in this way: at the output of the M-to-2 analysis channelizer all the signals have been downsampled and their spectra, or their spectral fragments, have been translated to the first Nyquist zone by the channelizing process. In order to reassemble the segments into a wider-bandwidth superchannel, the time series from each segment must be upsampled and frequency shifted to their appropriate positions so that they can be added together for forming the time series corresponding to the wider-bandwidth assembled signal.

At the end of the down conversion and up conversion processes, all the received spectra have been frequency-translated and, when necessary, recomposed in the first Nyquist zone. The analysis down converter channelizer shifted to baseband the signal spectra and their fragments that are exactly centered on its channels' center frequencies. This engine was not able to compensate the frequency offsets deriving from the arbitrary frequency positioning of the received signals.

The frequency offset compensation task is accomplished by the complex frequency rotators that follow the small up converter synthesizers. They are connected with the channel configuration block that provides them proper information about the signal center frequencies.

Once the signals are perfectly centered at DC, arbitrary interpolators are used to adjust their sampling rate to provide us exactly two samples per symbol needed for the further processing stages that are performed in the digital receiver.

7.9.1.1 Postanalysis block and synthesis up converter channelizers

At the output of the analysis channelizer all the channels have been aliased to baseband. As a consequence, the channelized segments of the received spectrum have been aliased to the first Nyquist zone (see Fig. 7.61).

The following three options, which cover all the possible cases of bandwidth positioning in the channelized spectrum, have to be considered and solved, by further processing the analysis channelizer outputs, in order to achieve, at the end of the analysis–synthesis chain, all the received bands shifted to baseband:

- 1. The aliased baseband channel could contain only one signal spectrum whose bandwidth is narrower than the channel bandwidth. The carrier frequency of the signal generally does not coincide with the center frequency of the channel.
- 2. The aliased baseband channel could contain two or more spectra, or also their fragments, belonging to different signals. These spectra are arbitrarily positioned in the channel bandwidth.
- 3. The baseband channel could contain only one spectral fragment belonging to one of the received signals which has bandwidth larger than the channel bandwidth.

In the first option, at the output of the analysis channelizer, the signal spectra that are entirely contained in one single channel reside in the first Nyquist zone. Eventually the receiver has to compensate the frequency offsets derived from their arbitrary center frequency positioning and resample them to obtain the desired output sampling rate of two samples per symbol. The complex frequency rotators and the arbitrary interpolators perform these tasks at the end of the receiver chain.

For the case in which two or more spectra are processed by one single channel, more tasks need to be performed before frequency offset compensation and arbitrary interpolation occur. We, in fact, have to separate, by filtering, the bands, or their fragments, belonging to different signals before processing all of them independently. The separation task is performed by the postanalysis block. It performs a further channelization process that filters, from every baseband aliased channel, the bands belonging to different signals. Note that some of the filtered signals, and precisely the ones with bandwidths narrower than the channel bandwidth and entirely contained in the channel, only have to be frequency shifted before being delivered to the arbitrary interpolator; else, the spectral fragments belonging to signals processed in more than one channel have to be passed through the up converter synthesizers before being frequency shifted and resampled.

In the third case, which concerns the signals with bandwidths wider than the channel spacing, the analysis channelizer partitioned the bandwidths into several fragments and aliased every fragment to baseband. In order to recombine them we must first upsample each input time series and second translate them to their proper spectral region. We can then form the sum to obtain the superchannel representation of the original signal bandwidth. Those are exactly the tasks performed by the synthesis channelizers.

7.9.1.2 High-quality arbitrary interpolator

After all the bands have been translated to zero frequency, we need to resample them to obtain exactly the two samples per symbol needed for the subsequent processing performed in the receiver.

In this subsection we present the high-quality interpolator structure used to obtain two samples per symbol needed for the second- and third-tier processing tasks in a digital receiver. It is well known that the dynamic range of an arbitrary interpolator should match the system's quantization noise level [8]. The error due to the linear interpolation process, in fact, is not observable if it is below the noise level attributed to the signal quantization process. Since the error due to the b-bit quantized signal is 2^{-b}, the interpolation error or, equivalently, the level of residual spectral artifacts has to be below this threshold. In other words, if the oversampling factor, N, satisfies Eq. (7.34), then the interpolation error will not be noticeable:

$$\left[\frac{1}{2N}\right]^2 \leq \frac{1}{2^b}. \tag{7.34}$$

Thus, if we interpolate a 16-bit dataset, we keep the spectral artifacts below the quantization noise level if N is greater than or equal to 128. To interpolate the signals by a factor of $N = 128$, we break the upsampling process of the baseband centered spectra in two stages.

As depicted in Fig. 7.67, we perform an initial 1-to-4 upsampling followed by a 1-to-32 upsampling obtained by using the high-quality arbitrary interpolator shown in Fig. 7.68. The initial four-times oversampling has the effect of strongly decreasing the length of the polyphase filters used in the interpolator, which significantly reduces the total workload of this structure. More details on this topic can be found in [8].

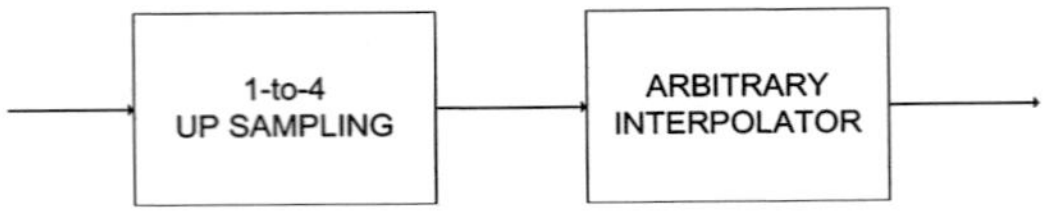

FIGURE 7.67

Two stage arbitrary interpolator.

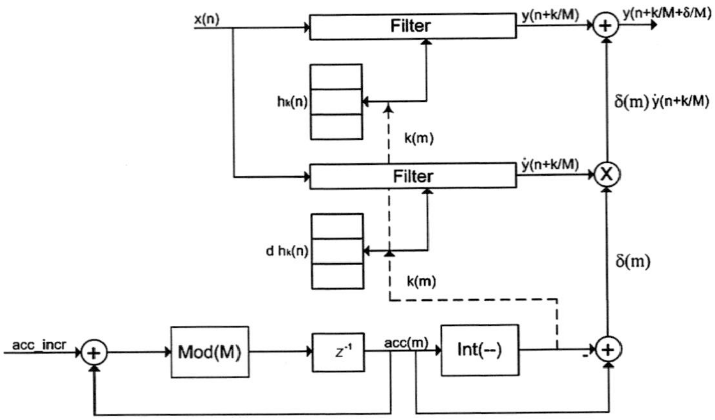

FIGURE 7.68

Block diagram of high-quality arbitrary interpolator.

Note that the initial 1-to-4 signal upsampling can either be achieved by using two half-band filters in cascade or we can previously upsample the signals by properly choosing P_n, the IFFT size of the synthesis channelizers. Also, at the output of the synthesizers some of the signals could be already upsampled by some factors because the IFFT size of the synthesizer has to be an even number [8] and it also has to satisfy Nyquist criteria for every signal band.

The arbitrary interpolator shown in Fig. 7.68 performs the linear interpolation between two available adjacent signal samples (two-neighbor interpolation). The interpolation process operates as follows: A new input data sample is delivered to the interpolation filter register on each accumulator overflow. The integer part of the accumulator content selects one of the M filter weights, $h_k(n)$, and one of the M derivative filter weights, $dh_k(n)$, while the filters compute the amplitude $y(n+k/M)$ and the first derivative $\dot{y}(n+k/M)$ at the kth interpolated point in the M-point interpolated output grid. Then the structure uses the local Taylor series to form the interpolated sample value between grid points. The output clock directs the accumulator process to add the desired increment Δ (acc_incr in Fig. 7.68) to the modulo M accumulator and increments the output index m. The increment Δ can be computed according to the following equation: $\Delta = M f_{in}/f_{out}$. This process continues until the accumulator overflows at one of the output clock increments. Upon this overflow the input index is incremented, a new input data point is shifted into the filter, and the process continues as before.

7.9.1.3 Nyquist filters

Particular attention has to be paid in designing the low-pass prototype filters used in the analysis channelizer. The signal spectra have to be randomly located in the frequency domain and their bandwidths can easily span and occupy more than one baseline channel; also, in the signal disassembling and reassembling processes, we need to collect all the energy corresponding to a single signal without losses.

The Nyquist filter presents the interesting property of having a band edge gain equal to 0.5 (or −6 dB). By using this filter as prototype in our channelizers, we place M of them across the whole spanned spectrum with each filter centered on kf_s/M. All adjacent filters exhibit −6 dB overlap at their band edges. The channelizer working under this configuration is able to collect all the signal energy across its full operating spectrum range even if signals occupy more than one adjacent channel or it resides in the channel's overlapping transition bandwidths. In the following paragraphs, for completeness, we provide a brief introduction on Nyquist filters.

Nyquist pulses is the name given to the wave shapes $f(n)$ required to communicate over band-limited channels with no intersymbol interference (ISI). When sampled at equally spaced time increments they have to verify the following requirement, which is known as the Nyquist pulse criterion for zero ISI:

$$f(n) = \begin{cases} 0 & \text{if} \quad n \neq 0, \\ 1 & \text{if} \quad n = 0. \end{cases} \tag{7.35}$$

There are infinite such functions that satisfy this set of restrictions. The one with minimum bandwidth is the ubiquitous $\sin(x)/x$, which is variously known as the cardinal pulse when used for band-limited interpolation and as the Nyquist pulse when used in pulse shaping. The transform of this wave shape, $R(f)$, is the unit-area rectangle with spectral support $1/T$ Hz. Unfortunately this waveform is noncausal and further it resides on an infinite support. If the pulse resided on a finite support we could delay the response sufficiently for it to be causal. We have to form finite support approximations to the $\sin(x)/x$ pulse. The first approximation to this pulse is obtained by convolving the rectangular spectrum, $R(f)$, with an even symmetric, continuous spectrum $W(f)$ with finite support α/T. The convolution between $R(f)$ and $W(f)$ in the frequency domain is equivalent to a product in the time domain between $r(t)$ and $w(t)$, where $w(t)$ is the inverse transform of $W(f)$. The effect of the spectral convolution is to increase the two-sided bandwidth from $1/T$ to $(1+\alpha)/T$. The excess bandwidth α/T is the cost we incur to form filters on finite support. The term α is called the roll-off factor and is typically on the order of 0.5–0.1, with many systems using values of $\alpha = 0.2$. The transition bandwidth caused by the convolution is seen to exhibit odd symmetry about the half amplitude point of the original rectangular spectrum. This is a desired consequence of requiring even symmetry for the convolving spectral mass function. When the windowed signal is sampled at the symbol rate $1/T$ Hz, the spectral component residing beyond the $1/T$ bandwidth folds about the frequency $\pm 1/2T$ into the original bandwidth. This folded spectral component supplies the additional amplitude required to bring the spectrum to the constant amplitude of $R(f)$.

Following this reasoning, we note that the significant amplitude of the windowed wave shape is confined to an interval of approximate width $4T/\alpha$ so that a filter with $\alpha = 0.2$ spans approximately $20T$, or 20 symbol durations. We can elect to simply truncate the windowed impulse response to obtain a finite support filter, and often choose the truncation points at $\pm 2T/\alpha$. A second window, a rectangle, performs this truncation. The result of this second windowing operation is a second spectral convolution with its transform. This second convolution induces passband ripple and out-of-band side lobes in the spectrum of the finite support Nyquist filter. The description of this band-limited spectrum normalized to unity passband gain is presented as

$$H(w) = \begin{cases} 1 & \text{for } \frac{|w|}{w_{\text{sym}}} \leq (1-\alpha), \\ 0.5\left\{1+\cos\left\{\frac{2\pi}{\alpha}\left[\frac{w}{w_{\text{sym}}}-(1-\alpha)\right]\right\}\right\} & \text{for } (1-\alpha) \leq \frac{|w|}{w_{\text{sym}}} \leq (1+\alpha), \\ 0 & \text{for } \frac{|w|}{w_{\text{sym}}} \geq (1+\alpha). \end{cases} \tag{7.36}$$

The continuous-time domain expression for the cosine-tapered Nyquist filter is shown in Eq. (7.37). Here we see the windowing operation of the Nyquist pulse as a product with the window that is the transform of the half-cosine spectrum:

$$h(t) = f_{\text{sym}} \frac{\sin(\pi f_{\text{sym}} t)}{(\pi f_{\text{sym}} t)} \frac{\cos(\pi \alpha f_{\text{sym}} t)}{[1-(2\alpha f_{\text{sym}} t)^2]}. \tag{7.37}$$

Since the Nyquist filter is band-limited, we can form the samples of a digital filter by sampling the impulse response of the continuous filter. Normally this involves two operations. The first is a scaling factor applied to the impulse response by dividing by the sample rate, and the second is the sampling process in which we replace t with nT_s or n/f_s. The sample rate must exceed the two-sided bandwidth of the filter that, due to the excess bandwidth, is wider than the symbol rate. It is standard to select the sample rate f_s to be an integer multiple of the symbol rate f_{SYM} so that the filter operates at M samples per symbol. It is common to operate the filter at four or eight samples per symbol.

In the design of the proposed analysis channelizer, the Nyquist prototype low-pass filter has to be designed with its two-sided 3 dB bandwidth equal to $1/M$th of the channelizer input sampling frequency. This is equivalent to the filter impulse response having approximately M samples between its peak and first zero crossing and having approximately M samples between its zero crossings. The integer M is also the size of the IFFT as well as the number of channels in the analysis channelizer. The prototype filter must also exhibit reasonable transition bandwidth and sufficient out-of-band attenuation or stopband level. We designed our system for a dynamic range of 80 dB, which is the dynamic range of a 16-bit processor.

7.9.2 Digital down converter simulation results

For simulation purposes, in this section we consider, at the input to the M-to-2 analysis channelizer, a composite spectrum that contains 12 QAM signals with 5 different bandwidths randomly located in the frequency domain. In particular, the signal constellations are 4-QAM, 16-QAM, 64-QAM, and 256-QAM, while the signal bandwidths are 1.572132 MHz, 3.892191 MHz, 5.056941 MHz, 5.360537 MHz, and 11.11302 MHz, respectively. It is clear that we used two different signal bandwidths for the 256-QAM constellation (5.360537 MHz and 11.11302 MHz).

The signals are shaped by square-root Nyquist filters with 20% excess bandwidth. At the input of the analysis channelizer all the signals are resampled to achieve a sample rate of 192 MHz. The spectrum is shown in Fig. 7.56. In particular, the upper subplot of Fig. 7.69 shows, superimposed on the composite received spectrum, the 61 channels of the analysis channelizer. It is easy to recognize that the received spectra are arbitrarily located. Their center frequencies do not coincide with the channel center frequencies. That is clearly shown in the lower subplot of Fig. 7.69, in which the enlarged view

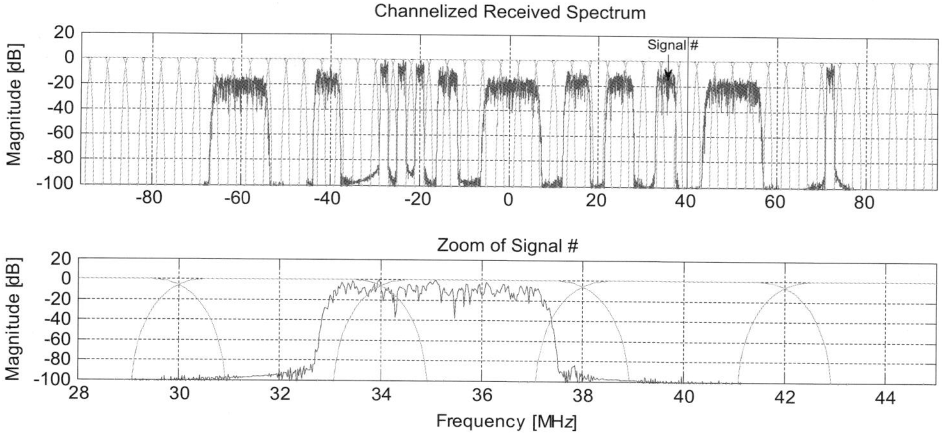

FIGURE 7.69

Channelized received spectrum and zoom of one of the received signals.

of one of the received signals is presented. The arbitrary signal positioning is the reason for which the polyphase down converter channelizer, by itself, is not able to directly shift them to DC. The IFFT size of the analysis M-to-2 down converter channelizer is $M = 48$ with an output sample rate of 8 MHz.

Fig. 7.70 shows the impulse response and the magnitude response of the designed prototype low-pass Nyquist filter. It is designed to have 48 samples per symbol. Its length is 1200 taps, while its stopband attenuation is −80 dB. Note that since this filter is M-path partitioned, the length of each filter in the M-path bank is only 25 taps.

The 61 spectra, at the output of the analysis channelizer, are depicted in Fig. 7.71. The down converter channelizer has partitioned the entire frequency range into 48 segments. It is easy to recognize in this figure the different spectra composing the received signal. Note that because of the arbitrary frequency positioning, it is possible that also the signals having bandwidths narrower than the channelizer channel bandwidth occupy more than one analysis channelizer channel. Also, in this case, before shifting these signals to zero frequency by using complex frequency rotators, we need to pass them through a synthesis channelizer that reassembles their fragments.

Before delivering the analysis channelizer outputs to the synthesis channelizers, we need to separate, if necessary, by filtering, those spectra that belong to different signals lying in the same channel. An example of this is represented by channel 30 in Fig. 7.71. It contains fragments of two spectra belonging to different signals. The filter design in the postanalysis block, of course, depends on the bands that have to be resolved. Their sample rate has to be the same as the analysis channelizer output rate (8 MHz). An example of postanalysis filters along with the filtered signals, for the 30th analysis channelizer channel, is shown in Fig. 7.72. In particular, the signal spectra and the filters used for separating them are shown in the upper subplot, while the separated spectra are shown in the lower subplots.

At the output of the postanalysis block, all the spectra with bandwidths narrower than the analysis channel bandwidth lying in a single analysis channel can be directly delivered to the complex heterodyne that translates them to DC. All the other signals, the ones with bandwidths wider than the analysis

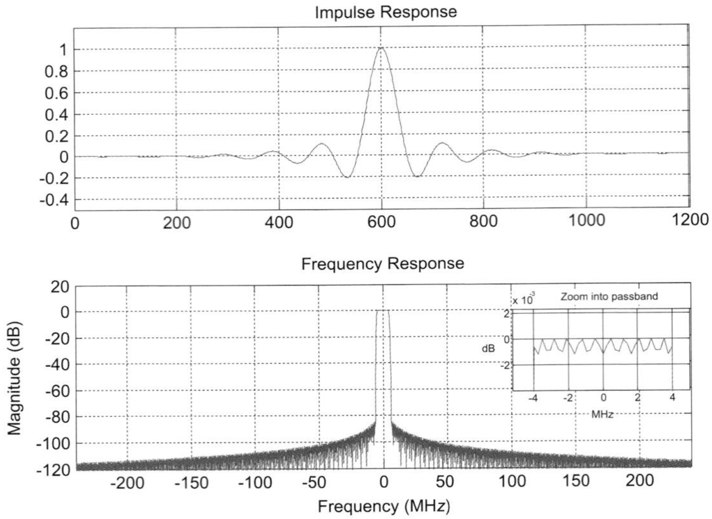

FIGURE 7.70

Nyquist prototype filter.

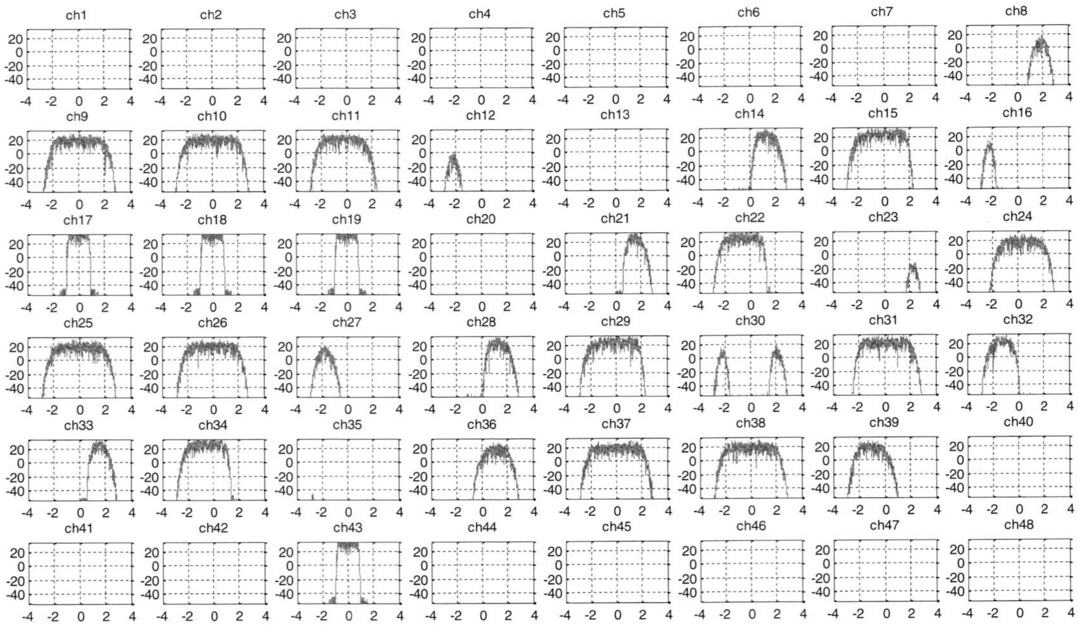

FIGURE 7.71

Analysis channelizer outputs.

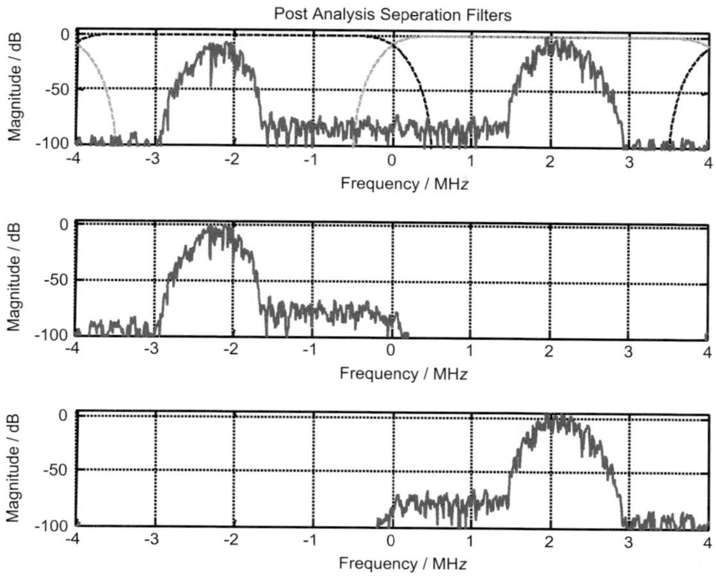

FIGURE 7.72

Postanalysis filters and filtered signals for channel 30 of the analysis channelizer.

channel bandwidth and the ones with bandwidths narrower than the analysis channelizer channel bandwidth, which as a consequence of the analysis process are delivered to two different channelizer outputs, need to be processed by the small synthesis channelizers. We have, in fact, to upsample, frequency shift, and recombine the time series from coupled analysis channelizer outputs. In the example of Fig. 7.69, four of the received spectra, the narrowest ones, are directly sent to the frequency rotators. Eight of them are processed through the synthesis channelizer. The IFFT sizes P_n, with $n = 0, 1, 2$, of the synthesizers we selected to process the three remaining bandwidths are $P_0 = 6$, $P_1 = 4$, and $P_2 = 2$ points. These sizes are the minimum possible chosen to satisfy the Nyquist sampling criterion for each output signal. Note that because of the structure of the synthesizers, we can only have an even number of IFFT points [8].

At the output of the synthesizer, all the signal fragments are recombined in baseband but still some of them have a residual frequency offset that needs to be compensated. The signal spectra before frequency offset compensation are shown in Fig. 7.73. Here it is clearly visible that many of the signals are not centered at DC (the red[1] line in Fig. 7.73 represents the signals' center frequency). We compensate these frequency offsets by using complex frequency rotators. Note that by estimating the signal energy in the synthesizer channels, we could easily recover the carrier offsets affecting the received spectra. The frequency offset recovery is another of the many applications of the polyphase channelizer. Other papers addressing this topic are being prepared.

[1] For interpretation of color in Fig. 7.73, the reader is referred to the web version of this book.

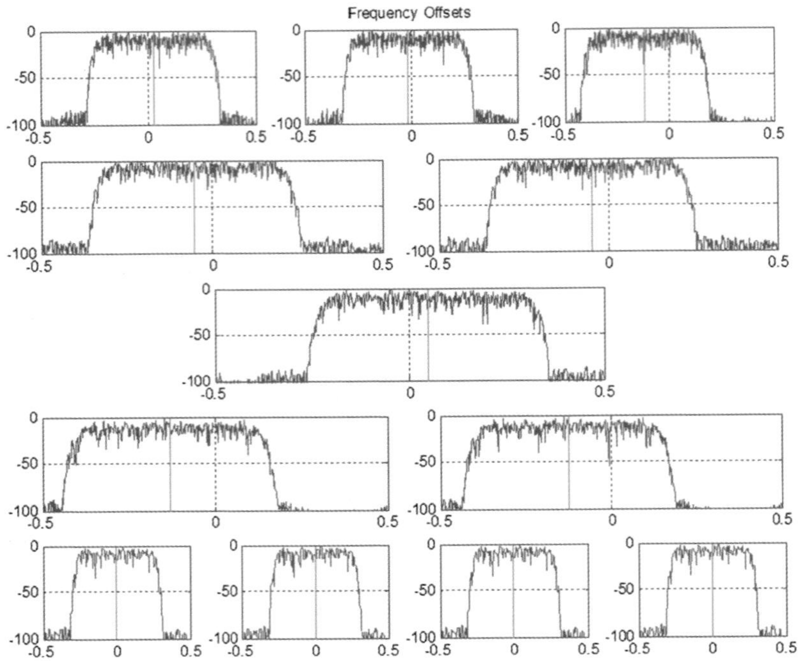

FIGURE 7.73

Log magnitude of synthesizer outputs with frequency offsets in the normalized frequency domain.

When they are DC centered, the signals need to be resampled. The signals at the outputs of the synthesis channelizers have different sampling frequencies. All that we need at this point is to interpolate them to obtain exactly two samples per symbol for each of them. We use an arbitrary interpolator to achieve the sample rate conversion. The interpolated, DC shifted signals are shown in Fig. 7.74. We also match filtered each of the eight channelized and reconstructed time signals and present their constellations in Fig. 7.75. Here we see that all of the QAM constellations are perfectly reconstructed, which demonstrates the correct functionality of the proposed receiver.

7.9.3 Proposed up converter architecture

Fig. 7.76 shows the complete block diagram of the proposed up converter, which has the structure that is a dual of the down converter structure presented in the previous sections. It is composed of a 2-to-M upsampler synthesis channelizer, which has been presented in Section 7.7.1, preceded by N preprocessing analysis blocks which are small P_n-to-2 analysis down converter channelizers. Their number, N, corresponds to the number of the signal spectra wider than the synthesis channelizer channel bandwidth. Their IFFT size, P_n, is chosen to completely span the bandwidth of the nth input signal spectrum. Their task is to decompose the wider input spectra into P_n partitions matching the bandwidth

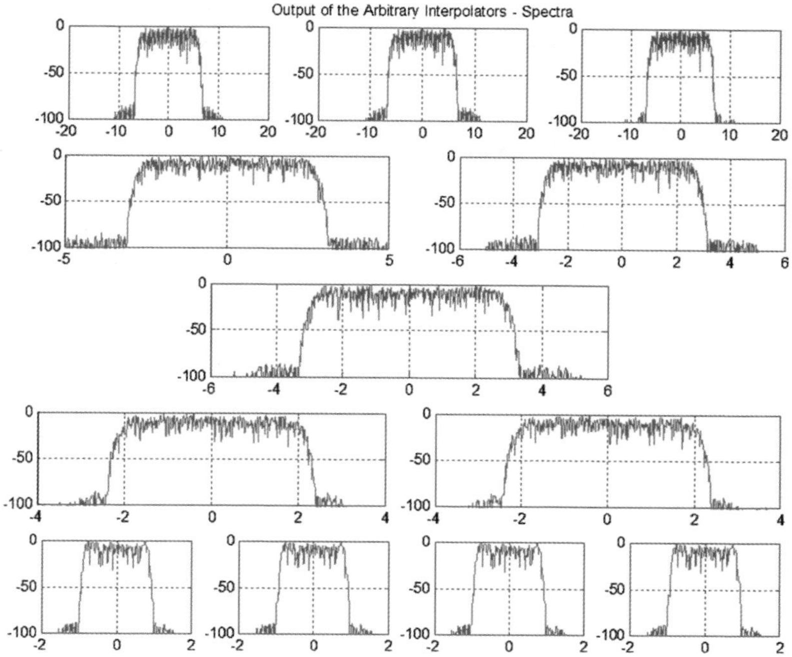

FIGURE 7.74

Log magnitude of the heterodyned and interpolated spectra in the frequency domain [MHz].

and sample rate of the baseline synthesis channelizer which will coherently recombine them in the receiver.

All the input signals to the 2-to-M synthesis channelizer with bandwidths less than or equal to the channel spacing are to be upsampled and translated from baseband to the selected center frequency by the channelizing process. For these signals we may only have to filter and resample to obtain the desired sampling rate of two samples per channel bandwidth. However, we expect that many of the spectra we presented to the synthesis channelizer have bandwidths that are wider than the synthesizer channel spacing. To accommodate these signals we need to use the small analysis channelizers that partition their bandwidths into several fragments, translate all of them to baseband, and reduce their sample rate to twice the channel bandwidth. The analysis channelizers are designed as P_n-to-2 polyphase down converter channelizers, where P_n is approximately twice the number of baseline channels spanned by the wide-band spectrum. They were presented in Section 7.5 of this chapter, and Fig. 7.55 shows their block diagram.

We briefly recall here that such a system accepts $N/2$ input samples and outputs time samples from N output channels. The N-point input buffer is fed by a dual input commutator with an $N/2$ sample offset. The N-path filter contains polynomials of the form $H_r(z^2)$ and $z^{-1}H_{r+N/2}(z^2)$ in its upper and lower halves, respectively. The circular output buffer performs the phase rotation alignment of the IFFT

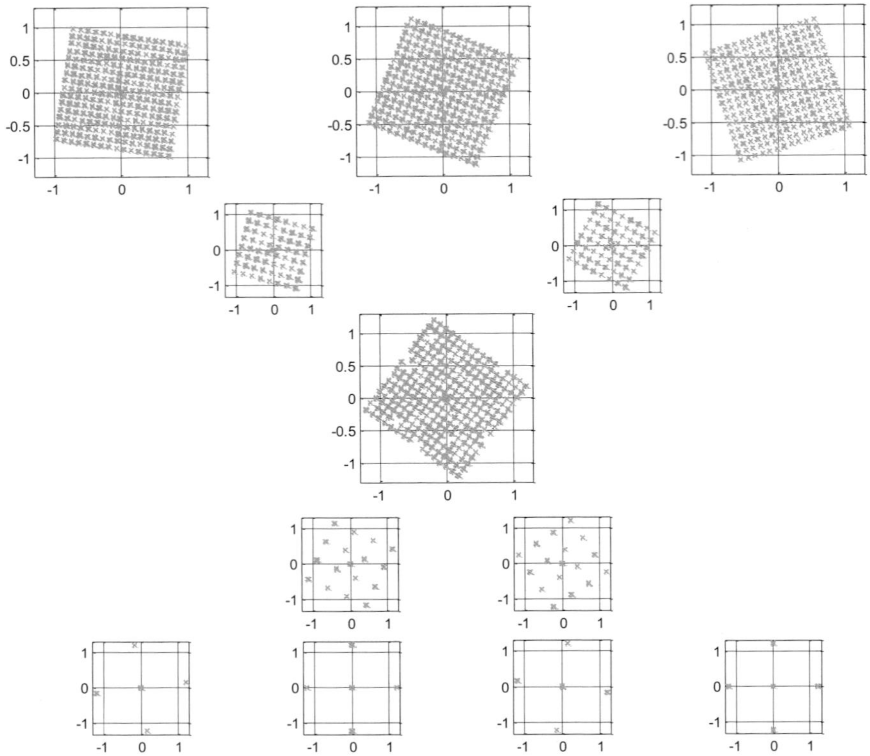

FIGURE 7.75

Heterodyned and interpolated constellations.

block with the $N/2$ stride shifting time origin of the N-path polyphase filter. The IFFT size for these analysis channelizers, P_n, is also the number of their filter outputs.

In order to recombine the segmented signal components we have to satisfy the Nyquist criteria for their sum. Since these are preprocessor analysis channelizers that feed the M-path synthesis channelizer we must have their output sample rate two times their channel bandwidth. We can achieve this by selecting P_n to be approximately twice the number of channels being merged in the synthesizer and setting its input sample rate to be $2f_s P_n$, so that the preprocessor output rate per channel is the required $2f_s$. Remember that the IFFT block sizes must be even to perform the 2-to-N resampling by the technique described in Section 7.5; also remember that an actual system may have a few standard size IFFTs to be used for the analysis channelizer and the user may have to choose from the small list of available block sizes.

The channel selector placed between the analysis channelizer bank and the synthesis channelizer also connected with the input channel configuration block provides the correct outputs from the preprocessor analysis channelizer to the input synthesizer, while the channel configuration block provides the necessary information to the selector block that is connected to the synthesis input series. The selector

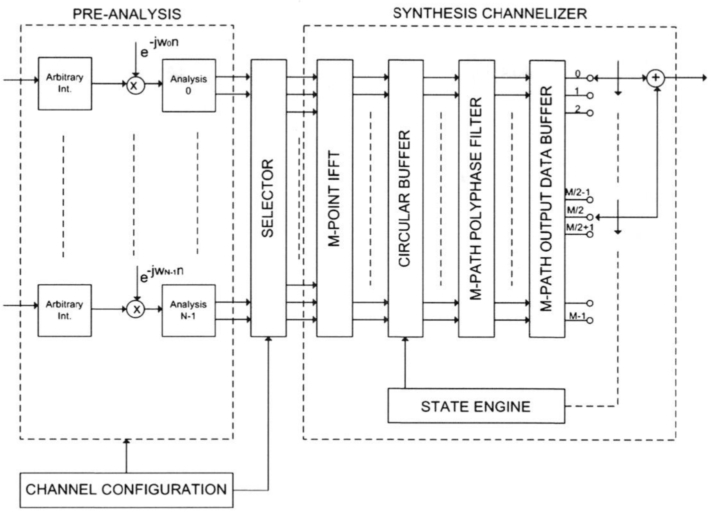

FIGURE 7.76

Block diagram of the proposed up converter; arbitrary interpolators, complex frequency rotators, analysis down converter channelizers, and synthesis up converter channelizer.

routes all the segments required to assemble the wider bandwidth channel to the synthesizer, which performs their frequency shift and reassembly.

Depending on the desired center frequency of the disassembled spectrum an earlier complex heterodyne may be required before the analyzers to shift the about to be disassembled signals with the proper frequency offset.

7.9.4 Digital up converter simulation results

In the simulation results shown in this section we consider a set of eight distinct baseband input signals to be delivered to the 2-to-M up converter synthesis channelizer. These are QPSK signals with three different bandwidths as shown in Fig. 7.77. The symbol rates chosen for the three signals denoted 1, 2, and 3 are 7.5, 15.0, and 30.0 MHz. These signals are shaped by square-root Nyquist filters with 25% excess bandwidth; hence, the two-sided bandwidths are 7.5·1.25, 15.0·1.25, and 30.0·1.25 MHz, respectively.

The IFFT size of the baseline 2-to-M up converter channelizer is $M = 48$ with 10 MHz channel spacing for which the required input sample rate per input signal is 20 MHz and for which the output sample rate will be 480 MHz. For ease of signal generation, all three signals were shaped and upsampled to a sample rate of 60 MHz with shaping filters designed for eight, four, and two samples per symbol. Signal 1, represented in the first line of Fig. 7.77, is downsampled 3-to-1 to obtain the desired sample rate of 20 MHz for the synthesis channelizer. Signals 2 and 3, on the second and third line of Fig. 7.77, respectively, are downsampled 6-to-2 in the six-point IFFT analysis channelizers which form 10 MHz channels at a sample rate of 20 MHz. Signal 2 is spanned by three 10 MHz channels which

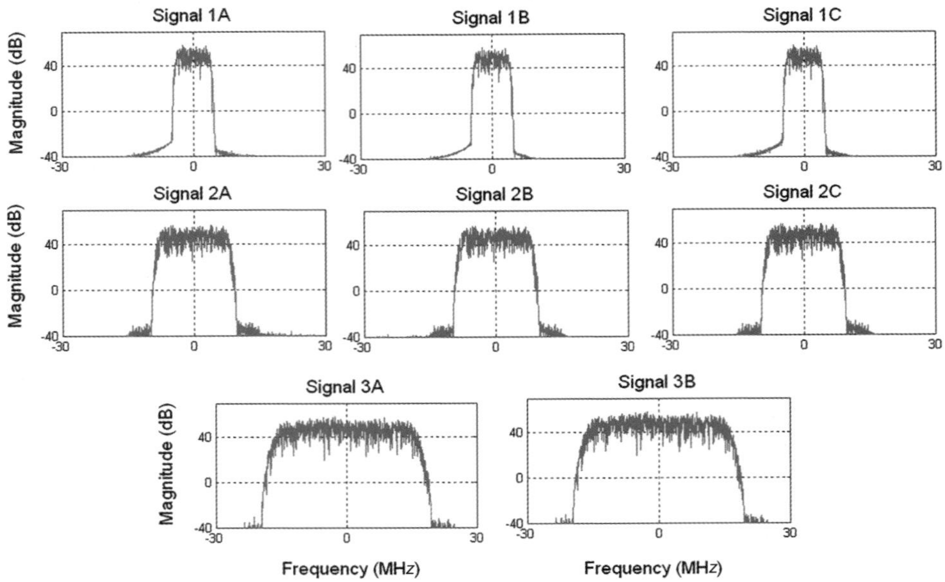

FIGURE 7.77

Example of baseband spectra to be up converted.

will feed three input ports of the 48-point synthesizer IFFT, while signal 3 is spanned by five 10 MHz channels which will feed five input ports of the 48-point IFFT. The IFFT inputs are controlled by a channel control block that routes them to the proper synthesis channelizer ports.

Any of the signals presented to the analysis channelizers is previously shifted to a desired frequency offset, by means of a complex heterodyne if required. The output channels that span the offset input bandwidth of the analysis channelizer are the channels passed on the synthesizer and these change due to a frequency offset. The inserted baseband frequency offset will survive the synthesis channelizer.

Fig. 7.78 shows all the spectra of the output time series from the six-channel polyphase 6-to-2 analysis channelizer engine processing signal 3, which is the signal with the widest band. Also seen in the same figure is the spectrum of signal 3 and the frequency response of all the six channels formed by one of the analysis channelizers. Note the spectra in the upper subplots have been filtered by the channelizer with the 10 MHz Nyquist passband frequency response, have been translated to baseband, and have been sampled at 20 MHz. Five of these segments are presented to the five input ports of the 2-to-48 synthesis channelizer centered on the desired frequency translation index.

The up converter polyphase synthesis channelizer accepts time sequences sampled at 20 MHz with bandwidths less than 10 MHz. We have delivered three signals that satisfy these constraints along with four signals that were conditioned by analysis channelizers that partitioned their bandwidths into segments that also satisfied the input signal constraints. The spectra of the separate components delivered to the synthesis channelizer are shown in Fig. 7.79. It is easy to recognize in this figure the different

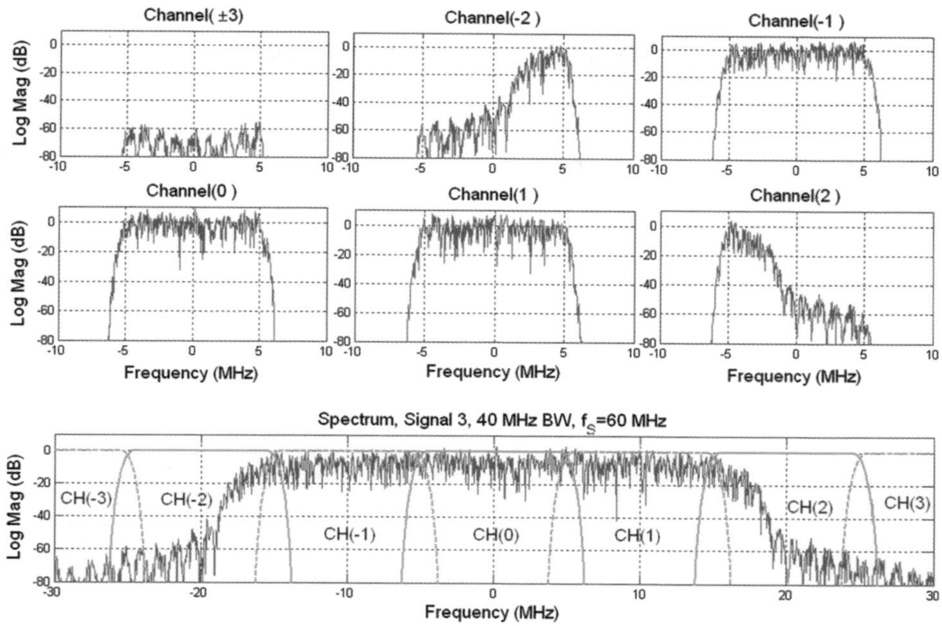

FIGURE 7.78

Spectral fragments formed by six-channel 6-to-2 downsample analysis channelizer processing signal 3.

spectra composing the received signal. Here we see that filters 4 to 8 are segments of a single frequency band fragmented in Fig. 7.78.

At this point, we upsample, frequency shift, and recombine the time series from coupled channel outputs using the synthesis channelizer. The frequency shifted spectra of the eight signals, including six that have been fragmented in analysis channelizers and then defragmented in the synthesis channelizer, are plotted in Fig. 7.80.

7.10 Case study: enabling automated signal detection, segregation, and classification

In this section, we will present a wireless radiofrequency use case where the polyphase channelizer can improve performance over standard techniques that rely on the FFT. Using the channelizer will allow us to differentiate nearby signals with very wide dynamic range to a finer degree than an equivalent short-time Fourier transform (STFT). Next, a channelizer will be used to revert the detected and localized signals back to the time domain for either demodulation or for classification algorithms that require time domain data. Finally, we present a machine learning technique that uses a modified strip spectral

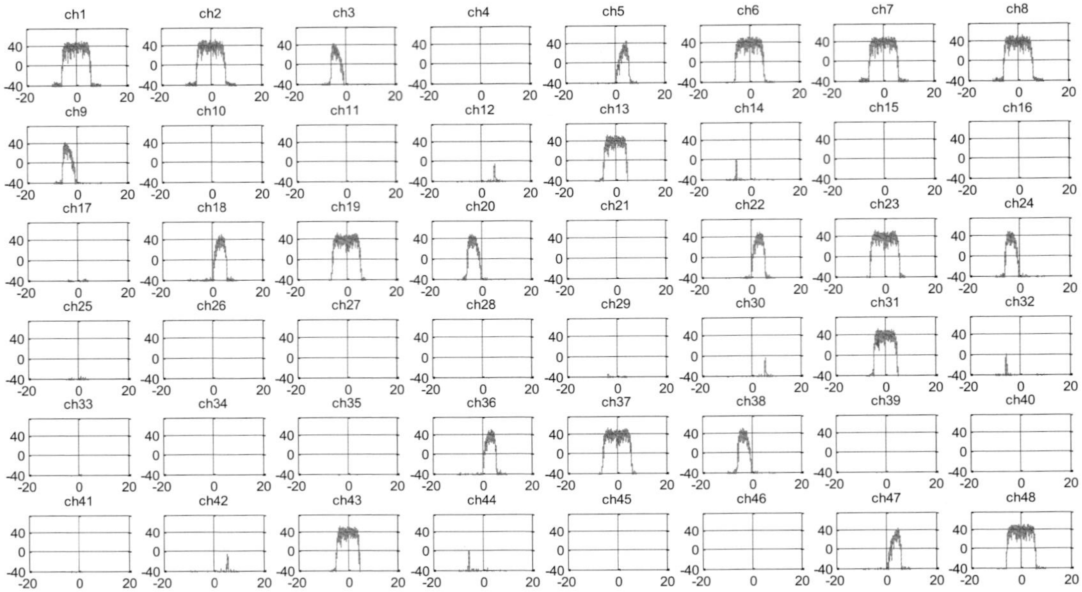

FIGURE 7.79

2-to-*M* up converter channelizer inputs.

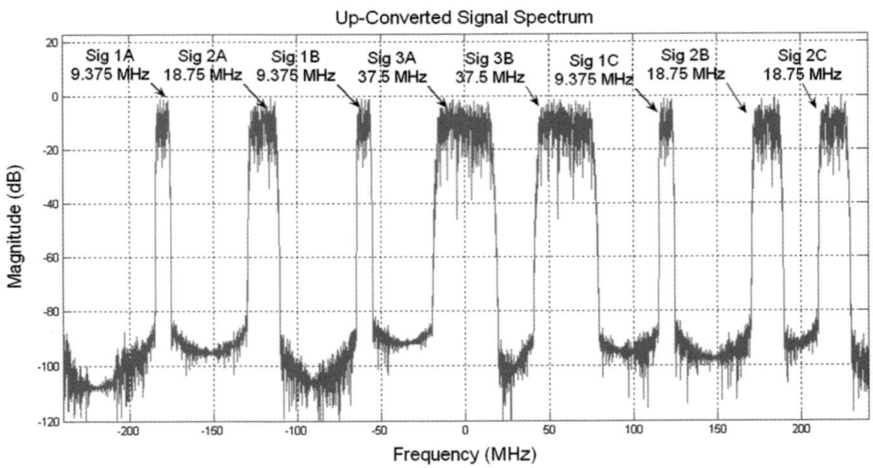

FIGURE 7.80

Up converted synthesized spectrum with unequal bandwidths fragmented and defragmented spectral components.

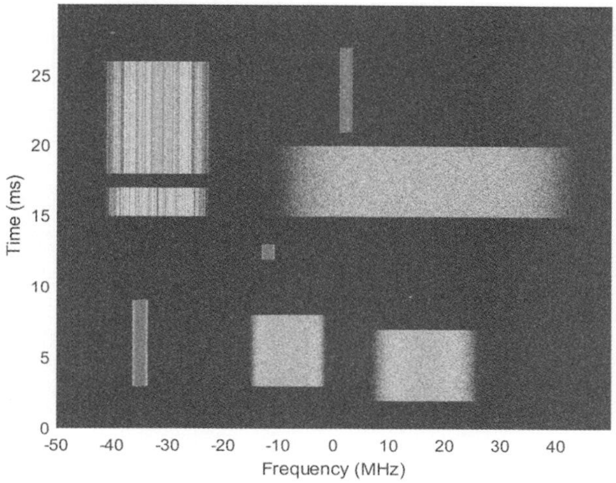

FIGURE 7.81

An example spectrogram defining the energy or power at a particular time and frequency. Several different signals with varying time and frequency extents are shown.

correlation analyzer (SSCA) incorporating a channelizer and a convolutional neural network to classify the signal that was detected.

7.10.1 Signal detection and segregation using the channelizer

Creating a view into the time/frequency plane is normally achieved using an STFT and referred to as a spectrogram, as shown in Fig. 7.81. The STFT relies on an FFT to estimate the amount of power at a particular time and frequency. Recall that a channelizer can also provide an estimation of the power at a particular frequency, and when successive estimates are stacked in a similar fashion used to construct the spectrogram, an estimate of the signal power over time and frequency is achieved. Fig. 7.82 is a spectrogram of the same signal used to produce Fig. 7.81 with the STFT replaced by the channelizer. We will refer to the spectrogram created using the channelizer as the channogram. In this scenario, the signals are sufficiently spread out that no clear gain was achieved by using the channelizer. In the examples to come, we will demonstrate the channelizer's ability to resolve signals at finer frequency resolutions than the STFT.

Estimating the power in time and frequency using an STFT requires the designer to choose several parameters:

1. window length,
2. window type,
3. FFT length,
4. window overlap.

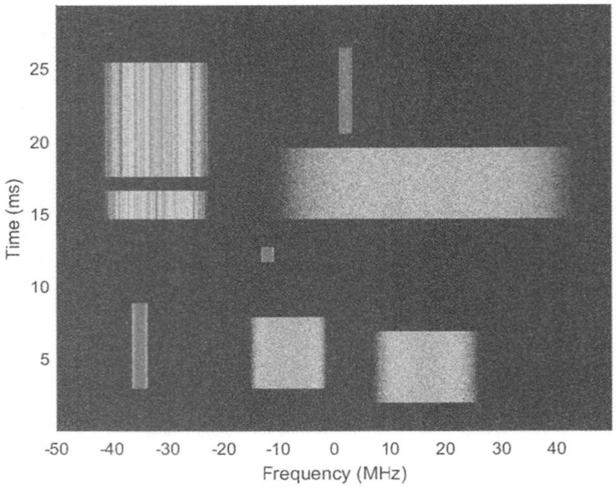

FIGURE 7.82

The channogram which is a spectrogram created using a polyphase channelizer.

The window length sets the fundamental frequency resolution that can be achieved. For example, given a sample rate of $fs = 100$ MHz and a window length of 1024, the fundamental frequency resolution that can be achieved is $\frac{f_s}{N} = 97.7$ kHz. This, however, will not be achievable in practice due to spectral leakage caused by the window type. If no window type is selected, by default you have chosen a box, or square, window. This window will produce the most narrow main lobe with very high side lobes. If the signals we wish to resolve are of equal power, then this is the best window to use. Other windows, such as the Kaiser or Hamming window, will increase the main lobe width while reducing the side lobe height. More information on the effects of windowing can be found in [9]. The FFT length cannot be used to increase the resolution between two independent signals; however, it does provide finer resolution when trying to identify the peak power content of a particular signal. Increasing the FFT length is the same as oversampling the frequency domain. The overlap amount determines the time domain resolution of the STFT and can not be used to improve the frequency resolution. The most important parameter, therefore, is the choice of window length if resolving signals is the highest priority. The spectrogram of Fig. 7.81 was made using a 1024-sample window length, An important point to note when using the STFT is that the window length and the fundamental frequency resolution are coupled. The number of FFT points used can be no smaller than the window length. This is not true for the channelizer, as will be explained next.

The power of the channelizer when using it to estimate power spectra is the ability to increase the effective window length without increasing the total number of output channels you need to compute [10]. For example, if we want a fundamental frequency resolution of 97.7 kHz, then we would choose a channelizer with 1024 output channels. Unlike with the STFT, we may now choose a prototype filter length that reduces the windowing spectral leakage effect down to any level we desire. This is equivalent to saying that we can set the window length to be any multiple of the number of samples selected as input to the channelizer. Fig. 7.83 and Fig. 7.84 show the improvement in frequency we can expect

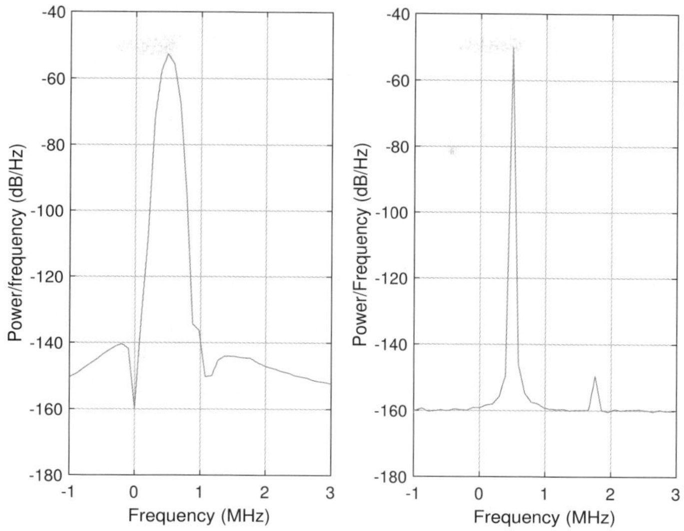

FIGURE 7.83

Power spectral density estimation using the STFT on the left and the channelizer on the right. The channelizer clearly reveals the existence of two sine waves, while the STFT approach does not.

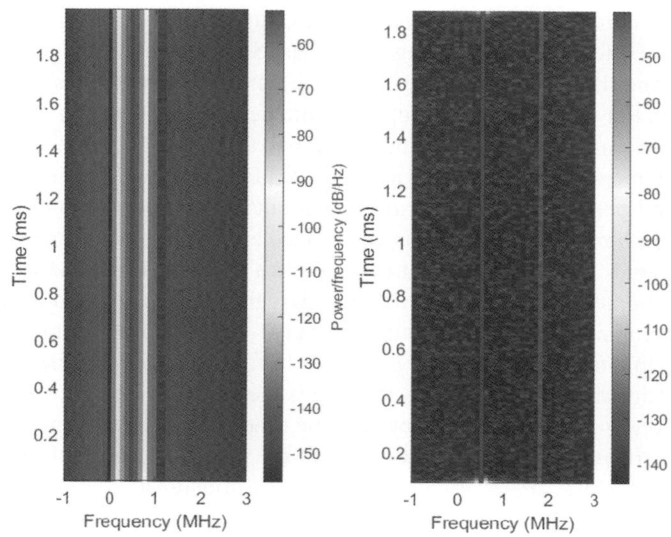

FIGURE 7.84

Spectrogram estimation using the STFT on the left and the channelizer on the right.

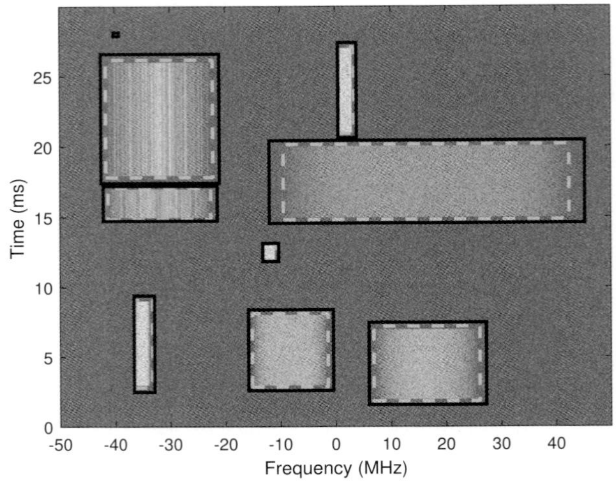

FIGURE 7.85

Example of energy detector output after being applied to the channogram.

given two sine waves that are separated in power by 100 dB. This is an extremely large dynamic range, yet the channelizer has no problem revealing the two tones, while the STFT approach hides the second tone in the skirt of the window.

With our improved ability to distinguish between signals that are very closely spaced, we will now discuss the process used to detect and segregate signals. Signal detection is the process of declaring when a signal exists and then localizing the signal in the time/frequency plane. If prior information is known about the signal or signals to be detected, a suitable detector that uses this information can be designed. If no prior information is known about the signal, then an energy detector can be used. Energy detection can be implemented by convolving a variety of 2D convolution kernels with the channogram shown in Fig. 7.82 and performing some form of thresholding or object/blob detection on the output. This leads to a series of boxes, as shown in Fig. 7.85, once energy detection has completed. The dashed boxes are the true bounds of the energy that rose above a predefined threshold while the solid lines represent a padded region that we will use as the boundary that will be reverted back to IQ time domain samples. The reason for padding is to provide additional context to algorithms that will attempt to classify these signals in later steps. In can be helpful to have access to the entire transition region from the noise floor to the signal region, which the padding is meant to ensure.

Now that the regions we wish to convert back to the time domain have been defined, we may apply a synthesis channelizer to revert these regions back to the time domain. First the vertical boundaries defined by the solid black lines are reverted back to the time domain and then the time domain is sliced according to the horizontal boundaries of the box to complete the process. The real and imaginary time domain samples that result are shown in Fig. 7.86 and Fig. 7.87. The spectrogram for each of these time domain samples is then shown in Fig. 7.88. At this point, each of the signals has been segregated and both the IQ files and the channogram are available for further classification of these signals. In the next subsection we will describe a technique for signal classification that begins with these outputs.

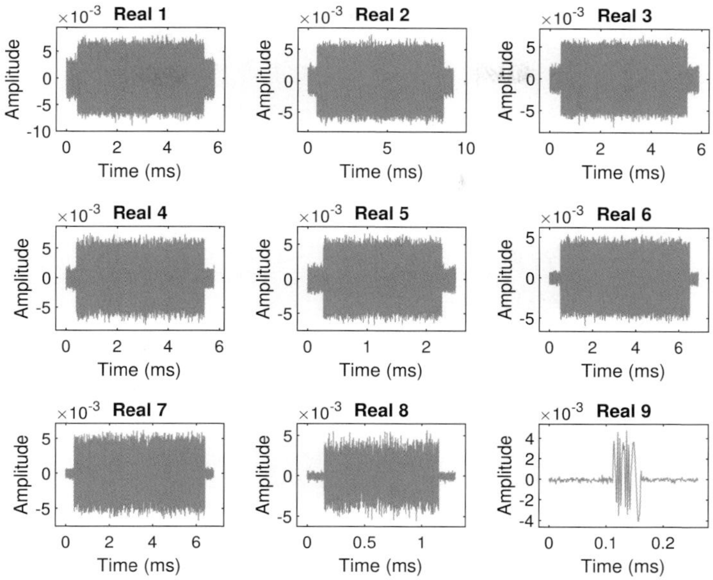

FIGURE 7.86

The real time domain samples for each boxed signal region.

7.10.2 Signal classification

Suppose we wish to determine whether each of the detected and segregated signals is single-carrier, multicarrier, direct sequence spread spectrum (DSSS), or frequency hopping. This type of classification will be referred to as modality classification, where each of the aforementioned choices represents a particular modality type. The approach we present for this classification task will take advantage of the inherent cyclostationarity in communication signals as a distinguishing feature. These features will be passed to a neural network to produce the modality type decision. Neural network approaches to modality classification that use the STFT of the signal as the distinguishing feature are significantly outperformed by neural networks that use cyclostationary features such as the spectral correlation density (SCD).

To define the SCD, we return to one of the most important quantities of a random process, $x(t)$, and the autocorrelation function $\mathcal{R}(t, \tau)$:

$$\mathcal{R}_x(t, \tau) = \mathbb{E}(x(t+\tau)x^*(t)). \tag{7.38}$$

For communication signals, this function is generally almost periodic. Functions that are almost periodic are a generalization of periodic functions in the sense that the frequencies in a Fourier decomposition of such a function are not harmonics of a fundamental tone. These frequencies can range over any

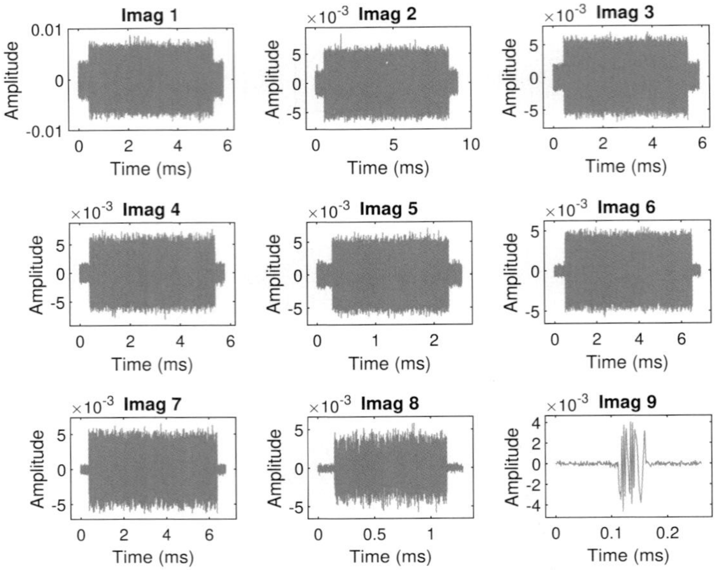

FIGURE 7.87

The imaginary time domain samples for each boxed signal region.

(countable) set. Accordingly, we have the decomposition

$$\mathcal{R}_x(t,\tau) = \sum_{\alpha} R_x^{\alpha}(\tau) e^{2i\pi\alpha t}\,. \tag{7.39}$$

The frequencies in the set α are called the cycle frequencies of the signal $x(t)$. The SCD is then the Fourier transform of the function $\mathcal{R}_x^{\alpha}(\tau)$ with respect to the delay τ,

$$\mathrm{SCD}(f,\alpha) = \int R_x^{\alpha}(\tau) e^{-2i\pi f\tau} d\tau\,. \tag{7.40}$$

We can form a cyclostationary feature function (CFF) using the SCD by computing

$$F(\alpha) = \max_{f}\, \mathrm{SCD}(f,\alpha)\,, \tag{7.41}$$

which is useful when attempting to characterize the cyclic frequencies of unknown signals. An example CFF plot using (7.41) is shown in Figs. 7.89 and 7.90. It is important to note that these features are less sensitive to noise, and are therefore very helpful in low-SNR signal classification tasks.

We now note the results in Fig. 7.91. This figure shows the accuracy of classification of signals versus their Es/No. To generate the plot, a number of signals of three classes, viz., bluetooth, ofdm wifi, and DSSS, were synthetically generated at various Es/No levels. These signals were then fed to

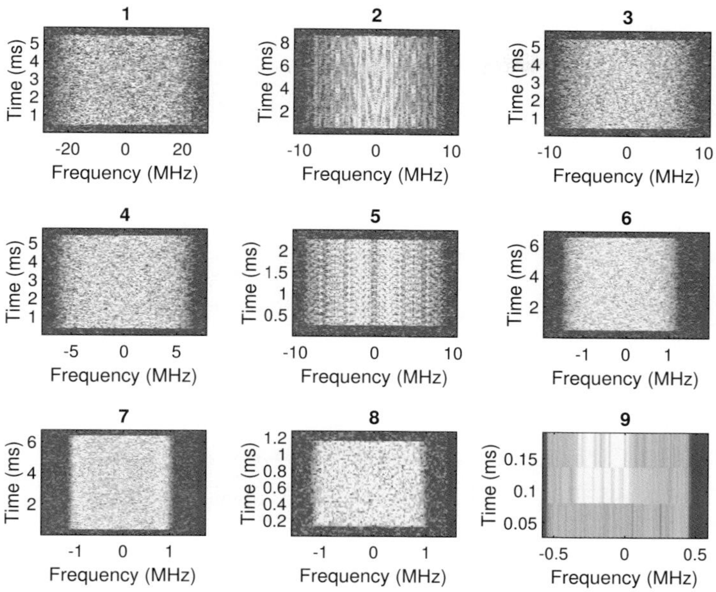

FIGURE 7.88

The spectrogram for each boxed region after segregating all the signals.

two different neural networks; one of these was trained using STFT features and the other was trained using SCD features of the input signals.

Any machine learning algorithm divides the input data it is given into three sets: training, validation, and testing. The accuracy of the output of the neural network on the testing set is recorded. The overall accuracy at each Es/No level for each signal label is then plotted in the figure. One notes that at every Es/No level, the network trained using the SCD features outperforms the network trained using the STFT features.

While (7.40) defines the SCD, it is not an efficient method for estimating the SCD from collected IQ samples. Both a reduction in computational complexity and an improvement in cycle frequency resolution can be achieved by using an SCD estimator known as the SSCA. The SSCA is the fastest known algorithm for estimating the SCD at a large number of points in the (f, α)-plane and will be used for feature extraction.

The SSCA algorithm is shown in the flow diagram depicted in Fig. 7.92. The input to the algorithm is raw IQ samples that are subsequently channelized using a windowed FFT-based channelizer. The polyphase channelizer could be incorporated into the SSCA as well. Each channel then contains a portion of the input spectrum centered at the Fourier frequencies of the first FFT. The contents of each channel are multiplied by the input signal once more, and the result is sent through another FFT. The number of channels (N_p) is chosen to be much smaller than the length of each channel (N) for reasons of statistical reliability.

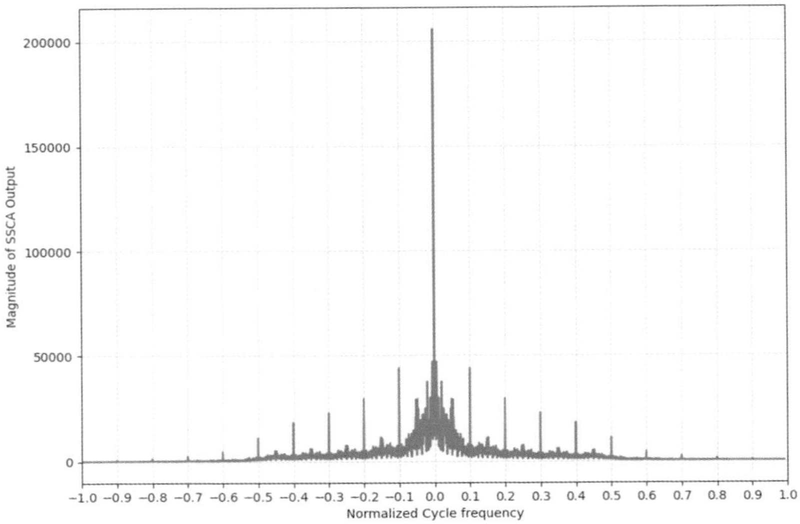

FIGURE 7.89

The cyclic features computed by the SSCA for a Type-1 DSSS-BPSK signal, where the same PN sequence is used to spread each data symbol. We can see sharp cyclic features at the chip rate of the signal.

We will now describe the neural network that was used to implement the signal modality classifier. As depicted in Fig. 7.93, the network consists of an input layer of dimensions 256×32. This means that the parameters of the SSCA algorithm are chosen as $N = 256$ and $N_p = 32$. The input then feeds into a convolutional layer followed by a pooling layer and another convolutional layer. A pooling layer takes a matrix and performs a pooling operation, which means that it reduces the dimension of the input by computing the maximum of subblocks of a given size. A convolutional layer computes the convolution of the input matrix with a given kernel that is slid along the matrix. The output of the convolutional and pooling layers is fed into fully connected layers. These layers compute the ReLU activation function on their input after applying proper weights and biases. Thus, if the input to the first fully connected layer is the matrix A, then this layer computes the output B for each output neuron such that

$$B = \text{ReLU}(W \cdot A + b)\,, \tag{7.42}$$

where the function ReLU is given by

$$\text{ReLU}(y) = \max(y, 0)\,. \tag{7.43}$$

The ReLU function has been shown to perform better for learning algorithms than sigmoid activation functions. The final layer in Fig. 7.93 computes probabilities of what it thinks the modality of a given test signal instance is, and its prediction for this signal would be the label with the highest computed probability. We note that the dimensions of these layers are hyperparameters whose choice was made

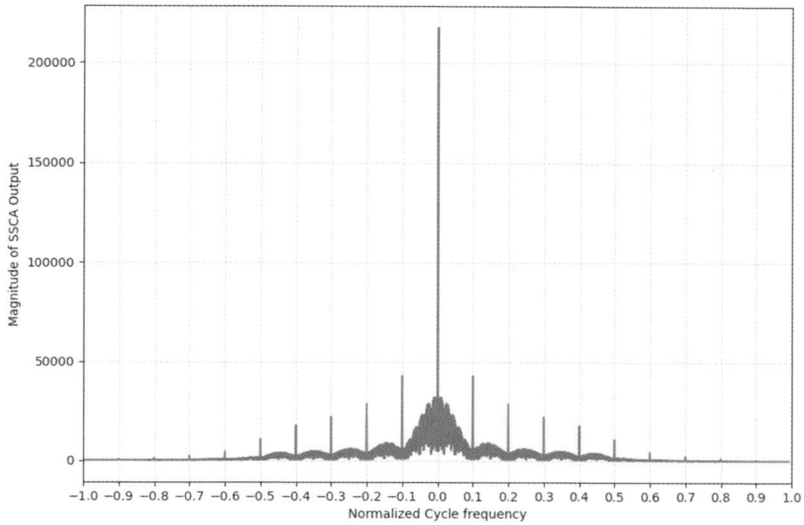

FIGURE 7.90

The cyclic features computed by the SSCA for a Type-2 DSSS-BPSK signal. Unlike the Type-1 DSSS, each data symbol uses a unique PN sequence for spreading, making it ideal for covert communications. Even in this case, the cyclostationarity is preserved, and we see sharp cyclic features at the chip rate.

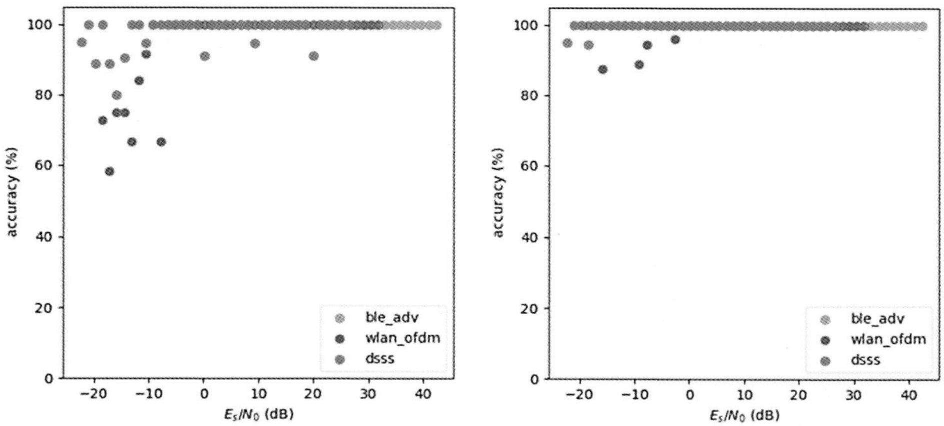

FIGURE 7.91

Comparison of STFT as input features to a neural network (left) against SCD (right).

through trial and error with the additional requirement that the whole architecture be as computationally lightweight as possible.

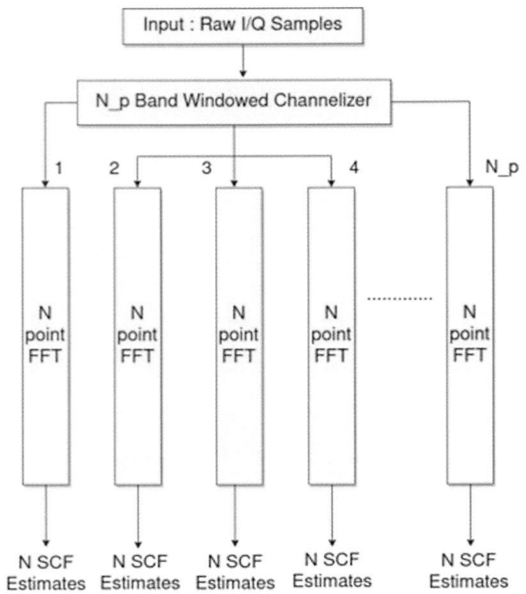

FIGURE 7.92

The SSCA algorithm.

The training algorithm proceeds by attempting to minimize the categorical cross-entropy I, which is defined as

$$I = -\sum_i y_i \log \hat{y}_i. \tag{7.44}$$

The sum is over all the signals in the training set, y_i is the true label of the ith signal, and $\hat{y}_i$ is the prediction which is a function of all the weights and biases introduced above. The Adam algorithm is used to find the optimal weights and biases in each layer, with a learning rate of 0.001, momenta of 0.9 and 0.999, and an epsilon parameter of 10^{-8}.

Finally, we comment on the interpretation of the output of the neural network, which is depicted in Fig. 7.94. The confusion matrix, as it is called, shows the fraction of signals of a given label that have been classified as being signals of the same label and that have been misclassified as signals of a different label. The confusion matrix of an ideal classifier should be the identity matrix. The output was constructed, as noted in Fig. 7.93, using synthetic overt signals at various SNR levels as well as real-time emanation data collected from a desktop monitor.

Acknowledgment

This section is based upon work supported in part by the Office of the Director of National Intelligence (ODNI), Intelligence Advanced Research Projects Activity (IARPA), via [2021-2106240007]. The views and conclusions contained herein are those of the authors and should not be interpreted as

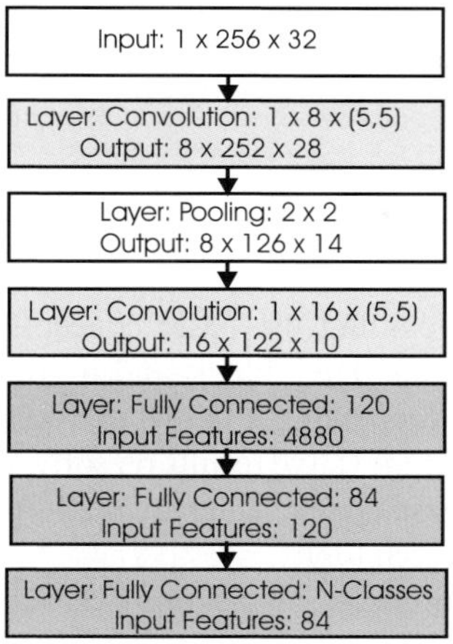

FIGURE 7.93

The neural network that is used in the modality classification.

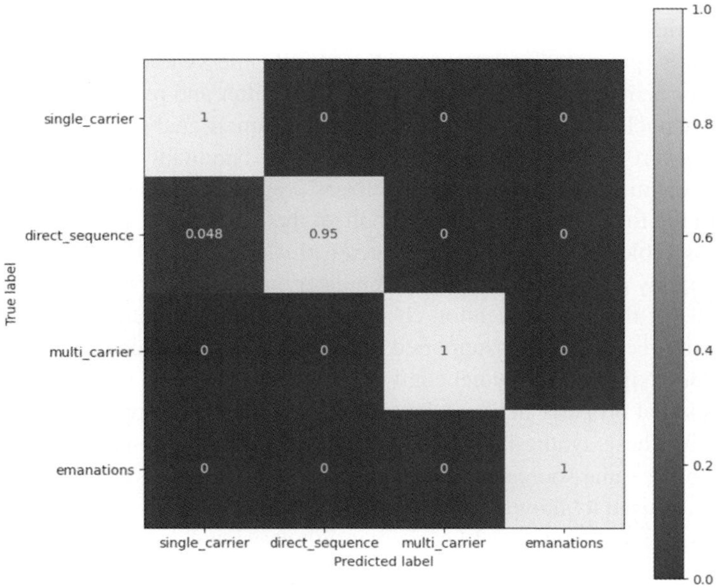

FIGURE 7.94

The output of the neural network in Fig. 7.93. This output was generated with synthetically generated overt signals and emanations captured from a desktop monitor.

necessarily representing the official policies, either expressed or implied, of ODNI, IARPA, or the US Government. The US Government is authorized to reproduce and distribute reprints for governmental purposes notwithstanding any copyright annotation therein.

7.11 Closing comments

This chapter has the intention of providing the basic concepts of multirate signal processing and polyphase filter banks which, when an IDFT block and a commutator are applied, are known as polyphase channelizer because of the processing task that they perform on the input signals. This chapter also has the intention of presenting innovative results on software and cognitive radio designs. While the polyphase filter banks and polyphase channelizers are well-known topics in the DSP area, the software radio design is still an open research topic.

The document started with preliminaries, in Section 7.2, on the resampling process of a digital signal as opposed to the sampling process of an analog signal. In Section 7.3 an introduction on digital filters has been provided and the differences with analog filters have been explained. In Section 7.4 an introduction on the window method for digital filter design has been provided. Multirate filters have been defined and explained in Section 7.5, while the polyphase decomposition of a prototype filter has been introduced in Section 7.6 along with the standard upsampler and downsampler polyphase channelizers. In Section 7.7 the modifications of the standard polyphase channelizer that allow us to change the output sampling rate have been presented. These engines are the basic components of the proposed architectures (presented in Sections 7.8 and 7.9): the synthesis and analysis channelizers, which are suitable for being used as software-defined transmitter and receiver, respectively.

The synthesis channelizer, in fact, when supported by small analysis channelizers, is able to simultaneously up convert multiple signals having arbitrary bandwidths to randomly located center frequencies. In this engine small analysis channelizers preprocess wider-bandwidth signals into segments acceptable to the following synthesis channelizer that upsamples, translates to the proper center frequency, and reassembles them. A channel selector block, connected with a channel configuration block, is used to properly deliver the analysis channelizer outputs to the synthesis channelizer.

On the other side of the communication chain, the analysis channelizer, which is embedded in a software-defined radio receiver, when supported by small synthesis channelizers, is able to simultaneously demodulate these signals. A channel configuration block, inserted in the receiver, communicates with a selector block that properly delivers the analysis channelizer outputs to the synthesizer up converters that follow it. These synthesizers upsample, translate, and reassemble the spectral fragments when they belong to the same source signal. Nyquist prototype low-pass filters, which are perfect reconstruction filters, are used to allow the reconstruction of the signal fragments without energy losses.

Complex frequency rotators are used for compensating residual frequency offsets and arbitrary interpolators provide us exactly two samples per symbol required for the following processing tasks of the receiver.

Theoretical reasoning and simulation results are presented, for both the receiver and the transmitter, in order to demonstrate their correct functionality.

Note that because in both of the proposed DUC and DDC structures, the IFFT sizes of the small synthesis and analysis channelizers are chosen in order to give us at least two samples per symbol for every processed bandwidth, they result to be very efficient, from a computational point of view, when the received signal is composed of spectra with widely varying bandwidths.

Slightly different versions of channelizers can be used based on different input signals and/or to add functionalities to the proposed architectures. Channelizers are, in fact, highly efficient and very flexible structures. Among their most common applications we find spectral analysis, modem synchronization (phase, frequency, and time), and channel equalization.

Glossary

- Filter calculation procedure that transforms the input signals into others
- Digital filter: filter that operates on digital signals
- Multirate filter: digital filter that contains a mechanism to increase or decrease the sampling rate while processing input signals
- Polyphase filter: digital filter partitioned following Eq. (7.3)
- Polyphase channelizer: flexible digital device which can arbitrarily change the sample rate and the bandwidth of the input signal and can also select randomly located Nyquist zones
- Software radio: radio device in which the digitization of the signal occurs before the intermediate frequency translation
- Cognitive radio: software radio with the capability to adapt its operating parameters according to the interactions with the surrounding radio environment

References

[1] M. Bellanger, G. Bonnerot, M. Coudreuse, Digital filtering by polyphase network: application to sample-rate alteration and filter bank, IEEE Transactions on Acoustics, Speech, and Signal Processing 24 (1976) 109–114.

[2] M. Bellanger, J. Daguet, TDM-FDM transmultiplexer: digital polyphase and FFT, IEEE Transactions on Communications 22 (1974) 1199–1205.

[3] R. Bagheri, A. Mirzaei, M.E. Heidari, Software-defined radio receiver: dream to reality, IEEE Communications Magazine 44 (2006).

[4] X. Chen, E. Venosa, F. Harris, Polyphase synthesis filter bank up-converts unequal channel bandwidths with arbitrary center frequencies-design II, in: SDR 2010, Washington, DC, December 2010.

[5] F.J. Harris, On detecting white space spectra for spectral scavenging in cognitive radios, Wireless Personal Communications: Special Issue on Cognitive Radio Technologies (Springer) 45 (3) (May 2008) 325–342.

[6] F. Harris, C. Dick, X. Chen, E. Venosa, M-path channelizer with arbitrary center frequency assignments, in: WPMC 2010, 2010.

[7] F. Harris, C. Dick, X. Chen, E. Venosa, Wideband 160 channel polyphase filter bank cable TV channelizer, IET Signal Processing (2010).

[8] F.J. Harris, Multirate Signal Processing for Communication Systems, Prentice Hall, Upper Saddle River, New Jersey, 2004.

[9] Fredric J. Harris, On the use of windows for harmonic analysis with the discrete Fourier transform, Proceedings of the IEEE 66 (1) (1978) 51–83.

[10] Fred Harris, On the relationship between multirate polyphase FIR filters and windowed, overlapped, FFT processing, in: Twenty-Third Asilomar Conference on Signals, Systems and Computers, 1989, vol. 1, IEEE Computer Society, 1989, pp. 485–486.

[11] F.J. Harris, C. Dick, M. Rice, Digital receivers and transmitters using polyphase filter banks for wireless communications, IEEE Transactions on Microwave Theory and Techniques, Special Issue of Microwave Theory and Techniques, MTT 51 (4) (April 2003) 1395–1412.

[12] F. Harris, W. Lowdermilk, Software defined radio, IEEE Instrumentation & Measurement Magazine (2010).

[13] E.I. Jury, F.J. Mullin, The analysis of sampled data control system with a periodically time varying sampling rate, IRE Transactions on Automatic Control AC-4 (1959) 15–21.

[14] E.I. Jury, F.J. Mullin, The analysis of sampled-data control systems with a periodically time-varying sampling rate, IRE Transactions on Automatic Control 4 (1959) 15–20.

[15] G.M. Kranc, Input-output analysis of multirate feedback systems, IRE Transactions on Automatic Control AC-2 (1956) 21–28.

[16] J. Mitola, Cognitive radio architecture evolution, Proceedings of the IEEE 7 (2009).

[17] Jun Ma, G. Li Ye, B.H. Juang, Signal processing in cognitive radio, Proceedings of the IEEE 97 (2009).

[18] M. Ribey, Exploration of transmultiplexers in telecommunication networks, IEEE Transactions on Communications C0M-30 (1982) 14931497.

[19] A. Sahai, S.M. Mishra, R. Tandra, K.A. Woyach, Cognitive radios for spectrum sharing, IEEE Signal Processing Magazine 26 (2009).

[20] R.W. Schafer, L.R. Rabiner, Design of digital filter banks for speech analysis, The Bell System Technical Journal 50 (1971) 3097–3115.

[21] M. Sherman, C. Rodriguez, R. Reddy, IEEE standards supporting cognitive radio and networks, dynamic spectrum access, and coexistence, IEEE Communications Magazine 46 (2008).

[22] R. Tandra, S.M. Mishra, A. Sahai, What is a spectrum hole and what does it take to recognize one?, Proceedings of the IEEE 97 (2009).

[23] Tore Ulversøy, Software defined radio: challenges and opportunities, IEEE Communications Surveys and Tutorials 12 (4) (2010) 531–548.

[24] E. Venosa, X. Chen, F. Harris, Polyphase analysis filter bank down-converts unequal channel bandwidths with arbitrary center frequencies-design II, in: SDR 2010, Washington, DC, December. 2010.

[25] P. Vary, U. Heute, A short-time spectrum analyzer with polyphase-network and DFT, Signal Processing 2 (1980) 55–65.

[26] B. Wang, K.J. Ray Liu, Advances in cognitive radio networks: a survey, IEEE Journal of Selected Topics in Signal Processing 5 (2011).

CHAPTER

Modern transform design for practical audio/image/video coding applications 8

Trac D. Tran
Department of Electrical and Computer Engineering, The Johns Hopkins University, Baltimore, MD, United States

8.1 Introduction

With the recent explosion in popularity of the Internet, wireless communication, and portable computing, the demands for and the interests in digital multimedia (digitized speech, audio, image, video, computer graphics, and their combinations) are growing exponentially. A high-performance filter bank is typically at the heart of every state-of-the-art digital multimedia system whose compression paradigm is illustrated in Fig. 8.1. Current popular international image/video compression standards such as JPEG [1], MPEG2 [2], and H.263 [3] are all based on the 8×8 discrete cosine transform (DCT), an eight-channel eight-tap orthogonal filter bank with fast computational algorithms based on sparse factorizations. The image compression JPEG2000 standard [4] employs the wavelet transform, an iterated two-channel biorthogonal filter bank, as its decorrelation engine. Of extreme importance is the ability to design a filter bank that can fully exploit (i) the statistical properties of a particular signal or class of signals; (ii) the goals of the applications; and (iii) the computational resources available.

Although the roots of signal transformations can be traced back to works in the 19th century by prominent mathematicians such as Laplace and Fourier, the exciting development of modern, practical digital transforms and filter banks only has a few years of history [5–8] as illustrated in the timeline of Fig. 8.2. These new systems provide more effective tools to represent digital signals not only for analysis but also for processing and compression purposes. For multimedia signals, representations by transform coefficients are usually more compact and efficient than the time representations, but are just as informative. Taking advantage of the normally sparse transform coefficient matrix, we can perform a majority of signal processing tasks directly in the transform domain at a lower level of computational complexity.

On the other hand, fast, efficient, and low-cost transforms have played a crucial role in revolutionary advances throughout the history of digital signal processing. The discrete Fourier transform (DFT) is an excellent signal analysis tool. However, its popularity in practical systems was not widespread until the discovery of the fast Fourier transform (FFT) by Cooley and Tukey in 1965 that reduces the complexity level from $\mathcal{O}(n^2)$ to $\mathcal{O}(n \log n)$ [9]. Another example, the 8×8 DCT [10], has a theoretically elegant closed-form expression. It was chosen as the transform-of-choice in most of current image and video coding standards mainly because it possesses various fast algorithms [11] that can reduce the transform complexity from 64 multiplications and 56 additions to as low as 5 multiplications and 29 additions per 8 transform coefficients.

Signal Processing and Machine Learning Theory. https://doi.org/10.1016/B978-0-32-391772-8.00014-4
Copyright © 2024 Elsevier Inc. All rights reserved.

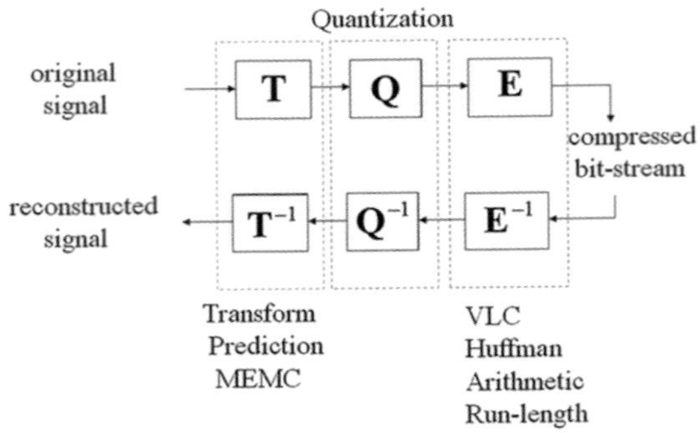

FIGURE 8.1

Standard compression paradigm in current digital multimedia systems.

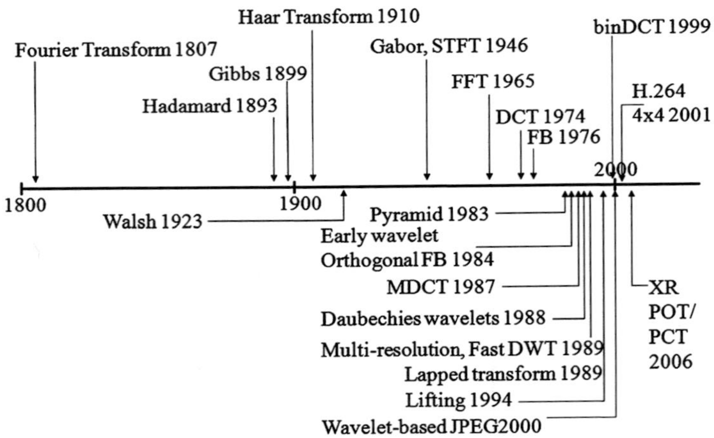

FIGURE 8.2

A historical timeline of significant developments in signal decomposition and transformation.

In this chapter, we review the design and development progress of modern transforms for application in digital image/video coding and processing. These new state-of-the-art transforms provide versatility and flexibility in time-frequency mapping, coding performance, cost of implementation as well as integration into modern digital multimedia processing/communication systems. We assert that the most successful philosophy in transform design is to construct highly complex systems from modular cascades of similar, simple building blocks, each propagating a set of desired transform properties. These novel transforms will be designed to have as many of the following features as possible:

- orthogonal or at least near-orthogonal,
- symmetric/antisymmetric (linear-phase) basis functions,

- good stopband attenuation,
- smoothness of the basis functions for perceptually pleasant reconstruction,
- integer-to-integer mapping capability with exact recovery for a unifying lossless/lossy coding framework,
- fast computation and efficient implementations in both software and hardware: multiplierless property for low-power real-time systems; in-place computation; low memory buffering; VLSI-friendliness with high modularity and regularity in construction; facilitating region-of-interest coding/decoding and computational parallelism.

The organization of the chapter is as follows. In Section 8.2, we offer a review of important background materials, concepts, motivations, and previous related works in transform design. Common design strategy and desirable cost functions are discussed in Section 8.3. Next, Section 8.4 describes the direct scaling-based design method for modern integer-coefficient transforms. Section 8.5 presents a totally different philosophy in transformation design via general parameterization and construction of polyphase matrices based on lifting steps (also known as ladder structures). The section also discusses in detail the subset of solutions that allows the construction and implementation of various dyadic-coefficient transforms purely from shift-and-add operations. The design procedure of spectral factorization of maxflat half-band filters that lead to popular wavelet filter pairs is described in Section 8.6. Advanced design mainly via optimization with a larger number of channels and/or longer filter lengths is covered in Section 8.7. Numerous design examples along with current practical applications will be demonstrated throughout along with discussions on each design's advantages as well as disadvantages and other interesting properties. Finally, we briefly summarize the chapter with a few concluding remarks in Section 8.8.

8.2 Background and fundamentals

We provide in this section the notation, common background, a brief description of the historical development of transforms (especially for compression applications), and a discussion on desirable properties of a state-of-the-art modern transform in multimedia coding and processing.

8.2.1 Notation

Let $\mathcal{R}$, $\mathcal{Q}$, and $\mathcal{Z}$ denote the sets of real numbers, rational numbers, and integers, respectively. Also, let $\mathcal{D}$ denote the set of dyadic rationals, i.e., all rational numbers that can be represented in the form of $\frac{k}{2^m}$ where $k, m \in \mathcal{Z}$. Bold-faced lowercase characters are used to denote vectors while bold-faced uppercase characters are used to denote matrices. Let $\mathbf{A}^T$, $\mathbf{A}^{-1}$, $|\mathbf{A}|$, and a_{ij} denote respectively the transpose, the inverse, the determinant, and the ith jth element of the matrix $\mathbf{A}$. Several special matrices with reserved symbols are: the polyphase matrix of the analysis bank $\mathbf{E}(z)$, the polyphase matrix of the synthesis bank $\mathbf{R}(z)$, the identity matrix $\mathbf{I}$, the reversal or antidiagonal matrix $\mathbf{J}$, the null matrix (or vector) $\mathbf{0}$, the unity vector $\mathbf{1}$, the permutation matrix $\mathbf{P}$, and the diagonal matrix $\mathbf{D}$. The letter M is usually reserved for the number of channels of the filter bank or the size of the transform. Finally, we denote the ℓ_p-norm of an N-point vector $\mathbf{x}$ as $\|\mathbf{x}\|_p = \left(\sum_{i=0}^{N-1} |x_i|^p\right)^{1/p}$, where the ℓ_2-norm $\|\mathbf{x}\|_2 = \sqrt{\mathbf{x}^T\mathbf{x}}$ is of special interest.

8.2.2 Transform fundamentals

A transform is most often thought of as a linear mapping which can be carried out by a simple matrix multiplication. A block transform $\mathbf{T}_A$ of size M is simply an $M \times M$ scalar matrix, mapping the signal vector $\mathbf{x}$ to its corresponding transform coefficients $\mathbf{c}$. Here, the subscript A is used to denote the analysis transform often employed in an encoder to analyze or to decorrelate input data samples. At the decoder, we need the inverse operator $\mathbf{T}_S$ to recover $\mathbf{x}$ back, where the subscript S denotes the synthesis or signal reconstruction operation. The series of mappings are depicted in Fig. 8.3. The operation $\mathbf{Q}$ in the middle of Fig. 8.3 represents approximation (in the form of quantization for instance), processing, coding, communication, or any combination thereof. Obviously, if the coefficients can be recovered losslessly or $\hat{\mathbf{c}} = \mathbf{c}$, we simply require $\mathbf{T}_S$ to be the exact inverse of $\mathbf{T}_A$ to reconstruct the input signal perfectly.

The columns $\boldsymbol{\phi}_i$ of $\mathbf{T}_A^T$ are often called the analysis basis functions while the columns $\hat{\boldsymbol{\phi}}_i$ of $\mathbf{T}_S$ are called the synthesis basis functions. $\mathbf{T}_A$ is said to be an invertible transform or a transform that has perfect reconstruction or biorthogonal when $\mathbf{T}_S\mathbf{T}_A = \mathbf{T}_A\mathbf{T}_S = \mathbf{I}$. In the special case that the synthesis transform has real coefficients and is simply the transpose of the analysis transform $\mathbf{T}_S = \mathbf{T}_A^{-1} = \mathbf{T}_A^T$, the transform $\mathbf{T}_A$ is said to be orthogonal. In this case, we can conceptually think of representing the N-point input signal as

$$\mathbf{x} = c_0\boldsymbol{\phi}_0 + c_1\boldsymbol{\phi}_1 + \cdots + c_{N-1}\boldsymbol{\phi}_{N-1} = \sum_{i=0}^{N-1} c_i\boldsymbol{\phi}_i, \quad \text{where } c_i = \langle \boldsymbol{\phi}_i, \mathbf{x} \rangle. \tag{8.1}$$

However, our goal is to design the transform that yields the sparsest and most compact representation

$$\mathbf{x} = \sum_{i \in \mathcal{S}; |\mathcal{S}| = K \ll N} c_i\boldsymbol{\phi}_i. \tag{8.2}$$

In (8.2), the set of significant coefficients are indexed by the set $\mathcal{S}$ whose cardinality K is much smaller than the signal dimension N.

For 2D signals such as still images or frames in video sequences, we typically employ the separable transformation approach where all rows are transformed and then all columns are transformed or vice

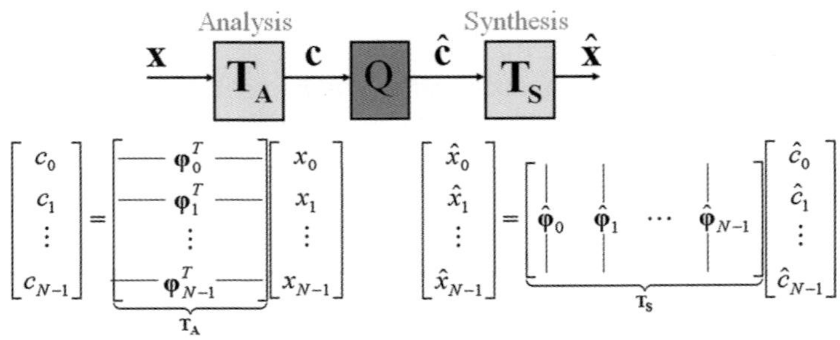

FIGURE 8.3

Signal decomposition and reconstruction as matrix operations.

versa. Mathematically, if we let $\mathbf{X}$ be the image of interest in the form of a matrix, then the transform coefficients $\mathbf{C}$ of $\mathbf{X}$ can be computed as

$$\mathbf{C} = \mathbf{T}_A \mathbf{X} \mathbf{T}_A^T, \tag{8.3}$$

while the inverse synthesis mapping can be computed as

$$\mathbf{X} = \mathbf{T}_S \mathbf{C} \mathbf{T}_S^T. \tag{8.4}$$

Note that the computation order of rows or columns in (8.3) and (8.4) is not critical and the strategy can be straightforwardly extended to signals of any higher dimension.

8.2.3 Optimal orthogonal transform

Obtaining the sparsest representation as in (8.2) above is certainly a signal-dependent task. Let $\mathbf{c}$ be the coefficients of the input signal $\mathbf{x}$, obtained from a certain linear transformation $\mathbf{T}$. Let us further assume that the input signal $\mathbf{x}$ can be modeled as a wide-sense stationary stochastic process. Then, the correlation matrix of the output signal $\mathbf{c}$ is

$$\mathbf{R}_{cc} = E\{\mathbf{T}\mathbf{x}\mathbf{T}^T\} = \mathbf{T}\mathbf{R}_{xx}\mathbf{T}^T.$$

The optimal orthogonal linear transform $\mathbf{T}$ here is the one that can fully decorrelate $\mathbf{x}$. In order to achieve this, $\mathbf{R}_{cc}$ must be a diagonal matrix with the eigenvalues of $\mathbf{R}_{xx}$ being its diagonal entries and each row of $\mathbf{T}$ being an eigenvector of $\mathbf{R}_{xx}$. This optimal transform is often called the Karhunen–Loève transform (KLT) of signal $\mathbf{x}$, also known as principal component analysis (PCA) or the Hotelling transform. Obviously, the KLT depends on the autocorrelation matrix $\mathbf{R}_{xx}$. Thus it is signal-dependent, computationally expensive, and costly to communicate to the decoder if the signal of interest is not stationary. To find fast algorithms and signal-independent transforms for practical applications, a typical approach is to start with a simplified stationary signal model, try to find the approximation of its KLT, and then develop fast computational algorithms of the resulting approximation. The DCT is developed following this approach [10,11].

8.2.4 Popular transforms in signal processing: DFT, WHT, DCT

The problem of transform design has undergone several important evolutions. In the beginning, transformation was mainly thought of as a continuous-time continuous-amplitude change of basis operation (and often in infinite-dimensional functional space) to solve complicated sets of equations. Discrete-time transforms are often obtained from sampling the continuous-time version. Four popular discrete-time transforms in signal processing are the DFT, the Walsh–Hadamard transform (WHT), the type-II DCT, and the type-IV DCT, shown below in matrix representation format, respectively:

$$[\mathbf{F}] = \frac{1}{\sqrt{M}}\left[W_M^{mn}\right];\ 0 \le m, n \le M-1;\ \ W_M = e^{-j\frac{2\pi}{M}}, \tag{8.5}$$

$$\mathbf{H}_2 = \begin{bmatrix} 1 & 1 \\ 1 & -1 \end{bmatrix};\quad \mathbf{H}_{2^m} = \mathbf{H}_2 \otimes \mathbf{H}_{2^{m-1}} = \begin{bmatrix} \mathbf{H}_{2^{m-1}} & \mathbf{H}_{2^{m-1}} \\ \mathbf{H}_{2^{m-1}} & -\mathbf{H}_{2^{m-1}} \end{bmatrix};\ m \in \mathcal{Z}, m \ge 2, \tag{8.6}$$

$$\left[\mathbf{C}_M^{II}\right] = \sqrt{\frac{2}{M}}\left[K_m \cos\left(\frac{m(n+1/2)\pi}{M}\right)\right];\ 0 \le m, n \le M-1;\ K_i = \begin{cases} \frac{1}{\sqrt{2}} & i = 0, \\ 1 & i > 0, \end{cases} \tag{8.7}$$

$$\left[\mathbf{C}_M^{IV}\right] = \sqrt{\frac{2}{M}}\left[\cos\left(\frac{(m+1/2)(n+1/2)\pi}{M}\right)\right];\ 0 \le m, n \le M-1. \tag{8.8}$$

Out of the four mentioned above, only the Hadamard transform has integer coefficients. Unfortunately, it is too simplistic to offer reasonable coding performance. On the other hand, the DCT offers very good compression performance but it has irrational coefficients. The DFT is not only irrational, but it also has complex-valued coefficients.

8.2.5 The filter bank connection

A typical discrete-time, maximally decimated M-channel filter bank is depicted in Fig. 8.4. At the analysis stage, the input signal $x[n]$ is passed through a bank of M analysis filters $H_i(z)$, $i = 0, 1, \ldots, M-1$, each of which preserves a frequency band. These M filtered signals are then decimated by M to preserve the system's overall sampling rate. The resulting subband signals can be coded, processed, and/or transmitted independently or jointly. At the synthesis stage, the subbands are combined by a set of upsamplers and M synthesis filters $F_i(z)$, $i = 0, 1, \ldots, M-1$, to form the reconstructed signal $\hat{x}[n]$. In the absence of processing errors (e.g., quantization), filter banks that yield the output $\hat{x}[n]$ as a pure delayed version of the input $x[n]$, i.e., $\hat{x}[n] = x[n-n_0]$, are called *perfect reconstruction* filter banks. From a traditional transform perspective, the analysis filter $h_i[n]$ is the ith analysis basis function $\boldsymbol{\phi}_i$. Similarly, the synthesis filter $f_i[n]$ yields the ith synthesis basis function $\hat{\boldsymbol{\phi}}_i$.

The filter bank in Fig. 8.4(a) can also be represented in terms of its polyphase matrices as shown in Fig. 8.4(b). $\mathbf{E}(z)$ is the analysis bank's polyphase matrix and $\mathbf{R}(z)$ is the synthesis bank's polyphase matrix. Note that both $\mathbf{E}(z)$ and $\mathbf{R}(z)$ are $M \times M$ matrices whose elements are polynomials in z [7]. If $\mathbf{E}(z)$ is invertible with a minimum-phase determinant (stable inverse), one can obtain perfect reconstruction by simply choosing $\mathbf{R}(z) = \mathbf{E}^{-1}(z)$. In other words, any choice of $\mathbf{R}(z)$ and $\mathbf{E}(z)$ that satisfies

$$\mathbf{R}(z)\mathbf{E}(z) = \mathbf{E}(z)\mathbf{R}(z) = z^{-l}\mathbf{I}, \quad l \ge 0, \tag{8.9}$$

yields perfect reconstruction. Since filter banks with finite impulse response (FIR) filters are very valuable in practice, we can restrict the determinants of both polyphase matrices to be monomials [6–8] as well:

$$|\mathbf{R}(z)| = z^{-m} \quad \text{and} \quad |\mathbf{E}(z)| = z^{-n} \quad m, n \in \mathcal{Z}. \tag{8.10}$$

A popular choice of $\mathbf{R}(z)$ yielding *paraunitary* or *orthogonal* systems is

$$\mathbf{R}(z) = z^{-K}\mathbf{E}^T(z^{-1}), \tag{8.11}$$

where K is the order of a properly designed $\mathbf{E}(z)$. In the case where $\mathbf{E}(z)$ may not be paraunitary but (8.9) still holds, the filter bank is said to be *biorthogonal*. Now, the design of a complicated M-band FIR perfect reconstruction filter bank is reduced to choosing appropriate polynomial matrices $\mathbf{E}(z)$ and $\mathbf{R}(z)$ such that (8.9) and (8.10) are satisfied.

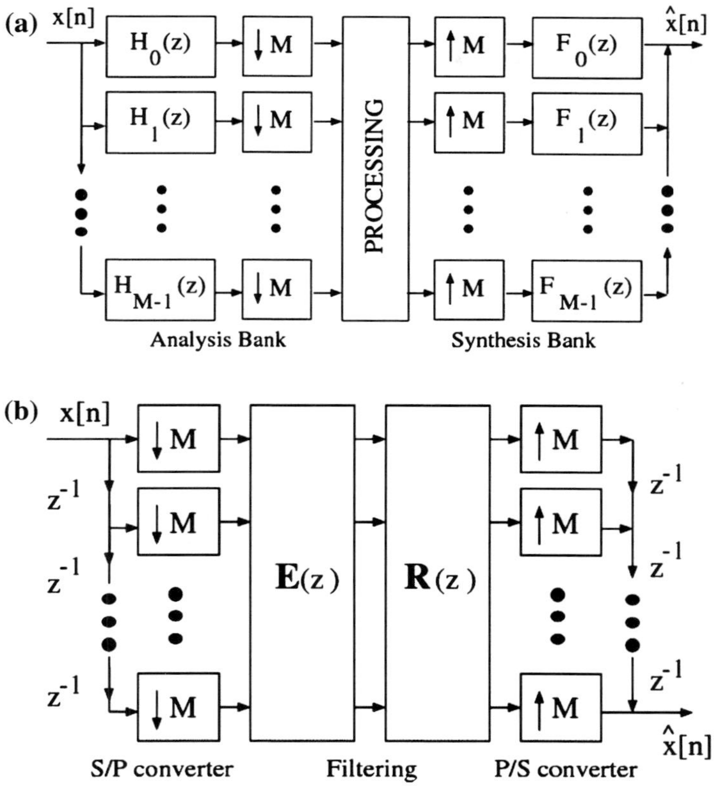

FIGURE 8.4

M-channel filter bank. (a) Conventional representation. (b) Polyphase representation.

When the filter length equals the number of channels M, the polyphase matrices $\mathbf{E}(z)$ and $\mathbf{R}(z)$ become scalar matrices of zero order and they become equivalent to the conventional transforms above, i.e., $\mathbf{E}(z) = \mathbf{T}_A$ and $\mathbf{R}(z) = \mathbf{T}_S$. When the filter length is higher than the number of channels, we have to deal with matrices whose entries are polynomials in z. In the wavelet transform, $M = 2$ and we almost never have a scalar polyphase matrix, except the Haar wavelet. When M equals the support of the input signal, we have a global transform and maximum frequency resolution becomes achievable. On the other hand, when M is much smaller than the input size (for example, most imaging applications have a transform size of 16×16 or less), we have a block transform coding framework where the downsamplers and the delay chain on the left of Fig. 8.4(b) serve as serial-to-parallel blocking mechanism whereas the upsamplers and the delay (or advance) chain on the right of Fig. 8.4(b) implement the parallel-to-serial unblocking mechanism. In this case, the global transform matrices $\mathbf{T}_A$ and $\mathbf{T}_S$ have block-diagonal structure and maximum time/space resolution becomes achievable.

8.2.6 The lapped transform connection

The lapped transform (LT) is defined as a linear transformation that partitions the input signal into small *overlapped* blocks and then processes each block independently. The equivalence between the LT, the filter bank in Section 8.2.5, and the general linear transformation in Section 8.2.2 has been well established in [12].

Consider the M-channel perfect reconstruction filter bank with FIR filters, all of length $L = KM$ for presentation elegance. The polyphase matrix $\mathbf{E}(z)$ as shown in Fig. 8.4(b) is of order $K-1$ and can be expressed as $\mathbf{E}(z) = \sum_{i=0}^{K-1} \mathbf{E}_i z^{-1}$. Similarly, the synthesis polyphase matrix $\mathbf{R}(z)$ can be represented as $\mathbf{R}(z) = \sum_{i=0}^{K-1} \mathbf{R}_i z^{-1}$. With $\mathbf{J}$ as the time-reversal matrix, if we define

$$\mathbf{E} = [\mathbf{E}_{K-1}\mathbf{J} \ \cdots \ \mathbf{E}_1\mathbf{J} \ \mathbf{E}_0\mathbf{J}] \quad \text{and} \quad \mathbf{R} = \begin{bmatrix} \mathbf{J}\mathbf{R}_0 \\ \mathbf{J}\mathbf{R}_1 \\ \vdots \\ \mathbf{J}\mathbf{R}_{K-1} \end{bmatrix},$$

then the global transform matrices can be shown to take the following forms:

$$\mathbf{T}_A = \begin{bmatrix} \ddots & & & & \mathbf{0} \\ & \mathbf{E} & & & \\ & & \mathbf{E} & & \\ & & & \mathbf{E} & \\ \mathbf{0} & & & & \ddots \end{bmatrix}$$

$$= \begin{bmatrix} \ddots & \ddots & & \ddots & & & & \mathbf{0} \\ & \mathbf{E}_{K-1}\mathbf{J} & \mathbf{E}_{K-2}\mathbf{J} & \cdots & \mathbf{E}_0\mathbf{J} & & & \\ & & \mathbf{E}_{K-1}\mathbf{J} & \mathbf{E}_{K-2}\mathbf{J} & \cdots & \mathbf{E}_0\mathbf{J} & & \\ & & & \mathbf{E}_{K-1}\mathbf{J} & \mathbf{E}_{K-2}\mathbf{J} & \cdots & \mathbf{E}_0\mathbf{J} & \\ \mathbf{0} & & & & \ddots & \ddots & & \ddots \end{bmatrix} \tag{8.12}$$

and

$$\mathbf{T}_S = \begin{bmatrix} \ddots & & & & \mathbf{0} \\ & \mathbf{R} & & & \\ & & \mathbf{R} & & \\ & & & \mathbf{R} & \\ \mathbf{0} & & & & \ddots \end{bmatrix}$$

$$
= \begin{bmatrix}
\ddots & & & & \mathbf{0} \\
\ddots & \mathbf{JR}_0 & & & \\
 & \mathbf{JR}_1 & \mathbf{JR}_0 & & \\
 & \vdots & \mathbf{JR}_1 & \mathbf{JR}_0 & \\
 & \mathbf{JR}_{K-1} & \vdots & \mathbf{JR}_1 & \\
 & & \mathbf{JR}_{K-1} & \vdots & \\
 & & & \mathbf{JR}_{K-1} & \ddots \\
\mathbf{0} & & & & \ddots
\end{bmatrix}. \tag{8.13}
$$

Although the blocking windows overlap, the total number of produced transform coefficients is the same as the total number of input samples. Hence, the LT does not introduce any redundancy. In traditional block transform processing, the number of input samples from each block is the same as the number of basis functions, i.e., $L = M$. In this case, the transform matrix $\mathbf{P}$ becomes square $(M \times M)$ and there is no overlapping between neighboring blocks. However, as the basis vectors of block transforms do not overlap, there may be discontinuities along the boundary regions of the blocks when heavy quantization is involved. Different approximations of those boundary regions in each side of the border may cause an artificial "edge" in between blocks. This is the so-called *blocking* effect. Allowing overlapping block boundary leads to coding improvement over nonoverlapped transforms such as the DCT on two counts: (i) from the analysis viewpoint, the overlapping basis takes into account interblock correlation, hence, provides better energy compaction; (ii) from the synthesis viewpoint, the overlapping basis either eliminates or significantly reduces blocking discontinuities.

However, the global transform matrices $\mathbf{T}_A$ and $\mathbf{T}_S$ remain to be very sparse, leading to fast computation of transform coefficients $\mathbf{c}$. The philosophy is that a few borrowed samples are enough to eliminate blocking artifacts and to improve coding performance.

8.3 Design strategy

All of the transforms presented in the previous sections are designed to have high practical value. They all have perfect reconstruction. Some of them even have real and symmetric basis functions. However, for the transforms to achieve high coding performance, several other properties are also needed. Transforms can be obtained using unconstrained nonlinear optimization where some of the popular cost criteria are: coding gain C_{CG}, DC leakage C_{DC}, attenuation around mirror frequencies C_M, and stopband attenuation in both analysis and synthesis bank – C_A and C_S. In the particular field of image compression, all of these criteria are well-known desired properties in yielding the best reconstructed image quality [6]. The cost function in the optimization process can be a weighted linear combination of these measures as follows:

$$
C_{\text{Overall}} = \alpha_1 C_{\text{CG}} + \alpha_2 C_{\text{DC}} + \alpha_3 C_M + \alpha_4 C_A + \alpha_5 C_S, \tag{8.14}
$$

where each of the individual cost functions will be further elaborated in the next section.

8.3.1 Desirable transform properties

- *Symmetry:* $\mathbf{J}\boldsymbol{\phi}_i = \pm\boldsymbol{\phi}_i$. Linear-phase basis functions are critical in image/video processing applications (but not so much in audio processing applications).
- *Orthogonality:* Orthogonality offers mathematical elegance and the norm preservation property. However, practical codecs often employ biorthogonal transforms (e.g., the 9/7-tap wavelet in JPEG2000). In other words, the relaxation of the tight orthogonal constraint can sometimes provide a much needed flexibility in transform design.
- *Energy compaction:* Energy compaction, also known as coding gain, is defined as

$$C_{\mathrm{CG}} = 10 \times \log_{10}\left(\frac{\sigma^2_{\text{input}}}{\left[\prod_{i=1}^{M}\sigma_i^2\|\hat{\boldsymbol{\phi}}_i\|^2\right]^{1/M}}\right), \tag{8.15}$$

where σ^2_{input} is the variance of the input signal; σ_i^2 is the variance of the ith subband (generated from the ith analysis basis function $\boldsymbol{\phi}_i$); and $\|\hat{\boldsymbol{\phi}}_i\|^2$ is the norm-squared of the ith synthesis basis function. Note that for orthogonal transforms, this term disappears and Eq. (8.15) reduces down to the ratio (in dB) of the arithmetic mean over the geometric mean of the subband variances. The signal model is the commonly used $AR(1)$ process with intersample autocorrelation coefficient $\rho = 0.95$. As mentioned above, the coding gain can be thought of as an approximate measure of the transform's energy compaction capability. Among the listed criteria, higher coding gain correlates most consistently with higher objective performance (measured in Mean Squared Error (MSE) or Peak Sinal-to-Noise Ratio (PSNR)) in coding applications. Transforms with higher coding gain compact more signal energy into a lower number of coefficients, leading to a sparser, more compact representation and more efficient entropy coding.
- *Stopband attenuation:* Stopband attenuation is defined as the summation of the stopband energy of all analysis and/or synthesis basis functions. It can be formulated as

$$\text{analysis stopband attenuation} \quad C_A = \sum_{i=0}^{M-1}\int_{\omega\in\Omega_i}|\boldsymbol{\phi}_i(e^{j\omega})|^2 d\omega, \tag{8.16}$$

$$\text{synthesis stopband attenuation} \quad C_S = \sum_{i=0}^{M-1}\int_{\omega\in\Omega_i}|\hat{\boldsymbol{\phi}}_i(e^{j\omega})|^2 d\omega, \tag{8.17}$$

where Ω_i denotes the stopband of the ith basis function.
Stopband attenuation is a classical performance criterion in filter design. On the analysis side, the stopband attenuation cost helps in improving the signal decorrelation and decreasing the amount of aliasing. In meaningful images, we know a priori that most of the energy is concentrated in the low-frequency region. Hence, besides the few basis functions that are designated to cover this low-frequency range, high stopband attenuation for the remainder in this part of the frequency spectrum becomes extremely desirable. On the contrary, at the synthesis side, the synthesis basis functions (or filters) covering low-frequency bands need to have high stopband attenuation near and/or at $\omega = \pi$ to enhance their smoothness.

- *Zero DC leakage:* The DC leakage cost function measures the amount of DC energy that leaks out to the bandpass and high-pass subbands. The main idea is to concentrate all signal energy at DC into the DC coefficients. This proves to be advantageous in both signal decorrelation and the prevention of discontinuities in the reconstructed signals. Low DC leakage can prevent the annoying checkerboard artifact that usually occurs when high-frequency bands are severely quantized. Zero DC leakage is equivalent to attaining one vanishing moment in wavelet design (a necessary condition for the convergence of the wavelet construction): $\langle \boldsymbol{\phi}_i, \mathbf{1} \rangle = 0;\ \forall i \neq 0$. Depending on whether the focus is on the transform coefficients or the filter bank coefficients, it can be stated as

$$C_{\mathrm{DC}} = \sum_{i \neq 0} \langle \boldsymbol{\phi}_i, \mathbf{1} \rangle \quad \text{or} \quad C_{\mathrm{DC}} = \sum_{i=1}^{M-1} \sum_{n=0}^{L-1} h_i[n]. \tag{8.18}$$

- *Attenuation at mirror frequencies:* The mirror frequency cost function is a generalization of C_{DC} above. The focus is now on all aliasing frequencies $\omega_m = \frac{2m\pi}{M};\ m \in \mathcal{Z},\ 1 \leq m \leq \frac{M}{2}$, instead of just at DC. Frequency attenuation at mirror frequencies is very important in the further reduction of blocking artifacts: the filter responses (except the only high-pass filter in the system) should vanish at these mirror frequencies as well. The corresponding cost function is

$$C_M = \sum_{i=0}^{M-2} |H_i(e^{j\omega_m})|^2; \quad \omega_m = \frac{2m\pi}{M}, \quad m \in \mathcal{Z},\ 1 \leq m \leq \frac{M}{2}. \tag{8.19}$$

  Low DC leakage and high attenuation near the mirror frequencies are not as essential to the coder's objective performance as coding gain. However, they do improve the visual quality of the reconstructed signal significantly.
- *Integer-to-integer mapping* with exact invertibility and tight dynamic range extension: These two properties are critical for lossless coding applications.
- *Practical considerations:* Integer or dyadic-rational coefficients; hardware- and software-friendly; low latency; small bus width; parallelizability; structural regularity; low memory buffering; and in-place computation.

8.4 Approximation approach via direct scaling

The simplest approach to design an integer transform is to approximate a well-known infinite precision (represented using floating-point precision) transform by a close finite precision (fixed-point) version. For instance, an integer-coefficient transform $\mathbf{T}_M$ can be obtained from a well-known irrational-coefficient transform $\mathbf{C}_M$ as follows:

$$\mathbf{T}_M(\alpha, \mathbf{C}_M) = \mathrm{round}(\alpha \mathbf{C}_M), \tag{8.20}$$

where α is a parameter to control the precision of the approximation. Obviously, the larger α is, the closer the approximation gets along with a higher implementation cost and a larger output dynamic range.

8.4.1 H.264 4 x 4 transform design

The most well-known integer transform successfully designed from this scaling approach is the four-point bit-exact DCT approximation in the video coding international standard H.264 or MPEG-4 Part 10 [13,14]. In this particular example, the target transform is the four-point type-II DCT in (8.7), which can be expressed as follows:

$$\mathbf{C}_4 = \begin{bmatrix} 1 & 1 & 1 & 1 \\ c & s & -s & -c \\ 1 & -1 & -1 & 1 \\ s & -c & c & -s \end{bmatrix}; \quad c = \sqrt{2}\cos\frac{\pi}{8}, \quad s = \sqrt{2}\sin\frac{\pi}{8}. \tag{8.21}$$

In the earlier design, the approximation $\mathbf{T}_4$ transform is designed with the scaling parameter $\alpha = 26$ yielding the following integer version:

$$\mathbf{T}_4(26) = \text{round}\,(26\mathbf{C}_4) = \begin{bmatrix} 13 & 13 & 13 & 13 \\ 17 & 7 & -7 & -17 \\ 13 & -13 & -13 & 13 \\ 7 & -17 & 17 & -7 \end{bmatrix}, \tag{8.22}$$

which happens to retain symmetry, orthogonality, and the coding gain of the DCT. The only drawback here is the extended dynamic range which requires the transformation stage to be implemented on a 32-bit architecture. The standardization committee later selected the following integer transform, proposed independently by Microsoft and Nokia [14]:

$$\mathbf{T}_4\left(\frac{5}{2}\right) = \text{round}\left(\frac{5}{2}\mathbf{C}_4\right) = \begin{bmatrix} 1 & 1 & 1 & 1 \\ 2 & 1 & -1 & -2 \\ 1 & -1 & -1 & 1 \\ 1 & -2 & 2 & -1 \end{bmatrix}, \tag{8.23}$$

whose inverse turns out to be

$$\mathbf{T}_4^{-1}\left(\frac{5}{2}\right) = \frac{1}{20}\begin{bmatrix} 5 & 4 & 5 & 2 \\ 5 & 2 & -5 & -4 \\ 5 & -2 & -5 & 4 \\ 5 & -4 & 5 & -2 \end{bmatrix}. \tag{8.24}$$

The current transform in H.264 is designed with the scaling parameter $\alpha = 2.5$ (actually, any parameter in the range $2.3 \le \alpha < 3$ also yields exactly the same result). Its coding gain as defined in (8.15) is only 0.02 dB lower than the original DCs. Similar to the earlier design, this transform retains symmetry as well as orthogonality, and offers an efficient multiplierless implementation as depicted in Fig. 8.5. Most importantly, the lower dynamic range allows the entire transformation stage to fit within a 16-bit architecture. It is interesting to note that when the binary shifts of 2 and $\frac{1}{2}$ in the structures of Fig. 8.5 are removed, we are left with the Walsh–Hadamard transform $\mathbf{H}_4$ as defined in (8.6). Despite the important improvements, several drawbacks still exist: (i) it is not an integer-to-integer transform with exact invertibility (its inverse involves a division by 5); and (ii) its dynamic range control is still not tight enough. Hence, H.264 loses the capability of lossy and lossless coding based on the same compression framework.

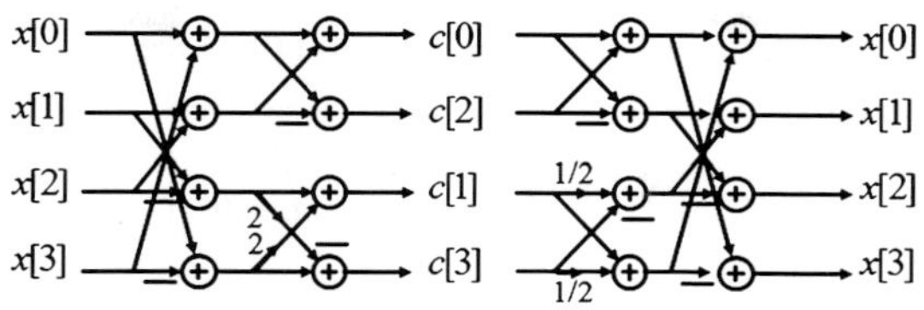

FIGURE 8.5

Fast implementation of the H.264 transform (left) and inverse transform (right). No multiplications are needed, only binary additions and shifts.

8.4.2 Integer DCT design via the principle of dyadic symmetry

The direct scaling approach is actually studied extensively by Cham, whose focus is on the two cases $M = 8$ and $M = 16$ [15–17]. Cham's approach can be characterized as direct scaling with orthogonal constraint. In other words, since there is no guarantee that scaling retains the orthogonality property, we seek additional condition(s) on the integer transform coefficients such that the scaled-up integer transform is always orthogonal. Let us illustrate this design procedure with the $M = 8$ case, in which the free integer parameters come from the set of $\{a, b, c, d, e, f\}$:

$$\mathbf{ICT}_8 = \begin{bmatrix} 1 & 1 & 1 & 1 & 1 & 1 & 1 & 1 \\ a & b & c & d & -d & -c & -b & -a \\ e & f & -f & -e & -e & -f & f & e \\ 1 & -1 & -1 & 1 & 1 & -1 & -1 & 1 \\ c & -a & d & b & -b & -d & a & -c \\ f & -e & e & -f & -f & e & -e & f \\ d & -c & b & -a & a & -b & c & -d \end{bmatrix}. \tag{8.25}$$

In this particular case, it can be proven that the only condition they have to satisfy in order for the resulting integer cosine transform (ICT) to be orthogonal [15] is

$$ab = ac + bd + cd. \tag{8.26}$$

An exhaustive search, where the cost function is set to be the transform coding gain C_{CG} as defined in (8.15), up to a certain maximum resolution is performed to identify a rather large family of ICTs. One example of a parameter set that yields a high-performance ICT is $a = 230$, $b = 201$, $c = 134$, $d = 46$, $e = 3$, and $f = 1$.

8.4.3 Direct scaling of a rotation angle

It is worth noting that although direct scaling of a given orthogonal matrix does not generally retain orthogonality, it always holds in the 2×2 case. In this special case, any orthogonal matrix can be characterized by a single rotation angle, and it is easy to see that orthogonality is robust under direct scaling. For any rotation angle $\mathbf{R}_\theta$, the direct scaling method produces the approximation

$$\hat{\mathbf{R}}_\theta = \text{round}\,(\alpha \mathbf{R}_\theta) = \text{round}\left(\alpha \begin{bmatrix} \cos\theta & \sin\theta \\ -\sin\theta & \cos\theta \end{bmatrix}\right) = \begin{bmatrix} \hat{c} & \hat{s} \\ -\hat{s} & \hat{c} \end{bmatrix}, \tag{8.27}$$

which always retains orthogonality regardless of the scaling parameter choice:

$$\hat{\mathbf{R}}_\theta^T \hat{\mathbf{R}}_\theta = \begin{bmatrix} \hat{c} & -\hat{s} \\ \hat{s} & \hat{c} \end{bmatrix} \begin{bmatrix} \hat{c} & \hat{s} \\ -\hat{s} & \hat{c} \end{bmatrix} = \begin{bmatrix} \hat{c}^2 + \hat{s}^2 & 0 \\ 0 & \hat{c}^2 + \hat{s}^2 \end{bmatrix}. \tag{8.28}$$

The observation above leads to the following systematic integer-approximating design procedure which does not require complicated orthogonality constraints on the parameter set:

- Obtain a rotation-based sparse factorization of the target transform.
- Apply direct scaling to each rotation.

Note that each rotation angle can be approximated independently with a different degree of accuracy. This independent approximation approach above is actually very well known in practical fixed-point implementations. Most fixed-point IDCT structures are obtained following this procedure [18].

The direct scaling design approach in this section offers a few practical advantages:

- Several desirable properties of the original transform can be robustly retained (most importantly, symmetry, orthogonality, and high coding gain).
- The design procedure leads to efficient integer implementations that are often equivalent to fixed-point hardware implementations.

However, this approach is only applicable to transform approximation, not construction of new ones. Furthermore, a few disadvantages still remains in this ad hoc trial-and-error method:

- There is in general no exact inverse (not useful for lossless coding).
- In fact, the inverse operation often involves division operations.
- It is difficult to control the dynamic range expansion and to apply when the transform size/complexity increases.

8.5 Approximation approach via structural design

The structural design approach seeks to capture structurally as many desirable properties as possible and then optimize the parameters of the structure for high coding gain (or any other desirable property for the targeted application). In other words, we use basic low-order building blocks (such as butterflies, lifting steps, scaling factors, and delay elements) to construct much more complicated high-order transforms with certain predetermined properties (such as integer-mapping, perfect invertibility, and symmetry). This design approach can be applied to transform approximation as well as new transform construction. Integer- or dyadic-rational-coefficient transforms can be easily designed by approximating the optimal structural parameters by (dyadic) rationals.

8.5.1 Lifting step

One critical component in modern transform design is the lifting step [19], which is simply an elementary upper triangular or a lower triangular matrix with unity diagonal elements, also known as a shear operation in matrix theory or the ladder structure in signal processing literature [20]. Fig. 8.6 depicts a typical lifting step on the left and its corresponding inverse on the right. The upper triangular matrix as shown in Fig. 8.6 is often referred to as the update lifting step while the lower triangular matrix is called the prediction lifting step in the literature. The operator P in Fig. 8.6 is very general as far as perfect reconstruction is concerned: **P** can certainly be any scalar value; **P** actually can be any polynomial with delays (as popularized in fast wavelet transform implementation); in fact, **P** can even be any nonlinear operator (to generate morphological wavelets for instance). Invertibility is structurally guaranteed since to invert a lifting step, we only need to subtract out (add in) what has been added in (subtracted out) at the forward transform.

One advantage that the lifting scheme offers is the versatility and the simplicity in constructing fast transforms that can map integers to integers. If a *floor* (or *round*, or *ceiling*, or any other quantization) operator is placed after **P** in the lifting step in Fig. 8.6, the transformation can now map integers to integers with perfect reconstruction regardless of whether the lifting step has integer coefficients or not. This property can be confirmed through a simple inspection as demonstrated in Fig. 8.7.

Moreover, if the lifting step is chosen to be dyadic, the nonlinear operation can be incorporated into the division using binary shift as illustrated in Fig. 8.8. A scalar lifting step of value $\frac{k}{2^m}$ can be implemented as a multiplication by k followed by a division by 2^m. Division by 2^m followed by a truncation is equivalent to a binary right-shift by m places. The numerator k can be easily implemented using bit-shift and add operations as well, or we can split the fraction $\frac{k}{2^m}$ into a cascade of pure power-of-two lifting steps, i.e., $\frac{k}{2^m} = \sum_i \frac{1}{2^i}$. The latter approach leads to an implementation with solely binary right-shifts, preventing the bit-depth expansion in intermediate results. With all scaling factors set to unity, multiplierless transforms can be easily constructed via a cascade of multiple lifting steps. If the scaling

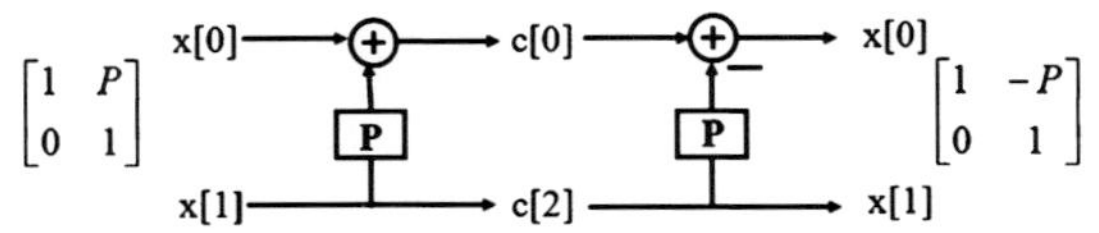

FIGURE 8.6

A lifting step (left) and its corresponding inverse (right).

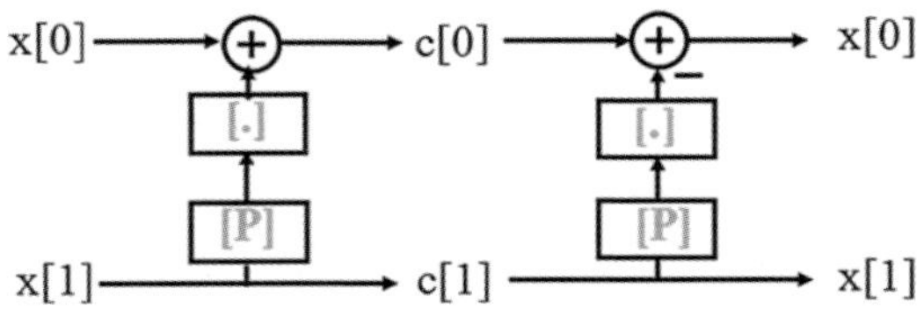

FIGURE 8.7

Integer-to-integer mapping with exact invertibility of a lifting step.

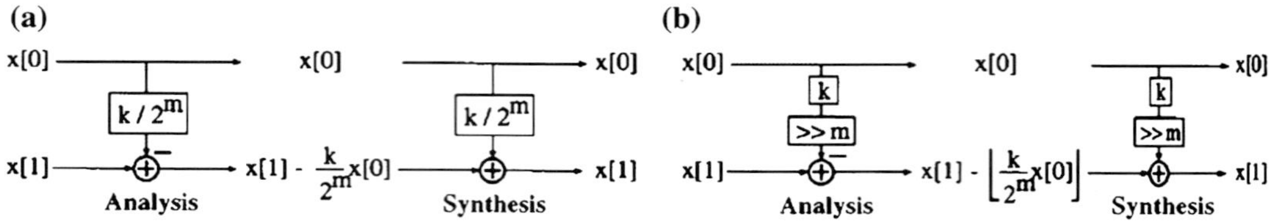

FIGURE 8.8

Lifting step with (a) dyadic-rational coefficient and (b) the resulting multiplierless implementation.

factors are powers of two, we can still retain the multiplierless feature on both the analysis and the synthesis side. Besides the integer-friendly property, the lifting scheme also offers several other interesting advantages: simplicity of wavelet design on irregular intervals, in-place computation, a connection to spatial prediction, and the capability of incorporating nonlinear filtering.

It is quite straightforward to extend the lifting result above to the higher-dimensional case of any arbitrary $M \times M$ matrix since it is well known from LDU matrix factorization that every $M \times M$ invertible matrix **V** can be completely characterized by $M(M-1)$ elementary matrices, M diagonal scaling factors, and a permutation matrix. In other words, any invertible matrix **V** can be factorized as $\mathbf{V} = \mathbf{PLDU}$ where the upper triangular matrix **U** can be constructed from $\frac{M(M-1)}{2}$ elementary update lifting steps labeled u_{ij}, the lower triangular matrix **V** can be constructed from $\frac{M(M-1)}{2}$ elementary prediction lifting steps labeled p_{ij}, and the diagonal matrix **D** contains the scaling factors α_i. The construction of **V** as described above is depicted in Fig. 8.9. Note that the parameterization can be accomplished with rotation angles and diagonal scaling factors as well (see Fig. 8.10).

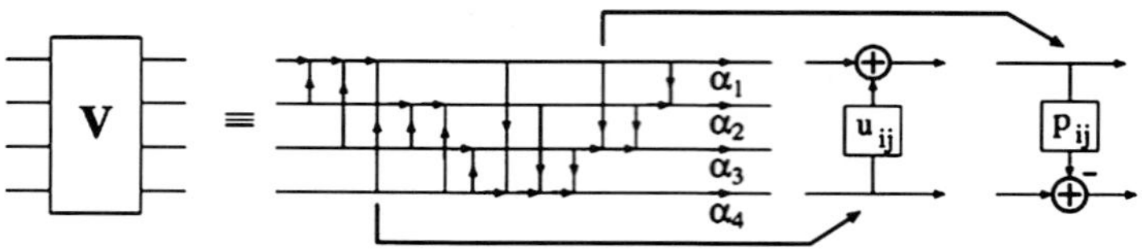

FIGURE 8.9

General parameterization of any invertible matrix via lifting steps.

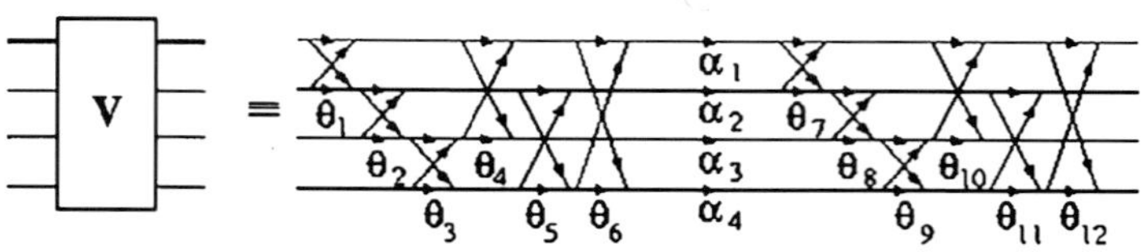

FIGURE 8.10

General parameterization of any invertible matrix via rotation angles.

8.5.2 Lifting-based approximation

Lifting steps are very convenient in transform approximation as well. Fig. 8.11 shows the equivalent representation of the popular rotation angle by three lifting steps or two lifting steps and two scaling factors. Sometimes, the latter representation can lead to a more efficient approximation since the two scaling factors can be absorbed into the quantization stage. Fig. 8.12 demonstrates a systematic approach to rotation approximation via multiplier-free lifting:

- first compute the theoretical values of the lifting steps,
- obtain dyadic-rational approximating values from theoretical ones (the resolution of the approximating values is the trade-off between computational complexity and coding performance),
- obtain the multiplierless implementation from the dyadic-rational lifting steps.

Since most existing fast transform algorithms are obtained from sparse factorization of the transform matrix as a cascade of rotation angles, we can systematically break down the design procedure one rotation independently at a time. An example of a successful application of this lifting-based approxi-

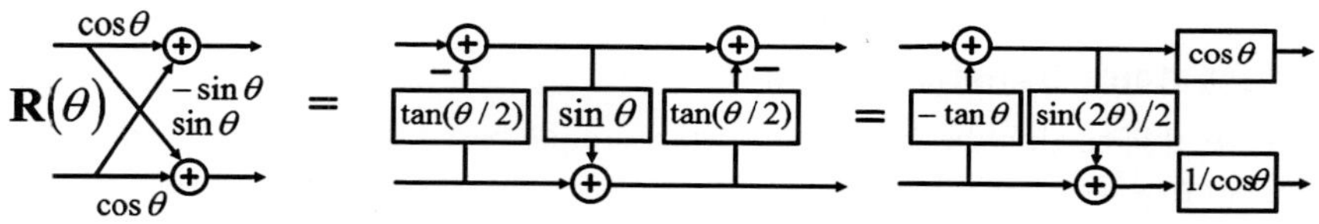

FIGURE 8.11

Representation of a rotation angle via either three lifting steps or two lifting steps and two scaling factors.

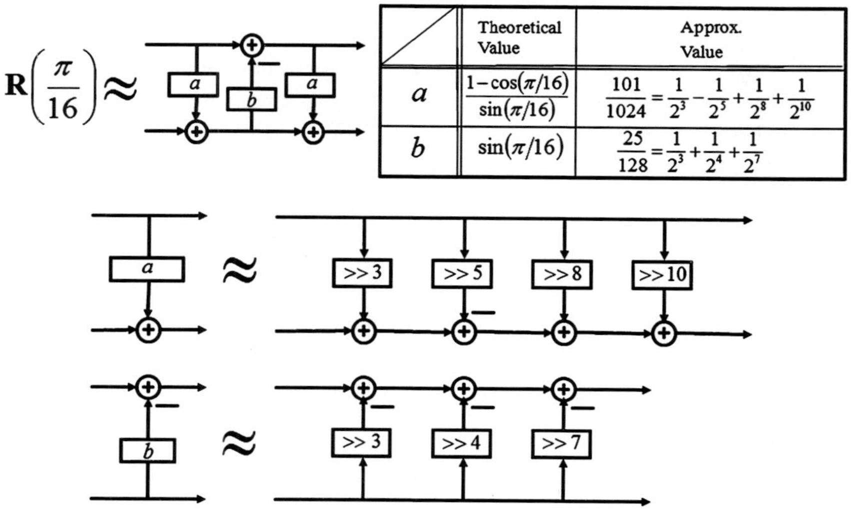

	Theoretical Value	Approx. Value
a	$\frac{1-\cos(\pi/16)}{\sin(\pi/16)}$	$\frac{101}{1024}=\frac{1}{2^3}-\frac{1}{2^5}+\frac{1}{2^8}+\frac{1}{2^{10}}$
b	$\sin(\pi/16)$	$\frac{25}{128}=\frac{1}{2^3}+\frac{1}{2^4}+\frac{1}{2^7}$

FIGURE 8.12

Example of a rotation approximation by dyadic-rational lifting steps.

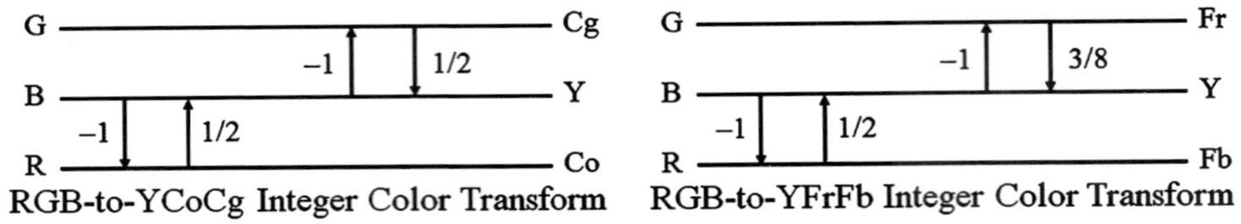

FIGURE 8.13

Multiplierless 3 × 3 color transforms with exact invertibility.

mation approach is the design of multiplierless IDCT for video coding in an effort to end the drifting effects from IDCT mismatch. Liu et al. showed that standard-compliance IDCT can be achieved on a 16-bit architecture whereas on a 32-bit architecture, approximation via lifting and butterfly can be so accurate that the reconstructed video sequences are virtually drifting-free when pairing with a 64-bit floating-point IDCT [21].

8.5.3 Lossless color transform design

We demonstrate in this section the simplest design example based on lifting construction for $M > 2$ – 3 × 3 color transforms. Fig. 8.13 depicts two different designs developed independently and submitted to the JVT standardization committee at the same time: the RGB-to-YCoCg transform (on the left of Fig. 8.13) by Malvar et al. [22] and the RGB-to-YFbFr transform (on the right of Fig. 8.13) by Topiwala et al. [23].

It is interesting to note that two designs are remarkably similar since they are both designed using our lifting approximation framework described in the previous sections. First, the optimal KLT is obtained where the autocorrelation matrix is derived from a set of high-quality test color images acquired by Kodak. Lifting-based approximation is then applied to obtain the dyadic lifting coefficients. Practical experiments show that both transforms decorrelate the three color planes more efficiently than existing popular color transforms such as the RGB-to-YUV and the RGB-to-YCbCr transform while additionally providing the pivotal feature of lossless integer mapping with low dynamic range (there is only a 1-bit expansion on the two chrominance components).

Finally, both transforms only utilize four lifting steps (instead of theoretically six) without any diagonal scaling. At the cost of one more addition and one more bit-shift operation per luminance sample, the RGB-to-YFbFr transform attains a coding gain of 4.812 dB while the RGB-to-YCoCg transform achieves $C_{CG} = 4.619$ dB (the optimal KLT in this case has a coding gain of 4.966 dB).

8.5.4 Integer DCT design

The DCT is the most widely used transform in current international image/video compression standards. This four-point dyadic-coefficient design example is a close approximation of the four-point DCT and it follows the DCT's factorization very closely. The resulting structure is depicted in Fig. 8.14.

In the sparse four-point DCT factorization case, there are three butterflies and only one irrational rotation angle of $\frac{\pi}{8}$ [24]. We can employ only two lifting steps for this rotation angle (since the scaling factors associated with the two-lifting step structure can be combined with the quantization step sizes to

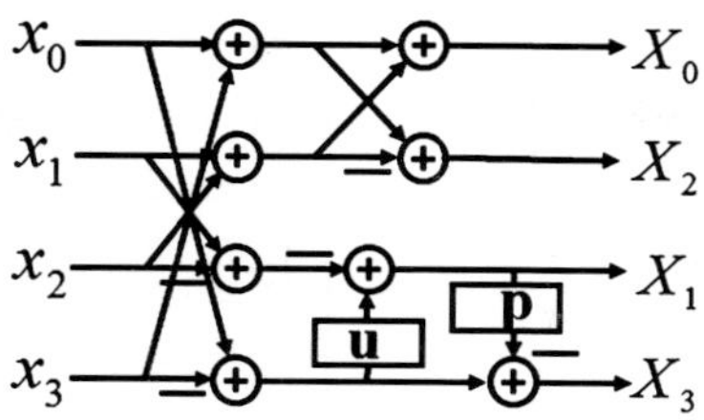

FIGURE 8.14

4×4 integer DCT structure.

save computational complexity further) while keeping the butterflies intact. Many different transforms can be represented using this structure; the difference between them lies only in the parameter choices as demonstrated below:

- $u = 1$; $p = 1/2$: essentially the Walsh–Hadamard transform,
- $u = 7/16$; $p = 3/8$: Liang et al. submission to H.264 [25,26],
- $u = 1/2$; $p = 2/5$: the current four-point H.264 transform (the last two basis functions are off by 2, 5/2 scaling),
- $u = 1/2$; $p = 1/2$: the current four-point Photo Core Transform (PCT) in HD-Photo/JPEG-XR [27, 28].

It is interesting to note that the last two options yield two transforms with an identical coding gain of 7.55 dB. The lifting implementation allows the PCT [27,28] to map integers to integers losslessly and to have a much tighter dynamic range control comparing to the rotation-based implementation of H.264 transform (this is where H.264 loses its lossless coding capability). The second option with $u = 7/16$ and $p = 3/8$ essentially achieves the same coding gain as the theoretical four-point DCT (7.57 dB for the $AR(1)$ signal model with $\rho = 0.95$). Note that all four options allow 16-bit implementation with 9-bit input data.

Integer DCTs of larger sizes ($M > 4$) can be easily designed using the same procedure. The rotation-factorized representation of the transform of interest is first obtained, and then each rotation angle can be approximated methodically as demonstrated in Section 8.5.2. With various degrees of approximation accuracy (more accuracy obviously requires more complexity), these integer DCTs can be tuned to cover the gap between the Walsh–Hadamard transform and the original DCT. The corresponding solution often allows a 16-bit implementation, enables lossless compression, and maintains satisfactory compatibility with the floating-point DCT. The 8×8 and 16×16 integer DCTs are depicted in Figs. 8.15 and 8.16, respectively.

Table 8.1 lists the analytical values of all the lifting parameters and some configurations of the integer DCT family (called binary DCT or binDCT in short) as shown in Fig. 8.15. The dyadic lifting values are obtained by truncating or rounding the corresponding analytical values with different accuracy levels. $C_g(8)$ and $C_g(4)$ are the coding gains of these 8×8 binDCTs and the 4×4 binDCTs embedded in them. Table 8.1 also shows the MSE between the binDCT output and the true DCT output for the popular 8×8 case. The MSE here is the expected ℓ_2-norm of the error in the produced coefficients and it is computed as follows. Assume that $\mathbf{C}$ is the true $M \times M$ DCT and $\hat{\mathbf{C}}$ is its approximated integer

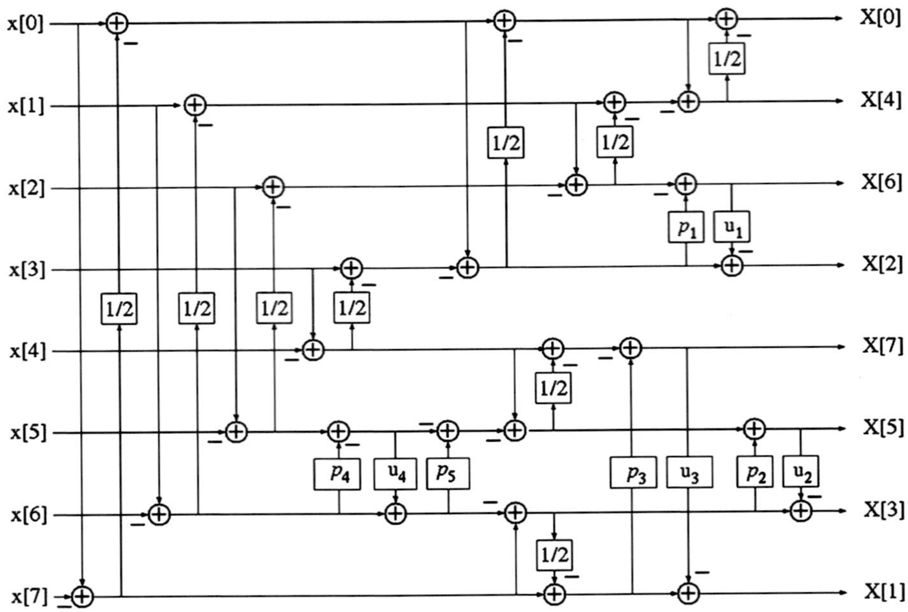

FIGURE 8.15

8 × 8 integer DCT.

Table 8.1 Different configurations of the 8 × 8 Integer DCT in Fig. 8.15.

	Floating-point	binDCT-C1	C2	C3	C4	C5
p_1	0.4142135623	13/32	7/16	3/8	1/2	1/2
u_1	0.3535533905	11/32	3/8	3/8	3/8	1/2
p_2	0.6681786379	11/16	5/8	7/8	7/8	1
u_2	0.4619397662	15/32	7/16	1/2	1/2	1/2
p_3	0.1989123673	3/16	3/16	3/16	3/16	1/4
u_3	0.1913417161	3/16	3/16	3/16	1/4	1/4
p_4	0.4142135623	13/32	7/16	7/16	7/16	1/2
u_4	0.7071067811	11/16	11/16	11/16	3/4	3/4
p_5	0.4142135623	13/32	3/8	3/8	3/8	1/2
Shifts	–	27	23	21	18	13
Adds	–	42	37	36	33	28
MSE	–	1.1E−5	8.5E−5	4.2E−4	5.8E−4	2.3E−3
$C_g(8)$ (dB)	–	8.8251	8.8220	8.8159	8.8033	8.7686
$C_g(4)$ (dB)	–	7.5697	7.5697	7.5566	7.5493	7.5485

version. For an input column vector **x**, the difference between the DCT coefficients and the binDCT coefficients is

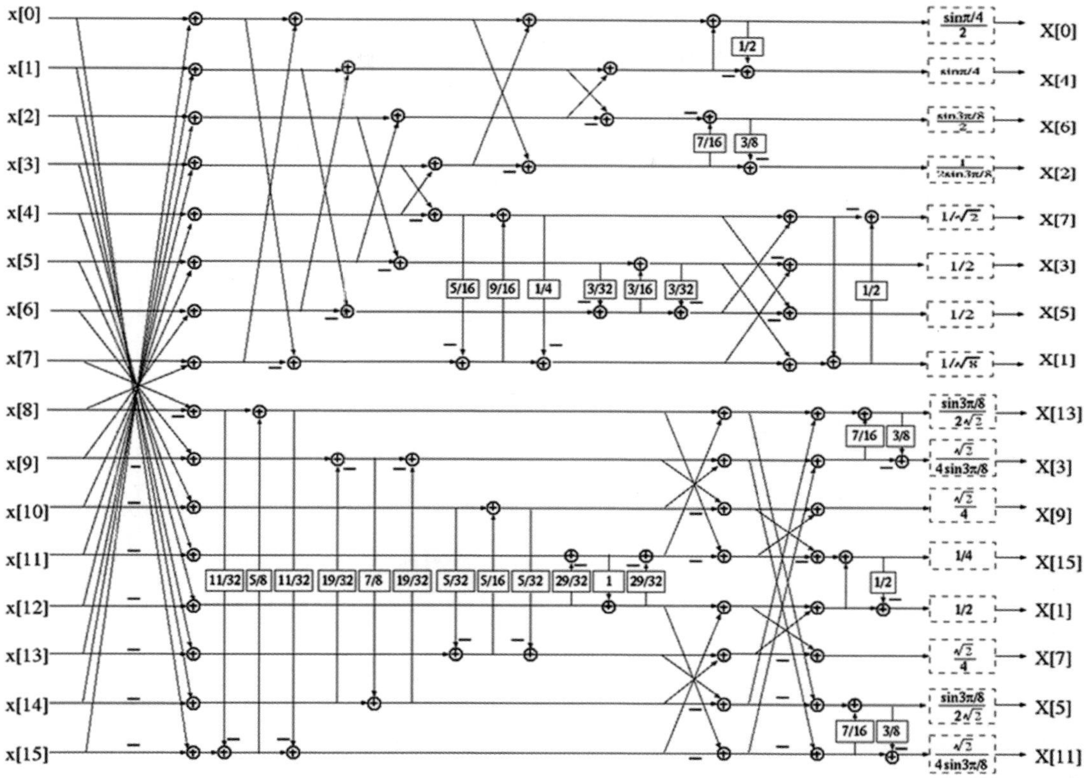

FIGURE 8.16

16 × 16 integer DCT.

$$\mathbf{e} = \mathbf{Cx} - \hat{\mathbf{C}}\mathbf{x} = (\mathbf{C} - \hat{\mathbf{C}})\mathbf{x} \triangleq \mathbf{Dx} \tag{8.29}$$

and the MSE of each coefficient is

$$\epsilon \triangleq \frac{1}{M} E[\mathbf{e}^T \mathbf{e}] = \frac{1}{M} \text{trace}\{\mathbf{ee}^T\} = \frac{1}{M} \text{trace}\{\mathbf{DR}_{xx}\mathbf{D}^T\}, \tag{8.30}$$

where $\mathbf{R_{xx}} \triangleq \mathbf{E}[\mathbf{xx^T}]$ is the autocorrelation matrix of the input signal. The low MSE level as depicted in Table 8.1 indicates that the integer transform approximation is very accurate. The interested reader is referred to [21,26] for more details on the design of this particular family of multiplierless cosine transforms.

8.5.5 Lifting for complex-coefficient transforms

So far, we have focused on real-coefficient transforms since they are mostly employed in image/video coding applications. For complex-coefficient transforms such as the FFT, it turns out that lifting steps

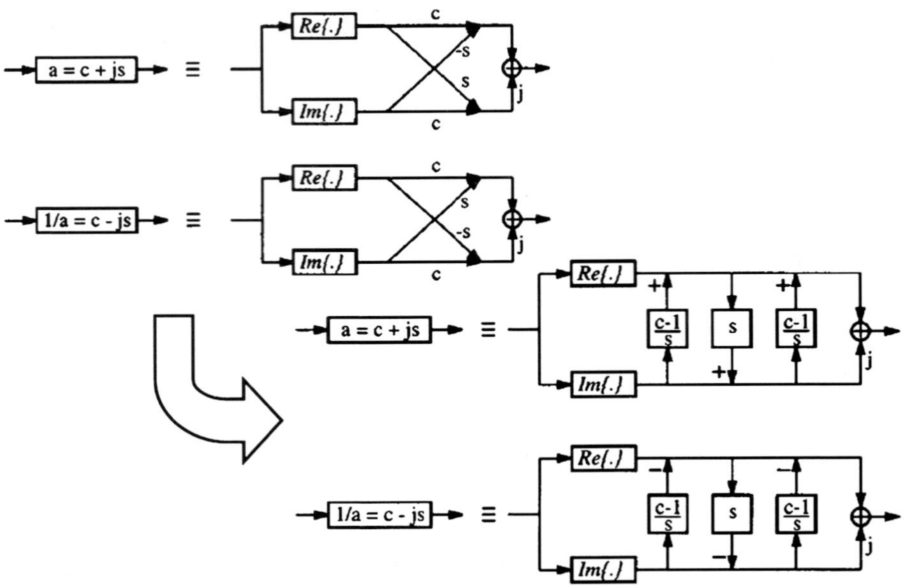

FIGURE 8.17

Implementation of a complex multiplication operation via lifting steps.

can be as easily applied as in the real-coefficient case. Fig. 8.17 illustrates how to model a complex multiplication operation by a rotation angle and the resulting equivalent lifting representation [20,29]. Each lifting step in Fig. 8.17 can then be approximated by a dyadic-rational value as previously shown. One particularly nice feature of the FFT is its regularity in factorized format: its implementation is recursive and all values of rotation angles involved are easily computed. Hence, complex-coefficient transforms such as the FFT can be easily constructed or approximated in multiplierless fashion as well. Again, the lifting coefficients appearing in the structures can be quantized directly to obtain different resolutions (resulting in trade-off in computational cost) while preserving the exact reversibility property.

8.6 Wavelet filter design via spectral factorization

8.6.1 Wavelets and filter banks

One of the newest additions to the transform field is the wavelet transform, which can be interpreted as an iteration of a two-channel filter bank with certain degrees of regularity on its low-pass output depicted in Fig. 8.18. Part of the beauty and power of wavelets is their elegance: complicated tiling of the time-frequency plane can be easily achieved by merely iterating a two-channel decomposition. Furthermore, the wavelet representation of signals as a coarse approximation and detailed components at different resolutions allows fast execution of many DSP applications such as image browsing, database retrieval, scalable multimedia delivery, etc.

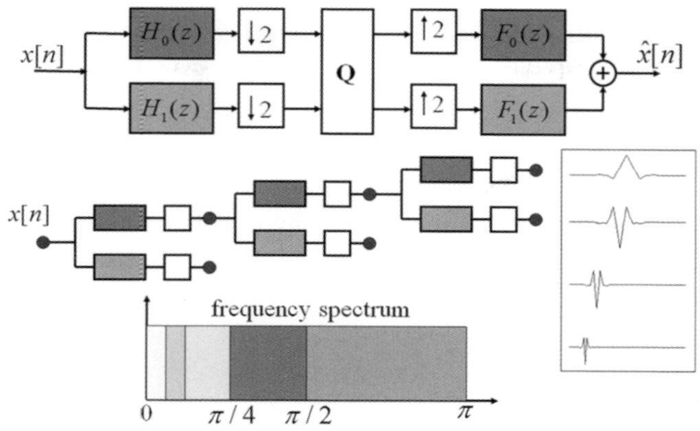

FIGURE 8.18

The discrete wavelet transform as iteration of a special two-channel filter bank on its low-pass output.

The theory and design of two-channel filter banks as well as two-band wavelets have been studied extensively and we refer the readers to several excellent text books on the topic [5–8,30]. The simplest design procedure is the spectral factorization of a low-pass half-band filter with some regularity constraints (vanishing moments). Unlike all other transforms mentioned in this chapter, all of the degrees of freedom in wavelet design are allocated to regularity. In the scope of this chapter, we will briefly cover this most successful wavelet filter design approach for completeness only.

8.6.2 Spectral factorization

It can be verified in a straightforward fashion that in order for the two-channel filter bank in Fig. 8.18 to have perfect reconstruction (i.e., $\hat{x}[n] = x[n - D]$, where D is a certain time delay), the four filters involved have to satisfy the following two conditions:

$$\text{aliasing cancelation:} \quad F_0(z)H_0(-z) + F_1(z)H_1(-z) = 0, \tag{8.31}$$

$$\text{distortion elimination:} \quad F_0(z)H_0(z) + F_1(z)H_1(z) = 2z^{-D}. \tag{8.32}$$

The following alternating-sign construction provides an elegant way to cancel aliasing and to reduce the design parameters involved by a factor of two

$$\begin{cases} F_0(z) = H_1(-z), \\ F_1(z) = -H_0(-z). \end{cases} \tag{8.33}$$

With the relationship between the analysis and synthesis filters set as in (8.33), the distortion elimination condition in (8.32) reduces to

$$F_0(z)H_0(z) - F_0(-z)H_0(-z) = 2z^{-D}. \tag{8.34}$$

If we define the product filter $P_0(z)$ as $P_0(z) \triangleq F_0(z)H_0(z)$, then it has to satisfy the following condition:

$$P_0(z) - P_0(-z) = 2z^{-D}, \tag{8.35}$$

which is well known in the signal processing community as the half-band condition (a filter that has one and only one odd power of z^{-1}). Hence, the design of a two-channel perfect reconstruction filter bank with all four FIR filters can follow the standard procedure listed below:

- Design a low-pass half-band filter $P_0(z)$.
- Factor $P_0(z)$ into $H_0(z)$ and $F_0(z)$.
- Use the aliasing cancelation condition in (8.33) to obtain $H_1(z)$ and $F_1(z)$.

The difference between any common filter pair designed from the approach above and a good wavelet filter pair is the degree of regularity or the level of smoothness of the resulting basis functions, called the scaling function and the wavelet function, after many levels of repeated iterations. As mentioned above, Daubechies decided to allocate all of the degrees of freedom in the design to the degree of regularity or the number of vanishing moments [30]. In other words, an additional constraint labeled *maxflat* is imposed on top of the half-band condition in (8.35), resulting in the maxflat half-band filter defined as follows:

$$P_0(z) = (1 + z^{-1})^{2P} Q_0(z). \tag{8.36}$$

This key filter has $2P$ zeros (roots) at $z = -1$ or $\omega = \pi$ and $Q_0(z)$ is the polynomial of minimum order (which turns out to be $2P - 2$) such that the half-band condition in (8.35) is met. The closed-form expression of this family of maxflat half-band filters can be found in [5,6,8,30].

The key step in the design process obviously lies in the second spectral factorization procedure. There are multiple ways to split the roots $\{z_i\}$ of $P_0(z)$ and the desirable properties of the resulting filters $\{H_0(z), H_1(z), F_0(z), F_1(z)\}$ dictate the constraints in the root assignment. Here are three popular desirable properties and their associated constraints in spectral factorization:

- For the filters to have real coefficients, we need z_i and z_i^* to stay together: assign both to the same filter.
- To obtain orthogonal solutions, we need to separate z_i and z_i^{-1} to different filters: we must assign z_i to $H_0(z)$ and z_i^{-1} to $F_0(z)$ or vice versa.
- On the contrary, to obtain linear-phase solutions, we need z_i and z_i^{-1} to stay together: both are either assigned to $H_0(z)$ or to $F_0(z)$.

Note that from the second and third conditions above, we cannot design a solution with both orthogonality and linear phase in the two-channel case (except the trivial 2-tap Haar solution). When the number of channels increases ($M > 2$), having both orthogonality and linear phase is actually quite easy to achieve – the type-II DCT in (8.7) and the WHT in (8.6) are two typical examples.

8.6.2.1 5-/3-tap symmetric and 4-tap orthogonal wavelet filters

Let us consider the following maxflat half-band filter with four roots at $z = -1$ and two real roots at $z^{-1} = 2 \pm \sqrt{3}$:

$$\begin{aligned} P_0(z) &= \frac{1}{16}\left(-1 + 9z^{-2} + 16z^{-3} + 9z^{-4} - z^{-6}\right) \\ &= \frac{1}{16}\left(1 + z^{-1}\right)^4 \left(-1 + 4z^{-1} - z^{-2}\right). \end{aligned} \tag{8.37}$$

Assigning two roots at $z = -1$ to $F_0(z)$ and the rest to $H_0(z)$, we obtain the following real-coefficient symmetric filter pair:

$$\begin{cases} H_0(z) = -\frac{1}{8} + \frac{1}{4}z^{-1} + \frac{3}{4}z^{-2} + \frac{1}{4}z^{-3} - \frac{1}{8}z^{-4}, \\ H_1(z) = -\frac{1}{2} + z^{-1} - \frac{1}{2}z^{-2}. \end{cases} \tag{8.38}$$

This particular solution was first published in [31], so they are often referred to as the Le Gall 5/3 filters. Not only do they have dyadic coefficients; their lifting multiplierless implementation also provides an efficient reversible integer-to-integer mapping as discussed in Section 8.6.3. The discrete wavelet transform with these 5-/3-tap filters is the default transformation option for the lossless mode in the international image coding standard JPEG2000 [4].

On the other hand, if we assign the roots $\{-1, -1, 2 - \sqrt{3}\}$ to $H_0(z)$ while giving the roots $\{-1, -1, 2 + \sqrt{3}\}$ to $F_0(z)$, we arrive at the celebrated compact-support orthogonal Daubechies D4 wavelet pair

$$\begin{cases} H_0(z) = \frac{1}{4\sqrt{2}}\left[(1 + \sqrt{3}) + (3 + \sqrt{3})z^{-1} + (3 - \sqrt{3})z^{-2} + (1 - \sqrt{3})z^{-3}\right], \\ H_1(z) = \frac{1}{4\sqrt{2}}\left[(1 - \sqrt{3}) - (3 - \sqrt{3})z^{-1} + (3 + \sqrt{3})z^{-2} - (1 + \sqrt{3})z^{-3}\right]. \end{cases} \tag{8.39}$$

With this particular maxflat half-band filter $P_0(z)$, there are only two possible orthogonal solutions. The pair in (8.39) above is called the minimum-phase solution. Switching the roles of $H_i(z)$ and $H_i(z)$ yields the other maximum-phase solution. There are quite a few more real symmetric solutions. In fact, the 2-/6-tap and the 4-/4-tap pair are also quite popular in the image processing community.

8.6.2.2 9-/7-tap symmetric and eight-tap orthogonal wavelet filters

Let us consider the higher-order symmetric maxflat half-band filter with eight roots at $z = -1$, two real roots, and four complex roots; the complex plane of $P_0(z)$ is shown on the top part of Fig. 8.19:

$$P_0(z) = \frac{(1 + z^{-1})^8}{2048}\left(-5 + 40z^{-1} - 131z^{-2} + 208z^{-3} - 131z^{-4} + 40z^{-5} - 5z^{-6}\right).$$

There are exactly four orthogonal solutions from spectral factorization in this case: one minimum-phase, one maximum-phase, and two mixed-phase solutions (they are time-reversals of each other just like the minimum-phase solution is the time-reversed version of the maximum-phase solution) as illustrated in Fig. 8.19.

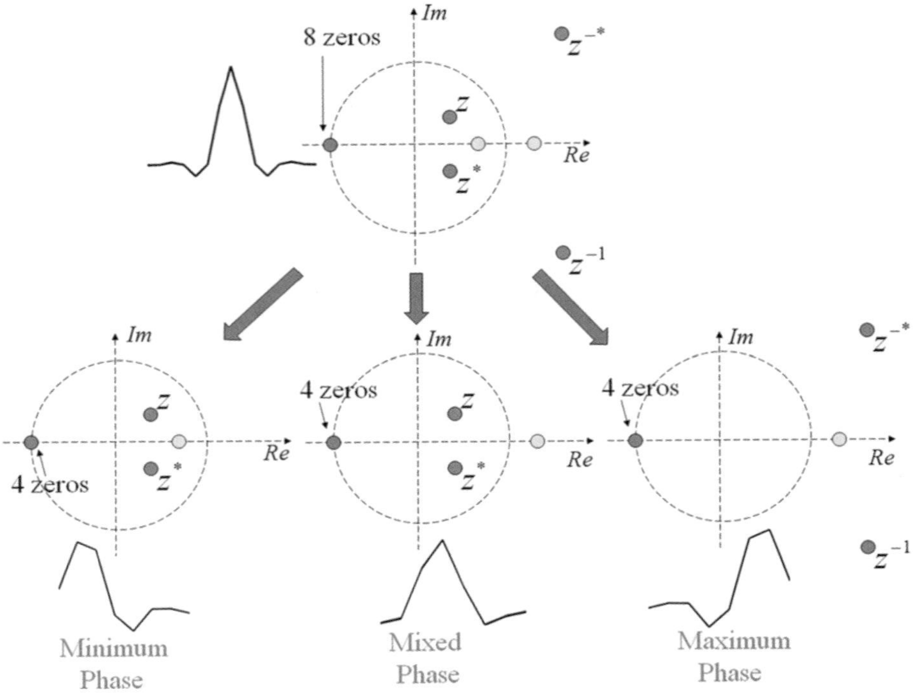

FIGURE 8.19

Orthogonal wavelet filter design via spectral factorization.

In the biorthogonal linear-phase case, there are many more nontrivial possibilities. The best solution in terms of attaining the highest theoretical coding gain as well as the highest consistent coding performance in practice is the 9-/7-tap filter pair with $H_0(z)$ having four roots at $z = -1$ and four complex roots as depicted in Fig. 8.20. This is often referred to as the Daubechies 9/7 wavelet filter pair [32], which is JPEG2000's default for the floating-point high-performance option [4]. Various interesting properties of these JPEG2000 wavelet filters have been explored in depth in [33].

It is worth noting that sacrificing a few vanishing moments (precisely, reducing the number of roots at $z = -1$ from eight to six in $P_0(z)$) opens up more flexibility in achieving other desirable properties such as dyadic coefficients. The following 9-/7-tap filter pair is called *binlet* – binary wavelet. The analysis bank now only has two vanishing moments (instead of four as in the Daubechies 9/7 JPEG2000 pair) whereas the synthesis bank still retains four vanishing moments since the vanishing moments – intimately related to the level of smoothness – are more critical in signal reconstruction than in signal decomposition. The 9-/7-tap binlet filter coefficients are

$$\begin{cases} h_0[n] = \frac{1}{64}[1\ 0\ -8\ 16\ 46\ 16\ -8\ 0\ 1], \\ h_1[n] = \frac{1}{16}[1\ 0\ -9\ 16\ -9\ 0\ 1]. \end{cases} \tag{8.40}$$

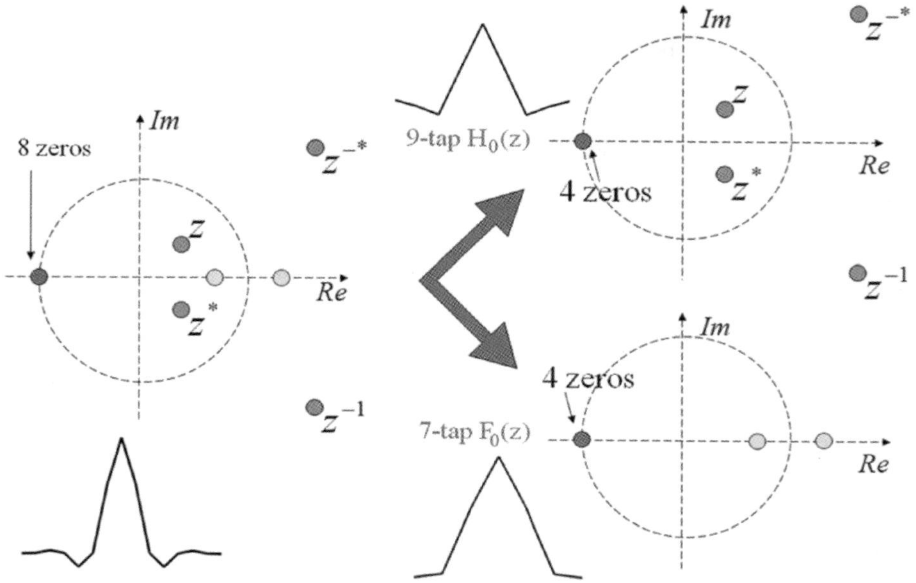

FIGURE 8.20

Linear-phase biorthogonal wavelet filter design via spectral factorization.

8.6.3 Lifting and the wavelet transform

The lifting scheme was originally designed for the wavelet transform and two-channel filter bank decomposition [19]. Two-channel filter banks implemented with the lifting scheme save roughly half of the computational complexity due to the exploitation of the intersubband relationship between coefficients. It has been proven that any 2×2 polyphase matrix $\mathbf{E}(z)$ with unity determinant can be factored into a cascade of prediction $\mathbf{P}(z)$, update $\mathbf{U}(z)$ lifting steps, and a diagonal scaling matrix as shown in Fig. 8.21 [34].

As previously mentioned, one of the most elegant as well as the most popular wavelet filter pair in practice is the Le Gall 5/3 [31] whose polyphase matrix can be constructed with two lifting steps:

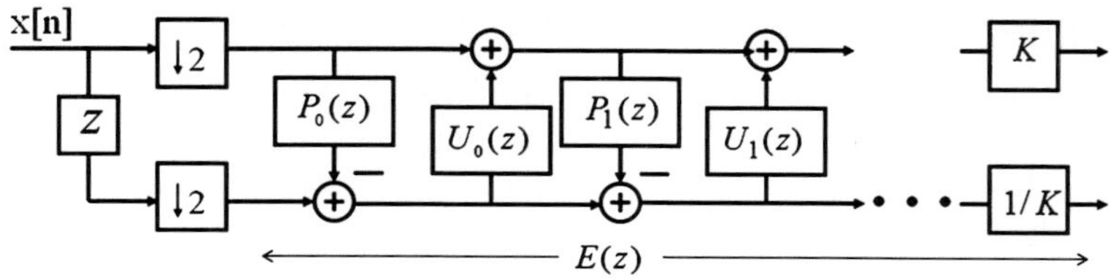

FIGURE 8.21

General wavelet construction and implementation with lifting steps.

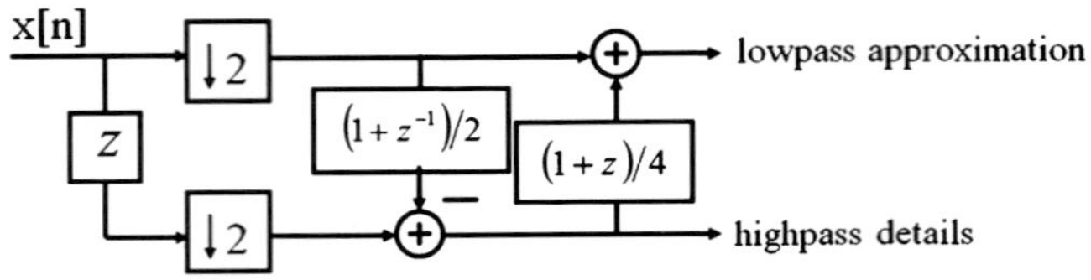

FIGURE 8.22

The 5/3-tap integer wavelet implementation in lifting.

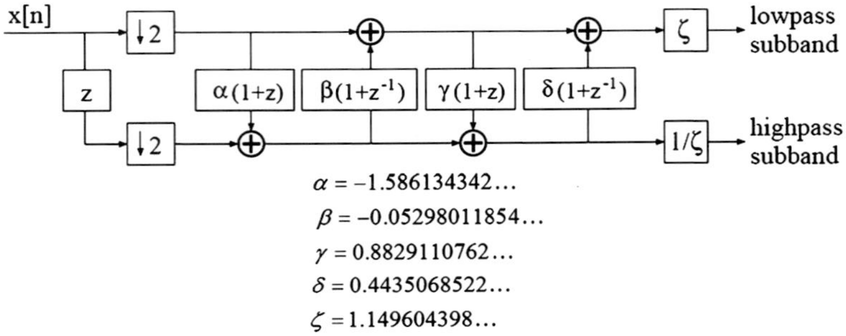

FIGURE 8.23

The 9/7-tap double-precision wavelet implementation in lifting.

$P(z) = \frac{1}{2}(1 + z^{-1})$ and $U(z) = \frac{1}{4}(1 + z)$. This transform depicted in Fig. 8.22 leads to a multiplierless implementation, and since it can map integers to integers with exact invertibility, it is naturally adopted by JPEG2000 for the lossless (or low-cost) compression mode [4]. It has a very intuitive time-domain interpretation: each high-pass detail wavelet coefficient is simply the prediction error between the average of two neighboring even-indexed samples and the in-between odd-indexed sample. Fig. 8.23 shows the lifting implementation of the irrational-coefficient 9/7 Daubechies wavelet filter pair [32,34] as previously discussed in Section 8.6.2.2 (see Figure 8.22).

8.7 Higher-order design approach via optimization

One major drawback with nonoverlapped block transforms such as the DCT and its integer versions as described in the previous section is the existence of annoying blocking artifacts from discontinuities at block boundaries in low-bitrate applications. Transforms with overlapping basis functions such as the wavelet transform, the wavelet packet, and the LT can elegantly solve the blocking problems, producing much more visually pleasant reconstructed images and video frames.

From a filter bank perspective, higher-order design lengthens the filters, leading to a drastic improvement in frequency resolution and hence decorrelation efficiency while still retaining time or space resolution. As mentioned above, from a filter bank perspective, the wavelet transform and LT are simply higher-order systems whose polyphase matrices contain polynomials in z whereas all examples discussed so far are block transforms whose polyphase matrices are regular scalar matrices of zero order.

8.7.1 General modular construction

In the wavelet case where the basic building block is a two-channel filter bank, the most general design approach is the spectral factorization of a low-pass half-band filter. The lifting scheme is used mainly for fast and efficient wavelet implementation. When the number of channels M increases, spectral factorization is not applicable and the most successful design approach is the modular construction from the lattice structure pioneered by Vaidyanathan [7] and Malvar [12]. The first filter bank design based on this philosophy is illustrated in Fig. 8.24 where each modular stage is a rotation angle (coupled with a delay element), propagating the orthogonality property. This simple structure has been proven to utilize the minimum number of delay elements in the resulting implementation since it shares the delays between the two channels. More importantly, the structure in Fig. 8.24 covers *all* possible orthogonal solutions in the two-channel case. Indeed, when $K = 2$, the particular selection of $\{\theta_0 = 60°, \theta_1 = -15°\}$ generates the D4 orthogonal Daubechies wavelet filters in (8.39). It is a simple exercise for the reader to verify that the resulting polyphase matrix here always satisfies the paraunitary condition in (8.11), i.e., $\mathbf{E}^{-1}(z) = z^{-(K-1)}\mathbf{E}^T(z^{-1})$.

Similarly, in the more general M-channel case, we can construct the high-order polyphase matrix $\mathbf{E}(z)$ as in Fig. 8.25 from a cascade of many low-order building blocks $\mathbf{G}_i(z)$, each propagating certain desirable properties such as symmetry, perfect reconstruction (or more elegantly, orthogonality), and zero DC leakage. The free parameters of the structure within each building block $\mathbf{G}_i(z)$ are then further optimized to achieve other remaining desirable properties such as coding gain and stopband attenuation

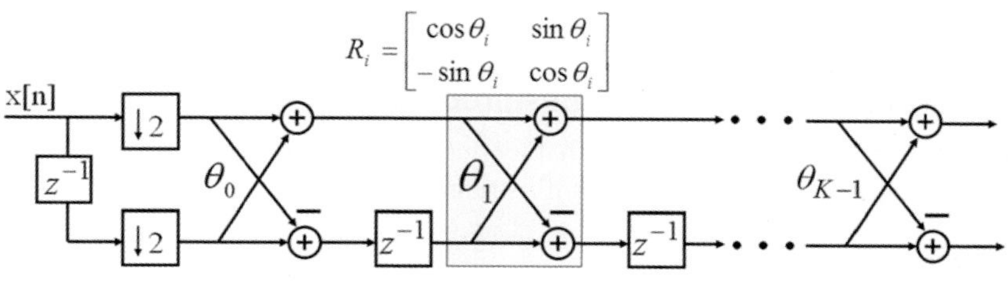

FIGURE 8.24

First two-channel $2K$-tap filter bank design via the lattice structure propagating orthogonality.

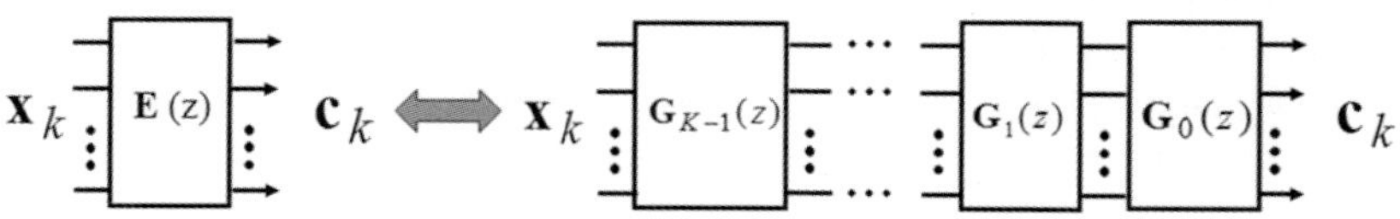

FIGURE 8.25

Construction of a high-order polyphase matrix **E**(z) from a cascade of low-order building blocks $\mathbf{G}_i(z)$.

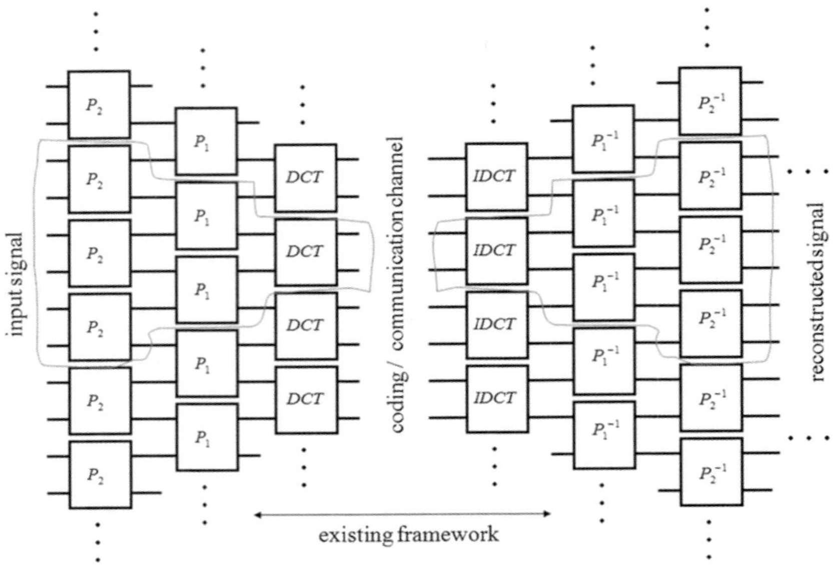

FIGURE 8.26

General transformation framework for block coding systems with pre- and postprocessing operators.

as discussed in Section 8.3. This modular construction design approach yields numerous interesting filter bank solutions, which the reader can learn about in much greater detail in two excellent textbooks [6,7].

8.7.2 Higher-order design with pre-/postprocessing operators

Within the more restrictive constraint of this chapter, let us present in depth a general structure that has an intuitive time-domain interpretation [35] as illustrated in Fig. 8.26.

8.7.2.1 *A simple example*

Several basic properties of this design approach are best explained through an almost trivial example – the addition of two-point pre- and postprocessing (or pre-/postfiltering) operators linking any two DCT blocks in traditional block coders such as JPEG [1] depicted in Fig. 8.27(a). This is a particular preprocessor in the general framework of Fig. 8.26 and the postprocessor has been chosen to be the exact inverse of the preprocessing operator.

Let $\{x_i\}$, $\{p_i\}$, $\{\hat{p}_i\}$, and $\{\hat{x}_i\}$ be the set of the preprocessor's input samples, the preprocessor's output samples, the postprocessor's input samples, and the postprocessor's output samples, respectively. We examine the postprocessor first since its operation is slightly more intuitive than the preprocessor's.

Consider $\hat{p}_i$ and $\hat{p}_{i+1}$ in Fig. 8.27(b) – two input samples of the postprocessor at the boundary of two neighboring IDCT blocks – and their corresponding output samples, $\hat{x}_i$ and $\hat{x}_{i+1}$. Define a crude measure of blocking artifact D as the absolute difference between two samples at the block boundary.

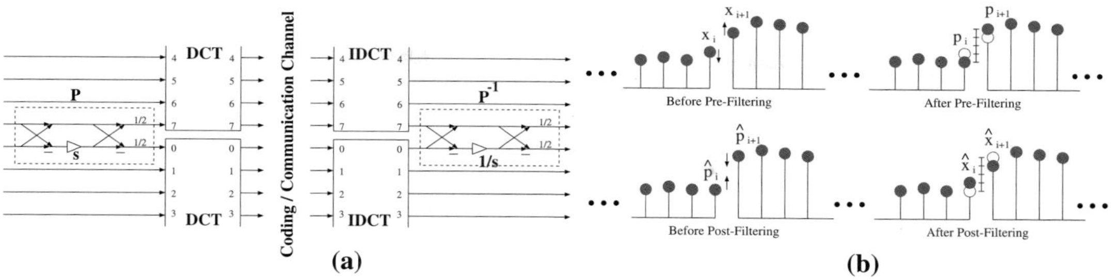

FIGURE 8.27

Basic demonstration of the pre-/postprocessing design. (a) Simple 2×2 pre-/postprocessing operator pair $\{\mathbf{P}, \mathbf{P}^{-1}\}$ linking two DCT blocks. (b) Preprocessed and postprocessed effects.

Without any postprocessing, $D = |\hat{p}_i - \hat{p}_{i+1}|$. It is intuitive that in order to reduce the blocking artifact D, we simply have to pull $\hat{p}_i$ and $\hat{p}_{i+1}$ closer (blurring out the discontinuity at the boundary) depending on how far they are apart. One systematic method to achieve this objective is presented in Fig. 8.27(a) where the scaling factor s is the single degree of freedom.

With $\mathbf{P}^{-1} = \frac{1}{2}\begin{bmatrix} 1 & 1 \\ 1 & -1 \end{bmatrix}\begin{bmatrix} 1 & 0 \\ 0 & 1/s \end{bmatrix}\begin{bmatrix} 1 & 1 \\ 1 & -1 \end{bmatrix}$, we can easily derive the following relationships:

$$\hat{x}_i = \hat{p}_i + \frac{1/s - 1}{2}(\hat{p}_i - \hat{p}_{i+1}) \quad \text{and} \quad \hat{x}_{i+1} = \hat{p}_{i+1} - \frac{1/s - 1}{2}(\hat{p}_i - \hat{p}_{i+1}).$$

Hence, after postprocessing,

$$\begin{aligned} \widehat{D} = |\hat{x}_i - \hat{x}_{i+1}| &= |(\hat{p}_i - \hat{p}_{i+1}) + (1/s - 1)(\hat{p}_i - \hat{p}_{i+1})| \\ &= |(\hat{p}_i - \hat{p}_{i+1})/s| = |1/s||\hat{p}_i - \hat{p}_{i+1}|. \end{aligned} \tag{8.41}$$

Choosing $|1/s| < 1$ (or equivalently $|s| > 1$) guarantees a decrease in blocking artifact. This postprocessing effect is demonstrated in Fig. 8.27(b). For example, if we choose $s = 2$, then the postprocessing operator partitions the distance $D = |\hat{p}_i - \hat{p}_{i+1}|$ into four segments of equal length, moves $\hat{x}_i$ up one segment length from $\hat{p}_i$, and moves $\hat{x}_{i+1}$ down one segment length from $\hat{p}_{i+1}$. So, postprocessing adaptively reduces the observed discrepancy D by a factor of two.

The preprocessor modifies the samples in the exact opposite direction. It attempts to "decorrelate" the boundary by lowering the smaller-value sample x_i while increasing the larger-value x_{i+1} based on a percentage of their difference $|x_i - x_{i+1}|$. Again, the choice $s > 1$ makes intuitive sense. If $s < 1$, samples are adjusted in the wrong direction. A good choice of s in the energy compaction sense for a smooth $AR(1)$ signal model is the *golden ratio* $\frac{1+\sqrt{5}}{2}$ or a close approximation such as $\frac{8}{5}$ [35] (note that this choice also yields a multiplierless postprocessor).

8.7.2.2 More general solutions

The seemingly trivial example above actually provides a valuable lesson. When more than two samples are involved, the preprocessing operator should be designed as a *flattening* operator. It should attempt to make the input to the DCT as homogeneous as possible, thus improving the overall energy compaction of the combined decomposition. This is quite consistent with most preprocessing schemes in practice: smoothing the input signal improves coding efficiency. However, in this framework, perfect reconstruction can be easily maintained (by choosing the postprocessor as the inverse of the preprocessor). High-frequency signal components are never eliminated; they are only slightly shifted spatially. In other words, we take full advantage of the block-based framework by *carefully aligning high-frequency components at block boundaries*. Discontinuities between DCT blocks, i.e., actual high-frequency contents, do not affect coding performance whereas, within each block, data samples are smoothened out, enhancing the core block transform's effectiveness in energy compaction.

The most general higher-order transform as shown in Fig. 8.27 is equivalent to an M-channel critically sampled filter bank with filters of length $L = KM$ whose analysis polyphase matrix turns out to be

$$\mathbf{E}(z) = \mathbf{G}_0\mathbf{G}_1(z)\mathbf{G}_1(z)\ldots\mathbf{G}_{K-1}(z), \tag{8.42}$$

where each propagating component takes on the form $\mathbf{G}_i(z) = \mathbf{\Lambda}(z)\,\mathbf{P}_i$ with

$$\mathbf{\Lambda}(z) \triangleq \begin{bmatrix} \mathbf{I} & \mathbf{0} \\ \mathbf{0} & z\mathbf{I} \end{bmatrix} \begin{bmatrix} \mathbf{0} & \mathbf{I} \\ \mathbf{I} & \mathbf{0} \end{bmatrix} = \begin{bmatrix} \mathbf{0} & \mathbf{I} \\ z\mathbf{I} & \mathbf{0} \end{bmatrix} \tag{8.43}$$

being the permuted advance chain. The synthesis polyphase matrix is then

$$\mathbf{R}(z) = \mathbf{G}_{K-1}^{-1}(z)\mathbf{G}_{K-2}^{-1}(z)\ldots\mathbf{G}_0^{-1}. \tag{8.44}$$

The more detailed polyphase representation is shown in Fig. 8.28 for the case of $M = 4$ and $K = 1$. By choosing the inverse $\mathbf{G}_i^{-1}(z) = \mathbf{P}_i^{-1}\,\mathbf{\Lambda}^{-1}(z)$, we ensure that the resulting filter bank or decomposition has perfect reconstruction. Enforcing the tighter constraint $\mathbf{P}^{-1} = \mathbf{P}^T$ leads to a paraunitary system, i.e., an orthogonal mapping.

To obtain linear-phase basis functions, a further structural constraint has to be imposed on the basic building block $\mathbf{P}$. A general closed-form M-point preprocessing operator $\mathbf{P}$ that works well with DCT of any size M, generating a linear-phase mapping, is presented in [35]. This general preprocessing block operator is shown in Fig. 8.29(a) and it should be placed squarely between two adjacent DCT blocks as illustrated in Fig. 8.28. The linear operator consists of two stages of butterflies and a matrix $\mathbf{V}$ between them:

$$\mathbf{P} = \frac{1}{2}\begin{bmatrix} \mathbf{I} & \mathbf{J} \\ \mathbf{J} & -\mathbf{I} \end{bmatrix} \begin{bmatrix} \mathbf{I} & \mathbf{0} \\ \mathbf{0} & \mathbf{V} \end{bmatrix} \begin{bmatrix} \mathbf{I} & \mathbf{J} \\ \mathbf{J} & -\mathbf{I} \end{bmatrix}, \tag{8.45}$$

where the reader is reminded that $\mathbf{I}$, $\mathbf{J}$, and $\mathbf{0}$ are the $\frac{M}{2} \times \frac{M}{2}$ identity matrix, the reversal matrix, and the null matrix, respectively. The matrix $\mathbf{V}$, controlling pre- and postprocessing behavior, can be further chosen as

$$\mathbf{V} = \mathbf{J}\mathbf{C}_{M/2}^{II^T}\mathbf{S}\mathbf{C}_{M/2}^{IV}\mathbf{J}, \tag{8.46}$$

where $\mathbf{C}_{M/2}^{II}$ is the $\frac{M}{2}$-point type-II DCT matrix in (8.7), $\mathbf{C}_{M/2}^{IV}$ is the $\frac{M}{2}$-point type-IV DCT matrix in (8.8), and $\mathbf{S}$ is a diagonal matrix that introduces biorthogonality. One example of $\mathbf{S}$ is diag$\{\frac{8}{5}, 1, \ldots, 1\}$. The structure produces an orthogonal preprocessing linear operator if $\mathbf{S}$ is chosen to be the identity

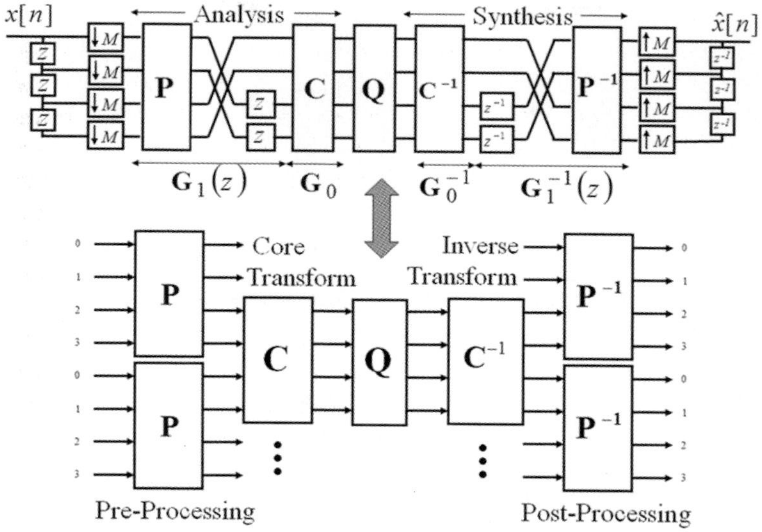

FIGURE 8.28

Construction of an M-channel filter bank from two $M \times M$ matrices **P** and **C** (top) and the equivalent time-domain interpretation of pre- and postprocessing (bottom).

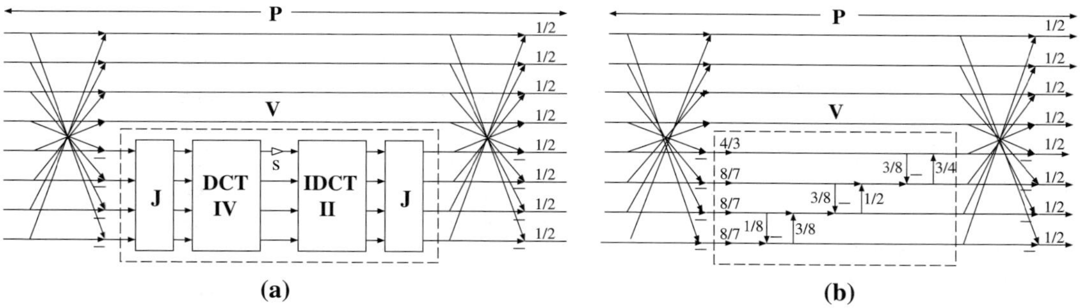

FIGURE 8.29

Preprocessors for the DCT. (a) General closed-form M-point prefilter. (b) Fast eight-point rational-coefficient preprocessing operator.

matrix. In general, it can be straightforwardly proven that any orthogonal matrix **V** would yield a linear-phase orthogonal transform. For the particular case of eight-point pre-/postprocessing, a fast rational-coefficient approximation of **P** is shown in Fig. 8.29(b).

It is interesting to note the flattening property of the preprocessing scheme as demonstrated in an image processing example shown in Fig. 8.30 where preprocessing is carried out in 2D separably. When the size of the preprocessing operator grows, the preprocessed image (which will then be the input of the

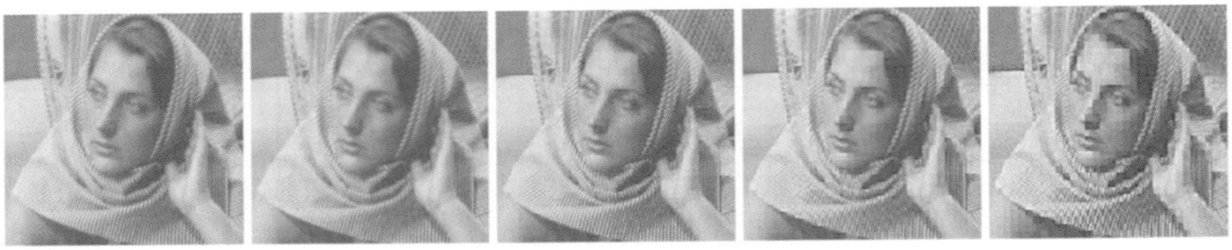

FIGURE 8.30

Preprocessing's blockwise flattening effect with 8×8 block size. From left to right: original image; after two-point, four-point, six-point, and eight-point preprocessing.

traditional block DCT decomposition) becomes more blocky since each 8×8 block becomes smoother and less correlated with its neighbors, and more high-frequency components are shifted to the block boundary. The closed-form preprocessor **P** of (8.45) depicted in Fig. 8.29(a) is used to generate this example. At the decoder side, the postprocessor – if chosen as the inverse of its counterpart – serves as a *smoothening interpolating operator* placed between block boundaries, mitigating blocking artifacts.

8.7.2.3 HD photo or JPEG-XR transform

Both concepts of multiplierless transform design via the lifting scheme as described in Section 8.5.4 and pre-/postprocessing of Section 8.7.2 are on full display in JPEG-XR [27,28]. The still-image compression algorithm/file format JPEG-XR has been widely distributed along with Windows Vista and Windows 7 operating systems and was officially approved as an ISO/IEC International standard in 2010. The entire transformation stage in JPEG-XR (formerly known as Microsoft HD-Photo) – including a base integer transform called photo core transform (PCT) **C** and an additional preprocessing operator called photo overlap transform (POT) **C** – is just one particular solution of the more general framework illustrated in Figs. 8.26 and 8.28.

Both operators are constructed from cascades of hardware-friendly dyadic-rational lifting steps. In fact, the choice of the two dyadic lifting steps $\frac{1}{2}$ in PCT comes from Configuration C5 as shown previously in Table 8.1. In Fig. 8.31, two pairs of scaling factors $\sqrt{2}$ and $\frac{1}{\sqrt{2}}$ are shown merely to

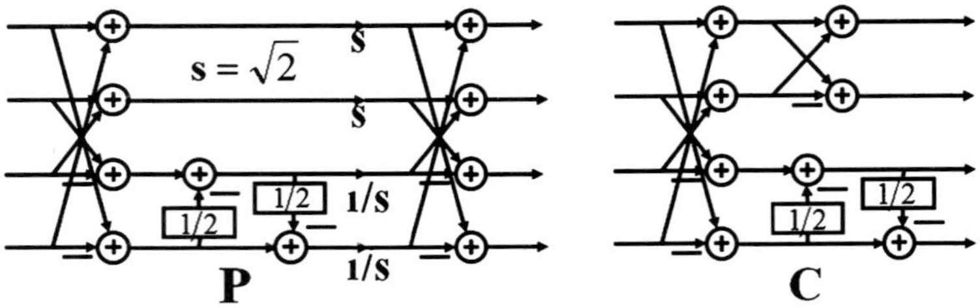

FIGURE 8.31

The photo overlap transform **P** (left) and the photo core transform **C** (right) in HD photo or JPEG-XR transformation stage.

illustrate the conceptual design. The actual implementation in JPEG-XR utilizes dyadic-rational lifting steps to approximate each pair of irrational scaling. Therefore, the entire transformation stage (including both PCT and POT) in JPEG-XR is constructed from a cascade of only dyadic-rational lifting steps [27,28].

8.7.3 Adaptive decomposition design

Here, by adaptivity, we refer to time-varying adaptable operators that the encoder or the decoder can select on-the-fly based on the local statistical behavior of the input. Adaptivity can lead to significant coding improvements if the amount of side information is cleverly avoided or kept to a minimum. It is clear that long basis functions associated with large operator size are best for representing smooth signal regions. On the other hand, transient signals, abrupt discontinuities, and strong edges are best captured with short high-time resolution basis. Based on this observation, various adaptive operators are developed to provide a powerful decomposition framework with strong time-frequency localization. Several options are under consideration:

- varying the support of pre- and postprocessing operator at each block boundary;
- varying the transform block size;
- a combination of both of the above;
- allowing every operator's parameters to be time-varying.

8.7.3.1 Adaptive pre-/postprocessing support

Within the pre-/postprocessing framework in Section 8.7.2, based on the energy of the transform coefficients generated, we can decide to turn on or off pre-/postprocessing. It is just as easy to vary the size of preprocessing support dynamically. This adaptive-borrowing signal decomposition is illustrated in Fig. 8.32(a) where the block size is fixed to $M = 4$ while the preprocessing operator can be chosen among: no processing, 2×2 processing, or 4×4 processing. In other words, from top to bottom, we are switching from a 4×8 to a 4×7 to a 4×6 to a 4×5 and finally to a 4×4 linear mapping. With the transform set to the DCT of a fixed block size (e.g., 8) and the size of the pre-/postprocessing operator chosen from the set $\{0, 2, 4, 8\}$, the side information for each block boundary is only 2 bits. The side information can be lowered with Huffman or context-based adaptive arithmetic coding.

A visual illustration of adaptive pre-/postprocessing is depicted in the leftmost binary image in Fig. 8.33. Here, dark pixels indicate blocks which the coder select the minimum-size 2×2 operator whereas light pixels indicate blocks where the maximum-size 8×8 preprocessing operator is preferred. The decision is based on the minimum bit length needed to encode each particular data block and we have employed an exhaustive search [36]. It is clear that pre- and postprocessing *should be avoided* at or near strong edges of the input image.

8.7.3.2 Adaptive transform block size

Another adaptive decomposition scheme is to employ variable block sizes. In slowly changing homogeneous parts of the signal, a large data block is desirable (a higher frequency resolution is needed). On the other hand, in fast-changing transient parts of the signal, it is more advantageous to switch to a small block size (a higher time resolution is needed). Such a signal-adaptive switching scheme has proven to be very effective in practice. For instance, MPEG-4's Advanced Audio Coder switches back-and-forth between a 256-point high-time resolution short window and a 2048-point high-frequency resolution

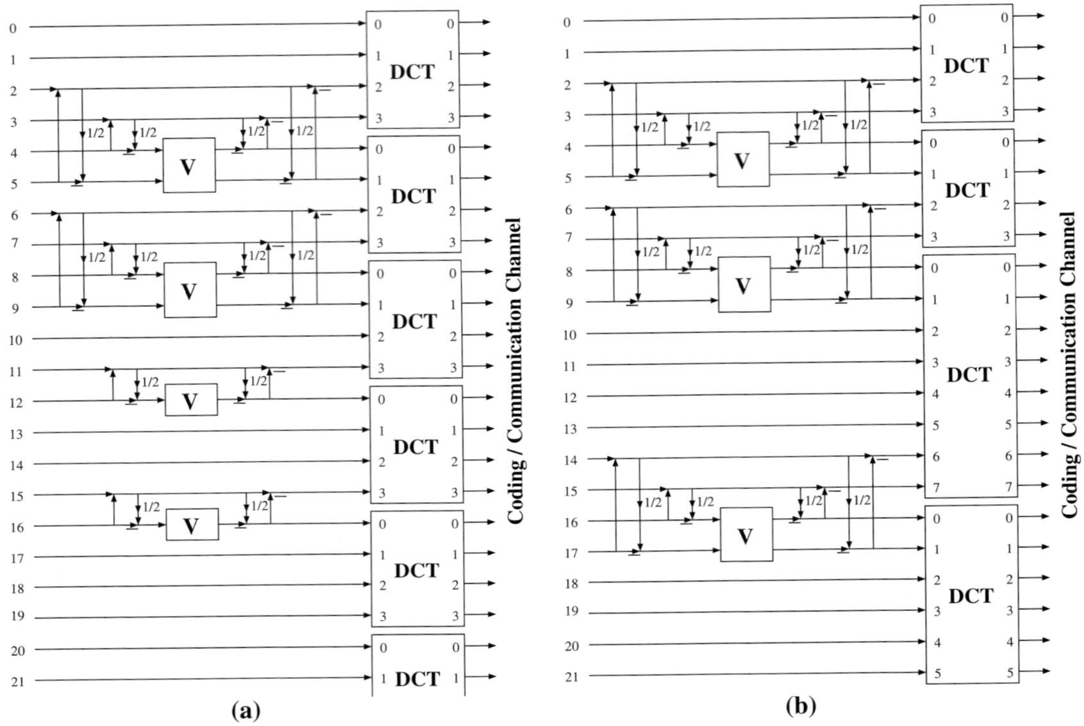

FIGURE 8.32

Examples of adaptive time-varying decomposition. (a) Adaptive preprocessing with fixed DCT block size. (b) Adaptive DCT block size with fixed preprocessing operators.

long window to avoid pre-echo and to improve coding efficiency [37]. A variable-block size decomposition scheme as depicted in Fig. 8.32(b) can be employed where the preprocessing operator is fixed whereas the DCT block size is allowed to vary. The side information, just like in the adaptive-processing case, can be kept manageable as well with a well-chosen set of block sizes; see, e.g., {4, 8, 16, 32} for image/video applications.

A demonstration of the adaptive block transform algorithm is shown in the right of Fig. 8.33. An expensive exhaustive search algorithm is set up to generate this example: pre- and postprocessing is fixed to the maximum size allowable based on the size of any two neighboring DCT blocks; block size selection is restricted to {4, 8, 16}; and the criterion is again the minimum amount of bits required to encode each local 16×16 image region. This example vividly confirms that the optimally adaptive decomposition algorithm has a tendency to latch onto strong local image features [36].

In the more general case, both adaptive preprocessing with different sizes, with multiple stages, and adaptive variable block size can be combined. This decomposition scheme generates a large library of basis functions that the encoder can choose from depending on the input signal behavior. How to

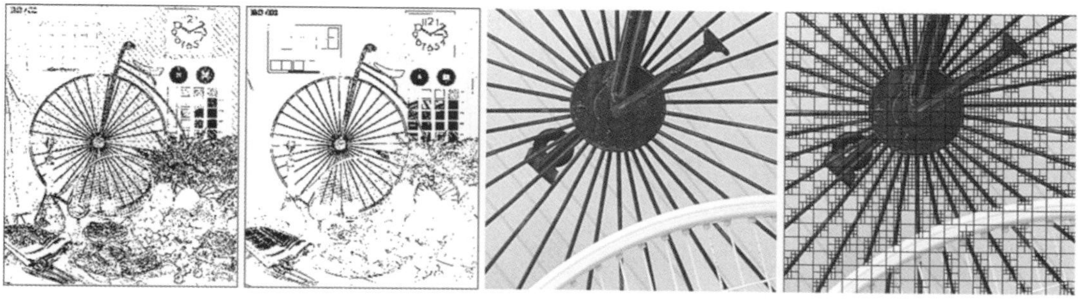

FIGURE 8.33

Visual demonstration of adaptive decomposition algorithms. From left to right: pre-/postprocessing mode selection for maximum local coding efficiency via exhaustive search; pre-/postprocessing mode selection for fast overflow/underflow criterion; portion of the original *Bike* image; optimal block size selection based on a maximum coding efficiency criterion.

make the right decision quickly and how to minimize the amount of side information involved are two important open research problems.

8.7.4 Modulated design

One drawback with the cascading structural approach in (8.42) and illustrated in Fig. 8.26 is that the optimization becomes problematic when the application demands or desires a larger number of channels M and/or a longer filter length $L = KM$. In speech and audio coding applications, for example, frequency resolution of the transform is often critical and the number M of frequency bins (or channels) is typically in the hundreds or even thousands [12,38].

Given that most of the cost function in Section 8.3.1 is not convex, the huge number of free parameters in the matrices $\{\mathbf{G}_0, \mathbf{P}_i\}$ – from $2K\binom{M/2}{2}$ in the linear-phase orthogonal case to $K\binom{M}{2}$ in the orthogonal case to KM^2 in the most general case [35] – poses a major challenge to any optimization approach. One particular elegant design that can efficiently deal with this obstacle is the cosine-modulated approach where a low-pass prototype filter is designed and the rest of the filters are obtained from cosine or sine modulation as illustrated in Fig. 8.34. Many different cosine-modulation approaches have been developed and the most significant difference among them is the low-pass prototype choice and the phase of the cosine/sine sequence [6,7,12].

For the case $K = 1$ ($M \times 2M$ transform), the modulated design with the type-IV DCT in (8.8) is often referred to as the modified DCT (MDCT) [39] or the modulated LT (MLT) [40]. The transform basis functions (or the filter coefficients) are given as

$$h_k[n] = \sqrt{\frac{2}{M}}\, h[n] \cos\left[\left(n + \frac{M+1}{2}\right)\left(k + \frac{1}{2}\right)\frac{\pi}{M}\right] \tag{8.47}$$

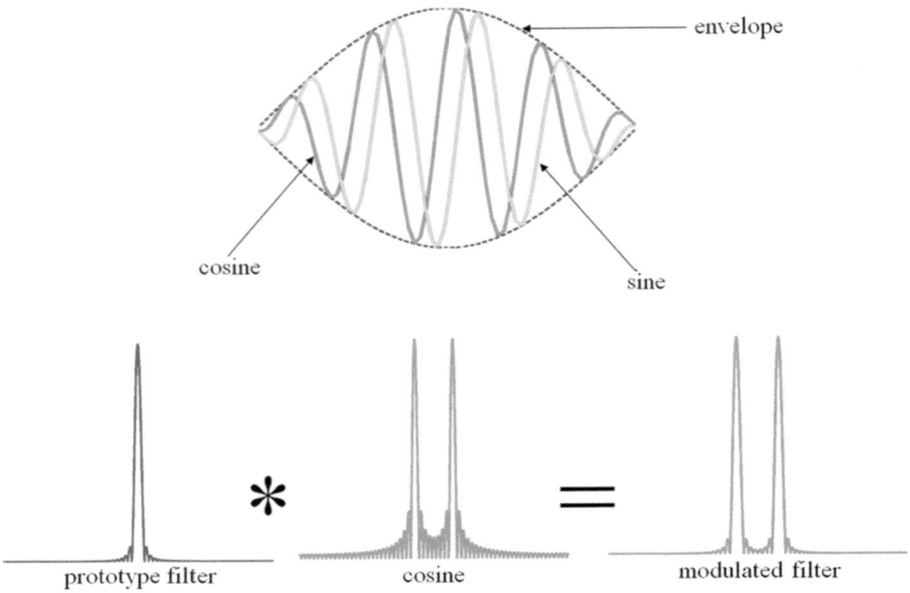

FIGURE 8.34

Construction via cosine/sine modulation.

for $0 \le k \le M-1, 0 \le n \le 2M-1$ and $h[n]$ can be thought of as a symmetric window modulating the cosine sequence or the impulse response of a low-pass prototype (with a cutoff frequency of $\frac{\pi}{2M}$). It has been shown that as far as perfect reconstruction is concerned, the choice of the prototype window $h[n]$ is rather flexible: all it has to satisfy is the following two constraints:

$$\begin{cases} h[n] = h[2M-1-n], \\ h^2[n] + h^2[M-1-n] = 1. \end{cases} \tag{8.48}$$

A nice choice with a closed-form expression for the prototype filter $h[n]$ is

$$h[n] = -\sin\left[\left(n+\frac{1}{2}\right)\frac{\pi}{2M}\right], \tag{8.49}$$

for which the reader can easily verify that it does not only satisfy both conditions in (8.48) but it also produces a smooth symmetric low-pass filter.

It is interesting to note that the MDCT/MLT as defined in (8.47) and (8.49) also fits the pre- and post-processing framework in Fig. 8.28. In this interpretation, the prototype window is completely governed by the preprocessing operator $\mathbf{P}$ while the orthogonal core transform $\mathbf{C}^{IV}$ takes care of the cosine-modulation part. The particular window choice in (8.49) leads to a series of $\frac{M}{2}$ pairwise rotation angles, which yields an overall orthogonal transform. Unfortunately, we did have to give up the linear-phase property. However, for speech and audio applications, that is not a critical property. The MDCT/MLT as

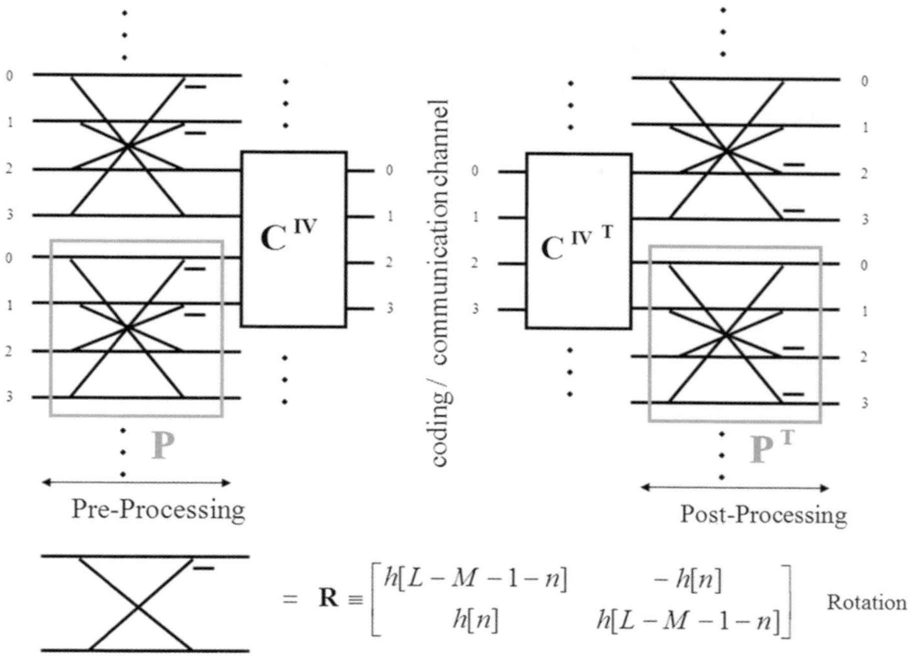

FIGURE 8.35

Modulated LT or MDCT implementation with pre-/postprocessing operators.

depicted in Fig. 8.35 still has a very efficient implementation. Finally, as discussed in Section 8.5.2, the MDCT can easily be approximated with dyadic lifting steps, rendering a multiplierless transformation with exact integer-to-integer mapping [41].

Both MP3 and MPEG-2 AAC employ the choice of window in (8.49). However, this is not the only choice. In fact, the open source Vorbis codec selects a slightly different sine window

$$h[n] = \sin\left\{\frac{\pi}{2}\sin^2\left[\left(n+\frac{1}{2}\right)\frac{\pi}{2M}\right]\right\}. \tag{8.50}$$

MP3 utilizes a hybrid adaptive decomposition approach: the MDCT/MLT is applied to the output of a 32-channel 512-tap near-perfect reconstruction filter bank. MP3's hybrid filter bank adapts to the signal characteristics via the block switching technique similarly to the adaptive mechanism shown in Fig. 8.32: the codec adaptively selects either a 6×12 MDCT (labeled short block for higher time resolution) or an 18×36 MDCT (labeled long block for higher frequency resolution). On the other hand, MPEG-4 AAC dynamically switches between a 128×256 MDCT (short blocks) and a 1024×2048 MDCT (long blocks) [38]. The long block's frequency resolution is rather high, offering improved coding efficiency for stationary signals whereas the shorter blocks provide optimized coding capability for transient signals. The transform choice in MP3 and AAC clearly indicates the importance of adaptivity in state-of-the-art compression systems.

Finally, if the application requires even higher frequency resolution, the reader should consider the extended version of the MDCT/MLT, namely the extended lapped transform (ELT) [12,42] whose basis functions are

$$h_k[n] = h[n] \cos\left\{\left[k + \frac{1}{2}\right]\left[\left(n - \frac{L-1}{2}\right)\frac{\pi}{M} + (N+1)\frac{\pi}{2}\right]\right\} \tag{8.51}$$

for $0 \le k \le M-1$, $0 \le n \le L-1$, and $h[n]$ is again the impulse response of the prototype low-pass window. A major advantage of the ELT is the existence of various fast implementation algorithms which still rely on the DCT type-IV as the modulation engine but employ more layers of preprocessing operators (similarly to the overall framework shown in Fig. 8.26).

8.8 Conclusion

In this chapter, we first briefly reviewed the deep and rich history of transform design and then presented a general systematic approach for modern transform design: the structural construction of high-order transforms from low-order basic building blocks: butterflies, lifting steps, and possibly scaling factors. We concentrated on the integer or dyadic-rational coefficient case which yields powerful integer invertible transforms. Numerous key examples in mainstream audio, image, and video applications were also presented.

References

[1] W.B. Pennebaker, J.L. Mitchell, JPEG: Still Image Compression Standard, Van Nostrand Reinhold, New York, NY, 1993.

[2] J.L. Mitchell, D. LeGall, C. Fogg, MPEG Video Compression Standard, Chapman & Hall, New York, NY, 1996.

[3] ITU Telecom, Standardization Sector of ITU, Video coding for low bit rate communication, ITU-T Recommendation H.263, March 1996.

[4] D.S. Taubman, M.W. Marcellin, JPEG2000: Image Compression Fundamentals, Standards, and Practice, Kluwer Academic Publishers, Norwell, MA, 2002.

[5] S.G. Mallat, A Wavelet Tour of Signal Processing, Academic Press, San Diego, CA, 1998.

[6] G. Strang, T.Q. Nguyen, Wavelets and Filter Banks, second ed., Wellesley-Cambridge Press, Boston, 1998.

[7] P.P. Vaidyanathan, Multirate Systems and Filter Banks, Prentice Hall, Englewood Cliffs, NJ, 1993.

[8] M. Vetterli, J. Kovacevic, Wavelets and Subband Coding, Prentice Hall, Englewood Cliffs, NJ, 1995.

[9] J.W. Cooley, J.W. Tukey, An algorithm for the machine calculation of complex Fourier series, Math. Comput. 19 (1965) 297–301.

[10] N. Ahmed, T. Natarajan, K.R. Rao, Discrete cosine transform, IEEE Trans. Comput. C 23 (1974) 90–93.

[11] K.R. Rao, P. Yip, Discrete Cosine Transform: Algorithms, Advantages, Applications, Academic Press, New York, NY, 1990.

[12] H.S. Malvar, Signal Processing with Lapped Transforms, Artech House, Boston, MA, 1992.

[13] I.E.G. Richardson, H.264 and MPEG-4 Video Compression, John Wiley and Sons, 2003.

[14] H. Malvar, A. Hallapuro, M. Karczewicz, L. Kerofsky, Low-complexity transform and quantization in H.264/AVC, IEEE Trans. Circuits Syst. Video Technol. 13 (2003) 598–603.

[15] W.K. Cham, Development of integer cosine transforms by the principle of dyadic symmetry, IEE Proc. 136 (1989) 276–282.

[16] W.K. Cham, Y.T. Chan, An order-16 integer cosine transform, IEEE Trans. Signal Process. 39 (1991) 1205–1208.

[17] K.T. Lo, W.-K. Cham, Development of simple orthogonal transforms for image compression, IEE Proc. Vis. Image Signal Process. 142 (1995) 22–26.

[18] I.D. Tun, S.U. Lee, On the fixed-point-error analysis of several fast DCT algorithms, IEEE Trans. Circuits Syst. Video Technol. 3 (1993) 27–41.

[19] W. Sweldens, The lifting scheme: a custom-design construction of bi-orthogonal wavelets, Appl. Comput. Harmon. Anal. 3 (1997) 185–200.

[20] A.A.M.L. Bruekens, A.W.M. vanden Enden, New networks for perfect inversion and perfect reconstruction, IEEE J. Sel. Areas Commun. 10 (1992) 130–137.

[21] L. Liu, T.D. Tran, P. Topiwala, From 16-bit to high-accuracy idct approximation: fruits of single architecture affiliation, in: Proceedings of SPIE Applications of Digital Image Processing XXX, August 2007.

[22] H.S. Malvar, G.J. Sullivan, YCoCg-R: a color space with RGB reversibility and low dynamic range, JVT ISO/IEC MPEG and ITU-T VCEG, Document No. JVT-I014r3, July 2003.

[23] P. Topiwala, C. Tu, New invertible integer color transforms based on lifting steps and coding of 4:4:4 video, JVT ISO/IEC MPEG and ITU-T VCEG, Document No. JVT-I014r8, July 2003.

[24] K.R. Rao, P. Yip, V. Britanak, Discrete Cosine Transform: Algorithms, Advantages, Applications, Academic Press, 2007.

[25] J. Liang, T.D. Tran, P. Topiwala, A 16-bit architecture for h.26l, treating DCT transforms and quantization, ITU-T Q.6/SG16, Document No. VCEG-M16, June 2001.

[26] J. Liang, T.D. Tran, Fast multiplier-less approximations of the DCT with the lifting scheme, IEEE Trans. Signal Process. 49 (2001) 3032–3044.

[27] S. Srinivasan, C. Tu, S.L. Regunathan, G.J. Sullivan, Hd photo: a new image coding technology for digitalphotography, in: Proceedings of SPIE Applications of Digital Image Processing XXX, vol. 6696, August 2007.

[28] C. Tu, S. Srinivasan, G.J. Sullivan, S. Regunathan, H.S. Malvar, Low-complexity hierarchical lapped transform for lossy-to-lossless image coding in jpeg xr/hd photo, in: Proceedings of SPIE Applications of Digital Image Processing XXXI, vol. 7073, August 2008.

[29] S. Oraintara, Y.J. Chen, T. Nguyen, Integer fast Fourier transform, IEEE Trans. Signal Process. 50 (2002) 607–618.

[30] I. Daubechies, Ten Lectures on Wavelets, CBMS-NSF Regional Conference Series in Applied Mathematics, SIAM, Philadelphia, PA, 1992.

[31] D. LeGall, A. Tabatabai, Subband coding of digital images using symmetric short kernel filters and arithmetic coding techniques, in: Proceedings of the IEEE International Conference on Acoustics, Speech, and Signal Processing (ICASSP), 1988, pp. 761–765.

[32] M. Antonini, M. Barlaud, P. Mathieu, I. Daubechies, Image coding using wavelet transform, IEEE Trans. Image Process. 1 (1992) 205–220.

[33] M. Under, T. Blu, Mathematical properties of the JPEG2000 wavelet filters, IEEE Trans. Image Process. 12 (2003) 1080–1090.

[34] I. Daubechies, W. Sweldens, Factoring wavelet transforms into lifting steps, J. Fourier Anal. Appl. 4 (3) (1998) 247–269.

[35] T.D. Tran, J. Liang, C. Tu, Lapped transform via time-domain pre- and post-filtering, IEEE Trans. Signal Process. 51 (2003) 1557–1571.

[36] W. Dai, L. Liu, T.D. Tran, Adaptive block-based image coding with pre-/post-filtering, in: Proceedings of the Data Compression Conference, March 2005, pp. 73–82.

[37] J.D. Johnston, S.R. Quackenbush, J. Herre, B. Grill, Review of MPEG-4 general audio coding, in: Multimedia Systems, Standards, and Networks, 2000, pp. 131–155.
[38] A. Spanias, T. Painter, V. Atti, Audio Signal Processing and Coding, John Wiley and Sons, Hoboken, NJ, 2007.
[39] J.P. Princen, A.W. Johnson, A.B. Bradley, Subband/transform coding using filter bank designs based on time domain aliasing cancellation, in: Proceedings of the IEEE International Conference on Acoustics, Speech, and Signal Processing (ICASSP), May 1987, pp. 2161–2164.
[40] H.S. Malvar, Lapped transforms for efficient transform/subband coding, IEEE Trans. Acoust. Speech Signal Process. 38 (1990) 969–978.
[41] Y. Yokotani, R. Geiger, G. Schuller, S. Oraintara, K.R. Rao, Lossless audio coding using the IntMDCT and rounding error shaping, IEEE Trans. Audio Speech Lang. Process. 14 (2006) 2201–2211.
[42] H.S. Malvar, Extended lapped transforms: properties, applications and fast algorithms, IEEE Trans. Signal Process. 40 (1992) 2703–2714.

CHAPTER

9 Data representation: from multiscale transforms to neural networks

Yufang Bao[a] and Hamid Krim[b]

[a]*Department of Mathematics and Computer Science, UNC Fayetteville State University, Fayetteville, NC, United States*

[b]*Electrical and Computer Engineering Department, North Carolina State University, Raleigh, NC, United States*

9.1 Introduction

9.1.1 An overview of multiscale transforms

One's first and most natural reflex when presented with an unfamiliar object is to carefully look it over and hold it up to the light to inspect its different facets in the hope of recognizing it, or of at least relating any of its aspects to a more familiar and well-known entity. This almost innate strategy pervades all science and engineering disciplines.

Physical phenomena (e.g., earth vibrations) are monitored and observed by way of measurements in a form of temporal and/or spatial data sequences. Analyzing such data is tantamount to extracting information which is useful to further our understanding of the underlying process (e.g., frequency and amplitude of vibrations may be an indicator for an imminent earthquake). Visual or manual analysis of typically massive amounts of acquired data (e.g., in remote sensing) is impractical, making one resort to adapted mathematical tools and analysis techniques to better cope with potential intricacies and complex structure of the data. Among these tools, a variety of functional transforms (e.g., Fourier transform) exist which, in many cases, may facilitate and simplify an analytical solution of a problem. More frequently (and just as importantly) they also provide an alternative view of, and a better insight into, the problem (this, in some sense, is analogous to exploring and inspecting data under a "different light"). An illustration of such a "simplification" is shown in Fig. 9.1.1, where a rather intricate signal $x(t)$ shown in the leftmost figure may be displayed/viewed in a different space as two elementary tones. In Fig. 9.1.2, a real bird chirp is similarly displayed as a fairly rich signal which, when considered in an appropriate space, is reduced and "summarized" to a few "atoms" in the time-frequency (T-F) representation. Transformed signals may formally be viewed as *convenient* representations in a different domain which is itself described by a set of vectors/functions $\{\phi_i(t)\}_{i=\{1,2,\cdots,N\}}$. A contribution of a signal $x(t)$ along a direction "$\phi_i(t)$" (its projection) is given by the following inner product:

$$C_i(x) = < x(t), \phi_i(t) > = \int_{-\infty}^{\infty} x(t)\phi_i(t)dt, \tag{9.1.1.1}$$

where the compact notation "$< \cdot, \cdot >$" for an inner product is used. The choice of functions in the set is usually intimately tied to the nature of information that we are seeking to extract from the signal of interest. A Fourier function, for example, is specified by a complex exponential "$e^{j\omega t}$" and reflects the

Signal Processing and Machine Learning Theory. https://doi.org/10.1016/B978-0-32-391772-8.00015-6
Copyright © 2024 Elsevier Inc. All rights reserved.

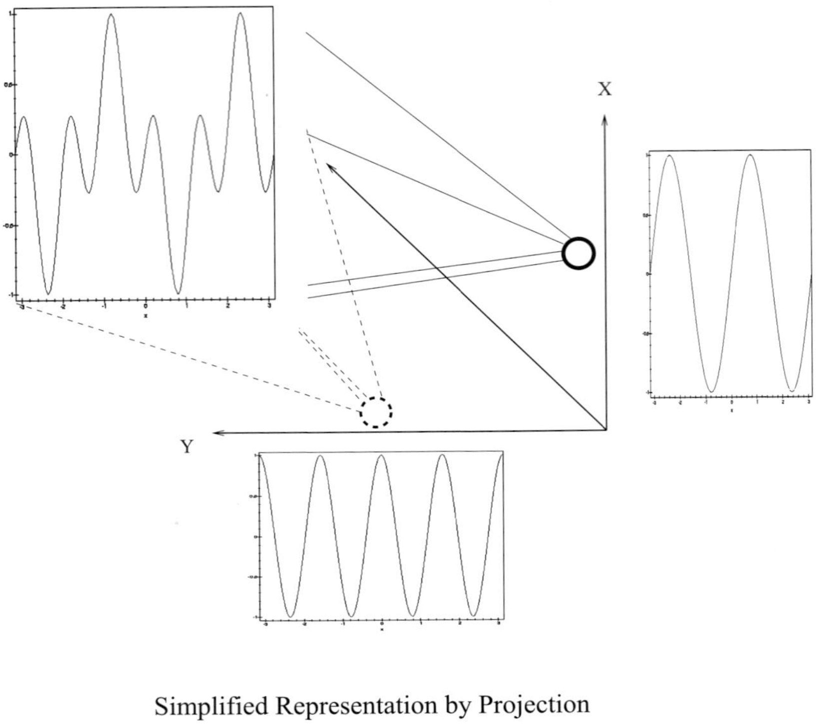

FIGURE 9.1.1

A canonical function-based projection to simplify a representation.

harmonic nature of a signal, and provides it with a simple and an intuitively appealing interpretation as a "tone." Its spectral (frequency) representation is a Dirac impulse which is well localized in frequency. A Fourier function is hence well adapted to represent a class of signals with a fairly well localized "spectrum," and is highly inadequate for a transient signal which, in contrast, tends to be temporally well localized. This thus calls for a different class of analyzing functions, namely those with good temporal compactness. The wavelet transform, by virtue of its mathematical properties, indeed provides such an analysis framework which in many ways is complementary to that of the Fourier transform.

The success of classical wavelets has motivated active research in the multiscale analysis field where multidisciplinary researchers pursue the best multiscale basis/frame in order to efficiently approximate a function that has singularities across scales. Although the classical wavelets can reach the optimal approximation for 1D functions, it has problems in generating sparse coefficients to approximate a 2D function with discontinuities. The classical wavelets locate singularities of a 2D function in isotropic and multiscale representations. The Fourier frequency plane is partitioned with scaled square windows that are only in the horizontal/vertical directions. The bandpass wavelet atoms, in general, are conceptually equivalent to applying differential operators (such as calculating the gradient difference) to an image to measure the regularities in localized areas along the horizontal/vertical directions. This

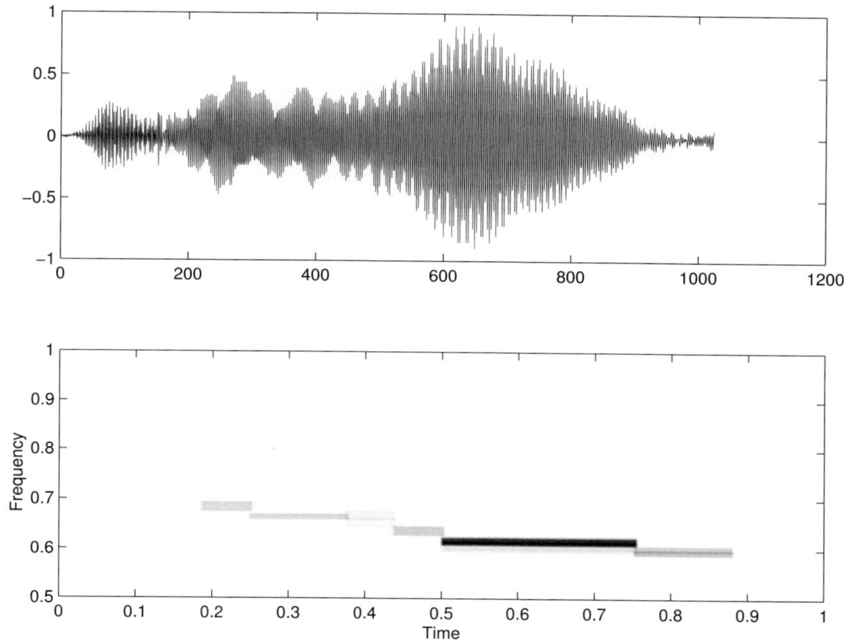

FIGURE 9.1.2

Bird chirp with a simplified representation in an appropriate basis.

isotropic measurement lacks flexibility, and therefore, although the classical wavelets can identify discontinuities efficiently, they cannot efficiently detect the direction-oriented curves in an image even at fine scales. Even though some improvements have been introduced to adopt anisotropic dilation in different directions, the problems still persist.

A 2D signal, visually represented as an image, typically contains spatially distributed geometrical shapes that can be characterized by edges that signal the transients from one object to other objects. The edges are the discontinuities in an image intensity function along curves, contours, or lines that are directional-oriented. Scientists have acknowledged [1,2,3] the instinct of the human eye in recognizing objects in a multiscale manner. The eye can quickly zoom in to the most important isotropic or anisotropic features. Researchers in mathematics and engineering fields have extended this natural vision to the new era of anisotropic wavelets. These new transformations have allowed researchers to selectively capture edges, which are important features in image processing.

The recently developed curvelets [4,5,6,7], contourlets [8,9], and shearlets [10,11,12,13] aim to mathematically animate the eye's function in order to catch the anisotropic curve features. We generally refer to them as anisotropic wavelets. The beauty with anisotropic wavelets is their flexibility in partitioning the Fourier plane into a collection of atoms with varied scales and orientations. One advantage of anisotropic wavelets is that when a 2D function is projected into these wedge-shaped atoms, the energy of singularities can be confined to only a sparse number of atoms, resulting in only a small amount of significant coefficients across scales. They are more efficient tools than the classical wavelets in capturing the direction-oriented curve features. Another advantage of anisotropic wavelets is that many nice features of classical wavelets are retained with anisotropic wavelets. Some essential results are parallel to those for classical wavelets, for example, the exact reconstruction and energy conservation formula. This has not only made anisotropic wavelets more attractive tools, but also allows fast development of anisotropic wavelets using classical wavelets as guidelines.

The goal of this chapter being of a tutorial nature, we cannot help but provide a somewhat high-level exposition of the theoretical frameworks of classical and anisotropic wavelets, while our focus will be to provide working knowledge of the tools as well as their potential for application in information sciences in general. Some of the MATLAB® code for the pictures used in this paper can be downloaded [14]. Before we get into the introduction of the various wavelets, we will introduce some classical Fourier transform-related topics in the remainder of this section.

9.1.2 Signal representation in functional bases

As noted above, a proper selection of a set of analyzing functions heavily impacts the resulting representation of an observed signal in the dual space, and hence the resulting insight as well.

When considering finite energy signals (these are also said to be $L^2(\mathcal{R}^d)$), i.e.,

$$\int_{\mathcal{R}^d} | x(\boldsymbol{t}) |^2 \, d\boldsymbol{t} < \infty, \tag{9.1.2.1}$$

a convenient functional vector space is that which is endowed with a norm inducing inner product, namely a Hilbert space, which will also be our assumed space throughout.

Definition 9.1. A vector space endowed with an inner product which in turn induces a norm such that every sequence of functions $x_n(\boldsymbol{t})$ whose elements are asymptotically close is convergent (this convergence is in the Cauchy sense whose technical details are deliberately left out to help the flow of the chapter) is called a Hilbert space.

Remark. An $L^2(\mathcal{R}^d), d = 1, 2$, space is a Hilbert space.

Definition 9.2. Given $\alpha > 0$, a function $x(\boldsymbol{t})$ is said to be Lipschitz-α at $\boldsymbol{t}_0 \in \mathcal{R}^d$ if there exists a positive constant K and a polynomial $p_{n,\boldsymbol{t}_0}$ with degree $n = [\alpha]$ such that for all $\boldsymbol{t}$ in a neighborhood of $\boldsymbol{t}_0$ we have

$$|x(\boldsymbol{t}) - p_{n,\boldsymbol{t}_0}(\boldsymbol{t})| \le K|\boldsymbol{t} - \boldsymbol{t}_0|^{\alpha}. \tag{9.1.2.2}$$

It is shown that the function $x(\boldsymbol{t})$ is necessarily m times differentiable at a point $\boldsymbol{s}_0$ if $x(\boldsymbol{t})$ is uniformly Lipschitz with $\alpha > m$ in a neighborhood of $\boldsymbol{s}_0$, namely, Eq. (9.1.2.2) is satisfied for all points $\boldsymbol{t}_0$ in the neighborhood of $\boldsymbol{s}_0$. Further, if $x(\boldsymbol{t})$ is Lipschitz-α with $\alpha < 1$ at $\boldsymbol{t}_0$, then f is not differentiable at $\boldsymbol{t}_0$.

Definition 9.3. A set of vectors or functions $\{\phi_i(\boldsymbol{t})\}, i \in \mathcal{N}$, which are linearly independent and span a vector space is called a *basis*. If in addition these elements are orthonormal, then the basis is said to be an orthonormal basis.

Among the desirable properties of a selected basis in such a space are *adaptivity* and a *capacity* to preserve and reflect some key signal characteristics (e.g., signal smoothness). Fourier bases have in the past been at the center of all science and engineering applications. They have, in addition, provided an efficient framework for analyzing periodic as well as nonperiodic signals with dominant modes.

9.1.3 The continuous Fourier transform

The Fourier transform is a powerful tool in understanding features of signals/images through a basis. A Fourier transform essentially measures the spectral content of a signal $x(\boldsymbol{t})$ across all frequencies "ω," viewed as a row vector, and is defined as the decomposition of the signal over a basis $\{e^{-j\omega \boldsymbol{t}}\}$.

Definition 9.4. For a signal $x(\boldsymbol{t}), \boldsymbol{t} \in \mathcal{R}^d$, viewed as a column vector that is an absolutely integrable function, its Fourier transform is defined as

$$X(\omega) = \int_{\mathcal{R}^d} x(\boldsymbol{t}) e^{-j\omega \boldsymbol{t}} d\boldsymbol{t}. \tag{9.1.3.1}$$

The formula is also referred to as the Fourier integral. The signal $x(\boldsymbol{t})$ can be recovered by its inverse Fourier transform, which is defined as

$$x(\boldsymbol{t}) = \frac{1}{(2\pi)^d} \int_{\mathcal{R}^d} X(\omega) e^{j\omega \boldsymbol{t}} d\omega. \tag{9.1.3.2}$$

As an orthonormal transform, the Fourier transform enjoys a number of interesting properties [15]. To simplify, we are constrained to the 1D function $f(t)$ in the following until we get to the sections on anisotropic wavelets. If we use $FT(f)(\omega) = F(\omega)$ to represent the Fourier transform of $f(t)$, we can list some of its important properties as follows:

- linearity:

$$FT(f+g)(\omega) = F(\omega) + G(\omega),$$

- convolution:

$$FT(f \star g)(\omega) = F(\omega)G(\omega),$$

- multiplication:

$$FT(fg)(\omega) = F(\omega) \star G(\omega),$$

- shifting:

$$FT(x(t-s))(\omega) = e^{-js\omega} X(\omega),$$

- scaling:

$$FT(x(at))(\omega) = |a|^{-1} X(a^{-1}\omega),$$

- differentiation:

$$FT(\frac{\partial^n x(t)}{\partial t^n})(\omega) = (j\omega)^n X(\omega).$$

Throughout, and unless otherwise specified, the $\star$ symbol between two functions usually indicates a convolution between the two functions, and "*" as a superscript denotes a complex conjugate. Furthermore, a useful property of the 1D Fourier transform is its energy preservation in the dual space (transform domain),

$$\int_{\mathcal{R}} | x(t) |^2 \, dt = \int_{\mathcal{R}} | X(\omega) |^2 \, d\omega.$$

This is also true for signals in $\mathcal{R}^d$, and is referred to as the Plancherel–Parseval property [15].

9.1.4 Fourier series

If a canonical function of a selected basis is $\phi(t) = e^{j\omega t}$, where $\omega \in \mathcal{R}$, then we speak of a Fourier basis which yields a Fourier series for periodic signals and a Fourier transform (FT) for aperiodic signals.

The Fourier transform is defined for a broad class of signals [16], and may also be specialized to derive the Fourier series of a periodic signal. This is easily achieved by noting that a periodic signal $\widetilde{x}(t)$ may be evaluated over a period T, which in turn leads to

$$\tilde{x}(t) = x(t+T) = \sum_{n=-\infty}^{\infty} \alpha_n^x e^{jn\omega_0 t}, \tag{9.1.4.1}$$

with $\omega_0 = 2\pi/T$ and $\alpha_n^x = 1/T \int_0^T x(t) e^{-jn\omega_0 t} dt$.

In real applications, however, $x(t)$ is measured and sampled at discrete times, requiring that the aforementioned transform be extended to obtain a spectral representation which closely approximates the theoretical truth and which remains just as informative. Towards that end, we proceed to define the discrete-time Fourier transform (DTFT) as

$$X(e^{j\omega}) = \sum_{i=-\infty}^{\infty} x(i) e^{-j\omega i}. \tag{9.1.4.2}$$

This expression may also be extended for finite observation (finite-dimensional) signals via the fast Fourier transform (FFT) [15]. While these transforms have been and remain crucial in many applications, they show limitations in problems requiring a good "time-frequency" localization, as often encountered in transient analysis. This may be easily understood by reinterpreting Eq. (9.1.4.2) as a weighted transform where each of the time samples is equally weighted in the summation for $X(e^{j\omega})$. Prior to introducing alternatives to uniform weighting in the Fourier space, we first discuss the practical implementation and exploitation of the DTFT.

9.1.5 The discrete Fourier transform

In applications, the discrete Fourier transform (DFT) is the computational surrogate for the continuous Fourier transform and a natural extension of the DTFT. It results in a periodic sequence of numbers that is mapped to discretely sampled values of an analog signal (viewed as a time function). The sampling is performed over an N-long window which is extended periodically. The DFT also views a finite length signal, $x(n)$, as a periodic signal, and is defined as

$$X(k) = \sum_{n=0}^{N-1} x(n) e^{\frac{-j2\pi kn}{N}}.$$

The original signal may be recovered by applying an inverse transform

$$x(n) = \frac{1}{N} \sum_{k=0}^{N-1} X(k) e^{\frac{j2\pi kn}{N}}.$$

The DFT decomposes a signal into an orthonormal basis. The energy is preserved by the Plancherel formula described earlier and written as

$$||f||^2 = \sum_{n=0}^{N-1} |x(k)|^2 = \frac{1}{N} \sum_{k=0}^{N-1} |X(k)|^2.$$

The DFT of a 2D signal (image) is correspondingly defined as

$$X(k_1, k_2) = \sum_{n_1, n_2=0}^{N-1} x(n_1, n_2) e^{\frac{-j2\pi(k_1 n_1 + k_2 n_2)}{N}}.$$

The original signal may also be recovered by applying an inverse transform

$$x(n_1, n_2) = \frac{1}{N^2} \sum_{k_1, k_2=0}^{N-1} X(k_1, k_2) e^{\frac{j2\pi(k_1 n_1 + k_2 n_2)}{N}}.$$

Since the DFT is based on the sampling taken over a certain length, the frequency content of the original function may not be adequately represented by the DFT when the sampling rate is insufficiently low. This will cause the high-frequency content to fold back into the lower-frequency content and yield aliasing.

To overcome some of the limitations of the uniform weighting of a function of a classical Fourier transform, Gabor [17] first proposed to use a different weighting window leading to the so-called windowed Fourier transform, which served well in many practical applications [18] and is discussed next.

9.1.6 The windowed Fourier transform

As just mentioned, when our goal is to analyze very local features, such as those present in transient signals, for instance, it makes sense to introduce a focusing window as follows:

$$W_{\mu,\omega}(t) = e^{j\omega t} W(t - \mu),$$

where $\| W_{(\mu,\omega)} \| = 1 \ \forall (\mu, \omega) \in \mathbb{R}^2$ and $W(\cdot)$ is typically a smooth function with a compact support. This yields the following parameterized transform:

$$\widehat{X}_W(\omega, \mu) = < x(t), W_{\mu,\omega}(t) > . \tag{9.1.6.1}$$

The selection of a proper window is problem-dependent and is ultimately resolved by the desired spectrotemporal trade-off, which is itself constrained by the Heisenberg uncertainty principle [18,19]. From

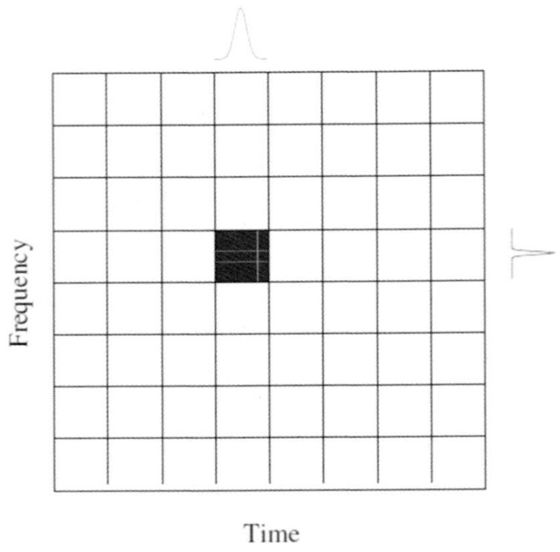

FIGURE 9.1.3

A Gaussian waveform results in a uniform analysis in the time-frequency plane.

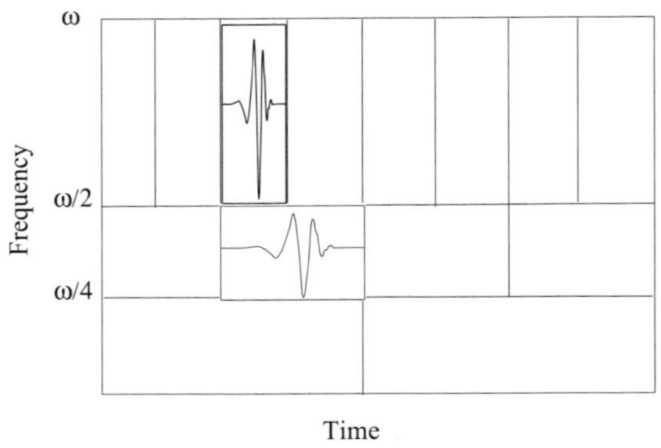

FIGURE 9.1.4

A time-frequency tiling of a dyadic wavelet basis by a proper subsampling of a wavelet frame.

the T-F distribution perspective, and as discussed by Gabor [17] and displayed in Fig. 9.1.3, the Gaussian window may be shown to have less temporal and spectral support than any other function. It hence represents the best compromise between the temporal and spectral resolution.

Its numerical implementation entails a discretization of the modulation and translation parameters and results in a uniform partitioning of the T-F plane, as illustrated in Fig. 9.1.4. Different windows result in various T-F distributions of elementary atoms, favoring either temporal or spectral resolution,

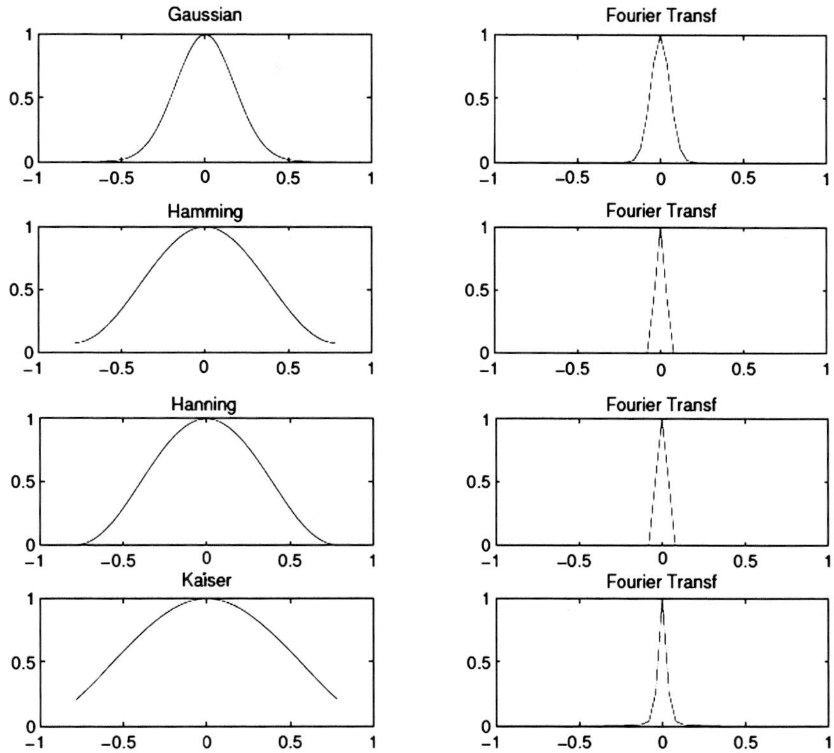

FIGURE 9.1.5

Trade-off resulting from windows.

as may be seen for the different windows in Fig. 9.1.5. While representing an optimal T-F compromise, the uniform coverage of the T-F plane by the Gabor transform falls short of adequately resolving a signal whose components are spectrally far apart. This may easily and convincingly be illustrated by the study case in Fig. 9.1.6, where we note the number of cycles which may be enumerated within a window of fixed time width. It is readily seen that while the selected window (shown grid) may be adequate for one fixed-frequency component, it is inadequate for another lower-frequency component. An analysis of a spectrum exhibiting such a time-varying behavior is ideally performed by way of a frequency-dependent time window in different designs, as we elaborate in the next several sections. The wavelet transform described next offers a highly adaptive window which is of compact support, and which, by virtue of its dilations and translations, covers different spectral bands at all instants of time. Anisotropic wavelets using different windows, such as the curvelet transform, the contourlet transform, and the shearlet transform, will be introduced following the discussion of the wavelet transform and wavelet decomposition.

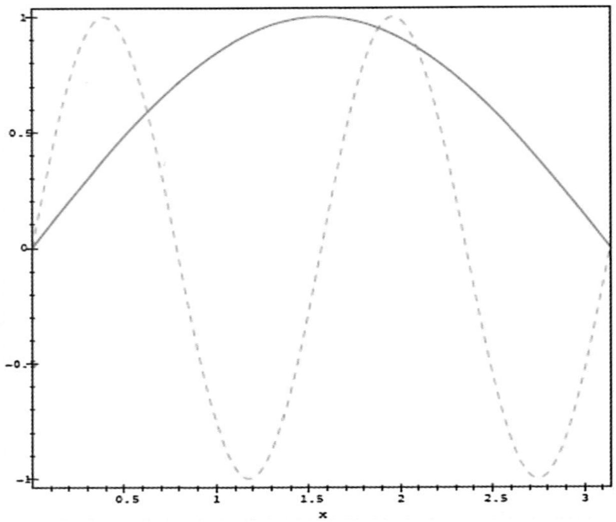

FIGURE 9.1.6

Time windows and frequency trade-off.

9.2 Wavelets: a multiscale analysis tool

This section is organized as follows. In Section 9.2.1 we present a multiscale analysis based on a wavelet basis and elaborate on its properties and implications, as well as its applications. In Section 9.2.2 we discuss a specific estimation technique which is believed to be sufficiently generic and general to be useful in a number of different applications of information sciences in general, in particular for signal processing and communications.

9.2.1 The wavelet transform

Much like the Fourier transform, the wavelet transform is based on an elementary function, which is well localized in time and frequency. In addition to a compactness property, this function has to satisfy a set of properties to be admissible as a wavelet. The first fundamental property is stated next.

Definition 9.5. [18,20,21] A wavelet is a finite energy function $\psi(\cdot)$ (i.e., $\psi(\cdot) \in L^2(\mathcal{R})$) with zero mean,

$$\int_{-\infty}^{\infty} \psi(t)dt = 0. \tag{9.2.1.1}$$

Wavelets are commonly normalized so that $\| \psi \| = \int | \psi |^2 dt = 1$. The wavelet also constitutes a fundamental building block in the construction of functions (atoms) spanning the time-scale plane by

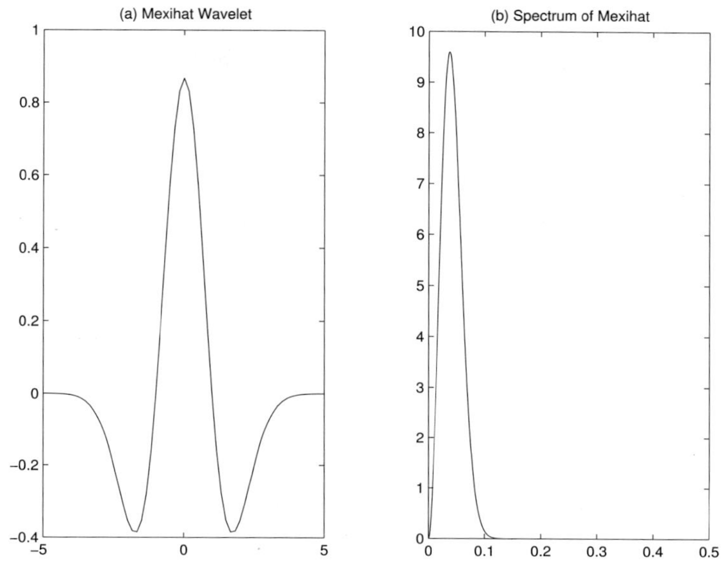

FIGURE 9.2.7

Admissible Mexican hat wavelet.

way of dilation and translation parameters. We hence write

$$\psi_{\mu,\xi}(t) = \frac{1}{\sqrt{\xi}}\psi\left(\frac{t-\mu}{\xi}\right),$$

where the scaling factor $\xi^{\frac{1}{2}}$ ensures an energy invariance of $\psi_{\mu,\xi}(t)$ over all dilations "ξ" $\in \mathcal{R}^+$ and translations "μ" $\in \mathcal{R}$. With such a function in hand, and towards mitigating the limitation of a windowed Fourier transform, we proceed to define a Wavelet transform with such a capacity. A scale-dependent window with good time localization properties as shown in Fig. 9.2.7 yields a transform for $x(t)$ given by

$$\mathcal{W}_x(\mu,\xi) = \int_{-\infty}^{\infty} x(t)\frac{1}{\sqrt{\xi}}\psi^*(\frac{t-\mu}{\xi})dt, \tag{9.2.1.2}$$

where "*" denotes complex conjugate. This is of course in contrast to the Gabor transform, whose window width remains constant throughout. A T-F plot of a continuous wavelet transform is shown in Fig. 9.2.8 for a corresponding $x(t)$.

9.2.2 Inverting the wavelet transform

Similarly to the weighted Fourier transform, the wavelet transform is a redundant representation, which with a proper normalization factor leads to the reconstruction formula

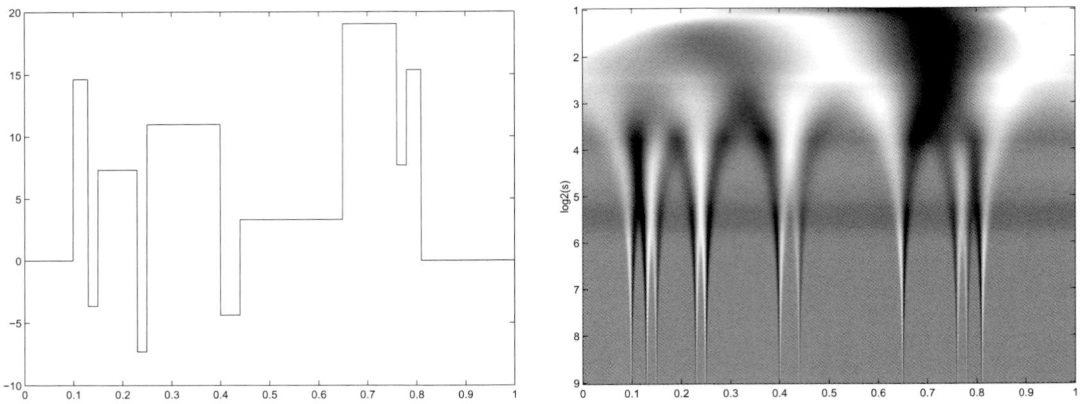

FIGURE 9.2.8

Continuous signal with corresponding wavelet transform with cones of influence around singularities.

$$x(t) = \frac{1}{C_\psi} \int_0^{\infty} \int_{-\infty}^{\infty} \mathcal{W}_x(\mu, \xi) \frac{1}{\sqrt{\xi}} \psi\left(\frac{t-\mu}{\xi}\right) \frac{d\xi}{\xi^2} d\mu, \tag{9.2.2.1}$$

with $C_\psi = \int_0^{+\infty} \frac{\hat{\Psi}^2(\omega)}{\omega} d\omega$.

While the direct and inverse wavelet transform have been successfully applied in a number of different problems [18], their computational cost due to their continuous/redundant nature is considered as a serious drawback. An immediate and natural question arises about a potential reduction in the computational complexity, and hence in the wavelet transform redundancy. This clearly has to be carefully carried out to guarantee a sufficient coverage of the T-F plane, and thereby ensure a proper reconstruction of the transformed signal. Upon discretizing the scale and dilation parameters, a dimension reduction relative to a continuous transform is achieved. The desired adaptivity of the window width with varying frequency naturally results by selecting a geometric variation of the dilation parameter $\xi = \xi_0^m$, $m \in \mathcal{Z}$. To obtain a systematic and consistent coverage of the T-F plane, our choice of the translation parameter "μ" should be in congruence with the fact that at any scale "m," the coverage of the whole line $\mathcal{R}$ (for instance) is complete and the translation parameter is in step with the chosen wavelet $\psi(t)$, i.e., $\mu = n\mu_0\xi_0^m$, with $n \in \mathcal{Z}$. This hence gives the following scale- and translation-adaptive wavelet:

$$\psi(t)_{m,n} = \xi_0^{-m/2} \psi\left(\frac{t}{\xi_0^m} - n\mu_0\right), \tag{9.2.2.2}$$

where the factor "$\xi_0^{-m/2}$" ensures a unit energy function. This reduction in dimensionality yields a redundant discrete wavelet transform endowed with a structure which lends itself to an iterative and fast inversion/reconstruction of a transformed signal $x(t)$.

Definition 9.6. The set of resulting wavelet coefficients $\{< x(t), \psi_{mn}(t) >\}_{(m,n)\in\mathcal{Z}^2}$ completely characterizes $x(t)$, and hence leads to a stable reconstruction [18,19] if $\forall x(t) \in L^2(\mathcal{R})$ (or finite energy signal) the following condition on the energy holds:

$$A \parallel x(t) \parallel^2 \leq \sum_{m,n} |< x(t), \psi_{mn}(t) >|^2 \leq B \parallel x(t) \parallel^2 . \tag{9.2.2.3}$$

Such a set of functions $\{\psi_{mn}(t)\}_{(m,n)\in\mathcal{Z}^2}$ then constitutes a frame.

The energy inequality condition intuitively suggests that the redundancy should be controlled to avoid instabilities in the reconstruction (i.e., too much redundancy makes it more likely that any perturbation of coefficients will yield an unstable/inconsistent reconstruction). If the frame bounds "A" and "B" are equal, the corresponding frame is said to be tight, and if furthermore $A = B = 1$, it is an orthonormal basis. Note, however, that for any $A \neq 1$ a frame is not a basis. In some cases, the frame bounds specify a redundancy ratio which may help guide one in inverting a frame representation of a signal, since a unique inverse does not exist. An efficient computation of an inverse (for reconstruction), or more precisely of a pseudoinverse, may only be obtained by a judicious manipulation of the reconstruction formula [18,19]. This is in a finite-dimensional setting, similar to a linear system of equations with a rank-deficient matrix whose column space has an orthogonal complement (the union of the two subspaces yields all of the Hilbert space), and hence whose inversion is ill-conditioned. The size of this space is determined by the order of the deficiency (number of linearly dependent columns) and explains the potential for instability in a frame projection-based signal reconstruction [18]. This type of "effective rank" utilization is encountered in numerical linear algebra applications (e.g., signal subspace methods [22], model order identification, etc.). The close connection between the redundancy of a frame and the rank deficiency of a matrix suggests that solutions available in linear algebra [23] may provide insight in solving our frame-based reconstruction problem. A well-known iterative solution to a linear system and based on a gradient search solution is described in [24] (and in fact coincides with a popular iterative solution to the frame algorithm for reconstructing a signal) for solving

$$\mathbf{f} = \mathbf{L}^{-1}\mathbf{b},$$

where "$\mathbf{L}$" is a matrix operating on $\mathbf{f}$ and $\mathbf{b}$ is the data vector. Starting with an initial guess $\mathbf{f}_0$ and iterating on it leads to

$$\mathbf{f}_n = \mathbf{f}_{n-1} + \alpha(\mathbf{b} - \mathbf{L}\mathbf{f}_{n-1}). \tag{9.2.2.4}$$

When α is appropriately selected, the latter iteration may be shown to yield [18]

$$\lim_{n\to+\infty} \mathbf{f}_n = \mathbf{f}. \tag{9.2.2.5}$$

Numerous other good solutions have also been proposed in the literature; their detailed descriptions are given in Refs. [18,19,25,26].

9.2.3 Wavelets and multiresolution analysis

While a frame representation of a signal is of lower dimension than that of its continuous counterpart, a complete elimination of redundancy is only possible by orthonormalizing a basis. A proper construction of such a basis ensures orthogonality within scales as well as across scales. The existence of such a

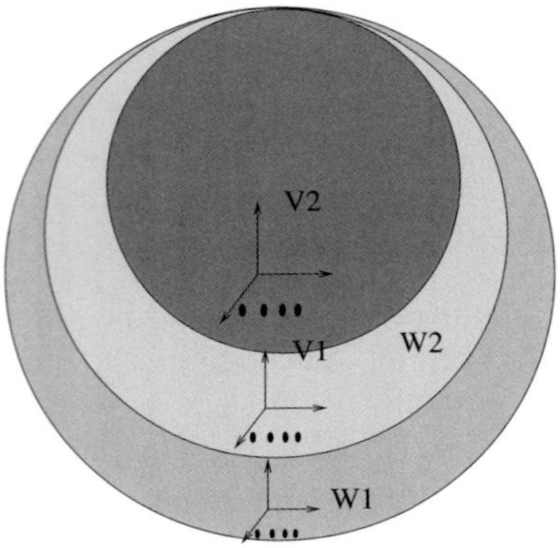

FIGURE 9.2.9

Hierarchy of wavelet bases.

basis with a dyadic scale progression was first shown, and an explicit construction was given by Meyer and Daubechies [19,27]. The connection to subband coding discovered by Mallat [28] resulted in a numerically stable and efficient implementation which helped propel wavelet analysis at the forefront of computational sciences (see Daubechies and Meyer [29] for a comprehensive historical as well as technical development [19], which also led to Daubechies' celebrated orthonormal wavelet bases).

To help maintain a smooth flow of this chapter and achieve the goal of endowing an advanced undergraduate with working knowledge of the multiresolution analysis (MRA) framework, we give a high-level introduction, albeit comprehensive, and we refer the interested reader for most of the technical details to the sources in the bibliography. For example, the trade-off in T-F resolution advocated earlier lies at the heart of the often technical wavelet design and construction. The balance between the spectral and temporal decay, which constitutes one of the key design criteria, has led to a wealth of new functional bases, and this upon the introduction of the now classical MRA framework [18,30]. The pursuit of a more refined analysis arsenal resulted in the recently introduced nonlinear MRA in [31,32,33].

9.2.3.1 Multiresolution analysis

The MRA theory developed by Mallat and Meyer [27,34] may be viewed as a functional analytic approach to subband signal analysis, which was previously introduced in and applied to engineering problems [35,36]. The clever connection between an orthonormal wavelet decomposition of a signal and its filtering by a bank of corresponding filters led to an axiomatic formalization of the theory and subsequent equivalence. This, as a result, opened up a new wide avenue of research in the development and construction of new functional bases as well as filter banks [21,37,38,39]. The construction of a telescopic set of nested approximation and detail subspaces $\{V_j\}_{j\in\mathcal{Z}}$ and $\{W_j\}_{j\in\mathcal{Z}}$, each endowed with an orthonormal basis, as shown in Fig. 9.2.9, is a key step of the analysis. An inter- and intrascale orthogonality of the wavelet functions, as noted above, is preserved, with the interscale orthogonality expressed as

$$W_i \perp W_j \ \forall i \neq j. \tag{9.2.3.1}$$

By replacing the discrete parameter wavelet $\psi_{ij}(t)$ in Eq. (9.2.1.2), where $(i, j) \in \mathcal{Z}^2$ (*set of all positive and negative 2-tuple integers*), respectively, denote the translation and the scale parameters (i.e., $\mu = 2^j i,\ \xi = 2^J$), we obtain the orthonormal wavelet coefficients as [18]

$$C_j^i(x) = < x(t), \psi_{ij}(t) > . \tag{9.2.3.2}$$

The orthogonal complementarity of the scaling subspace and that of the residuals/details amount to synthesizing the higher-resolution subspace

$$V_i \oplus W_i = V_{i-1}.$$

Iterating this property leads to a reconstruction of the original space where the observed signal lies, which is taken to be V_0 in practice, i.e., the observed signal at its first and finest resolution (this may be viewed as implicitly accepting the samples of a given signal as the coefficients in an approximation space with a scaling function corresponding to that which the subsequent analysis is based on). The dyadic scale progression has been thoroughly investigated, and its wide acceptance/popularity is due to its tight connection with subband coding, whose practical implementation is fast and simple. Other nondyadic developments have also been explored (see, e.g., Vetterli and Kovacevic [21]).

The qualitative characteristics of the MRA we have thus far discussed may be succinctly stated as follows.

Definition 9.7. A sequence $\{V_i\}_{i\in\mathcal{Z}}$ of closed subspaces of $L^2(\mathcal{R})$ is a multiresolution approximation if the following properties hold [18]:

- $\forall (j,k) \in \mathcal{Z}^2,\ x(t) \in V_j \leftrightarrow x(t - 2^j k) \in V_j,$
- $\forall j \in \mathcal{Z},\ V_{j+1} \subset V_j,$
- $\forall j \in \mathcal{Z},\ x(t) \in V_j \leftrightarrow x(\frac{t}{2}) \in V_{j+1},$
- $\lim_{j\to+\infty} V_j = \bigcap_{j=-\infty}^{\infty} V_j = \{0\},$
- $\lim_{j\to-\infty} V_j = \overline{\bigcup_{j=-\infty}^{\infty} V_j} = L^2(\mathcal{R}),$
- there exists a function $\phi(t)$ such that $\{\phi(t-n)\}_{n\in\mathcal{Z}}$ is a Riesz basis of V_0, where the "overbar" denotes closure of the space.

9.2.3.2 Properties of wavelets

A wavelet analysis of a signal assumes a judiciously preselected wavelet, and hence prior knowledge about the signal itself. As stated earlier, the properties of an analyzing wavelet have a direct impact on the resulting multiscale signal representation. Carrying out a useful and meaningful analysis is hence facilitated by a good understanding of some fundamental wavelet properties.

9.2.3.3 Vanishing moments

Recall that one of the fundamental admissibility conditions of a wavelet is that its first moment be zero. This is intuitively understood in the sense that a wavelet focuses on residuals or oscillating features of a signal. This property may in fact be further exploited by constructing a wavelet with an arbitrary number of vanishing moments. We say that a wavelet has n vanishing moments if

$$\int \psi(t)t^i dt = 0, \; i = \{0, 1, \cdots, n-1\}. \tag{9.2.3.3}$$

Reflecting a bit on the properties of a Fourier transform of a function [15], it is easy to note that the number of zero moments of a wavelet reflects the behavior of its Fourier transform around zero. This property is also useful in applications like compression, where it is highly desirable to maximize the number of small/negligible coefficients and only preserve a minimal number of large coefficients. The cost associated with an increased number of vanishing moments is that of an increased support size for the wavelet, hence that of the corresponding filter [18,19], and hence that of some of its localizing potential.

9.2.3.4 Regularity/smoothness

The smoothness of a wavelet "$\psi(t)$" is important for an accurate and parsimonious signal approximation. For a large class of wavelets (those relevant to applications), the smoothness (or regularity) property of a wavelet, which may also be measured by its degree of differentiability ($d^\alpha \psi(t)/dt^\alpha$) or equivalently by its Lipschitzity "γ," is also reflected by its number of vanishing moments [18]. The larger the number of vanishing moments, the smoother the function. In applications such as image coding, a smooth analyzing wavelet is useful for not only compressing the image, but also for controlling the visual distortion due to errors. The associated cost (i.e., some trade-off) is again a size increase in the wavelet support, which may in turn make it more difficult to capture local features, such as important transient phenomena.

9.2.3.5 Wavelet symmetry

At the exception of a Haar wavelet, compactly supported real wavelets are asymmetric around their center point. A symmetric wavelet clearly corresponds to a symmetric filter which is characterized by a linear phase. The symmetry property is important for certain applications where symmetric features are crucial (e.g., a symmetric error in image coding is better perceived). In many applications, however, it is generally viewed as a property secondary to those described above. When such a property is desired, truly symmetric biorthogonal wavelets have been proposed and constructed [19,40], with the slight disadvantage of using different analysis and synthesis mother wavelets. In Fig. 9.2.10, we show some illustrative cases of symmetric wavelets. Other nearly symmetric functions (referred to as symlets) have also been proposed, and for a detailed discussion of the pros and cons of each, the interested reader is deferred to Daubechies [19].

9.2.3.6 A filter bank view: implementation

As noted earlier, the connection between an MRA of a signal and its processing by a filter bank was not only of intellectual interest, but was of breakthrough proportion for general applications as well. It indeed provided a highly efficient numerical implementation for a theoretically very powerful methodology. Such a connection is most easily established by invoking the nestedness property of

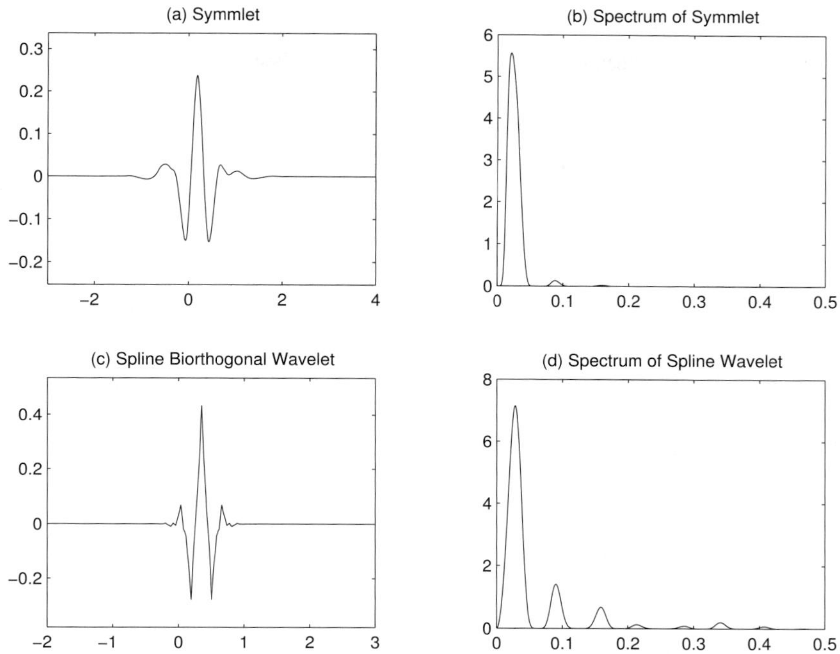

FIGURE 9.2.10

Biorthogonal wavelets preserve symmetry and symlets nearly do.

multiresolution subspaces. Specifically, it implies that if $\phi(2^{-j}t) \in V_j$ and $V_j \subset V_{j-1}$, we can write

$$\frac{1}{2^{j/2}}\phi(2^{-j}t) = \frac{1}{2^{\frac{j-1}{2}}}\sum_{k=-\infty}^{\infty} h(k)\phi(2^{-j+1}t-k), \tag{9.2.3.4}$$

where $h(k) =< \phi(2^{-j}t), \phi(2^{-j+1}t-k) >$, i.e., the expansion coefficient at time shift k. By taking the Fourier transform of Eq. (9.2.3.4), we obtain

$$\Phi(2^j\omega) = \frac{1}{2^{1/2}}H(2^{j-1}\omega)\Phi(2^{j-1}\omega), \tag{9.2.3.5}$$

which when iterated through scales leads to

$$\Phi(\omega) = \prod_{p=-\infty}^{\infty}\frac{h(2^{-p}\omega)}{\sqrt{2}}\Phi(0). \tag{9.2.3.6}$$

The complementarity of scaling and detail subspaces noted earlier stipulates that any function in subspace W_j may also be expressed in terms of V_{j-1}, where $V_{j-1} = \text{Span}\{\phi_{j-1,k}(t)\}_{(j,k)\in\mathcal{Z}^2}$, or

$$\frac{1}{2^{j/2}}\psi(2^{-j}t) = \sum_{i=-\infty}^{\infty} \frac{1}{2^{(j-1)/2}} g(i)\phi(2^{-j+1}t - i). \tag{9.2.3.7}$$

In the Fourier domain, this leads to

$$\Psi(2^j\omega) = \frac{1}{\sqrt{2}} G(2^{j-1}\omega)\Phi(2^{j-1}\omega), \tag{9.2.3.8}$$

whose iteration also leads to an expression of the wavelet Fourier transform in terms of the transfer function $G(\omega)$, as previously given for $\Phi(\omega)$ in Eq. (9.2.3.6). In light of these equations it is clear that the filters $\{G(\omega), H(\omega)\}$ may be used to compute functions at successive scales. This in turn implies that the coefficients of any function $x(t)$ may be similarly obtained as they are merely the result of an inner product of the signal of interest $x(t)$ with a basis function,

$$\begin{aligned} < x(t), \psi_{ij}(t) > &= \sum < x(t), g(k)\frac{1}{2^{(j-1)/2}}\phi(2^{-j+1}(t - 2^{-j}i) - k) > \\ &= m_k g(k)\mathcal{A}_{i-k}^{j+1}(x). \end{aligned} \tag{9.2.3.9}$$

To complete the construction of the filter pair $\{H(\cdot), G(\cdot)\}$, the properties between the approximation subspaces $\left(\{V_j\}_{j\in\mathcal{Z}}\right)$ and detail subspaces $\left(\{W_j\}_{j\in\mathcal{Z}}\right)$ are exploited to derive the design criteria for the discrete filters. Specifically, the same scale orthogonality property between the scaling and detail subspaces in

$$\sum_{k\in\mathcal{Z}} \Phi(2^j\omega + 2k\pi)\Psi(2^j\omega + 2k\pi) = 0, \ \forall\, j, \tag{9.2.3.10}$$

is the first property that the resulting filters should satisfy. For the sake of illustration, fix $j = 1$ and use

$$\sqrt{2}\Phi(2\omega) = H(\omega)\Phi(\omega). \tag{9.2.3.11}$$

Using Eq. (9.2.3.11) together with the orthonormality property of jth scale basis functions $\{\Phi_{jk}(t)\}$ [18], i.e., $\sum |\Phi(\omega + 2k\pi)|^2 = 1$, where we separate even and odd terms, yields the first property of one of the so-called conjugate mirror filters,

$$|H(\omega)|^2 + |H(\omega + \pi)|^2 = 1. \tag{9.2.3.12}$$

Using the nonoverlapping property expressed in Eq. (9.2.3.10) together with the evaluations of $\Psi(2\omega)$ and $\Phi(2\omega)$ and making a similar argument as in the preceding equation, we obtain the second property of conjugate mirror filters,

$$H(\omega)G^*(\omega) + H(\omega + \pi)G^*(\omega + \pi) = 0. \tag{9.2.3.13}$$

The combined structure of the two filters is referred to as "a conjugate mirror filter bank," see Fig. 9.2.11, and their respective impulse responses completely specify the corresponding wavelets, see

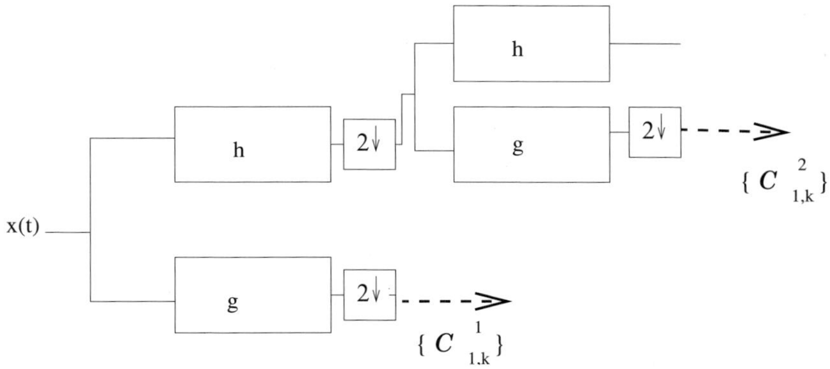

FIGURE 9.2.11

Filter bank implementation of a wavelet decomposition.

Fig. 9.2.12, (see numerous references, e.g., [20], for additional technical details). Note that the literature in the multiresolution studies tends to follow one of two possible strategies:

- a more functional analytic approach, which is mostly followed by applied mathematicians/mathematicians and scientists [18,19,30] (we could not possibly do justice to the numerous good texts now available, the author cites what he is most familiar with),
- a more filtering-oriented approach that is widely used among engineers and some applied mathematicians [21,38,41].

9.2.3.7 Refining a wavelet basis: wavelet packet and local cosine bases

A selected analysis wavelet is not necessarily well adapted to any observed signal. This is particularly the case when the signal is time-varying and has a rich spectral structure.

One approach is to then partition the signal into homogeneous spectral segments and by the same token find it an adapted basis. This is tantamount to further refining the wavelet basis and results in what is referred to as a wavelet packet basis. A similar adapted partitioning may be carried out in the time domain by way of an orthogonal local trigonometric basis (sine or cosine). The two formulations are very similar, and the solution to the search for an adapted basis in both cases is resolved in precisely the same way. In the interest of space, we focus in the following development only on wavelet packets.

9.2.3.8 Wavelet packets

We maintained above that the selection of an analysis wavelet function should be carried out with reference to the signal of interest. This of course assumes that we have some prior knowledge about the signal at hand. While plausible in some cases, this assumption is very unlikely in most practical cases. Yet it is highly desirable to select a basis which might still lead to an adapted representation of an a priori "unknown" signal. Coifman et al. [42] proposed to refine the standard wavelet decomposition. Intuitively, they proposed to further partition the spectral region of the details (i.e., wavelet coefficients) in addition to partitioning the coarse/scaling portion as ordinarily performed for a wavelet representa-

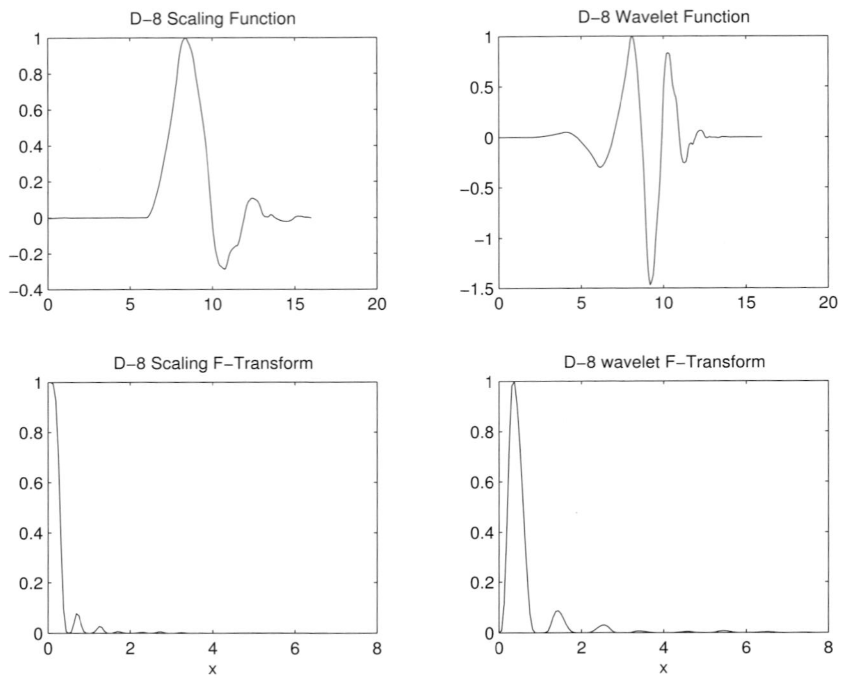

FIGURE 9.2.12

A Daubechies-8 function and its spectral properties.

tion. This, as shown in Fig. 9.2.13, yields an overcomplete representation of a signal, in other words a dictionary of bases with a tree structure. The binary nature of the tree, as further discussed below, affords a very efficient search for a basis which is best adapted to a given signal in the set.

Formally, we accomplish a partition refinement of say a subspace U_j by way of a new wavelet basis which includes both its approximation and its details, as shown in Fig. 9.2.13. This construction due to Coifman, Meyer, and Wickerhauser [42] may be simply understood as one's ability to find a basis for all the subspaces $\{V_j\}_{j\in\mathcal{Z}}$ and $\{W_j\}_{j\in\mathcal{Z}}$ by iterating the nestedness property. This then amounts to expressing two new functions $\widetilde{\psi}_j^0(t)$ and $\widetilde{\psi}_j^1(t)$ in terms of $\{\widetilde{\psi}_{j-1,k}\}_{(j,k)\in\mathcal{Z}^2}$, an orthonormal basis of a generic subspace U_{j-1} (which in our case may be either V_{j-1} or W_{j-1}),

$$\widetilde{\psi}_j^0(t) = \sum_{k=-\infty}^{\infty} h(k)\widetilde{\psi}_{j-1}(t-2^{j-1}k), \tag{9.2.3.14}$$

$$\widetilde{\psi}_j^1(t) = \sum_{k=-\infty}^{\infty} g(k)\widetilde{\psi}_{j-1}(t-2^{j-1}k), \tag{9.2.3.15}$$

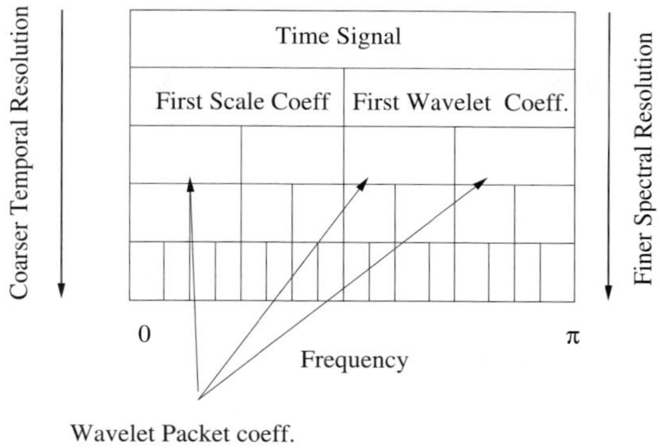

FIGURE 9.2.13

Wavelet packet basis structure.

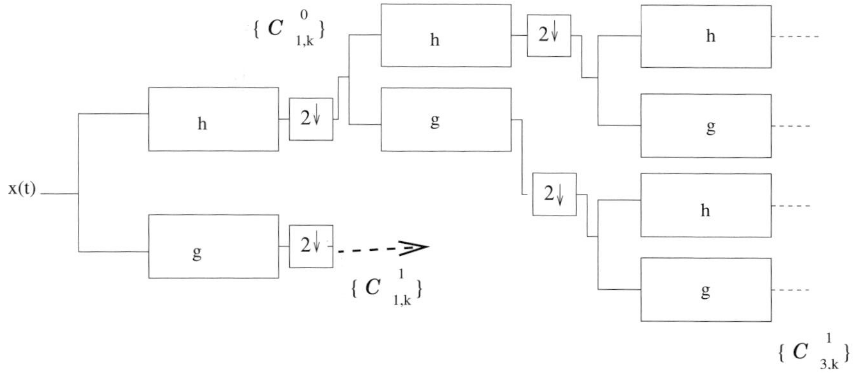

FIGURE 9.2.14

Wavelet packet filter bank realization.

where $h(\cdot)$ and $g(\cdot)$ are the impulse responses of the filters in a corresponding filter bank, and the combined family $\{\widetilde{\psi}_j^0(t-2^jk), \widetilde{\psi}_j^1(t-2^jk)\}_{(j,k)\in\mathcal{Z}^2}$ is an orthonormal basis of U_j. This, as illustrated in Fig. 9.2.13, is graphically represented by a binary tree where at each of the nodes the corresponding wavelet packet coefficients reside. The implementation of a wavelet packet decomposition is a straightforward extension of that of a wavelet decomposition, and consists of an iteration of both $H(\cdot)$ and $G(\cdot)$ bands (see Fig. 9.2.14) to naturally lead to an overcomplete representation. This is reflected on the tree by the fact that, at the exception of the root and bottom nodes which only have two children nodes and one ancestor node, respectively, each node bears two children nodes and one ancestor node.

9.2.3.9 Basis search

To identify the best-adapted basis in an overcomplete signal representation, as just noted above, we first construct a criterion which, when optimized, will reflect desired properties intrinsic to the signal being analyzed. The earliest proposed criterion applied to a wavelet packet basis search is the so-called entropy criterion [43]. Unlike Shannon's information theoretic criterion, this is additive and makes use of the coefficients residing at each node of the tree in lieu of computed probabilities. The presence of more complex features in a signal necessitates such an adapted basis to ultimately achieve an ideally more parsimonious/succinct representation. As pointed out earlier, when searching for a wavelet packet or local cosine best basis, we typically have a dictionary $\mathcal{D}$ of possible bases with a binary tree structure. Each node (j, j') (where $j \in \{0, \dots, J\}$ represents the depth and $j' \in \{0, \dots, 2^j - 1\}$ represents the branches on the jth level) of the tree then corresponds to a given orthonormal basis $\mathcal{B}_{j,j'}$ of a vector subspace of $\ell^2(\{1, \dots, N\})$ (Hilbert space of finite energy sequences). Since a particular partition $p \in \mathcal{P}$ of $[0, 1]$ is composed of intervals $I_{j,j'} = [2^{-j} j', 2^{-j}(j' + 1)[$, an orthonormal basis of $\ell^2(\{1, \dots, N\})$ is given by $\mathcal{B}^p = \cup_{\{(j,j')|I_{j,j'} \in \mathcal{P}\}} \mathcal{B}_{j,j'}$. By taking advantage of the property

$$\text{Span}\{\mathcal{B}_{j,j'}\} = \text{Span}\{\mathcal{B}_{j+1,2j'}\} \overset{\perp}{\oplus} \text{Span}\{\mathcal{B}_{j+1,2j'+1}\}, \tag{9.2.3.16}$$

where $\oplus$ denotes a subspace direct sum, we associate to each node a cost $\mathcal{C}(\cdot)$. We can then perform a bottom-up comparison of children versus parent costs (this in effect will eliminate the redundant/inadequate leaves of the tree) and ultimately prune the tree.

Our goal is to then choose the basis which leads to the *best* approximation of $\{x[t]\}$ among a collection of orthonormal bases $\{\mathcal{B}^p = \{\Psi x_i p\}_{1 \le i \le N} \mid p \in \mathcal{P}\}$, where the subscript x_i emphasizes that it is adapted to $\{x[t]\}$. Trees of wavelet packet bases studied by Coifman and Wickerhauser [44] are constructed by quadrature mirror filter (QMF) banks and comprise functions that are well localized in time and frequency. This family of orthonormal bases partitions the frequency axis into intervals of different sizes, with each set corresponding to a specific wavelet packet basis. Another family of orthonormal bases studied by Malvar [45] and Coifman and Meyer [42] can be constructed with a tree of windowed cosine functions and corresponds to a division of the time axis into intervals of dyadically varying sizes.

For a discrete signal of size N (e.g., the size of the wavelet packet tableau shown in Fig. 9.2.13), one can show that a tree of wavelet packet bases or local cosine bases has $P = N(1 + \log_2 N)$ distinct vectors but includes more than $2^{\frac{N}{2}}$ different orthogonal bases. One can also show that the signal expansion in these bases is computed with algorithms that require $O(N \log_2 N)$ operations. Wickerhauser and Coleman [44] proposed that for any signal $\{x[m]\}$ and an appropriate functional $\mathcal{K}(\cdot)$, one finds the best basis $\mathcal{B}^{p_0}$ by minimizing an "additive" cost function

$$\text{Cost}(\mathbf{x}, \mathcal{B}^{\text{p}}) = \sum_{i=1}^{N} \mathcal{K}(|\langle \mathbf{x}, \Psi x_i p \rangle|^2) \tag{9.2.3.17}$$

over all bases. As a result, the basis which results from minimizing this cost function corresponds to the "best" representation of the observed signal. The resulting pruned tree of Fig. 9.2.15 bears the coefficients at the remaining leaves/nodes.

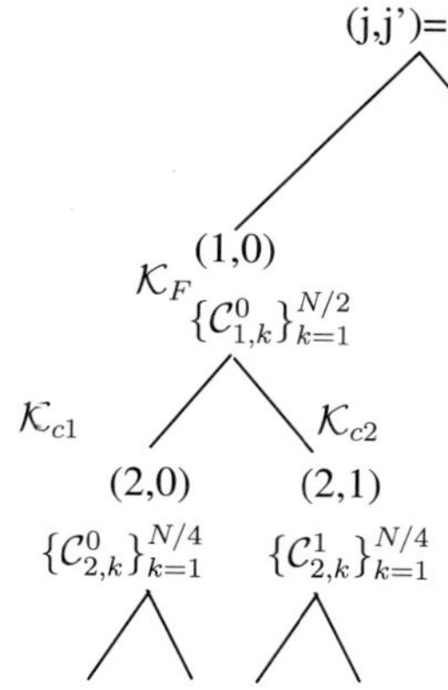

FIGURE 9.2.15

Tree pruning in search for a best basis.

9.2.4 Multiresolution applications in estimation problems

The computational efficiency of wavelet decomposition together with all of its properties has triggered unprecedented interest in its application in the area of information sciences (see, e.g., [46,47,48,49, 50,51]). Specific applications have ranged from compression [52,53,54,55] to signal/image modeling [56,57,58,59] and from signal/image enhancement to communications [60,61,62,63,64,65,66,67]. The literature in statistical applications as a whole has seen an explosive growth in recent years, and in the interest of space, we will focus our discussion on a somewhat broader perspective which may, in essence, be usefully reinterpreted in a number of different instances.

9.2.4.1 Signal estimation/denoising and modeling

Denoising may be heuristically interpreted as a quest for parsimony of a representation of a signal. Wavelets, as described above, have a great capacity for energy compaction, particularly at or near singularities. Given a particular wavelet, its corresponding basis as noted above is not universally optimal for all signals and particularly not for a noisy one; this difficulty may be lifted by adapting the representation to the signal in a "best" way possible and according to some criterion. The first of the two possible ways is to pick an optimal basis in a wavelet packet dictionary and discard the negligible coefficients [68]. The second, which focuses on reconstruction of a signal in noise, and which we discuss here, accounts for the underlying noise statistics in the multiscale domain to separate the noisy signal into the "mostly" signal part and the "mostly" noise part [69,70,71]. We opt here to discuss a more general setting which assumes unknown noise statistics and where a signal reconstruction is still sought in some optimal way, as we elaborate below. This approach is particularly appealing in that it may be reduced to the setting in earlier developments.

9.2.4.2 Problem statement

Consider an additive noise model

$$x(t) = s(t) + n(t), \tag{9.2.4.1}$$

where $s(t)$ is an unknown but deterministic signal corrupted by a zero-mean noise process $n(t)$ and $x(t)$ is the observed, i.e., noisy, signal. The objective is to recover the signal $\{s(t)\}$ based on the observations $\{x(t)\}$.

The underlying signal is modeled with an orthonormal basis representation,

$$s(t) = \sum_i C_i^s \psi_i(t),$$

and similarly the noise is represented as

$$n(t) = \sum_i C_i^n \psi_i(t).$$

By linearity, the observed signal can also be represented in the same fashion, with its coefficients given by

$$C_i^x = C_i^s + C_i^n.$$

A key assumption we make is that for certain values of i, $C_i^s = 0$. In other words, the corresponding observation coefficients C_i^x represent "pure noise," rather than signal corrupted by noise. As shown by Krim and Pesquet [72], this is a reasonable assumption in view of the spectral and structural differences between the underlying signal $s(t)$ and the noise $n(t)$ across scales. Given this assumption, wavelet-based denoising consists of determining which wavelet coefficients represent primarily signal and which mostly capture noise. The goal is to then localize and isolate the "mostly signal" coefficients. This may be achieved by defining an information measure as a function of the wavelet coefficients. It identifies the "useful" coefficients as those whose inclusion improves the data explanation. One such measure is Rissanen's information theoretic approach (or minimum description length [MDL]) [73]. In other words, the MDL criterion is utilized for resolving the trade-off between model complexity (each retained coefficient increases the number of model parameters) and goodness-of-fit (each truncated coefficient decreases the fit between the received, i.e., noisy, signal and its reconstruction).

9.2.4.3 The coding length criterion

Wavelet thresholding is essentially an order estimation problem, one of balancing model accuracy against overfitting and one of capturing as much of the "signal" as possible, while leaving out as much of the "noise" as possible. One approach to this estimation problem is to account for any prior knowledge available on the signal of interest, which usually is of probabilistic nature. This leads to a Bayesian estimation approach as developed in [74,75]. While generally more complex, it does provide a regularization capacity which is much needed in low-signal-to-noise ratio (SNR) environments.

A parsimony-driven strategy which we expound on here addresses the problem of modeling in general and that of compression in particular. In addition it provides a fairly general and interesting framework where the signal is assumed deterministic and unknown and results in some intuitively sensible and mathematically tractable techniques [68]. Specifically, it brings together Rissanen's work on stochastic complexity and coding length [73,76] and Huber's work on minimax statistical robustness [77,78].

Following Rissasen, we seek the data representation that results in the shortest encoding of both observations and complexity constraints. As a departure from the commonly assumed Gaussian likelihood, we rather assume that the noise distribution f of our observed sequence is a (possibly) scaled version of an unknown member of the family of ε-contaminated normal distributions,

$$\mathcal{P}_\varepsilon = \{(1-\varepsilon)\Phi + \varepsilon G : G \in \mathcal{F}\},$$

where Φ is the standard normal distribution, $\mathcal{F}$ is the set of all suitably smooth distribution functions, and $\varepsilon \in (0, 1)$ is the known fraction of contamination (this is no loss of generality, since ε may always be estimated if unknown). Note that this study straightforwardly reduces to the additive Gaussian noise case by setting the mixture parameter $\varepsilon = 0$ and is in that sense more general.

For fixed model order, the expectation of the MDL criterion is the entropy plus a penalty term which is independent of both the distribution and the functional form of the estimator. In accordance with the minimax principle, we seek the least favorable noise distribution and evaluate the MDL criterion for that distribution. In other words, we solve a minimax problem where the entropy is maximized over all distributions in $\mathcal{P}_\varepsilon$ and the description length is minimized over all estimators in $\mathcal{S}$. The saddle point (provided its existence) yields a minimax robust version of MDL, which we call the minimax description length (MMDL) criterion.

In [68], it is shown that the least favorable distribution in $\mathcal{P}_\varepsilon$, which also maximizes the entropy, is one which is Gaussian in the center and Laplacian ("double exponential") in the tails and switches from one to the other at a point whose value depends on the fraction of contamination ε.

Proposition 9.1. *The distribution $f_H \in \mathcal{P}_\varepsilon$ that minimizes the negentropy is*

$$f_H(c) = \begin{cases} (1-\varepsilon)\phi_\sigma(a)e^{\frac{1}{\sigma^2}(ac+a^2)}, & c \le -a, \\ (1-\varepsilon)\phi_\sigma(c), & -a \le c \le a, \\ (1-\varepsilon)\phi_\sigma(a)e^{\frac{1}{\sigma^2}(-ac+a^2)}, & a \le c, \end{cases} \tag{9.2.4.2}$$

where ϕ_σ is the normal density with mean zero and variance σ^2 and a is related to ε by the equation

$$2\left(\frac{\phi_\sigma(a)}{a/\sigma^2} - \Phi_\sigma(-a)\right) = \frac{\varepsilon}{1-\varepsilon}. \tag{9.2.4.3}$$

9.2.4.4 Coding for worst-case noise

Let the set of wavelet coefficients obtained from the observed signal be denoted by $\mathcal{C}^N = \{C_1^x, C_2^x, \ldots, C_N^x\}$ as a time series without regard to the scale and where the superscript indicates the corresponding process. Let exactly K of these coefficients contain signal information, while the remainder only contain noise. If necessary, we reindex these coefficients so that

$$C_i^x = \begin{cases} C_i^s + C_i^n, & i = 1, 2, \ldots, K, \\ C_i^n, & \text{otherwise.} \end{cases} \tag{9.2.4.4}$$

By assumption, the set of noise coefficients $\{C_i^n\}$ is a sample of independent, identically distributed (i.d.d.) random variates drawn from Huber's distribution f_H. It follows by Eq. (9.2.4.4) that the observed coefficients C_i^x obey the distribution $f_H(c - C_i^s)$ when $i = 1, 2, \ldots, K$ and $f_H(c)$ otherwise.

Thus, the likelihood function is given by

$$\ell(\mathcal{C}^N;\, K) \;=\; \prod_{i\le K} f_H(C_i^x - C_i^s) \prod_{i>K} f_H(C_i^x).$$

Since f_H is symmetric and unimodal with a maximum at the origin, the above expression is maximized (with respect to the signal coefficient estimates $\{\hat{C}_i^s\}$) by setting

$$\hat{C}_i^s \;=\; C_i^x,$$

for $i = 1, 2, \ldots, K$. It follows that the maximized likelihood (given K) is

$$\ell^*(\mathcal{C}^N;\, K) \;=\; \prod_{i\le K} f_H(0) \prod_{i>K} f_H(C_i^x).$$

Thus, the problem is reduced to choosing the optimal value of K, in the sense of minimizing the MDL criterion,

$$\begin{aligned}\mathcal{L}(\mathcal{C}^N;\, K) \;&=\; -\log \ell^*(\mathcal{C}^N;\, K) \;+\; K\log N \\ &=\; -\sum_{i\le K} \log f_H(0) \;-\; \sum_{i>K} \log f_H(C_i^x) \;+\; K\log N.\end{aligned} \tag{9.2.4.5}$$

Neglecting terms independent of K, this is equivalent to minimizing

$$\tilde{\mathcal{L}}(\mathcal{C}^N;\, K) \;=\; \frac{1}{2\sigma^2}\sum_{i>K} \eta(C_i^x) \;+\; K\log N,$$

where

$$\eta(c) \;=\; \begin{cases} c^2 & \text{if } |\, c\, | < a, \\ a\,|\, c\,| \;-\; a^2 & \text{otherwise} \end{cases}$$

is proportional to the exponent in Huber's distribution f_H. This can simply be achieved by a thresholding scheme [68].

Proposition 9.2. *The coefficient* $|\, C_i^x\, |$ *is truncated as follows.*

Case 1. *When* $\log N > \frac{a^2}{2\sigma^2}$, *the coefficient* $|\, C_i^x\, |$ *is truncated if*

$$|\, C_i^x\, | < \frac{a}{2} \;+\; \frac{\sigma^2}{a}\log N.$$

Case 2. *When* $\log N \le \frac{a^2}{2\sigma^2}$, *the coefficient* $|\, C_i^x\, |$ *is truncated if*

$$|\, C_i^x\, | < \sigma\sqrt{2\log N}.$$

Remarks. More ample details may be found in [68].

When $\sigma^2 \to 0$, the thresholding scheme reduces to Case 2, and C_i^x is never truncated; since this represents the no-noise case, it is reasonable that all coefficients should be retained in the reconstruction. On the other hand, for large σ^2, the thresholding scheme reduces to Case 1, which is more conservative. For $\sigma^2 \to \infty$, the SNR becomes zero and the best one can do is to estimate the signal as identically zero.

Similarly, when $a \to \infty$, the noise distribution becomes purely Gaussian, and the thresholding scheme reduces to Case 2, as expected. The resulting threshold of this particular noise case coincides with the results of [69,70] and is qualitatively similar to that derived in [71]. On the other hand, when $a \to 0$, the noise distribution becomes purely Laplacian, and the thresholding scheme reduces to Case 1.

Finally, when $N \to 1$, the thresholding scheme reduces to Case 2, suggesting that outliers are unlikely to occur in a small sample, and it is hence more reasonable to assume purely Gaussian noise. On the other hand, for large N, the thresholding scheme reduces to Case 1, since outliers are highly likely to occur in a large sample.

It is important to distinguish the minimax error result obtained in [69], which was achieved over a signal smoothness class, from those discussed here and derived in [68], which are obtained over a family of noise distributions.

9.2.4.5 Numerical experiments

In the examples that follow, we demonstrate the performance of the robust thresholding procedure described above and compare it with that of the thresholding scheme based upon the assumption of normally distributed noise.

Example I

Using WAVELAB (available from the Stanford Statistics Department, courtesy of D.L. Donoho and I.M. Johnstone), we synthesized a broken ramp signal of length $N = 1024$. This signal admits an efficient representation in a wavelet basis, i.e., one with very few nonzero coefficients. The noise was additive and i.i.d., obeying an $N(0, \sigma^2)$ distribution contaminated by a fraction $\varepsilon = 10\%$ of white Gaussian noise with distribution $N(0, 9\sigma^2)$. The overall SNR was maintained at 10 dB (see Fig. 9.2.16, top).

We implemented two estimators. The first one was based on a purely Gaussian noise assumption (i.e., $\varepsilon = 0$), where the thresholding scheme due to [72,79] was used. The second was the MMDL robust estimator described above. The reconstructions based on each estimator appear in Fig. 9.2.16. As may easily be observed, and in contrast to the MMDL technique, the Gaussian assumption induces a high susceptibility to outliers.

Monte Carlo simulations were carried out to evaluate the reconstruction performance over a range of SNRs. At each value of the SNR, 100 experiments were conducted, and the cumulative reconstruction error is displayed in Fig. 9.2.17. The robust estimator uniformly outperforms the *classic* estimator in both L_1- and L_2-errors over a wide range of SNRs. Furthermore, the performances of the Gaussian and robust estimators become indistinguishable at high SNRs, i.e., small noise variance, showing that robustness does not come at the cost of reduced efficiency.

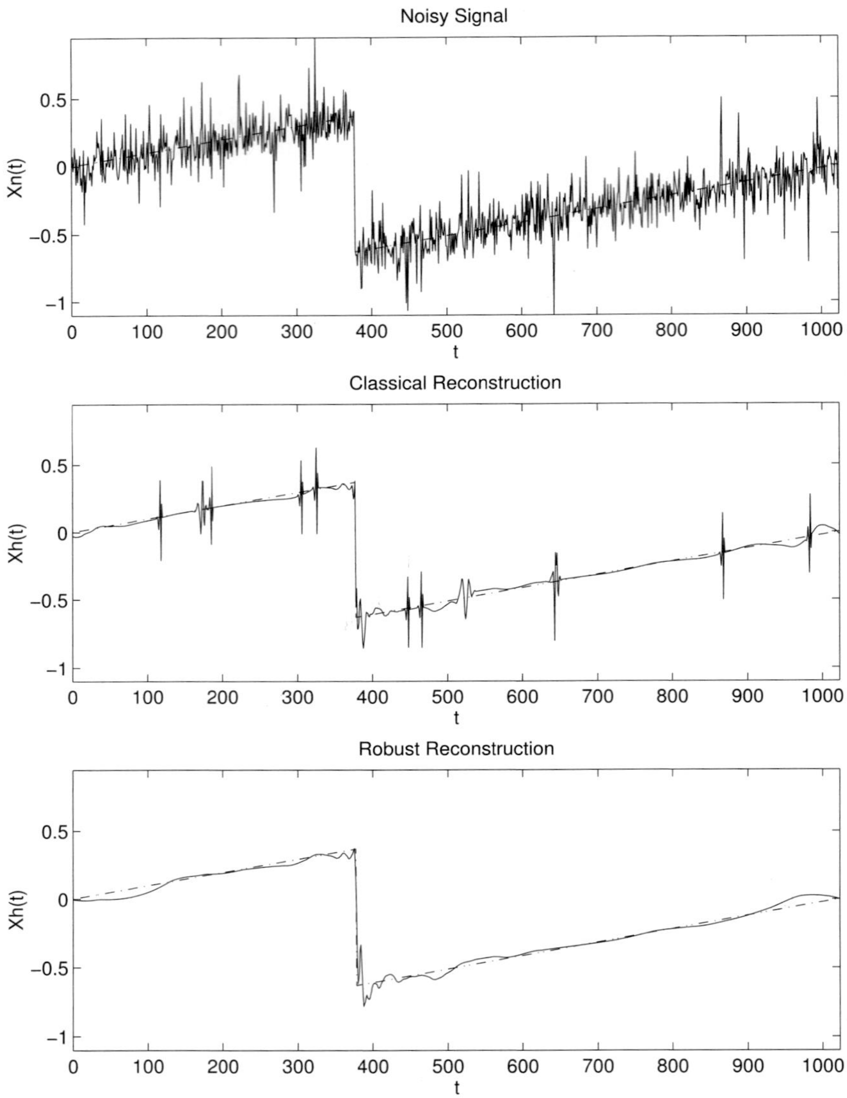

FIGURE 9.2.16

Noisy ramp signal and its Gaussian and robust reconstructions.

Bounding the reconstruction error.

Although the robust estimator-based reconstruction error is much improved, it is still potentially unbounded. As further discussed in [68], and due to the compactness of wavelets, unbounded

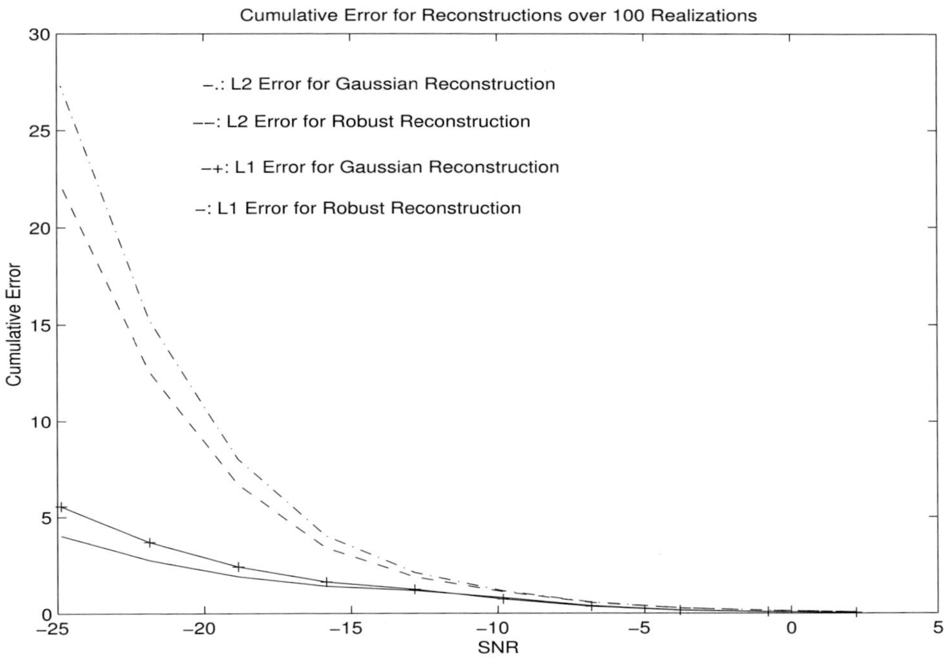

FIGURE 9.2.17

L_1- and L_2-error performance versus SNR for the Gaussian and robust estimators.

noise will still result in unbounded reconstruction error, a property that may be considered undesirable. This problem may be circumvented by making the assumption that the signal has bounded energy; in that case, one of at least two alternatives is possible:

1. In practice, the signal is known to be bounded, and prior knowledge of the physical properties of the signal may be used to determine the $\|\cdot\|_\infty$ of the sequence of signal wavelet coefficients $\{C_i^s\}$. This information may be used to truncate observed coefficients $\{C_i^x\}$ not only below, as discussed earlier, but also above.
2. In the absence of such prior knowledge, it may still be possible to bound the reconstruction error through an adaptive supremum–secondary thresholding scheme based upon some representation criterion, e.g., entropy.

The first of these approaches uses the following modified thresholding. Let $\beta > 0$ be an upper bound on the magnitude of the signal coefficients. Then

$$\tilde{C}_i^s = \begin{cases} 0, & \text{if } |C_i^x| \le \frac{a}{2} + \frac{\sigma^2}{a}\log N, \\ \hat{C}_i^s = C_i^x, & \text{if } \frac{a}{2} + \frac{\sigma^2}{a}\log K \le |C_i^x| \le \beta, \\ \beta\,\mathrm{sgn}(C_i^x), & \text{if } \beta \le |C_i^x|, \end{cases}$$

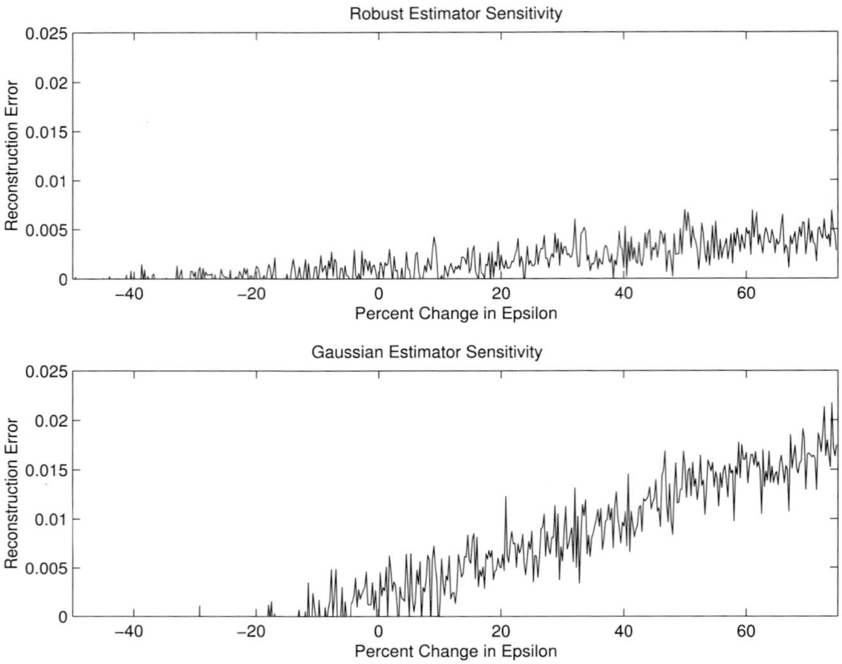

FIGURE 9.2.18

Error stability versus variation in the mixture parameter ε.

provided that $\log N > a^2/2\sigma^2$ and $\beta > (a/2) + (\sigma^2/a)\log N$. The graph shows that although the robust estimator's reconstruction error initially grows more slowly than that of the Gaussian estimator, the two errors soon converge as the variance of the outliers grows. The reconstruction error for the bounded-error estimator, however, levels off past a certain magnitude of the outlier, as expected.

Sensitivity analysis for the fraction of contamination.

Although such crucial assumptions as the normality of the noise or exact knowledge of its variance σ^2 usually go unremarked, it is often thought that Huber-like approaches are limited by the assumption of a known ε. We demonstrate the resilience and robustness of the approach by studying the sensitivity of the estimator to changes in the assumed value of ε.

Fig. 9.2.18 shows the total reconstruction error as a function of variation in the true fraction of contamination ε from white Gaussian noise with distribution $N(0, 9\sigma^2)$. In other words, an abscissa of "0" corresponds to an assumed true fraction of contamination equal to 0; larger abscissas correspond to outliers of larger magnitude than assumed by the robust estimator, and vice versa. Clearly, the Gaussian estimator assumes zero contamination throughout, namely, the estimator did not acknowledge the contamination from $N(0, 9\sigma^2)$. The figure in the bottom shows that the reconstruction error for the Gaussian estimator grows very rapidly as the true fraction of contamination increases, whereas that of the robust estimator is nearly flat over a broad range, as shown in the figure in the top. This should not come as a surprise; outliers are, by definition, rare events, and for a localized procedure such as wavelet expansion, the precise frequency of outliers is much less important than their existence at all.

Example II

The example above assumed a fixed wavelet basis. As discussed above, however, highly nonstationary signals can be represented most efficiently by using an adaptive basis that automatically chooses resolutions as the behavior of the signal varies over time. Using an L^2-error criterion we search for the best basis of a speech signal and a chirp signal shown in Fig. 9.2.19 and errors are further analyzed (see [80] for more details).

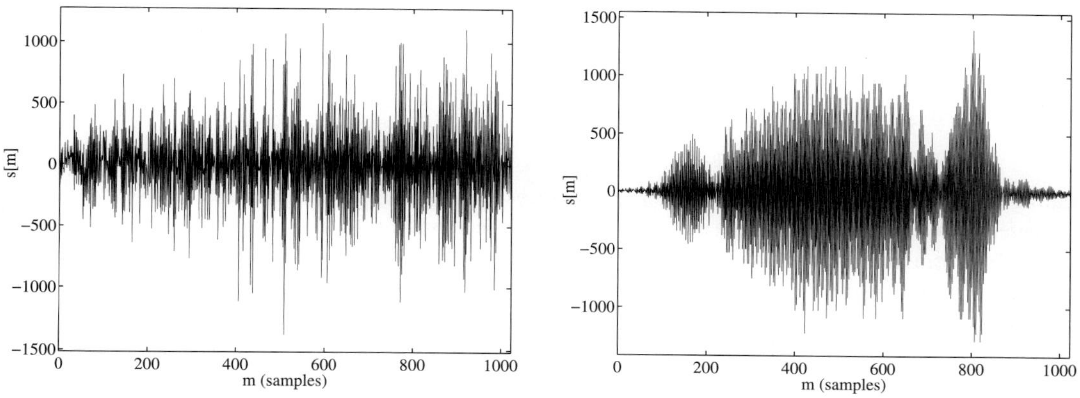

FIGURE 9.2.19

Two real signals, a voice (left) and a bird chirp (right), are used to analyze the performance of the best basis search by an L^2-error criterion.

The separability property of most carefully designed wavelets and their inability to achieve fine tuned directional analysis has led to a search for alternative wavelets which are more sensitive to arbitrary curves and image edges. These include the introduction of curvelets, contourlets (an efficient implementation of curvelets), and shearlets, which are discussed in more detail in the next three sections.

9.3 Curvelets and their applications

We first introduce the 2D curvelet transform developed by Candès and Donoho around 2000 [4]. Initially, they proposed the ridgelet transform in the Cartesian Fourier plane [81] in order to locate segmented curve features in an image. Soon, they discovered that multiscale partitioning of the Fourier plane using polar coordinate parameters can provide much simpler structures. This led to their development of the current continuous curvelet transform. This is closely related to Hart Smith's parabolic

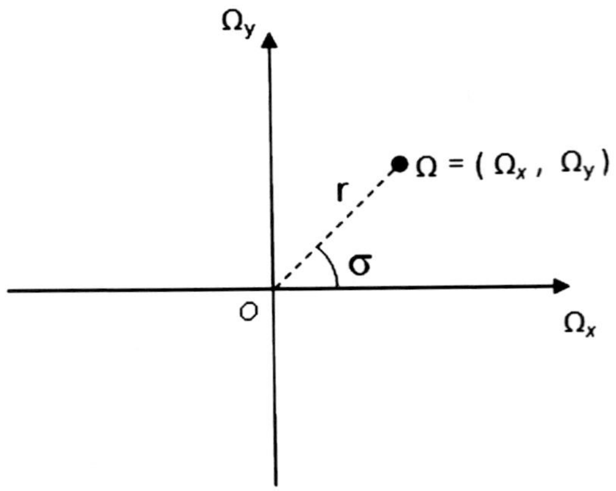

FIGURE 9.3.20

The polar and rectangle coordinates.

scaling-based transform using the Fourier integral operators [82]. The convenience in using polar coordinates to define the continuous curvelet transform indeed has a cost in implementing it in the digital world. Candès, Donoho, and their collaborators have also introduced the discretized curvelet transform, in which the DFT involving polar coordinates has been resampled using Cartesian coordinates [83].

In a 2D spatial plane, the curvelet atoms' effective impact is similar to applying differential operators to an image along selected directions to measure the regularities in localized areas that can be contained inside rotated rectangles. Large coefficients can be obtained when a differential operator is applied across the discontinuities along a curve, and thus can be used to characterize salient features in an image.

To start, we first review the conversion between polar coordinates and Cartesian coordinates, see Fig. 9.3.20. In the Fourier frequency plane, a point $\mathbf{\Omega} = (\Omega_x, \Omega_y)$ with Cartesian coordinates, can be converted to polar coordinates (r, σ) by

$$r = \pm|\mathbf{\Omega}| = \pm\sqrt{\Omega_x^2 + \Omega_y^2}, \quad \sigma = \arctan(\Omega_y/\Omega_x),$$

assuming we confine the angle σ to $(-\pi/2, \pi/2)$ (the angle between the polar axis and the ray from the pole to the point). Meanwhile, the polar coordinates (r, σ) can be converted to the Cartesian coordinates (Ω_x, Ω_y) by

$$\Omega_x = r\cos\sigma, \quad \Omega_y = r\sin\sigma. \tag{9.3.0.1}$$

Let R_θ be the operator that rotates a point $\mathbf{\Omega}$ in the counterclockwise (ccw) direction around the origin by an angle of θ rad. The end point in polar coordinates will be $R_\theta(\mathbf{\Omega}) = (r, \sigma + \theta)$. In Cartesian coordinates, the operator can be written as a matrix

$$R_\theta = \begin{pmatrix} \cos\theta & -\sin\theta \\ \sin\theta & \cos\theta \end{pmatrix},$$

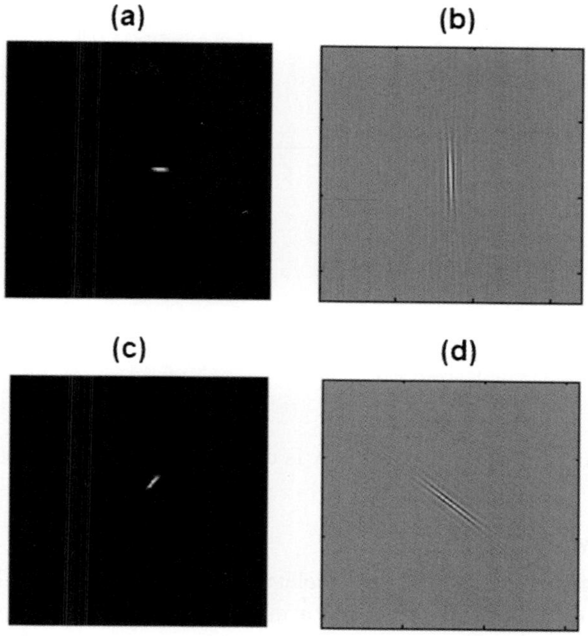

FIGURE 9.3.21

Fourier transform of an image and its rotated version. (a) Selected Fourier content of an image. (b) Its corresponding spatial waveform. (c) Fourier content rotated counterclockwise by 45 degrees. (d) Its corresponding spatial waveform is also rotated counterclockwise by 45 degrees.

and the end point of the rotation, $R_\theta(\mathbf{\Omega}) = R_\theta \cdot \mathbf{\Omega}$, is the multiplication of the rotating matrix and the vector $\mathbf{\Omega}$ ($\mathbf{\Omega}$ can be viewed as a column vector).

The following proposition states that when the input of a function is rotated by θ rad in the ccw direction, its Fourier transform is also rotated by θ rad in the Fourier plane (see Fig. 9.3.21).

Proposition 9.3. *Let the Fourier transform of $f(\boldsymbol{x}) \in L^2(\mathcal{R}^2)$ be $F(\mathbf{\Omega})$. We have*

$$F(R_\theta(\mathbf{\Omega})) = \int_{\mathcal{R}^2} f(R_\theta(\boldsymbol{x}))\, d\boldsymbol{x}. \tag{9.3.0.2}$$

9.3.1 The continuous curvelet transform

The continuous curvelet transform is defined as the inner product of a function with curvelet atoms at various dilating scales, rotating angles, and translating locations. It calculates the energy of the function correlated to each curvelet atom so that the frequency content of the function can be examined separately in each localized subband. At each scale, the curvelet atoms are generated by rotating a smooth 2D window with compact support in its frequency domain and shifting its waveform in the

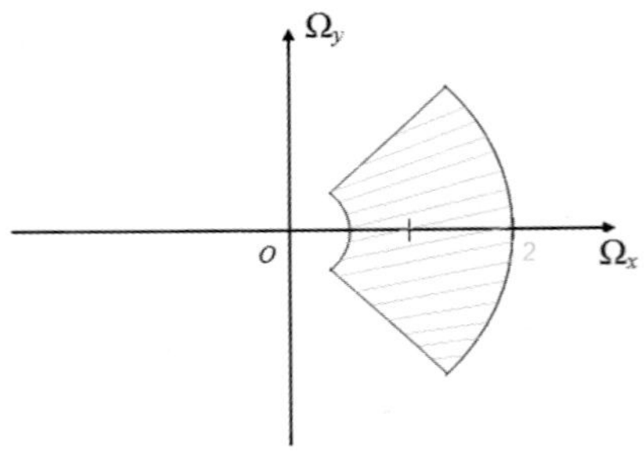

FIGURE 9.3.22

A mother curvelet window $\Psi_a(\Omega)$ at scale $a = 1$.

spatial domain to various locations. This window is called the mother curvelet at scale a, as shown in Fig. 9.3.22.

9.3.1.1 *Mother curvelets*

At scale a, the Fourier transform of a mother curvelet is written as $\Psi_a(\Omega_x, \Omega_y)$ in Cartesian coordinates or as $\Psi_a(r, \sigma)$ in polar coordinates. It is defined using two 1D windows, along the radial and angular directions, and is given by

$$\Psi_a(r,\sigma) = a^{3/4}\,\Psi_1(ar)\Psi_2(\frac{\sigma}{\sqrt{a}}), \quad 0 < a \le 1, \tag{9.3.1.1}$$

where the radial (or amplitude) window $\Psi_1(r)$ is a smooth, bandpass, and nonnegative real-valued function that has its support $r \in [1/2, 2]$. This interval is chosen for the convenience of using dyadic intervals for discretization. The angular (or phase) window $\Psi_2(\sigma)$ is a smooth, low-pass, and nonnegative real-valued even function that has its support $\sigma \in [-1, 1]$.

At scale a, the mother curvelet $\Psi_a(r, \sigma)$ is constructed by scaling the two window functions, Ψ_1 and Ψ_2, in two different ways. The support of this mother curvelet is a single polar wedge with an area $k \cdot a^{-3/2}$ (k is a constant), which is located inside $\left\{(r,\sigma) : \frac{1}{2a} < r < \frac{2}{a}, |\sigma| < \sqrt{a}\right\}$ and is symmetric with respect to the polar axis. The length of the wedge in the polar axis direction is at the same order as a^{-1}, and the larger arc length in the angular direction is at the same order as $a^{-1/2}$, as shown in Fig. 9.3.22. Approximately, the wedge has width $\approx$ length^2. It can be observed that the mother curvelets at two different scales a, a', with $a \neq a'$, have no scaling relation between them, so they differ slightly at various scales.

9.3.1.2 *The amplitude and angular windows*

The choice of the amplitude and angular windows not only decides the properties of the mother curvelets, but also decides the properties of the curvelet atoms. The above construction shows that the curvelet windows have compact supports in the Fourier plane because both Ψ_1 and Ψ_2 have compact supports. Further, the following proposition (derived in [84]) implies that their corresponding spatial waveforms are such that $\psi_1(x), \psi_2(t) \in C^\infty$. Hence the mother curvelets are C^∞ in the spatial plane.

Proposition 9.4. *A function $\psi(x)$ is bounded and n times continuously differentiable with bounded derivatives if its Fourier transform $\Psi(\Omega_x)$ satisfies*

$$\int_{-\infty}^{+\infty} |\Psi(\Omega_x)|(1+|\Omega_x|^n)\, d\Omega_x < +\infty. \quad (9.3.1.2)$$

The compact support of a mother curvelet in the Fourier plane indicates that the spatial waveform of the mother curvelet does not have a compact support in the spatial plane. Its waveform spreads out, which is known as the side robe effects. To allow the curvelets to focus on localized areas in the spatial domain for detecting the singularities, the windows Ψ_1 and Ψ_2 should be chosen as smooth as possible in the Fourier plane so that their spatial waveforms decay rapidly. A function is said to have a rapid decay if there exists a constant M_n, such that the following is true for all integers $n > 0$ for $f(\boldsymbol{x})$:

$$|f(\boldsymbol{x})| \leq \frac{M_n}{(1+|\boldsymbol{x}|^n)}, \quad \forall x. \quad (9.3.1.3)$$

Proposition 9.4 is also true for the inverse Fourier transform. If the spatial waveform of a mother curvelet decays fast, then it indicates that Fourier windows $\Psi_1(r)$ and $\Psi_2(\sigma)$ should be C^n. Therefore, the windows Ψ_1 and Ψ_2 should be smooth and n times continuously differentiable. In the literature, the windows are often required to be C^∞ to ensure the spatial functions decay rapidly.

At scale a, the mother waveform oscillates along the horizontal direction with most of its support located inside an ellipsoid-shaped area along a ridge in the vertical direction. The ridge is defined as the (line) locations where the amplitude of a function reaches its maximum amplitude. This ellipsoid becomes deformed and has a "needle-like" shape (the area becomes longer and narrower) when the scale a gets smaller. Finally, the amplitude and angular windows Ψ_1 and Ψ_2 should satisfy the following admissibility conditions:

$$\int_0^\infty \frac{\Psi_1^2(r)}{r}\, dr = C_{\Psi_1}, \quad (9.3.1.4)$$

$$\int_{-1}^{1} \Psi_2^2(\sigma)\, d\sigma = C_{\Psi_2}. \quad (9.3.1.5)$$

The admissibility conditions in Eqs. (9.3.1.4) and (9.3.1.5) can be normalized such that $C_{\Psi_1} = 1$ and $C_{\Psi_2} = 1$. Eq. (9.3.1.4) indicates that $\Psi_1(0) = 0$, namely, $\Psi_1(r)$ is the transfer function of a bandpass filter.

A continuously differentiable function, Ψ_1, that satisfies $\Psi_1(0) = 0$ is sufficient to ensure that Ψ_1 satisfies the admissibility condition in Eq. (9.3.1.4) [84]. The selection of the windows for constructing the curvelets will be further discussed in detail in the discrete case.

9.3.1.3 The curvelet atoms

Definition 9.8. A curvelet atom is an element in the set $\{\psi_{a,\theta,\boldsymbol{b}}(\boldsymbol{x}), \boldsymbol{x} \in \mathcal{R}^2, a \in (0,1], \theta \in [0, 2\pi), \boldsymbol{b} \in \mathcal{R}^2\}$. It can be defined as the inverse Fourier transform of a function $\Psi_{a,\theta,\boldsymbol{b}}(\boldsymbol{\Omega})$ as follows:

$$\Psi_{a,\theta,\boldsymbol{b}}(\boldsymbol{\Omega}) = exp(-j\, \langle \boldsymbol{b}, \boldsymbol{\Omega} \rangle)\Psi_a(R_\theta \boldsymbol{\Omega}), \quad (9.3.1.6)$$

where $\Psi_a(\Omega)$ is a mother curvelet. The polar coordinate format of $\Psi_a(R_\theta \Omega)$ is equal to $\Psi_a(r, \sigma + \theta)$.

Specifically, at scale a, the curvelet atom at $\theta = 0$, $\mathbf{b} = \mathbf{0}$ (define $\mathbf{0} = (0, 0)$) is the mother curvelet because

$$\Psi_{a,0,\mathbf{0}}(\mathbf{\Omega}) = \Psi_a(\mathbf{\Omega}). \tag{9.3.1.7}$$

The spatial waveform of the curvelet atom, $\psi_{a,\theta,\mathbf{b}}(\boldsymbol{x})$, takes a complex value and can be written in terms of a mother curvelet as

$$\psi_{a,\theta,\mathbf{b}}(\boldsymbol{x}) = \psi_{a,0,\mathbf{0}}(\boldsymbol{x})(R_\theta(\boldsymbol{x} - \mathbf{b})), \quad a \in (0, 1], \theta \in [0, 2\pi), \mathbf{b} \in \mathcal{R}^2. \tag{9.3.1.8}$$

Fig. 9.3.21 shows two curvelet atoms (in both Fourier and spatial planes) at different orientations. Although complex-valued curvelet atoms are capable of separating the amplitude and the phase information contained in a 2D function, in many situations the real-valued curvelet atoms can be built by modifying the mother curvelet windows, namely, replacing the radial window $\Psi_1(r)$ with $\Psi_{R,1}(r) = \Psi_1(r) + \Psi_1(-r)$ or replacing the angular window $\Psi_2(\sigma)$ with $\Psi_{R,2}(\sigma) = \Psi_2(\sigma) + \Psi_2(\sigma + \pi)$. It results in mother curvelets supported on two polar wedges that are symmetric with respect to the x, y-axes in the rectangle coordinates. Fig. 9.3.23(a) shows the frequency support (the gray area) of a mother curvelet that generates complex-valued curvelet atoms in the spatial domain, while Fig. 9.3.23(b) shows the frequency support of a mother curvelet that generates real-valued curvelet atoms.

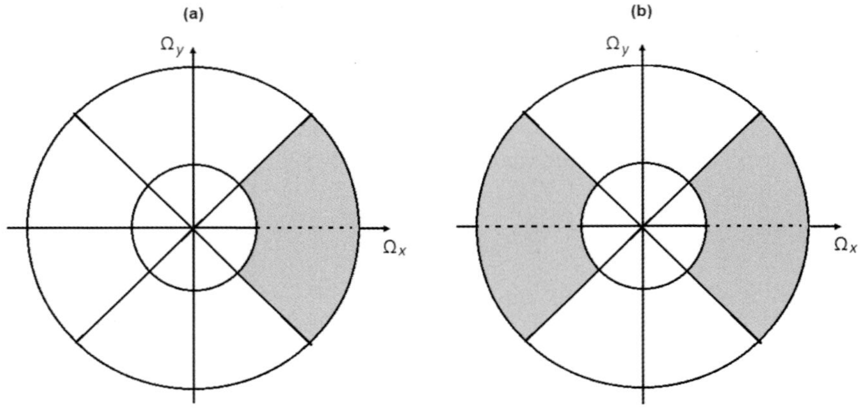

FIGURE 9.3.23

Examples of mother curvelets. (a) A mother curvelet that generates a family of complex-valued curvelets. The mother curvelet has its frequency content supported in a single wedge, represented by the gray area. (b) A mother curvelet that generates a family of real-valued curvelets. The mother curvelet has its frequency content supported in a pair of symmetric wedges, represented by the gray area.

The curvelet atoms generated by the mother curvelets dissect the Fourier plane into localized subbands so that the Fourier plane is packed redundantly with all the dilated and rotated curvelet atoms. Note that shifting in the spatial plane has no effect on the amplitude of the content in the Fourier plane. Fig. 9.3.24 shows pictures of a curvelet atom: (a) in the spatial space, (b) its Fourier transform in frequency space, (c) the graph of the oscillatory function along the θ-direction taken from the real part of the wavelet atom, and (d) a zoomed in view of the center 64×64 pixels for a better view of the content.

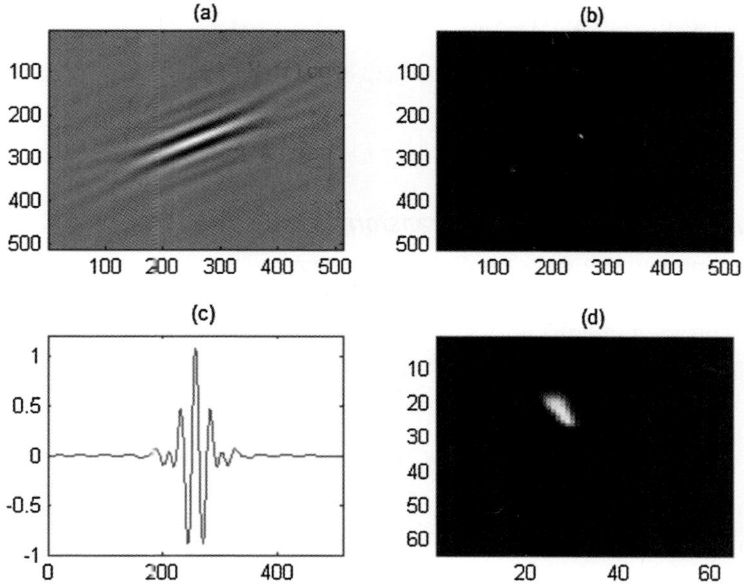

FIGURE 9.3.24

(a) The real part of a curvelet atom in the spatial space. (b) The Fourier transform of the curvelet atom. (c) The graph of the oscillatory function along the θ-direction taken from the real part of the curvelet atom. (d) Zoom-in view of the central region of the Fourier transform shown in (b).

9.3.1.4 The formal definition of the continuous curvelet transform

Definition 9.9. Let $f(\boldsymbol{x}) \in L^2(\mathcal{R}^2)$ and let $\{\psi_{a,\,\theta,\,\mathbf{b}}(\boldsymbol{x})\}_{a\in(0,1],\ \theta\in[0,2\pi),\ \mathbf{b}\in\mathcal{R}^2}$ be the families of curvelet atoms defined above. The continuous curvelet transform is defined as

$$C_f(a,\theta,\mathbf{b}) = \langle f, \psi_{a,\,\theta,\,\mathbf{b}}\rangle = \int_{\mathcal{R}^2} f(\boldsymbol{x})\psi^*_{a,\,\theta,\,\mathbf{b}}(\boldsymbol{x})\,d\boldsymbol{x}, \tag{9.3.1.9}$$

where $\{C_f(a,\theta,\mathbf{b}),\ a\in(0,1],\ \theta\in[0,2\pi),\ \mathbf{b}\in\mathcal{R}^2\}$ are called the curvelet coefficients of the function $f(\boldsymbol{x})$.

Note that here we mentioned the families of curvelet atoms. This is because each mother curvelet at scale a generates a family of curvelet atoms. The curvelet transform projects a 2D function into the families of curvelet atoms to receive a series of decomposed coefficients. Each coefficient reflects the correlation between the function and a curvelet atom focusing inside a localized area. Typically, the curvelet coefficients are calculated in the Fourier plane using the following result, which is the Parseval formula.

Proposition 9.5. *The curvelet transform can be calculated in the Fourier plane as*

$$
\begin{aligned}
C_f(a,\theta,\boldsymbol{b}) &= \langle F, \Psi_{a,\,\theta,\,\boldsymbol{b}} \rangle \\
&= \frac{1}{(2\pi)^2} \int_{\mathcal{R}^2} F(\Omega_x, \Omega_y) \Psi^*_{a,\,\theta,\,\boldsymbol{b}}(\Omega_x,\, \Omega_y)\, d\Omega_x\, d\Omega_y,
\end{aligned} \tag{9.3.1.10}
$$

for $a \in (0, 1]$, $\theta \in [0, 2\pi)$, $\boldsymbol{b} \in \mathcal{R}^2$.

9.3.1.5 The properties of the curvelet transform

The vanishing moments and the oscillatory behavior

In the classical wavelet theory, a wavelet with n vanishing moments has been interpreted as a multiscale differential operator of order n. A similar concept can be extended to the curvelets in parallel. A mother curvelet $\psi_{a,0,\mathbf{0}}$ is said to have q vanishing moments if its waveform in spatial plane satisfies

$$
\int_{-\infty}^{\infty} \psi_{a,0,\mathbf{0}}(x, y) x^n\, dx = 0, \text{ for all } 0 < n \le q. \tag{9.3.1.11}
$$

The curvelet windows we selected can have infinitely many vanishing moments as Eq. (9.3.1.11) is satisfied for any q.

The curvelet window that has infinitely many vanishing moments ensures that the coefficients of an image function tend to zero when a mother curvelet is centered at $\boldsymbol{x}$. In the spatial plane, a rotated curvelet atom with oscillatory direction at θ rad, if shifted to the spatial location $\boldsymbol{b} \in \mathcal{R}^2$, plays the role of an angular differential operator in a localized area centered at $\boldsymbol{b}$ to measure the regularity along the orientation, as shown in Fig. 9.3.25. This area is referred to as the ellipse support area. If an image function satisfies a localized Lipschitz regularity along the radial direction at location $\boldsymbol{x}$, then a polynomial can be used to approximate the image function in a localized area. The aforementioned infinitely many vanishing moments of the curvelet windows will therefore lead to many curvelet coefficients tending to zero.

The decay rate of the curvelet coefficients

As mentioned above, the requirement for infinitely many vanishing moments ensures that the curvelets decay rapidly in the spatial domain. A curvelet coefficient therefore measures the correlation between a function and a curvelet atom which has its support focused on a spatial ellipse area. The resulting coefficients may be classified into the following three cases [7]:

1. The magnitude of a curvelet coefficient tends to zero if the ellipse support of the curvelet atom does not intersect with a singularity of the function.
2. The magnitude of a curvelet coefficient tends to zero if the ellipse support of the curvelet atom intersects with a singularity, but its ridge direction is not oriented at the tangential direction of a curve.
3. The magnitude of a curvelet coefficient decays slowly if the ellipse support of the curvelet atom intersects with a discontinuity, and its ridge is oriented along the tangential direction of the curve where the singularities are located. More precisely, the curvelet oscillates along the normal direction of the curve to measure the regularities in a local area centered at $\boldsymbol{b}$.

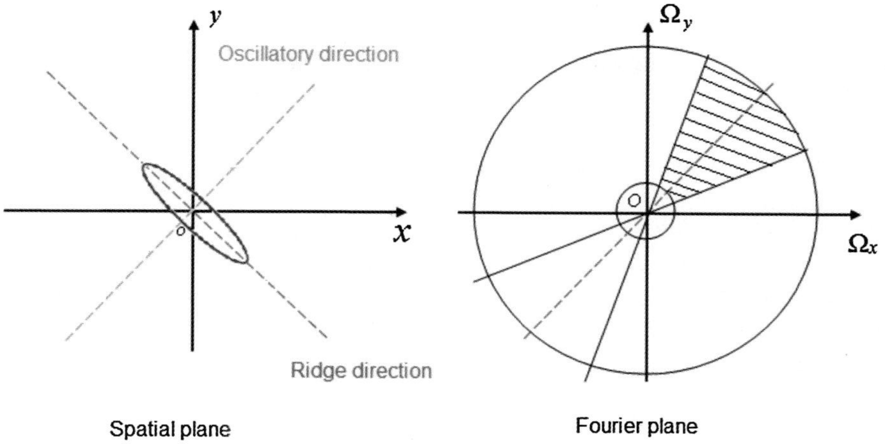

FIGURE 9.3.25

A curvelet atom in its spatial plane and Fourier plane.

9.3.1.6 The reproducing formula

Similarly to the classical wavelets, the admissibility conditions in Eqs. (9.3.1.4) and (9.3.1.5) ensure that all the curvelet atoms cover the Fourier plane completely, so the curvelet transform is a redundant representation of a function that can be perfectly reconstructed.

Theorem 9.1. *If the Fourier transform of a high-pass function $f \in L^2(\mathcal{R}^2)$ vanishes (is equal to 0) for $|\boldsymbol{\Omega}| < 2/a_0$, this function can be perfectly reconstructed from the curvelet coefficients according to the following Calderón-like reproducing formula:*

$$f(\boldsymbol{x}) = \int_0^{2\pi} \int_0^{a_0} \int_{\mathcal{R}^2} C_f(a, \theta, \boldsymbol{b}) \psi_{a,\,\theta,\,\boldsymbol{b}}(\boldsymbol{x}) \frac{d\boldsymbol{b}}{a^{1/2}} \frac{da}{a^{3/2}} \frac{d\theta}{a}. \tag{9.3.1.12}$$

For the proof of this theorem, the reader is referred to [7].

This reproducing formula can be extended to reconstruct any function $f \in L^2(\mathcal{R}^2)$ that includes both low-frequency and high-frequency components. To do so, we can introduce a scaling function, $\phi(\boldsymbol{x})$, the so-called father curvelet, through its Fourier transform, $\Phi(\boldsymbol{\Omega})$, given by

$$\Phi^2(\boldsymbol{\Omega}) = 1 - \Psi^2(\boldsymbol{\Omega}), \quad \text{where } \Psi^2(\boldsymbol{\Omega}) = \int_0^{a_0|\boldsymbol{\Omega}|} |\Psi_1(r)|^2 \frac{dr}{r}. \tag{9.3.1.13}$$

The function $\Phi(\boldsymbol{\Omega})$ can be viewed as an aggregation of curvelets at scales greater than a_0, written as

$$\Phi(\boldsymbol{\Omega}) = \int_{a_0}^{+\infty} \frac{|\Psi_1(r|\boldsymbol{\Omega}|)|^2}{r} dr = \int_{a_0|\boldsymbol{\Omega}|}^{+\infty} \frac{|\Psi_1(r)|^2}{r} dr, \tag{9.3.1.14}$$

and may further be written as

$$\Phi(\boldsymbol{\Omega}) = \begin{cases} 1, & |\boldsymbol{\Omega}| \le 1/(2a_0), \\ (1 - \Psi^2(\boldsymbol{\Omega}))^{1/2}, & 1/(2a_0) < |\boldsymbol{\Omega}| < 2/a_0, \\ 0, & |\boldsymbol{\Omega}| \ge 2/a_0. \end{cases}$$

Note that a_0 is limited to $0 < a_0 \le 1$, and is typically set to $a_0 = 1$.

The high-frequency content in $\Phi(\boldsymbol{\Omega})$ equaling zero indicates the father curvelet is the impulse response of a low-pass frequency filter. Shifting the father curvelet $\phi(\boldsymbol{x})$ builds the shifting atoms $\phi_{\boldsymbol{b}}(\boldsymbol{x}) = \phi(\boldsymbol{x} - \boldsymbol{b})$, $\boldsymbol{b} \in \mathcal{R}^2$. It can be seen that the father curvelet is isotropic and is not scaled. It only produces an atom that is shifted to spatial location $\boldsymbol{b}$.

Theorem 9.2. *Any $f \in L^2(\mathcal{R}^2)$ can be reproduced as*

$$\begin{aligned} f(\boldsymbol{x}) &= \int_{\mathcal{R}^2} < \phi_{\boldsymbol{b}}, f > \phi_{\boldsymbol{b}}(\boldsymbol{x})\, d\boldsymbol{b} \\ &+ \int_0^1 \int_0^{2\pi} \int_{\mathcal{R}^2} C_f(a, \theta, \boldsymbol{b}) \psi_{a,\theta,\boldsymbol{b}}(\boldsymbol{x}) \frac{d\boldsymbol{b}}{a^{3/2}} \frac{da}{a} \frac{d\theta}{a^{1/2}} \end{aligned} \tag{9.3.1.15}$$

and it satisfies the Plancherel formula

$$\begin{aligned} \|f(\boldsymbol{x})\|_2^2 &= \int_{\mathcal{R}^2} < \phi_{\boldsymbol{b}}, f >^2 d\boldsymbol{b} \\ &+ \int_0^1 \int_0^{2\pi} \int_{\mathcal{R}^2} |C_f(a, \theta, \boldsymbol{b})|^2 \frac{d\boldsymbol{b}}{a^{3/2}} \frac{da}{a} \frac{d\theta}{a^{1/2}}. \end{aligned} \tag{9.3.1.16}$$

The term $\frac{d\mathbf{b}}{a^{3/2}} \frac{da}{a} \frac{d\theta}{a^{1/2}}$ may be viewed as the reference measure in the parameter space of $(a, \theta, \boldsymbol{b})$. It suggests that the curvelets use anisotropic measuring of the unit cell in the space of $(a, \theta, \boldsymbol{b})$. This theorem ensures that both the curvelets and the classical wavelets can be decomposed into a low-frequency component and a combination of detail components. The polar wedge-shaped curvelet windows in the frequency domain outperform the classical wavelet in the 2D case, as they allow one to study 2D functions by measuring the energy inside the windows oriented along selective orientations to efficiently identify the regularities.

9.3.2 The discrete curvelet transform

The continuous curvelets can be sampled along their scale, rotation, and translation parameters to form a pyramid of discrete curvelets for numerical computation. A tight frame can be adopted for efficiency so that a 2D function may be projected onto the discrete curvelet frame with the lowest number of large coefficients. Here only one tight frame is introduced. First we introduce the window functions to be used in generating the curvelets.

9.3.2.1 The selection of windows

As we mentioned earlier, the mother curvelets are primarily defined by two window functions, Ψ_1 and Ψ_2. A good choice is the Meyer wavelet windows in the Fourier domain. Fig. 9.3.26 shows the

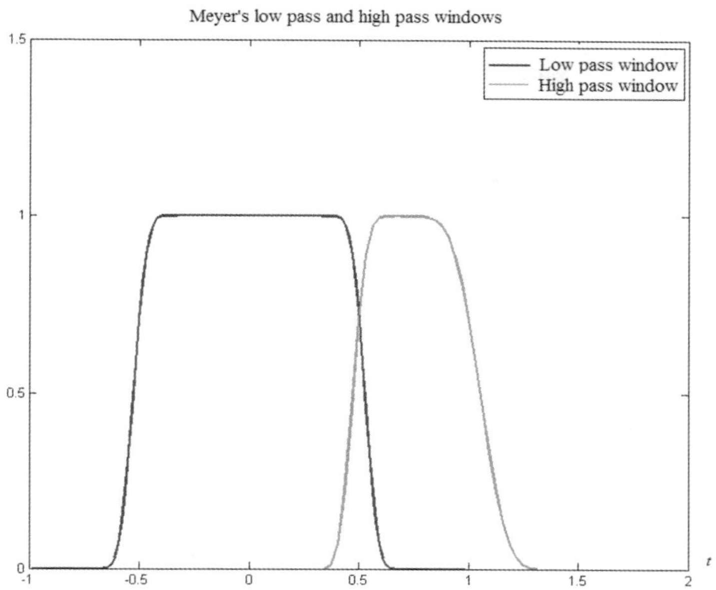

FIGURE 9.3.26

The Meyer's wavelet windows: low-pass $\Psi_2(t)$, in blue (dark gray in print version) color, and high-pass $\Psi_1(t)$, in red (light gray in print version) color. Only its component in the positive axis is shown here.

high-pass and low-pass windows that are orthogonal and overlapping. The low-pass window in the 1D Fourier domain may be used as the angular window, $\Psi_2(\sigma)$, and is defined as

$$\Psi_2(\sigma) = \begin{cases} 1, & |\sigma| \le 1/3, \\ \cos(\frac{\pi}{2}\beta(3|\sigma| - 1)), & 1/3 < |\sigma| \le 2/3, \\ 0, & |\sigma| > 2/3. \end{cases} \tag{9.3.2.1}$$

The high-pass window in the 1D Fourier domain may also be used as the radial window $\Psi_1(r)$, and is defined as

$$\Psi_1(r) = \begin{cases} 0, & |r| \le 2/3 \text{ and } |r| \ge 8/3, \\ \sin(\frac{\pi}{2}\beta(\frac{3|r|}{2} - 1)), & 2/3 < |r| \le 4/3, \\ \cos(\frac{\pi}{2}\beta(\frac{3|r|}{4} - 1)), & 4/3 < |r| < 8/3, \end{cases} \tag{9.3.2.2}$$

where $\beta(x)$ is any smooth function that defines the transition bands. It is defined on $[0, 1]$ and satisfies

$$\beta(x) + \beta(1 - x) = 1, \quad x \in \mathcal{R}. \tag{9.3.2.3}$$

The following admissibility conditions ensure that the discretized curvelet atoms generated from these windows can completely cover the Fourier plane.

Proposition 9.6. *On a discrete scale, the two windows selected above, Ψ_1 and Ψ_2, satisfy the following admissibility conditions:*

$$\sum_{k=-\infty}^{\infty} |\Psi_1(2^{-k}r)|^2 = 1, \quad r > 0, \tag{9.3.2.4}$$

$$\sum_{l=-\infty}^{\infty} |\Psi_2(\sigma - l)|^2 = 1, \quad \sigma \in \mathcal{R}. \tag{9.3.2.5}$$

The proof of the proposition only needs to be shown in the overlapping regions between two adjacent windows; for example, for $\Psi_1(r)$, when $r \in (2/3, 3/4)$, we have

$$\begin{aligned}\sum_{k=-\infty}^{\infty} |\Psi_1(2^{-k}r)|^2 &= |\Psi_1(r)|^2 + |\Psi_1(2r)|^2 \\ &= \sin^2(\frac{\pi}{2}\beta(\frac{3|r|}{2} - 1)) + \cos^2(\frac{\pi}{2}\beta(\frac{3|2r|}{4} - 1)) \\ &= 1.\end{aligned} \tag{9.3.2.6}$$

Refer to [85] for details of the proof for a different set of window functions. Note that Eq. (9.3.2.5) is the same as Eq. (9.3.1.5), while Eq. (9.3.2.4) is similar to Eq. (9.3.1.4) for the continuous case when $C_{\Psi_1} = \ln 2$. The selected windows can be scaled to fit $\text{supp}\{\Psi_1(r)\} = [1/2, 2]$ and $\text{supp}\{\Psi_2(\sigma)\} = [-1, 1]$.

The discretized windows are used in the following to build the discrete curvelet atoms. Meanwhile, the coarse-scale father curvelet window can be defined as the low-pass window $\Phi(r)$ in the radial direction which satisfies

$$|\Phi(r)|^2 + \sum_{k \geq 0} |\Psi_1(2^{-k}r)|^2 = 1. \tag{9.3.2.7}$$

The choice of $\beta(x)$ determines the transition band of the Meyer window. The definition of $\beta(x)$, for example, can be chosen as

$$\beta(x) = x^4(35 - 84x + 70x^2 - 20x^3), \tag{9.3.2.8}$$

which leads to the resulting windows Ψ_1 and Ψ_2 being in C^3. For $\beta(x)$ of the form

$$\beta(x) = x^3(5 - 5x + x^2), \tag{9.3.2.9}$$

the resulting windows Ψ_1 and Ψ_2 are C^2.

Finally, for

$$\beta(x) = x^2(3 - 2x), \tag{9.3.2.10}$$

the resulting windows Ψ_1 and Ψ_2 are C^1. The 1D window Ψ_1 has infinitely many vanishing moments in the parameter r because $\Psi_1(r) = 0$ for $|r| < 2/3$. The mother curvelets constructed from the windows Ψ_1 and Ψ_2 should have infinitely many moments too [83,85].

9.3.2.2 Constructing a tight frame of curvelet atoms

Now we are to construct a tight Parseval curvelet frame by sampling the continuous curvelets. With a tight frame, the whole Fourier plane can be covered with discrete curvelet atoms with minimum overlap, and the following equation is satisfied:

$$\sum_{n\in\Gamma} |\langle f, \psi_n\rangle|^2 = C \parallel f \parallel_2^2 . \tag{9.3.2.11}$$

The three parameters $(a, \theta, \boldsymbol{b})$ will be sampled in order to obtain a tight frame. First, at each scale k, the radial scale is sampled on a dyadic interval at $a_k = 2^{-k},\ k \geq 0$, resulting in the selection of the mother curvelet, $\psi_{k,0,\mathbf{0}}$, defined as the inverse Fourier transform of $\Psi_{k,0,\mathbf{0}}(r, \sigma)$, which is created using the polar coordinates as

$$\Psi_{k,0,\mathbf{0}}(r,\sigma) = \begin{cases} \Psi_1(2^{-k}\cdot r)\Psi_2(\dfrac{2^{(k+2)/2}\sigma}{\pi})2^{-3k/4}, & k \text{ is an even integer,} \\ \Psi_1(2^{-k}\cdot r)\Psi_2(\dfrac{2^{(k+1)/2}\sigma}{\pi})2^{-3k/4}, & k \text{ is an odd integer.} \end{cases}$$

Second, the rotation angular parameter is sampled at $\theta_{k,l}$, given as

$$\theta_{k,l} = \begin{cases} \pi l 2^{-(k+2)/2}, & k \text{ is an even integer, } 0 \leq l < 2^{\frac{(k+2)}{2}+1} - 1, \\ \pi l 2^{-(k+1)/2}, & k \text{ is an odd integer, } 0 \leq l < 2^{\frac{(k+1)}{2}+1} - 1. \end{cases}$$

For example, at $k = 0$ and $k = 1$, $\theta_{k,l} = \frac{\pi}{2}l,\ l = 0, 1, 2, 3$; at $k = 2$ and $k = 3$, $\theta_{k,l} = \frac{\pi}{4}l,\ l = 0, 1, 2, \cdots, 7$. It can be deduced that

$$\sum_{l=0}^{2^{\frac{(k+2)}{2}+1}-1} |\Psi_2(\frac{2^{(k+2)/2}(\sigma + \theta_{k,l})}{\pi})|^2 = 1 \quad \text{if } k \text{ is an even integer} \tag{9.3.2.12}$$

and

$$\sum_{l=0}^{2^{\frac{(k+1)}{2}+1}-1} |\Psi_2(\frac{2^{(k+1)/2}(\sigma + \theta_{k,l})}{\pi})|^2 = 1 \quad \text{if } k \text{ is an odd integer.} \tag{9.3.2.13}$$

This results in the discrete curvelet atoms

$$\psi_{k,l,\boldsymbol{b}}(\boldsymbol{x}) = \psi_{k,0,\mathbf{0}}(R_{\theta_{k,l}}(\boldsymbol{x} - \boldsymbol{b})). \tag{9.3.2.14}$$

Finally, the parameter $\boldsymbol{b}$ is sampled at

$$\boldsymbol{b}_s^{k,l} = R_{\theta_{k,l}}^{-1}(s_1 2^{-k}, s_2 2^{-k/2}), \tag{9.3.2.15}$$

where $\boldsymbol{b}_s^{k,l}$ denotes the grids defined by rotating the dyadic points by an angle $\theta_{k,l}$. The discrete curvelet atoms are given as

$$\psi_{k,l,s}(\boldsymbol{x}) = \psi_{k,0,\boldsymbol{0}}(R_{\theta_{k,l}}(\boldsymbol{x} - \boldsymbol{b}_s^{k,l})), \quad \boldsymbol{s} = (s_1, s_2). \tag{9.3.2.16}$$

This curvelet frame consists of interleaved curvelet atoms, separately selected from the samplings of two adjacent scales, the even and odd scales. Under this tight frame, the discrete curvelet transform coefficients are given by

$$Cf_{k,l,s} = \langle f, \psi_{k,l,s} \rangle. \tag{9.3.2.17}$$

The coefficients are calculated in the Fourier plane using the following formula based on the mother curvelet:

$$Cf_{k,l,s} = \frac{1}{(2\pi)^2} \int_{\mathcal{R}^2} F(\boldsymbol{\Omega}) \Psi_k^*(R_{\theta_{k,l}}(\boldsymbol{\Omega})) e^{j\langle \mathbf{b}_s^{k,l}, \boldsymbol{\Omega} \rangle} d\boldsymbol{\Omega}. \tag{9.3.2.18}$$

Theorem 9.3. *For a high-pass function $f \in L^2(\mathcal{R}^2)$, Eqs. (9.3.2.4), (9.3.2.12), and (9.3.2.13) guarantee that a reproducing formula can be obtained by*

$$f = \sum_k \sum_l \sum_{s \in \mathcal{R}^2} Cf_{k,l,s} \psi_{k,l,s}, \tag{9.3.2.19}$$

and the energy conservation is maintained:

$$\| f \|_2^2 = \sum_k \sum_l \sum_{s \in \mathcal{R}^2} |Cf_{k,l,s}|^2. \tag{9.3.2.20}$$

Meanwhile, if we add the father curvelet $\phi_{\boldsymbol{b}}(x)$ into the frame, calculate the father curvelet coefficients from the inner product between functions $\phi_{\boldsymbol{b}}(x)$ and f, put both with the curvelet coefficients, and still write them as $Cf_{k,l,s}$, then any function in $L^2(\mathcal{R}^2)$ can be reconstructed from the integrated families of the curvelet atoms, resulting in the same formulas of Eqs. (9.3.2.19) and (9.3.2.20).

Under this curvelet tight frame, Candès and Donoho [5] have proved the following theorem, which estimates the approximation error under the curvelet coefficients. It shows a faster convergence rate than that using the classical wavelets in 2D.

Theorem 9.4. *Let $f(x)$ be a twice differentiable function in $\mathcal{R}^2$ except at discontinuities along a C^2-curve. Let f_N^C be the N-term approximation of f reconstructed by using the n largest curvelet coefficients obtained by simple thresholding. The approximation error follows*

$$\| f - f_N^C \|_2^2 \leq C N^{-2} (log N)^3.$$

The proof can be found in [5].

9.3.2.3 Implementation procedure for the discrete curvelet transform

The discrete curvelet transform is implemented using a low-pass filter P_0 together with a set of passband filters $\Delta_1, \Delta_2, \cdots, \Delta_s$. This bank of filters partitions the whole frequency plane into $s + 1$ decompositions that are concentrated and are located near certain frequencies in a sequence of dyadic intervals.

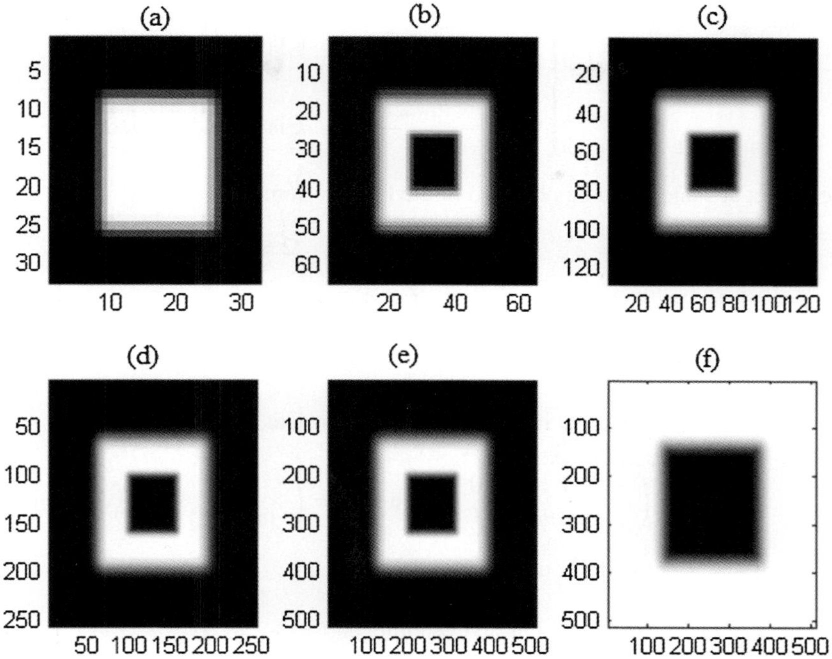

FIGURE 9.3.27

Examples of concentric square windows used in the discrete curvelet transform at several scales. (a) The low-frequency level ($k = 0$). (b) Level $k = 1$. (c) Level $k = 2$. (d) Level $k = 3$. (e) Level $k = 4$. (f) The remaining high-frequency level ($k = 5$). Note that the number of pixels in the x, y-direction is shown for each image because of the different size of the window at different scales.

Indeed, in the digital curvelet implementation, this bank of filters is generated using concentric squares instead of concentric circles.

To generate the concentric squares, as shown in Fig. 9.3.27, a low-pass filter at scale k is first generated using a smooth Meyer window $\Phi(r)$ following Eq. (9.3.2.1) as

$$\Phi_k(\Omega_x, \Omega_y) = \Phi(2^{-k}\Omega_x)\Phi(2^{-k}\Omega_y). \tag{9.3.2.21}$$

Here P_0 refers to the window when $k = 0$. Then the bandpass filters Δ_k corresponding to bandpass window functions $\Psi_{1,k}(\Omega_x, \Omega_y)$ are defined as

$$\Psi_{1,k}(\Omega_x, \Omega_y) = \sqrt{\Phi_{k+1}^2(\Omega_x, \Omega_y) - \Phi_k^2(\Omega_x, \Omega_y)}, \quad k \geq 0.$$

The angular window $\Psi_{2,k}(\Omega_x, \Omega_y)$ in the Cartesian plane is also defined using the function $\Psi_2(\sigma)$ in Eq. (9.3.2.1),

$$\Psi_{2,k}(\Omega_x, \Omega_y) = \begin{cases} \Psi_2(\frac{2^{k/2}\Omega_y}{\Omega_x}), & k \text{ is an even integer,} \\ \Psi_2(\frac{2^{(k-1)/2}\Omega_y}{\Omega_x}), & k \text{ is an odd integer.} \end{cases}$$

The mother curvelet at scale k can be obtained in the Fourier domain as

$$\Psi_k(\mathbf{\Omega}) = 2^{-3j/4}\Psi_{1,k}(\mathbf{\Omega})\Psi_{2,k}(\mathbf{\Omega}), \quad \mathbf{\Omega} = (\Omega_x, \Omega_y).$$

The curvelet atoms at $\mathbf{b} = \mathbf{0}$ can be obtained in the Fourier domain as

$$\Psi_{k,l}(\mathbf{\Omega}) = \Psi_k(S_{\theta_{k,l}}^{-1}\mathbf{\Omega}) = 2^{-3j/4}\Psi_{1,k}(\mathbf{\Omega})\Psi_{2,k}(S_{\theta_{k,l}}^{-1}(\mathbf{\Omega})), \quad \mathbf{\Omega} = (\Omega_x, \Omega_y),$$

where $S_{\theta_{k,l}}$ is the shear matrix with its inverse $S_{\theta_{k,l}}^{-1}$. Both matrices are given as

$$S_{\theta_{k,l}} = \begin{pmatrix} 1 & 0 \\ -\tan(\theta_{k,l}) & 1 \end{pmatrix} \text{ and } S_{\theta_{k,l}}^{-1} = \begin{pmatrix} 1 & 0 \\ \tan(\theta_{k,l}) & 1 \end{pmatrix}.$$

The selection of $\theta_{k,l}$ is defined through

$$\tan(\theta_{k,l}) = \begin{cases} l \cdot 2^{-k/2}, & k \text{ is an even integer, } l = -2^{k/2}, \cdots, 2^{k/2} - 1, \\ l \cdot 2^{-(k-1)/2}, & k \text{ is an odd integer, } l = -2^{(k-1)/2}, \cdots, 2^{(k-1)/2} - 1. \end{cases}$$

Here $\Psi_{k,l}(\Omega_x, \Omega_y)$ indeed defines the Cartesian approximation of the curvelet atoms in the polar Fourier plane along the horizontal cone direction. Rotating this system by $\pi/2$ defines the Cartesian approximation along the vertical cone direction. The two systems together define the Cartesian approximation of the curvelet atoms [83,86,87]. Using the discretized curvelet atoms, the Curvelab software [88] calculates the discrete curvelet coefficients using two different approaches, the unequispaced FFT (USFFT) and the wrapping method. It is believed that the wrapping method is simpler to understand. We summarize the two implementation procedures in the following.

The unequispaced FFT

The USFFT method is based on the operation of shearing a curvelet atom to an input function F in the frequency domain according to the following formula:

$$\begin{aligned} C_f(k, l, \boldsymbol{s}) &= \frac{1}{(2\pi)^2}\int_{\mathcal{R}^2} F(\mathbf{\Omega})\Psi_k(S_\theta^{-1}\mathbf{\Omega})exp(j\left\langle S_\theta^{-T}\boldsymbol{b}, \mathbf{\Omega}\right\rangle)\, d\Omega_x\, d\Omega_y \\ &= \frac{1}{(2\pi)^2}\int_{\mathcal{R}^2} F(S_\theta\mathbf{\Omega})\Psi_k(\mathbf{\Omega})exp(j\left\langle \boldsymbol{b}, \mathbf{\Omega}\right\rangle)\, d\Omega_x\, d\Omega_y, \end{aligned} \quad (9.3.2.22)$$

with $\theta = \theta_{k,l}$, $\boldsymbol{b} \simeq (s_1 2^{-k}, s_2 2^{-k/2})$. Note that $\Psi_k(\mathbf{\Omega})$ is a real-valued function; therefore, $\Psi_k^*(\mathbf{\Omega}) = \Psi_k(\mathbf{\Omega})$ is used in Eq. (9.3.2.22). The procedure for implementing the USFFT to calculate the discrete curvelet coefficients is summarized in the following. For more details, refer to [83,86].

1. Apply a 2D FFT to a function $f(\boldsymbol{x})$ and obtain its discrete Fourier samples $F(n_1, n_2)$, $0 < n_1, n_2 \leq N$, where N is the sample size.

2. For each scale/angle pair (k,l), resample (or interpolate) $F(n_1,n_2)$ to obtain sheared sample values, written as $F(n_1,n_2-n_1\tan\theta_{k,l})$ for $(n_1,n_2)\in R_k$. Here R_k is a rectangle of length $=2^k$ and width $=2^{k/2}$, which is located at the center of the bandpass content of the mother curvelet at scale k.
3. Multiplying the resampled F with the mother curvelet window $\Psi_{k,l}$, we have $Cf_{k,l}(n_1,n_2)=F(n_1,n_2-n_1\tan\theta_{k,l})\Psi_{k,l}(n_1,n_2)$.
4. Apply an inverse 2D FFT to each $Cf_{k,l}(n_1,n_2)$ obtained in the previous step to calculate the discrete curvelet coefficients $C_f(k,l,s)$.

In Curveletlab [88], the following package in MATALB is used to decompose the original image into curvelet coefficients:

```
%   is_real    Type of the transform
%                   0: complex-valued curvelets
%                   1: real-valued curvelets
Curveletcoef = fdct_usfft(X,is_real);
```

The inverse curvelet transform is simply computed by "reversing" all the operations of the direct transform with some adjustments.

1. For each scale/angle pair (k,l), perform a (properly normalized) 2D FFT of each curvelet coefficient array $C_f(k,l,s)$ to obtain $(Cf)_{k,l}(n_1,n_2)$ in $\Psi_{k,l}$.
2. Now $(Cf)_{k,l}(n_1,n_2)$ should be viewed as samples on the sheared grid. Resample them back to the standard Cartesian grid.
3. Integrate $(Cf)_{k,l}(n_1,n_2)$ from all scales and angles together. This recovers $F(n_1,n_2)$.
4. Take a 2D inverse FFT to obtain $f(x,y)$.

Fig. 9.3.28 provides the diagram for implementing the USFFT curvelet transform and its inverse transform.

In Curveletlab, the following MATLAB syntax is used to reconstruct the original image from the curvelet coefficients:

```
Y = ifdct_usfft(C,is_real);
```

The computational cost for the forward and inverse transform requires approximately $O(n^2\log n)$ operations and $O(n^2)$ memory storage.

The wrapping method

The wrapping method uses the following formula to calculate the curvelet coefficients:

$$C_f(k,l,s)=\frac{1}{(2\pi)^2}\int_{\mathcal{R}^2}F(\mathbf{\Omega})\Psi_k(S_\theta^{-1}(\mathbf{\Omega}))exp(j\,\langle \boldsymbol{b},\mathbf{\Omega}\rangle)\,d\Omega_x\,d\Omega_y. \tag{9.3.2.23}$$

Note that $S_\theta^{-1}\boldsymbol{b}$ in Eq. (9.3.2.22) is replaced by $\boldsymbol{b}$, which takes its value on the rectangle grid. In this method, the curvelet atoms $\Psi_{k,l}=\Psi_k(S_{\theta_{k,l}}^{-1}(\mathbf{\Omega}))$ should be obtained by applying the shear operation $S_{\theta_{k,l}}^{-1}$ to the rectangle R_k as in the USFFT method. The resulting window should be a trapezoidal-shaped region.

The following is the procedure for implementing the wrapping technique to calculate the discrete curvelet coefficients:

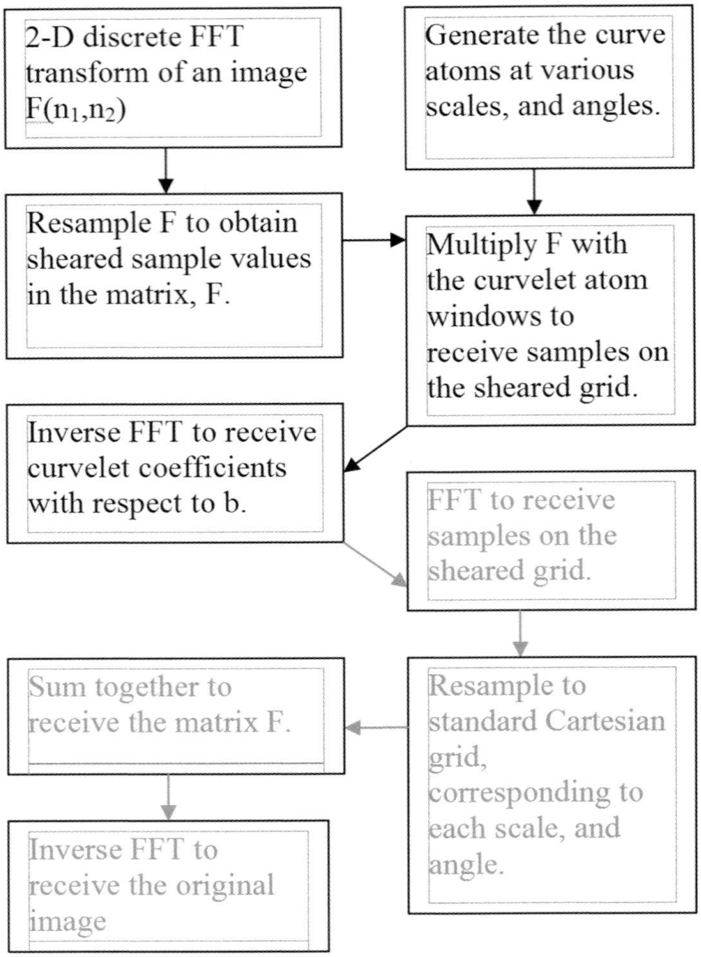

FIGURE 9.3.28

The USFFT curvelet transform diagram (in black color) and its inverse transform (in red (light gray in print version) color).

1. Calculate the 2D FFT of an image $f(\boldsymbol{x})$ to obtain the Fourier samples $F(n_1, n_2)$.
2. For (k, l), decompose $F(n_1, n_2)$ into each curvelet window by calculating $F(n_1, n_2)\Psi_{k,l}(n_1, n_2)$.
3. Wrap this product around the origin in the Fourier plane. This is done by relabeling the samples.
4. Apply the inverse 2D FFT to the wrapped $F(n_1, n_2)$ to calculate the discrete curvelet coefficients $C_f(k, l, s)$.

In Curveletlab, the following syntax is used to calculate the curvelet coefficients using the wrapping method:

$$\text{Curveletcoefwrp} = \text{fdct\_wrapping(X,0)}.$$

The inverse curvelet transform is computed as follows:

1. For each (k, l), perform a properly normalized 2D FFT of each curvelet coefficient array $C_f(k, l, s)$ to obtain $(Cf)_{k,l}(n_1, n_2)$ in $\Psi_{k,l}$.
2. For each (k, l), multiply the array of the coefficients by the corresponding wrapped curvelet $\Psi_{k,l}(n_1, n_2)$, which gives $(|\Psi_{k,l}|^2 F)(n_1, n_2)$.
3. Unwrap each array $(|\Psi_{k,l}|^2 F)[n_1, n_2]$ with $\Psi_{k,l}(n_1, n_2)$ on the frequency grid and integrate them all to form $F(n_1, n_2)$.
4. Finally, take a 2D inverse FFT of $F(n_1, n_2)$ to obtain $f(x, y)$.

In Curveletlab, the following syntax is used to reconstruct the original image from the curvelet coefficients:

$$\text{image} = \text{fdct\_wrapping\_dispcoef(Curveletcoefwrp)}.$$

In the wrapping approach, both the forward and inverse transforms are computed in $O(n^2 \log n)$ operations and require $O(n^2)$ storage.

9.3.3 Applications

Researchers have applied the curvelet transform in solving partial differential equations and in numerous other applications including image processing, seismic exploration, fluid mechanics, and data compression [87,89,90,91,92,93,94,95,96,97,98,99,100]. All applications take advantage of the flexibility of curvelets to selectively capture curve-based features. For illustration in this manuscript, we limit ourselves to the application of curvelets in image denoising.

The assumption about noisy signals in previous sections can be similarly made for 2D noisy images of interest herein. It is well known that a 2D isotropic wavelet transform of an image exhibits a large number of significant coefficients at every fine scale, making wavelet denoising more complex as many more wavelet coefficients are needed for reconstructing edges while smoothing out the noise [68,69,70, 101]. The development of curvelet theory has shown that discrete curvelet coefficients provide a near-optimal way to represent smooth objects with discontinuities along C^2 continuous curves in an image. The curvelet-based denoising method is similar to that of the wavelet. It consists of determining which curvelet coefficients represent the transient features in an image. It is shown that the high-amplitude curvelet coefficients usually indicate the position of edges and the low-amplitude curvelet coefficients usually capture noise present in the image. Therefore, the denoising is accomplished by applying a threshold to the curvelet coefficients.

A hard threshold estimator is defined as

$$I_T(x) = \begin{cases} x, & \text{if } |x| \geq T, \\ 0, & \text{if } |x| < T. \end{cases}$$

Applying the thresholds is equivalent to locally averaging the noise contribution present in the image around the local area the image is smooth over.

Here we assume the noise to have a normal distribution $N(0, \sigma^2)$, and $y(\boldsymbol{n})$ denotes the discrete curvelet coefficients from a noisy image. The following hard threshold is applied to estimate the curvelet

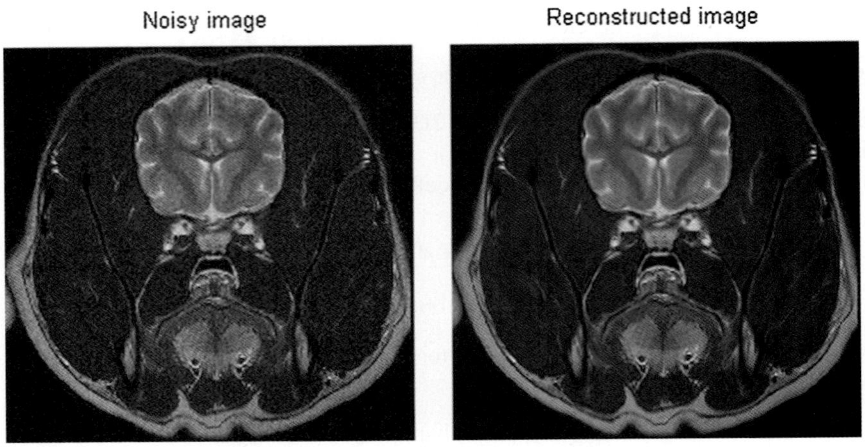

FIGURE 9.3.29

Shown in the left is a noisy image and shown in the right is the image after applying the hard threshold denoising method.

coefficients, $\tilde{y}(\boldsymbol{n})$, of the denoised image:

$$\tilde{y}(\boldsymbol{n}) = \begin{cases} y(\boldsymbol{n}), & \text{if } |y(\boldsymbol{n})|\sigma \geq T_a, \\ 0, & \text{if } |y(\boldsymbol{n})|\sigma < T_a. \end{cases}$$

The threshold T_a is chosen as scale-dependent and can be estimated from the noisy image.

It is shown that the curvelet-based denoising displays higher sensitivity than the wavelet-based denoising (see Fig. 9.3.29). In curvelet denoising, salient features in the noisy image can be clearly enhanced, while the noise can be selectively removed [102]. However, even though the pure discrete curvelet denoising has improved the image denoising method, it is interesting to see that a better denoising effect has been achieved by a hybrid approach [89] called the total variation (TV) method that combines the discrete curvelet transform with the classical wavelet method. The TV method has shown a higher capability in exploiting curve-like features.

9.4 Contourlets and their applications

9.4.1 The contourlet transform

In this section, we introduce the contourlet transform, proposed by Do and Vetterli [9]. The contourlet transform is a discrete filter bank implementation that expands multidimensional signals into multiresolution copies for analysis, approximation, and compression [9,103]. Here we consider only 2D signals, that is, images. A filter bank technique uses multirate signal processing techniques that have been successfully applied in many application areas [41,104]. It is a natural and practical approach to efficiently compute a set of good basis functions for an image [21,105]. The contourlet filter bank technique proceeds to decompose the frequency domain into angularly oriented directions from fine to coarse resolution. It implements similar features to curvelet analysis so that certain frequency regions in the coarse or detailed version can be focused on under a selected basis/frame. This advances the classical discrete filter bank technique that is used in classical wavelet analysis. It allows the representation of a 2D signal to be optimized using only a few significant coefficients corresponding to the geometrical features in an image, such as edges with continuous curves.

The multiresolution decomposition filter bank technique is represented by the Laplacian pyramid (LP) decomposition, originally developed by Burt and Adelson [106] for image coding. Its development has inspired the usage of discrete filter bank techniques to decompose a 2D image into images of multiresolution in discrete-time wavelet bases. This has further inspired the research for building the connection between the filter banks and classical wavelets led by Mallat [84] and Daubechies [20]. In turn, this has led to the development of the filter bank technique, which has played an active role in constructing various bases for classical wavelets. It has strengthened the connection between harmonic analysis and discrete signal processing. A filter bank with a directional-oriented decomposition capacity was later proposed by Bamberger [107]. The recently proposed contourlet filter banks have integrated the LP filter bank and the directional filter bank (DFB) for multiresolution anisotropic wavelet decomposition.

The contourlet transform is a tree-structured filter bank that cascades multiple LP and DFB filter banks iteratively. In general, the filter banks use maximally decimated filter banks similar to those in subband coding and invoke the usage of multirate signal processing to reduce the computational complexity. Each filter bank in the structure can be divided into two stages: an analysis filter bank and a synthesis filter bank. An analysis filter bank decomposes a signal into different subsignals. Each subsignal is downsampled to its frequency content concentrated in a segment in the Fourier plane. A synthesis filter bank upsamples and combines the output signals from the analysis filter bank into one signal. The basic structure of the contourlets is a two-channel maximally decimated filter bank, referred to as a QMF bank.

In synthesis, a perfect reconstruction is achieved when the output signal equals the original signal. An optimal approximation of the original signal is also of interest when perfect reconstruction is difficult to achieve. This is useful in signal compression, as an individual image of reduced size is efficient for transmission even though the total sampling size might be increased. To avoid a redundant expansion of a signal, filters that implement orthonormal bases are sought. However, filters that implement biorthogonal bases or tight frames are necessary in some cases in order to relax some constraints.

A digital contourlet filter bank inevitably inherits the distortion problem in digital signal processing; to avoid this problem, a nondownsampling version of the contourlets has been proposed [108], which is parallel to the nonsubsampled discrete wavelet transform (NSWT) with the à trous algorithm. Additionally, an M-band method that provided nonredundant DFB was proposed in [109].

In this section, we downplay the distortion problems with downsampling/upsampling to keep our focus on MRA. Polyphase decomposition is typically used to form a fast computing system for implementing filter banks [21,41,110]. In the following, we start by introducing some basic sampling operations in digital image processing. We then introduce the LP and DFB filter banks before we introduce the contourlets as the combination of the two types of filter banks.

9.4.1.1 Downsampling and upsampling operators

Downsampling is a decimation operator that reduces the original image size, while upsampling is an interpolation operator that increases the original image size. The two operations are the basic techniques that form the multirate digital system. In the 2D digital image processing, downsampling is used to generate multiple components with reduced size for the economy of analyzing an image, while upsampling is used to recover the parent image. In this section and the following, $\boldsymbol{n} \in \mathcal{Z}^2$ is viewed as a column vector $\boldsymbol{n} = (n_1, n_2)^T$ since matrix multiplication is involved. Here the superscript "T" refers to the transpose of a vector or a matrix.

Both sampling operations involve a 2×2 sampling matrix given by

$$\boldsymbol{M} = \begin{pmatrix} m_1 & 0 \\ 0 & m_2 \end{pmatrix}, \qquad \text{where } m_1, m_2 \in \mathcal{Z}.$$

Given a 2D image $x(\boldsymbol{n})$, $\boldsymbol{n} = (n_1, n_2) \in \mathcal{Z}^2$, with its z-transform $X(z)$ (see Appendix 9.A.1), an $\boldsymbol{M}$-fold downsampling (decimator) takes $x(\boldsymbol{n})$ as an input and produces an output, $x_d(\boldsymbol{n})$, given by

$$x_d(\boldsymbol{n}) = x(\boldsymbol{M}\boldsymbol{n}).$$

The z-transform of an $\boldsymbol{M}$-fold downsampled signal, $x_d(\boldsymbol{n})$, is given as

$$X_d(z) = |Det(\boldsymbol{M})|^{-1} \sum_{\boldsymbol{k}=(k_1,k_2)\in\mathcal{S}} X(z_1^{1/m_1} e^{-j2\pi k_1/m_1}, z_2^{1/m_2} e^{-j2\pi k_2/m_2}),$$

where $\mathcal{S}$ is the set of all integer points in $\boldsymbol{M}\boldsymbol{t}, \boldsymbol{t} \in [0, 1) \times [0, 1)$.

An $\boldsymbol{M}$-fold upsampling (expander or interpolator) takes $x(\boldsymbol{n})$ as an input and produces an output, $x_u(\boldsymbol{n})$, given by

$$x_u(\boldsymbol{n}) = \begin{cases} x(\boldsymbol{s}), & \text{if} \quad \boldsymbol{n} = \boldsymbol{M}\boldsymbol{s}, \ \boldsymbol{s} \in \mathcal{Z}^2, \\ 0, & \text{otherwise.} \end{cases}$$

The z-transform of an $\boldsymbol{M}$-fold upsampled signal $x_u(\boldsymbol{n})$ is given as

$$X_u(z) = X(z_1^{m_1}, z_2^{m_2}), \text{ with } z = (z_1, z_2).$$

The dyadic subsampling occurs in the x- or y-direction when $|det(\boldsymbol{M})| = 2$, where $det(\boldsymbol{M})$ is the determinant of a matrix $\boldsymbol{M}$. Dyadic subsampling in both the x- and y-directions occurs when $m_1 = 2$ and $m_2 = 2$ in the diagonal matrix $\boldsymbol{M}$. The dyadic sampling is usually used in the digital filter bank design of a discrete wavelet transform (DWT) or discrete anisotropic wavelet transform. Sampling operations that invoke nondiagonal matrices are to be introduced in the following section.

9.4.1.2 Polyphase decomposition

Polyphase decomposition involves subsampling and upsampling operations which use a nondiagonal matrix $\boldsymbol{M}$ of nonsingular integer values, given as

$$\boldsymbol{M} = \begin{pmatrix} m_{11} & m_{12} \\ m_{21} & m_{22} \end{pmatrix}.$$

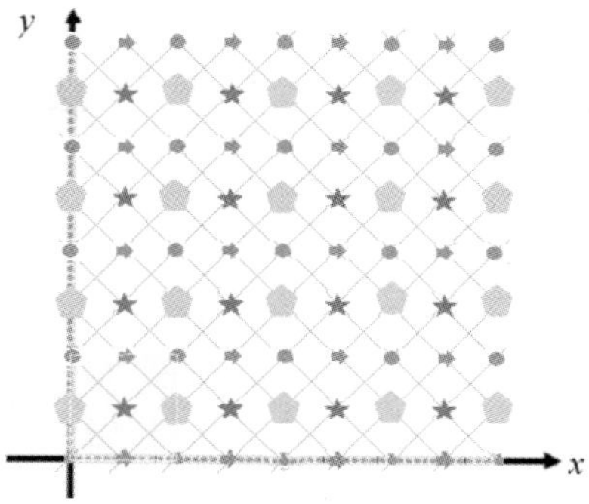

FIGURE 9.4.30

The resampling lattice generated by a dyadic sampling matrix.

This corresponds to a shearing (resampling) operation. The collection of all the subsampling vectors, $\boldsymbol{Mn}$, is called a lattice, given by

$$LAT(\boldsymbol{M}) = \{\boldsymbol{m} : \boldsymbol{m} = \boldsymbol{Mn}, \boldsymbol{n} \in \mathcal{Z}^2\},$$

which is indeed a vector space spanned by the linear combination of basis column vectors from $\boldsymbol{M}$, written as $n_1\boldsymbol{M}_1 + n_2\boldsymbol{M}_2$, with $\boldsymbol{M}_1 = (m_{11}, m_{21})^T$, $\boldsymbol{M}_2 = (m_{12}, m_{22})^T$. In our context, a nonsingular subsampling matrix $\boldsymbol{M}$ refers to the matrices defined in the section of DFB.

Let $\mathcal{S}$ be the set of points of $\boldsymbol{Mt}$, $\boldsymbol{t} \in [0, 1) \times [0, 1)$, called the fundamental parallelepiped of matrix $\boldsymbol{M}$. It consists of all real-valued vectors in an area of the parallelogram controlled by $\boldsymbol{M}_1$ and $\boldsymbol{M}_2$. An example of $\mathcal{S}$ is shown in the shadowed square area in Fig. 9.4.30. However, we are only interested in the integer-valued points in $\mathcal{S}$. According to the division theorem in number theory [41], any integer vector can be written as $\boldsymbol{Mn} + \boldsymbol{s}$ with $\boldsymbol{s} = (s_1, s_2) \in \mathcal{S}$, so there are $D = |det(\boldsymbol{M})|$ integer points in $\mathcal{S}$. We can label each integer vector $\boldsymbol{s} \in \mathcal{S}$ using index $k = 0, 1, \cdots, D - 1$. Note that k cannot be a natural ordering like the 1D case. The sampling lattice of $\boldsymbol{Mn} + \boldsymbol{s}$ results from shifting $LAT(\boldsymbol{M})$ to the location $\boldsymbol{s} \in \mathcal{S}$, yielding the kth coset of $LAT(\boldsymbol{M})$ generated by integer-valued points inside $\mathcal{S}$. Therefore, the original lattice of integer numbers in the coordinate system is the union of all the cosets.

For a 2D image, polyphase decomposition is the subsampling over the cosets defined above. The kth component subimage takes a value at the location vector in the kth coset associated with $\boldsymbol{s}$, and is defined as

$$x^{(k,s)}(\boldsymbol{n}) = x(\boldsymbol{Mn} + \boldsymbol{s}).$$

Denote by $x_u^{(k,s)}(\boldsymbol{n})$ the upsampled version of the kth polyphase component $x^{(k,s)}(\boldsymbol{n})$, with a shifting by $\boldsymbol{s} \in \mathcal{S}$, defined as

$$x_u^{(k,s)}(\boldsymbol{n}) = \begin{cases} x^{(k,s)}(\boldsymbol{m}), & \text{if } \boldsymbol{n} = \boldsymbol{Mm} + \boldsymbol{s},\ \boldsymbol{s} \in \mathcal{S}, \\ 0, & \text{otherwise.} \end{cases}$$

The parent image can then be reconstructed as the sum of $x_u^{(k,s)}(\boldsymbol{n})$, namely,

$$x(\boldsymbol{n}) = \sum_{k=0}^{D-1} x_u^{(k,s)}(\boldsymbol{n}).$$

Its z-transform can be written as

$$X(z) = \sum_{s} z_1^{-s_1} z_2^{-s_2} X^{(k,s)}(z^{\boldsymbol{M}}), \quad \text{with } z = (z_1, z_2), \boldsymbol{s} = (s_1, s_2) \in \mathcal{S}.$$

Here, for a nondiagonal matrix $\boldsymbol{M}$, $z^{\boldsymbol{M}}$ stands for $(z_1^{m_{11}} z_2^{m_{21}}, z_1^{m_{12}} z_2^{m_{22}})$, and $X^{(k,s)}(z)$ is the z-transform of $x^{(k,s)}(\boldsymbol{n})$ (see [39]).

Definition 9.10. To be consistent with the subsampling concepts, the polyphase representation of the z-transform $X(z)$ of an image $x(\boldsymbol{n})$ is defined as a vector $\boldsymbol{x}(z)$, written as

$$\boldsymbol{x}(z) = (X_0(z), X_1(z), \cdots, X_{D-1}(z)),$$

with $X_k(z) = X^{(k,s)}(z)$, for $k = 0, 1, \cdots, D-1$.

For example, a dyadic subsampling matrix $\boldsymbol{M}$ will have its cosets generated by points $(0,0)$, $(0,1)$, $(1,0)$, $(1,1)$ in S; in short, we write the points as (i, j), $i = 0, 1$, $j = 0, 1$, and the z-transform of $x(\boldsymbol{n})$ can be written as

$$X(z) = X_0(z_1^2, z_2^2) + z_1^{-1} X_1(z_1^2, z_2^2) + z_2^{-1} X_2(z_1^2, z_2^2) + z_1^{-1} z_2^{-1} X_3(z_1^2, z_2^2),$$

where $X_k(z_1, z_2) = \sum_{n_1} \sum_{n_2} z_1^{-n_1} z_2^{-n_2} x(2n_1 + i, 2n_2 + j)$, $k = 2 \times j + i$, $i = 0, 1$, $j = 0, 1$. The polyphase representation can then be written as

$$\boldsymbol{x}(z) = (X_0(z_1, z_2), X_1(z_1, z_2), X_2(z_1, z_2), X_3(z_1, z_2)).$$

The use of a polyphase decomposition in the design of a finite impulse response (FIR) filter bank reduces the computational complexity in the realization of a filter bank, in which the polyphase representation is frequently used. Note that $X(z)$ is a vector, the z-transfer function takes a scalar value, and the polyphase representation $\boldsymbol{x}(z)$ takes a vector value. We assume the signal is periodically expanded if it is of finite length.

A filter bank with a solo implementation of the polyphase decomposition is also termed as a lazy filter bank because of its simple reconstruction structure that involves only the basic operations, such as addition, delay, downsampling, and upsampling. An iterated lazy filter bank leads to the so-called lazy wavelet.

9.4.2 Laplacian pyramid frames

Introduced by Burt and Adelson [8,106], the LP was initially aimed at representing data in a multiresolution setting from fine to coarse levels for data compression purposes. The analysis filter bank of LP iteratively decomposes an image into multiple subimages of coarse approximation, namely, low-resolution subimages with low-pass content and detail images representing the difference between the

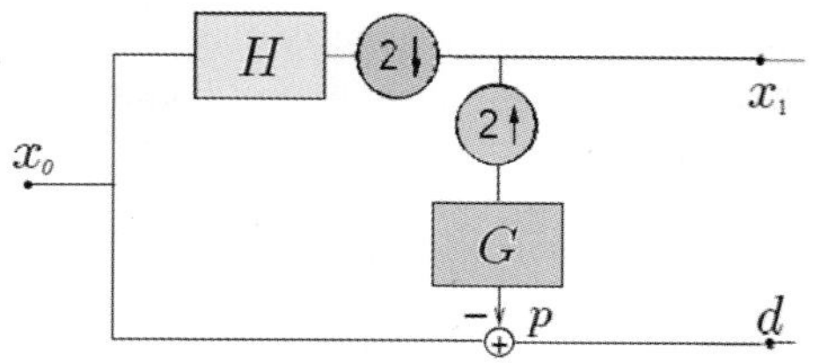

FIGURE 9.4.31

The LP filter bank in a single level.

low-pass filtered image and the input image. The synthesis filter bank integrates the outputs from the analysis filters into an estimate of the original image. Perfect reconstruction occurs when the estimate is a replica of the original image.

The analysis filter bank of the LP results in a sequence of bandpass images and a further upsampled low-resolution image in the end level. The synthesis filters are used to recover the original image from this sequence of subimages. The double-layered filter bank analyzes an image in various components of size-reduced images with their frequency contents limited to subbands. The pyramid design has inspired the connection of filter banks with the wavelet implementation and calculating linear expansions of multiresolution signals.

9.4.2.1 Analysis filter banks and synthesis filter banks of the Laplacian pyramid

The analysis filter bank is shown in Fig. 9.4.31. It computes the approximation and the detail image at one single level. Take $x_0(\boldsymbol{n})$ as the input image. The analysis LP filter bank decomposes the image using a low-pass filter H (analysis filter) with the impulse response $\bar{h}(\boldsymbol{n}) = h(-\boldsymbol{n})$. The output from the low-pass filter H is downsampled, resulting in $x_1(\boldsymbol{n})$, which can be written as

$$x_1(\boldsymbol{n}) = \sum_{\boldsymbol{s}\in \mathcal{Z}^2} x_0(\boldsymbol{s}) h(\boldsymbol{Mn} - \boldsymbol{s}) \equiv \langle x(\cdot), \bar{h}(\cdot - \boldsymbol{Mn})\rangle, \tag{9.4.2.1}$$

where $\boldsymbol{M}$ is a sampling matrix which defines the downsampling. Note that the "$\equiv$" sign in Eq. (9.4.2.1) is defined by slightly abusing the inner product notation for two vectors. The z-transform of $x_1(\boldsymbol{n})$ can be written in a polyphase representation as

$$X_1(z) = \boldsymbol{h}(z)\boldsymbol{x}_0^T(z), \tag{9.4.2.2}$$

with

$$\boldsymbol{x}_0(z) = (X_{00}(z), X_{01}(z), \ldots, X_{0(D-1)}(z))$$

and

$$\boldsymbol{h}(z) = (H_0(z), H_1(z), \ldots, H_{D-1}(z)),$$

which both are $1 \times D$ polyphase representation vectors.

The detail image is the difference between the approximation image $x_1(\boldsymbol{n})$ and the original image $x_0(\boldsymbol{n})$. It is calculated as the difference between a version of the approximation image $p(\boldsymbol{n})$, which is an upsampled version of $x_1(\boldsymbol{n})$, and the original image $x_0(\boldsymbol{n})$, and is written as

$$d(\boldsymbol{n}) = x_0(\boldsymbol{n}) - p(\boldsymbol{n}).$$

The low-pass image $p(\boldsymbol{n})$ of the original size is obtained as the following:

$$p(\boldsymbol{n}) = \sum_{s \in \mathcal{Z}^2} x_1(\boldsymbol{s}) g(\boldsymbol{n} - \boldsymbol{M}\boldsymbol{s}).$$

To calculate $p(\boldsymbol{n})$, the already downsampled image $x_1(\boldsymbol{n})$ is upsampled and filtered by another low-pass filter G (synthesis filter) with impulse response $g(\boldsymbol{n})$. The polyphase representation $\boldsymbol{p}(z)$ can then be written as

$$\boldsymbol{p}^T(z) = (P_0(z), P_1(z), \cdots, P_{D-1}(z))^T = \boldsymbol{g}^T(z) X_1(z), \tag{9.4.2.3}$$

where $\boldsymbol{g}(z) = (G_0(z), G_1(z), \ldots, G_{D-1}(z))$ is a $1 \times D$ vector and $X_1(z)$ is a scalar function. Thus the polyphase representation of $d(\boldsymbol{n})$ can be calculated using Eq. (9.4.2.3) and can be written as

$$\boldsymbol{d}^T(z) = (I - \boldsymbol{g}^T(z)\boldsymbol{h}(z))\boldsymbol{x}_0^T(z). \tag{9.4.2.4}$$

If we cascade Eqs. (9.4.2.2) and (9.4.2.4), we will have a vector equation

$$\begin{pmatrix} X_1(z) \\ \boldsymbol{d}^T(z) \end{pmatrix} = \begin{pmatrix} \boldsymbol{h}(z) \\ I - \boldsymbol{g}^T(z)\boldsymbol{h}(z) \end{pmatrix} \boldsymbol{x}_0^T(z) = \Phi(z)\boldsymbol{x}_0^T(z), \tag{9.4.2.5}$$

where $\Phi(z)$ is a matrix of size $(D+1) \times D$.

In the synthesis filter bank, a simple reconstruction of the original image can be computed by directly inverting the steps in the analysis filter. A more efficient reconstruction method is to search for the inverse operator under an orthogonal frame system to calculate the orthogonal projections that is defined as the pseudoinverse of matrix $\Phi(z)$. This approach provides the best approximation [111].

The LP filter bank can be realized through a polyphase filter bank structure, in which the low-pass filter H is determined by its transfer function $H(z)$ and the high-pass filter K_i, $i = 0, 1, 2, \cdots, D-1$, is determined by a transfer function whose polyphase representation is given as the ith row vector of the matrix $\Phi(z) = (I - \boldsymbol{g}^T(z))\boldsymbol{h}(z)$. Therefore the high-pass filters are derived from the low-pass filters H and G in the filter bank described in Fig. 9.4.31. Fig. 9.4.32 shows the polyphase structure filter bank for dyadic subsampling which is to be used in the contourlet filter bank.

Based on the structure of a single-level analysis filter bank shown in Fig. 9.4.31, the procedure of multiresolution decomposition using the LP filter bank is listed as follows:

- Input an image $x_0(\boldsymbol{n})$ into the one-level LP filter bank. The outputs are the low-resolution upsampled image $x_1(\boldsymbol{n})$ and a detailed difference image $d_1(\boldsymbol{n})$.
- Input $x_1(\boldsymbol{n})$ into the analysis filters again. The outputs are a low-resolution image $x_2(\boldsymbol{n})$ that is further upsampled and a detailed image $d_2(\boldsymbol{n})$ that is the difference between $x_1(\boldsymbol{n})$ and $x_2(\boldsymbol{n})$.
- Iterate the procedure K times.

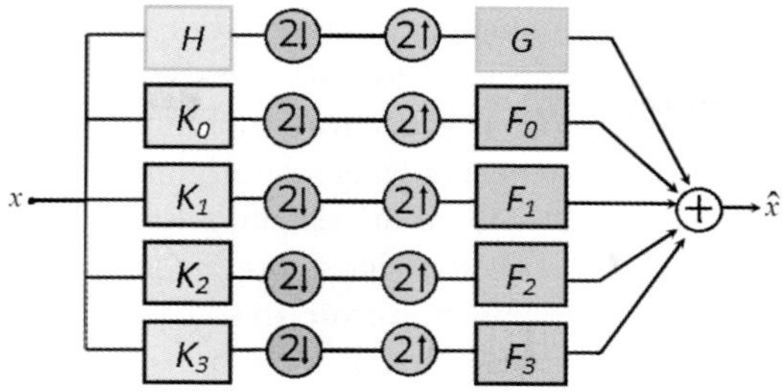

FIGURE 9.4.32

The LP filter implemented using polyphase decomposition with dyadic subsampling.

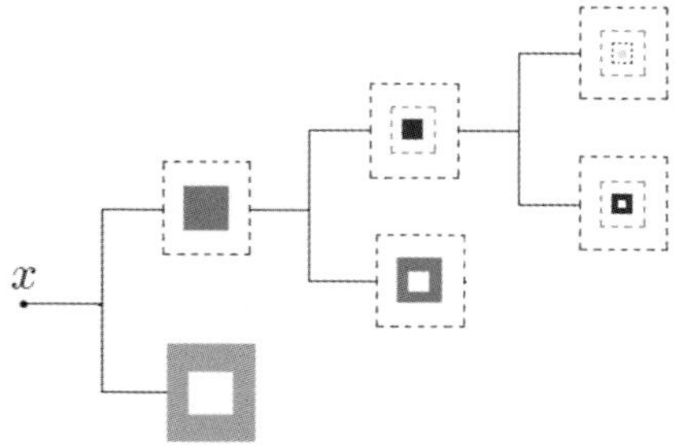

FIGURE 9.4.33

An LP filter bank that decomposes the frequency content of an image into three levels.

The output sequence $\{x_k\}_{k=1,2,\cdots,K}$ is called a Gaussian pyramid because Gaussian low-pass filters are typically used. The filtering structure for generating the bandpass sequence $\{d_k\}_{k=1,2,\cdots,K}$ is called the LP. As x_k is further decomposed, the final outputs are the sequence of detail images d_k, with only one low-resolution image at the end level, thus yielding a filter bank that is generally referred to as the LP. It is a cascade of a sequence of filter banks for a multiresolution approximation of an original image.

9.4.2.2 *The wavelet frame associated with the Laplacian pyramid*

A wavelet transform can be implemented by computing the fine to coarse multiresolution decomposition as shown in Fig. 9.4.33. The wavelet frame [8] defined in Eq. (9.2.2.3) for the decomposition can be specified by using the filters H and G of the LP filter bank to determine the mother/father wavelets.

In the LP filter bank, a frame operator $\hat{K}_i$ can be defined as $\hat{K}_i(x, \phi_i) = \langle x, \phi_i \rangle$, where each ϕ_i is the $(i+1)$th row of the matrix $\Phi(z)$ in Eq. (9.4.2.5). It characterizes the filter K_i in the polyphase filter bank. This shows that the LP, with bounded output for any bounded input, provides a frame expansion in $l^2(\mathcal{Z}^2)$ [9]. If the frame vector $\{\phi_i\}$ is normalized with $||\phi_i|| = 1$ and we have $A = B = 1$, then $\{\phi_i\}$ is an orthonormal basis by the discussion on Definition 9.6.

A nonorthonormal frame induces a redundant representation of an image, for which the reconstruction has to use a redundant frame operator that involves a dual frame of $\hat{K}_i$, which performs the inverse/pseudoinverse of the matrix Φ. It generates instability that affects the quality of the reconstructed image. Perfect reconstruction occurs when the original image is replicated. In numerical analysis, the research interests have been focused on searching for orthonormal bases for optimal approximations. With an orthogonal basis, when a signal $x(\boldsymbol{n})$ is projected onto a subspace V, given as $P_V(x)$, it generates the minimum approximation error. In some cases, due to the requirement of the basis functions, we have to relax the requirement for an orthonormal basis and replace it with an orthogonal basis, a biorthogonal basis, or even a tight frame for MRA. This is used to inject some flexibilities, such as minimizing the redundancy of the expansion.

The LP uses a pair of biorthogonal filters with a sampling matrix $\boldsymbol{M}$ and satisfies

$$\langle \bar{h}(\cdot - \boldsymbol{M}\boldsymbol{v}), g(\cdot - \boldsymbol{M}\boldsymbol{l})\rangle = \delta(\boldsymbol{v} - \boldsymbol{l}).$$

The filters become orthogonal when they satisfy

$$h(\boldsymbol{n}) = g(-\boldsymbol{n}) \text{ and } \langle g(\cdot), g(\cdot - \boldsymbol{M}\boldsymbol{n})\rangle = \delta(\boldsymbol{n}).$$

The orthogonal requirement on the filters H and G is equivalent to

$$H^*(z) = G(z) \text{ and } G^*(z)G(z) = 1.$$

Perfect reconstruction can then be achieved, thus providing a simple structure that connects the LP filter bank with the classical wavelets [84] for MRA. A similar implementation involving a directional decomposition that further connects the anisotropic curvelets with the discrete filter banks is discussed next.

9.4.3 Directional filter banks

Introduced by Bamberger and Smith [107] in 1992, a DFB is a maximally decimated QMF bank that partitions the 2D Fourier plane with 2^r directional-oriented wedge-shaped subbands at each level r. A perfect reconstruction of the original image may be obtained. The implementation introduced here is using a polyphase-like tree-structured FIR filter bank [9,112]. This DFB has a simplified structure and uses fan filters and resampling operators with dyadic downsampling in each level.

9.4.3.1 Resampling operators

The design of a DFB requires resampling operators controlled by sampling matrices, such as shearing operators. Here we first introduce the resampling matrices.

Definition 9.11. A quincunx sampling matrix is defined as one of the following two matrices:

$$\boldsymbol{Q}_0 = \begin{pmatrix} 1 & -1 \\ 1 & 1 \end{pmatrix}, \quad \boldsymbol{Q}_1 = \begin{pmatrix} 1 & 1 \\ -1 & 1 \end{pmatrix}.$$

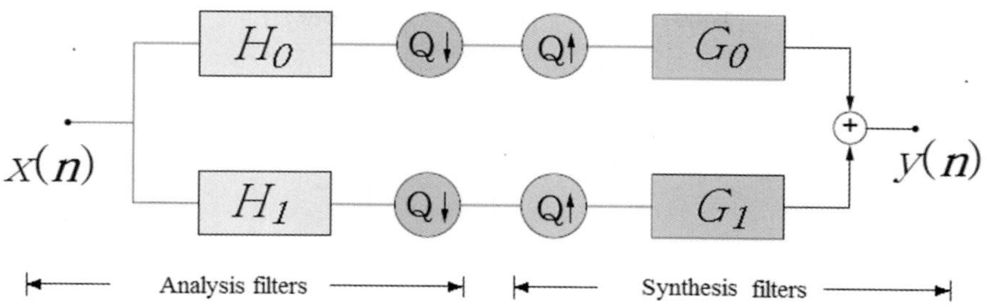

FIGURE 9.4.34

The quincunx filter bank with resampling matrix Q.

Definition 9.12. An integer matrix $\boldsymbol{M}$ is said to be a unimodular matrix if $|det(\boldsymbol{M})| = 1$. Here, we refer to the following four matrices $\boldsymbol{R}_i$, $i = 0, 1, 2, 3$:

$$\boldsymbol{R}_0 = \begin{pmatrix} 1 & 1 \\ 0 & 1 \end{pmatrix}, \quad \boldsymbol{R}_1 = \begin{pmatrix} 1 & -1 \\ 0 & 1 \end{pmatrix}, \quad \boldsymbol{R}_2 = \begin{pmatrix} 1 & 0 \\ 1 & 1 \end{pmatrix}, \quad \boldsymbol{R}_3 = \begin{pmatrix} 1 & 0 \\ -1 & 1 \end{pmatrix}.$$

The two quincunx matrices can be further decomposed as

$$\boldsymbol{Q}_0 = \boldsymbol{R}_1 \boldsymbol{D}_0 \boldsymbol{R}_2 = \boldsymbol{R}_2 \boldsymbol{D}_1 \boldsymbol{R}_1 \quad \text{and} \quad \boldsymbol{Q}_1 = \boldsymbol{R}_0 \boldsymbol{D}_0 \boldsymbol{R}_3 = \boldsymbol{R}_3 \boldsymbol{D}_1 \boldsymbol{R}_0,$$

where the two diagonal matrices $\boldsymbol{D}_0$ and $\boldsymbol{D}_1$ are the dyadic subsampling matrix along x- and y-axis respectively, and are given as

$$\boldsymbol{D}_0 = \begin{pmatrix} 2 & 0 \\ 0 & 1 \end{pmatrix}, \quad \boldsymbol{D}_1 = \begin{pmatrix} 1 & 0 \\ 0 & 2 \end{pmatrix}.$$

A filter that uses $\boldsymbol{Q}_0$ or $\boldsymbol{Q}_1$ carries out a downsampling operation that sheers an image along the orientation at 45 degrees or −45 degrees. In the DFB, $\boldsymbol{Q}_0$ and $\boldsymbol{Q}_1$ are typically cascaded together to result in dyadic downsampling in both the x- and y-directions. This is because $\boldsymbol{Q}_0\boldsymbol{Q}_1 = \boldsymbol{Q}_1\boldsymbol{Q}_0 = 2\boldsymbol{I}$, where $\boldsymbol{I}$ is the unit matrix.

9.4.3.2 The design of the quincunx filter banks

A quincunx filter bank (QFB) [112,113] is a two-channel QMF as shown in Fig. 9.4.34. It incorporates the analysis and synthesis filters with the decimators and expanders being the quincunx resampling matrices. A QFB with fan filters used in the analysis part is a typical example. The fan filters are nonseparable filters. In Fig. 9.4.35 is shown the support of one fan filter which can be designed starting from a 1D filter [41].

The single-level structure analysis filter bank of QFB decomposes the frequency plane of a 2D image into two wedge-shaped cones. The horizontal cone uses H_0, a horizontal pass fan filter. The vertical cone uses H_1, a vertical pass fan filter. The synthesis filters, denoted by G_0 and G_1, perfectly reconstruct the original image when the two dual filter frames satisfy the following conditions:

$$H_0(\boldsymbol{\omega})G_0(\boldsymbol{\omega}) + H_1(\boldsymbol{\omega})G_1(\boldsymbol{\omega}) = 2,$$

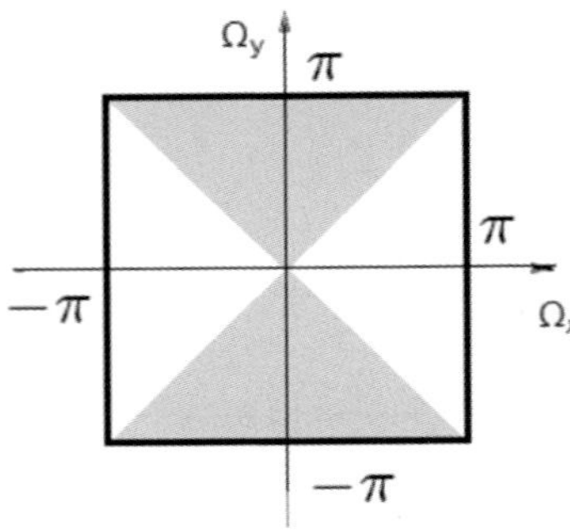

FIGURE 9.4.35

The support of a fan filter in the Fourier plane. The shaded area is the bandpass area.

$$H_0(\omega+\pi)G_0(\omega)+H_1(\omega+\pi)G_1(\omega)=0.$$

If the synthesis filters are the reversed versions of the analysis filters, then the frame generated by the QFB is an orthogonal frame. This simplifies the design of the synthesis filters. When the analysis and synthesis filters are restricted to be FIR, the perfect reconstruction conditions imply that the synthesis filters are specified by the analysis filters (up to a shift and scale factor i) as follows:

$$G_0(z)=z^i H_1(-z),\ G_1(z)=-z^i H_0(z).$$

9.4.3.3 *The design of the directional filter banks*

A QFB is the essential QMF bank in a tree-structured DFB filter bank. An iterative DFB decomposes the frequency space of a 2D image into 2^r small angularly oriented wedges at the rth level. The DFB is a cascade of multiple QFBs, together with selective resampling matrices to determine the shearing angle of the segmented wedge in the Fourier plane. The iterative DFB can be viewed as proceeding in multiple stages. According to the Nobel identity shown in Fig. 9.4.37(a), the overall output in two adjacent stages can be viewed as applying a decimation $M=Q_0Q_1$ using the overall analysis filter $H_1(z)H_2(z^{M^T})$. Fig. 9.4.37(b) shows the overall effect of a DFB in two stages.

The DFB decompositions in multiple levels are carried out in the following procedures:

- At the first level decomposition, the DFB uses a pair of fan filters with supports complementing each other. The output is resampled using $\boldsymbol{Q}_1$. The two fan filters split the Fourier plane into two cone-shaped regions: one along the vertical direction y and another along the horizontal direction x. The output is downsampled by 2 in the x-direction.
- At the second level, the same pair of fan filters are used; however, the output image is resampled using $\boldsymbol{Q}_1$, so that the four subimages correspond to the four directional subbands shown in Fig. 9.4.38. The filter structure for generating these four images is shown in Fig. 9.4.36. The outputs at this level are downsampled by 2 in both the x- and y-directions.
- At the level $r \geq 3$, the DFB partitions the Fourier plane into 2^r wedges. Fig. 9.4.39 shows the partition at the level $r=3$. This is realized by attaching to the QFB at the previous level one of the four unimodular resampling operations, $\boldsymbol{R}_i,\ i=0,1,2,3$, before and after the analysis filters. Fig. 9.4.40

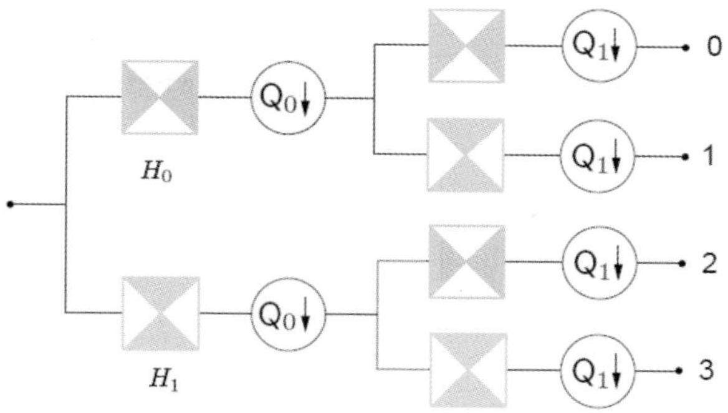

FIGURE 9.4.36

The partition of the Fourier plane using DFB at the first two levels.

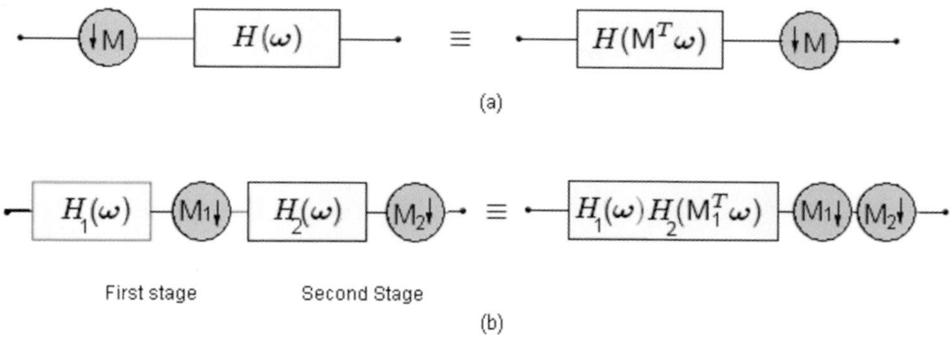

FIGURE 9.4.37

The effect of the Noble identity.

shows the filter bank structure for generating the zeroth wedge in the Fourier frequency plane at the third level.

To ensure that each channel of the filters uses a diagonal sampling matrix, a backsampling operation [114] is applied so that, at each level r, the vth channel subband can be equivalently viewed as being filtered by an analysis filter H_v, for $v = 0, 1, \cdots, 2^r - 1$. The diagonal sampling operator is represented by the sampling matrix

$$S_v^{(r)} = \begin{cases} diag(2^{r-1}, 2), & 0 \leq v < 2^{r-1}, \\ diag(2, 2^{r-1}), & 2^{v-1} \leq v < 2^r. \end{cases}$$

The DFB is a perfect reconstruction filter bank if and only if the QFB used at each level achieves perfect reconstruction.

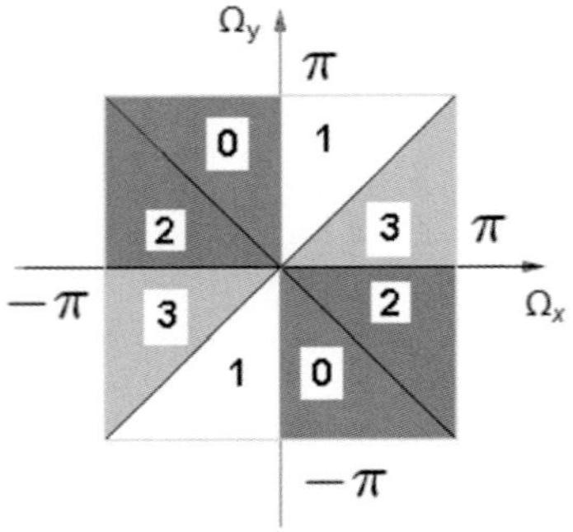

FIGURE 9.4.38

The four directional bands output from the first two-level partitions.

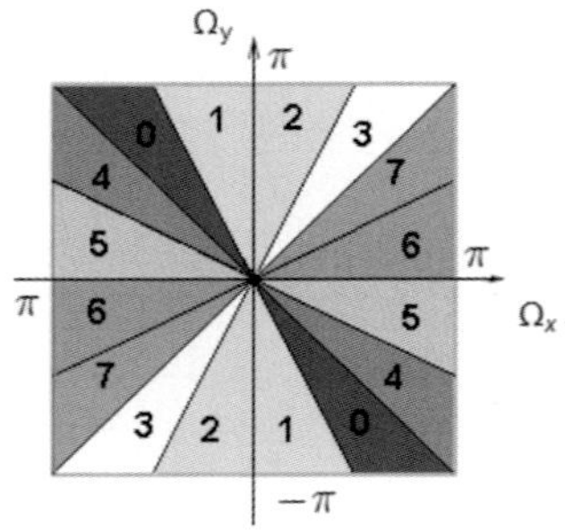

FIGURE 9.4.39

The eight directional bands output from a DFB at the level $r = 3$.

9.4.3.4 Constructing the basis for the directional filter bank

As shown in Fig. 9.4.41, in the DFB at the rth level, every analysis filter H_v with impulse response h_v is followed by a diagonal sampling $\mathbf{S}_v^{(r)}$, $v = 0, 1, \cdots, 2^r - 1$. The outputs are 2^r wedges in the Fourier plane. Each wedge is an angularly oriented segment of the Fourier content of the original image. Correspondingly, the original image can be reconstructed from applying the diagonal sampling matrix $\mathbf{S}_v^{(r)}$ to the outputs at the rth level, followed by applying the synthesis filters $\{G_v\}_{v=0,1,\cdots,2^r-1}$, with impulse response $\{g_v(\boldsymbol{n})\}_{v=0,1,\cdots,2^r-1}$. We also write $g_v(\boldsymbol{n}), h_v(\boldsymbol{n})$ as $g_v^{(r)}(\boldsymbol{n}), h_v^{(r)}(\boldsymbol{n})$ for the rth level.

Each filter pair, with H_v and G_v in the filter bank as the analysis and synthesis filters, is related to the angular wedge region $\boldsymbol{U}_v$, $v = 0, 1, \cdots, 2^r - 1$. The filters generate a pair of dual frames of $l^2(\mathcal{Z}^2)$. Therefore, a basis of $l^2(\mathcal{Z}^2)$ can be obtained by translating the impulse responses of synthesis filters $g_v(\boldsymbol{n})$ as $\{g_{v,\boldsymbol{m}}^{(r)}(\boldsymbol{n}) = g_v^{(r)}(\boldsymbol{n} - \mathbf{S}_v^{(r)}\boldsymbol{m}) : 0 \le v < 2^r, \boldsymbol{m} \in \mathcal{Z}^2\}$. Under this basis, an image $x(\boldsymbol{n}) \in l^2(\mathcal{Z}^2)$ can be uniquely reconstructed as

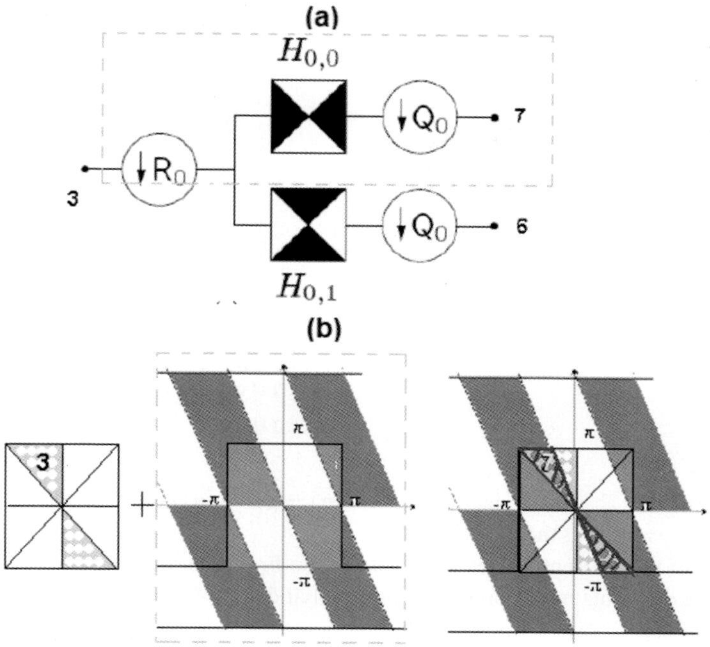

FIGURE 9.4.40

(a) A resampling operator is attaching to a QFB to implement the DFB. (b) Its effects on partitioning the Fourier plane: the third region in the four partitions is used as the input, the output is the wedge indicated in the last picture.

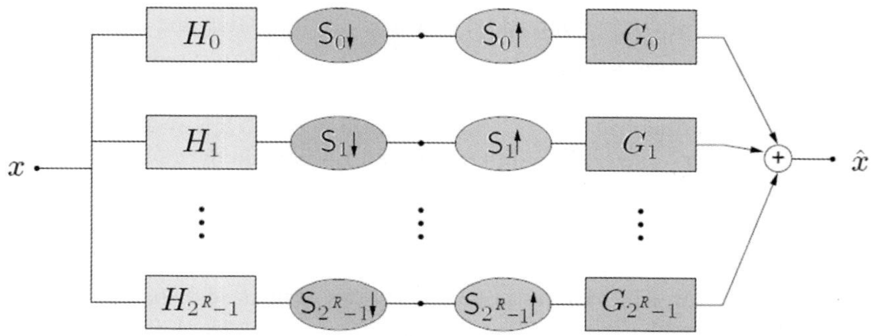

FIGURE 9.4.41

The analysis and synthesis filters in a DFB that partitions the Fourier plane into 2^R wedges.

$$x(\boldsymbol{n}) = \sum_{v=0}^{2^r-1} \sum_{\boldsymbol{m}\in\mathcal{Z}^2} c_v(\boldsymbol{m}) g_v^{(r)}(\boldsymbol{n} - \boldsymbol{S}_v^{(r)}\boldsymbol{m}),$$

where $c_v(\boldsymbol{m}) = \langle x(\cdot), h_v^{(r)}(\boldsymbol{S}_v^{(r)}\boldsymbol{m} - \cdot)\rangle$, $v = 0, 1, \cdots, 2^r - 1$, are the coefficients with respect to a basis $h_v(\boldsymbol{S}_v^{(r)}\boldsymbol{m} - \cdot)$ of $l^2(\mathcal{Z}^2)$ at the rth DFB decomposition level. The two dual bases $l^2(\mathcal{Z}^2)$ satisfy the following conditions when they are biorthogonal frames:

$$\langle g_{v'}^{(r)}(\cdot - \boldsymbol{S}_{v'}^{(r)}\boldsymbol{m}'), h_v^{(r)}(\boldsymbol{S}_v^{(r)}\boldsymbol{m} - \cdot)\rangle = \delta(v - v')\delta(\boldsymbol{m} - \boldsymbol{m}'),$$

with the Dirac function given as

$$\delta(a - a') = \begin{cases} 0, & \text{when } a \neq a', \\ 1, & \text{when } a = a'. \end{cases}$$

An orthogonal basis of $l^2(\mathcal{Z}^2)$ is obtained when $h_v^{(r)}(\boldsymbol{n}) = g_v^{(r)}(-\boldsymbol{n}))$, that is, when the synthesis filters of the DFB are time-reverses of the analysis filters.

The following are two extreme cases of DFBs. In the first case, the filter system generates a basis of the $l^2(\mathcal{Z}^2)$ whose elements have compact supports that do not overlap with each other in the spatial domain. The second case shows a filter system that generates a basis of $l^2(\mathcal{Z}^2)$ whose elements have compact supports that do not overlap with each other in the Fourier domain.

Case 1: In this case, the analysis DFB implements a polyphase decomposition with the vth polyphase component $x_v(\boldsymbol{n}) = x(p_v^T + \boldsymbol{S}_v^{(r)}\boldsymbol{n})$, where

$$\boldsymbol{p}_v = \begin{cases} (v, v), & \text{if } \ 0 \leq v < 2^{r-1}, \\ (v + 1 - 2^{r-1}, v - 2^{r-1}), & \text{if } \ 2^{r-1} \leq v < 2^r. \end{cases}$$

The polyphase decomposition is realized when each filter G_v takes the z-transform as $G_v(z) = z^{-\boldsymbol{p}_v}$, where the basis $g_v^{(r)}(\boldsymbol{n} - \boldsymbol{S}_v^{(r)}\boldsymbol{m}) = \delta(\boldsymbol{n} - p_v^T - \boldsymbol{S}_v^{(r)}\boldsymbol{m})$ extends a standard orthonormal basis of $l^2(\mathcal{Z}^2)$.

Case 2: Another orthonormal basis of $l^2(\mathcal{Z}^2)$ is obtained using a DFB with ideal fan filters at each level, resulting in the synthesis filters G_v with Fourier transform given by

$$G_v(\boldsymbol{\omega}) = 2^{v/2}\delta_{\boldsymbol{R}_v}(\boldsymbol{\omega}) exp\{j\langle\boldsymbol{\omega}, \boldsymbol{p}_v^T\rangle\}.$$

These two cases, however, will have noncompact support in their dual spaces. Namely, in the first case, the Fourier content of each element in the basis is supported in the whole Fourier plane, while in the second case, each element in the basis is a sinc function with support in the whole spatial domain.

In applications, a DFB realization that generates a basis of compact support in the Fourier plane and fast decay in the spatial plane works better to separate a signal into subsignals that are localized in various frequency regions. In the contourlets [9], a biorthogonal filter pair "9-7" is usually used.

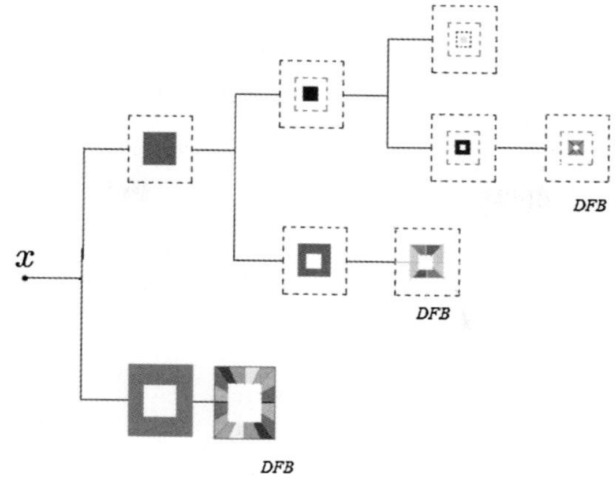

FIGURE 9.4.42

The contourlet filter bank partitions the Fourier plane into wedges for three levels.

9.4.4 Contourlet filter bank

Now we are ready to address the contourlet filter bank, which is a tree-structured double-layered filter bank that combines the LP and DFB to decompose an original image iteratively. Fig. 9.4.42 shows the multiscale decomposition result generated by the contourlet filter bank.

9.4.4.1 *The wavelet frames of Laplacian filter banks*

The LP filters in the contourlet filter bank are octave-band filters: The image is decomposed into a low-pass frequency content and a bandpass content. The low-pass filtered image is an approximation of the high-resolution image with dyadic downsampling applied in both directions. The bandpass filtered image is the detail difference between the original image and the approximation image. This decomposition proceeds iteratively. At level k, the original input image is decomposed into several subimages that are focused on subbands with a corona support as shown in Fig. 9.4.33, where the Fourier plane of an image is decomposed into three levels.

The LP filter provides a frame expansion in $L^2(\mathcal{R}^2)$. Under MRA, the LP filter bank is associated with a wavelet frame represented by nested subspaces $\{V_k\}_{k\in\mathcal{Z}}$ (refer to Section 9.2) that satisfies

$$V_k \subset V_{k-1}, \quad \text{for } \forall\, k \in \mathcal{Z},$$

$$\text{Closure}(\bigcup_{k=-\infty}^{\infty} V_k) = L^2(\mathcal{R}^2).$$

The low-pass filter G in Fig. 9.4.31 defines a wavelet frame of $L^2(\mathcal{R}^2)$. The filter structure uniquely defines $\phi(t)$ that satisfies the following equation (called an orthogonal scaling function):

$$\phi(t) = D^{\frac{1}{2}} \sum_{\boldsymbol{n}\in\mathcal{Z}^2} g(\boldsymbol{n})\phi(\boldsymbol{M}t - \boldsymbol{n}).$$

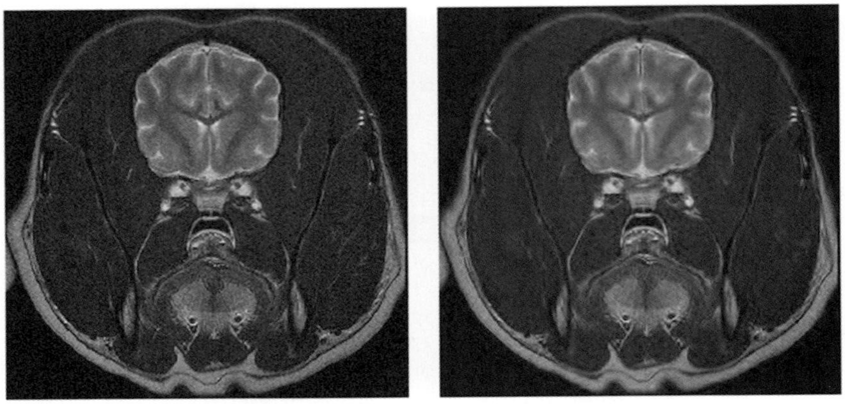

FIGURE 9.4.43

Shown on the left is a noisy image, shown on the right is the image after applying the hard threshold denoising method using the contourlets.

Recall that, in general, $D = |det(\boldsymbol{M})|$, where $\boldsymbol{M}$ is a diagonal matrix that corresponds to subsampling. In our LP filter bank introduced here, the subsampling is constrained to a dyadic subsampling only; therefore, we have $D = 4$, and

$$\boldsymbol{M} = \begin{pmatrix} 2 & 0 \\ 0 & 2 \end{pmatrix}.$$

The subspace V_k is spanned by an orthogonal basis $\{\phi_{k,\boldsymbol{n}}\}_{\boldsymbol{n}\in\mathcal{Z}^2}$ that is defined as

$$\phi_{k,\boldsymbol{n}}(t) = D^{-\frac{k}{2}}\phi(\boldsymbol{M}^{-k}t - \boldsymbol{n}).$$

Writing the orthogonal complement of V_k in V_{k-1} as W_k (see Section 9.2), we have

$$V_{k-1} = V_k \bigoplus W_k.$$

Now let $\{f_i\}_{i=0,1,2,3}$ represent the impulse responses of the high-pass synthesis filters $\{F_i\}_{i=0,1,2,3}$ in the reconstruction part of the LP polyphase structured filter bank, as shown in Fig. 9.4.32. It generates the mother wavelet function ψ_i for each $i = 0, 1, 2, 3$ as the following equation:

$$\psi_i(t) = D^{\frac{1}{2}} \sum_{\boldsymbol{s}\in\mathcal{Z}^2} f_i(\boldsymbol{s})\phi(\boldsymbol{M}t - \boldsymbol{s}), t \in \mathcal{R}^2, \quad i = 0, 1, 2, 3.$$

A tight frame of the detail space W_k can be obtained as a family of dilated and translated functions, $\psi_{k,i,\boldsymbol{n}}(t)$, given as

$$\psi_{k,i,\boldsymbol{n}}(t) = D^{-\frac{k}{2}}\psi_i(\boldsymbol{M}^{-k}t - \boldsymbol{n}), \text{ for } k \in \mathcal{Z}, \ \boldsymbol{n} \in \mathcal{Z}^2, \ t \in \mathcal{R}^2, \ i = 0, 1, 2, 3. \tag{9.4.4.1}$$

Here, $\psi_{k,i,\boldsymbol{n}}(t)$ corresponds to the ith polyphase component of the LP filter bank at level k. Under this set of tight frames, given an input image $x(\boldsymbol{n})$, the outputs of the LP filter bank at the kth level are the approximation wavelet coefficients $c_k(\boldsymbol{n})$, and the detail wavelet coefficients are $d_{k,i}(\boldsymbol{n})$. They can be obtained from the following formulas based on the approximation coefficients at the previous level:

$$c_k(\boldsymbol{n}) = \langle x, \phi_{k,\boldsymbol{n}}\rangle = \sum_{s\in\mathcal{Z}^2} c_{k-1}(\boldsymbol{s})g(\boldsymbol{s} - \boldsymbol{M}\boldsymbol{n}),$$

$$d_{k,i}(\boldsymbol{n}) = \langle x, \psi_{k,i,\boldsymbol{n}}\rangle = \sum_{s\in\mathcal{Z}^2} c_{k-1}(\boldsymbol{s})f_i(\boldsymbol{s} - \boldsymbol{M}\boldsymbol{n}).$$

Theorem 9.5. *It can be shown [111] that $L^2(\mathcal{R}^2)$ can be decomposed into detail spaces W_k, $k \in \mathcal{Z}$, that are mutually orthogonal, and can be written as*

$$L^2(\mathcal{R}^2) = \bigoplus_{k\in\mathcal{Z}} W_k.$$

At a scale level k, the family of functions $\{\psi_{k,i,\boldsymbol{n}}(t) :\ 0 \le i < D, \boldsymbol{n} \in \mathcal{Z}^2\}$ is a tight frame of W_k. In addition, the entire family $\{\psi_{k,i,\boldsymbol{n}}(t) : 0 \le i < D, k \in \mathcal{Z}, \boldsymbol{n} \in \mathcal{Z}^2\}$ forms a tight frame of $L^2(\mathcal{R}^2)$.

9.4.4.2 The wavelet frames of directional filter banks

In a contourlet filter bank, the high-pass content output from the LP filter bank is the input for the DFB filter bank. The DFB generates 2^r directional-oriented wedges that partition the Fourier plane. Typically, only the high-frequency content is decomposed since it corresponds to salient features in an image. The directional-oriented filter bank provides an anisotropic basis for an efficient decomposition.

As provided in Section 9.4.3.4, under the DFB, at level r, the family $\{g^{(r)}_{v,\boldsymbol{n}}(\boldsymbol{m}) = g^{(r)}_{v,\boldsymbol{n}}(\boldsymbol{m} - \boldsymbol{S}^{(r)}_v\boldsymbol{n}) :$ $0 \le v < 2^r, \boldsymbol{n} \in \mathcal{Z}^2\}$ is generated and is called a directional orthonormal basis of $l^2(\mathcal{Z}^2)$. Each $g^{(r)}_{v,\boldsymbol{n}}(\boldsymbol{m})$ is the impulse response of a directional filter. When $v = 0, \cdots, 2^{r-1} - 1$, it corresponds to an angle in $[-45°, 45°]$, and when $v = 2^{r-1}, \cdots, 2^r - 1$, it corresponds to an angle in $[45°, 135°]$.

This directional orthonormal basis $\{g^{(r)}_{v,\boldsymbol{n}}(\boldsymbol{m}) :\ 0 \le v < 2^r, \boldsymbol{n} \in \mathcal{Z}^2\}$ is used to construct the frames of the further decomposed detail space. Since the detail subspaces W_k from the LP correspond to four components of the polyphase decomposition, the following function is introduced using Eq. (9.4.4.1),

$$\mu_{k,2\boldsymbol{n}+s_i}(t) = \psi_{k,i,\boldsymbol{s}}(t),$$

where $s_0 = (0, 0)$, $s_1 = (1, 0)$, $s_2 = (0, 1)$, $s_3 = (1, 1)$. With a lattice of dyadic subsampling in both directions, the points $\boldsymbol{s}_i$, $i = 0, 1, 2, 3$, represent all the cosets with the dyadic subsampling; therefore, any integer vector $\boldsymbol{s}$ can be written as $\boldsymbol{n} + \boldsymbol{s}_i$, and we have the family $\mu_{k,\boldsymbol{s}}(t)$ constituting a tight frame of W_k at level k in the LP decomposition.

Now define

$$\psi^{(r)}_{k,v,\boldsymbol{n}}(t) = \sum_{s\in\mathcal{Z}^2} g^{(r)}_v(\boldsymbol{s} - \boldsymbol{S}^{(r)}_v\boldsymbol{n})\mu_{k,\boldsymbol{s}}(t).$$

The set of functions $\{\psi^{(r)}_{k,v,\boldsymbol{n}}\}_{\boldsymbol{n}\in\mathcal{Z}^2}$ constitutes a tight frame of the finer partition subspace of the detail space W_k, written as $W^{(r)}_{k,v}$, with v referring to the vth angular direction. The frame bounds are equal to 1 for each $k = 1, \cdots, K$ and $v = 0, \cdots, 2^r - 1$. Also note that the index v is referred to an angular direction in DFB, which is different from the index i, which refers to a polyphase component in the LP filter.

A directional subspace $W_{k,v}^{(r)}$ is defined on a wedge contained in a rectangular grid that has support in the spatial domain as an oval area, similar to what is shown in Fig. 9.3.25. These subspaces are orthogonal with

$$W_{k,v}^{(r)} = W_{k,2v}^{(r+1)} \bigoplus W_{k,2v+1}^{(r+1)}$$

and

$$W_k = \bigoplus_{v=0}^{2^r-1} W_{k,v}^{(r)}.$$

Further, combining $\psi_{k,v,\boldsymbol{n}}^{(r)}(t)$ together with $\phi_{k_0,\boldsymbol{n}}$ for $k \le k_0,\ 0 \le v < 2^r,\ \boldsymbol{n} \in \mathcal{Z}^2$, we obtain a tight frame of $l^2(\mathcal{R}^2)$, as in the following theorem [113].

Theorem 9.6. *$l^2(\mathcal{R}^2)$ can be decomposed into mutually orthogonal subspaces as*

$$l^2(\mathcal{R}^2) = V_{k_0} \bigoplus \left(\bigoplus_{k \le k_0} W_k \right).$$

For a sequence of finite positive integers $k \le k_0$, the family

$$\{\phi_{k_0,\boldsymbol{n}}(t), \psi_{k,v,\boldsymbol{n}}^{(r)}(t) :\ k \le k_0, 0 \le v < 2^r, \boldsymbol{n} \in \mathcal{Z}^2\}$$

is a tight frame of $L^2(\mathcal{R}^2)$.

Theorem 9.7. *$l^2(\mathcal{R}^2)$ can be decomposed into mutually orthogonal subspaces as*

$$l^2(\mathcal{R}^2) = \bigoplus_{k \in \mathcal{Z}} W_k.$$

The family

$$\{\psi_{k,v,\boldsymbol{n}}^{(r)}(t) : k \in \mathcal{Z}, 0 \le v \le 2^r - 1, \boldsymbol{n} \in \mathcal{Z}^2\}$$

is a directional wavelet tight frame of $L^2(\mathcal{R}^2)$. In each case, the frame bounds are equal to 1.

This combination of frames from both the LP and DFB filter banks yields frequency partitions similar to those achieved by curvelets. The support of the LP is reduced to one fourth of its original size while the number of directions of the DFB is doubled. Hence, it connects the directional anisotropic wavelets with the digital filter bank system.

In the Contourlet MALAB program [115], the syntax that is used for decomposition is

```
coeffs = pdfbdec( double(im), pfilter, dfilter, nlevels );
```

The syntax that is used for reconstructing the original image from the contourlet coefficients is

imrec = pdfbrec(coeffs, pfilter, dfilter);

Since the contourlets can be viewed as a filter bank implementation of the curvelets, here we only show in Fig. 9.4.43 an image that demonstrates the denoising outcome using the contourlets.

9.5 Shearlets and their applications

The sheartlet transform is an alternative anisotropic multiresolution system proposed by Guo, Labate, and their colleagues [11,13,12,116,117,118,119,120,121,122,123]. It was inspired by the composite wavelet theory, which takes advantage of geometric affine transforms, such as translation and dilation (or resampling) in a multidimensional Cartesian coordinate system (our focus is on 2D). In the spatial plane, the affine transforms generate a frame of functional elements (also called building blocks) that are oscillatory waveforms across various scales, locations, and orientations.

Similar to the curvelet transform, a shearlet transform provides a Parseval (normalized tight) frame equipped with well-localized functions at various scales, locations, and directions. When a 2D function is projected into this system, sparse significant coefficients, which correspond to discontinuities that occur along C^2 piecewise smooth curves, can be used to optimally approximate 2D functions. It overcomes a known problem with the classical 2D wavelet transform, in which a larger number of coefficients are needed in order to optimally represent a 2D function with discontinuities.

Meanwhile, the shearlet transform has its advantages over the curvelet transform. One advantage with the continuous shearlet transform is that it can be defined directly in a rectangular coordinate system using an affine transform. It avoids the usage of polar coordinates (as curvelets) and hence avoids the computationally complex numerical conversion from polar coordinates to rectangular coordinates. The second advantage with the shearlet transform is that it exhibits a better capability at distinguishing different types of corner (junction) points, which can be used to separate corner (junction) points from the regular edge points. In addition, the shearlet transform can be mathematically related to group theoretic methods, and can hence be interpreted as a natural building block in measuring the smoothness of a 2D function. It can also be related to the so-called co-orbit space theory [124].

In this subsection, we first describe the composite wavelets [10,125] before we introduce the shearlets, since the shearlets can be viewed as a special class of the composite wavelets. Applications of the shearlet transform in image processing enjoy some interesting properties in applications in edge detection, feature extraction, and denoising. We will also discuss more recent applications at the end of this section.

9.5.1 Composite wavelets

The composite wavelet system provides a general framework with angularly oriented oscillatory spatial waveforms that can be used for representing geometrical features of curves. Since it heavily relies on the affine transform, we proceed by first introducing the affine transform system $\mathcal{H}$, given as follows.

Definition 9.13. Given a function $\psi(\boldsymbol{x}) \in L^2(\mathcal{R}^2)$, a continuous affine system $\mathcal{H}$ generated by $\psi(\boldsymbol{x})$ is defined as

$$\mathcal{H} = \{\psi_{\boldsymbol{G},\boldsymbol{b}}(\boldsymbol{x}) = |det\boldsymbol{G}|^{1/2}\psi(\boldsymbol{G}(\boldsymbol{x}-\boldsymbol{b})) : \boldsymbol{b}, \boldsymbol{x} \in \mathcal{R}^2, \boldsymbol{G} \in \mathcal{G}\}, \tag{9.5.1.1}$$

where $\mathcal{G}$ is a set of invertible 2×2 matrices.

All elements in the affine system $\mathcal{H}$ are called composite wavelets when $\mathcal{H}$ forms a tight (Parseval) frame of $L^2(\mathcal{R}^2)$, namely, for any $f \in L^2(\mathcal{R}^2)$, the following equation holds:

$$\sum_{\boldsymbol{b},\boldsymbol{G}} |\langle f, \psi_{\boldsymbol{G},\boldsymbol{b}}\rangle|^2 = ||f||^2, \tag{9.5.1.2}$$

where $\boldsymbol{G}$ is an invertible 2×2 matrix in the set $\mathcal{G}$.

Under this general framework of composite wavelets, various wavelet systems can be introduced as special cases for decomposing an image. The 2D isotropic continuous (classical) wavelet transform is a special example. It can be obtained when $\mathcal{G}$, the set of matrices in Eq. (9.5.1.1), takes the form $\mathcal{G} = \{c\boldsymbol{I} : c > 0\}$, where $\boldsymbol{I}$ is the unit matrix. Furthermore, the separable classical wavelets can also be obtained when we further use a 2D separable function $\psi(\boldsymbol{x})$, given as $\psi(x, y) = \psi_1(x)\psi_2(y)$. The most exciting case of composite wavelets is the shearlet transform system obtained from using the shearing and translating operations. A shearlet transform system defines a new class of wavelets that produces a set of efficient anisotropic components that can flexibly be adapted to represent the geometrical shapes.

9.5.2 Shearlet atoms in the spatial plane

The shearlet atoms are defined by selecting the matrix set $\mathcal{G}$ in Eq. (9.5.1.1), in the affine system $\mathcal{H}$, as $\mathcal{G} = \{\boldsymbol{M}_{as} : a > 0, s \in \mathcal{R}\}$, with $\boldsymbol{M}_{as}$ given as

$$\boldsymbol{M}_{as} = \begin{pmatrix} a & -\sqrt{a}s \\ 0 & \sqrt{a} \end{pmatrix}. \tag{9.5.2.1}$$

It is useful to write $\boldsymbol{M}_{as} = \boldsymbol{B}_s\boldsymbol{A}_a$, where

$$\boldsymbol{A}_a = \begin{pmatrix} a & 0 \\ 0 & \sqrt{a} \end{pmatrix} \text{ and } \boldsymbol{B}_s = \begin{pmatrix} 1 & -s \\ 0 & 1 \end{pmatrix}.$$

Note that the inverse matrix $\boldsymbol{M}_{as}^{-1}$ can be written as $\boldsymbol{A}_a^{-1}\boldsymbol{B}_s^{-1}$, and is given as

$$\boldsymbol{M}_{as}^{-1} = \begin{pmatrix} a^{-1} & sa^{-1} \\ 0 & a^{-1/2} \end{pmatrix}. \tag{9.5.2.2}$$

Indeed, the matrix $\boldsymbol{M}_{as}$ is the composition of two affine operations in a plane: a parabolic scaling, $\boldsymbol{A}_a$, followed by a shearing $\boldsymbol{B}_s$. The shearlets are defined on the basis of the operator, $D_{\boldsymbol{M}_{as}}$, defined by the matrix $\boldsymbol{M}_{as}$ as

$$D_{\boldsymbol{M}_{as}}\psi(\boldsymbol{x}) = |det(\boldsymbol{M}_{as})|^{-\frac{1}{2}}\psi(\boldsymbol{M}_{as}^{-1}\boldsymbol{x}),$$

and the translation operator $T_{\boldsymbol{b}}$ is defined as

$$T_{\boldsymbol{b}}\psi(\boldsymbol{x}) = \psi(\boldsymbol{x} - \boldsymbol{b}),$$

yielding the shearlet atoms defined as

$$\psi_{a,s,\boldsymbol{b}}(\boldsymbol{x}) = T_{\boldsymbol{b}} D_{\boldsymbol{M}_{as}} \psi(\boldsymbol{x}). \tag{9.5.2.3}$$

Definition 9.14. The family of shearlet atoms in the spatial plane is defined as the collection of all atoms in the affine system generated by $\psi(\boldsymbol{x}) \in L^2(\mathcal{R}^2)$, given by

$$\psi_{a,s,\boldsymbol{b}}(\boldsymbol{x}) = |det(\boldsymbol{M}_{as})|^{-\frac{1}{2}} \psi(\boldsymbol{M}_{as}^{-1}(\boldsymbol{x} - \boldsymbol{b})), \quad a > 0,\ s \in \mathcal{R},\ \boldsymbol{x}, \boldsymbol{b} \in \mathcal{R}^2. \tag{9.5.2.4}$$

The function $\psi(\boldsymbol{x}) \in L^2(\mathcal{R}^2)$ is called the mother shearlet window.

Note that because $det(\boldsymbol{M}_{as}) = a^{3/2}$, Eq. (9.5.2.4) can also be written as

$$\psi_{a,s,\boldsymbol{b}}(\boldsymbol{x}) = a^{-\frac{3}{4}} \psi(\boldsymbol{M}_{as}^{-1}(\boldsymbol{x} - \boldsymbol{b})). \tag{9.5.2.5}$$

In the literature about shearlets, the matrix $\boldsymbol{M}_{as}$ can also be chosen as

$$\boldsymbol{M}_{as} = \begin{pmatrix} a & \sqrt{as} \\ 0 & \sqrt{a} \end{pmatrix}, \text{ with } \boldsymbol{M}_{as}^{-1} = \begin{pmatrix} a^{-1} & -sa^{-1} \\ 0 & a^{-1/2} \end{pmatrix}.$$

In this paper, we stick with the matrix $\boldsymbol{M}_{as}$ given by Eq. (9.5.2.1), which makes Eq. (9.5.2.4) achieve the shearing effects. Hence, the inverse matrix $\boldsymbol{M}_{as}^{-1}$ is used in its Fourier domain definition; however, one can alternatively use $\boldsymbol{M}_{as}^{-1}$ in the spatial domain and use $\boldsymbol{M}_{as}$ in the Fourier plane [118].

9.5.3 Shearlet atoms in the frequency domain

The mother shearlet window defines the properties of the shearlet atoms. Every shearlet atom is a result of an affine transform of the mother shearlet window. We first discuss the choice of the mother shearlet window in order to quickly address the geometrical features of an image using a shearlet transform.

9.5.3.1 *The mother shearlet window*

The rule of thumb for selecting a mother shearlet window in a shearlet transform is to generate shearlet atoms that are able to focus on localized frequency content to represent geometrical features. Under this shearlet system, the energy of a function $f(\boldsymbol{x})$ will be mostly focused on a sparse number of coefficients when the function is decomposed. In general, the mother shearlet window is chosen as a C^∞ function with a compact support in the Fourier plane, so that it decays rapidly in the spatial domain. The mother shearlet window function $\psi(x, y)$ is defined by its Fourier transform $\Psi(\boldsymbol{\Omega})$, $\boldsymbol{\Omega} = (\Omega_x, \Omega_y)$, as

$$\Psi(\boldsymbol{\Omega}) = \Psi_1(\Omega_x)\Psi_2(\frac{\Omega_y}{\Omega_x}),$$

where $\Psi_1(\Omega_x)$ and $\Psi_2(\Omega_y)$ satisfy the following two conditions:

- $\Psi_1(\Omega_x)$ is a 1D high-pass odd function that satisfies the regular 1D admissibility condition,

$$\int_0^\infty |\Psi_1(a\Omega_x)|^2 \frac{da}{a} = 1, \tag{9.5.3.1}$$

- $\Psi_2(\Omega_y)$ is a 1D positive low-pass even function that satisfies

$$\int_{\mathcal{R}} |\Psi_2^2(\Omega_y)|\, d\Omega_y = 1. \tag{9.5.3.2}$$

A good choice for the two 1D windows are the Meyer's wavelet windows defined in Section 9.3.

9.5.3.2 The Fourier content of shearlet atoms

Given the mother shearlet window defined above, the Fourier transform of the shearlet atoms defined in Eq. (9.5.2.4) can be written as

$$\begin{aligned}\Psi_{a,s,\boldsymbol{b}}(\boldsymbol{\Omega}) &= a^{\frac{3}{4}} e^{-j2\pi <\boldsymbol{\Omega},\boldsymbol{b}>} \Psi(\boldsymbol{M}_{as}^T \boldsymbol{\Omega}) \\ &= a^{\frac{3}{4}} e^{-j2\pi <\boldsymbol{\Omega},\boldsymbol{b}>} \Psi_1(a\Omega_x)\Psi_2(\frac{\sqrt{a}(\Omega_y - s\Omega_x)}{a\Omega_x}) \\ &= a^{\frac{3}{4}} e^{-j2\pi <\boldsymbol{\Omega},\boldsymbol{b}>} \Psi_1(a\Omega_x)\Psi_2(a^{-\frac{1}{2}}(\frac{\Omega_y}{\Omega_x} - s)), \quad a > 0,\ s \in \mathcal{R},\ \boldsymbol{b} \in \mathcal{R}^2.\end{aligned} \tag{9.5.3.3}$$

In addition to requiring that the functions Ψ_1 and Ψ_2 satisfy the conditions in Eqs. (9.5.3.1) and (9.5.3.2), we also assume $\text{supp}\{\Psi_1\} \subset [-2, -\frac{1}{2}] \cup [\frac{1}{2}, 2]$ and $\text{supp}\{\Psi_2\} \subset [-1, 1]$. The frequency content of each shearlet atom which is yielded from the mother shearlet window is supported on a pair of trapezoids that are symmetric with respect to the origin point. They are centered between the two lines, $\Omega_y = m\Omega_x$, with slopes $m = s \pm \sqrt{a}$. We refer to s as the orientation of this pair of trapezoids. Fig. 9.5.44 shows some examples of the trapezoid pairs. Each supporting region may be written as

$$\{(\Omega_x, \Omega_y) \in \mathcal{R}^2 : \Omega_x \in [-\frac{2}{a}, -\frac{1}{2a}] \cup [\frac{1}{2a}, \frac{2}{a}], -\sqrt{a} + s \le \frac{\Omega_y}{\Omega_x} \le \sqrt{a} + s\}.$$

So far, the mother shearlet window generates real-valued spatial shearlet atoms; however, the complex-valued spatial shearlet atoms can also be obtained if the support of Ψ_1 is confined to $[\frac{1}{2}, 2]$. In both cases, the mother shearlet reproduces the shearlet atoms that are angularly oriented waveforms in the spatial domain, with its shape controlled by the shearing parameter s and scaling parameter a (see Fig. 9.5.45).

9.5.4 The continuous shearlet transform

Definition 9.15. Let $f(\boldsymbol{x}) \in L^2(\mathcal{R}^2)$ and let $\{\psi_{a,s,\boldsymbol{b}}(\boldsymbol{x}) : a > 0, s \in \mathcal{R}, \boldsymbol{b} \in \mathcal{R}^2\}$ be the family of shearlet atoms defined in Eq. (9.5.2.4). The continuous shearlet transform is defined as

$$\gamma_f(a, s, \boldsymbol{b}) = \langle f, \psi_{a,s,\boldsymbol{b}} \rangle = \int_{\mathcal{R}^2} f(\boldsymbol{x})\psi_{a,s,\boldsymbol{b}}^*(\boldsymbol{x})\, d\boldsymbol{x}, \tag{9.5.4.1}$$

where $\{\gamma_f(a, s, \boldsymbol{b})\}_{a>0, s\in\mathcal{R}, \boldsymbol{b}\in\mathcal{R}^2}$ are called the shearlet coefficients of the function $f(\boldsymbol{x})$.

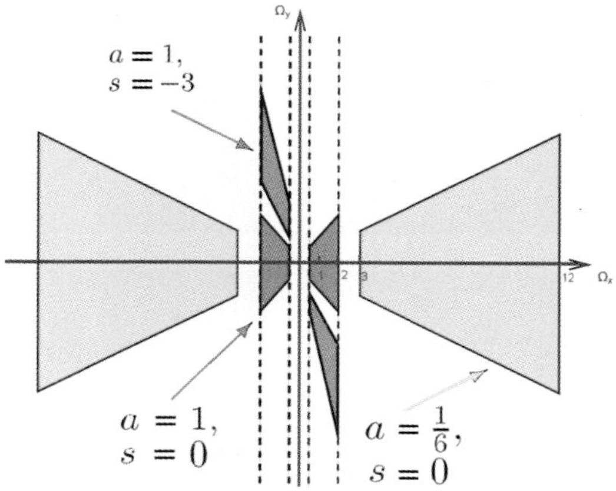

FIGURE 9.5.44

The frequency supports of some shearlet atoms in the Fourier plane, where a is the scale and s is the orientation of a pair of trapezoids.

Note that the shearlet coefficients of the function $f(\boldsymbol{x})$ can also be obtained in the Fourier domain as

$$\begin{aligned}\gamma_f(a,s,\boldsymbol{b}) &= \langle F, \Psi_{a,s,\boldsymbol{b}}\rangle \\ &= \int_{\mathcal{R}^2} F(\boldsymbol{\Omega})\Psi^*_{a,s,\boldsymbol{b}}(\boldsymbol{\Omega})\, d\boldsymbol{\Omega} \\ &= a^{3/4}\int_{\mathcal{R}^2} F(\boldsymbol{\Omega})\Psi_1(a\Omega_x)\Psi_2(a^{-\frac{1}{2}}(\frac{\Omega_y}{\Omega_x}-s))e^{j2\pi<\boldsymbol{\Omega},\boldsymbol{b}>}\, d\boldsymbol{\Omega}. \end{aligned} \tag{9.5.4.2}$$

It can be proved that the mother shearlet window function given in the previous section satisfies the following admissibility condition:

$$\int_{\mathcal{R}}\int_0^{\infty} |\Psi(\boldsymbol{M}_{as}^T\boldsymbol{\Omega})|^2 a^{-3/2}\, da\, ds = 1 \quad \text{for almost every } \boldsymbol{\Omega}\in\mathcal{R}^2.$$

This admissibility condition shows that the family of shearlets generates a reproducing system, which yields the following theorem.

Theorem 9.8. *For a function $f \in L^2(\mathcal{R}^2)$, the exact reproduction can be obtained by the following generalized Calderón reproducing formula:*

$$f(\boldsymbol{x}) = \int_{\mathcal{R}^2}\int_{-\infty}^{\infty}\int_0^{\infty} \gamma_f(a,s,\boldsymbol{b})\psi_{a,s,\boldsymbol{b}}(\boldsymbol{x})\,\frac{da}{a^3}\, ds\, d\boldsymbol{b}. \tag{9.5.4.3}$$

In addition, we have

$$||f(\boldsymbol{x})||^2 = \int_{\mathcal{R}^2}\int_{-\infty}^{\infty}\int_0^{\infty} |\gamma_f(a,s,\boldsymbol{b})|^2\,\frac{da}{a^3}\, ds\, d\boldsymbol{b}. \tag{9.5.4.4}$$

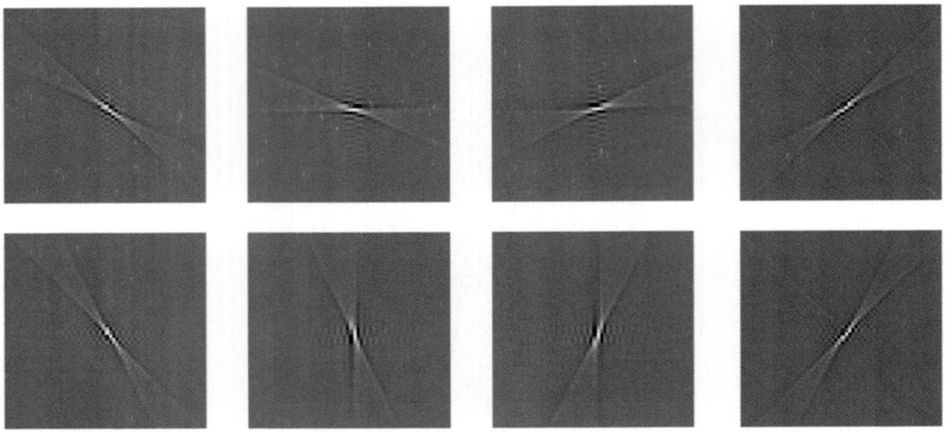

FIGURE 9.5.45

The eight spatial waveform images of shearlet atoms corresponding to different orientations at scale = 3. Note that the images are zoomed in to show only the center portions of the images.

9.5.4.1 Properties of a shearlet transform

A 2D function f that is smooth except for discontinuities along a curve Γ can be modeled as an image in $[0, 1]^2$ showing an object and a boundary, and is written as $O \cup \Gamma = \cup_{n=1}^{L}(O_n \cup C_n)$, namely, L small objects O_n, $n = 1, 2, \cdots, L$, form a connected open set O in the image with a boundary Γ, connected by piecewise smooth curves of finite length, with a finite number of corner (junction) points.

As $a \to 0$, a shearlet atom has a slim support in the spatial domain (see Fig. 9.5.45). If this slim support is centered at a point on an edge curve and is aligned along the normal direction of the curve, it results in slow decay or large value shearlet coefficients.

The following conclusions demonstrate that the location and orientation of singularities can be well characterized by the decay rate of continuous shearlet coefficients, which can be used to further classify the geometrical features at singularity points [120,121,126].

- If $\boldsymbol{b} \notin \Gamma$, then

$$\lim_{a \to 0+} a^{-N} \gamma_f(a, s, \boldsymbol{b}) = 0, \text{ for } \quad \forall s \in \mathcal{R}, \quad \forall N \in \mathcal{N},$$

  namely, the continuous shearlet coefficients satisfy

$$|\gamma_f(a, s, \boldsymbol{b})| \le k a^N, \quad \forall N \in \mathcal{N}, \quad \text{when } a \to 0+, \ k \text{ is a constant.}$$

- If $\boldsymbol{b} \in \Gamma$ is a regular edge point, namely, the curve is at least C^2 smooth near $\boldsymbol{b}$, but s is not at the normal direction of the curve at $\boldsymbol{b}$, then we have

$$\lim_{a \to 0+} a^{-N} \gamma_f(a, s, \boldsymbol{b}) = 0,$$

  namely, $\gamma_f(a, s, \boldsymbol{b})$ decays rapidly to 0 as $a \to 0$.

- If $\boldsymbol{b} \in \Gamma$ is a regular edge point, namely, the curve is at least C^2 smooth near $\boldsymbol{b}$, and s is at the normal direction of the curve at $\boldsymbol{b}$, that is, s is at the orientation that is perpendicular to Γ at $\boldsymbol{b}$, we have

$$\lim_{a \to 0+} a^{-3/4} \gamma_f(a, s, \boldsymbol{b}) = k f_{\boldsymbol{b}}, \quad k \text{ is a constant},$$

where $f_{\boldsymbol{b}}$ is the jump of a function f at $\boldsymbol{b}$ along the normal direction.

In regard to the case that $\boldsymbol{b} \in \Gamma$ is a corner (junction) point, the following results follow:

- If $\boldsymbol{b} \in \Gamma$ is a corner (junction) point, that is, more than two curves are merged together at $\boldsymbol{b}$, the curves are not smooth at $\boldsymbol{b}$, and s corresponds to one of the normal directions of the curves, we have

$$0 < \lim_{a \to 0+} a^{-3/4} \gamma_f(a, s, \boldsymbol{b}) < +\infty.$$

Namely, $\gamma_f(a, s, \boldsymbol{b})$ decays as $a^{3/4}$, as $a \to 0$.
- If $\boldsymbol{b} \in \Gamma$ is a corner (junction) point, but s does not correspond to any of the normal directions of the curves, the shearlet transform decays as $a^{9/4}$ except in some special case. Note that in the corner points, as $a \to 0+$, $\gamma_f(a, s, \boldsymbol{b})$ decays as $a^{3/4}$ in two or more different normal directions at $\boldsymbol{b}$ when two or more curves are merged together at $\boldsymbol{b}$.
- A spike type singularity point has a different behavior for the decay of the shearlet coefficients. For example, for a Dirac delta function centered at $\boldsymbol{b}_0$, we have $|\gamma_f(a, s, \boldsymbol{b})| \asymp a^{-3/4}$, as $a \to 0$ for all s when $\boldsymbol{b} = \boldsymbol{b}_0$. Therefore, the shearlet coefficients actually grow at fine scales for $\boldsymbol{b} = \boldsymbol{b}_0$. This is in contrast to the case that $|\gamma_f(a, s, \boldsymbol{b})|$ rapidly decays to zero if $\boldsymbol{b}$ is a regular edge point.

The Lipschitz regularity is commonly used to describe the local regularity of a function. In shearlet theory, it has been shown [120] that the decay rate of the shearlet coefficients of a function f depends on the local regularity of f. This further says that amplitudes of the shearlet coefficients are useful to describe the regularity points of a curve when the curve can be measured by Lipschitz regularity. If inequality (9.1.2.2) holds for all $\boldsymbol{b} \in O$ with constant $K > 0$ that is independent of $\boldsymbol{b}_0$, then the function f is uniformly Lipschitz-α over an open set O.

The following result [120] provides us with further insight into the shearlet coefficients. It takes advantage of the fact that the mother shearlet has an infinite number of vanishing moments, namely

$$\int x^n \psi(x, y)\, dx = 0, \quad \text{for} \quad n = 0, 1, \cdots, \infty.$$

Theorem 9.9. *If function $f(\boldsymbol{x}) \in L^2(\mathcal{R}^2)$ is Lipschitz-α, $\alpha > 0$ near a point $\boldsymbol{b}_0$, we will have*

$$|\gamma_f(a, s, \boldsymbol{b})| \le k a^{\frac{1}{2}(\alpha + \frac{3}{2})}(1 + |a^{-1/2}(\boldsymbol{b} - \boldsymbol{b}_0)|), \ \textit{for } a < 1, \ k \textit{ is a constant}.$$

9.5.4.2 Shearlets in the horizontal and vertical directions

In this section, we explain that the family of shearlet atoms can be split into two families, one in the horizontal direction and another in the vertical direction. The separation will ease the numerical implementation of the shearlet transform because the nonuniform angular covering of the frequency plane becomes a problem when the slope of a trapezoid is too steep, namely, when $|s|$ takes a large value, as shown in Fig. 9.5.44.

The following proposition ensures that reproducing a function $f \in L^2(\mathcal{R}^2)$ using shearlets can be separated into reconstruction under three systems: two (horizontal and vertical) shearlet systems, plus a coarse-scale isotropic system. First, we introduce the horizontal shearlet family, which can be defined by restricting the values of a and s to $0 < a < 1$ and $-2 \leq s \leq 2$. The shearlets are focused on the horizontal direction zone $C^{(h)}$ defined as

$$C^{(h)} = \{(\Omega_x, \Omega_y) \in \mathcal{R}^2 : |\Omega_x| \geq 2 \text{ and } |\frac{\Omega_y}{\Omega_x}| \leq 1\}.$$

Proposition 9.7. *Any function $f(\boldsymbol{x}) \in L^2(\mathcal{R}^2)$ that has its support in the horizontal zone $C^{(h)}$ can be exactly reconstructed by the following reproducing formula:*

$$f(\boldsymbol{x}) = \int_{\mathcal{R}^2} \int_{-2}^{2} \int_{0}^{1} \gamma_f(a, s, \boldsymbol{b}) \psi_{a,s,\boldsymbol{b}}(\boldsymbol{x}) \frac{da}{a^3} \, ds \, d\boldsymbol{b}, \tag{9.5.4.5}$$

assuming the mother shearlet $\psi(\boldsymbol{x})$ satisfies the conditions given in Section 9.5.3.1 and the shearlet atoms $\psi_{a,s,\boldsymbol{b}}(\boldsymbol{x})$ are defined as in Eq. (9.5.2.4).

The proof of this proposition can be found in [119]. The overall requirement for s is to indeed satisfy $-\sqrt{a} - 1 < s < \sqrt{a} + 1$. This proposition can be extended to show that any function $f(\boldsymbol{x}) \in L^2(\mathcal{R}^2)$ with support in the vertical zone defined as

$$C^{(v)} = \{(\Omega_x, \Omega_y) \in \mathcal{R}^2 : |\Omega_x| \geq 2 \text{ and } |\frac{\Omega_y}{\Omega_x}| > 1\}$$

can be exactly reproduced using the shearlet system defined by Eq. (9.5.2.4); however, the matrix $\boldsymbol{M}_{as}$ should be replaced by $\boldsymbol{M}_{as}^T$, and the mother shearlet window should be replaced by $\psi^{(v)}$ with its Fourier transform $\Psi^{(v)}(\boldsymbol{\Omega})$ being defined as

$$\Psi^{(v)}(\boldsymbol{\Omega}) = \Psi_1(\Omega_y)\Psi_2(\frac{\Omega_x}{\Omega_y}).$$

To distinguish the mother shearlet functions for the vertical/horizontal zones, we write them as $\psi^{(v)}$ (generating the vertical shearlets) and $\psi^{(h)}$ (generating the horizontal shearlets), respectively. In addition, we can also define an isotropic function $\phi(\boldsymbol{x})$ as a C^∞ window function in the spatial plane $\mathcal{R}^2$. Its Fourier transform is defined as $\Phi(\boldsymbol{\Omega}) = 1$ for $\boldsymbol{\Omega} \in [-1/2, 1/2]^2$ and as $\Phi(\boldsymbol{\Omega}) = 0$ for $\boldsymbol{\Omega} \notin [-2, 2]^2$. This function Φ can be called the scaling function, or the father shearlet function.

Now, any function $f(\boldsymbol{x}) \in L^2(\mathcal{R}^2)$ can be reconstructed using these three systems introduced above as

$$\begin{aligned} f(\boldsymbol{x}) = & \int_{\mathcal{R}^2} \langle f(\cdot), \phi(\cdot - \boldsymbol{b}) \rangle \phi(x - \boldsymbol{b}) \, d\boldsymbol{b} \\ & + \int_{\mathcal{R}^2} \int_{-2}^{2} \int_{0}^{1} \gamma_f^{(h)}(a, s, \boldsymbol{b}) \psi_{a,s,\boldsymbol{b}}^{(h)}(\boldsymbol{x}) \frac{da}{a^3} \, d\boldsymbol{b} \, ds \\ & + \int_{\mathcal{R}^2} \int_{-2}^{2} \int_{0}^{1} \gamma_f^{(v)}(a, s, \boldsymbol{b}) \psi_{a,s,\boldsymbol{b}}^{(v)}(\boldsymbol{x}) \frac{da}{a^3} \, d\boldsymbol{b} \, ds. \end{aligned} \tag{9.5.4.6}$$

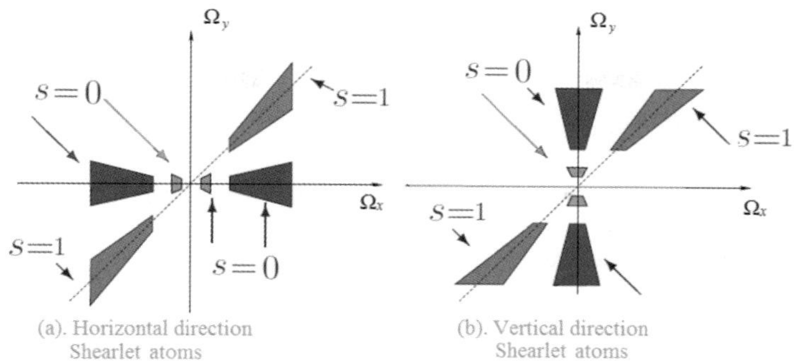

FIGURE 9.5.46

The frequency support of some horizontal and vertical shearlet atoms in the Fourier plane at two different scales.

Typically, in finding the salient features of an image, only the high-frequency content of a function in $L^2(\mathcal{R}^2)$ matters, and it corresponds to the curvelet atoms in the horizontal and vertical cones, written as $\gamma_f^{(h)}(a, s, \boldsymbol{b})$ and $\gamma_f^{(v)}(a, s, \boldsymbol{b})$. In general, we refer to the following two horizontal/vertical cones after we separate the low-pass content from the bandpass content:

1. the horizontal cone,

$$\mathcal{D}^{(h)} = \{(\Omega_x, \Omega_y) : |\frac{\Omega_y}{\Omega_x}| \leq 1\},$$

2. the vertical cone,

$$\mathcal{D}^{(v)} = \{(\Omega_x, \Omega_y) : |\frac{\Omega_x}{\Omega_y}| \geq 1\}.$$

Separating shearlet atoms into two families that cover the horizontal or vertical directions has made it easier for numerical implementation of the shearlet transform. Indeed, each "vertical" direction shearlet atom can be obtained by rotating the corresponding "horizontal" shearlet atom by $\frac{\pi}{2}$. Fig. 9.5.46 shows some examples of horizontal and vertical shearlet atoms for $s = 0,\ s = 1$ and some a. In the following, we focus on the numerical implementation of shearlets in these two directions.

9.5.5 Discrete shearlet atoms

Several discretization methods may be chosen to generate the discrete shearlet atoms along the horizontal cone and the vertical cone directions. We discretize the continuous shearlet transform by sampling the scale, shear, and translation parameters as follows: the scale parameter is sampled at $a = 2^{-2k},\ k \geq 0$; the shearing parameter is sampled as $s_{k,l} = l2^{-k},\ -2^k \leq l < 2^k$; and the translation parameter is sampled at $\boldsymbol{b} = B_l A_2^k \boldsymbol{m}$, with $\boldsymbol{m} \in \mathcal{Z}^2$,

$$A_2^k = \begin{pmatrix} 2^{-2k} & 0 \\ 0 & 2^{-k} \end{pmatrix}, \text{ and } B_l = \begin{pmatrix} 1 & -l \\ 0 & 1 \end{pmatrix}.$$

After sampling, Eq. (9.5.2.5) is rewritten as

$$\psi_{k,l,\boldsymbol{m}}(\boldsymbol{x}) = 2^{3k/2}\psi(B_l^{-1}A_2{-}k(\boldsymbol{x} - B_l A_2^k \boldsymbol{m})). \tag{9.5.5.1}$$

The following equation leads to the discretization method that we introduce in this section:

$$\{T_{A_2^k B_l}\boldsymbol{b}\}D_{A_2^k B_l}\psi(\boldsymbol{x}) = D_{A_2^k B_l}T_{\boldsymbol{b}}\psi(\boldsymbol{x}), \quad \boldsymbol{x} \in \mathcal{R}^2.$$

The discrete shearlet atoms in the horizontal direction are therefore given as the following set:

$$\{\psi_{k,l,\boldsymbol{m}}^{(h)}(\boldsymbol{x}) = D_{A_2^k B_l}T_{\boldsymbol{b}}\psi^{(h)}(\boldsymbol{x}) : k \geq 0, -2^k \leq l < 2^k, \boldsymbol{m} \in \mathcal{Z}^2\}.$$

The discrete shearlet atoms can be further written as:

$$\psi_{k,l,\boldsymbol{m}}^{(h)}(\boldsymbol{x}) = 2^{3k/2}\psi^{(h)}(\boldsymbol{B}_l^{-1}\boldsymbol{A}_2^{-k}\boldsymbol{x} - \boldsymbol{m}), \tag{9.5.5.2}$$

where

$$\boldsymbol{B}_l^{-1}\boldsymbol{A}_2^{-k} = \begin{pmatrix} 2^{2k} & l2^k \\ 0 & 2^k \end{pmatrix} \text{ and } \boldsymbol{A}_2^k\boldsymbol{B}_l = \begin{pmatrix} 2^{-2k} & -l2^{-2k} \\ 0 & 2^{-k} \end{pmatrix}.$$

The Fourier transform of $\psi_{k,l,\boldsymbol{m}}^{(h)}(\boldsymbol{x})$ can be written as

$$\begin{aligned}\Psi_{k,l,\boldsymbol{m}}^{(h)}(\boldsymbol{\Omega}) &= 2^{-3k/2}\Psi^{(h)}((\boldsymbol{A}_2^k\boldsymbol{B}_l)^T\boldsymbol{\Omega})e^{-j2\pi\langle \boldsymbol{A}_2^k\boldsymbol{B}_l\boldsymbol{m},\boldsymbol{\Omega}\rangle} \\ &= 2^{-3k/2}\Psi_1(2^{-2k}\Omega_x)\Psi_2(2^k\frac{\Omega_y}{\Omega_x} - l)e^{-j2\pi\langle \boldsymbol{A}_2^k\boldsymbol{B}_l\boldsymbol{m},\boldsymbol{\Omega}\rangle}.\end{aligned} \tag{9.5.5.3}$$

The mother shearlet function can be similarly chosen as in the continuous case. Recall that the functions $\Psi_1(\Omega_x)$ and $\Psi_2(\Omega_y)$ are $C^\infty(\mathcal{R})$, and shearlet atoms constructed from them are well localized in the Fourier plane. The window $\Psi_1(\Omega_x)$ is supported on $[-2, -1/2] \cup [1/2, 2]$ in the Fourier plane, and it satisfies

$$\sum_{k=0}^{\infty}|\Psi_1(2^{-2k}\Omega_x)|^2 = 1, \text{ for } |\Omega_x| > \frac{1}{2}. \tag{9.5.5.4}$$

The window $\Psi_2(\Omega_y)$ is supported on $[-1, 1]$ in the Fourier plane, and it satisfies

$$|\Psi_2(\Omega_y - 1)|^2 + |\Psi_2(\Omega_y)|^2 + |\Psi_2(\Omega_y + 1)|^2 = 1, \text{ for } |\Omega_y| \leq 1, \tag{9.5.5.5}$$

which leads to the following equation:

$$\sum_{l=-2^k}^{2^k-1}|\Psi_2(2^k\Omega_y + l)|^2 = 1, \text{ for } |\Omega_y| \leq 1, \ k \geq 0. \tag{9.5.5.6}$$

The Fourier content of a discrete shearlet atom $\Psi^{(h)}_{k,l,\boldsymbol{m}}$ is supported on a pair of trapezoids oriented along a line with slope $l2^{-k}$, of size about $2^{2k} \times 2^k$, written as

$$W_{k,l} = \{\boldsymbol{\Omega} = (\Omega_x, \Omega_y) : \Omega_x \in [-2^{2k+1}, -2^{2k-1}] \cup [2^{2k-1}, 2^{2k+1}], |\frac{\Omega_y}{\Omega_x} - l2^{-k}| \leq 2^{-k}\}.$$

This system of discrete shearlet atoms defines a Parseval frame on the horizontal cone $\mathcal{D}^{(h)}$. An exact reproducing theorem can then be obtained.

Theorem 9.10. *Let the horizontal shearlet atoms be defined as in Eq. (9.5.5.2), using the shearlet window that satisfies Eqs. (9.5.5.4) and (9.5.5.6). This shearlet system is a Parseval frame for $L^2(\mathcal{D}^{(h),2})$.*

The discrete shearlets on the vertical zone can be similarly defined as above. In addition, the low-pass isotropic scaling shearlets can be generated by the translations of a scaling function.

9.5.6 Numerical implementation of the shearlet transform

A numerically efficient implementation of a shearlet transform was previously introduced in [118,127] based on combining an LP with appropriate shearing filters. The problem with the implementation is the large side lobe effects around significant edges. Recently, an improved implementation based on separately calculating the vertical and horizontal shearlets was proposed by Yi et al. [13,122,123] and will be introduced here.

For convenience, the new implementation reformulates the shearlet transform by introducing the following functions in the horizontal and vertical zones separately:

$$W^{(h)}_{kl}(\boldsymbol{\Omega}) = 2^{k/2}\Psi_2(2^k\frac{\Omega_y}{\Omega_x} - l)\chi_{\mathcal{D}^{(h)}}(\boldsymbol{\Omega}), \tag{9.5.6.1}$$

$$W^{(v)}_{kl}(\boldsymbol{\Omega}) = 2^{k/2}\Psi_2(2^k\frac{\Omega_x}{\Omega_y} - l)\chi_{\mathcal{D}^{(v)}}(\boldsymbol{\Omega}). \tag{9.5.6.2}$$

The function $W^{(d)}_{kl}(\boldsymbol{\Omega})$, where $d = v$ or h, is a window function supported on a pair of trapezoids in the Fourier plane. It corresponds to a spatial function $w^{(d)}_{kl}(n_1, n_2)$. Now the Fourier content of the shearlets in Eq. (9.5.5.2) can be written as

$$\Psi^{(d)}_{k,l,\boldsymbol{m}}(\boldsymbol{\Omega}) = 2^{-2k}V^{(d)}(2^{-2k}\boldsymbol{\Omega})W^{(d)}_{kl}(\boldsymbol{\Omega})e^{-j2\pi\langle \boldsymbol{m},\boldsymbol{\Omega}\rangle}, \quad d = v, h,$$

where $V^{(d)}(\Omega_x, \Omega_y)$ is defined as

$$V^{(h)}(\Omega_x, \Omega_y) = \Psi_1(\Omega_x) \text{ and } V^{(v)}(\Omega_x, \Omega_y) = \Psi_1(\Omega_y).$$

The discrete shearlet transform of a discrete sampled function $f(n_1, n_2)$ can be computed as

$$\begin{aligned}\gamma^{(d)}_f(k, l, \boldsymbol{m}) &= \int_{\mathcal{R}^2} F(\boldsymbol{\Omega})\Psi^{(d*)}_{k,l,\boldsymbol{m}}(\boldsymbol{\Omega})\, d\boldsymbol{\Omega} \\ &= \int_{\mathcal{R}^2} 2^{-2k}F(\boldsymbol{\Omega})V^{(d*)}(2^{-2k}\boldsymbol{\Omega})W^{(d*)}_{kl}(\boldsymbol{\Omega})e^{j2\pi\langle \boldsymbol{m},\boldsymbol{\Omega}\rangle}\, d\boldsymbol{\Omega}\end{aligned}$$

$$= \int_{\mathcal{R}^2} V_k^{(d*)} f(\boldsymbol{\Omega}) W_{kl}^{(d*)}(\boldsymbol{\Omega}) e^{j2\pi\langle \boldsymbol{m},\boldsymbol{\Omega}\rangle}\, d\boldsymbol{\Omega} \quad \text{for } d = v, h. \tag{9.5.6.3}$$

Note that here the subscript "$*$" indicates complex conjugate, with

$$V_k^{(d)} f(\boldsymbol{\Omega}) = aF(\boldsymbol{\Omega}) V^{(d)}(a\boldsymbol{\Omega}), \; a = 2^{-2k},$$

which is the output from the high-pass filter in the horizontal/vertical zone. Eq. (9.5.6.3) can be viewed as the inverse Fourier transform, which is equivalent to a convolution in the spatial plane, and can be written as

$$\gamma_f^{(d)}(k, l, \boldsymbol{m}) = (v_k^{(d)} f \star w_{kl}^{(d)})(\boldsymbol{m}).$$

Here $v_k^{(d)} f$ is the inverse Fourier transform of the function $V_k^{(d*)} f(\boldsymbol{\Omega})$ and can be written as

$$v_k^{(d)} f(\boldsymbol{b}) = \int_{\mathcal{R}^2} V_k^{(d*)} f(\boldsymbol{\Omega}) e^{j2\pi\langle \boldsymbol{b},\boldsymbol{\Omega}\rangle}\, d\boldsymbol{\Omega}.$$

The value of $v_k^{(d)} f(\boldsymbol{b})$ is obtained through an iterative calculation as in the algorithm introduced later. The value of $w_{kl}^{(d)}(\boldsymbol{b})$ is the inverse Fourier transform of $W_{kl}^{(d*)}(\boldsymbol{\Omega})$, obtained through a mapping of the Fourier plane from the Cartesian grid to the pseudopolar grid, which is defined using the following pseudopolar coordinates $(\zeta_x, \zeta_y) \in \mathcal{R}^2$:

$$\begin{aligned} (\zeta_x, \zeta_y) &= (\Omega_x, \frac{\Omega_y}{\Omega_x}) \quad \text{if } (\Omega_x, \Omega_y) \in \mathcal{D}^{(v)}, \\ (\zeta_x, \zeta_y) &= (\Omega_y, \frac{\Omega_x}{\Omega_y}) \quad \text{if } (\Omega_x, \Omega_y) \in \mathcal{D}^{(h)}. \end{aligned} \tag{9.5.6.4}$$

Fig. 9.5.47 shows the mapping from the Fourier plane to the pseudopolar coordinates. Under this mapping, the following result is obtained:

$$\begin{aligned} V_a^{(d)} f(\zeta_x, \zeta_y) &= V_a^{(d)} f(\Omega_x, \Omega_y), \\ W^{(d)}(2^k(\zeta_y - l)) &= W_{kl}^{(d)}(\Omega_x, \Omega_y). \end{aligned}$$

Note that each directional window $W_{kl}^{(d)}$ in the Fourier plane can be obtained by translating the window $W^{(d)}$ in the pseudopolar grid using the function Ψ_2 in Eqs. (9.5.6.1) and (9.5.6.2), resulting in the directional localization supported in a wedge in the Fourier plane, as shown in Fig. 9.5.48. This results in the spatial waveform $w_{kl}^{(d)}(\boldsymbol{m})$ to be used in the following algorithm.

The digital implementation of the discrete shearlet transform in Eq. (9.5.6.3) can be realized through a filter bank as follows. Let H_k and G_k be, respectively, the low-pass and high-pass filters of a wavelet transform with $2k - 1$ zeros inserted between consecutive coefficients of a given 1D filter (H or G). Define $u \star H$ and $u \star G$ to be the separable convolution of the rows and the columns of u with H and G, respectively.

Note that G is the wavelet filter corresponding to a high-pass odd function ψ_1 and H is the filter corresponding to the coarse scale. The window $W^{(d)}$ is related to an even low-pass function ψ_2. ψ_1 and ψ_2 can be selected through the 1D Meyer wavelets.

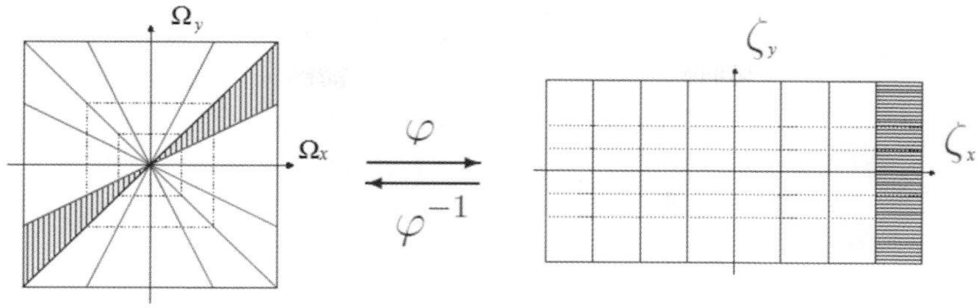

FIGURE 9.5.47

The mapping from a wedge in the Fourier plane to a rectangle in the pseudopolar coordinates according to Eq. (9.5.6.4).

FIGURE 9.5.48

Images of the 2^3 windows that partition the Fourier plane at various orientations.

The following is a cascade algorithm for implementing the discrete shearlet transform. Let $f \in l_2(\mathcal{Z}^2)$ be a 2D image. Define

$$S_0 f = f(\boldsymbol{x}),$$
$$S_k f = (S_{k-1} f) \star (H_k, H_k), \ k \geq 1.$$

For $d = v, h$, the discrete shearlet transform can be calculated as follows:

$$\gamma_f^{(d)}(k, l, \boldsymbol{m}) = (v_k^{(d)} f \star w_{kl}^{(d)})(\boldsymbol{m}), \qquad \text{for } k \geq 0, \ -2^k \leq l \leq 2^k - 1, \ \boldsymbol{m} \in \mathcal{Z}^2, \tag{9.5.6.5}$$

where

$$v_k^{(h)} f = (S_k f) \star (G_k, \delta), \ v_k^{(v)} f = (S_k f) \star (\delta, G_k).$$

Fig. 9.5.49 shows a representation of the discrete shearlet coefficients for different orientations at decomposition levels 2 and 3. For convenience, the vertical and horizontal shearlet coefficients are relabeled according to the parameter l, which represents the orientation:

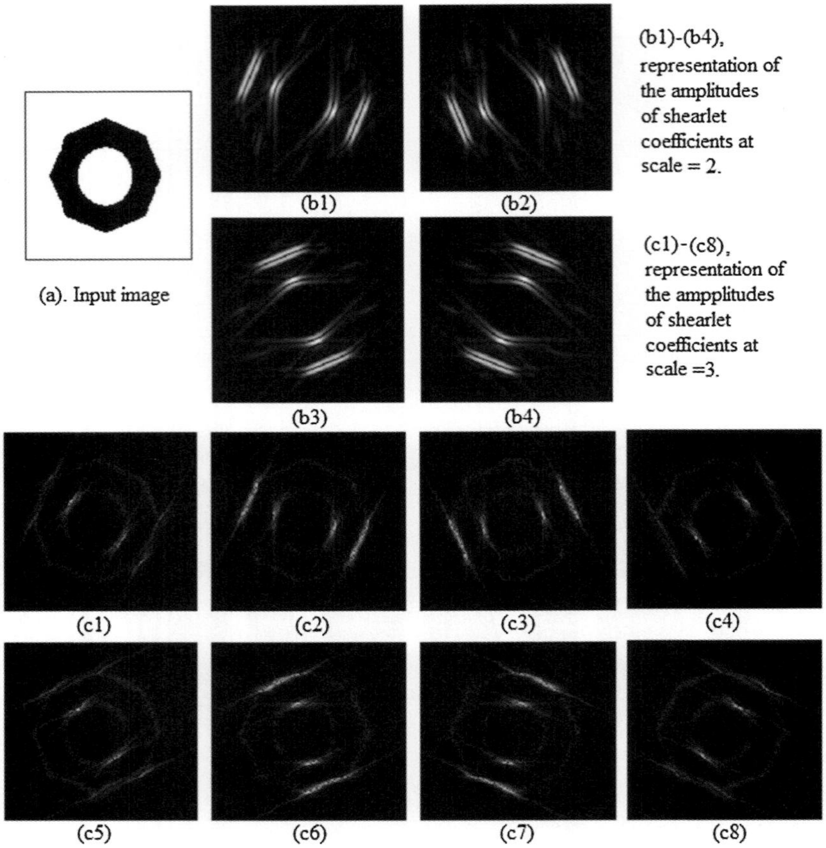

FIGURE 9.5.49

(a). The image of a disk is shown. (b1–b4) The representations of shearlet coefficients of the disk image at multiple scales for several values of the orientation index (l) for scale = 2. (c1–c8) The representations for scale = 3 are shown.

$$\gamma_f(k,l,\boldsymbol{m}) = \begin{cases} \gamma_f^{(h)}(k, l-1-2^k, \boldsymbol{m}), & 1 \le l \le 2^{k+1}, \\ \gamma_f^{(v)}(k, 3*2^k-1-l, \boldsymbol{m}), & 2^{k+1} < l \le 2^{k+2}. \end{cases} \tag{9.5.6.6}$$

9.5.7 Applications

In the following, we are to separately introduce the applications of shearlet transform in processing edge-related information, for example, estimation of edge orientation, edge detection, feature classification, and image denoising.

Edge-related tasks such as edge detection can be very difficult when noise, mixed with multiple edges, is presented in an image. The asymptotic decay property of the shearlet coefficients γ_f suggests that a shearlet transform can be a very efficient tool to determine the edges and their orientations inside an image f. A shearlet transform not only captures the singularities in a 2D function, but also identifies the geometrical feature of curves where the singularities are located. This leads to feature classification, namely, to determine whether a singularity point is a regular edge point or an irregular edge point, such as a corner or junction point.

Another difficult task in image processing is denoising. Noise is the most prevalent as spike singularities are classified as isolated points that have significantly higher gradients than their neighbors. The fact that the shearlet transform coefficients on spike points actually grow at fine scales shows that the shearlet transform can be used to distinguish noise points from true edge points and can therefore be an efficient tool for denoising.

9.5.7.1 Estimation of edge orientation

Detecting edge orientation using a shearlet transform takes advantage of geometrical information contributions at various scales, directions, and locations. Multiple orientations may be found when more than two curves are merged at a location. Recently, Yi and Krim [13,122,123] proposed to estimate the edge orientation using the following formula:

$$l(k, \boldsymbol{m}) = \text{argmax}_l |\gamma_f(k, l, \boldsymbol{m})|, \tag{9.5.7.1}$$

where $\boldsymbol{m}$ is an edge point and k is a fine scale. The orientation angle, $l(k, \boldsymbol{m})$, can then be directly calculated. Note that this angle is a discrete value associated with $l = 1, 2, 3, \cdots, 2^k$ possible orientations. To reduce the quantization error, the values of the orientations will be interpolated using a parabolic function of l to model the continuous shearlet transform $\gamma_f(k, l, \boldsymbol{m})$, and the orientation angle of the edge is associated with the maximum value in the parabolic function. The following procedure describes the algorithm that uses shearlet transform to detect the orientation of an edge.

Shearlet orientation detection algorithm:

- Compute $l_1 = l(k, \boldsymbol{m})$ using Eq. (9.5.7.1); let $l_0 = ((l_1 - 1) \mod N), l_2 = ((l_1 + 1) \mod N)$, where N is the size of the set of indices l.
- l_i, $i = 0, 1, 2$, are the slopes of three adjacent discretized directions. Let θ_i, $i = 0, 1, 2$, be the angles of orientations calculated from l_i.
- Let $S(\theta) = c_1\theta^2 + c_2\theta + c_3$. When $\theta = \theta_i$, $S(\theta_i)$ is identified with $\gamma_f(k, l_i, \boldsymbol{b}), i = 0, 1, 2,$ then c_1, c_2, c_3 is obtained by fitting the system $S(\theta_i) = c_1\theta_i^2 + c_2\theta_i + c_3$. The detected angle, $\theta_{max} = -\frac{c_2}{2c_1}$, $c_1 < 0$, is the value where the parabolic function $S(\theta)$ achieves its maximum.

Fig. 9.5.50 provides an error analysis of applying this algorithm to an image with a circle object.

9.5.7.2 Feature classification

We present in this section a computationally efficient shearlet approach for classifying several different feature points in an image. This method has advantages over wavelet-based methods, on account of the fact that the shearlet transform catches the directional features of an image.

In a typical image, there usually exist four classes of points, namely, junction and corner points, points near edges, points on smooth edges, and points inside a smooth region. The different types

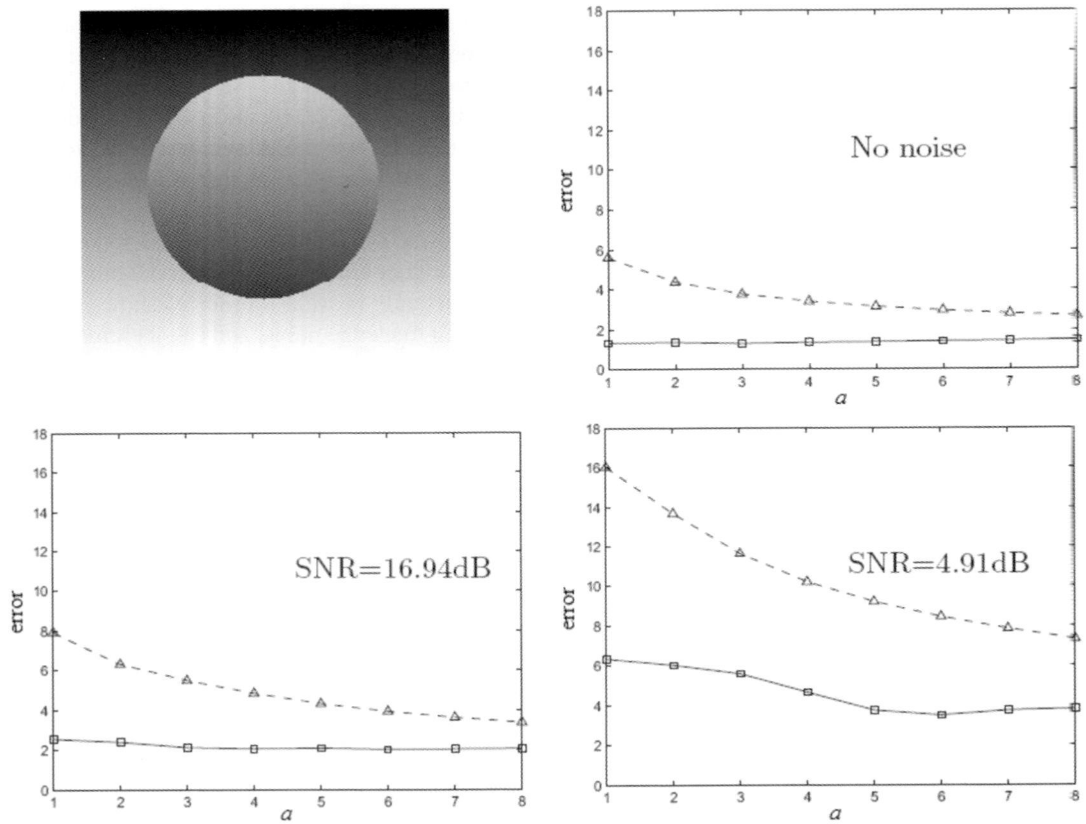

FIGURE 9.5.50

Comparison of the average error in the estimation of edge orientation (calculated using equation (3.4.9) in [122]) for the disk image shown on the left, using the wavelet method (dashed line) versus the shearlet method (solid line), as a function of the scale a, for various SNRs (additive white Gaussian noise). Courtesy from [122].

of points may be classified by examining the behavior of the discrete shearlet transform coefficients according to the shearlet properties at these points.

First, at a fixed scale k_0, we define the energy function at each spatial point $\boldsymbol{m}$ as

$$E_{k_0}(\boldsymbol{m}) = \sum_{l} |\gamma_f(k_0, l, \boldsymbol{m})|. \tag{9.5.7.2}$$

The points with large energy values $E(\boldsymbol{m})$ are classified as boundary points.

Second, we define the following function:

$$s_{\boldsymbol{m}_0}(l) = |\gamma_f(k_0, l, \boldsymbol{m}_0)|.$$

The function is tested on the boundary points, and the corner and junction points will be separated from the regular edge points by applying the following criteria:

- The point $\boldsymbol{m}_0$ is a corner/junction point if more than one peak is presented in the function $s_{\boldsymbol{m}_0}(l)$ at $\boldsymbol{m}_0$.
- The point $\boldsymbol{m}_0$ is a regular edge point if the function $s_{\boldsymbol{m}_0}(l)$ has only a single peak at $\boldsymbol{m}_0$.

In order to implement this algorithm, for each boundary point $\boldsymbol{m}$ at scale k_0, let $L_{\boldsymbol{m}}$ be the set that collects all the orientations at which the local maxima of the amplitudes of the discrete shearlet coefficients occur along parameter l, namely,

$$L_{\boldsymbol{m}} = \{l : |\gamma_f(k_0, l, \boldsymbol{m})| > |\gamma_f(k_0, l+1, \boldsymbol{m})| \text{ and } |\gamma_f(k_0, l, \boldsymbol{m})| > |\gamma_f(k_0, l-1, \boldsymbol{m})|\}.$$

To verify the significance of the peaks in $L_{\boldsymbol{m}}$ for each boundary point $\boldsymbol{m}$, in the following algorithm, we use normalized shearlet coefficients at the point $\boldsymbol{m}$ defined as follows, in order to find the local maximum points:

$$p_{\boldsymbol{m}}(l) = \begin{cases} \dfrac{|\gamma_f(k_0, l, \boldsymbol{m})|}{\sum_{l \in L_k} |\gamma_f(k_0, l, \boldsymbol{m})|}, & l \in L_k, \\ 0, & l \notin L_k. \end{cases} \tag{9.5.7.3}$$

Feature classification algorithm:

- Use the energy function $E_{k_0}(\boldsymbol{m})$ defined in Eq. (9.5.7.2) to cluster the image points into three mutually exclusive sets I_1, I_2, I_3. The K-means clustering algorithm with Euclidean metric is used as the clustering algorithm.
- Reorder the index of sets I_i, namely, let

$$i_{max} = \text{argmax}_i \frac{1}{|I_i|} \sum_{\boldsymbol{m} \in I_i} E_{k_0}(\boldsymbol{m}) \text{ and } i_{min} = \text{argmin}_i \frac{1}{|I_i|} \sum_{\boldsymbol{m} \in I_i} E_{k_0}(\boldsymbol{m}).$$

  The set $I_{i_{max}}$ is the set of boundary points, because the significant coefficients only occur at the boundary points. The set $I_{i_{min}}$ contains regular points inside a region because the coefficients rapidly decay to zero at these points. The set I_i between sets $I_{i_{min}}$ and $I_{i_{max}}$ contains the near-edge points.
- For a point $\boldsymbol{m} \in I_{imax}$, sort the entries of the normalized coefficients $p_{\boldsymbol{m}}(l)$ from the greatest to the smallest, and denote them as $[\tilde{p}_{\boldsymbol{m}}(1), \cdots, \tilde{p}_{\boldsymbol{m}}(N)]$. Using again the K-means clustering algorithm on $I_{i_{max}}$, this set can be further classified into one group of smooth edge points and another group of corner and junction points according to the number of peaks at each boundary point.

Note that corner points can be further separated from junction points, and junction points can be separated into different classes corresponding to different geometric features. Fig. 9.5.51 illustrates the result of applying the feature classification algorithm.

9.5.7.3 Edge detection

Edge detection is a difficult task when noise is present and when several edges intersect each other and are close together. The asymptotic decay rate of the continuous shearlet transform provides a better solution to precisely capture the geometry of edges, and it can be used to obtain both locations and

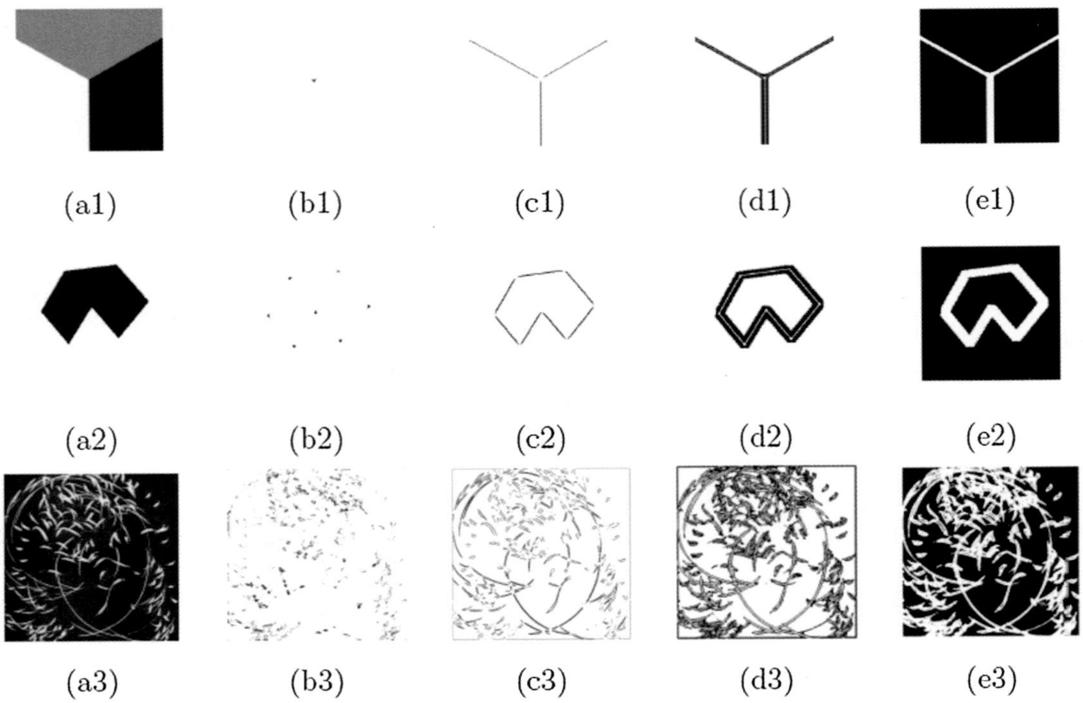

FIGURE 9.5.51

(a1–a3) Test images. (b1–b3) Identification of corners and junctions. (c1–c3) Identification of smooth edge points. (d1–d3) Identification of points near the edges. (e1–e3) Identification of regular points (smooth regions). Courtesy from [122].

orientations of edges in an image f. Here, we describe an effective edge detection scheme to precisely detect the shape of an edge using the shearlet transform that can be attributed to the anisotropic analytic and geometric properties of shearlets.

In this algorithm, the first step is to identify the edge candidates in an image f. They are selected by the shearlet transform modulus maxima at the kth scale level, that is, those points $\boldsymbol{m}$ that are the local maxima of the function

$$M_k f(\boldsymbol{m})^2 = \sum_l |\gamma_f(k, l, \boldsymbol{m})|^2.$$

The second step is to confirm edge points by excluding the noise points. According to the properties of shearlet the transform, if $\boldsymbol{m}$ is an edge point, the shearlet transform of f has the property that

$$|\gamma_f(k, l, \boldsymbol{m})| \sim C2^{-\beta k}, \ \beta > 0.$$

Note that points with $\beta < 0$ will be classified as noise; therefore, the edge points can be identified as those points that have $\beta > 0$, and the task of edge detection is transferred to the task of determining the sign of β.

Theoretically, this can be estimated by computing the best linear fit to data

$$\{\log|\gamma_f(k,l,\boldsymbol{m})|\}_{k=1,2,\cdots,L}.$$

The magnitude of the shearlet coefficients at each point is compared with the values of its neighbors along the gradient direction (this is obtained from the orientation map of the shearlet decomposition). If the magnitude is smaller, the point is discarded; if it is the largest, it is kept. This yields a set of possible edge points. This is the so-called nonmaximal suppression routine. Further, this edge candidate set will be filtered by a window filter, and a smooth edge set can be selected by a threshold. In practice, the following shearlet edge detection algorithm is implemented to simplify the computation complexity.

Shearlet edge detection algorithm:

- Let $f \in l_2(\mathcal{Z}^2)$ be a 2D image. Calculate $\gamma_f^{(d)}(k,l,\boldsymbol{m})$, which is given as follows:

$$\gamma_f^{(d)}(k,l,\boldsymbol{m}) = (v_k^{(d)} f \star w_{kl}^{(d)})(\boldsymbol{m}), \qquad \text{for } k \geq 0,\ -2^k \leq l \leq 2^k - 1,\ \boldsymbol{m} \in \mathcal{Z}^2, \tag{9.5.7.4}$$

where

$$\begin{aligned} S_0 f &= f(\boldsymbol{x}), \\ S_k f &= (S_{k-1} f) \star (H_k, H_k),\ k \geq 1, \\ v_k^{(h)} f &= (S_k f) \star (G_k, \delta), \\ v_k^{(v)} f &= (S_k f) \star (\delta, G_k). \end{aligned}$$

- Define

$$\chi_l^{(d)}(\boldsymbol{m}) = \begin{cases} 1, & \text{if } \gamma_f^{(d)}(k,l,\boldsymbol{m}) > \gamma_f^{(d)}(k-1,l,\boldsymbol{m}), \\ 0, & \text{otherwise.} \end{cases} \quad d = v, h. \tag{9.5.7.5}$$

  This defines a function to indicate the points that have negative Lipschitz regularity.
- Based on this indicator function, define

$$R_k^{(d)} f(\boldsymbol{m}) = \sum_l \gamma_f^{(d)}(k,l,\boldsymbol{m}) \chi_l^{(d)}(\boldsymbol{m}), \quad d = v, h.$$

- Modify $v_k^{(d)} f, d = v, h$, according to the following formula:

$$v_k^{(d)} f = \begin{cases} v_k^{(d)} f + R_k^{(d)} f, & \text{if } |v_k^{(d)} f| \leq |R_k^{(d)} f|, \\ R_k^{(d)} f, & \text{otherwise,} \end{cases} \quad d = v, h. \tag{9.5.7.6}$$

  This formula modifies $v_k^{(d)} f$ so that the value of $v_k^{(d)} f$ is kept small (unchanged) for locations with positive Lipschitz regularity and the value of $v_k^{(d)} f$ is increased over the scales for locations with negative Lipschitz regularity.

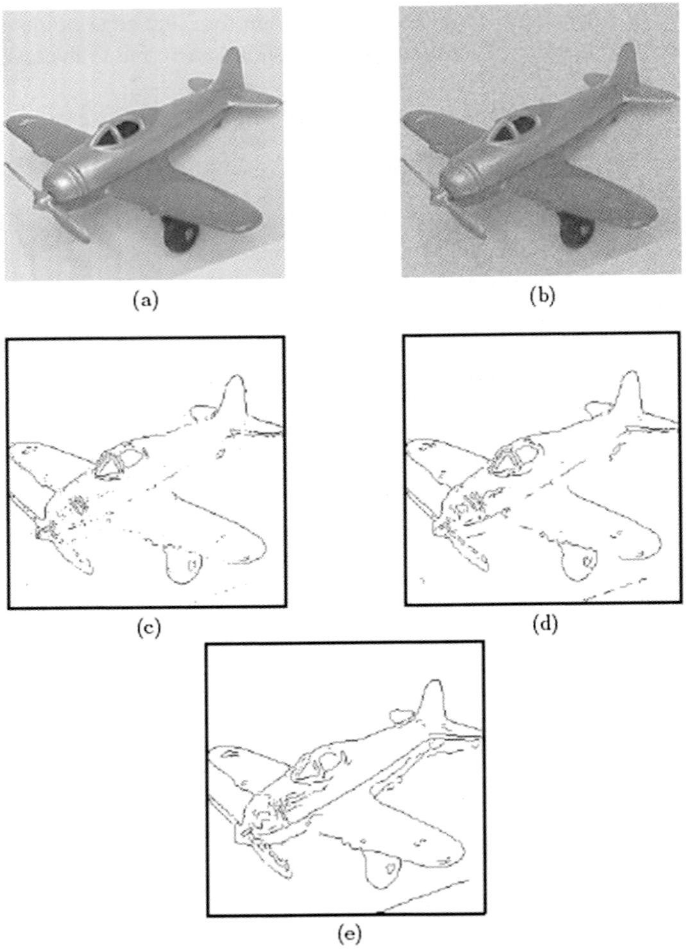

FIGURE 9.5.52

Results of edge detection methods. (a) Original image. (b) Noisy image (PSNR = 24.58 dB). (c) Sobel edge detector result (FOM = 0.54). (d) Wavelet result (FOM = 0.84). (e) Shearlet result (FOM = 0.94). Courtesy from [122].

- After the discrete shearlet $\gamma_f^{(d)}(k, l, \boldsymbol{m})$ is calculated using the updated $v_k^{(d)} f$, the edge at level k will be decided by checking the local maxima of the following function:

$$E_k(\boldsymbol{m}) = (\sum_l \gamma_f^{(v)}(k, l, \boldsymbol{m}))^2 + (\sum_l \gamma_f^{(h)}(k, l, \boldsymbol{m}))^2.$$

Fig. 9.5.52 shows the results of edge detection, in which the Pratt figure of merit (FOM) is used to measure the performance of an edge detector [128].

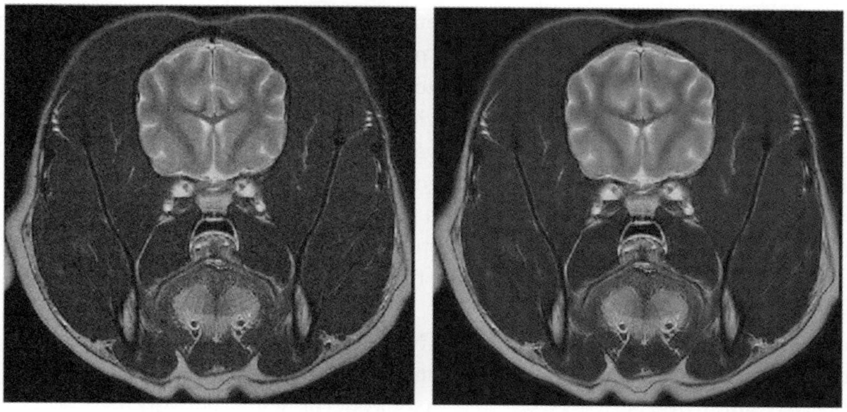

FIGURE 9.5.53

In this picture, the noisy brain image is shown to the left, and the filtered image using the shearlet denoising method is shown to the right.

9.5.7.4 Image denoising

The properties of shearlet coefficients at fine scales show that a 2D function f can be reproduced using the following shearlet representation (see [129]):

$$f(\boldsymbol{b}) = \sum_{k,l,\boldsymbol{m}\in M_1} \langle f, \psi_{k,l,\boldsymbol{m}}\rangle \psi_{k,l,\boldsymbol{m}}(\boldsymbol{b}) + \sum_{k,l,\boldsymbol{m}\in M_2} \langle f, \psi_{k,l,\boldsymbol{m}}\rangle \psi_{k,l,\boldsymbol{m}}(\boldsymbol{b}), \tag{9.5.7.7}$$

where M_1 is the set of coefficients associated with the smooth regions of f and M_2 is the set of coefficients associated with the edges of f.

A simple approach to achieve image denoising is carried out by using a shrinkage approach to remove shearlet coefficients below a threshold τ. A denoised image using the discrete shearlet transform can be expressed as

$$\tilde{f}(\boldsymbol{b}) \approx \sum_{k,l,\boldsymbol{m}\in M_1} \langle f, \psi_{k,l,\boldsymbol{m}}\rangle \psi_{k,l,\boldsymbol{m}}(\boldsymbol{b}) + \sum_{k,l,\boldsymbol{m}\in M_2^c} \langle f, \psi_{k,l,\boldsymbol{m}}\rangle \psi_{k,l,\boldsymbol{m}}(\boldsymbol{b}), \tag{9.5.7.8}$$

where M_2^c is the set of shearlet coefficients whose values are greater than τ. This approach is able to optimally capture directional features and reach the desired denoising effect. Fig. 9.5.53 shows a denoising result using the shrinkage method.

9.6 Incorporating wavelets into neural networks

Artificial neural network (ANN) originally gained researchers' attention on account of their promising performance as well as their structure, which is also reminiscent of the human brain functionality and its associated perception. Subsequent developments led to a refinement of ANN using wavelets to produce a more efficient analysis tool to seek nonlinear and time-varying features in signals/images. Unlike ANN, wavelets were capable of systematically accounting for the spectral information associated with features. Thus, a natural goal which arises is that of integrating wavelet analysis with ANN for a joint exploitation of their respective advantages in statistical inference applications. Specifically, researchers have focused on developing algorithms that integrate the advantages of ANN, i.e., self-adaptivity, fault tolerance, and robustness, with the advantages of the well-founded and established wavelet transformation in T-F localization for capturing scale-invariant features. This up-to-date, hybrid approach promises to leverage the strengths of deep learning, thereby providing a very powerful framework for supervised learning.

Numerous numerical stability challenges of ANN, hindering convergence, have necessitated the use of a compact and local kernel approach to yield convolutional neural network (CNN) [130] and, more recently, other activation-free networks referred to as Volterra neural network (VNN) [131,132]. The popularity of CNN may be attributed to a variety of properties including its self-learning and adaptive capacity, as well as its strong inference ability. It uses convolution as its basic representational atom and a bank of these atoms to build one representation layer in the so-called deep learning structure of many layers. Uniquely, the structure of CNN, by way of layers of convolutionally adapted filter banks, provides an efficient flexibility in the size of channels (number of filters) and the depth of the network (number of layers). This improved efficiency of CNN compared to ANN at capturing proper features has led to state-of-the-art performance in signal/image inference. Additionally, it has been followed by constant improvements and perspectives to address identification/annotation in video tracking [133] as well as superresolution [134], denoising [135], etc.

Early research efforts focused on 1D signals, through which researchers realized that the activation function was a key component in a hidden layer/unit which determined the success of a neural network. Further, it was found that the resulting output expression of the ANN bore a form similar to that of wavelet reconstruction. These discoveries led to the generation of a new idea: replacing the activation function in the ANN with wavelet atoms to subsequently yield wavelet neural network (WNN) [136].

Although WNN seemed to be a promising improvement on previous techniques, the computational complexity of multidimensional wavelets (even for 1D signals) makes the application of these tools to 2D images quite challenging. One of the key obstacles of this task is managing the cost of selection and storage of wavelet (orthonormal and nonorthonormal frame) representations, which can very quickly become prohibitively expensive. On the other hand, CNN can produce acceptable efficiency with large datasets and hold promise for joint exploitation of wavelets, which could form a solid groundwork for resolving the computational challenges in inference and processing problems.

In this section, we will introduce the neural network methods that use wavelets for both 1D signals and 2D images in a balanced manner. While the extension from 1D to 2D signals is algebraically tedious, it is theoretically and conceptually natural. For the sake of clarity, we separately detail the 1D and 2D cases.

9.6.1 Using wavelets in neural networks for signal processing

For a single-input single-output (SISO) system, the output from an ANN with a fully connected single hidden layer is analogous to that of a nonlinear regression model,

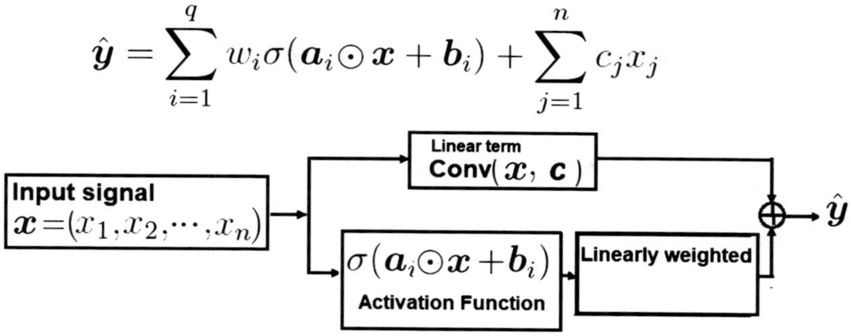

FIGURE 9.6.54

The diagram structure of RBFN with one single hidden layer.

$$\hat{y} = f(\boldsymbol{x}) = \sum_{i=1}^{q} w_i \sigma(\boldsymbol{a}_i^T \boldsymbol{x} + b_i) + \sum_{j=1}^{n} c_j x_j + w_0, \tag{9.6.1.1}$$

where $\boldsymbol{x} = (x_1, x_2, \cdots, x_n) \in \mathcal{R}^n$ is an input signal, q is the number of neurons or hidden units, w_i, c_j are linear coefficients, w_0 is a bias term from the hidden layer to the network output, and σ is an activation function that represents a hidden neuron controlled by the weights $\boldsymbol{a}_i$ and the hidden layer bias b_i. The choices of most popular activation functions are nonlinear functions such as:

- Rectified Linear Unit (ReLU) activation: $ReLU(x) = max(x, 0)$,
- sigmoidal activation: $\sigma(x) = \dfrac{1}{1+e^{-x}}$,
- hyperbolic tangent (Tanh) activation: $\mathrm{Tanh}(x) = \dfrac{e^x - e^{-x}}{e^x + e^{-x}}$.

Since there is no precise formula for ANN, alternatively, the output from an ANN can also yield the following form:

$$\hat{y} = f(\boldsymbol{x}) = \sum_{i=1}^{q} w_i \sigma(\boldsymbol{a}_i \odot \boldsymbol{x} + \boldsymbol{b}_i) + \sum_{j=1}^{n} c_j x_j + w_0, \tag{9.6.1.2}$$

where "$\odot$" is the componentwise multiplication operator and the activation function takes the format of a radial basis function (RBF); therefore, in this case, ANNs are also called RBF network (RBFN) [137].

Fig. 9.6.54 provides a diagram of RBFN while we view the linear term as a convolution and ignore the bias term. For a multiple-input multiple-output (MIMO) system, given a set of N input/output pairs, a training dataset, $(\boldsymbol{x}_k, y_k)$, $\boldsymbol{x}_k \in \mathcal{R}^n$, $k = 1, \cdots, N$, is the input signal with the corresponding true label $y_k \in \mathcal{R}$ for classification purposes. The output from an ANN is the estimated label $\hat{y}_k$. The network is trained via a backpropagation (BP) algorithm that uses a gradient decent method to select the weight parameters, $w_i, \boldsymbol{a}_i, \boldsymbol{b}_i, \boldsymbol{c}_j, w_0$, by minimizing a cost function (or loss function) such as an L_p-norm between the output of the network and the desired output. When $p = 2$, the cost function is the mean square error (MSE),

$$L_2 = \frac{1}{N}\sum_{k=1}^{N}(\hat{y}_k - y_k)^2.$$

The similarity between Eq. (9.6.1.2) and the wavelet reconstruction formula (see previous Section 9.2) is analyzed in [139,142]. In the following subsections, we will extend the RBFN to the ANN-based WNN, developed in the 1990s. Then, we will present the deep convolutional framelets neural network (FrameletsNN), a recent development, to compare the advantages associated with the two wavelet-based neural networks.

9.6.1.1 Wavelet neural networks

In a simple structured neural network with only one single hidden layer, WNN [136,138,139,140] replaces the activation function in RBFN to allow the algorithm to function similarly to an adaptive inverse wavelet transform while maintaining the advantages of the BP neural network. Without the bias (shifting) term, the output can be written as

$$\hat{y} = \sum_{k=1}^{q} w_i \psi(a_i(x - b_i)) + cx, \quad x \in \mathcal{R}, \tag{9.6.1.3}$$

where $\psi(a_i(x - b_i))$ are wavelet atoms derived from a 1D mother wavelet $\psi : \mathcal{R} \to \mathcal{R}$ with a_i, b_i as the scaling and translation parameters, respectively, and q is the number of wavelet neurons, also called wavelons.

WNN quickly becomes more cumbersome in the case of a multidimensional input signal $\boldsymbol{x} \in \mathcal{R}^n$ The output (with the bias term) can be written as

$$\hat{y} = \sum_{i=1}^{q} w_i \psi(\boldsymbol{a}_i \odot (\boldsymbol{x} - \boldsymbol{b}_i)) + \sum_{j=1}^{n} c_j x_j + w_0, \quad \boldsymbol{x}, \boldsymbol{a}_i, \boldsymbol{b}_i \in \mathcal{R}^n, \tag{9.6.1.4}$$

where ψ is the multidimensional mother wavelet $\psi : \mathcal{R}^n \to \mathcal{R}$ and $w_i, \boldsymbol{a}_i, \boldsymbol{b}_i, c_j, w_0$ are weights used in the hidden layer. This amounts to a total of $q + 2nq + n + 1$ parameters being estimated. Even though the number of weights can be reduced to $2q + qn + n + 1$ by using the scalar dilating parameter $\boldsymbol{a}_i = a$, it is still much more expensive to train a multidimensional WNN to estimate $\boldsymbol{x} \in \mathcal{R}^n$ than to train a 1D WNN. Therefore, researchers are more focused on multidimensional WNN in an effort to reduce the computational burden associated with higher-dimensional data.

The diagram of a multidimensional WNN in Fig. 9.6.55, in contrast to the diagram of an ANN in Fig. 9.6.54, shows that the localized spectral information from the wavelet transform ensures that useful feature information will be utilized in a more delicate manner than a general activation function in optimizing the weights by BP. This allows WNN to mitigate the problem of inaccurate prediction results due to the existence of local extrema. However, the fact that the linear term, if viewed as a convolution operator, is separated from the wavelet-related neurons is a limitation that can be addressed in FrameletsNN (to be introduced later in Section 9.6.1.2).

In training a WNN, the weights are randomly initialized, followed by iterative updates to minimize the cost function. Numerous viable solutions [139,141] have been proposed with the intent of obtaining faster convergence of the algorithm, thus strengthening the initialization of WNN. These involve

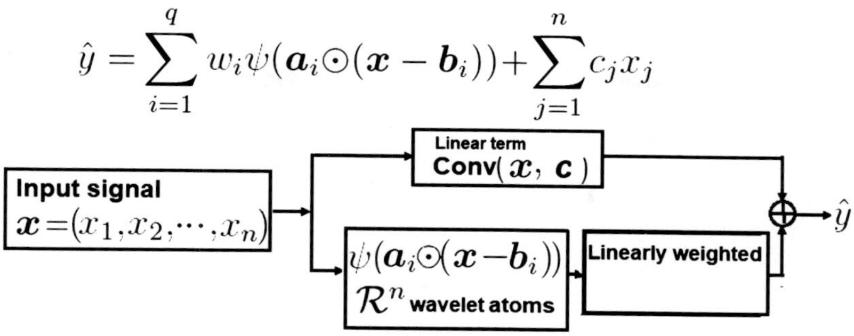

FIGURE 9.6.55

Structure of a WNN with one single hidden layer.

procedures to select a "good" mother wavelet and wavelet atoms in order to facilitate the training. To keep this section more focused on related topics, we will only briefly discuss how to initiate the steps for training the WNN using a BP algorithm. For details, readers are referred to [139,141].

Selecting a mother wavelet

Since $\boldsymbol{x} = (x_1, x_2, \cdots, x_n) \in \mathcal{R}^n$ is estimated as a function on a compact domain, usually the rule of thumb for selecting the mother wavelet is to ensure that it is compactly supported or rapidly vanishing to alleviate the computational burden due to the large number of weights that need to be trained. The following mother wavelets are commonly used:

$$\text{Gaussian derivative: } \psi(\boldsymbol{x}) = ||\boldsymbol{x}||e^{-||\boldsymbol{x}||^2/2},$$

$$\text{Mexican hat derivative: } \psi(\boldsymbol{x}) = (n - ||\boldsymbol{x}||^2)e^{-||\boldsymbol{x}||^2/2},$$

$$\text{Morlet wavelet: } \psi(\boldsymbol{x}) = \cos(5||\boldsymbol{x}||)e^{-||\boldsymbol{x}||^2/2}.$$

In the case where a multiplicative function is desired, a mother wavelet can be constructed as the product of 1D wavelets:

$$\psi(\boldsymbol{x}) = \prod_{i=1}^{n} \psi(x_i).$$

Selecting wavelet atoms

Once a multidimensional mother wavelet is selected, a library of wavelons will be created as dilated/translated wavelet atoms. The next step will be to select the regressors on account of the linear regression form in the first term of Eq. (9.6.1.4). The regressors are wavelets determined by the dilation and translation parameters (a_i, b_i). The neural network will be trained to optimize the parameters. However, in a WNN, the initialization of the parameters is a crux to reduce the training time for the neural network. Here, we summarize three ways to "best" initialize the parameters for the regressors. A detailed discussion can be found in Zhang [139].

- Residual-based selection (RBS): A simple method to build the "best set" is to iteratively select the wavelet atom from the library that best fits the output data at each step, adding new wavelets to the developing set one-by-one until high efficiency is reached while balancing it against maintaining low computational burden.
- Backward elimination: One of the most effective methods is to start with the library containing all wavelets and remove wavelets of least contribution to the fitting (small wavelet coefficients) from the library in an effective way that is similar to the wavelet shrinkage method [142,143].
- Stepwise selection by orthogonalization (SSO): SSO is an extension of the RBS method. In each step, a new wavelet atom will be selected through a modified Gram–Schmidt algorithm and normalized (see [139]). A similar modification was proposed by Oussar [141].

9.6.1.2 *Deep convolutional framelets neural networks*

Replacing the "$\odot$" operator in Eq. (9.6.1.2) with the convolution operator "$\star$" while ignoring the constant term, we can interpret the equation as a multichannel convolution (q-channel output) for one input signal $\boldsymbol{x}$, regulated by the activation function that can be viewed as a composite of nonlinearities and pooling, followed by a fully connected 1×1 convolution. This replacement is what led to the creation of the CNN, which is much more computationally efficient by virtue of needing only a small number of weights to be used as a convolution filter.

In this subsection, we introduce first the convolution framelets [144] (called Framelets) and then the FrameletsNN [145,146]. Framelets induce an energy-compact representation of a signal by way of local and nonlocal bases. This representation opened up a new avenue to incorporate wavelets into the deeper layers of a CNN, leading to the development of FrameletsNN, a CNN-based deep learning method with a wavelet-friendly architecture that has improved performance over the popular CNN-based U-Net reconstruction [134]. Similar to U-Net, FrameletsNN consists of a contracting path that involves pooling functions that carry downsampled features and an expansive path that involves unpooling for upsampled feature reconstruction. The critical role of pooling/unpooling in a neural network for feature extraction in signal/image reconstruction was discussed in [147,148].

9.6.1.2.1 Convolution framelets (Framelets)

Framelets [144] is a patch-based representation of a signal for energy compaction. A redundant representation consisting of patches that are wrapped around $\boldsymbol{x} = (x_1, x_2, \cdots, x_n)$ is used to yield the Hankel matrix $H_d(\boldsymbol{x})$ of $n \times d$, defined as

$$H_d(\boldsymbol{x}) = \begin{bmatrix} x_1 & x_2 & \cdots & x_d \\ x_2 & x_3 & \cdots & x_{d+1} \\ \vdots & \vdots & \ddots & \vdots \\ x_n & x_1 & \cdots & x_{d-1} \end{bmatrix}.$$

By way of interpreting an index $i \leq 0$ as $i = n + i$, namely an index wrapping around a signal, $\boldsymbol{x}$ can be reconstructed back as

$$x_k = \frac{1}{d} \sum_{i=0}^{d-1} [H_d(\boldsymbol{x})]_{k-i,i+1}, \; k = 1, \cdots, n. \tag{9.6.1.5}$$

For a SISO system, let $\overline{\boldsymbol{v}} = (v_d, v_{d-1}, \cdots, v_1)^T$ represent the index-reversed vector copied from a vector $\boldsymbol{v} = (v_1, v_2, \cdots, v_d)^T \in \mathcal{R}^d$. The (circular) convolution [145] of an input signal $\boldsymbol{x}$ with a convolutional filter $\overline{\boldsymbol{v}}$ generates the output $\boldsymbol{y}$,

$$\boldsymbol{y} = \boldsymbol{x} \star \overline{\boldsymbol{v}} = H_d(\boldsymbol{x})\boldsymbol{v}. \tag{9.6.1.6}$$

Suppose the Hankel matrix $H_d(\boldsymbol{x})$ with rank q, $1 \le q \le d$, has a singular value decomposition (SVD) with $H_d(\boldsymbol{x}) = U\Sigma V^T$, where $\Sigma = (\sigma_{ij}) \in \mathcal{R}^{q\times q}$ is a diagonal matrix, $U = (\boldsymbol{u}_1, \boldsymbol{u}_2, \cdots, \boldsymbol{u}_q) \in \mathcal{R}^{n\times q}$, and $V = (\boldsymbol{v}_1, \boldsymbol{v}_2, \cdots, \boldsymbol{v}_q) \in \mathcal{R}^{d\times q}$ are orthogonal matrices (column vectors are orthonormal to each other). Here U and V are referred to as nonlocal and local bases, respectively. Similarly, in Framelets, the Hankel matrix is decomposed as $H_d(\boldsymbol{x}) = \boldsymbol{\Phi}\boldsymbol{C}\boldsymbol{V}^T$. The nonlocal basis $\boldsymbol{\Phi}_{n\times n} = (\boldsymbol{\phi}_1, \boldsymbol{\phi}_2, \cdots, \boldsymbol{\phi}_n)$, $\boldsymbol{\phi}_i \in \mathcal{R}^{n\times 1}$, is designed to capture, as much as possible, the "variance" within the dataset. The local basis $\boldsymbol{V}_{d\times d} = (\boldsymbol{v}_1, \boldsymbol{v}_2, \cdots, \boldsymbol{v}_d)$, $\boldsymbol{v}_i \in \mathcal{R}^{d\times 1}$, is known to represent a local segment of a signal in an "energy compaction" fashion. As shown by Yin [144], when both local and nonlocal bases are selected with energy concentrated on their low-frequency components in the order of increased frequencies, they promote signal representations with "certain energy concentration patterns," and $\boldsymbol{C} = (c_{ij}) \in \mathcal{R}^{n\times d}$ will be optimized with nonzero entries only lying along the diagonal in the upper $d \times d$ block.

Define the outer product of a nonlocal basis (the column vectors of $\boldsymbol{\Phi}$) with a local basis (the column vectors of $\boldsymbol{V}$) as

$$\boldsymbol{\Psi}_{ij} = \boldsymbol{\phi}_i \boldsymbol{v}_j^T, \; i = 1, \cdots, n, \; j = 1, \cdots, d.$$

It can be shown [144] that $\{\boldsymbol{\Psi}_{ij}\}$ forms an orthonormal basis for the space of all $n \times d$ matrices, $\mathcal{R}^{n\times d}$, and the Hankel matrix can be written as

$$H_d(\boldsymbol{x}) = \sum_{i=1}^{n}\sum_{j=1}^{d} < \boldsymbol{x}, \boldsymbol{\phi}_i \star \boldsymbol{v}_j > \boldsymbol{\Psi}_{ij} = \sum_{i=1}^{n}\sum_{j=1}^{d} < \boldsymbol{x}, \boldsymbol{\phi}_i \star \boldsymbol{v}_j > \boldsymbol{\phi}_i \boldsymbol{v}_j^T,$$

where the inner product of two vectors is defined as $< \boldsymbol{\phi}, \boldsymbol{v} >= \boldsymbol{\phi}^T \boldsymbol{v}$. Hence, combining this equation with Eq. (9.6.1.5), we will have the representation of $\boldsymbol{x}$ as

$$\boldsymbol{x} = \frac{1}{d}\sum_{i=1}^{n}\sum_{j=1}^{d} < \boldsymbol{x}, \boldsymbol{\phi}_i \star \boldsymbol{v}_j > \boldsymbol{\phi}_i \star \boldsymbol{v}_j. \tag{9.6.1.7}$$

So we define the Framelets as

$$\boldsymbol{\psi}_{ij} = \boldsymbol{\phi}_i \star \boldsymbol{v}_j, \; i = 1, \cdots, n, \; j = 1, \cdots, d.$$

Subsequently, the framelet coefficients are defined as $c_{ij} =< \boldsymbol{x}, \boldsymbol{\psi}_{ij} >$ with all coefficients forming the elements of the framelet coefficient matrix C. Eq. (9.6.1.7) indicates that the framelets $\boldsymbol{\psi}_{ij}$, $i = 1, \cdots, n$, $j = 1, \cdots, d$, constitute a tight frame for signals in $\mathcal{R}^n$. To optimize the selection of the frame, for a given nonlocal basis $\boldsymbol{\Phi}$, the local basis $\boldsymbol{V}$ is optimally learned from the data with the goal of making the framelet coefficient matrix $\boldsymbol{C}$ sufficiently sparse so that its energy will concentrate only on the upper left block storing coefficients (along the diagonal direction of $\boldsymbol{C}$). A more parsimonious representation of the Framelets using dimension reduction can be found in Yin [144].

9.6.1.2.2 Deep convolutional framelets neural networks

Framelets provides a foundation for the development of deep convolutional framelets neural networks (FrameletsNN) [145]. If we relax the condition on matrices used in SVD, the following proposition will provide the formula for using framelets to represent a signal, thus allowing multistage implementation.

Proposition 9.8. *Let $\boldsymbol{\Phi}_{n\times m} = (\boldsymbol{\phi}_1, \boldsymbol{\phi}_2, \cdots, \boldsymbol{\phi}_m) \in \mathcal{R}^{n\times m}$ be a generalized nonlocal basis matrix paired with its dual matrix $\tilde{\boldsymbol{\Phi}} \in \mathcal{R}^{n\times m}$, and, similarly, let $\boldsymbol{V}_{d\times q} = (\boldsymbol{v}_1, \boldsymbol{v}_2, \cdots, \boldsymbol{v}_q)$ be a local basis matrix paired with its dual matrix $\tilde{\boldsymbol{V}} = (\tilde{\boldsymbol{v}}_1, \tilde{\boldsymbol{v}}_2, \cdots, \tilde{\boldsymbol{v}}_q) \in \mathcal{R}^{d\times q}$. Both satisfy the condition $\tilde{\boldsymbol{\Phi}}\boldsymbol{\Phi}^T = I_{n\times n}$ and $\tilde{\boldsymbol{V}}\boldsymbol{V}^T = I_{d\times d}$, namely,*

$$\tilde{\boldsymbol{\Phi}}\boldsymbol{\Phi}^T = \sum_{i=1}^{m} \tilde{\boldsymbol{\phi}}_i \boldsymbol{\phi}_i^T = I_{n\times n}, \tag{9.6.1.8}$$

$$\tilde{\boldsymbol{V}}\boldsymbol{V}^T = \sum_{j=1}^{q} \tilde{\boldsymbol{v}}_j \boldsymbol{v}_j^T = I_{d\times d}. \tag{9.6.1.9}$$

A signal $\boldsymbol{x}$ can be perfectly reconstructed as

$$\boldsymbol{x} = \frac{1}{d}\sum_{j=1}^{q} (\tilde{\Phi}\boldsymbol{c}_j) \star \tilde{\boldsymbol{v}}_j, \quad \boldsymbol{c}_j = \Phi^T(\boldsymbol{x} \star \overline{\boldsymbol{v}}_j), \tag{9.6.1.10}$$

and $\boldsymbol{x}$ can be further represented as

$$\boldsymbol{x} = \frac{1}{d}\sum_{j=1}^{q}\sum_{i=1}^{m} < \boldsymbol{x}, \boldsymbol{\phi}_i \star \boldsymbol{v}_j > \tilde{\boldsymbol{\phi}}_i \star \tilde{\boldsymbol{v}}_j. \tag{9.6.1.11}$$

Note that the frame conditions in Eqs. (9.6.1.8) and (9.6.1.9) can be further relaxed to $P_{R(\boldsymbol{\phi})}$ or $P_{R(V)}$, where $P_{R(S)}$ denotes the projection to the range space of S, namely, the set of possible linear combinations of its column vectors.

The encoder and decoder in a one-layer FrameletsNN

To elaborate further on perfect reconstruction, we can first write $H_d(\boldsymbol{x})$ as

$$H_d(\boldsymbol{x}) = \tilde{\Phi}\Phi^T H_d(\boldsymbol{x})\boldsymbol{V}\tilde{\boldsymbol{V}}^T = \tilde{\Phi}\boldsymbol{C}\tilde{\boldsymbol{V}}^T. \tag{9.6.1.12}$$

The middle term of Eq. (9.6.1.12), denoted as $\boldsymbol{C} = \Phi^T H_d(\boldsymbol{x})\boldsymbol{V}$, describes a one-layer encoder architecture as a single-input multiple-output (SIMO) system, and the reconstruction of $\boldsymbol{x}$ from $H_d(\boldsymbol{x}) = \tilde{\Phi}\boldsymbol{C}\tilde{\boldsymbol{V}}^T$ describes the decoder architecture as a multiple-input single-output (MISO) system. Together they form a one-layer FrameletsNN that will be described further as follows.

In the encoder layer, the convolution of one input signal $\boldsymbol{x}$ with a multichannel filter $\overline{\boldsymbol{V}}_{d\times q} = (\overline{\boldsymbol{v}}_1, \overline{\boldsymbol{v}}_2, \cdots, \overline{\boldsymbol{v}}_q)$ is a SIMO system that generates a multichannel output $\boldsymbol{Y}$ that can be written as

$$\begin{aligned} \boldsymbol{Y} &= (\boldsymbol{y}_1, \boldsymbol{y}_2, \cdots, \boldsymbol{y}_q) \\ &= (\boldsymbol{x} \star \overline{\boldsymbol{v}}_1, \boldsymbol{x} \star \overline{\boldsymbol{v}}_2, \cdots, \boldsymbol{x} \star \overline{\boldsymbol{v}}_q) \\ &= \boldsymbol{x} \star \overline{\boldsymbol{V}} = H_d(\boldsymbol{x})\boldsymbol{V}. \end{aligned} \tag{9.6.1.13}$$

Applying the nonlocal basis $\boldsymbol{\Phi}$ to the multichannel output yields the framelet coefficient matrix

$$\begin{aligned} \boldsymbol{C} &= (\boldsymbol{c}_1, \boldsymbol{c}_2, \cdots, \boldsymbol{c}_q) = \Phi^T \boldsymbol{Y} = \Phi^T \boldsymbol{x} \star \overline{\boldsymbol{V}} \\ &= (\Phi^T(\boldsymbol{x} \star \overline{\boldsymbol{v}}_1), \Phi^T(\boldsymbol{x} \star \overline{\boldsymbol{v}}_2), \cdots, \Phi^T(\boldsymbol{x} \star \overline{\boldsymbol{v}}_q)). \end{aligned} \tag{9.6.1.14}$$

The decoder architecture aims for perfect reconstruction of the original signal, $\boldsymbol{x}$, via a generalized inverse, $H_d^\dagger$, of H_d to yield $H_d^\dagger(H_d(\boldsymbol{x})) = \boldsymbol{x}$, which is a MISO system [144,145]. We can convert Eq. (9.6.1.12) back to

$$H_d(\boldsymbol{x}) = \tilde{\Phi}\boldsymbol{C}\tilde{\boldsymbol{V}}^T = \sum_{j=1}^{q} (\tilde{\Phi}\boldsymbol{c}_j)\tilde{\boldsymbol{v}}_j^T.$$

Further using Eq. (9.6.1.5), similar to Eq. (9.6.1.7) we have

$$\boldsymbol{x} = H_d^\dagger(H_d(\boldsymbol{x})) = \frac{1}{d}\sum_{j=1}^{q} (\tilde{\Phi}\boldsymbol{c}_j) \star \tilde{\boldsymbol{v}}_j = (\tilde{\Phi}\boldsymbol{C}) \star \tilde{\tilde{\boldsymbol{V}}},$$

where $\tilde{\tilde{\boldsymbol{V}}}$ is a vector defined from the matrix $\tilde{\boldsymbol{V}}$,

$$\tilde{\tilde{\boldsymbol{V}}} = \frac{1}{d}\begin{bmatrix} \tilde{\boldsymbol{v}}_1 \\ \tilde{\boldsymbol{v}}_2 \\ \vdots \\ \tilde{\boldsymbol{v}}_q \end{bmatrix} \in \mathcal{R}^{dq\times 1}.$$

In a single-stage frameletsNN, the search for a filter $\boldsymbol{V}$ is limited to the following space with positive framelet coefficients:

$$H = \{\boldsymbol{x} \in \mathcal{R}^n | \boldsymbol{x} = (\tilde{\Phi}\boldsymbol{C}) \star \tilde{\tilde{\boldsymbol{V}}}, \boldsymbol{C} = \Phi^T(\boldsymbol{x} \star \overline{\boldsymbol{V}}), [\boldsymbol{C}]_{ij} \geq 0\}.$$

Let $(\boldsymbol{x}_k; \boldsymbol{y}_k)$, $k = 1, \cdots, N$, be a training dataset, where $\boldsymbol{y}_k$ is the ground-truth signal corresponding to the kth input signal $\boldsymbol{x}_k$. The goal of FrameletsNN is to learn $(\boldsymbol{V}, \tilde{\boldsymbol{V}})$ from the training data through minimizing the cost function

$$\min_{(\boldsymbol{V},\tilde{\boldsymbol{V}})} \sum_{k=1}^{N} ||\boldsymbol{y}_k - F(\boldsymbol{x}_k; \boldsymbol{V}, \tilde{\boldsymbol{V}})||^2, \tag{9.6.1.15}$$

where $F(\boldsymbol{x}_k; \boldsymbol{V}, \tilde{\boldsymbol{V}}) = (\tilde{\Phi}\boldsymbol{C}) \star \tilde{\tilde{\boldsymbol{V}}}$ and $\boldsymbol{C} = \rho(\Phi^T(\boldsymbol{x} \star \overline{\boldsymbol{V}}))$. An activation function $\rho(\cdot)$, such as ReLU, is installed in the encoder to allow only positive framelet coefficients to participate in the reconstruction [145]. In the following, we will sometimes omit the activation functions in the equations to simplify notation, and only mark it out in the diagram figures.

The multiple-input multiple-output system

For p input signals $\boldsymbol{Z}_{n\times p} = (z_1, z_2, \cdots, z_p)$, each can be associated with a q-channel filter (each of length d), $\boldsymbol{V}_j = (\boldsymbol{v}_1^j, \boldsymbol{v}_2^j, \cdots, \boldsymbol{v}_q^j) \in \mathcal{R}^{d\times q}$, $j = 1, 2, \cdots, p$. Hence, the MIMO filter kernel is a block-structured matrix, or its index-reversed version; together they are defined as

$$\mathcal{V}_{pd\times q} = \begin{bmatrix} \boldsymbol{V}_1 \\ \boldsymbol{V}_2 \\ \vdots \\ \boldsymbol{V}_p \end{bmatrix}, \quad \overline{\mathcal{V}}_{pd\times q} = \begin{bmatrix} \overline{\boldsymbol{V}}_1 \\ \overline{\boldsymbol{V}}_2 \\ \vdots \\ \overline{\boldsymbol{V}}_p \end{bmatrix}, \tag{9.6.1.16}$$

where $\overline{\boldsymbol{V}}_j = [\overline{\boldsymbol{v}}_1^j, \overline{\boldsymbol{v}}_2^j, \cdots, \overline{\boldsymbol{v}}_q^j]$. The corresponding convolution can be written as

$$\boldsymbol{C} = \boldsymbol{Z} \star \overline{\mathcal{V}} = \sum_{j=1}^{p} H_d(\boldsymbol{z}_j)\boldsymbol{V}_j = H_{d|p}(\boldsymbol{Z})\mathcal{V}, \tag{9.6.1.17}$$

where $H_{d|p}(\boldsymbol{Z}) = (H_d(z_1), H_d(z_2), \cdots, H_d(z_p))$ is called an extended Hankel matrix. Each output vector of the MIMO system can be written as

$$\boldsymbol{c}_i = \sum_{j=1}^{p} (z_j \star \overline{\boldsymbol{v}}_i^j), \text{ for } i = 1, \cdots, q.$$

The encoder and decoder in the deeper layers for perfect reconstruction

The single-layer FrameletsNN described above can be expanded to add new layers (also called stages) of encoders and decoders in turn as MIMO systems to deepen the network with a varied number of channels in each layer. Write the output from the kth encoder layer as $\boldsymbol{C}^{(k)}$, $k = 1, 2, \cdots, K$. Also, denote the kth-layer local and nonlocal bases as $\mathcal{V}^{(k)}$ and $\boldsymbol{\Phi}^{(k)}$, respectively, with $\tilde{\mathcal{V}}^{(k)}$ and $\tilde{\boldsymbol{\Phi}}^{(k)}$ as their corresponding dual bases. $\boldsymbol{\Phi}^{(k)}$ and $\tilde{\boldsymbol{\Phi}}^{(k)}$ satisfy Eq. (9.6.1.8). $\mathcal{V}^{(k)}$ and $\tilde{\mathcal{V}}^{(k)}$ satisfy condition

$$\mathcal{V}^{(k)}\tilde{\mathcal{V}}^{(k)T} = I_{p_k d_k \times p_k d_k}, \quad k = 2, \cdots, K. \tag{9.6.1.18}$$

Note that the local basis of the first layer $\mathcal{V}^{(1)}$ is a matrix of size $d_1 \times q_1$ that satisfies Eq. (9.6.1.9) as we assume a single input at the beginning layer. The local basis of the kth layer ($k = 2, \cdots, K$) is a block matrix of size $d_k p_k \times q_k$ due to its nature as a MIMO system. For details with respect to the specification of parameters at each layer, the reader is referred to [145].

Under the new layered notations, the q_1-channel output from the encoder in the first layer can be rewritten as $\boldsymbol{C}^{(1)} = \boldsymbol{\Phi}^{(1)T}(\boldsymbol{x} \star \overline{\mathcal{V}}^{(1)})$. Similar to the first-layer FrameletsNN, $\boldsymbol{C}^{(1)}$ now becomes the p_2-channel input to a MIMO encoder; thus, $p_2 = q_1$. It generates a q_2-channel output of the second layer as $\boldsymbol{C}^{(2)} = \boldsymbol{\Phi}^{(2)T}(\boldsymbol{C}^{(1)} \star \overline{\mathcal{V}}^{(2)})$, and subsequently, the output from the encoder layers can be written as

$$\begin{aligned} \boldsymbol{C}^{(1)} &= \boldsymbol{\Phi}^{(1)T}(\boldsymbol{x} \star \overline{\mathcal{V}}^{(1)}), \\ \boldsymbol{C}^{(k)} &= \boldsymbol{\Phi}^{(k)T}(\boldsymbol{C}^{(k-1)} \star \overline{\mathcal{V}}^{(k)}), \quad \text{for } k > 1. \end{aligned} \tag{9.6.1.19}$$

Meanwhile, using the new layered notations, a MIMO decoder in the first layer can be rewritten as $\boldsymbol{x} = (\tilde{\boldsymbol{\Phi}}^{(1)T}\boldsymbol{C}^{(1)}) \star \tilde{\overline{\mathcal{V}}}^{(1)}$. The second-layer (the same for the following layers) reconstruction leads to the extended Hankel matrix

$$H_{d_2|p_2}(\boldsymbol{C}^{(1)}) = \tilde{\boldsymbol{\Phi}}^{(2)} \boldsymbol{C}^{(2)} \tilde{\mathcal{V}}^{(2)T}, \tag{9.6.1.20}$$

which, by way of Eq. (9.6.1.5), further leads to reconstructing its input $\boldsymbol{C}^{(1)}$ as

$$\begin{aligned} \boldsymbol{C}^{(1)} &= \frac{1}{d_2}\left[\sum_{j=1}^{q_2}(\tilde{\boldsymbol{\Phi}}^{(2)} \boldsymbol{c}_j^2) \star \tilde{\boldsymbol{v}}_j^{2,1}, \sum_{j=1}^{q_2}(\tilde{\boldsymbol{\Phi}}^{(2)} \boldsymbol{c}_j^2) \star \tilde{\boldsymbol{v}}_j^{2,2}, \cdots, \sum_{j=1}^{q_2}(\tilde{\boldsymbol{\Phi}}^{(2)} \boldsymbol{c}_{p_2}^2) \star \tilde{\boldsymbol{v}}_j^{2,p_2}\right] \\ &= (\tilde{\boldsymbol{\Phi}}^{(2)} \boldsymbol{C}^{(2)}) \star \tilde{\tilde{\mathcal{V}}}^{(2)}, \end{aligned} \tag{9.6.1.21}$$

where $\tilde{\tilde{\mathcal{V}}}^{(2)}$ is reshaped from $\tilde{\mathcal{V}}^{(2)}_{(p_2 d_2)\times q_2}$ as

$$\tilde{\tilde{\mathcal{V}}}^{(2)} = \frac{1}{d_2}\begin{bmatrix} \tilde{\boldsymbol{v}}_1^{2,1} & \cdots & \tilde{\boldsymbol{v}}_1^{2,p_2} \\ \tilde{\boldsymbol{v}}_2^{2,1} & \cdots & \tilde{\boldsymbol{v}}_2^{2,p_2} \\ \vdots & \vdots & \vdots \\ \tilde{\boldsymbol{v}}_{q_2}^{2,1} & \cdots & \tilde{\boldsymbol{v}}_{q_2}^{2,p_2} \end{bmatrix} \in \mathcal{R}^{(d_2 q_2)\times p_2}.$$

This operation will be continued in the deeper layers, so the output from each decoder layer yields

$$\begin{aligned} \boldsymbol{C}^{(k-1)} &= (\tilde{\boldsymbol{\Phi}}^{(k)T} \boldsymbol{C}^{(k)}) \star \tilde{\tilde{\mathcal{V}}}^{(k)}, \quad \text{for } k > 1, \\ \boldsymbol{x} &= (\tilde{\boldsymbol{\Phi}}^{(1)T} \boldsymbol{C}^{(1)}) \star \tilde{\tilde{\mathcal{V}}}^{(1)}. \end{aligned} \tag{9.6.1.22}$$

The diagram in Fig. 9.6.56 shows an encoder/decoder with three layers of perfect reconstruction ($K = 3$). The same module can be repeated after three stages in depth. Ideally, the reconstruction will be perfect; however, due to noise and the estimated local basis $\mathcal{V}$, the reconstruction is an approximate of the input, so we write it as $\hat{\boldsymbol{C}}^{(k)}$ for the kth-layer reconstruction.

The encoder and decoder layers for FrameletsNN

The framelet representation reveals that the nonlocal basis plays a key role in the efficiency of perfect reconstruction. The SVD of $H_d(\boldsymbol{x})$ exposes the limitation of using an average operator as a pooling operator in U-Nets as the basis does not satisfy the frame condition as required for a perfect reconstruction. This suggests the potential to replace the average pooling operator (and their corresponding unpooling operator) in each layer of CNN with the nonlocal basis $\boldsymbol{\Phi}$ ($\tilde{\boldsymbol{\Phi}}$) (layer is not specified) so that the condition in Definition 9.6 can be best satisfied. To do so, a nonlocal basis $\boldsymbol{\Phi}$ can be arranged as a composition of a tight filter bank in the order of the frequency content,

$$\boldsymbol{\Phi} = [\Phi_1 \quad \Phi_2 \quad \cdots \quad \Phi_L],$$

where $\Phi_l,\ l = 1, \cdots L$, is the lth subband operator, each of size $n \times n/L$ (assume the size n is divisible by L), such that

$$\Phi\Phi^T = \sum_{l=1}^{L} \Phi_l \Phi_l^T = c I_{n\times n}.$$

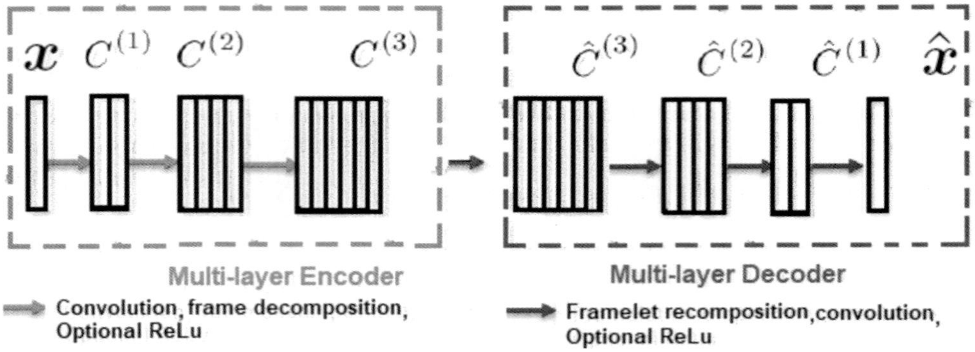

FIGURE 9.6.56

Multilayer encoder and decoder for FrameletsNN.

In the 1D case, let $L = 2$. We have the low-pass filter bank Φ_{low} and the high-pass filter Φ_{high} that satisfy $\Phi_{low}\Phi_{low}^T + \Phi_{high}\Phi_{high}^T = I$. If we use Haar wavelets, the matrices are given as

$$\boldsymbol{\Phi}_{low}^T = \frac{1}{\sqrt{2}} \begin{bmatrix} 1 & 1 & 0 & 0 & \cdots & 0 & 0 \\ 0 & 0 & 1 & 1 & \cdots & 0 & 0 \\ \vdots & \vdots & \vdots & \vdots & \ddots & \vdots & \vdots \\ 0 & 0 & 0 & 0 & \cdots & 1 & 1 \end{bmatrix}_{\frac{n}{2}\times n}, \tag{9.6.1.23}$$

$$\boldsymbol{\Phi}_{high}^T = \frac{1}{\sqrt{2}} \begin{bmatrix} 1 & -1 & 0 & 0 & \cdots & 0 & 0 \\ 0 & 0 & 1 & -1 & \cdots & 0 & 0 \\ \vdots & \vdots & \vdots & \vdots & \ddots & \vdots & \vdots \\ 0 & 0 & 0 & 0 & \cdots & 1 & -1 \end{bmatrix}_{\frac{n}{2}\times n}. \tag{9.6.1.24}$$

The following alternative form directly accounts for adding a bypass connection to fulfill the frame condition with the nonlocal basis, which provides high-pass content to compensate for the loss from the use of an average operator as a low-pass filter:

$$\boldsymbol{\Phi}_{ext} \equiv \begin{bmatrix} I & \boldsymbol{\Phi} \end{bmatrix} = \begin{bmatrix} I & \boldsymbol{\Phi}_{low} & \boldsymbol{\Phi}_{high} \end{bmatrix},$$

such that $\boldsymbol{\Phi}_{ext}\boldsymbol{\Phi}_{ext}^T = I + \boldsymbol{\Phi}\boldsymbol{\Phi}^T = I + \boldsymbol{\Phi}_{low}\boldsymbol{\Phi}_{low}^T + \boldsymbol{\Phi}_{high}\boldsymbol{\Phi}_{high}^T$. Then the framelet coefficients for a signal $\boldsymbol{x}$ can be expressed as

$$\begin{aligned} \boldsymbol{C}_{ext} &= \boldsymbol{\Phi}_{ext}^T H_d(\boldsymbol{x}) \boldsymbol{V} = \boldsymbol{\Phi}_{ext}^T(\boldsymbol{x} \star \overline{\boldsymbol{V}}) \\ &= \begin{bmatrix} (\boldsymbol{x} \star \overline{\boldsymbol{V}}) \\ \boldsymbol{\Phi}^T(\boldsymbol{x} \star \overline{\boldsymbol{V}}) \end{bmatrix} = \begin{bmatrix} (\boldsymbol{x} \star \overline{\boldsymbol{V}}) \\ \boldsymbol{\Phi}_{low}^T(\boldsymbol{x} \star \overline{\boldsymbol{V}}) \\ \boldsymbol{\Phi}_{high}^T(\boldsymbol{x} \star \overline{\boldsymbol{V}}) \end{bmatrix} = \begin{bmatrix} (\boldsymbol{x} \star \overline{\boldsymbol{V}}) \\ C_{low} \\ C_{high} \end{bmatrix}. \end{aligned}$$

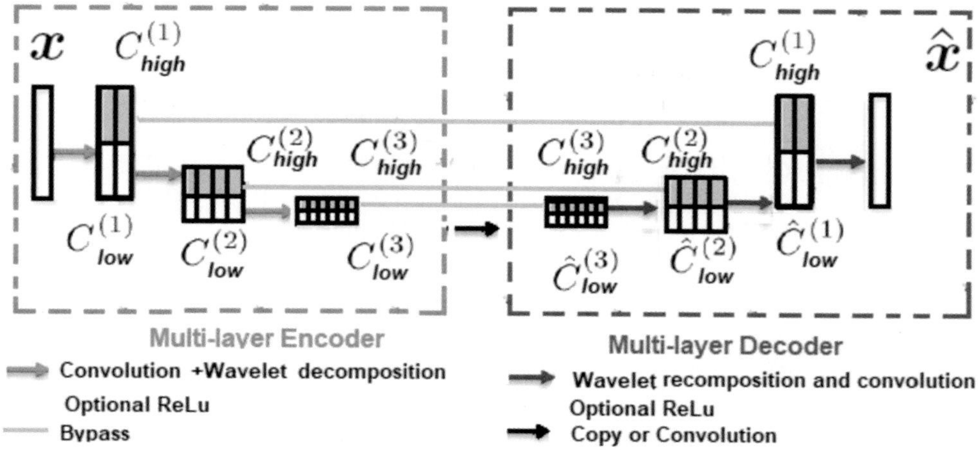

FIGURE 9.6.57

Multilayer encoder and decoder for FrameletsNN that combines frequency localization information.

This expression highlights three approaches participating in an encoder convolution layer in the FrameletsNN:

1. The bypass (skip) connection shown in the first row.
2. The low-pass filter shown in the second row that can be used as a pooling function. The output is much like the low-pass content from the wavelet decomposition; we write it as $\boldsymbol{C}_{low}$.
3. The high-pass filter in the third row that can be used in the unpooling function. The output is much like the high-pass content from the wavelet decomposition; we write it as $\boldsymbol{C}_{high}$.

In the decoder, the pooled data will be unpooled by the matrix $\tilde{\boldsymbol{\Phi}}$, which will be further concatenated to the bypass term for a perfect reconstruction. Thus, we have the concatenated output signal as

$$M_c = \begin{bmatrix} \boldsymbol{x} \star \overline{\boldsymbol{V}} & \tilde{\boldsymbol{\Phi}}\boldsymbol{\Phi}^T(\boldsymbol{x} \star \overline{\mathcal{V}}) \end{bmatrix} = \begin{bmatrix} \boldsymbol{x} \star \overline{\mathcal{V}} & \tilde{\Phi}\boldsymbol{C}_{low} & \tilde{\Phi}\boldsymbol{C}_{high} \end{bmatrix}.$$

The reconstructed signal is obtained as $\boldsymbol{y} = H_d^{\dagger}\left(M_c\tilde{\mathcal{V}}^T\right)$.

This analysis built the foundation for FrameletsNN to use residual networks as well as an MRA reconstruction method to deal with the problem encountered in perfect reconstruction. Namely, the estimated local basis hardly satisfies the required frame condition and typically only an approximation can be obtained. Hence, by combining frequency localization with added layers of encoder and decoder, FrameletsNN proceeds in an MRA manner to allow the low-pass frequency content to be further decomposed at the lower-level stage, concatenating the high-pass frequency component to the estimated low-pass content for reconstruction (see Fig. 9.6.57).

To write out the formulas for the MRA, we first decompose the kth-layer nonlocal orthonormal basis $\boldsymbol{\Phi}^{(k)} = [\boldsymbol{\Phi}_{low}^{(k)} \quad \boldsymbol{\Phi}_{high}^{(k)}]$, for which we use the Haar wavelet as in Eqs. (9.6.1.23) and (9.6.1.24);

hence, $\tilde{\boldsymbol{\Phi}}^{(k)} = \boldsymbol{\Phi}^{(k)}$. By way of Eq. (9.6.1.20), the following formulas provide the reconstruction of the input to each layer of the network:

$$H_{d_1}(\boldsymbol{x}) = \boldsymbol{\Phi}_{low}^{(1)} \boldsymbol{C}_{low}^{(1)} \tilde{\mathcal{V}}^{(1)T} + \boldsymbol{\Phi}_{high}^{(1)} \boldsymbol{C}_{high}^{(1)} \tilde{\mathcal{V}}^{(1)T},$$
$$H_{d_k|p_k}(\boldsymbol{C}_{low}^{(k-1)}) = \boldsymbol{\Phi}_{low}^{(k)} \boldsymbol{C}_{low}^{(k)} \tilde{\mathcal{V}}^{(k)T} + \boldsymbol{\Phi}_{high}^{(k)} \boldsymbol{C}_{high}^{(k)} \tilde{\mathcal{V}}^{(k)T}, \quad k = 2, \cdots, K.$$

The K layers of encoder of the FrameletsNN with the ReLU can be described as

$$\boldsymbol{C}_{low}^{(1)} = \rho(\boldsymbol{\Phi}_{low}^{(1)T} (\boldsymbol{x} \star \overline{\mathcal{V}}^{(1)})), \qquad \boldsymbol{C}_{high}^{(1)} = \boldsymbol{\Phi}_{high}^{(1)T} (\boldsymbol{x} \star \overline{\mathcal{V}}^{(1)}),$$
$$\boldsymbol{C}_{low}^{(k)} = \rho(\boldsymbol{\Phi}_{low}^{(k)T} (\boldsymbol{C}_{low}^{(k-1)} \star \overline{\mathcal{V}}^{(k)})), \quad \boldsymbol{C}_{high}^{(k)} = \boldsymbol{\Phi}_{high}^{(k)T} (\boldsymbol{C}_{low}^{(k-1)} \star \overline{\mathcal{V}}^{(k)}), \quad k = 2, \cdots, K.$$

Note that the resolution will be reduced by half in each layer of the encoder. To reverse to the original signal of higher-resolution content, at each stage, the low-frequency content is concatenated with the high-frequency one to yield the estimate $\hat{\boldsymbol{C}}^{(k)} = [\hat{\boldsymbol{C}}_{low}^{(k)} \quad \boldsymbol{C}_{high}^{(k)}]$, which will subsequently be used to reconstruct the low-frequency content in the higher layer. The K layers of decoder can be described as

$$\begin{aligned} \hat{\boldsymbol{C}}_{low}^{(k-1)} &= \rho((\tilde{\boldsymbol{\Phi}}^{(k)} \hat{\boldsymbol{C}}^{(k)}) \star \tilde{\tilde{\mathcal{V}}}^{(k)}), \quad k = 2, \cdots, K, \\ \hat{\boldsymbol{x}} &= \rho((\tilde{\boldsymbol{\Phi}}^{(1)} \hat{\boldsymbol{C}}^{(1)}) \star \tilde{\tilde{\mathcal{V}}}^{(1)}), \end{aligned} \tag{9.6.1.25}$$

where low-pass content estimated at the last layer can be directly bypassed to the decoder to concatenate it with the high-pass content from the higher layer for reconstruction, or a convolution filter can be applied to it before it is sent to the decoder at the same layer. Note the contrast with WNN, which uses wavelets to replace the activation layer that is not related to convolution. Using the nonlocal basis in the FrameletsNN allows it to incorporate high-frequency content into the reconstruction. The local basis can be trained by optimizing a lost function that is similar to Eq. (9.6.1.15) to account for the optimization over all the stages. For further details the reader is referred to [145].

9.6.2 Using wavelets in neural networks for image processing

The success of LeNet [130] launched CNN-based algorithms onto the center stage as the dominant technique in image processing. Capturing image features with localized weights made CNN significantly more efficient than its counterpart, the fully connected ANN. Further adding to CNN's growing success were the development of related algorithms and technological advancement. The availability of high-performance GPU computing systems stimulated the development of AlexNet [149] with deeper CNN architecture for image classification. VGG networks [150] went deeper, using as many as 19 layers of convolution while reducing a convolution field to a notably small size of 3 × 3. The batch normalization layers [151] extended its support for generic convolution blocks that include nonlinearities/pooling as well. Researchers have used generic convolution blocks as the backbone of CNN to construct a deeper network of increasing complexity with the intent of improving performance as the increasing number of network layers exponentially enhances the representation power [152]. More algorithms were further designed in order to fine tune the CNN with well-known improvements such as the network in network (NiN) [153], GoogleNet [154], residual network (ResNet) [155], densely connected network (DenseNet) [157], principled hierarchical (deep) dictionary learning[156], and metalearning [158,159,160,161], just to name a few.

Researchers further extended the utility of CNN-based object classification tasks to include image restoration. The task of retrieving a superresolution (SR) image from its lower-resolution (LR) version or denoising an image at the same resolution has found a wide variety of applications in many research fields. One such example is U-Net [134,162], a remarkable CNN-based algorithm, which has found much use in the medical imaging field among others. This algorithm includes a contracting path (left side) and an expansive path (right side) for image restoration, and many improvements and applications have since built upon U-Net.

Still, wavelets hold a distinct advantage over CNN in that they are able to capture scale-invariant features. Meanwhile, the CNN-based ResNet or DenseNet are able to account for the long short-term memory [155]. Researchers combined the advantage of wavelet transformation in T-F localization with the advantage of ResNet by adding a bypass to circumvent the problem that gradients are likely to disappear during an error in BP in deeper networks. An even bolder attempt by researchers was to combine the advantage of DenseNet with wavelets, further adding bypasses to connect the input from each layer to all the outputs of the subsequent layers to make the last layer densely connected to all preceding layers.

In the following subsections, we will discuss some research efforts that sought to embed wavelets into deeper neural networks for image restoration only. Readers interested in using wavelets for image registration/annotation are referred to [163,164].

9.6.2.1 Cascading wavelets with CNN-based algorithms

One straightforward solution for integrating the wavelets with CNN is to cascade them in a parallel fashion, namely, we first decompose the input signals/images into a selected wavelet basis/frame and, in turn, feed the coefficients into a CNN-based algorithm. To this end, two algorithms [165,166,167,168] were proposed almost simultaneously.

9.6.2.1.1 Deep wavelet superresolution

Deep wavelet superresolution (DWSR) is a CNN-based ResNet algorithm [165] to capture the residuals of wavelet coefficients between an LR image and its corresponding true SR version. The 2D DWT (Haar wavelets are used) is applied to the upsampled LR image to generate four low-resolution subband (LRSB) wavelet coefficient images: LA (average), LV (vertical), LH (horizontal), and LD (diagonal). The LSRBs are fed into a sequence of convolution filters. The first layer has a 64-channel filter of size $4 \times 3 \times 3$, the last layer has a four-channel filter of size $64 \times 3 \times 3$, and the middle N layers ($N = 10$ was empirically selected) all have a 64-channel filter of size $64 \times 3 \times 3$. Zero padding at each layer is applied for the output to have the same image size as the subband wavelet coefficients, and no pooling/unpooling is applied. In the algorithm, the 2D DWT is applied to the corresponding high-resolution true image in parallel to receive the true superresolution subbands (SRSBs), labeled as SA, SV, SH, and SD.

The algorithm learns the four subband residuals, $\Delta\hat{S}B$, from the true version, ΔSB, the difference between the true SRSBs and the LRSBs, in an iterative manner. The parameters (weights and bias) of the network, $(\boldsymbol{\Theta}, \boldsymbol{b})$, minimizing the cost function are obtained as,

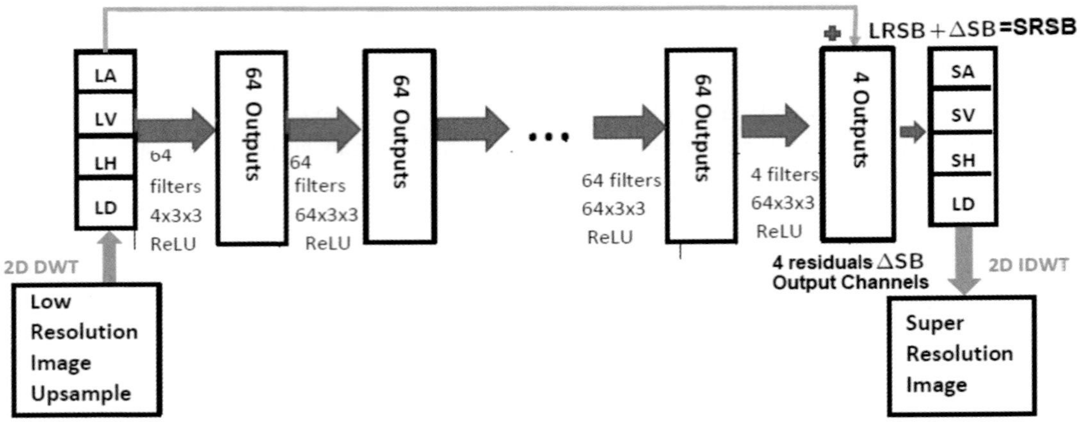

FIGURE 9.6.58

The network structure of DWSR. The algorithm used about 10 convolution layers.

$$(\boldsymbol{\Theta}, \boldsymbol{b}) = \underset{(\boldsymbol{\Theta}, \boldsymbol{b})}{\operatorname{argmin}} \frac{1}{2} ||\Delta SB - \Delta \hat{S} B||_2^2 + \lambda ||\boldsymbol{\Theta}||_2^2,$$

for selected $\lambda > 0$. Fig. 9.6.58 provides the details for the DWSR network architecture. The estimated residual SRSBs are further added back to the LRSBs to yield the estimated SRSBs as follows:

$$\begin{aligned} & \text{Estimated SRSBs } \{\hat{S}A, \hat{S}V, \hat{S}H, \hat{S}D\} \\ = \; & LRSB + \Delta\hat{S}B \\ = \; & \{LA + \hat{\Delta S}A, LV + \hat{\Delta S}V, LH + \hat{\Delta S}H, LD + \hat{\Delta S}D\}. \end{aligned} \quad (9.6.2.1)$$

9.6.2.1.2 Wavelet residual network

The wavelet residual network (WavResNet) [166,167,168] algorithm is another CNN-based algorithm that cascades wavelet coefficients with a CNN-based network in a parallel fashion. It utilizes more layers and bypass paths than DWSR. Because CNN learning can benefit from the deep learning on features in wavelet subbands, WavResNet uses multiple convolution layers together with five convolution module units for denoising, which amount to a total of 20 convolution layers for training, making it significantly deeper than DWSR. In addition, WavResNet uses a 320-channel filter while DWSR used a 64-channel filter in each convolution layer, except in the last layer, where only a four-channel filter is used to output the subbands. Thus, WavResNet is designed to capture more salient features.

For consistency, we use the same notations for the subbands as used in DWSR. WavResNet also uses the four subband wavelet coefficients as input, labeled as $X = (LA, LH, SV, LD)$. It estimates the wavelet coefficients of the noise ϵ, namely the difference between the subbands of the noisy input and those of the clean version for the task of image denoising. The wavelet coefficients of a denoised image are inverted as $Y = X - \epsilon$. This reduces the topological complexity of the label manifolds.

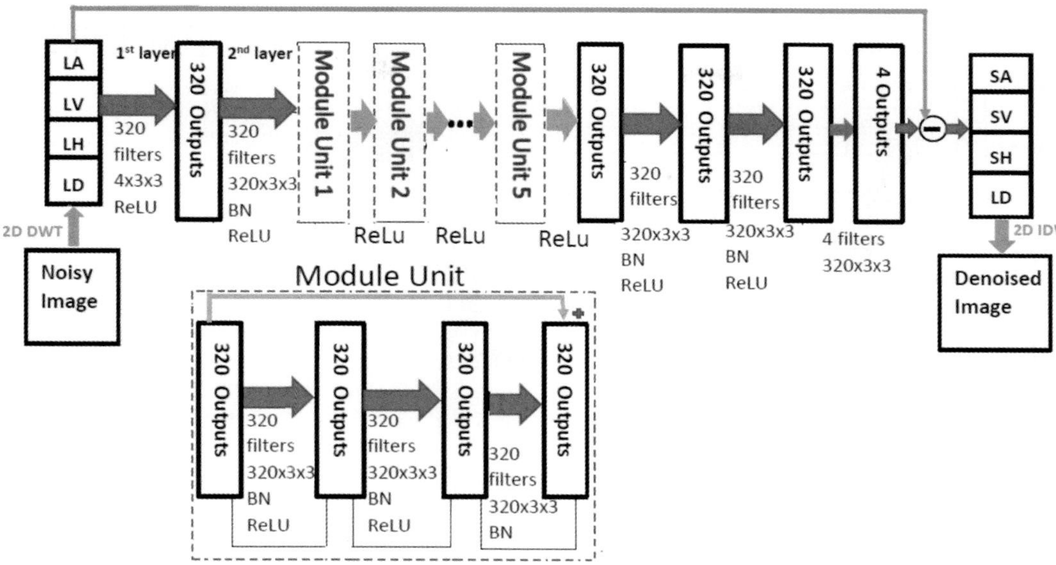

FIGURE 9.6.59

The first row shows the network structure of WavResNet. The second row shows the details inside a module unit that is used in WaveResNet.

WavResNet can also be used to reconstruct SR images. To this end, an LR image needs to be upsampled by bicubic interpolation. WavResNet increases the number of module units used in the network for SR reconstruction. There are two versions of WavResNet. One version adds the bypass link from the original input to the end output while each convolution module unit also utilizes a bypass link adding the input of each module to its output. Fig. 9.6.59 shows the structure diagram of this version for image denoising. The other version uses a network that is more similar to DenseNet. It concatenates all outputs from the modules in the intermediate layers to the end output to be further processed by convolution filters (diagram can be found in [166,167]).

In both DWSR and WavResNet, the wavelet coefficients are obtained first and fed into the neural networks as the input. Namely, the neural networks use the preprocessed wavelet coefficients as their input. Hence they are compatible with flexible use of wavelets. As an example, directional wavelets are also used [167] as an input to WavResNet for low-dose X-ray CT image reconstruction. However, it is worth to mention that the pooling/unpooling layer is missing in both DWSR and WavResNet, which makes them somewhat less effective.

9.6.2.2 Deep convolutional framelets neural networks

As mentioned earlier, with some tedious algebraic calculation, 1D FrameletsNN [145,169] can be extended to process 2D images. For 1D FrameletsNN, if the nonlocal and local bases satisfy the given frame conditions in Eqs. (9.6.1.8) and (9.6.1.9), in a single-layer FrameletsNN architecture, the perfect

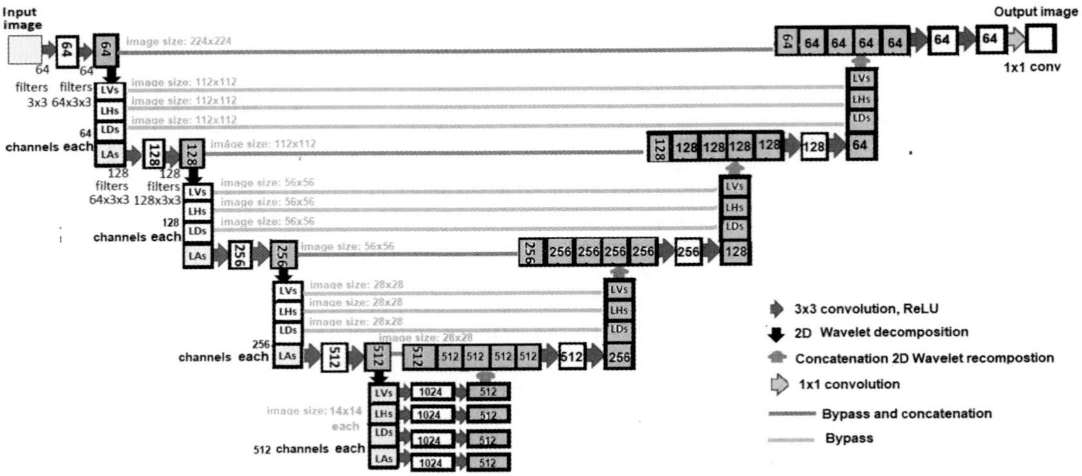

FIGURE 9.6.60

The network structure of FrameletsNN. In the contracting path, the input to each layer in depth is highlighted in yellow (light gray in print version), while the images to be bypassed to the decoder layers are highlighted in gray. Assume one input image of 224 × 224. Image size in each depth stage is marked in orange (mid gray in print version). The number inside each box indicates the number of output channels.

reconstruction of a signal, combining the encoder and decoder, can be written as

$$x = H_d^{\dagger}\left(\tilde{\boldsymbol{\Phi}}\boldsymbol{\Phi}^T H_d(x) V \tilde{V}^T\right).$$

Similarly, when the nonlocal and local bases satisfy the given frame conditions in Eqs. (9.6.1.8) and (9.6.1.18), the perfect reconstruction of a 2D image $X \in \mathcal{R}^{n \times p}$ yields

$$X = H_{d|p}^{\dagger}\left(\tilde{\boldsymbol{\Phi}}\boldsymbol{\Phi}^T H_{d|p}(X) V \tilde{V}^T\right).$$

FrameletsNN for the 2D case, shown in Fig. 9.6.60, is an extension of its 1D version shown in Fig. 9.6.58. Deeper layers are added in a similar fashion as in the 1D case. The 2D version intuitively visualizes this U-Net-based algorithm. In the contracting path (encoder), the nonlocal basis as a generalized pooling in each stage reduces both the length and width of an image by half to arrive to the four subbands in the next stage. The low-frequency content (LA) is further decomposed using the nonlocal basis that is specified to the layer. This works as a generalized pooling operation following an MRA manner. Meanwhile, the high-pass wavelet coefficients (LV, LH, LD) are bypassed to the expansive path (decoder) to be concatenated to the estimated low-pass content. However, the perfect reconstruction is not guaranteed, making the reconstruction similar to the wavelet reconstruction while taking advantage of the CNN to search for the optimal local basis when training the network.

FrameletsNN makes it essential to select a nonlocal basis $\boldsymbol{\Phi}$ as the basis for SVD to enable energy compaction. This procedure allows it to naturally substitute the pooling layer with wavelets including

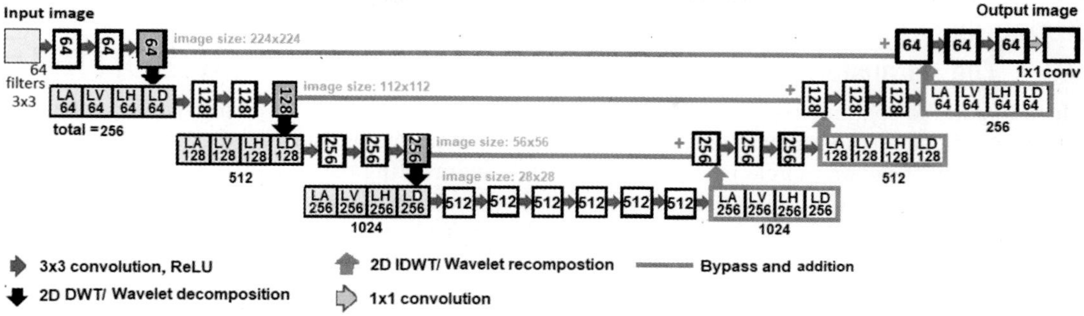

FIGURE 9.6.61

The network structure of MWNN. Assume one input image of 224 × 224. Image size in each depth stage is marked in orange (mid gray in print version). The number of output images is shown inside or under a box.

both low-pass and high-pass spectral spatial information, and this procedure is repeated following an MRA manner. This, when compared to the average pooling/unpooling operation, further addresses the limitations of U-Net and provides a significantly improved image reconstruction method.

9.6.2.3 *Multilevel wavelet CNN*

Multilevel wavelet CNN (MWCNN) [170] utilizes DWT to replace the pooling operations in the CNN. The weights Θ used in MWCNN are optimized by minimizing the cost function defined as

$$L(\Theta) = \frac{1}{2N}\sum_{i=1}^{N} ||F(\mathbf{y}_i;\Theta) - (\mathbf{y}_i - \mathbf{x}_i)^2]||.$$

Much like the FrameletsNN, the algorithm has a structure based on the U-Net architecture (see Fig. 9.6.61) and uses 2D DWT to replace the average pooling function as an improvement over U-Net. What makes it significantly different from FrameletsNN is that while FrameletsNN only applies the convolution to the low-pass content and skips the high-pass content to the decoder layer, MWCNN concatenates all four bands of the wavelet coefficients and applies the convolution to the concatenated block. Later, the output from the encoder is added to the wavelet recomposition. It is worthwhile to mention that this operation is used only in the last layer (stage) in FrameletsNN for dealing with image denoising.

The overall effectiveness of MWCNN, especially in preserving detailed texture, provides evidence further proving the advantages of infusing wavelets in all the deep stages (layers) of a CNN-based algorithm. The deeply embedded wavelets along the contracting/expansive path are tantamount to tentative interactions between convolution and wavelets, leading to a reduction in the resolution of feature maps while also increasing the size of the receptive field in the encoder/decoder layers. Ultimately these benefit computational efficiency, making such techniques by far the most efficient way to utilize the spatial frequency content captured by wavelets together with CNN. This coincides with the most recently developed VNN [131,132], which addresses the pooling/unpooling layers using Volterra filters directly in a convolution layer, pointing to a new direction for neural network development.

Appendix 9.A

9.A.1 The z-transform

Much like in the continuous-time domain, we can cover the whole complex plane by using a Laplace transform in contrast to the harmonic Fourier transform along the imaginary axis. We may cover the whole interior of the unit circle in the discrete domain by invoking the so-called z-transform. Due to the direct relevance to the derivations of the various multiscale transforms in this chapter, we define the z-transform in the 2D plane. Given a 2D discretized signal $x(\boldsymbol{n}),\ \boldsymbol{n} \in \mathcal{Z}^2$, its z-transform $X(z),\ z \in \mathcal{Z}^2$, is defined as

$$X(z) = \sum_{n \in \mathcal{Z}^2} = x(\boldsymbol{n}) z^{-\boldsymbol{n}} = \sum_{n \in \mathcal{Z}^2} x(\boldsymbol{n}) z_1^{-n_1} z_2^{-n_2}, \text{ with } \boldsymbol{n} = (n_1, n_2), z = (z_1, z_2).$$

Note that the z-transform of an image $x(\boldsymbol{n})$ is reduced to its discrete-time Fourier series when we evaluate its z-transform function at $z = e^{j\omega}$.

Given a discrete 2D image $x(\boldsymbol{n})$ as an input to a 2D linear discrete invariant system with an impulse response of $h(\boldsymbol{n})$, the resulting output is the following convolution sum:

$$y(\boldsymbol{n}) = x(\boldsymbol{n}) * h(\boldsymbol{n}) = \sum_{s \in \mathcal{Z}^2} x(s_1, s_2) h(n_1 - s_1, n_2 - s_2).$$

The z-transform of the output $y(\boldsymbol{n})$ of the system is (similarly to the continuous analog) $Y(z) = H(z)X(z)$, where the z-transform of the impulse response $h(\boldsymbol{n})$ is written as $H(z)$ and called the transfer function of the filter system. The z-transform is a very widely used tool in the analysis and design of discrete filter systems.

References

[1] D.H. Hubel, T.N. Wiesel, Receptive fields binocular interaction and functional architecture in the cats visual cortex, J. Physiol. 160 (1962) 106–154.

[2] B. Olshausen, D. Field, Emergence of simple-cell receptive led properties by learning a sparse code for natural images, Nature 381 (1996) 607–609.

[3] B. Olshausen, D. Field, Sparse coding of sensory inputs, Curr. Opin. Neurobiol. 14 (2004) 481–487.

[4] E.J. Candès, D. Donoho, Curvelets - a surprisingly effective nonadaptive representation for objects with edges, in: A. Cohen, C. Rabut, L. Schumaker (Eds.), Curves and Surface Fitting: Saint-Malo 1999, Vanderbilt Univ. Press, Nashville, 2000, pp. 105–120.

[5] E.J. Candès, D.L. Donoho, New tight frames of curvelets and optimal representations of objects with c2 singularities, Commun. Pure Appl. Math. 57 (2) (2004) 219–266.

[6] E. Candès, D. Donoho, Continuous curvelet transform: I. Resolution of the wavefront set, Appl. Comput. Harmon. Anal. 19 (2003) 162–197.

[7] E. Candès, D. Donoho, Continuous curvelet transform: II. Discretization and frames, Appl. Comput. Harmon. Anal. 19 (2003) 198–222.

[8] M.N. Do, M. Vetterli, in: G.V. Welland (Ed.), Contourlets in Beyond Wavelets, Academic Press, 2003.

[9] M.N. Do, M. Vetterli, The contourlet transform an efficient directional multiresolution image representation, IEEE Trans. Image Process. 14 (2005) 2091–2106.

[10] K. Guo, D. Labate, W. Lim, G. Weiss, E. Wilson, Wavelets with composite dilations, Electron. Res. Announc. Am. Math. Soc. 10 (2004) 78–87.
[11] K. Guo, D. Labate, W. Lim, G. Weiss, E. Wilson, Wavelets with composite dilations and their MRA properties, Appl. Comput. Harmon. Anal. 20 (2006) 202–236.
[12] D. Labate, W. Lim, G. Kutyniok, G. Weiss, Sparse multidimensional representation using shearlets, in: SPIE Conf. Wavelets XI, San Diego, CA, USA, 2005, pp. 254–262.
[13] S. Yi, D. Labate, G.R. Easley, H. Krim, A shearlet approach to edge analysis and detection, IEEE Trans. Image Process. 18 (2009) 929–941.
[14] http://www.vissta.ncsu.edu/.
[15] A. Papoulis, Signal Analysis, Mc Graw Hill, 1977.
[16] E. Brigham, The Fast Fourier Transform, Prentice Hall, 1974.
[17] D. Gabor, Theory of communication, J. IEE 93 (1946) 429–457.
[18] S. Mallat, A Wavelet Tour of Signal Processing, Academic Press, 1997.
[19] I. Daubechies, S. Mallat, A. Willsky, Special edition on wavelets and applications, IEEE Trans. Inf. Theory (1992).
[20] I. Daubechies, Ten Lectures on Wavelets, SIAM, Philadelphia, PA, 1992.
[21] M. Vetterli, J. Kovacevic, Wavelets and Subband Coding, Prentice Hall, Englewood Cliffs, NJ, 1995.
[22] H. Krim, On the distribution of optimized multiscale representations, in: Proc. IEEE Int. Conf. on Acoustics Speech & Signal Processing ICASSP, vol. V, Munich, Germany, May 1997.
[23] G.H. Golub, C.F. VanLoan, Matrix Computations, The Johns Hopkins University Press, Baltimore, Maryland, 1984.
[24] D. Luenberger, Optimization by Vector Space Methods, J. Wiley, New York, London, 1968.
[25] K. Gröchenig, Acceleration of the frame algorithm, IEEE Trans. Signal Process. 41 (Dec. 1993) 3331–3340.
[26] A.B. Hamza, H. Krim, Image denoising: a nonlinear robust statistical approach, IEEE Trans. Signal Process. 49 (2001) 3045–3054.
[27] Y. Meyer, Ondelettes et opérateur, vol. 1, Hermann, Paris, 1990.
[28] S. Mallat, A theory for multiresolution signal decomposition: the wavelet representation, IEEE Trans. Pattern Anal. Mach. Intell. PAMI-11 (Jul. 1989) 674–693.
[29] Y. Meyer, Wavelets and Applications, first ed., SIAM, Philadelphia, 1992.
[30] Y. Meyer, Wavelets and Operators, Cambridge University Press, Cambridge, 1993.
[31] F.J. Hampson, J.C. Pesquet, A nonlinear decomposition with perfect reconstruction, preprint, 1998.
[32] P.L. Combettes, J.C. Pesquet, Convex multiresolution analysis, IEEE Trans. Pattern Anal. Mach. Intell. 20 (Dec. 1998) 1308–1318.
[33] W. Sweldens, The lifting scheme: a construction of second generation wavelets, SIAM J. Math. Anal. 29 (9) (1997) 511–546.
[34] S. Mallat, Multiresolution approximation and wavelet orthonormal bases of $L^2(\mathbb{R})$, Trans. Am. Math. Soc. 315 (Sep. 1989) 69–87.
[35] J. Woods, S. O'Neil, Sub-band coding of images, IEEE Trans. Acoust. Speech Signal Process. 34 (May 1986) 1278–1288.
[36] M. Vetterli, Multidimensional subband coding some theory and algorithms, Signal Process. 6 (April 1984) 97–112.
[37] F. Meyer, R. Coifman, Brushlet a tool for directional image analysis and image compression, Appl. Comput. Harmon. Anal. (1997) 147–187.
[38] G. Strang, T. Nguyen, Wavelets and Filter Banks, first ed., Wellesley-Cambridge Press, Boston, 1996.
[39] P.P. Vaidyanathan, Multirate digital filters filter banks polyphase networks and applications a tutorial, Proc. IEEE 78 (Jan. 1990) 56–93.
[40] A. Cohen, I. Daubechies, J.C. Feauveau, Biorthogonal bases of compactly supported wavelets, 1992.
[41] P.P. Vaidyanathan, Multirate Systems and Filter Banks, Prentice Hall, New Jersey, 1992.

[42] R. Coifman, Y. Meyer, Remarques sur l'analyse de Fourier à fenêtre, C. R. Acad. Sci., Sér. I (1991) 259–261.
[43] R.R. Coifman, M.V. Wickerhauser, Entropy-based algorithms for best basis selection, IEEE Trans. Inf. Theory IT-38 (Mar. 1992) 713–718.
[44] M.V. Wickerhauser, INRIA lectures on wavelet packet algorithms, in: Ondelettes et paquets d'ondelettes, Roquencourt, Jun. 1991, pp. 31–99.
[45] H. Malvar, Lapped transforms for efficient transform subband coding, IEEE Trans. Acoust. Speech Signal Process. ASSP-38 (Jun. 1990) 969–978.
[46] A. Aldroubi, E.M. Unser, Wavelets in Medicine and Biology, CRC Press, 1996.
[47] A. Akansu, E.M.J. Smith, Subband and Wavelet Transforms, Kluwer, 1995.
[48] A. Arneodo, F. Argoul, J.E.E. Bacry, J. Muzy, Ondelettes Multifractales et Turbulence, Diderot, Paris, France, 1995.
[49] A. Antoniadis, E.G. Oppenheim, Wavelets and Statistics, Lecture Notes in Statistics, Springer Verlag, 1995.
[50] P. Mueller, B. Vidakovic (Eds.), Bayesian Inference in Wavelet Based Models, first ed., Lecture Notes in Statistics, vol. 141, Springer-Verlag, 1999.
[51] B. Vidakovic, Statistical Modeling by Wavelets, John Wiley, New York, 1999.
[52] K. Ramchandran, M. Vetterli, Best wavelet packet bases in a rate-distorsion sense, IEEE Trans. Image Process. 2 (April 1993).
[53] J. Shapiro, Embedded image coding using zerotrees of wavelet coefficients, IEEE Trans. Signal Process. 41 (1993) 3445–3462.
[54] P. Cosman, R. Gray, M. Vetterli, Vector quantization of image subbands: a survey, IEEE Trans. Image Process. 5 (Feb. 1996) 202–225.
[55] A. Kim, H. Krim, Hierarchical stochastic modeling of SAR imagery for segmentation/compression, IEEE Trans. Signal Process. 47 (Feb. 1999) 458–468.
[56] M. Basseville, M. Benveniste, A. Chou, K. Golden, R. Nikoukhah, A.S. Willsky, Modeling and estimation of multiresolution stochastic processes, IEEE Trans. Inf. Theory IT-38 (Mar. 1992) 529–532.
[57] M. Luettgen, W. Karl, A. Willsky, Likelihood calculation for multiscale image models with applications in texture discrimination, IEEE Trans. Image Process. (July 1994).
[58] E. Fabre, New fast smoothers for multiscale systems, IEEE Trans. Signal Process. 44 (Aug. 1995) 1893–1911.
[59] P. Fieguth, W. Karl, A. Willsky, C. Wunsch, Multiresolution optimal interpolation and statistical analysis of topex/Poseidon satellite altimetry, IEEE Trans. Geosci. Remote Sens. 33 (March 1995) 280–292.
[60] M.K. Tsatsanis, G.B. Giannakis, Principal component filter banks for optimal multiresolution analysis, IEEE Trans. Signal Process. 43 (Aug. 1995) 1766–1777.
[61] M.K. Tsatsanis, G.B. Giannakis, Optimal linear receivers for DS-CDMA systems a signal processing approach, IEEE Trans. Signal Process. 44 (Dec. 1996) 3044–3055.
[62] A. Scaglione, G.B. Giannakis, S. Barbarossa, Redundant filterbank precoders and equalizers, parts I and II, IEEE Trans. Signal Process. 47 (July 1999) 1988–2022.
[63] A. Scaglione, S. Barbarossa, G.B. Giannakis, Filterbank transceivers optimizing information rate in block transmissions over dispersive channels, IEEE Trans. Inf. Theory 45 (April 1999) 1019–1032.
[64] R. Learned, H. Krim, A. Willsky, W. Karl, Wavelet packet based multiple access communication, in: Wavelet Applications in Signal and Image Processing, vol. II, Oct. 1994, pp. 246–259.
[65] R. Learned, A. Willsky, D. Boroson, Low complexity optimal joint detection for oversaturated multiple access communications, IEEE Trans. Signal Process. 45 (Jan 1997).
[66] A. Lindsey, Wavelet packet modulation for orthogonally multiplexed communication, IEEE Trans. Signal Process. 45 (May 1997).
[67] K. Wong, J. Wu, T. Davidson, Q. Jin, Wavelet packet division multiplexing and wavelet packet design under timing error effects, IEEE Trans. Signal Process. 45 (Dec. 1997) 2877–2886.
[68] H. Krim, I. Schick, Minimax description length for signal denoising and optimized representation, IEEE Trans. Inf. Theory (April 1999) 809–908, Special Issue.

[69] D.L. Donoho, I.M. Johnstone, Ideal spatial adaptation by wavelet shrinkage, preprint Dept. Stat., Stanford Univ., Jun. 1992.

[70] H. Krim, S. Mallat, D. Donoho, A. Willsky, Best basis algorithm for signal enhancement, in: Proc. IEEE Int. Conf. on Acoustics Speech & Signal Processing, ICASSP'95, Detroit, MI, IEEE, May 1995.

[71] N. Saito, Local feature extraction and its applications using a library of bases, PhD thesis, Yale University, Dec. 1994.

[72] H. Krim, J.C. Pesquet, On the statistics of best bases criteria, in: Wavelets in Statistics, in: Lecture Notes in Statistics, Springer-Verlag, July 1995.

[73] J. Rissanen, Modeling by shortest data description, Automatica 14 (1978) 465–471.

[74] B. Vidakovic, Nonlinear wavelet shrinkage with Bayes rules and Bayes, J. Am. Stat. Assoc. 93 (1998) 173–179.

[75] D. Leporini, J.C. Pesquet, H. Krim, Best basis representation with prior statistical models, in: Bayesian Inference in Wavelet-Based Models, in: Lecture Notes in Statistics, Springer-Verlag, 1999, ch. 11.

[76] J. Rissanen, Stochastic complexity and modeling, Ann. Stat. 14 (1986) 1080–1100.

[77] P. Huber, Robust estimation of a location parameter, Ann. Math. Stat. 35 (1964) 1753–1758.

[78] P. Huber, Théorie de l'inférence statistique robuste, tech. rep., Univ. de Montreal, Montreal, Quebec, Canada, 1969.

[79] D. Donoho, I. Johnstone, Adapting to unknown smoothness via wavelet shrinkage, J. Am. Stat. Assoc. 90 (Dec. 1995) 1200–1223.

[80] H. Krim, D. Tucker, S. Mallat, D. Donoho, Near-optimal risk for best basis search, IEEE Trans. Inf. Theory (Nov. 1999).

[81] E.J. Candès, D. Donoho, Ridgelets: a key to higher-dimensional intermittency?, R. Soc. Lond. Philos. Trans. Ser. A Math. Phys. Eng. Sci. 357 (1999) 2495–2509.

[82] H.A. Smith, A parametrix construction for wave equations with $c^{1,1}$ coefficients, Ann. Inst. Fourier (Grenoble) 48 (1998) 797–835.

[83] E. Candès, L. Demanet, D. Donoho, L. Ying, Fast discrete curvelet transforms, Multiscale Model. Simul. 5 (2006) 861–899.

[84] S. Mallat, Adaptive Wavelet Collocation for Nonlinear A Wavelet Tour of Signal Processing, Academic Press, San Diego, CA, USA, 1998.

[85] J. Ma, G. Plonka, Curvelets transform: a review of recent applications, IEEE Signal Process. Mag. (2010) 118–133.

[86] M. Fadili, J. Starck, Curvelets and ridgelets, Encycl. Complex. Syst. Sci. 3 (2009) 1718–1738.

[87] F.J. Herrmann, P.P. Moghaddam, C.C. Stolk, Sparsity- and continuity-promoting seismic image recovery with curvelet frames, Appl. Comput. Harmon. Anal. 24 (2008) 150–173.

[88] http://www.curvelet.org/software.html.

[89] E.J. Candès, F. Guo, New multiscale transforms, minimum total variation synthesis: applications to edge-preserving image reconstruction, Signal Process. 82 (11) (2002) 1519–1543, special issue on Image and Video Coding Beyond Standards.

[90] E.J. Candès, D.L. Donoho, Recovering edges in ill-posed inverse problems: optimality of curvelet frames, Ann. Stat. 30 (2002) 784–842.

[91] E.J. Candès, L. Demanet, Curvelets and Fourier integral operators, C. R. Math. Acad. Sci. Paris, Ser. C. R. Math. 336 (5) (2003) 395–398.

[92] E.J. Candès, D.L. Donoho, Curvelets: new tools for limited-angle tomography, manuscript, 2004.

[93] E. Candès, J. Romberg, T. Tao, Stable signal recovery from incomplete and inaccurate information, Commun. Pure Appl. Math. 59 (2005) 1207–1233.

[94] L. Demanet, Curvelets wave atoms and wave equations, https://resolver.caltech.edu/CaltechETD:etd-05262006-133555f, 2006.

[95] J. Ma, A. Antoniadis, F.-X.L. Dimet, Curvelet-based multiscale detection and tracking for geophysical fluids, IEEE Trans. Geosci. 44 (12) (2006) 3626–3637.

[96] J. Ma, Curvelets for surface characterization, Appl. Phys. Lett. 90 (2007) 054109.
[97] J. Ma, Deblurring using singular integrals and curvelet shrinkage, Phys. Lett. A 368 (2007) 245–250.
[98] J.L. Starck, E.J. Candès, D.L. Donoho, The curvelet transform for image denoising, IEEE Trans. Image Process. 11 (6) (2002) 670–684.
[99] J. Starck, M. Elad, D. Donoho, Image decomposition via the combination of sparse representation and a variational approach, IEEE Trans. Image Process. 14 (10) (2005) 1570–1582.
[100] C. Zhang, L. Cheng, Z. Qiu, L. Cheng, Multipurpose watermarking based on multiscale curvelet transform, IEEE Trans. Inf. Forensics Secur. 3 (4) (2008) 611–619.
[101] Y. Bao, H. Krim, Upon bridging scalespace to multiscale frame analysis, in: Wavelets in Signal and Image Analysis from Theory to Practice, Computational Imaging and Vision, vol. 19, 2001, Chapter 6.
[102] M. Elad, Why simple shrinkage is still relevant for redundant representations, IEEE Trans. Inf. Theory 52 (2006) 5559–5569.
[103] Y. Lu, M.N. Do, Multidimensional directional filter banks and surfacelets, IEEE Trans. Image Process. 16 (2007) 918–931.
[104] H.S. Malvar, Signal Processing with Lapped Transforms, Artech House, Norwood, MA, 1992.
[105] K.-O. Cheng, N.-F. Law, W.-C. Siu, Multiscale directional filter bank with applications to structured and random texture retrieval, Pattern Recognit. 40 (2006) 1182–1194.
[106] P.J. Burt, E.H. Adelson, The Laplacian pyramid as a compact image code, IEEE Trans. Commun. 31 (1983) 532–540.
[107] R.H. Bamberger, M.J.T. Smith, A filter bank for the directional decomposition of images: theory and design, in: Proc. IEEE Int. Conf. Acoust. Speech and Signal Proc., 1999, pp. 1417–1420.
[108] A.L. Cunha, M.N. Do, A new directional filterbank for image analysis and classification, IEEE Trans. Image Process. 48 (2005).
[109] S. Durand, M-band filtering and nonredundant directional wavelets, Appl. Comput. Harmon. Anal. 22 (2006) 124–139.
[110] S.K. Mitra, Digital Signal Processing: A Computer-Based Approach, Mc Graw Hill Higher Education, 2004.
[111] M.N. Do, M. Vetterli, Framing pyramids, IEEE Trans. Signal Process. 51 (2003) 2329–2342.
[112] S. Park, M.J.T. Smith, R.M. Mersereau, The contourlet transform: an efficient directional multiresolution image representation, IEEE Trans. Image Process. 48 (2005) 797–835.
[113] M.N. Do, Directional multiresolution image representations, PhD thesis, Swiss Federal Institute of Technology Lausanne, November 2001.
[114] S. Park, M.J.T. Smith, R.M. Mersereau, A new directional filterbank for image analysis and classification, in: Proc. IEEE Int. Conf. Acoust. Speech and Signal Proc., 1999, pp. 1417–1420.
[115] http://www.ifp.illinois.edu/~minhdo/software/.
[116] K. Guo, G. Kutyniok, D. Labate, Sparse multidimensional representations using anisotropic dilation and shear operators, in: G. Chen, M. Lai (Eds.), Wavelets and Splines, Nashboro Press, Nashville, TN, 2006, pp. 189–201.
[117] K. Guo, D. Labate, Optimally sparse multidimensional representation using shearlets, SIAM J. Math. Anal. (2007) 298–318.
[118] G. Easley, D. Labate, W.-Q. Lim, Sparse directional image representations using the discrete shearlet transform, Appl. Comput. Harmon. Anal. 25 (2008) 25–46.
[119] G. Kutyniok, D. Labate, Resolution of the wavefront set using continuous shearlets, Trans. Am. Math. Soc. 361 (2009) 2719–2754.
[120] G. Easley, K. Guo, D. Labate, Analysis of singularities and edge detection using the shearlet transform, in: Proc. Int. Conf. on SAMPling Theory and Applications SAMPTA, 2009.
[121] K. Guo, D. Labate, W. Lim, Edge analysis and identification using the continuous shearlet transform, Appl. Comput. Harmon. Anal. 27 (2009) 24–46.

[122] S. Yi, D. Labate, G.R. Easley, H. Krim, Edge detection and processing using shearlets, in: IEEE Int. Conf. on Image Processing (ICIP), 2008, pp. 1148–1151.
[123] S. Yi, Shape Dynamic Analysis, PhD thesis, North Carolina State University, Dept. of EECS Raleigh NC, 2011.
[124] S. Dahlke, G. Kutyniok, G. Steidl, G. Teschke, Shearlet coorbit spaces and associated Banach frames, Appl. Comput. Harmon. Anal. 27 (2009) 195–214.
[125] K. Guo, W. Lim, D. Labate, G. Weiss, E. Wilson, The theory of wavelets with composite dilations, in: C. Heil (Ed.), Harmonic Analysis and Applications, Birkhäuser Boston, 2006, pp. 231–249.
[126] G. Easley, D. Labate, F. Colonna, Shearlet based total variation for denoising, IEEE Trans. Image Process. 18 (2009) 260–268.
[127] http://www.math.uh.edu/~dlabate/software.html.
[128] W. Pratt, Digital Image Processing, Wiley Interscience Publications, 1978.
[129] K. Guo, D. Labate, Characterization and analysis of edges using the continuous shearlet transform, SIAM Imaging Sci. 2 (2009) 959–986.
[130] Y. LeCun, L. Bottou, Y. Bengio, P. Haffner, Gradient-based learning applied to document recognition, http://yann.lecun.com/exdb/lenet/, 1998.
[131] S. Roheda, H. Krim, Conquering the CNN over-parameterization dilemma: a Volterra filtering approach for action recognition, in: Proc. the Thirty-Fourth AAAI Conf. on Artificial Intelligence AAAI-20, 2020.
[132] S. Ghanem, S. Roheda, H. Krim, Latent code-based fusion: a Volterra neural network approach, https://arxiv.org/pdf/2104.04829.pdf, 2021.
[133] L. Wang, W. Ouyang, X. Wang, L. Huchuan, STCT: sequentially training convolutional networks for visual tracking, in: Conf. on Computer Vision and Pattern Recognition, 2016.
[134] O. Ronneberger, P. Fischer, T. Brox, Unet convolutional networks for biomedical image segmentation, in: Proc. Int. Conf. on Medical Image Computing and Computer-Assisted Intervention, 2015, pp. 234–241.
[135] C. Tian, Y. Xu, W. Zuo, Image denoising using deep CNN with batch renormalization, Neural Netw. 121 (2020) 461–471.
[136] Q. Zhang, A. Benveniste, Wavelet networks, IEEE Trans. Neural Netw. 3 (1992) 889–898.
[137] S. Chen, C. Cowan, P. Grant, Orthogonal least squares learning algorithm for radical basis function networks, IEEE Trans. Neural Netw. 2 (1991) 302–309.
[138] Q. Zhang, Regressor selection and wavelet network construction, Technical Report 709, vol. INTRIA, 1993.
[139] Q. Zhang, Using wavelet network in nonparametric estimation, IEEE Trans. Neural Netw. 82 (1997) 227–236.
[140] Y. Pati, P. Krishnaprasad, Analysis and synthesis of feedforward neural networks using discrete affine wavelet transformations, IEEE Trans. Neural Netw. 41 (1993) 73–85.
[141] Y. Oussar, G. Dreyfus, Initialization by selection for wavelet network training, Neurocomputing 34 (2000) 131–143.
[142] A. Benveniste, A. Juditsky, B. Delyon, Q. Zhang, P.-Y. Glorennec, Wavelets in identification, in: SYSID'94 10th IFAC Symposium on System Identification, Copenhagen, 1994.
[143] A. Juditsky, Wavelet estimators: Adapting to unknown smoothness, Technical Report IRISA, vol. 815, 1994.
[144] R. Yin, T. Gao, Y. Lu, I. Daubechies, A tale of two bases: local-nonlocal regularization on image patches with convolution framelets, SIAM J. Imaging Sci. 10 (2017) 711–750.
[145] J. Ye, Y. Han, E. Cha, Deep convolutional framelets: a general deep learning framework for inverse problems, SIAM J. Imaging Sci. 11 (2) (2018) 991–1048.
[146] https://github.com/jongcye/FramingUNet.
[147] T. Wiatowski, H. Bölcskei, A mathematical theory of deep convolutional neural networks for feature extraction, arXiv:1512.06293, 2015.
[148] V. Papyan, Y. Romano, M. Elad, Convolutional neural networks analyzed via convolutional sparse coding, J. Mach. Learn. Res. 18 (2017) 1–52.

[149] A. Krizhevsky, I. Sutskever, G.E. Hinton, Imagenet classification with deep convolutional neural networks, Adv. Neural Inf. Process. Syst. 8611 (2012) 1097–1105.
[150] K. Simonyan, A. Zisserman, Very deep convolutional networks for large-scale image recognition, arXiv preprint, arXiv:1409.1556.
[151] S. Santurkar, D. Tsipras, A. Ilyas, A. Madry, How does batch normalization help optimization?, in: Advances in Neural Information Processing Systems, 2018, pp. 2483–2493.
[152] M. Telgarsky, Benefits of depth in neural networks, https://arxiv.org/abs/1602.04485, 2016.
[153] M. Lin, Q. Chen, S. Yan, Network in network, https://arxiv.org/abs/1312.4400, 2013.
[154] C. Szegedy, W. Liu, Y. Jia, P. Sermanet, S. Reed, D. Anguelov, A. Rabinovich, Going deeper with convolutions, in: Proc. IEEE Conf. on Computer Vision and Pattern Recognition, 2015, pp. 1–9.
[155] K. He, X. Zhang, S. Ren, J. Sun, Deep residual learning for image recognition, in: Proc. IEEE Conf. on Computer Vision and Pattern Recognition, 2016, pp. 770–778.
[156] S. Mahdizadehaghdam, A. Panahi, H. Krim, Deep dictionary learning: a PARametric NETwork approach, IEEE Trans. Image Process. 28 (2019) 4790–4802.
[157] G. Huang, Z. Liu, L. Van Der Maaten, K.Q. Weinberger, Densely connected convolutional networks, in: Proc. IEEE Conf. on Computer Vision and Pattern Recognition, 2017, pp. 4700–4708.
[158] A. Santoro, S. Bartunov, M. Botvinick, D. Wierstra, T. Lillicrap, Meta-learning with memory-augmented neural networks, in: Int. Conf. on Machine Learning (ICML), 2016.
[159] B. Lake, R. Salakhutdinov, J. Tenenbaum, Human-level concept learning through probabilistic program induction, Science 350 (6266) (2015) 1332–1338.
[160] F. Sung, L. Zhang, T. Xiang, P.H.S. Torr, T.M. Hospedales, Learning to compare: relation network for few-shot learning, in: Proc. IEEE Conf. on Computer Vision and Pattern Recognition, 2018.
[161] Y. Bao, Varied few-shot relation network for non-uniformly distributed source, in: International Conf. on Image Processing Computer Vision and Pattern Recognition, 2021.
[162] https://github.com/kilgore92/PyTorch-UNet.
[163] S. Fujieda, K. Akayama, T. Hachisuka, Wavelet convolutional neural networks, https://arxiv.org/abs/1707.07394, 2018.
[164] N.S. Suhaimi, Z. Othman, M.R. Yaakub, Analyzing prediction performance between wavelet neural network and product-unit neural network, J. Phys.: Conf. Ser. 1432 (2020).
[165] T. Guo, H.S. Mousavi, T. Vu, V. Monga, Deep wavelet prediction for image super-resolution, in: IEEE Conf on Computer Vision and Pattern Recognition Workshops, 2017, http://signal.ee.psu.edu/research/DWSR.html.
[166] E. Kang, J. Min, J. Ye, Wavelet domain residual network *wavresnet* for low-dose x-ray CT reconstruction, Med. Phys. 44 (Oct. 2016).
[167] E. Kang, J. Min, J. Ye, Wavenet: a deep convolutional neural network using directional wavelets for low-dose x-ray CT reconstruction, Med. Phys. 44 (Oct. 2016).
[168] W. Bae, J. Yoo, J. Ye, Beyond deep residual learning for image restoration: persistent homology-guided manifold simplification, arXiv:1611.06345v4 [cs.CV], 8 Jun 2017.
[169] Y. Han, J. Ye, Framing U-Net via deep convolutional framelets: application to sparse-view CT, https://arxiv.org/abs/1805.08620, 2018.
[170] P. Liu, H. Zhang, W. Lian, W. Zuo, Multilevel wavelet convolutional neural network, arXiv:1907.03128v1, 2019.

CHAPTER

Frames in signal processing

10

Lisandro Lovisolo[a], Eduardo A.B. da Silva[b], and Allan Freitas da Silva[b]

[a]*PROSAICO – DETEL/PEL – UERJ, Rio de Janeiro, Brazil*
[b]*SMT – DEL/PEE – COPPE/UFRJ, Rio de Janeiro, Brazil*

10.1 Introduction

Frames were introduced by Duffin and Schaeffer [1] in 1952 for the study of nonharmonic Fourier series [2,3]. The central idea was to represent a signal by its projections on a sequence of elements $\{e^{\,j2\pi\lambda_n t}\}_n$, $n \in \mathbb{Z}$, not restricting λ_n to be multiples nF of a fundamental frequency F as in harmonic (traditional) Fourier series. One can readily see that without such restriction the set $\{e^{\,j2\pi\lambda_n t}\}_n$ tends to be highly redundant, or, equivalently, overcomplete. For instance, in a space $L^2(-T,T)$, it may consist of a sequence of elements or functions that are greater in number than a basis for $L^2(-T,T)$, the latter being given for example by a harmonic Fourier series, among others.

Overcompleteness may make it hard to find a representation of a function $f(t) \in L^2(-T,T)$ using the set $\{e^{\,j2\pi\lambda_n t}\}_n$, in the form

$$f(t) = \sum_n c_n e^{\,j2\pi\lambda_n t}. \tag{10.1}$$

This is so because due to overcompleteness, the weights or coefficients c_n in Eq. (10.1) cannot be computed by the simple approach of projecting $f(t)$ into each element of $\{e^{\,j2\pi\lambda_n t}\}_n$. Such projection would be equivalent to using c_n proportional to $\langle f(t), e^{\,j2\pi\lambda_n t}\rangle$, where $\langle f(t), g(t)\rangle$ represents the inner product $\int f(t)g^*(t)dt$.[1] One should note that when using an orthonormal basis for representing a signal (as is the sequence of elements defining harmonic Fourier series), computing the inner product is the standard approach for finding c_n.

Despite this difficulty, overcompleteness may bring some advantages, as one may obtain more compact, robust, or stable signal representations than using bases [4,5,6,7]. This is one of the reasons why frames have been widely researched and employed since the 1990s in mathematics, statistics, computer science, and engineering [4].

The basic idea leading to the definition of frames is the representation of signals from a given space using more coefficients than the minimum necessary – the essential concept behind overcompleteness. This idea is behind modern analog-to-digital converters (ADCs) that sample signals at rates that are higher than the Nyquist rate with low resolution [8]. When the signal is sampled at a rate highly exceeding the one of Nyquist, the requirement on the sample precision can be relaxed [6]. This is a

[1] $g^*(t)$ denotes the complex conjugate of $g(t)$.

Signal Processing and Machine Learning Theory. https://doi.org/10.1016/B978-0-32-391772-8.00016-8
Copyright © 2024 Elsevier Inc. All rights reserved.

different perspective as compared to traditional Nyquist sampling, when in general one requires that a good quantization precision is employed so that signal reconstruction errors can be neglected.

In addition to the above, there is the motivation of achieving robustness to nonuniform and/or irregular sampling, which is useful in the cases where there is clock jitter. This has led to a large set of results and some applications of frame theory [9]. These ideas were further extrapolated, leading to the definition of signal sampling based on their "rate of innovation" [10], which takes into account the number of degrees of freedom per unit of time when sampling a given class of signals. In some way, these concepts have led to the so-called "compressed sensing" (CS) perspective [11].

Although frames provide the mathematical justification for several signal processing methods and are at the heart of the flow of ideas and achievements briefly described above, it took a long time until frames came definitely into play. The first book concerning the topic was written by Young in 1980 [2]. In [12] Daubechies, Grossman, and Meyer realized that frames can somehow be understood in a way very similar to expansions using bases. This established the link among frames and wavelets [5,13, 14], and new breath was given to the research on frames. Applications of the "frame theory" have broadly appeared. Frames have been an active area of research, allowing the production of a plethora of theoretical results and applications. Good material on frames can be found in [4,5,13], and [6] is a good book dedicated to this topic. We show a few of the relevant results regarding frames in this chapter.

10.1.1 Notation

In this chapter we use $\mathbf{x}$ to denote a signal in a discrete space, which can be finite-dimensional or not. The value of the nth coordinate or sample of $\mathbf{x}$ is referred to as $\mathrm{x}[n]$. A bold capital is used for matrices, as $\mathbf{G}$. Continuous functions or signals are explicitly denoted with reference to the independent variable, as in $x(t)$. When just a nonbold letter such as x is employed it is because the definition or result applies in both cases – discrete and continuous spaces. Moreover, $\|\mathrm{x}\|$ refers to $\langle \mathrm{x}, \mathrm{x}\rangle$. The inner product $\langle \mathrm{g}, \mathrm{h}\rangle$ is to be understood as $\langle g(t), h(t)\rangle = \int g(t)h^*(t)dt$ in continuous spaces, and $\langle \mathbf{g}, \mathbf{h}\rangle = \sum_n \mathrm{g}[n]\mathrm{h}^*[n]$ in discrete spaces.

10.2 Basic concepts

Let us briefly introduce what are frames. A frame of a space $\mathbb{H}$ is a set of elements that spans $\mathbb{H}$. Therefore, a frame $\mathcal{G} = \{\mathrm{g}_k\}_{k\in\mathcal{K}}$ (a set of elements g_k which are indexed by $k \in \mathcal{K}$) can be used to express any $x \in \mathbb{H}$ by means of a linear combination of g_k, that is,

$$\mathrm{x} = \sum_{k\in\mathcal{K}} c_k \mathrm{g}_k, \tag{10.2}$$

where c_k are called the frame coefficients.

The only restriction that we have imposed on $\mathcal{G} = \{\mathrm{g}_k\}_{k\in\mathcal{K}}$ for it being a frame is that it should span $\mathbb{H}$. Therefore, $\mathcal{G}$ is in general overcomplete, meaning that the set of frame coefficients $\{c_k\}_{k\in\mathcal{K}}$ that can express x by means of Eq. (10.2) is not unique, as the overcompleteness of $\mathcal{G}$ means that its elements are not linearly independent.

The above means that there may exist different ways to compute the frame coefficients c_k used in Eq. (10.2) to synthesize x. We employ interchangeably both terms "expansion" and "representation" since all the frame elements (generally in larger number than the space dimension) may a priori be employed to express the signal. Due to overcompleteness one usually employs more elements than the support or dimension of the space $\mathbb{H}$ to represent the signal. This is equivalent to saying that the frame coefficients for a given signal may be in larger number than the coefficients employed in the canonical signal representation.

Example 10.1. Consider the set of elements $\mathcal{G}$ in $\mathbb{R}^2$, $\mathcal{G} = \{[0 \quad 1/2], [1/4 \quad -1/4], [-1/2 \quad 0]\}$. Any vector $\mathbf{x} = [\mathrm{x}[0], \mathrm{x}[1]]$ projected into these elements gives a set of coefficients:

$$\tilde{\mathbf{c}} = \begin{bmatrix} \tilde{c}_0 \\ \tilde{c}_1 \\ \tilde{c}_2 \end{bmatrix} = \begin{bmatrix} 0 & 1/2 \\ 1/4 & -1/4 \\ -1/2 & 0 \end{bmatrix} \begin{bmatrix} \mathrm{x}[0] \\ \mathrm{x}[1] \end{bmatrix} = \begin{bmatrix} \mathrm{x}[1]/2 \\ (\mathrm{x}[0] - \mathrm{x}[1])/4 \\ -\mathrm{x}[0]/2 \end{bmatrix}. \tag{10.3}$$

The original vector can be obtained from $\tilde{\mathbf{c}}$ using, for example,

$$\begin{bmatrix} \mathrm{x}[0] \\ \mathrm{x}[1] \end{bmatrix} = \tilde{\mathbf{G}}_0^{\mathrm{T}} \tilde{\mathbf{c}} = \begin{bmatrix} 0 & 0 & -2 \\ 0 & -4 & -2 \end{bmatrix} \begin{bmatrix} \tilde{c}_0 \\ \tilde{c}_1 \\ \tilde{c}_2 \end{bmatrix}, \tag{10.4}$$

$$\begin{bmatrix} \mathrm{x}[0] \\ \mathrm{x}[1] \end{bmatrix} = \tilde{\mathbf{G}}_1^{\mathrm{T}} \tilde{\mathbf{c}} = \begin{bmatrix} 0 & 0 & -2 \\ 2 & 0 & 0 \end{bmatrix} \begin{bmatrix} \tilde{c}_0 \\ \tilde{c}_1 \\ \tilde{c}_2 \end{bmatrix}, \tag{10.5}$$

or

$$\begin{bmatrix} \mathrm{x}[0] \\ \mathrm{x}[1] \end{bmatrix} = \tilde{\mathbf{G}}_2^{\mathrm{T}} \tilde{\mathbf{c}} = \begin{bmatrix} 2 & 4 & 0 \\ 2 & 0 & 0 \end{bmatrix} \begin{bmatrix} \tilde{c}_0 \\ \tilde{c}_1 \\ \tilde{c}_2 \end{bmatrix}, \tag{10.6}$$

with $\tilde{\mathbf{G}}_0$, $\tilde{\mathbf{G}}_1$, and $\tilde{\mathbf{G}}_2$ being only a few among infinite other possibilities. That is, the matrix $\tilde{\mathbf{G}}^{\mathrm{T}}$ that reconstructs $\mathbf{x}$ from its projection on a set $\mathcal{G}$ is not unique.

Now consider that $\mathbf{G}$ is the matrix constructed such that its rows are the frame elements. One can use $\mathbf{G}$ to compute the coefficients vector $\tilde{\mathbf{c}}$ with

$$\tilde{\mathbf{c}} = \mathbf{G}\mathbf{x}. \tag{10.7}$$

Note that in this example, if $\dim(\mathbf{x})$ represents the dimensions of element $\mathbf{x}$, then $\dim(\mathbf{x}) = 2 \times 1$, $\dim(\mathbf{G}) = 3 \times 2$, $\dim(\tilde{\mathbf{c}}) = 3 \times 1$, and $\dim(\tilde{\mathbf{G}}^{\mathrm{T}}) = 2 \times 3$. More generally, if a frame for $\mathbb{R}^N$ has $K > N$ elements, then $\dim(\mathbf{x}) = N \times 1$, $\dim(\mathbf{G}) = K \times N$, $\dim(\tilde{\mathbf{c}}) = K \times 1$, and $\dim(\tilde{\mathbf{G}}^{\mathrm{T}}) = N \times K$. From Eq. (10.7), the matrix $\tilde{\mathbf{G}}^{\mathrm{T}}$ used to obtain $\mathbf{x} = \tilde{\mathbf{G}}^{\mathrm{T}}\tilde{\mathbf{c}}$ must be such that $\tilde{\mathbf{G}}^{\mathrm{T}}\mathbf{G} = \mathbf{I}_N$ ($\mathbf{I}_N$ is the identity matrix of dimension equating the dimension of the vectors in $\mathcal{G}$, which is 2 in the present example).

Naturally, the roles of the matrices $\mathbf{G}$ and $\tilde{\mathbf{G}}$ above can be interchanged. Since $\tilde{\mathbf{G}}^{\mathrm{T}}\mathbf{G} = \mathbf{I}_N$, we have $\mathbf{G}^{\mathrm{T}}\tilde{\mathbf{G}} = \mathbf{I}_N$, and thus one can employ $\tilde{\mathbf{G}}$ as a projection operator on $\mathbf{x}$ for finding a set of coefficients $\mathbf{c}$, and reconstruct $\mathbf{x}$ using

$$\mathbf{x} = \mathbf{G}\mathbf{c}. \tag{10.8}$$

In general, the best choice for $\tilde{\mathbf{G}}$ depends on the application. For instance, consider the case where the coefficient $\tilde{c}_0$ has been corrupted to e and one must reconstruct the original source from the corrupted coefficient vector $\hat{\mathbf{c}} = \begin{bmatrix} e & \tilde{c}_1 & \tilde{c}_2 \end{bmatrix}^T$. It is possible to see that using the matrix $\tilde{\mathbf{G}}_0$ defined in Eq. (10.4) one can still reconstruct the original vector $\mathbf{x}$ from the corrupted $\hat{\mathbf{c}}$. A similar argument can be made for a corruption in the coefficients $\tilde{c}_1$ and $\tilde{c}_2$: a perfect reconstruction is still possible using the matrices $\tilde{\mathbf{G}}_1$ from Eq. (10.5) and $\tilde{\mathbf{G}}_2$ from Eq. (10.6). However, if one used the matrix

$$\tilde{\mathbf{G}}^{\mathrm{T}} = \begin{bmatrix} 1 & 2 & -1 \\ 1 & -2 & -1 \end{bmatrix} \tag{10.9}$$

to perform the recovery, then the recovered signal would also be corrupted, although any of the matrices in Eqs. (10.4), (10.5), (10.6), and (10.9) would provide an exact reconstruction if no error occurred. □

As highlighted in the example above, the set $\mathbf{c}$ is redundant in the sense that the vectors employed to compute $\{c_k\}_{k\in\mathcal{K}}$ (onto which $\mathbf{x}$ is projected) may be linearly dependent. This has the advantage that the wrong computation of some coefficients of the expansion or even their losses may not prevent the recovery of the original signal from the remaining coefficients within an acceptable error margin. This is so because, as one has different possible signal reconstruction operators provided by the frame concept, one might choose, for example, those that provide a reconstruction that is independent of the value of the corrupted coefficient. Ideas along these lines have led to applications of frame expansions for signal sampling with low quantization requirements [4], and also to the development of transmission systems that are robust to erasures (when data are corrupted or lost but one knows which piece of data has been corrupted or lost) over a network, which corresponds to the deletion of the correspondent frame elements [15,16,17].

10.2.1 The dual frame

The above example links to the strategy for computing frame coefficients using the inverse or dual frame [5,6,4,7], which provides the dual reconstruction formulas

$$\mathrm{x} = \sum_{k\in\mathcal{K}} \langle \mathrm{x}, \mathrm{g}_k \rangle \tilde{\mathrm{g}}_k = \sum_{k\in\mathcal{K}} \langle \mathrm{x}, \tilde{\mathrm{g}}_k \rangle \mathrm{g}_k . \tag{10.10}$$

From Eq. (10.10) one may define $\tilde{\mathcal{G}} = \{\tilde{\mathrm{g}}_k\}_{k\in\mathcal{K}}$ as the inverse or dual frame to the frame $\mathcal{G} = \{\mathrm{g}_k\}_{k\in\mathcal{K}}$. For a given signal x, since $\mathcal{G}$ is overcomplete, $\tilde{\mathcal{G}}$ is, in general, not unique [6], as shown in Example 10.1.

Each coefficient c_k of a frame expansion can be viewed as a measure of "how much" the signal "has" of frame element $\{\mathrm{g}_k\}_{k\in\mathcal{K}}$. Naturally, this property can be used to infer or analyze signal characteristics. Since in an overcomplete frame its elements are linearly dependent, at least two of its elements are not orthogonal. This implies that signal characteristics that are different, but similar to one another, can be effectively observed by projecting the signal into different frame elements that are similar to these characteristics. If an orthogonal basis was used instead, such characteristics would not be as evident from the projections, since each characteristic would be "partitioned" among the projections, with each partition being orthogonal to one another. Therefore, the similar characteristics would be spread along different projections, and would therefore not be as evident as they might in the case of a carefully chosen redundant frame.

The reasoning in the above paragraph has been largely employed to justify the use of the so-called Gabor frames, which are discussed in Section 10.5.4. Gabor frames are constructed by modulation and translation of a prototype function, thus providing a time-frequency analysis capability [4,18,19,20,21, 22]. Gabor frames employ sets of time- and frequency-translated versions of a predefined function to analyze or synthesize a signal. These frames are named after Gabor due their origin in [23]. The main idea is to explore functions that have some desired energy concentration in the time and frequency domains to analyze the time-frequency content that is represented in a signal. Several tools have been developed for achieving this, using the decomposition of a signal into a Gabor frame [24,25,26]. As the example regarding the Gaborgram in Section 10.6.3 shows, the inverse frame elements may not resemble the frame elements and some care must be taken when using frame expansions to capture specific signal features.

Although the elements of a frame that are used to express a signal x are in general selected a priori, frame expansions allow for the use of elements with special properties. The Balian–Low theorem [13] shows that it is not possible to construct a basis with elements that are well localized in both time and frequency domains simultaneously. Frames do not impose uniqueness to the signal representation; therefore, the definition of the frame elements is less restrictive than it is for basis elements. Hence, for a frame we can achieve a better localization of its elements simultaneously in the time and frequency domains than for a basis, better dealing with the limitations imposed by the uncertainty principle [4,19, 21].

The property of frames discussed above has been largely employed by the so-called windowed Fourier analysis schemes [4,27]. In addition, since the frame coefficients may not be unique for a given signal, the set of frame coefficients can be altered or improved – considering the linear dependencies among frame elements – in order to highlight desired signal features. Moreover, frames can be designed for specific applications depending on the features that one desires to extract or analyze in a signal. Examples of such specific designs are ones using wavelets [4,5,7], and other analysis techniques such as time-frequency [4,28] or structurally/geometrically oriented techniques [29,30].

In general, the computational burden of decomposing discrete signals into frames surpasses the computational demands of traditional expansions into bases, as, for example, the ones of the fast Fourier transform (FFT) [31]. However, despite the higher computational complexity, with frame expansions, as mentioned above, we can obtain a separation of signal structures that is much better than the one that is possible using traditional approaches. Mallat [4] explores this in his proposed superresolution techniques.

Due to the characteristics above, frame applications encompass, among others, signal coding [4], signal sampling [1,4,6], signal analysis [4,5], transient detection [24,25], and communication systems design [6,32]. Frames have also been related to the analysis and design of filter banks [33,34].

10.2.2 Signal analysis and synthesis using frames

A frame or its dual can be employed to analyze a signal and obtain the signal expansion, i.e., the frame coefficients, and to synthesize it from the coefficients obtained in the analysis process. This is illustrated in Fig. 10.1. As Eq. (10.24) implies, the roles of $\{g_k\}_{k\in\mathcal{K}}$ and $\{\tilde{g}_k\}_{k\in\mathcal{K}}$ can be interchanged in Fig. 10.1, providing different perspectives for signal analysis.

We note that the "system model" of Fig. 10.1 is very similar to the one employed in several signal processing tasks – for example in filter banks [31]. This resemblance between frames and several signal

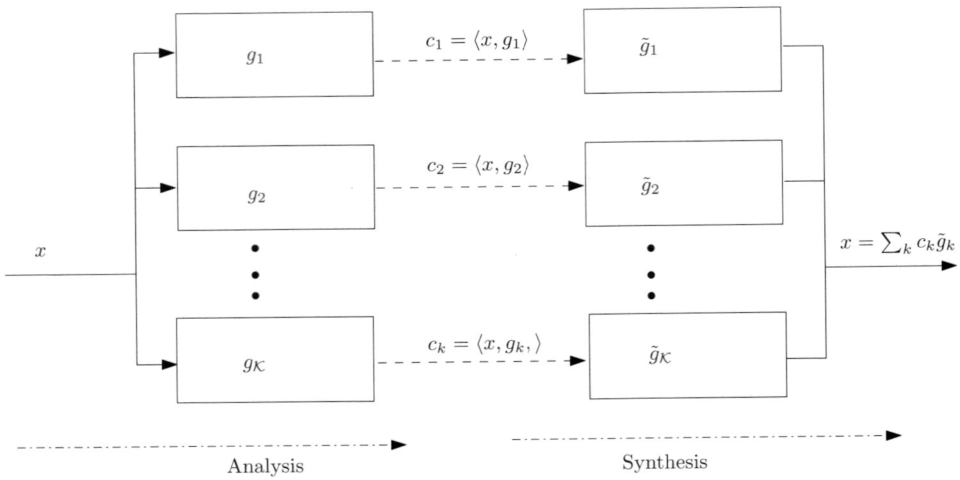

FIGURE 10.1

Analysis and synthesis of a signal using a frame and its dual – as noted in the text, the roles can be interchanged.

processing tasks provides a large range of potential applications for frames. To highlight an important feature, the frame elements may be designed to highlight or analyze desired signal features that may be "overlapping" or linearly dependent.

In the scenario of filter banks, a relevant and special case is the one of a "perfect reconstruction filter bank" [31]. In this case the analysis and reconstruction filter banks correspond to a pair of dual frames. A more detailed analysis of the input–output mapping in this case is provided in Section 10.4.1, where we discuss frames of discrete spaces.

Some examples of common frame constructions for signal analysis purposes are Gabor and wavelet frames. These provide expansion systems or representations that have different supports on the time-frequency plane. We will discuss them in Sections 10.5.4 and 10.5.5, respectively.

10.3 Relevant definitions

Now that we have introduced frames and have outlined some of their basic properties, we are ready to present a more formal definition.

Definition 10.1. A sequence of elements $\mathcal{G} = \{\mathrm{g}_k\}_{k\in\mathcal{K}}$ in a space $\mathbb{H}$ is a *frame* for $\mathbb{H}$ if there exist constants A and B, $0 < A \le B < \infty$, such that [4,5,7,6]

$$A\|\mathrm{x}\|^2 \le \sum_{k\in\mathcal{K}} |\langle \mathrm{x}, \mathrm{g}_k\rangle|^2 \le B\|\mathrm{x}\|^2, \quad \forall \mathrm{x} \in \mathbb{H}. \tag{10.11}$$

(i) The numbers A and B are called the lower and upper *frame bounds*, respectively, and are not unique. The optimal lower frame bound is the supremum on all A and the optimal upper frame bound is the infimum on all B [6].
(ii) It is said that the frame is normalized [5] if $||g_k|| = 1, \ \forall k \in \mathcal{K}$. □

A commonly addressed problem related to frames is how to determine if a given sequence of elements $\{g_k\}_{k\in\mathcal{K}}$ in $\mathbb{H}$ is a frame of $\mathbb{H}$. This is often referred to as the frame characterization problem [6]. This characterization is commonly accomplished by the frame bounds. From Eq. (10.11), we note that if $A = 0$, then there is a signal x that is orthogonal to all the elements $\{g_k\}_{k\in\mathcal{K}}$ and, therefore, the signal cannot be represented as their linear combination.

That is the reason for the requirement that $A > 0$. The requirement for $B < \infty$ can be understood considering that if for a given signal there is no upper bound in Eq. (10.11). This means that the set of elements is too "dense" for that signal.

Given a set of elements in a space $\mathbb{H}$, in order to verify if it is or is not a frame one may compute the frame bounds. It is in general much easier to find the upper frame bound B than the lower frame bound A.

Example 10.2. In $\mathbb{R}^N$ any set of $M < \infty$ finite norm vectors $\{g_k\}_{k=1,\dots,M}$ necessarily implies an upper bound $B < \infty$. On the other hand, for the lower bound to be $A > 0$, the set should span $\mathbb{H}$, and such a verification requires a more careful analysis. □

10.3.1 The frame operator

Let us now define the frame operator and state the frame decomposition result.

Definition 10.2. The *frame operator* $S\{\cdot\}$ is defined as

$$S : \mathbb{H} \to \mathbb{H}, \ S\{\mathrm{x}\} = \sum_{k\in\mathcal{K}} \langle \mathrm{x}, g_k \rangle g_k. \tag{10.12}$$

From this definition some properties are derived [6]:

(i) The frame operator is bounded, invertible, self-adjoint, and positive. Therefore, it assumes an inverse $S^{-1}\{\cdot\}$. This is referred to as the *inverse frame operator*, which should be such that

$$\mathrm{x} = S\{S^{-1}\{\mathrm{x}\}\} = \sum_{k\in\mathcal{K}} \langle S^{-1}\{\mathrm{x}\}, g_k \rangle g_k. \tag{10.13}$$

(ii) If the lower and upper frame bounds of $\{g_k\}_{k\in\mathcal{K}}$ are A and B, respectively, then $\left\{S^{-1}\{g_k\}\right\}_{k\in\mathcal{K}}$ is a frame with lower and upper bounds B^{-1} and A^{-1}, respectively, and the frame operator for $\left\{S^{-1}\{g_k\}\right\}_{k\in\mathcal{K}}$ is S^{-1}. □

In the following example we show a numerical illustration of the computation of the frame operator.

Example 10.3. We return to Example 10.1 and find the frame operator for the set of elements $\mathcal{G} = \{[0 \quad 1/2], [1/4 \quad -1/4], [-1/2 \quad 0]\}$. According to the definition contained in Eq. (10.12), we find

$$
\begin{aligned}
S\{\mathrm{x}\} &= \sum_{k\in\mathcal{K}} \langle \mathrm{x}, \mathrm{g}_k\rangle \mathrm{g}_k \\
&= \mathrm{x}[1]/2\begin{bmatrix} 0 \\ 1/2 \end{bmatrix} + (\mathrm{x}[0]/4 - \mathrm{x}[1]/4)\begin{bmatrix} 1/4 \\ -1/4 \end{bmatrix} - \mathrm{x}[0]/2\begin{bmatrix} -1/2 \\ 0 \end{bmatrix} \\
&= \begin{bmatrix} 5/16 & -1/16 \\ -1/16 & 5/16 \end{bmatrix}\begin{bmatrix} \mathrm{x}[0] \\ \mathrm{x}[1] \end{bmatrix}.
\end{aligned}
$$

□

From the definition of the frame operator one has the *frame decomposition* result that provides the reconstruction formulas [6]

$$\mathrm{x} = \sum_{k\in\mathcal{K}} \langle \mathrm{x}, \mathrm{g}_k\rangle S^{-1}\{\mathrm{g}_k\} = \sum_{k\in\mathcal{K}} \langle \mathrm{x}, S^{-1}\{\mathrm{g}_k\}\rangle \mathrm{g}_k. \tag{10.14}$$

Note that this provides a way to compute $\tilde{\mathrm{g}}_k = S^{-1}\{\mathrm{g}_k\}$, which are the elements of the inverse frame or the so-called canonical dual.

Example 10.4. Suppose that a normalized frame $\{\mathrm{g}_k\}_{k\in\mathcal{K}}$ is used to express the information of a signal x, in the form of its coefficients c_k, which are transmitted or stored in order to later reconstruct x, that is,

$$c_k = \langle \mathrm{x}, S^{-1}\{\mathrm{g}_k\}\rangle, \tag{10.15}$$

where $S^{-1}\{x\}$ is the inverse frame operator. From Definition 10.2 we also have

$$\mathrm{x} = \sum_k c_k \mathrm{g}_k. \tag{10.16}$$

However, due to quantization the coefficients cannot be recovered exactly, and this can be modeled by a noise term w_k added to coefficient c_k, $k \in \mathcal{K}$. In this case, instead of x we will reconstruct $\hat{\mathrm{x}}$ given by

$$\hat{\mathrm{x}} = \sum_{k\in\mathcal{K}} (c_k + w_k)\,\mathrm{g}_k = \sum_{k\in\mathcal{K}} \left(\left\langle \mathrm{x}, S^{-1}\{\mathrm{g}_k\}\right\rangle + w_k\right) \mathrm{g}_k. \tag{10.17}$$

Then, due to the frame reconstruction equation above, we have

$$\hat{\mathrm{x}} = \mathrm{x} + \sum_{k\in\mathcal{K}} w_k \mathrm{g}_k, \tag{10.18}$$

that is, the reconstructed signal is corrupted by a noise term.

We should note that in this scenario, we know that c_k (obtained by means of Eq. (10.15)) are the expansion of a signal over a frame. However, this may not be the case for the set of coefficients

$$\{c_k + w_k\}_{k\in\mathcal{K}} = \left\{\langle \mathrm{x}, S^{-1}\{\mathrm{g}_k\}\rangle + w_k\right\}_{k\in\mathcal{K}} \tag{10.19}$$

used to reconstruct $\hat{\mathrm{x}}$ in Eq. (10.17). However, there may not exist any $\tilde{\mathrm{x}} \in \mathbb{H}$ that generates the quantized coefficients $c_k + w_k$ when decomposed in the frame $\{\mathrm{g}_k\}_{k\in\mathcal{K}}$, and this fact can be exploited to generate a better approximation $\hat{x}$ from the quantized coefficients. Assuming that $w_k \in \ell^2(\mathbb{N})$, i.e., the noise is bounded, we can project $\left\{\langle x, S^{-1}\{g_k\}\rangle + w_k\right\}_{k\in\mathcal{K}}$ using the operator [4,6]

$$Q\{c_k\} = \left\{\left\langle \sum_k c_k S^{-1}\{g_k\}, g_j \right\rangle\right\}_{j\in\mathcal{K}}. \tag{10.20}$$

Using this operator in the projected-quantized frame coefficients scenario above, that is, applying apply $Q\{\cdot\}$ to Eq. (10.19), we have

$$Q\left\{\langle x, S^{-1}\{g_k\}\rangle + w_k\right\}_{k\in\mathcal{K}} = \{c_k\}_{k\in\mathcal{K}} + Q\{w_k\}_{k\in\mathcal{K}}. \tag{10.21}$$

Using the above, the signal can be reconstructed by means of

$$\hat{x} = \sum_{k\in\mathcal{K}} (c_k + Q\{w_k\})\, g_k = x + \sum_{k\in\mathcal{K}} Q\{w\}_k\, g_k. \tag{10.22}$$

Supposing that the frame $\{g_k\}_{k\in\mathcal{K}}$ is normalized, considering the quantization noise in Eq. (10.17) to be white and zero-mean, and considering $E[\cdot]$ to denote the expected value, in [4] it is shown that

$$E\left[|Q\{w\}|^2\right] \leq \frac{1}{A} E\left[|w|^2\right]. \tag{10.23}$$

That is, when reconstructing the signal by means of the processed coefficients as in Eq. (10.21), the energy of the resulting quantization noise is reduced if A (the lower frame bound) is greater than one. □

Example 10.5. We evaluate the noise resilience property evidenced by Eq. (10.23) with a simple experiment. Consider a grayscale version of the image *astronaut* in Fig. 10.2. It is encoded using Gabor frames (they are formalized in Section 10.5.4) having increasing lower bounds A and frame redundancy. For each case, we corrupt the frame coefficients by a white Gaussian noise with variance σ^2 and reconstruct the original image. Fig. 10.3(b–d) shows the reconstructed image using each frame. For comparison, Fig. 10.3(a) shows the original image corrupted by noise with the same variance σ^2. The images have better quality as the lower frame bound A increases. However, it should be noted that this robustness is achieved by increasing the redundancy in the frame representation. We compute the mean square error (MSE) obtained for each reconstructed image, which corresponds to the term $E\left[|Q\{w\}|^2\right]$ in Eq. (10.23), and compare it to the upper bound $\frac{1}{A}E\left[|w|^2\right] = \frac{\sigma^2}{A}$, plotting the values in Fig. 10.4. Note that the reconstruction error in all cases was smaller than $\frac{\sigma^2}{A}$. □

10.3.2 The inverse frame

Definition 10.3. The *inverse frame* $\tilde{\mathcal{G}} = \{\tilde{g}_k\}_{k\in\mathcal{K}}$ to a frame $\mathcal{G} = \{g_k\}_{k\in\mathcal{K}}$ is the one that provides the reconstruction formulas

$$x = \sum_{k\in\mathcal{K}} \langle x, \tilde{g}_k\rangle g_k = \sum_{k\in\mathcal{K}} \langle x, g_k\rangle \tilde{g}_k. \tag{10.24}$$

□

FIGURE 10.2

Grayscale version of the image *astronaut*, obtained from the *scikit-image* library [35].

Example 10.6. Let us go back to Example 10.1, where we have seen two different possibilities for "inversion" of the frame given by the rows of the matrix

$$\mathbf{G} = \begin{bmatrix} 0 & 1/2 \\ 1/4 & -1/4 \\ -1/2 & 0 \end{bmatrix}. \tag{10.25}$$

In this example we show that the canonical dual is associated with the inverse frame operator. Using a pseudoinversion algorithm [4] we find the pseudoinverse $\mathbf{G}^+$ to $\mathbf{G}$ ($\mathbf{G}^+\mathbf{G} = \mathbf{I}_2$):

$$\mathbf{G}^+ = \begin{bmatrix} 1/3 & 2/3 & -5/3 \\ 5/3 & -2/3 & -1/3 \end{bmatrix}. \tag{10.26}$$

We should note that this is just another example of possible frame inversion, as shown in [4], since $\mathbf{G}$ has an infinite number of left inverses. □

Definition 10.4. When $A = B$ the frame is said to be *tight* [6] and in this case

$$S^{-1}\{\cdot\} = S\{\cdot\}/A. \tag{10.27}$$

In addition, frames for which $A \approx B$ are said to be *snug* [34] and for these $S^{-1}\{\cdot\} \approx S\{\cdot\}/A$. □

Definition 10.5. A frame $\mathcal{G} = \{g_k\}_{k \in \mathcal{K}}$ is said to be *normalized* when

$$\|g_k\| = c, \ \forall\, k \in \mathcal{K}, \quad \text{and } c > 0 \text{ is a real constant.} \tag{10.28}$$

□

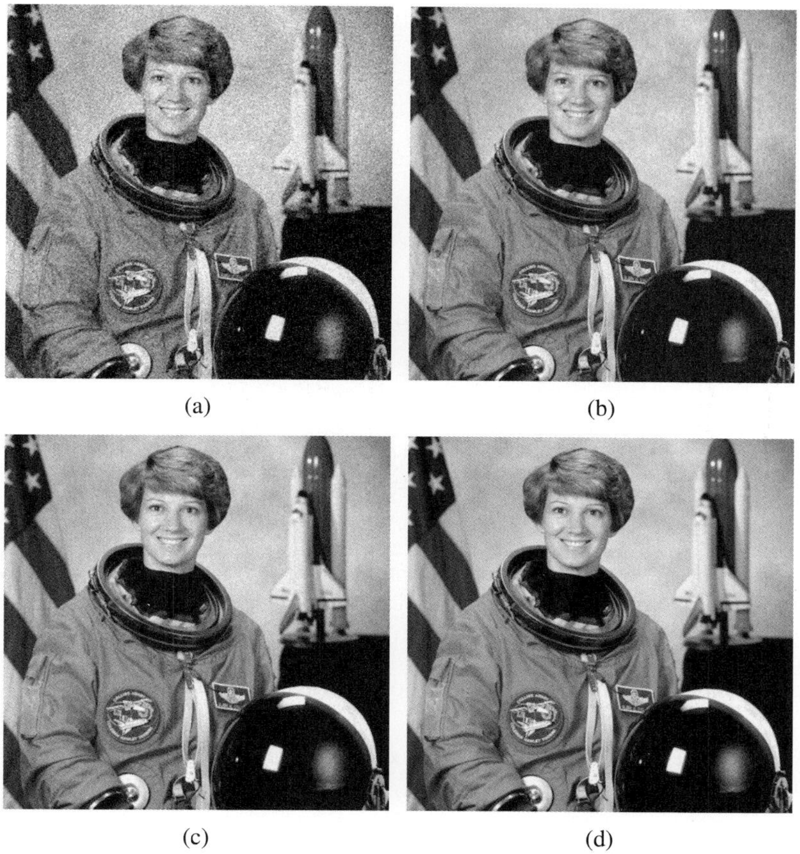

FIGURE 10.3

Effect of the noise on the frame coefficients. In (a), the original image *astronaut* is corrupted by noise with variance $\sigma^2 = 1000$. In (b–d), the original image is decomposed using different Gabor frames, the frame coefficients are corrupted by noise with variance σ^2, and the image is reconstructed using (b) a Gabor frame with $A \approx 1.83$ and two times the number of coefficients, (c) a Gabor frame with $A \approx 3.99$ and four times the number of coefficients, and (d) a Gabor frame with $A \approx 8$ and eight times the number of coefficients.

10.3.3 Characterization of frames: basic property

Frame bounds provide limits for the energy scaling of signals when they are represented using the projections into the frame elements. Due to that, frames are often characterized in terms of their frame bounds. It is very common to also use the *frame bounds ratio* A/B [13] to characterize a frame. It is also possible to define the "tightness" of a frame from its frame bounds ratio: the closer that A and B are, the tighter the frame is.

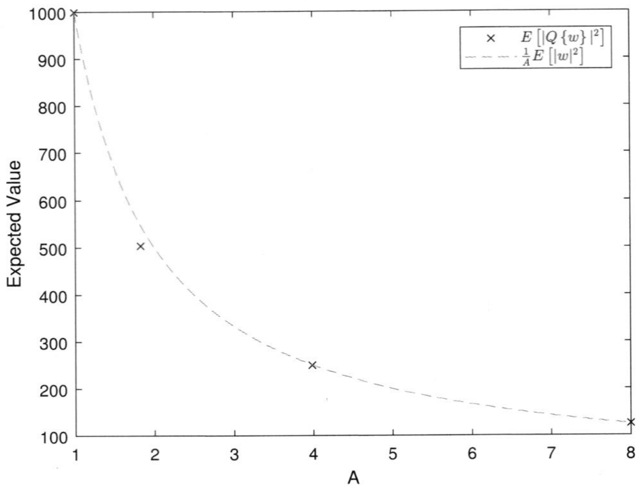

FIGURE 10.4

Comparison between the MSE obtained for each reconstructed image in Fig. 10.3 (black cross), which correspond to $E\left[|Q\{w\}|^2\right]$, and the upper bound $\frac{1}{A}E\left[|w|^2\right] = \frac{\sigma^2}{A}$ (red (light gray in print version) dashed line).

Daubechies [5] showed the following property of the frame operator defined in Eq. (10.12):

$$S\{\mathrm{x}\} = \frac{A+B}{2}\left(\mathrm{x} - R\{\mathrm{x}\}\right), \text{ i.e., } \mathrm{x} = \frac{2}{A+B}\sum_{k\in\mathcal{K}}\langle \mathrm{x}, \mathrm{g}_k\rangle \mathrm{g}_k + R\{\mathrm{x}\}, \tag{10.29}$$

where $R\{\mathrm{x}\}$ can be understood as the error incurred in reconstructing x from the projections of x into the frame elements instead of into the inverse frame elements.

In addition, she also showed that

$$R\{\mathrm{x}\} = I\{\mathrm{x}\} - \frac{2}{A+B}S\{\mathrm{x}\}, \tag{10.30}$$

where $I\{\mathrm{x}\}$ is the identity operator. Therefore, we can conclude that

$$-\frac{B-A}{B+A}I\{\mathrm{x}\} \leq R\{\mathrm{x}\} \leq \frac{B-A}{B+A}I\{\mathrm{x}\} \quad \equiv \quad \|R\{\mathrm{x}\}\| \leq \frac{B-A}{B+A}\|\mathrm{x}\|. \tag{10.31}$$

The above implies that if B/A is close to one, then the error incurred in the reconstruction by using the frame itself instead of its inverse frame is small. This error gets smaller as A and B become closer. Therefore, from an analysis–synthesis perspective, if the frame is tight there is no need to find its inverse. Tight frames are self-dual [36], that is, the inverse frame to a tight frame is a scaled version of the frame itself.

For normalized frames, Daubechies refers to $\frac{A+B}{2}$ as the frame redundancy. Several studies have been developed on normalized tight frames. For example, in [37] it is shown that these frames have some noise reduction property. That is, if a given reconstruction distortion is required, one can use fewer bits to represent the frame coefficients in comparison to the precision required to represent coefficients derived from the representation in a basis. This is related to the acquisition–quantization problem we have discussed in Example 10.4 above.

Example 10.7. Let $\{\mathbf{e}_1, \mathbf{e}_2, \mathbf{e}_3\}$ be an orthonormal basis of the 3D space. Define the following vectors:

$$\mathbf{g}_1 = \frac{\sqrt{2}}{\sqrt{5}}\mathbf{e}_1 - \frac{\sqrt{3}}{\sqrt{5}}\mathbf{e}_2, \qquad \mathbf{g}_2 = \frac{\sqrt{2}}{\sqrt{5}}\mathbf{e}_2 + \frac{\sqrt{3}}{\sqrt{5}}\mathbf{e}_3, \qquad \mathbf{g}_3 = \frac{\sqrt{3}}{\sqrt{5}}\mathbf{e}_1 + \frac{\sqrt{2}}{\sqrt{5}}\mathbf{e}_3,$$
$$\mathbf{g}_4 = \frac{\sqrt{3}}{\sqrt{5}}\mathbf{e}_1 + \frac{\sqrt{2}}{\sqrt{5}}\mathbf{e}_2, \qquad \mathbf{g}_5 = -\frac{\sqrt{3}}{\sqrt{5}}\mathbf{e}_2 + \frac{\sqrt{2}}{\sqrt{5}}\mathbf{e}_3, \qquad \mathbf{g}_6 = \frac{\sqrt{2}}{\sqrt{5}}\mathbf{e}_1 - \frac{3}{\sqrt{5}}\mathbf{e}_3.$$

For any $\mathbf{x} \in \mathbb{R}^3$ we have

$$\begin{aligned}
\sum_k |\langle \mathbf{x}, \mathbf{g}_k\rangle|^2 = &\left|\left\langle \mathbf{x}, \frac{\sqrt{2}}{\sqrt{5}}\mathbf{e}_1\right\rangle + \left\langle \mathbf{x}, -\frac{\sqrt{3}}{\sqrt{5}}\mathbf{e}_2\right\rangle\right|^2 + \left|\left\langle \mathbf{x}, \frac{\sqrt{2}}{\sqrt{5}}\mathbf{e}_2\right\rangle + \left\langle \mathbf{x}, \frac{\sqrt{3}}{\sqrt{5}}\mathbf{e}_3\right\rangle\right|^2 + \\
&\left|\left\langle \mathbf{x}, \frac{\sqrt{3}}{\sqrt{5}}\mathbf{e}_1\right\rangle + \left\langle \mathbf{x}, \frac{\sqrt{2}}{\sqrt{5}}\mathbf{e}_3\right\rangle\right|^2 + \left|\left\langle \mathbf{x}, \frac{\sqrt{3}}{\sqrt{5}}\mathbf{e}_1\right\rangle + \left\langle \mathbf{x}, \frac{\sqrt{2}}{\sqrt{5}}\mathbf{e}_2\right\rangle\right|^2 + \\
&\left|\left\langle \mathbf{x}, -\frac{\sqrt{3}}{\sqrt{5}}\mathbf{e}_2\right\rangle + \left\langle \mathbf{x}, \frac{\sqrt{2}}{\sqrt{5}}\mathbf{e}_3\right\rangle\right|^2 + \left|\left\langle \mathbf{x}, \frac{\sqrt{2}}{\sqrt{5}}\mathbf{e}_1\right\rangle + \left\langle \mathbf{x}, -\frac{\sqrt{3}}{\sqrt{5}}\mathbf{e}_3\right\rangle\right|^2 = \\
&= 2\left(|\langle \mathbf{x}, \mathbf{e}_1\rangle|^2 + |\langle \mathbf{x}, \mathbf{e}_2\rangle|^2 + |\langle \mathbf{x}, \mathbf{e}_3\rangle|^2\right) \\
&= 2\|\mathbf{x}\|^2.
\end{aligned}$$

Therefore the six vectors define a tight frame with $A = B = 2$. This number can in some sense be understood as a measure of redundancy of the frame expansion [4]. □

Tight frames, due to their operational simplicity, are relevant to many applications. For example, tight frames were shown to be the most resistant to coefficient erasures [15]. An erasure occurs when transiting information over a network one knows that a given piece of information is unreliable. If a frame is used to compute the information that is transmitted over the network, then the reconstruction error under erasures is lower for a tight frame than for nontight ones. Tight frames can also provide sparse signal decompositions [38]. Due to these and other interesting properties a lot of attention has been given to the construction of tight frames [39,40,41,16,38].

Example 10.8. Based on Example 10.7, define the new set of vectors:

$$\mathbf{r}_1 = 2\mathbf{g}_1, \qquad \mathbf{r}_2 = \mathbf{g}_2, \qquad \mathbf{r}_3 = \mathbf{g}_3,$$
$$\mathbf{r}_4 = \mathbf{g}_4, \qquad \mathbf{r}_5 = \mathbf{g}_5, \qquad \mathbf{r}_6 = \mathbf{g}_6.$$

For any $\mathbf{x} \in \mathbb{R}^3$ we have

$$\sum_k |\langle \mathbf{x}, \mathbf{r}_k\rangle|^2 = |\langle \mathbf{x}, \mathbf{r}_1\rangle|^2 + \sum_{k\neq 1} |\langle \mathbf{x}, \mathbf{r}_k\rangle|^2$$

$$
\begin{aligned}
&= |\langle \mathbf{x}, 2\mathbf{g}_1\rangle|^2 + \sum_{k\neq 1} |\langle \mathbf{x}, \mathbf{g}_k\rangle|^2 \\
&= 4\,|\langle \mathbf{x}, \mathbf{g}_1\rangle|^2 + \sum_{k\neq 1} |\langle \mathbf{x}, \mathbf{g}_k\rangle|^2 \\
&= 3\,|\langle \mathbf{x}, \mathbf{g}_1\rangle|^2 + \sum_{k} |\langle \mathbf{x}, \mathbf{g}_k\rangle|^2 \\
&= 3\,|\langle \mathbf{x}, \mathbf{g}_1\rangle|^2 + 2\|\mathbf{x}\|^2.
\end{aligned}
$$

From the above, one can see that what prevents the frame from being tight is the term $3\,|\langle \mathbf{x}, \mathbf{g}_1\rangle|^2$, which makes the value of $\sum_k |\langle \mathbf{x}, \mathbf{r}_k\rangle|^2$ depend on the direction of $\mathbf{x}$. If $\mathbf{x}$ is orthogonal to $\mathbf{g}_1$, $|\langle \mathbf{x}, \mathbf{g}_1\rangle|^2 = 0$, and thus $\sum_k |\langle \mathbf{x}, \mathbf{r}_k\rangle|^2 = 2\|\mathbf{x}\|^2$. On the other hand, the maximum value of $|\langle \mathbf{x}, \mathbf{g}_1\rangle|^2$ occurs when $\mathbf{x}$ is proportional to $\mathbf{g}_1$, and since $\|\mathbf{g}_1\| = 1$, its maximum value is $|\langle \mathbf{x}, \mathbf{g}_1\rangle|^2 = \|\mathbf{x}\|^2$, and in this case $\sum_k |\langle \mathbf{x}, \mathbf{r}_k\rangle|^2 = 5\|\mathbf{x}\|^2$. Therefore, this set of vectors defines a frame with bounds $A = 2$ and $B = 5$. We can infer from this analysis that the frame bounds are related to the directions of highest and lowest variation in space. In Section 10.4.2 we show that for finite vector spaces, they can be obtained by computing the eigenvalues of the so-called Gram matrix. □

10.4 Some computational remarks

In general, signal processing algorithms are implemented in digital processors. Therefore, the signals are invariably discrete and some simplifications take place when frames are concerned.

10.4.1 Frames in discrete spaces

Consider the case of discrete spaces such as $\ell^2(\mathbb{Z})$, that is, the space of square-summable sequences, i.e., $\mathbf{x} \in \ell^2(\mathbb{Z})$ if $\sum_{n\in\mathbb{Z}} |\mathrm{x}[n]|^2 < \infty$, where $\mathrm{x}[n]$ is the value of the nth coordinate/sample of $\mathbf{x}$. In general, in such spaces, signal analysis is accomplished by using sliding windows. This is actually the case for filter banks [31], where according to the structure depicted in Fig. 10.1, the synthesis equation is [34,33]

$$
\mathrm{x}[n] = \sum_{k=0}^{K-1} \sum_{m=-\infty}^{\infty} c_{k,m} g_{k,m}[n]. \tag{10.32}
$$

In the above equation each signal's sample $\mathrm{x}[n]$, $n \in \mathbb{Z}$, is synthesized as a sum of the samples $g_{k,m}[n] = g_k[n - mN]$, $N \le K$, where K is the cardinality of the frame. Here $\mathbf{g}_{k,m}$ are translated versions of $\mathbf{g}_k$, referring to the mth length-N signal block.

Similar concepts can be brought to more general cases of frames in $\ell^2(\mathbb{Z})$, bearing in mind the correct usage of a frame and its dual. In this case, Eq. (10.32) can be written as

$$
\mathrm{x}[n] = \sum_{k=0}^{K-1} \sum_{m=-\infty}^{\infty} \langle \mathbf{x}, \tilde{\mathbf{g}}_{k,m}\rangle g_{k,m}[n]. \tag{10.33}
$$

In other words, as discussed in Section 10.2.2, the frame and its dual should be employed in the two distinct tasks of analysis (decomposition) and synthesis (reconstruction) of **x**.

Up to this point, the definitions involving frames were valid for any Hilbert space $\mathbb{H}$. When considering a discrete Hilbert space such as $\ell^2(\mathbb{Z})$, a space in which digital signal processing applications may exist (they may also exist in generic finite vector spaces – see Section 10.4.2), using finite support vectors may be handy, as in the case of finite impulse response (FIR) filter banks [31].

A signal $\mathbf{x} \in \ell^2(\mathbb{Z})$ can be expressed using Eq. (10.32) if the family of functions [33]

$$\mathcal{G} = \left\{\mathbf{g}_{k,m} : g_{k,m}[n] = g_k[n - mN],\ k = 0,\ 1, \ldots, K - 1,\ m \in \mathbb{Z}\right\} \tag{10.34}$$

is a frame for $\ell^2(\mathbb{Z})$. This result provides an important feature of frames for $\ell^2(\mathbb{Z})$.

Definition 10.6. A set of vectors $\mathcal{G}$ as defined in Eq. (10.34) (i.e., block-translated) is a *frame* of $\ell^2(\mathbb{Z})$ if there exist constants $0 < A \leq B < \infty$ such that for any $\mathbf{x} \in \ell^2(\mathbb{Z})$ we have

$$A\|\mathbf{x}\|^2 \leq \sum_{k=0}^{\mathcal{K}-1} \sum_{m=-\infty}^{\infty} \left|\langle \mathbf{x}, \mathbf{g}_{k,m}\rangle\right|^2 \leq B\|\mathbf{x}\|^2. \tag{10.35}$$

□

We should note that if Eq. (10.35) holds for $\mathcal{G}$ as in Eq. (10.34), then there exists another frame

$$\tilde{\mathcal{G}} = \left\{\tilde{\mathbf{g}}_{k,m} : \tilde{g}_{k,m}[n] = \tilde{g}_k[n - mN],\ k = 0,\ 1, \ldots, K - 1,\ m \in \mathbb{Z}\right\} \tag{10.36}$$

that can be employed for obtaining the coefficients $c_{k,m}$ in Eq. (10.33).

Naturally, the roles of $\mathcal{G}$ and $\tilde{\mathcal{G}}$ can be interchanged, providing

$$\mathrm{x}[n] = \sum_{k=0}^{K-1} \sum_{m=-\infty}^{\infty} \langle \mathbf{x}, \mathbf{g}_{k,m}\rangle \tilde{g}_{k,m}[n]. \tag{10.37}$$

As can be noted, given $\mathcal{G}$ as in Eq. (10.34), $\tilde{\mathcal{G}}$ as in (10.36) satisfying Eqs. (10.33) and (10.37) is not unique.

10.4.2 Finite vector spaces

In a finite vector space $\mathbb{H}^N$ one in general restricts the frame to have a finite number K of elements, and then Eq. (10.11) becomes

$$A\|\mathbf{x}\|^2 \leq \sum_{k=1}^{K} |\langle \mathbf{x}, \mathbf{g}_k\rangle|^2 \leq B\|\mathbf{x}\|^2, \qquad \mathbf{x} \in \mathbb{H}^N. \tag{10.38}$$

In this case, as illustrated in examples above, some computational simplifications may take place.

Definition 10.7. *Analysis* and *synthesis* operators are defined as follows.

(i) The *synthesis operator* is defined as

$$T\{\cdot\}:\mathbb{C}^K \to \mathbb{H}^N, \quad T\{c_k\}_{k=1}^{k=K} = \sum_{k=1}^{K} c_k \mathbf{g}_k, \tag{10.39}$$

where $\mathbb{C}^K$ is a K-dimensional complex vector space.

(ii) The *analysis operator* $T^*\{\cdot\}$ is defined as

$$T^*\{\cdot\}:\mathbb{H}^N \to \mathbb{C}^K, \quad T^*\{\mathbf{x}\} = \{\langle \mathbf{x}, \mathbf{g}_k\rangle\}_{k=1}^{k=K}. \tag{10.40}$$

It is also known as the adjunct operator of $T\{\cdot\}$.

(iii) The operator $T\{\cdot\}$ synthesizes $\mathbf{x}$ from the frame coefficients c_k. These frame coefficients are obtained when $\tilde{T}^*\{\cdot\}$, the analysis operator of a dual frame, is applied to $\mathbf{x}$. Such operator is given by

$$\tilde{T}^*\{\cdot\}:\mathbb{H}^N \to \mathbb{C}^K, \quad \tilde{T}^*\{\mathbf{x}\} = \{\langle \mathbf{x}, \tilde{\mathbf{g}}_k\rangle\}_{k=1}^{k=K}. \tag{10.41}$$

(iv) The frame operator can be expressed using the analysis and synthesis operators as

$$S:\mathbb{H}^N \to \mathbb{H}^N, \quad S\{\mathbf{x}\} = T\{T^*\{\mathbf{x}\}\} = \sum_{k=1}^{K} \langle \mathbf{x}, \mathbf{g}_k\rangle \mathbf{g}_k. \tag{10.42}$$

Note that the frame operator does not perform the frame reconstruction depicted in Eq. (10.24) since it does not use $\tilde{\mathbf{g}}_k$. □

In vector spaces the operators $T\{\cdot\}$ and $T^*\{\cdot\}$ can be represented by matrices [6], as in the examples above.

Assuming that $\mathbf{x}$ and the frame elements $\mathbf{g}_k$ are organized as column vectors with dimensions $N \times 1$ and the coefficients are organized as a column vector $\mathbf{c}$ of dimensions $K \times 1$, the following properties apply:

$$\mathbf{T} = \begin{bmatrix} \mathbf{g}_1^{\mathrm{T}} \\ \vdots \\ \mathbf{g}_K^{\mathrm{T}} \end{bmatrix} \qquad \dim(\mathbf{T}) = K \times N, \tag{10.43}$$

$$T\{\mathbf{c}\} = \mathbf{T}^{\mathrm{T}}\mathbf{c} \qquad \dim(\mathbf{T}^{\mathrm{T}}) = N \times K, \tag{10.44}$$

$$T^*\{\mathbf{x}\} = \mathbf{T}\mathbf{x} \qquad \dim(\mathbf{T}^*) = K \times N, \tag{10.45}$$

$$S\{\mathbf{x}\} = T\{\mathbf{T}\{\mathbf{x}\}\} = \mathbf{T}^{\mathrm{T}}\mathbf{T}\mathbf{x} = \mathbf{S}\mathbf{x} \qquad \dim(\mathbf{S}) = N \times N, \tag{10.46}$$

$$S^{-1}\{\mathbf{x}\} = \left(\mathbf{T}^{\mathrm{T}}\mathbf{T}\right)^{-1}\mathbf{x} = \mathbf{S}^{-1}\mathbf{x} \qquad \dim\left(\mathbf{S}^{-1}\right) = N \times N. \tag{10.47}$$

Note that in Eq. (10.44) we show how to use the frame elements $\mathbf{g}_k$ to perform a synthesis operation. If one wants to reconstruct a signal from the coefficients obtained from the analysis operation in Eq. (10.45), the dual frame must be used.

The dual frame elements can be computed by means of

$$\tilde{\mathbf{g}}_k = S^{-1}\{\mathbf{g}_k\} = \left(\mathbf{T}^{\mathrm{T}}\mathbf{T}\right)^{-1}\mathbf{g}_k, \tag{10.48}$$

and we have

$$\tilde{T}\{\mathbf{c}\} = \tilde{\mathbf{T}}^{\mathrm{T}}\mathbf{c} = \mathbf{T}\left(\mathbf{T}^{T}\mathbf{T}\right)^{-1}\mathbf{c} \qquad \dim(\tilde{\mathbf{T}}^{\mathrm{T}}) = N \times K, \tag{10.49}$$

$$\tilde{T}^*\{\mathbf{x}\} = \tilde{\mathbf{T}}^*\mathbf{x} = \left(\mathbf{T}^{T}\mathbf{T}\right)^{-1}\mathbf{T}\mathbf{x} \qquad \dim(\tilde{\mathbf{T}}^*) = K \times N. \tag{10.50}$$

Example 10.9. As we have seen above, the frame operator $\mathbf{S} = \mathbf{T}^T\mathbf{T}$. If ρ_i are the eigenvalues of $\mathbf{S}$, then the frame bounds are given by $A = \min_i\{\rho_i\}$ and $B = \max_i\{\rho_i\}$ [6]. Thus, if $\mathbf{T}^T\mathbf{T} = A\mathbf{I}_N$ ($\mathbf{I}_N$ is the identity matrix of size N), then the frame is tight ($A = B$). Hence, for a tight frame $\mathbf{S}^{-1}\mathbf{S} = \mathbf{I}_N$. □

Example 10.10. The Gram matrix or Gramian of a sequence of vectors $\{\mathbf{g}_k\}_{k\in\mathcal{K}}$ is the matrix G whose elements are defined by $g_{i,j} = \left\{\langle \mathbf{g}_i, \mathbf{g}_j\rangle\right\}_{i,j\in\mathcal{K}}$. It can be employed to analyze different frame characteristics. For example, in [42], it is employed to investigate the symmetry properties of tight frames, providing tight frame design procedures. In [38] the Gramian is also employed for analyzing and designing frames with desired features.

Up to this subsection we have dealt with the analysis, synthesis, and frame operators, as well as their counterparts corresponding to an inverse frame. As one may notice, a great deal of frame characteristics can be extracted from them. The Gram matrix G also carries useful information about frames. We readily note that $G = \mathbf{T}\mathbf{T}^{\mathrm{T}}$ and thus $\dim(G) = K \times K$. Since each of its entries $g_{i,j}$ represents the inner product between two frame elements, the Gram matrix provides information on how similar the frame elements are to each other; therefore, it can be understood as containing indicatives of a sort of redundancy between frame elements, and thus of connections among the different c_k. □

Example 10.11. We highlight some consequences of the frame bounds in finite vector spaces. (i) If the lower frame bound A is zero, then there is a signal $\mathbf{x}$ such that $\sum_{k\in\mathcal{K}} |\langle \mathrm{x}, g_k\rangle|^2 = 0$. Consequently, the frame elements are all orthogonal to $\mathbf{x}$ and the frame would not be capable of representing it. (ii) If the frame has a finite set of elements, then, necessarily, due to the Cauchy–Schwartz inequality, the upper frame bound condition is automatically satisfied. □

10.5 Construction of frames from a prototype signal

It is common to process signals in order to search for specific patterns in time, frequency, or even scale. In this case, one often a priori specifies a set of desired behaviors or patterns to be searched for in signals to be analyzed. These are usually specified as a set of parameterized waveforms. If synthesis is not required, the restrictions on such a set are milder than in the case when synthesis is required. In the latter case, one should guarantee the set of parameterized functions or waveforms to be a frame. This is what is discussed in the sequel. We suppose a signal processing scenario where a given signal has to be analyzed and synthesized using a set of parameterized waveforms that are derived from operations such as change of position in time (translation), change of frequency (modulation), and change of support (dilation) of a prototype signal. In this section we discuss some conditions under which a frame is generated when these operations are applied to a prototype waveform.

10.5.1 Translation, modulation, and dilation operators

There are several ways to construct frames from operations on a prototype signal. Some of them are based on translations, modulations, and dilations of a function $g(t) \in L^2(\mathbb{R})$ (the space of square-integrable functions) [6].

Definition 10.8. *Translation* by $a \in \mathbb{R}$ is defined as

$$T_a : L^2(\mathbb{R}) \to L^2(\mathbb{R}), \quad T_a g(t) = g(t-a). \tag{10.51}$$

□

Definition 10.9. *Modulation* by $b \in \mathbb{R}$ is defined as

$$E_b : L^2(\mathbb{R}) \to L^2(\mathbb{R}), \quad E_b g(t) = g(t) e^{2\pi jbt}. \tag{10.52}$$

□

Definition 10.10. *Dilation* by $c \in \mathbb{R} - \{0\}$ is defined as

$$D_c : L^2(\mathbb{R}) \to L^2(\mathbb{R}), \quad D_c g(t) = \frac{1}{\sqrt{c}} g\left(\frac{t}{c}\right). \tag{10.53}$$

□

10.5.2 Common frame constructions

The most common approaches to construct frames from a prototype are:

- A frame in $L^2(\mathbb{R})$ constructed by translations of a given function $g(t) \in L^2(\mathbb{R})$, through

$$\{T_{na} g(t)\}_{n \in \mathbb{Z}}, \text{ with } a > 0, \tag{10.54}$$

  is called a *frame of translates* [6].
- A *Weyl–Heisenberg* or *Gabor frame* is a frame in $L^2(\mathbb{R})$ constructed from a fixed function $g(t) \in L^2(\mathbb{R})$, by means of

$$\{E_{mb} T_{na} g(t)\}_{m,n \in \mathbb{Z}}, \text{ with } a, b > 0. \tag{10.55}$$

  Such frame is also called a *Gabor system* or a *windowed Fourier frame* [4,5,6,43] due to its connection to time-frequency analysis techniques.
- A frame constructed by dilations and translations of a prototype (mother) function $g(t) \in L^2(\mathbb{R})$ by

$$\{T_{nac^j} D_{c^j} g(t)\}_{j,n \in \mathbb{Z}} = \{c^{\frac{-j}{2}} g\left(c^{-j} t - na\right)\}_{j,n \in \mathbb{Z}}, \text{ with } c > 1 \text{ and } a > 0, \tag{10.56}$$

  is called a *wavelet frame* [4,5,6,7].

In the following subsections we discuss some conditions for obtaining frames using the above approaches and some examples of their applications and relevance.

10.5.3 Frames of translates

Digital signal processing is heavily based on signal sampling, that is, the capability of sampling a function at regular intervals and reconstructing the signal from these samples [4,6,31,44]. In a broad sense a sample corresponds to a value that is the outcome of an evaluation/measurement of the function/signal. The sampling operation can often be modeled as the projection of the function/signal over a predefined element. Regular sampling corresponds to translating such element to regularly spaced positions.

Definition 10.11. A *frame of translates* is a frame for $L^2(\mathbb{R})$ obtained through operations $\{T_{na}g(t)\}_{n\in\mathbb{Z}}$ on a fixed function $g(t)$ with $a > 0$. □

Which conditions should hold on $g(t)$ in order for $\{T_{na}g(t)\}_{n\in\mathbb{Z}}$ to be a frame? In this case it can be shown that [6] $\{T_{na}g(t)\}_{n\in\mathbb{Z}}$ is a frame with bounds A and B if and only if

$$aA \leq \sum_{k\in\mathbb{Z}} \left| \hat{g}\left(\frac{f+k}{a}\right) \right| \leq aB, \tag{10.57}$$

for $f \in \mathbb{R} - \left\{ f \in \mathbb{R} : \sum_{k\in\mathbb{Z}} \left| \hat{g}\left(\frac{f+k}{a}\right) \right| = 0 \right\}$, where $\hat{g}(f) = \int_{-\infty}^{\infty} g(t)e^{\mathrm{j}2\pi f t}dt$ is the Fourier transform of $g(t)$.

Frames of translates are intimately related with sampling. The example below highlights their connection with the Shannon sampling theorem [44].

Example 10.12. In this example one explores the connection between the Shannon sampling theorem [44,45] and frames. Let a band-limited function of time $x(t)$ be sampled using a train of impulses at a rate $1/T$. Assume that the signal bandwidth is W. If $T < \frac{1}{2W}$, we can reconstruct the signal using

$$x(t) = \sum_n x(nT) \frac{\sin\left[(t-nT)\frac{\pi}{T}\right]}{(t-nT)\frac{\pi}{T}}. \tag{10.58}$$

This is so because the infinite set of sinc pulses centered at nT provides a basis/frame of the W Hz band-limited space.

Define R as the "amount of oversampling"

$$R = \frac{1}{2WT}, \text{ for } R \geq 1. \tag{10.59}$$

Since at the Nyquist rate [31] $2WT = 1$, $R \geq 1$ gives $1/T \geq 2W$. Under these conditions we have [45]

$$x(t) = \frac{1}{R} \sum_n x(nT) \frac{\sin\left[(t-nT)\frac{\pi}{RT}\right]}{(t-nT)\frac{\pi}{RT}}. \tag{10.60}$$

Note that for $R \geq 1$, two sinc pulses centered at kT and lT $(k, l \in \mathbb{Z})$ are actually not orthogonal. Indeed, when $R \geq 1$, the set of T temporally spaced sinc pulses is a frame in the W-band-limited signal space, because although they expand the space, the sinc pulses centered at nT with $R > 1$ are not linearly independent. In addition, the infinite set of sinc pulses above is shown to be a tight frame and R is interpreted as a redundancy factor for the frame [10,45]. We note that for $R = 1$, Eq. (10.60) is the same as Eq. (10.58) and the signal samples occur where the sinc pulses equate either one (for $t = nT$) or zero (other cases). However, as R increases, the sinc pulse centered at nT gets wider and is not necessarily zero for $t - nT = kT$, $k \in \mathbb{Z}^*$. □

10.5.4 Gabor frames

While the introduction of the frame concept is traced back to 1952 [1], according to Christensen [6] the first mention of what are now called Gabor or Weyl–Heisenberg frames can be traced back to even before that, since they appear in the work of Gabor on communications [23] in 1946 and in the book of von Neumann on quantum mechanics originally published in 1932 [46].

Let us simplify the idea of Gabor and simply state that its vision was to represent a signal $x(t)$ by means of

$$x(t) = \sum_m \sum_n \tilde{c}_{m,n} E_{mb} T_{na} g(t). \tag{10.61}$$

The function $g(t)$ is chosen to be a Gaussian due to its time-frequency concentration [19,21,47]. This representation can provide an analysis of the time-frequency content around the points (na, mb) in the time-frequency plane. The resemblance between this idea and the frame definition sparks, and the following straightforward definition emerges.

Definition 10.12. A *Gabor frame* is a frame for $L^2(\mathbb{R})$ obtained through translation and modulation operations on a fixed function $g(t)$, $\{E_{mb} T_{na} g(t)\}_{m,n\in\mathbb{Z}}$, with $a, b > 0$. □

Now, one has the problem of determining conditions for the set $\{E_{mb} T_{na} g(t)\}_{n,m\in\mathbb{Z}}$ to be capable of representing any signal $x(t) \in L^2(\mathbb{R})$, i.e., to be a frame of $L^2(\mathbb{R})$. In addition, there is the question of how one computes the set $\{c_{m,n}\}_{n,m\in\mathbb{Z}^2}$ that allows the synthesis of a signal using Eq. (10.61). The answer to the latter question lies in the concept of inverse frame. We will start, however, from the first question: for a given $g(t)$, when is $\{E_{mb} T_{na} g(t)\}_{m,n\in\mathbb{Z}}$ a frame? The answer depends on a complicated interplay between $g(t)$, a, and b. For example, in [48] a set of nonintuitive conditions was shown to hold on a and b in order for a Gabor system to be generated. In what follows, several results collected from the frame literature are presented, which provide either sufficient or necessary conditions for $\{E_{mb} T_{na} g(t)\}_{m,n\in\mathbb{Z}}$ to constitute a frame.

For $\{E_{mb} T_{na} g(t)\}_{m,n\in\mathbb{Z}}$ to constitute a frame it is necessary that the product $ab \leq 1$ [4,5,6]. However, the assumption $ab \leq 1$ does not guarantee the generation of a frame for every $g(t)$ (see for example [48]). It should be observed that ab is a measure of the density of the frame in the time-frequency plane [4,5,19,21,20,43]; the smaller ab is, the denser is the frame. Fig. 10.5 illustrates this concept.

If $\{E_{mb} T_{na} g(t)\}_{m,n\in\mathbb{Z}}$ constitutes a frame, then the frame bounds necessarily satisfy [6]

$$\forall t \in \mathbb{R}, \quad A \leq \tfrac{1}{b} \textstyle\sum_{n\in\mathbb{Z}} |g(t - na)|^2 \leq B, \tag{10.62}$$

$$\forall f \in \mathbb{R}, \quad A \leq \tfrac{1}{a} \textstyle\sum_{k\in\mathbb{Z}} |\hat{g}(f - kb)|^2 \leq B. \tag{10.63}$$

A well-known sufficient condition for $\{E_{mb} T_{na} g(t)\}_{m,n\in\mathbb{Z}}$ to be a frame is presented in [14]. Let $a, b > 0$, let $g(t) \in L^2(\mathbb{R})$, let $g^*(t)$ denote the complex conjugate of $g(t)$, and suppose that $\exists\, A, B > 0$ such

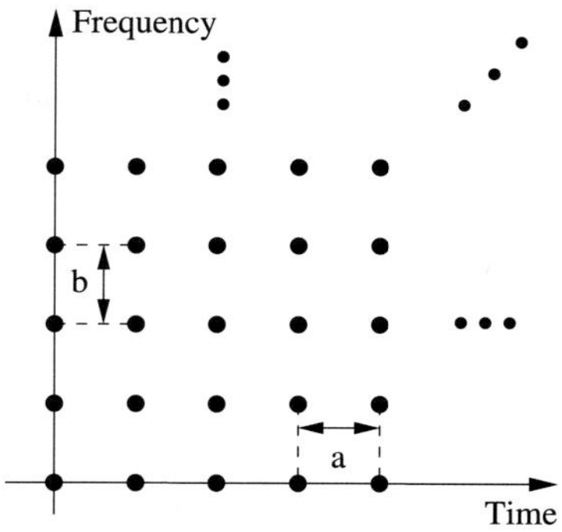

FIGURE 10.5

Gabor system elements location in the time-frequency plane.

that

$$A \leq \sum_{n\in\mathbb{Z}} |g(t-na)|^2 \leq B \;\forall t \in \mathbb{R} \text{ and} \tag{10.64}$$

$$\sum_{k\neq 0} \left\| \sum_{n\in\mathbb{Z}} T_{na} g(t) T_{na+\frac{k}{b}} g^*(t) \right\|_\infty < A. \tag{10.65}$$

Then $\{E_{mb}T_{na}g(t)\}_{m,n\in\mathbb{Z}}$ is a Gabor frame for $L^2(\mathbb{R})$.

A more general sufficient condition [6,49] for the generation of a frame $\{E_{mb}T_{na}g(t)\}_{m,n\in\mathbb{Z}}$ for $a, b > 0$ given and $g(t) \in L^2(\mathbb{R})$ is that if

$$B := \frac{1}{b} \sup_{t\in[0,a]} \sum_{k\in\mathbb{Z}} \left| \sum_{n\in\mathbb{Z}} g(t-na)\, g^*\left(t - na - \frac{k}{b}\right) \right| < \infty, \quad \text{and} \tag{10.66}$$

$$A := \frac{1}{b} \inf_{t\in[0,a]} \left[\sum_{n\in\mathbb{Z}} |g(t-na)|^2 - \sum_{k\neq 0} \left| \sum_{n\in\mathbb{Z}} g(t-na)\, g^*\left(t - na - \frac{k}{b}\right) \right| \right] > 0, \tag{10.67}$$

then $\{E_{mb}T_{na}g(t)\}_{m,n\in\mathbb{Z}}$ is a frame for $L^2(\mathbb{R})$ with frame bounds A, B. Note that this result shows that if $g(t)$ has a limited support $[0, 1/d]$, then for any set $ab \leq 1$ and $b < d$ a Gabor frame is obtained.

Other conditions for the generation of Gabor frames exist (see for example [4,5,6,27]). Reference [50] provides an extension of the results in Eqs. (10.66) and (10.67) for irregularly sampled time-frequency parameters, that is, when the set (an, bm) is replaced by more generic pairs $(a_{n,m}, b_{n,m}) \in [na, (n+1)a] \times [mb, (m+1)b]$.

In Section 10.6.3 we discuss the use of such frames for time-frequency analysis by means of the Gaborgram, and in Section 10.6.4 we present an example of "how to find" the inverse frame to a Gabor frame. In Section 10.6.5 a condition for constructing a Gabor frame in discrete space from its continuous counterpart is stated.

10.5.5 Wavelet frames

In wavelet analysis, translated and scaled versions of function $\psi(t)$ are employed for representing a signal $x(t) \in L(\mathbb{R})$. In this sense, the signal is to be represented using functions like

$$\psi_{c,a}(t) = \frac{1}{\sqrt{c}}\psi\left(\frac{t-a}{c}\right). \tag{10.68}$$

If one thinks about the signal projections over different $\psi_{c,a}(t)$ to represent the signal, the similarity with the Gabor "time-frequency" approach is unavoidable. We apply operations to a fixed prototype function and use the resulting modified versions of the prototype to express the signal. Discrete pairs (a, c) define how the prototype function is modified and thus provide what signal characteristics are being analyzed.

If we employ a continuum of (a, c) pairs, we have the so-called *continuous wavelet transform* [4,5,7,51]

$$W_\psi(a,c) = \langle x(t), \psi_{c,a}(t)\rangle = \int_{-\infty}^{\infty} x(t)\frac{1}{\sqrt{c}}\psi\left(\frac{t-a}{c}\right)dt. \tag{10.69}$$

We can compute the inverse wavelet transform using

$$x(t) = K_\psi \int_0^{\infty}\int_{-\infty}^{\infty} W_\psi(a,c)\frac{1}{\sqrt{c}}\psi\left(\frac{t-a}{c}\right)da\frac{dc}{c^2}. \tag{10.70}$$

In the equations above, we have considered that $\psi(t)$ is real and K_ψ is a constant that depends on $\psi(t)$ [4,52,53]. It is worthy to say that, similarly, the Gabor frame can be linked to the so-called *short time Fourier transform* [4,21].

In the case of wavelet frames, we are concerned with the case of a discrete set of pairs (a, c). In this context, in 1910 [54] Haar has shown that for the function

$$\psi(t) = \begin{cases} 1, & 0 \le t < 1/2, \\ -1, & 1/2 \le t < 1, \\ 0, & \text{otherwise}, \end{cases} \tag{10.71}$$

the set of dilated and translated versions

$$\psi_{j,k}(t) = T_k D_{2^{-j}}\{f(t)\} = 2^{j/2}\psi\left(2^j t - k\right), \quad j,k \in \mathbb{Z},$$

is a basis for $L^2(\mathbb{R})$.

A more thorough study of such possibilities occurred in the beginning of the 1980s. Calderón in 1964 [52] and Grossman and Morlet in 1985 [53] introduced the *wavelet transform*. In continuation, the first *wavelet frame* construction appeared in 1986 in [12]. Great effort was dedicated to the study of wavelets due to their capability of performing multiresolution analysis [4,5,7,13,51]. The book [55] collects some relevant papers for wavelet analysis theory.

We should note that in Eqs. (10.68)–(10.70), the order in which the translation and dilation operations are applied differs from the one previously presented in Section 10.5.2. Therefore, for clarity, we now define what is meant by a wavelet frame.

Definition 10.13. For $c > 1$, $a > 0$, and $g \in L^2(\mathbb{R})$, a frame constructed by dilations and translations as

$$\{T_{nac^j} D_{c^j} g(t)\}_{j,n\in\mathbb{Z}} = \{c^{-\frac{j}{2}} g\left(c^{-j}t - na\right)\}_{j,n\in\mathbb{Z}} \tag{10.72}$$

is called a wavelet frame, as defined in Section 10.5.2. □

In [5] both necessary and sufficient conditions are provided to construct wavelet frames. For example, if $\{T_{nac^j} D_{c^j} g(t)\}_{j,n\in\mathbb{Z}} = \left\{c^{\frac{j}{2}} g\left(c^j t - na\right)\right\}_{j,n\in\mathbb{Z}}$ is a frame with frame bounds A and B, then necessarily [56]

$$A \le \frac{1}{a}\sum_{j\in\mathbb{Z}} \left|\hat{g}\left(c^j f\right)\right|^2 \le B. \tag{10.73}$$

A sufficient condition to generate a wavelet frame [6] is the following. Suppose that $c > 1$, $a > 0$, and $g(t) \in L^2(\mathbb{R})$ are given. If

$$B := \frac{1}{b} \sup_{|f|\in[1,c]} \sum_{j,n\in\mathbb{Z}} \left|\hat{g}\left(c^j f\right)\hat{g}\left(c^j f + \frac{n}{a}\right)\right| < \infty, \quad \text{and} \tag{10.74}$$

$$A := \frac{1}{b} \inf_{|f|\in[1,c]} \left[\sum_{n\in\mathbb{Z}} \left|\hat{g}\left(c^j f\right)\right|^2 - \sum_{n\neq 0}\sum_{j\in\mathbb{Z}} \left|\hat{g}\left(c^j f\right)\hat{g}\left(c^j f + \frac{n}{a}\right)\right|\right] > 0, \tag{10.75}$$

then $\{T_{nac^j} D_{c^j} g(t)\}_{j,n\in\mathbb{Z}}$ is a frame for $L^2(\mathbb{R})$ with bounds A, B given by the expressions above.

Reference [50] provides an extension of this condition for irregularly sampled time-scale parameters, when the set $\left(an, c^j\right)$ is replaced by any pair $\left(a_{n,j}, c_{n,j}\right) \in [c^j an, c^j a(n+1)] \times [c^j, c^{j+1}]$.

10.5.6 Finite vector spaces

When considering finite vector spaces $\mathbb{H}^N$ the simpler solution is to truncate or box-window the elements of the discrete space frame; however, this simple approach alters the frame bounds [57]. An alternative is to consider that the vector space is generated from an N-length section of an N-periodic $l^2(\mathbb{Z})$ space, where the translation operator is a circular shift. The circular shift of a signal does not change the vector norm; this way the frame in $\mathbb{H}^N$ has the same frame bounds as the frame in the N-periodic $l^2(\mathbb{Z})$.

10.6 Some remarks and highlights on applications

10.6.1 Signal analysis

We have discussed the construction of different frames from a fixed prototype function. We have also presented examples that span the possibility of signal representation, analysis, and synthesis using functions with prescribed characteristics. These are some of the main features of frames that are relevant for signal processing applications: detection of signal features and analyzing and synthesizing signals using a set of predefined signals. Frames bring freedom when compared to bases as the number of elements to be employed in the analysis–synthesis process can be larger than the signal space dimension.

In the literature, there is a large number of examples of different frame constructions using a fixed prototype signal; we have presented some in Section 10.5. Fig. 10.6 illustrates different functions with different characteristics that can be used in this framework. It illustrates possible fixed functions to be employed to detect localized phenomena in a signal.

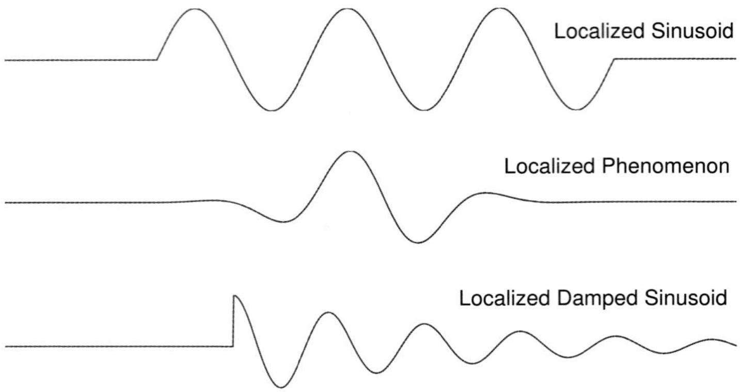

FIGURE 10.6

Examples of fixed functions that can be used to construct frames.

The Gabor and wavelet approaches commonly used for building frames from a fixed function were discussed. In some cases, the density in time, frequency, or scale of the phenomena to be analyzed or detected in the signal [58,24,25] may be such that techniques for reducing the number of frame elements may be handy [59].

Example 10.13. As the reader may have already noticed, wavelet frames can be used in a similar way to Gabor frames for time-frequency analysis. However, in this case a better phrasing would be *time-scale analysis*. Fig. 10.7 depicts the different tilings of the time-frequency plane obtained by these approaches. These tilings depict the Heisenberg boxes [4] of the frame elements – these boxes roughly correspond to the time-frequency plane area where most of the elements' energy is concentrated. □

10.6.2 Robust data transmission

An advantage of frames is that they can increase the robustness of data transmission. Assume that at some point the system experiences losses, that is, a certain number of coefficients are lost or delayed, and the receiver decides to reconstruct the source with the available information. Due to the redundancy introduced by the frames, it may still be possible to reconstruct the original source. In order to do so, however, we must analyze the information contained in the remaining coefficients.

Example 10.14. Consider the set of frame elements $\mathcal{G} = \{[1 \quad 0], [0 \quad 1], [-1 \quad 0], [0 \quad -1]\}$. We can see that this frame is robust to one erasure, since if any element is missing, the remaining elements

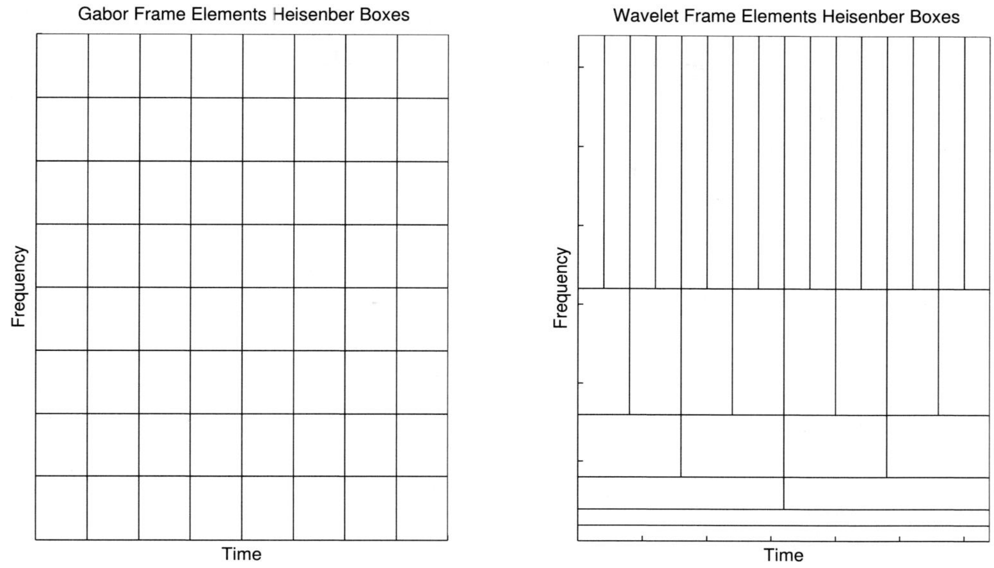

FIGURE 10.7

Heisenberg boxes in the time-frequency plane for Gabor (left) and wavelet (right) frames.

form a frame that spans $\mathbb{R}^2$. On the other hand, for two erasures, it may not be possible to recover the original information. For instance, if the coefficients corresponding to the elements $[1 \quad 0]$ and $[-1 \quad 0]$ are lost, the remaining coefficients do not contain information about one of the signal dimensions.

Consider now the set of frame elements $\mathcal{G} = \{[1 \quad 0], [0 \quad 1], [1 \quad 1], [1 \quad -1]\}$. For this new set of frame elements, under any two erasures, since the remaining elements still span $\mathbb{R}^2$, it is still possible to reconstruct the original vector. □

The above exemplifies that, for a frame with N elements to be resilient to K erasures, any subset of $N - K$ elements must also be a frame. Consider Σ as a diagonal matrix containing 0's in the rows corresponding to the elements that were lost. We can see that the received signal is $\Sigma\mathbf{G}\mathbf{x}$. If Σ is known, the receiver may compute a left inverse for $\Sigma\mathbf{G}$ and use it to reconstruct the original source. Note that under K erasures, if the remaining elements span $\mathbb{R}^N$, the matrix $\Sigma\mathbf{G}$ has a left inverse and a perfect reconstruction can be found. For a large amount of erasures, the matrix $\Sigma\mathbf{G}$ may not have a left inverse and only a partial reconstruction of $\mathbf{x}$ is possible.

Alternatively, if the erasures are not known, the receiver may use the original inverse frame $\tilde{\mathbf{G}}$ and assume that only an approximation of $\mathbf{x}$ can be found. This motivates several studies [16,60] to define "optimal frames," in the sense that they minimize the reconstruction error independently of the elements that were deleted. It can be shown [60], for instance, that Grassmannian frames [32] are optimal for up to two erasures.

Example 10.15. We illustrate the resilience to erasures provided by the frame theory. Consider the image *astronaut* in Fig. 10.2. We split the image in blocks of size 16×16 and encode each block into the coefficients of a Gabor frame with double the number of coefficients. To simulate the loss of information during the transmission, we randomly delete a fraction of the frame coefficients and reconstruct the original image using the remaining ones. For comparison, we perform a similar experiment using instead the discrete cosine transform (DCT) coefficients of the image, which are transmitted under similar conditions. We use the same 16×16 blocks, compute the DCT transform for each block, and randomly delete the same ratio of DCT coefficients as in the frame experiment.

Fig. 10.8(a and c) shows the resulting images obtained when 40% and 60% of the frame coefficients, respectively, are randomly set to zero, while Fig. 10.8(b and d) shows the images obtained when 40% and 60% of the DCT coefficients, respectively, are randomly set to zero.

Note that for the DCT case, even with a low ratio of erasures, some blocks can be completely lost. This behavior is due to the fact that the DCT transform concentrates information in a small number of coefficients. Therefore, if an erasure occurs in one of these coefficients, almost no information for the block can be retrieved. On the other hand, the Gabor coefficients allow the reconstruction of the original image with no noticeable error when the number of erasures is small enough.

This experiment was repeated for several erasure ratios, with multiple realizations being performed for each erasure ratio. In each realization, the MSE between the original and reconstructed images using either the frame or the DCT coefficients were computed. For each erasure ratio, the average MSE was taken among all realizations. The results are shown in Fig. 10.9, where it is possible to see that this Gabor frame provides immunity to erasures of up to 50% of the coefficients. This ratio is compatible with the fact that the employed frame contains twice as many elements as the space dimension. This redundancy can be regarded as the price that has to be paid for having robustness to this level of erasures. In addition, under a large erasure ratio that implies a reconstruction error using the remaining frame coefficients, the reconstruction error is smaller than the one generated by the DCT under the same condition. □

10.6.3 Gaborgram

In the beginning of Section 10.5.4 we have presented the idea of Gabor, that is, of representing a signal $x(t)$ by means of

$$x(t) = \sum_m \sum_n c_{m,n} E_{mb} T_{na} g(t) = \sum_m \sum_n c_{m,n} e^{jmb} g(t - na). \tag{10.76}$$

As can be noticed, this equation assumes that there is a representation for any $x(t)$ using several $g_{mb,na}(t) = E_{mb} T_{na} g(t)$. We have seen that for these to be true it is enough for the set $\{E_{mb} T_{na} g(t)\}_{m,n}$ to be a frame. This is commonly referred to as the Gaborgram, and consists in the plot of the coefficients $c_{m,n}$ using the lattice presented in Fig. 10.5. As discussed above a similar concept can be used for wavelet frames.

We have also seen that to compute the frame coefficients $c_{m,n}$, the inverse frame to $\{E_{mb} T_{na} g(t)\}_{m,n}$ must be used. The next section illustrates how to obtain the inverse frame in the case of an exponential function. This gains special relevance if one notes that a Gabor frame of damped sinusoids is inspired by the connection among transients and damped sinusoids in physical systems.

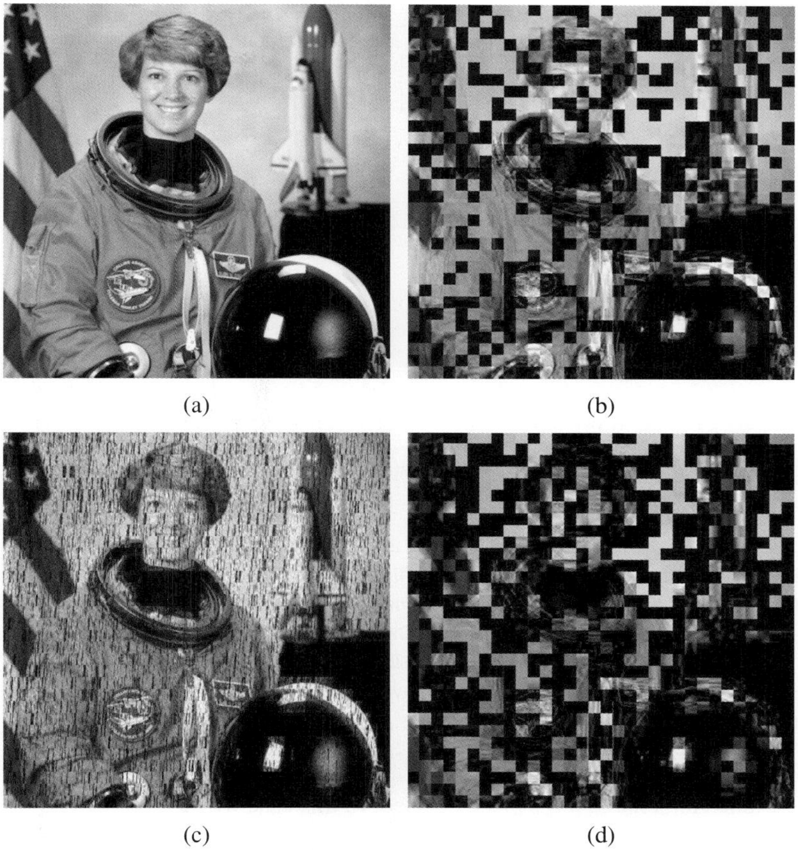

FIGURE 10.8

Impact of frame coefficient erasures in image reconstruction. (a) Image *astronaut* recovered after it was split in blocks of size 16×16, each block being encoded into the coefficients of a Gabor frame, 40% of which were randomly set to zero. (b) A similar case in which instead of the Gabor frame a DCT transform is employed. Subfigures (c) and (d) present results equivalent to (a) and (b), respectively, when 60% of coefficients are set to zero.

10.6.4 Inverse Gabor frame

The power of the Gabor system for signal analysis and synthesis is not difficult to perceive. The problem is how to find the coefficients $c_{m,n}$ that provide the reconstruction

$$x(t) = \sum_m \sum_n c_{m,n} E_{mb} T_{na} g(t) = \sum_m \sum_n c_{m,n} e^{jmb} g(t - na). \tag{10.77}$$

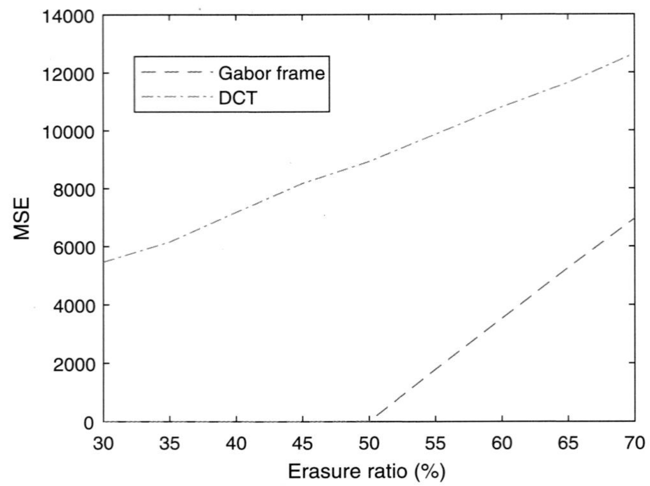

FIGURE 10.9

Reconstruction error (MSE) with respect to the ratio of erasures. The blue (mid gray in print version) dashed curve shows the error obtained when the image *astronaut* is split in 16×16 blocks, each block is converted to the coefficients of a Gabor frame, a proportion of coefficients are set to zero, and the image is reconstructed from the remaining coefficients. For comparison purposes, the red (light gray in print version) dash-dotted curve shows the case where the 16×16 blocks are converted to the coefficients of a DCT transform and the same proportion of erasures is applied to its coefficients.

The inverse frame elements $\{\tilde{g}(t)\}_{m,n}$ can be used. In this case we have

$$x(t) = \sum_m \sum_n \langle x(t), \tilde{g}_{m,n}(t)\rangle E_{mb} T_{na} g(t). \tag{10.78}$$

Daubechies [13] has shown that, for Gabor frames, the inverse frame must be such that

$$\tilde{g}_{m,n}(t) = E_{mb} T_{na} \tilde{g}(t). \tag{10.79}$$

We have

$$x(t) = \sum_m \sum_n \langle x(t), E_{mb} T_{na} \tilde{g}(t)\rangle E_{mb} T_{na} g(t). \tag{10.80}$$

From the above equations, we notice that $\tilde{g}(t)$ must be known in order to compute the inverse frame. Eq. (10.80) provides a biorthogonality condition between $g(t)$ and $\tilde{g}(t)$ (the “frame decomposition” formulas discussed in Section 10.3.1). This biorthogonality condition is derived in [61] from Eq. (10.80), where it becomes

$$\frac{1}{ab} \int g(t) \tilde{g}\left(t - \frac{n}{b}\right) e^{-\mathrm{j}2\pi m \frac{t}{a}} dt = \delta(n)\delta(m). \tag{10.81}$$

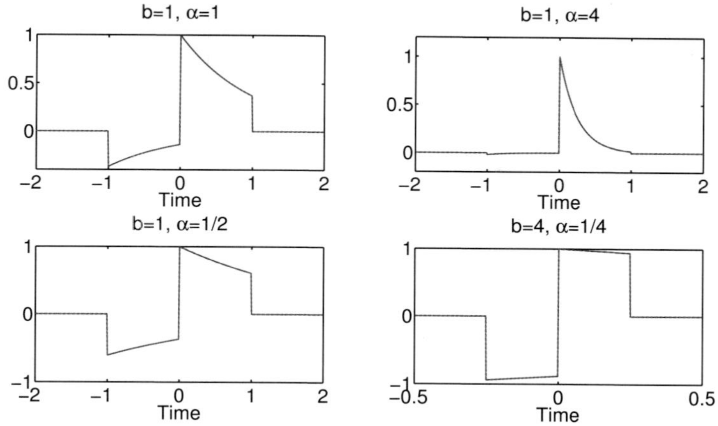

FIGURE 10.10

Examples of inverse Gabor frame prototype function of the one-sided exponential function (the set $\tilde{g}(t)$ biorthogonal to $g(t) = \sqrt{2\alpha}e^{-\alpha t}u(t)$) for different values of α.

Here $\delta(x)$ denotes the impulse of x. In [61] it is also discussed that Eq. (10.81) accepts more than one solution in the oversampled case ($ab < 1$) [27].

Example 10.16. In [24] the problem of building a Gabor system (or Gabor frame) based on the one-sided exponential function is presented. In this case, we have

$$g(t) = \begin{cases} \sqrt{2\alpha}e^{-\alpha t}, & t \geq 0, \\ 0, & \text{otherwise.} \end{cases} \tag{10.82}$$

Using the Zak transform [62,48], Ref. [24] shows that for large oversampling ratio ($ab < 1$),

$$\tilde{g}(t) \approx \begin{cases} -ke^{-\alpha\left(t+\frac{2}{b}\right)}, & -1 \leq tb < 0, \\ ke^{-\alpha t}, & 0 \leq tb < 1, \\ 0, & \text{otherwise,} \end{cases} \tag{10.83}$$

where k is a constant that depends on a, b, and α [24].

Fig. 10.10 illustrates $\tilde{g}(t)$ for different cases of a and b. In order to just focus on the waveform of $\tilde{g}(t)$, we ignore here the value of the constant k and make all functions scaled so that $\max_t \tilde{g}(t) = 1$. □

10.6.5 Gabor frames in discrete spaces

Due to the time-shift frequency-shift structure of Gabor frames, these have a broad range of applications for signal processing. Since most algorithms take place in a discrete space we should evaluate how to construct Gabor frames in this space.

Frames in discrete spaces ($l^2(\mathbb{Z})$) can be obtained by time-sampling the elements of frames in $L^2(\mathbb{R})$ [6]. If $g(t) \in L^2(\mathbb{R})$ is such that

$$\lim_{\epsilon \to 0} \sum_{k \in \mathbb{Z}} \frac{1}{\epsilon} \int_{-\frac{1}{2}\epsilon}^{\frac{1}{2}\epsilon} |g(k+t) - g(k)|^2 dt = 0 \tag{10.84}$$

and $\{E_{m/Q} T_{n/P} g(t)\}_{m,n \in \mathbb{Z}}$ with $Q, P \in \mathbb{N}$ is a frame for $L^2(\mathbb{R})$ with frame bounds A and B, then $\{E_{m/Q} T_{n/P} \mathbf{g}_D\}_{n \in \mathbb{Z},\ m=0,1,\ldots,M-1}$ is a frame for $l^2(\mathbb{Z})$ with frame bounds A and B [6]. The vector $\mathbf{g}_D$ is the discretized version of $g(t)$ with one sample per time unit, i.e., $\mathbf{g}_D = \{\mathrm{g}(j)\}_{j \in \mathbb{Z}}$.

10.6.6 Fast analysis and synthesis operators for Gabor frames

Gabor or Weyl–Heisenberg frames in N-dimensional spaces are also interesting because one can find fast algorithms to compute their analysis and synthesis operators. Now, we consider that the number of "points" of the Weyl–Heisenberg frame in the frequency axis is such that $Q = N/r$, $Q, r \in \mathbb{N}$, and also that the number of points in the time axis $P \in \mathbb{N}$ (the same restrictions were employed in Section 10.6.5). These restrictions actually lead to $a = 1/Q$ and $b = N/P$ in Eq. (10.61). In this case, $c_{n,m}$ are defined for $n \in \{0 \ \ldots \ P-1\}$ and $m \in \{0 \ \ldots \ Q-1\}$ and are given by

$$c_{n,m} = \langle \mathbf{x}, E_{m\frac{1}{Q}} T_{n\frac{N}{P}} \mathbf{g} \rangle = \sum_{l=0}^{N-1} \mathrm{x}[l] e^{\frac{-\mathrm{j}2\pi l m}{Q}} T_{n\frac{N}{P}} \mathrm{g}[l]. \tag{10.85}$$

Defining

$$\mathrm{f}_{n\frac{N}{P}}[k] = \mathrm{x}[k] T_{n\frac{N}{P}} \mathrm{g}[k] = \mathrm{x}[k] \mathrm{g}\left[\left(k - n\frac{N}{P}\right) \mathrm{mod} N\right], \tag{10.86}$$

$$\mathbf{f}_{n\frac{N}{P}} = \left[\mathrm{f}_{n\frac{N}{P}}[0], \ldots, \mathrm{f}_{n\frac{N}{P}}[N-1]\right], \tag{10.87}$$

we obtain

$$c_{n,m} = \langle \mathbf{x}, E_{m\frac{1}{Q}} T_{n\frac{N}{P}} \mathbf{g} \rangle = \sum_{l=0}^{N-1} \mathrm{f}_{n\frac{N}{P}}[l] e^{\frac{-\mathrm{j}2\pi l m}{Q}}. \tag{10.88}$$

For $Q = N/r$, $r \in \mathbb{N}$, we obtain

$$c_{n,m} = \langle \mathbf{x}, E_{m\frac{1}{Q}} T_{n\frac{N}{P}} \mathbf{g} \rangle = \sum_{l=0}^{N-1} \mathrm{f}_{n\frac{N}{P}}[l] e^{\frac{-\mathrm{j}2\pi l}{N} mr} = DFT\{\mathbf{f}_{n\frac{N}{P}}\}[mr], \tag{10.89}$$

where $DFT\{\mathbf{x}\}[k]$ is the kth sample of the discrete Fourier transform of $\mathbf{x}$, which can be computed using fast algorithms (FFT) [31].

Using the result in Eq. (10.89), the analysis operator (which was discussed in Section 10.4.2) of such Weyl–Heisenberg frames is given by

$$\mathbf{T}^*\{\cdot\}: \mathbb{H}^N \to \mathbb{C}^{PQ}, \tag{10.90}$$

$$\mathbf{T}^*\{\mathbf{x}\}: c_{n,m} = \{\langle \mathbf{x}, E_{m\frac{1}{Q}} T_{n\frac{N}{P}} \mathbf{g} \rangle\},\ n \in [0, P-1],\ m \in [0, Q-1], \tag{10.91}$$

$$c_{n,m} = FFT\{\mathbf{f}_{n\frac{N}{P}}\}[mr],\ \mathrm{f}_{n\frac{N}{P}}[k] = \mathrm{x}[k]\mathrm{g}\left[\left(k - n\frac{N}{P}\right) \bmod N\right]. \tag{10.92}$$

Similarly, for the synthesis operator of such frames we have

$$\mathbf{T}\{\cdot\}: \mathbb{C}^{PQ} \to \mathbb{H}^N,$$

$$\mathbf{T}\{c_{n,m}\} = [\mathrm{x}[0], \ldots, \mathrm{x}[N-1]] = \sum_{n=0}^{P-1}\sum_{m=0}^{Q-1} c_{n,m} E_{m\frac{1}{Q}} T_{n\frac{N}{P}} \mathbf{g}. \tag{10.93}$$

The last equation is the sum of PQ length-N vectors and can be computed using the inverse discrete Fourier transform as follows:

$$\mathrm{x}[l] = \sum_{n=0}^{P-1}\sum_{m=0}^{Q-1} c_{n,m} e^{\mathrm{j}\frac{2\pi ml}{Q}} \mathrm{f}_{n\frac{N}{P}}[l] = \sum_{n=0}^{P-1} \mathrm{f}_{n\frac{N}{P}}[l] \sum_{m=0}^{Q-1} c_{n,m} e^{\mathrm{j}\frac{2\pi}{N} lmr}. \tag{10.94}$$

Denoting $c_n(m) = c_{n,m}$ we can define

$$\mathbf{c}_n = [\mathrm{c}_n[0], \mathrm{c}_n[1], \ldots, \mathrm{c}_n[Q-1]] \tag{10.95}$$

and its upsampled version

$$U(\mathbf{c}_n)[k] = \begin{cases} \mathrm{c}_n[k/r],\ k/r \in \{0, 1, \ldots, Q-1\}, \\ 0,\ \text{otherwise}. \end{cases} \tag{10.96}$$

Thus, the synthesis operator is such that

$$\mathrm{x}[l] = \sum_{n=0}^{P-1} \mathrm{f}_{n\frac{N}{P}}[l] \sum_{m=0}^{Q-1} U(\mathbf{c}_n)[mr] e^{\mathrm{j}\frac{2\pi}{N} lmr} = \sum_{n=0}^{P-1} \mathrm{f}_{n\frac{N}{P}}[l] IFFT\{U(\mathbf{c}_n)\}[l]. \tag{10.97}$$

Fig. 10.11 presents a way to compute the expansion coefficients or the reconstructed signal when a frame is used either as an analysis or as a synthesis operator. Note that in the outputs of each FFT block in Fig. 10.11, there is a serial-to-parallel converter, while at the inverse FFT (IFFT) inputs a parallel-to-serial converter exists. In Fig. 10.11 all the frame coefficients can be obtained using $PN(1 + \log_2(N))$ operations, which can be reduced if one takes into account that just Q FFT/IFFT coefficients need to be computed at each FFT/IFFT branch.

10.6.7 Time-frequency content analysis using frame expansions

As we have seen in several cases, frames provide a powerful tool for signal analysis. This is specially true when one considers Gabor and wavelet frames. These provide a natural tiling of the time-frequency plane. We now discuss how a signal expansion can be mapped into a time-frequency plot.

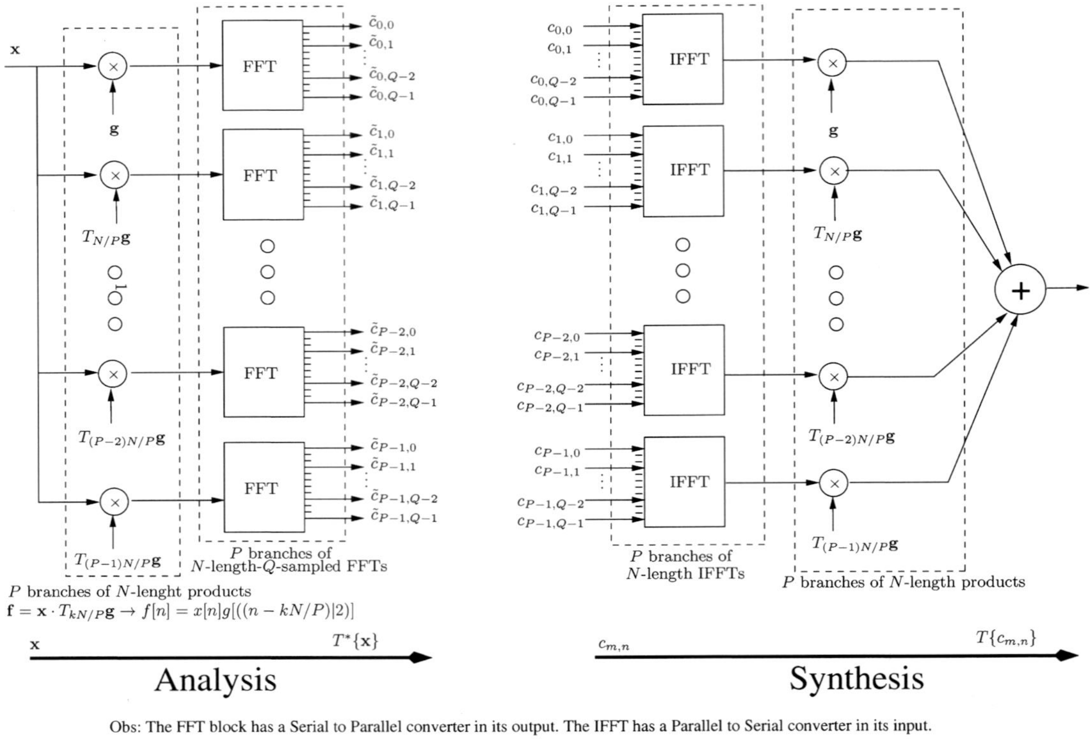

FIGURE 10.11

Weyl–Heisenberg fast analysis and synthesis using FFTs and IFFTs.

The Wigner–Ville distribution (WD) is probably one of the most well-known and widely used tools for time-frequency analysis of signals. The WD of a signal $x(t)$ is defined as [4,20,63,18]

$$WD_x(t, f) = \int_{-\infty}^{+\infty} x\left(t + \frac{\tau}{2}\right) x^*\left(t - \frac{\tau}{2}\right) e^{-2\pi jf\tau} d\tau. \tag{10.98}$$

In some applications the WDs of signal decompositions have been employed in order to analyze the time-frequency content of signals [4,20]. The WD of a signal $x(t)$, $WD_x(t, f)$, is a "measure" of the energy density in the signal in both time (t) and frequency (f) simultaneously. However, it is just meaningful when regions of the time-frequency plane are considered as a whole (that is, as local averages) – due to the uncertainty principle a single time-frequency point (t', f') cannot be properly analyzed [4,20]. In addition, the WD of a signal has the drawback of not being restricted to be positive, a mandatory characteristic for an energy density.

In Section 10.3, we have seen that a signal x can be decomposed into a frame $\{g_k\}_{k\in\mathcal{K}}$ or into its dual $\{\tilde{g}_k\}_{k\in\mathcal{K}}$ and reconstructed by means of

$$x = \sum_{k\in\mathcal{K}} \langle x, \tilde{g}_k\rangle g_k = \sum_{k\in\mathcal{K}} \langle x, g_k\rangle \tilde{g}_k. \tag{10.99}$$

Once x is decomposed into the frame coefficients $\langle x, \tilde{g}_k\rangle$ as a linear combination of the frame elements, its time-frequency content can be analyzed using the linear combination of the time-frequency content of the frame elements as follows:

$$WD_{\tilde{x}}(t, f) \quad = \sum_{k\in\mathcal{K}} \langle x(t), \tilde{g}_k(t)\rangle \, WD_{g_k}(t, f). \tag{10.100}$$

Approaches such as the one above that infer signal characteristics from signal decompositions are common and powerful tools for signal analysis, detection, and estimation. Specifically, in the case of time-frequency analysis, such approaches avoid the cross-terms that appear when the original signal is directly analyzed [4,20,43,63,18,19,21,22].

It is important to note that there are cases where $\tilde{g}_k(t)$ do not, neither physically nor visually or in its time-frequency content, resemble the associated $g_k(t)$. The coefficients used for obtaining the time-frequency content are computed as $\langle x(t), \tilde{g}_k(t)\rangle$, which is a measure of similarity between $x(t)$ and $\tilde{g}_k(t)$. This is inconvenient in the case that $g_k(t)$ and $\tilde{g}_k(t)$ are too dissimilar; therefore, in general this approach is left aside.

For instance, in [43] the time-frequency content of signals is estimated using the matching pursuit (MP) signal decomposition algorithm, a step-based decomposition algorithm that iterates to successively obtain better signal approximations [64]. The approximation is accomplished by selecting the dictionary element that best matches the signal. This match is evaluated by projecting the signal on the dictionary elements. The best-matching element is scaled by its corresponding projection and subtracted from the signal generating a residual error. The process is then iteratively applied to the residual error. This algorithm is greedy [65,66,67] in the sense that it minimizes the approximation error at each iteration. However, it is not optimal as a sequence of greedy iterations occur, and due to this fact several variants for MP exist [67,68,69,70].

Although the dictionary employed in MP is not required to be a frame, it is a common procedure to employ a frame as a dictionary [59,71,72], because this choice guarantees that any signal can be analyzed and synthesized. It is common to use a mix of Gabor and wavelet frames to build the dictionary used in MP [43]. Doing this, the dictionary elements are placed in different locations in the time-frequency plane and the several scales provide different concentrations for the energy of the elements in time and frequency. A widely used dictionary in this context is one formed by Gaussian functions in different scales with different time and frequency shifts. In this case one obtains the MP expansion of a signal $x(t)$ into the Gabor dictionary as

$$x(t) \approx \sum_{n=1}^{M} \gamma_n g_{i(n)}(t), \tag{10.101}$$

where $i(n)$ indexes the element selected for approximating the residual error in the decomposition step n. From this representation the time-frequency content of $x(t)$ is analyzed by means of

$$WD_x(t, f) = \sum_{n=1}^{M} \gamma_n WD_{g_{i(n)}}(t, f). \tag{10.102}$$

Note that the time-frequency analysis approach using MP decomposition, associated with Eqs. (10.101) and (10.102), resembles the one associated with Eqs. (10.99) and (10.100). However, in the MP algorithm the coefficients γ_n are obtained by a similarity calculation with the function $g_{i(n)}(t)$ itself, whose time-frequency analysis is used in Eq. (10.102). This circumvents the problem of the dissimilarity between $g_k(t)$ and $\tilde{g}_k(t)$ discussed above in relation to Eq. (10.100). In [73] a similar approach is used, but using a modified version of MP to obtain the signal expansion. Note the resemblance between the approach in Eq. (10.100) and the ones in Eqs. (10.101) and (10.102).

10.7 Conclusion

In this chapter we have provided an overview of frames and their applications in signal processing. We started by introducing the concept of overcomplete decompositions, highlighting their main characteristics and showing their advantages using practical examples. We then commented on the dual frames and on how frames can be employed to perform analysis and synthesis of signals. After this, we provided formal definitions of the frame operator, inverse frames, and frame bounds and illustrated their usage. We then formalized the use of frames in discrete and finite-dimensional spaces. Shifting into more practical matters, we showed how to generate frames from a prototype function by using translations, modulations, and dilations. From this, we analyzed the widely used frames of translates, Gabor frames, and wavelet frames. We then described selected applications in signal analysis, where we showed that frame representations have intrinsic properties that can increase the resilience to noise and losses in robust data transmission systems. We formalized the Gaborgram and presented a way to compute it more efficiently using the FFT. We finished the chapter by discussing how to compute time-frequency analysis using decompositions and the MP algorithm.

References

[1] R.J. Duffin, A.C. Schaeffer, A class of nonharmonic Fourier series, Transactions of the American Mathematical Society 72 (1952) 341–366.
[2] R. Young, An Introduction to Nonharmonic Fourier Series, Academic Press, New York, 1980.
[3] R. Paley, N. Wiener, The Fourier Transforms in the Complex Domain, American Mathematical Society Colloquium Publ. Ser., vol. 19, 1934.
[4] S. Mallat, A Wavelet Tour of Signal Processing, 1st edition, Academic Press, San Diego, California, USA, 1998.
[5] I. Daubechies, Ten Lectures on Wavelets, Society for Industrial and Applied Mathematics, Philadelphia, Pennsylvania, USA, 1991.
[6] O. Christensen, An Introduction to Frames and Riesz Bases, 2nd edition, Applied and Numerical Harmonic Analysis, Birkhäuser, Boston, Basel, Berlin, 2016.

[7] C.S. Burrus, R.A. Gopinath, H. Guo, Introduction to Wavelets and Wavelets Transforms a Primer, 1st edition, Prentice Hall, Upper Saddle River, New Jersey 07458, USA, 1998.
[8] Z. Cvetkovic, M. Vetterli, On simple oversampled A/D conversion in $l^2(\mathbb{R})$, IEEE Transactions on Information Theory 47 (1) (2001) 146–154.
[9] F. Marvasti (Ed.), Nonuniform Sampling: Theory and Practice, 1st edition, Springer, 2001.
[10] M. Vetterli, P. Marziliano, T. Blu, Sampling signals with finite rate of innovation, IEEE Transactions on Signal Processing 50 (6) (June 2002) 1417–1428.
[11] R.G. Baraniuk, Compressive sensing [lecture notes], IEEE Signal Processing Magazine 24 (4) (2007) 118–121.
[12] I. Daubechies, A. Grossman, Y. Meyer, Painless nonorthogonal expansions, Journal Mathematical Physics 27 (1986) 1271–1283.
[13] I. Daubechies, The wavelet transform, time-frequency localization and signal analysis, IEEE Transactions on Information Theory 36 (5) (1990) 961–1005.
[14] C. Heil, D. Walnut, Continuous and discrete wavelet transforms, SIAM Review 31 (1989) 628–666.
[15] V.K. Goyal, J. Kovacevic, J.A. Kelner, Quantized frame expansions with erasures, Applied Computational Harmonic Analysis 10 (2001) 203–233.
[16] P.G. Casazza, J. Kovacevic, Equal-norm tight frames with erasures, Advances in Computational Mathematics–Especial ISSUE on Frames (2002) 387–430.
[17] J. Kovacevic, P.L. Dragotti, V.K. Goyal, Filter bank frame expansions with erasures, IEEE Transactions on Information Theory 48 (6) (June 2002) 1439–1450.
[18] G. Matz, F. Hlawatsch, Wigner distributions (nearly) everywhere: time-frequency analysis of signals, systems, random processes, signal spaces, and frames, Elsevier Signal Processing 83 (2003) 1355–1378.
[19] L. Cohen, Time-frequency distributions - a review, Proceeding of the IEEE 77 (7) (July 1989) 941–981.
[20] F. Hlawatsch, G.F. Boudreaux-Bartels, Linear and quadratic time-frequency signal representations, IEEE Signal Processing Magazine 9 (April 1992) 21–67.
[21] Leon Cohen, Time-Frequency Analysis, Prentice Hall Signal Processing Series, Prentice Hall, Prentice Hall PTR, Englewood Cliffs, New Jersey 07632, 1995.
[22] Kalheiz Gröchenig, Foundations of Time-Frequency Analysis, Birkhäuser, New York, USA, 2001.
[23] D. Gabor, Theory of communications, IEE Journal 93 (1946) 429–457.
[24] B. Friedlander, A. Zeira, Oversampled Gabor representation for transient signals, IEEE Transactions on Signal Processing 43 (9) (September 1995) 2088–2094.
[25] B. Friedlander, B. Porat, Detection of transient signals by the Gabor representation, IEEE Transactions on Acoustics, Speech, and Signal Processing 37 (2) (February 1989) 169–180.
[26] M. Zibulski, Y.Y. Zeevi, Discrete multiwindow Gabor-type transforms, IEEE Transactions on Signal Processing 45 (1997) 1428–1442.
[27] M. Zibulski, Y.Y. Zeevi, Oversampling in the Gabor scheme, IEEE Transactions on Signal Processing 43 (9) (August 1993) 2679–2687.
[28] D.L. Donoho, M. Vetterli, R.A. DeVore, I. Daubechies, Data compression and harmonic analysis, IEEE Transactions on Information Theory 44 (6) (October 1998) 2435–2476.
[29] E.J. Candes, D.L. Donoho, Ridgelets: a key to higher-dimensional intermittency?, Philosophical Transactions: Mathematical, Physical and Engineering Sciences 357 (1999) 2495–2509.
[30] D.L. Donoho, A.G. Flesia, Can recent innovations in harmonic analysis 'explain' key findings in natural image statistics?, Network: Computation in Neural Systems, Taylor & Francis 12 (2001) 371–393.
[31] P.S.R. Diniz, E.A.B. da Silva, S. Lima Netto, Digital Signal Processing: System Analysis and Design, second edition, Cambridge University Press, 2010.
[32] T. Strohmer, R.W. Heath Jr., Grassmannian frames with applications to coding and communication, Applied Computational Harmonic Analysis 14 (3) (2003) 257–275.
[33] Z. Cvetkovic, Martin Vetterli, Oversampled filter banks, IEEE Transactions on Signal Processing 46 (5) (1998) 1245–1255.

[34] H. Bölcksei, F. Hlawatsh, H.G. Feichtinger, Frame-theoretic analysis of oversampled filter banks, IEEE Transactions on Signal Processing 46 (12) (August 1998) 3256–3268.

[35] Stéfan van der Walt, Johannes L. Schönberger, Juan Nunez-Iglesias, François Boulogne, Joshua D. Warner, Neil Yager, Emmanuelle Gouillart, Tony Yu, the scikit-image contributors, scikit-image: image processing in Python, PeerJ 2 (6 2014) e453.

[36] V.K. Goyal, M. Vetterli, N.T. Thao, Quantized overcomplete expansions in $\mathbb{R}^N$: analysis, synthesis, and algorithms, IEEE Transactions on Information Theory 44 (1) (January 1998) 16–31.

[37] N.J. Munch, Noise reduction in tight Weyl-Heisenberg frames, IEEE Transactions on Information Theory 38 (2) (March 1992) 608–616.

[38] J.A. Tropp, I.S. Dhillon, R.W. Heath Jr., T. Strohmer, Designing structured tight frames via an alternating projection method, IEEE Transactions on Information Theory 51 (2005) 188–209.

[39] Y.C. Eldar, G.D. Forney Jr., Optimal tight frames and quantum measurement, IEEE Transactions on Information Theory 48 (3) (2002) 599–610.

[40] Y.C. Eldar, H. Bolcskei, Geometrically uniform frames, IEEE Transactions on Information Theory 49 (4) (April 2003) 993–1006.

[41] J.J. Benedetto, M. Fickus, Frame potentials, Advances in Computational Mathematics 18 (2003) 357–385.

[42] R. Vale, S. Waldron, Tight frames and their symmetries, Constructive Approximation 21 (2005) 83–112.

[43] S. Mallat, Z. Zhang, Matching pursuits with time-frequency dictionaries, IEEE Transactions on Signal Processing 41 (12) (December 1993) 3397–3415.

[44] A.J. Jerri, The Shannon sampling theorem – its various extensions and applications: a tutorial review, Proceedings of the IEEE 65 (11) (Nov. 1977) 1565–1596.

[45] R.J. Marks II, Introduction to Shannon Sampling and Interpolation Theory, Springer-Verlag, 1991.

[46] J. von Neumann, Mathematische Grundlagen der Quantenmechanik, Springer, Berlin, 1932, English version: Mathematical Foundations of Quantum Mechanics, Princeton Univ. Press, 1955.

[47] Shie Qian, Dapang Chen, Joint time-frequency analysis, IEEE Signal Processing Magazine (March 1999) 52–67.

[48] A.J.E.M. Janssen, Zak transforms with few zeros and the tie, in: H.G. Feichtinger, T. Strohmer (Eds.), Advances in Gabor Analysis, Birkhäuser, Boston, 2002.

[49] P.G. Casazza, O. Christensen, Weyl-Heisenberg frames for subspaces of $L^2(\mathbb{R})$, Proceedings American Mathematical Society 129 (2001) 145–154.

[50] W. Sun, X. Zhou, Irregular wavelet/Gabor frames, Applied Computational and Harmonic Analysis 13 (2002) 63–76.

[51] Mladen Victor Wickerhauser, Adapted Wavelet Analysis from Theory to Software, IEEE Press, Piscataway, NJ, USA, 1998.

[52] A.P. Calderón, Intermediate spaces and interpolation, the complex method, Studia Mathematica 24 (1964) 113–190.

[53] A. Grossmann, J. Morlet, Decomposition of Hardy functions into square integrable wavelets of constant shape, SIAM Journal on Mathematical Analysis 15 (1984) 723–736.

[54] A. Haar, Zur theorie der orthogonalen funktionensysteme, Mathematische Annalen 69 (1910) 331–371, Translated version: on the theory of orthogonal function systems, by G. Zimmermann.

[55] Christopher Heil, David F. Walnut, Fundamental Papers in Wavelet Theory, Princeton University Press, 2006.

[56] C. Chui, X. Shi, Inequalities of Littlewod-Paley type for frames and wavelets, SIAM Journal on Mathematical Analysis 24 (1993) 263–277.

[57] T. Strohmer, Numerical analysis of the non-uniform sampling problem, Computational Applied Mathematics 122 (2000) 297–316.

[58] R. Balan, I. Daubechies, V. Vaishampayan, The analysis and design of windowed Fourier frame based multiple description source coding schemes, IEEE Transactions on Information Theory 46 (November 2000) 2491–2536.

[59] K. Engan, S.O. Aase, J.H. Husoy, Designing frames for matching pursuits algorithms, in: IEEE International Conference on Acoustics, Speech, and Signal Processing, May 1998, pp. 1817–1820.
[60] Roderick B. Holmes, Vern I. Paulsen, Optimal frames for erasures, Linear Algebra and Its Applications 377 (2004) 31–51.
[61] J. Wexler, S. Raz, Discrete Gabor expansions, Signal Processing 21 (3) (1990) 207–221.
[62] J. Zak, Finite translations in solid-state physics, Physics Review Letters 19 (1967) 1385.
[63] T. Claasen, W. Mecklenbrauker, The aliasing problem in discrete-time Wigner distributions, IEEE Transactions on Acoustics, Speech, and Signal Processing 31 (1983) 1067–1072.
[64] Lisandro Lovisolo, Eduardo A.B. da Silva, Paulo S.R. Diniz, On the statistics of matching pursuit angles, Signal Processing 90 (2010).
[65] R.A. DeVore, V.N. Temlyakov, Some remarks in greedy algorithms, Advances in Computational Mathematics 5 (1996) 173–187.
[66] G. Davis, S. Mallat, M. Avellaneda, Adaptive greedy approximations, Journal of Constructive Approximation 13 (1997) 57–98.
[67] J.A. Tropp, Greed is good: algorithmic results for sparse approximation, IEEE Transactions on Information Theory 50 (2004) 2231–2241.
[68] Y.C. Pati, R. Rezaiifar, P.S. Krishnaprasad, Orthogonal matching pursuit: recursive function approximation with applications to wavelet decomposition, in: IEEE Twenty-Seventh Asilomar Conference on Signals, Systems and Computers, 1993.
[69] J.A. Tropp, A.C. Gilbert, Signal recovery from random measurements via orthogonal matching pursuit, IEEE Transactions on Information Theory 53 (2007) 4655–4666.
[70] L. Rebollo-Neira, D. Lowe, Optimized orthogonal matching pursuit approach, IEEE Signal Processing Letters 9 (2002) 137–140.
[71] L. Lovisolo, M.A.M. Rodrigues, E.A.B. da Silva, P.S.R. Diniz, Efficient coherent decompositions of power systems signals using damped sinusoids, IEEE Transactions on Signal Processing 53 (2005) 3831–3846.
[72] P.J. Durka, D. Ircha, K.J. Blinowska, Stochastic time-frequency dictionaries for matching pursuit, IEEE Transactions on Signal Processing 49 (2001) 507–510.
[73] A. Papandreou-Suppappola, S.B. Suppappola, Analysis and classification of time-varying signals with multiple time-frequency structures, IEEE Signal Processing Letters 9 (2002) 92–95.

CHAPTER

11 Parametric estimation

Ryan M. Corey[a], Suleyman Serdar Kozat[b], and Andrew C. Singer[a]

[a] *University of Illinois, Urbana, IL, United States*

[b] *Bilkent University, Ankara, Turkey*

11.1 Introduction

In signal processing, we often wish to recover a small amount of useful information from a large amount of data, such as the words spoken in a sound recording, the binary message encoded in a cellular broadcast, or positional coordinates from a set of readings from a sensor network. We can infer these parameters of interest from their effects on the structure and content of the signal that we observe. Similarly, we might wish to analyze or control a complex system, such as an electronic circuit or a communication channel, that generated or modified an observed signal. We can often approximate such an unknown system using a model with a tractable number of parameters that we can infer from the observed signal.

11.1.1 What is parametric estimation?

Parametric estimation is the practice of modeling an unknown signal or system by a finite number of free parameters. Using prior knowledge or informed assumptions about the signal or system of interest, we choose a model with known structure but unknown parameter values. Then, we use optimization methods to select the parameter values that best explain the observed signal. For example, if we are given an infinite length signal but we know in advance that the signal is a sinusoid, then we can fully represent it with only three parameters: its frequency, amplitude, and phase. More often, a parametric model can capture only part of the information in a signal; for example, a transcript of a speech recording reflects the words spoken but not the tone, cadence, or identity of the speaker. In other cases, we know little about the signal itself but can model the system that produced or acted on it. For example, if the same sound is recorded by a compact array of several identical microphones, then the signal in each microphone will differ from the signal in each other microphone by a time delay that depends on the direction of the source. By estimating those time delays, we can infer the location of the sound source, which is itself a parameter of the acoustic system.

It is useful to compare parametric and nonparametric methods for analyzing data. Suppose that we are given a set of unordered data, such as an opinion poll, and asked to analyze its distribution. We might generate a histogram of the data, which can capture fine detail but is difficult to summarize. If we have reason to believe that the responses are normally distributed, however, we can use a parametric model, estimating the mean and variance of the normal distribution from our dataset. For ordered data, or signals, we are often interested in their spectral content. We can analyze the spectrum nonparametrically

Signal Processing and Machine Learning Theory. https://doi.org/10.1016/B978-0-32-391772-8.00017-X
Copyright © 2024 Elsevier Inc. All rights reserved.

by taking a discrete Fourier transform. Because the transform has the same number of samples as the original sequence, this analysis might not be useful for a long signal. If we know that the signal was generated by a linear system with a tractable number of poles and zeros, such as a human vocal tract, then we can instead fit the observed data to a linear system model whose parameters are its rational transfer function coefficients. By estimating these parameters, we can approximate the system and study its frequency response. In pattern recognition applications, such as image classification, a nonparametric system might use a nearest-neighbor algorithm to search for the closest match to the observed signal in a database, while a neural network uses a high-dimensional parametric model that captures the structure of the images.

Parametric estimation can be used to transform and analyze signals with known structure. Because models generally have fewer free parameters than unprocessed signals, parametric models are often used for data compression. For example, the autoregressive (AR) models introduced later in this chapter have been used for speech compression. If a model can easily predict a given signal, then the signal can be represented using only the prediction error, or unpredictable components, and the parameters of the model used for prediction. Parametric models can also be used to analyze the content of a signal, such as the symbols encoded into a communication broadcast or the words in a speech recording. Parametric estimation is especially useful if we know how the signal was generated. In those cases, the parameters of interest control the behavior of the *system* rather than the signal itself.

Indeed, one of the most common applications of parametric estimation in signal processing is to find an approximate model of a complex or unknown system, which is often a physical channel or device. We can then analyze the properties of these systems, such as their stability, predict their effects on other signals, and even control their behavior. For example, in an audio conferencing system, we can find a parametric model for the acoustic path between a loudspeaker and a microphone, then use that model to predict the signal at the microphone, and subtract it from the recorded signal, thereby canceling the echo. Sometimes the predictions alone may be all that is desired, such as in forecasting or recommendation systems. In machine learning or adaptive systems, parametric estimation can be used to find a differentiable model for the system of interest so that it can be used as part of a larger optimization framework. In other cases, it is the system parameters themselves that are of interest to an application. Time differences of arrival between sensors can be used to localize a signal source, while the Doppler shift applied to a signal can be used to estimate the relative speed of an object.

11.1.2 Parametric models for signals and systems

The art of parametric estimation lies in choosing a good model. Ideally, we should choose a model based on prior knowledge about the phenomenon of interest, for example based on physics. Sometimes we have direct knowledge of the system and can readily identify the relevant parameters, such as the weighting of an unfair coin, the carrier frequency of a broadcast, or the gain applied by a linear amplifier. In other cases, we know what kind of model would be reasonable, such as a pole-zero model for a linear time-invariant (LTI) system, but we do not know exactly how many parameters we should use. Communication channels, vocal tracts, and room acoustics can all be well modeled as linear systems but may be too complex to characterize exactly. We can control the complexity of the model by choosing the number of parameters: It should be rich enough to enable insights into the salient characteristics of the signal or system, but parsimonious enough to be computationally tractable and avoid overfitting. If we have no prior knowledge about the structure of the problem, then a parametric model may be inappropriate. For example, if our data follow a bimodal distribution, we should not try to model them by a normal distribution. Likewise, if an LTI system uses feedback, it cannot be effectively approximated by a feedforward model, even if it has many coefficient parameters.

Once we have selected an appropriate model, we must choose a set of parameters that best explain the observed data. Parameter estimation is typically framed as an optimization problem, but there are different objectives that might be optimized depending on the application. In some cases, such as direction-of-arrival estimation, we are concerned with the accuracy of the parameter itself, and we seek to minimize its deviation from the true value. In other applications, such as linear prediction for echo cancelation, performance depends only indirectly on the parameter values. Instead, we are concerned with modeling error, that is, the ability of the parameters to explain the observed data.

Consider the following motivational example. A sequence $x[n]$ is generated using the following linear constant-coefficient difference equation (LCCDE):

$$x[n] = -0.8x[n-1] + 0.33x[n-2] + 0.73x[n-3] + 0.315x[n-4] + w[n], \tag{11.1}$$

where $w[n]$ is a sequence of independent, identically distributed, zero-mean random values. We will formally introduce AR models in Section 11.2, but from (11.1) we can see that the sequence $x[n]$ is determined by its four most recent values to within the unknown and unpredictable white noise sequence $w[n]$.

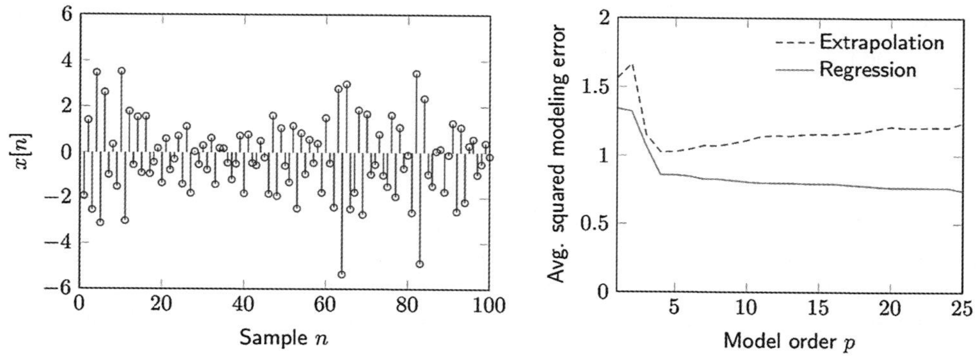

FIGURE 11.1

Model order and diminishing returns. Left is a sequence generated by a fourth-order signal model and right is the regression error and extrapolation error as a function of model order. The regression error applies the fitted parameters to the same 100 samples used to find them, while the extrapolation error uses these parameters to fit 100,000 data samples generated with the same model, but outside the 100 training samples.

Suppose that we are given 100 samples from this sequence, shown on the left in Fig. 11.1, and told that the signal was generated by a difference equation of the form (11.1), but we do not know the number of coefficients or their values. We try several linear models with different numbers of coefficients, fitting each model to the system by minimizing the regression error, that is, the error between the observed data and the data predicted by the model. The right plot of Fig. 11.1 shows the average squared modeling error for models of order 1 through 25, that is, with 1 to 25 coefficients. The regression error decreases monotonically with increasing model order but has diminishing returns after just four coefficients. On the other hand, the extrapolation error, which is evaluated on samples outside the training range, achieves its minimum at the true model order and increases for higher orders. We can see that although more parameters always provide a better fit to the training data, too many parameters can harm extrapolation performance by overfitting the parameters to the data. In practice, it can be challenging to determine the number of parameters to use for a given parametric model. Model order is often chosen based on prior knowledge of the system of interest, computational complexity constraints, or the amount of available training data. It can also be determined in a data-driven manner. We will revisit model order selection in Section 11.5.

In this chapter, we describe several parametric models that are commonly used to analyze signals and systems and the techniques used to estimate their parameters based on observed data. We present both stochastic and data-driven formulations for parameter estimation. We focus on rational transfer function models, that is, on systems with a finite number of poles and zeros, which can be used in batch, segmented, and on-line recursive formulations. Finally, we discuss advanced techniques for model order selection.

The background of the reader assumed in this chapter is a working knowledge of discrete-time signals and systems, some basic probability theory including knowledge of stationary random processes, and the mathematical skills of a senior level undergraduate student in electrical engineering or a similar discipline. An exemplary treatment of these topics includes the texts by [1], [2], and [3]. These texts cover discrete-time signal and system theory, with the latter covering some random processes. Additional material on probability and stochastic processes as applied to the problems in this chapter can be found in [4] and [5].

11.2 Preliminaries

11.2.1 Signals

Parametric estimation techniques can be used with unordered data, for example by fitting a histogram to a parametric distribution, but in signal processing we are generally more concerned with ordered data, also known as signals. A signal is a set of values indexed by an independent variable which we will call time, although the index could also some other quantity, such as space (for an image), storage locations in a memory, positions on a magnetic tape, or any other integer-valued index set. In this chapter we restrict our attention to discrete-time signals of the form $x[n]$, where the time index n is an integer. Such a sequence could be sampled from a continuous-time signal, such as an acoustic waveform recorded onto a computer using an analog-to-digital converter, or it could be intrinsically discrete, such as a sequence of symbols in a digital transmission.

Mathematically, we allow the index n to take on any integer value, so that the sequence $x[n]$ contains infinitely many samples. Indeed, one of the benefits of parametric modeling is that we can characterize an infinite length sequence using a finite number of parameters. For example, if $x[n]$ is periodic with period P, then it is fully characterized by P samples. If it is a sinusoid, then it is determined by its frequency, amplitude, and phase parameters. In real-world applications, we must perform parametric estimation using a finite number of samples. A sequence of N samples can of course be fully represented by N parameters, but to analyze the signal or the system that produced it, it is useful to select a model with $p \ll N$ parameters. As in the motivating example above, the N samples used for parameter estimation are often a small subset of the overall signal.

The methods in this chapter can be applied to real- or complex-valued signals and to scalar- or vector-valued signals, but we will assume for convenience that $x[n]$ is a sequence of real-valued scalars. In many engineering applications, signal values are drawn from a finite alphabet, or set of possible values. For example, communication signals comprise values from a modulation alphabet indexed by groups of bits from a transmitted digital message. These message bits can be thought of as parameters that control the signal observed by the recipient. Similarly, the digital signals processed by logic circuits and computers are stored as sequences of bits corresponding to quantized signal values. The finite precision of a digital signal can introduce nonlinear distortion effects into signal processing operations that are difficult to model. For simplicity, in this chapter we assume that the observed sequence $x[n]$ can take on a continuous range of values, even if it was originally generated from a finite alphabet, and that signal processing operations have perfect precision.

11.2.2 Data-driven framework

Because parametric estimation can be applied in many different engineering contexts, in this chapter we will consider several frameworks for analyzing signals and systems. In particular, we will show how to estimate the parameters of a model using both data-driven and stochastic optimization methods. The first and perhaps simplest formulation is to assume that the signals of interest are deterministic, that is, that the signal values $x[n]$ can take on any finite real value (or complex value, in the case of complex signals) and that we have no prior information, such as a probability distribution, to restrict their values. We often refer to these methods as "data-driven" because they rely on the data alone, rather than other prior knowledge or assumptions, to select model parameters.

In many problems for which parametric estimation is useful, such as linear prediction or echo cancelation, we want to find parameter values that best explain the observed signal. Thus, the optimization problem is not posed in terms of the parameter values directly. Instead, we measure the error between the signal data that we observe and the signal data that the model predicts. A convenient and popular performance measure is the square error between signals. The accumulated square error E_{LS} between an estimate $\hat{x}[n]$ and a desired signal $x_d[n]$ over a block of samples, $n = 0, \ldots, N$, is given by

$$E_{LS} = \sum_{n=0}^{N} \left(x_d[n] - \hat{x}[n]\right)^2. \tag{11.2}$$

We can also write the cost function as an integral square error measured in the frequency domain:

$$E_{IS} = \frac{1}{2\pi} \int_{-\pi}^{\pi} \left|X_d(\omega) - \hat{X}(\omega)\right|^2 d\omega. \tag{11.3}$$

Both E_{LS} and E_{IS} capture the degree to which the approximating time domain signal $\hat{x}[n]$ or frequency domain estimate $\hat{X}(\omega)$ matches the deterministic signal of interest, $x_d[n]$ or $X_d(\omega)$. Assuming that the desired signal $x_d[n]$ is zero outside the range $0 \leq n \leq N$ and that $X_d(\omega) = \sum_{n=0}^{N} x_d[n] e^{j\omega n}$, i.e., $X_d(\omega)$ is the discrete-time Fourier transform of $x_d[n]$, it can be shown that $E_{LS} = E_{IS}$ using Parseval's relation. Thus, a parametric model that accurately predicts the signal will also accurately reflect its frequency content.

As a simple example, suppose that we have measured a signal $x[n]$ that corresponds to the daily temperature of the water in a local pond. If we wanted to compute a good approximating signal, $\hat{x}[n] = c$, i.e., a signal whose value remains constant over the entire interval, and we used the accumulated square error over the collected data signal as a measure of performance, we could write

$$E_{LS} = \sum_{n=0}^{N} (x[n] - c)^2 \tag{11.4}$$

and seek to find the value of c that minimizes the total accumulated square error over the observation interval. Differentiating E_{LS} with respect to c and setting the result equal to zero, we obtain

$$\frac{\partial E_{LS}}{\partial c} = -2 \sum_{n=0}^{N} (x[n] - c), \tag{11.5}$$

$$0 = -2 \sum_{n=0}^{N} (x[n] - c), \tag{11.6}$$

which can be solved for the minimizing value of c, yielding

$$c = \frac{1}{N+1} \sum_{n=0}^{N} x[n]. \tag{11.7}$$

This is the well-known result that the best constant-valued approximation of a given sequence, measured by square error, is the sample average of that sequence. This is one simple way in which a parameter representing the sequence can be estimated from the data directly.

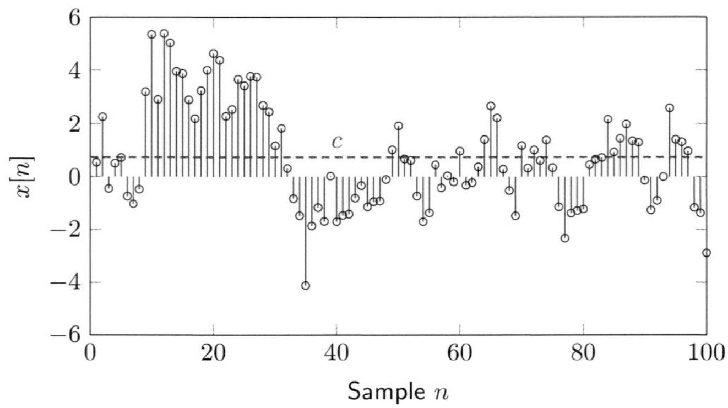

FIGURE 11.2

Approximating a signal with a constant value.

Fig. 11.2 shows the result of approximating a signal $x[n]$ with another signal that takes on only a single constant value. The signal in this example has an average value of c over the interval shown. As a result, the best approximating constant-valued signal, in terms of minimizing the squared approximation error over the interval shown, is the mean of the signal, c.

11.2.3 Stochastic model and criteria

Parametric estimation can also be approached using stochastic, that is, random, signal models, which can help to account for some of the uncertainty in a signal or system of interest. Such models can also incorporate prior knowledge about the signals, such as the correlation between samples. When we use a stochastic framework, we choose measures of performance that are similar in spirit to those of our deterministic analysis, but which are defined in terms of probabilities or statistical averages. In statistical signal processing, performance is often measured in terms of the mean square error (MSE), which is the stochastic average of the data-driven square error measure from the previous section. Like that measure, the MSE has desirable mathematical properties that make optimization problems more tractable.

The MSE performance measure is closely related to the second-order statistics of a random process, which allow us summarize the behavior of a signal with a relatively small number of parameters [6]. In general, to fully statistically characterize a discrete-time random signal $x[n]$, we would need to find the joint probability density function between any combination of samples $f_x(x[n_1], x[n_2], x[n_3], \ldots, x[n_k])$ for all possible k-tuples $n_1, n_2, n_3, \ldots, n_k$ and for all possible k. This complete statistical knowledge may be available for certain signal or system models, but in general, such a complete characterization is neither available nor appropriate. Instead, we will often use second-order statistical models for random signals of interest. Second-order models are parameterized by two statistical functions, the mean and the covariance. The mean value of a random signal $x[n]$ is defined as

$$m_x[n] = E\{x[n]\}, \tag{11.8}$$

where $E\{\}$ denotes the statistical expectation operation which is taken with respect to the complete stochastic characterization of the signal. The covariance function of the signal is defined as

$$\begin{aligned} C_x[m,n] &= E\left\{(x[m] - m_x[m])(x[n] - m_x[n])^*\right\} \\ &= E\left\{(x[m]x[n]^*)\right\} - m_x[m]m_x[n]^*, \end{aligned} \tag{11.9}$$

where, again, the expectation is taken with respect to the complete stochastic characterization of the random signal. Note that if the random signal is complex-valued, then by convention the second term in the definition is conjugated. If the signal is real-valued, then there is no conjugation.

In general, the mean and covariance of a process are both functions of time. However, when the random signal $x[n]$ is *wide-sense stationary* (WSS), then both the mean and covariance take on a special form [6] [1]. For WSS random signals, the mean value of the signal is a constant, $m_x[n] = m_x$, and the covariance function is only a function of the difference in times between the two samples,

$$C_x[m,n] = C_x[0, n-m] \triangleq C_x[n-m], \tag{11.10}$$

where, with a slight abuse of notation, we use C_x to denote both the general and the WSS covariance functions. Wide-sense stationarity is a generalization of strict stationarity, which requires that none of the statistical properties of the signal vary over time (or location in space, or another independent variable used as the index to the sequence). The second-order statistics of WSS signals do not depend on the specific time but are functions of how close in time two samples are taken with respect to one another.

Returning to our earlier example of representing a signal, now a random signal, by a single constant estimate of that signal, we might seek to minimize the following MSE rather than the accumulated square error as before:

$$E_{MS} = E\left\{\left(x[n] - \hat{x}[n]\right)^2\right\},$$
$$E_{MS} = E\left\{\left(x[n] - c\right)^2\right\}, \tag{11.11}$$

where once again we seek a single value of the constant c to represent the signal over the region of interest. If the signal $x[n]$ is stationary, then we can find a simple solution to this estimation problem by differentiating with respect to the parameter as follows:

$$\frac{\partial E_{MS}}{\partial c} = -E\left\{2\left(x[n] - c\right)\right\},$$
$$0 = -2E\left\{x[n]\right\} + 2c,$$
$$c = m_x. \tag{11.12}$$

This solution is the stochastic equivalent of the deterministic least-squares estimation problem; the minimum MSE (MMSE) constant estimate of a WSS random signal is given by the mean of the signal.

11.3 Parametric models for linear time-invariant systems

In the field of signal processing, parametric estimation is often used to model the systems that generate or act on a signal of interest. We begin our study of parametric system models with LTI systems, which are mathematically tractable and represent many real-world systems of interest, such as electronic circuits, acoustic propagation in a stationary environment, and digital signal processing filters. Furthermore, LTI systems are often a good small-scale or short-term approximation for more complex nonlinear and time-varying systems. When a system is LTI, the output of the system $y[n]$ for a given input $x[n]$ can always be written through the convolution sum,

$$y[n] = \sum_{m=-\infty}^{\infty} h[m]\, x[n-m], \tag{11.13}$$

where $h[n]$ is the *impulse response* of the LTI system, that is, its output when the input is a discrete-time impulse $\delta[n]$.

LTI systems can be classified as having either finite impulse response (FIR) or infinite impulse response (IIR). For an FIR system of order N, the impulse response $h[n]$ has N nonzero terms, so that for a causal system the output is given by

$$y[n] = \sum_{m=0}^{N} h[m]\, x[n-m]. \tag{11.14}$$

It is straightforward to model FIR systems because the coefficients of $h[n]$ form a complete set of parameters that fully determine the behavior of the system.

Not all LTI systems have FIRs, but some IIR systems can still be represented by a finite set of parameters. These systems are defined recursively using a finite-order LCCDE:

$$\sum_{k=0}^{M} a_k y[n-k] = \sum_{k=0}^{N} b_k x[n-k]. \tag{11.15}$$

The coefficients $a_0, \ldots, a_M$ and $b_0, \ldots, b_N$ of the LCCDE comprise a finite set of parameters that capture the behavior of such a system. The z-transform relationship between the input and output of such a system can be written as a rational transfer function, $H(z)$,

$$H(z) = \frac{\sum_{k=0}^{N} b_k z^{-k}}{\sum_{k=0}^{N} a_k z^{-k}}, \tag{11.16}$$

which, together with the region of convergence of the system function, determines the behavior of the LTI system. Although rational transfer functions describe systems whose impulse responses may be infinite in length, they can still be described through a finite number of parameters, namely the coefficients of the LCCDE or equivalently the coefficients of the numerator and denominator polynomials in its transfer function.

In this section, we will consider methods for estimating the coefficients of different kinds of rational transfer function LTI models. The simplest such models are *autoregressive* (AR) systems, whose transfer functions have only denominator coefficients, and *moving average* (MA) systems, whose transfer functions have only numerator coefficients. When both numerator and denominator coefficients are present, an ARMA model is the result. In both data-driven and stochastic formulations, there exist efficient algorithms to find coefficients that minimize the (mean) square error of the signal estimated by the model.

11.3.1 Autoregressive models

A discrete-time signal $x[n]$ is an AR process of order p if it can be written in the form [6]

$$x[n] + a_1 x[n-1] + \cdots + a_p x[n-p] = w[n], \tag{11.17}$$

where $a_1, \ldots, a_p$ are parameters of the AR model and the signal $w[n]$ is a zero-mean white noise process of variance σ^2, i.e., we have $C_w[m] = \sigma^2 \delta[m]$. This formulation can be viewed as an LCCDE with input $w[n]$ and output $x[n]$, as illustrated as a flow graph for the system in Fig. 11.3. By rearranging the order of the terms in the definition, we can write the AR process recursively as

$$x[n] = -a_1 x[n-1] - a_2 x[n-2] - \cdots - a_p x[n-p] + w[n]. \tag{11.18}$$

Thus, a pth-order AR process can be written as a linear combination of the p most recent values of the signal plus a term that is statistically independent of the past values of the process; this is the so-called *innovations sequence*, and represents the unpredictable components in the random signal.

We can interpret this difference equation in two ways. First, we can treat $w[n]$ as the input and $x[n]$ as the output of an all-pole filter. Ignoring that the sequences $x[n]$ and $w[n]$ may be random for now

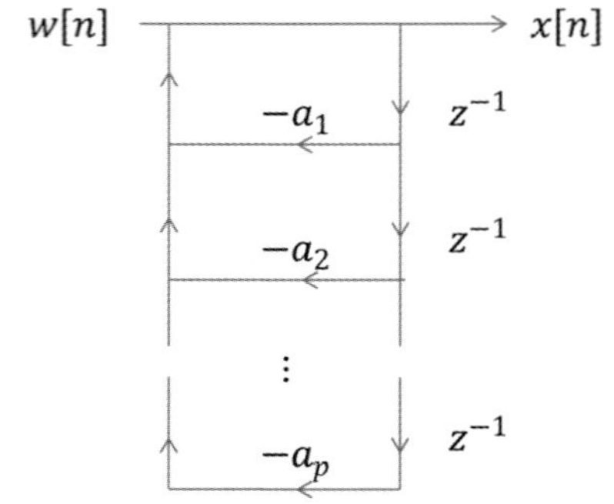

FIGURE 11.3

Autoregressive process flow diagram.

and considering only the input–output relationship between them, we can take the z-transform of both sides to find

$$X(z) + \sum_{k=1}^{p} a_k z^{-k} X(z) = W(z) \tag{11.19}$$

$$X(z)\left(1 + \sum_{k=1}^{p} a_k z^{-k}\right) = W(z) \tag{11.20}$$

$$X(z) = \left(\frac{1}{1 + \sum_{k=1}^{p} a_k z^{-k}}\right) W(z). \tag{11.21}$$

The term in parentheses is the rational transfer function $H(z) = \frac{X(z)}{W(z)}$ that relates $X(z)$ and $W(z)$. Because the numerator polynomial is $B(z) = 1$, this transfer function corresponds to an "all-pole" filter with denominator polynomial $A(z) = 1 + \sum_{k=1}^{p} a_k z^{-k}$. We can view the sequence $x[n]$ as being generated by filtering the sequence $w[n]$ with this all-pole filter, as illustrated in Fig. 11.4. This input–output interpretation of an AR process provides us with a simple way to generate random processes (or deterministic signals) with a spectral shape that can be parametrically controlled through the coefficients of $A(z)$.

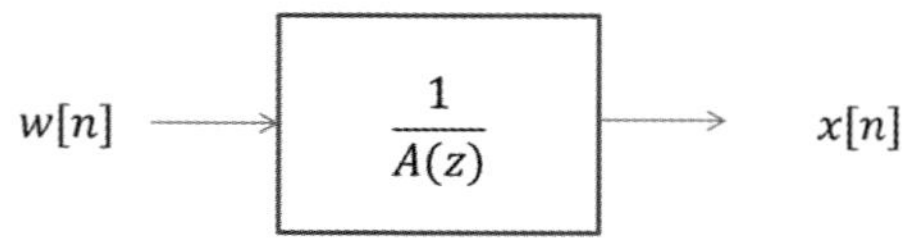

FIGURE 11.4

Input–output transfer function perspective of the autoregressive model.

We can also interpret the AR model as a form of feedforward prediction. Note that the definition of the process represents a convolution of the input sequence $x[n]$ with the AR parameters a_k, such that the output signal is the innovations sequence, i.e.,

$$\sum_{k=0}^{p} a_k x[n-k] = w[n], \tag{11.22}$$

where we set $a_0 = 1$. The significance of this can be seen if we slightly rearrange terms, yielding

$$x[n] - \underbrace{\left(-\sum_{k=1}^{p} a_k x[n-k]\right)}_{\hat{x}[n]} = w[n]. \tag{11.23}$$

The term in parentheses can be interpreted as an estimate of the signal $x[n]$ based on its most recent p samples, so that

$$x[n] - \hat{x}[n] = w[n]. \tag{11.24}$$

By writing the process in this form, we see that the sequence $x[n]$ is rather unique in that it can be best estimated by its most recent past values and that after this *predictable component* is subtracted, all that remains is a white noise sequence that is independent of the past values. The term "autoregressive" comes from the ability to write the sequence $x[n]$ using a "regression model" over its own past values, i.e., using a finite linear combination of its own past values. This way, we can view $A(z)$ as the transfer function of an FIR linear prediction filter that produces an estimate $\hat{x}[n]$ of $x[n]$ based on past samples. Then the innovations sequence $w[n]$ is the output of a prediction error filter, which removes the predictable component of $x[n]$.

Because they can be expressed in this way, AR models are frequently used in prediction applications. For example, one might attempt to predict the next value of a colored noise signal with an AR model. Fig. 11.5 shows the input $x[n]$ and output $\hat{x}[n]$ of an FIR prediction filter that is designed to track the input using only past values of the observed signal. The signal used in Fig. 11.2 is reused in Fig. 11.5.

Since colored noise can be generated by filtering a white noise signal, we can view the signal of interest as if it were originally generated by filtering such a white noise signal. Thus, we expect that the error between the predicted and measured signals will be a white noise sequence. Fig. 11.6 shows the prediction error sequence, which indeed appears uncorrelated from sample to sample, i.e., white.

11.3.2 Moving average models

A discrete-time signal $x[n]$ is called an MA process of order q if it can be written in the form [6] [3]

$$x[n] = b_0 w[n] + b_1 w[n-1] + \cdots + b_q w[n-q], \tag{11.25}$$

where $b_0, \ldots, b_q$ are parameters of the MA model and the signal $w[n]$ is a white noise process, i.e., we have that $C_w[m] = \sigma^2 \delta[m]$. Since the sequence $x[n]$ can be viewed as a linear combination of the

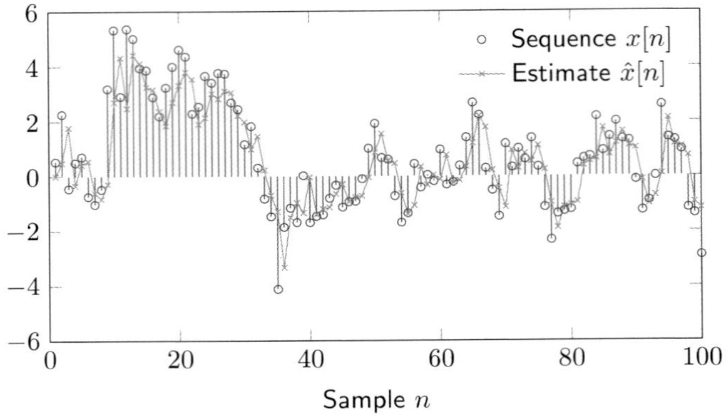

FIGURE 11.5

Tracking colored noise with an AR model. The signal $x[n]$ is illustrated using circles and its linear prediction $\hat{x}[n]$ is represented with cross marks.

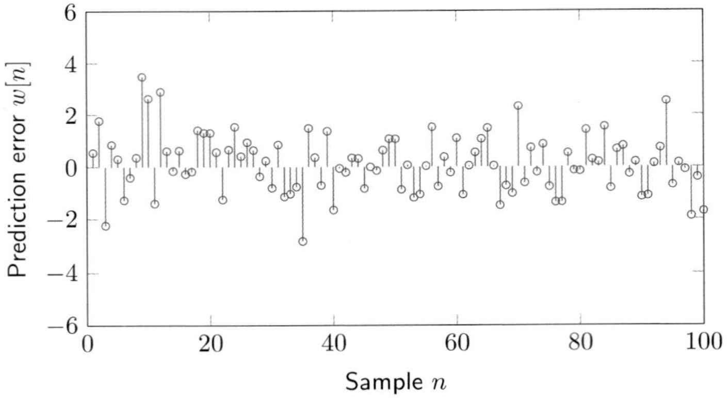

FIGURE 11.6

Prediction error, $w[n] = x[n] - \hat{x}[n]$, from estimating colored noise using an AR model.

$q+1$ most recent values of the sequence $w[n]$, it is called a "moving average." From this structure, whenever the sequence $x[n]$ is given as the output of an FIR filter whose input is a white noise sequence, then we can refer to $x[n]$ as an MA sequence. More generally, even when the input is not a white noise sequence, the process of filtering with an FIR filter can be viewed as taking a "moving average" of the input. Financial models often use 50-day and 100-day MAs as indicators of trends in various stock prices and indices. In such a context, q and b_k are usually taken such that $b_k = \frac{1}{(q+1)}$, $k = 0, \ldots, q$, so that $x[n] = \frac{1}{q+1}\sum_{k=0}^{q} w[n-k]$, i.e., the arithmetic mean of the most recent $q+1$ values of $w[n]$.

Viewing the relationship between the sequences $x[n]$ and $w[n]$ with a signal flowgraph model, in Fig. 11.7a, we see that $x[n]$ can be viewed as an FIR filtered version of the sequence $w[n]$. Similarly, we can view this in Fig. 11.7b via an input–output transfer function relationship between the two sequences.

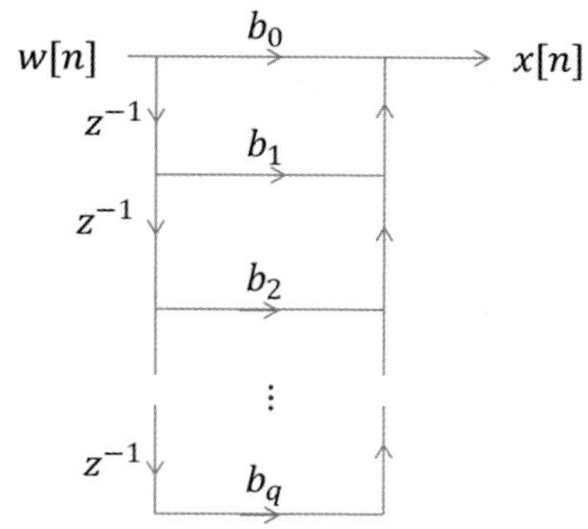

FIGURE 11.7a

Moving average flowgraph model.

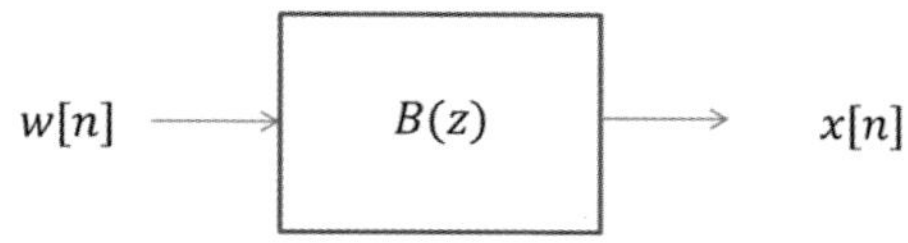

FIGURE 11.7b

Transfer function perspective of the MA model relating $x[n]$ and $w[n]$.

11.3.3 Autoregressive moving average models

A discrete-time signal $x[n]$ is an autoregressive moving average (ARMA) process of order (p, q) if it can be written in the form [6] [3]

$$x[n] + a_1 x[n-1] + \cdots + a_p x[n-p] = b_0 w[n] + \cdots + b_q w[n-q], \tag{11.26}$$

where $a_1, \ldots, a_p, b_0, \ldots, b_q$ are parameters of the ARMA model and the signal $w[n]$ is a white noise process, i.e., we have $C_w[m] = \sigma^2 \delta[m]$. We can once again relate the sequences $x[n]$ and $w[n]$ through a transfer function relationship, such as

$$X(z) = \left(\frac{\sum_{k=0}^{q} b_k z^{-k}}{1 + \sum_{k=1}^{p} a_k z^{-k}} \right) W(z). \tag{11.27}$$

This input–output relationship implies that we can view the sequence $x[n]$ as being generated by filtering the sequence $w[n]$ with a rational transfer function, i.e., a "pole-zero" filter with finite-order numerator and denominator polynomials, where the numerator polynomial is given by $B(z) = \sum_{k=0}^{q} b_k z^{-k}$ and the denominator polynomial is given by $A(z) = 1 + \sum_{k=1}^{p} a_k z^{-k}$. It is usually assumed that the rational function is "strictly proper," which means that the numerator order is strictly less than that of the denominator, i.e., $q < p$. For this reason, we may refer to the sequence as an ARMA(p) process, rather than ARMA(p, q). Figs. 11.8a and 11.8b illustrate a flowgraph and transfer function perspective of the ARMA model relating the sequences $x[n]$ and $w[n]$.

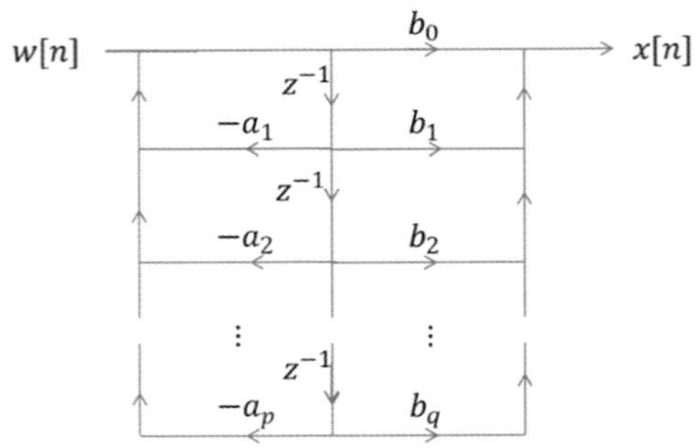

FIGURE 11.8a

ARMA flowgraph model using a direct form II structure.

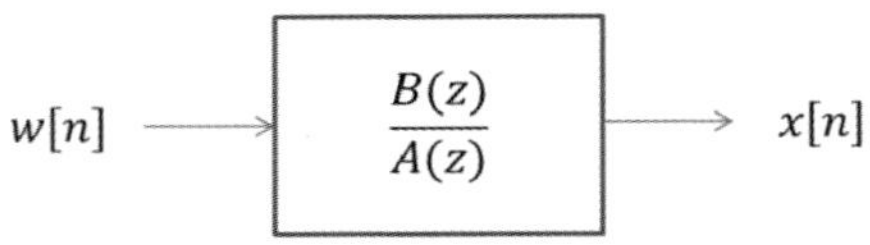

FIGURE 11.8b

Transfer function perspective of the ARMA model relating $x[n]$ and $w[n]$.

11.3.4 Parametric modeling for system function approximation

Once we have selected an appropriate LTI system model, we must estimate its parameters. Generally, our goal is to match the impulse response or frequency response of the model to that of the observed system, which can be derived from a physical model or measured directly using probe signals. Although these three LTI system models – AR, MA, and ARMA – are all parameterized by their rational transfer function coefficients, the methods used to estimate their parameters vary considerably.

The parameter estimation problem can be posed in either the time domain or the frequency domain, and we can move freely between these domains using the Fourier transform. The frequency response $H\left(e^{j\omega}\right)$ of a discrete-time system is related to its impulse response $h[n]$ by the discrete-time Fourier transform [1],

$$H\left(e^{j\omega}\right) = \sum_{n=-\infty}^{\infty} h[n] e^{-j\omega n}, \tag{11.28}$$

$$h[n] = \frac{1}{2\pi}\int_{-\pi}^{\pi} H\left(e^{j\omega}\right) e^{j\omega n} d\omega. \tag{11.29}$$

Suppose that we wish to choose the coefficients of an ARMA(p, q) model such that the frequency response of the model system closely approximates a given frequency response $H\left(e^{j\omega}\right)$. Using least-squares optimization, the coefficient vectors are the solution to the minimization

$$\{\boldsymbol{a}, \boldsymbol{b}\}_{LS} = \underset{\{a_k\}_{k=1}^{p}, \{b_k\}_{k=0}^{q}}{\operatorname{argmin}} \frac{1}{2\pi}\int_{-\pi}^{\pi} \left| H\left(e^{j\omega}\right) - \frac{\sum_{k=0}^{q} b_k e^{-\mathrm{j}k\omega}}{1+\sum_{k=1}^{p} a_k e^{-\mathrm{j}k\omega}} \right|^2 d\omega. \tag{11.30}$$

We can also write the optimization problem in the time domain using the desired impulse response $h[n]$. Applying Parseval's relation [1], which equates energy calculated in the time domain with that calculated in the frequency domain, we have

$$\{\boldsymbol{a}, \boldsymbol{b}\}_{LS} = \underset{\{a_k\}_{k=1}^{p}, \{b_k\}_{k=0}^{q}}{\operatorname{argmin}} \sum_{n=-\infty}^{\infty} \left| h[n] - \frac{1}{2\pi}\int_{-\pi}^{\pi} \frac{\sum_{k=0}^{q} b_k e^{-\mathrm{j}k\omega}}{1+\sum_{k=1}^{p} a_k e^{-\mathrm{j}k\omega}} e^{j\omega n} d\omega \right|^2. \tag{11.31}$$

Both minimization problems are quite cumbersome. The ARMA(p, q) model is the most general, but it is also the most difficult to use because the cost function is highly nonlinear in the parameters. Thus, we will start with AR and MA parameter estimation before returning to the more general ARMA parameter estimation problem.

The ARMA problem formulation is difficult because the AR parameters, that is, the feedback components, are nonlinearly related to the optimization criterion. If we were only concerned with an MA estimate, that is, a feedforward model, and we wanted to use this least-squares criterion, then the time-domain formulation can provide a simple solution. Without the AR coefficients, (11.31) becomes

$$\{\boldsymbol{b}\}_{LS} = \underset{\{b_k\}_{k=0}^{q}}{\operatorname{argmin}} \sum_{n=-\infty}^{\infty} \left| h[n] - \frac{1}{2\pi}\int_{-\pi}^{\pi} \sum_{k=0}^{q} b_k e^{-\mathrm{j}k\omega} e^{j\omega n} d\omega \right|^2, \tag{11.32}$$

which can be rewritten

$$\{\boldsymbol{b}\}_{LS} = \underset{\{b_k\}_{k=0}^{q}}{\operatorname{argmin}} \sum_{n=-\infty}^{\infty} \left| h[n] - \sum_{k=0}^{q} b_k \delta[n-k] \right|^2, \tag{11.33}$$

yielding the simple, though somewhat unsatisfying, solution

$$b_k = h[k], k = 0, \ldots, q. \tag{11.34}$$

This is nothing more than a simple truncation approximation to the desired impulse response for the system of interest.

Impulse response truncation, while simple, cannot efficiently model systems that contain feedback. Such systems have IIRs and must be modeled using poles. First, consider an all-pole transfer function, that is, an AR model. The optimization problem (11.31) is still difficult to solve directly, but we can transform the problem into a more tractable one. Recall the form of the AR model for the transfer function:

$$H(z) = \frac{1}{1 + \sum_{k=1}^{p} a_k z^{-k}}. \tag{11.35}$$

Beginning with this expression and taking the inverse z-transform, we see that the impulse response of the modeled system must satisfy

$$h[n] + \sum_{k=1}^{p} a_k h[n-k] = \delta[n]. \tag{11.36}$$

Now given a desired (or measured) impulse response, which we will denote $h_d[n]$, we attempt to minimize the deviation between $h_d[n]$ and the impulse response $h[n]$ of the approximate system:

$$\{\boldsymbol{a}\}_{LS} = \underset{\{a_k\}_{k=1}^{p}}{\operatorname{argmin}} \sum_{n=-\infty}^{\infty} \left| h_d[n] + \sum_{k=1}^{p} a_k h_d[n-k] \right|^2. \tag{11.37}$$

We now have a quadratic minimization in the parameters of interest, which admits a closed-form solution. Specifically, denoting the quadratic error term in the minimization above as E, we can write

$$0 = \frac{\partial E}{\partial a_\ell} = 2 \sum_{n=-\infty}^{\infty} \left(h_d[n] + \sum_{k=1}^{p} a_k h_d[n-k] \right) h_d[n-\ell], \ell = 1, \dots, p, \tag{11.38}$$

which comprise a set of linear equations in a_ℓ, $\ell = 1, \dots, p$. We can write these equations more compactly by exchanging the order of summation:

$$\sum_{k=1}^{p} a_k \underbrace{\sum_{n=-\infty}^{\infty} h_d[n-k] h_d[n-\ell]}_{\hat{R}_{h_d h_d}[k-\ell]} = - \underbrace{\sum_{n=-\infty}^{\infty} h_d[n] h_d[n-\ell]}_{\hat{R}_{h_d h_d}[\ell]}, \ell = 1, \dots, p. \tag{11.39}$$

We have made use of the definition $\hat{R}_{h_d h_d}[\ell] = \sum_{n=-\infty}^{\infty} h_d[n] h_d[n-\ell] = \hat{R}_{h_d h_d}[-\ell]$, which is often called the deterministic autocorrelation sequence for $h_d[n]$. We can then summarize the set of equations in matrix form, sometimes referred to as the normal equations [6] [3],

$$\hat{R}\vec{a} = -\vec{\hat{r}}, \tag{11.40}$$

or

$$\begin{bmatrix} \hat{R}_{h_d h_d}[0] & \cdots & \hat{R}_{h_d h_d}[p-1] \\ \vdots & \ddots & \vdots \\ \hat{R}_{h_d h_d}[p-1] & \cdots & \hat{R}_{h_d h_d}[0] \end{bmatrix} \begin{bmatrix} a_1 \\ \vdots \\ a_p \end{bmatrix} = - \begin{bmatrix} \hat{R}_{h_d h_d}[1] \\ \vdots \\ \hat{R}_{h_d h_d}[p] \end{bmatrix}, \tag{11.41}$$

which has the closed-form solution

$$\vec{a} = -\hat{R}^{-1}\vec{\hat{r}}. \tag{11.42}$$

The matrix $\hat{R}$ has a special structure that allows the system of linear equations to be solved efficiently. Because it is symmetric and constant along its diagonals, $\hat{R}$ is known as a symmetric Toeplitz matrix, for which fast linear equation solvers can be readily obtained [3] [6]. Furthermore, the values of the correlation matrix also appear on the right-hand side of the equation. Such a system can be solved quickly using an algorithm known as Levinson–Durbin recursion [3].

The optimization problem can also be posed stochastically. If we were to replace the accumulated square error criterion with an MSE criterion, i.e., we sought to minimize the MSE

$$\{\boldsymbol{a}\}_{LS} = \underset{\{a_k\}_{k=1}^{p}}{\operatorname{argmin}}\, E\left\{\left|h_d[n] + \sum_{k=1}^{p} a_k h_d[n-k]\right|^2\right\}, \tag{11.43}$$

we would naturally arrive at a similar form for its solution:

$$R\vec{a} = -\vec{r}, \tag{11.44}$$

or

$$\begin{bmatrix} R_{h_d h_d}[0] & \cdots & R_{h_d h_d}[p-1] \\ \vdots & \ddots & \vdots \\ R_{h_d h_d}[p-1] & \cdots & R_{h_d h_d}[0] \end{bmatrix} \begin{bmatrix} a_1 \\ \vdots \\ a_p \end{bmatrix} = -\begin{bmatrix} R_{h_d h_d}[1] \\ \vdots \\ R_{h_d h_d}[p] \end{bmatrix}, \tag{11.45}$$

which has the analytical solution

$$\vec{a} = -R^{-1}\vec{r}. \tag{11.46}$$

The only difference in our solution lies in replacing the deterministic autocorrelation $\hat{R}$ with the statistical autocorrelation, $R_{h_d h_d}[\ell] = E\{h_d[n]h_d[n-\ell]\}$, where the expectation is taken with respect to the distribution of the random sequence $h_d[n]$. This approach to AR modeling is often called "linear prediction" or "linear predictive modeling" due to the use of the prediction error form that is used to make the optimization problem linear in the parameters of interest (see [6], [3], and [7]).

We can now return to the problem of estimating an ARMA model with both poles and zeros. There are several ways of incorporating zeros into the modeling problem while still using the much simpler prediction-error formulation. Perhaps the simplest is to use the zeros to match the first q values of the impulse response exactly, that is, to first solve for the denominator coefficients as above and then select

$$b_\ell = h_d[\ell] + \sum_{k=1}^{p} a_k h_d[\ell-k],\, k = 0, \ldots q, \tag{11.47}$$

which follows from the system function relation

$$H_d(z) = \left(\frac{\sum_{k=0}^{q} b_k z^{-k}}{1 + \sum_{k=1}^{p} a_k z^{-k}}\right). \tag{11.48}$$

This is a two-step process: First, the AR parameters are determined using linear predictive modeling, and then the MA parameters are selected by exactly matching the first q samples of the impulse response. This approach is often called Prony's method [7] [3]. In general, this method provides a reasonable set of AR parameters; however, the MA parameters are not particularly well modeled. That is, if a system were known to have a given ARMA form and its parameters were to be estimated from noisy observations using Prony's method, the a_k coefficients would closely match those of the true underlying system, but the b_k parameters would likely be a poor match.

We can improve the fit of the MA coefficients by separating the ARMA model into a cascade of AR and MA systems and using least-squares parameter estimation for each component. As before, we use the predictive least-squares approach to find the AR parameters according to the normal equations. Next, we define the impulse response $v[n]$ of the AR-only system:

$$v[n] = -\sum_{k=1}^{p} a_k v[n-k] + \delta[n]. \tag{11.49}$$

This sequence becomes the input to an MA-only system with transfer function $B(z)$. Now, the goal is to minimize the discrepancy between the output of that MA system and the desired impulse response $h_d[n]$. Specifically, we select the MA coefficients to minimize the square error criterion

$$\{\boldsymbol{b}\} = \underset{\{b_k\}_{k=0}^{q}}{\text{argmin}} \sum_{n=-\infty}^{\infty} \left| h_d[n] + \sum_{k=0}^{q} b_k v[n-k] \right|^2, \tag{11.50}$$

which is once again a quadratic minimization admitting a closed-form solution. The linear equations reduce to

$$\sum_{k=0}^{q} b_k \underbrace{\sum_{n=-\infty}^{\infty} v[n-k]\, v[n-\ell]}_{\hat{R}_{vv}[k-\ell]} = -\underbrace{\sum_{n=-\infty}^{\infty} h_d[n]\, v[n-\ell]}_{\hat{R}_{h_d v}[\ell]}, \ell = 0, \ldots, q, \tag{11.51}$$

which admits the solution

$$\vec{b} = -\begin{bmatrix} \hat{R}_{vv}[0] & \cdots & \hat{R}_{vv}[q] \\ \vdots & \ddots & \vdots \\ \hat{R}_{vv}[q] & \cdots & \hat{R}_{vv}[0] \end{bmatrix}^{-1} \begin{bmatrix} \hat{R}_{h_d v}[0] \\ \vdots \\ \hat{R}_{h_d v}[q] \end{bmatrix}, \tag{11.52}$$

$$\vec{b} = -(\hat{R}_{vv})^{-1} \vec{r}_{hv}, \tag{11.53}$$

where we denote

$$\hat{R}_{vv} = \begin{bmatrix} \hat{R}_{vv}[0] & \cdots & \hat{R}_{vv}[q] \\ \vdots & \ddots & \vdots \\ \hat{R}_{vv}[q] & \cdots & \hat{R}_{vv}[0] \end{bmatrix}, \quad \vec{r}_{hv} = \begin{bmatrix} \hat{R}_{h_d v}[0] \\ \vdots \\ \hat{R}_{h_d v}[q] \end{bmatrix}. \tag{11.54}$$

This system can be solved using similar techniques to the normal equations used for the AR coefficients. The cascaded AR and MA systems can now closely approximate the desired ARMA system. Selection of the parameters p and q in AR and ARMA models is known as the "model order selection" problem. In some applications these parameters can be determined from knowledge of the underlying physics or the structure of the problem. In others, these parameters are often determined empirically. A common selection method is to use a small segment of data to compute the modeling or extrapolation error, as in Fig. 11.1, and select the model order p in the region where there is diminishing return in modeling error reduction. Typically, q is then selected to be the same value for model symmetry.

11.4 Joint process estimation and sequential modeling

In many signal processing problems, we are tasked with estimating or predicting the output of a system based on its input. If the system is unknown but we have access to past input and output signals, then we can use parametric estimation to create an approximate model of the system and apply that model to future input signals. The process of estimating a desired output signal that is correlated with a known input signal is referred to as *joint process estimation* [2]. The problem is closely associated with *adaptive filtering* algorithms, which iteratively update the parameters of the model as new data are received. Fig. 11.9 depicts a generic joint process estimation or adaptive filtering problem.

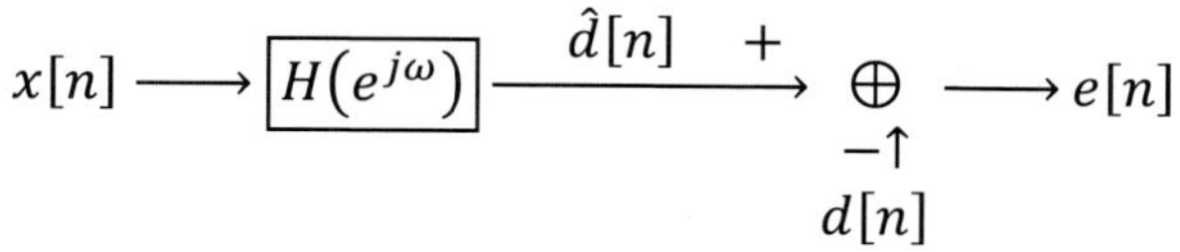

FIGURE 11.9

A system-level diagram of the adaptive filtering problem, where the parameters of a filter with frequency response $H\left(e^{j\omega}\right)$ are adjusted such that the output $\hat{d}[n]$ approximates the desired signal $d[n]$.

In Fig. 11.9, $d[n]$ is a desired signal, such as the observed output of an unknown system, and $\hat{d}[n]$ is the output of the model system $H(e^{j\omega})$, which we assume to be an order-p FIR system model with impulse response $\{h[k]\}_{k=0}^{p}$. In joint process estimation, the parameters $\{h[k]\}_{k=0}^{p}$ are adjusted to minimize the discrepancy $e^2[n] = \left(d[n] - \hat{d}[n]\right)^2$ between the output of the model system and the target signal [8]. Joint process estimation can be used for system identification, where H is designed to approximate the response of an unknown system. In that case, $d[n]$ is the observed output of the unknown system in response to a known input $x[n]$. If the parametric model perfectly matched the unknown system, then $e[n]$ would be zero. Such a model can be used to predict the output of the system of interest in response to future input signals. In many applications, such as acoustic echo cancelation, we cannot observe the system output directly, only a mixture of the output with other signals, but we can still apply the same estimation techniques.

In this section, we consider three variations on the joint process estimation problem. First, when a batch of data $\{x[k]\}_{k=0}^{N}$ is observed and a single time-invariant set of parameters $\{h[k]\}_{k=0}^{p}$ is selected based on this entire set of observations, the problem is referred to as *batch* or *block* processing. If instead the system of interest varies over time, then the model must also be adjusted every so often; we will show how to use the segmented least-squares method to process the data in blocks. Finally, when the model is adjusted in an on-line (sequential) manner, such that the filter parameters $\{h[k]\}_{k=0}^{p}$ are updated after each sample, we can use adaptive filtering algorithms [8].

11.4.1 Batch joint process estimation

We begin with the batch case, where the system is assumed to be time-invariant and we use all available data to estimate its parameters. We will first consider the stochastic case, where the input and output signals are assumed to be jointly WSS with known second-order statistics. The discrepancy to be minimized takes the form

$$\{h_{MMSE}[k]\}_{k=0}^{p} = \underset{\{h[k]\}_{k=0}^{p}}{\operatorname{argmin}}\, E\left\{\left(d[n] - \hat{d}[n]\right)^2\right\}. \tag{11.55}$$

Writing out this minimization using the parameters of the FIR system model, we have

$$\{h_{MMSE}[k]\}_{k=0}^{p} = \underset{\{h[k]\}_{k=0}^{p}}{\operatorname{argmin}}\, \epsilon_{MMSE}, \tag{11.56}$$

$$\epsilon_{MMSE} = E\left\{\left(d[n] - \sum_{k=0}^{p} h[k]\, x[n-k]\right)^2\right\}. \tag{11.57}$$

To proceed, we assume that the sequences $x[n]$ and $d[n]$ are WSS stochastic processes with zero mean and covariance functions [6]

$$C_x[n, n-m] = E\{x[n]\, x[n-m]\} = R_x[m] = R_x[-m], \tag{11.58}$$

$$C_d[n, n-m] = E\{d[n]\, d[n-m]\} = R_d[m] = R_d[-m], \tag{11.59}$$

$$C_{xd}[n, n-m] = E\{x[n]\, d[n-m]\} = R_{xd}[m] = R_{dx}[-m]. \tag{11.60}$$

We can now solve for the minimizing set of parameters by differentiating this quadratic expression with respect to each of the unknown parameters, which leads to

$$0 = \frac{\partial \epsilon_{MMSE}}{\partial h[\ell]} = -2E\left\{\left(d[n] - \sum_{k=0}^{p} h[k]\, x[n-k]\right) x[n-\ell]\right\}, \ell = 0, \ldots, p. \tag{11.61}$$

Rearranging terms, we arrive at

$$0 = E\{d[n]\, x[n-\ell]\} - \sum_{k=0}^{p} h[k]\, E\{x[n-k]\, x[n-\ell]\}, \ell = 0, \ldots, p, \tag{11.62}$$

$$E\{d[n]\, x[n-\ell]\} = \sum_{k=0}^{p} h[k]\, E\{x[n-k]\, x[n-\ell]\}, \ell = 0, \ldots, p, \tag{11.63}$$

$$R_{dx}[\ell] = \sum_{k=0}^{p} h[k] R_{xx}[k-\ell], \ell = 0, \ldots, p. \tag{11.64}$$

This system of equations is another form of the autocorrelation normal equations we have seen previously for AR modeling. Expressing (11.64) as a matrix equation, we have

$$\begin{bmatrix} R_{xx}[0] & \cdots & R_{xx}[p] \\ \vdots & \ddots & \vdots \\ R_{xx}[p] & \cdots & R_{xx}[0] \end{bmatrix} \begin{bmatrix} h[0] \\ \vdots \\ h[p] \end{bmatrix} = \begin{bmatrix} R_{dx}[0] \\ \vdots \\ R_{dx}[p] \end{bmatrix}, \tag{11.65}$$

which has the solution

$$\begin{bmatrix} h_{MMSE}[0] \\ \vdots \\ h_{MMSE}[p] \end{bmatrix} = \begin{bmatrix} R_{xx}[0] & \cdots & R_{xx}[p] \\ \vdots & \ddots & \vdots \\ R_{xx}[p] & \cdots & R_{xx}[0] \end{bmatrix}^{-1} \begin{bmatrix} R_{dx}[0] \\ \vdots \\ R_{dx}[p] \end{bmatrix}, \tag{11.66}$$

$$\vec{h}_{MMSE} = R^{-1}\vec{p}. \tag{11.67}$$

As in the AR modeling problem, the matrix R is an autocorrelation matrix and is therefore symmetric, positive semidefinite, and Toeplitz. We can apply fast and numerically stable algorithms to solve for the MMSE-optimal parameters [8] [6], which minimize the MSE. The filter defined by (11.67) is often referred to as the Wiener solution or the Wiener filter [8].

Since the second-order statistics of the inputs and outputs may not be available, we can also formulate joint process estimation as a deterministic least-squares problem. Using the least-squares criterion, the squared error discrepancy is given by

$$\{h_{LS}[k]\}_{k=0}^{p} = \underset{\{h[k]\}_{k=0}^{p}}{\operatorname{argmin}} \sum_{n=-\infty}^{\infty} \left(d[n] - \sum_{k=0}^{p} h[k] x[n-k] \right)^2. \tag{11.68}$$

Solving this minimization, we find the system of equations

$$\sum_{n=-\infty}^{\infty} d[n] x[n-\ell] = \sum_{k=0}^{p} h[k] \sum_{n=-\infty}^{\infty} x[n-k] x[n-\ell], \ell = 0, \ldots, p, \tag{11.69}$$

$$\hat{R}_{dx}[\ell] = \sum_{k=0}^{p} h[k] \hat{R}_{xx}[k-\ell], \ell = 0, \ldots, p, \tag{11.70}$$

where $\hat{R}_{dx}[\ell]$ and $\hat{R}_{xx}[\ell]$ are defined using summations over the observed data rather than taking an expectation with respect to an underlying stochastic model. The resulting set of normal equations are equivalent to (11.66) with the stochastic auto- and cross-correlation functions replaced by their deterministic counterparts:

$$\begin{bmatrix} h_{LS}[0] \\ \vdots \\ h_{LS}[p] \end{bmatrix} = \begin{bmatrix} \hat{R}_{xx}[0] & \cdots & \hat{R}_{xx}[p] \\ \vdots & \ddots & \vdots \\ \hat{R}_{xx}[p] & \cdots & \hat{R}_{xx}[0] \end{bmatrix}^{-1} \begin{bmatrix} \hat{R}_{dx}[0] \\ \vdots \\ \hat{R}_{dx}[p] \end{bmatrix}, \quad (11.71)$$

$$\vec{h}_{LS} = \hat{R}^{-1}\hat{\vec{p}}. \quad (11.72)$$

As in the stochastic formulation, this system of linear equations can be solved efficiently making use of the special structure of the matrix.

11.4.2 Segmented least-squares

In the previous section, we considered a batch formulation of the parameter estimation problem, where we find a single set of parameters that explain the entire set of observed data. However, there are many practical applications where the mechanism that gives rise to the observed signals varies over time. In prediction-based speech coding, for example, a new set of parameters is estimated every few milliseconds as the shape of the vocal tract changes. In this section, we show how to partition the data into segments such that each segment has a different LTI system model. In the next section, we will consider models that vary continuously over time.

In the FIR modeling problem from the previous section, we sought a single set of filter coefficients that minimized modeling error over the entire dataset:

$$\{h_{LS}[k]\}_{k=0}^{p} = \underset{\{h[k]\}_{k=0}^{p}}{\operatorname{argmin}} \sum_{n=0}^{N} \left(d[n] - \sum_{k=0}^{p} h[k]\, x[n-k] \right)^2 . \quad (11.73)$$

However, there may be a better fit of the data using one set of parameters from $0 \le n \le L$ and another set of parameters from $L < n \le N$. There may in fact be three or more distinct regions over which a different set of parameters would fit the data well. This problem formulation was first posed by R. Bellman in [9]. We denote by $e_{i,j}$ the square error within a segment from time index i to time index j:

$$e_{i,j} = \min_{\{h[k]\}_{k=0}^{p}} \sum_{n=i}^{j} \left(d[n] - \sum_{k=0}^{p} h[k]\, x[n-k] \right)^2 . \quad (11.74)$$

The total square error is the sum of these errors over each segment of the data. To discourage overfitting, we also add a penalty term proportional to the number of segments. Thus, the segmented least-squares (SLS) minimization problem is given by

$$E = \min_{M,\{i_1,\ldots,i_M\}} MC + \sum_{j=1}^{M} e_{i_j, i_{j+1}}, \quad (11.75)$$

where $i_1 = 0$, $i_M = N$, and $C > 0$ is a positive constant. This formulation seeks to minimize the total modeling error, while at the same time adding a penalty of C to the modeling error for each new segment added to the model. This penalty term encourages the algorithm to find a small number of segments that fit the data well. It can be verified that if $C = 0$, the system could choose a different set of parameters for each sample to achieve $E = 0$, which would not be a useful model. We can solve the optimization problem using a recursion known as the Bellman equation. First, we recursively compute the optimal SLS loss function for each partition:

$$E(j) = \min_{0 \le i \le j} \left\{ e_{i,j} + C + E(i-1) \right\}, \tag{11.76}$$

where $E(j)$ is the SLS objective function (penalized total square error) up through sample j. Next, the optimal partitioning of the data sequence can be obtained by tracing $E(j)$ from $N \ge j \ge 0$.

To illustrate the SLS algorithm, a sequence $x[n]$ was generated by changing the parameters of a second-order AR process every 1000 samples. The sequence is plotted in Fig. 11.10. Also shown in the figure, depicted using vertical lines, are the regions selected by the SLS algorithm over which the parameters of a second-order model ($p = 2$) remain constant. As seen in Fig. 11.10, the SLS algorithm can identify the change in parameters nearly perfectly. For this example, the constant C from (11.75) above was selected to be twice the variance of the signal $x[n]$.

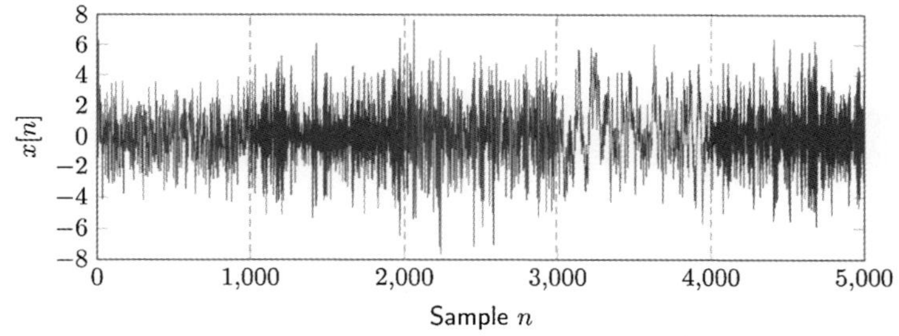

FIGURE 11.10

A sequence generated by changing the parameters of a second-order autoregressive process every 1000 samples. Also shown are the regions selected by the segmented least-squares algorithm.

11.4.3 Recursive estimation with RLS and LMS

In certain on-line signal processing applications, we must provide predictions immediately without waiting to observe a complete dataset or segment. In other applications, the system of interest varies continuously in time, for example due to motion, and the parametric model must be regularly updated. In either case, we must provide estimates sequentially, using only current and past observations to infer the parameters. That is, unlike the previous approaches where we have the batch data $\{x[k]\}_{k=0}^{N}$, the parameters $\{h[k]\}_{k=0}^{p}$ must be selected at each time index n based only on $\{x[k]\}_{k=-\infty}^{n}$.

Let us first consider a deterministic formulation of the problem using the least-squares criterion. Using all available data up to time index n, the optimization problem is given by

$$\{h_{LS}[k,n]\}_{k=0}^{p} = \underset{\{h[k]\}_{k=0}^{p}}{\operatorname{argmin}} \sum_{t=-\infty}^{n} \left(d[t] - \sum_{k=0}^{p} h[k]\, x[t-k] \right)^2 . \tag{11.77}$$

The estimated parameters are now functions of time and are updated as new observations arrive. In time-varying problems, older data are less useful than newer data. To emphasize the most recent data in the estimate, the least-squares criterion is often defined over a window of observations or weighted to emphasize more recent samples:

$$\{h_{LS}[k,n]\}_{k=0}^{p} = \underset{\{h[k]\}_{k=0}^{p}}{\text{argmin}} \sum_{t=-\infty}^{n} \lambda^{n-t} \left(d[t] - \sum_{k=0}^{p} h[k]\, x[t-k] \right)^2 . \tag{11.78}$$

With this weighted criterion, the effective window length is set by the weighting parameter $0 < \lambda \leq 1$. For example, when $\lambda = 0.9$, any samples that are more than $(1-\lambda)^{-1} = 10$ samples old are weighted by e^{-1} or less. The weighted least-squares criterion leads to a set of normal equations,

$$\begin{bmatrix} h_{LS}[0,n] \\ \vdots \\ h_{LS}[p,n] \end{bmatrix} = \begin{bmatrix} \hat{R}_{xx}[0,n] & \cdots & \hat{R}_{xx}[p,n] \\ \vdots & \ddots & \vdots \\ \hat{R}_{xx}[p,n] & \cdots & \hat{R}_{xx}[0,n] \end{bmatrix}^{-1} \begin{bmatrix} \hat{R}_{dx}[0,n] \\ \vdots \\ \hat{R}_{dx}[p,n] \end{bmatrix}, \tag{11.79}$$

$$\vec{h}_{LS}[n] = \hat{R}[n]^{-1}\, \hat{\vec{p}}[n]. \tag{11.80}$$

These equations are similar to those in the batch problem except that the sample autocorrelation matrix and cross-correlation vector are time-varying and are calculated using only the available data:

$$\hat{R}_{xx}[k-\ell, n] = \sum_{t=-\infty}^{n} \lambda^{n-t} x[t-k]\, x[t-\ell],\ \ell = 0, \ldots, p, \tag{11.81}$$

$$\hat{R}_{dx}[\ell, n] = \sum_{t=-\infty}^{n} \lambda^{n-t} d[t]\, x[t-\ell],\ \ell = 0, \ldots, p. \tag{11.82}$$

Although the normal equations can be solved more efficiently than a general linear system of equations, it is still computationally prohibitive to calculate a new sample correlation matrix and solve the system at every sample index. We can update the filter more efficiently by writing the weighted sample correlation matrices recursively as

$$\hat{R}_{xx}[k-\ell, n+1] = \lambda \hat{R}_{xx}[k-\ell, n] + x[n+1-k]\, x[n+1-\ell],\ \ell = 0, \ldots, p, \tag{11.83}$$

$$\hat{R}_{dx}[\ell, n+1] = \lambda \hat{R}_{dx}[\ell, n] + d[n+1]\, x[n+1-\ell],\ \ell = 0, \ldots, p. \tag{11.84}$$

We can then apply the matrix inversion lemma [8] to write the estimated model parameters at time $n+1$ in terms of the estimated parameters at time n:

$$e[n] = d[n] - \hat{d}[n], \tag{11.85}$$

$$\vec{g}[n] = P[n]\vec{x}[n] \left\{ \lambda + \vec{x}[n]^T P[n] \vec{x}[n] \right\}^{-1}, \tag{11.86}$$

$$P[n+1] = \lambda^{-1} P[n] - \vec{g}[n]\vec{x}[n]^T \lambda^{-1} P[n], \tag{11.87}$$

$$\vec{h}[n+1] = \vec{h}[n] + \vec{g}[n]\, e[n], \tag{11.88}$$

where $P[n] = \hat{R}[n]^{-1}$ is updated directly. This sequential update is known as the "recursive least-squares" (RLS) algorithm [8]. While there exist more computationally efficient lattice implementations of this update without the matrix inversion lemma, the RLS algorithm is shown to be numerically more stable [8] than such lattice recursions, owing to the additional redundant computations. Many variations of the RLS algorithm have been developed, including approaches specifically designed to improve performance when numerical precision is limited.

We observe that the estimated parameter vector $\vec{h}[n]$ at time n is updated by adding a vector, called the gain vector, times the estimation error $e[n]$:

$$\vec{h}[n+1] = \vec{h}[n] + \vec{g}[n]e[n], \tag{11.89}$$

where the gain vector $\vec{g}[n]$ is the inverse correlation matrix $P[n]$ multiplied by the data vector $\vec{x}[n]$ with certain power scaling. It is common to reduce the computational complexity of this update by replacing the scaled inverse correlation matrix by the inverse of the trace of the correlation matrix

$$\vec{h}[n+1] = \vec{h}[n] + \frac{1}{Trace\left(\hat{R}[n]\right)}\vec{x}[n]e[n]. \tag{11.90}$$

Subsequently, the inverse of the trace, which corresponds to the power in the stochastic process, is often further replaced by a small positive constant $\mu > 0$,

$$\vec{h}[n+1] = \vec{h}[n] + \mu\vec{x}[n]e[n], \tag{11.91}$$

yielding the celebrated least-mean squares (LMS) adaptive algorithm [10].

The LMS recursion can also be derived based using a stochastic formulation. When the underlying stochastic model is known, i.e., the cross- and autocorrelation functions are known, the parameters that minimize the MSE are the solution to the normal equations

$$\vec{h}_{MMSE} = R^{-1}\vec{p}. \tag{11.92}$$

The same normal equations can be solved iteratively by applying a gradient descent algorithm [8] to the MSE cost function, yielding a recursive update

$$\vec{h}^{(k+1)} = \vec{h}^{(k)} - \mu\nabla_{\vec{h}^{(k)}}E\left[e^2[n]\right], \tag{11.93}$$

$$\vec{h}^{(k+1)} = \vec{h}^{(k)} + \mu\left(\vec{p} - R\vec{h}^{(k)}\right), \tag{11.94}$$

which converges to $\vec{h}_{MMSE}$ as k increases provided that μ is selected appropriately [8] [10]. The notation in (11.93) and (11.94) uses the parameter k to denote an iteration of the algorithm, rather than a time index. As such, the superscript (k) is chosen to highlight this difference. If the underlying R and p are not known, then the same iteration can be carried out in the time domain by replacing the expectations over variables by their instantaneous temporal values, in other words by replacing $E\left[e^2[n]\right]$ with $e^2[n]$, to yield

$$\vec{h}[n+1] = \vec{h}[n] - \mu\nabla_{\vec{h}[n]}e^2[n], \tag{11.95}$$

$$\vec{h}[n+1] = \vec{h}[n] + \mu\vec{x}[n]e[n], \tag{11.96}$$

which yields the stochastic interpretation of the LMS algorithm. Hence, the LMS algorithm can be considered a gradient descent algorithm, where a "stochastic" gradient is used in the update instead of the true gradient [8]. Fig. 11.11 shows an example of the path traced by gradient descent for a two-parameter filter. In this example, a signal $y[n]$ is generated as the output of a filter with impulse response $h[n] = 0.5\delta[n] + 0.25\delta[n-1]$, with an input $x[n]$ generated as a sequence of independent, identically distributed Gaussian random values. In this example, the LMS algorithm was initialized with $\vec{h}[0] = [0, 0]^\top$, and iterations of (11.95) and (11.96) through 1000 iterations were carried out, where the vector $\vec{x}[n] = [x[n], x[n-1]]^\top$, and $e[n] = \left(y[n] - \vec{h}[n]^\top \vec{x}[n]\right)$, and $\mu = 0.02$ are used. The algorithm can be seen to converge to the vicinity of the correct values and remain within a small "misadjustment" error of these values.

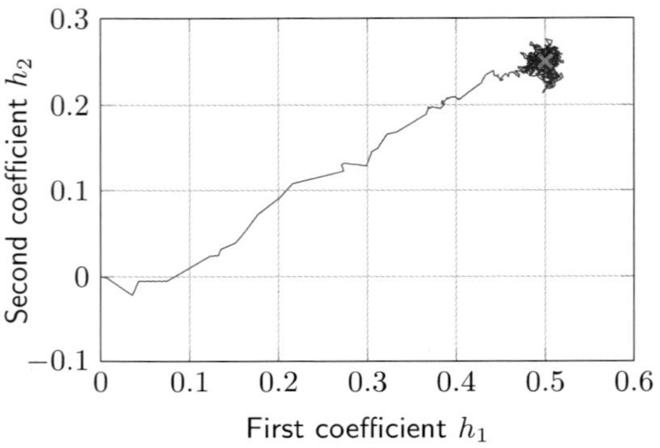

FIGURE 11.11

Graphical depiction of the convergence of the stochastic gradient (LMS) algorithm, depicted for a two-parameter filter initialized at $\vec{h}[0] = [0, 0]^\top$.

11.5 Model order estimation

Even when a parametric model is known to be appropriate for a particular application, e.g., an AR, ARMA, or MA model, we still need to decide on the number of parameters in that model, that is, the order of the model. Determination of the proper number of parameters to use for a parametric model is a difficult problem and has a rich history in many applications in signal processing, machine learning, and communications [8] [11]. If the complete statistical properties of the underlying signals were known, then increasing the number of parameters would always increase the modeling power. However, since the model parameters must be learned from a limited amount of training data, a high-order model can cause overfitting. As we saw in the example from Section 11.1.2, a large model can exhibit poor extrapolation outside of the training set; that is, too many parameters can be detrimental to performance.

To avoid such overfitting problems, much of the early work in the model order determination literature focused on probabilistic methods, such as the minimum description length (MDL) or the Akaike information criterion (AIC) [12] [6]. These methods establish a balance between model order and modeling power, quantified by certain maximum likelihood or information theoretic measures, using regularization terms. For example, for AR modeling with white Gaussian noise, the leading term of the penalty term in the MDL for a model of order p and signal length n is given by $\left(\frac{p}{2}\right)\log(n)$. Hence, for a signal $d[t]$ according to the original definition of MDL, the model order is selected by minimizing

$$p_{MDL} = \underset{p}{\operatorname{argmin}} \left\{ \min_{\{h[k]\}_{k=0}^{p}} \sum_{t=1}^{n} \left(d[t] - \sum_{k=1}^{p} h[k]\, d[t-k] \right)^2 + \frac{p}{2}\log(n) \right\}. \tag{11.97}$$

While these statistical approaches have been widely used in different signal processing applications, they require considerable statistical information on the signals of interest and can be fragile in regimes in which the signals do not behave according to the assumed statistical model.

However, the machine learning and information theory communities have recently introduced new approaches to the model order selection problem [13] [14]. In the universal model order selection approach, instead of making hard decisions at each instant based on a single assumed statistical model, one uses a performance-based mixture of all models of different orders. The goal is to perform as well as or better than the best model in the mixture. The resulting algorithms rely on lower-order models in the short-data record regime and emphasize higher-order models as the data record grows to accommodate them. The elegance of the universal mixture methods is that this transition occurs naturally: By monitoring the performance of each model on the data observed so far and placing more emphasis on better-performing models, the algorithm changes the mixture over time to leverage the respective strengths of both low- and high-order models.

For example, consider a linear prediction application using an AR model. Without prior information about the signal structure, the choice of order of the linear predictor depends heavily on the amount of data available. Instead of fixing a particular model order, one can use a convex combination of order-p predictors $\hat{x}_p[n]$, for $p = 1, \ldots M$. The combined prediction is given by

$$\hat{x}[n] = \sum_{p=1}^{M} \mu_p[n]\, \hat{x}_p[n], \tag{11.98}$$

where the combination weights $\mu_p[n]$ are calculated as $\mu_p[n] \sim \exp\left(-\sum_{t=1}^{n}(x(t) - \hat{x}_p[t])^2\right)$, i.e., based on the performance of the pth predictor on the observed data. The output is therefore a performance-weighted mixture.

Because the filters in the mixture differ from each other only in their order, they can be efficiently implemented using a lattice structure. The resulting universal predictor has computational complexity that is linear in the largest model order [10]. Such a model combination is shown to achieve the following square error performance for any N [13]:

$$\frac{1}{N}\sum_{n=1}^{N}\left(x[n] - \hat{x}[n]\right)^2 \le \min_{p<M} \frac{1}{N}\sum_{n=1}^{N}\left(x[n] - \hat{x}_p[n]\right)^2 + \frac{1}{N}\ln(M). \tag{11.99}$$

This filter asymptotically achieves the performance of the best model order in the set of models. Therefore, these filters are called "universal filters" and are part of a larger family of "universal" methods, which are explored in more detail in [13] [14] [15] [16].

References

[1] A. Oppenheim, R.W. Schafer, J.R. Buck, Discrete-Time Signal Processing, 2nd ed., Prentice Hall, 1999.

[2] L. Rabiner, B. Gold, Theory and Application of Digital Signal Processing, Prentice-Hall, Inc., Englewood Cliffs, 1975.

[3] L.R. Rabiner, R.W. Schafer, Introduction to Digital Speech Processing, Prentice Hall, Inc., 1978.

[4] W. Feller, An Introduction to Probability Theory and Its Applications, Wiley, 1968.

[5] B. Hajek, An exploration of random processes for engineers [online], available: http://www.ifp.illinois.edu/~hajek/Papers/randomprocesses.html, 2011.

[6] S. Kay, Fundamentals of Statistical Signal Processing, Volume I: Estimation Theory (v. 1), Prentice Hall, 1993.

[7] J. Mahkoul, Linear prediction: a tutorial review, in: Proceedings of the IEEE, 1975, pp. 561–580.

[8] A.H. Sayed, Fundamentals of Adaptive Filtering, Wiley-IEEE Press, 2003.

[9] R. Bellman, On the approximation of curves by line segments using dynamic programming, Communications of the ACM 4 (6) (1961) 284.

[10] M.L. Honig, D.G. Messerschmitt, Adaptive Filters: Structures, Algorithms and Applications, Springer, 1984.

[11] J.G. Proakis, Digital Communications, McGraw-Hill, 1983.

[12] H.V. Poor, An Introduction to Signal Detection and Estimation, Springer, New York, 1988.

[13] A.C. Singer, M. Feder, Universal linear prediction by model order weighting, IEEE Transactions on Signal Processing 47 (10) (October 1999) 2685–2699.

[14] N. Cesa-Bianchi, G. Lugosi, Prediction Learning and Games, Cambridge University Press, 2006.

[15] A.C. Singer, S.S. Kozat, M. Feder, Universal linear least squares prediction upper and lower bounds, IEEE Transactions on Information Theory 48 (8) (August 2002) 2354–2362.

[16] S.S. Kozat, A.C. Singer, G.C. Zeitler, Universal piecewise linear prediction via context trees, IEEE Transactions on Signal Processing 55 (7) (July 2007) 3730–3745.

CHAPTER

Adaptive filters

12

Vítor H. Nascimento and Magno T.M. Silva

Dept. of Electronic Systems Engineering, University of São Paulo, São Paulo, SP, Brazil

12.1 Introduction

Adaptive filters are employed in situations in which the environment is constantly changing, so that a fixed system would not have adequate performance. As they are usually applied in real-time applications, they must be based on algorithms that require a small number of computations per input sample.

These algorithms can be understood in two complementary ways. The most straightforward way follows directly from their name: an adaptive filter uses information from the environment and from the very signal it is processing to change itself, so as to optimally perform its task. The information from the environment may be sensed in real time (in the form of a so-called desired signal), or may be provided a priori, in the form of previous knowledge of the statistical properties of the input signal (as in blind equalization).

On the other hand, we can think of an adaptive filter also as an algorithm to separate a mixture of two signals. The filter must have some information about the signals to be able to separate them; usually this is given in the form of a reference signal, related to only one of the two terms in the mixture. The filter then has two outputs, corresponding to each signal in the mixture (see Fig. 12.1). As a byproduct of separating the signals, the reference signal is processed in useful ways. This way of viewing an adaptive filter is very useful, particularly when one is learning the subject.

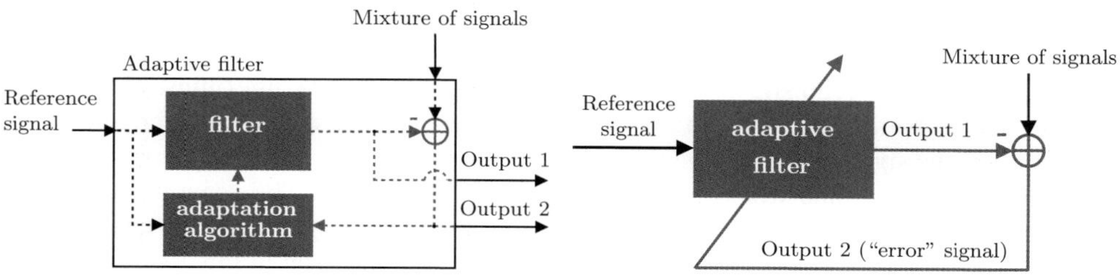

FIGURE 12.1

Inputs and outputs of an adaptive filter. Left: Detailed diagram showing inputs, outputs, and internal variables. Right: Simplified diagram.

In the following sections we give an introduction to adaptive filters, covering subjects ranging from basic principles to some of the most important developments. Since the adaptive filtering literature is

Signal Processing and Machine Learning Theory. https://doi.org/10.1016/B978-0-32-391772-8.00018-1
Copyright © 2024 Elsevier Inc. All rights reserved.

vast, we are not able to treat all interesting topics. Along the text, we reference important textbooks, classic papers, and a sample of the most promising literature. Among the many very good textbooks on the subject, four of the most popular are [117,172,317,388].

We start with an application example in the next section. In it, we show how an adaptive filter is applied to cancel acoustic echo in hands-free telephony. In the following sections we return frequently to this example, to illustrate new concepts and methods on a practical situation. The example is completed by an overview of how adaptive filters work, in which we use only deterministic arguments and concepts from basic linear systems theory, in Section 12.1.2. This overview shows many of the compromises that arise in adaptive filter design, without the more complicated math. Section 12.1.3 gives a small glimpse of the wide variety of applications of adaptive filters, describing a few other examples. Of course, in order to fully understand the behavior of adaptive filters, one must use estimation theory and stochastic processes, which is the subject of Sections 12.2 and 12.3. The remaining sections are devoted to extensions and improvements to the basic algorithms.

Adaptive filter theory brings together results from several fields: signal processing, matrix analysis, control and systems theory, stochastic processes, and optimization. We provide short reviews for most of the necessary material in separate links ("boxes"), which the reader may consult if needed.

12.1.1 Motivation – acoustic echo cancelation

Suppose you are in a virtual meeting, using your computer to talk to friends. Imagine your friend is not using headphones, but the computer microphone and loudspeakers. Let us assume that your friend's computer does not have an acoustic echo canceler. In this case, the sound of your voice will leave the loudspeaker, propagate inside your friend's room, and find its way to your friend's microphone, resulting in an echo signal. The echo will be composed of a direct signal from the loudspeaker to the microphone and also from reflexions from the walls and ceiling. This signal is fed back to the microphone and you will hear your own voice with delay and attenuation, as shown in Fig. 12.2.

The echo signal depends on the acoustic characteristics of each room and also on the location of the loudspeaker and microphone. Furthermore, the acoustic characteristics can change during a phone call, since your friend may open or close a door, for example. In order to cancel the acoustic echo effectively, the variations of the environment must be tracked by an adaptive filter, which performs on-line updating of its parameters through an algorithm. Fig. 12.3 shows the insertion of an adaptive filter in the system of Fig. 12.2 to cancel the echo signal. In this configuration, the adaptive filter must identify the echo path and provide at its output an estimate of the echo signal. Thus, a synthesized version of the echo is subtracted from the signal picked up by the computer microphone. In an ideal scenario, the resulting signal will be free of echo and will contain only the signal of interest. The general setup of an echo canceler is shown in Fig. 12.4. The *near-end* signal (near with respect to the echo canceler) may be simply noise when the person inside the room is in silence, as considered in Figs. 12.2 and 12.3, or it can be a signal of interest, that is, the voice of the user in the room.

You most likely have already heard the output of an adaptive echo canceler. However, you probably did not pay attention to the initial period, when you can hear the adaptive filter learning the echo impulse response, if you listen carefully. We illustrate this through a simulation, using a measured impulse response of 256 samples (sampled at 8 kHz) and an echo signal produced from a recorded voice signal. The measured impulse response is shown in Fig. 12.5. In the first example, the near-end signal is low-power noise (more specifically, white Gaussian noise with zero mean and variance

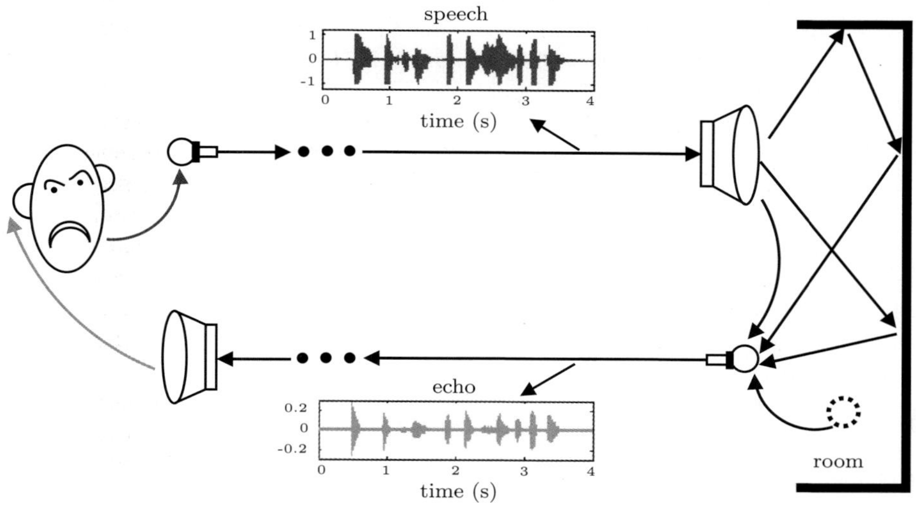

FIGURE 12.2

Echo in hands-free telephony. Go to http://www.lps.usp.br/FAs/echo/speech.wav to listen to a speech signal and to http://www.lps.usp.br/FAs/echo/echo.wav to listen to the corresponding echo signal.

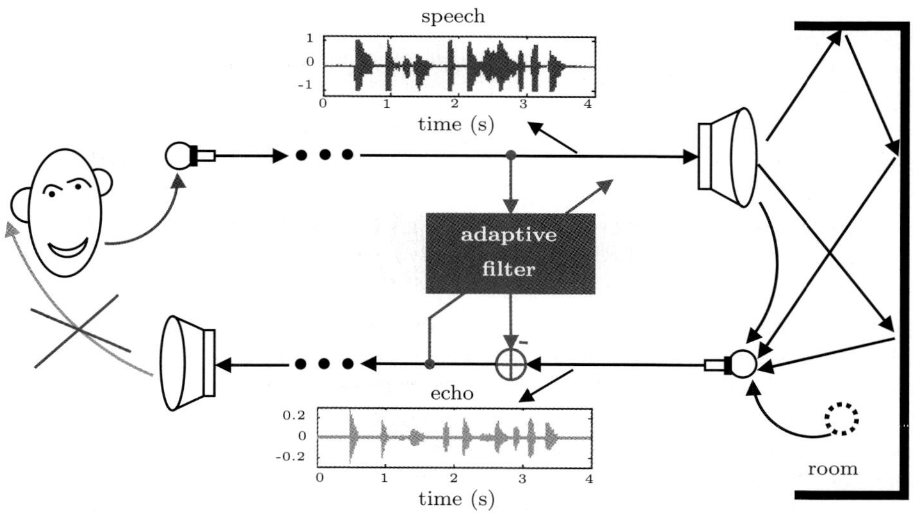

FIGURE 12.3

Acoustic echo cancelation using an adaptive filter.

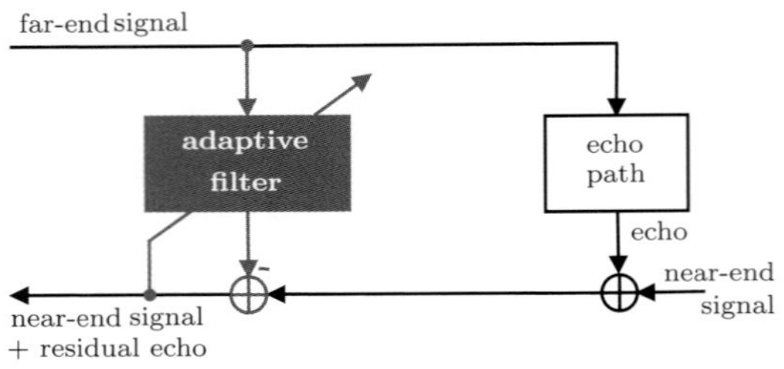

FIGURE 12.4

General echo canceler configuration.

$\sigma^2 = 10^{-4}$). Listen to `e_nlms.wav`, and pay attention to the beginning. You will notice that the voice (the echo) is fading away, while the adaptive filter is learning the echo impulse response. This file is the output of an echo canceler adapted using the *normalized least-mean squares* (NLMS) algorithm, which is described in Section 12.3.2.

Listen now to `e_lsl.wav`. In this case the filter was adapted using the modified *error-feedback least-squares lattice* (EF-LSL) algorithm [248], which is a low-cost and stable version of the *recursive least-squares* (RLS) algorithm described in Section 12.3.3. Now you will notice that the echo fades away much faster. This fast convergence is characteristic of the EF-LSL algorithm, and is of course a very desirable property. On the other hand, the NLMS algorithm is simpler to implement and more robust against imperfections in the implementation. An adaptive filtering algorithm must satisfy several requirements, such as convergence rate, tracking capability, noise rejection, computational complexity, and robustness. These requirements are conflicting, so that there is no best algorithm that will outperform all the others in every situation. The many algorithms available in the literature represent different compromises between these requirements [117,172,317].

Fig. 12.6 shows 4 seconds of the echo signal prior to cancelation (bottom) and a measure of the quality of echo cancelation, the so-called *echo return loss enhancement* (ERLE) (top). The ERLE is defined as the following ratio:

$$\text{ERLE} = \frac{\text{power of original echo}}{\text{power of residual echo}}.$$

The higher the ERLE, the better the performance of the canceler. The ERLE drops when the original echo is weak (there is nothing to cancel!) and increases again when the echo is strong. The figure shows that EF-LSL performs better than NLMS (but at a higher computational cost).

Assuming now that the near-end signal is another recorded voice signal (click here to listen to this signal), we applied again the EF-LSL algorithm to cancel the echo. This case simulates a person talking on the other side. Since the speech of this person is not *correlated*[1] to the far-end signal, the output of the adaptive filter converges again to an estimate of the echo signal. Thus, a synthesized version of the echo is subtracted from the signal picked up by the computer microphone and the resulting signal

[1] For more on this, see Eq. (12.4) and Section 12.2.

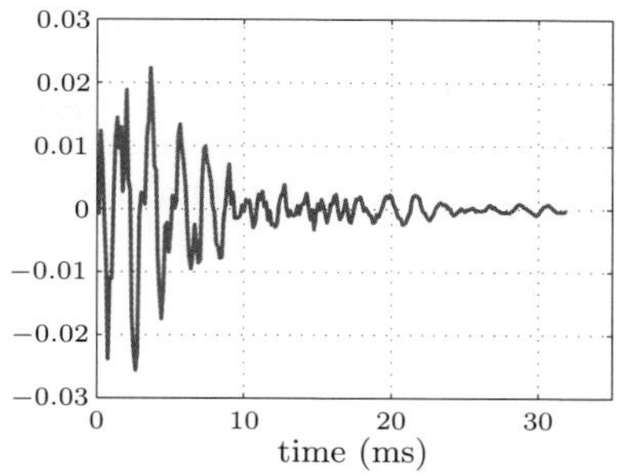

FIGURE 12.5

Measured impulse response (256 samples).

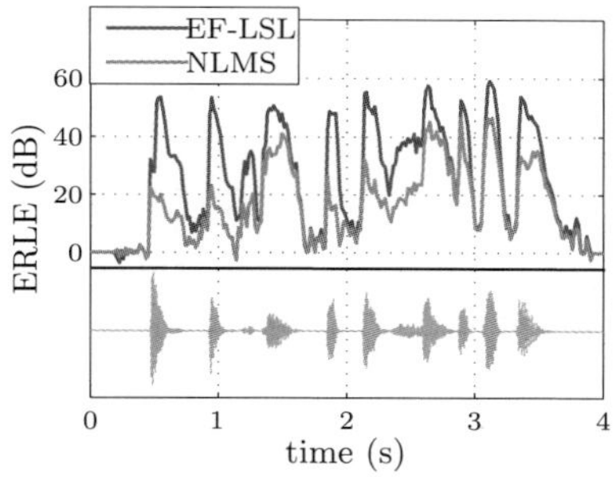

FIGURE 12.6

Echo return loss enhancement for the NLMS ($\tilde{\mu} = 0.5$, $\delta = 10^{-5}$, $M = 256$) and EF-LSL ($\lambda = 0.999$, $\delta = 10^{-5}$) algorithms. Bottom trace: Echo signal prior to cancelation. The meaning of the parameters is explained in Sections 12.3.1, 12.3.2, and 12.3.3.

converges to the speech of the person using the computer. This result can be verified by listening to the file `e_1s1DT.wav`.

This is a typical example of use of an adaptive filter. Although in our simulation we assumed the environment (the echo impulse response) is fixed, in practice it changes constantly, requiring a constant retuning of the canceling filter. Although the echo itself is not known, we have access to a reference signal (the clean speech of the far-end user). This is the signal that is processed to obtain an estimate of the echo. Finally, the signal captured by the near-end microphone (the so-called *desired* signal) is a mixture of two signals: the echo and the near-end speech (or ambient noise, when the near-end user is silent). The task of the adaptive filter can be seen as to separate these two signals.

We should mention that the name desired signal is somewhat misleading, although widespread in the literature. Usually we are interested in separating the "desired" signal into two parts, one of which is of no interest at all!

We proceed now to show how an algorithm can perform this task, modeling all signals as periodic to simplify the arguments.

12.1.2 A quick tour of adaptive filtering

Before embarking on a detailed explanation of adaptive filtering and its theory, it is a good idea to present the main ideas in a simple setting, to help the reader create intuition. In this section, we introduce the least-mean squares (LMS) algorithm and some of its properties using only deterministic arguments and basic tools from linear systems theory.

12.1.2.1 *Posing the problem*

Consider again the echo cancelation problem, seen in the previous section (see Fig. 12.7). Let $y(n)$ represent the echo, $v(n)$ the voice of the near-end user, and $x(n)$ the voice of the far-end user (the reference); $d(n)$ represents the signal that would be received by the far-end user without an echo canceler (the mixture, or desired signal).

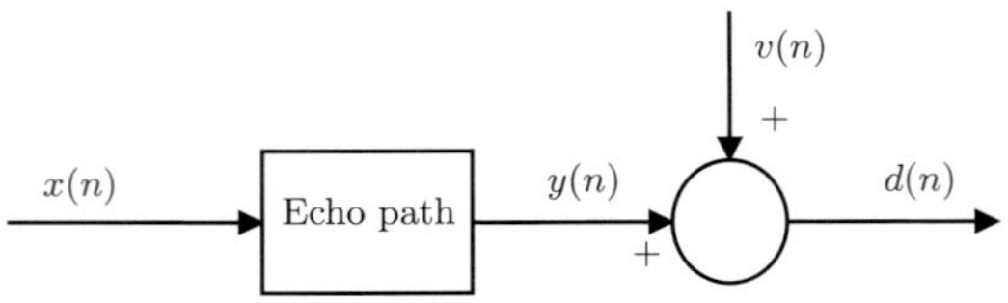

FIGURE 12.7

A model for the echo.

The echo path represents the changes that the far-end signal suffers when it goes through the digital-to-analog (D/A) converter, the loudspeaker, the air path between the loudspeaker and the microphone (including all reflections), the microphone itself and its amplifier, and the analog-to-digital (A/D) converter. The microphone signal $d(n)$ will always be a mixture of the echo $y(n)$ and an uncorrelated signal $v(n)$, which is composed of the near-end speech plus noise. Our goal is to remove $y(n)$ from $d(n)$. The challenge is that we have no means to measure $y(n)$ directly, and the echo path is constantly changing.

12.1.2.2 *Measuring how far we are from the solution*

How is the problem solved, then? Since we cannot measure $y(n)$, the solution is to estimate it from the measurable signals $x(n)$ and $d(n)$. This is done as follows: imagine for now that all signals are periodic, so they can be decomposed as Fourier series,

$$x(n) = \sum_{k=1}^{K_0} A_k \cos(k\omega_0 n + \varphi_k), \qquad v(n) = \sum_{\ell=1}^{K_1} B_\ell \cos(\ell\omega_1 n + \theta_\ell), \tag{12.1}$$

for certain amplitudes A_k and B_ℓ, phases φ_k and θ_ℓ, and frequencies ω_0 and ω_1. The highest frequencies appearing in $x(n)$ and $v(n)$ must satisfy $K_0\omega_0 < \pi$, $K_1\omega_1 < \pi$, since we are dealing with sampled signals (Nyquist criterion). We assume for now that the echo path is fixed (not changing) and can be modeled by an unknown linear transfer function $H(z)$. In this case, after the transients vanish, $y(n)$ would also be a periodic sequence with fundamental frequency ω_0,

$$y(n) = \sum_{k=1}^{K_0} C_k \cos(k\omega_0 n + \psi_k),$$

with $C_k e^{j\psi_k} = H(e^{jk\omega_0}) A_k e^{j\varphi_k}$. The signal picked up by the microphone is then

$$d(n) = y(n) + v(n) = \sum_{k=1}^{K_0} C_k \cos(k\omega_0 n + \psi_k) + \sum_{\ell=1}^{K_1} B_\ell \cos(\ell\omega_1 n + \theta_\ell).$$

Since we know $x(n)$, if we had a good approximation $\hat{H}(z)$ of $H(z)$, we could easily obtain an approximation $\hat{y}(n)$ of $y(n)$ and subtract it from $d(n)$, as in Fig. 12.8.

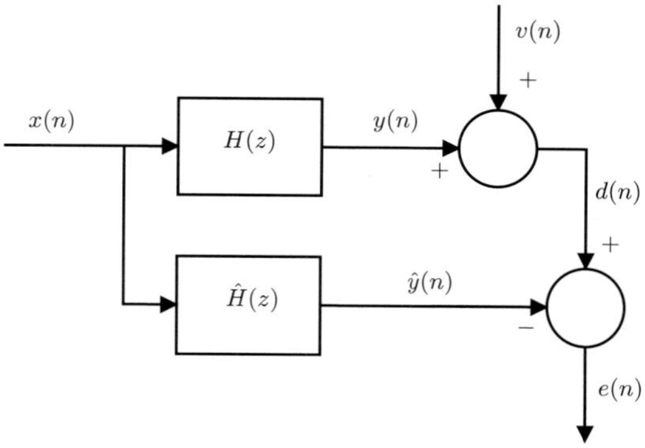

FIGURE 12.8

Echo canceler.

How could we find $\hat{H}(z)$ in real time? Recall that only $d(n)$, $x(n)$, $\hat{y}(n)$, and $e(n)$ can be observed. For example, how could we know that we have the exact filter, that is, $\hat{H}(z) = H(z)$, by looking only at these signals? To answer this question, take a closer look at $e(n)$. Let the output of $\hat{H}(z)$ be

$$\hat{y}(n) = \sum_{k=1}^{K_0} \hat{C}_k \cos(k\omega_0 n + \hat{\psi}_k).$$

Then

$$\begin{aligned} e(n) &= d(n) - \hat{y}(n) = y(n) - \hat{y}(n) + v(n) \\ &= \sum_{k=1}^{K_0} \left[C_k \cos(k\omega_0 n + \psi_k) - \hat{C}_k \cos(k\omega_0 n + \hat{\psi}_k) \right] \\ &\quad + \sum_{\ell=1}^{K_1} B_\ell \cos(\ell\omega_1 n + \theta_\ell) \\ &= \sum_{k=1}^{K_0} \tilde{C}_k \cos(k\omega_0 n + \tilde{\psi}_k) + \sum_{\ell=1}^{K_1} B_\ell \cos(\ell\omega_1 n + \theta_\ell), \end{aligned} \tag{12.2}$$

where $\tilde{C}_k e^{j\tilde{\psi}_k} = C_k e^{j\psi_k} - \hat{C}_k e^{j\hat{\psi}_k}$, $k = 1 \ldots K_0$.

Assuming that the cosines in $x(n)$ and $v(n)$ have no frequencies in common, the simplest approach is to measure the average power of $e(n)$. Indeed, if all frequencies $k\omega_0$ and $\ell\omega_1$ appearing in $e(n)$ are different from one another, then the average power of $e(n)$ is

$$P = \sum_{k=1}^{K_0} \frac{\tilde{C}_k^2}{2} + \sum_{\ell=1}^{K_1} \frac{B_\ell^2}{2}. \tag{12.3}$$

If $\hat{y}(n) = y(n)$, then P is at its minimum,

$$P_o = \min_{\hat{H}(z)} P = \sum_{\ell=1}^{K_1} \frac{B_\ell^2}{2},$$

which is the average power of $v(n)$. In this case, $e(n)$ is at its optimum, $e(n) = e_o(n) = v(n)$, and the echo is completely canceled.

It is important to see that this works because the two signals we want to separate, $y(n)$ and $v(n)$, are *uncorrelated*, that is, they are such that

$$\lim_{N\to\infty} \frac{1}{N+1} \sum_{-N/2}^{N/2} y(n)v(n) = \lim_{N\to\infty} \frac{1}{N+1} \sum_{-N/2}^{N/2} y(n)e_o(n) = 0. \tag{12.4}$$

This property is known as the *orthogonality condition*. A form of orthogonality condition will appear in general whenever one tries to minimize a quadratic function, as is the case here. This is discussed in more detail in Section 12.2.

The average power P could then be used as a measure of how good our approximation $\hat{y}(n)$ is. The important point is that it is very easy to find a good approximation to P. It suffices to low-pass filter $e^2(n)$, for example,

$$\hat{P}(n) = \frac{1}{N} \sum_{k=0}^{N-1} e^2(n-k) \tag{12.5}$$

is a good approximation if the window length N is large enough.

The adaptive filtering problem then becomes an optimization problem, with the particularity that the cost function that is being minimized is not known exactly, but only through an approximation, as in (12.5). In the following sections, we discuss in detail the consequences of this fact.

It is also important to remark that the average error power is not the only possible choice for the cost function. Although it is the most popular choice for a number of reasons, in some applications other choices are more adequate. Even posing the adaptive filtering problem as an optimization problem is not the only alternative, as shown by recent methods based on projections onto convex sets [365].

12.1.2.3 Choosing a structure for the filter

Our approximation $\hat{y}(n)$ will be built by minimizing the estimated average error power (12.5) as a function of the echo path model $\hat{H}(z)$, that is, our estimated echo path will be the solution of

$$\hat{H}_o(z) = \arg\min_{\hat{H}(z)} \hat{P}(n). \tag{12.6}$$

Keep in mind that we are looking for a way of solving (12.6) in real time, since $\hat{P}(n)$ can only be obtained by measuring $e(n)$ for a certain amount of time. We must then be able to implement the current approximation $\hat{H}(z)$, so that $\hat{y}(n)$ and $e(n)$ may be computed.

This will impose practical restrictions on the kind of function that $\hat{H}(z)$ may be: since memory and processing power are limited, we must choose beforehand a structure for $\hat{H}(z)$. Memory and processing power will impose a constraint on the maximum order the filter may have. In addition, in order to write the code for the filter, we must decide if it will depend on past outputs (infinite impulse response [IIR] filters) or only on past inputs (finite impulse response [FIR] filters). These choices must be made based on some knowledge of the kind of echo path that our system is likely to encounter.

To explain how these choices can be made, let us simplify the problem a little more and restrict the far-end signal $x(n)$ to a simple sinusoid (i.e., assume that $K_0 = 1$). In this case,

$$y(n) = C_1 \cos(\omega_0 n + \varphi_1),$$

with $C_1 e^{j\psi_1} = H(e^{j\omega_0}) A_1 e^{j\varphi_1}$. Therefore, $\hat{H}(z)$ must satisfy

$$\left[\hat{H}(e^{j\omega}) = H(e^{j\omega})\right]_{\omega=\omega_0} \tag{12.7}$$

only for $\omega = \omega_0$; the value of $\hat{H}(e^{j\omega})$ for other frequencies is irrelevant, since the input $x(n)$ has only one frequency. Expanding (12.7), we obtain

$$\mathrm{Re}\{\hat{H}(e^{j\omega_0})\} = \mathrm{Re}\{H(e^{j\omega_0})\}, \qquad \mathrm{Im}\{\hat{H}(e^{j\omega_0})\} = \mathrm{Im}\{H(e^{j\omega_0})\}. \tag{12.8}$$

The optimum $\hat{H}(z)$ is defined through two conditions. Therefore an FIR filter with just two coefficients would be able to satisfy both conditions for any value of $H(e^{j\omega_0})$, so we could define

$$\hat{H}(z) = w_0 + w_1 z^{-1},$$

and the values of w_0 and w_1 would be chosen to solve (12.6). Note that the argument generalizes for $K_0 > 1$: in general, if $x(n)$ has K_0 harmonics, we could choose $\hat{H}(z)$ as an FIR filter with length $2K_0$, two coefficients per input harmonic.

This choice is usually not so simple: in practice, we would not know K_0, and in the presence of noise (as we shall see later), a filter with fewer coefficients might give better performance, even though it would not completely cancel the echo. The structure of the filter also affects the dynamic behavior of the adaptive filter, so simply looking at equations such as (12.7) does not tell the whole story. Choosing the structure for $\hat{H}(z)$ is one of the most difficult steps in adaptive filter design. In general, the designer must test different options, initially perhaps using recorded signals, but ultimately building a prototype and performing some tests.

12.1.2.4 Searching for the solution

Returning to our problem, assume that we have decided that an FIR filter with length M is adequate to model the echo path, that is,

$$\hat{H}(z) = \sum_{k=0}^{M-1} w_k z^{-k}.$$

Our minimization problem (12.6) now reduces to finding the coefficients $w_0 \ldots w_{M-1}$ that solve

$$\min_{w_0 \,\ldots\, w_{M-1}} \hat{P}(n). \tag{12.9}$$

Since $\hat{P}(n)$ must be computed from measurements, we must use an iterative method to solve (12.9). Many algorithms could be used, as long as they depend only on measurable quantities ($x(n)$, $d(n)$, $\hat{y}(n)$, and $e(n)$). We will use the steepest descent algorithm (also known as gradient algorithm) as an example now. Later we show other methods that also could be used. If you are not familiar with the gradient algorithm, see Box 12.1 for an introduction.

The cost function in (12.9) is

$$\hat{P}(n) = \frac{1}{N}\sum_{k=0}^{N-1} e^2(n-k) = \frac{1}{N}\sum_{k=0}^{N-1}\left[d(n-k) - \hat{y}(n-k)\right]^2.$$

We need to rewrite this equation so that the steepest descent algorithm can be applied. Recall also that now the filter coefficients will change. Hence we define the vectors

$$\boldsymbol{w}(n) = \begin{bmatrix} w_0(n) & w_1(n) & \ldots & w_{M-1}(n) \end{bmatrix}^T$$

and

$$\boldsymbol{x}(n) = \begin{bmatrix} x(n) & x(n-1) & \ldots & x(n-M+1) \end{bmatrix}^T,$$

where $(\cdot)^T$ denotes transposition. At each instant, we have $\hat{y}(n) = \boldsymbol{w}^T(n)\boldsymbol{x}(n)$, and our cost function becomes

$$\hat{P}(n) = \frac{1}{N}\sum_{k=0}^{N-1}\left[d(n-k) - \boldsymbol{w}^T(n-k)\boldsymbol{x}(n-k)\right]^2, \tag{12.10}$$

which depends on $\boldsymbol{w}(n) \ldots \boldsymbol{w}(n-N+1)$. In order to apply the steepest descent algorithm, let us keep the coefficient vector $\boldsymbol{w}(n)$ constant during the evaluation of $\hat{P}$, as follows. Starting from an initial condition $\boldsymbol{w}(0)$, compute for $n = 0, N, 2N, \ldots$ (the notation is explained in Boxes 12.2 and 12.6)

$$\boldsymbol{w}(n+N) = \boldsymbol{w}(n) - \alpha \left.\frac{\partial \hat{P}(n+N-1)}{\partial \boldsymbol{w}^T}\right|_{\boldsymbol{w}=\boldsymbol{w}(n)}, \tag{12.11}$$

where α is a positive constant and $\boldsymbol{w}(n+N-1) = \boldsymbol{w}(n+N-2) = \cdots = \boldsymbol{w}(n)$. Our cost function now depends on only one $\boldsymbol{w}(n)$ (compare with (12.10)):

$$\begin{aligned}\hat{P}(n+N-1) &= \frac{1}{N}\sum_{k=0}^{N-1}\left[d(n+N-1-k) - \boldsymbol{w}^T(n)\boldsymbol{x}(n+N-1-k)\right]^2 \\ &= \frac{1}{N}\sum_{k=0}^{N-1}\left[d(n+k) - \boldsymbol{w}^T(n)\boldsymbol{x}(n+k)\right]^2, \quad n = 0, N, 2N, \ldots.\end{aligned} \tag{12.12}$$

We now need to evaluate the gradient of $\hat{P}$. Expanding the expression above, we obtain

$$\hat{P}(n+N-1) = \frac{1}{N}\sum_{k=0}^{N-1}\left[d(n+k) - \sum_{\ell=0}^{M-1} w_\ell(n)x(n+k-\ell)\right]^2,$$

so

$$\frac{\partial \hat{P}(n-N+1)}{\partial w_m(n)} = -\frac{2}{N}\sum_{k=0}^{N-1}\left[d(n+k) - \boldsymbol{w}^T(n)\boldsymbol{x}(n+k)\right]x(n+k-m)$$
$$= -\frac{2}{N}\sum_{k=0}^{N-1} e(n+k)x(n+k-m)$$

and

$$\frac{\partial \hat{P}(n-N+1)}{\partial \boldsymbol{w}^T} = -\frac{2}{N}\sum_{k=0}^{N-1} e(n+k)\boldsymbol{x}(n+k),\; n=0, N, 2N, \ldots. \tag{12.13}$$

As we needed, this gradient depends only on measurable quantities, $e(n+k)$ and $\boldsymbol{x}(n+k)$, for $k=0\ldots N-1$. Our algorithm for updating the filter coefficients then becomes

$$\boldsymbol{w}(n+N) = \boldsymbol{w}(n) + \mu\frac{1}{N}\sum_{k=0}^{N-1} e(n+k)\boldsymbol{x}(n+k),\; n=0, N, 2N, \ldots, \tag{12.14}$$

where we introduced the overall step size $\mu = 2\alpha$.

We still must choose μ and N. The choice of μ is more complicated and will be treated in Section 12.1.2.5. The value usually employed for N may be a surprise: in almost all cases, one uses $N=1$, resulting in the so-called LMS algorithm

$$\boldsymbol{w}(n+1) = \boldsymbol{w}(n) + \mu e(n)\boldsymbol{x}(n), \tag{12.15}$$

proposed initially by Widrow and Hoff in 1960 [387] (Ref. [386] describes the history of the creation of the LMS algorithm). The question is, how can this work if no average is being used for estimating the error power? An intuitive answer is not complicated: assume that μ is a very small number so that $\mu = \mu_0/N$ for a large N. In this case, we can approximate $e(n+k)$ from (12.15) as follows:

$$\boldsymbol{w}(n+1) = \boldsymbol{w}(n) + \frac{\mu_0}{N}e(n)\boldsymbol{x}(n) \approx \boldsymbol{w}(n), \quad \text{if } N \text{ is large.}$$

Since $\boldsymbol{w}(n+1) \approx \boldsymbol{w}(n)$, we have $e(n+1) = d(n+1) - \boldsymbol{w}^T(n+1)\boldsymbol{x}(n+1) \approx d(n+1) - \boldsymbol{w}^T(n)\boldsymbol{x}(n+1)$. Therefore, we could approximate

$$e(n+k) \approx d(n+k) - \boldsymbol{w}^T(n)\boldsymbol{x}(n+k), \quad k=0\ldots N-1,$$

so N steps of the LMS recursion (12.15) would result in

$$\boldsymbol{w}(n+N) \approx \boldsymbol{w}(n) + \frac{\mu_0}{N}\sum_{k=0}^{N-1}\left[d(n+k) - \boldsymbol{w}^T(n)\boldsymbol{x}(n+k)\right]\boldsymbol{x}(n+k),$$

just what would be obtained from (12.14). The conclusion is that although there is no explicit average being taken in the LMS algorithm (12.15), the algorithm in fact computes an implicit, approximate average if the step size is small. This is exactly what happens, as can be seen in the animations in Fig. 12.9. These simulations were prepared with

$$x(n) = \cos(0.4\pi n + \varphi_1), \qquad v(n) = 0.2\cos(0.2\pi n + \theta_1),$$

where φ_1 and θ_1 were chosen randomly in the interval $[0, 2\pi)$. Several step sizes were used. The estimates computed with the LMS algorithm are marked by the crosses (in red in the web version). The initial condition is shown at the left, and the theoretical optimum is shown at the right end of each figure. In the simulations, the true echo path was modeled by the fixed filter

$$H(z) = 1.5 + 0.5z^{-1}.$$

For comparison, we also plotted the estimates that would be obtained by the gradient algorithm using the exact value of the error power at each instant. These estimates are marked with small dots (in blue in the web version).

Note that when a small step size is used, such as $\mu = 0.1$ in Fig. 12.9(a), the LMS filter stays close to the dots obtained assuming exact knowledge of the error power. However, as the step size increases, the LMS estimates move further away from the dots. Although convergence is faster, the LMS estimates do not reach the optimum and stay there: instead, they hover around the optimum. For larger step sizes, the LMS estimates can go quite far from the optimum (Fig. 12.9(b and c)). Finally, if the step size is too large, the algorithm will diverge, that is, the filter coefficients grow without bounds (Fig. 12.9(d)).

12.1.2.5 *Trade-off between speed and precision*

The animations in Fig. 12.9 illustrate an important problem in adaptive filters: even if one could magically choose as initial conditions the exact optimum solutions, the adaptive filter coefficients would not stay there! This happens because the filter does not know the exact value of the cost function it is trying to minimize (in this case, the error power): since $\hat{P}(n)$ is an approximation, noise (and the near-end signal in our echo cancelation example) would make the estimated gradient nonzero, and the filter would wander away from the optimum solution. This effect keeps the minimum error power obtainable using the adaptive filter always a little higher than the optimum value. The difference between the (theoretical, unattainable without perfect information) optimum and the actual error power is known as *excess mean square error* (EMSE), and the ratio between the EMSE and the optimum error is known as *misadjustment* (see Eqs. (12.172) and (12.175)).

For small step sizes the misadjustment is small, since the LMS estimates stay close to the estimates that would be obtained with exact knowledge of the error power. However, a small step size also means slow convergence. This trade-off exists in all adaptive filtering algorithms, and has been an intense topic of research: many algorithms have been proposed to allow faster convergence without increasing the misadjustment. We will see some of them in the next sections.

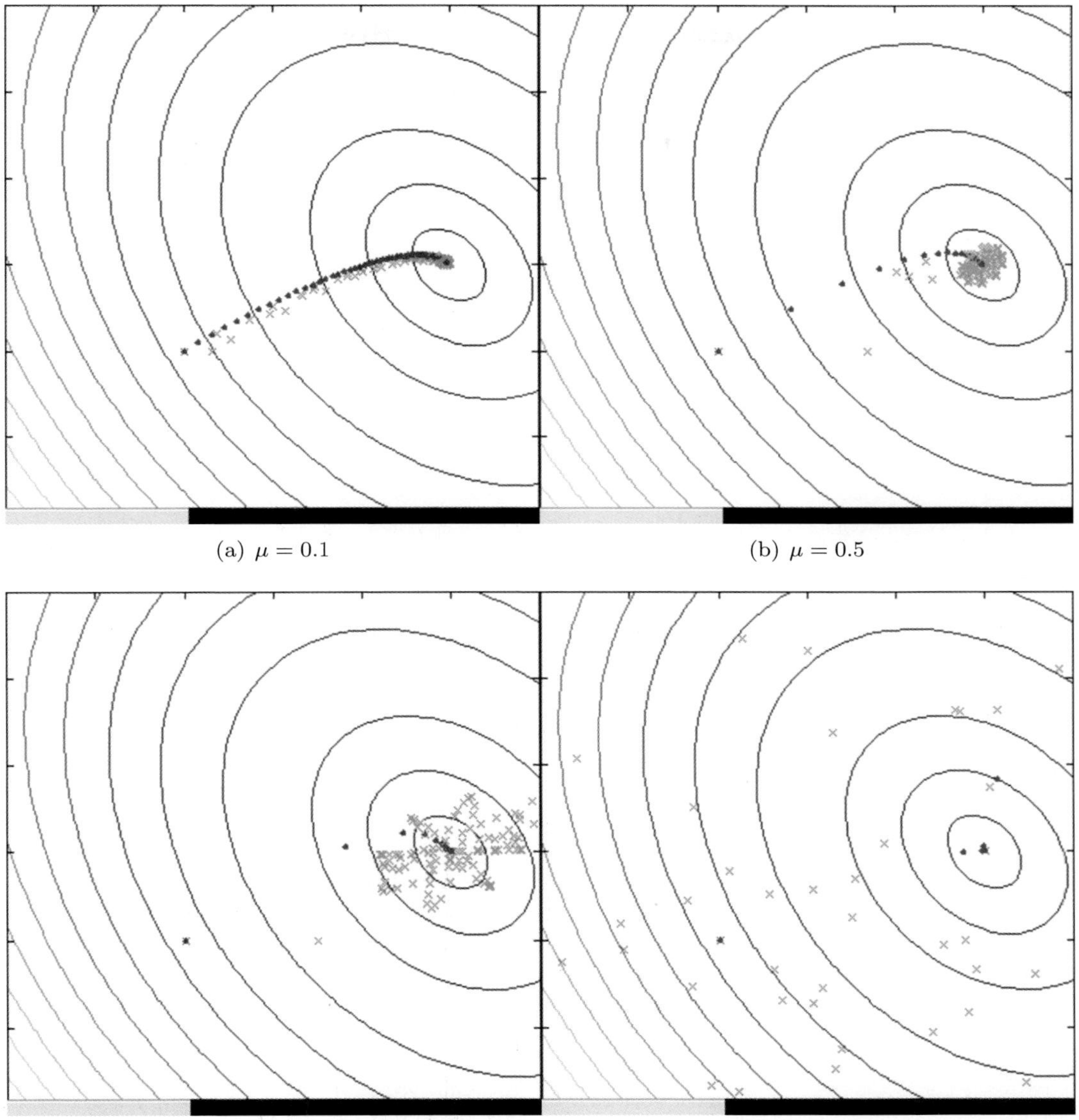

(a) $\mu = 0.1$ (b) $\mu = 0.5$

(c) $\mu = 1.1$ (d) $\mu = 1.9$

FIGURE 12.9

LMS algorithm for echo cancelation with sinusoidal input. Click on each caption to see animations in your browser or go to http://www.lps.usp.br/FAs/quick_introduction/.

The misadjustment is central to evaluation of the performance of an adaptive filter. In fact, if we want to eliminate the echo from the near-end signal, we would like that after convergence, $e(n) \approx v(n)$. This is indeed what happens when the step size is small (see Fig. 12.10(a)). However, when the step size is increased, although the algorithm converges more quickly, the performance after convergence is not very good, because of the wandering of the filter weights around the optimum (see Fig. 12.10(b–d) – in the last case, for $\mu = 1.9$, the algorithm is diverging).

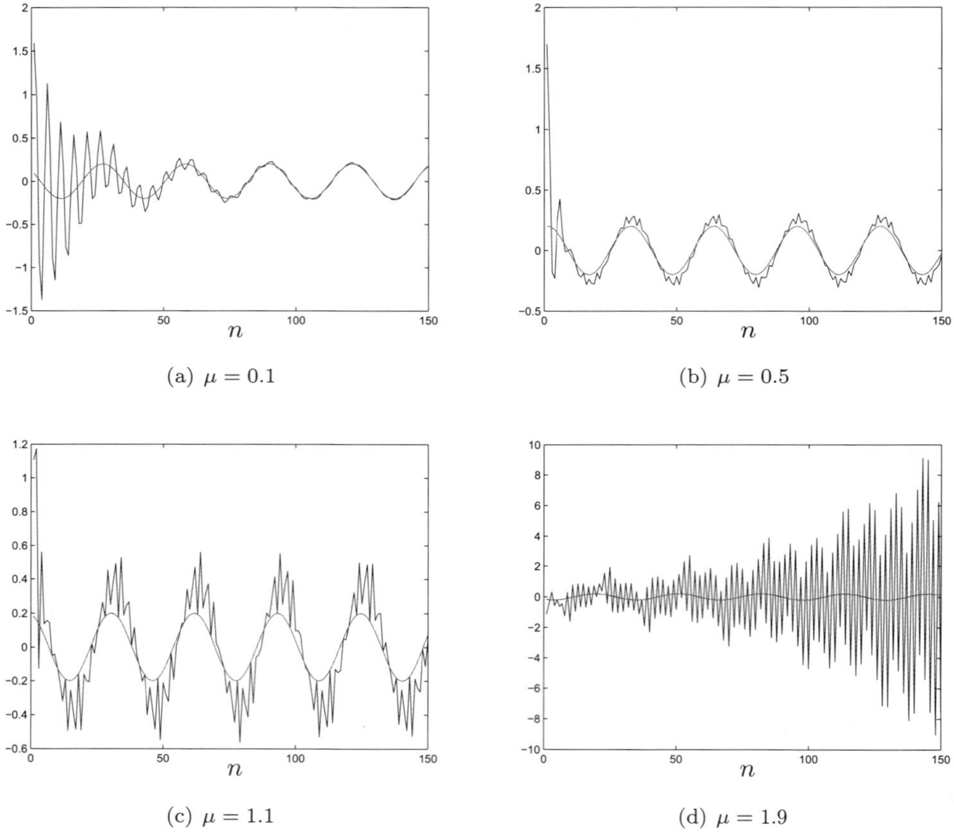

(a) $\mu = 0.1$ (b) $\mu = 0.5$ (c) $\mu = 1.1$ (d) $\mu = 1.9$

FIGURE 12.10

Compromise between convergence rate and misadjustment in LMS. Plots of $e(n) \times n$ for several values of μ.

We will stop this example here for the moment, and move forward to a description of adaptive filters using probability theory and stochastic processes. We will return to the example during discussions of stability and model order selection.

The important message from this section is that an adaptive filter is able to separate two signals in a mixture by minimizing a well-chosen cost function. The minimization must be performed iteratively, since the true value of the cost function is not known: only an approximation to it may be computed at each step. When the cost function is quadratic, as in the example shown in this section, the components of the mixture are separated such that they are in some sense orthogonal to each other.

Box 12.1: Steepest descent algorithm

Given a cost function

$$J(\boldsymbol{w}) = J(w_0, \ldots, w_{M-1}),$$

with $\boldsymbol{w} = \begin{bmatrix} w_0 & \ldots & w_{M-1} \end{bmatrix}^T$ ($(\cdot)^T$ denotes transposition), we want to find the minimum of J, that is, we want to find $\boldsymbol{w}_\mathrm{o}$ such that

$$\boldsymbol{w}_\mathrm{o} = \arg\min_{\boldsymbol{w}} J(\boldsymbol{w}).$$

The solution can be found using the gradient of $J(\boldsymbol{w})$, which is defined by

$$\nabla_{\boldsymbol{w}} J(\boldsymbol{w}) = \frac{\partial J}{\partial \boldsymbol{w}^T} = \begin{bmatrix} \frac{\partial J}{\partial w_0} \\ \frac{\partial J}{\partial w_1} \\ \vdots \\ \frac{\partial J}{\partial w_{M-1}} \end{bmatrix}.$$

See Box 12.2 for a brief explanation about gradients and Hessians, that is, derivatives of functions of several variables. The notation $\partial J/\partial \boldsymbol{w}^T$ is most convenient when we deal with complex variables, as in Box 12.6. We use it for real variables for consistency.

Since the gradient always points to the direction in which $J(\boldsymbol{w})$ increases most quickly, the steepest descent algorithm searches iteratively for the minimum of $J(\boldsymbol{w})$ taking at each iteration a small step in the opposite direction, i.e., towards $-\nabla_{\boldsymbol{w}} J(\boldsymbol{w})$:

$$\boldsymbol{w}(n+1) = \boldsymbol{w}(n) - \frac{\mu}{2} \nabla_{\boldsymbol{w}} J(\boldsymbol{w}(n)), \tag{12.16}$$

where μ is the step size, i.e., a constant controlling the speed of the algorithm. As we shall see, this constant should not be too small (or the algorithm will converge too slowly) neither too large (or the recursion will diverge).

As an example, consider the quadratic cost function with $M = 2$

$$J(\boldsymbol{w}) = 1 - 2\boldsymbol{w}^T \boldsymbol{r} + \boldsymbol{w}^T \boldsymbol{R} \boldsymbol{w}, \tag{12.17}$$

with

$$\boldsymbol{r} = \begin{bmatrix} 1 \\ 0 \end{bmatrix}, \qquad \boldsymbol{R} = \begin{bmatrix} \frac{4}{3} & -\frac{2}{3} \\ -\frac{2}{3} & \frac{4}{3} \end{bmatrix}. \tag{12.18}$$

The gradient of $J(\boldsymbol{w})$ is

$$\nabla_{\boldsymbol{w}} J(\boldsymbol{w}) = \frac{\partial J}{\partial \boldsymbol{w}^T} = -2\boldsymbol{r} + 2\boldsymbol{R}\boldsymbol{w}. \tag{12.19}$$

Of course, for this example we can find the optimum $\boldsymbol{w}_o$ by equating the gradient to zero:

$$\nabla_{\boldsymbol{w}} J(\boldsymbol{w}_o) = \mathbf{0} \Longrightarrow \boldsymbol{R}\boldsymbol{w}_o = \boldsymbol{r} \Longrightarrow \boldsymbol{w}_o = \boldsymbol{R}^{-1}\boldsymbol{r} = \begin{bmatrix} 0.9877 \\ 0.4902 \end{bmatrix}. \tag{12.20}$$

Note that in this case the *Hessian* of $J(\boldsymbol{w})$ (the matrix of second derivatives, see Box 12.2) is simply $\nabla^2_{\boldsymbol{w}} J = 2\boldsymbol{R}$. As $\boldsymbol{R}$ is symmetric with positive eigenvalues ($\lambda_1 = 2$ and $\lambda_2 = \frac{2}{3}$), it is positive definite, and we can conclude that $\boldsymbol{w}_o$ is indeed a minimum of $J(\boldsymbol{w})$. (See Boxes 12.2 and 12.3. Box 12.3 lists several useful properties of matrices.)

Even though in this case we could compute the optimum solution through (12.20), we will apply the gradient algorithm with the intention of understanding how it works and which are its main limitations. From (12.16) and (12.17), we obtain the recursion

$$\boldsymbol{w}(n+1) = \boldsymbol{w}(n) + \mu\big(\boldsymbol{r} - \boldsymbol{R}\boldsymbol{w}(n)\big). \tag{12.21}$$

The user must choose an initial condition $\boldsymbol{w}(0)$.

In Fig. 12.11 the evolution of the approximations computed through (12.21) is plotted against the level sets (i.e., curves of constant value) of the cost function (12.17). Different choices of step size μ are shown. In Fig. 12.11(a), a small step size is used. The algorithm converges to the correct solution, but slowly. In Fig. 12.11(b), we used a larger step size – convergence is now faster, but the algorithm still needs several iterations to reach the solution. If we try to increase the step size even further, convergence at first becomes slower again, with the algorithm oscillating towards the solution (Fig. 12.11(c)). For even larger step sizes, such as that in Fig. 12.11(d), the algorithm diverges: the filter coefficients get further and further away from the solution.

It is important to find the maximum step size for which the algorithm (12.21) remains stable. For this, we need concepts from linear systems and linear algebra, in particular eigenvalues, their relation to stability of linear systems, and the fact that symmetric matrices always have real eigenvalues (see Box 12.3).

The range of allowed step sizes is found as follows. Rewrite (12.21) as

$$\boldsymbol{w}(n+1) = [\boldsymbol{I} - \mu\boldsymbol{R}]\,\boldsymbol{w}(n) + \mu\boldsymbol{r},$$

which is a linear recursion in state-space form ($\boldsymbol{I}$ is the identity matrix).

Linear systems theory [274] tells us that this recursion converges as long as the largest eigenvalue of $\boldsymbol{A} = \boldsymbol{I} - \mu\boldsymbol{R}$ has an absolute value of less than one. The eigenvalues of $\boldsymbol{A}$ are the roots of

$$\det(\beta\boldsymbol{I} - (\boldsymbol{I} - \mu\boldsymbol{R})) = \det((\beta - 1)\boldsymbol{I} + \mu\boldsymbol{R}) = -\det((1-\beta)\boldsymbol{I} - \mu\boldsymbol{R}) = 0.$$

Let $1 - \beta = \nu$. Then β is an eigenvalue of $\boldsymbol{A}$ if and only if $\nu = 1 - \beta$ is an eigenvalue of $\mu\boldsymbol{R}$. Therefore, if we denote by λ_i the eigenvalues of $\boldsymbol{R}$, the stability condition is

$$-1 < 1 - \mu\lambda_i < 1\ \forall i \Leftrightarrow 0 < \mu < \frac{2}{\lambda_i}\ \forall i \Leftrightarrow 0 < \mu < \frac{2}{\max\{\lambda_i\}}.$$

The stability condition for our example is thus $0 < \mu < 1$, in agreement with what we saw in Fig. 12.11.

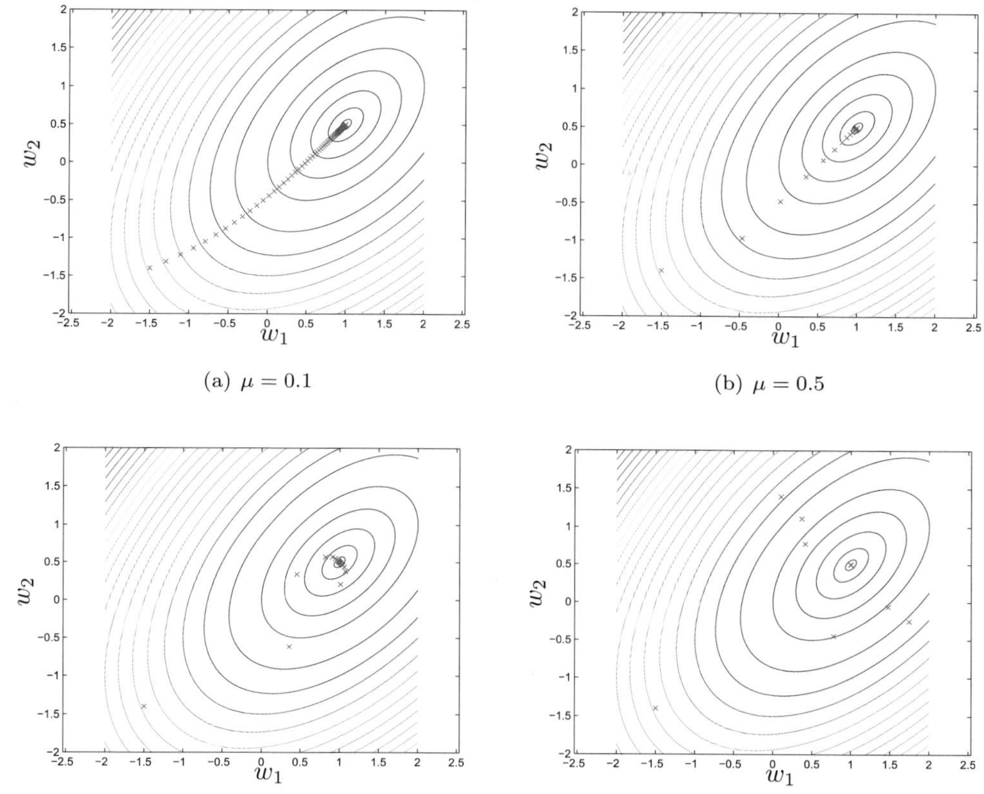

(a) $\mu = 0.1$ (b) $\mu = 0.5$

(c) $\mu = 0.9$ (d) $\mu = 1.1$

FIGURE 12.11

Performance of the steepest descent algorithm for different step sizes. The crosses (x) represent the successive approximations $\boldsymbol{w}(n)$, plotted against the level curves of (12.17). The initial condition is the point in the lower left.

The gradient algorithm leads to relatively simple adaptive filtering algorithms; however, it has an important drawback. As one can see in Fig. 12.11(a), the gradient does not point directly to the direction of the optimum. This effect is heightened when the level sets of the cost function are very elongated, as in Fig. 12.12. This case corresponds to a quadratic function in which $\boldsymbol{R}$ has one eigenvalue much smaller than the other. In this example, we replaced the matrix $\boldsymbol{R}$ in (12.18) by

$$\boldsymbol{R}' = \frac{1}{9}\begin{bmatrix} 25 & 20 \\ 20 & 25 \end{bmatrix}.$$

This matrix has eigenvalues $\lambda_1' = 5$ and $\lambda_2' = \frac{5}{9}$.

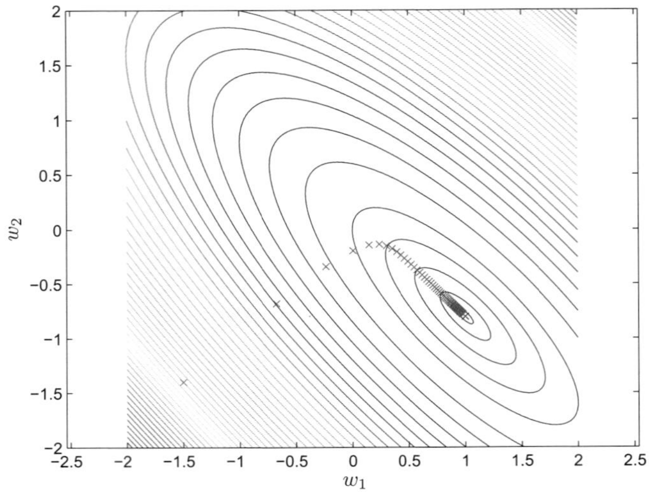

FIGURE 12.12

Performance of the steepest descent algorithm for a problem with large ratio of eigenvalues. The crosses (x) represent the successive approximations $\boldsymbol{w}(n)$, plotted against the level curves of (12.17). The initial condition is the point in the lower left. Step size $\mu = 0.1$.

Let us see why this is so. Recall from Box 12.3 that for every symmetric matrix there exists an orthogonal matrix $\boldsymbol{U}$ (that is, $\boldsymbol{U}^{-1} = \boldsymbol{U}^T$) such that

$$\boldsymbol{U}^T \boldsymbol{R} \boldsymbol{U} = \begin{bmatrix} \lambda_1 & 0 \\ 0 & \lambda_2 \end{bmatrix} \triangleq \boldsymbol{\Lambda}, \tag{12.22}$$

where we defined the diagonal eigenvalue matrix $\boldsymbol{\Lambda}$.

Let us apply a change of coordinates to (12.19). The optimum solution to (12.17) is

$$\boldsymbol{w}_o = \boldsymbol{R}^{-1}\boldsymbol{r}.$$

Our first change of coordinates is to replace $\boldsymbol{w}(n)$ by $\tilde{\boldsymbol{w}}(n) = \boldsymbol{w}_o - \boldsymbol{w}(n)$ in (12.19). We subtract (12.19) from $\boldsymbol{w}_o$ and replace $\boldsymbol{r}$ in (12.19) by $\boldsymbol{r} = \boldsymbol{R}\boldsymbol{w}_o$ to obtain

$$\tilde{\boldsymbol{w}}(n+1) = \tilde{\boldsymbol{w}}(n) - \mu \boldsymbol{R} \tilde{\boldsymbol{w}}(n) = [\boldsymbol{I} - \mu \boldsymbol{R}]\, \tilde{\boldsymbol{w}}(n). \tag{12.23}$$

Next, we multiply this recursion from both sides by $\boldsymbol{U}^T = \boldsymbol{U}^{-1}$:

$$\boldsymbol{U}^T \tilde{\boldsymbol{w}}(n+1) = \boldsymbol{U}^T [\boldsymbol{I} - \mu \boldsymbol{R}] \underbrace{\boldsymbol{U}\boldsymbol{U}^T}_{=\boldsymbol{I}} \tilde{\boldsymbol{w}}(n) = [\boldsymbol{I} - \mu \boldsymbol{\Lambda}] \boldsymbol{U}^T \tilde{\boldsymbol{w}}(n).$$

Defining $\bar{\boldsymbol{w}}(n) = \boldsymbol{U}^T \tilde{\boldsymbol{w}}(n)$, we have rewritten the gradient equation in a new set of coordinates, such that now the equations are uncoupled. Given that $\boldsymbol{\Lambda}$ is diagonal, the recursion is simply

$$\begin{bmatrix} \bar{w}_1(n+1) \\ \bar{w}_2(n+1) \end{bmatrix} = \begin{bmatrix} (1-\mu\lambda_1)\bar{w}_1(n) \\ (1-\mu\lambda_2)\bar{w}_2(n) \end{bmatrix}. \tag{12.24}$$

Note that as long as $|1-\mu\lambda_i| < 1$ for $i = 1, 2$, both entries of $\bar{\boldsymbol{w}}(n)$ will converge to zero, and consequently $\boldsymbol{w}(n)$ will converge to $\boldsymbol{w}_{\mathrm{o}}$.

The stability condition for this recursion is

$$|1-\mu\lambda_i| < 1, \; i = 1, 2 \Rightarrow \mu < \frac{2}{\max\{\lambda_1, \lambda_2\}}.$$

When one of the eigenvalues is much larger than the other, say, when $\lambda_1 \gg \lambda_2$, the rate of convergence for the direction relative to the smaller eigenvalue becomes

$$1 > 1 - \mu\lambda_2 > 1 - 2\frac{\lambda_{\min}}{\lambda_{\max}} \approx 1,$$

and even if one of the coordinates in $\bar{\boldsymbol{w}}(n)$ converges quickly to zero, the other will converge very slowly. This is what we saw in Fig. 12.12. The ratio $\lambda_{\max}/\lambda_{\min}$ is known as the *eigenvalue spread* of a matrix.

In general, the gradient algorithm converges slowly when the eigenvalue spread of the Hessian is large. One way of solving this problem is to use a different optimization algorithm, such as *Newton* or *quasi-Newton* algorithms, which use the inverse of the Hessian matrix, or an approximation to it, to improve the search direction used by the gradient algorithm.

Although these algorithms converge very quickly, they require more computational power than the gradient algorithm. We will see more about them when we describe the RLS algorithm in Section 12.3.3.

For example, for the quadratic cost function (12.17), the Hessian is equal to $2\boldsymbol{R}$. If we had a good approximation $\hat{\boldsymbol{R}}$ to $\boldsymbol{R}$, we could use the recursion

$$\boldsymbol{w}(n+1) = \boldsymbol{w}(n) + \mu\hat{\boldsymbol{R}}^{-1}\left[\boldsymbol{r} - \boldsymbol{R}\boldsymbol{w}(n)\right]. \tag{12.25}$$

This class of algorithms is known as *quasi-Newton* (if $\hat{\boldsymbol{R}}^{-1}$ is an approximation) or *Newton* (if $\hat{\boldsymbol{R}}^{-1}$ is exact). In the Newton case, we would have

$$\boldsymbol{w}(n+1) = (1-\mu)\boldsymbol{w}(n) + \boldsymbol{w}_{\mathrm{o}},$$

that is, the algorithm moves at each step precisely in the direction of the optimum solution. Of course, this is not a very interesting algorithm for a quadratic cost function as in this example (the optimum solution is used in the recursion!). However, this method is very useful when the cost function is not quadratic and in adaptive filtering, when the cost function is not known exactly.

Box 12.2: Gradients

Consider a function of several variables $J(\boldsymbol{w})$, where

$$\boldsymbol{w} = \begin{bmatrix} w_0 \\ w_1 \\ \vdots \\ w_{M-1} \end{bmatrix}.$$

Its gradient is defined by

$$\nabla_{\boldsymbol{w}} J(\boldsymbol{w}) = \frac{\partial J}{\partial \boldsymbol{w}^T} = \begin{bmatrix} \frac{\partial J}{\partial w_0} \\ \frac{\partial J}{\partial w_1} \\ \vdots \\ \frac{\partial J}{\partial w_{M-1}} \end{bmatrix}.$$

The real value of the notation $\partial J/\partial \boldsymbol{w}^T$ will only become apparent when working with functions of complex variables, as shown in Box 12.6. We also define, for consistency,

$$\frac{\partial J}{\partial \boldsymbol{w}} = \begin{bmatrix} \frac{\partial J}{\partial w_0} & \frac{\partial J}{\partial w_1} & \cdots & \frac{\partial J}{\partial w_{M-1}} \end{bmatrix}.$$

As an example, consider the quadratic cost function

$$J(\boldsymbol{w}) = b - 2\boldsymbol{w}^T \boldsymbol{r} + \boldsymbol{w}^T \boldsymbol{R} \boldsymbol{w}. \tag{12.26}$$

The gradient of $J(\boldsymbol{w})$ is (verify!)

$$\nabla_{\boldsymbol{w}} J(\boldsymbol{w}) = \frac{\partial J}{\partial \boldsymbol{w}^T} = -2\boldsymbol{r} + 2\boldsymbol{R}\boldsymbol{w}. \tag{12.27}$$

If we use the gradient to find the minimum of $J(\boldsymbol{w})$, it would be necessary to also check the second-order derivative, to make sure the solution is not a maximum or a saddle point. The second-order derivative of a function of several variables is a matrix, the *Hessian*. It is defined by

$$\nabla^2_{\boldsymbol{w}} J = \frac{\partial^2 J}{\partial \boldsymbol{w} \partial \boldsymbol{w}^T} = \begin{bmatrix} \frac{\partial}{\partial \boldsymbol{w}} \left\{ \frac{\partial J}{\partial w_0} \right\} \\ \vdots \\ \frac{\partial}{\partial \boldsymbol{w}} \left\{ \frac{\partial J}{\partial w_{M-1}} \right\} \end{bmatrix} = \begin{bmatrix} \frac{\partial^2 J}{\partial w_0^2} & \cdots & \frac{\partial^2 J}{\partial w_0 \partial w_{M-1}} \\ \vdots & \ddots & \vdots \\ \frac{\partial^2 J}{\partial w_{M-1} \partial w_0} & \cdots & \frac{\partial^2 J}{\partial w_{M-1}^2} \end{bmatrix}. \tag{12.28}$$

Note that for the quadratic cost function (12.26), the Hessian is equal to $\boldsymbol{R} + \boldsymbol{R}^T$. It is equal to $2\boldsymbol{R}$ if $\boldsymbol{R}$ is symmetric, that is, if $\boldsymbol{R}^T = \boldsymbol{R}$. This will usually be the case here. Note that $\boldsymbol{R} + \boldsymbol{R}^T$ is always symmetric: in fact, since for well-behaved functions J we have

$$\frac{\partial^2 J}{\partial w_i \partial w_j} = \frac{\partial^2 J}{\partial w_j \partial w_i},$$

the Hessian is usually symmetric.

Assume that we want to find the minimum of $J(\boldsymbol{w})$. The first-order conditions are then

$$\left.\frac{\partial J}{\partial \boldsymbol{w}}\right|_{\boldsymbol{w}_o} = \mathbf{0}.$$

The solution $\boldsymbol{w}_o$ is a minimum of $J(\boldsymbol{w})$ if the Hessian at $\boldsymbol{w}_o$ is a positive semidefinite matrix, that is, if

$$\boldsymbol{x}^T \nabla^2_{\boldsymbol{w}} J(\boldsymbol{w}_o)\boldsymbol{x} \geq 0$$

for all directions $\boldsymbol{x}$.

Box 12.3: Useful results from matrix analysis

Here we list without proof a few useful results from matrix analysis. Detailed explanations can be found, for example, in [176,177,197,242].

Fact 12.1 (Traces). The *trace* of a matrix $\boldsymbol{A} \in \mathbb{C}^{M\times M}$ is the sum of its diagonal elements:

$$\operatorname{Tr}(\boldsymbol{A}) = \sum_{i=1}^{M} a_{ii}, \tag{12.29}$$

in which a_{ii} are the entries of $\boldsymbol{A}$. For any two matrices $\boldsymbol{B} \in \mathbb{C}^{M\times N}$ and $\boldsymbol{C} \in \mathbb{C}^{N\times M}$, we have

$$\operatorname{Tr}(\boldsymbol{BC}) = \operatorname{Tr}(\boldsymbol{CB}). \tag{12.30}$$

In addition, if λ_i, $i = 1 \ldots M$, are the eigenvalues of a square matrix $\boldsymbol{A} \in \mathbb{C}^{M\times M}$, then

$$\operatorname{Tr}(\boldsymbol{A}) = \sum_{i=1}^{M} \lambda_i. \tag{12.31}$$

Fact 12.2 (Singular matrices). A matrix $\boldsymbol{A} \in \mathbb{C}^{M\times M}$ is singular if and only if there exists a nonzero vector $\boldsymbol{u}$ such that $\boldsymbol{Au} = \mathbf{0}$.

The *null space* or *kernel* $N(\boldsymbol{A})$ of a matrix $\boldsymbol{A} \in \mathbb{C}^{M\times N}$ is the set of vectors $\boldsymbol{u} \in \mathbb{C}^N$ such that $\boldsymbol{Au} = \mathbf{0}$. A square matrix $\boldsymbol{A} \in \mathbb{C}^{M\times M}$ is nonsingular if and only if $N(\boldsymbol{A}) = \{\mathbf{0}\}$, that is, if its null space contains only the null vector.

Note that if $\boldsymbol{A} = \sum_{k=1}^{K} \boldsymbol{u}_k \boldsymbol{v}_k^H$, with $\boldsymbol{u}_k, \boldsymbol{v}_k \in \mathbb{C}^M$, then $\boldsymbol{A}$ cannot be invertible if $K < M$. This is because when $K < M$, there always exists a nonzero vector $\boldsymbol{x} \in \mathbb{C}^M$ such that $\boldsymbol{v}_k^H \boldsymbol{x} = 0$ for $1 \leq k \leq K$, and thus $\boldsymbol{Ax} = \mathbf{0}$ with $\boldsymbol{x} \neq \mathbf{0}$.

Fact 12.3 (Inverses of block matrices). Assume the square matrix $\boldsymbol{A}$ is partitioned such that

$$\boldsymbol{A} = \begin{bmatrix} \boldsymbol{A}_{11} & \boldsymbol{A}_{12} \\ \boldsymbol{A}_{21} & \boldsymbol{A}_{22} \end{bmatrix},$$

with $\boldsymbol{A}_{11}$ and $\boldsymbol{A}_{22}$ square. If $\boldsymbol{A}$ and $\boldsymbol{A}_{11}$ are invertible (nonsingular), then

$$\boldsymbol{A}^{-1} = \begin{bmatrix} \boldsymbol{I} & -\boldsymbol{A}_{11}^{-1}\boldsymbol{A}_{12} \\ \boldsymbol{0} & \boldsymbol{I} \end{bmatrix} \begin{bmatrix} \boldsymbol{A}_{11}^{-1} & \boldsymbol{0} \\ \boldsymbol{0} & \boldsymbol{\Delta}^{-1} \end{bmatrix} \begin{bmatrix} \boldsymbol{I} & \boldsymbol{0} \\ -\boldsymbol{A}_{21}\boldsymbol{A}_{11}^{-1} & \boldsymbol{I} \end{bmatrix},$$

where $\boldsymbol{\Delta} = \boldsymbol{A}_{22} - \boldsymbol{A}_{21}\boldsymbol{A}_{11}^{-1}\boldsymbol{A}_{12}$ is the *Schur complement* of $\boldsymbol{A}_{11}$ in $\boldsymbol{A}$. If $\boldsymbol{A}_{22}$ and $\boldsymbol{A}$ are invertible, then

$$\boldsymbol{A}^{-1} = \begin{bmatrix} \boldsymbol{I} & \boldsymbol{0} \\ -\boldsymbol{A}_{22}^{-1}\boldsymbol{A}_{21} & \boldsymbol{I} \end{bmatrix} \begin{bmatrix} \boldsymbol{\Delta}'^{-1} & \boldsymbol{0} \\ \boldsymbol{0} & \boldsymbol{A}_{22}^{-1} \end{bmatrix} \begin{bmatrix} \boldsymbol{I} & -\boldsymbol{A}_{12}\boldsymbol{A}_{22}^{-1} \\ \boldsymbol{0} & \boldsymbol{I} \end{bmatrix},$$

where $\boldsymbol{\Delta}' = \boldsymbol{A}_{11} - \boldsymbol{A}_{12}\boldsymbol{A}_{22}^{-1}\boldsymbol{A}_{21}$.

Fact 12.4 (Matrix inversion lemma). Let $\boldsymbol{A}$ and $\boldsymbol{B}$ be two nonsingular $M \times M$ matrices, let $\boldsymbol{D} \in \mathbb{C}^{K \times K}$ also be nonsingular, and let $\boldsymbol{C}, \boldsymbol{E} \in \mathbb{C}^{M \times K}$ be such that

$$\boldsymbol{A} = \boldsymbol{B} + \boldsymbol{C}\boldsymbol{D}\boldsymbol{E}^H. \tag{12.32}$$

The inverse of $\boldsymbol{A}$ is then given by

$$\boldsymbol{A}^{-1} = \boldsymbol{B}^{-1} - \boldsymbol{B}^{-1}\boldsymbol{C}(\boldsymbol{D}^{-1} + \boldsymbol{E}^H\boldsymbol{B}^{-1}\boldsymbol{C})^{-1}\boldsymbol{E}^H\boldsymbol{B}^{-1}. \tag{12.33}$$

The lemma is most useful when $K \ll M$. In particular, when $K = 1$, $\boldsymbol{D} + \boldsymbol{E}^H\boldsymbol{B}^{-1}\boldsymbol{C}$ is a scalar, and we have

$$\boldsymbol{A}^{-1} = \boldsymbol{B}^{-1} - \frac{\boldsymbol{B}^{-1}\boldsymbol{C}\boldsymbol{E}^H\boldsymbol{B}^{-1}}{\boldsymbol{D}^{-1} + \boldsymbol{E}^H\boldsymbol{B}^{-1}\boldsymbol{C}}.$$

Fact 12.5 (Symmetric matrices). Let $\boldsymbol{A} \in \mathbb{C}^{M \times M}$ be *Hermitian symmetric* (or simply *Hermitian*), that is, $\boldsymbol{A}^H = (\boldsymbol{A}^T)^* = \boldsymbol{A}$. Then we have:

1. All eigenvalues λ_i, $i = 1 \dots M$, of $\boldsymbol{A}$ are real.
2. $\boldsymbol{A}$ has a complete set of orthonormal eigenvectors $\boldsymbol{u}_i$, $i = 1 \dots M$. Since the eigenvectors are orthonormal, $\boldsymbol{u}_i^H\boldsymbol{u}_j = \delta_{ij}$, where $\delta_{ij} = 1$ if $i = j$ and 0 otherwise.
3. Arranging the eigenvectors in a matrix $\boldsymbol{U} = \begin{bmatrix} \boldsymbol{u}_1 & \dots & \boldsymbol{u}_M \end{bmatrix}$, we find that $\boldsymbol{U}$ is *unitary*, that is, $\boldsymbol{U}^H\boldsymbol{U} = \boldsymbol{U}\boldsymbol{U}^H = \boldsymbol{I}$. In addition,

$$\boldsymbol{U}^H\boldsymbol{A}\boldsymbol{U} = \boldsymbol{\Lambda} = \begin{bmatrix} \lambda_1 & 0 & \dots & 0 \\ 0 & \lambda_2 & \dots & 0 \\ \vdots & \vdots & \ddots & \vdots \\ 0 & 0 & \dots & \lambda_M \end{bmatrix}.$$

4. If $A = A^T \in \mathbb{R}^{M\times M}$, it is called simply *symmetric*. In this case, not only the eigenvalues, but also the eigenvectors are real.
5. If A is symmetric and invertible, then A^{-1} is also symmetric.

Fact 12.6 (Positive definite matrices). If $A \in \mathbb{C}^{M\times M}$ is Hermitian symmetric and moreover for all $u \in \mathbb{C}^M$ we have

$$u^H A u \geq 0, \tag{12.34}$$

then A is *positive semidefinite*. Positive semidefinite matrices have nonnegative eigenvalues: $\lambda_i \geq 0$, $i = 1 \dots M$. They may be singular if one or more eigenvalues are zero.

If inequality (12.34) is strict, that is, if

$$u^H A u > 0 \tag{12.35}$$

for all $u \neq 0$, then A is *positive definite*. All eigenvalues λ_i of positive definite matrices satisfy $\lambda_i > 0$. Positive definite matrices are always nonsingular.

Finally, all positive definite matrices admit a Cholesky factorization, i.e., if A is positive definite, there is a lower triangular matrix L such that $A = LL^H$ [165].

Fact 12.7 (Vector norms). Norms measure the length of a vector. The usual *Euclidean* norm (also known as ℓ_2-norm) of a vector $x \in \mathbb{C}^M$ is defined by $\|x\|_2 = \sqrt{\sum_{k=1}^M |x_k|^2}$. Since we use this norm more frequently, we denote it simply by $\|x\|$.

Other common options are the ℓ_1-norm, $\|x\|_1 \triangleq \sum_{k=1}^M |x_k|$, and the ℓ_∞-norm, $\|x\|_\infty \triangleq \max_{1\leq k\leq M} |x_k|$. It can be shown that for any vector $x \in \mathbb{C}^M$, we have [304]

$$\|x\|_2 \leq \|x\|_1 \leq \sqrt{M}\|x\|_1. \tag{12.36}$$

Fact 12.8 (Schwarz's inequality). Let $x, y \in \mathbb{C}^M$. Then it follows that $|x^H y| \leq \|x\|_2 \|y\|_2$.

Fact 12.9 (Norms and spectral radius). The *spectral radius* $\rho(A)$ of a matrix $A \in \mathbb{C}^{M\times M}$ is

$$\rho(A) = \max_{1\leq i\leq M} |\lambda_i|, \tag{12.37}$$

in which λ_i are the eigenvalues of A. The spectral radius is important, because a linear system

$$x(n+1) = Ax(n)$$

is stable if and only if $\rho(A) \leq 1$, as is well known. A useful inequality is that for any matrix norm $\|\cdot\|$ such that for any $A, B \in \mathbb{C}^{M\times M}$

$$\|AB\| \leq \|A\|\|B\|,$$

we have

$$\rho(A) \leq \|A\|. \tag{12.38}$$

This property is most useful when used with norms that are easy to evaluate, such as the 1-norm,

$$\|\boldsymbol{A}\|_1 = \max_{1\le i\le M} \sum_{j=1}^{M} |a_{ij}|,$$

where a_{ij} are the entries of $\boldsymbol{A}$. We could also use the infinity norm,

$$\|\boldsymbol{A}\|_\infty = \max_{1\le j\le M} \sum_{i=1}^{M} |a_{ij}|.$$

12.1.3 Applications

Due to the ability to adjust themselves to different environments, adaptive filters can be used in different signal processing and control applications. Thus, they have been used as a powerful device in several fields, such as communications, radar, sonar, biomedical engineering, active noise control, modeling, etc. It is common to divide these applications into four groups:

1. interference cancelation;
2. system identification;
3. prediction; and
4. inverse system identification.

In the first three cases, the goal of the adaptive filter is to find an approximation $\hat{y}(n)$ for the signal $y(n)$, which is contained in the signal $d(n) = y(n) + v(n)$. Thus, as $\hat{y}(n)$ approaches $y(n)$, the signal $e(n) = d(n) - \hat{y}(n)$ approaches $v(n)$. The difference between these applications is what we are interested in. In interference cancelation, we are interested in the signal $v(n)$, as is the case of acoustic echo cancelation, where $v(n)$ is the speech of a person using a hands-free telephone (go back to Figs. 12.4 and 12.7). In system identification we are interested in the filter parameters, and in prediction we may be interested in the signal $v(n)$ and/or in the filter parameters. Note that in all these applications (interference cancelation, system identification, and prediction), the signal $e(n)$ is an approximation for $v(n)$ and should not converge to zero, except when $v(n) \equiv 0$. In inverse system identification, differently from the other applications, the signal at the output of the adaptive filter must be as close as possible to the signal $d(n)$ and thus, ideally, the signal $e(n)$ should be zeroed. In the sequel, we give an overview of these four groups in order to arrive at a common formulation that will simplify our analysis of adaptive filtering algorithms in further sections.

12.1.3.1 Interference cancelation

In a general interference cancelation problem, we have access to a signal $d(n)$, which is a mixture of two other signals, $v(n)$ and $y(n)$. We are interested in one of these signals, and want to separate it from the other (the interference) (see Fig. 12.13). Even though we do not know $v(n)$ or $y(n)$, we have some information about $y(n)$, usually in the form of a reference signal $x(n)$ that is related to $y(n)$ through a filtering operation $\mathcal{H}$. We may know the general form of this operation; for example, we may know that the relation is linear and well approximated by an FIR filter with 200 coefficients. However, we do not know the parameters (the filter coefficients) necessary to reproduce it. The goal of the adaptive filter is to find an approximation $\hat{y}(n)$ to $y(n)$, given only $x(n)$ and $d(n)$. In the process of finding this $\hat{y}$, the adaptive filter will construct an approximation for the relation $\mathcal{H}$. A typical example of an interference cancelation problem is the echo cancelation example we gave in Section 12.1.1. In that example, $x(n)$ is the far-end voice signal, $v(n)$ is the near-end voice signal, and $y(n)$ is the echo. The relation $\mathcal{H}$ is usually well approximated by a linear FIR filter with a few hundred taps.

The approximation for $\mathcal{H}$ is a byproduct, which does not need to be very accurate as long as it leads to a $\hat{y}(n)$ close to $y(n)$. This configuration is shown in Fig. 12.13 and is called *interference cancelation*, since the interference $y(n)$ should be canceled. Note that we do not want to make $e(n) \equiv 0$; otherwise, we would not only be killing the interference $y(n)$, but also the signal of interest $v(n)$.

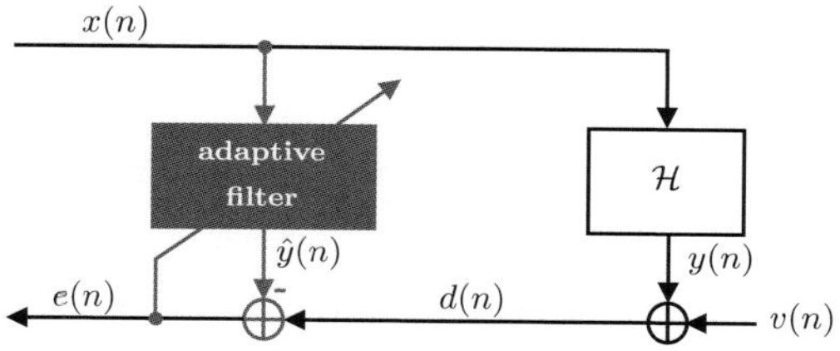

FIGURE 12.13

Scheme for interference cancelation and system identification. $x(n)$ and $y(n)$ are correlated to each other.

There are many applications of interference cancelation. In addition to acoustic and line echo cancelation, we can mention, for example, adaptive notch filters for cancelation of sinusoidal interference (a common application is removal of 50 or 60 Hz interference from the mains line), cancelation of the maternal electrocardiography in fetal electrocardiography, cancelation of echoes in long-distance telephone circuits, active noise control (as in noise-canceling headphones), and active vibration control.

12.1.3.2 System identification

In interference cancelation, the coefficients of the adaptive filter converge to an approximation for the true relation $\mathcal{H}$, but as mentioned before, this approximation is a byproduct that does not need to be very accurate. However, there are some applications in which the goal is to construct an approximation as accurate as possible for the unknown relation $\mathcal{H}$ between $x(n)$ and $y(n)$, thus obtaining a model for the unknown system. This is called *system identification* or *modeling* (the diagram in Fig. 12.13 applies also for this case). The signal $d(n)$ is now composed of the output $y(n)$ of an unknown system $\mathcal{H}$, plus noise $v(n)$. The reference signal $x(n)$ is the input to the system which, when possible, is chosen to be white noise. In general, this problem is harder than interference cancelation, as we shall see in Section 12.2.1.5, due to some conditions that the reference signal $x(n)$ must satisfy. However, one point does not change: in the ideal case, the signal $e(n)$ will be equal to the noise $v(n)$. Again, we do not want to make $e(n) \equiv 0$; otherwise, we would be trying to model the noise (however, the smaller the noise, the easier the task, as one would expect).

In many control systems, an unknown dynamic system (also called *plant* in control system terminology) is identified on-line and the result is used in a self-tuning controller, as depicted in Fig. 12.14. The plant and the adaptive filter have the same input $x(n)$. In practical situations, the plant to be modeled is noisy, which is represented by the signal $v(n)$ added to the plant output. The noise $v(n)$ is generally uncorrelated with the plant input. The task of the adaptive filter is to minimize the error model $e(n)$ and track the time variations in the dynamics of the plant. The model parameters are continually fed back to the controller to obtain the controller parameters used in the self-tuning regulator loop [30].

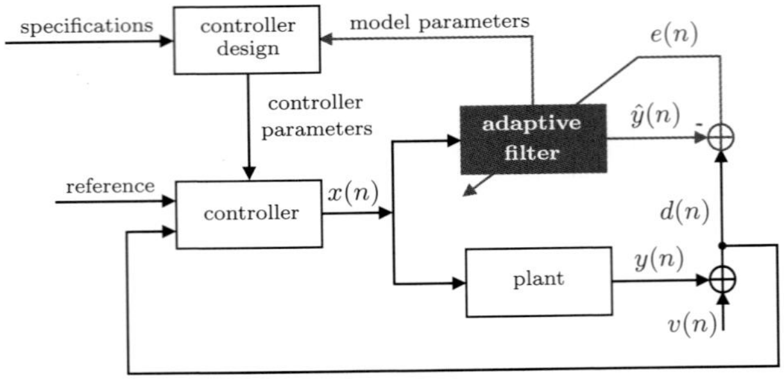

FIGURE 12.14

Scheme for plant identification in a control system.

There are many applications that require the identification of an unknown system. In addition to control systems, we can mention, for example, the identification of the room impulse response, used to study the sound quality in concert rooms, and the estimation of the communication channel impulse response, required by maximum likelihood detectors and blind equalization techniques based on second-order statistics.

12.1.3.3 Prediction

In *prediction*, the goal is to find a relation between the current sample $d(n)$ and previous samples $d(n-L)\ldots d(n-L-M+1)$, as shown in Fig. 12.15. Therefore, we want to model $d(n)$ as a part $y(n)$ that depends only on $d(n-L)\ldots d(n-L-M+1)$ and a part $v(n)$ that represents new information. For example, for a linear model, we would have

$$d(n) = \underbrace{\sum_{k=0}^{M-1} h_k d(n-L-k)}_{y(n)} + v(n).$$

Thus, we in fact have again the same problem as in Fig. 12.13. The difference is that now the reference signal is a delayed version of $d(n)$, $x(n) = d(n-L)$, and the adaptive filter will try to find an approximation $\hat{y}(n)$ for $y(n)$, thereby separating the "predictable" part $y(n)$ of $d(n)$ from the "new information" $v(n)$, to which the signal $e(n) = d(n) - \hat{y}(n)$ should converge. The system $\mathcal{H}$ in Fig. 12.13 now represents the relation between $d(n-L)$ and the predictable part of $d(n)$.

Prediction finds application in many fields. We can mention, for example, linear predictive coding (LPC) and adaptive differential pulse code modulation (ADPCM) used in speech coding, adaptive line enhancement (ALE), autoregressive spectral analysis, etc.

For example, ALE seeks the solution for a classical detection problem, whose objective is to separate a narrow-band signal $y(n)$ from a wide-band signal $v(n)$. This is the case, for example, of finding a low-level sine wave (predictable narrow-band signal) in noise (nonpredictable wide-band signal). The signal

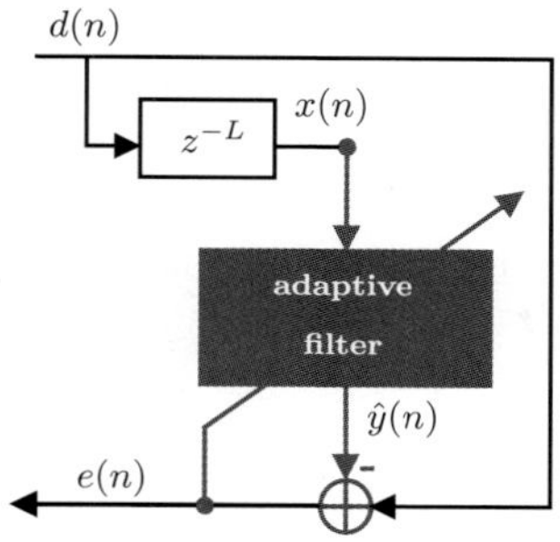

FIGURE 12.15

Scheme for prediction. $x(n) = d(n - L)$ is a delayed version of $d(n)$.

$d(n)$ is constituted by the sum of these two signals. Using $d(n)$ as the reference signal and a delayed replica of it, i.e., $d(n - L)$, as input of the adaptive filter as in Fig. 12.15, the output $\hat{y}(n)$ provides an estimate of the (predictable) narrow-band signal $y(n)$, while the error signal $e(n)$ provides an estimate of the wide-band signal $v(n)$. The delay L is also known as *decorrelation parameter* of the ALE, since its main function is to remove the correlation between the wide-band signal $v(n)$ present in the reference $d(n)$ and in the delayed predictor input $d(n - L)$.

12.1.3.4 Inverse system identification

Inverse system identification, also known as *deconvolution*, has been widely used in different fields such as communications, acoustics, optics, image processing, and control, among others. In communications, it is also known as *channel equalization* and the adaptive filter is commonly called *equalizer*. Adaptive equalizers play an important role in digital communications systems, since they are used to mitigate the intersymbol interference (ISI) introduced by dispersive channels. Due to its importance, we focus on the equalization application, which is explained in the sequel.

A simplified baseband communications system is depicted in Fig. 12.16. The signal $s(n)$ is transmitted through an unknown channel, whose model is constituted by an FIR filter with transfer function $H(z) = h_0 + h_1 z^{-1} + \cdots + h_{K-1} z^{K-1}$ and additive noise $\eta(n)$. Due to the channel memory, the signal at the receiver contains contributions not only from $s(n)$, but also from the previous symbols $s(n-1), s(n-2), \ldots, s(n-K+1)$, i.e.,

$$x(n) = \underbrace{\sum_{k=0}^{L-1} h_k s(n-k)}_{\text{pre-ISI}} + h_L s(n-L) + \underbrace{\sum_{k=L+1}^{K-1} h_k s(n-k)}_{\text{post-ISI}} + \eta(n). \tag{12.39}$$

Assuming that the overall channel equalizer system imposes a delay of L samples, the adaptive filter will try to find an approximation $\hat{y}(n)$ for $d(n) = s(n - L)$, and for this purpose the two summations in (12.39), which constitute the ISI, must be mitigated. Of course, when one is transmitting information the receiver will not have access to $d(n - L)$. The filter will adapt during a *training* phase, in which the transmitter sends a preagreed signal. We can see that in this case, the role of $x(n)$ and $d(n)$ is the reverse of that in system identification. In this case, one would indeed like to have $e(n) \equiv 0$. The problem is that this is not possible, given the presence of noise in $x(n)$. The role of the adaptive filter is to approximately invert the effect of the channel, at the same time trying to suppress the noise (or at least, not to amplify it too much). In one sense, we are back to the problem of separating two signals, but now the mixture is in the reference signal $x(n)$, so what can and what cannot be done is considerably different from the other cases.

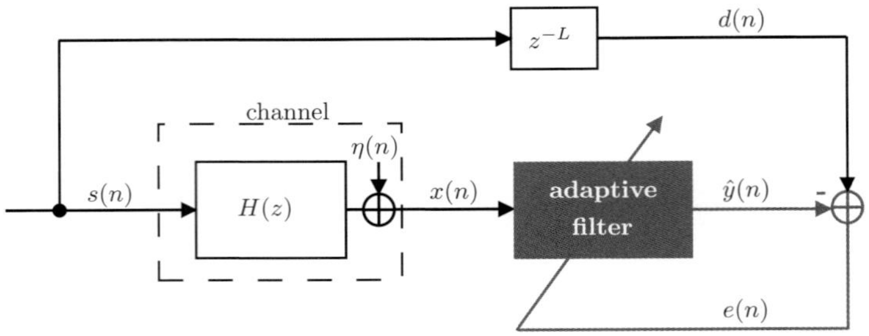

FIGURE 12.16

Simplified communications system with an adaptive equalizer in training mode.

The scheme of Fig. 12.16 is also called training mode, since the delayed version of the transmitted sequence $d(n) = s(n - L)$ (training sequence) is known at the receiver. After the convergence of the filter, the signal $d(n)$ is changed to the estimate $\hat{s}(n - L)$ obtained at the output of a decision device, as shown in Fig. 12.17. In this case, the equalizer works in the so-called decision-directed mode. The decision device depends on the signal constellation; for example, if $s(n) = \pm 1$, the decision device returns $+1$ for $\hat{y}(n) \geq 0$ and -1 for $\hat{y}(n) < 0$.

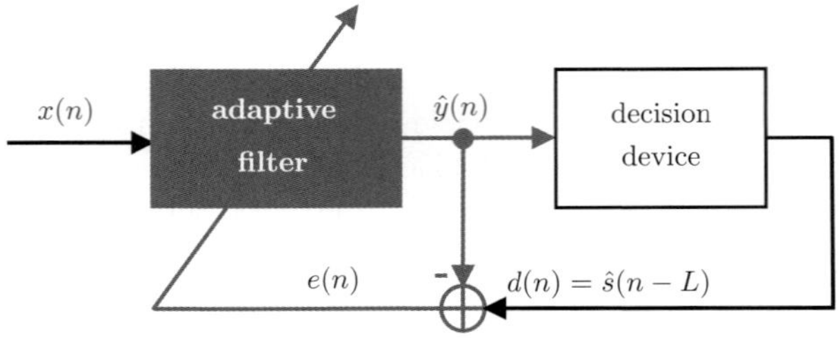

FIGURE 12.17

Adaptive equalizer in decision-directed mode.

12.1.3.5 A common formulation

In all the four groups of applications, the inputs of the adaptive filter are given by the signals $x(n)$ and $d(n)$ and the output by $\hat{y}(n)$. Note that the input $d(n)$ appears effectively in the signal $e(n) = d(n) - \hat{y}(n)$, which is computed and fed back at each time instant n, as shown in Fig. 12.18. In the literature, the signal $d(n)$ is referred to as desired signal, $x(n)$ as reference signal, and $e(n)$ as error signal. These names unfortunately are somewhat misleading, since they give the impression that our goal is to recover exactly $d(n)$ by filtering (possibly in a nonlinear way) the reference $x(n)$. In almost all applications, this is far from the truth, as previously discussed. In fact, except in the case of channel equalization, exactly zeroing the error would result in very poor performance.

The only application in which $d(n)$ is indeed a "desired signal" is channel equalization, in which $d(n) = s(n - L)$. Despite this particularity in channel equalization, the common feature in all adaptive filtering applications is that the filter must learn a relation between the reference $x(n)$ and the desired signal $d(n)$ (we will use this name to be consistent with the literature). In the process of building this relation, the adaptive filter is able to perform useful tasks, such as separating two mixed signals or recovering a distorted signal. Section 12.2 explains this idea in more detail.

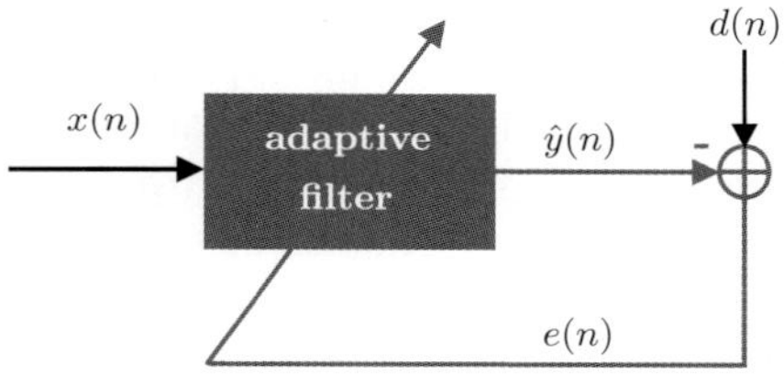

FIGURE 12.18

Inputs and output of an adaptive filter.

12.2 Optimum filtering

We now need to extend the ideas of Section 12.1.2, which applied only to periodic (deterministic) signals, to more general classes of signals. For this, we will use tools from the theory of stochastic processes. The main ideas are very similar: as before, we need to choose a structure for our filter and a means of measuring how far or how near we are to the solution. This is done by choosing a convenient cost function, whose minimum we must search iteratively, based only on measurable signals.

In the case of periodic signals, we saw that minimizing the error power (12.3), repeated below,

$$P = \sum_{k=1}^{K_0} \frac{\tilde{C}_k^2}{2} + \sum_{\ell=1}^{K_1} \frac{B_\ell^2}{2},$$

would be equivalent to zeroing the echo. However, we were not able to compute the error power exactly – we used an approximation through a time average, as in (12.5), repeated here,

$$\hat{P}(n) = \frac{1}{N} \sum_{k=0}^{N-1} e^2(n-k),$$

where N is a convenient window length (in Section 12.1.2, we saw that choosing a large window length is approximately equivalent to choosing $N = 1$ and a small step size). Since we used an approximated estimate of the cost, our solution was also an approximation: the estimated coefficients did not converge exactly to their optimum values, but instead hovered around the optimum (Figs. 12.9 and 12.10).

The minimization of the time average $\hat{P}(n)$ also works in the more general case in which the echo is modeled as a random signal – but what is the corresponding exact cost function? The answer is given by the property of *ergodicity* that some random signals possess. If a random signal $s(n)$ is such that its mean $\mathrm{E}\{s(n)\}$ does not depend on n and its autocorrelation $\mathrm{E}\{s(n)s(k)\}$ depends only on $n-k$, it is called wide-sense stationary (WSS) [201]. If $s(n)$ is also (mean square) ergodic, then

$$\text{average power of } s(n) \triangleq \lim_{N\to\infty} \frac{1}{N+1} \sum_{n=0}^{N} s^2(n) \underbrace{= \mathrm{E}\{s^2(n)\}}_{\text{if } s(t) \text{ is ergodic}} . \tag{12.40}$$

Note that $\mathrm{E}\{s^2(n)\}$ is an ensemble average, that is, the expected value is computed over all possible realizations of the random signal. Relation (12.40) means that for an ergodic signal, the time average of a single realization of the process is equal to the ensemble average of all possible realizations of the process. This ensemble average is the exact cost function that we would like to minimize.

In the case of adaptive filters, (12.40) holds approximately for $e(n)$ for a finite value of N if the environment does not change too fast and if the filter adapts slowly. Therefore, for random variables we will still use the average error power as a measure of how well the adaptive filter is doing. The difference is that in the case of periodic signals, we could understand the effect of minimizing the average error power in terms of the amplitudes of each harmonic in the signal, but now the interpretation will be in terms of ensemble averages (means and variances).

Although the error power is not the only possible choice for cost function, it is useful to study this choice in detail. Quadratic cost functions such as the error power have a number of properties that make them popular. For example, they are differentiable, and hence it is relatively easy to find a closed-form solution for the optimum, and in the important case where all signals are Gaussian, the optimum filters are linear. Of course, quadratic cost functions are not the best option in all situations: the use of other cost functions, or even of adaptive filters not based on the minimization of a cost function, is becoming more common. We will talk about some of the most important alternatives in Sections 12.5.2 and 12.5.4.

Given the importance of quadratic cost functions, in this section we study them in detail. Our focus is on what could and what could not be done if the filter were able to measure $\mathrm{E}\{e^2(n)\}$ perfectly at each instant. This discussion has two main goals. The first is to learn what is feasible, so we do not expect more from a filter than it can deliver. The second is to enable us to make more knowledgeable choices when designing a filter, for example, which filter order should be used, and for system identification, which input signal would result in better performance. In Section 12.4.4.3 we will study the performance of adaptive filters taking into consideration their imperfect knowledge of the environment.

12.2.1 Linear least-mean squares estimation

In the examples we gave in previous sections, a generic adaptive filtering problem boils down to the following. We are given two sequences, $\{x(n)\}$ and $\{d(n)\}$, as depicted in Fig. 12.18. It is known from physical arguments that $d(n)$ has two parts, one that is in some sense related to $x(n), x(n-1), \ldots$ and another that is not, and that both parts are combined additively, that is,

$$d(n) = \underbrace{\mathcal{H}(x(n), x(n-1), \dots)}_{y(n)} + v(n), \tag{12.41}$$

where $\mathcal{H}(\cdot)$ is an unknown relation and $v(n)$ is the part "unrelated" to $\{x(n)\}$. In our examples so far, $\mathcal{H}(\cdot)$ was linear, but we do not need to restrict ourselves to this case. Our objective is to extract $y(n)$ and $v(n)$ from $d(n)$, based only on observations of $\{x(n)\}$ and $\{d(n)\}$. We saw in Section 12.1.2 that the average power of the difference $e(n)$ between $d(n)$ and our current approximation $\hat{y}(n)$ could be used as a measure of how close we are to the solution. In general, $\mathcal{H}(\cdot)$ is time-varying, i.e., it depends directly on n. We will not write this dependence explicitly to simplify the notation.

In this section we ask in some sense the inverse problem, that is, given two sequences $\{x(n)\}$ and $\{d(n)\}$, what sort of relation will be found between them if we use $\mathrm{E}\{e^2(n)\}$ as a standard? The difference is that we now do not assume a model such as (12.41); instead, we want to know what kind of model results from the exact minimization of $\mathrm{E}\{e^2(n)\}$, where $e(n)$ is defined as the difference between $d(n)$ and a function of $x(n), x(n-1), \dots$.

We can answer this question in an entirely general way, without specifying beforehand the form of $\mathcal{H}$. This is done in Box 12.4. However, if we have information about the physics of a problem and know that the relation $\mathcal{H}(x(n), x(n-1), \dots)$ is of a certain kind, we can restrict the search for the solution $\hat{\mathcal{H}}$ to this class. This usually reduces the complexity of the filter and increases its convergence rate (because less data is necessary to estimate a model with fewer unknowns, as we will see in Section 12.3). In this section we focus on this second option.

The first task is to describe the class $\mathcal{F}$ of allowed functions $\hat{\mathcal{H}}$. This may be done by choosing a relation that depends on a few parameters, such as (recall that the adaptive filter output is $\hat{y}(n) = \hat{\mathcal{H}}(x(n), x(n-1), \dots)$)

$$\begin{aligned}
&\mathcal{F}_{\text{FIR}} \text{ (FIR filter):} && \hat{y}(n) = w_0 x(n) + \dots + w_{M-1} x(n-M+1), && (12.42)\\
&\mathcal{F}_{\text{IIR}} \text{ (IIR filter):} && \hat{y}(n) = -a_1 \hat{y}(n-1) + b_0 x(n), && (12.43)\\
&\mathcal{F}_{\text{V}} \text{ (Volterra filter):} && \hat{y}(n) = w_0 x(n) + w_1 x(n-1) + w_{0,0} x^2(n) + && \\
& && \quad + w_{0,1} x(n) x(n-1) + w_{1,1} x(n-1)^2, && (12.44)\\
&\mathcal{F}_{\text{S}} \text{ (saturation):} && \hat{y}(n) = \arctan(a x(n)). && (12.45)
\end{aligned}$$

In each of these cases, the relation between the input sequence $\{x(n)\}$ and $d(n)$ is constrained to a certain class, for example, linear length-M FIR filters in (12.42), first-order IIR filters in (12.43), and second-order Volterra filters in (12.44). Each class is described by a certain number of parameters: the filter coefficients $w_0, \dots, w_{M-1}$ in the case of FIR filters, a_1 and b_0 in the case of IIR filters, and so on. The task of the adaptive filter will then be to choose the values of the parameters that best fit the data.

For several practical reasons, it is convenient if we make the filter output depend linearly on the parameters, as happens in (12.42) and in (12.44). It is important to distinguish linear in the parameters from input–output linear: (12.44) is linear in the parameters, but the relation between the input sequence $\{x(n)\}$ and the output $\hat{y}(n)$ is nonlinear. What may come as a surprise is that the IIR filter of Eq. (12.43) is *not* linear in the parameters; in fact, $\hat{y}(n) = -a_1 \hat{y}(n-1) + b_0 x(n) = a_1^2 \hat{y}(n-2) - a_1 b_0 x(n-1) + b_0 x(n) = \dots$ – one can see that $\hat{y}(n)$ depends nonlinearly on a_1 and b_0.

Linearly parameterized classes $\mathcal{F}$ such as $\mathcal{F}_{\text{FIR}}$ (12.42) and $\mathcal{F}_{\text{V}}$ (12.44) are popular because in general it is easier to find the optimum parameters, both theoretically and in real time. In fact, when the filter output depends linearly on the parameters and the cost function is a *convex* function of the error, it can be shown that the optimal solution is unique (see Box 12.5). This is a very desirable property, since it simplifies the search for the optimal solution.

In the remainder of this section we will concentrate on classes of relations that are linear in the parameters. As $\mathcal{F}_{\text{V}}$ shows, this does not imply that we are restricting ourselves to linear models. There are, however, adaptive filtering algorithms that use classes of relations that are not linear in the parameters, such as IIR adaptive filters. On the other hand, blind equalization algorithms are based on nonconvex cost functions.

Assume then that we have chosen a convenient class of relations $\mathcal{F}$ that depends linearly on its parameters. That is, we want to solve the problem

$$\min_{\hat{\mathcal{H}}\in\mathcal{F}} \mathrm{E}\left\{\left[d(n) - \hat{\mathcal{H}}(x(n), x(n-1), \dots)\right]^2\right\}. \tag{12.46}$$

We want to know which properties the solution to this problem will have when $\mathcal{F}$ depends linearly on a finite number of parameters. In this case, $\mathcal{F}$ is a linear combinations of certain functions ϕ_i of $x(n), x(n-1), \dots,$

$$\hat{y}(n) = w_0\phi_0 + w_1\phi_1 + \cdots + w_{M-1}\phi_{M-1} \triangleq \boldsymbol{w}^T\boldsymbol{\phi}, \tag{12.47}$$

where in general $\phi_i = \phi_i(x(n), x(n-1), \dots)$, $0 \le i \le M-1$. The vector $\boldsymbol{\phi}$ is known as *regressor*. In the case of length-M FIR filters, we would have

$$\phi_0 = x(n), \qquad \phi_1 = x(n-1), \qquad \dots \qquad \phi_{M-1} = x(n-M+1),$$

whereas in the case of second-order Volterra filters with memory 1, we would have

$$\phi_0 = x(n), \quad \phi_1 = x(n-1), \quad \phi_2 = x^2(n), \quad \phi_3 = x(n)x(n-1), \quad \phi_4 = x^2(n-1).$$

Our problem can then be written in general terms as follows. Find $\boldsymbol{w}_{\text{o}}$ such that

$$\boldsymbol{w}_{\text{o}} = \arg\min_{\boldsymbol{w}} \mathrm{E}\left\{\left(d - \boldsymbol{w}^T\boldsymbol{\phi}\right)^2\right\}. \tag{12.48}$$

We omitted the dependence of the variables on n to lighten the notation. Note that, in general, $\boldsymbol{w}_{\text{o}}$ will also depend on n.

To solve this problem, we use the fact that $\boldsymbol{w}^T\boldsymbol{\phi} = \boldsymbol{\phi}^T\boldsymbol{w}$ to expand the expected value

$$J(\boldsymbol{w}) \triangleq \mathrm{E}\left\{\left(d - \boldsymbol{w}^T\boldsymbol{\phi}\right)^2\right\} = \mathrm{E}\{d^2\} - 2\boldsymbol{w}^T\,\mathrm{E}\{d\boldsymbol{\phi}\} + \boldsymbol{w}^T\,\mathrm{E}\{\boldsymbol{\phi}\boldsymbol{\phi}^T\}\boldsymbol{w}.$$

Recall that the weight vector $\boldsymbol{w}$ is *not* random, so we can take it out of the expectations.

Define r_d (the autocorrelation of d), $\boldsymbol{r}_{d\boldsymbol{\phi}}$ (the cross-correlation vector), and $\boldsymbol{R}_{\boldsymbol{\phi}}$ (the autocorrelation matrix) as follows:

$$r_d = \mathrm{E}\{d^2\}, \qquad \boldsymbol{r}_{d\boldsymbol{\phi}} = \mathrm{E}\{d\boldsymbol{\phi}\}, \qquad \boldsymbol{R}_{\boldsymbol{\phi}} = \mathrm{E}\{\boldsymbol{\phi}\boldsymbol{\phi}^T\}. \tag{12.49}$$

The cost function then becomes

$$J(\boldsymbol{w}) = r_d - 2\boldsymbol{w}^T \boldsymbol{r}_{d\boldsymbol{\phi}} + \boldsymbol{w}^T \boldsymbol{R}_{\boldsymbol{\phi}} \boldsymbol{w}. \tag{12.50}$$

Differentiating $J(\boldsymbol{w})$ with respect to $\boldsymbol{w}$, we obtain

$$\frac{\partial J}{\partial \boldsymbol{w}^T} = -2\boldsymbol{r}_{d\boldsymbol{\phi}} + 2\boldsymbol{R}_{\boldsymbol{\phi}} \boldsymbol{w},$$

and equating the result to zero, we see that the optimal solution must satisfy

$$\boldsymbol{R}_{\boldsymbol{\phi}} \boldsymbol{w}_{\mathrm{o}} = \boldsymbol{r}_{d\boldsymbol{\phi}}. \tag{12.51}$$

These are known as the *normal*, or *Wiener–Hopf*, equations. The solution, $\boldsymbol{w}_{\mathrm{o}}$, is known as the *Wiener solution*. A note of caution: the Wiener solution is not the same thing as the Wiener filter. The Wiener filter is the linear filter that minimizes the mean square error (MSE), without restriction of filter order [183]. The difference is that the Wiener solution has the filter order prespecified and is not restricted to linear filters, for example, if we choose $\mathcal{F}_V$ for our class of allowed functions.

When the autocorrelation matrix is nonsingular (which is usually the case), the Wiener solution is

$$\boldsymbol{w}_{\mathrm{o}} = \boldsymbol{R}_{\boldsymbol{\phi}}^{-1} \boldsymbol{r}_{d\boldsymbol{\phi}}. \tag{12.52}$$

Given $\boldsymbol{w}_{\mathrm{o}}$, the optimum error will be

$$v_{\mathrm{o}} = d - \boldsymbol{w}_{\mathrm{o}}^T \boldsymbol{\phi}. \tag{12.53}$$

Note that the expected value of v_{o} is not necessarily zero:

$$\mathrm{E}\{v_{\mathrm{o}}\} = \mathrm{E}\{d\} - \boldsymbol{w}_{\mathrm{o}}^T \, \mathrm{E}\{\boldsymbol{\phi}\}. \tag{12.54}$$

If, for example, $\mathrm{E}\{\boldsymbol{\phi}\} = \boldsymbol{0}$ and $\mathrm{E}\{d\} \neq 0$, then $\mathrm{E}\{v_{\mathrm{o}}\} = \mathrm{E}\{d\} \neq 0$. In practice, it is good to keep $\mathrm{E}\{v_{\mathrm{o}}\} = 0$, because we usually know that v_{o} should approximate a zero-mean signal, such as speech or noise. We will show shortly, in Section 12.2.1.4, how to guarantee that v_{o} has zero mean.

12.2.1.1 Orthogonality condition

A key property of the Wiener solution is that the optimum error is orthogonal to the regressor $\boldsymbol{\phi}$, that is,

$$\mathrm{E}\{v_{\mathrm{o}}\boldsymbol{\phi}\} = \mathrm{E}\{\boldsymbol{\phi}(d - \underbrace{\boldsymbol{w}_{\mathrm{o}}^T \boldsymbol{\phi}}_{=\boldsymbol{\phi}^T \boldsymbol{w}_{\mathrm{o}}})\} = \mathrm{E}\{d\boldsymbol{\phi}\} - \mathrm{E}\{\boldsymbol{\phi}\boldsymbol{\phi}^T\}\boldsymbol{w}_{\mathrm{o}} = \boldsymbol{r}_{d\boldsymbol{\phi}} - \boldsymbol{R}_{\boldsymbol{\phi}} \boldsymbol{w}_{\mathrm{o}} = \boldsymbol{0}. \tag{12.55}$$

We saw a similar condition in Section 12.1.2. It is very useful to remember this result, known as the *orthogonality condition*: from it, we will find when to apply the cost function (12.48) to design an adaptive filter. Note that when v_{o} has zero mean, (12.55) also implies that v_{o} and $\boldsymbol{\phi}$ are uncorrelated.

Remark 12.1. You should not confuse orthogonal with uncorrelated. Two random variables x and y are orthogonal if $\mathrm{E}\{xy\} = 0$, whereas they are uncorrelated if $\mathrm{E}\{xy\} = \mathrm{E}\{x\}\,\mathrm{E}\{y\}$. The two concepts coincide only if either x or y has zero mean.

Remark 12.2. The orthogonality condition is an intuitive result if we remember that we can think of the space of random variables with finite variance as a vector space. In fact, define the cross-correlation $r_{xy} = \mathrm{E}\{xy\}$ between two random variables x and y as the inner product between x and y. Then the autocorrelation of a random variable x, $\mathrm{E}\{x^2\}$, would be interpreted as the square of the "length" of x. In this case, our problem (12.48) is equivalent to finding the vector in the subspace spanned by $\phi_0, \dots, \phi_{M-1}$ that is closest to d. As is well known, the solution is such that the error is orthogonal (or *normal* — this is the reason for the name *normal* equations to (12.51)) to the subspace. So, the orthogonality condition results from the vector space structure of our problem and the quadratic nature of our cost function (see Fig. 12.19).

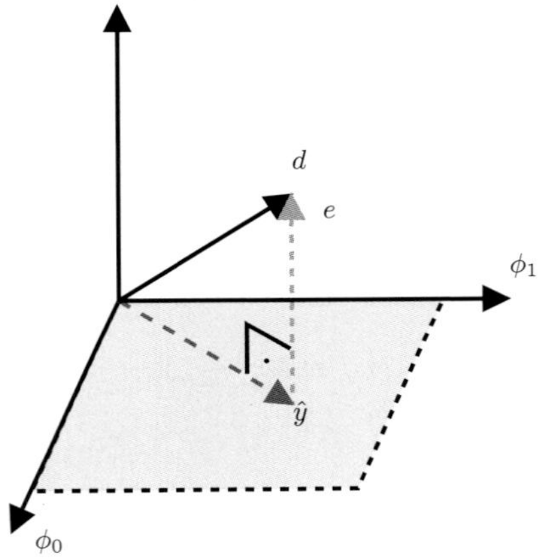

FIGURE 12.19

Orthogonality in vector spaces with inner products.

Remark 12.3. The difference between (12.55) and the corresponding result obtained in Box 12.4, Eq. (12.106), is that here the optimum error is orthogonal only to the functions ϕ_i of $x(n), x(n-1), \dots$ that were included in the regressor (and their linear combinations), not to any function, as in (12.106). This is not surprising: in this section the optimization was made only over the functions in class $\mathcal{F}$, and in Box 12.4 we allow any function. Both solutions, (12.107) and (12.52), will be equal if the general solution according to Box 12.4 is indeed in $\mathcal{F}$.

Remark 12.4. The optimal MSE is (recall that $\mathrm{E}\{v_{\mathrm{o}}\boldsymbol{\phi}\} = \mathbf{0}$)

$$\begin{aligned} J_{\min} = r_{v,\mathrm{o}} = \mathrm{E}\{v_{\mathrm{o}}^2\} = \mathrm{E}\{v_{\mathrm{o}}(d - \boldsymbol{w}_{\mathrm{o}}^T\boldsymbol{\phi})\} = \mathrm{E}\{v_{\mathrm{o}}d\} = \\ = \mathrm{E}\{(d - \boldsymbol{w}_{\mathrm{o}}^T\boldsymbol{\phi})d\} = \mathrm{E}\{d^2\} - \boldsymbol{w}_{\mathrm{o}}^T\boldsymbol{r}_{d\boldsymbol{\phi}} = r_d - \boldsymbol{r}_{d\boldsymbol{\phi}}^T\boldsymbol{R}_{\boldsymbol{\phi}}^{-1}\boldsymbol{r}_{d\boldsymbol{\phi}}. \end{aligned} \tag{12.56}$$

Now that we have the general solution to (12.48), we turn to the following question: for which class of problems is the quadratic cost function adequate?

12.2.1.2 Implicit versus physical models

From (12.55), we see that the fact that we are minimizing the MSE between d and a linear combination of the regressor $\boldsymbol{\phi}$ induces a model

$$d = \mathbf{w}_o^T \boldsymbol{\phi} + v_o, \tag{12.57}$$

in which v_o is orthogonal to $\boldsymbol{\phi}$. This is not an assumption, as we sometimes see in the literature, but a consequence of the quadratic cost function we chose. In other words, there is always a relation such as (12.57) between any pair $(d, \boldsymbol{\phi})$, as long as both have finite second-order moments. This relation may make sense from a physical analysis of the problem, as we saw in the echo cancelation example, or be simply a consequence of solving (12.48) (see Section 12.2.1.3 for examples).

Consider the following example. Assume that $\phi \sim \text{Uniform}(-1, 1)$ is a uniform scalar random variable and $d = \sin(\phi)$. Then

$$R_\phi = \mathrm{E}\{\phi^2\} = \int_{-1}^{1} \frac{1}{2}\phi^2 \, \mathrm{d}\phi = \frac{1}{3}$$

and

$$r_{d\phi} = \int_{-1}^{1} \frac{1}{2}\phi \sin(\phi) \, \mathrm{d}\phi = -\frac{1}{2}\phi\cos(\phi)\Big|_{-1}^{1} + \int_{-1}^{1} \frac{1}{2}\cos(\phi) \, \mathrm{d}\phi = -\cos(1) + \sin(1).$$

Therefore, in this case we have $w_o = 3(\sin(1) - \cos(1)) \approx 0.9035$ and $v_o = \sin(\phi) - 3(\sin(1) - \cos(1))\phi$. One can check that in this case $\mathrm{E}\{v_o\} = 0$ and

$$\mathrm{E}\{v_o\phi\} = \mathrm{E}\{\phi\sin(\phi) - 3(\sin(1) - \cos(1))\phi^2\} = 0,$$

that is, we can say that there exist w_o and v_o such that

$$d = w_o\phi + v_o,$$

with v_o orthogonal to ϕ, even though the definitions of ϕ and d do not involve w_o or v_o. Fig. 12.20 shows the relation between d and $w_o\phi$.

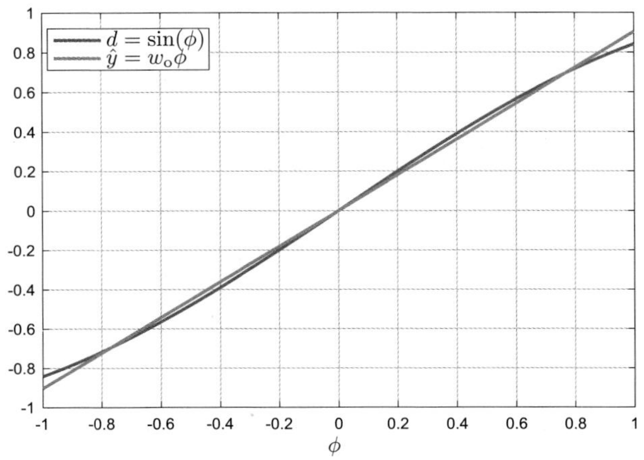

FIGURE 12.20

Comparison between d and $w_o\phi$.

Relation (12.57) is never an assumption; it is a consequence of the quadratic cost function we chose to minimize. Note that all we can say is that $\boldsymbol{\phi}$ and v_{o} are *orthogonal*, not *independent*. The analysis of adaptive filters usually makes the *assumption* that $\boldsymbol{\phi}$ and v_{o} are independent, in order to simplify the derivations (see Section 12.4.3).

On the other hand, the orthogonality condition allows us to find out when the solution of (12.48) will be able to successfully solve a problem: assume now that we know beforehand, by physical arguments about the problem, that d and $\boldsymbol{\phi}$ must be related through

$$d = \boldsymbol{w}_*^T \boldsymbol{\phi} + v, \tag{12.58}$$

where $\boldsymbol{w}_*$ is a certain coefficient vector and v is orthogonal to $\boldsymbol{\phi}$. Will the solution to (12.48) be such that $\boldsymbol{w}_{\text{o}} = \boldsymbol{w}_*$ and $v_{\text{o}} = v$?

To check if this is the case, we only need to evaluate the cross-correlation vector $\boldsymbol{r}_{d\boldsymbol{\phi}}$ assuming the model (12.58):

$$\boldsymbol{r}_{d\boldsymbol{\phi}} = \mathrm{E}\{d\boldsymbol{\phi}\} = \mathrm{E}\{\boldsymbol{\phi}(\underbrace{\boldsymbol{w}_*^T\boldsymbol{\phi}}_{=\boldsymbol{\phi}^T\boldsymbol{w}_*} + v)\} = \mathrm{E}\{\boldsymbol{\phi}\boldsymbol{\phi}^T\}\boldsymbol{w}_* + \underbrace{\mathrm{E}\{\boldsymbol{\phi}v\}}_{=\boldsymbol{0}} = \boldsymbol{R}_{\boldsymbol{\phi}}\boldsymbol{w}_*. \tag{12.59}$$

Therefore, $\boldsymbol{w}_*$ also obeys the normal equations, and we can conclude that $\boldsymbol{w}_{\text{o}} = \boldsymbol{w}_*$, $v_{\text{o}} = v$, as long as $\boldsymbol{R}_{\boldsymbol{\phi}}$ is nonsingular.

Even if $\boldsymbol{R}_{\boldsymbol{\phi}}$ is singular, the optimal error v_{o} will always be equal to v (in a certain sense). In fact, if $\boldsymbol{R}_{\boldsymbol{\phi}}$ is singular, then there exists a vector $\boldsymbol{a}$ such that $\boldsymbol{R}_{\boldsymbol{\phi}}\boldsymbol{a} = \boldsymbol{0}$ (see Fact 12.2 in Box 12.3), and thus any solution $\boldsymbol{w}_{\text{o}}$ of the normal equations (we know from (12.59) that there is at least one, $\boldsymbol{w}_*$) will have the form

$$\boldsymbol{w}_{\text{o}} = \boldsymbol{w}_* + \boldsymbol{a}, \qquad \boldsymbol{a} \in N(\boldsymbol{R}_{\boldsymbol{\phi}}), \tag{12.60}$$

where $N(\boldsymbol{R}_{\boldsymbol{\phi}})$ is the null space of $\boldsymbol{R}_{\boldsymbol{\phi}}$. In addition,

$$0 = \boldsymbol{a}^T \boldsymbol{R}_{\boldsymbol{\phi}}\boldsymbol{a} = \boldsymbol{a}^T \mathrm{E}\{\boldsymbol{\phi}\boldsymbol{\phi}^T\}\boldsymbol{a} = \mathrm{E}\{\boldsymbol{a}^T\boldsymbol{\phi}\boldsymbol{\phi}^T\boldsymbol{a}\} = \mathrm{E}\left\{(\boldsymbol{a}^T\boldsymbol{\phi})^2\right\}.$$

Therefore, $\boldsymbol{a}^T\boldsymbol{\phi} = 0$ with probability one. Then,

$$v_{\text{o}} = d - \boldsymbol{w}_{\text{o}}^T\boldsymbol{\phi} = d - (\boldsymbol{w}_* + \boldsymbol{a})^T\boldsymbol{\phi} = d - \boldsymbol{w}_*^T\boldsymbol{\phi}$$

with probability one. Therefore, although we cannot say that $v = v_{\text{o}}$ always, we can say that they are equal with probability one. In other words, when $\boldsymbol{R}_{\boldsymbol{\phi}}$ is singular we may not be able to identify $\boldsymbol{w}_*$, but (with probability one) we are still able to separate d into its two components. More about this can be found in Section 12.2.1.5.

12.2.1.3 Undermodeling

We just saw that the solution to the quadratic cost function (12.48) is indeed able to separate the two components of d when the regressor that appears in the physical model (12.58) is the same $\boldsymbol{\phi}$ that we used to describe our class $\mathcal{F}$, that is, when our choice of $\mathcal{F}$ includes the general solution from Box 12.4. Let us check now what happens when this is not the case.

Assume there is a relation

$$d = \boldsymbol{w}_*^T \boldsymbol{\phi}_\mathrm{e} + v, \tag{12.61}$$

in which v is orthogonal to $\boldsymbol{\phi}_\mathrm{e}$, but our class $\mathcal{F}$ is defined through a regressor $\boldsymbol{\phi}$ that is a subset of the "correct" one, $\boldsymbol{\phi}_\mathrm{e}$. This situation is called *undermodeling*: our class is not rich enough to describe the true relation between $\{x(n)\}$ and $\{d(n)\}$.

Assume then that

$$\boldsymbol{\phi}_\mathrm{e} = \begin{bmatrix} \boldsymbol{\phi} \\ \boldsymbol{\theta} \end{bmatrix}, \tag{12.62}$$

where $\boldsymbol{\phi} \in \mathbb{R}^M$, $\boldsymbol{\theta} \in \mathbb{R}^{K-M}$, with $K > M$. The autocorrelation of $\boldsymbol{\phi}_\mathrm{e}$ is

$$\boldsymbol{R}_{\boldsymbol{\phi}_\mathrm{e}} = \mathrm{E}\{\boldsymbol{\phi}_\mathrm{e}\boldsymbol{\phi}_\mathrm{e}^T\} = \begin{bmatrix} \boldsymbol{R}_{\boldsymbol{\phi}} & \boldsymbol{R}_{\boldsymbol{\phi}\boldsymbol{\theta}} \\ \boldsymbol{R}_{\boldsymbol{\phi}\boldsymbol{\theta}}^T & \boldsymbol{R}_{\boldsymbol{\theta}} \end{bmatrix},$$

where $\boldsymbol{R}_{\boldsymbol{\phi}\boldsymbol{\theta}} = \mathrm{E}\{\boldsymbol{\phi}\boldsymbol{\theta}^T\}$. From our hypothesis that $\boldsymbol{\phi}_\mathrm{e}$ is orthogonal to v, that is,

$$\boldsymbol{0} = \mathrm{E}\{v\boldsymbol{\phi}_\mathrm{e}\} = \begin{bmatrix} \mathrm{E}\{v\boldsymbol{\phi}\} \\ \mathrm{E}\{v\boldsymbol{\theta}\} \end{bmatrix},$$

we conclude that $\boldsymbol{\phi}$ is also orthogonal to v, so

$$\boldsymbol{r}_{d\boldsymbol{\phi}} = \mathrm{E}\left\{\boldsymbol{\phi}(\boldsymbol{\phi}_\mathrm{e}^T \boldsymbol{w}_* + v)\right\} = \mathrm{E}\{\boldsymbol{\phi}\boldsymbol{\phi}_\mathrm{e}^T\}\boldsymbol{w}_* + \underbrace{\mathrm{E}\{\boldsymbol{\phi}v\}}_{=\boldsymbol{0}} = \begin{bmatrix} \boldsymbol{R}_{\boldsymbol{\phi}} & \boldsymbol{R}_{\boldsymbol{\phi}\boldsymbol{\theta}} \end{bmatrix} \boldsymbol{w}_*.$$

Assuming that $\boldsymbol{R}_{\boldsymbol{\phi}}$ is nonsingular, solving the normal equations (12.51) we obtain

$$\boldsymbol{w}_\mathrm{o} = \boldsymbol{R}_{\boldsymbol{\phi}}^{-1} \begin{bmatrix} \boldsymbol{R}_{\boldsymbol{\phi}} & \boldsymbol{R}_{\boldsymbol{\phi}\boldsymbol{\theta}} \end{bmatrix} \boldsymbol{w}_* = \begin{bmatrix} \boldsymbol{I} & \boldsymbol{R}_{\boldsymbol{\phi}}^{-1}\boldsymbol{R}_{\boldsymbol{\phi}\boldsymbol{\theta}} \end{bmatrix} \boldsymbol{w}_*. \tag{12.63}$$

Now we have two interesting special cases: first, if $\boldsymbol{\phi}$ and $\boldsymbol{\theta}$ are orthogonal, i.e., if $\boldsymbol{R}_{\boldsymbol{\phi}\boldsymbol{\theta}} = \boldsymbol{0}$, then $\boldsymbol{w}_\mathrm{o}$ contains the first M elements of $\boldsymbol{w}_*$. Let us consider a specific example: assume that

$$\boldsymbol{\phi} = \begin{bmatrix} x(n) \\ x(n-1) \\ \vdots \\ x(n-M+1) \end{bmatrix}, \quad \text{but} \quad \boldsymbol{\phi}_\mathrm{e} = \begin{bmatrix} x(n) \\ x(n-1) \\ \vdots \\ x(n-K+1) \end{bmatrix}, \tag{12.64}$$

with $M < K$. Then $\boldsymbol{\theta} = \begin{bmatrix} x(n-M) & \dots & x(n-K+1) \end{bmatrix}^T$. This situation is very common in practice.

In this case, $\boldsymbol{R}_{\boldsymbol{\phi}\boldsymbol{\theta}} = \boldsymbol{0}$ when $\{x(n)\}$ is zero-mean white noise. This is one reason why white noise is the preferred input to be used in system identification: even in the (very likely in practice) case in which $M < K$, at least the first elements of $\boldsymbol{w}_*$ are identified without bias.

On the other hand, if $\boldsymbol{R}_{\boldsymbol{\phi}\boldsymbol{\theta}} \neq \boldsymbol{0}$, then $\boldsymbol{w}_\mathrm{o}$ is a mixture of the elements of $\boldsymbol{w}_*$. This happens in (12.64) when the input sequence $\{x(n)\}$ is not white. In this case, the optimum filter $\boldsymbol{w}_\mathrm{o}$ takes advantage of the correlation between the entries of $\boldsymbol{\phi}$ and $\boldsymbol{\theta}$ to estimate $\boldsymbol{\theta}$ given $\boldsymbol{\phi}$, and uses these estimated values to obtain a better approximation for d. Sure enough, if we write down the problem of approximating $\boldsymbol{\theta}$ linearly from $\boldsymbol{\phi}$, namely,

$$\boldsymbol{W}_{\mathrm{o}} = \arg\min_{\boldsymbol{W}} \mathrm{E}\left\{ \left\| \boldsymbol{\theta} - \boldsymbol{W}^T \boldsymbol{\phi} \right\|^2 \right\},$$

we will see that the solution is exactly $\boldsymbol{W}_{\mathrm{o}} = \boldsymbol{R}_{\boldsymbol{\phi}}^{-1} \boldsymbol{R}_{\boldsymbol{\phi\theta}}$.

What is important to notice is that in both cases the optimum error v_{o} is not equal to v:

$$v_{\mathrm{o}} = d - \boldsymbol{w}_{\mathrm{o}}^T \boldsymbol{\phi} = \boldsymbol{w}_*^T \boldsymbol{\phi}_{\mathrm{e}} + v - \boldsymbol{w}_{\mathrm{o}}^T \boldsymbol{\phi}.$$

Partitioning $\boldsymbol{w}_* = \begin{bmatrix} \boldsymbol{w}_{*,1}^T & \boldsymbol{w}_{*,2}^T \end{bmatrix}^T$, with $\boldsymbol{w}_{*,1} \in \mathbb{R}^M$, $\boldsymbol{w}_{*,2} \in \mathbb{R}^{K-M}$, we have

$$\begin{aligned} v_{\mathrm{o}} &= \left(\boldsymbol{w}_{*,1} - \boldsymbol{w}_{\mathrm{o}}\right)^T \boldsymbol{\phi} + \boldsymbol{w}_{*,2}^T \boldsymbol{\theta} + v = -\boldsymbol{w}_{*,2}^T \boldsymbol{R}_{\boldsymbol{\phi\theta}}^T \boldsymbol{R}_{\boldsymbol{\phi}}^{-1} \boldsymbol{\phi} + \boldsymbol{w}_{*,2}^T \boldsymbol{\theta} + v \\ &= \boldsymbol{w}_{*,2}^T \left(\boldsymbol{\theta} - \boldsymbol{R}_{\boldsymbol{\phi\theta}}^T \boldsymbol{R}_{\boldsymbol{\phi}}^{-1} \boldsymbol{\phi} \right) + v. \end{aligned}$$

Although $v_{\mathrm{o}} \neq v$, the reader may check that indeed $\mathrm{E}\{v_{\mathrm{o}} \boldsymbol{\phi}\} = \mathbf{0}$.

As another example, assume that we have two variables such that $d = ax + bx^3 + v$, in which v has zero mean and is independent of x, and a and b are constants. Assume that we choose as regressor simply $\phi = x$, so we try to solve

$$w_{\mathrm{o}} = \arg\min_{w} \mathrm{E}\left\{ (d - wx)^2 \right\}.$$

In this case we have

$$R_\phi = \mathrm{E}\{x^2\}, \qquad r_{d\phi} = \mathrm{E}\{xd\} = \mathrm{E}\{ax^2 + bx^4\},$$

since $\mathrm{E}\{x^k v\} = \mathrm{E}\{x^k\}\,\mathrm{E}\{v\} = 0$. The optimum solution is thus

$$w_{\mathrm{o}} = \frac{a\,\mathrm{E}\{x^2\} + b\,\mathrm{E}\{x^4\}}{\mathrm{E}\{x^2\}},$$

and the optimum error is

$$v_{\mathrm{o}} = d - w_{\mathrm{o}} x = ax + bx^3 + v - \frac{a\,\mathrm{E}\{x^2\} + b\,\mathrm{E}\{x^4\}}{\mathrm{E}\{x^2\}} x = v + b\left(x^2 - \frac{\mathrm{E}\{x^4\}}{\mathrm{E}\{x^2\}} \right) x \neq v.$$

However, v_{o} is again orthogonal to x (but not to x^3, which we did not include in the regressor).

What happens in the opposite situation, when we include more entries in $\boldsymbol{\phi}$ than necessary? As one would expect, in this case the parameters related to these unnecessary entries will be zeroed in the optimum solution $\boldsymbol{w}_{\mathrm{o}}$. This does not mean, however, that we should hasten to increase $\boldsymbol{\phi}$ to make $\mathcal{F}$ as large as possible and leave to the filter the task of zeroing whatever was not necessary. This is because increasing the number of parameters to be estimated has a cost. The first and more obvious cost is in terms of memory and the number of computations. However, we saw in Section 12.1.2 that an actual adaptive filter usually does not converge to the exact optimum solution, but rather hovers around it. Because of this, increasing the number of parameters may in fact decrease the quality of the solution. We explain this in more detail in Section 12.4.

12.2.1.4 Zero and nonzero mean variables

Up to now we did not assume anything about the means of d and $\boldsymbol{\phi}$. If both d and $\boldsymbol{\phi}$ have zero mean, we can interpret all autocorrelations as variances or covariance matrices, according to the case. To see this, assume that

$$\mathrm{E}\{d\} = 0, \qquad \mathrm{E}\{\boldsymbol{\phi}\} = \mathbf{0}. \tag{12.65}$$

Then $\mathrm{E}\{\hat{y}\} = \mathrm{E}\{\boldsymbol{w}_{\mathrm{o}}^T\boldsymbol{\phi}\} = 0$, and thus

$$\mathrm{E}\{v_{\mathrm{o}}\} = \mathrm{E}\{d\} - \boldsymbol{w}_{\mathrm{o}}^T\,\mathrm{E}\{\boldsymbol{\phi}\} = 0 + \boldsymbol{w}_{\mathrm{o}}^T\mathbf{0} = 0,$$

so the optimum error has zero mean. In this case, $r_d = \mathrm{E}\{d^2\} = \sigma_d^2$ is the variance of d, and $\mathrm{E}\{v_{\mathrm{o}}^2\} = \sigma_{v,\mathrm{o}}^2$ is the variance of v_{o}, so (12.56) becomes

$$J_{\min} = \sigma_{v,\mathrm{o}}^2 = \mathrm{E}\{v_{\mathrm{o}}^2\} = \sigma_d^2 - \boldsymbol{r}_{d\boldsymbol{\phi}}^T\boldsymbol{R}_{\boldsymbol{\phi}}^{-1}\boldsymbol{r}_{d\boldsymbol{\phi}}. \tag{12.66}$$

Since $\boldsymbol{R}_{\boldsymbol{\phi}}$ is positive definite, we conclude that the uncertainty in v_{o} (its variance) is never larger than the uncertainty in d (Box 12.3).

However, v_{o} may have a nonzero mean if either d or $\boldsymbol{\phi}$ has nonzero mean. This is verified as follows. Define

$$\bar{d} = \mathrm{E}\{d\}, \qquad \bar{\boldsymbol{\phi}} = \mathrm{E}\{\boldsymbol{\phi}\}. \tag{12.67}$$

Then

$$\mathrm{E}\{v_{\mathrm{o}}\} = \mathrm{E}\{d - \boldsymbol{r}_{d\boldsymbol{\phi}}^T\boldsymbol{R}_{\boldsymbol{\phi}}^{-1}\boldsymbol{\phi}\} = \bar{d} - \boldsymbol{r}_{d\boldsymbol{\phi}}^T\boldsymbol{R}_{\boldsymbol{\phi}}^{-1}\bar{\boldsymbol{\phi}} \neq 0$$

in general (for example, if $\bar{\boldsymbol{\phi}} = \mathbf{0}$, but $\bar{d} \neq 0$).

In many applications, v_{o} should approximate some signal that is constrained to have zero mean, such as speech or noise, or v_{o} will be passed through a system with a high DC gain (such as an integrator), so it may be useful to guarantee that v_{o} has zero mean. How can this be done if the means of d and $\boldsymbol{\phi}$ are not zero? This problem is quite common, particularly when $\boldsymbol{\phi}$ includes nonlinear functions of $x(n)$. Fortunately, there are two easy solutions. First, if one knows the means of d and $\boldsymbol{\phi}$, one could define zero-mean variables

$$d_{\mathrm{c}} = d - \mathrm{E}\{d\}, \qquad \boldsymbol{\phi}_{\mathrm{c}} = \boldsymbol{\phi} - \mathrm{E}\{\boldsymbol{\phi}\} \tag{12.68}$$

and use these instead of the original variables in (12.48). This is the simplest solution, but we often do not know the means $\bar{d}$ and $\bar{\boldsymbol{\phi}}$.

The second solution is used when the means are not known: simply add an extra term to $\boldsymbol{\phi}$, defining an extended regressor (not to be confused with the extended regressor from Section 12.2.1.3)

$$\boldsymbol{\phi}_{\mathrm{e}} = \begin{bmatrix} 1 \\ \boldsymbol{\phi} \end{bmatrix}, \tag{12.69}$$

and an extended weight vector $\boldsymbol{w}_e \in \mathbb{R}^{M+1}$. The first coefficient of $\boldsymbol{w}_e$ will take care of the means, as we show next. The extended cross-correlation vector and autocorrelation matrix are

$$\boldsymbol{r}_{d\phi_e} = \begin{bmatrix} \bar{d} \\ \boldsymbol{r}_{d\phi} \end{bmatrix}, \qquad \boldsymbol{R}_{\phi_e} = \begin{bmatrix} 1 & \bar{\boldsymbol{\phi}}^T \\ \bar{\boldsymbol{\phi}} & \boldsymbol{R}_{\phi} \end{bmatrix}.$$

Assuming that $\boldsymbol{R}_{\phi_e}$ and $\boldsymbol{R}_{\phi}$ are nonsingular, we can compute $\boldsymbol{w}_{o,e}$ if we evaluate the inverse of $\boldsymbol{R}_{\phi_e}$ using Schur complements (see Fact 12.3 in Box 12.3) as follows:

$$\begin{aligned}
\boldsymbol{w}_{o,e} &= \boldsymbol{R}_{\phi_e}^{-1}\boldsymbol{r}_{d\phi_e} \\
&= \begin{bmatrix} 1 & \boldsymbol{0}^T \\ -\boldsymbol{R}_{\phi}^{-1}\bar{\boldsymbol{\phi}} & \boldsymbol{I} \end{bmatrix} \begin{bmatrix} (1-\bar{\boldsymbol{\phi}}^T\boldsymbol{R}_{\phi}^{-1}\bar{\boldsymbol{\phi}})^{-1} & \boldsymbol{0}^T \\ \boldsymbol{0} & \boldsymbol{R}_{\phi}^{-1} \end{bmatrix} \begin{bmatrix} 1 & -\bar{\boldsymbol{\phi}}^T\boldsymbol{R}_{\phi}^{-1} \\ \boldsymbol{0} & \boldsymbol{I} \end{bmatrix} \begin{bmatrix} \bar{d} \\ \boldsymbol{r}_{d\phi} \end{bmatrix} \\
&= \begin{bmatrix} 1 & \boldsymbol{0}^T \\ -\boldsymbol{R}_{\phi}^{-1}\bar{\boldsymbol{\phi}} & \boldsymbol{I} \end{bmatrix} \begin{bmatrix} (1-\bar{\boldsymbol{\phi}}^T\boldsymbol{R}_{\phi}^{-1}\bar{\boldsymbol{\phi}})^{-1} & \boldsymbol{0}^T \\ \boldsymbol{0} & \boldsymbol{R}_{\phi}^{-1} \end{bmatrix} \begin{bmatrix} \bar{d}-\bar{\boldsymbol{\phi}}^T\boldsymbol{R}_{\phi}^{-1}\boldsymbol{r}_{d\phi} \\ \boldsymbol{r}_{d\phi} \end{bmatrix} \\
&= \begin{bmatrix} 1 & \boldsymbol{0}^T \\ -\boldsymbol{R}_{\phi}^{-1}\bar{\boldsymbol{\phi}} & \boldsymbol{I} \end{bmatrix} \begin{bmatrix} \frac{\bar{d}-\bar{\boldsymbol{\phi}}^T\boldsymbol{R}_{\phi}^{-1}\boldsymbol{r}_{d\phi}}{1-\bar{\boldsymbol{\phi}}^T\boldsymbol{R}_{\phi}^{-1}\bar{\boldsymbol{\phi}}} \\ \boldsymbol{R}_{\phi}^{-1}\boldsymbol{r}_{d\phi} \end{bmatrix} = \begin{bmatrix} \frac{\bar{d}-\bar{\boldsymbol{\phi}}^T\boldsymbol{R}_{\phi}^{-1}\boldsymbol{r}_{d\phi}}{1-\bar{\boldsymbol{\phi}}^T\boldsymbol{R}_{\phi}^{-1}\bar{\boldsymbol{\phi}}} \\ \boldsymbol{R}_{\phi}^{-1}\boldsymbol{r}_{d\phi} - \frac{\bar{d}-\bar{\boldsymbol{\phi}}^T\boldsymbol{R}_{\phi}^{-1}\boldsymbol{r}_{d\phi}}{1-\bar{\boldsymbol{\phi}}^T\boldsymbol{R}_{\phi}^{-1}\bar{\boldsymbol{\phi}}}\boldsymbol{R}_{\phi}^{-1}\bar{\boldsymbol{\phi}} \end{bmatrix}.
\end{aligned}$$

Using this result, we can evaluate now $\mathrm{E}\{v_{o,e}\} = \mathrm{E}\{d\} - \boldsymbol{w}_{o,e}^T\mathrm{E}\{\boldsymbol{\phi}_e\}$ as follows (we used the fact that the inverse of a symmetric matrix is also symmetric from Box 12.3):

$$\begin{aligned}
\mathrm{E}\{v_{o,e}\} &= \bar{d} - \frac{\bar{d}-\bar{\boldsymbol{\phi}}^T\boldsymbol{R}_{\phi}^{-1}\boldsymbol{r}_{d\phi}}{1-\bar{\boldsymbol{\phi}}^T\boldsymbol{R}_{\phi}^{-1}\bar{\boldsymbol{\phi}}} - \boldsymbol{r}_{d\phi}^T\boldsymbol{R}_{\phi}^{-1}\bar{\boldsymbol{\phi}} + \frac{\bar{d}-\bar{\boldsymbol{\phi}}^T\boldsymbol{R}_{\phi}^{-1}\boldsymbol{r}_{d\phi}}{1-\bar{\boldsymbol{\phi}}^T\boldsymbol{R}_{\phi}^{-1}\bar{\boldsymbol{\phi}}}\bar{\boldsymbol{\phi}}^T\boldsymbol{R}_{\phi}^{-1}\bar{\boldsymbol{\phi}} \\
&= \bar{d} - \boldsymbol{r}_{d\phi}^T\boldsymbol{R}_{\phi}^{-1}\bar{\boldsymbol{\phi}} - \frac{\bar{d}-\bar{\boldsymbol{\phi}}^T\boldsymbol{R}_{\phi}^{-1}\boldsymbol{r}_{d\phi}}{1-\bar{\boldsymbol{\phi}}^T\boldsymbol{R}_{\phi}^{-1}\bar{\boldsymbol{\phi}}}\left(1-\bar{\boldsymbol{\phi}}^T\boldsymbol{R}_{\phi}^{-1}\bar{\boldsymbol{\phi}}\right) = 0.
\end{aligned}$$

12.2.1.5 *Sufficiently rich signals*

We saw in Section 12.2.1.2 that even when $\boldsymbol{R}_{\phi}$ is singular, we can recover the optimum v_o (with probability one), but not the optimum $\boldsymbol{w}_*$. Let us study this problem in more detail, since it is important for system identification. So, what does it mean in practice to have a singular autocorrelation function $\boldsymbol{R}_{\phi}$? The answer to this question can be quite complicated when the input signals are nonstationary (i.e., when $\boldsymbol{R}_{\phi}$ is not constant), so in this case we refer the interested reader to texts in adaptive control (such as [179]), which treat this problem in detail (although usually from a deterministic point of view). We consider here only the case in which the signals $\{d(n), x(n)\}$ are jointly WSS.

In this case, $\boldsymbol{R}_{\phi}$ may be singular because some entries of $\boldsymbol{\phi}$ are linear combinations of the others. For example, if we choose $\phi_0 = x(n)$ and $\phi_1 = 2x(n)$, then

$$\boldsymbol{R}_{\boldsymbol{\phi}} = \mathrm{E}\{x^2(n)\}\begin{bmatrix} 1 & 2 \\ 2 & 4 \end{bmatrix},$$

which is singular.

However, there are more subtle ways in which $\boldsymbol{R}_{\boldsymbol{\phi}}$ may become singular. Assume for example that our input sequence is given by

$$x(n) = \cos(\omega_0 n + \varphi), \tag{12.70}$$

where $0 < \omega_0 \le \pi$ is the frequency and φ is a random phase, uniformly distributed in the interval $[0, 2\pi)$. This kind of model (sinusoidal signals with random phases), by the way, is the bridge between the periodic signals from Section 12.1.2 and the stochastic models discussed in this section. In this case, consider a vector $\boldsymbol{\phi}$ formed from a delay line with $x(n)$ as input,

$$\boldsymbol{\phi} = \begin{bmatrix} x(n) & x(n-1) & \ldots & x(n-M+1) \end{bmatrix}^T.$$

If $M = 2$, we have

$$\boldsymbol{R}_{\boldsymbol{\phi}} = \begin{bmatrix} \mathrm{E}\{x^2(n)\} & \mathrm{E}\{x(n)x(n-1)\} \\ \mathrm{E}\{x(n)x(n-1)\} & \mathrm{E}\{x^2(n-1)\} \end{bmatrix} = \frac{1}{2}\begin{bmatrix} 1 & \cos(\omega_0) \\ \cos(\omega_0) & 1 \end{bmatrix}.$$

The determinant of $\boldsymbol{R}_{\boldsymbol{\phi}}$ is $1 - \cos^2(\omega_0) = \sin^2(\omega_0) \neq 0$ if $\omega_0 \neq 0$ and $\omega \neq \pi$.

However, for any $M \ge 3$, the resulting $\boldsymbol{R}_{\boldsymbol{\phi}}$ will be singular: we can write

$$\boldsymbol{\phi} = \frac{1}{2}e^{j(\omega_0 n+\varphi)}\begin{bmatrix} 1 \\ e^{-j\omega_0} \\ \vdots \\ e^{-j(M-1)\omega_0} \end{bmatrix} + \frac{1}{2}e^{-j(\omega_0 n+\varphi)}\begin{bmatrix} 1 \\ e^{j\omega_0} \\ \vdots \\ e^{j(M-1)\omega_0} \end{bmatrix},$$

and since $\mathrm{E}\{(e^{j\omega_0 n+j\varphi})^2\} = \mathrm{E}\{e^{j2(\omega_0 n+\varphi)}\} = 0$, $\mathrm{E}\{e^{j(\omega_0 n+\varphi)}e^{-j(\omega_0 n+\varphi)}\} = 1$, we have

$$\boldsymbol{R}_{\boldsymbol{\phi}} = \frac{1}{4}\begin{bmatrix} 1 \\ e^{-j\omega_0} \\ \vdots \\ e^{-j(M-1)\omega_0} \end{bmatrix}\begin{bmatrix} 1 & e^{j\omega_0} & \ldots & e^{j(M-1)\omega_0} \end{bmatrix} + \frac{1}{4}\begin{bmatrix} 1 \\ e^{j\omega_0} \\ \vdots \\ e^{j(M-1)\omega_0} \end{bmatrix}\begin{bmatrix} 1 & e^{-j\omega_0} & \ldots & e^{-j(M-1)\omega_0} \end{bmatrix}.$$

Since $\boldsymbol{R}_{\boldsymbol{\phi}}$ is the sum of two rank-one matrices, its rank is at most two. It will be two when $0 < \omega_0 < \pi$, since in this case the vectors are linearly independent. We conclude that when the input signal is a sinusoid as in (12.70), the optimum solution to (12.48) will never be able to identify more than two coefficients. This argument can be generalized: if $x(n)$ is composed of K sinusoids of different frequencies, then $\boldsymbol{R}_\phi$ can be nonsingular only if $M \leq 2K$.

The general conclusion is that $\boldsymbol{R}_\phi$ will be singular or not, depending on which functions are chosen as inputs, and on the input signal. We say that the input is *sufficiently rich* for a given problem if it is such that $\boldsymbol{R}_\phi$ is nonsingular. We show a few examples next.

12.2.1.6 Examples

The concepts we saw in Sections 12.2.1.2–12.2.1.5 can be understood intuitively if we return to our example in Section 12.1.2. In that example, the true echo was obtained by filtering the input $x(n)$ by an unknown $H(z)$, so we had

$$d(n) = \sum_{k=0}^{K} h_k x(n-k) + v(n),$$

where K is the true order of the echo path, $v(n) = B\cos(\omega_1 n + \theta)$, and $x(n) = A\cos(\omega_0 n + \varphi)$. We considered only one harmonic for simplicity.

We are trying to cancel the echo by constructing an approximated transfer function $\hat{H}(z)$, which is modeled as an FIR filter with M coefficients. As we saw in Section 12.1.2, if $x(n)$ is a simple sinusoid, the echo will be completely canceled if we satisfy (12.8), reproduced below:

$$\mathrm{Re}\{\hat{H}(e^{j\omega_0})\} = \mathrm{Re}\{H(e^{j\omega_0})\}, \qquad \mathrm{Im}\{\hat{H}(e^{j\omega_0})\} = \mathrm{Im}\{H(e^{j\omega_0})\}.$$

If $K = M = 1$, so $H(z) = h_0 + h_1 z^{-1}$ and $\hat{H}(z) = w_0 + w_1 z^{-1}$, this condition becomes

$$\begin{aligned} w_0 + w_1\cos(-\omega_0) &= h_0 + h_1\cos(-\omega_0), \\ w_1\sin(-\omega_0) &= h_1\sin(-\omega_0), \end{aligned}$$

with a single solution $w_0 = h_0$, $w_1 = h_1$.

On the other hand, if the input were as before (a single sinusoid), but $K = 2$, so $H(z) = h_0 + h_1 z^{-1} + h_2 z^{-2}$, the equations would be

$$\begin{aligned} w_0 + w_1\cos(-\omega_0) &= h_0 + h_1\cos(-\omega_0) + h_2\cos(-2\omega_0), \\ w_1\sin(-\omega_0) &= h_1\sin(-\omega_0) + h_2\sin(-2\omega_0), \end{aligned}$$

and the solution would be unique (since $\boldsymbol{R}_\phi$ is nonsingular), but would not approximate h_0 and h_1 in general.

This is an example of an undermodeled system, as we saw in Section 12.2.1.3. Fig. 12.21 shows the LMS algorithm derived in Section 12.1.2 applied to this problem. Note that even though $w_0(n)$, $w_1(n)$ do not converge to h_0, h_1, in this example the error $e(n)$ does indeed approximate well $v(n)$. This happens because in this case $x(n)$ is periodic and the term $x(n-2)$ can be obtained exactly as a linear combination of $x(n)$ and $x(n-1)$ (check that using the expressions for v_o from Section 12.2.1.2, we obtain $v_o = v$ in this case).

If we tried to identify $H(z)$ by adding an extra coefficient to $\hat{H}(z)$, the equations would be

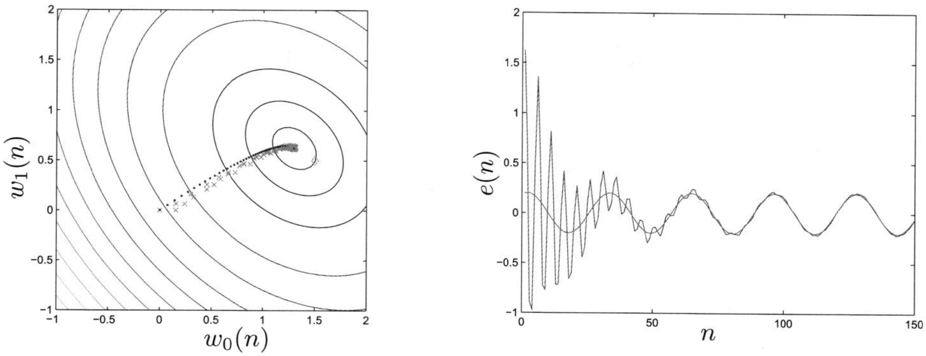

(a) LMS trajectory in (w_0, w_1)-space. The diamond marks the true value of the first two terms in $H(z)$.

(b) Output error

FIGURE 12.21

LMS in the undermodeled case.

$$\begin{aligned} w_0 + w_1 \cos(-\omega_0) + w_2 \cos(-2\omega_0) &= h_0 + h_1 \cos(-\omega_0) + h_2 \cos(-2\omega_0), \\ w_1 \sin(-\omega_0) + w_2 \sin(-2\omega_0) &= h_1 \sin(-\omega_0) + h_2 \sin(-2\omega_0). \end{aligned} \tag{12.71}$$

Now $w_0 = h_0$, $w_1 = h_1$, and $w_2 = h_2$ is a solution, but not the only one. Depending on the initial condition, the filter would converge to a different solution, as can be seen in Fig. 12.22 (the level sets are not shown, because they are not closed curves in this case). The input is not rich enough (in harmonics) to excite the unknown system $H(z)$ so that the adaptive filter might identify it, as we saw in Section 12.2.1.5. However, for all solutions the error converges to $v(n)$.

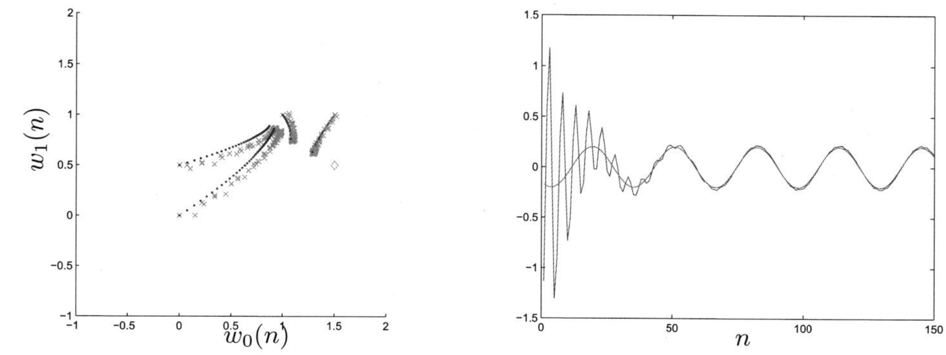

(a) LMS trajectory in (w_0, w_1)-space. The diamond marks the true value of the first two terms in $H(z)$.

(b) Output error

FIGURE 12.22

LMS under nonsufficiently rich input.

If the input had a second harmonic ($K_0 = 2$), we would add two more equations to (12.71), corresponding to $\hat{H}(e^{2j\omega_0}) = H(e^{2j\omega_0})$, and the filter would be able to identify systems with up to four coefficients.

The important message is that the adaptive filter only "sees" the error $d(n) - \hat{y}(n)$. If a mismatch between the true echo path $H(z)$ and the estimated one $\hat{H}(z)$ is not observable through looking only at the error signal, the filter will not converge to $H(z)$. Whether a mismatch is observable or not through the error depends on both $H(z)$ and the input signal being used: the input must excite a large enough number of frequencies so that $H(z)$ can be estimated correctly. For this reason, in system identification a good input would be white noise, which contains all frequencies.

12.2.2 Complex variables and multichannel filtering

Adaptive filters used in important applications such as communications and radar have complex variables as input signals. In this section, we extend the results of Section 12.2.1 to consider complex signals. We start with the general solution and then restrict the result to the more usual case of circular signals.

We show next that the general solution for complex signals is equivalent to an adaptive filter with two inputs and two outputs. Indeed, if all input variables are complex, then the error $e(n) = d(n) - \hat{y}(n)$ will also be complex, that is, $e(n) = e_{\mathrm{r}}(n) + je_{\mathrm{i}}(n)$, where $e_{\mathrm{r}}(n), e_{\mathrm{i}}(n) \in \mathbb{R}$ are the real and imaginary parts of $e(n)$. We must then minimize $\mathrm{E}\{|e(n)|^2\} = \mathrm{E}\{e_{\mathrm{r}}^2(n)\} + \mathrm{E}\{e_{\mathrm{i}}^2(n)\}$, the total power of the complex signal $e(n)$. The quadratic cost function (12.46) must be changed to

$$J(\boldsymbol{w}) = \mathrm{E}\{|e(n)|^2\} = \mathrm{E}\{e(n)e^*(n)\} = \mathrm{E}\{e_{\mathrm{r}}^2(n)\} + \mathrm{E}\{e_{\mathrm{i}}^2(n)\},$$

where $e^*(n)$ is the complex conjugate of $e(n)$. Now, we need to be careful about what exactly is the coefficient vector $\boldsymbol{w}$ in this case: if the regressor is $\boldsymbol{\phi} = \boldsymbol{\phi}_{\mathrm{r}} + j\boldsymbol{\phi}_i$ and $d = d_{\mathrm{r}} + jd_{\mathrm{i}}$, in principle $\hat{y} = \hat{y}_{\mathrm{r}} + j\hat{y}_{\mathrm{i}}$ should be such that both its real and imaginary parts depend on both the real and imaginary parts of $\boldsymbol{\phi}$, that is,

$$\begin{aligned} \hat{y}_{\mathrm{r}} &= \boldsymbol{w}_{\mathrm{rr}}^T\boldsymbol{\phi}_{\mathrm{r}} + \boldsymbol{w}_{\mathrm{ri}}^T\boldsymbol{\phi}_{\mathrm{i}}, \\ \hat{y}_{\mathrm{i}} &= \boldsymbol{w}_{\mathrm{ir}}^T\boldsymbol{\phi}_{\mathrm{r}} + \boldsymbol{w}_{\mathrm{ii}}^T\boldsymbol{\phi}_{\mathrm{i}}. \end{aligned} \tag{12.72}$$

Note that there is no a priori reason for us to imagine that there should be a special relation between the pairs $(\boldsymbol{w}_{\mathrm{rr}}, \boldsymbol{w}_{\mathrm{ri}})$ and $(\boldsymbol{w}_{\mathrm{ir}}, \boldsymbol{w}_{\mathrm{ii}})$. That is, we are dealing with two different coefficient vectors and one extended regressor:

$$\boldsymbol{w}_{\mathrm{r}} = \begin{bmatrix} \boldsymbol{w}_{\mathrm{rr}} \\ \boldsymbol{w}_{\mathrm{ri}} \end{bmatrix}, \qquad \boldsymbol{w}_{\mathrm{i}} = \begin{bmatrix} \boldsymbol{w}_{\mathrm{ir}} \\ \boldsymbol{w}_{\mathrm{ii}} \end{bmatrix}, \qquad \boldsymbol{\theta} = \begin{bmatrix} \boldsymbol{\phi}_{\mathrm{r}} \\ \boldsymbol{\phi}_{\mathrm{i}} \end{bmatrix}. \tag{12.73}$$

Let us proceed for the time being separating the real and imaginary channels of our filter. Later, in Section 12.2.2.1, we return to complex algebra. With these definitions, we can write

$$\hat{y}_{\mathrm{r}} = \boldsymbol{w}_{\mathrm{r}}^T\boldsymbol{\theta}, \qquad \hat{y}_{\mathrm{i}} = \boldsymbol{w}_{\mathrm{i}}^T\boldsymbol{\theta}. \tag{12.74}$$

The cost function then becomes

$$J(\boldsymbol{w}_{\mathrm{r}}, \boldsymbol{w}_{\mathrm{i}}) = \mathrm{E}\{(d_{\mathrm{r}} - \boldsymbol{w}_{\mathrm{r}}^T\boldsymbol{\theta})^2\} + \mathrm{E}\{(d_{\mathrm{i}} - \boldsymbol{w}_{\mathrm{i}}^T\boldsymbol{\theta})^2\}$$
$$= \mathrm{E}\{d_{\mathrm{r}}^2 - 2\boldsymbol{w}_{\mathrm{r}}^T\boldsymbol{\theta} d_{\mathrm{r}} + \boldsymbol{w}_{\mathrm{r}}^T\boldsymbol{\theta}\boldsymbol{\theta}^T\boldsymbol{w}_{\mathrm{r}}\} + \mathrm{E}\{d_{\mathrm{i}}^2 - 2\boldsymbol{w}_{\mathrm{i}}^T\boldsymbol{\theta} d_{\mathrm{i}} + \boldsymbol{w}_{\mathrm{i}}^T\boldsymbol{\theta}\boldsymbol{\theta}^T\boldsymbol{w}_{\mathrm{i}}\}$$
$$= r_{d_{\mathrm{r}}} - 2\boldsymbol{w}_{\mathrm{r}}^T\boldsymbol{r}_{d_{\mathrm{r}}\boldsymbol{\theta}} + \boldsymbol{w}_{\mathrm{r}}^T\boldsymbol{R}_{\boldsymbol{\theta}}\boldsymbol{w}_{\mathrm{r}} + r_{d_{\mathrm{i}}} - 2\boldsymbol{w}_{\mathrm{i}}^T\boldsymbol{r}_{d_{\mathrm{i}}\boldsymbol{\theta}} + \boldsymbol{w}_{\mathrm{i}}^T\boldsymbol{R}_{\boldsymbol{\theta}}\boldsymbol{w}_{\mathrm{i}}, \quad (12.75)$$

where the autocorrelation matrix and cross-correlation vectors are given by

$$\boldsymbol{R}_{\boldsymbol{\theta}} = \begin{bmatrix} \boldsymbol{R}_{\mathrm{r}} & \boldsymbol{R}_{\mathrm{ri}} \\ \boldsymbol{R}_{\mathrm{ir}} & \boldsymbol{R}_{\mathrm{i}} \end{bmatrix}, \qquad \boldsymbol{r}_{d_{\mathrm{r}}\boldsymbol{\theta}} = \begin{bmatrix} \boldsymbol{r}_{\mathrm{rr}} \\ \boldsymbol{r}_{\mathrm{ri}} \end{bmatrix}, \qquad \boldsymbol{r}_{d_{\mathrm{i}}\boldsymbol{\theta}} = \begin{bmatrix} \boldsymbol{r}_{\mathrm{ir}} \\ \boldsymbol{r}_{\mathrm{ii}} \end{bmatrix}, \quad (12.76)$$

where we simplified the notation, writing $\boldsymbol{R}_{\mathrm{r}}$ instead of $\boldsymbol{R}_{\boldsymbol{\phi}_{\mathrm{r}}}$, $\boldsymbol{R}_{\mathrm{ri}}$ instead of $\boldsymbol{R}_{\boldsymbol{\phi}_{\mathrm{r}}\boldsymbol{\phi}_{\mathrm{i}}}$, $\boldsymbol{r}_{\mathrm{rr}}$ instead of $\boldsymbol{r}_{d_{\mathrm{r}}\boldsymbol{\phi}_{\mathrm{r}}}$, and so on.

Differentiating (12.75) to find the minimum, we obtain

$$\frac{\partial J}{\partial \boldsymbol{w}_{\mathrm{r}}^T} = -2\boldsymbol{r}_{d_{\mathrm{r}}\boldsymbol{\theta}} + 2\boldsymbol{R}_{\boldsymbol{\theta}}\boldsymbol{w}_{\mathrm{r}}, \qquad \frac{\partial J}{\partial \boldsymbol{w}_{\mathrm{i}}^T} = -2\boldsymbol{r}_{d_{\mathrm{i}}\boldsymbol{\theta}} + 2\boldsymbol{R}_{\boldsymbol{\theta}}\boldsymbol{w}_{\mathrm{i}}. \quad (12.77)$$

Therefore, the optimum filters satisfy

$$\begin{bmatrix} \boldsymbol{R}_{\mathrm{r}} & \boldsymbol{R}_{\mathrm{ri}} \\ \boldsymbol{R}_{\mathrm{ir}} & \boldsymbol{R}_{\mathrm{i}} \end{bmatrix}\begin{bmatrix} \boldsymbol{w}_{\mathrm{rr,o}} \\ \boldsymbol{w}_{\mathrm{ri,o}} \end{bmatrix} = \begin{bmatrix} \boldsymbol{r}_{\mathrm{rr}} \\ \boldsymbol{r}_{\mathrm{ri}} \end{bmatrix}, \qquad \begin{bmatrix} \boldsymbol{R}_{\mathrm{r}} & \boldsymbol{R}_{\mathrm{ri}} \\ \boldsymbol{R}_{\mathrm{ir}} & \boldsymbol{R}_{\mathrm{i}} \end{bmatrix}\begin{bmatrix} \boldsymbol{w}_{\mathrm{ir,o}} \\ \boldsymbol{w}_{\mathrm{ii,o}} \end{bmatrix} = \begin{bmatrix} \boldsymbol{r}_{\mathrm{ir}} \\ \boldsymbol{r}_{\mathrm{ii}} \end{bmatrix}. \quad (12.78)$$

The optimum error is then

$$v_{r,\mathrm{o}} = d_r - \begin{bmatrix} \boldsymbol{w}_{\mathrm{rr,o}}^T & \boldsymbol{w}_{\mathrm{ri,o}}^T \end{bmatrix}\boldsymbol{\theta}, \qquad v_{i,\mathrm{o}} = d_i - \begin{bmatrix} \boldsymbol{w}_{\mathrm{ir,o}}^T & \boldsymbol{w}_{\mathrm{ii,o}}^T \end{bmatrix}\boldsymbol{\theta}. \quad (12.79)$$

There is also an orthogonality condition for this case: using (12.78), we obtain

$$\mathrm{E}\{v_{r,\mathrm{o}}\boldsymbol{\theta}\} = \mathrm{E}\{d_r\boldsymbol{\theta}\} - \mathrm{E}\{\boldsymbol{\theta}\boldsymbol{\theta}^T\}\begin{bmatrix} \boldsymbol{w}_{\mathrm{rr,o}} \\ \boldsymbol{w}_{\mathrm{ri,o}} \end{bmatrix} = \boldsymbol{0}, \quad (12.80a)$$

$$\mathrm{E}\{v_{i,\mathrm{o}}\boldsymbol{\theta}\} = \mathrm{E}\{d_i\boldsymbol{\theta}\} - \mathrm{E}\{\boldsymbol{\theta}\boldsymbol{\theta}^T\}\begin{bmatrix} \boldsymbol{w}_{\mathrm{ir,o}} \\ \boldsymbol{w}_{\mathrm{ii,o}} \end{bmatrix} = \boldsymbol{0}. \quad (12.80b)$$

The value of the cost function at the minimum can be obtained using this result. As in the case of real variables,

$$\mathrm{E}\{v_{r,\mathrm{o}}^2\} = \mathrm{E}\left\{v_{r,\mathrm{o}}\left(d_r - \begin{bmatrix} \boldsymbol{w}_{\mathrm{rr,o}}^T & \boldsymbol{w}_{\mathrm{ri,o}}^T \end{bmatrix}\boldsymbol{\theta}\right)\right\}$$
$$= \mathrm{E}\{v_{r,\mathrm{o}}d_r\} = \mathrm{E}\left\{d_r^2\right\} - \begin{bmatrix} \boldsymbol{w}_{\mathrm{rr,o}}^T & \boldsymbol{w}_{\mathrm{ri,o}}^T \end{bmatrix}\mathrm{E}\{d_r\boldsymbol{\theta}\},$$

where we used (12.80) in the second equality. Doing similarly for $v_{\mathrm{i,o}}$, we obtain

$$r_{v_{\mathrm{r,o}}} = \mathrm{E}\{v_{r,\mathrm{o}}^2\} = r_{d_{\mathrm{r}}} - \begin{bmatrix} \boldsymbol{w}_{\mathrm{rr,o}}^T & \boldsymbol{w}_{\mathrm{ri,o}}^T \end{bmatrix}\boldsymbol{r}_{d_{\mathrm{r}}\boldsymbol{\theta}}, \quad (12.81a)$$

$$r_{v_{\mathrm{i,o}}} = \mathrm{E}\{v_{\mathrm{i,o}}^2\} = r_{d_{\mathrm{i}}} - \begin{bmatrix} \boldsymbol{w}_{\mathrm{ir,o}}^T & \boldsymbol{w}_{\mathrm{ii,o}}^T \end{bmatrix}\boldsymbol{r}_{d_{\mathrm{i}}\boldsymbol{\theta}}, \quad (12.81b)$$

$$J_{\min} = r_{v_{\mathrm{r,o}}} + r_{v_{\mathrm{i,o}}}. \quad (12.81c)$$

What we have just described is a multichannel (two-input two-output) optimal filter. In the next section, we rederive it using complex algebra. Note that multichannel filters have applications that are unrelated to complex variables, particularly in control [179], stereo echo cancelation [153,295], and active noise control [169,194,268,352].

12.2.2.1 *Widely linear complex least-mean squares*

This general solution can be obtained in a more straightforward way if we use the complex gradients described in Box 12.6. Keeping the regressor $\boldsymbol{\theta}$ as in (12.73) and defining the extended complex weight vector as $\boldsymbol{\omega} = \boldsymbol{w}_{\mathrm{r}} + j\boldsymbol{w}_{\mathrm{i}}$, we can write the cost function as

$$J(\boldsymbol{\omega}) = \mathrm{E}\left\{(d - \boldsymbol{\omega}^H\boldsymbol{\theta})(d - \boldsymbol{\omega}^H\boldsymbol{\theta})^*\right\} = r_d - \boldsymbol{\omega}^H\boldsymbol{r}_{d\theta} - \boldsymbol{r}_{d\theta}^H\boldsymbol{\omega} + \boldsymbol{\omega}^H\boldsymbol{R}_{\theta}\boldsymbol{\omega},$$

where we defined $r_d = \mathrm{E}\{|d|^2\}$ and $\boldsymbol{r}_{d\theta} = \mathrm{E}\{d^*\boldsymbol{\theta}\}$. Note that $(\cdot)^H$ represents the Hermitian transpose, that is, transpose followed by conjugation. Differentiating $J(\boldsymbol{\omega})$ with respect to ω^H, as described in Box 12.6, we obtain

$$\frac{\partial J}{\partial \boldsymbol{\omega}^H} = -\boldsymbol{r}_{d\theta} + \boldsymbol{R}_{\theta}\boldsymbol{\omega}. \tag{12.82}$$

The normal equations then become

$$\boldsymbol{R}_{\theta}\boldsymbol{\omega}_{\mathrm{o}} = \boldsymbol{r}_{d\theta}. \tag{12.83}$$

Expanding the real and imaginary parts of (12.83), we recover (12.78). Similarly, the optimum error and the minimum value of the cost are

$$\nu_{\mathrm{o}} = d - \boldsymbol{\omega}_{\mathrm{o}}^H\boldsymbol{\theta}, \qquad J_{\min} = \mathrm{E}\{|\nu_{\mathrm{o}}|^2\} = \mathrm{E}\{|d|^2\} - \boldsymbol{\omega}_{\mathrm{o}}^H\boldsymbol{r}_{d\theta}. \tag{12.84}$$

Comparing with (12.81), we see that $\nu_{\mathrm{o}} = v_{\mathrm{r,o}} + jv_{\mathrm{i,o}}$, and the value of $J_{\min}$ is the same as before. However, as the reader probably noticed, using a complex weight vector $\boldsymbol{\omega}$ and the complex derivatives of Box 12.6, all expressions are much simpler to derive.

The filter using a complex weight vector $\boldsymbol{\omega}$ and the real regressor $\boldsymbol{\theta}$ constitutes a version of what is known as *widely linear* complex filter. This version was recently proposed in [13] as a reduced-complexity alternative to the widely linear complex filter originally proposed in [291,292], which uses as regressor the extended vector

$$\boldsymbol{\phi}_{\mathrm{e}} = \begin{bmatrix} \boldsymbol{\phi} \\ \boldsymbol{\phi}^* \end{bmatrix}, \tag{12.85}$$

composed of the original regressor $\boldsymbol{\phi}$ and its complex conjugate $\boldsymbol{\phi}^*$ (computational complexity is reduced because the number of operations required to compute $\boldsymbol{\omega}^H\boldsymbol{\theta}$ is about half of what is necessary to evaluate $\boldsymbol{\omega}^H\boldsymbol{\phi}_{\mathrm{e}}$). Widely linear estimation has received much attention, due to a number of different applications, such as those described in [4,5,87,123,223,404].

But why the name *widely* linear? The reason for this name is that the complex regressor $\boldsymbol{\phi}$ does not appear linearly in the expressions, but only by separating its real and imaginary parts (as in $\boldsymbol{\theta}$) or by including its complex conjugate (as in $\boldsymbol{\phi}_{\mathrm{e}}$). Both these expressions, from the point of view of a complex variable, are nonlinear.

What would be a *linear* complex filter, then? This is the subject of the next section.

12.2.2.2 Linear complex least-mean squares

There are many applications in which the data and regressor are such that their real and imaginary parts are similarly distributed, so that

$$\boldsymbol{R}_{\mathrm{r}} = \boldsymbol{R}_{\mathrm{i}}, \qquad r_{d_{\mathrm{r}}} = r_{d_{\mathrm{i}}}, \qquad \boldsymbol{R}_{\mathrm{ri}} = -\boldsymbol{R}_{\mathrm{ir}}, \qquad \boldsymbol{r}_{\mathrm{rr}} = \boldsymbol{r}_{\mathrm{ii}}, \qquad \boldsymbol{r}_{\mathrm{ri}} = -\boldsymbol{r}_{\mathrm{ir}}. \tag{12.86}$$

Under these conditions, we say that the pair $(d, \boldsymbol{\phi})$ is *complex circular*, or simply *circular*.

In this case, we can rewrite (12.78) as

$$\begin{bmatrix} \boldsymbol{R}_{\mathrm{r}} & \boldsymbol{R}_{\mathrm{ri}} \\ -\boldsymbol{R}_{\mathrm{ri}} & \boldsymbol{R}_{\mathrm{r}} \end{bmatrix} \begin{bmatrix} \boldsymbol{w}_{\mathrm{rr}} \\ \boldsymbol{w}_{\mathrm{ri}} \end{bmatrix} = \begin{bmatrix} \boldsymbol{r}_{\mathrm{rr}} \\ \boldsymbol{r}_{\mathrm{ri}} \end{bmatrix}, \qquad \begin{bmatrix} \boldsymbol{R}_{\mathrm{r}} & \boldsymbol{R}_{\mathrm{ri}} \\ -\boldsymbol{R}_{\mathrm{ri}} & \boldsymbol{R}_{\mathrm{r}} \end{bmatrix} \begin{bmatrix} \boldsymbol{w}_{\mathrm{ir}} \\ \boldsymbol{w}_{\mathrm{ii}} \end{bmatrix} = \begin{bmatrix} -\boldsymbol{r}_{\mathrm{ri}} \\ \boldsymbol{r}_{\mathrm{rr}} \end{bmatrix}.$$

Expand these equations as follows (notice the change in order in (12.87b)):

$$\boldsymbol{R}_{\mathrm{r}}\boldsymbol{w}_{\mathrm{rr}} + \boldsymbol{R}_{\mathrm{ri}}\boldsymbol{w}_{\mathrm{ri}} = \boldsymbol{r}_{\mathrm{rr}}, \qquad -\boldsymbol{R}_{\mathrm{ri}}\boldsymbol{w}_{\mathrm{rr}} + \boldsymbol{R}_{\mathrm{r}}\boldsymbol{w}_{\mathrm{ri}} = \boldsymbol{r}_{\mathrm{ri}}, \tag{12.87a}$$

$$\boldsymbol{R}_{\mathrm{r}}\boldsymbol{w}_{\mathrm{ii}} - \boldsymbol{R}_{\mathrm{ri}}\boldsymbol{w}_{\mathrm{ir}} = \boldsymbol{r}_{\mathrm{rr}}, \qquad \boldsymbol{R}_{\mathrm{ri}}\boldsymbol{w}_{\mathrm{ii}} + \boldsymbol{R}_{\mathrm{r}}\boldsymbol{w}_{\mathrm{ir}} = -\boldsymbol{r}_{\mathrm{ri}}. \tag{12.87b}$$

Now we can see that if $(\boldsymbol{w}_{\mathrm{rr}}, \boldsymbol{w}_{\mathrm{ri}})$ is a solution to (12.87a), then

$$(\boldsymbol{w}_{\mathrm{ii}} = \boldsymbol{w}_{\mathrm{rr}}, \boldsymbol{w}_{\mathrm{ir}} = -\boldsymbol{w}_{\mathrm{ri}}) \tag{12.88}$$

is a solution to (12.87b). Therefore, when the input signals are circular, the number of degrees of freedom in the problem is actually smaller. In this case, there is a very nice way of recovering the solutions, working only with complex algebra. Indeed, defining $\boldsymbol{w} = \boldsymbol{w}_{\mathrm{rr}} + j\boldsymbol{w}_{\mathrm{ri}}$, we have

$$\begin{aligned} \boldsymbol{w}^H\boldsymbol{\phi} &= (\boldsymbol{w}_{\mathrm{rr}} - j\boldsymbol{w}_{\mathrm{ri}})^T(\boldsymbol{\phi}_{\mathrm{r}} + j\boldsymbol{\phi}_{\mathrm{i}}) \\ &= \left(\boldsymbol{w}_{\mathrm{rr}}^T\boldsymbol{\phi}_{\mathrm{r}} + \boldsymbol{w}_{\mathrm{ri}}^T\boldsymbol{\phi}_{\mathrm{i}}\right) + j\left(\boldsymbol{w}_{\mathrm{rr}}^T\boldsymbol{\phi}_{\mathrm{i}} - \boldsymbol{w}_{\mathrm{ri}}^T\boldsymbol{\phi}_{\mathrm{r}}\right), \end{aligned} \tag{12.89}$$

which is equivalent to (12.72) with the identities (12.88). Using this definition, it is possible to obtain the optimum solution (12.87a), (12.87b), (12.88) using only complex algebra, following exactly the same steps as our derivation for the real case in the previous section.

This path is very useful; we can derive almost all results for real or complex circular adaptive filters using the same arguments. Let us see how it goes. First, we define $\hat{y} = \boldsymbol{w}^H\boldsymbol{\phi}$, so our problem becomes

$$\min_{\boldsymbol{w}\in\mathbb{C}^M} \mathrm{E}\left\{\left|d - \boldsymbol{w}^H\boldsymbol{\phi}\right|^2\right\}. \tag{12.90}$$

Expanding the cost, recalling that now $\left|d - \boldsymbol{w}^H\boldsymbol{\phi}\right|^2 = (d - \boldsymbol{w}^H\boldsymbol{\phi})(d - \boldsymbol{w}^H\boldsymbol{\phi})^* = (d - \boldsymbol{w}^H\boldsymbol{\phi})(d^* - \boldsymbol{\phi}^H\boldsymbol{w})$, we obtain

$$\begin{aligned} J(\boldsymbol{w}) &= \mathrm{E}\left\{|d|^2 - \boldsymbol{w}^H\boldsymbol{\phi}d^* - d\boldsymbol{\phi}^H\boldsymbol{w} + \boldsymbol{w}^H\boldsymbol{\phi}\boldsymbol{\phi}^H\boldsymbol{w}\right\} \\ &= r_d - \boldsymbol{w}^H\boldsymbol{r}_{d\boldsymbol{\phi}} - \boldsymbol{r}_{d\boldsymbol{\phi}}^H\boldsymbol{w} + \boldsymbol{w}^H\boldsymbol{R}_{\boldsymbol{\phi}}\boldsymbol{w}, \end{aligned} \tag{12.91}$$

where we defined $r_d = \mathrm{E}\{|d|^2\}$, $\boldsymbol{r}_{d\boldsymbol{\phi}} = \mathrm{E}\{d^*\boldsymbol{\phi}\}$, $\boldsymbol{R}_{\boldsymbol{\phi}} = \mathrm{E}\{\boldsymbol{\phi}\boldsymbol{\phi}^H\}$.

We now need to find the minimum of this cost. One easy solution is to use the rules for differentiation of real functions of complex variables, as described in Box 12.6. These rules are very useful and surprisingly easy to use: basically, we can treat $\boldsymbol{w}$ and its conjugate $\boldsymbol{w}^H$ as if they were independent variables, that is,

$$\frac{\partial J}{\partial \boldsymbol{w}^H} = -\boldsymbol{r}_{d\phi} + \boldsymbol{R}_{\phi}\boldsymbol{w}. \tag{12.92}$$

Equating the gradient to zero, the solution is (assuming that $\boldsymbol{R}_{\phi}$ is nonsingular)

$$\boldsymbol{w}_{\mathrm{o}} = \boldsymbol{R}_{\phi}^{-1}\boldsymbol{r}_{d\phi}, \tag{12.93}$$

which is exactly equal to (12.52). This is the advantage of this approach: the expressions for complex and real problems are *almost* equal. The important differences are the conjugates that appear in the definitions of $\boldsymbol{R}_{\phi}$ and $\boldsymbol{r}_{d\phi}$ and the absence of a factor of 2 in the gradient (12.92). This follows from the definition of complex derivative we used (see Box 12.6). Here this difference is canceled out in the solution (12.93), but further on we will find situations in which there will be a different factor for the case of real and for the case of complex variables.

Let us check that (12.93) is really equivalent to (12.87a). Expanding $\boldsymbol{R}_{\phi}$ and $\boldsymbol{r}_{d\phi}$ and using the circularity conditions (12.86), we obtain

$$\boldsymbol{R}_{\phi} = \mathrm{E}\{\boldsymbol{\phi}\boldsymbol{\phi}^H\} = \mathrm{E}\{(\boldsymbol{\phi}_{\mathrm{r}} + j\boldsymbol{\phi}_{\mathrm{i}})(\boldsymbol{\phi}_{\mathrm{r}} - j\boldsymbol{\phi}_{\mathrm{i}})^T\} = \boldsymbol{R}_{\mathrm{r}} + \boldsymbol{R}_{\mathrm{i}} + j\,(\boldsymbol{R}_{\mathrm{ir}} - \boldsymbol{R}_{\mathrm{ri}}) \tag{12.94a}$$

$$= 2\boldsymbol{R}_{\mathrm{r}} - 2j\,\boldsymbol{R}_{\mathrm{ri}} \tag{12.94b}$$

and

$$\begin{aligned}\boldsymbol{r}_{d\phi} &= \mathrm{E}\{d^*\boldsymbol{\phi}\} = \mathrm{E}\{(d_{\mathrm{r}} - jd_{\mathrm{i}})(\boldsymbol{\phi}_{\mathrm{r}} + j\boldsymbol{\phi}_{\mathrm{i}})\} = \boldsymbol{r}_{\mathrm{rr}} + \boldsymbol{r}_{\mathrm{ii}} + j\,(\boldsymbol{r}_{\mathrm{ri}} - \boldsymbol{r}_{\mathrm{ir}})\\ &= 2\boldsymbol{r}_{\mathrm{rr}} + 2j\boldsymbol{r}_{\mathrm{ri}},\end{aligned} \tag{12.95}$$

so, equating the gradient (12.92) to zero, $\boldsymbol{w}_{\mathrm{o}} = \boldsymbol{w}_{\mathrm{r,o}} + j\boldsymbol{w}_{\mathrm{i,o}}$ must satisfy

$$\begin{aligned}2\,(\boldsymbol{R}_{\mathrm{r}} - j\boldsymbol{R}_{\mathrm{ri}})\left(\boldsymbol{w}_{\mathrm{r,o}} + j\boldsymbol{w}_{\mathrm{i,o}}\right) &= 2\,(\boldsymbol{r}_{\mathrm{rr}} + j\boldsymbol{r}_{\mathrm{ri}}),\\ \boldsymbol{R}_{\mathrm{r}}\boldsymbol{w}_{\mathrm{r,o}} + \boldsymbol{R}_{\mathrm{ri}}\boldsymbol{w}_{\mathrm{i,o}} + j\left(\boldsymbol{R}_{\mathrm{r}}\boldsymbol{w}_{\mathrm{i,o}} - \boldsymbol{R}_{\mathrm{ri}}\boldsymbol{w}_{\mathrm{r,o}}\right) &= \boldsymbol{r}_{\mathrm{rr}} + j\boldsymbol{r}_{\mathrm{ri}}.\end{aligned} \tag{12.96}$$

Comparing the real and imaginary parts of (12.96) with (12.87a), we see that they indeed correspond to the same set of equations.

We now make a few important remarks.

Remark 12.5. The orthogonality condition still holds in the complex circular case, but with a small difference: defining the optimum error as before,

$$v_{\mathrm{o}} = d - \boldsymbol{w}_{\mathrm{o}}^H\boldsymbol{\phi},$$

we obtain, noting that $(\boldsymbol{w}_{\mathrm{o}}^H\boldsymbol{\phi})^* = \boldsymbol{\phi}^H\boldsymbol{w}_{\mathrm{o}}$,

$$\mathrm{E}\{v_{\mathrm{o}}^*\boldsymbol{\phi}\} = \mathrm{E}\{d^*\boldsymbol{\phi} - \boldsymbol{\phi}\boldsymbol{\phi}^H\boldsymbol{w}_{\mathrm{o}}\} = \boldsymbol{r}_{d\phi} - \boldsymbol{R}_{\phi}\boldsymbol{w}_{\mathrm{o}} = \boldsymbol{0}. \tag{12.97}$$

Remark 12.6. The complex gradient (12.92) seems strange because it does not include the factors of 2 that would appear in the real case. This difference will appear in many expressions in the following sections, whenever we need to differentiate between the real and complex cases. However, the difference is illusory: if you go back to (12.94b) and (12.95), you will see that the factor of 2 is actually there, just hidden by the definition of autocorrelation and cross-correlation under the circularity conditions.

Remark 12.7. The optimum error power can be obtained as in the real case, and is equal to

$$\mathrm{E}\{|v_\mathrm{o}|^2\} = r_d - \boldsymbol{w}_\mathrm{o}^H \boldsymbol{r}_{d\boldsymbol{\phi}} = r_d - \boldsymbol{r}_{d\boldsymbol{\phi}}^H \boldsymbol{R}_{\boldsymbol{\phi}}^{-1} \boldsymbol{r}_{d\boldsymbol{\phi}}. \tag{12.98}$$

Box 12.4: Least-mean squares estimation

Here we consider the general problem of finding the optimum relation $\hat{\mathcal{H}}_\mathrm{o}$ between $d(n)$ and $x(n)$, such that

$$\hat{\mathcal{H}}_\mathrm{o} = \arg\min_{\hat{\mathcal{H}}} \mathrm{E}\left\{d(n) - \hat{\mathcal{H}}(x(n), x(n-1), \dots)\right\}, \tag{12.99}$$

without restricting $\mathcal{H}$ to a particular class.

In order to understand this problem well, let us consider first the simpler case where the relation between $x(n)$ and $d(n)$ is memoryless, that is, our function depends only on the current value of $x(n)$. Dropping the time index to simplify the notation, our problem then becomes how to find a function $\hat{\mathcal{H}}$ that, applied to a certain random variable x, approximates a certain random variable d so that the MSE is minimized, that is, we want to solve

$$\min_{\hat{\mathcal{H}}} \mathrm{E}\{[d - \hat{\mathcal{H}}(x)]^2\}. \tag{12.100}$$

Assume that the joint probability density function (pdf) $p_{dx}(d, x)$ of d and x is known (we are assuming that d and x are continuous random variables, but the final result in (12.101) also holds if they are discrete). Then,

$$\begin{aligned}
\mathrm{E}\{[d - \hat{\mathcal{H}}(x)]^2\} &= \int\int [d - \hat{\mathcal{H}}(x)]^2 p_{dx}(d, x)\,\mathrm{d}d\,\mathrm{d}x \\
&= \int \underbrace{\left\{\int [d - \hat{\mathcal{H}}(x)]^2 p_{d|x}(d|x)\,\mathrm{d}d\right\}}_{\mathrm{E}_d\{[d-\hat{\mathcal{H}}(x)]^2|x\}} p_x(x)\,\mathrm{d}x \\
&= \mathrm{E}_x\left\{\mathrm{E}_d\left\{[d - \hat{\mathcal{H}}(x)]^2|x\right\}\right\},
\end{aligned} \tag{12.101}$$

where we use the subscripts d and x in E_d and E_x to highlight that the expectations are taken with respect to d or x only.

Note that the inner integrand $\mathrm{E}_d\left\{[d - \hat{\mathcal{H}}(x)]^2|x\right\}$ in (12.101) is nonnegative. Therefore, if we minimize it for each value of x, we will obtain the minimum of the full expression. That is, the optimum

$\hat{\mathcal{H}}_o$ is the function defined by

$$\hat{\mathcal{H}}_o(x) = \arg\min_{\hat{\mathcal{H}}(x)} \mathrm{E}_d\left\{[d - \hat{\mathcal{H}}(x)]^2|x\right\}. \quad (12.102)$$

It is important to remember that, for fixed x, $\hat{\mathcal{H}}(x)$ is simply a number. This can be seen more easily expanding the expectations in (12.102):

$$\begin{aligned}\mathrm{E}_d\{[d - \hat{\mathcal{H}}(x)]^2|x\} &= \int \left(d^2 - 2d\hat{\mathcal{H}}(x) + \hat{\mathcal{H}}^2(x)\right) p_{d|x}(d|x)\,\mathrm{d}d \\ &= \int d^2 p_{d|x}(d|x)\,\mathrm{d}d - 2\hat{\mathcal{H}}(x)\int d p_{d|x}(d|x)\,\mathrm{d}d + \hat{\mathcal{H}}^2(x)\int p_{d|x}(d|x)\,\mathrm{d}d \\ &= \mathrm{E}_d\{d^2|x\} - 2\,\mathrm{E}\{d|x\}\hat{\mathcal{H}}(x) + \hat{\mathcal{H}}^2(x).\end{aligned}$$

Differentiating with respect to $\hat{\mathcal{H}}(x)$ and equating to zero, we obtain the solution

$$\hat{\mathcal{H}}_o(x) = \mathrm{E}_d\{d|x\} \triangleq \hat{y}(n), \quad (12.103)$$

which is indeed a function of x, as desired.

Let us study the properties of this solution. First, define

$$v_o = d - \hat{\mathcal{H}}_o(x).$$

Note that v_o and $\hat{\mathcal{H}}_o$ are not necessarily equal to v and $\mathcal{H}$ from (12.41). Evaluating the average of $\hat{\mathcal{H}}_o(x)$, we obtain

$$\mathrm{E}_x\{\hat{\mathcal{H}}_o(x)\} = \mathrm{E}_x\{\mathrm{E}_d\{d|x\}\} = \mathrm{E}\{d\},$$

by an argument similar to that used to obtain (12.101). We conclude that v_o has zero mean. In addition, for any function $f(x)$ of x for which the expected values exist, we have

$$\mathrm{E}\{v_o f(x)\} = \mathrm{E}_x\{\mathrm{E}_d\{d\,f(x)\}\} - \mathrm{E}\{\mathrm{E}_d\{d|x\}f(x)\} = 0 = \mathrm{E}\{v_o\}\,\mathrm{E}\{f(x)\}, \quad (12.104)$$

since $\mathrm{E}_d\{d\,f(x)|x\} = f(x)\,\mathrm{E}_d\{d|x\}$. This result means that the error v_o is uncorrelated to any function of the reference data x, which is a very strong condition. It basically means that we have extracted all second-order information available in x about d.

Table 12.1 Counterexample: (12.104) does not imply independence.

r \ s	**−1**	**0**	**1**
−1	0	1/2	0
1	1/4	0	1/4

We remark that (12.104) does not imply that x and v_o are independent. A simple counterexample is the following. Consider two discrete-valued random variables r and s with joint probabilities given by Table 12.1. Then $\mathrm{E}\{sf(r)\} = 1/4 \cdot (-1) \cdot f(+1) + 1/4 \cdot (+1) \cdot f(+1) + 1/2 \cdot 0 \cdot f(-1) = 0$, $\forall f(\cdot)$, but r and s are not independent.

In the more general case in which $\hat{\mathcal{H}}$ is allowed to depend on previous samples of $x(n)$, the arguments and the final conclusion are similar,

$$\arg\min_{\hat{\mathcal{H}}} \mathrm{E}\left\{\left[d - \hat{\mathcal{H}}(x(n), x(n-1), \dots)\right]^2\right\} = \mathrm{E}\big\{d \big| x(n), x(n-1), \dots\big\}, \tag{12.105}$$

and defining

$$v_o(n) = d(n) - \mathrm{E}\big\{d \big| x(n), x(n-1), \dots\big\},$$

we know that $v_o(n)$ has zero mean and is uncorrelated with any function of $x(n), x(n-1), \dots,$

$$\mathrm{E}\{v_o(n)\} = 0, \qquad \mathrm{E}\{v_o(n) f(x(n), x(n-1), \dots\} = 0. \tag{12.106}$$

Note that, in general, the optimal solution of (12.105) depends on n, so we should write

$$\hat{\mathcal{H}}_{o,n}(x(n), x(n-1), \dots) = \mathrm{E}\big\{d \big| x(n), x(n-1), \dots\big\} \tag{12.107}$$

to make the time dependence explicit.

We can now return to the question at the start of this section. From our results thus far, we see that minimizing $\mathrm{E}\{e^2(n)\}$ leads to a model which relates $d(n)$ and the sequence $\{x(n)\}$ such that

$$d(n) = \hat{\mathcal{H}}_{o,n}(x(n), x(n-1), \dots) + v_o(n),$$

in which $v_o(n)$ is uncorrelated with any function $f(\cdot)$ of $x(n), x(n-1), \dots$. In other words, the solution of (12.105) imposes such a model on our data, even if this model makes no physical sense at all. This is an important point: such a model will always result from the minimization of the MSE, at least as long as the statistics of $d(n)$ and $\{x(n)\}$ are such that the expected values in (12.105) are finite.

Now, what would happen if we knew beforehand, by physical arguments about the data, that a model such as (12.41), repeated here for convenience,

$$d(n) = \underbrace{\mathcal{H}(x(n), x(n-1), \dots)}_{y(n)} + v(n),$$

holds between $d(n)$ and $\{x(n)\}$, such that $v(n)$ has zero mean and is uncorrelated to any function of $x(n), x(n-1), \dots$? Will the solution of (12.105) be such that

$$\hat{\mathcal{H}}_o(x(n), x(n-1), \dots) = \mathcal{H}(x(n), x(n-1), \dots), \qquad v_o(n) = v(n)?$$

To answer this question, let us go back to (12.100) and use (12.41) to find the solution. We thus have (we omit the arguments of $d(n)$, $v(n)$, and $\mathcal{H}(\cdot)$ for simplicity)

$$\min_{\hat{\mathcal{H}}} \mathrm{E}\{[d - \hat{\mathcal{H}}(x)]^2\} = \min_{\hat{\mathcal{H}}} \mathrm{E}\left\{[\mathcal{H} + v - \hat{\mathcal{H}}]^2\right\}.$$

Since v is orthogonal to any function of $x(n), x(n-1), \dots$ by assumption, it will be orthogonal to $\mathcal{H}$ and $\hat{\mathcal{H}}$ in particular. Therefore, we can simplify the cost as

$$\min_{\hat{\mathcal{H}}} \mathrm{E}\{[d-\hat{\mathcal{H}}(x)]^2\} = \min_{\hat{\mathcal{H}}} \left[\mathrm{E}\left\{(\mathcal{H}-\hat{\mathcal{H}})^2\right\} + \mathrm{E}\{v^2\}\right]. \tag{12.108}$$

Both terms in (12.108) are positive, and only the first depends on $\hat{\mathcal{H}}$. Therefore, the solution is indeed $\hat{\mathcal{H}} = \mathcal{H}$ and $v_o = v$, as one would wish (to be precise, $\mathcal{H}$ and $\hat{\mathcal{H}}$ could differ on a set of probability zero, which can be seen as a generalization of our discussion on sufficiently rich signals in Section 12.2.1.5).

We conclude that we can use (12.103) as a criterion for separating the two parts $y(n)$ and $v(n)$ of our model for $d(n)$ if and only if $v(n)$ satisfies (12.106). This holds, for example, if $v(n)$ has zero mean and is independent of $x(n-k)$ for all $k \geq 0$.

Note that since $\hat{y}(n) = \hat{\mathcal{H}}_{o,n}(x(n), x(n-1), \ldots)$ is itself a function of $x(n), x(n-1), \ldots$, Eq. (12.106) implies that

$$\mathrm{E}\{v_o(n)\hat{y}(n)\} = 0,$$

and we have

$$\begin{aligned}
\mathrm{E}\{v_o^2(n)\} &= \mathrm{E}\left\{v_o\left[d(n)-\hat{y}(n)\right]\right\} = \mathrm{E}\{v_o d(n)\} = \\
&= \mathrm{E}\left\{\left[d(n)-\hat{y}(n)\right]d(n)\right\} = \\
&= \mathrm{E}\{d^2(n)\} - \mathrm{E}\left\{d(n)\hat{y}(n)\right\} = \\
&= \mathrm{E}\{d^2(n)\} - \mathrm{E}\left\{\left[\hat{y}(n)+v_o(n)\right]\hat{y}(n)\right\} = \\
&= \mathrm{E}\{d^2(n)\} - \mathrm{E}\left\{\hat{y}^2(n)\right\}.
\end{aligned}$$

Recalling that $\mathrm{E}\{\hat{y}(n)\} = \mathrm{E}\{d(n)\}$, if we add and subtract $\mathrm{E}^2\{d(n)\}$ to the last result, we conclude that

$$\mathrm{E}\{v_o^2(n)\} = \sigma_d^2 - \sigma_{\hat{y}}^2 \leq \sigma_d^2, \tag{12.109}$$

where σ_d^2 and $\sigma_{\hat{y}}^2$ are the variances of $d(n)$ and of $\hat{y}(n)$.

The solution we just found is very general (it assumes little about $\mathcal{H}$), and there are ways of computing it adaptively from the data, using for example algorithms known as particle filters [79,88,120]. The solution requires, however, that one builds approximations for the joint probability distribution of $d(n)$ and $x(n), x(n-1), \ldots$, which in general requires a large amount of data, resulting in relatively high complexity and relatively slow convergence. If you know more about the relation between $\{x(n)\}$ and $d(n)$, this information might be used to constrain the search to a smaller class of problems, potentially leading to simpler algorithms with faster convergence. That is the goal of adaptive filters, and the reason why we restricted ourselves to this case in the main text.

Box 12.5: Convex functions

Convex sets are such that lines between any two points in the set always stay entirely in the set (see Fig. 12.23). More formally, a set $\mathcal{S} \in \mathbb{R}^n$ is convex if

$$s_1, s_2 \in \mathcal{S} \Rightarrow \lambda s_1 + (1-\lambda)s_2 \in \mathcal{S}, \quad \text{for all } \lambda \in [0, 1].$$

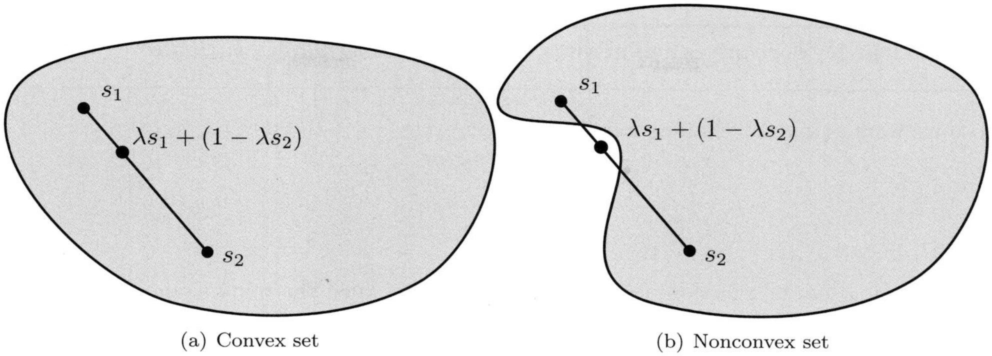

FIGURE 12.23

Convexity.

Convex *functions*, on the other hand, are functions $f(\boldsymbol{x}) : \mathbb{R}^n \to \mathbb{R}$ such that for all $\boldsymbol{x}_1, \boldsymbol{x}_2 \in \mathbb{R}^n$ and $0 < \lambda < 1$,

$$f(\lambda \boldsymbol{x}_1 + (1 - \lambda)\boldsymbol{x}_2) \leq \lambda f(\boldsymbol{x}_1) + (1 - \lambda) f(\boldsymbol{x}_2).$$

The function is called *strictly convex* if the inequality is strict ($<$ instead of $\leq$). For example, $f(x) = x^2$ is strictly convex. Fig. 12.24 shows examples of convex and nonconvex functions. An important property of strictly convex functions is that they always have a unique minimum, that is, they never have suboptimal local minima.

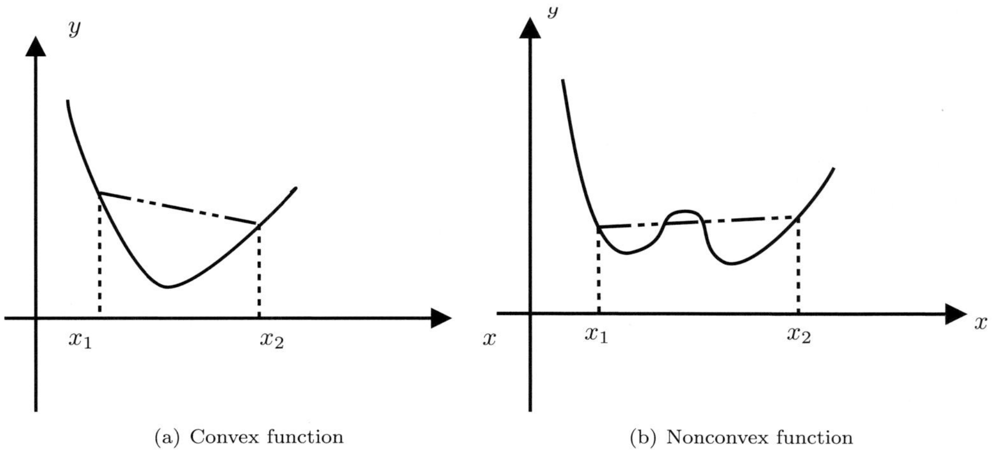

FIGURE 12.24

Convex functions.

Box 12.6: Wirtinger derivatives and minima of functions of a complex variable

Suppose we want to minimize a function

$$f : \mathbb{C} \to \mathbb{R}, \\ w \in \mathbb{C} \mapsto f(w) \in \mathbb{R}, \tag{12.110}$$

that is, a function that takes a complex argument to a real value. The minimization problem is well defined, since the image of f is real. However, we cannot use standard arguments about derivatives and stationary points to find the minimum, because a function like this is never analytic, that is, the standard derivative used in complex analysis does not exist.

To see why, we need to think about f as a function of two variables, x and y, the real and imaginary parts of w. Take $f(w) = |w|^2 = ww^*$ for example. We can write it as a function of x and y as

$$\hat{f}(x, y) = f(x + jy) = x^2 + y^2. \tag{12.111}$$

Since it is a function of two variables, the derivatives in principle depend on the direction we are moving along. Consider the derivative at a point w along the x-axis, that is, keeping y constant:

$$\lim_{\epsilon \to 0} \frac{(x+\epsilon)^2 + y^2 - (x^2 + y^2)}{\epsilon} = 2x.$$

On the other hand, along the y-axis, we have

$$\lim_{\epsilon \to 0} \frac{x^2 + (y+\epsilon)^2 - (x^2 + y^2)}{\epsilon} = 2y,$$

that is, the derivative at point w is different according to the direction we are moving. This is essentially what is meant by saying that the function is not analytic at w (see for example [202] for a more complete discussion about analytic functions).

Consider now the analytic function $g = w^2$. Its derivative is the same, no matter which direction we take:

$$\lim_{\epsilon \to 0} \frac{(x + \epsilon + jy)^2 - (x + jy)^2}{\epsilon} \\ = \lim_{\epsilon \to 0} \frac{(x+\epsilon)^2 - y^2 + 2j(xy + \epsilon y) - (x^2 - y^2 + 2jxy)}{\epsilon} = 2w.$$

Similarly, differentiating along the y-axis we obtain again the same result

$$\lim_{\epsilon \to 0} \frac{(x + j(y+\epsilon))^2 - (x + jy)^2}{\epsilon} = 2w.$$

When the derivative at a certain point w_0 is the same along every direction, we can define derivatives of complex functions writing simply

$$\frac{\mathrm{d}g(w_0)}{\mathrm{d}w} = \lim_{\Delta w \to 0} \frac{g(w_0 + \Delta w) - g(w_0)}{\Delta w}, \tag{12.112}$$

since we know the limit does not depend on the direction along which Δw goes to zero.

It can be shown that a function is analytic at a given point only if the *Cauchy–Riemann conditions* hold at that point. Separating a generic complex function g into its real and imaginary parts $g(w) = u(x, y) + jv(x, y)$, the Cauchy–Riemann conditions are [202]

$$\frac{\partial u}{\partial x} = \frac{\partial v}{\partial y}, \qquad \frac{\partial u}{\partial y} = -\frac{\partial v}{\partial x}.$$

In the case of a *real* function of a complex variable, such as f defined as in (12.110), $v \equiv 0$, and of course the conditions could hold only for points in which the real part has zero partial derivatives.

This means that we cannot search for the stationary points of a function such as (12.110) using the derivative defined as in (12.112). We could of course always go back and interpret the function as a function of two variables, as we did in (12.111). In order to find the minimum of $f(w)$, we would need to find the stationary points of $\hat{f}(x, y)$,

$$\frac{\partial \hat{f}}{\partial x} = 0, \qquad \frac{\partial \hat{f}}{\partial y} = 0.$$

This works just fine, of course, but there is a nice trick, which allows us to keep working with complex variables in a simple way. The idea is to define derivatives differently, such that optimization problems become easy to solve, using what is known as *Wirtinger calculus*.

Real functions of complex variables are usually defined in terms of variables and its conjugates, since $w + w^*$ and ww^* are both real. The idea is to define derivatives such that w and w^* can be treated as if they were independent variables, as follows. Despite the motivation, the new derivatives are defined for general functions $g : \mathbb{C} \to \mathbb{C}$.

Define partial derivatives of a function $g : \mathbb{C} \to \mathbb{C}$ with respect to w and w^* as

$$\frac{\partial g}{\partial w} = \frac{1}{2}\left\{\frac{\partial g}{\partial x} - j\frac{\partial g}{\partial y}\right\}, \qquad \frac{\partial g}{\partial w^*} = \frac{1}{2}\left\{\frac{\partial g}{\partial x} + j\frac{\partial g}{\partial y}\right\}. \tag{12.113}$$

One can check that if g is analytic at a given point (so g satisfies the Cauchy–Riemann conditions at that point), then $\partial g/\partial w^* = 0$ at that point.

Consider, for example, $g(w) = w^m (w^*)^n = (x + jy)^m (x - jy)^n$. Then,

$$\begin{aligned}
\frac{\partial g}{\partial w} &= \frac{1}{2}\Big\{m(x + jy)^{m-1}(x - jy)^n + n(x + jy)^m (x - jy)^{n-1} \\
&\quad - j\left(jm(x + jy)^{m-1}(x - jy)^n - jn(x + jy)^m (x - jy)^{n-1}\right)\Big\} \\
&= \frac{1}{2}\Big\{mw^{m-1}(w^*)^n + nw^m (w^*)^{n-1} + mw^{m-1}(w^*)^n - nw^m (w^*)^{n-1}\Big\} \\
&= mw^{m-1}(w^*)^n.
\end{aligned}$$

As advertised, the final result is what we would obtain if we had treated w and w^* as two independent variables. Similarly, one can check that

$$\frac{\partial g}{\partial w^*} = n w^m (w^*)^{n-1}.$$

An important special case is the quadratic function $f(w) = d - w^* r - r^* w + w^* w R$, which assumes only real values if d and R are real. Using the previous result, we see that

$$\frac{\partial f}{\partial w} = -r^* + w^* R, \qquad \frac{\partial f}{\partial w^*} = -r + wR.$$

Note that there is no factor of 2 in this case. This difference between Wirtinger derivatives and standard real derivatives will propagate to many expressions in this text.

Now, the definitions (12.113) are useful for minimizing real functions of complex variables as in (12.110), since from (12.113)

$$\frac{\partial f}{\partial w^*} = 0 \Leftrightarrow \frac{\partial \hat{f}}{\partial x} = 0 \text{ and } \frac{\partial \hat{f}}{\partial y} = 0.$$

Applying this idea to our quadratic function, we see that

$$\frac{\partial f}{\partial w^*} = 0 \Rightarrow R w_o = r \Rightarrow w_o = r/R,$$

if $R \neq 0$. In addition,

$$\frac{\partial^2 f}{\partial w \partial w^*} = R,$$

and we can conclude that w_o is a minimum as long as $R > 0$.

The same conclusions could easily be obtained working directly with $\hat{f}$. The Wirtinger derivatives are merely a shortcut that makes working with complex variables very similar to working with real variables. A good detailed description of Wirtinger derivatives can be found in [190].

In the vector case, the definitions and results are very similar. Given a function

$$g : \mathbb{C}^M \to \mathbb{R},$$

$$\boldsymbol{w} = \begin{bmatrix} w_0 \\ w_1 \\ \vdots \\ w_{M-1} \end{bmatrix} \in \mathbb{C}^M \mapsto g(\boldsymbol{w}),$$

with $w_i = x_i + j y_i$, $i = 0 \dots M-1$, we define

$$\frac{\partial g}{\partial \boldsymbol{w}} = \begin{bmatrix} \frac{\partial g}{\partial w_0} & \cdots & \frac{\partial g}{\partial w_{M-1}} \end{bmatrix} = \frac{1}{2} \begin{bmatrix} \frac{\partial g}{\partial x_0} - j \frac{\partial g}{\partial y_0} & \cdots & \frac{\partial g}{\partial x_{M-1}} - j \frac{\partial g}{\partial y_{M-1}} \end{bmatrix},$$

$$\frac{\partial g}{\partial \boldsymbol{w}^H} = \begin{bmatrix} \frac{\partial g}{\partial w_0^*} \\ \vdots \\ \frac{\partial g}{\partial w_{M-1}^*} \end{bmatrix} = \frac{1}{2} \begin{bmatrix} \frac{\partial g}{\partial x_0} + j \frac{\partial g}{\partial y_0} \\ \vdots \\ \frac{\partial g}{\partial x_{M-1}} + j \frac{\partial g}{\partial y_{M-1}} \end{bmatrix}.$$

Using our definition of gradients, the second derivative of f (its *Hessian*) is

$$\frac{\partial^2 f}{\partial \boldsymbol{w} \partial \boldsymbol{w}^H} = \begin{bmatrix} \frac{\partial}{\partial \boldsymbol{w}} \left\{ \frac{\partial f}{\partial w_0^*} \right\} \\ \vdots \\ \frac{\partial}{\partial \boldsymbol{w}} \left\{ \frac{\partial f}{\partial w_{M-1}^*} \right\} \end{bmatrix} = \begin{bmatrix} \frac{\partial^2 f}{\partial w_0 \partial w_0^*} & \cdots & \frac{\partial^2 f}{\partial w_{M-1} \partial w_0^*} \\ \vdots & \ddots & \vdots \\ \frac{\partial^2 f}{\partial w_0 \partial w_{M-1}^*} & \cdots & \frac{\partial^2 f}{\partial w_{M-1} \partial w_{M-1}^*} \end{bmatrix}.$$

The most important example is the quadratic function

$$f(\boldsymbol{w}) = r_d - \boldsymbol{w}^H \boldsymbol{r}_{d\phi} - \boldsymbol{r}_{d\phi}^H \boldsymbol{w} + \boldsymbol{w}^H \boldsymbol{R}_\phi \boldsymbol{w},$$

where r_d is a real variable and $\boldsymbol{R}_\phi$ is Hermitian symmetric. Then, we have

$$\frac{\partial f}{\partial \boldsymbol{w}^H} = -\boldsymbol{r}_{d\phi} + \boldsymbol{R}_\phi \boldsymbol{w}.$$

This example is the reason for the notation $\partial f / \partial \boldsymbol{w}^H$ that we have been using for the gradient. We see that a stationary point of f is such that $\boldsymbol{R}_\phi \boldsymbol{w}_o = \boldsymbol{r}_{d\phi}$. In addition,

$$\frac{\partial^2 f}{\partial \boldsymbol{w} \partial \boldsymbol{w}^H} = \boldsymbol{R}_\phi,$$

and we conclude that $\boldsymbol{w}_o$ is the single point of minimum of f if and only if $\boldsymbol{R}_\phi$ is positive definite.

Note that in the complex case, the gradient and the Hessian do not have the factors of 2 that we saw in the real case (see Box 12.1.)

12.3 Stochastic algorithms

In this section, we focus on three of the most important adaptive filters: the LMS algorithm, the NLMS algorithm, and the RLS algorithm, which serve as the basis of many others. These three stochastic algorithms are derived in detail and analyzed using different methods. We present different approaches for analyzing an adaptive filter, because the variety and complexity of their behavior makes it difficult for a single technique to be able to handle all situations. The analyses provided here allow us to predict the behavior of the algorithms in transient and steady state, and choose values of parameters (such as the step size) for which the filters operate adequately. In particular, we study conditions under which the filters will remain stable.

We assume linearly parameterized classes, such as FIR ($\mathcal{F}_{\text{FIR}}$) and Volterra ($\mathcal{F}_{\text{V}}$) filters (see Section 12.2.1). In these classes, each element of the input regressor vector is given by a certain function $\phi_i, i = 0, 1, \ldots, M-1$, of $x(n),\ x(n-1),\ \ldots,\ x(n-N)$. In the case of length-$M$ FIR filters, we have

$$\begin{aligned}\boldsymbol{\phi}(n) &= \begin{bmatrix} \phi_0(n) & \phi_1(n) & \cdots & \phi_{M-1}(n) \end{bmatrix}^T \\ &= \begin{bmatrix} x(n) & x(n-1) & \cdots & x(n-M+1) \end{bmatrix}^T,\end{aligned} \tag{12.114}$$

whereas in the case of second-order Volterra filters with memory $N = 1$ and real-valued signals, we have

$$\begin{aligned}\boldsymbol{\phi}(n) &= \begin{bmatrix} \phi_0(n) & \phi_1(n) & \phi_2(n) & \phi_3(n) & \phi_4(n) \end{bmatrix}^T \\ &= \begin{bmatrix} x(n) & x(n-1) & x^2(n) & x(n)x(n-1) & x^2(n-1) \end{bmatrix}^T.\end{aligned} \tag{12.115}$$

As much as possible, our presentation in this section is independent of the choices of $\mathcal{F}$ and $\phi(\cdot)$. However, some algorithms are designed for a specific class $\mathcal{F}$ (the most common case are fast algorithms designed for FIR filters (12.114)). In these cases, we will alert the reader of the restriction.

Using the results of Box 12.6, the derivation of algorithms for real or circular complex inputs is very similar, the only differences being the need of conjugating a few variables in some expressions and the appearance of a factor $\beta = 2$ in the expressions of gradients of functions of real variables and $\beta = 1$ in the complex case.

In general, the output of a length-M adaptive filter is given by

$$\hat{y}(n) = \boldsymbol{w}^H(n)\boldsymbol{\phi}(n) = w_0^*(n)\phi_0(n) + w_1^*(n)\phi_1(n) + \cdots + w_{M-1}^*(n)\phi_{M-1}(n). \tag{12.116}$$

In the case of Volterra filters, it is common to use the notation $w_{k,\ell}$ for the weight related to the input $x(n-k)x(n-\ell)$ (see Eq. (12.44)). However, to obtain a common formulation independent of the class $\mathcal{F}$, we denote the weight vector as

$$\boldsymbol{w}(n) = \begin{bmatrix} w_0(n) & w_1(n) & w_2(n) & \cdots & w_{M-1}(n) \end{bmatrix}^T. \tag{12.117}$$

The task of the adaptive filter will be to choose in a recursive form the values of the parameters $w_0,\ w_1,\ \ldots,\ w_{M-1}$ that best fit the data at each time instant n.

Fig. 12.25 shows the scheme of a length-M adaptive filter for linearly parameterized classes, where Φ denotes the set of functions $\phi_i(\cdot)$, $i = 0, 1, \ldots, M-1$, of $x(n),\ x(n-1),\ \ldots,\ x(n-N)$ that generate the elements $\phi_0(n), \phi_1(n), \ldots, \phi_{M-1}(n)$ of the input regressor vector $\boldsymbol{\phi}(n)$. The number of samples of the signal $x(n)$ needed to form $\phi_m(n)$ for $0 \le m < M$ (the *memory* of the filter) depends on the class $\mathcal{F}$ and is given by $N+1$. An FIR filter with length M requires $N+1 = M$ samples. On the other hand, a Volterra filter with memory N requires $N+1$ samples, but the length of the corresponding adaptive filter depends on the order of the Volterra series expansion – in the example above in (12.115), $N = 1$ and $M = 5$. At each time instant, the output of the adaptive filter $\hat{y}(n)$ is compared with the desired signal $d(n)$ to compute the error $e(n) = d(n) - \hat{y}(n)$. In order to minimize a cost function of this error, a stochastic algorithm uses $e(n)$ to adjust the filter weights.

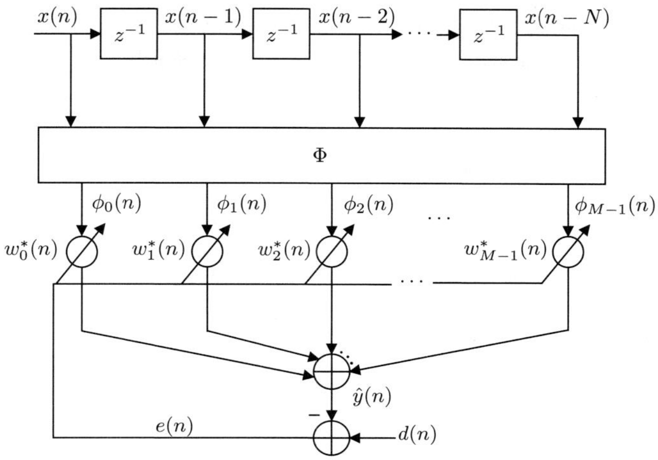

FIGURE 12.25

Scheme for a length-M adaptive filter assuming linearly parameterized classes.

12.3.1 LMS algorithm

The update equation of the steepest descent algorithm is given by

$$\boldsymbol{w}(n+1) = \boldsymbol{w}(n) - \frac{\mu}{\beta}\nabla_{\boldsymbol{w}} J(n), \tag{12.118}$$

where μ is a step size and $\beta = 1$ (resp. $\beta = 2$) for complex (resp. real) data (see Box 12.6). This algorithm requires exact measurements of the gradient vector. For the case of minimization of the MSE, that is, if $J(n) = \mathrm{E}\{|e(n)|^2\}$,

$$\nabla_{\boldsymbol{w}} J(n) = -\beta \boldsymbol{r}_{d\boldsymbol{\phi}} + \beta \boldsymbol{R}_{\boldsymbol{\phi}} \boldsymbol{w}(n) \tag{12.119}$$

at each iteration n (see Box 12.1 for an introduction to gradient algorithms). Note that the exact gradient requires prior knowledge of the correlation matrix $\boldsymbol{R}_{\boldsymbol{\phi}}$ and of the cross-correlation vector $\boldsymbol{r}_{d\boldsymbol{\phi}}$, which is not feasible in real-time applications. For example, in the hands-free virtual meeting application described in Section 12.1.1, the input signal to the adaptive filter is the far-end speech (see Figs. 12.3 and 12.4). A speech signal is nonstationary, and therefore it is not possible to obtain exact estimates for $\boldsymbol{R}_{\boldsymbol{\phi}}$ and $\boldsymbol{r}_{d\boldsymbol{\phi}}$ at each time instant. Furthermore, good estimates would demand much processing time, which could insert an inadmissible delay in the system, making this approach not feasible, except in off-line applications. Similar restrictions appear in virtually all adaptive filtering applications.

To circumvent this problem, a stochastic version of the steepest descent algorithm was proposed, the LMS algorithm [387]. Instead of minimizing the exact MSE, the LMS algorithm minimizes a straightforward approximation to it: $|e(n)|^2$. The idea, as we saw in Section 12.1.2.4, is that the algorithm will approximately average the cost function if adaptation is slow (small step size). The estimation error is (in the equations that follow, note that if x and $\boldsymbol{y}$ are real, then $x^* = x$ and $\boldsymbol{y}^H = \boldsymbol{y}^T$, so the equations are valid for both real and complex signals)

$$e(n) = d(n) - \boldsymbol{w}^H(n)\boldsymbol{\phi}(n), \tag{12.120}$$

so the gradient of the instantaneous cost function is

$$\widehat{\nabla}_{\boldsymbol{w}} J(n) = -\beta\boldsymbol{\phi}(n)d^*(n) + \beta\boldsymbol{\phi}(n)\boldsymbol{\phi}^H(n)\boldsymbol{w}(n) = -\beta\boldsymbol{\phi}(n)e^*(n). \tag{12.121}$$

Since $e(n)$ depends on the old estimate of the weight vector, it is called an a priori estimation error, as opposed to the a posteriori error, which will be introduced in (12.130). Instead of minimizing the instantaneous cost function, the stochastic gradient defined in (12.121) can be obtained by replacing the instantaneous estimates

$$\widehat{\boldsymbol{R}}_{\boldsymbol{\phi}}(n) = \boldsymbol{\phi}(n)\boldsymbol{\phi}^H(n) \tag{12.122}$$

and

$$\widehat{\boldsymbol{r}}_{d\boldsymbol{\phi}}(n) = \boldsymbol{\phi}(n)d^*(n) \tag{12.123}$$

in (12.119).

Replacing (12.121) in (12.118), we arrive at the update equation for LMS, i.e.,

$$\boldsymbol{w}(n+1) = \boldsymbol{w}(n) + \mu\boldsymbol{\phi}(n)e^*(n). \tag{12.124}$$

As explained in Section 12.1.2.4 and illustrated in the animations of Fig. 12.9, although there is no explicit average being taken in the LMS algorithm, it in fact computes an implicit approximate average if the step size is small.

Remark 12.8. We introduced LMS through instantaneous approximations (12.119) to the autocorrelation matrix $\boldsymbol{R}_{\boldsymbol{\phi}}$ and the cross-correlation vector $\boldsymbol{r}_{d\boldsymbol{\phi}}$, but we can check that the same result is obtained if we apply the gradient algorithm using as cost function the instantaneous square error, $|e(n)|^2$, instead of the average square error $\mathrm{E}\{|e(n)|^2\}$.

The LMS algorithm is summarized in Table 12.2. Its computational cost increases linearly with the length of the filter, i.e., the computational cost is $\mathcal{O}(M)$. Table 12.3 shows the number of real multiplications and real additions per iteration for real- and complex-valued signals. The quantities inside brackets were computed in previous steps. We assume that a complex multiplication requires four real multiplications and two real additions, and a complex addition requires two real additions (it is possible to use fewer multiplications and more additions). Different ways of carrying out the calculations may result in a different computational cost. For example, if we first computed $\mu \times \boldsymbol{\phi}(n)$ followed by $[\mu\boldsymbol{\phi}(n)] \times e^*(n)$, LMS would require $M-1$ more real multiplications per iteration than the ordering of Table 12.3, considering real-valued signals. For complex-valued signals, the cost would be even higher, since LMS would require $4M-2$ more real multiplications and $2M$ more real additions.

Table 12.2 Summary of the LMS algorithm.

Initialization:
Set $w(0) = 0$
For $n = 0, 1, 2, \ldots$, compute
$\hat{y}(n) = \boldsymbol{w}^H(n)\boldsymbol{\phi}(n)$
$e(n) = d(n) - y(n)$
$\boldsymbol{w}(n+1) = \boldsymbol{w}(n) + \mu e^*(n)\boldsymbol{\phi}(n)$

Table 12.3 Computational cost of the LMS algorithm for complex- and real-valued signals in terms of real multiplications and real additions per iteration.

Term	**Real signals**		**Complex signals**	
	×	+	×	+
$\hat{y}^*(n) = \boldsymbol{\phi}^H(n)\,[\boldsymbol{w}(n)]$	M	$M-1$	$4M$	$4M-2$
$e^*(n) = d^*(n) - \left[\hat{y}^*(n)\right]$		1		2
$\mu\left[e^*(n)\right]$	1		2	
$\left[\mu e^*(n)\right]\boldsymbol{\phi}(n)$	M		$4M$	$2M$
$\boldsymbol{w}(n+1) = [\boldsymbol{w}(n)] + \left[\mu e^*(n)\boldsymbol{\phi}(n)\right]$		M		$2M$
Total per iteration	$2M+1$	$2M$	$8M+2$	$8M$

12.3.1.1 A deterministic approach for the stability of LMS

Finding the range of step sizes such that the recursion (12.124) converges is a complicated task. There are two main approaches to finding the maximum step size: an average approach and a deterministic (worst-case) approach. The average approach will be treated in Section 12.4.3. Since the deterministic approach is easy to explain intuitively, we address it in the sequel (although a precise proof is not so simple; see Refs. [328,353] for a full analysis).

For simplicity, assume that our filter is such that $d(n) \equiv 0$. Although this is not a very likely situation to encounter in practice, it allows us to arrive at the important conclusions, but avoiding technical details that would considerably complicate the discussion.

Let us rewrite the LMS recursion, expanding the error with $d(n) = 0$ as $e^*(n) = -(\boldsymbol{w}^H(n)\boldsymbol{\phi}(n))^* = -\boldsymbol{\phi}^H(n)\boldsymbol{w}(n)$ to obtain

$$\boldsymbol{w}(n+1) = \left[\boldsymbol{I} - \mu\boldsymbol{\phi}(n)\boldsymbol{\phi}^H(n)\right]\boldsymbol{w}(n). \tag{12.125}$$

This is a linear system, but not a time-invariant one. It will be stable if the eigenvalues $\lambda_i(n)$ of the matrix $\boldsymbol{A}(n) = \boldsymbol{I} - \mu\boldsymbol{\phi}(n)\boldsymbol{\phi}^H(n)$ satisfy

$$\max_i |\lambda_i(n)| \le 1, \quad \text{for all } n \ge 0. \tag{12.126}$$

We have to be careful; if $d(n)$ were not identically zero, condition (12.126) would have to be slightly modified to guarantee stability (see, e.g., [14,353]).

The eigenvalues of $\boldsymbol{A}(n)$ are easy to find if we note that $\boldsymbol{\phi}(n)$ is an eigenvector of $\boldsymbol{A}(n)$ (the case $\boldsymbol{\phi}(n) = \mathbf{0}$ is trivial):

$$\boldsymbol{A}(n)\boldsymbol{\phi}(n) = \boldsymbol{\phi}(n) - \mu\boldsymbol{\phi}(n)\boldsymbol{\phi}^H(n)\boldsymbol{\phi}(n) = (1 - \mu\|\boldsymbol{\phi}(n)\|^2)\boldsymbol{\phi}(n),$$

where $\|\boldsymbol{\phi}(n)\|^2 = \boldsymbol{\phi}^H(n)\boldsymbol{\phi}(n)$ is the square of the Euclidean norm. The corresponding eigenvalue is $\lambda_1 = 1 - \mu\|\boldsymbol{\phi}(n)\|^2$. All other eigenvalues are equal to one (the eigenvectors are all vectors orthogonal to $\boldsymbol{\phi}(n)$). Therefore, the LMS algorithm (12.125) will be stable whenever

$$-1 \leq 1 - \mu\|\boldsymbol{\phi}(n)\|^2 \leq 1 \Longleftrightarrow 0 \leq \mu \leq \frac{2}{\sup\limits_{n\geq 0} \|\boldsymbol{\phi}(n)\|^2}. \tag{12.127}$$

In this relation we used the *supremum* (sup) of a sequence, which is a generalization of the maximum, defined to avoid a technicality. Consider the sequence $f(n) = \{1 - 0.5^n\}$, $n \geq 0$. Strictly speaking, it has no maximum, since there is no n_0 for which $f(n_0) \geq f(n)$ for all $n \geq 0$. The supremum is defined so we can avoid this problem: $\sup_{n\geq 0} f(n) = 1$. Therefore, the supremum is equal to the maximum whenever the latter exists, but is also defined for some sequences that do not have a maximum. The *infimum* (inf) has a similar relation to the minimum.

This condition is sufficient for stability, but is too conservative, since $\|\boldsymbol{\phi}(n)\|^2$ is not at its highest value at all instants. A popular solution is to use *normalization*. The idea is simple: adopt a time-varying step size $\mu(n)$ such that

$$\mu(n) = \frac{\tilde{\mu}}{\varepsilon + \|\boldsymbol{\phi}(n)\|^2}, \tag{12.128}$$

where $0 < \tilde{\mu} < 2$ and $\varepsilon > 0$ is used to avoid division by zero. The resulting algorithm is known as *normalized LMS* (NLMS). We will talk more about it in the next section.

12.3.2 Normalized LMS algorithm

One problem with the LMS algorithm is how to choose the step size. How large should it be to enable a high convergence rate, provide an acceptable misadjustment, and even ensure the stability of LMS? More importantly, how can this choice be made so that the filter will work correctly even for very different input signals? One answer to this problem is the NLMS algorithm, which we describe now. It uses a time-varying step size, which is particularly useful when the statistics of the input signals change quickly.

Thus, replacing the fixed step size μ by the time-varying step size $\mu(n)$ in (12.124), we obtain

$$\boldsymbol{w}(n+1) = \boldsymbol{w}(n) + \mu(n)\boldsymbol{\phi}(n)e^*(n). \tag{12.129}$$

In order to increase the convergence speed, we could try to find $\mu(n)$ such that the *a posteriori estimation error*,

$$\xi(n) = d(n) - \boldsymbol{w}^H(n+1)\boldsymbol{\phi}(n), \tag{12.130}$$

is zeroed. Note that in contrast to the a priori estimation error defined in (12.120), $\xi(n)$ depends on the updated estimate of the weight vector, hence the name *a posteriori* estimation error. Replacing (12.129) in (12.130), we obtain

$$\xi(n) = \underbrace{d(n) - \boldsymbol{w}^H(n)\boldsymbol{\phi}(n)}_{e(n)} - \mu(n)\boldsymbol{\phi}^H(n)\boldsymbol{\phi}(n)e(n)$$
$$= \left[1 - \mu(n)\boldsymbol{\phi}^H(n)\boldsymbol{\phi}(n)\right] e(n). \tag{12.131}$$

In order to enforce $\xi(n) = 0$ at each time instant n, we must select

$$\mu(n) = \frac{1}{\boldsymbol{\phi}^H(n)\boldsymbol{\phi}(n)} = \frac{1}{\|\boldsymbol{\phi}(n)\|^2}. \tag{12.132}$$

Replacing (12.132) in (12.129), we obtain the update equation of the normalized least-squares algorithm, i.e.,

$$\boldsymbol{w}(n+1) = \boldsymbol{w}(n) + \frac{1}{\|\boldsymbol{\phi}(n)\|^2}\boldsymbol{\phi}(n)e^*(n). \tag{12.133}$$

Note that the time-varying step size $\mu(n)$ depends inversely on the instantaneous power of the input vector $\boldsymbol{\phi}(n)$. Indeed, we will show in Section 12.4.3 that the maximum value of the fixed step size μ to ensure the convergence and mean square stability of LMS depends inversely on the power of the input signal.

In order to make (12.133) more reliable for practical implementations, it is common to introduce two positive constants: a fixed step size $\tilde{\mu}$ to control the rate of convergence (and the misadjustment) and the regularization factor ε to prevent division by a small value when $\|\boldsymbol{\phi}(n)\|^2$ is small. With these constants, the NLMS update equation reduces to

$$\boldsymbol{w}(n+1) = \boldsymbol{w}(n) + \frac{\tilde{\mu}}{\varepsilon + \|\boldsymbol{\phi}(n)\|^2}\boldsymbol{\phi}(n)e^*(n). \tag{12.134}$$

Some authors refer to (12.133) as NLMS and to (12.134) as ε-NLMS. However, for simplicity we will refer to both as NLMS. The name "normalized" is due to a most useful property of NLMS: the range of values of $\tilde{\mu}$ for which the algorithm remains stable is independent of the statistics of $x(n)$, i.e., $0 < \tilde{\mu} < 2$, as can be observed in Section 12.3.1.1.

As pointed out in [317], the NLMS algorithm can also be interpreted as a quasi-Newton algorithm. As briefly explained in Box 12.1, a quasi-Newton algorithm updates the weights using approximations for the Hessian matrix and gradient vector, i.e.,

$$\boldsymbol{w}(n+1) = \boldsymbol{w}(n) - \frac{\mu}{\beta}\left[\widehat{\nabla}^2_{\boldsymbol{w}} J(n)\right]^{-1} \widehat{\nabla}_{\boldsymbol{w}} J(n). \tag{12.135}$$

Assuming the following instantaneous approximations,

$$\widehat{\nabla}_{\boldsymbol{w}} J(n) = -\beta\boldsymbol{\phi}(n)d^*(n) + \beta\boldsymbol{\phi}(n)\boldsymbol{\phi}^H(n)\boldsymbol{w}(n) = -\beta\boldsymbol{\phi}(n)e^*(n) \tag{12.136}$$

and

$$\widehat{\nabla}^2_{\boldsymbol{w}} J(n) = \varepsilon\boldsymbol{I} + \boldsymbol{\phi}(n)\boldsymbol{\phi}^H(n), \tag{12.137}$$

and replacing them in (12.135), we arrive at the stochastic recursion

$$\boldsymbol{w}(n+1) = \boldsymbol{w}(n) + \mu\left[\varepsilon\boldsymbol{I} + \boldsymbol{\phi}(n)\boldsymbol{\phi}^H(n)\right]^{-1}\boldsymbol{\phi}(n)e^*(n). \tag{12.138}$$

The term $\varepsilon\boldsymbol{I}$ guarantees that (12.137) is invertible (a rank-one matrix such as $\boldsymbol{\phi}(n)\boldsymbol{\phi}^H(n)$ is never invertible).

To compute $\left[\varepsilon\boldsymbol{I} + \boldsymbol{\phi}(n)\boldsymbol{\phi}^H(n)\right]^{-1}$, we can use the *matrix inversion lemma* (Box 12.3), which gives us an expression for the inverse of an $M \times M$ matrix of the form

$$\boldsymbol{A} = \boldsymbol{B} + \boldsymbol{C}\boldsymbol{D}\boldsymbol{C}^H, \tag{12.139}$$

where $\boldsymbol{B}$ and $\boldsymbol{D}$ are two square matrices with dimensions M and N, respectively, and $\boldsymbol{C}$ is an $M \times N$ matrix. According to the matrix inversion lemma, the inverse of $\boldsymbol{A}$ is given by

$$\boldsymbol{A}^{-1} = \boldsymbol{B}^{-1} - \boldsymbol{B}^{-1}\boldsymbol{C}(\boldsymbol{D}^{-1} + \boldsymbol{C}^H\boldsymbol{B}^{-1}\boldsymbol{C})^{-1}\boldsymbol{C}^H\boldsymbol{B}^{-1}. \tag{12.140}$$

Thus, identifying

$$\boldsymbol{A} = \varepsilon\boldsymbol{I} + \boldsymbol{\phi}(n)\boldsymbol{\phi}^H(n), \qquad \boldsymbol{B} = \varepsilon\boldsymbol{I}, \qquad \boldsymbol{C} = \boldsymbol{\phi}(n), \quad \text{and} \quad \boldsymbol{D} = 1,$$

we obtain

$$\left[\varepsilon\boldsymbol{I} + \boldsymbol{\phi}(n)\boldsymbol{\phi}^H(n)\right]^{-1} = \varepsilon^{-1}\boldsymbol{I} - \frac{\varepsilon^{-2}}{1 + \varepsilon^{-1}\boldsymbol{\phi}^H(n)\boldsymbol{\phi}(n)}\boldsymbol{\phi}(n)\boldsymbol{\phi}^H(n). \tag{12.141}$$

Multiplying (12.141) from the right by $\boldsymbol{\phi}(n)$, after some algebraic manipulations, we get

$$\left[\varepsilon\boldsymbol{I} + \boldsymbol{\phi}(n)\boldsymbol{\phi}^H(n)\right]^{-1}\boldsymbol{\phi}(n) = \frac{\boldsymbol{\phi}(n)}{\varepsilon + \boldsymbol{\phi}^H(n)\boldsymbol{\phi}(n)} = \frac{\boldsymbol{\phi}(n)}{\varepsilon + \|\boldsymbol{\phi}(n)\|^2}. \tag{12.142}$$

Replacing (12.142) in (12.138), we arrive at (12.134). Since Newton-based algorithms converge faster than gradient-based algorithms, it is expected that NLMS exhibits a higher convergence rate than that of LMS, which is indeed the case in general.

The NLMS algorithm is summarized in Table 12.4. To obtain its computational cost, we can use the same procedure considered in the computation of the number of operations of LMS (see Table 12.3). For complex-valued signals, NLMS requires $10M + 2$ real multiplications, $10M$ real additions, and one real division per iteration. For real-valued signals, it requires $3M + 1$ real multiplications and $3M$ real additions per iteration. Note that for input vectors with a shift structure such as the FIR filters of Eq. (12.114), its computational cost can be reduced if $\|\boldsymbol{\phi}(n)\|^2$ is computed by the following recursion

$$\|\boldsymbol{\phi}(n)\|^2 = \|\boldsymbol{\phi}(n-1)\|^2 - |x(n-M)|^2 + |x(n)|^2 \tag{12.143}$$

with initialization $\|\boldsymbol{\phi}(-1)\| = 0$.

Table 12.4 Summary of the NLMS algorithm.

Initialization:
Set $\boldsymbol{w}(0) = \mathbf{0}$ and $0 < \tilde{\mu} < 2$
For $n = 0, 1, 2, \ldots$, compute
$\hat{y}(n) = \boldsymbol{w}^H(n)\boldsymbol{\phi}(n)$
$e(n) = d(n) - \hat{y}(n)$
$\boldsymbol{w}(n+1) = \boldsymbol{w}(n) + \dfrac{\tilde{\mu}}{\varepsilon + \Vert\boldsymbol{\phi}(n)\Vert^2} e^*(n)\boldsymbol{\phi}(n)$

12.3.3 RLS algorithm

The convergence rate of LMS type algorithms varies considerably with the input signal $x(n)$, since they are based on steepest descent optimization algorithms. If $x(n)$ is a white noise, the convergence rate is high. On the other hand, if the correlation between successive samples of $x(n)$ is high, LMS type algorithms converge slowly. The RLS algorithm does not have this drawback. However, this advantage has a price: a considerable increase in the computational cost. Furthermore, numerical errors play a more important role in RLS and may even lead to divergence of the weight estimates. Different versions of RLS were proposed to circumvent these problems. In the sequel, we obtain the equations of the conventional RLS algorithm.

The RLS can be derived in different ways (much as NLMS). Each form of arriving at the algorithm highlights a different set of its properties: one that leads to many new algorithms and insights is the connection between RLS and Kalman filters described in [319].

We present here shorter routes, starting with a derivation based on deterministic arguments. In fact, RLS solves a deterministic least-squares problem, which is based on the weighted least-squares cost function

$$J_{\text{LS}}(n) = \sum_{\ell=0}^{n} \lambda^{n-\ell} |\xi(\ell)|^2, \tag{12.144}$$

where $0 \ll \lambda < 1$ is a constant known as *forgetting factor*,

$$\xi(\ell) = d(\ell) - \boldsymbol{w}^H \boldsymbol{\phi}(\ell) \tag{12.145}$$

are a posteriori errors, and

$$\boldsymbol{w}(n+1) = \arg\min_{\boldsymbol{w}} J_{\text{LS}}(n). \tag{12.146}$$

The factor $\lambda^{n-\ell}$ emphasizes the most recent errors, forgetting the errors from the remote past. This is important; otherwise, the algorithm would give too much weight to the remote past and would not be able to track variations in the input signals. Note that $\boldsymbol{w}(n+1)$ minimizes $J_{\text{LS}}(n)$ at each time instant. Thus, RLS provides an exact solution for the least-squares problem at each time instant. As the NLMS algorithm, RLS seeks to minimize the a posteriori error $|\xi(n)|^2$, searching for an exact solution to an optimization problem. The difference is that RLS searches for a weight vector $\boldsymbol{w}(n+1)$ that takes into consideration the whole past history of inputs, minimizing $\lambda^n|\xi(0)|^2 + \lambda^{n-1}|\xi(1)|^2 + \cdots + \lambda|\xi(n-1)|^2 + |\xi(n)|^2$ at each time instant n, not only the current value $|\xi(n)|^2$ as in NLMS.

The error sequence $\xi(\ell)$ weighted by $\lambda^{(n-\ell)/2}$ in the interval $0 \le \ell \le n$ can be written in vectorial form, i.e.,

$$\boldsymbol{\xi}(n) = \boldsymbol{d}(n) - \boldsymbol{\Phi}(n)\boldsymbol{w}^*, \tag{12.147}$$

where

$$\boldsymbol{\xi}(n) = \begin{bmatrix} \xi(n) & \lambda^{1/2}\xi(n-1) & \cdots & \lambda^{n/2}\xi(0) \end{bmatrix}^T, \tag{12.148}$$

$$\boldsymbol{d}(n) = \begin{bmatrix} d(n) & \lambda^{1/2}d(n-1) & \cdots & \lambda^{n/2}d(0) \end{bmatrix}^T, \tag{12.149}$$

and

$$\boldsymbol{\Phi}(n) = \begin{bmatrix} \phi_0(n) & \phi_1(n) & \cdots & \phi_{M-1}(n) \\ \lambda^{1/2}\phi_0(n-1) & \lambda^{1/2}\phi_1(n-1) & \cdots & \lambda^{1/2}\phi_{M-1}(n-1) \\ \lambda\phi_0(n-2) & \lambda\phi_1(n-2) & \cdots & \lambda\phi_{M-1}(n-2) \\ \vdots & \ddots & \ddots & \vdots \\ \lambda^{n/2}\phi_0(0) & \lambda^{n/2}\phi_1(0) & \cdots & \lambda^{n/2}\phi_{M-1}(0) \end{bmatrix}. \tag{12.150}$$

Note that the weighted least-squares cost function can be rewritten in terms of the vector $\boldsymbol{\xi}(n)$, i.e.,

$$J_{\text{LS}}(n) = \boldsymbol{\xi}^H(n)\boldsymbol{\xi}(n) = \|\boldsymbol{\xi}(n)\|^2 = \sum_{\ell=0}^{n} \lambda^{n-\ell}|\xi(\ell)|^2. \tag{12.151}$$

Now, we have to find the weight vector $\boldsymbol{w}(n+1)$ that minimizes (12.151). The gradient of $J_{\text{LS}}(n)$ with relation to $\boldsymbol{w}^H$ is given by

$$\nabla_{\boldsymbol{w}} J_{\text{LS}}(n) = -\beta\boldsymbol{\Phi}^T(n)\boldsymbol{d}^*(n) + \beta\boldsymbol{\Phi}^T(n)\boldsymbol{\Phi}^*(n)\boldsymbol{w}, \tag{12.152}$$

where as usual, $\beta = 1$ (resp. $\beta = 2$) for complex (resp. real) data. Defining

$$\widehat{\boldsymbol{R}}_{\boldsymbol{\phi}}(n) = \boldsymbol{\Phi}^T(n)\boldsymbol{\Phi}^*(n) = \sum_{\ell=0}^{n} \lambda^{n-\ell}\boldsymbol{\phi}(\ell)\boldsymbol{\phi}^H(\ell) \tag{12.153}$$

and

$$\widehat{\boldsymbol{r}}_{d\boldsymbol{\phi}}(n) = \boldsymbol{\Phi}^T(n)\boldsymbol{d}^*(n) = \sum_{\ell=0}^{n} \lambda^{n-\ell}d^*(\ell)\boldsymbol{\phi}(\ell), \tag{12.154}$$

in which $\boldsymbol{\phi}(n)$ is defined as before, (12.152) can be rewritten as

$$\nabla_{\boldsymbol{w}} J_{\text{LS}}(n) = -\beta\widehat{\boldsymbol{r}}_{d\boldsymbol{\phi}}(n) + \beta\widehat{\boldsymbol{R}}_{\boldsymbol{\phi}}(n)\boldsymbol{w}. \tag{12.155}$$

Equating $\nabla_{\boldsymbol{w}} J_{\mathrm{LS}}(n)$ to the null vector, we get the normal equations

$$\widehat{\boldsymbol{R}}_{\boldsymbol{\phi}}(n)\boldsymbol{w}(n+1) = \widehat{\boldsymbol{r}}_{d\boldsymbol{\phi}}(n). \tag{12.156}$$

Note that these are a deterministic version of the normal equations we saw in Section 12.2.1. Therefore, to minimize $J_{\mathrm{LS}}(n)$, the weight vector $\boldsymbol{w}(n+1)$ must satisfy

$$\boldsymbol{w}(n+1) = \widehat{\boldsymbol{R}}_{\boldsymbol{\phi}}^{-1}(n)\widehat{\boldsymbol{r}}_{d\boldsymbol{\phi}}(n), \tag{12.157}$$

which requires the solution of the linear system of Eqs. (12.156) at each time instant. Note that the solution is guaranteed to exist and to be unique only if $\widehat{\boldsymbol{R}}_{\boldsymbol{\phi}}(n)$ is nonsingular. In particular, the inverse does not exist if $n < M$ (Box 12.3). Assuming $\lambda = 1$ and that the signals $x(n)$ and $d(n)$ are jointly stationary and ergodic, we obtain the following steady-state relations with probability one:

$$\lim_{n\to\infty} \frac{1}{n+1} \widehat{\boldsymbol{R}}_{\boldsymbol{\phi}}(n)\big|_{\lambda=1} = \boldsymbol{R}_{\boldsymbol{\phi}} \quad \text{and} \quad \lim_{n\to\infty} \frac{1}{n+1} \widehat{\boldsymbol{r}}_{d\boldsymbol{\phi}}(n)\big|_{\lambda=1} = \boldsymbol{r}_{d\boldsymbol{\phi}}.$$

Therefore, the deterministic normal equations converge to the stochastic normal equations when $\lambda = 1$, and the weight vector obtained from (12.157) converges to the Wiener solution, i.e.,

$$\lim_{n\to\infty} \boldsymbol{w}(n+1)\big|_{\lambda=1} = \boldsymbol{R}_{\boldsymbol{\phi}}^{-1}\boldsymbol{r}_{d\boldsymbol{\phi}} = \boldsymbol{w}_{\mathrm{o}}. \tag{12.158}$$

We should mention that with $\lambda = 1$ the convergence rate of RLS goes to zero, i.e., it loses the ability to follow the statistical variations of the observed data. An equivalent result is obtained for LMS or NLMS when the fixed step size μ or $\tilde{\mu}$ is replaced by $1/n$ [195].

The computational cost of (12.157) is high due to the solution of the linear system, which requires $\approx M^3/3$ operations to solve at each instant if we use Gauss elimination – incidentally, note that computing the inverse and using (12.157) requires approximately three times as many computations. In order to solve the normal equations in real time and with less cost, we should rewrite (12.153) and (12.154) as recursions, i.e.,

$$\widehat{\boldsymbol{R}}_{\boldsymbol{\phi}}(n) = \lambda\widehat{\boldsymbol{R}}_{\boldsymbol{\phi}}(n-1) + \boldsymbol{\phi}(n)\boldsymbol{\phi}^H(n) \tag{12.159}$$

and

$$\widehat{\boldsymbol{r}}_{d\boldsymbol{\phi}}(n) = \lambda\widehat{\boldsymbol{r}}_{d\boldsymbol{\phi}}(n-1) + d^*(n)\boldsymbol{\phi}(n), \tag{12.160}$$

with initializations $\widehat{\boldsymbol{R}}_{\boldsymbol{\phi}}(-1) = \delta\boldsymbol{I}$ and $\widehat{\boldsymbol{r}}_{d\boldsymbol{\phi}}(-1) = \boldsymbol{0}$, where δ is a small positive constant. Note that (12.160) is identical to (12.154), but (12.159) leads to the estimate

$$\widehat{\boldsymbol{R}}_{\boldsymbol{\phi}}(n) = \lambda^{n+1}\delta\boldsymbol{I} + \sum_{\ell=0}^{n} \lambda^{n-\ell}\boldsymbol{\phi}(\ell)\boldsymbol{\phi}^H(\ell), \tag{12.161}$$

and therefore the initialization guarantees the existence of the inverse at all instants. For $0 \ll \lambda < 1$, the initialization $\widehat{\boldsymbol{R}}_{\boldsymbol{\phi}}(-1) = \delta\boldsymbol{I}$ is forgotten and (12.159) becomes close to (12.153) as n increases. Furthermore, if one wants the algorithm to forget quickly the initialization, besides using $0 \ll \lambda < 1$, one should choose δ small (δ^{-1} large). Note however that a certain amount of regularization may be useful in practice, to avoid problems when the input signal has long stretches with low power.

In order to obtain a recursion for the inverse matrix $\widehat{\boldsymbol{R}}_{\boldsymbol{\phi}}^{-1}(n)$, we can use the matrix inversion lemma (Box 12.3) by comparing (12.139) with (12.159) and identifying

$$\boldsymbol{A} = \widehat{\boldsymbol{R}}_{\boldsymbol{\phi}}(n), \qquad \boldsymbol{B} = \lambda \widehat{\boldsymbol{R}}_{\boldsymbol{\phi}}(n-1), \qquad \boldsymbol{C} = \boldsymbol{\phi}(n), \quad \text{and} \quad \boldsymbol{D} = 1.$$

Thus, using (12.140), we obtain

$$\widehat{\boldsymbol{R}}_{\boldsymbol{\phi}}^{-1}(n) = \lambda^{-1}\widehat{\boldsymbol{R}}_{\boldsymbol{\phi}}^{-1}(n-1) - \frac{\lambda^{-2}\widehat{\boldsymbol{R}}_{\boldsymbol{\phi}}^{-1}(n-1)\boldsymbol{\phi}(n)\boldsymbol{\phi}^H(n)\widehat{\boldsymbol{R}}_{\boldsymbol{\phi}}^{-1}(n-1)}{1+\lambda^{-1}\boldsymbol{\phi}^H(n)\widehat{\boldsymbol{R}}_{\boldsymbol{\phi}}^{-1}(n-1)\boldsymbol{\phi}(n)}. \tag{12.162}$$

Defining $\boldsymbol{P}(n) \triangleq \widehat{\boldsymbol{R}}_{\boldsymbol{\phi}}^{-1}(n)$, with some algebraic manipulations in (12.162), we arrive at

$$\boldsymbol{P}(n) = \frac{1}{\lambda}\left[\boldsymbol{P}(n-1) - \frac{\boldsymbol{P}(n-1)\boldsymbol{\phi}(n)\boldsymbol{\phi}^H(n)\boldsymbol{P}(n-1)}{\lambda + \boldsymbol{\phi}^H\boldsymbol{P}(n-1)\boldsymbol{\phi}(n)}\right] \tag{12.163}$$

with initialization $\boldsymbol{P}(-1) = \delta^{-1}\boldsymbol{I}$. This equation allows us to compute the inverse correlation matrix in a recurrent form, avoiding the explicit computation of a matrix inverse. However, the implementation of (12.163) in finite precision arithmetic requires some care to avoid numerical problems, as explained in Section 12.3.3.1.

To simplify the following equations, it is convenient to define the column vector

$$\boldsymbol{k}(n) = \frac{\boldsymbol{P}(n-1)\boldsymbol{\phi}(n)}{\lambda + \boldsymbol{\phi}^H(n)\boldsymbol{P}(n-1)\boldsymbol{\phi}(n)}, \tag{12.164}$$

which is the so-called Kalman gain. Thus, (12.163) can be rewritten as

$$\boldsymbol{P}(n) = \lambda^{-1}\left[\boldsymbol{I} - \boldsymbol{k}(n)\boldsymbol{\phi}^H(n)\right]\boldsymbol{P}(n-1). \tag{12.165}$$

Using (12.165) and some algebraic manipulations in (12.164), we arrive at

$$\begin{aligned}\boldsymbol{k}(n) &= \lambda^{-1}\boldsymbol{P}(n-1)\boldsymbol{\phi}(n) - \lambda^{-1}\boldsymbol{k}(n)\boldsymbol{\phi}^H(n)\boldsymbol{P}(n-1)\boldsymbol{\phi}(n)\\ &= \lambda^{-1}\left[\boldsymbol{I} - \boldsymbol{k}(n)\boldsymbol{\phi}^H(n)\right]\boldsymbol{P}(n-1)\boldsymbol{\phi}(n)\\ &= \boldsymbol{P}(n)\boldsymbol{\phi}(n).\end{aligned} \tag{12.166}$$

Thus, using (12.160) and (12.166) in (12.157), the weight vector $\boldsymbol{w}(n+1)$ can also be computed recursively, i.e.,

$$\begin{aligned}\boldsymbol{w}(n+1) &= \boldsymbol{P}(n)\widehat{\boldsymbol{r}}_{d\boldsymbol{\phi}}(n)\\ &= \lambda\boldsymbol{P}(n)\widehat{\boldsymbol{r}}_{d\boldsymbol{\phi}}(n-1) + \boldsymbol{P}(n)\boldsymbol{\phi}(n)d^*(n)\\ &= \lambda\boldsymbol{P}(n)\widehat{\boldsymbol{r}}_{d\boldsymbol{\phi}}(n-1) + \boldsymbol{k}(n)d^*(n).\end{aligned} \tag{12.167}$$

Replacing (12.165) in (12.167), we arrive at

$$\begin{aligned} \boldsymbol{w}(n+1) &= \boldsymbol{P}(n-1)\widehat{\boldsymbol{r}}_{d\boldsymbol{\phi}}(n-1) - \boldsymbol{k}(n)\boldsymbol{\phi}^H(n)\boldsymbol{P}(n-1)\widehat{\boldsymbol{r}}_{d\boldsymbol{\phi}}(n-1) + \boldsymbol{k}(n)d^*(n) \\ &= \boldsymbol{w}(n) - \boldsymbol{k}(n)\boldsymbol{\phi}^H(n)\boldsymbol{w}(n) + \boldsymbol{k}(n)d^*(n) \\ &= \boldsymbol{w}(n) + \boldsymbol{k}(n)[\underbrace{d(n) - \boldsymbol{w}^H(n)\boldsymbol{\phi}(n)}_{e(n)}]^*. \end{aligned} \tag{12.168}$$

The RLS algorithm was derived here as an exact solution for the least-squares problem. However, it also can be interpreted as a *quasi*-Newton algorithm assuming the instantaneous approximation for the gradient vector as in (12.136) and considering $\boldsymbol{P}(n)$ as an approximation for the inverse Hessian matrix. Replacing these approximations in (12.135), we arrive at the same stochastic recursion (12.168).

The RLS algorithm is summarized in Table 12.5. Its computational cost increases with M^2, much faster than those of LMS and NLMS. Different ways of carrying out the calculations may result in a slightly different computational cost, but all with the same order of magnitude, i.e., $\mathcal{O}(M^2)$. Performing the calculations in the following order [317]:

$$\begin{aligned}
&\boldsymbol{\phi}^H(n) \times [\boldsymbol{w}(n)], \\
&d^*(n) - \left[\boldsymbol{\phi}^H(n)\boldsymbol{w}(n)\right], \\
&\lambda^{-1}\boldsymbol{\phi}(n), \\
&\boldsymbol{P}(n-1) \times \left[\lambda^{-1}\boldsymbol{\phi}(n)\right], \\
&\boldsymbol{\phi}^H(n) \times \left[\lambda^{-1}\boldsymbol{P}(n-1)\boldsymbol{\phi}(n)\right], \\
&1 + \left[\lambda^{-1}\boldsymbol{\phi}^H(n)\boldsymbol{P}(n-1)\boldsymbol{\phi}(n)\right], \\
&1/\left[1 + \lambda^{-1}\boldsymbol{\phi}^H(n)\boldsymbol{P}(n-1)\boldsymbol{\phi}(n)\right], \\
&\left[\lambda^{-1}\boldsymbol{\phi}^H(n)\boldsymbol{P}(n-1)\boldsymbol{\phi}(n)\right] \times \left[\frac{1}{1 + \lambda^{-1}\boldsymbol{\phi}^H(n)\boldsymbol{P}(n-1)\boldsymbol{\phi}(n)}\right], \\
&\left[\lambda^{-1}\boldsymbol{P}(n-1)\boldsymbol{\phi}(n)\right] \times \left[\frac{\lambda^{-1}\boldsymbol{\phi}^H(n)\boldsymbol{P}(n-1)\boldsymbol{\phi}(n)}{1 + \lambda^{-1}\boldsymbol{\phi}^H(n)\boldsymbol{P}(n-1)\boldsymbol{\phi}(n)}\right], \\
&\boldsymbol{P}(n)\boldsymbol{\phi}(n) = \left[\lambda^{-1}\boldsymbol{P}(n-1)\boldsymbol{\phi}(n)\right] - \left[\frac{\lambda^{-2}\boldsymbol{P}(n-1)\boldsymbol{\phi}(n)\boldsymbol{\phi}^H(n)\boldsymbol{P}(n-1)\boldsymbol{\phi}(n)}{1 + \lambda^{-1}\boldsymbol{\phi}^H(n)\boldsymbol{P}(n-1)\boldsymbol{\phi}(n)}\right], \\
&[\boldsymbol{P}(n)\boldsymbol{\phi}(n)] \times \left[d^*(n) - \boldsymbol{\phi}^H(n)\boldsymbol{w}(n)\right], \\
&\boldsymbol{w}(n+1) = [\boldsymbol{w}(n)] + \left[\boldsymbol{P}(n)\boldsymbol{\phi}(n)\left(d^*(n) - \boldsymbol{\phi}^H(n)\boldsymbol{w}(n)\right)\right],
\end{aligned}$$

the conventional RLS requires $4M^2 + 16M + 1$ real multiplications, $4M^2 + 12M - 1$ real additions, and one real division per iteration for complex-valued signals. For real-valued data, it requires $M^2 + 5M + 1$ real multiplications, $M^2 + 3M$ real additions, and one real division per iteration. Note that the quantities inside brackets were computed in previous steps.

Table 12.5 Summary of the conventional RLS algorithm.

Initialization:
Set $P(-1) = \delta^{-1}I$; δ = small positive constant
$w(0) = 0$ and $0 \ll \lambda < 1$
For $n = 0, 1, 2, \ldots$, compute
$k(n) = \dfrac{P(n-1)\phi(n)}{\lambda + \phi^H(n)P(n-1)\phi(n)}$
$\hat{y}(n) = w^H(n)\phi(n)$
$e(n) = d(n) - \hat{y}(n)$
$w(n+1) = w(n) + k(n)e^*(n)$
$P(n) = \lambda^{-1}\left[I - k(n)\phi^H(n)\right]P(n-1)$

12.3.3.1 *Practical implementation of RLS*

The RLS algorithm must be implemented carefully, because it is sensitive to numerical errors, which may accumulate and make the algorithm unstable. The problem can be understood intuitively as follows. In the conventional RLS algorithm, the inverse of the autocorrelation matrix is computed recursively by (expanding (12.162))

$$P(n) = \lambda^{-1}P(n-1) - \frac{\lambda^{-2}P(n-1)\phi(n)\phi^H(n)P(n-1)}{1 + \lambda^{-1}\phi^H(n)P(n-1)\phi(n)},$$

i.e., as the difference between two positive (semi)definite matrices. Note that since $\lambda \leq 1$, the factor multiplying the current $P(n-1)$ is $\lambda^{-1} \geq 1$. In infinite precision arithmetic, any growth due to this factor will be compensated by the second term. However, in finite precision arithmetic this compensation may not take place, and the factor λ^{-1} may make the recursion numerically unstable – usually, what happens is that $P(n)$ loses its positive definite character, so the matrix will have a negative eigenvalue, which in its turn leads to the divergence of the coefficients. Thus, to avoid problems one might begin by guaranteeing that $P(n)$ stays symmetric and positive definite for all time instants n (these properties are known as *backward consistency* of RLS [298,348]). Symmetry can be guaranteed by computing only the upper or lower triangular part of $P(n)$ and copying the result to obtain the rest of elements. However, it is also necessary to guarantee positive definiteness. Reference [57] describes how this can be achieved.

Due to the fast convergence rate of RLS, a large body of literature is devoted to finding numerically stable and low-cost (i.e., $\mathcal{O}(M)$) versions of RLS. This is not an easy problem, which required the work of numerous researchers to be solved. Nowadays the user has a very large number of different versions of RLS to choose from. There are versions with formal proofs of stability, versions that work well in practice but without formal proofs, versions with $\mathcal{O}(M^2)$ complexity, and versions with $\mathcal{O}(M)$ complexity. The interested reader can find more references in Section 12.5.1.2.

One approach to ensure the consistency of $P(n)$ uses a property of symmetric and positive definite matrices (see Box 12.3): Cholesky factorizations. A symmetric and positive definite matrix $P(n)$ may always be factored as

$$\boldsymbol{P}(n) = \mathcal{P}(n)\mathcal{P}^H(n), \tag{12.169}$$

where $\mathcal{P}(n)$ is a lower triangular $M \times M$ matrix called the Cholesky factor of $\boldsymbol{P}(n)$. Thus, an algorithm that computes $\mathcal{P}(n)$ instead of $\boldsymbol{P}(n)$ can avoid the numerical instability of the conventional RLS algorithm. There are many algorithms based on this main idea; for some of them a precise stability proof is available [298]. A very complete survey of QR-RLS algorithms is available in [19].

Another approach is available when the regressor sequence $\{\boldsymbol{\phi}(n)\}$ has shift structure. In this case, it is possible to derive *lattice*-based RLS filters that are in practice stable, although no formal proof is available, as described, for example, in [248].

A more practical solution for the implementation of the RLS algorithm was proposed in [402]. This approach avoids propagation of the inverse covariance matrix, using an iterative method to solve (12.156) at each step. The computational complexity is kept low by using the previous solution as initialization and by restricting the majority of the multiplications to multiplications by powers of two (which can be implemented as simple shifts) (see Section 12.5.1.3).

12.3.4 Comparing convergence rates

To compare the convergence of LMS, NLMS, and RLS, we next show two examples. In Example 12.E-3.1 we consider a system identification application, and in Example 12.E-3.2 we use these algorithms to update the weights of an equalizer.

As we will see in more detail in Section 12.4, adaptive filters are compared based on how well they handle different classes of signals. For example, we could be interested in seeing how two filters handle white noise or how they handle a certain level of correlation at the input. Given the stochastic nature of the inputs, the performance measures are defined in terms of averaged cost functions, most commonly the MSE, that is, the average of $|e(n)|^2$ over all possible inputs in a given class. These averages can be computed theoretically, as we show in Section 12.4, or via simulations.

For example, the MSE as a function of time, that is, the curve $\mathrm{E}\{|e(n)|^2\} \times n$, is the *learning curve* of the filter. A comparison of learning curves obtained theoretically and from simulations can be seen in Fig. 12.31 further ahead. When simulations are used, one runs a filter L times for an interval of N samples, starting always from the same conditions. For each run ℓ, the error $e^{(\ell)}(n)$ is computed for $1 \le n \le N$, and one obtains the so-called *ensemble average learning curve*

$$\hat{E}(n) \triangleq \frac{1}{L}\sum_{\ell=1}^{L} |e^{(\ell)}(n)|^2.$$

Usually the ensemble average learning curve will be a reasonable approximation of the true learning curve for a reasonably small value of L (say, from 50 to 1000), but in some situations this may not be true (see Refs. [259,262] for examples and a detailed explanation).

Example 12.E-3.1. The aim of this example is to identify the system $\boldsymbol{w}_o = \begin{bmatrix}1 & -0.8\end{bmatrix}^T$, assuming that the regressor $\boldsymbol{\phi}(n)$ is obtained from a process $x(n)$ generated with a first-order autoregressive model, whose transfer function is $1/(1-0.8z^{-1})$. This model is fed with an independent and identically distributed (i.i.d.) Gaussian random process with unitary variance. Moreover, additive i.i.d. noise $v(n)$ with variance $\sigma_v^2 = 0.01$ is added to form the desired signal.

Considering an adaptive filter with $M = 2$ coefficients, we can obtain a contour plot of the MSE cost as a function of w_0 and w_1 as shown in Fig. 12.26. The animations in Fig. 12.26 illustrate the behavior of LMS ($\mu = 0.01$), NLMS ($\tilde{\mu} = 0.05$, $\varepsilon = 0.5$), and RLS ($\lambda = 0.99$, $\delta = 1$) initialized at $\boldsymbol{w}(0) = [-1.5 \quad -1.4]^T$. The correspondent curves of MSE along the iterations, estimated from the ensemble average of 1000 independent runs, are shown in Fig. 12.27. As expected for a colored input signal, RLS converges much faster than NLMS and LMS. NLMS converges faster than LMS and achieves the solution through a different path. To obtain a good behavior of NLMS in this example, we had to choose a relatively large regularization factor ε. This is always necessary when NLMS is used with few coefficients, since in this case $\|\boldsymbol{\phi}(n)\|^2$ has a high probability of becoming too close to zero and destabilize the algorithm. This is shown in the analysis of Section 12.4.

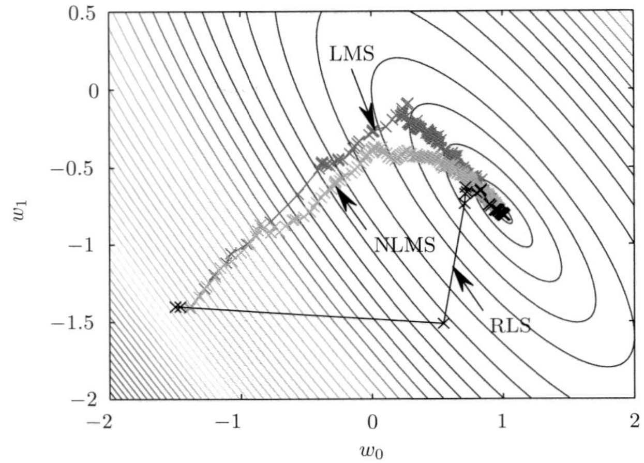

FIGURE 12.26

LMS ($\mu = 0.01$), NLMS ($\tilde{\mu} = 0.05$, $\varepsilon = 0.5$), and RLS ($\lambda = 0.99$, $\delta = 1$) initialized at $\boldsymbol{w}(0) = [-1.5 \quad -1.4]^T$, $\sigma_v^2 = 0.01$. An animation is also available.

Example 12.E-3.2. Now, we assume the transmission of a quadrature phase shift-keying (QPSK) signal, whose symbols belong to the alphabet $\{\pm 1 \pm j1\}$. This signal suffers the effect of the channel (taken from [85])

$$H(z) = (-0.2 + j0.3) + (-0.5 + j0.4)\,z^{-1} + (0.7 - j0.6)\,z^{-2} + (0.4 + j0.3)\,z^{-3} + (0.2 + j0.1)\,z^{-4} \tag{12.170}$$

and of additive white Gaussian noise (AWGN) under a signal-to-noise ratio (SNR) of 30 dB. The equalizer is an adaptive filter with $M = 15$ coefficients initialized with zero and the desired signal is the transmitted sequence delayed by $L=9$ samples (go back to Fig. 12.16 to see the equalization scheme). The adaptation parameters (μ, $\tilde{\mu}$, ε, λ) were chosen in order to obtain the same steady-state performance of LMS, NLMS, and RLS (note that with a longer filter, the regularization parameter ε can be

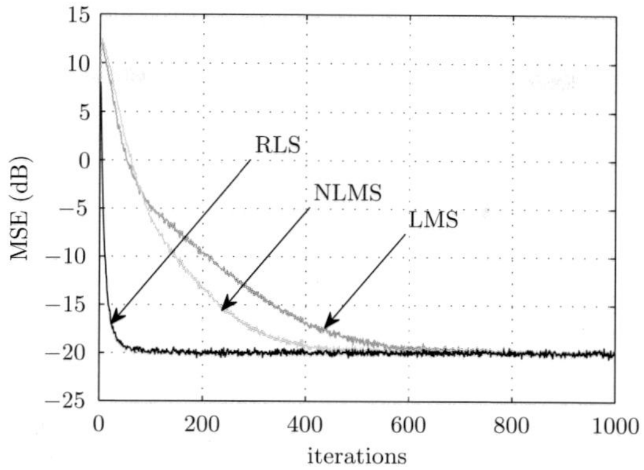

FIGURE 12.27

MSE along the iterations for LMS ($\mu = 0.01$), NLMS ($\tilde{\mu} = 0.05$, $\varepsilon = 0.5$), and RLS ($\lambda = 0.99$, $\delta = 1$) initialized at $\boldsymbol{w}(0) = [-1.5 \quad -1.4]^T$, $\sigma_v^2 = 0.01$, and ensemble average of 1000 independent runs.

much smaller than in the previous example). Fig. 12.28 shows the MSE along the iterations, estimated from the ensemble average of 1000 independent runs. Again, RLS presents the fastest convergence rate, followed by NLMS and LMS. These algorithms achieve the same steady-state MSE and therefore present the same performance after convergence, which can be observed in Fig. 12.29, where the equalizer output signal constellations are shown. The MATLAB® file used in this simulation is available at http://www.lps.usp.br.

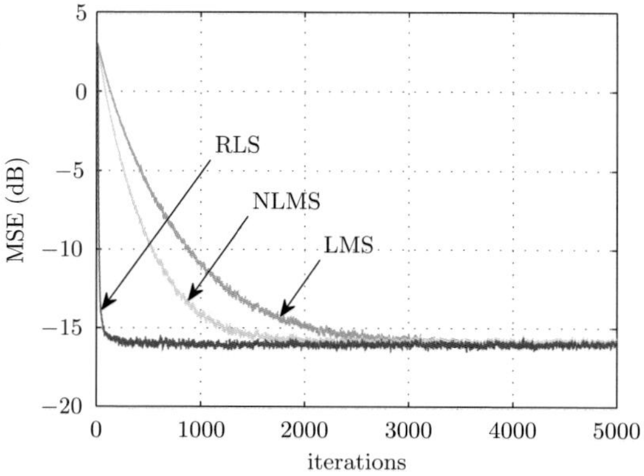

FIGURE 12.28

MSE along the iterations for LMS ($\mu = 10^{-3}$), NLMS ($\tilde{\mu} = 0.08$, $\varepsilon = 10^{-5}$), and RLS ($\lambda = 0.997$, $\delta = 1$) initialized at $\boldsymbol{w}(0) = \boldsymbol{0}$; QPSK, $L = 9$, $M = 15$, and ensemble average of 1000 independent runs; equalization of the channel (12.170).

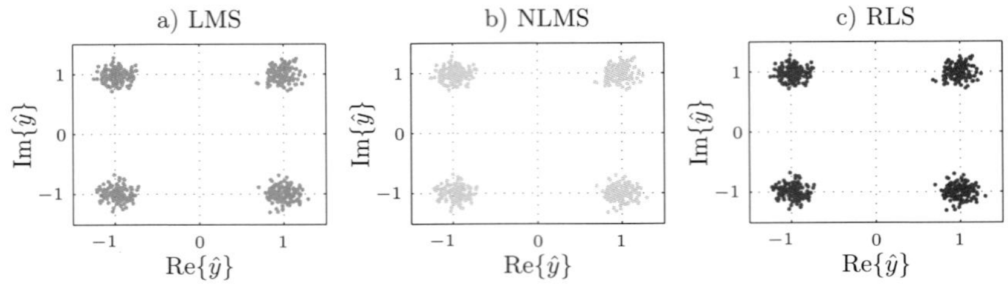

FIGURE 12.29

Equalizer output signal constellations after convergence for (a) LMS ($\mu = 10^{-3}$), (b) NLMS ($\tilde{\mu} = 0.08$, $\varepsilon = 10^{-5}$), and (c) RLS ($\lambda = 0.997$, $\delta = 1$); initialized at $\boldsymbol{w}(0) = \boldsymbol{0}$; QPSK, $L = 9$, $M = 15$; equalization of the channel (12.170).

12.4 Statistical analysis

Closed-form expressions for the mean square performance of adaptive filters facilitate comparisons between different algorithms and provide information about stability, convergence rates, MSE, and tracking capability. Since adaptive algorithms are obtained from stochastic approximations of the cost functions, gradients, and Hessians, their performance will degrade in comparison with the performance of the exact algorithms. Thus, stochastic analyses are also important to measure the performance degradation in relation to the exact algorithms.

The analysis of an adaptive algorithm constitutes a very hard problem, since adaptive filters are time-varying, stochastic, and nonlinear systems. Therefore, it is common to introduce simplifying approximations. They consist in most cases of disregarding the dependence between variables, for example, approximating the expected value of the product of two random variables, say, $\mathrm{E}\{xy\}$, by the product of the expected values, $\mathrm{E}\{x\}\,\mathrm{E}\{y\}$. For historical reasons, these approximations are usually referred to as *assumptions* in the literature.

In some situations, in particular for LMS and some of its variants, the approximations are proved to converge to the exact solution in the limiting case of slow adaptation ($\mu \to 0$ in the case of LMS), and their predictions tend to be reasonable for the range of step sizes used in practice. We will not reproduce these proofs here, since they are in general very technical, but they can be found in [63,64,134,221,234, 354] for the case of LMS and in [195] for more general filters.

The literature contains a few exact analytical results, but only for LMS, under very particular conditions [124,144,149,344]. In addition, the complexity of these exact models grows very quickly with the number of coefficients, so they are only useful for short filters (up to about five coefficients).

The performance of an adaptive filter is usually described by evaluating the mean and the mean square (variance) of the weight error vector and of the error signal $e(n)$. This usually works very well for slow adaptation (i.e., small step size). However, for fast adaptation, the behavior of an adaptive filter is considerably more complicated than can be seen by looking only at means and variances. Some of the phenomena that appear in LMS with large step sizes are described in [262,320]. References [105,178,259,265] study algorithms with strong nonlinearities, for which the usual approximations are not always valid.

The analysis of NLMS was considered in [6,8,48,49,99,258,349,364]. Many works consider only tapped-delay line regressors; the case of Volterra filters is considered for example in [71,192,230,338, 358]. For RLS filters, important models for infinite precision arithmetic can be found in [126,253]. The case of finite precision arithmetic is discussed in Section 12.5.1.2.

The real importance of the approximations used in the adaptive filtering literature is that they lead to reasonably simple models that capture well the overall behavior of an adaptive filter. Simple models provide intuition about how a filter will behave if a given parameter is changed, which is important for a designer to make good decisions. Some approximations are made because without them the mathematical analysis would not be feasible, but others are used mainly because they lead to simpler models. Of course, there are situations in which it is important to study a filter in more detail, reducing the number of assumptions used in the analysis, as in the references just cited.

In this section we analyze three of the most common adaptive filters – LMS, NLMS, and RLS – starting with LMS in Section 12.4.3, and then showing how to extend these results to the other algorithms in Section 12.4.4. To start, we need to define some useful performance measures. The MSE is simply the expected value of $|e(n)|^2$:

$$J(n) \triangleq \mathrm{E}\{|e(n)|^2\}. \tag{12.171}$$

The most widely measure of performance used in the literature of adaptive filtering is the *excess MSE* (EMSE), which is defined as

$$\zeta(n) \triangleq \mathrm{E}\{|e_a(n)|^2\}, \tag{12.172}$$

where the *excess a priori error* is given by

$$e_a(n) = \tilde{\boldsymbol{w}}^H(n)\boldsymbol{\phi}(n) \tag{12.173}$$

and the weight error vector by

$$\tilde{\boldsymbol{w}}(n) = \boldsymbol{w}_{\mathrm{o}}(n) - \boldsymbol{w}(n). \tag{12.174}$$

Note that the optimal solution $\boldsymbol{w}_{\mathrm{o}}(n)$ is now allowed to be time-varying. Note also the difference between $e_a(n)$ and $e(n)$, defined in (12.120). Both are a priori errors, but $e_a(n)$ is computed with $\tilde{\boldsymbol{w}}(n)$, whereas $e(n)$ is computed with $\boldsymbol{w}(n)$. The EMSE measures how much $J(n)$ exceeds its minimum value $J_{\min}(n)$ due to adaptation: if we were able to find the optimum filter coefficients $\boldsymbol{w}_{\mathrm{o}}(n)$ at each time, the EMSE would be zero. Since the actual filter coefficients are never exactly equal to the optimum values, the EMSE measures the effect of this difference on the error variance. Closely related to the EMSE, the *misadjustment* is defined as

$$\mathcal{M}(n) \triangleq \zeta(n)/\sigma_v^2, \tag{12.175}$$

where $\sigma_v^2 = \mathrm{E}\{|v_{\mathrm{o}}|^2\}$.

The *excess a posteriori error* $e_p(n)$ is mostly used in the analysis and derivations of algorithms. It is the error obtained *after* using the current data $\boldsymbol{\phi}(n)$ to update the coefficient vector (hence the names *a priori* and *a posteriori*), i.e.,

$$e_p(n) = \tilde{\boldsymbol{w}}^H(n+1)\boldsymbol{\phi}(n). \tag{12.176}$$

Again, note the difference with respect to $\xi(n)$, defined in (12.130).

Another common performance measure is the *mean square deviation* (MSD). It measures how far $\boldsymbol{w}(n)$ is from $\boldsymbol{w}_o(n)$ on average:

$$\chi(n) \triangleq \mathrm{E}\{\|\tilde{\boldsymbol{w}}(n)\|^2\}. \tag{12.177}$$

The EMSE is the most frequently performance measure used for comparisons between adaptive filters; however, in some applications, particularly for system identification, the MSD is more adequate.

We now will discuss a general model for the input signals, considering a set of approximations that leads to reasonable, yet simple models.

12.4.1 Data model

From the orthogonality principle (recall Section 12.2), the optimal solution $\boldsymbol{w}_o(n)$ satisfies the property

$$\mathrm{E}\left\{\boldsymbol{\phi}(n)\left[d(n) - \boldsymbol{w}_o^H(n)\boldsymbol{\phi}(n)\right]\right\} = \mathbf{0}, \tag{12.178}$$

i.e., the input regressor vector $\boldsymbol{\phi}(n)$ is uncorrelated with the optimal estimation error

$$v_o(n) \triangleq d(n) - \boldsymbol{w}_o^H(n)\boldsymbol{\phi}(n), \tag{12.179}$$

whose mean is zero and whose variance $\sigma_v^2(n) = \mathrm{E}\{|v_o(n)|^2\}$ is equal to the minimum cost $J_{\min}(n)$. Thus, a linear regression model for $d(n)$ holds, i.e.,

$$d(n) = \boldsymbol{w}_o^H(n)\boldsymbol{\phi}(n) + v_o(n). \tag{12.180}$$

It then follows that

$$e(n) = d(n) - \boldsymbol{w}^H(n)\boldsymbol{\phi}(n) = \tilde{\boldsymbol{w}}^H(n)\boldsymbol{\phi}(n) + v_o(n) = e_a(n) + v_o(n), \tag{12.181}$$

and the MSE is given by

$$\begin{aligned} J(n) = \mathrm{E}\{|e(n)|^2\} &= \mathrm{E}\{|e_a(n)|^2\} + 2\,\mathrm{E}\{e_a(n)v_o(n)\} + \mathrm{E}\{|v_o(n)|^2\} \\ &= \zeta(n) + 2\,\mathrm{E}\{e_a(n)v_o(n)\} + \sigma_v^2(n). \end{aligned}$$

The first approximation (or assumption) used in adaptive filtering analysis is to disregard the cross-term $\mathrm{E}\{e_a(n)v_o(n)\}$. Since $v_o(n)$ and $\boldsymbol{\phi}(n)$ are uncorrelated, but not necessarily independent, the cross-term does not vanish in general. From observations, however, one notes that the cross-term is usually negligible, compared to the other terms. In order to obtain a simple model for the adaptive filter, we will assume that $\boldsymbol{\phi}(n)$ and $v_o(n)$ are independent of each other and of previous samples $\boldsymbol{\phi}(n-k)$, $v(n-k)$ for $k > 0$.

12.A-1. The sequences $\{\boldsymbol{\phi}(n)\}$ and $\{v_o(n)\}$ are i.i.d. and independent of each other (not only orthogonal). In addition, the sequence $\{\boldsymbol{\phi}(n)\}$ is assumed stationary and $\{v_o(n)\}$ is zero-mean.

The last two assumptions are not in fact necessary, but they simplify the analysis. In general the sequence $\{\boldsymbol{\phi}(n)\}$ is not i.i.d.: for example, if $\boldsymbol{\phi}(n)$ is formed from a tapped-delay line or from a Volterra series expansion (see Eqs. (12.114) and (12.115)), $\boldsymbol{\phi}(n)$ and $\boldsymbol{\phi}(n-1)$ will have terms in common. However, assuming that the sequences are i.i.d. leads to simple models that in general are good approximations for the range of step sizes usually found in practice, in particular for small step sizes. The analysis based on Assumption 12.A-1 is known in the literature as *independence theory* (see [244] for an alternative justification for the approximations).

Going back to the adaptive filter recursions (see Tables 12.2–12.5 and Eq. (12.223) further ahead) we see that the current weight error $\tilde{\boldsymbol{w}}(n)$ is a function of past inputs $\boldsymbol{\phi}(n-1)$, $\boldsymbol{\phi}(n-2), \dots$ and noise samples $v(n-1)$, $v(n-2), \dots$, but *not* of the current samples $\boldsymbol{\phi}(n)$ and $v_o(n)$. Therefore, under Assumption 12.A-1, $\tilde{\boldsymbol{w}}(n)$ is independent of $v_o(n)$ and $\boldsymbol{\phi}(n)$, so that

$$\mathrm{E}\{e_a(n)v_o(n)\} = \mathrm{E}\{\tilde{\boldsymbol{w}}^H(n)\}\,\mathrm{E}\{\boldsymbol{\phi}(n)\}\,\mathrm{E}\{v_o(n)\} = 0.$$

We conclude that under Assumption 12.A-1, the MSE $J(n) = \mathrm{E}\{|e(n)|^2\}$ is related to the EMSE via

$$J(n) = \zeta(n) + J_{\min}(n) = \zeta(n) + \sigma_v^2, \tag{12.182}$$

since Assumption 12.A-1 implies that $J_{\min}(n)$ is a constant.

We also need a model for the variation of the optimal solution. Again, our choice is a model that captures the main consequences of a nonconstant $\boldsymbol{w}_o(n)$, without unnecessarily complicating the analysis.

12.A-2. Random-walk model. In a nonstationary environment, $\boldsymbol{w}_o$ varies as

$$\boldsymbol{w}_o(n+1) = \boldsymbol{w}_o(n) + \boldsymbol{q}(n), \tag{12.183}$$

where $\boldsymbol{q}(n)$ is a zero-mean i.i.d. vector with positive definite autocorrelation matrix $\boldsymbol{Q} = \mathrm{E}\{\boldsymbol{q}(n)\boldsymbol{q}^H(n)\}$, independent of the initial condition $\boldsymbol{w}(0)$ and of $\{\boldsymbol{\phi}(\ell), v(\ell)\}$ for all ℓ.

The covariance matrix of $\boldsymbol{w}_o$, according to (12.183), grows slowly to infinity. A more general model for the variation of $\boldsymbol{w}_o$ uses the recursion

$$\begin{cases} \boldsymbol{w}_o(n+1) = \boldsymbol{w}_o(n) + \boldsymbol{\theta}(n), \\ \boldsymbol{\theta}(n) = \alpha\boldsymbol{\theta}(n-1) + \boldsymbol{q}(n), \qquad 0 \le |\alpha| < 1, \end{cases} \tag{12.184}$$

whose covariance matrix remains bounded, but the analysis becomes more complicated [97,317]. Therefore, it is common in the literature to adopt the model (12.183). Note that both models assume that $\{\boldsymbol{x}(n)\}$ is a stationary random sequence, although $\boldsymbol{w}_o(n)$ is allowed to vary with n. A model considering nonstationary regressors (in fact, cyclostationary regressors) can be found in [133].

We remark that the performance in tracking of different adaptive filtering algorithms is usually studied by direct comparison. However, after a theoretical analysis for the Kalman filter applied to models (12.183) and (12.184) was developed in [97], it became possible to make performance comparisons against a fixed standard, since the Kalman filter is optimum for (12.183) and (12.184) if $\boldsymbol{Q}$ and α are known.

12.4.2 Relating the autocorrelation matrix of the weight error vector to the EMSE and the MSD

The autocorrelation matrix of the weight error vector is defined as

$$\boldsymbol{S}(n) \triangleq \mathrm{E}\left\{\tilde{\boldsymbol{w}}(n)\tilde{\boldsymbol{w}}^H(n)\right\}. \tag{12.185}$$

This matrix is important in the statistical analysis of adaptive filters since it is related to the MSD and the EMSE as we show next.

Recall that the MSD is given by

$$\chi(n) = \mathrm{E}\{\|\tilde{\boldsymbol{w}}(n)\|^2\} = \sum_{i=0}^{M-1} \mathrm{E}\{\tilde{w}_i^2(n)\}.$$

Now, the terms $\mathrm{E}\{\tilde{w}_i^2(n)\}$ are exactly the terms that appear on the diagonal of $\boldsymbol{S}(n)$; therefore, recalling that the *trace* of a square matrix is the sum of its main-diagonal elements (see Box 12.3), we obtain

$$\chi(n) = \sum_{i=0}^{M-1} \mathrm{E}\{\tilde{w}_i^2(n)\} = \sum_{m=0}^{M-1} s_{mm}(n) = \mathrm{Tr}(\boldsymbol{S}(n)). \tag{12.186}$$

Under Assumption 12.A-1, $\boldsymbol{\phi}(n)$ and $\tilde{\boldsymbol{w}}(n)$ are independent, as we just saw. Therefore,

$$\begin{aligned}
\zeta(n) &= \mathrm{E}\{|\tilde{\boldsymbol{w}}^H(n)\boldsymbol{\phi}(n)|^2\} = \mathrm{E}\{\tilde{\boldsymbol{w}}^H(n)\boldsymbol{\phi}(n)(\tilde{\boldsymbol{w}}^H(n)\boldsymbol{\phi}(n))^*\} \\
&= \mathrm{E}\{\tilde{\boldsymbol{w}}^H(n)\boldsymbol{\phi}(n)\boldsymbol{\phi}^H(n)\tilde{\boldsymbol{w}}(n)\} = \mathrm{E}\left\{\mathrm{E}\{\tilde{\boldsymbol{w}}^H(n)\boldsymbol{\phi}(n)\boldsymbol{\phi}^H(n)\tilde{\boldsymbol{w}}(n)|\tilde{\boldsymbol{w}}(n)\}\right\} \\
&= \mathrm{E}\left\{\tilde{\boldsymbol{w}}^H(n)\,\mathrm{E}\{\boldsymbol{\phi}(n)\boldsymbol{\phi}^H(n)|\tilde{\boldsymbol{w}}(n)\}\tilde{\boldsymbol{w}}(n)\right\} = \mathrm{E}\{\tilde{\boldsymbol{w}}^H(n)\boldsymbol{R}_\phi\tilde{\boldsymbol{w}}(n)\},
\end{aligned}$$

where we used Assumption 12.A-1 to write $\mathrm{E}\{\boldsymbol{\phi}(n)\boldsymbol{\phi}^H(n)\} = \boldsymbol{R}_\phi$ (constant) and two properties of expected values. The first property is that for any two random variables a, b,

$$\mathrm{E}\{ab\} = \mathrm{E}\{\mathrm{E}\{ab|b\}\} = \mathrm{E}\{\mathrm{E}\{a|b\}b\},$$

where the inner expected value is taken with respect to a and the second with respect to b. The second property is that if a and b are independent, then $\mathrm{E}\{a|b\}$ does not depend on b, i.e., $\mathrm{E}\{a|b\} = \mathrm{E}\{a\}$.

To proceed, we apply a useful trick from matrix theory. One can check by direct computation that if $\boldsymbol{A} \in \mathbb{C}^{K\times L}$ and $\boldsymbol{B} \in \mathbb{C}^{L\times K}$, then $\mathrm{Tr}(\boldsymbol{AB}) = \mathrm{Tr}(\boldsymbol{BA})$. Applying this property with $\boldsymbol{A} \leftarrow \tilde{\boldsymbol{w}}^H(n)$ and $\boldsymbol{B} \leftarrow \boldsymbol{R}_\phi\tilde{\boldsymbol{w}}(n)$, we obtain

$$\zeta(n) = \mathrm{Tr}\left(\boldsymbol{R}_\phi\,\mathrm{E}\{\tilde{\boldsymbol{w}}(n)\tilde{\boldsymbol{w}}^H(n)\}\right) = \mathrm{Tr}\big(\boldsymbol{R}_\phi\boldsymbol{S}(n)\big). \tag{12.187}$$

Since $\boldsymbol{R}_{\phi}$ is symmetric, there always exists an orthogonal matrix $\boldsymbol{U}$ (i.e., $\boldsymbol{U}^{-1} = \boldsymbol{U}^H$) that diagonalizes $\boldsymbol{R}_{\phi}$, that is,

$$\boldsymbol{U}^H \boldsymbol{R}_{\phi} \boldsymbol{U} = \begin{bmatrix} \lambda_1 & 0 & \cdots & 0 \\ 0 & \lambda_2 & \cdots & 0 \\ \vdots & \vdots & \ddots & \vdots \\ 0 & 0 & 0 & \lambda_M \end{bmatrix} \triangleq \boldsymbol{\Lambda}, \tag{12.188}$$

where $\lambda_1, \lambda_2, \ldots, \lambda_M$ are the eigenvalues of $\boldsymbol{R}_{\phi}$ (see Box 12.3). Defining the rotated matrix

$$\boldsymbol{S}'(n) \triangleq \boldsymbol{U}^H \boldsymbol{S}(n) \boldsymbol{U}$$

and recalling that $\boldsymbol{U}\boldsymbol{U}^H = \boldsymbol{I}$, (12.186) and (12.187) can be rewritten respectively as

$$\chi(n) = \mathrm{Tr}(\boldsymbol{U} \underbrace{\boldsymbol{U}^H \boldsymbol{S}(n) \boldsymbol{U}}_{\boldsymbol{S}'(n)} \boldsymbol{U}^H) = \mathrm{Tr}(\boldsymbol{U} \boldsymbol{S}'(n) \boldsymbol{U}^H) \tag{12.189}$$

and

$$\zeta(n) = \mathrm{Tr}(\boldsymbol{U} \underbrace{\boldsymbol{U}^H \boldsymbol{R}_{\phi} \boldsymbol{U}}_{\boldsymbol{\Lambda}} \underbrace{\boldsymbol{U}^H \boldsymbol{S}(n) \boldsymbol{U}}_{\boldsymbol{S}'(n)} \boldsymbol{U}^H) = \mathrm{Tr}(\boldsymbol{U} \boldsymbol{\Lambda} \boldsymbol{S}'(n) \boldsymbol{U}^H). \tag{12.190}$$

Using again the fact that $\mathrm{Tr}(\boldsymbol{A}\boldsymbol{B}) = \mathrm{Tr}(\boldsymbol{B}\boldsymbol{A})$, with $\boldsymbol{A} \leftarrow \boldsymbol{U}$, we obtain

$$\chi(n) = \mathrm{Tr}\big(\boldsymbol{S}'(n)\big) \tag{12.191}$$

and

$$\zeta(n) = \mathrm{Tr}\big(\boldsymbol{\Lambda} \boldsymbol{S}'(n)\big). \tag{12.192}$$

Note that both the MSD (12.191) and the EMSE (12.192) depend only on the diagonal entries of $\boldsymbol{S}'(n)$. We can work therefore only with these diagonal entries, defining the vectors

$$\boldsymbol{s}'(n) = \mathrm{diag}(\boldsymbol{S}'(n)) = \left[\, s'_0(n) \; s'_1(n) \; \cdots \; s'_{M-1}(n) \,\right]^T \tag{12.193}$$

and

$$\boldsymbol{l} = [\, \lambda_0 \; \ldots \; \lambda_{M-1} \,]^T, \tag{12.194}$$

where $\mathrm{diag}(\boldsymbol{A})$ represents a column vector with the diagonal elements of $\boldsymbol{A}$, and evaluating the MSD and EMSE respectively as

$$\chi(n) = \sum_{k=0}^{M-1} s'_k(n) \tag{12.195}$$

and

$$\zeta(n) = \boldsymbol{l}^T \boldsymbol{s}'(n) = \sum_{k=0}^{M-1} \lambda_k s'_k(n). \tag{12.196}$$

Again, the MSD and EMSE depend only on the elements of the diagonal of the rotated matrix $\boldsymbol{S}'(n)$. In the case of EMSE, these elements are weighted by the eigenvalues of the autocorrelation matrix $\boldsymbol{R}_{\boldsymbol{\phi}}$. As we shall see in Section 12.4.3, the stability of LMS can be studied through a recursion for the vector $\boldsymbol{s}'(n)$.

Thus, according to the previous results, in order to obtain closed-form expressions for the MSD and the EMSE of an adaptive algorithm, one path is to find a recursion for the covariance matrix $\boldsymbol{S}(n)$ or for the vector $\boldsymbol{s}'(n)$. However, this is not the only method to analyze adaptive algorithms, as we shall see in Section 12.4.4.

12.4.3 Statistical analysis of the LMS algorithm

In this section, we present a statistical analysis for the LMS algorithm, assuming that the optimal solution $\boldsymbol{w}_o$ remains constant ($\boldsymbol{q}(n) \equiv \boldsymbol{0}$, for all n). This analysis predicts the behavior of LMS in steady state and in transient, and provides a range of the step size μ for which LMS operates adequately. If you are familiar with the LMS analysis, go to Section 12.4.4, where we present a unified analysis for LMS, NLMS, and RLS algorithms in a nonstationary environment.

We first rewrite (12.124) in terms of the weight error vector $\tilde{\boldsymbol{w}}$. Thus, subtracting both sides of (12.124) from $\boldsymbol{w}_o$, we obtain

$$\begin{aligned} \boldsymbol{w}_o - \boldsymbol{w}(n+1) &= \boldsymbol{w}_o - \boldsymbol{w}(n) - \mu\boldsymbol{\phi}(n)e^*(n), \\ \tilde{\boldsymbol{w}}(n+1) &= \tilde{\boldsymbol{w}}(n) - \mu\boldsymbol{\phi}(n)e^*(n). \end{aligned} \tag{12.197}$$

Replacing $e^*(n) = e_a^*(n) + v_o^*(n) = \boldsymbol{\phi}^H(n)\tilde{\boldsymbol{w}}(n) + v_o^*(n)$ in (12.197), we arrive at

$$\tilde{\boldsymbol{w}}(n+1) = \left[\boldsymbol{I} - \mu\boldsymbol{\phi}(n)\boldsymbol{\phi}^H(n)\right]\tilde{\boldsymbol{w}}(n) - \mu\boldsymbol{\phi}(n)v_o^*(n). \tag{12.198}$$

First-order analysis

Taking expectations on both sides of (12.198), we obtain

$$\begin{aligned} \mathrm{E}\{\tilde{\boldsymbol{w}}(n+1)\} &= \mathrm{E}\left\{\left[\boldsymbol{I} - \mu\boldsymbol{\phi}(n)\boldsymbol{\phi}^H(n)\right]\tilde{\boldsymbol{w}}(n)\right\} - \mu\,\mathrm{E}\{\boldsymbol{\phi}(n)v_o^*(n)\} \\ &= \mathrm{E}\left\{\left[\boldsymbol{I} - \mu\boldsymbol{\phi}(n)\boldsymbol{\phi}^H(n)\right]\tilde{\boldsymbol{w}}(n)\right\}, \end{aligned} \tag{12.199}$$

where the last equality follows from the fact that $\boldsymbol{\phi}(n)$ is orthogonal to $v_o^*(n)$, according to the orthogonality condition (see Eq. (12.97)). Since $\tilde{\boldsymbol{w}}(n)$ is assumed to be independent of $\boldsymbol{\phi}(n)$ (see Assumption 12.A-1), we arrive at

$$\mathrm{E}\{\tilde{\boldsymbol{w}}(n+1)\} = \left[\boldsymbol{I} - \mu\boldsymbol{R}_{\boldsymbol{\phi}}\right]\mathrm{E}\{\tilde{\boldsymbol{w}}(n)\}. \tag{12.200}$$

This is the same recursion we saw in Box 12.1. According to linear systems theory, this recursion converges as long as the largest eigenvalue of $\boldsymbol{A} = \boldsymbol{I} - \mu \boldsymbol{R}_{\phi}$ has absolute value less than one, which implies (see Box 12.1)

$$\begin{aligned} &|1-\mu\lambda| < 1 \Leftrightarrow 0 < \mu < \frac{2}{\lambda_k}, \forall k \Leftrightarrow \\ &0 < \mu < \frac{2}{\lambda_{\max}}, \end{aligned} \tag{12.201}$$

where λ_k, $k = 1, 2, \ldots, M$, are the eigenvalues of $\boldsymbol{R}_{\phi}$ and $\lambda_{\max}$ is its maximum eigenvalue. We should notice that this range for μ does not ensure the stability of LMS: $0 < \mu < 2/\lambda_{\max}$ guarantees the convergence of the mean of $\tilde{\boldsymbol{w}}(n) = \boldsymbol{w}_{\mathrm{o}} - \boldsymbol{w}(n)$ towards $\boldsymbol{0}$, but the autocorrelation of $\tilde{\boldsymbol{w}}(n)$ may still be diverging. We show next that the range for μ based on a second-order analysis is indeed more restrictive.

Second-order analysis

Now, we use (12.198) to find a recursion for $\boldsymbol{S}(n)$ and (12.187) to evaluate the EMSE of LMS. Multiplying (12.198) by its Hermitian, taking the expectations of both sides, we arrive at

$$\begin{aligned} \mathrm{E}\{\tilde{\boldsymbol{w}}(n+1)\tilde{\boldsymbol{w}}^H(n+1)\} &= \mathrm{E}\{\tilde{\boldsymbol{w}}(n)\tilde{\boldsymbol{w}}^H(n)\} \\ &- \mu\, \overbrace{\mathrm{E}\{\tilde{\boldsymbol{w}}(n)\tilde{\boldsymbol{w}}^H(n)\boldsymbol{\phi}(n)\boldsymbol{\phi}^H(n)\}}^{\mathcal{A}} \\ &- \mu\, \overbrace{\mathrm{E}\{\boldsymbol{\phi}(n)\boldsymbol{\phi}^H(n)\tilde{\boldsymbol{w}}(n)\tilde{\boldsymbol{w}}^H(n)\}}^{\mathcal{B}} \\ &+ \mu^2\, \overbrace{\mathrm{E}\{\boldsymbol{\phi}(n)\boldsymbol{\phi}^H(n)\tilde{\boldsymbol{w}}(n)\tilde{\boldsymbol{w}}^H(n)\boldsymbol{\phi}(n)\boldsymbol{\phi}^H(n)\}}^{\mathcal{C}} \\ &+ \mu^2\, \overbrace{\mathrm{E}\{|v_{\mathrm{o}}(n)|^2\boldsymbol{\phi}(n)\boldsymbol{\phi}^H(n)\}}^{\mathcal{D}} \\ &- \mu\, \overbrace{\mathrm{E}\{v_{\mathrm{o}}(n)\tilde{\boldsymbol{w}}(n)\boldsymbol{\phi}^H(n)\}}^{\mathcal{E}} \\ &- \mu\, \overbrace{\mathrm{E}\{v_{\mathrm{o}}^*(n)\boldsymbol{\phi}(n)\tilde{\boldsymbol{w}}^H(n)\}}^{\mathcal{F}} \\ &+ \mu^2\, \overbrace{\mathrm{E}\{v_{\mathrm{o}}(n)\boldsymbol{\phi}(n)\boldsymbol{\phi}^H(n)\tilde{\boldsymbol{w}}(n)\boldsymbol{\phi}^H(n)\}}^{\mathcal{G}} \\ &+ \mu^2\, \overbrace{\mathrm{E}\{v_{\mathrm{o}}^*(n)\boldsymbol{\phi}(n)\tilde{\boldsymbol{w}}^H(n)\boldsymbol{\phi}(n)\boldsymbol{\phi}^H(n)\}}^{\mathcal{H}}. \end{aligned} \tag{12.202}$$

Now, using Assumption 12.A-1, we can evaluate the terms $\mathcal{A}$–$\mathcal{H}$ of (12.202):

$\mathcal{A}$ Recalling that Assumption 12.A-1 implies that $\tilde{\boldsymbol{w}}(n)$ is independent of $\boldsymbol{\phi}(n)$, $\mathcal{A}$ can be approximated by

$$\begin{aligned}\mathcal{A} &= \mu \,\mathrm{E}\left\{\mathrm{E}\left\{\tilde{\boldsymbol{w}}(n)\tilde{\boldsymbol{w}}^H(n)\boldsymbol{\phi}(n)\boldsymbol{\phi}^H(n)|\boldsymbol{\phi}(n)\right\}\right\} \\ &\approx \mu \,\mathrm{E}\left\{\mathrm{E}\left\{\tilde{\boldsymbol{w}}(n)\tilde{\boldsymbol{w}}^H(n)\right\}\boldsymbol{\phi}(n)\boldsymbol{\phi}^H(n)\right\} \\ &= \mu \boldsymbol{S}(n)\,\mathrm{E}\left\{\boldsymbol{\phi}(n)\boldsymbol{\phi}^H(n)\right\} = \mu \boldsymbol{S}(n)\boldsymbol{R}_{\boldsymbol{\phi}}.\end{aligned} \tag{12.203}$$

$\mathcal{B}$ Analogously, we obtain for $\mathcal{B}$

$$\mathcal{B} \approx \mu \boldsymbol{R}_{\boldsymbol{\phi}}\boldsymbol{S}(n). \tag{12.204}$$

$\mathcal{C}$ Using Assumption 12.A-1, $\mathcal{C}$ can be rewritten as

$$\begin{aligned}\mathcal{C} &= \mu^2 \,\mathrm{E}\left\{\boldsymbol{\phi}(n)\boldsymbol{\phi}^H(n)\tilde{\boldsymbol{w}}(n)\tilde{\boldsymbol{w}}^H(n)\boldsymbol{\phi}(n)\boldsymbol{\phi}^H(n)\right\} \\ &\approx \mu^2 \,\mathrm{E}\left\{\boldsymbol{\phi}(n)\boldsymbol{\phi}^H(n)\boldsymbol{S}(n)\boldsymbol{\phi}(n)\boldsymbol{\phi}^H(n)\right\}.\end{aligned} \tag{12.205}$$

We return to this term further on.

$\mathcal{D}$ Using the assumption that $v_o(n)$ is independent of $\boldsymbol{\phi}(n)$ and $\tilde{\boldsymbol{w}}(n)$, $\mathcal{D}$ reduces to

$$\mathcal{D} = \mu^2 \,\mathrm{E}\{|v_o(n)|^2\boldsymbol{\phi}(n)\boldsymbol{\phi}^H(n)\} \approx \mu^2\sigma_v^2 \boldsymbol{R}_{\boldsymbol{\phi}}. \tag{12.206}$$

$\mathcal{E}$ Using the fact that $v_o(n)$ is a zero-mean random variable and the assumption that it is independent of $\boldsymbol{\phi}(n)$ and $\tilde{\boldsymbol{w}}(n)$, $\mathcal{E}$ is an $M \times M$ null matrix. Using the same arguments, $\mathcal{F}$, $\mathcal{G}$, and $\mathcal{H}$ are also null matrices. We should point out that the zero-mean assumption for $v_o(n)$ is not necessary, but it really facilitates the analysis. Furthermore, as we saw in Section 12.2.1.4, we can enforce $\mathrm{E}\{v_o(n)\} = 0$ using simple methods.

From the previous results, (12.202) reduces to

$$\begin{aligned}\mathbf{S}(n+1) \approx\; & \mathbf{S}(n) - \mu \boldsymbol{S}(n)\boldsymbol{R}_{\boldsymbol{\phi}} - \mu \boldsymbol{R}_{\boldsymbol{\phi}}\boldsymbol{S}(n) \\ & + \underbrace{\mu^2 \,\mathrm{E}\left\{\boldsymbol{\phi}(n)\boldsymbol{\phi}^H(n)\boldsymbol{S}(n)\boldsymbol{\phi}(n)\boldsymbol{\phi}^H(n)\right\}}_{\mathcal{C}} + \mu^2\sigma_v^2 \boldsymbol{R}_{\boldsymbol{\phi}}.\end{aligned} \tag{12.207}$$

This recursion is convenient to obtain closed-form expressions for the EMSE and the MSD of LMS. Observe that the four first terms of the right-hand side of (12.207) are linear in $\boldsymbol{S}(n)$. Assuming that μ is sufficiently small, $\mathcal{C}$ can be neglected with respect to the three first terms on the right-hand side of (12.207). Thus, for small step sizes, (12.207) reduces to

$$\mathbf{S}(n+1) \approx \mathbf{S}(n) - \mu\left[\boldsymbol{S}(n)\boldsymbol{R}_{\boldsymbol{\phi}} + \boldsymbol{R}_{\boldsymbol{\phi}}\boldsymbol{S}(n)\right] + \mu^2\sigma_v^2 \boldsymbol{R}_{\boldsymbol{\phi}}, \tag{12.208}$$

with initialization $\boldsymbol{S}(0) = [\boldsymbol{w}_o - \boldsymbol{w}(0)][\boldsymbol{w}_o - \boldsymbol{w}(0)]^H$.

Since we can obtain the EMSE and the MSD from $\boldsymbol{S}(n)$ and $\boldsymbol{R}_{\phi}$ through (12.187) and (12.186), we should use the recursion (12.208) to compute $\boldsymbol{S}(n)$ and then evaluate $\zeta(n) = \text{Tr}(\boldsymbol{S}(n)\boldsymbol{R}_{\phi})$ and $\chi(n) = \text{Tr}(\boldsymbol{S}(n))$. Particularly, for $n \to \infty$, since $\boldsymbol{S}(n+1) \approx \boldsymbol{S}(n)$, we obtain

$$\boldsymbol{S}(\infty)\boldsymbol{R}_{\phi} + \boldsymbol{R}_{\phi}\boldsymbol{S}(\infty) = \mu\sigma_v^2\boldsymbol{R}_{\phi}. \tag{12.209}$$

Taking the trace of both sides of (12.209), we arrive at an analytical expression for the steady-state EMSE of LMS, i.e.,

$$\zeta^{\text{LMS}}(\infty) \approx \frac{\mu\sigma_v^2\,\text{Tr}(\boldsymbol{R}_{\phi})}{2} \quad \text{(for sufficiently small } \mu\text{)}. \tag{12.210}$$

This expression indicates that the EMSE of LMS increases with μ and with the length of the filter. The dependence on the filter length is through the trace of $\boldsymbol{R}_{\phi}$. It is more evident when the class of filters we are considering is FIR, that is, when $\boldsymbol{\phi}(n) = \begin{bmatrix} x(n) & \ldots x(n-M+1) \end{bmatrix}^T$ (a tapped-delay line). In this case, if $\{x(n)\}$ is stationary, then $\text{Tr}(\boldsymbol{R}_{\phi}) = M\,\text{E}\{x^2(n)\}$, M times the average power of $x(n)$.

Turning now to the filter coefficients, if we multiply both sides of (12.209) from the right by $\boldsymbol{R}_{\phi}^{-1}$, taking the trace of both sides, and recalling that $\text{Tr}(\boldsymbol{AB}) = \text{Tr}(\boldsymbol{BA})$, we obtain an expression for the steady-state MSD of LMS, i.e.,

$$\chi^{\text{LMS}}(\infty) \approx \frac{\mu\sigma_v^2 M}{2} \quad \text{(for sufficiently small } \mu\text{)}. \tag{12.211}$$

Again, the smaller the step size μ and the filter length M, the smaller the MSD of LMS.

These results explain what we observed in the simulations in Section 12.1.2 (Figs. 12.9 and 12.10): for larger values of μ the filter converges faster, but hovers around the optimum solution in a larger region. The size of this region is what the MSD attempts to measure – the EMSE measures how much this hovering affects the error.

These quantities are important in practical situations. For example, in an actual implementation of an acoustic echo canceler, when the near-end user is speaking, his or her voice is part of $v(n)$. Since the near-end speech is normally stronger than the echo, σ_v^2 tends to be quite large when the near-end speaker is talking. If the adaptive filter were allowed to keep adapting at these moments, (12.210) shows that the filter estimates would wander too far from the optimum. To avoid this problem, the step size is reduced to zero whenever the near-end user is speaking. A *double-talk detector* is used to decide when to stop adaptation.

It can be shown that (12.208) is stable for sufficiently small μ. However, (12.208) cannot be used to find a range of values of μ that guarantee stability, due to the approximations made, particularly the discarding of term $\boldsymbol{C}$. Thus, in order to study the stability of LMS, we consider next an approximation for this term, based the assumption that the input vector is Gaussian. This assumption allows us to approximate $\boldsymbol{C}$ as

$$\boldsymbol{C} \approx \boldsymbol{R}_{\phi}\,\text{Tr}(\boldsymbol{S}(n)\boldsymbol{R}_{\phi}) + \beta\boldsymbol{R}_{\phi}\boldsymbol{S}(n)\boldsymbol{R}_{\phi}, \tag{12.212}$$

where as usual, $\beta = 1$ (resp. $\beta = 2$) for complex (resp. real) data. See Box 12.7 for the derivation of (12.212).

Replacing (12.212) in (12.207), we arrive at

$$\begin{aligned}\boldsymbol{S}(n+1) \approx\ & \boldsymbol{S}(n) - \mu\left[\boldsymbol{S}(n)\boldsymbol{R}_\phi + \boldsymbol{R}_\phi\boldsymbol{S}(n)\right] \\ & + \mu^2\left[\boldsymbol{R}_\phi \operatorname{Tr}\left(\boldsymbol{S}(n)\boldsymbol{R}_\phi\right) + \beta\boldsymbol{R}_\phi\boldsymbol{S}(n)\boldsymbol{R}_\phi + \sigma_v^2\boldsymbol{R}_\phi\right].\end{aligned} \tag{12.213}$$

The stability of (12.213) is determined through a recursion for the vector $\boldsymbol{s}'(n)$, which contains the elements of the diagonal of the rotated matrix $\boldsymbol{S}'(n) = \boldsymbol{U}^H\boldsymbol{S}(n)\boldsymbol{U}$ (see Section 12.4.2). A recursion for $\boldsymbol{S}'(n)$ can be obtained by multiplying (12.213) from the left by $\boldsymbol{U}^H$ and from the right by $\boldsymbol{U}$ and recalling that $\boldsymbol{U}\boldsymbol{U}^H = \boldsymbol{I}$, which leads to

$$\begin{aligned}\boldsymbol{S}'(n+1) \approx\ & \boldsymbol{S}'(n) - \mu\left[\boldsymbol{S}'(n)\boldsymbol{\Lambda} + \boldsymbol{\Lambda}\boldsymbol{S}'(n)\right] \\ & + \mu^2\left[\boldsymbol{\Lambda}\operatorname{Tr}(\boldsymbol{\Lambda}\boldsymbol{S}'(n)) + \beta\boldsymbol{\Lambda}\boldsymbol{S}'(n)\boldsymbol{\Lambda}\right] + \mu^2\sigma_v^2\boldsymbol{\Lambda},\end{aligned} \tag{12.214}$$

where $\boldsymbol{\Lambda}$ is defined as in (12.188). Thus, a recursion for $\boldsymbol{s}'(n+1)$ is given by

$$\boldsymbol{s}'(n+1) = \left[\boldsymbol{I} - 2\mu\boldsymbol{\Lambda} + \mu^2\left(\beta\boldsymbol{\Lambda}^2 + \boldsymbol{l}\boldsymbol{l}^T\right)\right]\boldsymbol{s}'(n) + \mu^2\sigma_v^2\boldsymbol{l}, \tag{12.215}$$

where $\boldsymbol{l}$ is defined in (12.194).

The system matrix $\boldsymbol{A}$ of (12.215) is given by

$$\boldsymbol{A} = \boldsymbol{I} - 2\mu\boldsymbol{\Lambda} + \mu^2\left(\beta\boldsymbol{\Lambda}^2 + \boldsymbol{l}\boldsymbol{l}^T\right) = (\boldsymbol{I} - \mu\boldsymbol{\Lambda})^2 + \mu^2(\beta-1)\boldsymbol{\Lambda}^2 + \mu^2\boldsymbol{l}\boldsymbol{l}^T.$$

Although it is possible to obtain the exact values of all eigenvalues of $\boldsymbol{A}$, as done in [353], we follow here a much simpler, approximate route. We can use Fact 12.9 in Box 12.3 to find an upper bound for the largest eigenvalue ν_i of $\boldsymbol{A}$, using the 1-norm of $\boldsymbol{A}$:

$$\max_{0\le i\le M-1} |\nu_i| \le \|\boldsymbol{A}\|_1 = \max_{0\le i\le M-1} \sum_{j=0}^{M-1} |a_{ij}|,$$

where a_{ij} are the entries of $\boldsymbol{A}$. If we find a range of μ for which $\|\boldsymbol{A}\|_1 \le 1$, then the recursion will be stable. This will be of course a conservative condition (sufficient, but not necessary).

Now note that the ith column of $\boldsymbol{A}$ has entries $\lambda_i\lambda_j$ if $i \ne j$ and $(1-\mu\lambda_i)^2 + \mu^2(\beta-1)\lambda_i^2 + \mu^2\lambda_i^2$ in the term corresponding to the diagonal. Since all terms are positive, the 1-norm is

$$\begin{aligned}& \max_{0\le i\le M-1}\left[(1-\mu\lambda_i)^2 + \mu^2(\beta-1)\lambda_i^2 + \mu^2\lambda_i\sum_{j=0}^{M-1}\lambda_j\right] \\ &\quad = \max_{0\le i\le M-1}\left[(1-\mu\lambda_i)^2 + \mu^2(\beta-1)\lambda_i^2 + \mu^2\lambda_i\operatorname{Tr}(\boldsymbol{\Lambda})\right].\end{aligned}$$

Recall that $\operatorname{Tr}(\boldsymbol{\Lambda}) = \operatorname{Tr}(\boldsymbol{R}_\phi)$ (Fact 12.1 from Box 12.3). Therefore, the recursion will be stable if

$$1 - 2\mu\lambda_i + \mu^2\lambda_i^2 + \mu^2(\beta-1)\lambda_i^2 + \mu^2\lambda_i\operatorname{Tr}(\boldsymbol{R}_\phi) \le 1, \quad 0 \le i \le M-1.$$

Simplifying, we obtain the condition

$$\mu \leq \frac{2}{\beta\lambda_i + \mathrm{Tr}(\boldsymbol{R}_\phi)}, \quad 0 \leq i \leq M-1.$$

The smallest bound is for $\lambda_i = \lambda_{\max}$:

$$\mu \leq \frac{2}{\beta\lambda_{\max} + \mathrm{Tr}(\boldsymbol{R}_\phi)}. \tag{12.216}$$

If we replace $\lambda_{\max}$ by $\mathrm{Tr}(\boldsymbol{R}_\phi) \geq \lambda_{\max}$ in the denominator, we obtain a simpler, but more conservative condition:

$$\mu \leq \frac{2}{(\beta+1)\,\mathrm{Tr}(\boldsymbol{R}_\phi)}. \tag{12.217}$$

For real data, this range reduces to

$$0 < \mu < \frac{2}{3\,\mathrm{Tr}(\boldsymbol{R}_\phi)} \quad \text{(real data)}, \tag{12.218}$$

and for complex data, this range reduces to

$$0 < \mu < \frac{1}{\mathrm{Tr}(\boldsymbol{R}_\phi)} \quad \text{(complex data)}. \tag{12.219}$$

If the input is a tapped-delay line as in (12.114), $\mathrm{Tr}(\boldsymbol{R}_\phi) = \sum_{k=0}^{M-1} \mathrm{E}\{|x(n-k)|^2\}$. If $\{x(n)\}$ is a stationary sequence, the stability condition for LMS is easy to obtain: we need to choose μ in the range

$$0 < \mu < \frac{2}{(\beta+1) \times M \times (\text{average power of } x(n))}. \tag{12.220}$$

This concludes our stability analysis for LMS. We now show how to extend these results to NLMS and RLS in a unified way. At the end of the next section, we give examples comparing the models to simulated results.

12.4.4 A unified statistical analysis

In this section, we assume a nonstationary environment, where the optimal solution varies according to the random walk model (see Assumption 12.A-2). We start by obtaining a general update equation for the LMS, NLMS, and RLS algorithms, in order to provide a unified statistical analysis for these algorithms.

12.4.4.1 A general update equation

In order to provide a unified analysis for the LMS, NLMS, and RLS algorithms, we consider the following general update equation:

$$\boldsymbol{w}(n+1) = \boldsymbol{w}(n) + \rho(n)\,\boldsymbol{M}(n)\boldsymbol{\phi}(n)e^*(n), \tag{12.221}$$

where $\rho(n)$ is a step size, $\boldsymbol{M}(n)$ is a nonsingular matrix with Hermitian symmetry ($\boldsymbol{M}^H(n) = \boldsymbol{M}(n)$), and $e(n) = d(n) - \boldsymbol{w}^H(n)\boldsymbol{\phi}(n)$ is the estimation error. Many adaptive algorithms can be written as in (12.221), with proper choices of $\rho(n)$ and $\boldsymbol{M}(n)$. The LMS, NLMS, and RLS algorithms employ the step sizes $\rho(n)$ and the matrices $\boldsymbol{M}(n)$ as in Table 12.6, where $\boldsymbol{I}$ is the $M \times M$ identity matrix, μ and $0 < \tilde{\mu} < 2$ are step sizes, and $0 \ll \lambda < 1$ is a forgetting factor. For RLS, $\boldsymbol{M}(n) = (1-\lambda)^{-1}\boldsymbol{P}(n)$ (see Eq. (12.163)). Note that since $(1-\lambda)$ plays the role of step size in RLS, we multiply the matrix $\boldsymbol{P}(n)$ by $(1-\lambda)^{-1}$, keeping the update equation of the RLS weights as in (12.168).

We must rewrite (12.221) in terms of the weight error vector $\tilde{\boldsymbol{w}}$. Thus, subtracting both sides of (12.221) from $\boldsymbol{w}_o(n+1) = \boldsymbol{w}_o(n) + \boldsymbol{q}(n)$ (Eq. (12.183)), we obtain

$$\begin{aligned} \boldsymbol{w}_o(n+1) - \boldsymbol{w}(n+1) &= \boldsymbol{w}_o(n) + \boldsymbol{q}(n) - \boldsymbol{w}(n) - \rho(n)\boldsymbol{M}(n)\boldsymbol{\phi}(n)e^*(n), \\ \tilde{\boldsymbol{w}}(n+1) &= \tilde{\boldsymbol{w}}(n) - \rho(n)\boldsymbol{M}(n)\boldsymbol{\phi}(n)e^*(n) + \boldsymbol{q}(n). \end{aligned} \tag{12.222}$$

Replacing $e^*(n) = e_a^*(n) + v_o^*(n) = \boldsymbol{\phi}^H(n)\tilde{\boldsymbol{w}}(n) + v_o^*(n)$ in (12.222), we arrive at

$$\tilde{\boldsymbol{w}}(n+1) = \left[\boldsymbol{I} - \rho(n)\boldsymbol{M}(n)\boldsymbol{\phi}(n)\boldsymbol{\phi}^H(n)\right]\tilde{\boldsymbol{w}}(n) - \rho(n)\boldsymbol{M}(n)\boldsymbol{\phi}(n)v_o^*(n) + \boldsymbol{q}(n). \tag{12.223}$$

Table 12.6 Parameters of LMS, NLMS, and RLS algorithms.

Alg.	$\rho(n)$	$\boldsymbol{M}^{-1}(n)$
LMS	μ	$\boldsymbol{I}$
NLMS	$\dfrac{\tilde{\mu}}{\varepsilon + \|\boldsymbol{\phi}(n)\|^2}$	$\boldsymbol{I}$
RLS	$1-\lambda$	$(1-\lambda)\sum_{\ell=1}^{n}\lambda^{n-\ell}\boldsymbol{\phi}(\ell)\boldsymbol{\phi}^H(\ell)$

We next use recursion (12.223) to evaluate the EMSE and MSD of LMS, NLMS, and RLS.

12.4.4.2 *Alternative analysis methods*

Closed-form expressions for the EMSE and MSD can be obtained in two ways:

(i) Finding a recursion for the correlation matrix of the weight error vector, denoted by $\boldsymbol{S}(n)$. The EMSE and MSE can be easily computed given $\boldsymbol{S}(n)$, as we saw in Section 12.4.2. This is the earlier approach in the literature [117,172] and therefore, we refer to it here as the *traditional* method.

(ii) Using the energy conservation analysis of [313,317,321,396]. The energy conservation approach relies on an energy conservation relation that holds for a large class of adaptive filters. The energy conservation method presents some advantages when compared to the traditional one. For example, using energy conservation, one can obtain steady-state results directly, bypassing several of the difficulties encountered in obtaining them as a limiting case of a transient analysis. This method can also be used in transient analyses of adaptive algorithms (see, e.g., Part V of [317]).

It is important to understand that both methods involve different sets of approximations (assumptions), so they do not always arrive at the same expressions. Generally, they tend to be equal in the limit of slow adaptation (i.e., if $\rho(n) \to 0$ in (12.223)). For larger values of $\rho(n)$, which method will provide the best approximation depends on the algorithm. As for the work necessary to arrive at the final answers, the energy conservation approach is almost always easier to use if one is interested only in steady-state results. If one is interested in transient performance and stability studies, then the traditional method may be simpler, depending on the algorithm under study.

To give an overview of both methods, we first use the traditional method to obtain the transient and steady-state performance of the LMS, NLMS, and RLS algorithms. This is an extension of what we did for LMS in Section 12.4.3, taking advantage of the similarities between the algorithms. In the sequel, we employ the energy conservation method to obtain closed-form expressions for the steady-state EMSE of the same algorithms.

In this section we allow the environment to be nonstationary, i.e., the optimal solution $\boldsymbol{w}_o$ is time-varying. The results for stationary environments can be obtained simply by making $\boldsymbol{Q} = \boldsymbol{0}$ in the equations below. In order to simplify the arguments, we require that the regressor sequence $\{\boldsymbol{\phi}(n)\}$ be WSS, that is, we assume that $\boldsymbol{R}_{\boldsymbol{\phi}}$ is constant.

12.4.4.3 Analysis with the traditional method

The traditional analysis method consists of finding a recursion for $\boldsymbol{S}(n)$ and using (12.186) and (12.187) to evaluate the MSD and the EMSE, respectively. A recursion for $\boldsymbol{S}(n)$ can be obtained by multiplying (12.223) by its Hermitian, taking the expectations of both sides, and recalling that Assumption 12.A-1 implies that $\tilde{\boldsymbol{w}}(n)$ is independent of $\boldsymbol{\phi}(n)$ and of the zero-mean $v_o(n)$. Furthermore, using the fact that $\mathrm{E}\{\tilde{\boldsymbol{w}}(n)\boldsymbol{q}^H(n)\} = \boldsymbol{0}$ since the sequence $\{\boldsymbol{q}(n)\}$ is i.i.d. (Assumption 12.A-2), we arrive at

$$\begin{aligned}
\mathrm{E}\{\tilde{\boldsymbol{w}}(n+1)\tilde{\boldsymbol{w}}^H(n+1)\} = {} & \mathrm{E}\{\boldsymbol{q}(n)\boldsymbol{q}^H(n)\} + \mathrm{E}\{\tilde{\boldsymbol{w}}(n)\tilde{\boldsymbol{w}}^H(n)\} \\
& - \overbrace{\mathrm{E}\{\rho(n)\tilde{\boldsymbol{w}}(n)\tilde{\boldsymbol{w}}^H(n)\boldsymbol{\phi}(n)\boldsymbol{\phi}^H(n)\boldsymbol{M}(n)\}}^{\mathcal{A}} \\
& - \overbrace{\mathrm{E}\{\rho(n)\boldsymbol{M}(n)\boldsymbol{\phi}(n)\boldsymbol{\phi}^H(n)\tilde{\boldsymbol{w}}(n)\tilde{\boldsymbol{w}}^H(n)\}}^{\mathcal{B}} \\
& + \overbrace{\mathrm{E}\{\rho^2(n)\boldsymbol{M}(n)\boldsymbol{\phi}(n)\boldsymbol{\phi}^H(n)\tilde{\boldsymbol{w}}(n)\tilde{\boldsymbol{w}}^H(n)\boldsymbol{\phi}(n)\boldsymbol{\phi}^H(n)\boldsymbol{M}(n)\}}^{\mathcal{C}} \\
& + \overbrace{\mathrm{E}\{\rho^2(n)|v_o(n)|^2\boldsymbol{M}(n)\boldsymbol{\phi}(n)\boldsymbol{\phi}^H(n)\boldsymbol{M}(n)\}}^{\mathcal{D}} .
\end{aligned} \tag{12.224}$$

To simplify (12.224), we also need two additional assumptions.

12.A-3. Matrix $\boldsymbol{M}(n)$ varies slowly in relation to $\tilde{\boldsymbol{w}}(n)$. Thus, when $\boldsymbol{M}(n)$ appears inside the expectations of (12.224), we simply replace it by its mean. For LMS and NLMS, this assumption is not necessary, since $\boldsymbol{M}(n) = \boldsymbol{I}$. For RLS, $\boldsymbol{M}(n) = (1-\lambda)^{-1}\boldsymbol{P}(n) \approx (1-\lambda)^{-1}\mathrm{E}\{\boldsymbol{P}(n)\}$ if $\lambda \approx 1$.

12.A-4. The input regressor vector $\boldsymbol{\phi}(n)$ is assumed to be Gaussian. This assumption makes the analysis more tractable, since Gaussianity facilitates the computation of the expected values in $\mathcal{C}$. Note however that this assumption will never hold in the case of Volterra filters, since (for example) $x(n)$ and $x^2(n)$ cannot both be Gaussian.

Now, using Assumptions 12.A-1–12.A-4, we can evaluate the terms $\mathcal{A}$–$\mathcal{D}$ of (12.224):

$\mathcal{A}$ The term $\mathcal{A}$ can be approximated by

$$\begin{aligned}\mathcal{A} &= \mathrm{E}\left\{\mathrm{E}\left\{\rho(n)\tilde{\boldsymbol{w}}(n)\tilde{\boldsymbol{w}}^H(n)\boldsymbol{\phi}(n)\boldsymbol{\phi}^H(n)\boldsymbol{M}(n)|\boldsymbol{\phi}(n)\right\}\right\}\\ &\approx \mathrm{E}\left\{\rho(n)\,\mathrm{E}\left\{\tilde{\boldsymbol{w}}(n)\tilde{\boldsymbol{w}}^H(n)\right\}\boldsymbol{\phi}(n)\boldsymbol{\phi}^H(n)\right\}\mathrm{E}\{\boldsymbol{M}(n)\}\\ &= \boldsymbol{S}(n)\,\mathrm{E}\left\{\rho(n)\boldsymbol{\phi}(n)\boldsymbol{\phi}^H(n)\right\}\mathrm{E}\{\boldsymbol{M}(n)\}.\end{aligned} \tag{12.225}$$

For LMS, this term reduces to

$$\mathcal{A}^{\mathrm{LMS}} \approx \mu \boldsymbol{S}(n)\boldsymbol{R}_{\boldsymbol{\phi}}. \tag{12.226}$$

For NLMS, we must obtain an approximation for

$$\mathrm{E}\left\{\rho(n)\boldsymbol{\phi}(n)\boldsymbol{\phi}^H(n)\right\} = \tilde{\mu}\,\mathrm{E}\left\{\frac{\boldsymbol{\phi}(n)\boldsymbol{\phi}^H(n)}{\varepsilon + \boldsymbol{\phi}^H(n)\boldsymbol{\phi}(n)}\right\}. \tag{12.227}$$

It is possible to evaluate this expected value exactly (see [6,7,48,49]), but our intention here is to obtain the simplest possible model. Therefore, we will use two additional approximations.

12.A-5. The number of coefficients M is large enough for each element $\boldsymbol{\phi}(n)\boldsymbol{\phi}^H(n)$ in the numerator to be approximately independent of the denominator $\sum_{k=0}^{M-1}|\phi(n-k)|^2$ and for the following approximation to hold:

$$\mathrm{E}\left\{\frac{1}{(\varepsilon + \boldsymbol{\phi}^H(n)\boldsymbol{\phi}(n))^k}\right\} \approx \frac{1}{\mathrm{E}\left\{(\varepsilon + \boldsymbol{\phi}^H(n)\boldsymbol{\phi}(n))^k\right\}}. \tag{12.228}$$

This is equivalent to applying the averaging principle proposed in [314], since for large M, $\|\boldsymbol{\phi}(n)\|^2$ tends to vary slowly compared to the individual entries of $\boldsymbol{\phi}(n)\boldsymbol{\phi}^H(n)$.
Thus, (12.227) can be approximated as

$$\mathrm{E}\left\{\frac{\boldsymbol{\phi}(n)\boldsymbol{\phi}^H(n)}{\varepsilon + \boldsymbol{\phi}^H(n)\boldsymbol{\phi}(n)}\right\} \approx \mathrm{E}\left\{\frac{1}{\varepsilon + \boldsymbol{\phi}^H(n)\boldsymbol{\phi}(n)}\right\}\mathrm{E}\left\{\boldsymbol{\phi}(n)\boldsymbol{\phi}^H(n)\right\}. \tag{12.229}$$

Under Assumption 12.A-5, the first expected value in (12.229) can be approximated by

$$\mathrm{E}\left\{\frac{1}{\varepsilon + \boldsymbol{\phi}^H(n)\boldsymbol{\phi}(n)}\right\} \approx \frac{1}{\mathrm{E}\{\varepsilon + \boldsymbol{\phi}^H(n)\boldsymbol{\phi}(n)\}} = \frac{1}{\varepsilon + \mathrm{Tr}(\boldsymbol{R}_{\boldsymbol{\phi}})}. \tag{12.230}$$

An alternative, more accurate approximation is obtained when $\boldsymbol{\phi}(n)$ is Gaussian and a tapped-delay line.

12.A-6. The regressor $\boldsymbol{\phi}(n)$ is formed by a tapped-delay line with Gaussian entries as in (12.114), and $\varepsilon = 0$.

Under Assumptions 12.A-4, 12.A-5, and 12.A-6 it can be shown (see [99] for details) that

$$\mathrm{E}\left\{\frac{1}{\varepsilon+\boldsymbol{\phi}^H(n)\boldsymbol{\phi}(n)}\right\}\approx\frac{1}{\sigma_x^2(M-2)}. \tag{12.231}$$

Both approximations are only valid for long filters. However, (12.231) shows that NLMS needs a large regularization constant if $M < 3$, which is verified in practice [48]. We will present the final results for NLMS in terms of Assumption 12.A-6, but recall that a less precise, but more general option is available using only Assumption 12.A-4 and approximations such as (12.230).

Thus, the term $\boldsymbol{\mathcal{A}}$ for NLMS reduces to (using (12.231))

$$\boldsymbol{\mathcal{A}}^{\mathrm{NLMS}}\approx\frac{\tilde{\mu}}{\sigma_x^2(M-2)}\boldsymbol{S}(n)\boldsymbol{R}_{\boldsymbol{\phi}}. \tag{12.232}$$

Finally, for RLS, we obtain

$$\boldsymbol{\mathcal{A}}^{\mathrm{RLS}}\approx\boldsymbol{S}(n)\boldsymbol{R}_{\boldsymbol{\phi}}\overline{\boldsymbol{P}}(n), \tag{12.233}$$

where $\overline{\boldsymbol{P}}(n)\triangleq\mathrm{E}\{\boldsymbol{R}_{\boldsymbol{\phi}}^{-1}(n)\}$, which we will evaluate later on.

$\boldsymbol{\mathcal{B}}$ Analogously, we obtain for $\boldsymbol{\mathcal{B}}$

$$\boldsymbol{\mathcal{B}}\approx\mathrm{E}\{\boldsymbol{M}(n)\}\,\mathrm{E}\left\{\rho(n)\boldsymbol{\phi}(n)\boldsymbol{\phi}^H(n)\right\}\boldsymbol{S}(n). \tag{12.234}$$

Particularizing for each algorithm, we have

$$\boldsymbol{\mathcal{B}}^{\mathrm{LMS}}\approx\mu\boldsymbol{R}_{\boldsymbol{\phi}}\boldsymbol{S}(n), \tag{12.235}$$

$$\boldsymbol{\mathcal{B}}^{\mathrm{NLMS}}\approx\frac{\tilde{\mu}}{\sigma_x^2(M-2)}\boldsymbol{R}_{\boldsymbol{\phi}}\boldsymbol{S}(n), \tag{12.236}$$

and

$$\boldsymbol{\mathcal{B}}^{\mathrm{RLS}}\approx\overline{\boldsymbol{P}}(n)\boldsymbol{R}_{\boldsymbol{\phi}}\boldsymbol{S}(n). \tag{12.237}$$

$\boldsymbol{\mathcal{C}}$ The term $\boldsymbol{\mathcal{C}}$ can be approximated by

$$\begin{aligned}\boldsymbol{\mathcal{C}}&=\mathrm{E}\left\{\rho^2(n)\boldsymbol{M}(n)\boldsymbol{\phi}(n)\boldsymbol{\phi}^H(n)\tilde{\boldsymbol{w}}(n)\tilde{\boldsymbol{w}}^H(n)\boldsymbol{\phi}(n)\boldsymbol{\phi}^H(n)\boldsymbol{M}(n)\right\}\\&\approx\mathrm{E}\{\boldsymbol{M}(n)\}\,\mathrm{E}\left\{\rho^2(n)\boldsymbol{\phi}(n)\boldsymbol{\phi}^H(n)\boldsymbol{S}(n)\boldsymbol{\phi}(n)\boldsymbol{\phi}^H(n)\right\}\mathrm{E}\{\boldsymbol{M}(n)\}.\end{aligned} \tag{12.238}$$

Under the Gaussianity assumption, Assumption 12.A-4, we have (see Box 12.7)

$$\mathrm{E}\left\{\boldsymbol{\phi}(n)\boldsymbol{\phi}^H(n)\boldsymbol{S}(n)\boldsymbol{\phi}(n)\boldsymbol{\phi}^H(n)\right\}=\boldsymbol{R}_{\boldsymbol{\phi}}\,\mathrm{Tr}(\boldsymbol{S}(n)\boldsymbol{R}_{\boldsymbol{\phi}})+\beta\boldsymbol{R}_{\boldsymbol{\phi}}\boldsymbol{S}(n)\boldsymbol{R}_{\boldsymbol{\phi}}, \tag{12.239}$$

where $\beta=1$ (resp. $\beta=2$) for complex (resp. real) data. Thus, for LMS we have

$$\boldsymbol{\mathcal{C}}^{\mathrm{LMS}}\approx\mu^2\left[\boldsymbol{R}_{\boldsymbol{\phi}}\,\mathrm{Tr}(\boldsymbol{S}(n)\boldsymbol{R}_{\boldsymbol{\phi}})+\beta\boldsymbol{R}_{\boldsymbol{\phi}}\boldsymbol{S}(n)\boldsymbol{R}_{\boldsymbol{\phi}}\right]. \tag{12.240}$$

For NLMS, using the same arguments to get (12.229), the following approximation holds for large M:

$$\begin{aligned}
&\mathrm{E}\left\{\rho^2(n)\boldsymbol{\phi}(n)\boldsymbol{\phi}^H(n)\boldsymbol{S}(n)\boldsymbol{\phi}(n)\boldsymbol{\phi}^H(n)\right\} \\
&\quad\approx \tilde{\mu}^2\,\mathrm{E}\left\{\frac{1}{\left[\varepsilon+\boldsymbol{\phi}^H(n)\boldsymbol{\phi}(n)\right]^2}\right\}\mathrm{E}\left\{\boldsymbol{\phi}(n)\boldsymbol{\phi}^H(n)\boldsymbol{S}(n)\boldsymbol{\phi}(n)\boldsymbol{\phi}^H(n)\right\}.
\end{aligned} \tag{12.241}$$

Under Assumptions 12.A-4 and 12.A-5, using (12.281) and (12.282) from Box 12.7 with $\boldsymbol{S}=\boldsymbol{I}$, the first expectation on the right-hand side of (12.241) can be approximated for large M by [46]

$$\mathrm{E}\left\{\frac{1}{\left[\varepsilon+\boldsymbol{\phi}^H(n)\boldsymbol{\phi}(n)\right]^2}\right\} \approx \frac{1}{\varepsilon^2+2\varepsilon\,\mathrm{Tr}(\boldsymbol{R}_{\boldsymbol{\phi}})+\beta\,\mathrm{Tr}(\boldsymbol{R}_{\boldsymbol{\phi}}^2)+\mathrm{Tr}^2(\boldsymbol{R}_{\boldsymbol{\phi}})}. \tag{12.242}$$

On the other hand, under Assumption 12.A-6, we obtain [99]

$$\mathrm{E}\left\{\frac{1}{\left[\boldsymbol{\phi}^H(n)\boldsymbol{\phi}(n)\right]^2}\right\} \approx \frac{1}{\sigma_x^4(M-2)(M-4)}. \tag{12.243}$$

Replacing (12.243) and (12.239) in (12.241), we arrive at

$$\mathcal{C}^{\mathrm{NLMS}} \approx \frac{\tilde{\mu}^2}{\sigma_x^4(M-2)(M-4)}\left[\boldsymbol{R}_{\boldsymbol{\phi}}\,\mathrm{Tr}(\boldsymbol{S}(n)\boldsymbol{R}_{\boldsymbol{\phi}})+\beta\boldsymbol{R}_{\boldsymbol{\phi}}\boldsymbol{S}(n)\boldsymbol{R}_{\boldsymbol{\phi}}\right]. \tag{12.244}$$

Finally, for RLS we obtain

$$\mathcal{C}^{\mathrm{RLS}} \approx \overline{\boldsymbol{P}}(n)\left[\boldsymbol{R}_{\boldsymbol{\phi}}\,\mathrm{Tr}(\boldsymbol{S}(n)\boldsymbol{R}_{\boldsymbol{\phi}})+\beta\boldsymbol{R}_{\boldsymbol{\phi}}\boldsymbol{S}(n)\boldsymbol{R}_{\boldsymbol{\phi}}\right]\overline{\boldsymbol{P}}(n). \tag{12.245}$$

$\mathcal{D}$ Using the assumption that $v_o(n)$ is independent of $\boldsymbol{\phi}(n)$ and $\tilde{\boldsymbol{w}}(n)$, $\mathcal{D}$ reduces to

$$\begin{aligned}
\mathcal{D} &= \mathrm{E}\{\rho^2(n)|v_o(n)|^2\boldsymbol{M}(n)\boldsymbol{\phi}(n)\boldsymbol{\phi}^H(n)\boldsymbol{M}(n)\} \\
&\approx \sigma_v^2\,\mathrm{E}\{\boldsymbol{M}(n)\}\,\mathrm{E}\{\rho^2(n)\boldsymbol{\phi}(n)\boldsymbol{\phi}^H(n)\}\,\mathrm{E}\{\boldsymbol{M}(n)\}.
\end{aligned} \tag{12.246}$$

Particularizing for each algorithm, we obtain

$$\mathcal{D}^{\mathrm{LMS}} \approx \mu^2\sigma_v^2\,\boldsymbol{R}_{\boldsymbol{\phi}}(n), \tag{12.247}$$

$$\mathcal{D}^{\mathrm{NLMS}} \approx \frac{\tilde{\mu}^2}{\sigma_x^4(M-2)(M-4)}\sigma_v^2\,\boldsymbol{R}_{\boldsymbol{\phi}}(n), \tag{12.248}$$

where we could also have used (12.242), and

$$\mathcal{D}^{\mathrm{RLS}} \approx \sigma_v^2\,\overline{\boldsymbol{P}}(n)\boldsymbol{R}_{\boldsymbol{\phi}}\overline{\boldsymbol{P}}(n). \tag{12.249}$$

From the previous results, (12.224) reduces to

$$\begin{aligned}
\mathbf{S}(n+1) \approx \mathbf{S}(n) &- \mathrm{E}\{\rho(n)\}\left[\boldsymbol{S}(n)\boldsymbol{R}_{\boldsymbol{\phi}}\,\mathrm{E}\{\boldsymbol{M}(n)\}+\mathrm{E}\{\boldsymbol{M}(n)\}\boldsymbol{R}_{\boldsymbol{\phi}}\boldsymbol{S}(n)\right] \\
&+\mathrm{E}\{\rho^2(n)\}\,\mathrm{E}\{\boldsymbol{M}(n)\}\left[\boldsymbol{R}_{\boldsymbol{\phi}}\,\mathrm{Tr}(\boldsymbol{S}(n)\boldsymbol{R}_{\boldsymbol{\phi}})+\beta\boldsymbol{R}_{\boldsymbol{\phi}}\boldsymbol{S}(n)\boldsymbol{R}_{\boldsymbol{\phi}}\right]\mathrm{E}\{\boldsymbol{M}(n)\}
\end{aligned}$$

$$+\sigma_v^2\,\mathrm{E}\{\rho^2(n)\}\,\mathrm{E}\{\boldsymbol{M}(n)\}\boldsymbol{R}_{\boldsymbol{\phi}}\,\mathrm{E}\{\boldsymbol{M}(n)\}+\boldsymbol{Q} \tag{12.250}$$

with initialization $\boldsymbol{S}(0)=[\boldsymbol{w}_{\mathrm{o}}(0)-\boldsymbol{w}(0)][\boldsymbol{w}_{\mathrm{o}}(0)-\boldsymbol{w}(0)]^H$.

For RLS, we still need an approximation for $\overline{\boldsymbol{P}}(n)=\mathrm{E}\{\boldsymbol{P}(n)\}$. Initializing the estimate of the autocorrelation matrix with $\widehat{\boldsymbol{R}}_{\boldsymbol{\phi}}(-1)=\delta\boldsymbol{I}$, we will have at iteration n

$$\widehat{\boldsymbol{R}}_{\boldsymbol{\phi}}(n)=\lambda^n\delta\boldsymbol{I}+\sum_{\ell=1}^{n}\lambda^{n-\ell}\boldsymbol{\phi}(\ell)\boldsymbol{\phi}^H(\ell). \tag{12.251}$$

Taking the expectations of both sides of (12.251), we obtain

$$\mathrm{E}\left\{\widehat{\boldsymbol{R}}_{\boldsymbol{\phi}}(n)\right\}=\lambda^n\delta\boldsymbol{I}+\boldsymbol{R}_{\boldsymbol{\phi}}\frac{1-\lambda^n}{1-\lambda}. \tag{12.252}$$

For $\lambda\approx 1$, the variance of $\widehat{\boldsymbol{R}}_{\boldsymbol{\phi}}(n)$ is small, and we can approximate $\overline{\boldsymbol{P}}(n)\approx\left[\mathrm{E}\left\{\widehat{\boldsymbol{R}}_{\boldsymbol{\phi}}(n)\right\}\right]^{-1}$, so that

$$\overline{\boldsymbol{P}}(n)\approx\left[\lambda^n\delta\boldsymbol{I}+\boldsymbol{R}_{\boldsymbol{\phi}}\frac{1-\lambda^n}{1-\lambda}\right]^{-1}. \tag{12.253}$$

This approximation is good for large n and in steady state. During the initial phases of the transient, however, the approximation will only be reasonable if δ is large enough to make the last term in (12.252) small compared to $\lambda^n\delta\boldsymbol{I}$. This is because for $n<M$, the term $\sum_{\ell=1}^{n}\lambda^{n-\ell}\boldsymbol{\phi}(\ell)\boldsymbol{\phi}^H(\ell)$ in (12.251) will be singular, and thus it will not be well approximated by $(1-\lambda^n)\boldsymbol{R}_{\boldsymbol{\phi}}/(1-\lambda)$.

Using the previous approximations, (12.250) can be particularized for the LMS, NLMS, and RLS algorithms as shown in Table 12.7. Since we can obtain the EMSE and the MSD from $\boldsymbol{S}(n)$ and $\boldsymbol{R}_{\boldsymbol{\phi}}$ through (12.187) and (12.186), we should use the recursions of Table 12.7 to compute $\boldsymbol{S}(n)$ and then evaluate $\zeta(n)=\mathrm{Tr}(\boldsymbol{S}(n)\boldsymbol{R}_{\boldsymbol{\phi}})$ and $\chi(n)=\mathrm{Tr}(\boldsymbol{S}(n))$.

Table 12.7 Recursions for covariance matrix $\boldsymbol{S}(n+1)$. The expressions for NLMS assume that the regressor is a tapped-delay line. Alternative expressions for NLMS are available using (12.230) **and** (12.242).

Alg.	$\boldsymbol{S}(n+1)$
LMS	$\boldsymbol{S}(n+1)\approx\boldsymbol{S}(n)-\mu\left[\boldsymbol{S}(n)\boldsymbol{R}_{\boldsymbol{\phi}}+\boldsymbol{R}_{\boldsymbol{\phi}}\boldsymbol{S}(n)\right]+\mu^2\left[\boldsymbol{R}_{\boldsymbol{\phi}}\,\mathrm{Tr}\left(\boldsymbol{S}(n)\boldsymbol{R}_{\boldsymbol{\phi}}\right)+\beta\boldsymbol{R}_{\boldsymbol{\phi}}\boldsymbol{S}(n)\boldsymbol{R}_{\boldsymbol{\phi}}+\sigma_v^2\boldsymbol{R}_{\boldsymbol{\phi}}\right]+\mathbf{Q}$
NLMS	$\boldsymbol{S}(n+1)\approx\boldsymbol{S}(n)-\frac{\tilde{\mu}}{\sigma_x^2(M-2)}\left[\boldsymbol{S}(n)\boldsymbol{R}_{\boldsymbol{\phi}}+\boldsymbol{R}_{\boldsymbol{\phi}}\boldsymbol{S}(n)\right]+\frac{\tilde{\mu}^2}{\sigma_x^4(M-2)(M-4)}\left[\boldsymbol{R}_{\boldsymbol{\phi}}\,\mathrm{Tr}\left(\boldsymbol{S}(n)\boldsymbol{R}_{\boldsymbol{\phi}}\right)+\beta\boldsymbol{R}_{\boldsymbol{\phi}}\boldsymbol{S}(n)\boldsymbol{R}_{\boldsymbol{\phi}}+\sigma_v^2\boldsymbol{R}_{\boldsymbol{\phi}}\right]+\mathbf{Q}$
RLS	$\boldsymbol{S}(n+1)\approx\boldsymbol{S}(n)-\left[\boldsymbol{S}(n)\boldsymbol{R}_{\boldsymbol{\phi}}\overline{\boldsymbol{P}}(n)+\overline{\boldsymbol{P}}(n)\boldsymbol{R}_{\boldsymbol{\phi}}\boldsymbol{S}(n)\right]+\overline{\boldsymbol{P}}(n)\left[\boldsymbol{R}_{\boldsymbol{\phi}}\,\mathrm{Tr}\left(\boldsymbol{S}(n)\boldsymbol{R}_{\boldsymbol{\phi}}\right)+\beta\boldsymbol{R}_{\boldsymbol{\phi}}\boldsymbol{S}(n)\boldsymbol{R}_{\boldsymbol{\phi}}+\sigma_v^2\boldsymbol{R}_{\boldsymbol{\phi}}\right]\overline{\boldsymbol{P}}(n)+\mathbf{Q}$

Example 12.E-4.3. To verify the accuracy of the expressions of Table 12.7, we consider a system identification application in a stationary environment ($\boldsymbol{Q}=\mathbf{0}$). The optimal solution is a low-pass

FIR filter with linear phase, whose coefficients are shown in Fig. 12.30. The regressor $\boldsymbol{\phi}(n)$ is obtained from a process $x(n)$ generated with a first-order autoregressive model, whose transfer function is $\sqrt{1-b^2}/(1-bz^{-1})$. Note that the sequence $\{\boldsymbol{\phi}(n)\}$ is not i.i.d. This model is fed with an i.i.d. Gaussian random process, whose variance is such that $\text{Tr}(\boldsymbol{R}_{\boldsymbol{\phi}}) = 1$. Moreover, additive i.i.d. noise $v_o(n)$ with variance $\sigma_v^2 = 0.01$ is added to form the desired signal.

Fig. 12.31 shows the EMSE for RLS ($\lambda = 0.995$, $\delta = 1$), NLMS ($\tilde{\mu} = 0.1$, $\varepsilon = 10^{-5}$), and LMS ($\mu = 0.05$), estimated from the ensemble average of 2000 independent runs. The dashed lines represent the theoretical EMSE computed as $\zeta(n) = \text{Tr}(\boldsymbol{S}(n)\boldsymbol{R}_{\boldsymbol{\phi}})$. We can observe a good agreement between theory and simulation.

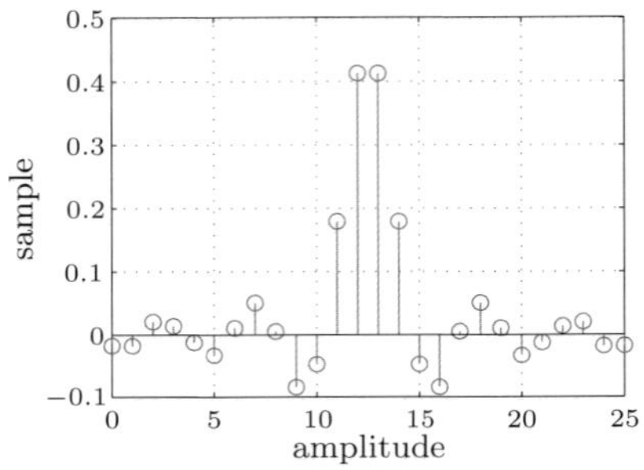

FIGURE 12.30

Impulse response $\boldsymbol{w}_o$ of the unknown low-pass filter.

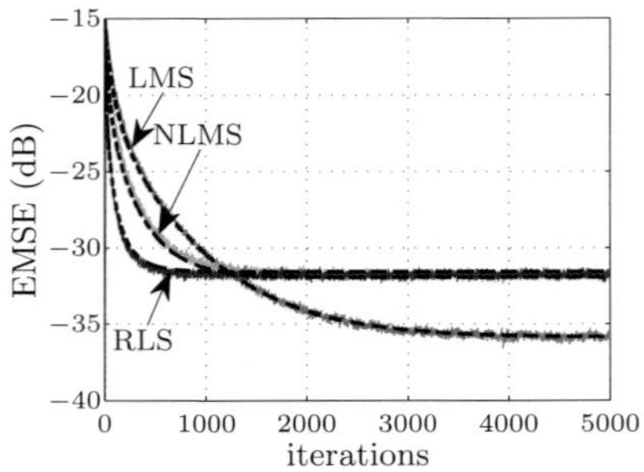

FIGURE 12.31

EMSE for RLS ($\lambda = 0.995$, $\delta = 1$), NLMS ($\tilde{\mu} = 0.1$, $\varepsilon = 10^{-5}$), and LMS ($\mu = 0.05$); $b = 0.8$, $\sigma_v^2 = 10^{-2}$, $\sigma_q^2 = 0$; mean of 2000 independent runs. The dashed lines represent the predicted values of $\zeta(n)$ for each algorithm.

Fig. 12.32 shows the EMSE of RLS ($\lambda = 0.995$), assuming different values for δ. We can observe that the theoretical EMSE of RLS presents a good agreement with simulation during the initial phases of the transient, but only for large values of δ due to the approximation (12.253).

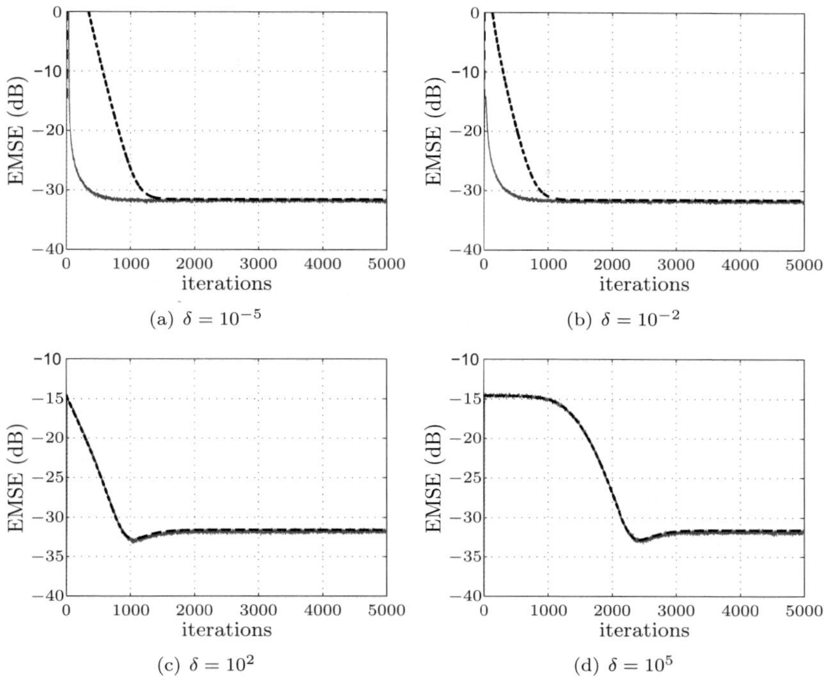

FIGURE 12.32

EMSE for RLS considering $\lambda = 0.995$ and different values of δ; $b = 0.8$, $\sigma_v^2 = 10^{-2}$, $\sigma_q^2 = 0$; mean of 2000 independent runs. The dashed lines represent the predicted values of $\zeta(n)$ for each algorithm.

12.4.4.4 Steady-state analysis with the energy conservation method

This method (also known as the *feedback* approach [222,313]) is based on an energy conservation relation that holds for a large class of adaptive filters. To obtain this relation for the algorithms of the form (12.221), we first assume a stationary environment ($\boldsymbol{q}(n) \equiv \boldsymbol{0}$).

Recall that the excess a posteriori error is given by

$$e_p(n) = \tilde{\boldsymbol{w}}^H(n+1)\boldsymbol{\phi}(n). \tag{12.254}$$

Then, we multiply both sides of the Hermitian of (12.222) (with $\boldsymbol{q}(n) \equiv \boldsymbol{0}$) from the right by $\boldsymbol{\phi}(n)$ to obtain

$$\tilde{\boldsymbol{w}}^H(n+1)\boldsymbol{\phi}(n) = \tilde{\boldsymbol{w}}^H(n)\boldsymbol{\phi}(n) - \rho(n)\boldsymbol{\phi}^H(n)\boldsymbol{M}(n)\boldsymbol{\phi}(n)e(n). \tag{12.255}$$

Using (12.173) and (12.254), (12.255) reduces to

$$e_p(n) = e_a(n) - \rho(n)\|\boldsymbol{\phi}(n)\|^2_{\boldsymbol{M}(n)}e(n), \tag{12.256}$$

where $\|\boldsymbol{u}\|_{\boldsymbol{A}}^2 = \boldsymbol{u}^H \boldsymbol{A}\boldsymbol{u}$ stands for the Euclidean norm of $\boldsymbol{u}$ weighted by the matrix $\boldsymbol{A}$. Using (12.256) to eliminate $e(n)$ in (12.222) (with $\boldsymbol{q}(n) \equiv \boldsymbol{0}$) and assuming that $\boldsymbol{\phi}(n) \neq \boldsymbol{0}$, we get

$$\tilde{\boldsymbol{w}}(n+1) + \frac{\boldsymbol{M}(n)\boldsymbol{\phi}(n)}{\|\boldsymbol{\phi}(n)\|_{\boldsymbol{M}(n)}^2} e_a^*(n) = \tilde{\boldsymbol{w}}(n) + \frac{\boldsymbol{M}(n)\boldsymbol{\phi}(n)}{\|\boldsymbol{\phi}(n)\|_{\boldsymbol{M}(n)}^2} e_p^*(n). \tag{12.257}$$

Defining

$$\bar{\rho}(n) \triangleq \begin{cases} 1/\|\boldsymbol{\phi}(n)\|_{\boldsymbol{M}(n)}^2, & \text{if } \boldsymbol{\phi}(n) \neq \boldsymbol{0}, \\ 0, & \text{otherwise,} \end{cases} \tag{12.258}$$

to include the case of zero regressor and equating the squared weighted norms on both sides of (12.257) with $\boldsymbol{M}^{-1}(n)$ as a weighting matrix, the cross-terms cancel and we obtain

$$\|\tilde{\boldsymbol{w}}(n+1)\|_{\boldsymbol{M}^{-1}(n)}^2 + \bar{\rho}(n)|e_a(n)|^2 = \|\tilde{\boldsymbol{w}}(n)\|_{\boldsymbol{M}^{-1}(n)}^2 + \bar{\rho}(n)|e_p(n)|^2. \tag{12.259}$$

This relation shows how the weighted energies of the weight error vectors at two successive time instants are related to the energies of the excess a priori and excess a posteriori errors. This relation is the reason why this approach is known as the energy conservation method – (12.259) relates the (weighted) coefficient error energies at instants n and $n+1$ to the a priori and a posteriori errors. In [317], this energy conservation relation is used extensively to find models for a wide class of adaptive filters. We follow here a shorter route, however, and derive in a more direct way the relations necessary for the study of just LMS, NLMS, and RLS.

We can obtain the steady-state EMSE by computing the squared weighted norms on both sides of (12.222) with $\boldsymbol{M}^{-1}(n)$ as a weighting matrix, which leads to

$$\begin{aligned}\|\tilde{\boldsymbol{w}}(n+1)\|_{\boldsymbol{M}^{-1}(n)}^2 &= \|\tilde{\boldsymbol{w}}(n)\|_{\boldsymbol{M}^{-1}(n)}^2 + \rho^2(n)|e(n)|^2\|\boldsymbol{\phi}(n)\|_{\boldsymbol{M}(n)}^2 \\ &\quad - \rho(n)\left[e_a(n)e^*(n) + e_a^*(n)e(n)\right] + \|\boldsymbol{q}(n)\|_{\boldsymbol{M}^{-1}(n)}.\end{aligned} \tag{12.260}$$

By taking expectations on both sides of (12.260), we arrive at (we already used Assumption 12.A-2 to eliminate cross-terms with $\boldsymbol{q}(n)$ on the right-hand side)

$$\begin{aligned}\overbrace{\mathrm{E}\left\{\|\tilde{\boldsymbol{w}}(n+1)\|_{\boldsymbol{M}^{-1}(n)}^2\right\}}^{\mathcal{A}} &= \overbrace{\mathrm{E}\left\{\|\tilde{\boldsymbol{w}}(n)\|_{\boldsymbol{M}^{-1}(n)}^2\right\}}^{\mathcal{B}} \\ &\quad + \overbrace{\mathrm{E}\left\{\rho^2(n)|e(n)|^2\|\boldsymbol{\phi}(n)\|_{\boldsymbol{M}(n)}^2\right\}}^{\mathcal{C}} \\ &\quad - \overbrace{\mathrm{E}\left\{\rho(n)\left[e_a(n)e^*(n) + e_a^*(n)e(n)\right]\right\}}^{\mathcal{D}} + \\ &\quad + \overbrace{\|\boldsymbol{q}(n)\|_{\boldsymbol{M}^{-1}(n)}^2}^{\mathcal{E}}.\end{aligned} \tag{12.261}$$

To proceed, we have to assume again that matrix $\boldsymbol{M}^{-1}(n)$ varies slowly in relation to $\tilde{\boldsymbol{w}}(n)$ and $\boldsymbol{q}(n)$, as in Assumption 12.A-3. Since now we are dealing only with the steady state, this approximation is in general good. Instead of Assumption 12.A-1, in energy conservation analysis the following assumptions are used.

12.A-7. $\|\boldsymbol{\phi}(n)\|^2_{\boldsymbol{M}(n)}$ is independent of $e_a(n)$ at the steady state.

12.A-8. The noise sequence $\{v_o(n)\}$ is i.i.d. and independent of the input sequence $\{\boldsymbol{\phi}(n)\}$.

Assumption 12.A-7 is referred to in the literature as the *separation principle* and is reasonable in steady state since for large n, $e_a(n)$ tends to be less sensitive to the regressor data, in particular for long filters. Assumption 12.A-8 is a weakened version of Assumption 12.A-1, which is necessary to relate the MSE with the EMSE through (12.182).

Under Assumptions 12.A-2, 12.A-3, and 12.A-7, we can now evaluate the terms $\mathcal{A}$–$\mathcal{E}$:

$\mathcal{A}$ Using Assumption 12.A-3, we obtain

$$\mathcal{A} \approx \mathrm{E}\left\{\|\tilde{\boldsymbol{w}}(n+1)\|^2_{\mathrm{E}\{\boldsymbol{M}^{-1}(n)\}}\right\}. \tag{12.262}$$

For LMS and NLMS, we have

$$\mathcal{A}^{\mathrm{LMS}} = \mathcal{A}^{\mathrm{NLMS}} \approx \mathrm{E}\left\{\|\tilde{\boldsymbol{w}}(n+1)\|^2\right\}. \tag{12.263}$$

For RLS,

$$\mathcal{A}^{\mathrm{RLS}} \approx \mathrm{E}\left\{\|\tilde{\boldsymbol{w}}(n+1)\|^2_{\boldsymbol{R}_{\boldsymbol{\phi}}}\right\}. \tag{12.264}$$

$\mathcal{B}$ Assuming that the filter is stable and reaches a steady state, we have

$$\mathcal{B} = \mathrm{E}\left\{\|\tilde{\boldsymbol{w}}(n)\|^2_{\boldsymbol{M}^{-1}(n)}\right\} \approx \mathrm{E}\left\{\|\tilde{\boldsymbol{w}}(n+1)\|^2_{\boldsymbol{M}^{-1}(n)}\right\}. \tag{12.265}$$

$\mathcal{C}$ Given that $e(n) = e_a(n) + v_o(n)$, since $v_o(n)$ is assumed to be independent of $e_a(n)$ (Assumption 12.A-8), we obtain

$$\mathrm{E}\{|e(n)|^2\} \approx \mathrm{E}\{|e_a(n)|^2\} + \sigma_v^2.$$

Then, using Assumptions 12.A-3 and 12.A-8, we arrive at

$$\mathcal{C} \approx \mathrm{E}\left\{\rho^2(n)\boldsymbol{\phi}^H(n)\,\mathrm{E}\{\boldsymbol{M}(n)\}\boldsymbol{\phi}(n)\right\}\left(\mathrm{E}\{|e_a(n)|^2\} + \sigma_v^2\right). \tag{12.266}$$

For LMS, this term reduces to

$$\mathcal{C}^{\mathrm{LMS}} \approx \mu^2\,\mathrm{Tr}(\boldsymbol{R}_{\boldsymbol{\phi}})\left(\mathrm{E}\{|e_a(n)|^2\} + \sigma_v^2\right). \tag{12.267}$$

For NLMS,

$$\mathcal{C}^{\mathrm{NLMS}} \approx \tilde{\mu}^2\,\mathrm{E}\left\{\frac{1}{\varepsilon + \boldsymbol{\phi}^H(n)\boldsymbol{\phi}(n)}\right\}\left(\mathrm{E}\{|e_a(n)|^2\} + \sigma_v^2\right), \tag{12.268}$$

which for long filters and Gaussian tapped-delay line regressors can be approximated as (see Eq. (12.231) and the alternative (12.230) for general regressors)

$$\mathcal{C}^{\text{NLMS}} \approx \frac{\tilde{\mu}^2}{\sigma_x^2(M-2)}\left(\mathrm{E}\{|e_a(n)|^2\}+\sigma_v^2\right). \tag{12.269}$$

Finally, for RLS we obtain

$$\mathcal{C}^{\text{RLS}} \approx (1-\lambda)^2 M\left(\mathrm{E}\{|e_a(n)|^2\}+\sigma_v^2\right). \tag{12.270}$$

$\mathcal{D}$ Using the same arguments, $\mathcal{D}$ reduces to

$$\mathcal{D} \approx 2\,\mathrm{E}\{\rho(n)\}\,\mathrm{E}\{|e_a(n)|^2\}. \tag{12.271}$$

For LMS, $\mathcal{D}^{\text{LMS}} \approx 2\mu\,\mathrm{E}\{|e_a(n)|^2\}$. Using Eq. (12.231) for NLMS, we get

$$\mathcal{D}^{\text{NLMS}} \approx \frac{2\tilde{\mu}}{\sigma_x^2(M-2)}\,\mathrm{E}\{|e_a(n)|^2\}.$$

For RLS, this term is given by $\mathcal{D}^{\text{RLS}} \approx 2(1-\lambda)\,\mathrm{E}\{|e_a(n)|^2\}$.

$\mathcal{E}$ Using Assumption 12.A-3 and recalling that $\mathrm{Tr}(\boldsymbol{AB})=\mathrm{Tr}(\boldsymbol{BA})$, we obtain

$$\mathcal{E} \approx \mathrm{E}\left\{\boldsymbol{q}^H(n)\,\mathrm{E}\{\boldsymbol{M}^{-1}(n)\}\boldsymbol{q}(n)\right\} = \mathrm{Tr}\left(\mathrm{E}\left\{\boldsymbol{M}^{-1}(n)\right\}\mathrm{E}\{\boldsymbol{q}(n)\boldsymbol{q}^H(n)\}\right).$$

For LMS and NLMS this reduces to

$$\mathcal{E}^{\text{LMS}} = \mathcal{E}^{\text{NLMS}} \approx \mathrm{Tr}(\boldsymbol{Q}). \tag{12.272}$$

For RLS,

$$\mathcal{E}^{\text{RLS}} \approx \mathrm{Tr}(\boldsymbol{R}_{\boldsymbol{\phi}}\boldsymbol{Q}). \tag{12.273}$$

Using these approximations in (12.261), we arrive at the following steady-state approximation:

$$\mathrm{E}\{|e_a(n)|^2\} \approx \frac{\sigma_v^2\,\mathrm{E}\left\{\rho^2(n)\boldsymbol{\phi}^H(n)\,\mathrm{E}\{\boldsymbol{M}(n)\}\boldsymbol{\phi}(n)\right\}+\mathrm{Tr}(\boldsymbol{Q}\,\mathrm{E}\{\boldsymbol{M}^{-1}(n)\})}{2\,\mathrm{E}\{\rho(n)\}-\mathrm{E}\left\{\rho^2(n)\boldsymbol{\phi}^H(n)\,\mathrm{E}\{\boldsymbol{M}(n)\}\boldsymbol{\phi}(n)\right\}}. \tag{12.274}$$

Particularizing (12.274) for LMS, NLMS, and RLS, we obtain the results of the second column of Table 12.8, which hold over a wide range of step sizes and forgetting factors.

12.4.4.5 ***Relation between the results obtained with both analysis methods***

As we explained before, each method is based on a different set of approximations. We now show how they are related.

The energy conservation method is capable of obtaining closed-form expressions for the EMSE using less restrictive assumptions; in particular, $\boldsymbol{\phi}(n)$ is not required to be Gaussian. These results are equivalent to those obtained for LMS and RLS using the traditional analysis, taking the recursions of Table 12.7 to the limit as $n \to \infty$ and assuming $\beta = 0$ (recall that the values of β used for the traditional analysis are only valid if $\boldsymbol{\phi}(n)$ is Gaussian). The assumption of $\beta = 0$ implies the following approximation:

$$\mathrm{E}\left\{\boldsymbol{\phi}(n)\boldsymbol{\phi}^H(n)\boldsymbol{S}(n)\boldsymbol{\phi}(n)\boldsymbol{\phi}^H(n)\right\} \approx \boldsymbol{R}_{\boldsymbol{\phi}}\,\mathrm{Tr}(\boldsymbol{S}(n)\boldsymbol{R}_{\boldsymbol{\phi}}).$$

For small step sizes, this approximation does not have a large impact, since we are disregarding a term of $\mathcal{O}(\mu^2)$, which is small compared to other $\mathcal{O}(\mu)$ terms. For NLMS, in order to recover (12.274) using the traditional method, we would choose $\beta = 0$ and make

$$\mathrm{E}\left\{\frac{1}{\left[\varepsilon + \boldsymbol{\phi}^H(n)\boldsymbol{\phi}(n)\right]^2}\right\} \approx \left[\mathrm{E}\left\{\frac{1}{\varepsilon + \boldsymbol{\phi}^H(n)\boldsymbol{\phi}(n)}\right\}\right]^2,$$

which leads to reasonable results at the steady state for long filters.

For small step sizes (or forgetting factors close to one), the denominator of (12.274) can be approximated by $2\,\mathrm{E}\{\rho(n)\}$. In this case, the expressions of the second column reduce to those of the third column of Table 12.8. Again, these results can be obtained from the recursions of Table 12.7, disregarding the term $\boldsymbol{\mathcal{C}}$ (Eq. (12.238)), which will be negligible compared to $\mathcal{B}$ in (12.224) if $\rho \approx 0$.

The results for steady-state analysis using the traditional method are commonly obtained as the limiting case of a transient analysis. Although this procedure is adequate for understanding both the steady-state and the transient behavior of an adaptive algorithm, it tends to be more laborious than using the energy conservation method to analyze the steady-state behavior. This is one of the largest advantages of this last method.

Table 12.8 Expressions for the steady-state EMSE obtained from the energy conservation method for LMS, NLMS, and RLS.

Alg.	$\zeta(\infty)$ for a wider range of μ, $\tilde{\mu}$, and λ	$\zeta(\infty)$ for small μ and $\tilde{\mu}$, and $\lambda \approx 1$
LMS	$\dfrac{\mu\sigma_v^2\,\mathrm{Tr}(\boldsymbol{R}_{\boldsymbol{\phi}}) + \mu^{-1}\,\mathrm{Tr}(\boldsymbol{Q})}{2 - \mu\,\mathrm{Tr}(\boldsymbol{R}_{\boldsymbol{\phi}})}$	$\dfrac{\mu\sigma_v^2\,\mathrm{Tr}(\boldsymbol{R}_{\boldsymbol{\phi}}) + \mu^{-1}\,\mathrm{Tr}(\boldsymbol{Q})}{2}$
NLMS	$\dfrac{\tilde{\mu}\sigma_v^2 + \tilde{\mu}^{-1}\sigma_x^2(M-2)\,\mathrm{Tr}(\boldsymbol{Q})}{2 - \tilde{\mu}}$	$\dfrac{\tilde{\mu}\sigma_v^2 + \tilde{\mu}^{-1}\sigma_x^2(M-2)\,\mathrm{Tr}(\boldsymbol{Q})}{2}$
RLS	$\dfrac{\sigma_v^2(1-\lambda)M + \dfrac{\mathrm{Tr}(\boldsymbol{Q}\boldsymbol{R}_{\boldsymbol{\phi}})}{1-\lambda}}{2-(1-\lambda)M}$	$\dfrac{\sigma_v^2(1-\lambda)M + \dfrac{\mathrm{Tr}(\boldsymbol{Q}\boldsymbol{R}_{\boldsymbol{\phi}})}{1-\lambda}}{2}$

12.4.4.6 Optimal step size for tracking

From the results of the third column of Table 12.8, we can observe that the expressions for the steady-state EMSE have two terms: one for the stationary environment, which increases as the step size increases, and another that appears in the nonstationary case, which increases as the step size decreases. Therefore, there exist optimal adaptation parameters ρ_{o} that minimize the EMSE. These parameters are μ_{o}, $\tilde{\mu}_{\mathrm{o}}$, and $1 - \lambda_{\mathrm{o}}$ for LMS, NLMS, and RLS, respectively. The corresponding minima, denoted by $\zeta_{\min}(\infty)$, can also be evaluated. All these results are summarized in Table 12.9. The results for a wider range of adaptation parameters lead to more complicated expressions for ρ_o and $\zeta_o(\infty)$ (see, e.g., [317] and [97] for an extension to the more general model (12.184) for time variations).

Table 12.9 Expressions for the optimal adaptation parameters (ρ_o) and minimum steady-state EMSE ($\zeta_{\min}(\infty)$) obtained from the expressions of the third column of Table 12.8.

Alg.	ρ_o	$\zeta_{\min}(\infty)$
LMS	$\mu_o = \sqrt{\dfrac{\mathrm{Tr}(\boldsymbol{Q})}{\sigma_v^2 \mathrm{Tr}(\boldsymbol{R}_\phi)}}$	$\sqrt{\sigma_v^2 \mathrm{Tr}(\boldsymbol{R}_\phi)\mathrm{Tr}(\boldsymbol{Q})}$
NLMS	$\tilde{\mu}_o = \sqrt{\dfrac{(M-2)\sigma_x^2 \mathrm{Tr}(\boldsymbol{Q})}{\sigma_v^2}}$	$\sqrt{\sigma_v^2\sigma_x^2(M-2)\mathrm{Tr}(\boldsymbol{Q})}$
RLS	$1-\lambda_o = \sqrt{\dfrac{\mathrm{Tr}(\boldsymbol{Q}\boldsymbol{R}_\phi)}{M\sigma_v^2}}$	$\sqrt{\sigma_v^2 M \mathrm{Tr}(\boldsymbol{Q}\boldsymbol{R}_\phi)}$

We can use the results of Table 12.9 to compare the tracking performance of the algorithms. For instance, comparing the minimum EMSE for LMS to that of RLS, we obtain the ratio

$$\frac{\zeta_{\min}^{\mathrm{RLS}}}{\zeta_{\min}^{\mathrm{LMS}}} = \sqrt{\frac{M\,\mathrm{Tr}(\boldsymbol{Q}\boldsymbol{R}_\phi)}{\mathrm{Tr}(\boldsymbol{R}_\phi)\mathrm{Tr}(\boldsymbol{Q})}}. \tag{12.275}$$

Clearly, the results of this comparison depend on the environment. There are situations where RLS has superior tracking capability compared to LMS, and vice versa. This is highlighted considering three different choices for matrix $\boldsymbol{Q}$. Consider the following three situations [131]:

(i) If $\boldsymbol{Q}$ is a multiple of $\boldsymbol{I}$, we have

$$\frac{\zeta_{\min}^{\mathrm{RLS}}}{\zeta_{\min}^{\mathrm{LMS}}} = \sqrt{\frac{M\,\mathrm{Tr}(\boldsymbol{R}_\phi)}{\mathrm{Tr}(\boldsymbol{R}_\phi)\mathrm{Tr}(\boldsymbol{I})}} = 1, \tag{12.276}$$

and thus the performance of LMS is similar to that of RLS.

(ii) If $\boldsymbol{Q}$ is a multiple of $\boldsymbol{R}_\phi$, we have

$$\frac{\zeta_{\min}^{\mathrm{RLS}}}{\zeta_{\min}^{\mathrm{LMS}}} = \sqrt{\frac{M\,\mathrm{Tr}(\alpha\boldsymbol{R}_\phi^2)}{\alpha\,\mathrm{Tr}(\boldsymbol{R}_\phi)\mathrm{Tr}(\boldsymbol{R}_\phi)}} \geq 1. \tag{12.277}$$

This result can be verified by noting that if λ_k are the eigenvalues of $\boldsymbol{R}_\phi$ and defining $\boldsymbol{\lambda} = \begin{bmatrix}\lambda_1 & \dots & \lambda_M\end{bmatrix}^T$, we can use matrix diagonalization (see Box 12.3) to show that $\mathrm{Tr}(\boldsymbol{R}_\phi) = \|\boldsymbol{\lambda}\|_1$ (the ℓ_1-norm) and $\mathrm{Tr}(\boldsymbol{R}_\phi^2) = \|\boldsymbol{\lambda}\|_2^2$ (the Euclidean norm). Then, applying inequality (12.36) from Box 12.3, we obtain (12.277). We conclude that in this case LMS is superior.

(iii) Finally, if $\boldsymbol{Q}$ is a multiple of $\boldsymbol{R}_{\boldsymbol{\phi}}^{-1}$, we obtain

$$\frac{\zeta_{\min}^{\mathrm{RLS}}}{\zeta_{\min}^{\mathrm{LMS}}} = \sqrt{\frac{M\,\mathrm{Tr}(\boldsymbol{I})}{\mathrm{Tr}(\boldsymbol{R}_{\boldsymbol{\phi}})\mathrm{Tr}(\boldsymbol{R}_{\boldsymbol{\phi}}^{-1})}} = \frac{M}{\sqrt{\mathrm{Tr}(\boldsymbol{R}_{\boldsymbol{\phi}})\mathrm{Tr}(\boldsymbol{R}_{\boldsymbol{\phi}}^{-1})}} \le 1. \tag{12.278}$$

This inequality can be derived as follows. Define $\boldsymbol{\nu} = \begin{bmatrix}\lambda_1^{-1/2} & \dots & \lambda_M^{-1/2}\end{bmatrix}^T$ and $\boldsymbol{\lambda}_{1/2} = \begin{bmatrix}\lambda_1^{1/2} & \dots & \lambda_M^{1/2}\end{bmatrix}^T$ for λ_k as before. Then $\mathrm{Tr}(R_{\boldsymbol{\phi}}) = \|\boldsymbol{\lambda}_{1/2}\|_2^2$, $\mathrm{Tr}(R_{\boldsymbol{\phi}}^{-1}) = \|\boldsymbol{\nu}\|_2^2$, and applying Schwarz's inequality (see Box 12.3) we obtain

$$M = \boldsymbol{\nu}^T\boldsymbol{\lambda}_{1/2} \le \|\boldsymbol{\nu}\|_2\|\boldsymbol{\lambda}_{1/2}\|_2 = \sqrt{\mathrm{Tr}(R_{\boldsymbol{\phi}})\,\mathrm{Tr}(R_{\boldsymbol{\phi}}^{-1})},$$

and (12.278) follows. We conclude that in this case RLS is superior.

These choices for $\boldsymbol{Q}$ do not model practical situations, but they highlight that even though the convergence rate of RLS is in general much higher than that of LMS, this advantage, unexpectedly, does not necessarily follow to the problem of tracking a slowly varying parameter vector.

Example 12.E-4.4. To verify the accuracy of the expressions in Tables 12.8 and 12.9, we use the LMS, NLMS, and RLS filters to identify the same low-pass FIR system of Fig. 12.30. Now, the environment is assumed nonstationary with $\boldsymbol{Q} = 10^{-6}\boldsymbol{I}$. The input signal and minimum MSE (σ_v^2) are the same of Example 12.E-4.3.

Fig. 12.33 shows the measured steady-state EMSE with varying adaptation factors considering the theoretical and experimental results for LMS, NLMS, and RLS. Each value of experimental EMSE was obtained from the ensemble average of 100 independent runs. The theoretical minimum EMSEs predicted by the expressions of Table 12.9 are $\zeta_o^{\mathrm{LMS}}(\infty) = \zeta_o^{\mathrm{RLS}}(\infty) = -32.9251$ dB and $\zeta_o^{\mathrm{NLMS}}(\infty) = -33.0989$ dB, which correspond to the optimal adaptation parameters $\mu_o = 5.1 \times 10^{-2}$, $\tilde{\mu}_o = 4.9 \times 10^{-2}$, and $1 - \lambda_o = 2 \times 10^{-3}$ ($\lambda_o = 0.998$). The experimental values for $\zeta_{\min}(\infty)$ and ρ_o are close to the values predicted by the expressions of Table 12.9 as shown in Fig. 12.33.

Example 12.E-4.5. Again, we consider a system identification application but with the initial optimal solution given by

$$\boldsymbol{w}_o^T(0) = \begin{bmatrix}0.5349 & 0.9527 & -0.9620 & -0.0158 & -0.1254\end{bmatrix}.$$

The input signal is assumed to be colored Gaussian noise with variance $\sigma_x^2 = 0.2$. This signal is obtained from the filtering of a white Gaussian noise by a first-order autoregressive model ($b = 0.8$). We also assume that the minimum MSE is $\sigma_v^2 = 0.01$.

As predicted by (12.275), the performance of RLS and LMS is similar when $\boldsymbol{Q} = \sigma_q^2\boldsymbol{I}$. Thus, assuming $\boldsymbol{Q} = 10^{-6}\boldsymbol{I}$, we computed μ_o and λ_o from the expressions of Table 12.9 and used these parameters in the adaptation, such that the same minimum EMSE could be achieved by both algorithms in this situation. Fig. 12.34 shows the EMSE estimated from the ensemble average of 5000 independent runs for RLS ($\lambda = \lambda_o = 0.9955$, $\delta = 4.5 \times 10^{-3}$) and LMS ($\mu = \mu_o = 0.0224$). At every 7×10^4 iterations, the nonstationary environment is changed. During the first 7×10^4 iterations, $\boldsymbol{Q} = 10^{-6}\boldsymbol{R}_{\boldsymbol{\phi}}$ and LMS

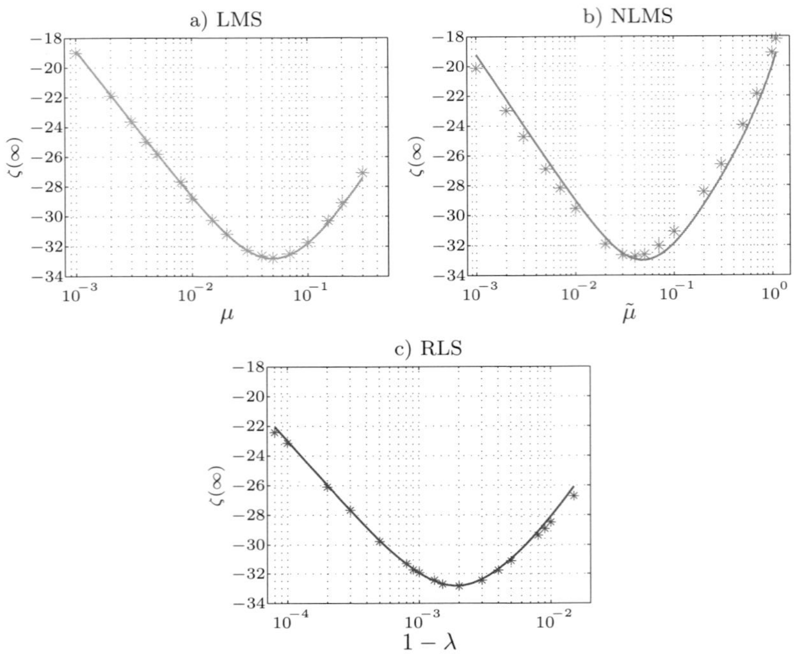

FIGURE 12.33

Theoretical and experimental EMSE as a function of the adaptation parameter for (a) LMS, (b) NLMS, and (c) RLS. The asterisks represent simulation results and the solid lines represent the theoretical models.

presents a slightly better tracking performance than RLS. When $\boldsymbol{Q} = 10^{-6}\boldsymbol{R}_{\boldsymbol{\phi}}^{-1}$, this behavior changes: RLS becomes better than LMS. Finally, for $\mathbf{Q} = 10^{-6}\boldsymbol{I}$, RLS and LMS present similar performance. The dashed lines represent the theoretical values of $\zeta(\infty)$ for each algorithm, predicted from the expressions of the second column of Table 12.8.

Box 12.7: Fourth-order moments of Gaussian vectors

The second-order moments of a random sequence are related to its average power (of course, this is an approximation unless the sequence is stationary and ergodic). This makes second-order moments intuitive to work with and relatively easy to evaluate in practice. Therefore, when studying an adaptive filter, it is common to try to describe its performance in terms of the autocorrelation of the input sequence $\boldsymbol{x}(n)$.

During the analysis of LMS, one encounters a term that contains fourth-order powers of the input vector $\boldsymbol{x}(n)$:

$$\boldsymbol{F} \triangleq \mathrm{E}\{\boldsymbol{x}(n)\boldsymbol{x}^H(n)\tilde{\boldsymbol{w}}(n)\tilde{\boldsymbol{w}}^H(n)\boldsymbol{x}(n)\boldsymbol{x}^H(n)\}. \tag{12.279}$$

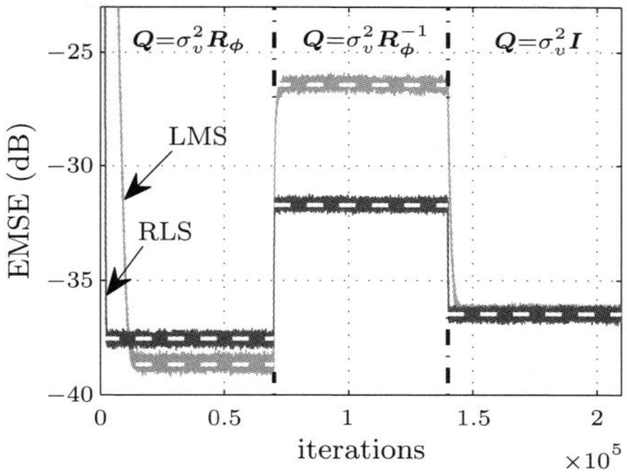

FIGURE 12.34

EMSE for RLS ($\lambda = 0.9955$, $\delta = 4.5 \times 10^{-3}$) and LMS ($\mu = 0.0224$); $b = 0.8$, $\sigma_v^2 = 10^{-2}$, $\sigma_q^2 = 10^{-6}$; mean of 5000 independent runs. The dashed lines represent the predicted values of $\zeta(\infty)$ for each algorithm.

In order to evaluate a term such as this, one needs to know much more information about the statistics of $\boldsymbol{x}(n)$ than is provided simply by its mean and autocorrelation. However, in general this additional information is not available, and one assumes that the signals are Gaussian in order to proceed.

Why not any other distribution? First, because many processes in Nature are indeed approximately Gaussian. Second, because uncorrelated Gaussian variables are also independent, which is of great help when evaluating fourth-order terms, such as $\mathrm{E}\{x^2(n)x^2(n-1)\}$ or $\mathrm{E}\{x^3(n)x(n-2)\}$.

Although in many cases the input sequence is not Gaussian (for example, speech does not follow a Gaussian distribution), this approximation describes the most important features of adaptive filter learning with reasonable accuracy and leads to a relatively simple model.

In order to simplify the notation, we will write simply $\boldsymbol{x}$ and $\tilde{\boldsymbol{w}}$ instead of $\boldsymbol{x}(n)$ and $\tilde{\boldsymbol{w}}(n)$ in the following. Assume then that $\boldsymbol{x}$ is a Gaussian random vector, with autocorrelation $\mathrm{E}\{\boldsymbol{x}\boldsymbol{x}\} = \boldsymbol{R}$, and that $\tilde{\boldsymbol{w}}$ is another vector independent of $\boldsymbol{x}$, with autocorrelation $\mathrm{E}\{\tilde{\boldsymbol{w}}\tilde{\boldsymbol{w}}^H\} = \boldsymbol{S}$ (we do not need to know the exact distribution of $\tilde{\boldsymbol{w}}$). Assume also that all variables have zero mean. Our goal is to evaluate

$$\boldsymbol{F} = \mathrm{E}\{\boldsymbol{x}\boldsymbol{x}^H\tilde{\boldsymbol{w}}\tilde{\boldsymbol{w}}^H\boldsymbol{x}\boldsymbol{x}^H\}.$$

Since $\tilde{\boldsymbol{w}}$ is independent of $\boldsymbol{x}$, we can write

$$\boldsymbol{F} = \mathrm{E}\{\boldsymbol{x}\boldsymbol{x}^H\tilde{\boldsymbol{w}}\tilde{\boldsymbol{w}}^H\boldsymbol{x}\boldsymbol{x}^H\} = \mathrm{E}\{\boldsymbol{x}\boldsymbol{x}^H\,\mathrm{E}\{\tilde{\boldsymbol{w}}\tilde{\boldsymbol{w}}^H\}\boldsymbol{x}\boldsymbol{x}^H\} = \mathrm{E}\{\boldsymbol{x}\boldsymbol{x}^H\boldsymbol{S}\boldsymbol{x}\boldsymbol{x}^H\}. \tag{12.280}$$

The autocorrelation of $\boldsymbol{x}$ is in general a full matrix,

$$\boldsymbol{R} = \begin{bmatrix} r_{11} & r_{12} & \cdots & r_{1M} \\ r_{12}^* & r_{22} & \cdots & r_{2M} \\ \vdots & \vdots & \ddots & \vdots \\ r_{1M}^* & r_{2M}^* & \cdots & r_{MM} \end{bmatrix},$$

that is, in general the elements of $\boldsymbol{x}$ are correlated; however, $\boldsymbol{R}$ has Hermitian symmetry and is nonnegative definite, which implies that there exists a unitary matrix $\boldsymbol{U}$ such that

$$\boldsymbol{U}^H\boldsymbol{R}\boldsymbol{U} = \boldsymbol{\Lambda} = \begin{bmatrix} \lambda_1 & 0 & \dots & 0 \\ 0 & \lambda_2 & \dots & 0 \\ \vdots & \vdots & \ddots & \vdots \\ 0 & 0 & \dots & \lambda_M \end{bmatrix},$$

where $\lambda_i \geq 0$ for $i = 1 \dots M$. Define $\boldsymbol{x}' = \boldsymbol{U}^H\boldsymbol{x}$. Then $\mathrm{E}\{\boldsymbol{x}'\boldsymbol{x}'^H\} = \mathrm{E}\{\boldsymbol{U}^H\boldsymbol{x}\boldsymbol{x}^H\boldsymbol{U}\} = \boldsymbol{U}^H\,\mathrm{E}\{\boldsymbol{x}\boldsymbol{x}^H\}\boldsymbol{U} = \boldsymbol{\Lambda}$, and we see that the entries of $\boldsymbol{x}'$ are uncorrelated. Since $\boldsymbol{x}$ is Gaussian and a linear transformation of a Gaussian vector is also Gaussian, the entries of $\boldsymbol{x}'$ are independent of each other. Multiplying (12.280) by $\boldsymbol{U}^H$ on the left and by $\boldsymbol{U}$ on the right and recalling that $\boldsymbol{U}^H\boldsymbol{U} = \boldsymbol{U}\boldsymbol{U}^H = \boldsymbol{I}$, we obtain

$$\boldsymbol{F}' \triangleq \boldsymbol{U}^H\boldsymbol{F}\boldsymbol{U} = \mathrm{E}\{\boldsymbol{U}^H\boldsymbol{x}\boldsymbol{x}^H\underbrace{\boldsymbol{U}\boldsymbol{U}^H}_{=\boldsymbol{I}}\boldsymbol{S}\underbrace{\boldsymbol{U}\boldsymbol{U}^H}_{=\boldsymbol{I}}\boldsymbol{x}\boldsymbol{x}^H\boldsymbol{U}\} = \mathrm{E}\{\boldsymbol{x}'\boldsymbol{x}'^H\boldsymbol{U}^H\boldsymbol{S}\boldsymbol{U}\boldsymbol{x}'\boldsymbol{x}'^H\}.$$

Defining $\bar{\boldsymbol{S}} = \boldsymbol{U}^H\boldsymbol{S}\boldsymbol{U}$, we can expand $\boldsymbol{F}$ in terms of the elements s'_{ij} of $\bar{\boldsymbol{S}}$ and x'_k of $\boldsymbol{x}'$. Each element of $\boldsymbol{F}$ will have terms of the form

$$\mathrm{E}\{x'_{k_1}x'_{k_2}x'^{*}_{k_3}x'^{*}_{k_4}\}s'_{ij}.$$

If the signals are real, then, as x'_{k_1} is independent of x'_{k_2} if $k_1 \neq k_2$, these expected values will be nonzero only if $k_1 = k_2 = k_3 = k_4$, if $k_1 = k_3$ and $k_2 = k_4$, if $k_1 = k_4$ and $k_2 = k_3$, or if $k_1 = k_2$ and $k_3 = k_4$. Evaluation of all such terms that appear in the expression for $\boldsymbol{F}'$ will result in

$$\boldsymbol{F}|_{\text{real signals}} = \boldsymbol{R}\,\mathrm{Tr}(\boldsymbol{S}\boldsymbol{R}) + 2\boldsymbol{R}\boldsymbol{S}\boldsymbol{R}. \tag{12.281}$$

If, on the other hand, the signals are complex-valued, then we need to know how the real and imaginary parts of each entry of $\boldsymbol{x}$ relate to each other. One common assumption is that $\boldsymbol{x}$ is *circularly* Gaussian. We explain more about this in Section 12.2.2.1. For now, we only remark that if this is the case, the real and imaginary parts of each entry of $\boldsymbol{x}$ and $\boldsymbol{x}'$ are such that $\mathrm{E}\{x_k^2\} = \mathrm{E}\{x'^2_k\} = 0$ for all k (note that $\mathrm{E}\{|x_k|^2\} = r_{kk}$ and $\mathrm{E}\{|x'_k|^2\} = \lambda_k$ are still nonzero). This will result in a different expression for $\boldsymbol{F}$:

$$\boldsymbol{F}|_{\text{complex signals}} = \boldsymbol{R}\,\mathrm{Tr}(\boldsymbol{S}\boldsymbol{R}) + \boldsymbol{R}\boldsymbol{S}\boldsymbol{R}. \tag{12.282}$$

12.5 Extensions and current research

In this section we briefly describe some important extensions to the basic adaptive filtering algorithms, as well as some promising research topics. The intention is not to describe in any detail all these techniques, but rather to give a list of key references where the interested reader may find more information.

It is usual when starting such lists to put a disclaimer that the authors do not claim to have included all the important contributions, nor that their list is really of the most important contributions to an area.

This is not intended for legal purposes, nor false modesty: the size of the literature is indeed so large (a search on "adaptive filter" on Google gives over one million hits) that we cannot claim to know with certainty the best of it all. We included therefore the techniques that we found useful during our own research and experience, which is perforce limited.

12.5.1 Finite precision arithmetic

Any operation made using finite precision arithmetic may involve errors. For example, the product of $a = 0.957$ and $b = 0.542$ would result in $c = 0.518694$. However, if all numbers were stored keeping only three digits, the final result would be $\bar{c} = 0.519$, with an error of $c - \bar{c} = -3.06 \times 10^{-4}$ [389,390]. These *quantization* errors may accumulate and modify the behavior of an adaptive filter in important ways. A precise analysis of the effect of quantization errors is quite complicated, since these errors are highly nonlinear functions of the variables.

When finite precision arithmetic effects are studied, it is usually important to know exactly how and in which order the arithmetic operations are performed. In particular, the exact numerical representation used is important: fixed-point versus floating-point, truncation versus rounding, etc. For this reason, it is more difficult to perform analyses that are equally valid for a large class of filters. The most important differences are between gradient-based algorithms (i.e., algorithms similar to LMS and NLMS) and Hessian-based algorithms (i.e., algorithms similar to RLS). We will give references to works treating both cases, starting with LMS.

12.5.1.1 Finite precision effects in LMS

The simplest models for adaptive filters in finite precision arithmetic treat the quantization error as random noise, uniformly distributed and independent of all variables [12,24,70,160]. These models show that the quantization errors will add another term to the EMSE. However, this term does not converge to zero when the step size is reduced to zero, even in a stationary environment. Quite the opposite; it is inversely proportional to the step size (similar to the EMSE in a nonstationary environment).

Another important undesirable effect is that the product $\mu e(n)$ may be rounded down to zero when $e(n)$ becomes small (underflow), virtually stopping the adaptation. When the step size is small, this may happen when the filter is rather far from the optimum solution. Therefore, there is an optimum value for the step size – neither too large, to avoid increasing the misadjustment, neither too small, to avoid underflow and excessive growth of the EMSE. A similar effect happens in RLS [117, p. 595].

This underflow problem, which is usually known as the *stopping phenomenon*, was studied in more detail in [42,43,50,51], using a nonlinear model. These works show that the adaptation does not really stop, but rather convergence is reduced to a very low rate. Note that this phenomenon is less important in floating-point arithmetic, since floating-point arithmetic is able to represent much smaller numbers without underflow than fixed-point arithmetic.

Finite precision effects may have worse consequences; in particular, they may make the adaptive filter unstable. In the case of LMS, this phenomenon is rare, and will only happen when the regressor sequence $\{\boldsymbol{\phi}\}$ has a very ill-conditioned covariance matrix and the noise v_o has a nonzero mean (this nonzero mean may itself appear due to quantization errors) [95,161,261]. This problem is not difficult to solve, using regularization as described in Section 12.5.2. The problem of numerical stability in RLS is much more serious and harder to solve, and has received much attention, as we see next.

In addition to studying the effect of quantization errors in the performance of an adaptive filter, another important topic is how to reduce the computational cost of a filter, taking advantage of implementation details. One such approach is to replace the error $e(n)$ in the LMS recursion by its sign, i.e., to use the recursion

$$\boldsymbol{w}(n+1) = \boldsymbol{w}(n) + \mu \operatorname{sign}(e(n))\, \boldsymbol{\phi}(n), \tag{12.283}$$

where sign$(\cdot)$ is the sign function, defined by

$$\operatorname{sign}(e) = \begin{cases} 1, & \text{if } e > 0, \\ 0, & \text{if } e = 0, \\ -1, & \text{if } e < 0. \end{cases} \tag{12.284}$$

The resulting algorithm, known as *sign-LMS*, has a smaller computational cost if μ is restricted to be a power of two, that is, if $\mu = 2^k$ for an integer (usually negative) k. In this case, $\mu \operatorname{sign}(e(n))$ is always equal to $\pm 2^k$ (or zero), and the product $[\mu \operatorname{sign}(e(n))]\, \boldsymbol{\phi}(n)$ can be implemented in fixed-point arithmetic as M shifts, instead of M multiplications. In floating-point arithmetic, the multiplications are replaced by sums, since we would need to add k to the exponent terms of all entries of $\boldsymbol{\phi}(n)$. This algorithm is very robust (see also Section 12.5.4), although its convergence rate under Gaussian inputs is considerably slower than that of LMS. Sign-LMS was thoroughly studied in [129–132,158].

Another possibility that works surprisingly well, reducing the computational cost with only a slight decrease in convergence speed, is the *power-of-two LMS* algorithm [392]. In it, the error is rounded to the nearest power of two, using the function

$$f(e) = \begin{cases} \operatorname{sign}(e), & |e| \geq 1, \\ \operatorname{sign}(e) 2^{\lfloor \log_2(|e|) \rfloor}, & 2^{-B+1} \leq |e| < 1, \\ 0, & |e| < 2^{-B+1}, \end{cases}$$

where B is the number of bits used to represent the error and $\lfloor x \rfloor$ returns the largest integer smaller than x. In this case, μ should also be a power of two.

Yet another approach consists in using sum-of-powers-of-two (SPOT) representations [104] for coefficients. Although one may argue that any number in binary form is in some sense a sum of powers-of-two, SPOT numbers are a modification, in which a number x is represented as

$$x = \sum_{k=0}^{K-1} x_k 2^{b_k}, \tag{12.285}$$

in which $x_k = \pm 1$ (it is also possible to define x_k assuming values in the set $\{-1, 0, 1\}$) and b_k are the exponents. This class of numbers is used in [103] to derive a version of the RLS algorithm.

12.5.1.2 Finite precision effects in RLS

As we briefly described in Section 12.3.3.1, the conventional RLS is numerically unstable in finite precision arithmetic and has $\mathcal{O}(M^2)$ complexity. To solve these problems, different versions of RLS were proposed in the literature. Many of these versions are based on coordinate transformations of the state-space representation of the conventional RLS algorithm. This representation can be obtained by replacing $\boldsymbol{k}(n)$ by $\boldsymbol{P}(n)\boldsymbol{\phi}(n)$ in (12.167), i.e.,

$$\boldsymbol{w}(n+1) = \lambda \boldsymbol{P}(n) \widehat{\boldsymbol{r}}_{d\boldsymbol{\phi}}(n-1) + \boldsymbol{P}(n)\boldsymbol{\phi}(n) d^*(n). \tag{12.286}$$

Recalling that $\boldsymbol{w}(n) = \boldsymbol{P}(n-1)\widehat{\boldsymbol{r}}_{d\boldsymbol{\phi}}(n-1)$, we can replace $\widehat{\boldsymbol{r}}_{d\boldsymbol{\phi}}(n-1)$ in (12.286) as a function of $\boldsymbol{P}^{-1}(n-1)$ and $\boldsymbol{w}(n)$, which leads to

$$\boldsymbol{w}(n+1) = \lambda \boldsymbol{P}(n)\boldsymbol{P}^{-1}(n-1)\boldsymbol{w}(n) + \boldsymbol{P}(n)\boldsymbol{\phi}(n)d^*(n). \tag{12.287}$$

This equation and the definition of the a priori error $e(n)$ characterize the adaptation and filtering operations of the RLS algorithm and constitute its state-space representation, i.e.,

$$\begin{bmatrix} \boldsymbol{w}(n+1) \\ e^*(n) \end{bmatrix} = \begin{bmatrix} \lambda \boldsymbol{P}(n)\boldsymbol{P}^{-1}(n-1) & \boldsymbol{P}(n)\boldsymbol{\phi}(n) \\ -\boldsymbol{\phi}^H(n) & 1 \end{bmatrix} \begin{bmatrix} \boldsymbol{w}(n) \\ d^*(n) \end{bmatrix}. \tag{12.288}$$

In this representation, $\boldsymbol{w}(n)$ is the state, $e^*(n)$ is the output, and $d^*(n)$ is the input.

Performing a coordinate transformation, (12.288) can be transformed to an unlimited number of systems with the same input–output relation, hence solving the same least-squares problem (that is, the output error is always the same). Although all these realizations are equivalent in infinite precision arithmetic, the numerical behavior will vary from one coordinate system to another. Thus, the numerical instability of the conventional RLS can be avoided by choosing a convenient transformation of (12.288) [214,298].

An alternative method of implementing RLS is based on the QR decomposition (QRD) [165] to triangularize the input data matrix. The main advantages of QRD-RLS-based algorithms are the possibility of implementation in systolic arrays (e.g., [92]) and the improved numerical behavior in finite precision arithmetic. Some versions require $\mathcal{O}(M^2)$ operations per iteration, but others have a reduced computational cost of $\mathcal{O}(M)$. The low-cost versions of QRD-RLS algorithms are known as *fast*-QR algorithms. Some important references on this subject are [11,21,92,245–247,294,300,306–310,319,334, 335]. A good tutorial on QRD-RLS-based algorithms can be found in [19], where some more advanced developments as well as the basic concepts are covered.

Another important class of fast algorithms solves the least-squares problem in a recursive form based on lattice realization. These algorithms are very attractive due to the possible modular implementation and low cost ($\mathcal{O}(M)$). Lattice-RLS-based algorithms are derived by solving the forward and backward linear prediction problems and require time-update and order-update recursions. They explore the time-shift property of the input signal vector (such as in (12.114)) and do not compute explicitly the filter weights. Recall that many applications do not require the computation of the filter weights and have regressors that satisfy the time-shift property as, for example, channel equalization and interference cancelation (see Section 12.1.3). In these applications, a fast RLS algorithm can be a reasonable choice since it presents a good trade-off between convergence rate and computational cost. One of these algorithms is the modified EF-LSL algorithm, which presents reliable numerical properties when implemented in finite precision arithmetic [248]. This algorithm was proposed for echo cancelation applications and was used in the example of Section 12.1.1. Some lattice-RLS-based algorithms can be found in Refs. [150,199,203,208,231,248,273,296,314].

Lattice-RLS algorithms have been generalized to filter structures other than tapped-delay lines, in particular to Laguerre filters [237,240,241].

Other fast algorithms that solve the least-squares problem in a recursive form are the fast transversal RLS (FTRLS) algorithms. Unlike the lattice-based algorithms, the FTRLS algorithms require only time-recursive equations and therefore compute explicitly the filter weights. The main drawback of these algorithms is that they are very sensitive to quantization effects and become unstable if certain actions are not taken [117]. The list of FTRLS algorithms is vast. Some of the most important versions can be found in Refs. [56,93,94,135,254,350,351].

12.5.1.3 DCD-RLS

The DCD-RLS algorithm, proposed in [402], is a low-complexity alternative to the RLS algorithm, based on the dichotomous coordinate descent (DCD) algorithm for function minimization proposed in [401]. The DCD algorithm is designed to be easily implementable in hardware (such as FPGAs, for example) and thus avoids multiplications, divisions, and other operations that are costly to implement in hardware, replacing them as much as possible by additions and comparisons.

The RLS weight vector estimates are given by the solution of the following set of linear equations, which we repeat from Section 12.3.3:

$$\widehat{\boldsymbol{R}}_{\boldsymbol{\phi}}(n)\boldsymbol{w}(n+1)=\widehat{\boldsymbol{r}}_{d\boldsymbol{\phi}}(n), \tag{12.289}$$

where $\widehat{\boldsymbol{R}}_{\boldsymbol{\phi}}(n)$ and $\widehat{\boldsymbol{r}}_{d\boldsymbol{\phi}}(n)$ are given by the recursions

$$\widehat{\boldsymbol{R}}_{\boldsymbol{\phi}}(n)=\lambda\widehat{\boldsymbol{R}}_{\boldsymbol{\phi}}(n-1)+\boldsymbol{\phi}(n)\boldsymbol{\phi}^H(n) \tag{12.290}$$

and

$$\widehat{\boldsymbol{r}}_{d\boldsymbol{\phi}}(n)=\lambda\widehat{\boldsymbol{r}}_{d\boldsymbol{\phi}}(n-1)+d^*(n)\boldsymbol{\phi}(n). \tag{12.291}$$

The difficulty in solving these expressions is that a general solution for (12.289) involves a number of operations of the order of M^3 ($\mathcal{O}(M^3)$) for a filter with M coefficients. This is usually too costly for practical applications, except perhaps for very short filters. The classical RLS algorithm described in Section 12.3.3 solves this problem partially by using the matrix inversion lemma to compute the inverse of $\widehat{\boldsymbol{R}}_{\boldsymbol{\phi}}(n)$ recursively using the already computed inverse of $\widehat{\boldsymbol{R}}_{\boldsymbol{\phi}}(n-1)$. This reduces the total number of computations to $\mathcal{O}(M^2)$. However, the resulting algorithm is still difficult to implement in practice, because numerical errors may easily cause divergence.

The DCD-RLS algorithm avoids both these problems by using a couple of clever tricks. First, assume that we have a good approximation $\hat{\boldsymbol{w}}(n)$ for $\boldsymbol{w}(n)$. Then we could use a recursive algorithm to find an approximation $\hat{\boldsymbol{w}}(n+1)$ to the solution of (12.289), using our current approximation $\hat{\boldsymbol{w}}(n)$ as an initial condition. Since the initial condition is already close to the solution, we would need only a few iterations of our recursive algorithm to obtain $\hat{\boldsymbol{w}}(n+1)$. However, this idea only helps if each iteration of the recursive algorithm is very cheap to compute. We explain next how this can be done using DCD.

DCD minimization of quadratic functions

Let us consider first a 1D quadratic problem

$$\min_{w}\left\{f(w)=\frac{1}{2}aw^2-bw+c\right\},$$

where $a>0$, b, and c are constants. Assume in addition that we want to implement our algorithm in hardware, using fixed-point arithmetic.

In order to solve this problem, we could proceed as follows. Assume we have an initial approximation $\hat{w}(0)$. We choose a step size $H > 0$ and compute

$$\Delta f_+ = f(\hat{w}(0) + H) - f(\hat{w}(0)), \qquad \Delta f_- = f(\hat{w}(0) - H) - f(\hat{w}(0)).$$

Then, if $\Delta f_+ < 0$, we choose $\hat{w}(1) = \hat{w}(0) + H$, and if $\Delta f_- < 0$, we choose $\hat{w}(1) = \hat{w}(0) - H$. If both are positive, we choose $\hat{w}(1) = \hat{w}(0)$ and reduce the step size by half, i.e., $H \leftarrow H/2$ (note that if $f(\cdot)$ is convex, the case of both $\Delta f_+ < 0$ and $\Delta f_- < 0$ can never occur). With the new estimate $\hat{w}(1)$, the procedure can be repeated to find a new improved estimate $\hat{w}(2)$, and so on. It can be shown that this algorithm will converge to the minimum of $f(\cdot)$ whenever $f(\cdot)$ is a convex function (see Fig. 12.35).

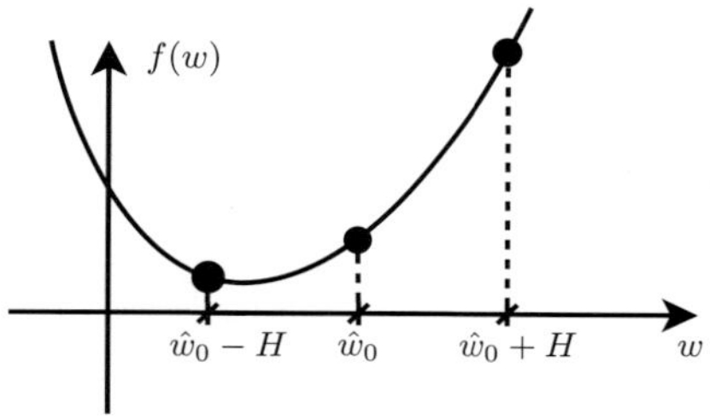

FIGURE 12.35

DCD applied to 1D minimization.

Although the description given so far explains the general working of the DCD algorithm, it does not show how it can be efficiently implemented, using as few operations as possible. We will turn to this point now, but already extending the algorithm to minimize a quadratic function of several variables. Consider then the problem of finding the solution $\boldsymbol{w}_o$ to

$$\min_{\boldsymbol{w}} \left\{ f(\boldsymbol{w}) = \frac{1}{2}\boldsymbol{w}^T \boldsymbol{R} \boldsymbol{w} - \boldsymbol{b}^T \boldsymbol{w} + c \right\},$$

where $\boldsymbol{R}$ is a positive definite matrix, $\boldsymbol{b}$ is a vector, and c is a constant (note that the minimum $\boldsymbol{w}_o$ of $f(\boldsymbol{w})$ is also the solution $\boldsymbol{w}_o$ to $\boldsymbol{R}\boldsymbol{w}_o = \boldsymbol{b}$, that is, any algorithm to find $\boldsymbol{w}_o$ can also be used to solve the normal equations (12.289)).

Assume that we have an initial approximation $\hat{\boldsymbol{w}}(0)$ to the solution and we want to find an improved approximation, changing one entry of $\hat{\boldsymbol{w}}(0)$ at a time. In order to differentiate each new approximation, we use the following notation. We define $\hat{\boldsymbol{w}}^{(0)}(0) = \hat{\boldsymbol{w}}(0)$. We shall first seek an improved estimate to the first entry of $\hat{\boldsymbol{w}}^{(0)}(0)$. The resulting vector will be denoted by $\hat{\boldsymbol{w}}^{(1)}(0)$. Continuing, we find an improved estimate to the second entry of $\hat{\boldsymbol{w}}^{(1)}(0)$, resulting in vector $\hat{\boldsymbol{w}}^{(2)}(0)$, and so on. After updating all entries of the initial estimate, we have $\hat{\boldsymbol{w}}^{(M)}(0)$. We then let $\hat{\boldsymbol{w}}^{(0)}(1) = \hat{\boldsymbol{w}}^{(M)}(0)$ and repeat the procedure.

Then we need to check if $f(\hat{\boldsymbol{w}}^{(0)}(0) \pm H\boldsymbol{e}_1) < f(\hat{\boldsymbol{w}}(0))$ if we add or subtract H to the first entry of $\hat{\boldsymbol{w}}^{(0)}(0)$ ($\boldsymbol{e}_1$ is the vector $\begin{bmatrix}1 & 0 & \dots & 0\end{bmatrix}^T$). The change Δf_+ in the cost function obtained by adding

H to the first entry of $\hat{\boldsymbol{w}}^{(0)}(0)$ is

$$\Delta f_+ = \frac{1}{2}\left(\hat{\boldsymbol{w}}^{(0)}(0) + H\boldsymbol{e}_1\right)^T \boldsymbol{R}\left(\hat{\boldsymbol{w}}^{(0)}(0) + H\boldsymbol{e}_1\right) - \boldsymbol{b}^T\left(\hat{\boldsymbol{w}}^{(0)}(0) + H\boldsymbol{e}_1\right) + c \tag{12.292}$$

$$- \left(\frac{1}{2}\hat{\boldsymbol{w}}^{(0)T}(0)\boldsymbol{R}\hat{\boldsymbol{w}}^{(0)}(0) - \boldsymbol{b}^T\hat{\boldsymbol{w}}^{(0)}(0) + c\right)$$

$$= H\boldsymbol{e}_1^T\boldsymbol{R}\hat{\boldsymbol{w}}^{(0)}(0) + \frac{1}{2}H^2\boldsymbol{e}_1^T\boldsymbol{R}\boldsymbol{e}_1 - H\boldsymbol{b}^T\boldsymbol{e}_1. \tag{12.293}$$

Similarly, the variation Δf_- obtained from subtracting H from the first entry of $\hat{\boldsymbol{w}}(0)$ is

$$\Delta f_- = -H\boldsymbol{e}_1^T\boldsymbol{R}\hat{\boldsymbol{w}}^{(0)}(0) + \frac{1}{2}H^2\boldsymbol{e}_1^T\boldsymbol{R}\boldsymbol{e}_1 + H\boldsymbol{b}^T\boldsymbol{e}_1. \tag{12.294}$$

Therefore, we should make

$$\hat{\boldsymbol{w}}^{(1)}(0) = \begin{cases} \hat{\boldsymbol{w}}^{(0)}(0) + H\boldsymbol{e}_1 & \text{if } \Delta f_+ < 0, \\ \hat{\boldsymbol{w}}^{(0)}(0) - H\boldsymbol{e}_1 & \text{if } \Delta f_- < 0, \\ \hat{\boldsymbol{w}}^{(0)}(0), & \text{otherwise.} \end{cases} \tag{12.295}$$

The same steps can be repeated for the second entry of $\hat{\boldsymbol{w}}^{(1)}(0)$, and so on, that is, we can update the $(m+1)$th entry of $\hat{\boldsymbol{w}}^{(m)}(0)$ to obtain a new approximation $\hat{\boldsymbol{w}}^{(m+1)}(0)$ for $m = 0 \ldots M-1$. After obtaining $\hat{\boldsymbol{w}}^{(M)}(0)$, we have updated all entries of $\hat{\boldsymbol{w}}(0)$. We can then make $\hat{\boldsymbol{w}}^{(0)}(1) = \hat{\boldsymbol{w}}^{(M)}(0)$ and repeat the procedure. However, if no update is made for $m = 1 \ldots M$, before repeating we decrease the step H by a factor of two (i.e., $H \leftarrow H/2$). By reducing H only when no update is necessary for any entry of the vector, we avoid problems that would appear if the initial value of H were too small.

Implemented as described so far, the algorithm has a high computational cost. For example, for the evaluation of Δf_+ at each step we must:

- Evaluate $\boldsymbol{e}_m^T\boldsymbol{R}\hat{\boldsymbol{w}}^{(m-1)}(i)$. This is equivalent to multiplying the mth row of $\boldsymbol{R}$ by $\hat{\boldsymbol{w}}^{(m-1)}(i)$, involving M multiplications and $M-1$ additions.
- Multiply $-\boldsymbol{e}_m^T\boldsymbol{R}\hat{\boldsymbol{w}}^{(m-1)}(i) + \boldsymbol{b}^T\boldsymbol{e}_m$ by H and add the result to $(H^2/2)R_{m,m}$ ($R_{m,m}$ is the element (m,m) of $\boldsymbol{R}$). This requires two multiplications and two additions, assuming $H^2/2$ is precomputed.

The same steps would be necessary for the computation of Δf_-. Note that all these computations would have to be repeated at each step; therefore, the update of all entries of $\hat{\boldsymbol{w}}^{(0)}(i)$ would require more than M^2 multiplications.

However, the number of computations can be reduced substantially if we take advantage of some properties of hardware implementations of arithmetic operations, as we describe next. Assume in the following that all variables are fixed-point binary numbers (see Box 12.8).

First, we can avoid the computation of $\boldsymbol{R}\hat{\boldsymbol{w}}^{(m-1)}(i)$ by introducing a residue vector that is updated at each step. Let $\boldsymbol{r}^{(0)}(0) = \boldsymbol{b} - \boldsymbol{R}\hat{\boldsymbol{w}}^{(0)}(0)$ be the initial residue vector. Note that this is the residue of solving $\boldsymbol{R}\boldsymbol{w} = \boldsymbol{b}$ using $\hat{\boldsymbol{w}}^{(0)}(0)$ as an approximation to the solution.

At the end of the first step, $\hat{\boldsymbol{w}}^{(0)}(0)$ is updated as in (12.295). Since only one entry of $\hat{\boldsymbol{w}}^{(0)}(0)$ is changed, the modification in the residue is simple to evaluate:

$$\boldsymbol{r}^{(1)}(0) = \boldsymbol{b} - \boldsymbol{R}\hat{\boldsymbol{w}}^{(1)}(0) = \begin{cases} \boldsymbol{b} - \boldsymbol{R}(\hat{\boldsymbol{w}}^{(0)}(0) + H\boldsymbol{e}_1) = \boldsymbol{r}^{(0)}(0) - H\boldsymbol{R}\boldsymbol{e}_1, & \text{if } \Delta f_+ < 0, \\ \boldsymbol{b} - \boldsymbol{R}(\hat{\boldsymbol{w}}^{(0)}(0) - H\boldsymbol{e}_1) = \boldsymbol{r}^{(0)}(0) + H\boldsymbol{R}\boldsymbol{e}_1, & \text{if } \Delta f_- < 0, \\ \boldsymbol{r}^{(0)}(0), & \text{otherwise.} \end{cases} \tag{12.296}$$

This still requires M multiplications (H multiplied by the first column of $\boldsymbol{R}$). However, if H is a power of two, that is, if $H = 2^k$ for some integer k, then the computation of $\boldsymbol{r}^{(1)}(0)$ in fixed-point arithmetic requires only M shifts and M additions, which are much easier to implement in hardware than multiplications. Moreover, if both Δf_+ and Δf_- are positive, no update is necessary.

Next, we note that it is not really necessary to compute both Δf_+ and Δf_-: from (12.293) and (12.294), when updating the mth element of $\hat{\boldsymbol{w}}^{(m-1)}(i)$ at step i, we have

$$\Delta f_+ = -H\boldsymbol{e}_m^T \boldsymbol{r}^{(m-1)}(i) + \frac{1}{2}H^2 \boldsymbol{e}_m^T \boldsymbol{R}\boldsymbol{e}_m = -H r_m^{(m-1)}(i) + \frac{1}{2}H^2 R_{m,m}, \tag{12.297}$$

where $r_m^{(m-1)}(i)$ is the mth entry of the residue vector $\boldsymbol{r}^{(m-1)}(i)$, $\boldsymbol{r}^{(0)}(i+1) \triangleq \boldsymbol{r}^{(M)}(i)$, and $\boldsymbol{e}_m$ is the mth column of the $M \times M$ identity matrix. Similarly,

$$\Delta f_- = H r_m^{(m-1)}(i) + \frac{1}{2}H^2 R_{m,m}. \tag{12.298}$$

Then, recalling that H and $R_{m,m}$ are positive ($\boldsymbol{R}$ is positive definite), $\Delta f_+ < 0$ only if

$$r_m^{(m-1)}(i) > \frac{1}{2} H R_{m,m} \geq 0.$$

Repeating the arguments for Δf_-, we see that the mth entry of $\hat{\boldsymbol{w}}^{(m-1)}(i)$ will be updated if and only if

$$|r_m^{(m-1)}(i)| > \frac{1}{2} H R_{m,m}, \tag{12.299}$$

with updates

$$\hat{\boldsymbol{w}}^{(m)}(i) = \hat{\boldsymbol{w}}^{(m-1)}(i) + H \operatorname{sign}(r_m^{(m-1)}(i))\boldsymbol{e}_m, \tag{12.300}$$

$$\boldsymbol{r}^{(m)}(i) = \boldsymbol{r}^{(m-1)}(i) - H \operatorname{sign}(r_m^{(m-1)}(i))\boldsymbol{R}\boldsymbol{e}_m. \tag{12.301}$$

Eqs. (12.299)–(12.301) require one comparison, $M+1$ shifts, and $M+1$ additions. Table 12.10 summarizes the DCD algorithm for minimization of quadratic cost functions. Note that the algorithm will update the entries of $\hat{\boldsymbol{w}}(i)$ bit by bit. It therefore makes sense to choose for the initial step size H the power-of-two value represented by the most significant bit in the fixed-point representation being used. The conditional jump in step 10 prevents problems in case H is chosen too small.

As seen in Table 12.10, the total number of operations never exceeds $(2M+2)(B-1) + N_u(4M+3)$ (note that here we are counting additions, comparisons, and shifts as operations – the algorithm does

Table 12.10 Summary of the DCD algorithm for minimization of a quadratic cost function.

Step	Initialization: Choose step $H = 2^k$, number of bits B, and maximum update count N_u. Let $\hat{w} = 0$, $r = b$, $h = H$, and $c = 0$.	Number of additions, shifts, and comparisons	multiplications
1	for $b = 0 : B - 1$		
2	$h \leftarrow h/2$	1	0
3	flag $\leftarrow 0$	0	0
4	for $m = 1 : M$		
5	if $\lvert r_m \rvert > \frac{1}{2} h R_{m,m}$	2	0
6	$\hat{w}_m \leftarrow \hat{w}_m + h\,\text{sign}(r_m)$	1	0
7	$\boldsymbol{r} \leftarrow \boldsymbol{r} - h\,\text{sign}(r_m)\boldsymbol{R}\boldsymbol{e}_m$	2M	0
8	$c \leftarrow c + 1$, flag $\leftarrow 1$	1	0
9	if $c > N_u$, stop algorithm.	1	0
10	if flag $= 1$, go to step 2	1	0
	Total: (worst case)	$\leq (2M + 2)(B - 1) + N_u(2M + 3)$	0

not involve multiplications). In DCD-RLS, the maximum number of updates N_u can be chosen between one and eight with good performance, while keeping the cost low.

When applied to RLS, one final simplification can be used when implementing DCD: instead of updating all entries of vector $\hat{\boldsymbol{w}}(i)$, we can only update the entry corresponding to the largest residue. As explained in [402], the resulting algorithm (called *leading DCD*) is still guaranteed to converge to the optimum solution. This algorithm is described in Table 12.12.

DCD-RLS

As we mentioned before, DCD-RLS uses the DCD algorithm just described to solve the normal equations as each new sample arrives. Since DCD is an iterative algorithm, we can use the current solution $\boldsymbol{w}(n)$ as the initial value for computation of $\boldsymbol{w}(n+1)$. In this way, the number of updates N_u of DCD can be restricted to a small number (in many situations one to four updates are enough for performance close to that of exact RLS). This further reduces the computational cost of the algorithm.

In the following, we will use $\boldsymbol{w}(n+1)$ to denote the exact solution of (12.289) (i.e., the exact RLS estimate) and $\hat{\boldsymbol{w}}(n+1)$ to denote the approximation computed by DCD-RLS (both would be equal if we let $N_u \to \infty$). Assume that at time instant n, we have available both $\hat{\boldsymbol{w}}(n)$ and the residue

$$\boldsymbol{r}(n) = \widehat{\boldsymbol{r}}_{d\boldsymbol{\phi}}(n-1) - \widehat{\boldsymbol{R}}_{\boldsymbol{\phi}}(n-1)\hat{\boldsymbol{w}}(n). \tag{12.302}$$

In order to apply the DCD algorithm to solve for $\hat{\boldsymbol{w}}(n+1)$, it is convenient to update the difference

$$\boldsymbol{\Delta w}(n+1) = \boldsymbol{w}(n+1) - \hat{\boldsymbol{w}}(n). \tag{12.303}$$

Define also

$$\boldsymbol{\Delta}\widehat{\boldsymbol{R}}_{\boldsymbol{\phi}}(n) = \widehat{\boldsymbol{R}}_{\boldsymbol{\phi}}(n) - \widehat{\boldsymbol{R}}_{\boldsymbol{\phi}}(n-1), \tag{12.304}$$

$$\Delta\widehat{\boldsymbol{r}}_{d\phi}(n) = \widehat{\boldsymbol{r}}_{d\phi}(n) - \widehat{\boldsymbol{r}}_{d\phi}(n-1). \tag{12.305}$$

The normal equations (12.289) then become

$$\widehat{\boldsymbol{R}}_{\phi}(n)\left[\hat{\boldsymbol{w}}(n) + \Delta\boldsymbol{w}(n+1)\right] = \widehat{\boldsymbol{r}}_{d\phi}(n),$$

and thus

$$\begin{aligned}\widehat{\boldsymbol{R}}_{\phi}(n)\Delta\boldsymbol{w}(n+1) &= \widehat{\boldsymbol{r}}_{d\phi}(n-1) + \Delta\widehat{\boldsymbol{r}}_{d\phi}(n) - \widehat{\boldsymbol{R}}_{\phi}(n-1)\hat{\boldsymbol{w}}(n) - \Delta\widehat{\boldsymbol{R}}_{\phi}(n)\hat{\boldsymbol{w}}(n)\\ &= \boldsymbol{r}(n) + \Delta\widehat{\boldsymbol{r}}_{d\phi}(n) - \Delta\widehat{\boldsymbol{R}}_{\phi}(n)\hat{\boldsymbol{w}}(n) \stackrel{\Delta}{=} \boldsymbol{\beta}(n).\end{aligned} \tag{12.306}$$

Given an approximate solution $\Delta\hat{\boldsymbol{w}}(n+1)$ to these equations, we obtain an updated approximate solution $\hat{\boldsymbol{w}}(n+1) = \hat{\boldsymbol{w}}(n) + \Delta\hat{\boldsymbol{w}}(n+1)$. Note that the residue of solving (12.289) using $\hat{\boldsymbol{w}}(n+1)$,

$$\begin{aligned}\boldsymbol{r}(n+1) &= \widehat{\boldsymbol{r}}_{d\phi}(n) - \widehat{\boldsymbol{R}}_{\phi}(n)\widehat{\boldsymbol{w}}(n+1)\\ &= \widehat{\boldsymbol{r}}_{d\phi}(n-1) + \Delta\widehat{\boldsymbol{r}}_{d\phi}(n) - \widehat{\boldsymbol{R}}_{\phi}(n)\left[\hat{\boldsymbol{w}}(n) + \Delta\hat{\boldsymbol{w}}(n+1)\right]\\ &= \widehat{\boldsymbol{r}}_{d\phi}(n-1) + \Delta\widehat{\boldsymbol{r}}_{d\phi}(n) - \widehat{\boldsymbol{R}}_{\phi}(n)\Delta\hat{\boldsymbol{w}}(n+1) - \left[\widehat{\boldsymbol{R}}_{\phi}(n-1) + \Delta\widehat{\boldsymbol{R}}_{\phi}(n)\right]\hat{\boldsymbol{w}}(n)\\ &= \boldsymbol{r}(n) + \Delta\widehat{\boldsymbol{r}}_{d\phi}(n) - \widehat{\boldsymbol{R}}_{\phi}(n)\Delta\hat{\boldsymbol{w}}(n+1) - \Delta\widehat{\boldsymbol{R}}_{\phi}(n)\hat{\boldsymbol{w}}(n)\\ &= \boldsymbol{\beta}(n) - \widehat{\boldsymbol{R}}_{\phi}(n)\Delta\hat{\boldsymbol{w}}(n+1),\end{aligned} \tag{12.307}$$

is exactly the residue of approximately solving (12.306) by $\Delta\hat{\boldsymbol{w}}(n+1)$.

To complete, we only need to evaluate $\Delta\widehat{\boldsymbol{R}}_{\phi}(n)$ and $\Delta\widehat{\boldsymbol{r}}_{d\phi}(n)$. From (12.304), (12.305), (12.290), and (12.291), we obtain

$$\Delta\widehat{\boldsymbol{R}}_{\phi}(n) = (\lambda - 1)\widehat{\boldsymbol{R}}_{\phi}(n-1) + \boldsymbol{\phi}(n)\boldsymbol{\phi}^T(n), \tag{12.308}$$

$$\Delta\widehat{\boldsymbol{r}}_{d\phi}(n) = (\lambda - 1)\widehat{\boldsymbol{d}}_{d\phi}(n-1) + d(n)\boldsymbol{\phi}(n). \tag{12.309}$$

Substituting these results in the definition of $\boldsymbol{\beta}(n)$, we obtain

$$\begin{aligned}\boldsymbol{\beta}(n) &= \boldsymbol{r}(n) + (\lambda - 1)\widehat{\boldsymbol{r}}_{d\phi}(n-1) + d(n)\boldsymbol{\phi}(n) - \left[(\lambda - 1)\widehat{\boldsymbol{R}}_{\phi}(n-1) + \boldsymbol{\phi}(n)\boldsymbol{\phi}^T(n)\right]\hat{\boldsymbol{w}}(n)\\ &= \boldsymbol{r}(n) + (\lambda - 1)\boldsymbol{r}(n) + e(n)\boldsymbol{\phi}(n) = \lambda\boldsymbol{r}(n) + e(n)\boldsymbol{\phi}(n).\end{aligned} \tag{12.310}$$

We can therefore use the DCD algorithm to compute an approximate solution to (12.306) and update the residue defined by (12.307), as described in Table 12.11.

As mentioned before, the number of operations of the DCD algorithm can be reduced if we modify the algorithm in Table 12.10 to update only the entries of the weight vector corresponding to the largest residue entry, as described in Table 12.12.

The total number of operations of DCD-RLS as described in Tables 12.11 and 12.12 is $M^2 + 4M$ multiplications and $M^2/2 + 3.5M + (2M+1)N_u + B$ additions. Most of these operations are due to the updating of $\widehat{\boldsymbol{R}}_{\phi}$ in Table 12.11, which is responsible for the terms in M^2. While this cannot be helped for general regressors $\boldsymbol{\phi}(n)$, the operation count can be substantially reduced when our filter is modeling an FIR relation as in (12.42) (i.e., $\boldsymbol{\phi}(n)$ is a tapped-delay line), as we show next.

Table 12.11 Summary of the DCD-RLS algorithm for general regressors.

Initialization:
$\hat{\boldsymbol{w}}(0) = \mathbf{0}$, $\boldsymbol{r}(0) = \mathbf{0}$, $\widehat{\boldsymbol{R}}_{\phi}(-1) = \boldsymbol{\Pi} > 0$
for $n = 0, 1, 2, \ldots$
$\widehat{\boldsymbol{R}}_{\phi}(n) = \lambda \widehat{\boldsymbol{R}}(n-1) + \boldsymbol{\phi}(n)\boldsymbol{\phi}^T(n)$ (use (12.313) for tapped-delay lines)
$\hat{y}(n) = \hat{\boldsymbol{w}}^T(n)\boldsymbol{\phi}(n)$,
$e(n) = d(n) - \hat{y}(n)$,
$\boldsymbol{\beta}(n) = \lambda \boldsymbol{r}(n) + e(n)\boldsymbol{\phi}(n)$,
Use the DCD algorithm to compute new $\boldsymbol{\Delta}\hat{\boldsymbol{w}}(n+1)$ and $\boldsymbol{r}(n+1)$ from $\widehat{\boldsymbol{R}}_{\phi}(n)\boldsymbol{\Delta}\hat{\boldsymbol{w}}(n+1) = \boldsymbol{\beta}(n)$,
$\hat{\boldsymbol{w}}(n+1) = \hat{\boldsymbol{w}}(n) + \boldsymbol{\Delta}\hat{\boldsymbol{w}}(n+1)$.

Table 12.12 Summary of leading-element DCD algorithm for use in DCD-RLS, from [402].

Step	**Initialization:** Choose initial condition step $H = 2^k$, number of bits B, and maximum update count N_u. Let $\Delta\hat{w} = 0$, $r = \beta(n)$, $h = H/2$, $b = 1$.
	for $k = 1 : N_u$
1	$p = \arg \max_{1 \le m \le M} \{\lvert r_m \rvert\}$, go to step 4
2	$b \leftarrow b + 1$, $h \leftarrow h/2$
3	Stop if $b > B$
4	Go to step 2 if $\lvert r_p \rvert \le (h/2) R_{p,p}$
5	$\Delta\hat{w}_p = \Delta\hat{w}_p + \text{sign}(r_p)h$
6	$\boldsymbol{r} = \boldsymbol{r} - \text{sign}(r_p)h\widehat{\boldsymbol{R}}_{\phi}\boldsymbol{e}_p$

If

$$\boldsymbol{\phi}(n) = \begin{bmatrix} x(n) & x(n-1) & \ldots & x(n-M+1) \end{bmatrix}^T, \tag{12.311}$$

then $\widehat{\boldsymbol{R}}(n)$ and $\widehat{\boldsymbol{R}}(n-1)$ share a common structure. Assume that the initial condition $\widehat{\boldsymbol{R}}(-1) = \boldsymbol{\Pi} = \text{diag}(\lambda^{M-1}, \lambda^{M-2}, \ldots, 1)$. Then

$$\widehat{\boldsymbol{R}}(0) = \lambda \begin{bmatrix} \lambda^{M-1} & 0 & \ldots & 0 \\ 0 & \lambda^{M-2} & \ldots & 0 \\ \vdots & \vdots & \ddots & \vdots \\ 0 & 0 & \ldots & 1 \end{bmatrix} + \begin{bmatrix} x(0) \\ 0 \\ \vdots \\ 0 \end{bmatrix} \begin{bmatrix} x(0) & 0 & \ldots & 0 \end{bmatrix}$$

$$
= \begin{bmatrix} \lambda^M + x^2(0) & 0 & \dots & 0 \\ 0 & \lambda^{M-1} & \dots & 0 \\ \vdots & \vdots & \ddots & \vdots \\ 0 & 0 & \dots & \lambda \end{bmatrix} = \left[\begin{array}{c|c} \lambda^M + x^2(0) & \mathbf{0}^T \\ \hline \mathbf{0} & [\widehat{\boldsymbol{R}}_{\boldsymbol{\phi}}(-1)]_{1:M-1,1:M-1} \end{array}\right],
$$

where $[\widehat{\boldsymbol{R}}_{\boldsymbol{\phi}}(-1)]_{1:M-1,1:M-1}$ denotes the top-left $(M-1)\times(M-1)$ block of $\widehat{\boldsymbol{R}}_{\boldsymbol{\phi}}(-1)$. We note that except for its first row and column, $\widehat{\boldsymbol{R}}(0)$ can be obtained directly from $\widehat{\boldsymbol{R}}(-1)$.

Following this observation, we claim that for tapped-delay line regressors, the computation of $\widehat{\boldsymbol{R}}_{\boldsymbol{\phi}}(n)$ in Table 12.11 can be replaced by an update of only its first column $[\widehat{\boldsymbol{R}}_{\boldsymbol{\phi}}(n)]_1$, that is,

$$
[\widehat{\boldsymbol{R}}_{\boldsymbol{\phi}}(n)]_1 = \lambda[\widehat{\boldsymbol{R}}_{\boldsymbol{\phi}}(n-1)]_1 + x(n)\boldsymbol{\phi}(n). \tag{12.312}
$$

Since $\widehat{\boldsymbol{R}}_{\boldsymbol{\phi}}(n)$ is symmetric, its first row is the transpose of (12.313) and does not need to be evaluated again.

We show next by induction that this pattern holds for all n. We just showed that it is true for $n = 0$. Assume then that for a certain n, we have

$$
\widehat{\boldsymbol{R}}_{\boldsymbol{\phi}}(n) = \left[\begin{array}{c|c} \lambda[\widehat{\boldsymbol{R}}_{\boldsymbol{\phi}}(n-1)]_1 + x(n)\boldsymbol{\phi}(n) & \begin{array}{c} (\cdot)^T \\ \\ [\widehat{\boldsymbol{R}}_{\boldsymbol{\phi}}(n-1)]_{1:M-1,1:M-1} \end{array} \end{array}\right], \tag{12.313}
$$

where we use $(\cdot)^T$ to indicate that the entries of the first row are the same as the entries of the first column. We show next that $\widehat{\boldsymbol{R}}_{\boldsymbol{\phi}}(n+1)$ must follow the same pattern. Indeed,

$$
\begin{aligned}
\widehat{\boldsymbol{R}}_{\boldsymbol{\phi}}(n+1) &= \lambda\widehat{\boldsymbol{R}}_{\boldsymbol{\phi}}(n) + \begin{bmatrix} x(n+1) \\ x(n) \\ \vdots \\ n(n-M+2) \end{bmatrix} \begin{bmatrix} x(n+1) & x(n) & \dots & x(n-M+2) \end{bmatrix} \\
&= \left[\begin{array}{c|c} \lambda[\widehat{\boldsymbol{R}}_{\boldsymbol{\phi}}(n)]_1 & \begin{array}{c} (\cdot)^T \\ \\ \lambda[\widehat{\boldsymbol{R}}_{\boldsymbol{\phi}}(n-1)]_{1:M-1,1:M-1} \end{array} \end{array}\right] \\
&+ \left[\begin{array}{c|c} x(n+1)\boldsymbol{\phi}(n+1) & \begin{array}{c} (\cdot)^T \\ \begin{bmatrix} x^2(n) & \dots & x(n)x(n-M+2) \\ \vdots & \ddots & \vdots \\ x(n)x(n-M+2) & \dots & x^2(n-M+2) \end{bmatrix} \end{array} \end{array}\right] \\
&= \left[\begin{array}{c|c} \lambda[\widehat{\boldsymbol{R}}_{\boldsymbol{\phi}}(n)]_1 + x(n+1)\boldsymbol{\phi}(n+1) & \begin{array}{c} (\cdot)^T \\ \\ [\widehat{\boldsymbol{R}}_{\boldsymbol{\phi}}(n)]_{1:M-1,1:M-1} \end{array} \end{array}\right],
\end{aligned}
$$

where in the last step we identified the (2, 2) block of the intermediate result with the first $M-1$ rows and columns of $\widehat{\boldsymbol{R}}_{\phi}(n)$, from (12.290).

Note that in practice we should not actually move the entries of $\widehat{\boldsymbol{R}}_{\phi}(n-1)$ to their new positions; rather, we should use an indexing system to access the entries of $\widehat{\boldsymbol{R}}_{\phi}(n)$ to avoid this costly moving operation. The total number of operations of DCD-RLS using (12.313) is $3M$ multiplications and $2MN_u + 6M$ additions. This is not much larger than the complexity of NLMS. The DCD-RLS algorithm has been implemented in FPGAs and runs for long periods of time without observation of divergence [402].

Fig. 12.36 shows a comparison of RLS and DCD-RLS in a system identification problem. We are trying to identify an $M = 10$ FIR filter, so we use (12.313) to update $\widehat{\boldsymbol{R}}_{\phi}(n)$ with lower cost. The input signal $x(n)$ is white Gaussian noise with zero mean and unit variance. The optimum coefficient vector $\boldsymbol{w}_{\text{o}}(n)$ follows a random-walk model as in (12.183) with $\boldsymbol{Q} = 10^{-5}\boldsymbol{I}$ (the initial value $\boldsymbol{w}_{\text{o}}(0)$ is random, taken from a Gaussian distribution with covariance matrix equal to $\boldsymbol{I}$). Both RLS and RLS-DCD use a forgetting factor of $\lambda = 0.95$. RLS-DCD is implemented with leading DCD (Table 12.12), $B = 16$, and $H = 2$. The maximum number of updates N_u is varied from one to eight. The optimum vector $\boldsymbol{w}_{\text{o}}(n)$ suffers an abrupt change at $n = 400$. The learning curves were obtained from the average of $L = 500$ realizations. As the figure shows, even for $N_u = 1$, the performance of RLS-DCD is very close to that of

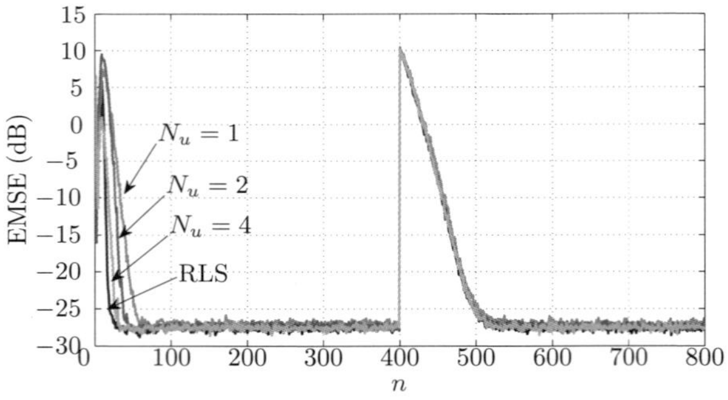

FIGURE 12.36

EMSE for RLS and DCD-RLS for different values of N_u, with $\lambda = 0.95$, $M = 10$, and for DCD, $H = 2$ and $B = 16$. Average of $L = 500$ realizations.

RLS. The only difference is a smaller convergence rate at the start of the algorithm. However, after the abrupt modification of $\boldsymbol{w}_{\text{o}}(n)$ at instant $n = 400$, all filters converge equally quickly to the new vector. This shows that RLS-DCD does not lose in tracking capability when compared to standard RLS.

12.5.2 Regularization

There are applications in which the autocorrelation matrix $\boldsymbol{R}_{\phi}$ may become very ill-conditioned (i.e., nearly singular). In these cases, it can be shown that the LMS weight estimates may slowly drift to quite large values, even though the error remains small. The problem arises when the filter is implemented in finite precision arithmetic, because the variables holding the filter weights may overflow, thereby making the error grow suddenly very large.

This can happen, for example, for certain kinds of equalizers in communications [95,160,161]. The mechanism through which the LMS coefficients may slowly drift is explained in [261,329].

The usual solution to this problem is to modify the cost function to add a term proportional to $\|\boldsymbol{w}\|^2$, so that large values of the weight vector are penalized:

$$J_{\text{leaky}} = \mathrm{E}\left\{\left|d(n) - \boldsymbol{w}^H\boldsymbol{\phi}(n)\right|^2\right\} + \alpha\|\boldsymbol{w}\|^2, \tag{12.314}$$

where $\|\cdot\|$ is the Euclidean norm and $\alpha > 0$ is a small constant. With this cost function, the corresponding algorithm is

$$\boldsymbol{w}(n+1) = (1-\mu\alpha)\boldsymbol{w}(n) + \mu e^*(n)\boldsymbol{\phi}(n). \tag{12.315}$$

This is known as the *leaky*-LMS algorithm [388]. It can be shown that the introduction of leakage will bias the solution away from the Wiener solution, but will prevent any excessive growth of the weight coefficients. A detailed analysis of the leaky-LMS algorithm is available in [233]. Reference [261] proposes a modified version of the algorithm without the bias.

Difficulties with singular or near-singular autocorrelation matrices also appears with RLS. For example, if the input is a periodic signal with little noise, the autocorrelation matrix will become nearly singular, and matrix $\boldsymbol{P}(n)$ in RLS will diverge. This problem, as well as a solution to it using a variable forgetting factor, is described in [220]. A constant regularization method based on DCD can be found in [209]. Another approach, valid only for short filters, but which allows constant regularization, is described in [370].

The leaky-LMS algorithm has found other important applications, most particularly in beamforming, in which it is used to make the beamformer robust against array imperfections [180]. The algorithms used in beamforming are examples of constrained adaptive filters, in which the cost function incorporates equality conditions that must be satisfied at all times (see Section 12.5.11).

Another algorithm closely related to leaky-LMS, in which the extra term in the cost function penalizes large values of the filter output (that is, $\alpha|\boldsymbol{w}^H(n)\boldsymbol{\phi}(n)|^2$ instead of $\alpha\|\boldsymbol{w}(n)\|^2$ in (12.314)), was recently employed to improve the performance of FIR adaptive filters in the presence of saturation nonlinearities [45].

More recent applications of regularization involve the use of the ℓ_1-norm,

$$\|\boldsymbol{w}\|_1 = |w_0| + |w_1| + \cdots + |w_{M-1}|,$$

instead of the Euclidean norm. This is because the ℓ_1-norm promotes solutions that are sparse, that is, a filter that minimizes the cost function

$$J_{\text{sparse}} = \mathrm{E}\left\{\left|d(n) - \boldsymbol{w}^H\boldsymbol{\phi}(n)\right|^2\right\} + \alpha\|\boldsymbol{w}\|_1, \tag{12.316}$$

will favor solutions in which the vector $\boldsymbol{w}$ has only a few nonzero entries (an intuitive explanation for this can be found in [67], which is a good introduction to compressive sensing). We should note that the cost function in (12.316) is still convex. Sparser solutions can also be achieved when the ℓ_1-norm is replaced by the reweighted ℓ_1-norm [68] or by an ℓ_0-norm approximation [166,206,207]. However, the use of these pseudo-norms makes the problem nonconvex. Algorithms for promoting sparsity have received much attention, due to the interest in compressive sensing applications, and also in echo cancelation applications [17,18,86,186,243,279,403]. See also the references for the proportionate NLMS (PNLMS) algorithm in Section 12.5.3 and the adaptive filters with constraints in Section 12.5.11.

12.5.3 Variable step size

The steady-state analysis of Section 12.4.4 shows that the step size μ plays an import role in controlling the performance of the LMS algorithm, since it is one of the main factors affecting the convergence rate and the misadjustment. To obtain a better trade-off between convergence speed and steady-sate EMSE, several authors proposed different forms of adjusting the step size of LMS. Variable step size algorithms allow the filters to dynamically adjust their performance in response to conditions in the input data and error signals.

In this context, one of the most popular algorithms is the variable step size LMS (VSLMS) proposed in [170]. In the VSLMS algorithm, each coefficient $w_k(n)$, $k = 0, \ldots, M-1$, is updated with a time-varying step size $\mu_k(n)$, whose adjustment is given by

$$\mu_k(n) = \mu_k(n-1) + \rho \operatorname{sign}[e(n)\phi_k(n)] \operatorname{sign}[e(n-1)\phi_k(n-1)], \tag{12.317}$$

where ρ is a small positive constant. This algorithm operates in a very intuitive way. When the algorithm has not yet converged, the gradient term $e(n)\phi_k(n)$ shows a consistently positive or negative direction in successive iterations and the step size $\mu_k(n)$ increases up to an allowed maximum value. On the other hand, near steady state $e(n)\phi_k(n)$ approaches zero and its sign changes in successive iterations, which reduces $\mu_k(n)$ (the step size is lower-bounded by a small minimum value). Therefore, during the initial convergence VSLMS uses a large step size, which leads to a high convergence rate. When its coefficients are close to the optimum solution, the algorithm is updated with a small step size, which allows VSLMS to achieve a small EMSE. A variant of this algorithm is obtained by dropping the sign functions in (12.317), i.e.,

$$\mu_k(n) = \mu_k(n-1) + \rho\, e(n)e(n-1)\phi_k(n)\phi_k(n-1). \tag{12.318}$$

The choice between (12.317) and (12.318) depends on the application [139].

The NLMS algorithm, proposed independently by Nagumo and Noda [255] and Albert and Gardner [10], can be interpreted as a VSLMS with a particular choice for $\mu(n)$ (see Section 12.3.2). An important variant of NLMS is the so-called *power-normalized* LMS (PN-LMS) algorithm, which uses the following variable step size:

$$\mu(n) = \frac{\tilde{\mu}}{\varepsilon + p(n)}, \tag{12.319}$$

where

$$p(n) = \gamma p(n-1) + (1-\gamma)|\phi_0(n)|^2, \tag{12.320}$$

with $p(-1) = 0$ and γ being a positive scalar chosen from within the interval $0 < \gamma \leq 1$. This algorithm is useful when the regressor is a tapped-delay line, that is, when $\phi_i(n) = x(n-i)$, so that the scalar $p(n)$ is an estimate for the power of the input sequence $\{x(n)\}$. The range of $\tilde{\mu}$ in (12.319) is $0 \leq \tilde{\mu} \leq \frac{2}{M}$ [317].

Over the years, several variable step size algorithms were proposed in the literature. Some of these algorithms can be found in Refs. [1,16,40,47,54,128,137,167,196,232,277,281,287,333,379] and the references therein. Besides the variable step size approaches, an alternative scheme to improve the trade-off between convergence speed and steady-sate EMSE is constituted by combinations of adaptive algorithms, described in Section 12.5.8.1.

Another class of variable step size algorithms was developed specifically to accelerate the convergence of filters whose optimum weight vector $\boldsymbol{w}_{\text{o}}$ is sparse, that is, has only a small fraction of nonzero elements. The primary application for this is echo cancelation, mainly for echo due to reflections in the telephone connection. The algorithms are known as *proportionate* NLMS (PNLMS) algorithms and usually have very good performance for approximating sparse weight vectors [38,39,106,107,125,154].

12.5.4 Non-Gaussian noise and robust filters

Consider a problem in which we know that the regressor $\boldsymbol{\phi}(n)$ and the desired signal $d(n)$ are related through a model (we assume for simplicity that all variables are real)

$$d(n) = \boldsymbol{w}_{\text{o}}^T \boldsymbol{\phi}(n) + v(n), \tag{12.321}$$

in which $v(n)$ is independent of $\boldsymbol{\phi}(n)$. It is shown in [382] that minimization of the *fourth* power of the error may lead to an adaptive filter with better compromise between convergence rate and EMSE than LMS if $v(n)$ is *sub-Gaussian*. Let us explain what this means.

Recall that for any random variable x we have $0 \le \sigma_x^2 = \mathrm{E}\{x^2\} - (\mathrm{E}\{x\})^2$. Similarly, if we define $\mu_x = \mathrm{E}\{x\}$,

$$\mathrm{E}\left\{(x-\mu_x)^4\right\} \ge \left(\mathrm{E}\left\{(x-\mu_x)^2\right\}\right)^2 = \sigma_x^4, \qquad \mathrm{E}\left\{(x-\mu_x)^6\right\} \ge \left(\mathrm{E}\left\{(x-\mu_x)^2\right\}\right)^3 = \sigma_x^6.$$

In general, we can write $\mathrm{E}\{(x-\mu_x)^4\} = \alpha\sigma_x^4$, with $\alpha \ge 1$. For example, consider the distributions in Table 12.13. Distributions for which $\frac{\mathrm{E}\{(x-\mu_x)^4\}}{\sigma_x^4} < 3$, such as binary and uniform, are called *sub-Gaussian*,

Table 12.13 Fourth- and sixth-order moments of different distributions.

Distribution	$\frac{\mathrm{E}\{(x-\mu_x)^4\}}{\sigma_x^4}$	$\frac{\mathrm{E}\{(x-\mu_x)^6\}}{\sigma_x^6}$
Binary	1	1
Uniform	9/5	27/7
Gaussian	3	15
Exponential	6	90

and those for which $\frac{\mathrm{E}\{x^4\}}{\left(\mathrm{E}\{(x-\mu_x)^2\}\right)^2} > 3$, such as the exponential, are called *super-Gaussian*. These relations are usually given in terms of the *kurtosis* γ_4 of a distribution

$$\gamma_4 = \frac{\mathrm{E}\left\{(x-\mu_x)^4\right\}}{\sigma_x^4} - 3.$$

Therefore, a distribution is sub-Gaussian if its kurtosis is negative and super-Gaussian if its kurtosis is positive. A distribution with positive kurtosis is such that its pdf $f(x)$ decays more slowly than the Gaussian as $|x| \to \infty$.

The authors of [382] show that when the noise $v(n)$ is sub-Gaussian, an algorithm based on the minimization of

$$J_{LMP} = \mathrm{E}\{|e(n)|^p\} \tag{12.322}$$

with $p > 2$ results in lower steady-state EMSE than LMS. They study particularly the case $p = 4$, for which the stochastic gradient algorithm is of the form

$$\boldsymbol{w}(n+1) = \boldsymbol{w}(n) + \mu e^3(n)\boldsymbol{\phi}(n). \tag{12.323}$$

This algorithm is known as the *least-mean fourth* (LMF) algorithm. Analyses of general algorithms with error nonlinearities can be found in [331], and a unified analysis is available in [9].

The LMF algorithm can be understood intuitively as a variable step size version of LMS:

$$\boldsymbol{w}(n+1) = \boldsymbol{w}(n) + \left(\mu e^2(n)\right) e(n)\boldsymbol{\phi}(n).$$

When the error is large, the "equivalent" step size $\mu e^2(n)$ is large, and the filter converges quickly. On the other hand, when the error is small, the "equivalent" step size is reduced, which consequently reduces the EMSE.

This idea works well if the distributions of $\boldsymbol{\phi}(n)$ and $v(n)$ are indeed sub-Gaussian, but it was soon noticed that the algorithm becomes sensitive to outliers (large but rare values of the input signals). This is also easy to understand: if the error is too large, the "equivalent" step size may be such that the recursion becomes unstable. References [80,363] proposed changing the cost function to a mixture of quadratic and quartic, in an attempt to reduce this sensitivity. In [316, p. 313], a different modification of the cost function is proposed, with the same purpose. The new cost function is the following:

$$J_{\text{LMF,mod}} = \mathrm{E}\{f(e(n))\}, \qquad f(e) = \begin{cases} e^4, & \text{if } |e| \le \gamma, \\ e^2, & \text{if } |e| > \gamma, \end{cases} \tag{12.324}$$

in which γ is a constant. References [178,259] show that when the regressor is Gaussian, LMF has a nonzero probability of diverging. From these results one can see that the solution proposed in [316] indeed stabilizes the algorithm.

Since the LMF recursion contains a cubic nonlinearity, it can be used as a simpler platform to understand the behavior of blind equalization algorithms such as the constant modulus algorithm (CMA) (see Section 12.5.5), which also contains a cubic nonlinearity.

Set-membership algorithms [117], which are explained in more detail in Section 12.5.10, can be seen as an extreme version of (12.324), in which $f(e)$ is reduced to simply 0 when $|e| \le \gamma$. The corresponding cost function is

$$J_{\text{SM}} = \mathrm{E}\{f(e(n))\}, \qquad f(e) = \begin{cases} 0, & \text{if } |e| \le \gamma, \\ e^2, & \text{if } |e| > \gamma. \end{cases} \tag{12.325}$$

Set-membership algorithms are particularly well suited to bounded noise, i.e., to the case in which one knows that $|v| \le \gamma$.

All these methods are suited to the case in which the noise $v(n)$ is sub-Gaussian. When the noise is super-Gaussian, there is a small but significant probability that $v(n)$ will assume a very large value. This may make the performance of even LMS become poor. The solution is to give less weight to large errors, and one choice is to use $p < 2$ in (12.322). This will give rise to several *robust* adaptive filters, that is, algorithms that are insensitive to outliers. It should be noted that LMS itself is quite robust in terms of energy (quadratic) relations: in fact, [171] shows that LMS is optimal in the $\mathcal{H}_\infty$ sense (see also [316, Chapter 17]). However, for noise distributions with higher kurtosis, it is convenient to use a smaller value of p [31,185,227,330]. A variable step size algorithm robust against high-kurtosis noise is described in [407].

One important algorithm of this class is the *sign-LMS* algorithm [158,174], which is obtained from (12.322) with $p = 1$, and which we already saw in Section 12.5.1.1.

For very impulsive signals, methods based on *order statistics* (such as the median, for example) may produce better results, although their complexity tends to be higher [147].

12.5.5 Blind equalization

As described in Section 12.1.3, the objective of equalization is to mitigate the ISI introduced by dispersive channels in order to recover the transmitted sequence. Assuming that the equalizer is an adaptive FIR filter, its coefficients may be updated in two different ways:

(i) using a supervised algorithm in the training and decision-directed modes, as explained in Section 12.1.3; or

(ii) using a blind equalization algorithm, which uses higher-order statistics (HOS) of the transmitted signal instead of a training sequence, as shown in Fig. 12.37. In this case, the available bandwidth is used in a more efficient manner due to the absence of a training sequence.

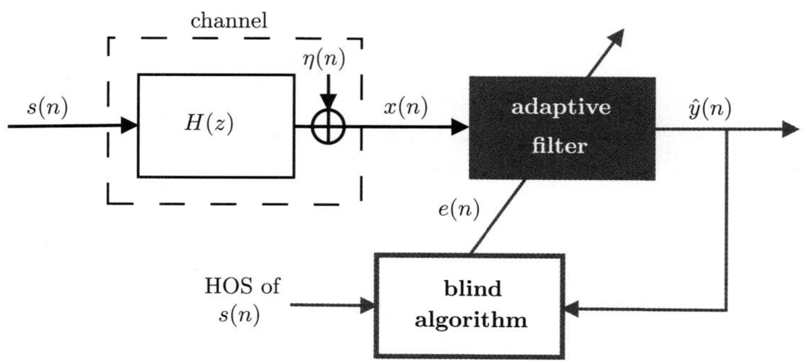

FIGURE 12.37

Simplified communications system with a blind adaptive equalizer.

The literature contains many blind equalization algorithms based on HOS, but the most popular is the CMA, proposed independently by Godard [163] and Treichler and Agee [369] in the 1980s. CMA seeks to minimize the constant modulus cost function defined as

$$J_{\text{CM}} = \mathrm{E}\left\{(r - |\hat{y}(n)|^2)^2\right\}, \tag{12.326}$$

where $r = \mathrm{E}\{|s(n)|^4\}/\mathrm{E}\{|s(n)|^2\}$ is a dispersion constant which contains information about the HOS of the transmitted signal $s(n)$ and $\hat{y}(n) = \boldsymbol{w}^H \boldsymbol{x}(n)$ is the equalizer output. This function penalizes deviations in the modulus of the equalized signal away from the dispersion constant r. Different from the MSE cost function used in supervised adaptive filtering, J_{CM} is not convex in relation to the coefficients of the equalizer (see the definition of convex functions in Box 12.5). It has local minima, and constant modulus-based algorithms can get stuck at these suboptimal solutions.

CMA is obtained from an instantaneous approximation for the gradient of J_{CM} in relation to $\boldsymbol{w}$. Defining the estimation error

$$e(n) = (r - |\hat{y}(n)|^2)\hat{y}(n), \tag{12.327}$$

the update equation of CMA can be written as

$$\boldsymbol{w}(n+1) = \boldsymbol{w}(n) + \mu e^*(n)\boldsymbol{x}(n), \tag{12.328}$$

where μ stands for a step size. Since CMA is a stochastic gradient-type algorithm, the similarity of (12.328) to the LMS update equation is not surprising. Due to this similarity, CMA is sometimes interpreted as a blind version of the LMS algorithm [282]. Thus, as in the LMS, small step sizes lead to a small misadjustment and slow convergence. However, the similarity to LMS stops here. The multimodality of the constant modulus cost function makes the prediction of the behavior of CMA a hard task.

To illustrate the constant modulus cost function, we assume the transmission of a binary signal $\{-1, +1\}$ through the IIR channel $H(z) = 1/(1 + 0.6z^{-1})$ in the absence of noise [116]. Considering an FIR equalizer with $M = 2$ coefficients, we can obtain a 3D plot of J_{CM} as a function of w_0 and w_1 as shown in Fig. 12.38. As indicated in the figure, J_{CM} contains two global minima at $\pm[1\ \ 0.6]^T$ and two local minima. Note that if $\boldsymbol{w}$ converges to one of the global minima, the transmitted sequence or its inverse will be recovered. These two possibilities occur due to the fact that CMA seeks to recover only the modulus of the transmitted signal and does not solve phase ambiguities introduced by the channel. The phase ambiguities can be corrected by using differential code, which is based on the change of sign between two successive samples.

The animations in Fig. 12.39 illustrate the behavior of CMA initialized at the different points. Initializing at $\boldsymbol{w}(0) = [1\ \ 0]^T$, the weight vector converges to $[+1\ \ +0.6]^T$. Now, considering $\boldsymbol{w}(0) = [-0.4\ \ 0.05]^T$, we get $\boldsymbol{w}(n) \approx [-1\ \ -0.6]^T$ at the steady state. On the other hand, if CMA is initialized at $[0.05\ \ -0.4]^T$, the algorithm gets stuck at a local minimum. We should note that in the presence of noise or depending on the step size, CMA also can escape from a minimum and drift to another solution.

Despite the name of the algorithm, CMA also works for nonconstant modulus signals. However, as shown analytically in [222,342], its best performance occurs for constant modulus signals since the variability in the transmitted signal modulus plays a role similar to the measurement noise in supervised algorithms.

The stochastic analysis of constant modulus-based algorithms is a hard task. Due to the nonlinearity and multimodality of the constant modulus cost function, additional assumptions are necessary. A good review of the main results in the analysis of CMA can be found in [182], but many results were obtained after its publication. These results address different issues of constant modulus-based algorithms, such

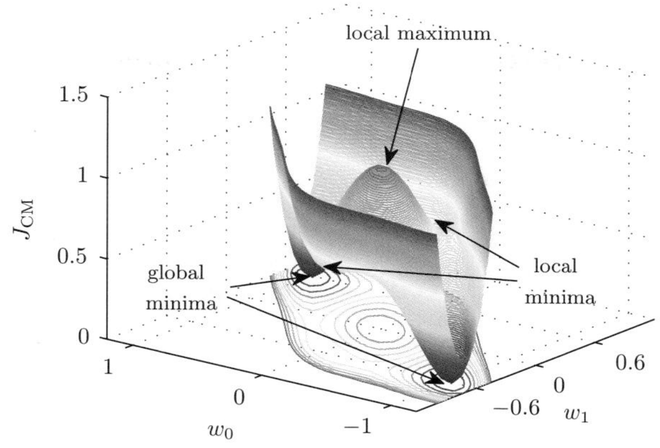

FIGURE 12.38

Constant modulus cost function as a function of the equalizer weights.

as the steady-state and transient behavior [65,66,145,222,286,341–343,396], convergence and stability [105,249,250,265,284,285,301,359], phase rotation [394,397,398], and solutions for nonconstant modulus signals [3,35,146,235,236,339].

12.5.6 Subband and transform-domain adaptive filters

In transform-domain adaptive filters, the input signals are transformed (using the fast Fourier transform, the discrete cosine transform, or other convenient transform), and the adaptation is performed in the transform domain. The gains are twofold. First, using block processing, one can obtain large reductions in computational cost. Second, by using the decorrelation properties of the transforms, it is possible to use different step sizes for each tap in the transformed domain, thereby reducing the equivalent eigenvalue spread and increasing the convergence rate [36,96,138,143,239,256,257,337,356].

A related approach is that of subband adaptive filters, which are also useful to both reduce the computational cost and increase the convergence speed of long filters, in particular for acoustic echo. The main idea is to split the input signal in several parts using an *analysis filter bank*. Each part, relative to a particular band of the full spectrum, is processed separately, and finally all parts are joined together again by a *synthesis* filter bank [375]. Since each subsignal has a smaller bandwidth, the sampling rate can be reduced. The whole process must be performed carefully to avoid as much as possible a deterioration of the signals due to aliasing, while still gaining in computational cost, in comparison with standard full-band filters [152,159,184,288–290,383]. It is important to point out that it is possible to avoid the introduction of extra delay (compared with a full-band filter) [173,238,252,272].

Good introductions to the subject can be found in [117,139,317].

12.5.7 Affine projections algorithm

The *affine projections* algorithm (APA) is an extension of NLMS, designed to obtain a convergence rate closer to that of RLS, but with a low computational cost, as close as possible to that of NLMS. We saw

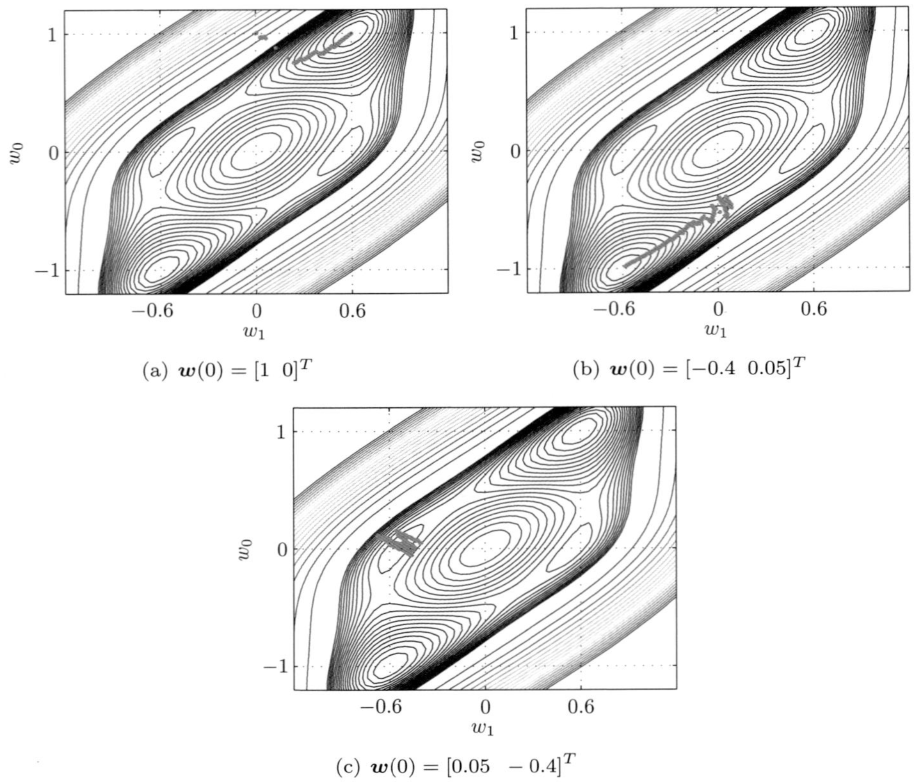

FIGURE 12.39

CMA initialized at different points, $\mu = 0.016$, channel $H(z) = 1/(1 + 0.6z^{-1})$ in the absence of noise. You can find animations for each simulation at http://www.lps.usp.br/FAs/CMA/.

in Section 12.3.2 that NLMS can be derived as a method for choosing the step size so that the current a posteriori estimation error is zeroed (12.131). The APA extends this idea by canceling a vector of a posteriori errors obtained from data from times $n - K$ to n and the weight vector at time $n + 1$:

$$\boldsymbol{\xi}(n) = \boldsymbol{d}(n) - \boldsymbol{\Phi}(n)\boldsymbol{w}^*(n+1), \tag{12.329}$$

in which

$$\boldsymbol{d}(n) = \begin{bmatrix} d(n) \\ d(n-1) \\ \vdots \\ d(n-K-1) \end{bmatrix}, \qquad \boldsymbol{\Phi}(n) = \begin{bmatrix} \boldsymbol{\phi}^T(n) \\ \boldsymbol{\phi}^T(n-1) \\ \vdots \\ \boldsymbol{\phi}^T(n-K-1) \end{bmatrix}. \tag{12.330}$$

The APA chooses as next weight estimate the vector $\boldsymbol{w}(n+1)$ closest to the current estimate $\boldsymbol{w}(n)$ such that $\boldsymbol{\xi}(n) = \boldsymbol{0}$ [20,189,251,275,311,312]:

$$\boldsymbol{w}(n+1) = \arg \min_{\substack{\boldsymbol{w} \\ \text{s.t. } \boldsymbol{\xi}(n)=\boldsymbol{0}}} \|\boldsymbol{w} - \boldsymbol{w}(n)\|^2. \tag{12.331}$$

The solution to this problem is a recursion,

$$\boldsymbol{w}(n+1) = \boldsymbol{w}(n) + \boldsymbol{\Phi}^T(n)\left(\boldsymbol{\Phi}^*(n)\boldsymbol{\Phi}^T(n)\right)^{-1}\boldsymbol{e}(n), \tag{12.332}$$

where

$$\boldsymbol{e}(n) = \boldsymbol{d}(n) - \boldsymbol{\Phi}(n)\boldsymbol{w}^*(n). \tag{12.333}$$

When $K = 1$ and $K = 2$, (12.332) reduces to NLMS and binormalized LMS (BNLMS) [20,117], respectively. Similarly to what is done for NLMS, it is convenient to introduce a regularization term $\varepsilon \boldsymbol{I}$, with $\varepsilon > 0$ and a step size $0 < \mu \leq 1$ in (12.332), leading to

$$\boldsymbol{w}(n+1) = \boldsymbol{w}(n) + \mu\boldsymbol{\Phi}^T(n)\left(\varepsilon\boldsymbol{I} + \boldsymbol{\Phi}^*(n)\boldsymbol{\Phi}^T(n)\right)^{-1}\boldsymbol{e}(n). \tag{12.334}$$

The advantage of this algorithm is that $\boldsymbol{\Phi}^*(n)\boldsymbol{\Phi}^T(n)$ is $K \times K$ (as opposed to the $M \times M$ matrix that must be recursively inverted for RLS), and the value of K does not need to be large, which keeps the computational cost low. In general, a value of two or three already gives a good improvement to the convergence rate even for regressors whose covariance matrix has large spreading factor. Furthermore, for $K \in \{1, 2, 3\}$, which are the most common choices, the matrix inversion has a closed-form solution [117].

In order to implement (12.334), it is useful to remember that computing the inverse explicitly is never a good idea: it is much better to solve the linear system

$$\left(\varepsilon\boldsymbol{I} + \boldsymbol{\Phi}^*(n)\boldsymbol{\Phi}^T(n)\right)\boldsymbol{a} = \boldsymbol{e}(n) \tag{12.335}$$

and then compute $\boldsymbol{\Phi}^T(n)\boldsymbol{a}$. From a numerical point of view, the best option would be to use the QRD of $\boldsymbol{\Phi}(n)$ and avoid forming $\boldsymbol{\Phi}^*(n)\boldsymbol{\Phi}^T(n)$, but the computational cost would be larger. When the regressor is a tapped-delay line and K is large, one can use a fast version of APA, in which the solution of (12.335) is obtained in $\mathcal{O}(K)$ operations, instead of the usual $\mathcal{O}(K^3)$ [157,305,362].

Recent works focusing on the analysis of the APA are [109,110,315,332]. The problem of regularization (choice of ε) is considered in [280,302]. Applications to echo cancelation can be found in [37,278].

A related approach is the sliding-window RLS algorithm, a version of RLS in which only a finite memory is considered, instead of a forgetting factor as used in Section 12.3.3 [224,225].

12.5.8 Cooperative estimation

In this section, we briefly describe important advances in adaptive filtering through cooperative estimation. We first describe combinations of adaptive filters and in the sequel, we focus on distributed adaptive filtering.

12.5.8.1 Combinations of adaptive filters

Combinations of adaptive filters have received considerable attention, since they decrease the sensitivity of the filter to choices of parameters such as the step size, forgetting factor, or filter length. The idea is to combine the outputs of two (or several) different independently run adaptive algorithms to achieve better performance than that of a single filter. In general, this approach is more robust than variable parameter schemes [25].

The first combined scheme that attracted attention was the convex combination of adaptive filters due to its relative simplicity and the proof that the optimum combination is universal, i.e., the combined estimate is at least as good as the best of the component filters in steady state, for stationary inputs [27]. This scheme was proposed in [226] and further extended and analyzed in [27,28]. The original idea was to combine one fast and one slow LMS filter to obtain an overall filter with fast convergence and low misadjustment, but it was extended to other algorithms to take advantage of different characteristics, as explained below.

Fig. 12.40 shows the convex combination of two adaptive filters, in which the output of the overall filter is computed as

$$\hat{y}(n) = \eta(n)\hat{y}_1(n) + [1 - \eta(n)]\hat{y}_2(n), \tag{12.336}$$

where $\eta(n) \in [0, 1]$ is the mixing parameter, $\hat{y}_i(n) = \boldsymbol{w}_i^H(n)\boldsymbol{\phi}(n)$, $i = 1, 2$, are the outputs of the transversal filters, $\boldsymbol{\phi}(n)$ is the common regressor vector, and $\boldsymbol{w}_i(n)$ are the weight vectors of the component filters. Note that the weight vector and the estimation error of the overall filter are given respectively by

$$\boldsymbol{w}(n) = \eta(n)\boldsymbol{w}_1(n) + [1 - \eta(n)]\boldsymbol{w}_2(n) \tag{12.337}$$

and

$$e(n) = d(n) - \hat{y}(n) = \eta(n)e_1(n) + [1 - \eta(n)]e_2(n), \tag{12.338}$$

where $e_i(n) = d(n) - \hat{y}_i(n)$ are the estimation errors of each component filter.

In order to restrict the mixing parameter to the interval $[0, 1]$ and to reduce gradient noise when $\eta(n) \approx 0$ or $\eta(n) \approx 1$, a nonlinear transformation and an auxiliary variable $a(n)$ are used, i.e.,

$$\eta(n) = \text{sgm}[a(n)] \triangleq \frac{1}{1 + \mathrm{e}^{-a(n)}}, \tag{12.339}$$

where $a(n)$ is updated to minimize the overall squared error $e^2(n)$, i.e.,

$$a(n+1) = a(n) - \frac{\mu_a}{2}\nabla_a e^2(n) = a(n) + \mu_a\big[\hat{y}_1(n) - \hat{y}_2(n)\big]e(n)\eta(n)\big[1 - \eta(n)\big]. \tag{12.340}$$

In practice, $a(n)$ must be restricted by saturation of (12.340) to an interval $[-a_+, a_+]$, since the factor $\eta(n)\big[1 - \eta(n)\big]$ would virtually stop adaptation if $a(n)$ were allowed to grow too much. The correct adjustment of the step size μ_a depends on the input signal and additive noise powers and also on the step sizes of the component filters. To make the choice of μ_a independent of the filtering scenario, a normalized scheme was proposed in [32].

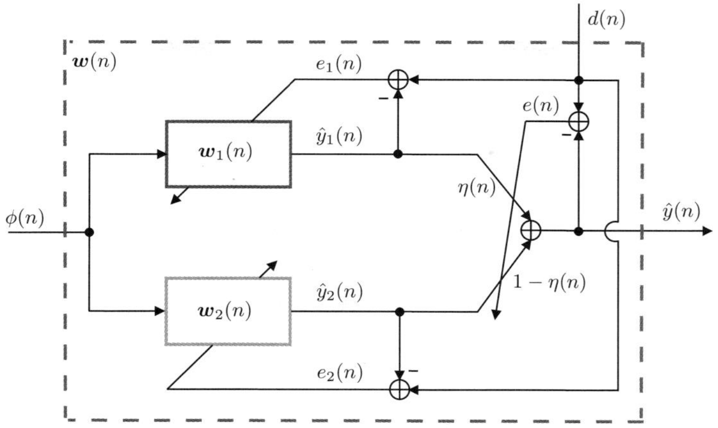

FIGURE 12.40

Convex combination of two transversal adaptive filters.

The combination of one fast and one slow LMS (with $\mu_1 > \mu_2$) operates in a very intuitive way. When fast changes appear, the fast filter outperforms the slow one, making $\eta(n) \approx 1$. In stationary situations, the slow filter gets a lower quadratic error and $\eta(n) \approx 0$, which allows the overall filter to achieve the misadjustment of the slow LMS filter. There are situations in which the combination outperforms each component filter and in these cases, $0 < \eta(n) < 1$. This behavior can be observed in the simulation results shown in Fig. 12.41, where the convex combination of two LMS filters with different step sizes ($\mu_1 = 0.1$ and $\mu_2 = 0.01$) was used to identity the system (from [27])

$$\boldsymbol{w}_{\rm o} = \begin{bmatrix} 0.9003 & -0.5377 & 0.2137 & -0.0280 & 0.7826 & 0.5242 & -0.0871 \end{bmatrix}^T.$$

The regressor $\boldsymbol{\phi}(n)$ is obtained from a process $x(n)$ generated with a first-order autoregressive model, whose transfer function is $\sqrt{1-b^2}/(1-bz^{-1})$. This model is fed with an i.i.d. Gaussian random process, whose variance is such that $\mathrm{Tr}(\boldsymbol{R}_{\boldsymbol{\phi}}) = 1$. Moreover, additive i.i.d. noise $v(n)$ with variance $\sigma_v^2 = 0.01$ is added to form the desired signal. Fig. 12.41(a) shows the EMSE curves estimated from the ensemble average of 500 independent runs and filtered by a moving average filter with 128 coefficients to facilitate the visualization. The combined filter acquires the faster convergence of μ_1-LMS and attains the EMSE of μ_2-LMS. Fig. 12.41(b) shows the time evolution of the mixing parameter. We can observe that it rapidly changes toward $\mathrm{sgm}[a_+]$ during the initial convergence, while its steady-state value is $1 - \mathrm{sgm}[a_+]$.

After the publication of [226], many papers on combinations of adaptive filters appeared in the literature. Apparently, the idea of combining the outputs of several different independently run adaptive algorithms was first proposed in [15] and later improved in [270,271]. The algorithms proposed in [15,270,271] are based on a Bayesian argument and construct an overall (combined) filter through a linear combination of the outputs of several independent adaptive filters. The weights are the a posteriori probabilities that the underlying models used to describe each individual algorithm are "true." Since the weights add up to one, in a sense these first papers also proposed "convex" combinations of algorithms.

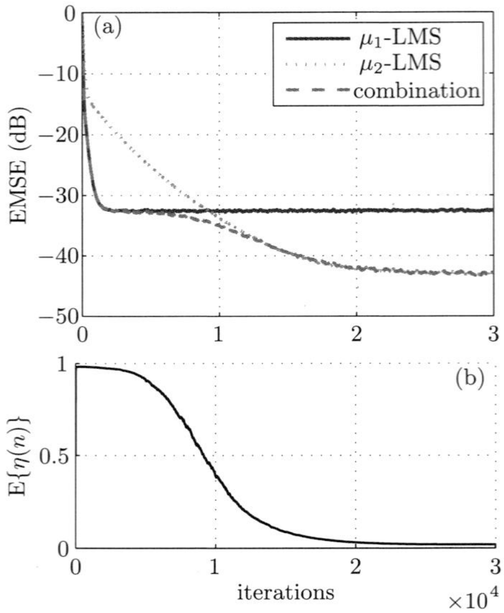

FIGURE 12.41

(a) EMSE for μ_1-LMS, μ_2-LMS, and their convex combination. (b) Ensemble average of $\eta(n)$. $\mu_1 = 0.1$, $\mu_2 = 0.01$, $\mu_a = 100$, $a^+ = 4$, $b = 0.8$; mean of 500 independent runs.

Unconstrained linear combinations of adaptive filters were also proposed in [188]. However, the method of [27,226] has received more attention due to its relative simplicity and the proof that the optimum combination is universal. By this, we mean that if the combination uses at every instant the optimum value of η, then the combined filter is always at least as good as the best component filter. However, in practice η must also be estimated, so the performance of the combination may be slightly worse than that of the best filter.

In the sequel, we briefly describe some of the most important contributions in this area. The key ideas and principles behind these combination schemes and a variety of examples can be found in the overview article [25].

In [28], the convergence of the overall filter is greatly improved by transferring a part of the fast filter $\boldsymbol{w}_1$ to the slow filter $\boldsymbol{w}_2$ when $\lambda(n) \geq \gamma$, that is,

$$\boldsymbol{w}_2(n+1) = \alpha\left[\boldsymbol{w}_2(n) + \mu_2 e_2(n)\boldsymbol{x}(n)\right] + (1-\alpha)\boldsymbol{w}_1(n), \tag{12.341}$$

where α is a parameter close to 1 and γ is a threshold close to the maximum value that can be reached by $\lambda(n)$. The computational cost can be reduced as suggested in [260], by simply copying the fast filter coefficients to the slow filter, instead of the linear combination of (12.341).

A steady-state analysis of the combined filter was presented in [27]. This analysis shows that the convex combination is universal if the optimum η is used, a property that was exploited to design filters with improved tracking performance. Due to the nonlinear function (12.339), a transient analysis of the convex combination needs additional simplifications. This issue was addressed in [264], which provide models for the transient of the convex combination based on first-order Taylor series approximations.

The convex combination was used in [405] to improve the performance of a variable tap length LMS algorithm in a low-SNR environment (SNR ≤ 0 dB). The adaptation of the tap length in the variable tap length LMS algorithm is highly affected by the parameter choice and the noise level. Combination approaches improve such adaptation by exploiting advantages of parallel adaptive filters with different parameters.

The convex combination was exploited in [342] to improve the tracking performance of adaptive filters by combining filters with different tracking capabilities, as in the cases of LMS and RLS (see Section 12.4.4). The combination of algorithms of different families was also addressed in [266], where it was shown that a combination of two filters from the same family (i.e., two LMS or two RLS filters) cannot improve the performance over that of a single filter of the same type with optimal selection of the step size (or forgetting factor). However, combining LMS and RLS filters, it is possible to simultaneously outperform the optimum LMS and RLS filters. In other words, combination schemes can achieve smaller errors than optimally adjusted individual filters. This result was extended in [97] to a larger class of models for the variation of the optimum weight vector following (12.184) instead of the random-walk model of (12.183). It is shown in [97] that the combination of RLS and LMS filters achieves a performance close to that of the Kalman filter, which is optimum for tracking linear models such as (12.184), but requires knowledge of the environment and $\mathcal{O}(M^2)$ operations, while combinations of LMS and RLS filters can be implemented in $\mathcal{O}(M)$ operations and do not require a priori knowledge of the environment.

Convex combinations were used for sparse echo cancelation in [26], considering the combination of two improved proportionate normalized LMS (IPNLMS) filters. The combined scheme increases the IPNLMS robustness to channels with different degrees of sparsity and also alleviates the trade-off between rate of convergence and steady-state misadjustment imposed by the selection of the step size. Still considering the echo cancelation application, [34] proposes a combination of adaptive Volterra kernels that presents a similar behavior to that of the complete Volterra filters, but with a reduction in the computational load. This scheme is robust regardless of the level of nonlinear distortion, which is a desired property for nonlinear echo cancelation.

In [198], the convex combination was also used as a scheme for adaptively biasing the weights of adaptive filters using an output multiplicative factor. Interpreting the biased estimator as the combination of the original filter and a filter with constant output equal to zero, practical schemes to adaptively adjust the multiplicative factor were proposed. This scheme provides a convenient bias versus variance trade-off, leading to reductions in the filter MSE, especially in situations with a low SNR.

Extending [27], the authors of [52] proposed an affine combination of two LMS algorithms, where the condition on the mixing parameter is relaxed, allowing it to be negative. Thus, this scheme can be interpreted as a generalization of the convex combination since the mixing parameter is not restricted to the interval [0, 1]. This approach allows for smaller EMSE in theory, but suffers from larger gradient noise in some situations. Under certain conditions, the optimum mixing parameter was proved to be negative in steady state. Although the optimal linear combiner is unrealizable, two realizable algorithms were introduced. One is based on a stochastic gradient search and the other is based on the ratio of the average error powers from each individual adaptive filter. Under some circumstances, both algorithms present performance close to the optimum.

Similarly to the convex combination, the correct adjustment of the step size for the updating of the mixing parameter in the affine combination (using the stochastic algorithm of [52]) depends on characteristics of the filtering scenario. This issue was addressed in [66], where a transient analysis for the affine combination of two adaptive filters was presented. The results of this analysis were used to facilitate the adjustment of the free parameters of the scheme and to propose two normalized algorithms to update the mixing parameter. The scheme based on the ratio of error powers of the two filters proposed in [52] to update the mixing parameter was analyzed in [44].

Finally, [187] studied different mixing strategies in which the overall output is formed as the weighted linear combination (not necessarily constrained to convex or affine, as before) of the outputs of several component algorithms for stationary and certain nonstationary data.

12.5.8.2 Distributed adaptive filtering

Adaptive networks use the information from data collected at nodes distributed over a certain region. Each node collects noisy observations related to a parameter of interest and interacts with its neighbors considering a certain network topology. The objective is to obtain an estimate of the parameter of interest as accurate as the one that would be obtained if each node had access to the information across the entire network [216,218,318]. Distributed estimation algorithms are useful in several contexts, including wireless and sensor networks, where scalability, robustness, and low power consumption are desirable. They also find applications in precision agriculture and environmental monitoring and transportation [73,318].

Fig. 12.42 shows a network composed by N nodes distributed over some geographical area. At time instant n, every node k takes a measurement $\{d_k(n), \boldsymbol{\phi}_k(n)\}$ to estimate some parameter vector $\boldsymbol{w}_o$. The scalar measurement $d_k(n)$ represents the desired signal and $\boldsymbol{\phi}_k(n)$ denotes the length-M input regressor vector for node k. There are different solutions to the problem of estimating $\boldsymbol{w}_o$. In the *centralized* solution, every node transmits its data $\{d_k(n), \boldsymbol{\phi}_k(n)\}$ to a fusion center for processing. This solution is nonrobust to failure in the fusion center. In *distributed* solutions, every node exchanges information with a subset of its neighboring nodes, and the processing is distributed among all nodes in the network. The set of nodes connected to node k (including k itself) is denoted by $\mathcal{N}_k$ and is called the neighborhood of node k.

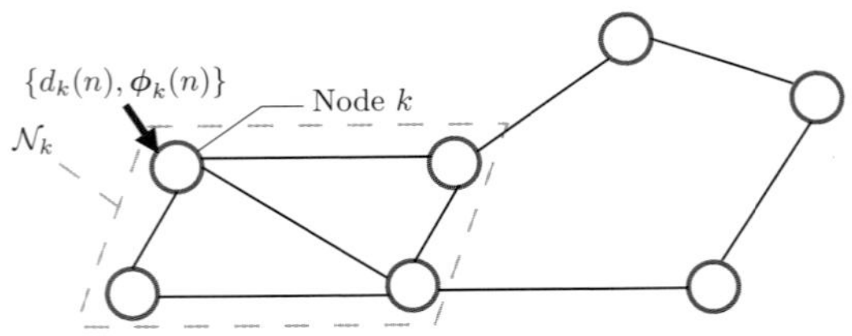

FIGURE 12.42

Diffusion network with $N = 7$ nodes. At time n, every node k takes a measurement $\{d_k(n), \boldsymbol{\phi}_k(n)\}$.

There are three main network topologies used in distributed estimation: incremental, diffusion, and probabilistic diffusion [218,318]. In incremental networks, the nodes are connected so that information flows sequentially from node to node, in a cyclic manner (Fig. 12.43(a)). In diffusion networks, each node communicates with its entire neighborhood at each instant (Fig. 12.43(b)). Finally, in probabilistic diffusion networks each node communicates with a (possibly random) subset of its neighbors at each instant (Fig. 12.43(c)).

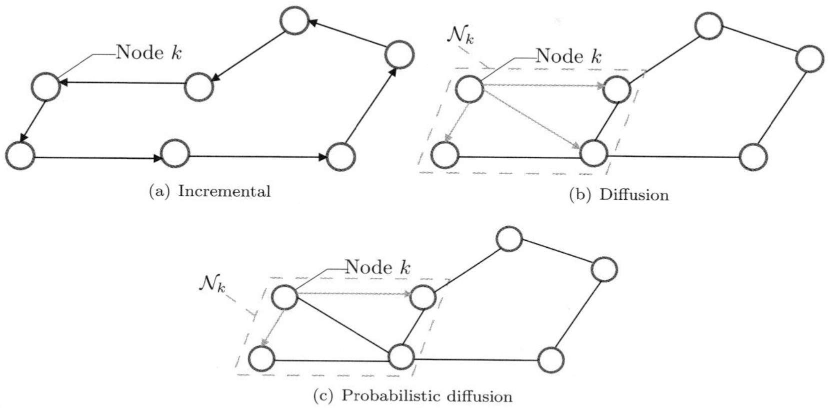

FIGURE 12.43

Different network topologies.

Several estimation algorithms have been proposed in the context of distributed adaptive filtering, such as incremental LMS, incremental RLS, diffusion LMS, diffusion RLS, diffusion Kalman filtering, and smoothing algorithms. In these algorithms, the nodes share information with their neighbors and perform local adaptation and merging of information using convex combinations of available estimates. Distributed algorithms only require local communications between neighboring nodes and can attain good estimation performance compared to centralized solutions [72–76,78,90,140,142,204,215,228, 229,263,318,326,361].

Much effort has been devoted to the development of diffusion strategies that are able to learn and adapt from continuous streaming data and exhibit fast convergence, good tracking capability, and low computational cost. When these strategies are implemented in wireless sensor networks, for example, energy consumption is the most critical constraint [29,141,360]. As a result, several selective transmission mechanisms have been proposed to reduce the energy consumption associated with the communication processes [23,29,41,91,141,217,360,366,391,395,406]. Some of these solutions seek to censor the nodes by avoiding the transmission of information to any of their neighbors when the information is not sufficiently new to significantly improve the current network estimates. This allows the censored nodes to turn their transmitters off, thus saving energy, and reduces the amount of information used in the processing [41,141,366].

Adaptive distributed algorithms have also been used to model biological systems due to their self-organization property. Examples include fish joining together in schools, bacterial motility, and bird flight formations [77,83,84,371–373].

12.5.9 Adaptive IIR filters

Adaptive IIR filters represent an advantageous alternative in relation to adaptive FIR filters, due to their capacity to provide long impulse responses with a small number of coefficients. Their use may be desirable in applications requiring hundreds or thousands of taps, such as satellite channel and mobile radio equalizers. The advantages of adaptive IIR filters have a price: slow convergence, possible instability, error surfaces with local minima, or biased global minima [62,117,269,299,336]. In addition, not all long impulse responses will necessarily be well approximated by an IIR filter with low order.

The nonlinear relation between the adaptive filter coefficients and the internal signals makes the gradient computation and convergence analysis more complicated than those of adaptive FIR filters [117]. Over the years, problems related to local minima, stability, and the effect of poles close to the unit circle have been addressed by several authors, leading to different adaptive algorithms and realization structures [60,61,100,101,118,136,181,219,297,357].

Another difficulty with adaptive IIR filters is that the performance of simple constant gain algorithms may rapidly degrade as the order of the filter grows. Even in the absence of local minima, the adaptation of filters with order greater than two can remain almost stopped in regions where the MSE is relatively high. This issue was addressed in [62], where a large acceleration in the convergence rate of IIR filters was obtained.

12.5.10 Set membership and projections onto convex sets

When the desired signal and the regressor are related as

$$d(n) = \boldsymbol{w}_{\mathrm{o}}^H \boldsymbol{\phi}(n) + v(n),$$

where $v(n)$ is *bounded*, that is, $|v(n)| \leq \gamma$ for all n, it is natural to consider that each pair $(d(n), \boldsymbol{\phi}(n))$ defines a region in which the true solution $\boldsymbol{w}_*$ must lie:

$$\boldsymbol{w}_{\mathrm{o}} \in \mathcal{S}(n) = \left\{ \boldsymbol{w} \in \mathbb{C}^M : \left| d(n) - \boldsymbol{w}^H \boldsymbol{\phi}(n) \right| \leq \gamma \right\}. \tag{12.342}$$

For example, if all vectors are 2D and real, the region $\mathcal{S}(n)$ will be as in Fig. 12.44.

Set membership algorithms aim to find the set of possible solutions by tracking the intersection

$$\mathcal{T}(n) = \bigcap_{k=0}^{n} \mathcal{S}(k) \tag{12.343}$$

of all $\mathcal{S}(n)$ [53]. In the control literature and earlier algorithms, this is done by approximating $\mathcal{T}(n)$ by ellipsoids [53,108,115]. However, very good performance is obtained if one uses just any point contained in $\mathcal{T}(n)$ as a point estimate of $\boldsymbol{w}_{\mathrm{o}}$ at each instant. In fact, it is more interesting to restrict the memory, so that the filter can track variations of $\boldsymbol{w}_{\mathrm{o}}$. In this case, only the last K constraint sets $\mathcal{S}(n)$ are used:

$$\mathcal{T}_K(n) = \bigcap_{k=n-K+1}^{n} \mathcal{S}(k). \tag{12.344}$$

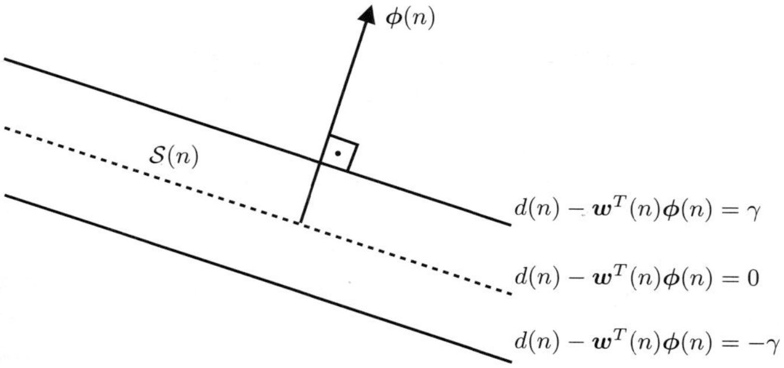

FIGURE 12.44

Constraint imposed by data at time n.

This approach has a very nice property: close to steady state, if the weight estimates are already inside $\mathcal{T}_K(n)$, it is not necessary to perform any update. This may considerably reduce the average complexity of the filter [2,111,119,121,122,205,384,385].

This approach has also been extended to a more general setting, using projections onto convex sets and subgradients. This allows the extension of the results to nondifferentiable cost functions, with different descriptions for the constraints (not restricted to (12.342)), also allowing the design of kernel adaptive filters [346,347,365,393] (see Section 12.5.13).

We should note that the assumption of bounded noise was used in the earlier works, especially in control theory. However, this assumption is not really necessary and in adaptive filtering literature, it is common to assume $v(n)$ is unbounded (e.g., Gaussian), but with finite variance [117].

12.5.11 Adaptive filters with constraints

In some applications, one wishes to minimize a cost function given certain constraints; a typical case is beamforming. In the simplest example, an array of antennas or microphones receives a signal coming from a certain (known) direction and interference from other (unknown) directions. The output of the array is a linear combination of the signals $\phi_i(n)$ received at each antenna, e.g.,

$$\hat{y}(n) = w_0^*\phi_0(n) + w_1^*\phi_1(n) + \cdots + w_{M-1}^*\phi_{M-1}(n).$$

One way of improving the SNR at the output is to minimize the output power, under the restriction that the gain of the array in the direction of the desired signal is one. This is described by a constraint of the kind

$$\boldsymbol{c}^H\boldsymbol{w} = 1,$$

where $\boldsymbol{w} = \begin{bmatrix} w_0 & \dots & w_{M-1} \end{bmatrix}^T$ is the weight vector and $\boldsymbol{c} \in \mathbb{C}^M$ is a vector describing the constraint. The goal of the filter would then be to find the optimum $\boldsymbol{w}_o$ that is the solution of

$$\boldsymbol{w}_o = \arg \min_{\substack{\boldsymbol{w} \\ \text{s.t. } \boldsymbol{c}^H\boldsymbol{w}=1}} |\hat{y}(n)|^2.$$

This solution is known as the *minimum-variance distortionless response* (MVDR), or Capon beamformer, since it was originally proposed in [69]. There is an extensive body of literature on this subject [22,55,102,151,162,200,347,376,377,380,381]. Introductions to the topic can be found in [191,368,378].

Other kinds of constraints can be useful to design specialized filters for identification of sparse models or to guarantee that the weight estimates remain in a bounded region (box constraints), such as

$$a_m \le w_m(n) \le b_m, \quad \text{for } 1 \le m \le M, \tag{12.345}$$

where $a_m \le b_m$ are known constants, or norm constraints such as

$$\|\boldsymbol{D}\boldsymbol{w}(i)\|_p \le \tau, \tag{12.346}$$

where $\boldsymbol{D}$ is a diagonal matrix with nonnegative entries, τ is a constant, and $\|\boldsymbol{x}\|_p$ is a vector norm, such as ℓ_1, ℓ_2, or ℓ_∞. Box constraints can be used to keep the estimates inside a known region where the coefficients should be (for example, if the coefficients must be nonnegative, one can choose $a_m = 0$, $b_m \to \infty$). The nonnegative LMS algorithm described in [82] imposes exactly this kind of constraints. Norm constraints, on the other hand, can be used to impose sparsity constrains (if the ℓ_1-norm is used). An extension to the RLS-DCD algorithm considering box and norm constraints is described in [267].

12.5.12 Reduced-rank adaptive filters

In applications in which large filters are necessary, such as acoustic echo cancelation and nonlinear models using Volterra expansions, it may be useful to reduce the order of the problem by working in a smaller subspace of the original problem. This reduces the number of parameters to be estimated, increasing the convergence speed and reducing the EMSE [324]. Several methods have been proposed, such as those based on eigendecompositions [168,355], the multistage Wiener filter [164,175,325], the auxiliary vector filtering algorithm [276], and the joint iterative optimization method [112–114,400]. This last method has the advantage of having low computational cost.

12.5.13 Kernel adaptive filtering

Kernel adaptive filters are able to solve nonlinear problems by mapping the input vectors into a high-dimensional space where a standard linear adaptive filter is employed (see, e.g., [81,98,212]). They have been used in different applications such as time series prediction [212,323,367,399], system identification [283], echo cancelation [374], and channel equalization [211,212].

These filters use a Mercer kernel to map the input column vector $\boldsymbol{\phi} \in \mathbb{R}^M$ into a high-dimensional feature space $\mathcal{H}$ as the column vector $\boldsymbol{\varphi}(\boldsymbol{\phi}) \in \mathcal{H}$. A Mercer kernel is a continuous, symmetric, and positive definite function $\kappa : \mathbb{R}^M \times \mathbb{R}^M \to \mathbb{R}$, such that [98,193,212,327]

$$\kappa(\boldsymbol{\phi}, \boldsymbol{\phi}') = \boldsymbol{\varphi}(\boldsymbol{\phi})^T \boldsymbol{\varphi}(\boldsymbol{\phi}') \tag{12.347}$$

for all M-length column vectors $\boldsymbol{\phi}, \boldsymbol{\phi}' \in \mathbb{R}^M$.[2] Eq. (12.347) is known as the kernel trick since the inner product $\boldsymbol{\varphi}(\boldsymbol{\phi})^T \boldsymbol{\varphi}(\boldsymbol{\phi}')$ can be computed without knowing explicitly the feature map $\boldsymbol{\varphi}(\cdot)$.

[2] In this section, in order to simplify the arguments, we assume that all quantities are real.

The polynomial and Gaussian are two typical examples of kernel functions used in the literature. The polynomial kernel induces a finite-dimensional Hilbert space and is defined as

$$\kappa(\boldsymbol{\phi}, \boldsymbol{\phi}') = \left(\boldsymbol{\phi}^T \boldsymbol{\phi}' + 1\right)^q, \tag{12.348}$$

where the integer $q \geq 1$ is the degree of the polynomial. The Gaussian kernel induces an infinite-dimensional Hilbert space and is defined as

$$\kappa(\boldsymbol{\phi}, \boldsymbol{\phi}') = \exp\left(-\frac{\|\boldsymbol{\phi} - \boldsymbol{\phi}'\|^2}{2\sigma^2}\right), \tag{12.349}$$

where $\|\cdot\|$ denotes the Euclidean norm and $\sigma > 0$ is the kernel width [212]. Selection of an appropriate width remains an important issue to achieve good performance [33,340]. Unless the target function can be well approximated by a polynomial function, the Gaussian kernel is usually the default choice since it is numerically stable and capable of universal approximation [212].

Eq. (12.347) makes it possible to obtain different kernelized versions of the standard adaptive filters, as is the case of the kernel LMS (KLMS) algorithm [211]. In the space $\mathcal{H}$, the LMS algorithm is used to update the filter weight column vector $\boldsymbol{\Omega}(n-1)$ in order to estimate the desired signal $d(n)$, i.e.,

$$y(n) = \boldsymbol{\varphi}(\boldsymbol{\phi}(n))^T \boldsymbol{\Omega}(n-1), \tag{12.350}$$

$$e(n) = d(n) - y(n), \tag{12.351}$$

$$\boldsymbol{\Omega}(n) = \boldsymbol{\Omega}(n-1) + \mu e(n) \boldsymbol{\varphi}(\boldsymbol{\phi}(n)), \tag{12.352}$$

where $\boldsymbol{\Omega}(0) = \mathbf{0}$, $\boldsymbol{\phi}(n)$ is the input vector, $y(n)$ is the filter output, $e(n)$ represents the estimation error, and μ is the step size. Fig. 12.45 shows a scheme for a kernel adaptive filter applied to nonlinear system identification.

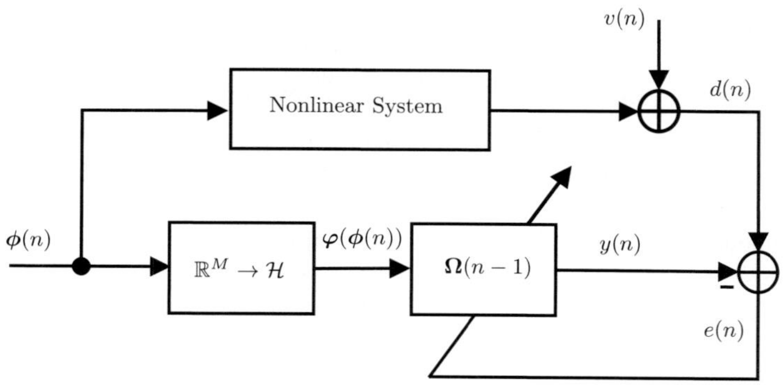

FIGURE 12.45

KLMS applied for nonlinear system identification.

Using the Representer Theorem [212], $\boldsymbol{\Omega}(n)$ can be expressed as a linear combination of past samples, i.e.,

$$\boldsymbol{\Omega}(n) = \sum_{i=1}^{n} \mu e(i)\varphi(\boldsymbol{\phi}(i)), \tag{12.353}$$

which allows us to rewrite (12.350) as [211]

$$y(n) = \mu \sum_{i=1}^{n-1} e(i)\kappa(\boldsymbol{\phi}(n), \boldsymbol{\phi}(i)). \tag{12.354}$$

In spite of the high-dimensionality of the feature space, $y(n)$ can be computed as a linear combination of the kernel function evaluated over the present input vector and all previous input vectors of the filter. These vectors are usually collected into a set called dictionary $\mathcal{D}(n) = \{\boldsymbol{\phi}_1 = \boldsymbol{\phi}(1), \boldsymbol{\phi}_2 = \boldsymbol{\phi}(2), \cdots, \boldsymbol{\phi}_{n-1} = \boldsymbol{\phi}(n-1)\}$, whose cardinality grows linearly with the incoming vectors, i.e., $N(n) = n-1$. This implies high computational cost and requires much memory.

To avoid this linear growth, many sparsification techniques have been proposed to include in the dictionary only informative data (e.g., [58,81,127,148,156,210,213,293,303,345,367]). Among these techniques, the novelty criterion (NC) is the most used [293]. At time instant n, NC computes the distance from $\boldsymbol{\phi}(n)$ to all dictionary vectors $\boldsymbol{\phi}_i$, $i = 1, 2, \cdots, N(n)$. If the minimum distance is greater than a threshold, $\boldsymbol{\phi}(n)$ is considered to carry important information and must be included in the dictionary as member $\boldsymbol{\phi}_{N(n+1)}$ with $N(n+1) = N(n) + 1$. Otherwise, the dictionary must remain unchanged, i.e., $\mathcal{D}(n+1) \leftarrow \mathcal{D}(n)$ and $N(n+1) = N(n)$. An alternative to NC used in conjunction with KLMS is the quantized KLMS (QKLMS) algorithm [81]. QKLMS uses NC and updates the weight of the closest vector when the vector $\boldsymbol{\phi}(n)$ is not included in the dictionary. The use of "redundant" data in this update leads to an algorithm with better accuracy and a more compact dictionary [81]. Besides NC, we should also mention some other interesting techniques, such as the coherence criterion (CC) [303], random Fourier feature (RFF) [58], approximate linear dependence (ALD) [127], and Gram–Schmidt KLMS (GS-KLMS) [59]. CC is a variant of NC and presents similar behavior. RFF uses Bochner's theorem to approximate the kernel by a sum of cosines whose parameters are drawn from a pdf related to the Fourier transform of the kernel [58]. ALD was proposed in [127] as a sparsification technique for the kernel RLS (KRLS) algorithm. It computes $\min_{\mathbf{b}} \| \sum_i^{N(n)} b_i \varphi(\boldsymbol{\phi}_i) - \varphi(\boldsymbol{\phi}(n)) \|$, where $\mathbf{b} = \begin{bmatrix} b_1 & b_2 & \cdots & b_{N(n)} \end{bmatrix}^T$, and uses vector $\mathbf{b}$ as a representation of the mapped vector $\varphi(\boldsymbol{\phi}(n))$. The input vector $\boldsymbol{\phi}(n)$ is included in the dictionary if the computed norm is lower than a threshold. Thus, ALD computes a basis (not necessarily orthogonal) that spans the vector space of the mapped vectors contained in the dictionary. Finally, GS-KLMS computes an orthonormal basis using the Gram–Schmidt process to span the vector space of the mapped vectors contained in the dictionary. At every time instant n, this technique verifies if the norm of the projection of $\varphi(\boldsymbol{\phi}(n))$ onto that space is smaller than a threshold. If this occurs, $\varphi(\boldsymbol{\phi}(n))$ cannot be well represented in the basis. Thus, $\boldsymbol{\phi}(n)$ must be included in the dictionary and a new vector for the basis is computed. Otherwise the dictionary and the basis remain unchanged [59]. All these sparsification techniques provide a reduction in the computational burden and memory of kernel-based algorithms. However, this reduction is not sufficient to use these methods in real-time problems, which remains an open question in the literature.

Kernel adaptive filters have also been used in nonlinear system modeling with diffusion networks [89,155]. Potential applications include monitoring of complex natural phenomena or infrastructure. A naive distributed implementation requires the algorithm to exchange all the local datasets between the agents, which can become unfeasible. Therefore, it is common to assume a shared dictionary among nodes, whose selection is an open research question [322].

Box 12.8: Fixed-point arithmetic

In fixed-point arithmetic, all variables are restricted to lie in a fixed interval $[-a, a)$ and are stored as a sequence of B bits, $b_0 b_1 \dots b_{B-1}$. Assuming that two's-complement arithmetic is used and that $a = 1$, b_0 represents the signal of the variable and the sequence of bits represents the number

$$-b_0 + \sum_{m=1}^{B-1} b_m 2^{-m}.$$

References

[1] T. Aboulnasr, K. Mayyas, A robust variable step-size LMS-type algorithm, IEEE Trans. Signal Process. 45 (3) (Mar. 1997) 631–639.
[2] T. Aboulnasr, K. Mayyas, Complexity reduction of the NLMS algorithm via selective coefficient update, IEEE Trans. Signal Process. 47 (5) (1999) 1421–1424.
[3] S. Abrar, A. Nandi, Blind equalization of square-QAM signals: a multimodulus approach, IEEE Trans. Commun. 58 (6) (Jun. 2010) 1674–1685.
[4] T. Adali, S. Haykin (Eds.), Adaptive Signal Processing, John Wiley & Sons, Inc., Mar 2010.
[5] T. Adali, H. Li, R. Aloysius, On properties of the widely linear MSE filter and its LMS implementation, in: 43rd Annual Conference on Information Sciences and Systems, Mar. 2009, pp. 876–881.
[6] T.Y. Al-Naffouri, M. Moinuddin, Exact performance analysis of the ϵ-NLMS algorithm for colored circular Gaussian inputs, IEEE Trans. Signal Process. 58 (10) (Oct. 2010) 5080–5090.
[7] T.Y. Al-Naffouri, M. Moinuddin, N. Ajeeb, B. Hassibi, A.L. Moustakas, On the distribution of indefinite quadratic forms in Gaussian random variables, IEEE Trans. Commun. 64 (1) (Jan 2016) 153–165.
[8] T.Y. Al-Naffouri, M. Moinuddin, M.S. Sohail, Mean weight behavior of the NLMS algorithm for correlated Gaussian inputs, IEEE Signal Process. Lett. 18 (1) (2011) 7–10.
[9] T.Y. Al-Naffouri, A.H. Sayed, Transient analysis of adaptive filters with error nonlinearities, IEEE Trans. Signal Process. 51 (3) (Mar. 2003) 653–663.
[10] A.E. Albert, L.S. Gardner Jr., Stochastic Approximation and Nonlinear Regression, MIT Press, 1967.
[11] S. Alexander, A. Ghimikar, A method for recursive least squares filtering based upon an inverse QR decomposition, IEEE Trans. Signal Process. 41 (1) (Jan. 1993) 20.
[12] S.T. Alexander, Transient weight misadjustment properties for the finite precision LMS algorithm, IEEE Trans. Acoust. Speech Signal Process. ASSP-35 (9) (Sept. 1987) 1250–1258.
[13] F.G. Almeida Neto, V.H. Nascimento, M.T.M. Silva, Reduced-complexity widely linear adaptive estimation, in: 7th International Symposium on Wireless Communication Systems, Sept. 2010, pp. 399–403.
[14] B.D.O. Anderson, R.R. Bitmead, C.R. Johnson Jr., P.V. Kokotovic, R.L. Kosut, I.M.Y. Mareels, L. Praly, B.D. Riedle, Stability of Adaptive Systems: Passivity and Averaging Analysis, MIT Press, Cambridge, MA, 1986.

[15] P. Andersson, Adaptive forgetting in recursive identification through multiple models, Int. J. Control 42 (Nov. 1985) 1175–1193.
[16] W.-P. Ang, B. Farhang-Boroujeny, A new class of gradient adaptive step-size LMS algorithms, IEEE Trans. Signal Process. 49 (4) (Apr 2001) 805–810.
[17] D. Angelosante, J.A. Bazerque, G.B. Giannakis, Online adaptive estimation of sparse signals: where RLS meets the ℓ_1-norm, IEEE Trans. Signal Process. 58 (7) (2010) 3436–3447.
[18] D. Angelosante, G.B. Giannakis, RLS-weighted Lasso for adaptive estimation of sparse signals, in: Proc. IEEE Int. Conf. Acoustics, Speech and Signal Processing ICASSP 2009, 2009, pp. 3245–3248.
[19] J.A. Apolinário Jr. (Ed.), QRD-RLS Adaptive Filtering, Springer, NY, 2009.
[20] J.A. Apolinário Jr., M.L.R. Campos, P.S.R. Diniz, Convergence analysis of the binormalized data-reusing LMS algorithm, IEEE Trans. Signal Process. 48 (11) (2000) 3235–3242.
[21] J.A. Apolinário Jr., P.S.R. Diniz, A new fast QR algorithm based on a priori errors, IEEE Signal Process. Lett. 4 (11) (Nov. 1997) 307–309.
[22] S. Applebaum, Adaptive arrays, IEEE Trans. Antennas Propag. 24 (5) (1976) 585–598.
[23] R. Arablouei, S. Werner, Y.-F. Huang, K. Doğançay, Distributed least mean-square estimation with partial diffusion, IEEE Trans. Signal Process. 62 (2) (2013) 472–484.
[24] S. Ardalan, Floating-point error analysis of recursive least-squares and least-mean-squares adaptive filters, IEEE Trans. Circuits Syst. 33 (12) (1986) 1192–1208.
[25] J. Arenas-García, L.A. Azpicueta-Ruiz, M.T.M. Silva, V.H. Nascimento, A.H. Sayed, Combinations of adaptive filters: performance and convergence properties, IEEE Signal Process. Mag. 33 (1) (2016) 120–140.
[26] J. Arenas-García, A.R. Figueiras-Vidal, Adaptive combination of proportionate filters for sparse echo cancellation, IEEE Trans. Audio Speech Lang. Process. 17 (Aug. 2009) 1087–1098.
[27] J. Arenas-García, A.R. Figueiras-Vidal, A.H. Sayed, Mean-square performance of a convex combination of two adaptive filters, IEEE Trans. Signal Process. 54 (3) (Mar. 2006) 1078–1090.
[28] J. Arenas-García, M. Martínez-Ramón, A. Navia-Vázquez, A.R. Figueiras-Vidal, Plant identification via adaptive combination of transversal filters, Signal Process. 86 (2006) 2430–2438.
[29] R. Arroyo-Valles, S. Maleki, G. Leus, A censoring strategy for decentralized estimation in energy-constrained adaptive diffusion networks, in: 2013 IEEE 14th Workshop on Signal Processing Advances in Wireless Communications (SPAWC), IEEE, 2013, pp. 155–159.
[30] K.J. Åström, B. Wittenmark, Adaptive Control, Addison-Wesley, 1995.
[31] G. Aydin, O. Arikan, A.E. Cetin, Robust adaptive filtering algorithms for α-stable random processes, IEEE Trans. Circuits Syst. II 46 (2) (1999) 198–202.
[32] L. Azpicueta-Ruiz, A. Figueiras-Vidal, J. Arenas-García, A normalized adaptation scheme for the convex combination of two adaptive filters, in: Proc. IEEE Int. Conf. Acoustics, Speech and Signal Processing ICASSP 2008, Las Vegas, NV, USA, 2008, pp. 3301–3304.
[33] L.A. Azpicueta-Ruiz, J. Arenas-García, M.T.M. Silva, R. Candido, Combined filtering architectures for complex nonlinear systems, in: D. Comminiello, J. Principe (Eds.), Adaptive Learning Methods for Nonlinear System Modeling, Butterworth-Heinemann/Elsevier, Oxford, 2018, pp. 243–264, chapter 11.
[34] L.A. Azpicueta-Ruiz, M. Zeller, A.R. Figueiras-Vidal, J. Arenas-García, W. Kellermann, Adaptive combination of Volterra kernels and its application to nonlinear acoustic echo cancellation, IEEE Trans. Acoust. Speech Signal Process. 19 (Jan. 2011) 97–110.
[35] K. Banovic, E. Abdel-Raheem, M.A.S. Khalid, A novel radius-adjusted approach for blind adaptive equalization, IEEE Signal Process. Lett. 13 (Jan. 2006) 37–40.
[36] F. Beaufays, Transform-domain adaptive filters: an analytical approach, IEEE Trans. Signal Process. 43 (2) (1995) 422–431.
[37] J. Benesty, P. Duhamel, Y. Grenier, A multichannel affine projection algorithm with applications to multichannel acoustic echo cancellation, IEEE Signal Process. Lett. 3 (2) (1996) 35–37.
[38] J. Benesty, S.L. Gay, An improved PNLMS algorithm, in: International Conference on Acoustics, Speech and Signal Processing, vol. 2, IEEE, 2002, pp. 1881–1884.

[39] J. Benesty, C. Paleologu, S. Ciochina, Proportionate adaptive filters from a basis pursuit perspective, IEEE Signal Process. Lett. 17 (12) (2010) 985–988.
[40] J. Benesty, H. Rey, L. Vega, S. Tressens, A nonparametric VSS NLMS algorithm, IEEE Signal Process. Lett. 13 (10) (Oct. 2006) 581–584.
[41] D.K. Berberidis, V. Kekatos, G. Wang, G.B. Giannakis, Adaptive censoring for large-scale regressions, in: Proc. IEEE Int. Conf. Acoustics, Speech and Signal Processing ICASSP 2015, IEEE, 2015, pp. 5475–5479.
[42] J.C.M. Bermudez, N.J. Bershad, A nonlinear analytical model for the quantized LMS algorithm — the arbitrary step size case, IEEE Trans. Signal Process. 44 (5) (May 1996) 1175–1183.
[43] J.C.M. Bermudez, N.J. Bershad, Transient and tracking performance analysis of the quantized LMS algorithm for time-varying system identification, IEEE Trans. Signal Process. 44 (8) (Aug. 1996) 1990–1997.
[44] J.C.M. Bermudez, N.J. Bershad, J.-Y. Tourneret, Stochastic analysis of an error power ratio scheme applied to the affine combination of two LMS adaptive filters, Signal Process. 91 (11) (2011) 2615–2622.
[45] J.C.M. Bermudez, M.H. Costa, Optimum leakage factor for the MOV-LMS algorithm in nonlinear modeling and control systems, in: Proc. IEEE Int. Conf. Acoustics, Speech and Signal Processing ICASSP 2002, vol. 2, 2002.
[46] J.C.M. Bermudez, M.H. Costa, A statistical analysis of the ϵ-NLMS and NLMS algorithms for correlated Gaussian signals, in: Anais do XX Simpósio Brasileiro de Telecomunicações, 2003, pp. 1–5, may be obtained at http://www.lps.usp.br/FAs/bermudez.pdf.
[47] N. Bershad, On the optimum gain parameter in LMS adaptation, IEEE Trans. Acoust. Speech Signal Process. 35 (7) (Jul. 1987) 1065–1068.
[48] N.J. Bershad, Analysis of the normalized LMS algorithm with Gaussian inputs, IEEE Trans. Acoust. Speech Signal Process. ASSP-34 (1986) 793–806.
[49] N.J. Bershad, Behavior of the ϵ-normalized LMS algorithm with Gaussian inputs, IEEE Trans. Acoust. Speech Signal Process. ASSP-35 (May 1987) 636–644.
[50] N.J. Bershad, J.C. Bermudez, New insights on the transient and steady-state behavior of the quantized LMS algorithm, IEEE Trans. Signal Process. 44 (10) (Oct. 1996) 2623–2625.
[51] N.J. Bershad, J.C.M. Bermudez, A nonlinear analytical model for the quantized LMS algorithm-the power-of-two step size case, IEEE Trans. Signal Process. 44 (11) (1996) 2895–2900.
[52] N.J. Bershad, J.C.M. Bermudez, J.-Y. Tourneret, An affine combination of two LMS adaptive filters—transient mean-square analysis, IEEE Trans. Signal Process. 56 (5) (May 2008) 1853–1864.
[53] D.P. Bertsekas, I.B. Rhodes, Recursive state estimation for a set-membership description of uncertainty, IEEE Trans. Autom. Control AC-16 (2) (Apr. 1971) 117–128.
[54] R.C. Bilcu, P. Kuosmanen, K. Egiazarian, A transform domain LMS adaptive filter with variable step-size, IEEE Signal Process. Lett. 9 (2) (Feb. 2002) 51–53.
[55] N. Bornhorst, M. Pesavento, A.B. Gershman, Distributed beamforming for multi-group multicasting relay networks, IEEE Trans. Signal Process. 60 (1) (2012) 221–232.
[56] J.-L. Botto, G.V. Moustakides, Stabilizing the fast Kalman algorithms, IEEE Trans. Acoust. Speech Signal Process. 37 (9) (1989) 1342–1348.
[57] G.E. Bottomley, S.T. Alexander, A novel approach for stabilizing recursive least squares filters, IEEE Trans. Signal Process. 39 (8) (1991) 1770–1779.
[58] P. Bouboulis, S. Pougkakiotis, S. Theodoridis, Efficient KLMS and KRLS algorithms: a random Fourier feature perspective, in: IEEE Statistical Signal Processing Workshop (SSP), June 2016, pp. 1–5.
[59] A.A. Bueno, M.T.M. Silva, Gram-Schmidt-based sparsification for kernel dictionary, IEEE Signal Process. Lett. 27 (2020) 1130–1134.
[60] P.M.S. Burt, M. Gerken, A polyphase IIR adaptive filter: error surface analysis and application, in: Proc. IEEE Int. Conf. Acoustics, Speech, and Signal Process, vol. 3, Apr. 1997, pp. 2285–2288.
[61] P.M.S. Burt, M. Gerken, A polyphase IIR adaptive filter, IEEE Trans. Circuits Syst. II 49 (5) (May 2002) 356–359.

[62] P.M.S. Burt, P. Regalia, A new framework for convergence analysis and algorithm development of adaptive IIR filters, IEEE Trans. Signal Process. 53 (Aug. 2005) 3129–3140.
[63] H.J. Butterweck, Iterative analysis of the steady-state weight fluctuations in LMS-type adaptive filters, Technical Report EUT 96-E-299, Eindhoven University of Technology, Netherlands, June 1996.
[64] H.J. Butterweck, A wave theory of long adaptive filters, IEEE Trans. Circuits Syst. I 48 (6) (June 2001) 739–747.
[65] R. Candido, M.T.M. Silva, M.D. Miranda, V.H. Nascimento, A statistical analysis of the dual-mode CMA, in: Proc. IEEE Int Circuits and Systems (ISCAS) Symp., 2010, pp. 2510–2513.
[66] R. Candido, M.T.M. Silva, V.H. Nascimento, Transient and steady-state analysis of the affine combination of two adaptive filters, IEEE Trans. Signal Process. 58 (8) (2010) 4064–4078.
[67] E.J. Candès, M.B. Wakin, People hearing without listening: an introduction to compressive sampling, IEEE Signal Process. Mag. 25 (2) (2008) 21–30.
[68] E.J. Candès, M.B. Wakin, S.P. Boyd, Enhancing sparsity by reweighted ℓ_1 minimization, J. Fourier Anal. Appl. 14 (2008) 877–905.
[69] J. Capon, High-resolution frequency-wavenumber spectrum analysis, Proc. IEEE 57 (8) (1969) 1408–1418.
[70] C. Caraiscos, B. Liu, A roundoff error analysis of the LMS adaptive algorithm, IEEE Trans. Acoust. Speech Signal Process. ASSP-32 (1) (Feb. 1984) 34–41.
[71] A. Carini, E. Mumolo, G.L. Sicuranza, V-vector algebra and its application to Volterra-adaptive filtering, IEEE Trans. Circuits Syst. II 46 (5) (May 1999) 585–598.
[72] F.S. Cattivelli, C.G. Lopes, A.H. Sayed, Diffusion recursive least-squares for distributed estimation over adaptive networks, IEEE Trans. Signal Process. 56 (5) (2008) 1865–1877.
[73] F.S. Cattivelli, A.H. Sayed, Diffusion LMS strategies for distributed estimation, IEEE Trans. Signal Process. 58 (3) (2010) 1035–1048.
[74] F.S. Cattivelli, A.H. Sayed, Diffusion strategies for distributed Kalman filtering and smoothing, IEEE Trans. Autom. Control 55 (9) (2010) 2069–2084.
[75] F.S. Cattivelli, A.H. Sayed, Analysis of spatial and incremental LMS processing for distributed estimation, IEEE Trans. Signal Process. 59 (4) (2011) 1465–1480.
[76] F.S. Cattivelli, A.H. Sayed, Distributed detection over adaptive networks using diffusion adaptation, IEEE Trans. Signal Process. 59 (5) (2011) 1917–1932.
[77] F.S. Cattivelli, A.H. Sayed, Modeling bird flight formations using diffusion adaptation, IEEE Trans. Signal Process. 59 (5) (2011) 2038–2051.
[78] R.L.G. Cavalcante, I. Yamada, B. Mulgrew, An adaptive projected subgradient approach to learning in diffusion networks, IEEE Trans. Signal Process. 57 (7) (2009) 2762–2774.
[79] A.T. Cemgil, A tutorial introduction to Monte Carlo methods, Markov chain Monte Carlo and particle filtering, in: Academic Press Library in Signal Processing, Elsevier, 2014, pp. 1065–1114.
[80] J.A. Chambers, O. Tanrikulu, A.G. Constantinides, Least mean mixed-norm adaptive filtering, Electron. Lett. 30 (19) (Sept. 15, 1994).
[81] B. Chen, S. Zhao, P. Zhu, J.C. Principe, Quantized kernel least mean square algorithm, IEEE Trans. Neural Netw. Learn. Syst. 23 (1) (Jan 2012) 22–32.
[82] J. Chen, C. Richard, J.C.M. Bermudez, P. Honeine, Nonnegative least-mean-square algorithm, IEEE Trans. Signal Process. 59 (11) (Nov 2011) 5225–5235.
[83] J. Chen, A.H. Sayed, Bio-inspired cooperative optimization with application to bacteria motility, in: Proc. IEEE Int. Conf. Acoustics, Speech and Signal Processing ICASSP 2011, 2011, pp. 5788–5791.
[84] J. Chen, X. Zhao, A.H. Sayed, Bacterial motility via diffusion adaptation, in: Proc. Conf Signals, Systems and Computers (ASILOMAR) Record of the Forty Fourth Asilomar Conf., 2010, pp. 1930–1934.
[85] S. Chen, Low complexity concurrent constant modulus algorithm and soft directed scheme for blind equalization, IEE Proc., Vis. Image Signal Process. 150 (Oct. 2003) 312–320.
[86] Y. Chen, Y. Gu, A.O. Hero, Sparse LMS for system identification, in: Proc. IEEE Int. Conf. Acoustics, Speech and Signal Processing ICASSP 2009, Apr. 2009, pp. 3125–3128.

[87] A. Chinatto, C. Junqueira, J.M.T. Romano, Interference mitigation using widely linear arrays, in: 17th European Signal Processing Conference, Sept. 2009, pp. 353–357.

[88] N. Chopin, O. Papaspiliopoulos, An Introduction to Sequential Monte Carlo, Springer International Publishing, 2020.

[89] S. Chouvardas, M. Draief, A diffusion kernel LMS algorithm for nonlinear adaptive networks, in: Proc. IEEE Int. Conf. Acoustics, Speech and Signal Processing ICASSP 2016, 2016, pp. 4164–4168.

[90] S. Chouvardas, K. Slavakis, S. Theodoridis, Adaptive robust distributed learning in diffusion sensor networks, IEEE Trans. Signal Process. 59 (10) (2011) 4692–4707.

[91] S. Chouvardas, K. Slavakis, S. Theodoridis, Trading off complexity with communication costs in distributed adaptive learning via Krylov subspaces for dimensionality reduction, IEEE J. Sel. Top. Signal Process. 7 (2) (2013) 257–273.

[92] J. Cioffi, The fast adaptive ROTOR's RLS algorithm, IEEE Trans. Acoust. Speech Signal Process. 38 (4) (Apr 1990) 631–653.

[93] J. Cioffi, T. Kailath, Fast, recursive-least-squares transversal filters for adaptive filtering, IEEE Trans. Acoust. Speech Signal Process. 32 (2) (1984) 304–337.

[94] J. Cioffi, T. Kailath, Windowed fast transversal filters adaptive algorithms with normalization, IEEE Trans. Acoust. Speech Signal Process. 33 (3) (1985) 607–625.

[95] J.M. Cioffi, J.J. Werner, The tap-drifting problem in digitally implemented data-driven echo cancellers, Bell Syst. Tech. J. 64 (1) (Jan. 1985) 115–138.

[96] G. Clark, S. Mitra, S. Parker, Block implementation of adaptive digital filters, IEEE Trans. Acoust. Speech Signal Process. 29 (3) (1981) 744–752.

[97] R. Claser, V.H. Nascimento, On the tracking performance of adaptive filters and their combinations, IEEE Trans. Signal Process. 69 (2021) 3104–3116.

[98] D. Comminiello, J.C. Príncipe (Eds.), Adaptive Learning Methods for Nonlinear System Modeling, Butterworth-Heinemann - Elsevier, UK, 2018.

[99] M.H. Costa, J.C.M. Bermudez, An improved model for the normalized LMS algorithm with Gaussian inputs and large number of coefficients, in: Proc. IEEE Int. Conf. Acoustics, Speech and Signal Processing ICASSP 2002, vol. 2, 2002, pp. 1385–1388.

[100] J.E. Cousseau, P.S.R. Diniz, A general consistent equation-error algorithm for adaptive IIR filtering, Signal Process. 56 (2) (1997) 121–134.

[101] J.E. Cousseau, P.S.R. Diniz, New adaptive IIR filtering algorithms based on the Steiglitz-McBride method, IEEE Trans. Signal Process. 45 (5) (May 1997) 1367–1371.

[102] H. Cox, R. Zeskind, M. Owen, Robust adaptive beamforming, IEEE Trans. Acoust. Speech Signal Process. 35 (10) (1987) 1365–1376.

[103] L.F. da S. Coelho, L. Lovisolo, M.P. Tcheou, A proposal for multiplierless RLS adaptive filters with implementation complexity constraint, in: 2020 IEEE International Symposium on Circuits and Systems (ISCAS), IEEE, Oct 2020.

[104] E.A.B. da Silva, L. Lovisolo, A.J.S. Dutra, P.S.R. Diniz, FIR filter design based on successive approximation of vectors, IEEE Trans. Signal Process. 62 (15) (Aug 2014) 3833–3848.

[105] O. Dabeer, E. Masry, Convergence analysis of the constant modulus algorithm, IEEE Trans. Inf. Theory 49 (6) (June 2003) 1447–1464.

[106] F. das Chagas de Souza, O.J. Tobias, R. Seara, D.R. Morgan, A PNLMS algorithm with individual activation factors, IEEE Trans. Signal Process. 58 (4) (2010) 2036–2047.

[107] F. das Chagas de Souza, O.J. Tobias, R. Seara, D.R. Morgan, Stochastic model for the mean weight evolution of the IAF-PNLMS algorithm, IEEE Trans. Signal Process. 58 (11) (2010) 5895–5901.

[108] S. Dasgupta, Y.-F. Huang, Asymptotically convergent modified recursive least-squares with data-dependent updating and forgetting factor for systems with bounded noise, IEEE Trans. Inf. Theory 33 (3) (1987) 383–392.

[109] S.J.M. de Almeida, J.C.M. Bermudez, N.J. Bershad, A stochastic model for a pseudo affine projection algorithm, IEEE Trans. Signal Process. 57 (1) (2009) 107–118.
[110] S.J.M. de Almeida, J.C.M. Bermudez, N.J. Bershad, M.H. Costa, A statistical analysis of the affine projection algorithm for unity step size and autoregressive inputs, IEEE Trans. Circuits Syst. I 52 (7) (2005) 1394–1405.
[111] R.C. de Lamare, P.S.R. Diniz, Set-membership adaptive algorithms based on time-varying error bounds for CDMA interference suppression, IEEE Trans. Veh. Technol. 58 (2) (2009) 644–654.
[112] R.C. de Lamare, R. Sampaio-Neto, Reduced-rank interference suppression for DS-CDMA based on interpolated fir filters, IEEE Commun. Lett. 9 (3) (Mar. 2005) 213–215.
[113] R.C. de Lamare, R. Sampaio-Neto, Reduced-rank adaptive filtering based on joint iterative optimization of adaptive filters, IEEE Signal Process. Lett. 14 (12) (Dec. 2007) 980–983.
[114] R.C. de Lamare, R. Sampaio-Neto, Adaptive reduced-rank processing based on joint and iterative interpolation, decimation, and filtering, IEEE Trans. Signal Process. 57 (7) (July 2009) 2503–2514.
[115] J.R. Deller, M. Nayeri, S.F. Odeh, Least-square identification with error bounds for real-time signal processing and control, Proc. IEEE 81 (6) (1993) 815–849.
[116] Z. Ding, Y. Li, Blind Equalization and Identification, Marcel Dekker, 2001.
[117] P.S.R. Diniz, Adaptive Filtering: Algorithms and Practical Implementation, 5th edition, Springer, 2020.
[118] P.S.R. Diniz, J.E. Cousseau, A. Antoniou, Fast parallel realization of IIR adaptive filters, IEEE Trans. Circuits Syst. II 41 (8) (Aug. 1994) 561–567.
[119] P.S.R. Diniz, S. Werner, Set-membership binormalized data-reusing LMS algorithms, IEEE Trans. Signal Process. 51 (1) (Jan. 2003) 124–134.
[120] P.M. Djuric, J.H. Kotecha, J. Zhang, Y. Huang, T. Ghirmai, M.F. Bugallo, J. Miguez, Particle filtering, IEEE Signal Process. Mag. 20 (5) (2003) 19–38.
[121] K. Dogancay, O. Tanrikulu, Adaptive filtering algorithms with selective partial updates, IEEE Trans. Circuits Syst. II 48 (8) (2001) 762–769.
[122] S.C. Douglas, Adaptive filters employing partial updates, IEEE Trans. Circuits Syst. II 44 (3) (1997) 209–216.
[123] S.C. Douglas, D.P. Mandic, Performance analysis of the conventional complex LMS and augmented complex LMS algorithms, in: Proc. IEEE Int. Conf. Acoustics, Speech and Signal Processing ICASSP 2010, 2010, pp. 3794–3797.
[124] S.C. Douglas, W. Pan, Exact expectation analysis of the LMS adaptive filter, IEEE Trans. Signal Process. 43 (12) (Dec. 1995) 2863–2871.
[125] D.L. Duttweiler, Proportionate normalized least-mean-squares adaptation in echo cancellers, IEEE Trans. Speech Audio Process. 8 (5) (Sept. 2000) 508–518.
[126] E. Eleftheriou, D. Falconer, Tracking properties and steady-state performance of RLS adaptive filter algorithms, IEEE Trans. Acoust. Speech Signal Process. 34 (5) (1986) 1097–1110.
[127] Y. Engel, S. Mannor, R. Meir, The kernel recursive least-squares algorithm, IEEE Trans. Signal Process. 52 (8) (Aug. 2004) 2275–2285.
[128] J.B. Evans, P. Xue, B. Liu, Analysis and implementation of variable step size adaptive algorithms, IEEE Trans. Signal Process. 41 (8) (Aug. 1993) 2517–2535.
[129] E. Eweda, Almost sure convergence of a decreasing gain sign algorithm for adaptive filtering, IEEE Trans. Acoust. Speech Signal Process. 36 (10) (1988) 1669–1671.
[130] E. Eweda, A tight upper bound of the average absolute error in a constant step-size sign algorithm, IEEE Trans. Acoust. Speech Signal Process. 37 (11) (1989) 1774–1776.
[131] E. Eweda, Comparison of RLS, LMS, and sign algorithms for tracking randomly time-varying channels, IEEE Trans. Signal Process. 42 (11) (Nov. 1994) 2937–2944.
[132] E. Eweda, Convergence analysis of the sign algorithm without the independence and Gaussian assumptions, IEEE Trans. Signal Process. 48 (9) (2000) 2535–2544.

[133] E. Eweda, N.J. Bershad, J.C.M. Bermudez, Stochastic analysis of the LMS and NLMS algorithms for cyclostationary white Gaussian and non-Gaussian inputs, IEEE Trans. Signal Process. 66 (18) (Sep 2018) 4753–4765.

[134] E. Eweda, O. Macchi, Convergence of an adaptive linear estimation algorithm, IEEE Trans. Autom. Control AC-19 (2) (Feb. 1984) 119–127.

[135] D. Falconer, L. Ljung, Application of fast Kalman estimation to adaptive equalization, IEEE Trans. Commun. 26 (10) (1978) 1439–1446.

[136] H. Fan, A structural view of asymptotic convergence speed of adaptive IIR filtering algorithms. I. Infinite precision implementation, IEEE Trans. Signal Process. 41 (4) (Apr. 1993) 1493–1517.

[137] B. Farhang-Boroujeny, Variable-step-size LMS algorithm: new developments and experiments, IEE Proc., Vis. Image Signal Process. 141 (5) (Oct. 1994) 311–317.

[138] B. Farhang-Boroujeny, Analysis and efficient implementation of partitioned block LMS adaptive filters, IEEE Trans. Signal Process. 44 (11) (1996) 2865–2868.

[139] B. Farhang-Boroujeny, Adaptive Filters - Theory and Applications, John Wiley & Sons, West Sussex, 1998.

[140] J. Fernandez-Bes, J. Arenas-García, M.T.M. Silva, L.A. Azpicueta-Ruiz, Adaptive diffusion schemes for heterogeneous networks, IEEE Trans. Signal Process. 65 (Nov. 2017) 5661–5674.

[141] J. Fernandez-Bes, R. Arroyo-Valles, J. Arenas-García, J. Cid-Sueiro, Censoring diffusion for harvesting WSNs, in: 2015 IEEE 6th International Workshop on Computational Advances in Multi-Sensor Adaptive Processing (CAMSAP), IEEE, 2015, pp. 237–240.

[142] J. Fernandez-Bes, L.A. Azpicueta-Ruiz, J. Arenas-García, M.T.M. Silva, Distributed estimation in diffusion networks using affine least-squares combiners, Digit. Signal Process. 36 (Jan. 2015) 1–14.

[143] E. Ferrara, Fast implementations of LMS adaptive filters, IEEE Trans. Acoust. Speech Signal Process. 28 (4) (1980) 474–475.

[144] A. Feuer, E. Weinstein, Convergence analysis of LMS filters with uncorrelated Gaussian data, IEEE Trans. Acoust. Speech Signal Process. ASSP-33 (1) (Feb. 1985) 222–229.

[145] I. Fijalkow, C.E. Manlove, C.R. Johnson Jr., Adaptive fractionally spaced blind CMA equalization: excess MSE, IEEE Trans. Signal Process. 46 (1) (Jan. 1998) 227–231.

[146] J.M. Filho, M.D. Miranda, M.T.M. Silva, A regional multimodulus algorithm for blind equalization of QAM signals: introduction and steady-state analysis, Signal Process. 92 (11) (Nov. 2012) 2643–2656.

[147] A. Flaig, G.R. Arce, K.E. Barner, Affine order-statistic filters: "medianization" of linear FIR filters, IEEE Trans. Signal Process. 46 (8) (1998) 2101–2112.

[148] A. Flores, R.C. de Lamare, Set-membership adaptive kernel NLMS algorithms: design and analysis, Signal Process. 154 (2019) 1–14.

[149] S. Florian, A. Feuer, Performance analysis of the LMS algorithm with a tapped delay line (two-dimensional case), IEEE Trans. Acoust. Speech Signal Process. ASSP-34 (6) (Dec. 1986) 1542–1549.

[150] B. Friedlander, Lattice filters for adaptive processing, IEEE Trans. Signal Process. 70 (8) (Aug. 1982) 829–867.

[151] O.L. Frost, An algorithm for linearly constrained adaptive array processing, Proc. IEEE 60 (8) (1972) 926–935.

[152] I. Furukawa, A design of canceller of broadband acoustic echo, in: Proc. Int. Teleconference Symposium, Tokyo, Japan, 1984, pp. 1–8.

[153] T. Gansler, J. Benesty, New insights into the stereophonic acoustic echo cancellation problem and an adaptive nonlinearity solution, IEEE Trans. Speech Audio Process. 10 (5) (2002) 257–267.

[154] T. Gansler, S.L. Gay, M.M. Sondhi, J. Benesty, Double-talk robust fast converging algorithms for network echo cancellation, IEEE Trans. Speech Audio Process. 8 (6) (2000) 656–663.

[155] W. Gao, J. Chen, C. Richard, J. Huang, Diffusion adaptation over networks with kernel least-mean-square, in: Proc. of IEEE 6th International Workshop on Computational Advances in Multi-Sensor Adaptive Processing (CAMSAP), 2015, pp. 217–220.

[156] S. Garcia-Vega, X.-J. Zeng, J. Keane, Learning from data streams using kernel least-mean-square with multiple kernel-sizes and adaptive step-size, Neurocomputing 339 (2019) 105–115.

[157] S.L. Gay, S. Tavathia, The fast affine projection algorithm, in: Proc. IEEE Int. Conf. Acoustics, Speech and Signal Processing ICASSP 1995, vol. 5, 1995, pp. 3023–3026.

[158] A. Gersho, Adaptive filtering with binary reinforcement, IEEE Trans. Inf. Theory 30 (2) (1984) 191–199.

[159] A. Gilloire, M. Vetterli, Adaptive filtering in subbands with critical sampling: analysis, experiments, and application to acoustic echo cancellation, IEEE Trans. Signal Process. 40 (8) (1992) 1862–1875.

[160] R.D. Gitlin, J.E. Mazo, M.G. Taylor, On the design of gradient algorithms for digitally implemented adaptive filters, IEEE Trans. Circuit Theory CT-20 (2) (Mar. 1973) 125–136.

[161] R.D. Gitlin, H.C. Meadors Jr., S.B. Weinstein, The tap-leakage algorithm: an algorithm for the stable operation of a digitally implemented, fractionally spaced adaptive equalizer, Bell Syst. Tech. J. 61 (8) (Oct. 1982) 1817–1839.

[162] G.-O. Glentis, K. Berberidis, S. Theodoridis, Efficient least squares adaptive algorithms for FIR transversal filtering, IEEE Signal Process. Mag. 16 (4) (1999) 13–41.

[163] D.N. Godard, Self-recovering equalization and carrier tracking in two-dimensional data communication systems, IEEE Trans. Commun. 28 (11) (Nov. 1980) 1867–1875.

[164] J.S. Goldstein, I.S. Reed, L.L. Scharf, A multistage representation of the Wiener filter based on orthogonal projections, IEEE Trans. Inf. Theory 44 (7) (1998) 2943–2959.

[165] G.H. Golub, C.F.V. Loan, Matrix Computations, 3rd edition, Johns Hopkins, 1996.

[166] Y. Gu, J. Jin, S. Mei, ℓ_0 norm constraint LMS algorithm for sparse system identification, IEEE Signal Process. Lett. 16 (9) (Sept. 2009) 774–777.

[167] A. Gupta, S. Joshi, Variable step-size LMS algorithm for fractal signals, IEEE Trans. Signal Process. 56 (4) (April 2008) 1411–1420.

[168] A.M. Haimovich, Y. Bar-Ness, An eigenanalysis interference canceler, IEEE Trans. Signal Process. 39 (1) (1991) 76–84.

[169] C.H. Hansen, S.D. Snyder, Active Control of Noise and Vibration, Taylor & Francis, 1997.

[170] R. Harris, D. Chabries, F. Bishop, A variable step (VS) adaptive filter algorithm, IEEE Trans. Acoust. Speech Signal Process. 34 (2) (Apr. 1986) 309–316.

[171] B. Hassibi, A.H. Sayed, T. Kailath, LMS is H^∞-optimal, in: Proc. Conference on Decision and Control, vol. 1, San Antonio, TX, Dec. 1993, pp. 74–79.

[172] S. Haykin, Adaptive Filter Theory, 5th edition, Prentice Hall, 2014.

[173] N. Hirayama, H. Sakai, S. Miyagi, Delayless subband adaptive filtering using the Hadamard transform, IEEE Trans. Signal Process. 47 (6) (1999) 1731–1734.

[174] D. Hirsch, W. Wolf, A simple adaptive equalizer for efficient data transmission, IEEE Trans. Commun. Technol. 18 (1) (1970) 5–12.

[175] M.L. Honig, J.S. Goldstein, Adaptive reduced-rank interference suppression based on the multistage Wiener filter, IEEE Trans. Commun. 50 (6) (2002) 986–994.

[176] R.A. Horn, C.R. Johnson, Matrix Analysis, Cambridge University Press, 1987.

[177] R.A. Horn, C.R. Johnson, Topics in Matrix Analysis, Cambridge University Press, Cambridge, MA, 1991.

[178] P.I. Hübscher, J.C.M. Bermudez, V.H. Nascimento, A mean-square stability analysis of the least mean fourth (LMF) adaptive algorithm, IEEE Trans. Signal Process. 55 (8) (Aug. 2007) 4018–4028.

[179] P. Ioannou, B. Fidan, Adaptive Control Tutorial, SIAM, PA, Dec. 2006.

[180] N. Jablon, Adaptive beamforming with the generalized sidelobe canceller in the presence of array imperfections, IEEE Trans. Antennas Propag. 34 (8) (1986) 996–1012.

[181] C.R. Johnson Jr., Adaptive IIR filtering: current results and open issues, IEEE Trans. Inf. Theory 30 (2) (Mar. 1984) 237–250.

[182] C.R. Johnson Jr., P. Schniter, T.J. Endres, J.D. Behm, D.R. Brown, R.A. Casas, Blind equalization using the constant modulus criterion: a review, Proc. IEEE 86 (10) (Oct. 1998) 1927–1950.

[183] T. Kailath, A.H. Sayed, B. Hassibi, Linear Estimation, Prentice Hall, NJ, 2000.
[184] W. Kellermann, Kompensation akustischer echos in frequenzteilbändern, Frequenz 39 (1985) 209–215.
[185] S.R. Kim, A. Efron, Adaptive robust impulse noise filtering, IEEE Trans. Signal Process. 43 (8) (1995) 1855–1866.
[186] Y. Kopsinis, K. Slavakis, S. Theodoridis, Online sparse system identification and signal reconstruction using projections onto weighted ℓ_1 balls, IEEE Trans. Signal Process. 59 (3) (2011) 936–952.
[187] S.S. Kozat, A.T. Erdogan, A.C. Singer, A.H. Sayed, Steady-state MSE performance analysis of mixture approaches to adaptive filtering, IEEE Trans. Signal Process. 58 (8) (Aug. 2010) 4050–4063.
[188] S.S. Kozat, A.C. Singer, Multi-stage adaptive signal processing algorithms, in: Proc. SAM Signal Process. Workshop, 2000, pp. 380–384.
[189] S.G. Kratzer, D.R. Morgan, The partial-rank algorithm for adaptive beamforming, Proc. SPIE 564 (1985) 9–14.
[190] K. Kreutz-Delgado, The complex gradient operator and the CR-calculus, arXiv:0906.4835, June 2009.
[191] H. Krim, M. Viberg, Two decades of array signal processing research: the parametric approach, IEEE Signal Process. Mag. 13 (4) (1996) 67–94.
[192] F. Kuech, W. Kellermann, Orthogonalized power filters for nonlinear acoustic echo cancellation, Signal Process. 86 (6) (2006) 1168–1181.
[193] S.Y. Kung, Kernel Methods and Machine Learning, Cambridge University Press, 2014.
[194] S.M. Kuo, D.R. Morgan, Active Noise Control Systems – Algorithms and DSP Implementations, Wiley Series in Telecommunications and Signal Processing, Wiley-Interscience, 1996.
[195] H.J. Kushner, G.G. Yin, Stochastic Approximation Algorithms and Applications, Applications of Mathematics, Springer, 1997.
[196] R.H. Kwong, E.W. Johnston, A variable step size LMS algorithm, IEEE Trans. Signal Process. 40 (7) (July 1992) 1633–1642.
[197] A.J. Laub, Matrix Analysis for Scientists and Engineers, SIAM, PA, 2005.
[198] M. Lázaro-Gredilla, L.A. Azpicueta-Ruiz, A.R. Figueiras-Vidal, J. Arenas-García, Adaptively biasing the weights of adaptive filters, IEEE Trans. Signal Process. 58 (7) (2010) 3890–3895.
[199] D. Lee, M. Morf, B. Friedlander, Recursive least squares ladder estimation algorithms, IEEE Trans. Circuits Syst. 28 (6) (1981) 467–481.
[200] L. Lei, J.P. Lie, A.B. Gershman, C.M.S. See, Robust adaptive beamforming in partly calibrated sparse sensor arrays, IEEE Trans. Signal Process. 58 (3) (2010) 1661–1667.
[201] A. Leon-Garcia, Probability, Statistics, and Random Processes for Electrical Engineering, 3rd edition, Prentice Hall, Jan. 2008.
[202] W.R. LePage, Complex Variables and the Laplace Transform for Engineers, Dover, 1980.
[203] H. Lev-Ari, T. Kailath, J. Cioffi, Least-squares adaptive lattice and transversal filters: a unified geometric theory, IEEE Trans. Inf. Theory 30 (2) (1984) 222–236.
[204] L. Li, J. Chambers, A new incremental affine projection-based adaptive algorithm for distributed networks, Signal Process. 88 (10) (2008) 2599–2603.
[205] M.V.S. Lima, P.S.R. Diniz, H. Yazdanpanah, Set-membership constrained frequency-domain algorithm, IEEE Trans. Circuits Syst. II, Express Briefs 68 (2) (Feb. 2021) 797–801.
[206] M.V.S. Lima, T.N. Ferreira, W.A. Martins, P.S.R. Diniz, Sparsity-aware data-selective adaptive filters, IEEE Trans. Signal Process. 62 (17) (Sept. 2014) 4557–4572.
[207] M.V.S. Lima, W.A. Martins, P.S.R. Diniz, Affine projection algorithms for sparse system identification, in: Proc. IEEE Int. Conf. Acoustics, Speech and Signal Processing ICASSP 2013, 2013, pp. 5666–5670.
[208] F. Ling, D. Manolakis, J. Proakis, Numerically robust least-squares lattice-ladder algorithms with direct updating of the reflection coefficients, IEEE Trans. Acoust. Speech Signal Process. 34 (4) (1986) 837–845.
[209] J. Liu, Y. Zakharov, Low complexity dynamically regularised RLS algorithm, Electron. Lett. 44 (14) (2008) 886–888.

[210] W. Liu, I. Park, J.C. Principe, An information theoretic approach of designing sparse kernel adaptive filters, IEEE Trans. Neural Netw. 20 (12) (Dec. 2009) 1950–1961.
[211] W. Liu, P.P. Pokharel, J.C. Principe, The kernel least-mean-square algorithm, IEEE Trans. Signal Process. 56 (2) (Feb 2008) 543–554.
[212] W. Liu, J.C. Príncipe, S. Haykin, Kernel Adaptive Filtering: A Comprehensive Introduction, Wiley, 2010.
[213] Y. Liu, C. Sun, S. Jiang, A reduced Gaussian kernel least-mean-square algorithm for nonlinear adaptive signal processing, Circuits Syst. Signal Process. 38 (Jan. 2019) 371–394.
[214] S. Ljung, L. Ljung, Error propagation properties of recursive least-squares adaptation algorithms, Automatica 21 (2) (Mar. 1985) 157–167.
[215] C.G. Lopes, L.F.O. Chamon, V.H. Nascimento, Towards spatially universal adaptive diffusion networks, in: 2014 IEEE Global Conference on Signal and Information Processing (GlobalSIP), IEEE, Dec 2014.
[216] C.G. Lopes, A.H. Sayed, Incremental adaptive strategies over distributed networks, IEEE Trans. Signal Process. 55 (Aug. 2007) 4064–4077.
[217] C.G. Lopes, A.H. Sayed, Diffusion adaptive networks with changing topologies, in: 2008 IEEE International Conference on Acoustics, Speech and Signal Processing, IEEE, 2008, pp. 3285–3288.
[218] C.G. Lopes, A.H. Sayed, Diffusion least-mean squares over adaptive networks: formulation and performance analysis, IEEE Trans. Signal Process. 56 (7) (2008) 3122–3136.
[219] R. Lopez-Valcarce, F. Perez-Gonzalez, Adaptive lattice iir filtering revisited: convergence issues and new algorithms with improved stability properties, IEEE Trans. Signal Process. 49 (4) (Apr. 2001) 811–821.
[220] C.S. Ludovico, J.C.M. Bermudez, A recursive least squares algorithm robust to low-power excitation, in: Proc. IEEE Int. Conf. on Acoustics, Speech, and Signal Processing ICASSP 2004, 2004.
[221] O. Macchi, E. Eweda, Convergence analysis of self-adaptive equalizers, IEEE Trans. Inf. Theory IT-30 (2) (Mar. 1984) 161–176.
[222] J.Y. Mai, A.H. Sayed, A feedback approach to the steady-state performance of fractionally spaced blind adaptive equalizers, IEEE Trans. Signal Process. 48 (1) (Jan. 2000) 80–91.
[223] D.P. Mandic, S. Still, S.C. Douglas, Duality between widely linear and dual channel adaptive filtering, in: Proc. IEEE Int. Conf. Acoustics, Speech and Signal Processing ICASSP 2009, 2009, pp. 1729–1732.
[224] D. Manolakis, F. Ling, J. Proakis, Efficient time-recursive least-squares algorithms for finite-memory adaptive filtering, IEEE Trans. Circuits Syst. 34 (4) (1987) 400–408.
[225] K. Maouche, D.T.M. Slock, Performance analysis and FTF version of the generalized sliding window recursive least-squares (GSWRLS) algorithm, in: Proc. Conf Signals, Systems and Computers Record of the Twenty-Ninth Asilomar Conf., vol. 1, 1995, pp. 685–689.
[226] M. Martínez-Ramón, J. Arenas-García, A. Navia-Vázquez, A.R. Figueiras-Vidal, An adaptive combination of adaptive filters for plant identification, in: 14th International Conference on Digital Signal Processing (DSP 2002), 2002, pp. 1195–1198.
[227] E. Masry, Alpha-stable signals and adaptive filtering, IEEE Trans. Signal Process. 48 (11) (2000) 3011–3016.
[228] G. Mateos, I.D. Schizas, G.B. Giannakis, Closed-form MSE performance of the distributed LMS algorithm, in: Proc. IEEE 13th Digital Signal Processing Workshop and 5th IEEE Signal Processing Education Workshop DSP/SPE 2009, 2009, pp. 66–71.
[229] G. Mateos, I.D. Schizas, G.B. Giannakis, Distributed recursive least-squares for consensus-based in-network adaptive estimation, IEEE Trans. Signal Process. 57 (11) (2009) 4583–4588.
[230] V.J. Mathews, G.L. Sicuranza, Polynomial Signal Processing, Wiley-Interscience, New York, 2000.
[231] V.J. Mathews, Z. Xie, Fixed-point error analysis of stochastic gradient adaptive lattice filters, IEEE Trans. Acoust. Speech Signal Process. 38 (1) (1990) 70–80.
[232] V.J. Mathews, Z. Xie, A stochastic gradient adaptive filter with gradient adaptive step size, IEEE Trans. Signal Process. 41 (6) (1993) 2075–2087.
[233] K. Mayyas, T. Aboulnasr, The leaky LMS algorithm: MSE analysis for Gaussian data, IEEE Trans. Signal Process. 45 (4) (Apr. 1997) 927–934.

[234] J.E. Mazo, On the independence theory of equalizer convergence, Bell Syst. Tech. J. 58 (May–June 1979) 963–993.
[235] J. Mendes Filho, M.T.M. Silva, M.D. Miranda, A family of algorithms for blind equalization of QAM signals, in: Proc. IEEE Int. Conf. Acoustics, Speech, and Signal Process. ICASSP 2011, Prague, Czech Republic, 2011, pp. 3388–3391.
[236] J. Mendes Filho, M.T.M. Silva, M.D. Miranda, V.H. Nascimento, A region-based algorithm for blind equalization of QAM signals, in: Proc. of the IEEE/SP 15th Workshop on Statistical Signal Processing, Cardiff, UK, 2009, pp. 685–688.
[237] R. Merched, Extended RLS lattice adaptive filters, IEEE Trans. Signal Process. 51 (9) (Sept. 2003) 2294–2309.
[238] R. Merched, M. Petraglia, P.S.R. Diniz, A delayless alias-free subband adaptive filter structure, IEEE Trans. Signal Process. 47 (6) (June 1999) 1580–1591.
[239] R. Merched, A.H. Sayed, An embedding approach to frequency-domain and subband adaptive filtering, IEEE Trans. Signal Process. 48 (9) (2000) 2607–2619.
[240] R. Merched, A.H. Sayed, Order-recursive RLS Laguerre adaptive filtering, IEEE Trans. Signal Process. 48 (11) (2000) 3000–3010.
[241] R. Merched, A.H. Sayed, RLS-Laguerre lattice adaptive filtering: error-feedback, normalized, and array-based algorithms, IEEE Trans. Signal Process. 49 (11) (2001) 2565–2576.
[242] C.D. Meyer, Matrix Analysis and Applied Linear Algebra, SIAM, Philadelphia, USA, 2000.
[243] G. Mileounis, B. Babadi, N. Kalouptsidis, V. Tarokh, An adaptive greedy algorithm with application to nonlinear communications, IEEE Trans. Signal Process. 58 (6) (2010) 2998–3007.
[244] J. Minkoff, Comment: on the unnecessary assumption of statistical independence between reference signal and filter weights in feedforward adaptive systems, IEEE Trans. Signal Process. 49 (May 2001) 1109.
[245] M.D. Miranda, L. Aguayo, M. Gerken, Performance of the a priori and a posteriori QR-LSL algorithms in a limited precision environment, in: Proc. IEEE Int. Conf. Acoustics, Speech, and Signal Processing ICASSP 1997, vol. 3, 1997, pp. 2337–2340.
[246] M.D. Miranda, M. Gerken, A hybrid QR-lattice least squares algorithm using a priori errors, in: Proc. 38th Midwest Symp. Circuits and Systems, Proceedings, vol. 2, 1995, pp. 983–986.
[247] M.D. Miranda, M. Gerken, A hybrid least squares QR-lattice algorithm using a priori errors, IEEE Trans. Signal Process. 45 (12) (Dec. 1997) 2900–2911.
[248] M.D. Miranda, M. Gerken, M.T.M. Silva, Efficient implementation of error-feedback LSL algorithm, Electron. Lett. 35 (16) (1999) 1308–1309.
[249] M.D. Miranda, M.T.M. Silva, V.H. Nascimento, Avoiding divergence in the constant modulus algorithm, in: Proc. IEEE Int. Conf. Acoustics, Speech and Signal Processing ICASSP 2008, Las Vegas, NV, USA, 2008, pp. 3565–3568.
[250] M.D. Miranda, M.T.M. Silva, V.H. Nascimento, Avoiding divergence in the Shalvi-Weinstein algorithm, IEEE Trans. Signal Process. 56 (11) (Nov. 2008) 5403–5413.
[251] D.R. Morgan, S.G. Kratzer, On a class of computationally efficient, rapidly converging, generalized NLMS algorithms, IEEE Signal Process. Lett. 3 (8) (1996) 245–247.
[252] D.R. Morgan, J.C. Thi, A delayless subband adaptive filter architecture, IEEE Trans. Signal Process. 43 (8) (1995) 1819–1830.
[253] G.V. Moustakides, Study of the transient phase of the forgetting factor RLS, IEEE Trans. Signal Process. 45 (10) (1997) 2468–2476.
[254] G.V. Moustakides, S. Theodorides, Fast Newton transversal filters — a new class of adaptive estimation algorithms, IEEE Trans. Signal Process. 39 (10) (Oct. 1991) 2184–2193.
[255] J. Nagumo, A. Noda, A learning method for system identification, IEEE Trans. Autom. Control 12 (3) (Jun. 1967) 282–287.
[256] S. Narayan, A. Peterson, M. Narasimha, Transform domain LMS algorithm, IEEE Trans. Acoust. Speech Signal Process. 31 (3) (1983) 609–615.

[257] S.S. Narayan, A.M. Peterson, Frequency domain least-mean-square algorithm, Proc. IEEE 69 (1) (1981) 124–126.

[258] V.H. Nascimento, A simple model for the effect of normalization on the convergence rate of adaptive filters, in: Proc. of the 2004 IEEE International Conference on Acoustics, Speech and Signal Processing, 2004, pp. II-453–II-456.

[259] V.H. Nascimento, J.C.M. Bermudez, Probability of divergence for the least-mean fourth (LMF) algorithm, IEEE Trans. Signal Process. 54 (4) (Apr. 2006) 1376–1385.

[260] V.H. Nascimento, R.C. de Lamare, A low-complexity strategy for speeding up the convergence of convex combinations of adaptive filters, in: Proc. IEEE Int. Conf. Acoustics, Speech and Signal Processing ICASSP 2012, 2012, pp. 3553–3556.

[261] V.H. Nascimento, A.H. Sayed, Unbiased and stable leakage-based adaptive filters, IEEE Trans. Signal Process. 47 (12) (Dec. 1999) 3261–3276.

[262] V.H. Nascimento, A.H. Sayed, On the learning mechanism of adaptive filters, IEEE Trans. Signal Process. 48 (6) (June 2000) 1609–1625.

[263] V.H. Nascimento, A.H. Sayed, Continuous-time distributed estimation, in: Proc. 45th Asilomar Conference on Signals, Systems, and Computers, Pacific Grove, CA, Nov. 2011, pp. 1–5.

[264] V.H. Nascimento, M.T. Silva, R. Candido, J. Arenas-García, A transient analysis for the convex combination of adaptive filters, in: Proc. IEEE Workshop on Statistical Signal Process., Cardiff, UK, 2009, pp. 53–56.

[265] V.H. Nascimento, M.T.M. Silva, Stochastic stability analysis for the constant-modulus algorithm, IEEE Trans. Signal Process. 56 (10) (Oct. 2008) 4984–4989.

[266] V.H. Nascimento, M.T.M. Silva, L.A. Azpicueta-Ruiz, J. Arenas-García, On the tracking performance of combinations of least mean squares and recursive least squares adaptive filters, in: Proc. IEEE Int. Conf. Acoustics, Speech and Signal Processing ICASSP 2010, Dallas, USA, 2010, pp. 3710–3713.

[267] V.H. Nascimento, Y.V. Zakharov, RLS adaptive filter with inequality constraints, IEEE Signal Process. Lett. 23 (5) (May 2016) 752–756.

[268] P.A. Nelson, S.J. Elliot, Active Control of Sound, Academic Press, 1992.

[269] S.L. Netto, P.S.R. Diniz, P. Agathoklis, Adaptive IIR filtering algorithms for system identification: a general framework, IEEE Trans. Ed. 38 (1) (Feb. 1995) 54–66.

[270] M. Niedźwiecki, Identification of nonstationary stochastic systems using parallel estimation schemes, IEEE Trans. Autom. Control 35 (3) (Mar. 1990) 329–334.

[271] M. Niedźwiecki, Multiple-model approach to finite memory adaptive filtering, IEEE Trans. Signal Process. 40 (2) (Feb. 1992) 470–473.

[272] K. Nishikawa, H. Kiya, Conditions for convergence of a delayless subband adaptive filter and its efficient implementation, IEEE Trans. Signal Process. 46 (4) (1998) 1158–1167.

[273] R.C. North, J.R. Zeidler, W.H. Ku, T.R. Albert, A floating-point arithmetic error analysis of direct and indirect coefficient updating techniques for adaptive lattice filters, IEEE Trans. Signal Process. 41 (5) (1993) 1809–1823.

[274] A.V. Oppenheim, A.S. Willsky, Signals and Systems, Prentice-Hall, 1996.

[275] K. Ozeki, T. Umeda, An adaptive filtering algorithm using an orthogonal projection to an affine subspace and its properties, Electron. Commun. Jpn. 67-A (1984) 19–27.

[276] D.A. Pados, G.N. Karystinos, An iterative algorithm for the computation of the MVDR filter, IEEE Trans. Signal Process. 49 (2) (2001) 290–300.

[277] C. Paleologu, J. Benesty, S. Ciochina, A robust variable forgetting factor recursive least-squares algorithm for system identification, IEEE Signal Process. Lett. 15 (2008) 597–600.

[278] C. Paleologu, J. Benesty, S. Ciochina, A variable step-size affine projection algorithm designed for acoustic echo cancellation, IEEE Trans. Audio Speech Lang. Process. 16 (8) (2008) 1466–1478.

[279] C. Paleologu, J. Benesty, S. Ciochina, Sparse adaptive filters for echo cancellation, Synth. Lect. Speech Audio Process. 6 (1) (Jun 2010) 1–124.

[280] C. Paleologu, J. Benesty, S. Ciochina, Regularization of the affine projection algorithm, IEEE Trans. Circuits Syst. II 58 (6) (2011) 366–370.
[281] C. Paleologu, S. Ciochina, J. Benesty, Variable step-size NLMS algorithm for under-modeling acoustic echo cancellation, IEEE Signal Process. Lett. 15 (2008) 5–8.
[282] C.B. Papadias, D.T.M. Slock, Normalized sliding window constant modulus and decision-direct algorithms: a link between blind equalization and classical adaptive filtering, IEEE Trans. Signal Process. 45 (Jan. 1997) 231–235.
[283] W.D. Parreira, J.C.M. Bermudez, C. Richard, J.Y. Tourneret, Stochastic behavior analysis of the Gaussian kernel least-mean-square algorithm, IEEE Trans. Signal Process. 60 (5) (May 2012) 2208–2222.
[284] F.R.M. Pavan, M.T.M. Silva, M.D. Miranda, Avoiding divergence in the constant modulus algorithm for blind equalization of MIMO systems, in: IEEE Sensor Array and Multichannel Signal Processing Workshop (SAM), Rio de Janeiro, Brazil, 2016, pp. 1–5.
[285] F.R.M. Pavan, M.T.M. Silva, M.D. Miranda, A numerically robust blind equalization scheme applied to MIMO communication systems, J. Franklin Inst. 1 (Jan. 2018) 596–624.
[286] F.R.M. Pavan, M.T.M. Silva, M.D. Miranda, Performance analysis of the multiuser Shalvi-Weinstein algorithm, Signal Process. 163 (2019) 153–165.
[287] D.I. Pazaitis, A.G. Constantinides, A novel kurtosis driven variable step-size adaptive algorithm, IEEE Trans. Signal Process. 47 (3) (1999) 864–872.
[288] M.R. Petraglia, R.G. Alves, P.S.R. Diniz, New structures for adaptive filtering in subbands with critical sampling, IEEE Trans. Signal Process. 48 (12) (2000) 3316–3327.
[289] M.R. Petraglia, P.B. Batalheiro, Nonuniform subband adaptive filtering with critical sampling, IEEE Trans. Signal Process. 56 (2) (2008) 565–575.
[290] M.R. Petraglia, S.K. Mitra, Adaptive FIR filter structure based on the generalized subband decomposition of fir filters, IEEE Trans. Circuits Syst. II 40 (6) (1993) 354–362.
[291] B. Picinbono, Wide-sense linear mean square estimation and prediction, in: Proc. IEEE Int. Conf. Acoustics, Speech and Signal Processing ICASSP 1995, vol. 3, 1995, pp. 2032–2035.
[292] B. Picinbono, P. Bondon, Second-order statistics of complex signals, IEEE Trans. Signal Process. 45 (2) (Feb. 1997) 411–420.
[293] J. Platt, A resource-allocating network for function interpolation, Neural Comput. 2 (1991) 213–225.
[294] I.K. Proudler, J.G. McWhirter, T.J. Shepherd, Computationally efficient QR decomposition approach to least squares adaptive filtering, IEE Proc., F, Radar Signal Process. 138 (4) (1991) 341–353.
[295] H.I.K. Rao, B. Farhang-Boroujeny, Fast LMS/Newton algorithms for stereophonic acoustic echo cancelation, IEEE Trans. Signal Process. 57 (8) (2009) 2919–2930.
[296] M.J. Reed, B. Liu, Analysis of simplified gradient adaptive lattice algorithms using power-of-two quantization, in: Proc. IEEE Int Circuits and Systems Symp., 1990, pp. 792–795.
[297] P.A. Regalia, Stable and efficient lattice algorithms for adaptive IIR filtering, IEEE Trans. Signal Process. 40 (2) (Feb. 1992) 375–388.
[298] P.A. Regalia, Numerical stability properties of a QR-based fast least squares algorithm, IEEE Trans. Signal Process. 41 (6) (June 1993) 2096–2109.
[299] P.A. Regalia, Adaptive IIR Filtering in Signal Processing and Control, Marcel Dekker, 1995.
[300] P.A. Regalia, M.G. Bellanger, On the duality between fast QR methods and lattice methods in least-squares adaptive filtering, IEEE Trans. Signal Process. 39 (Apr. 1991) 879–891.
[301] P.A. Regalia, M. Mboup, Undermodeled equalization: a characterization of stationary points for a family of blind criteria, IEEE Trans. Signal Process. 47 (Mar. 1999) 760–770.
[302] H. Rey, L.R. Vega, S. Tressens, J. Benesty, Variable explicit regularization in affine projection algorithm: robustness issues and optimal choice, IEEE Trans. Signal Process. 55 (5) (2007) 2096–2109.
[303] C. Richard, J.C.M. Bermudez, P. Honeine, Online prediction of time series data with kernels, IEEE Trans. Signal Process. 57 (3) (Mar. 2009) 1058–1067.

[304] Roger A. Horn, C.R. Johnson, Matrix Analysis, 2nd edition, Cambridge University Press, 2013.

[305] G. Rombouts, M. Moonen, A fast exact frequency domain implementation of the exponentially windowed affine projection algorithm, in: Proc. and Control Symp. Adaptive Systems for Signal Processing, Communications 2000, AS-SPCC, The IEEE, 2000, pp. 342–346.

[306] A. Rontogiannis, S. Theodoridis, On inverse factorization adaptive LS algorithms, Signal Process. 52 (1997) 35–47.

[307] A.A. Rontogiannis, S. Theodoridis, New fast inverse QR least squares adaptive algorithms, in: Proc. Int. Conf. Acoustics, Speech, and Signal Processing, vol. 2, 1995, pp. 1412–1415.

[308] A.A. Rontogiannis, S. Theodoridis, New multichannel fast QRD-LS adaptive algorithms, in: Proc. 13th Int. Conf. Digital Signal Processing, vol. 1, 1997, pp. 45–48.

[309] A.A. Rontogiannis, S. Theodoridis, Multichannel fast QRD-LS adaptive filtering: new technique and algorithms, IEEE Trans. Signal Process. 46 (11) (1998) 2862–2876.

[310] A.A. Rontogiannis, S. Theodoridis, New fast QR decomposition least squares adaptive algorithms, IEEE Trans. Signal Process. 46 (8) (1998) 2113–2121.

[311] S. Roy, J.J. Shynk, Analysis of the data-reusing LMS algorithm, in: Proc. 32nd Midwest Symp. Circuits and Systems, 1989, pp. 1127–1130.

[312] M. Rupp, A family of adaptive filter algorithms with decorrelating properties, IEEE Trans. Signal Process. 46 (3) (1998) 771–775.

[313] M. Rupp, A.H. Sayed, A time-domain feedback analysis of filtered-error adaptive gradient algorithms, IEEE Trans. Signal Process. 44 (1996) 1428–1439.

[314] C. Samson, V.U. Reddy, Fixed point error analysis of the normalized ladder algorithm, IEEE Trans. Acoust. Speech Signal Process. 31 (5) (Oct. 1983) 1177–1191.

[315] S.G. Sankaran, A.A.L. Beex, Convergence behavior of affine projection algorithms, IEEE Trans. Signal Process. 48 (4) (2000) 1086–1096.

[316] A.H. Sayed, Fundamentals of Adaptive Filtering, Wiley-Interscience, 2003.

[317] A.H. Sayed, Adaptive Filters, Wiley, NJ, 2008.

[318] A.H. Sayed, Adaptation, Learning, and Optimization over Networks, Foundations and Trends in Machine Learning, vol. 7, Now Publishers Inc., Hanover, MA, 2014.

[319] A.H. Sayed, T. Kailath, A state-space approach to adaptive RLS filtering, IEEE Signal Process. Mag. 11 (3) (July 1994) 18–60.

[320] A.H. Sayed, V.H. Nascimento, Energy conservation and the learning ability of LMS adaptive filters, in: S. Haykin, B. Widrow (Eds.), Least-Mean-Square Adaptive Filters, Wiley, 2003.

[321] A.H. Sayed, M. Rupp, Error-energy bounds for adaptive gradient algorithms, IEEE Trans. Signal Process. 44 (8) (Aug. 1996) 1982–1989.

[322] S. Scardapane, J. Chen, C. Richard, Adaptation and learning over networks for nonlinear system modeling, in: D. Comminiello, J. Principe (Eds.), Adaptive Learning Methods for Nonlinear System Modeling, Butterworth-Heinemann/Elsevier, Oxford, 2018, pp. 223–242, chapter 10.

[323] M. Scarpiniti, D. Comminiello, R. Parisi, A. Uncini, A collaborative approach to time-series prediction, in: B. Apolloni, S. Bassis, A. Esposito, F.C. Morabito (Eds.), Neural Nets WIRN11, IOS Press, 2011, pp. 178–185.

[324] L. Scharf, D. Tufts, Rank reduction for modeling stationary signals, IEEE Trans. Acoust. Speech Signal Process. 35 (3) (1987) 350–355.

[325] L.L. Scharf, E.K.P. Chong, M.D. Zoltowski, J.S. Goldstein, I.S. Reed, Subspace expansion and the equivalence of conjugate direction and multistage Wiener filters, IEEE Trans. Signal Process. 56 (10) (2008) 5013–5019.

[326] I.D. Schizas, G. Mateos, G.B. Giannakis, Distributed LMS for consensus-based in-network adaptive processing, IEEE Trans. Signal Process. 57 (6) (2009) 2365–2382.

[327] B. Scholkopf, A.J. Smola, Learning with Kernels: Support Vector Machines, Regularization, Optimization, and Beyond, MIT Press, Cambridge, 2002.

[328] W.A. Sethares, The least mean square family, in: N. Kalouptsidis, S. Theodoridis (Eds.), Adaptive System Identification and Signal Processing Algorithms, Prentice Hall, 1993.

[329] W.A. Sethares, D.A. Lawrence, C.R. Johnson Jr., R.R. Bitmead, Parameter drift in LMS adaptive filters, IEEE Trans. Acoust. Speech Signal Process. ASSP-34 (Aug. 1986) 868–878.

[330] M. Shao, C.L. Nikias, Signal processing with fractional lower order moments: stable processes and their applications, Proc. IEEE 81 (7) (1993) 986–1010.

[331] R. Sharma, W.A. Sethares, J.A. Bucklew, Asymptotic analysis of stochastic gradient-based adaptive filtering algorithms with general cost functions, IEEE Trans. Signal Process. 44 (9) (Sept. 1996) 2186–2194.

[332] H.-C. Shin, A.H. Sayed, Mean-square performance of a family of affine projection algorithms, IEEE Trans. Signal Process. 52 (1) (2004) 90–102.

[333] H.-C. Shin, A.H. Sayed, W.-J. Song, Variable step-size NLMS and affine projection algorithms, IEEE Signal Process. Lett. 11 (2) (Feb. 2004) 132–135.

[334] M. Shoaib, S. Werner, J.A. Apolinário, Multichannel fast QR-decomposition algorithms: weight extraction method and its applications, IEEE Trans. Signal Process. 58 (1) (2010) 175–188.

[335] M. Shoaib, S. Werner, J.A. Apolinário, T.I. Laakso, Solution to the weight extraction problem in fast QR-decomposition RLS algorithms, in: Proc. IEEE Int Acoustics, Speech and Signal Processing Conf. ICASSP 2006, vol. 3, 2006.

[336] J.J. Shynk, Adaptive IIR filtering, IEEE ASSP Mag. 6 (2) (Apr. 1989) 4–21.

[337] J.J. Shynk, Frequency-domain and multirate adaptive filtering, IEEE Signal Process. Mag. 9 (1) (1992) 14–37.

[338] G.L. Sicuranza, A. Carini, A. Fermo, Nonlinear adaptive filters for acoustic echo cancellation in mobile terminals, in: K.E. Barner, G.R. Arce (Eds.), Nonlinear Signal and Image Processing: Theory, Methods, and Applications, in: The Electrical Engineering and Applied Signal Processing Series, CRC Press, 2003, pp. 223–255, chapter 7.

[339] M.T.M. Silva, J. Arenas-García, A soft-switching blind equalization scheme via convex combination of adaptive filters, IEEE Trans. Signal Process. 61 (5) (Mar. 2013) 1171–1182.

[340] M.T.M. Silva, R. Candido, J. Arenas-García, J.A. Azpicueta-Ruiz, Improving multikernel adaptive filtering with selective bias, in: Proc. IEEE Int. Conf. Acoustics, Speech, and Signal Process. ICASSP 2018, Calgary, Canada, 2018, pp. 4529–4533.

[341] M.T.M. Silva, M.D. Miranda, Tracking issues of some blind equalization algorithms, IEEE Signal Process. Lett. 11 (9) (Sept. 2004) 760–763.

[342] M.T.M. Silva, V.H. Nascimento, Improving the tracking capability of adaptive filters via convex combination, IEEE Trans. Signal Process. 56 (7) (July 2008) 3137–3149.

[343] M.T.M. Silva, V.H. Nascimento, Tracking analysis of the constant modulus algorithm, in: Proc. IEEE Int. Conf. Acoustics, Speech and Signal Processing ICASSP 2008, Las Vegas, NV, USA, 2008, pp. 3561–3564.

[344] T.T.P. Silva, P. Lara, F. Igreja, F.D.V.R. Oliveira, L. Tarrataca, D.B. Haddad, An exact expectation model for the LMS tracking abilities, IEEE Trans. Signal Process. 68 (2020) 5882–5893.

[345] K. Slavakis, P. Bouboulis, S. Theodoridis, Online learning in reproducing kernel Hilbert spaces, in: R. Chellapa, S. Theodoridis (Eds.), Academic Press Library in Signal Processing: Signal Processing Theory and Machine Learning, vol. 1, Academic Press, Chennai, 2014, pp. 883–987, chapter 17.

[346] K. Slavakis, S. Theodoridis, I. Yamada, Online kernel-based classification using adaptive projection algorithms, IEEE Trans. Signal Process. 56 (7) (2008) 2781–2796.

[347] K. Slavakis, S. Theodoridis, I. Yamada, Adaptive constrained learning in reproducing kernel Hilbert spaces: the robust beamforming case, IEEE Trans. Signal Process. 57 (12) (2009) 4744–4764.

[348] D.T.M. Slock, The backward consistency concept and round-off error propagation dynamics in recursive least-squares filtering algorithms, Opt. Eng. 31 (Jun. 1992) 1153–1169.

[349] D.T.M. Slock, On the convergence behavior of the LMS and the normalized LMS algorithms, IEEE Trans. Signal Process. 41 (9) (Sept. 1993) 2811–2825.

[350] D.T.M. Slock, T. Kailath, Fast transversal filters with data sequence weighting, IEEE Trans. Acoust. Speech Signal Process. 37 (3) (1989) 346–359.

[351] D.T.M. Slock, T. Kailath, Numerically stable fast transversal filters for recursive least squares adaptive filtering, IEEE Trans. Signal Process. 39 (1) (1991) 92–114.

[352] S.D. Snyder, Active Noise Control Primer, Springer-Verlag, 2000.

[353] V. Solo, Averaging analysis of the LMS algorithm, in: C.T. Leondes (Ed.), Control and Dynamic Systems, in: Stochastic Techniques in Digital Signal Processing Systems, Part 2 of 2, vol. 65, Academic Press, 1994, pp. 379–397.

[354] V. Solo, X. Kong, Adaptive Signal Processing Algorithms, Prentice Hall, Englewood Cliffs, NJ, 1995.

[355] Y. Song, S. Roy, Blind adaptive reduced-rank detection for DS-CDMA signals in multipath channels, IEEE J. Sel. Areas Commun. 17 (11) (1999) 1960–1970.

[356] J.-S. Soo, K.K. Pang, Multidelay block frequency domain adaptive filter, IEEE Trans. Acoust. Speech Signal Process. 38 (2) (1990) 373–376.

[357] S. Stearns, Error surfaces of recursive adaptive filters, IEEE Trans. Acoust. Speech Signal Process. 29 (3) (Jun. 1981) 763–766.

[358] A. Stenger, W. Kellermann, Adaptation of a memoryless preprocessor for nonlinear acoustic echo cancelling, Signal Process. 80 (9) (2000) 1747–1760.

[359] R. Suyama, R.R.F. Attux, J.M.T. Romano, M. Bellanger, On the relationship between least squares and constant modulus criteria for adaptive filtering, in: The 37th Asilomar Conference on Signals, Systems and Computers, vol. 2, 2003, pp. 1293–1297.

[360] N. Takahashi, I. Yamada, Link probability control for probabilistic diffusion least-mean squares over resource-constrained networks, in: 2010 IEEE International Conference on Acoustics, Speech and Signal Processing, IEEE, 2010, pp. 3518–3521.

[361] N. Takahashi, I. Yamada, A.H. Sayed, Diffusion least-mean squares with adaptive combiners: formulation and performance analysis, IEEE Trans. Signal Process. 58 (9) (2010) 4795–4810.

[362] M. Tanaka, S. Makino, J. Kojima, A block exact fast affine projection algorithm, IEEE Trans. Speech Audio Process. 7 (1) (1999) 79–86.

[363] O. Tanrikulu, J.A. Chambers, Convergence and steady-state properties of the least-mean mixed-norm (LMMN) adaptive algorithm, IEEE Trans. Signal Process. 143 (3) (June 1996) 137–142.

[364] M. Tarrab, A. Feuer, Convergence and performance analysis of the normalized LMS algorithm with uncorrelated Gaussian data, IEEE Trans. Inf. Theory 34 (4) (July 1988) 680–691.

[365] S. Theodoridis, K. Slavakis, I. Yamada, Adaptive learning in a world of projections, IEEE Signal Process. Mag. 28 (1) (2011) 97–123.

[366] D.G. Tiglea, R. Candido, M.T.M. Silva, A low-cost algorithm for adaptive sampling and censoring in diffusion networks, IEEE Trans. Signal Process. 69 (Jan. 2021) 58–72.

[367] F.A. Tobar, S.Y. Kung, D.P. Mandic, Multikernel least mean square algorithm, IEEE Trans. Neural Netw. Learn. Syst. 25 (2) (Feb. 2014) 265–277.

[368] H.L.V. Trees, Optimum Array Processing, Wiley, 2002.

[369] J.R. Treichler, B. Agee, A new approach to multipath correction of constant modulus signals, IEEE Trans. Acoust. Speech Signal Process. ASSP-28 (Apr. 1983) 334–358.

[370] M.C. Tsakiris, C.G. Lopes, V.H. Nascimento, An array recursive least-squares algorithm with generic nonfading regularization matrix, IEEE Signal Process. Lett. 17 (12) (Dec. 2010) 1001–1004.

[371] S.-Y. Tu, A.H. Sayed, Foraging behavior of fish schools via diffusion adaptation, in: Proc. 2nd Int Cognitive Information Processing (CIP) Workshop, 2010, pp. 63–68.

[372] S.-Y. Tu, A.H. Sayed, Tracking behavior of mobile adaptive networks, in: Proc. Conf Signals, Systems and Computers (ASILOMAR) Record of the Forty Fourth Asilomar Conf., 2010, pp. 698–702.

[373] S.-Y. Tu, A.H. Sayed, Cooperative prey herding based on diffusion adaptation, in: Proc. IEEE Int. Conf. Acoustics, Speech and Signal Processing ICASSP 2011, 2011, pp. 3752–3755.

[374] S.V. Vaerenbergh, L.A. Azpicueta-Ruiz, D. Comminiello, A split kernel adaptive filtering architecture for nonlinear acoustic echo cancellation, in: Proc. of the European Signal Processing Conference (EUSIPCO), Budapest, Hungary, 2016, pp. 1768–1772.
[375] P.P. Vaidyanathan, Multirate Systems and Filter Banks, Prentice Hall, Englewood Cliffs, NJ, 1993.
[376] B.D. Van Veen, An analysis of several partially adaptive beamformer designs, IEEE Trans. Acoust. Speech Signal Process. 37 (2) (1989) 192–203.
[377] B.D. van Veen, Minimum variance beamforming with soft response constraints, IEEE Trans. Signal Process. 39 (9) (1991) 1964–1972.
[378] B.D. Van Veen, K.M. Buckley, Beamforming: a versatile approach to spatial filtering, IEEE ASSP Mag. 5 (2) (1988) 4–24.
[379] L. Vega, H. Rey, J. Benesty, S. Tressens, A new robust variable step-size NLMS algorithm, IEEE Trans. Signal Process. 56 (5) (May 2008) 1878–1893.
[380] S.A. Vorobyov, H. Chen, A.B. Gershman, On the relationship between robust minimum variance beamformers with probabilistic and worst-case distortionless response constraints, IEEE Trans. Signal Process. 56 (11) (2008) 5719–5724.
[381] S.A. Vorobyov, A.B. Gershman, Z.-Q. Luo, N. Ma, Adaptive beamforming with joint robustness against mismatched signal steering vector and interference nonstationarity, IEEE Signal Process. Lett. 11 (2) (2004) 108–111.
[382] E. Walach, B. Widrow, The least mean fourth (LMF) adaptive algorithm and its family, IEEE Trans. Inf. Theory IT-30 (2) (Mar. 1984) 275–283.
[383] S. Weiss, R.W. Stewart, R. Rabenstein, Steady-state performance limitations of subband adaptive filters, IEEE Trans. Signal Process. 49 (9) (Sept. 2001) 1982–1991.
[384] S. Werner, J.A. Apolinário, M.L. de Campos, P.S.R. Diniz, Low-complexity constrained affine-projection algorithms, IEEE Trans. Signal Process. 53 (12) (Dec. 2005) 4545–4555.
[385] S. Werner, M.L.R. de Campos, P.S.R. Diniz, Partial-update NLMS algorithms with data-selective updating, IEEE Trans. Signal Process. 52 (4) (Apr. 2004) 938–949.
[386] B. Widrow, Thinking about thinking: the discovery of the LMS algorithm, IEEE Signal Process. Mag. 22 (1) (2005) 100–106.
[387] B. Widrow, M. Hoff, Adaptive switching circuits, in: IRE WESCON Conv. Rec., vol. 4, 1960, pp. 96–104.
[388] B. Widrow, S.D. Stearns, Adaptive Signal Processing, Prentice Hall, Englewood Cliffs, 1985.
[389] J.H. Wilkinson, Rounding Errors in Algebraic Processes, Her Majesty's Stationery Office, 1963.
[390] J.H. Wilkinson, The Algebraic Eigenvalue Problem, Oxford University Press, London, UK, 1965.
[391] S. Xu, R.C. de Lamare, H.V. Poor, Adaptive link selection algorithms for distributed estimation, EURASIP J. Adv. Signal Process. 2015 (1) (2015) 86.
[392] P. Xue, B. Liu, Adaptive equalizer using finite-bit power-of-two quantizer, IEEE Trans. Acoust. Speech Signal Process. 34 (6) (1986) 1603–1611.
[393] I. Yamada, K. Slavakis, K. Yamada, An efficient robust adaptive filtering algorithm based on parallel subgradient projection techniques, IEEE Trans. Signal Process. 50 (5) (2002) 1091–1101.
[394] J. Yang, J.-J. Werner, G.A. Dumont, The multimodulus blind equalization and its generalized algorithms, IEEE J. Sel. Areas Commun. 20 (Jun. 2002) 997–1015.
[395] L. Yang, H. Zhu, K. Kang, X. Luo, H. Qian, Y. Yang, Distributed censoring with energy constraint in wireless sensor networks, in: Proc. IEEE Int. Conf. Acoustics, Speech and Signal Processing ICASSP 2018, IEEE, 2018, pp. 6428–6432.
[396] N. Yousef, A.H. Sayed, A unified approach to the steady-state and tracking analysis of adaptive filters, IEEE Trans. Signal Process. 49 (2) (Feb. 2001) 314–324.
[397] J.-T. Yuan, T.-C. Lin, Equalization and carrier phase recovery of CMA and MMA in blind adaptive receivers, IEEE Trans. Signal Process. 58 (6) (June 2010) 3206–3217.
[398] J.-T. Yuan, K.-D. Tsai, Analysis of the multimodulus blind equalization algorithm in QAM communication systems, IEEE Trans. Commun. 53 (Sept. 2005) 1427–1431.

[399] M. Yukawa, Multikernel adaptive filtering, IEEE Trans. Signal Process. 60 (9) (Sept. 2012) 4672–4682.

[400] M. Yukawa, R. de Lamare, R. Sampaio-Neto, Efficient acoustic echo cancellation with reduced-rank adaptive filtering based on selective decimation and adaptive interpolation, IEEE Trans. Speech Audio Process. 16 (4) (2008) 696–710.

[401] Y. Zakharov, T. Tozer, Multiplication-free iterative algorithm for LS problem, Electron. Lett. 40 (9) (2004) 567–569.

[402] Y. Zakharov, G. White, J. Liu, Low-complexity RLS algorithms using dichotomous coordinate descent iterations, IEEE Trans. Signal Process. 56 (7) (2008) 3150–3161.

[403] Y.V. Zakharov, V.H. Nascimento, DCD-RLS adaptive filters with penalties for sparse identification, IEEE Trans. Signal Process. 61 (12) (Jun 2013) 3198–3213.

[404] X. Zhang, Y. Xia, C. Li, L. Yang, D.P. Mandic, Complex properness inspired blind adaptive frequency-dependent I/Q imbalance compensation for wideband direct-conversion receivers, IEEE Trans. Wirel. Commun. 19 (9) (Sep 2020) 5982–5992.

[405] Y. Zhang, J. Chambers, Convex combination of adaptive filters for a variable tap-length LMS algorithm, IEEE Signal Process. Lett. 13 (10) (Oct. 2006) 628–631.

[406] X. Zhao, A.H. Sayed, Single-link diffusion strategies over adaptive networks, in: Proc. IEEE Int. Conf. Acoustics, Speech and Signal Processing ICASSP 2012, 2012, pp. 3749–3752.

[407] Y.R. Zheng, V.H. Nascimento, Two variable step-size adaptive algorithms for non-Gaussian interference environment using fractionally lower-order moment minimization, Digit. Signal Process. 23 (3) (May 2013) 831–844.

CHAPTER

Machine learning

Review and trends

13

Marcele O.K. Mendonça[a], **Sergio L. Netto**[a], **Paulo S.R. Diniz**[a], **and Sergios Theodoridis**[b,c]

[a]*Program of Electrical Engineering and Department of Electronics & Computer Engineering, COPPE/Poli/Universidade Federal do Rio de Janeiro, Rio de Janeiro, Brazil*

[b]*Department of Informatics and Telecommunications, National and Kapodistrian University of Athens, Athens, Greece*

[c]*Electronic Systems Department, Aalborg University, Aalborg, Denmark*

13.1 Introduction

Machine learning (ML) is a pervasive technology finding more applications on a daily basis. ML is the critical tool in methods designed to provide prediction and classification in a wide variety of fields. One should bear in mind that despite taking part in the world of artificial intelligence (AI), ML has its limitations. There are many tasks in which natural intelligence cannot be matched by ML solutions relying on data. In many cases, for ML, it is hard to adapt or generalize the solution to new problems, since it relies on the data available and ignores temporal and domain changes [53]. The broader view of AI includes ML as an action or execution of distinct components that would get closer to an AI benchmark if combined. In the long run, the idea contains features like cognition, perception, and actions entirely controlled by a computer.

In summary, it is possible to predict that there is a long road ahead for ML researchers; for instance, one can incorporate some form of reasoning to approach the goal of proper AI solutions. While AI encompasses all aspects of intelligence, ML can be considered as only one component of AI.

This chapter covers some classical methods of ML and delves into some current problems being addressed by the ML community [48], [8].

The adaptive filtering algorithms covered in this book fall into the class of learning algorithms, mainly utilizing linear parametric models, that are widely used in real-time applications with inherent streaming data. The algorithms are efficient computationally and can be applied to deal with some particular nonlinear models as well [13]. From now on, we aim at presenting ways of learning from data beyond the realm and without the constraints akin to classical online adaptive filtering.

In this chapter, we present the basic methods of ML that hopefully can motivate the reader to learn further concepts of this fascinating field and set the stage for the many aspects of the ML discussed in the remaining chapters of this book. The chapter describes the basic concepts of ML balancing the presentation between established methods and more recent challenges in the field.

Many textbooks cover ML, statistical inference, neural networks (NNs), and related topics [2,3,10, 15,21,24,30,45,58,62–65]. Here we selected some topics that we found more relevant to develop new tools in this area and complement the remaining topics of this book. Furthermore, wherever appropriate, we discuss the connections with the signal processing methods discussed in the previous chapter.

Signal Processing and Machine Learning Theory. https://doi.org/10.1016/B978-0-32-391772-8.00019-3
Copyright © 2024 Elsevier Inc. All rights reserved.

We should have two approaches to describe ML theory and algorithms. Some topics are covered in more detailed form, whereas others are covered in a more informative and intuitive format. The ML field is ample and encompasses many branches of knowledge, so it is only possible to describe a narrow range of tools.

In the detailed cases, we should include the following information when it applies:

- cost function,
- network structure,
- learning algorithm,
- regularization (if applicable),
- universal approximation property (if applicable),
- an example (where and when it applies).

The presentation starts with a broad introduction to the basic concepts related to the learning process inherent to any ML algorithm. In there, we discuss some issues akin to the data use in an ML application, such as how to prepare the data; how to separate the datasets for training, validation, and testing; and how to define a minibatch. Then, we discuss some learning modes highlighting unsupervised, semisupervised, and supervised learning.

13.2 Learning concepts

Learning is the ability to acquire knowledge, enabling one to perform new and more complex tasks. Although not restricted to humans, our amazing learning capacity distinguishes us from other animal species. Such intelligence is enlarged by our acute sense of the surroundings, expanded memory (due to writing), developed communication skills, and amazing processing capability (thanks to a disproportionate and highly evolved brain). In the last few decades, however, a new in-lab bred *species* has also developed a large ability to acquire, store, communicate, and process information, enabling it to learn human-like skills such as language translation, music composition, image interpretation, and data classification.

This chapter deals with this amazing process where we humans can teach this new species, the computer, how to analyze data and extract information from it in order to develop new abilities and perform ever more advanced tasks. To do so in a successful manner, we need to dissect the learning process and use this analysis to develop algorithms that are able to mimic this ability.

Nature observation has become the fundamental cornerstone in science ever since Galileo pointed his telescope to Jupiter and took note of the peculiar movements of some of the planet's moons. From then on, recording data from nature and interpreting them has led to many scientific discoveries and fundamental laws.

Nowadays, we use the same approach to make computers learn new tasks. Given enough data, we have now developed algorithms that can teach a computer an underlying pattern within a given process. Once this pattern is mathematically modeled or, as we say, learned by the computer, the machine can reproduce it on newly acquired data. The most important aspects in this entire process are the data, the process at hand, and the training framework: One must have enough data to represent a process and have an effective learning strategy to identify and mathematically represent a given pattern within the process.

13.2.1 Data concepts

As mentioned above, data play a major role in the ML process. Consider a given process $\mathcal{P}$ that maps the m-dimensional data space $\mathcal{U}^m$ into an output space $\mathcal{Y}$, as represented in Fig. 13.1. Due to practical issues, one only has access to a limited amount of data $\{\mathbf{u}^m\} \subset \mathcal{U}^m$ and $\{y\} \subset \mathcal{Y}$ about $\mathcal{P}$. During the learning process, datasets $\{\mathbf{u}^m\}$ and $\{y\}$ are our only windows into the soul of $\mathcal{P}$. Therefore, these sets must represent as faithfully and completely as possible the process one is attempting to model.

FIGURE 13.1

Block diagram representation of a given pattern $\mathcal{P}$ that maps the data space $\mathcal{W}^m$ into an output space $\mathcal{Y}$.

Any noise component in either $\{\mathbf{u}^m\}$ or $\{y\}$ shall cause a suboptimal learning process, leading to a biased representation $\hat{\mathcal{P}}$ of $\mathcal{P}$. Noisy $\mathbf{u}^m$ data result from practical measurements, where a given sensor may be even off due to some operational failure, whereas a noisy y is commonly associated with improper mislabeling, which often results from imperfect human intervention. In lab conditions, such impairments can be controlled down to a certain level, which often guarantees a successful learning procedure. In several practical situations, however, these data imperfections are strong enough to hinder the learning process, which shall require some additional data preprocessing stages to become feasible.

As brilliantly demonstrated by Claude E. Shannon, not all bits of data represent bits of information. Therefore, a crucial stage in the learning framework is to perform information extraction from the available data, where the concept of information is pretty much dependent on the nature of the underlying pattern one is trying to represent. In this stage, each piece of information extracted from $\{\mathbf{u}^m\}$ is commonly referred to as a process attribute or feature.

As will be later assessed, system training is a somewhat intricate process, where one searches for the best model $\hat{\mathcal{P}}^*$ for $\mathcal{P}$ within a subset of candidates $\hat{\mathcal{P}}$. This search may become impractical in cases of high values of data dimension m. This issue is commonly addressed by some form of dimensionality reduction, leading to an input mapping of the form $\mathbf{u}^m \to \mathbf{x}^n$, where $\mathbf{x}$ denotes the feature vector and commonly $n << m$, indicating a significant reduction in the data dimension and in the complexity of the associated learning process.

Other important aspect arises when the available data do not completely represent the original sets $\mathcal{U}^m$ and $\mathcal{Y}$. In such a case, our best trained model $\hat{\mathcal{P}}^*$ may seem to map well $\{\mathbf{u}^m\}$ into $\{y\}$, despite being significantly distinct from $\mathcal{P}$. When this happens, our estimated model $\hat{\mathcal{P}}^*$ will have poor generalization capability and very little practical usage. Such impairment may be detected by incorporating an extra step into the learning process which evaluates the degree of generalization achieved by the model. This is accomplished by separating the available data into two independent subsets $(\{\mathbf{u}^m\}_{\text{tr}}, \{y\}_{\text{tr}})$ and $(\{\mathbf{u}^m\}_{\text{te}}, \{y\}_{\text{te}})$, where the first one is used for model training and the second one is used for testing its performance in a different dataset. It is very important to ensure that there is no contamination between these two datasets, meaning that no piece of information employed in system training is used during the system testing, guaranteeing a truly independent and reliable performance assessment. This dataset partition reduces the amount of training data, probably affecting a little bit the final estimate $\hat{\mathcal{P}}^*$ yielded by the learning process. By allowing the evaluation of the final trained model into completely independent data, however, one is able to evaluate the model's ability to represent new data, thus measuring the process generalization capability, which is critical in most practical situations.

Following the discussion presented in this section, we may incorporate the data preprocessing, feature extraction, dimensionality reduction, and training/testing partition stages into the overall learning process, leading to a more detailed framework, as depicted in Fig. 13.2.

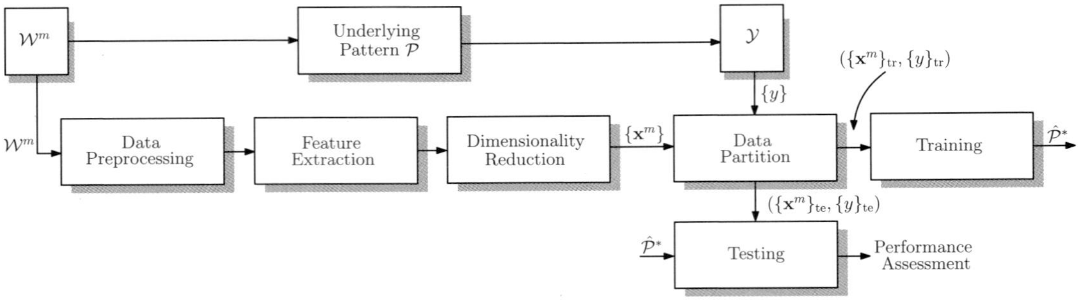

FIGURE 13.2

Block diagram of a more detailed machine learning pipeline.

In this procedure, the "Training" step includes the search, as performed by a given parameter adjustment algorithm, for a best model $\mathcal{P}^*$, according to a given performance criterion, as detailed in Fig. 13.3.

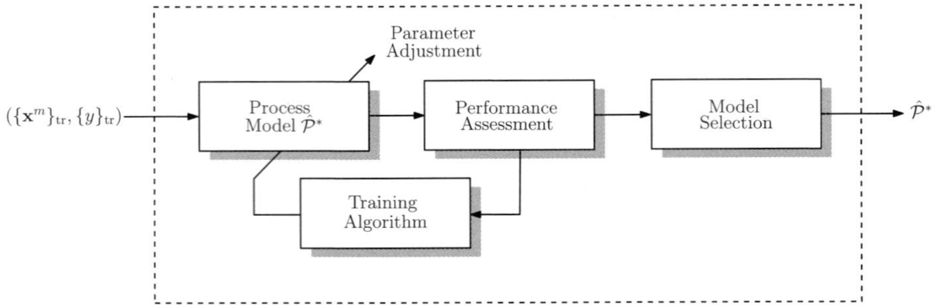

FIGURE 13.3

Detailed block diagram for the model training procedure.

Although still soon in this section, we can say, from past practical experience, that data quality is usually the bottleneck in the whole learning process: The quality of your final system will be affected most by the quantity and most importantly by the quality of the available data. As we are still early in our learning journey, some of these aspects may seem a bit hermetic to the reader at this stage. All of them, however, will be revisited along this chapter from a practical point of view, which shall clarify their most important concepts in due time.

13.2.2 Kinds of learning

At this point, we hope the reader is convinced that ML is a very important and interesting field. We now proceed to give indications that it is also a very broad field quite challenging to cover in a single book chapter. In fact, the learning process is very diverse and rich enough to fill a large bookshelf or even an entire library section. Among the several types of learning processes, we may cite:

- Classification × regression: Classification is a learning process where the expected output belongs to a discrete and finite set of values (such as "class 1," "class 2," …, "class N"), whereas regression shall output a real number.
- Binary × multiclass: A binary problem is a particular classification problem where the output set has only $N = 2$ classes, whereas the multiclass problem has $N > 2$. (See Fig. 13.4.)

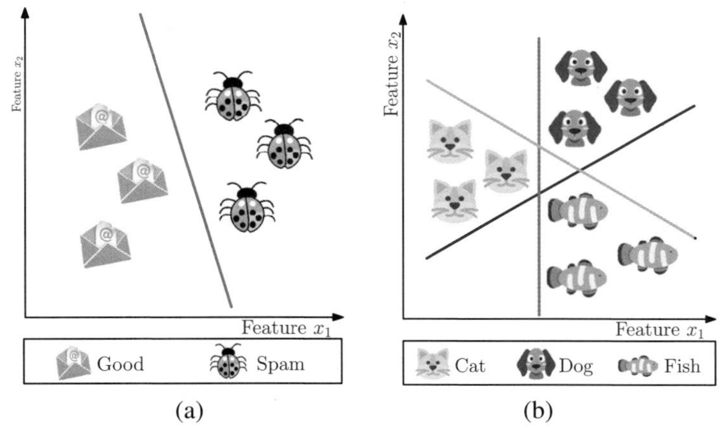

FIGURE 13.4

Illustration of (a) binary and (b) multiclass classification problems.

- Unsupervised × supervised: The unsupervised classification problem, also known as clustering, groups the input data into subsets (clusters) without labeling them, whereas the supervised problem uses prior information about the data (commonly obtained through painful manual annotation) to train an algorithm that partitions the input data into a predefined set of classes. (See Fig. 13.5.)
- Online × active (learning by asking) × batch learning: In the online scheme, the input data are acquired in real time, during the training process. In the active scheme, the learning process asks for more data until some performance criterion is satisfied. Finally, batch learning (possibly the most common form among these three) uses an entire precollected set of input samples. If the entire set is divided into several smaller subsets, the strategy is also referred to as minibatch learning.
- Structured learning: This strategy refers to situations where the underlying pattern presents some form of structure, like speech recognition systems predicting practical sentences that must follow a formal grammatical set of rules (structure).
- Reinforcement learning (RL): In this type of learning, an algorithm models a given process by performing it over and over again until reaching good performance levels in several occasions.

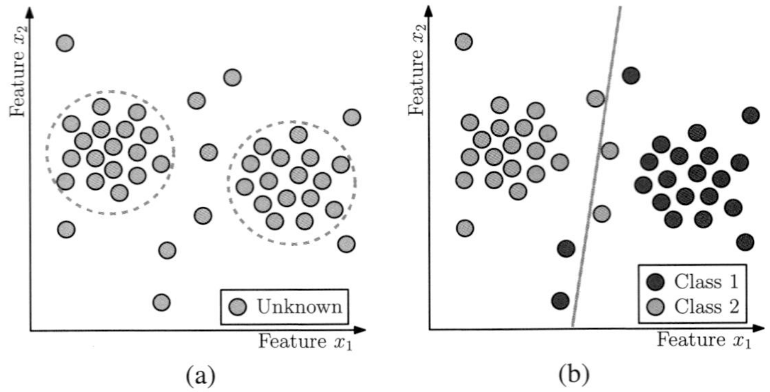

FIGURE 13.5

Illustration of (a) unsupervised and (b) supervised classification problems.

- Adversarial learning: In this strategy, two systems are used in tandem to model a given process. One system tries to learn about the process whereas the other system attempts to deceive the first one. In this manner, the second system reinforces the learning process performed by the first system, thus providing stronger and more robust results.
- Continuous learning: Practical situations where the target system dynamically changes along time require some form of (continuous) learning, where the trained model must adjust itself to reflect the time-varying nature of the underlying pattern.

This chapter covers several aspects associated with some of these learning strategies, mostly focusing on the batch learning approach. In particular, unsupervised learning is addressed in Section 13.3, the most common supervised algorithms are presented in Section 13.4, adversarial learning is introduced in Section 13.6.9, and RL is treated in Section 13.8.

13.2.3 Underfitting and overfitting

Overfitting or overtraining is literally the condition of excessively adjusting our model to spurious underlying patterns within the training data. When this happens, our model adjusts itself to data subtleties that have no direct counterpart along the original target process. Such a faulty training condition may lead to biased models $\hat{\mathcal{P}}^*$ with excellent performance along the training dataset, but with a very bad generalization capability when analyzing new data generated by $\mathcal{P}$. This process misrepresentation commonly occurs when one uses an excessively complex model and the training data present one or more of the following issues:

- a stochastic noise component in the measured data;
- missing data samples due to temporary sensor failures;
- data mislabeling caused by faulty human intervention.

The opposite condition to overfitting is called underfitting, which occurs when the model is not able to represent completely the true nature of $\mathcal{P}$. This situation often occurs when there are too few training data available and/or our model is not sufficiently complex to represent fully the target process. Such a faulty training condition often generates biased models $\hat{\mathcal{P}}^*$ with poor performance in both the training and testing datasets. (See Figs. 13.6 and 13.7.)

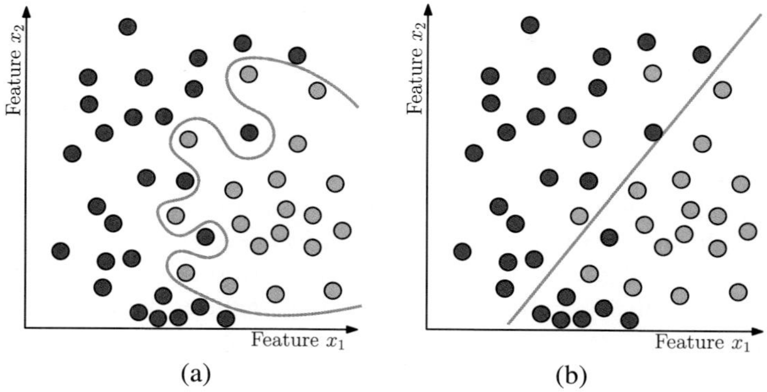

FIGURE 13.6

Illustration of (a) overfitting and (b) underfitting in a classification problem.

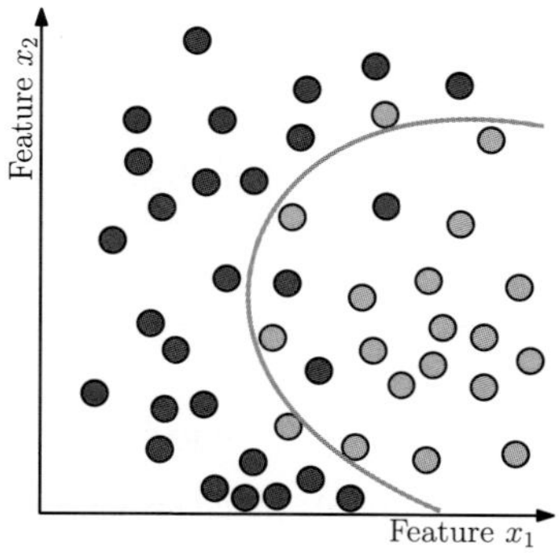

FIGURE 13.7

Illustration of a good fit for the classification problem in Fig. 13.6.

As the entire training process is based only on the available data, meaning that one does not have direct access to the true process, any significant discrepancy between $(\{\mathbf{u}^m\}_{\text{tr}}, \{y\}_{\text{tr}})$, and $\mathcal{P}$ leads the training algorithm astray thus converging to suboptimal solutions $\hat{\mathcal{P}}^*$ in both overfitting and underfitting situations.

In practice, overfitting can be detected by excellent levels of performance in the training stage followed by poor performance in the test dataset. When this happens, our first impulse is to retrain the algorithm in order to improve its performance in the testing stage. In this strategy, however, the test dataset is somehow guiding the training process, violating the independence condition between these two learning stages.

Dealing with overfitting is perhaps one of the biggest challenges in the field of ML: how can someone be rattled by amazing performance levels during training? A bad generalization capability, however, means that our selected model has little applicability in practice. Therefore, one may ask, "How can we detect and mitigate overtraining?" The answer to this question lies in two distinct and complementary strategies, namely validation and regularization.

In order to measure the generalization capability of our model, one cannot base this analysis on the testing dataset to avoid data snooping along the training procedure. We may, however, separate the original training dataset into two new subsets: one for the effective system training and the other, referred to as the validation set, to verify the performance of our model with nontraining and nontesting data. This training–validation partition can be performed in several different ways, allowing one to estimate the average and standard deviation of our system's performance over different validation sets.

Regularization is basically a training strategy where one reduces the number of degrees of freedom of the model by restricting the domain of its parameters. Common parameter constraints include a maximum value for their absolute values (ℓ_1 constraint) or total energy (ℓ_2 constraint), for instance. By imposing such limitations during the system training our model loses the ability to represent complicated patterns which are often associated with the process misrepresentation along the available data.

13.3 Unsupervised learning

13.3.1 Introduction

Unsupervised learning attempts to divide the whole mass of available data into subgroups, where all elements of any given partition share similar characteristics and elements from distinct partitions are as different as possible. If properly performed, this process allows us to further analyze each subgroup separately, thus breaking a large and complex problem into several smaller and simpler tasks, as preconized by Descartes. Each data partition is commonly referred to as a cluster and, therefore, unsupervised learning is also known as clusterization.

Clustering algorithms often involve the following steps [67]:

1. Feature selection and extraction: Each available piece of data must first be characterized by some of its specific attributes or features. Proper features shall be immune to noise and easy to extract and interpret. Choosing the proper feature set decreases the subsequent workload but usually requires great knowledge about the problem at hand.
2. Clustering algorithm selection and tuning: There is a plethora of clustering algorithms in the literature [3,63,67]. Different clustering algorithms result from different data similarity or proximity functions. Once such a function is selected, clustering becomes an optimization problem, which also has a vast associated literature. The proper clustering algorithm is often chosen based on the user's familiarity and on the algorithm's robustness and convergence speed.
3. Cluster validation: Different data clusters often result from different algorithms or even from the same algorithm with different execution parameters. Therefore, evaluating the final result with effective standards and criteria is an important step in the learning process. A typical assessment, for instance, analyzes the final number of clusters and verifies if there is no over- or underpartition;

4. Result interpretation: As mentioned above, once clustering is performed on a given dataset, one may extract information in an easier and more effective way by interpreting the data at both the intra- and intercluster levels.

In the three sections that follow we consider some clustering algorithms that altogether, far from exhausting the field, provide a good coverage of the matter in both historical and practical aspects.

13.3.2 K-Means algorithm

In the K-means algorithm, we assume the existence of K clusters in a given dataset and initially select at random K points, $\boldsymbol{\mu}_k$, for $k = 1, 2, \ldots, K$, not necessarily belonging to our dataset, to act as the prototypes of each cluster.

We then assign each datapoint $\mathbf{x}$ to the closest prototype following some distance definition, such as the Euclidean one, creating a set of K clusters.

For each cluster, a new prototype is defined as the point that minimizes the total distance to all its points. If the Euclidean distance is employed, these prototypes are called centroids and they are determined as the mean of all points belonging to that cluster, justifying the algorithm nomenclature.

The previous two steps (data assignment and prototype calculation) are repeated until some convergence criterion is reached. Possible rules may consider a given number of iterations, total displacement of all prototypes in consecutive iterations, and so on. Convergence of the K-means algorithm may occur to a locally optimal solution and tends to be highly dependent on the initial set of prototype points and nonrobust to outliers [3].

13.3.3 Self-organizing map

A self-organizing map (SOM) or a Kohonen map [36] is the numerical or visual representation of a one-layer fully connected 2D NN, where neighboring cells respond to specific but similar input signal patterns. In this process, each input activates a cell or a group of cells in a particular region of the network. Similar inputs will tend to activate close by cells and vice versa. After convergence, the SOM provides a 2D representation of the input domain, where each network region corresponds to a particular data pattern.

The SOM training process is based on the concept of competitive learning, where the cells compete with each other to decide which one will respond or be activated by a given input. Once the winner cell is determined, the network weights are adjusted in such a way that the cells become more similar to the input the closer they are to the winning cell.

For an n-dimensional input vector $\mathbf{x}$, each cell $c_{i,j}$ of an $I \times J$ SOM has an n-dimensional weight vector $\mathbf{w}_{i,j}$ associated with it, with $i = 1, 2, \ldots, I$ and $j = 1, 2, \ldots, J$. We may assume that all weights are initialized with random values. For every input $\mathbf{x}$, a distance is determined between it and all weight vectors $\mathbf{w}_{i,j}$, and the network element with minimum distance becomes the best-matching cell (BMC) for that particular input. Once the BMC is determined, the weight vectors of all neighboring cells are adjusted in such a way that they become more similar to the input the closer the corresponding cell is to the BMC.

Algebraically speaking, for any input $\mathbf{x}$, we determine its distance to all network weight vectors $\mathbf{w}_{i,j}$, as given by

$$d(i, j) = \|\mathbf{x} - \mathbf{w}_{i,j}\|, \tag{13.1}$$

with i, j as before. The minimum value of $d(i, j)$, say $d(i^*, j^*)$, identifies the network cell with a weight vector most similar to the input. Once the BMC position is determined, (i^*, j^*) in this case, we adjust the network weight vectors such that [37]

$$\mathbf{w}_{i,j} = \mathbf{w}_{i,j} + \eta T(i, j)(\mathbf{x} - \mathbf{w}_{i,j}), \tag{13.2}$$

where η is a learning rate parameter (that may vary along the convergence process) and $T(i, j)$ is a function of the distance between the (i, j) cell and the BMC for that particular input. In order to enforce larger adjustments for cells closer to the BMC, $T(i, i)$ may be defined, for instance, as

$$T(i, j) = e^{-\frac{\|(i,j)-(i^*,j^*)\|}{2\sigma^2}}, \tag{13.3}$$

where σ is the number of cells being adjusted, as one may opt to restrict the learning process to a close neighborhood around the BMC to reduce computational complexity. The entire weight adjustment process is performed for all available inputs, over and over again, until the SOM converges to a dataset match.

13.3.4 Mean-shift clustering

The mean-shift algorithm (MSA) is a member of the hierarchical clustering family, where the algorithm itself figures out how many clusters there ought to be, as opposed for instance to K-means, where that number must be specified a priori.

The MSA starts by considering that a given n-dimensional datapoint $\mathbf{x}$ is the initial value of a cluster prototype $\boldsymbol{\mu}_{\mathbf{x}}(0)$. The MSA then determines all datapoints within a neighborhood $\mathcal{N}$ of radius or bandwidth R around that prototype and obtains a new prototype by shifting the previous one in the direction of the average point $\boldsymbol{\mu}_{\mathcal{N}}$ within $\mathcal{N}$, that is,

$$\boldsymbol{\mu}_{\mathbf{x}}(t+1) = \boldsymbol{\mu}_{\mathbf{x}}(t) + \eta(\boldsymbol{\mu}_{\mathcal{N}} - \boldsymbol{\mu}_{\mathbf{x}}(t)). \tag{13.4}$$

In this iterative procedure, η sets the convergence learning rate, which can be made variable, as well as the value of R.

In every iteration, a new cluster is defined within the new neighborhood of the shifted prototype, and subsequently a new prototype displacement is performed, and so on. Once the procedure converges for a given datapoint, the same algorithm is replicated to the other datapoints outside the clusters that have already been determined.

13.3.5 MNIST example

We consider the Modified National Institute of Standards and Technology (MNIST) [39] database, which is a large dataset containing digits handwritten by students and employees of the United States Census Bureau. This dataset consists of 60,000 and 10,000 training and test examples, respectively. The input of this set is a 28×28 matrix, where each value represents a pixel of the image. The input signal is normalized to the range 0 to 1. The output dataset has integer values between 0 and 9. Fig. 13.8 illustrates some entries of this dataset. For the unsupervised learning example, we only consider digits 0, 1, and 4 from the training set. Each 28×28 matrix example representing a digit is reshaped into a vector

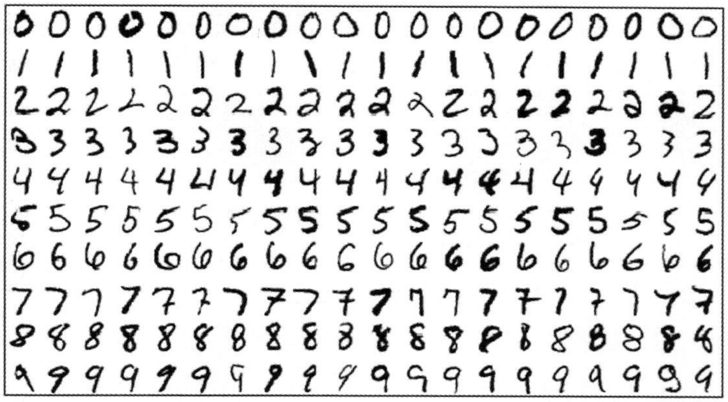

FIGURE 13.8

Sample images from the MNIST dataset.

with 784 elements. To facilitate the visualization, principal component analysis (PCA) is employed to reduce the original 784 features to only two features [56]. The obtained features are illustrated in Fig. 13.9(d), where red (mid gray in print version), blue (dark gray in print version), and green (light gray in print version) points represent digits 0, 1, and 4, respectively. After repeatedly computing the centroids, both K-means and mean-shift algorithms obtain similar clusters as shown in Fig. 13.9(a) and Fig. 13.9(c). Although the performance is similar for this example, the K-means algorithm is less computational intensive, whereas the MSA is more automatic when deciding the number of clusters to be formed. On the other hand, the SOM training process obtains a map of activated neurons in an attempt to reflect the data pattern. The resulting map of neurons can then be used to determine the clusters, as depicted in Fig. 13.10. We can visualize the clusters back in the feature space in Fig. 13.9(b). Indeed, the SOM training method is appropriate to mimic the data behavior, and it is an adequate tool to obtain and visualize clusters when the data dimension is high.

13.4 Supervised learning

This section presents some classical and more recent methods of supervised learning that benefit from the availability of annotated data. With the reference signal it is expected that the training period will be more informative to the network, resulting in improved classification or regression performance.

13.4.1 Perceptron

One of the earliest and simplest NN tools is the so-called perceptron, enabling pattern classification, which is linearly separable. As such, the perceptron finds the hyperplane that divides the original space into two regions. The original sources related to the perceptron concept date back to the 1940s. Some

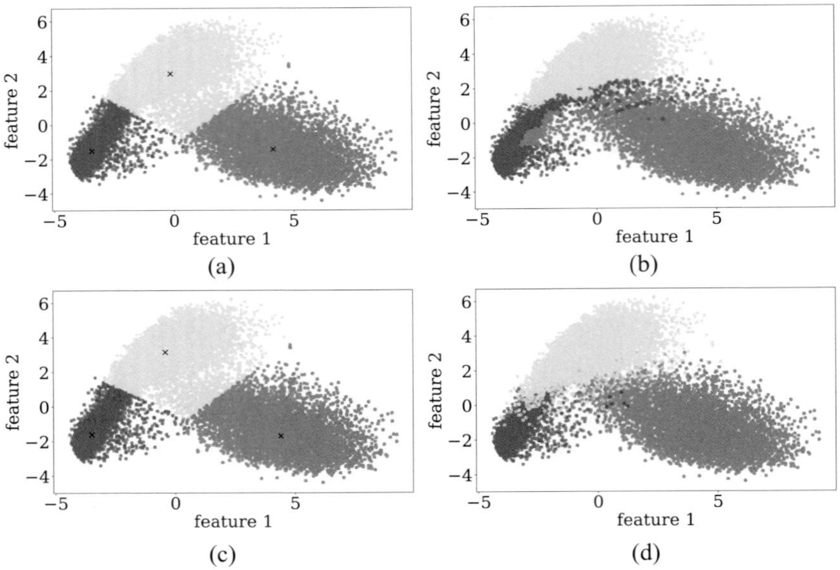

FIGURE 13.9

Clusters formed using (a) K-means, (b), SOM and (c) mean-shift algorithms for three classes from the MNIST dataset (d).

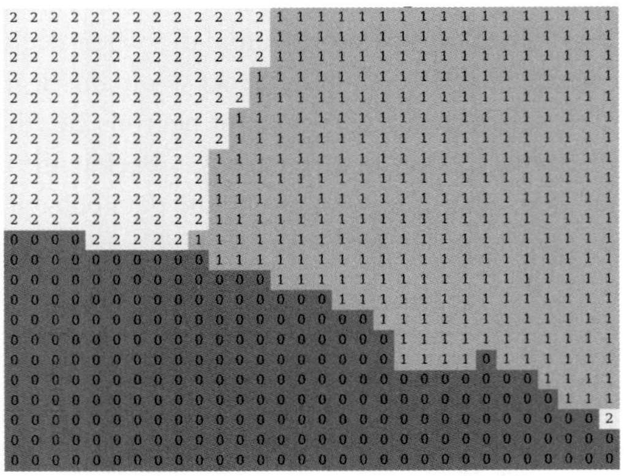

FIGURE 13.10

Neuron map obtained using SOM.

highlights should be given to [26,43,49]. An excellent review of the origin of the perceptron can be found in [24].

Consider a dataset $\{(\mathbf{x}(1), y(1)), (\mathbf{x}(2), y(2)), \cdots, (\mathbf{x}(Q), y(Q))\}$ such that $\mathbf{x}(q) \in \mathbb{R}^{n\times 1}$ belongs to one of the classes (ω_1, ω_2). If the two classes are linearly separable, at least one hyperplane can classify the set of training samples correctly. The hyperplane is defined by the normal vector $\mathbf{w}_o$ such that $\mathbf{w}_o^T \mathbf{x}(q) = 0$ and the classification is

$$y(q) = \text{sign}\left[\mathbf{w}^T \mathbf{x}(q)\right] = \begin{cases} -1, & \text{if } \mathbf{x}(q) \in \omega_1, \\ +1, & \text{if } \mathbf{x}(q) \in \omega_2, \end{cases} \tag{13.5}$$

where sign $[\cdot]$ is the sign function.

The bias term of the hyperplane is included in the weight vector $\mathbf{w}$ to simplify the notation. From the definition of the sign function, we have $y(q)(\mathbf{x}^T(q)\mathbf{w}) > 0$ if only if $\{\mathbf{x}(q), y(q)\}$ is correctly classified. A popular example of the application of this algorithm is the credit scoring model [66], whose goal is to identify if a person can receive credit based on the information stored in a database.

The perceptron learning algorithm aims at finding the previously mentioned hyperplane. The algorithm is initialized with parameter $\mathbf{w}$ as a random or zero vector. The update rule per iteration is

$$\mathbf{w} = \begin{cases} \mathbf{w} + \mu y(q)\mathbf{x}(q) & \text{if } \mathbf{x}(q) \text{ is misclassified,} \\ \mathbf{w} & \text{otherwise,} \end{cases} \tag{13.6}$$

where $\mu > 0$ is the step size parameter, which controls the algorithm's convergence. It is possible to interpret this operation by employing a geometric interpretation, as depicted in Fig. 13.11. Since $\mathbf{x}(q)$ is misclassified by $\mathbf{w}$, the term $\mu y(q)\mathbf{x}(q)$ is added and the hyperplane moves so that the updated coefficient $\mathbf{w}$ classifies $\mathbf{x}(q)$ correctly.

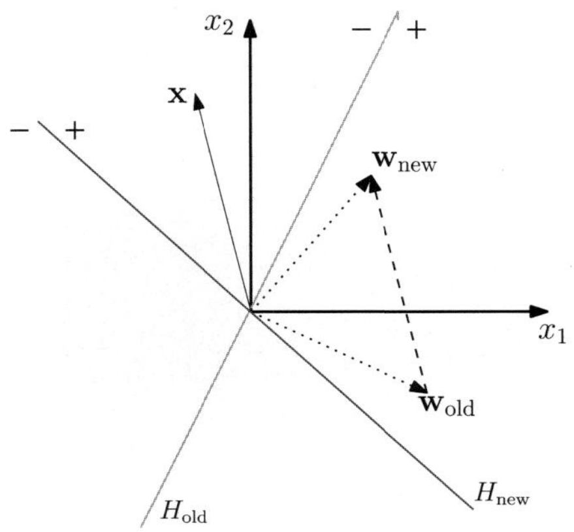

FIGURE 13.11

The vector $\mathbf{x}$ is misclassified by the hyperplane H_{old} (in red; light gray in print version). The update rule adds $\mathbf{x}$ to the coefficient vector, $\mathbf{w}_{\text{new}} = \mathbf{w}_{\text{old}} + \mathbf{x}$, where in this case $\mu = 1$. The new hyperplane H_{new} (in blue; dark gray in print version) classifies correctly the vector $\mathbf{x}$.

After a finite number of iterations, an optimal solution $\mathbf{w}_o$ is reached. This result holds regardless of which criterion is chosen to select the data at each iteration and also how the weight vector in the algorithm is initialized [47]. The perceptron algorithm is outlined in Table 13.1.

Table 13.1 Perceptron algorithm.

Perceptron algorithm
Initialize
$\mathbf{w}$ = random vectors or zero vectors,
set $\mu > 0$,
Do for $t > 0$ (epoch)
count = 0
Do for $q = 1 : Q$ (for iteration q, select the data $\{\mathbf{x}(q), y(q)\}$)
if $y(q)(\mathbf{x}^T(q)\mathbf{w}) < 0$
$\mathbf{w} = \mathbf{w} + \mu y(q)\mathbf{x}(q)$
count = count + 1
end if
end
if count = 0
break; (If count = 0, then all data is classified correctly)
end if
end

The perceptron can be extended to generate a network with multiple layers called multilayer perceptron (MLP). The latter is a universal approximator able to represent any nonlinear function [29]. The classification illustrated in Fig. 13.12 cannot be implemented by the perceptron, even though this function is considered simple. Therefore, it is suitable to apply an MLP algorithm, a version of the perceptron with multiple layers, for this problem. The added layers between input and output are called hidden layers. Each layer is comprised of neurons, also known as nodes in this kind of problem. A sign function is calculated in each of the nodes belonging to the hidden and output layers, as in the perceptron algorithm.

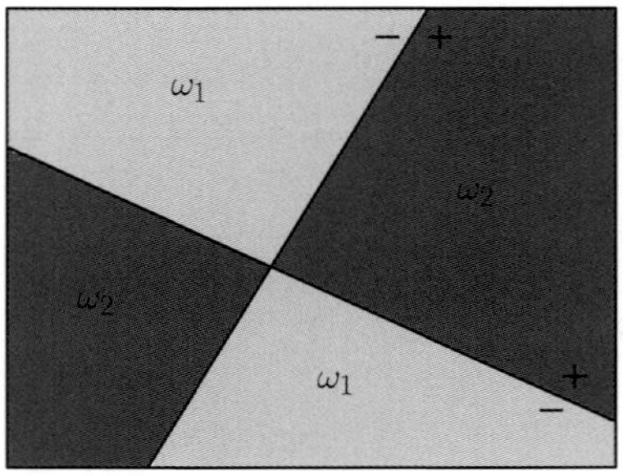

FIGURE 13.12

Two classes (ω_1, ω_2) are formed by the union of polyhedral regions. The configuration in this image is related to the Boolean XOR function.

In the following, we present more complex problems which require NNs such as the feedforward multilayer NN. The concepts and definitions of these NNs are following described.

13.4.2 Fully connected neural networks

The structure of the feedforward multilayer NN includes an input layer, several internal layers, and an output layer as defined in the following. Let us start by considering a dataset of input–output pairs,

$$\mathcal{D} = \{(\mathbf{x}(1), y(1)), (\mathbf{x}(2), y(2)), \cdots, (\mathbf{x}(Q), y(Q))\}, \tag{13.7}$$

where $\mathbf{x}(q) \in \mathbb{R}^{n\times 1}$ for $q = 1, \cdots Q$. The inherent quantities belonging to each layer l are the input vector denoted as $\mathbf{h}^{(l)}$ and the output vector denoted as $\mathbf{a}^{(l)}$. The connection between the data vectors of layers $l-1$ and l is performed by applying the weight matrix $\mathbf{W}^{(l)}$. For reference the NN main variables are organized in Table 13.2.

Table 13.2 Configuration of an arbitrary neural network.

Input vector of layer l	$\mathbf{h}^{(l)} \in \mathbb{R}^{d^{(l)}\times 1}$
Output vector of layer l	$\mathbf{a}^{(l)} \in \mathbb{R}^{(d^{(l)}+1)\times 1}$
Weight matrix between layers $l-1$ and l	$\mathbf{W}^{(l)} \in \mathbb{R}^{(d^{(l)}+1)\times(d^{(l+1)})}$

As previously mentioned, a powerful NN used in supervised learning to approximate the nonlinear mapping between the input–output pairs is the MLP. The MLP is composed of fully connected neurons or nodes that are organized in several layers, as illustrated in Fig. 13.13. The input layer denoted by $l=0$ acquires the signal $\mathbf{x}(q)$ and the output layer $l=L$ takes care of the prescribed task, i.e., makes a decision or prediction $\hat{y}(q)$ related to the input. The hidden layers lie between them, indexed by $0<l<L$. These models of networks are known as feedforward since the data move forward from the input layer to the output layer [24,54].

Each layer l consists of $d^{(l)}$ neurons. For the input and hidden layers, there are bias parameters as part of the network. The weight matrix $\mathbf{W}^{(l)}$ connects the layers $l-1$ and l, so that $w_{ij}^{(l)}$ links the ith node of the previous layer $l-1$ to the jth node of layer l. Each node of the hidden layers, except for the bias nodes, includes an activation function f whose argument is the weighted input of the layer. For example, at the output layer, the activation function is f_L. Some of the most widely used activation functions are the ReLU function, the sigmoid logistic function, and the hyperbolic tangent function [20,52].

In regression problems, the number of nodes in the output layer is $d^{(L)}=1$, producing a continuous numeric value $\hat{y}$ to be compared with the reference value y. In classification problems, $d^{(L)}$ represents the number of classes. The desired signal vector $\mathbf{y}$ is one-hot-encoded, meaning that if c is the correct class, $y_c=1$ and $y_i=0$ for $i\neq c$. In this case, utilizing the softmax activation function at the output layer is useful since the output signal $\hat{\mathbf{y}}$ returns a probability distribution of the classes, that is, $\hat{y}_i = P(c=i)$, and the class with greater probability is chosen.

Training the MLP involves adapting the weights and biases of the model to minimize a cost function of the error in prediction or classification related to the input signal. In the forward pass, the input signal vector $\mathbf{x}(q)$ flows forward through the network and produces the estimate $\hat{y}(q)$, which is then

Table 13.3 Gradient descent backpropagation algorithm.

Gradient descent backpropagation algorithm

Initialize
$\{(\mathbf{x}(1), \mathbf{y}(1)), (\mathbf{x}(2), \mathbf{y}(2)), \cdots, (\mathbf{x}(Q), \mathbf{y}(Q))\}$,
$\mathcal{W} = \{\mathbf{W}^{(1)}, \mathbf{W}^{(2)}, \cdots, \mathbf{W}^{(L)}\}$ (random),
Select: step size $\mu > 0$, number of epoch T, minibatch size b, number of layers $L+1$,
number of nodes $(d^{(1)}, \cdots, d^{(L)})$, activation function f, output activation function f_L,
objective function J
$I = Q/b$
Do for $t = 1 : T$
  Do for $i = 1 : I$ (for each iteration, randomly select b examples
in training dataset $\rightarrow \mathbf{X}_i^t = [\mathbf{x}_1, \mathbf{x}_2, \cdots, \mathbf{x}_b]$, $\mathbf{Y}_i^t = [\mathbf{y}_1, \mathbf{y}_2, \cdots, \mathbf{y}_b]$)
[Forward Propagation]:
    $\mathbf{S}^0 = [\mathbf{s}_1^0, \cdots, \mathbf{s}_b^0] = [ones(1, b); \mathbf{X}_i^t]$ ($ones(1, b)$ are the bias term)
    Do for $l = 1 : L-1$
      $\mathbf{E}^{(l)} = (\mathbf{W}^{(l)})^T \mathbf{S}^{(l-1)}$
      $\mathbf{S}^{(l)} = [ones(1, b); f(\mathbf{E}^{(l)})]$
    end
    $\mathbf{E}^{(L)} = (\mathbf{W}^{(L)})^T \mathbf{S}^{(L-1)}$
    $\hat{\mathbf{Y}} = [\hat{\mathbf{y}}_1, \cdots, \hat{\mathbf{y}}_b] = \mathbf{S}^{(L)} = g(\mathbf{E}^{(L)})$
[Backpropagation]:
    $\mathbf{\Delta}^{(L)} = [\delta_1^{(L)}, \cdots, \delta_b^{(L)}] = \frac{\partial J}{\partial \mathbf{E}^{(L)}}$
    Do for $l = L-1 : -1 : 1$
      $\mathbf{\Delta}^{(l)} = f'(\mathbf{E}^{(l)}) \otimes [\mathbf{W}^{(l+1)} \mathbf{\Delta}^{(l+1)}]_1^{d^{(l)}}$
    end
[Updating the weights]:
    Do for $l = 1 : L$
      $\mathbf{W}^{(l)} = \mathbf{W}^{(l)} - \frac{\mu}{b} \mathbf{S}^{(l-1)} (\mathbf{\Delta}^{(l)})^T$
    end
  end
  $J_{\text{train}}(\mathbf{W}) = \frac{1}{Q} \sum_{q=1}^{Q} \sum_{p=1}^{d^{(L)}} J_q^p(\mathbf{W})$ (and J_{val} if this set is previously defined)
end

saturation in the tails and centering on a nonzero value. Currently, the tangent hyperbolic (tanh) and ReLU represent the standard.

The standard way to select the number of hidden layers and neurons in a training network is to add layers and neurons until the validation error shows no improvement. The step size μ is a crucial parameter in any learning algorithm based on the gradient descent search method, including NNs. However, the selection of μ requires some moderation, given that large values might lead to divergence, whereas low values might stick the cost function to a local minimum. A possible solution is to decrease the value of μ at each iteration.

There are many other minimizers available in the literature. Some of them are more flexible optimizers, such as Adagrad, adaptive moment (Adam), and RMSProp [1,11,33,51,68]. Some of these methods employ adaptive step sizes resulting in improved and smoother convergence in comparison with the gradient descent algorithm.

If the weights are too large or too small, the activation function will be saturated, resulting in low gradients and slower convergence.

The weights $\mathcal{W}$ are randomly initialized by default. If the weights are too large or too small, the activation function will be saturated, resulting in low gradients and slower convergence. One can deal with this drawback by initializing the weights with a uniform or normal distribution.

13.4.2.1 Local error features

In supervised learning of NNs, each output neuron produces an estimate of a target signal and their subtraction forms an error measure defined according to the cost function (13.12). As a rule, the closer to zero the error $e(k)$ is in principle, the less informative or relevant will be the contribution of the pair $(\mathbf{x}(k), \mathbf{y}(k))$ to the parameter update at iteration k. Considering all the output neurons, the error vector for the data pair $(\mathbf{x}, \mathbf{y})$ is

$$\mathbf{e}(\hat{\mathbf{y}}, \mathbf{y}) = [e(\hat{y}_1, y_1), e(\hat{y}_2, y_2), \cdots, e(\hat{y}_{d^{(L)}}, y_{d^{(L)}})], \tag{13.17}$$

where $\mathbf{y} = [y_1, y_2, \cdots, y_{d^{(L)}}]$ is the target signal and $\hat{\mathbf{y}} = [\hat{y}_1, \hat{y}_2, \cdots, \hat{y}_{d^{(L)}}]$ is the signal estimated by the NN. The total sum of this error is then expressed as

$$\mathbb{E}\left[\hat{\mathbf{y}}, \mathbf{y}\right] = \sum_{n=1}^{d^{(L)}} e\left(\hat{y}_n, y_n\right). \tag{13.18}$$

Fig. 13.14 illustrates a possible behavior of the error signals in regression or classification NN problems. A closer analysis of the behavior of the error along the learning process is directly related to a measure of how informative the current data are.

The next step is to describe in detail the algorithms to solve regression and classification problems.

13.4.3 Regression

The chosen objective function is the mean square error (MSE),

$$J(\mathbf{W}) = \frac{1}{Q}\sum_{q=1}^{Q}\left[\frac{1}{2}(\hat{y}(q) - y(q))^2\right], \tag{13.19}$$

where $\hat{y}(q) = y_1^{(L)}(q) = x_1^{(L)}(q)$ and the output activation function is a linear function. Since the output has only one dimension in regression problems, Eq. (13.18) can be rewritten as

$$E(\hat{y}, y) = e(\hat{y}, y) = y - \hat{y}, \tag{13.20}$$

where y is the desired value and $\hat{y}$ is the target value.

At each epoch t, it has been observed in many examples that the error signal approaches a Gaussian distribution,

$$e \sim \mathcal{N}(0, (\sigma_e^t)^2), \tag{13.21}$$

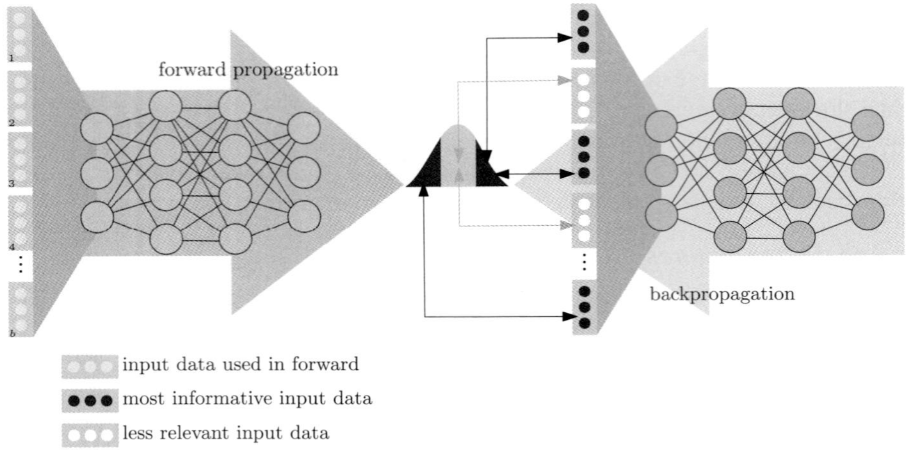

FIGURE 13.14

Neural network feedforward and backpropagation illustration.

where $(\sigma_e^t)^2$ is the error variance. By normalizing this error distribution, we obtain

$$\frac{\mathrm{e}}{\sigma_e^t} \sim \mathcal{N}(0, 1). \tag{13.22}$$

To corroborate the discussion, Fig. 13.15 illustrates the observed error statistics in an ML example employing a deep NN (DNN).

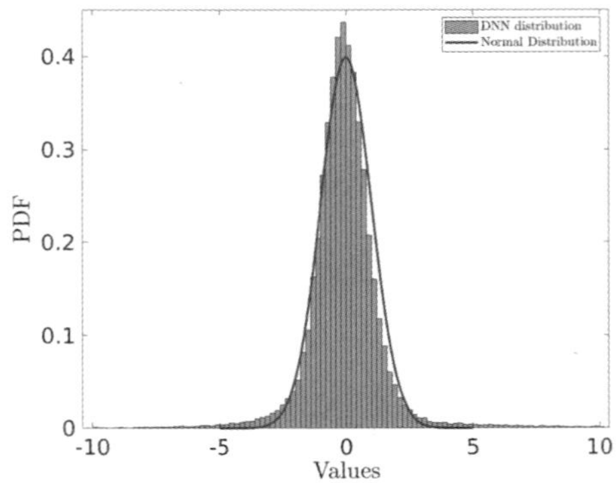

FIGURE 13.15

Histogram of a DNN error signal.

At iteration i in epoch t, the estimated error variance is calculated by

$$(\sigma_e^t(i))^2 = (1 - \lambda_e)\sigma_e^2 + (\lambda_e)(\sigma_e^t(i-1))^2, \tag{13.23}$$

where σ_e^2 is the error variance related to the data in the ith iteration and λ_e is a forgetting factor.

At each start of epoch t, the estimated error variance depends on the last error variance $(\sigma_e^{t-1}(b))^2$ from the previous epoch; thus, the equation for this dependence is established by

$$(\sigma_e^t(0))^2 = (\sigma_e^{t-1}(b))^2. \tag{13.24}$$

The NN algorithm for a regression problem is outlined in Table 13.4.

13.4.4 A regression example

In regression problems, the goal is to approximate a mapping function from input variables to a continuous output variable. Since the desired output is a continuous quantity, the activation function at the output layer is linear.

Consider the communication between a base station (BS) and a mobile phone user in Fig. 13.16. The medium between the transmitter and the receiver, known as channel in wireless communications, transforms the transmitted signal and generates interfering signals that affect the quality of the received signal.

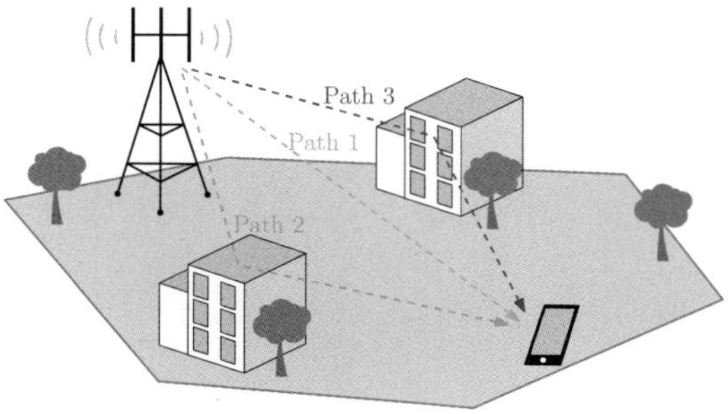

FIGURE 13.16

Illustration of the wireless communication between the base station and a mobile user.

For such environments, orthogonal frequency division multiplexing (OFDM) systems are quite suitable as they can mitigate the interference by adding redundant elements to the transmitted signal. Nevertheless, to undo the channel effect, we need to estimate the channel response. To do so, pilot symbols, which are known at the receiver, are transmitted and used to invert the received signal and obtain an estimate for the channel response.

The channel models a moving pedestrian in urban scenarios. For this example, we can build our training dataset by collecting previously obtained channel estimates. With the training dataset, the NN

Table 13.4 Feedforward multilayer NN algorithm for a regression problem.

Feedforward Multilayer NN algorithm for regression

Initialize

$\{(\mathbf{x}(1), \mathbf{y}(1)), (\mathbf{x}(2), \mathbf{y}(2)), \cdots, (\mathbf{x}(Q), \mathbf{y}(Q))\}$, $\mathcal{W} = \{\mathbf{W}^{(1)}, \mathbf{W}^{(2)}, \cdots, \mathbf{W}^{(L)}\}$ (random),

Select: step size $\mu > 0$, number of epoch T, minibatch size b, number of layers L, number of nodes $(d^{(1)}, \cdots, d^{(L)})$, activation function f, output function f_L, forgetting factor λ_e

Objective function J = Mean Squared Error

Define $I = M/b$, $\sigma_e^0(b) = 0$

Do for $t = 1 : T$

$(\sigma_e^t(0))^2 = (\sigma_e^{t-1}(b))^2$

Do for $i = 1 : I$ (for each iteration, randomly select b examples in the training dataset $\rightarrow \mathbf{X}_{(t,i)} = [\bar{\mathbf{x}}(1), \bar{\mathbf{x}}(2), \cdots, \bar{\mathbf{x}}(b)]$, $\mathbf{Y}_{(t,i)} = [\bar{\mathbf{y}}(1), \bar{\mathbf{y}}(2), \cdots, \bar{\mathbf{y}}(b)]$)

[Forward Propagation]:

$\mathbf{Y}^0 = [\bar{y}^0(1), \bar{y}^0(2), \cdots, \bar{y}^0(b)] = [ones(1, b); \mathbf{X}_i^t]$ ($ones(1, b)$ are the bias term)

Do for $l = 1 : L - 1$

$\mathbf{X}^{(l)} = (\mathbf{W}^{(l)})^T \mathbf{Y}^{(l-1)}$

$\mathbf{Y}^{(l)} = [ones(1, b);\ f(\mathbf{X}^{(l)})]$

end

$\mathbf{X}^{(L)} = (\mathbf{W}^{(L)})^T \mathbf{Y}^{(L-1)}$

$\hat{\mathbf{Y}}_{(t,i)} = [\hat{y}(1), \hat{y}(2), \cdots, \hat{y}(b)] = \mathbf{Y}^{(L)} = g(\mathbf{X}^{(L)})$

[Backpropagation]:

$\boldsymbol{\Delta}_{\mathcal{R}}^L = [\delta^{(L)}(k_1), \delta^{(L)}(k_2), \cdots, \delta^{(L)}(k_p)] = g'(\mathbf{X}_{\mathcal{R}}^L) \otimes (\hat{\mathbf{Y}}_{\mathcal{R}} - \mathbf{Y}_{\mathcal{R}})$

Do for $l = L - 1 : -1 : 1$

$\boldsymbol{\Delta}_{\mathcal{R}}^l = f'(\mathbf{X}_{\mathcal{R}}^l) \otimes [\mathbf{W}^{(l+1)} \boldsymbol{\Delta}_{\mathcal{R}}^{l+1}]_1^{d^{(l)}}$

end

[Updating the weights]:

Do for $l = 1 : L$

$\mathbf{W}^{(l)} = \mathbf{W}^{(l)} - \frac{\mu}{b} \mathbf{Y}_{\mathcal{R}}^{(l-1)} (\boldsymbol{\Delta}_{\mathcal{R}}^l)^T$

end

end

$J_{\text{train}}(\mathbf{W}) = \frac{1}{Q} \sum_{q=1}^{Q} J_q^{(1)}(\mathbf{W})$ (and J_{val} if this set is previously defined)

end

will be able to learn the problem of obtaining an improved channel estimate. The input of the NN is the estimated channel via least squares (LS) and the desired output is the refined channel estimate. The improved channel estimator (ICE) net is illustrated in Fig. 13.17, where it has one hidden layer with hyperbolic tangent as the activation function. In Fig. 13.18, we present the MSE between the channel estimate and the true channel response using the traditional method LS and the NN as function of the signal-to-noise ratio (SNR). Fig. 13.19 presents an example of estimated channel frequency response at the data subcarriers using the ICE subnet. Since the ICE has access to typical channel examples during training, it is natural that the ICE can outperform the traditional channel estimator in this example.

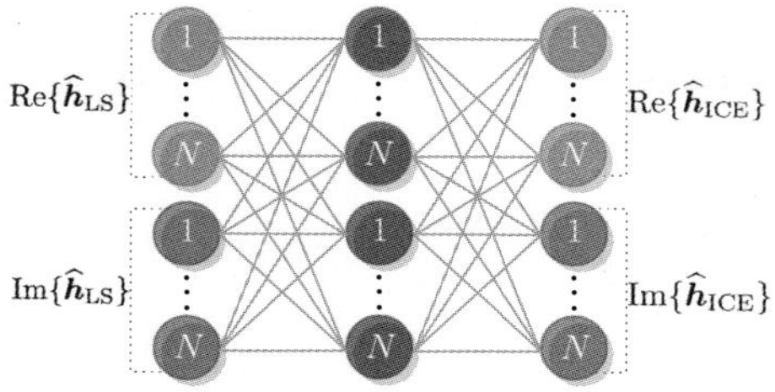

FIGURE 13.17

ICE neural network used to estimate the channel response.

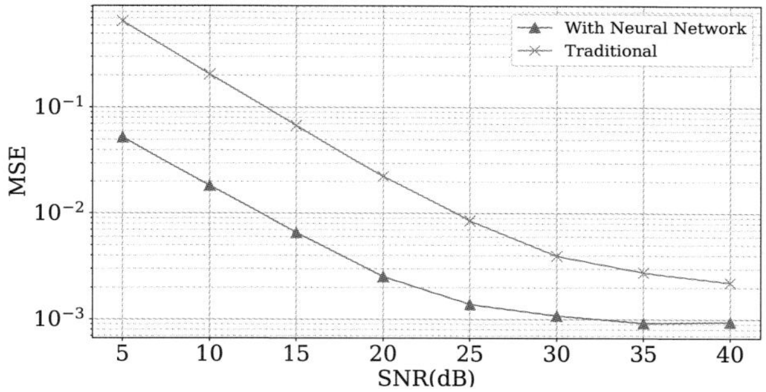

FIGURE 13.18

MSE between estimated and true channel frequency responses.

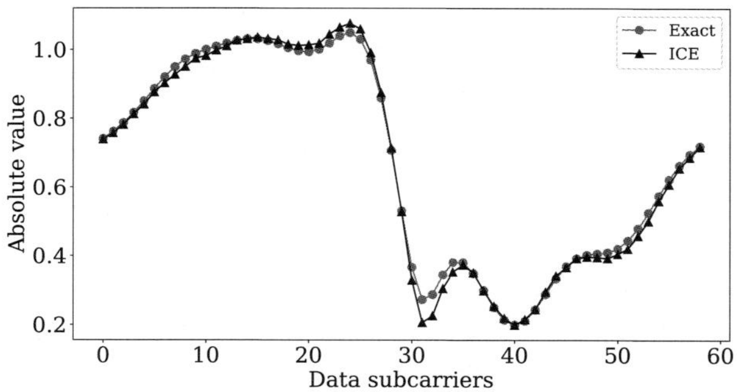

FIGURE 13.19

Estimated channel frequency response using ICE subnet for SNR = 20 dB.

13.4.5 Classification

In the classification problem, we consider two objective functions: the MSE,

$$J(\mathcal{W}) = \frac{1}{Q}\sum_{q=1}^{Q}\sum_{p=1}^{d^{(L)}}\left[\frac{1}{2}(\hat{y}_p(q) - y_p(q))^2\right], \tag{13.25}$$

and the CE (binary),

$$J(\mathcal{W}) = \frac{1}{Q}\sum_{q=1}^{Q}\sum_{p=1}^{d^{(L)}}\left[-y_p(q)\ln(\hat{y}_p(q)) - (1 - y_p(q))\ln(1 - \hat{y}_p(q))\right], \tag{13.26}$$

where the adopted output values are 0, 1. Eq. (13.25) is one of the most widely used in signal processing and NNs, while the relation in (13.26) obtains the best results, according to more recent research, when combined with the softmax activation function in the output layer,

$$\hat{y}_p(q) = y_p^{(L)}(q) = \frac{\exp(x_p^{(L)}(q))}{\sum_{j=1}^{d^{(L)}}\exp(x_p^{(L)}(q))}, \quad \text{for } p = 1, \cdots, d^{(L)}. \tag{13.27}$$

The error value $e(\cdot)$ in Eq. (13.17) is defined for the MSE and CE, respectively, as

$$e(\hat{y}_p, y_p) = (\hat{y}_p - y_p)^2, \tag{13.28}$$

$$e(\hat{y}_p, y_p) = -y_p\ln(\hat{y}_p) - (1 - y_p)\ln(1 - \hat{y}_p), \tag{13.29}$$

where $\hat{y}_p$ is the estimated output for the pth class and y_p is the pth desired value for the output $\mathbf{y}$.

In classification problems, since the last layer has multiple outputs corresponding to multiple classes, it is difficult to infer a distribution for the error signal, as it was viable in the regression problem, Eq. (13.21). In this case, it is not feasible to discuss the error distribution properties as we did in Eq. (13.18).

The complete procedure for feedforward multilayer NN is described in Table 13.5.

13.4.6 A classification example

In this chapter, we utilize some standard examples based on openly available data. Let us consider the MNIST [39] database, which is a large dataset containing digits handwritten by students and employees of the United States Census Bureau. This dataset consists of 60,000 and 10,000 training and test examples, respectively. The input of this set is a 28×28 matrix, where each value represents a pixel of the image. The input signal is normalized to the range 0 to 1. The output dataset has integer values between 0 and 9. Fig. 13.8 in Section 13.3.5 illustrates some entries of this dataset.

This dataset has been previously preprocessed and therefore has no missing values. Since we are modeling an image classification problem, there are no categorical variables, and we have not eliminated any variables before the learning process.

Table 13.5 Feedforward multilayer NN algorithm for classification problems.

Feedforward multilayer NN algorithm for classification
Initialize
$\{(\mathbf{x}(1), \mathbf{y}(1)), (\mathbf{x}(2), \mathbf{y}(2)), \cdots, (\mathbf{x}(Q), \mathbf{y}(Q))\}$, $\mathcal{W} = \{\mathbf{W}^{(1)}, \mathbf{W}^{(2)}, \cdots, \mathbf{W}^{(L)}\}$ (random),
Select: step size $\mu > 0$, number of epoch I, minibatch size b, number of layers L,
number of nodes $(d^{(1)}, \cdots, d^{(L)})$, activation function f, forgetting factor λ_e
Option 1: Objective function J = Mean Squared Error, output function f_L = Linear
or Hyperbolic tangent;
Option 2: Objective function J = Cross-Entropy, output function f_L = Softmax;
Do for $t = 1 : T$
Do for $i = 1 : I$ (for each iteration, randomly select b examples
in training dataset $\rightarrow \mathbf{X}_{(t,i)} = [\bar{\mathbf{x}}(1), \bar{\mathbf{x}}(2), \cdots, \bar{\mathbf{x}}(b)]$, $\mathbf{Y}_{(t,i)} = [\bar{\mathbf{y}}(1), \bar{\mathbf{y}}(2), \cdots, \bar{\mathbf{y}}(b)]$)
[Forward Propagation]:
$\mathbf{Y}^0 = [\bar{\mathbf{y}}^0(1), \bar{\mathbf{y}}^0(2), \cdots, \bar{\mathbf{y}}^0(b)] = [ones(1, b); \mathbf{X}_{(t,i)}]$ ($ones(1, b)$ are the bias term)
Do for $l = 1 : L - 1$
$\mathbf{X}^{(l)} = (\mathbf{W}^{(l)})^T \mathbf{Y}^{(l-1)}$
$\mathbf{Y}^{(l)} = [ones(1, b); f(\mathbf{X}^{(l)})]$
end
$\mathbf{X}^{(L)} = (\mathbf{W}^{(L)})^T \mathbf{Y}^{(L-1)}$
$\hat{\mathbf{Y}}_{(t,i)} = [\hat{\mathbf{y}}(1), \hat{\mathbf{y}}(2), \cdots, \hat{\mathbf{y}}(b)] = \mathbf{Y}^{(L)} = g(\mathbf{X}^{(L)})$
[Backpropagation]:
$\boldsymbol{\Delta}_C(L)L = [\boldsymbol{\delta}^{(L)}(k_1), \boldsymbol{\delta}^{(L)}(k_2), \cdots, \boldsymbol{\delta}^{(L)}(t_{\text{bin}})] = \begin{cases} g'(\mathbf{X}_C^L) \otimes (\hat{\mathbf{Y}}_C - \mathbf{Y}_C), & \text{if } \textit{option 1} \text{ is chosen} \\ (\hat{\mathbf{Y}}_C - \mathbf{Y}_C), & \text{if } \textit{option 2} \text{ is chosen} \end{cases}$
Do for $l = L - 1 : -1 : 1$
$\boldsymbol{\Delta}_C^l = f'(\mathbf{X}_C^l) \otimes [\mathbf{W}^{(l+1)} \boldsymbol{\Delta}_C^{l+1}]_1^{d^{(l)}}$
end
[Updating the weights]:
Do for $l = 1 : L$
$\mathbf{W}^{(l)} = \mathbf{W}^{(l)} - \frac{\mu}{b} \mathbf{Y}_C^{(l-1)} (\boldsymbol{\Delta}_C^{(l)})^T$
end
end
$J_{\text{train}}(\mathcal{W}) = \frac{1}{Q} \sum_{q=1}^{Q} \sum_{p=1}^{d^{(L)}} J_q^p(\mathcal{W})$ (and J_{val} if this set is previously defined)
end

When the cost function is the CE error and the softmax function is the output activation function, the step size parameter is $\mu = 0.1$. For MSE in the objective function and linear function as output activation, the step size is equal to 0.01. The other parameters are $b = 256$ and $T = 100$ in both cases. In addition, the number of nodes in each hidden layer is equal to 1024.

Figs. 13.20(a) and 13.20(b) show the results with CE as the objective function and softmax as the output activation function and results with MSE as the objective function and a linear function as the output activation function, respectively. Also, when compared with simulation $b = 32$, the result is improved in comparison with the case $b = 256$.

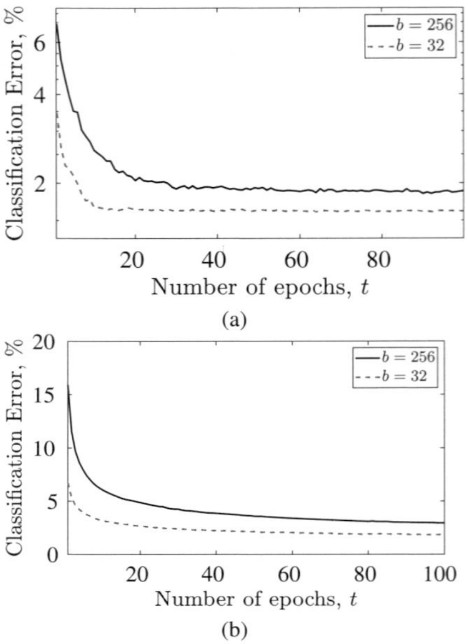

FIGURE 13.20

MNIST handwritten digit recognition simulation: Test classification error (%) when (a) the output activation function is softmax and the objective function is the CE error and (b) the output activation function is a linear function and the objective function is MSE.

13.4.7 EMNIST letter dataset

The EMNIST letter dataset considered now is provided by the National Institute of Standards and Technology (NIST) [55]. This dataset can represent the 26 letters of the alphabet at the output. The letters EMNIST contain 120,000 training examples and 25,000 test examples and the data have been normalized to values between 0 and 1. The input is a 28×28 matrix, where each input represents a pixel in the image. It has no categorical variables, so it does not need any transformation in the variables. This dataset had already been preprocessed [6]. Fig. 13.21 depicts some entries of this dataset.

The choice of the objective function and the output activation function affects the choice μ; thus, if we choose MSE as the cost function and a linear function as the output function, the parameters in this NN will be $\mu = 0.01$, $b = 256$, and $T = 100$. On the other hand, if the objective function is CE and the output function is softmax, the parameters will be $\mu = 0.1$, $b = 256$, and $T = 100$. Moreover, the number of nodes in the hidden layers in this simulation is 1024.

The results regarding the test classification error are shown in Figs. 13.22(a) and 13.22(b) for CE and MSE, respectively. The performance of simulation by setting $b = 32$ in CE as objective function and $b = 128$ in MSE as objective function shows a slight difference in the accuracy of results.

In Table 13.6, the averaged MSE over the last 10 epochs is presented for the classification problem. We can observe that CE as an objective function combined with softmax as output activation function

FIGURE 13.21

Sample images from the EMNIST letter dataset.

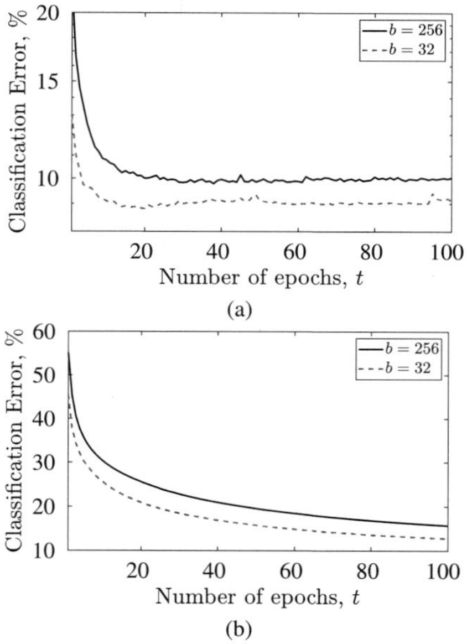

FIGURE 13.22

EMNIST letter simulation: Test classification error (%) when (a) the output activation function is softmax and the objective function is CE error and (b) the output activation function is a linear function and the objective function is MSE.

achieves better results. In Table 13.7, we conclude that the minibatch requires equivalent computational cost.

Table 13.6 Comparison of test errors between classification problems (blue [dark gray in print version] is the best and red [mid gray in print version] is the worst for each problem).

	MNIST (CE)	MNIST (MSE)	EMNIST (CE)	EMNIST (MSE)
$b = 256$	1.87±0.023	2.97±0.037	9.99±0.069	15.95±0.163
$b = 32$	1.60±0.009	1.84±0.031	9.13±0.142	12.96±0.118

Table 13.7 Approximated number of flops in one epoch.

	MNIST	EMNIST
$b = 256$	566×10^9	1145×10^9
$b = 32$	566×10^9	1145×10^9

13.4.8 Dropout

Dropout is a stochastic technique to avoid overfitting during the training, providing a way of combining exponentially many different NN architectures. Dropout is a computationally efficient regularization procedure and can be applied to a wide range of networks. Dropout consists of dropping out nodes in the network, including all its connections, as shown in Fig. 13.23. In many cases, a node is retained with probability p, according to a Bernoulli distribution, independent of other nodes; that is, its value is temporarily set to zero in each iteration. This procedure is applied at each layer. In addition, the derivatives of the loss function are backpropagated through the network.

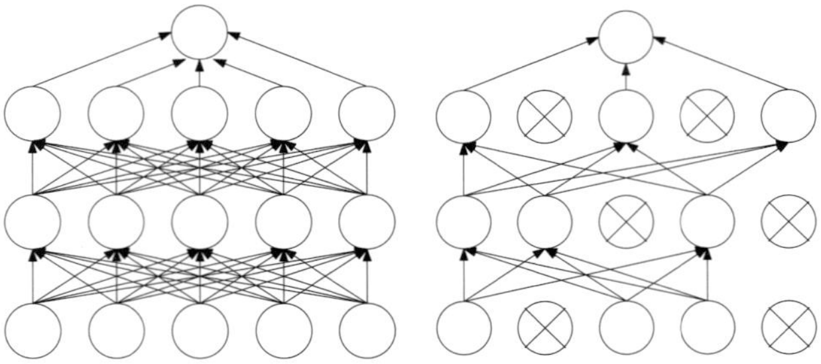

FIGURE 13.23

(a) A neural network with two hidden layers. (b) Dropout applied in the network producing a thinned net.

At the testing procedure, it is unfeasible to average all combinations of many thinned models. A solution is to employ an NN without dropout and utilizing a reduced version of the trained weights. If a node is dropped with probability p during training, the weights are multiplied by p at the test, as depicted in Fig. 13.24.

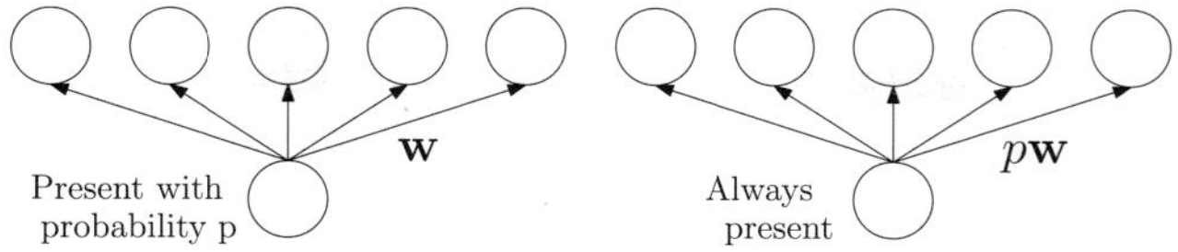

FIGURE 13.24

(a) The training node present in the process with probability p. (b) The test phase with node always present and weight multiplied by p.

13.4.9 Number of flops

A flop is a short description for a floating-point operation [23]. There are several difficulties associated with calculating a precise count of floating points operations. First, addition and subtraction are always counted as a flop. Second, multiplication is usually counted as a flop. Finally, divisions and exponents are more complicated because we must consider each one as some flops. The values are defined as four flops for divisions and eight flops for exponents. In this chapter, each operation is considered a flop, including divisions and exponents. We express the number of flops in two steps for forward and backpropagation algorithms. For instance, an NN with four layers contains two hidden layers and has i nodes at the input layer, j nodes at the second layer, k nodes at the third layer, and l nodes at the output layer. The data-related parameters are denoted as T (epoch), b (minibatch size), and I (iteration).

At each iteration for a certain epoch, we have b training examples. In the forward propagation, the following operations are performed from one layer to the next layer:

$$\mathbf{X}^{(l)} = \mathbf{W}^{(l)^T}\mathbf{Y}^{(l-1)}, \tag{13.30}$$

$$\mathbf{Y}^{(l)} = [ones(1, b);\ f(\mathbf{X}^{(l)})]. \tag{13.31}$$

From the first layer to the second layer, the number of flops for a first operation, i.e., the product between two matrices $\mathbf{W}^{(1)} \in \mathbb{R}^{i\times j}$ and $\mathbf{Y}^0 \in \mathbb{R}^{i\times b}$, entails jb inner products between vectors of size i. This inner product involves $i-1$ additions and i multiplications. Hence, the resulting number of flops is $(2i-1)jb$. The activation function ReLU applied in the second equation has 0 flops. Then, we have $(2i-1)jb$ flops.

Using the same rationale from the second layer to the third layer, we have $(2j-1)kb$ flops. For the propagation from the third layer to the last layer, the matrix multiplication requires $(2k-1)lb$ flops. In the last part, we obtain the estimated value in NN. In the regression problem, the output activation function is linear, resulting in 0 flops. In the classification problem, the softmax function has $l-1$ additions, 1 division, and $l+1$ exponents for b training examples, resulting in a total of $(2l+1)b$.

Therefore, the total number of flops for feedforward propagation depends on the problem. For the regression problem it is $(2ij + 2jk + 2kl - j - k - l)b$ flops, whereas for the classification problem it is $(2ij + 2jk + 2kl + l + 1)b$ flops.

Starting from the last layer to the third layer, we compute the sensitivity vector in the fourth layer. To obtain $\mathbf{\Delta}^{(3)} = \hat{\mathbf{Y}}_{(t,i)} - \mathbf{Y}^{(L)}$, we require $l(b)$ flops. In the remaining layers, we compute the sensitive weights from the previously obtained vectors. For example, from the third layer to the second layer, we have $\mathbf{\Delta}^{(2)} = f'(\mathbf{X}^{(2)}) \otimes [\mathbf{W}^{(3)}\mathbf{\Delta}^{(3)}]_1^{\varrho^2}$. The number of flops in the derivative of the activation function is zero. The multiplication matrix $\mathbf{W}^{(3)}\boldsymbol{\Delta}^{(3)}$ requires $k(b)$ inner products involving $l-1$ additions and l multiplications. The elementwise operation has $k(b)$ flops. Then, the resulting number of flops is $2lk(b)$ in this propagation. Finally, we have the same procedure from the second to the first layer, and the number of flops required is $2kj(b)$.

The last step is the weight update; for example, from the fourth to the third layer, we have the update $\mathbf{W}^{(3)} = \mathbf{W}^{(3)} - \frac{\mu}{b}\mathbf{Y}^{(2)}(\boldsymbol{\Delta}^{(2)})^T$. The multiplication matrix results in a total of $(2(b)-1)kl$ flops. On top of that, there are sums of the matrices, with a number of flops equal to kl. By accumulating the number of multiplications and additions, we end up with $2(b)kl$ flops. We employ the same rationale from the third to the second layer, and we obtain $2(b)jk$, and from the second to the first layer, and we obtain $2(b)ij$. Then, the total number of flops in backpropagation is

$$\begin{aligned}&2(b)ij + 2(b)jk + 2(b)kl + 2kj(b) + 2kl(b) + l(b) = \\ &(b)(2ij + 2jk + 2kl + 2kj + 2kl + l).\end{aligned} \tag{13.32}$$

13.4.10 Convolutional neural networks

A convolutional NN (CNN) is a deep learning (DL) algorithm especially suitable for approximating an unknown mapping function when the input is a 2D signal, as an image for example. As discussed in Section 13.4.2, fully connected neural networks (FCNNs) can be used to solve tasks with images as input. Nevertheless, the image dimension and its complexity play a crucial role in how computationally intensive the task can become.

CNNs are composed of a sequence of convolutional, pooling, and fully connected layers, designed to transform the image and extract low-dimensional features, which are then easier to process. Fig. 13.25 illustrates an input image flowing through the CNN, generating feature maps which are further processed until the last layer is reached, where the image is finally predicted as a dog.

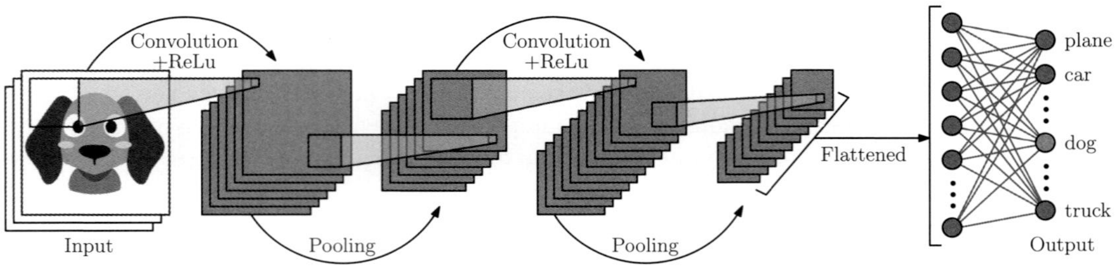

FIGURE 13.25

A CNN example.

In the convolutional layer, a set of filters with learnable weights are convolved across the width and height of the input, as shown in Fig. 13.26(a), and the nonlinear activation function ReLu is applied. Different feature maps are obtained from each filter, and they are stacked to form the input of the next layer. To further reduce the spatial size of the feature map, a downsampling is performed by the pooling layer, as the max pooling operation illustrated in Fig. 13.26(b). In this way, the amount of network parameters is also reduced and overfitting can be prevented.

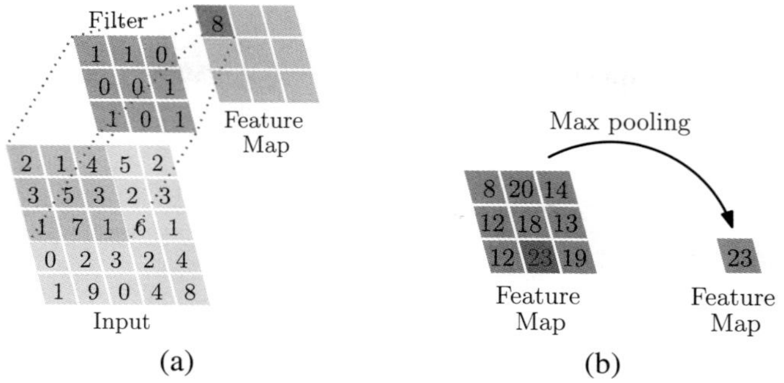

FIGURE 13.26

Illustration of (a) convolutional and (b) max pooling layers.

The final set of feature maps is flattened to form the input of a fully connected layer, and they are processed as in the FCNN case. The predicted class is the one with greatest probability after the softmax output function.

Among the popular datasets in computer vision, we can mention the CIFAR10 dataset, which is composed of 60,000 32×32 color images labeled in 10 classes. Fig. 13.27 depicts 10 samples of each class from the CIFAR10 dataset. As an example, we perform a classification task using CIFAR10 as the dataset and Resnet18 (cite) as the model. The classification error for the test dataset is shown in Fig. 13.28 at each epoch.

13.4.11 Recurrent neural networks

Again let us denote the dataset $\{(\mathbf{x}(1), y(1)), (\mathbf{x}(2), y(2)), \cdots, (\mathbf{x}(K), y(K))\}$ such that $\mathbf{x}(k) \in \mathbb{R}^{n\times 1}$. In the present discussion, our aim is to map a sequential[1] data.

13.4.11.1 Standard RNN

Fig. 13.29(a) shows the compact representation of the RNN network. The unfolded representation of the RNN, shown in Fig. 13.29(b), is also known as vanilla RNN.

The RNN entails generating a hidden state $\mathbf{h}(k)$ based on the input vector $\mathbf{x}(k)$, the previous $\mathbf{h}(k-1)$, and a bias vector $\mathbf{b}_h$ as follows:

$$\begin{aligned}\mathbf{h}(k) &= f\left(\mathbf{h}(k-1), \mathbf{x}(k), \mathbf{b}_h\right)\\ &= \tanh\left(\mathbf{W}_f\mathbf{h}(k-1) + \mathbf{W}_i\mathbf{x}(k) + \mathbf{b}_h\right),\end{aligned} \tag{13.33}$$

where $\mathbf{x}(k)$ is the input of the node generation $\mathbf{h}(k)$, $\mathbf{W}_f$ is the feedback matrix, $\mathbf{W}_i$ is the input matrix, and $\mathbf{b}_h$ is the hidden-state bias vector. The output of the RNN is

[1] In some cases, it consists of streaming data.

FIGURE 13.27

Sample images from the CIFAR10 dataset.

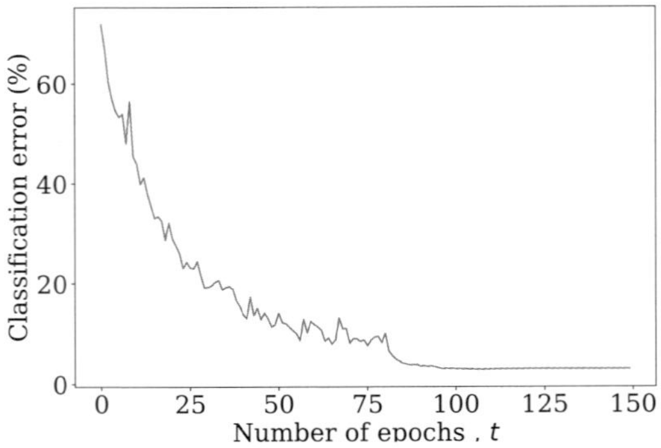

FIGURE 13.28

Classification error of the Resnet18 model for the CIFAR10 dataset.

$$\begin{aligned}\hat{\mathbf{y}}(k) &= g\left(\mathbf{W}_o\mathbf{h}(k) + \mathbf{W}_i\mathbf{x}(k) + \mathbf{b}_o\right)\\ &= \text{softmax}\left(\mathbf{W}_o\mathbf{h}(k) + \mathbf{W}_i\mathbf{x}(k) + \mathbf{b}_o\right)\\ &= \sigma\left(\mathbf{W}_o\mathbf{h}(k) + \mathbf{W}_i\mathbf{x}(k) + \mathbf{b}_o\right).\end{aligned} \tag{13.34}$$

The functions $f(\cdot)$ and $g(\cdot)$ are applied individually to each entry of the argument vector.

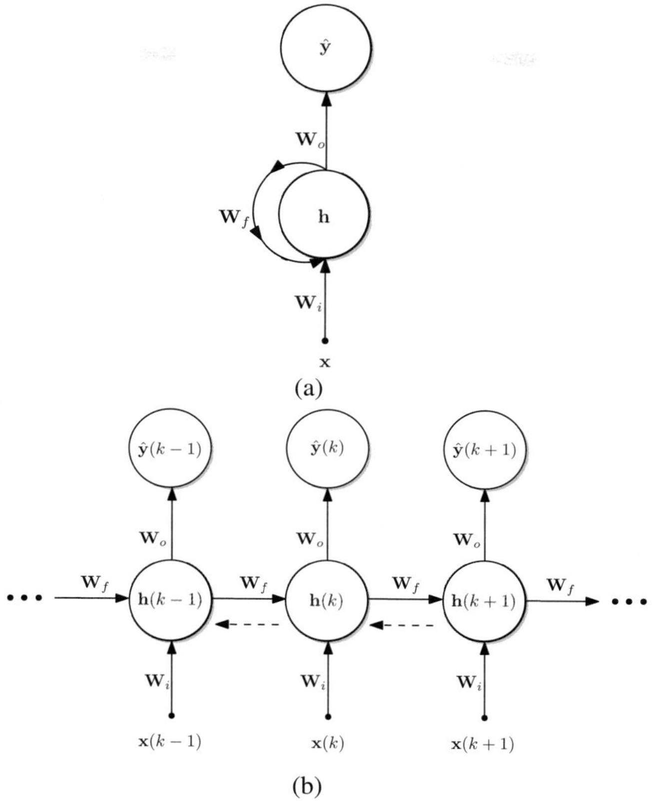

FIGURE 13.29

(a) RNN configuration. (b) Its unfolded representation. The parameters are initialized randomly.

The generic cost function is

$$\bar{J}(\bar{\mathbf{y}}, \mathbf{y}) = \sum_{k=1}^{K} \left[J(\hat{\mathbf{y}}(k), \mathbf{y}(k)) \right], \tag{13.35}$$

where $\mathbf{y}$ is the label vector and $\hat{\mathbf{y}}$ is the RNN output vector generated through the state vector. In the following text, we will denote $J(\hat{\mathbf{y}}(k), \mathbf{y}(k))$ as $J(k)$, for simplification.

In the RNN structure, the state vector $\mathbf{h}(k)$ connects to the cost function directly at instant k, and in the following instants $k+1$ up to instant K. With the chain rule, we can compute the partial derivative with respect to the state vector as

$$\frac{\partial \bar{J}(\hat{\mathbf{y}}, \mathbf{y})}{\partial \mathbf{h}(k)} = \left[\frac{\partial \mathbf{h}(k+1)}{\partial \mathbf{h}(k)} \right]^T \frac{\partial \bar{J}(\hat{\mathbf{y}}, \mathbf{y})}{\partial \mathbf{h}(k+1)} + \left[\frac{\partial \hat{\mathbf{y}}(k)}{\partial \mathbf{h}(k)} \right]^T \frac{\partial \bar{J}(\hat{\mathbf{y}}, \mathbf{y})}{\partial \hat{\mathbf{y}}(k)}, \tag{13.36}$$

where the dependency between $\mathbf{h}(k+1)$ and $\mathbf{h}(k)$ is taken into account. Eq. (13.36) entails a recursive expression that propagates backwards; see [62]. Further discussions on gradients are presented in the Appendix, Section 13.11.2.

In the forward step, the computational flow related to the state vector can be represented as

$$\left(\hat{\mathbf{y}}(1);\mathbf{h}(1)\right) \to \cdots \to \left(\hat{\mathbf{y}}(K-1);\mathbf{h}(K-1)\right) \to \left(\hat{\mathbf{y}}(K);\mathbf{h}(K)\right). \tag{13.37}$$

Similarly, in the backward step, the computational flow related to the state vector can be represented as

$$\frac{\partial \bar{J}(\hat{\mathbf{y}},\mathbf{y})}{\partial \mathbf{h}(1)} \leftarrow \cdots \leftarrow \frac{\partial \bar{J}(\hat{\mathbf{y}},\mathbf{y})}{\partial \mathbf{h}(K-1)} \leftarrow \frac{\partial \bar{J}(\hat{\mathbf{y}},\mathbf{y})}{\partial \mathbf{h}(K)}. \tag{13.38}$$

Note that the RNN is a kind of feedforward network with K layers. The unique feature of the RNN is that the inputs are applied to all layers, and the layers themselves produce outputs. As such, the gradients require a distinct backpropagation strategy.

13.4.11.2 Activation functions

One of the most widely used activation functions in RNN networks is the hyperbolic tangent, defined as

$$\tanh(z) = \frac{\exp^{2z} - 1}{\exp^{2z} + 1}. \tag{13.39}$$

Its partial derivative is

$$\begin{aligned}
\frac{\partial \tanh(z)}{\partial z} &= \frac{2\exp^{2z}}{\exp^{2z}+1} - 2\exp^{2z}\frac{\exp^{2z}-1}{(\exp^{2z}+1)^2} \\
&= \frac{2\exp^{4z}+2\exp^{2z}-2\exp^{4z}+2\exp^{2z}}{(\exp^{2z}+1)^2} = \frac{(\exp^{2z}+1)^2-(\exp^{2z}-1)^2}{(\exp^{2z}+1)^2} \\
&= 1-\tanh^2(z).
\end{aligned} \tag{13.40}$$

Another activation function widely used in RNN networks is the softmax function, redefined here for convenience:

$$\text{softmax}_i(\mathbf{z}) = \sigma_i(\mathbf{z}) = \frac{\exp^{z_i}}{\sum_{l=1}^{L}\exp^{z_l}}. \tag{13.41}$$

Its corresponding partial derivative is

$$\begin{aligned}
\frac{\partial \text{softmax}_i(\mathbf{z})}{\partial z_j} &= \text{softmax}_i(\mathbf{z})\left[\delta_{ij} - \text{softmax}_j(\mathbf{z})\right] \\
&= \sigma_i(\mathbf{z})\left[\delta_{ij} - \sigma_j(\mathbf{z})\right],
\end{aligned} \tag{13.42}$$

where δ_{ij} is the Kronecker delta, which means $\delta_{ii} = 1$ and $\delta_{ij} = 0$ for $i \neq j$. The proof follows from the definition in Eq. (13.41).

13.4.11.3 **Long short-term memory**

The long short-term memory (LSTM) NN replaces the standard RNN in the hidden layers in order to cope with the numerical issues inherent to the latter caused by vanishing or exploding values of recursive gradient computations. Unlike the standard RNN whose internal hidden layers have two inputs, the LSTM cells have three inputs, namely, the cell of the previous layer, the vector originating from the input layer, and the hidden state of the previous layer; see for instance [5] and [28]. The LSTM structure includes a forget gate, an input gate, an output gate, and a memory cell, denoted as $\mathbf{f}(k)$, $\mathbf{i}(k)$, $\mathbf{o}(k)$, and $\mathbf{c}(k)$, respectively. The gates are logical units that control the error in selected quantities to halt possible gradient anomalies. The input forget gate decides if the memory is set to zero; the input gate determines if the memory cell is updated; and the output gate decides the visibility of the current state information. These gates use either a softmax or sigmoid activation, and the latter case is employed here due to its differentiability. Indeed, the forget gates avert the extreme increase or decrease of the backpropagated errors. Usually LSTM-based networks include long memory to preserve valuable events required to provide meaningful outcomes. Fig. 13.30 details the main components of an LSTM cell.

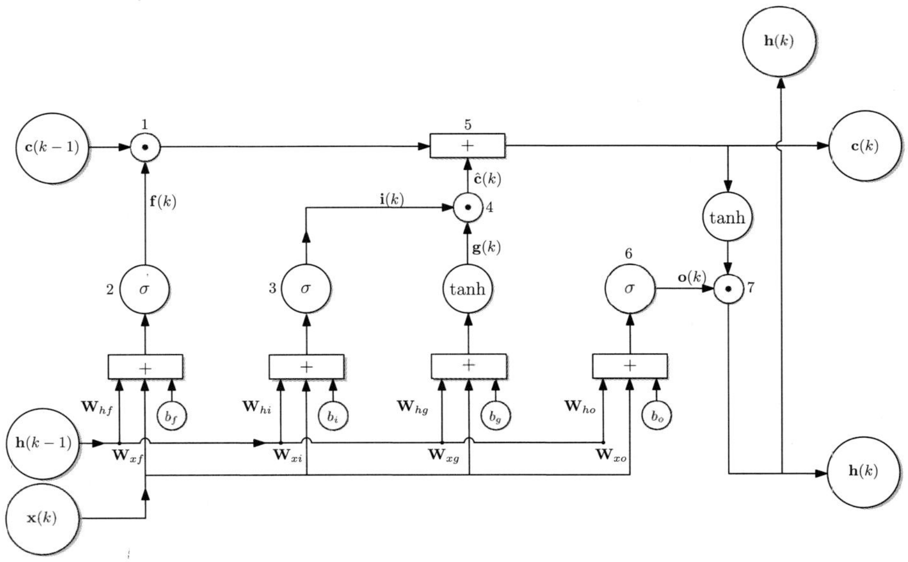

FIGURE 13.30

LSTM unit representation. The numbered nodes represent the following tasks: (1) forget part of the cell contents; (2) forget gate computation; (3) input gate computation; (4) compute cell content; (5) add new content to the cell; (6) compute output gate; and (7) include to the hidden state some cell information.

As can be observed, the LSTM scheme consists of some gates and an internal cell denoted as $\mathbf{c}(k)$, defined as follows. In the LSTM solution, the hidden state $\mathbf{h}(k)$ is generated from the input vector $\mathbf{x}(k)$, the previous $\mathbf{h}(k-1)$, and several internal operations including the internal cell. The following equations describe the mathematical operations:

$$\mathbf{f}(k) = \sigma\left(\mathbf{W}_{hf}\mathbf{h}(k-1) + \mathbf{W}_{xf}\mathbf{x}(k) + \mathbf{b}_f\right), \tag{13.43}$$

$$\mathbf{i}(k) = \sigma\left(\mathbf{W}_{hi}\mathbf{h}(k-1) + \mathbf{W}_{xi}\mathbf{x}(k) + \mathbf{b}_i\right), \tag{13.44}$$

$$\begin{aligned}\mathbf{g}(k) &= \tanh\left(\mathbf{h}(k-1), \mathbf{x}(k), \mathbf{b}_g\right)\\ &= \tanh\left(\mathbf{W}_{hg}\mathbf{h}(k-1) + \mathbf{W}_{xg}\mathbf{x}(k) + \mathbf{b}_g\right),\end{aligned} \tag{13.45}$$

$$\mathbf{o}(k) = \sigma\left(\mathbf{W}_{ho}\mathbf{h}(k-1) + \mathbf{W}_{xo}\mathbf{x}(k) + \mathbf{b}_o\right), \tag{13.46}$$

where $\mathbf{x}(k)$ is the input of the node generation $\mathbf{h}(k)$. The vectors $\mathbf{f}(k)$, $\mathbf{i}(k)$, and $\mathbf{o}(k)$ represent the forget, input, and output gates, respectively. The symbol $\mathbf{g}(k)$ represents an internal signal.

The softmax function, represented by $\sigma(\cdot)$, is suitable to provide a gating effect to the quantities involved, imposed by its gradient. Here $\tanh(\cdot)$ plays the normal role of activation function, distributing the gradient values to prevent exploding or vanishing values.

The internal information of the LSTM block is

$$\mathbf{c}(k) = \mathbf{f}(k) \odot \mathbf{c}(k-1) + \mathbf{i}(k) \odot \mathbf{g}(k), \tag{13.47}$$

$$\mathbf{h}(k) = \mathbf{o}(k) \odot \tanh\left(\mathbf{c}(k)\right). \tag{13.48}$$

The output of the LSTM layers might be given by

$$\hat{\mathbf{y}}(k) = \sigma\left(\mathbf{W}_{hy}\mathbf{h}(k) + \mathbf{b}_y\right). \tag{13.49}$$

This output is not shown in Fig. 13.30 for simplicity.

The gradient derivations for the LSTM configuration are presented in Section 13.11.2 as an illustration and for completeness since they are rarely found in the literature.

The feedforward process is as follows:

$$\mathbf{x}(1), \mathbf{h}(1), \mathbf{c}(1) \xrightarrow{\mathcal{W},\boldsymbol{\beta}} \mathbf{x}(2), \mathbf{h}(2), \mathbf{c}(2) \xrightarrow{\mathcal{W},\boldsymbol{\beta}} \cdots \mathbf{x}(K), \mathbf{h}(K), \mathbf{c}(K) \xrightarrow{\mathcal{W},\boldsymbol{\beta}} \tag{13.50}$$

where the symbols $\mathcal{W}$ and $\boldsymbol{\beta}$ congregate the coefficients and the bias terms of the LSTM cells, respectively. Note that for all cells the coefficients and bias terms are the same; otherwise, the LSTM updates become too complex.

The objective function should be reduced by changing the weights $\mathcal{W}$ and $\boldsymbol{\beta}$. By using the gradient descent algorithm, the weights are iteratively updated following the negative gradient direction,

$$\mathcal{W}(k+1) = \mathcal{W}(k) - \mu \frac{\partial \bar{J}(k)}{\partial \mathcal{W}}, \tag{13.51}$$

$$\boldsymbol{\beta}(k+1) = \boldsymbol{\beta}(k) - \mu \frac{\partial \bar{J}(k)}{\partial \boldsymbol{\beta}}, \tag{13.52}$$

where μ is the step size. As the standard, the coefficient update requires a set of partial derivatives.

The backpropagation entails the calculation of the derivatives, starting from the error signal at the network output. The following chain shows how the procedure works:

$$\delta(K) \xrightarrow{\mathcal{W},\boldsymbol{\beta}} \frac{\partial \bar{J}(K)}{\partial \mathcal{W}}, \frac{\partial \bar{J}(K)}{\partial \boldsymbol{\beta}} \xrightarrow{\mathcal{W},\boldsymbol{\beta}} \frac{\partial \bar{J}(K-1)}{\partial \mathcal{W}}, \frac{\partial \bar{J}(K-1)}{\partial \boldsymbol{\beta}} \xrightarrow{\mathcal{W},\boldsymbol{\beta}} \frac{\partial \bar{J}(1)}{\partial \mathcal{W}}, \frac{\partial \bar{J}(1)}{\partial \boldsymbol{\beta}}. \tag{13.53}$$

Table 13.8 Multilayer RNN algorithm using LSTM.
Feedforward multilayer RNN algorithm
Initialize
$\{(\mathbf{x}(1), \mathbf{y}(1)), (\mathbf{x}(2), \mathbf{y}(2)), \cdots, (\mathbf{x}(K), \mathbf{y}(K))\}$,
$\mathcal{W} = \{\mathbf{W}_{hf}, \mathbf{W}_{xf}, \mathbf{W}_{hi}, \mathbf{W}_{xi}, \mathbf{W}_{hg}, \mathbf{W}_{xg}, \mathbf{W}_{ho}, \mathbf{W}_{xo}\}$ (random),
$\boldsymbol{\beta} = \{\mathbf{b}_f, \mathbf{b}_i, \mathbf{b}_g, \mathbf{b}_o\}$ (random),
Select: step size $\mu > 0$
Objective function $\bar{J}$ = Mean Squared Error
[Forward Propagation]:
Compute the forward path as per the procedure in (13.50)
[Backpropagation]:
Compute the partial derivatives according to the procedure described in (13.53)
[Updating the weights]:
Update the coefficients according to Eqs. (13.51) and (13.52)
end

The RNN algorithm utilizing the LSTM strategy is outlined in Table 13.8. Note that the table description is compact for simplicity.

A popular and more compact alternative to the LSTM is the gate recurrent unit (GRU). The GRU contains only two gates, the update and reset gates; see [14] for further discussions.

13.4.11.4 Types of RNNs

There are several configurations for the RNN structure that are application-specific. The simplest standard form of RNN is the one-to-one configuration depicted in Fig. 13.31.

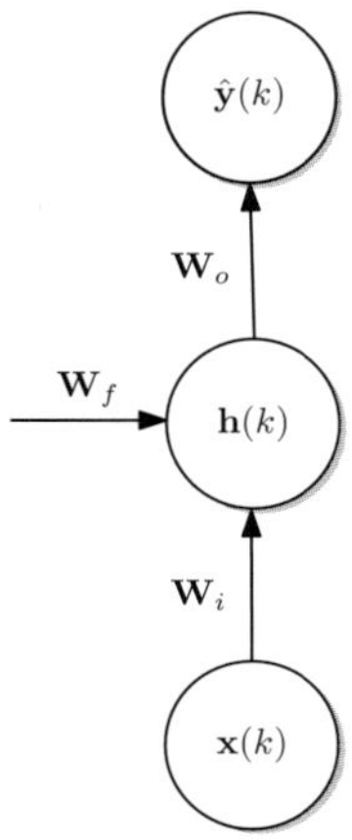

FIGURE 13.31

One-to-one structure.

The many-to-one configuration, shown in Fig. 13.32, corresponds to a sequence of inputs to one output, where for instance a sentence is analyzed by the network to classify the inherent sentiment. The sentiment classification can be positive or negative. The many-to-one configuration is also known as sequence-to-vector model.

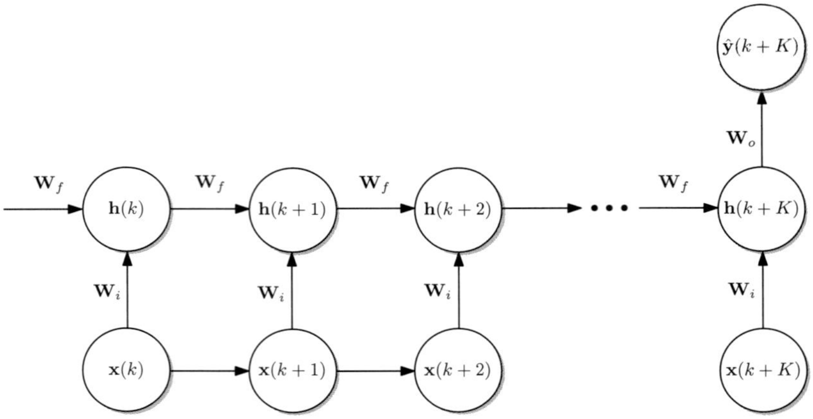

FIGURE 13.32

Many-to-one structure.

Fig. 13.33 illustrates the one-to-many configuration, which finds application in image captioning producing a series of words forming a sentence. This configuration, slightly modified by concatenating the outputs with the hidden states, is used in music generation. The one-to-many configuration is also known as vector-to-sequence model.

The first many-to-many configuration of Fig. 13.34, also known as sequence-to-sequence model, is employed in name entity recognition, word prediction, sentence generation, and video frame labeling and classification. This structure has equally sized inputs and outputs.

The second many-to-many configuration, depicted in Fig. 13.35, finds application in machine translation applications where one can feed the network with a sentence in one language and the application outputs the sentence in the target language. This configuration also applies to the automatic summarizing of sentences. This structure has unequally sized inputs and outputs. In some cases, we can configure this model to perform and encoder–decoder task where the input vectors are mapped into an internal representation, usually not amenable to human interpretation, which in turn is mapped into an output sequence.

13.4.11.5 ***Using RNN***

The basic RNN and any other forms of RNN structures, such as LSTM, can be part of many ML configurations and applications. Let us consider the scheme illustrated in Fig. 13.36 where the input data consisting, for instance, of K images are applied to a feature extraction block. Then, the set of features is applied to an LSTM classifier, which, in turn, will produce the decisions related to classes of the input images.

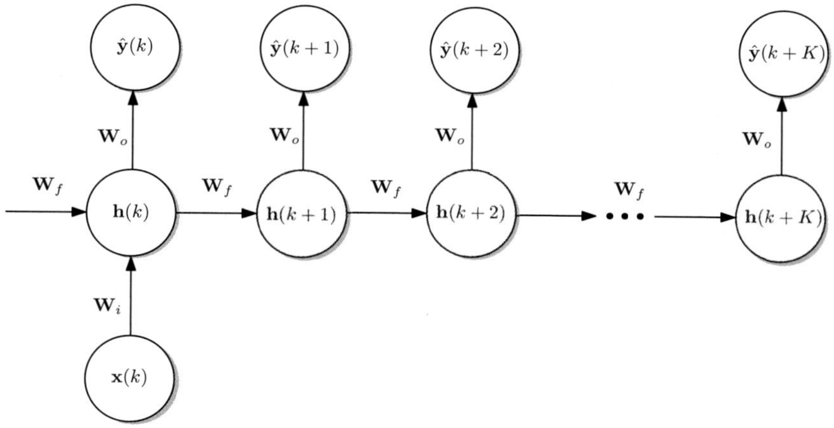

FIGURE 13.33

One-to-many structure.

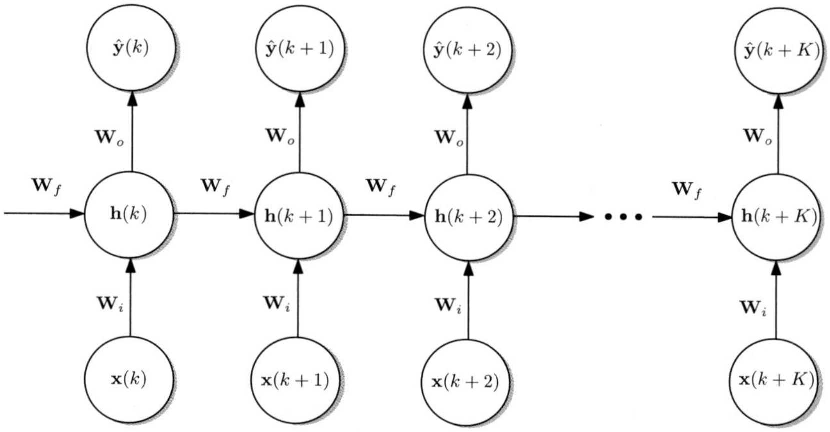

FIGURE 13.34

Many-to-many structure.

13.4.11.6 An example with RNN using LSTM

The RNN employing LSTM has found applications in many fields, where each one relies on the choice of the configuration and the amount of data available. For long time series, the solution falls in the DNN class, although what is considered DNN is debatable since some classical CNNs, such as the so-called LeNet, had six layers. NNs with so many layers are considered deep by some definitions; however, here we call an NN with more than six layers a DNN.

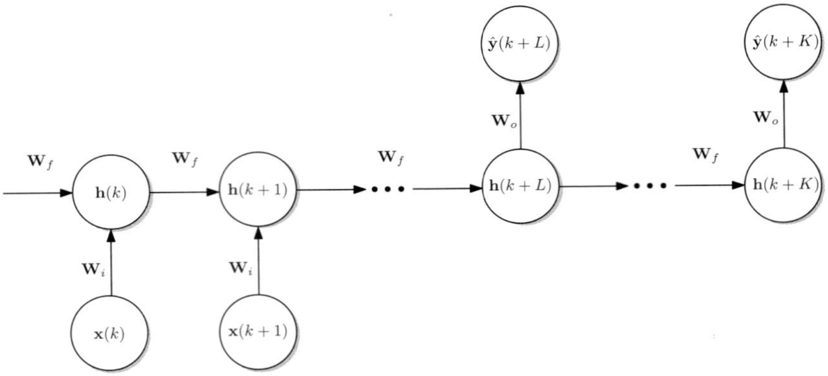

FIGURE 13.35

Many-to-many structure.

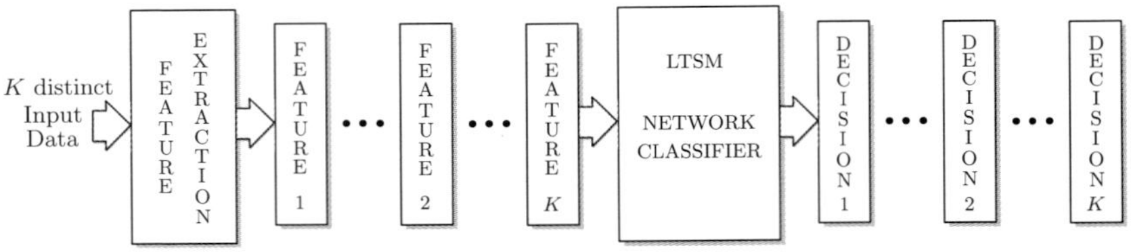

FIGURE 13.36

A possible classification scheme using RNN.

To illustrate the use of an LSTM configuration, we utilize a prediction example inspired by the following site:

https://machinelearningmastery.com/time-series-prediction-lstm-recurrent-neural-networks-python-keras/.

In this example we build an LSTM model to predict Amazon's future stock prices. We use one previous time step as input variable to predict the next observation. Fig. 13.37 shows the original data in blue (dark gray in print version), the predictions during training in orange, and the predictions on the unseen test dataset in green (light gray in print version). For this illustrative example, the model was able to fit both the training and test datasets quite well.

13.4.12 Support vector machines

Support vector machines (SVMs) are ML tools conceived to perform novelty detection, prediction, and classification. The SVMs belong to the class of supervised learning ML algorithms, and their solutions rely on convex optimization, generating global optimum solutions. Unlike many ML methods based on a probabilistic model, the SVM utilizes geometric reasoning. In SVM, we define a cost function that is minimized during the training process. In some ways, the SVM approach provides some geometric

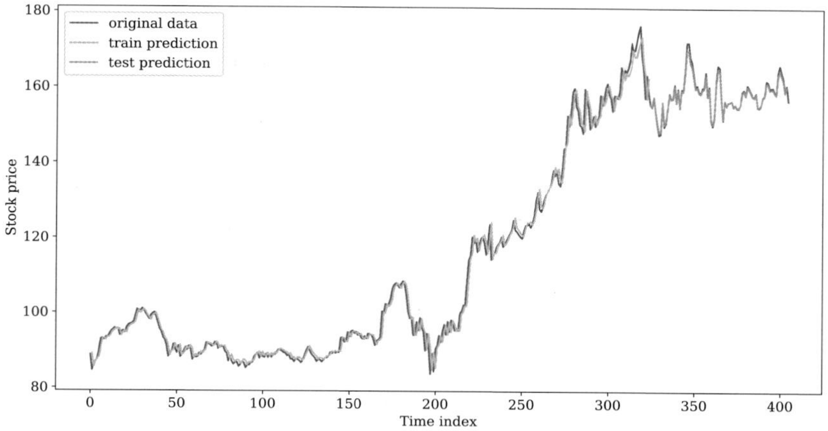

FIGURE 13.37

LSTM trained on a regression formulation of a stock price prediction problem.

interpretation to supervised learning as quotes [10], and it is basically meant to implement a two-class classifier. The presentation starts with the classification task of separating datapoints into two distinct classes.

Consider a classification problem where we separate two classes of datapoints that live in the $\mathbb{R}^n$-dimensional space using a hyperplane. For example, for $n = 2$, the hyperplane is a line, and we can choose three different ones, for example, to separate the datapoints in Fig. 13.38. The observed hyperplanes in Fig. 13.38 can differentiate the two classes. However, changing the positions of the points that are closer to the hyperplanes might change the classification. Thus, the greater the variation needed to misclassify these datapoints, the better is the hyperplane.

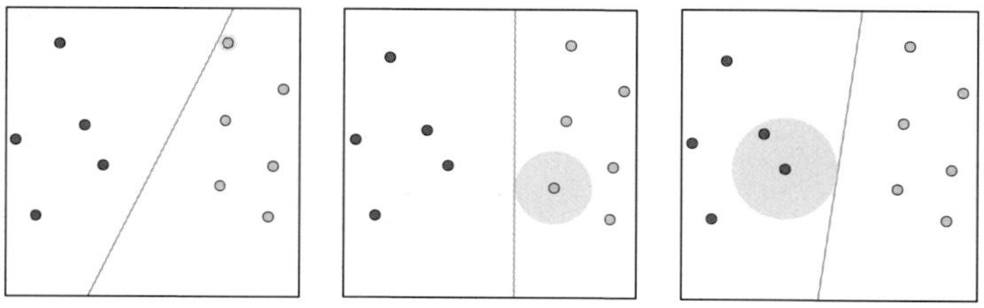

FIGURE 13.38

Better linear separation.

The optimal hyperplane is the one that separates the data with maximum margin, as illustrated in Fig. 13.39. The position and orientation of the optimal hyperplane are determined by the support vectors.

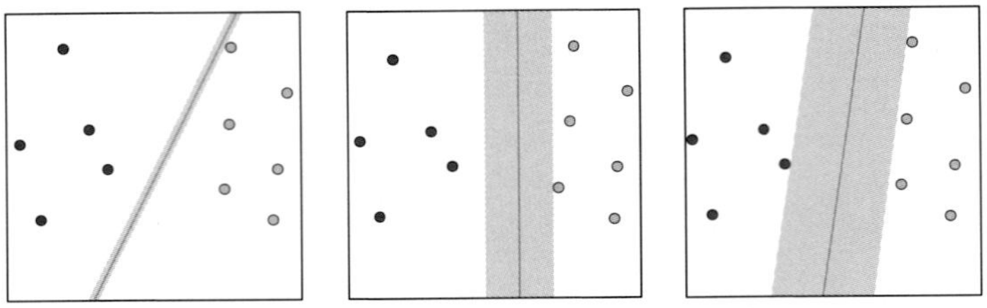

FIGURE 13.39

Margins.

Let $\mathbf{x}(q)$ in Fig. 13.40 be the nearest datapoint to the plane $\mathcal{S}$,

$$\mathbf{w}^{\mathrm{T}}\mathbf{x} + b = 0, \tag{13.54}$$

where $\mathbf{w} = [w_1, w_2, \cdots w_Q]$, $\mathbf{x} = [x_1, x_2, \cdots x_Q]$, and b is bias. By normalizing the weights so that $|\mathbf{w}^T\mathbf{x}| = 1$, the planes $\mathcal{S}_1$ and $\mathcal{S}_2$ are

$$\mathcal{S}_1 : \mathbf{w}^{\mathrm{T}}\mathbf{x} + b = 1,$$
$$\mathcal{S}_2 : \mathbf{w}^{\mathrm{T}}\mathbf{x} + b = -1.$$

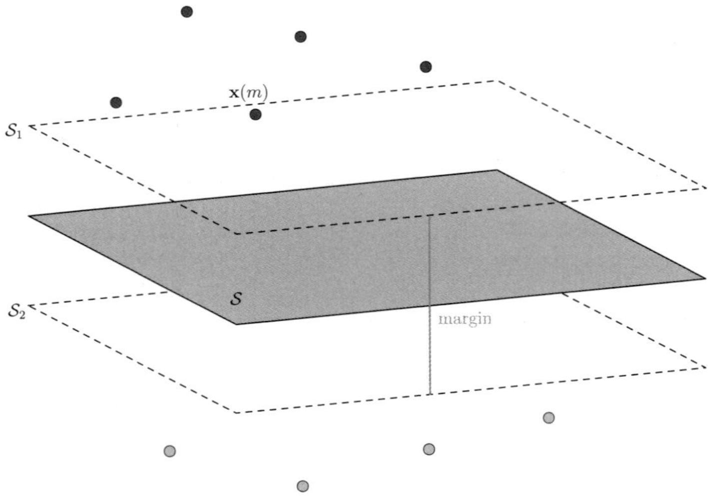

FIGURE 13.40

Finding the fattest margin.

As $y(q) \in \{+1, -1\}$ is the classification label, $y(q)(\mathbf{w}^{\mathrm{T}}\mathbf{x}(q) + b) \geq 1$, for $q = 1, \cdots Q$. Then, the maximum margin is

$$\lambda = \mathrm{d}(x(q), \mathcal{S}), \tag{13.55}$$

where $\mathrm{d}(\cdot)$ is the Euclidean distance as illustrated in Fig. 13.41. The decision rule can be defined as

$$\begin{cases} \mathbf{w}^{\mathrm{T}}\mathbf{x}(q) + b \geq 1, & \text{if data are +,} \\ \mathbf{w}^{\mathrm{T}}\mathbf{x}(q) + b \leq -1, & \text{if data are } -. \end{cases}$$

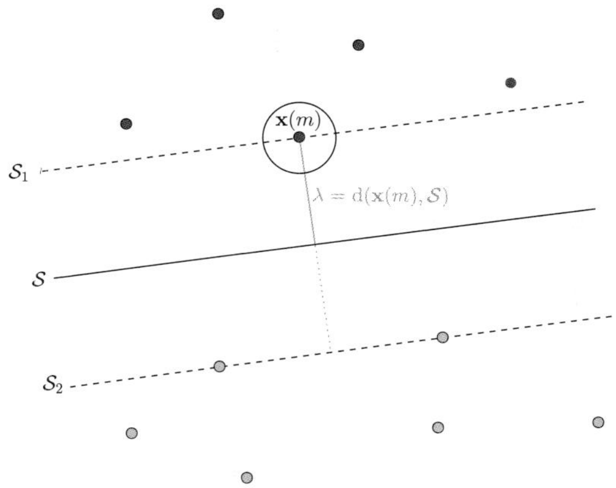

FIGURE 13.41

Decision rule.

Let us compute the distance between $\mathbf{x}(q)$ and the plane $\mathbf{w}^{\mathrm{T}}\mathbf{x} + b = 0$. If we take any two points $\mathbf{x}'$ and $\mathbf{x}''$ on the plane, then $\mathbf{w}^{\mathrm{T}}\mathbf{x}' + b = 0$ and $\mathbf{w}^{\mathrm{T}}\mathbf{x}'' + b = 0$. Since $\mathbf{w}^{\mathrm{T}}(\mathbf{x}' - \mathbf{x}'') = 0$, the vector $\mathbf{w}$ is perpendicular to the hyperplane, as illustrated in Fig. 13.42.

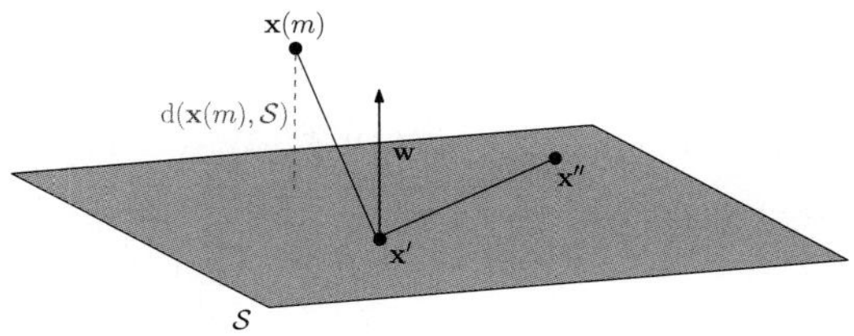

FIGURE 13.42

Distance between $x(m)$ and the hyperplane.

As $\mathbf{w}^T\mathbf{x}(q)+b=1$ and $\mathbf{w}^T\mathbf{x}'+b=0$, $\mathbf{w}^T(\mathbf{x}(q)-\mathbf{x}')=1$. As also shown in Fig. 13.42, the distance is the projection of vector $(\mathbf{x}(q)-\mathbf{x}')$ on $\mathbf{w}$:

$$\mathrm{d}(\mathbf{x}(q),\mathcal{S})=\frac{|\mathbf{w}^T(\mathbf{x}(q)-\mathbf{x}')|}{\|\mathbf{w}\|}=\frac{1}{\|\mathbf{w}\|}. \tag{13.56}$$

13.4.12.1 The optimization problem

To obtain the optimal hyperplane, we need to maximize the margin. The resulting optimization problem is

$$\begin{aligned}&\text{Maximize } \frac{1}{\|\mathbf{w}\|}\\ &\quad \text{s.t. } y(q)\left(\mathbf{w}^T\mathbf{x}(q)+b\right)\geq 1,\ \text{for } q=1,\ldots,Q,\end{aligned} \tag{13.57}$$

which is equivalent to

$$\begin{aligned}&\text{Minimize } \frac{1}{2}\|\mathbf{w}\|^2\\ &\quad \text{s.t. } y(q)\left(\mathbf{w}^T\mathbf{x}(q)+b\right)\geq 1,\ \text{for } q=1,\ldots,Q.\end{aligned} \tag{13.58}$$

13.4.12.2 Hard-margin SVM

We can use the Lagrangian to solve the optimization problem in Eq. (13.58). The Lagrangian is defined as

$$\begin{aligned}\xi^{\mathrm{L}}(\mathbf{w},b,\boldsymbol{\alpha})&=\frac{1}{2}\|\mathbf{w}\|^2+\sum_{q=1}^{Q}\alpha_q\left[1-y(q)\left(\mathbf{w}^T\mathbf{x}(q)+b\right)\right]\\ &=\frac{1}{2}\|\mathbf{w}\|^2-\sum_{q=1}^{Q}\alpha_q y(q)\left(\mathbf{w}^T\mathbf{x}(q)\right)-b\sum_{q=1}^{Q}\alpha_m y(q)+\sum_{q=1}^{Q}\alpha_q,\end{aligned} \tag{13.59}$$

where $\alpha_q, q=1,\cdots Q$, are the Lagrange multipliers.

Minimizing for $\mathbf{w}$ and b, we have

$$\nabla_{\mathbf{w}}\xi^{\mathrm{L}}=\mathbf{w}-\sum_{q=1}^{Q}\alpha_q y(q)\mathbf{x}(Q) \tag{13.60}$$

and

$$\frac{\partial\xi^{\mathrm{L}}}{\partial b}=\sum_{q=1}^{Q}\alpha_q y(q). \tag{13.61}$$

By setting $\nabla_{\mathbf{w}}\xi^{\mathrm{L}}$ and $\frac{\partial \xi^{\mathrm{L}}}{\partial b}$ to zero, we obtain

$$\mathbf{w} = \sum_{q=1}^{Q} \alpha_q y(q)\mathbf{x}(q) \tag{13.62}$$

and

$$\sum_{q=1}^{Q} \alpha_q y(q) = 0. \tag{13.63}$$

By replacing **w** in the Lagrangian in Eq. (13.59), we get

$$\xi^{\mathrm{L}}(\boldsymbol{\alpha}) = -\frac{1}{2}\sum_{n=1}^{Q}\sum_{q=1}^{Q} y(n)y(q)\alpha_n\alpha_q \mathbf{x}^T(n)\mathbf{x}(q) + \sum_{q=1}^{Q}\alpha_q \tag{13.64}$$

and we can maximize $\xi^{\mathrm{L}}(\boldsymbol{\alpha})$ such that $\boldsymbol{\alpha} \geq 0$, or equivalently minimize $\xi^{\mathrm{L}}(\boldsymbol{\alpha}) = -\xi^{\mathrm{L}}(\boldsymbol{\alpha})$. To solve this, we can use quadratic programming (QP), defined as

$$\begin{aligned} \min \xi^{\mathrm{L}}(\boldsymbol{\alpha}) &= \frac{1}{2}\boldsymbol{\alpha}^T \tilde{\mathbf{R}}\boldsymbol{\alpha} - \mathbf{1}^T\boldsymbol{\alpha}, \\ \text{s.t. } \mathbf{y}^T\boldsymbol{\alpha} &= 0, \\ \boldsymbol{\alpha} &\geq 0, \end{aligned} \tag{13.65}$$

where

$$\tilde{\mathbf{R}} = \begin{bmatrix} y(1)y(1)\mathbf{x}^T(1)\mathbf{x}(1) & y(1)y(2)\mathbf{x}^T(1)\mathbf{x}(2) & \dots & y(1)y(Q)\mathbf{x}^T(1)\mathbf{x}(Q) \\ y(2)y(1)\mathbf{x}^T(2)\mathbf{x}(1) & y(2)y(2)\mathbf{x}^T(2)\mathbf{x}(2) & \dots & y(2)y(Q)\mathbf{x}^T(2)\mathbf{x}(Q) \\ \vdots & \vdots & \ddots & \vdots \\ y(Q)y(1)\mathbf{x}^T(Q)\mathbf{x}(1) & y(Q)y(2)\mathbf{x}^T(Q)\mathbf{x}(2) & \dots & y(Q)y(Q)\mathbf{x}^T(Q)\mathbf{x}(Q) \end{bmatrix} \tag{13.66}$$

is the matrix of quadratic coefficients.

Example: Suppose that points (1, 3) and (0, 1) belong to class 1 and points (2, −1) and (3, −2) belong to class 2. Then

$$\mathbf{x} = \begin{bmatrix} 1 & 3 \\ 0 & 1 \\ 2 & -1 \\ 3 & -2 \end{bmatrix} \mathbf{y} = \begin{bmatrix} 1 \\ 1 \\ -1 \\ -1 \end{bmatrix}. \tag{13.67}$$

By using Eq. (13.66),

$$\tilde{\mathbf{R}} = \begin{bmatrix} 10 & 3 & 1 & 3 \\ 3 & 1 & 1 & 2 \\ 1 & 1 & 5 & 8 \\ 3 & 2 & 8 & 13 \end{bmatrix}. \tag{13.68}$$

The MATLAB® formulation for solving a QP is

$$\text{Min } L_D(\alpha) = \frac{1}{2}(\alpha^T R\alpha) + f^T\alpha, \tag{13.69}$$

$$\text{s.t. } A\alpha \le a, \tag{13.70}$$

$$B\alpha = b, \tag{13.71}$$

where

$$A\alpha \le a \rightarrow \alpha \ge 0, \tag{13.72}$$

$$B\alpha = b \rightarrow y^{\mathrm{T}}\alpha = 0. \tag{13.73}$$

The QP solution is $\boldsymbol{\alpha} = [\alpha_1, \ldots, \alpha_Q]$ and it can be obtained via MATLAB command $\boldsymbol{\alpha} = quadprog(\tilde{\mathbf{R}}, \mathbf{f}, \mathbf{A}, \mathbf{a}, \mathbf{B}, \mathbf{b})$, where

$$\mathbf{A} = \begin{bmatrix} -1 & 0 & 0 & 0 \\ 0 & -1 & 0 & 0 \\ 0 & 0 & -1 & 0 \\ 0 & 0 & 0 & -1 \end{bmatrix}, \ \mathbf{a} = \begin{bmatrix} 0 \\ 0 \\ 0 \\ 0 \end{bmatrix}, \ \mathbf{B} = \begin{bmatrix} 1 & 1 & -1 & -1 \\ 0 & 0 & 0 & 0 \\ 0 & 0 & 0 & 0 \\ 0 & 0 & 0 & 0 \end{bmatrix}, \ \mathbf{b} = \begin{bmatrix} 0 \\ 0 \\ 0 \\ 0 \end{bmatrix}, \tag{13.74}$$

and

$$\mathbf{f} = \begin{bmatrix} -1 \\ -1 \\ -1 \\ -1 \end{bmatrix}. \tag{13.75}$$

For this example,

$$\boldsymbol{\alpha} = \begin{bmatrix} 0 \\ 0.25 \\ 0.25 \\ 0 \end{bmatrix}. \tag{13.76}$$

Then, the coefficient vector $\mathbf{w}$ can be recovered by

$$\begin{aligned} \mathbf{w} = \sum_{q=1}^{Q} \alpha_q y(q)\mathbf{x}(q) &= 0.25 \times 1 \times [0 \ \ 1]^{\mathrm{T}} + 0.25 \times (-1) \times [2 \ \ -1]^{\mathrm{T}} \\ &= [-0.5 \ \ 0.5]^{\mathrm{T}}, \end{aligned} \tag{13.77}$$

and b can be found by using any data with $\alpha > 0$,

$$b = y(q) - \mathbf{w}^T\mathbf{x}(q) = 1 - [-0.5 \ \ 0.5] \cdot [0 \ \ 1]^{\mathrm{T}} = 0.5. \tag{13.78}$$

Many of the α_qs are zero due to the constrained optimization condition

$$\alpha_q \left[1 - y(q)\left(\mathbf{w}^T\mathbf{x}(q) + b\right)\right] = 0. \tag{13.79}$$

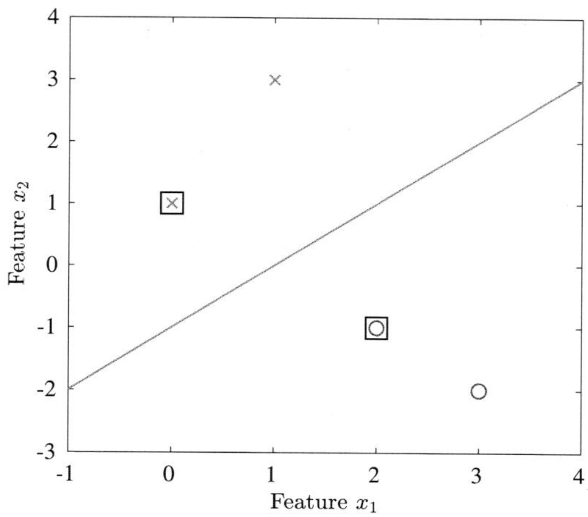

FIGURE 13.43

Example.

However, the equality $y(q)\left(\mathbf{w}^T\mathbf{x}(q)+b\right)=1$ is true only for the nearest points to $\mathbf{w}$. For the remaining internal points, we have $\alpha_q=0$. The points with $\alpha_q>0$ are the *support vectors*. One can observe that the support vectors, highlighted in Fig. 13.43, are indeed the nearest points to the separator plane $\mathbf{w}$ in red (light gray in print version).

13.4.12.3 Support vectors

Support vectors are the closest $x(q)$ to the plane. They live on the margin ($y(q)\left(\mathbf{w}^T\mathbf{x}(k)+b\right)=1$), as illustrated in Fig. 13.44. The decision boundary is determined only by the support vectors as

$$\mathbf{w}=\sum_{q=1}^{N_{sv}}\alpha_q y(q)\mathbf{x}(q), \tag{13.80}$$

where N_{sv} is the number of support vectors. We can solve for b using any support vector k,

$$y(k)\left(\mathbf{w}^T\mathbf{x}(k)+b\right)=1. \tag{13.81}$$

For testing new data, we can compute

$$\mathbf{w}^{\mathrm{T}}\mathbf{x}(p)+b=\sum_{q=1}^{k}\alpha_q y(q)\mathbf{x}^T(k)\mathbf{x}(p). \tag{13.82}$$

Then, we classify $\mathbf{x}(p)$ as class 1 if the sum is positive and as class 2 otherwise.

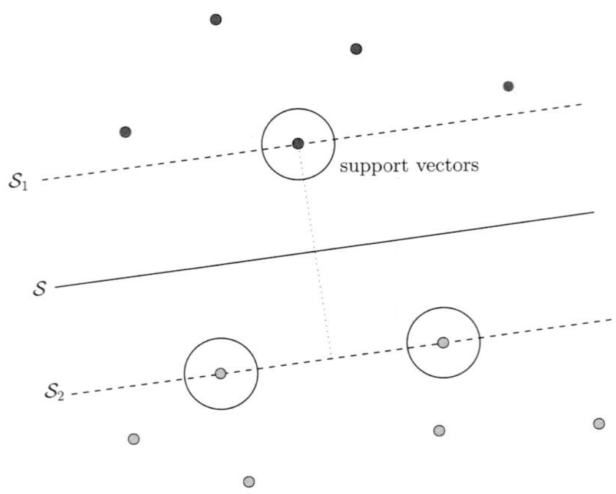

FIGURE 13.44

Support vectors.

As discussed in [2], an upper bound in the cross-validation error relates to the number of support vectors, while nonsupport vector data are not taken into account. The bound result is

$$\mathbb{E}[E_{\mathrm{o}}] \leq \frac{\mathbb{E}[\text{number of support vectors}]}{M-1}. \tag{13.83}$$

13.4.12.4 Nonlinear transformations

Suppose we apply a nonlinear function $\boldsymbol{\upsilon}$ to the points that live in $\mathcal{X}$ space and obtain

$$\boldsymbol{\upsilon} : \mathbb{R}^Q \rightarrow \mathcal{H}, \tag{13.84}$$

so that

$$\mathbf{x}(q) \mapsto \mathbf{h}(q) = \boldsymbol{\upsilon}\left(\mathbf{x}(q)\right), \tag{13.85}$$

which live in an alternative Hilbert space denoted as $\mathcal{H}$.

The freedom to choose the mapping function $\boldsymbol{\upsilon}(\cdot)$ allows effective similarity measures between vectors after mapping into high-dimensional space, called the feature space. The SVM can be performed at the feature space as follows:

$$\langle \boldsymbol{\upsilon}\left(\mathbf{x}(k)\right), \boldsymbol{\upsilon}\left(\mathbf{x}(l)\right)\rangle = \langle \mathbf{h}(k), \mathbf{h}(l)\rangle = \kappa\left(\mathbf{x}(k), \mathbf{x}(l)\right), \tag{13.86}$$

where the feature space can have a much higher dimension than $\mathbb{R}^N$. The so-called *kernel trick* allows the computation of inner products via functional evaluations.

Define a kernel function $\kappa(\cdot,\cdot)$ such that $\forall \mathbf{x} \in \mathbb{R}^Q$, $\kappa(\mathbf{x},\cdot)$ in a Hilbert space $\mathcal{H}$. A reproducing kernel Hilbert space (RKHS) is a Hilbert space in which the value of a function $f(\mathbf{x}) \in \mathcal{H}$ can be evaluated as $f(\mathbf{x}) = \langle f(\cdot), \kappa(\mathbf{x},\cdot)\rangle$. This reproduction property of a function via the kernel $\kappa(\mathbf{x},\cdot)$ can be realized via functional evaluations by means of what is called the kernel trick. Fig. 13.45 illustrates the mapping of data vectors into the feature space, where the mathematical representation is so that if

$$f(\cdot) = \sum_{r=0}^{R} \alpha_r \kappa(\mathbf{x}(r), \cdot) \in \mathcal{H}, \tag{13.87}$$

then

$$f(\mathbf{x}) = \sum_{r=0}^{R} \alpha_r \kappa(\mathbf{x}(r), \mathbf{x}) \in \mathbb{R}. \tag{13.88}$$

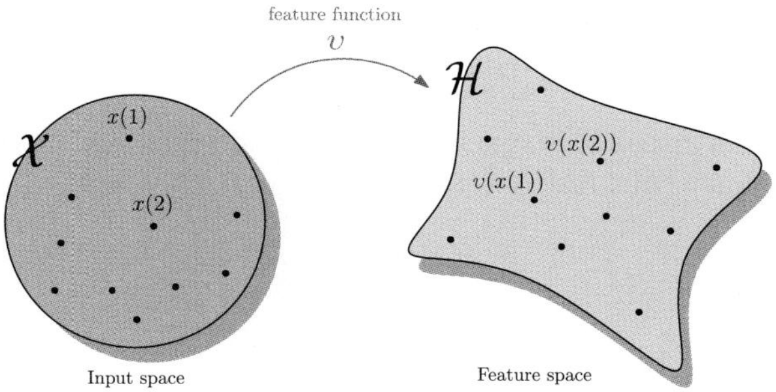

FIGURE 13.45

Kernel mapping representation.

With the application of the kernel mapping, the Lagrangian becomes

$$\xi^{\mathrm{L}}(\boldsymbol{\alpha}) = \sum_{r=1}^{Q} \alpha_n - \frac{1}{2}\sum_{r=1}^{Q}\sum_{q=1}^{Q} y(r)y(q)\alpha_n\alpha_m \mathbf{h}^{\mathrm{T}}(r)\mathbf{h}(q). \tag{13.89}$$

In this case, the support vectors live in $\mathcal{H}$ space. In the $\mathcal{X}$ space, we have the *preimages* of the support vectors, as illustrated in Fig. 13.46. The margin remains the same in $\mathcal{H}$ space.

In the following, we discuss the properties of the kernels adequate to build RKHS feature spaces.

13.4.12.5 Mercer kernels

One can form a similarity matrix $\mathbf{K}$ called Gram matrix and described as

$$\mathbf{K} = \begin{bmatrix} \kappa(\mathbf{x}(1), \mathbf{x}(1)) & \kappa(\mathbf{x}(1), \mathbf{x}(2)) & \cdots & \kappa(\mathbf{x}(1), \mathbf{x}(Q)) \\ \kappa(\mathbf{x}(2), \mathbf{x}(1)) & \kappa(\mathbf{x}(2), \mathbf{x}(2)) & \cdots & \kappa(\mathbf{x}(2), \mathbf{x}(Q)) \\ \vdots & \vdots & \ddots & \vdots \\ \kappa(\mathbf{x}(Q), \mathbf{x}(1)) & \kappa(\mathbf{x}(Q), \mathbf{x}(1)) & \cdots & \kappa(\mathbf{x}(Q), \mathbf{x}(Q)) \end{bmatrix}. \tag{13.90}$$

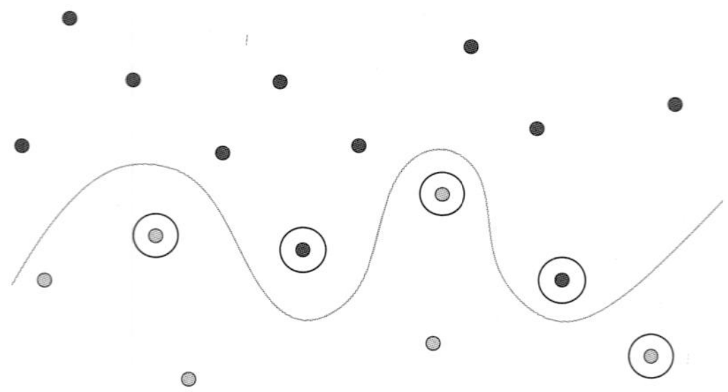

FIGURE 13.46

Support vectors in a nonlinear problem.

The Gram matrix is inherently positive semidefinite, and those kernel matrices that satisfy this condition are called *Mercer kernels*.

Assuming that the Gram matrix $\mathbf{K}$ is symmetric, it can be decomposed in M orthonormal eigenvectors $\mathbf{g}_i$ composing the columns of a matrix $\mathbf{G}$ so that

$$\mathbf{G}^{\mathrm{T}}\mathbf{K}\mathbf{G} = \begin{bmatrix} \lambda_1 & 0 & \cdots & 0 \\ 0 & \lambda_2 & & \vdots \\ \vdots & 0 & \cdots & \vdots \\ \vdots & \vdots & & 0 \\ 0 & 0 & \cdots & \lambda_Q \end{bmatrix} = \mathbf{\Lambda}, \tag{13.91}$$

where the eigenvalues $\lambda_i \geq 0$.

The kernel-based SVM entails replacing the inner product performed in the original data space to measure similarity by the kernel function evaluation in the feature space.

13.4.12.6 Some kernel functions

A pivotal step to perform the data transformation into $\mathcal{H}$ is the choice of the kernel function. A proper choice is beneficial in terms of the SVM performance and relates to the type of data. A widely used kernel function is the polynomial kernel,

$$\kappa\left(\mathbf{x}(k), \mathbf{x}(l)\right) = (c\mathbf{x}^{\mathrm{T}}(k)\mathbf{x}(l) + d)^p, \tag{13.92}$$

where $p \in \mathbb{N}$ and $d \geq 0$. If $d \neq 0$, this is an inhomogeneous polynomial kernel. Also, the nonnegative value of d is required to guarantee that the kernel matrix is positive definite, and in any case, the choice of parameters requires that the diagonal entries of the kernel matrix be positive. Let us take a simple example: The data vectors have two entries, i.e., $Q = 1$, and the polynomial power is $p = 2$. In this case

$$\begin{aligned}\kappa\left(\mathbf{x}(k), \mathbf{x}(l)\right) &= (c\mathbf{x}^{\mathrm{T}}(k)\mathbf{x}(l) + d)^p \\ &= d^2 + 2cd(x_0(k)x_0(l) + x_1(k)x_1(l))\end{aligned}$$

$$
\begin{aligned}
&+c^2\left[x_0^2(k)x_0^2(l)+x_1^2(k)x_1^2(l)+2x_0(k)x_0(l)x_1(k)x_1(l)\right] \\
&=\left[\begin{array}{cccccc} d & \sqrt{2cd}x_0(k) & \sqrt{2cd}x_1(k) & cx_0^2(k) & cx_1^2(k) & \sqrt{2}cx_0(k)x_1(k) \end{array}\right] \\
&\times\left[\begin{array}{cccccc} d & \sqrt{2cd}x_0(k) & \sqrt{2cd}x_1(k) & cx_0^2(k) & cx_1^2(k) & \sqrt{2}cx_0(k)x_1(k) \end{array}\right]^{\mathrm{T}} \\
&=\langle \boldsymbol{v}\left(\mathbf{x}(k)\right), \boldsymbol{v}\left(\mathbf{x}(l)\right)\rangle \\
&=\boldsymbol{v}^{\mathrm{T}}\left(\mathbf{x}(k)\right)\boldsymbol{v}\left(\mathbf{x}(l)\right).
\end{aligned} \tag{13.93}
$$

Some stability issues might show up for large values of p; therefore, this value is usually chosen to be less than 10.

There are many types of kernel functions available in the literature. A few of them are listed in Table 13.9.

Table 13.9 Kernel functions $\kappa\left(\mathbf{x}(k), \mathbf{x}(l)\right)$.

Polynomial	Cosine	Sigmoid	Laplacian	Gaussian
$(c\mathbf{x}^{\mathrm{T}}(k)\mathbf{x}(l)+d)^p$	$\frac{\mathbf{x}^{\mathrm{T}}(k)\mathbf{x}(l)}{\\|\mathbf{x}(k)\\|_2\\|\mathbf{x}(l)\\|_2}$	$\tanh(c\mathbf{x}^{\mathrm{T}}(k)\mathbf{x}(l)+d)$	$\mathrm{e}^{-\frac{\\|\mathbf{x}(k)-\mathbf{x}(l)\\|}{\sigma}}$	$\mathrm{e}^{-\frac{1}{2}(\mathbf{x}(k)-\mathbf{x}(l))^{\mathrm{T}}\boldsymbol{\Sigma}^{-1}(\mathbf{x}(k)-\mathbf{x}(l))}$
$c, d\in\mathbb{R}$, and $p\in\mathbb{N}$	$c, d\in\mathbb{R}$	$c, d\in\mathbb{R}$	$\sigma>0$	$\boldsymbol{\Sigma}$ diagonal, entries σ_i^2

A few comments are appropriate. The sigmoid kernel does not generate a positive semidefinite Gram matrix and is widely used in MLP. The Gram matrix must be positive semidefinite, and symmetric so that the kernel function relates to an inner product in the feature space, benefiting from the fact that the kernel function is more efficient to compute than the inner product. In the Gaussian case, the free parameter choice is often $\sigma>1$. In all cases, the choice of the kernel parameters strongly affects the SVM performance, and it is not a trivial task.

Another family of kernel mapping relies on replacing the kernel evaluation by an approximation entailing the inner product of two vectors, each of them representing the outcome of a randomized feature map; see [50] for details. It entails the approximation of a kernel function through an inner product as

$$
\kappa\left(\mathbf{x}(l), \mathbf{x}(k)\right)\approx\mathbf{z}^{\mathrm{T}}\left(\mathbf{x}(k)\right)\mathbf{z}\left(\mathbf{x}(l)\right), \tag{13.94}
$$

where $\mathbf{z}(\cdot):\mathbb{R}^Q\rightarrow\mathbb{R}^D$, for $M\gg D$.

In terms of the kernel matrix the random Fourier feature achieves the following approximation:

$$
\mathbf{K}\approx\left[\begin{array}{cccc}
\mathbf{z}^T\left(\mathbf{x}(1)\right)\mathbf{z}\left(\mathbf{x}(1)\right) & \mathbf{z}^T\left(\mathbf{x}(1)\right)\mathbf{z}\left(\mathbf{x}(2)\right) & \cdots & \mathbf{z}^T\left(\mathbf{x}(1)\right)\mathbf{z}\left(\mathbf{x}(M)\right) \\
\mathbf{z}^T\left(\mathbf{x}(k-1)\right)\mathbf{z}\left(\mathbf{x}(1)\right) & \mathbf{z}^T\left(\mathbf{x}(2)\right)\mathbf{z}\left(\mathbf{x}(2)\right) & \cdots & \mathbf{z}^T\left(\mathbf{x}(2)\right)\mathbf{z}\left(\mathbf{x}(Q)\right) \\
\vdots & \vdots & \ddots & \vdots \\
\mathbf{z}^T\left(\mathbf{x}(Q)\right)\mathbf{z}\left(\mathbf{x}(1)\right) & \mathbf{z}^T\left(\mathbf{x}(Q)\right)\mathbf{z}\left(\mathbf{x}(2)\right) & \cdots & \mathbf{z}^T\left(\mathbf{x}(Q)\right)\mathbf{z}\left(\mathbf{x}(Q)\right)
\end{array}\right]. \tag{13.95}
$$

In the kernel case, we also need to employ QP to find the SVM solution as

$$\begin{aligned} \min \xi^{\mathrm{L}}(\boldsymbol{\alpha}) &= \frac{1}{2}\boldsymbol{\alpha}^T \tilde{\mathbf{K}} \boldsymbol{\alpha} - \mathbf{1}^T \boldsymbol{\alpha}, \\ \text{s.t. } \mathbf{y}^T \boldsymbol{\alpha} &= 0, \\ \boldsymbol{\alpha} &\geq 0, \end{aligned} \tag{13.96}$$

where

$$\tilde{\mathbf{K}} = \begin{bmatrix} y(1)y(1)\kappa(1,1) & y(1)y(2)\kappa(1,2) & \dots & y(1)y(Q)\kappa(1,Q) \\ y(2)y(1)\kappa(2,1) & y(2)y(2)\kappa(2,2) & \dots & y(2)y(Q)\kappa(2,Q) \\ \vdots & \vdots & \ddots & \vdots \\ y(Q)y(1)\kappa(Q,1) & y(Q)y(2)\kappa(Q,2) & \dots & y(Q)y(Q)\kappa(Q,Q) \end{bmatrix} \tag{13.97}$$

is the matrix of quadratic coefficients.

Example: In a toy example, 40 points consisting of two classes live in a 2D input space as depicted in Fig. 13.47. After applying a polynomial kernel, the datapoints are mapped to a 3D feature space, where performing the binary classification is easier as depicted in Fig. 13.48(a). The support vectors are highlighted inside the red circles (light gray in print version) in Fig. 13.48(a).

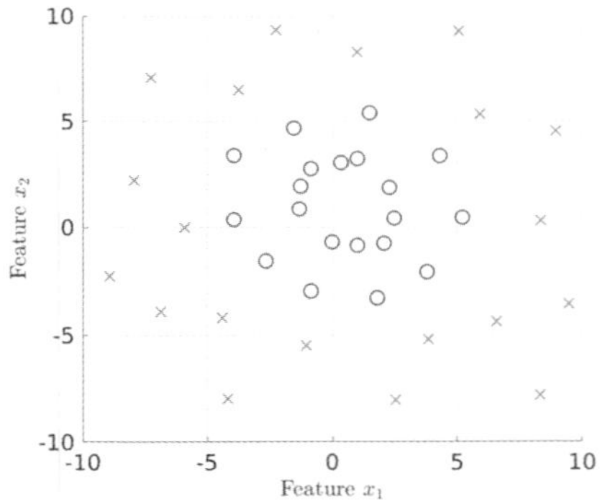

FIGURE 13.47

Data samples in input space.

The boundaries cross the support vectors which can be seen highlighted inside the red circles (light gray in print version). It is also possible to classify the data applying a Gaussian kernel, as shown in Fig. 13.48(b). In this case, however, the Gaussian kernel leads to a complex separator hyperplane, and all the datapoints become support vectors. We should then remind ourselves that when the model is too complex for our dataset, we are at risk of overfitting.

As can be observed, with the proper choice of the kernel parameters, it is possible to determine an adequate nonlinear boundary for correct classification in this example.

As discussed here, the SVM provides a decision solution without any a posteriori probability information. However, the so-called relevance vector machines (RLVs), which are related to SVMs, utilize

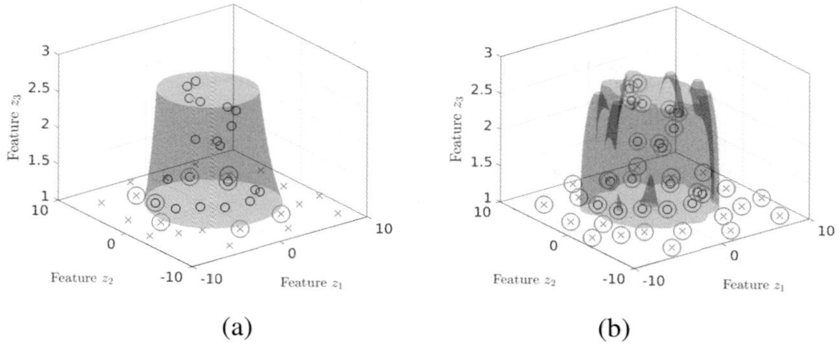

FIGURE 13.48

Illustration of (a) polynomial and (b) Gaussian kernels for the toy example in Fig. 13.47.

the Bayesian sparse kernel concepts and can be applied with benefits to address classification and regression problems [3]. Although the SMV is originally a two-class classifier, there are attempts to extend it to multiple classes with some drawbacks [3].

13.5 Ensemble learning

A good measure of a classifier performance is its average accuracy over a training-independent dataset. In many situations, however, one must also consider the accuracy variance over different conditions in order to infer how trustworthy the average statistics are: most of the times, a large variance reduction more than compensates for a small decrease in average accuracy. Ensemble learning combines several distinct classifiers to reduce overall performance variance, exploring the so-called wisdom of the crowd concept. In this combination, strategies must be employed to guarantee that all separate classifiers contribute significantly to the final system, without any classification trend dominating over the others. Two examples of ensemble learning algorithms are briefly described below.

13.5.1 Random forest

When playing a game of "Who is it?," one must guess a hidden card by asking yes-or-no questions about the unknown character: Is he/she blond? Does he/she wear glasses? Usually our next question depends on the answers to the previous ones. Therefore, if we know that our hidden character is male, we may later want to find out if he has a mustache or not; something that would not make much sense for female characters. A good strategy that maximizes long-run winning chances is to use questions that eliminate the largest number of possibilities independently of the possible outcomes, that is, we want questions that divide our current dataset in two subsets of about the same size.

This whole process can be represented by what is called a binary decision tree, which is a directed graph as illustrated in Fig. 13.49. In this model, each question is referred to as a node and each possible answer as a branch. The first node is called the root node and the final guesses at the bottom of the tree are the so-called leaves, indicating that our tree is somewhat upside down.

Given a preannotated dataset, one may grow or, more exactly, train a particular decision tree by following the "Who is it?" metaphor: Each question (node) shall attempt to split the corresponding dataset into two equally sized subsets and each subsequent question (child node) depends on the answer given to the previous one (parent node), leading most likely to an asymmetrical tree. This whole process, where a particular data characteristic (attribute) must be selected along its value that better divides the current dataset, is often implemented on a trial-and-error basis, leading to a very computationally intensive procedure. In the end, the tree leaves must indicate our final guess about which class a particular datapoint belongs to.

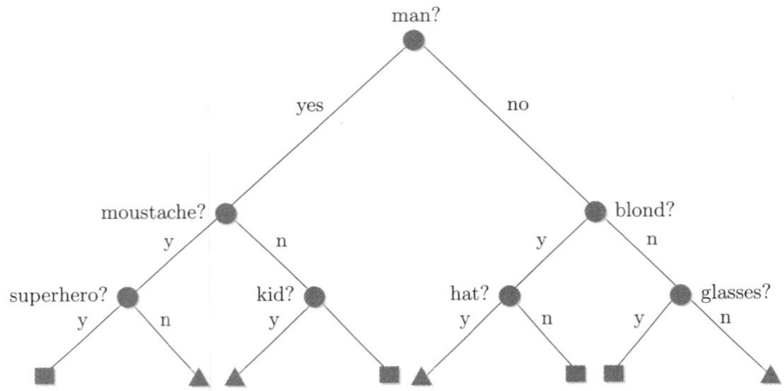

FIGURE 13.49

Example of decision tree applied to a possible "Who is it?" game.

After the decision tree is trained, using it to classify a given sample is straightforward, as one only needs to follow the path defined by the correct answers/branches to each question/node along the tree for that particular datapoint.

Decision trees are characterized by being easy-to-interpret models, which work equally well with both numerical and categorical data attributes and are robust to outliers or mislabeled data. However, optimal decision trees require a very computationally intensive training process and practical approximations cannot guarantee global optima are reached. They are also prone to overfitting, as 100% accuracy in training datasets can always be achieved if the tree is allowed to grow indefinitely. This can be mitigated by a tree pruning process, where a maximum depth or performance threshold is enforced. Since small data changes may lead to completely different structures, decision trees also suffer from instability, which ultimately corresponds to high levels of performance variance.

This last characteristic motivated the development of an ensemble of decision trees, as illustrated in Fig. 13.50, appropriately referred to as a random forest.

Decorrelation among individual classifiers may be achieved by enforcing a random component during training of each tree. Some common strategies for this include:

- splitting the initial training dataset in several subsets following a bootstrap aggregation approach to guarantee a similar statistical distribution to each subset as in the original one;

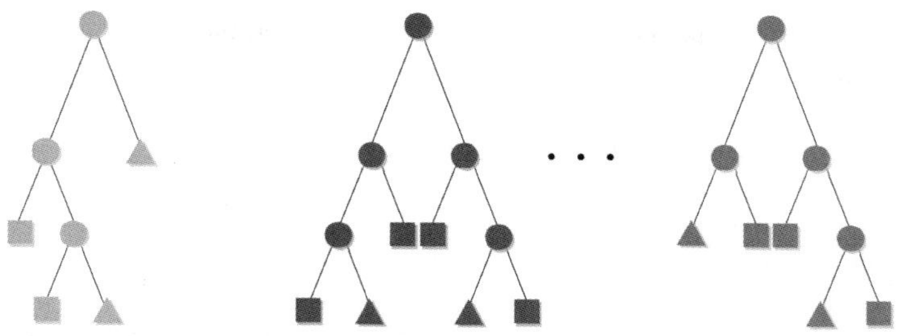

FIGURE 13.50

Illustration of an ensemble of decision trees.

- splitting the set of attributes extracted from the initial data in several distinct subsets, with possible nonnull intersection, and use each subset to train a different tree.

These two strategies are often combined to train distinct decision trees, which are then combined to form a random forest, whose output is taken as the mode of all individual tree outputs.

Random forests have few hyperparameters, the main ones being maximum tree depth and total number of trees. Adjusting these variables, however, is not a complicated task, as the final performance is quite robust to wide a range of their values. Training random forests is quite computationally intensive, but the process can be performed in parallel, greatly reducing the overall duration.

Generally speaking, a random forest is characterized by easy handling of heterogeneous (numerical and categorical) data, intrinsic implementation of feature selection, and very little performance variance, leading to a very good generalization capability.

13.5.2 Boosting

Training individual classifiers in parallel, although advantageous from a computation time point of view, may not be optimal with respect to final performance. In boosting, classification units are trained in sequence and the performances of previously trained ones drive the training process of the following units. The main idea here is to "boost" the performance of several weak learners, each one performing slightly better than a random decision-making scheme, into a strong and accurate ensemble algorithm. New classifiers are added in an attempt to correct the classification errors generated by the initial ones, in such a way that each new ensemble member becomes an expert on the errors of all its predecessors.

In boosting, one initializes the entire process by giving the same importance to all training datapoints. After developing a first weak classifier, using any model, such as decision tree, perceptron, and so on, one evaluates its performance in the entire training dataset. Importance is then decreased for the correctly identified points and increased for all misclassified points, and a new classifier is trained based on the dataset with modified weights. This process of error evaluation, dataset reweighting, and new unit training is repeated until the overall performance remains unchanged after a few consecutive iterations.

Adaptive boosting (AdaBoost) was the first successful boosting algorithm [17]. In it, if a given individual $h_i(y)$ classifier has error probability ϵ_i, its ensemble weight α_i is given by

$$\alpha_i = \frac{1}{2}\ln\left(\frac{1-\epsilon_i}{\epsilon_i}\right), \tag{13.98}$$

and the training datapoints must be reweighted in such a way that

$$w(i+1) = \frac{w(i)}{2} \times \begin{cases} \dfrac{1}{1-\epsilon_i}, & \text{if it is a hit,} \\ \dfrac{1}{\epsilon_i}, & \text{otherwise.} \end{cases} \tag{13.99}$$

After iterating $i = 1, 2, \ldots, I$, the final ensemble output for the AdaBoost algorithm becomes

$$H(y) = \text{sign}\left(\sum_{i=1}^{I} \alpha_i h_i(y)\right). \tag{13.100}$$

Other variations with improved implementations and performances, such as the real AdaBoost, gradient boosting [16], and extreme gradient boosting (XGBoost) [4] algorithms, can be found in the related literature.

13.6 Deep learning

DL refers to NNs with a large number of layers. In the initial works, one would consider any network with more than three layers DL; nowadays, however, this number has grown to many thousands of layers, and it is still increasing in more demanding applications. In order for these systems to work properly, mainly avoiding gradient fading or divergence during the backpropagation process, several adjustments needed to be made on the network itself and its training procedure. In this section, we analyze several of these combined improvements that have led to a true scientific revolution, the consequences of which increase on a daily basis and we are still trying to understand.

13.6.1 Large dataset corpora

Networks with many layers starve for huge amounts of data to adjust their many coefficients. Therefore, one of the main technological advances that boosted the DL breakthrough was the appearance of large and properly annotated datasets. Some entries of this dataset are shown in Fig. 13.8 in Section 13.3.5.

The MNIST database includes 70,000 examples of handwritten digits which have been size-normalized and centered in a (28×28)-pixel image, where each pixel is described by its 8-bit gray level. This database was widely explored by Yann LeCun's research group, who developed the so-called LeNet-5 DL network for the automatic recognition of postal codes [40].

Another widely known dataset, which also played a major role in early DL development and expansion, is the ImageNet. This database is comprised of 14 million images categorized into 22,000 distinct classes [9]. Some samples from ImageNet are shown in Fig. 13.51. Starting in 2010, a challenge ran yearly on an ImageNet subset (with about 1.4 million images divided into *only* 1000 classes) to showcase image classification algorithms based on computer vision techniques. In its 2012 version, the challenge was won by a large margin by the so-called AlexNet DL architecture developed by Geoffrey Hinton's research group at the University of Toronto, Canada [38]. After that year, all versions of the challenge were won by DL algorithms with ever increasing numbers of layers. Following AlexNet's success, in what seems today as a blink of an eye, researchers all over the world started applying DL to all kinds of problems, strangely enough with great success.

FIGURE 13.51

Illustrative samples from the ImageNet dataset.

Nowadays, the importance of these data corpora is summarized into the overused refrain *data is the new oil*. New challenges appear almost on a daily basis. Dataset repositories and even specialized technical journals have appeared on this topic. Despite all this, in our experience, good data, particularly properly annotated data, are almost always the bottleneck for the development of a good DL system. Any resource spent in this direction is paid tenfold in final system performance and most importantly in overall problem understanding.

13.6.2 Nonlinear activation function

Traditional NNs were mostly limited to three layers due to numerical problems during the network training stage. Those issues resulted mainly from gradient fading along the backpropagation process across several layers. Any attempt to avoid such fading caused the opposite effect of gradient divergence. Balancing between fading and divergence was a difficult task which limited the number of network layers down to three for several decades. Such impairment restricted the use of NNs in several practical situations, which ultimately led to financial cuts in research funding on this topic.

Thanks to the groundbreaking work of a few obstinate and brilliant researchers, we now know that a few adjustments in the training procedure and the network structure were required to enable stacking of a large number of layers. In this section, we list some of these adjustments that altogether have led to a true revolution in several research areas.

The most common activation functions are the sigmoid and tanh ones, respectively described by

$$\begin{aligned}\phi_\sigma(x) &= \frac{1}{1+e^{-x}},\\ \phi_t(x) &= \tanh(x) = \frac{e^x - e^{-x}}{e^x + e^{-x}}.\end{aligned} \tag{13.101}$$

As depicted in Fig. 13.52, the sigmoid function maps the real variable x into the $0 \leq \sigma(x) \leq 1$ range, with a possible interpretation of probability of a given event. However, the nonzero average value of the sigmoid leads to a bias that propagates along the network training process. The tanh function can be seen as a scaled zero-mean version of the sigmoid function,

$$\phi_t(x) = 2\phi_\sigma(2x) - 1. \tag{13.102}$$

Those two functions, however, as also shown in Fig. 13.52, have strong saturation regions for large values of $|x|$. This characteristic corresponds to very low values of their derivatives in this range, which ultimately leads to gradient fading, slowing down or even halting the training of deeper networks.

In order to avoid this last issue, the ReLU, defined as

$$\phi_r(x) = \max(0, x), \tag{13.103}$$

was introduced and found great success in several DL applications [31,38,46]. Besides its evident simplicity, the ReLU allows for a much simpler gradient computation and avoids saturation in its positive part, speeding up the training process.

Nowadays, among the plethora of activation functions that can be found in the related literature, the ReLU is the most widely used one, although the tanh can be a solid choice and the sigmoid is employed whenever the application calls for a probability-like output within the [0, 1] interval.

13.6.3 Network weight initialization

A proper weight initialization procedure also constitutes a useful tool for speeding up the learning process of DNNs.

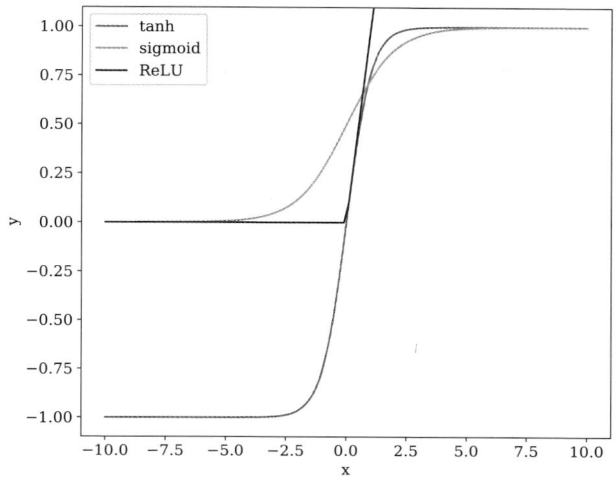

FIGURE 13.52

Most common activation functions for neural networks: sigmoid, tanh, and ReLU.

On the one hand, very small weights, with values close to zero, consistently shrink the signals in the forward pass across the network, turning it useless. On the other hand, large weights tend to generate large inputs to the activation function which saturate the corresponding neurons, thus hindering their learning ability, as seen in Fig. 13.52. Therefore, one should aim for an in-between situation with the weights properly distributed for both the network backward training and forward operation.

Since the saturation intervals are quite spread along the input range, a uniform weight distribution leads to an output distribution with large peaks in the two (low and high) saturation regions, which is obviously undesired. Therefore, a proper initialization strategy should distribute the network weights randomly in order to achieve a more uniform distribution for the activation function output. To do so, one may use the so-called Xavier or Glorot distributions [19]. The Glorot normal distribution is a truncated normal distribution with zero mean and standard deviation $\sqrt{\frac{2}{N_i+N_o}}$, where N_i and N_o are the numbers of neurons in the preceding and subsequent layers, respectively. An alternative distribution is the Glorot uniform with range $[-\ell, \ell]$, with $\ell = \sqrt{\frac{6}{N_i+N_o}}$.

When using the ReLU activation function, one may consider the He initialization scheme [25], which uses a standard deviation of $\sqrt{\frac{2}{N_i}}$. This value compensates for the fact that the ReLU output is null for half of its inputs, thus doubling the weight variance to keep the output signal variance constant.

13.6.4 Loss function

The loss function attempts to evaluate how much the output y of an NN resembles the target or desired output value $\hat{y}$ for each different input x.

Perhaps the simplest cost function is given by the MSE, defined as

$$\mathcal{L}_{\mathrm{MSE}}(\mathbf{y}, \hat{\mathbf{y}}) = \frac{1}{n}\sum_{i=1}^{n}(y_i - \hat{y}_i)^2, \tag{13.104}$$

where n is the total number of inputs. The squared operation in this function equally penalizes positive and negative differences between y and $\hat{y}$ and allows for an easy computation of its gradient with respect to a network coefficient w as required by the backpropagation algorithm:

$$\frac{\partial \mathcal{L}_{\mathrm{MSE}}}{\partial w} = -\frac{1}{n}\sum_{i=1}^{n} 2(y_i - \hat{y}_i)\frac{\partial \hat{y}_i}{\partial w}. \tag{13.105}$$

The main drawback of the MSE function is that when the activation function becomes saturated, any coefficient change causes little to no impact in the overall cost function, that is, if $\frac{\partial \hat{y}_i}{\partial w} \approx 0$, then $\frac{\partial \mathcal{L}_{\mathrm{MSE}}}{\partial w} \approx 0$. This is associated with the fading gradient effect, which greatly slows down the network learning process.

The effect of saturated neurons on the network learning speed may be diminished by substituting the MSE by the CE cost function defined as

$$\mathcal{L}_{\mathrm{CE}}(\mathbf{y}, \hat{\mathbf{y}}) = -\frac{1}{n}\sum_{i=1}^{n}\left[y_i \ln \hat{y}_i + (1 - y_i)\ln(1 - \hat{y}_i)\right]. \tag{13.106}$$

Once again, similar to the MSE function, any difference between y and $\hat{y}$ increases the total CE cost. In the CE case, however, the gradient with respect to a network coefficient w, after a few straightforward calculations, is given by

$$\frac{\partial \mathcal{L}_{\mathrm{CE}}}{\partial w} = -\frac{1}{n}\sum_{i=1}^{n}\frac{y_i - \hat{y}_i}{\hat{y}_i(1 - \hat{y}_i)}\frac{\partial \hat{y}_i}{\partial w}. \tag{13.107}$$

In this case, in the neuron saturation regions where $\frac{\partial \hat{y}_i}{\partial w} \approx 0$, the denominator factors $\hat{y}_i$ and $(1 - \hat{y}_i)$ refrain the CE gradient from approaching zero, thus avoiding the dreadful fading effect.

In Fig. 13.53, we present an illustrative example when CE and MSE losses are employed in a classification problem. By considering $\hat{y}_1 = 1$ and $\hat{y}_2 = 0$ in Eqs. (13.106) and (13.104), we vary the classifier output in order to obtain the loss surfaces as in Fig. 13.53. Observe that the loss magnitude of CE is much higher than that of MSE, especially for $\mathbf{y} = [1, 1]$, $\mathbf{y} = [0, 1]$, and $\mathbf{y} = [0, 0]$, which would represent different classes. Since CE penalizes the error better for different categories, it is more reasonable for classification problems. In regression problems, there is no need for this abrupt behavior, so MSE is more adequate.

13.6.5 Learning algorithm

Network training consists of finding the best set of network coefficients w (weights and bias) that minimize a given loss function $\mathcal{L}$. As seen in Section 13.4.2, this procedure usually follows a gradient-

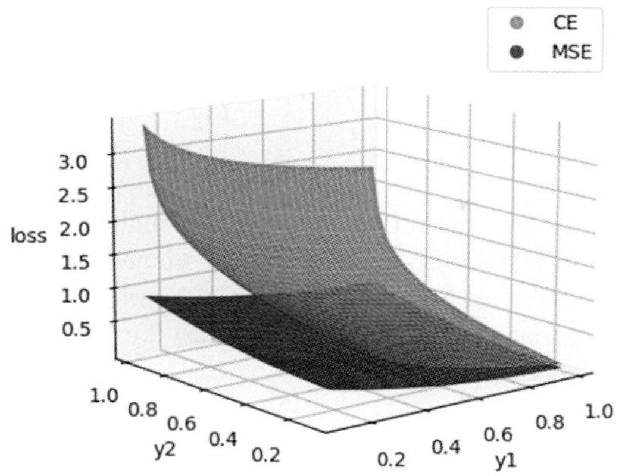

FIGURE 13.53

Loss surfaces for CE and MSE.

based weight adjustment of the form

$$w(k+1) = w(k) - \mu \left. \frac{\partial \mathcal{L}}{\partial w} \right|_{w=w(k)}, \tag{13.108}$$

where μ controls the learning rate.

Determining the partial derivatives of $\mathcal{L}$ in this equation is a very costly operation, especially for large datasets. Therefore, training data are often partitioned in smaller subsets, reducing the computational cost in each iteration. Data correlation provides some stability in the gradient estimation for each minibatch. This strategy allows one to perform a larger number of updates for a given computational cost, thus speeding up the learning process.

Finding a good value for μ in Eq. (13.108) is a very difficult task in most practical problems: Small values of μ make learning too slow, requiring a large number of iterations (and a large amount of time) to converge, whereas large values of μ may lead to algorithm divergence. In addition, the highly nonconvex nature of $\mathcal{L}$ often leads to suboptimal convergence due to the ever-present local minima of its surface in the parameter space. Momentum-based algorithms adaptively adjust the learning rate along the network training and avoid the weight vector getting trapped in the local minima valleys by averaging the partial derivatives over several consecutive iterations. Among the several momentum-based strategies perhaps the most currently employed are the RMSProp and Adam algorithms described below.

The RMSProp algorithm uses a weight update procedure of the form

$$w(k+1) = w(k) - \frac{1}{\sqrt{m(k+1)}} \left. \frac{\partial \mathcal{L}}{\partial w} \right|_{w=w(k)}, \tag{13.109}$$

where $m(k+1)$ is determined by

$$m(k+1)=\alpha m(k)+(1-\alpha)\left(\left.\frac{\partial \mathcal{L}}{\partial w}\right|_{w=w(k)}\right)^2. \tag{13.110}$$

In this process, RMSProp works as some form of gradient normalization, which avoids extreme cases of learning rate, where $m(k)$ is a smoothed version of the squared gradient for each weight. In this scheme α is a hyperparameter often set to $\alpha = 0.9$ or adjusted during validation.

Adam estimation uses a similar gradient normalization to the RMSProp but employs a smoothed gradient version in the weight update step [34]:

$$\begin{aligned} v(k+1) &= \beta v(k)+(1-\beta)\left.\frac{\partial \mathcal{L}}{\partial w}\right|_{w=w(k)}, \\ m(k+1) &= \alpha m(k)+(1-\alpha)\left(\left.\frac{\partial \mathcal{L}}{\partial w}\right|_{w=w(k)}\right)^2, \\ w(k+1) &= w(k)-\frac{1}{\sqrt{m(k+1)}}v(k+1). \end{aligned} \tag{13.111}$$

RMSProp and Adam can be used interchangeably with the designer's choice mostly based on his or her previous experience with those (or any other) momentum algorithms. In some situations, one may even opt to the standard stochastic steepest descent procedure, which has fewer parameters to set.

13.6.6 Network regularization

Deeper models are prone to overfitting as their extremely large number of parameters can model very complicated data idiosyncrasies. This undesired behavior may be mitigated by increasing training data, applying early stop during training as guided by validation, or using network regularization. In this last approach, constraints are enforced in the parameter values, restricting their respective domain, to avoid modeling very complex and probably undesired patterns. Among the most used regularization strategies, we consider the weight decay and dropout techniques described below.

Weight decay or ℓ_1/ℓ_2 regularization constrains the ℓ_1- (sum of absolute values) or ℓ_2-norms (sum of squared values) of the network weight matrix $\mathbf{W}$. These strategies penalize large weight values, favor smooth networks, and are implemented by incorporating a penalty to the loss function definition, such as

$$\begin{aligned} \ell_1 &: \mathcal{L}_1(\mathbf{y},\hat{\mathbf{y}}) = \mathcal{L}(\mathbf{y},\hat{\mathbf{y}})+\lambda\sum_i\sum_j |w_{i,j}|, \\ \ell_2 &: \mathcal{L}_2(\mathbf{y},\hat{\mathbf{y}}) = \mathcal{L}(\mathbf{y},\hat{\mathbf{y}})+\lambda\sqrt{\sum_i\sum_j |w_{i,j}|^2}, \end{aligned} \tag{13.112}$$

where $w_{i,j}$ denotes the (i,j) entry of $\mathbf{W}$. In this process, λ is a system hyperparameter usually initialized as $\lambda = 0$ and slowly increased along the training process in order to enforce network regularization. In practice, ℓ_1 regularization reduces the number of nonzero coefficients, whereas ℓ_2 regularization tends to remove larger parameters.

Weight decay is the standard regularization technique in several statistical modeling fields. In the DL community, however, a different approach is often employed due to its great success in the AlexNet architecture. The dropout regularization [57] randomly removes or ignores different neurons in each training step. This can be interpreted as the elementwise multiplication of a binary masking matrix to the original weight matrix, leading to the new weight matrix

$$\mathbf{W}_d = \mathbf{M} \odot \mathbf{W}, \tag{13.113}$$

with $\mathbf{M}$ generated according to a Bernoulli(p) distribution. In practice, as each network behaves differently, finding an appropriate value of p often requires great experimentation. In that sense, one may start with $p = 0$ and increase its value up to $p = 0.5$, particularly for the last hidden layers of the network, and analyze the effect of these changes along the training process. Dropout can be interpreted as an extreme case of ensemble learning (see Section 13.5), where several distinct but correlated models are trained for a given task and the final model is given by their output average or mode values. As a final note on the dropout approach, one must keep in mind that the final network will have all active weights, as opposed to the incomplete networks generated along training. Therefore, one must adjust the weight values in the complete network to compensate for the dropout removal.

13.6.7 Deep learning frameworks

Among the several platforms for DL development, perhaps the most used ones are TensorFlow and PyTorch [35].

TensorFlow is an open source library developed by Google first made available to the general public in November 2015. TensorFlow allows for optimized model training and execution in several hardware configurations, including multiple CPU and GPU servers. As of its 2.0 version released in 2019, TensorFlow has a built-in mode for simplified network design. Using TensorFlow is highly facilitated by a high-level front-end library such as Keras calling the TensorFlow platform in the background.

PyTorch was first released in January 2017, almost one year later than TensorFlow. PyTorch has a very dynamic interface facilitating user iteration. The nature of PyTorch makes for a very flexible design strategy, as the user can easily modify the entire network during its training process. Other platforms often require the designer to specify all network characteristics before executing the training procedure. PyTorch also simplifies script debugging as one can choose desired outputs on the fly.

Along the years, TensorFlow and PyTorch have learned with each other, and each platform has incorporated advantages initially associated with the other. Nowadays, their most recent versions are somewhat equivalent in terms of usage and performance. All in all, as often occurs in practice, the best framework for us is the one that our lab staff is more familiar with. We just need to find the right material (book, video, class notes, etc.) or the right person to assist us along the initial "Hello world" designs and enjoy the rest of the ride.

13.6.8 Autoencoders

An autoencoder is a type of artificial NN that reduces its high-dimensional input, obtains a compressed and meaningful representation of it, and then uses this compressed version of the information to reconstruct the input. As autoencoders create their own labels from the training data, they are considered self-supervised algorithms. The autoencoder consists of three parts: encoder, code, and decoder, as shown in Fig. 13.54.

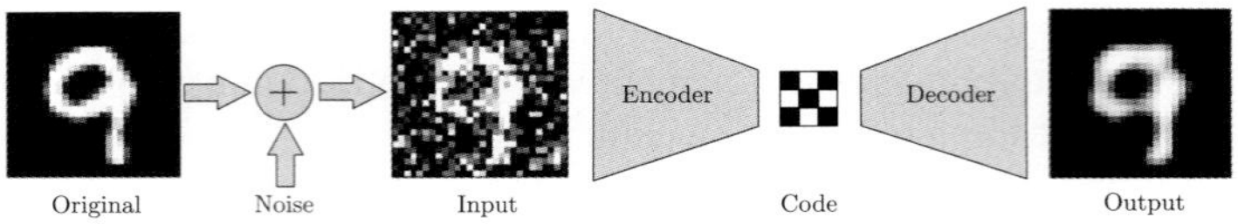

FIGURE 13.54

A denoising autoencoder example.

The encoder compresses the input and, as a result, produces the code. The code is a reduced feature representation of the input, also called the latent-space representation. Then, by minimizing the loss function, the decoder reconstructs the input using the code. Therefore, the autoencoder is particularly helpful for compression, feature extraction, and denoising tasks. For instance, in denoising autoencoders, the input is corrupted by noise, as illustrated in Fig. 13.54. The autoencoder is then responsible for reconstructing the clean version of the input. In other words, denoising autoencoders aim at making the encoder resistant to perturbations in the input data. For example, consider a denoising example using the MNIST [39] database: an extensive dataset containing digits handwritten by students and employees of the United States Census Bureau. This dataset was first detailed in Section 13.3.5. Some samples of the original MNIST dataset are shown in the first row of images in Fig. 13.55. The samples are corrupted by noise, resulting in the second row of images illustrated in Fig. 13.55. The noisy images are used as input to the autoencoder. The autoencoder processes the input and obtains as output the denoised images in the third row of Fig. 13.55.

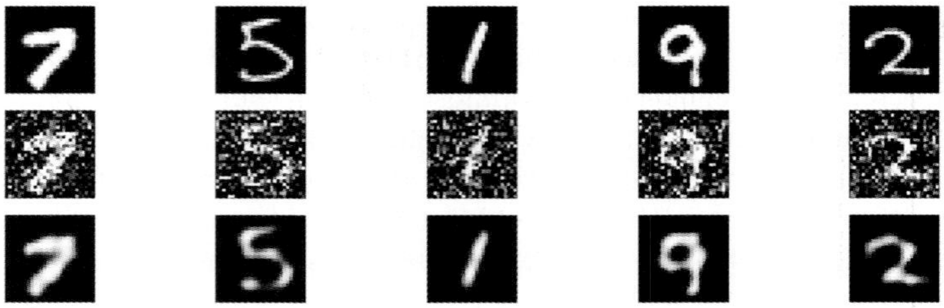

FIGURE 13.55

Results obtained using a denoising autoencoder. The first row consists of the original images, the second row is the corrupted images used as input to the autoencoder, and the third row is the output of the autoencoder.

13.6.9 Adversarial training

In Section 13.4.10, we observed how powerful the deep CNNs can be at solving classification tasks even better than us, humans. However, the NNs are very susceptible to small perturbations in the input image, barely unseen by the human eyes [22,60]. These malicious perturbations of the network input are known as adversarial examples. They can easily fool the trained model with high confidence, as illustrated in Fig. 13.56.

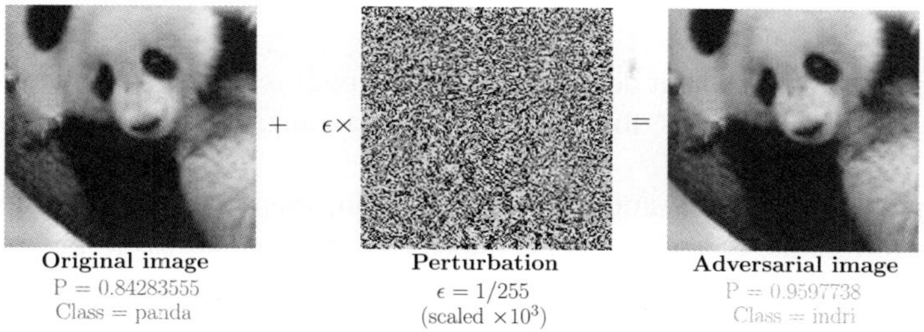

FIGURE 13.56

Example of adversarial image generation.

Adversarial examples are created by finding a small perturbation $\boldsymbol{\eta}$ to be added to the clean image $\mathbf{x}$ so that the new image $\mathbf{x}' = \mathbf{x} + \boldsymbol{\eta}$ is misclassified by the model. So far, adversarial training is the state-of-the-art defense approach to mitigate the effects of adversarial attacks and hence improve the model robustness. In adversarial training, the adversarial examples are continually produced and injected during the training process of a DNN classifier,

$$f_{\mathbf{w}}(\mathbf{x}) : \mathbb{R}^N \rightarrow \{1 \cdots C\}, \tag{13.114}$$

with $\mathbf{w}$ weights, which maps an input image $\mathbf{x}$ to a label y from a dataset $\mathcal{D}$ with C possible classes. Its main objective consists of solving the min-max optimization problem

$$\begin{aligned} &\min_{\mathbf{w}} \frac{1}{|\mathcal{D}|} \sum_{\mathbf{x}, y \in \mathcal{D}} \max_{\boldsymbol{\eta}} J(f_{\mathbf{w}}(\mathbf{x} + \boldsymbol{\eta}), y), \\ &\text{s.t. } \|\boldsymbol{\eta}\|_p \leq \epsilon, \end{aligned} \tag{13.115}$$

where $J(f_{\mathbf{w}}(\mathbf{x} + \boldsymbol{\eta}), y)$ is the loss function on the adversarial sample and $\boldsymbol{\eta}$ is a small perturbation constrained by ϵ.

The inner maximization problem in Eq. (13.115) is responsible for creating the adversarial examples by maximizing the loss function J in an effort to change the classification so that $f_{\mathbf{w}}(\mathbf{x} + \boldsymbol{\eta}) \neq f_{\mathbf{w}}(\mathbf{x})$. To ensure that the adversarial example keeps similar to the clean sample, the distance between them ($\|\boldsymbol{\eta}\|_p$) is constrained to be less than ϵ under a particular norm. The norms aim to quantify how imperceptible to humans an adversarial example is. Some examples of norms are the l_0-norm, l_2-norm, and l_∞-norm. In the l_0-norm, ϵ bounds the total number of pixels in $\mathbf{x}'$ that can be modified with respect to $\mathbf{x}$. For the l_2-norm, ϵ bounds the total squared distance between the values of pixels in $\mathbf{x}'$ and the corresponding pixels in $\mathbf{x}$. And in the l_∞-norm, also known as max norm, ϵ bounds the maximum difference between any pixel in $\mathbf{x}'$ and the corresponding pixel in $\mathbf{x}$.

After the inner maximization problem is addressed, it is time to solve the outer minimization problem in Eq. (13.115) and find the model parameters that minimize the loss on the generated adversarial examples. The model parameters

$$\mathbf{w}_t = \mathbf{w}_{t-1} + \mu \frac{1}{|\mathcal{B}|} \sum_{\mathbf{x}, y \in \mathcal{B}} \nabla_{\mathbf{w}} J(f_{\mathbf{w}}(\mathbf{x} + \eta^*), y) \tag{13.116}$$

are updated using stochastic gradient descent (SGD), where $\mathcal{B}$ is the current minibatch. The gradient is evaluated at the maximum point η^* found in the inner maximization problem, thanks to the Danskin theorem [7].

When creating the adversarial example, the amount of information available to the attacker regarding the model categorizes the type of attack. If the attacker has full access to the model, including its gradients, it is known as a white-box attack. On the other hand, when the attacker has no access to the model, but it has some prior information, for example, the classes' probabilities, it is considered as a black-box attack.

13.6.9.1 White-box attack

(A) Fast gradient sign method (FGSM)

FGSM is an attack introduced by [22] that creates adversarial examples by differentiating the loss with respect to the input and updating it so that the loss J would increase on that sample,

$$\mathbf{x}' = \mathbf{x} + \epsilon \operatorname{sign}(\nabla_{\mathbf{x}} J(\mathbf{w}, \mathbf{x}, y)), \tag{13.117}$$

where $\operatorname{sign}(\cdot)$ is the sign function, $\nabla_{\mathbf{x}} J(\mathbf{w}, \mathbf{x}, y)$ is the loss gradient with respect to $\mathbf{x}$, and ϵ controls the perturbation of the adversarial examples.

We can see the impact on the classification when an adversarial image is presented to a standard trained network in Fig. 13.57. The classification is changed from car to plane with a probability of $P = 0.917503$, but the difference between the adversarial and the clean images is imperceptible to the human eye. The adversarial image is obtained via a FGSM attack with $\epsilon = 4/255$. The original image is obtained from the CIFAR10 dataset. The images in Fig. 13.57 are then presented to an adversarially trained network via FGSM attacks with $\epsilon = 4/255$. The classification results are shown in Fig. 13.58, where both the clean and the adversarial images are correctly classified.

(B) Projected gradient descent (PGD)

Proposed by [42], the PGD attack attempts to solve the inner maximization problem in Eq. (13.115) following an iterative procedure. In each iteration i, PGD employs the update rule

$$\mathbf{x}_i' = \operatorname{clip}_{\mathbf{x}+\epsilon}(\mathbf{x}_{i-1} + \alpha \operatorname{sign}(\nabla_{\mathbf{x}} J(\mathbf{w}, \mathbf{x}, y))), \tag{13.118}$$

where function $\operatorname{clip}_{\mathbf{x}+\epsilon}(\cdot)$ clips the input at the positions around the predefined perturbation range and α is a step size.

An illustrative example is shown in Fig. 13.59, in which both clean and adversarial images are presented to a standard trained model. For the considered input image, a perturbation with $\epsilon = 1/255$ is sufficient to fool the Resnet18 model with a probability of $P = 0.9297474$. Nevertheless, the adversarially trained Resnet18 is able to correctly classify the same images from Fig. 13.59 as shown in Fig. 13.60.

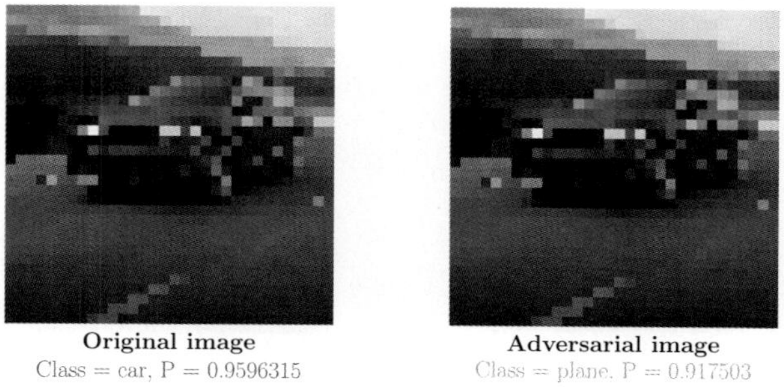

FIGURE 13.57

Classification results for original and adversarial examples via a standard trained Resnet18 network using the CIFAR10 dataset. The adversarial image is obtained via a FGSM attack with $\epsilon = 4/255$.

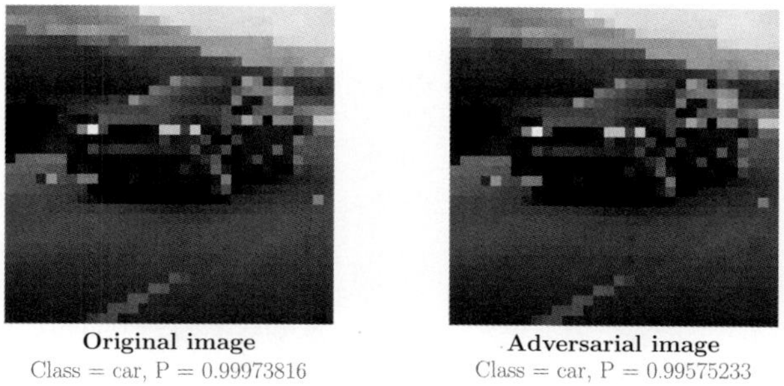

FIGURE 13.58

Classification results for original and adversarial examples via an adversarially trained Resnet18 network using the CIFAR10 dataset. The adversarial image is obtained via an FGSM attack with $\epsilon = 4/255$.

13.6.9.2 Black-box attack

(A) Pixel attack (sparsity-aware)

Originally, the pixel attack was proposed in [59] to perturb only one pixel with differential evolution (DE). This black-box attack aims at solving the sparse optimization problem

$$\begin{aligned} &\max_{\eta} P_{\text{adv}}(\mathbf{x}+\boldsymbol{\eta}), \\ &\text{s.t. } \|\boldsymbol{\eta}\|_0 \le d, \end{aligned} \tag{13.119}$$

where only the probabilities of the classes are available to perform the attack. The method tries to maximize the probability of the class being the adversarial one P_{adv} while the number of pixels being modified is constrained to d. The input image is represented by vector $\mathbf{x}$, where each element of $\mathbf{x}$ is a pixel of the image, and vector $\boldsymbol{\eta}$ has d nonzero elements. The adversarial class can be target or

FIGURE 13.59

Classification results for original and adversarial examples via a standard trained Resnet18 network using the ImageNet dataset. The adversarial image is obtained via a PGD attack with $\epsilon = 1/255$ and 20 iterations.

FIGURE 13.60

Classification results for original and adversarial examples via an adversarially trained Resnet18 network using the ImageNet dataset. The adversarial image is obtained via a PGD attack with $\epsilon = 1/255$ and 20 iterations.

untargeted. The perturbation is encoded into an array composed of five elements: xy-coordinates and RGB values of the perturbation. By using DE, a population of candidates or children for the perturbation vector is created at each iteration based on past candidates or parents. The children compete with their parents in terms of fitness and the winners survive for the next generation.

As shown in Figs. 13.61 and 13.62, $d = 3$ pixels are sufficient to fool Resnet18 for the CIFAR10 and ImageNet datasets, respectively. The perturbed pixels are highlighted in red (light gray in print version). The maximum number of iterations is set to 100 and the early-stop criterion will be triggered when the probability of the target class exceeds 90% or when the label of true class is lower than 5% in the case of nontargeted attacks.

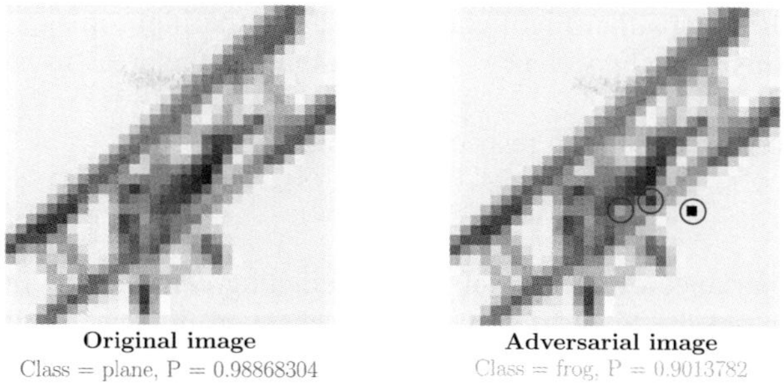

FIGURE 13.61

Classification results for original and adversarial examples via a standard trained Resnet18 network using the CIFAR10 dataset. The adversarial image is obtained via pixel attack with $d = 3$ pixels.

FIGURE 13.62

Classification results for original and adversarial examples via a standard trained Resnet18 network using the ImageNet dataset. The adversarial image is obtained via pixel attack with $d = 3$ pixels.

13.7 CNN visualization

Wondering how to interpret the results of some sophisticated DL methods is legitimate. This issue is critical in applications where high risks are involved. This section briefly discusses an approach related to CNN visualization.

13.7.1 Saliency map (vanilla and guided backpropagation)

By composing multiple layers, DNNs are able to process a big amount of data and hence learn complex tasks, as presented in Section 13.4.10. Nevertheless, we are still far from understanding how NNs work inside. What elements in the image lead to a particular prediction? How can small imperceptible perturbations be enough to make the NNs misclassify an image with high confidence? What makes an NN robust? We want to discuss these open questions with a helpful tool known as a saliency map.

Saliency maps are heatmaps that highlight the pixels of the input image that affected the output classification most. In this way, the saliency maps help unravel the areas in the image that impact the NN prediction.

By computing the loss gradient with respect to the input pixels,

$$\nabla_{\mathbf{x}} J(\mathbf{x}, y_c, \mathbf{w}), \tag{13.120}$$

we can analyze the importance of each pixel when predicting class c. The pixels for which the loss gradient would be large are the pixels that need to be changed the least to affect the class score the most.

Fig. 13.63 depicts the saliency map for a correctly classified gorilla when the network is trained in a standard manner. Observe that the highlighted pixels have a gorilla shape, indicating that the background is less important to the prediction. On the other hand, the adversarial version in Fig. 13.64 has a slightly different saliency map. In this case, we notice that some parts of the background influence the wrong classification. This dynamic behavior of the saliency maps suggests that the standard trained network is prone to fail when small perturbations are added to the clean image. By adversarially training the network, as presented in Section 13.6.9, we can obtain the saliency maps shown in Figs. 13.65 and 13.66 for clean and adversarial examples, respectively. Observe, however, that the saliency maps are the same in this case, indicating that the adversarially trained network is robust to small changes in the pixels.

FIGURE 13.63

Classification result of a clean image and its saliency map for a standard trained Resnet18 using the ImageNet dataset.

13.8 Deep reinforcement learning

RL is an ML algorithm in which an AI agent learns from the environment by interacting with it. The agent acts on the environment and it receives as feedback the reward and the state of the environment, as illustrated in Fig. 13.67. The agent aims at finding the action that maximizes its total reward.

In Q-learning, the Q-value or total reward $Q_t(s, a)$ represents how good action a for a particular state s at time t is. As training progresses, the algorithm stores the current Q-values for all possible

FIGURE 13.64

Classification result of an adversarial image and its saliency map for a standard trained Resnet18 using the ImageNet dataset.

FIGURE 13.65

Classification result of a clean image and its saliency map for an adversarially PGD trained Resnet18 using the ImageNet dataset.

combinations of s and a in the Q-table. The best action is then chosen based on the maximum Q-value for a given state. The Q-learning algorithm updates the Q-value $Q_t(s, a)$ as follows:

$$Q_{t+1}(s,a) = (1-\alpha)Q_t(s,a) + \alpha(r + \gamma \max_{a'} Q_t(s', a')), \tag{13.121}$$

where s' is the state reached from state s when performing action a, r is the immediate reward received when moving from s to s', α is the learning rate, and γ is the discount factor. The maximum Q-value $\max_{a'} Q_t(s', a')$ in the next state when considering all the available actions that lead to the next state is computed using the Q-table. After some iterations, the Q-table becomes a helpful tool to obtain the best action for a given state.

Nevertheless, the Q-table might become too large, exceeding the computer memory. A deep Q-network (DQN) can be used to approximate $Q_t(s, a)$, avoiding the use of a Q-table to remember the solutions. Instead of storing all the Q-values in the Q-table, an NN is trained to produce different Q-values for each action. The action related with the maximum Q-value is then chosen.

FIGURE 13.66

Classification result of an adversarial image and its saliency map for an adversarially PGD trained Resnet18 using the ImageNet dataset.

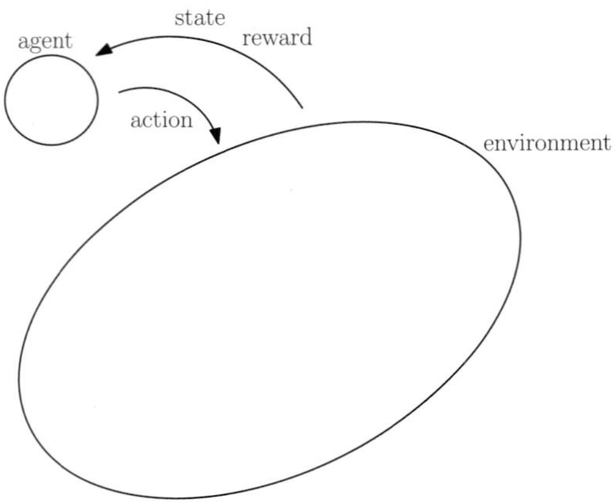

FIGURE 13.67

Illustration of reinforcement learning.

The NN is trained by minimizing the loss function

$$\mathcal{L} = (r + \gamma \max_{a'} \widehat{Q}(s', a') - Q(s, a))^2, \tag{13.122}$$

where $\max_{a'} \widehat{Q}(s', a')$ is the prediction using the current model on the next state s'.

The whole RL procedure is illustrated in Fig. 13.68, in which the ϵ-greedy policy dictates how to learn the problem. In the beginning, the exploration rate ϵ employed by the ϵ-greedy policy is high, and hence we uniformly select possible actions. The actions and associated states are stored in the memory replay, which is then used to train the NN. As the training progresses, the exploration rate ϵ decays and the action is chosen using the NN more often.

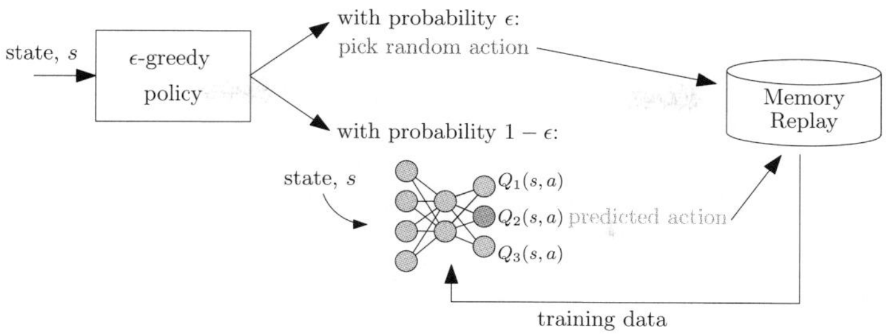

FIGURE 13.68

Illustration of ϵ-greedy policy.

For instance, consider a maze game illustrated in Fig. 13.69(a). The black squares represent walls, the pink squares (mid gray in print version) are starting and ending positions, and the yellow (light gray in print version) and white squares are possible positions the player can choose. Here, the agent is the player, the environment is the maze itself, the state is the player's position, the action is the direction the player can take (left, right, up, and down), and the reward is based on the player's position in the maze. For this example, the reward is -1 when the player reaches yellow squares (light gray in print version), whereas 0 is the reward for white squares. After 54 steps, our agent can reach the ending position, as shown in Fig. 13.69(b).

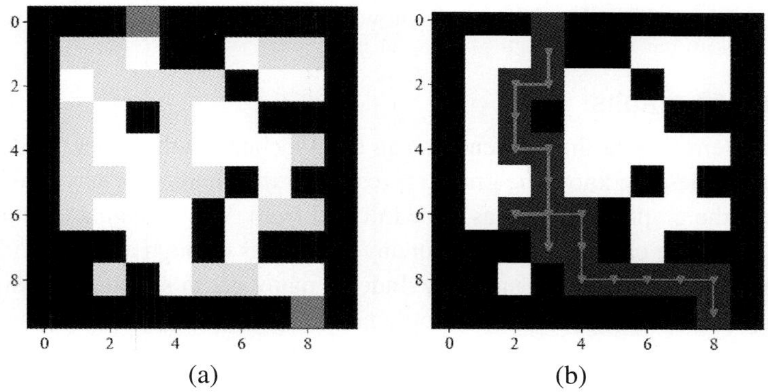

FIGURE 13.69

Illustration of (a) starting maze and (b) learned path.

13.9 Current trends

This section describes some topics in the ML field which are currently widely investigated. The descriptions are brief but aim to attract readers to pursue further research.

13.9.1 Transfer learning

DL architectures starve for data in order to adjust their parameters along their computationally intensive training procedures. However, getting more labeled data is very costly and time consuming, which motivates us to find data-efficient strategies to train such structures in new problems. One possible way to reduce this data dependency is to avoid starting the network training from scratch and use a pretrained network on a similar problem as our initial condition.

For instance, initial layers tend to extract basic features such as lines, edges, and simple shapes when dealing with images. Deeper layers, in contrast, extract high-level features such as textures and image content. The closer the layer is to the network output, the more tuned it becomes to the problem at hand. Therefore, when dealing with image-related problems, one may use a deep network that was previously trained on the ImageNet challenge as a starting point. Although the specific aim of the two problems may be different, the basic image processing aspects may be similar in a way that the second problem can take advantage of the knowledge acquired when processing the ImageNet dataset. The network is then retrained using our limited dataset specially devised for our particular new task.

A similar approach can be used in the speech recognition problem. When dealing with languages with limited annotated data, one may opt to start from an English-language recognizer, developed based on thousands of hours of annotated English speech, and then further adjust or retrain the network for recognizing another language with a limited dataset.

This strategy is often referred to as transfer learning, as all information acquired in the previous problem is transferred to the new one via the network parameters used as initial conditions in the training procedure. In this manner, the network does not start learning a new problem from scratch but as we humans often do, it takes advantage of all knowledge acquired from previous similar situations.

13.9.2 Learning on graphs

There are many problems where the data entries can be associated with graphs, as described in Chapters 14 and 16. Sometimes, we know the graph representing information or physical networks beforehand. In other cases, the graph structure has to be inferred from data through a learning process. From the ML point of view, if we can relate the data in any two nodes through an edge with a weight value, then a graph learning problem follows naturally. Indeed, many graph signal processing tools apply to ML, such as in graph NN problems where we apply nonlinearities to the data employed as graph data.

A graph learning framework naturally represents the connectivity experienced by devices that are the sources of big datasets, falling within the domains of optimization methods required to solve many ML problems.

13.9.3 Interpretable artificial intelligence

Despite their remarkable achievements in many applications, many ML and AI solutions often lack interpretability or explanations to corroborate their reliability. This issue is particularly relevant in the context of the widely used DL solution. The aim is to incorporate interpretability to classical ML models exposing the relationships among the domains that produced the observed data [61]. As advocated in [44], to be helpful, interpretability needs a precise definition in the context of ML, involving the prediction and descriptive accuracies, as well as relevancy.

In the prediction stage of the ML process, we have to ensure the model is adequate to produce high prediction accuracy when employing the available test data. Then, the concern is if the interpretation of the learned relations by a model is faithful; this issue relates to descriptive accuracy. Finally, the overall interpretation is relevant only if it offers insight into the problem at hand and for a targeted audience; see [44] for details.

In many cases, traditional ML-based decisions are not understandable, motivating a booming interest of the research community in interpretable or explainable AI (XAI). This challenging research area has advanced; for instance, it is possible to search for a meaningful dictionary matrix whose columns and their combinations have semantic, and conceptual meaning is a successful approach for creating practical XAI algorithms, as illustrated in Chapter 17.

Some tools and directions applicable to interpretable ML are discussed in Chapters 15, 16, and 17 of this book.

13.9.4 Federated learning

Federated learning (FL) is a new paradigm in the ML field for admitting decentralized training performed on edge devices. FL allows smartphones, drones, medical monitoring devices, vehicles, etc., to learn using data available in the corresponding edge devices. The local learning algorithms rely on the processing power and storage capability of each device. Let us suppose a device has access to a downloaded model, and it is able to learn further by using the locally available data. Instead of uploading the data for processing in a cloud infrastructure, it is beneficial to perform a limited and focused update with the local infrastructure. For further details, refer to the review article on the subject [18] and the references therein.

Fig. 13.70 illustrates a possible configuration including edge devices with varying storage and computational capabilities, where each device might contribute to the FL process constrained by its respective limitation.

The FL concept entails new ML tools to deal with heterogeneous edge devices, with distinct computational and storage capabilities, each device with its inherent privacy and security requirements. The full implementation of the learning system using FL benefits from the data available in millions of devices, possibly alleviating the burden of the technical requirements of learning from large datasets which are typically processed in homogeneous networks placed at distinct servers. The referred burden consists of high-throughput and low-latency connections to training data required by solo cloud-based computing learning systems.

The edge devices learn from their data and perform focused updates, and only the outcomes are sent to the cloud. The edge devices can immediately use the information they learn, whereas at the clouds the encrypted contributions of several devices are properly averaged. The information of the individual device should not be decrypted separately for privacy purposes.

The FL configuration might include several access points, such as in an urban environment as depicted in Fig. 13.71, where the edge devices communicate with the closer access point and all data updates are congregated in a central server.

The structure of the FL system consists of an individual multinode learning process, where each node i has available N_i dataset input–output pairs denoted as

$$\mathcal{D}_i = \{(\mathbf{x}_i(1), y_i(1)), (\mathbf{x}_i(2), y_i(2)), \cdots, (\mathbf{x}_i(N_i), y_i(N_i))\}, \tag{13.123}$$

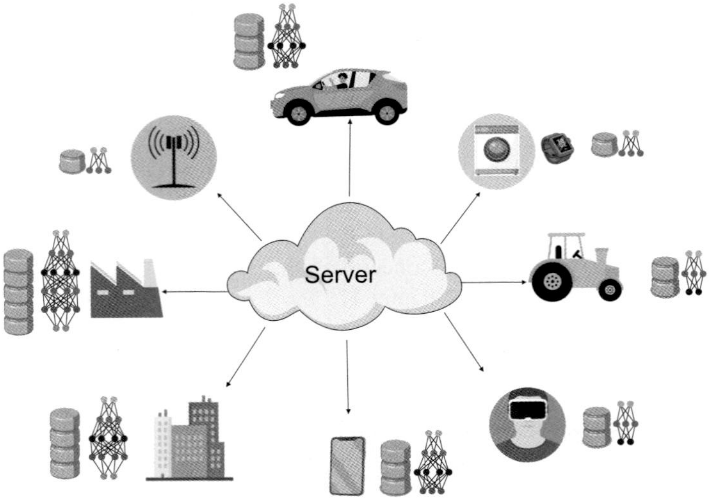

FIGURE 13.70

Federated learning illustration.

where $\mathbf{x}_i(q) \in \mathbb{R}^{n_i} \times 1$ for $q = 1, \cdots N_i$.

At the edge device, the objective is the minimization of the local function denoted as

$$\tilde{J}_i(\mathbf{W}_i; \mathcal{D}_i)\big|_{q=1}^{N_i} = \tilde{J}_i\left(\mathbf{W}_i; \mathbf{x}_i(q), y_i(q)\right)\big|_{q=1}^{N_i}, \tag{13.124}$$

where for the ith device, the aim is to minimize the local cost function utilizing the currently available data,

$$J_i(\mathbf{W}_i; \mathcal{D}_i)\big|_{l=1}^{N_i} = \frac{1}{N_i}\sum_{q=1}^{N_i} \tilde{J}_i\left(\mathbf{W}_i; \mathbf{x}_i(q), y_i(q)\right)\big|_{l=1}^{N_i}, \tag{13.125}$$

where J_i represents the objective function related to the ith edge device. The edge updated coefficient is then given by

$$\mathbf{W}_i(k+1) = \min_{\mathbf{W}_i} J_i(\mathbf{W}_i; \mathcal{D}_i)\big|_{l=1}^{N_i}. \tag{13.126}$$

The minimization of the local cost function at the edge devices uses the available data at node i. The set of data available is usually small to grasp an accurate model. In addition, it is possible that the chosen subset of parameters updated at node i, $\mathbf{W}_i(k+1)$, does not encompass the full set of learning coefficients.

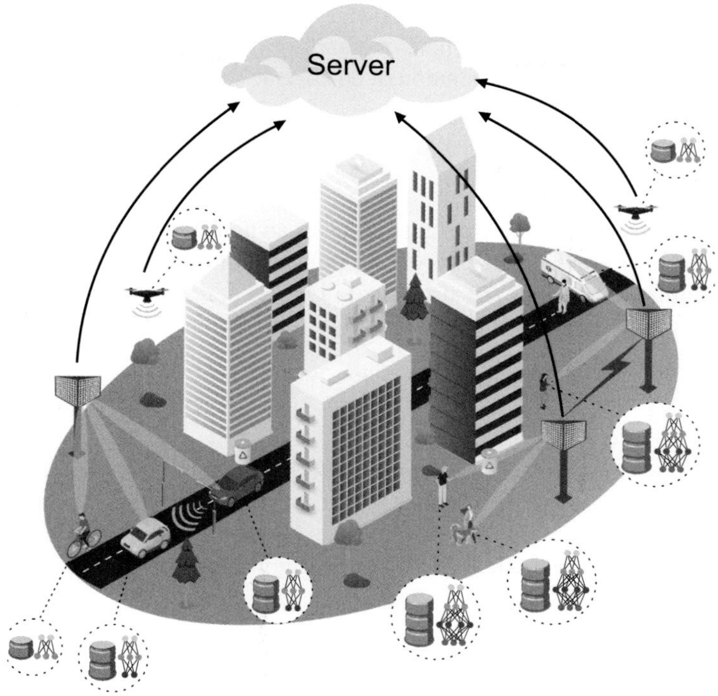

FIGURE 13.71

Federated learning system including several access points in an urban area.

In a centralized configuration, the overall cost function to be minimized is

$$J(\mathcal{W}) = \sum_{i=1}^{N} \alpha_i J_i \left(\mathbf{W}_i; \mathcal{D}_i\right), \tag{13.127}$$

where $\alpha_i = \frac{N_i}{\sum_{q=1}^{N} N_q}$, where N is the number of edge devices. The symbol $\mathcal{W}(k)$ represents the set of central parameters. The updated coefficient is then given by

$$\mathcal{W}(k+1) = \min_{\mathcal{W}(k)} J\left(\mathcal{W}(k); \mathcal{D}(k)\right), \tag{13.128}$$

where $\mathcal{D}(k)$, encompasses $\mathcal{D}_i$, $\forall i$.

At each edge device, a gradient descent algorithm updates the weights iteratively following the negative gradient direction,

$$\mathbf{W}_i(l+1) = \mathbf{W}_i(l) - \mu_i \frac{\partial J_i(\mathbf{W}_i(l); \mathcal{D}_i(l))}{\partial \mathbf{W}_i(l)}\bigg|_{l=1,\ldots,E}, \tag{13.129}$$

where μ_i is the step size employed in the ith device, $\mathcal{D}_i(k)$ represents a randomly selected minibatch at the edge, and $\mathbf{W}_i(1) = \mathcal{W}(k)$. The E samples are uniformly drawn from the set of N_i local samples.

A possible way to aggregate the contribution of all edge devices to the parameter update is given by

$$\mathcal{W}(k+1) = \frac{N}{s} \sum_{i \in \mathcal{S}} \alpha_i \mathbf{W}_i(k+1), \tag{13.130}$$

where $\mathcal{S}$ represents the set of s edge devices that provides an update at the current iteration.

It is now convenient to provide some very brief information related to FL convergence following the work of [41]. Similar discussions are also found in [12], but details of the convergence issues fall beyond our scope here. The convergence of the FL algorithm requires that

$$\mathbb{E}[J(\mathcal{W})] - J(\mathcal{W}_o) \leq \frac{L}{\epsilon + KE - 1} \left(\frac{2B}{\nu^2} + \frac{\epsilon}{2} \mathbb{E}\left[\|\mathcal{W}(KE) - \mathcal{W}_o\|^2 \right] \right), \tag{13.131}$$

where the cost functions of the edge devices are considered L-smooth,[2] ν-strongly convex.[3] They also have stochastic gradient with bounded variance σ_i^2, and the $\|\cdot\|^2$ of their stochastic gradient is uniformly bounded by G^2; see [41]. There is a degree of not independent and identically distributed (i.i.d.) heterogeneity among the users given by Ψ [41]. The FL algorithms stops after KE iterations providing as solution $\mathcal{W}(T)$. By selecting $\epsilon = \max\left(\frac{8L}{\nu}; E\right)$ and $\mu_i = \frac{2}{\nu(\epsilon + kE)}$, we get the convergence where $B = \sum_{i=1}^{N} \alpha_i \sigma_i^2 + 6L\Psi + 8(E-1)^2 G^2$. This discussion mainly illustrates that the proof of convergence available for the FL algorithm requires a diminishing step size, a function of k.

13.9.5 Causal machine learning

ML methods perform well when assuming that the available data are i.i.d. However, the classical ML methods need to generalize better when data distribution changes occur. As a result, a current challenge in ML is to investigate tools to improve the generalization performance, particularly in prediction and classification involving large datasets originating from natural signals. A solution is formally proposing a model approach to the data generation process through the so-called structural causal model (SCM). The aim is to estimate the data behavior after an intervention is made in the data generation process. The end result is a causal inference procedure that might close the gap between ML and natural intelligence.

Classical ML inherently assumes that the dataset is i.i.d. Let us consider the supervised case with the representation repeated here for convenience:

$$\mathcal{D} = \{(\mathbf{x}(1), y(1)), (\mathbf{x}(2), y(2)), \cdots, (\mathbf{x}(Q), y(Q))\}. \tag{13.132}$$

The main goal is to acquire information about the conditional probability distribution denoted by $p(\mathbf{y}|\mathbf{X})$. The i.i.d. assumption is crucial to justify the division of the dataset into training, validation, and testing subsets. In a nutshell, for successful learning, it is expected that at least the training dataset

---

2 L-Smooth means $J(\mathcal{W}) \leq J(\mathcal{W}') + (\mathcal{W} - \mathcal{W}')^T \frac{\partial [J(\mathcal{W})]}{\partial \mathcal{W})} |_{\mathcal{W}'} + \frac{L}{2} \|\mathcal{W} - \mathcal{W}'\|_2^2$.

3 ν-Strongly convex means $J(\mathcal{W}) \geq J(\mathcal{W}') + (\mathcal{W} - \mathcal{W}')^T \frac{\partial [J(\mathcal{W})]}{\partial \mathcal{W})} |_{\mathcal{W}'} + \frac{\nu}{2} \|\mathcal{W} - \mathcal{W}'\|_2^2$.

follows the same distribution as the test dataset. A simple, intuitive example consists of a situation where in the whole training dataset, the object is placed in a given class of backgrounds, and in the testing phase, unexpected types of backgrounds surround the object. Broad discussions related to causal learning are available in the tutorial articles [53], [32].

A replacement for the i.i.d. assumption is SCM, where the dataset originates from samples of interventional distributions. The basic idea is to generate several datasets relying on distinct distributions represented as

$$\mathcal{D}_i = \{(\mathbf{x}_i(1), y_i(1)), (\mathbf{x}_i(2), y_i(2)), \cdots, (\mathbf{x}_i(Q), y_i(Q))\}, \tag{13.133}$$

so that we deal with a set of distributions $p(\mathbf{y}_i|\mathbf{X}_i)$, where $i \in I$ and I represents the set of instances. A distinct distribution generates a separate set of data samples for each instance.

A directed acyclic graph (DAG) is a valuable tool to represent well a priori assumptions about the causal relations in data generation. For instance, a parent node connects to the children's nodes through directed edges. In addition, the graphs help express how to factorize the distribution of the set of random variables leading to a substantial reduction in the complexity inherent to joint distributions.

Let us assume the joint distribution

$$p(a, b, c, d) = p(a|b, c, d)p(b|c, d)p(c|d)p(d), \tag{13.134}$$

where a, b, c, d represent the random variables and we utilized the probability rule to factorize the referred distribution.

If, for instance, a is independent of all the remaining variables, as per the DAG of Fig. 13.72, we can use the marginalized model $p(a)$ and simplify the conditionals as follows:

$$p(a, b, c, d) = p(a)p(b|a)p(c|a)p(d|b, c). \tag{13.135}$$

Note that the latter joint distribution involves four random variables, whereas six are present in Eq. (13.134).

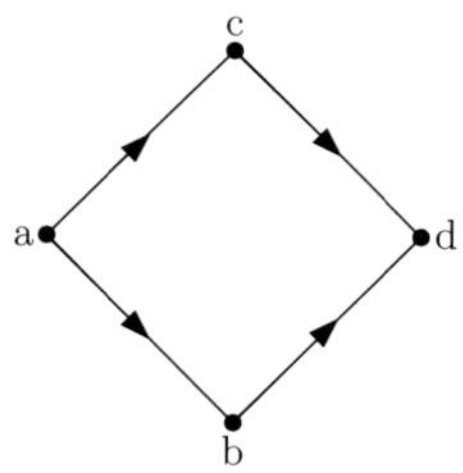

FIGURE 13.72

DAG of a Bayesian network.

Consider a graph whose nodes are connected with directed edges, where each edge connects a parent node to a child node. Let the ith node host a random variable x_i with the corresponding probability density function represented by $p(x_i)$. The parents of x_i are denoted as $\text{pa}(x_i)$, whereas if x_i is an ancestor of x_j, then $x_i \in \text{an}(x_j)$. Similarly, for a descendant of x_i represented by x_l we utilize the notation $x_l \in \text{de}(x_i)$.

If the graph contains the node x_m, $\forall m$ with joint distribution $p(\mathbf{x})$, the so-called Markov condition states that

$$p(x_m|\text{pa}(x_m)) = p(x_m|\mathbf{x} \notin \text{de}(x_m)), \tag{13.136}$$

where every node x_m is considered independent of the nondescendant nodes denoted as $\notin \text{de}(x_m)$. This condition leads to the following factorization of the joint distribution:

$$p(\mathbf{x}) = \prod_m p(x_m|\text{pa}(x_m)). \tag{13.137}$$

The Bayesian network representation aids the decomposition of complex joint distributions into smaller conditional distributions. In addition, we can benefit from the DAG structure to perform interventions.

The DAG of a Bayesian network is a starting point for providing interventions in the distribution of interest by acting on some variable denoted by do(·). Let us consider the case

$$p(a, c, d|b) = p(a)p(b|a)p(c|a)p(d|b, c). \tag{13.138}$$

By taking action on b, it is possible to observe its effect on the overall distribution according to the graph configuration as depicted in Fig. 13.73. The action is an intervention that can be described as

$$p(a, c, d|\text{do}(b)) = p(a)p(c|a)p(d|b, c) \neq p(a, c, d|b), \tag{13.139}$$

since that by taking an action, it follows that $p(b|a) = 1$. After performing this action, the variable a has no bearing on the behavior of b.

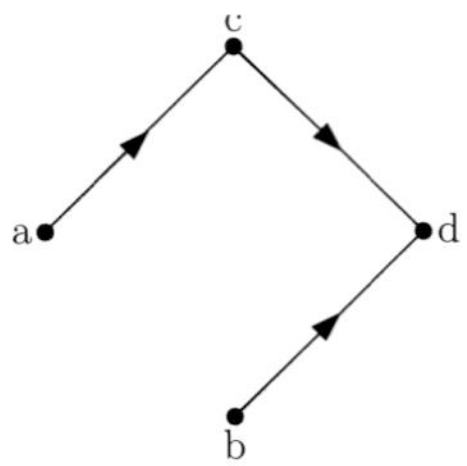

FIGURE 13.73

Causal Bayesian network as a result of intervention.

If we take a set $\mathcal{L}$ of interventions, it follows that

$$p(\mathbf{x}) = \prod_{m \notin \mathcal{L}} p(x_m|\text{pa}(x_m)). \tag{13.140}$$

This expression represents a truncated factorization of the distribution.

In ML strategies, the causal assumption is not often imposed; therefore, it is an extra ingredient to widen the generalization power of the ML solutions.

A missing tool for expanding the concept of causal ML is a counterfactual probe related to the possibility of imagining a situation where a particular process a has caused b, that is, a counterfactual account for a hypothetical backward-looking intervention related to the observed outcome. In order to allow a wider range of applications, in addition to employing the causal factorization of Eq. (13.140), it is possible to assume that each node signal is generated as

$$x_i = f_i(\mathrm{pa}(x_i), u_i) \tag{13.141}$$

for all nodes in the graph, giving rise to an SCM, where the random signal u_i accounts for an explained contribution to x_i and $f_i(\cdot)$, $\forall i$ are deterministic functions. In typical cases, the noise terms $u_1, u_2, \ldots, u_i, \ldots$ are assumed jointly independent. Through human actions or learning methods, these tools allow the analysis of contrafactuals, interventions, and observations. The SCM brings about the flexibility of computing intervention-based distributions, like the causal DGM, in addition to implementing counterfactuals by fixing the values of the noise signals.

The described tools form the basis for developing causal ML solutions, a subject with growing research interest; see [53], [32], and the references therein.

13.10 Concluding remarks

This chapter presented a review of ML, attempting to cover a wide range of branches of the field and provide the background required to endeavor in the topics presented in the chapters ahead. Since ML is poised to take part in many products under development, research efforts continue to widen the capabilities of ML-based solutions. The chapter illustrated many signal processing tools that are somehow incorporated or adapted to the existing ML algorithms; however, many new ideas flourished by the growing availability of large datasets and computer resources. Much more is yet to come!

13.11 Appendix: RNN's gradient derivations

This appendix shows the derivation of the gradients related to the RNNs.

13.11.1 Gradient derivations of the standard RNN

Further development of a gradient-based learning algorithm requires a number of partial derivatives of the chosen cost function with respect to the network parameters. All expressions apply the chain rule, and they are discussed below.

Let us first consider the expression of the gradient with respect to $\mathbf{W}_o$ and $\mathbf{b}_o$ as

$$\frac{\partial \bar{J}(\hat{\mathbf{y}}, \mathbf{y})}{\partial \mathbf{W}_o} = \sum_{k=1}^{K} \left[\frac{\partial J(k)}{\partial \hat{\mathbf{y}}(k)}\right]^T \frac{\partial \hat{\mathbf{y}}(k)}{\partial \mathbf{W}_o},$$

$$\frac{\partial \bar{J}(\hat{\mathbf{y}}, \mathbf{y})}{\partial \mathbf{b}_o} = \sum_{k=1}^{K} \left[\frac{\partial J(k)}{\partial \hat{\mathbf{y}}(k)} \right]^T \frac{\partial \hat{\mathbf{y}}(k)}{\partial \mathbf{b}_o}. \tag{13.142}$$

We now continue by deriving the expression of the gradient with respect to $\mathbf{W}_i$ by accessing the transition between the instances k and $k+1$:

$$\begin{aligned}\frac{\partial J(k+1)}{\partial \mathbf{W}_i} &= \left[\frac{\partial J(k+1)}{\partial \mathbf{h}(k+1)} \right]^T \frac{\partial \mathbf{h}(k+1)}{\partial \mathbf{W}_i} + \left[\frac{\partial J(k+1)}{\partial \mathbf{h}(k)} \right]^T \frac{\partial \mathbf{h}(k)}{\partial \mathbf{W}_i} \\ &= \left[\frac{\partial J(k+1)}{\partial \mathbf{h}(k+1)} \right]^T \frac{\partial \mathbf{h}(k+1)}{\partial \mathbf{W}_i} \\ &+ \left[\frac{\partial J(k+1)}{\partial \mathbf{h}(k+1)} \right]^T \frac{\partial \mathbf{h}(k+1)}{\partial \mathbf{h}(k)} \frac{\partial \mathbf{h}(k)}{\partial \mathbf{W}_i}. \end{aligned} \tag{13.143}$$

The above equation only deals with the transition $k \rightarrow k+1$, where we consider that the layer output $\hat{\mathbf{y}}(k+1)$ feels the effect of $\mathbf{W}_i$ through $\mathbf{h}(k+1)$ and $\mathbf{x}(k+1)$. The second equality originates from the fact that $\mathbf{h}(k+1)$ is in some way a function of $\mathbf{h}(k)$; therefore, some sort of backpropagation scheme is required to compute the partial derivative. The dashed line in Fig. 13.29 represents the backpropagation path required to update the network parameters.

When considering the data at $k+1$, we need the so-called backpropagation through time (BPTT) to calculate the partial derivative of the cost function with respect to coefficients of the matrix $\mathbf{W}_i$. In Eq. (13.143), we can expand the summation to include all instants k, resulting in

$$\frac{\partial J(k+1)}{\partial \mathbf{W}_i} = \sum_{l=1}^{k+1} \left[\frac{\partial J(k+1)}{\partial \mathbf{h}(k+1)} \right]^T \frac{\partial \mathbf{h}(k+1)}{\partial \mathbf{h}(l)} \frac{\partial \mathbf{h}(l)}{\partial \mathbf{W}_i}. \tag{13.144}$$

By taking into account the entire time sequence and highlighting the output signal, we obtain

$$\frac{\partial \bar{J}(\hat{\mathbf{y}}, \mathbf{y})}{\partial \mathbf{W}_i} = \sum_{k=0}^{K-1} \sum_{l=1}^{k+1} \left[\frac{\partial J(k+1)}{\partial \hat{\mathbf{y}}(k+1)} \right]^T \frac{\partial \hat{\mathbf{y}}(k+1)}{\partial \mathbf{h}(k+1)} \frac{\partial \mathbf{h}(k+1)}{\partial \mathbf{h}(l)} \frac{\partial \mathbf{h}(l)}{\partial \mathbf{W}_i}. \tag{13.145}$$

The partial derivative of the partial cost function with respect to $\mathbf{W}_f$, considering only the transition from k to $k+1$, is

$$\begin{aligned}\frac{\partial J(k+1)}{\partial \mathbf{W}_f} &= \left[\frac{\partial J(k+1)}{\partial \hat{\mathbf{y}}(k+1)} \right]^T \frac{\partial \hat{\mathbf{y}}(k+1)}{\partial \mathbf{h}(k+1)} \frac{\partial \mathbf{h}(k+1)}{\partial \mathbf{W}_f} \\ &= \left[\frac{\partial J(k+1)}{\partial \hat{\mathbf{y}}(k+1)} \right]^T \frac{\partial \hat{\mathbf{y}}(k+1)}{\partial \mathbf{h}(k+1)} \frac{\partial \mathbf{h}(k+1)}{\partial \mathbf{h}(k)} \frac{\partial \mathbf{h}(k)}{\partial \mathbf{W}_f}. \end{aligned} \tag{13.146}$$

The second equality originates from the fact that $\mathbf{h}(k+1)$ depends on $\mathbf{h}(k)$.

For $k+1$, we apply the BPTT to compute the partial derivative of the cost function with respect to coefficients of the matrix $\mathbf{W}_f$:

$$\frac{\partial J(k+1)}{\partial \mathbf{W}_f} = \sum_{l=1}^{k+1} \left[\frac{\partial J(k+1)}{\partial \hat{\mathbf{y}}(k)} \right]^T \frac{\partial \hat{\mathbf{y}}(k)}{\partial \mathbf{h}(k+1)} \frac{\partial \mathbf{h}(k+1)}{\partial \mathbf{h}(l)} \frac{\partial \mathbf{h}(l)}{\partial \mathbf{W}_f}. \tag{13.147}$$

By considering the entire time sequence we obtain

$$\frac{\partial \bar{J}(\hat{\mathbf{y}}, \mathbf{y})}{\partial \mathbf{W}_f} = \sum_{k=0}^{K-1} \sum_{l=1}^{k+1} \left[\frac{\partial J(k+1)}{\partial \hat{\mathbf{y}}(k)} \right]^T \frac{\partial \hat{\mathbf{y}}(k)}{\partial \mathbf{h}(k+1)} \frac{\partial \mathbf{h}(k+1)}{\partial \mathbf{h}(l)} \frac{\partial \mathbf{h}(l)}{\partial \mathbf{W}_f}. \tag{13.148}$$

In the same vein, a specific and widely used cost function in RNNs is the relative entropy function given by

$$\begin{aligned} \mathcal{L}(\hat{\mathbf{y}}, \mathbf{y}) &= \sum_{k=1}^{K} \left[\mathcal{L}(\hat{\mathbf{y}}(k), \mathbf{y}(k)) \right] \\ &= -\sum_{i=1}^{L} \sum_{k=1}^{K} \left[\mathbf{y}_i(k) \ln\left(\frac{\hat{\mathbf{y}}_i(k)}{\mathbf{y}_i(k)} \right) \right], \end{aligned} \tag{13.149}$$

where L is the length of vectors $\hat{\mathbf{y}}(k)$ and $\mathbf{y}(k)$, so that

$$\frac{\partial \mathcal{L}(\hat{\mathbf{y}}, \mathbf{y})}{\partial \hat{\mathbf{y}}(k)} = \big(\hat{\mathbf{y}}(k) - \mathbf{y}(k) \big). \tag{13.150}$$

Eq. (13.150) is valid if the cost function is the logistic function and the activation function is the softmax. This combination is crucial to avoid undesirable behavior in the network's learning process [62], since it generates gradients depend only on the error vector. A brief proof of this statement is worth pursuing. Since

$$\mathcal{L}(\hat{\mathbf{y}}_i, \mathbf{y}_i) = \sum_{k=1}^{K} \mathbf{y}_i(k) \ln\left(\frac{\hat{\mathbf{y}}_i(k)}{\mathbf{y}_i(k)} \right), \tag{13.151}$$

assuming that $\hat{\mathbf{y}}_i(k)$ is generated by applying the softmax to $\mathbf{h}_i(k)$, we have

$$\begin{aligned} \frac{\partial \mathcal{L}(\hat{\mathbf{y}}_i, \mathbf{y}_i)}{\partial \mathbf{h}_i(k)} &= -\frac{\partial}{\partial \mathbf{h}_i(k)} \left\{ \sum_{k=1}^{K} \mathbf{y}_i(k) \mathbf{h}_i(k) - \sum_{k=1}^{K} \mathbf{y}_i(k) \ln\left[\sum_{i=1}^{L} \exp(\mathbf{h}_i(k)) \right] \right\} \\ &= -\mathbf{y}_i(k) + \sum_{k=1}^{K} \mathbf{y}_i(k) \frac{\exp(\mathbf{h}_i(k))}{\sum_{i=1}^{L} \exp(\mathbf{h}_i(k))} \\ &= -\mathbf{y}_i(k) + \hat{\mathbf{y}}_i(k) \sum_{i=1}^{L} \mathbf{y}_i(k) = -(\mathbf{y}_i(k) - \hat{\mathbf{y}}_i(k)). \end{aligned} \tag{13.152}$$

The RNN uses the input layer vector $\mathbf{x}(k)$, for instance at time k, and the feedforward process representation is

$$\cdots \mathbf{h}(k-1) \parallel \hat{\mathbf{y}}(k-1) \xrightarrow[\mathbf{x}(k);\tanh\|\text{softmax}]{\mathbf{W}_i;\mathbf{W}_f;b_h\|\mathbf{W}_i;\mathbf{W}_o;b_o} \mathbf{h}(k) \parallel \hat{\mathbf{y}}(k) \xrightarrow[\mathbf{x}(k+1);\tanh\|\text{softmax}]{\mathbf{W}_i;\mathbf{W}_f;b_h\|\mathbf{W}_i;\mathbf{W}_o;b_o} \cdots, \tag{13.153}$$

where the symbol $\|$ in the present context represents simultaneous operations involving the network parameters, the input data, and the corresponding nonlinear functions.

An important issue affecting RNN algorithms is the vanishing and exploding behavior of some gradient components. We provide a brief illustration, inspired by [62], coming from the partial derivative recursion and applied to the scalar state, given by

$$\frac{\partial h(k+1)}{\partial h(k)} = w_f\left(1-h^2(k+1)\right). \tag{13.154}$$

The above equality results from the partial derivative of tanh as per Eq. (13.40). In a sequence of iterations, one gets

$$\frac{\partial J(k+1)}{\partial h(k)} = w_f^2\left(1-h^2(k+1)\right)\left(1-h^2(k+2)\right)\frac{\partial J(k+1)}{\partial h(k+1)} + \cdots. \tag{13.155}$$

This expression derives from Eqs. (13.33) and (13.40). As can be observed, since the terms $h^2(\cdot)$ result from the tanh operation, $|h(\cdot)| < 1$. Therefore, it is likely that the first term in the above equation will vanish in the long run. On the other hand, the powers of w_f will increase and if $w_f > 1$, Eq. (13.155) is prone to explode. This behavior requires alternative solutions for RNN, particularly in deep networks such as clipping or other structures.

The same conclusion follows if we consider a simple network mapping an input vector to an output vector, through many layers with the same weight matrix $\mathbf{W}_f$ being applied. This situation occurs in RNNs but not in the standard NNs and DNNs, and the RNN dynamic range is highly dependent on the weight matrix eigenvalues. This behavior also affects the gradients, leading to potential explosion or vanishing of their values.

13.11.2 Gradient derivations for the LSTM scheme

In the following, we present the gradient-related quantities required to implement the BPTT to update the LSTM-based network [5], [28]. As expected, the gradient calculations and the backpropagation procedures are more involved than the standard NN cases. Fortunately, these expressions present some structure that can benefit the implementation.

Define the output error as $J(k) = \frac{1}{2}\left(\mathbf{y}(k) - \hat{\mathbf{y}}(k)\right)^2$ to derive the variations in the quantities associated with the output of the network. For this case, the error in the network at k is

$$\boldsymbol{\delta}(k) = \mathbf{y}(k) - \hat{\mathbf{y}}(k). \tag{13.156}$$

The partial variation in the output signal with respect to the output coefficients is given by

$$\frac{\partial J(k)}{\mathbf{W}_{hy}} = \sum_{k=1}^{K}\left(\mathbf{h}(k) \odot \boldsymbol{\delta}(k)\right), \tag{13.157}$$

and the partial derivative of the cost function with respect to the hidden LSTM state is

$$\frac{\partial J(k)}{\partial \mathbf{h}(k)} = \mathbf{W}_{hy}\boldsymbol{\delta}(k), \tag{13.158}$$

where we will assume $\mathbf{W}_{hy}$ is an identity matrix from now on to simplify the derivations.

The internal variations in the LSTM building block are quantified as discussed below. The gradient of the cost function with respect to their output gate is

$$\begin{aligned}\frac{\partial J(k)}{\partial \mathbf{o}(k)} &= \left[\frac{\partial J(k)}{\partial \mathbf{h}(k)}\right]^T \frac{\partial \mathbf{h}(k)}{\partial \mathbf{o}(k)} = \left[\frac{\partial \mathbf{h}(k)}{\partial \mathbf{o}(k)}\right]^T \frac{\partial J(k)}{\partial \mathbf{h}(k)} \\ &= \tanh\left(\mathbf{c}(k)\right) \odot \boldsymbol{\delta}(k)\end{aligned} \tag{13.159}$$

by utilizing Eqs. (13.48) and (13.158). Using the same equations, the gradient with respect to the internal cell coefficients is

$$\begin{aligned}\frac{\partial J(k)}{\partial \mathbf{c}(k)} &= \left[\frac{\partial J(k)}{\partial \mathbf{h}(k)}\right]^T \frac{\partial \mathbf{h}(k)}{\partial \mathbf{c}(k)} = \left[\frac{\partial \mathbf{h}(k)}{\partial \mathbf{c}(k)}\right]^T \frac{\partial J(k)}{\partial \mathbf{h}(k)} \\ &= \left[\mathbf{1} - \tanh^2\left(\mathbf{c}(k)\right)\right] \odot \mathbf{o}(k) \odot \boldsymbol{\delta}(k),\end{aligned} \tag{13.160}$$

where $\mathbf{1}$ represents a column vector with entries equal to one. By taking into consideration Eq. (13.160), the gradient with respect to the forget gate is

$$\begin{aligned}\frac{\partial J(k)}{\partial \mathbf{f}(k)} &= \mathbf{c}(k-1) \odot \frac{\partial J(k)}{\partial \mathbf{c}(k)} \\ &= \mathbf{c}(k-1) \odot \left[\mathbf{1} - \tanh^2\left(\mathbf{c}(k)\right)\right] \odot \mathbf{o}(k) \odot \boldsymbol{\delta}(k),\end{aligned} \tag{13.161}$$

which derives from Eqs. (13.40) and (13.48). Also from Eqs. (13.47) and (13.161), it is possible to show that

$$\begin{aligned}\frac{\partial J(k)}{\partial \mathbf{c}(k-1)} &= \left[\frac{\partial J(k)}{\partial \mathbf{c}(k)}\right]^T \frac{\partial \mathbf{c}(k)}{\partial \mathbf{c}(k-1)} \\ &= \mathbf{f}(k) \odot \left[\mathbf{1} - \tanh^2\left(\mathbf{c}(k)\right)\right] \odot \mathbf{o}(k) \odot \boldsymbol{\delta}(k),\end{aligned} \tag{13.162}$$

whereas from Eqs. (13.45), (13.48), and (13.160), it follows that

$$\begin{aligned}\frac{\partial J(k)}{\partial \mathbf{g}(k)} &= \mathbf{i}(k) \odot \frac{\partial J(k)}{\partial \mathbf{c}(k)} \\ &= \mathbf{i}(k) \odot \left[\mathbf{1} - \tanh^2\left(\mathbf{c}(k)\right)\right] \odot \mathbf{o}(k) \odot \boldsymbol{\delta}(k),\end{aligned} \tag{13.163}$$

$$\begin{aligned}\frac{\partial J(k)}{\partial \mathbf{i}(k)} &= \mathbf{g}(k) \odot \frac{\partial J(k)}{\partial \mathbf{c}(k)} \\ &= \mathbf{g}(k) \odot \left[\mathbf{1} - \tanh^2\left(\mathbf{c}(k)\right)\right] \odot \mathbf{o}(k) \odot \boldsymbol{\delta}(k).\end{aligned} \tag{13.164}$$

By employing the internal variations, we can determine the coefficient matrix updates as

$$\begin{aligned}
\frac{\partial J(k)}{\partial \mathbf{W}_{ho}} &= \frac{\partial \mathbf{o}(k)}{\partial \mathbf{W}_{ho}} \frac{\partial J(k)}{\partial \mathbf{o}(k)} \\
&= \mathbf{o}(k) \odot (\mathbf{1} - \mathbf{o}(k)) \odot \mathbf{h}(k-1) \odot \frac{\partial J(k)}{\partial \mathbf{o}(k)} \\
&= \mathbf{o}(k) \odot (\mathbf{1} - \mathbf{o}(k)) \odot \mathbf{h}(k-1) \odot \tanh(\mathbf{c}(k)) \odot \boldsymbol{\delta}(k), \qquad (13.165)\\
\frac{\partial J(k)}{\partial \mathbf{W}_{hi}} &= \frac{\partial \mathbf{i}(k)}{\partial \mathbf{W}_{hi}} \frac{\partial J(k)}{\partial \mathbf{i}(k)} \\
&= \mathbf{i}(k) \odot (\mathbf{1} - \mathbf{i}(k)) \odot \mathbf{h}(k-1) \odot \frac{\partial J(k)}{\partial \mathbf{i}(k)} \\
&= \mathbf{i}(k) \odot (\mathbf{1} - \mathbf{i}(k)) \odot \mathbf{h}(k-1) \odot \mathbf{g}(k) \odot \left[\mathbf{1} - \tanh^2(\mathbf{c}(k))\right] \odot \mathbf{o}(k) \odot \boldsymbol{\delta}(k), \qquad (13.166)\\
\frac{\partial J(k)}{\partial \mathbf{W}_{hf}} &= \frac{\partial \mathbf{f}(k)}{\partial \mathbf{W}_{hf}} \frac{\partial J(k)}{\partial \mathbf{f}(k)} \\
&= \mathbf{f}(k) \odot (\mathbf{1} - \mathbf{f}(k)) \odot \mathbf{h}(k-1) \odot \frac{\partial J(k)}{\partial \mathbf{f}(k)} \\
&= \mathbf{f}(k) \odot (\mathbf{1} - \mathbf{f}(k)) \odot \mathbf{h}(k-1) \odot \mathbf{c}(k-1) \odot \left[\mathbf{1} - \tanh^2(\mathbf{c}(k))\right] \odot \mathbf{o}(k) \odot \boldsymbol{\delta}(k), \qquad (13.167)\\
\frac{\partial J(k)}{\partial \mathbf{W}_{hg}} &= \frac{\partial \mathbf{g}(k)}{\partial \mathbf{W}_{hg}} \frac{\partial J(k)}{\partial \mathbf{g}(k)} \\
&= \mathbf{g}(k) \odot \left(\mathbf{1} - \mathbf{g}^2(k)\right) \odot \mathbf{h}(k-1) \odot \frac{\partial J(k)}{\partial \mathbf{g}(k)} \\
&= \mathbf{g}(k) \odot \left(\mathbf{1} - \mathbf{g}^2(k)\right) \odot \mathbf{h}(k-1) \odot \mathbf{i}(k) \odot \left[\mathbf{1} - \tanh^2(\mathbf{c}(k))\right] \odot \mathbf{o}(k) \odot \boldsymbol{\delta}(k), \qquad (13.168)\\
\frac{\partial J(k)}{\partial \mathbf{W}_{xo}} &= \frac{\partial \mathbf{o}(k)}{\partial \mathbf{W}_{ho}} \frac{\partial J(k)}{\partial \mathbf{o}(k)} \\
&= \mathbf{o}(k) \odot (\mathbf{1} - \mathbf{o}(k)) \odot \mathbf{x}(k) \odot \frac{\partial J(k)}{\partial \mathbf{o}(k)} \\
&= \mathbf{o}(k) \odot (\mathbf{1} - \mathbf{o}(k)) \odot \mathbf{x}(k) \odot \tanh(\mathbf{c}(k)) \odot \boldsymbol{\delta}(k), \qquad (13.169)\\
\frac{\partial J(k)}{\partial \mathbf{W}_{xi}} &= \frac{\partial \mathbf{i}(k)}{\partial \mathbf{W}_{hi}} \frac{\partial J(k)}{\partial \mathbf{i}(k)} \\
&= \mathbf{i}(k) \odot (\mathbf{1} - \mathbf{i}(k)) \odot \mathbf{x}(k) \odot \frac{\partial J(k)}{\partial \mathbf{i}(k)} \\
&= \mathbf{i}(k) \odot (\mathbf{1} - \mathbf{i}(k)) \odot \mathbf{x}(k) \odot \mathbf{g}(k) \odot \left[\mathbf{1} - \tanh^2(\mathbf{c}(k))\right] \odot \mathbf{o}(k) \odot \boldsymbol{\delta}(k), \qquad (13.170)\\
\frac{\partial J(k)}{\partial \mathbf{W}_{xf}} &= \frac{\partial \mathbf{f}(k)}{\partial \mathbf{W}_{xf}} \frac{\partial J(k)}{\partial \mathbf{f}(k)} \\
&= \mathbf{f}(k) \odot (\mathbf{1} - \mathbf{f}(k)) \odot \mathbf{x}(k) \odot \frac{\partial J(k)}{\partial \mathbf{f}(k)} \\
&= \mathbf{f}(k) \odot (\mathbf{1} - \mathbf{f}(k)) \odot \mathbf{x}(k) \odot \mathbf{c}(k-1) \odot \left[\mathbf{1} - \tanh^2(\mathbf{c}(k))\right] \odot \mathbf{o}(k) \odot \boldsymbol{\delta}(k), \qquad (13.171)
\end{aligned}$$

$$\begin{aligned}\frac{\partial J(k)}{\partial \mathbf{W}_{xg}} &= \frac{\partial \mathbf{g}(k)}{\partial \mathbf{W}_{xg}} \frac{\partial J(k)}{\partial \mathbf{g}(k)} \\ &= \mathbf{g}(k) \odot \left(\mathbf{1} - \mathbf{g}^2(k)\right) \odot \mathbf{x}(k) \odot \frac{\partial J(k)}{\partial \mathbf{g}(k)} \\ &= \mathbf{g}(k) \odot \left(\mathbf{1} - \mathbf{g}^2(k)\right) \odot \mathbf{x}(k) \odot \mathbf{i}(k) \odot \left[\mathbf{1} - \tanh^2\left(\mathbf{c}(k)\right)\right] \odot \mathbf{o}(k) \odot \boldsymbol{\delta}(k),\end{aligned} \tag{13.172}$$

using Eqs. (13.159), (13.164), (13.161), (13.163), and (13.164), respectively. Therefore, the overall contributions to the coefficient matrix gradients related to the state vector $\mathbf{h}(k-1)$ are given by

$$\frac{\partial \bar{J}(k)}{\partial \mathbf{W}_{ho}} = \sum_{k=1}^{K} \frac{\partial J(k)}{\partial \mathbf{W}_{ho}}, \tag{13.173}$$

$$\frac{\partial \bar{J}(k)}{\partial \mathbf{W}_{hi}} = \sum_{k=1}^{K} \frac{\partial J(k)}{\partial \mathbf{W}_{hi}}, \tag{13.174}$$

$$\frac{\partial \bar{J}(k)}{\partial \mathbf{W}_{hf}} = \sum_{k=1}^{K} \frac{\partial J(k)}{\partial \mathbf{W}_{hf}}, \tag{13.175}$$

$$\frac{\partial \bar{J}(k)}{\partial \mathbf{W}_{hg}} = \sum_{k=1}^{K} \frac{\partial J(k)}{\partial \mathbf{W}_{hg}}, \tag{13.176}$$

$$\frac{\partial \bar{J}(k)}{\partial \mathbf{W}_{xo}} = \sum_{k=1}^{K} \frac{\partial J(k)}{\partial \mathbf{W}_{xo}}, \tag{13.177}$$

$$\frac{\partial \bar{J}(k)}{\partial \mathbf{W}_{xi}} = \sum_{k=1}^{K} \frac{\partial J(k)}{\partial \mathbf{W}_{xi}}, \tag{13.178}$$

$$\frac{\partial \bar{J}(k)}{\partial \mathbf{W}_{xf}} = \sum_{k=1}^{K} \frac{\partial J(k)}{\partial \mathbf{W}_{xf}}, \tag{13.179}$$

$$\frac{\partial \bar{J}(k)}{\partial \mathbf{W}_{xg}} = \sum_{k=1}^{K} \frac{\partial J(k)}{\partial \mathbf{W}_{xg}}. \tag{13.180}$$

The state update, required by the backpropagation procedure, has the following form:

$$\begin{aligned}\frac{\partial J(k)}{\partial \mathbf{h}(k-1)} &= \frac{\partial \mathbf{o}(k)}{\partial \mathbf{h}(k-1)} \frac{\partial J(k)}{\partial \mathbf{o}(k)} + \frac{\partial \mathbf{i}(k)}{\partial \mathbf{h}(k-1)} \frac{\partial J(k)}{\partial \mathbf{i}(k)} \\ &+ \frac{\partial \mathbf{f}(k)}{\partial \mathbf{h}(k-1)} \frac{\partial J(k)}{\partial \mathbf{f}(k)} + \frac{\partial \mathbf{g}(k)}{\partial \mathbf{h}(k-1)} \frac{\partial J(k)}{\partial \mathbf{g}(k)} \\ &= \mathbf{o}(k) \odot (\mathbf{1} - \mathbf{o}(k)) \odot \mathbf{W}_{ho} \frac{\partial J(k)}{\partial \mathbf{o}(k)} + \mathbf{i}(k) \odot (\mathbf{1} - \mathbf{i}(k)) \odot \mathbf{W}_{hi} \frac{\partial J(k)}{\partial \mathbf{i}(k)} \\ &+ \mathbf{f}(k) \odot (\mathbf{1} - \mathbf{f}(k)) \odot \mathbf{W}_{hf} \frac{\partial J(k)}{\partial \mathbf{f}(k)} + \mathbf{g}(k) \odot \left(\mathbf{1} - \mathbf{g}^2(k)\right) \odot \mathbf{W}_{hg} \frac{\partial J(k)}{\partial \mathbf{g}(k)}.\end{aligned} \tag{13.181}$$

The partial derivatives of the cost function with respect to the bias coefficients follow the same type of derivations, resulting in

$$\frac{\partial J(k)}{\partial \mathbf{b}_f(k)} = \left[\frac{\partial J(k)}{\partial \mathbf{f}(k)}\right]^T \frac{\partial \mathbf{f}(k)}{\partial \mathbf{b}_f(k)}$$
$$= \mathbf{f}(k) \odot [\mathbf{1} - \mathbf{f}(k)] \odot \mathbf{c}(k-1) \odot \left[\mathbf{1} - \tanh^2(\mathbf{c}(k))\right] \odot \mathbf{o}(k) \odot \boldsymbol{\delta}(k), \tag{13.182}$$

$$\frac{\partial J(k)}{\partial \mathbf{b}_i(k)} = \left[\frac{\partial J(k)}{\partial \mathbf{i}(k)}\right]^T \frac{\partial \mathbf{i}(k)}{\mathbf{b}_i(k)}$$
$$= \mathbf{i}(k) \odot [\mathbf{1} - \mathbf{i}(k)] \odot \mathbf{g}(k) \odot \left[\mathbf{1} - \tanh^2(\mathbf{c}(k))\right] \odot \mathbf{o}(k) \odot \boldsymbol{\delta}(k), \tag{13.183}$$

$$\frac{\partial J(k)}{\partial \mathbf{b}_o(k)} = \left[\frac{\partial J(k)}{\partial \mathbf{o}(k)}\right]^T \frac{\partial \mathbf{i}(k)}{\partial \mathbf{b}_o(k)}$$
$$= \mathbf{o}(k) \odot [\mathbf{1} - \mathbf{o}(k)] \odot \tanh(\mathbf{c}(k)) \odot \boldsymbol{\delta}(k), \tag{13.184}$$

$$\frac{\partial J(k)}{\partial \mathbf{b}_g(k)} = \left[\frac{\partial J(k)}{\partial \mathbf{g}(k)}\right]^T \frac{\partial \mathbf{i}(k)}{\partial \mathbf{b}_g(k)}$$
$$= \mathbf{i}(k) \odot \left[\mathbf{1} - \tanh^2(\mathbf{c}(k))\right] \odot \mathbf{o}(k) \odot \left[\mathbf{1} - \mathbf{g}^2(k)\right] \odot \boldsymbol{\delta}(k). \tag{13.185}$$

The reader should bear in mind that in all the expressions presented so far, due to the Kronecker products, the vector orderings are irrelevant.

References

[1] A. Antoniou, W.-S. Lu, Practical Optimization - Algorithms and Engineering Applications, 2nd edition, Springer, New York, NY, 2021.

[2] Y.S. Abu-Mostafa, M. Magdon-Ismail, H.-T. Lin, Learning from Data, AMLbook.com, Pasadena, CA, 2012.

[3] C.M. Bishop, Pattern Recognition and Machine Learning, Springer, New York, NY, 2006.

[4] T. Chen, C. Guestrin, XGBoost: A scalable tree boosting system, in: Proceedings of the 22nd ACM SIGKDD International Conference on Knowledge Discovery and Data Mining, 2016, p. 785794.

[5] G. Chen, A gentle tutorial of recurrent neural network with error backpropagation, ArXiv, arXiv:1610.02583v3 [abs], 2018.

[6] G. Cohen, S. Afshar, J. Tapson, A. van Schaik, EMNIST: an extension of MNIST to handwritten letters, arXiv preprint, arXiv:1702.05373, 2017.

[7] J.M. Danskin, The theory of max-min, with applications, SIAM Journal on Applied Mathematics 14 (4) (1966) 641–664.

[8] H. Dbouk, H. Geng, C.M. Vineyard, N.R. Shanbhag, Low-complexity fixed-point convolutional neural networks for automatic target recognition, in: International Conference on Acoustics, Speech, and Signal Processing (ICASSP-2020), 2020, pp. 1598–1602.

[9] J. Deng, W. Dong, R. Socher, L.-J. Li, K. Li, L. Fei-Fei, ImageNet: A large-scale hierarchical image dataset, in: Proc. Conference on Computer Vision and Pattern Recognition, 2009.

[10] M.P. Deisenroth, A.A. Faisal, C.S. Ong, Mathematics for Machine Learning, Cambridge University Press, Cambridge, UK, 2020.

[11] J.C. Duchi, E. Hazan, Y. Singer, Adaptive subgradient methods for online learning and stochastic optimization, Journal of Machine Learning Research 12 (2011) 2121–2159.
[12] C.T. Dinh, N.H. Tran, M.N.H. Nguyen, C.S. Hong, W. Bao, A.Y. Zomaya, V. Gamoli, Federated learning over wireless networks: convergence analysis and resource allocation, arXiv:1910.13067v4, 2020, pp. 1–15.
[13] P.S.R. Diniz, Adaptive Filtering: Algorithms and Practical Implementation, 5th edition, Springer, Cham, Switzerland, 2020.
[14] R. Dey, F.M. Salem, Gate-variants of Gated Recurrent Unit (GRU) neural networks, in: 2017 IEEE 60th International Midwest Symposium on Circuits and Systems (MWSCAS), 2017, pp. 1597–1600, https://doi.org/10.1109/MWSCAS.2017.8053243.
[15] B. Efron, T. Hastie, Computer Age Statistical Inference, Cambridge University Press, Cambridge, UK, 2016.
[16] J.H. Friedman, Greedy function approximation: A gradient boosting machine, The Annals of Statistics 29 (5) (2001) 11891232.
[17] Y. Freund, R.E. Schapire, A decision-theoretic generalization of on-line learning and an application to boosting, Journal of Computer and System Sciences 55 (1) (1997) 119–139.
[18] T. Gafni, N. Shlezinger, K. Cohen, Y.C. Eldar, H.V. Poor, Federated learning: an signal processing perspective, IEEE Signal Processing Magazine (2022) 14–41, https://doi.org/10.1109/MSP.2021.3125282.
[19] X. Glorot, Y. Bengio, Understanding the difficulty of training deep feedforward neural networks, in: Proc. International Conference on Artificial Intelligence and Statistics, Sardinia, Italy, 2010, pp. 249–256.
[20] X. Glorot, A. Bordes, Y. Bengio, Deep sparse rectifier neural networks, in: Proceedings of the Fourteenth International Conference on Artificial Intelligence and Statistics, 2011, pp. 315–323.
[21] I. Goodfellow, Y. Bengio, A. Courville, Deep Learning, MIT Press, Cambridge, MA, 2016.
[22] I.J. Goodfellow, J. Shlens, C. Szegedy, Explaining and harnessing adversarial examples, arXiv preprint, arXiv:1412.6572, 2014.
[23] G.H. Golub, C.F. Van Loan, Matrix Computations, 4th ed., John Hopkins University Press, Baltimore, ISBN 978-1-421407944, 2013.
[24] S. Haykin, Neural Networks and Learning Machines, Prentice-Hall, Englewood Cliffs, NJ, 2009.
[25] K. He, X. Zhang, S. Ren, J. Sun, Delving deep into rectifiers: surpassing human-level performance on ImageNet classification, in: Proc. International Conference on Artificial Intelligence and Statistics, Santiago, Chile, 2015, pp. 1026–1034.
[26] D.O. Hebb, The Organization of Behavior: A Neural Psychological Theory, John Wiley & Sons, New York, ISBN 9780415654531, 1949.
[27] R. Hecht-Nielsen, Theory of the backpropagation neural network, in: Neural Networks for Perception, Elsevier, 1992, pp. 65–93.
[28] S. Hochreiter, J. Schmidhuber, Long short-term memory, Neural Computation 9 (1997) 1735–1780.
[29] K. Hornik, M. Stinchcombe, H. White, Multilayer feedforward networks are universal approximators, Neural Networks 2 (1989) 359–366.
[30] T. Hastie, R. Tibshirani, J. Friedman, Elements of Statistical Learning: Data Mining, Inference, and Prediction, Springer, New York, NY, 2016.
[31] K. Jarrett, K. Kavukcuoglu, M.A. Ranzato, Y. LeCun, What is the best multi-stage architecture for object recognition?, in: Proc. International Conference on Computer Vision, Kyoto, Japan, 2009, pp. 2146–2153.
[32] J. Kaddour, A. Lynch, Q. Liu, M.J. Kusner, R. Silva, Causal machine learning: A survey and open problems, ArXiv, https://arxiv.org/abs/2206.15475, 2022.
[33] D. Kingma, J. Ba, Adam: A method for stochastic optimization, in: International Conference on Learning Representations, 2017, pp. 1–15, https://arxiv.org/abs/1412.6980.
[34] D.P. Kingma, J. Ba, Adam: A method for stochastic optimization, in: Proc. International Conference for Learning Representations, San Diego, USA, 2015, pp. 1–15.
[35] J. Krohn, G. Beyleveld, A. Bassens, Deep Learning Illustrated, Pearson Addison-Wesley, USA, 2020.
[36] T. Kohonen, Self-organized formation of topologically correct feature maps, Biological Cybernetics 43 (1982) 59–69.

[37] T. Kohonen, The self-organizing map, Proceedings of the IEEE 78 (9) (1990) 1464–1480.
[38] A. Krizhevsky, I. Sutskever, G.E. Hinton, ImageNet classification with deep convolutional networks, Advances in Neural Information Processing Systems 25 (2) (2012).
[39] Y. LeCun, C. Cortes, C.J.C. Burges, MNIST dataset of handwritten digit, http://yann.lecun.com/exdb/mnist/.
[40] Y. LeCun, L. Bottou, Y. Bengio, P. Haffner, Gradient-based learning applied to document recognition, Proceedings of the IEEE 86 (11) (1998) 2278–2324.
[41] X. Li, W. Yang, Z. Zhang, K. Huang, S. Wang, On the convergence of FedAvg on non-iid data, arXiv: 1907.02189v4, 2020, pp. 1–26.
[42] A. Madry, A. Makelov, L. Schmidt, D. Tsipras, A. Vladu, Towards deep learning models resistant to adversarial attacks, arXiv preprint, arXiv:1706.06083, 2017.
[43] W.S. McCulloch, W. Pitts, A logical calculus of the activities immanent in nervous activities, Bulletin of Mathematical Biophysics 5 (1943) 115–133.
[44] W.J. Murdoch, C. Singh, R. Abbasi-Asl, B. Yu, Definitions, methods, and applications in interpretable machine learning, Proceedings of the National Academy of Science (PNAS) 116 (44) (2019) 22071–22080.
[45] K.P. Murphy, Machine Learning: A Probabilistic Perspective, MIT Press, Cambridge, MA, 2012.
[46] V. Nair, G.E. Hinton, Rectified linear units improve restricted Boltzmann machines, in: Proc. International Conference on Machine Learning, Haifa, Israel, 2010.
[47] A.B. Novikoff, On convergence proofs for perceptrons, Tech. rep., Stanford Research Institute, 1963, https://apps.dtic.mil/sti/pdfs/AD0298258.pdf.
[48] G. Ortiz-Jiménez, A. Modas, S.M. Moosavi-Dezfooli, P. Frossard, Optimism in the face of adversity: Optimism in the face of adversity: Understanding and improving deep learning through adversarial robustness, Proceedings of the IEEE 109 (2021) 635–659, https://doi.org/10.1109/JPROC.2021.3050042.
[49] F. Rosenblatt, The perceptron: A probabilistic model for information storage and organization in the brain, Psychological Review 65 (1958) 386–408.
[50] A. Rahimi, B. Recht, Random features for large-scale kernel machines, in: NIPS'07: Proceedings of the 20th International Conference on Neural Information Processing Systems, Curran Associates, Inc., 2008, pp. 1–8.
[51] S. Ruder, An overview of gradient descent optimization algorithms, ArXiv, arXiv:1609.04747 [abs], 2016.
[52] P. Ramachandran, B. Zoph, Q.V. Le, Searching for activation functions, ArXiv, arXiv:1710.05941 [abs], 2017.
[53] B. Schölkopf, F. Locatello, S. Bauer, N.R. Ke, N. Kalchbrenner, Y. Bengio, Towards causal representation learning, Proceedings of the IEEE 109 (2021) 612–634, https://doi.org/10.1109/JPROC.2021.3058954.
[54] J. Schmidhuber, Deep learning in neural networks: An overview, Neural Networks 61 (2015) 85–117.
[55] National Institute of standards and Technology (NIST), Emnist Letters dataset, https://www.nist.gov/itl/products-and-services/emnist-dataset.
[56] J.B.O. Souza Filho, L.-D. Van, T.-P. Jung, P.S.R. Diniz, Online component analysis, architectures and applications, Foundations and Trends in Signal Processing 16 (3–4) (2022) 224–429, https://doi.org/10.1561/2000000112, NOW Publishers, Boston, MA, USA.
[57] N. Srivastava, G.E. Hinton, A. Krizhevsky, I. Sutskever, R. Salakhutdinov, Dropout: A simple way to prevent neural networks from overfitting, Journal of Machine Learning Research 15 (2014) 1929–1958.
[58] B. Scholkopf, A.J. Smola, Learning with Kernels: State Vector Machines, Regularization, Optimization, and Beyond, MIT Press, Cambridge, MA, 2002.
[59] J. Su, D.V. Vargas, K. Sakurai, One pixel attack for fooling deep neural networks, IEEE Transactions on Evolutionary Computation 23 (5) (2019) 828–841.
[60] C. Szegedy, W. Zaremba, I. Sutskever, J. Bruna, D. Erhan, I. Goodfellow, R. Fergus, Intriguing properties of neural networks, arXiv preprint, arXiv:1312.6199, 2013.
[61] E. Tjoa, C. Guan, A survey on explainable artificial intelligence (XAI): Toward medical XAI, IEEE Transactions on Neural Networks and Learning Systems 32 (11) (2021) 4793–4813.
[62] S. Theodoridis, Machine Learning: A Bayesian Optimization Perspective, 2nd edition, Academic Press, Oxford, UK, 2020.

[63] S. Theodoridis, K. Koutroumbas, Patter Recognition, Academic Press, London, UK, 2009.

[64] V.N. Vapnik, Statistical Learning Theory, Wiley Interscience, New Delhi, India, 1998.

[65] J. Watt, R. Borhani, A.K. Katsaggelos, Machine Learning Refined: Foundations, Algorithms, and Applications, 2nd edition, Cambridge University Press, Cambridge, UK, 2020.

[66] D. West, Neural network credit scoring models, Computers & Operations Research 27 (11–12) (2000) 1131–1152.

[67] R. Xu, D. Wunsch II, Survey of clustering algorithms, IEEE Transactions on Neural Networks 16 (3) (2005) 645–678.

[68] M.D. Zeiler, ADADELTA: An adaptive learning rate method, arXiv preprint, arXiv:1212.5701, 2012.

CHAPTER

A primer on graph signal processing

14

Wallace Alves Martins[a], Juliano Bandeira Lima[b], Cédric Richard[c], and Symeon Chatzinotas[a]
[a]Interdisciplinary Centre for Security, Reliability and Trust (SnT), University of Luxembourg, Luxembourg, Luxembourg
[b]Department of Electronics and Systems, Federal University of Pernambuco (UFPE), Recife, Brazil
[c]Department of Electrical Engineering, University Côte d'Azur, Nice, France

14.1 The case for GSP

Over the last decades, an enormous amount of data associated with network structures has been constantly generated, stored, and processed in the most diverse systems related to engineering and technology. This includes, for instance, measurements of some climatological variable acquired by sensors arbitrarily placed at a geographic region, attributes of points of a 3D cloud representing some virtual object, and opinion scores of people grouped according to their interests and proximity relationships in a social network. The structures on which the data or signals lie in those applications can be modeled as graphs whose vertices are connected by edges inferred from a variety of influence or dependency criteria. To be more specific, in a sensor network, for instance, the vertices would represent the sensors (respecting their geographic positions), and the signal samples would assume the values of the corresponding measurements; each sensor pair could be connected by an edge weighted according to the geographical distance between them.

The aforementioned sensor network graph-based model contrasts with what is usually adopted to process discrete-time signals, in which the samples are equidistantly placed throughout an axis (the signals' domain) and have left- and right-side immediate neighbors only; something similar happens in the case of digital images, where the pixels are arranged, also equidistantly, over a rectangular grid on a plane. Although not always apparent, the uniform structure of the signals' domain plays a key role in the development of classical signal processing tools. Thus, how can we analyze and process signals lying over arbitrary networks, taking into account the structure of such networks? The search for an answer to this question has driven the emergence of the so-called *graph signal processing* (GSP). In short, GSP aims to extend concepts and operations of classical *digital signal processing* (DSP) to scenarios in which the signals lie over *irregular* domains. In the present context, the referred *irregularity* means that the signal samples can lie over the vertices of an arbitrary graph.

Extending DSP concepts to the GSP framework usually involves complicated tasks and requires the creation of new strategies and paradigms. This is mainly due to the fact that several obvious notions in their original DSP context are missing or at least they become unclear in GSP. The idea of origin (i.e., the reference point of the domain, usually denoted by 0) of time or space, for example, is not extensible to graph domains, but it is also not always relevant, since the entire structure of the graph is commonly considered when establishing the corresponding analysis and processing tools. This means

Signal Processing and Machine Learning Theory. https://doi.org/10.1016/B978-0-32-391772-8.00020-X
Copyright © 2024 Elsevier Inc. All rights reserved.

that for many GSP tools, if the graph vertices are identified by numbered labels, it does not matter whether we employ a particular labeling or any of its permuted versions. On the other hand, deleting an edge from a graph may have a relevant impact on the spectral content of the corresponding signal. The reason is that the idea of high- or low-frequency components remains respectively related to more or less abrupt value changes from a sample to an adjacent one, that is, between samples on vertices connected by an edge.

Another issue that constitutes in itself a GSP subarea is graph inference. Of course there are practical scenarios in which the corresponding graph model is directly provided due to the nature of the application. For example, a citation graph can be easily constructed using a set of papers as vertices and the respective numbers of citations as signal samples. Directional unweighted edges connect each paper to the papers cited by it. On the other hand, there are contexts in which the construction of an adequate corresponding graph-based model, albeit greatly simplified, may involve complex problems and choices with subjective character. A graph with vertices representing brain regions connected as a function of their interdependence, on which a signal of brain activity level is defined, for instance, is a good illustration of the abovementioned possibility. In any case, it is clear that the results achieved by employing GSP tools highly depend on the underlying graph we construct during the system modeling.

In order to make a didactic presentation, we chose to build an example graph to be used to illustrate most of the GSP concepts discussed in the next sections. Such a graph is the collaboration graph shown in Fig. 14.1. It represents a network of 89 authors (vertices) of scientific papers, where an edge connects a pair of vertices if they are related to researchers coauthoring at least one paper. The graph signal (GS) represents the h-index[1] of each author. In the figure, the color of each vertex is defined according to the value of the corresponding GS sample and employing the sidebar colormap. This graph was constructed using as *seed* vertices the authors of this chapter and the organizer of this book. Some of the main coauthors of these researchers were also included, as well as their main coauthors, and so on. The size and the connectivity of the graph were limited by means of an arbitrary exclusion of some vertices and edges. All information used to construct the graph was freely obtained from the Digital Bibliography & Library Project (DBLP), Scopus, and Google Scholar websites.

Considering the aspect of Fig. 14.1, it is natural to assume that the edges simply indicate a mutual coauthorship relation. In this sense, the edges can be viewed as *undirected* and *unweighted*, that is, having unitary weight.[2] Otherwise, it is possible to consider a *directed* and *weighted* version of the same graph. In this case, the connection between two researchers is made by two edges, one in each direction. Taking into account the meaning of a coauthorship relation, it seems reasonable that the weight of the edge going from Martins to Lima, for example, should encode the (scientific) influence exerted by Martins on Lima, and vice versa. Accordingly, we propose to weigh the edge going from Martins to Lima (resp. from Lima to Martins) using the ratio of the number of works coauthored by them to the total number of works authored by Lima (resp. Martins). Throughout our text, we clearly indicate which version – either undirected or directed – of this graph is being considered. This choice basically depends on the concept or operation we want to highlight.

[1] The h-index is an attempt to quantify both the author's productivity and impact in terms of publications. It is defined as the maximum integer number h such that the author has published at least h papers that have each been cited at least h times.
[2] More precise definitions related to fundamentals of graph theory are given in the next section.

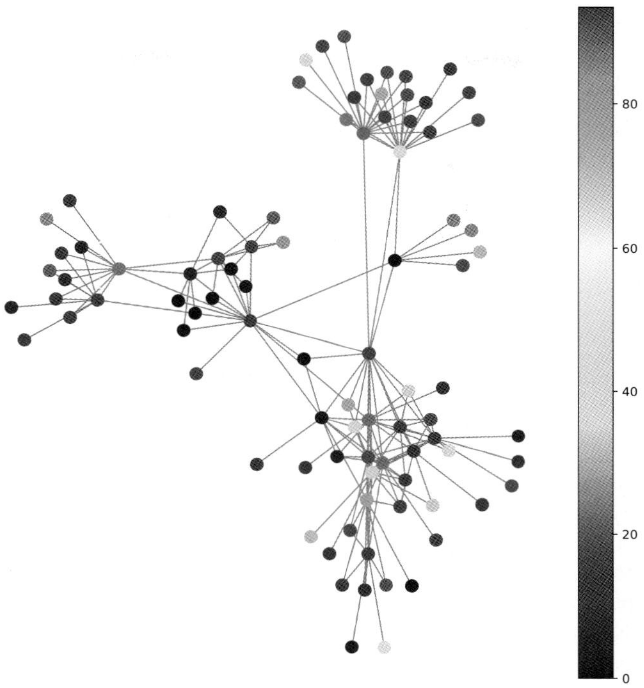

FIGURE 14.1

Collaboration graph connecting a set of 89 authors of scientific papers. An edge connects a pair of vertices if they represent researchers coauthoring at least one paper. The graph signal corresponds to the h-index of each author.

14.2 Fundamentals of graph theory

A graph is commonly defined as the ordered pair $(\mathcal{V}, \mathcal{E})$, in which the set $\mathcal{V}$ contains the so-called graph *vertices* and the set of *edges* $\mathcal{E}$ is a subset of $\mathcal{V}^2$ [1]. We will usually indicate by $|\mathcal{V}| = N$ and $|\mathcal{E}| = E$ the number of vertices and the number of edges of a graph, respectively, with $|\cdot|$ denoting the cardinality, or number of elements, of the (finite) set. For our purposes it is convenient to represent a graph as the structure $\mathcal{G} = \{\mathcal{V}, \mathbf{A}\}$, endowed with the (weighted) *adjacency matrix* $\mathbf{A}$ which captures the vertex-to-vertex relations: if $A_{i,j} \neq 0$, then there is an edge of weight $A_{i,j}$ from vertex v_j to v_i. By d_i^- we denote the *indegree* of vertex v_i, consisting of the sum of weights of all incoming edges to vertex v_i. Likewise, the *outdegree* d_i^+ is the sum of weights of edges departing from v_i.

A graph is called *undirected* if and only if its adjacency matrix is symmetric, in which case the *degree* of vertex v_i is defined as $d_i^- = d_i^+ = d_i$. In this case, a graph is said to be *d-regular* whenever all graph vertices have degree d. If $\mathbf{A}$ is asymmetric, however, the respective graph is *directed* and its pictorial representation depicts the edges as arrows to account for the unidirectional relation between adjacent vertices. Examples of directed and undirected graphs are shown in Fig. 14.2. Such graphs

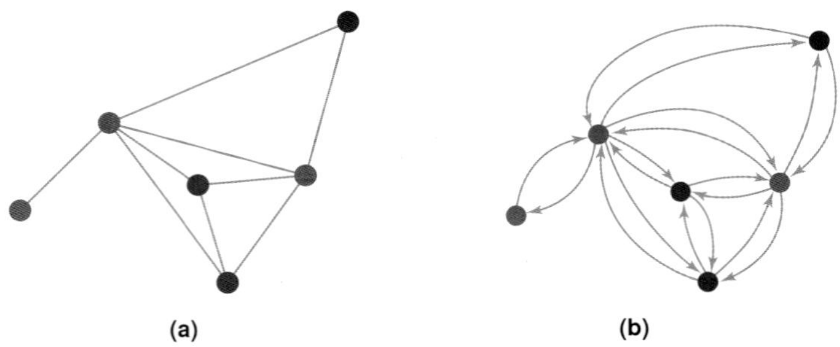

FIGURE 14.2

Examples of (a) undirected and (b) directed graphs, defined over the same vertex set.

are obtained after an arbitrary pruning of the collaboration graph shown in Fig. 14.1. In particular, in Fig. 14.2b, we always draw two edges with opposite directions to connect each pair of adjacent vertices to emphasize that these edges can have different weights, as explained in the previous section. Naturally, in directed graphs, the influence may not be mutual, in which case two adjacent vertices are connected by an edge with only one direction.

The adjacency matrix is a possible building block for one of the main frameworks of GSP, which will be covered soon, but another matrix of great importance, mainly in the branch of GSP originated from spectral graph theory, is the *Laplacian matrix*

$$\boldsymbol{L} \triangleq \boldsymbol{D} - \boldsymbol{A}, \tag{14.1}$$

with the *degree matrix* $\boldsymbol{D}$ being a diagonal matrix with degree d_i as its ith entry. Depending on the context, $\boldsymbol{D}$ may be taken as the indegree or outdegree matrix, although when the Laplacian matrix is used the graphs considered are more often undirected.

A *path* is a set of distinct edges (with the same orientation, if the graph is directed) linking distinct vertices. A *cycle* is a path with equal starting and end points/vertices, and if a graph has a cycle it is called *cyclic* (otherwise it is called *acyclic*). If the cycle has only one edge, it is called a *loop*. One refers to *multiple edges* whenever a single pair of vertices is connected by two or more edges. An undirected graph is called *simple* if it has no loops or multiple edges.

A graph is said to be *complete* if any two of its vertices are adjacent. GSP over such graphs may be extremely cumbersome, for the computational complexity of many of its techniques depends heavily on the number of graph edges. For most applications, it is desirable to have a small number of edges while keeping the graph *connected*, i.e., for any pair of vertices there exists a set of distinct edges (with the same orientation, if directed) connecting them without making a cycle.

A graph is said to be *unweighted* if all its edges have unit weight. A *subgraph* of $\mathcal{G}$ is a graph $\mathcal{G}' = (\mathcal{V}', \boldsymbol{A}')$ with edge set $\mathcal{E}'$, in which $\mathcal{V}' \subset \mathcal{V}$ and $\mathcal{E}' \subset \mathcal{E}$. A *connected component* of $\mathcal{G}$ is a connected subgraph $\mathcal{G}' = (\mathcal{V}', \boldsymbol{A}')$ in which any vertex in $\mathcal{V}'$ is linked exclusively to another vertex also in $\mathcal{V}'$.

The *neighborhood* of a vertex v_i is the set $\mathcal{N}_i$ of all vertices adjacent to v_i. Sometimes it is useful as well to denote by $\mathcal{N}(i, K)$ the set of vertices connected to v_i through a path of length K or less. This notion is represented in Fig. 14.3. The graph shown in the figure is, again, obtained after an arbitrary pruning of the collaboration graph shown in Fig. 14.1.

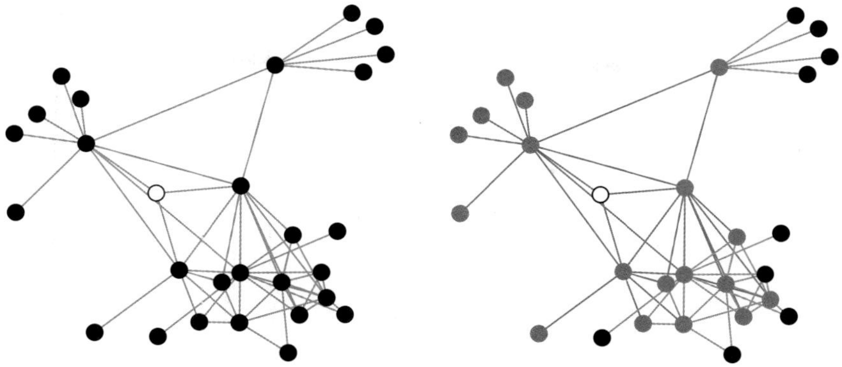

FIGURE 14.3

(a) A graph and (b) the set of vertices $\mathcal{N}(i, 2)$ shown in red (dark gray in print version), with v_i being depicted in white. The edges linking v_i to the elements in $\mathcal{N}(i, 2)$ are also highlighted in red (dark gray in print version).

The reader is encouraged to refer to this section whenever necessary. For a broader glossary with a solid introduction to graph theory, the authors recommend [2,3].

14.3 Graph signals and systems

14.3.1 Graph signals

In DSP, a signal is mathematically represented by a numerical sequence. In this context, the ith number of a sequence x is denoted by x_i, $i \in \mathbb{Z}$. A signal $\boldsymbol{x}$ defined over $\mathcal{G} = \{\mathcal{V}, \mathbf{A}\}$, with $|\mathcal{V}| = N$, is a discrete-domain function mapping the graph vertex set to a scalar set, usually the complex or real numbers,

$$x : \mathcal{V} \rightarrow \mathbb{C} \mid x(v_n) = x_n, \tag{14.2}$$

so that $\boldsymbol{x}$, called *graph signal* (GS), can be seen as a vector in $\mathbb{C}^N$ *with entries indexed by the vertices of* $\mathcal{G}$. In this context, the space of such GSs is denoted as $\mathbb{C}^{\mathcal{V}}$. Once the vertices $\mathcal{V} = \{v_0, \ldots, v_{N-1}\}$ are clearly labeled, it is not ambiguous to represent the signal as the column vector $\boldsymbol{x} = [x_0\ x_1\ \ldots\ x_{N-1}]^{\top}$, $x_n \in \mathbb{C}$, with $n \in \{0, \ldots, N-1\}$.

There are several ways to provide illustrative representations of GSs. In general, given the usual graph drawing, the vertex labeling is omitted for the sake of simplicity, as it will be assumed that the signal sample x_n is assigned to vertex v_n. The signal values can be indicated in three manners: by writing down its numerical value next to the respective vertex, by drawing over each vertex a vertical line whose height is related to the corresponding sample value, or by using a pseudocolor scale, the latter of which is the scheme adopted throughout this chapter, considering $x_n \in \mathbb{R}$.

In the present context, the directed ring graph is of particular importance and will be revisited a few times throughout this chapter. It can be viewed as a link between GSP and classical DSP theory, since it

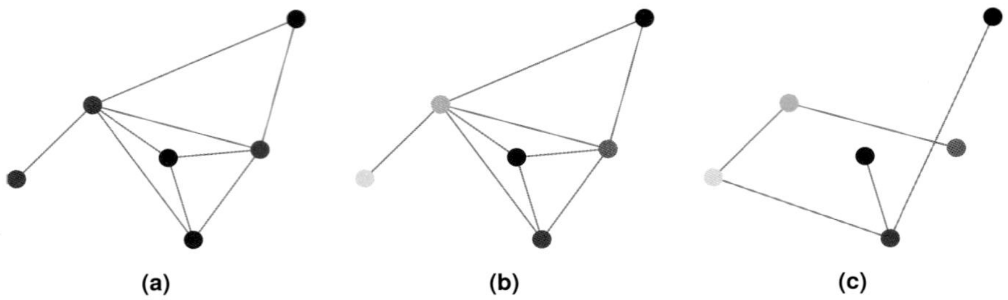

FIGURE 14.4

Graphs with the same set of vertices, but with (a) low-pass, (b) high-pass, and (c) low-pass spectra, depending on the signal and the corresponding set of edges.

can be used to model the finite length discrete-time domain. Its directed edges model the causality of the time domain, whereas the feedback edge accounts for the boundary condition of periodicity imposed by the *discrete Fourier transform* (DFT) analysis. Other signals that arise in practical applications have the respective graphs easily identified. The rectangular grid, for example, models the digital image domain [4].

The spectral characteristics of a signal depend heavily on the domain over which it is defined, but one does not need to acknowledge this in the context of DSP, since in this case the domains are always regular and uniform.[3] From the classical theory, the common understanding states that a signal has mostly low frequencies if adjacent samples have similar values, and high frequencies otherwise. When dealing with signals defined over graphs, it is clear that the adjacency relations depend on the graph topology, and therefore one may foresee that *the same signal values may present different spectra when defined over different graphs*.

A preliminary and intuitive discussion regarding this issue can be carried out by considering, again, the graph shown in Fig. 14.2a, which is also presented in Fig. 14.4a. Taking into account the colormap employed throughout this chapter (see Fig. 14.1), we observe that in Fig. 14.2a, the signal values have low variation when transitioning from a vertex to a corresponding adjacent vertex. This characteristic can be associated with the predominance of low frequencies in the graph spectrum. On the other hand, if we preserve the graph topology, but arbitrarily change the signal, so that samples with high values are adjacent to the ones with small values, high-frequency components should appear. This would be the case shown in Fig. 14.4b, where we can see a red (mid gray in print version) vertex directly connected to dark blue (black in print version) vertices. Finally, we consider Fig. 14.4c, which shows a graph having the same set of vertices of the two previous subfigures and the same signal values of Fig. 14.4b, but a different set of edges. In this case, the edges connect only vertices that have slightly different values, which gives a smooth characteristic to the GS and, therefore, a low-pass spectrum. We remark that in the context of the collaboration graph we have been considering, a low-pass GS can be related

[3] One could argue that in the theory of *nonuniform sampling*, the signal is defined over an irregular domain, since the samples may be randomly spaced. Even in this case, however, the classical techniques still aim to *recover* the signal so as to represent it in its usual – and *uniform* – domain.

to a scenario in which coauthors who collaborate intensively have low disparity in their productivity (in terms of the h-index); correspondingly, a high-pass signal suggests a scenario in which there may be a very productive researcher, surrounded by coauthors with relatively lower productivity.

14.3.2 Inferring graphs

Some contexts in which GSP is to be applied, to perform whatever signal processing technique is necessary, provide clear information on *how the underlying graph is structured*. This is the case with the example we introduced at the end of Section 14.1. Specifically in this example, we can choose to consider the graph as being undirected or directed and, in the latter case, we need to make decisions about how to weigh the edges. On the other hand, the connections between the vertices, that is, the authors, are well defined. As another example, let us reconsider a graph modeling the brain regions and their interdependencies. One might suppose that each region influences or is influenced, in terms of functional activity, by the regions physically adjacent to it. However, based on specific experiments, one actually observes that regions without physical connectivity can be functionally interdependent. Therefore, a graph associated with this scenario must somehow be able to capture these interrelationships.

As an additional example, let us consider a sensor network that may be covering, for instance, the entire territory of a large country and measuring some climatological variable. If we desire to model this scenario as a graph, in order to use GSP tools, how is one supposed to weigh the respective graph edges, and before this, how does one decide which vertices to connect? Would representing each sensor as a vertex of the graph and connecting it to all other sensors that were within a certain radius (in terms of geographic distance) be the only option? Clearly it is not. Although generally the problem of graph inference is complex, this type of geography-based graph has an adequate topology for estimation methods.

The general idea is that if there is a clear metric to evaluate the *expected* similarity between samples as a function of the available information regarding the respective vertices, then this metric may be used as the edge weight and a threshold is set so that any weight below this value causes the respective edge to be eliminated. In the case of vertices which have geodesic positions, the Euclidean distance may be used as the metric because vertices that are closer together are *expected* to have similar signal samples, and therefore the adjacency matrix of the underlying graph may have entries given by

$$A_{ij} = \begin{cases} \exp\left(-\dfrac{\text{dist}^2(v_i, v_j)}{2\theta^2}\right) & \text{if dist}(v_i, v_j) < T, \\ 0 & \text{otherwise,} \end{cases} \tag{14.3}$$

as used in [5]. The choice of the parameters T and[4] θ (standard deviation of the distribution) and the choice of how to use the metric (in this case, inside a Gaussian distribution) are dictated by the application and by the analyst's experience.

However, if there is an isolated vertex, far from the others, the use of (14.3) may lead to a compromise between keeping the graph connected and obtaining a sparse adjacency matrix, since imposing

[4] T indicates a distance threshold above which we set the edge weight to zero, effectively leaving the vertices unlinked. This means that the distance between them is assumed to be too high for any significant interdependence to exist.

connectivity to the graph in this case implies increasing T, and therefore having many edges. To deal with this problem and still have a good representation of the underlying graph, one alternative is to connect a vertex to its K closest neighbors (setting K to an appropriate value, according to the context) and weight the edges using the Gaussian distribution in (14.3).

As previously discussed, these methods require an adequate metric to evaluate the expected similarity between samples in the graph vertex, but given the diverse areas in which GSs may arise, estimating the topology of the underlying graph constitutes a challenge of its own [6–9].

14.3.3 Graph shift operators

In the study of systems for processing GSs, a path naturally analogous to that taken by the classical DSP is to define, as a starting point, elementary operators that can be used to design more elaborate systems for performing filtering. When the purpose is to consider linear operators in this context, it can be seen that the role of the referred elementary operator is played by any matrix, which, multiplied by a GS represented as a vector, produces another signal whose samples result from a linear combination of the samples of the original signal. In this context, the first approaches that systematized the basic GSP concepts could have chosen any well-established matrix operator from graph theory as building blocks of the new theory; in fact, these approaches did so by focusing on operators that allowed some meaningful physical interpretation or some parallel with the classical DSP, as described below.

The first among the prominent approaches in this scenario comes from the so-called *algebraic signal processing* (ASP), in which case the adjacency matrix is used as the abovementioned elementary operator. The most natural motivation for such a choice comes to light by observing the way the adjacency matrix

$$\boldsymbol{C} = \begin{bmatrix} & & & 1 \\ 1 & & & \\ & \ddots & & \\ & & 1 & \end{bmatrix} \tag{14.4}$$

of the directed ring graph acts on a signal over the same graph. Denoting this signal by $\boldsymbol{x} = [x_0\ x_1\ \dots\ x_{N-1}]^{\top}$, one has

$$\boldsymbol{C}\boldsymbol{x} = \begin{bmatrix} & & & 1 \\ 1 & & & \\ & \ddots & & \\ & & 1 & \end{bmatrix} \begin{bmatrix} x_0 \\ x_1 \\ \vdots \\ x_{N-1} \end{bmatrix} = \begin{bmatrix} x_{N-1} \\ x_0 \\ \vdots \\ x_{N-2} \end{bmatrix} \triangleq \boldsymbol{x}^{\langle 1 \rangle}. \tag{14.5}$$

Clearly, the signal resulting from the operation above corresponds to the unit (circular) shift of $\boldsymbol{x}$, which also confirms that the directed ring graph can be viewed as a graph model for the discrete-time domain. This suggests a generalization that consists in designating the adjacency matrix of an arbitrary graph as a *graph shift operator* (GSO), when it is chosen to play the role of elementary operator in this context. In other words, for a signal $\boldsymbol{x}$ defined over the graph $\mathcal{G} = \{\mathcal{V}, \boldsymbol{A}\}$, the adjacency matrix $\boldsymbol{A}$ acts as a *filter* which "delays" $\boldsymbol{x}$ by one unit, producing the delayed version represented hereinafter by $\boldsymbol{x}^{\langle 1 \rangle} = \boldsymbol{A}\boldsymbol{x}$.

In the other remarkable GSP framework, the basic concepts are developed from the graph spectral theory [5]. In this case, the matrix adopted as elementary operator to process GSs is the Laplacian matrix. As previously presented in this chapter, such a matrix has the expression

$$\boldsymbol{L} = \boldsymbol{D} - \boldsymbol{A}, \tag{14.6}$$

where $\boldsymbol{D}$ is a diagonal matrix and $\boldsymbol{L}$ is usually identified as a combinatorial or unnormalized graph Laplacian. Working on (14.6) it is possible to see $\boldsymbol{L}$ as a difference operator acting upon a signal $\boldsymbol{x}$, updating each sample with the difference between the value on a vertex and its neighbors. That is,

$$\boldsymbol{L}\boldsymbol{x} = \boldsymbol{D}\boldsymbol{x} - \boldsymbol{A}\boldsymbol{x}, \tag{14.7}$$

implying that

$$\begin{aligned} [\boldsymbol{L}\boldsymbol{x}]_n &= d_n x_n - \sum_{m \mid v_m \in \mathcal{N}_n} A_{nm} x_m \\ &= \sum_m A_{nm} x_n - \sum_{m \mid v_m \in \mathcal{N}_n} A_{nm} x_m \\ &= \sum_{m \mid v_m \in \mathcal{N}_n} A_{nm}(x_n - x_m). \end{aligned} \tag{14.8}$$

In fact, it can be shown that the graph Laplacian operator is the discrete counterpart to the continuous Laplace–Beltrami (second-order) derivative operator on a manifold [3,10]. In this sense, $\boldsymbol{L}$ can be interpreted as a second-order difference operator. However, in order to be more general and to emphasize the possibility of other choices for the graph elementary operator, we will refer to this matrix as being a GSO as well. Note that from a terminology point of view, the key point here is the use of the term *shift*, which naturally translates the idea of a basic building block in a signal processing framework.

Variations of the Laplacian that could also be used as GSO include the normalized Laplacian, given by

$$\boldsymbol{L}_{\mathrm{n}} = \boldsymbol{D}^{-1/2}\boldsymbol{L}\boldsymbol{D}^{-1/2}, \tag{14.9}$$

and the random walk Laplacian, given by

$$\boldsymbol{L}_{\mathrm{rw}} = \boldsymbol{D}^{-1}\boldsymbol{L}. \tag{14.10}$$

Each GSO allows specific interpretations regarding its action in the vertex domain and also with respect to the corresponding spectral domain. Naturally, the latter depends on the eigendecomposition or the Jordan canonical form of the respective operator, which will be discussed in the next sections. From this point forward, whenever we want to refer to a GSO without necessarily specifying it, it will be denoted by $\boldsymbol{S}$ (instead of being denoted by $\boldsymbol{A}$ or $\boldsymbol{L}$, for instance).

14.4 Graph Fourier transform

The topic of spectral analysis is key in signal processing, and since the introduction of the first GSP concepts, many researchers have spent some time reflecting upon how this would fit into their theory.

The starting point was to look at the classical Fourier transform as the signal decomposition over a basis of eigenfunctions of linear time-invariant systems [11,12], as in the case of the basis comprised of time-domain complex exponential functions. Thus, the *graph Fourier transform* (GFT) could be defined as the decomposition over a basis of eigenvectors of *linear shift-invariant* (LSI) systems.

Let us take the graph $\mathcal{G} = \{\mathcal{V}, \boldsymbol{A}\}$, $|\mathcal{V}| = N$, and the corresponding GSO $\boldsymbol{S}$. Note that at this point, it is not necessary to specify which matrix has been chosen as GSO; in any case, $\boldsymbol{S}$ admits the Jordan decomposition

$$\boldsymbol{S} = \boldsymbol{V}\boldsymbol{J}\boldsymbol{V}^{-1}, \tag{14.11}$$

where $\boldsymbol{V}$ contains the N Jordan (generalized) eigenvectors of $\boldsymbol{S}$ in its columns,

$$\boldsymbol{V} \triangleq [\boldsymbol{v}_0 \; \boldsymbol{v}_1 \; \dots \; \boldsymbol{v}_{N-1}], \tag{14.12}$$

and $\boldsymbol{J}$ is a block-diagonal matrix formed of the so-called Jordan blocks. In particular, if $\boldsymbol{S}$ is diagonalizable, (14.11) coincides with its eigendecomposition, so that $\boldsymbol{J}$ reduces to a diagonal matrix whose entries are the eigenvalues of $\boldsymbol{S}$.

When we mentioned that the role to be played by the GSO would be that of an elementary operator for GSs, this was associated with the idea that more elaborate graph systems could be designed from $\boldsymbol{S}$. In this sense, let us also assume that the operator related to the referred systems can be written as polynomials in $\boldsymbol{S}$.[5] Since a matrix and its powers share the same set of eigenvectors, the columns of $\boldsymbol{V}$ form a basis of vectors invariant to LSI systems. Besides, given that the subspaces generated by the linearly independent eigenvectors of an eigenvalue of $\boldsymbol{S}$ are irreducible and have null intersection and the dimensions of all subspaces add to N [13], $\boldsymbol{V}$ provides a basis which is invariant to LSI systems for the space of signals defined over $\mathcal{G}$.

Therefore, a signal $\boldsymbol{x}$ may be decomposed into its components with respect to $\boldsymbol{V}$ as

$$\begin{aligned} \boldsymbol{x} &= \widehat{x}_0 \boldsymbol{v}_0 + \dots + \widehat{x}_{N-1} \boldsymbol{v}_{N-1} \\ &= \boldsymbol{V} [\widehat{x}_0 \; \widehat{x}_1 \; \dots \; \widehat{x}_{N-1}]^\top \\ &= \boldsymbol{V}\widehat{\boldsymbol{x}}, \end{aligned} \tag{14.13}$$

and this is the synthesis equation of the GFT. Hence, the analysis equation is

$$\widehat{\boldsymbol{x}} = \boldsymbol{V}^{-1}\boldsymbol{x}. \tag{14.14}$$

Again, the notion of what a Fourier transform should be, derived from the analogy described at the beginning of this section and expressed in the analysis and synthesis equations above, is independent of the matrix chosen as GSO. In this context, it is clear that the eigenvectors $\boldsymbol{v}_i$ of $\boldsymbol{S}$ can be interpreted as "frequency components" associated with the "graph frequencies" given by the respective eigenvalues λ_i, as the Fourier component $\mathrm{e}^{-\mathrm{j}\Omega t}$, in the continuous-time domain t, is associated with the frequency Ω.

[5] In fact, LSI graph systems, or simply LSI filters, are polynomials in $\boldsymbol{S}$. This will be covered in detail in Section 14.5.

If the graph $\mathcal{G}$ is undirected, the Fourier basis is orthonormal, and then Parseval's identity holds for adjacency- and Laplacian-based GFT, for instance. That is,

$$\boldsymbol{x}^{\mathrm{H}}\boldsymbol{y} = \widehat{\boldsymbol{x}}^{\mathrm{H}}\widehat{\boldsymbol{y}}, \tag{14.15}$$

for GSs $\boldsymbol{x}$ and $\boldsymbol{y}$ on $\mathcal{G}$. Since $\boldsymbol{V}$ is unitary, $\boldsymbol{x}^{\mathrm{H}}\boldsymbol{y} = \langle \boldsymbol{V}\widehat{\boldsymbol{x}}, \boldsymbol{V}\widehat{\boldsymbol{y}}\rangle = \langle \boldsymbol{V}^{\mathrm{H}}\boldsymbol{V}\widehat{\boldsymbol{x}}, \widehat{\boldsymbol{y}}\rangle = \widehat{\boldsymbol{x}}^{\mathrm{H}}\widehat{\boldsymbol{y}}$. The interpretation of Parseval's identity is that the energy of the GS is conserved when passing from the vertex to the frequency domain.

There are other important aspects to be considered which suggest specific interpretations depending on $\boldsymbol{S}$. Such aspects are mainly related to the way we order the eigenvalues of $\boldsymbol{S}$ and, consequently, the respective spectral components. Clarifying this aspect is necessary to allow the distinction between low and high graph frequencies, for instance. The next subsections are devoted to dealing with this question, which is discussed by considering the adjacency matrix and the Laplacian as GSOs.

14.4.1 Adjacency-based graph Fourier transform

It has been emphasized that the directed ring graph can be viewed as a link between GSP and DSP, because it models the discrete-time domain (more specifically, the domain of periodic extensions of discrete-time signals). If we choose the adjacency matrix $\boldsymbol{A}$ as GSO, this provides a way of checking how consistent with the classical theory the GSP tools are. When investigating how the GFT would act upon discrete-time signals, one should first diagonalize the adjacency matrix $\boldsymbol{C}$ of the directed ring graph, given by (14.4). Since it is circulant, it is known to be diagonalized by the DFT matrix $\boldsymbol{F}$, with entries

$$F_{n,k} = \frac{\mathrm{e}^{-\mathrm{j}\frac{2\pi}{N}nk}}{\sqrt{N}}, \tag{14.16}$$

which contains in its *rows* the DFT eigenvectors. The calculation of the characteristic polynomial of $\boldsymbol{C}$,

$$p_{\boldsymbol{C}}(\lambda) = \det(\lambda \boldsymbol{I} - \boldsymbol{C}) = \begin{vmatrix} \lambda & & & -1 \\ -1 & \lambda & & \\ & \ddots & \ddots & \\ & & -1 & \lambda \end{vmatrix} = \lambda^N - 1, \tag{14.17}$$

shows that its eigenvalues are the N complex roots of unity. Setting these eigenvalues as the entries of a diagonal matrix $\boldsymbol{\Lambda}_{\boldsymbol{C}}$, the eigendecomposition of $\boldsymbol{C}$ may be written as

$$\boldsymbol{C} = \boldsymbol{F}^{-1}\boldsymbol{\Lambda}_{\boldsymbol{C}}\boldsymbol{F}, \tag{14.18}$$

and one can see that, in the case of directed ring graphs, the GFT and the DFT matrices *coincide*, since $\boldsymbol{V}^{-1} = \boldsymbol{F}$. This equivalence indicates a consistency with the classical theory.

From this point forward, let us turn specifically to the question of ordering the graph frequencies. For this purpose, the following fact deserves to be highlighted: whenever one has at least one eigenvalue of $\boldsymbol{A}$ with geometric multiplicity greater than one, the same frequency may be associated with two or more linearly independent frequency components, as indeed was the case in the example of Fig. 14.5.

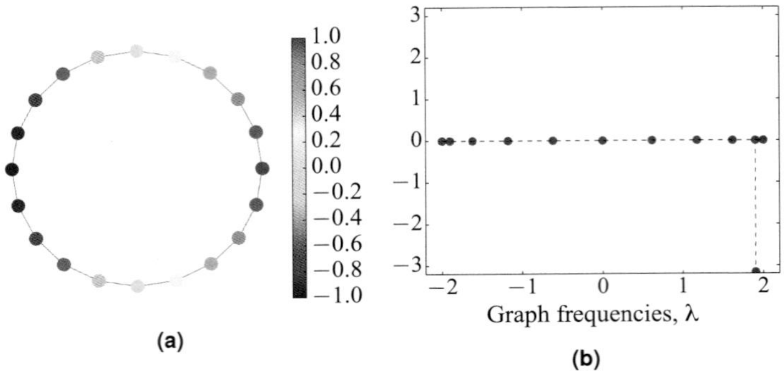

FIGURE 14.5

(a) Signal defined over an undirected ring graph and (b) its spectrum in the GSP_A sense, in which the graph frequencies correspond to the eigenvalues of the adjacency matrix.

Furthermore, this figure shows that although the signal seems to be smooth, its frequency components are mostly associated with eigenvalues of high magnitude,[6] which is counterintuitive and provides a motivation to define a clear criterion to distinguish high and low graph frequencies.

The following mathematical reasoning consists of taking a metric which quantifies the expected signal smoothness and using it to propose or confirm a notion of graph frequency. The metric used by Sandryhaila and Moura was the *total variation* (TV), taken from classical real analysis and defined for differentiable functions as [14,15]

$$\|f\|_V = \int_{-\infty}^{\infty} |f'(t)|\mathrm{d}t, \tag{14.19}$$

where f' denotes the first derivative of f.

For discrete domain functions $f_N[n]$, the Riemann integral is replaced by first-order differences,

$$\|f_N\|_V = \sum_p |f_N[n_p + 1] - f_N[n_p]|, \tag{14.20}$$

which clearly quantifies the dissimilarity between contiguous values of the function f_N. With this in mind, it was natural for Sandryhaila and Moura to use this metric in their mathematical formulation of frequency in GSP, wherein they represented the TV of a finite length *discrete-time signal* $\boldsymbol{x}$ as

$$\mathrm{TV}(\boldsymbol{x}) = \sum_n |x_n - x_{n-1 \bmod N}|. \tag{14.21}$$

[6] The spectral magnitude associated with the eigenvalue with the highest magnitude is null because the GS in the example has null DC value; this will become clearer later.

From (14.5), one can see that (14.21) may be written in terms of the ℓ_1-norm[7] as $\mathrm{TV}(\boldsymbol{x}) = \|\boldsymbol{x} - \boldsymbol{C}\boldsymbol{x}\|_1$, by using the directed ring graph adjacency matrix to perform the cyclic shift. From that point, the generalization consisted of using this expression and defining the *TV on graphs* of a signal $\boldsymbol{x}$ defined over the graph $\mathcal{G} = \{\mathcal{V}, \boldsymbol{A}\}$ as

$$\mathrm{TV_G}(\boldsymbol{x}) \triangleq \|\boldsymbol{x} - \boldsymbol{A}^{\mathrm{norm}}\boldsymbol{x}\|_1, \tag{14.22}$$

with $\boldsymbol{A}^{\mathrm{norm}} = |\lambda_{\max}|^{-1}\boldsymbol{A}$ and with $\lambda_{\max}$ being the eigenvalue of $\boldsymbol{A}$ having the largest absolute value. The normalization of the adjacency matrix avoids the excessive magnification of the shifted signal [16].

Let the (possibly complex) eigenvalues of $\boldsymbol{A}$ be ordered according to

$$|\lambda_0| \le |\lambda_1| \le \cdots \le |\lambda_{N-1}| \triangleq |\lambda_{\max}|. \tag{14.23}$$

The eigenvalues above are respectively associated with the eigenvectors $(\boldsymbol{v}_i)_{i=0}^{N-1}$, scaled so that $\|\boldsymbol{v}_i\|_1 = 1\ \forall i$. Taking the TV (on graphs) of the eigenvector $\boldsymbol{v}_k$, one has

$$\begin{aligned}
\mathrm{TV_G}(\boldsymbol{v}_k) &= \|\boldsymbol{v}_k - \boldsymbol{A}\boldsymbol{v}_k\|_1 \\
&= \left\| \boldsymbol{v}_k - \frac{1}{|\lambda_{\max}|}\lambda_k \boldsymbol{v}_k \right\|_1 \\
&= \left| 1 - \frac{\lambda_k}{|\lambda_{\max}|} \right| \|\boldsymbol{v}_k\|_1 \\
&= |\lambda_k - |\lambda_{\max}|| \frac{\|\boldsymbol{v}_k\|_1}{|\lambda_{\max}|}
\end{aligned}$$

so that, since $\|\boldsymbol{v}_k\|_1 = 1$, one has

$$\Big|\lambda_i - |\lambda_{\max}|\Big| \le \Big|\lambda_j - |\lambda_{\max}|\Big| \Longleftrightarrow \mathrm{TV_G}(\boldsymbol{v}_i) \le \mathrm{TV_G}(\boldsymbol{v}_j), \tag{14.24}$$

i.e., frequency components associated with eigenvalues closer to the real point $|\lambda_{\max}|$ in the complex plane are *smoother* (because they have lower TV), and therefore are said to be of *low frequency*. Fig. 14.6 illustrates this ordering for graph frequencies. Additionally, the fact that in the spectrum of the signal in Fig. 14.5a the frequency components are mostly associated with eigenvalues of high magnitude is clarified (note that since the graph is undirected, its adjacency matrix is symmetric and the eigenvalues are real-valued).

Let us take the undirected version of the graph in Fig. 14.1 to verify the consistency of the notion of frequency just derived. For this, along with the TV on graphs, also the number of zero crossings (i.e., the number of edges connecting vertices with signal samples of different sign) will be used to *quantify* frequency [5]. This quantity is also related to frequency in classical theory: the more a discrete signal has contiguous samples with different sign, the higher are its frequency components. These two functions, the TV on graphs and the number of zero crossings, were calculated for each of the adjacency matrix eigenvectors, and the result is shown in Fig. 14.7, in which the eigenvectors $\boldsymbol{v}_k$ are

[7] The ℓ_1-norm and the ℓ_2-norm are particular cases of the ℓ_p-norm of a vector $\boldsymbol{x} \in \mathbb{C}^N$, defined as $\|\boldsymbol{x}\|_p \triangleq \left(\sum_{n=0}^{N-1} |x_n|^p\right)^{1/p}$.

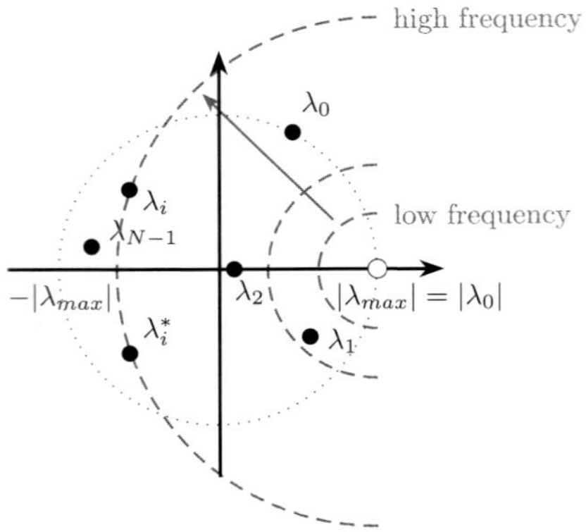

FIGURE 14.6

Frequency ordering of graph signals, from low to high frequencies, in the complex plane [16].

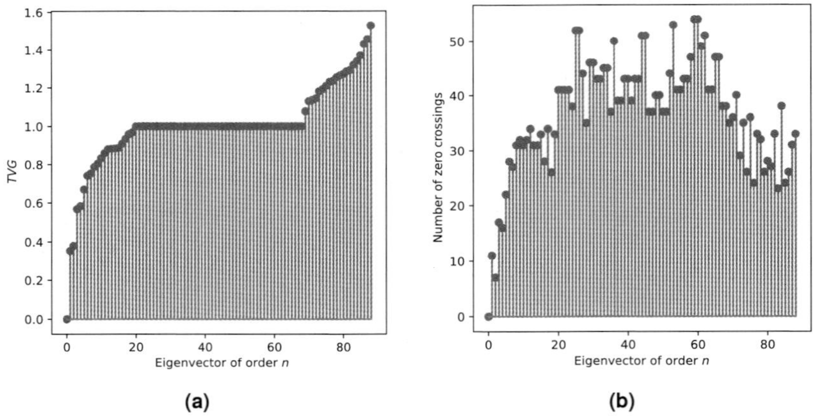

FIGURE 14.7

(a) Total variation and (b) number of zero crossings of the eigenvectors $(\boldsymbol{v}_i)_{i=0,\ldots,N-1}$ of the adjacency matrix $\boldsymbol{A}$ of the graph in Fig. 14.1, ordered so that the respective eigenvalues appear from the closest to the furthest from the real point $|\lambda_{\max}|$ in the complex plane. That is, according to (14.24) and Fig. 14.6, the eigenvectors are arranged in ascending order of frequency.

ordered in such a way that the respective eigenvalues λ_k appear from the closest to the furthest from the real point $|\lambda_{\max}|$ in the complex plane. Along almost the entire axis representing the indices of the ordered eigenvectors, the behavior of both metrics is similar. More specifically, both the TV and the number of zero crossings have increasing values up to the eigenvector of index close to 20. Then, up to the eigenvector of index close to 70, the values of both metrics show little or no fluctuation. When we consider the eigenvectors associated with higher frequencies, we observe a downward trend in the number of zero crossings, while the TV shows the opposite behavior. Since the number of zero crossings is indifferent to GS variations which do not change sign, it was already expected to be less accurate as a figure of merit for frequency. Likely, the aforementioned downward trend stems from this inaccuracy. Despite this disparity, it is worth highlighting how the adopted eigenvector ordering indeed implies an ascending frequency order, since both functions in Figs. 14.7a and 14.7b, in general, agree on the tendency of growth. More than that, $\mathrm{TV}_{\mathrm{G}}(\boldsymbol{v}_k)$ grows *monotonically*, as it should do according to (14.24).

14.4.2 Laplacian-based graph Fourier transform

When the Laplacian matrix is chosen as GSO, the definition of the respective GFT brings another interesting interpretation. First, we recover the fact that the continuous-time Fourier transform consists of a decomposition into a basis of eigenfunctions of the continuous Laplacian operator (second derivative), which is denoted here by Δ. Therefore, since $\boldsymbol{L}$ can be viewed as an approximate version of Δ, using an eigenbasis of $\boldsymbol{L}$ to define the GFT as in (14.14) is somehow consistent with the classical theory. More specifically, we have

$$\Delta e^{j\Omega t} = \frac{\partial^2}{\partial t^2} e^{j\Omega t} = -\Omega^2 e^{j\Omega t}, \tag{14.25}$$

so that the eigenvalues of $\boldsymbol{L}$, that is, the graph frequencies, would be related to $-\Omega^2$.

In general, the graphs for which $\boldsymbol{L}$ is taken as GSO are undirected and have edges with real-valued weights, so that such a matrix is always real-valued, symmetric, and therefore diagonalizable. Moreover, if the referred weights are all nonnegative, for any real-valued signal $\boldsymbol{x} = [x_0\ x_1\ \ldots\ x_{N-1}]^\top$, one may write

$$\boldsymbol{x}^\top \boldsymbol{L}\boldsymbol{x} = \frac{1}{2} \sum_{(i,j)\in\mathcal{E}} A_{ij}(x_i - x_j)^2 \geq 0. \tag{14.26}$$

In other words, $\boldsymbol{L}$ is positive semidefinite and, equivalently, its eigenvalues are all real and nonnegative. Additionally, it has been shown that zero is an eigenvalue with multiplicity equal to the number of connected components of the graph (see [5] and references therein), and therefore connected graphs have only one null eigenvalue of $\boldsymbol{L}$. Gathering all these facts, the eigenvalues of the Laplacian matrix are real and can be ordered as $\lambda_0 = 0 < \lambda_1 \leq \cdots \leq \lambda_{N-1}$.

The ordering criterion established above allows us to return to the possible parallel between the current GFT approach and the classical case, observing what follows. In (14.25), the frequency information is hidden in the eigenvalues $-\Omega^2$ associated with each eigenfunction $e^{j\Omega t}$, and the closer to zero is the eigenvalue, the smoother is its respective harmonic component. The same occurs with the eigenvalues and eigenvectors of the Laplacian matrix, and a way of verifying this is using the norm $\|\boldsymbol{L}\boldsymbol{x}\|_2$: since $\boldsymbol{L}$ is a difference operator, $\|\boldsymbol{L}\boldsymbol{x}\|_2$ acts as a metric similar to the TV on graphs in (14.21). If $\boldsymbol{v}_i$ is the eigenvector of $\boldsymbol{L}$ associated with the eigenvalue λ_i,

$$\|\boldsymbol{L}\boldsymbol{v}_i\| = \|\lambda_i \boldsymbol{v}_i\| = \lambda_i \|\boldsymbol{v}_i\| \quad (\text{since } \lambda_i \geq 0), \tag{14.27}$$

so that if $\|\boldsymbol{v}_i\|_2 = 1$ (normalized magnitude), then

$$\lambda_i \leq \lambda_j \Rightarrow \|\boldsymbol{L}\boldsymbol{v}_i\|_2 \leq \|\boldsymbol{L}\boldsymbol{v}_j\|_2, \tag{14.28}$$

in which case $\boldsymbol{v}_i$ is smoother than $\boldsymbol{v}_j$.

Besides, looking at the unique null eigenvalue of $\boldsymbol{L}$ (for connected undirected graphs), one has

$$\boldsymbol{L}\boldsymbol{v}_0 = \lambda_0 \boldsymbol{v}_0 = \boldsymbol{0}, \tag{14.29}$$

which consists of a system of homogeneous linear equations on the entries of vector $\boldsymbol{v}_0$. The solutions for such a system form the so-called *null space* of $\boldsymbol{L}$, and since its dimension equals the geometric multiplicity of $\lambda_0 = 0$, it follows that this null space is 1D. Therefore, it suffices to find a single nontrivial solution of (14.29) to obtain a basis for the possible eigenvectors $\boldsymbol{v}_0$. From (14.8), one may see that

$$[\boldsymbol{L}\boldsymbol{v}_0]_n = 0 \Rightarrow \sum_{m \mid v_m \in \mathcal{N}_n} A_{nm}(v_{0n} - v_{0m}) = 0, \tag{14.30}$$

which shows that any *constant* vector is a solution for (14.29). This leads to a quite satisfactory conclusion: the eigenvector associated with the null eigenvalue (frequency) is constant, as it is known to occur in the classical signal processing theory. If the eigenvectors are normalized, then each entry of $\boldsymbol{v}_0$ equals $1/\sqrt{N}$.

Fig. 14.8 visually confirms the notion of frequency based on the eigenvalues of the Laplacian matrix, using the example we described at the end of Section 14.1. In general, the smoothness of the eigenvectors decreases as the frequency (eigenvalue) rises (see Fig. 14.8e). Specifically considering the network modeled by the graph in this example, additional inferences can be carried out. Roughly speaking, it can be observed that the graph has three main branches: one at the left, one at the top, and one at the bottom of the image. Obviously, when we plot the eigenvector $\boldsymbol{v}_0$ as a signal over the graph in Fig. 14.8a, its constancy is verified by the fact that we have all the vertices painted with the same color. The eigenvector $\boldsymbol{v}_1$ corresponds to the first nonconstant spectral component, which has a remarkably smooth behavior. When $\boldsymbol{v}_1$ is plotted over the graph in Fig. 14.8b, such a characteristic is observed through gradual variations in the colors of the vertices, starting from the (left, upper, and lower) ends of each of the mentioned three branches and meeting in the central part of the graph. Finally, in Fig. 14.8c, we plot the eigenvector $\boldsymbol{v}_{88}$, the one representing the highest-frequency component. The highlight of the respective image is the presence of at least one vertex to which a high-value (red (mid gray in print version)) sample is associated, connected to other vertices on which there are samples with relatively low amplitudes (light (dark gray in print version) and dark blue (black in print version)). This is clearly related to high-frequency spectral content. Fig. 14.8d is an alternative way of visualizing $\boldsymbol{v}_{88}$, after its positive and negative samples are respectively replaced with 1 and -1; this allows us to get a better visual sense of the relatively large number of zero crossings such an eigenvector has. Fig. 14.8f shows the number of zero crossings of each eigenvector, meaning the number of graph edges connecting samples of the eigenvector with different sign.

14.5 Graph filtering

In the collaboration graph illustrated in Fig. 14.1, one can clearly see that a few nodes have a much higher h-index value than their corresponding neighbors. These nodes tend to exert greater influence on their coauthors in terms of research topics than the other way round. Detecting which nodes are under such influence is a matter of combining knowledge coming from the GS itself, i.e., the vector comprised

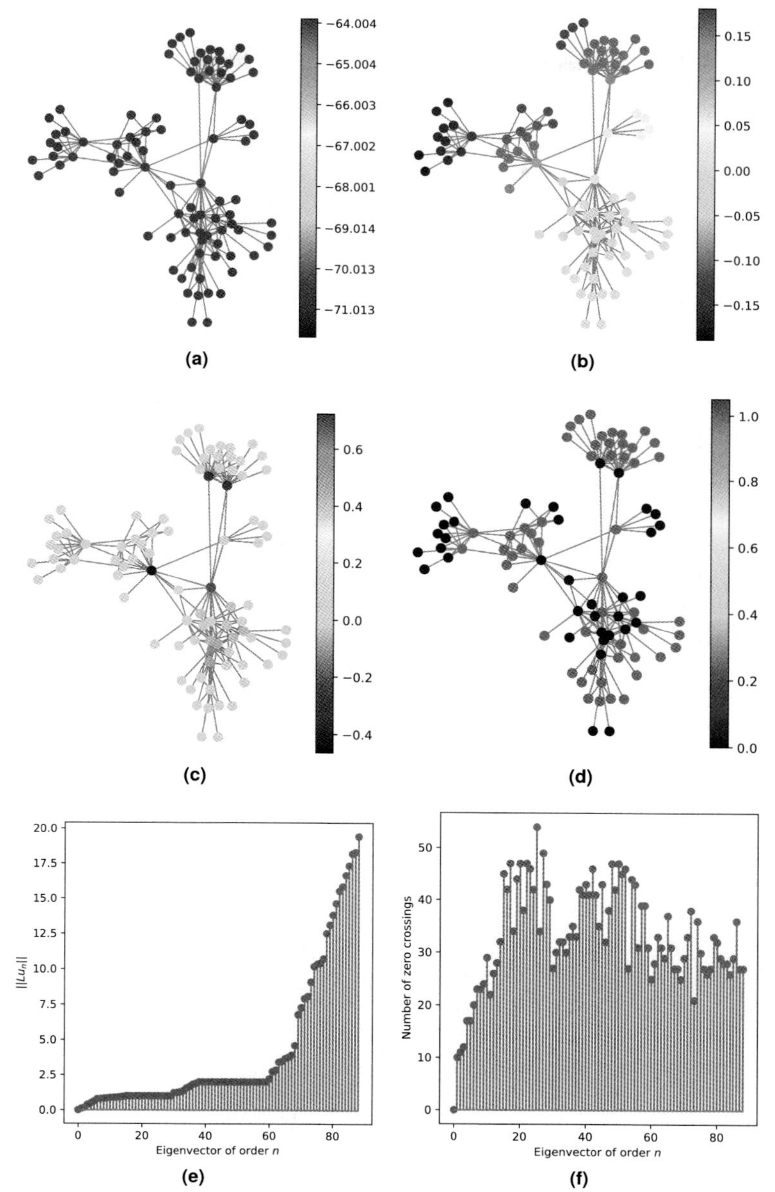

FIGURE 14.8

Some eigenvectors of the Laplacian matrix of the collaboration graph with 89 vertices shown in Fig. 14.1: (a) $\boldsymbol{v}_0$, (b) $\boldsymbol{v}_1$, and (d) $\boldsymbol{v}_{88}$. Alternatively, in (e) a binarized version of $\boldsymbol{v}_{88}$ is shown, with positive samples set to 1 and negative samples set to -1. In (c) and (f) two measures of signal smoothness are depicted: the norm $\|\boldsymbol{L}\boldsymbol{v}_i\|_2$ and the number of zero crossings of the eigenvectors.

of all h-index values, and from the graph, i.e., the set of nodes corresponding to authors/researchers, and vertices corresponding to established coauthorship/collaboration. On the other hand, some other nodes are connected only to neighbors with rather similar h-index values. These nodes tend to work on topics of common affinity without any particular node having a clearly greater influence over the others. Again, it is challenging, maybe even impossible considering a single GS, to draw conclusions from this type of observation by inspecting only either the GS, i.e., following a classical DSP approach, or the graph, i.e., following a conventional graph theoretic approach. Rather, the combination of the knowledge about the graph structure with the knowledge about the signal values on top of the graph nodes is the key to infer proper information, and perhaps even to confirm our intuition, regarding the influence on the particular research topics of the published works.

This section details the concept of graph filtering, which refers to the processing of GSs that takes into account the graph structure (domain of the signal).

14.5.1 Convolution and filtering

The natural approach to generate signals that make the task of extracting information relevant to the aforementioned discussion easier is filtering. In particular, linear filtering is the simplest approach, which in classical DSP is mathematically modeled via a (linear) convolution operation.

A linear filtering operator $\mathcal{L}\{\cdot\}$ in classical DSP acts over a discrete-time input signal $x[n]$ to produce an output $y[n]$ as follows:

$$\begin{aligned} y[n] &= \mathcal{L}\{x[n]\} \\ &\overset{\text{(a)}}{=} \mathcal{L}\left\{\sum_{m\in\mathbb{Z}} x[m]\delta[n-m]\right\} \\ &\overset{\text{(b)}}{=} \sum_{m\in\mathbb{Z}} x[m]\underbrace{\mathcal{L}\{\delta[n-m]\}}_{\triangleq h_m[n]} \\ &\overset{\text{(c)}}{=} \sum_{m\in\mathbb{Z}} x[m]\cdot\underbrace{h_m[n]}_{\triangleq h[n-m]} \\ &= \sum_{m\in\mathbb{Z}} x[m]h[n-m] \\ &\triangleq (x*h)[n], \end{aligned} \tag{14.31}$$

which is the well-known convolution sum. In (a) we rewrote the input signal as a function of the Kronecker delta

$$\delta[n] \triangleq \begin{cases} 1, & n=0, \\ 0, & n\in\mathbb{Z}\setminus\{0\}, \end{cases} \tag{14.32}$$

using the well-known sampling property of $\delta[n]$. In (b) we used the linearity of the linear filtering operator and defined the response of the filter to an m-sample shifted discrete-time impulse. In (c) we assumed that the linear filter is also shift-invariant so that $h_m[n] = h_0[n-m] \triangleq h[n-m]$.

Similarly, considering the sampling property of the Dirac delta $\delta(t)$, the continuous-time output signal $y(t)$ of a shift-invariant linear filter processing an input $x(t)$ is the convolution integral

$$y(t) = \int_{-\infty}^{\infty} x(\tau)h(t-\tau)\mathrm{d}\tau \triangleq (x * h)(t). \tag{14.33}$$

The filter's output signal h is called *impulse response* and carries the main pieces of information of a shift-invariant linear system.

One of the main properties of the convolution operation in signal processing is that it can be implemented either in the original domain (e.g., time) or in the transformed domain (e.g., frequency) via pointwise multiplications. Indeed, the convolution theorem establishes the equivalence, under certain assumptions as always, between the convolution of two signals in the original domain and the elementwise multiplication of their transformed-domain representations.

In classical DSP, the (circular[8]) convolution theorem may be exemplified as follows:

$$(x \circledast h)[n] = \text{IDFT}\left\{\text{DFT}\left\{x[n]\right\} \odot \text{DFT}\left\{h[n]\right\}\right\}, \tag{14.34}$$

wherein DFT $\{\cdot\}$ denotes the DFT of its argument (with a conveniently chosen length), IDFT $\{\cdot\}$ denotes the inverse DFT, and $\odot$ denotes the Hadamard product (i.e., elementwise product) between the related sequences. In this case, DFT $\{h[n]\}$ is called the frequency response of the filter.

Convolution between graph signals

Inspired by the above DSP example, we shall use the frequency domain as the starting point for the definition of the convolution operation. More specifically, we can define the convolution between two GSs $\boldsymbol{x}$ and $\boldsymbol{h}$ as

$$\boldsymbol{x} * \boldsymbol{h} \triangleq \boldsymbol{V}(\widehat{\boldsymbol{x}} \odot \widehat{\boldsymbol{h}}), \tag{14.35}$$

where $\widehat{\boldsymbol{x}} = \boldsymbol{V}^{-1}\boldsymbol{x}$, $\widehat{\boldsymbol{h}} = \boldsymbol{V}^{-1}\boldsymbol{h}$, and $\boldsymbol{V}$ is the linear transformation associated with the IGFT. It is worth noting that, by definition, $\boldsymbol{x} * \boldsymbol{h} = \boldsymbol{h} * \boldsymbol{x}$ since the Hadamard product is commutative.

Note that if we want to keep using the convolution operation as the basic processing action associated with linear filtering, we shall consider that $\boldsymbol{h}$ in (14.35) plays the role of impulse response, whereas $\widehat{\boldsymbol{h}}$ plays the role of frequency response. In this context, let us define $\widehat{\boldsymbol{H}} \triangleq \text{diag}\{\widehat{\boldsymbol{h}}\}$. We can then write $\widehat{\boldsymbol{x}} \odot \widehat{\boldsymbol{h}} = \widehat{\boldsymbol{H}} \cdot \widehat{\boldsymbol{x}}$, so that

$$\begin{aligned}\boldsymbol{x} * \boldsymbol{h} &= \underbrace{\boldsymbol{V}\widehat{\boldsymbol{H}}\boldsymbol{V}^{-1}}_{\triangleq \boldsymbol{H}} \cdot \boldsymbol{x} \\ &= \boldsymbol{H}\boldsymbol{x},\end{aligned} \tag{14.36}$$

in which $\boldsymbol{H}$ is the vertex-domain matrix associated with the linear filtering.

[8] The classical DSP approach assumes finite length sequences which can be extended periodically. Under suitable conditions on the signals' lengths, the circular convolution is equivalent to the linear convolution associated with linear filtering.

Impulse graph signal

In order to justify the aforementioned nomenclature of impulse and frequency responses, let us again trace a parallel with DSP. In classical DSP, an impulse is defined as in (14.32). For this type of signal, the ideas of origin ($n = 0$) and regularity of the domain play an important role, bringing some difficulties for the corresponding GSP generalization.[9] Alternatively, an impulse can be thought of as the signal that has a perfectly flat frequency-domain representation, expressing the fact that all frequency components contribute equally to the signal's composition. In this text, we shall adopt this property as the basic one to be generalized for the impulse GS, which is defined as

$$\boldsymbol{\delta}_0 \triangleq \boldsymbol{V}\mathbf{1}_N, \tag{14.37}$$

with $\mathbf{1}_N$ denoting a column vector with all its N entries equal to 1. Note that $\widehat{\boldsymbol{\delta}}_0 = \mathbf{1}_N$. Further, observe that the vertex-domain representation of the impulse $\boldsymbol{\delta}_0$ depends on the particular graph structure or, more specifically, on the particular IGFT matrix $\boldsymbol{V}$.

Considering the above definition for the impulse, the underlying impulse response GS $\boldsymbol{h}_0$ of a linear filter parameterized by matrix $\boldsymbol{H}$ is

$$\boldsymbol{h}_0 \triangleq \boldsymbol{H}\boldsymbol{\delta}_0, \tag{14.38}$$

whose graph-frequency representation is

$$\begin{aligned}\widehat{\boldsymbol{h}}_0 &\triangleq \boldsymbol{V}^{-1}\boldsymbol{h}_0 \\ &= \widehat{\boldsymbol{H}}\boldsymbol{V}^{-1}\boldsymbol{\delta}_0 \\ &= \widehat{\boldsymbol{H}}\mathbf{1}_N.\end{aligned} \tag{14.39}$$

In other words, considering the definition of GS convolution in (14.35), we note that $\boldsymbol{h} = \boldsymbol{h}_0$ is indeed the impulse response and $\widehat{\boldsymbol{h}} = \widehat{\boldsymbol{h}}_0$ is the frequency response.

As a final remark on the impulse GS definition, note that it preserves the following nice property: the impulse signal plays the role of an identity element in the space of GSs for the convolution operation. Indeed, we have

$$\begin{aligned}\boldsymbol{x} * \boldsymbol{\delta}_0 &= \boldsymbol{V}(\widehat{\boldsymbol{x}} \odot \widehat{\boldsymbol{\delta}}_0) \\ &= \boldsymbol{V}(\widehat{\boldsymbol{x}} \odot \mathbf{1}_N) \\ &= \boldsymbol{V}\widehat{\boldsymbol{x}} \\ &= \boldsymbol{x}.\end{aligned} \tag{14.40}$$

[9] One could define an impulse GS centered at the nth vertex as a vector whose entries are all null but the nth entry, which would be equal to 1. Whereas such a definition preserves some natural properties of the impulse in the vertex domain, it makes things more difficult when considering the convolution operation in (14.35).

Graph filters

When the graph frequencies (i.e., the eigenvalues of the GSO $\boldsymbol{S}$), $\lambda_0, \ldots, \lambda_{N-1}$, are such that we can find a solution $\tilde{\boldsymbol{h}}$ to the linear system

$$\underbrace{\begin{bmatrix} 1 & \lambda_0 & \cdots & \lambda_0^L \\ 1 & \lambda_1 & \cdots & \lambda_1^L \\ \vdots & \vdots & \cdots & \vdots \\ 1 & \lambda_{N-1} & \cdots & \lambda_{N-1}^L \end{bmatrix}}_{\triangleq \boldsymbol{U}} \underbrace{\begin{bmatrix} \tilde{h}_0 \\ \tilde{h}_1 \\ \vdots \\ \tilde{h}_{L-1} \end{bmatrix}}_{=\tilde{\boldsymbol{h}}} = \underbrace{\begin{bmatrix} \widehat{h}_0 \\ \widehat{h}_1 \\ \vdots \\ \widehat{h}_{N-1} \end{bmatrix}}_{=\widehat{\boldsymbol{h}}}, \tag{14.41}$$

then we can replace $\widehat{h}_n$ with $\sum_{\ell=0}^{L} \tilde{h}_\ell \lambda_n^\ell$ and, therefore, obtain

$$\widehat{\boldsymbol{H}} = \sum_{\ell=0}^{L} \tilde{h}_\ell \boldsymbol{\Lambda}^\ell, \tag{14.42}$$

thus yielding the *graph filter*

$$\begin{aligned} \boldsymbol{H} &= \sum_{\ell=0}^{L} \tilde{h}_\ell \underbrace{\boldsymbol{V}\boldsymbol{\Lambda}^\ell \boldsymbol{V}^{-1}}_{=\boldsymbol{S}^\ell} \\ &= \sum_{\ell=0}^{L} \tilde{h}_\ell \boldsymbol{S}^\ell. \end{aligned} \tag{14.43}$$

By definition, *linearity* always holds for such graph filters, since the distributivity of matrix multiplication with respect to vector addition guarantees that

$$\boldsymbol{H}(\alpha_1 \boldsymbol{x}_1 + \alpha_2 \boldsymbol{x}_2) = \alpha_1 \boldsymbol{H}\boldsymbol{x}_1 + \alpha_2 \boldsymbol{H}\boldsymbol{x}_2. \tag{14.44}$$

In addition to linearity, another useful desirable property is *shift-invariance*, which refers to the possibility to commute filtering (i.e., multiplication by $\boldsymbol{H}$) and shifting (i.e., multiplication by $\boldsymbol{S}$) operations. In other words, in order for a graph filter $\boldsymbol{H}$ to be LSI it is required that $\boldsymbol{S}\boldsymbol{H}\boldsymbol{x} = \boldsymbol{H}\boldsymbol{S}\boldsymbol{x}\ \forall \boldsymbol{x}$, and therefore $\boldsymbol{S}\boldsymbol{H} = \boldsymbol{H}\boldsymbol{S}$. Once again, given the expression in (14.43), the corresponding graph filter is LSI, for $\boldsymbol{H}$ is a polynomial in $\boldsymbol{S}$, exactly as happens in DSP, in which LSI filters have polynomial representations in z^{-1}.

Let us take the filter in (14.43) and its output to the input $\boldsymbol{x}$ given by

$$\begin{aligned} \boldsymbol{H}\boldsymbol{x} &= \sum_{\ell=0}^{L} \tilde{h}_\ell \boldsymbol{S}^\ell \boldsymbol{x} \\ &= \boldsymbol{V} \underbrace{\left(\sum_{\ell=0}^{L} \tilde{h}_\ell \boldsymbol{\Lambda}^\ell \right)}_{\triangleq \tilde{h}(\boldsymbol{\Lambda})} \boldsymbol{V}^{-1} \boldsymbol{x}. \end{aligned} \tag{14.45}$$

Taking the GFT of both sides of the last equation yields

$$V^{-1}Hx = \tilde{h}(\Lambda)\widehat{x}, \tag{14.46}$$

which confirms that left-multiplication by H (action of the filter in the vertex domain) is equivalent to the left-multiplication in the frequency domain by the matrix $\tilde{h}(\Lambda)$. In other words, $\tilde{h}(\Lambda)$ represents the frequency response of H.

As S is an $N \times N$ matrix, we shall assume that $L \leq N-1$ without loss of generality – indeed, if $L > N-1$, then we could rewrite H as a polynomial in S of degree at most $N-1$ due to, for instance, the Cayley–Hamilton theorem. On the other hand, guaranteeing the existence of a solution $\tilde{h}$ in general requires $L \geq N-1$. Thus, let us assume $L = N-1$ and $\lambda_n \neq \lambda_m$ for all $n \neq m$, so that the (square) Vandermonde matrix U is nonsingular. In this case, we have

$$\tilde{h} = U^{-1}\widehat{h} \tag{14.47a}$$

$$= (VU)^{-1}h. \tag{14.47b}$$

Eq. (14.47a) shows that the filter taps can be obtained from the filter's frequency response using the inverse of the Vandermonde matrix containing the underlying graph frequencies. Eq. (14.47b) shows the relation between the filter's taps and impulse response, since they are not identical in general. This contrasts with what happens in classical DSP, wherein a finite duration impulse response (FIR) filter is characterized by its taps, which also comprise the impulse response. This happens because both the basis functions and the frequencies themselves can be written as complex exponential functions in classical DSP, so that $VU = I_N$; in GSP, one may have $VU \neq I_N$ in general.

14.5.2 Graph filter design

The most common way to design a graph filter is to start with the definition of a desired frequency-domain response to be approximated. For instance, consider the function $f : [\lambda_{\min}, \lambda_{\max}] \to \mathbb{C}$ and let $\widehat{h}_n = f(\lambda_n)$. When filtering is performed in the graph-frequency domain, then one could directly use $\widehat{h}_0, \widehat{h}_1, \ldots, \widehat{h}_{N-1}$ for this operation; in this case, if we consider that the notation $f(\Lambda)$ represents a diagonal matrix with entries $f(\lambda_n)$, then the corresponding graph filter is $Vf(\Lambda)V^{-1}$.

Although filters can be designed regardless of the graph spectrum, developing graph filters without taking into account the actual graph spectrum may lead to undesirable results due to the irregular (i.e., nonuniform) distribution of eigenvalues. For example, consider a graph with $N = 4$ nodes and Laplacian-based eigenvalues $\lambda_0 = 0,\ \lambda_1 = 2.0,\ \lambda_2 = 2.5,\ \lambda_3 = 3.0$ and define the following scalar indicator function:

$$\chi_{[\lambda<\lambda_c]}(\lambda) = \begin{cases} 1, & \lambda < \lambda_c, \\ 0, & \text{otherwise.} \end{cases} \tag{14.48}$$

The half-band low-pass filter $\chi_{[\lambda<1.5]}(\lambda)$ will remove all non-DC components of a GS, which might not be intended originally. Indeed, the only eigenvalue in the set $\{\lambda_0, \lambda_1, \lambda_2, \lambda_3\}$ that is smaller than 1.5 is $\lambda_0 = 0$ (DC), eventually implying that all the other components (corresponding to the frequencies $\lambda_1 = 2.0,\ \lambda_2 = 2.5,\ \lambda_3 = 3.0$) will be zeroed by this low-pass filtering process. The interesting part of this example is that $\chi_{[\lambda<1.5]}(\lambda)$ represents an ideal low-pass filter with cutoff frequency exactly in the middle of the underlying Laplacian spectrum, since $\frac{\lambda_0+\lambda_3}{2} = 1.5$.

Note that if the degree L in (14.43) is not sufficiently large, say $L < N-1$, then we cannot guarantee the existence of a solution to the linear system of equations in (14.41). This means that we would only be able to approximate the ideal graph filter $\boldsymbol{V} f(\boldsymbol{\Lambda})\boldsymbol{V}^{-1}$ using the LSI graph filter $\boldsymbol{H} = \sum_{\ell=0}^{L} \tilde{h}_\ell \boldsymbol{S}^\ell$. Yet, this approximation might be interesting due to some good additional properties that polynomial (in $\boldsymbol{S}$) filters enjoy, viz.:

- $\boldsymbol{H}$ can be computed without accessing the entire eigendecomposition of the GSO $\boldsymbol{S}$.
- Depending on the sparsity level of the GSO and also on the polynomial degree L, polynomial filters are localized in the vertex domain, which might be handy for big data-related graphs as well as for distributed processing.

Least-squares approximation

The approximation of an ideal graph filter can be implemented, for instance, via least-squares approximation. Indeed, starting from the ideal function f and its specific values on the graph spectrum $\widehat{h}_n = f(\lambda_n)$, we consider the following least-squares problem related to the linear system in (14.41):

$$\underset{\tilde{\boldsymbol{h}}}{\text{minimize}} \left\| \widehat{\boldsymbol{h}} - \boldsymbol{U}\tilde{\boldsymbol{h}} \right\|_2, \tag{14.49}$$

whose solution is

$$\tilde{\boldsymbol{h}} = (\boldsymbol{U}^{\mathrm{H}}\boldsymbol{U})^{-1}\boldsymbol{U}^{\mathrm{H}}\widehat{\boldsymbol{h}}. \tag{14.50}$$

Chebyshev polynomial approximation

An alternative approach that might be useful when the GSO is the Laplacian matrix is to use Chebyshev polynomials $C_\ell : [-1, 1] \to \mathbb{R}$, which satisfy the following recursion:

$$C_\ell(\xi) = 2\xi C_{\ell-1}(\xi) - C_{\ell-2}(\xi). \tag{14.51}$$

Given a function f in the space of square-integrable functions in the interval $[-1, 1]$, denoted by $L_2\left([-1, 1], \frac{1}{\sqrt{1+\xi^2}}\right)$, with respect to the measure $\frac{1}{\sqrt{1+\xi^2}}$, its Lth-order Chebyshev approximation is given by $\frac{1}{2}\tilde{h}_0 + \sum_{\ell=1}^{L} \tilde{h}_\ell C_\ell(\xi)$, where

$$\tilde{h}_\ell = \int_{-1}^{1} \frac{C_\ell(\xi) f(\xi)}{\sqrt{1+\xi^2}} \mathrm{d}\xi = \frac{2}{\pi} \int_{0}^{\pi} \cos(\ell\theta) f(\cos(\theta)) \mathrm{d}\theta. \tag{14.52}$$

The matrix $f(\boldsymbol{L}) \triangleq \boldsymbol{V} f(\boldsymbol{\Lambda})\boldsymbol{V}^{-1}$ can be approximated by a Chebyshev polynomial in the Laplacian matrix. First, in order to encompass the entire Laplacian spectrum, it is necessary to map the interval $[0, \lambda_{\max}]$ into $[-1, 1]$, giving the following construction of Chebyshev polynomials in $\boldsymbol{L}$:

$$C_\ell(\boldsymbol{L}) = 2\overline{\boldsymbol{L}} C_{\ell-1}(\boldsymbol{L}) - C_{\ell-2}(\boldsymbol{L}), \tag{14.53}$$

wherein

$$\overline{\boldsymbol{L}} \triangleq \frac{2}{\lambda_{\max}}\boldsymbol{L} - \boldsymbol{I}, \tag{14.54a}$$

$$C_0(\boldsymbol{L}) \triangleq \boldsymbol{I}, \tag{14.54b}$$

$$C_1(\boldsymbol{L}) \triangleq \overline{\boldsymbol{L}}, \tag{14.54c}$$

so that the resulting graph filter is

$$\boldsymbol{H} \triangleq \frac{1}{2}\tilde{h}_0 + \sum_{\ell=1}^{L} \tilde{h}_\ell C_\ell(\boldsymbol{L}), \tag{14.55}$$

with coefficients defined by

$$\tilde{h}_\ell \triangleq \frac{2}{\pi} \int_0^\pi \cos(\ell\theta) f\left(\frac{\lambda_{\max}}{2}(\cos\theta + 1)\right) d\theta. \tag{14.56}$$

This polynomial approximation is still valid for any real symmetric matrix with $\lambda_{\min}$ not necessarily zero, such as in the case of an adjacency matrix of an undirected graph. In this case, $\overline{\boldsymbol{L}} \triangleq \frac{2}{\lambda_{\max}-\lambda_{\min}}(\boldsymbol{L} - \lambda_{\min}\boldsymbol{I}) - \boldsymbol{I}$ and

$$\tilde{h}_\ell \triangleq \frac{2}{\pi} \int_0^\pi \cos(\ell\theta) f\left(\frac{\lambda_{\max}}{2}(\cos\theta + 1) + \lambda_{\min}\right) d\theta.$$

Jackson–Chebyshev polynomial approximation

In order to cancel the ripples associated with Chebyshev polynomials as well as to enhance the stopband attenuation we can use the Jackson–Chebyshev polynomials [17]. Indeed, the coefficients of the Chebyshev polynomial can be adapted to approximate the graph filter $f(\boldsymbol{L})$ as $\sum_{\ell=0}^{L} \check{h}_\ell \boldsymbol{L}^\ell$ with $\check{h}_0 = \frac{1}{2}\tilde{h}_0$ and $\check{h}_\ell = \gamma_{\ell,L}\tilde{h}_\ell$ for $\ell \geq 1$, with

$$\gamma_{\ell,L} = \frac{\left(1 - \frac{\ell}{L+2}\right)\sin\left(\frac{\pi}{L+2}\right)\cos\left(\frac{\ell\pi}{L+2}\right) + \frac{\ell}{L+2}\cos\left(\frac{\pi}{L+2}\right)\sin\left(\frac{\ell\pi}{L+2}\right)}{\sin\left(\frac{\pi}{L+2}\right)}. \tag{14.57}$$

It is worth highlighting that the benefits of employing the Jackson–Chebyshev approximation come at a cost of expanding the transition band.

The following example illustrates the implementation of an ideal half-band low-pass graph filter using least-squares, Chebyshev, and Jackson–Chebyshev polynomial approximations. Consider a randomly connected sensor graph $\mathcal{G}$ with $N = 200$ nodes whose Laplacian matrix is normalized by its largest eigenvalue. In addition, consider an ideal half-band low-pass filter $f(\lambda) = \chi_{[\lambda<\lambda_{\max}/2]}(\lambda)$. Fig. 14.9 shows an ideal low-pass filter $f(\lambda)$ approximated by polynomials of degrees $L = 10$ and $L = 50$ (top and bottom, respectively) for least-squares, Chebyshev, and Jackson–Chebyshev polynomial approximations. In comparison with the Chebyshev polynomial approximation, the Jackson–Chebyshev polynomial approximation of degree $L = 10$ has a larger transition band which is narrowed as long as the polynomial degree increases.

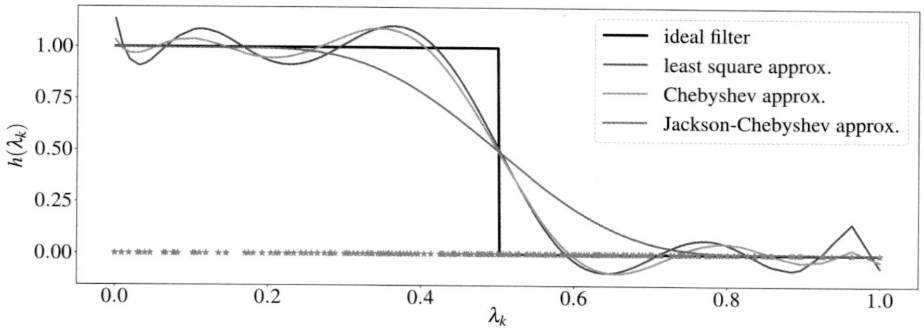

(a) $Q = 10$

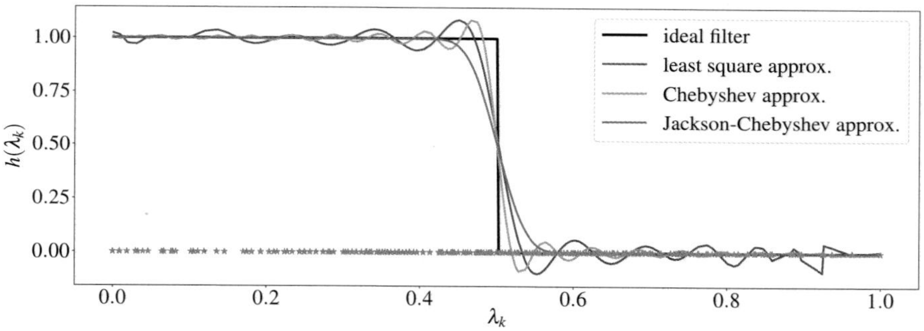

(b) $Q = 50$

FIGURE 14.9

Ideal low pass filter and polynomial approximations. The stars represent the Laplacian eigenvalues.

14.6 Down- and upsampling graph signals

In the collaboration graph illustrated in Fig. 14.1, we could have practical difficulties in obtaining the h-index values of some authors. This might be the case when the authors have no profile listed on the Digital Bibliography & Library Project (DBLP), Scopus, and Google Scholar websites. Yet, given the nature of the h-index and considering the collaboration structure captured by the graph model, we could infer the h-index values that would be achieved in steady state by these authors. This inference problem can be cast as a problem of recovering a GS from one of its sampled versions. Alternatively, one can think of the nodes with available h-index values as a graph on its own that will serve as the basic building block for a denser (virtual) graph comprised of the original nodes with available h-index values as well as upcoming collaborating nodes with interpolated h-index values. This interpolation/forecasting problem can be cast as a problem of upsampling GSs. This discussion clarifies the importance of developing a sampling theory for GSP that can answer questions like: For which conditions can one obtain perfect reconstruction of GSs from their sampled versions? Or what are the samples that favor a more accurate/robust reconstruction of the full GS?

14.6.1 Band-limited graph signals

In classical DSP, it is quite common to have the need to work with multirate signals and filters. As one would expect, there are many GSP applications where the underlying abstract problems can be cast as some sort of multirate manipulation of signals or multirate design of systems. Although the generality of the graph domain brings many advantages in terms of modeling, it also raises several challenges, since the domain loses some interesting properties on which classical DSP tools strongly rely. Consider, for instance, the downsampling of a discrete-time signal by a factor of 2. Such operation can be straightforwardly implemented by dropping every odd sample of the signal. The generalization of this operation for GSP is not straightforward, since an "odd node" is not a well-defined concept for a general graph. Due to this difficulty, a plethora of graph sampling approaches have been developed [18] focusing on many different applications, such as filter banks, active learning, data interpolation, etc.

In classical signal processing, the Shannon–Nyquist sampling theorem states that a sampled band-limited signal can be perfectly recovered by a sinc[10] interpolation if the sampling rate is at least twice the bandwidth. More specifically, if $f(t)$ is a continuous square-integrable function such that $\widehat{f}(\xi)=0$ for all $|\xi|>\xi_b$ and the sampling frequency $\xi_s \geq 2\xi_b$, then $f(t)$ can be recovered by the following interpolation formula:

$$f(t)=\sum_{n\in\mathbb{Z}} f\left(\frac{n\pi}{\xi_s}\right)\frac{\sin(\xi_s t)}{\xi_s t}. \tag{14.58}$$

The set of square-integrable[11] band-limited functions

$$\mathrm{PW}_{\xi_s}(\mathbb{R}) \triangleq \left\{f(t)\in L^2(\mathbb{R})\;;\;\widehat{f}(\xi)=0,\,|\xi|>\xi_s\right\} \tag{14.59}$$

is called a *Paley–Wiener space*.

Sampling and interpolation operators

In the following we shall describe several attempts to extend sampling theory to GSP. Let us first start by defining $\mathcal{S}\triangleq\{s_0,\ldots,s_{M-1}\}\subset\mathcal{V}$ as a subset of vertices with $M\leq N$ nodes; the vector of measurements $\boldsymbol{x}_{\mathcal{S}}\in\mathbb{R}^M$ is given by $\boldsymbol{x}_{\mathcal{S}}=\boldsymbol{\Psi}_{\mathcal{S}}\boldsymbol{x}$, where the sampling operator

$$[\boldsymbol{\Psi}_{\mathcal{S}}]_{mn}\triangleq\begin{cases}1, & \text{if } v_n=s_m,\\ 0, & \text{otherwise}\end{cases} \tag{14.60}$$

selects from $\mathcal{V}$ the nodes in $\mathcal{S}$ that will have the signals' samples retained.

---

[10] The sinc function is defined as $\mathrm{sinc}(x)=\frac{\sin(x)}{x}$ for any $x\in\mathbb{R}\setminus\{0\}$ and $\mathrm{sinc}(0)=1$.
[11] The set of square-integrable real functions is denoted by $L^2(\mathbb{R})$. That is, $f(t)\in L^2(\mathbb{R})\Leftrightarrow\int_{\mathbb{R}}f^2(t)\mathrm{d}t<\infty$.

On the other hand, the interpolation operator $\mathbf{\Phi}$ is an $N \times M$ matrix such that the recovered signal is

$$\tilde{\boldsymbol{x}} \triangleq \mathbf{\Phi}\mathbf{\Psi}_{\mathcal{S}}\boldsymbol{x}. \tag{14.61}$$

If $\tilde{\boldsymbol{x}} = \boldsymbol{x}$, the pair of sampling and interpolation operators $(\mathbf{\Phi}, \mathbf{\Psi}_{\mathcal{S}})$ can perfectly recover the signal $\boldsymbol{x}$ from its sampled version.

As the rank of $\mathbf{\Phi}\mathbf{\Psi}_{\mathcal{S}}$ is smaller than or equal to M, this perfect reconstruction is not possible for all $\boldsymbol{x} \in \mathbb{R}^N$ when $M < N$. This fact makes clear the necessity to constrain the class of possible GSs of interest that can be perfectly reconstructed by their sampled versions. Once gain, we draw inspiration from DSP and look for a subclass of signals with a certain degree of "sparsity" in its frequency-domain representation.

Band-limitedness

The GS $\boldsymbol{x}_{\mathrm{b}}$ is said to be $\mathcal{F}$-band-limited if $[\widehat{\boldsymbol{x}}_{\mathrm{b}}]_n = 0$ for all indices n such that the corresponding frequency $\lambda_n \notin \mathcal{F} \subset \{\lambda_0, \ldots, \lambda_{N-1}\}$, that is, the frequency content of $\boldsymbol{x}_{\mathrm{b}}$ is restricted to the set of frequencies $\mathcal{F}$. The space of $\mathcal{F}$-band-limited GSs on the graph $\mathcal{G}$ with GFT $\boldsymbol{V}^{-1}$ will be denoted as $\mathrm{BL}_{\mathcal{F}}(\boldsymbol{V})$.

Some works also restrict the support of the frequency content and consider that a GS $\boldsymbol{x}_{\mathrm{b}}$ is K-band-limited if $[\widehat{\boldsymbol{x}}_{\mathrm{b}}]_k = 0$ for $k \geq K$, wherein it is assumed that the columns of $\boldsymbol{V}$ are ordered so that the associated eigenvalues are properly sorted to reflect the notion of high/lower frequencies – for instance, in the case of the GSO being the Laplacian matrix, the eigenvalues (real numbers) are sorted in ascending order. When $\boldsymbol{S} = \boldsymbol{L}$, one can also define a ξ-band-limited GS $\boldsymbol{x}_{\mathrm{b}}$ when $[\widehat{\boldsymbol{x}}_{\mathrm{b}}]_k = 0$ for all indices k such that $\lambda_k > \xi$ [19]. In this case, the space of ξ-band-limited GSs, $\mathrm{BL}_{\xi}(\boldsymbol{V})$, is the graph equivalent of the $\mathrm{PW}_{\xi}(\mathbb{R})$ from classical signal processing.

Fig. 14.10 exemplifies some of the aforementioned concepts. In this example, we have three different signals defined over the same unweighted graph with $N = 10$ vertices. More specifically, Fig. 14.10a depicts a realization of a random GS that is full-band, as shown in Fig. 14.10b, in which it is clear that the signal has nonzero components across all frequencies $\lambda_0, \lambda_1, \ldots, \lambda_9$. In this example, the adopted GSO for computing the GFT was the Laplacian matrix. Starting from the GS in Figs. 14.10a and 14.10b, we generated two different band-limited GSs. In Figs. 14.10c and 14.10d we have a K-band-limited GS, with $K = 5$, obtained from the original GS by zeroing the frequency-domain components associated with the frequencies λ_k with $k \geq K$. In Figs. 14.10e and 14.10f we have an $\mathcal{F}$-band-limited GS, obtained from the original GS by zeroing the frequency-domain components associated with the frequencies λ_k with $k \in \{1, 3, 5, 7, 9\}$. By comparing Figs. 14.10a and 14.10c, we notice that the K-band-limited GS is a much smoother version of the original GS. On the other hand, however the $\mathcal{F}$-band-limited GS in Fig. 14.10e is smoother than the original GS, it clearly has stronger signal variations than the K-band-limited GS. This example also makes clear that band-limitedness might not be apparent directly from a quick look at the vertex-domain representation of a GS. Indeed, one may have an $\mathcal{F}$-band-limited GS with strong signal variations across adjacent nodes. Yet, the band-limitedness is an important property for achieving perfect reconstruction, as we shall detail now.

Conditions for perfect reconstruction

Given the set of band-limited GSs $\mathrm{BL}_{\mathcal{F}}(\boldsymbol{V})$, designing a pair of sampling and interpolation operators $(\mathbf{\Phi}, \mathbf{\Psi}_{\mathcal{S}})$ boils down to finding a subset of nodes $\mathcal{S} \subset \mathcal{V}$ whose signal values capture all the information

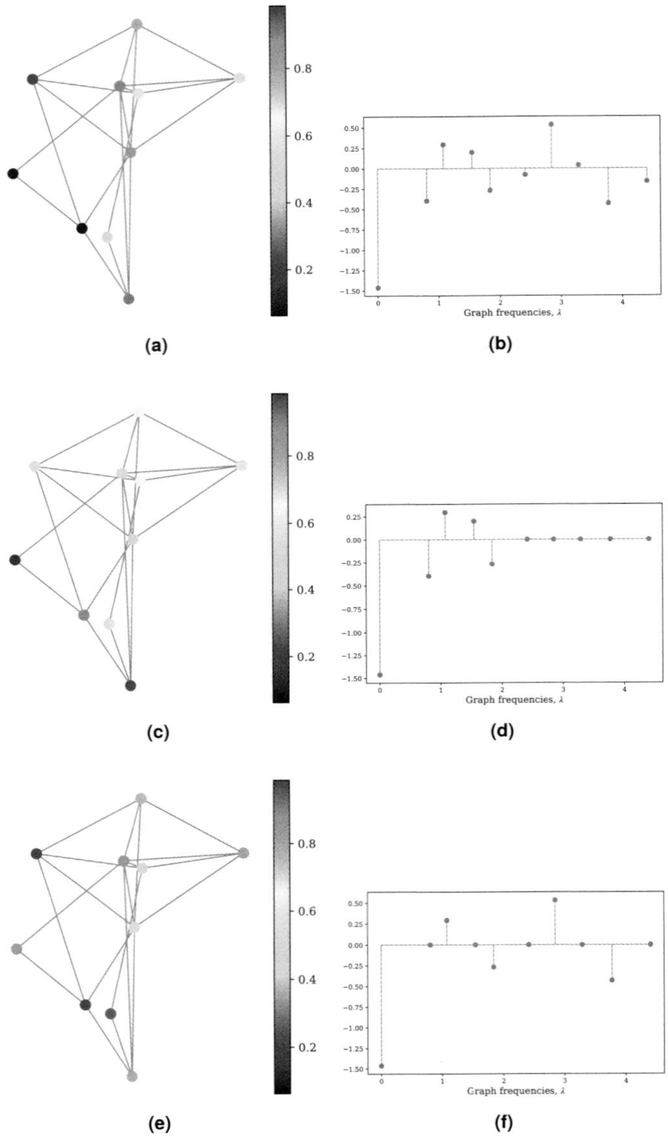

FIGURE 14.10

Example of signals defined over an unweighted graph with $N = 10$ vertices. (a) Full-band graph signal in the vertex domain. (b) Graph-frequency representation of the GS in (a). (c) K-band-limited graph signal in the vertex domain, with $K = 5$. (d) Graph-frequency representation of the GS in (c). (e) $\mathcal{F}$-band-limited graph signal in the vertex domain, with $\mathcal{F} = \{\lambda_0, \lambda_2, \lambda_4, \lambda_6, \lambda_8\}$. (f) Graph-frequency representation of the GS in (e). The adopted GSO was the Laplacian matrix.

about any GS $\boldsymbol{x} \in \mathrm{BL}_{\mathcal{F}}(\boldsymbol{V})$. In this context, $\mathcal{S}$ is called a *uniqueness set* for $\mathrm{BL}_{\mathcal{F}}(\boldsymbol{V})$ when $\boldsymbol{x}_{\mathcal{S}} = \boldsymbol{y}_{\mathcal{S}}$ implies $\boldsymbol{x} = \boldsymbol{y}$ for any two GSs $\boldsymbol{x}, \boldsymbol{y} \in \mathrm{BL}_{\mathcal{F}}(\boldsymbol{V})$.

The following *admissibility condition* guarantees the perfect reconstruction of an $\mathcal{F}$-band-limited GS for some sampling sets.

If the sampling operator $\boldsymbol{\Psi}_{\mathcal{S}}$ satisfies

$$\mathrm{rank}(\boldsymbol{\Psi}_{\mathcal{S}} \boldsymbol{V}_{:,\mathcal{F}}) = |\mathcal{F}| = K, \tag{14.62}$$

then $\boldsymbol{x}_{\mathrm{b}} = \boldsymbol{\Phi}\boldsymbol{\Psi}_{\mathcal{S}}\boldsymbol{x}_{\mathrm{b}}$ as long as $\boldsymbol{\Phi} = \boldsymbol{V}_{:,\mathcal{F}}\boldsymbol{\Sigma}$, with $\boldsymbol{\Sigma}$ satisfying $\boldsymbol{\Sigma}\boldsymbol{\Psi}_{\mathcal{S}}\boldsymbol{V}_{:,\mathcal{F}} = \boldsymbol{I}_K$ and with $\boldsymbol{V}_{:,\mathcal{F}}$ being a submatrix of $\boldsymbol{V}$ with columns restricted to the indices associated with the frequencies in $\mathcal{F}$.

Note that we omitted the dependency of $\boldsymbol{\Phi}$ on both $\mathcal{S}$ and $\mathcal{F}$ in order to simplify the notation. The result above can be justified by noting that:

(i) the columns of $\boldsymbol{\Phi}$ span $\mathrm{BL}_{\mathcal{F}}(\boldsymbol{V})$, and
(ii) $\boldsymbol{x} = \boldsymbol{\Phi}\boldsymbol{\Psi}_{\mathcal{S}}\boldsymbol{x}$ for all $\boldsymbol{x} \in \mathrm{BL}_{\mathcal{F}}(\boldsymbol{V})$, meaning that $\boldsymbol{\Phi}\boldsymbol{\Psi}_{\mathcal{S}}$ is a projection operator on $\mathrm{BL}_{\mathcal{F}}(\boldsymbol{V})$; thus, if the original GS $\boldsymbol{x}$ already belongs to $\mathrm{BL}_{\mathcal{F}}(\boldsymbol{V})$, then $\boldsymbol{\Phi}\boldsymbol{\Psi}_{\mathcal{S}}$ is an identity operator.

In order to prove (i), we start with the condition in (14.62) and recall that, by definition of $\boldsymbol{\Sigma}$, $\mathrm{rank}(\boldsymbol{\Sigma}\boldsymbol{\Psi}_{\mathcal{S}}\boldsymbol{V}_{:,\mathcal{F}}) = K$, which means that matrix $\boldsymbol{\Sigma}$ spans $\mathrm{BL}_{\mathcal{F}}(\boldsymbol{V})$. Therefore, $\boldsymbol{\Phi} = \boldsymbol{V}_{:,\mathcal{F}}\boldsymbol{\Sigma}$ also spans $\mathrm{BL}_{\mathcal{F}}(\boldsymbol{V})$. To further prove (ii), i.e., that $\boldsymbol{\Phi}\boldsymbol{\Psi}_{\mathcal{S}}$ is a projection operator, it is sufficient to show that it is an idempotent matrix as follows:

$$(\boldsymbol{\Phi}\boldsymbol{\Psi}_{\mathcal{S}})^2 = \boldsymbol{\Phi}\boldsymbol{\Psi}_{\mathcal{S}}\boldsymbol{\Phi}\boldsymbol{\Psi}_{\mathcal{S}} = \boldsymbol{V}_{:,\mathcal{F}}\boldsymbol{\Sigma}\boldsymbol{\Psi}_{\mathcal{S}}\boldsymbol{V}_{:,\mathcal{F}}\boldsymbol{\Sigma}\boldsymbol{\Psi}_{\mathcal{S}} = \boldsymbol{V}_{:,\mathcal{F}}\boldsymbol{I}_K\boldsymbol{\Sigma}\boldsymbol{\Psi}_{\mathcal{S}} = \boldsymbol{\Phi}\boldsymbol{\Psi}_{\mathcal{S}}. \tag{14.63}$$

The condition in (14.62) is also equivalent to [20]

$$\mathrm{SV}_{\max}(\boldsymbol{V}_{\overline{\mathcal{S}},\mathcal{F}}) \leq 1, \tag{14.64}$$

where $\mathrm{SV}_{\max}(.)$ stands for the largest singular value and $\overline{\mathcal{S}} = \mathcal{V} \setminus \mathcal{S}$ is the complement of $\mathcal{S}$. This means that no $\mathcal{F}$-band-limited signal over the graph $\mathcal{G}$ is supported on $\overline{\mathcal{S}}$.

Still focused on this rank characterization of the sampling operator, note that in order to have $\boldsymbol{\Sigma}\boldsymbol{\Psi}_{\mathcal{S}}\boldsymbol{V}_{:,\mathcal{F}} = \boldsymbol{I}_K$, we must have $|\mathcal{S}| = M \geq K$, since $\mathrm{rank}(\boldsymbol{V}_{:,\mathcal{F}}) = K$. If $M \geq K$, $\boldsymbol{\Sigma}$ is the pseudoinverse of $\boldsymbol{V}_{\mathcal{S},\mathcal{F}} \triangleq \boldsymbol{\Psi}_{\mathcal{S}}\boldsymbol{V}_{:,\mathcal{F}}$ and the interpolation operator is

$$\boldsymbol{\Phi} = \boldsymbol{V}_{:,\mathcal{F}}\left(\boldsymbol{V}_{\mathcal{S},\mathcal{F}}^{\mathrm{H}}\boldsymbol{V}_{\mathcal{S},\mathcal{F}}\right)^{-1}\boldsymbol{V}_{\mathcal{S},\mathcal{F}}^{\mathrm{H}}. \tag{14.65}$$

Note that as $\boldsymbol{V}$ is nonsingular, there is always at least one subset $\mathcal{S}$ such that the condition in (14.62) is satisfied. Nonetheless, for many choices of $\mathcal{S}$, $\boldsymbol{V}_{\mathcal{S},\mathcal{F}}$ can be full rank but ill-conditioned, leading to large reconstruction errors, especially in the presence of noisy measurements or in the case of approximately band-limited GSs, which is addressed in the next section. To overcome this issue, optimal sampling strategies, in the sense of minimizing reconstruction error, can be employed [21].

14.6.2 Approximately band-limited GSs

In practice, many GSs that we will encounter do not satisfy the aforementioned strict definitions of band-limitedness, thus calling for an alternative definition of approximately band-limitedness. A GS is $(\mathcal{F}, \epsilon)$-band-limited if [20]

$$\boldsymbol{x} = \boldsymbol{x}_{\mathrm{b}} + \boldsymbol{\eta}, \tag{14.66}$$

where $\boldsymbol{x}_{\mathrm{b}}$ is an $\mathcal{F}$-band-limited GS and $\boldsymbol{\eta}$ is an $\overline{\mathcal{F}}$-band-limited GS such that $\|\boldsymbol{\eta}\|_2 < \epsilon$, with $\overline{\mathcal{F}} \triangleq \{\lambda_0, \ldots, \lambda_{N-1}\} \setminus \mathcal{F}$.

When the signal $\boldsymbol{x}$ is sampled on the subset $\mathcal{S}$ and recovered by the interpolator in (14.65), it is possible to show that the error energy of the reconstructed signal is upper-bounded by [20]

$$\|\tilde{\boldsymbol{x}} - \boldsymbol{x}\|_2 \leq \frac{\|\boldsymbol{\eta}\|_2}{\cos(\theta_{\mathcal{S},\mathcal{F}})}, \tag{14.67}$$

where $\theta_{\mathcal{S},\mathcal{F}}$ is the maximum angle between the subspace of signals supported on $\mathcal{S}$ and the subspace of $\mathcal{F}$-band-limited GS; more specifically,

$$\cos(\theta_{\mathcal{S},\mathcal{F}}) = \inf_{\|z\|=1} \left\{ \|\boldsymbol{\Psi}_{\mathcal{S}} z\|_2 \ : \ \boldsymbol{V}_{:,\mathcal{F}} \boldsymbol{V}^{\mathrm{H}}_{:,\mathcal{F}} z = z \right\}. \tag{14.68}$$

Therefore, the error bound (14.67) is minimized when the subspace of signals in $\mathrm{BL}_{\mathcal{F}}(\boldsymbol{V})$ is as much as possible aligned with the subspace of signals supported on $\mathcal{S}$, so that $\theta_{\mathcal{S},\mathcal{F}}$ is minimized. From (14.68), it is clear that $\cos(\theta_{\mathcal{S},\mathcal{F}}) = \mathrm{SV}_{\min}(\boldsymbol{V}_{\mathcal{S},\mathcal{F}})$; therefore, in order to minimize the upper bound of the reconstruction error in (14.67), the set $\mathcal{S}$ should maximize $\mathrm{SV}_{\min}(\boldsymbol{V}_{\mathcal{S},\mathcal{F}})$. This optimal sampling strategy will be further discussed in the next section.

14.6.3 Optimal sampling strategies

Due to the irregular topology of graphs, the sample size can influence not only the reconstruction error when sampling and interpolating a GS by $\boldsymbol{\Psi}_{\mathcal{S}}$ and $\boldsymbol{\Phi}$, respectively, but also the nodes in $\mathcal{S}$ themselves. Therefore, whenever possible, it is favorable to select *where* to sample in addition to *how many* nodes. It is also worth remembering that not every set $\mathcal{S}$ with cardinality $|\mathcal{S}| = M \geq K$ is a uniqueness set for $\mathrm{BL}_{\mathcal{F}}(\boldsymbol{V})$. In the previous section, it was shown that the reconstruction ℓ_2-error of an approximately band-limited GS is minimized when the chosen sampling set $\mathcal{S}$ maximizes $\mathrm{SV}_{\min}(\boldsymbol{V}_{\mathcal{S},\mathcal{F}})$. Therefore, the optimal sampling set $\mathcal{S}^{\mathrm{opt}}$ in the presence of model mismatching is

$$\mathcal{S}^{\mathrm{opt}} \triangleq \arg \max_{\mathcal{S} \in \mathcal{A}_M} \ \mathrm{SV}_{\min}(\boldsymbol{V}_{\mathcal{S},\mathcal{F}}), \tag{14.69}$$

where $\mathcal{A}_M$ is the class of all sampling sets $\mathcal{S}$ with $|\mathcal{S}| = M$ and satisfying the admissibility condition described in Section 14.6.1.

Besides model mismatching, reconstruction errors can also arise due to measurement noise. Consider a zero-mean and uncorrelated Gaussian vector $\boldsymbol{\eta}$ that is added to the sampled $\mathcal{F}$-band-limited GS, $\boldsymbol{x}$, as follows:

$$x_{\mathcal{S}} = \boldsymbol{\Psi}_{\mathcal{S}} x + \eta. \tag{14.70}$$

Then the recovered signal $\tilde{x}$ is

$$\tilde{x} = \boldsymbol{\Phi} x_{\mathcal{S}} = \boldsymbol{\Phi}\boldsymbol{\Psi}_{\mathcal{S}} x + \boldsymbol{\Phi}\eta = x + \boldsymbol{\Phi}\eta.$$

Since $\|\boldsymbol{\Phi}\eta\|_2 = \|\boldsymbol{V}_{:,\mathcal{F}}\boldsymbol{\Sigma}\eta\|_2 \leq \|\boldsymbol{V}_{:,\mathcal{F}}\|_2\|\boldsymbol{\Sigma}\|_2\|\eta\|_2$ and $\|\boldsymbol{V}_{:,\mathcal{F}}\|_2$ and $\|\eta\|_2$ are fixed, to minimize the upper bound for the error $\|\boldsymbol{\Phi}\eta\|_2$, one must minimize $\|\boldsymbol{\Sigma}\|_2$. Since $\boldsymbol{\Sigma}\boldsymbol{V}_{\mathcal{S},\mathcal{F}} = \boldsymbol{I}_K$, minimizing the largest singular value of $\boldsymbol{\Sigma}$, its spectral norm, is equivalent to maximizing the smallest singular value of $\boldsymbol{V}_{\mathcal{S},\mathcal{F}}$, leading to the optimization problem in (14.69). This problem is equivalent to the E-optimal design problem from experimental design theory.

Finding the optimal set $\mathcal{S}^{\text{opt}}$ is a combinatorial optimization problem that may require an exhaustive search in all possible subsets of $\mathcal{V}$ with size M. An alternative approach can be obtained by the following E-optimal greedy sampling algorithm [21][12]:

- **Input**: $\boldsymbol{V}_{:,\mathcal{F}}$ and M, the size of the sampling set
- **Output**: a subset of nodes $\mathcal{S}$
- **Procedure**:

  1. *while* $|\mathcal{S}| < M$:
  2. $\quad m \leftarrow \underset{n}{\text{argmax}}\, f(\mathcal{S}) = \text{SV}_{\min}(\boldsymbol{V}_{[\mathcal{S}\cup\{n\}],\mathcal{F}})$
  3. $\quad \mathcal{S} \leftarrow \mathcal{S} \cup \{m\}$
  4. *end*

- **return** $\mathcal{S}$

As the set function $f(\mathcal{S}) = \text{SV}_{\min}(\boldsymbol{V}_{[\mathcal{S}\cup\{n\}],\mathcal{F}})$ is not submodular,[13] a greedy algorithm, such as the above E-optimal greedy sampling algorithm, does not guarantee that the resulting sampling set $\mathcal{S}$ is close to a solution $\mathcal{S}^{\text{opt}}$ from (14.69). There are other sampling strategies derived from experimental design that can also be used in this context. For instance, if the covariance matrix of the random vector η in (14.70) is known, the interpolation operator can take this knowledge into account. The interested reader is referred to [20] for further details regarding other sampling criteria for GS.

A common concern in the extension of classical DSP methods to GSP is having to compute the GFT, a key building block in GSP that has to be computed globally (i.e., considering the whole graph), for very large graphs. Therefore, the aforementioned sampling method, which requires the computation of at least K Laplacian eigenvectors, becomes prohibitive as the size of the graph increases. To deal with the computational cost problem, inspired by *compressed sensing*, different sampling strategies based on random sampling have been proposed. The interested reader is referred to [22] for further information.

[12] A greedy algorithm searches a local optimal set at each stage of the algorithm. For a general greedy algorithm, there is no guarantee of finding a global optimal set or even near-optimal set.

[13] A submodular set function has the property that the difference in the incremental value of the function that a single element makes when added to an input set decreases as the size of the input set increases. Mathematically, $f : 2^{\mathcal{V}} \rightarrow \mathbb{R}$ is submodular when for every $\mathcal{S}_1, \mathcal{S}_2 \subset \mathcal{V}$, with $\mathcal{S}_1 \subset \mathcal{S}_2$, and for every $v \in \mathcal{V} \setminus \mathcal{S}_2$ we have $f(\mathcal{S}_1 \cup \{v\}) - f(\mathcal{S}_1) \geq f(\mathcal{S}_2 \cup \{v\}) - f(\mathcal{S}_2)$.

14.6.4 Interpolation of band-limited GSs

Up to now we have discussed the joint design of sampling and interpolation operators. In this section, we briefly introduce an interpolation approach that does not depend on the particular sampling method and relies on a different definition for band-limited signals that takes into account which frequencies are nonzero. In the forthcoming discussion, we shall consider that the GSO is the Laplacian matrix ($\boldsymbol{S} = \boldsymbol{L}$), so that the frequencies (eigenvalues) are nonnegative real numbers.

Let us first recall that the signal $\boldsymbol{x}$ is said to be ξ-band-limited if $\widehat{x}_k = 0$ for all k such that $\lambda_k > \xi$. The space of ξ-band-limited signals on the graph $\mathcal{G}$ with GFT $\boldsymbol{V}^\top$ will be analogous to the Paley–Wiener space and denoted by $\mathrm{BL}_\xi(\boldsymbol{V})$.

Unlike the $\mathrm{BL}_\mathcal{F}(\boldsymbol{V})$ space, $\mathrm{BL}_\xi(\boldsymbol{V})$ takes the values of the Laplacian eigenvalues into account. Given an ξ-band-limited GS, we aim to recover the full original GS $\boldsymbol{x}$ from few samples [23]. Note that $\xi_1 < \xi_2 \implies \mathrm{BL}_{\xi_1}(\boldsymbol{V}) \subset \mathrm{BL}_{\xi_2}(\boldsymbol{V})$. The following definition of a Λ-set will be useful for the forthcoming results.

Given $\Lambda > 0$, the subset $\mathcal{S} \subset \mathcal{V}$ is a Λ-set if

$$\|\boldsymbol{x}\|_2 \leq \Lambda \|\boldsymbol{L}\boldsymbol{x}\|_2 \tag{14.71}$$

holds for all GSs $\boldsymbol{x}$ supported on $\mathcal{S}$.

Considering the former definition of a Λ-set, the following result [24] provides a condition on a subset $\mathcal{S}$ to be a uniqueness set for the subspace $\mathrm{BL}_\xi(\boldsymbol{V})$.

The set $\mathcal{S} \subset \mathcal{V}$ is a uniqueness set for $\mathrm{BL}_\xi(\boldsymbol{V})$ *when its complement $\overline{\mathcal{S}}$ is a Λ-set with $0 < \xi < \frac{1}{\Lambda}$.*

In order to justify this result, let $\boldsymbol{x}, \boldsymbol{y} \in \mathrm{BL}_\xi(\boldsymbol{V})$; we need to show that if $\boldsymbol{x}_\mathcal{S} = \boldsymbol{y}_\mathcal{S}$, then $\boldsymbol{x} = \boldsymbol{y}$ in order to prove that $\mathcal{S}$ is a uniqueness set. Note that if $\boldsymbol{x} \in \mathrm{BL}_\xi(\boldsymbol{V})$, then $\boldsymbol{v}_k^\top \boldsymbol{x} = 0\ \forall k$ such that $\lambda_k > \xi$. Thus, we have

$$\begin{aligned}
\|\boldsymbol{L}\boldsymbol{x}\|_2 &= \|\boldsymbol{V}\boldsymbol{\Lambda}\widehat{\boldsymbol{x}}\|_2 \\
&\leq \underbrace{\|\boldsymbol{V}\|_2}_{=1} \cdot \underbrace{\|\boldsymbol{\Lambda}\widehat{\boldsymbol{x}}\|_2}_{=\sqrt{\sum_k (\lambda_k \widehat{x}_k)^2}} \\
&\leq \left(\max_{\lambda_k \leq \xi} \lambda_k\right) \|\widehat{\boldsymbol{x}}\|_2 \\
&\leq \xi \|\boldsymbol{x}\|_2.
\end{aligned} \tag{14.72}$$

Suppose $\boldsymbol{x}_\mathcal{S} = \boldsymbol{y}_\mathcal{S}$ but $\boldsymbol{y} \neq \boldsymbol{x}$. Then $\boldsymbol{x} - \boldsymbol{y} \in \mathrm{BL}_\xi(\boldsymbol{V})$ is supported on $\overline{\mathbf{S}}$, and using the assumption that $\overline{\mathbf{S}}$ is a Λ-set with $0 < \xi < \frac{1}{\Lambda}$, we can write

$$\|\boldsymbol{x} - \boldsymbol{y}\| \leq \Lambda \|\boldsymbol{L}(\boldsymbol{x} - \boldsymbol{y})\| \leq \Lambda\xi \|\boldsymbol{x} - \boldsymbol{y}\| < \|\boldsymbol{x} - \boldsymbol{y}\|, \tag{14.73}$$

where the second inequality follows from (14.72), thus arriving at a contradiction. Hence, we must have $\boldsymbol{x} = \boldsymbol{y}$, and the result is proved.

The following result [23] provides a cutoff frequency ξ^* for GSs supported on $\mathcal{S}$, that is, the maximum ξ such that any signal in $\mathrm{BL}_\xi(\boldsymbol{V})$ can be perfectly recovered given a known subset of samples $\mathcal{S}$.

Given a proper subset $\mathcal{S} \subsetneq \mathcal{V}$ of vertices of a connected graph (with only one connected component), let $[\boldsymbol{L}^2]_{\overline{\mathcal{S}}}$ denote the submatrix of $\boldsymbol{L}^2 = \boldsymbol{L} \cdot \boldsymbol{L}$ containing only rows and columns corresponding to nodes in $\overline{\mathcal{S}}$. Then, the set $\mathcal{S}$ is a uniqueness set for $\mathrm{BL}_\xi(\boldsymbol{V})$ *with $0 < \xi \le \xi^* = \overline{\lambda}_{\min}$, where $\overline{\lambda}^2_{\min} > 0$ denotes the smallest eigenvalue of $[\boldsymbol{L}^2]_{\overline{\mathcal{S}}}$.*

Indeed, let z be a GS supported on $\overline{\mathcal{S}}$. Then

$$\frac{\|\boldsymbol{L}z\|_2^2}{\|z\|_2^2} = \frac{z_{\overline{\mathcal{S}}}^\top ([\boldsymbol{L}^2]_{\overline{\mathcal{S}}})\, z_{\overline{\mathcal{S}}}}{\|z_{\overline{\mathcal{S}}}\|_2^2} \triangleq R(z_{\overline{\mathcal{S}}}, \left[\boldsymbol{L}^2\right]_{\overline{\mathcal{S}}}) \tag{14.74}$$

is the Rayleigh quotient of $[\boldsymbol{L}^2]_{\overline{\mathcal{S}}}$. Since $\overline{\lambda}^2_{\min} = \min\limits_{z_{\overline{\mathcal{S}}} \in \mathbb{R}^{N-M}} R(z_{\overline{\mathcal{S}}}, [\boldsymbol{L}^2]_{\overline{\mathcal{S}}})$,

$$\frac{\|\boldsymbol{L}z\|_2^2}{\|z\|_2^2} \ge \overline{\lambda}^2_{\min} \iff \frac{1}{\overline{\lambda}_{\min}}\|\boldsymbol{L}z\|_2 \ge \|z\|_2, \tag{14.75}$$

which means that $\overline{\mathcal{S}}$ is a $\frac{1}{\sqrt{\overline{\lambda}_{\min}}}$-set and, therefore, it follows from a previous result that $\mathcal{S}$ is a uniqueness set for $\mathrm{BL}_\xi(\boldsymbol{V})$ with $0 < \xi \le \xi^* = \overline{\lambda}_{\min}$.

On the guarantee of existence of a reconstruction formula for a sampled GS, [24, Theorem 4.1] states that if $\mathcal{S}$ is a uniqueness set for $\mathrm{BL}_\xi(\boldsymbol{V})$, then there exists a frame $\{\boldsymbol{v}_m\}_{m\in\mathcal{S}} \subset \mathrm{BL}_\xi(\boldsymbol{V})$ such that

$$x_n = \sum_{m\in\mathcal{S}} x_m v_{mn}, \quad \forall \boldsymbol{x} \in \mathrm{BL}_\xi(\boldsymbol{V}).$$

Therefore, once a cutoff frequency ξ^* is obtained, the reconstruction of any signal restricted to a uniqueness set is obtained by least-squares projection. Any band-limited signal in $\mathrm{BL}_\xi(\boldsymbol{V})$, $\xi < \xi^*$, can be written as a linear combination of the Laplacian eigenvectors corresponding to eigenvalues smaller than ξ^*. Let us denote by $\mathcal{F}^*$ the index set of the Laplacian eigenvectors corresponding to eigenvalues smaller than ξ^*. Then $\boldsymbol{x} = \boldsymbol{V}_{\mathcal{F}^*}\widehat{\boldsymbol{x}}\ \forall \boldsymbol{x} \in \mathrm{PW}_{\xi^*}(\boldsymbol{V})$. Because $\mathcal{S}$ is a uniqueness set for $\mathrm{BL}_\xi(\boldsymbol{V})$, it contains all the spectral information of any GS in $\boldsymbol{x} \in \mathrm{BL}_\xi(\boldsymbol{V})$ and then $\boldsymbol{x} = \boldsymbol{V}_{\mathcal{S}^*}\hat{\boldsymbol{x}}_{\mathcal{F}^*}$. Therefore, given a subset of nodes $\mathcal{S}$, with $\overline{\mathcal{S}}$ being the remaining nodes whose samples are unknown, let $(\boldsymbol{L}^2)_{\overline{\mathcal{S}}}$ be the submatrix of $\boldsymbol{L}^2$ with rows and columns corresponding to nodes in $\overline{\mathcal{S}}$. Then $\mathcal{S}$ is a uniqueness set for $\mathrm{BL}_{\overline{\lambda}^*_{\min}}(\boldsymbol{V})$, where $\overline{\lambda}^2_{\min}$ is the smallest singular value of $(\boldsymbol{L}^2)_{\overline{\mathcal{S}}}$. $\mathrm{SV}^*_{\min}$ is the largest bandwidth such that a GS can be perfectly recovered from $\mathcal{S}$. The unknown GS supported on $\overline{\mathcal{S}}$ can be recovered by

$$\boldsymbol{x}_{\overline{\mathcal{S}}} = \boldsymbol{V}_{\overline{\mathcal{S}},\mathcal{F}^*}(\boldsymbol{V}_{\mathcal{S},\mathcal{F}^*}^\top \boldsymbol{V}_{\mathcal{S},\mathcal{F}^*})^{-1}\boldsymbol{V}_{\mathcal{S},\mathcal{F}^*}^\top \boldsymbol{x}_{\mathcal{S}}. \tag{14.76}$$

The interpolator (14.76) is equivalent to (14.65), except for the definition of the set $\mathcal{F}$.

14.7 Examples and applications

This section is dedicated to presenting some practical applications in which GSP tools have been successfully employed. More specifically, we describe how GSP has supported the creation of new techniques for on-line filtering of network signals, image and video processing, 3D point cloud processing, and spatiotemporal analysis and visualization. In our text, we opted for a more qualitative description of the possibilities allowed by the GSP in these contexts. In this sense, we consider some pioneering works in each of the aforementioned fields of application, connecting them to other relevant publications.

14.7.1 On-line filtering of network signals

One of the main difficulties in translating classical DSP techniques and algorithms into the GSP framework is to mold the technique in such a way that a generalization becomes apparent. This is true of filtering – which, as we have seen in Section 14.5, is more easily handled in theory in the frequency domain – and even more so when dealing with adaptive filters, which rely heavily on the regularity of the signal domain. To this date, the path for designing adaptive filtering techniques for GSP is not crystal clear. Since the field of signal processing on graphs is still incipient, we verify that many works are still ad hoc in nature.

A common feature of the majority of works in this field is the fact that they bypass some of the difficulties of dealing with the transformation of the time domain to the graph domain by working with both domains simultaneously. This approach allows us to consider a linear and uniform time dimension, while modeling the irregular dimension using graphs. In this way, GSP theory can be applied to the *snapshots* of the resulting multivariable signal

$$\boldsymbol{x}[\cdot] : \mathbb{Z} \to \mathbb{C}^{\mathcal{V}} \tag{14.77}$$

since each sample in time – when thinking in terms of classical adaptive filtering – has now become a whole GS $\boldsymbol{x}[k] \in \mathbb{C}^{\mathcal{V}}$, instead of just a scalar value.

The most widely used framework in the literature for developing adaptive graph filters tackles the problem of interpolating sampled GSs (Section 14.6). As mentioned before, graphs are particularly useful for modeling networks. Real-world networks can be huge, and attributes of a number of nodes may be unavailable, or deliberately undisclosed due to privacy matters [25]. Further, some practical networks change their set of nodes over time, so that new nodes may come to exist without enough time to sharing their node attributes with the central unit for processing. The interpolation problem, therefore, naturally arises in these and many other practical applications, including semisupervised learning of categorical data, ranking problems, and matrix completion problems [26]. All of these applications rely on predicting the property of some nodes (e.g., class, ranking, or function value) considering the knowledge of signal values at neighboring nodes, or in other words, *interpolating* those values. The quality of the interpolation depends on the smoothness of the GS, which can be assessed taking into account the similarity between nodes captured by link weights in the graph – there is an underlying assumption that the graph model is accurate. For instance, in social networks, people tend to rate movies similar to their friends, and in financial networks, companies that trade with each other usually belong to the same category.

The appeal for using adaptive filtering ideas in the GS interpolation problem arises from the expected benefits of enabling on-line reconstruction and tracking time-varying GSs in noisy environments [27]. Although traditional interpolation methods like kriging [28] can be used for handling signal inference in irregular domains with static reference values, adaptive algorithms seem more suitable for dealing with dynamic GSs due to their reduced complexity and the benefits of on-line estimation.

The adaptive estimation of GSs deals with a scenario where one intends to recover a *band-limited* – or approximately band-limited in practice – GS $\boldsymbol{x}[k] \in \mathbb{C}^{\mathcal{V}}$, with frequency set $\mathcal{F}$, from a reduced set $\mathcal{S}[k] \subseteq \mathcal{V}$ of noisy measurements drawn from $\boldsymbol{d}[k] = \boldsymbol{x}[k] + \boldsymbol{\eta}[k]$ – recall the definitions in Section 14.6. Since only the node signals indexed by $\mathcal{S}[k]$ are acquired at iteration k, the reference measurements at time instant k are in fact $\boldsymbol{\Psi}_{\mathcal{S}[k]}\boldsymbol{d}[k]$.

We define the (a priori) error vector $\boldsymbol{e}[k] \in \mathbb{C}^{\mathcal{S}[k]}$ as the difference between the measured reference GS $\boldsymbol{\Psi}_{\mathcal{S}[k]}\boldsymbol{d}[k]$ and the respective positions of the current estimate $\boldsymbol{y}[k]$, that is,

$$\begin{aligned}\boldsymbol{e}[k] &= \boldsymbol{\Psi}_{\mathcal{S}[k]}\boldsymbol{d}[k] - \boldsymbol{\Psi}_{\mathcal{S}[k]}\boldsymbol{y}[k]\\ &= \boldsymbol{\Psi}_{\mathcal{S}[k]}\left(\boldsymbol{d}[k] - \boldsymbol{V}_{:,\mathcal{F}}\hat{\boldsymbol{y}}_{\mathcal{F}}[k]\right),\end{aligned} \tag{14.78}$$

where $\hat{\boldsymbol{y}}_{\mathcal{F}}[k] \in \mathbb{C}^{|\mathcal{F}|}$ is the frequency-domain estimate of $\boldsymbol{y}[k]$ considering only the frequency support $\mathcal{F}$.

Unlike adaptive filters from classical DSP, the adaptive graph filter here is not explicitly parameterized. In addition, the graph filtering algorithm employs only some *ancillary information* related to the GS properties, such as the knowledge of the sampling and frequency sets $\mathcal{S}[k]$ and $\mathcal{F}$, respectively. In other words, the filtering process does not act upon a given *input* GS and the only external driving force of the algorithm is dictated by the reference GS $\boldsymbol{\Psi}_{\mathcal{S}[k]}\boldsymbol{d}[k]$.

As an example, the interpolation-based graph counterpart of the well-known least-mean squares (LMS) algorithm [29] aims to solve the problem

$$\underset{\hat{\boldsymbol{y}}_{\mathcal{F}}}{\text{minimize}}\ \mathbb{E}\left\{ \|\boldsymbol{\Psi}_{\mathcal{S}[k]}\left(\boldsymbol{d}[k] - \boldsymbol{V}_{:,\mathcal{F}}\hat{\boldsymbol{y}}_{\mathcal{F}}\right)\|_2^2 \right\} \tag{14.79}$$

via a stochastic gradient approach, giving rise to the graph-frequency domain update equation

$$\hat{\boldsymbol{y}}_{\mathcal{F}}[k+1] = \hat{\boldsymbol{y}}_{\mathcal{F}}[k] + \mu \boldsymbol{V}_{:,\mathcal{F}}^{\top}\boldsymbol{e}[k], \tag{14.80}$$

or, in the vertex domain,

$$\boldsymbol{y}[k+1] = \boldsymbol{y}[k] + \mu \boldsymbol{V}_{:,\mathcal{F}}\boldsymbol{V}_{:,\mathcal{F}}^{\top}\boldsymbol{e}[k], \tag{14.81}$$

in which $\mu > 0$ is the convergence factor. This parameter can be chosen so as to achieve an asymptotically unbiased estimator. It can be proved that this can be attained when

$$0 < \mu < \frac{2}{\lambda_{\max}\left(\boldsymbol{V}_{:,\mathcal{F}}^{\top}\boldsymbol{\Psi}_{\mathcal{S}}\boldsymbol{V}_{:,\mathcal{F}}\right)}. \tag{14.82}$$

In addition to the graph LMS algorithm [29] briefly described above, other works have extended the classical adaptive filtering framework to the interpolation of GSs via normalized LMS (NLMS) [30] and recursive least-squares (RLS) [31] algorithms. A totally different class of adaptive graph filters have also appeared in [32–35]. These works propose LMS-based graph diffusion strategies to obtain adaptive graph filters that can be used in many applications analogous to the traditional system identification, signal enhancement, deconvolution/equalization, and signal prediction, but now considering GSs. These LMS-based adaptive graph filters along with their NLMS and RLS extensions are described in great detail in [36]. Further extensions of graph filters to nonlinear settings have also appeared in [37–40].

14.7.2 Image processing

The fields of image and video processing are among those in which GSP tools have been most applied. Clearly, the basic structure in which a digital image is organized can be viewed as a rectangular grid graph on which the pixels (vertices) lie and interconnect with their immediate vertical and horizontal neighbors by means of edges with unit weight. The GS is then related to the attributes of each pixel (color, transparency, depth, etc.) and, naturally, the numerical value of the corresponding sample depends on the code employed to represent the image. Consequently, a video can be interpreted as a time-varying GS or even as a signal over a multidimensional graph. The fact is that, considering this basic model as reference, the GSP allows us to adjust the referred underlying graphs so that they can better reflect the structure of a given image. This has enabled the development of GSP-based techniques devoted to image compression, restoration, filtering, and segmentation, for instance, that achieve results very competitive with those of classical methods.

In this context, the GFT plays an essential role and provides high flexibility, especially if we are interested in compressing digital images. In fact, as shown in [41], the GFT is optimal for decorrelation of an image following a Gaussian Markov random field (GMRF) model. Suboptimal GFT-like transforms can therefore be obtained by adjusting the graph employed to model an image. This is usually performed so that a reasonable trade-off between the amount of side information (graph structure) to be sent to the decoder and the amount of energy compaction provided by the transform is reached. Such a strategy seems to be more intuitive than others using, for instance, the Karhunen–Loève transform (KLT), which takes a statistical approach, describing correlations among image pixels through estimates of their linear correlation coefficients [42].

As a first example of what was explained above, we observe that if the rectangular grid graph model is adopted, the GFT can be defined so that it coincides with the 2D discrete cosine transform (DCT) [43,44]. A possibility arising from this finding is to define a 2D-DCT generalization identified as steerable DCT (SDCT). Such a transform is obtained by applying rotations to specific pairs of 2D-DCT basis vectors or, employing the GFT terminology, rotating pairs of eigenvectors associated with the same eigenvalue of the Laplacian of the corresponding graph. In terms of energy compaction, the SDCT has potential to be more effective than the 2D-DCT. This is due to the SDCT particularity of employing a basis whose vectors are better aligned with the directions of the discontinuities of a given image block if the corresponding rotation angles are properly chosen. On the other hand, if a complete SDCT-based codec is designed, it is necessary to take into account the need to inform the decoder about the angles used to transform each image block. Details regarding steerable transforms and its use in image compression can be found in several works [42–48].

Another proposal to GFT design aiming at image compression is that based on symmetries [49]. In this case, each image block is modeled as a totally or partially symmetric grid graph, which is motivated by the fact that symmetries are often present in real-world data. Experimental tests have demonstrated that symmetry-based GFTs outperform the 2D-DCT in the representation of both natural images and residual signals obtained from intraprediction based on the directional modes adopted in H.265 standard. In addition, we highlight the method proposed in [50], which introduces a novel algorithm for graph estimation. Such an algorithm takes into account the coding of both the signal and the graph topology in rate-distortion terms. The cost of the graph description is considered in the formulation of an optimization problem which also involves the minimization of the sparsity of the GFT coefficients on the dual graph. Experimental results show that the proposed technique outperforms classical fixed transforms for both natural and piecewise smooth images; it is also comparable to state-of-the-art graph-based coding methods designed for depth map images. Furthermore, GFT-based compression has been applied to light-field imaging, which is a technology for capturing images, building upon conventional digital photography, that presents challenging tasks related to the amount of data generated. The works in [51,52] propose methods for modeling the blocks of pixels from a light field as graphs and use the GFT to compress the images, leveraging the energy concentration provided by the Fourier analysis of the corresponding GSs. Other GSP-based methods devoted to image and video compression can be found in [53], for example.

Even before the systematic development of GSP tools, its concepts were already used in image filtering as well. Graph spectral image smoothing, which appears to have been documented for the first time in [54], is a good example to illustrate such a possibility. In this case, the low-pass filtering can be performed by employing the heat kernel filter, whose expression in the graph spectral domain is

$$\widetilde{\boldsymbol{h}}(\lambda) = \mathrm{e}^{-\tau\lambda}. \tag{14.83}$$

The last equation is related to the computation of the τth power of the heat (discrete) diffusion operator $\boldsymbol{R} = \mathrm{e}^{-\boldsymbol{L}}$. If the signal $\boldsymbol{f}$ represents the initial amount of heat at each vertex, $\boldsymbol{R}^{\tau}\boldsymbol{f}$ represents the amount of heat at each vertex after time τ. Such a parameter also provides a notion of scale, as clearly explained in [54] and [5], and, naturally, it determines the decaying speed of $\widetilde{\boldsymbol{h}}(\lambda)$. In fact, the heat kernel has been employed to perform low-pass filtering of GSs in several works (see, for example, [55–57]).

In general, graph spectral filtering can be used to implement discrete versions of several continuous filtering techniques. This is applicable, for instance, to image denoising, where the noisy GS $\boldsymbol{y} = \boldsymbol{f}_0 + \eta$ is given by the sum of the clean signal $\boldsymbol{f}_0$ and the Gaussian noise η. Assuming that $\boldsymbol{f}_0$ is smooth, the term $\boldsymbol{f}^{\top}\boldsymbol{L}\boldsymbol{f}$ is used to regularize the corresponding ill-posed inverse problem, which, for a fixed $\gamma > 0$, can be written as

$$\underset{\boldsymbol{f}}{\operatorname{argmin}}\{\|\boldsymbol{f} - \boldsymbol{y}\|_2^2 + \gamma \boldsymbol{f}^{\top}\boldsymbol{L}\boldsymbol{f}\}. \tag{14.84}$$

After solving (14.84), one obtains the optimal reconstruction

$$f_*(i) = \sum_{\ell=0}^{N-1} \left[\frac{1}{1+\gamma\lambda_\ell}\right] \widehat{y}(\lambda_\ell) u_\ell(i), \tag{14.85}$$

or, equivalently, $\boldsymbol{f} = \widehat{h}(\boldsymbol{L})\boldsymbol{y}$, where $\widehat{h}(\lambda) := 1/(1+\gamma\lambda)$ can be interpreted as a low-pass filter [5,58]. It has been demonstrated that using the above described strategy has potential to provide results that are better than those produced by classical Wiener and Gaussian filtering, for example. The graph model employed to carry out the referred denoising method can be constructed by connecting each pixel to its eight immediate neighbors and setting the edge weights according to a similarity criterion.

Finally, we can mention some more specific and sophisticated GSP-based techniques devoted to image filtering. In [59], the authors extend the GSP context to the trilateral filter, which is a data-dependent filter commonly used as an edge-preserving smoothing method for image processing. They introduce a spectral graph representation of such a filter, which enables them to design an effective GS denoising filter with a Tikhonov regularization. In [60], a spectral graph-based vertex-varying Wiener filtering framework in the joint vertex-frequency domain for denoising of signals is proposed. The authors demonstrate that their method provides very competitive results when compared with high-performance denoising methods. An iterative graph-based filtering for image abstraction and stylization is proposed in [61]. The respective framework is designed so that iterative filtering without requiring any weight updates can be performed. The authors demonstrate that the proposed method can yield significantly improved visual quality for stylized images as compared to other existing methods.

14.7.3 3D point cloud processing

Point clouds are digital virtual objects formed by points lying in a 3D space for which a specific coordinate system is defined. In contrast to digital images and light fields, whose pixels lie in fixed and regular grids, in a 3D cloud, the points are arbitrarily positioned in the referred space, so that they can be viewed as vertices of a graph possibly connected according to a specific criterion. The feasibility of using GSP tools in this context is related to the fact that one or more attributes can be associated with the points of a 3D cloud. Such an attribute can be a color, for example, which can be numerically encoded and, therefore, represent the value of the corresponding sample in the respective GS [62].

The usual strategy for establishing edges connecting the points of a 3D cloud (vertices of the graph) employs the well-known k-nearest neighbor (k-NN) approach. The weight $W_{i,j}$ of the edge connecting the ith and jth vertices can be computed by combining some metric and a weighting function. In most works on this theme, the thresholded Gaussian kernel

$$W_{i,j} = \begin{cases} \exp\left(-\frac{\|p_i - p_j\|^2}{2\theta^2}\right), & \text{if } p_j \in C_k(i) \text{ or } p_i \in C_k(j), \\ 0, & \text{otherwise} \end{cases} \tag{14.86}$$

has been used; in the last equation, θ is a variance parameter, $C_k(i)$ is the set containing the k closest points to p_i, and $\|p_i - p_j\|$ denotes the Euclidean distance between points p_i and p_j.

In this context, GSP-based techniques have been developed, for example, with the purpose of compressing attributes of 3D point clouds [63–68]. In general, these methods assume the point cloud is organized into an octree, where each occupied leaf node is a voxel that represents a point from the cloud. After that, the respective graph is created as explained earlier in this section. This allows the structure of the graph, that is, the octree, to be efficiently encoded and equally recovered in the decoder. The attribute compression process itself usually has the GFT as its centerpiece, since such a transform is optimum for decorrelating a signal following a GMRF model. Details regarding this fact can be found in [41,63]. To be more specific, the GFT is applied in a block basis to the corresponding GS, which can represent, for example, the components of a certain color channel. The resulting signal is then quantized and entropy coded; this can be performed using classical techniques for these purposes, but adapted to the statistical behavior of the GFT coefficients for the signal being processed.

The final result of the compression process includes the data coming from the graph structure (octree) encoding and the data resulting from the corresponding attribute compression. Due to quantization, the 3D point cloud decoding is lossy, so that the performance of the described technique can be evaluated by computing the reconstruction error for a given bitrate. It has been demonstrated that, in general, such a performance is better than that obtained by methods based on conventional transforms such as the DCT.[14] Basically, this is due to the fact that correlations in the 3D space are better captured in the graph transform.

In the context of 3D point cloud compression, we can also specifically mention GSP-based schemes for estimating and compensating motion in dynamic point clouds [64,65,68], techniques that employ subdivisional meshes, graph wavelet transforms and variations of the GFT [67,69,70], and schemes that perform progressive graph-signal sampling and encoding for static 3D geometry representation [71].

GSP-based techniques with the aim of denoising point clouds have also been proposed. Such a task is recurrently necessary, since the occurrence of errors is common when creating a 3D model from images. In [72], for example, the authors present a technique that uses the structure of the graph created from a point cloud and convex optimization methods to perform the respective denoising. In this case, the focus is on removing outliers and correcting the position of the remaining vertices. In [73], the authors introduce an extension to GSP of a deconvolution algorithm based on the so-called Stein's unbiased risk estimate-linear expansion threshold (SURE-LET) approach [74]. The idea is then used to perform deblurring of point cloud attributes. The method proposed in [75] carries out local 3D point cloud denoising via bipartite graph approximation and TV. The authors show that their algorithm achieves the best overall denoising performance when compared to state-of-the-art non-GSP schemes with similar complexity. In [76], the authors employ the spectral graph wavelet transform (SGWT) to jointly perform denoising of geometry and color attributes of 3D point clouds. They remove the noise from the SGWT coefficients by applying data-driven adaptive soft-thresholding.

Segmentation is another task that can benefit from GSP tools in the context of 3D point clouds. This has been demonstrated in previous works, which employ, for instance, graph convolutional networks and hypergraph spectral clustering [77–79]. Finally, we should mention the possibility of performing 3D point cloud watermarking in the GFT domain [80,81], which has shown promising results when compared to those obtained by non-GSP methods.

14.7.4 Spatiotemporal analysis and visualization

Another interesting GSP application scenario is related to analysis and visualization of time-varying data on graphs. The possibility of considering a time axis along which each sample of a GS has its value modified enhances the use of GSP tools and allows their crossing with consolidated classical DSP techniques. Naturally, a wide range of real-world problems can be modeled by a time-varying graph, including traffic data on street intersections in a given region along a day and according to the day of the week, the evolution of crime data over the years in the neighborhoods of a city, the changes in the attributes of a dynamic 3D point cloud, and the fluctuation of climatological variables measured by sensors in a network, to mention a few. In this scope, some GSP-based techniques have been developed and demonstrated to have potential to provide relevant results.

In [82], for instance, the problem of spatiotemporal prediction is addressed. This problem boils down to forecasting (temporal prediction) and interpolation (spatial prediction). The former refers to

[14] We remark that the DCT is a particular case of the GFT. If the DCT is used in the current application, instead of inferring an arbitrary graph from the points in the cloud, a 1D chain (or path graph) is formed.

predicting some physical phenomenon using historical data acquired by a network of spatially distributed sensors. The latter refers to predicting the phenomenon with a higher spatial resolution. In this context, spatiotemporal data can be seen as a network signal in which a time series is associated with each network element; the dynamics (time-domain evolution) of the time series depends on the network structure (spatial domain), rather than on the isolated network elements only. The interpolation is useful to generate a denser (virtual) network. Classical predictive models assume independence of data samples and disregard relevant spatial information. In [82], a neural network (NN) architecture is proposed to handle spatiotemporal correlations by employing GSP in conjunction with a recurrent NN (RNN). Thus, the inherent nature of spatiotemporal data is addressed by jointly forecasting and interpolating the underlying network signals. A global interpolation approach is adopted as it provides accurate results when the signal is smooth in the GSP sense, whereas an RNN forecasting model is adopted given its prior success in network prediction. Herein, not only the sampled GS is input to a predictive model but also its spectral components, which carry spatial information on the underlying graph. The major contribution of the model is, therefore, the ability to learn spatiotemporal features by observing a few nodes of the entire graph.

In [83], a new batch reconstruction method of time-varying GSs is proposed by exploiting the smoothness of the temporal difference signals. The approach also includes an on-line distributed technique that has been shown to be suitable for practical applications faced with real-world requirements. In [84], the authors investigate the problem of recovering spatiotemporal signals from partially known entries. They propose a low rank- and differential smoothness-based recovery method, which introduces the differential smooth prior of time-varying GSs to the field of spatiotemporal signal analysis. In another previous work [85], the problem of reconstruction of time-varying GSs is used to model the number of new COVID-19 cases. The authors propose a method based on the minimization of the Sobolev norm in GSP and obtain results that outperform state-of-the-art algorithms in two COVID-19 databases.

The issue of visualizing time-varying GSs for the purpose of performing specific inferences has also been addressed in the literature. In [86], for example, a visualization method that relies on graph wavelet theory and a stacked graph metaphor to enable the visual analysis of time-varying data defined on the nodes of a graph is proposed. The technique allows to identify regions where data present abrupt and mild spatial and/or temporal variation while still being able to show how such changes evolve over time, making the identification of events an easier task. In [87], the same authors propose a methodology that employs a fast approximation of a graph wavelet transform to derive a set of wavelet coefficients suitable to identify activity patterns on large networks (including their temporal recurrence).

In this section, we illustrate the possibility of visualizing and performing spatial pattern detection in spatiotemporal data by briefly describing the GSP-based technique introduced in [88]. In that paper, the authors first introduce a graph Laplacian of Gaussian (GLoG) filter devoted to the detection of node pairs to which abrupt variations of the corresponding GS are related. This is equivalent to a boundary detection procedure, usually employed in the context of image processing. More specifically, the approach consists in pointwise multiplying the GFT of the signal by the graph spectral filter

$$\widetilde{\nabla^2 G}(\lambda) = -4\pi^2\lambda^2 e^{-\sigma^2\lambda^2}, \tag{14.87}$$

where σ is a user-defined parameter and λ is the independent variable in the graph spectral domain. The inverse GFT of the filtered signal is then computed, and in the vertex domain a thresholding-based procedure is applied in order to decide which are the strongest boundary nodes.

In a time-varying GS, the GLoG filter (14.87) can be applied to each time slice t_j, $j \in \{1, 2, \ldots, m\}$, and the probability $p_e(\tau_i)$ of a node τ_i being a boundary node can be estimated by

$$p_e(\tau_i) = \frac{1}{m} \sum_{k=1}^{m} I(\tau_i, t_k), \tag{14.88}$$

where

$$I(\tau_i, t_k) = \begin{cases} 1, & \text{if } \tau_i \text{ is a boundary node in time slice } t_k, \\ 0, & \text{otherwise.} \end{cases} \tag{14.89}$$

In this context, the function

$$p(t_i, t_k) = \begin{cases} p_e(\tau_i), & \text{if } I(\tau_i, t_k) = 1, \\ 1 - p_e(\tau_i), & \text{if } I(\tau_i, t_k) = 0 \end{cases} \tag{14.90}$$

computes how probable the observed configuration of a node τ_i (in t_k) is in terms of it being a boundary node or not. This allows to define the (graph) entropy $E(t_k)$, which corresponds to a measurement of the degree of randomness of a time slice. It is given by

$$E(t_k) = -\sum_{i=1}^{n} p(\tau_i, t_k) \log p(\tau_i, t_k). \tag{14.91}$$

The larger the entropy, the more unpredictable the time slice is in terms of its boundary node configuration. This means that if we plot the entropy over time and produce an *entropy diagram*, time slices where the signal presents unexpected (high entropy) or predictable (low entropy) boundary node configurations can be visualized.

The described approach has been evaluated by using synthetic and real data. Among the case studies considered by the authors, the analysis of downtown Manhattan taxi data deserves to be highlighted; the respective graph has 4694 street intersections as its nodes and 6350 street blocks connecting the intersections as its edges. The time-varying GS corresponds to the number of taxi pickups in each intersection, aggregated into intervals of fixed duration, along the period of one week. The generated entropy diagram reveals, for instance, that taxi pickup follows a certain (expected) pattern in the mornings and afternoons, becoming more random at dawn, specially in the weekend. In this case, by employing the referred diagram as well as visualizing boundary node configuration patterns, the analyst can draw a picture of the dynamics of the city with regard to the regions to which people move and at what times this happens. This kind of interpretation can be extended to other application scenarios involving time-varying data on graphs.

14.8 A short history of GSP

In the last years, theory and applications related to GSP have been widely developed and attracted the attention of several scholars. In fact, it seems that the number of researchers working on GSP is still increasing at the time this chapter is being written. This is reflected in the increasing quantity, quality, and interdisciplinarity of the works that keep appearing to this date addressing such a subject. Although the systematization of the knowledge related to GSP has started in the beginning of the 2010s, the use of graph theoretic mathematical tools to solve signal processing problems predates this period. Throughout the previous sections of this chapter, some works responsible for such an *embryonal* stage of the GSP were mentioned. In these publications, one identifies the presence, albeit not always very evidently, of a first opening to the new paradigm of employing a graph to characterize the domain on which a signal lies. From another point of view, this is equivalent to looking at a graph beyond its structure, allowing its nodes to be associated with values that are not simple labels, but have a physical meaning or interpretation as a function of their magnitude.

In this line, the authors of [54] represent the image pixel lattice using a weighted undirected graph. They propose a specific methodology to determine the edge weights, after which the Laplacian is computed and the diffusion across the graph structure with time is captured by using the heat equation (diffusion kernels on graphs had already been studied in [89], for example). This is then employed to perform grayscale and color image smoothing, but more striking than this is the fact that the developed technique and the theoretical justifications presented by the authors remain very useful and support several GSP concepts and methods.

Within the same scope, it is also important to highlight the emergence of transforms that had connections with what would come to be identified as the GFT. This started in 2010 and was leveraged by researchers working mainly in the area of image coding. At that moment, the bases for the respective transforms already corresponded to eigenbases of the Laplacian of a graph that could have been generated from the edges of the image, if the purpose was to efficiently encode some depth map, or from more arbitrary graph models connecting the pixels, if the purpose was to exploit redundancies and compress image blocks. In this framework, we would like to mention as a standout result the demonstration that the GFT is optimal for decorrelation of an image following a GMRF model [41]. Other results that supported advances in this context can be found in [63,90–92], for instance.

Another series of papers related to the beginnings of GSP is the one dealing with graph wavelets. A work that seems to have pioneered this is the one in which the authors propose a novel method for constructing wavelet transforms of functions defined on the vertices of an arbitrary finite weighted graph [93]. Their approach is based on defining scaling using the spectral decomposition of the graph Laplacian, which, at that time, was already pointed out as being the graph analog of the Fourier domain. In fact, many other developments followed the aforementioned work, providing theoretical advances and covering application scenarios. In this context, concepts related to wavelet filter banks for graph structured data [94], wavelet frames on multislice graphs [95], the graph wavelet transform applied to image segmentation [96], graph wavelets for multiscale community mining [97], and wavelet-based visualization of time-varying data on graphs [86], for instance, have been investigated.

Despite all the abovementioned original contributions, when we think about the systematization of the foundations of the GSP and its emergence as a well-established research line, it seems natural to refer to two approaches that have moved more clearly in this direction. The first of these approaches has as its early stage the *algebraic signal processing* (ASP) theory, published by Püschel and Moura in 2008 [98,99]. The ASP expands DSP by moving to an algebraic point of view: each signal processing theory is studied as a triple $(\mathscr{A}, \mathscr{M}, \Phi)$ consisting of an algebra $\mathscr{A}$ (a vector space endowed with multiplication between vectors), an $\mathscr{A}$-module $\mathscr{M}$ (a vector space over the same base field as $\mathscr{A}$ which admits left-multiplication by elements of $\mathscr{A}$), and a linear transformation Φ. Here $\mathscr{A}$ is called the filter space, $\mathscr{M}$ is the signal space, and Φ is the Fourier transform (homomorphism over $\mathscr{M}$) associated with the structure.

When these authors drew inspiration from ASP to develop their GSP theory, the starting point was necessarily to define the unit-shift operator of GSs, the reason being that such an operator in ASP is the building block of the algebra $\mathscr{A}$ (as, for example, the unit delay z^{-1} is the building block for filters of discrete-time and finite length signals $\mathscr{A} = \{\sum_{\ell=0}^{N-1} h_\ell z^{-\ell} | h_\ell \in \mathbb{C}\}$). To do so, the shift of discrete-time signals, defined over directed ring graphs, was investigated.

The other approach is based on the graph spectral theory, which is a branch of graph theory concerned with the eigendecomposition of graph characteristic matrices, such as the adjacency and Laplacian matrices, and the properties derived from such a decomposition. Developed from matrix theory and linear algebra, it finds applications in quantum mechanics, communications theory, chemistry, and other fields [3]. It has been used by Ortega, Shuman, and many others as foundation for the proposal and growth of a theoretical framework for GSP based on the spectral properties of the Laplacian matrix. In this branch of GSP, it is worth mentioning that the authors use to restrict the graph space in which their tools are applied to undirected graphs with nonnegative real edge weights, as Sandryhaila and Moura point out [100].

In short, we can state that the structure built based on the aforementioned approaches has served as the main reference for what has been done in GSP. Throughout its still short trajectory, GSP has brought together researchers from the most diverse areas, being established as a multidisciplinary research field with potential for application in modern practical scenarios. Currently, judging by the community's efforts to propose solutions to problems using GSP, this theme seems to be in a maturing process driven by the possibility for creativity it gives to the authors. The expectation is, therefore, that the representativeness of GSP within signal processing will continue to grow and that its effective use in real-world applications will become increasingly concrete.

References

[1] P. Feofiloff, Y. Kohayakawa, Y. Wakabayashi, Uma Introdução Sucinta à Teoria dos Grafos, https://www.ime.usp.br/~pf/teoriadosgrafos/, 2011.

[2] J.A. Bondy, U.S.R. Murty, Graph Theory, Graduate Texts in Mathematics, vol. 244, 2008.

[3] F.R.K. Chung, Spectral Graph Theory, vol. 92, American Mathematical Soc., 1997.

[4] A. Sandryhaila, J.M. Moura, Nearest-neighbor image model, in: Image Processing (ICIP), 2012 19th IEEE International Conference on, IEEE, 2012, pp. 2521–2524.

[5] D.I. Shuman, S.K. Narang, P. Frossard, A. Ortega, P. Vandergheynst, The emerging field of signal processing on graphs: extending high-dimensional data analysis to networks and other irregular domains, IEEE Signal Processing Magazine 30 (3) (2013) 83–98.

[6] J. Mei, J.M. Moura, Signal processing on graphs: performance of graph structure estimation, in: Acoustics, Speech and Signal Processing (ICASSP), 2016 IEEE International Conference on, IEEE, 2016, pp. 6165–6169.

[7] S. Sardellitti, S. Barbarossa, P.D. Lorenzo, Graph topology inference based on transform learning, in: 2016 IEEE Global Conference on Signal and Information Processing (GlobalSIP), IEEE, 2016.

[8] V.R.M. Elias, W.A. Martins, S. Werner, Diffusion-based virtual graph adjacency for Fourier analysis of network signals, in: XXXVIII Simpósio Brasileiro de Telecomunicações e Processamento de Sinais, SBrT'20, SBrT, 2020, pp. 1–5.

[9] V.R.M. Elias, W.A. Martins, S. Werner, Extended adjacency and scale-dependent graph Fourier transform via diffusion distances, IEEE Transactions on Signal and Information Processing over Networks 6 (2020) 592–604.
[10] R.R. Coifman, S. Lafon, A.B. Lee, M. Maggioni, B. Nadler, F. Warner, S.W. Zucker, Geometric diffusions as a tool for harmonic analysis and structure definition of data: diffusion maps, Proceedings of the National Academy of Sciences 102 (21) (2005) 7426–7431.
[11] A.V. Oppenheim, A.S. Willsky, S.H. Nawab, Signals and Systems, Prentice-Hall Signal Processing Series, Prentice Hall, 1997, https://books.google.com.br/books?id=LwQqAQAAMAAJ.
[12] P.S.R. Diniz, E.A.B. da Silva, S.L. Netto, Digital Signal Processing: System Analysis and Design, Cambridge University Press, 2010.
[13] A. Sandryhaila, J.M.F. Moura, Discrete signal processing on graphs: graph Fourier transform, in: 2013 IEEE International Conference on Acoustics, Speech and Signal Processing, Institute of Electrical and Electronics Engineers (IEEE), 2013.
[14] W. Rudin, Real and Complex Analysis, Tata McGraw-Hill Education, 1987.
[15] S. Mallat, A Wavelet Tour of Signal Processing, Academic Press, 1999.
[16] A. Sandryhaila, J.M.F. Moura, Discrete signal processing on graphs: frequency analysis, IEEE Transactions on Signal Processing 62 (12) (2014) 3042–3054.
[17] S. Li, Y. Jin, D.I. Shuman, Scalable M-channel critically sampled filter banks for graph signals, IEEE Transactions on Signal Processing 67 (15) (2018) 3954–3969.
[18] A. Ortega, P. Frossard, J. Kovačević, J.M. Moura, P. Vandergheynst, Graph signal processing: overview, challenges, and applications, Proceedings of the IEEE 106 (5) (2018) 808–828.
[19] S.K. Narang, A. Ortega, Downsampling graphs using spectral theory, in: IEEE International Conference on Acoustics, Speech and Signal Processing (ICASSP), 2011, pp. 4208–4211.
[20] P. Lorenzo, S. Barbarossa, P. Banelli, Sampling and recovery of graph signals, in: Cooperative and Graph Signal Processing, Academic Press, Italy, 2018, pp. 261–282.
[21] S. Chen, R. Varma, A. Sandryhaila, J. Kovačević, Discrete signal processing on graphs: sampling theory, IEEE Transactions on Signal Processing 63 (24) (2015) 6510–6523.
[22] G. Puy, N. Tremblay, R. Gribonval, P. Vandergheynst, Random sampling of bandlimited signals on graphs, Applied and Computational Harmonic Analysis 44 (2) (2018) 446–475.
[23] S.K. Narang, A. Gadde, A. Ortega, Signal processing techniques for interpolation in graph structured data, in: 2013 IEEE International Conference on Acoustics, Speech and Signal Processing, 2013, pp. 5445–5449.
[24] I. Pesenson, Sampling in Paley-Wiener spaces on combinatorial graphs, Transactions of the American Mathematical Society 360 (10) (2008) 5603–5627.
[25] Y. Shen, G. Leus, G.B. Giannakis, Online graph-adaptive learning with scalability and privacy, IEEE Transactions on Signal Processing 67 (9) (2019) 2471–2483.
[26] S.K. Narang, A. Ortega, Compact support biorthogonal wavelet filterbanks for arbitrary undirected graphs, IEEE Transactions on Signal Processing 61 (19) (2013) 4673–4685.
[27] M.J.M. Spelta, W.A. Martins, Online temperature estimation using graph signals, in: XXXVI Simpósio Brasileiro de Telecomunicações e Processamento de Sinais, SBrT'18, SBrT, 2018, pp. 1–5.
[28] N. Cressie, The origins of kriging, Mathematical Geology 22 (1990) 239–252.
[29] P.D. Lorenzo, S. Barbarossa, P. Banelli, S. Sardellitti, Adaptive least mean squares estimation of graph signals, IEEE Transactions on Signal and Information Processing over Networks 2 (4) (2016) 555–568.
[30] M.J.M. Spelta, W.A. Martins, Normalized LMS algorithm and data-selective strategies for adaptive graph signal estimation, Signal Processing 167 (2020) 107326.
[31] P.D. Lorenzo, E. Isufi, P. Banelli, S. Barbarossa, G. Leus, Distributed recursive least squares strategies for adaptive reconstruction of graph signals, in: 2017 25th European Signal Processing Conference (EUSIPCO), 2017, pp. 2289–2293.
[32] R. Nassif, C. Richard, J. Chen, A.H. Sayed, A graph diffusion LMS strategy for adaptive graph signal processing, in: 2017 51st Asilomar Conference on Signals, Systems, and Computers, 2017, pp. 1973–1976.

[33] R. Nassif, C. Richard, J. Chen, A.H. Sayed, Distributed diffusion adaptation over graph signals, in: 2018 IEEE International Conference on Acoustics, Speech and Signal Processing (ICASSP), 2018, pp. 4129–4133.

[34] F. Hua, R. Nassif, C. Richard, H. Wang, A.H. Sayed, A preconditioned graph diffusion LMS for adaptive graph signal processing, in: 2018 26th European Signal Processing Conference (EUSIPCO), 2018, pp. 111–115.

[35] F. Hua, R. Nassif, C. Richard, H. Wang, A.H. Sayed, Decentralized clustering for node-variant graph filtering with graph diffusion LMS, in: 2018 52nd Asilomar Conference on Signals, Systems, and Computers, 2018, pp. 1418–1422.

[36] P.S.R. Diniz, M.L.R. Campos, W.A. Martins, M.V.S. Lima, J.A. Apolinário Jr., Online Learning and Adaptive Filters, Cambridge University Press, 2022.

[37] V.C. Gogineni, V.R.M. Elias, W.A. Martins, S. Werner, Graph diffusion kernel LMS using random Fourier features, in: 2020 54th Asilomar Conference on Signals, Systems, and Computers, 2020, pp. 1528–1532.

[38] V.R.M. Elias, V.C. Gogineni, W.A. Martins, S. Werner, Adaptive graph filters in reproducing kernel Hilbert spaces: design and performance analysis, IEEE Transactions on Signal and Information Processing over Networks 7 (2021) 62–74.

[39] V.R.M. Elias, V.C. Gogineni, W.A. Martins, S. Werner, Kernel regression on graphs in random Fourier features space, in: ICASSP 2021 - 2021 IEEE International Conference on Acoustics, Speech and Signal Processing (ICASSP), 2021, pp. 5235–5239.

[40] V.R.M. Elias, V.C. Gogineni, W.A. Martins, S. Werner, Kernel regression over graphs using random Fourier features, IEEE Transactions on Signal Processing (2022) 1–14.

[41] C. Zhang, D. Florencio, Analyzing the optimality of predictive transform coding using graph-based models, IEEE Signal Processing Letters 20 (1) (2013) 106–109.

[42] G. Cheung, E. Magli, Y. Tanaka, M.K. Ng, Graph spectral image processing, Proceedings of the IEEE (2018) 1–24.

[43] G. Fracastoro, E. Magli, Subspace-sparsifying steerable discrete cosine transform from graph Fourier transform, in: 2016 IEEE International Conference on Image Processing (ICIP), 2016, pp. 1534–1538.

[44] G. Fracastoro, S.M. Fosson, E. Magli, Steerable discrete cosine transform, IEEE Transactions on Image Processing 26 (1) (2017) 303–314.

[45] G. Fracastoro, E. Magli, Steerable discrete Fourier transform, IEEE Signal Processing Letters 24 (3) (2017) 319–323.

[46] M. Masera, G. Fracastoro, M. Martina, E. Magli, A novel framework for designing directional linear transforms with application to video compression, in: ICASSP 2019 - 2019 IEEE International Conference on Acoustics, Speech and Signal Processing (ICASSP), 2019, pp. 1812–1816.

[47] R. Peloso, M. Capra, L. Sole, M. Ruo Roch, G. Masera, M. Martina, Steerable-discrete-cosine-transform (sdct): hardware implementation and performance analysis, Sensors 20 (5) (2020).

[48] V.S. Lima, F. Madeiro, J.B. Lima, Three-dimensional steerable discrete cosine transform with application to 3d image compression, Multidimensional Systems and Signal Processing 32 (2) (2021) 491–519.

[49] A. Gnutti, F. Guerrini, R. Leonardi, A. Ortega, Symmetry-based graph Fourier transforms for image representation, in: 2018 25th IEEE International Conference on Image Processing (ICIP), 2018, pp. 2575–2579.

[50] G. Fracastoro, D. Thanou, P. Frossard, Graph transform optimization with application to image compression, IEEE Transactions on Image Processing 29 (2020) 419–432.

[51] V.R.M. Elias, W.A. Martins, Graph Fourier transform for light field compression, in: XXXV Simpósio Brasileiro de Telecomunicações e Processamento de Sinais, SBrT'17, SBrT, 2017, pp. 881–885.

[52] V.R.M. Elias, W.A. Martins, On the use of graph Fourier transform for light-field compression, Journal of Communication and Information Systems 33 (1) (2018) 92–103.

[53] H.E. Egilmez, Y.-H. Chao, A. Ortega, Graph-based transforms for video coding, IEEE Transactions on Image Processing 29 (2020) 9330–9344.

[54] F. Zhang, E.R. Hancock, Graph spectral image smoothing using the heat kernel, Pattern Recognition 41 (11) (2008) 3328–3342.

[55] D. Thanou, X. Dong, D. Kressner, P. Frossard, Learning heat diffusion graphs, IEEE Transactions on Signal and Information Processing over Networks 3 (3) (2017) 484–499.

[56] H.E. Egilmez, E. Pavez, A. Ortega, Graph learning from filtered signals: graph system and diffusion kernel identification, IEEE Transactions on Signal and Information Processing over Networks 5 (2) (2019) 360–374.

[57] D. Abramian, M. Larsson, A. Eklund, I. Aganj, C.-F. Westin, H. Behjat, Diffusion-informed spatial smoothing of fMRI data in white matter using spectral graph filters, NeuroImage 237 (2021) 118095.

[58] M. Liu, Y. Wei, Image denoising using graph-based frequency domain low-pass filtering, in: 2019 IEEE 4th International Conference on Image, Vision and Computing (ICIVC), 2019, pp. 118–122.

[59] M. Onuki, S. Ono, M. Yamagishi, Y. Tanaka, Graph signal denoising via trilateral filter on graph spectral domain, IEEE Transactions on Signal and Information Processing over Networks 2 (2) (2016) 137–148.

[60] A.C. Yagan, M.T. Özgen, Spectral graph based vertex-frequency Wiener filtering for image and graph signal denoising, IEEE Transactions on Signal and Information Processing over Networks 6 (2020) 226–240.

[61] H. Sadreazami, A. Asif, A. Mohammadi, Iterative graph-based filtering for image abstraction and stylization, IEEE Transactions on Circuits and Systems II: Express Briefs 65 (2) (2018) 251–255.

[62] W. Hu, J. Pang, X. Liu, D. Tian, C.-W. Lin, A. Vetro, Graph signal processing for geometric data and beyond: theory and applications, IEEE Transactions on Multimedia (2021) 1.

[63] C. Zhang, D. Florêncio, C. Loop, Point cloud attribute compression with graph transform, in: 2014 IEEE International Conference on Image Processing (ICIP), 2014, pp. 2066–2070.

[64] D. Thanou, P.A. Chou, P. Frossard, Graph-based motion estimation and compensation for dynamic 3d point cloud compression, in: 2015 IEEE International Conference on Image Processing (ICIP), 2015, pp. 3235–3239.

[65] D. Thanou, P.A. Chou, P. Frossard, Graph-based compression of dynamic 3d point cloud sequences, IEEE Transactions on Image Processing 25 (4) (2016) 1765–1778.

[66] R.A. Cohen, D. Tian, A. Vetro, Attribute compression for sparse point clouds using graph transforms, in: 2016 IEEE International Conference on Image Processing (ICIP), 2016, pp. 1374–1378.

[67] A. Anis, P.A. Chou, A. Ortega, Compression of dynamic 3d point clouds using subdivisional meshes and graph wavelet transforms, in: 2016 IEEE International Conference on Acoustics, Speech and Signal Processing (ICASSP), 2016, pp. 6360–6364.

[68] Y. Xu, W. Hu, S. Wang, X. Zhang, S. Wang, S. Ma, Z. Guo, W. Gao, Predictive generalized graph Fourier transform for attribute compression of dynamic point clouds, IEEE Transactions on Circuits and Systems for Video Technology 31 (5) (2021) 1968–1982.

[69] E. Pavez, B. Girault, A. Ortega, P.A. Chou, Region adaptive graph Fourier transform for 3d point clouds, in: 2020 IEEE International Conference on Image Processing (ICIP), 2020, pp. 2726–2730.

[70] E. Pavez, A.L. Souto, R.L.D. Queiroz, A. Ortega, Multi-resolution intra-predictive coding of 3d point cloud attributes, in: 2021 IEEE International Conference on Image Processing (ICIP), 2021, pp. 3393–3397.

[71] M. Zhao, G. Cheung, D. Florencio, X. Ji, Progressive graph-signal sampling and encoding for static 3d geometry representation, in: 2017 IEEE International Conference on Image Processing (ICIP), 2017, pp. 735–739.

[72] Y. Schoenenberger, J. Paratte, P. Vandergheynst, Graph-based denoising for time-varying point clouds, in: 2015 3DTV-Conference: The True Vision - Capture, Transmission and Display of 3D Video (3DTV-CON), 2015, pp. 1–4.

[73] K. Yamamoto, M. Onuki, Y. Tanaka, Deblurring of point cloud attributes in graph spectral domain, in: 2016 IEEE International Conference on Image Processing (ICIP), 2016, pp. 1559–1563.

[74] F. Xue, F. Luisier, T. Blu, Multi-Wiener sure-let deconvolution, IEEE Transactions on Image Processing 22 (5) (2013) 1954–1968.

[75] C. Dinesh, G. Cheung, I.V. Bajić, C. Yang, Local 3d point cloud denoising via bipartite graph approximation amp; total variation, in: 2018 IEEE 20th International Workshop on Multimedia Signal Processing (MMSP), 2018, pp. 1–6.
[76] M.A. Irfan, E. Magli, Joint geometry and color point cloud denoising based on graph wavelets, IEEE Access 9 (2021) 21149–21166.
[77] Q. Lu, C. Chen, W. Xie, Y. Luo, Pointngcnn: deep convolutional networks on 3d point clouds with neighborhood graph filters, Computers & Graphics 86 (2020) 42–51.
[78] S. Zhang, S. Cui, Z. Ding, Hypergraph spectral clustering for point cloud segmentation, IEEE Signal Processing Letters 27 (2020) 1655–1659.
[79] W. Li, Q. Ma, W. Tian, X. Na, Graph convolution network with double filter for point cloud segmentation, in: 2020 5th International Conference on Intelligent Informatics and Biomedical Sciences (ICIIBMS), 2020, pp. 168–173.
[80] H. Al-Khafaji, C. Abhayaratne, Graph spectral domain blind watermarking, in: ICASSP 2019 - 2019 IEEE International Conference on Acoustics, Speech and Signal Processing (ICASSP), 2019, pp. 2492–2496.
[81] F.A.B.S. Ferreira, J.B. Lima, A robust 3D point cloud watermarking method based on the graph Fourier transform, Multimedia Tools and Applications 79 (3) (2020) 1921–1950.
[82] G. Lewenfus, W.A. Martins, S. Chatzinotas, B. Ottersten, Joint forecasting and interpolation of time-varying graph signals using deep learning, IEEE Transactions on Signal and Information Processing over Networks 6 (2020) 761–773.
[83] K. Qiu, X. Mao, X. Shen, X. Wang, T. Li, Y. Gu, Time-varying graph signal reconstruction, IEEE Journal of Selected Topics in Signal Processing 11 (6) (2017) 870–883.
[84] X. Mao, K. Qiu, T. Li, Y. Gu, Spatio-temporal signal recovery based on low rank and differential smoothness, IEEE Transactions on Signal Processing 66 (23) (2018) 6281–6296.
[85] J.H. Giraldo, T. Bouwmans, On the minimization of Sobolev norms of time-varying graph signals: estimation of new coronavirus disease 2019 cases, in: 2020 IEEE 30th International Workshop on Machine Learning for Signal Processing (MLSP), 2020, pp. 1–6.
[86] P. Valdivia, F. Dias, F. Petronetto, C.T. Silva, L.G. Nonato, Wavelet-based visualization of time-varying data on graphs, in: 2015 IEEE Conference on Visual Analytics Science and Technology (VAST), IEEE, 2015, pp. 1–8.
[87] A.D. Col, P. Valdivia, F. Petronetto, F. Dias, C.T. Silva, L.G. Nonato, Wavelet-based visual analysis of dynamic networks, IEEE Transactions on Visualization and Computer Graphics 24 (8) (2018) 2456–2469.
[88] L.G. Nonato, F.P. do Carmo, C.T. Silva, GLoG: Laplacian of Gaussian for spatial pattern detection in spatio-temporal data, IEEE Transactions on Visualization and Computer Graphics 27 (8) (2021) 3481–3492.
[89] R. Kondor, J. Lafferty, Diffusion kernels on graphs and other discrete structures, in: Proceedings of the 19th International Conference on Machine Learning, 2002, pp. 315–322.
[90] G. Shen, W.-S. Kim, S.K. Narang, A. Ortega, J. Lee, H. Wey, Edge-adaptive transforms for efficient depth map coding, in: 28th Picture Coding Symposium, IEEE, 2010.
[91] G. Cheung, W.-S. Kim, A. Ortega, J. Ishida, A. Kubota, Depth map coding using graph based transform and transform domain sparsification, in: 2011 IEEE 13th International Workshop on Multimedia Signal Processing, 2011, pp. 1–6.
[92] W. Hu, G. Cheung, X. Li, O. Au, Depth map compression using multi-resolution graph-based transform for depth-image-based rendering, in: 2012 19th IEEE International Conference on Image Processing, 2012, pp. 1297–1300.
[93] D.K. Hammond, P. Vandergheynst, R. Gribonval, Wavelets on graphs via spectral graph theory, Applied and Computational Harmonic Analysis 30 (2) (2011) 129–150.
[94] S.K. Narang, A. Ortega, Perfect reconstruction two-channel wavelet filter banks for graph structured data, IEEE Transactions on Signal Processing 60 (6) (2012) 2786–2799.
[95] N. Leonardi, D. Van De Ville, Tight wavelet frames on multislice graphs, IEEE Transactions on Signal Processing 61 (13) (2013) 3357–3367.

[96] A. Ozdemir, S. Aviyente, Graph wavelet transform: application to image segmentation, in: 2014 48th Asilomar Conference on Signals, Systems and Computers, 2014, pp. 496–499.
[97] N. Tremblay, P. Borgnat, Graph wavelets for multiscale community mining, IEEE Transactions on Signal Processing 62 (20) (2014) 5227–5239.
[98] M. Püschel, J.M.F. Moura, Algebraic signal processing theory: foundation and 1-D time, IEEE Transactions on Signal Processing 56 (8) (2008) 3572–3585.
[99] M. Püschel, J.M.F. Moura, Algebraic signal processing theory: 1-D space, IEEE Transactions on Signal Processing 56 (8) (2008) 3586–3599.
[100] A. Sandryhaila, J.M.F. Moura, Big data analysis with signal processing on graphs: representation and processing of massive data sets with irregular structure, IEEE Signal Processing Magazine 31 (5) (2014) 80–90.

CHAPTER

Tensor methods in deep learning 15

Yannis Panagakis[a], Jean Kossaifi[b], Grigorios G. Chrysos[c], James Oldfield[d], Taylor Patti[b], Mihalis A. Nicolaou[e], Anima Anandkumar[b,f], and Stefanos Zafeiriou[g]

[a]*Department of Informatics and Telecommunications, National and Kapodistrian University of Athens, Athens, Greece*
[b]*NVIDIA, Santa Clara, CA, United States*
[c]*Department of Electrical Engineering, Ecole Polytechnique Federale de Lausanne (EPFL), Lausanne, Switzerland*
[d]*Queen Mary University of London, London, United Kingdom*
[e]*Computation-based Science and Technology Research Center at The Cyprus Institute, Nicosia, Cyprus*
[f]*Caltech, Pasadena, CA, United States*
[g]*Department of Computing, Imperial College London, London, United Kingdom*

15.1 Introduction

Deep learning architectures such as convolutional, recurrent, and self-attention networks have led to qualitative breakthroughs in a wide variety of artificial intelligence application realms, notably computer vision, speech, and natural language processing, among many other machine learning tasks [1]. The key ingredient of their success is that they effectively leverage statistical properties of data such as stationarity (e.g., shift- and time-invariance) and compositionality (e.g., the hierarchical structure of images and natural language) to learn approximations of high-dimensional, nonlinear functions that can successfully describe complex real-world data.

At the core of such learning architectures are multidimensional blocks of parameters, naturally represented by higher-order tensors. Indeed, deep convolutional neural networks (DCNNs) [2,3] consist of alternating multidimensional convolutional layers, pointwise nonlinear functions (e.g., ReLU), and downsampling (pooling) layers, while also containing tensor-structured fully connected layers. The use of multidimensional convolutions in convolutional neural network (CNNs) allows extracting local features shared across the data domain. In turn, this greatly reduces the number of learnable parameters and consequently alleviates the effect of the curse of dimensionality, without sacrificing the ability to approximate the target function. When dealing with sequential data such as text, deep recurrent networks [1] and more recently deep self-attention networks [4] have been proved effective in capturing long-range temporal dependencies and higher-order correlations, by modeling higher-order interactions among data and parameters through a series of multidimensional parametric blocks. Such tensors of parameters, representing for example convolutional kernels or attention blocks, can be viewed as multilinear mappings generalizing the notion of linear mappings represented by matrices.

Concretely, tensors are a multidimensional generalization of matrices, and a core mathematical object in multilinear algebra – just like vectors and matrices are in linear algebra. The *order* of a tensor is the number of indices needed to address its elements. A matrix is a second-order tensor of two *modes*,

Signal Processing and Machine Learning Theory. https://doi.org/10.1016/B978-0-32-391772-8.00021-1
Copyright © 2024 Elsevier Inc. All rights reserved.

namely rows and columns, requiring two indices to access its elements. In analogy, an Nth-order tensor has N modes, and N indices index it.

Tensors have found applications in a wide range of disciplines in science and engineering, ranging from many-body quantum systems [5] to algebraic geometry and theoretical computer science [6], to mention a few. In data science and related fields such as machine learning, signal processing, computer vision, and statistics, tensors have been used to represent and analyze information hidden in multidimensional data, such as images and videos, or to capture and exploit higher-order similarities or dependencies among vector-valued variables. Just as pairwise (i.e., second-order) similarities of vector samples are represented by covariance or correlation matrices, a third- (or higher)-order tensor can represent similarities among three (or more) samples in a set. Such tensors are often referred to as higher-order statistics. In this context, tensor methods mainly focus on extending matrix-based learning models such as component analysis [7], dictionary learning (e.g., [8]), and regression models (e.g., [9]), to handle data represented by higher-order tensors or estimating parameters of latent variable models by analyzing higher-order statistics [10].

Such developments were for a long time independent of the rise of deep learning, where currently tensors and their decompositions play a pivotal role. Deep learning models are usually overparameterized involving an enormous number (typically tens of millions or even billions) of unknown parameters. Even though overparameterization of deep neural networks allows finding good local minima, particularly when they are trained with gradient descent [11,12], such a large number of trainable parameters makes deep networks prone to overfitting and noise [13], hindering the analysis of their generalization properties (i.e., how well they perform on unseen data) [14]. Tensor decompositions can significantly reduce the number of unknown parameters of deep models and further mitigate the curse of dimensionality. Indeed, tensor decompositions can be applied to the weights of neural network layers to compress them and in some cases speed them up [15,16,17]. In addition, using tensor-based operations within deep neural networks allows to preserve and leverage the topological structure in the data while resulting in parsimonious models [18,19]. In CNNs, (nonseparable) multichannel convolution kernels can be decomposed into a sum of mode-separable convolutions using low-rank tensor decompositions, resulting in network compression and a reduction in the number of parameters [20,21]. Deep network compression is significant in the deployment of deep learning models in resource-limited devices, while reducing the number of parameters acts as implicit regularization that improves generalization across tasks and domains [22,23,24]. Furthermore, randomized tensor decomposition of neural networks' building blocks has been proved effective in robustifying neural networks against adversarial attacks [25,26] and to various types of random noise, such as noise occurring naturally when capturing data such as MRI data [27].

Quantum computing (QC) aims to provide significant improvements to classical computing, leading to the so-called quantum advantage. The field of quantum machine learning (QML) aims to bring similar advantages to machine and deep learning algorithms, for which an enabling factor is quantum simulation. However, it also suffers significantly from the curse of dimensionality, scaling exponentially fast with either the number of data (qubits) or the depth of the quantum circuit. In this context, tensor methods have been shown to mitigate this curse of dimensionality by reducing the memory footprint of QC systems and broadly accelerating simulations of such systems.

In this chapter, besides the fundamentals of tensors and their decompositions, we focus on recent developments that have gradually increased tensor-related methods, especially in deep learning architectures. We particularly focus on tensor methods for (1) compressing deep architectures, (2) modeling higher-order interactions among data points and among model parameters, (3) interpreting the latent space of generative adversarial networks (GANs), and (4) making deep networks robust to noise and adversarial attacks. Finally, we introduce tensor methods in the context of the emerging field of QML.

15.2 Preliminaries on matrices and tensors

To make the chapter self-contained, we first introduce the notational conventions used in this chapter and then review some of the fundamental concepts of linear and multilinear algebra, i.e., the algebra of higher-order tensors.

To ensure consistency in terminology and notations that would be familiar to machine learning researchers, we have adopted that employed in widely cited overview papers [28,29,10]. However, there is one point where we differentiate. We employ a different ordering of elements when tensors are flattened to matrices than that used in [29] which may be familiar to some of the readers (cf. Definition 15.2.1).

15.2.1 Notation

Scalars are denoted by plain letters, i.e., x, I, J. First-order tensors are *vectors*, denoted by bold lowercase letters, e.g., $\mathbf{x}$. Second-order tensors are *matrices*, denoted by bold uppercase letters, e.g., $\mathbf{X}$. $\mathbf{I}$ denotes the identity matrix of compatible dimensions. The transpose of $\mathbf{X}$ is denoted $\mathbf{X}^\top$. The ith column of $\mathbf{X}$ is denoted as $\mathbf{x}_i$.

The sets of real and integer numbers are denoted by $\mathbb{R}$ and $\mathbb{Z}$, respectively. A set of M real matrices (vectors) of varying dimensions is denoted by $\{\mathbf{X}^{(m)} \in \mathbb{R}^{I_m \times N}\}_{m=1}^{M}$ ($\{\mathbf{x}^{(m)} \in \mathbb{R}^{I_m}\}_{m=1}^{M}$). Finally, for any $i, j \in \mathbb{Z}$, with $i < j$, $[i\,..\,j]$ denotes the set of integers $\{i, i+1, \cdots, j-1, j\}$.

Tensors of order three or higher are denoted by $\mathcal{X}$. The reader is reminded that the order of a tensor is the number of indices (dimensions) needed to address its elements. Each dimension is called a mode. Specifically, an Nth-order tensor has N indices, with each index addressing a mode of $\mathcal{X}$. Assuming $\mathcal{X}$ is real-valued, it is defined over the tensor space $\mathbb{R}^{I_1 \times I_2 \times \cdots \times I_N}$, where $I_n \in \mathbb{Z}$ for $n = 1, 2, \ldots, N$. An element $(i_1, i_2, \cdots, i_N)$ of tensor $\mathcal{X} \in \mathbb{R}^{I_1 \times I_2 \times \cdots \times I_N}$ is denoted as $\mathcal{X}_{i_1, i_2, \cdots, i_N}$ or $\mathcal{X}(i_1, i_2, \cdots, i_N)$. This corresponds to viewing a tensor as a multidimensional array in $\mathbb{R}^{I_1 \times I_2 \times \cdots \times I_N}$. Furthermore, given a set of N vectors (or matrices) that correspond to each mode of $\mathcal{X}$, the nth vector (or matrix) is denoted as $\mathbf{u}^{(n)}$ (or $\mathbf{U}^{(n)}$).

The elementwise ℓ_p of $\mathcal{X}$ is

$$\|\mathcal{X}\|_p = \|\text{vec}(\mathcal{X})\|_p = \left(\sum_{i_1, i_2, \cdots, i_N} |\mathcal{X}_{i_1, i_2, \cdots, i_N}|^p \right)^{1/p}.$$

When $p = 1$, the ℓ_1-norm is denoted by $\|\mathcal{X}\|_1$ and is defined as the sum of the absolute values of the tensor's elements. The ℓ_1-norm is used as a measure of sparsity in practice. For $p = 2$, the *Frobenius norm* is denoted as $\|\mathbf{X}\|_F$ for matrices and it is generalized to higher-order tensors by the *tensor norm*, denoted $\|\mathcal{X}\|$.

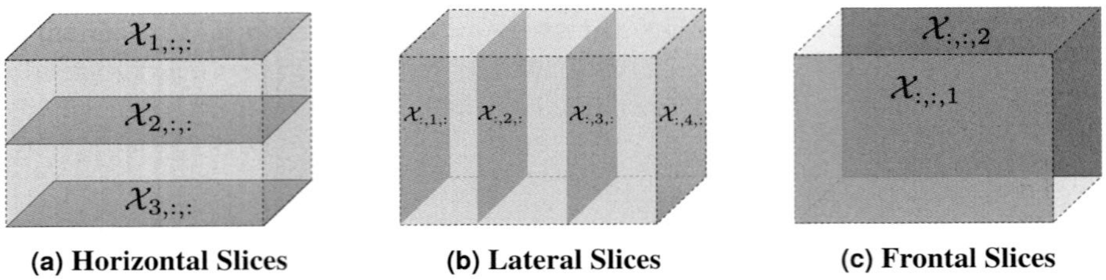

FIGURE 15.1

Illustration of the slices of a third-order tensor of size $3 \times 4 \times 2$.

Fibers are a generalization of the concept of rows and columns of matrices to tensors. They are obtained by fixing all indices but one. A colon is used to denote all elements of a mode, e.g., the mode-1 fibers of $\mathcal{X}$ are denoted as $\mathcal{X}_{:,i_2,i_3}$.

Slices (Fig. 15.1) are obtained, for a third-order tensor, by fixing one of the indices.

Table 15.1 Illustration of the fibers and unfolding of a third-order tensor.

Fibers	**mode-0 fibers** 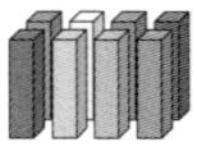	**mode-1 fibers** 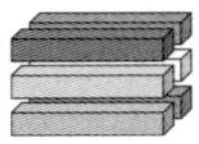	**mode-2 fibers** 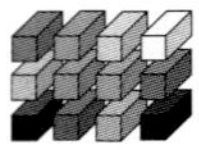
TensorLy	**mode-0 unfolding** 	**mode-1 unfolding** 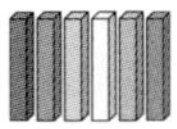	**mode-2 unfolding** 
Matlab	**mode-0 unfolding** 	**mode-1 unfolding** 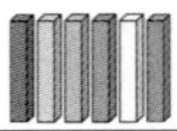	**mode-2 unfolding** 

15.2.2 Transforming tensors into matrices and vectors

Definition 15.2.1 (Tensor unfolding). Given a tensor $\mathcal{X} \in \mathbb{R}^{I_1 \times I_2 \times \cdots \times I_N}$, its mode-$n$ unfolding is a matrix $\mathbf{X}_{[n]} \in \mathbb{R}^{I_n \times I_M}$ with $M = \prod_{\substack{k=1,\\ k \neq n}}^{N} I_k$ and is defined by the mapping from element $(i_1, i_2, \cdots, i_N)$ to (i_n, j), with $j = 1 + \sum_{\substack{k=1,\\ k \neq n}}^{N} (i_k - 1) \times \prod_{\substack{m=k+1,\\ m \neq n}}^{N} I_m$.

Tensor unfolding corresponds to a reordering of the fibers of the tensor as the columns of a matrix. There are several definitions of tensor unfolding, which vary by the ordering of the fibers. In particular, our definition corresponds to a rowwise underlying ordering (or C-ordering) of the elements. The main other definition, popularized by Kolda and Bader [29], corresponds to a columnwise (or Fortran) ordering of the elements, which is more efficient for libraries using that ordering, such as MATLAB®. Conversely, our definition naturally leads to better performance for implementation with C-ordering of the elements, which is the case with most Python/C++ software libraries as well as most GPU libraries [30]. Table 15.1 illustrates the link between the fibers and the unfoldings, as well as the difference between our definition (rowwise, or C-ordering) and the columnwise, or Fortran ordering [29].

Definition 15.2.2 (Tensor vectorization). Given a tensor $\mathcal{X} \in \mathbb{R}^{I_1 \times I_2 \times \cdots \times I_N}$, we can flatten it into a vector vec($\mathcal{X}$) of size $I_1 \cdot I_2 \cdot \ldots \cdot I_N$ defined by the mapping from element $(i_1, i_2, \cdots, i_N)$ of $\mathcal{X}$ to element j of vec($\mathcal{X}$), with $j = 1 + \sum_{k=1}^{N}(i_k - 1) \times \prod_{m=k+1}^{N} I_m$.

15.2.3 Matrix and tensor products

Here, we will define several matrix and tensor products employed by the surveyed methods.

Definition 15.2.3 (Kronecker product). Given two matrices $\mathbf{A} \in \mathbb{R}^{I_1 \times I_2}$ and $\mathbf{B} \in \mathbb{R}^{J_1 \times J_2}$, their Kronecker product is defined as

$$\mathbf{A} \otimes \mathbf{B} = \begin{bmatrix} a_{11}\mathbf{B} & \cdots & a_{1I_2}\mathbf{B} \\ \vdots & \ddots & \vdots \\ a_{I_1 1}\mathbf{B} & \cdots & a_{I_1 I_2}\mathbf{B} \end{bmatrix} \in \mathbb{R}^{I_1 \cdot J_1 \times I_2 \cdot J_2}.$$

Definition 15.2.4 (Khatri–Rao product). Given two matrices $\mathbf{A} \in \mathbb{R}^{I \times R}$ and $\mathbf{B} \in \mathbb{R}^{J \times R}$ with the same number of columns, their Khatri–Rao product, also known as columnwise Kronecker product, is defined as $\mathbf{A} \odot \mathbf{B} \in \mathbb{R}^{I \cdot J \times R}$:

$$\mathbf{A} \odot \mathbf{B} = \left[\mathbf{A}_{:,1} \otimes \mathbf{B}_{:,1}, \quad \mathbf{A}_{:,2} \otimes \mathbf{B}_{:,2}, \quad \cdots, \quad \mathbf{A}_{:,R} \otimes \mathbf{B}_{:,R}\right].$$

Furthermore, the Khatri–Rao product of a set of M matrices $\{\mathbf{X}^{(m)} \in \mathbb{R}^{I_m \times N}\}_{m=1}^{M}$ is denoted by $\mathbf{X}^{(1)} \odot \mathbf{X}^{(2)} \odot \cdots \odot \mathbf{X}^{(M)}$.

Definition 15.2.5 (Hadamard product). The Hadamard product of $\mathbf{A}, \mathbf{B} \in \mathbb{R}^{I \times J}$ is the elementwise product symbolized as $\mathbf{A} * \mathbf{B}$ and is defined elementwise as $\mathbf{A}_{(i,j)}\mathbf{B}_{(i,j)}$.

Definition 15.2.6 (Outer product). The symbol $\circ$ denotes the vector outer product. Given a set of N vectors $\{\mathbf{x}^{(n)} \in \mathbb{R}^{I_n}\}_{n=1}^{N}$, their outer product is denoted as $\mathcal{X} = \mathbf{x}^{(1)} \circ \mathbf{x}^{(2)} \circ \cdots \circ \mathbf{x}^{(N)} \in \mathbb{R}^{I_1 \times I_2 \times \cdots \times I_N}$ and defines a *rank*-one Nth-order tensor.

Definition 15.2.7 (n-mode product). For a tensor $\mathcal{X} \in \mathbb{R}^{I_1 \times I_2 \times \cdots \times I_N}$ and a matrix $\mathbf{M} \in \mathbb{R}^{R \times I_n}$, the n-mode product of a tensor is a tensor of size $(I_1 \times \cdots \times I_{n-1} \times R \times I_{n+1} \times \cdots \times I_N)$ and can be expressed using unfolding of $\mathcal{X}$ and the classical dot product as follows:

$$\mathcal{X} \times_n \mathbf{M} = \mathbf{M}\mathbf{X}_{[n]} \in \mathbb{R}^{I_1 \times \cdots \times I_{n-1} \times R \times I_{n+1} \times \cdots \times I_N}.$$

The *n-mode vector product* of a tensor $\mathcal{X} \in \mathbb{R}^{I_1 \times I_2 \times \ldots \times I_N}$ with a vector $\mathbf{x} \in \mathbb{R}^{I_n}$ is denoted by $\mathcal{X} \times_n \mathbf{x} \in \mathbb{R}^{I_1 \times I_2 \times \cdots \times I_{n-1} \times I_{n+1} \times \cdots \times I_N}$. The resulting tensor is of order $N-1$ and is defined elementwise as

$$(\mathcal{X} \times_n \mathbf{x})_{i_1,\ldots,i_{n-1},i_{n+1},\ldots,i_N} = \sum_{i_n=1}^{I_n} x_{i_1,i_2,\ldots,i_N} x_{i_n}. \tag{15.1}$$

In order to simplify the notation, we denote $\mathcal{X} \times_1 \mathbf{x}^{(1)} \times_2 \mathbf{x}^{(2)} \times_3 \cdots \times_N \mathbf{x}^{(N)} = \mathcal{X} \prod_{n=1}^{N} \times_n \mathbf{x}^{(n)}$.

Definition 15.2.8 (Generalized inner product). For two tensors $\mathcal{X}, \mathcal{Y} \in \mathbb{R}^{I_1 \times I_2 \times \cdots \times I_N}$ of the same size, their inner product is defined as $\langle \mathcal{X}, \mathcal{Y} \rangle = \sum_{i_1=1}^{I_1} \sum_{i_2=1}^{I_2} \cdots \sum_{i_n=1}^{I_N} \mathcal{X}_{i_1,i_2,\cdots,i_n} \mathcal{Y}_{i_1,i_2,\cdots,i_n}$. For two tensors $\mathcal{X} \in \mathbb{R}^{D_x \times I_1 \times I_2 \times \cdots \times I_N}$ and $\mathcal{Y} \in \mathbb{R}^{I_1 \times I_2 \times \cdots \times I_N \times D_y}$ sharing N modes of the same size, we similarly define the generalized inner product along the N last (respectively first) modes of $\mathcal{X}$ (respectively $\mathcal{Y}$) as

$$\langle \mathcal{X}, \mathcal{Y} \rangle_N = \sum_{i_1=1}^{I_1} \sum_{i_2=1}^{I_1} \cdots \sum_{i_n=1}^{I_N} \mathcal{X}_{:,i_2,i_3,\cdots,i_n} \mathcal{Y}_{i_1,i_2,\cdots,i_{n-1},:},$$

with $\langle \mathcal{X}, \mathcal{Y} \rangle_N \in \mathbb{R}^{I_x \times I_y}$.

Definition 15.2.9 (Convolution). We denote a regular convolution of $\mathcal{X}$ with $\mathbf{W}$ as $\mathcal{X} \star \mathbf{W}$. For 1D convolutions, we write the convolution of a tensor $\mathcal{X} \in \mathbb{R}^{I_1 \times I_2 \times \cdots \times I_N}$ with a vector $\mathbf{v} \in \mathbb{R}^K$ along the nth mode as $\mathcal{X} \star_n \mathbf{v}$. In practice, as done in current deep learning frameworks [31], we use cross-correlation, which differs from a convolution by a flip of the kernel. This does not impact the results since the weights are learned end-to-end. In other words, $(\mathcal{X} \star_n \mathbf{v})_{i_1,\cdots,i_N} = \sum_{k=1}^{K} \mathbf{v}_k \mathcal{X}_{i_1,\cdots,i_{n-1},i_n+k,i_{n+1},\cdots,I_N}$.

15.2.4 Tensor diagrams

Most tensor methods involve a series of tensor operators, such as tensor contractions,[1] resulting in several sums over various modes and tensors, with as many indices, which can be cumbersome to read and write.

Tensor diagrams are pictorial representations of tensor algebraic operations in the form of undirected graphs offering a more intuitive way to read and write tensor operators and design tensor methods. The vertices (circles) represent tensors, and edges represent the modes of the tensors. The degree of each vertex (the number of edges coming out of it) thus represents the order of the corresponding tensor. Consequently, using tensor diagrams a scalar (Fig. 15.2a) would simply be represented by a vertex while a vector (i.e., a tensor with one index) would have one edge. A matrix (Fig. 15.2b) would be a vertex with two edges, and so on for third- (Fig. 15.2c) and fourth-order (Fig. 15.2d) tensors.

By employing tensor diagrams, tensor contraction over a common dimension of two tensors is represented by connecting the two corresponding edges. For instance, given matrices $\mathbf{A} \in \mathbb{R}^{I \times J}$, $\mathbf{B} \in \mathbb{R}^{J \times K}$, we can simply express their product $\mathbf{C} = \mathbf{AB} = \sum_{j=1}^{J} \mathbf{a}_{:,j} \mathbf{b}_{j,:}^{\top}$, as illustrated in Fig. 15.3a.

[1] The n-mode product of a tensor with a matrix is a tensor contraction between the nth mode of the tensor and the second mode of the matrix.

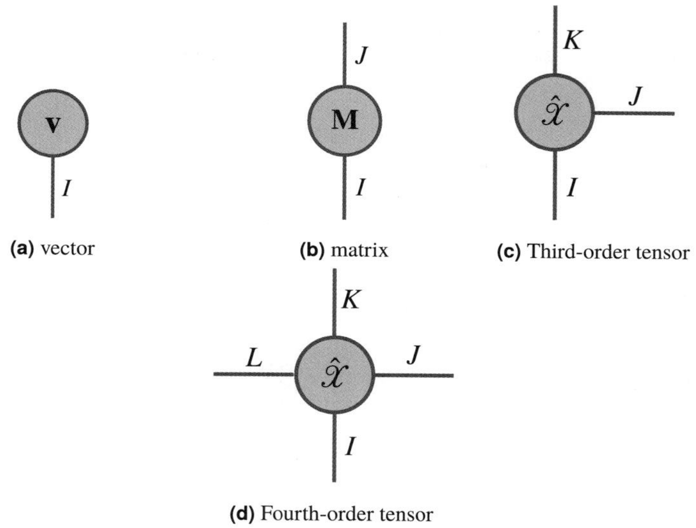

FIGURE 15.2

Representation of tensors of various orders using tensor diagrams. The vertices represent tensors and their degree represents the order of the tensors.

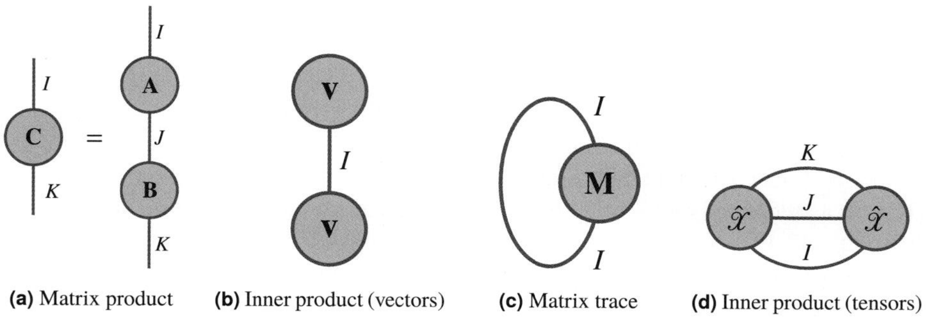

FIGURE 15.3

Representation of some tensor contractions using tensor diagrams.

15.3 Tensor decomposition

Matrix decomposition forms the mathematical backbone in learning parsimonious, low-dimensional representations from high-dimensional vector data samples, such as images flattened into vectors. *Component analysis* (e.g., [32]) and *sparse coding* (e.g., [33,34]) are prominent examples of representation learning methods that extract simple data representations by means of low-rank matrix decompositions and sparse, possibly overcomplete representations, respectively. Given M vector data samples $\{\mathbf{x}_m\}_{m=1}^{M}$, where each sample $\mathbf{x}_m$ is an I-dimensional vector, that is, $\mathbf{x}_m \in \mathbb{R}^I_m$, such a dataset can be represented by a data matrix $\mathbf{X} \in \mathbb{R}^{I \times J}$ which contains in its columns data vectors. Matrix decompositions seek to factorize $\mathbf{X}$ as a product of two factor matrices $\mathbf{U}^{(1)} \in \mathbb{R}^{I \times R}$ and $\mathbf{U}^{(2)} \in \mathbb{R}^{J \times R}$, namely

$$\mathbf{X} = \mathbf{U}^{(1)} \mathbf{U}^{(2)\top}. \tag{15.2}$$

Assuming $\mathbf{X}$ is of low rank $R < \min\{I, J\}$, Eq. (15.2) is referred to as *low-rank matrix decomposition*. Hence, $\mathbf{X}$ can be expressed as a sum of R rank-one matrices: $\mathbf{X} = \mathbf{U}^{(1)}\mathbf{U}^{(2)\top} = \sum_{r=1}^{R} \mathbf{u}_r^{(1)} \circ \mathbf{u}_i^{(2)} = \sum_{r=1}^{R} \mathbf{u}_r^{(1)} \mathbf{u}_r^{(2)\top}$. The concept of low-rank matrix decomposition is generalized for higher-order tensors by low-rank tensor decomposition, which will be discussed next. Interestingly, as opposed to matrix decomposition, which is *not unique*,[2] low-rank tensor decomposition, such as canonical-polyadic (CP) decomposition, also known as parallel factor analysis (PARAFAC) [35,36,37], is unique under mild algebraic assumptions (see [28] and references therein).

In multivariate linear regression models, enforcing low-rank constraints on the coefficient matrix offers an effective reduction of unknown parameters, which facilitates reliable parameter estimation and model interpretation. Likewise, low-rank structures can be imposed in tensor regression models such as those presented in Section 15.3.4.

Tensor decomposition extends the concept of matrix decomposition to higher-order tensors. Here, we give a brief overview of the three most fundamental tensor decomposition models that are most useful in the context of this chapter. For more details on these models, the interested reader is referred to [38,29,28,39], where algorithmic matters are also covered in detail. We introduce all three decompositions for an Nth-order tensor $\mathcal{X} \in \mathbb{R}^{I_1 \times I_2 \times \cdots \times I_N}$.

15.3.1 Canonical-polyadic decomposition

CP decomposition, also referred to as PARAFAC [35,36,37], decomposes $\mathcal{X}$ into a sum of R rank-one tensors. The rank of a tensor $\mathcal{X}$ is the minimum number of rank-one tensors that sum to $\mathcal{X}$ and generalizes the notion of matrix rank to high-order tensors. However, computing the tensor rank is NP-hard [40,41]. Hence, a predefined rank should be known in advance or estimated from data using Bayesian approaches (e.g., [42,43]) or deep neural networks [44]. When directly learning the decomposition end-to-end with stochastic gradient descent, the rank parameter is validated along with other parameters.

Formally, CP decomposition seeks to find the vectors $\mathbf{u}_r^{(1)}, \mathbf{u}_r^{(2)}, \cdots, \mathbf{u}_r^{(N)}$ such that

$$\mathcal{X} = \sum_{r=1}^{R} \underbrace{\mathbf{u}_r^{(1)} \circ \mathbf{u}_r^{(2)} \circ \cdots \circ \mathbf{u}_r^{(N)}}_{\text{rank-one components (tensors)}}. \tag{15.3}$$

These vectors can be collected into N matrices, $\{\mathbf{U}^{(n)} \in \mathbb{R}^{I_n \times R}\}_{n=1}^N$, with each matrix defined as

$$\mathbf{U}^{(n)} = \left[\mathbf{u}_1^{(n)}, \mathbf{u}_2^{(n)}, \cdots, \mathbf{u}_R^{(n)}\right]. \tag{15.4}$$

---

[2] To verify this, consider any invertible matrix $\mathbf{Q} \in \mathbb{R}^{R \times R}$. Then we have $\mathbf{X} = \mathbf{U}^{(1)}\mathbf{U}^{(2)\top} = \mathbf{U}^{(1)}\mathbf{Q}\mathbf{Q}^{-1}\mathbf{U}^{(2)\top}$.

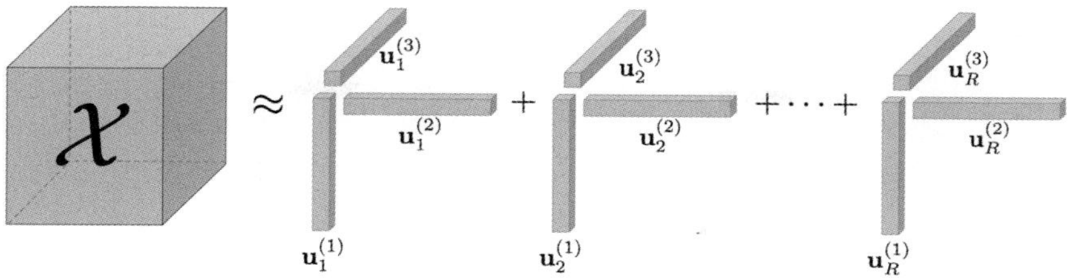

FIGURE 15.4

Illustration of a CP decomposition of third-order tensor $\mathcal{X}$ into a sum of rank-one tensors.

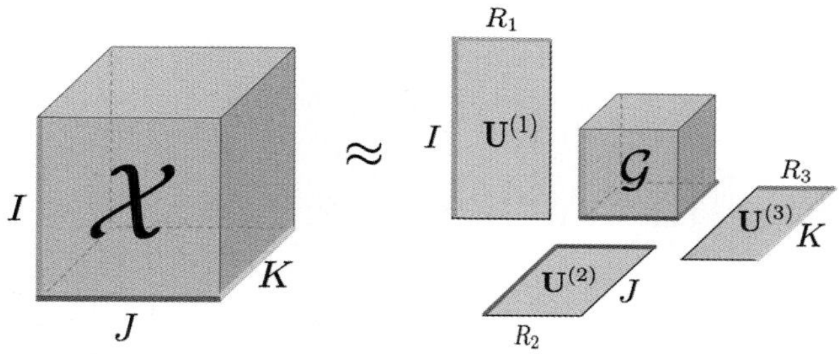

FIGURE 15.5

Illustration of a Tucker decomposition of a third-order tensor $\mathcal{X}$ into a core tensor and three matrices.

CP decomposition is denoted more compactly as $\mathcal{X} = [\![\mathbf{U}^{(1)}, \mathbf{U}^{(2)}, \cdots, \mathbf{U}^{(N)}]\!]$ [29]. A representation of a tensor using its CP decomposition is sometimes referred to as its Kruskal format. A pictorial representation of CP decomposition is shown in Fig. 15.4.

15.3.2 Tucker decomposition

Tucker decomposition [45] decomposes, nonuniquely, $\mathcal{X}$ into a core tensor $\mathcal{G} \in \mathbb{R}^{R_1 \times R_2 \times \cdots \times R_N}$ and a set of factor matrices $\left(\mathbf{U}^{(1)}, \mathbf{U}^{(2)}, \cdots, \mathbf{U}^{(N)}\right)$, with $\mathbf{U}^{(n)} \in \mathbb{R}^{R_n \times I_n}$, $n = 1, 2, \ldots, N$, which are multiplied with $\mathcal{G}$ along its modes as follows:

$$\mathcal{X} = \mathcal{G} \times_1 \mathbf{U}^{(1)} \times_2 \mathbf{U}^{(2)} \times_3 \cdots \times_N \mathbf{U}^{(N)}. \tag{15.5}$$

The core tensor captures interactions between the columns of factor matrices. If $R_n \ll I_n$, $\forall n$, then the core tensor can be viewed as a compressed version of $\mathcal{X}$. It is also worth noting that CP decomposition can be expressed as Tucker decomposition, with $R_n = R$, $\forall n \in \{1, 2, \ldots N\}$, and the core tensor being superdiagonal. Tucker decomposition is denoted compactly as $[\![\mathcal{G}; \mathbf{U}^{(1)}, \mathbf{U}^{(2)}, \cdots, \mathbf{U}^{(N)}]\!]$ [29], and it is shown pictorially in Fig. 15.5. By imposing the factor matrices to be orthonormal, the Tucker model is known as higher-order singular value decomposition (HOSVD) [46].

In practice, data are corrupted by noise, and as in the matrix case, both the Tucker and CP decompositions are not exact. Therefore, they are approximated by optimizing a suitable criterion representing

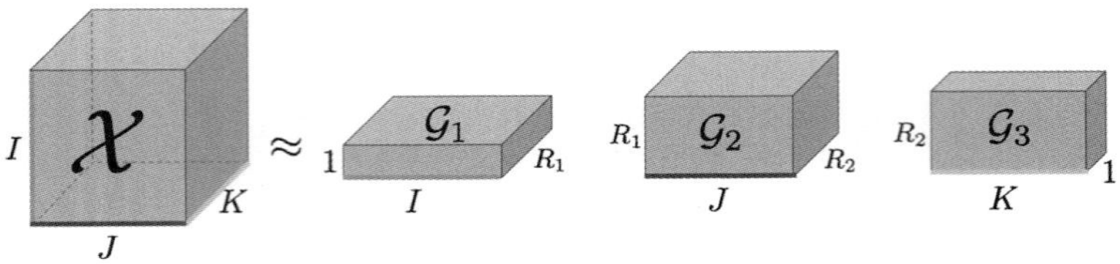

FIGURE 15.6

Illustration of a tensor-train decomposition of a third-order tensor $\mathcal{X}$ into a series of third-order cores. The boundary conditions dictate $R_1 = R_{N+1} = 1$.

fitting loss. Typically, the least-squares criterion is employed. In this case, assuming that $\{R_n\}_{n=1}^N$ are known, the Tucker decomposition is computed by solving the following nonconvex minimization problem:

$$\min_{\mathcal{G}, \{\mathbf{U}^{(n)}\}_{n=1}^N} \quad \|\mathcal{X} - [\![\mathcal{G}; \mathbf{U}^{(1)}, \mathbf{U}^{(2)}, \cdots, \mathbf{U}^{(N)}]\!]\|^2. \tag{15.6}$$

The most common optimization algorithm for the Tucker and CP decompositions is the alternating least-square (ALS) algorithm [47], where the unknowns are estimated iteratively. More specifically, the ALS algorithm fixes, sequentially, all but one factor, which is then updated by solving a linear least-squares problem. However, the ALS algorithm is not guaranteed to converge towards a minimum; see [48] for example.

15.3.3 Tensor-train

Tensor-train (TT) decomposition [49], also known as the matrix product state (MPS) [50] in quantum physics, expresses $\mathcal{X}$ as a product of third-order tensors $\mathcal{G}_1 \in \mathbb{R}^{R_1 \times I_1 \times R_2}, \cdots, \mathcal{G}_N \in \mathbb{R}^{R_N \times I_N \times R_{N+1}}$ (the cores of the decomposition), such that

$$\mathcal{X}(i_1, i_2, \cdots, i_N) = \underbrace{\mathcal{G}_1[i_1]}_{R_1 \times R_2} \times \underbrace{\mathcal{G}_2[i_2]}_{R_2 \times R_3} \times \cdots \times \underbrace{\mathcal{G}_N[i_N]}_{R_N \times R_{N+1}} .$$

The boundary conditions of the TT (open boundary conditions) decomposition dictate $R_1 = R_{N+1} = 1$. The decomposition is illustrated in Fig. 15.6. The open boundary conditions can be replaced by periodic boundary conditions (the tensor-ring) [51,52] which instead contracts together the first and last cores along R_1 and R_{N+1}.

Let us now depict the aforementioned tensor decompositions using tensor diagrams. While representing tensor decompositions visually (e.g., Figs. 15.4 and 15.5), we are inherently limited to a maximum of third-order tensors, tensor diagrams conveniently represent decompositions of higher-order tensors. For instance, let $\mathcal{X} \in \mathbb{R}^{I_1 \times I_2 \times I_3 \times I_4 \times I_5}$ be a fifth-order tensor. Its CP decomposition is shown in Fig. 15.7, its Tucker decomposition is shown in Fig. 15.8, and Fig. 15.9 depicts its TT decomposition.

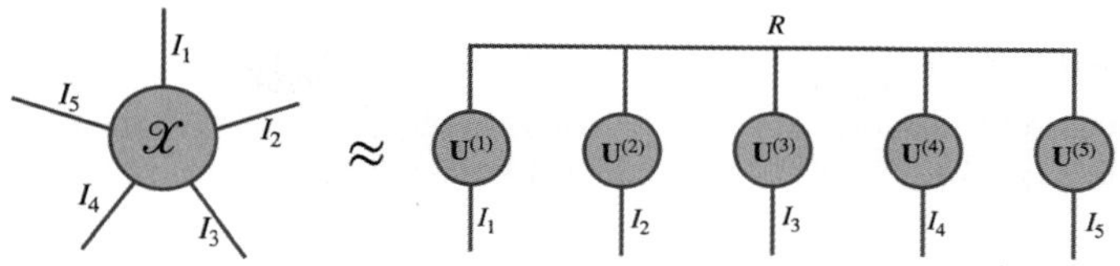

FIGURE 15.7

Representation of a CP decomposition using tensor diagrams.

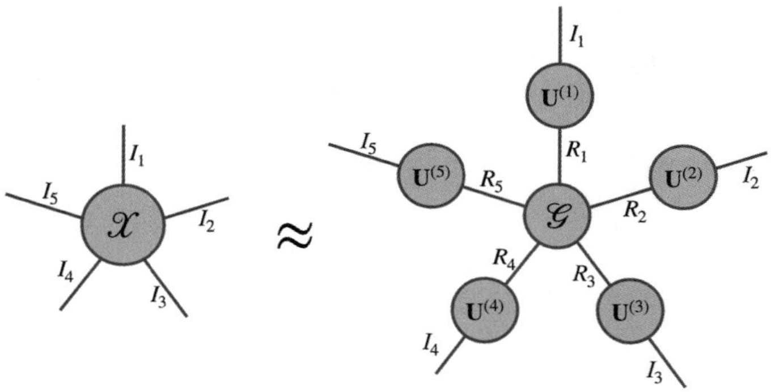

FIGURE 15.8

Representation of a Tucker decomposition using tensor diagrams.

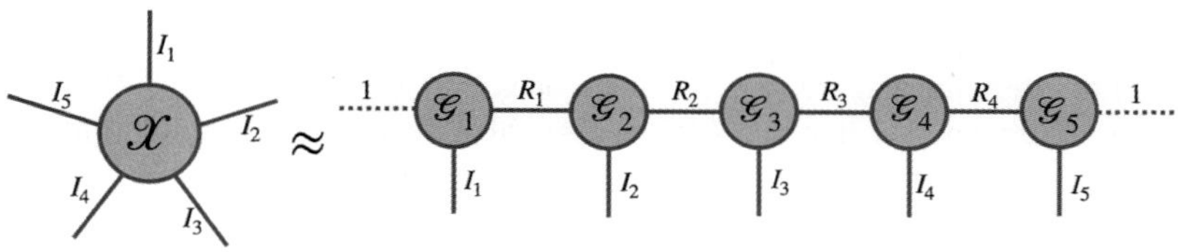

FIGURE 15.9

Representation of a tensor-train decomposition using tensor diagrams.

15.3.4 Tensor regression

Besides learning representations, low-rank tensor decomposition facilitates reducing the parameters of tensor regression models and prevents overfitting. Tensor regression models generalize linear regression to higher-order tensors. More specifically, in tensor regression a regression output $y \in \mathbb{R}$ is expressed as the inner product between the observation tensor $\mathcal{X}$ and a weight tensor $\mathcal{W}$ with the same dimension as $\mathcal{X}$: $y = \langle \mathcal{X}, \mathcal{W} \rangle + b$. Several low-rank tensor regression models have been proposed [9,53,54,55,56,57]. These methods share the assumption that the unknown regression weight tensor is low-rank. This is enforced by expressing the weight tensor in CP [9,54] or the Tucker form [58,56,59].

15.4 Tensor methods in deep learning architectures

Tensors play a central role in modern deep learning architectures. The basic building blocks of deep neural networks, such as multichannel convolutional kernels and attention blocks, are essentially tensor mappings, represented by tensors. Deep neural networks, as mentioned earlier, operate in a regime where the number of parameters is typically tens of millions or even billions, and is thus much larger than the number of training data. In this overparameterized regime, incorporating tensor decompositions into deep learning architectures leads to a significant reduction in the number of unknown parameters. This allows the design of networks that further preserve the topological structure in the data, being more parsimonious in terms of parameters, more data-efficient, and more robust to various types of noise and domain shifts. Besides the design of deep networks, tensor analysis can be beneficial in demystifying the success behind neural networks, having already been used to prove universal approximation properties of neural networks [60] and to understand the inductive bias leveraged by networks in computer vision [61]. This section overviews recent developments in the use of tensor methods to design efficient and robust deep neural networks and the theoretical understanding of their properties.

15.4.1 Parameterizing fully connected layers

Leveraging tensor decompositions within neural networks by reparameterizing fully connected layers was first proposed in [15]. Since the weights of fully connected layers are represented as matrices, tensor decompositions cannot be applied directly. Therefore, the weight matrix needs to be reshaped to obtain a higher-order tensor.

Specifically, consider an input matrix $\mathbf{W}$ of size $I \times J$, whose dimensions can be expressed as $I = I_1 \times I_2 \times \cdots \times I_N$ and $J = J_1 \times J_2 \times \cdots \times J_N$. Hence, $\mathbf{W}$ is tensorized first by reshaping it to a higher-order tensor of size $I_1 \times I_2 \times \cdots \times I_N \times J_1 \times J_2 \times \cdots \times J_N$. Next, by permuting the dimensions and reshaping it again, an Nth-order tensor of $I_1 J_1 \times I_2 J_2 \times \cdots \times I_N J_N$ is obtained. This tensor is then compressed in the TT format. The approximated weights are reconstructed from that low-rank TT factorization during inference and are then reshaped back into a matrix.

In this setting, other types of decompositions have also been explored, including tensor-ring [62] and block term decomposition [63]. This strategy has also been extended to parameterize other types of layers [64].

In the context of two-layer fully connected neural networks, tensors have also been used for deriving alternatives to (stochastic) gradient descent methods for training, with guaranteed generalization bounds. Specifically, in [65], CP decomposition is applied on the cross-moment tensor, which encodes the correlation between the third-order score function (i.e., the normalized derivative of the input probability density function) to obtain the parameters of the first layer. The parameters of the second layer are obtained next via regression. Under mild assumptions, this approach comes with guaranteed risk bounds in both the realizable setting (i.e., the target function can be approximated with zero error from the neural network under consideration) and the nonrealizable one, where the target function is arbitrary.

15.4.2 Tensor contraction layers

In modern deep neural networks such as CNNs, each layer's output is an (activation) tensor. However, activation tensors are usually flattened and passed to subsequent fully connected layers, resulting in structural information loss. A natural way to preserve the multilinear structure of activation tensors is to incorporate tensor operations into deep networks. For instance, tensor contraction can be applied to an activation tensor to obtain a low-dimensional representation of it [18]. In other words, an input activation tensor is contracted along each mode with a projection matrix (an n-mode product as defined earlier).

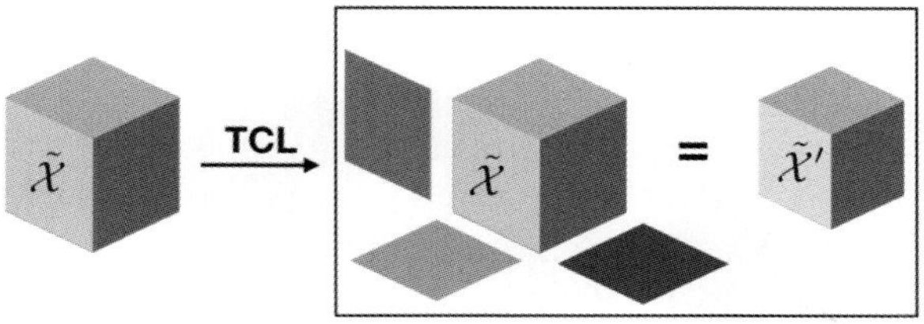

FIGURE 15.10

An input tensor $\mathcal{X}$ is contracted into a smaller tensor $\mathcal{X}'$ through a tensor contraction layer [18].

Let us assume a network involves an input activation tensor $\mathcal{X}$ of size $S \times I_1 \times \cdots \times I_N$, where S is the batch size. A tensor contraction layer (TCL) contracts the input activation tensor along each of its dimensions except the batch size with a series of factor matrices $\{\mathbf{V}^{(n)}\}_{n=2}^{N}$ (Fig. 15.10). The result is a compact core tensor $\mathcal{X}'$ of smaller size $S \times R_1 \times \cdots \times R_N$ defined as

$$\mathcal{X}' = \mathcal{X} \times_2 \mathbf{V}^{(2)} \times_3 \mathbf{V}^{(3)} \times \cdots \times_N \mathbf{V}^{(N)}, \tag{15.7}$$

with $\mathbf{V}^{(n)} \in \mathbb{R}^{R_n \times I_n}$, $n \in [2\,..\,N]$. Note that the projections start at the second mode as we do not contract along the first mode S, which corresponds to the minibatch size. These factors are learned in an end-to-end manner concurrently with all the other network parameters by gradient backpropagation. A TCL that produces a compact core of smaller size is denoted as size-$(R_2, \cdots, R_N)$ TCL, or TCL-$(R_2, \cdots, R_N)$.

In addition to preserving the multilinear structure in the input, TCLs have much fewer parameters than a corresponding fully connected layer. We can see this by establishing the link between a fully connected layer with structured weights and a tensor regression layer (TRL). Let us assume an input activation tensor $\mathcal{X}$ of size $S_1 \times I_1 \times \cdots \times I_N$. A size-$(R_2, \cdots, R_N)$ TCL parameterized by weight factors $\mathbf{V}^{(2)}, \cdots, \mathbf{V}^{(N)}$ has a total of $\sum_{n=2}^{N} I_n \times R_n$ parameters. By unfolding the input tensor (i.e., vectorizing each sample of the batch), TCL can be equivalently expressed as a fully connected layer parameterized by the weight matrix $\mathbf{W} = \left(\mathbf{V}^{(2)} \otimes \cdots \otimes \mathbf{V}^{(N)}\right)^{\top}$, which computes

$$\mathcal{X}_{[1]}\mathbf{W} = \mathcal{X}_{[1]}\left(\mathbf{V}^{(2)} \otimes \cdots \otimes \mathbf{V}^{(N)}\right)^{\top}.$$

However, since fully connected layers have no structure on their weights, the corresponding fully connected layer parameterized by a weight matrix $\mathbf{W}$ of the same size as above would have a total of $\prod_{n=2}^{N} I_n \times R_n$ parameters, compared to the $\sum_{n=2}^{N} I_n \times R_n$ parameters of the TCL.

15.4.3 Tensor regression layers

In CNNs, after an activation tensor has been obtained through a series of convolutional layers, predictions (outputs) are typically generated by flattening this activation tensor or applying a spatial pooling, before using a fully connected output layer. However, both approaches discard the multilinear structure of the activation tensor. To alleviate this, tensor regression models can be incorporated within deep neural networks and trained end-to-end jointly with the other parameters of the network [19].

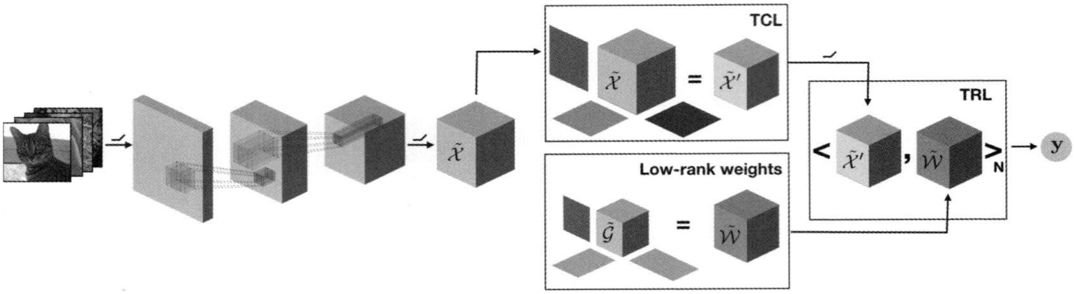

FIGURE 15.11

Illustration of a convolutional deep network where the activation tensor is processed through a TCL followed by a TRL to produce the outputs while preserving the topological structure [19]. Note that y, here, is a scalar.

Specifically, let us assume we have a set of S input activation tensors $\mathcal{X}^{(s)} \in \mathbb{R}^{I_1 \times I_2 \times \cdots \times I_N}$ and corresponding target labels $y^{(s)} \in \mathbb{R}$, with $s \in [1 \,..\, S]$. A TRL estimates the regression weight tensor $\mathcal{W} \in \mathbb{R}^{I_1 \times I_2 \cdots \times I_N}$, by further assuming that it admits a low-rank decomposition (Fig. 15.11). Originally, TRL has been introduced by assuming a Tucker structure on the regression weights. That is, for a rank-$(R_1, \cdots, R_N)$ Tucker decomposition, we have

$$y^{(s)} = \langle \mathcal{X}^{(s)}, \mathcal{W} \rangle + b$$
$$\text{with } \mathcal{W} = \mathcal{G} \times_1 \mathbf{U}^{(1)} \times_2 \mathbf{U}^{(2)} \cdots \times_N \mathbf{U}^{(N)} \tag{15.8}$$

with $\mathcal{G} \in \mathbb{R}^{R_1 \times \cdots \times R_N}$, $\mathbf{U}^{(n)} \in \mathbb{R}^{I_n \times R_n}$ for each n in $[1 \,..\, N]$ and $\mathbf{U}^{(N)} \in \mathbb{R}^{1 \times R_N}$. For simplicity, here, we introduce the problem for a scalar output, but it is established, in general, for a tensor-valued response. The inner product between the activation tensor and the regression weight is then replaced by a tensor contraction along matching modes.

In addition to the implicit regularization induced by the low-rank structure of the regression weights, TRLs result in a much smaller number of parameters. For an input activation tensor $\mathcal{X} \in \mathbb{R}^{S \times I_1 \times I_2 \times \cdots \times I_N}$, a rank-$(R_1, R_2, \cdots, R_N, R_{N+1})$ TRL, and a d-dimensional output, a fully connected layer taking the same input $\mathcal{X}$ (after a flattening layer) would have n_{FC} parameters, with $n_{\text{FC}} = d \times \prod_{n=1}^{N} I_n$. By comparison, a rank-$(R_1, R_2, \cdots, R_N, R_{N+1})$ TRL taking $\mathcal{X}$ as input has n_{TRL} parameters, with

$$n_{\text{TRL}} = \prod_{n=1}^{N+1} R_n + \sum_{n=1}^{N} R_n \times I_n + R_{N+1} \times d.$$

As previously observed for the TCL, TRL is much more parsimonious in terms of the number of parameters. Note that while we introduced the TRL here with a Tucker structure on the regression weight tensor, in general, any low-rank format can be used, including CP and TT [66].

15.4.4 Parameterizing convolutional layers

The weights of convolutional layers are naturally represented as tensors. For instance, a 2D convolution is represented as a fourth-order tensor. Hence convolutional layers can be directly compressed by employing tensor decomposition. In fact, there is a close link between deep neural networks and efficient convolutional blocks, which we will highlight in this section.

15.4.4.1 1 × 1 *convolutions*

The first thing to notice is that 1×1 convolutions are tensor contractions. This type of convolution is frequently used within deep neural networks in order to create a data bottleneck. For a 1×1 convolution Φ, defined by kernel $\mathcal{W} \in \mathbb{R}^{T \times C \times 1 \times 1}$ and applied to an activation tensor $\mathcal{X} \in \mathbb{R}^{C \times H \times W}$, we denote the squeezed version of $\mathcal{W}$ along the first mode as $\mathbf{W} \in \mathbb{R}^{T \times C}$. We then have

$$\Phi(\mathcal{X})_{t,y,x} = \mathcal{X} \star \mathcal{W} = \sum_{k=1}^{C} \mathcal{W}_{t,k,y,x} \mathcal{X}_{k,y,x} = \mathcal{X} \times_1 \mathbf{W}.$$

Note that, here, we have $x = y = 1$ as we are considering a 1×1 convolution.

15.4.4.2 *Kruskal convolutions*

Let us now focus on how tensor decomposition can be applied to the convolutional kernels of CNNs and how this results not only in fewer parameters but also in faster operations. CP decomposition, for instance, allows separating the modes of a convolutional kernel, resulting in a Kruskal form. Although the application of tensor decomposition in the context of CNNs is relatively new and of practical interest nowadays, the idea of separable representation of linear operators by means of tensor decomposition is not new (see [67]) and in fact was one of the original motivations for tensor rank decomposition of the matrix multiplication tensor (see [68] for example).

In computer vision, the use of separable convolutions was proposed in [69], in the context of learnable filter banks. In the context of deep learning, [70] proposed to leverage redundancies across channels by exploiting separable convolutions. Furthermore, it is possible to start from a pretrained convolutional kernel and apply CP decomposition to it in order to obtain a separable convolution. This was proposed for 2D convolutions in [16], where CP decomposition of the kernel was achieved by minimizing the reconstruction error between the pretrained weights and the corresponding CP approximation. The authors demonstrated both space savings and computational speedups. Similar results can be obtained using different optimization strategies such as the tensor power method [17].

An advantage of factorizing the kernel using CP decomposition is that the resulting factors can be used to parameterize a series of efficient depthwise separable convolutions, replacing the original convolution [16], as illustrated in Fig. 15.12. This can be seen by considering the expression of a regular convolution:

$$\mathcal{F}_{t,y,x} = \sum_{k=1}^{C} \sum_{j=1}^{H} \sum_{i=1}^{W} \mathcal{W}(t, k, j, i) \mathcal{X}(k, j + y, i + x). \tag{15.9}$$

A Kruskal convolution is obtained by expanding the kernel $\mathcal{W}$ in the CP form as introduced in Section 15.3.1. By reordering the terms, we can then obtain an efficient reparameterization [16]:

$$\mathcal{F}_{t,y,x} = \underbrace{\sum_{r=1}^{R} \mathbf{U}^{(T)}_{t,r} \underbrace{\left[\sum_{i=1}^{W} \mathbf{U}^{(W)}_{i,r} \underbrace{\left(\sum_{j=1}^{H} \mathbf{U}^{(H)}_{j,r} \underbrace{\left[\sum_{k=1}^{C} \mathbf{U}^{(C)}_{k,r} \mathcal{X}(k, j + y, i + x) \right]}_{1\times 1 \text{ conv}} \right)}_{\text{depthwise conv}} \right]}_{\text{depthwise conv}}}_{1\times 1 \text{ convolution}}. \tag{15.10}$$

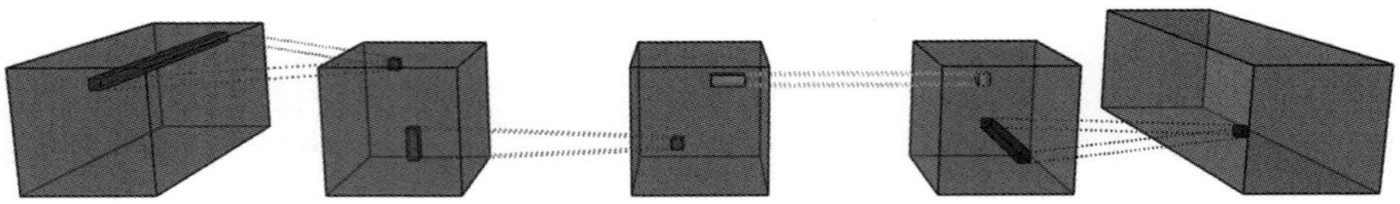

FIGURE 15.12

Illustration of a 2D Kruskal convolution, as expressed in Eq. (15.10), with matching colors. First, a 1×1 convolution reduces the number of input channels to the rank (blue (dark gray in print version)). Two depthwise convolutions are then applied on the height and width of the activation tensor (red (mid gray in print version) and green (light gray in print version)). Finally, a second 1×1 convolution restores the number of channels from the rank of the CP decomposition to the desired number of output channels (black).

15.4.4.3 Tucker convolutions

As previously, we consider the convolution $\mathcal{F} = \mathcal{X} \star \mathcal{W}$ described in (15.9). However, instead of a Kruskal structure, the convolution kernel $\mathcal{W}$ is assumed to admit a Tucker (Fig. 15.13) decomposition as follows:

$$\mathcal{W}(t, s, j, i) = \sum_{r_1=0}^{R_1} \sum_{r_2=0}^{R_2} \sum_{r_3=0}^{R_3} \sum_{r_4=0}^{R_4} \mathcal{G}_{r_1,r_2,r_3,r_4} \mathbf{U}^{(T)}_{t,r_1} \mathbf{U}^{(C)}_{s,r_2} \mathbf{U}^{(H)}_{j,r_3} \mathbf{U}^{(W)}_{i,r_4}.$$

This approach allows for an efficient reformulation [71]: first, the factors along the spatial dimensions are absorbed into the core by writing $\mathcal{H} = \mathcal{G} \times_3 \mathbf{U}^{(H)}_{j,r_3} \times_4 \mathbf{U}^{(W)}_{i,r_4}$. By reordering the term, we can now see that a Tucker convolution is equivalent to first transforming the number of channels and then applying a (small) convolution before restoring the channel dimension from the rank to the target number of channels:

$$\mathcal{F}_{t,y,x} = \underbrace{\sum_{r_1=1}^{R_1} \mathbf{U}^{(T)}_{t,r_1} \left[\underbrace{\sum_{j=1}^{H}\sum_{i=1}^{W}\sum_{r_2=1}^{R_2} \mathcal{H}_{r_1,r_2,j,i} \left[\underbrace{\sum_{k=1}^{C} \mathbf{U}^{(C)}_{k,r_2} \mathcal{X}(k, j+y, i+x)}_{1\times 1 \text{ conv}} \right]}_{H\times W \text{ conv}} \right]}_{1\times 1 \text{ conv}} . \tag{15.11}$$

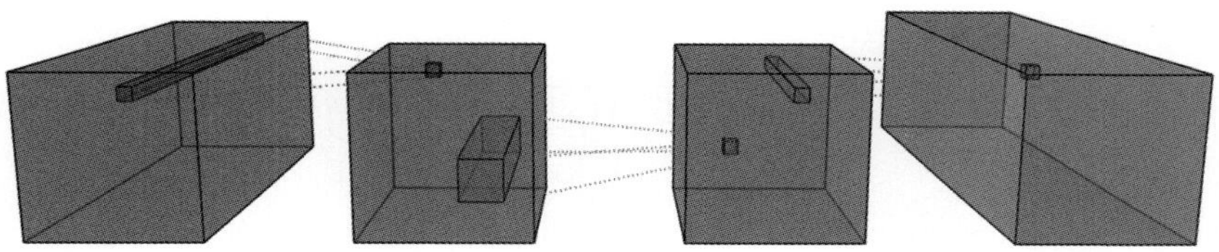

FIGURE 15.13

Illustration of a Tucker convolution expressed as a series of small, efficient convolutions, as illustrated in Eq. (15.11). Note that this is the approach taken by ResNet for the bottleneck blocks. After the full kernel has been decomposed, the factors along the input and output channels are used to parameterize 1×1 convolutions, applied respectively first (blue (dark gray in print version)) and last (green (mid gray in print version)). The remaining two factors are absorbed into the core and used to parameterize a regular 2D convolution (red (mid gray in print version)).

Besides the aforementioned types of convolutions (i.e., bottleneck layers, separable convolutions), [72] identified maximally resource-efficient convolutional layers utilizing tensor networks and decompositions. Tensor decompositions have also been used for unifying different successful architectures. Particularly, in [23], the block term decomposition [73] is used to unify different architectures that employ residual connections, such as ResNet [74], while proposing a new architecture that improves parameter efficiency. This implies that searching over the space of tensor decompositions and their parameters might be an excellent proxy task to limit the search space for architecture search.

From a theoretical point of view, using CP decomposition to replace convolutional with low-rank layers results in deriving the upper bound for the generalization error of the compressed network [75]. This bound improves previous results on the compressibility and generalizability of neural networks [76]. It is also related to the layers' rank, providing insights into the design of neural networks (with low-rank tensors as layers).

15.4.4.4 Multistage compression

The factorization strategies mentioned above allow for efficient compression of pretrained, fully connected, and convolutional layers of deep neural networks by applying tensor decomposition to the learned weights. Fine-tuning allows recovery for loss of performance when increasing the compression ratio. However, it is also possible to apply multistate compression [77] to further compress without loss of performance. Instead of compressing and then fine-tuning, the idea is to adopt an iterative approach by alternating low-rank factorization with rank selection and fine-tuning.

15.4.4.5 General N-dimensional separable convolutions and transduction

The design of general (N-dimensional) fully separable convolution has been introduced in the context of deep neural networks in [21]. Let us assume an $(N+1)$th-order input activation tensor $\mathcal{X} \in \mathbb{R}^{C \times D_1 \times \cdots \times D_N}$ corresponding to N dimensions with C channels. The higher-order separable convolution, which we denote as Φ, is defined by a kernel $\mathcal{W} \in \mathbb{R}^{T \times C \times K_1 \times \cdots \times K_N}$. By expressing this kernel in a factorized Kruskal form as $\mathcal{W} = [\![\lambda;\ \mathbf{U}^{(T)}, \mathbf{U}^{(C)}, \mathbf{U}^{(K_1)}, \cdots, \mathbf{U}^{(K_N)}]\!]$, we obtain

$$\Phi(\mathcal{X})_{t,i_1,\cdots,i_N} = \sum_{r=1}^{R} \sum_{s=1}^{C} \sum_{i_1=1}^{K_1} \cdots$$

$$\cdots \sum_{i_N=1}^{K_N} \lambda_r \left[\mathbf{U}^{(T)}_{t,r} \mathbf{U}^{(C)}_{s,r} \mathbf{U}^{(K_1)}_{i_1,r} \cdots \mathbf{U}^{(K_N)}_{i_N,r} \mathcal{X}_{s,i_1,\cdots,i_N}\right].$$

The separable higher-order convolution can be obtained by reordering the term into

$$\mathcal{F} = \left(\rho\left(\mathcal{X} \times_1 \mathbf{U}^{(T)}\right)\right) \times_1 \left(\operatorname{diag}(\lambda)\mathbf{U}^{(C)}\right), \tag{15.12}$$

where ρ applies the 1D spatial convolutions:

$$\rho(\mathcal{X}) = \left(\mathcal{X} \star_1 \mathbf{U}^{(K_1)} \star_2 \mathbf{U}^{(K_2)} \star \cdots \star_N \mathbf{U}^{(K_N)}\right). \tag{15.13}$$

A factorized separable convolution can be extended from N dimensions to $N+K$ ($N, K \in \mathbb{N}$) by means of transduction [21]. Since the convolution is fully separable, it is possible to add additional factors. In particular, to apply transduction from N to $N+1$, it is sufficient to add an additional factor $\mathbf{U}^{K_{N+1}}$ and update ρ in Eq. (15.13) to

$$\rho(\mathcal{X}) = \left(\mathcal{X} \star_1 \mathbf{U}^{(K_1)} \star_2 \mathbf{U}^{(K_2)} \star \cdots \star_N \mathbf{U}^{(K_N)} \star_{N+1} \mathbf{U}^{(K_{N+1})}\right).$$

15.4.4.6 Parameterizing full networks

So far, we reviewed methods focusing on parameterizing a single layer. It is also possible to jointly parameterize a whole CNN with a single low-rank tensor. This is the idea of "T-Net" [20], where all convolutional layers of a stacked hourglass [78] (a type of U-Net [79]) are parameterized by a single eighth-order tensor. Each of the modes of this tensor models a modality of the network, allowing to jointly model and regularize all the layers. Hence correlations across channels, layers, convolutional blocks, and subnetworks of the model can be leveraged. Imposing low-rank constraints on this model tensor results in large parameter space savings with little to no impact on performance, or even improved performance, depending on the compression ratio. That is, small compression ratios result in improved performance, while it is possible to reach substantial compression ratios with only minor decreases in performance. In addition, it is possible to partially reconstruct the full tensor to obtain a Tucker factorization of the kernel of each layer, thus allowing for speedups as described in Section 15.4.4.3.

15.4.4.7 Parameterizing recurrent neural networks

Higher-order structures occur naturally when we try to generalize existing architectures. One example is long short-term memory (LSTM), which is based on Markovian models that traditionally incorporate the previous time step through a transition matrix. Extending the history to consider more time steps and their interactions results in a higher-order tensor, allowing to model non-Markovian dynamics. While a naive version would result in an exponential number of parameters, it can be made tractable by using a low-rank structure on this higher-order tensor, such as TT [80]. To further improve the efficiency of the model and enable efficient learning of convolutional LSTM in higher orders while preserving the spatial structure, a convolutional TT decomposition has been successfully used [81].

15.4.4.8 Preventing degeneracy in factorized convolutions

As discussed above, when parameterizing DCNNs, one can either train the factorized layers from scratch in an end-to-end fashion via stochastic gradient descent or decompose existing convolutional kernels to obtain the factorized version. In the latter case, the authors of [82] showed that when applying decomposition to obtain a factorized convolution, degeneracy can appear in the decomposition due to diverging components. They tackle this issue and enable stable factorization of convolutional layers through sensitivity minimization.

15.4.5 Tensor structures in polynomial networks and attention mechanisms

Apart from CNNs, tensor structures arise in polynomial networks and attention mechanisms, capturing higher-order (e.g., multiplicative) interactions among data or parameters. Here, we provide an overview of how tensor methods are incorporated within such models.

Multiplicative interactions and polynomial function approximation have long been used in the domain of machine learning. The Group Method of Data Handling (GMDH) [83] proposes building quadratic polynomials from input elements. The elements in each quadratic polynomial are predetermined, which does not allow the method to capture data-driven correlations among different elements. The method was later extended to higher-degree polynomials [84]. Low-degree polynomial networks were also used in pi-sigma [85] and sigma-pi-sigma networks [86]. The pi-sigma network includes a single hidden layer, and the output is the product of all hidden representations with fixed weights equal to 1. The idea of sigma-pi-sigma is to sum multiple pi-sigma networks. In the context of kernel methods, factorization machines [87] and their higher-order extension [88] model both low- and higher-order interactions between variables, using factorized parameters via matrix/tensor decompositions. By employing factorized parameters, such models are able to estimate higher-order interactions even in tasks where data are sparse (e.g., recommender systems), unlike other kernel methods such as support vector machines.

More recently, deeper neural networks have been proposed to capture multiplicative interactions [89,90]. For instance, highway networks [89] introduce an architecture that enables second-order interactions between the inputs. In [91,92,93], the authors augment the original convolution with a multiplicative interaction to increase the representational power. In [90], the authors group several such multiplicative interactions as a second-degree polynomial and prove that the second-degree polynomials extend the class of functions that can be represented with zero error.

Recent empirical [94,95] and theoretical [90,96] results indicate that multivariate higher-degree polynomials expand the classes of functions that can be approximated with deep neural networks. Approximating a function using a multivariate higher-degree polynomial involves learning the multivariate polynomial expansion parameters using input–output training data. Higher-order tensors naturally represent the parameters of such a multivariate polynomial [97], whose dimensionality increases exponentially to the degree of the polynomial expansion. This is another instance of the curse of dimensionality that tensor decompositions can mitigate. Moreover, in this setting, choosing different tensor decompositions results in different recursive equations that compose the polynomial function approximation and are hence linked to different neural network architectures.

Particularly, let $z \in \mathbb{R}^d$ be the input, e.g., the latent noise in a generative model or an image in a discriminative model, and let $\boldsymbol{x} \in \mathbb{R}^o$ be the target, e.g., an output image or a class label. A polynomial expansion[3] of the input z can be considered for approximating the target $\boldsymbol{x}$. That is, a vector-valued function $\boldsymbol{G}(z) : \mathbb{R}^d \to \mathbb{R}^o$ expresses the high-degree multivariate polynomial expansion:

$$\boldsymbol{x} = G(z) = \sum_{n=1}^{N} \left(\mathcal{W}^{[n]} \prod_{j=2}^{n+1} \times_j z \right) + \boldsymbol{\beta}, \tag{15.14}$$

where $\boldsymbol{\beta} \in \mathbb{R}^o$ and $\left\{ \mathcal{W}^{[n]} \in \mathbb{R}^{o \times \prod_{m=1}^{n} \times_m d} \right\}_{n=1}^{N}$ are the learnable parameters. The form of (15.14) can approximate any smooth function (for large N). As aforementioned the number of parameters required to accommodate all higher-order interactions of the input increases exponentially with the desired degree of the polynomial, which is impractical for high-dimensional data. To make this model practical, its parameters can be reduced by (1) making low-rank assumptions and (2) sharing factors between the decompositions of each parameter tensor, which amounts to jointly factorizing all the parameter tensors $\mathcal{W}^{[n]}$. To illustrate this, a certain form of CP decomposition with shared factors among low-rank decompositions of weight tensors [95] is selected. Concretely, the parameter tensor $\mathcal{W}^{[n]}$ is factorized using n factor matrices. The first and last factor matrices (called $\boldsymbol{U}_{[1]}$ and $\boldsymbol{C}$, respectively) are shared across all parameter tensors. Then, depending on the interactions between the layers we want to forge, we can share the corresponding factor matrices. That results (see [95,100] for a detailed derivation) in a simple recursive relationship, which can be expressed as

$$\boldsymbol{x}_n = \left(\boldsymbol{U}_{[n]}^T z \right) * \boldsymbol{x}_{n-1} + \boldsymbol{x}_{n-1}, \tag{15.15}$$

for $n = 2, \ldots, N$ with $\boldsymbol{x}_1 = \boldsymbol{U}_{[1]}^T z$ and $\boldsymbol{x} = \boldsymbol{C}\boldsymbol{x}_N + \boldsymbol{\beta}$. The parameters $\boldsymbol{C} \in \mathbb{R}^{o \times k}$ and $\boldsymbol{U}_{[n]} \in \mathbb{R}^{d \times k}$ for $n = 1, \ldots, N$ are learnable with $k \in \mathbb{N}$ being the rank of the CP decomposition.

Each recursive call in (15.15) expands the approximation degree by one. Then, the final output $\boldsymbol{x}$ is a polynomial of degree N. The recursive relationship of (15.15) enables us to implement the polynomial using a hierarchical neural network, where the approximation degree defines the depth of the network. A schematic assuming a third-degree expansion ($N = 3$) is illustrated in Fig. 15.14. The tensor decomposition selected and the sharing of the factors have a crucial role in the resulting architecture. As aforementioned, by selecting different decompositions, new architectures can emerge [95], while additional constraints can be placed on the structure (e.g., shift-invariance) by allowing factor matrices to implement convolutions (e.g., via the im2col operator or a circulant operator).

---

[3] The Stone–Weierstrass theorem [98] guarantees that any smooth function can be approximated by a polynomial. Multivariate function approximation is covered by an extension of the Weierstrass theorem, e.g., in [99] (p. 19).

Higher-degree polynomials can be represented as a product of lower-degree polynomials, which have been extensively studied for both univariate and multivariate cases. Berlekamp's algorithm [101] was one of the first to successfully factor univariate polynomials over finite fields; the Cantor–Zassenhaus algorithm [102] has since become more popular. Factorizing multivariate polynomials as a linear combination of univariate polynomials has recently been studied using tensor decompositions [103]. The work of [104] demonstrates that the aforementioned factorization of multivariate polynomials is related to the joint CP decomposition.

In deep polynomial networks, instead of using a single polynomial, also products of polynomials can be used [95]. That is, the higher-degree multivariate polynomial function is expressed as a product of low-degree multivariate polynomials. This approach enforces variable separability in two levels: first in the parameters of low-degree polynomials via tensor decomposition as aforementioned and second by allowing approximation as a product of simpler functions. The product of polynomials increases the expansion degree without increasing the number of layers significantly.

The aforementioned expansions have been utilized in nonadversarial generation [105] and conditional generation [106], where multiple input variables are defined. An upper bound on the Lipschitz constant on this class of models has been derived in [107], while [108] proved that when multiplicative interactions are added to the network, it can learn higher-frequency functions faster.

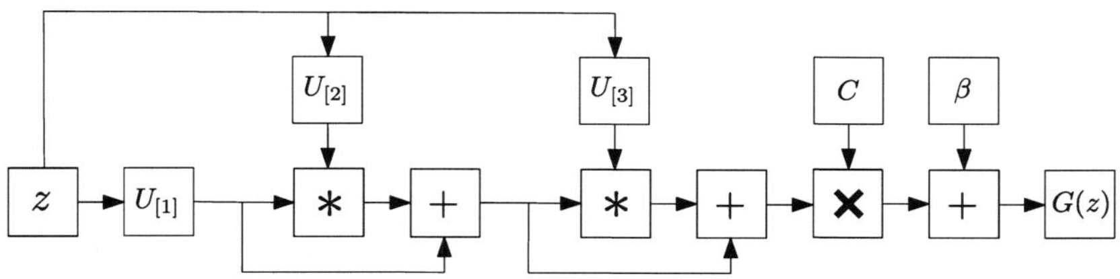

FIGURE 15.14

Schematic illustration of the polynomial expansion for third-degree approximation of (15.15).

Furthermore, the original attention mechanism [109] is focused on neural machine translation, yet it expresses multiplicative interactions. The idea is to capture long-range dependencies that might exist in a temporal sequence. This was later extended to attention mechanisms for image and video data [110]. Multihead attention learns different matrices per head; concatenating those matrices, a third-order tensor is obtained. The authors utilize a CP decomposition to reduce both the parameters and the redundancy in the learned matrices [111]. In [112], the authors extend self-attention to capture not only inter- but also intrachannel correlations by constructing an appropriate tensor such that the variance of spatial- and channelwise features is regularized.

15.4.6 Tensor structures in generative adversarial networks

In addition to their utility for discriminative tasks, tensor methods have also been shown to be useful for generative tasks [113,114,115,116,117], particularly for GANs [25]. For example, multilinear factorizations have been applied to the GAN latent or feature space, providing further insights and enabling a new set of downstream tasks. Some recent advances in this direction are detailed below.

15.4.6.1 *Multilinear factorization of latent space*

Modern image generators map a latent code $\mathbf{z} \sim \mathcal{N}(\mathbf{0}, \mathbf{I})$ to a synthetic image through a sequence of 2D transpose convolutional layers. Thus, the intermediate representations of a batch of images are fourth-order tensors $\mathcal{Z} \in \mathbb{R}^{N \times H \times W \times C}$, each sample of which has height, width, and channel modes. The style of the image at each layer is often modulated with AdaIN [94,118] via a location-scale transformation parameterized by latent "style code" $\mathbf{w} \in \mathbb{R}^d$. Latent representations of GANs have been shown to contain rich semantic structure [119], wherein interpretable directions can be found that affect high-level changes to the semantic content of the synthetic images [120,121,122,123,124,125,126,127]. Recent work has found that tensor-based factorization can be used to uncover such semantic directions in pre-trained GAN generators. These methods can be considered to fall into one of two categories – those finding directions in the latent space of style codes [115,116] and those operating on the convolutional feature maps [128,117]. Interestingly enough, all the aforementioned methods can be viewed as employing Tucker style decompositions of various forms on two different stages of intermediate generator representations. They are summarized in Table 15.2 and detailed below.

Table 15.2 Summary of recent methods using Tucker style decompositions to locate interpretable directions in pretrained GANs. $\mathcal{G}_i$ are the method-specific core tensors.

Method	Latent space	Decomposition	Generator features factorized	Additional constraints
τGAN [115,116]	Style codes	$[\![\mathcal{G}_1; \mathbf{U}^{(d)}, \mathbf{U}^{(P_1)}, \ldots, \mathbf{U}^{(P_k)}]\!]$	$\mathcal{T} \in \mathbb{R}^{d \times P_1 \times \cdots \times P_k}$	None
TCA [117]	Feature maps	$[\![\mathcal{G}_2; \mathbf{I}_N, \mathbf{U}^{(H)}, \mathbf{U}^{(W)}, \mathbf{U}^{(C)}]\!]$	$\mathcal{Z} \in \mathbb{R}^{N \times H \times W \times C}$	None
PandA [128]	Feature maps	$[\![\mathcal{G}_3; \mathbf{I}_N, \mathbf{U}^{(C)}, \mathbf{U}^{(S)}]\!]$	$\mathcal{Z} \in \mathbb{R}^{N \times C \times H \cdot W}$	$\mathbf{U}^{(S)} \geq \mathbf{0}$

15.4.6.1.1 Style code factorization

Operating on style codes $\mathbf{w} \in \mathbb{R}^d$, the authors of [115,116] propose embedding [129] a structured dataset of faces in the target modes of variation into the latent space. For example, given images of faces with P identities, E expressions, and R viewpoints, a structured tensor of style codes $\mathcal{T} \in \mathbb{R}^{d \times P \times E \times R}$ is formed. The HOSVD [130] is then applied (assuming zero-mean $\mathcal{T}$) to learn a basis $\mathbf{U}^{(n)}$ for each of the subspaces $\mathcal{T} \approx \mathcal{G} \times_1 \mathbf{U}^{(d)} \times_2 \mathbf{U}^{(E)} \times_3 \mathbf{U}^{(P)} \times_4 \mathbf{U}^{(R)}$, where $\mathcal{G} \in \mathbb{R}^{d' \times E' \times P' \times R'}$ is the core tensor. The authors show that interpretable directions corresponding to a target mode of variation n can be generated from the rows of the respective factor matrix $\mathbf{U}^{(n)}$, producing compelling results for emotion editing [116] and attribute transfer [115].

15.4.6.1.2 Feature map factorization

Tensor methods have also proven to be useful for finding interpretable directions in the intermediate convolutional features directly [117,128]. Given N samples' intermediate generator feature maps $\mathcal{Z} \in \mathbb{R}^{N \times H \times W \times C}$, a basis for the subspace spanned by each mode's fibers $\mathbf{U}^{(H)}, \mathbf{U}^{(W)}, \mathbf{U}^{(C)}$ is learned in [117] to better separate style-based variation by applying a Tucker decomposition [131] $\mathcal{Z} \approx \mathcal{Z} \times_2 \mathbf{U}^{(H)}\mathbf{U}^{(H)\top} \times_3 \mathbf{U}^{(W)}\mathbf{U}^{(W)\top} \times_4 \mathbf{U}^{(C)}\mathbf{U}^{(C)\top}$, where the columns of $\mathbf{U}^{(C)}$ are found to correspond to high-level attributes relating to the style of the synthetic images [117] (e.g., "hair color"). Motivated by this finding, the authors of [128] further consider a joint seminonnegative factorization of a dataset of images' *parts* in addition to their appearances. Where [117] can be seen as a multilinear decomposition of the raw activation tensors $\mathcal{Z} \in \mathbb{R}^{N\times H\times W\times C}$, [128] instead factorizes the N feature maps' channel-mode unfoldings $\mathbf{Z}_{i(3)} \in \mathbb{R}^{C\times S}$ with $S \triangleq H \cdot W$. The former imposes a strict separable structure between the two spatial modes, whilst the latter combines the height and width modes to learn factors that can better span semantic parts of an image. Nonnegative parts' factors $\mathbf{P} \in \mathbb{R}^{S\times R_S}$ are learned for the combined spatial mode, which control *whereabouts* in the feature maps the various learned appearance factors $\mathbf{A} \in \mathbb{R}^{C\times R_C}$ are present through the factors' multiplicative interactions [90]. This is shown schematically in Fig. 15.15.

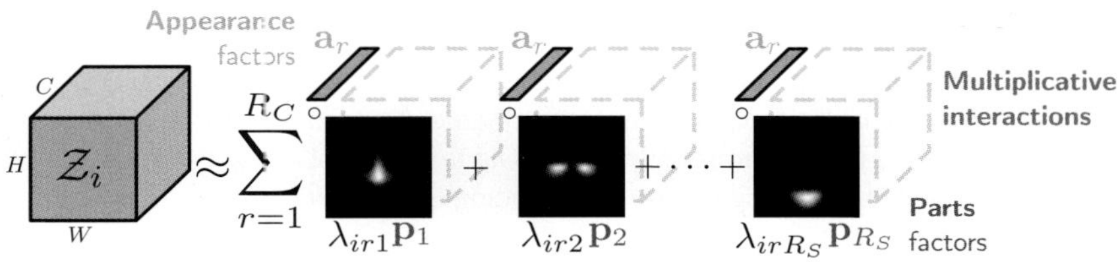

FIGURE 15.15

An overview of PandA [128]. A seminonnegative Tucker decomposition is applied to a dataset of generator activations $\mathcal{Z}_i \in \mathbb{R}^{H\times W\times C}$. Each factor has an intuitive interpretation: the factors for the spatial modes $\mathbf{p}_j$ control the parts, determining at *which spatial locations* in the feature maps the various appearances $\mathbf{a}_k$ are present, through their multiplicative interactions.

Concretely, let $\hat{\mathcal{Z}} \in \mathbb{R}^{N\times C\times S}$ be a batch of the mode-3 unfoldings of the N samples' convolutional features (i.e., with the spatial modes combined). The following constrained optimization problem is posed:

$$\min_{\mathbf{A},\mathbf{P}\geq\mathbf{0}} \sum_{i=1}^{N} ||\hat{\mathbf{Z}}_i - \mathbf{A}\left(\mathbf{A}^\top \hat{\mathbf{Z}}_i \mathbf{P}\right)\mathbf{P}^\top||_F^2, \tag{15.16}$$

$$= \min_{\mathbf{A},\mathbf{P}\geq\mathbf{0}} \sum_{i=1}^{N} ||(\hat{\mathcal{Z}} - \hat{\mathcal{Z}} \times_2 \mathbf{A}\mathbf{A}^\top \times_3 \mathbf{P}\mathbf{P}^\top)_i||_F^2, \tag{15.17}$$

which we rewrite here in Eq. (15.17) with the mode-n product to make the connection to tensor decompositions explicit. We see that this is a special form of seminonnegative Tucker decomposition which can also be expressed concisely as $\hat{\mathcal{Z}} \approx [\![(\hat{\mathcal{Z}} \times_2 \mathbf{A}^\top \times_3 \mathbf{P}^\top); \mathbf{I}_N, \mathbf{A}, \mathbf{P}]\!]$ using the notation introduced in [132].

The learned factors for parts and appearances can be used for a range of tasks. For example, the change of basis $\mathbf{A}^\top \mathbf{Z}_i$ specifies how much of each appearance vector is present at each spatial location, localizing the learned appearance concepts in the images. Additionally, local image editing with appearance $\mathbf{a}_j$ can be achieved at a target part $\mathbf{p}_k$ in the image by adding the rank-one matrix $\mathbf{a}_j \circ \mathbf{p}_k$ (refolded back to a third-order tensor) to the original feature maps.

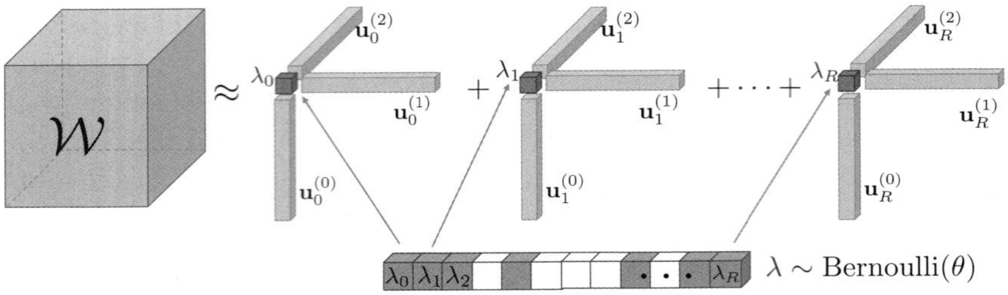

FIGURE 15.16

Tensor dropout on a CP decomposition. Visualization of tensor dropout applied to a CP factorization. Here, a Bernoulli dropout is applied by first sampling a multivariate Bernoulli random vector λ. Its elements have value 1 (red (dark gray in print version)) with probability θ and 0 (white) with probability $1-\theta$. Each rank-one CP component is kept or discarded according to the corresponding Bernoulli variable, thus inducing stochasticity on the rank of the decomposition.

15.4.7 Robustness of deep networks

Deep neural networks produce remarkable results in computer vision applications but are typically very vulnerable to noise, such as capture noise, corruption, and adversarial attacks of domain shift. Below, we provide a brief review of tensor methods that robustify deep networks.

15.4.7.1 Tensor dropout

One known way to increase robustness is to incorporate randomization during training. For example, dropout [133] randomly drops activations, while DropConnect [134] applies the randomization to the weights of fully connected layers. However, randomizing directly in the original space results in sparse weights or activations (with zeroed values), which affects the high-order statistics.

Tensor dropout [27] is a principled way to incorporate randomization to robustify the model without altering its statistics. It consists of applying sketching matrices in the latent subspace spanned by a tensor decomposition, either CP (see Fig. 15.16) or Tucker (see Fig. 15.17), leading to significantly improved robustness to random noise (e.g., during capture) or adversarial attacks.

15.4.7.2 Defensive tensorization

By combining the reparameterization approach described in Section 15.4.4.5 with tensor dropout, one can drastically improve the robustness of a DCNN to adversarial attacks. When combined with adversarial training, this approach proved successful against a wide range of attacks [135].

15.4.7.3 Tensor-Shield

In addition to the line of research that consists in modifying the weights or activations of the network to increase robustness to adversarial attacks, it is possible to modify the input image directly. For instance, Shield [136] uses JPEG encoding to remove any potential adversarial noise. However, the defensive capacity of this approach is limited. Instead of JPEG encoding, in [137] tensor decomposition techniques are utilized as a preprocessing step to find a low-rank approximation of images which can significantly discard high-frequency perturbations.

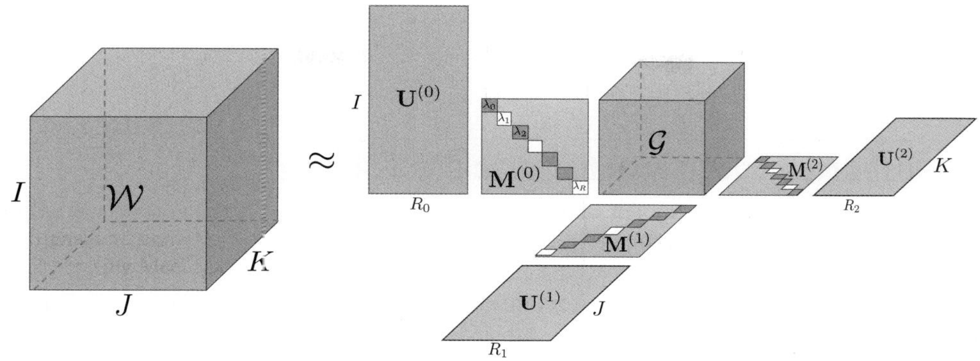

FIGURE 15.17

Tensor dropout on a Tucker decomposition. Visualization of tensor dropout applied to a Tucker decomposition. Several types of randomization can be used. Here, a Bernoulli dropout is represented, applied in the latent subspace of the decomposition, by contracting the core along each mode with a series of (diagonal) Bernoulli sketching matrices, $\mathbf{M}^{(k)} = \text{diag}(\lambda^{(k)})$.

15.4.8 Applications

Tensor-based deep learning architectures have found a wide range of applications, some of which have been mentioned already. Some additional applications, mostly related to the technical content of the chapter, are discussed next. Due to the prevalence of deep learning methods in computer vision, many have been evaluated in related applications, in terms of both generative modeling and predictive analysis (e.g., image classification). This has also led to rigorous evaluations in terms of performance and effectiveness in large-scale standard benchmark datasets such as ImageNet, applications to large-scale problems in earth observation such as land-cover classification [138], and many other related tasks. Furthermore, CNNs with tensor regression layers have been used for a multitude of applications, including affect estimation from faces [139,21], while randomized TRL has been utilized for age prediction from MRI scans [27], outperforming compared methods. A clear example of a seminal lightweight architecture that facilitates mobile and embedded applications are the Mobilenet V1 [140] and Mobilenet V2 [141] architectures, which can be seen as special cases of Kruskal and Tucker convolutions (Section 15.4.4).

Integrating tensors in deep learning architectures, for example with low-rank tensor factorization, can lead to many advantageous properties such as robustness and regularization, efficiency, and parameter reduction. Furthermore, one can straightforwardly leverage the structure induced by a factorization to adapt a model to different domains or facilitate information sharing between multiple tasks. For example, Tucker decomposition leads to a core tensor that represents a latent subspace which we can project from and onto. In this way, applications that include cross-task and cross-domain incremental learning are enabled, usually by facilitating weight sharing through joint factorization [22] or sharing residual units [23]. Approaches that propose learning a task-agnostic core that can be shared between tasks and domains have also been proposed. In more detail, the shared subspace represented by the core tensor can be specialized to new tasks and domains via task-specific projections, leading to much higher efficiency in computer vision [24] and natural language processing tasks [142].

15.5 Tensor methods in quantum machine learning

QML seeks to demonstrate a significant improvement over classical machine and deep learning methods, leading to a quantum advantage. Similarly to the vast majority of optimization algorithms used in deep learning, QML also employs gradient-based optimization, for example to optimize quantum networks. QC is though still at its infancy, and hence quantum simulation remains essential. Simulations on informative scales nevertheless come with a significant computational cost, which overwhelms classical computing resources very quickly. In a similar fashion to deep learning, quantum tensor operations can leverage GPU acceleration. For example, approaches such as networks of factorized tensors have led to an exponential reduction in computational overhead. In this section, we provide an overview of recent research in this direction.

15.5.1 Quantum mechanics

The behavior of small-scale particles is not likely to be intuitive for the macroscopic reader. Quantum mechanics, the branch of physics that describes the dynamics of electrons, photons, atoms, and other microscopic objects, is ripe with bizarre features that are quite distinct from human perception of reality. One such characteristic of quantum mechanics is its affinity for discrete quantities. For instance, quantum light is not allowed to come in any quantity, but rather in discrete packets that contain an integer number of photons [143]. In this chapter, we will focus on quantum systems with a discrete set of two-levels or "states." Such a discrete quantum system is known as a "qubit," the fundamental unit of quantum computation.

The status of a qubit is known as its "wavefunction," which is most commonly represented as the "bra" $|\psi\rangle$ in the Dirac notation favored by most physicists [144]. Mathematically, $|\psi\rangle$ can be represented as a column vector,

$$|\psi\rangle = \alpha|0\rangle + \beta|1\rangle, \quad |0\rangle = \begin{bmatrix} 1 \\ 0 \end{bmatrix}, \quad |1\rangle = \begin{bmatrix} 0 \\ 1 \end{bmatrix}. \tag{15.18}$$

Here, α and β are complex coefficients and $|\alpha|^2 + |\beta|^2 = 1$ such that $|\psi\rangle$ is a normal vector. Note that the wavefunction $|\psi\rangle$ is generally both the states $|0\rangle$ and $|1\rangle$ at the same time, in contrast to classical systems for which only a single discrete outcome is possible. We call this phenomenon "superposition" and observe quantum wavefunctions like $|\psi\rangle$ through measurement. Measurement of $|\psi\rangle$ will yield *either* the state $|0\rangle$ or $|1\rangle$, not both, and will transform our qubit completely into the resultant state, collapsing the wavefunction. In particular, this leads to the interpretation of $|\alpha|^2$ ($|\beta|^2$) as the probability of observing the state $|0\rangle$ ($|1\rangle$) in measurement.

As exotic as the qubit formalism may seem, it has remained computationally tractable until this point. However, the scale of the wavefunction becomes much more demanding when we consider multiqubit systems. We can describe the wavefunction of an ensemble of n qubits through the multiqubit wavefunction

$$|\Psi\rangle = \sum_{i_1=0}^{1}\sum_{i_2=0}^{1}\cdots\sum_{i_n=0}^{1}\alpha_{i_1 i_2 \cdots i_n}|i_1\rangle \otimes |i_2\rangle \otimes \cdots \otimes |i_n\rangle, \tag{15.19}$$

where $\otimes$ is the Kronecker product and $\alpha_{i_1 i_2 \cdots i_n}$ is the coefficient of each tensor-product state [145,146]. As each qubit can take one of two states and all qubit combinations are possible, the wavefunction $|\Psi\rangle$ has 2^n different states and is traditionally represented as a 2^n-dimensional vector. As the form of Eq. (15.19) suggests, however, quantum states that are not full-rank can often be represented more compactly in some form of tensor decomposition [147].

The exponential scaling of quantum wavefunctions with the number of qubits n has both complicated the study of quantum systems and motivated the quest for quantum computation. As for the former, many branches of physics, from quantum chemistry [148,149] to condensed matter [150,151], are intractable to fully simulate, as a mere 30-electron wavefunction requires over a billion coefficients to simulate. As a result, the first quantum computers to be proposed were quantum simulators [151,152,153], which seek to reproduce the dynamics of difficult to study quantum systems with those of controllable quantum systems [154]. Nevertheless, due to the low entanglement [155], high locality [156], and considerable decoherence [157,158,159] of many naturally occurring quantum states, tensor network decompositions have proven indispensable in modern quantum research. Similarly, since the 1990s, quantum computation for classical applications, ranging from integer factoring [160] to image classification [161,162] and countless algorithms in between, have been proposed. While these QC algorithms often rely on the exponentially large state space of quantum systems, decoherence due to noise and the potential irrelevance of certain subspaces can make tensor decompositions useful.

15.5.2 Matrix product states for quantum simulation

The most common form of tensor networks for quantum simulation are the so-called "matrix product state (MPS)" representations of quantum wavefunctions [145,146], which is the quantum community's term for the TT and tensor-ring formalisms. The factorization was popularized in the 1990s by pioneering research on the density matrix renormalization group for identifying quantum ground states in condensed matter physics simulations [163,164,165].

In the MPS formalism, we write $|\Psi\rangle$ as the contraction of K tensor cores,

$$|\Psi\rangle = \sum_{j_1}\sum_{j_2}\cdots\sum_{j_{K-1}} M_{j_1}^{p_1} M_{j_1 j_2}^{p_2} \cdots M_{j_{K-1}}^{p_K}. \tag{15.20}$$

In this formalism, the j indices are the bonds connecting adjacent cores. Their order determines the rank of the tensor network. The p indices are the physical or visible indices, which can be used to express the elements of $|\Psi\rangle$ or can be contracted with other quantum operators, e.g., a distinct quantum wavefunction $|\Psi'\rangle$ or some quantum observable $\hat{O}$. We here assume that each p index represents a single qubit and thus has dimension $|p| = 2$; however, multiple qubits can be encoded into larger tensor cores. As TT is general, it can be used to express any quantum state $|\Psi\rangle$, as long as the virtual bond dimensions are large enough. For example, for general n-qubit states, we require each j index to have dimension $|j| = 2^{n-1}$, which is commensurate with the 2^n elements of a standard vector representation of $|\Psi\rangle$. For useful implementations of MPS, however, $|j|$ is much smaller than 2^{n-1} and the $2n|j|^2$-dimensionality of Eq. (15.20) offers a significant compression.

MPS can be used for compact quantum calculations. As a canonical example, let us consider the inner product between two wavefunctions $\langle\Psi'|\Psi\rangle$, where $\langle\Psi'|$ is some distinct MPS factorized wavefunction

$$\langle\Psi'| = \sum_{j_1'}\sum_{j_2'}\cdots\sum_{j_{K-1}'} M_{j_1'}^{p_1'\,\prime} M_{j_1' j_2'}^{p_2'\,\prime} \cdots M_{j_{K-1}'}^{p_K'}. \tag{15.21}$$

In the standard vector representation of quantum wavefunctions, this operation would be the matrix multiplication of 1×2^n and $2^n \times 1$ vectors, producing a single scalar. In MPS format, this operation is described by the contraction

$$\langle\Psi'|\Psi\rangle = \sum_{\{p\}}\sum_{j_1' j_1}\cdots\sum_{j_{K-1}' j_{K-1}} M_{j_1}^{p_1} M_{j_1'}^{p_1'\,\prime} M_{j_1' j_2'}^{p_2'\,\prime} \cdots M_{j_{K-1}'}^{p_K'\,\prime} M_{j_1 j_2}^{p_2} \cdots M_{j_{K-1}}^{p_K}. \tag{15.22}$$

15.5.3 Factorized quantum circuit simulation with TensorLy-Quantum

TT factorization can be extended to quantum observables and unitaries to produce matrix product operators (MPOs) [145,146,166]:

$$\hat{O} = \sum_{j_1}\sum_{j_2}\cdots\sum_{j_K} M_{j_1}^{p_1 p^1} M_{j_1 j_2}^{p_2 p^2} \cdots M_{j_{K-1}}^{p_K p^K}, \tag{15.23}$$

where each tensor core M_{j_l} now has two physical indices p_l and p^l. In quantum circuit simulation, factorized operators such as $\hat{O}$ are used as both unitary operators (which represent the action of quantum logic gates on the input quantum wavefunction) and observables (which represent the measurements of quantum wavefunctions after the circuits and serve as the computations result).

Several classical simulators for quantum computation offer end-to-end implementation of quantum circuits in MPS/MPO formalism [167,168,169], including TensorLy-Quantum [170], a quantum extension of the above discussed TensorLy software package. As the bond dimension of quantum tensor networks grows with the intricacy of the wavefunction, or the portion of Hilbert space that contains occupied states, the memory and FLOP requirements of the simulation grow quickly with circuit depth. However, as the computational complexity of MPS representations grows only linearly in the number of tensor cores, large numbers of qubits can be simulated for shallow circuits.

When simulating quantum circuits using MPS/MPO, it is typical to combine both the wavefunction evolution and the measurement calculation as a single large tensor calculation. As an example, consider that the action of a quantum circuit on some initial input quantum state $|\Psi_{in}\rangle$ is captured by the contraction of $|\Psi_{\text{in}}\rangle$ with a unitary U,

$$|\Psi_{\text{out}}\rangle = U|\Psi_{\text{in}}\rangle. \tag{15.24}$$

Likewise, the energy expectation value E_{out} of a quantum wavefunction $|\Psi_{\text{out}}\rangle$ is its inner product

$$E_{\text{out}} = \langle\Psi_{\text{out}}|H|\Psi_{\text{out}}\rangle \tag{15.25}$$

with the so-called "Hamiltonian" operator H. That is, the Hamiltonian is the sum of all energy-relevant observables of a quantum system. Combining these two elements, we can study the dynamics of a quantum circuit as the single contraction

$$E_{\text{out}} = \langle \Psi_{\text{in}} | U^{\dagger} H U | \Psi_{\text{in}} \rangle. \tag{15.26}$$

We note that in the regime in which QML will be useful, it will not be simulatable with tensor networks nor any other kind of classical representation. This is in part by definition, as useful quantum computation is said to have a "quantum advantage" or offer some capacity of speedup not accessible through classical computation [171,172,173], and it is also due to the exponentially large kernel spaces that useful QML would offer [174,175].

In addition to QML techniques for classical applications, progress has also been made in classical machine learning for quantum applications. Research areas include state classification [176,177,178], quantum phase prediction [179,180], and quantum compilation [181,182]. While the impact of classical machine learning on quantum science is potentially large, classical state spaces are fundamentally limited in their ability to represent exponential quantum Hilbert spaces. Notwithstanding, in low-rank cases, tensor networks may play a major role in the advance of these applications.

15.6 Software

In this section, to accelerate the learning curve and enable researchers and practitioners to get started in the field of tensor methods quickly, we review available software packages for tensor algebra and introduce the TensorLy library [30].

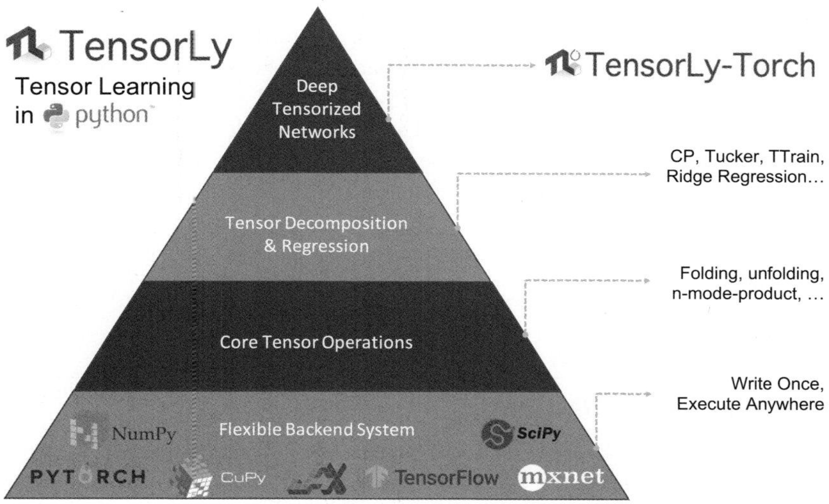

FIGURE 15.18

Overview of the TensorLy and TensorLy-Torch libraries.

TensorLy is a Python library that aims at making tensor learning simple and accessible. It provides a high-level API for tensor methods, including core tensor operations, tensor decomposition, regression, and deep tensorized architectures. It has a flexible backend that allows running operations seamlessly using NumPy, PyTorch, TensorFlow, etc. The project is also open-sourced under the BSD license, making it suitable for both academic and commercial use. TensorLy is also optimized for Python and its ecosystem. For instance, the unfolding is redefined as introduced in Section 15.2 to match the underlying C-ordering of the elements in memory, resulting in better performance. An overview of the functionalities of TensorLy can be seen in Fig. 15.18. By default, it uses NumPy as its backend. However, this can be set to any of the deep learning frameworks, optimized for running large-scale methods, and in the companion, we also use the popular PyTorch library. Once PyTorch has been installed, it can be easily used as a backend for TensorLy, allowing all the operations to run transparently on multiple machines and GPUs. Finally, TensorLy has received several extensions, including TensorLy-Torch[4] [183], which builds on top of TensorLy and PyTorch and implements out-of-the-box layers for tensor operations within deep neural networks. In addition, TensorLy-Quantum [170] builds upon TensorLy and provides a library for tensor-based QML, with efficient implementations of tensor methods and networks that can provide significant speedups in quantum simulations and computation.

Other libraries exist for tensor operations in Python. In particular, t3f [184] is a TensorFlow library for working with TT decomposition on CPUs and GPUs. It also supports Riemannian optimization operations. Tntorch [185] is a library for tensor networks in PyTorch, while TensorNetwork [186] additionally supports JAX, TensorFlow, and NumPy. Scikit-Tensor [187] is a NumPy library for tensor learning, supporting several tensor decompositions.

In MATLAB, packages such as the Tensor Toolbox [188] or TensorLab [189] provide extensive support for tensor decomposition. In C++, ITensor [169] provides efficient support for tensor network calculations.

Finally, efficient operations on sparse tensors require specialized implementations. Indeed, while libraries such as NumPy are highly optimized for matrix and tensor contractions, they do not support operations on sparse tensors. Explicitly representing sparse tensors as dense arrays is very inefficient in terms of memory use and prohibitive for very large tensors. Instead, these can be represented efficiently, e.g., using the COOrdinate format, which instead of storing all the elements only stores the coordinates and values of nonzero elements. As a result, implementing them efficiently can be tricky. While there is a large body of work focusing on sparse matrix support, support for sparse tensor algebra is scarcer. Libraries such as PyData Sparse[5] for CPUs and the Minkowski Engine [190] for GPUs are being developed and provide an efficient way to leverage large, sparse multidimensional data. TensorLy also supports sparse tensor operations on CPUs through the PyData Sparse library.

15.7 Limitations and guidelines

Leveraging tensor methods in deep learning comes with a set of desirable properties. However, several practical and scientific challenges exist that may hinder wider adoption from new researchers in the

[4] http://tensorly.org/torch/dev/.

[5] https://sparse.pydata.org/.

area. In this section, we discuss some of the most important ones and provide some practical guidelines to address them.

15.7.1 Choosing the appropriate decomposition

While fundamental to the design of tensorized deep networks, choosing the appropriate decomposition for a given problem setting relies on heuristics. Despite the lack of principled algorithms to determine the right decomposition, some intuition can be leveraged when approaching this task. For example, CP decomposition (sum of rank-one tensors) is considered appropriate for learning interpretable latent factors. Due to separability properties, CP can be a good choice when combined with transduction for generalization. In this context, one can, e.g., train on a static domain (such as images) and generalize to videos [21] during inference. In fact, CP factorization can be seen as a MobileNet-v2 block, with rank equal to the product of the expansion factor with the number of input channels. Thus, for achieving speedup one should choose the rank accordingly, i.e., to be a power of 2. As a downside, it is ill-advised to use CP when a tensor has both very large and small modes, since the same rank is used in all dimensions.

On the other hand, Tucker decomposition can be considered more flexible, allowing each mode to have a different rank. It is deemed suitable for subspace learning (e.g., applied on the activation space of deep networks), since it facilitates projecting to and from the latent subspace represented by its core. A Tucker convolution gives more flexibility by allowing each mode to have a different rank, but the rank chosen along the input and output channels should also be selected with speed in mind.

Finally, TT decomposition favors higher compression ratios due to disentangling of the dimensions. However, it may be desirable to jointly model the dimensions if intercorrelations need to be preserved [15]. For efficient learning of convolutional LSTMs in higher orders, convolutional TT decompositions have proven successful [81].

We note that many tensor decompositions exist that can be suitable for tensorizing deep learning architectures. For example, if one seeks a compromise between CP and Tucker in terms of core tensor rank, then turning to the block term decomposition [191] is likely to be a good choice. Albeit with some new considerations, the intuition and empirical rules regarding decompositions we do not discuss in detail are largely inherited from the deep learning context.

15.7.2 Rank selection

Most tensor methods require some estimate (or knowledge) of the decomposition rank. Unfortunately, determining the tensor rank is considered an NP-hard [40,41] problem. Similarly to the choice of decomposition, heuristics are also applied in practice to estimate tensor rank. Practically, it is usually beneficial to initially select a rank that preserves 90% of the parameters and then gradually reduce the rank, e.g., through multistage compression [77], while monitoring performance. Other methods have also been proposed, such as Bayesian matrix factorization [192] or introducing a Lasso type penalty that can automatically set components to zero [21]. It should be noted that in case of end-to-end training within a deep framework, all parameters are trained jointly and thus rank selection becomes less critical. However, several tasks still require solving appropriate tensor-based problems that include rank determination, for example in the context of interpreting generative networks.

Clearly, the problem of <u>rank selection</u> and the <u>choice of decomposition</u> depend on tensor structure, and thus are clearly interdependent. A lack of theory can be observed in terms of guiding optimal choices and enabling efficient search over model architectures. However, research into autotensorization can have an important impact, where the goal is to automatically choose the rank and decomposition in order to optimally approximate a pretrained network in a compressed form.

15.7.3 Training pitfalls and numerical issues

Most difficulties in training tensor methods in a deep learning context are due to numerical issues and instabilities. For example, the composition of tensor contractions is highly prone to instability (gradient vanishing or exploding). This issue becomes even more prominent when using lower precision (e.g., int8) and necessitates the development of more principled initialization approaches that also account for structure that can mitigate the problem. While using full precision (i.e., "float64") can understandably alleviate the majority of such issues, modern deep learning frameworks are focused on speeding up learning and inference, and thus often make sacrifices in terms of precision. For example, reduced precision (e.g., float32) is commonly employed, or quantization to int16, int8, or binary is often used, amplifying issues that relate to numerical stability. Mixed-precision tools (available for instance through NVIDIA's Amp) can help in this regard. Concretely, reduced precision is used when this does not impact performance, and automatic loss scaling is applied to prevent small gradient values from vanishing. If the setting allows using only positive values, then working in log-space is also a feasible avenue that can improve numerical stability. Finally, initialization of the decomposition factors is critical to ensure that the reconstruction values are in the expected range. This is similar to the case of deep convolutional networks in general, where careful initialization needs to take place (e.g., using He [193] or Glorot [194] initialization). Typically, these methods use a Gaussian distribution with zero mean and a small variance. Finally, it is possible to perform variance analysis and thus initialize factors directly, in such way that the reconstruction has the desired statistics.

References

[1] Y. LeCun, Y. Bengio, G. Hinton, Deep learning, Nature 521 (7553) (2015) 436–444, https://doi.org/10.1038/nature14539.
[2] Y. LeCun, Y. Bengio, Convolutional networks for images, speech, and time series, in: The Handbook of Brain Theory and Neural Networks, vol. 3361 (10), 1995, p. 1995.
[3] Y. LeCun, L. Bottou, Y. Bengio, P. Haffner, Gradient-based learning applied to document recognition, Proc. IEEE 86 (11) (1998) 2278–2324.
[4] A. Vaswani, N. Shazeer, N. Parmar, J. Uszkoreit, L. Jones, A.N. Gomez, Ł. Kaiser, I. Polosukhin, Attention is all you need, Adv. Neural Inf. Process. Syst. 30 (2017).
[5] Simone Montangero, Introduction to Tensor Network Methods, Springer, 2018.
[6] J.M. Landsberg, Tensors: geometry and applications, Represent. Theory 381 (402) (2012) 3.
[7] H. Lu, K.N. Plataniotis, A. Venetsanopoulos, Multilinear Subspace Learning: Dimensionality Reduction of Multidimensional Data, 1st edition, Chapman & Hall/CRC, 2013.
[8] M. Bahri, Y. Panagakis, S. Zafeiriou, Robust Kronecker component analysis, IEEE Trans. Pattern Anal. Mach. Intell. 41 (2019) 2365–2379.
[9] W. Guo, I. Kotsia, I. Patras, Tensor learning for regression, IEEE Trans. Image Process. 21 (2) (2012) 816–827.
[10] A. Anandkumar, R. Ge, D. Hsu, S.M. Kakade, M. Telgarsky, Tensor decompositions for learning latent variable models, J. Mach. Learn. Res. 15 (2014) 2773–2832.

[11] S.S. Du, J.D. Lee, On the power of over-parametrization in neural networks with quadratic activation, in: Proc. Int. Conf. Mach. Learn. (ICML), 2018.
[12] M. Soltanolkotabi, A. Javanmard, J.D. Lee, Theoretical insights into the optimization landscape of over-parameterized shallow neural networks, IEEE Trans. Inf. Theory (2018).
[13] R. Caruana, S. Lawrence, L. Giles, Overfitting in neural nets: backpropagation, conjugate gradient, and early stopping, in: Advances in Neural Inf. Process. Syst. (NeurIPS), 2001.
[14] H. Mhaskar, Q. Liao, T. Poggio, Learning functions: when is deep better than shallow, arXiv preprint, arXiv:1603.00988, 2016.
[15] A. Novikov, D. Podoprikhin, A. Osokin, D. Vetrov, Tensorizing neural networks, in: Advances in Neural Inf. Process. Syst. (NeurIPS), 2015, pp. 442–450.
[16] V. Lebedev, Y. Ganin, M. Rakhuba, I.V. Oseledets, V.S. Lempitsky, Speeding-up convolutional neural networks using fine-tuned CP-decomposition, in: Proc. Int. Conf. Learn. Representations (ICLR), 2015.
[17] M. Astrid, S. Lee, CP-decomposition with tensor power method for convolutional neural networks compression, CoRR, arXiv:1701.07148 [abs], 2017.
[18] J. Kossaifi, A. Khanna, Z. Lipton, T. Furlanello, A. Anandkumar, Tensor contraction layers for parsimonious deep nets, in: Proc. IEEE Conf. Comput. Vis. Pattern Recognit. (CVPR), 2017, pp. 1940–1946.
[19] J. Kossaifi, Z.C. Lipton, A. Kolbeinsson, A. Khanna, T. Furlanello, A. Anandkumar, Tensor regression networks, J. Mach. Learn. Res. 21 (123) (2020) 1–21.
[20] J. Kossaifi, A. Bulat, G. Tzimiropoulos, M. Pantic, T-Net: parametrizing fully convolutional nets with a single high-order tensor, in: Proc. IEEE Conf. Comput. Vis. Pattern Recognit. (CVPR), 2019, pp. 7814–7823.
[21] J. Kossaifi, A. Toisoul, A. Bulat, Y. Panagakis, T.M. Hospedales, M. Pantic, Factorized higher-order CNNs with an application to spatio-temporal emotion estimation, in: Proc. IEEE Conf. Comput. Vis. Pattern Recognit. (CVPR), 2020, pp. 6060–6069.
[22] Y. Yang, T.M. Hospedales, Deep multi-task representation learning: a tensor factorisation approach, in: Proc. Int. Conf. Learn. Representations (ICLR), 2017.
[23] Y. Chen, X. Jin, B. Kang, J. Feng, S. Yan, Sharing residual units through collective tensor factorization to improve deep neural networks, in: Proc. of the Twenty-Seventh Int. Joint Conf. on Artif. Intell., IJCAI-18, Int. Joint Conf. on Artif. Intell. Org., 2018, pp. 635–641.
[24] A. Bulat, J. Kossaifi, G. Tzimiropoulos, M. Pantic, Incremental multi-domain learning with network latent tensor factorization, in: Proc. AAAI Conf. Artif. Intell. (AAAI), 2020.
[25] I. Goodfellow, J. Pouget-Abadie, M. Mirza, B. Xu, D. Warde-Farley, S. Ozair, A. Courville, Y. Bengio, Generative adversarial nets, in: Advances in Neural Inf. Process. Syst. (NeurIPS), 2014.
[26] A. Madry, A. Makelov, L. Schmidt, D. Tsipras, A. Vladu, Towards deep learning models resistant to adversarial attacks, in: Proc. Int. Conf. Learn. Representations (ICLR), 2018.
[27] A. Kolbeinsson, J. Kossaifi, Y. Panagakis, A. Anandkumar, I. Tzoulaki, P. Matthews, Tensor dropout for robust learning, CoRR, arXiv:1902.10758 [abs], 2020.
[28] N.D. Sidiropoulos, L. De Lathauwer, X. Fu, K. Huang, E.E. Papalexakis, C. Faloutsos, Tensor decomposition for signal processing and machine learning, IEEE Trans. Signal Process. 65 (13) (2017) 3551–3582.
[29] T.G. Kolda, B.W. Bader, Tensor decompositions and applications, SIAM Rev. 51 (3) (2009) 455–500.
[30] J. Kossaifi, Y. Panagakis, A. Anandkumar, M. Pantic, Tensorly: tensor learning in python, J. Mach. Learn. Res. 20 (26) (2019) 1–6.
[31] A. Paszke, S. Gross, S. Chintala, G. Chanan, E. Yang, Z. DeVito, Z. Lin, A. Desmaison, L. Antiga, A. Lerer, Automatic differentiation in PyTorch, in: Advances in Neural Inf. Process. Syst. (NeurIPS), 2017.
[32] F. De la Torre, A least-squares framework for component analysis, IEEE Trans. Pattern Anal. Mach. Intell. 34 (6) (2012) 1041–1055.
[33] Z. Zhang, Y. Xu, J. Yang, X. Li, D. Zhang, A survey of sparse representation: algorithms and applications, IEEE Access 3 (2015) 490–530, https://doi.org/10.1109/access.2015.2430359.
[34] J. Mairal, F. Bach, J. Ponce, G. Sapiro, Online learning for matrix factorization and sparse coding, J. Mach. Learn. Res. 11 (2010) 19–60.

[35] F.L. Hitchcock, The expression of a tensor or a polyadic as a sum of products, J. Math. Phys. 6 (1–4) (1927) 164–189, https://doi.org/10.1002/sapm192761164.
[36] J.D. Carroll, J.-J. Chang, Analysis of individual differences in multidimensional scaling via an n-way generalization of "Eckart-Young" decomposition, Psychometrika 35 (1970) 283–319.
[37] R.A. Harshman, Foundations of the PARAFAC procedure: models and conditions for an "explanatory" multi-modal factor analysis, UCLA Work. Pap. Phon. 16 (1970) 1–84.
[38] E.E. Papalexakis, C. Faloutsos, N.D. Sidiropoulos, Tensors for data mining and data fusion: models, applications, and scalable algorithms, ACM Trans. Intell. Syst. Technol. (TIST) 8 (2) (2016) 1–44.
[39] A. Cichocki, D. Mandic, L. De Lathauwer, G. Zhou, Q. Zhao, C. Caiafa, H.A. Phan, Tensor decompositions for signal processing applications: from two-way to multiway component analysis, IEEE Signal Process. Mag. 32 (2) (2015) 145–163.
[40] J. Håstad, Tensor rank is NP-complete, J. Algorithms (Print) 11 (4) (1990) 644–654.
[41] C.J. Hillar, L.-H. Lim, Most tensor problems are NP-hard, J. ACM 60 (6) (Nov. 2013), https://doi.org/10.1145/2512329.
[42] Q. Zhao, L. Zhang, A. Cichocki, Bayesian CP factorization of incomplete tensors with automatic rank determination, IEEE Trans. Pattern Anal. Mach. Intell. 37 (9) (2015) 1751–1763, https://doi.org/10.1109/TPAMI.2015.2392756.
[43] P. Rai, Y. Wang, S. Guo, G. Chen, D. Dunson, L. Carin, Scalable Bayesian low-rank decomposition of incomplete multiway tensors, in: Proc. Int. Conf. Mach. Learn. (ICML), vol. 32, PMLR, Bejing, China, 2014, pp. 1800–1808.
[44] M. Zhou, Y. Liu, Z. Long, L. Chen, C. Zhu, Tensor rank learning in CP decomposition via convolutional neural network, Signal Process. Image Commun. 73 (2019) 12–21.
[45] L.R. Tucker, Some mathematical notes on three-mode factor analysis, Psychometrika 31 (1966) 279–311.
[46] L. De Lathauwer, B. De Moor, J. Vandewalle, A multilinear singular value decomposition, SIAM J. Matrix Anal. Appl. 21 (4) (2000) 1253–1278, https://doi.org/10.1137/S0895479896305696.
[47] M.J. Mohlenkamp, Musings on multilinear fitting, Linear Algebra Appl. 438 (2) (2013) 834–852, https://doi.org/10.1016/j.laa.2011.04.019, Tensors and Multilinear Algebra.
[48] P. Comon, Tensors: a brief introduction, IEEE Signal Process. Mag. 31 (2014) 44–53, https://doi.org/10.1109/MSP.2014.2298533.
[49] I.V. Oseledets, Tensor-train decomposition, SIAM J. Sci. Comput. 33 (5) (2011) 2295–2317.
[50] S. Rommer, S. Östlund, Class of ansatz wave functions for one-dimensional spin systems and their relation to the density matrix renormalization group, Phys. Rev. B 55 (1997) 2164–2181, https://doi.org/10.1103/PhysRevB.55.2164.
[51] M. Espig, K.K. Naraparaju, J. Schneider, A note on tensor chain approximation, Comput. Vis. Sci. 15 (6) (2012) 331–344, https://doi.org/10.1007/s00791-014-0218-7.
[52] Q. Zhao, G. Zhou, S. Xie, L. Zhang, A. Cichocki, Tensor ring decomposition, CoRR, arXiv:1606.05535 [abs], 2016.
[53] G. Rabusseau, H. Kadri, Low-rank regression with tensor responses, in: D.D. Lee, M. Sugiyama, U.V. Luxburg, I. Guyon, R. Garnett (Eds.), Advances in Neural Inf. Process. Syst. (NeurIPS), Curran Associates, Inc., 2016, pp. 1867–1875.
[54] H. Zhou, L. Li, H. Zhu, Tensor regression with applications in neuroimaging data analysis, J. Am. Stat. Assoc. 108 (502) (2013) 540–552.
[55] N. Batmanghelich, A. Dong, B. Taskar, C. Davatzikos, Regularized tensor factorization for multi-modality medical image classification, in: Int. Conf. Medical Image Comput. and Comput.-Assisted Intervention, Springer, 2011, pp. 17–24.
[56] R. Yu, Y. Liu, Learning from multiway data: simple and efficient tensor regression, in: Proc. Int. Conf. Mach. Learn. (ICML), vol. 48, 2016, pp. 373–381.
[57] X. Song, H. Lu, Multilinear regression for embedded feature selection with application to fMRI analysis, in: Proc. AAAI Conf. Artif. Intell. (AAAI), 2017.

[58] B. Romera-paredes, H. Aung, N. Bianchi-berthouze, M. Pontil, Multilinear multitask learning, in: S. Dasgupta, D. Mcallester (Eds.), Proc. Int. Conf. Mach. Learn. (ICML), vol. 28, JMLR Workshop and Conf. Proc., 2013, pp. 1444–1452.
[59] X. Li, D. Xu, H. Zhou, L. Li, Tucker tensor regression and neuroimaging analysis, Stat. Biosci. 10 (3) (2018) 520–545, https://doi.org/10.1007/s12561-018-9215-6.
[60] N. Cohen, A. Shashua, Convolutional rectifier networks as generalized tensor decompositions, in: Proc. Int. Conf. Mach. Learn. (ICML), 2016, pp. 955–963.
[61] Y. Levine, D. Yakira, N. Cohen, A. Shashua, Deep learning and quantum entanglement: fundamental connections with implications to network design, in: Proc. Int. Conf. Learn. Representations (ICLR), 2018.
[62] Q. Zhao, M. Sugiyama, L. Yuan, A. Cichocki, Learning efficient tensor representations with ring structure networks, in: ICLR Workshops, 2018.
[63] J. Ye, G. Li, D. Chen, H. Yang, S. Zhe, Z. Xu, Block-term tensor neural networks, Neural Netw. 130 (2020) 11–21, https://doi.org/10.1016/j.neunet.2020.05.034.
[64] T. Garipov, D. Podoprikhin, A. Novikov, D. Vetrov, Ultimate tensorization: compressing convolutional and FC layers alike, in: NeurIPS Workshop: Learning with Tensors: Why Now and How?, 2016.
[65] M. Janzamin, H. Sedghi, A. Anandkumar, Beating the perils of non-convexity: guaranteed training of neural networks using tensor methods, arXiv preprint, arXiv:1506.08473, 2015.
[66] X. Cao, G. Rabusseau, J. Pineau, Tensor regression networks with various low-rank tensor approximations, CoRR, arXiv:1712.09520 [abs], 2017.
[67] A. Cichocki, A.-H. Phan, Q. Zhao, N. Lee, I. Oseledets, M. Sugiyama, D.P. Mandic, Tensor networks for dimensionality reduction and large-scale optimization: part 2 applications and future perspectives, Found. Trends Mach. Learn. 9 (6) (2017) 431–673, https://doi.org/10.1561/2200000067.
[68] G.O. Berger, P. Absil, L.D. Lathauwer, R.M. Jungers, M.V. Barel, Equivalent polyadic decompositions of matrix multiplication tensors, CoRR, arXiv:1902.03950 [abs], 2019.
[69] R. Rigamonti, A. Sironi, V. Lepetit, P. Fua, Learning separable filters, in: Proc. IEEE Conf. Comput. Vis. Pattern Recognit. (CVPR), 2013.
[70] M. Jaderberg, A. Vedaldi, A. Zisserman, Speeding up convolutional neural networks with low rank expansions, in: Proc. Brit. Mach. Vis. Conf. (BMVC), 2014.
[71] Y. Kim, E. Park, S. Yoo, T. Choi, L. Yang, D. Shin, Compression of deep convolutional neural networks for fast and low power mobile applications, in: Proc. Int. Conf. Learn. Representations (ICLR), 2016.
[72] K. Hayashi, T. Yamaguchi, Y. Sugawara, S.-i. Maeda, Exploring unexplored tensor network decompositions for convolutional neural networks, in: Advances in Neural Inf. Process. Syst. (NeurIPS), 2019, pp. 5552–5562.
[73] L. De Lathauwer, Decompositions of a higher-order tensor in block terms—part II: definitions and uniqueness, SIAM J. Matrix Anal. Appl. 30 (3) (2008) 1033–1066, https://doi.org/10.1137/070690729.
[74] K. He, X. Zhang, S. Ren, J. Sun, Deep residual learning for image recognition, in: Proc. IEEE Conf. Comput. Vis. Pattern Recognit. (CVPR), 2016.
[75] J. Li, Y. Sun, J. Su, T. Suzuki, F. Huang, Understanding generalization in deep learning via tensor methods, in: Proc. Int. Conf. Artif. Intell. Statist. (AISTATS), 2020.
[76] S. Arora, N. Cohen, W. Hu, Y. Luo, Implicit regularization in deep matrix factorization, in: H. Wallach, H. Larochelle, A. Beygelzimer, F. d'Alché-Buc, E. Fox, R. Garnett (Eds.), Advances in Neural Inf. Process. Syst. (NeurIPS), vol. 32, 2019, pp. 7413–7424.
[77] J. Gusak, M. Kholiavchenko, E. Ponomarev, L. Markeeva, P. Blagoveschensky, A. Cichocki, I. Oseledets, Automated multi-stage compression of neural networks, in: Proc. IEEE Int. Conf. Comput. Vis. (ICCV), 2019.
[78] A. Newell, K. Yang, J. Deng, Stacked hourglass networks for human pose estimation, in: Proc. Eur. Conf. Comput. Vis. (ECCV), 2016.
[79] O. Ronneberger, P. Fischer, T. Brox, U-Net: convolutional networks for biomedical image segmentation, in: Medical Image Comput. Comput.-Assisted Intervention (MICCAI), 2015, pp. 234–241.

[80] R. Yu, S. Zheng, A. Anandkumar, Y. Yue, Long-term forecasting using higher order tensor rnns, Learning, arXiv:1711.00073 [cs.LG], 2017.

[81] J. Su, W. Byeon, J. Kossaifi, F. Huang, J. Kautz, A. Anandkumar, Convolutional tensor-train LSTM for spatio-temporal learning, in: Advances in Neural Information Processing Systems, vol. 33, 2020, pp. 13714–13726.

[82] A.-H. Phan, K. Sobolev, K. Sozykin, D. Ermilov, J. Gusak, P. Tichavský, V. Glukhov, I. Oseledets, A. Cichocki, Stable low-rank tensor decomposition for compression of convolutional neural network, in: A. Vedaldi, H. Bischof, T. Brox, J.-M. Frahm (Eds.), ECCV, Springer International Publishing, Cham, 2020, pp. 522–539.

[83] A.G. Ivakhnenko, Polynomial theory of complex systems, IEEE Trans. Syst. Man Cybern. Syst. (4) (1971) 364–378.

[84] S.-K. Oh, W. Pedrycz, B.-J. Park, Polynomial neural networks architecture: analysis and design, Comput. Electr. Eng. 29 (6) (2003) 703–725.

[85] Y. Shin, J. Ghosh, The pi-sigma network: an efficient higher-order neural network for pattern classification and function approximation, in: Int. Joint Conf. Neural Netw., vol. 1, 1991, pp. 13–18.

[86] C.-K. Li, A sigma-pi-sigma neural network (SPSNN), Neural Process. Lett. 17 (1) (2003) 1–19.

[87] S. Rendle, Factorization machines, in: 2010 IEEE International Conference on Data Mining, 2010, pp. 995–1000.

[88] M. Blondel, A. Fujino, N. Ueda, M. Ishihata, Higher-order factorization machines, in: Advances in Neural Inf. Process. Syst. (NeurIPS), vol. 29, 2016.

[89] R.K. Srivastava, K. Greff, J. Schmidhuber, Highway networks, arXiv preprint, arXiv:1505.00387, 2015.

[90] S.M. Jayakumar, W.M. Czarnecki, J. Menick, J. Schwarz, J. Rae, S. Osindero, Y.W. Teh, T. Harley, R. Pascanu, Multiplicative interactions and where to find them, in: Proc. Int. Conf. Learn. Representations (ICLR), 2020.

[91] X. Dong, J. Huang, Y. Yang, S. Yan, More is less: a more complicated network with less inference complexity, in: Proc. IEEE Conf. Comput. Vis. Pattern Recognit. (CVPR), 2017, pp. 5840–5848.

[92] Y. Wang, L. Xie, C. Liu, S. Qiao, Y. Zhang, W. Zhang, Q. Tian, A. Yuille, Sort: second-order response transform for visual recognition, in: Proc. IEEE Int. Conf. Comput. Vis. (ICCV), 2017, pp. 1359–1368.

[93] G. Zoumpourlis, A. Doumanoglou, N. Vretos, P. Daras, Non-linear convolution filters for CNN-based learning, in: Proc. IEEE Int. Conf. Comput. Vis. (ICCV), 2017, pp. 4761–4769.

[94] T. Karras, S. Laine, T. Aila, A style-based generator architecture for generative adversarial networks, in: Proc. IEEE Conf. Comput. Vis. Pattern Recognit. (CVPR), 2019.

[95] G. Chrysos, S. Moschoglou, G. Bouritsas, Y. Panagakis, J. Deng, S. Zafeiriou, π-nets: deep polynomial neural networks, in: Proc. IEEE Conf. Comput. Vis. Pattern Recognit. (CVPR), 2020.

[96] F.-L. Fan, M. Li, F. Wang, R. Lai, G. Wang, Expressivity and trainability of quadratic networks, arXiv preprint, arXiv:2110.06081, 2021.

[97] L. Sorber, M.V. Barel, L.D. Lathauwer, Numerical solution of bivariate and polyanalytic polynomial systems, SIAM J. Numer. Anal. 52 (4) (2014) 1551–1572.

[98] M.H. Stone, The generalized Weierstrass approximation theorem, Math. Mag. 21 (5) (1948) 237–254.

[99] S. Nikol'skii, Analysis III: Spaces of Differentiable Functions, Encyclopaedia of Mathematical Sciences, Springer Berlin Heidelberg, 2013.

[100] G.G. Chrysos, S. Moschoglou, G. Bouritsas, J. Deng, Y. Panagakis, S.P. Zafeiriou, Deep polynomial neural networks, https://doi.org/10.1109/TPAMI.2021.3058891, 2021.

[101] E.R. Berlekamp, Factoring polynomials over finite fields, Bell Syst. Tech. J. 46 (8) (1967) 1853–1859.

[102] D.G. Cantor, H. Zassenhaus, A new algorithm for factoring polynomials over finite fields, Math. Comput. (1981) 587–592.

[103] P. Dreesen, M. Ishteva, J. Schoukens, Decoupling multivariate polynomials using first-order information and tensor decompositions, SIAM J. Matrix Anal. Appl. 36 (2) (2015) 864–879.

[104] P. Comon, Y. Qi, K. Usevich, A polynomial formulation for joint decomposition of symmetric tensors of different orders, in: Int. Conf. Latent Variable Anal. Signal Separation, Springer, 2015, pp. 22–30.
[105] G.G. Chrysos, Y. Panagakis, Naps: non-adversarial polynomial synthesis, Pattern Recognit. Lett. 140 (2020) 318–324.
[106] G. Chrysos, M. Georgopoulos, Y. Panagakis, Conditional generation using polynomial expansions, in: Advances in Neural Inf. Process. Syst. (NeurIPS), 2021.
[107] Z. Zhu, F. Latorre, G. Chrysos, V. Cevher, Controlling the complexity and Lipschitz constant improves polynomial nets, in: Proc. Int. Conf. Learn. Representations (ICLR), 2022.
[108] M. Choraria, L.T. Dadi, G. Chrysos, J. Mairal, V. Cevher, The spectral bias of polynomial neural networks, in: Proc. Int. Conf. Learn. Representations (ICLR), 2022.
[109] D. Bahdanau, K. Cho, Y. Bengio, Neural machine translation by jointly learning to align and translate, in: Proc. Int. Conf. Learn. Representations (ICLR), 2015.
[110] X. Wang, R. Girshick, A. Gupta, K. He, Non-local neural networks, in: Proc. IEEE Conf. Comput. Vis. Pattern Recognit. (CVPR), 2018, pp. 7794–7803.
[111] J.-B. Cordonnier, A. Loukas, M. Jaggi, Multi-head attention: collaborate instead of concatenate, arXiv preprint, arXiv:2006.16362, 2020.
[112] F. Babiloni, I. Marras, G. Slabaugh, S. Zafeiriou, TESA: tensor element self-attention via matricization, in: Proc. IEEE Conf. Comput. Vis. Pattern Recognit. (CVPR), 2020, pp. 13945–13954.
[113] G.G. Chrysos, J. Kossaifi, Z. Yu, A. Anandkumar, Unsupervised controllable generation with self-training, in: Int. Joint Conf. on Neural Netw. (IJCNN), Jul 2021.
[114] M. Georgopoulos, J. Oldfield, M.A. Nicolaou, Y. Panagakis, M. Pantic, Mitigating demographic bias in facial datasets with style-based multi-attribute transfer, Int. J. Comput. Vis. (IJCV) 129 (7) (2021) 2288–2307.
[115] R. Haas, S. Graßhof, S.S. Brandt, Tensor-based subspace factorization for StyleGAN, in: Proc. Int. Conf. Automatic Face and Gesture Recognit. (FG), 2021.
[116] R. Haas, S. Graßhof, S.S. Brandt, Tensor-based emotion editing in the StyleGAN latent space, in: Proc. Conf. Comput. Vis. Pattern Recognit. Workshops (CVPR'W), 2022.
[117] J. Oldfield, M. Georgopoulos, Y. Panagakis, M.A. Nicolaou, P. Ioannis, Tensor component analysis for interpreting the latent space of GANs, in: Proc. Brit. Mach. Vis. Conf. (BMVC), 2021.
[118] X. Huang, S. Belongie, Arbitrary style transfer in real-time with adaptive instance normalization, in: Proc. IEEE Int. Conf. Comput. Vis. (ICCV), 2017, pp. 1501–1510.
[119] A. Radford, L. Metz, S. Chintala, Unsupervised representation learning with deep convolutional generative adversarial networks, in: Proc. Int. Conf. Learn. Representations (ICLR), 2016.
[120] Y. Shen, C. Yang, X. Tang, B. Zhou, InterFaceGAN: interpreting the disentangled face representation learned by GANs, IEEE Trans. Pattern Anal. Mach. Intell. (2020).
[121] Y. Shen, B. Zhou, Closed-form factorization of latent semantics in GANs, in: Proc. IEEE Conf. Comput. Vis. Pattern Recognit. (CVPR), 2021.
[122] E. Härkönen, A. Hertzmann, J. Lehtinen, S. Paris, GANSpace: discovering interpretable GAN controls, in: Advances in Neural Inf. Process. Syst. (NeurIPS), vol. 33, 2020, pp. 9841–9850.
[123] A. Voynov, A. Babenko, Unsupervised discovery of interpretable directions in the GAN latent space, in: Proc. Int. Conf. Mach. Learn. (ICML), 2020, pp. 9786–9796.
[124] C. Tzelepis, G. Tzimiropoulos, I. Patras, WarpedGANSpace: finding non-linear RBF paths in GAN latent space, in: Proc. IEEE Int. Conf. Comput. Vis. (ICCV), 2021, pp. 6393–6402.
[125] J. Zhu, R. Feng, Y. Shen, D. Zhao, Z. Zha, J. Zhou, Q. Chen, Low-rank subspaces in GANs, in: Advances in Neural Inf. Process. Syst. (NeurIPS), 2021.
[126] Z. Wu, D. Lischinski, E. Shechtman, StyleSpace analysis: disentangled controls for stylegan image generation, in: Proc. IEEE Conf. Comput. Vis. Pattern Recognit. (CVPR), 2021.
[127] R. Abdal, P. Zhu, N.J. Mitra, P. Wonka, StyleFlow: attribute-conditioned exploration of StyleGAN-generated images using conditional continuous normalizing flows, ACM Trans. Graph. (TOG) 40 (3) (May 2021).

[128] J. Oldfield, C. Tzelepis, Y. Panagakis, M.A. Nicolaou, I. Patras, PandA: unsupervised learning of parts and appearances in the feature maps of GANs, arXiv:2206.00048, 2022.
[129] O. Tov, Y. Alaluf, Y. Nitzan, O. Patashnik, D. Cohen-Or, Designing an encoder for stylegan image manipulation, ACM Trans. Graph. (TOG) 40 (4) (2021) 1–14.
[130] M.A.O. Vasilescu, D. Terzopoulos, Multilinear analysis of image ensembles: tensorfaces, in: Proc. Eur. Conf. Comput. Vis. (ECCV), 2002.
[131] H. Lu, K.N. Plataniotis, A.N. Venetsanopoulos, MPCA: multilinear principal component analysis of tensor objects, IEEE Trans. Neural Netw. 19 (1) (2008) 18–39.
[132] T.G. Kolda, Multilinear operators for higher-order decompositions, Tech. Rep. SAND2006-2081, Sandia National Laboratories, April 2006.
[133] N. Srivastava, G. Hinton, A. Krizhevsky, I. Sutskever, R. Salakhutdinov, Dropout: a simple way to prevent neural networks from overfitting, J. Mach. Learn. Res. 15 (56) (2014) 1929–1958.
[134] L. Wan, M. Zeiler, S. Zhang, Y.L. Cun, R. Fergus, Regularization of neural networks using DropConnect, in: S. Dasgupta, D. McAllester (Eds.), Proc. Int. Conf. Mach. Learn. (ICML), vol. 28, 2013, pp. 1058–1066.
[135] A. Bulat, J. Kossaifi, S. Bhattacharya, Y. Panagakis, G. Tzimiropoulos, N.D. Lane, M. Pantic, Defensive tensorization: randomized tensor parametrization for robust neural networks, https://openreview.net/forum?id=r1gEXgBYDH, 2020.
[136] N. Das, M. Shanbhogue, S.-T. Chen, F. Hohman, S. Li, L. Chen, M.E. Kounavis, D.H. Chau, SHIELD, in: Proc. 24th ACM SIGKDD Int. Conf. Knowl. Discovery & Data Mining, Jul 2018.
[137] N. Entezari, E.E. Papalexakis, TensorShield: tensor-based defense against adversarial attacks on images, ArXiv, arXiv:2002.10252 [abs], 2020.
[138] K. Makantasis, A.D. Doulamis, N.D. Doulamis, A. Nikitakis, Tensor-based classification models for hyperspectral data analysis, IEEE Trans. Geosci. Remote Sens. 56 (12) (2018) 6884–6898.
[139] A. Mitenkova, J. Kossaifi, Y. Panagakis, M. Pantic, Valence and arousal estimation in-the-wild with tensor methods, in: IEEE Int. Conf. on Autom. Face & Gesture Recognit. (FG), IEEE, 2019, pp. 1–7.
[140] A.G. Howard, M. Zhu, B. Chen, D. Kalenichenko, W. Wang, T. Weyand, M. Andreetto, H. Adam, MobileNets: efficient convolutional neural networks for mobile vision applications, CoRR, arXiv:1704.04861 [abs], 2017.
[141] M. Sandler, A. Howard, M. Zhu, A. Zhmoginov, L.-C. Chen, MobileNetV2: inverted residuals and linear bottlenecks, in: Proc. IEEE Conf. Comput. Vis. Pattern Recognit. (CVPR), 2018.
[142] C.C. Papadopoulos, Y. Panagakis, M. Koubarakis, M.A. Nicolaou, Efficient learning of multiple nlp tasks via collective weight factorization on bert, in: 2022 Annual Conference of the North American Chapter of the Association for Computational Linguistics Findings (NAACL 2022), 2022.
[143] Eleanor G. Rieffel, Wolfgang H. Polak, Quantum Computing: A Gentle Introduction, The MIT Press, 2014.
[144] R. Shankar, Principles of Quantum Mechanics, Springer, 2013.
[145] R. Orús, A practical introduction to tensor networks: matrix product states and projected entangled pair states, Ann. Phys. 349 (2014) 117–158, https://doi.org/10.1016/j.aop.2014.06.013.
[146] J.C. Bridgeman, C.T. Chubb, Hand-waving and interpretive dance: an introductory course on tensor networks, J. Phys. A, Math. Theor. 50 (2017) 223001, https://doi.org/10.1088/1751-8121/aa6dc3.
[147] D. Perez-Garcia, F. Verstraete, M.M. Wolf, J.I. Cirac, Matrix product state representations, Quantum Inf. Comput. 7 (5) (2007) 401–430.
[148] Y. Cao, J. Romero, J.P. Olson, M. Degroote, P.D. Johnson, M. Kieferová, I.D. Kivlichan, T. Menke, B. Peropadre, N.P.D. Sawaya, S. Sim, L. Veis, A. Aspuru-Guzik, Quantum chemistry in the age of quantum computing, Chem. Rev. 119 (19) (2019) 10856–10915.
[149] T. Helgaker, W. Klopper, D.P. Tew, Quantitative quantum chemistry, Mol. Phys. 106 (16–18) (2008) 2107–2143.
[150] N. Ashcroft, N.D. Mermin, Solid State Physics, Cengage Learning, 1976.
[151] R. Feynman, T. Ising, T. Ising, Quantum Simulations of Magnetism, 1982.

[152] H. Weimer, M. Müller, I. Lesanovsky, P. Zoller, H.P. Büchler, A Rydberg quantum simulator, Nat. Phys. 6 (5) (2010) 382–388.
[153] R. Islam, E. Edwards, K. Kim, S. Korenblit, C. Noh, H. Carmichael, G.-D. Lin, L.-M. Duan, C.-C. Joseph Wang, J. Freericks, C. Monroe, Onset of a quantum phase transition with a trapped ion quantum simulator, Nat. Commun. 2 (1) (2011) 1–6.
[154] S. Ebadi, T.T. Wang, H. Levine, A. Keesling, G. Semeghini, A. Omran, D. Bluvstein, R. Samajdar, H. Pichler, W.W. Ho, S. Choi, S. Sachdev, M. Greiner, V. Vuletić, M.D. Lukin, Quantum phases of matter on a 256-atom programmable quantum simulator, Nature 595 (7866) (2021) 227–232.
[155] G. Vidal, Efficient classical simulation of slightly entangled quantum computations, Phys. Rev. Lett. 91 (2003) 147902, https://doi.org/10.1103/PhysRevLett.91.147902.
[156] M. Fannes, B. Nachtergaele, R.F. Werner, Finitely correlated states on quantum spin chains, Commun. Math. Phys. 144 (1992) 443–490, https://doi.org/10.1007/BF02099178.
[157] C. Huang, F. Zhang, M. Newman, J. Cai, X. Gao, Z. Tian, J. Wu, H. Xu, H. Yu, B. Yuan, M. Szegedy, Y. Shi, J. Chen, Classical simulation of quantum supremacy circuits, https://arxiv.org/abs/2005.06787, 2020.
[158] B. Villalonga, D. Lyakh, S. Boixo, H. Neven, T.S. Humble, R. Biswas, E.G. Rieffel, A. Ho, S. Mandrà, Establishing the quantum supremacy frontier with a 281 pflop/s simulation, Quantum Sci. Technol. 5 (3) (2020) 034003.
[159] F. Pan, P. Zhang, Simulation of quantum circuits using the big-batch tensor network method, Phys. Rev. Lett. 128 (2022) 030501, https://doi.org/10.1103/PhysRevLett.128.030501.
[160] P.W. Shor, Algorithms for quantum computation: discrete logarithms and factoring, in: Proceedings 35th Annual Symposium on Foundations of Computer Science, Ieee, 1994, pp. 124–134.
[161] M. Ostaszewski, P. Sadowski, P. Gawron, Quantum image classification using principal component analysis, arXiv preprint, arXiv:1504.00580, 2015.
[162] Y. Dang, N. Jiang, H. Hu, Z. Ji, W. Zhang, Image classification based on quantum k-nearest-neighbor algorithm, Quantum Inf. Process. 17 (9) (2018) 1–18.
[163] S.R. White, Density matrix formulation for quantum renormalization groups, Phys. Rev. Lett. 69 (1992) 2863–2866, https://doi.org/10.1103/PhysRevLett.69.2863.
[164] I.P. McCulloch, From density-matrix renormalization group to matrix product states, J. Stat. Mech. Theory Exp. 2007 (10) (2007) P10014, https://doi.org/10.1088/1742-5468/2007/10/p10014.
[165] U. Schollwöck, The density-matrix renormalization group in the age of matrix product states, Ann. Phys. 326 (1) (2011) 96–192, https://doi.org/10.1016/j.aop.2010.09.012, January 2011 Special Issue.
[166] C. Hubig, I.P. McCulloch, U. Schollwöck, Generic construction of efficient matrix product operators, Phys. Rev. B 95 (2017) 035129, https://doi.org/10.1103/PhysRevB.95.035129.
[167] J. Gray, quimb: a python package for quantum information and many-body calculations, J. Open Sour. Softw. 3 (29) (2018) 819, https://doi.org/10.21105/joss.00819.
[168] T. Nguyen, D. Lyakh, E. Dumitrescu, D. Clark, J. Larkin, A. McCaskey, Tensor network quantum virtual machine for simulating quantum circuits at exascale, https://arxiv.org/abs/2104.10523, 2021.
[169] M. Fishman, S. White, E.M. Stoudenmire, The ITensor software library for tensor network calculations, ArXiv, arXiv:2007.14822 [abs], 2020.
[170] T.L. Patti, J. Kossaifi, S.F. Yelin, A. Anandkumar, Tensorly-quantum: quantum machine learning with tensor methods, https://arxiv.org/abs/2112.10239, 2021.
[171] D. Riste, M.P. da Silva, C.A. Ryan, A.W. Cross, A.D. Córcoles, J.A. Smolin, J.M. Gambetta, J.M. Chow, B.R. Johnson, Demonstration of quantum advantage in machine learning, npj Quantum Inf. 3 (1) (2017) 1–5.
[172] S. Bravyi, D. Gosset, R. Koenig, M. Tomamichel, Quantum advantage with noisy shallow circuits, Nat. Phys. 16 (10) (2020) 1040–1045.
[173] H.-Y. Huang, R. Kueng, J. Preskill, Information-theoretic bounds on quantum advantage in machine learning, Phys. Rev. Lett. 126 (19) (2021) 190505.

[174] M. Schuld, N. Killoran, Quantum machine learning in feature Hilbert spaces, Phys. Rev. Lett. 122 (4) (2019) 040504.
[175] Maria Schuld, Supervised quantum machine learning models are kernel methods, https://arxiv.org/abs/2101.11020.
[176] J. Gao, L.-F. Qiao, Z.-Q. Jiao, Y.-C. Ma, C.-Q. Hu, R.-J. Ren, A.-L. Yang, H. Tang, M.-H. Yung, X.-M. Jin, Experimental machine learning of quantum states, Phys. Rev. Lett. 120 (24) (2018) 240501.
[177] Z. Abohashima, M. Elhosen, E.H. Houssein, W.M. Mohamed, Classification with quantum machine learning: a survey, arXiv preprint, arXiv:2006.12270, 2020.
[178] M. Schuld, I. Sinayskiy, F. Petruccione, An introduction to quantum machine learning, Contemp. Phys. 56 (2) (2015) 172–185.
[179] P. Broecker, J. Carrasquilla, R.G. Melko, S. Trebst, Machine learning quantum phases of matter beyond the fermion sign problem, Sci. Rep. 7 (1) (2017) 1–10.
[180] J. Carrasquilla, Machine learning for quantum matter, Adv. Phys. X 5 (1) (2020) 1797528.
[181] F.T. Chong, D. Franklin, M. Martonosi, Programming languages and compiler design for realistic quantum hardware, Nature 549 (7671) (2017) 180–187.
[182] Z. He, L. Li, S. Zheng, Y. Li, H. Situ, Variational quantum compiling with double q-learning, New J. Phys. 23 (3) (2021) 033002.
[183] Tensorly-torch, https://github.com/tensorly/torch, 2020.
[184] A. Novikov, P. Izmailov, V. Khrulkov, M. Figurnov, I. Oseledets, Tensor train decomposition on tensorflow (t3f), J. Mach. Learn. Res. 21 (30) (2020) 1–7.
[185] R. Ballester-Ripoll, tntorch – tensor network learning with pytorch, https://github.com/rballester/tntorch, 2018.
[186] Tensor network, https://github.com/google/TensorNetwork, 2019.
[187] M. Nickel, scikit-tensor, https://github.com/mnick/scikit-tensor/, 2007.
[188] B.W. Bader, T.G. Kolda, MATLAB tensor toolbox version 2.6, http://www.sandia.gov/~tgkolda/TensorToolbox/, February 2015.
[189] N. Vervliet, O. Debals, L. Sorber, M. Van Barel, L. De Lathauwer, Tensorlab 3.0, https://www.tensorlab.net, Mar. 2016.
[190] C. Choy, J. Gwak, S. Savarese, 4D spatio-temporal convnets: Minkowski convolutional neural networks, in: Proc. IEEE Conf. Comput. Vis. Pattern Recognit. (CVPR), 2019, pp. 3075–3084.
[191] C. Chatzichristos, E. Kofidis, L. De Lathauwer, S. Theodoridis, S. Van Huffel, Early soft and flexible fusion of EEG and fMRI via tensor decompositions, arXiv preprint, arXiv:2005.07134, 2020.
[192] S. Nakajima, R. Tomioka, M. Sugiyama, S. Babacan, Perfect dimensionality recovery by variational Bayesian pca, in: F. Pereira, C.J.C. Burges, L. Bottou, K.Q. Weinberger (Eds.), Advances in Neural Information Processing Systems, vol. 25, Curran Associates, Inc., 2012, https://proceedings.neurips.cc/paper/2012/file/26337353b7962f533d78c762373b3318-Paper.pdf.
[193] K. He, X. Zhang, S. Ren, J. Sun, Delving deep into rectifiers: surpassing human-level performance on imagenet classification, in: Proceedings of the IEEE International Conference on Computer Vision (ICCV), 2015.
[194] X. Glorot, Y. Bengio, Understanding the difficulty of training deep feedforward neural networks, in: Proceedings of the Thirteenth International Conference on Artificial Intelligence and Statistics, vol. 9, JMLR Workshop and Conference Proceedings, 2010, pp. 249–256.

CHAPTER

Nonconvex graph learning: sparsity, heavy tails, and clustering

16

José Vinícius de M. Cardoso, Jiaxi Ying, and Daniel P. Palomar
The Hong Kong University of Science and Technology, Hong Kong SAR, China

16.1 Introduction

The current big data era has been enriched with seamless connectivity experienced by devices that generate unprecedented amounts of data at an ever-increasing pace. This interconnected scenario clearly demands robust models that can accurately estimate, detect, and predict relationships among entities that are part of this network.

One way to model such interconnections is via *graphical models* [30]. A particular class of graphical models, *undirected* Gaussian graphical models, has found extreme success across a number of practical fields, including biology and finance. Such success is related to its ℓ_1-norm penalized *convex* formulation and the development of the graphical lasso algorithm via iterative block coordinate descent methods [4,19,54]. Graphical lasso is a sparse estimator for the *unconstrained* precision (inverse covariance) matrix of a multivariate Gaussian distribution. Mathematically, it can be expressed as the following *convex* program:

$$\begin{array}{ll} \underset{\boldsymbol{\Theta}}{\text{minimize}} & \operatorname{tr}(\boldsymbol{\Theta}\boldsymbol{S}) - \log\det(\boldsymbol{\Theta}) + \lambda\|\boldsymbol{\Theta}\|_1, \\ \text{subject to} & \boldsymbol{\Theta} \succ \mathbf{0}, \end{array} \tag{16.1}$$

where $\boldsymbol{S}$ is the sample covariance matrix and $\|\cdot\|_1$ denotes the ℓ_1-norm.

Efficient optimization formulations along with scalable optimization algorithms have been key for performing inference in graphical models. More recently, there has been a growing interest in *constraining* graphical models into a particular family or structure for a variety of practical reasons, including (i) introducing prior information available from the task at hand and (ii) testing assumptions on the data generating process. Most notably, constraints on the spectral decomposition of the adjacency and Laplacian matrices of a graph have been applied to learn bipartite and k-component graphs for data clustering [26,28,40], and sign constraints on the elements of the precision matrix of a Gaussian Markov random field (GMRF) have been employed to learn total positivity models for stock markets [2,46,51]. From an optimization standpoint, however, such spectral constraints are typically *nonconvex*, which demands the usage of powerful yet scalable optimization algorithms.

While learning the structure of general graphical models is an NP-hard task [3], its importance is critical towards understanding and leveraging the information contained in such structures.

Signal Processing and Machine Learning Theory. https://doi.org/10.1016/B978-0-32-391772-8.00022-3
Copyright © 2024 Elsevier Inc. All rights reserved.

Learning graphs from data is a fundamental problem in the statistical graph learning and signal processing fields [16,17,19,29,42,52,62], having a direct impact on applications such as clustering [21,22,26,28,40,47,49], finance [13,33,35], network topology inference [11,36,45], graph neural nets [56], geometric deep learning [6], and graph variational autoencoders [24,32].

In this chapter, we discuss recent advancements in graph learning estimation methods that were only possible via nonconvex formulations and the subsequent application of optimization frameworks that can handle such nonconvexities. In addition, the algorithms discussed in this chapter are implemented in the R programming language and they are available in the open source packages: https://github.com/mirca/sparseGraph and https://github.com/dppalomar/spectralGraphTopology.

16.1.1 Learning undirected graphs

An undirected, weighted graph is denoted as a triple $\mathcal{G} = (\mathcal{V}, \mathcal{E}, \boldsymbol{W})$, where $\mathcal{V} = \{1, 2, \ldots, p\}$ is the node set, $\mathcal{E} \subseteq \{\{u, v\} : u, v \in \mathcal{V}, u \neq v\}$ is the edge set, that is, a subset of the set of all possible unordered pairs of nodes such that $\{u, v\} \in \mathcal{E}$ if and only if there exists a link between nodes u and v, and $\boldsymbol{W} \in \mathbb{R}_+^{p\times p}$ is the symmetric weighted adjacency matrix that satisfies $W_{ii} = 0$, $W_{ij} > 0$ if and only if $\{i, j\} \in \mathcal{E}$ and $W_{ij} = 0$ otherwise. The combinatorial, unnormalized graph Laplacian matrix $\boldsymbol{L}$ is defined as $\boldsymbol{L} \triangleq \boldsymbol{D} - \boldsymbol{W}$, where $\boldsymbol{D} \triangleq \mathsf{Diag}(\boldsymbol{W}\mathbf{1})$ is the degree matrix. A graph is said to be connected if and only if $D_{ii} > 0\ \forall i$; otherwise, the graph is called disconnected.

The basic idea behind learning a graph is to answer the following question: given a data matrix whose columns represent signals (observations) measured at the graph nodes, how can one design a graph that "best" fits such data matrix without possibly any (or with at most partial) knowledge of the underlying graph structure? By "graph structure," the Laplacian, adjacency, or incidence matrices of the graph is often understood, or even a more general graph shift operator [34]. In addition, the observed signals do not need to live in regular, ordered spaces and can take arbitrary values, such as categorical and numerical values, and hence the probability distribution of the data can be highly unknown. Fig. 16.1 illustrates such setting.

A p-dimensional, real-valued, Gaussian random vector $\boldsymbol{x}$ with mean vector $\mathbb{E}[\boldsymbol{x}] \triangleq \boldsymbol{\mu}$ and rank-deficient precision matrix $\boldsymbol{L}$ is said to form a Laplacian-constrained GMRF (LGMRF) [28,57–59] of rank $p - k$, $k \geq 1$, with respect to a graph $\mathcal{G}$, when its probability density function is given as

$$p(\boldsymbol{x}) \propto \sqrt{\det{}^*(\boldsymbol{L})} \exp\left\{-\frac{1}{2}(\boldsymbol{x} - \boldsymbol{\mu})^\top \boldsymbol{L}(\boldsymbol{x} - \boldsymbol{\mu})\right\}, \tag{16.2}$$

where $\det^*(\boldsymbol{L})$ is the pseudodeterminant of $\boldsymbol{L}$, i.e., the product of its positive eigenvalues [25].

Assume we are given n observations of $\boldsymbol{x}$, i.e., $\boldsymbol{X} = [\boldsymbol{x}_1, \boldsymbol{x}_2, \ldots, \boldsymbol{x}_n]^\top$, $\boldsymbol{X} \in \mathbb{R}^{n\times p}$, $\boldsymbol{x}_i \in \mathbb{R}^{p\times 1}$. The goal of graph learning algorithms is to learn a Laplacian matrix, or equivalently an adjacency matrix, given only the data matrix $\boldsymbol{X}$, i.e., often without any knowledge of the edge set $\mathcal{E}$.

To that end, the penalized maximum likelihood estimator (MLE) of the Laplacian-constrained precision matrix, on the basis of the observed data $\boldsymbol{X}$, may be formulated as the following optimization program:

$$\begin{aligned} &\underset{\boldsymbol{L}\succeq\boldsymbol{0}}{\mathsf{minimize}} && \mathsf{tr}(\boldsymbol{L}\boldsymbol{S}) - \log\det{}^*(\boldsymbol{L}) + h(\boldsymbol{L}), \\ &\mathsf{subject\ to} && \boldsymbol{L}\mathbf{1} = \mathbf{0},\ L_{ij} = L_{ji} \leq 0, \end{aligned} \tag{16.3}$$

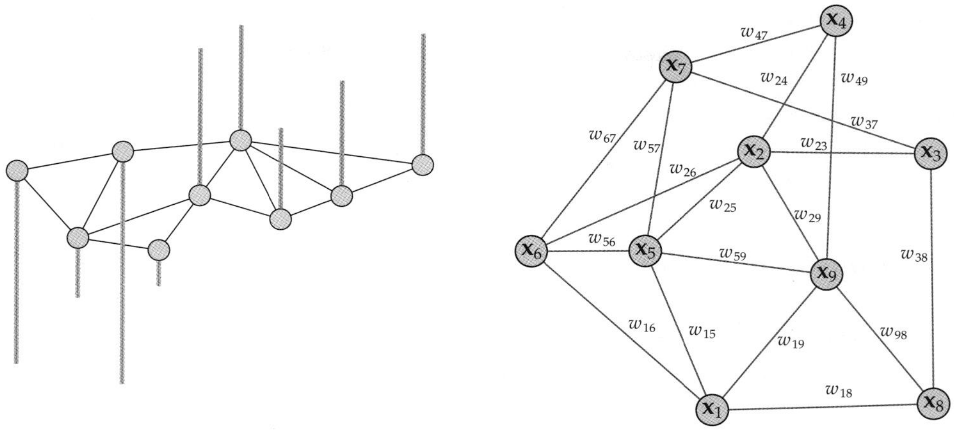

(a) A graph signal

(b) A graph to be estimated on the basis of its graph signals

FIGURE 16.1

Illustration of a hypothetical signal observed in a graph (Fig. 16.1a). The circles represent the graph nodes, the thin black lines denote the graph edges, indicating the relationships among nodes, and the vertical bars denote the signal intensities measured at each node. Graph learning techniques seek to estimate the underlying graph structure (edge weights W_{ij} in Fig. 16.1b) through the graph signal measurements.

where $\boldsymbol{S}$ is a similarity matrix, e.g., the sample covariance (or correlation) matrix $\boldsymbol{S} \propto \boldsymbol{X}^\top \boldsymbol{X}$, and h is a regularization function to promote certain properties on $\boldsymbol{L}$ such as sparsity or low-rankness. We note that we can express the Laplacian matrix via its linear operator, i.e., $\boldsymbol{L} = \mathcal{L}\boldsymbol{w}$ [28], where $\boldsymbol{w} \in \mathbb{R}^{p(p-1)/2}$ is the vectorized form of the upper triangular part of the adjacency matrix, also known as the vector of graph weights. In addition, for connected graphs, it follows that $\det^*(\mathcal{L}\boldsymbol{w}) = \det(\mathcal{L}\boldsymbol{w} + \boldsymbol{J})$, $\boldsymbol{J} \triangleq \frac{1}{p}\mathbf{1}\mathbf{1}^\top$ [17].

Problem (16.3) is fundamental in the graph signal processing and statistical machine learning fields and has served as a cornerstone for many extensions, primarily those involving the inference of structure onto the Laplacian matrix $\boldsymbol{L}$ [17,28,42]. Even though problem (16.3) is convex, provided we assume a convex choice for h, it is not adequate to be solved by disciplined convex programming languages, such as cvxpy [14], particularly due to scalability issues related to the computation of the term $\log\det^*(\boldsymbol{L})$ [17,62]. Indeed, recently, considerable efforts have been directed towards the design of scalable, iterative algorithms based on block coordinate descent [54], majorization-minimization (MM) [48], and the alternating direction method of multipliers (ADMM) [5] to solve problem (16.3) in an efficient fashion; see, e.g., [17] and [62].

Estimators based on Gaussian assumptions have been proposed for connected graphs [15,17,23,29,62]. Some of their properties, such as sparsity, are yet being investigated [58,59]. The authors in [28] and [40] proposed optimization programs for learning the class of k-component graphs, as such class is an appealing model for clustering tasks due to the spectral properties of the Laplacian matrix. However, a major shortcoming in their formulations is the lack of constraints on the degrees of the nodes, which allows for trivial solutions, i.e., graphs with isolated nodes.

In the following sections, we discuss optimization formulations for connected sparse graphs, heavy-tailed graphs, and k-component graphs. The latter is particular appealing for machine learning applications such as clustering. Before diving into the specific formulations, we briefly discuss the optimization frameworks leveraged for the design of optimization algorithms.

16.1.2 Majorization-minimization: a brief visit

MM is one of the primary tools to tackle optimization problems, in particular nonconvex ones, due to its flexible framework. In this short section, we briefly revisit the MM recipe.

The MM framework seeks to solve the following general optimization problem:

$$\begin{array}{ll} \underset{\boldsymbol{x}}{\text{minimize}} & f(\boldsymbol{x}) \\ \text{subject to} & \boldsymbol{x} \in \mathcal{X}, \end{array} \tag{16.4}$$

where we consider f a continuously differentiable, possibly nonconvex function and $\mathcal{X}$ is a nonempty closed set.

The general idea behind MM is to find a sequence of feasible points $\{\boldsymbol{x}^i\}_{i\in\mathbb{N}}$ by minimizing a sequence of carefully constructed global upper bounds of f. The popular expectation-maximization (EM) algorithm is a special case of MM [55].

At point $\boldsymbol{x}^i$, we design a continuous global upper bound function $g\left(\cdot, \boldsymbol{x}^i\right) : \mathcal{X} \to \mathbb{R}$ such that

$$g\left(\boldsymbol{x}, \boldsymbol{x}^i\right) \geq f(\boldsymbol{x}), \ \forall\, \boldsymbol{x} \in \mathcal{X}. \tag{16.5}$$

Then, in the minimization step we update $\boldsymbol{x}$ as

$$\boldsymbol{x}^{i+1} \in \underset{\boldsymbol{x}\in\mathcal{X}}{\arg\min}\; g(\boldsymbol{x}, \boldsymbol{x}^i). \tag{16.6}$$

The global upper bound function $g(\cdot, \boldsymbol{x}^i)$ must satisfy the following conditions in order to guarantee theoretical convergence:

1. $g\left(\boldsymbol{x}, \boldsymbol{x}^i\right) \geq f(\boldsymbol{x})\ \forall\, \boldsymbol{x} \in \mathcal{X}$,
2. $g\left(\boldsymbol{x}^i, \boldsymbol{x}^i\right) = f\left(\boldsymbol{x}^i\right)$,
3. $\nabla g\left(\boldsymbol{x}^i, \boldsymbol{x}^i\right) = \nabla f\left(\boldsymbol{x}^i\right)$,
4. $g(\boldsymbol{x}, \boldsymbol{x}^i)$ is continuous on both $\boldsymbol{x}$ and $\boldsymbol{x}^i$.

A thorough discussion about MM along with a significant number of its extensions with practical examples and comparisons to other optimization frameworks can be found in [48].

16.1.3 Alternating direction method of multipliers: a brief visit

ADMM is a primal-dual framework designed to solve the following class of optimization problems:

$$\begin{aligned} \underset{x,z}{\text{minimize}} \quad & f(\boldsymbol{x}) + g(\boldsymbol{z}) \\ \text{subject to} \quad & \boldsymbol{A}\boldsymbol{x} + \boldsymbol{B}\boldsymbol{z} = \boldsymbol{c}, \end{aligned} \tag{16.7}$$

where $\boldsymbol{x} \in \mathbb{R}^n$ and $\boldsymbol{z} \in \mathbb{R}^m$ are the optimization variables, $\boldsymbol{A} \in \mathbb{R}^{p \times n}$, $\boldsymbol{B} \in \mathbb{R}^{p \times m}$, and, $\boldsymbol{c} \in \mathbb{R}^p$ are parameters, and f and g are convex, proper, closed, possibly nondifferentiable functions.

The central object in the ADMM framework is the augmented Lagrangian function, which is given as

$$L_\rho(\boldsymbol{x}, \boldsymbol{z}, \boldsymbol{y}) = f(\boldsymbol{x}) + g(\boldsymbol{z}) + \boldsymbol{y}^\top(\boldsymbol{A}\boldsymbol{x} + \boldsymbol{B}\boldsymbol{z} - \boldsymbol{c}) + \frac{\rho}{2}\|\boldsymbol{A}\boldsymbol{x} + \boldsymbol{B}\boldsymbol{z} - \boldsymbol{c}\|_2^2, \tag{16.8}$$

where ρ is a penalty parameter.

The basic workflow of the ADMM algorithm is summarized in Algorithm 16.1.

Algorithm 16.1: ADMM framework.

Data: $\boldsymbol{z}^0$, $\boldsymbol{y}^0$, $\boldsymbol{A}$, $\boldsymbol{B}$, $\boldsymbol{c}$, $\rho > 0$
Result: $\boldsymbol{x}^\star$, $\boldsymbol{z}^\star$, $\boldsymbol{y}^\star$

$l \leftarrow 0$
while *not converged* **do**
  $\boldsymbol{x}^{l+1} \leftarrow \underset{\boldsymbol{x} \in \mathcal{X}}{\text{argmin}}\ L_\rho\left(\boldsymbol{x}, \boldsymbol{z}^l, \boldsymbol{y}^l\right)$
  $\boldsymbol{z}^{l+1} \leftarrow \underset{\boldsymbol{z} \in \mathcal{Z}}{\text{argmin}}\ L_\rho\left(\boldsymbol{x}^{l+1}, \boldsymbol{z}, \boldsymbol{y}^l\right)$
  $\boldsymbol{y}^{l+1} \leftarrow \boldsymbol{y}^l + \rho\left(\boldsymbol{A}\boldsymbol{x}^{l+1} + \boldsymbol{B}\boldsymbol{z}^{l+1} - \boldsymbol{c}\right)$
  $l \leftarrow l + 1$
end

The convergence of ADMM algorithms is attained provided that the following conditions are met:

1. $\text{epi}(f) = \{(\boldsymbol{x}, t) \in \mathbb{R}^n \times \mathbb{R} : f(\boldsymbol{x}) \le t\}$ and $\text{epi}(g) = \{(\boldsymbol{z}, s) \in \mathbb{R}^m \times \mathbb{R} : g(\boldsymbol{z}) \le s\}$ are both closed nonempty convex sets;
2. the unaugmented Lagrangian function L_0 has a saddle point.

We refer readers to [5], where elaborate convergence results are discussed.

16.2 Sparse graphs

Sparse graph structures are of great applicability in practical scenarios because more often than not real-world graphs are sparse. One alternative to generate sparse graphs is to conduct postprocessing pruning of edges whose weights are below a specified threshold. However, this task may not be adequate because such threshold may not be known a priori and it may also lead to disconnected graphs, which may be undesirable. Therefore, it is paramount to include sparsity-promoting regularizers directly into the objective function or constraints of the optimization formulation. The most common sparse-inducing function is the ℓ_1-norm, which has been a key part of graphical lasso [19,52].

However, the ℓ_1-norm may not always promote sparsity. To see that, consider optimization of a loss function f that attains its minimum at $\boldsymbol{x}^\star$,

$$\begin{array}{ll} \underset{\boldsymbol{x}\in\mathbb{R}^p}{\text{minimize}} & f(\boldsymbol{x}), \\ \text{subject to} & \boldsymbol{x}^\top \mathbf{1} = 1, \ \boldsymbol{x} \geq \mathbf{0}. \end{array} \tag{16.9}$$

It is straightforward to see that changing the objective function to $f(\boldsymbol{x}) + \lambda\|\boldsymbol{x}\|_1$, $\lambda > 0$, does not change the solution of the optimization problem. This toy problem is used to illustrate that careful design choices must be considered before utilizing regularizations in optimization formulations.

Fig. 16.2 illustrates the effect of the ℓ_1-norm on the estimated graphs. We generate $n = 300$ data samples from a tree graph with $p = 50$ nodes, as shown in Fig. 16.2a, as in $\boldsymbol{X} \sim \mathcal{N}(\mathbf{0}, \boldsymbol{L}^\dagger)$, where the edge weights in $\boldsymbol{L}$ are uniformly sampled from $U[2, 5]$. We use the data matrix $\boldsymbol{X}$ to recover the graph $\boldsymbol{L}$ by solving problem (16.3) with $h(\cdot)$ set as the ℓ_1-norm with regularization hyperparameter λ. It can be observed that in this scenario, the estimator defined as in problem (16.3) yields denser graphs as the hyperparameter λ increases. More details can be found in [57,58].

Recently, theoretical and empirical studies on sparse formulations for undirected, weighted graphs have been conducted [57–59,61]. The main conclusion of these works is that the ℓ_1-norm is unable to learn sparse structures in the settings of problem (16.3). Therefore, they rely on concave regularizers such as the minimax concave penalty (MCP) [60], smoothly clipped absolute deviation (SCAD) [18], or the reweighted ℓ_1-norm [7].

The primary shortcoming of these regularizers is that they make the optimization formulation nonconvex. However, efficient optimization frameworks, such as MM [48], can be used to generate a sequence of easy-to-solve convex problems with convergence guarantees. For instance, Ying et al. [58,59] directly used MM on the optimization formulations with MCP, SCAD, and the logarithm approximation to the ℓ_0-norm to design an estimator that turns out to be a sequence of constrained quadratic programs. Zhang et al. [61] used a difference-of-convex algorithm, which can be interpreted as an instance of MM [48], to design an estimator for connected graphs. These formulations are identical to that of problem (16.3), except that the function $h(\cdot)$ is MCP, SCAD, or the logarithm approximation to the ℓ_0-norm.

As a practical example, we illustrate the design an MM-based estimator that solves a sequence of majorized *convex* problems to obtain a sparse graph. The algorithm presented here is analogous to that of Ying et al. [58]. Recall that the general formulation for graph learning is stated in problem (16.3). Assuming that the graph is connected, we have the following formulation:

$$\begin{array}{ll} \underset{\boldsymbol{L}\succeq\mathbf{0}}{\text{minimize}} & \operatorname{tr}(\mathcal{L}\boldsymbol{w}\boldsymbol{S}) - \log\det(\mathcal{L}\boldsymbol{w} + \boldsymbol{J}) + h(\boldsymbol{w}), \\ \text{subject to} & \boldsymbol{w} \geq \mathbf{0}, \end{array} \tag{16.10}$$

where $h(\cdot)$ is a concave regularizer such as MCP, SCAD [58], or approximation to the ℓ_0-norm [26].

As briefly revisited in Section 16.1.2, the MM recipe consists of two steps: (1) the majorization step and the (2) minimization step. For the majorization step, a simple but often effective rule of thumb to construct a majorizer of an objective function is to a take the first-order Taylor approximation of its

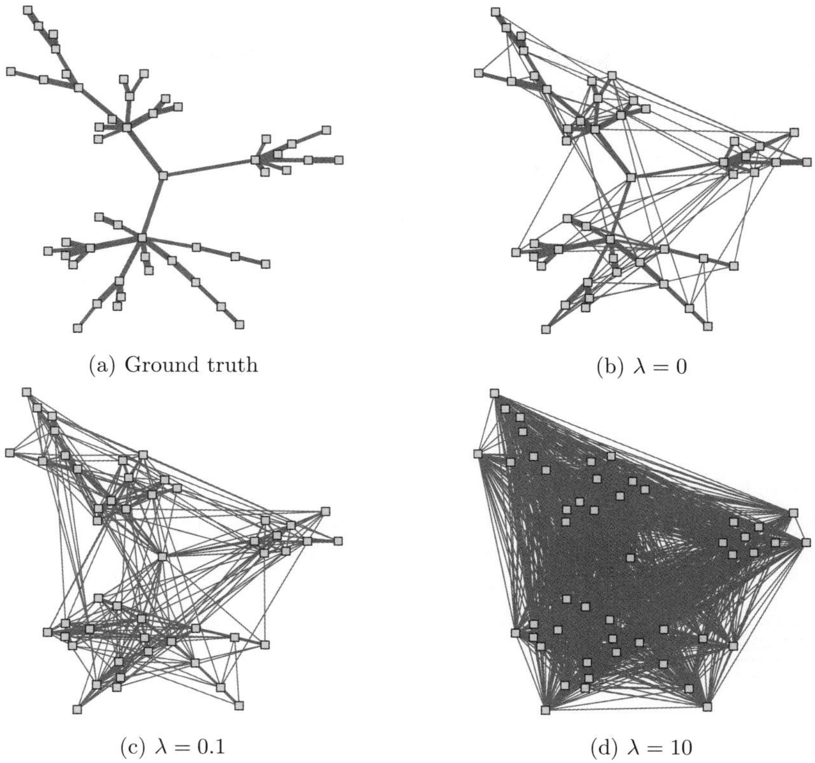

FIGURE 16.2

Estimating Laplacian matrices via ℓ_1-norm regularization for different values of the regularization hyperparameter. (a) Ground truth graph. (b) $\lambda = 0$. (c) $\lambda = 0.1$. (d) $\lambda = 10$. We note that increasing the regularization intensity actually makes the estimate graph more dense. The numbers of nonzero edges in (b), (c), and (d) are 135, 286, and 1225, respectively. The true graph in (a) has 49 edges and the graph in (d) is fully connected. The relative errors of the learned graphs in (b), (c), and (d) are 0.14, 0.64, and 0.99, respectively.

concave components [48]. Hence, for the setting of problem (16.10), we are required to find the first-order Taylor expansion of $h(\cdot)$, as the other terms are convex. Denoting by $g(\boldsymbol{w}, \boldsymbol{w}^i)$ the majorizer of the objective function in problem (16.10) at a point $\boldsymbol{w}^i$, we construct g as follows:

$$g(\boldsymbol{w}, \boldsymbol{w}^i) = \mathrm{tr}\,(\mathcal{L}\boldsymbol{w}\boldsymbol{S}) - \log\det(\mathcal{L}\boldsymbol{w} + \boldsymbol{J}) + \underbrace{h(\boldsymbol{w}^i) + \langle \boldsymbol{w} - \boldsymbol{w}^i, \nabla_{\boldsymbol{w}} h(\boldsymbol{w}^i)\rangle}_{\text{first-order Taylor expansion of } h(\cdot)}. \tag{16.11}$$

The next step is to find a minimizer of $g(\cdot, \boldsymbol{w}^i)$, i.e.,

$$\boldsymbol{w}^{i+1} \in \underset{\boldsymbol{w} \geq \boldsymbol{0}}{\arg\min}\ \mathrm{tr}\,(\mathcal{L}\boldsymbol{w}\boldsymbol{S}) - \log\det(\mathcal{L}\boldsymbol{w} + \boldsymbol{J}) + \langle \boldsymbol{w}, \nabla_{\boldsymbol{w}} h(\boldsymbol{w}^i)\rangle. \tag{16.12}$$

Note that problem (16.12) is convex; therefore, it can be solved by several optimization algorithms such as block coordinate descent and projected gradient descent [58]. We can also rely on computational convex frameworks, e.g., cvxpy [14] for low-dimensional cases. Algorithm 16.2 summarizes the implementation of an MM algorithm for sparse graph learning similar to that of [58].

In addition, note that Algorithm 16.2 only depends on the regularizer h through its gradient ∇h. For reference, we list below the gradients of MCP and SCAD:

$$h'_{\mathsf{MCP}}(x) = \begin{cases} \lambda - \dfrac{x}{\gamma} & x \in [0, \gamma\lambda], \\ 0 & x \in [\gamma\lambda, \infty], \end{cases} \tag{16.13}$$

$$h'_{\mathsf{SCAD}}(x) = \begin{cases} \lambda & x \in [0, \lambda], \\ \dfrac{(\gamma\lambda - x)}{\gamma} - 1 & x \in [\lambda, \gamma\lambda], \\ 0 & x \in [\gamma\lambda, \infty], \end{cases} \tag{16.14}$$

where γ and λ are positive hyperparameters.

Algorithm 16.2: Sparse graph learning.

Data: Similarity matrix $\boldsymbol{S} \in \mathbb{R}^{p \times p}$, initial estimate of the graph weights $\boldsymbol{w}^0$.
Result: Graph Laplacian estimation: $\mathcal{L}\boldsymbol{w}^\star$.

$i \leftarrow 0$
while *convergence criteria* ***not*** *met* **do**
  $\triangleright$ update $\boldsymbol{w}^{i+1} \in \underset{\boldsymbol{w} \geq \boldsymbol{0}}{\mathsf{arg\,min}}\ \mathsf{tr}\,(\mathcal{L}\boldsymbol{w}\boldsymbol{S}) - \log\det(\mathcal{L}\boldsymbol{w} + \boldsymbol{J}) + \langle \boldsymbol{w}, \nabla_{\boldsymbol{w}} h(\boldsymbol{w}^i)\rangle$.
  $i \leftarrow i + 1$
end

Now, we illustrate the advantages of Algorithm 16.2 for sparse graph learning in a practical experiment involving financial time-series data. More precisely, we perform an experiment with data from returns of stocks belonging to the S&P500 in order to showcase the importance of appropriate sparsity-promoting regularizations while learning graphs. To that end, we collect price data of $p = 85$ stocks belonging to three sectors, namely, Communication Services (red), Utilities (blue), and Real Estate (green), from April 22, 2019, to December 30, 2020, resulting in $n = 429$ observations.

Fig. 16.3 shows the estimated graphs by problem (16.3) using the ℓ_1-norm and MCP, which we respectively denote as "Gaussian MLE" and "Gaussian MLE with MCP." We observe that the modularity value of the MCP-based graph is twice as much as the one based on the ℓ_1-norm. In addition, the learned graph by MCP is sparser, which can aid interpretability of the financial network. However, improved results can be obtained if we take into account that the data distribution is actually heavy-tailed, which we will discuss in the next section.

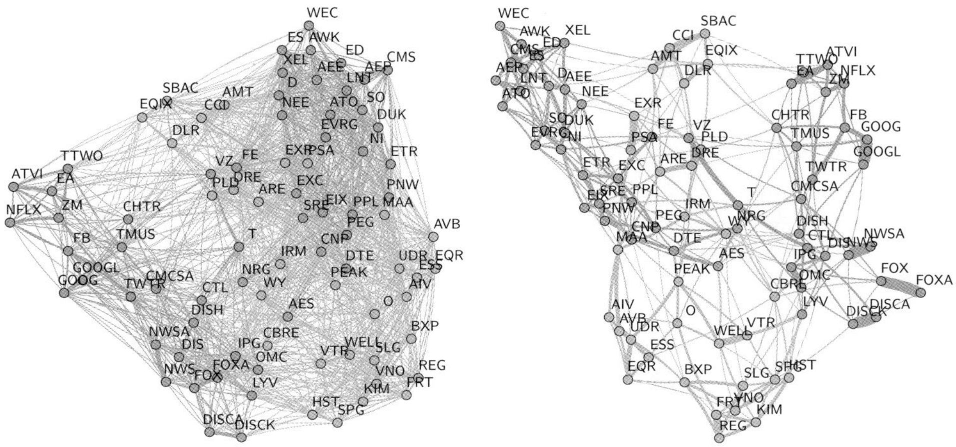

(a) Gaussian MLE, modularity $= 0.23$

(b) Gaussian MLE with MCP, modularity $= 0.46$

FIGURE 16.3

Learned graphs of S&P500 stocks during the COVID-19 pandemic.

16.3 Heavy-tail graphs

Heavy-tail events and outliers are prevalent in contemporaneous datasets. The appropriate modeling of such events is critical to design performant graph estimators in practical scenarios [43].

Enforcing sparsity is one possible way to remove spurious conditional correlations between nodes in the presence of data with outliers. However, assuming a principled, heavy-tailed statistical distribution directly brings more benefits, rather than simply imposing arbitrary, nonconvex regularizations, because they are often cumbersome to deal with from a theoretical perspective and, in practice, they bring the additional task of tuning hyperparameters, which is often repetitive.

Data from financial instruments, e.g., returns of equities, currencies, and cryptocurrencies, can be extremely heavy-tailed. We illustrate that phenomenon empirically using returns data from the S&P500 index from different time periods. Fig. 16.4 shows histograms of the log-returns of the S&P500 index in different periods along with a fit of a Gaussian probability density function. As can be observed, there exists a significant discrepancy between the fitted density and its empirical counterpart, especially around the tails of the distribution.

In order to address the inherently heavy-tailed nature of such datasets, the authors in [8] considered the Student t-distribution under the improper Markov random field assumption [44] with Laplacian structural constraints, that is, they assume the data generating process can be modeled with a multivariate zero-mean Student t-distribution, whose probability density function can be written as

$$p(\boldsymbol{x}) \propto \sqrt{\det^*(\boldsymbol{\Theta})}\left(1+\frac{\boldsymbol{x}^\top \boldsymbol{\Theta}\boldsymbol{x}}{\nu}\right)^{-\frac{\nu+p}{2}}, \quad \nu > 2, \tag{16.15}$$

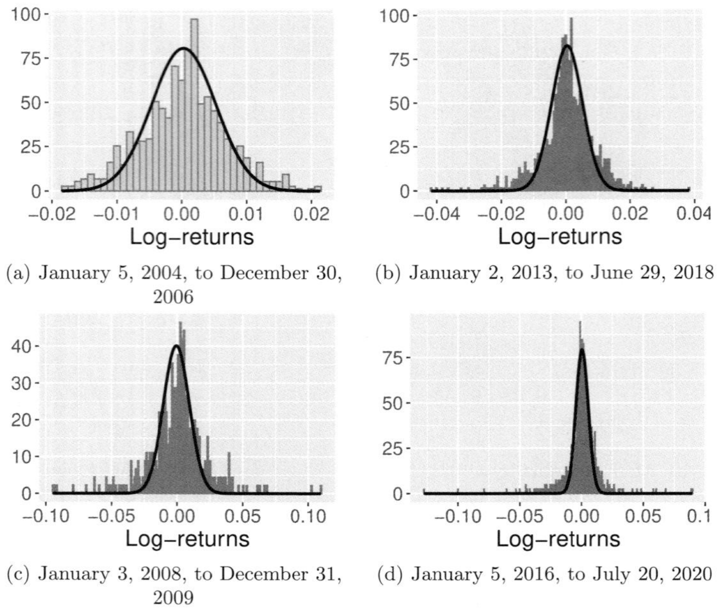

FIGURE 16.4

Histograms of the S&P500 log-returns during the mentioned time periods, where the solid curve represents a Gaussian fit. It can be noticed that the tails of the Gaussian decay much faster than the tails of the empirical histograms, indicating the presence of heavy tails or outliers.

where $\boldsymbol{\Theta}$ is a positive semidefinite inverse scatter matrix modeled as a combinatorial graph Laplacian matrix. This results in a robustified version of the MLE for connected graph learning, i.e.,

$$\begin{array}{ll}\underset{\boldsymbol{w}\geq\mathbf{0},\boldsymbol{\Theta}\succeq\mathbf{0}}{\text{minimize}} & \dfrac{p+\nu}{n}\displaystyle\sum_{i=1}^{n}\log\left(1+\frac{\boldsymbol{x}_i^\top\mathcal{L}\boldsymbol{w}\boldsymbol{x}_i}{\nu}\right)-\log\det\left(\boldsymbol{\Theta}+\boldsymbol{J}\right),\\ \text{subject to} & \boldsymbol{\Theta}=\mathcal{L}\boldsymbol{w},\ \mathfrak{d}\boldsymbol{w}=\boldsymbol{d},\end{array} \tag{16.16}$$

where $\mathfrak{d}:\mathbb{R}^{p(p-1)/2}\to\mathbb{R}^p$ is the degree operator defined as $\mathfrak{d}\boldsymbol{w}\triangleq\mathsf{diag}(\mathcal{L}\boldsymbol{w})$. The constraint $\mathfrak{d}\boldsymbol{w}=\boldsymbol{d}$ enables the learning of additional graph structures such as regular graphs and it is crucial for k-component graphs to avoid isolated nodes.

From a theoretical perspective, the Student t model naturally yields sparse graphs. Comparing the objective function in problem (16.16) to that of problem (16.3), we note that the Student t contains a $\log(\cdot)$ term in place of a linear term of the graph weights. The usage of a log function to promote sparsity is closely related to the iteratively reweighted ℓ_1-norm as an approximation for the ℓ_0-norm problem [7]. Problem (16.16) is, in general, nonconvex due to the summation of log terms and hence it is challenging to handle directly. Therefore, the authors in [8] relied on optimization frameworks such as the ADMM and MM to come up with a convergent algorithm for this problem.

In what follows, we illustrate how to design an ADMM algorithm for problem (16.16). Since problem (16.16) is nonconvex, elaborate convergence results are required; however, for the sake of simplicity we refer interested readers to the supplementary material of [8].

We start by following the ADMM exposition in Section 16.1.3. Then the partial augmented Lagrangian function of problem (16.16) can be written as

$$\begin{aligned} L_\rho(\boldsymbol{\Theta}, \boldsymbol{w}, \boldsymbol{Y}, \boldsymbol{y}) = {} & \frac{p+\nu}{n}\sum_{i=1}^{n}\log\left(1+\frac{\boldsymbol{x}_{i,*}^\top \mathcal{L}\boldsymbol{w}\boldsymbol{x}_{i,*}}{\nu}\right) - \log\det(\boldsymbol{\Theta}+\boldsymbol{J}) + \langle \boldsymbol{y}, \mathfrak{d}\boldsymbol{w}-\boldsymbol{d}\rangle \\ & + \frac{\rho}{2}\|\mathfrak{d}\boldsymbol{w}-\boldsymbol{d}\|_2^2 + \langle \boldsymbol{Y}, \boldsymbol{\Theta}-\mathcal{L}\boldsymbol{w}\rangle + \frac{\rho}{2}\|\boldsymbol{\Theta}-\mathcal{L}\boldsymbol{w}\|_{\mathrm{F}}^2. \end{aligned} \tag{16.17}$$

The subproblem for $\boldsymbol{\Theta}$ can be written as

$$\boldsymbol{\Theta}^{l+1} = \underset{\boldsymbol{\Theta}\succeq \boldsymbol{0}}{\arg\min} - \log\det(\boldsymbol{\Theta}+\boldsymbol{J}) + \langle \boldsymbol{\Theta}, \boldsymbol{Y}^l\rangle + \frac{\rho}{2}\left\|\boldsymbol{\Theta}-\mathcal{L}\boldsymbol{w}^l\right\|_{\mathrm{F}}^2. \tag{16.18}$$

Now, making the simple affine transformation $\boldsymbol{\Omega}^{l+1} = \boldsymbol{\Theta}^{l+1} + \boldsymbol{J}$, we have

$$\boldsymbol{\Omega}^{l+1} = \underset{\boldsymbol{\Omega}\succ \boldsymbol{0}}{\arg\min} - \log\det(\boldsymbol{\Omega}) + \langle \boldsymbol{\Omega}, \boldsymbol{Y}^l\rangle + \frac{\rho}{2}\left\|\boldsymbol{\Omega}-\mathcal{L}\boldsymbol{w}^l-\boldsymbol{J}\right\|_{\mathrm{F}}^2, \tag{16.19}$$

which can be expressed as a proximal operator [41],

$$\boldsymbol{\Omega}^{l+1} = \mathsf{prox}_{\rho^{-1}(-\log\,\det(\cdot)+\langle \boldsymbol{Y}^l,\cdot\rangle)}\left(\mathcal{L}\boldsymbol{w}^l+\boldsymbol{J}\right), \tag{16.20}$$

whose closed-form solution is given by Lemma 16.1.

Lemma 16.1. *The global minimizer of problem* (16.20) *is [12,53]*

$$\boldsymbol{\Omega}^{l+1} = \frac{1}{2\rho}\boldsymbol{U}\left(\boldsymbol{\Gamma}+\sqrt{\boldsymbol{\Gamma}^2+4\rho\boldsymbol{I}}\right)\boldsymbol{U}^\top, \tag{16.21}$$

where $\boldsymbol{U}\boldsymbol{\Gamma}\boldsymbol{U}^\top$ *is the eigenvalue decomposition of* $\rho\left(\mathcal{L}\boldsymbol{w}^l+\boldsymbol{J}\right)-\boldsymbol{Y}^l$.

Hence the closed-form solution for (16.18) is

$$\boldsymbol{\Theta}^{l+1} = \boldsymbol{\Omega}^{l+1} - \boldsymbol{J}. \tag{16.22}$$

The subproblem for $\boldsymbol{w}$ can be written as

$$\begin{aligned} \underset{\boldsymbol{w}\geq\boldsymbol{0}}{\mathsf{minimize}}\ & \frac{\rho}{2}\boldsymbol{w}^\top\left(\mathfrak{d}^*\mathfrak{d}+\mathcal{L}^*\mathcal{L}\right)\boldsymbol{w} - \left\langle \boldsymbol{w}, \mathcal{L}^*\left(\boldsymbol{Y}^l+\rho\boldsymbol{\Theta}^{l+1}\right) - \mathfrak{d}^*\left(\boldsymbol{y}^l-\rho\boldsymbol{d}\right)\right\rangle \\ & + \frac{p+\nu}{n}\sum_{i=1}^{n}\log\left(1+\frac{\boldsymbol{x}_{i,*}^\top\mathcal{L}\boldsymbol{w}\boldsymbol{x}_{i,*}}{\nu}\right), \end{aligned} \tag{16.23}$$

where $\mathcal{L}^*$ and $\mathfrak{d}^*$ are the adjoint operators of the Laplacian and degree operators, respectively [8].

We employ the MM framework to formulate an efficient iterative algorithm to obtain a stationary point of problem (16.23). We proceed by constructing a global upper bound of problem (16.23). Using the fact that the logarithm is globally upper-bounded by its first-order Taylor expansion, we have

$$\log\left(1+\frac{t}{b}\right) \leq \log\left(1+\frac{a}{b}\right)+\frac{t-a}{a+b}, \quad \forall a \geq 0,\ t \geq 0,\ b > 0, \tag{16.24}$$

which results in the following upper bound:

$$\log\left(1+\frac{\langle \boldsymbol{w}, \mathcal{L}^* \boldsymbol{x}_{i,*}\boldsymbol{x}_{i,*}^\top\rangle}{\nu}\right) \leq \frac{\langle \boldsymbol{w}, \mathcal{L}^* \boldsymbol{x}_{i,*}\boldsymbol{x}_{i,*}^\top\rangle}{\langle \boldsymbol{w}^j, \mathcal{L}^* \boldsymbol{x}_{i,*}\boldsymbol{x}_{i,*}^\top\rangle+\nu}+c_1, \tag{16.25}$$

where $c_1 = \log\left(1+\frac{\langle \boldsymbol{w}^j, \mathcal{L}^* \boldsymbol{x}_{i,*}\boldsymbol{x}_{i,*}^\top\rangle}{\nu}\right) - \frac{\langle \boldsymbol{w}^j, \mathcal{L}^* \boldsymbol{x}_{i,*}\boldsymbol{x}_{i,*}^\top\rangle}{\langle \boldsymbol{w}^j, \mathcal{L}^\star \boldsymbol{x}_{i,*}\boldsymbol{x}_{i,*}^\top\rangle+\nu}$ is a constant.

By upper-bounding the objective function of problem (16.23), at point $\boldsymbol{w}^j = \boldsymbol{w}^l$, with (16.25), the vector of graph weights $\boldsymbol{w}$ can then be updated by solving the following nonnegative, quadratic-constrained, strictly convex problem:

$$\begin{aligned}
\boldsymbol{w}^{j+1} &= \underset{\boldsymbol{w}\geq\boldsymbol{0}}{\arg\min}\ \frac{\rho}{2}\boldsymbol{w}^\top\left(\mathfrak{d}^*\mathfrak{d}+\mathcal{L}^*\mathcal{L}\right)\boldsymbol{w} - \left\langle \boldsymbol{w}, \mathcal{L}^*\left(\boldsymbol{Y}^l+\rho\boldsymbol{\Theta}^{l+1}\right) - \mathfrak{d}^*\left(\boldsymbol{y}^l-\rho\boldsymbol{d}\right)\right\rangle \\
&\quad + \frac{p+\nu}{n}\sum_{i=1}^{n}\frac{\langle \boldsymbol{w}, \mathcal{L}^* \boldsymbol{x}_{i,*}\boldsymbol{x}_{i,*}^\top\rangle}{\langle \boldsymbol{w}^j, \mathcal{L}^* \boldsymbol{x}_{i,*}\boldsymbol{x}_{i,*}^\top\rangle+\nu} \\
&= \underset{\boldsymbol{w}\geq\boldsymbol{0}}{\arg\min}\ \frac{\rho}{2}\boldsymbol{w}^\top\left(\mathfrak{d}^*\mathfrak{d}+\mathcal{L}^*\mathcal{L}\right)\boldsymbol{w} + \left\langle \boldsymbol{w}, \mathcal{L}^*\left(\tilde{\boldsymbol{S}}^j-\boldsymbol{Y}^l-\rho\boldsymbol{\Theta}^{l+1}\right) + \mathfrak{d}^*\left(\boldsymbol{y}^l-\rho\boldsymbol{d}\right)\right\rangle,
\end{aligned} \tag{16.26}$$

where $\tilde{\boldsymbol{S}}^j \triangleq \frac{1}{n}\sum_{i=1}^{n}\frac{(p+\nu)}{\langle \boldsymbol{w}^j, \mathcal{L}^*(\boldsymbol{x}_{i,*}\boldsymbol{x}_{i,*}^\top)\rangle+\nu}\boldsymbol{x}_{i,*}\boldsymbol{x}_{i,*}^\top$ is a weighted sample covariance matrix.

Problem (16.26) is a convex quadratic program, and hence it can be solved efficiently.

The dual variables $\boldsymbol{Y}$ and $\boldsymbol{y}$ are updated as

$$\boldsymbol{Y}^{l+1} = \boldsymbol{Y}^l + \rho\left(\boldsymbol{\Theta}^{l+1} - \mathcal{L}\boldsymbol{w}^{l+1}\right) \tag{16.27}$$

and

$$\boldsymbol{y}^{l+1} = \boldsymbol{y}^l + \rho\left(\mathfrak{d}\boldsymbol{w}^{l+1} - \boldsymbol{d}\right). \tag{16.28}$$

Algorithm 16.3 summarizes the implementation of an ADMM algorithm to solve problem (16.16).

To illustrate the benefits of the robustified MLE, we consider learning a financial stock market network comprised of S&P500 stocks belonging to three sectors, namely, Communication Services (red), Utilities (blue), and Real Estate (green), totaling $p = 82$ stocks, during the time period from January 3, 2014, to December 29, 2017, resulting in $n = 1006$ observations. In order to obtain descriptive insights

Algorithm 16.3: Connected Student t graph learning.

Data: Data matrix $\boldsymbol{X} \in \mathbb{R}^{n \times p}$, initial estimate of the graph weights $\boldsymbol{w}^0$, desired degree vector $\boldsymbol{d}$, penalty parameter $\rho > 0$, degrees of freedom ν, tolerance $\epsilon > 0$

Result: Laplacian estimation: $\mathcal{L}\boldsymbol{w}^\star$

initialize $\boldsymbol{Y} = \boldsymbol{0}$, $\boldsymbol{y} = \boldsymbol{0}$
$l \leftarrow 0$
while $\max\left(|\boldsymbol{r}^l|\right) > \epsilon$ *or* $\max\left(|\boldsymbol{s}^l|\right) > \epsilon$ **do**
  ▷ update $\boldsymbol{\Theta}^{l+1}$ via (16.22)
  ▷ update $\boldsymbol{w}^{l+1}$ by iterating the solution of (16.26)
  ▷ update $\boldsymbol{Y}^{l+1}$ as in (16.27)
  ▷ update $\boldsymbol{y}^{l+1}$ as in (16.28)
  ▷ compute residual $\boldsymbol{r}^{l+1} = \boldsymbol{\Theta}^{l+1} - \mathcal{L}\boldsymbol{w}^{l+1}$
  ▷ compute residual $\boldsymbol{s}^{l+1} = \mathfrak{d}\boldsymbol{w}^{l+1} - \boldsymbol{d}$
  $l \leftarrow l + 1$
end

into this dataset, we measure its degree of heavy-tailedness and annualized volatility.[1] The former is obtained by fitting the degrees of freedom of a Student t-distribution to the matrix of log-returns, whereby we obtain $\nu \approx 5.5$ and $\sigma \approx 21\%$. This scenario can be considered as having a moderate amount of heavy-tailedness.

Fig. 16.5 depicts the learned connected graphs on the aforementioned time periods. It can be readily noticed that the graph learned with the Student t-distribution (Fig. 16.5c) is sparser than those learned with the Gaussian assumption (Figs. 16.5a and 16.5b), which results from the fact that the Gaussian distribution is more sensitive to outliers. Moreover, the Student t graph presents a higher degree of interpretability as measured by its modularity value. In addition, a larger number of intersector connections, as indicated by gray-colored edges, which are often spurious from a practical perspective, are present in the graphs learned using the Gaussian MLE and Gaussian MLE with MCP regularization. Sparsity regularization provides a means to remove edges between nodes in the presence of data with outliers and possibly increase the modularity of the resulting graph. However, they bring the additional task of tuning hyperparameters, which is often repetitive and impractical for real-time applications. A cleaner graph, without the need for postprocessing or additional regularization, is obtained directly by using the Student t assumption. More details can be found in [8].

16.4 Clustering

Recently, Kumar et al. [26,28], Nie et al. [40] proposed optimization programs for learning the class of k-component graphs, as such class is an appealing model for clustering tasks due to the spectral

[1] The annualized volatility is computed as $\sigma = \frac{\sqrt{252}}{p}\sum_{i=1}^{p}\sigma_i$, where σ_i is the daily sample standard deviation of the ith stock.

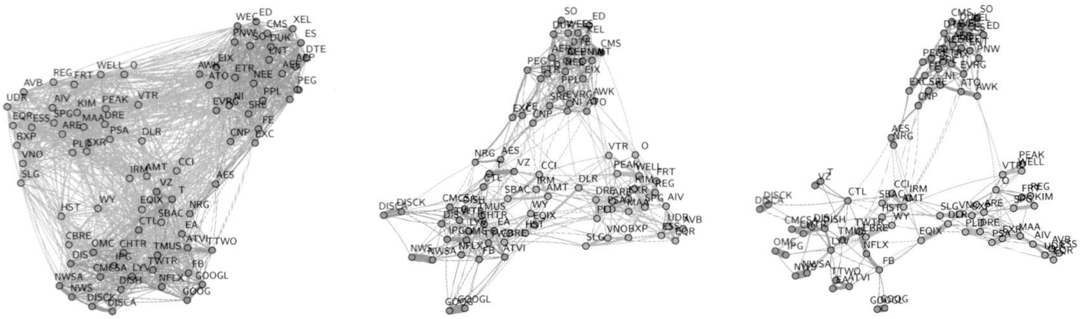

(a) Gaussian MLE, modularity $= 0.31$

(b) Gaussian MLE with MCP, modularity $= 0.49$

(c) Student t MLE, modularity $= 0.54$

FIGURE 16.5

Learned graphs of S&P500 stocks.

properties of the Laplacian matrix. Clustering may be accomplished through graphs by directly taking advantage of the fact that a graph that has k disconnected components must satisfy $\mathsf{rank}(\boldsymbol{L}) = p - k$ [9].

More precisely, Nie et al. [40] proposed the constrained Laplacian-rank (CLR) algorithm, which works in two stages. In the first stage it estimates a connected graph and in the second stage it heuristically projects the graph onto the set of Laplacian matrices of dimension p with rank $p - k$, where k is the given number of graph components. This approach is summarized as follows:

1. Obtain an initial affinity matrix $\boldsymbol{A}^\star$ as the optimal solution of

$$\begin{array}{ll} \underset{\boldsymbol{A}}{\mathsf{minimize}} & \frac{1}{2}\mathsf{tr}\left(\boldsymbol{A}\boldsymbol{Z}\right) + \frac{\eta}{2}\left\|\boldsymbol{A}\right\|_\mathrm{F}^2, \\ \mathsf{subject\ to} & \mathsf{diag}\left(\boldsymbol{A}\right) = \boldsymbol{0}, \;\; \boldsymbol{A}\boldsymbol{1} = \boldsymbol{1}, \;\; A_{ij} \geq 0 \;\forall i, j. \end{array} \tag{16.29}$$

   Note that problem (16.29) is a convex quadratic program; hence, it can be readily solved computationally.

2. Find a projection of $\boldsymbol{A}^\star$ such that $\boldsymbol{L}^\star = \mathsf{Diag}(\frac{\boldsymbol{B}^{\star\top} + \boldsymbol{B}^\star}{2}) - \frac{\boldsymbol{B}^{\star\top} + \boldsymbol{B}^\star}{2}$ has rank $p - k$:

$$\begin{array}{ll} \underset{\boldsymbol{B}, \boldsymbol{L}\succeq\boldsymbol{0}}{\mathsf{minimize}} & \left\|\boldsymbol{B} - \boldsymbol{A}^\star\right\|_\mathrm{F}^2, \\ \mathsf{subject\ to} & \boldsymbol{B}\boldsymbol{1} = \boldsymbol{1}, \;\; \mathsf{rank}(\boldsymbol{L}) = p - k, \\ & \boldsymbol{L} = \mathsf{Diag}(\frac{\boldsymbol{B}^\top + \boldsymbol{B}}{2}) - \frac{\boldsymbol{B}^\top + \boldsymbol{B}}{2}, \end{array} \tag{16.30}$$

   where k is the desired number of graph components.
   Note that problem (16.30) is nonconvex due to the rank constraint. However, a suitable approximation can be made, which leads to the design of an alternate optimization scheme. We point interested readers to [40] for the derivation.

Spectral constraints on the Laplacian matrix are an intuitive way to recover k-component graphs as the multiplicity of its zero eigenvalue, i.e., the nullity of $\boldsymbol{L}$, dictates the number of components of a graph. The first framework to impose structures on the estimated Laplacian matrix under a multivariate Gaussian setting was proposed by Kumar et al. [26,28], through the use of spectral constraints, as follows:

$$\begin{aligned}
\underset{\boldsymbol{w}\geq\boldsymbol{0},\boldsymbol{U},\boldsymbol{\lambda}}{\text{minimize}} \quad & \operatorname{tr}(\mathcal{L}\boldsymbol{w}\boldsymbol{S}) - \sum_{i=1}^{p-k}\log(\lambda_i) + \frac{\eta}{2}\left\|\mathcal{L}\boldsymbol{w} - \boldsymbol{U}\mathsf{Diag}(\boldsymbol{\lambda})\boldsymbol{U}^\top\right\|_{\mathrm{F}}^2, \\
\text{subject to} \quad & \boldsymbol{U}^\top\boldsymbol{U} = \boldsymbol{I},\ \boldsymbol{U} \in \mathbb{R}^{p\times(p-k)}, \\
& \boldsymbol{\lambda} \in \mathbb{R}_+^{p-k},\ c_1 < \lambda_1 < \cdots < \lambda_{p-k} < c_2,
\end{aligned} \tag{16.31}$$

where the term $\frac{\eta}{2}\left\|\boldsymbol{L} - \boldsymbol{U}\mathsf{Diag}(\boldsymbol{\lambda})\boldsymbol{U}^\top\right\|_{\mathrm{F}}^2$, often called spectral regularization, is added as a penalty term to indirectly promote $\boldsymbol{L}$ to have the same rank as $\boldsymbol{U}\mathsf{Diag}(\boldsymbol{\lambda})\boldsymbol{U}^\top$, i.e., $p-k$, k is the number of components of the graph to be chosen a priori, $\eta > 0$ is a hyperparameter that controls the penalization on the spectral factorization of $\boldsymbol{L}$, and c_1 and c_2 are positive, real-valued constants employed to promote bounds on the eigenvalues of $\boldsymbol{L}$. While problem (16.31) is nonconvex, the authors in [28] employed the block MM framework [48] to obtain a stationary point. Note that the block MM framework is a natural extension of the MM principle for multiple blocks of variables in a coordinate descent fashion, i.e., the variables are partitioned into blocks and MM is applied to one block while keeping the value of the other blocks fixed. This framework provides a clear flexibility benefit when designing majorization functions. The convergence for the algorithm proposed in [28] is quite involved; therefore, we refer interested readers to their supplementary material. We denote the estimator in problem (16.31) as SGL.

We now elaborate on the algorithm proposed by Kumar et al. [28]. The subproblem for $\boldsymbol{w}$ can be written as

$$\underset{\boldsymbol{w}\geq\boldsymbol{0}}{\text{minimize}} \quad \operatorname{tr}(\mathcal{L}\boldsymbol{w}\boldsymbol{S}) + \tfrac{\eta}{2}\left\|\mathcal{L}\boldsymbol{w} - \boldsymbol{U}\mathsf{Diag}(\boldsymbol{\lambda})\boldsymbol{U}^\top\right\|_{\mathrm{F}}^2, \tag{16.32}$$

which is a convex QP without closed form; nonetheless, it can be solved via standard QP solvers or projected gradient descent methods.

The subproblem for $\boldsymbol{U}$ can be written as

$$\begin{aligned}
\underset{\boldsymbol{U}}{\text{minimize}} \quad & \left\|\mathcal{L}\boldsymbol{w} - \boldsymbol{U}\mathsf{Diag}(\boldsymbol{\lambda})\boldsymbol{U}^\top\right\|_{\mathrm{F}}^2, \\
\text{subject to} \quad & \boldsymbol{U}^\top\boldsymbol{U} = \boldsymbol{I},\ \boldsymbol{U} \in \mathbb{R}^{p\times(p-k)},
\end{aligned} \tag{16.33}$$

which can be further simplified as

$$\begin{aligned}
\underset{\boldsymbol{U}}{\text{maximize}} \quad & \operatorname{tr}\left(\boldsymbol{U}^\top\mathcal{L}\boldsymbol{w}\boldsymbol{U}\mathsf{Diag}(\boldsymbol{\lambda})\right), \\
\text{subject to} \quad & \boldsymbol{U}^\top\boldsymbol{U} = \boldsymbol{I},\ \boldsymbol{U} \in \mathbb{R}^{p\times(p-k)}.
\end{aligned} \tag{16.34}$$

Problem (16.34) is an optimization on the orthogonal Stiefel manifold $\mathrm{St}(p, p-k) = \{\boldsymbol{U} \in \mathbb{R}^{p\times p-k} : \boldsymbol{U}^\top\boldsymbol{U} = \boldsymbol{I}\}$. From [1] the optimizer of (16.34) is the $p-k$ eigenvectors of $\mathcal{L}\boldsymbol{w}$ associated with the $p-k$ largest eigenvalues of $\mathcal{L}\boldsymbol{w}$.

The subproblem for $\boldsymbol{\lambda}$ is given as

$$\begin{array}{ll} \underset{\boldsymbol{\lambda}}{\mathsf{minimize}} & -\sum_{i=1}^{p-k} \log(\lambda_i) + \frac{\eta}{2} \left\| \mathcal{L}\boldsymbol{w} - \boldsymbol{U} \mathsf{Diag}(\boldsymbol{\lambda}) \boldsymbol{U}^\top \right\|_F^2, \\ \mathsf{subject\ to} & \boldsymbol{\lambda} \in \mathbb{R}_+^{p-k},\ c_1 < \lambda_1 < \cdots < \lambda_{p-k} < c_2, \end{array} \tag{16.35}$$

which can be simplified as

$$\begin{array}{ll} \underset{\boldsymbol{\lambda}}{\mathsf{minimize}} & -\sum_{i=1}^{p-k} \log(\lambda_i) + \frac{\eta}{2} \left\| \boldsymbol{u} - \boldsymbol{\lambda} \right\|_2^2, \\ \mathsf{subject\ to} & \boldsymbol{\lambda} \in \mathbb{R}_+^{p-k},\ c_1 < \lambda_1 < \cdots < \lambda_{p-k} < c_2, \end{array} \tag{16.36}$$

where $\boldsymbol{u} = \mathsf{diag}(\boldsymbol{U}^\top \mathcal{L}\boldsymbol{w}\boldsymbol{U})$.

Problem (16.36) is a convex optimization problem with isotonic constraints. While it can be solved via general-purpose convex solvers, dedicated algorithms are needed for high-dimensional cases [50].

Algorithm 16.4 summarizes the scheme of Kumar et al. [28] for k-component graph learning.

Algorithm 16.4: k-component graph learning (SGL).

Data: Similarity matrix $\boldsymbol{S} \in \mathbb{R}^{p \times p}$, initial estimate of the graph weights $\boldsymbol{w}^0$, integer number of graph components k.

Result: Graph Laplacian estimation: $\mathcal{L}\boldsymbol{w}^\star$.

```
i ← 0
while convergence criteria not met do
    ▷ update w^{i+1} via (16.32)
    ▷ update U^{i+1} via (16.34)
    ▷ update λ^{i+1} via (16.36)
    i ← i + 1
end
```

Note that problem (16.31) learns a k-component graph without the need for a two-stage algorithm. However, a clear caveat of this formulation is that it does not control the degrees of the nodes in the graph, which may result in a trivial solution that contains isolated nodes, turning out not to be useful for clustering tasks, especially when applied to noisy datasets or to datasets that are not significantly Gaussian-distributed. In addition, choosing values for hyperparameters η, c_1, and c_2 is often an intricate task.

To showcase the capabilities of the estimators CLR and SGL, we perform experiments with both synthetic and real datasets. Fig. 16.6 illustrates the results of clustering synthetic structures via SGL using the spatial coordinates of the nodes as features, with 100 nodes per cluster, i.e., the data matrix is $\boldsymbol{X} \in \mathbb{R}^{100 \times n}$, where n is the number of spatial dimensions and each row of $\boldsymbol{X}$ is associated with a node's coordinates. More precisely, we have $n = 2$ in Figs. 16.6a–16.6e and $n = 3$ in Fig. 16.6f. We observe that SGL succeeds in correctly clustering the nodes according to their cluster membership.

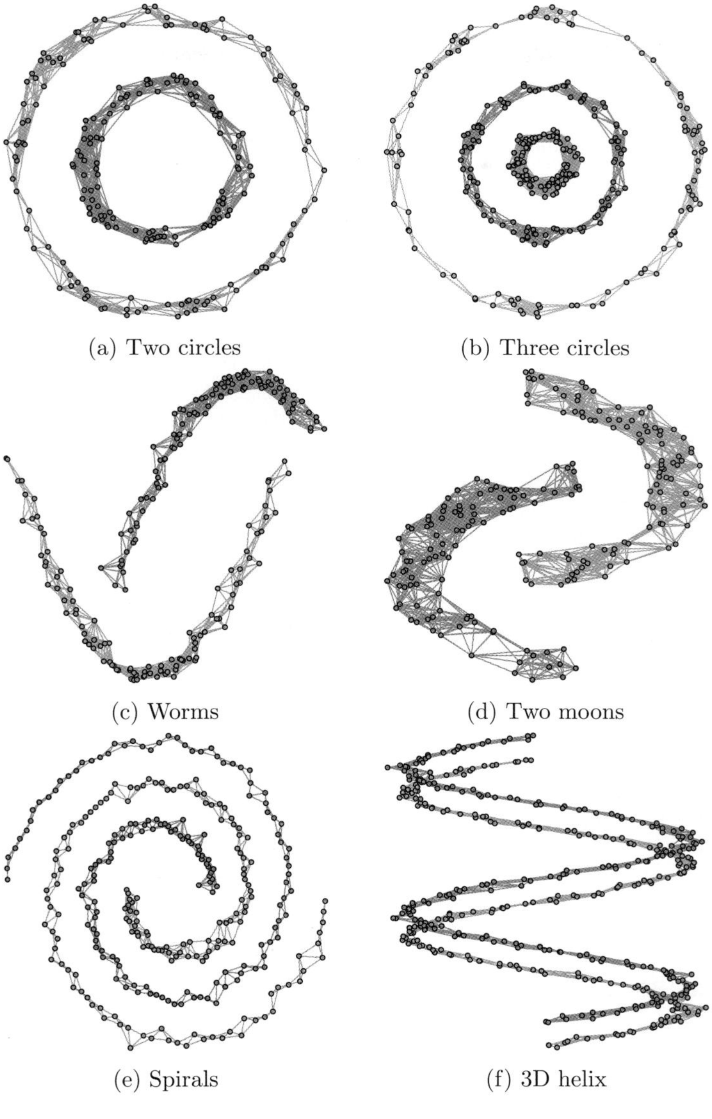

(a) Two circles (b) Three circles
(c) Worms (d) Two moons
(e) Spirals (f) 3D helix

FIGURE 16.6

The estimator defined in problem (16.31) is able to perfectly cluster the data points according to their cluster membership for synthetic structures.

As for a real-world dataset, we consider foreign exchange data from the 34 most traded currencies in the period from January 2, 2019, to December 31, 2020, totaling $n = 522$ observations. The data matrix is composed by the log-returns of the currencies' prices with respect to the United States dollar. Unlike in the experiment involving S&P500 stocks, there is no classification standard for currencies; hence, we use a community detection algorithm [10] in order to create classes within the learned graph. In particular, the algorithm in [10] takes as input the learned Laplacian matrix of the graph and outputs a membership assignment that maximizes the modularity of the graph.

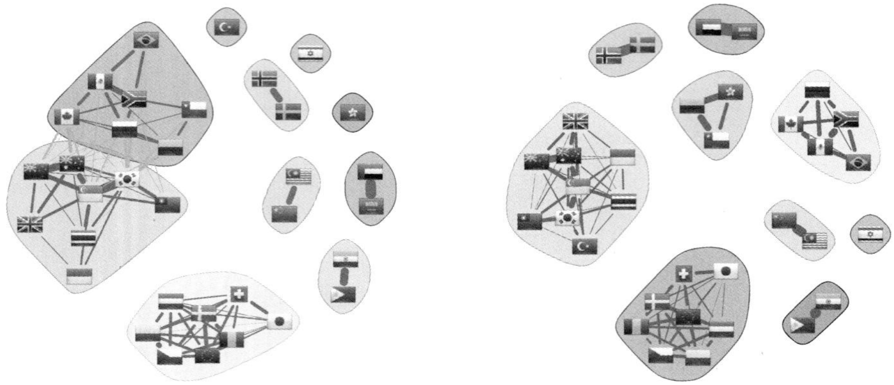

(a) SGL, modularity = 0.62 (b) CLR, modularity = 0.79

FIGURE 16.7

Learned nine-component graphs of currencies.

Fig. 16.7 depicts the learned nine-component graphs of currencies during the time window from January 2, 2019, to December 31, 2020, where we can readily observe a few meaningful connections between currencies of countries that are geographically close to each other. We can observe that SGL and CLR allow the existence of isolated nodes in the learned graphs, which may not be ideal if we would like to cluster a particular node. More details can be found in [8].

16.4.1 Soft-clustering via bipartite graphs

Bipartite graphs are graphs whose node set can be partitioned in two disjoint groups [9]. More formally, an undirected, weighted, bipartite graph can be defined as a 4-tuple $\mathcal{G} \triangleq \{\mathcal{V}_r, \mathcal{V}_q, \mathcal{E}, \boldsymbol{W}\}$, where $\mathcal{V}_r \triangleq \{1, 2, \ldots, r\}$ and $\mathcal{V}_q \triangleq \{r+1, r+2, \ldots, r+q\}$ are the node sets associated with a group of objects and classes, respectively, $\mathcal{E} \subseteq \{\{u, v\} : u \in \mathcal{V}_r, v \in \mathcal{V}_q\}$ is the edge set, that is, a subset of the set of all possible unordered pairs of $r+q$ nodes such that $\{u, v\} \in \mathcal{E}$ if and only if nodes u and v are connected, and $\boldsymbol{W} \in \mathbb{R}_+^{p\times p}$ is the symmetric weighted adjacency matrix that satisfies $W_{ii} = 0$, $W_{ij} > 0$ if and only if $\{i, j\} \in \mathcal{E}$ and $W_{ij} = 0$ otherwise. Fig. 16.8 displays an instance of such model.

Properties associated with the spectral decomposition of graph matrices have demonstrated prolific advantages that enable learning graphs with specific structures, such as bipartite, k-component, and bipartite k-component graphs [26–28,38–40]. Let $\boldsymbol{W} = \boldsymbol{V}\mathsf{Diag}(\boldsymbol{\psi})\boldsymbol{V}^\top$ be the spectral decomposition

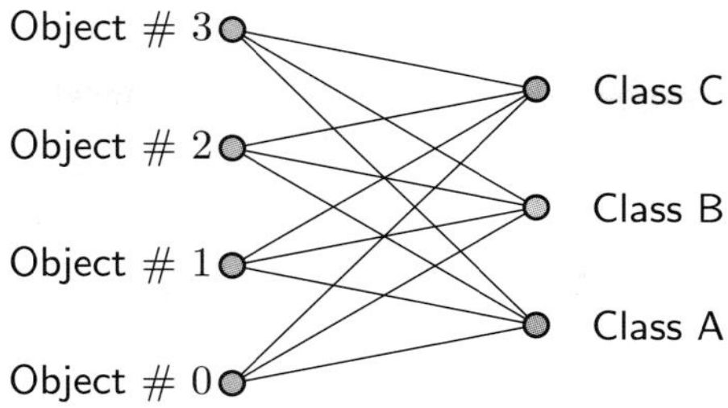

FIGURE 16.8

A bipartite graph illustrating the modeling of dependencies between a collection of objects and their classes. The edges of the graph correspond to nonzero elements in its adjacency matrix $\boldsymbol{W}$.

of the adjacency matrix. Then the adjacency matrix of a bipartite graph satisfies the following spectral properties [20]:

(P1) $\boldsymbol{\psi}$ contains exactly $r - q$ zero elements;
(P2) $\boldsymbol{\psi}$ is antisymmetric, i.e., $\boldsymbol{\psi} \in C_{\boldsymbol{\psi}}$, where

$$\begin{aligned} C_{\boldsymbol{\psi}} = \{ & \boldsymbol{\psi} \in \mathbb{R}^p : \boldsymbol{\psi}_i = -\boldsymbol{\psi}_{2q+1-i}, \; c_1 \geq \boldsymbol{\psi}_1 \geq \boldsymbol{\psi}_2 \geq \cdots \geq \boldsymbol{\psi}_q \geq c_2, \\ & i = 1, 2, \ldots, q \}, \end{aligned} \tag{16.37}$$

where c_1 and c_2 are positive constants.

Leveraging (P1) and (P2), Kumar et al. [26,27] proposed the following *nonconvex* program to learn a bipartite graph:

$$\begin{array}{ll} \underset{\boldsymbol{w}, \boldsymbol{V}, \boldsymbol{\psi}}{\text{minimize}} & \operatorname{tr}(\mathcal{L}\boldsymbol{w}\boldsymbol{S}) - \log\det(\mathcal{L}\boldsymbol{w} + \boldsymbol{J}) + \frac{\gamma}{2} \left\| \mathcal{A}\boldsymbol{w} - \boldsymbol{V}\mathsf{Diag}(\boldsymbol{\psi})\boldsymbol{V}^\top \right\|_\mathrm{F}^2, \\ \text{subject to} & \boldsymbol{w} \in \mathbb{R}_+^{p(p-1)/2}, \; \boldsymbol{V}^\top \boldsymbol{V} = \boldsymbol{I}_{2q}, \; \boldsymbol{V} \in \mathbb{R}^{p \times 2q}, \; \boldsymbol{\psi} \in C_{\boldsymbol{\psi}}, \end{array} \tag{16.38}$$

where $\mathcal{L}$ and $\mathcal{A}$ [26] are the Laplacian and adjacency operators, respectively, and $\boldsymbol{w}$ is the vector of graph weights. While the estimator proposed in Kumar et al. [26] does not directly assume knowledge of the partition of the node set, it does require the availability of the number of nodes in each group. We denote the estimator defined by problem (16.38) as SGA. The development of a block MM optimization algorithm to solve problem (16.38) is substantially similar to that of problem (16.31); hence, we omit the mathematical derivations.

Assuming a connected bipartite graph, Nie et al. [38] proposed a heuristic approach that relies on the availability of a graph similarity matrix $\boldsymbol{B}^0 \in \mathbb{R}^{r \times q}$ and used the fact that the adjacency matrix of a bipartite graph can be written as $\boldsymbol{W} = [\boldsymbol{0}\ \boldsymbol{B};\ \boldsymbol{B}^\top\ \boldsymbol{0}]$. More precisely, their estimator is obtained as the solution of the following optimization problem:

$$\begin{array}{ll} \underset{\boldsymbol{B} \in \mathbb{R}_+^{r \times q}}{\text{minimize}} & \left\| \boldsymbol{B} - \boldsymbol{B}^0 \right\|_\mathrm{F}^2, \\ \text{subject to} & \boldsymbol{B} \geq \boldsymbol{0}, \; \boldsymbol{B}\boldsymbol{1}_q = \boldsymbol{1}_r. \end{array} \tag{16.39}$$

Problem (16.39) is a Euclidean projection of the rows of $\boldsymbol{B}$ onto the probability simplex. While it can be solved efficiently via standard quadratic programming solvers, this approach lacks statistical support and its performance in practical applications heavily depends on the quality of the initial graph similarity matrix. We denote the estimator defined by problem (16.39) as SOBG.

We illustrate the performance of the state-of-the-art algorithms in terms of quantitative measures such as node label accuracy and modularity. The accuracy is computed as the ratio between the number of correctly predicted labels and the number of nodes in the object set, whereas the modularity of a graph is defined as in [37]. The modularity measures the strength of division of a graph into groups. A high modularity value means that the nodes from the same group are more likely to be connected.

We conduct a soft-clustering experiment using the MNIST image data [31]. While this dataset has been widely investigated in supervised machine (deep) learning settings, which achieved human-like accuracy, we instead use it as a proof of concept for unsupervised soft-clustering via graphical models. MNIST provides a collection of 28×28 grayscaled images of handwritten digits from 0 to 9. The nodes in the classes set are defined such that every node corresponds to a unique digit, i.e., $\mathcal{V}_q = \{0, 1, \dots, 9\}$. We assign 500 nodes for the object set, each of which represents an image randomly selected from the testing set. We randomly select the images in a stratified way, such that every label (digit) appears 50 times. The signal for a node $v_i \in \mathcal{V}_q$, associated with digit i, is constructed by averaging 1000 randomly selected images from the training set whose label corresponds to i. We construct the data matrix $\boldsymbol{X}$ by vectorizing and stacking the images in a columnwise manner. The quantitative measures are then computed as an average of 50 realizations of this randomized experiment.

Table 16.1 Performance of the algorithms in soft-clustering handwritten digits.

Algorithm	Accuracy	Modularity
SOBG	0.76 +/- 0.01	0.29 +/- 0.01
SGA	0.62 +/- 0.02	0.35 +/- 0.05

We then proceed to learn graphs by SOBG and SGA. The final label assigned to the ith image in the object set corresponds to $\arg\max_{j \in 1,\dots,q} B_{ij} - 1$. Fig. 16.9 shows the estimated graphs, while Table 16.1 provides quantitative results, where we observe that SGA outputs a sparser graph that turns out to have a higher modularity value than that of SOBG; on the other hand, SOBG yields a graph that is more accurate.

16.5 Conclusion

In this chapter, we reviewed state-of-the-art formulations and numerical optimization algorithms for graph learning under different practical requirements, e.g., sparsity and heavy tails, and different structures, such as k-components and bipartite. We illustrated the power of modern graph learning estimators via experiments that included practical real-world datasets such as returns from financial equities and currencies as well as images of handwritten digits.

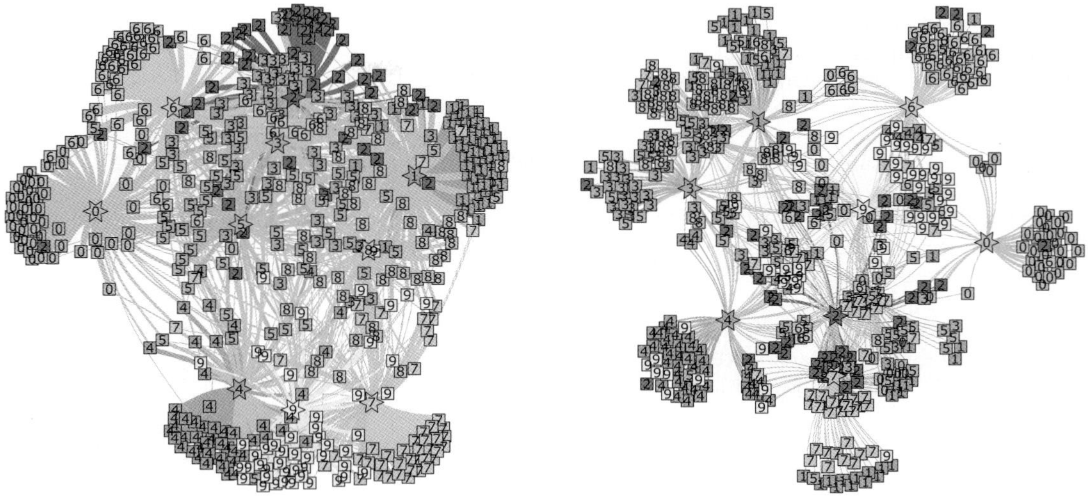

(a) SOBG (b) SGA

FIGURE 16.9

Learned bipartite graphs of handwritten digital images in the MNIST dataset. Each color represents a digit. Gray-colored edges represent connections between images of distinct digits.

Acknowledgments

This work was supported by the Hong Kong GRF 16207019 research grant.

References

[1] P.-A. Absil, R. Mahony, R. Sepulchre, Optimization Algorithms on Matrix Manifolds, Princeton University Press, Princeton, NJ, 2007.

[2] R. Agrawal, U. Roy, C. Uhler, Covariance matrix estimation under total positivity for portfolio selection, Journal of Financial Econometrics (2020, 09).

[3] A. Anandkumar, V.Y.F. Tan, F. Huang, A.S. Willsky, High-dimensional Gaussian graphical model selection: walk summability and local separation criterion, Journal of Machine Learning Research 13 (1) (2012) 2293–2337.

[4] O. Banerjee, L.E. Ghaoui, A. d'Aspremont, Model selection through sparse maximum likelihood estimation for multivariate Gaussian or binary data, Journal of Machine Learning Research 9 (15) (2008) 485–516.

[5] S. Boyd, N. Parikh, E. Chu, B. Peleato, J. Eckstein, Distributed optimization and statistical learning via the alternating direction method of multipliers, Foundations and Trends in Machine Learning 3 (1) (2011) 1–122.

[6] M.M. Bronstein, J. Bruna, Y. LeCun, A. Szlam, P. Vandergheynst, Geometric deep learning: going beyond Euclidean data, IEEE Signal Processing Magazine 34 (4) (2017) 18–42.

[7] E.J. Candès, M.B. Wakin, S.P. Boyd, Enhancing sparsity by reweighted ℓ_1 minimization, Journal of Fourier Analysis and Applications 14 (5) (2008) 877–905.

[8] J.V.M. Cardoso, J. Ying, D.P. Palomar, Graphical models in heavy-tailed markets, in: Advances in Neural Information Processing Systems (NeurIPS), 2021.
[9] F.R.K. Chung, Spectral Graph Theory, CBMS Regional Conference Series in Mathematics, vol. 92, 1997.
[10] A. Clauset, M.E.J. Newman, C. Moore, Finding community structure in very large networks, Physical Review E 70 (2004) 066111.
[11] M. Coutino, E. Isufi, T. Maehara, G. Leus, State-space network topology identification from partial observations, arXiv:1906.10471, 2019.
[12] P. Danaher, P. Wang, D.M. Witten, The joint graphical lasso for inverse covariance estimation across multiple classes, Journal of the Royal Statistical Society Series B 76 (2) (2014) 373–397.
[13] M.L. de Prado, Building diversified portfolios that outperform out of sample, The Journal of Portfolio Management 42 (4) (2016) 59–69.
[14] S. Diamond, S. Boyd, CVXPY: a Python-embedded modeling language for convex optimization, Journal of Machine Learning Research 17 (83) (2016) 1–5.
[15] X. Dong, D. Thanou, P. Frossard, P. Vandergheynst, Learning Laplacian matrix in smooth graph signal representations, IEEE Transactions on Signal Processing 64 (23) (2016) 6160–6173.
[16] X. Dong, D. Thanou, M. Rabbat, P. Frossard, Learning graphs from data: a signal representation perspective, IEEE Signal Processing Magazine 36 (3) (2019) 44–63.
[17] H.E. Egilmez, E. Pavez, A. Ortega, Graph learning from data under Laplacian and structural constraints, IEEE Journal of Selected Topics in Signal Processing 11 (6) (2017) 825–841.
[18] J. Fan, R. Li, Variable selection via nonconcave penalized likelihood and its oracle properties, Journal of the American Statistical Association 96 (456) (2001) 1348–1360.
[19] J. Friedman, T. Hastie, R. Tibshirani, Sparse inverse covariance estimation with the graphical lasso, Biostatistics 9 (2008) 432–441.
[20] C. Godsil, G. Royle, Algebraic Graph Theory, Graduate Texts in Mathematics, Springer Science, 2001.
[21] B. Hao, W.W. Sun, Y. Liu, G. Cheng, Simultaneous clustering and estimation of heterogeneous graphical models, Journal of Machine Learning Research 18 (217) (2018) 1–58.
[22] C. Hsieh, A. Banerjee, I.S. Dhillon, P.K. Ravikumar, A divide-and-conquer method for sparse inverse covariance estimation, in: Advances in Neural Information Processing Systems (NeurIPS'12), 2012, pp. 2330–2338.
[23] V. Kalofolias, How to learn a graph from smooth signals, in: Proceedings of the 19th International Conference on Artificial Intelligence and Statistics, vol. 51, 2016, pp. 920–929.
[24] T.N. Kipf, M. Welling, Variational graph auto-encoders, in: Bayesian Deep Learning Workshop, Advances in Neural Information Processing Systems (NeurIPS), 2016.
[25] O. Knill, Cauchy–Binet for pseudo-determinants, Linear Algebra and Its Applications 459 (2014) 522–547.
[26] S. Kumar, J. Ying, J.V.M. Cardoso, D.P. Palomar, A unified framework for structured graph learning via spectral constraints, Journal of Machine Learning Research 21 (2020) 1–60.
[27] S. Kumar, J. Ying, J.V. de M. Cardoso, D.P. Palomar, Bipartite structured Gaussian graphical modeling via adjacency spectral priors, in: 53rd Annual Asilomar Conference on Signals, Systems, and Computers, 2019.
[28] S. Kumar, J. Ying, J.V. de, M. Cardoso, D.P. Palomar, Structured graph learning via Laplacian spectral constraints, in: Advances in Neural Information Processing Systems (NeurIPS), 2019.
[29] B.M. Lake, J.B. Tenenbaum, Discovering structure by learning sparse graph, in: Proceedings of the 33rd Annual Cognitive Science Conference, 2010.
[30] S.L. Lauritzen, Graphical Models, vol. 17, Clarendon Press, 1996.
[31] Y. LeCun, C. Cortes, C. Burges, MNIST handwritten digit database, in: ATT Labs [Online], 2010, https://yann.lecun.com/exdb/mnist.
[32] Q. Liu, M. Allamanis, M. Brockschmidt, A.L. Gaunt, Constrained graph variational autoencoders for molecule design, in: Advances in Neural Information Processing Systems (NeurIPS), 2018.
[33] R.N. Mantegna, Hierarchical structure in financial markets, The European Physical Journal B 11 (1) (1999) 193–197.

[34] A.G. Marques, S. Segarra, G. Leus, A. Ribeiro, Sampling of graph signals with successive local aggregations, IEEE Transactions on Signal Processing 64 (7) (2016) 1832–1843.
[35] G. Marti, F. Nielsen, M. Bińkowski, P. Donnat, A review of two decades of correlations, hierarchies, networks and clustering in financial markets, arXiv:1703.00485, 2017.
[36] G. Mateos, S. Segarra, A.G. Marques, A. Ribeiro, Connecting the dots: identifying network structure via graph signal processing, IEEE Signal Processing Magazine 36 (3) (2019) 16–43.
[37] M.E.J. Newman, Modularity and community structure in networks, Proceedings of the National Academy of Sciences of the United States of America 103 (2006) 8577–8582.
[38] F. Nie, X. Wang, C. Deng, H. Huang, Learning a structured optimal bipartite graph for co-clustering, in: Proceedings of the 31st International Conference on Neural Information Processing Systems, Neurips'17, 2017, pp. 4132–4141.
[39] F. Nie, X. Wang, H. Huang, Clustering and projected clustering with adaptive neighbors, in: Proceedings of the 20th ACM SIGKDD International Conference on Knowledge Discovery & Data Mining, KDD'14, 2014.
[40] F. Nie, X. Wang, M.I. Jordan, H. Huang, The constrained Laplacian rank algorithm for graph-based clustering, in: Proceedings of the Thirtieth AAAI Conference on Artificial Intelligence, AAAI'16, 2016, pp. 1969–1976.
[41] N. Parikh, S. Boyd, Proximal algorithms, Foundations and Trends in Optimization 1 (3) (2014) 127–239.
[42] E. Pavez, H.E. Egilmez, A. Ortega, Learning graphs with monotone topology properties and multiple connected components, IEEE Transactions on Signal Processing 66 (9) (2018) 2399–2413.
[43] S.I. Resnick, Heavy-Tail Phenomena: Probabilistic and Statistical Modeling, Springer-Verlag, New York, 2007.
[44] H. Rue, L. Held, Gaussian Markov Random Fields: Theory and Applications, Chapman & Hall/CRC, 2005.
[45] S. Segarra, A.G. Marques, G. Mateos, A. Ribeiro, Network topology inference from spectral templates, IEEE Transactions on Signal and Information Processing over Networks 3 (3) (2017) 467–483.
[46] M. Slawski, M. Hein, Estimation of positive definite M-matrices and structure learning for attractive Gaussian Markov random fields, Linear Algebra and Its Applications 473 (2015) 145–179.
[47] S. Sun, Y. Zhu, J. Xu, Adaptive variable clustering in Gaussian graphical models, in: Proceedings of the Seventeenth International Conference on Artificial Intelligence and Statistics, vol. 33, 2014, pp. 931–939.
[48] Y. Sun, P. Babu, D.P. Palomar, Majorization-minimization algorithms in signal processing, communications, and machine learning, IEEE Transactions on Signal Processing 65 (3) (2017) 794–816.
[49] K.M. Tan, D. Witten, A. Shojaie, The cluster graphical lasso for improved estimation of Gaussian graphical models, Computational Statistics & Data Analysis 85 (2015) 23–36.
[50] X. Wang, J. Ying, J.V.M. Cardoso, D.P. Palomar, Efficient algorithms for general isotone optimization, in: The Thirty-Sixth AAAI Conference on Artificial Intelligence (AAAI), 2022.
[51] Y. Wang, U. Roy, C. Uhler, Learning high-dimensional Gaussian graphical models under total positivity without adjustment of tuning parameters, in: Proceedings of the Twenty Third International Conference on Artificial Intelligence and Statistics, vol. 108, 2020, pp. 2698–2708.
[52] D.M. Witten, J.H. Friedman, N. Simon, New insights and faster computations for the graphical lasso, Journal of Computational and Graphical Statistics 20 (4) (2011) 892–900.
[53] D.M. Witten, R. Tibshirani, Covariance-regularized regression and classification for high dimensional problems, Journal of the Royal Statistical Society. Series B (Statistical Methodology) 71 (3) (2009) 615–636.
[54] S.J. Wright, Coordinate descent algorithms, Mathematical Programming 151 (2015) 3–34.
[55] T.T. Wu, K. Lange, The MM alternative to EM, Statistical Science 25 (4) (2010) 492–505.
[56] Z. Wu, S. Pan, F. Chen, G. Long, C. Zhang, P.S. Yu, A comprehensive survey on graph neural networks, arXiv:1901.00596, 2019.
[57] J. Ying, J.V.M. Cardoso, D.P. Palomar, Does the ℓ_1-norm learn a sparse graph under Laplacian constrained graphical models?, arXiv:2006.14925, 2020.
[58] J. Ying, J.V.M. Cardoso, D.P. Palomar, Nonconvex sparse graph learning under Laplacian-structured graphical model, in: Advances in Neural Information Processing Systems (NeurIPS), 2020.

[59] J. Ying, J.V.M. Cardoso, D.P. Palomar, Minimax estimation of Laplacian constrained precision matrices, in: Proceedings of the 24th International Conference on Artificial Intelligence and Statistics, in: Proceedings of Machine Learning Research, vol. 130, PMLR, 2021, pp. 3736–3744.
[60] C.-H. Zhang, Nearly unbiased variable selection under minimax concave penalty, The Annals of Statistics 38 (2) (2010) 894–942.
[61] Y. Zhang, K.-C. Toh, D. Sun, 2020, Learning graph Laplacian with MCP.
[62] L. Zhao, Y. Wang, S. Kumar, D.P. Palomar, Optimization algorithms for graph Laplacian estimation via ADMM and MM, IEEE Transactions on Signal Processing 67 (16) (2019) 4231–4244.

CHAPTER

Dictionaries in machine learning

17

Kenneth Kreutz-Delgado[a,c], **Bhaskar Rao**[a], **Igor Fedorov**[d], **and Srinjoy Das**[b]

[a] *University of California San Diego, San Diego, CA, United States*
[b] *West Virginia University, Morgantown, WV, United States*
[c] *Pattern Computer Inc., Friday Harbor, WA, United States*
[d] *Meta, Menlo Park, CA, United States*

17.1 Data-driven AI via dictionary learning for sparse signal processing

In the beginning there was information. The word came later. The transition was achieved by the development of organisms with the capacity for selectively exploiting this information in order to survive and perpetuate their kind.

– Fred I. Dretske, 1981 [46]

What use can the brain make of the massive flow of sensory information that occurs without any associated rewards or punishments?

– Horace Barlow, 1989 [12]

How does one process information-laden streams of sensory data and extract thereof "dictionaries of words" whose subsequent recognition can lead to actionable consequences? The answer to this question is of immense importance given that the ability to learn representation vectors from observed signals increasingly is at the foundation of much of our modern autonomous and decision making technologies. The combination of machine learning and contemporary 21st-century statistical signal processing is at the frontier of research and development for sensor-driven artificial intelligence (AI). One only needs to search an academic internet database using the combination of the strings "dictionary learning" and "machine learning" to obtain thousands of hits and thereby quickly discover that dictionary learning has become a primary implement in the contemporary toolkit of modern machine learning in sensor-driven environments. This fact has influenced a new generation of machine learning textbooks written to ensure that the next generation of data scientists will have a requisite mastery of advanced tools of contemporary statistical signal processing *and* the techniques of machine learning [166,105,168,90]; a symbiosis critical for creating algorithms capable of interfacing with the physical world in an intelligent and purposeful manner.

As well outlined and described in [153], the approximately 20-year period between 1990 and 2010 witnessed a sea change from a mature basis-based (Fourier, Laplace, etc.) approach to signal processing to a rigorous frame-based (wavelets, dictionaries, etc.) perspective that has greatly enhanced our ability

Signal Processing and Machine Learning Theory. https://doi.org/10.1016/B978-0-32-391772-8.00023-5
Copyright © 2024 Elsevier Inc. All rights reserved.

to represent, process, and extract information from signals detected in the world via a variety of sensory modalities.[1] This 21st-century ("not your grandfather's") signal processing science is beginning to find its way into the basic curricula of undergraduate education [166] and has become an important, and even central, theme in some introductory-level graduate textbooks [168,114].

In this chapter we emphasize the fact that *the natural sensory world has sparse structure that can be learned* in a data-driven manner. This learned structure can be used not only for applications in classical signal processing (such as sensing, channel equalization, compression, denoising, interpolation/infilling, etc.), but also for AI applications involving discrimination, classification, regression, and explanation of complex events sensed in the real physical world. The natural world is the world that all living organisms have evolved to survive in, and within which intelligent robots, autonomous vehicles, continuous glucose monitoring sensors, earthquake warning systems, greenhouse gas detectors, and all other sensor-based reasoning systems must effectively function. The appropriate AI tools for such applications cannot merely only be applicable to simple webpage click-count contingency tables stored in a spreadsheet, but must be able to interface nicely with and process features extracted from sensory data, often streaming and often in real time. Any successfully autonomous agent, real or synthetic, that navigates or interfaces with the natural, physical world must be able to sense that world in a manner that allows for the extraction of actionable, and even interpretable, information. Data scientists who can function at the interface of the sensory world and AI systems must have a sophisticated knowledge of how to combine the analysis and synthesis tools of both machine learning and statistical signal processing. For this reason, we present dictionary learning as an important data-driven methodology for learning and encoding structure extracted from observations of domains encountered in the natural world that enhances our understanding of those domains in support of machine learning applications.

The section topics are as follows. Sections 17.2 and 17.3 state the sparse representation problem; define terminology; discuss the difference between designed and learned dictionaries; and describe some of the commonly used algorithms to obtained a sparse representation vector $\boldsymbol{x} \in \mathbb{X} = \mathbb{R}^m$ for a given observation vector $\boldsymbol{y} \in \mathbb{Y} = \mathbb{R}^n$. Section 17.4 derives the two most commonly used dictionary learning algorithms, the Method of Moments (MOM) and K-singular value decomposition (K-SVD) algorithms, and describes their relationship to K-means clustering, PCA, and factor analysis (FA). Section 17.5 describes the Bayesian approach to dictionary learning, derives the Sparse Bayesian Learning (SBL) dictionary learning algorithm, and discusses the SBL algorithm from an information theory perspective. Section 17.6 describes the nonnegative matrix factorization (NMF) approach to dictionary learning within the SBL framework. Section 17.7 discusses some connections between manifold learning and dictionary learning. Section 17.8 discusses the use of dictionary learning for clustering and classification. In Section 17.9 we discuss how the stacking of dictionaries can be used for feature extraction at multiple scales of resolution and can be combined with tasks such as classification or regression. In Section 17.10 we discuss an approach to kernelizing dictionary learning. Finally, in Section 17.11 we close with some final comments.

[1] The survey references [153,169,93] discuss the important, path-breaking leap in the rigor of the formalism of sparse and compressive sensing that was provided by the seminal papers [25,44]. The contemporary sparse signal processing approach is described at length (in roughly chronological order) in the following books and specialist monographs: [157,49,52,63,139,111,150,164,176].

This chapter's goals are primarily pedagogical. For this purpose, notationally and conceptually, we have been greatly influenced by reference [168] for describing the statement of the sparse inverse problem[2] and by [153,174] when motivating the dictionary learning problem. Indeed, these are excellent first-encounter resources the reader might read before, while, or after exploring the content of this chapter. For going beyond these introductory resources there are many followup references, many of which are cited herein.

17.2 Sparsity and sparse representations

17.2.1 *"Something rather than anything"* – sparse structure in the natural world

That the world impinges on our sensory modalities in a statistically succinct and structured manner that can be exploited via a process of "redundancy removal" is an empirical fact that has been noted by scientists concerned with animal sensory systems for several decades [9,11,132,133].

Suppose we were to encounter a two-dimensional sensory world having completely uniform independent statistical structure at all scales, i.e., an unconstrained, maximum-entropy sensory world. A "typical" image drawn from such a distribution[3] would look like Fig. 17.1(A) in the limit that the spatial scale goes to zero, corresponding to spatially and temporally independent noise that has equal power at all frequencies in all spatial directions. In this limit, Fig. 17.1(A) becomes a white blank with no discernable structure at any scale; i.e., seeing "anything" yields "nothing of interest". In this unconstrained maximum-entropy sensory world, an image like Fig. 17.1(B) is *not* typical and perceiving such an image would be a probability zero event.

Yet perception of the natural world reveals quite the opposite [9,11,132,133]. There is structure in the natural world [45,104], and under the pressure of survival selection the sensory organs of animals have evolved to efficiently exploit this structure [183,22,23,182]. In the natural world the image shown in Fig. 17.1(b) is a realized *something* that is of interest precisely because it differs from other realized somethings in a distributionally nontrivial way. Data scientists and practitioners of modern 21st-century statistical signal processing have come to understand that we have much to learn about nature's solutions (witness, e.g., the highly evolved primate visual system) to the problem of understanding the structured natural world from sensory data [62,45,200,175,82,174,201,104].

The structure that exists in the natural world can be learned and codified in dictionaries whose elements (a.k.a. *atoms*) serve as "words" that each succinctly capture an atomic piece of that structure. A picturesque metaphor is to think of an overcomplete dictionary as a huge lexicon of words that enables one to find just the right succinct expression of a few words eloquently combined. Rather than struggle to convey a complicated idea using a small dictionary with a restricted vocabulary, with an appropriately large dictionary one has access to the single, most apt word that perfectly describes the situation. Thus in a large array of possible words (all contained in an overcomplete dictionary) we search for a *few* highly *meaningful* ones. Thus a quest for sparsity is often a quest for meaning.

2 Specifically we highly recommend Chapters 9 and 10 of [168] for descriptive introductions and explanations of concepts and algorithms that are too lengthy to provide here.

3 For a rigorous definition of "typicality" see [37].

A) Color Television Noise B) Camera Image of La Jolla, California

FIGURE 17.1

Natural images. The natural world generates data with low-dimensional structure [62,133,45].

17.2.2 Sparsity: "*meaning*" versus "*information as choice of possibilities*"

It is important to distinguish between "meaning" and "information" in the modern *engineering technical sense* of the word "information." There is an unfortunate confusion between the use of the word "information" in the vernacular sense and that used in its technical information theoretic sense. In its vernacular sense the denotation of the word "information" is conflated with that of the word "meaning" (as in the text message string "FYI"), whereas its technical sense it stands for "range of choice of possible outcome" or "range of choice of possibility."[4] The separation of the concepts of "meaning" and "choice" in the modern theory of information founded by Shannon [159,160] was a key step in the development of a practically useful mathematical theory of information [144,165].

"Entropy" is a scalar measure of how much choice exists[5] and a technical measure of information in the sense that once one knows what choice of possibilities occurred, then all other possibilities are ruled out. We stress again that this definition is independent of the possibility of the existence of any "meaning" associated with that choice (which could be due to, e.g., instantiation of a purely random noise). Any meaning that does happen to be attached to the occurrence of that particular choice is either apparent to some human (or animal) that becomes aware of that choice, or else is "meaningful" in an actionable sense if its occurrence triggers subsequent automatic/mechanical consequences. Because of

[4] As expressed by Shannon and Weaver in [160]: "Information must not be confused with meaning," and "Information is a measure of one's freedom of *choice* when one selects a message," [emphasis added]. This can be stated using the axiomatic setup of a probability space $(\Omega, \mathcal{F}, \mu)$ with a finite sample space $|\Omega| < \infty$. The primary outcomes of interest are the probability atoms of the σ-algebra $\mathcal{F}$; the (finite in this case) number of such possible atomic outcomes and knowledge of which of these atoms occurs (or could occur) as an experimental outcome is directly related to information in the technical sense ("which outcome did nature *choose*"). For excellent pedagogic discussions of these matters, see [144,165]. For a detailed, mathematically rigorous discussion, see [117].

[5] Shannon entropy is actually a measure of average "surprise." But since surprise directly relates to what choice of outcomes occurred, our choice of wording should be harmless enough.

the definition of entropy as a measure of "information as choice of outcome" one needs to be mindful of the fact that noise and an intentional signal both have information, and that generally the sum of them has even more information, and that therefore removing the noise from the sum to retrieve the signal is a loss of ("bad") information. Because of difficulties associated with "filtering" out noise, it is common to work with noiseless models[6] or, if that is not possible, to distinguish between "signal" and "noise" by assuming that "signals" have statistical structure (i.e., correlations among the signals) while "noise" does not (i.e., noise is uncorrelated, a.k.a. "white noise").[7] We assume in this chapter that the reader has some familiarity with information theory; if that is not the case, then the brief summary given in [8] is highly recommended.[8]

To maximize entropy is to maximize choice; to minimize entropy is to minimize choice. Note that extracting sparse structure from sensory data generally corresponds to a minimization of choice (since many possibility entities that do not fit the structure are "zeroed out"). Often a search for sparse structure is a search for *meaning*, i.e., for "information" in the vernacular sense, *not for information in the technical sense*. And to look for structured patterns is to ask what is irrelevant and can be thrown away. This is the reason that in the pattern recognition literature one encounters papers with titles like "Pattern Recognition as a Quest for Minimum Entropy" [180]. To minimize entropy is to minimize choice, often a chaotic range of choices, and it is common, especially in the theory of thermodynamics, to associate low entropy with order and high entropy with disorder. Thus a search for a sparse (low-entropy) representation $\boldsymbol{x}$ of an observation $\boldsymbol{y}$ is a search for order that denotes a pattern and a pattern can be a signifier of a meaningful situation to a human being or AI system. However, one must *not* do this in a manner that throws away "good" information (such as contained in a statistically structured signal of interest), but preferentially (nearly only) gets rid of "bad" information (such as white noise). This can be done by entropy minimization of *factorial representations* that are simultaneously constrained to faithfully represent observed data. These points will be discussed later below once an appropriate mathematical framework has been set up.

17.2.3 The linear sparse generative model

First consider the noiseless model case

$$\boldsymbol{y} = A\boldsymbol{x} \quad \text{with} \quad A \in \mathbb{R}^{n\times m} \quad \text{and} \quad \mathsf{rank}(A) = n. \tag{17.1}$$

We say that the full row-rank matrix A is *fat* when $m > n$ and *square* when $m = n$. We refer to either A or the collection of its m columns as a *dictionary*, where the distinction between these two cases should be clear from the context. In this chapter we are primarily concerned with the situation when A is *fat and full-rank*, in which case the system is underdetermined and the dictionary A (equivalently, its set of columns) is said to be *overcomplete*.[9]

[6] Even this is not true, it may lead to acceptable results.

[7] The problem of dealing with correlated noise is generally nontrivial and, in this case, it is not unusual, at least initially, to distinguish between noise and signal.

[8] An excellent and very accessible introduction to information theory is [165]. The very readable modern classic account is [37].

[9] Note that the noiseless overcomplete dictionary model (17.1) is different from the standard principal component analysis (PCA) model which assumes a (possibly rank-deficient) square matrix, $m = n$, that can be derived from an estimate of the covariance matrix of $\boldsymbol{y}$ when $\boldsymbol{y}$ is random [115,88].

To know the model (17.1) means that we know the dictionary A and any statistical assumptions that might hold on the vector $\boldsymbol{x}$ if it is assumed to be random.[10] For any observed $\boldsymbol{y} \in \mathbb{Y} = \mathbb{R}^n$ we assume that it was *generated* from some vector $\boldsymbol{x} \in \mathbb{X} = \mathbb{R}^m$ via the above model.[11] Given the assumption that (17.1) serves as a *generative model* of $\boldsymbol{y}$, then given an observation $\boldsymbol{y}$ we can ask when there exists a unique generator $\boldsymbol{x}$ and how it can be found. These and other questions are the subject matter of sparse signal processing and compressive sensing.

More realistically, one can consider the model-with-error case,

$$\boldsymbol{y} = A\boldsymbol{x} + \boldsymbol{\epsilon}, \tag{17.2}$$

where a variety of stochastic and/or nonstochastic modeling assumptions can be placed on the noise error term $\boldsymbol{\epsilon}$. For example, one can have a stochastic "signal-plus-noise" model

$$\boldsymbol{y} = \boldsymbol{s} + \boldsymbol{n}, \quad \boldsymbol{s} = A\boldsymbol{x}, \tag{17.3}$$

where $\boldsymbol{s} = A\boldsymbol{x}$ satisfies the modeling assumptions of (17.1) and $\boldsymbol{n} = \boldsymbol{\epsilon}$ is a random measurement noise, a situation that corresponds to a noisily observed signal of interest that is sparsely generated. Or $\boldsymbol{\epsilon}$ can be a deterministic modeling error,

$$\boldsymbol{\epsilon} \stackrel{\text{def}}{=} \boldsymbol{y} - \hat{\boldsymbol{y}}, \quad \hat{\boldsymbol{y}} = A\boldsymbol{x},$$

where $\hat{\boldsymbol{y}}$ is an approximation to $\boldsymbol{y}$ that has a sparse representation, in which case we say that $\boldsymbol{x}$ provides a *sparse approximation* of $\boldsymbol{y}$, a situation that is encountered when one is interested in a compressed representation $\boldsymbol{x}$ of an observation (say an image) $\boldsymbol{y}$. And, of course, one can combine these two scenarios.[12]

In all cases considered above, noisily observed or not, our interest is in solving an inverse problem of the form $\boldsymbol{\eta} = A\boldsymbol{x}$ for $\boldsymbol{\eta} = \boldsymbol{y}, \boldsymbol{s}$, or $\hat{\boldsymbol{y}}$. To a large extent we focus on the paradigmatic error-free model (17.1)[13] and we interchangeably refer to $\boldsymbol{x}$ as either a *solution* (to $\boldsymbol{y} = A\boldsymbol{x}$) or as a *representation* (of $\boldsymbol{y}$). We define the *sparsity* $\mathsf{sp}(\boldsymbol{x})$ of the n-vector $\boldsymbol{x}$ to be its number of nonzero elements.[14] Evidently, because $\mathsf{rank}(A) = n$ means that A must have n linearly independent columns, $0 \leq \mathsf{sp}(\boldsymbol{x}) \leq n$. It is the case that $\mathsf{sp}(\boldsymbol{x}) = \|\boldsymbol{x}\|_0$, where

$$\|\boldsymbol{x}\|_0 \stackrel{\text{def}}{=} \lim_{p \to 0+} \left(\sum_{j=1}^{m} |x_j|^p \right)^{\frac{1}{p}}. \tag{17.4}$$

Here $\|\boldsymbol{x}\|_0$ gives the count of the nonzero components of $\boldsymbol{x}$ and is often called (as an usually harmless abuse of terminology) the "zero-norm." By definition, $\boldsymbol{x}$ is said to be *k-sparse* if its sparsity is *at most* k, $\mathsf{sp}(\boldsymbol{x}) \leq k$.

---

[10] Unless specified otherwise, the default assumption is that $\boldsymbol{x}$ is deterministic or conditioned on.

[11] Unless otherwise stated, we assume $\mathbb{Y} = \mathbb{R}^n$ and $\mathbb{X} = \mathbb{R}^m$.

[12] Analogously to Footnote 9, note that the noisy overcomplete dictionary model (17.2) is different from the standard FA model that applies when $\boldsymbol{y}$ is random and A is assumed to be *tall*, $n > m$, and one-to-one [115,88].

[13] Which has great practical utility even in the case of modeling error provided the error is "negligible."

[14] The sparsity of $\boldsymbol{x}$ should not be confused with the *spark* of the dictionary matrix A, $\mathsf{spark}(A)$, to be mentioned below.

17.2.4 Sufficient conditions for a unique sparse solution

So far the only conditions placed on the $n \times m$ dictionary A are that it is fat and has full (row) rank, $\mathsf{rank}(A) = n \leq m$. Subject to only these conditions, which imply that A must have at least one subset of n linearly independent columns, every observed $\boldsymbol{y} \in \mathbb{R}^n = \mathsf{range}(A)$ is guaranteed to have some (not necessarily unique) k-sparse representation $\boldsymbol{x}$ with $k \leq n$. Additional conditions that one can place on the dictionary A to guarantee *uniqueness* of a sufficiently sparse solution to $\boldsymbol{y} = A\boldsymbol{x}$ are naturally of keen interest. A sufficient condition based on a unique representation property (URP) (defined below) was given in [75]. This was followed a few years later by the path-breaking papers [25,27,44], which led to a variety of sufficient conditions based on the concepts of *spark*, *mutual coherence*, and the *restricted isometry property* (RIP). In the interest of space we briefly describe the spark of a matrix followed by some general comments.[15]

Following [44], for a full row-rank matrix A one defines the *spark* of A as

$$\mathsf{spark}(A) = \text{smallest number of columns of } A \text{ which are linearly } \textit{dependent} \leq n + 1. \tag{17.5}$$

Note that by definition any $k < \mathsf{spark}(A)$ columns selected from A are linearly *independent*. By definition we say that the *unique representation property* (URP) holds for an $n \times m$ full row-rank matrix A when $\mathsf{spark}(A) = n + 1$ [75]. A key result is the following sufficient condition for uniqueness:

- Given $\boldsymbol{y}$, a full row-rank matrix A, and a solution $\boldsymbol{x}$ such that $\boldsymbol{y} = A\boldsymbol{x}$, $\boldsymbol{x}$ is unique if

$$\mathsf{sp}(\boldsymbol{x}) < \frac{1}{2}\mathsf{spark}(A).$$

As a corollary we have that if A satisfies the URP, then a sparse solution $\boldsymbol{x}$ is unique if $\mathsf{sp}(\boldsymbol{x}) < (n + 1)/2$.

For arbitrary matrices of even moderate size, the direct verification of the sufficient conditions based on spark, mutual coherence, or the URP property is computationally intractable. For this reason one often works with structured families of *designed matrices*[16] that provably satisfy the sufficient conditions. Alternatively, as discussed in this chapter, one can *learn dictionaries* from observational data that, at least empirically, provide stable sparse representations of the data.[17] One way to build designed matrices is to randomly and independently select the elements of the dictionaries as such dictionaries can be shown to satisfy the desired sufficient conditions (and in particular the URP) with high probability. In particular, such a randomly selected matrix ensures that the *uniform uncertainty principle* (UUP) holds [26], which is sufficient to guarantee the URP-related sufficient conditions for uniqueness of a sufficiently sparse solution. This fact enhances the belief that learning dictionaries from noisy natural data will also satisfy the sufficient conditions with high probability and that sufficiently sparse solutions (in particular those that satisfy the URP-justified condition $\mathsf{sp}(\boldsymbol{x}) < n/2$ [75]) are likely to be unique with high probability.

[15] For more details see the very accessible discussion of these results given in Chapter 9 of [168].

[16] By a "designed matrix" we mean a matrix that has been built ("designed") to satisfy specified properties. Note that this should not be confused with a *design* matrix (a.k.a. *data matrix)* whose columns or rows are comprised of observed data vectors.

[17] Somewhat confusingly, if one learns (estimates) dictionaries from data for specific application purposes, say for classification, such dictionaries are also sometimes referred to as "designed" in the literature. We eschew this latter terminology: In this chapter, by "designed" we mean something akin to "handcrafted to meet specified constraints."

17.2.5 Sparsity classes

Each representation $\boldsymbol{x} \in \mathbb{R}^m$ of $\boldsymbol{y} = A\boldsymbol{x}$ has a distinct *sparsity pattern* or *sparsity class* c_x defined by $j \in c_x$ if and only if $x_j \neq 0$. The cardinality of the set c_x is the sparsity $|c_x| = \mathsf{sp}(\boldsymbol{x})$. For $|c_x| = k$, c_x has the form

$$c_x = (j_1, \quad \cdots \quad , j_k)\,.$$

The collection of all possible sparsity patterns is denoted by $\mathsf{S}(m)$, $c_x \in \mathsf{S}(m)$, and the number of distinct possible sparsity classes is $|\mathsf{S}(m)| = 2^m$. In particular, the number of sparsity classes c_x with sparsity $k = |c_x|$ is $\binom{m}{k}$ and the sum of these numbers is 2^m.

If we assume that we are working with a unique sparse representation of $\boldsymbol{y}$, then we can reference the sparsity pattern as $c_y = c_x$. Often we will make this assumption, in which case the sparsity class will be denoted as c_y, or even just c when $\boldsymbol{y}$ is tacitly understood.

Knowledge of the sparsity class c_x allows us to write

$$\boldsymbol{y} = A\boldsymbol{x} = \sum_{j=1}^{m} x_j \boldsymbol{a}_j = \sum_{j \in c_x} x_j \boldsymbol{a}_j = \sum_{j=1}^{m} s_{jc_x} x_j \boldsymbol{a}_j \quad \text{for} \quad s_{jc_x} = \mathbf{1}\{j \in c_x\}, \tag{17.6}$$

where x_j is the jth component of $\boldsymbol{x}$, $\boldsymbol{a}_j$ is the jth column of A, and $\mathbf{1}\{\cdot\}$ denotes the 0–1 indicator function where the value 1 indicates that the argument is a true statement. The 0–1 indicators s_{jc_x} are elements of an m-vector representation of the class c_x denoted by $\boldsymbol{s}_{c_x}$ such that

$$x_j \neq 0 \iff s_{jc_x} = 1 \text{ and } x_j = 0 \iff s_{jc_x} = 0.$$

Given an observation $\boldsymbol{y}$, assume that it has a unique *maximally sparse*[18] representation $c = c_y = c_x$. Given the $n \times m$ full rank, fat dictionary matrix $A = (\boldsymbol{a}_1 \quad \cdots \quad \boldsymbol{a}_m)$, define the $m \times k$ *subdictionary* (submatrix) A_c by

$$A_c = (\boldsymbol{a}_{i_1} \quad \cdots \quad \boldsymbol{a}_{i_k}) \in \mathbb{R}^{n \times k} \quad \text{and}$$
$$\boldsymbol{x}_c^T = (x_{i_1} \quad \cdots \quad x_{i_k}) \in \mathbb{R}^k \quad \text{for distinct} \quad i_j \in c,\ |c| = k \leq \frac{n+1}{2}.$$

We then have

$$\boldsymbol{y} = A\boldsymbol{x} = A_c \boldsymbol{x}_c.$$

Conceptually, this shows that one way to solve the underdetermined inverse problem $y = Ax$ for the unique maximally sparse solution x is via a two-step procedure:

S1: Given $\boldsymbol{y}$ determine its maximal sparsity class $c = c_y$.

S2: Compute $\boldsymbol{x}_c = A_c^+ \boldsymbol{y} = \left(A_c^T A_c\right)^{-1} A_c^T \boldsymbol{y}$.

---

[18] By this we mean that $k = |c|$ is a minimum. Note that to be *maximally sparse* is to have a *minimum* value for the *sparsity* k. For this reason, it would seem better to refer to k as the *diversity* [140], but the current terminology seems entrenched in the sparse signal processing literature. However, below we will refer to functions that we minimize to maximize the "sparsity" k as *diversity measures*.

The classification step S1 is generally a nonlinear one. Step S2 is the standard (linear) pseudoinverse solution $x_c = A_c^+ y$ for a full-rank overdetermined linear system of equations $y = A_c x_c$.

Note that the classification step S1 provides pointers to each of the $k = |c|$ columns that (when weighted) sum up to provide a sparse representation of $y \in \mathsf{range}(A_c)$. If we force k to take the value $k = 1$, the two-step procedure corresponds to vector quantization (VQ), which is a coding algorithm that selects a single vector (in this case, a weighted single column of A) to serve as an exemplar for measurements in the vicinity of the measurement y [65]. Thus the two-step procedure S1 and S2, and more generally the sparse inverse problem itself, can be viewed as a generalization of VQ.[19]

17.2.6 Bases, frames, dictionaries, and sparsity classes

Given an $n \times m$ dictionary matrix $A = (a_1 \quad \cdots \quad a_m)$ which is onto we have that for all $y \in \mathbb{R}^n$,[20]

$$y \in \mathbb{R}^n = \mathsf{range}(A) = \mathsf{span}(\mathcal{D}) \text{ with } \mathcal{D} = \{a_1, \cdots, a_m\} \text{ and } m \geq n.$$

The *dictionary* $\mathcal{D}$ is also known as a *frame*. If $m = n$, then $\mathcal{D}$ is a *basis*, also called a *tight frame* or a *complete frame*. If $m = n$ and the elements of $\mathcal{D}$ are orthonormal, then $\mathcal{D}$ is an *orthonormal basis*.

Associated with the basis vectors of a general (not necessarily orthonormal) tight frame is a *dual basis* that is also a tight frame such that the basis and dual basis are biorthonormal.[21] An orthonormal basis is a tight frame for which the basis and the dual basis are identical. If $m > n$ the dictionary (frame) is said to be *overcomplete* and if $m < n$ the frame is *undercomplete*. Consider the system $y = Ax$ for arbitrary y: When A is overcomplete the system is underdetermined; when A is complete the system is determined; and when A is undercomplete the system is overdetermined.

The column vector a_j is referred to as a *word* or *atom* of the dictionary/frame $\mathcal{D}$. At times, we also refer to a_j as a *component* by analogy with PCA. As nicely described in [153], as one moves from an orthonormal basis to a tight frame to an overcomplete dictionary one increases the expressive power of a solution to the inverse problem $y = Ax$, whereas as one moves in the reverse direction from overcomplete dictionary to tight frame to orthonormal basis one increases the computational ease of solving the inverse problem, i.e., there is a trade-off between expressiveness and computational ease. Given an overcomplete dictionary $\mathcal{D}$, one might be interested in extracting n elements to serve as a basis (tight frame) that appears to be very good for subsequent use for finding representations of new observations y; such a methodology falls under the rubric of "best basis selection."

For a large enough dictionary $m \gg n$ the probability that there is a single word a_j, or a very small number of words, that with high accuracy expresses y up to a constant multiple α, $y \approx \alpha a_j$, becomes high. When a single word can always do the job, we have sparsity $k = 1$, in which case the problem becomes equal to VQ, and when $k > 1$ is the case, then the problem is a generalization of VQ as briefly mentioned at the end of the previous Subsection 17.2.5. From the perspective of the discussion given in that subsection, we can view a particular selection of words to represent y as a choice of a sparsity

[19] This perspective has proven to be very fruitful for creating dictionary learning algorithms [54,53,2,4], as discussed below.

[20] The *span* of a set of vectors is the set of all linear combinations of the vectors.

[21] Every Banach space (i.e., normed linear vector space) $\mathcal{Y}$ with a countable basis (where in our case we have $\mathcal{Y} = \mathbb{R}^n$) has a dual Banach space $\mathcal{Y}^*$ of bounded linear operators with a countable dual basis. If $\mathcal{Y}$ is also a Hilbert space (in our case $\mathbb{R}^n$ with the standard Euclidean metric), then it is self-dual so that the space and its dual can be treated as the same space $\mathcal{Y} \equiv \mathcal{Y}^*$. In this latter case, the dual basis, also in this case called the *reciprocal basis*, is an alternative basis for the space $\mathcal{Y}$. See [106,92].

class $c_y \in \mathsf{S}(m)$ where as m gets larger we have a greater chance of finding the perfect set of a very few words that can express $\boldsymbol{y}$ succinctly and even, if the dictionary is chosen appropriately, meaningfully.

17.3 Sparse signal processing

Here we briefly mention some of the primary algorithms used to solve the sparse representation problem given a dictionary A. More details on sparse signal processing and compressive sensing can be found in [24,169,168] and a succinct survey is given in [93].

17.3.1 Measures of sparsity and diversity

The most commonly used procedures for searching for sparse representations minimize explicitly defined scalar measures of *diversity*, $d(\boldsymbol{x})$. Minimizing diversity is equivalent to maximizing concentration (in particular, sparsity) and for this reason the same measure can be referred to both as a concentration measure and as a diversity measure. Thus one speaks of maximizing sparsity (a type of concentration) while typically actually implementing a diversity minimization algorithm.

There are many measures of diversity that have been considered for use in ecology and economics [140,80,181] and described in the sparse signal processing literature [81]. In this literature, there are a variety of properties that have been proposed as good ones for diversity measures to satisfy with a particularly useful one being concavity.[22,23] Unfortunately when constraints are operative strictly concave measures generally have a multitude of local minima which is problematic when utilizing optimization procedures to find sparse representations. One strictly concave diversity measure, and therefore subject to the problem of multiple local minima, often described in the literature is the Shannon entropy.[24] Perhaps the most commonly described diversity measures are the *p-diversity measures* $d(\boldsymbol{x}) = \mathsf{div}_p(\boldsymbol{x})$, where

$$\mathsf{div}_p(\boldsymbol{x}) \stackrel{\mathsf{def}}{=} \left(\sum_{j=1}^{m} |x_j|^p \right)^{\frac{1}{p}} \quad \text{for } 0 \le p \le 1. \tag{17.7}$$

The value of $\mathsf{div}_0(\boldsymbol{x})$ is defined by

$$\mathsf{div}_0(\boldsymbol{x}) = \lim_{p \to 0^+} \mathsf{div}_p(\boldsymbol{x}) = \mathsf{sp}(\boldsymbol{x}) = \|\boldsymbol{x}\|_0 = \sum_{j=1}^{m} \mathbf{1}\{x_j \neq 0\},$$

which is the count of the nonzero components of $\boldsymbol{x}$ described earlier. For $p < 1$, the p-diversity measure is strictly concave, while for $p = 1$ we have $\mathsf{div}_p(\boldsymbol{x}) = \|\boldsymbol{x}\|_1$, where the ℓ_1-norm $\|\cdot\|_1$, which is not *strictly* concave, is both concave and convex and satisfies the triangle inequality. Even for $p < 1$ the

[22] See [95,148,94] for brief comments and [81,116,7] for rigorous discussions of desirable properties of concentration measures.
[23] However, while being a desirable property, concavity is not a necessarily condition for being a useful diversity measure. For example, reference [132] uses the measure $\mathsf{d}(x_i) = \log(1 + x_i^2)$, which is not everywhere concave.
[24] For example, see reference [180].

p-diversity measures are referred to (erroneously, but generally harmlessly) as the "p-norm" diversity measures $\|x\|_p = \mathsf{div}_p(x)$.[25] A diversity measure $d(x)$ is said to be *additive* if[26]

$$d(x) = \alpha_1 \mathsf{d}_1(x_1) + \cdots + \alpha_m \mathsf{d}_m(x_m) \tag{17.8}$$

with $\alpha_j > 0$, $\mathsf{d}_j(x_j) \geq 0$, and $\mathsf{d}_j(x_j) = 0 \iff x_j = 0$ for $j = 1, \cdots, m$. A particularly popular diversity measure is the 1-norm diversity,

$$d(x) = \mathsf{div}_1(x) = \|x\|_1 = |x_1| + \cdots + |x_n|.$$

This diversity measure is convex (and concave) and almost everywhere differentiable.

The so-called "zero-norm" $\|\cdot\|_0 = \mathsf{div}_0(\cdot) = \mathsf{sp}(\cdot)$ and the 1-norm $\|\cdot\|_1 = \mathsf{div}_1(\cdot)$ measures are the diversity measures most frequently encountered in the contemporary sparse signal processing literature. Note that $\|x\|_0$ and $\|x\|_1$ are both additive. The (nondifferentiable) zero-norm measure $\|\cdot\|_0$ provides the conceptual ideal: Its value *is* the sparsity and is a minimum when sparsity is a maximum. On the other hand, the 1-norm measure $\|\cdot\|_1$ is differentiable and (nonstrictly) convex, and minimizing it can yield sparse solutions [168], though generally this solution is not maximally sparse. However, as noted in Subsection 17.2.4 above, sufficient conditions on the dictionary matrix can ensure that if minimizing $\|\cdot\|_1$ yields a "sparse enough" solution, then with high confidence that solution is unique and maximally sparse.

17.3.2 Sparsity-inducing optimizations of diversity measures

Ideally, sparse solutions are found by exactly solving the optimization problem

$$\min_x \mathsf{sp}(x) = \|x\|_0 \quad \text{subject to} \quad y = Ax. \tag{17.9}$$

Because this is a combinatorial problem which is NP-hard, solutions are commonly determined using heuristic greedy search algorithms where one greedily collects atoms one-at-a-time until a sparse, though generally not necessarily maximally sparse, solution is obtained. Commonly used sequential search greedy algorithms are *Matching Pursuit* (MP), *Orthogonal Matching Pursuit* (OMP), and *Compressive Sampling Matching Pursuit* (CoSaMP), among others [168]. If one assumes that the matrix A is UUP, or well matched to the data via learning, then a sufficiently sparse solution (e.g., $\mathsf{sp}(x) < n/2$) is assumed to be an optimal solution. One can try a variety of greedy selection orderings to increase the chance that such a solution will be found.

To handle the model-with-error case discussed in Eq. (17.2) *et seq.*, where $\epsilon = y - Ax \neq 0$, it is common to consider the optimization

$$\min_x \|y - Ax\|_2^2 \quad \text{subject to} \quad \mathsf{sp}(x) = \|x\|_0 \leq k \tag{17.10}$$

for a specified sparsity level k. This is a deterministic optimization problem that is useful in both the deterministic and stochastic error cases.

---

[25] As is well known, for $p \geq 0$ the right-hand side of Eq. (17.7) defines a true norm, the so-called p-norm $\|x\|_p$. However, the p-diversity measure $\mathsf{div}_p(x)$, which is defined for $p < 1$, does not satisfy the triangle inequality required of a true norm. Rather, because of the strict concavity, it satisfies the reverse triangle inequality.

[26] As discussed in Section 17.3.3 below, the additivity assumption corresponds to assuming a factorial Bayesian prior on x.

Because the above optimizations are typically difficult to perform, it is a common procedure to replace the nondifferentiable measure $\mathsf{sp}(\boldsymbol{x})$ by a differentiable diversity measure $d(\boldsymbol{x})$. Thus in the noiseless case one commonly relaxes (17.10) to the problem

$$\min_{\boldsymbol{x}} d(\boldsymbol{x}) \quad \text{subject to} \quad \boldsymbol{y} = A\boldsymbol{x}, \tag{17.11}$$

with $d(\cdot)$ being a differentiable diversity measure. Since a general diversity measure is nonconvex,[27] in general this will have many local minima. But as previously mentioned, if a sufficiently sparse solution can be found, it is not necessarily unreasonable to assume that it is an acceptable answer. For general, smooth concave measures one can determine local solutions to (17.11) via the use of *iteratively reweighted least-squares* [75,148,28,187].

A particularly nice choice of diversity measure is the convex/concave 1-norm $d(\boldsymbol{x}) = \mathsf{div}_1(\boldsymbol{x}) = \|\boldsymbol{x}\|_1$ for which (17.11) is a convex optimization problem that has a generically unique sparse solution that is amenable to solution via standard convex optimization procedures. Furthermore, the use of the 1-norm diversity measure leads to computational tractable optimization problems for handling the noisy model equation (17.2). In the noisy case, one looks for a solution $\boldsymbol{x}$ to one or another of the following three *equivalent* regularized optimization problems:

Sparse representations via regularized optimizations

Tikhonov [21,32]	**LASSO** [171]	**Basis pursuit (BP)** [32,24]	
$\min_{\boldsymbol{x}} \|\boldsymbol{y} - A\boldsymbol{x}\|_2^2 + \lambda\|\boldsymbol{x}\|_1$	$\min_{\boldsymbol{x}} \|\boldsymbol{y} - A\boldsymbol{x}\|_2^2$	$\min_{\boldsymbol{x}} \|\boldsymbol{x}\|_1$	(17.12)
	subject to $\|\boldsymbol{x}\|_1 \le \rho$	subject to $\|\boldsymbol{y} - A\boldsymbol{x}\|_2^2 \le \epsilon$	

which yield equivalent sparse solutions for appropriate choices of the nonnegative regularization parameters λ, ρ, ϵ [21]. These convex optimization problems have unique sparse solutions that can be obtained using a variety of convex optimization algorithms [21,32]. Further, if appropriate conditions exist on A to ensure that a sufficiently sparse solution is unique (see Section 17.2.4 above), then for appropriately selected values of the hyperparameters λ, ρ, and ϵ the unique solution to the ideal optimization problem (17.10) will be found [24,26,28,49,52,63,114,139,150,157,176].

17.3.3 The stochastic model and sparse Bayesian learning

Considering the stochastic noisy model (17.3),

$$\boldsymbol{y} = A\boldsymbol{x} + \boldsymbol{n}.$$

The sparsity algorithms described in the previous subset can be obtained within a Bayesian framework by assuming that $\boldsymbol{x}$ is random and taking

$$P(\boldsymbol{x}) = \exp(-\beta d(\boldsymbol{x}))/Z \quad \text{and} \quad P(\boldsymbol{n}) = \mathsf{N}(\boldsymbol{n}; 0, \sigma^2 \mathsf{I}) \tag{17.13}$$

[27] For example, if we consider a strictly concave measure $d(x) = \mathsf{div}_p(\boldsymbol{x})$ for $p < 1$.

with $\beta > 0$ and normalization factor ("partition function")

$$Z = \int \exp(-\beta d(\boldsymbol{x}))\, dx.$$

A maximum a posteriori (MAP) estimate of $\boldsymbol{x}$ is then given by

$$\hat{\boldsymbol{x}}_\lambda = \arg\max_{\boldsymbol{x}} P(\boldsymbol{x}|\boldsymbol{y}) = \arg\max_{\boldsymbol{x}} P(\boldsymbol{y}|\boldsymbol{x})P(\boldsymbol{x}) = \arg\min_{\boldsymbol{x}} \left(\|\boldsymbol{y} - A\boldsymbol{x}\|_2^2 + \lambda d(\boldsymbol{x}) \right) \tag{17.14}$$

with $\lambda = \beta/\sigma^2$. This provides a statistical interpretation of the Tikhonov regularization form of an inverse problem optimization,

$$\min_{\boldsymbol{x}} \|\boldsymbol{y} - A\boldsymbol{x}\|_2^2 + \lambda d(\boldsymbol{x}). \tag{17.15}$$

Note that taking $d(\boldsymbol{x}) = \|\boldsymbol{x}\|_1$ gives the LASSO/BP optimization discussed in the previous Subsection 17.3.2, while in the limit $\lambda \to 0^+$ we obtain $\hat{\boldsymbol{x}}_\lambda = \hat{\boldsymbol{x}}_0$, which is a solution to the problem (17.11). In the MAP formalism, the sparsity-inducing mechanism is due to the *explicit* presence of the diversity measure $d(\boldsymbol{x})$ in the resulting optimizations. Such a choice for $d(\boldsymbol{x})$ makes $P(\boldsymbol{x})$ a *sparsity-inducing prior* for $\boldsymbol{x}$. As we will discuss below, there are good computational and statistical reasons for taking $d(\boldsymbol{x})$ to be of the additive form[28]

$$d(\boldsymbol{x}) = \alpha_1 d_1(x_1) + \cdots + \alpha_2 \mathsf{d}_m(x_m), \tag{17.16}$$

which corresponds to the assumption of a *factorial prior distribution*,

$$P(\boldsymbol{x}) = \prod_{j=1}^{m} P(x_j), \quad P(x_j) = \exp\left(-\beta_j d_j(x_j)\right)/Z_j, \quad \beta_j = \beta\alpha_j, \quad Z = Z_1 \cdots Z_m. \tag{17.17}$$

As discussed further in Subsection 17.5.1 below, this framework is often referred to as a "Type I" estimation formalism. A richer, more general sparsity-enforcing hierarchical framework is described in Subsection 17.5.1 which in the machine learning literature is commonly referred to as *sparse Bayesian learning* (SBL) framework [172,56,173]. SBL is based on a stochastic model that induces sparsity via a different mechanism than the direct use of sparsity-inducing functions $\mathsf{d}(x_j)$. SBL is an (empirical) hierarchical Bayes formalism that takes the prior on $\boldsymbol{x}$ to be of a Gaussian factorial form with $P(x_j; \gamma_j) = \mathsf{N}(x_j; 0, \gamma_j)$,[29]

$$P(\boldsymbol{x}; \boldsymbol{\gamma}) = \mathsf{N}(\boldsymbol{x}; 0, \Gamma) = \mathsf{N}(x_1; 0, \gamma_1) \cdots \mathsf{N}(x_n; 0, \gamma_m) \quad \text{with} \quad \Gamma = \operatorname{diag}(\boldsymbol{\gamma}) = \operatorname{diag}(\gamma_1 \quad \cdots \quad \gamma_m),$$

where the "hyperparameters" γ_j themselves are also random variables.

Note that the factorizability of the prior is equivalent to the assumption that $\Gamma = \mathsf{diag}(\boldsymbol{\gamma}) = \mathsf{diag}(\gamma_1 \quad \cdots \quad \gamma_m)$ is diagonal and that each component x_j is assigned its own variance γ_j. SBL does

---

28 See Eq. (17.8).

29 Much of the SBL literature parameterizes the shown prior by the precision $1/\gamma_j$. We prefer to work with the variance γ_j.

not directly force the components of $\boldsymbol{x}$ to be sparse during the learning process. Instead, remarkably, it is some of the diagonal elements γ_j of Γ that become zero as they are learned, a learning process that is referred to as "Type II" estimation. *Note that it is the resulting sparsity of* $\operatorname{diag}\Gamma$ *that enforces sparsity of* $\boldsymbol{x}$ because $\gamma_j \equiv 0 \iff x_j = 0$ almost surely.

When $\gamma_j \neq 0$, we say that x_j is *relevant*, so SBL is a search for *relevance* [128,172,20,189]. The prior variances γ_j and the noise variance σ^2 can be learned by maximizing the *evidence* [107,20,168] (a.k.a. *marginal likelihood*) for the model parameters $\boldsymbol{\gamma}$ and σ^2,

$$\begin{aligned} P(\boldsymbol{y}|\gamma_1, \cdots, \gamma_m, \sigma^2) &= \int P(\boldsymbol{y}, \boldsymbol{x}|\gamma_1, \cdots, \gamma_m, \sigma^2) d\boldsymbol{x} \\ &= \int P(\boldsymbol{y}|\boldsymbol{x}, \gamma_1, \cdots, \gamma_m, \sigma^2) P(\boldsymbol{x}|\gamma_1, \cdots, \gamma_m, \sigma^2) d\boldsymbol{x}, \end{aligned}$$

which is a Gaussian distribution under our model assumptions. It is a remarkable[30] fact that maximizing this Gaussian likelihood with respect to the variances γ_j and σ^2 can yield sparsity as a result of many variance values γ_j becoming arbitrarily small so that $\gamma_j \to 0$ for those values [56,188]. Optimizing the evidence for every new observation $\boldsymbol{y}$ results in *automatic relevance determination* (ARD) of the components of its representation vector $\boldsymbol{x}$ [128,189]. Beyond the further discussion given below in Subsection 17.5.1, detailed introductions to the SBL formalism can be found in [20,168]. For deeper theoretical and algorithmic details see the foundational papers [172,56,173,190,188,191,189,192].

17.4 Dictionary learning I – basic models and algorithms

As noted above, within the theoretical foundation that now exists it is possible to build "designed dictionaries" that have desirable efficient sparsity encoding and decoding properties. However, the natural world presents us with signals that are structured according to principles that are still being elucidated and for which dictionaries can be learned that provide excellent encoding and representation capabilities even if it is computationally intractable to determine if they satisfy known sufficient conditions, or even when they likely do not satisfy such conditions. By learning dictionaries that match to the statistical structure of natural signals, one can obtain efficient sparse encoding/decoding capability while often even gaining some understanding of the nature of the phenomenon at hand or of the natural sensory mechanisms that are seen in the animal kingdom, such as the mammalian vision system [62,132,133,82,201].

For our purposes, dictionary learning is defined to be any *natural data*-driven procedure that leads to the parametric statistical estimation ("learning") of an overcomplete set of atoms (a.k.a. words) from which one can determine a sparse representation of any future signal encountered in the natural world of interest. In this section we detail some of the most commonly used dictionary methods.[31] Two of the most commonly used dictionary learning algorithms, the Method of Directions (MOD) and KSVD, can

[30] Remarkable because sparsity of the variances $\boldsymbol{\gamma}$ is generally not explicitly enforced. In particular this "unforced" behavior occurs if the sensory world satisfies the "Sparseland" assumption [50,49] as discussed in Subsection 17.5.3.

[31] These are discussed in a variety of textbooks [47,33,49,168,199]. In particular, an excellent way for one to learn about the subject is to use references [168] and [47] as an entry path into the subject.

be viewed as generalizations of VQ. In the next subsection we set up the general dictionary learning framework and in the subsequent subsection we briefly describe VQ.

17.4.1 Dictionary learning from observational data – the setup

Assume that we have a collection of independently observed data vectors $\boldsymbol{y}_\ell$, $\ell = 1, \cdots, L$, each of which is generated by the representation model (17.2):

$$\boldsymbol{y}_\ell = A\boldsymbol{x}_\ell + \boldsymbol{\epsilon}_\ell, \quad \ell = 1, \cdots, L \quad \Longleftrightarrow \quad \underbrace{\begin{pmatrix} \boldsymbol{y}_1 & \cdots & \boldsymbol{y}_L \end{pmatrix}}_{\boldsymbol{Y}} = A \underbrace{\begin{pmatrix} \boldsymbol{x}_1 & \cdots & \boldsymbol{x}_L \end{pmatrix}}_{\boldsymbol{X}} + \underbrace{\begin{pmatrix} \boldsymbol{\epsilon}_1 & \cdots & \boldsymbol{\epsilon}_L \end{pmatrix}}_{\boldsymbol{E}}.$$

It is assumed that any observation $\boldsymbol{y}_\ell$ is generated by a unique k-sparse representation vector $\boldsymbol{x}_\ell$, for some specified level, k, of sparseness that is independent of ℓ. Therefore, by assumption, $\mathsf{sp}(\boldsymbol{x}_\ell) \le k$ for all $\ell = 1, \cdots, L$. Subject to this sparseness constraint, we call the model (17.2) a *k-sparse representation model*, or simply a *sparse representation model* if k is understood. In practice some lower bound on the value of k will be observed to hold for the entire set of estimated representation vectors $\boldsymbol{x}_\ell$, $\ell = 1, \cdots, L$, and/or it may be reasonable to assume that with high likelihood the dictionary matrix A satisfies the URP so that one can take $n/2$ to be a reasonable upper bound on k.

Although the $n \times L$ *data matrix* $\boldsymbol{Y}$ is available,[32,33] the dictionary A is unknown except for its size, which is $n \times m$, and it is assumed that A is onto. Each of the unobservable sparse representation vectors $\boldsymbol{x}_\ell$, $\ell = 1, \cdots, L$, are also unknown, which means we have a so-called *blind estimation problem*. To facilitate estimability of A we will require that our algorithms converge to estimates of A that belong to a set $\mathcal{A}$ of matrices that have specified structural properties.[34] Even then, as evident from (17.6), identifiability of A is ascertainable only up to the ordering of its columns and up to the "signs" of the columns, meaning up to $\pm\boldsymbol{a}_j$ due to the ambiguity induced by $\boldsymbol{a}_j x_j = (-\boldsymbol{a}_j)(-x_j)$.

Summarizing, we have the following,

Dictionary learning problem
Given an $n \times L$ data matrix $\boldsymbol{Y}$ and a k-sparse representation model

$$\boldsymbol{Y} = A\boldsymbol{X} + \boldsymbol{E} \tag{17.18}$$

with A $n \times m$, onto, belonging to $\mathcal{A}$, and with specified (stochastic and/or nonstochastic) assumptions on $\boldsymbol{E}$, estimate the dictionary matrix A up to the ordering and "signs" of its columns.

[32] Some readers, particularly those with a statistics and data science background, may be used to working with the $L \times n$ data matrix $\mathcal{Y} = \boldsymbol{Y}^T$ whose ith *row* is given by $\boldsymbol{y}_i^T$, i.e., whose ith row is equal to the components ("covariates") of $\boldsymbol{y}_i$ which is the ith *column* of $\boldsymbol{Y}$ (e.g., see [115,86]). Such readers should note that in this chapter the data matrix is the $n \times L$ matrix $\boldsymbol{Y} = \mathcal{Y}^T$.

[33] We will indicate that $\boldsymbol{y}_\ell$ is a column of $\boldsymbol{Y}$, and therefore one of our data samples, by the shorthand notation $\boldsymbol{y}_\ell \in \boldsymbol{Y}$. Similarly $C \subset \boldsymbol{Y}$ will denote a subset of the data.

[34] Typically a matrix $A \in \mathcal{A}$ is required to have either a normalized Frobenius norm $\|A\|_F^2 = \mathsf{trace}(A^T A)$ or normalized columns (i.e., atoms) $\|\boldsymbol{a}_j\|^2 = \boldsymbol{a}_j^T \boldsymbol{a}_j = 1$.

Given a sparsity measure $d(\boldsymbol{x})$, define its *matrix extension*, $D(\cdot)$, to the $m \times L$ matrix $\boldsymbol{X} = (\boldsymbol{x}_1 \ \cdots \ \boldsymbol{x}_L)$, and the (cumulative) *sparsity*, $\mathsf{sp}(\cdot)$- and *max-sparsity*, $\overline{\mathsf{sp}}(\cdot)$, of $\boldsymbol{X}$ as[35]

$$D(\boldsymbol{X}) = \sum_{\ell=1}^{L} d(\boldsymbol{x}_\ell), \quad \mathsf{sp}(\boldsymbol{X}) = \sum_{\ell=1}^{L} \mathsf{sp}(\boldsymbol{x}_\ell), \quad \text{and} \quad \overline{\mathsf{sp}}(\boldsymbol{X}) = \max_{1 \le \ell \le L} \mathsf{sp}(\boldsymbol{x}_\ell).$$

A standard approach to solving the blind dictionary learning problem is to simultaneously solve for A and $\boldsymbol{X}$ as solutions to the Tikhonov optimization problem

$$\min_{\boldsymbol{X}, A} \|\boldsymbol{Y} - A\boldsymbol{X}\|_F^2 + \lambda D(\boldsymbol{X}) \tag{17.19}$$

subject to normalization constraints placed on A to ensure its identifiability. This optimization was proposed and numerically solved in the foundational dictionary learning papers [132,133].

The regularization parameter λ is chosen to ensure that $\boldsymbol{X}$ is sufficiently sparse, $\mathsf{sp}(\boldsymbol{X}) < k$, for a specified sparsity k. Note that

$$\|\boldsymbol{Y} - A\boldsymbol{X}\|_F^2 + \lambda D(\boldsymbol{X}) = \sum_{\ell=1}^{L} \left(\|\boldsymbol{y}_\ell - A\boldsymbol{x}_\ell\|^2 + \lambda d(\boldsymbol{x}_\ell) \right).$$

If we assume the stochastic model (17.3) and (17.13), then with the independent sampling assumption on $\boldsymbol{y}_\ell$ we have[36]

$$P(\boldsymbol{Y}, \boldsymbol{X}; A) = \prod_{\ell=1}^{L} P(\boldsymbol{y}_\ell, \boldsymbol{x}_\ell; A) = \prod_{\ell=1}^{L} P(\boldsymbol{y}_\ell | \boldsymbol{x}_\ell; A) P(\boldsymbol{x}_\ell).$$

Note, then, that

$$\arg\max_{\boldsymbol{X}, A} P(\boldsymbol{Y}, \boldsymbol{X}; A) = \arg\min_{\boldsymbol{X}, A} \sum_{\ell=1}^{L} \left(\|\boldsymbol{y}_\ell - A\boldsymbol{x}_\ell\|^2 + \lambda d(\boldsymbol{x}_\ell) \right) = \arg\min_{\boldsymbol{X}, A} \|\boldsymbol{Y} - A\boldsymbol{X}\|_F^2 + \lambda D(\boldsymbol{X}) \tag{17.20}$$

with $\lambda = \beta / \sigma^2$. This gives a statistical interpretation to the optimization problem (17.19). Note however that

$$\arg\max_{\boldsymbol{X}, A} P(\boldsymbol{Y}, \boldsymbol{X}; A) \neq \arg\max_{\boldsymbol{X}, A} \frac{P(\boldsymbol{Y}, \boldsymbol{X}; A)}{P(\boldsymbol{Y}; A)} = \arg\max_{\boldsymbol{X}, A} P(\boldsymbol{X} | \boldsymbol{Y}; A).$$

By optimizing the joint likelihood $P(\boldsymbol{Y}, \boldsymbol{X}; A)$ with respect to A rather than the conditional likelihood $P(\boldsymbol{X} | \boldsymbol{Y}; A)$, we are side-stepping the difficult computational issue arising from the A dependence of the model evidence $P(\boldsymbol{Y}; A)$.

[35] See Footnote 18. Note that $\overline{\mathsf{sp}}(\boldsymbol{X}) \le k$ means that *all* L columns of $\boldsymbol{X}$ are k-sparse.

[36] Note that deterministic parameters are set apart via use of a semicolon. Also note that for the stochastic model (17.3) and (17.13) we have $P(\boldsymbol{y}_\ell | \boldsymbol{x}_\ell; A) = \mathsf{N}(\boldsymbol{y}_\ell; A\boldsymbol{x}_\ell, \sigma^2 I)$ for all ℓ with σ^2 not dependent on A.

Reference [133] notes that ideally one can determine a maximum likelihood estimate (MLE) of A by maximizing

$$P(\boldsymbol{Y};A) = \int_{\boldsymbol{X}} P(\boldsymbol{Y},\boldsymbol{X};A)d\boldsymbol{X} = \int_{\boldsymbol{X}} P(\boldsymbol{Y}|\boldsymbol{X};A)P(\boldsymbol{X})d\boldsymbol{X}, \tag{17.21}$$

but since the integration is generally intractable, one can optimistically assume that $P(\boldsymbol{X})$ is sufficiently tightly peaked enough to allow the approximation

$$P(\boldsymbol{Y};A) = \int_{\boldsymbol{X}} P(\boldsymbol{Y}|\boldsymbol{X};A)P(\boldsymbol{X})d\boldsymbol{X} \approx P(\boldsymbol{Y}|\hat{\boldsymbol{X}};A)P(\hat{\boldsymbol{X}}) = P(\boldsymbol{Y},\hat{\boldsymbol{X}};A),$$

where $\hat{\boldsymbol{X}}$ is the value of $\boldsymbol{X}$ that maximizes $P(\boldsymbol{Y},\boldsymbol{X};A)$ for the current estimate of A.[37] This gives the problem (17.20) shown above, showing that it can be viewed as a procedure for obtaining an approximate MLE for the dictionary A given the observed data $\boldsymbol{Y}$.

Here is another way to look at problem (17.19) from a stochastic perspective. Given the data $\boldsymbol{Y}$, note that for large L and *known* $\boldsymbol{X}$,

$$\mathbb{E}\left\{\|\boldsymbol{y}-A\boldsymbol{x}\|^2+\lambda d(\boldsymbol{x})\right\} \approx \frac{1}{L}\sum_{\ell=1}^{L}\left(\|\boldsymbol{y}_\ell - A\boldsymbol{x}_\ell\|^2+\lambda d(\boldsymbol{x}_\ell)\right) = \frac{1}{L}\left(\|\boldsymbol{Y}-A\boldsymbol{X}\|_F^2+\lambda D(\boldsymbol{X})\right). \tag{17.22}$$

Thus if sparse solutions $\boldsymbol{X}$ *were* known, we would be able to consider the solution

$$\arg\min_A \mathbb{E}\left\{\|\boldsymbol{y}-A\boldsymbol{x}\|^2+\lambda d(\boldsymbol{x})\right\} = \arg\min_A \mathbb{E}\left\{\|\boldsymbol{y}-A\boldsymbol{x}\|^2\right\} \approx \arg\min_A \frac{1}{L}\sum_{\ell=1}^{L}\|\boldsymbol{y}_\ell - A\boldsymbol{x}_\ell\|^2. \tag{17.23}$$

However, since we have a blind estimation problem where $\boldsymbol{X}$ is unknown, and therefore cannot be averaged out, we can instead attempt to simultaneously estimate A and $\boldsymbol{X}$ given $\boldsymbol{Y}$ via the optimization shown in (17.19), which, in turn, is a relaxed procedure for attempting to find a solution to the following ideal joint dictionary/sparse representation problem:

Joint dictionary/representation learning problem

$$(A,\boldsymbol{X}) \leftarrow \arg\min_{A,\boldsymbol{X}} \|\boldsymbol{Y}-A\boldsymbol{X}\|_F^2 \;\text{ subject to }\; \overline{\mathsf{sp}}(\boldsymbol{X}) \le k \;\text{ and }\; A \in \mathcal{A}. \tag{17.24}$$

Unfortunately, the simultaneous optimization of A and $\boldsymbol{X}$ is generally too difficult to perform in combination and instead it is standard to cycle between the $\boldsymbol{X}$ and A optimizations as follows.

---

[37] This is known as the *Laplace approximation*.

A1. Cyclic optimization-based dictionary learning algorithm
Data: $n \times L$ data matrix $\boldsymbol{Y}$; λ; m; k; ϵ; dictionary constraint set $\mathcal{A}$
Initialize: $n \times m$ dictionary matrix A; unconverged $\leftarrow$ TRUE;
While unconverged
C1. $\boldsymbol{X} \leftarrow \arg\min_{\boldsymbol{X}} \|\boldsymbol{Y} - A\boldsymbol{X}\|_F^2$ subject to $\overline{\mathsf{sp}}(\boldsymbol{X}) \leq k$;
or
C1′. $\boldsymbol{X} \leftarrow \arg\min_{\boldsymbol{X}} \|\boldsymbol{Y} - A\boldsymbol{X}\|_F^2 + \lambda D(\boldsymbol{X})$;
C2. $A \leftarrow \arg\min_A \|\boldsymbol{Y} - A\boldsymbol{X}\|_F^2$ subject to $A \in \mathcal{A}$;
C3. *If* $\overline{\mathsf{sp}}(\boldsymbol{X}) \leq k$ *and* $\|\boldsymbol{Y} - A\boldsymbol{X}\|_F < \epsilon$ *Then* unconverged $\leftarrow$ FALSE;
End While;
Return A, $\boldsymbol{X}$;

The sparse-solutions update step **C1** of cyclic optimization algorithm **A1** can be performed using any of the sparse representation algorithms mentioned in Section 17.3. It is important to note that the dictionary update step **C2** is performed *after* step **C1**, which means that step **C2** is performed conditioned on the fact that $\boldsymbol{X}$ has sparse structure.[38]

Reference [133] performs step **C1′** via gradient descent on $\boldsymbol{X}$ and step **C2** of Algorithm **A1** via the gradient descent algorithm shown below in (17.42). Below, we describe several algorithms specific for performing step **C2**. These dictionary update algorithms, which solve

$$\textbf{C2}: \quad A \leftarrow \arg\min_A \|\boldsymbol{Y} - A\boldsymbol{X}\|_F^2 \text{ subject to } A \in \mathcal{A}, \tag{17.25}$$

can learn atoms (columns of A) that match well to the ability to find sparse solutions using standard sparse representation algorithms.

However, before we look at such algorithms beginning in Subsection 17.4.4 we discuss how some dictionary learning algorithms designed to solve (17.25) can be viewed as generalizations of vector quantization (VQ) algorithms [65] and how they relate to principal component analysis (PCA) and factor analysis (FA).

17.4.2 Dictionary learning as a generalization of K-means clustering and vector quantization

Vector quantization (VQ) is a procedure based on a disjoint partitioning of the observation signal space $\mathbb{Y} = \mathbb{R}^n$ into "cluster regions" that then allows assignment to any subsequent observation $\boldsymbol{y} \in \mathbb{Y}$ of a "one-hot" vector $\boldsymbol{x}$ that points to the centroid (representative member) of the cluster that contains $\boldsymbol{y}$. VQ and hard clustering share the same goal of finding an exhaustive partitioning $\mathcal{P} = \{C_1, \cdots, C_K\}$ of a signal space $\mathbb{Y}$ into K disjoint sets (a collection of *equivalence classes* or *clusters*),

$$\mathbb{Y} = C_1 \cup \cdots \cup C_K \quad \text{with} \quad C_i \cap C_j = \emptyset \text{ for } i \neq j,$$

and *determining a unique representative member of each partition region* $\boldsymbol{a}_j \in C_j$. Every observation $\boldsymbol{y}$ is then uniquely assigned to some cluster C_j (which we can refer to as the $\boldsymbol{y}$-cluster). Depending on the

[38] This property will be exploited in the K-SVD algorithm described below.

application, the uniquely specified representative member $\boldsymbol{a}_j$ of a cluster C_j, $\boldsymbol{a}_j \iff C_j$, is variously called the *codeword*, *centroid*, *prototype*, *archetype*, or *pattern*, among other terms, for that cluster.

In particular, in VQ a cluster representative $\boldsymbol{a}_j$ is generally called a *codeword* and the collection of codewords is called a *codebook*, denoted as[39]

$$\mathbf{C} = \{\boldsymbol{a}_1, \cdots, \boldsymbol{a}_K\} \simeq \mathcal{P} = \{C_1, \cdots, C_K\},$$

where the K-element partitioning $\mathcal{P}$ and corresponding K-codeword codebook $\mathbf{C}$ are created with the goal of minimizing, on average, the coding error that arises from replacing any observation $\boldsymbol{y}$ by a finite precision ("quantized") codeword representative [65] according to

$$\boldsymbol{a}_j \leftarrow \boldsymbol{y} \iff \boldsymbol{y} \in C_j.$$

In the *clustering* literature, the term *centroid* is used to refer to $\boldsymbol{a}_j \in C_j$ because one commonly searches for a representative in C_j that corresponds to some definition of centrality (commonly the mean, median, or mode) [170,158], whereas in the *pattern recognition* literature, it is the terms *pattern*, *exemplar*, *archetype*, *prototype*, and *template* that are commonly used to refer to the $\boldsymbol{y}$-cluster representative that serves as an indicator of which class is most likely to be the case when performing classification of an observation $\boldsymbol{y}$ [170,20]. Going forward, for a given C_j we define the *centroid function*

$$\boldsymbol{a}_j = \mathsf{centroid}(C_j)$$

to be the operation of producing or extracting a unique representative element $\boldsymbol{a}_j \in C_j$. Note that given a centroid function, knowledge of the partition $\mathcal{P} = \{C_1, \cdots, C_K\}$ is sufficient to construct an associated codebook $\mathbf{C} = \{\boldsymbol{a}_1, \cdots, \boldsymbol{a}_K\}$.

The primary distinction between VQ used for coding purposes and hard clustering in general is that the cluster representative in VQ, when it is used for coding purposes, is restricted to being a finite precision object, whereas no such restriction is generally imposed in hard clustering. Further, whereas VQ can be analyzed and performed assuming a known distribution over $\mathbb{Y}$, in clustering the underlying generative distribution is generally unknown and only a data matrix $\boldsymbol{Y}$ of observations is available. Given a known distribution and a fixed value for K, the *generalized Lloyd algorithm*[40] (GLA) is a well-known greedy algorithm for learning a signal space partitioning and corresponding codebook. It can be used to empirically perform VQ when only a data matrix $\boldsymbol{Y}$ is at hand, in which case the GLA is equivalent to the well-known *K-means algorithm*[41] (KMA) used in pattern recognition for learning a K-element clustering partitioning $\mathcal{P}$ and the K corresponding cluster centroids $\mathbf{C}$. We use the terminology "GLA" and "KMA" interchangeably in this chapter, with a bias towards "KMA" because of its well-established machine learning connotations.

---

[39] It is no accident that we are using the same symbols $\boldsymbol{a}_j$ to denote a codeword as we used to denote an atom/word of a dictionary. Indeed, soon we will explain why we take *codebook*, *dictionary*, and *dictionary matrix* to be essentially synonymous entities.

[40] "Generalized" because it applies to vector data. The original Lloyd algorithm applies to scalar data [65].

[41] The name refers to the fact that one is looking for K clusters, with K prespecified, and that each cluster representative $\boldsymbol{a}_j$ is taken to be the sample *mean* over all observations $\boldsymbol{y} \in C_j$. Although the sample mean is still the most common choice for each cluster centroid, other measures are also used, for example as in the *K-Medoids* algorithm [170].

Given a data matrix of L observations $\boldsymbol{Y} = \begin{pmatrix} \boldsymbol{y}_1 & \cdots & \boldsymbol{y}_L \end{pmatrix}$ the goal is to determine a value for K, a K-element partition $\mathcal{P} = \{C_1, \cdots, C_K\}$, and a corresponding K-element set of cluster representatives (codebook) $\mathbf{C} = \{\boldsymbol{a}_1, \cdots, \boldsymbol{a}_K\}$ such that given a new observation $\boldsymbol{y} \in C_j$ and corresponding representative $\boldsymbol{a}_j \in C_j$, a *discrepancy* measure $d(\boldsymbol{y}, \boldsymbol{a}_j) \geq 0$ is minimized on average, where $d(\boldsymbol{y}, \boldsymbol{a}_j) = 0 \iff \boldsymbol{y} = \boldsymbol{a}_j$. A straightforward, but not the only, choice of a discrepancy measure is the squared Euclidean distance

$$\mathsf{dis}(\boldsymbol{y}, \boldsymbol{a}_j) = \|\boldsymbol{y} - \boldsymbol{a}_j\|_2^2.$$

Conceptually, one can solve this problem for each possible value of K and then afterwards select an optimal value for K.[42] Fixing K and given data $\boldsymbol{Y}$, the optimization to be solved is

$$(\hat{\mathcal{P}}, \hat{\mathbf{C}}) = \arg\min_{\mathcal{P}, \mathbf{C}} \frac{1}{L} \sum_{j=1}^{K} \sum_{\ell=1}^{L} \mathsf{dis}(\boldsymbol{y}_\ell, \boldsymbol{a}_j) \mathbf{1}(\boldsymbol{y}_\ell \in C_j) = \arg\min_{\mathcal{P}, \mathbf{C}} \frac{1}{L} \sum_{j=1}^{K} \sum_{\boldsymbol{y}_\ell \in C_j} \mathsf{dis}(\boldsymbol{y}_\ell, \boldsymbol{a}_j), \tag{17.26}$$

where $\mathbf{1}(\cdot)$ denotes the indicator function and $\boldsymbol{y}_\ell$ is an observation (column of $\boldsymbol{Y}$). This is a combinatorial optimization problem that is NP-hard and which is commonly attacked in a suboptimal manner using the KMA, a greedy algorithm that can be designed to be guaranteed to find a locally (but not globally) optimal solution when applied to a set of data.

To implement the KMA, in addition to fixing a value of K one also specifies a centroid function

$$\boldsymbol{a}_j = \mathsf{centroid}(C_j) \tag{17.27}$$

and defines C_j to be the *neighborhood* of $\boldsymbol{a}_j$ given by

$$C_j = \mathsf{nbh}(\boldsymbol{a}_j) \stackrel{\mathsf{def}}{=} \left\{ \boldsymbol{y} \in \mathbb{Y} \,\middle|\, \mathsf{dis}(\boldsymbol{y}, \boldsymbol{a}_j) < \mathsf{dis}(\boldsymbol{y}, \boldsymbol{a}_k) \text{ for all } k \neq j \right\} \subset \mathbb{Y}. \tag{17.28}$$

In practice, we are given a data matrix $\boldsymbol{Y}$ and define the *empirical neighborhood* on the data by[43]

$$\tilde{C}_j = \mathsf{enbh}(\boldsymbol{a}_j) \stackrel{\mathsf{def}}{=} \left\{ \boldsymbol{y}_\ell \in \boldsymbol{Y} \,\middle|\, \mathsf{dis}(\boldsymbol{y}_\ell, \boldsymbol{a}_j) < \mathsf{dis}(\boldsymbol{y}_\ell, \boldsymbol{a}_k) \text{ for all } k \neq j \right\} \subset \boldsymbol{Y}. \tag{17.29}$$

Further, we denote an *empirical* centroid function that applies to empirical neighborhoods by "$\mathsf{ecentroid}(\tilde{C}_j)$."

At convergence of the KMA we require the following consistency condition to hold:

$$\boldsymbol{a}_j = \mathsf{ecentroid}(\tilde{C}_j) \quad \text{and} \quad \tilde{C}_j = \mathsf{enbh}(\boldsymbol{a}_j), \tag{17.30}$$

[42] A crucial difference between VQ and general clustering of empirical data is that the number of codewords (respectively, cluster centers), K, usually can be readily ascertained for VQ lossy coding purposes, whereas the number of clusters that "actually" exists in the world from which data are collected is generally unknown. The determination of the number of clusters K from the data itself is a difficult problem and a variety of methods have been proposed, including the elbow method, the silhouette index, and the gap statistic, among others. See Chapter 16 of [170] for a thorough discussion of this issue.

[43] $\boldsymbol{y}_\ell \in \boldsymbol{Y}$ denotes the fact that $\boldsymbol{y}_\ell$ is a column of $\boldsymbol{Y}$, i.e., that $\boldsymbol{y}_\ell$ is an empirical data sample. Similarly, $\tilde{C}_j = \mathsf{enbh}(\boldsymbol{a}_j) \subset \boldsymbol{Y}$ means that all of the elements of $\tilde{C}_j$ are columns of $\boldsymbol{Y}$.

or equivalently,

$$\tilde{C}_j = \mathsf{enbh}(\mathsf{ecentroid}(\tilde{C}_j)) \text{ and } \tilde{\boldsymbol{a}}_j = \mathsf{ecentroid}\left(\mathsf{enbh}(\tilde{\boldsymbol{a}}_j)\right), \tag{17.31}$$

for all $j = 1, \cdots, K$.

With the above structure in place, the KMA attempts to solve the problem

$$\widehat{\mathcal{P}} = \arg\min_{\widetilde{\mathcal{P}}} \underbrace{\frac{1}{L}\sum_{j=1}^{K} \sum_{\boldsymbol{y}_\ell \in \tilde{C}_j = \mathsf{enbh}(\boldsymbol{a}_j) \subset \boldsymbol{Y}} \mathsf{dis}(\boldsymbol{y}_\ell, \boldsymbol{a}_j)}_{\mathcal{L}(\widetilde{\mathcal{P}}|\mathbf{Y})} \text{ subject to } \boldsymbol{a}_j = \mathsf{ecentroid}(\tilde{C}_j), \tag{17.32}$$

where $\mathcal{L}(\widetilde{\mathcal{P}}|\mathbf{Y})$ is the empirical loss function to be optimized over an empirical partitioning $\widetilde{\mathcal{P}} = \{\tilde{C}_1, \cdots, \tilde{C}_K\}$ over the data $\mathbf{Y}$. This optimization is suboptimally done via the following iteration, which for the standard choices of $\mathsf{dis}(\cdot,\cdot)$ and $\mathsf{ecentroid}(\cdot)$ (see below) is provably convergent to a local minimum.

A2. K-means clustering algorithm (KMA)

Data: $n \times L$ data matrix $\boldsymbol{Y}$; K; $\mathsf{dis}(\cdot)$; $\mathsf{ecentroid}(\cdot)$; $\epsilon > 0$;
Initialize: Codebook $\mathbf{C} = \{\boldsymbol{a}_1, \cdots, \boldsymbol{a}_K\}$; $\mathsf{unconverged} \leftarrow \mathsf{TRUE}$;
  $\tilde{C}_j = \mathsf{enbh}(\boldsymbol{a}_j)$ for $j = 1, \cdots, K$;
While $\mathsf{unconverged}$
  K1. $\boldsymbol{a}_j \leftarrow \mathsf{ecentroid}(\tilde{C}_j)$ for $j = 1, \cdots, K$;
  K2. $\tilde{C}_j \leftarrow \mathsf{enbh}(\boldsymbol{a}_j)$ for $j = 1, \cdots, K$;
  K3. *If* $\|\boldsymbol{a}_j - \mathsf{ecentroid}(\tilde{C}_j)\|_2 < \epsilon$ for all $j = 1, \cdots, K$,
    Then $\mathsf{unconverged} \leftarrow \mathsf{FALSE}$;
End While;
Return $\mathbf{C} = \{\boldsymbol{a}_1, \cdots, \boldsymbol{a}_K\}$;

We remind the reader that Algorithm **A2** is also known as the generalized Lloyd algorithm (GLA). Note that at each pass through the algorithm, step **K1** rebuilds the codebook $\mathbf{C} = \{\boldsymbol{a}_1, \cdots, \boldsymbol{a}_K\}$; step **K2** performs greedy local optimizations to rebuild the partition of the empirical data $\widetilde{\mathcal{P}} = \{\tilde{C}_1, \cdots, \tilde{C}_K\}$ by optimally assigning observations that are closest to $\boldsymbol{a}_j$ to class $\tilde{C}_j$; and step **K3** tests whether the codebook and partitioning of the data adequately satisfy (to within an $\epsilon > 0$) the consistency condition (17.30). Once the KMA has converged we can associate an exemplar $\boldsymbol{a}_i$ (equivalently, a neighborhood C_i) to a new observation $\boldsymbol{y} \in \mathbb{Y}$ via

$$\boldsymbol{a}_i = \arg\min_{\boldsymbol{a}_j \in \mathbf{C}} \mathsf{dis}(\boldsymbol{y}, \boldsymbol{a}_j).$$

If we interpret the codebook as defining an (unnormalized) dictionary $A \in \mathbb{R}^{n \times K}$ by

$$A = (\boldsymbol{a}_1 \ \cdots \ \boldsymbol{a}_K) \iff \mathbf{C} = \{\boldsymbol{a}_1, \cdots, \boldsymbol{a}_K\},$$

then we can interpret the above procedure as solving the inverse problem

$$y = Ax + n,$$

where $x \in \mathbb{R}^K$ is a sparsity 1 ($k = 1$) "one-hot" indicator unit vector x that picks out a single column of A to represent the observation y. As noted in [54,53,2,4],[44] general dictionary learning can be interpreted as a powerful extension of this idea. For example, suppose the components of x are still constrained to be 0 or 1 in value but are allowed to have up to six nonzero components (i.e., they are allowed to be k-sparse for $k = 6$) and take $K = 100$. It is evident that sparsity 1, the case considered above, only allows $K = 100$ prototype codewords ("concepts") to be expressed by y, whereas if up to six codewords can be combined, we have

$$\binom{100}{1} + \binom{100}{2} + \binom{100}{3} + \binom{100}{4} + \binom{100}{5} + \binom{100}{6} = 1{,}271{,}427{,}895 \approx 1.27 \times 10^9 \quad (17.33)$$

possible compound concepts that y can express, which is a huge, combinatorial boost in expressive power. Further, in our more general model (17.2) we allow the nonzero components of x to take any real value[45]; thus it is evident that the expressive power of the overdetermined dictionary model is quite remarkable.

The standard, most commonly encountered, version of the KMA takes

$$\mathsf{dis}(y, a_j) = \|y - a_j\|_2^2 = \|y - Ae_j\|_2^2 \quad \text{and} \quad a_j = \mathsf{ecentroid}(\tilde{C}_j) = \frac{1}{|\tilde{C}_j|} \sum_{y_\ell \in \tilde{C}_j \subset Y} y_\ell, \quad (17.34)$$

where e_j is the canonical unit vector whose coordinate value is 1 in the jth position and 0 elsewhere. The empirical loss function of (17.32) then becomes

$$\mathcal{L}(\widetilde{\mathcal{P}}|\mathbf{Y}) = \frac{1}{L} \sum_{j=1}^{K} \sum_{\ell=1}^{L} \underbrace{\mathbf{1}(y_\ell \in \tilde{C}_j)}_{\overset{\mathsf{def}}{=} \gamma_{\ell j}} \|y_\ell - Ae_j\|_2^2 = \frac{1}{L} \sum_{\ell=1}^{L} \|y_\ell - Ax_\ell\|_2^2, \quad (17.35)$$

where we define the one-hot vectors $x_\ell \in \mathbb{R}^K$ by

$$x_\ell = \sum_{j=1}^{K} \gamma_{\ell j} e_j \quad \text{for } \ell = 1, \cdots, L.$$

Because e_j are the canonical unit vectors for $\mathbb{R}^K$, it is the case that $\gamma_{\ell j}$ is the jth component of x_ℓ, $(x_\ell)_j = \gamma_{\ell j}$, and the fact that x_ℓ is a one-hot vector corresponds to the fact that, as an indicator function,

[44] The discussion in [2] is particularly recommended. See also Section 17.4.5 and Appendix 17.A below.

[45] This fact allows us to normalize A to ensure identifiability without any loss of generality.

$\gamma_{\ell j} = 1 \iff \boldsymbol{y}_\ell \in \tilde{C}_j$. Thus,[46]

$$\widetilde{\mathcal{P}} = \{\tilde{C}_1, \cdots, \tilde{C}_K\} \iff X = (\boldsymbol{x}_1 \quad \cdots \quad \boldsymbol{x}_L) = \boldsymbol{\Gamma}^T \quad \text{with} \quad (\boldsymbol{\Gamma})_{\ell j} = \gamma_{\ell j}, \tag{17.36}$$

as a consequence of $(\boldsymbol{x}_\ell)_j = \gamma_{\ell j}$ and the fact that

$$\boldsymbol{y}_\ell \in \tilde{C}_j \iff \gamma_{\ell j} = 1.$$

This allows us to make the identification $\mathcal{L}(\widetilde{\mathcal{P}}|\mathbf{Y}) = \mathcal{L}(X|\mathbf{Y})$ for purposes of optimization. Making this notational change and continuing with Eq. (17.35), we determine that the empirical loss function of (17.32) is equivalent to

$$\mathcal{L}(X|Y) = \frac{1}{L}\text{trace}\left\{(Y - AX)^T(Y - AX)\right\} = \frac{1}{L}\|Y - AX\|_F^2 \propto \|Y - AX\|_F^2.$$

The matrix $X = \Gamma^T$ is a 0–1 patterned matrix such that no two rows can share the value 1 in the same column position. Denoting the set of all such matrices by $\boldsymbol{\mathcal{X}}$, we indicate this constraint on the form of X by $X \in \boldsymbol{\mathcal{X}}$. As a member of $\boldsymbol{\mathcal{X}}$ the rows of the $K \times L$ matrix X are necessarily orthogonal; in particular the matrix XX^T is diagonal with the jth diagonal value equal to the cardinality of $\tilde{C}_j$, $|\tilde{C}_j| = (XX^T)_{jj}$. Furthermore, $X \in \boldsymbol{\mathcal{X}}$ implies $\mathsf{sp}(X) \le K$ and $\overline{\mathsf{sp}}(X) \le 1$.[47] If we weaken the constraint set to $\bar{\boldsymbol{\mathcal{X}}} = \{\text{matrices with elements 0 or 1}\}$, then

$$\boldsymbol{\mathcal{X}} = \left\{X \in \bar{\boldsymbol{\mathcal{X}}} \;\middle|\; \overline{\mathsf{sp}}(X) \le 1\right\}.$$

Summarizing, the KMA, a well-known, important algorithm in signal processing and in machine learning, provides a locally optimal solution to the 1-sparse dictionary learning problem.[48]

VQ problem

$$\min_{A,X} \|Y - AX\|_F^2 \quad \text{subject to} \quad \overline{\mathsf{sp}}(X) \le 1 \text{ and } X \in \bar{\boldsymbol{\mathcal{X}}}. \tag{17.37}$$

The minimization with respect to X (corresponding to step **C1** in the cyclic optimization dictionary learning algorithm **A1**) is equivalent to step **K2** of the KMA algorithm **A2**, while the minimization with respect to A (corresponding to step C2 of cyclic optimization) is equivalent to step **K1** of KMA and yields the estimate for $A\boldsymbol{e}_j = \boldsymbol{a}_j$ shown in (17.34). This is a case where each step of a cyclic

[46] Note that $(\boldsymbol{x}_\ell)_j = \gamma_{\ell j}$ means that the ℓth column, $\boldsymbol{x}_\ell$, of the Boolean matrix $X \in \mathbb{R}^{K\times L}$ is equal to the jth row of the Boolean matrix $\boldsymbol{\Gamma} \in \mathbb{R}^{L\times K}$.

[47] If at least one column of the data matrix Y is nonzero, then $\overline{\mathsf{sp}}(X) = 1$, and if all columns of Y are nonzero, then $\mathsf{sp}(X) = K$.

[48] Note that for known A and a single measurement vector $\boldsymbol{y}$, this is similar to the inverse problem (17.10), provided that we set $k = 1$ and impose the constraint that the solution is a one-hot vector in (17.10).

descent optimization strategy never increases the cost and is therefore guaranteed to converge to a locally optimal solution. Note that A is $n \times m$ with $m = K$.[49]

The constraint $X \in \bar{\mathcal{X}}$ ensures identifiability, i.e., that the VQ problem has a unique solution. Generally, to ensure unique identifiability some form of constraints is placed on X – which is henceforth denoted by $X \in \mathcal{X}$ for some specified constraint set $\mathcal{X}$ – and/or on the dictionary matrix A, denoted by $A \in \mathcal{A}$ for some specified constraint set $\mathcal{A}$. Given specified constraint sets $\mathcal{X}$ and $\mathcal{A}$, the general dictionary learning problem is a generalization of VQ (i.e., of the K-means clustering problem).

Dictionary learning problem as generalized VQ

$$\min_{A,X} \|Y - AX\|_F^2 \quad \text{subject to} \quad \overline{\text{sp}}(X) \le k, \quad X \in \mathcal{X}, \quad \text{and} \quad A \in \mathcal{A}. \tag{17.38}$$

This should be compared to Eq. (17.37) above. It is quite common to impose no constraints on X except for the ever-present sparsity constraint $\overline{\text{sp}}(X) \le k$, in which case it is tacitly understood that $\mathcal{X} = \mathbb{X} = \mathbb{R}^n$ and that uniqueness of a resulting solution will then result from the sparsity constraint and uniqueness-inducing properties of A as discussed in Section 17.2.4.

Below, in Subsection 17.4.4 and Subsection 17.4.5, we will discuss the MOD and K-SVD dictionary learning algorithms. As described in the original literature, both of these algorithms were designed to function as algorithmic extensions of the KMA to the problem of dictionary learning [53,4]. The degree to which this indeed is the case is discussed in [154].[50]

17.4.3 Dictionary learning as a generalization of PCA and FA

The stochastic overcomplete dictionary model (17.3) can be viewed as a generalization of the *factor analysis* (FA) model, which is an important model in multivariate statistics and machine learning [97, 115,15,88,86,14]. The standard FA model is given by

$$\boldsymbol{y} = A\boldsymbol{x} + \boldsymbol{n}$$

with the assumptions that the unknown $n \times m$ matrix of *factor loadings* A is *undercomplete* ($n \ge m$) with full column rank m; the vector of latent variables (the *factors*) $\boldsymbol{x}$ is Gaussian with zero mean and covariance I; $\boldsymbol{n}$ is Gaussian with zero mean and covariance $\sigma^2 I$; and $\boldsymbol{x}$ and $\boldsymbol{n}$ are independent, $\boldsymbol{x} \perp\!\!\!\perp \boldsymbol{n}$. A critical point to note is that the necessarily independent columns of the "tall" matrix A, which we refer to as the *factor directions* or *factor components*, are generally not orthogonal or normalized.

For the FA model, the covariance matrix of $\boldsymbol{y}$ is given by

$$\Sigma_y = AA^T + \sigma^2 I.$$

[49] Thus a better terminology when the number of cluster centers is m would be m-means algorithm. However, historically it was once very common to use "K" to refer to the number of clusters and as a consequence KMA has become standardized. Although the alternative name of "generalized Lloyd algorithm" (GLA) is notationally noncommittal regarding "m" or "K," the machine learning connotations of KMA better serve the themes of this chapter.

[50] A point of connection between K-SVD is also described in Appendix 17.A.

PCA is based on an eigendecomposition of Σ_y to determine its dominant *principal directions* (eigenvectors, a.k.a. *principal components*) which are necessarily orthogonal and typically taken to be normalized. FA can be interpreted as a generalization of PCA where the orthogonal principal directions of PCA are shifted to new, generally nonorthogonal, factor directions (the columns of A), a process generally referred to as "factor rotation" even through the loss of orthogonality means that the transformation is not generally orthogonal.[51] In essence, the move from PCA to FA is a move from an orthogonal basis (the principal directions of $\boldsymbol{y}$) of $\mathbb{Y}$ to a tight frame (that includes the factor directions of $\boldsymbol{y}$) that still spans $\mathbb{Y}$ yet has greater interpretative power.

This last point is critical for understanding the distinction between PCA and FA. The orthogonal principal directions provided by PCA are mathematically convenient, but generally interpretability of the principal directions is lost. FA takes interpretability of the factor directions to be more important than mathematical convenience. In essence, in FA the columns of A are desired to have conceptual or semantic meaning or interpretability. Similarly, learning and solving the overcomplete dictionary model (17.2) can also be viewed as a search for columns of A that have interpretable meaning, a fact consistent with our often referring to them as *words*, except that now we are willing to grow our dictionary until it is highly *overcomplete* and consider arbitrary (albeit "sparse") combinations of words in order to tremendously increase its expressive power as discussed in the previous Subsection 17.4.2. This is important and topical in machine learning: currently, *explainable AI* (XAI) is understood to be an important and very difficult area of research [184]. In a very important sense, a search for a highly expressible dictionary matrix whose columns (words) and combinations of those columns have semantic and conceptual meaning is a powerful approach for creating practical XAI algorithms. We will return to this point when we consider an interesting example taken from [174] in Section 17.5.4 below.

17.4.4 Dictionary learning using the Method of Directions

Motivated by the fact that the dictionary learning problem can be viewed as a generalization of the VQ problem, as is evident from a comparison of problems (17.37) and (17.38), the authors of [54,53] proposed the cyclic optimization approach of iteratively performing an optimization to update $\boldsymbol{X}$ (analogously to performing step **K1** of the KMA **A1**) followed by an optimization to update the dictionary matrix A (analogously to performing step **K3** of the KMA).

Obtaining a sparse estimate of $\boldsymbol{X}$ given a current estimate of A to perform step **C1** of a cyclic optimization **A1** can be done using a variety of possible algorithms, as described in Section 17.3 above. After an updated sparse estimate of $\boldsymbol{X}$ is at hand, the next problem is to obtain an updated estimate of A (step **C2** of a cyclic optimization **A1**). If A' is the current estimate we can write

$$\boldsymbol{\Delta} = A - A' \iff A = A' + \boldsymbol{\Delta},$$

where $\boldsymbol{\Delta}$ is the "direction" in A-space one moves to go from A' to a new estimate A. One looks for an "optimal" direction $\boldsymbol{\Delta}$ that yields a good estimate A and for this reason the approach is called the *Method of Directions* (MOD) [54,53]. Note that defining the residual $\boldsymbol{R} = \boldsymbol{Y} - A'\boldsymbol{X}$ we have

$$\|\boldsymbol{Y} - A\boldsymbol{X}\|_F^2 = \|(\boldsymbol{Y} - A'\boldsymbol{X}) - \boldsymbol{\Delta}\boldsymbol{X}\|_F^2 = \|\boldsymbol{R} - \boldsymbol{\Delta}\boldsymbol{X}\|_F^2, \tag{17.39}$$

[51] Often such transformations are referred to as "oblique rotations."

which shows that, given A', one minimizes over the update direction $\mathbf{\Delta}$ to obtain an optimal direction $\hat{\mathbf{\Delta}}$ and then form the update $A = A' + \hat{\mathbf{\Delta}}$. Alternatively, one can obtain an update by solving problem (17.25). Solving the latter problem yields the update

$$A = \boldsymbol{Y}\boldsymbol{X}^{+}, \tag{17.40}$$

where $\boldsymbol{X}^{+} : \mathbb{R}^{n\times L} \to \mathbb{R}^{n\times m}$ is the pseudoinverse of $\boldsymbol{X} : \mathbb{R}^{n\times m} \to \mathbb{R}^{n\times L}$ viewed as an operator whose action is $A \to A\boldsymbol{X}$.[52] For $L \geq m$ and $\mathsf{rank}(\boldsymbol{X}) = m$ we have

$$\boldsymbol{X}^{+} = \boldsymbol{X}^{T}\left(\boldsymbol{X}\boldsymbol{X}^{T}\right)^{-1}. \tag{17.41}$$

With $\boldsymbol{X}$ being an $m \times L$ matrix where typically m and L are quite large in value, the computation of the pseudoinverse is generally very computationally demanding. If the computation of an SVD $X = U\Sigma V^{T} = U_1 S V_1^{T}$ is computationally feasible, then $\boldsymbol{X}^{+} = V_1 S^{-1} U_1^{T}$ [166]. A modification of MOD to a sequential form that is more computationally tractable is given in [154].

It is critical that $\boldsymbol{X}$ be adequately sparse, $\overline{\mathsf{sp}}(\boldsymbol{X}) \leq k$, in order to force the minimizing solution A for the least-squares loss $\|\boldsymbol{Y} - A\boldsymbol{X}\|_F^2$ to match to that fact. To ensure identifiability, typically after each update (17.40) the dictionary A is then normalized,[53] a constraint we denote by $A \in \boldsymbol{\mathcal{A}}$.

A3. Method of Directions (MOD) dictionary learning

Data: $n \times L$ data matrix $\boldsymbol{Y}$; m; k; ϵ; conditions for belonging to $\boldsymbol{\mathcal{A}}$;
Initialize: $n \times m$ dictionary matrix $A \in \boldsymbol{\mathcal{A}}$; unconverged $\leftarrow$ TRUE;
$\boldsymbol{X} \leftarrow \arg\min_{\boldsymbol{X}} \|\boldsymbol{Y} - A\boldsymbol{X}\|_F^2$ subject to $\overline{\mathsf{sp}}(\boldsymbol{X}) \leq k$;
While unconverged
M1. $A \leftarrow \boldsymbol{Y}\boldsymbol{X}^{+}$;
M2. Project A into $\boldsymbol{\mathcal{A}}$;
M3. $\boldsymbol{X} \leftarrow \arg\min_{\boldsymbol{X}} \|\boldsymbol{Y} - A\boldsymbol{X}\|_F^2$ subject to $\overline{\mathsf{sp}}(\boldsymbol{X}) \leq k$;
M4. *If* $\|\boldsymbol{Y} - A\boldsymbol{X}\|_F^2 < \epsilon$ *Then* unconverged $\leftarrow$ FALSE;
End While;
Return A;

Step **M3** of the MOD algorithm **A3** (which corresponds to step **C1** of a cyclic optimization **A1**) can be performed using any effective sparse solution algorithm. Unfortunately, unlike the KMA, the MOD algorithm is not provably convergent. The problem is that step **M3** is not guaranteed to produce a value for $\|\boldsymbol{Y} - A\boldsymbol{X}\|_F^2$ that is reduced, or unchanged, from its previous value computed for the previous values of A and $\boldsymbol{X}$. However, in practice convergence is generally seen despite the lack of a rigorous convergence proof. Reference [154] points out the desirable property of step **M1** that A is updated in a manner that does not require a disruption of the (sparse and otherwise) structure that currently exists

[52] Interestingly, this is equal to the standard matrix pseudoinverse of the $n \times m$ matrix $\boldsymbol{X}$ viewed as an operator from $\mathbb{X} = \mathbb{R}^m$ to $\mathbb{Y} = \mathbb{R}^n$.
[53] Typically by normalizing its columns to one, $\|\boldsymbol{a}_j\|_2 = 1$, $j = 1, \cdots, n$, or forcing $\|A\|_F = 1$.

in X, and which is independent of any additional structural constraints on X that might be added to the loss function minimized in step **M3**. Reference [154] also presents a sequential version of MOD that minimizes its computational cost.

An alternative to the explicit solution $A = YX^+$ shown in step **M1** is to determine A via gradient descent on the loss function $\|Y - AX\|_F^2$. Using the matrix gradient of the loss [149,42],

$$\nabla_A \|Y - AX\|_F^2 = -2(Y - AX)X^T,$$

one estimates A via iteration on

$$A \leftarrow A + \lambda(Y - AX)X^T, \tag{17.42}$$

for $\lambda > 0$. Note that this is essentially the dictionary update algorithm independently proposed in [133].

17.4.5 Dictionary learning using K-SVD

The K-SVD dictionary learning algorithm [2,51,49] is one that is quite commonly used to handle the dictionary learning step **C2** of the cyclic optimization dictionary learning procedure **A1**. The K-SVD algorithm minimizes $\| Y - AX \|_F^2$ sequentially by updating one column a_j of A at a time and doing so in a manner that preserves the sparse structure that exists in a current solution. To begin, let r_j^T denote the jth row of the $m \times L$ matrix X,

$$X = \begin{pmatrix} x_1 & \cdots & x_L \end{pmatrix} = \begin{pmatrix} r_1^T \\ \vdots \\ r_m^T \end{pmatrix} = R^T \in \mathbb{R}^{m\times L}. \tag{17.43}$$

Then, to isolate a single column a_j of A, write

$$\| Y - AX \|_F^2 = \left\| Y - \sum_{i=1}^{m} a_i r_i^T \right\|_F^2 = \left\| \left(Y - \sum_{i\neq j}^{m} a_i r_i^T \right) - a_j r_j^T \right\|_F^2 = \left\| Y_j - a_j r_j^T \right\|_F^2 \tag{17.44}$$

$$\text{with} \quad Y_j \overset{\text{def}}{=} Y - \sum_{i\neq j}^{m} a_i r_i^T \in \mathbb{R}^{n\times L}, \quad a_i \in \mathbb{R}^n, \quad \text{and} \quad r_i \in \mathbb{R}^L. \tag{17.45}$$

Note that the isolated outer product $a_j r_j^T$ is an $n \times L$ rank-1 matrix. The goal of the K-SVD algorithm is to sequentially minimize $\left\| Y_j - a_j r_j^T \right\|_F^2$ with respect to both a_j and r_j^T for $j = 1, \cdots, m$, and to do so in a manner that preserves the sparse structure existing in $R = X^T \in \mathbb{R}^{L\times m}$. Note that once this has been done for $j = 1, \cdots, m$, we will have performed a complete update of the dictionary A, thereby completing step **C2** of a cyclic optimization **A1**. The fact that we focus on a single column a_i at a time in (17.44) is similar in spirit to the KMA, which is concerned with the updating of individual centroids; see Appendix 17.A for an elaboration of this point.

To address the general problem of minimizing (17.44), we show that we can build and minimize a "sparsity-enforced" version of the "isolated" loss function

$$\left\| \boldsymbol{Y}_j - \boldsymbol{a}_j \boldsymbol{r}_j^T \right\|_F^2 \tag{17.46}$$

that: (1) will be tractable to optimize and (2) will allow that optimization to preserve the sparse structure that exists in the current estimate $\boldsymbol{X}$. A brief outline of the procedure is given here with details of the derivation provided in Appendix 17.A.

We define the *row sparsity* $s_j = \mathsf{sp}(\boldsymbol{r}_j)$ to be the number of nonzero components of the current estimate of the L-element row vector $\boldsymbol{r}_j^T$ and identify the locations of those s_j nonzero components[54] by the following sets of integers and canonical unit L-vectors:

$$\{j_1, \cdots, j_{s_j}\} \quad \text{and} \quad \{\boldsymbol{e}_{j_1}, \cdots, \boldsymbol{e}_{j_{s_j}}\},$$

each integer j_{i_k} taking values between 1 and L and each $\boldsymbol{e}_{i_k} \in \mathbb{R}^L$ having the value 1 in the kth position. From these data we can construct the *sparse downsampling matrix*

$$\boldsymbol{S}_j \stackrel{\text{def}}{=} \boldsymbol{S}(\boldsymbol{r}_j) \stackrel{\text{def}}{=} \begin{pmatrix} \boldsymbol{e}_{j_1} & \cdots & \boldsymbol{e}_{j_{s_j}} \end{pmatrix} \in \mathbb{R}^{L \times s_j}. \tag{17.47}$$

We then form the *downsampled* entities,

$$\tilde{\boldsymbol{r}}_j^T = \boldsymbol{r}_j^T \boldsymbol{S}_j \in \mathbb{R}^{1 \times s_j} \quad \text{and} \quad \tilde{\boldsymbol{Y}}_j \stackrel{\text{def}}{=} \boldsymbol{Y}_j \boldsymbol{S}_j \in \mathbb{R}^{n \times s_j}, \tag{17.48}$$

to finally arrive at the *downsampled loss function*

$$\left\| \tilde{\boldsymbol{Y}}_j - \boldsymbol{a}_j \tilde{\boldsymbol{r}}_j^T \right\|_F^2. \tag{17.49}$$

As discussed in Appendix 17.A, minimizing the downsampled loss (17.49) is equivalent to minimizing the part of the loss (17.46) that is consistent with the goal of preserving the sparse structure of $\boldsymbol{r}_j$.

Minimization of (17.49) with respect to $\boldsymbol{a}_j$ and $\tilde{\boldsymbol{r}}_j$ is easily done by extracting the principal (maximum) singular value and associated singular vectors, $(\sigma_1, \boldsymbol{u}_1, \boldsymbol{v}_1)$, after which the answer is determined from[55]

$$\boldsymbol{a}_j = \boldsymbol{u}_1, \quad \tilde{\boldsymbol{r}}_j = \sigma_1 \boldsymbol{v}_1, \quad \text{and} \quad \boldsymbol{r}_j = \boldsymbol{S}_j \tilde{\boldsymbol{r}}_j. \tag{17.50}$$

This procedure is performed for each value of $j = 1, \cdots, m$ to complete step **C2** of the cyclic optimization learning algorithm **A1**. Summarizing, the K-SVD algorithm is given as follows.

---

[54] It is important to remember that we are assuming that step **C1** of cyclic optimization algorithm **A1** has been performed so that $\boldsymbol{R} = \boldsymbol{X}^T$ indeed has sparse structure. Thus it is the case that generally one can have $s_j < L$.

[55] As discussed in [154], although the K-SVD procedure preserves the sparse structure in $\boldsymbol{X}$, during the sequential dictionary procedure there effectively is a disruption of the other structural properties of $\boldsymbol{X}$ so that the resulting updated dictionary is not consistent with the current sparse solution $\boldsymbol{X}$. Because of this disruption, reference [154] claims that K-SVD is not, strictly speaking, a generalization of the KMA. However, the K-SVD algorithm is motivated by the properties of the KMA [4] and a point of connection is described in Appendix 17.A.

A4. K-SVD dictionary learning algorithm

Data: $n \times L$ data matrix $\boldsymbol{Y}$; m; k; ϵ;

$\boldsymbol{\mathcal{A}} =$ column normalized matrices;

Initialize: $n \times m$ dictionary matrix $A \in \boldsymbol{\mathcal{A}}$; unconverged $\leftarrow$ TRUE;

$\boldsymbol{R}^T = \boldsymbol{X} \leftarrow \arg\min_{\boldsymbol{X}} \|\boldsymbol{Y} - A\boldsymbol{X}\|_F^2$ subject to $\overline{\mathsf{sp}}(\boldsymbol{X}) \leq k$;

While unconverged

K1. **K_SVD**($\boldsymbol{Y}, A, \boldsymbol{X} = \boldsymbol{R}^T$);

Initialize: $j \leftarrow 1$;

While $j < m + 1$

$\boldsymbol{S}_j \leftarrow \boldsymbol{S}(\boldsymbol{r}_j^T)$;

$\boldsymbol{Y}_j \leftarrow \boldsymbol{Y} - \sum_{i=1, i \neq j}^{m} \boldsymbol{a}_i \boldsymbol{r}_i^T$;

$\tilde{\boldsymbol{Y}}_j \leftarrow \boldsymbol{Y}_j \boldsymbol{S}_j$;

$(\sigma_1, \boldsymbol{u}_1, \boldsymbol{v}_1) \leftarrow$ principal-SVD($\tilde{\boldsymbol{Y}}_j$);

$\boldsymbol{a}_j \leftarrow \boldsymbol{u}_1$;

$\tilde{\boldsymbol{r}}_j^T \leftarrow \sigma_1 \boldsymbol{v}_1^T$;

$\boldsymbol{r}_j^T \leftarrow \tilde{\boldsymbol{r}}_j^T \boldsymbol{S}_j^T$;

$j \leftarrow j + 1$;

End While;

Return A;

K2. $\boldsymbol{X} \leftarrow \arg\min_{\boldsymbol{X}} \|\boldsymbol{Y} - A\boldsymbol{X}\|_F^2$ subject to $\overline{\mathsf{sp}}(\boldsymbol{X}) \leq k$;

K3. *If* $\|\boldsymbol{Y} - A\boldsymbol{X}\|_F^2 < \epsilon$ *Then* unconverged $\leftarrow$ FALSE;

End While;

Return A;

In the above, the function principal-SVD(M) extracts the principal (dominant) singular value σ_1 and its associated left and right singular vectors, respectively $\boldsymbol{u}_1$ and $\boldsymbol{v}_1$, from the matrix $\tilde{\boldsymbol{Y}}_j$. As mentioned in Appendix 17.A, this does not necessarily involve the cost of performing a full SVD. Note that the update of the column $\boldsymbol{a}_j$ in the inner loop means that A has been updated; and the update of the row $\boldsymbol{r}_j^T$ means that the matrix $\boldsymbol{X} = \boldsymbol{R}^T$ has been updated. These updates affect later iterations of the inner loop. If the sparsity level is set to $k = 1$, K-SVD learns atoms that when scaled serve as single exemplars of observations; for example, given any new observation $\boldsymbol{y}$, we obtain a $k = 1$ representation $\boldsymbol{y} \approx A\boldsymbol{x} = x_j \boldsymbol{a}_j$ for some "shape" $\boldsymbol{a}_j$ and some "gain" x_j, j, $1 \leq j \leq m$. This conceptually corresponds to performing a scaled version of VQ known as *gain-shape VQ* [65].

17.5 Dictionary learning II – the hierarchical/empirical Bayes approach

17.5.1 Hierarchical model for sparse Bayesian learning – Type I versus Type II estimation

We have already encountered a simplified version of the model we describe here in Section 17.3.3 above. Here, our model is

$$y = Ax + n, \tag{17.51}$$

with $A \in \mathbb{R}^{n\times m}$, $m \geq n$, $\text{rank}(A) = m$, $x \perp\!\!\!\perp n$,

$$n \sim p(n; \Lambda) = \mathsf{N}(n; 0, \Lambda), \tag{17.52}$$

$$x \,\big|\, \gamma \sim p(x|\gamma) = \mathsf{N}(x; 0, \Gamma), \quad \Gamma = \text{diag}(\gamma), \tag{17.53}$$

and

$$p(\gamma) = p(\gamma_1)\cdots p(\gamma_m) \quad \text{with} \quad \gamma \perp\!\!\!\perp n. \tag{17.54}$$

Note that Eq. (17.53) implies

$$p(x|\gamma) = p(x_1|\gamma_1)\cdots p(x_m|\gamma_m). \tag{17.55}$$

The covariance Λ is an arbitrary, deterministic full-rank $n \times n$ matrix that may be known or unknown, where in the latter case it must be estimated.[56,57] The dictionary A is a deterministic unknown subject to the constraint that $A \in \mathcal{A}$, where $\mathcal{A}$ represents a set of appropriately constrained dictionary matrices.[58]

Note that with this model we have

$$y \,\big|\, x \sim p(y|x; A, \Lambda) = \mathsf{N}(y; Ax, \Lambda) \tag{17.56}$$

and, because when conditioned on γ the variable y is a linear combination of Gaussian variables,

$$y \,\big|\, \gamma \sim p(y|\gamma; A, \Lambda) = \mathsf{N}(y; 0, \Sigma_y) \quad \text{with} \quad \Sigma_y = \Lambda + A\Gamma A^T \quad \text{and} \quad \Gamma = \text{diag}(\gamma). \tag{17.57}$$

Although the diagonal matrix Γ can be (or become) singular, the matrix Σ_y is always invertible. For notational convenience we will often suppress the notational dependence of probability distributions on the unknown parameters A and Λ.

For the model defined above, we have the distributional factorization,

$$p(y, x, \gamma) = p(y|x)p(x|\gamma)p(\gamma). \tag{17.58}$$

This corresponds to the two-step ("two-level") Markov chain shown in Fig. 17.2(a). Note that the latent variable x is a primary variable of interest as its determination provides a sparse representation for the observed signal y. In the Bayesian literature it is common to call the vector x in $p(y|x)$ a "Level I parameter" that determines the behavior of y, and the vector γ in $p(x|\gamma)$ a "hyperparameter" (a "Level II parameter") that determines the behavior of the Level I "parameter" x. In the parlance of [73,74,18] the estimation of the Level II parameter (hyperparameter) γ from observations y, followed by a subsequent estimation of x, is a "Type II" estimation problem.

56 Note that heretofore we have made the simplifying assumption that $\Lambda = \lambda I$.
57 A rationale for the factorial forms of $p(x|\gamma)$ and $p(\gamma)$ is mentioned in Section 17.5.3 below.
58 Such as the set of Frobenius or column normalized matrices.

If one marginalizes out the hyperparameters $\boldsymbol{\gamma}$ from $p(\boldsymbol{y}, \boldsymbol{x}, \boldsymbol{\gamma})$,[59] then one has the factorization $p(\boldsymbol{y}, \boldsymbol{x}) = p(\boldsymbol{y}|\boldsymbol{x})p(\boldsymbol{x})$, which corresponds to the one-step ("one-level") Markov chain shown in Fig. 17.2(b). The use of this one-step factorization to estimate the "Level I parameter" $\boldsymbol{x}$ given an observation $\boldsymbol{y}$ is called a "Type I" estimation problem.[60]

Our model naturally lends itself to "Type I" sparse signal processing, which includes the MAP approaches described earlier in Section 17.3.3. Note that with

$$p(x, \gamma) = p(x|\gamma)p(\gamma) = \mathsf{N}(x; 0, \Gamma)p(\gamma) = \prod_{j=1}^{m} \mathsf{N}(x_j; 0, \gamma_j)p(\gamma_j)$$

we have

$$p(\boldsymbol{x}) = \int p(\boldsymbol{x}, \boldsymbol{\gamma})d\gamma = \prod_{j=1}^{m} p(x_j) \quad \text{with} \quad p(x_j) = \int N(x_j; 0, \gamma_j)p(\gamma_j)d\gamma_j.$$

It is the case that under reasonable conditions on $p(\gamma_j)$, the marginal distribution $p(x_j)$ is a Gaussian scale mixture (GSM) density that has desirable sparsity-inducing properties.[61] Indeed, as detailed in references [135,134,70,69], with the right choices of $p(\gamma_j)$ the marginal $p(x_j)$ takes on the form shown in Eq. (17.17),

$$p(x_j) = \int N(x_j; 0, \gamma_j)p(\gamma_j)d\gamma_j = \frac{1}{Z_j}\exp\left(-\beta_j d_j(x_j)\right),$$

where $d_j(\cdot)$ is a diversity measure which enforces sparsity of realizations drawn from $p(x_j)$. With

$$p(\boldsymbol{y}, \boldsymbol{x}) = p(\boldsymbol{y}|\boldsymbol{x})p(\boldsymbol{x}) = \mathsf{N}(\boldsymbol{y}; A\boldsymbol{x}, \Sigma_y)p(\boldsymbol{x}) = \mathsf{N}(\boldsymbol{y}; A\boldsymbol{x}, \Sigma_y)\prod_{j=1}^{m} p(x_j), \tag{17.59}$$

we see that many of the sparse representation learning algorithms described above in Section 17.3 (applicable when both A and Λ are known) fit within the framework of Type I MAP estimation of $\boldsymbol{x}$ given $\boldsymbol{y}$.

Continuing with the assumption (for now) that A and Λ are known, we turn to the estimation of $\boldsymbol{x}$ within the Type II framework. Here we note that $\boldsymbol{x} \,\big|\, (\boldsymbol{\gamma}, \boldsymbol{y})$ is Gaussian [172,191],

$$p(\boldsymbol{x}|\boldsymbol{y}, \boldsymbol{\gamma}) = \frac{p(\boldsymbol{y}, \boldsymbol{x}|\boldsymbol{\gamma})}{p(\boldsymbol{y}|\boldsymbol{\gamma})} = \frac{p(\boldsymbol{x}|\boldsymbol{y})p(\boldsymbol{y}|\boldsymbol{\gamma})}{p(\boldsymbol{y}|\boldsymbol{\gamma})} = \mathsf{N}(\boldsymbol{x}; \hat{\boldsymbol{\mu}}, \hat{\Sigma}), \tag{17.60}$$

with (using the definitions given above in Eq. (17.57))

[59] When doing so it is common to say that one has integrated out the "nuisance parameters" $\boldsymbol{\gamma}$.

[60] Although this terminology appears to be rarely used, discussion or estimation of parameters χ relevant for the specification of $p(\boldsymbol{\gamma}) = p(\boldsymbol{\gamma}; \chi)$ is referred to as a "Level III" or "Type III" consideration.

[61] More commonly, for mathematical convenience, one writes the shown marginalization integration resulting in $p(x_j)$ in terms of the precision $\beta_j = 1/\gamma_j$.

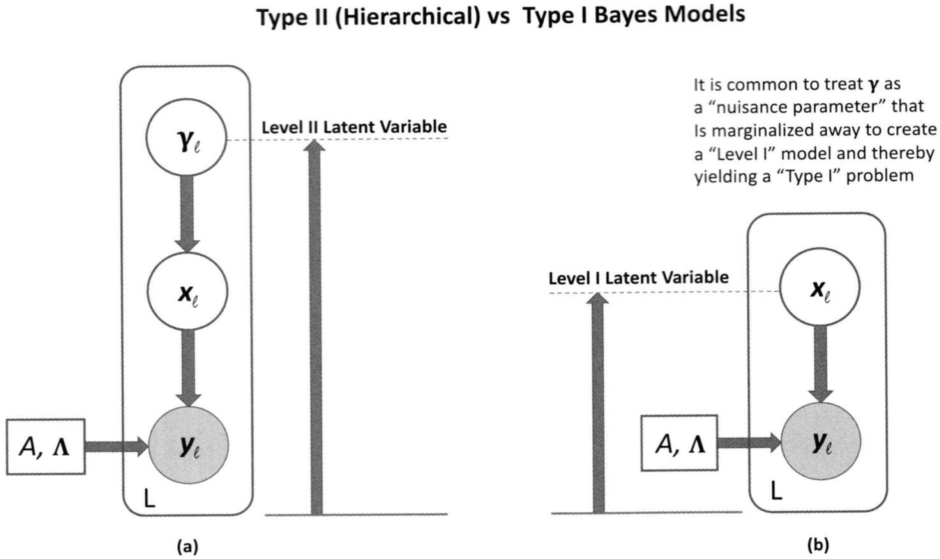

FIGURE 17.2

Bayesian latent variable models. Independent and identically distributed (i.i.d.) samples of latent and observed random variables are shown for the model described in the text using "plate notation." "Plate notation" means that random variables indexed by ℓ (shown contained within the "plates") are i.i.d. sampled L times, $\ell = 1, \cdots, L$ [124]. Only the random variable $\boldsymbol{y}$ is observable, a fact denoted by the shading of its node, while the remaining random variables $\boldsymbol{x}$ and $\boldsymbol{\gamma}$ are unobservable latent variables, as denoted by their unshaded nodes. Deterministic parameters are encased in squares which are unshaded if known and shaded if unknown. Here we show the dictionary matrix A and the covariance matrix Λ as deterministic unknowns. Note that the realization values of all the variables $\boldsymbol{y}$, $\boldsymbol{x}$, and $\boldsymbol{\gamma}$ change with every observation $\ell = 1, \cdots, L$. In particular, it is important to note that the instantiation value of γ, which is treated as a random *hyperparameter* for the conditional density $p(\boldsymbol{x}|\boldsymbol{\gamma})$, changes with *each* observed data instance ℓ. The random variable $\boldsymbol{x}$ is a "Level I" random variable (relative to the observed variable $\boldsymbol{y}$) while $\boldsymbol{\gamma}$ is a "Level II" variable (again, relative to the observation $\boldsymbol{y}$). Estimation of a Level I latent variable is a "Type I" or "Level I" problem, while estimation of a Level II latent variable is a "Type II" or "Level II" problem. In the model shown in panel (a) on the left the Level II variable is $\boldsymbol{\gamma}$. The "Type I" and "Type II" nomenclature was first proposed in [73,74]. This nomenclature emphasizes the fact that although it is common to integrate out the hyperparameter $\boldsymbol{\gamma}$ in the joint distribution $P(\boldsymbol{y}, \boldsymbol{x}, \boldsymbol{\gamma})$ to form $p(\boldsymbol{y}, \boldsymbol{x})$ and then deal with the estimation of the "Level I" latent variable $\boldsymbol{x}$, which is generally the variable of direct interest (see panel (b) on the right), it instead can be advantageous, conceptually and practically, to estimate the hyperparameter $\boldsymbol{\gamma}$ located one step higher in the shown Bayesian hierarchy (see panel (a) on the left), after which an estimate of $\boldsymbol{x}$ can be computed [18,173,193,141]. This latter approach often involves marginalizing out the Level I latent variable $\boldsymbol{x}$ from the joint distribution $P(\boldsymbol{y}, \boldsymbol{x}, \boldsymbol{\gamma})$ to obtain $P(\boldsymbol{y}, \boldsymbol{\gamma}) = P(\boldsymbol{\gamma}|\boldsymbol{y})P(\boldsymbol{y})$ and then estimating γ from $P(\boldsymbol{\gamma}|\boldsymbol{y})$.

$$\hat{\mu} = \hat{\mu}(\boldsymbol{y}, \boldsymbol{\gamma}) = \Gamma A^T \Sigma_y^{-1} \boldsymbol{y} \quad \text{and} \quad \hat{\boldsymbol{\Sigma}} = \hat{\boldsymbol{\Sigma}}(\boldsymbol{\gamma}) = \Gamma - \Gamma A^T \Sigma_y^{-1} A \Gamma = \left(\Gamma^{-1} + A^T \Lambda^{-1} A\right)^{-1}. \tag{17.61}$$

Given this conditionally Gaussian model, the minimum mean square error (MMSE) and MAP estimates of $\boldsymbol{x}$ given $\boldsymbol{y}$ and $\boldsymbol{\gamma}$ are identical and given by [172,191]

$$\boxed{\hat{\boldsymbol{x}} = \hat{\boldsymbol{x}}(\boldsymbol{y}, \boldsymbol{\gamma}) = \hat{\boldsymbol{\mu}}(\boldsymbol{y}, \boldsymbol{\gamma}) = \Gamma A^T \Sigma_y^{-1} \boldsymbol{y},} \tag{17.62}$$

where

$$\Gamma = \text{diag}(\boldsymbol{\gamma}) \quad \text{and} \quad \Sigma_y = \Lambda + A\Gamma A^T. \tag{17.63}$$

This is a remarkable result which shows that $\hat{x}_j$ is estimated to be "irrelevant" if and only if $\gamma_j = 0$ and that sparsity is attained by the solution $\hat{\boldsymbol{x}}$ only if $\mathsf{sp}(\boldsymbol{\gamma}) \leq n$.

This fact motivates the SBL Type II *empirical Bayes* methodology of "automatic relevance determination" (ARD) first proposed in [172]. A purely Bayesian method would directly estimate $\boldsymbol{x}$ given an observation $\boldsymbol{y}$ from $p(\boldsymbol{x}, \boldsymbol{y}) = p(\boldsymbol{x}|\boldsymbol{y})p(\boldsymbol{y})$ [70], which would be obtained by marginalization of $p(\boldsymbol{y}, \boldsymbol{x}, \boldsymbol{\gamma})$ over $\boldsymbol{\gamma}$ to thereby produce an appropriate "Type I" model (see Fig. 17.2(b) and Eq. (17.59)). "Empirical Bayes" is a quasi-Bayesian methodology where the Level II hyperparameter $\boldsymbol{\gamma}$ is first estimated as indicated in Fig. 17.3(c) (often using point-estimate approaches from classical, non-Bayesian statistics) and then used to produce a Bayesian estimate (in this case, a Bayesian conditional mean estimate given $\boldsymbol{y}$ and $\hat{\boldsymbol{\gamma}}$) of the Level I variable $\boldsymbol{x}$ as indicated in Fig. 17.3(d).

What is quite surprising is that in practice Type II estimation often is superior to Type I estimation for obtaining sparse solutions to the overcomplete inverse problem $\boldsymbol{y} = A\boldsymbol{x}$ [189]. This superiority in performance also is seen in the Type II dictionary learning algorithms to be discussed shortly. Reference [172] suggests the determination of the MAP estimate[62]

$$\hat{\boldsymbol{\gamma}} = \arg\max_{\boldsymbol{\gamma}} p(\boldsymbol{\gamma}|\boldsymbol{y}) = \arg\max_{\boldsymbol{\gamma}} p(\boldsymbol{\gamma}, \boldsymbol{y}) = \arg\max_{\boldsymbol{\gamma}} p(\boldsymbol{y}|\boldsymbol{\gamma})p(\boldsymbol{\gamma}), \tag{17.64}$$

with $p(\boldsymbol{y}|\boldsymbol{\gamma})$ and $p(\boldsymbol{\gamma})$ given by Eqs. (17.57) and (17.54), respectively.[63] This optimization is equivalent to

$$\hat{\boldsymbol{\gamma}} = \arg\min_{\Gamma = \text{diag}(\boldsymbol{\gamma}) \geq 0} \left(\boldsymbol{y}^T \left(\Lambda + A\Gamma A^T\right)^{-1} \boldsymbol{y} + \ln\det\left(\Lambda + A\Gamma A^T\right) - 2\sum_{j=1}^{m} \ln p(\gamma_j) \right). \tag{17.65}$$

With γ_j being a scale parameter, it is common to use the *Jeffreys prior*

$$p(\gamma_j) = \frac{1}{\gamma_j} \iff \ln p(\gamma_j) = -\ln \gamma_j, \tag{17.66}$$

which is an "uninformative" improper prior that is agnostic regarding scale [87,18,107]. The rationale for using an uninformative prior in estimating a sparse representation is that unless very strong prior

---

[62] If, as we shall be concerned with in Section 17.5.2 below, the optimization is also over unknown parameters, then $\text{argmax}\, p(\boldsymbol{\gamma}, \boldsymbol{y}) \neq \text{argmax}\, p(\boldsymbol{\gamma}|\boldsymbol{y})$ (see Footnote 64).

[63] This optimization choice will be further discussed in the context of dictionary learning in the following three sections. In particular, it will be given an information theoretic interpretation in Section 17.5.3.

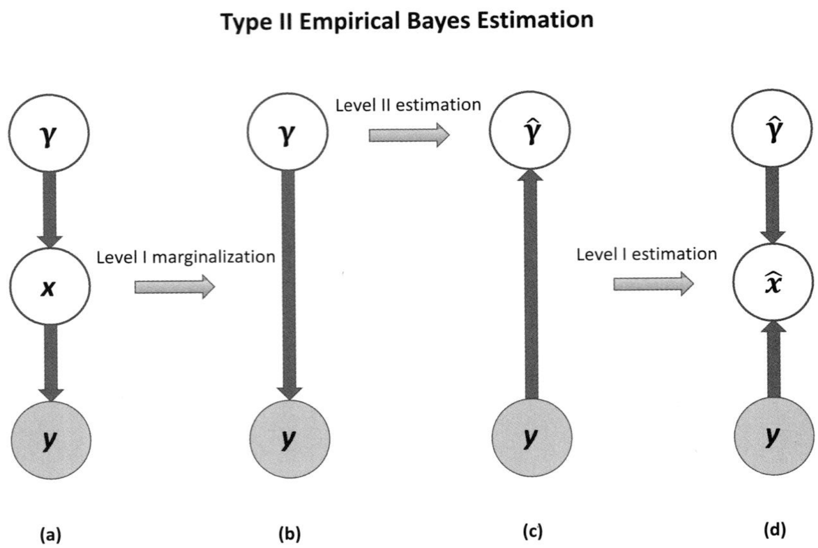

FIGURE 17.3

Empirical Bayesian approach to learning. Here we assume that A and Λ are known and consider the problem of estimating the latent variable $\boldsymbol{x}$ given an observation $\boldsymbol{y}$. A purely Bayesian approach would marginalize ("average out") the latent variable (hyperparameter) $\boldsymbol{\gamma}$ and then solve the Type I estimation problem shown in Fig. 17.2(b). Empirical Bayes is a quasi-Bayesian approach that estimates the Level II hyperparameter $\boldsymbol{\gamma}$ from the distribution $p(\boldsymbol{\gamma}, \boldsymbol{y}) = p(\boldsymbol{\gamma} | \boldsymbol{y}) p(\boldsymbol{y})$ (formed via the marginalization of $p(\boldsymbol{y}, \boldsymbol{x}, \boldsymbol{\gamma})$ shown above in the move from Fig. 17.3(a) to Fig. 17.3(b)) followed by an estimation of $\boldsymbol{x}$ from $p(\boldsymbol{x} | \boldsymbol{y}, \hat{\boldsymbol{\gamma}})$.

beliefs exist to suggest otherwise, the observation $\boldsymbol{y}$ itself should force the degree of sparsity appropriate for the particular realization at hand. This heuristic works well in practice and it is the assumption on the form of $p(\gamma_j)$ that we henceforth take to be the case. Note that assuming the uninformative Jeffreys prior results in the last term on the right-hand side of Eq. (17.65) becoming

$$-2 \sum_{j=1}^{m} \ln p(\gamma_j) = 2 \ln \det \Gamma \iff -\ln p(\boldsymbol{\gamma}) = \ln \det \Gamma. \tag{17.67}$$

A variety of iterative procedures have been proposed to obtain the estimate in Eq. (17.65). In particular, rather than tackle the difficult optimization in (17.65) directly, one can treat the (latent) marginalization variable $\boldsymbol{x}$ as a missing variable and apply the EM algorithm to obtain the required estimate of $\boldsymbol{\gamma}$ [20]. This involves taking the expectation of the log-likelihood of the factorization equation (17.58) conditioning on the observation $\boldsymbol{y}$ and assuming the current estimate $\hat{\boldsymbol{\gamma}}$ for the purpose of taking the required expectations. This procedure, which is detailed in Appendix 17.B, results in the following. Given a current estimate $\hat{\boldsymbol{\gamma}}$, the EM updates of its components are given by

$$\gamma_j^+ = 3\left(\hat{\mu}_j^2(\boldsymbol{y}, \hat{\boldsymbol{\gamma}}) + \hat{\Sigma}_{jj}(\hat{\boldsymbol{\gamma}})\right) \quad \text{for} \quad j = 1, \cdots, m, \tag{17.68}$$

where $\hat{\mu}_j(\boldsymbol{y}, \hat{\boldsymbol{\gamma}})$ is the jth component of $\hat{\boldsymbol{\mu}}(\boldsymbol{y}, \hat{\boldsymbol{\gamma}})$ and $\hat{\Sigma}_{jj}(\hat{\boldsymbol{\gamma}})$ is the jth diagonal element of $\hat{\Sigma}(\hat{\boldsymbol{\gamma}})$ (see Eq. (17.61)). Given the EM update $\hat{\boldsymbol{\gamma}}^+$, one performs the reassignment $\boldsymbol{\gamma} \leftarrow \hat{\boldsymbol{\gamma}}^+$ and iterates the EM procedure until convergence. Note from Eq. (17.62) that the EM algorithm iteratively updates the desired sparse estimate of $\boldsymbol{x}$ via $\hat{\boldsymbol{x}} = \hat{\boldsymbol{\mu}}(\boldsymbol{y}, \hat{\boldsymbol{\gamma}})$.

Summarizing the EM algorithm-based SBL procedure for determining a sparse representation $\boldsymbol{x}$ for an observation $\boldsymbol{y}$ assuming the applicability of the model (17.51)–(17.57) and the Jeffreys prior (17.66), we have the following.

A5. SBL algorithm for sparse representation learning

Data: n-dimensional observation vector $\boldsymbol{y}$;
$n \times m$ dictionary matrix A, $n \times n$ noise covariance matrix Λ;
Initialization values for m-dimensional positive hyperparameter $\boldsymbol{\gamma}$;
Initialization values for m-dimensional representation vector $\boldsymbol{x}$;
Boolean convergence function CONV($\boldsymbol{x}, \boldsymbol{\gamma}$);

While CONV($\boldsymbol{x}, \boldsymbol{\gamma}$) $\neq$ TRUE

$\Gamma \leftarrow \text{diag}(\boldsymbol{\gamma})$;
$\Sigma_y \leftarrow \Lambda + A\Gamma A^T$;
$\hat{\boldsymbol{\Sigma}} \leftarrow \Gamma - \Gamma A^T \Sigma_y^{-1} A\Gamma$;
$\hat{\boldsymbol{\mu}} \leftarrow \Gamma A^T \Sigma_y^{-1} \boldsymbol{y}$;
$\gamma_j^+ \leftarrow 3\left(\hat{\mu}_j^2 + \hat{\Sigma}_{jj}\right)$ for $j = 1, \cdots, m$;
$\boldsymbol{\gamma} \leftarrow \boldsymbol{\gamma}^+$;
$\boldsymbol{x} \leftarrow \hat{\boldsymbol{\mu}}$;

End While;

Return $\boldsymbol{x}$; $\hat{\boldsymbol{\Sigma}}$; $\boldsymbol{\gamma}$;

17.5.2 Dictionary learning in the SBL framework

In the previous section we assumed that the dictionary A and the noise covariance matrix are known. Now we turn to learning these two quantities given independent and identically distributed (i.i.d.) samples $\boldsymbol{y}_\ell$, $\ell = 1, \cdots, L$, which is the situation shown, using "plate notation," in Fig. 17.2 [60,71,57,59,89]. Similarly to the discussion surrounding Eq. (17.18) we represent this multiobservation situation as

$$\boldsymbol{Y} = A\boldsymbol{X} + \boldsymbol{N}, \tag{17.69}$$

with $\boldsymbol{Y} \in \mathbb{R}^{n \times L}$ and $\boldsymbol{X} \in \mathbb{R}^{m \times L}$. Similarly we define the $m \times L$ data matrix $\boldsymbol{G}$ to be the matrix whose columns are comprised of $\boldsymbol{\gamma}_1 \cdots \boldsymbol{\gamma}_L$, in that order,

$$\boldsymbol{G} \stackrel{\text{def}}{=} (\boldsymbol{\gamma}_1 \quad \cdots \quad \boldsymbol{\gamma}_L) \in \mathbb{R}^{m \times L}. \tag{17.70}$$

Given the i.i.d. sampling assumption we have

$$p(\boldsymbol{Y}, \boldsymbol{X}, \boldsymbol{G}; A, \Lambda) = \prod_{\ell=1}^{L} p(\boldsymbol{y}_\ell, \boldsymbol{x}_\ell, \boldsymbol{\gamma}_\ell; A, \Lambda) = \prod_{\ell=1}^{L} p(\boldsymbol{y}_\ell|\boldsymbol{x}_\ell; A, \Lambda) p(\boldsymbol{x}_\ell|\boldsymbol{\gamma}_\ell) p(\boldsymbol{\gamma}_\ell)$$

and variations thereof. Although in this section we take A and Λ to be unknown parameters of interest, we will often suppress them notationally in order to reduce notational clutter.

In principle, one can perform the marginalizations

$$p(\boldsymbol{Y}; A, \Lambda) = \int p(\boldsymbol{Y}, \boldsymbol{X}, \boldsymbol{G}; A, \Lambda) d\boldsymbol{X} d\boldsymbol{G}$$

and then minimize the negative log-likelihood function

$$-\ln p(\boldsymbol{Y}; A, \Lambda)$$

with respect to A and Λ to obtain MLEs of the dictionary A and the noise covariance matrix Λ. In practice, this is difficult to do, so instead we treat $\boldsymbol{X}$ as missing data and obtain simultaneous estimates of $\boldsymbol{G}$, $\boldsymbol{A}$, and Λ from the optimization[64,65]

$$\max_{\boldsymbol{G}, A, \Lambda} p(\boldsymbol{Y}, \boldsymbol{G}; A, \Lambda), \tag{17.71}$$

with

$$p(\boldsymbol{Y}, \boldsymbol{G}; A, \Lambda) = \int p(\boldsymbol{Y}, \boldsymbol{X}, \boldsymbol{G}; A, \Lambda) d\boldsymbol{X}.$$

The optimization (17.71) is equivalent to minimizing

$$-\ln p(\boldsymbol{Y}, \boldsymbol{G}; A, \Lambda). \tag{17.72}$$

The procedure used to do so is a generalization of the EM algorithm-based approach described in Section 17.5.1. However, in addition to estimating the variances $\boldsymbol{G}$, we now also desire to learn the unknown parameters A and Λ. The steps involved in deriving an algorithm for determining estimates of A, Λ, and G (i.e., $\hat{\boldsymbol{\gamma}}_\ell$ for $\ell = 1, \cdots, L$) are rather detailed and have been relegated to Appendix 17.C.

The resulting EM algorithm-based dictionary learning procedure is given as follows:

---

64 Unlike the optimization in the previous section, note that here $\arg\max p(\boldsymbol{Y}, \boldsymbol{G}; A, \Lambda) \neq \arg\max p(\boldsymbol{Y} \mid \boldsymbol{G}; A, \Lambda)$ as a consequence of the dependence of the optimization on the unknown parameters A and Λ.

65 Some discussion regarding possible information theoretic justifications and interpretations of the use of this optimization is given in the following Section 17.5.3.

A6. D-SBL algorithm for sparse representation dictionary learning

Data: $n \times L$-dimensional observation data matrix $\boldsymbol{Y} = (\boldsymbol{y}_1 \ \cdots \ \boldsymbol{y}_L)$;
Initialization values for $n \times m$ dictionary matrix A;
Initialization values for $n \times n$ noise covariance matrix Λ;
Initialization values for $m \times L$ hyperparameter matrix $\boldsymbol{G} = (\boldsymbol{\gamma}_1 \ \cdots \ \boldsymbol{\gamma}_L)$;
Initialization values for $m \times L$ representation matrix $\boldsymbol{X} = (\boldsymbol{x}_1 \ \cdots \ \boldsymbol{x}_L)$;
Boolean convergence function $\mathsf{CONV}(\boldsymbol{\mu}, \boldsymbol{\gamma}, A, \Lambda)$;

While $\mathsf{CONV}(\boldsymbol{X}, \boldsymbol{G}, A, \Lambda) \neq$ TRUE
Initialize: $\ell \leftarrow 1$; $\mathbf{R}_{yy} \leftarrow 0$; $\mathbf{R}_{y\mu} \leftarrow 0$; $\mathbf{R}_{\mu\mu} \leftarrow 0$; $\boldsymbol{\Sigma}_{xx} \leftarrow 0$;
While $\ell < L + 1$
$\Gamma_\ell \leftarrow \mathrm{diag}(\boldsymbol{\gamma}_\ell)$;
$\Sigma_{y,\ell} \leftarrow \Lambda + A\Gamma_\ell A^T$;
$\hat{\boldsymbol{\Sigma}}_\ell \leftarrow \Gamma_\ell - \Gamma_\ell A^T \Sigma_{y,\ell}^{-1} A \Gamma_\ell$;
$\hat{\boldsymbol{\mu}}_\ell \leftarrow \Gamma_\ell A^T \Sigma_{y,\ell}^{-1} \boldsymbol{y}_\ell$;
$\gamma_{j,\ell}^{+} \leftarrow 3\left(\hat{\mu}_{j,\ell}^2 + \hat{\Sigma}_{jj,\ell}\right)$ for $j = 1, \cdots, m$;
$\boldsymbol{\Sigma}_{xx} \leftarrow \boldsymbol{\Sigma}_{xx} + \hat{\boldsymbol{\Sigma}}_\ell$;
$\mathbf{R}_{yy} \leftarrow \mathbf{R}_{yy} + \boldsymbol{y}_\ell \boldsymbol{y}_\ell^T$;
$\mathbf{R}_{y\mu} \leftarrow \mathbf{R}_{y\mu} + \boldsymbol{y}_\ell \hat{\boldsymbol{\mu}}_\ell^T$;
$\mathbf{R}_{\mu\mu} \leftarrow \mathbf{R}_{\mu\mu} + \hat{\boldsymbol{\mu}}_\ell \hat{\boldsymbol{\mu}}_\ell^T$;
$\boldsymbol{x}_\ell \leftarrow \hat{\boldsymbol{\mu}}_\ell$;
$\ell \leftarrow \ell + 1$;
End While;
$\mathbf{R}_{xx} \leftarrow \mathbf{R}_{\mu\mu} + \boldsymbol{\Sigma}_{xx}$;
$A \leftarrow \mathbf{R}_{y\mu} \mathbf{R}_{xx}^{-1}$;
$\Lambda \leftarrow \left(\mathbf{R}_{yy} - \mathbf{R}_{y\mu} A^T - A \mathbf{R}_{y\mu}^T + A \mathbf{R}_{xx} A^T\right)/L$;
$\boldsymbol{G} \leftarrow \boldsymbol{G}^{+}$;
End While;
Return A, Λ;

17.5.2.1 ***Scalable SBL dictionary learning***

Algorithm A6 may scale poorly in practice as the size of the dataset grows (i.e., as L becomes large) and as the size of the dictionary grows (i.e., as m becomes large). More concretely, scalability refers to the memory cost of running A6, as measured by the number of parameters which need to be stored, and the computational cost, as measured by the number of per-iteration multiply-and-accumulate (MAC) operations. A detailed analysis of the memory and computational costs of A6 is given in [59], along with improvements which lead to dramatically faster and less memory-intensive algorithms. The high computational cost of A6 results from iterating over all L data points and computing $\hat{\boldsymbol{\Sigma}}_\ell^{-1}$. One way in which the computational cost can be decreased is by only updating the posterior statistics for a batch of data instead of all data points. This is referred to as incremental EM and has a similar flavor to stochastic gradient descent, whereby only a batch of data is processed at each algorithm iteration [129]. The per-iteration computational complexity is dominated by the matrix inversion in $\hat{\Sigma}_\ell^{-1}$. If the covariance matrix Λ is diagonal, i.e., $\sigma^2 I$, one alternative is to approximate $\hat{\Sigma}_\ell^{-1}$ using $\left(\text{diag}(\Gamma_l^{-1} + \sigma^{-2} A^T A)\right)^{-1}$ [127] such that all matrix inversions are done on diagonal matrices. In terms of memory cost, incremental EM requires the storage of Lm^2 parameters, corresponding to the sufficient statistic $\hat{\Sigma}_\ell$ for all L data points. Using the diagonal approximation, the memory cost can be reduced to Lm.

17.5.3 Information theory, generative models, and "Sparseland"

For convenience, set $\boldsymbol{\theta} = (A, \Lambda)$. We have (at least) two interesting ways that we can interpret the distribution $p(\boldsymbol{y}; \boldsymbol{\theta})$ obtained from the hierarchical Bayes model described in Section 17.5.1:

1. We can interpret it as a generative model that provides a faithful model of the *actual* statistical structure of the world. In this case we assume that the true (but unknown) distribution $p(\boldsymbol{y}) = p_{\text{true}}(\boldsymbol{y})$ of sensory observations $\boldsymbol{y}$ drawn for the natural world corresponds to a marginalization of the distribution $p(\boldsymbol{y}, \boldsymbol{x}, \boldsymbol{\gamma}) = p_{\text{true}}(\boldsymbol{y}, \boldsymbol{x}, \boldsymbol{\gamma})$ given by the model shown in Eqs. (17.51)–(17.55) for true (but unknown) values of the size of the dictionary, m, the parameters, $\boldsymbol{\theta} = (A, \Lambda)$, and the prior probabilities $p(\gamma_j)$.
2. We can view it as an *engineering* imposition, i.e., an almost certainly mismatched model *imposed* on the data to obtain a tractable procedure for obtaining a desirable representation $\boldsymbol{x}$ of an observation $\boldsymbol{y}$, for which a resulting learned representation $\boldsymbol{x}$ may or may not be sparse depending on the functional form selected for the imposed model priors $p(\gamma_j)$, $j = 1, \cdots m$.

In either case we are interested in determining the model parameters m, $\boldsymbol{\theta} = (A, \Lambda)$, and the prior $p(\boldsymbol{\gamma})$ that ensure $p(\boldsymbol{y}; \boldsymbol{\theta}) \approx p(\boldsymbol{y}) = p_{\text{true}}(\boldsymbol{y})$ with sufficient accuracy over a sufficiently wide range of possible observation values. Thus it is desirable to have a learning procedure such that given L i.i.d. observational samples and reasonable assumptions on $p(\boldsymbol{y}) = p_{\text{true}}(\boldsymbol{y})$, the procedure yields a model distribution $p(\boldsymbol{y}; \boldsymbol{\theta})$ for which the Kullback–Leibler divergence

$$\mathsf{D}\left(p(\boldsymbol{y}) \| p(\boldsymbol{y}; \boldsymbol{\theta})\right) = \mathbb{E}_{p(\boldsymbol{y})}\left\{\ln \frac{p(\boldsymbol{y})}{p(\boldsymbol{y}; \boldsymbol{\theta})}\right\} = \int p(\boldsymbol{y}) \ln \frac{p(\boldsymbol{y})}{p(\boldsymbol{y}; \boldsymbol{\theta})} d\boldsymbol{y} \geq 0$$

is a minimum and ideally such that $\mathsf{D}\left(p(\boldsymbol{y}) \| p(\boldsymbol{y}; \boldsymbol{\theta})\right) \approx 0$.[66]

To arrive at a mathematically tractable procedure for doing so, consider the related problem

$$\hat{\boldsymbol{\theta}} = \arg\min_{\boldsymbol{\theta}} \mathsf{D}\left(p(\boldsymbol{y}, \boldsymbol{\gamma}) \| p(\boldsymbol{y}, \boldsymbol{\gamma}; \boldsymbol{\theta})\right), \tag{17.73}$$

with $p(\boldsymbol{y}, \boldsymbol{\gamma}) = p_{\text{true}}(\boldsymbol{y}, \boldsymbol{\gamma})$, noting that

$$p(\boldsymbol{y}, \boldsymbol{\gamma}) = p(\boldsymbol{y}, \boldsymbol{\gamma}; \boldsymbol{\theta}) \implies p(\boldsymbol{y}) = p(\boldsymbol{y}; \boldsymbol{\theta})$$

via marginalization.

---

[66] Recall that the divergence is zero if and only if $p(\boldsymbol{y}) = p(\boldsymbol{y}; \boldsymbol{\theta})$ and that the divergence has a variety of very useful properties and is related to several important information theoretic and statistical quantities [83,39,37,165].

A well-known relationship between the cross-entropy[67]

$$\mathsf{C}\,(p(\boldsymbol{y},\boldsymbol{\gamma})||p(\boldsymbol{y},\boldsymbol{\gamma};\boldsymbol{\theta})) = \mathbb{E}_{p(\boldsymbol{y},\boldsymbol{\gamma})}\left\{\ln \frac{1}{p(\boldsymbol{y},\boldsymbol{\gamma};\boldsymbol{\theta})}\right\} = -\int p(\boldsymbol{y},\boldsymbol{\gamma}) \ln p(\boldsymbol{y},\boldsymbol{\gamma};\boldsymbol{\theta}) d\boldsymbol{y} d\boldsymbol{\gamma}$$

and divergence from the true distribution $p(\boldsymbol{y},\boldsymbol{\gamma})$ to the model distribution $p(\boldsymbol{y},\boldsymbol{\gamma};\boldsymbol{\theta})$ is

$$\mathsf{C}\,(p(\boldsymbol{y},\boldsymbol{\gamma})||p(\boldsymbol{y},\boldsymbol{\gamma};\boldsymbol{\theta})) = \mathsf{H}(p(\boldsymbol{y},\boldsymbol{\gamma})) + \mathsf{D}\,(p(\boldsymbol{y},\gamma)\|p(\boldsymbol{y},\gamma;\boldsymbol{\theta}))\,.$$

The quantity[68]

$$\mathsf{H}(p(\boldsymbol{y},\boldsymbol{\gamma})) = \mathbb{E}_{p(\boldsymbol{y},\boldsymbol{\gamma})}\left\{\ln \frac{1}{p(\boldsymbol{y},\boldsymbol{\gamma})}\right\} = -\int p(\boldsymbol{y},\boldsymbol{\gamma}) \ln p(\boldsymbol{y},\boldsymbol{\gamma}) d\boldsymbol{y} d\boldsymbol{\gamma},$$

which is the entropy of the true distribution $p(\boldsymbol{y},\boldsymbol{\gamma})$, is an unknown (assumed finite) constant which we can ignore. Thus the solution to (17.73) can be obtained from the cross-entropy optimization

$$\hat{\boldsymbol{\theta}} = \arg\min_{\boldsymbol{\theta}} \mathsf{C}\,(p(\boldsymbol{y},\gamma)\|p(\boldsymbol{y},\gamma;\boldsymbol{\theta}))\,. \tag{17.74}$$

Now exploit the fact that for large values of L we have

$$\mathsf{C}\,(p(\boldsymbol{y},\gamma)\|p(\boldsymbol{y},\gamma;\boldsymbol{\theta})) \approx \widehat{\mathsf{C}}\,(\boldsymbol{Y},\boldsymbol{G};\boldsymbol{\theta})) \stackrel{\mathsf{def}}{=} -\frac{1}{L}\sum_{\ell=1}^{L} \ln p(\boldsymbol{y}_\ell,\boldsymbol{\gamma}_\ell;\boldsymbol{\theta}) = -\frac{1}{L}\ln p(\boldsymbol{Y},\boldsymbol{G};A,\Lambda),$$

where $\boldsymbol{Y}$ and $\boldsymbol{G}$ are the data matrices defined at the outset of Section 17.5.2 and the last step corresponds to the expression shown in Eq. (17.72). It is now evident that the EM algorithm procedure for solving Eq. (17.72) described in the previous Subsection 17.5.2[69] provides an approximate solution $\hat{\boldsymbol{\theta}}$ to Eq. (17.73)/(17.74). After marginalization, this yields

$$p(\boldsymbol{y};\hat{\boldsymbol{\theta}}) \approx p(\boldsymbol{y}).$$

The existence of *Sparseland* – Actual vs. enforced sparsity. Consider the first case described at the outset of this section. If indeed our model provides a true representation of the stochastic behavior of observations taken in the actual world, then it is reasonable to aspire to the outcome $\mathsf{D}(p(\boldsymbol{y})\|p(\boldsymbol{y};\hat{\boldsymbol{\theta}})) \approx 0$. Further, if the world *actually does* have *sparse structure*, an assumption referred to as the *Sparseland assumption* in [50,49], we expect that using an *uninformative* prior for $p(\boldsymbol{\gamma})$, e.g., in our case the Jeffreys prior, will *allow* our model to naturally (i.e., in an unforced manner) obtain "true" sparse representations of observations $\boldsymbol{y}$ as then the likelihood of the model parameters $\boldsymbol{\theta} = (A,\Lambda)$ will be consistent

[67] The double stroke symbol "||" serves as a reminder of the essentially nonsymmetry of both the cross-entropy and the divergence.

[68] Note that $\mathsf{H}(p(\boldsymbol{y},\boldsymbol{\gamma})) = \mathsf{C}\,(p(\boldsymbol{y},\boldsymbol{\gamma})||p(\boldsymbol{y},\boldsymbol{\gamma}))$, i.e., $\mathsf{H}(p(\boldsymbol{y},\boldsymbol{\gamma}))$ is the cross-entropy of $p(\boldsymbol{y},\boldsymbol{\gamma})$ with itself. For this reason, the entropy $\mathsf{H}\,(p(\boldsymbol{y},\boldsymbol{\gamma}))$ is sometimes called the *self-entropy.*

[69] The derivation given above in Sections 17.5.1 and 17.5.2 assumed the noninformative Jeffreys prior. However, it is straightforward to incorporate a variety of alternative, sparsity-inducing priors [70].

with the world of sparsely generated observations, i.e., will be consistent with the Sparseland assumption. This indeed is what synthetic data simulations given in the literature essentially demonstrate; data "generated" by a sparse truth model allow that model to be learned and the generating sparse signal realizations $\boldsymbol{x}$ to be accurately estimated. In a sense, the hierarchical Bayes framework clarifies the Sparseland assumption if we define Sparseland to be a world generated by a Type II hierarchical Bayes model for which the condition $\mathsf{sp}(\boldsymbol{\gamma}) < n$ is a probability 1 event (or equivalently, that $\mathsf{sp}(\boldsymbol{\gamma}) \geq n$ is a probability 0 event).[70]

However, if data observed in the world are *not* truly generated sparsely, then when using an uninformative prior – a prior that does *not* impose any sparsity assumptions – we should not be surprised to obtain solutions no sparser than $\mathsf{sp}(\boldsymbol{x}) = n$, which is equivalent to the (unsurprising) fact that a basis for $\mathbb{Y} = \mathbb{R}^n$ is n-dimensional.[71]

Nonetheless, if the data are not truly generated sparsely, and yet one *insists* on obtaining sparse representations, one can *impose* this assumption on the model by an appropriate choice of an "informative" sparsity-inducing prior $p(\boldsymbol{\gamma})$ [134,135]. This corresponds to the second interpretation of the hierarchical generative model mentioned at the outset of this section. The result of this forced imposition of sparsity can be viewed as a form of "lossy compression" and here we expect that we will be content if $\mathsf{D}(p(\boldsymbol{y}) \| p(\boldsymbol{y}; \hat{\boldsymbol{\theta}}))$ is "small enough" for some requisite level of compression (i.e., sparseness). The choice of a sparsity-inducing prior imposes an *inductive bias* in favor of sparse representations on our model and resulting model-based algorithm.

This suggest that the "uninformative prior" hierarchical Bayesian model, where we eschew an a priori bias towards sparsity, can be used as a type of "Sparseland detector," where if the use of the uninformative prior gives a sparsity-inducing model we judge that the world indeed has natural sparse structure. One also can imagine the degree of sparsity attained by the use of a "sliding sparsity"-enforcing prior to be a measure of the "nearness to Sparseland" of the data.

Generative modeling of signals in the natural world. Let $p(\boldsymbol{y}) = p_{\text{true}}(\boldsymbol{y})$ denote the true distribution of $\boldsymbol{y}$ as observed in the natural world. What does it mean for an information processing system, such as an animal's nervous system, that $p(\boldsymbol{y})$ can be well represented by a distribution $p(\boldsymbol{y}; \boldsymbol{\theta})$, $\boldsymbol{\theta} = (A, \Lambda)$, according to the model given in Eqs. (17.51)–(17.55)? It has long been argued that under evolutionary pressure natural sensory processing systems will have evolved into being which are capable of *efficiently* extracting relevant *information* about the world [11,12,13,8,62] and that it is the case that a sensory system based on the model given in Eqs. (17.51)–(17.54) can satisfy those desiderata assuming that the external, natural sensory world indeed has sparse structure. It has been experimentally verified that sparse structure of natural signals is ubiquitous [132,34,82] and argued from physical principles that the physical behavior of the world encourages sparsely structured signal generation on the part of the external, physical environment [45,104], so we assume that natural world signals are indeed capable of being sparsely represented.

The overcomplete model $\boldsymbol{y} = A\boldsymbol{x} + \boldsymbol{n}$ provides a very large increase in expressive power over linear methods such as PCA and FA due to its ability to utilize a general frame/dictionary of very large size that contains words (component vectors) that are generally nonorthogonal and from which one can

[70] When generating synthetic sparse data this probability 1 condition is virtually always enforced.

[71] However, the representation for each observation $\boldsymbol{y}$ will generally be based on a different set of m columns drawn from the n columns of A, which reflects the fact that the model will still allow for *best basis selection*.

nonlinearly select representation vectors using sparse signal processing algorithms. Furthermore, as discussed earlier in Section 17.4.2, its expressive power is far greater than that of VQ due to the ability to combinatorially sum up different sets of weighted component vectors. Thus our model can serve to create a sensor capable of detecting a vast amount of information presented by the natural world.

Further, it has been argued that an organism can efficiently and quickly, i.e., in a computationally reasonable manner, detect when the world has changed in some statistically significant way via the use of a *factorial code* [13,8]. This criterion is captured in the factorized joint distributions shown in Eqs. (17.54)–(17.55). Note that if $\boldsymbol{\gamma}$ were not factorial, then neither would be the marginal distribution $p(\boldsymbol{x})$ of $p(\boldsymbol{x}, \boldsymbol{\gamma})$. Independence of the components x_i, $i = 1, \cdots, m$, means that one has statistical independence of the x_i-weighted words $\{x_i \boldsymbol{a}_i,\ i = 1, \cdots, m\}$ even though the unweighted words (dictionary elements) $\boldsymbol{a}_i \in \mathbb{Y} = \mathbb{R}^n$ themselves are generally nonorthogonal as vectors in $\mathbb{R}^n$, so that $\boldsymbol{a}_i^T \boldsymbol{a}_j \neq 0$. Note that each weighted word is a statistically independent representation of a complex, patterned situation in the world corresponding to an "atomic" (possibly noisy) observation of the world $\boldsymbol{y} = x_i \boldsymbol{a}_i + n$, and therefore changes in the observed frequency of occurrence of this patterned situation – which can be detected as a changed relative to its prediction occurrence as given by the current learned model – can be done in isolation of the other atomic patterns of the world which are encoded in the other weighted words. This *statistical decoupling* of the ability to detect changes in the learned atomic aspects of the world (aspects which enable great representational power) computationally bounds the problem of monitoring those changes in the environment which matter for the survival of a biological organism.

An additional aspect of the hierarchical Bayesian model is that the choice of the prior for $\boldsymbol{\gamma}$ provides a higher-level procedure for "deploying" the dictionary words in order to best extract information about the world contained in an observation $\boldsymbol{y}$. The nonzero elements, x_j, of the resulting sparse representation (which correspond to the deployed words) then provide coded information about the state of the world provided by $\boldsymbol{y}$. Speaking metaphorically, we interpret $\boldsymbol{\gamma}$ as the "boss" of the representation-extracting procedure and conclude that *it is desirable to maximize the mutual information between* $\boldsymbol{\gamma}$ *and the observation* $\boldsymbol{y}$ so that $\boldsymbol{\gamma}$ can give the best possible "deployment orders" to the weighting factors x_j, $j = 1, \cdots, m$. From this perspective, as a "knowledgeable boss," $\boldsymbol{\gamma}$ has enough knowledge about the world to even be able to *generate* synthetic observations $\boldsymbol{y}$. And a "good boss" who is pretty confident in their workers' abilities to work in a world that everyone agrees already has sparse structure does not need to overly pressure subordinates, so we can take the Jeffreys prior of Eq. (17.66) to hold for $\boldsymbol{\gamma}$ which allows the boss to also be a good listener of their workers.

To make these metaphorical comments more concrete, consider the generative model Markov chain[72]

72 Note that $\boldsymbol{y} \xrightarrow{\ \boldsymbol{\theta}\ } \boldsymbol{x} \longrightarrow \boldsymbol{\gamma}$ is Markov equivalent to $\boldsymbol{y} \xleftarrow{\ \boldsymbol{\theta}\ } \boldsymbol{x} \longleftarrow \boldsymbol{\gamma}$ *only if* the constraints $p(\boldsymbol{\gamma}; \boldsymbol{\theta}) = p(\boldsymbol{\gamma})$ and $p(\boldsymbol{y}; \boldsymbol{\theta}) = p(\boldsymbol{y})$ hold. This is because the serial Markov chain on the left corresponds to the factorization $p(\boldsymbol{y}, \boldsymbol{x}, \boldsymbol{\gamma}; \boldsymbol{\theta}) = p(\boldsymbol{\gamma}|\boldsymbol{x}) p(\boldsymbol{x}|\boldsymbol{y}; \boldsymbol{\theta}) p(\boldsymbol{y})$ while the chain on the right corresponds to $p(\boldsymbol{y}, \boldsymbol{x}, \boldsymbol{\gamma}; \boldsymbol{\theta}) = p(\boldsymbol{y}|\boldsymbol{x}; \boldsymbol{\theta}) p(\boldsymbol{x}|\boldsymbol{\gamma}) p(\boldsymbol{\gamma})$, so that marginalization forces these constraints. To remove these constraints, one can forgo the assumption that the root nodes of the serial chains correspond to marginals which are independent of the model parameters. Note that in this section we essentially *do* assume that the marginals are independent of the model parameters because we assume at the outset that $p(\boldsymbol{\gamma})$ is independent of $\boldsymbol{\theta}$, so that $H(\boldsymbol{\gamma})$ is independent of $\boldsymbol{\theta}$, and then we make $p(\boldsymbol{y})$ approximately independent of $\boldsymbol{\theta}$, so that as a consequence $H(\boldsymbol{y})$ is approximately independent of $\boldsymbol{\theta}$.

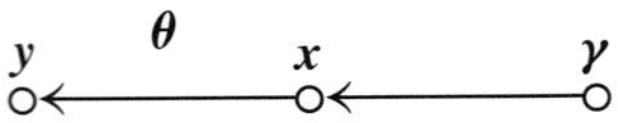

as a graphical representation of the factorial model (17.54)–(17.55) with the Jeffreys prior of Eq. (17.66), and consider the optimization of the mutual information between $\boldsymbol{y}$ and $\boldsymbol{\gamma}$,

$$\hat{\boldsymbol{\theta}} = \arg\max_{\boldsymbol{\theta}} \mathsf{I}(\boldsymbol{y}, \boldsymbol{\gamma}; \boldsymbol{\theta}). \tag{17.75}$$

The optimization is performed to select a model for which $\boldsymbol{\gamma}$ has maximal knowledge about the world as embodied in observations $\boldsymbol{y}$. For the shown model we have [37]

$$\mathsf{I}(\boldsymbol{y}, \boldsymbol{\gamma}; \boldsymbol{\theta}) = \mathsf{H}(\boldsymbol{\gamma}) + \mathsf{H}(\boldsymbol{y}; \boldsymbol{\theta}) - \mathsf{H}(\boldsymbol{\gamma}, \boldsymbol{y}; \boldsymbol{\theta}),$$

where we take $\mathsf{H}(\boldsymbol{\gamma})$ to be independent of $\boldsymbol{\theta}$. Now one can see that the use of a factorization distribution for $\boldsymbol{\gamma}$ can serve to maximize mutual information (the information that $\boldsymbol{\gamma}$ "knows" about $\boldsymbol{y}$) as a consequence of the inequality

$$\mathsf{H}(\boldsymbol{\gamma}) \leq \mathsf{H}(\gamma_1) + \cdots + \mathsf{H}(\gamma_m),$$

where equality holds (in which case the maximum is attained) if and only if the distribution for $\boldsymbol{\gamma}$ is factorial [8].

However, rather than directly maximize $\mathsf{I}(\boldsymbol{y}, \boldsymbol{\gamma}; \boldsymbol{\theta})$, we make the two further approximations that

$$\mathsf{H}(\boldsymbol{y}; \boldsymbol{\theta}) \approx \mathsf{H}(\boldsymbol{y})$$

and as θ gets close to its optimum value we have $p(\boldsymbol{y}, \boldsymbol{\gamma}; \boldsymbol{\theta}) \approx P(\boldsymbol{y}, \boldsymbol{\gamma})$ so that[73]

$$\mathsf{H}(\boldsymbol{\gamma}, \boldsymbol{y}; \boldsymbol{\theta}) \approx -\mathbb{E}_{\boldsymbol{\gamma}, \boldsymbol{y}} \{\ln p(\boldsymbol{\gamma}, \boldsymbol{y}; \boldsymbol{\theta})\} = \mathsf{C}\left(p(\boldsymbol{y}, \gamma) \| p(\boldsymbol{y}, \gamma; \boldsymbol{\theta})\right).$$

This yields the approximation

$$\mathsf{I}(\boldsymbol{y}, \boldsymbol{\gamma}; \boldsymbol{\theta}) \approx \mathsf{H}(\boldsymbol{y}) + \mathsf{H}(\boldsymbol{\gamma}) - \mathsf{C}\left(p(\boldsymbol{y}, \gamma) \| p(\boldsymbol{y}, \gamma; \boldsymbol{\theta})\right), \tag{17.76}$$

where $\mathsf{H}(y)$ and $\mathsf{H}(\gamma)$ are constants. This shows that the goal of maximizing the mutual information $\mathsf{I}(\boldsymbol{y}, \boldsymbol{\gamma}; \boldsymbol{\theta})$ with respect to the model parameters to ensure that $\boldsymbol{\gamma}$ is maximally informative about $\boldsymbol{y}$ is approximately accomplished via the minimization shown in equation (17.74), which was discussed at the outset of this subsection and which, as described there, can be done using the Jeffreys prior-based algorithm described at length in Subsection 17.5.2.

17.5.4 Dictionary learning for conceptual understanding of the world

The interpretation of dictionary learning and subsequent sparse feature extraction as a generalization of K-means clustering is potentially a very powerful one. Whereas K-means provides a mechanism to

[73] Recall that the entropy (a.k.a. self-entropy) is related to the cross-entropy as $\mathsf{C}\left(p(\boldsymbol{y}, \boldsymbol{\gamma}; \boldsymbol{\theta}) \| p(\boldsymbol{y}, \gamma; \boldsymbol{\theta})\right) = \mathsf{H}(\boldsymbol{\gamma}, \boldsymbol{y}; \boldsymbol{\theta})$. Thus we are invoking our approximation that the joint distribution of $\boldsymbol{y}$ and $\boldsymbol{\gamma}$ is independent of $\boldsymbol{\theta}$ in the first argument of the cross-entropy.

learn and point to a number of single conceptual, atomic situations (or patterns or events or influences or $\cdots$) perceived in the world, dictionary learning provides a way to identify compound situations that are comprised of a multitude of atomic situations. However, even single atomic situations are complex enough that a mere basis representation is inadequate to allow them to be expressed in an easily interpretable and distinct manner. To make this point clearer, let us consider an illuminating example from [174].

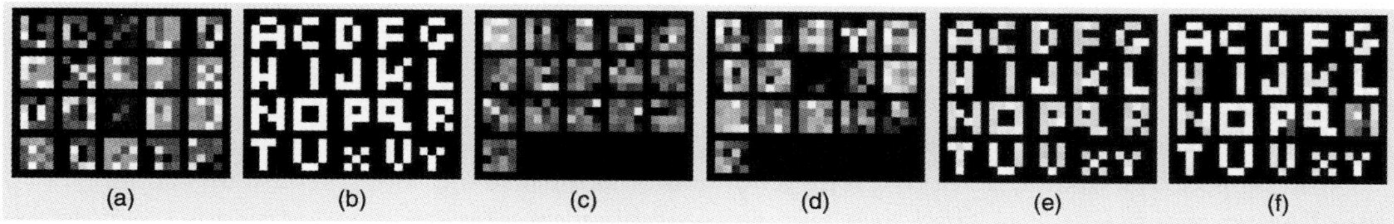

FIGURE 17.4

Towards XAI? – Learned dictionaries can capture interpretable structure. Courtesy of [174].

Consider an image space of $4 \times 4 = 16$-pixel grayscale images. The image space is $n = 16$-dimensional and therefore any image $\mathcal{I}$ can be represented and reconstructed from a 16-element basis set, a.k.a. a tight frame. It is common to represent the image $\mathcal{I}$ as a 16-element column vector $\boldsymbol{y} \in \mathbb{R}^n$, $n = 16$, obtained from stacking the pixel columns, so that $\mathcal{I} \leftrightarrow \boldsymbol{y}$. Note that there are 20 such 16-pixel images shown in Fig. 17.4(b); they are $4 \times 4 = 16$-pixel exemplars of 20 English letters, which we take to be represented by the $n = 16$-dimensional column vectors $\underline{\boldsymbol{a}}_j$, $j = 1, \cdots, \underline{m} = 20$. Assume that the presence of the exemplar images is not known and can only be detected indirectly via observations arising from the noisy generative model

$$\boldsymbol{y} = \underline{A}\boldsymbol{x} + \boldsymbol{n} \quad \text{with} \quad \underline{A} = \begin{pmatrix} \underline{\boldsymbol{a}}_1 & \cdots & \underline{\boldsymbol{a}}_m \end{pmatrix} \in \mathbb{R}^{n \times \underline{m}}, \quad n = 16, \quad \underline{m} = 20, \quad \text{and} \quad \mathsf{sp}(\boldsymbol{x}) = \underline{k} = 2.$$

This model says that observations are generated by a noisy sum of $\underline{k} = 2 = \mathsf{sp}(\boldsymbol{x})$ exemplar letters. Twenty representative samples drawn from this generative model are shown in Fig. 17.4(a). There are a total of $L = 5000$ such samples, $\boldsymbol{Y} = \begin{pmatrix} \boldsymbol{y}_1 & \cdots & \boldsymbol{y}_L \end{pmatrix}$, which can be used to learn a hypothesized generative model presumed to be of the form

$$\boldsymbol{Y} = A\boldsymbol{X} + \boldsymbol{N}, \quad A \in \mathbb{R}^{16 \times m}.$$

In general the unknowns of the hypothesized model include the size m of the frame, the $n \times m$ dictionary A, and the $m \times L$ representation vectors $\boldsymbol{X}$. The only thing known is that $n = 16$ and the assumption of a reasonably tight upper bound k on the sparsity of solutions, $\overline{\mathsf{sp}}(\boldsymbol{X}) \leq k$, which can be determined during the dictionary learning process.

Because $\boldsymbol{y} \in \mathbb{R}^n$ with $n = 16$, a representation for any dataset $\boldsymbol{Y}$ is possible if A is a tight frame, i.e., if it is 16×16 and nonsingular. However, note that there are 20 letters of interest as shown in Fig. 17.4(b). Although, of course, these 20 exemplar images can each be represented as a linear combination of 16 basis vectors that span the image space, such tight frame representations generally will not be sparse. Reference [174] shows that by using PCA [88] or independent component analysis (ICA) [83], one can learn 16 basis vectors, as shown in Fig. 17.4(c) and Fig. 17.4(d), respectively. By using either one of these bases one can uniquely solve $\boldsymbol{y} = A\boldsymbol{x}$ given an image $\boldsymbol{y}$ for an image representation vector via a simple inversion, $\boldsymbol{x} = A^{-1}\boldsymbol{y}$. However, in either of these tight frame cases the derived representation $\boldsymbol{x}$ generally will not be sparse and/or easily interpretable. This will almost certainly be the situation if the atoms themselves are noninterpretable, which is the case for the PCA and ICA solutions as shown in Fig. 17.4(c and d).[74]

If we know that there are $m = 20$ atomic entities of interest (in this case, letters), then it is natural to use a dictionary of $m = 20$ atoms, which for our example corresponds to a 16×20 dictionary matrix A. Fig. 17.4(e) shows the 20 16-dimensional columns (laid out as 4×4 pixel grids) of the dictionary matrix A obtained using the gradient descent dictionary learning algorithm shown in Eq. (17.42) [133], while Fig. 17.4(f) shows the columns from finding A using the K-SVD algorithm. It is evident that taking $m = 20 > 16 = n$ greatly increases the interpretability of the dictionary atoms.

Now suppose we want to detect overlaid letter combinations. In the simplest case this could be two-letter English words like "to," "in," "it," "as," etc., combined in a layered "bag-of-words" fashion. For example, in Fig. 17.4(a) representative noisy images made up of the sum of two letters plus additive white Gaussian noise are shown. For a well-learned dictionary, like the one shown in Fig. 17.4(f), it will be possible to decompose the image  as the sum of the letters T and O – "to." And sometimes, as for this simple example, this procedure can have meaning. And because dictionary learning has the ability to extract meaning, it can be used in conjunction with other machine learning methods; for example in [125] dictionary learning is used at the input layer of a deep neural network to good effect.

For an $n \times m$ dictionary matrix A, the fact that one knows, or can detect, the truth that $m = \underline{m} = 20$ is generally nontrivial. In the example shown here, this corresponds to knowing that there are 20 atomic classes (letters in this case) in play. Determining the value of m is equivalent to the very difficult problem of model order determination, which in the case of classification requires determining the number of distinct classes. From the perspective of dictionary learning viewed as a powerful generalization of clustering that allows us to deal with the case where multiple atomic classes get admixed to form complex compound classes, this problem is seemingly amplified. However, it may well be the case that dictionary learning is robust in the limit that our choice of m gets large in that there may be a propensity in "Sparseland" for meaningful atoms to pop up out of a sea of nonsense atoms because the "meaningful atoms" match to the structure in the world, while the remaining atoms just match to the noise. If this latter fact is true, after a large dictionary is learned one can then keep a count of which atoms are constantly used to represent situations encountered in the world and which ones are never in play, and in this way eventually (over evolutionary time?) learn an appropriate model order.

17.6 Nonnegative matrix factorization in dictionary learning

As emphasized in the previous section, an increasingly important concern of contemporary research in machine learning is the development of algorithms capable of providing semantic understanding of sit-

[74] PCA results in orthonormal frames while, similarly to FA, ICA can find nonorthonormal tight frames. This allows ICA to generally have greater expressive power than PCA, which in this case appears to be manifested in the fact that the learned ICA atoms seem to show some hints of "letter-like" structure. Recall that a blind inverse problem at best can produce a solution for A that is unique only up to the sign and ordering of the columns of A, so it is not clear in these two examples which atoms, if any, are comparable.

uations in the world in addition to having predictive or discriminative capabilities and thereby provide explanations of how or why such predictions were arrived at. Such a semantic or interpretative capability falls under the rubric of *explainable AI* (XAI) [184,155,145]. An important approach to developing multivariate statistical learning algorithms capable of both explainability and dimensionality reduction is *nonnegative matrix factorization* (NMF), which is applicable when observed data only take nonnegative values [99,66,68,35,67]. Popular applications of NMF include cases where the underlying data are nonnegative and a decomposition into constituent parts serves some application-specific benefit. One prime example of this is speech denoising, where denoising is performed in the spectrogram domain and the observations are decomposed (additively) into signal and noise [186]. Other applications of NMF include document clustering [196], hyperspectral image denoising [103], and recommendation systems [64].

When interpreting NMF, it is standard to consider the low-noise case

$$\boldsymbol{Y} = \boldsymbol{A}\boldsymbol{X} + \boldsymbol{N} \approx \boldsymbol{A}\boldsymbol{X},$$

which is a realistic assumption or approximation in many scenarios.[75] We denote componentwise nonnegativity of a matrix M by $M \geq 0$ or $0 \leq M$.[76] Thus, in the low-noise limit we are interested in the problem

$$\boldsymbol{Y} \approx \boldsymbol{A}\boldsymbol{X} \quad \text{with} \quad 0 \leq \boldsymbol{Y} \in \mathbb{R}^{n \times L}, \;\; 0 \leq \boldsymbol{A} \in \mathbb{R}^{n \times m}, \;\; 0 \leq \boldsymbol{X} \in \mathbb{R}^{m \times L}. \tag{17.77}$$

Classically NMF assumes that $m \leq n$ and quite frequently that $m \ll n$. We refer to this classical case as *undercomplete* NMF (uNMF). Our interest is in the overcomplete case when $m > n$, which we refer to as *overcomplete* NMF (oNMF) or, equivalently, as *nonnegative dictionary learning* (NDL). It is important to distinguish between uNMF (i.e., "usual" NMF) and oNMF (NDL) as applications and simulations suggest that qualitatively they can behave quite differently.

As noted in the foundational paper [99], VQ, PCA, and (standard undercomplete) NMF can be interpreted as variants of a matrix factorization procedure resulting in components (a.k.a. atoms or features) providing different types of explainability. The case where $m \ll n$ can yield a *parts-based* representation [99]. This arises from the fact that

$$\boldsymbol{y}_\ell \approx \boldsymbol{A}\boldsymbol{x}_\ell = x_{1\ell}\boldsymbol{a}_1 + \cdots + x_{m\ell}\boldsymbol{a}_m \quad \text{with} \quad m \ll n \ll L$$

for *every* positive observation n-vector $\boldsymbol{y} \geq 0$ and associated positive representation m-vector $\boldsymbol{x}_\ell \geq 0$, $\ell = 1, \cdots, L$. The fact that *all* L positive observations $\boldsymbol{y}_\ell$ have to be additively assembled as a positive sum of the *same* small set of positive atoms $\boldsymbol{a}_j \geq 0$ forces the atoms to have a parts-like structure in contrast to techniques such as VQ and PCA which yields atoms having holistic structure [99]. In essence, when $m \ll n$ NMF performs a type of FA subject to nonnegativity constraints which produces parts-like factor components.

NDL (oNMF) generalizes standard NMF (uNMF) to the overcomplete case $m > n$. Assuming that the URP condition holds whenever $m > n$,[77] any n columns of the $n \times m$ dictionary matrix A are lin-

---

[75] Such as when one works with high-quality images or word count data extracted from documents. It has also been argued that this assumption makes sense for data sensed by animals in the natural world [16].

[76] Be careful not to confuse this with the positive definiteness partial ordering on square matrix which typically uses the same notation. The meaning of $M \geq 0$ should be clear from the context.

[77] See Sections 17.2.4 and 17.2.5.

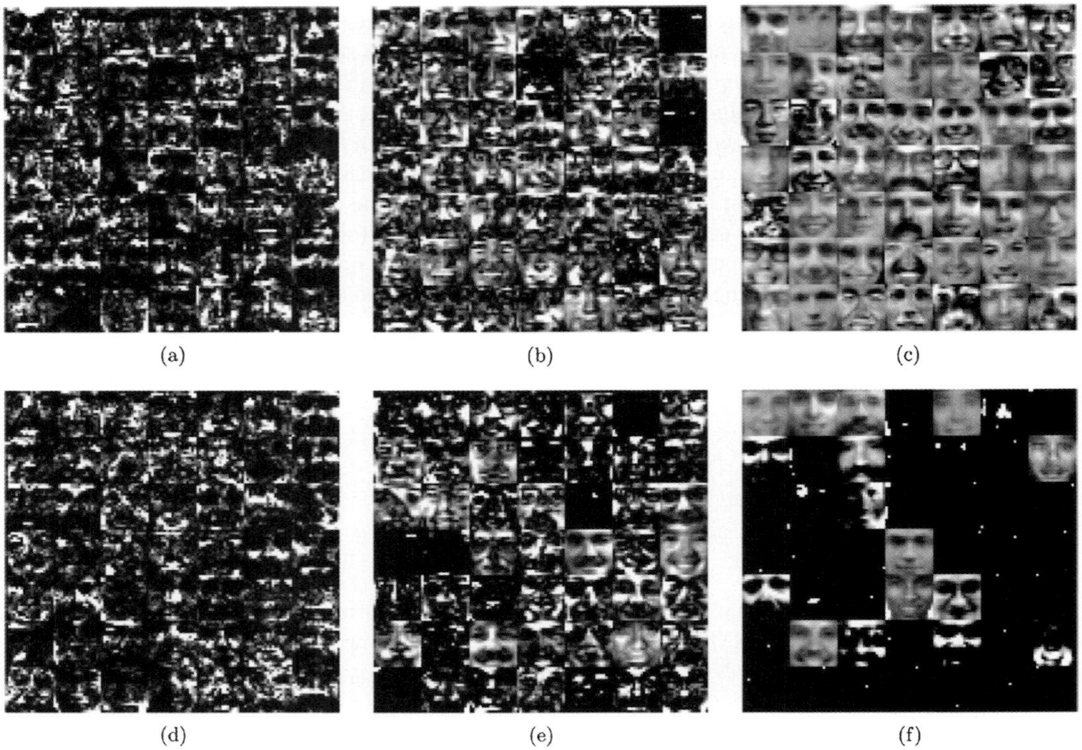

FIGURE 17.5

Nonnegative dictionaries can capture whole-part relationships and discover structure. Visualization of atoms for dictionaries A used for representing faces that are learned using the sparsity-regularized, overcomplete NMF "Type I" algorithms given in [58]. These algorithms learn a nonnegative factorization $\boldsymbol{Y} \approx \boldsymbol{A}\boldsymbol{X}$ given a nonnegative data matrix $\boldsymbol{Y}$ of database images. The algorithm used to develop the atoms shown in the top row enforces sparsity in the solutions $\boldsymbol{X}$. The algorithm in the bottom row enforces sparsity both in the dictionary A and in the solutions $\boldsymbol{X}$. Both algorithms have a sparsity regularization parameter that increases in strength from left to right so atoms shown on the right yield sparser representations than the atoms shown on the left, while the atoms in the middle yield an intermediate level of sparsity. In both rows, as one moves from left to right one can see the atoms change from being "parts" that must be assembled to present an image (yielding less sparse representations) to atoms that can "holistically" represent an entire image in the sense of vector quantization (which are maximally sparse sparsity-1 representations). In the bottom row, moving from left to right one sees the development of holistic atoms that can point to holistic template atoms and parts-based atoms that serve to "clean up" the representation adding missing "parts." Courtesy of [58].

early independent. The existence (or enforcement) of sparse solutions to the oNMF problem allows for the existence of holistic, nonparts-based atoms. For example, if a single column of A can (necessarily holistically) explain an observation $\boldsymbol{y}$, then a sparsity $k = 1$ representation for $\boldsymbol{y}$ exists. Indeed, in the limit that $m \to \infty$, essentially any $\boldsymbol{y}$ can be assigned a holistic, sparsity $k = 1$ representation.[78] Thus, our heuristic discussion suggests that the $m \ll n$ limit tends to enforce the creation of parts-based representations, while taking $m \gg n$ and enforcing sparsity tends to yield holistic representations. One can conjecture that taking m only moderately greater than n will yield holistic, maximally sparse representations for frequently occurring values of $\boldsymbol{y}$ and parts-based, less sparse representations for rarely occurring values of $\boldsymbol{y}$. Such behavior is experimentally shown in Fig. 17.5 taken from reference [58].

A popular approach used in standard uNMF to determine a positive factorization $\boldsymbol{Y} \approx A\boldsymbol{X}$ is through the use of multiplicative update rules, whereby updates to A and $\boldsymbol{X}$ are generated by elementwise multiplication with nonnegative matrices [61]. Generalizations of these multiplicative update rules to oNMF that minimizes a sparsity-regularized objective for a general class of sparsity-promoting regularizers is given in [58,57], and earlier approaches to solving the oNMF problem can be found in [79,3].

Both Bayesian Type I methods [58,57] and Type II methods [127,126] can be used to develop oNMF algorithms. When compared to the Type I approach, the Type II method exhibits superior performance despite requiring additional implementation approximations. Fig. 17.5 gives a visualization of the atoms learned from a database of human faces using the Type I algorithm developed in [58]. As described in the figure caption, this visualization shows the ability of the oNMF algorithm given to learn atoms capable of creating parts-based, holistic, and mixed representations as a function of parametric and algorithmic choices made by the user.

17.7 Dictionaries, data manifold learning, and geometric multiresolution analysis

There are situations where one can argue that it is reasonable to model high n-dimensional ($n \gg 1$) signals measured in the world as being constrained to lie on intrinsically low d-dimensional manifolds ($d \ll n$) embedded in the signal space corrupted by some degree of noise in the ambient signal space $\mathbb{R}^n$. Such is the case, for example, when imaging a world that is well modeled by articulated interconnected rigid body objects [45,40] and/or by localized, texturally distinct image patches [142]. Multiscale (a.k.a. multiresolution) modeling of such situations by the use of multiresolution dictionaries was explored in the papers [179,10,178], and these investigations were amplified and extended in the works [31,123,30, 5,85,108,102]. In this section we briefly review the work of this latter literature, noting that references [85,108,102] in particular provide a good survey of the prior art in this area.

We begin by motivating the problem using terminology already established in this chapter. Suppose the observed data $\boldsymbol{y} \in \mathbb{R}^n$ actually lie on a smooth manifold $\mathbb{Y} \subset \mathbb{R}^n$ of constant intrinsic dimension $d \ll n$. Then *locally* around any point in $\boldsymbol{y}_0 \in \mathbb{Y}$, i.e., in a small enough neighborhood $\mathsf{N}(\boldsymbol{y}_0)$, we have (see Section 17.2.5 for the definition of a sparsity class c_y)

$$\boldsymbol{y} = \boldsymbol{y}_0 + \sum_{j \in c_y} x_j \boldsymbol{a}_j + \boldsymbol{\epsilon} \quad \text{with} \quad |c_y| = d, \tag{17.78}$$

[78] In this limit, the problem becomes one of vector quantization.

where the d vectors $\boldsymbol{a}_j$, $j \in c_y$, span the local tangent plane $\mathsf{T}_{y_0}\mathbb{Y}$ of the d-dimensional manifold $\mathbb{Y}$ located at the point y_0 and the *accuracy* ϵ is appropriately "small." Note that determining (17.78) for any $\boldsymbol{y}$, which requires the determination of its associated frame c_y and frame location $\boldsymbol{y}_0$, can be viewed as a form of *localized best-basis selection* where the number of basis vectors $d = |c_y|$ in the local frame c_y is equal to the intrinsic dimensionality of the data manifold $d = \dim \mathsf{T}_{y_0}\mathbb{Y}$. It is convenient to take each $\boldsymbol{a}_j$ to be a unit vector and to define the frame location $\boldsymbol{y}_0$ by

$$\boldsymbol{y}_0 = x_{y_0}\boldsymbol{a}_{y_0},$$

with $\boldsymbol{a}_{y_0}$ being a unit vector. We represent each tangent space by the $k \overset{\text{def}}{=} d+1$ element frame

$$\tilde{\mathbf{c}}_y = \{\boldsymbol{a}_{y_0}\} \cup \{\boldsymbol{a}_j\}_{j \in c_y}, \qquad |\tilde{c}_y| = k = d+1,$$

where it is understood that $\boldsymbol{a}_{y_0}$ must be appropriately scaled by x_{y_0} in order to site the frame c_y at the point $\boldsymbol{y}_0 = x_{y_0}\boldsymbol{a}_{y_0}$.

Assume that a collection of τ tangent planes is sufficient to cover the data manifold $\mathbb{Y}$.[79] Then the collection of vectors

$$\mathcal{D} = \bigcup_{\substack{y \in \mathbb{Y} \\ |\tilde{\mathbf{c}}_y| = k}} \tilde{\mathbf{c}}_y$$

forms an overcomplete dictionary that can adequately (up to accuracy ϵ) represent any observation $\boldsymbol{y} \in \mathbb{Y}$. Note that

$$|\mathcal{D}| = m \leq k\tau = (d+1)\tau$$

with strict inequality $m < k\tau$ if neighboring tangent planes share vectors $\boldsymbol{a}_j$. Note that this strict inequality $m < k\tau$ can occur if we impose the condition that m, the size of the frame $\mathcal{D}$, must be as small as possible subject to bounds on the size of the error ϵ. Given the dictionary $\mathcal{D}$, we arrange its elements into the columns of the $n \times m$ dictionary matrix A and thereby attain the *best-localized k-basis model of accuracy* ϵ

$$\boldsymbol{y} = A\boldsymbol{x} + \epsilon = A_{\tilde{\mathbf{c}}}\boldsymbol{x}_{\tilde{\mathbf{c}}} + \epsilon \quad \text{with} \quad \tilde{\mathbf{c}} = \tilde{\mathbf{c}}_y \ \text{ and } \ |\tilde{\mathbf{c}}| = k = d+1, \tag{17.79}$$

which is just a rewriting of Eq. (17.78). This model provides a representation of the k-dimensional data manifold $\mathbb{Y}$ which is accurate up to the approximation error ϵ.

Taking stock, with $A \in \mathbb{R}^{n \times m}$ and $A_{\tilde{c}} \in \mathbb{R}^{n \times k}$, it is evident that (up to the size of the error ϵ) a solution to Eq. (17.79) exists with sparsity no more than $k = d+1$.[80] It will often be the case that the construction (17.79) yields a model that is overcomplete relative to the ambient space $\mathbb{R}^n$, $m > n$. However, if the data manifold is of sufficiently low intrinsic dimension, d, and a relatively low number, τ, of tangent planes is sufficient to cover the data manifold to a requisite degree of precision, it is possible

[79] By this, we mean that the error ϵ in (17.78) is bounded to within some acceptable level uniformly for all $\boldsymbol{y}$.

[80] Recall that $\boldsymbol{x}$ is said to be k-sparse if at most k of its components are nonzero.

(a) S-manifold (b) scale 6 (c) scale 10

FIGURE 17.6

Data manifold learning using localized scale-dependent dictionaries. Visual demonstration of dictionary learning of an unknown data manifold of data sampled from an intrinsically low-dimensional manifold embedded in high-dimensional real Euclidean space (example and images taken from reference [102]). Data are (appropriately uniformly) sampled from the S-shaped surface shown in (a) and then isometrically embedded, with a randomly chosen position and pose, into $n = 100$-dimensional Euclidean space. Small-variance uniformly sampled random noise is then independently added to all $n = 100$ components to create noisy sampled observation vectors $\boldsymbol{y}_\ell \in \mathbb{R}^{100}$, $\ell = 1, \cdots, L = 50{,}000$. From these data alone, one is to learn a scale-dependent dictionary that provides an acceptable approximation to the unknown manifold. After the intrinsic dimension, $d = 2$, has been learned (see text), the multiscale procedures described in [102] can be applied to learn a collection of flats (affine subspaces) that in the aggregate provide an approximation to the unknown manifold to a specified level of accuracy. In (b) a moderate degree of approximation accuracy is obtained by learning and adaptively fitting on the order of $2^6 = 64$ (i.e., for scale $s = 6$) flats to the unknown data manifold. In (a) a more accurate fit is obtained by learning on the order of $2^{10} = 1024$ flats (corresponding to scale $s = 10$). Courtesy of [102].

to have $n > m = k\tau = (d + 1)\tau$. This emphasizes the point that redundancy (overcompleteness) is actually relative to the intrinsic dimensionality of the data (in this case, d) and not to the dimensionality, n, of the ambient space within which data are collected.

Some thinking leads to the realization that the size of the dictionary $m = |\mathcal{D}|$ should be dependent on the desired level of local accuracy $m = m(\epsilon)$. This is a supposition which can be formulated in a rigorous manner within the framework of *geometric multiresolution analysis* (GMRA) [31,123,30,5, 85,108,102]. Furthermore, as described in these references, this can be done in a manner that results in a tree structure hierarchy of *multiscale dictionaries* (with the atoms of these multiscale dictionaries referred to as *multiscale geometric wavelets*[81]). Finite sample nonasymptotic results and approximation error bounds are obtained by assuming that the data are generated by a distribution that is supported on or near a manifold of intrinsic dimension $d \ll n$, subject to conditions on the data generation process and appropriately constructed multiscale data structures that are rigorously presented in [5,108,102].

[81] This terminology is somewhat misleading as the dictionary atoms do not generally obey the properties (e.g., translation and dilation properties) often demanded of wavelets [31].

A key contribution of the work reported in [5,108,102] is the elucidation of the importance of constructing data structures from the observed data that capture multiscale neighborhood structure, in particular the construction of tree structures that encode data coverings (*cover trees*) and partitions (*partition trees*) at different scales so that one can "zoom in" and "zoom out" to various data scales in a consistent manner [105,19]. This enables one to create scaled neighborhood-conditioned data covariance matrices from which the intrinsic dimension of the data can be estimated as the number of dominant eigenvalues.[82] By looking for stability of this number over space and scale (a form of *topological persistence* [48]), one can obtain a reliable and accurate estimate of the intrinsic dimension [105]. Having an accurate estimate of the intrinsic dimension and with a multiscale cover/partition tree at hand, one can then turn to the multiscale determination of dictionaries that can be used to represent the data at a variety of scales, with finer scales corresponding to more accurate data representations. For a given scale, s, the data manifold will be approximated by a collection of local, approximate tangent spaces (*flats* a.k.a. affine subspaces) of the form (17.78) where the number of such flats is of order 2^s. Fig. 17.6 shows an example taken from [102].

17.8 Hard clustering and classification in dictionary learning

17.8.1 Clustering

Clustering in its hard sense carves up the world of observations into disjoint subsets that collectively cover the set of observations and the range of possible future observations. This is clustering as partitioning [121,122,118,161]. Because there are an infinity of ways to partition the world, and hence many different types of algorithms available to do so, it is not surprising that clustering in the abstract – i.e., as a partitioning decoupled from any specified goal – is ill-posed as it stands [55,185]. Any clustering algorithm used for such a purpose has a built-in "inductive bias" that favors some kinds of partitioning over others and thus it is unwise to ascribe any intrinsic meaning to the outcome of the algorithm without some deep reflection [17].

At the end of Section 17.5.3, we discussed how the imposition of a sparsity-enforcing prior imposes an inductive bias on the hierarchical/empirical Bayes model that favors selection of low-dimensional affine subspaces (flats). Similarly to how standard clustering techniques are used in spectral clustering after a spectral graph theory-based dimensionality reduction has been performed [119,177] to produce representation vectors or how clustering can be performed after a PCA dimensionality reduction step has first produced representation directions in data space [43], we can perform clustering on the representation vectors we compute. In particular, one possible way to do this is in three steps:

C0 From the data learn a *single* overcomplete dictionary and use this learned dictionary to obtain K distinct *sparsity classes*. This will result in the data being disjointly partitioned into the K sparsity classes according to the sparsity pattern of their representations.[83]

[82] Equivalently, as the number of dominant singular values of the data matrix [105].

[83] See Section 17.2.5. The sparsity classes are comprised of the atoms defining the subspaces of interest. Generally K, or an upper bound on K, will be prespecified. Note that this is a simplified approach in that the dictionary is learned for sparse representation purposes only, without regard to the fact that it will subsequently be used for clustering purposes. We will comment on this point later below.

C1 Within each sparsity class, perform further "second-level" clustering of the data.
C2 Assign a new observation to a cluster by determining its sparsity class from its sparse representation and then performing a second-level clustering of its representation within that sparsity class.

Note that clusters detected in this manner are naturally identified using two indices: one to indicate the sparsity class found in step **C0**, the second to indicate the second-level cluster found in step **C1**. The first step **C0** requires the commitment to a procedure to select a low number of subspaces (i.e., K sparsity classes) that occur most of the time or approximately all of the time. The second step **C1** requires the selection and deployment of a second-level clustering algorithm to be applied to the representations occurring within those subspaces. Note that these procedural choices add additional aspects to our inductive bias.

Continuing with the ideal formalism **C0–C1**, we can interpret the selection of a subspace in step **C0** as the detection of a set of "atoms" (or "words") capable of describing a "domain of discourse." The second step **C1** detects the particular way that those atoms have been weighted and then combined to form an isolated, identifiable statement within that domain. This is a particularly compelling way to think about this clustering procedure if the nonnegativity constraints described in Section 17.6 are imposed.

Interestingly, however, the procedure **C0–C1** just described does not seem to have been fully explored in the literature. The proposed procedure **C0–C1**, which uses K class-dependent sparsity ("local") subspaces determined from the columns of a *single* learned dictionary, is fundamentally different from the multidictionary approaches followed in [177,43,163,147] where global spaces ("global" because, in essence, several cluster-dependent dictionaries are learned) are determined (one for each cluster) based on the data followed by which clustering is performed on a new data sample by selecting the *entire* cluster-dependent dictionary that best matches the sample. In particular, in [163,147] the two steps **C0** and **C1** have been combined into a single step (we call this single step "step **CD1**") and relaxed to the problem of creating full dictionaries (each one, of course, comprised of columns which are atoms) that are each entirely specific to a single cluster[84] and the remaining clustering step **C2** (which we now call "step **CD2**") is performed simply by selecting the class-specific dictionary that gives the best-fitted sparse representation. In this simplified approach we have the following two-step procedure:

CD1 Given data matrix $\boldsymbol{Y}$, specify an integer K. Then learn K classes $\mathcal{P} = \{C_1, \cdots, C_K\}$ with partition $\mathbb{Y}$ with each class having a distinct dictionary which we can assign to the set $\mathbf{A} = \{A^{(1)}, \cdots, A^{(K)}\}$. Note that the K classes (equivalently, the K distinct class-dependent dictionaries) represent the K clusters into one of which a new data instance will be assigned as per the next step.
CD2 Perform clustering of a new observation $\boldsymbol{y}$ by selecting the best-fitting class-specific dictionary $A^{(j)}$ and assigning $\boldsymbol{y} \in C_j$.

This is the procedure, for example, described in [163,147] where, for a specified number K of clusters, step **CD1** is performed by solving (for specified values of k, K, and m)[85]

[84] Note that a single dictionary will generally correspond to many distinct sparsity classes of the type defined in Section 17.2.5.

[85] As noted in [147,47], one can increase cluster separability by adding a penalty term of the form $\gamma \sum_{i \neq j} \|A^{(i)^T} A^{(j)}\|_F^2$ to the loss function that is being minimized on the right-hand side of (17.80). This term penalizes the coherence between dictionaries.

$$(\hat{\mathcal{P}}, \hat{\mathbf{A}}) = \arg\min_{\mathcal{P},\mathbf{A}} \frac{1}{L} \sum_{j=1}^{K} \sum_{\ell=1}^{L} \mathsf{dis}(\boldsymbol{y}_\ell, A^{(j)}) \mathbf{1}(\boldsymbol{y}_\ell \in C_j) \tag{17.80}$$

with $A^{(j)} \in \mathbb{R}^{n \times m}$, for $j = 1, \cdots, K$, and

$$\mathsf{dis}(\boldsymbol{y}, A) \stackrel{\text{def}}{=} \min_{\mathbf{sp}(\boldsymbol{x}) \leq k} \|\boldsymbol{y} - A\boldsymbol{x}\|_2^2. \tag{17.81}$$

It is illuminating to compare Eq. (17.80) with Eq. (17.26). Note that in step **CD1** the goal is to simultaneously learn K classes for the data $\boldsymbol{Y}$ and, at the same time, the K dictionaries $A^{(i)}$, $i = 1, \cdots, K$. In general this is quite hard and – in the spirit of the K-means clustering algorithm – one will often resort to a greedy algorithm that cycles between dictionary learning and clustering.

Once Eq. (17.80) is solved, step **CD2** is given by

$$\boldsymbol{y} \in C_j \quad \text{for} \quad j = \arg\min_{1 \leq i \leq K} \mathsf{dis}(\boldsymbol{y}, A^{(i)}). \tag{17.82}$$

We emphasize again that the procedure **C0–C2** described at the outset of this section involves learning a *single dictionary* for which we determine sparsity classes as a first level of clustering and then searching further within each sparsity class for additional subclusters. The second procedure **CD1–CD2** involves learning *multiple dictionaries*, one for each class. Note that procedure **C0–C2** can be simplified by dropping step **C1** and contenting ourselves with merely clustering according to sparsity class alone. But note that this simplification still uses only a *single* learned dictionary – we will further discuss this simplification below.[86]

Let us discuss a possible program for attacking the problem of solving step **C0** discussed at the outset of this section. Having learned a *single* overcomplete dictionary $A \in \mathbb{R}^{n \times m}$, $n < m$, from the data $\boldsymbol{Y}$ in any one of the manners described earlier means that for each data vector $\boldsymbol{y}_\ell$, $\ell = 1, \cdots, L$, we have the ability to solve the k-sparse inverse problem

$$\boldsymbol{y}_\ell \approx A\boldsymbol{x}_\ell$$

to obtain a k-sparse representation $\boldsymbol{x}_\ell \in \mathbb{R}^m$ having a specific 0–1 sparsity c_{c_ℓ}. We can encode this sparsity pattern as a Boolean *sparsity-pattern vector* $\boldsymbol{c}_\ell \in \mathbb{B}^m \subset \mathbb{R}^m$ such that the nonzero elements of $\boldsymbol{c}_\ell$ correspond to the nonzero elements of $\boldsymbol{x}_\ell$.[87] Note that $\mathsf{sp}(\boldsymbol{x}_\ell) = \boldsymbol{c}_\ell^T \boldsymbol{c}_\ell = |c_{x_\ell}| \leq k$ by construction of $\boldsymbol{c}_\ell$ and the requirement that $\boldsymbol{x}_\ell$ be k-sparse. It is convenient to construct the sparsity-data Boolean matrix $\mathbf{C} = [\boldsymbol{c}_1 \ \cdots \ \boldsymbol{c}_L] \in \mathbb{B}^{m \times L}$. Now note that *the goal of step **C0** is to cluster the binary vectors* $\boldsymbol{c}_\ell$, $\ell = 1, \cdots, L$ (i.e., the columns of $\mathbf{C}$) *into K distinct classes.* Fig. 17.5(c) and Fig. 17.5(f) shown above suggest that (at least for a benign world admitting appropriately parameterized nonnegative models) this can be done for reasonably bounded values of K.

For large values of m and L, clustering (the columns of) $\mathbf{C}$ can be a computationally onerous task. For example, if $m = 100$ and the representations $\boldsymbol{x}_\ell$ are k-sparse for $k = 6$, the computation of

---

86 This distinction between classification using multiple, per class, dictionaries and classification by sparsity class within a single dictionary is discussed in [47].

87 The 2^m elements of the finite set $\mathbb{B}$ correspond to the corners of the unit hypercube in $\mathbb{R}^m$.

Eq. (17.33) shows that there are 1.27×10^9 possible sparsity patterns to consider as (sparsity class) cluster centroids. Therefore heuristic approaches often must be used in practice. A variety of algorithms exist that can be used to cluster Boolean vectors [91,6,195]. However, especially for high-dimensional data, the convergence of such algorithms to acceptable solutions can be tricky and depends sensitively on the nature of the distance measure.[88] And, of course, the difficult problem of how to choose K, the number of clusters, must also be addressed.[89] Assuming this clustering procedure tractably leads to good solutions, the end result will be a collection of K cluster representatives $\hat{\boldsymbol{c}}_\kappa \in \mathbb{B}^m$, $\kappa = 1, \cdots, K$.

Note that even if step **C0** can be successfully performed, one would then need to perform step **C1** by performing second-level clustering of the data within each learned sparsity class $\hat{\boldsymbol{c}}_\kappa$. Thus the procedure **C0-C2**, which requires clustering into sparsity classes followed by clustering within each sparsity class, appears to be computationally daunting. Thus we note, as mentioned above, that a simpler (though potentially suboptimal) way to proceed is to only assign a new observation to a single sparsity class and *not* expend the effort to find second-level clusters within each sparsity class.[90]

This simplification of procedure **C0–C2** means that once we have determined the cluster representatives $\hat{\boldsymbol{c}}_\kappa$ in step **C0**, we forgo step **C1** and instead immediately perform step **C2** (i.e., cluster a new data vector $\boldsymbol{y}$) via the following procedure:

$$\text{solve for } k\text{-sparse representation vector } \boldsymbol{x} \text{ such that } \|\boldsymbol{y} - A\boldsymbol{x}\|_2^2 \text{ is minimized;} \tag{17.83}$$

$$\text{then assign } \boldsymbol{y} \in C_j \text{ for } j = \arg\min_{\kappa} \|\boldsymbol{y} - A(\hat{\boldsymbol{c}}_\kappa \circ \boldsymbol{x})\|_2^2, \tag{17.84}$$

where A is the *single* overcomplete dictionary learned in step **C0** and $\hat{\boldsymbol{c}}_\kappa \circ \boldsymbol{x}$ denotes the elementwise Hadamard product, which here serves to project $\boldsymbol{x}$ onto the sparsity pattern represented by $\hat{\boldsymbol{c}}_\kappa$. Note that *all* sparsity classes are involved when solving for $\boldsymbol{x}$ in the first step, which allows the K different sparsity classes to "compete" in explaining $\boldsymbol{y}$. The second step would then select the class that is most effective in explaining $\boldsymbol{y}$ after the "competition" is resolved.

Note that (17.82) and (17.84) show that after clusters have been discovered from a sample of unlabeled data, then a new observation is assigned to the cluster that results the representation having the minimum residual error. This procedure also can be applied for learning to classify labeled data as we discuss next.

17.8.2 Classification using residual errors

What differentiates classification from clustering? Rather than relying solely on the built-in inductive bias of a clustering procedure, to perform classification one further requires a partition of the observation space into regions corresponding to values of some utility, loss, or classification function. From this perspective, clustering is classification with a purpose, i.e., the inductive bias is imposed by the user or external environment (such as when an animal in its natural ecosystem learns to classify predator from nonpredator). Now when a new observation is assigned to one cluster versus another, different,

[88] And, partially for this reason, there have been a variety of distance measures proposed for categorical data [195,41].

[89] Thus, perhaps it is not surprising that the more straightforward procedure of learning cluster- and class-dependent dictionaries appears to be a not unpopular approach to clustering and classification [163,147,77,162,143,101].

[90] Note that this is philosophically similar to procedure **CD1–CD2**. The difference is that **CD1–CD2** learns a different dictionary for each cluster, whereas here we learn a single dictionary and from that dictionary learn a different sparsity class for each cluster.

cluster, we have a *difference that makes a difference*. Different requirements/goals result in different partitionings (clusterings) of the world into disjoint cells and one can ask what relationship one such clustering has to another.

Similarly to the multidictionary **CD1–CD2** clustering procedure described in the previous Subsection 17.8.1, it is natural to ask if learning a bank of (in this case, label-specific) dictionaries and corresponding sparse representations can be used to enhance the ability to classify measured situations, and indeed early on such investigations were begun. Classification can be view as "contextualized" clustering where knowledge that an observation belongs to a particular cluster is relevant for determining if a particular context (i.e., "class") is the case. And this does not have to be done by straightforwardly matching to categorical target-label values during training; for example the research reported in [77,162,143] trained multiple dictionaries, each one trained for a specific context, and thereafter made a discrimination between the contexts for a future observation by determining which context dictionary produces a sparse representation that best explained that observation in the squared-error sense. For example, suppose that context-1 data result in a learned dictionary $A^{(1)}$ and context-2 data in a learned dictionary $A^{(2)}$. Then in a new, unknown context, observation $\boldsymbol{y}$ can be determined to have a sparse representation $\boldsymbol{x}^{(1)}$ using the context-1 dictionary and a sparse representation $\boldsymbol{x}^{(2)}$ using the context-2 dictionary, which allows for the following simple rule for classifying $\boldsymbol{y}$ [77,162]:

$$\|\boldsymbol{y} - A^{(1)}\boldsymbol{x}^{(1)}\|^2 \underset{\text{decide 2}}{\overset{\text{decide 1}}{\lessgtr}} \|\boldsymbol{y} - A^{(2)}\boldsymbol{x}^{(2)}\|^2, \tag{17.85}$$

which is just the decision rule equation (17.82) for the binary classification case $K = 2$. It is evident that this binary dictionary procedure readily extends to the general, nonbinary, multiclass case.[91] Generalizations of this approach that incorporate nonnegativity constraints and allow for competition between the class-specific learned dictionaries $A^{(j)}$ when learning the class representations $\boldsymbol{x}^{(j)}$ are described in [101].

To modify the single-dictionary clustering procedure that leads to Eqs. (17.83)–(17.84) in order to simultaneously learn dictionary A and sparsity classes $\boldsymbol{c}_\kappa$ appears to be computationally nontrivial. In general, a straightforward combinatorial search for K k-sparse sparsity-pattern vectors $\boldsymbol{c}_\kappa$ would be a difficult computational endeavor for large values of K. One way to approach the problem would be to solve the following problem for $\hat{A} \in \mathbb{R}^{m\times n}$ and $\hat{\boldsymbol{c}}_\kappa \in \mathbb{B}^m$, $\kappa = 1, \cdots, K$, given labeled data $\boldsymbol{y}_\ell$, $\ell = 1, \cdots, L$, for K classes[92]:

$$(\hat{A}, \hat{\boldsymbol{c}}_\kappa, \kappa = 1, \cdots, K) = \arg \min_{\substack{A, k\text{-sparse } \boldsymbol{c}_\kappa \in \mathbb{B}^m \\ \kappa=1,\cdots,K}} \frac{1}{L} \sum_{\kappa=1}^{K} \sum_{\ell=1}^{L} \mathsf{dis}(\boldsymbol{y}_\ell, \boldsymbol{c}_\kappa; A) \mathbf{1}(\boldsymbol{y}_\ell \in C_\kappa), \tag{17.86}$$

$$\text{where } \mathsf{dis}(\boldsymbol{y}, \boldsymbol{c}_\kappa; A) = \|\boldsymbol{y} - A(\boldsymbol{c}_\kappa \circ \hat{\boldsymbol{x}})\|_2^2 \text{ subject to } \hat{\boldsymbol{x}} = \arg \min_{k\text{-sparse}\,\boldsymbol{x}} \|\boldsymbol{y} - A\boldsymbol{x}\|_2^2. \tag{17.87}$$

91 As discussed, for example, in references [112,194,163]. Note that the decision rule in this case is precisely Eq. (17.82). This approach is used to great effect for face recognition in [194], where the algorithm is referred to as *Sparse Representation-based Classification* (SRC), a terminology that is commonly encountered in the literature. The SRC approach to classification is enhanced in [197,198] by the use of a loss function that directly encourages class separation.

92 As mentioned in Footnote 85, a penalization term can be added to the loss function that forces class separation.

The procedure in Eq. (17.86) consists of learning a single dictionary *at the same time* that one learns the sparsity-pattern vectors $\mathbf{c}_K$. Here a dictionary matrix A is learned that enhances classification *and* facilitates classification of the data.[93] This should provide superior performance compared to separating the clustering and classification steps, particularly in a world that is not likely to be accurately modeled because of the need to work with reasonable model parameter values (e.g., of m and K).

An alternative and generally more powerful approach to classification based on layering dictionaries, which also allows the use of loss functions more general than residual error, is described in the next section. It is often the case that the concern with solving a difficult problem like (17.83)–(17.84) or (17.86)–(17.87) is strongly coupled with a desire to have a decision making process that is *explainable*. A simple rule like (17.85) hides the inner workings of the decision rule, whereas approaches bringing structural features to light, like the class sparsities encoded in Boolean vectors like $\hat{\mathbf{c}}_K$ can elucidate structure in the world that perhaps one can attach an interpretable conceptual framework to. Other procedures have been proposed to perform classification using dictionaries. A particularly accessible discussion of classification using residual errors can be found in [47].

17.9 Multilayer dictionary learning and classification

Perhaps the most straightforward way to combine dictionary learning with a classifier such as a support vector machine or a neural network is to prepend a dictionary-based sparse feature extraction preprocessing stage to the network. This approach involves first independently learning the sparse codes and dictionary elements for application to downstream tasks such as classification or regression, followed by learning the network weights using the sparse codes provided by the dictionary in the manner shown in Eqs. (17.88) and (17.89) and in Fig. 17.7(a) [110]. Here $\mathcal{L}$ denotes the loss function used in classification, $(\boldsymbol{y}_\ell, z_\ell)$ are respectively the input data and expected or ground truth labels for $l = 1, \dots, \ell$ sets of data, $\boldsymbol{D} = A$ is the learned dictionary with the sparse codes $\boldsymbol{x}_\ell$ for $\ell = 1, \dots, L$, and the classifier weights are denoted as W. We have

$$\min_{\boldsymbol{x}_l, A} \frac{1}{\ell} \sum_{\ell=1}^{L} \frac{1}{2} \|\boldsymbol{y}_\ell - A\boldsymbol{x}_\ell\|_2^2 + \lambda \|\boldsymbol{x}_\ell\|_1 \tag{17.88}$$

and

$$\min_{W} \frac{1}{\ell} \sum_{\ell=1}^{L} \mathcal{L}(z_\ell, W\boldsymbol{x}_\ell) + \frac{\gamma}{2} \|W\|_F^2. \tag{17.89}$$

A version of this procedure is used to good effect in [125].[94] As a simple example, consider the case where a prelearned dictionary is used to preprocess the data which are then used to train a simple single hidden-layer multilayer perceptron (MLP) with softmax classier outputs. The steps are as follows:

---

[93] This indicates that one can strengthen the sophistication of clustering procedure **C0–C2** which, as mentioned in Footnote 83, decouples dictionary learning from clustering. One way to do this would be to search over possible clustering during the optimization shown in (17.86). Of course this would greatly increase the complexity of an already nontrivial problem.

[94] See Fig. 3 of that reference.

a)

b)

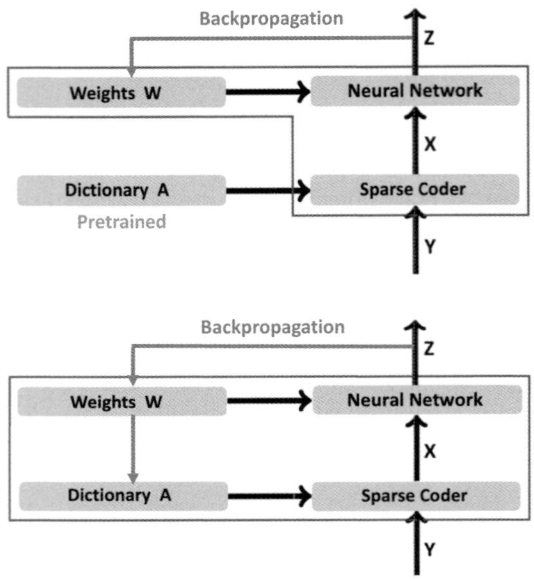

FIGURE 17.7

Combining a dictionary with a neural network. Learning the dictionary and associated weights for sparse classification/regression tasks [110]. **(a)** Backpropagation is used to learn the neural network weights W from a pretrained dictionary A. **(b)** Backpropagation is used to learn both the weights W of the neural network and the dictionary A. This combined learning procedure is nontrivial, as detailed in [110].

1. Collect labeled data $\boldsymbol{y}_\ell \in \mathbb{R}^n$, $\ell = 1, \cdots, L$, for K classes. Select the dimension m of the corresponding representation vectors $\boldsymbol{x}_\ell$ and the sparsity level k. Select the number of neurons, ν, in the perceptron (single-layer network). Encode the K classes as one-hot unit vectors $\boldsymbol{e}_j$, $j = 1, \cdots, K$.
2. Using all the data, $\boldsymbol{Y} \in \mathbb{R}^{n\times L}$, learn an overcomplete dictionary $A \in \mathbb{R}^{n\times m}$ trained to provide a k-sparse representation vector $\boldsymbol{x}_\ell$ for each data sample $\boldsymbol{y}_\ell$.
3. Transform the labeled data into labeled k-sparse representation vectors $\boldsymbol{y}_\ell \to \boldsymbol{x}_\ell$.
4. Train the weights $\boldsymbol{W} \in \mathbb{R}^{K\times m}$ of the perceptron (e.g., via backpropagation) using a loss function that forces the output of the softmax function to match to the K one-hot class encoding vectors.

The example given above is one of several possible architectures. This particular example decouples dictionary learning from learning weights of the network performing classification or regression, as shown in Fig. 17.7(a).

Learning the dictionary independently of classification as done in Eqs. (17.88) and (17.89) does not give optimal performance for such tasks. A more sophisticated approach is to simultaneously learn the dictionary and the classifier weights. Towards that end a more comprehensive approach for tasks such as regression and classification was proposed in [110,36]. The learning architecture for such task-driven dictionary learning is shown in Fig. 17.7(b), where learning the dictionary-based features used for supervised learning is performed in a fully supervised manner. In this case the objective function for learning is as follows:

$$\min_{W,A} \frac{1}{\ell}\sum_{\ell=1}^{L} \mathcal{L}(z_\ell, W\boldsymbol{x}_\ell^*(\boldsymbol{y}_l; A)) + \frac{\gamma}{2}\|W\|_F^2, \tag{17.90}$$

where

$$\boldsymbol{x}^*(\boldsymbol{y}, A) = \arg\min_{\boldsymbol{x}} \frac{1}{2}\|\boldsymbol{y} - A\boldsymbol{x}\|_2^2 + \lambda_1\|\boldsymbol{x}\|_1 + \frac{\lambda_2}{2}\|\boldsymbol{x}\|_2^2. \tag{17.91}$$

Under a few technical assumptions on the probability distributions of $(\boldsymbol{y}, z)$ it is possible to compute the gradient of $\mathcal{L}$ with derived learning rules for both W and A. The interested reader is referred to [110] for more details on the gradient updates for W and A.

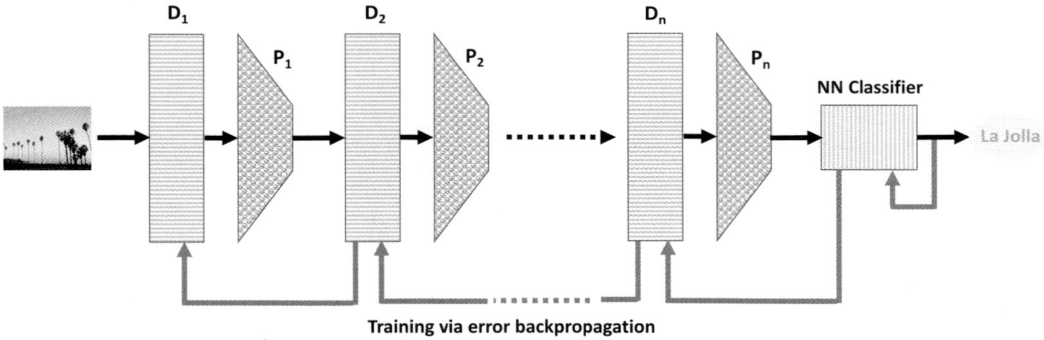

FIGURE 17.8

Deep dictionary learning. An architecture for learning dictionaries $\boldsymbol{D}_i = A_i$, $i = 1, \cdots, n$, in a deep dictionary network used for image classification. The backend classifier can be any shallow or deep neural network. This architecture is a generalization of the architecture shown in Fig. 17.7(b), created by layering multiple dictionaries prior to the neural network. Because the dictionaries D_i are overcomplete, the pooling stages P_i also serve the function of preventing the dimensionality of each layer from growing to an unwieldy size [29,109].

This formulation is extended to an architecture consisting of multilevel dictionaries with an image classifier in [29] as indicated in Fig. 17.8. In this case simultaneous learning of dictionary elements, sparse codes, and network weights is performed using backpropagation. In such architectures every layer l consisting of a sparse code has to be overcomplete with respect to its previous one $(l-1)$. Therefore it is required to perform dimensionality reduction on layer l before dictionary learning is performed on it to generate the sparse codes which form layer $l+1$. This bottleneck problem leading to the growth of overly fat dictionaries as highlighted in [167] can be resolved by using pooling layers as described in [29] and [109]. In these cases the pooling functions are parameter-free and differentiable, which allows them to be used in backpropagation in order to learn the dictionaries at different levels. Using similar mechanisms it is also possible to learn more general parameterized smooth pooling functions in addition to the classifier weights and the dictionaries. As shown in Fig. 17.9, such deep architectures extract features at multiple resolutions in the stack of dictionaries and thereby provide a

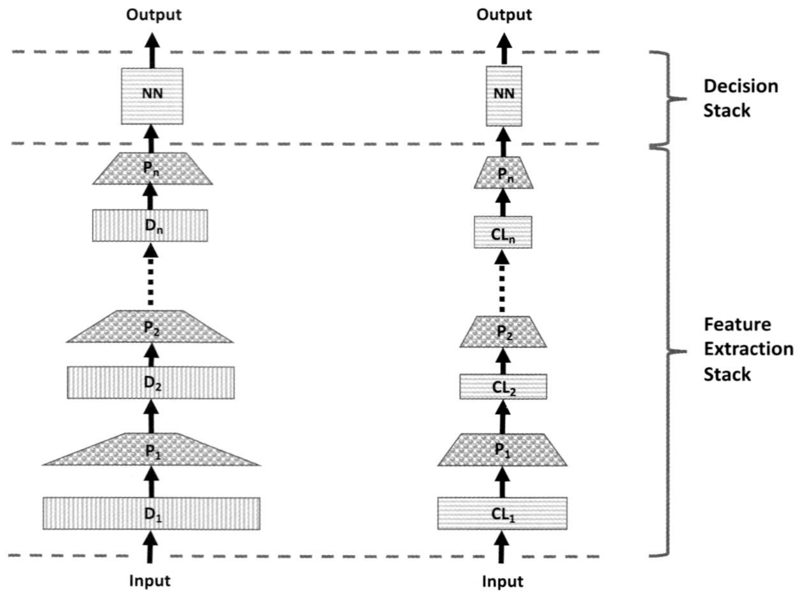

FIGURE 17.9

Two alternative deep feature extraction architectures: dictionaries vs. convolutional networks. (A) On the right is shown an archetypical deep convolutional neural network (DCNN) used in the feature extraction stack. The stacked coding elements are convolutional layers (CLs) with pooling elements (P) interposed between the CLs. Note that setting a pooling layer to the identity map "turns off" pooling. A trained DCNN can provide highly useful features for further processing in a "decision stack" using a variety of decision-oriented or regression-oriented neural networks. This combination is known to be very powerful and highly effective [96]. Essentially each CL/P combination extracts information from the input at a higher (generally coarser and/or more abstract) level of resolution. (B) Shown on the left is an architecture where the feature extraction stack is comprised of an interleaving of overcomplete dictionary (D) and pooling (P) layers. Here, the pooling not only provides control over the level of resolution of the features extracted at each layer, they are also critical for ensuring that the dimensions internal to the feature extraction stack remain reasonably bounded. This is because pooling ameliorates the explosion of dimensionality that can occur for deep dictionary architectures due to the fact each dictionary is taken to be overcomplete, thereby requiring that the domain of each dictionary be of higher dimension than the range. Recent work has shown indications that this architecture can be competitive with DCNNs while also providing the possibility of extracting features at each layer that are interpretable by humans.

powerful alternative to similar feature extraction capabilities of deep architectures such as convolutional neural networks [98]. In this context there has been some recent work focusing on the connections between architectures such as CNNs and sparse coding to understand important considerations such as uniqueness of representation in the network. For more details the reader is referred to [136,137,138].

Several other proposals have incorporated frameworks for jointly updating classifier weights and dictionary elements using backpropagation. These include applications for texture classification [162] and image analysis for histopathology [76]. In addition using the method outlined in [197] it is also possible to learn separate dictionaries for each class along with the network weights as described in [152]. However, all these proposals focus on undercomplete dictionaries which are jointly trained with the classifier owing to the problem with dimensionality expansion using sparse codes at every layer as described before.

17.10 Kernel dictionary learning

In this section we only cursorily describe the use of kernel methods in dictionary learning. Fortunately, an overview and discussion of this topic is given in the very accessible textbook [47], which in particular showcases an important central idea proposed in [130,131] that allows for kernelization of many well-known dictionary algorithms. Here, and in Appendix 17.D, we describe the rationale behind the kernelization procedure of [130,131] in some depth and only allude to the resulting kernelized dictionary learning algorithms in passing. For specific details on these algorithms see [47] and [72,1,84]. We end this section with a brief discussion of the meaningfulness of *overcompleteness* and *sparsity* within the context of the kernelized dictionary learning problem and mention the relevance of the work in [72] to this discussion.

The many uses of kernel methods in machine learning are motivated by the fact that a finite collection of data which are not linearly separable in the original data space are linearly separable when transformed in a one-to-one manner to a sufficiently high-dimensional space [168]. Further, many multivariate statistical learning and inference algorithms depend only on the inner products between data samples while performing well on data that are linearly separable. Kernel methods work by using a "kernel function," $K(\boldsymbol{y}_i, \boldsymbol{y}_j)$, the evaluation of which corresponds to an inner product $\langle \cdot, \cdot \rangle$ in some (generally unknown) high-dimensional (often infinite-dimensional) Hilbert space $\mathcal{M}$ (which, for simplicity, is taken to be over real-valued scalars) to which (at least conceptually) the data are mapped via some (generally unknown) function $\phi(\boldsymbol{y}) \in \mathcal{M}$. That is,

$$K(\boldsymbol{y}_i, \boldsymbol{y}_j) = \langle \phi(\boldsymbol{y}_i), \phi(\boldsymbol{y}_j) \rangle, \tag{17.92}$$

for data $\boldsymbol{y}_i$, $i = 1, \cdots, L$. The fact that one can evaluate the left-hand side of (17.92) in lieu of the right-hand side is known as the "kernel trick" [156,38,78]. If the computational cost of computing the value of a known kernel function is reasonable, then one potentially can get both the benefit of high-dimensional separability and the ability to use multivariate statistical techniques that can be expressed in terms of inner products and to do so without the need to explicitly produce the mapping $\phi(\boldsymbol{y})$ needed to explicitly perform the inner product shown on the right-hand side of (17.92).

A binary function $K(\cdot, \cdot)$ is guaranteed to have the property (17.92) (for some generally unknown space $\mathcal{M}$ and mapping $\phi(\cdot)$) if it satisfies Mercer's theorem,[95] a consequence of which is that the *kernel matrix*

$$\boldsymbol{K} = \Big(K(\boldsymbol{y}_i, \boldsymbol{y}_j) \Big) \in \mathbf{R}^{L \times L}$$

[95] See the discussions in [156,38,78]. It is also the case that more complex kernel functions can be systematically built from multiplying and adding simpler kernel functions [78].

is symmetric and positive semidefinite for all L and data y_i, $i = 1, \cdots n$. Given training data Y, note that the kernel matrix can be computed prior to subsequent use in a kernel-dependent algorithm so that this computational overhead is not consequential for the cost of subsequent algorithm deployment. If $K(\cdot, \cdot)$ satisfies (17.92), we call it a *Mercer kernel*. Further, we call $\mathcal{M}$ a *Mercer space*, $\phi(\cdot)$ a *Mercer mapping* (or *Mercer transformation*), and a vector $\mu \in \mathcal{M}$ (e.g., $\mu = \phi(y)$) a *Mercer vector* (or *Mercer feature*). There are many kernel functions described in the literature, with two very commonly used ones being the *radial basis function* (RBF) and the *polynomial* kernels which are respectively given by

$$K_\sigma(y_i, y_j) = \exp\left\{-\left(\frac{\|y_i - y_j\|}{\sigma}\right)^2\right\} \quad \text{and} \quad K_{c,d} = \left(y_i^T y_j + c\right)^d$$

for σ real and strictly positive, c real-valued, and d a nonzero, finite positive integer. The Mercer space for the RBF kernel is infinite-dimensional, while the Mercer space for the polynomial kernel has finite dimension.

As well motivated in [47], the merit of transforming the observations to Mercer space is that the ability to separate classes and/or discern fine structure in the data can be greatly enhanced. It is true that a deterministic mapping at best can only preserve information in the Shannon sense; nonetheless, non-stochastic (deterministic) *structural distinctions* between situations or classes (for which the Shannon theory can even be irrelevant – recall, for example, that mean shifts leave the Shannon entropy invariant) can be enhanced in a manner that rises above the noise level (which, of course, is also affected by the transformation) to thereby enhance the ability to perform classification and prediction. A classic toy example of this is shown and described in Fig. 17.10 for data transformed by $\phi : \mathbb{R}^2 \to \mathbb{R}^3$, where for $y = (\eta_1, \eta_2)^T \in \mathbb{R}^2$,

$$\phi(y) = \left(\eta_1, \eta_2, \exp\left(-(\eta_1^2 + \eta_2^2)/\sigma^2\right)\right)^T = \left(y^T, \exp\left(-\|y\|^2/\sigma^2\right)\right)^T \in \mathbb{R}^3 v, \tag{17.93}$$

with corresponding kernel function

$$K(y_i, y_j) = \phi^T(y_i)\phi(y_j) = y_i^T y_j + \exp\left(-\left(\|y_i\|^2 + \|y_j\|^2\right)/\sigma^2\right). \tag{17.94}$$

Reference [47] discusses some simple motivating examples that suggests why dictionary learning can benefit from mapping the observations $Y \in \mathbb{R}^{n \times L}$ via transformations,

$$\Phi(Y) \stackrel{\text{def}}{=} \left[\phi(y_1) \quad \cdots \quad \phi(y_L)\right] \in \mathbb{R}^{n' \times L}, \quad \phi(y_i) \in \mathcal{M}, \quad n' = \dim(\mathcal{M}) \leq \infty,$$

where $\boldsymbol{\phi}(\cdot) : \mathbb{Y} \to \mathcal{M}$ is a (generally unknown) Mercer transformation associated with some known (specified) Mercer kernel function $K(\cdot, \cdot)$. For instance, the example of Fig. 17.10 suggests that in the Mercer space one can learn directions (i.e., atoms/columns of a dictionary in the Mercer space) that are of great utility for making decisions in the original data space. Note that in general $\Phi(Y)$ has L columns and a potentially infinite number of "rows" $n' = \dim(\mathcal{M}) \leq \infty$. Formally we have (see Appendix 17.D)

$$\Phi^*(Y)\Phi(Y) = \left(\langle\phi(y_i), \boldsymbol{\phi}(y_j)\rangle\right) = \left(K(y_i, y_j)\right) = K \in R^{L \times L}.$$

Conceptually, once a transformation of the data $Y \in \mathbb{R}^{n \times L}$ has been performed, one would then proceed to learn a dictionary $D \in \mathbb{R}^{n' \times m}$ capable of providing sparse representations $X \in \mathbb{R}^{m \times L}$ of the

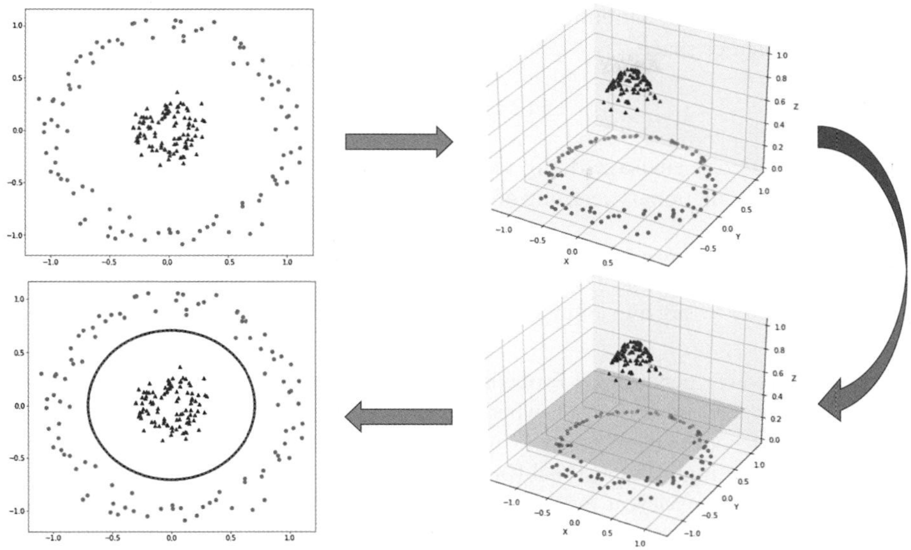

FIGURE 17.10

Transforming to higher dimensions can facilitate data-driven machine learning. Consider the two-class sample data shown in the upper left-hand side (LHS). It is evident that there is low-entropy, almost deterministic, class separation structure (cf. the discussion in [180]) that an omniscient Genie would recognize and can be captured using polar coordinates to create a radius threshold classifier as shown in the lower LHS image. However, mere mortal data scientists are not Genies possessing all-knowing insight; they somehow must effectively handle large amounts of data situated in very high-dimensional spaces that have been collected using the Cartesian (and other) coordinate systems associated with standard collection devices and modalities. What is needed are general approaches capable of deployment in highly complex, difficult to understand data environments, and the use of kernel functions provides a rich family of such approaches [156,38,78]. In the example shown above, the kernel function of Eq. (17.94) corresponds to the Mercer transformation equation (17.93), whose application to the data shown in the top LHS image results in the top right-hand side (RHS) image. It is evident that the Mercer transformation "stretches" the structural difference between the two classes along a new direction in a manner that keeps the transformed noisiness in the data reasonably bounded (i.e., so that the stochastic signal-to-noise behavior remains favorable to the detection process). The move from the upper RHS to the lower RHS shows the fact that the differential stretching of the two classes along the new direction allows for discrimination using a simple separating hyperplane in the high-dimensional feature space. The so-called "kernel trick" is used to perform this classification in the original space via use of the kernel function (17.94). Determining a class by thresholding-transformed data values in the z-direction as shown in the lower RHS is equivalent to thresholding along the radial direction of the original data as shown in the lower LHS, the approach that a benevolent Genie would have suggested at the outset, but that here has been functionally replicated without any such global awareness of the data structure. Note that in principle the z-direction of the transformed space can be learned as a column of a dictionary in that space, showing that learning an overcomplete linear model in the transformed space can be quite powerful for performing inference in the original data space. See reference [131] for a comparison of dictionary learning for nontransformed data versus Mercer-transformed data for this toy example using a polynomial kernel.

transformed data $\boldsymbol{\Phi}(\boldsymbol{Y}) \in \mathbb{R}^{n' \times L}$ assuming the validity of a linear generative model in the (possibly) infinite-dimensional Mercer space,

$$\boldsymbol{\Phi}(\boldsymbol{Y}) = D\boldsymbol{X} + \boldsymbol{E}. \tag{17.95}$$

Again conceptually, one can envisage doing so via some implementation of steps **C1** and **C2** of the cyclic-coordinate descent dictionary learning algorithm **A1** discussed above in Section 17.4.1:

$$\boldsymbol{X} \leftarrow \arg\min_{\boldsymbol{X}} \|\boldsymbol{\Phi}(\boldsymbol{Y}) - D\boldsymbol{X}\|_{GF}^2 \;\text{ subject to }\; \overline{\mathsf{sp}}(\boldsymbol{X}) \leq k, \tag{17.96}$$

$$D \leftarrow \arg\min_{D} \|\boldsymbol{\Phi}(\boldsymbol{Y}) - D\boldsymbol{X}\|_{GF}^2 \;\text{ subject to }\; D \in \mathcal{D}. \tag{17.97}$$

The norms shown here are "generalized Frobenius norms" that are approximate for an operator norm on bounded linear operators $\boldsymbol{\Phi} : \mathbb{R}^L \to \mathcal{M}$, where $\mathcal{M}$ is possibly infinite-dimensional (see Eq. (17.123) in Appendix 17.D). Of course it generally is not tractable to solve these optimizations directly due to the typically extremely high-dimensional (indeed, often infinite-dimensional) nature of the Mercer space $\mathcal{M}$ and the fact that the Mercer mapping $\phi(\cdot)$ is not known. Thus, as is usual in the kernel-based learning literature, one is led to consider variants of the kernel trick that correspond to solving the problem in a more indirect manner. Following an approach that is a key contribution of [130,131], we briefly describe a way to do this for dictionary learning.[96]

Recall (see Section 17.4.1) that the cyclic-coordinate descent algorithm is a procedure to solve the joint dictionary/representation learning problem,

$$(D, \boldsymbol{X}) \leftarrow \arg\min_{D,\boldsymbol{X}} \|\boldsymbol{\Phi}(\boldsymbol{Y}) - D\boldsymbol{X}\|_{GF}^2 \;\text{ subject to }\; \overline{\mathsf{sp}}(\boldsymbol{X}) \leq k \;\text{ and }\; D \in \mathcal{D}, \tag{17.98}$$

where the generalized Frobenius norm is described in Appendix 17.D. An important and useful result, first presented and proved in [130,131], is the fact that for the purposes of solving the optimization problem (17.98) it suffices to assume the dictionary form[97]

$$D = \boldsymbol{\Phi}(\boldsymbol{Y})A, \quad A \in \mathbb{R}^{L \times m}. \tag{17.99}$$

This key insight enables us to switch from the optimization (17.98) to the problem

$$(A, \boldsymbol{X}) \leftarrow \arg\min_{A,\boldsymbol{X}} \|\boldsymbol{\Phi}(\boldsymbol{Y})(I - A\boldsymbol{X})\|_{GF}^2 \;\text{ subject to }\; \overline{\mathsf{sp}}(\boldsymbol{X}) \leq k \;\text{ and }\; A \in \mathcal{A}. \tag{17.100}$$

---

[96] Further details are given in Appendix 17.D.

[97] As discussed in Appendix 17.D, the primary requirement is that the structural constraints imposed on D are "columnwise" so that D can be "projected" into the constraint set $\mathcal{D}$ without disrupting the existing sparsity structure of $\boldsymbol{X}$. Specifically, given $D\boldsymbol{X}$ with $\overline{\mathsf{sp}}(\boldsymbol{X}) \leq k$ and $D \notin \mathcal{D}$, let S be an $L \times L$ diagonal matrix such that $DS \in \mathcal{D}$. For example, such a postmultiplication of D can perform column normalization or Frobenius normalization. Then $D\boldsymbol{X} = DSS^{-1}\boldsymbol{X} = D'\boldsymbol{X}'$ with $D' \in \mathcal{D}$ and $\overline{\mathsf{sp}}(\boldsymbol{X}) \leq k$. With $D = \boldsymbol{\Phi}A$, $\boldsymbol{\Phi} = \boldsymbol{\Phi}(\boldsymbol{X})$, we have $D\boldsymbol{X} = \boldsymbol{\Phi}A\boldsymbol{X} = \boldsymbol{\Phi}\,(AS)\,(S^{-1}\boldsymbol{X}) = \boldsymbol{\Phi}A'\boldsymbol{X}'$, with $A' = AS$ and $\boldsymbol{X}' = S^{-1}\boldsymbol{X}$, showing that we are free to impose structural constraints on the columns of A in lieu of directly imposing constraints on D if that is our preference. Since a (often the) primary reason to impose constraints on the dictionary is to ensure identifiability of the separate factors $D = \boldsymbol{\Phi}A$ and $\boldsymbol{X}$ in the product expression $D\boldsymbol{X} = \boldsymbol{\Phi}A\boldsymbol{X}$, for this purpose it is entirely reasonable to opt to constrain A.

As shown in Appendix 17.D, this form can be immediately recast in terms of the kernel matrix $\boldsymbol{K}$,

$$(A, \boldsymbol{X}) \leftarrow \arg\min_{A,\boldsymbol{X}} \text{trace}\,(I - A\boldsymbol{X})^T \boldsymbol{K}(I - A\boldsymbol{X}) \text{ subject to } \overline{\mathsf{sp}}(\boldsymbol{X}) \leq k \text{ and } A \in \mathcal{A}. \tag{17.101}$$

In this optimization, the identity matrix I is $L \times L$, A is $L \times m$, and $\boldsymbol{X}$ is $m \times L$. Thus we are looking for an approximation to the identity given by the factorization $A\boldsymbol{X} \approx I$ subject to the constraint of sparsity on $\boldsymbol{X}$ and the identifiability constraint $A \in \mathcal{A}$. If the amount of data L is very large, then it is also very likely the case that $m \ll L$ even if the model is overcomplete in the sense that $m \geq n$. Note that the optimization in (17.101) is a weighted Frobenius norm on the "error" $I - A\boldsymbol{X}$ with weighting given by the kernel matrix $\boldsymbol{K}$. In the ideal case that $A\boldsymbol{X} = I$, there is zero error because the model perfectly fits the data; otherwise the finite-dimensional kernel matrix-weighted optimization (17.101) finds the best fit $A\boldsymbol{X} \approx I$ in the (generally) infinite-dimensional Mercer space.

As described in [130,131,47], one can now proceed to derive kernelized variants of many of the standard dictionary learning algorithms, including Method of Directions (MOD) and K-SVD, and the reader is invited to peruse that literature. More advanced aspects can be found in [72,1,84].

Some last comments should be given that pertain to the fact that there are difficulties attendant to the nature of "overcompleteness" and "sparsity" in the kernel dictionary learning context. In the kernelized model

$$\boldsymbol{\Phi} = D\boldsymbol{X} + \boldsymbol{N} = \boldsymbol{\Phi} A\boldsymbol{X} + \boldsymbol{E},$$

the columns of $\boldsymbol{\Phi}$ and D are (generally) infinite-dimensional vectors in $\mathcal{M}$; the number of columns (dictionary elements) of D and A is m; and the number of rows of A is L. Thus, the dictionary A is $L \times m$, and if we set $n' = \dim \mathcal{M} = \dim \phi(\mathbb{Y})$, the dictionary D is $n' \times m$. In the kernel framework we generally have

$$m \ll L \ll n' = \dim \phi(\mathbb{Y}) = \infty.$$

This shows that the dictionaries D and A are very "tall" indeed, which certainly does not conform to the concept of dictionary overcompleteness generally encountered in the literature where one demands that a dictionary be a "fat" matrix.

Further, the concept of sparseness becomes muddled: Is sparseness defined relative to the dimension of the original data space $n = \dim(\mathbb{Y})$ or relative to the dimension of the transformed data space $n' = \dim(\mathcal{M})$? From the perspective of obtaining a sparse representation for $\boldsymbol{y}$, one would like to continue to demand that the first criterion holds, $\mathsf{sp}(\boldsymbol{x}) = k \ll n = \dim \mathbb{Y}$. However, the conditions that guarantee the existence of an exact sparse solution do not generally apply to tall matrices such as D and A.

One way to proceed would be to replace the high-dimensional Mercer data operator Φ by a finite-dimensional $\hat{n} \times L$ approximation such that

$$\boldsymbol{K} \approx \hat{\boldsymbol{K}} = \hat{\boldsymbol{\Phi}}^T \hat{\boldsymbol{\Phi}}$$

and for which we ensure that $n < \hat{n} < m \ll \dim \mathcal{M}$. Having done this, one could use any of the standard dictionary learning approaches to learn a dictionary for the model,

$$\hat{\boldsymbol{\Phi}} = D\boldsymbol{X} + \boldsymbol{E},$$

where now the dictionary D is fat, $\hat{n} \times m$, so that the theory of compressive sensing using overcomplete dictionaries applies.

Having set the stage, the reader is invited to read [72] and the subsequent discussion in [47], which proposes exactly the methodology described in the previous paragraph. The algorithm proposed in [72] uses a Nyström sampling procedure for determining a low rank-$\hat{n}$ approximation of the symmetric, positive semidefinite kernel matrix $\boldsymbol{K}$.

17.11 Conclusion

The seminal paper [96] ignited an intense concentrated effort by the theoretical and practical engineering community to develop and deploy the powerful deep neural networks algorithms and systems. Deep neural networks and related technologies are now so well developed that turn-key systems exist which have been democratized to the point that university undergraduates can use them to good effect and are hired immediately into industries that are keen to apply these technologies.

The research community is now focusing increasingly on critical domain-specific applications and the increasing need for understanding the "why" of machine learning algorithms. As controls engineers well know, it is important to know, learn or and detect comprehensible features *in the real physical world* in order to create *robustly stable* feedback control laws applicable to real-world dynamical systems. This is true for many types of complex systems in the world: energy distribution grids, embedded biomedical control devices, next-generation supersonic passenger jets, robotic microsurgery – the list of such safety critical applications is almost endless.

The data needed to meet the needs of real-world, real-time applications do not arise from simple link-click counts, or monitoring of moving-watching preferences, or collection of answers to questionnaires. To collect and process the myriad signals afforded us by the material and natural worlds we interact with is highly nontrivial and we have developed sophisticated sensors, mathematical theories, and algorithms to facilitate these interactions and the extraction of interpretable and useful actionable features and information [146,107,113,151,105,168]. It generally requires concentrated effort and a significant intellectual investment, whether in graduate school or on the job, to gain mastery and deep understanding of the theories of modern signal processing and data science to a level where one can solve, and deploy the solutions to, the difficult problems encountered when dealing with the natural, physical world.

Several books exist that well describe the theory and methodology of dictionary learning as it is commonly utilized [33,47,49,139,150], and some of the highlights of that theory are described in this chapter, where we have endeavored to describe how the standard methodologies for dictionary learning, specifically the MOD algorithm and the K-SVD algorithm, can be related to the standard multivariable statistical analysis techniques of K-means clustering, PCA, and FA. We also discuss points of fruitful interactions between dictionary learning and machine learning techniques such as data manifold learning, clustering, classification, kernelization, and neural networks. It is clear that there is a lot of promise in these interactions – and a lot of work that remains to be done! Now that many areas of machine learning and deep neural networks are sufficiently mature, researchers can pay more attention to the many unresolved questions that remain to be solved and it is hoped that this chapter will be an impetus to such investigations.

One point of separation between the presentation given here and standard references such as [33, 47,49,139,150] is our emphasis on the fact that the natural signals encountered in the world often have sparse structure (i.e., "live" in what has been referred to as "Sparseland" [49,50]) and that, as argued by many, are well modeled by a hierarchical/empirical Bayesian generative model framework. In this chapter we have taken some pains to showcase this latter approach to dictionary learning and to provide references to the relevant literature. It is a general belief amongst experimental biologists and cognitive scientists that obtaining patterns relevant for succinct representations of natural signals provides an enhanced ability for a biological agent to recognize situations in the world, detect changes in those situations, and predict likely outcomes as a result of acting on such detections [8,13,12,62,100,82,132]. It seems reasonable that, to the extent this belief about the world is true, such benefits can also accrue to man-made artificial agents acting in the natural word. This opens up many other areas of research for aspiring researchers to look at.

Further, scientists and engineers – and users! – want and need *understanding*; they want/need to know what a pattern discerned in the world *means*, and *why* that particular pattern is important for providing a solution to the particular problem at hand. Dictionary learning can have a particularly important and powerful role to play in a future where *explainable* AI (XAI) comes to the foreground as a consequence of decision makers, politicians, legal systems, medical professionals, scientists, and engineers asking, and demanding answers to, important "why" questions. This is much research to be done in this regard, but, as is tantalizingly evident from a glance at Figs. 17.4 and 17.5, there is strong evidence that semantically meaningful (interpretable) patterns can be learned from natural data. This is a difficult research area that likely lies at the interface of information theory, signal processing, semiotics, computer science ontology theory, and psychology. A risky area of research no doubt, but one of huge potential payoff for those researchers who can accept the risk and cost and are willing and able to take the long view.

Appendix 17.A Derivation and properties of the K-SVD algorithm

Here, we use the notation and variables established in Sections 17.4.2 and 17.4.5.

Relationship between K-SVD and K-means

Reference [154] argues that, strictly speaking, K-SVD is not a generalization of the KMA because when restricted to performing the steps of the KMA it does not fully correctly do so.[98] Nonetheless, the K-SVD algorithm does have an insightful point of contact with the KMA, provided certain constraints are imposed, as we now show.

Assume that all columns of $\boldsymbol{X}$ have sparsity 1. This imposes the constraint that $\boldsymbol{r}_j^T = \begin{pmatrix}\gamma_{1j} & \cdots & \gamma_{Lj}\end{pmatrix}$ for $r_{\ell j} \equiv \gamma_{\ell j} \in \{0, 1\}$, with $\gamma_{\ell j}$ as defined in (17.35).[99] To show a correspondence with step **K1** of the KMA **A2**, assume that A is known and that the only nonzero columns of $\boldsymbol{Y}_j$ of (17.45) are in

[98] As reference [154] notes, this is a consequence of using the SVD to perform the minimization step of the K-SVD algorithm.

[99] This corresponds to the K-means constraint $\boldsymbol{X} = \boldsymbol{\Gamma}^T$ of Eq. (17.36), which is here equivalent to the constraint $\boldsymbol{R} = \boldsymbol{\Gamma}$.

$\tilde{C}_j = \text{enbh}(\boldsymbol{a}_j)$,[100] which is equivalent to constraining the columns of $\boldsymbol{Y}_j$ to be $\gamma_{\ell j}\boldsymbol{y}_\ell = r_{\ell j}\boldsymbol{y}_\ell$, for $\ell = 1, \cdots, L$. In this case,

$$\boldsymbol{a}_j \leftarrow \arg\min_{\boldsymbol{a}_j} \left\| \boldsymbol{Y}_j - \boldsymbol{a}_j \boldsymbol{r}_j^T \right\|_F^2 = \arg\min_{\boldsymbol{a}_j} \sum_{\ell=1}^{L} r_{\ell j} \|\boldsymbol{y}_\ell - \boldsymbol{a}_j\|^2 = \frac{1}{|\tilde{C}_j|} \sum_{\boldsymbol{y}_\ell \in \tilde{C}_j \subset \boldsymbol{Y}} \boldsymbol{y}_\ell,$$

which is the optimal update shown in (17.34). To show a correspondence with step **K2** of the KMA **A2**, note that for fixed A minimizing

$$\| \boldsymbol{Y} - A\boldsymbol{X} \|_F^2 = \| \boldsymbol{Y} - \begin{pmatrix} A\boldsymbol{x}_1 & \cdots & A\boldsymbol{x}_L \end{pmatrix} \|_F^2 = \left\| \begin{pmatrix} \cdots & \boldsymbol{y}_\ell - \sum_{i=1}^{m} \boldsymbol{a}_i \underbrace{(\boldsymbol{x}_\ell)_i}_{r_{\ell i}} & \cdots \end{pmatrix} \right\|_F^2$$

$$= \left\| \begin{pmatrix} \cdots & \boldsymbol{y}_\ell - \sum_{i=1}^{m} \boldsymbol{a}_i r_{\ell i} & \cdots \end{pmatrix} \right\|_F^2$$

subject to the constraints that $r_{\ell i} = \gamma_{\ell i} \in \{0, 1\}$ forces a competition between hard class assignments $\boldsymbol{y}_\ell \in \tilde{C}_i$ via the optimization

$$\boldsymbol{X}^T = \boldsymbol{\Gamma} = \boldsymbol{R} \leftarrow \arg\min_{\boldsymbol{R}} \sum_{\ell=1}^{L} \left\| \boldsymbol{y}_\ell - \sum_{i=1}^{m} r_{\ell i} \boldsymbol{a}_i \right\|^2,$$

whose solution is given by step **K2** of the KMA. With updates for A and $\boldsymbol{R}$ in hand, one approximately recomputes $\boldsymbol{Y}_j$ by ignoring the presumably small residual "quantization errors" in

$$\boldsymbol{Y} - \sum_{i \neq j}^{m} \boldsymbol{a}_i \boldsymbol{r}_i^T = \boldsymbol{Y}_j + \text{matrix of residual columns.} \tag{17.102}$$

A residual column is of the form $(\boldsymbol{y}_\ell - \boldsymbol{a}_i)$ (presumably ≈ 0) with $\boldsymbol{y}_\ell \in \tilde{C}_i = \text{enbh}(\boldsymbol{a}_i)$ for some $i \neq j$.

Derivation of the K-SVD Algorithm

Recalling the definition of the construction of the downsampling matrix

$$\boldsymbol{S}_j = \boldsymbol{S}(\boldsymbol{r}_j) = \begin{pmatrix} \boldsymbol{e}_{j_1} & \cdots & \boldsymbol{e}_{j_{s_j}} \end{pmatrix} \in \mathbb{R}^{L \times s_j},$$

we have

$$\boldsymbol{r}_j^T \boldsymbol{e}_{j_i} \neq 0 \quad \text{for} \quad i = 1, \cdots, s_j \quad \text{and} \quad \boldsymbol{r}_j^T \boldsymbol{e}_p = 0 \quad \text{otherwise.}$$

[100] See step **K2** of the KMA **A2**. Of course this single-column "quantization" (a.k.a. VQ-based "lossy compression" of the data $\boldsymbol{Y}$) creates a model that only approximately explains $\boldsymbol{Y}_j$, as indicated in the line after Eq. (17.102).

We also have the following, easily verified properties[101,102]:

$$S_j = \begin{pmatrix} e_{j_1} & \cdots & e_{j_{s_j}} \end{pmatrix} \in \mathbb{R}^{L \times s_j}; \quad \tilde{r}_j^T \overset{\text{def}}{=} r_j^T S_j = \begin{pmatrix} (r_j)_{j_1} & \cdots & (r_j)_{j_{s_j}} \end{pmatrix}; \quad \tilde{r}_j^T S_j^T = r_j^T;$$

$$S_j^T S_j = I \ (s_j \times s_j); \quad S_j S_j^T = I_j \ (L \times L); \quad \bar{I}_j = I - I_j; \quad \bar{I}_j I_j = I_j \bar{I}_j = 0;$$

$$r_j^T I_j = r_j^T; \quad r_j^T \bar{I}_j = 0;$$

$I_j =$ diagonal matrix whose only nonzero elements are ones in the (j_i, j_i) positions for $i = 1, \cdots, s_j$;

$$I_j = S_j S_j^T \text{ and}$$

$\bar{I}_j$ are symmetric and idempotent and thus orthogonal projection operators with $I = I_j + \bar{I}_j$;

$$\langle M I_j, N \bar{I}_j \rangle_F = \text{trace}\left(M I_j \bar{I}_j N\right) = 0 \implies \|M I_j + N \bar{I}_j\|_F^2 = \|M I_j\|_F^2 + \|N \bar{I}_j\|_F^2.$$

Note that postmultiplication of r_j^T by S_j downsamples the sparse row vector r_j^T to the dense row vector $\tilde{r}_j^T$ by removing zeros, while postmultiplication of $\tilde{r}_j^T$ by S_j^T upsamples the dense row vector $\tilde{r}_j^T$ to the sparse row vector $\tilde{r}_j^T$ by adding (padding) zeros. These facts are captured succinctly by the statements

$$r_j^T = r_j^T I_j = r_j^T S_j S_j^T = \tilde{r}_j^T S_j^T = r_j^T \quad \text{and} \quad \tilde{r}_j^T = \tilde{r}_j^T I = \tilde{r}_j^T S_j^T S_j = r_j^T S_j = \tilde{r}_j^T.$$

Now note that

$$Y_j - a_j r_j^T = \left(Y_j - a_j r_j^T\right) I = \left(Y_j - a_j r_j^T\right)(I_j + \bar{I}_j) = Y_j \bar{I}_j + \left(Y_j - a_j r_j^T\right) I_j.$$

This gives

$$\left\| Y_j - a_j r_j^T \right\|_F^2 = \| Y_j \bar{I}_j \|_F^2 + \left\| \left(Y_j - a_j r_j^T\right) I_j \right\|_F^2.$$

But

$$\begin{aligned} \left\| \left(Y_j - a_j r_j^T\right) I_j \right\|_F^2 &= \text{trace}\left(Y_j - a_j r_j^T\right) \underbrace{I_j I_j}_{= I_j = S_j S_j^T} \left(Y_j - a_j r_j^T\right)^T \\ &= \text{trace}\left(\tilde{Y}_j - a_j \tilde{r}_j^T\right)\left(\tilde{Y}_j - a_j \tilde{r}_j^T\right)^T, \end{aligned}$$

where

$$\tilde{r}_j^T = r_j^T S_j \quad \text{and} \quad \tilde{Y}_j = Y_j S_j. \tag{17.103}$$

[101] From an algorithmic perspective, the operator $S(\cdot)$ that creates the downsampling matrix $S_j = S(r_j^T)$ automatically detects the length, L, of its argument, r_j^T, and the number, s_j, and locations, $1 \le j_i \le L$, $i = 1, \cdots, s_j$, of the nonlinear elements of r_j^T and from this information builds the matrix S_j.

[102] Note that the identity matrix symbol I is overloaded and, depending on the context, denotes either $s_j \times s_j$ or $L \times L$, where in the latter case we have $I = I_j + \bar{I}_j$ with $I_j \bar{I}_j = 0$.

Thus,

$$\left\| Y_j - a_j r_j^T \right\|_F^2 = \| Y_j \bar{I}_j \|_F^2 + \left\| \tilde{Y}_j - a_j \tilde{r}_j^T \right\|_F^2 . \tag{17.104}$$

Note that $\tilde{Y}_j S_j^T \neq Y_j$; the columns of Y_j that have been zeroed out by postmultiplication by S_j to form $\tilde{Y}_j = Y_j S_j$ cannot be recovered from $\tilde{Y}_j$ and instead are contained in $Y_j \bar{I}_j$. The error shown on the left-hand side of (17.104) decomposes into the sum of the two error terms shown on the right-hand side. The first term is the cost due to there being no "explanation" by a_j of the data in $Y_j \bar{I}_j$ given the current estimate of r_j^T, while the second term is due to the mismatch in the explanation by a_j of the data in $\tilde{Y}_j$ given the current estimate of r_j^T.

Unconstrained optimization of the left-hand side with respect to r_j^T and of the right-hand side of (17.104) with respect to $\tilde{r}_j^T$ attain different goals. Note that on the right-hand side we are structurally enforcing a sparsity patterning constraint that exists in the current estimate of r_j^T. Therefore, by optimizing the expression on the right-hand side one is preserving whatever sparse structure already exists in r_j^T *without* a need to impose that condition as an explicit constraint. To obtain the *same* answer using the loss function on the left-hand side, one would have to explicitly impose a patterning constraint since otherwise an unconstrained optimization over all values of r_j^T might destroy the sparsity structure that exists in the current value of r_j^T.

Since an *unconstrained* solution to minimizing a loss function of the form $\left\| Y_j - a_j r_j^T \right\|_F^2$ can be obtained in a very straightforward manner via use of SVD, as we discuss in the next paragraph, the loss function on the right-hand side of (17.104) is preferred if the preservation of sparsity during the dictionary update stage is deemed important. Furthermore, practically there are computational advantages to working with the latter rather than the former, and significant computational advantages when compared to MOD.

To preserve the sparsity structure of r_j^T, we optimize the right-hand side of (17.104) to obtain the problem

$$\min_{a_j, \tilde{r}_j^T} \left\| \tilde{Y}_j - a_j \tilde{r}_j^T \right\|_F^2 . \tag{17.105}$$

Once this has been solved for a_j and $\tilde{r}_j^T$ we determine r_j^T by zero-padded upsampling: $r_j^T = \tilde{r}_j^T S_j^T$. It is well known that the optimal solution to the rank-1 approximation problem

$$\min_{M, \, \mathrm{rank}(M)=1} \| \tilde{Y}_j - M \|_F^2$$

is given by $M = \sigma_1 u_1 v_1^T$, where σ_1 is the largest singular value of $\tilde{Y}_j$ and u_1 and v_1 are the associated left and right singular vectors, respectively.[103] Thus one is led to consider solutions for a_j and $\tilde{r}_j^T$ that

[103] This is a consequence of the *Eckart–Young theorem* [166]. Indeed, the rank 1 approximation is quite strong and is known to be optimal for the Frobenius norm, the spectral norm, and the nuclear norm, among others. By convention u_1 and v_1 are normalized to 1.

satisfy

$$\boldsymbol{a}_j \tilde{\boldsymbol{r}}_j^T = \sigma_1 \boldsymbol{u}_1 \boldsymbol{v}_1^T.$$

Unit normalization of the columns of the dictionary to ensure identifiability is easily ensured by taking $\boldsymbol{a}_j = \boldsymbol{u}_1$, which then results in $\tilde{\boldsymbol{r}}_j^T = \sigma_1 \boldsymbol{v}_1^T$ followed by the zero-padded upsampling step $\boldsymbol{r}_j^T = \tilde{\boldsymbol{r}}_j^T \boldsymbol{S}_j^T$. This procedure preserves the sparsity structure of the jth row of $\boldsymbol{X}$, $\boldsymbol{r}_j^T$.

Note that to minimize $\left\| \boldsymbol{Y}_j - \boldsymbol{a}_j \boldsymbol{r}_j^T \right\|_F^2$ via this procedure requires extracting the dominant singular value, left singular vector, and right singular vector from the $n \times L$ matrix $\boldsymbol{Y}_j$, whereas to perform this procedure to solve (17.105) we instead work with the $n \times s_j$ matrix $\tilde{\boldsymbol{Y}}_j$, where often $s_j \ll L$. After the joint estimation of $\boldsymbol{a}_j$ and $\boldsymbol{r}_j^T$ we update both A and X and then move to the estimation of $\boldsymbol{a}_{j+1}$ and $\boldsymbol{r}_{j+1}^T$, proceeding in this manner until all pairs $(\boldsymbol{a}_j, \boldsymbol{r}_j^T)$, $j = 1, \cdots, m$, have been processed. At this point step **C2** of the cyclic dictionary learning procedure has been completed and we return to step **C1** to recompute from anew a sparse solution matrix $\boldsymbol{X}$ in preparation for the next step **C2** iteration.

To perform step **M1** (which is equivalent to step **C2** for a general cyclic optimization algorithm **A1**) of the MOD dictionary learning algorithm **A3** requires the computation of the pseudoinverse $\boldsymbol{X}^+$, which, in turn, can be found from the SVD of the $m \times L$ matrix $\boldsymbol{X}$. To perform cyclic optimization algorithm step **C2** for K-SVD requires the computation of the dominant singular value, σ_1, associated left singular vector, $\boldsymbol{u}_1$, and associated right singular vector, $\boldsymbol{v}_1$, of the $n \times s_j$ matrix $\tilde{\boldsymbol{Y}}_j$ for $j = 1, \cdots, m$. Since it is not uncommon to encounter situations with $n \ll m$ and $s_j \ll L$, the computational edge often is in favor of K-SVD. In addition the enforcement of the constraint that the space structure of $\boldsymbol{X} = \boldsymbol{R}^T$ is preserved throughout the matrix update step enhances the convergence behavior of K-SVD. Further, as pointed out in [4,49] one can iteratively compute the SVD solution for $\boldsymbol{a}_j$ and $\tilde{\boldsymbol{r}}_j^T$ by noting that the separate least squares solutions for each of them that minimizes $\left\| \tilde{\boldsymbol{Y}}_j - \boldsymbol{a}_j \tilde{\boldsymbol{r}}_j^T \right\|_F^2$ are easily computed as

$$\hat{\boldsymbol{a}}_j = \tilde{\boldsymbol{Y}}_j \left(\tilde{\boldsymbol{r}}_j^+\right)^T = \frac{\tilde{\boldsymbol{Y}}_j \tilde{\boldsymbol{r}}_j}{\|\tilde{\boldsymbol{r}}_j\|^2} \quad \text{and} \quad \hat{\tilde{\boldsymbol{r}}}_j^T = \boldsymbol{a}_j^+ \boldsymbol{Y}_j = \frac{\boldsymbol{a}_j^T \boldsymbol{Y}_j}{\|\boldsymbol{a}_j\|^2}$$

and that iterating between these two solutions only a few times leads to a good approximation,[104]

$$\hat{\boldsymbol{a}}_j \hat{\tilde{\boldsymbol{r}}}_j^T \approx \sigma_1 \boldsymbol{u}_1 \boldsymbol{v}_1^T.$$

Once this procedure has converged, our final estimates for $\hat{\boldsymbol{a}}_j$ and $\hat{\tilde{\boldsymbol{r}}}_j^T$ are rescaled to preserve the value of their rank-1 matrix product while ensuring that $\hat{\boldsymbol{a}}_j$ is normalized to 1.

---

104 This corresponds to iterative coordinate descent of $\left\| \tilde{\boldsymbol{Y}}_j - \boldsymbol{a}_j \tilde{\boldsymbol{r}}_j^T \right\|_F^2$ with respect to $\boldsymbol{a}_j$ and $\tilde{\boldsymbol{r}}_j^T$. Since this procedure converges to values of $\boldsymbol{a}_j$ and $\tilde{\boldsymbol{r}}_j^T$ which minimize the loss and the loss function explicitly depends on their rank-1 outer product $\boldsymbol{a}_j \tilde{\boldsymbol{r}}_j^T$, their outer product must be equal to the optimal rank-1 approximation $\sigma_1 \boldsymbol{u}_1 \boldsymbol{v}_1^T$ which also minimizes the loss and is unique.

Appendix 17.B Derivation of the SBL EM update equation

An outline of the derivation of Eq. (17.68) is given in this appendix, where the notations used here are developed in the discussion preceding that equation. Ignoring irrelevant coefficients and terms that do not depend on $\boldsymbol{\gamma}$, the negative Q-function is effectively given by[105]

$$
\begin{aligned}
-2\,Q(\boldsymbol{\gamma},\hat{\boldsymbol{\gamma}};\boldsymbol{y}) &= -\mathbb{E}_{\boldsymbol{x}\mid\boldsymbol{y},\hat{\boldsymbol{\gamma}}}\{\ln p(\boldsymbol{y},\boldsymbol{x},\boldsymbol{\gamma})\}\\
&= -\mathbb{E}_{\boldsymbol{x}\mid\boldsymbol{y},\hat{\boldsymbol{\gamma}}}\{\ln p(\boldsymbol{y}|\boldsymbol{x})\} - \mathbb{E}_{\boldsymbol{x}\mid\boldsymbol{y},\hat{\boldsymbol{\gamma}}}\{\ln p(\boldsymbol{x}|\boldsymbol{\gamma})\} - \mathbb{E}_{\boldsymbol{x}\mid\boldsymbol{y},\hat{\boldsymbol{\gamma}}}\{\ln p(\boldsymbol{\gamma})\}\\
&\doteq -\mathbb{E}_{\boldsymbol{x}\mid\boldsymbol{y},\hat{\boldsymbol{\gamma}}}\{\ln p(\boldsymbol{x}|\boldsymbol{\gamma})\} - \ln p(\boldsymbol{\gamma}) \doteq \operatorname{trace}\left(\mathbb{E}_{\boldsymbol{x}\mid\boldsymbol{y},\hat{\boldsymbol{\gamma}}}\left\{\boldsymbol{x}\boldsymbol{x}^T\right\}\Gamma^{-1}\right) + \ln\det\Gamma + 2\ln\det\Gamma\\
&= \operatorname{trace}\left(\hat{\boldsymbol{\mu}}(\boldsymbol{y},\hat{\boldsymbol{\gamma}})\hat{\boldsymbol{\mu}}(\boldsymbol{y},\hat{\boldsymbol{\gamma}})^T + \hat{\boldsymbol{\Sigma}}(\hat{\boldsymbol{\gamma}})\right)\Gamma^{-1} + 3\ln\det\Gamma, \qquad (17.106)
\end{aligned}
$$

where this derivation uses the conditional probability (17.60)–(17.62), the definitions given in (17.61), and the Jeffreys prior assumption (17.67). Maximizing the Q-function is equivalent to minimizing this function with respect to the elements of the diagonal matrix $\Gamma = \operatorname{diag}(\boldsymbol{\gamma})$, and doing so results in the EM updates of Eq. (17.68)

$$
\gamma_j^+ = 3\boldsymbol{e}_j^T\left(\hat{\boldsymbol{\mu}}(\boldsymbol{y},\hat{\boldsymbol{\gamma}})\hat{\boldsymbol{\mu}}(\boldsymbol{y},\hat{\boldsymbol{\gamma}})^T + \hat{\boldsymbol{\Sigma}}(\hat{\boldsymbol{\gamma}})\right)\boldsymbol{e}_j = 3\left(\hat{\mu}_j^2(\boldsymbol{y},\hat{\boldsymbol{\gamma}}) + \hat{\Sigma}_{jj}(\hat{\boldsymbol{\gamma}})\right) \quad \text{for} \quad j = 1,\cdots,m, \qquad (17.107)
$$

where $\hat{\mu}_j(\boldsymbol{y},\hat{\boldsymbol{\gamma}})$ is the jth component of $\hat{\boldsymbol{\mu}}(\boldsymbol{y},\hat{\boldsymbol{\gamma}})$ and $\hat{\Sigma}_{jj}(\hat{\boldsymbol{\gamma}})$ is the jth diagonal element of $\hat{\Sigma}_{jj}(\hat{\boldsymbol{\gamma}})$.

Appendix 17.C SBL dictionary learning algorithm

The optimization (17.71) is determined by use of the EM algorithm to minimize

$$
-\ln p(\boldsymbol{Y},\boldsymbol{G};A,\Lambda) \qquad (17.108)
$$

via iterative minimization of the negative Q-function,

$$
-Q(\boldsymbol{G},A,\Lambda,\hat{\boldsymbol{G}},\hat{A},\hat{\Lambda};\boldsymbol{Y}) = -\mathbb{E}_{\boldsymbol{X}\mid\boldsymbol{Y},\hat{\boldsymbol{G}},\hat{A},\hat{\Lambda}}\{\ln p(\boldsymbol{Y},\boldsymbol{X},\boldsymbol{G};A,\Lambda)\} \qquad (17.109)
$$

given current estimates $\hat{\boldsymbol{G}}$, $\hat{A}$, and $\hat{\Lambda}$.

This procedure is a generalization of the approach described in the previous section where here, in addition to estimating the variances $\boldsymbol{G}$, we also want to learn the unknown parameters A and Λ. The goal is to determine estimates of A, Λ, and G (i.e., $\hat{\boldsymbol{\gamma}}_\ell$ for $\ell = 1,\cdots,L$) via the EM procedure. To do so, we work with the negative Q-function,

$$
-Q(\boldsymbol{G},A,\Lambda,\hat{\boldsymbol{G}},\hat{A},\hat{\Lambda};\boldsymbol{Y}) = -\mathbb{E}_{\boldsymbol{X}\mid\boldsymbol{Y},\hat{\boldsymbol{G}},\hat{A},\hat{\Lambda}}\{\ln p(\boldsymbol{Y},\boldsymbol{X},\boldsymbol{G};A,\Lambda)\}
$$

[105] "Effective equality" for the purpose of extracting the arg-max is denoted by the symbol "$\doteq$." Note we are still assuming, for the time being, that the parameters A and Λ are known.

$$= \underbrace{-\mathbb{E}_{X|Y,\hat{G},\hat{A},\hat{\Lambda}}\{\ln p(Y|X;A,\Lambda)\}}_{T_1} \underbrace{-\mathbb{E}_{X|Y,\hat{G},\hat{A},\hat{\Lambda}}\{\ln p(X|G)\}}_{T_2}$$
$$\underbrace{-\ln p(G)}_{T_3},$$

which we proceed to evaluate term-by-term. Note that the free parameters to be optimized are G, A, and Λ.

The last two terms T_2 and T_3 depend on the free parameter G and are handled in the same manner as in the previous section. Assuming the Jeffreys prior, the third term is straightforward to evaluate,

$$T_3 = -\sum_{\ell=1}^{L} \ln p(\boldsymbol{\gamma}_\ell) = \sum_{\ell=1}^{L} \ln \det \Gamma_\ell, \quad \Gamma_\ell = \mathrm{diag}(\boldsymbol{\gamma}_\ell).$$

With $\ln p(X|G) = \sum_{\ell=1}^{L} \ln p(\boldsymbol{x}_\ell|\boldsymbol{\gamma}_\ell)$ and the i.i.d. sampling assumption the second term is

$$\begin{aligned} T_2 &\doteq \frac{1}{2}\sum_{\ell=1}^{L}\left(\mathrm{trace}\left(\mathbb{E}_{\boldsymbol{x}_\ell|\boldsymbol{y}_\ell,\hat{\boldsymbol{\gamma}}_\ell,\hat{A},\hat{\Lambda}}\left\{\boldsymbol{x}_\ell\boldsymbol{x}_\ell^T\right\}\Gamma_\ell^{-1}\right) + \ln\det\Gamma_\ell\right) \\ &= \frac{1}{2}\sum_{\ell=1}^{L}\left(\left(\mathrm{trace}\left(\hat{\boldsymbol{\mu}}(\boldsymbol{y}_\ell,\hat{\boldsymbol{\gamma}}_\ell)\hat{\boldsymbol{\mu}}(\boldsymbol{y}_\ell,\hat{\boldsymbol{\gamma}}_\ell)^T + \hat{\boldsymbol{\Sigma}}(\hat{\boldsymbol{\gamma}}_\ell)\right)\Gamma_\ell^{-1}\right) + \ln\det\Gamma_\ell\right). \end{aligned}$$

The terms $\hat{\boldsymbol{\mu}}(\boldsymbol{y}_\ell,\hat{\boldsymbol{\gamma}}_\ell)$ and $\hat{\boldsymbol{\Sigma}}(\hat{\boldsymbol{\gamma}}_\ell)$ have dependencies on $\hat{A}$ and $\hat{\Lambda}_\ell = \mathrm{diag}(\hat{\boldsymbol{\gamma}}_\ell)$ which here are notationally suppressed (they are shown in detail in Eqs. (17.114)–(17.117) below). Term T_1 has no dependency on the free parameter G while terms T_2 and T_3 do not depend on the free parameters A and Λ. Thus the optimizations decouple into an optimization over the sum $T_2 + T_3$ with respect to G and an optimization over T_1 with respect to A and Λ.

We have

$$T_2 + T_3 \doteq \sum_{\ell=1}^{L}\left(\left(\mathrm{trace}\left(\hat{\boldsymbol{\mu}}(\boldsymbol{y}_\ell,\hat{\boldsymbol{\gamma}}_\ell)\hat{\boldsymbol{\mu}}(\boldsymbol{y}_\ell,\hat{\boldsymbol{\gamma}}_\ell)^T + \hat{\boldsymbol{\Sigma}}(\hat{\boldsymbol{\gamma}}_\ell)\right)\Gamma_\ell^{-1}\right) + 3\ln\det\Gamma_\ell\right),$$

which decouples so that each term in the sum indexed by ℓ can be optimized independently,

$$\gamma_{j,\ell}^{+} = \arg\min_{\gamma_{j,\ell}}\left[\left(\mathrm{trace}\left(\hat{\boldsymbol{\mu}}(\boldsymbol{y}_\ell,\hat{\boldsymbol{\gamma}}_\ell)\hat{\boldsymbol{\mu}}(\boldsymbol{y}_\ell,\hat{\boldsymbol{\gamma}}_\ell)^T + \hat{\boldsymbol{\Sigma}}(\hat{\boldsymbol{\gamma}}_\ell)\right)\Gamma_\ell^{-1}\right) + 3\ln\det\Gamma_\ell\right].$$

This single-sample optimization was performed in the previous section and we can immediately take over the answer determined there,

$$\begin{aligned} \gamma_{j,\ell}^{+} &= 3\boldsymbol{e}_j^T\left(\hat{\boldsymbol{\mu}}(\boldsymbol{y}_\ell,\hat{\boldsymbol{\gamma}}_\ell)\hat{\boldsymbol{\mu}}(\boldsymbol{y}_\ell,\hat{\boldsymbol{\gamma}}_\ell)^T + \hat{\boldsymbol{\Sigma}}(\hat{\boldsymbol{\gamma}}_\ell)\right)\boldsymbol{e}_j \\ &= 3\left(\hat{\mu}_j^2(\boldsymbol{y},\hat{\boldsymbol{\gamma}}_\ell) + \hat{\Sigma}_{jj}(\hat{\boldsymbol{\gamma}}_\ell)\right) \quad \text{for} \quad j = 1,\cdots,m, \quad \ell = 1,\cdots,L. \end{aligned} \tag{17.110}$$

We now examine term T_1, which contains the information needed to perform recursive updates of A and Λ. We have

$$T_1 \doteq \frac{1}{L}\sum_{\ell=1}^{L}\left(\text{trace}\left(\underbrace{\mathbb{E}_{\boldsymbol{x}_\ell \mid \boldsymbol{y}_\ell, \hat{\boldsymbol{\gamma}}_\ell, \hat{A}, \hat{\Lambda}}\left\{(\boldsymbol{y}_\ell - A\boldsymbol{x}_\ell)(\boldsymbol{y}_\ell - A\boldsymbol{x}_\ell)^T\right\}}_{C_\ell(A)}\Lambda^{-1}\right) + \log\det\Lambda\right)$$
$$= \text{trace}\left(\mathbf{C}(A)\Lambda^{-1}\right) + \ln\det\Lambda,$$

with

$$\mathbf{C}(A) \stackrel{\text{def}}{=} \frac{1}{L}\sum_{\ell=1}^{L} C_\ell(A) \quad \text{and} \quad C_\ell(A) = \mathbb{E}_{\boldsymbol{x}_\ell \mid \boldsymbol{y}_\ell, \hat{\boldsymbol{\gamma}}_\ell, \hat{A}, \hat{\Lambda}}\left\{(\boldsymbol{y}_\ell - A\boldsymbol{x}_\ell)(\boldsymbol{y}_\ell - A\boldsymbol{x}_\ell)^T\right\}. \tag{17.111}$$

For any value of A we can find a minimum of T_1 with respect to Λ by taking the matrix derivative with respect to Λ to determine a unique stationary point followed by demonstrating that point gives a global minimum. This can be done using the matrix derivative formulas[106]

$$\frac{\partial}{\partial X}\text{trace}\left(YX^{-1}\right) = -X^{-1}YX^{-1} \quad \text{and} \quad \frac{\partial}{\partial X}\ln\det X = X^{-1}.$$

The result is that T_1 is minimized with respect to Λ for any value of A when Λ takes the value

$$\hat{\Lambda}^+(A) = \mathbf{C}(A) = \frac{1}{L}\sum_{\ell=1}^{L} C_\ell(A), \tag{17.112}$$

which is computable once the functional form of $C_\ell(A)$ has been determined. Also note that once the function $\mathbf{C}(A)$ is available, the determination of an optimal update value for A is obtained by minimizing

$$\text{trace}\left(\mathbf{C}(A)\Lambda^{-1}\right). \tag{17.113}$$

We now turn to the determination of the functional form of $C_\ell(A)$. Note that

$$\mathbb{E}_{\boldsymbol{x}_\ell \mid \boldsymbol{y}_\ell, \hat{\boldsymbol{\gamma}}_\ell, \hat{A}, \hat{\Lambda}}\left\{\boldsymbol{x}_\ell\right\} = \hat{\boldsymbol{\mu}}\left(\boldsymbol{y}_\ell, \hat{\boldsymbol{\gamma}}_\ell\right) = \hat{\Gamma}_\ell\,\hat{A}^T\,\hat{\Sigma}_{y,\ell}^{-1}\,\boldsymbol{y}_\ell \tag{17.114}$$

with

$$\hat{\Gamma}_\ell = \text{diag}\left(\hat{\boldsymbol{\gamma}}_\ell\right) \quad \text{and} \quad \hat{\Sigma}_{y,\ell} = \hat{\Lambda} + \hat{A}\,\hat{\Gamma}_\ell\,\hat{A}^T, \tag{17.115}$$

[106] These can be found in [120], which is recommended because it emphasizes the fact that such matrix *derivative* identities often (but not always) can be easily found by *first* exploiting nice properties of the matrix *differential* and the trace and determinant operators.

and note that

$$\mathbb{E}_{\boldsymbol{x}_\ell \mid \boldsymbol{y}_\ell, \hat{\boldsymbol{\gamma}}_\ell, \hat{A}, \hat{\Lambda}}\left\{\boldsymbol{x}_\ell \boldsymbol{x}_\ell^T\right\} = \hat{\boldsymbol{\mu}}\left(\boldsymbol{y}_\ell, \hat{\boldsymbol{\gamma}}_\ell\right) \hat{\boldsymbol{\mu}}\left(\boldsymbol{y}_\ell, \hat{\boldsymbol{\gamma}}_\ell\right)^T + \hat{\boldsymbol{\Sigma}}(\hat{\boldsymbol{\gamma}}_\ell) \tag{17.116}$$

with

$$\hat{\boldsymbol{\Sigma}}(\hat{\boldsymbol{\gamma}}_\ell) = \hat{\Gamma}_\ell - \hat{\Gamma}_\ell \hat{A}^T \hat{\Sigma}_{y,\ell}^{-1} \hat{A} \hat{\Gamma}_\ell. \tag{17.117}$$

This yields, setting $\hat{\boldsymbol{\mu}}_\ell = \hat{\boldsymbol{\mu}}\left(\boldsymbol{y}_\ell, \hat{\boldsymbol{\gamma}}_\ell\right)$ and $\hat{\boldsymbol{\Sigma}}_\ell = \hat{\boldsymbol{\Sigma}}(\hat{\boldsymbol{\gamma}}_\ell)$,

$$C_\ell(A) = \boldsymbol{y}_\ell \boldsymbol{y}_\ell^T - A\,\hat{\boldsymbol{\mu}}_\ell \boldsymbol{y}_\ell^T - \boldsymbol{y}_\ell \hat{\boldsymbol{\mu}}_\ell^T A^T + A\left(\hat{\boldsymbol{\mu}}_\ell \hat{\boldsymbol{\mu}}_\ell^T + \hat{\boldsymbol{\Sigma}}_\ell\right) A^T = (\boldsymbol{y}_\ell - A\,\hat{\boldsymbol{\mu}}_\ell)(\boldsymbol{y}_\ell - A\,\hat{\boldsymbol{\mu}}_\ell)^T + A\,\hat{\boldsymbol{\Sigma}}_\ell\, A^T$$

and

$$\operatorname{trace}\left(C_\ell(A)\Lambda^{-1}\right) = \operatorname{trace}\left(\left[\boldsymbol{y}_\ell \boldsymbol{y}_\ell^T - 2A\hat{\boldsymbol{\mu}}_\ell \boldsymbol{y}_\ell^T + A\left(\hat{\boldsymbol{\mu}}_\ell \hat{\boldsymbol{\mu}}_\ell^T + \hat{\boldsymbol{\Sigma}}_\ell\right) A^T\right]\Lambda^{-1}\right).$$

It is clear that $C_\ell(A)$, and therefore $\mathbf{C}(A)$, depends on the current estimates of $\boldsymbol{\theta} = (\boldsymbol{G}, A, \Lambda)$. Henceforth we make this manifestly evident by writing $C_\ell(A) = C_\ell(A; \hat{\boldsymbol{\theta}})$ and $\mathbf{C}(A) = \mathbf{C}(A; \hat{\boldsymbol{\theta}})$. We also define

$$\mathbf{R}_{yy} = \frac{1}{L}\sum_{\ell=1}^{L} \boldsymbol{y}_\ell \boldsymbol{y}_\ell^T, \qquad \widehat{\mathbf{R}}_{\mu\mu}(\hat{\boldsymbol{\theta}}) = \frac{1}{L}\sum_{\ell=1}^{L} \hat{\boldsymbol{\mu}}_\ell \hat{\boldsymbol{\mu}}_\ell^T,$$

$$\widehat{\mathbf{R}}_{\mu y}(\hat{\boldsymbol{\theta}}) = \frac{1}{L}\sum_{\ell=1}^{L} \hat{\boldsymbol{\mu}}_\ell \boldsymbol{y}_\ell^T, \quad \widehat{\mathbf{R}}_{y\mu}(\hat{\boldsymbol{\theta}}) = \widehat{\mathbf{R}}_{\mu y}^T(\hat{\boldsymbol{\theta}}) = \frac{1}{L}\sum_{\ell=1}^{L} \boldsymbol{y}_\ell \hat{\boldsymbol{\mu}}_\ell^T,$$

$$\widehat{\mathbf{R}}_{\tilde{y}\tilde{y}}(A; \hat{\boldsymbol{\theta}}) = \frac{1}{L}\sum_{\ell=1}^{L} \tilde{\boldsymbol{y}}_\ell \tilde{\boldsymbol{y}}_\ell^T, \quad \text{where} \quad \tilde{\boldsymbol{y}}_\ell \overset{\text{def}}{=} \boldsymbol{y}_\ell - A\,\hat{\boldsymbol{\mu}}_\ell,$$

$$\widehat{\boldsymbol{\Sigma}}_{xx}(\hat{\boldsymbol{\theta}}) = \frac{1}{L}\sum_{\ell=1}^{L} \hat{\boldsymbol{\Sigma}}_\ell, \quad \text{and} \quad \widehat{\mathbf{R}}_{xx}(\hat{\boldsymbol{\theta}}) = \widehat{\mathbf{R}}_{\mu\mu}(\hat{\boldsymbol{\theta}}) + \widehat{\boldsymbol{\Sigma}}_{xx}(\hat{\boldsymbol{\theta}}).$$

With these definitions we have

$$\mathbf{C}(A; \hat{\boldsymbol{\theta}}) = \mathbf{R}_{yy} - \left(\widehat{\mathbf{R}}_{y\mu}(\hat{\boldsymbol{\theta}})\, A^T + A\,\widehat{\mathbf{R}}_{\mu y}(\hat{\boldsymbol{\theta}})\right) + A\,\widehat{\mathbf{R}}_{xx}(\hat{\boldsymbol{\theta}})\, A^T = \widehat{\mathbf{R}}_{\tilde{y}\tilde{y}}(A; \hat{\boldsymbol{\theta}}) + A\,\widehat{\boldsymbol{\Sigma}}_{xx}(\hat{\boldsymbol{\theta}})\, A^T, \tag{17.118}$$

$$\operatorname{trace}\left(\mathbf{C}(A; \hat{\boldsymbol{\theta}})\Lambda^{-1}\right) = \operatorname{trace}\left(\left[\mathbf{R}_{yy} - 2\,A\,\widehat{\mathbf{R}}_{\mu y}(\hat{\boldsymbol{\theta}}) + A\,\widehat{\mathbf{R}}_{xx}(\hat{\boldsymbol{\theta}})\, A^T\right]\Lambda^{-1}\right). \tag{17.119}$$

We can minimize this latter function with respect to A using the matrix derivative formulas [120]

$$\frac{\partial}{\partial X}\operatorname{trace}(XY) = Y \quad \text{and} \quad \frac{\partial}{\partial X}\operatorname{trace}\left(XYX^T Z\right) = YX^T Z + Y^T X^T Z^T$$

to obtain the EM update

$$\hat{A}^{+} = \widehat{\mathbf{R}}_{y\mu}(\hat{\boldsymbol{\theta}})\widehat{\mathbf{R}}_{xx}^{-1}(\hat{\boldsymbol{\theta}}), \tag{17.120}$$

which then yields the update to Λ as

$$\hat{\Lambda}^{+} = \mathbf{C}(\hat{A}^{+}; \hat{\boldsymbol{\theta}}). \tag{17.121}$$

Appendix 17.D Mathematical background for kernelizing dictionary learning

In this appendix we use the notation established in Section 17.10. The exposition given here is somewhat more detailed than that found in the original literature [130,131] and in the textbook [47]. This is done in order to more carefully deal with the (generally) nonfinite dimensionality of the Mercer space $\mathcal{M}$. For more on the mathematical techniques used in this section, see [106]. For convenience, we set $\phi_i = \phi(\boldsymbol{y}_i) \in \mathcal{M}$ and $\boldsymbol{\Phi} = \boldsymbol{\Phi}(Y)$ for $i = 1, \cdots, L$, and simply write

$$\boldsymbol{\Phi} = [\phi_1 \quad \cdots \quad \phi_L].$$

Given the data $\boldsymbol{\Phi} = [\phi_1 \quad \cdots \quad \phi_L]$, define a linear operator $\boldsymbol{\Phi} : \mathbb{R}^L \to \mathcal{M}$ by

$$\boldsymbol{\Phi} v = \sum_{i=1}^{L} v_i \phi_i \in \mathcal{M} \quad \text{for every} \quad v \in \mathbb{R}^L.$$

This linear operator is bounded because $\phi(\cdot)$ is assumed to be a well-defined function on its domain; $\mathbb{R}^L$ is a finite-dimensional real Hilbert space with the standard Euclidean inner product and norm given by

$$\langle v, v' \rangle = v^T v' \quad \text{and} \quad \|v\|^2 = v^T v.$$

Recall that the Mercer space $\mathcal{M}$ is assumed to be a real Hilbert space with an inner product $\langle \mu, \mu' \rangle$ and norm $\|\mu\|^2 = \langle \mu, \mu \rangle$. The distinction between $\langle \cdot, \cdot \rangle$ (respectively, $\| \cdot \|$) as an inner product (respectively, a norm) on $\mathbb{R}^L$ or $\mathcal{M}$ is made clear by the type of vectors ($\mathbb{R}^L$-vectors v or Mercer space vectors $\mu \in \mathcal{M}$) in its arguments.[107] Recall that real-valued inner products are linear in each of their arguments.

The adjoint linear operator $\Phi^* : \mathcal{M} \to \mathbb{R}^L$ is defined by the requirement that for every $\mu \in \mathcal{M}$ and every $v \in \mathbb{R}^L$,[108]

$$\langle \boldsymbol{\Phi}^* \mu, v \rangle = \langle \mu, \boldsymbol{\Phi} v \rangle.$$

Now note that $\boldsymbol{\Phi}^* \boldsymbol{\Phi} : \mathbb{R}^L \to \mathbb{R}^L$ must be an $L \times L$ matrix and that

[107] Both types of inner products can occur within a single equation so care must be taken when reading the discussion in this appendix.

[108] If $\mathcal{M}$ is finite-dimensional and its norm is also the simple Euclidean norm $\langle \mu, \mu \rangle = \mu^T \mu'$, then it is straightforward to show from the definition of the adjoint operator that $\boldsymbol{\Phi}^* = \boldsymbol{\Phi}^T$. Note also in this case that $\langle \phi_i, \phi_j \rangle = \phi_i^T \phi_j$.

$$v^T \left(\Phi^* \Phi\right) v' = \langle v, \Phi^* \Phi v' \rangle = \langle \Phi^* v, \Phi v' \rangle = \langle \sum_i v_i \phi_i, \sum_j v'_j \phi_j \rangle = \sum_i \sum_j v_i v'_j \langle \phi_i, \phi_j \rangle.$$

This shows that

$$\left(\Phi^* \Phi\right)_{ij} = \langle \phi_i, \phi_j \rangle = K(y_i, y_j),$$

where $K(y_i, y_j)$ is the ijth element of the $L \times L$ kernel matrix K. The kernel matrix K is positive semidefinite since for all $v \in \mathbb{R}^L$ we have

$$v^T K v = \langle v, K v \rangle = \langle v, \Phi^* \Phi v \rangle = \langle \Phi v, \Phi v \rangle = \|\Phi v\|^2 \geq 0.$$

As an $L \times L$ matrix on a finite-dimensional space $\mathbb{R}^L$, the kernel matrix is a mapping $K : \mathbb{R}^L \to \mathbb{R}^L$. With $\mathbb{R}^L$ having the standard inner product $\langle v, v' \rangle = v^T v'$, the adjoint of K is equal to its transpose: $K^* = K^T$. This means that if the kernel matrix can be shown to be self-adjoint, $K^* = K$, then it must be symmetric, $K^T = K$. The kernel matrix is indeed self-adjoint because

$$K^* = (\Phi^* \Phi)^* = \Phi^* \Phi = K$$

using the facts that $(\Phi^*)^* = \Phi$ and $(BF)^* = F^* B^*$ for linear operators $F : \mathbb{R}^L \to \mathcal{M}$ and $B : \mathcal{M} \to \mathbb{R}^L$.

Note that we have shown that *given* a Mercer transformation $\phi(\cdot)$ from a finite-dimensional Hilbert space to a general Hilbert space (Mercer space) we can construct and evaluate a kernel function $K(y_i, y_j) = \langle \phi(y_i), \phi(y_j) \rangle$ on the data Y, and assemble the resulting values into a kernel matrix K which is symmetric and positive semidefinite. That is, we have shown that[109]

$$\Phi \to K = K_{\Phi} = \Phi^* \Phi.$$

The hard problem is the reverse direction:

$$K \overset{?}{\to} \Phi = \Phi_K \quad \text{such that} \quad K = \Phi^* \Phi,$$

i.e., how to determine when a given, arbitrary binary function $K(\cdot, \cdot)$ is a kernel function for some (generally unknown) Mercer transformation $\phi(\cdot)$ to a (generally unknown) Mercer space. Fortunately Mercer's theorem can be used to answer this question and, further, given a set of kernel functions more complicated ones can be built from them [78,168].

As a linear operator, $\Phi : \mathbb{R}^L \to \mathcal{M}$ has a finite-dimensional range space[110]

$$\mathcal{R}(\Phi) = \Phi\left(\mathbb{R}^L\right) \subseteq \mathcal{M},$$

---

[109] Thus given $\Phi : \mathbb{R}^L \to \mathcal{M}$, the operator composition $\Phi^* \Phi$ is *always* a kernel and denoting it by the symbol K is an optional notational convenience.

[110] Note that the range space $\mathcal{R}(\Phi)$, which is a Hilbert subspace arising from the *linear operator* Φ acting on its entire domain space $\mathbb{R}^L$, is *not* the image arising from the *nonlinear operator* $\phi(\cdot)$ acting on the data space $\mathbb{Y} = \mathbb{R}^n$. In the latter case, $\phi(\mathbb{Y})$ is (at best) an embedded n-dimensional nonlinear manifold in a (possibly) infinite-dimensional Mercer space, in which case there exist Mercer vectors $\mu \notin \phi(\mathbb{Y})$. The image of the L data points y_i, $i = 1, \cdots, L$, comprising the columns of $\mathbb{Y}$ under the action of $\phi(\cdot)$ are a finite collection ("cloud") of L Mercer-transformed data points $\phi(y_i)$, $i = 1, \cdots, n$, in $\mathcal{M}$ that are "sprinkled" on the manifold $\phi(\mathbb{Y})$, i.e., $\phi(Y) \subset \phi(\mathbb{Y})$ is a strict inclusion. Further, the cloud of such points is collected into the L columns of Φ

which is the embedding under the action $\mathbf{\Phi}$ of the finite-dimensional space $\boldsymbol{R}^L$ into the (generally infinite-dimensional) Mercer space $\mathcal{M}$. The range space is a Hilbert subspace of $\mathcal{M}$ and its dimension is the *rank* r of $\mathbf{\Phi}$. We have

$$r = \dim \mathcal{R}(\mathbf{\Phi}) \leq \dim \mathbb{R}^L = L.$$

Because the range space $\mathcal{R}(\mathbf{\Phi})$ is finite-dimensional, it is topologically closed, $\overline{\mathcal{R}(\mathbf{\Phi})} = \mathcal{R}(\mathbf{\Phi})$.[111]

Given the linear operator $\mathbf{\Phi} : \mathbb{R}^L \to \mathcal{M}$, we have the following orthogonal decomposition of the Mercer space[112]:

$$\mathcal{M} = \mathcal{R}(\mathbf{\Phi}) \oplus \mathcal{R}(\mathbf{\Phi})^{\perp}. \tag{17.122}$$

Let P and $P_{\perp}$ respectively represent the orthogonal project operators onto the range space $\mathcal{R}(\mathbf{\Phi})$ and its orthogonal complement $\mathcal{R}(\mathbf{\Phi})^{\perp}$. As orthogonal projection operators they are both self-adjoint (e.g., $P^* = P$) and idempotent (e.g., $P^2 = P$). We further have $PP^{\perp} = P^{\perp}P = 0$. Also note that

$$P\mathbf{\Phi} = \mathbf{\Phi} \quad \text{and} \quad P^{\perp}\mathbf{\Phi} = 0.$$

Further, for any operator $\boldsymbol{H} : \mathbb{R}^L \to \mathcal{M}$ we must have

$$P\boldsymbol{H} = \mathbf{\Phi} M$$

for some $L \times L$ matrix M. That is, because we have $P\boldsymbol{H} : \mathbb{R}^L \to \mathcal{R}(\mathbf{\Phi}) \subseteq \mathcal{M}$, the operator $P\boldsymbol{H}$ must be representable in terms of the columns of $\mathbf{\Phi}$.

Denote the space of bounded linear operators from $\mathbb{R}^L \to \mathcal{M}$ by $\mathcal{B} = \mathcal{B}(\mathbb{R}^L, \mathcal{M})$ and the space of bounded linear operators in the reverse direction from $\mathcal{M} \to \mathbb{R}^L$ by $\mathcal{B}(\mathcal{M}, \mathbb{R}^L)$. We have seen that for every $\mathbf{\Phi} \in \mathcal{B}(\mathbb{R}^L, \mathcal{M})$ there exists an adjoint operator $\mathbf{\Phi}^* \in \mathcal{B}(\mathcal{M}, \mathbb{R}^L)$ and that the $L \times L$ square matrix $\mathbf{\Phi}^*\mathbf{\Phi}$ is symmetric and positive semidefinite. Thus $\mathbf{\Phi}^*\mathbf{\Phi}$ has L real and nonnegative eigenvalues $\sigma_i^2 \geq 0$, $i = 1, \cdots, L$, which we order as

$$\sigma_1^2 \geq \cdots \geq \sigma_L^2 \geq 0.$$

We now *tentatively* define a *generalized* Frobenius inner product on $B(\mathbb{R}^L, \mathcal{M})$, i.e., between any two linear operators $\mathbf{\Phi}, \mathbf{\Psi} \in \mathcal{B}(\mathbb{R}^L, \mathcal{M})$, by

$$\langle \mathbf{\Phi}, \mathbf{\Psi} \rangle = \operatorname{trace}\left(\mathbf{\Phi}^* \mathbf{\Psi}\right).$$

---

and the linear combination of all such points comprises the range space $\mathcal{R}(\mathbf{\Phi})$. That is, the range space $\mathcal{R}(\mathbf{\Phi})$ is the linear span of the Mercer-transformed data points $\phi(\mathbb{Y})$: $\mathcal{R}(\mathbf{\Phi}) = \operatorname{span} \phi(\mathbb{Y})$. The L weights used to form these linear combinations "live" in $\mathbb{R}^L$ and therefore one might call $\mathbb{R}^L$ the *weight space*, denoted, say, by $\mathbb{W} = \mathbb{R}^L$, in which case we can write $\mathbf{\Phi} : \mathbb{W} \to \mathcal{M}$. Note that $\phi(\mathbb{Y}) \subset \mathcal{R}(\mathbf{\Phi})$ denotes the fact that we have a strict inclusion of the columns of $\mathbf{\Phi}$ into $\mathcal{R}(\mathbf{\Phi})$.

111 In the infinite-dimensional case, this is not guaranteed to hold.

112 Although we do not use this fact, it is also the case that $\mathcal{R}(\mathbf{\Phi})^{\perp} = \mathcal{N}(\mathbf{\Phi}^*)$. Further, if the range space is not finite-dimensional, we more generally have $\mathcal{M} = \overline{\mathcal{R}(\mathbf{\Phi})} \oplus \mathcal{R}(\mathbf{\Phi})^{\perp} = \overline{\mathcal{R}(\mathbf{\Phi})} \oplus \mathcal{N}(\mathbf{\Phi}^*)$.

This is a *generalization* of the standard Frobenius inner product on matrices because $\mathbf{\Phi}$ and $\mathbf{\Psi}$ are not finite matrices.[113] This is a *tentative* definition because we have to demonstrate that the proposed norm is symmetric, linear in each argument, and positive definite when we take $\mathbf{\Phi} = \mathbf{\Psi}$.

It is obvious that we have linearity in each argument. If $\mathbf{\Phi} = \mathbf{\Psi}$ we have

$$\langle \mathbf{\Phi}, \mathbf{\Phi} \rangle = \text{trace}\left(\mathbf{\Phi}^* \mathbf{\Phi}\right) = \sigma_1^2 + \cdots + \sigma_L^2 \geq 0,$$

showing nonnegativity. To show definiteness, note that

$$\langle \mathbf{\Phi}, \mathbf{\Phi} \rangle = 0 \iff \sigma_1^2 = \cdots = \sigma_L^2 = 0 \iff \mathbf{\Phi}^* \mathbf{\Phi} = 0.$$

Thus for all $v \in \mathbb{R}^L$ we have

$$0 = v^T \mathbf{\Phi}^* \mathbf{\Phi} v = \langle v, \mathbf{\Phi}^* \mathbf{\Phi} v \rangle = \langle \mathbf{\Phi} v, \mathbf{\Phi} v \rangle = \|\mathbf{\Phi} v\|^2,$$

which can only be true for all v if $\mathbf{\Phi}$ is the zero operator, $\mathbf{\Phi} = \mathbf{0}$. To show symmetry, recall that for a real square $L \times L$ matrix $M : \mathbb{R}^L \to \mathbb{R}^L$, with $\mathbb{R}^L$ having inner product $\langle v, v' \rangle = v^T v'$, the adjoint of M is equal to its transpose $M^* = M^T$. In particular, this will be true for the real $L \times L$ matrix $M = \mathbf{\Phi}^* \mathbf{\Psi}$. Also recall that $(\mathbf{\Phi}^* \mathbf{\Psi})^* = \mathbf{\Psi}^* \mathbf{\Phi}$. Therefore, further recalling that trace M = trace M^T,

$$\langle \mathbf{\Phi}, \mathbf{\Psi} \rangle = \text{trace}\left[(\mathbf{\Phi}^* \mathbf{\Psi})\right] = \text{trace}\left[(\mathbf{\Phi}^* \mathbf{\Psi})^T\right] = \text{trace}\left[(\mathbf{\Phi}^* \mathbf{\Psi})^*\right] = \text{trace}\left[(\mathbf{\Psi}^* \mathbf{\Phi})\right] = \langle \mathbf{\Psi}, \mathbf{\Phi} \rangle.$$

The generalized Frobenius inner product on $\mathcal{B}(\mathbb{R}^L, \mathcal{M})$ yields the induced *generalized Frobenius norm*,

$$\|\mathbf{\Phi}\|_{GF}^2 = \langle \mathbf{\Phi}, \mathbf{\Phi} \rangle = \text{trace}\left(\mathbf{\Phi}^* \mathbf{\Phi}\right) = \sigma_1^2 + \cdots + \sigma_L^2 \geq 0, \tag{17.123}$$

where σ_i^2, $i = 1, \cdots, L$, are the real and necessarily nonnegative eigenvalues of the symmetric, positive semidefinite $L \times L$ kernel matrix $\mathbf{\Phi}^* \mathbf{\Phi}$. Let $I = P + P_\perp$ be the identity operator on $\mathcal{M}$ for P and let $P_\perp$ be the orthogonal projection operators defined above. It is straightforward to show the following.

Pythagorean theorem for complementary subspaces:

$$\|P\mathbf{\Phi} + P_\perp \mathbf{\Psi}\|_{GF}^2 = \|P\mathbf{\Phi}\|_{GF}^2 + \|P_\perp \mathbf{\Psi}\|_{GF}^2. \tag{17.124}$$

Indeed, we have

$$\begin{aligned} \|P\mathbf{\Phi} + P_\perp \mathbf{\Psi}\|_{GF}^2 &= \langle P\mathbf{\Phi} + P_\perp \mathbf{\Psi}, P\mathbf{\Phi} + P_\perp \mathbf{\Psi} \rangle = \langle P\mathbf{\Phi}, P\mathbf{\Phi} \rangle + \langle P_\perp \mathbf{\Psi}, P_\perp \mathbf{\Psi} \rangle \\ &= \|P\mathbf{\Phi}\|_{GF}^2 + \|P_\perp \mathbf{\Psi}\|_{GF}^2 \end{aligned}$$

as a consequence of the symmetry of the inner product, the linearity of each argument of the inner product, and the fact that all the cross-terms vanish since

$$\langle P\mathbf{\Phi}, P_\perp \mathbf{\Psi} \rangle = \langle P_\perp \mathbf{\Phi}, P\mathbf{\Psi} \rangle = \langle P P_\perp \mathbf{\Phi}, \mathbf{\Psi} \rangle = \langle P P_\perp \mathbf{\Psi}, \mathbf{\Phi} \rangle = 0$$

[113] However, they do have finite-dimensional domains, and hence range spaces, which is why the simple generalization to the standard Frobenius theory given here is straightforward to develop and verify.

due to the definition of the adjoint operator, P and $P_\perp$ being self-adjoint, and $PP_\perp = P_\perp P = 0$.

Consider the optimization required to solve problem (17.98):

$$\min_{D,X} \|\mathbf{\Phi} - DX\|_{GF}^2 \text{ subject to } \overline{\mathsf{sp}}(X) \le k \text{ and } D \in \mathcal{D}. \tag{17.125}$$

Using the identity decomposition $I = P + P_\perp$, we can write

$$DX = (I)DX = (P + P_\perp)DX = PDX + P_\perp DX \tag{17.126}$$

$$\text{with } PDX = \mathbf{\Phi}M \text{ for some } L \times L \text{ matrix } M. \tag{17.127}$$

The last step is true because $PDX \in \mathcal{R}(\mathbf{\Phi})$ and therefore must be of the form $PDX = \mathbf{\Phi}M$ for some matrix $M \in \mathbb{R}^{L\times L}$.[114] As a consequence of the Pythagorean theorem (17.124) we have (note that $\mathbf{\Phi} = P\mathbf{\Phi}$)

$$\|\mathbf{\Phi} - DX\|_{GF}^2 = \|\mathbf{\Phi} - PDX + P_\perp DX\|_{GF}^2 = \|\mathbf{\Phi} - PDX\|_{GF}^2 + \|P_\perp DX\|_{GF}^2 \ge \|\mathbf{\Phi} - PDX\|_{GF}^2.$$

Now consider D and X given a solution to (17.125). It must be the case that $D \in \mathcal{D}$ and $\underline{\mathsf{sp}}(X) \le k$. We have

$$\|\mathbf{\Phi} - DX\|_{GF}^2 = \|\mathbf{\Phi} - PDX\|_{GF}^2 + \|P_\perp DX\|_{GF}^2 \ge \|\mathbf{\Phi} - PDX\|_{GF}^2. \tag{17.128}$$

We ask: Does the (presumed) optimal solution on the far left of (17.128) attain the lower bound on the far right? Ignoring the optimization constraints, this is true if $PDX = \mathbf{\Phi}M$. But can this be done while preserving the constraints? Note that the value of X has not changed, so its sparsity structure is unaffected. Let S be a diagonal matrix such that $PDS \in \mathcal{D}$.[115] Then

$$\mathbf{\Phi} - PDX = \mathbf{\Phi} - (PDS)(S^{-1}X) = \mathbf{\Phi} - D'X',$$

where $D' = PDS \in \mathcal{D}$ and the transformation $X' = S^{-1}X$ preserves the sparsity constraint, $\overline{\mathsf{sp}}(X') \le k$. Thus the optimal solution can be represented as PDX and the lower bound shown on the far right-hand side of (17.128) is indeed attained.

As noted in Footnote 114, the orthogonal projection operator onto $\mathcal{R}(\mathbf{\Phi})$ has the form

$$P = \mathbf{\Phi}\left(\mathbf{\Phi}^*\mathbf{\Phi}\right)^+\mathbf{\Phi}^*.$$

This gives

$$PDX = \mathbf{\Phi}\left(\mathbf{\Phi}^*\mathbf{\Phi}\right)^+\mathbf{\Phi}^*DX = \mathbf{\Phi}AX,$$

where we have set

$$A = \left(\mathbf{\Phi}^*\mathbf{\Phi}\right)^+\mathbf{\Phi}^*D = \mathbf{\Phi}^+D.$$

---

[114] The projection operator P has the form $P = \mathbf{\Phi}(\mathbf{\Phi}^*\mathbf{\Phi})^+\mathbf{\Phi}^*$, where $(\mathbf{\Phi}^*\mathbf{\Phi})^+$ denotes the Moore–Penrose pseudoinverse of the kernel matrix $\mathbf{\Phi}^*\mathbf{\Phi}$. Thus $PDX = \mathbf{\Phi}M$ for $M = (\mathbf{\Phi}^*\mathbf{\Phi})^+\mathbf{\Phi}^*DX \in \mathbb{R}^{L\times L}$ and $PDX = \mathbf{\Phi}AX$ for $A = (\mathbf{\Phi}^*\mathbf{\Phi})^+\mathbf{\Phi}^*D \in \mathbb{R}^{L\times m}$.

[115] This can be done for constraints imposed for identifiability, such as column normalization or Frobenius normalization.

Here, $\mathbf{\Phi}^+$ is the pseudoinverse of $\mathbf{\Phi}$ and A can be interpreted as a pulling back of an m-element dictionary in the (potentially) infinite-dimensional Mercer space $\mathcal{M}$ to an m-element dictionary A in the finite-dimensional space $\mathbb{R}^L$. This is consistent with the fact that $PD = \mathbf{\Phi}A$, i.e., that $\mathbf{\Phi}$ maps A to the orthogonal projection of D into the finite-dimensional $\mathcal{R}(\mathbf{\Phi})$. Since we have seen that the optimal solution for D exists in this space, there is no harm in working with the $L \times m$ matrix A. This recognition that one equivalently can work with a dictionary on a finite-dimensional space in lieu of a dictionary on an infinite-dimensional space is a key contribution of [130,131].

References

[1] Vinayak Abrol, Pulkit Sharma, Anil Kumar Sao, Greedy dictionary learning for kernel sparse representation based classifier, Pattern Recognition Letters 78 (2016) 64–69.

[2] Michal Aharon, Michael Elad, Alfred Bruckstein, K-SVD: an algorithm for designing overcomplete dictionaries for sparse representation, IEEE Transactions on Signal Processing 54 (11) (2006) 4311–4322.

[3] Michal Aharon, Michael Elad, Alfred M. Bruckstein, K-SVD and its non-negative variant for dictionary design, in: Wavelets XI, vol. 5914, SPIE, 2005, pp. 327–339.

[4] Michel Aharon, Overcomplete Dictionaries for Sparse Representation of Signals, PhD thesis, Technion (Israel Institute of Technology), Haifa, Israel, 2006.

[5] William K. Allard, Guangliang Chen, Mauro Maggioni, Multi-scale geometric methods for data sets II: geometric multi-resolution analysis, Applied and Computational Harmonic Analysis 32 (3) (2012) 435–462.

[6] Bill Andreopoulos, Clustering categorical data, in: C.C. Aggarwal, C.K. Reddy (Eds.), Data Clustering, Chapman and Hall/CRC, 2018, pp. 277–304.

[7] Barry C. Arnold, José María Sarabia, Majorization and the Lorenz Order with Applications in Applied Mathematics and Economics, 2nd edition, Springer, 2018.

[8] Joseph J. Atick, Could information theory provide an ecological theory of sensory processing?, Network 3 (1992) 213–251.

[9] Fred Attneave, Some informational aspects of visual perception, Psychological Review 61 (3) (1954) 183–193.

[10] Richard G. Baraniuk, Michael B. Wakin, Random projections of smooth manifolds, Foundations of Computational Mathematics 9 (1) (2009) 51–77.

[11] Horace B. Barlow, Possible principles underlying the transformations of sensory messages, in: Walter A. Rosenblith (Ed.), Sensory Communication, MIT Press, 1961, pp. 217–234, chapter 13.

[12] Horace B. Barlow, Unsupervised learning, Neural Computation 1 (3) (1989) 295–311.

[13] Horace B. Barlow, Tej P. Kaushal, Graeme J. Mitchison, Finding minimum entropy codes, Neural Computation 1 (3) (1989) 412–423.

[14] David Bartholomew, Martin Knott, Irini Moustaki, Latent Variable Models and Factor Analysis: A Unified Approach, 3rd edition, Wiley, 2011.

[15] Alexander Basilevsky, Statistical Factor Analysis and Related Methods: Theory and Applications, Wiley–Interscience, 1994.

[16] Anthony J. Bell, Terrence J. Sejnowski, The 'independent components' of natural scenes are edge filters, Vision Research 37 (23) (1997) 3327–3338.

[17] Shai Ben-David, Clustering – what both theoreticians and practitioners are doing wrong, in: Proceedings of the AAAI Conference on Artificial Intelligence, vol. 32, 2018.

[18] James O. Berger, Statistical Decision Theory and Bayesian Analysis, 2nd edition, Springer, 1985.

[19] Alina Beygelzimer, Sham Kakade, John Langford, Cover trees for nearest neighbor, in: Proceedings of the 23rd International Conference on Machine Learning, 2006, pp. 97–104.

[20] Christopher M. Bishop, Pattern Recognition and Machine Learning, Springer, 2006.
[21] Stephen Boyd, Stephen P. Boyd, Lieven Vandenberghe, Convex Optimization, Cambridge University Press, 2004.
[22] Michael Bradie, Assessing evolutionary epistemology, Biology and Philosophy 6 (1986) 401–459.
[23] Michael Bradie, William Harms, Evolutionary epistemology, in: Edward N. Zalta (Ed.), The Stanford Encyclopedia of Philosophy, The Metaphysics Research Lab, Philosophy Department, Stanford University, 2020, Spring 2020 edition, available at https://plato.stanford.edu/archives/spr2020/entries/epistemology-evolutionary.
[24] Alfred M. Bruckstein, David L. Donoho, Michael Elad, From sparse solutions of systems of equations to sparse modeling of signals and images, SIAM Review 51 (1) (2009) 34–81.
[25] Emmanuel J. Candès, Justin Romberg, Terence Tao, Robust uncertainty principles: exact signal reconstruction from highly incomplete frequency information, IEEE Transactions on Information Theory 52 (2) (2006) 489–509.
[26] Emmanuel J. Candes, Terence Tao, Decoding by linear programming, IEEE Transactions on Information Theory 51 (12) (2005) 4203–4215.
[27] Emmanuel J. Candes, Terence Tao, Near-optimal signal recovery from random projections: universal encoding strategies?, IEEE Transactions on Information Theory 52 (12) (2006) 5406–5425.
[28] Emmanuel J. Candes, Michael B. Wakin, Stephen P. Boyd, Enhancing sparsity by reweighted ℓ_1 minimization, Journal of Fourier Analysis and Applications 14 (5) (2008) 877–905.
[29] Stefen Chan Wai Tim, Michèle Rombaut, Denis Pellerin, Multi-layer dictionary learning for image classification, in: International Conference on Advanced Concepts for Intelligent Vision Systems, Springer, 2016, pp. 522–533.
[30] Guangliang Chen, Anna V. Little, Mauro Maggioni, Lorenzo Rosasco, Some recent advances in multiscale geometric analysis of point clouds, Wavelets and Multiscale Analysis (2011) 199–225.
[31] Guangliang Chen, Mauro Maggioni, Multiscale geometric wavelets for the analysis of point clouds, in: 2010 44th Annual Conference on Information Sciences and Systems (CISS), IEEE, 2010, pp. 1–6.
[32] Scott Shaobing Chen, David L. Donoho, Michael A. Saunders, Atomic decomposition by basis pursuit, SIAM Review 43 (1) (2001) 129–159.
[33] Hong Cheng, Sparse Representation, Modeling and Learning in Visual Representation, Springer, 2015.
[34] Mads Græsbøll Christensen, Jan Østergaard, Søren Holdt Jensen, On compressed sensing and its application to speech and audio signals, in: 2009 Conference Record of the Forty-Third Asilomar Conference on Signals, Systems and Computers, IEEE, 2009, pp. 356–360.
[35] Andrzej Cichocki, Rafal Zdunek, Anh Huy Phan, Shun-ichi Amari, Nonnegative Matrix and Tensor Factorizations: Applications to Exploratory Multi-Way Data Analysis and Blind Source Separation, John Wiley & Sons, 2009.
[36] Florent Couzinie-Devy, Julien Mairal, Francis Bach, Jean Ponce, Dictionary learning for deblurring and digital zoom, arXiv preprint, arXiv:1110.0957, 2011.
[37] Thomas M. Cover, Joy A. Thomas, Elements of Information Theory, 2nd edition, Wiley-Interscience, 2006.
[38] John Shawe-Taylor, Nello Cristianini, Kernel Methods for Pattern Analysis, Cambridge University Press, 2004.
[39] Imre Csiszár, Paul C. Shields, Information Theory and Statistics: A Tutorial, Now Publishers Inc, 2004.
[40] Mark A. Davenport, Chinmay Hegde, Marco F. Duarte, Richard G. Baraniuk, Joint manifolds for data fusion, IEEE Transactions on Image Processing 19 (10) (2010) 2580–2594.
[41] M.M. Deza, E. Deza, Encyclopedia of Distances, 4th edition, Springer, 2016.
[42] Phoebus J. Dhrymes, Mathematics for Econometrics, 4th edition, Springer, 2013.
[43] Chris Ding, Xiaofeng He, K-means clustering via principal component analysis, in: Proceedings of the Twenty-First International Conference on Machine Learning, 2004, p. 29.
[44] David L. Donoho, Compressed sensing, IEEE Transactions on Information Theory 52 (4) (2006) 1289–1306.

[45] David L. Donoho, Carrie Grimes, Image manifolds which are isometric to Euclidean space, Journal of Mathematical Imaging and Vision 23 (1) (2005) 5–24.
[46] Fred I. Dretske, Knowledge and the Flow of Information, MIT Press, 1981.
[47] Bogdan Dumitrescu, Paul Irofti, Dictionary Learning Algorithms and Applications, Springer, 2018.
[48] Herbert Edelsbrunner, A Short Course in Computational Geometry and Topology, Springer, 2014.
[49] Michael Elad, Sparse and Redundant Representations: From Theory to Applications in Signal and Image Processing, Springer, 2010.
[50] Michael Elad, Welcome to Sparseland – Sparse and Redundant Representations and their Applications in Signal and Image Processing, September 4–8, 2017, Lectures given at the 2017 Summer School on Signal Processing Meets Deep Learning.
[51] Michael Elad, Michal Aharon, Image denoising via sparse and redundant representations over learned dictionaries, IEEE Transactions on Image Processing 15 (12) (2006) 3736–3745.
[52] Yonina C. Eldar, Gitta Kutyniok (Eds.), Compressed Sensing: Theory and Applications, Cambridge University Press, 2012.
[53] Kjersti Engan, Frame Based Signal Representation and Compression, PhD thesis, University of Stavanger, Stavanger, Norway, 2000.
[54] Kjersti Engan, Sven Ole Aase, J. Hakon Husoy, Method of optimal directions for frame design, in: 1999 IEEE International Conference on Acoustics, Speech, and Signal Processing. Proceedings. ICASSP99 (Cat. No. 99CH36258), vol. 5, IEEE, 1999, pp. 2443–2446.
[55] Vladimir Estivill-Castro, Why so many clustering algorithms: a position paper, ACM SIGKDD Explorations Newsletter 4 (1) (2002) 65–75.
[56] Anita C. Faul, Michael E. Tipping, Analysis of sparse Bayesian learning, Advances in Neural Information Processing Systems 14 (2002) 383–389.
[57] Igor Fedorov, Structured Learning with Scale Mixture Priors, PhD thesis, University of California, San Diego, La Jolla, California, USA, 2018.
[58] Igor Fedorov, Alican Nalci, Ritwik Giri, Bhaskar D. Rao, Truong Q. Nguyen, Harinath Garudadri, A unified framework for sparse non-negative least squares using multiplicative updates and the non-negative matrix factorization problem, Signal Processing 146 (2018) 79–91.
[59] Igor Fedorov, Bhaskar D. Rao, Multimodal sparse Bayesian dictionary learning, arXiv preprint, arXiv:1804.03740, 2018.
[60] Igor Fedorov, Bhaskar D. Rao, Truong Q. Nguyen, Multimodal sparse Bayesian dictionary learning applied to multimodal data classification, in: 2017 IEEE International Conference on Acoustics, Speech and Signal Processing (ICASSP), IEEE, 2017, pp. 2237–2241.
[61] Cédric Févotte, Jérôme Idier, Algorithms for nonnegative matrix factorization with the β-divergence, Neural Computation 23 (9) (2011) 2421–2456.
[62] David J. Field, What is the goal of sensory coding?, Neural Computation 6 (4) (1994) 559–601.
[63] Simon Foucart, Holger Rauhut, A Mathematical Introduction to Compressive Sensing, Birkhäuser, 2013.
[64] Rainer Gemulla, Erik Nijkamp, Peter J. Haas, Yannis Sismanis, Large-scale matrix factorization with distributed stochastic gradient descent, in: Proceedings of the 17th ACM SIGKDD International Conference on Knowledge Discovery and Data Mining, 2011, pp. 69–77.
[65] Allen Gersho, Robert M. Gray, Vector Quantization and Signal Compression, Springer Science & Business Media, 2012.
[66] Nicolas Gillis, The why and how of nonnegative matrix factorization, arXiv:1401.5226, 2014.
[67] Nicolas Gillis, Introduction to nonnegative matrix factorization, arXiv:1703.00663, 2017.
[68] Nicolas Gillis, Nonnegative Matrix Factorization, SIAM, 2020.
[69] Rikwik Giri, Bayesian Sparse Signal Recovery using Scale Mixtures with Applications to Speech, PhD thesis, University of California, San Diego, La Jolla, California, USA, 2016.
[70] Ritwik Giri, Bhaskar Rao, Type I and type II Bayesian methods for sparse signal recovery using scale mixtures, IEEE Transactions on Signal Processing 64 (13) (2016) 3418–3428.

[71] Mark Girolami, A variational method for learning sparse and overcomplete representations, Neural Computation 13 (11) (2001) 2517–2532.
[72] Alona Golts, Michael Elad, Linearized kernel dictionary learning, IEEE Journal of Selected Topics in Signal Processing 10 (4) (2016) 726–739.
[73] Irving J. Good, The Estimation of Probabilities: An Essay on Modern Bayesian Methods, MIT Press, 1965.
[74] Irving J. Good, How to estimate probabilities, IMA Journal of Applied Mathematics 2 (4) (12 1966) 364–383.
[75] Irina F. Gorodnitsky, Bhaskar D. Rao, Sparse signal reconstruction from limited data using FOCUSS: a re-weighted minimum norm algorithm, IEEE Transactions on Signal Processing 45 (3) (1997) 600–616.
[76] Nima Hatami, Mohsin Bilal, Nasir Rajpoot, Deep multi-resolution dictionary learning for histopathology image analysis, arXiv preprint, arXiv:2104.00669, 2021.
[77] Jostein Herredsvela, Kjersti Engan, Thor Ole Gulsrud, Karl Skretting, Detection of masses in mammograms by watershed segmentation and sparse representations using learned dictionaries, in: Proc of NORSIG, 2005.
[78] Thomas Hofmann, Bernhard Schölkopf, Alexander J. Smola, Kernel methods in machine learning, The Annals of Statistics 36 (3) (2008) 1171–1220.
[79] Patrik O. Hoyer, Non-negative sparse coding, in: Proceedings of the 12th IEEE Workshop on Neural Networks for Signal Processing, IEEE, 2002, pp. 557–565.
[80] Zdenûk Hubalek, Measures of species diversity in ecology: an evaluation, Folia Zoologica 49 (4) (2000) 241–260.
[81] Niall Hurley, Scott Rickard, Comparing measures of sparsity, IEEE Transactions on Information Theory 55 (10) (2009) 4723–4741.
[82] Aapo Hyvärinen, Jarmo Hurri, Patrick O. Hoyer, Natural Image Statistics: A Probabilistic Approach to Early Computational Vision, Springer Science & Business Media, 2009.
[83] Aapo Hyvärinen, Juha Karhunen, Erkki Oja, Independent Component Analysis, Wiley-Interscience, 2001.
[84] Denis C. Ilie-Ablachim, Bogdan Dumitrescu, Classification with incoherent kernel dictionary learning, in: 2021 23rd International Conference on Control Systems and Computer Science (CSCS), IEEE, 2021, pp. 106–111.
[85] Mark A. Iwen, Mauro Maggioni, Approximation of points on low-dimensional manifolds via random linear projections, Information and Inference: A Journal of the IMA 2 (1) (2013) 1–31.
[86] Alan Julian Izenman, Modern Multivariate Statistical Techniques: Regression, Classification, and Manifold Learning, Springer, 2008.
[87] Edwin T. Jaynes, Prior probabilities, IEEE Transactions on Systems Science and Cybernetics 4 (3) (1968) 227–241.
[88] Ian T. Jolliffe, Principal Component Analysis, 2nd edition, Springer, 2002.
[89] Geethu Joseph, Chandra R. Murthy, On the convergence of a Bayesian algorithm for joint dictionary learning and sparse recovery, IEEE Transactions on Signal Processing 68 (2019) 343–358.
[90] Anatoli Juditsky, Arkadi Nemirovski, Statistical Inference via Convex Optimization, Princeton University Press, 2020.
[91] Leonard Kaufman, Peter J. Rousseeuw, Finding Groups in Data: An Introduction to Cluster Analysis, Wiley-Interscience, 1990.
[92] Granino A. Korn, Theresa M. Korn, Mathematical Handbook for Scientists and Engineers, 2nd edition, Dover Publications, 1968.
[93] Ioannis A. Kougioumtzoglou, Ioannis Petromichelakis, Apostolos F. Psaros, Sparse representations and compressive sampling approaches in engineering mechanics: a review of theoretical concepts and diverse applications, Probabilistic Engineering Mechanics 61 (2020) 103082.
[94] Kenneth Kreutz-Delgado, Joseph F. Murray, Bhaskar D. Rao, Kjersti Engan, Te-Won Lee, Terrence J. Sejnowski, Dictionary learning algorithms for sparse representation, Neural Computation 15 (2) (2003) 349–396.

[95] Kenneth Kreutz-Delgado, Bhaskar D. Rao, Measures and algorithms for best basis selection, in: Proceedings of the 1998 IEEE International Conference on Acoustics, Speech and Signal Processing, ICASSP'98 (Cat. No. 98CH36181), vol. 3, IEEE, 1998, pp. 1881–1884.
[96] Alex Krizhevsky, Ilya Studkever, Georffrey E. Hinton, ImageNet classification with deep convolutional neural networks, Advances in Neural Information Processing Systems 25 (2012).
[97] D.N. Lawley, A.E. Maxwell, Factor Analysis as a Statistical Method, 2nd edition, London Butterworths, 1971.
[98] Yann LeCun, Yoshua Bengio, Geoffrey Hinton, Deep learning, Nature 521 (7553) (2015) 436–444.
[99] Daniel D. Lee, H. Sebastian Seung, Learning the parts of objects by non-negative matrix factorization, Nature 401 (6755) (1999) 788–791.
[100] Michael S. Lewicki, Terrence J. Sejnowski, Learning overcomplete representations, Neural Computation 12 (2000) 337–365.
[101] Yifeng Li, Alioune Ngom, Supervised dictionary learning via non-negative matrix factorization for classification, in: 2012 11th International Conference on Machine Learning and Applications, vol. 1, IEEE, 2012, pp. 439–443.
[102] Wenjing Liao, Mauro Maggioni, Adaptive geometric multiscale approximations for intrinsically low-dimensional data, Journal of Machine Learning Research 20 (2019) 1.
[103] Baihong Lin, Xiaoming Tao, Jianhua Lu, Hyperspectral image denoising via matrix factorization and deep prior regularization, IEEE Transactions on Image Processing 29 (2019) 565–578.
[104] Henry W. Lin, Max Tegmark, David Rolnick, Why does deep and cheap learning work so well?, Journal of Statistical Physics 168 (6) (2017) 1223–1247.
[105] Max A. Little, Machine Learning for Signal Processing: Data Science, Algorithms, and Computational Statistics, Oxford University Press, 2019.
[106] David G. Luenberger, Optimization by Vector Space Methods, Wiley-Interscience, 1969.
[107] David J.C. MacKay, Information Theory, Inference, and Learning Algorithms, Cambridge University Press, 2003.
[108] Mauro Maggioni, Stanislav Minsker, Nate Strawn, Multiscale dictionary learning: non-asymptotic bounds and robustness, The Journal of Machine Learning Research 17 (1) (2016) 43–93.
[109] Shahin Mahdizadehaghdam, Ashkan Panahi, Hamid Krim, Liyi Dai, Deep dictionary learning: a parametric network approach, IEEE Transactions on Image Processing 28 (10) (2019) 4790–4802.
[110] Julien Mairal, Francis Bach, Jean Ponce, Task-driven dictionary learning, IEEE Transactions on Pattern Analysis and Machine Intelligence 34 (4) (2012) 791–804.
[111] Julien Mairal, Francis Bach, Jean Ponce, Sparse Modeling for Image and Vision Processing, NOW Publishers, 2014.
[112] Julien Mairal, Francis Bach, Jean Ponce, Guillermo Sapiro, Andrew Zisserman, Discriminative learned dictionaries for local image analysis, in: 2008 IEEE Conference on Computer Vision and Pattern Recognition, IEEE, 2008, pp. 1–8.
[113] Stéphane Mallat, A Wavelet Tour of Signal Processing: The Sparse Way, 3rd edition, Academic Press, 2009.
[114] Angshul Manjubar, Compressed Sensing for Engineers, CRC Press – Taylor & Francis Group, 2019.
[115] Kanti V. Mardia, John T. Kent, John M. Bibby, Multivariate Analysis, Academic Press, 1979.
[116] Albert W. Marshall, Ingram Olkin, Barry C. Arnold, Inequalities: Theory of Majorization and Its Applications, 2nd edition, Springer, 2011.
[117] Nathaniel F.G. Martin, James W. England, Mathematical Theory of Entropy, Cambridge University Press, 1984.
[118] Marina Meilă, Comparing clusterings—-an information based distance, Journal of Multivariate Analysis 98 (5) (2007) 873–895.
[119] Marina Meilă, Spectral clustering, in: Handbook of Cluster Analysis, 2016, pp. 125–141.
[120] Thomas P. Minka, Old and new matrix algebra useful for statistics, https://www.microsoft.com/en-us/research/publication/old-new-matrix-algebra-useful-statistics/, December 2000.

[121] B.G. Mirkin, L.B. Cherny, Measurement of the distance between distinct partitions of a finite set of objects [in Russian], Avtomatika i Telemekhanika [Automation and Remote Control] 5 (1970) 120–127.
[122] Boris Mirkin, Mathematical Classification and Clustering, Kluwer Academic Publishing, 1996.
[123] Eric E. Monson, Guangliang Chen, Rachael Brady, Mauro Maggioni, Data representation and exploration with geometric wavelets, in: 2010 IEEE Symposium on Visual Analytics Science and Technology, IEEE, 2010, pp. 243–244.
[124] Kevin P. Murphy, Machine Learning: A Probabilistic Perspective, MIT Press, 2012.
[125] Joseph F. Murray, Kenneth Kreutz-Delgado, Visual recognition and inference using dynamic overcomplete sparse learning, Neural Computation 19 (9) (2007) 2301–2352.
[126] Alican Nalci, Rectified Sparse Bayesian Learning and Effects and Limitations of Nuisance Regression in Functional MRI, PhD thesis, University of California San Diego, La Jolla, California, 2019.
[127] Alican Nalci, Igor Fedorov, Maher Al-Shoukairi, Thomas T. Liu, Bhaskar D. Rao, Rectified Gaussian scale mixtures and the sparse non-negative least squares problem, IEEE Transactions on Signal Processing 66 (12) (2018) 3124–3139.
[128] Radford Neal, Bayesian Learning for Neural Networks, Springer, 1996.
[129] Radford M. Neal, Geoffrey E. Hinton, A view of the EM algorithm that justifies incremental, sparse, and other variants, in: Learning in Graphical Models, Springer, 1998, pp. 355–368.
[130] Hien Van Nguyen, Vishal M. Patel, Nasser M. Nasrabadi, Rama Chellappa, Kernel dictionary learning, in: 2012 IEEE International Conference on Acoustics, Speech and Signal Processing (ICASSP), IEEE, 2012, pp. 2021–2024.
[131] Hien Van Nguyen, Vishal M. Patel, Nasser M. Nasrabadi, Rama Chellappa, Design of non-linear kernel dictionaries for object recognition, IEEE Transactions on Image Processing 22 (12) (2013) 5123–5135.
[132] Bruno A. Olshausen, David J. Field, Emergence of simple-cell receptive field properties by learning a sparse code for natural images, Nature 381 (6583) (1996) 607–609.
[133] Bruno A. Olshausen, David J. Field, Sparse coding with an overcomplete basis set: a strategy employed by V1?, Vision Research 37 (23) (1997) 3311–3325.
[134] Jason Palmer, David Wipf, Kenneth Kreutz-Delgado, Bhaskar Rao, Variational EM algorithms for non-Gaussian latent variable models, Advances in Neural Information Processing Systems 18 (2006) 1059.
[135] Jason A. Palmer, Variational and Scale Mixture Representations of Non-Gaussian Densities for Estimation in the Bayesian Linear Model: Sparse Coding, Independent Component Analysis, and Minimum Entropy Segmentation, PhD thesis, University of California San Diego, La Jolla, California USA, 2006.
[136] Vardan Papyan, Yaniv Romano, Michael Elad, Convolutional neural networks analyzed via convolutional sparse coding, The Journal of Machine Learning Research 18 (1) (2017) 2887–2938.
[137] Vardan Papyan, Yaniv Romano, Jeremias Sulam, Michael Elad, Theoretical foundations of deep learning via sparse representations: a multilayer sparse model and its connection to convolutional neural networks, IEEE Signal Processing Magazine 35 (4) (2018) 72–89.
[138] Vardan Papyan, Jeremias Sulam, Michael Elad, Working locally thinking globally: theoretical guarantees for convolutional sparse coding, IEEE Transactions on Signal Processing 65 (21) (2017) 5687–5701.
[139] Vishal M. Patel, Rama Chellappa, Sparse Representations and Compressive Sensing for Imaging and Vision, Springer Science & Business Media, 2013.
[140] G.P. Patil, Charles Taillie, Diversity as a concept and its measurement, Journal of the American Statistical Association 77 (379) (1982) 548–561.
[141] Victor Peña, James O. Berger, Restricted type II maximum likelihood priors on regression coefficients, Bayesian Analysis 15 (4) (2020) 1281–1297.
[142] Gabriel Peyré, Manifold models for signals and images, Computer Vision and Image Understanding 113 (2) (2009) 249–260.
[143] Gabriel Peyré, Sparse modeling of textures, Journal of Mathematical Imaging and Vision 34 (1) (2009) 17–31.

[144] John R. Pierce, An Introduction to Information Theory: Symbols, Signals and Noise, 2nd revised edition, Dover Publications, 1980.

[145] Luca Pion-Tonachini, Kristofer Bouchard, Hector Garcia Martin, Sean Peisert, W. Bradley Holtz, Anil Aswani, Dipankar Dwivedi, Haruko Wainwright, Ghanshyam Pilania, Benjamin Nachman, Babetta L. Marrone, Nicola Falco, Prabhat, Daniel Arnold, Alejandro Wolf-Yadlin, Sarah Powers, Sharlee Climer, Quinn Jackson, Ty Carlson, Michael Sohn, Petrus Zwart, Neeraj Kumar, Amy Justice, Claire Tomlin, Daniel Jacobson, Gos Micklem, Georgios V. Gkoutos, Peter J. Bickel, Jean-Baptiste Cazier, Juliane Müller, Bobbie-Jo Webb-Robertson, Rick Stevens, Mark Anderson, Ken Kreutz-Delgado, Michael W. Mahoney, James B. Brown, Learning from learning machines: a new generation of AI technology to meet the needs of science, arXiv:2111.13786, 2021.

[146] Boaz Porat, Digital Processing of Random Signals: Theory and Methods, Prentice-Hall, 1994.

[147] Ignacio Ramirez, Pablo Sprechmann, Guillermo Sapiro, Classification and clustering via dictionary learning with structured incoherence and shared features, in: 2010 IEEE Computer Society Conference on Computer Vision and Pattern Recognition, IEEE, 2010, pp. 3501–3508.

[148] Bhaskar D. Rao, Kenneth Kreutz-Delgado, An affine scaling methodology for best basis selection, IEEE Transactions on Signal Processing 47 (1) (1999) 187–200.

[149] Calyampudi Radhakrishna Rao, Mareppalli Bhaskara Rao, Matrix Algebra and Its Applications to Statistics and Econometrics, World Scientific, 1998.

[150] Irina Rish, Genady Grabarnik, Sparse Modeling: Theory, Algorithms, and Applications, CRC Press, 2015.

[151] Michael Robinson, Topological Signal Processing, Springer, 2014.

[152] Ulises Rodríguez-Domínguez, Oscar Dalmau, Hierarchical discriminative deep dictionary learning, IEEE Access 8 (2020) 142680–142690.

[153] Ron Rubinstein, Alfred M. Bruckstein, Michael Elad, Dictionaries for sparse representation modeling, Proceedings of the IEEE 98 (6) (2010) 1045–1057.

[154] Sujit Kumar Sahoo, Anamitra Makur, Dictionary training for sparse representation as generalization of k-means clustering, IEEE Signal Processing Letters 20 (6) (2013) 587–590.

[155] Wojciech Samek, Grégoire Montavon, Andrea Vedaldi, Lars Kai Hansen, Klaus-Robert Müller (Eds.), Explainable AI: Interpreting, Explaining and Visualizing Deep Learning, vol. 11700, Springer Nature, 2019.

[156] Bernhard Schölkopf, Alexander J. Smola, Learning with Kernels: Support Vector Machines, Regularization, Optimization, and Beyond, MIT Press, 2002.

[157] Adriana Schulz, Eduardo Antônio Barros Da Silva, Luiz Velho, Compressive Sensing, IMPA, 2009.

[158] Rudolf Scitovski, Kristian Sabo, Francisco Martínez-Álvarez, Šime Ungar, Cluster Analysis and Applications, Springer, 2021.

[159] Claude Elwood Shannon, A mathematical theory of communication, The Bell System Technical Journal 27 (3) (1948) 379–423.

[160] Claude Elwood Shannon, Warren Weaver, The Mathematical Theory of Communication, The University of Illinois Press, 1949.

[161] Dan A. Simovici, Clustering: Theoretical and Practical Aspects, World Scientific, 2022.

[162] Karl Skretting, John Håkon Husøy, Texture classification using sparse frame-based representations, EURASIP Journal on Advances in Signal Processing 2006 (2006) 1–11.

[163] Pablo Sprechmann, Guillermo Sapiro, Dictionary learning and sparse coding for unsupervised clustering, in: 2010 IEEE International Conference on Acoustics, Speech and Signal Processing, IEEE, 2010, pp. 2042–2045.

[164] Jean-Luc Starck, Fionn Murtagh, Jalal Fadili, Sparse Image and Signal Processing: Wavelets and Related Geometric Multiscale Analysis, 2nd edition, Cambridge University Press, 2015.

[165] James V. Stone, Information Theory: A Tutorial Introduction, Sebtel Press, 2015.

[166] Gilbert Strang, Linear Algebra and Learning from Data, Wellesley–Cambridge Press, 2019.

[167] Xiaoxia Sun, Nasser M. Nasrabadi, Trac D. Tran, Supervised deep sparse coding networks for image classification, IEEE Transactions on Image Processing 29 (2019) 405–418.

[168] Sergio Theodoridis, Machine Learning: A Bayesian and Optimization Perspective, 2nd edition, Academic Press, 2020.
[169] Sergio Theodoridis, Yannis Kopsinis, Konstantinos Slavakis, Sparsity-aware learning and compressed sensing: an overview, in: Paulo Diniz, Johan Suykens, Rama Chellappa, Sergios Theodoridis (Eds.), Academic Press Library in Signal Processing Volume 1: Signal Processing Theory and Machine Learning, Academic Press, 2014, pp. 1271–1377, chapter 23.
[170] Sergio Theodoridis, Konstantinos Koutroumbas, Pattern Recognition, 4th edition, Academic Press, 2009.
[171] Robert Tibshirani, Regression shrinkage and selection via the LASSO, Journal of the Royal Statistical Society: Series B (Methodological) 58 (1) (1996) 267–288.
[172] Michael E. Tipping, Sparse Bayesian learning and the relevance vector machine, Journal of Machine Learning Research 1 (Jun) (2001) 211–244.
[173] Michael E. Tipping, Bayesian inference: an introduction to principles and practice in machine learning, in: O. Bousquet, U. von Luxberg, G. Rätsch (Eds.), Advanced Lectures on Machine Learning, Springer, 2004, pp. 41–62.
[174] Ivana Tošić, Pascal Frossard, Dictionary learning, IEEE Signal Processing Magazine 28 (2) (2011) 27–38.
[175] Manuela Vasconcelos, Nuno Vasconcelos, Natural image statistics and low-complexity feature selection, IEEE Transactions on Pattern Analysis and Machine Intelligence 31 (2) (2008) 228–244.
[176] Mathukumalli Vidyasagar, An Introduction to Compressed Sensing, SIAM, 2019.
[177] Ulrike Von Luxburg, A tutorial on spectral clustering, Statistics and Computing 17 (4) (2007) 395–416.
[178] Michael B. Wakin, Manifold-based signal recovery and parameter estimation from compressive measurements, arXiv preprint, arXiv:1002.1247, 2010.
[179] Michael B. Wakin, David L. Donoho, Hyeokho Choi, Richard G. Baraniuk, The multiscale structure of non-differentiable image manifolds, in: Wavelets XI, vol. 5914, International Society for Optics and Photonics, 2005, p. 59141B.
[180] Satosi Watanabe, Pattern recognition as a quest for minimum entropy, Pattern Recognition 13 (5) (1981) 381–387.
[181] Martin L. Weitzman, On diversity, The Quarterly Journal of Economics 107 (2) (1992) 363–405.
[182] Wikipedia – The Free Encyclopedia, Efficient coding hypothesis, downloaded August 2021, https://en.wikipedia.org/wiki/Efficient_coding_hypothesis.
[183] Wikipedia – The Free Encyclopedia, Evolutionary epistemology, downloaded August 2021, https://en.wikipedia.org/wiki/Evolutionary_epistemology.
[184] Wikipedia – The Free Encyclopedia, Explainable artificial intelligence, downloaded August 2021, https://en.wikipedia.org/wiki/Explainable_artificial_intelligence.
[185] Wikipedia – The Free Encyclopedia, Cluster analysis, downloaded May 2022, https://en.wikipedia.org/wiki/Cluster_analysis.
[186] Kevin W. Wilson, Bhiksha Raj, Paris Smaragdis, Ajay Divakaran, Speech denoising using nonnegative matrix factorization with priors, in: 2008 IEEE International Conference on Acoustics, Speech and Signal Processing, IEEE, 2008, pp. 4029–4032.
[187] David Wipf, Srikantan Nagarajan, Iterative reweighted ℓ_1 and ℓ_2 methods for finding sparse solutions, IEEE Journal of Selected Topics in Signal Processing 4 (2) (2010) 317–329.
[188] David Wipf, Jason Palmer, Bhaskar Rao, Perspectives on sparse Bayesian learning, Computer Engineering 16 (1) (2004) 249.
[189] David P. Wipf, Srikantan S. Nagarajan, A new view of automatic relevance determination, in: NIPS, 2007, pp. 1625–1632.
[190] David P. Wipf, Bhaskar D. Rao, Sparse Bayesian learning for basis selection, IEEE Transactions on Signal Processing 52 (8) (2004) 2153–2164.
[191] David P. Wipf, Bhaskar D. Rao, An empirical Bayesian strategy for solving the simultaneous sparse approximation problem, IEEE Transactions on Signal Processing 55 (7) (2007) 3704–3716.

[192] David P. Wipf, Bhaskar D. Rao, Srikantan Nagarajan, Latent variable Bayesian models for promoting sparsity, IEEE Transactions on Information Theory 57 (9) (2011) 6236–6255.

[193] David Paul Wipf, Bayesian Methods for Finding Sparse Representations, PhD thesis, University of California San Diego, La Jolla, California USA, 2006.

[194] John Wright, Allen Y. Yang, Arvind Ganesh, S. Shankar Sastry, Yi Ma, Robust face recognition via sparse representation, IEEE Transactions on Pattern Analysis and Machine Intelligence 31 (2) (2009) 210–227.

[195] Rui Xu, Donald Wunsch II, Clustering, IEEE Press/Wiley, 2009.

[196] Wei Xu, Xin Liu, Yihong Gong, Document clustering based on non-negative matrix factorization, in: Proceedings of the 26th Annual International ACM SIGIR Conference on Research and Development in Information Retrieval, 2003, pp. 267–273.

[197] Meng Yang, Lei Zhang, Xiangchu Feng, David Zhang, Fisher discrimination dictionary learning for sparse representation, in: 2011 International Conference on Computer Vision, IEEE, 2011, pp. 543–550.

[198] Meng Yang, Lei Zhang, Xiangchu Feng, David Zhang, Sparse representation based Fisher discrimination dictionary learning for image classification, International Journal of Computer Vision 109 (3) (2014) 209–232.

[199] Qiang Zhang, Baoxin Li, Dictionary Learning in Visual Computing, Morgan & Claypool Publishers, 2015.

[200] Li Zhaoping, Theoretical understanding of the early visual processes by data compression and data selection, Network: Computation in Neural Systems 17 (4) (2006) 301–334.

[201] Li Zhaoping, Understanding Vision: Theory, Models, and Data, Oxford University Press, 2014.

Index

A

A posteriori
errors, 776, 781, 810, 838
estimation error, 778, 838
Acceptable error margin, 654
Accuracy measure, 152
Acoustic echo cancelation, 707, 718, 740, 848
Acoustic surface wave (SAW) devices, 416
Activation function, 626–628, 630, 633, 883–885, 887, 889, 890, 893, 894, 897, 902, 904, 926–928, 951
nonlinear, 926
output, 927
Adaptive
filter, 21, 22, 89, 179, 270, 285, 717, 718, 720, 721, 725, 728, 730, 740–747, 749, 752, 755, 759, 760, 763, 768, 773, 774, 994, 995
coefficients, 22, 728, 846
combinations, 840
design, 718, 725
learning, 718, 817
linear combinations, 842
recursions, 793
theory, 718
filtering, 21, 707, 717, 722, 724, 735, 745, 746, 791, 792, 836, 839, 844, 845, 847, 848, 994, 995
algorithms, 707, 708, 720, 728, 733, 740, 748, 793, 818
applications, 775
FIR filters, 846
graph filters, 995, 996
IIR filters, 846
notch filters, 741
Adaptive boosting (AdaBoost), 924
Adaptive differential pulse code modulation (ADPCM), 742
Adaptive line enhancement (ALE), 742, 743
Adaptive moment (Adam), 886
Adaptivity, 540, 548
Adaptor
coefficients, 348, 349, 351, 353–355, 362, 363, 366, 368, 370, 372, 376
quantized, 366
network, 346, 358, 359
Additive white Gaussian noise (AWGN), 788
Adjacency matrix, 963, 964, 967, 968, 971–973
Admissibility condition, 989, 990
Adversarial
examples, 24, 933, 934, 938
learning, 874
noise, 1032
Adversarially trained network, 934, 938
Affine projections algorithm (APA), 837–839
Affinity matrix, 1062
Akaike information criterion (AIC), 715
Algebraic signal processing (ASP), 968, 1002, 1003
theory, 1002
Aliased
baseband channel, 466
center frequencies, 435
filter spectral component experiences, 405
frequency components, 436
frequency spans, 423
signal, 430, 436
Aliasing
DDC, 419, 421
distortion, 198, 199, 208, 245
effects, 15
functions, 225, 226
illustration, 199
process, 412
spurs, 232
Alternating direction method of multipliers (ADMM), 1051, 1053, 1058
algorithm, 1053, 1059, 1060
exposition, 1059
framework, 1053
Alternating least-square (ALS) algorithm, 1018
Ambient
noise, 721
signal space, 1119
space, 1120, 1121
Amplitude modulation (AM), 174
Analog
antialiasing filter, 191, 222, 245, 247, 248
approximation problem, 269, 316
bridge structure, 361
Cauer filter, 367
channel, 194
channel mismatches, 186
circuits, 83, 268
components, 83, 186, 198, 325, 462
counterparts, 340, 414

delay, 416
design, 83
domain, 340, 341, 415, 416, 429
errors, 187
filter, 6, 10, 205, 223, 244, 245, 248, 268–271, 317, 327, 339, 340, 367, 413–416, 428, 492
designs, 416
networks, 269
synthesis, 268
functions, 223
IF filter, 462
input signal, 2
lattice filter, 330
lattice structure, 330, 334, 335
low-pass filter, 316, 317
matching errors, 231
networks, 339, 358
processing, 462
processing techniques, 404
prototype, 10, 429
quadrature filters, 429
reconstruction filter, 190, 192, 204, 248
reconstruction functions, 223
recursive filter, 414
reference filter, 268, 339
samples, 6
signal, 7, 36, 83, 187, 188, 190–193, 222, 404, 405, 407, 409, 492, 542
signal acquisition systems, 2
stage, 461
system, 1
technologies, 83, 404
transfer function, 269, 339
waveform, 407, 413
world, 416, 429
Analog signal processing (ASP), 185, 194
Analog-to-digital (A/D) converter, 6, 185, 410, 722
Analysis
channelizer, 406, 464–467, 469–471, 475–479, 492
bank, 476
baseline, 407, 464
channel, 471
channel bandwidth, 407, 464, 473
engine, 478
output rate, 471
outputs, 466, 471, 492
filter, 14–16, 500, 591, 592, 596–598, 600
filter bank, 837
framework, 538
operator, 666
signal, 20, 48, 73, 655, 664, 673, 677, 681, 683, 684
Ancillary information, 995
Antialiasing filter, 199, 201, 245, 248
analog, 191, 222, 245, 247, 248
Aperiodic signals, 542
Application-specific integrated circuit (ASIC), 461
Approximate linear dependence (ALD), 850
Approximating signal, 693
Artificial intelligence (AI), 869, 938, 1073, 1074
Artificial neural network (ANN), 24, 625–627
Ascending frequency, 975
Audio signals, 183
Autocorrelation, 155–158, 160, 163, 166, 167, 170, 172–174, 181, 182, 709, 748, 750, 753, 765, 797, 816, 817
deterministic, 705
functions, 713, 756
matrix, 120, 512, 515, 709, 748, 749, 756, 761, 786, 794, 807, 831
sequence, 212
Automatic relevance determination (ARD), 1086, 1105
Autoregressive (AR)
models, 690, 697, 698
process, 183, 711
process flow diagram, 698
systems, 697
Autoregressive moving average (ARMA), 101, 182
models, 701
process, 701
Average
cross-power, 159, 160
power, 159, 160
Axiomatic approach, 125

B

Background noise, 162
Backpropagated errors, 903
Backpropagation (BP), 628, 639
algorithm, 627, 629
neural network, 628
Backpropagation through time (BPTT), 950–952
Backsampling operation, 597
Backward consistency, 786
Band filter, 302, 307, 308
Band-limited
approximation, 173
baseband WSS random process, 175
processes, 174
random signals, 174
Bandpass, 174, 423, 570
content, 583, 601, 613
filter, 14, 179, 212, 267, 275, 279, 283, 317, 321, 359, 363, 364, 377, 419, 421, 429, 432, 443, 444, 571, 581
filtering, 177, 429
frequency region, 212
images, 591

lattice, 364
lattice filters, 363
polyphase partition, 433
process, 175, 427
random process, 180
sampling, 211
sequence, 593
signals, 212
system, 179
wavelet, 538
window functions, 581
WSS random processes, 180
Bandstop filters, 268, 275, 279, 317, 321, 339, 363, 364, 375
Bandwidth
arbitrary, 312, 403
channel, 428, 435, 436, 444–446, 454, 464, 466, 467, 471, 473, 475, 476
filter, 429
frequency, 179
input signal, 465
minimum, 469
mismatch, 232
noise, 179
reduction, 427, 437
synthesis channelizer, 406, 464
transition, 301
wider, 406, 464, 467, 471
Base station (BS), 889
Baseband
channels, 406, 466
converter, 430
copy, 443
frequency, 461
input signals, 459, 477
mixer, 430
modulation bandwidth, 462
process, 174, 443
shifted signals, 464
shifting, 463
signal, 462
translation, 462
Baseline synthesis channelizer bandwidths, 464
Basis, 540
Batch learning, 873, 874
Bayes Rule, 126, 127
Bayesian
approach, 1074, 1106
estimate, 1105
estimators, 151
framework, 1084
generative model framework, 1137
learning, 1084
method, 1105
model, 1112, 1113
network, 948
Bernoulli trials, 129
Best-matching cell (BMC), 877, 878
Bias
coefficients, 956
component, 885
nodes, 883
parameters, 883
terms, 881, 904
vector, 899
Binary
classification, 920, 1126
dictionary procedure, 1126
representations, 191, 236
signal, 836
Binomial
distribution, 136
random variables, 136
Binormalized LMS (BNLMS), 839
Biorthogonal
bases, 587, 594
filter, 594
filter pair, 600
frames, 600
wavelet, 552, 553
wavelet transform, 18
Bireciprocal
filters, 369
property, 320
Bode
diagram, 70
phase plot, 74
plot, 70, 71
Bounded-input bounded-output (BIBO), 35, 64, 65
stability, 63, 65
Bucket-brigade device (BBD), 85
Butterworth filters, 318, 319, 325, 414

C

Cable channelizer, 446
Canceling filter, 721
Canonic signed-digit code (CSDC), 395
Canonical-polyadic (CP) decomposition, 1016, 1017
Canonical-polyadic decomposition, 1016
Capacity, 540
Carrier
frequency, 443, 460, 461, 466, 690
signal, 174
Cascade
algorithm, 617
connection, 16
form, 284, 285, 423

linear operators, 421
operators, 420
wavelet coefficients, 640
Cascade Integrator Comb (CIC) filter, 105, 106
Cascaded
AR systems, 707
transmission lines, 341, 359
Cauchy–Riemann conditions, 771
Cauer filters, 311, 313, 318–320, 325, 327, 328, 330
Cauer networks, 362
Causal
learning, 947
LTI system, 64, 65
signals, 52
Causal machine learning, 946
Central limit theorem (CLT), 148, 149
Central moment, 138
Centralized solutions, 844, 845
Chain rule, 168
Channel
analysis channelizer, 471
bandwidth, 428, 435, 436, 444–446, 454, 464, 466, 467, 471, 473, 475, 476
channelizer, 464
filter, 639, 640
frequency responses, 223
mismatch errors, 221
Channelized spectrum, 466
Channelizer
analysis, 406, 464–467, 469–471, 475–479, 492
channel, 464
channel bandwidth, 464, 471
converter, 403, 405–407, 427, 428, 432, 436–438, 444–446, 453, 454, 459, 464–466, 471
filters, 446
input sampling frequency, 470
output, 473
output sampling frequency, 464
output vector, 459
polyphase, 403, 406, 407, 427, 428, 445, 446, 457, 473, 479, 487, 492
synthesis, 406, 460, 464, 466–468, 471, 473–479
Chaotic signal component, 272
Chapman–Kolmogorov equation, 168
Characteristic function, 141, 148
Charge-coupled device (CCD), 85
Chebyshev II
approximation, 318, 319
filters, 318, 319, 325
Chirp signal, 567
Circular
buffer, 454
convolution, 425
data buffer, 453
output buffer, 459
shift, 453, 459
signals, 760
time shift, 453
Circulator, 345
Circulator filters, 341
Classic estimator, 563
Classical
discrete filter bank technique, 587
Fourier transform, 543
MRA framework, 550
wavelet, 538–540, 575, 576, 580, 587, 594
wavelet method, 586
wavelet theory, 574
Classification
algorithms, 479
binary, 920, 1126
data, 870
error, 894, 899, 923
example, 892
in dictionary learning, 1122
label, 911
NN problems, 887
problem, 873, 883, 885, 887, 892, 894, 897, 909, 928
results, 934
signal, 484, 485
step, 1081
task, 485, 486, 899, 909, 932
Clean signal, 997
Clock frequency, 384, 394–396
Clock signal, 85
Codewords, 1091, 1092
prototype, 1094
Coefficients
adaptive filter, 22, 728, 846
curvelet, 573–575, 580, 583–586
DC, 505
discrete curvelet, 582, 583, 585
discrete shearlet, 617, 621
dyadic, 519, 520
errors, 236, 810
filter, 229, 271, 282, 325, 326, 329, 389, 425, 426, 429, 531, 710, 726–728, 732, 740
frame, 652–655, 659, 663, 666, 676, 681
framelets, 631, 633, 636
IFFT, 681
learning, 944
linear, 627
LMS, 831
quantized, 377, 658, 659
shearlet, 608–611, 619, 621, 623, 625
signal, 565

wavelet, 548, 555, 560, 561, 585, 603, 630, 639–641, 643
weight, 831
Coherence criterion (CC), 850
Colored noise, 172, 173
signal, 699
Columnwise Kronecker product, 1013
Comb filters, 105
Combinatorial graph Laplacian matrix, 1058
Commensurate length transmission
line, 268, 336
line filters, 269, 324, 337
line network, 268, 269, 336, 339, 340
lossless, 332, 336
Communication
networks, 380
signals, 485, 693
Complete
filter, 386
frame, 1081
network, 931
Component filter, 840, 842
Compressed
network, 1025
passbands, 313
representation, 1078
sampling, 187
Compressive Sampling Matching Pursuit (CoSaMP), 1083
Concatenated output signal, 637
Conditional distribution, 137, 139, 145, 149
Conditional probabilities, 127, 137
Conjugate mirror filter, 554
bank, 554
Consistence, 152
Constant coefficients, 44, 45, 62, 387
Constant modulus algorithm (CMA), 834–836
Constant vector, 976
Continuous
analog domain, 462
learning, 874
random variable, 131
shearlet coefficients, 610
shearlet transform, 605, 608, 613, 619, 621
spectrum, 469
Continuous-time random processes, 154
Continuous-valued random processes, 154
Contourlet, 567, 586
coefficients, 605
filter bank, 587, 592, 601, 603
transform, 545, 586, 587
Conventional
analog receiver, 418
filters, 276
polyphase technique, 304
Convergence learning rate, 878
Converter
architecture, 474
baseband, 430
channelizer, 403, 405–407, 427, 428, 432, 436–438, 444–446, 453, 454, 459, 464–466, 471
output, 8
synthesis channelizer, 477
synthesizers, 466, 467
Convex
function, 748
problems, 1054
program, 1049
Convolution, 54, 175, 978, 1014
field, 638
filters, 630, 638, 639, 641
framelets, 630
integral, 41, 42, 61, 62, 979
kernels, 484, 1024
layers, 640, 643
linear, 408
module, 640, 641
multichannel, 630
operation, 81, 978–980
operator, 628, 630
process, 429
property, 81
spectral, 469
sum, 644, 978
theorem, 979
Convolutional
blocks, 1023, 1026
feature, 1031
feature maps, 1030
filter, 631
kernels, 1009, 1023, 1027
layers, 488, 1009, 1022, 1023, 1025–1027
LSTMs, 1027, 1039
networks, 999
TT decompositions, 1027, 1039
Convolutional layer (CL), 1130
Convolutional neural network (CNN), 481, 626, 630, 898, 1009
Correlation
coefficient, 147, 148, 153
matrix, 775, 784
Cost function, 21, 23, 497, 503, 505, 558, 627, 628, 693, 703, 724, 726, 728, 730–733, 735, 745, 746, 748–750, 761, 762, 774–776, 781, 782, 870, 883, 886, 887, 893, 894, 901, 908, 928
Covariance matrix, 149
Cross-entropy (CE), 885, 892, 893
Cross-power spectral density, 171, 172
Crossover frequency, 284

Cumulative probability distribution function (CDF), 131, 132
Curvelet
 atoms, 568–575, 579, 580, 582, 583
 coefficients, 573–575, 580, 583–586
 frame, 580
 mother, 570–574, 576, 578–580, 582, 583
 tight frame, 580
Cutoff frequency, 71, 72, 231, 324, 982, 992, 993
Cycle frequencies, 486
Cyclic frequencies, 486
Cyclostationary feature function (CFF), 486

D

d-regular graph, 963
Data
 classification, 870
 converters, 186
 matrix, 1087, 1090, 1093, 1098, 1101, 1107
 representations, 560, 1122
 signal, 693
DC
 coefficients, 505
 shifted signals, 474
 signal, 226
Decimated
 analysis filter, 15
 signal, 13
 subband signals, 16
Decimating filter, 462
Decimation, 12, 13, 105, 208, 245, 285, 307–309
 factor, 13, 15
 filters, 280, 303, 308
 operation, 13
Decoder
 architecture, 632, 633
 layer, 635, 643
 side, 528
Deconvolution, 743, 996
 algorithm, 999
Decorrelation parameter, 743
Deep
 convolutional
  framelets neural networks, 628, 630, 632, 641
  networks, 1040
 networks, 942, 952, 1010, 1011, 1020, 1021, 1032, 1039
 neural networks, 25, 1010, 1016, 1020, 1022, 1023, 1025–1027, 1032, 1038, 1136
 polynomial networks, 1029
Deep convolutional neural network (DCNN), 1009, 1027, 1032, 1130
Deep learning (DL), 25, 626, 640, 898, 924, 1009, 1010, 1020, 1023, 1034, 1038, 1039, 1050
 algorithms, 898, 1010
 architectures, 1010, 1020, 1033
 context, 1040
 frameworks, 931, 1014, 1038, 1040
 methods, 1033, 1034
Deep NN (DNN), 888
Deep Q-network (DQN), 939
Deep wavelet superresolution (DWSR), 639–641
 network architecture, 640
Degree matrix, 964
Delay elements, 90, 98, 230, 311, 359, 377–379, 508, 523
Demodulated signal, 115
Denoised image, 586, 625, 640
Denominator coefficients, 697, 705
Densely connected network (DenseNet), 639, 641
Deployed words, 1113
Derivative filter, 420
 operator, 421
 weights, 468
Designing
 adaptive filtering techniques, 994
 equiripple DC, 278
 FIR filters, 276, 426
 flexible radio architectures, 446
 frames, 667
 software radio architectures, 406
Desired signal, 721, 722, 745
Deterministic
 autocorrelation, 705
 autocorrelation sequence, 704
 signals, 161, 186, 194, 212, 214, 698
 signals sampling, 194
Dichotomous coordinate descent (DCD) algorithm, 822, 823, 825, 826
Dictionary
 algorithms, 1131
 atoms, 1116
 components, 30
 elements, 30, 683, 1113, 1127, 1129, 1130, 1135
 Gabor, 683
 learning, 30, 1010, 1073–1075, 1086–1088, 1090, 1094, 1096–1099, 1101
  algorithms, 1074, 1086, 1090, 1095, 1105, 1134, 1135, 1141
  framework, 1087
  problem, 1087, 1088, 1096
  procedure, 1099, 1108
  process, 1115
 matrix, 1081, 1083, 1087, 1090, 1096–1098, 1101, 1107, 1109, 1116, 1117, 1120
 methods, 1086
 overcomplete, 1075, 1081, 1096, 1097, 1120
 single, 1124, 1127

update
algorithms, 1090, 1099
stage, 1140
step, 1090
vectors, 850
Dictionary learning
nonnegative, 1117
Differential evolution (DE), 935, 936
Digital down converter (DDC), 403, 405, 418
Digital signal processing (DSP), 1, 185, 961
algorithms, 271
area, 403
equivalents, 414
filter design, 414
Digital up converter (DUC), 403, 443, 462
Digital-to-analog (D/A) converter, 6, 185, 722
Digitally assisted analog circuits, 187
Direct sequence spread spectrum (DSSS), 485, 489
Directed acyclic graph (DAG), 947, 948
Directional filter bank (DFB), 587, 589, 594, 603
Discrete
curvelet
atoms, 578–580
coefficients, 582–585
frame, 576
equivalent, 87
filters, 554, 644
frequencies, 73
random variable, 131
shearlet
atoms, 613–615
coefficients, 617, 621
transform, 615–617, 625
transform coefficients, 620
signal, 32, 86, 187, 193, 558, 655, 973
signal values, 187
Discrete cosine transform (DCT), 495, 499, 996, 999
coefficients, 514, 676
Discrete Fourier transform (DFT), 4, 120, 232, 411, 495, 542, 680
analysis, 966
filter banks, 405
matrix, 971
Discrete wavelet transform (DWT), 519, 548, 588, 639, 643
Discrete-time
random processes, 154
signal, 972
Discrete-time Fourier transform (DTFT), 4, 542
Discrete-valued random processes, 154
Discretized curvelet atoms, 578, 582
Distorted signal, 199, 745
Distributed solutions, 844
Distribution-ergodic, 162
Diversity measures, 1080, 1082, 1083
Double-talk detector, 799
Downconverter channelizer, 444
Downsampler polyphase channelizers, 492
Downsampling
digital signals, 407
dyadic, 594, 595, 601
operation, 405, 435, 446, 595
operators, 588
ratios, 453
Dyadic
coefficients, 519, 520
downsampling, 594, 595, 601
lifting coefficients, 512
sampling, 588
subsampling, 588, 595, 602, 603
subsampling matrix, 590

E

Echo
cancelation, 691, 693, 720, 722, 728, 741, 751, 762, 821, 832, 833, 839, 843, 848
impulse response, 718, 720, 721
signal, 718, 720
Echo return loss enhancement (ERLE), 720
Eigensignals, 36, 49, 62, 63, 66, 68, 78
Eigenvalue spread, 735, 837
Elaborated signal model, 111, 113
Electrical networks, 330
Embryonal stage, 1002
Encoder convolution layer, 637
Energy
compaction, 503, 504, 525, 526, 559, 631, 642, 996
compaction capability, 504
density spectrum, 81
signal, 210, 469, 505
spectrum, 81
Ensemble
average learning curve, 787
learning, 921, 931
learning algorithms, 921
Equalized signal, 836
Equalizer output signal constellations, 789
Equiripple
filters, 301
FIR filter, 276
Equivalent
circuit, 57
discrete, 87
forms, 75
representations, 76, 261
systems, 93
Ergodicity, 161, 746

Error
classification, 894, 899, 923
distribution, 888, 892
energy, 210
evaluation, 923
level, 230
metrics, 225
minimum, 728
model, 741
nonlinearities, 834
optimum, 749, 750, 754, 755, 761, 762, 764
power, 727, 728, 745, 746, 844
probability, 924
sequence, 238, 241, 782
signal, 22, 743, 745, 760, 790, 832, 887, 892, 904
signal component, 359
sources, 194
statistics, 888
surfaces, 846
value, 892
variance, 791, 888, 889
vector, 887, 951
weight, 793
Error-feedback least-squares lattice (EF-LSL) algorithm, 720
Euclidean norm, 739, 778, 810, 814, 831, 849
Evolution model, 168
Excess a posteriori error, 791, 809, 810
Excess a priori error, 791
Excess bandwidth, 446, 469, 470, 477
Excess mean square error (EMSE), 728, 791–799, 802, 803, 807, 808, 811–813, 819, 832
curves, 841
minimum, 814, 815
Expansion coefficients, 681
Expectation-maximization (EM) algorithm, 1052
Expected value, 138, 146
Explainable AI (XAI), 1097, 1117, 1137
Exponentially
damped sinusoids, 414
decaying sinusoids, 413, 414
Expressible dictionary matrix, 1097
Extended lapped transform (ELT), 534
Extrapolation error, 692, 707
Extreme gradient boosting (XGBoost), 924

F

Factor analysis (FA), 1074, 1090, 1096, 1097, 1112
model, 1078, 1096
Factorial representations, 1077
Factorized
convolutions, 1027
separable convolution, 1026
Fan filters, 594–596, 600
Fast Fourier transform (FFT), 404, 495, 542, 655
Fast gradient sign method (FGSM), 934, 935
Fast transversal RLS (FTRLS) algorithms, 821, 822
Fast-QR algorithms, 821
Fat dictionary matrix, 1080
Father curvelet coefficients, 580
Feature representations, 30
Federated learning (FL), 943–945
Feedback approach, 809
Feedforward network, 902
Field programmable gate array (FPGA), 267, 461
Figure of merit (FOM), 624
Filter
adaptive, 21, 22, 89, 179, 270, 285, 717, 718, 720, 721, 725, 728, 730, 740–747, 749, 752, 755, 759, 760, 763, 768, 773, 774, 994, 995
algorithms, 239, 378
approximations, 317
bandpass, 14, 179, 212, 267, 283, 363, 364, 419, 421, 429, 432, 443, 444, 571
bandwidth, 429
bank, 14, 16, 267, 270, 278, 283, 285, 287, 308, 309, 444, 447, 454, 495, 497, 500, 502, 516, 550, 552, 557, 586, 587, 590, 591, 655, 656, 664, 837, 986
coefficients, 505
connection, 500
decomposition, 521
design, 523
solutions, 524
channel, 639, 640
coefficients, 229, 271, 282, 325, 326, 329, 389, 425, 426, 429, 531, 710, 726–728, 732, 740, 747
convolutional, 631
design, 271, 307, 313, 424, 426, 471, 504
design process, 316
estimates, 799
frequency response, 426, 437
FRM, 286, 289, 290, 293, 296, 311, 313, 315
gain, 424, 439
implementation, 10
impulse response, 229, 423, 429, 470
input, 430
length, 296, 424, 501, 531, 799, 840
linear, 749, 978, 980
LMS, 728, 840, 841, 843
memory, 457, 459
MSE, 843
network, 315
nonrecursive, 280, 415, 423
optimal, 762
order, 9, 206, 270, 273–277, 283–285, 290, 293, 295–297, 309, 313, 317, 318, 324, 326, 327, 746, 749

output, 283, 284, 421, 423, 426, 430, 432, 439, 476, 747, 748, 831, 849
pairs, 289, 293, 429, 554, 598
parameters, 424, 708, 740
passband, 426
poles, 414
polyphase, 435, 436, 439, 444, 450, 451, 459, 467, 493
prototype, 286, 287, 289, 290, 295, 309, 310, 312, 405, 424, 425, 427, 470, 482, 492, 532
realization, 12, 311
requirements, 248
response, 6, 426, 505
segments, 434, 441
set, 457
space, 1003
specification, 290, 294, 297, 424
spectral, 438, 1000
stage outputs, 436
structures, 270–272, 279, 281, 296, 303, 319, 322, 323, 358, 388, 395, 596, 601, 821
system, 600, 644
taps, 423, 982
transition bandwidth, 425
wavelet, 497, 518–521, 616
weights, 437, 468, 730, 774, 821, 830
Filtering, 176
adaptive, 21, 707, 717, 722, 724, 735, 745, 746, 791, 792, 836, 839, 844, 845, 847, 848, 994, 995
algorithm, 848, 995
bandpass, 177, 429
effect, 176
Gaussian, 997
graph, 978
linear, 21, 978, 979
multichannel, 760
multirate, 427
nonlinear, 510
operations, 740, 821
optimum, 177, 745
process, 465, 995
scenario, 840, 844
spectral, 997
step, 405
structure, 593
task, 423, 424
techniques, 997
Finite impulse response (FIR), 5, 696
adaptive filters, 831
filter, 8, 243, 244, 267, 272–274, 276–281, 283, 284, 286, 289, 290, 297, 298, 416, 423, 424, 426, 500, 502, 518, 700, 725, 740, 741, 743, 747, 758, 774, 982
filter bank, 590, 594, 665
filtered version, 701
prediction filter, 699
solution, 178
subfilters, 230
systems, 96, 224
transfer function, 8
Finite word length, 375, 378
representations, 85
Forcing function, 44
Forgetting factor, 781, 802, 812, 813, 830
Fourier
frequency, 487, 597
frequency plane, 538, 568
plane, 539, 567, 569–573, 575, 578, 580, 582, 584, 587, 594, 596, 598, 600, 601, 603, 607, 614–616
representations, 77, 81
series coefficients, 78
synthesis, 76
Fourier transform (FT), 2, 3, 18, 48, 50, 62, 72, 78, 120, 194–196, 201–203, 408, 409, 414, 424, 425, 537, 538, 541, 542, 669, 690, 693, 702, 969–971, 975
Fourth power, 833
Fractional bandwidth, 438
Frame
algorithm, 549
applications, 655
bounds, 20, 549, 603, 604, 657, 661, 664, 667, 670, 673
bounds ratio, 661
characteristics, 667
characterization problem, 657
coefficients, 652–655, 659, 663, 666, 676, 681
complete, 1081
concept, 654, 670
condition, 632, 635, 636
constructions, 674
curvelet, 580
decomposition, 657, 658, 678
discrete curvelet, 576
elements, 653–656, 661, 667, 674, 675, 683
expansion, 593, 601, 654, 655, 663, 681
experiment, 676
index, 86
inverse, 658, 659, 662, 667, 670, 671, 675, 676, 678, 684
inversion, 660
literature, 670
location, 1120
operator, 593, 657, 658, 666, 667, 684
overcomplete, 654
reconstruction, 658, 666
redundancy, 659, 662
representation, 20, 549, 659, 684
theory, 652, 676
vector, 593

Framelets, 630–632
  coefficient, 631, 633, 636
  coefficient matrix, 631, 633
  convolution, 630
  representation, 635
FrameletsNN, 628, 630, 632, 633, 635, 637, 638, 641, 642
Framework
  analysis, 538
  dictionary learning, 1087
  kernel, 1135
  learning, 27, 871, 942
Frequency
  alias, 435
  axis, 69, 80, 558, 680
  bands, 198, 232, 276, 278, 282, 283, 286, 317, 428
  bandwidth, 179
  baseband, 461
  bins, 457
  components, 77, 198, 225, 245, 284, 407, 637, 970, 971, 973, 980
  content, 77, 543, 569, 587, 591, 608, 635, 693, 987
  dependency, 230
  domain, 3, 4, 57, 77, 79, 80, 94, 95, 169, 176, 208, 221, 225, 316, 317, 343, 403, 406, 408, 409, 411, 414, 569, 576, 582, 587, 607, 655, 693, 702, 971, 979, 982
  Fourier, 487, 597
  hopping, 485
  index, 459
  localization, 637
  masking techniques, 396
  Nyquist, 111, 198, 226, 286
  Nyquist sampling, 198
  offset, 464, 466, 467, 473, 477, 478, 492
  offset compensation, 464, 466, 467, 473
  partitions, 604
  passband, 317
  plane, 481
  positioning, 465
  range, 86, 284, 300, 301, 406
  region, 212, 245, 600
  representation, 78
  resolution, 482, 487
  response, 66, 176, 267, 269, 270, 979
    mismatch errors, 221
  rotators, 407, 473
  sampling, 189, 194, 196, 198, 199, 206, 208, 270, 283, 301, 311, 407, 409, 410, 446, 986
  sampling methods, 8
  scaling, 80
  selectivity, 307, 324
  shift, 77, 418, 473, 477, 479, 683
  shifted versions, 309
  shifting, 53
  shifting property, 81
  signal, 25
  spectrum, 314
  transformation, 268, 269, 279, 317, 339
Frequency division multiplexed (FDM)
  bandwidth, 428
  input signal, 428, 429
Frequency response masking (FRM)
  filter, 285, 286, 289, 290, 293, 296, 311, 313, 315
  FIR filter, 286
  interpolation filter structure, 306
  multirate filters, 301
  structure, 280, 286, 289, 293, 301, 302, 305, 307–310, 313, 314
  techniques, 267, 273
Full adder (FA), 382–384, 394
Fully connected neural network (FCNN), 898
Fundamental frequency, 73–76, 651
  resolution, 482

G

Gabor
  coefficients, 676
  dictionary, 683
  frames, 655, 659, 668, 670–672, 674, 676, 678–680
  system, 668
Gate recurrent unit (GRU), 905
Gaussian
  distribution, 134
  filtering, 997
  inputs, 820
  noise, 563, 815
Gaussian Markov random field (GMRF), 1049
  model, 996, 998, 1002
Gaussian scale mixture (GSM), 1103
Generalized Lloyd algorithm (GLA), 1091, 1093, 1096
Generative adversarial network (GAN), 1011, 1029, 1030
Generative model framework, 1137
Generic convolution blocks, 638
Geometric multiresolution analysis (GMRA), 1119, 1121
Gradient descent, 714, 1010, 1020, 1090, 1099
  algorithm, 713, 714, 884, 887, 904, 945, 1090
  backpropagation algorithm, 885
  dictionary learning algorithm, 1116
  search method, 886
  stochastic, 934, 1016, 1027, 1109
Gradient noise, 840
Gram matrix, 664, 667
Gram–Schmidt KLMS (GS-KLMS), 850
Graph filter, 981, 982
  design, 982
Graph filtering, 978
Graph Fourier transform (GFT), 970, 971

Graph Laplacian of Gaussian (GLoG) filter, 1000, 1001
Graph shift operator (GSO), 968
Graph signal (GS), 962, 965
convolution, 980
denoising filter, 998
Graph signal processing (GSP), 24, 961
Graphical
models, 1049
representation, 1114
Grassmannian frames, 675
Group Method of Data Handling (GMDH), 1027

H

Half-band
circulator-tree, 377
filter, 279, 282, 294, 295
FIR filters, 278, 282
Hankel matrix, 630, 631
Harmonically related signals, 76
Hermitian symmetric, 738, 739, 773
Hessian, 732, 735, 736, 773
Hessian matrix, 735, 779
High frequency selectivity, 309
Higher-order singular value decomposition (HOSVD), 1017, 1030
Higher-order statistics (HOS), 835, 836
Hilbert transform, 181
Holistic representations, 1119
Homogeneous
Markov chain, 168, 169
Markov sequences, 168
networks, 943

I

Improved channel estimator (ICE) net, 890
Improved proportionate normalized LMS (IPNLMS)
filters, 843
robustness, 843
Impulse response, 2, 4, 13, 36, 38, 40–42, 86, 92, 97, 98, 175, 176, 223, 224, 267, 273, 274, 423, 424, 429, 532, 534, 554, 557, 576, 591, 696, 697, 702, 718, 742, 979, 980, 982
antisymmetric, 10
coefficients, 224
echo, 718, 720, 721
filter, 229, 423, 429, 470
finite duration, 96
infinite duration, 98
truncation, 703
values, 244
windowed, 425, 469
Impulse signal, 36, 40, 980
Impulsive signals, 835
Incremental learning, 1033
Indegree, 963, 964
Independence theory, 793
Independent
noise, 1075
signals, 482
Independent and identically distributed (i.i.d.) samples, 1107
Independent component analysis (ICA), 1115, 1116
Infinite impulse response (IIR), 5, 99, 230, 696
adaptive filters, 748
filter, 9, 10, 12, 239, 267, 272, 279, 284, 311, 339, 414, 423, 725, 747, 846
filter design, 8
Infinite word length, 85
Information signal, 462
Input
Gaussian, 820
IFFT, 478
matrix, 48
signal bandwidths, 465
stability, 103, 104
Input–output stability, 105
criterion, 108
Instantaneous cost function, 776
Insufficient sampling frequency, 405
Integer
coefficients, 500, 509
transform coefficients, 507
Integer cosine transform (ICT), 507
Intelligence Advanced Research Projects Activity (IARPA), 490
Intelligible speech signal, 284
Interconnection networks, 343, 346, 352, 392
Interest
deterministic signal, 693
filters, 337, 344
noisily observed signal, 1078
random signals, 695
signal, 499, 537, 554, 555, 560, 696, 699, 718, 741
Interference cancelation, 740, 741
Interleaved curvelet atoms, 580
Intermediate frequency (IF) signals, 405
Internal stability, 103
Interpolated signal, 13, 443
Interpolation filter, 13, 304, 305, 307, 309, 405, 468
Interpretable artificial intelligence, 942
Interpreting generative networks, 1039
Intersymbol interference (ISI), 743, 835
Interval
estimate, 152
sampling, 126, 131, 133

Inverse
- 2D FFT, 583, 584
- correlation matrix, 713, 784
- covariance matrix, 787
- curvelet transform, 583, 585
- Fourier transform, 78, 195, 196, 203, 213, 222, 224, 229, 541, 571, 579, 616
- frame, 658, 659, 662, 667, 670, 671, 675, 676, 678, 684
  - elements, 655, 662
  - operator, 657, 658, 660
- Gabor frame, 677
- Hessian matrix, 785
- Laplace transform, 257, 259
- matrix, 606, 607, 784
- operation, 508
- operator, 498, 592
- preprocessor, 526
- problem, 747
- synthesis mapping, 499
- system identification, 740, 743
- transform, 543, 583, 585
- transformation, 241
- wavelet transform, 548, 628, 672

Inverse discrete Fourier transform (IDFT), 403, 681
Inverse fast Fourier transform (IFFT), 444, 450, 451, 453, 457, 459, 583, 585, 681
- algorithm, 403, 436
- block, 437, 454, 465, 476
- block sizes, 476
- coefficients, 681
- inputs, 478
- output vector, 459
- points, 473
- size, 466, 468, 471, 473, 474, 476, 477, 492

Irrational coefficients, 500
Irregular
- domains, 961
- sampling, 652

Irregularity, 961

J

Joint
- CDF, 143
- central moment, 147
- dictionary, 1089
- moment, 147
- probability, 126
- probability density function, 144

Jointly wide-sense ergodic, 162

K

K-means clustering algorithm (KMA), 1093
K-Medoids algorithm (KMA), 1091
k-nearest neighbor (*k*-NN) approach, 998
K-singular value decomposition (K-SVD) algorithms, 1074
Kaiser–Bessel window, 426
Kalman
- filter, 781, 793, 843
- filtering, 845

Karhunen–Loève transform (KLT), 499, 996
Kernel
- adaptive filters, 848, 851
- framework, 1135
- matrix, 1135, 1136, 1147
- trick, 916, 917

Kernel LMS (KLMS) algorithm, 849
Kernel RLS (KRLS) algorithm, 850
Kernelized dictionary learning, 1131
- algorithms, 1131

Kernelizing dictionary learning, 1074
Kirchhoff's voltage law (KVL), 39, 59
Kronecker product, 1013, 1035
Kruskal convolutions, 1023, 1024
Kullback–Leibler divergence, 1110
Kurtosis, 139, 833

L

Ladder filters synthesis, 329
Laguerre filters, 821
Laplace transform
- one-sided, 49, 52, 56
- pairs, 51, 53
- two-sided, 49

Laplacian
- matrix, 25, 27, 964, 969, 975, 976, 983, 984, 987, 992, 1003, 1050, 1051, 1062, 1063, 1066
- spectrum, 982, 983

Laplacian pyramid (LP), 590
- decomposition, 587
- filter, 601, 604
- filter bank, 587, 591, 592, 602

Laplacian-constrained GMRF (LGMRF), 1050
Lapped transform (LT), 502, 503, 522
Latent signal, 30
LC
- filter, 269, 270, 317, 319, 321, 323–326, 339
- network, 269, 336, 339, 340
- structure, 337

Leading DCD, 826, 830
Leaky-LMS algorithm, 831
Learned
- dictionary, 1122–1124, 1126, 1127
- representation, 1110

Learning
ability, 927
adaptive filter, 718, 817
adversarial, 874
algorithms, 27, 869, 870, 886, 928, 943, 1050, 1100, 1117
architectures, 1009, 1128
causal, 947
coefficients, 944
concepts, 870
continuous, 874
curve, 787, 830, 1037
deep, 25, 626, 640, 898, 924, 1009, 1010, 1020, 1023, 1034, 1038, 1039, 1050
dictionary, 30, 1010, 1073–1075, 1079, 1086–1088, 1090, 1094, 1096–1099, 1101
framework, 27, 871, 942
graphs, 1050
methods, 27, 949, 1015
model parameters, 33
modes, 870
problem, 942
procedure, 1110
process, 870–874, 876, 878, 887, 892, 926, 929, 942, 951, 1086
rate, 878, 929, 930, 939
representations, 1019
rules, 1129
settings, 1068
sparse representation, 1107
strategy, 870, 874
system, 943
tools, 33
unsupervised, 23, 874, 876, 878
Least squares (LS), 890
Least-mean fourth (LMF) algorithm, 834
Least-mean squares (LMS)
adaptive algorithm, 713
algorithm, 714, 722, 727, 728, 758, 773, 775, 776, 778, 796, 995
coefficients, 831
estimates, 728
filter, 728, 840, 841, 843
recursion, 713, 728
Likelihood, 128
Linear
coefficients, 627
combination, 38, 44, 45, 61, 64, 95, 269, 330, 339, 341, 652, 657, 683, 697, 699, 748, 750, 756, 758, 841
complex filter, 762
convolution, 408
filter, 749, 978, 980
filtering, 21, 978, 979
filtering operator, 978
Linear constant-coefficient difference equation (LCCDE), 691, 697
Linear predictive coding (LPC), 742
Linear shift-invariant (LSI)
filters, 981
graph filter, 983
graph systems, 970
system, 970
systems, 970
Linear time-invariant (LTI) system, 2, 36, 38, 41, 54, 55, 61–65, 73, 90, 101, 175, 176, 179, 181, 252, 253, 255, 690, 696, 697
causal, 64, 65
eigensignals, 61
model, 702, 710
response, 61, 78
Local
error features, 887
frame, 1120
Localized
frequency content, 607
signals, 479
Logarithmic frequency scale, 70
Logic signals, 394
Long short-term memory (LSTM), 903, 904, 1027
Loop filters, 253
Lossless
commensurate length transmission, 332, 336
network, 321, 322
sampling, 175
Low-resolution subband (LRSB), 639, 640
Lower-resolution (LR)
image, 639, 641
version, 639

M

Machine learning (ML), 22, 869
algorithms, 1, 23, 30, 1136
applications, 19, 27, 1052, 1074
causal, 946
classical, 1037
clustering, 27
connotations, 1091
dictionaries, 30
fields, 1051
methods, 1116
model, 23
problems, 30
quantum, 1010, 1034
researchers, 1011
solutions, 22, 24
systems, 153

tasks, 1009
techniques, 1136
theory, 1, 31, 32
tools, 1
unsupervised, 23
Magnitude spectrum, 78
Majorization-minimization (MM), 1051, 1052
Manifold learning, 1074, 1119, 1121, 1136
Marginal
CDF, 143
PDF, 144
Markov
chain, 168
homogeneous, 168, 169
process, 167
random process, 167
sequence, 167, 168
Masking filters, 286, 288, 289, 292, 293, 295, 296, 302, 304, 307–313, 315
Mass probability function, 134
Matching Pursuit (MP), 683, 684, 1083
algorithm, 684
decomposition, 684
expansion, 683
Mathematical
framework, 1077
representation, 62
Matrix
autocorrelation, 120, 512, 515, 709, 748, 749, 756, 761, 786, 794, 807, 831
dictionary, 1083, 1087, 1090, 1098, 1101, 1109
inverse, 784
inversion lemma, 780
representation, 1109
sampling, 591
trace, 737
Matrix product operator (MPO), 1036
Matrix product state (MPS), 1018, 1035, 1036
Maximally
decimated filter banks, 587
sparse representations, 1080, 1119
Maximum
frequency, 460
frequency component, 409
signal value, 247
Maximum a posteriori (MAP)
approaches, 1103
estimate, 1085, 1105
estimation, 1103
estimator, 151, 152
formalism, 1085
Maximum likelihood estimate (MLE), 1089
Maximum likelihood estimator (MLE), 1050
Maximum likelihood (ML) estimator, 151
McClellan–Parks–Rabiner (MPR), 426
Mean
instantaneous
cross-power, 156, 157, 167
power, 155, 156, 161, 166, 167
square value, 138, 155, 156
vector, 149, 150, 165
Mean square deviation (MSD), 792
Mean square error (MSE), 152, 628, 659, 695, 749, 887
Mean Squared Error (MSE), 504
Mean-shift algorithm (MSA), 878
Measurement noise, 836, 1078
Memristance, 83
Memristor, 83
Mercer kernels, 917, 918
Mere basis representation, 1115
Method of Directions (MOD), 1086, 1096–1099, 1135
Method of Moments (MOM), 1074
Microphone signal, 722
Minibatch learning, 873
Minimax concave penalty (MCP), 1054, 1056
Minimum
bandwidth, 469
EMSE, 814, 815
error, 728
mean quadratic error, 178
Minimum description length (MDL), 715
Minimum mean square error (MMSE), 1105
Minimum mean square (MMS) approximation, 425
Minimum MSE (MMSE) constant estimate, 696
Minimum-variance distortionless response (MVDR), 848
Misadjustment, 728, 791
Mismatch errors, 221
channel, 221
linear, 221, 230
nonlinear, 221
Mixed variable, 131
Modality classification, 485
Modular cascades, 496
Modulated
signal component, 113
sinusoid, 113
Modulation frequency, 113
Moment-generating function, 141, 148
Mother
curvelet, 570–574, 576, 578–580, 582, 583
curvelet windows, 572, 583
shearlet window, 607, 608, 612
shearlet window function, 609
Moving average (MA), 98, 182
systems, 697

Multichannel
convolution, 630
convolution kernels, 1010
convolutional kernels, 1020
filtering, 760
filters, 632, 762
Multidimensional signals, 86, 187
Multilayer perceptron (MLP), 882, 883, 919, 1127
Multimedia signals, 495
Multiple
experiment case, 129
signals, 403, 406
Multiple-input multiple-output (MIMO)
decoder, 634
encoder, 634
filter kernel, 633
system, 186, 627, 634
Multiple-input single-output (MISO) system, 250
Multiplier coefficients, 89
Multiply-and-accumulate (MAC) operations, 1109
Multirate
filter, 267, 283, 285, 286, 302, 316, 405, 406, 413, 416, 423, 427
banks, 225
realizations, 273
structures, 395
filtering, 427
FIR filters, 285
signal, 986
processing, 12, 14, 16, 407, 411, 492, 586, 587
Multiresolution analysis (MRA), 551, 552, 587, 594
framework, 550
reconstruction method, 637
theory, 550
Multiscale
representations, 538
signal representation, 551
Multistage Wiener filter, 848
Multivariable signal, 994
Multivariate statistical learning, 1131
Myriad signals, 1136

N

n-mode product, 1013, 1014, 1021
Natural
frequencies, 63
signals succinct representations, 1137
Near-end signal, 718, 720
Negative frequency, 429
Network
backward training, 927
characteristics, 931
compressed, 1025
design, 931
element, 1000
filter, 315
input, 933
layers, 638, 926
learning process, 928
lossless, 321, 322
output, 627, 904, 942
parameters, 898, 949, 950, 952, 1021
regularization, 930
signal, 1000
structure, 870, 926, 961, 1000
tensor, 1038
topology, 348, 1050
training, 928, 929, 942
training process, 926
training stage, 926
weight initialization, 926
Network in network (NiN), 639
Neural network (NN), 19, 21, 24, 30, 485, 487, 488, 625, 626, 628–630, 638, 641, 690, 869, 883, 1020
architecture, 1000, 1028
building blocks, 1010
classifier, 30
deep, 25, 1010, 1016, 1020, 1022, 1023, 1025–1027, 1032, 1038, 1136
development, 643
hierarchical, 1028
layers, 1010
methods, 626
Volterra, 626, 643
wavelet, 626, 628
weights, 24
Newton algorithm, 735
Noise
adversarial, 1032
bandwidth, 179
coefficients, 561
contribution, 585
covariance matrix, 1107–1109
distribution, 561, 563, 835
frequency response, 256
gains, 242
Gaussian, 563, 815
level, 843, 1132
models, 135
power, 161, 239, 247, 251
random, 161, 173, 819, 1010, 1032, 1076
random process, 172
reduction, 247, 251, 663
rejection, 720
resilience property, 659
samples, 793

sequence, 811
shaping, 186
signals, 172
sources, 239, 241, 243
statistics, 559
term, 658
variance, 243, 247, 563, 1086
Noiseless
case, 1084
models, 1077
Noisy signal, 21, 162, 559, 585
Noncausal filter, 230
Nonconvex, 1049
cost functions, 748
formulations, 27, 1050
function, 1052
graph learning, 27
problems, 27
program, 1067
regularizations, 1057
Nondiagonal matrix, 588, 590
Nondifferentiable cost functions, 847
Nondownsampling version, 587
Nonlinear
activation function, 898, 926
filter, 21
filtering, 510
Nonnegative dictionary learning (NDL), 1117
Nonnegative matrix factorization (NMF), 1074, 1117
Nonorthonormal frame, 594, 626
Nonoverlapping passbands, 15
Nonperiodic signals, 72, 77–79, 540
frequency spectrum, 77
Nonrecursive
analog filters, 416
digital filters, 358
filter, 280, 415, 423
FIR filters, 267, 272, 273, 358
Nonseparable filters, 595
Nonsingular subsampling matrix, 589
Nonstationary signals, 19
Nonuniform
filter banks, 309
sampling, 186, 220–223, 966
sampling grid, 220
Normal distribution, 134
Normalized least-mean squares (NLMS)
algorithm, 720, 773, 778–781, 832, 995
update equation, 779
Notch
filters, 277, 278
filters adaptive, 741
FIR filters, 277
Novelty criterion (NC), 850
Null space, 737, 752, 976
Numerator coefficients, 697
Numeric representation, 6
Numerical errors, 185, 781, 786, 822
Nyquist
criterion, 175
filter, 282, 446, 465, 469, 470
filter design, 465
frequency, 111, 198, 226, 286
passband frequency response, 478
sampling, 473, 652
criterion, 407
frequency, 198
rate, 407
zone, 212, 437, 444, 466, 467, 493

O

Objectionable stopband, 425
Offset errors, 230
One-sided
Laplace transform, 49, 52, 56
Laplace transform properties, 56
real exponential function, 50
spectrum, 76
Open boundary conditions, 1018
Optimal
error, 752
filter, 762
frames, 675
sampling, 990
sampling strategies, 989, 990
Optimum
error, 749, 750, 754, 755, 761, 762, 764
error power, 765
filter, 229, 746, 753, 761
filter coefficients, 791
filtering, 177, 745
Order statistics, 835
Ordinary differential equation (ODE), 44–46, 61, 63
Originator signal, 14
Orthogonal frequency division multiplexing (OFDM) system, 889
Orthogonal Matching Pursuit (OMP), 1083
Orthogonality condition, 724, 749, 750, 752, 761, 764
Orthonormal wavelet coefficients, 551
Outdegree, 963
Outdegree matrix, 964
Output
analysis channelizer, 466, 471, 492
channelizer, 473
equation, 47
matrix, 48

radio frequency amplitude, 462
stability, 103, 104
time, 475
Overcomplete
dictionary, 1075, 1081, 1096, 1097, 1120, 1128
frame, 654
representation, 20, 556, 557
signal representation, 558
Overcomplete NMF (oNMF), 1117
Overdetermined dictionary, 1094
Overflow errors, 236
Oversampling, 186, 206, 222, 226, 247, 482, 669
factor, 247, 467
ratio, 679
techniques, 186, 243, 244, 247

P

p-diversity measures, 1082, 1083
Paley–Wiener space, 986, 992
Parallel adaptor network, 355
Parallel factor analysis (PARAFAC), 1016
Parameter
estimation, 151
vector, 844
Parameterizing convolutional layers, 1023
Parks–McClellan (P-M), 426
Parseval
curvelet frame, 579
formula, 573
frame, 605, 615
Parsimonious
representation, 631
signal, 552
Partial differential equation, 44
Particle filters, 768
Partitioned filter, 422, 423
polyphase, 444
Pass fan filter, 595
Passband
attenuation, 320, 324
cutoff frequencies, 426
edge, 289, 290, 294, 304, 317, 319, 324, 326, 329, 360, 366
filters, 426, 580
frequency, 317
gain, 359, 470
performance, 309
regions, 276
requirement, 283
responses, 313, 317
ripple, 283–285, 289, 290, 300, 301, 312, 319, 324, 326, 327, 361, 366, 371, 373, 423, 425
sensitivity, 268, 322, 328, 329, 396
variations, 319, 363
width, 247
Passive LC filters, 321, 341
Pattern classification, 879
Peak Sinal-to-Noise Ratio (PSNR), 504
Perceptron learning algorithm, 881
Perfect reconstruction (PR), 199, 215, 224, 225
Periodic signal, 36, 73, 75–78, 542, 745, 746, 757, 831
Periodically sampled signal spectrum, 409
Phase
rotator outputs, 450
spectrum, 78
Photo core transform (PCT), 513, 528
Photo overlap transform (POT), 528, 529
Plancherel–Parseval property, 542
Point estimator, 151
Poisson
distribution, 136
process, 166
random process, 166
random variables, 136
Pole
filter, 414
radius, 319, 320, 366, 367, 369, 377
Pole-zero
diagram, 63, 64
plot, 63
Polynomial filters, 983
Polyphase
channelizer, 403, 406, 407, 427, 428, 445, 446, 457, 473, 479, 487, 492
decomposition, 405, 406, 427, 432, 433, 442, 444, 492, 587–590, 600, 603
filter, 435, 436, 439, 444, 450, 451, 459, 467, 493
filter bank, 405, 434, 446, 492, 592, 593
filter output data, 453
partitioned filter, 444
representation, 590–592
Positive definite, 739
character, 786
matrix, 739, 786
Positive semidefinite, 739
matrix, 737, 739
Postanalysis filters, 471
Posterior mean estimator, 152
Postfiltering, 524
Postprocessing channelizers, 464
Power spectral density (PSD), 20, 170, 171, 174, 176, 179, 181
Power spectrum, 182, 183, 212, 214, 216, 218, 219, 225
Power-normalized LMS (PN-LMS) algorithm, 832
Power-of-two LMS algorithm, 820
Preagreed signal, 743
Precision measure, 152

Prediction, 742
error, 690, 699, 705
error filter, 699
filter, 699
filter FIR, 699
Preimages, 917
Prelearned dictionary, 1127
Preprocessor analysis channelizers, 476
Pretrained
convolutional kernel, 1023
network, 942, 1040
Principal component analysis (PCA), 499, 879, 1077, 1081, 1090, 1096, 1097
Probability density function (PDF), 132–134, 765, 834, 850
Probe signals, 702
Processing element (PE), 379, 380, 383
Projected gradient descent (PGD), 934, 936, 939, 940
Proportionate NLMS (PNLMS) algorithms, 832, 833
Prototype
analog, 10, 429
Cauer filter, 311–313
channelizer filter, 454
codewords, 1094
filter, 286–290, 295, 309, 310, 312, 405, 424, 425, 427, 470, 482, 492, 532
masking filter, 303, 304
signal, 20, 667, 668, 674
Pseudoinverse, 549, 592, 594, 660
Pulse amplitude modulator (PAM), 194, 200, 201, 204
PyTorch, 931, 1038

Q

QPSK signals, 477
Quadratic programming (QP), 913, 914, 920
Quadrature amplitude modulation (QAM)
channels, 446
constellations, 474
signals, 470
Quadrature mirror filter (QMF), 558, 587, 594–596
Quadrature phase shift-keying (QPSK) signal, 788
Quality of service (QoS), 461
Quantization
errors, 6, 232, 236, 238, 239, 241, 250, 251, 819
noise, 207, 241, 242, 250, 659
noise variance, 241
Quantized
adaptor coefficients, 366
coefficients, 377, 658, 659
signal, 190, 693
Quantized KLMS (QKLMS) algorithm, 850
Quantum computing (QC), 1010, 1034, 1035
Quantum machine learning (QML), 1010, 1034
Quantum networks, 1034
Quasi-Newton algorithms, 735
Quasiperiodic signals, 183
Quincunx
resampling, 595
sampling, 594
Quincunx filter bank (QFB), 595, 596

R

Radial basis function (RBF), 627, 1132
Radio frequency (RF), 112, 429, 461
Radio spectrum, 406, 462, 463
Random
matrix, 119
noise, 161, 173, 819, 1010, 1032, 1076
process, 154
sample, 151
sampling, 991
sequence, 154
signal, 5, 123, 161, 175, 695, 697, 746
variable, 129
Random Fourier feature (RFF), 850
Rank selection, 1026, 1039
RBF network (RBFN), 627
Reactance network, 322
Receiving signals, 460
Recovering spatiotemporal signals, 1000
Rectangular spectrum, 469
Rectified Linear Unit (ReLU) activation, 627
Recurrent networks, 1009
Recurrent neural network (RNN), 899, 902, 1000, 1027
Recursive filters, 286, 308, 316
frequency, 311
Recursive least-squares (RLS)
algorithm, 713, 720, 735, 773, 781, 995
filters, 791, 815, 843
Redesigning, 464
Redundant
frame, 654
frame operator, 594
representation, 547, 575, 594, 630
Reflectance functions, 332, 335, 342, 356, 357
Region of convergence (ROC), 49–52
Regression error, 691
Regressor, 748
Reinforcement learning (RL), 23, 873, 874, 938, 940
Relative frequency, 125
Relevance vector machine (RLV), 920
Representation
capabilities, 1086
frame, 20, 549, 659, 684
framelets, 635
frequency, 78
learning problem, 1089, 1134

matrix, 1109
model, 1087
overcomplete, 20, 556, 557
polyphase, 590–592
power, 638
shearlet, 625
signal, 540, 631, 651, 655, 673
sparse, 25, 33, 1074, 1075, 1078–1082, 1084, 1086
spectral, 542
vectors, 1073, 1113, 1116
wavelet, 556
Reproducible filters, 270
Reproducing kernel Hilbert space (RKHS), 916, 917
Resampling
filters, 412
matrices, 594
operation, 430, 435
operator, 423
process, 403, 407, 411, 413, 492
quincunx, 595
switch, 426
Research interests, 594
Residual network (ResNet), 639
Resolving signals, 482
Resonance frequency, 67–69
Resonator filters, 329
Restricted isometry property (RIP), 1079
Ripple
passband, 414
stopband, 414
stopband elliptic filters, 414
Robust adaptive filters, 835
Root mean squared (RMS) bandwidth, 174
Rotated curvelet atoms, 572, 574

S

Sample
functions, 154
mean, 151
space, 124–126
Sample-and-hold (S/H) circuit, 191, 192, 231
Sampling
approaches, 986
bandpass, 211
clock, 416
compressed, 187
criteria, 991
different signals, 199
dyadic, 588
frequency, 189, 194, 196, 198, 199, 206, 208, 270, 283, 301, 311, 407, 409, 410, 445, 446, 986
grid, 221
instances, 189, 194
interval, 85
lossless, 175
matrix, 591, 594
method, 991, 992
Nyquist, 473, 652
operations, 587, 669
operator, 597, 989
optimal, 990
paradigm, 187
period, 3, 12, 109, 120, 189, 194, 196, 199, 220, 222, 245, 259, 408, 409
process, 214, 236, 403, 405, 407, 409, 411, 424, 470, 492
quincunx, 594
random, 991
rate, 12–14, 89, 105, 187, 221, 245, 405, 406, 409, 410, 444, 446, 543, 986
change, 12, 13, 302, 444, 447
changes, 33
increase, 105
sets, 989–991
signal, 652, 655, 669
size, 587
sparse, 187
strategies, 991
theorem, 6, 12, 194, 196, 198, 199, 201, 202, 206, 209, 212, 214
theory, 212, 985, 986
Schur complement, 738, 756
Seed vertices, 962
Segmented least-squares (SLS)
algorithm, 711
loss function, 711
minimization problem, 710
objective function, 711
Segmented signal components, 476
Segregated signals, 485
Selective resampling, 596
Self-adaptivity, 626
Self-organizing map (SOM), 877–880
Semisupervised learning, 23
Sentiment classification, 906
Separation principle, 811
Set-membership algorithms, 834, 846
Shannon sampling, 669
Shaping filter, 286–290, 292–294, 296, 302, 304, 307–311, 313, 315, 462, 615
Shearlet
atoms, 606–608, 611–614
coefficients, 608–611, 619, 621, 623, 625
representation, 625
transform, 545, 605, 607, 610, 611, 613, 615, 618, 619, 622
coefficients, 619
system, 606

Shift-invariance, 981
Shifted signal, 973
Short-time Fourier transform (STFT), 479, 481–483, 672
Sign-LMS algorithm, 820, 835
Signal
aliased, 430, 436
analog, 7, 36, 83, 187, 188, 190–193, 222, 404, 405, 407, 409, 492, 542
analysis, 20, 48, 73, 655, 664, 673, 677, 681, 683, 684
analysis purposes, 656
analysis tool, 495
approximations, 683
band, 468
bandwidths, 407, 410, 427, 461, 464, 470, 669
baseband, 462
binary, 836
cancelation, 329
carrier, 322
characteristics, 533, 540, 654, 672, 683
classification, 484, 485
coding, 655
coefficients, 565
compression, 587
constellations, 470, 744
data, 693
DC, 226
decomposition, 498, 520, 682
decomposition algorithm, 683
decorrelation, 504, 505
detection, 16, 462, 479, 481, 484
discrete, 32, 86, 187, 193, 558, 655, 973
echo, 718, 720
energy, 210, 469, 505
enhancement, 22, 996
error, 22, 743, 745, 760, 790, 832, 887, 892, 904
estimation, 559
flow graphs, 389
flowgraph model, 701
frequency, 25
frequency response, 256, 259
generation, 477, 1112
interest, 499, 537, 554, 555, 560, 696, 699, 718, 741
levels, 236, 244, 282, 355, 360
model, 525
network, 1000
power, 239, 481
prediction, 996
processing, 1, 4, 22, 24, 36, 79, 85, 175, 186, 301, 404, 430, 436, 499, 509, 546, 587, 689, 690, 692, 696, 714, 718, 740, 892, 969, 1051, 1073–1075
algorithms, 664
applications, 20, 194, 673, 711, 715
architecture, 405
classical, 986
community, 518
discrete-time, 252
fields, 1050
framework, 969
methods, 652, 869
network structures, 25
operations, 6, 185, 693
path, 430
problems, 14, 707, 1002
scenario, 667
sparse, 1073
statistical, 695
systems, 85, 186, 190
tasks, 460, 495, 655, 656
technique, 967
theory, 19, 976, 1002
toolbox, 109
tools, 1, 14, 20, 942, 949, 961
processors, 271
prototype, 20, 667, 668, 674
quantization process, 467
quantized, 190, 693
random, 5, 123, 161, 175, 695, 697, 746
range, 272
reconstruction, 185, 520, 559
errors, 652
methods, 186
operation, 498
operators, 654
regions, 484, 529
replicas, 409
representation, 540, 631, 651, 655, 673
samples, 412, 468, 670, 961, 962, 965, 967, 973
sampling, 652, 655, 669
sampling rate, 405, 410
source, 322, 345, 690
space, 1119
spectrum, 211, 277, 278, 405, 466, 478
structure, 715
subspace methods, 549
upsampling, 468
values, 5, 85, 94, 187, 188, 965, 966
vector, 389
wavelet coefficients, 565
windowed, 469
word length, 271, 272, 387
Signal-to-noise ratio (SNR), 161, 462, 490, 788, 843, 847, 890, 891
Single
atomic situations, 1115
cluster, 1123
conceptual atomic situations, 1115

DAC, 443
dictionary, 1124, 1127
filter structure, 423
frequency band, 428, 479
learned dictionary, 1123, 1124
overcomplete dictionary, 1122, 1124
Single-input multiple-output (SIMO) system, 250
Single-input single-output (SISO) system, 90, 626, 631
Singular value decomposition (SVD), 631, 632, 635, 642
Sinusoid, 21, 73, 77, 86, 109–112, 416, 417, 430, 452, 459, 689, 692
center frequency, 417
frequency, 111
Sinusoidal
basis functions, 73, 77
components, 22, 75
functions, 36, 73, 77
input, 73
input signal, 322
signals, 22, 72, 87, 177, 198, 199, 757
Skewness, 139
Smoothly clipped absolute deviation (SCAD), 1054, 1056
Social networks, 961, 994
Source signal, 492
Spam filters, 1
Sparse
adjacency matrix, 967
codes, 30, 31, 1127, 1129, 1131
graph learning, 1056
representation, 25, 33, 1074, 1075, 1078–1082, 1084, 1086
algorithms, 1090
dictionary learning, 1109
learning, 1107
learning algorithms, 1103
model, 1087
problem, 1089
vector, 1074, 1087
row vector, 1139
sampling, 187
signal
decompositions, 663
processing, 1078, 1082, 1083, 1103, 1113
realizations, 1112
solution matrix, 1141
Sparse Bayesian Learning (SBL), 1084, 1101
dictionary learning algorithm, 1074
dictionary learning scalable, 1109
framework, 1074, 1085, 1107
Sparse Representation-based Classification (SRC), 1126
Spatiotemporal signal analysis, 1000
Spectral
convolution, 469
filter, 438, 1000
filtering, 997
radius, 739
representation, 542
Spectral correlation density (SCD), 485–487
Spectral graph wavelet transform (SGWT), 999
Spectrum
analysis applications, 229
band, 15
combination, 15
continuous, 469
energy, 81
frequency, 314
signal, 211, 277, 278, 405, 466, 478
Speech signal, 36, 187, 283, 394, 567, 775
Stable
filter, 8
inverse, 500
Standard
channelizer configuration, 450
deviation, 139
State
equation, 47
matrix, 48
probabilities, 168
vector, 95, 955
Static
error, 236
offset mismatch errors, 221
Stationary signal, 533
model, 499
Statistical
correlation, 147
covariance, 147
independence, 146
mean, 138, 146
Steady-state probability vector, 169
Steerable DCT (SDCT), 996
Stein's unbiased risk estimate-linear expansion threshold (SURE-LET) approach, 999
Step response, 40, 42
Stochastic gradient descent (SGD), 934, 1016, 1027, 1109
Stochastic signals, 194
Stopband
attenuation, 9, 245, 274, 283–285, 320, 363, 366, 373, 471, 497, 503, 504, 523
edges, 286, 289, 290, 295, 296, 360, 366, 372, 423
filters, 321, 334, 359, 377
frequency responses, 426
peak ripple magnitude, 304
regions, 276
requirements, 283, 317
ripple, 9, 289, 290, 300, 366, 425, 426
sensitivity, 284, 328, 329, 361, 363

Stopping phenomenon, 819
Strict-sense stationary (SSS) random process, 157, 162
Strictly convex function, 769
Strip spectral correlation analyzer (SSCA), 481, 487–489
Structural causal model (SCM), 946, 947, 949
Sub-Gaussian, 833, 834
Subband
  adaptive filters, 837
  signal, 14–16, 500
  signal analysis, 550
  wavelet coefficients, 639, 640
Subclassifications, 423
Subdictionary, 1080
Subfilter, 230, 284, 289, 294, 296, 300, 301
  network, 286
Subnetwork, 279, 1026
Suboptimum filters, 229
Subsampling, 588, 589, 602
  concepts, 590
  dyadic, 588, 595, 602, 603
  vectors, 589
Subsignal, 587, 600, 837
Subspace learning, 1039
Successively delayed subfilters, 439
Succinct representation, 558
Sum-of-powers-of-two (SPOT) representations, 820
Super-Gaussian, 833–835
Superheterodyne frequency, 461
Superresolution (SR) image, 639
Superresolution subband (SRSB), 639
Supervised learning, 23, 870, 879, 883, 887, 908, 1128
Support vector, 915, 916
Support vector machine (SVM), 908, 912, 916, 918–920
Symbolic representations, 90
Synthesis
  bank, 497, 500, 503, 520
  basis functions, 498, 504
  channelizer, 406, 460, 464, 466–468, 471, 473–479
    bandwidth, 406, 464
    channel bandwidth, 474
    ports, 478
  digital FIR filters, 272
  equation, 78, 664
  filter, 15, 229, 308, 309, 500, 517, 591, 592, 595, 596, 598, 600
    bank, 15, 16, 587, 591, 592, 837
    impulse responses, 598
  Fourier, 76
  input series, 476
  mother wavelets, 552
  operation, 20, 666
  operator, 665, 666, 680, 681
  polyphase matrix, 502, 526
  problem, 286
  procedures, 337
  side, 504, 510
  stage, 500
  transform, 498
  vectors, 20
  viewpoint, 503
  wavelet, 18
Synthetic overt signals, 490

T

Target signal, 707
Temporal
  representation, 79
  signal, 79, 81
Tensor contraction layer (TCL), 1020, 1021
Tensor diagrams, 1014
Tensor network, 1025, 1037, 1038
  decompositions, 1035
Tensor regression layer (TRL), 1021, 1022
Tensor-ring, 1018, 1020, 1035
Tensor-train (TT) decomposition, 1018, 1019
TensorFlow, 931, 1038
  library, 1038
  platform, 931
Tensorized deep networks, 1039
Tensorizing deep learning architectures, 1039
Texture classification, 1130
Time
  autocorrelation, 159–162
  average, 159–161, 163
  cross-correlation, 159, 160
Time-frequency (T-F)
  distribution, 544
  distribution perspective, 544
  plane, 544, 548
  representation, 537
Time-invariant system, 175
Time-scale analysis, 674
Total probability, 126, 127
Total variation (TV), 586, 972, 973
Trace
  matrix, 737
  square matrix, 794
Trained network, 934, 938
Training phase, 743
Transduction, 1026, 1039
Transfer function, 62–64
Transition
  band, 206, 245, 248, 267, 273, 275, 280, 286, 287, 295, 984
  bandwidth, 301, 425, 469, 470
    control, 425
    filter, 425

PDF, 168
probability, 128, 168, 169
Transmitted
sequence, 744
signal, 460, 462, 835, 836
Transmultiplexer (TMUX), 267, 309
Transversal filters, 98, 840
Trilateral filter, 998
Tucker convolutions, 1024, 1025, 1033, 1039
Tuned radio frequency (TRF), 429, 430
Two-channel
decomposition, 516
filter bank, 517, 523
filter bank decomposition, 521
perfect reconstruction filter bank, 518
Two-sided spectrum, 76, 78

U

Unbiased sample variance, 152
Unconstrained precision, 1049
Uncorrelated
errors, 251
noise, 178
noise sources, 251
signal, 722
Undercomplete NMF (uNMF), 1117
Undersampling, 208
Undesired frequency components, 199, 230, 244
Undetermined coefficients, 45
Undirected
Gaussian graphical models, 1049
graph, 963, 964
Unequispaced FFT (USFFT), 582
Uniform distribution, 134
Uniform uncertainty principle (UUP), 1079, 1083
Unimodular resampling operations, 596
Unique representation property (URP), 1079
Uniqueness set, 989, 992, 993
Unitary, 738
Universal filters, 716
Unnormalized graph Laplacian matrix, 1050
Unprocessed signals, 690
Unrecoverable aliasing effects, 15
Unsupervised
classification problem, 873
learning, 23, 874, 876, 878
machine learning, 23
Unwanted signals, 317, 461
Unweighted words, 1113
Upsampled signals, 13, 15
Upsampling
digital signals, 407
graph signals, 985
operations, 13, 588
operators, 588
process, 467
signal, 468

V

Validation error, 886
Vandermonde matrix, 982
Variable bandwidth, 367
Variable step size LMS (VSLMS), 832
Variance, 139
Vector
error, 887, 951
random variable, 143, 145, 146, 148
signal, 389
Vector quantization (VQ), 1090, 1091
algorithms, 1090
Very large-scale integration (VLSI) implementations, 235
Voltage signal source, 345
Volterra filters, 643, 774, 791, 803, 843
Volterra neural network (VNN), 626, 643

W

Wanted signal, 317
Wavelet
bandpass, 538
coefficients, 548, 555, 560, 561, 585, 603, 630, 639–641, 643
filter, 497, 518–521, 616
banks, 1002
design, 516, 517
frames, 593, 601, 603, 656, 668, 672–674, 676, 681, 683, 684
representation, 556
series representation, 19
synthesis, 18
transform, 18, 19, 495, 501, 509, 516, 521, 522, 538, 545–548, 585, 587, 588, 593, 605, 606, 672, 999, 1000, 1002
transformation, 626, 639
Wavelet neural network (WNN), 626, 628
Wavelet residual network (WavResNet), 640, 641
Weight
coefficients, 831
error, 793
error vector, 790, 791, 794, 796, 802, 810
vector, 748, 756, 762, 781, 782, 784, 833, 881
Weighted least-squares (WLS) solutions, 8, 9
White Gaussian noise (WGN), 87, 111–113
White noise, 173
Wide-sense ergodic process, 162

Wide-sense stationary (WSS)
covariance functions, 695
process, 158, 159
random
process, 158, 161
sequence, 181
signal, 695, 696, 746
signals, 695
stochastic processes, 212, 708
Widely linear complex filter, 762
Wiener
filter, 177–179, 709, 749
solution, 709, 749, 783
Wiener–Hopf equations, 749
Wigner–Ville distribution (WD), 682
Window filter, 623
Windowed
Fourier frame, 668
Fourier transform, 543, 547
impulse response, 425, 469
signal, 469
Wireless radiofrequency, 479
Wirtinger
calculus, 771
derivatives, 770, 772
Word length, 191, 236, 239, 242, 244, 270, 272, 273, 280, 284, 349, 360, 361, 367, 377, 378
representation, 85